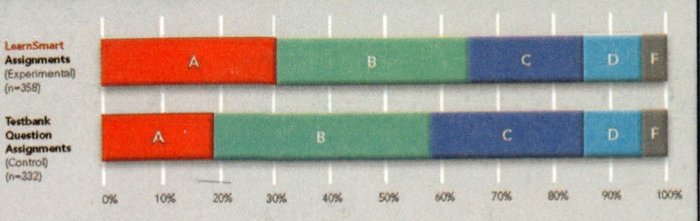

SMARTBOOK™

The first—and only—adaptive reading experience designed to transform the way students read.

> Engages students with a personalized reading experience
> Ensures students retain knowledge

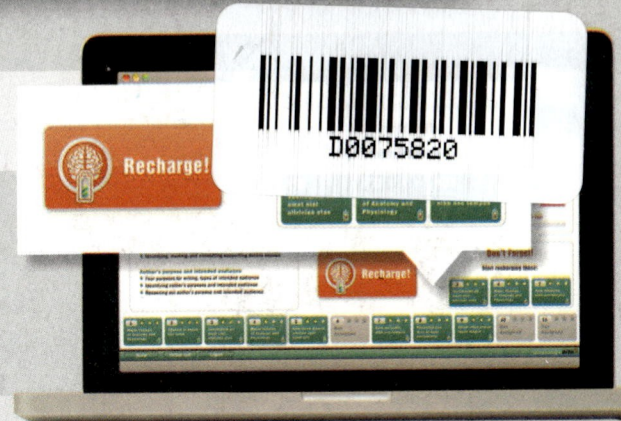

LEARNSMART®

The market leading **adaptive study tool** proven to strengthen memory recall, increase class retention and boost grades.

> Moves students beyond memorizing
> Allows instructors to align content with their goals
> Allows instructors to spend more time teaching higher level concepts

LEARNSMART PREP™

An adaptive course preparation tool that quickly and efficiently helps students prepare for college level work.

> Levels out student knowledge
> Keeps students on track

LEARNSMART ACHIEVE™

A learning system that continually adapts and provides learning tools to teach students the concepts they don't know.

> Adaptively provides learning resources
> A time management feature ensures students master course material to complete their assignments by the due date

LEARNSMART LABS™

LearnSmart Labs is a super-adaptive simulated lab experience that brings meaningful scientific exploration to students. Through a series of adaptive questions, LearnSmart Labs identifies a student's knowledge gaps and provides resources to quickly and efficiently close those gaps. Once the student has mastered the necessary basic skills and concepts, they engage in a highly realistic simulated lab experience that allows for mistakes and the execution of the scientific method.

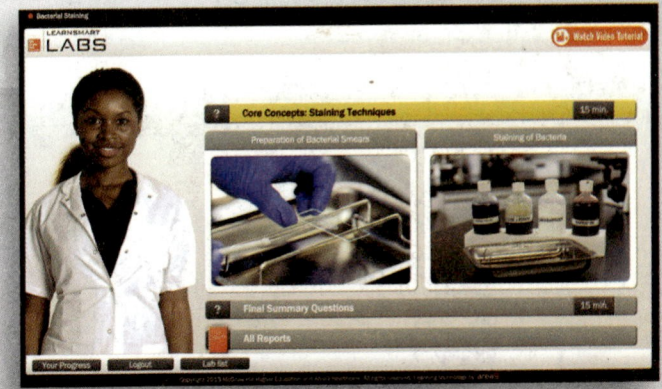

FOURTH EDITION

Microbiology
A Systems Approach

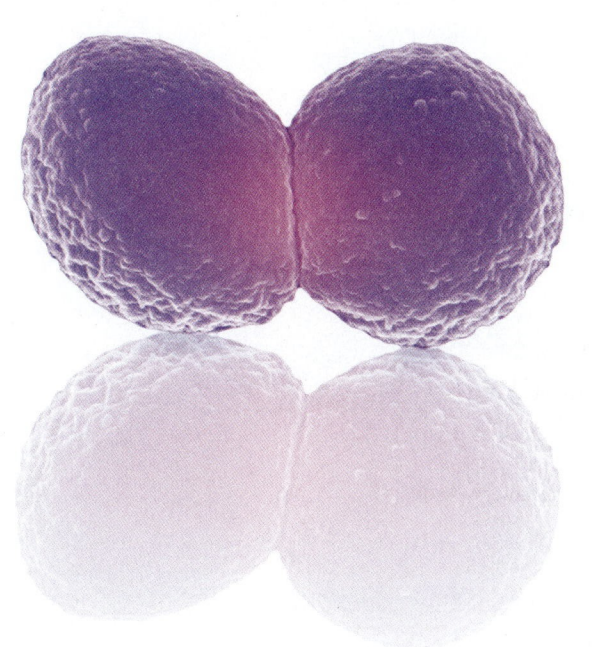
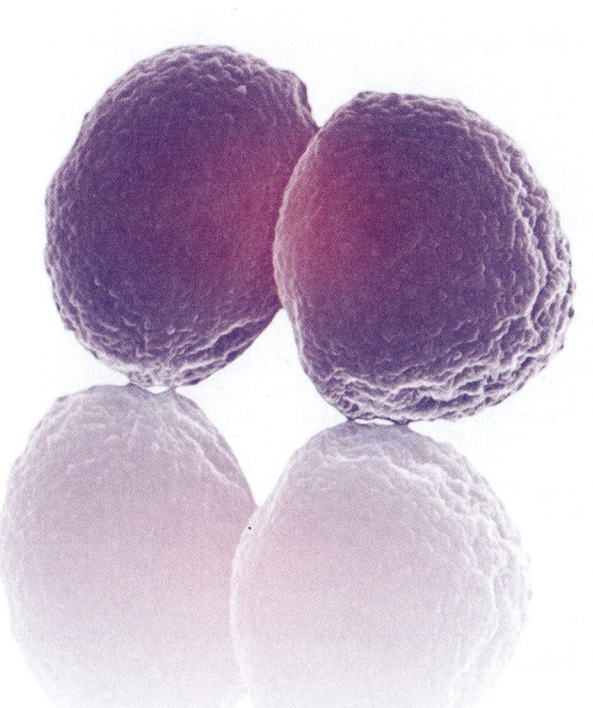

Marjorie Kelly Cowan
Miami University

Mc
Graw
Hill
Education

MICROBIOLOGY: A SYSTEMS APPROACH, FOURTH EDITION

Published by McGraw-Hill Education, 2 Penn Plaza, New York, NY 10121. Copyright © 2015 by McGraw-Hill Education. All rights reserved. Printed in the United States of America. Previous editions © 2012, 2009, and 2006. No part of this publication may be reproduced or distributed in any form or by any means, or stored in a database or retrieval system, without the prior written consent of McGraw-Hill Education, including, but not limited to, in any network or other electronic storage or transmission, or broadcast for distance learning.

Some ancillaries, including electronic and print components, may not be available to customers outside the United States.

This book is printed on acid-free paper.

2 3 4 5 6 7 8 9 0 DOW/DOW 1 0 9 8 7 6 5 4

ISBN 978–0–07–340243–7
MHID 0–07–340243–5

Senior Vice President, Products & Markets: *Kurt L. Strand*
Vice President, General Manager, Products & Markets: *Marty Lange*
Vice President, Content Production & Technology Services: *Kimberly Meriwether David*
Managing Director: *Michael S. Hackett*
Brand Manager: *Amy Reed*
Director of Development: *Rose M. Koos*
Product Developer: *Darlene Schueller*
Digital Product Analyst: *Jake Theobald*
Executive Marketing Manager: *Patrick E. Reidy*
Director, Content Production: *Terri Schiesl*
Content Project Manager: *Sherry Kane*
Senior Buyer: *Laura Fuller*
Senior Designer: *Laurie B. Janssen*
Cover Image: *Visuals Unlimited, Inc./Dr. Stanley Flegler/Gettyimages*
Senior Content Licensing Specialist: *John Leland*
Art Studio and Compositor: *Electronic Publishing Services Inc., NYC*
Typeface: *10/12 Palatino LT Std*
Printer: *R. R. Donnelley*

All credits appearing on page or at the end of the book are considered to be an extension of the copyright page.

Library of Congress Cataloging-in-Publication Data

Cowan, M. Kelly, author.
  Microbiology : a systems approach. – Fourth edition / Marjorie Kelly Cowan, Miami University–Middletown.
      pages cm
  Includes index.
  ISBN 978–0–07–340243–7 — ISBN 0–07–340243–5 (hard copy : alk. paper) 1. Microbiology–Textbooks.
I. Title.
  QR41.2.C69 2015
  579–dc23

                    2013039844

The Internet addresses listed in the text were accurate at the time of publication. The inclusion of a website does not indicate an endorsement by the authors or McGraw-Hill Education, and McGraw-Hill Education does not guarantee the accuracy of the information presented at these sites.

www.mhhe.com

Brief Contents

About the Authors

Kelly Cowan just celebrated her 20th anniversary at Miami University Middletown, an open admissions campus in Ohio. She received her Ph.D. at the University of Louisville, and later worked at the University of Maryland and the University of Groningen in the Netherlands. She specializes in teaching microbiology to nonmajors, and especially to pre-nursing and allied health students. She herself fell in love with microbiology while pursuing an undergraduate degree in dental hygiene. She has made it her personal mission to hear nurses and dental hygienists she encounters in everyday situations exclaim, "I *loved* my microbiology class!"

Having a *proven* educator as a digital author makes a *proven* learning system even better.

With this fourth edition, we are pleased to continue to have Jennifer Herzog on the team. Jen works hand-in-hand with the textbook author, creating online tools that truly complement and enhance the book's content. Because of Jen we now offer you a robust digital learning program, tied to Learning Outcomes, to enhance your lecture and lab, whether you run a traditional, hybrid, or fully online course.

Jennifer Herzog, M.S., M. Phil., is an assistant professor of biology at Herkimer County Community College, Herkimer, New York, where she regularly teaches biology and microbiology to nonmajors and allied health students. She has been an active member of the American Society for Microbiology for nearly 20 years, most recently serving as Chair of the ASM Conference for Undergraduate Education and serving as Chair-Elect for the ASM's Education Division. In addition, she currently authors the "Journal Watch" section of the ASM's *Journal of Microbiology & Biology Education* and serves on the ASM's Microbe Library Editorial Review Board.

Students:

Welcome to the microbial world! I think you will find it fascinating to understand how microbes interact with us, and with our environment. The interesting thing is that each of you has already had a lot of experience with microbiology. For one thing, you are thoroughly populated with microbes right now, and much of your own genetic material actually came from viruses and other microbes. And while you have probably had some bad experiences with quite a few microbes in the form of diseases, you have certainly been greatly benefited by them as well.

This book is suited for all kinds of students and doesn't require any prerequisite knowledge of biology or chemistry. If you are interested in entering the health care profession in some way, this book will give you a strong background in the biology of microorganisms, without overwhelming you with unnecessary details. Don't worry if you're not in the health professions. A grasp of this topic is important for everyone—and can be attained with this book.

—Kelly Cowan

I dedicate this book to all public health workers who devote their lives to bringing the advances and medicines enjoyed by the industrialized world to *all* humans.

McGraw Hill Education | LEARNSMART™
ADVANTAGE

LearnSmart® is one of the most effective and successful adaptive learning resources available on the market today. More than 2 million students have answered more than 1.3 billion questions in LearnSmart since 2009, making it the most widely used and intelligent adaptive study tool that's proven to strengthen memory recall, keep students in class, and boost grades. Students using LearnSmart are 13% more likely to pass their classes, and 35% less likely to drop out.

LearnSmart continuously adapts to each student's needs by building an individual learning path so students study smarter and retain more knowledge. Turnkey reports provide valuable insight to instructors, so precious class time can be spent on higher-level concepts and discussion.

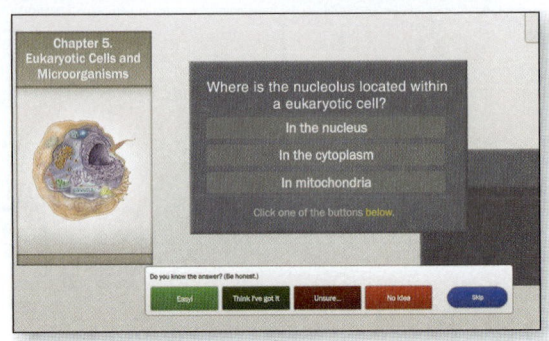

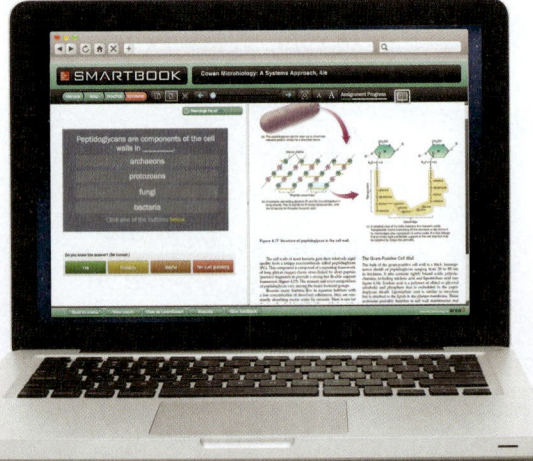

Fueled by LearnSmart—the most widely used and intelligent adaptive learning resource—**SmartBook™** is the first and only adaptive reading experience available today.

Distinguishing what a student knows from what they don't, and honing in on concepts they are most likely to forget, SmartBook personalizes content for each student in a continuously adapting reading experience. Reading is no longer a passive and linear experience, but an engaging and dynamic one where students are more likely to master and retain important concepts, coming to class better prepared.

As a result of the adaptive reading experience found in SmartBook, students are more likely to retain knowledge, stay in class, and get better grades.

LearnSmart Labs™ is a super-adaptive simulated lab experience that brings meaningful scientific exploration to students. Through a series of adaptive questions, LearnSmart Labs identifies a student's knowledge gaps and provides resources to quickly and efficiently close those gaps. Once the student has mastered the necessary basic skills and concepts, they engage in a highly realistic simulated lab experience that allows for mistakes and the execution of the scientific method.

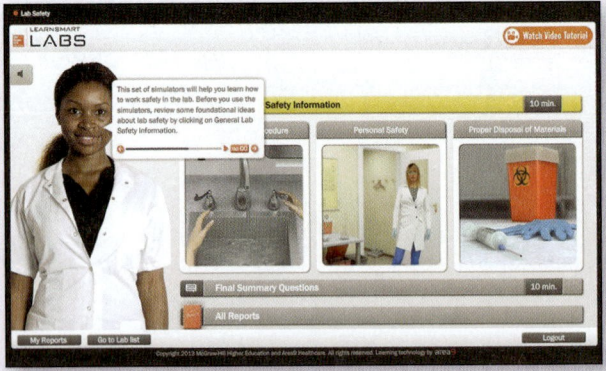

LearnSmart Prep™ The primary goal of LearnSmart Prep is to help students who are unprepared to take college-level courses. Using super-adaptive technology, the program identifies what a student doesn't know, and then provides "teachable moments" designed to mimic the office hour experience. When combined with a personalized learning plan, an unprepared or struggling student has all the tools they need to quickly and effectively learn the foundational knowledge and skills necessary to be successful in a college-level course.

Digital efficacy study shows results!

Digital efficacy study final analysis shows students experience higher success rates when required to use LearnSmart.

- Passing rates increased by an average of **11.5%** across the schools and by a weighted average of **7%** across all students.
- Retention rates increased an average of **10%** across the schools and by a weighted average of **8%** across all students.

Study details:

- Included two state universities and four community colleges.
- Control sections assigned chapter assignments consisting of testbank questions and the experimental sections assigned LearnSmart, both through McGraw-Hill Connect®.
- Both types of assignments were counted as a portion of the grade, and all other course materials and assessments were consistent.
- 358 students opted into the LearnSmart sections and 332 into the sections where testbank questions were assigned.

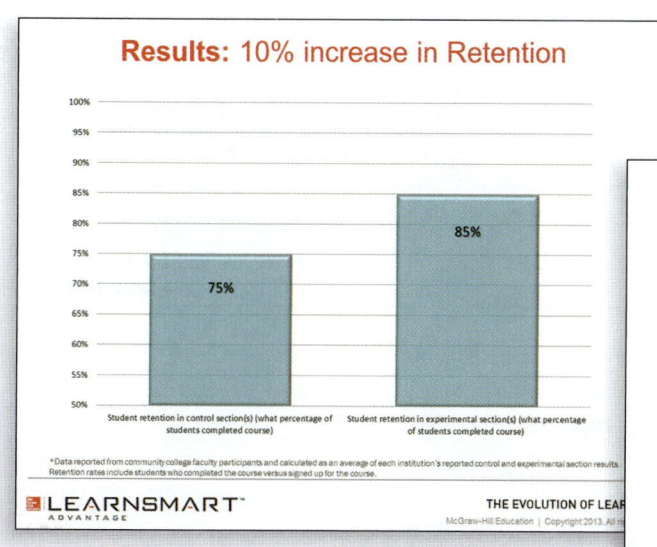

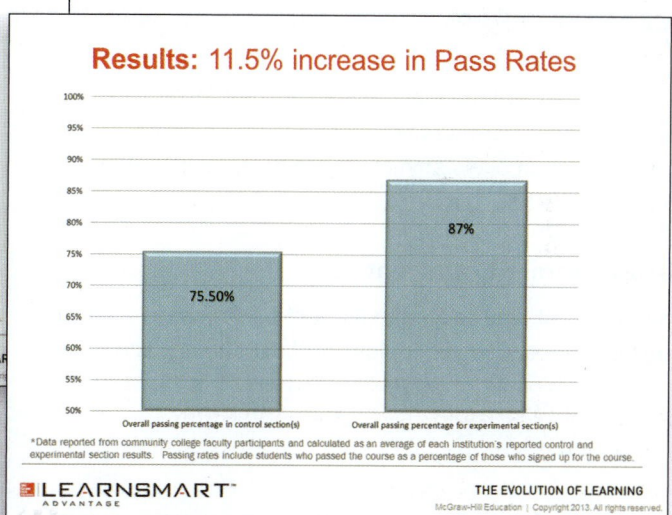

"LearnSmart has helped me to understand exactly what concepts I do not yet understand. I feel like after I complete a module I have a deeper understanding of the material and a stronger base to then build on to apply the material to more challenging concepts."

—Student

"After collecting data for five semesters, including two 8-week intensive courses, the trend was very clear: students who used LearnSmart scored higher on exams and tended to achieve a letter grade higher than those who did not."

—Gabriel Guzman, Triton College

"LearnSmart is intuitive and analyzes where the students' strengths and weaknesses are and develops a strategy to properly tutor the student. Connect Microbiology gives the students examples of test questions in several different formats and provides other materials to help them study and review the chapters."

—Stephen Wagner, Stephen F. Austin State University

Connecting to Core Concepts

McGraw-Hill ConnectPlus® Microbiology

McGraw-Hill Connect Microbiology is a digital teaching and learning environment that saves students and instructors time while improving performance over a variety of critical outcomes.

- From in-site tutorials to tips and best practices, to live help from colleagues and specialists—you're never left alone to maximize Connect's potential.
- Instructors have access to a variety of resources including assignable and gradable interactive questions based on textbook images, case study activities, tutorial videos, and more.
- Digital images, PowerPoint slides, and instructor resources are also available through Connect.
- Digital Lecture Capture: Get Connected. Get McGraw-Hill Tegrity Campus™. Capture your lectures for students. Easy access outside of class anytime, anywhere, on just about any device.

Gather assessment information

Generate powerful data related to student performance against Learning Outcomes, specific topics, level of difficulty, and more.

Visit www.mcgrawhillconnect.com.

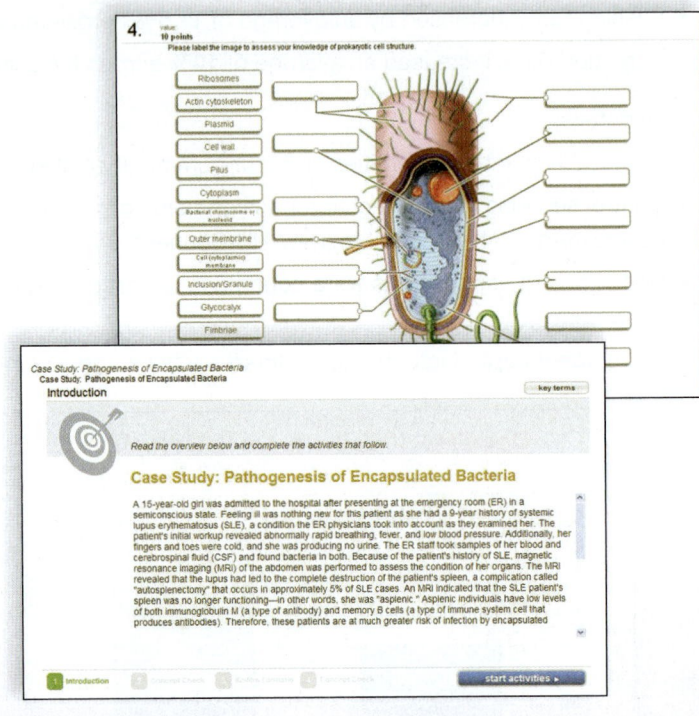

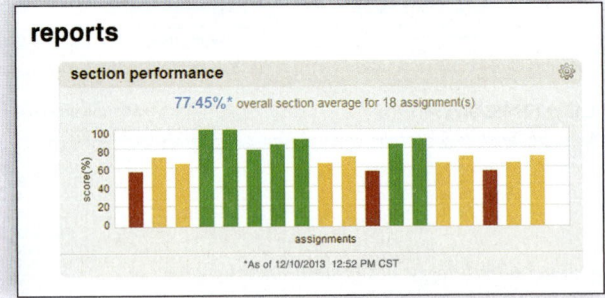

 Connect seamlessly integrates with every learning management system out there through McGraw-Hill Campus, so you can easily combine your course resources into a single platform. Instructors and students benefit from universal single sign-on, automatic registration, and gradebook synchronization.

Self-study resources are also available at **www.mhhe.com/cowan4e.**

Unique Interactive Question Types in Connect Tagged to ASM's Curriculum Guidelines for Undergraduate Microbiology

1 **Case Study:** Case studies come to life in a learning activity that is interactive, self-grading, and assessable. The integration of the cases with videos and animations adds depth to the content, and the use of integrated questions forces students to stop, think, and evaluate their understanding. Pre- and post-testing allow instructors and students to assess their overall comprehension of the activity.

2 **Concept Maps:** Concept maps allow students to manipulate terms in a hands-on manner in order to assess their understanding of chapter-wide topics. Students become actively engaged and are given immediate feedback, enhancing their understanding of important concepts within each chapter.

3 **What's the Diagnosis:** Specifically designed for the disease chapters of the text, this is an integrated learning experience designed to assess the student's ability to utilize information learned in the preceding chapters to successfully culture, identify, and treat a disease-causing microbe in a simulated patient scenario. This question type is true experiential learning and allows the students to think critically through a real-life clinical situation.

4 **Animations:** Animation quizzes pair our high-quality animations with questions designed to probe student understanding of the illustrated concepts.

5 **Tutorial Animation Learning Modules:** Making use of McGraw-Hill's collection of videos and animations, this question type presents an interactive, self-grading, and assessable activity. Pre- and post-testing is used to assess shifts in student comprehension. Integrated questions force students to stop, think, and evaluate their understanding of the process being presented. These tutorials take a stand-alone, static animation and turn it into an interactive learning experience for your students with real-time remediation.

6 **Labeling:** Using the high-quality art from the textbook, check your students' visual understanding as they practice interpreting figures and learning structures and relationships. Easily edit or remove any label you wish!

7 **Classification:** Ask students to organize concepts or structures into categories by placing them in the correct "bucket."

8 **Sequencing:** Challenge students to place the steps of a complex process in the correct order.

9 **Composition:** Fill in the blanks to practice vocabulary, and then reorder the sentences to form a logical paragraph (these exercises may qualify as "writing across the curriculum" activities!).

All McGraw-Hill ConnectPlus content is tagged to Learning Outcomes for each chapter as well as topic, section, Bloom's Level, and ASM Curriculum Guidelines to assist you in customizing assignments and in reporting on your students' performance against these points. This will enhance your ability to assess student learning in your courses by allowing you to align your learning activities to peer-reviewed standards from an international organization.

Presentation Tools Allow You to Customize Your Lectures

Enhanced Lecture Presentations contain lecture outlines, art, photos, tables, and animations embedded where appropriate. Fully customizable, but complete and ready to use, these presentations will enable you to spend less time preparing for lecture!

Animations Over 100 animations bring key concepts to life, available for instructors and students.

Animation PPTs Animations are truly embedded in PowerPoint® for ultimate ease of use! Just copy and paste into your custom slide show and you're done!

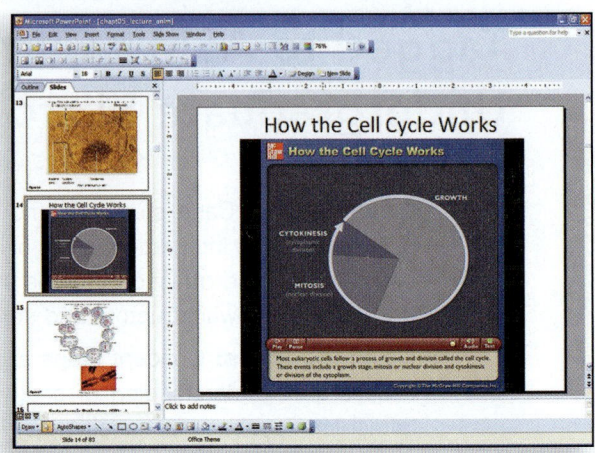

Take your course online—easily—with one-click Digital Lecture Capture

McGraw-Hill Tegrity Campus™ records and distributes your lectures with just a click of a button. Students can view them anytime/anywhere via computer, iPod, or mobile device. Tegrity Campus indexes as it records your slideshow presentations and anything shown on your computer so **students can use keywords to find exactly what they want to study.**

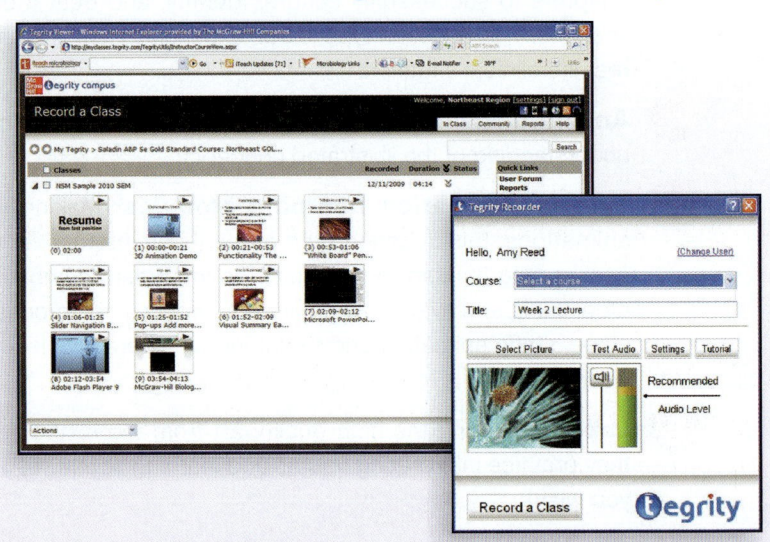

"This text is the complete package. It is well written and is supplemented with superior digital content."
— Nahel W. Awadallah, Johnston Community College

Be sure to visit Kelly's blog, www.microbiologymaven.com, where she and her guest bloggers tackle science and science teaching, as well as the occasional off-the-wall topic. If you subscribe (for free) you'll get emails once or twice a week with new entries: just enough to relieve stress and renew your sense of camaraderie with fellow instructors around the country.

Customize Your Course Materials to Your Learning Outcomes!

Create what you've only imagined.

Introducing **McGraw-Hill Create™**—a new, self-service website that allows you to create custom course materials—print and eBooks—by drawing upon McGraw-Hill's comprehensive, cross-disciplinary content. Add your own content quickly and easily. Tap into other rights-secured third-party sources as well. Then, arrange the content in a way that makes the most sense for your course. Even personalize your book with your course name and information! Choose the best format for your course: color print, black and white print, or eBook. The eBook is now even viewable on an iPad! And, when you are done you will receive a free PDF review copy in just minutes!

Finally, a way to quickly and easily create the course materials you've always wanted.

Imagine that.

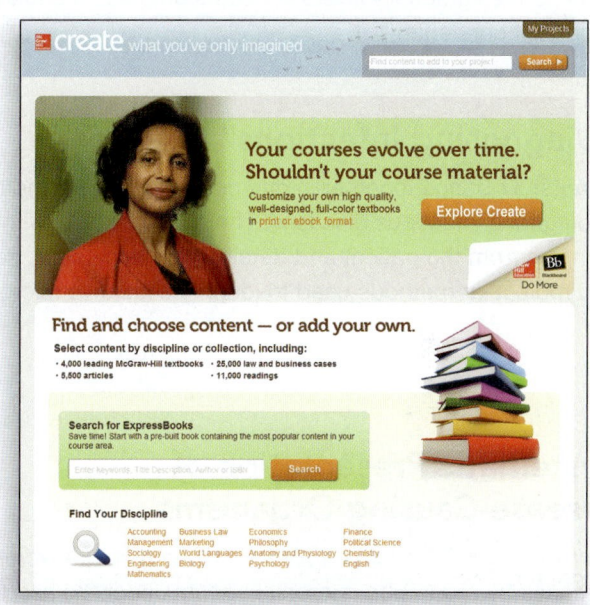

Visit McGraw-Hill Create—www.mcgrawhillcreate.com—today and begin building your perfect book.

Need a lab manual for your microbiology course? Customize any of these manuals—add your text material—and *Create* your perfect solution!

McGraw-Hill offers several lab manuals for the microbiology course. Contact your McGraw-Hill representative for packaging options with any of our lab manuals.

Brown/Smith: *Benson's Microbiological Applications: Laboratory Manual in General Microbiology,* 13th edition
Short Version (978-0-07-340241-3)
Complete Version (978-0-07-766802-0)

Chess: *Laboratory Applications in Microbiology: A Case Study Approach,* 3rd edition (978-0-07-340242-0)

Chess: *Photographic Atlas for Laboratory Applications in Microbiology* (978-0-07-737159-3)

Harley: *Laboratory Exercises in Microbiology,* 9th edition (978-0-07-751055-8)

Kleyn et al.: *Microbiology Experiments: A Health Science Perspective,* 7th edition (978-0-07-731554-2)

Morello: *Laboratory Manual and Workbook in Microbiology: Applications to Patient Care,* 11th edition (978-0-07-340239-0)

Connecting Students to Their Future Careers

Many students taking this course will be entering the health care field in some way, and it is absolutely critical that they have a good background in the biology of microorganisms. Author Kelly Cowan has made it her goal to help all students make the connections between microbiology and the world they see around them. Her textbooks have become known for their engaging writing style, instructional art program, and focus on active learning. The "building blocks" approach establishes the big picture first and then gradually layers concepts onto this foundation. This logical structure helps students build knowledge and *connect* important concepts.

"Diagnosing Infections" Chapter

Chapter 17 brings together in one place the current methods used to diagnose infectious diseases. The chapter starts with collecting samples from the patient and details the biochemical, serological, and molecular methods used to identify causative microbes.

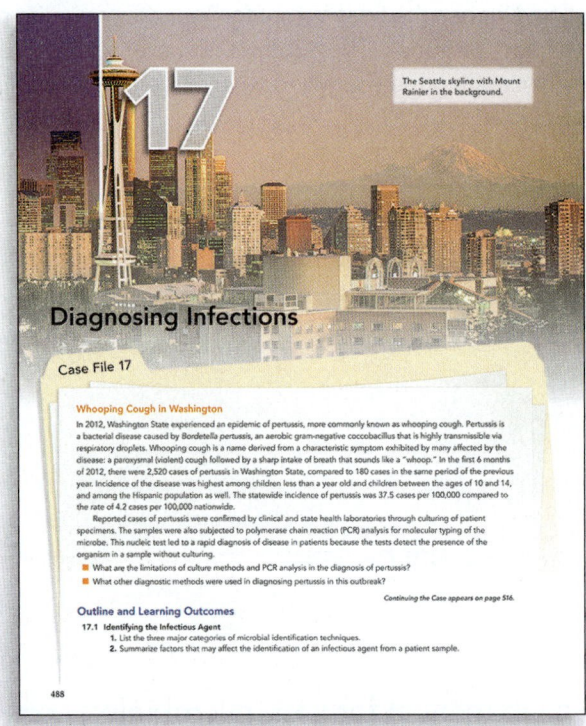

Systematic Presentation of Disease-Causing Organisms

Microbiology: A Systems Approach takes a unique approach to diseases by organizing microbial agents under the heading of the disease condition they cause. After all of them are covered the agents are summarized in a comparative table. Every condition gets a table, whether there is one possible cause or a dozen. Through this approach, students study how diseases affect patients—the way future health care professionals will encounter them in their jobs. A summary table follows the textual discussion of each disease and summarizes the characteristics of agents that can cause that disease. New to this edition: **Every disease table now contains national and worldwide epidemiological information for each causative agent.**

This approach is logical, systematic, and intuitive, as it encourages clinical and critical thinking in students—the type of thinking they will be using if their eventual careers are in health care. Students learn to examine multiple possibilities for a given condition and grow accustomed to looking for commonalities and differences among the various organisms that cause a given condition.

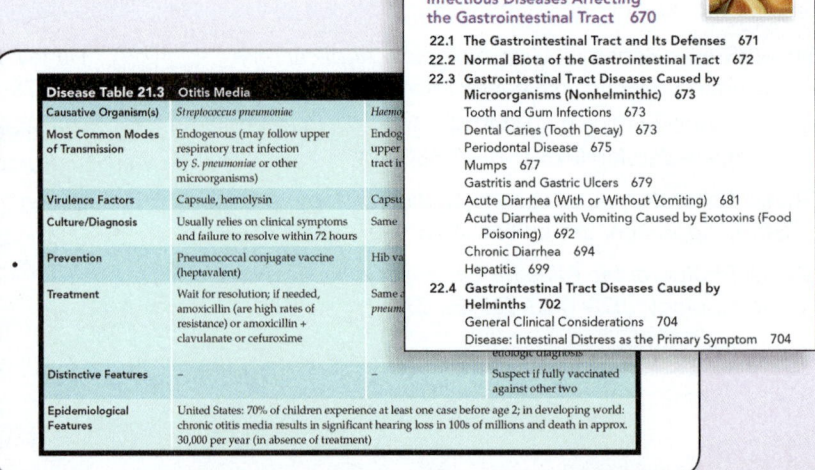

Chapter Opening Case Files!

Each chapter opens with a Case File, which helps students grasp the relevance of the material they're about to learn. The questions that directly follow the Case File challenge students to begin to think critically about what they are going to read, expecting that they'll be able to answer them once they've worked through the chapter. The Continuing the Case feature appears within the chapter where relevant, to help students follow the real-world application of the case. The Case File Wrap-Up summarizes the case at the end of the chapter, pulling together the applicable content and the chapter's topics. All of the case files are new in the fourth edition, including hot microbiological topics that are making news headlines today.

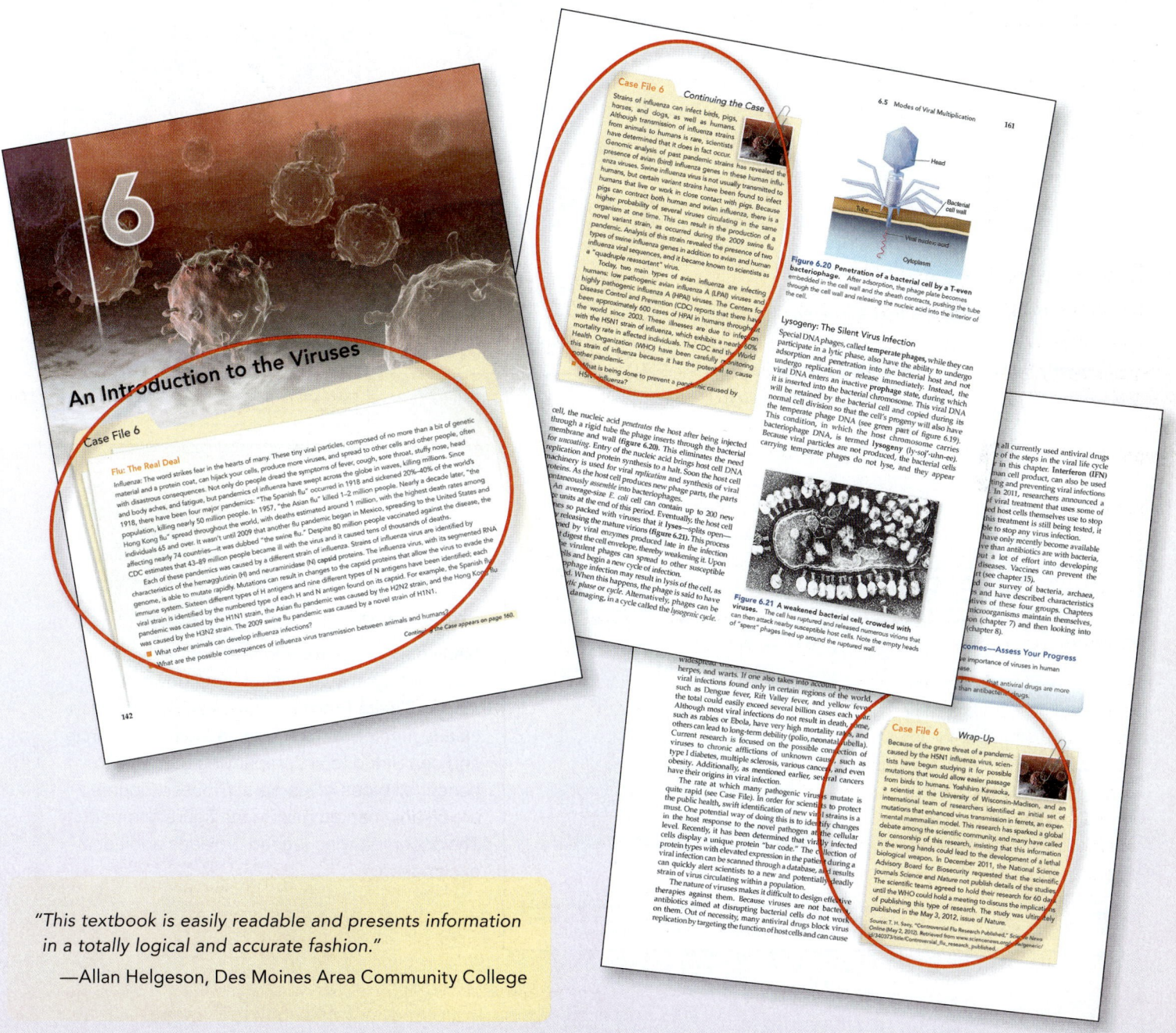

"This textbook is easily readable and presents information in a totally logical and accurate fashion."
—Allan Helgeson, Des Moines Area Community College

Connecting Students to the Content with Truly Instructional Art

Effective science illustrations not only look pretty, but help students visualize complex concepts and processes and paints a conceptual picture for them. The art combines vivid colors, multi-dimensionality, and self-contained narrative to help students study the challenging concepts of microbiology from a visual perspective. Drawings are often paired with photographs or micrographs to enhance comprehension.

"The readabililty makes this text a winner. Excellent text!"

—Kimberly Harding, Colorado Mountain College

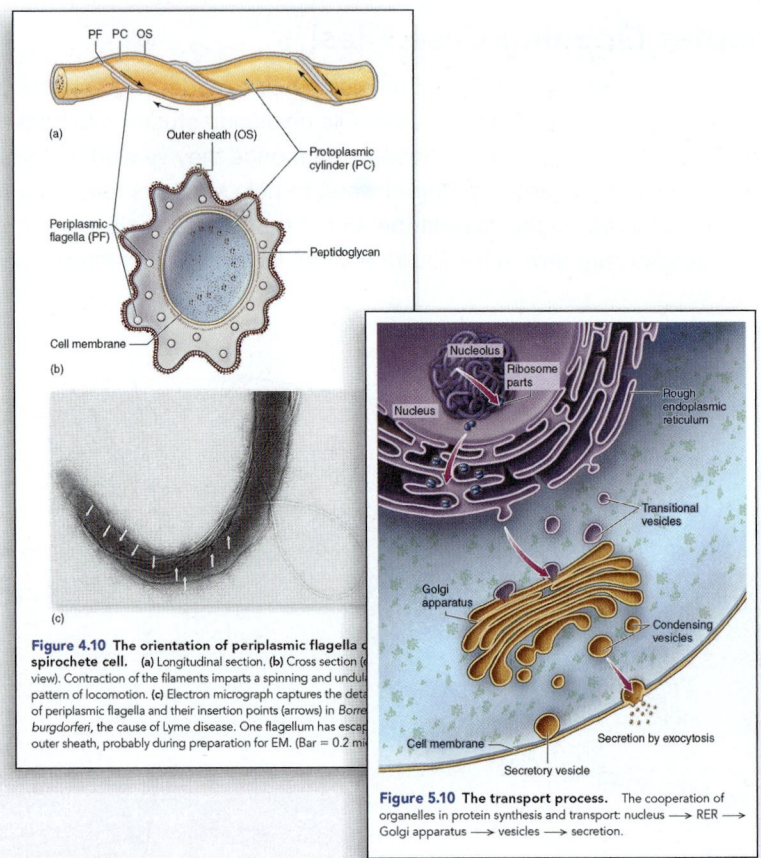

Figure 4.10 The orientation of periplasmic flagella of spirochete cell. **(a)** Longitudinal section. **(b)** Cross section (end view). Contraction of the filaments imparts a spinning and undula[...] pattern of locomotion. **(c)** Electron micrograph captures the deta[...] of periplasmic flagella and their insertion points (arrows) in *Borre[...] burgdorferi*, the cause of Lyme disease. One flagellum has escap[...] outer sheath, probably during preparation for EM. (Bar = 0.2 mi[...]

Figure 5.10 The transport process. The cooperation of organelles in protein synthesis and transport: nucleus ⟶ RER ⟶ Golgi apparatus ⟶ vesicles ⟶ secretion.

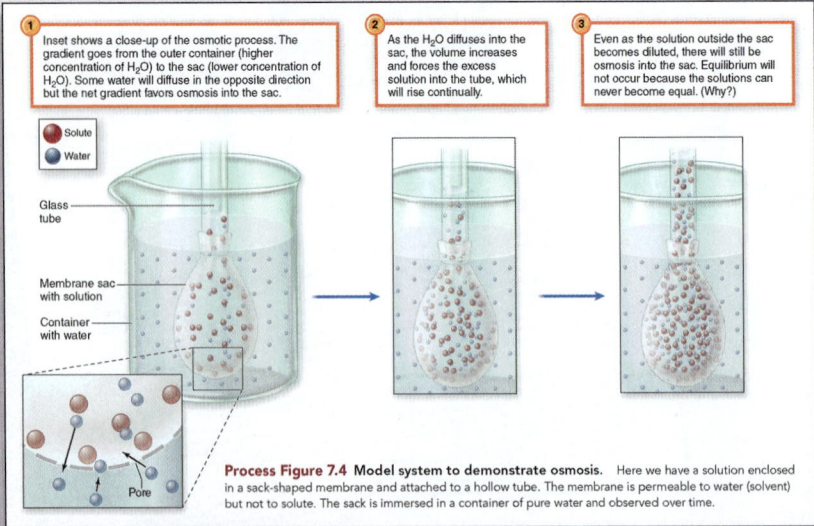

Process Figure 7.4 Model system to demonstrate osmosis. Here we have a solution enclosed in a sack-shaped membrane and attached to a hollow tube. The membrane is permeable to water (solvent) but not to solute. The sack is immersed in a container of pure water and observed over time.

Process Figures

Many difficult microbiological concepts are best portrayed by breaking them down into stages. These Process Figures show each step clearly marked with an orange, numbered circle and correlated to accompanying narrative to benefit all types of learners. Process Figures are clearly marked next to the figure number. The accompanying legend provides additional explanation.

Connecting Students to Microbiology with Relevant Examples

Real Clinical Photos Help Students Visualize Diseases

Clinical Photos

Color photos of individuals affected by disease provide students with a real life, clinical view of how microorganisms manifest themselves in the human body.

Combination Figures

Line drawings combined with photos give students two perspectives: the realism of photos and the explanatory clarity of illustrations. The authors chose this method of presentation often to help students comprehend difficult concepts.

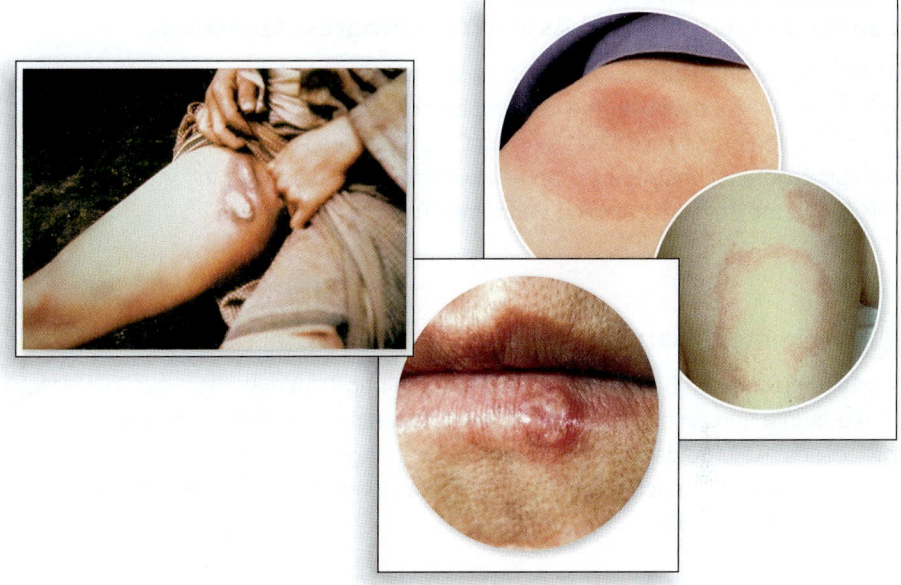

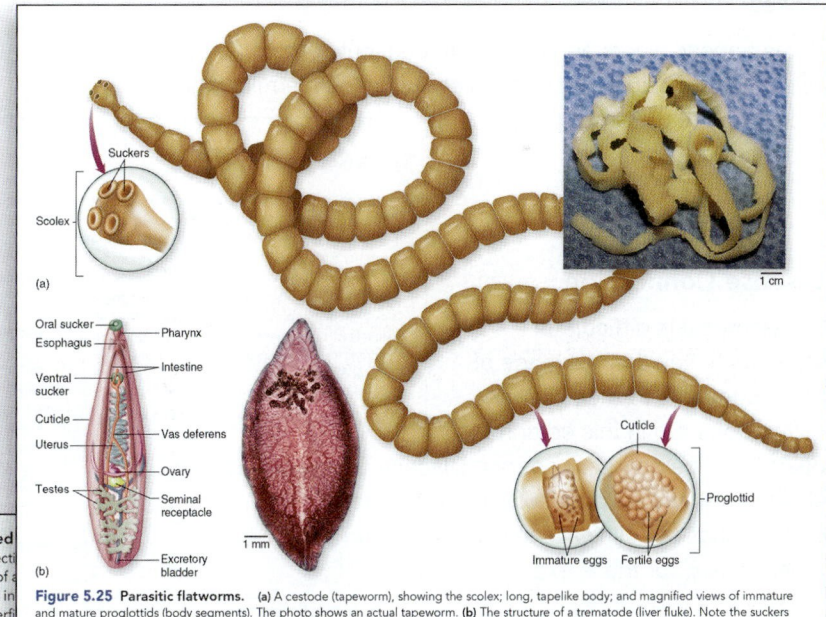

Figure 18.6 Staphylococcal scalded **(a)** Exfoliative toxin produced in local infecti outer layer of skin. **(b)** Photomicrograph of a epidermal shedding, or desquamation, is in because the level of separation is so superfi

Figure 5.25 Parasitic flatworms. **(a)** A cestode (tapeworm), showing the scolex; long, tapelike body; and magnified views of immature and mature proglottids (body segments). The photo shows an actual tapeworm. **(b)** The structure of a trematode (liver fluke). Note the suckers that attach to host tissue and the dominance of reproductive and digestive organs. The photo shows an actual liver fluke.

Connecting Students to Microbiology Through Student-Centered Pedagogy

Pedagogy Created to Promote Active Learning

Learning Outcomes and Assess Your Progress Questions

Every chapter in the book now opens with an outline—which is a list of Learning Outcomes. Assess Your Progress with the learning outcome questions conclude each major section of the text. The Learning Outcomes are tightly correlated to digital material. Instructors can easily measure student learning in relation to the specific Learning Outcomes used in their course.

Animated Learning Modules

Certain topics need help to come to life off the page. Animations, video, audio and text all combine to help students understand complex processes. Key topics have an Animated Learning Module assignable through Connect. An icon in the text indicates when these learning modules are available.

Notes

Notes appear, where appropriate, throughout the text. They give students helpful information about various terminologies, exceptions to the rule, or important clarifications.

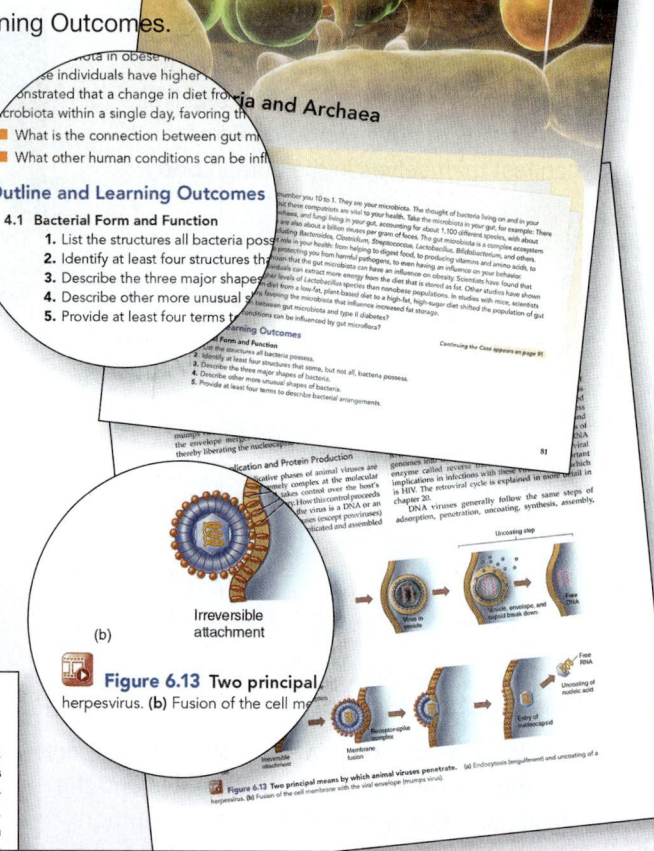

A Note on Terminology

The word *spore* can have more than one usage in microbiology. It is a generic term that refers to any tiny compact cell that is produced by vegetative or reproductive structures of microorganisms. Fungi have spores that serve as reproductive structures. The bacterial type discussed here is most accurately called an **endospore**, because it is produced inside a cell. Their function in survival, not in reproduction, because no increase is involved in their formation. In contrast, the fungi different types of spores for both survival and re chapter 5).

Disease Connection

Sometimes it is difficult for students to see the relevance of basic concepts to their chosen professions. So in this edition the basic science chapters contain Disease Connections, very short boxes that relate esoteric topics such as pH and growth phase to clinical situations (*H. pylori* and *M. tuberculosis,* for these examples).

Disease Connection

The fact that the poliovirus has tropisms for both neural and intestinal cells explains how it wreaks havoc on humans. Most people know that it causes paralysis; this is because it affects the neurons that make muscles work. But most people have no idea how you "catch" it. You catch it by ingesting water or food that is contaminated with the virus because it attaches to intestinal cells, and from there invades the nervous system. Polio is gone in the Western Hemisphere but still hangs on in three developing countries (as of 2013), despite the world health community's best efforts.

Tables

This edition contains numerous illustrated tables. Horizontal contrasting lines set off each entry, making it easy to read.

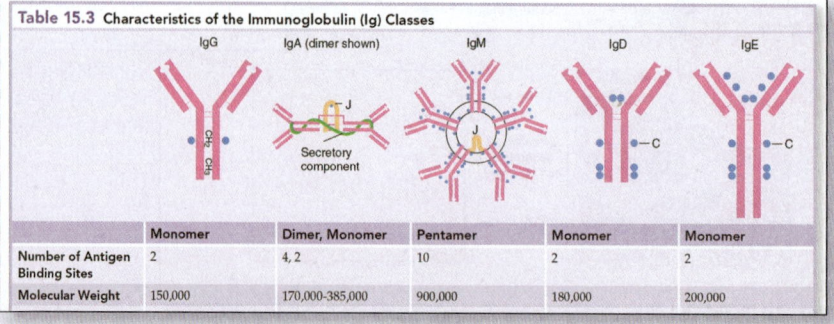

Table 15.3 Characteristics of the Immunoglobulin (Ig) Classes

	IgG	IgA (dimer shown)	IgM	IgD	IgE
	Monomer	Dimer, Monomer	Pentamer	Monomer	Monomer
Number of Antigen Binding Sites	2	4, 2	10	2	2
Molecular Weight	150,000	170,000–385,000	900,000	180,000	200,000

Insight Readings

Found throughout each chapter, current, real-world readings allow students to see an interesting application of the concepts they're studying.

System Summary Figures

"Glass body" figures at the end of each disease chapter highlight the affected organs and list the diseases that were presented in the chapter. In addition, the microbes are color coded by type of microorganism.

INFECTIOUS DISEASES AFFECTING
The Gastrointestinal Tract

Mumps
Mumps virus

Gastritis and Gastric Ulcer
Helicobacter pylori

Schistosomiasis
Schistosoma mansoni
Schistosoma japonicum

Acute Diarrhea
Salmonella
Shigella
E. coli O157:H7
Other E. coli
Campylobacter
Clostridium difficile
Vibrio cholerae
Cryptosporidium
Rotavirus
Norovirus

Chronic Diarrhea
EAEC
Cyclospora cayetanensis
Giardia lamblia
Entamoeba histolytica

Acute Diarrhea and/or Vomiting (Food Poisoning)
Staphylococcus aureus
Bacillus cereus
Clostridium perfringens

Helminthic Infections with Neurological and Muscular Symptoms
Trichinella spiralis

Dental Caries
Streptococcus mutans
Streptococcus sobrinus
Other bacteria

Periodontitis and Necrotizing Ulcerative Diseases
Tannerella forsythia
Aggregatibacter actinomycetemcomitans
Porphyromonas gingivalis
Treponema vincentii
Prevotella intermedia
Fusobacterium

Helminthic Infections with Intestinal and Migratory Symptoms
Ascaris lumbricoides
Necator americanus
Ancylostoma duodenale

Helminthic Infections with Liver and Intestinal Symptoms

Tract Infections Causing Intestinal Distress
Trichuris trichiura
Enterobius vermicularis
Taenia solium
Diphyllobothrium latum

Hepatitis
Hepatitis A or E
Hepatitis B or C

733

Taxonomic List of Organisms

A taxonomic list of organisms is presented at the end of each disease chapter so students can see the taxonomic position of microbes causing diseases in that body system.

> "I appreciate the organization in the way the topics are broken up so students can easily maintain their focus while reading. The Disease Tables, Insight Readings, and System Summary Figures are a great way for them to review and apply what they have learned."
>
> —Alicia D. Carley, Northwest Technical College

▶ Summing Up

Taxonomic Organization Microorganisms Causing Diseases in the Cardiovascular and Lymphatic System

Microorganism	Disease	Chapter Location
Gram-positive endospore-forming bacteria		
Bacillus anthracis	Anthrax	Anthrax, p. 622
Gram-positive bacteria		
Staphylococcus aureus	Acute endocarditis	Endocarditis, p. 611
Streptococcus pyogenes	Acute endocarditis	Endocarditis, p. 612
Streptococcus pneumoniae	Acute endocarditis	Endocarditis, p. 612
Gram-negative bacteria		
Yersinia pestis	Plague	Plague, p. 614
Francisella tularensis	Tularemia	Tularemia, p. 617
Borrelia burgdorferi	Lyme disease	Lyme disease, p. 618
Brucella abortus, B. suis	Brucellosis	Nonhemorrhagic fever diseases, p. 626
Coxiella burnetii	Q fever	Nonhemorrhagic fever diseases, p. 627
Bartonella henselae	Cat-scratch disease	Nonhemorrhagic fever diseases, p. 628
Bartonella quintana	Trench fever	Nonhemorrhagic fever diseases, p. 628
Ehrlichia chaffeensis, E. phagocytophila, E. ewingii	Ehrlichiosis	Nonhemorrhagic fever diseases, p. 629
Neisseria gonorrhoeae	Acute endocarditis	Endocarditis, p. 612
Rickettsia rickettsii	Rocky Mountain spotted fever	Nonhemorrhagic fever diseases, p. 629
DNA viruses		
Epstein-Barr virus	Infectious mononucleosis	Infectious mononucleosis, p. 621
RNA viruses		
Yellow fever viruses	Yellow fever	Hemorrhagic fevers, p. 624
Dengue fever viruses	Dengue fever	Hemorrhagic fevers, p. 624
Ebola and Marburg viruses	Ebola and Marburg hemorrhagic fevers	Hemorrhagic fevers, p. 625
Lassa fever virus	Lassa fever	Hemorrhagic fevers, p. 625
Chikungunya virus	Hemorrhagic fever	Hemorrhagic fevers, p. 624
Retroviruses		
Human immunodeficiency virus 1 and 2	HIV infection and AIDS	HIV infection and AIDS, p. 636
Human T-cell lymphotropic virus I	Adult T-cell leukemia	Leukemias, p. 637
Protozoa		
Plasmodium falciparum, P. vivax, P. ovale, P. malariae	Malaria	Malaria, p. 632
Trypanosoma cruzi	Chagas disease	Chagas disease, p. 630

Connecting Learning to Bloom's Taxonomy

The end-of-chapter material is linked to Bloom's Taxonomy. It has been carefully planned to promote active learning and provide review for different learning styles and levels of difficulty. Multiple-Choice and True-False Questions (Remember and Understand) precede the Critical Thinking, Concept Connections, Visual Connections Questions and Concept Mapping Exercises, which take the student through the Apply, Analyze, Evaluate, and Create levels. The consistent layout of each chapter allows students to develop a learning strategy and gain confidence in their ability to master the concepts, leading to success in the class!

Chapter Summary

A brief outline of the main chapter concepts is provided for students with important terms highlighted. Key terms are also included in the glossary at the end of the book. The chapter summary is now tagged with new American Society for Microbiology curricular guidelines.

Multiple Choice and True-False Questions

Students can assess their knowledge of basic concepts by answering these questions. Other types of questions and activities that follow build on this foundational knowledge. The ConnectPlus eBook allows students to quiz themselves interactively using these questions! Bloom's Levels for all questions are provided.

Critical Thinking Questions

Students use higher-order Bloom's skills (Apply, Analyze, Evaluate) with these questions. There is no single correct answer; this can open doors to discussion and application. New critical thinking questions have been added for the fourth edition.

Chapter Summary

6.1 The Search for the Elusive Viruses (ASM Guideline* 2.2)
- Viruses are noncellular entities whose properties have been identified through microscopy, tissue culture, and molecular biology.

6.2 The Position of Viruses in the Biological Spectrum (ASM Guidelines 1.5, 3.3, 4.4, 5.4)
- Viruses are infectious particles that invade every known type of cell. They are not alive, yet they are able to redirect the metabolism of living cells to reproduce virus particles.
- Viruses have a profound influence on the genetic makeup of the biosphere.
- Viral replication inside a cell usually causes death or loss of function of that cell.

6.3 The General Structure of Viruses (ASM Guidelines 2.3, 2.4, 4.4)
- Virus size range is from 20 nm to 1000 nm (diameter). Viruses are composed of an outer protein capsid containing either DNA or RNA plus a

Mimivirus

- Animal viruses can cause acute infections or can persist in host tissues as chronic latent infections that can reactivate periodically throughout the host's life. Some persistent animal viruses are oncogenic.
- Bacteriophages vary significantly from animal viruses in their methods of adsorption, penetration, site of replication, and method of exit from host cells.
- Lysogeny is a condition in which viral DNA is inserted into the bacterial chromosome and remains inactive for an extended period. It is replicated right along with the chromosome every time the bacterium divides.
- Some bacteria express virulence traits that are coded for by the bacteriophage DNA in their chromosomes. This phenomenon is called *lysogenic conversion*.

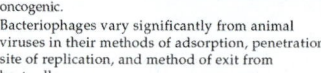

Multiple-Choice and True-False Questions | Bloom's Levels 1 and 2: Remember and Understand

Multiple-Choice Questions. Select the correct answer from the options provided.

1. A virus is a tiny infectious
   a. cell.   c. particle.
   b. living thing.   d. nucleic acid.

2. Viruses are known to infect
   a. plants.   c. fungi.
   b. bacteria. d. all organisms.

3. The nucleic acid of a virus is
   a. DNA only.   c. both DNA and RNA.
   b. RNA only.   d. either DNA or RNA.

4. The general steps in a viral multiplication cycle are
   a. adsorption, penetration, synthesis, assembly, and release.
   b. endocytosis, uncoating, replication, assembly, and budding.
   c. adsorption, uncoating, duplication, assembly, and lysis.
   d. endocytosis, penetration, replication, maturation, and exocytosis.

5. A prophage is a stage in the development of a/an
   a. bacterial virus.   c. lytic virus.
   b. poxvirus.   d. enveloped virus.

6. In general, RNA viruses multiply in the cell _____, and DNA viruses multiply in the cell _____.
   a. nucleus, cytoplasm   c. vesicles, ribosomes
   b. cytoplasm, nucleus   d. endoplasmic reticulum

7. Viruses
   a. tissu

9. Label the parts of this virus. Identify the capsid, nucleic acid, and other features of this virus.

10. Circle the viral infections from this list: cholera, rabies, plague, cold sores, whooping cough, tetanus, genital warts, gonorrhea, mumps, Rocky Mountain spotted fever, syphilis, rubella.

Critical Thinking Questions | Bloom's Levels 3, 4, and 5: Apply, Analyze, and Evaluate

Critical thinking is the ability to reason and solve problems using facts and concepts. These questions can be approached from a number of angles and, in most cases, they do not have a single correct answer.

1. Provide evidence in support of or refuting the following statement: Viruses are simple cellular agents of disease.

2. Summarize the unique properties of viruses and explain which of these characteristics allow them to function as "parasites."

3. a. Sketch the basic structure of both a nonenveloped and an enveloped virus, labeling all parts.
   b. Discuss the validity of the following statement: The viral capsid and envelope only provide functions that enhance the pathogenicity of a virus.

4. a. You identify a novel microbe in your laboratory and find that it possesses two types of nucleic acid. Explain why you immediately rule out the fact that this microbe is a virus.
   b. Describe the nucleic acid configuration of a positive-sense RNA virus and explain why its multiplication cycle is less complex than that of a retrovirus.

5. Define the term *tropism*, and provide at least one example illustrating how viral structure determines this property of a virus.

6. a. Provide one example of an oncogenic virus and explain the unique properties of its multiplication cycle that allow it to trigger the development of cancer.
   b. Compare and contrast the processes of latency and lysogeny, providing examples of latent viruses and lysogenic viruses.

7. Summarize the method used by most companies to manufacture influenza vaccine today, providing one clear advantage and one disadvantage of this process.

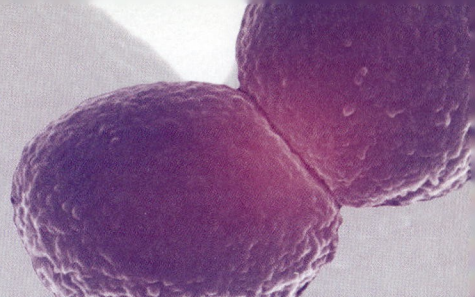

Concept Connections

A new feature that ties together topics in a visual manner, and calls on students' ability to Analyze and Create while connecting material from the chapter.

Visual Connections

Visual Connections questions take images and concepts learned in previous chapters and ask students to apply that knowledge to concepts newly learned in the current chapter. This helps students Evaluate information in new contexts and enhances learning.

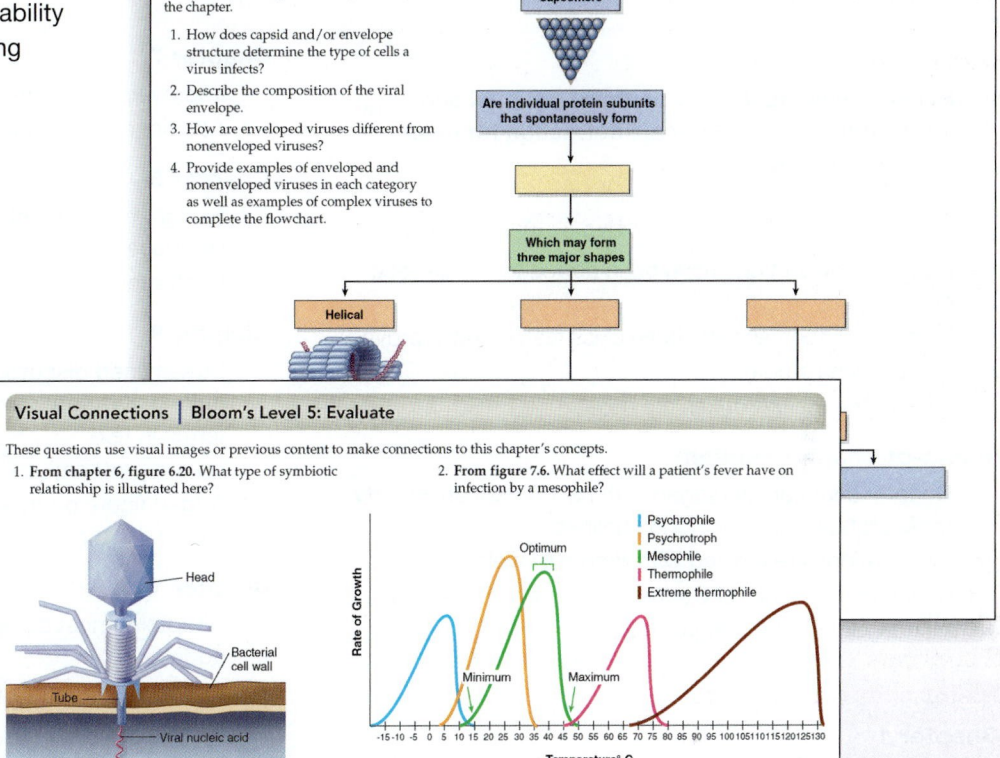

Concept Mapping

Every chapter contains a list of terms from which students are asked to construct (Create) a concept map. ConnectPlus expands this activity with interactive concept maps.

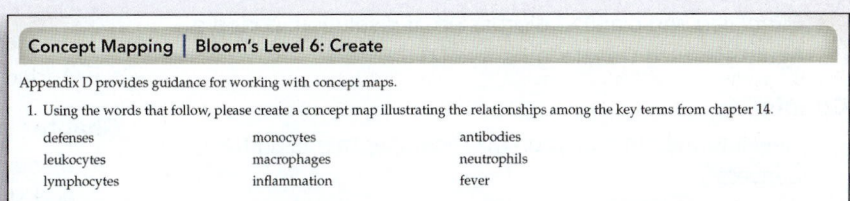

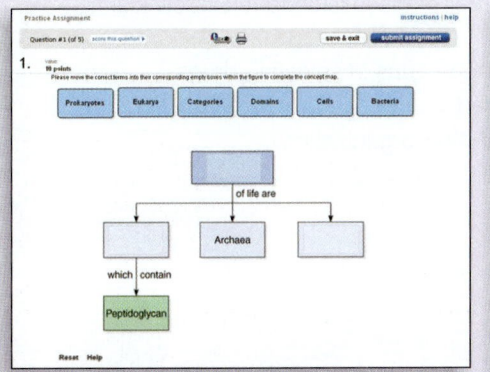

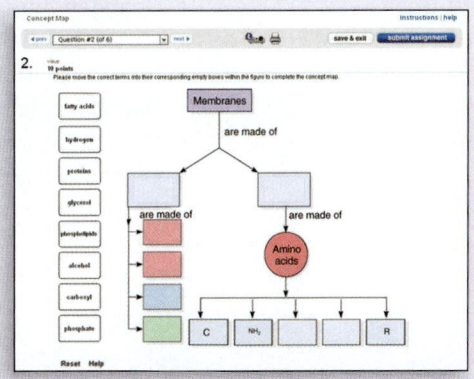

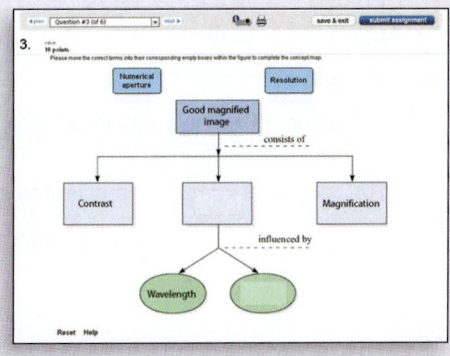

Changes to the Fourth Edition

New to Microbiology, A Systems Approach

Brand new

- **Every disease table now contains national and worldwide epidemiological information for each causative agent**

Global changes throughout the fourth edition

- Disease Connections have been added to nondisease chapters
- Learning Outcomes have been class tested and improved
- All new case studies
- 75% of the Insight boxes are new

In end-of-chapter section:

- Chapter summary is tagged with new American Society for Microbiology curricular guidelines
- All questions are labeled with Bloom's levels
- New feature: Concept Connections in each chapter
- All new Critical Thinking Questions

Major chapter changes

Chapter 1

- Revised discussion of history of cellular life on earth and the three domains

Chapter 3

- Simplified and clarified discussion of resolution; added a figure showing wavelengths

Chapter 4

- New information added on microcompartments and S layers

Chapter 5

- Updated protist classification
- Added O & P testing

Chapter 6

- Discussion of the new proposed viral domain
- Virus phage introduced
- DRACO broad-spectrum antiviral treatment described

Chapter 7

- Improved presentation of molecular transport
- Additional information on biofilms

Chapter 8

- Explanations of metabolic processes written in simpler language
- Illustrations greatly improved

Chapter 9

- Streamlined discussions of replication and translation by putting text right next to visuals and highlighting important terms in text
- Added proteomics
- Added figure on transformation so that there are three figures for three processes of horizontal gene transfer

Chapter 10

- Chapter almost completely new! Topics rewritten/updated/added
- Cloning, synthetic biology, miRNA strategies, sequencing and proteomics
- Many new figures

Chapter 11

- New figures and tables to make content more manageable
- Description of critical, semicritical, and noncritical medical devices
- Added discussion of disinfecting biofilms

Chapter 12

- Discussion of how new drugs may target host cell factors and still be selectively toxic
- More discussion of treating biofilm bacteria
- Changed the order of discussion to reflect clinical sequence
- Role of smartphone apps in selecting drugs
- New drugs added
- New tables for better organization
- Fecal therapy described

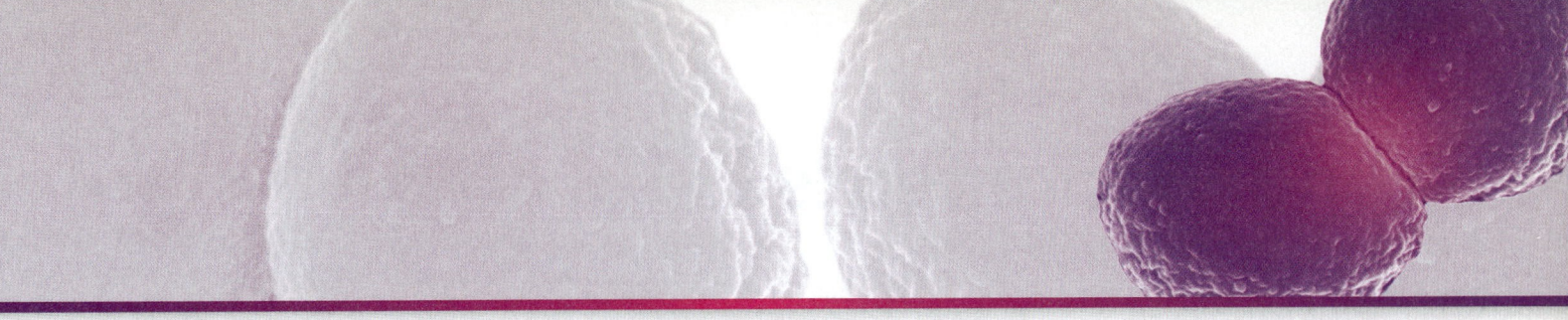

Chapter 13

- All new human microbiome section added
- Lots of information on gut microbiome
- New figure and information on newborn colonization
- More information of quorum sensing
- Added "the built environment" to reservoirs
- Updated section on healthcare-associated infection
- Added molecular Koch's postulates
- Added use of technology and social media in disease tracking
- Improved section on emerging diseases

Chapter 14

- Discussion of microbiome's role as first line of defense
- New research on platelets being involved in immunity
- Added information on collectins

Chapter 15

- Changed order of presentation: T cells first
- Much updated art
- Added information on CD80/CD28
- More emphasis on adult vaccines

Chapter 16

- More emphasis on hygiene hypothesis
- New research on autoimmunity

Chapter 17

- New section: "Breakthrough Methodologies" (deep sequencing, imaging, etc.)

Chapter 18

- MRSA, VRSA updated
- Vaccine information (i.e., MMRV) updated
- Leishmaniasis' creep into the United States discussed

Chapter 19

- Information on gut-brain axis added

Chapter 20

- Chagas disease added

Chapter 21

- Normal biota radically updated due to Human Microbiome Project
- Whooping cough epidemic addressed
- Much new information on both influenza and TB
- Causes of community-acquired pneumonia ranked as to frequency

Chapter 22

- Added information on 2010 mumps outbreak
- New information on non-O157:H7 STECs
- Added figure on most common causes of food-borne disease

Chapter 23

- Normal biota radically updated due to Human Microbiome Project
- Added information about head and neck cancers in males from HPV
- Added a figure summarizing incidence of *all* STIs

Chapter 24

- More bioremediation information
- More information on extreme environments
- Named *Prochlorococcus* as responsible for massive amounts of photosynthesis
- Two new figures: distribution of water on earth's surface and CO_2 levels over time

Chapter 25

- More emphasis on the transition from early biotech to genetically engineered organisms
- More detail about how coliform tests are not optimal
- More detail on HACCP, and new information about the Food Safety Modernization Act
- Updates on biofuels

Acknowledgments

I am most grateful to my students who have tried to teach me how to more effectively communicate this subject. All the professors (listed below) who reviewed manuscript for me were my close allies as well, especially when they were liberal in their criticism. Thanks to my co-author Jen Herzog for her great work on the digital side of things, and to Andrea Rediske for her writing assistance. Jill Kolodsick provided detailed reviews and I sincerely appreciated this feedback. My minders at McGraw-Hill are paragons of patience and professionalism: Darlene Schueller, Amy Reed, and Sherry Kane especially. Jeanne Patterson is the best copy editor west of the Mississippi. (Are you east of the Mississippi? Well, in the vicinity of the Mississippi....) Lastly, thanks to the thick-and-thin crew, my family: Taylor, Sam, Suzanne, Aaron, and Ted.

—Kelly Cowan

Reviewers

Abiodun Adibi, *Hampton University*

Cynthia Alonzo, *Community Colleges of Colorado*

Nahel W. Awadallah, *Johnston Community College*

John Bacheller, *Hillsborough Community College*

Michelle L. Badon, *University of Texas at Arlington*

Farah Bennani, *Front Range Community College—Westminster Campus*

Dena Berg, *Tarrant County College, NW*

Jennifer Bess, *Hillsborough Community College*

Cliff Boucher, *Tyler Junior College*

David Brady, *Southwestern Community College*

Chantae Calhoun, *Lawson State Community College*

Alicia D. Carley, *Northwest Technical College*

Sharron Crane, *Rutgers University*

Smruti Desai, *LoneStar College—CyFair*

Nichol Dolby, *Amarillo College*

Gillian Edwards, *University of California—Berkeley Extension (UNEX)*

Melissa Elliott, *Butler Community College*

Tracey Emmons, *Sandhills Community College*

Clifton Franklund, *Ferris State University*

Brinda Govindan, *San Francisco State University*

Kimberly Harding, *Colorado Mountain College*

Allan Helgeson, *Des Moines Area Community College*

Vida Irani, *Indiana University of Pennsylvania*

Sergei Markov, *Austin Peay State University*

Fernando P. Monroy, *Northern Arizona University*

Bethanye Branch Morgan, *Tarrant County College—Southeast*

Rita Moyes, *Texas A&M University*

Marcia Pierce, *Eastern Kentucky University*

Ines Rauschenbach, *Rutgers University* and *Union County College*

Luis A. Rodriguez, *San Antonio College*

Gene Scalarone, *Idaho State University*

Melissa Schreiber, *Valencia Community College*

Steven Scott, *Merritt College*

Jacqueline Spencer, *Thomas Nelson Community College*

Stephen Wagner, *Stephen F. Austin State University*

Holly Walters, *Cape Fear Community College*

Digital Reviewers

Cindy B. Anderson, *Mt. San Antonio College*

Jennifer Bess, *Hillsborough Community College*

Clifton Franklund, *Ferris State University*

Judy Haber, *California State University Fresno*

Ingrid Herrmann, *Santa Fe College*

Suzanne Long, *Monroe Community College*

Marty Lowe, *Bergen Community College*

Amee Mehta, *Seminole State College of Florida*

Amy Miller, *University of Cincinnati—Raymond Walters College*

Rita Moyes, *Texas A&M University*

Julie Oliver, *Cosumnes River College*

Jaime Parman-Ryans, *Walters State Community College*

Luis A. Rodriguez, *San Antonio College*

John R. Stevenson, *Miami University*

Janice Webster, *Ivy Tech Community College*

Van Wheat, *South Texas College*

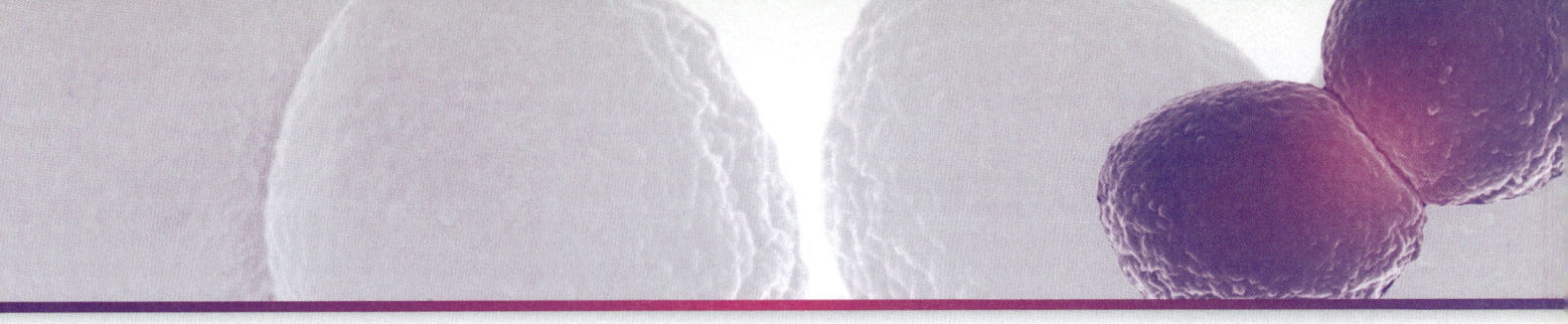

Symposium Participants

Cindy B. Anderson, *Mt. San Antonio College*

John Bacheller, *Hillsborough Community College*

Michelle L. Badon, *University of Texas at Arlington*

David Battigelli, *University of North Carolina—Greensboro*

Dena Berg, *Tarrant County College, NW*

Carroll Weaver Bottoms, *Collin College*

Nancy Boury, *Iowa State University*

Lance D. Bowen, *Truckee Meadows Community College*

William L. Boyko, *Sinclair Community College*

Toni Brem, *Wayne County Community College District—Northwest Campus*

Chad Brooks, *Austin Peay State University*

Linda D. Bruslind, *Oregon State University*

Lisa Burgess, *Broward College*

Elizabeth A. Carrington, *Tarrant County College District*

Joseph P. Caruso, *Florida Atlantic University*

Erin A. Christensen, *Middlesex County College*

James K. Collins, *University of Arizona*

David Daniel, *Weatherford College*

Elizabeth Emmert, *Salisbury University*

Susan Finazzo, *Georgia Perimeter College*

Teresa Fischer, *Indian River State College*

Carey Fox, *Brookdale Community College*

Clifton Franklund, *Ferris State University*

Jason Furrer, *University of Missouri*

Chris Gan, *Highline Community College*

Edwin Gines-Candelaria, *Miami Dade College*

Zaida M. Gomez-Kramer, *University of Central Arkansas*

Amy Goode, *Illinois Central College*

Todd Gordon, *Kansas City Kansas Community College*

Brinda Govindan, *San Francisco State University*

Julianne Grose, *Brigham Young University*

Gabriel E. Guzman, *Triton College*

Judy Haber, *California State University Fresno*

James B. Herrick, *James Madison University*

Dawn Janich, *Community College of Philadelphia*

James E. Johnson, *Central Washington University*

Kim Jones, *Suffolk County Community College—Ammerman Campus*

Eunice Kamunge, *Essex County College*

Angelo Kolokithas, *Northeast Wisconsin Technical College*

Terri J. Lindsey, *Tarrant County College District South*

Suzanne Long, *Monroe Community College*

Caroline H. McNutt, *Schoolcraft College*

Elizabeth F. McPherson, *The University of Tennessee*

Amee Mehta, *Seminole State College of Florida*

Sharon Miles, *Itawamba Community College*

Tracey Mills, *University of Indianapolis*

Pamela Monaco, *Molloy College*

Steven Moore, *Harding University*

Bethanye Branch Morgan, *Tarrant County College—Southeast*

Rita B. Moyes, *Texas A&M University*

Ruth A. Negley, *Harrisburg Area Community College—Gettysburg Campus*

Steven D. Obenauf, *Broward College—Central Campus*

Julie A. Oliver, *Cosumnes River College*

Janis Pace, *Navarro College, Midlothian Campus*

Todd P. Primm, *Sam Houston State University*

Jean Revie, *South Mountain Community College*

Jackie Reynolds, *Richland College*

Beverly A. Roe, *Erie Community College—South Campus*

Silvia Rossbach, *Western Michigan University*

Ben Rowley, *University of Central Arkansas*

Donald L. Rubbelke, *Lakeland Community College*

Mark A. Schneegurt, *Wichita State University*

Teri Shors, *University of Wisconsin Oshkosh*

Sasha A. Showsh, *University of Wisconsin—Eau Claire*

Heidi Smith, *Front Range Community College*

Sherry Stewart, *Navarro College*

Debby Sutton, *Mountain View College*

Steven J. Thurlow, *Jackson College*

Sanjay Tiwary, *Hinds Community College*

Stephen Wagner, *Stephen F. Austin State University*

Delon Washo-Krupps, *Arizona State University*

George Wawrzyniak, *Milwaukee Area Technical College*

Janice Webster, *Ivy Tech Community College*

Jim White, *Prairie State College*

John Whitlock, *Hillsborough Community College*

Table of Contents

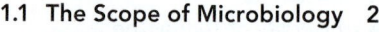
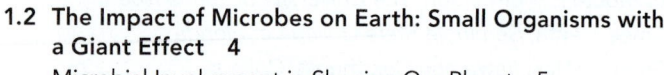
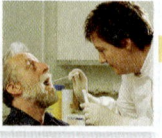

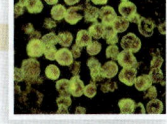

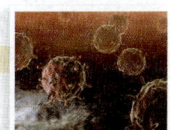

CHAPTER **7**

Microbial Nutrition, Ecology, and Growth   171

CHAPTER **8**

Microbial Metabolism: The Chemical Crossroads of Life   201

CHAPTER **9**

Microbial Genetics   233

CHAPTER **10**

Genetic Engineering and Recombinant DNA 268

CHAPTER **11**

Physical and Chemical Control of Microbes 292

CHAPTER **12**

Drugs, Microbes, Host— The Elements of Chemotherapy 323

CHAPTER **13**

Microbe-Human Interactions: Infection and Disease 355

CHAPTER **14**

Host Defenses I: Overview and Nonspecific Defenses 393

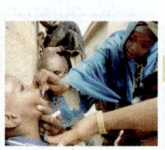

The Main Themes of Microbiology

Case File 1

It's Raining Bacteria

Bacteria are ubiquitous on the planet, but how profound is their impact on our lives? In addition to their impact on the earth's temperature, weathering, mineral extraction, and soil formation, recent studies have shown that bacteria have a major influence over another aspect of the earth's ecosystem: the weather. For years, scientists have believed that dust particles or minerals in clouds caused water droplets to coalesce into larger droplets and form rain, snow, or hail. However, recent research shows that bacteria are the predominant particles that induce the formation of precipitation.

After a hailstorm hit the Montana State University campus in Bozeman, Montana, Alexander Michaud and his collaborators gathered hailstones larger than 5 cm in diameter, separated them into four layers, and analyzed them as they melted. They were surprised to find that *Pseudomonas syringae*, a species of bacteria that is commonly implicated in infections of plants and as the cause of postharvest rots, grew from the water in the hailstones.

Michaud explains that bacteria found in the embryo—the first part of the hailstone to develop—initiate the growth of a hailstone. "In order for precipitation to occur, a nucleating particle must be present to allow for aggregation of water molecules," he states. "There is growing evidence that these nuclei can be bacteria or other biological particles."

- Why do you think that climate scientists never realized that microbes actually caused nucleation of water droplets in clouds?
- How does *P. syringae* make it rain at warm temperatures?

Continuing the Case appears on page 14.

Outline and Learning Outcomes

1.1 The Scope of Microbiology
   1. List the various types of microorganisms.
   2. Identify multiple professions using microbiology.

1.2 The Impact of Microbes on Earth: Small Organisms with a Giant Effect
   3. Describe the role and impact of microbes on the earth.
   4. Explain the theory of evolution and why it is called a theory.

1

1.3 Human Use of Microorganisms
   5. Explain one old way and one new way that humans manipulate organisms for their own uses.

1.4 Infectious Diseases and the Human Condition
   6. Summarize the relative burden of human disease caused by microbes, emphasizing the differences between developed countries and developing countries.

1.5 The General Characteristics of Microorganisms
   7. Differentiate among bacteria, archaea, and eukaryotic microorganisms.
   8. Identify a fourth type of microorganism.
   9. Compare and contrast the relative sizes of the different microbes.

1.6 The Historical Foundations of Microbiology
   10. Make a time line of the development of microbiology from the 1600s to today.
   11. List some recent microbiological discoveries of great impact.
   12. Explain what is important about the scientific method.

1.7 Naming, Classifying, and Identifying Microorganisms
   13. Differentiate among the terms *nomenclature*, *taxonomy*, and *classification*.
   14. Create a mnemonic device for remembering the taxonomic categories.
   15. Correctly write the binomial name for a microorganism.
   16. Draw a diagram of the three major domains.
   17. Explain the difference between traditional and molecular approaches to taxonomy.

1.1 The Scope of Microbiology

Microbiology is a specialized area of biology that deals with living things ordinarily too small to be seen without magnification. Such **microscopic** organisms are collectively referred to as **microorganisms** (my"-kroh-or'-gun-izms), **microbes,** or several other terms depending on the kind of microbe or the purpose. In the context of infection and disease, some people call them germs, viruses, or agents; others even call them "bugs"; but none of these terms are clear. In addition, some of these terms place undue emphasis on the disagreeable reputation of microorganisms. But, as we will learn throughout the course of this book, only a small minority of microorganisms are implicated in causing harm to other living beings. There are several major groups of microorganisms that we'll be studying. They are **bacteria, algae, protozoa, helminths** (parasitic invertebrate animals such as worms), and **fungi.** All of these microbes—just like plants and animals—can be infected by **viruses,** which are noncellular, **parasitic,** protein-coated genetic elements, dependent on their infected host. They can cause harm to the host they infect. Their evolutionary history and impact are intimately connected with the evolution of microbes and with all living organisms, including humans. As we will see in subsequent chapters, each group of microbes exhibits a distinct collection of biological characteristics.

The nature of microorganisms makes them both very easy and very difficult to study—easy because they reproduce so rapidly and we can quickly grow large populations in the laboratory and difficult because we usually can't see them directly. We rely on a variety of indirect means of analyzing them in addition to using microscopes.

Microbiologists study every aspect of microbes—their cell structure and function, their growth and physiology, their genetics, their taxonomy and evolutionary history, and their interactions with the living and nonliving environment. The latter includes their uses in industry and agriculture and the way they interact with mammalian hosts, in particular, their properties that may cause disease or lead to benefits.

Some descriptions of different branches of study appear in **table 1.1.** Studies in microbiology have led to greater understanding of many general biological principles. For example, the study of microorganisms established universal concepts concerning the chemistry of life (see chapters 2 and 8); systems of inheritance (see chapter 9); and the global cycles of nutrients, minerals, and gases (see chapter 24).

1.1 Learning Outcomes—Assess Your Progress

   1. List the various types of microorganisms.
   2. Identify multiple professions using microbiology.

Table 1.1 Microbiology—A Sampler

A. Medical Microbiology

This branch deals with microbes that cause diseases in humans and animals. Researchers examine factors that make the microbes virulent and mechanisms for inhibiting them.

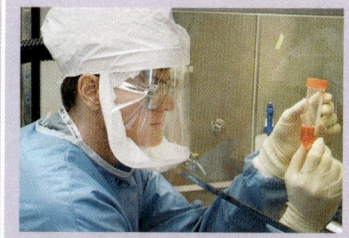

Figure A. A staff microbiologist at the Centers for Disease Control and Prevention (CDC) examines a culture of influenza virus identical to one that circulated in 1918. The lab is researching why this form of the virus was so deadly and how to develop vaccines and other treatments. Handling such deadly pathogens requires a high level of protection with special headgear and hoods.

B. Public Health Microbiology and Epidemiology

These branches monitor and control the spread of diseases in communities. Institutions involved in this concern are the U.S. Public Health Service (USPHS) with its main agency, the Centers for Disease Control and Prevention (CDC) located in Atlanta, Georgia, and the World Health Organization (WHO), the medical limb of the United Nations.

Figure B. Epidemiologists from the CDC employ an unusual method for microbial sampling. They are collecting grass clippings to find the source of an outbreak of tularemia in Massachusetts.

C. Immunology

This branch studies the complex web of protective substances and cells produced in response to infection. It includes such diverse areas as vaccination, blood testing, and allergy (see chapters 15, 16, and 17).

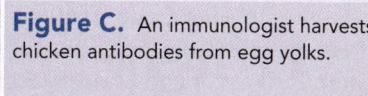

Figure C. An immunologist harvests chicken antibodies from egg yolks.

D. Industrial Microbiology

This branch safeguards our food and water, and also includes biotechnology, the use of microbial metabolism to arrive at a desired product, ranging from bread making to gene therapy. Microbes can be used to create large quantities of substances such as amino acids, beer, drugs, enzymes, and vitamins.

Figure D. Food inspectors sample a beef carcass for potential infectious agents. The safety of the food supply has wide-ranging importance.

E. Agricultural Microbiology

This branch is concerned with the relationships between microbes and domesticated plants and animals.

Plant specialists focus on plant diseases, soil fertility, and nutritional interactions.

Animal specialists work with infectious diseases and other associations animals have with microorganisms.

Figure E. Plant microbiologists examine images of alfalfa sprouts to see how microbial growth affects plant roots.

F. Environmental Microbiology

These microbiologists study the effect of microbes on the earth's diverse habitats. Whether the microbes are in freshwater or saltwater, topsoil or the earth's crust, they have profound effects on our planet. Subdisciplines of environmental microbiology are

Aquatic microbiology—the study of microbes in the earth's surface water;

Soil microbiology—the study of microbes in terrestrial parts of the planet;

Geomicrobiology—the study of microbes in the earth's crust; and

Astrobiology (also known as exobiology)—the search for/study of microbial and other life in places off of our planet.

Figure F. Researchers collect samples and data in Lake Erie.

1.2 The Impact of Microbes on Earth: Small Organisms with a Giant Effect

The most important knowledge that should emerge from a microbiology course is the profound influence microorganisms have on all aspects of the earth and its residents. For billions of years, microbes have extensively shaped the development of the earth's habitats and the evolution of other life forms. It is understandable that scientists searching for life on other planets first look for signs of microorganisms.

Single-celled organisms that preceded our current cell types arose on this planet about 3.5 billion years ago, according to the fossil record. It appears that they were the only living inhabitants until about 2.9 billion years ago. At that time, three types of cells arose from that original cell type: two were bacteria and archaea, and a more complex type of single-celled organism arose, the **eukaryote** (yoo"-kar-ee-ote). Eu-kary means *true nucleus,* because these were the only cells containing a nucleus. Bacteria and archaea have no true nucleus. For that reason, they have traditionally been called **prokaryotes** (meaning *prenucleus*). But researchers are suggesting we no longer use the term *prokaryote* because archaea and bacteria are so distant genetically.

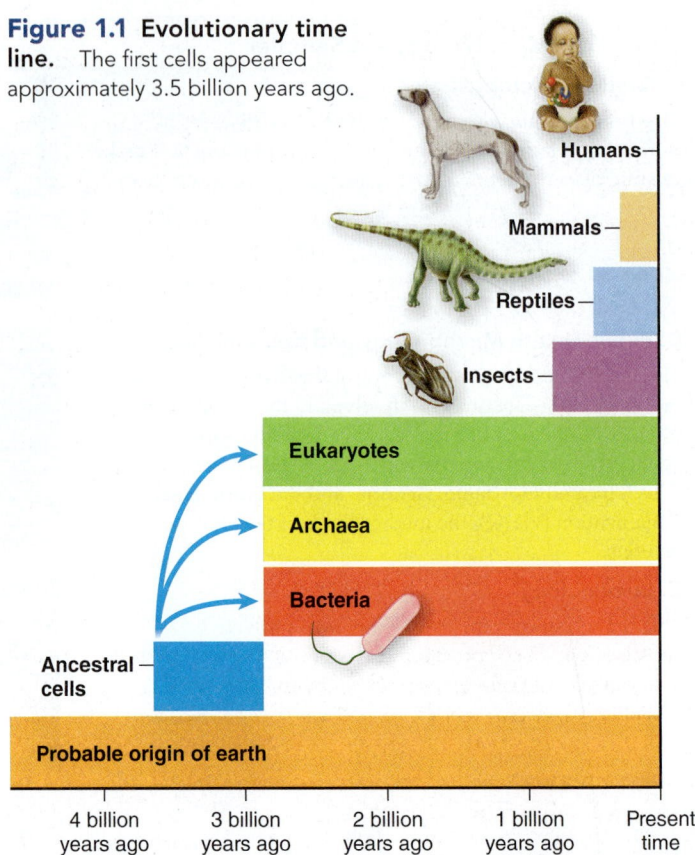

Figure 1.1 Evolutionary time line. The first cells appeared approximately 3.5 billion years ago.

A Note About Bacteria and Archaea

Microbiologists used to have it so easy, in the sense that we could use two terms to define all cell types: **prokaryote** and **eukaryote**. Prokaryotes referred to bacteria and archaea, that is, until genetic studies showed us that they are not closely related so we couldn't group them into a single category. Archaea seem to be genetically more related to eukaryotes, although structurally they resemble bacteria: thus the source of the prior confusion. So now we have three cell types: eukaryotes, bacteria, and archaea. In this book, we are going to focus on bacteria and the eukaryotes, because as far as we know these groups are responsible for the majority of human disease. We will address archaea in various sections of the book where the distinction is useful (for example, in this chapter), but mainly we will refer to bacteria, even when the description might also refer to archaea. It just might get confusing if we continue to say "bacteria and archaea" when the information you need is about bacteria.

Figure 1.1 illustrates the history of life on earth. On the scale pictured in the figure, humans seem to have just appeared. Bacteria preceded even the earliest animals by more than 2 billion years. This is a good indication that humans are not likely to—nor should we try to—eliminate bacteria from our environment. They've survived and adapted to many catastrophic changes over the course of their geologic history.

Another indication of the huge influence bacteria exert is how **ubiquitous** they are. Microbes can be found nearly everywhere, from deep in the earth's crust to the polar ice caps and oceans to inside the bodies of plants and animals. Being mostly invisible, the actions of microorganisms are usually not as obvious or familiar as those of larger plants and animals. They make up for their small size by occurring in large numbers and living in places that many other organisms cannot survive. Above all, they play central roles that are essential to life in the earth's landscape.

A Note About "Karyote" Versus "Caryote"

You will see the terms *prokaryote* and *eukaryote* spelled with c (*procaryote* and *eucaryote*) as well as k. Both spellings are accurate. This book uses the k spelling.

When we point out that single-celled organisms have adapted to a wide range of conditions over the 2.9 billion years of their presence on this planet, we are talking about **evolution.** Life in its present form would not be possible if the earliest life forms had not changed constantly, adapting to their environment and circumstances. Getting from the far left in figure 1.1 to the far right where humans appeared involved billions and billions of tiny changes, starting with the first cell that appeared about a billion years after the planet itself was formed.

You have no doubt heard this concept described as the "theory of evolution." Let's clarify some terms. **Evolution** is the accumulation of changes that occur in organisms as they adapt to their environments. It is documented every day in all corners of the planet, an observable phenomenon testable by science. It is often referred to as the **theory of evolution.** This has led to great confusion among the public. As we will explain in section 1.6, scientists use the term "theory" in a different way than the general public does. By the time a principle has been labeled a theory in science, it has undergone years and years of testing and not been disproven. This is much different than the common usage, as in "My theory is that he overslept and that's why he was late." The theory of evolution, like the germ theory and many other scientific theories, are labels for well-studied and well-established natural phenomena.

Microbial Involvement in Shaping Our Planet

Microbes are deeply involved in the flow of energy and food through the earth's ecosystems.[1] Most people are aware that plants carry out **photosynthesis,** which is the light-fueled conversion of carbon dioxide to organic material, accompanied by the formation of oxygen (called oxygenic photosynthesis). However, bacteria invented photosynthesis long before first plants appeared, first as a process that did not produce oxygen (*anoxygenic photosynthesis*). This anoxygenic

---

1. Ecosystems are communities of living organisms and their surrounding environment.

photosynthesis later evolved into oxygenic photosynthesis, which not only produced oxygen but also was much more efficient in extracting energy from sunlight. Hence, bacteria were responsible for changing the atmosphere of the earth from one without oxygen to one with oxygen. The production of oxygen also led to the use of oxygen for aerobic respiration and the formation of ozone, both of which set off an explosion in species diversification. Today, photosynthetic microorganisms (bacteria and algae) account for more than 70% of the earth's photosynthesis, contributing the majority of the oxygen to the atmosphere **(figure 1.2*a*).**

Another process that helps keep the earth in balance is the process of biological **decomposition** and nutrient recycling. Decomposition involves the breakdown of dead matter and wastes into simple compounds that can be directed back into the natural cycles of living things **(figure 1.2*b*).** If it were not for multitudes of bacteria and fungi, many chemical elements would become locked up and unavailable to organisms; we humans would drown in our own industrial and personal wastes! In the long-term scheme of things, microorganisms are the main forces that drive the structure and content of the soil, water, and atmosphere. For example:

- The very temperature of the earth is regulated by gases, such as carbon dioxide, nitrous oxide, and methane, which create an insulation layer in the atmosphere and help retain heat. Many of these gases are produced by microbes living in the environment and the digestive tracts of animals.

(a)

(b)

Figure 1.2 Examples of microbial habitats. **(a)** Summer pond with a thick mat of algae—a rich photosynthetic community. **(b)** Microbes play a large role in decomposing dead animal and plant matter.

- Recent estimates propose that large numbers of organisms exist within and beneath the earth's crust in sediments, rocks, and even volcanoes. It is increasingly evident that this enormous underground community of microbes is a significant influence on weathering, mineral extraction, and soil formation.
- Bacteria and fungi live in complex associations with plants that assist the plants in obtaining nutrients and water and may protect them against disease. Microbes form similar interrelationships with animals, notably, in the stomach of cattle, where a rich assortment of bacteria digest the complex carbohydrates of the animals' diets and cause the release of methane into the atmosphere.

1.2 Learning Outcomes—Assess Your Progress

3. Describe the role and impact of microbes on the earth.
4. Explain the theory of evolution and why it is called a theory.

1.3 Human Use of Microorganisms

Microorganisms clearly have monumental importance to the earth's operation. Their diversity and versatility make them excellent candidates for solving human problems. By accident or choice, humans have been using microorganisms for thousands of years to improve life and even to shape civilizations. Baker's and brewer's yeast, types of single-celled fungi, cause bread to rise and ferment sugar into alcohol to make wine and beers. Other fungi are used to make special cheeses such as Roquefort or Camembert. These and other "home" uses of microbes have been in use for thousands of years. For example, historical records show that households in ancient Egypt kept moldy loaves of bread to apply directly to wounds and lesions. When humans manipulate microorganisms to make products in an industrial setting, it is called biotechnology. For example, some specialized bacteria have unique capacities to mine precious metals or to clean up human-created contamination **(figure 1.3).**

Genetic engineering is an area of biotechnology that manipulates the genetics of microbes, plants, and animals for the purpose of creating new products and genetically modified organisms (GMOs). One powerful technique for designing GMOs is termed **recombinant DNA technology.** This technology makes it possible to transfer genetic material from one organism to another and to deliberately alter DNA.[2] Bacteria and fungi were some of the first organisms to be genetically engineered. This was possible because they are single-celled organisms and they are so adaptable to changes in their genetic makeup. Recombinant DNA technology has unlimited potential in terms of

(a)

(b)

(c)

Figure 1.3 Microbes at work. **(a)** An aerial view of a copper mine looks like a giant quilt pattern. The colored patches are bacteria in various stages of extracting metals from the ore. **(b)** Microbes as synthesizers. Fermenting tanks at a winery. **(c)** Members of a biohazard team from the National Oceanic and Atmospheric Agency (NOAA) participate in the removal and detoxification of 63,000 tons of crude oil released by a wrecked oil tanker on the coast of Spain. The bioremediation of this massive spill made use of naturally occurring soil and water microbes as well as commercially prepared oil-eating species of bacteria and fungi.

---

2. DNA, or deoxyribonucleic acid, is the chemical substance that comprises the genetic material of organisms.

medical, industrial, and agricultural uses. Microbes can be engineered to synthesize desirable products such as drugs, hormones, and enzymes.

Among the genetically unique organisms that have been designed by bioengineers are bacteria that mass produce antibiotic-like substances, yeasts that produce human insulin, pigs that produce human hemoglobin, and plants that contain natural pesticides or fruits that do not ripen too rapidly. Genetic engineering has also provided important human vaccines and therapies.

Another way of tapping into the unlimited potential of microorganisms is the science of **bioremediation** (by'-oh-ree-mee-dee-ay"-shun). This process involves the introduction of microbes into the environment to restore stability or to clean up toxic pollutants. Microbes have a surprising capacity to break down chemicals that would be harmful to other organisms. This includes even human-made chemicals that scientists have developed and for which there are no natural counterparts.

Agencies and companies have developed microbes to handle oil spills and detoxify sites contaminated with heavy metals, pesticides, and other chemical wastes **(figure 1.3c)**. One form of bioremediation that has been in use for some time is the treatment of water and sewage. Because clean freshwater supplies are dwindling worldwide, it will become even more important to find ways to reclaim polluted water.

1.3 Learning Outcomes—Assess Your Progress

5. Explain one old way and one new way that humans manipulate organisms for their own uses.

1.4 Infectious Diseases and the Human Condition

One of the most fascinating aspects of the microorganisms with which we share the earth is that, despite all of the benefits they provide, they also contribute significantly to human misery as **pathogens** (path'-oh-jenz). The vast majority of microorganisms that associate with humans cause no harm. In fact, they provide many benefits to their human hosts. It is important to note that a diverse microbial biota living in and on humans is an important part of human well-being. However, humankind is also plagued by nearly 2,000 different microbes that can cause various types of disease. Infectious diseases still devastate human populations worldwide, despite significant strides in understanding and treating them. The World Health Organization (WHO) estimates there are a total of 10 billion new infections across the world every year. Infectious diseases are also among the most common causes of death in much of humankind, and they still kill a significant percentage of the U.S. population. **Table 1.2** depicts the 10 top causes of death per year (by all causes, infectious and noninfectious) in the United States and worldwide. The worldwide death toll from infections is about 13 million people per year. For example, the World Health Organization reports that every 30 seconds a child dies from malaria.

Disease Connection

The most deadly lower respiratory tract infections are influenza and pneumonia. Seasonal influenza is generally hardest on the very young and very old, although during years when pandemic strains of the influenza virus are circulating young healthy adults can be severely affected. Influenza infections put you at risk for developing pneumonia, caused either by the influenza virus itself or by secondary viruses or bacteria. Of course, you can also develop pneumonia without first being infected by the influenza virus. Both of these diseases are thoroughly discussed in chapter 21.

Table 1.2 Top Causes of Death—All Diseases

United States	No. of Deaths	Worldwide	No. of Deaths
1. Heart disease	617,000	1. Heart disease	7.3 million
2. Cancer	565,000	2. Stroke	6.2 million
3. Chronic lower-respiratory disease	141,000	3. Lower-respiratory infections (influenza and pneumonia)*	3.5 million
4. Cerebrovascular disease	134,000	4. Chronic obstructive pulmonary disease	3.3 million
5. Accidents (unintentional injuries)	122,000	5. Diarrheal diseases	2.5 million
6. Alzheimer's disease	82,000	6. HIV/AIDS	1.8 million
7. Diabetes	71,000	7. Trachea, bronchus, lung cancers	1.4 million
8. Influenza and pneumonia	56,000	8. Tuberculosis	1.3 million
9. Kidney disease	48,000	9. Diabetes	1.3 million
10. Suicide	36,000	10. Road traffic accidents	1.2 million

*Diseases in red are those most clearly caused by microorganisms.

Source: Data from the World Health Organization and the Centers for Disease Control and Prevention. Data published in 2011 representing final figures for the year 2008.

In **figure 1.4,** you can see that high-income countries like ours see many more deaths caused by chronic, noninfectious diseases (heart disease, cancer, stroke) than those caused by infections. Low-income countries (on the left on the graph) suffer high rates of death from these diseases but even higher rates of deaths from infections. Economics is closely tied to survival in these countries.

Malaria, which kills between 700,000 and 1.2 million people every year worldwide, is caused by a microorganism transmitted by mosquitoes (see chapter 20). Currently, the most effective way for citizens of developing countries to avoid infection with the causal agent of malaria is to sleep under a bed net, because the mosquitoes are most active in the evening. Yet even this inexpensive solution is beyond the reach of many. Mothers in Southeast Asia and elsewhere have to make nightly decisions about which of their children will sleep under the single family bed net, because a second one, priced at about $5, is too expensive for them.

Adding to the overload of infectious diseases, we are also witnessing an increase in the number of new (emerging) and older (reemerging) diseases. AIDS, hepatitis C, and

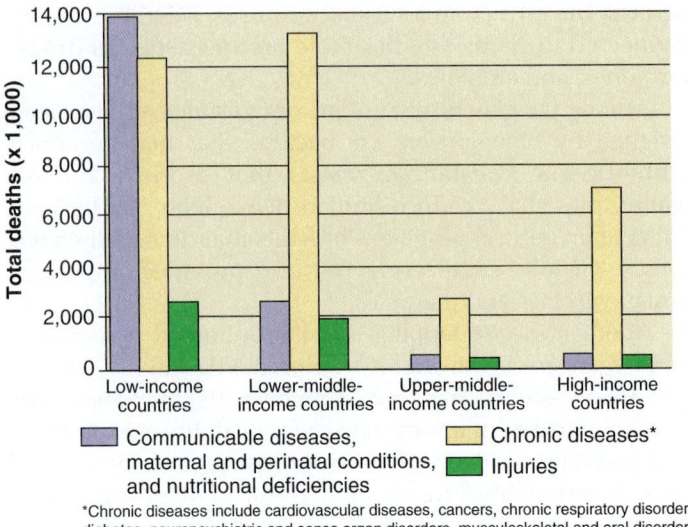

Figure 1.4 **The role of infectious diseases versus other causes of death in countries of varying income.**

INSIGHT 1.1   **The War Is Far from Over**

In 1964, the surgeon general of the United States told Congress, "It is time to close the book on infectious diseases. The war against pestilence is over." Recently discovered antibiotics and newly introduced vaccines were extremely effective against diseases that had haunted humankind for centuries. It was easy to think that humans had won the war over the microbes. What the surgeon general didn't realize was that the microbes that have inhabited this planet for millennia were slowly and quietly evolving to address the new threat.

Research on novel antibiotics slowed in the 1960s and 1970s, due in part to the sentiment among scientists and the medical community that once dangerous microbes were no longer a threat. Doctors regularly prescribed antibiotics for infections that were viral in origin, sometimes due to patient demand. Patients were careless in taking antibiotics, often not finishing a full prescription. Suddenly, drug-resistant strains such as methicillin-resistant *Staphylococcus aureus* (MRSA) and multidrug-resistant *Mycobacterium tuberculosis* (MDR-TB) began appearing worldwide.

In 2007, American Andrew Speaker left the country for his wedding. He had been previously diagnosed with MDR-TB, but preliminary tests showed that he was not a threat to others and had been cleared to travel by his doctors. *Mycobacterium tuberculosis* is a notoriously slow-growing organism, and after he left the country, further tests revealed that he harbored a strain of extensively drug-resistant tuberculosis (XDR-TB). During his travels throughout Europe and the Mediterranean, along with connecting flights in Canada and the United States, he unwittingly exposed thousands of fellow travelers to XDR-TB. Once he returned, Speaker submitted to voluntary quarantine for treatment, but the incident sparked an international firestorm.

This is only one example of the ability of microbes to adapt to the ever-changing world in which we live. Increased glo-

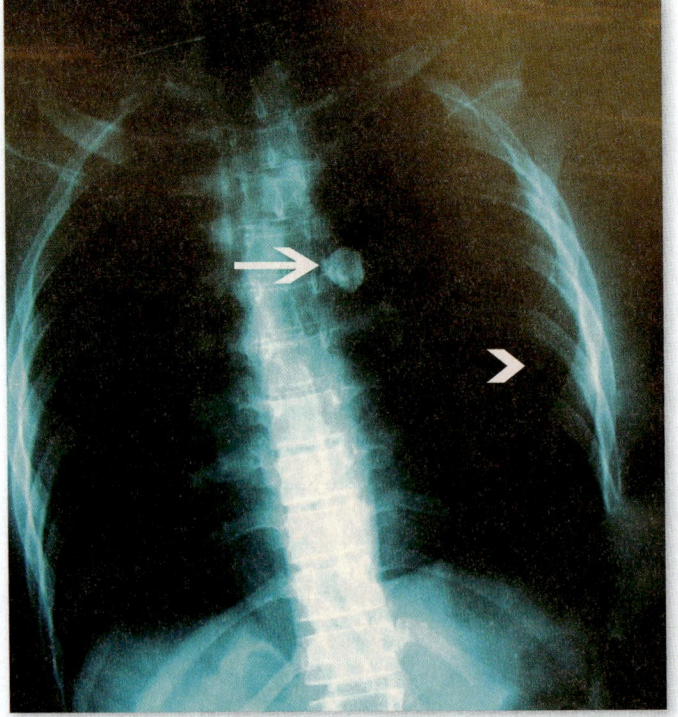

X ray showing a tubercle in a tuberculosis patient.

balization and travel, growing populations of susceptible and immune-suppressed individuals, emerging infectious diseases from previously unexplored areas of the world, and reemerging infectious diseases like tuberculosis show that not only is the war against pestilence not over, it's only just beginning.

Source: 2007. *J. Am. Med. Assoc.* vol. 298, no. 1, p. 83.

viral encephalitis are examples of diseases that cause severe mortality and morbidity. To somewhat balance this trend, there have also been some advances in eradication of diseases such as polio and leprosy and diseases caused by certain parasitic worms.

One of the most eye-opening discoveries in recent years is that many diseases that used to be considered noninfectious probably do involve microbial infection. The most famous of these is gastric ulcers, now known to be caused by a bacterium called *Helicobacter*. But there are more. An association has been established between certain cancers and bacteria and viruses, between diabetes and the coxsackievirus, and between schizophrenia and a virus called the Borna agent. Diseases as different as multiple sclerosis, obsessive compulsive disorder, coronary artery disease, and even obesity have been linked to chronic infections with microbes. It seems that the golden age of microbiological discovery, during which all of the "obvious" diseases were characterized and cures or preventions were devised for them, should more accurately be referred to as the *first* golden age. We're now discovering the subtler side of microorganisms. Their roles in quiet but slowly destructive diseases are now well known. These include female infertility, caused by *Chlamydia* infection, and malignancies such as liver cancer (hepatitis viruses) and cervical cancer (human papillomavirus). Here, again, low-income countries differ from high-income countries. It seems that up to 26% of cancers in low-income countries are caused by viruses or bacteria, while less than 7% of malignancies in the developed world are microbially induced.

As mentioned earlier, another important development in infectious disease trends is the increasing number of patients with weakened defenses that are kept alive for extended periods. They are subject to infections by common microbes that are not pathogenic to healthy people. There is also an increase in microbes that are resistant to drugs **(Insight 1.1)**. It appears that even with the most modern technology available to us, microbes still have the "last word," as the great French scientist Louis Pasteur observed.

1.5 The General Characteristics of Microorganisms

Cellular Organization

As discussed earlier, three basic cell lines appeared during evolutionary history. These lines—**Archaea, Eukarya,** and **Bacteria**—differ not only in the complexity of their cell structure **(figure 1.5a)** but also in contents and function.

A Note About Viruses

Viruses are subject to intense study by microbiologists. As mentioned before, they are not independently living cellular organisms. Instead, they are small particles that exist at the level of complexity somewhere between large molecules and cells **(figure 1.5b)**. Viruses are much simpler than cells; outside their host, they are composed essentially of a small amount of hereditary material (either DNA or RNA but never both) wrapped up in a protein covering that is sometimes enveloped by a protein-containing lipid membrane. In this extracellular state, they are individually referred to as a **virus particle** or **virion**. When inside their host organism, in the intracellular state, viruses usually exist only in the form of genetic material that confers a partial genetic program on the host organisms. That is why many microbiologists refer to viruses as parasitic particles; however, a few consider them to be very primitive organisms. Nevertheless, all biologists agree that viruses are completely dependent on an infected host cell's machinery for their multiplication and dispersal.

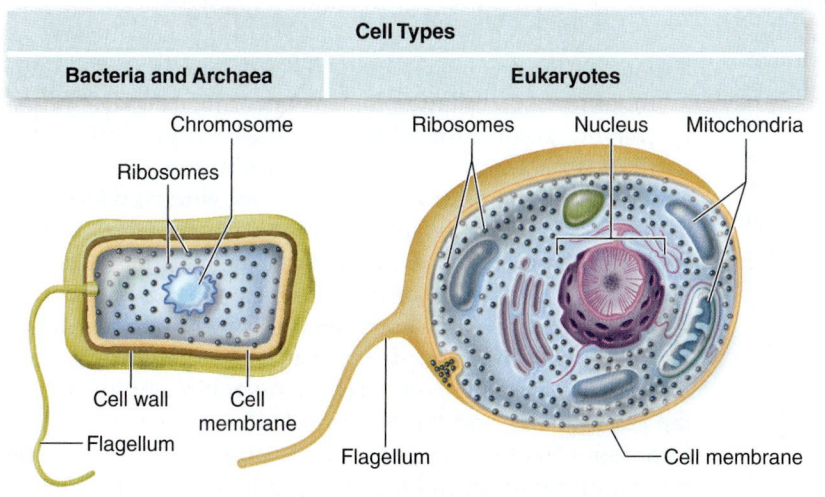

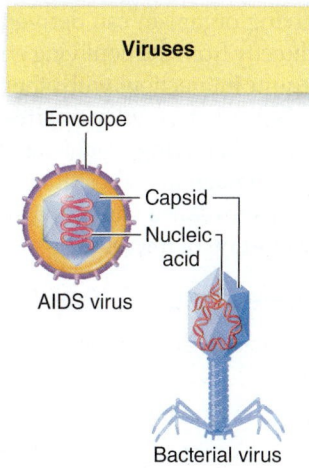

(a)

(b)

Figure 1.5
Cell structure.
(a) Comparison of a bacterial/archaeal cell and a eukaryotic cell (not to scale). (b) Two examples of viruses. These cell types and viruses are discussed in more detail in chapters 4, 5, and 6.

In general, bacterial and archaeal cells are about 10 times smaller than eukaryotic cells, and they generally lack many of the eukaryotic cell structures such as **organelles.** Organelles are small, double-membrane-bound structures in the eukaryotic cell that perform specific functions and include the nucleus, mitochondria, and chloroplasts. Bacteria and archaea are covered in more detail in chapter 4 and eukaryotes in chapter 5. All bacteria and archaea are microorganisms, but only some eukaryotes are microorganisms. The majority of microorganisms are single-celled (all bacteria and archaea and some eukaryotes), but some consist of a few cells **(figure 1.6).** Certain invertebrate animals—such as helminths (worms), many of which can be seen with the naked eye, are also included in the study of infectious diseases because of the way they are transmitted and the way the body responds to them, though they are not microorganisms.

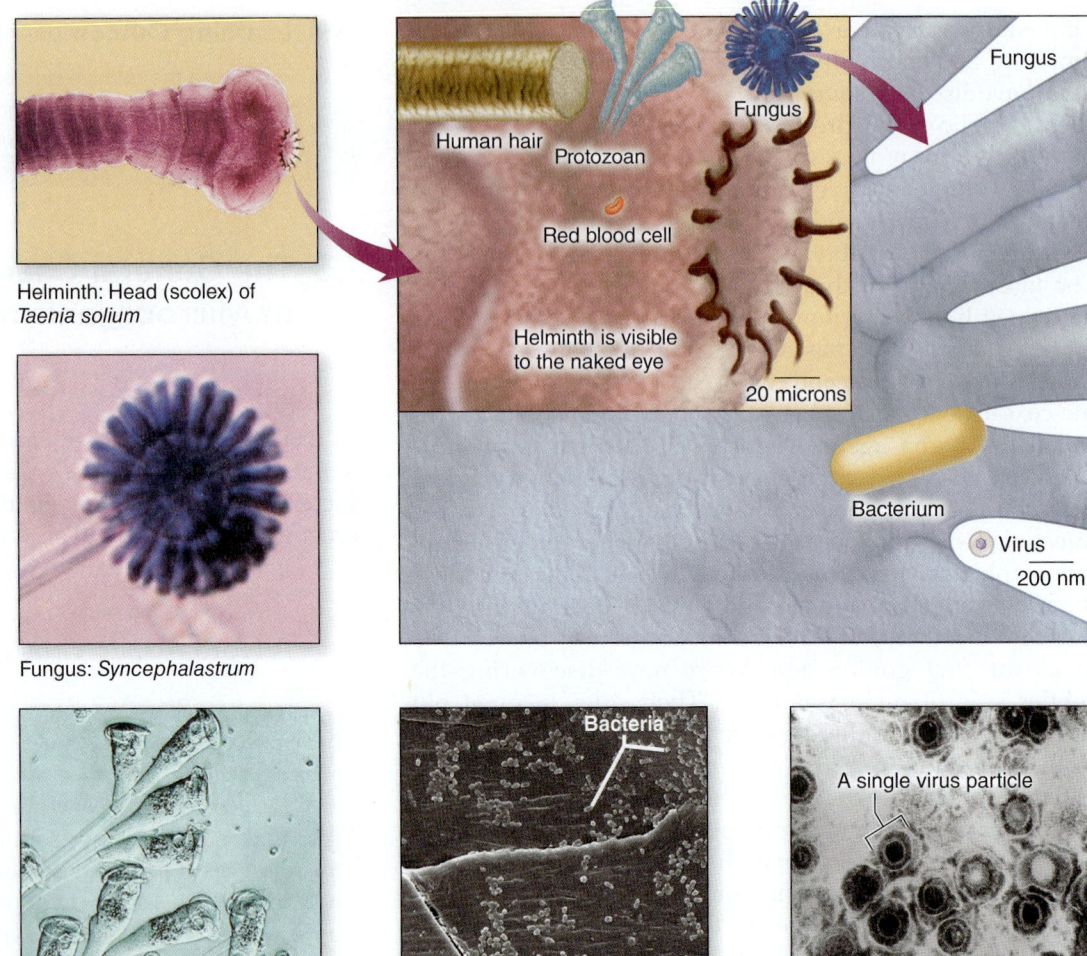

Helminth: Head (scolex) of *Taenia solium*

Fungus: *Syncephalastrum*

Protozoan: *Vorticella*

Bacterium: *E. coli*

Virus: Herpes simplex

Figure 1.6 Five types of microorganisms.   The drawing at top right shows relative size differences. The photos of organisms around the drawing are pictured at different magnifications in order to show their details.

Lifestyles of Microorganisms

The majority of microorganisms live a free existence in habitats such as soil and water, where they are relatively harmless and often beneficial. A free-living organism can derive all required foods and other factors directly from the nonliving environment. Some microorganisms require interactions with other organisms. Sometimes these microbes are termed **parasites.** They are harbored and nourished by other living organisms called **hosts.** A parasite's actions cause damage to its host through infection and disease. Although parasites cause important diseases, they make up only a small proportion of microbes.

1.5 Learning Outcomes—Assess Your Progress

7. Differentiate among bacteria, archaea, and eukaryotic microorganisms.
8. Identify a fourth type of microorganism.
9. Compare and contrast the relative sizes of the different microbes.

1.6 The Historical Foundations of Microbiology

If not for the extensive interest, curiosity, and devotion of thousands of microbiologists over the last 300 years, we would know little about the microscopic realm that surrounds us. Many of the discoveries in this science have resulted from the prior work of men and women who toiled long hours in dimly lit laboratories with the crudest of tools. Each additional insight, whether large or small, has added to our current knowledge of living things and processes. This section summarizes the prominent discoveries made in the past 300 years: microscopy; the rise of the scientific method; and the development of medical microbiology, including the germ theory and the origins of modern microbiological techniques. Table B.1 in appendix B summarizes some of the pivotal events in microbiology, from its earliest beginnings to the present.

The Development of the Microscope: "Seeing Is Believing"

From very earliest history, humans noticed that when certain foods spoiled they became inedible or caused illness, and yet other "spoiled" foods did no harm and even had enhanced flavor. Indeed, several centuries ago, there was already a sense that diseases such as the black plague and smallpox were caused by some sort of transmissible matter. But the causes of such phenomena were vague and obscure because the technology to study them was lacking. Consequently, they remained cloaked in mystery and regarded with superstition—a trend that led even well-educated scientists to believe in something called spontaneous generation **(Insight 1.2)**.

True awareness of the widespread distribution of microorganisms and some of their characteristics was finally made possible by the development of the first microscopes. These devices revealed microbes as discrete entities sharing many of the characteristics of larger, visible plants and animals. Several early scientists fashioned magnifying lenses, but their microscopes lacked the optical clarity needed for examining bacteria and other small, single-celled organisms. The likely earliest record of microbes is in the works of Englishman Robert Hooke. In the 1660s, Hooke studied a great diversity of material from household objects,

---

INSIGHT 1.2 Spontaneous Generation: A Hard Habit to Break

Most people have a vague idea of what microbes are. Even though many have never seen a microbe under the microscope, people often say, "I must be coming down with a bug," at the first sign of a cold or sore throat, but this hasn't always been so. For thousands of years, people thought that diseases were caused by a curse from God or miasmas in the air, and believed that plants, animals, and even people came from an invisible life-giving force. This theory was known as **spontaneous generation.**

Even after Robert Hooke and Antonie van Leeuwenhoek observed cells using primitive microscopes in the mid-1600s and 1700s, spontaneous generation was still a widely held belief by laypeople and scientists alike. It took over 200 years and a series of convincing experiments to disprove **abiogenesis** (a = without, bio = life, genesis = beginning, or *beginning in the absence of life*) in support of **biogenesis** (*beginning with life*).

One of the first scientists to test the theory of spontaneous generation was Francesco Redi using a simple experiment. He placed meat into two jars and covered one with fine gauze, preventing flies from landing on it, leaving the other one uncovered. Flies surrounded both jars, but maggots only appeared on the meat in the uncovered jar. The flies laid eggs on the gauze covering the second jar, and maggots appeared on the gauze but not the meat, proving that the meat was not the source of the maggots. This and other similar experiments put to rest any ideas that maggots, mice, and other complex organisms arose spontaneously.

The scientific community was not easily convinced. Franz Shultze and Theodor Schwann of Germany maintained that forces in the air were the source of life. They conducted a series of experiments in which they passed air through strong chemicals or glass tubes into heat-treated infusions in flasks. When the infusions remained devoid of life, they claimed that it was the treatment of the air that made it incapable of producing life.

It wasn't until the acclaimed chemist and biologist Louis Pasteur designed a series of elegant experiments that it was definitively shown that microbes in the dust in the air were the source of growth in infusions and broth. He filled flasks with broth and

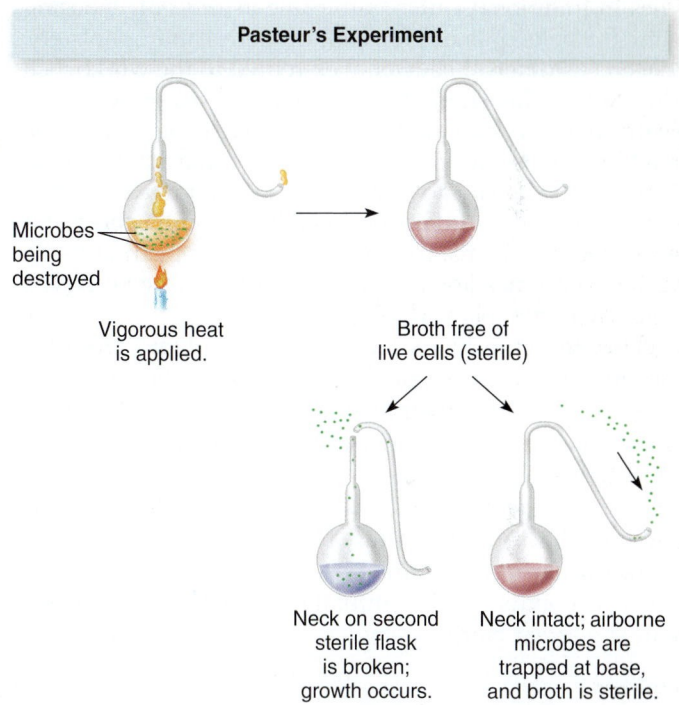

Pasteur's Experiment

Microbes being destroyed

Vigorous heat is applied.

Broth free of live cells (sterile)

Neck on second sterile flask is broken; growth occurs.

Neck intact; airborne microbes are trapped at base, and broth is sterile.

fashioned their openings into long, swan-neck-shaped tubes. The broth in the flasks was still open to the air but curved so that gravity would deposit any contaminants in the neck of the flask. He boiled the flasks to sterilize the broth and incubated them to encourage the growth of microbes. As long as the neck of the flask was intact, no microbes grew. If the swan-necked flask was broken off so that dust could fall into the container, microbes grew and the broth became cloudy. In reference to this compelling experiment, Pasteur said, "Never again shall the doctrine of spontaneous generation recover from the mortal blow that this one simple experiment has dealt it." Some of these original sterile flasks are still on display at the Pasteur Institute today.

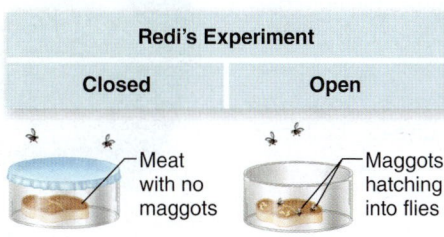

Redi's Experiment

Closed	Open
Meat with no maggots	Maggots hatching into flies

plants, and trees; described for the first time cellular structures in tree bark; and drew sketches of "little structures" that seemed to be alive. Using a single-lens microscope he made himself, Hooke described spots of mold he found on the sheepskin cover of a book:

> These spots appear'd, through a good Microscope, to be a very pretty shap'd vagetative body, which, from almost the same part of the Leather, shot out multitudes of small long cylindrical and transparent stalks, not exactly straight, but a little bended with the weight of a round and white knob that grew on the top of each of them. . . .

Figure 1.7a is a reproduction of the drawing he made to accompany his written observations. Hooke paved the way for even more exacting observations of microbes by Antonie van Leeuwenhoek (pronounced "Lay'-oow-un-hook"), a Dutch linen merchant and self-made microbiologist.

Imagine a dusty linen shop in Holland in the late 1600s. Ladies in traditional Dutch garb came in and out, choosing among the bolts of linens for their draperies and upholstery. Between customers, Leeuwenhoek retired to the workbench in the back of his shop, grinding glass lenses to ever-finer specifications so he could see with increasing clarity the threads in his fabrics. Eventually, he became interested in things other than thread counts. He took rainwater from a clay pot, smeared it on his specimen holder, and peered at it through his finest lens. He found "animals appearing to me ten thousand times less than those which may be perceived in the water with the naked eye."

He didn't stop there. He scraped the plaque from his teeth, and from the teeth of some volunteers who had never cleaned their teeth in their lives, and took a good close look at that. He recorded: "In the said matter there were many very little living animalcules, very prettily a-moving. . . . Moreover, the other animalcules were in such enormous numbers, that all the water . . . seemed to be alive." Leeuwenhoek started sending his observations to the Royal Society of London, and eventually he was recognized as a scientist of great merit.

(a)

(b)

Figure 1.7 The first depiction of microorganisms.
(a) Drawing of "hairy mould" colony made by Robert Hooke in 1665.
(b) Photomicrograph of the fungus probably depicted by Hooke. It is a species of *Mucor*, a common indoor mold.

in science, his descriptions of bacteria and protozoa (which he called "animalcules") were astute and precise. Because of Leeuwenhoek's extraordinary contributions to microbiology, he is known as the father of bacteriology and protozoology.

Disease Connection

The teeth are a perfect surface for accumulating a large assortment of bacteria. The clean tooth surface (immediately after a visit to the dental hygienist, for instance) immediately begins accumulating proteins from the saliva. This coated surface is then colonized by streptococcal bacteria, which are then colonized by other species of bacteria, which are then colonized by more bacteria, and so on. This creates a thick community of bacteria that eventually becomes visible as plaque—especially if you never brush your teeth, as with Leeuwenhoek's subjects. This plaque can lead to cavities (known as *caries*) or gum disease, both covered in chapter 22.

Leeuwenhoek constructed more than 250 small, powerful microscopes that could magnify up to 300 times **(figure 1.8)**. Considering that he had no formal training

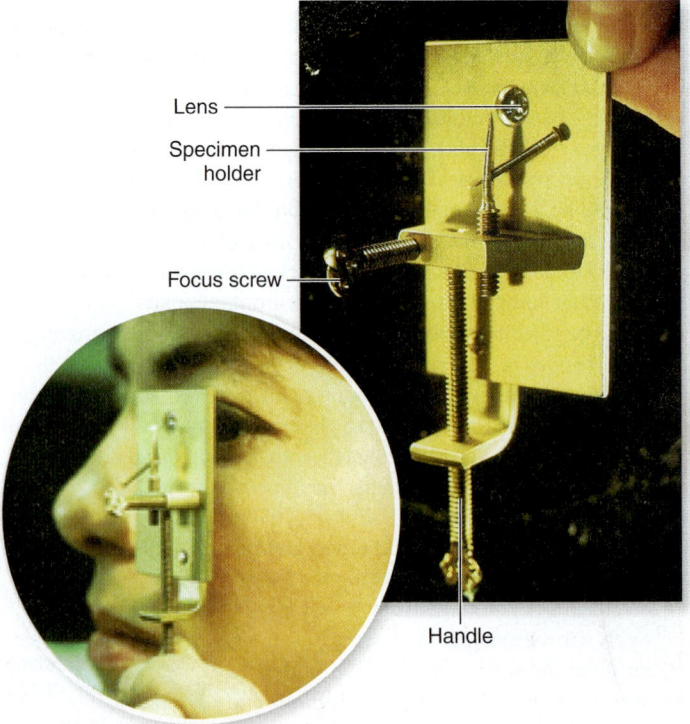

Lens
Specimen holder
Focus screw
Handle

Figure 1.8 Leeuwenhoek's microscope.   A brass replica of a Leeuwenhoek microscope.

From the time of Hooke and Leeuwenhoek, microscopes became more complex and improved with the addition of refined lenses, a condenser, finer focusing devices, and built-in light sources. The prototype of the modern compound microscope, in use from about the mid-1800s, was capable of magnifications of 1,000 times or more. Our modern student microscopes are not greatly different in basic structure and function from those early microscopes. The technical characteristics of microscopes and microscopy are a major focus of chapter 3.

These events marked the beginning of our understanding of microbes and the diseases they can cause. Discoveries continue at a breakneck pace, however. In fact, the 2000s are being widely called the Century of Biology, fueled by our new abilities to study genomes and harness biological processes. Microbes have led the way in these discoveries and continue to play a large role in the new research.

Of course, between the "Golden Age of Microbiology" and the "Century of Biology," there have been thousands of important discoveries. But to give you a feel for what has happened most recently, let's take a glimpse of some very recent discoveries that have had huge impacts on our understanding of microbiology.

Discovery of restriction enzymes—1970s. Three scientists, Daniel Nathans, Werner Arber, and Hamilton Smith, discovered these little molecular "scissors" inside bacteria. They chop up DNA in specific ways. Their job in the bacteria is to destroy invading (viral) DNA. The reason their discovery was such a major event in biology is that these enzymes can be harvested from the bacteria and then utilized in research labs to cut up DNA in a controlled way that then allows us to splice the DNA pieces into vehicles that can carry them into other cells. This opened the floodgates to genetic engineering—and all that has meant for the treatment of diseases, the investigation into biological processes, and the biological "revolution" of the 21st century.

The invention of the PCR technique—1980s. The **polymerase chain reaction (PCR)** was a breakthrough in our ability to detect tiny amounts of DNA and then amplify them into quantities sufficient for studying. It has provided a new and powerful method for discovering new organisms and diagnosing infectious diseases and for forensic work such as crime scene investigation. Its inventor is Kary Mullis, a scientist working at a company in California at the time. He won the Nobel Prize for this invention in 1993.

The importance of biofilms in infectious diseases—1980s and beyond. Biofilms are accumulations of bacteria and other microbes on surfaces. Often there are multiple species in a single biofilm and often they are several layers thick **(figure 1.9).** They have been recognized in environmental microbiology for a long time. Biofilms on rocks, biofilms on ship hulls, and even biofilms on ancient paintings have been well studied. We now understand that biofilms are relatively common in the human body (dental plaque is a great example) and may be responsible for infections that are tough to conquer, such as some ear infections and recalcitrant infections of the prostate. Biofilms are also a big danger to the

Channel

Biofilm material

Figure 1.9 A biofilm made of three different bacterial species. This biofilm was artificially grown in the lab by adding three bacterial species to a flowing chamber. The film is several bacterial layers thick and mimics the kinds of biofilms found in industrial settings, such as in water lines, and also in human infections.

success of any foreign body implanted in the body. Artificial hips, hearts, and even IUDs (intrauterine devices) have all been seen to fail due to biofilm colonization.

Disease Connection

Biofilms! That is exactly what the accumulation of plaque discussed in the previous Disease Connection is.

The importance of small RNAs—2000s. Once we were able to sequence entire genomes (another big move forward), scientists discovered something that turned a concept we literally used to call "dogma" on its head. You will learn in chapter 9 that DNA leads to the creation of proteins, the workhorses of all cells. The previously held "Central Dogma of Biology" was that RNA (a molecule related to DNA) was the go-between molecule. DNA was made into RNA, which dictated the creation of proteins. Genome sequencing has revealed that perhaps only 2% of DNA leads to a resulting protein. There is a lot of RNA that is being made that doesn't end up with a protein counterpart. These pieces of RNA are usually small. It now appears that they have absolutely critical roles in regulating what happens in the cell. This is important not just to correct scientific assumptions but to realize their practical potential as well. This discovery has led to new approaches to how diseases are treated. For example, if the small RNAs are in bacteria that infect humans, they can be new targets for antimicrobial therapy.

The preceding example highlights a feature of biology—and all of science—that is perhaps underappreciated. Because we have thick textbooks containing all kinds of assertions and "facts," many people think science is an iron-clad collection of facts. Wrong! Science is an ever-evolving collection of new information, gleaned from observable phenomena and synthesized with old information to come up with the current understandings of nature. Some of these observations have been confirmed so many times over such a long period of time that they are, if not "fact," very close to fact. Many other observations will be altered over and over again as new findings emerge. And that is the beauty of science.

The Establishment of the Scientific Method

A serious impediment to the development of true scientific reasoning and testing was the tendency of early scientists to explain natural phenomena by a mixture of belief, superstition, and argument. The development of an experimental system that answered questions objectively and was not based on prejudice marked the beginning of true scientific thinking. These ideas gradually crept into the consciousness of the scientific community during the 1600s. The general approach taken by scientists to explain a certain natural phenomenon is called the **scientific method.** A primary aim of this method is to formulate a **hypothesis,** a tentative explanation to account for what has been observed or measured. A good hypothesis should be in the form of a statement. It must be capable of being either supported or discredited by careful observation or experimentation. For example, the statement that "microorganisms cause diseases" can be experimentally determined by the tools of science, but the statement "diseases are caused by evil spirits" cannot.

Case File 1   *Continuing the Case*

Brent Christner, a scientist who worked with Michaud, has studied snow samples from around the world, and has observed that Michaud's findings are not unique to the hailstorm in Montana. Christner has found that bacteria are the most common warm-temperature rain and snow nucleators. He noted that researchers may have never realized that bacteria were so widespread because when samples were taken from clouds looking for nucleators, they used filters that trapped fine dust but ignored the much smaller microbe-sized particles.

P. syringae isn't the only biological nucleator of ice—other bacteria, fungi, diatoms, and algae can serve as nucleators as well. Water vapors in clouds freeze at temperatures below −35°C, but nucleators such as *P. syringae* can cause this to happen at much warmer temperatures. The bacteria possess a protein structure that provides a framework where free-floating water molecules can attach. When the water vapor clings to the bacteria and to other water molecules, it can freeze and fall back to the earth. Until this discovery, atmospheric scientists never realized the impact that biological nucleators had on the formation of rain, snow, and hail. It turns out that bacteria have been making it rain all along.

■ Can you think of some other applications for low-temperature ice nucleators?

Deductive and Inductive Reasoning

Science is a process of investigation using observation, experimentation, and reasoning. In some investigations, you make individual decisions by using accepted general principles as a guide. This is called deductive reasoning. Deductive reasoning, using general principles to explain specific observations, is the reasoning of mathematics, philosophy, politics, and ethics; deductive reasoning is also the way a computer works. All of us rely on deductive reasoning as a way to make everyday decisions—like whether you should open attachments in e-mails from unknown senders **(figure 1.10).** We use general principles as the basis for examining and evaluating these decisions.

Inductive Reasoning

Where do general principles come from? Religious and ethical principles often have a religious foundation; political principles reflect social systems. Some general principles, however, such as those behind the deductive reasoning example just given, are derived not from religion or politics but from observation of the physical world around us. If you drop an apple, it will fall whether or not you wish it to and despite any laws you may pass that forbid it to do so. Science is devoted to discovering the general principles that govern the operation of the physical world.

How do scientists discover such general principles? Scientists are, above all, observers: They look at the world to understand how it works. It is from observations that scientists determine the principles that govern our physical world.

The process of discovering general principles by careful examination of specific cases is termed *inductive reasoning.* This way of thought first became popular about 400 years ago, when Isaac Newton, Francis Bacon, and others began to conduct experiments and from the results infer general principles about how the world operates. Their experiments were sometimes quite simple. Newton's consisted simply of releasing an apple from his hand and watching it fall to the ground. From a host of particular observations, each no more complicated than the falling of an apple, Newton inferred a general principle—that all objects fall toward the center of the earth. This principle was a possible explanation, or hypothesis, about how the world works. You also make observations and formulate general principles based on your observations, like forming a general principle about the reliability of unknown e-mail attachments in figure 1.10. Like Newton, scientists work by forming and testing hypotheses, and observations are the materials on which they build them.

As you can see, the deductive process is used when a general principle has already been established; induction is a discovery process and leads to the creation of a general principle.

A lengthy process of experimentation, analysis, and testing eventually leads to conclusions that either support or refute the hypothesis. If experiments do not uphold the hypothesis—that is, if it is found to be flawed—the hypothesis

Deductive reasoning

Knowing that opening attachments from unknown senders can introduce viruses or other bad things to your computer, you choose the specific action of not opening the attachment.

General principle

Inductive reasoning

You have performed the specific action of clicking on unknown attachments three different times and each time your computer crashed. This leads you to conclude that opening unknown attachments can be damaging to your computer.

Figure 1.10 Deductive and inductive reasoning.

or some part of it is rejected; it is either discarded or modified to fit the results of the experiment. If the hypothesis is supported by the results from the experiment, it is not (or should not be) immediately accepted as fact. It then must be tested and retested. Indeed, this is an important guideline in the acceptance of a hypothesis. The results of the experiment must be published and then repeated by other investigators.

In time, as each hypothesis is supported by a growing body of data and survives rigorous scrutiny, it moves to the next level of acceptance—the **theory.** A theory is a collection of statements, propositions, or concepts that explains or accounts for a natural event. A theory is not the result of a single experiment repeated over and over again but is an entire body of ideas that expresses or explains many aspects of a phenomenon. It is not a fuzzy or weak speculation, as is sometimes the popular notion, but a viable declaration that has stood the test of time and has yet to be disproved by serious scientific endeavors. Often, theories develop and progress through decades of research and are added to and modified by new findings. At some point, evidence of the accuracy and predictability of a theory is so compelling that the next level of confidence is reached and the theory becomes a law, or principle. For example, although we still refer to the germ *theory* of disease, so little question remains that microbes can cause disease that it has clearly passed into the realm of law. The theory of evolution falls in this category as well.

Science and its hypotheses and theories must progress along with technology. As advances in instrumentation allow new, more detailed views of living phenomena, old theories may be reexamined and altered and new ones proposed. But scientists do not take the stance that theories or even "laws" are ever absolutely proved.

The characteristics that make scientists most effective in their work are curiosity, open-mindedness, skepticism, creativity, cooperation, and readiness to revise their views of natural processes as new discoveries are made. The events described in Insight 1.2 provide important examples.

The Development of Medical Microbiology

Early experiments on the sources of microorganisms led to the profound realization that microbes are everywhere: not only are air and dust full of them, but the entire surface of the earth, its waters, and all objects are inhabited by them. This discovery led to immediate applications in medicine. Thus, the seeds of medical microbiology were sown in the mid to latter half of the 19th century with the introduction of the germ theory of disease and the resulting use of sterile, aseptic, and pure culture techniques.

The Discovery of Spores and Sterilization

Following Pasteur's inventive work with infusions (see Insight 1.2), it was not long before English physicist John Tyndall provided the initial evidence that some of the microbes in dust and air have very high heat resistance and that particularly vigorous treatment is required to destroy them. Later, the discovery and detailed description of heat-resistant bacterial endospores by Ferdinand Cohn, a German botanist, clarified the reason that heat would sometimes fail to completely eliminate all microorganisms. The modern sense of the word **sterile,** meaning completely free of all life forms (including spores) and virus particles, was established from that point on (see chapter 11). The capacity to sterilize objects and materials is an absolutely essential part of microbiology, medicine, dentistry, and some industries.

The Development of Aseptic Techniques

From earliest history, humans experienced a vague sense that "unseen forces" or "poisonous vapors" emanating from decomposing matter could cause disease. As the study of microbiology became more scientific and the invisible was made visible, the fear of such mysterious vapors was replaced by the knowledge and sometimes even the fear of "germs." About 125 years ago, the first studies by Robert Koch clearly linked a microscopic organism with a specific disease. Since that time, microbiologists have conducted a continuous search for disease-causing agents.

At the same time that abiogenesis was being hotly debated, a few physicians began to suspect that microorganisms could cause not only spoilage and decay but also infectious diseases. It occurred to these rugged individualists that even the human body itself was a source of infection. Dr. Oliver Wendell Holmes, an American physician, observed that mothers who gave birth at home experienced fewer infections than did mothers who gave birth in the hospital; and the Hungarian Dr. Ignaz Semmelweis showed quite clearly that women became infected in the maternity ward after examinations by physicians coming directly from the autopsy room.

The English surgeon Joseph Lister took notice of these observations and was the first to introduce **aseptic** (ay-sep'-tik) **techniques** aimed at reducing microbes in a medical setting and preventing wound infections. Lister's concept of asepsis was much more limited than our modern

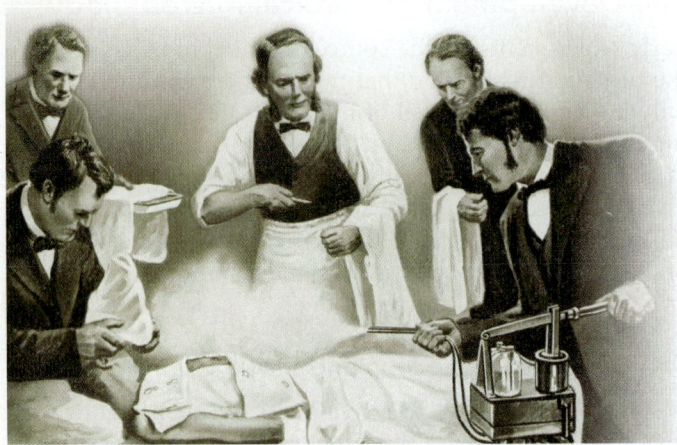

Figure 1.11 Joseph Lister's operating theater in the mid-1800s. This misting machine is releasing phenol.

precautions. It mainly involved disinfecting the hands and the air with strong antiseptic chemicals, such as phenol, prior to surgery **(figure 1.11).** It is hard for us to believe, but as recently as the late 1800s surgeons wore street clothes in the operating room and had little idea that hand washing was important. Lister's techniques and the application of heat for sterilization became the foundations for microbial control by physical and chemical methods, which are still in use today.

The Discovery of Pathogens and the Germ Theory of Disease

Louis Pasteur of France introduced techniques that are still used today. Pasteur made enormous contributions to our understanding of the microbial role in wine and beer formation. He invented pasteurization and completed some of the first studies showing that human diseases could arise from infection. These studies, supported by the work of other scientists, led to the **germ theory of disease.** Pasteur's contemporary, Robert Koch, established *Koch's postulates,* a series of proofs that verified the germ theory and could establish whether an organism was pathogenic and which disease it caused (see chapter 13). About 1875, Koch used this experimental system to show that anthrax was caused by a bacterium called *Bacillus anthracis.* So useful were his postulates that the causative agents of 20 other diseases were discovered between 1875 and 1900, and even today, they are the standard for identifying pathogens of plants and animals.

Numerous exciting technologies emerged from Koch's prolific and probing laboratory work. During this golden age of the 1880s, he realized that study of the microbial world would require separating microbes from each other and growing them in culture. It is not an overstatement to say that he and his colleagues invented most of the techniques that are described in chapter 3: inoculation, isolation, media, maintenance of pure cultures, and preparation of specimens

for microscopic examination. Other highlights in this era of discovery are presented in later chapters on microbial control (see chapter 11) and vaccination (see chapter 15).

1.6 Learning Outcomes—Assess Your Progress

10. Make a time line of the development of microbiology from the 1600s to today.

11. List some recent microbiological discoveries of great impact.

12. Explain what is important about the scientific method.

1.7 Naming, Classifying, and Identifying Microorganisms

Students just beginning their microbiology studies are often dismayed by the seemingly endless array of new, unusual, and sometimes confusing names for microorganisms. Learning microbial **nomenclature** is very much like learning a new language, and occasionally it may feel a bit overwhelming. But paying attention to proper microbial names is just like following a baseball game or a theater production: You cannot tell the players apart without a program! Your understanding and appreciation of microorganisms will be greatly improved by learning a few general rules about how they are named.

The science of classifying living beings is **taxonomy.** It originated more than 250 years ago when Carl von Linné (also known as Linnaeus; 1701–1778), a Swedish botanist, laid down the basic rules for *classification* and established taxonomic categories, or **taxa** (singular, *taxon*).

Von Linné realized early on that a system for recognizing and defining the properties of living beings would prevent chaos in scientific studies by providing each organism with a unique name and an exact "slot" in which to catalog it. This classification would then serve as a means for future identification of that same organism and permit workers in many biological fields to know if they were indeed discussing the same organism. The von Linné system has served well in categorizing the 2 million or more different kinds of organisms that have been discovered since that time, including organisms that have gone extinct.

The primary concerns of modern taxonomy are still naming, classifying, and identifying. These three areas are interrelated and play a vital role in keeping a dynamic inventory of the extensive array of living and extinct beings. In general,

Nomenclature is the assignment of scientific names to the various taxonomic categories and individual organisms.

Classification attempts the orderly arrangement of organisms into a hierarchy of taxa.

Identification is the process of discovering and recording the traits or organisms so that they may be recognized or named and placed in an overall taxonomic scheme.

With the rapid increase in knowledge largely due to the mind-boggling pace of improvement in scientific instrumentation and analysis, taxonomy has never stood still. Instead, it has evolved from a science that artificially classified organisms from a viewpoint of the organism's usefulness, danger, or esthetic appeal to humans to a science that devised a system of natural relationships between organisms. A survey of some general methods of identification appears in chapters 3 and 17. Discovery of present or extinct life forms in space would certainly provide an ultimate test for our existing taxonomy and shed light on the origins of life on our planet earth.

Assigning Specific Names

Many macroorganisms are known by a common name suggested by certain dominant features. For example, a bird species might be called a red-headed blackbird or a flowering plant species a black-eyed Susan. Some species of microorganisms are also called by informal names, including human pathogens such as "gonococcus" (*Neisseria gonorrhoeae*) or fermenters such as "brewer's yeast" (*Saccharomyces cerevisiae*), or the recent "Iraqabacter" (*Acinetobacter baumannii*), but this is not the usual practice. If we were to adopt common names such as the "little yellow coccus" the terminology would become even more cumbersome and challenging than scientific names. Even worse, common names are notorious for varying from region to region, even within the same country. A decided advantage of standardized nomenclature is that it provides a universal language, thereby enabling scientists from all countries to accurately exchange information.

The method of assigning a scientific or specific name is called the **binomial** (two-name) **system** of nomenclature. The scientific name is always a combination of the generic (genus) name followed by the species name. The generic part of the scientific name is capitalized, and the species part begins with a lowercase letter. Both should be italicized (or underlined if using handwriting), as follows:

Staphylococcus aureus

The two-part name of an organism is sometimes abbreviated to save space, as in *S. aureus*, but only if the genus name has already been stated. The source for nomenclature is usually Latin or Greek. If other languages such as English or French are used, the endings of these words are revised to have Latin endings. An international group oversees the naming of every new organism discovered, making sure that standard procedures have been followed and that there is not already an earlier name for the organism or another organism with that same name. The inspiration for names is extremely varied and often rather imaginative. Some species have been named in honor of a microbiologist who originally discovered the microbe or who has made outstanding contributions to the field, as seen in **Insight 1.3.** Other names may designate a characteristic of the microbe (shape, color), a location where it was found, or a disease it causes. Some examples of specific names, their pronunciations, and their origins are

- *Staphylococcus aureus* (staf'-i-lo-kok'-us ah'-ree-us) Gr. *staphule*, bunch of grapes, *kokkus*, berry, and Gr. *aureus*, golden. A common bacterial pathogen of humans.
- *Campylobacter jejuni* (cam'-peh-loh-bak-ter jee-joo'-neye) Gr. *kampylos*, curved, *bakterion*, little rod, and *jejunum*, a section of intestine. One of the most important causes of intestinal infection worldwide.
- *Lactobacillus sanfrancisco* (lak'-toh-bass-ill'-us san-fran-siss'-koh) L. *lacto*, milk, and *bacillus*, little rod. A bacterial species used to make sourdough bread.
- *Vampirovibrio chlorellavorus* (vam-py'-roh-vib-ree-oh klor-ell-ah'-vor-us) Fr. *vampire*; L. *vibrio*, curved cell; *Chlorella*, a genus of green algae; and *vorus*, to devour. A small, curved bacterium that sucks out the cell juices of *Chlorella*.
- *Giardia lamblia* (jee-ar'-dee-uh lam'-blee-uh) for Alfred Giard, a French microbiologist, and Vilem Lambl, a Bohemian physician, both of whom worked on the organism, a protozoan that causes a severe intestinal infection.

Here's a helpful hint: These names may seem difficult to pronounce and the temptation is to simply "slur over them." But when you encounter the names of microorganisms in the chapters ahead, it will be extremely useful to take the time to sound them out and repeat them until they seem familiar. You are much more likely to remember them that way—and they are less likely to end up in a tangled heap with all of the new language you will be learning.

The Levels of Classification

The main units of a classification scheme are organized into several descending ranks, beginning with a most general all-inclusive taxonomic category as a common denominator for organisms to exclude all others, and ending with the smallest and most specific category. This means that all members of the highest category share only one or a few general characteristics, whereas members of the lowest category are essentially the same kind of organism—that is, they share the majority of their characteristics. The taxonomic categories from top to bottom are **domain, kingdom, phylum** or **division,**[3] **class, order, family, genus,** and **species.** Thus, each kingdom can be subdivided into a series of phyla or divisions, each phylum is made up of several classes, each class contains several orders, and so on. Because taxonomic schemes are to some extent artificial, certain groups of organisms may not exactly fit into the main categories. In such a case, additional taxonomic levels can be imposed above (super) or below (sub) a taxon, giving us such categories as "superphylum" and "subclass."

---

3. The term *phylum* is used for bacteria, protozoa, and animals; the term *division* is used for algae, plants, and fungi.

INSIGHT 1.3   What's In a Name?

If you made an important discovery, wouldn't you want the world to know about it? People have a long history of naming things after themselves—cities, countries, mountains, rivers, and other landmarks are named after the people who founded or discovered them. Microbiologists and other scientists are no exception. A large number of organisms that you will encounter in this text have been named for their discoverer or in honor of a great scientist, although to be fair, most were named by others in their honor. Would *Escherichia coli* by any other name smell as sweet?

The following table lists some examples:

Theodor Escherich
(1857–1911).

Organism Name	Named For or In Honor Of
Bartonella	Alberto Barton, a Peruvian microbiologist who first discovered it in 1905.
Bordetella	Jules Bordet, a French microbiologist who isolated it as the cause of pertussis in 1906. He was awarded the Nobel Prize in Physiology or Medicine in 1919.
Borrelia	Amédée Borrel, a French biologist who also developed one of the earliest known gas masks.
Coxiella burnetti	Herald Rea Cox and Macfarlane Burnet, who worked together to discover the organism as the cause of Q fever in the 1930s.
Escherichia coli	Theodor Escherich, who discovered the organism in 1886.
Francisella tularensis	Edward Francis, an American bacteriologist in the early 1900s who contracted tularemia from a patient who had it and kept a careful record of his illness. He subsequently contracted it four more times while studying the organism and the disease. The species is named for Tulare County, California, where it was first discovered in ground squirrels.
Giardia lamblia	Alfred Mathieu Giard and Vilem Dusan Lambl, who are both credited with describing the organism in the late 1800s.
Klebsiella	Edwin Klebs, a German-Swiss pathologist.
Listeria	Discovered in 1926 and named in honor of Joseph Lister, the English surgeon who pioneered sterile surgical techniques.
Malassezia	Louis-Charles Malassez, a French anatomist and histologist who discovered it in 1904. He is also credited for the design of the hemocytometer.
Neisseria	Albert Ludwig Sigesmund Neisser, who discovered the cause of gonorrhea in the 1800s. He is also credited as the co-discoverer of the causative agent of leprosy.
Pasteurella multocida	Discovered in 1878 and named in honor of Louis Pasteur.
Rickettsia rickettsii	Howard Taylor Ricketts, who discovered that the carrier of Rocky Mountain spotted fever was a tick. He often injected himself with the pathogen causing the disease he was studying to measure its effects. He died of a strain of typhus that he isolated causing an outbreak in Mexico City in 1909.
Salmonella	Daniel Elmer Salmon, the administrator of the USDA in 1885. The organism was actually discovered by his assistant Theobald Smith. Since then, over 2,000 subtypes of *Salmonella* have been identified.
Shigella	Kiyoshi Shiga discovered the organism as the cause of dysentery in 1897.

Note: This is only a short list of a few of the organisms named for their discoverers or in honor of famous scientists. Obviously, organism names derived from the names of people are not limited strictly to bacteria. *Strigiphilus garylarsoni* is the scientific name of a louse found in owls named after Gary Larson, the cartoonist famous for "The Far Side" comic strip. Even laboratory instruments and tests are named for their inventors. Julius Richard Petri, a scientist working with Robert Koch in the 1880s, created the Petri dish, still used ubiquitously to grow bacteria on solid media in the laboratory.

Let's compare the taxonomic breakdowns of a human and a protozoan (proh'-tuh-zoh'-uhn) to illustrate the fine points of this system **(figure 1.12).** Humans and protozoa are both organisms with nucleated cells (eukaryotes); therefore, they are in the same domain but they are in different kingdoms. Humans are multicellular animals (Kingdom Animalia), whereas protozoa are single-celled organisms that, together with algae, belong to the Kingdom Protozoa. To emphasize just how broad

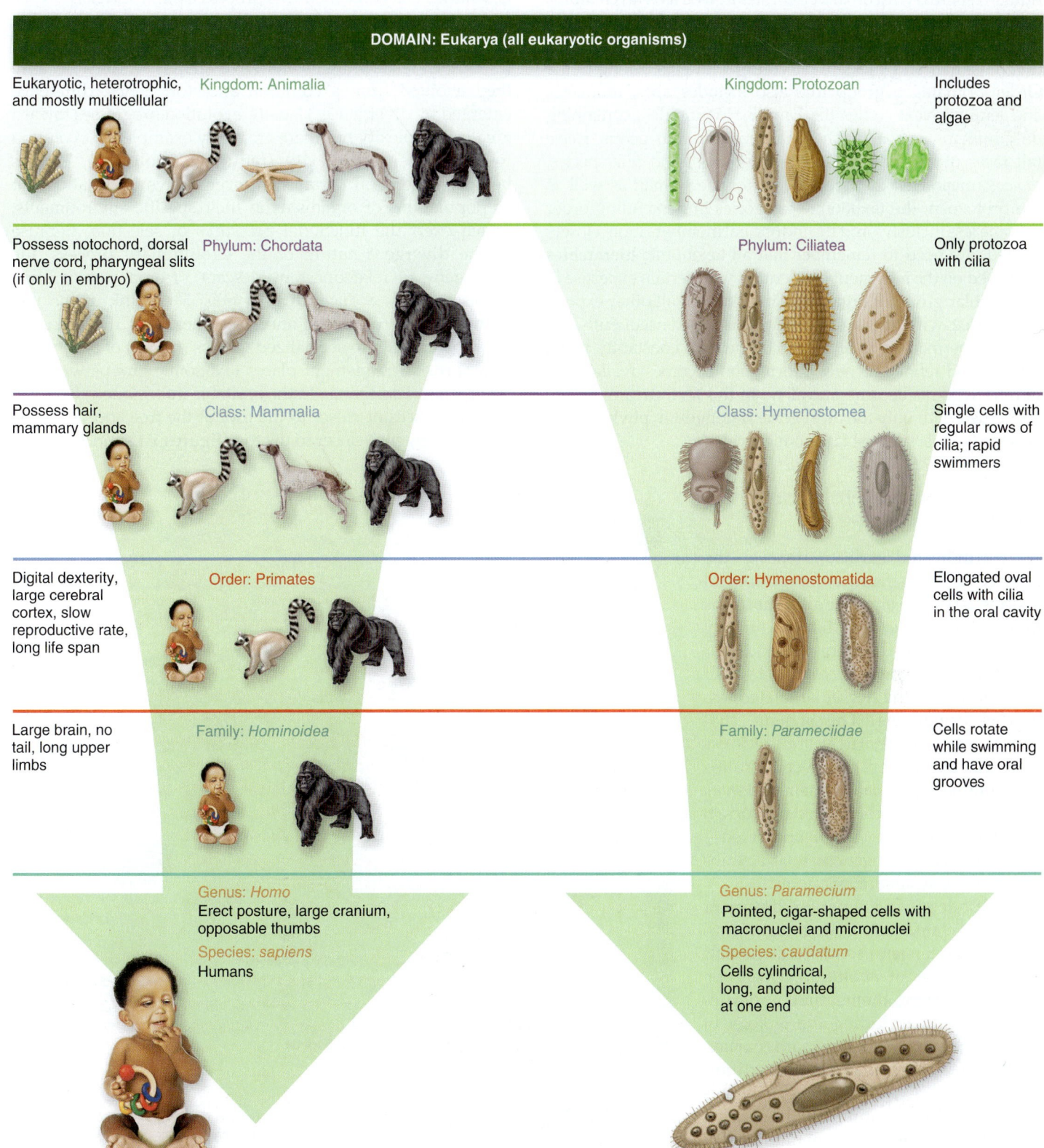

Figure 1.12 Sample taxonomy.   Two organisms belonging to the Eukarya domain, traced through their taxonomic series; on the left, a modern human, *Homo sapiens*; on the right, a common protozoan, *Paramecium caudatum.*

the category "kingdom" is, ponder the fact that we humans belong to the same kingdom as jellyfish. Of the several phyla within this kingdom, humans belong to the Phylum Chordata, but even a phylum is rather all-inclusive, considering that humans share it with other vertebrates as well as with creatures called sea squirts. The next level, Class Mammalia, narrows the field considerably by grouping only those vertebrates that have hair and suckle their young. Humans belong to the Order Primates, a group that also includes apes, monkeys, and lemurs. Next comes the Family Hominoidea, containing only humans and apes. The final levels are our genus, *Homo* (all races of modern and ancient humans), and our species, *sapiens* (meaning *wise*). Notice that for the human as well as the protozoan, the taxonomic categories in descending order become less inclusive and the individual members more closely related. We need to remember that all taxonomic **hierarchies** are based on the judgment of scientists with certain expertise in a particular group of organisms and that not all other experts may agree with the system being used. Consequently, no taxa are permanent to any degree; they are constantly being revised and refined as new information becomes available or new viewpoints become prevalent. In this text, we are usually concerned with only the most general (kingdom, phylum) and specific (genus, species) taxonomic levels.

The Origin and Evolution of Microorganisms

As we indicated earlier, *taxonomy,* the science of classification of biological species, is used to organize all of the forms of modern and extinct life. In biology today, there are different methods for deciding on taxonomic categories, but they all rely on the degree of relatedness among organisms. The scheme that represents the natural relatedness (relation by descent) between groups of living beings is called their *phylogeny* (Gr. *phylon,* race or class; L. *genesis,* origin or beginning), and—when unraveled— biologists use phylogenetic relationships to refine the system of taxonomy.

To understand the natural history of and the relatedness among organisms, we must understand some fundamentals of the process of evolution. Evolution is an important theme that underlies all of biology, including the biology of microorganisms. As we said earlier, evolution states that the hereditary information in living beings changes gradually through time (in humans it usually takes hundreds of millions of years) and that these changes result in various structural and functional changes through many generations. The process

of evolution is selective in that those changes that most favor the survival of a particular organism or group of organisms tend to be retained, whereas those that are less beneficial to survival tend to be lost. This is not always the case but it often is. Charles Darwin called this process *natural selection.*

Evolution is founded on the two preconceptions that (1) all new species originate from preexisting species and (2) closely related organisms have similar features because they evolved from a common ancestor; hence, difference emerged by divergence. Usually, evolution progresses toward greater complexity but there are many examples of evolution toward lesser complexity (reductive evolution). This is because individual organisms never evolve in isolation but as populations of organisms in their specific environments, which exert the functional pressures of selection. Because of the divergent nature of the evolutionary process, the phylogeny, or relatedness by descent, of organisms is often represented by a diagram of a tree. The trunk of the tree represents the origin of ancestral lines, and the branches show offshoots into specialized groups (clades) of organisms. This sort of arrangement places taxonomic groups with less divergence (less change in the heritable information) from the common ancestor closer to the root of the tree and taxa with lots of divergence closer to the top (**figures 1.13** and **1.14**).

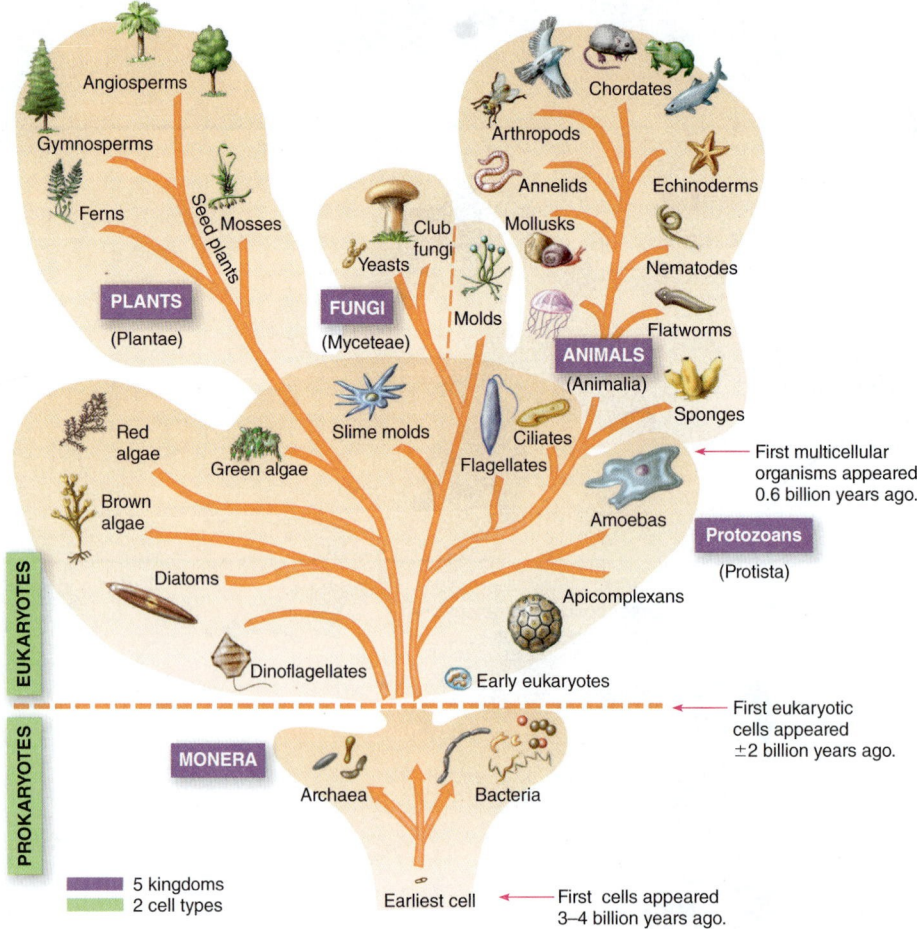

Figure 1.13 Traditional Whittaker system of classification.   In this system, kingdoms are based on cell structure and type, the nature of body organization, and nutritional type.

Systems of Presenting a Universal Tree of Life

The first trees of life were constructed a long time ago on the basis of just two kingdoms, plants and animals, by Charles Darwin and Ernst Haeckel. These trees were chiefly based on visible morphological characteristics. It became clear that certain (micro)organisms such as algae and protozoa, which only existed as single cells, did not truly fit either of those categories, so a third kingdom was recognized by Haeckel for these simpler organisms. It was named Protista (now called Protozoa). Eventually, when significant differences became evident among even the unicellular organisms, a fourth kingdom was established in the 1870s by Haeckel and named Monera. Almost a century passed before Robert Whittaker extended this work and added a fifth kingdom for fungi during the period of 1959 to 1969. The relationships that were used in Whittaker's tree were those based on structural similarities and differences, such as prokaryotic and eukaryotic cellular organization, and the way these organisms obtained their nutrition. These criteria indicated that there were five major taxonomic units, or kingdoms: the monera, protists, plants, fungi, and animals, all of which consisted of one of the two cell types, then known as prokaryotic and eukaryotic. Whittaker's five-kingdom system quickly became the standard (see figure 1.13).

With the rise of genetics as a molecular science, newer methods for determining phylogeny have led to the development of a differently shaped tree—with important implications for our understanding of evolutionary relatedness. Molecular genetics allowed an in-depth study of the structure and function of the genetic material at the molecular level. These studies have revealed that two of the four macromolecules that contribute to cellular structure and function, the proteins and nucleic acids, are very well suited to study how organisms differ from one another because their sequences can be aligned and compared. In 1975, Carl Woese discovered that one particular macromolecule, the ribonucleic acid in the small subunit of the ribosome (ssu rRNA), was highly conserved—meaning that it was nearly identical in organisms within the smallest taxonomic category, the species. Based on a vast amount of experimental data and the knowledge that protein synthesis proceeds in all organisms facilitated by the ribosome, Woese hypothesized that ssu rRNA provides a "biological chronometer" or a "living record" of the evolutionary history of a given organism. Extended analysis of this molecule in prokaryotic and eukaryotic cells indicated that all members in a certain group of bacteria, then known as archaeobacteria, had ssu rRNA with a sequence that was significantly different from the ssu rRNA found in other bacteria and in eukaryotes. This discovery led Carl Woese and collaborator George Fox to propose a separate taxonomic unit for the archaeobacteria, which they named **Archaea.** Under the microscope, they resembled the structure of bacteria, but molecular biology has revealed that the archaea, though seeming to be prokaryotic in nature, were actually more closely related to eukaryotic cells than to bacterial cells (see table 4.1). To reflect these relationships, Carl Woese and George Fox proposed an entirely new system that assigned all known organisms to one of the three major taxonomic units, the **domains,** each being a different type of cell (see figure 1.14).

The domains are the highest level in hierarchy and can contain many kingdoms. The prokaryotic cell types are represented by the domains Archaea and **Bacteria,** whereas

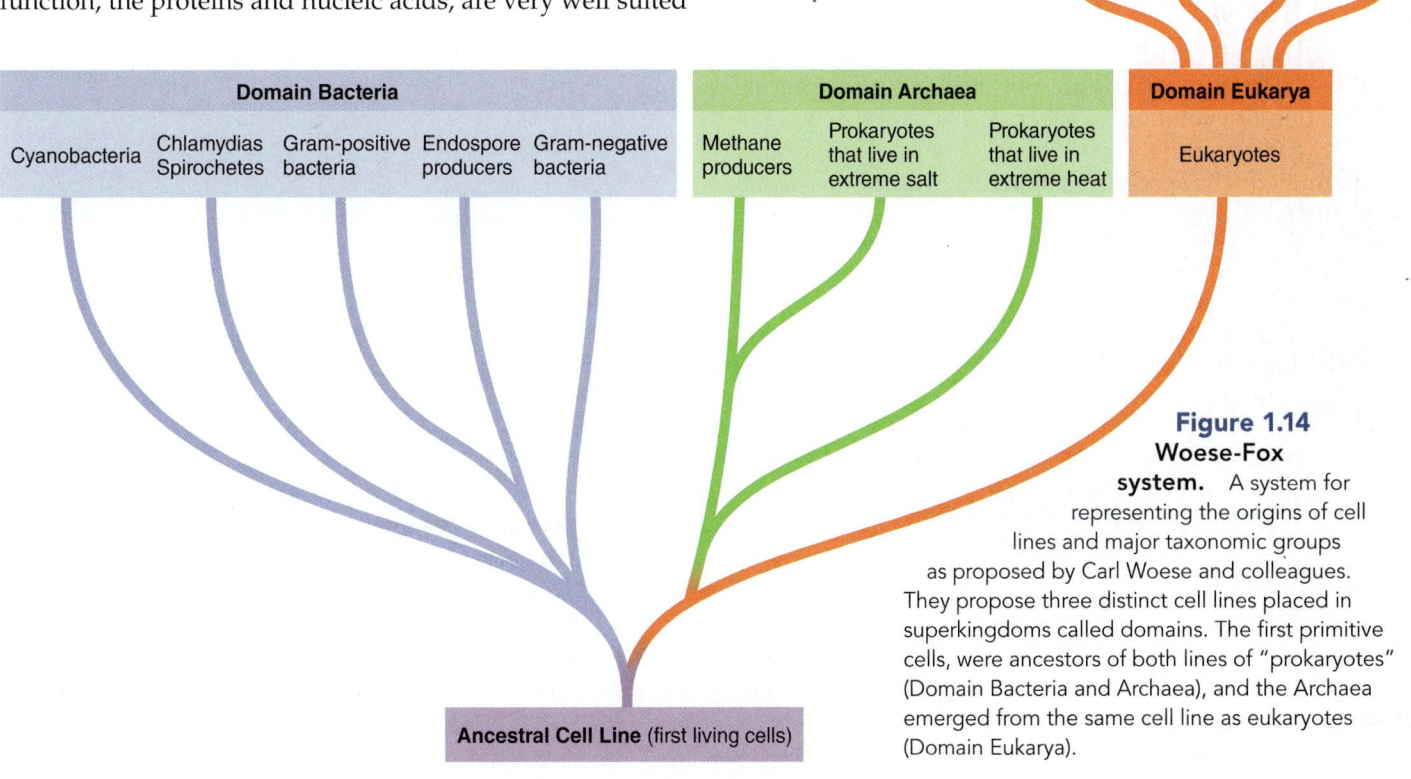

Figure 1.14
Woese-Fox system.   A system for representing the origins of cell lines and major taxonomic groups as proposed by Carl Woese and colleagues. They propose three distinct cell lines placed in superkingdoms called domains. The first primitive cells, were ancestors of both lines of "prokaryotes" (Domain Bacteria and Archaea), and the Archaea emerged from the same cell line as eukaryotes (Domain Eukarya).

eukaryotes are all placed in the domain **Eukarya.** Analysis of the ssu rRNAs from all organisms in these three domains suggests that all modern and extinct organisms on earth arose from a common ancestor. Therefore, eukaryotes did not emerge from prokaryotes. Both types of cells emerged separately from a different, now extinct, cell type.

To add another level of complexity, the most current data suggests that "trees" of life do not truly represent the relatedness—and evolution—of organisms at all. It has become obvious that genes travel horizontally—meaning from one species to another in nonreproductive ways—and that the neat generation-to-generation changes are combined with neighbor-to-neighbor exchanges of DNA. For example, it is estimated that 40% to 50% of human DNA has been carried to humans from other species (by viruses). Another example: The genome of the cow contains a piece of snake DNA. For these reasons, most scientists like to think of a *web* as the proper representation of life these days. Nevertheless, this new scheme does not greatly affect our presentation of most microbes, because we will discuss them at the genus or species level. But be aware that biological taxonomy and, more important, our view of how organisms evolved on earth are in a period of transition. Keep in mind that our methods of classification or evolutionary schemes reflect our current understanding and will change as new information is uncovered.

Disease Connection

Your experience of diarrhea may be affected by the phenomenon of gene sharing among species. One of the most serious causes of diarrhea, *E. coli* O157:H7, carries a toxin that it most likely picked up by incorporating a piece of DNA from *Shigella*, another bacterium, in its genome by accident.

Please note that viruses are not included in any of the classification or evolutionary schemes, because they are not cells or organisms and their position in a "tree of life" cannot be determined. The special taxonomy of viruses is discussed in chapter 6.

Case File 1   *Wrap-Up*

Have you ever found ice chunks in your ice cream? Have you ever traveled to the ski slopes only to find bare rocks and dirt instead of powder? *P. syringae* to the rescue! Scientists are now studying how *P. syringae* can prevent ice recrystallization in ice cream to prevent it from becoming crunchy. The same study showed that bacteria can aid in the formation of ice crystals at lower temperatures to aid in the production of artificial snow. Dr. Virginia Walker of Queen's University conducted the study and said, "Our findings will help to decrease the costs involved in further study of such [cold-tolerant] bacteria, since scientists will no longer need expeditions to the poles in order to isolate the bugs. Now they can find them in their own backyards." Further research has shown that *P. syringae* has no potential to cause disease in humans, so this organism is perfect for the prevention of chunky ice cream and the production of powdery, white slopes.

Source: 2011. Atmos. Chem. Phys. vol. 11, p. 9643.

1.7 Learning Outcomes—Assess Your Progress

13. Differentiate among the terms *nomenclature*, *taxonomy*, and *classification*.

14. Create a mnemonic device for remembering the taxonomic categories.

15. Correctly write the binomial name for a microorganism.

16. Draw a diagram of the three major domains.

17. Explain the difference between traditional and molecular approaches to taxonomy.

Chapter Summary

1.1 The Scope of Microbiology (ASM Guidelines* 2.4, 5.1)

- Microorganisms are defined as "living organisms too small to be seen with the naked eye." Among the members of this huge group of organisms are bacteria, algae, protozoa, fungi, parasitic worms (helminths), and viruses.
- Microorganisms live nearly everywhere and influence many biological and physical activities on earth.

- There are many kinds of relationships between microorganisms and humans; most are beneficial, but some are harmful.

1.2 The Impact of Microbes on Earth: Small Organisms with a Giant Effect (ASM Guidelines 1.1, 3.1, 5.1, 6.1)

- Groups of organisms are constantly evolving to produce new forms of life.
- Microbes are crucial to the cycling of nutrients and energy that are necessary for all life on earth.

ASM Curriculum Guidelines (American Society for Microbiology, 2012). Complete guidelines in appendix B of this book.

1.3 Human Use of Microorganisms (ASM Guidelines 4.5, 6.3, 6.4)

- Humans have learned how to manipulate microbes to do important work for them in industry, medicine, and in caring for the environment.

1.4 Infectious Diseases and the Human Condition (ASM Guideline 5.4)

- In the last 160 years, microbiologists have identified the causative agents for many infectious diseases. In addition, they have discovered distinct connections between microorganisms and diseases whose causes were previously unknown.
- While microbial diseases continue to cause disease worldwide, low-income countries are much harder hit by them directly and indirectly.

1.5 The General Characteristics of Microorganisms (ASM Guidelines 1.1, 2.4, 5.4)

- Excluding the viruses, there are three types of microorganisms: bacteria and archaea, which are small and lack a nucleus and (usually) organelles, and eukaryotes, which are larger and have both a nucleus and organelles.
- Viruses are not cellular and are therefore sometimes called particles rather than organisms. They are included in microbiology because of their small size and close relationship with cells.

1.6 The Historical Foundations of Microbiology (ASM Guidelines 2.4, 7.1)

- The microscope made it possible to see microorganisms and thus to identify their widespread presence, particularly as agents of disease.

- The theory of spontaneous generation of living organisms from "vital forces" in the air was disproved once and for all by Louis Pasteur.
- The scientific method is a process by which scientists seek to explain natural phenomena. It is characterized by specific procedures that either support or discredit an initial hypothesis.
- Knowledge acquired through the scientific method is rigorously tested by repeated experiments by many scientists to verify its validity. A collection of valid hypotheses is called a theory. A theory supported by much data collected over time is sometimes called a law, but the term *theory* may remain associated with it.
- Scientific dogma or theory changes through time as new research brings new information.
- Medical microbiologists developed the germ theory of disease and introduced the critically important concept of aseptic technique to control the spread of disease agents.

1.7 Naming, Classifying, and Identifying Microorganisms (ASM Guidelines 1.1, 1.5)

- The taxonomic system has three primary functions: naming, classifying, and identifying species.
- The major groups in the most advanced taxonomic system are (in descending order) domain, kingdom, phylum or division, class, order, family, genus, and species.
- Evolutionary patterns show a treelike or weblike branching thereby describing the diverging evolution of all life forms from the gene pool of a common ancestor.

- The Woese-Fox classification system places all eukaryotes in the domain Eukarya and subdivides the prokaryotes into the two domains Archaea and Bacteria.

Multiple-Choice and True-False Questions | Bloom's Levels 1 and 2: Remember and Understand

Multiple-Choice Questions. Select the correct answer from the options provided.

1. Which of the following is not considered a microorganism?
   a. alga
   b. bacterium
   c. protozoan
   d. mushroom

2. Which process involves the deliberate alteration of an organism's genetic material?
   a. bioremediation
   b. biotechnology
   c. decomposition
   d. recombinant DNA technology

3. Which of the following parts was absent from Leeuwenhoek's microscopes?
   a. focusing screw
   b. lens
   c. specimen holder
   d. condenser

4. Abiogenesis refers to the
   a. spontaneous generation of organisms from nonliving matter.
   b. development of life forms from preexisting life forms.
   c. development of aseptic technique.
   d. germ theory of disease.

5. A hypothesis can be defined as
   a. a belief based on knowledge.
   b. knowledge based on belief.
   c. a scientific explanation that is subject to testing.
   d. a theory that has been thoroughly tested.

6. When a hypothesis has been thoroughly supported by long-term study and data, it is considered
   a. a law.   c. a theory.
   b. a speculation.   d. proved.

7. Which is the correct order of the taxonomic categories, going from most specific to most general?
   a. domain, kingdom, phylum, class, order, family, genus, species
   b. division, domain, kingdom, class, family, genus, species
   c. species, genus, family, order, class, phylum, kingdom, domain
   d. species, family, class, order, phylum, kingdom

8. Which of the following are not eukaryotic?
   a. bacteria   c. protozoa
   b. archaea   d. both a and b

9. Order the following items by size, using numbers: 1 = smallest through 8 = largest.
   ___ adenovirus   ___ helminths
   ___ amoeba   ___ coccus-shaped bacterium
   ___ rickettsia   ___ white blood cell
   ___ protein   ___ atom

10. How would you classify a virus?
    a. prokaryotic
    b. eukaryotic
    c. neither a nor b

True-False Questions. If the statement is true, leave as is. If it is false, correct it by rewriting the sentence.

11. Organisms in the same order are more closely related than those in the same family.

12. Eukaryotes evolved from prokaryotes.

13. Archaea have no nucleus.

14. In order to be called a theory, a scientific idea has to undergo a great deal of testing.

15. Microbes are ubiquitous.

Critical Thinking Questions | Bloom's Levels 3, 4, and 5: Apply, Analyze, and Evaluate

Critical thinking is the ability to reason and solve problems using facts and concepts. These questions can be approached from a number of angles and, in most cases, they do not have a single correct answer.

1. Develop one argument in support of or refuting the following statement: "Viruses are living microorganisms."

2. Define the term *ubiquitous,* and provide examples illustrating why it is an appropriate term to use to describe microbes.

3. The following is an excerpt from a news article:

   "Dr. Miller suggests that the germicidal effect may be due to inhibition of protein synthesis. He and his research associates are now set to test this **theory** in their laboratory over the next two months."

   Explain whether or not the bolded term was used correctly in this example.

4. Describe how bacteria and archaea are ultimately credited for the evolution of aerobic respiration and the diversity of organisms seen today.

5. Discuss three examples of how humans utilize microbes and their products today.

6. Compare and contrast the Woese-Fox system with the Whittaker system of classification, discussing how genetic analysis provided information leading to the identification of a third cell lineage.

7. Differentiate the terms *emerging disease* and *reemerging disease,* providing examples of each.

8. Discuss how the findings of Louis Pasteur may have inspired Joseph Lister's development of aseptic techniques in surgical settings.

9. Based upon your reading of Insight 1.1, conduct additional research and discuss the current policy on infectious disease quarantine and how this process is administered in the United States.

10. You are a scientist researching West Nile virus, a mosquito-borne pathogen. You note that the number of cases of West Nile disease in your county skyrocketed to their highest levels ever this past summer, which also was the wettest summer in 100 years. Using the scientific method, develop a sound hypothesis explaining the increase in disease cases last summer and a method for testing this hypothesis.

Concept Connections | Bloom's Levels 4 and 6: Analyze and Create

This activity ties together multiple concepts in the chapter.

1. Describe the basic differences between bacterial and eukaryotic cell types.

2. Fill in the empty boxes to describe how microbes are helpful to humans.

3. Fill in the empty boxes to describe how microbes are harmful to humans.

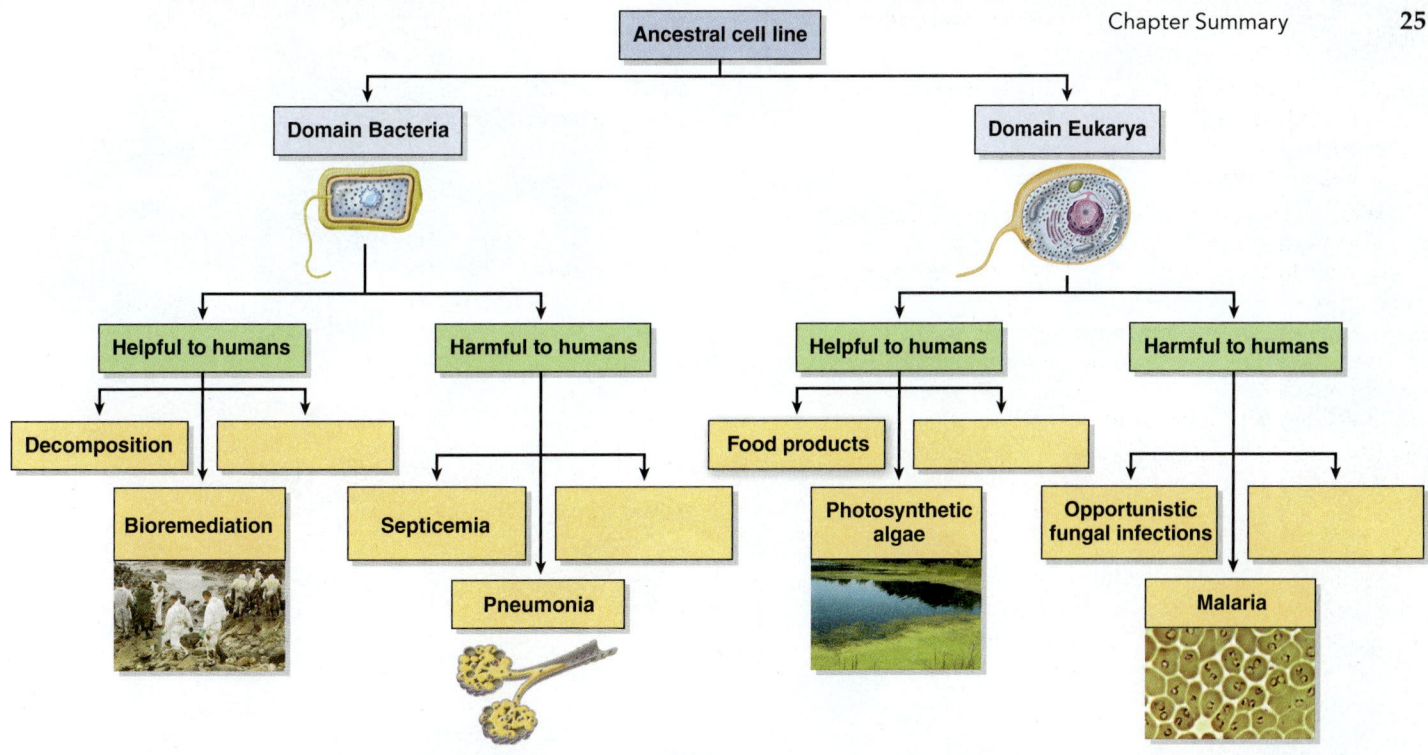

Visual Connections | Bloom's Level 5: Evaluate

These questions use visual images or previous content to make connections to this chapter's concepts.

1. **Figure 1.1.** Look at the yellow, green, and red bars and at the icon indicating the time that humans appeared. Speculate on the probability that we will be able to completely disinfect our planet or prevent all microbial diseases.

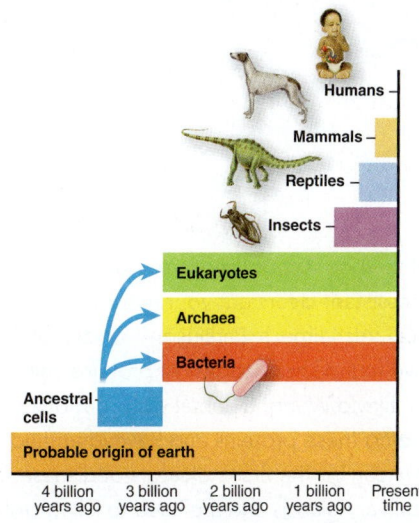

Concept Mapping | Bloom's Level 6: Create

Appendix D provides guidance for working with concept maps.

1. Using the words that follow, please create a concept map illustrating the relationships among these key terms from chapter 1.

microorganisms	helminths	viruses	decomposition	Bacteria
bacteria	protozoa	prokaryote	pathogens	Archaea
algae	fungi	eukaryote	organelles	Eukarya

www.mcgrawhillconnect.com

Enhance your study of this chapter with study tools and practice tests. Also ask your instructor about the resources available through ConnectPlus, including the media-rich eBook, interactive learning tools, and animations.

The Chemistry of Biology

Your Spit: It Might Save Your Life

You've heard it at your dentist's office dozens of times: the whooshing sound of your saliva being sucked down the vacuum system when the assistant clears out your mouth with that plastic tube. What if that saliva could be the key to diagnosing diseases like cancer, Alzheimer's, Parkinson's, or diabetes? According to recent research, it can.

Scientists have found short lengths of proteins called **biomarkers** excreted in the saliva whose presence can indicate disease status. Dr. David Wong, a dental expert at UCLA, along with groups at UCSF and the Scripps Research Institute have been cataloging these proteins and have identified over 2,200 proteins called the "salivary proteome." Many of these protein biomarkers can be the first indicators of disease. Wong says that about two-thirds of the proteins are produced normally by the salivary glands and are used to kill microbes and other pathogens, digest foods, or heal wounds within the mouth. The other one-third of the proteins come from other parts of the body and are then filtered through the blood and exit via the salivary glands. These proteins from the heart, liver, muscles, or other organs have no role in the mouth and elevated levels of these proteins can indicate disease.

Timothy Griffin, a biochemist at the University of Minnesota, has been studying the differences in salivary proteins between healthy people and those with cancer. He has found that the saliva of patients with cancer has unexpected proteins that occur at elevated levels or have been chemically modified when compared to patients who are cancer-free. Griffin has found biomarkers for oral cancer in patients who have not been diagnosed with the disease but only have an oral lesion or an elevated risk for cancer. Griffith's group has also been able to find biomarkers for breast cancer.

- What was the first biomarker discovered by scientists?
- Other than in the saliva, where can biomarkers be detected?

Continuing the Case appears on page 34.

Outline and Learning Outcomes

2.1 Atoms, Bonds, and Molecules: Fundamental Building Blocks
1. Explain the relationship between atoms and elements.
2. List and define four types of chemical bonds.

2.1 Atoms, Bonds, and Molecules: Fundamental Building Blocks

The universe is composed of an infinite variety of substances existing in the gaseous, liquid, and solid states. All such tangible materials that occupy space and have mass are called **matter.** The organization of matter—whether air, rocks, or bacteria—begins with individual building blocks called atoms. An **atom** is defined as a tiny particle that cannot be subdivided into smaller substances without losing its properties. Even in a science dealing with very small things, an atom's minute size is striking; for example, an oxygen atom is only 0.0000000013 mm (0.0013 nm) in diameter, and 1 million of them in a cluster would barely be visible to the naked eye.

The exact composition of atoms has been well established by extensive physical analysis using sophisticated instruments. In general, an atom derives its properties from a combination of subatomic particles called **protons** (p^+), which are positively charged; **neutrons** (n^0), which have no charge (are neutral); and **electrons** (e^-), which are negatively charged. The relatively larger protons and neutrons make up a central core, or *nucleus,*[1] that is surrounded by one or more electrons **(figure 2.1).** The nucleus makes up the larger mass (weight) of the atom, whereas the electron region accounts for the greater volume. To get a perspective on proportions,

---

1. Be careful not to confuse the nucleus of an atom with the nucleus of a cell (discussed later).

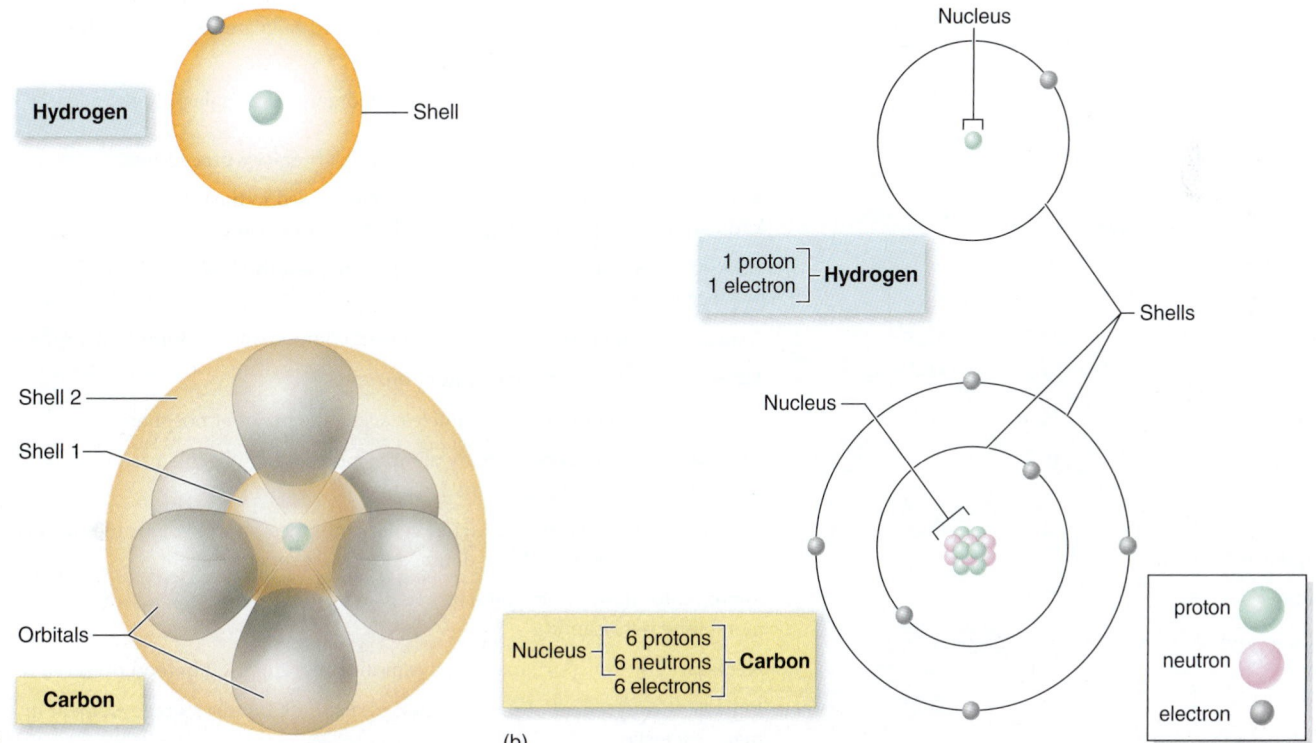

Figure 2.1 Models of atomic structure. **(a)** Three-dimensional models of hydrogen and carbon that approximate their actual structure. The nucleus is surrounded by electrons in orbitals that occur in levels called shells. Hydrogen has just one shell and one orbital. Carbon has two shells and four orbitals; the shape of the outermost orbitals is paired lobes rather than circles or spheres. **(b)** Simple models of the same atoms make it easier to show the numbers and arrangements of shells and electrons and the numbers of protons and neutrons in the nucleus. (Not to accurate scale.)

consider this: If an atom were the size of a baseball stadium, the nucleus would be about the size of a marble! The stability of atomic structure is largely maintained by (1) the mutual attraction of the protons and electrons (opposite charges attract each other); and (2) the exact balance of proton number and electron number, which causes the opposing charges to cancel each other out. At least in theory, then, isolated intact atoms do not carry a charge.

Different Types of Atoms: Elements and Their Properties

All atoms share the same fundamental structure. All protons are identical, all neutrons are identical, and all electrons are identical. But when these subatomic particles come together in specific, varied combinations, unique types of atoms called **elements** result. Each element has a characteristic atomic structure and predictable chemical behavior. To date, about 118 elements, both naturally occurring and artificially produced by physicists, have been described. By convention, an element is assigned a distinctive name with an abbreviated shorthand symbol. The elements are often depicted in a periodic table. **Table 2.1** lists some of the elements common to biological systems, their atomic characteristics, and some of the natural and applied roles they play.

The Major Elements of Life and Their Primary Characteristics

The unique properties of each element result from the numbers of protons, neutrons, and electrons it contains, and each element can be identified by certain physical measurements.

Table 2.1 The Major Elements of Life and Their Primary Characteristics

Element	Atomic Symbol*	Atomic Mass**	Examples of Ionized Forms	Significance in Microbiology
Calcium	Ca	40.1	Ca^{2+}	Part of outer covering of certain shelled amoebas; stored within bacterial spores
Carbon Carbon●	C C-14	12.0 14.0	CO_3^{2-}	Principal structural component of biological molecules Radioactive isotope used in dating fossils
Chlorine	Cl	35.5	Cl^-	Component of disinfectants, used in water purification
Cobalt Cobalt●	Co Co-60	58.9 60	Co^{2+}, Co^{3+}	Trace element needed by some bacteria to synthesize vitamins An emitter of gamma rays; used in food sterilization; used to treat cancer
Copper	Cu	63.5	Cu^+, Cu^{2+}	Necessary to the function of some enzymes; Cu salts are used to treat fungal and worm infections
Hydrogen Hydrogen●	H H3	1 3	H^+	Necessary component of water and many organic molecules; H_2 gas released by bacterial metabolism Has 2 neutrons; radioactive; used in clinical laboratory procedures
Iodine Iodine●	I I-131, I-125	126.9 131, 125	I^-	A component of antiseptics and disinfectants; used in the Gram stain Radioactive isotopes for diagnosis and treatment of cancers
Iron	Fe	55.8	Fe^{2+}, Fe^{3+}	Necessary component of respiratory enzymes; required by some microbes to produce toxin
Magnesium	Mg	24.3	Mg^{2+}	A trace element needed for some enzymes; component of chlorophyll pigment
Manganese	Mn	54.9	Mn^{2+}, Mn^{3+}	Trace element for certain respiratory enzymes
Nitrogen	N	14.0	NO^{3-}	Component of all proteins and nucleic acids; the major atmospheric gas
Oxygen	O	16.0		An essential component of many organic molecules; molecule used in metabolism by many organisms
Phosphorus● Phosphorus●	P P-32	31 32	PO_4^{3-}	A component of ATP, nucleic acids, cell membranes; stored in granules in cells Radioactive isotope used as a diagnostic and therapeutic agent
Potassium	K	39.1	K^+	Required for normal ribosome function and protein synthesis; essential for cell membrane permeability
Sodium	Na	23.0	Na^+	Necessary for transport; maintains osmotic pressure; used in food preservation
Sulfur	S	32.1	SO_4^{2-}	Important component of proteins; makes disulfide bonds; storage element in many bacteria
Zinc	Zn	65.4	Zn^{2+}	An enzyme cofactor; required for protein synthesis and cell division; important in regulating DNA

Notes: *Based on the Latin name of the element. The first letter is always capitalized; if there is a second letter, it is always lowercased.

**The atomic mass or weight is equal to the average mass number for the isotopes of that element.

Isotopes are variant forms of the same element that differ in the number of neutrons. These multiple forms occur naturally in certain proportions. Carbon, for example, exists primarily as carbon 12 with 6 neutrons; but a small amount (about 1%) is carbon 13 with 7 neutrons or carbon 14 with 8 neutrons. Although isotopes have virtually the same chemical properties, some of them have unstable nuclei that spontaneously release energy in the form of radiation. Such *radioactive isotopes* play a role in a number of research and medical applications. Because they emit detectable signs, they can be used to trace the position of key atoms or molecules in chemical reactions, they are tools in diagnosis and treatment, and they are even applied in sterilization procedures (see discussion of ionizing radiation in chapter 11). Another application of isotopes is in dating fossils and other ancient materials.

Electron Orbitals and Shells

The structure of an atom can be envisioned as a central nucleus surrounded by a "cloud" of electrons that constantly rotate about the nucleus in pathways (see figure 2.1). The pathways, called **orbitals,** are not actual objects or exact locations but represent volumes of space in which an electron is likely to be found. Electrons occupy energy shells, proceeding from the lower-level energy electrons nearest the nucleus to the higher-level energy electrons in the farthest orbitals.

Electrons fill the orbitals and shells in *pairs,* starting with the shell nearest the nucleus. The first shell contains one orbital and a maximum of 2 electrons; the second shell has four orbitals and up to 8 electrons; the third shell with nine orbitals can hold up to 18 electrons; and the fourth shell with 16 orbitals contains up to 32 electrons. The number of orbitals and shells and how completely they are filled depend on the numbers of electrons, so that each element

A Note About Mass, Weight, and Related Terms

Mass refers to the quantity of matter that an atomic particle contains. The proton and neutron have almost exactly the same mass, which is about 1.66×10^{-24} grams, a unit of mass known as a Dalton (Da) or unified atomic mass unit (U). All elements can be measured in these units. The terms *mass* and *weight* are often used interchangeably in biology, even though they apply to two different but related aspects of matter. Weight is a measurement of the gravitational pull on the mass of a particle, atom, or object. Consequently, it is possible for something with the same mass to have different weights. For example, an astronaut on the earth (normal gravity) would weigh more than the same astronaut on the moon (weak gravity). Atomic weight has been the traditional usage for biologists, because most chemical reactions and biological activities occur within the normal gravitational conditions on earth. This permits use of the atomic weight as a standard of comparison. You will also see the terms *formula weight* and *molecular weight* used interchangeably, and they are indeed synonyms. They both mean the sum of atomic weights of all atoms in a molecule.

INSIGHT 2.1	Na, Na, Na, Na, Sodium?

Looking at the periodic table, the atomic symbols all seem to make sense. H stands for hydrogen, He is helium, Li is lithium, Be is beryllium, and so on, until we get to Na. Sodium? How do you get Na from sodium? It turns out that the chemical symbol for sodium comes from the Latin word for sodium, *natrium.*

Sodium isn't the only atomic symbol that doesn't match its name. K for potassium comes from *kalium,* the Latin word for alkali. Ag and Au for silver and gold also come from their Latin roots—*argentum* and *aurum,* respectively. The atomic symbol for tungsten is W, from the Germanic name *wolfram,* which is derived from the word *wolframite,* meaning "the devourer of tin," since tungsten interferes with the smelting of tin. The chemical symbol for mercury comes from the Greek word *hydrargyrum,* meaning "liquid silver."

A number of elements are named for famous scientists: Curium (Cm) is named for Marie Curie, Einsteinium (Es) is named for Albert Einstein, and Mendelevium (Md) is named after the Russian chemist Dmitri Mendeleyev, to name a few. Other elements are named for locations (Americium, Berkelium, Californium), and some are named for planets (Plutonium, Uranium, Neptunium). Elementary!

Table salt: Sodium chloride crystals (magnification ×400).

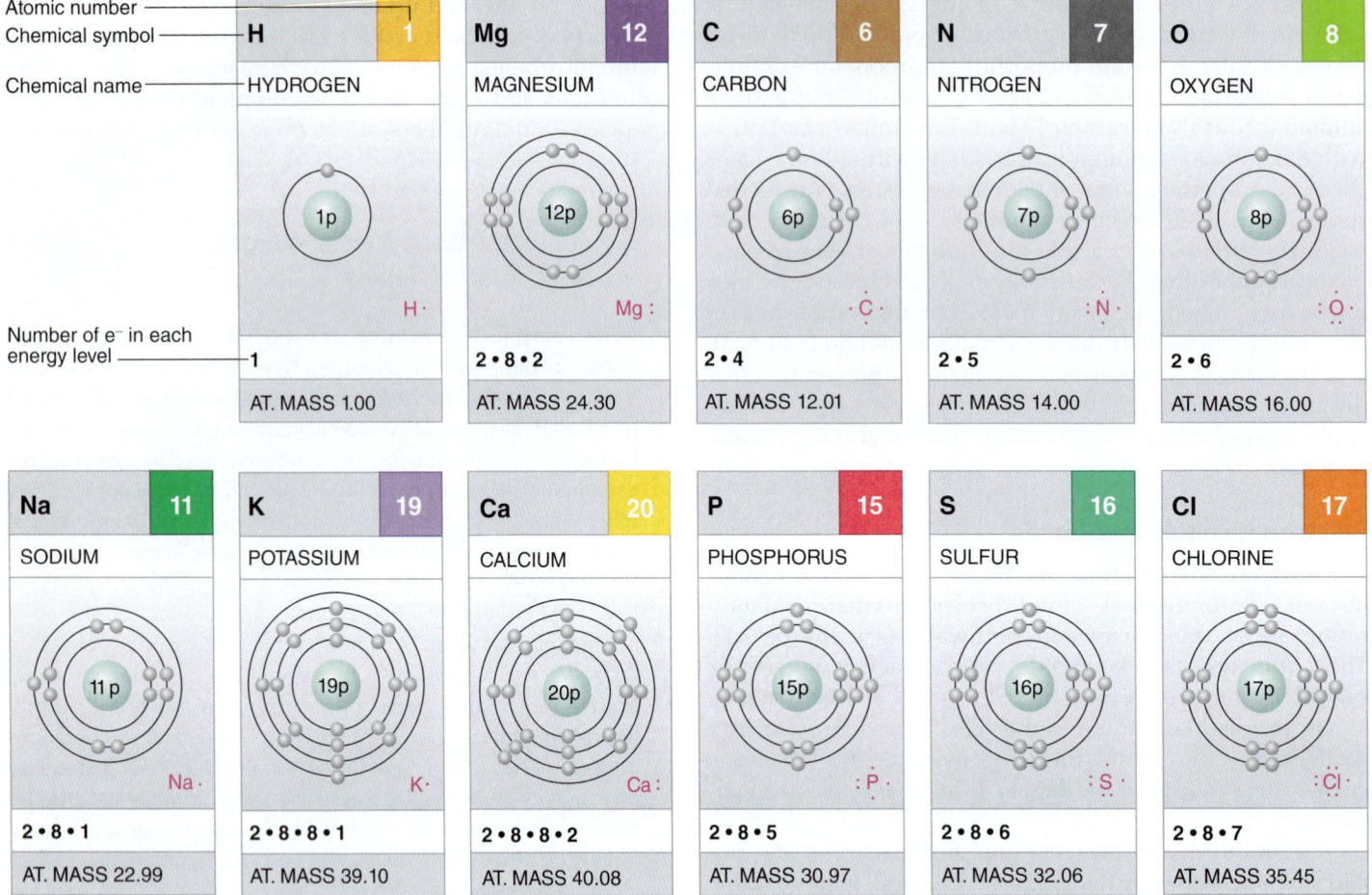

Atomic number
Chemical symbol
Chemical name
Number of e⁻ in each energy level

H	1
HYDROGEN	

H·

1

AT. MASS 1.00

Mg	12
MAGNESIUM	

Mg :

2 • 8 • 2

AT. MASS 24.30

C	6
CARBON	

·Ċ·

2 • 4

AT. MASS 12.01

N	7
NITROGEN	

:Ṅ·

2 • 5

AT. MASS 14.00

O	8
OXYGEN	

:Ö·

2 • 6

AT. MASS 16.00

Na	11
SODIUM	

Na·

2 • 8 • 1

AT. MASS 22.99

K	19
POTASSIUM	

K·

2 • 8 • 8 • 1

AT. MASS 39.10

Ca	20
CALCIUM	

Ca :

2 • 8 • 8 • 2

AT. MASS 40.08

P	15
PHOSPHORUS	

:Ṗ·

2 • 8 • 5

AT. MASS 30.97

S	16
SULFUR	

:S̈·

2 • 8 • 6

AT. MASS 32.06

Cl	17
CHLORINE	

:C̈l·

2 • 8 • 7

AT. MASS 35.45

Figure 2.2 Examples of biologically important atoms.   Simple models show how the shells are filled by electrons as the atomic numbers increase. Notice that these elements have incompletely filled outer shells since they have fewer than 8 electrons.

will have a unique pattern. For example, helium has only a filled first shell of 2 electrons; oxygen has a filled first shell and a partially filled second shell of 6 electrons; and magnesium has a filled first shell, a filled second one, and a third shell that fills only one orbital, so is nearly empty. As we will see, the chemical properties of an element are controlled mainly by the distribution of electrons in the outermost shell. Figure 2.1 and **figure 2.2** present various simplified models of atomic structure and electron maps. **Figure 2.3** presents all the elements in the familiar periodic table. The name of an element or its atomic symbol often reflects its chemical characteristics or a historical background, as seen in **Insight 2.1.**

Bonds and Molecules

Most elements do not exist naturally in pure, uncombined form but are bound together as molecules and compounds. A **molecule** is a distinct chemical substance that results from the combination of two or more atoms. Some molecules such as oxygen (O_2) and nitrogen gas (N_2) consist of atoms

of the same element. Molecules that are combinations of two or more *different* elements are termed **compounds.** Compounds such as water (H_2O) and biological molecules (proteins, sugars, fats) are the predominant substances in living systems. When atoms bind together in molecules, they lose the properties of the atom and take on the properties of the combined substance.

The **chemical bonds** of molecules and compounds result when two or more atoms share, donate (lose), or accept (gain) electrons **(figure 2.4).** The number of electrons in the outermost shell of an element is known as its **valence.** The valence determines the degree of reactivity and the types of bonds an element can make. Elements with a filled outer orbital are relatively stable because they have no extra electrons to share with or donate to other atoms. For example, helium has one filled shell, with no tendency either to give up electrons or to take them from other elements, making it a stable, inert (nonreactive) gas. Elements with partially filled outer orbitals are less stable and are more apt to form some sort of bond. Many chemical reactions are based on the

Periodic Table of the Elements

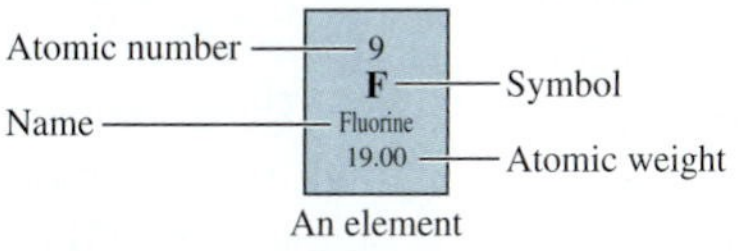

Group number —1A

Period number

Atomic number — 9
F — Symbol
Name — Fluorine
19.00 — Atomic weight
An element

1A	2A	3B	4B	5B	6B	7B	8B	8B	8B	1B	2B	3A	4A	5A	6A	7A	8A
1 **H** Hydrogen 1.0079																	2 **He** Helium 4.0026
3 **Li** Lithium 6.941	4 **Be** Beryllium 9.0122											5 **B** Boron 10.811	6 **C** Carbon 12.011	7 **N** Nitrogen 14.0067	8 **O** Oxygen 15.9994	9 **F** Fluorine 18.9984	10 **Ne** Neon 20.1797
11 **Na** Sodium 22.9898	12 **Mg** Magnesium 24.3050											13 **Al** Aluminum 26.9815	14 **Si** Silicon 28.0855	15 **P** Phosphorus 30.9738	16 **S** Sulfur 32.066	17 **Cl** Chlorine 35.4527	18 **Ar** Argon 39.948
19 **K** Potassium 39.0983	20 **Ca** Calcium 40.078	21 **Sc** Scandium 44.9559	22 **Ti** Titanium 47.88	23 **V** Vanadium 50.9415	24 **Cr** Chromium 51.9961	25 **Mn** Manganese 54.9380	26 **Fe** Iron 55.845	27 **Co** Cobalt 58.9332	28 **Ni** Nickel 58.693	29 **Cu** Copper 63.546	30 **Zn** Zinc 65.41	31 **Ga** Gallium 69.723	32 **Ge** Germanium 72.64	33 **As** Arsenic 74.9216	34 **Se** Selenium 78.96	35 **Br** Bromine 79.904	36 **Kr** Krypton 83.80
37 **Rb** Rubidium 85.4678	38 **Sr** Strontium 87.62	39 **Y** Yttrium 88.9059	40 **Zr** Zirconium 91.224	41 **Nb** Niobium 92.9064	42 **Mo** Molybdenum 95.94	43 **Tc** Technetium (98)	44 **Ru** Ruthenium 101.07	45 **Rh** Rhodium 102.9055	46 **Pd** Palladium 106.42	47 **Ag** Silver 107.8682	48 **Cd** Cadmium 112.411	49 **In** Indium 114.82	50 **Sn** Tin 118.710	51 **Sb** Antimony 121.760	52 **Te** Tellurium 127.60	53 **I** Iodine 126.9045	54 **Xe** Xenon 131.29
55 **Cs** Cesium 132.9054	56 **Ba** Barium 137.327	57 **La** Lanthanum 138.9055	72 **Hf** Hafnium 178.49	73 **Ta** Tantalum 180.9479	74 **W** Tungsten 183.84	75 **Re** Rhenium 186.207	76 **Os** Osmium 190.2	77 **Ir** Iridium 192.22	78 **Pt** Platinum 195.08	79 **Au** Gold 196.9665	80 **Hg** Mercury 200.59	81 **Tl** Thallium 204.3833	82 **Pb** Lead 207.2	83 **Bi** Bismuth 208.9804	84 **Po** Polonium (209)	85 **At** Astatine (210)	86 **Rn** Radon (222)
87 **Fr** Francium (223)	88 **Ra** Radium (226)	89 **Ac** Actinium (227)	104 **Rf** Rutherfordium (267)	105 **Db** Dubnium (268)	106 **Sg** Seaborgium (271)	107 **Bh** Bohrium (272)	108 **Hs** Hassium (270)	109 **Mt** Meitnerium (276)	110 **Ds** Darmstadtium (281)	111 **Rg** Roentgenium (280)	112 **Cn** Copernicium (280)	113 -- -- (284)	114 **Fl** Flerovium (284)	115 -- -- (288)	116 **Lv** Livermorium (280)	117 -- -- (294)	118 -- -- (294)

Lanthanides 6

58 **Ce** Cerium 140.115	59 **Pr** Praseodymium 140.9076	60 **Nd** Neodymium 144.24	61 **Pm** Promethium (145)	62 **Sm** Samarium 150.36	63 **Eu** Europium 151.964	64 **Gd** Gadolinium 157.25	65 **Tb** Terbium 158.9253	66 **Dy** Dysprosium 162.50	67 **Ho** Holmium 164.9303	68 **Er** Erbium 167.26	69 **Tm** Thulium 168.9342	70 **Yb** Ytterbium 173.04	71 **Lu** Lutetium 174.967
90 **Th** Thorium 232.0381	91 **Pa** Protactinium 231.0359	92 **U** Uranium 238.0289	93 **Np** Neptunium (237)	94 **Pu** Plutonium (244)	95 **Am** Americium (243)	96 **Cm** Curium (247)	97 **Bk** Berkelium (247)	98 **Cf** Californium (251)	99 **Es** Einsteinium (252)	100 **Fm** Fermium (257)	101 **Md** Mendelevium (258)	102 **No** Nobelium (259)	103 **Lr** Lawrencium (260)

Actinides 7

Figure 2.3
The periodic table.

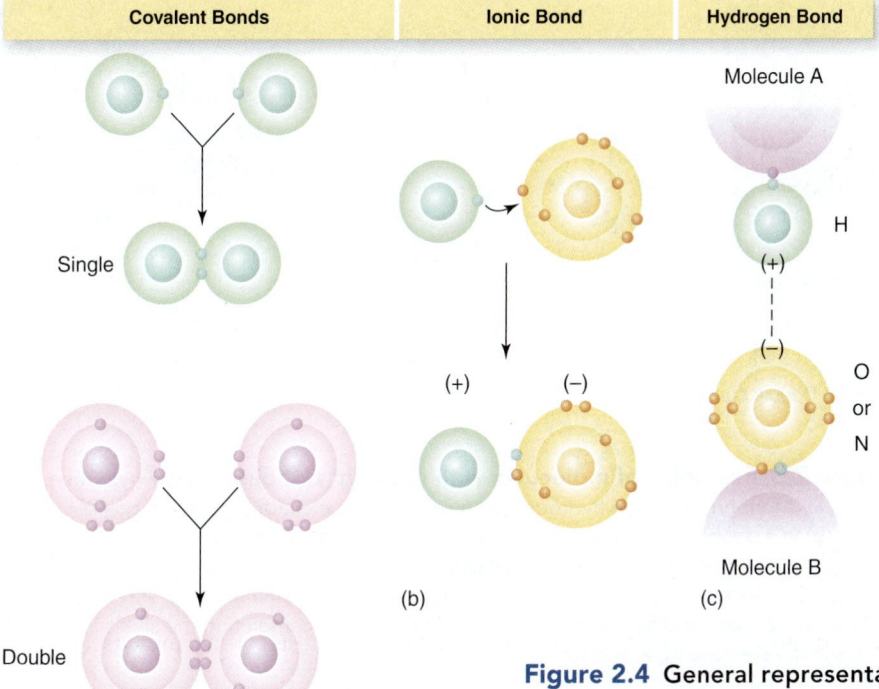

Covalent Bonds	Ionic Bond	Hydrogen Bond

Single

Double

(a)

(b)

(c)

Molecule A

H
(+)

(−)

O
or
N

Molecule B

Figure 2.4 General representation of three types of bonding.   **(a)** Covalent bonds, both single and double. **(b)** Ionic bond. **(c)** Hydrogen bond. Note that hydrogen bonds are represented in models and formulas by dotted lines, as shown in (c).

tendency of atoms with unfilled outer shells to gain greater stability by achieving, or at least approximating, a filled outer shell. For example, an atom such as oxygen that can accept 2 additional electrons will bond readily with atoms (such as hydrogen) that can share or donate electrons. We explore some additional examples of the basic types of bonding in the following section.

In addition to reactivity, the number of electrons in the outer shell also dictates the number of chemical bonds an atom can make. For instance, hydrogen can bind with one other atom, oxygen can bind with up to two other atoms, and carbon can bind with four.

Covalent Bonds and Polarity: Molecules with Shared Electrons

Covalent (cooperative valence) **bonds** form between atoms that share electrons rather than donating or receiving them. A simple example is hydrogen gas (H_2), which consists of two hydrogen atoms. A hydrogen atom has only a single electron, but when two of them combine, each will bring its electron to orbit about both nuclei, thereby approaching a filled orbital (2 electrons) for both atoms and thus creating a single covalent bond **(figure 2.5a).** Covalent bonding also occurs in oxygen gas (O_2) but with a difference. Because each atom has 2 electrons to share in this molecule, the combination creates two pairs of shared electrons, also known as a double covalent bond **(figure 2.5b).** The majority of the molecules associated with living things are composed of single and double covalent bonds between the most common biological elements (carbon, hydrogen, oxygen, nitrogen, sulfur, and phosphorus), which are discussed in more depth in chapter 7. Double bonds in molecules and compounds introduce more rigidity than single bonds. A slightly more complex pattern of covalent bonding is shown for methane gas (CH_4) in **figure 2.5c.**

Other effects of bonding result in differences in polarity. When atoms of different electronegativity[2] form covalent bonds, the electrons are not shared equally and may be pulled more toward one atom than another. This pull causes one end of a molecule to assume a partial negative charge and the other end to assume a partial positive charge. A molecule with such an asymmetrical distribution of charges is termed **polar** and has positive and negative poles. Observe the water

2. Electronegativity—the ability to attract electrons.

A Note About Diatomic Elements

You will notice that hydrogen, oxygen, nitrogen, chlorine, and iodine are often shown in notation with a 2 subscript—H_2 or O_2. These elements are diatomic (two atoms), meaning that in their pure elemental state, they exist in pairs, rather than as a single atom. The reason for this phenomenon has to do with their valences. The electrons in the outer shell are configured so as to complete a full outer shell for both atoms when they bind. You can see this for yourself in figures 2.3 and 2.5. Most of the diatomic elements are gases.

molecule shown in **figure 2.6** and note that, because the oxygen atom is larger and has more protons than the hydrogen atoms, it will tend to draw the shared electrons with greater force toward its nucleus. This unequal force causes the oxygen part of the molecule to express a negative charge (due to the electrons being attracted there) and the hydrogens to express a positive charge (due to the protons). The polar nature of water plays an extensive role in a number of biological reactions, which are discussed later.

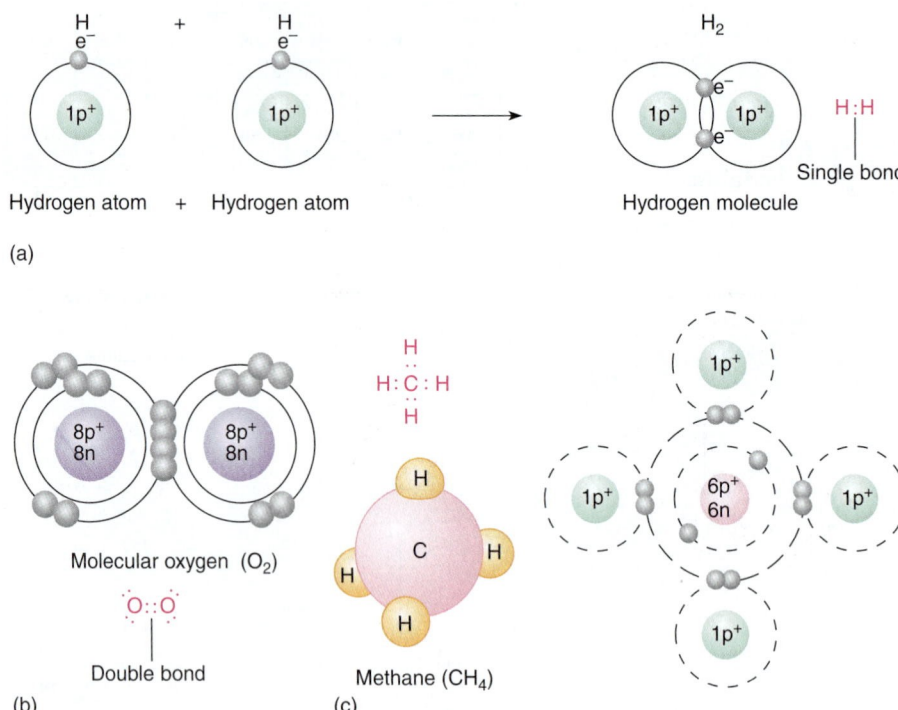

Figure 2.5 Examples of molecules with covalent bonding.   **(a)** A hydrogen molecule is formed when two hydrogen atoms share their electrons and form a single bond. **(b)** In a double bond, the outer orbitals of two oxygen atoms overlap and permit the sharing of 4 electrons (one pair from each) and the saturation of the outer orbital for both. **(c)** Simple, three-dimensional, and working models of methane. Note that carbon has 4 electrons to share and hydrogens each have one, thereby completing the shells for all atoms in the compound, and creating 4 single bonds.

Polarity is a significant property of many large molecules in living systems and greatly influences both their reactivity and their structure.

When covalent bonds are formed between atoms that have the same or similar electronegativity, the electrons are shared equally between the two atoms. Because of this balanced distribution, no part of the molecule has a greater attraction for the electrons. This sort of electrically neutral molecule is termed **nonpolar.**

Ionic Bonds: Electron Transfer Among Atoms

In reactions that form **ionic bonds,** electrons are transferred completely from one atom to another and are not shared. These reactions invariably occur between atoms with valences that complement each other, meaning that one atom has an unfilled shell that will readily accept electrons and the other atom has an unfilled shell that will readily lose electrons. A striking example is the reaction that occurs between sodium (Na) and chlorine (Cl). Elemental sodium is a soft, lustrous metal so reactive that it can burn flesh, and molecular chlorine is a very poisonous yellow gas. But when the two are combined, they form sodium chloride[3] (NaCl)— the familiar nontoxic table salt—a compound with properties quite different from either parent element **(figure 2.7).**

How does this transformation occur? Sodium has 11 electrons (2 in shell one, 8 in shell two, and only 1 in shell three), so it is 7 short of having a complete outer shell. Chlorine has 17 electrons (2 in shell one, 8 in shell two, and 7 in shell three), making it 1 short of a complete outer shell. These two atoms are very reactive with one another, because a sodium atom will readily donate its single electron and a chlorine atom will avidly receive it. (The reaction is slightly more involved than a single sodium atom's combining with a

---

3. In general, when a salt is formed, the ending of the name of the negatively charged ion is changed to *-ide.*

single chloride atom, but this complexity does not detract from the fundamental reaction as described here.) The outcome of this reaction is not many single, isolated molecules of NaCl but rather a solid crystal complex that interlinks millions of sodium and chloride ions **(figure 2.7c,d).**

Ionization: Formation of Charged Particles   Molecules with intact ionic bonds are electrically neutral, but they can produce charged particles when dissolved in a liquid called

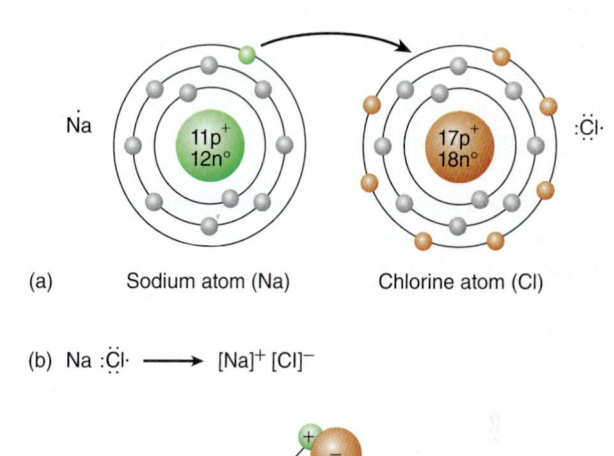

(a)   Sodium atom (Na)   Chlorine atom (Cl)

(b)   Na $:\ddot{C}l\cdot$ ⟶ $[Na]^+ [Cl]^-$

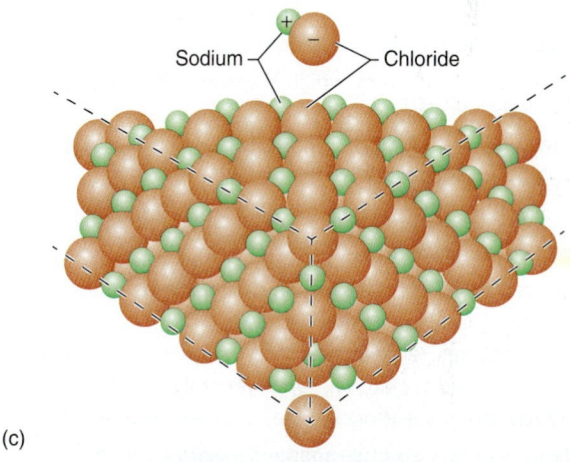

(c)

(d)

Figure 2.7 Ionic bonding between sodium and chlorine.
(a) When the two elements are placed together, sodium loses its single outer orbital electron to chlorine, thereby filling chlorine's outer shell. **(b)** Simple model of ionic bonding. **(c)** Sodium and chloride ions form large molecules, or crystals, in which the two atoms alternate in a definite, regular, geometric pattern. **(d)** Note the cubic nature of NaCl crystals at the macroscopic level.

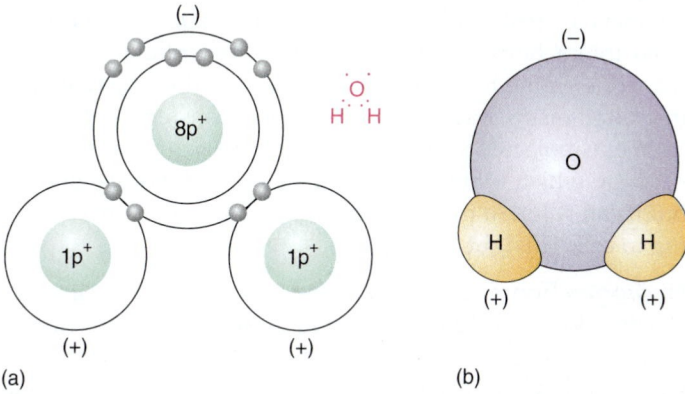

(a)   (b)

Figure 2.6 Polar molecule.   **(a)** A simple model and **(b)** a three-dimensional model of a water molecule indicate the polarity, or unequal distribution, of electrical charge, which is caused by the pull of the shared electrons toward the oxygen side of the molecule.

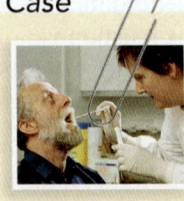
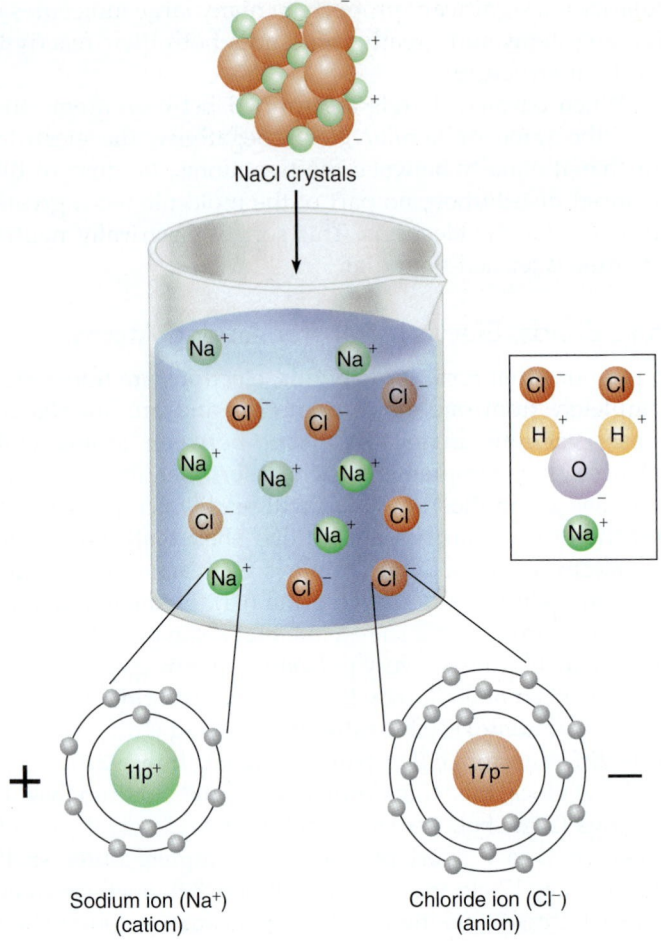

Figure 2.8 Ionization. When NaCl in the crystalline form is added to water, the ions are released from the crystal as separate charged particles (cations and anions) into solution. (See also figure 2.12.) In this solution, Cl^- ions are attracted to the hydrogen component of water, and Na^+ ions are attracted to the oxygen (box).

a solvent. This phenomenon, called **ionization,** occurs when the ionic bond is broken and the atoms dissociate (separate) into unattached, charged particles called **ions (figure 2.8).** To illustrate what gives a charge to ions, let us look again at the reaction between sodium and chlorine. When a sodium atom reacts with chlorine and loses 1 electron, the sodium is left with one more proton than electrons. This imbalance produces a positively charged sodium ion (Na^+). Chlorine, on the other hand, has gained 1 electron and now has 1 more electron than protons, producing a negatively charged ion (Cl^-). Positively charged ions are termed **cations,** and negatively charged ions are termed **anions.** (A good mnemonic device is to think of the "t" in cation as a plus (+) sign and the first "n" in anion as a negative (–) sign.) Substances such as salts, acids, and bases that release ions when dissolved in water are termed **electrolytes** because their charges enable

them to conduct an electrical current. Owing to the general rule that particles of like charge repel each other and those of opposite charge attract each other, we can expect ions to interact electrostatically with other ions and polar molecules. Such interactions are important in many cellular chemical reactions, in the formation of solutions, and in the reactions microorganisms have with dyes. The transfer of electrons from one molecule to another constitutes a significant mechanism by which biological systems store and release energy, but it can result in the formation of products that are damaging to cells **(Insight 2.2).**

Hydrogen Bonding Some types of bonding do not involve sharing, losing, or gaining electrons but instead are due to attractive forces between nearby molecules or atoms. One such bond is a **hydrogen bond,** a weak type of bond that forms between a hydrogen covalently bonded to one molecule and an oxygen or nitrogen atom on the same molecule or on a different molecule. Because hydrogen in a covalent bond tends to be positively charged, it will attract a nearby negatively charged atom and form an easily disrupted bridge with it. This type of bonding is usually represented in molecular models with a dotted line. A simple example of hydrogen bonding occurs between water molecules **(figure 2.9).** More extensive hydrogen bonding is partly responsible for the structure and stability of proteins and nucleic acids, as you will see later on.

Other similar noncovalent associations between molecules are the **van der Waals forces.** These weak attractions occur between molecules that demonstrate low levels of polarity. Neighboring groups with slight attractions will interact and remain associated. These forces are an essential factor in maintaining the cohesiveness of large molecules with many packed atoms.

It is safe to say that though each of these two types of bonds, hydrogen bonds and van der Waals forces, are relatively weak on their own, they provide great stability to molecules because there are often many of them in one area. The weakness of each individual bond also provides flexibility, allowing molecules to change their shapes and also to bind and unbind to other objects relatively easily. The fundamental processes of life involve bonding and unbonding (for example, the DNA helix has to "unbond" or unwind in order for replication to occur; enzymatic reactions require proteins to bind to other molecules and then be released), and hydrogen bonds and van der Waals forces are custom made for doing just that.

Chemical Shorthand: Formulas, Models, and Equations

The atomic content of molecules can be represented by a few convenient formulas. We have already been using the molecular formula, which concisely gives the atomic symbols and the number of the elements involved in subscript (CO_2, H_2O). More complex molecules such as glucose ($C_6H_{12}O_6$) can also be symbolized this way, but this formula is not unique, because fructose and galactose also share it.

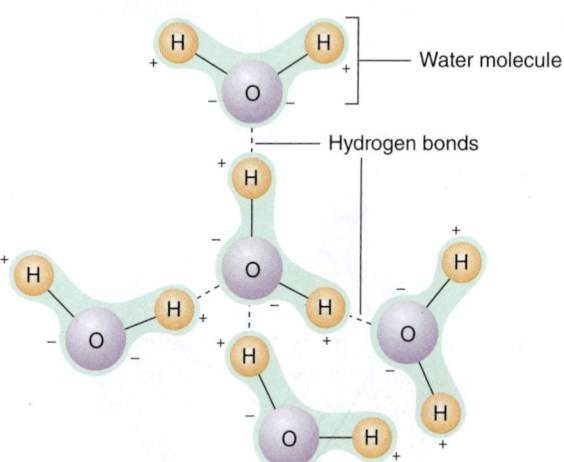

Figure 2.9 Hydrogen bonding in water.   Because of the polarity of water molecules, the negatively charged oxygen end of one water molecule is weakly attracted to the positively charged hydrogen end of an adjacent water molecule.

Molecular formulas are useful, but they only summarize the atoms in a compound; they do not show the position of bonds between atoms. For this purpose, chemists use structural formulas illustrating the relationships of the atoms and the number and types of bonds **(figure 2.10)**. Other structural models present the three-dimensional appearance of a molecule, illustrating the orientation of atoms (differentiated by color) and the molecule's overall shape **(figure 2.11)**. These are often called space-filling models, as you can get an idea of how the molecule actually occupies its space. The spheres surrounding each atom indicate how far the atom's influence can be felt, let's say. Sometimes it is also referred to as the atom's volume.

The printed page tends to make molecules appear static, but this picture is far from correct, because molecules are capable of changing through chemical reactions. For ease in tracing chemical exchanges between atoms or molecules, and to provide some sense of the dynamic character of reactions, chemists use shorthand equations containing symbols, numbers, and arrows to simplify or summarize the major characteristics of a reaction. Molecules entering or starting a reaction are called **reactants,** and substances left by a reaction are called **products.** In most instances, summary chemical reactions do not give the details of the exchange, in order to keep the expression simple and to save space.

In a *synthesis reaction,* the reactants bond together in a manner that produces an entirely new molecule (reactant A plus reactant B yields product AB). An example is the production of sulfur dioxide, a by-product of burning sulfur fuels and an important component of smog:

$$S + O_2 \rightarrow SO_2$$

Some synthesis reactions are not such simple combinations. When water is synthesized, for example, the reaction does not really involve one oxygen atom combining with two hydrogen atoms, because elemental oxygen exists as O_2 and elemental hydrogen exists as H_2. A more accurate equation for this reaction is

$$2H_2 + O_2 \rightarrow 2H_2O$$

The equation for reactions must be balanced—that is, the number of atoms on one side of the arrow must equal the number on the other side to reflect all of the participants in the reaction. To arrive at the total number of atoms in the reaction, multiply the prefix number by the subscript number; if no number is given, it is assumed to be 1.

In *decomposition reactions,* the bonds on a single reactant molecule are permanently broken to release two or more product molecules. One example is the resulting molecules when large nutrient molecules are digested into smaller units; a simpler example can be shown for the common chemical hydrogen peroxide:

$$2H_2O_2 \rightarrow 2H_2O + O_2$$

During *exchange reactions,* the reactants trade portions between each other and release products that are combinations

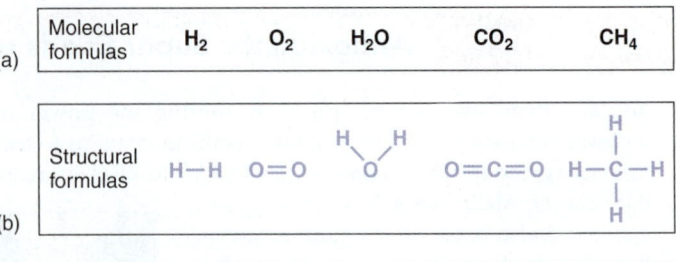

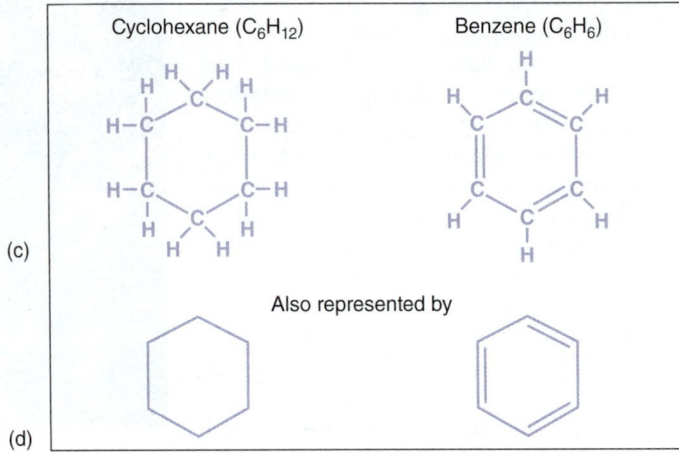

Figure 2.10 Comparison of molecular and structural formulas. **(a)** Molecular formulas provide a brief summary of the elements in a compound. **(b)** Structural formulas clarify the exact relationships of the atoms in the molecule, depicting single bonds by a single line and double bonds by two lines. **(c)** In structural formulas of organic compounds, cyclic or ringed compounds may be completely labeled, or **(d)** they may be presented in a shorthand form in which carbons are assumed to be at the angles and attached to hydrogens. See figure 2.15 for structural formulas of three sugars with the same molecular formula, $C_6H_{12}O_6$.

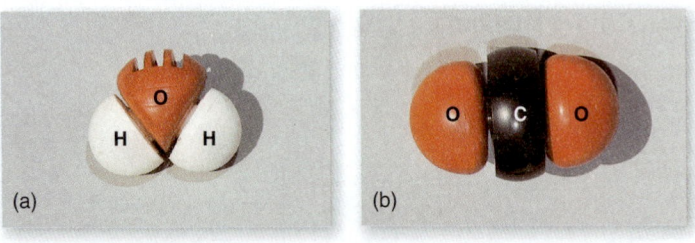

Figure 2.11 Three-dimensional, or space-filling, models of (a) water, (b) carbon dioxide, and (c) glucose. The red atoms are oxygen, the white ones hydrogen, and the black ones carbon.

of the two. This type of reaction occurs between acids and bases when they form water and a salt:

$$AB + XY \rightleftharpoons AX + BY$$

The reactions in biological systems can be reversible, meaning that reactants and products can be converted back and forth. These reversible reactions are symbolized with a double arrow, each pointing in opposite directions, as in the preceding exchange reaction. Whether a reaction is reversible depends on the proportions of these compounds, the difference in energy state of the reactants and products, and the presence of **catalysts** (substances that increase the rate of a reaction). Additional reactants coming from another reaction can also be indicated by arrows that enter or leave at the main arrow:

$$CD \quad C$$
$$X + XY \longrightarrow XYD$$

Solutions: Homogeneous Mixtures of Molecules

A **solution** is a mixture of one or more substances called **solutes** uniformly dispersed in a dissolving medium called a **solvent.** An important characteristic of a solution is that the solute cannot be separated by filtration or ordinary settling. The solute can be gaseous, liquid, or solid, and the solvent is usually a liquid. Examples of solutions are salt or sugar dissolved in water and iodine dissolved in alcohol. In general, a solvent will dissolve a solute only if it has similar electrical characteristics as indicated by the rule of solubility, expressed simply as "like dissolves like." For example, water is a polar molecule and will readily dissolve an ionic solute such as NaCl, yet a nonpolar solvent such as benzene will not dissolve NaCl.

Water is the most common solvent in natural systems, having several characteristics that suit it to this role. The polarity of the water molecule causes it to form hydrogen bonds with other water molecules, but it can also interact readily with charged or polar molecules. When an ionic solute such as NaCl crystals is added to water, it is dissolved, thereby releasing Na^+ and Cl^- into solution. Dissolution occurs because Na^+ is attracted to the negative pole of the water molecule and Cl^- is attracted to the positive pole; in this way, they are drawn away from the crystal separately into solution. As it leaves, each ion becomes **hydrated,** which means that it is surrounded by a sphere of water molecules **(figure 2.12).** Molecules such as salt or sugar that attract water to their surface are termed **hydrophilic.** Nonpolar molecules, such as benzene, that repel water are considered **hydrophobic.** A third class of molecules, which includes the phospholipids in cell membranes, is considered **amphipathic** because these molecules have both hydrophilic and hydrophobic properties.

Because most biological activities take place in aqueous (water-based) solutions, the concentration of these solutions can be very important (see chapter 7). The **concentration** of a solution expresses the amount of solute dissolved in a certain amount of solvent. It can be calculated by weight, volume, or percentage. A common way to calculate percentage of concentration is to use the weight of the solute, measured in grams (g), dissolved in a specified volume of solvent, measured in milliliters (ml). For example, dissolving 3 g of NaCl in 100 ml of water produces a 3% solution; dissolving 30 g in 100 ml produces a 30% solution; and dissolving 3 g in 1,000 ml (1 liter) produces a 0.3% solution.

A common way to express concentration of biological solutions is by its molar concentration, or *molarity* (M). A standard molar solution is obtained by dissolving one *mole,* defined as the molecular weight of the compound in grams, in 1 liter (1,000 ml) of solution. To make a 1 M solution of sodium chloride, we would dissolve 58 g of NaCl in 1 liter of solvent; a 0.1 M solution would require dissolving 5.8 g of NaCl in 1 liter of solvent.

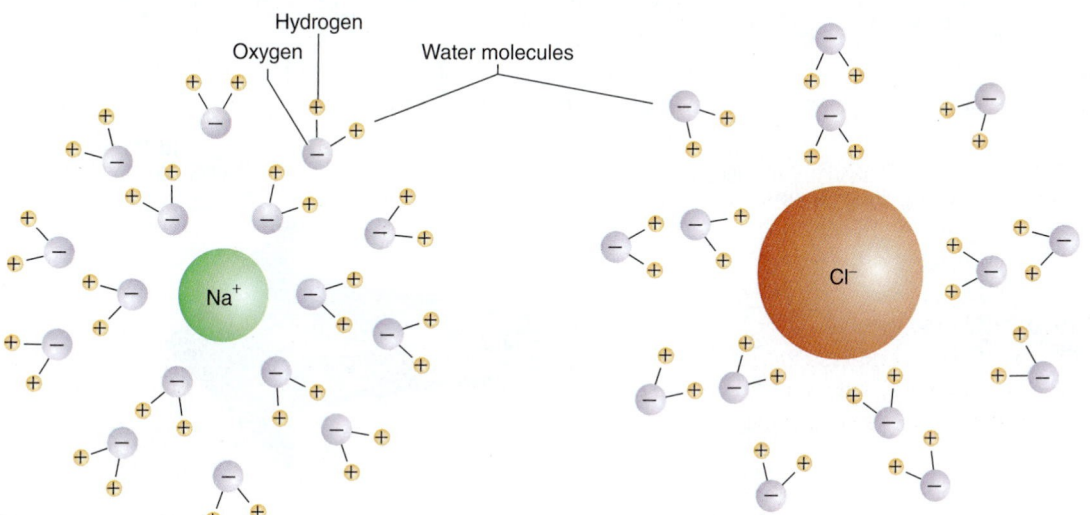

Hydrogen

Oxygen

Water molecules

Na^+

Cl^-

Figure 2.12 Hydration spheres formed around ions in solution. In this example, a sodium cation attracts the negatively charged region of water molecules, and a chloride anion attracts the positively charged region of water molecules. In both cases, the ions become covered with spherical layers of specific numbers and arrangements of water molecules.

Acidity, Alkalinity, and the pH Scale

Another factor with far-reaching impact on living things is their environment's relative degree of acidity or basicity (also called alkalinity). To understand how solutions become acidic or basic, we must look again at the behavior of water molecules. Hydrogens and oxygen tend to remain bonded by covalent bonds, but in certain instances, a single hydrogen can break away as the ionic form (H^+), leaving the remainder of the molecule in the form of an OH^- ion. The H^+ ion is positively charged because it is essentially a hydrogen ion that has lost its electron; the OH^- is negatively charged because it remains in possession of that electron. Ionization of water is constantly occurring, but in pure water containing no other ions, H^+ and OH^- are produced in equal amounts, and the solution remains neutral. By one definition, a solution is considered **acidic** when a component dissolved in water (acid) releases excess hydrogen ions[4] (H^+); a solution is **basic** when a component releases excess hydroxyl ions (OH^-), so that there is no longer a balance between the two ions.

To measure the acid and base concentrations of solutions, scientists use the **pH** scale, a graduated numerical scale that ranges from 0 (the most acidic) to 14 (the most basic). This scale is a useful standard for rating relative acidity and basicity; use **figure 2.13** to familiarize yourself with the pH readings of some common substances. Because the pH scale is a logarithmic scale, each increment (from pH 2.0 to pH 3.0) represents a tenfold change in concentration of ions. (Take a moment to glance at appendix A to review logarithms and exponents.)

---

4. Actually, it forms a hydronium ion (H_3O^+), but for simplicity's sake, we will use the notation of H^+.

Disease Connection

While many bacteria living in or on humans require a neutral pH (around 7.0) to thrive, the bacterium responsible for gastric ulcers, *Helicobacter pylori*, is a notable exception. It lives in the stomach, which has an acidic pH of around 2. It counters this hostile environment by secreting an enzyme called urease, which cleaves the common chemical urea into carbon dioxide and ammonia. The ammonia neutralizes stomach acid, allowing the bacterium to grow and thrive.

More precisely, the pH is based on the negative logarithm of the concentration of H^+ ions (symbolized as $[H^+]$) in a solution, represented as

$$pH = -\log[H^+]$$

The quantity is expressed in moles per liter. Recall that a mole is simply a standard unit of measurement and refers to the amount of substance containing 6×10^{23} atoms.

Acidic solutions have a greater concentration of H^+ than OH^-, starting with pH 0, which contains 1.0 mole H^+/liter. Each of the subsequent whole-number readings in the scale changes in $[H^+]$ by a tenfold reduction, so that pH 1 contains [0.1 mole H^+/liter], pH 2 contains [0.01 mole H^+/liter], and so on, continuing in the same manner up to pH 14, which contains [0.00000000000001 mole H^+/liter]. These same concentrations can be represented more manageably by exponents: pH 2 has an $[H^+]$ of 10^{-2} mole, and pH 14 has an $[H^+]$ of 10^{-14} mole **(table 2.2)**. It is evident that the pH units are derived from the exponent itself. Even

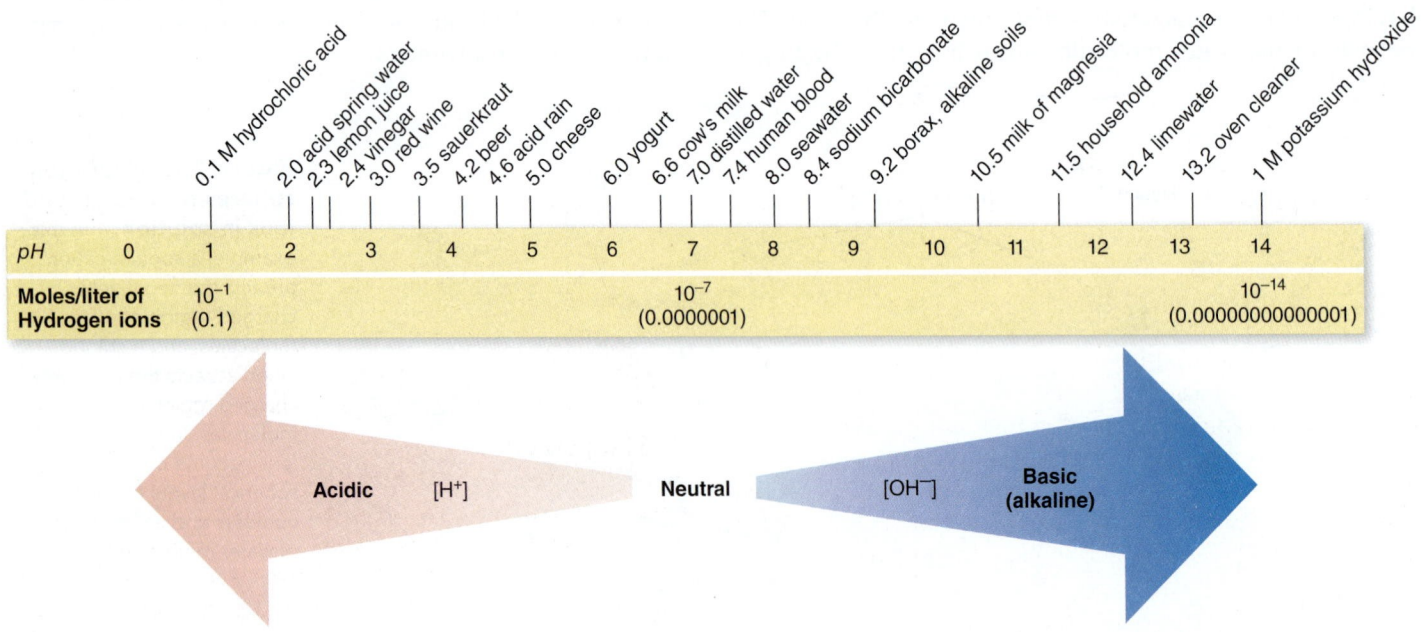

Figure 2.13 The pH scale. Shown are the relative degrees of acidity and basicity and the approximate pH readings for various substances.

Table 2.2 Hydrogen Ion and Hydroxide Ion Concentrations at a Given pH

Moles/Liter of Hydrogen Ions	Logarithm	pH	Moles/Liter of OH⁻
1.0	10^{-0}	0	10^{-14}
0.1	10^{-1}	1	10^{-13}
0.01	10^{-2}	2	10^{-12}
0.001	10^{-3}	3	10^{-11}
0.0001	10^{-4}	4	10^{-10}
0.00001	10^{-5}	5	10^{-9}
0.000001	10^{-6}	6	10^{-8}
0.0000001	10^{-7}	7	10^{-7}
0.00000001	10^{-8}	8	10^{-6}
0.000000001	10^{-9}	9	10^{-5}
0.0000000001	10^{-10}	10	10^{-4}
0.00000000001	10^{-11}	11	10^{-3}
0.000000000001	10^{-12}	12	10^{-2}
0.0000000000001	10^{-13}	13	10^{-1}
0.00000000000001	10^{-14}	14	10^{-0}

though the basis for the pH scale is [H^+], it is important to note that, as the [H^+] in a solution decreases, the [OH^-] increases in direct proportion. At midpoint—pH 7, or neutrality—the concentrations are exactly equal and neither predominates, this being the pH of pure water previously mentioned.

In summary, the pH scale can be used to rate or determine the degree of acidity or basicity of a solution. On this scale, a pH below 7 is acidic, and the lower the pH, the greater the acidity. A pH above 7 is basic, and the higher the pH, the greater the basicity. Incidentally, although pHs are given here in whole numbers, more often, a pH reading exists in decimal form, for example, pH 4.5 or 6.8 (acidic) and pH 7.4 or 10.2 (basic). Because of the damaging effects of very concentrated acids or bases, most cells operate best under neutral, weakly acidic, or weakly basic conditions (see chapter 7).

Aqueous solutions containing both acids and bases may be involved in **neutralization** reactions, which give rise to water and other neutral by-products. For example, when equal molar solutions of hydrochloric acid (HCl) and sodium hydroxide (NaOH, a base) are mixed, the reaction proceeds as follows:

$$HCl + NaOH \rightarrow H_2O + NaCl$$

Here the acid and base ionize to H^+ and OH^- ions, which form water, and other ions, Na^+ and Cl^-, which form sodium chloride. Any product other than water that arises when acids and bases react is called a salt. Many of the organic acids (such as lactic and succinic acids) that function in **metabolism** are available as the acid and the salt form (such as lactate, succinate), depending on the conditions in the cell (see chapter 8).

The Chemistry of Carbon and Organic Compounds

So far, our main focus has been on the characteristics of atoms; ions; and small, simple substances that play diverse roles in the structure and function of living things. These substances are often lumped together in a category called **inorganic chemicals.** A chemical is usually inorganic if it does not contain both carbon and hydrogen. Examples of inorganic chemicals include NaCl (sodium chloride), $Mg_3(PO_4)_2$ (magnesium phosphate), $CaCO_3$ (calcium carbonate), and CO_2 (carbon dioxide). In reality, however, most of the chemical reactions and structures of living things involve more complex molecules, termed **organic chemicals.** These are carbon compounds with a basic framework of the element carbon bonded to other atoms. Organic molecules vary in complexity from the simplest, methane (CH_4; see figure 2.5c), which has a molecular weight of 16, to certain antibody molecules (part of our immune systems) that have a molecular weight of nearly 1,000,000 and are among the most complex molecules on earth.

The role of carbon as the fundamental element of life can best be understood if we look at its chemistry and bonding patterns. The valence of carbon makes it an ideal atomic building block to form the backbone of organic molecules; it has 4 electrons in its outer orbital to be shared with other atoms (including other carbons) through covalent bonding. As a result, it can form stable chains containing thousands of carbon atoms and still has bonding sites available for forming covalent bonds with numerous other atoms. The bonds that carbon forms are linear, branched, or ringed; and it can form four single bonds, two double bonds, or one triple bond **(figure 2.14).** The atoms with which carbon is most often associated in organic compounds are hydrogen, oxygen, nitrogen, sulfur, and phosphorus.

Functional Groups of Organic Compounds

One important advantage of carbon's serving as the molecular skeleton for living things is that it is free to bind with an unending array of other molecules. These special molecular groups or accessory molecules that bind to organic compounds are called **functional groups.** Functional groups help define the chemical class of certain groups of organic compounds and confer unique reactive properties on the

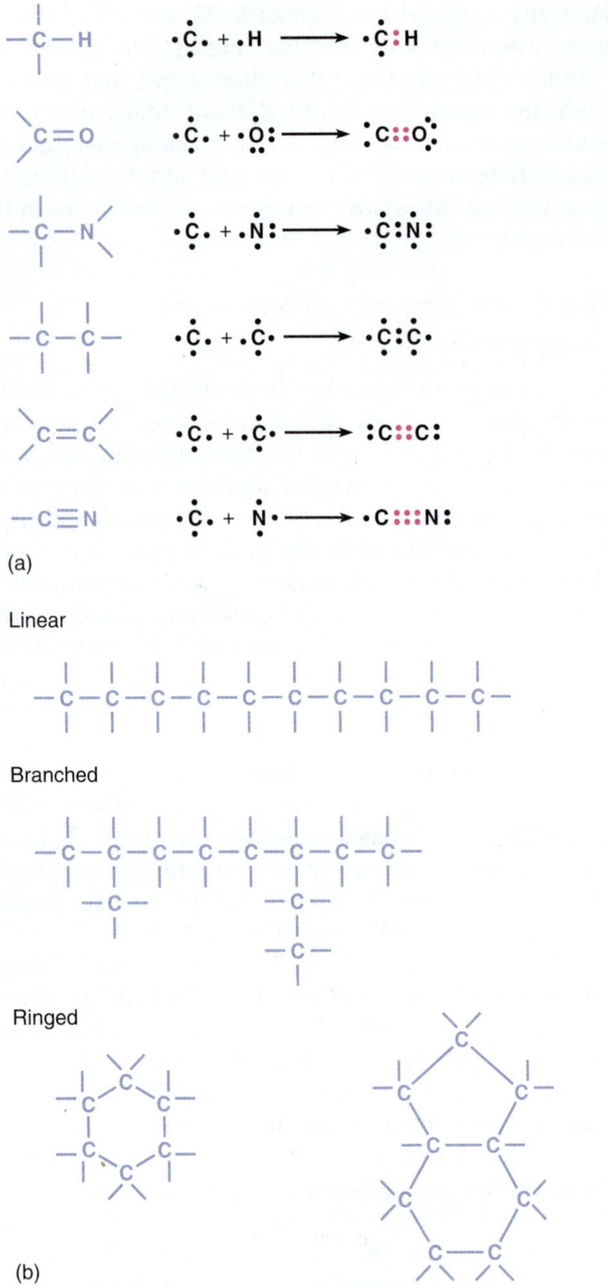

(a)

Linear

Branched

Ringed

(b)

Figure 2.14 The versatility of bonding in carbon.
In most compounds, each carbon makes a total of four bonds.
(a) Both single and double bonds can be made with other carbons,
oxygen, and nitrogen; single bonds are made with hydrogen.
Simple electron models show how the electrons are shared in these
bonds. **(b)** Multiple bonding of carbons can give rise to long chains,
branched compounds, and ringed compounds, many of which are
extraordinarily large and complex.

Table 2.3 Representative Functional Groups and Classes of Organic Compounds

Formula of Functional Group	Name	Class of Compounds
$R^* - O - H$	Hydroxyl	Alcohols, carbohydrates
$R - C \overset{O}{\underset{OH}{<}}$	Carboxyl	Fatty acids, proteins, organic acids
$R - C(H) - NH_2$ (with H below)	Amino	Proteins, nucleic acids
$R - C \overset{O}{\underset{O-R}{<}}$	Ester	Lipids
$R - C(H) - SH$ (with H below)	Sulfhydryl	Cysteine (amino acid), proteins
$R - C \overset{O}{\underset{H}{<}}$	Carbonyl, terminal end	Aldehydes, polysaccharides
$R - \overset{O}{\overset{\|}{C}} - C -$	Carbonyl, internal	Ketones, polysaccharides
$R - O - \overset{O}{\overset{\|}{P}} - OH$ (with OH below)	Phosphate	DNA, RNA, ATP

Note: The R designation on a molecule is shorthand for *residue*, and it indicates
that what is attached at that site varies from one compound to another.

designation on a molecule is shorthand for residue, and its
placement in a formula indicates that the residue (functional
group) varies from one compound to another.

2.1 Learning Outcomes—Assess Your Progress

1. Explain the relationship between atoms and elements.
2. List and define four types of chemical bonds.
3. Differentiate between a solute and a solvent.
4. Provide a brief definition of *pH*.

whole molecule **(table 2.3).** Because each type of functional
group behaves in a distinctive manner, reactions of an
organic compound can be predicted by knowing the kind
of functional group or groups it carries. Many reactions rely
upon functional groups such as R—OH or R—NH₂. The —R

2.2 Macromolecules: Superstructures of Life

The compounds of life fall into the realm of **biochemistry.** Biochemicals are organic compounds produced by (or components of) living things, and they include four main families: carbohydrates, lipids, proteins, and nucleic acids **(table 2.4).** The compounds in these groups are assembled from smaller molecular subunits, or building blocks, and because they are often very large compounds, they are termed **macromolecules.** All macromolecules except lipids are formed by polymerization, a process in which repeating subunits termed **monomers** are bound into chains of various lengths termed **polymers.** For example, proteins (polymers) are composed of a chain of amino acids (monomers). The large size and complex, three-dimensional shape of macromolecules enables them to function as structural components, molecular messengers, energy sources, enzymes (biochemical catalysts), nutrient stores, and sources of genetic information. In the following section and in later chapters, we consider numerous concepts relating to the roles of macromolecules in cells. Table 2.4 will also be a useful reference when you study metabolism in chapter 8.

Carbohydrates: Sugars and Polysaccharides

The term **carbohydrate** originates from the composition of members of this class: they are combinations of carbon (carbo-) and water (-hydrate). Although carbohydrates can be generally represented by the formula $(CH_2O)_n$, in which n indicates the number of units of this combination of atoms **(figure 2.15a),** some carbohydrates contain additional atoms of sulfur or nitrogen.

Carbohydrates exist in a great variety of configurations. The common term *sugar* **(saccharide)** refers to a simple carbohydrate such as a monosaccharide or a disaccharide. A **monosaccharide** is a simple sugar containing from 3 to 7 carbons; a **disaccharide** is a combination of two monosaccharides; and a **polysaccharide** is a polymer of five or more monosaccharides bound in linear or branched-chain

Table 2.4 Macromolecules and Their Functions

Macromolecule	Description/Basic Structure	Examples	Notes About the Examples
Carbohydrates			
Monosaccharides	3– to 7–carbon sugars	Glucose, fructose	Sugars involved in metabolic reactions; building block of disaccharides and polysaccharides
Disaccharides	Two monosaccharides	Maltose (malt sugar)	Composed of two glucoses; an important breakdown product of starch
		Lactose (milk sugar)	Composed of glucose and galactose
		Sucrose (table sugar)	Composed of glucose and fructose
Polysaccharides	Chains of monosaccharides	Starch, cellulose, glycogen	Cell wall, food storage
Lipids			
Triglycerides	Fatty acids + glycerol	Fats, oils	Major component of cell membranes; storage
Phospholipids	Fatty acids + glycerol + phosphate	Membrane components	
Waxes	Fatty acids, alcohols	Mycolic acid	Cell wall of mycobacteria
Steroids	Ringed structure	Cholesterol, ergosterol	In membranes of eukaryotes and some bacteria
Proteins			
	Amino acids	Enzymes, part of cell membrane, cell wall, ribosomes, antibodies	Serve as structural components and perform metabolic reactions
Nucleic Acids			
	Nucleotides Purines: adenine, guanine Pyrimidines: cytosine, thymine, uracil		
Deoxyribonucleic acid (DNA)	Contains deoxyribose sugar and thymine, not uracil	Chromosomes; genetic material of viruses	Mediate inheritance
Ribonucleic acid (RNA)	Contains ribose sugar and uracil, not thymine	Ribosomes; mRNA, tRNA	Facilitate expression of genetic traits

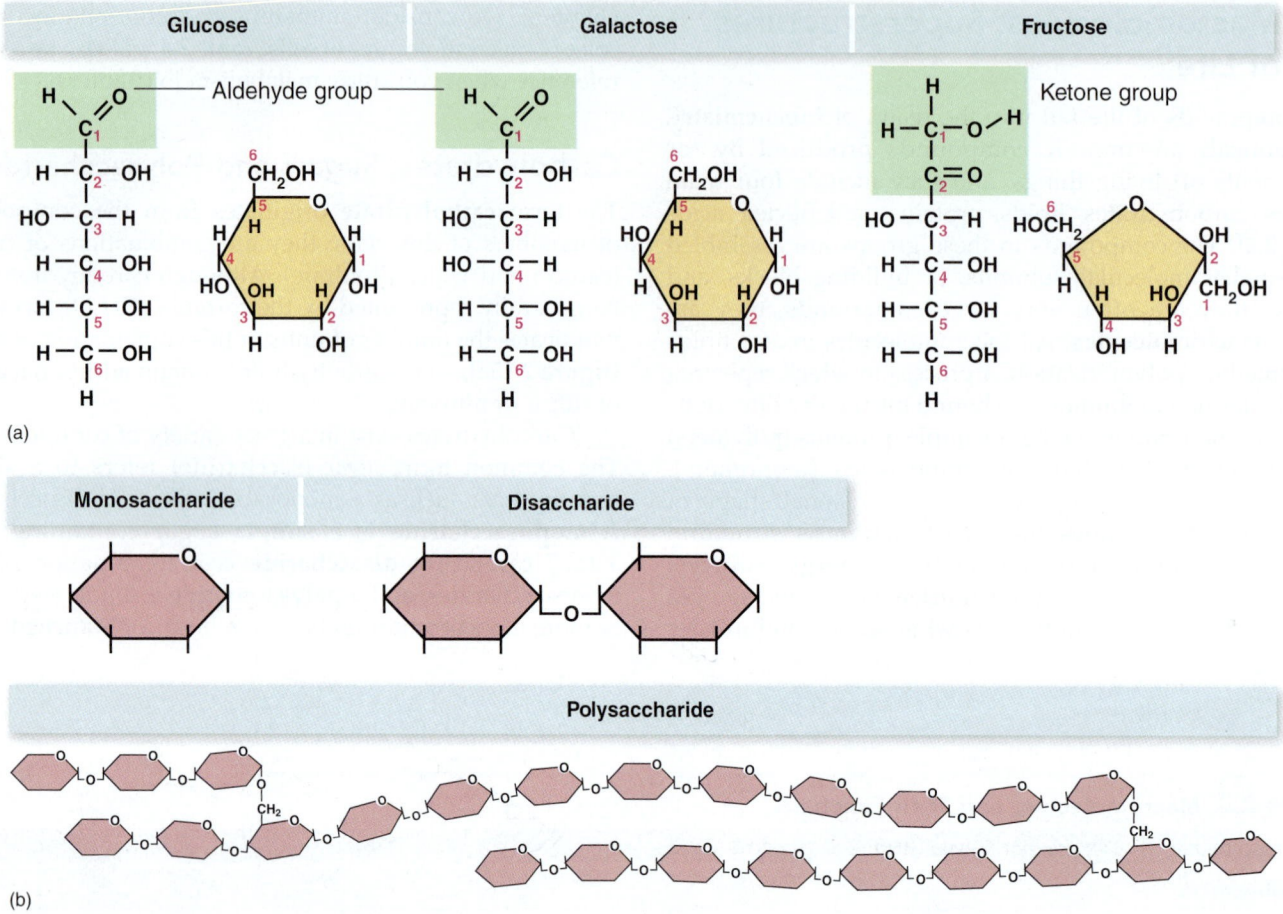

(a)

(b)

Figure 2.15 Common classes of carbohydrates. (a) Three hexoses with the same molecular formula and different structural formulas. Both linear and ring models are given. The linear form emphasizes aldehyde and ketone groups, although in solution the sugars exist in the ring form. Note that the carbons are numbered so as to keep track of reactions within and between monosaccharides. (b) Major saccharide groups, named for the number of sugar units each contains.

patterns **(figure 2.15b).** Monosaccharides and disaccharides are specified by combining a prefix that describes some characteristic of the sugar with the suffix *-ose.* For example, **hexoses** are composed of 6 carbons, and **pentoses** contain 5 carbons. **Glucose** (Gr. *glyko,* sweet) is the most common and universally important hexose; **fructose** is named for fruit (one place where it is found); and xylose, a pentose, derives its name from the Greek word for wood. Disaccharides are named similarly: **lactose** (L. *lacteus,* milk) is an important component of milk; **maltose** means malt sugar; and **sucrose** (Fr. *sucre,* sugar) is common table sugar or cane sugar.

The Nature of Carbohydrate Bonds

The subunits of disaccharides and polysaccharides are linked by means of **glycosidic bonds,** in which carbons (each is assigned a number) on adjacent sugar units are bonded to the same oxygen atom like links in a chain **(figure 2.16).** For example, maltose is formed when the number 1 carbon on a glucose bonds to the oxygen on the number 4 carbon on a second glucose; sucrose is formed when glucose and fructose bind oxygen between their number 1 and number 2 carbons; and lactose is formed when glucose and galactose connect by

their number 1 and number 4 carbons. In order to form this bond, 1 carbon gives up its OH group and the other (the one contributing the oxygen to the bond) loses the H from its OH group. Because a water molecule is produced, this reaction is known as **dehydration synthesis,** a process common to most polymerization reactions. Three polysaccharides (starch, cellulose, and glycogen) are structurally and biochemically distinct, even though all are polymers of the same monosaccharide—glucose. The basis for their differences lies primarily in the exact way the glucoses are bound together, which greatly affects the characteristics of the end product **(figure 2.17).** The synthesis and breakage of each type of bond require a specialized catalyst called an enzyme (see chapter 8).

The Functions of Polysaccharides

Polysaccharides typically contribute to structural support and protection and serve as nutrient and energy stores. The cell walls in plants and many microscopic algae derive their strength and rigidity from **cellulose,** a long, fibrous polymer **(figure 2.17a).** Because of this role, cellulose is probably one of the most common organic substances on the earth, yet it

(a)

(b)

Galactose   +   Glucose   ⟶   **Lactose**

is digestible only by certain bacteria, fungi, and protozoa. These microbes, called *decomposers,* play an essential role in breaking down and recycling plant materials. Some bacteria secrete slime layers of a glucose polymer called *dextran.* This substance causes a sticky layer to develop on teeth that leads to plaque, described later in chapter 22.

Other structural polysaccharides can be conjugated (chemically bonded) to amino acids, nitrogen bases, lipids, or proteins. **Agar,** an indispensable polysaccharide in preparing solid culture media, is a natural component of certain seaweeds. It is a complex polymer of galactose and sulfur-containing carbohydrates. The exoskeletons of certain fungi contain **chitin** (ky-tun), a polymer of glucosamine (a sugar with an amino functional group). **Peptidoglycan** (pep-tih-doh-gly'-kan) is one special class of compounds in which polysaccharides (glycans) are linked to peptide fragments (a short chain of amino acids). This molecule provides the main source of structural support to the bacterial cell wall. The cell wall of gram-negative bacteria also contains **lipopolysaccharide,** a complex of lipid and polysaccharide responsible for symptoms such as fever and shock (see chapters 4 and 13).

The outer surface of many cells has a "sugar coating" composed of polysaccharides bound in various ways to proteins (the combination is a glycoprotein). This structure, called the **glycocalyx,** functions in attachment to other cells or as a site for *receptors*—surface molecules that receive external stimuli or act as binding sites. Small sugar molecules account for the differences in human blood types, and carbohydrates are a component of large protein molecules called *antibodies.* Viruses also have glycoproteins on their surface with which they bind to and invade their host cells.

Polysaccharides are usually stored by cells in the form of glucose polymers such as starch **(figure 2.17*b*)** or **glycogen,** but only organisms with the appropriate digestive enzymes can break them down and use them as a nutrient source. Because a water molecule is required for breaking the bond between two glucose molecules, digestion is also termed **hydrolysis.** Starch is the primary storage food of green plants, microscopic algae, and some fungi; glycogen (animal starch) is a stored carbohydrate for animals and certain groups of bacteria and protozoa.

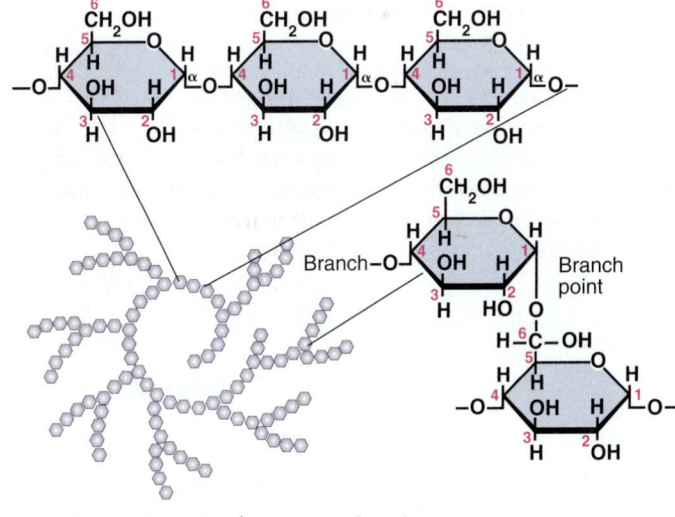

H bonds

(a) **Cellulose**

Branch–O

Branch point

(b) **Starch**

Figure 2.17 Polysaccharides. (a) Cellulose is composed of β glucose bonded in 1,4 bonds that produce linear, lengthy chains of polysaccharides that are H-bonded along their length. This is the typical structure of wood and cotton fibers. (b) Starch is also composed of glucose polymers, in this case α glucose. The main structure is amylose bonded in a 1,4 pattern, with side branches of amylopectin bonded by 1,6 bonds. The entire molecule is compact and granular.

Lipids: Fats, Phospholipids, and Waxes

The term **lipid,** derived from the Greek word *lipos,* meaning fat, is not a chemical designation but an operational term for a variety of substances that are not soluble in polar solvents such as water (recall that oil and water do not mix) but will dissolve in nonpolar solvents such as benzene and chloroform. This property occurs because the substances we call lipids contain relatively long or complex C—H (hydrocarbon) chains that are nonpolar and thus hydrophobic. The main groups of compounds classified as lipids are triglycerides, phospholipids, steroids, and waxes.

Important storage lipids are the **triglycerides,** a category that includes fats and oils. Triglycerides are composed of a single molecule of glycerol bound to three fatty acids **(figure 2.18). Glycerol** is a 3-carbon alcohol[5] with three OH groups that serve as binding sites, and fatty acids are long-chain hydrocarbon molecules with a carboxyl group (COOH) at one end that is free to bind to the glycerol. The hydrocarbon portion of a fatty acid can vary in length from 4 to 24 carbons; and, depending on the fat, it may be saturated or unsaturated. If all carbons in the chain are single-bonded to 2 other carbons and 2 hydrogens, the fat is saturated; if there is at least one C=C double bond in the

chain, it is unsaturated. The structure of fatty acids is what gives fats and oils (liquid fats) their greasy, insoluble nature. In general, solid fats (such as butter) are more saturated, and liquid fats (such as oils) are more unsaturated. In recent years there has been a realization that a type of triglyceride, called popularly *trans fat,* is harmful to the health of those who consume it. A trans fat is an unsaturated triglyceride with one or more of its fatty acids in a position (trans) that is not often found in nature, but it is a common occurrence in processed foods.

In most cells, triglycerides are stored in long-term concentrated form as droplets or globules. When they are acted on by digestive enzymes called lipases, the fatty acids and glycerol are freed to be used in metabolism. Fatty acids are a superior source of energy, yielding twice as much per gram as other storage molecules (starch). Soaps are K^+ or Na^+ salts of fatty acids whose qualities make them excellent grease removers and cleaners (see chapter 11).

Membrane Lipids

A class of lipids that serves as a major structural component of cell membranes is the **phospholipids.** Although phospholipids also contain glycerol and fatty acids, they have some significant differences from triglycerides. Phospholipids contain only two fatty acids attached to the glycerol, and the third glycerol binding site holds a

---

5. Alcohols are carbon compounds containing OH groups.

Figure 2.18 Synthesis and structure of a triglyceride. **(a)** Because a water molecule is released at each ester bond, this is another example of dehydration synthesis. The jagged lines and R symbol represent the hydrocarbon chains of the fatty acids, which are commonly very long. **(b)** Structural and three-dimensional models of fatty acids and triglycerides. (1) A saturated fatty acid has long, straight chains that readily pack together and form solid fats. (2) An unsaturated fatty acid—here a polyunsaturated one with 3 double bonds— has bends in the chain that prevent packing and produce oils (right).

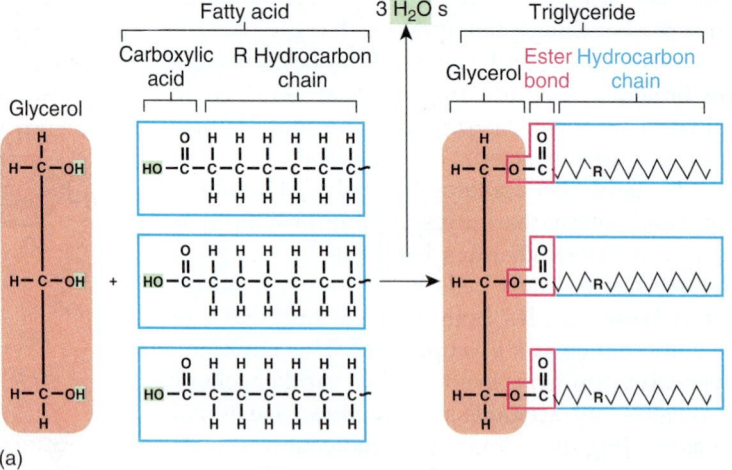

(a)

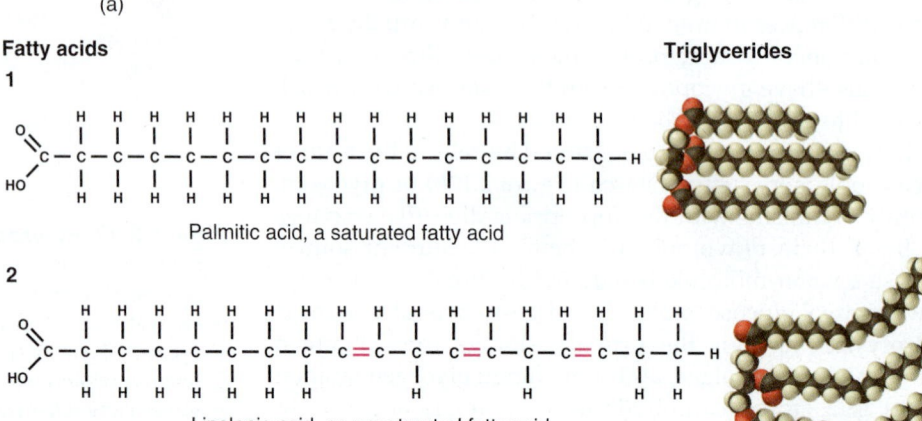

Fatty acids

1

Palmitic acid, a saturated fatty acid

2

Linolenic acid, an unsaturated fatty acid

(b)

Triglycerides

phosphate group. The phosphate is in turn bonded to an alcohol, which varies from one phospholipid to another **(figure 2.19a)**. These lipids have a hydrophilic region from the charge on the phosphoric acid–alcohol "head" of the molecule and a hydrophobic region that corresponds to the long, uncharged "tail" (formed by the fatty acids). When exposed to an aqueous solution, the charged heads are attracted to the water phase, and the nonpolar tails are repelled from the water phase **(figure 2.19b)**. This property causes lipids to naturally assume single and double layers (bilayers), which contribute to their biological significance in membranes. When two single layers of polar lipids come together to form a double layer, the outer hydrophilic face of each single layer will orient itself toward the solution, and the hydrophobic portions will become immersed in the core of the bilayer. The structure of lipid bilayers confers characteristics on membranes such as selective permeability and fluid nature **(Insight 2.3)**.

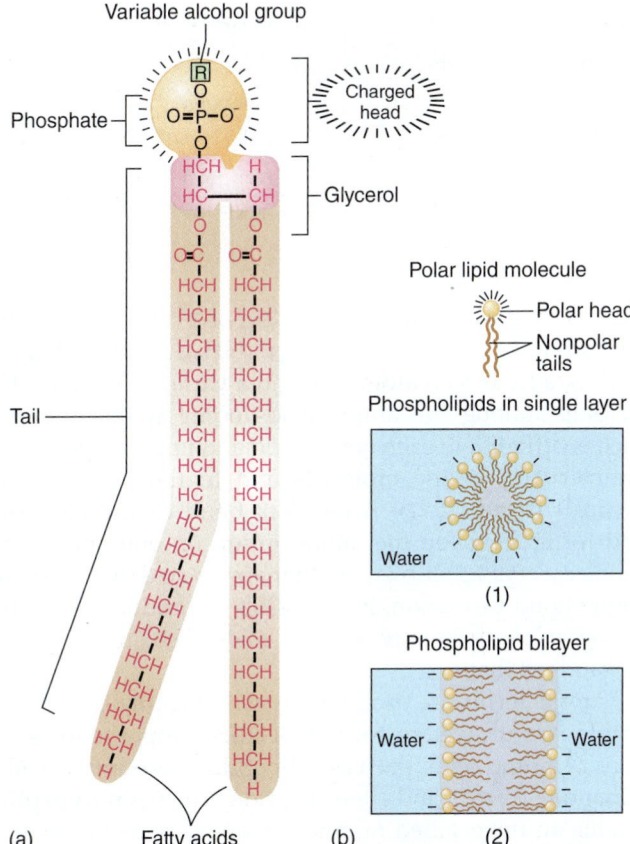

Figure 2.19 Phospholipids—membrane molecules. **(a)** A model of a single molecule of a phospholipid. The phosphate-alcohol head lends a charge to one end of the molecule; its long, trailing hydrocarbon chain is uncharged. **(b)** The behavior of phospholipids in water-based solutions causes them to become arranged (1) in single layers called micelles, with the charged head oriented toward the water phase and the hydrophobic nonpolar tail buried away from the water phase, or (2) in double-layered phospholipid systems with the hydrophobic tails sandwiched between two hydrophilic layers.

Labels in figure: Variable alcohol group; Phosphate; Charged head; Glycerol; Tail; Fatty acids; (a); (b); Polar lipid molecule; Polar head; Nonpolar tails; Phospholipids in single layer; Water; (1); Phospholipid bilayer; Water; Water; (2)

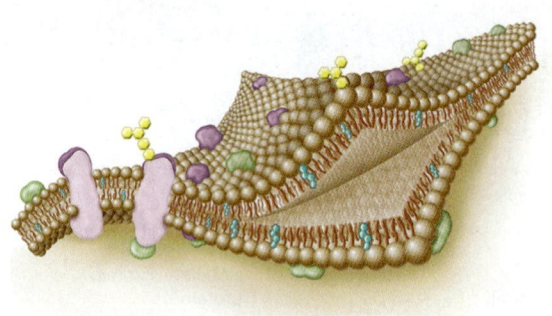

INSIGHT 2.3   Membranes: Cellular Skins

The word **membrane** appears frequently in descriptions of cells in this chapter and in chapters 4 and 5. The word itself describes any lining or covering, including such multicellular structures as the mucous membranes of the body. From the perspective of a single cell, however, a membrane is a thin, double-layered sheet composed of lipids such as phospholipids and sterols (averaging about 40% of membrane content) and protein molecules (averaging about 60%). The primary role of this structure is to completely encase the cytoplasm. Membranes are also components of eukaryotic organelles such as nuclei, mitochondria, and chloroplasts; and they appear in internal pockets of certain bacterial and archaeal cells. Even some viruses, which are not cells at all, can have a membranous protective covering.

Cell membranes are so thin—on the average, just 0.0070 μm (7 nm) thick—that they cannot actually be seen with an optical microscope. Even at magnifications made possible by electron microscopy (500,000×), very little of the precise architecture can be visualized, and a cross-sectional view has the appearance of railroad tracks. Following detailed microscopic and chemical analysis, S. J. Singer and C. K. Nicholson proposed a simple and elegant description of membrane structure called the **fluid mosaic model.** According to this model, a membrane is a continuous bilayer formed by lipids that are oriented with the polar lipid heads toward the outside and the nonpolar tails toward the center of the membrane. Embedded at numerous sites in this bilayer are various-size globular proteins. Some proteins are situated only at the surface; others extend fully through the entire membrane. The configuration of the inner and outer sides of the membrane can be quite different because of the variations in protein shape and position.

Membranes are dynamic and constantly changing because the lipid phase is in motion and many proteins can migrate freely about, somewhat as icebergs do in the ocean. This fluidity is essential to such activities as engulfment of food and discharge or secretion by cells. The structure of the lipid phase provides an impenetrable barrier to many substances. This property accounts for the selective permeability and capacity to regulate transport of molecules. It also serves to segregate activities within the cell's cytoplasm. Membrane proteins function in receiving molecular signals (receptors), in binding and transporting nutrients, and in acting as enzymes, topics to be discussed in chapters 7 and 8.

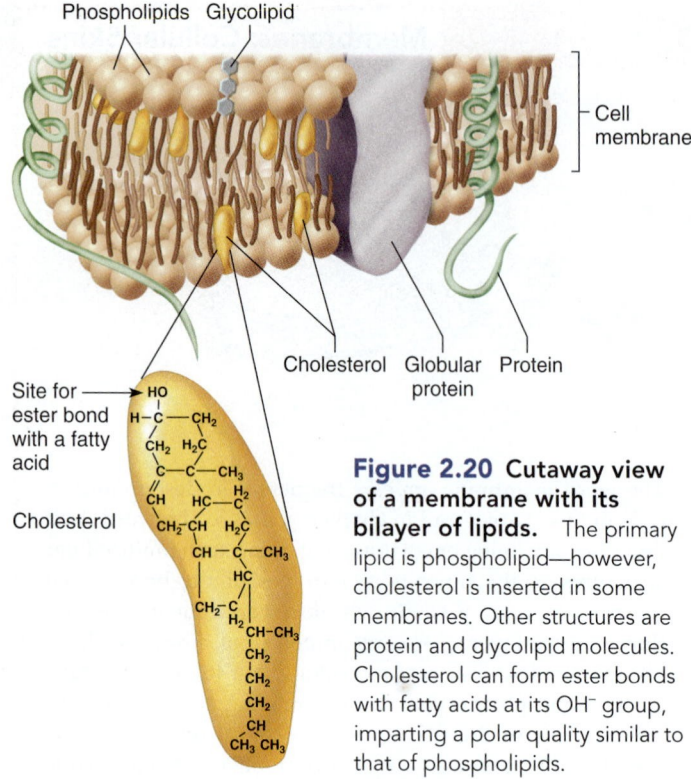

Phospholipids Glycolipid

Cell membrane

Site for ester bond with a fatty acid

Cholesterol

Cholesterol Globular protein Protein

Figure 2.20 Cutaway view of a membrane with its bilayer of lipids. The primary lipid is phospholipid—however, cholesterol is inserted in some membranes. Other structures are protein and glycolipid molecules. Cholesterol can form ester bonds with fatty acids at its OH⁻ group, imparting a polar quality similar to that of phospholipids.

Table 2.5 Twenty Amino Acids and Their Abbreviations

Acid	Abbreviation	Characteristic of R Groups
Alanine	Ala	Nonpolar
Arginine	Arg	+
Asparagine	Asn	Polar
Aspartic acid	Asp	−
Cysteine	Cys	Polar
Glutamic acid	Glu	−
Glutamine	Gln	Polar
Glycine	Gly	Nonpolar
Histidine	His	+
Isoleucine	Ile	Nonpolar
Leucine	Leu	Nonpolar
Lysine	Lys	+
Methionine	Met	Nonpolar
Phenylalanine	Phe	Nonpolar
Proline	Pro	Nonpolar
Serine	Ser	Polar
Threonine	Thr	Polar
Tryptophan	Trp	Nonpolar
Tyrosine	Tyr	Polar
Valine	Val	Nonpolar

Note: + = positively charged; − = negatively charged.

Steroids and Waxes

Steroids are complex ringed compounds commonly found in cell membranes and animal hormones. The best known of these is the sterol (meaning a steroid with an OH group) called **cholesterol (figure 2.20).** Cholesterol reinforces the structure of the cell membrane in animal cells and in an unusual group of cell-wall-deficient bacteria called the mycoplasmas (see chapter 4). The cell membranes of fungi also contain a sterol, called ergosterol.

Chemically, a *wax* is an ester formed between a long-chain alcohol and a saturated fatty acid. The resulting material is typically pliable and soft when warmed but hard and water resistant when cold (paraffin, for example). Among living things, fur, feathers, fruits, leaves, human skin, and insect exoskeletons are naturally waterproofed with a coating of wax. Bacteria that cause tuberculosis and leprosy produce a wax that repels ordinary laboratory stains and contributes to their pathogenicity.

Proteins: Shapers of Life

The predominant organic molecules in cells are **proteins,** a fitting term adopted from the Greek word *proteios,* meaning first or prime. To a large extent, the structure, behavior, and unique qualities of each living thing are a consequence of the proteins they contain. To best explain the origin of the special properties and versatility of proteins, we must examine their general structure. The building blocks of proteins are **amino acids,** which exist in 20 different naturally occurring forms **(table 2.5).** Various combinations of these amino acids account for the nearly infinite variety of proteins. Amino

acids have a basic skeleton consisting of a carbon (called the α carbon) linked to an amino group (NH_2), a carboxyl group (COOH), a hydrogen atom (H), and a variable R group. The variations among the amino acids occur at the R group, which is different in each amino acid and imparts the unique characteristics to the molecule and to the proteins that contain it **(figure 2.21).** A covalent bond called a **peptide bond** forms between the amino group on one amino acid and the carboxyl group on another amino acid. As a result of peptide bond formation, it is possible to produce molecules varying in length from two amino acids to chains containing thousands of them.

Various terms are used to denote the nature of proteins. **Peptide** usually refers to a molecule composed of short chains of amino acids, such as a dipeptide (two amino acids), a tripeptide (three), and a tetrapeptide (four). A **polypeptide** contains an unspecified number of amino acids but usually has more than 20 and is often a smaller subunit of a protein. A protein is the largest of this class of compounds and usually contains a minimum of 50 amino acids. It is common for the term *protein* to be used to describe all of these molecules; we used it in its general sense in the first sentence of this paragraph. But not all polypeptides are large enough to be considered proteins. In chapter 9, we see that protein synthesis is not just a random connection of amino acids; it is directed by information provided in DNA.

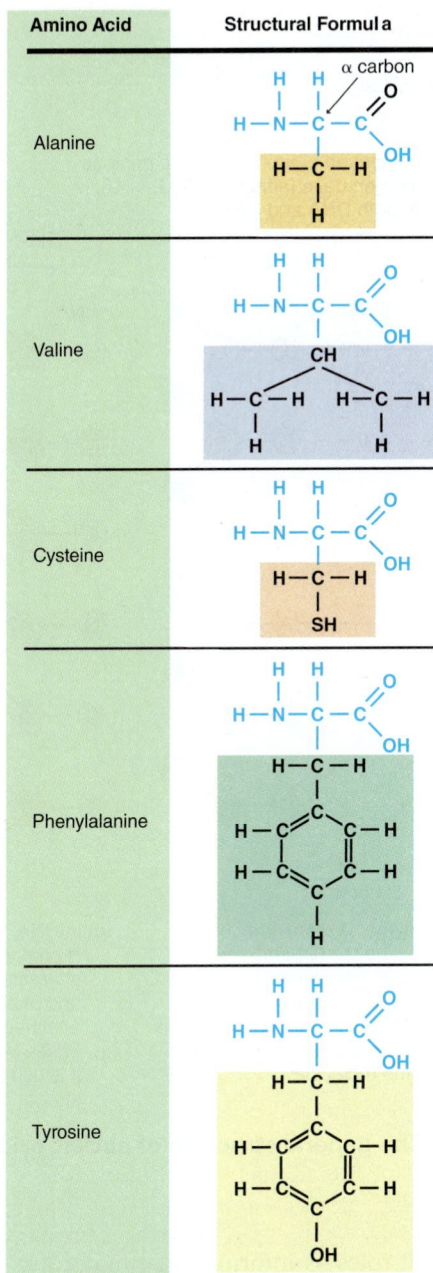

Amino Acid	Structural Formula

Figure 2.21 Structural formulas of selected amino acids. The basic structure common to all amino acids is shown in blue type; and the variable group, or R group, is placed in a colored box. Note the variations in structure of the R group.

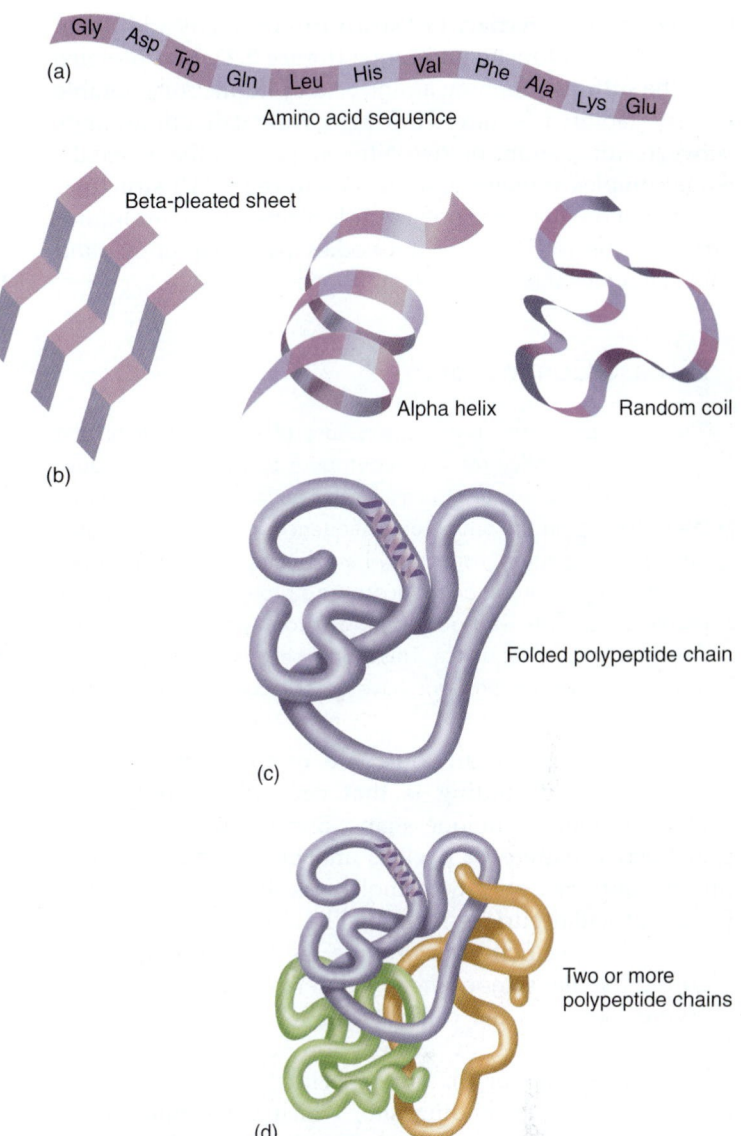

Figure 2.22 Stages in the formation of a functioning protein. **(a)** Its primary structure is a series of amino acids bound in a chain. **(b)** Its secondary structure develops when the chain forms hydrogen bonds that fold it into one of several configurations such as an alpha helix or beta-pleated sheet. Some proteins have several configurations in the same molecule. **(c)** A protein's tertiary structure is due to further folding of the molecule into a three-dimensional mass that is stabilized by hydrogen, ionic, and disulfide bonds between functional groups. **(d)** The quaternary structure exists only in proteins that consist of more than one polypeptide chain. The chains in this protein each have a different color.

Protein Structure and Diversity

The reason that proteins are so varied and specific is that they do not function in the form of a simple straight chain of amino acids (called the *primary structure*). A protein has a natural tendency to assume more complex levels of organization, called the secondary, tertiary, and quaternary structures **(figure 2.22).** The **primary (1°) structure** is the type, number, and order of amino acids in the chain, which varies extensively from protein to protein. The **secondary (2°) structure** arises when various functional groups exposed on the outer surface of the molecule interact by forming hydrogen bonds. This interaction causes the amino acid chain to twist into a coiled configuration called the α *helix* or to fold into an accordion pattern called a β-*pleated sheet*. Many proteins contain both types of secondary configurations. Proteins at the secondary level undergo a third degree of torsion called the **tertiary (3°) structure** created by additional bonds between functional groups **(figure 2.22c).** In proteins with the sulfur-containing amino acid **cysteine,** considerable tertiary stability is achieved through covalent disulfide bonds between sulfur atoms on two different parts of the molecule. Some complex proteins assume a **quaternary (4°) structure,** in which more than one polypeptide forms a large, multiunit protein. This is typical of antibodies (see chapter 15) and some enzymes that act in cell synthesis.

Disease Connection

Tiny changes in the primary structure of one protein on the surface of the influenza virus contribute to its ability to cause disease in people year after year. This surface protein accumulates changes in the amino acid sequence on a continual basis. The immune system, which may have been primed by previous infection or immunization to recognize a protein with a particular primary sequence, will not respond as strongly to a protein with a changed primary sequence. This is one reason you can get the flu more than once—even if you have been vaccinated in the past.

The most important outcome of the various forms of bonding and folding is that each different type of protein develops a unique shape, and its surface displays a distinctive pattern of pockets and bulges. As a result, a protein can react only with molecules that complement or fit its particular surface features like a lock and key. Such a degree of specificity can provide the functional diversity required for many thousands of different cellular activities. **Enzymes** serve as the catalysts for all chemical reactions in cells, and nearly every reaction requires a different enzyme (see chapter 8). This specificity comes from the architecture of the binding site, which determines which molecules fit it. The same is true of antibodies; **antibodies** are complex glycoproteins with specific regions of attachment for bacteria, viruses, and other microorganisms. Certain bacterial toxins (poisonous products) react with only one specific organ or tissue; and proteins embedded in the cell membrane have reactive sites restricted to a certain nutrient. The functional three-dimensional form of a protein is termed the *native state;* and if it is disrupted by some means, the protein is said to be *denatured.* Agents such as heat, acid, alcohol, and some disinfectants disrupt (and thus denature) the stabilizing intrachain bonds and cause the molecule to become nonfunctional, as described in chapter 11.

The Nucleic Acids: A Cell Computer and Its Programs

The nucleic acids, **deoxyribonucleic acid (DNA)** and **ribonucleic acid (RNA),** were originally isolated from the cell nucleus. Shortly thereafter, they were also found in other parts of nucleated cells, in cells with no nuclei (bacteria and archaea), and in viruses. The universal occurrence of nucleic acids in all known cells and viruses emphasizes

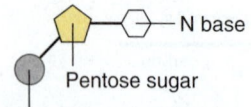

(a) A nucleotide, composed of a phosphate, a pentose sugar, and a nitrogen base (either A,T,C,G, or U), is the monomer of both DNA and RNA.

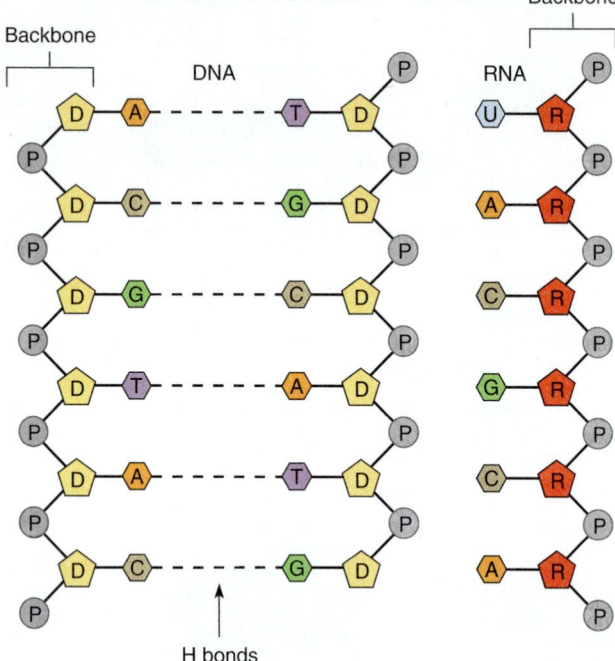

(b) In DNA, the polymer is composed of alternating deoxyribose (D) and phosphate (P) with nitrogen bases (A,T,C,G) attached to the deoxyribose. DNA almost always exists in pairs of strands, oriented so that the bases are paired across the central axis of the molecule.

(c) In RNA, the polymer is composed of alternating ribose (R) and phosphate (P) attached to nitrogen bases (A,U,C,G), but it is usually a single strand.

Figure 2.23 The general structure of nucleic acids.

their important roles as informational molecules. DNA, the master computer of cells, contains a special coded genetic program with detailed and specific instructions for each organism's heredity. It transfers the details of its program to RNA, "helper" molecules responsible for carrying out DNA's instructions and translating the DNA program into proteins that can perform life functions. For now, let us briefly consider the structure and some functions of DNA, RNA, and a close relative, adenosine triphosphate (ATP).

Both DNA and RNA are polymers of repeating units called **nucleotides,** each of which is composed of three smaller units: a **nitrogen base,** a **pentose** (5-carbon) sugar, and a **phosphate** (figure 2.23a).[6] The nitrogen base is a cyclic compound that comes in two forms: **purines** (two rings) and **pyrimidines** (one ring). There are two types of purines—**adenine (A)** and **guanine (G)**—and three types of pyrimidines—**thymine (T),**

---

6. The nitrogen base plus the pentose is called a *nucleoside.*

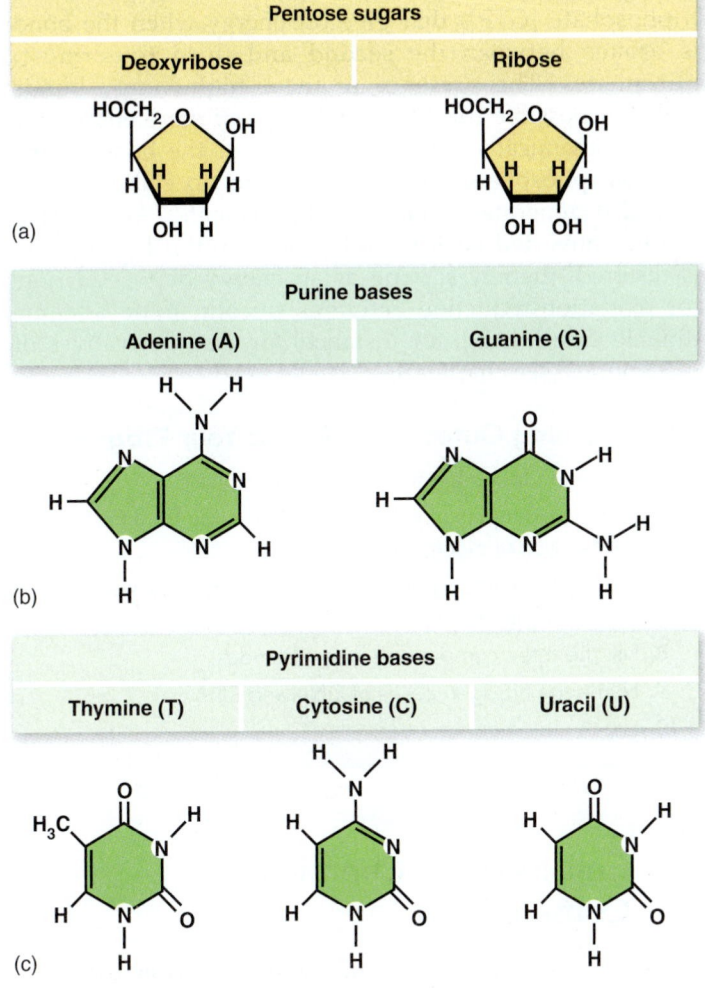

(a)

Pentose sugars

Deoxyribose	Ribose

(b)

Purine bases

Adenine (A)	Guanine (G)

(c)

Pyrimidine bases

Thymine (T)	Cytosine (C)	Uracil (U)

Figure 2.24 The sugars and nitrogen bases that make up DNA and RNA. **(a)** DNA contains deoxyribose, and RNA contains ribose. **(b)** A and G purine bases are found in both DNA and RNA. **(c)** Pyrimidine bases are found in both DNA and RNA, but T is found only in DNA, and U is found only in RNA.

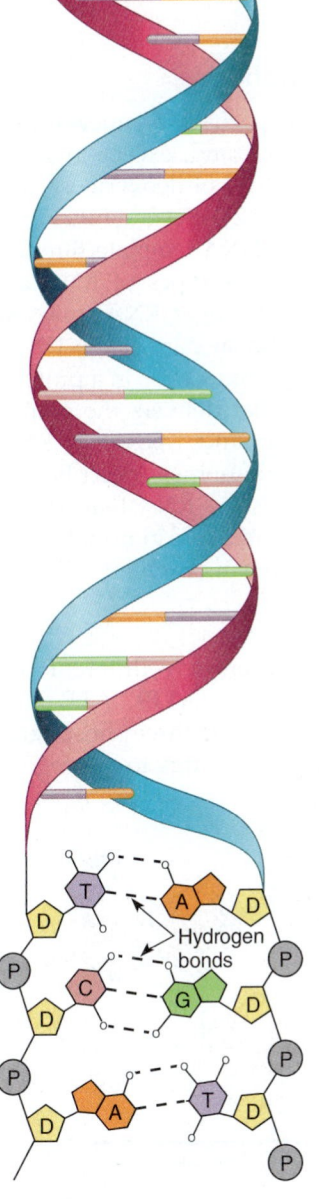

Figure 2.25 A structural representation of the double helix of DNA. Shown are the details of hydrogen bonds between the nitrogen bases of the two strands.

cytosine (C), and uracil (U) (figure 2.24). A characteristic that differentiates DNA from RNA is that DNA contains all of the nitrogen bases except uracil, and RNA contains all of the nitrogen bases except thymine. The nitrogen base is covalently bonded to the sugar **ribose** in RNA and **deoxyribose** (because it has one less oxygen than ribose) in DNA. Phosphate provides the final covalent bridge that connects sugars in series. Thus, the backbone of a nucleic acid strand is a chain of alternating phosphate-sugar-phosphate-sugar molecules, and the nitrogen bases branch off the side of this backbone (figure 2.23b,c).

The Double Helix of DNA

DNA is a huge molecule formed by two very long polynucleotide strands linked along their length by hydrogen bonds between complementary pairs of nitrogen bases.

The pairing of the nitrogen bases occurs according to a predictable pattern: Adenine always pairs with thymine, and cytosine with guanine. The bases are attracted in this way because each pair shares oxygen, nitrogen, and hydrogen atoms exactly positioned to align perfectly for hydrogen bonds **(figure 2.25)**.

For ease in understanding the structure of DNA, it is sometimes compared to a ladder, with the sugar-phosphate backbone representing the rails and the paired nitrogen bases representing the steps. Owing to the manner of nucleotide pairing and stacking of the bases, the actual configuration of DNA is a *double helix* that looks somewhat like a spiral staircase. As is true of protein, the structure of DNA is intimately related to its function. DNA molecules are usually extremely long. The hydrogen bonds between pairs break apart when DNA is being copied, and the fixed complementary base-pairing is essential to maintain the genetic code.

RNA: Organizer of Protein Synthesis

Like DNA, RNA consists of a long chain of nucleotides. However, RNA is usually a single strand, except in some viruses. It contains ribose sugar instead of deoxyribose and uracil instead of thymine (see figure 2.23). Several functional types of RNA are formed using the DNA template through a replication-like process. Three major types of RNA are important for protein synthesis. Messenger RNA (mRNA) is a copy of a gene (a single functional part of the DNA) that provides the order and type of amino acids in a protein; transfer RNA (tRNA) is a carrier that delivers the correct amino acids for protein assembly; and ribosomal RNA (rRNA) is a major component of ribosomes (described in chapter 4). A fourth type of RNA is the RNA that acts to regulate the genes and gene expression. More information on these important processes is presented in chapter 9.

ATP: The Energy Molecule of Cells

A relative of RNA involved in an entirely different cell activity is **adenosine triphosphate (ATP)**. ATP is a nucleotide containing adenine, ribose, and three phosphates rather than just one **(figure 2.26)**. It belongs to a category of high-energy compounds (also including guanosine triphosphate [GTP]) that give off energy when the bond is broken between the second and third (outermost) phosphates. The presence of these high-energy bonds makes it possible for ATP to release and store energy for cellular chemical reactions. Breakage of the bond of the terminal phosphate releases energy to do cellular work and also generates adenosine diphosphate (ADP). ADP can be converted back to ATP when the third phosphate is restored, thereby serving as an energy depot. Carriers for oxidation-reduction activities (nicotinamide adenine dinucleotide [NAD], for instance) are also derivatives of nucleotides (see chapter 8).

2.2 Learning Outcomes—Assess Your Progress

5. Name the four main families of biochemicals.
6. Provide examples of cell components made from each of the families of biochemicals.
7. Differentiate among primary, secondary, tertiary, and quaternary levels of protein structure.
8. List the three components of nucleotides.
9. Name the nitrogen bases of DNA and RNA.
10. List the three components of ATP.

2.3 Cells: Where Chemicals Come to Life

As we proceed in this chemical survey from the level of simple molecules to increasingly complex levels of macromolecules, at some point we cross a line from the realm of lifeless molecules and arrive at the fundamental unit of life called a **cell.**[7] A cell is indeed a huge aggregate of carbon, hydrogen, oxygen, nitrogen, and many other atoms; and it follows the basic laws of chemistry and physics, but it is much more. The combination of these atoms produces characteristics, reactions, and products that can only be described as *living.*

Fundamental Characteristics of Cells

The bodies of living things such as bacteria and protozoa consist of only a single cell, whereas those of animals and plants contain trillions of cells. Regardless of the organism, all cells have a few common characteristics. They tend to be spherical, polygonal, cubical, or cylindrical; and their protoplasm (internal cell contents) is encased in a cell or cytoplasmic membrane (see Insight 2.3). They have chromosomes containing DNA and ribosomes for protein synthesis, and they are exceedingly complex in function. Aside from these few similarities, most cell types fall into one of three fundamentally different lines (discussed

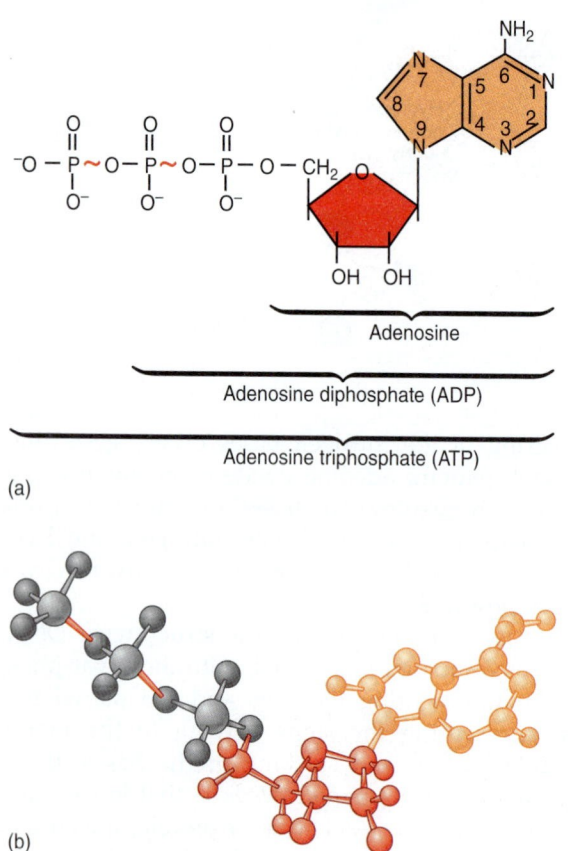

(a)

(b)

Figure 2.26 An ATP molecule.   **(a)** The structural formula. Wavy lines connecting the phosphates represent bonds that release large amounts of energy. **(b)** A ball-and-stick model.

7. The word *cell* was originally coined from an Old English term meaning "small room" because of the way plant cells looked to early microscopists.

in chapter 1): the small, seemingly simple bacterial and archaeal cells and the larger, structurally more complicated eukaryotic cells.

Animals, plants, fungi, and protozoans are composed of eukaryotic cells. Such cells contain a number of complex internal parts called organelles that perform useful functions for the cell involving growth, nutrition, or metabolism. By convention, organelles are defined as cell components that perform specific functions and are enclosed by membranes. Organelles also partition the eukaryotic cell into smaller compartments. The most visible organelle is the nucleus, a roughly ball-shaped mass surrounded by a double membrane that contains the DNA of the cell. Other organelles include the Golgi apparatus, endoplasmic reticulum, vacuoles, and mitochondria.

Bacterial and archaeal cells may seem to be the cellular "have nots" because, for the sake of comparison, they are described by what they lack. They have no nucleus and generally no other organelles. This apparent simplicity is misleading, however, because the fine structure of these cells is actually complex. Overall, bacterial and archaeal cells can engage in nearly every activity that eukaryotic cells can, and many can function in ways that eukaryotes cannot. Chapters 4 and 5 delve deeply into the properties of bacterial and eukaryotic cells.

2.3 Learning Outcome—Assess Your Progress

11. Recall three characteristics common to all cells.

Case File 2   *Wrap-Up*

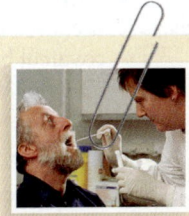

Blood work is required for diagnosis of many diseases and this involves needles, trained phlebotomists, and special equipment for testing and storage of blood. Other diagnostic tests are more invasive and painful, such as a lumbar puncture, and there is often a long wait before the doctor and the patient have the results. Saliva sampling requires no needles or special equipment—all patients have to do is spit into a cup or swipe a swab across their gums, and results can be available much more rapidly.

Dr. Wong's group developed a desktop device capable of analyzing protein and mRNA in saliva samples within minutes. Other groups are also working on rapid screening devices. John McDevitt of Rice University in Houston developed a biometer that can analyze saliva and other body samples that can be used onboard emergency vehicles. Individual test cards customized for specific health conditions can identify biomarkers found in body samples, making it easier to diagnose a patient having chest pain. Fluorescent markers on a test card will glow if a patient's saliva shows troponin T, a protein characteristic of a heart attack, which will allow doctors to diagnose a heart attack much more rapidly. So go ahead and spit—it might save your life.

Source: 2009. Clin. Cancer Res. *vol. 15, no. 17, p. 5743.*

Chapter Summary

2.1 Atoms, Bonds, and Molecules: Fundamental Building Blocks (ASM Guideline* 3.3)

- Protons (p^+) and neutrons (n^0) make up the nucleus of an atom. Electrons (e^-) orbit the nucleus.
- All elements are composed of atoms but differ in the numbers of protons, neutrons, and electrons they possess.
- Isotopes are varieties of one element that contain the same number of protons but different numbers of neutrons.
- The number of electrons in an element's outermost orbital (compared with the total number possible) determines the element's chemical properties and reactivity.
- Covalent bonds are chemical bonds in which electrons are shared between atoms. Equally distributed electrons form nonpolar covalent bonds, whereas unequally distributed electrons form polar covalent bonds.

- Ionic bonds are chemical bonds resulting from opposite charges. The outer electron shell either donates or receives electrons from another atom so that the outer shell of each atom is completely filled.
- Hydrogen bonds are weak chemical attractions that form between covalently bonded hydrogens and either oxygens or nitrogens on different molecules. These as well as van der Waals forces are critically important in biological processes.
- Chemical equations express the chemical exchanges between atoms or molecules.
- Solutions are mixtures of solutes and solvents that cannot be separated by filtration or settling.
- The pH, ranging from a highly *acidic* solution to a highly *basic* solution, refers to the concentration of hydrogen ions. It is expressed as a number from 0 to 14.
- Biologists define *organic molecules* as those containing both carbon and hydrogen.

*ASM Curriculum Guidelines (American Society for Microbiology, 2012). Complete guidelines in appendix B.

- Carbon is the backbone of biological compounds because of its ability to form single, double, or triple covalent bonds with itself and many different elements.
- Functional (R) groups are specific arrangements of organic molecules that confer distinct properties, including chemical reactivity, to organic compounds.

2.2 Macromolecules: Superstructures of Life (ASM Guidelines 2.1, 3.1, 4.2)

- Macromolecules are very large organic molecules (polymers) built up by polymerization of smaller molecular subunits (monomers).
- Carbohydrates are biological molecules whose polymers are monomers linked together by glycosidic bonds. Their main functions are protection and support (in organisms with cell walls) and also nutrient and energy stores.
- Lipids are biological molecules such as fats that are insoluble in water. Their main functions are as cell components, cell secretions, and nutrient and energy stores.
- Proteins are biological molecules whose polymers are chains of amino acid monomers linked together by peptide bonds.

- Proteins are called the "shapers of life" because of the many biological roles they play in cell structure and cell metabolism.
- Protein structure determines protein function. Structure and shape are dictated by amino acid composition and by the pH and temperature of the protein's immediate environment.
- Nucleic acids are biological molecules whose polymers are chains of nucleotide monomers linked together by phosphate–pentose sugar covalent bonds. Double-stranded nucleic acids are linked together by hydrogen bonds. Nucleic acids are information molecules that direct cell metabolism and reproduction. Nucleotides such as ATP also serve as energy-transfer molecules in cells.

2.3 Cells: Where Chemicals Come to Life (ASM Guideline 1.1)

- As the atom is the fundamental unit of matter, so is the cell the fundamental unit of life.

Multiple-Choice and True-False Questions | Bloom's Levels 1 and 2: Remember and Understand

Multiple-Choice Questions. Select the correct answer from the options provided.

1. The smallest unit of matter with unique characteristics is
   a. an electron.
   c. an atom.
   b. a molecule.
   d. a proton.

2. The _____ charge of a proton is exactly balanced by the _____ charge of a(an) _____.
   a. negative, positive, electron
   b. positive, neutral, neutron
   c. positive, negative, electron
   d. neutral, negative, electron

3. Electrons move around the nucleus of an atom in pathways called
   a. shells.
   c. circles.
   b. orbitals.
   d. rings.

4. Bonds in which atoms share electrons are defined as _____ bonds.
   a. hydrogen
   c. double
   b. ionic
   d. covalent

5. Hydrogen bonds can form between _____ adjacent to each other.
   a. two hydrogen atoms
   b. two oxygen atoms
   c. a hydrogen atom and an oxygen atom
   d. negative charges

6. An atom that can donate electrons during a reaction is called
   a. an oxidizing agent.
   c. an ionic agent.
   b. a reducing agent.
   d. an electrolyte.

7. A solution with a pH of 2 _____ than a solution with a pH of 8.
   a. has less H^+
   c. has more OH^-
   b. has more H^+
   d. is less concentrated

8. Proteins are synthesized by linking amino acids with _____ bonds.
   a. disulfide
   c. peptide
   b. glycosidic
   d. ester

9. DNA is a hereditary molecule that is composed of
   a. deoxyribose, phosphate, and nitrogen bases.
   b. deoxyribose, a pentose, and nucleic acids.
   c. sugar, proteins, and thymine.
   d. adenine, phosphate, and ribose.

10. RNA plays an important role in what biological process?
    a. replication
    c. lipid metabolism
    b. protein synthesis
    d. water transport

True-False Questions. If the statement is true, leave as is. If it is false, correct it by rewriting the sentence.

11. Elements have varying numbers of protons, neutrons, and electrons.

12. Covalent bonds are those that are made between two different elements.

13. A compound is called "organic" if it is made of all-natural elements.

14. Cysteine is the amino acid that participates in disulfide bonds in proteins.

15. Membranes are mainly composed of macromolecules called carbohydrates.

Critical Thinking Questions | Bloom's Levels 3, 4, and 5: Apply, Analyze, and Evaluate

Critical thinking is the ability to reason and solve problems using facts and concepts. These questions can be approached from a number of angles and, in most cases, they do not have a single correct answer.

1. Compare and contrast the characteristics of elemental iodine and iodine-125, and explain which form of iodine has practical applications in medicine today.

2. Support or refute the following statement: "Double bonding provides the plasma membrane with flexibility."

3. A small-diameter capillary tube is used to transfer blood from the fingertip to a hemocytometer during a complete blood count. Describe both the inter- and intramolecular forces that allow the blood to travel up the capillary tube without the use of suction.

4. The burning of coal imparts a large amount of sulfur dioxide into the atmosphere, which, when combined with water, creates sulfuric acid. Explain how such "acid rain" affects the water chemistry of inland lakes and streams.

5. Provide a definition of the term *organic molecule*, and provide two reasons why carbon is considered the "fundamental element of life."

6. Plant cell walls are composed of cellulose, a complex carbohydrate exhibiting a unique bond between its glucose subunits. Provide an explanation for the fact that humans cannot digest fruits and vegetables at an efficient level.

7. Draw a simple model of the following molecules: CH_4, NaCl, H_2O. Provide a definition of *polar molecule*, and identify the polar molecule listed here.

8. Compare and contrast the chemical bonding exhibited by secondary and tertiary levels of protein structure.

9. A new microbe was recently discovered that utilizes arsenic in place of phosphate in its DNA double helix. Provide a sound reason for whether or not this change will alter the information encoded by this organism's genetic material.

10. Microscopic analysis of a cell reveals the presence of ribosomes. Identify whether this is a eukaryotic, archaeal or bacterial cell type and explain your reasoning.

Concept Connections | Bloom's Levels 4 and 6: Analyze and Create

This activity ties together multiple concepts in the chapter.

1. What are two examples of polysaccharides in a cell?
2. What are four functions of proteins?
3. Where are phospholipids found?
4. What carbon-based molecules combine to form DNA?

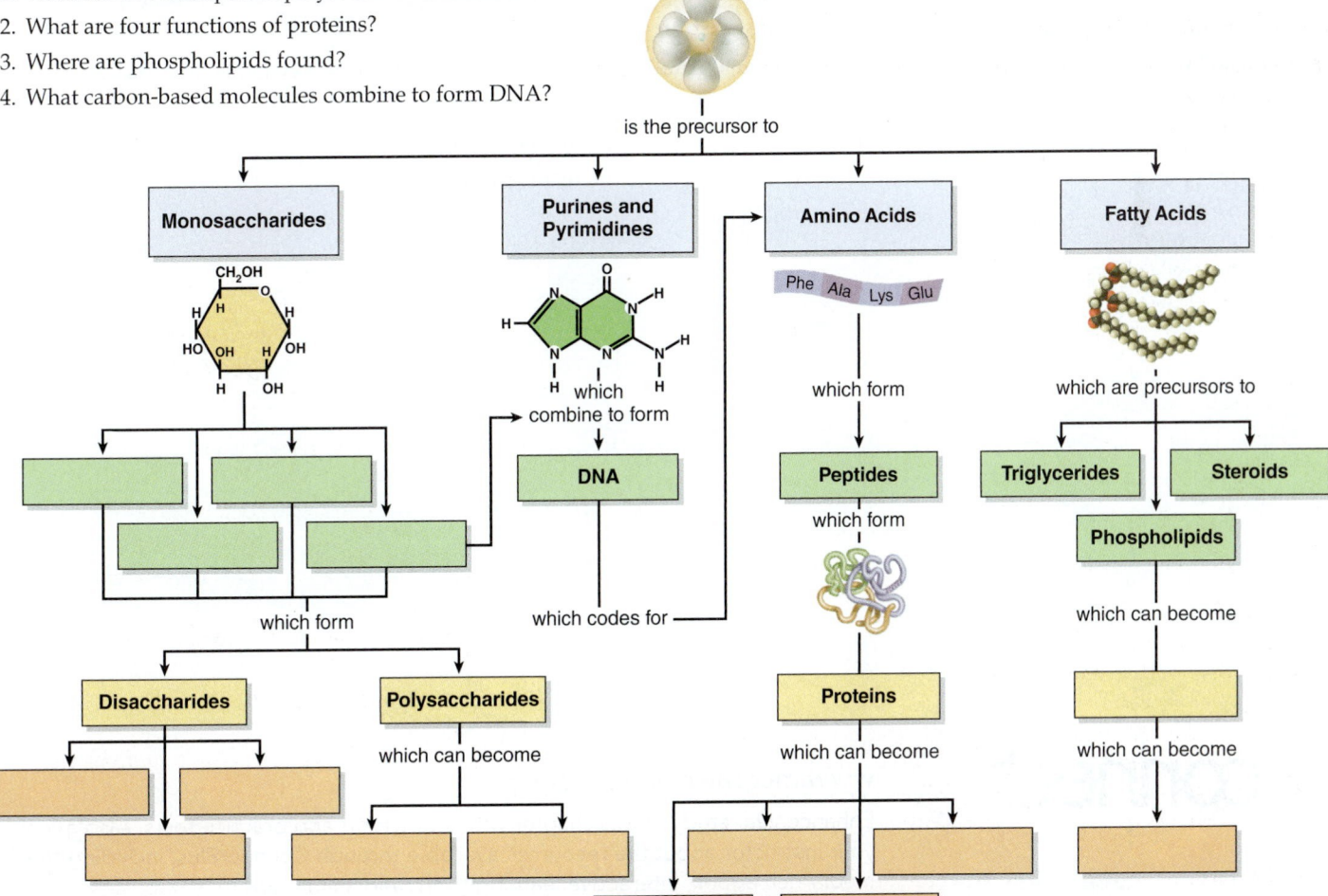

Visual Connections | Bloom's Level 5: Evaluate

This question uses visual images or previous content to make connections to this chapter's concepts.

1. **Figure 2.19a** and **Figure 2.20.** Speculate on why sterols like cholesterol can add "stiffness" to membranes that contain them.

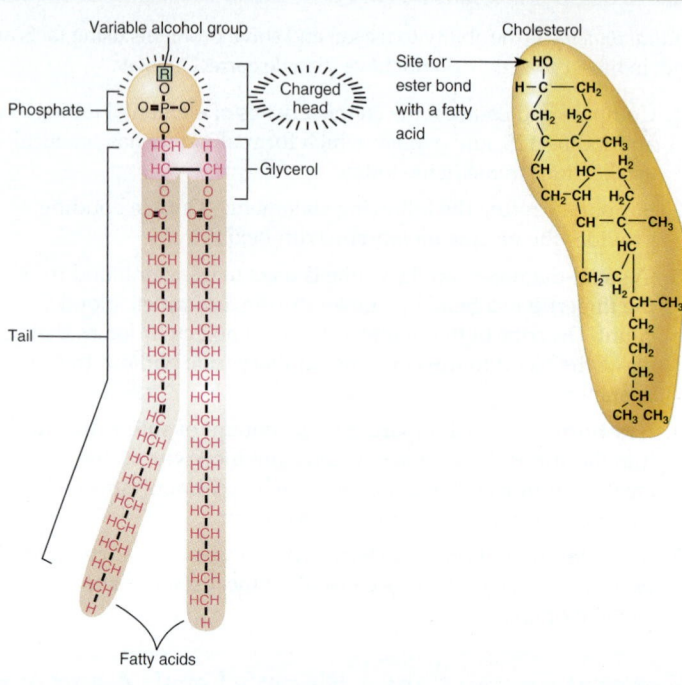

Concept Mapping | Bloom's Level 6: Create

Appendix D provides guidance for working with concept maps.

1. Using the words that follow, please create a concept map illustrating the relationships among these key terms from chapter 2.

elements	compounds	pH
atoms	chemical bonds	organic chemicals
molecules	ions	inorganic chemicals

www.mcgrawhillconnect.com

Enhance your study of this chapter with study tools and practice tests. Also ask your instructor about the resources available through ConnectPlus, including the media-rich eBook, interactive learning tools, and animations.

Tools of the Laboratory
Methods for the Culturing and Microscopic Analysis of Microorganisms

Case File 3

E. coli O104:H4: The Ugly Stepsister of *E. coli* O157:H7

In May 2011, disturbing reports of outbreaks of enterohemorrhagic *E. coli* (EHEC) began to emerge from Germany. Generally, strains of *E. coli* cause relatively mild cases of diarrhea, also known as traveler's diarrhea, or a mild form of dysentery, both of which resolve with minimal treatment. When the reporting of illness subsided in mid-June, there were a total of 855 cases of hemolytic uremic syndrome (HUS) and 2,987 cases of acute gastroenteritis associated with the outbreak. Thirty-five deaths were attributed to HUS, while 18 were due to EHEC gastroenteritis, making this the largest outbreak of food-borne disease ever recorded in Germany. Cases were reported throughout all of the states of Germany, and a majority of the cases were tied to consumption of contaminated bean sprouts. Epidemiologists linked several cases from the German outbreak to other cases found in Bordeaux, France. Six cases of illness and one death in the United States were linked to the outbreak, and all patients were determined to be recent travelers to Germany.

■ What are the steps that should be taken to identify the specific cause of this outbreak?

■ How is a case of EHEC different than other food-borne infections caused by *E. coli*?

Continuing the Case appears on page 66.

Outline and Learning Outcomes

3.1 Methods of Culturing Microorganisms: The Five I's
1. Explain what the Five I's mean and what each step entails.
2. Discuss three physical states of media and when each is useful.
3. Compare and contrast selective and differential media, and give an example of each.
4. Provide brief definitions for *defined* and *complex media*.

3.2 The Microscope: Window on an Invisible Realm
5. Convert among different lengths within the metric system.
6. Describe the earliest microscopes.
7. List and describe the three elements of good microscopy.
8. Differentiate between the principles of light and electron microscopy.
9. Compare and contrast the three main categories of stains, and provide examples of each.

3.1 Methods of Culturing Microorganisms: The Five I's

Biologists studying large organisms such as animals and plants can, for the most part, use their senses of sight, smell, hearing, and even touch to detect and evaluate identifying characteristics and to keep track of growth and developmental changes. Microbiologists, however, are confronted by some unique problems. First, most habitats harbor microbes in complex associations, so it is often necessary to separate the species from one another. Second, to maintain and keep track of such small research subjects, microbiologists often have to grow them under artificial (and thus distorting) conditions. It should be noted that in the last decade, advances in molecular techniques mean that "The Five I's" are no longer necessary in all cases. We'll see how this works in chapter 8. But even with the new advances, sometimes the best strategy is to grow the organism (in other words, go through the Five I's). A third difficulty in working with microbes is that they are invisible and widely distributed, and undesirable ones can be introduced into an experiment and cause misleading results.

The "Five I's" represent five basic techniques to manipulate, grow, examine, and characterize microorganisms in the laboratory: inoculation, incubation, isolation, inspection, and identification (the Five I's; **process figure 3.1**). These procedures make it possible to handle and maintain microorganisms as discrete entities whose detailed biology can be studied and recorded.

Inoculation: Producing a Culture

To cultivate, or **culture,** microorganisms, one introduces a tiny sample (the inoculum) into a container of nutrient **medium** (plural, *media*), which provides an environment in which they multiply. This process is called **inoculation.** Any instrument used for sampling and inoculation must initially be **sterile.** The observable growth that appears in or on the medium after **incubation** is known as a culture. The nature

---

Major Techniques Performed by Microbiologists to Locate, Grow, Observe, and Characterize Microorganisms

Specimen Collection:
Nearly any object or material can serve as a source of microbes. Common ones are body fluids and tissues, foods, water, or soil. Specimens are removed by some form of sampling device: a sterile swab, syringe, or a special transport system that holds, maintains, and preserves the microbes in the sample.

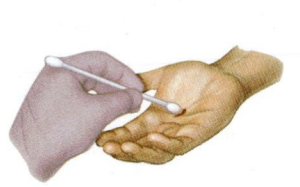  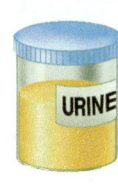

A GUIDE TO THE FIVE I's: How the Sample Is Processed and Profiled

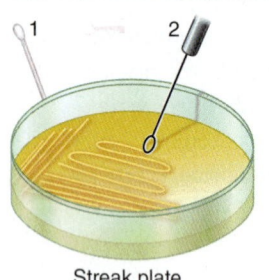

Streak plate

Syringe

Blood bottle

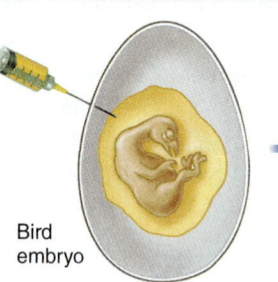

Bird embryo

Incubator

① Inoculation:
The sample is placed into a container of sterile **medium** containing appropriate nutrients to sustain growth. Inoculation involves spreading the sample on the surface of a solid medium or introducing the sample into a flask or tube. Selection of media with specialized functions can improve later steps of isolation and identification. Some microbes may require a live organism (animal, egg) as the growth medium.

② Incubation:
An incubator creates the proper growth temperature and other conditions. This promotes multiplication of the microbes over a period of hours, days, and even weeks. Incubation produces a culture—the visible growth of the microbe in or on the medium.

Process Figure 3.1 A summary of the general laboratory techniques carried out by microbiologists.   It is not necessary to perform all the steps shown or to perform them exactly in this order, but all microbiologists participate in at least some of these activities. In some cases, one may proceed right from the sample to inspection, and in others, only inoculation and incubation on special media are required.

of the sample being cultured depends on the objectives of the analysis. Clinical specimens for determining the cause of an infectious disease are obtained from body fluids (blood, cerebrospinal fluid), discharges (sputum, urine, feces), anatomical sites (throat, nose, ear, eye, genital tract), or diseased tissue such as an abscess or wound. Other samples subject to microbiological analysis are soil, water, sewage, foods, air, and inanimate objects. Procedures for proper specimen collection are discussed in chapter 17.

Incubation

Once a container of medium has been inoculated, it is **incubated,** which means it is placed in a temperature-controlled chamber (incubator) to encourage multiplication. Although microbes have adapted to growth at temperatures ranging from freezing to boiling, the usual temperatures used in laboratory propagation fall between 20°C and 45°C. Incubators can also control the content of atmospheric gases such as oxygen and carbon dioxide that may be required for the growth of certain microbes. During the incubation period (ranging from a day to several weeks), the microbe multiplies and produces growth that is observable macroscopically **(table 3.1).** Microbial growth in a liquid medium materializes as cloudiness, sediment, scum, or color. The most common manifestation of growth on solid media is the appearance of colonies, especially with bacteria and fungi. Colonies are actually large masses of piled-up cells (see chapter 7).

Before we continue to cover information on the Five I's, we will take a side trip to look at media in more detail.

Table 3.1 Incubation Conditions of Various Bacterial Pathogens

Organism	Optimum Growth Temperature (°C)	Culturing Time (Days)
Listeria monocytogenes	25–30	1–2
Pseudomonas fluorescens	25–30	1–2
Streptococcus pyogenes	37	1–2
Mycobacterium tuberculosis	37	28
Mycobacterium leprae	27–30	*

*Although *M. leprae* cannot be cultured *in vitro,* in its animal model it exhibits one of the longest doubling times (14 days) of any known bacterium.

Media: Providing Nutrients in the Laboratory

A major stimulus to the rise of microbiology in the late 1800s was the development of techniques for growing microbes out of their natural habitats and in pure form in the laboratory. This milestone enabled the close examination of a microbe and its morphology, physiology, and genetics. It was evident from the very first that for successful cultivation, each microorganism had to be provided with all of its required nutrients in an artificial medium.

Some microbes require only a very few simple inorganic compounds for growth; others need a complex list of specific inorganic and organic compounds. This tremendous diversity is evident in the types of media that can be prepared **(Insight 3.1).** More than 500 different types of media

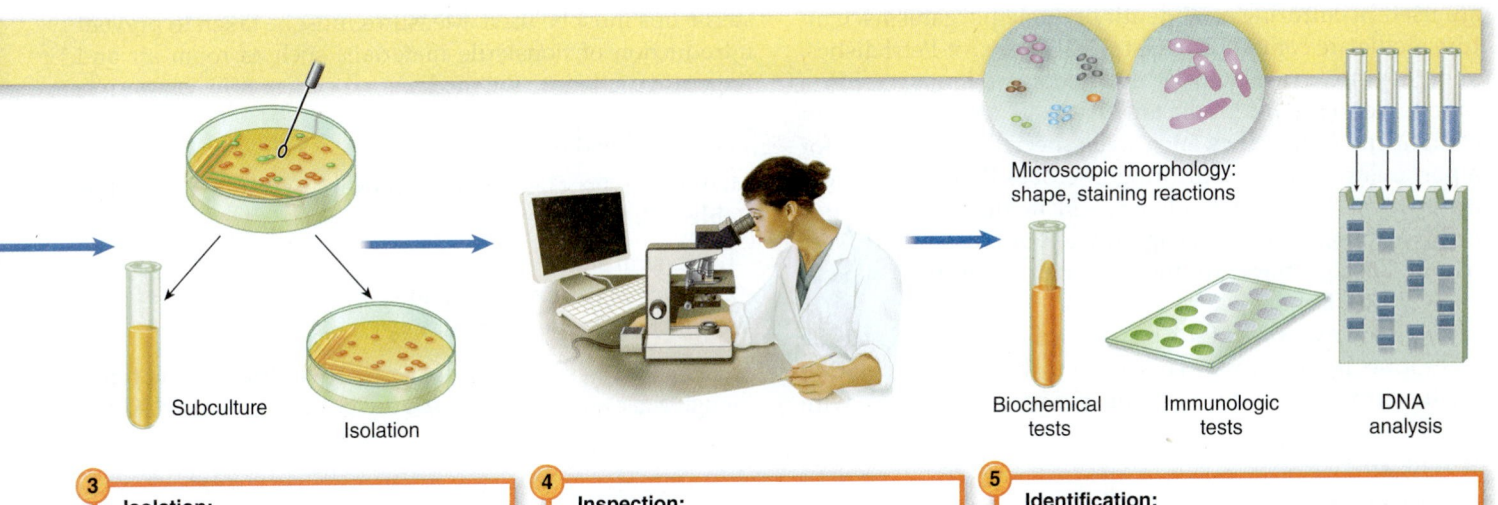

Microscopic morphology: shape, staining reactions

Subculture
Isolation

Biochemical tests   Immunologic tests   DNA analysis

3 Isolation:
One result of inoculation and incubation is **isolation** of the microbe. Isolated microbes may take the form of separate colonies (discrete mounds of cells) on solid media, or turbidity (free-floating cells) in broths. Further isolation by subculturing involves taking a bit of growth from an isolated colony and inoculating a separate medium. This is one way to make a pure culture that contains only a single species of microbe.

4 Inspection:
The colonies or broth cultures are observed macroscopically for growth characteristics (color, texture, size) that could be useful in analyzing the specimen contents. Slides are made to assess microscopic details such as cell shape, size, and motility. Staining techniques may be used to gather specific information on microscopic morphology.

5 Identification:
A major purpose of the Five I's is to determine the type of microbe, usually to the level of species. Information used in identification can include relevant data already taken during initial inspection and additional tests that further describe and differentiate the microbes. Specialized tests include biochemical tests to determine metabolic activities specific to the microbe, immunologic tests, and genetic analysis.

INSIGHT 3.1   Fanny's Fabulous Finding

Isolating pure cultures of bacteria was problematic in the early days of microbiology. Scientists were able to grow bacteria in what was essentially meat broth, but they had no way of separating out the different species of bacteria in the liquid mixed culture. Serial dilutions were used to isolate cells, but frequently the organism they were looking for was "lost" in the dilution series.

Some type of solid surface to encourage microbial growth was needed. Robert Koch was able to isolate Bacillus anthracis on thin slices of potatoes, but other species of bacteria didn't thrive in this medium. Koch had some success with using gelatin, but when left out in the warm laboratory, gelatin quickly liquefied, ruining the isolation experiment. Additionally, it was discovered that microbes produced an enzyme called gelatinase that cleaved the bonds between protein molecules, also liquefying the medium.

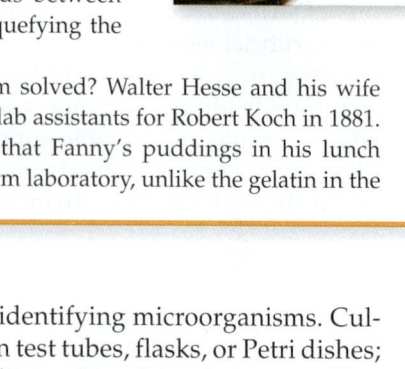

How was the problem solved? Walter Hesse and his wife Fanny started working as lab assistants for Robert Koch in 1881. One day, Walter noticed that Fanny's puddings in his lunch weren't melting in the warm laboratory, unlike the gelatin in the Petri dishes. When asked about this, Fanny told him that she'd learned the trick to perfect puddings was using a product made from seaweed called "agar-agar." Knowing the trouble Walter was having with finding solid microbial media, she suggested that they combine the agar-agar with the nutrient broth already used in the lab. Suddenly, Robert Koch's lab had the perfect microbial media for isolating bacterial colonies. The agar mixed with nutrient broth stayed firm at incubator temperatures and was impervious to bacterial enzymes. After this seemingly simple modification, Koch and his laboratory assistants were able to isolate the microbial cause of tuberculosis and numerous other pathogens.

So, the next time you pick up a Petri dish and look at the colorful varieties of bacterial colonies growing there, think of Fanny and Walter. Koch's lab assistants never received any compensation for the simple ingredient in Fanny's puddings that revolutionized microbiology at the time and is still used universally today.

Source: 2008. http://hardydiagnostics.com/articles/Agar-and-Fanny-Hesse01.pdf, Retrieved 7/3/2013.

are used in culturing and identifying microorganisms. Culture media are contained in test tubes, flasks, or Petri dishes; and they are inoculated by such tools as loops, needles, pipettes, and swabs. Media are extremely varied in nutrient content and consistency and can be specially formulated for a particular purpose. Culturing microbes that cannot grow on artificial media (i.e., all viruses) requires cell cultures or host animals. In this chapter, we will focus on artificial media, because these are the most frequently used type in clinical situations.

For an experiment to be properly controlled, sterile technique is necessary. This means that the inoculation must start with a sterile medium, and inoculating tools with sterile tips must be used. Measures must be taken to prevent introduction of nonsterile materials, such as room air and fingers, directly into the media.

Types of Media

Media can be classified according to three properties **(table 3.2):**

1. physical state,
2. chemical composition, and
3. functional type (purpose).

Most media discussed here are designed for bacteria and fungi, though algae and some protozoa can be propagated in media.

Table 3.2 Three Categories of Media Classification

Physical State*		Chemical Composition		Functional Type	
1.	Liquid	1.	Chemically defined (synthetic)	1. General purpose	5. Anaerobic growth
2.	Semisolid	2.	Complex; not chemically defined	2. Enriched	6. Specimen transport
3.	Solid (can be converted to liquid)			3. Selective	7. Assay
4.	Solid (cannot be liquefied)			4. Differential	8. Enumeration

*Some media can serve more than one function. For example, a medium such as brain-heart infusion is general purpose and enriched; mannitol salt agar is both selective and differential; and blood agar is both enriched and differential.

Physical States of Media

Liquid media are water-based solutions that do not solidify at temperatures above freezing and that tend to flow freely when the container is tilted **(process figure 3.2a)**. These media, termed *broths, milks,* or *infusions,* are made by dissolving various solutes in distilled water. A common laboratory medium, *nutrient broth,* contains beef extract and peptone dissolved in water. Methylene blue milk and litmus milk are opaque liquids containing whole milk and dyes. Fluid thioglycollate is a slightly viscous broth used for determining the oxygen requirement of different microbes.

At ordinary room temperature, **semisolid media** exhibit a clotlike consistency **(process figure 3.2b)** because they contain an amount of solidifying agent (agar or gelatin) that thickens them but does not produce a firm surface. Semisolid media are used to determine the motility of bacteria and to localize a reaction at a specific site.

Solid media provide a firm surface on which cells can form discrete colonies **(process figure 3.2c)** and are advantageous for isolating and culturing bacteria and fungi. Liquefiable solid media, sometimes called reversible solid media, contain a solidifying agent that changes their physical properties in response to temperature. By far the most widely used and effective of these agents is **agar,** a complex polysaccharide isolated from the red alga *Gelidium.* Agar is solid at room temperature, and it melts (liquefies) at the boiling temperature of water (100°C). Once liquefied, agar does not resolidify until it cools to 42°C.

Any medium containing 1% to 5% agar usually has the word *agar* in its name, the most common being nutrient agar. Like nutrient broth, it contains beef extract and peptone, as well as 1.5% agar by weight. Many of the examples covered in the section on functional categories of media contain agar. Although gelatin is not nearly as satisfactory as agar, it will create a reasonably solid surface in concentrations of 10% to 15%.

Chemical Content of Media

Media whose exact chemical compositions are known are termed *defined* (also known as *synthetic*). Such media contain pure organic and inorganic compounds that vary little from one source to another and have a molecular content specified by means of an exact formula. Defined media may contain nothing more than a few essential compounds such as salts and amino acids dissolved in water, or may be composed

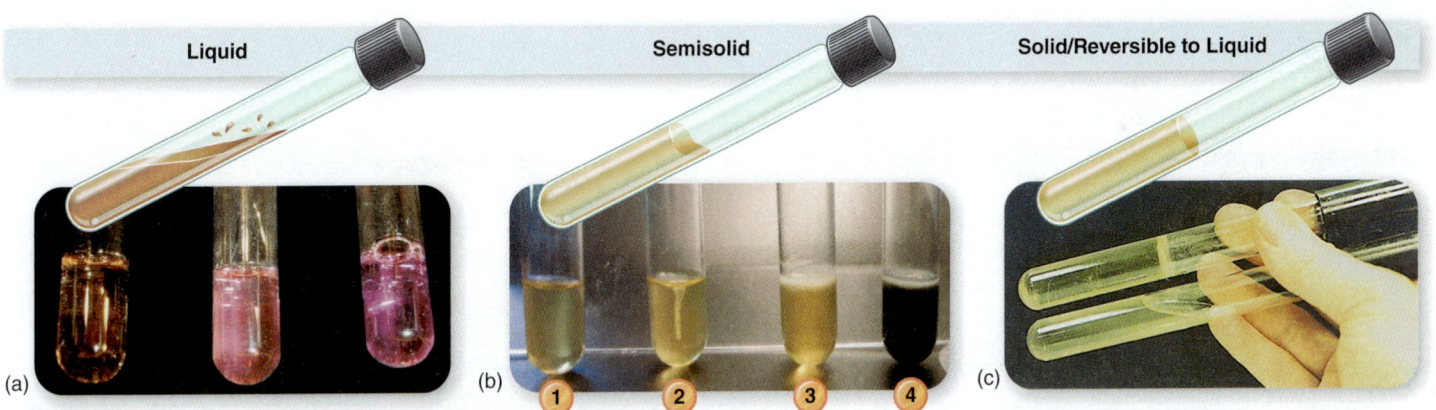

Process Figure 3.2 Media in different physical forms.   **(a)** Liquid media are water-based solutions that do not solidify at temperatures above freezing and that tend to flow freely when the container is tilted. Growth occurs throughout the container and can then present a dispersed, cloudy, or particulate appearance. Urea broth is used to show a biochemical reaction in which the enzyme urease digests urea and releases ammonium. This raises the pH of the solution and causes the dye to become increasingly pink. *Left:* uninoculated broth, pH 7; *middle:* weak positive, pH 7.5; *right:* strong positive, pH 8.0. **(b)** Semisolid media have more body than liquid media but less body than solid media. They do not flow freely and have a soft, clotlike consistency at room temperature. Semisolid media are used to determine the motility of bacteria and to localize a reaction at a specific site. Here, sulfur indole motility (SIM) medium is pictured. (1) The medium is stabbed with an inoculum and incubated. Location of growth indicates nonmotility (2) or motility (3). If H₂S gas is released, a black precipitate forms (4). **(c)** Media containing 1%–5% agar are solid enough to remain in place when containers are tilted or inverted. They are reversibly solid and can be liquefied with heat, poured into a different container, and resolidified. Solid media provide a firm surface on which cells can form discrete colonies. Nutrient gelatin contains enough gelatin (12%) to take on a solid consistency. The top tube shows it as a solid. The bottom tube indicates what happens when it is warmed or when microbial enzymes digest the gelatin and liquefy it.

of a variety of defined organic and inorganic chemicals (**table 3.3**). Such standardized and reproducible media are most useful in research and cell culture when the exact nutritional needs of the test organisms are known.

If even one component of a given medium is not chemically definable, the medium belongs in the *complex* category. Complex media contain extracts of animals, plants, or yeasts, including such materials as ground-up cells, tissues, and secretions. Examples are blood, serum, and meat extracts or infusions. Other nonsynthetic ingredients are milk, yeast extract, soybean digests, and peptone. Nutrient broth, blood agar, and MacConkey agar, though different in function and appearance, are all complex nonsynthetic media that present a rich mixture of nutrients for microbes that have complex nutritional needs.

Table 3.3 provides a practical application of the preceding two categories by comparing two different media for the growth of *Staphylococcus* species.

Media for Different Purposes

Until recently, microbiologists knew of only a few species of bacteria or fungi that could not be cultivated artificially. However, newer DNA-detection technologies have shown us that there are many times more microbes that we do not know how to cultivate in the lab than those that we do. The race is on to create media for the growth of these microbes out of their natural habitat!

General-purpose media are designed to grow as broad a spectrum of microbes as possible. As a rule, they are of the complex variety and contain a mixture of nutrients that could support the growth of a variety of microbial life. Examples include nutrient agar and broth, brain-heart infusion, and trypticase soy agar (TSA). An **enriched medium** contains complex organic substances such as blood, serum, hemoglobin, or special **growth factors** (specific vitamins, amino acids) that certain species require in order to grow. Bacteria that require growth factors and complex nutrients are termed **fastidious**. Blood agar, which is made by adding sterile sheep, horse, or rabbit blood to a sterile agar base **(figure 3.3a)**, is widely employed to grow fastidious streptococci and other pathogens. Pathogenic *Neisseria* (one

Table 3.3A	Chemically Defined Synthetic Medium for Growth and Maintenance of Pathogenic *Staphylococcus aureus*		
0.25 Grams Each of These Amino Acids	**0.5 Grams Each of These Amino Acids**	**0.12 Grams Each of These Amino Acids**
Cystine	Arginine	Aspartic acid
Histidine	Glycine	Glutamic acid
Leucine	Isoleucine	
Phenylalanine	Lysine	
Proline	Methionine	
Tryptophan	Serine	
Tyrosine	Threonine	
	Valine	

Additional ingredients

0.005   mole nicotinamide ⎤
0.005   mole thiamine   ⎥ Vitamins
0.005   mole pyridoxine   ⎥
0.5   micrograms biotin ⎦
1.25   grams magnesium sulfate ⎤
1.25   grams dipotassium hydrogen phosphate ⎥ Salts
1.25   grams sodium chloride ⎥
0.125   grams iron chloride ⎦

Ingredients dissolved in 1,000 milliliters of distilled water and buffered to a final pH of 7.0.

Table 3.3B	Brain-Heart Infusion Broth: A Complex, Nonsynthetic Medium for Growth and Maintenance of Pathogenic *Staphylococcus aureus*

27.5   grams brain, heart extract, peptone extract
2   grams glucose
5   grams sodium chloride
2.5   grams disodium hydrogen phosphate

Ingredients dissolved in 1,000 milliliters of distilled water and buffered to a final pH of 7.0.

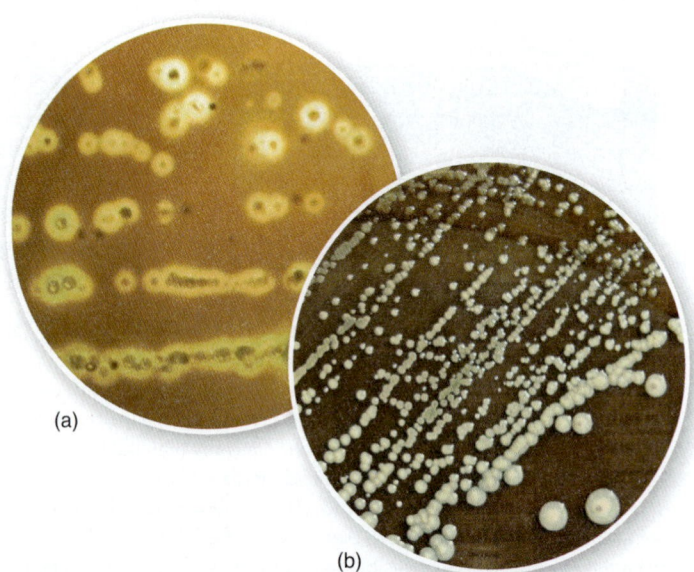

(a)

(b)

Figure 3.3 Examples of enriched media.   (a) Blood agar plate growing bacteria from the human throat. Note that this medium also differentiates among colonies by the zones of hemolysis (clear areas) they may show. **(b)** Culture of *Neisseria* sp. on chocolate agar. Chocolate agar gets its brownish color from cooked blood (not chocolate) and does not produce hemolysis.

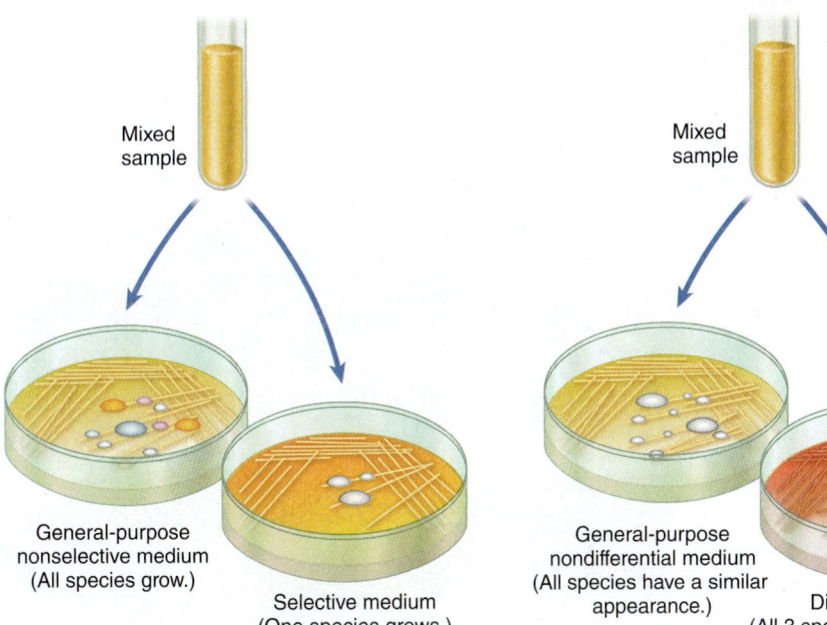

Figure 3.4 Comparison of selective and differential media with general-purpose media. **(a)** A mixed sample containing three different species is streaked onto plates of general-purpose nonselective medium and selective medium. Note the results. **(b)** Another mixed sample containing three different species is streaked onto plates of general-purpose nondifferential medium and differential medium.

Mixed sample

Mixed sample

General-purpose nonselective medium (All species grow.)

Selective medium (One species grows.)

(a)

General-purpose nondifferential medium (All species have a similar appearance.)

Differential medium (All 3 species grow but may have different appearances in some way.)

(b)

species causes gonorrhea) are grown on either Thayer-Martin medium or "chocolate" agar, a blood agar with added components **(figure 3.3b).**

Selective and Differential Media Some of the most inventive media recipes belong to the categories of selective and differential media **(figure 3.4).** These media are designed for special microbial groups, and they have extensive applications in isolation and identification. In a single step, they can permit the preliminary identification of a genus or even a species.

A **selective medium (table 3.4)** contains one or more agents that inhibit the growth of a certain microbe or microbes (call them A, B, and C) but not others (D) and thereby encourage, or *select*, microbe D and allow it to grow. Selective media are very important in primary isolation of a specific type of microorganism from samples containing dozens of different species—for example, feces, saliva, skin, water, and soil. They speed up isolation by suppressing the unwanted background organisms and favoring growth of the desired ones.

Disease Connection

Fastidious organisms may be missed during diagnosis if there is not an "index of suspicion" on the part of the health care provider. Even when provided with the correct nutrients, fastidious bacteria may grow more slowly than nonfastidious ones. For example, the bacterium that causes whooping cough, *Bordetella pertussis*, is strictly aerobic and requires certain nutrients to grow. If a patient presents with a painful upper throat or a cough, a practitioner may plate samples on blood agar and incubate them in 5% CO_2 to look for *Streptococcus* and examine them the next day. If no pathogens appear on the agar after a day or two, a diagnosis of viral infection may be made. But the real culprit might have been *B. pertussis*. This bacterium grows better on specialized media containing charcoal, and they form pin-prick-size colonies only several days after inoculation.

Table 3.4 Selective Media, Agents, and Functions		
Medium	**Selective Agent**	**Used For**
Mueller tellurite	Potassium tellurite	Isolation of *Corynebacterium diphtheriae*
Enterococcus faecalis broth	Sodium azide, tetrazolium	Isolation of fecal enterococci
Phenylethanol agar	Phenylethanol chloride	Isolation of staphylococci and streptococci
Tomato juice agar	Tomato juice, acid	Isolation of lactobacilli from saliva
MacConkey agar	Bile, crystal violet	Isolation of gram-negative enterics
Salmonella/Shigella (SS) agar	Bile, citrate, brilliant green	Isolation of *Salmonella* and *Shigella*
Lowenstein-Jensen	Malachite green dye	Isolation and maintenance of *Mycobacterium*
Mannitol salt agar	Sodium chloride	Isolation of *Staphylococcus* species
Sabouraud's agar	pH of 5.6 (acid)	Isolation of fungi—inhibits bacteria

Figure 3.5 Examples of media that are both selective and differential. (a) Mannitol salt agar is used to isolate members of the genus *Staphylococcus*. It is selective because *Staphylococcus* can grow in the presence of 7.5% sodium chloride, whereas many other species are inhibited by this high concentration. It contains a dye that also differentiates those species of *Staphylococcus* that produce acid from the fermentation of mannitol and turn the phenol red dye to a bright yellow. (b) MacConkey agar selects against gram-positive bacteria. It also differentiates between lactose-fermenting bacteria (indicated by a pink-red reaction in the center of the colony) and lactose-negative bacteria (indicated by an off-white colony with no dye reaction).

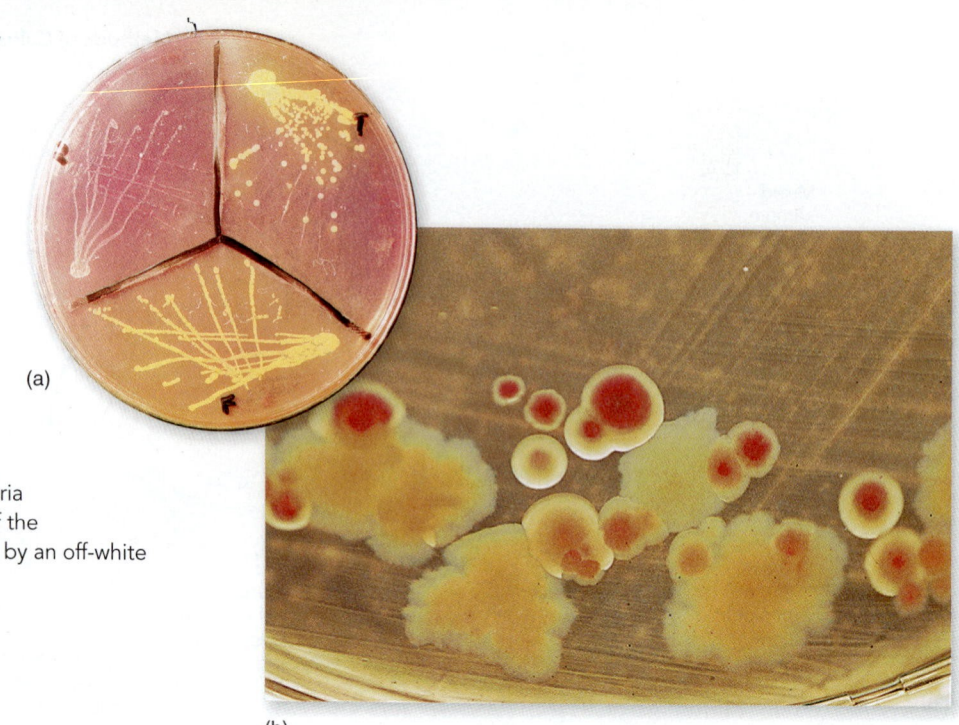

(a)

(b)

Mannitol salt agar (MSA) **(figure 3.5*a*)** contains a high concentration of NaCl (7.5%) that inhibits most human pathogens. One exception is the genus *Staphylococcus*, which grows well in this medium and consequently can be amplified in mixed samples. Media for isolating gram-negative intestinal pathogens (MacConkey agar, Hektoen enteric [HE] agar) contain bile salts, a component of feces, as a selective agent to inhibit most gram-positive bacteria **(figure 3.5*b*).** Other agents that have selective properties are dyes, such as methylene blue and crystal violet, acid, and antimicrobial drugs. Some selective media contain strongly inhibitory agents, such as selenite or sodium azide, to favor the growth of a pathogen that would otherwise be overlooked because of its low numbers in a specimen.

Differential media allow multiple types of microorganisms to grow but are designed to display visible differences among their colonies. Differentiation shows up as variations in colony size or color, in media color changes, or in the formation of gas bubbles and precipitates **(table 3.5).**

These variations often come from the type of chemicals these media contain and the ways that microbes react to them. For example, when microbe X metabolizes a certain substance not used by organism Y, then X will cause a visible change in the color of the colony or the medium and Y will not. The simplest differential media show two reaction types such as the use or nonuse of a particular nutrient or a color change in some colonies but not in others. Some media are sufficiently complex to show three or four different reactions **(figure 3.6).**

Dyes are frequently used as differential agents because many of them are pH indicators that change color in response to the production of an acid or a base. For example, mannitol salt agar contains phenol red, a dye that turns yellow when microbes acidify the medium by fermenting mannitol. MacConkey agar contains neutral red, a dye that turns pink or red when microbes metabolize lactose in the medium.

Although blood agar is a type of enriched medium used for the growth of fastidious microbes, the presence of intact

Table 3.5 Differential Media		
Medium	**Substances That Facilitate Differentiation**	**Differentiates Between**
Blood agar	Intact red blood cells	Types of hemolysis displayed by different species of *Streptococcus*
Mannitol salt agar	Mannitol, phenol red	Species of *Staphylococcus*
Hektoen enteric (HE) agar	Brom thymol blue, acid fuchsin, sucrose, salicin, thiosulfate, ferric ammonium citrate	*Salmonella, Shigella*, other lactose fermenters from nonfermenters
MacConkey agar	Lactose, neutral red	Bacteria that ferment lactose (lowering the pH) from those that do not
Urea broth	Urea, phenol red	Bacteria that hydrolyze urea to ammonia
Sulfur indole motility (SIM)	Thiosulfate, iron	H_2S gas producers from nonproducers
Triple-sugar iron agar (TSIA)	Triple sugars, iron, and phenol red dye	Fermentation of sugars, H_2S production
Birdseed agar	Seeds from thistle plant	*Cryptococcus neoformans* and other fungi

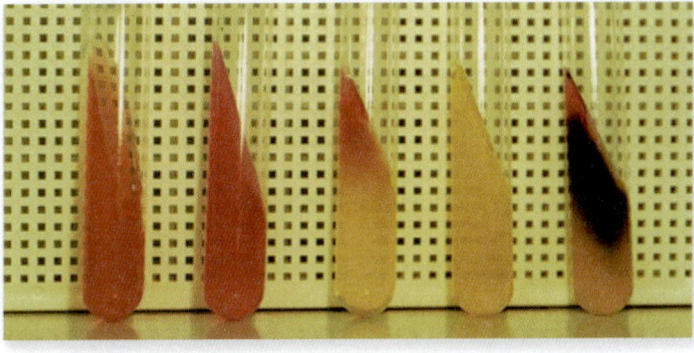

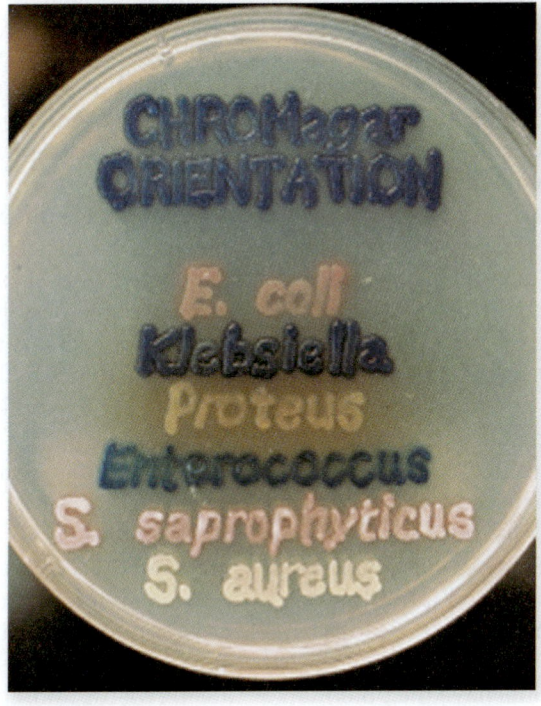

(b)

Figure 3.6   Media that differentiate characteristics.
(a) Triple-sugar iron agar (TSIA) in a slant tube. This medium contains three fermentable carbohydrates, phenol red to indicate pH changes, and a chemical (iron) that indicates H_2S gas production. Reactions (from left to right) are no growth; growth with no acid production; acid production in the bottom (butt) only; acid production all through the medium; and acid production in the butt with H_2S gas formation (black). **(b)** A state-of-the-art medium developed for culturing and identifying the most common urinary pathogens. CHROMagar Orientation uses color-forming reactions to distinguish at least seven species and permits rapid identification and treatment. In the example, the bacteria were streaked so as to spell their own names.

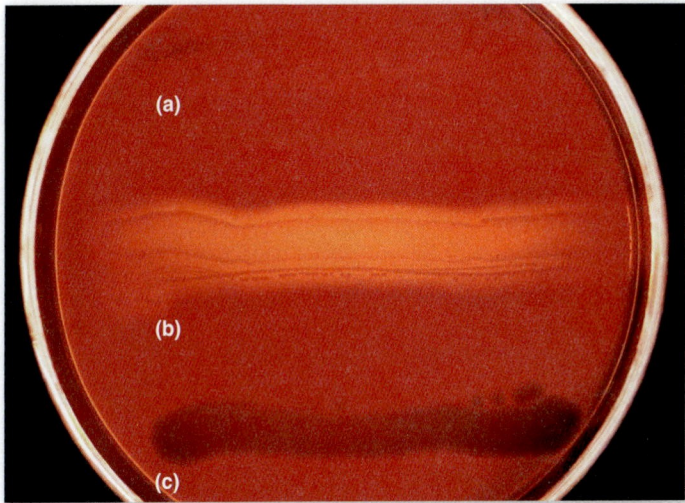

Figure 3.7   Types of hemolysis on blood agar.   (a) Microbe exhibiting gamma-hemolysis, or no change in the blood agar due to the lack of hemolysin production. **(b)** Microbe exhibiting beta-hemolysis, or a clearing of the blood agar due to complete red blood cell lysis. **(c)** Microbe exhibiting alpha-hemolysis, or greening of the blood agar due to incomplete red blood cell lysis.

red blood cells allows it to function as a differential medium as well. **Hemolysins** are enzymes that function to lyse red blood cells for the purpose of releasing iron-rich hemoglobin for growth. When grown on blood agar, some hemolysin-producing species completely lyse all red blood cells in the adjacent media resulting in beta-hemolysis, a clearing around the bacterial colony **(figure 3.7b)**. Other species only partially lyse the red blood cells producing alpha-hemolysis,

which appears as a greening of the agar around the colony **(figure 3.7c)**. Bacteria having no hemolysins result in no reaction in the agar, which is termed *gamma hemolysis* **(figure 3.7a)**.

A single medium can be both selective and differential, owing to its different ingredients. MacConkey agar and mannitol salt agar, for example, both appear in table 3.4 (selective media) and table 3.5 (differential media) due to their ability to simultaneously suppress the growth of some organisms while producing a visual distinction among the ones that do grow.

Miscellaneous Media   A **reducing medium** contains a substance (thioglycolic acid or cystine) that absorbs oxygen or slows the penetration of oxygen in a medium, thus reducing its availability. Reducing media are important for growing anaerobic bacteria or for determining oxygen requirements of isolates (described in chapter 7). **Carbohydrate fermentation media** contain sugars that can be fermented (converted to acids) and a pH indicator to show this reaction **(figure 3.8)**.

Transport media are used to maintain and preserve specimens that have to be held for a period of time before clinical analysis or to sustain delicate species that die rapidly if not held under stable conditions. Transport media contain salts, buffers, and absorbents to prevent cell destruction by enzymes, pH changes, and toxic substances but will not support growth. **Assay media** are used by technologists to test the effectiveness of antimicrobial drugs (see chapter 12) and by drug manufacturers to assess the effect of disinfectants, antiseptics, cosmetics, and preservatives on the growth of microorganisms. **Enumeration media** are used by industrial and environmental microbiologists to count the numbers of organisms in milk, water, food, soil, and other samples.

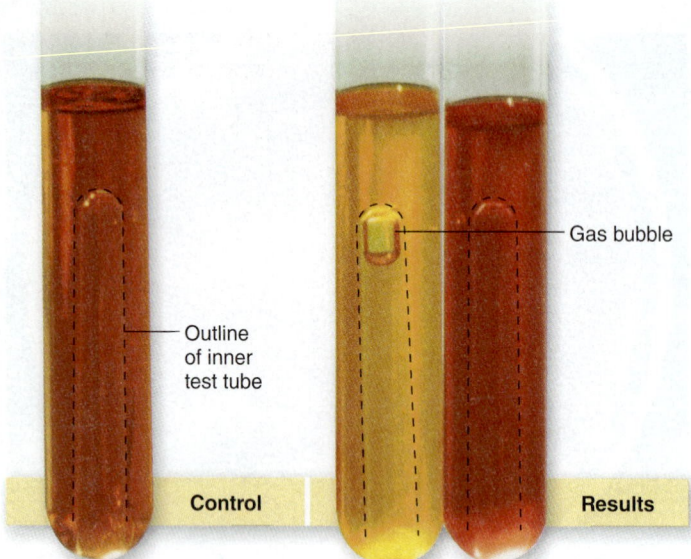

— Gas bubble

— Outline of inner test tube

Control   **Results**

Figure 3.8 Carbohydrate fermentation in broth.   This medium is designed to show fermentation (acid production) and gas formation by means of a small, inverted test tube for collecting gas bubbles. The tube on the left is an uninoculated negative control; the center tube is positive for acid (yellow) and gas (open space); the tube on the right shows growth but neither acid nor gas.

Isolation: Separating One Species from Another

Certain **isolation** techniques are based on the concept that if an individual bacterial cell is separated from other cells and provided adequate space on a nutrient surface, it will grow into a discrete mound of cells called a **colony (figure 3.9).** If formed from a single cell, a colony consists of just that one species and no other. Proper isolation requires that a small number of cells be inoculated into a relatively large volume or over an expansive area of medium selected to encourage the growth of the desired microbe. It generally requires the following materials: a medium that has a relatively firm surface (see agar in "Physical States of Media," page 59), a Petri dish (a clear, flat dish with a cover), and inoculating tools. In the streak plate method, a small droplet of culture or sample is spread over the surface of the medium with an **inoculating loop** in a pattern that gradually thins out the sample and separates the cells spatially over several sections of the plate **(process figure 3.10a).**

In the loop dilution, or pour plate, technique, the sample is inoculated serially into a series of cooled but still liquid agar tubes so as to dilute the number of cells in each successive tube in the series **(process figure 3.10b).** Inoculated tubes are then plated out (poured) into sterile Petri dishes and are allowed to solidify (harden). The end result (usually in the second or third plate) is that the number of cells per volume is so decreased that cells have ample space to grow into separate colonies. One difference between this and the streak plate method is that in this technique some of the colonies will develop deep in the medium itself and not just on the surface.

With the spread plate technique, a small volume of liquid, diluted sample is pipetted onto the surface of the medium and spread around evenly by a sterile spreading tool (sometimes called a "hockey stick"). Like the streak plate, cells are pushed onto separate areas on the surface so that they can form individual colonies **(process figure 3.10c).**

In some ways, culturing microbes is analogous to gardening. Cultures are formed by "seeding" tiny plots (media) with microbial cells. Extreme care is taken to exclude weeds (contaminants). Once microbes have grown after incubation, the clinician must *inspect* the container (Petri dish, test tube, etc.). A **pure culture** is a container of medium that grows only a single known species or type of microorganism **(figure 3.11a).** This type of culture is most frequently used for laboratory study, because it allows the systematic examination and control of one microorganism by itself. Instead of the term *pure culture,* some microbiologists prefer the term **axenic,** meaning that the culture is free of other living things except for the one being studied. A standard method for preparing a pure culture is to **subculture,** or make a second-level culture from a well-isolated colony. A tiny bit of cells is transferred into a separate container of media and incubated (see "Isolation" in process figure 3.1). Sometimes growing microbes in pure culture can tell you very little about how they act in a mixed-species environment. Being able to isolate and study them in this manner can be valuable, though, as long as you keep in mind that it is an unnatural state for them.

A **mixed culture (figure 3.11b)** is a container that holds two or more *identified,* easily differentiated species of microorganisms, not unlike a garden plot containing both carrots and onions. A **contaminated culture (figure 3.11c)** was once pure or mixed (and thus a known entity) but has since had **contaminants** (unwanted microbes of uncertain identity) introduced into it, like weeds into a garden. Because contaminants have the potential for causing disruption, constant vigilance is required to exclude them from microbiology

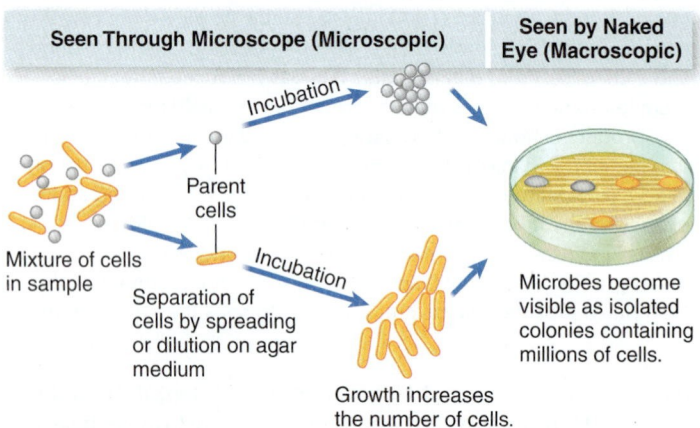

Seen Through Microscope (Microscopic)

Seen by Naked Eye (Macroscopic)

Incubation

Parent cells

Mixture of cells in sample

Separation of cells by spreading or dilution on agar medium

Incubation

Growth increases the number of cells.

Microbes become visible as isolated colonies containing millions of cells.

Figure 3.9 Isolation technique.   Stages in the formation of an isolated colony, showing the microscopic events and the macroscopic result. Separation techniques such as streaking can be used to isolate single cells. After numerous cell divisions, a macroscopic mound of cells, or a colony, will be formed. This is a relatively simple yet successful way to separate different types of bacteria in a mixed sample.

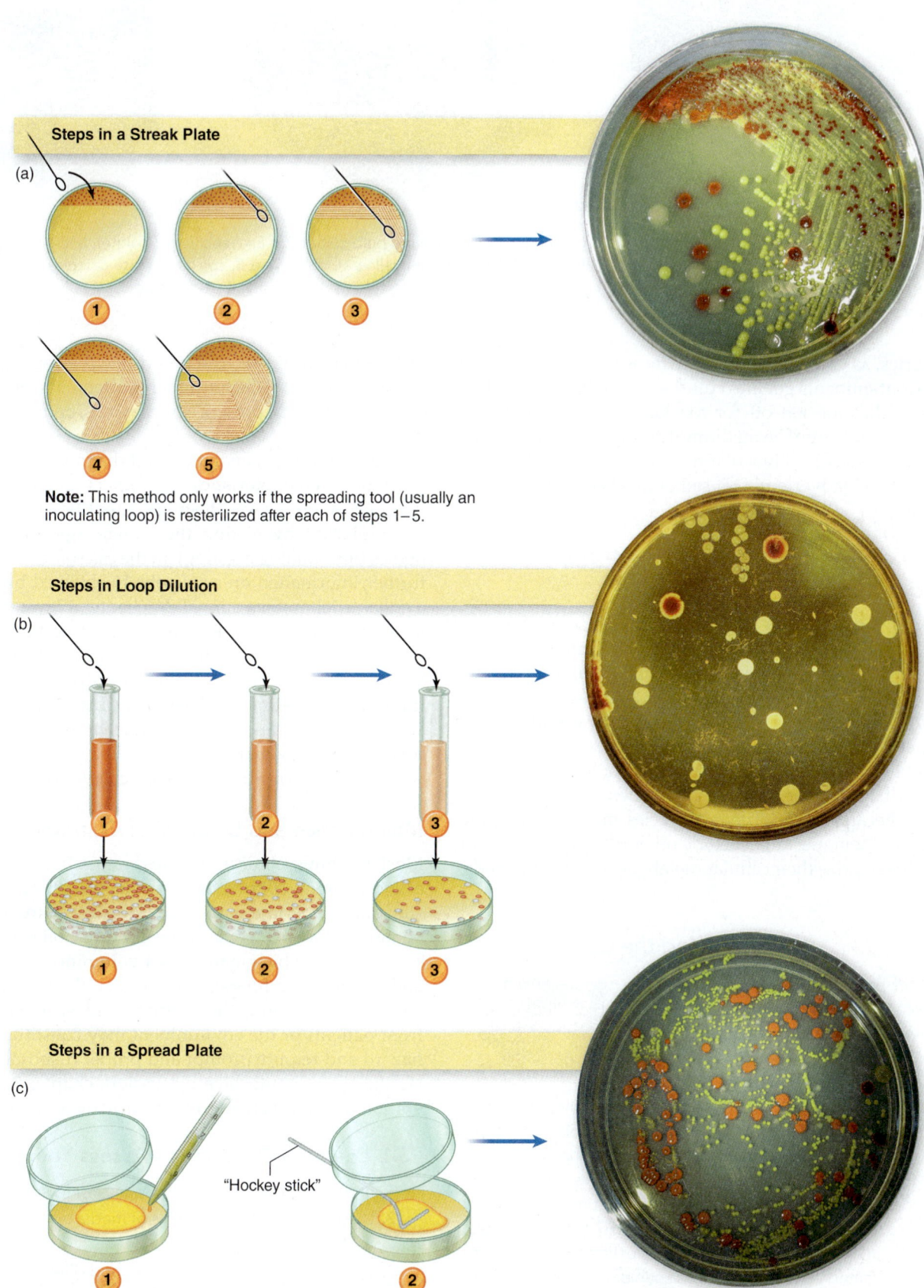

Steps in a Streak Plate

(a)

1 2 3

4 5

Note: This method only works if the spreading tool (usually an inoculating loop) is resterilized after each of steps 1–5.

Steps in Loop Dilution

(b)

1 2 3

1 2 3

Steps in a Spread Plate

(c)

"Hockey stick"

1 2

Process Figure 3.10 Methods for isolating bacteria.   **(a)** Steps in a quadrant streak plate and resulting isolated colonies of bacteria. Microbiologists use from 3–5 steps in this technique. **(b)** Steps in the loop dilution method and the appearance of plate 3. **(c)** Spread plate and its result.

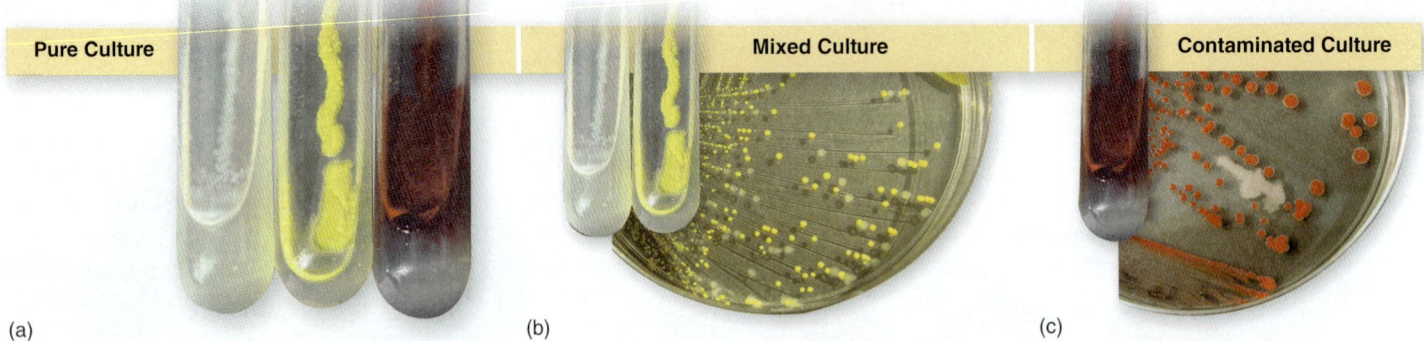

| Pure Culture | Mixed Culture | Contaminated Culture |

(a)   (b)   (c)

Figure 3.11 Various conditions of cultures.   **(a)** Three tubes containing pure cultures of *Escherichia coli* (white), *Micrococcus luteus* (yellow), and *Serratia marcescens* (red). **(b)** A mixed culture of *M. luteus* (bright yellow colonies) and *E. coli* (faint white colonies). **(c)** This plate of *S. marcescens* was overexposed to room air, and it has developed a large, white colony. Because this intruder is not desirable and not identified, the culture is now contaminated.

laboratories, as you will no doubt witness in your own experience. Contaminants get into cultures when the lids of tubes or Petri dishes are left off for too long, allowing airborne microbes to settle into the medium. They can also enter on an incompletely sterilized inoculating loop or on an instrument that you have inadvertently reused or touched to the table or your skin.

Rounding Out the Five I's: Inspection and Identification

How does one determine (i.e., identify) what sorts of microorganisms have been isolated in cultures? Certainly, microscopic appearance can be valuable in differentiating the smaller, simpler bacterial cells from the larger, more complex eukaryotic cells. Appearance can be especially useful in identifying eukaryotic microorganisms to the level of genus or species because of their distinctive morphological features; however, bacteria are generally not identifiable by these methods because very different species may appear quite similar. For them, we must include other techniques, some of which characterize their cellular metabolism. These methods,

called *biochemical tests,* can determine fundamental chemical characteristics such as nutrient requirements, products given off during growth, presence of enzymes, and mechanisms for deriving energy.

Several modern analytical and diagnostic tools that focus on genetic characteristics can detect microbes based on their DNA (genotypic testing). Identification can also be accomplished by testing the isolate against known antibodies (immunologic testing). In the case of certain pathogens, further information on a microbe is obtained by inoculating a suitable laboratory animal. In chapter 17, we present more detailed examples of these modern identification methods.

It is important to understand that a microbial profile can be prepared only by combining phenotypic, genotypic, and immunologic testing results with the macroscopic and microscopic analysis described in this chapter. The profile then becomes the raw material used in final identification as you will see in the concluding disease chapters of this textbook.

Maintenance and Disposal of Cultures

Most teaching, clinical, and research laboratories maintain a line of stock cultures that represent "living catalogs" for study and experimentation. The largest culture collection can be found at the American Type Culture Collection in Manassas, Virginia, which maintains a voluminous array of frozen and freeze-dried fungal, bacterial, viral, and algal cultures. On the other hand, the cultures and specimens collected from patients or the environment may constitute a potential hazard and require prompt and proper disposal. Both steam sterilizing (see autoclave discussion, chapter 11) and incineration (burning) are used to destroy microorganisms.

Case File 3   *Continuing the Case*

Cases of EHEC are usually associated with the notorious strain *E. coli* O157:H7, but this outbreak was caused by the lesser-known strain *E. coli* O104:H4. Both strains of *E. coli* produce the Shiga toxin that causes bloody diarrhea and HUS, but the O104:H4 strain had previously not been found in animals and only rarely in humans. Scientists and doctors analyzed stool samples of patients showing symptoms of EHEC and HUS in an attempt to isolate the Shiga-toxin-producing pathogen. Genomic analysis of specimens revealed that the O104:H4 strain rather than the O157:H7 strain was causing illness in the patients. Moreover, the scientists confirmed that it was resistant to at least 14 different antibiotics.

■ What type of microbial media might have been used to identify *E. coli* O104:H4?

3.1 Learning Outcomes—Assess Your Progress

1. Explain what the Five I's mean and what each step entails.
2. Discuss three physical states of media and when each is useful.
3. Compare and contrast selective and differential media, and give an example of each.
4. Provide brief definitions for *defined* and *complex media*.

3.2 The Microscope: Window on an Invisible Realm

Imagine Leeuwenhoek's excitement and wonder when he first viewed a drop of rainwater and glimpsed an amazing microscopic world teeming with unearthly creatures. Beginning microbiology students still experience this sensation, and even experienced microbiologists remember their first view. Before we examine microscopes, let's consider how small microbes actually are.

Microbial Dimensions: How Small Is Small?

When we say that microbes are too small to be seen with the unaided eye, what sorts of dimensions are we talking about? The concept of thinking small is best visualized by comparing microbes with the larger organisms of the macroscopic world and also with the atoms and molecules of the molecular world (**figure 3.12**). Whereas the dimensions of macroscopic organisms are usually given in centimeters (cm) and meters (m), those of microorganisms fall within the range of millimeters (mm) to micrometers (μm) to nanometers (nm). The size range of most microbes extends from the smallest bacteria, measuring around 200 nm, to protozoa and algae that measure 3 to 4 mm and are visible with the naked eye. Viruses, which can infect all organisms including microbes, measure between 20 nm and 800 nm, and some of them are thus not much bigger than large molecules, whereas others are just a tad larger than the smallest bacteria.

Disease Connection

Mycoplasma species are the smallest organisms able to grow on artificial medium. They are between 0.15 and 0.3 microns in size. These bacteria have no proper cell walls and therefore have varied shapes. The bacterium *Mycoplasma pneumoniae* is a common cause of pneumonia of a mild nature, which is often called walking pneumonia. We will learn more about this disease in chapter 21.

The microbial existence is indeed another world, but it would remain largely uncharted without an essential tool: the microscope. Your efforts in exploring microbes will be more meaningful if you understand some essentials of **microscopy** and specimen preparation.

Magnification and Microscope Design

A discovery by early microscopists that spurred the advancement of microbiology was that a clear, glass sphere could act as a lens to magnify small objects. Magnification in most microscopes results from a complex interaction between visible light waves and the curvature of the lens. When a beam or ray of light transmitted through air strikes and passes through the convex surface of glass, it experiences some

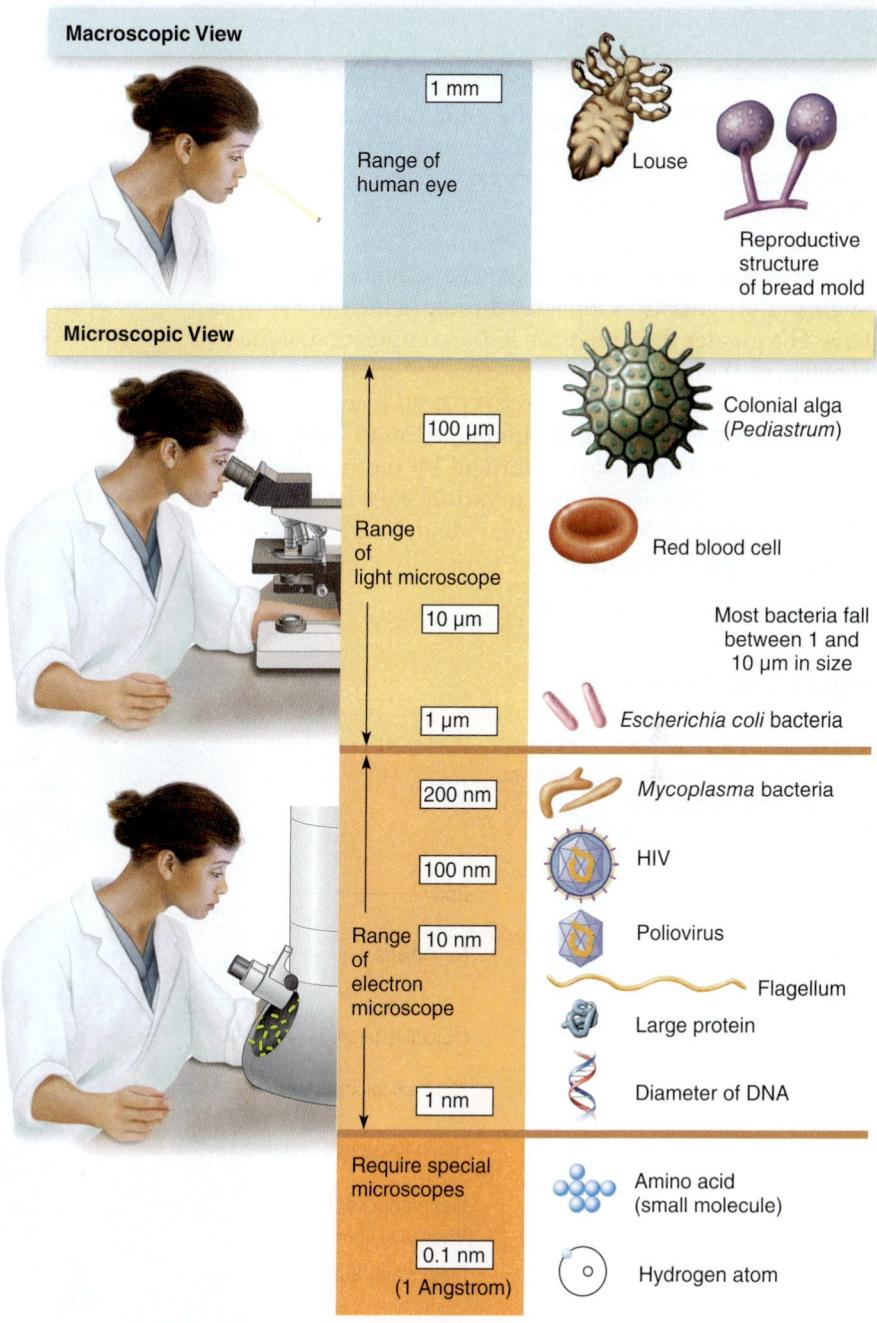

Figure 3.12 The size of things. Common measurements encountered in microbiology and a scale of comparison from the macroscopic to the microscopic, molecular, and atomic. Most microbes encountered in our studies will fall between 100 μm and 10 nm in overall dimensions. The microbes shown are more or less to scale within size zone but not *between* size zones.

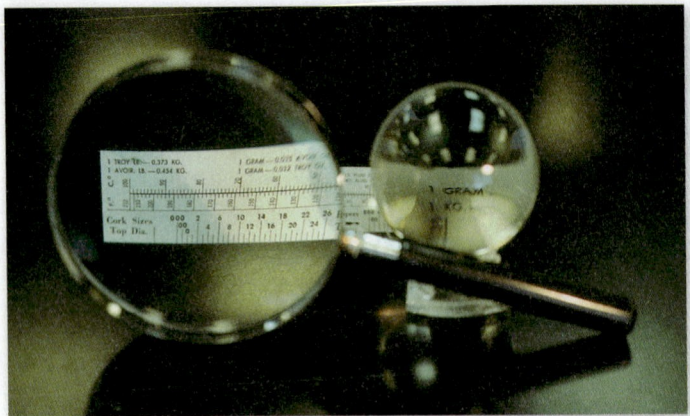

Figure 3.13 Effects of magnification. Demonstration of the magnification and image-forming capacity of clear glass "lenses." Given a proper source of illumination, this magnifying glass and crystal ball magnify a ruler two to three times.

degree of **refraction,** defined as the bending or change in the angle of the light ray as it passes through a medium such as a lens. The greater the difference in the composition of the two substances the light passes between, the more pronounced is the refraction. When an object is placed a certain distance from the spherical lens and illuminated with light, an optical replica, or image, of it is formed by the refracted light. Depending upon the size and curvature of the lens, the image appears enlarged to a particular degree, which is called its power of magnification and is usually identified with a

number combined with × (read "times"). This behavior of light is evident if one looks through an everyday object such as a glass ball or a magnifying glass **(figure 3.13).** It is basic to the function of all optical, or light, microscopes, though many of them have additional features that define, refine, and increase the size of the image.

The first microscopes were simple, meaning they contained just a single magnifying lens and a few working parts. Examples of this type of microscope are a magnifying glass, a hand lens, and Leeuwenhoek's basic little tool shown earlier in figure 1.8a. Among the refinements that led to the development of today's compound microscope were the addition of a second magnifying lens system, a lamp in the base to give off visible light and illuminate the specimen, and a special lens called the condenser that converges or focuses the rays of light to a single point on the object. The fundamental parts of a modern compound light microscope are illustrated in **figure 3.14.**

Principles of Light Microscopy

To be most effective, a microscope should provide three properties: magnification, resolution, and good contrast. Magnification of the object or specimen by a compound microscope occurs in two phases. The first lens in this system (the one closest to the specimen) is the objective lens, and the second (the one closest to the eye) is the ocular lens, or eyepiece **(figure 3.15).** The objective forms the initial image of the specimen, called the **real image.** When this image is projected up through the microscope body to the plane of the eyepiece,

Figure 3.14 The parts of a student laboratory microscope. This microscope is a compound light microscope with two oculars (called binocular).

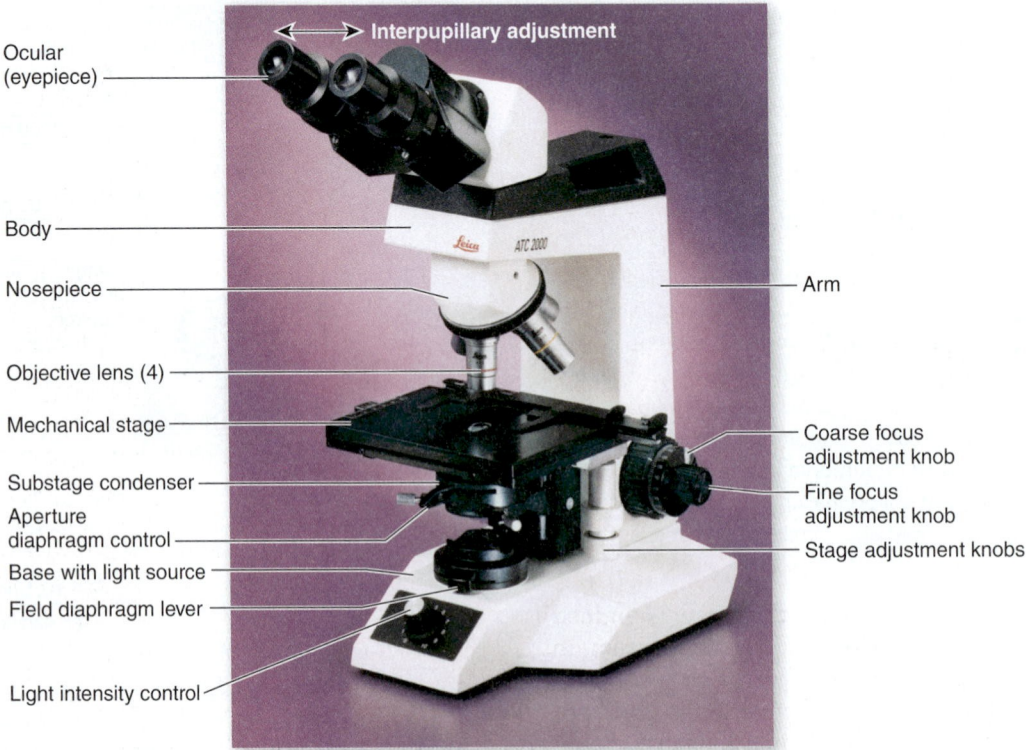

Interpupillary adjustment

Ocular (eyepiece)

Body

Nosepiece

Objective lens (4)

Mechanical stage

Substage condenser

Aperture diaphragm control

Base with light source

Field diaphragm lever

Light intensity control

Arm

Coarse focus adjustment knob

Fine focus adjustment knob

Stage adjustment knobs

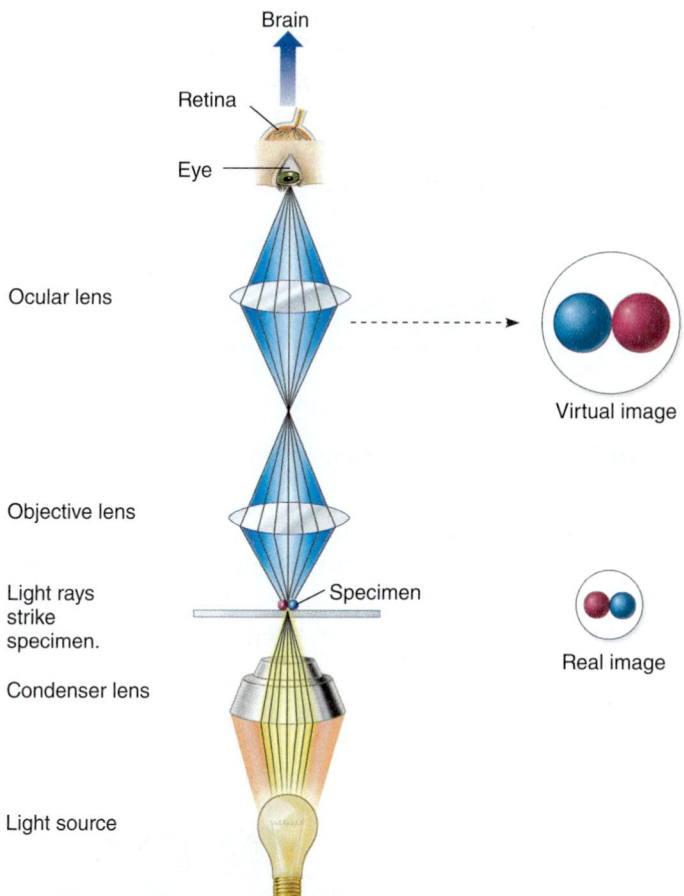

Figure 3.15 The pathway of light and the two stages in magnification of a compound microscope. As light passes through the condenser, it forms a solid beam that is focused on the specimen. Light leaving the specimen that enters the objective lens is refracted so that an enlarged primary image, the real image, is formed. One does not see this image, but its degree of magnification is represented by the lower circle. The real image is projected through the ocular, and a second image, the virtual image, is formed by a similar process. The virtual image is the final magnified image that is received by the retina and perceived by the brain. Notice that the lens systems cause the image to be reversed.

Microscopes are equipped with a nosepiece holding three or more objectives that can be rotated into position as needed. The power of the ocular usually remains constant for a given microscope. Depending on the power of the ocular, the total magnification of standard light microscopes can vary from 40× with a 4× objective (called the scanning objective) to 2,000× with the highest power objective (the oil immersion objective).

A Note About Oil Immersion Lenses

In order for the oil immersion lens to provide maximum resolution, a drop of oil must be inserted between the tip of the lens and the specimen on the glass slide. Because oil has the same optical qualities as glass, the light rays will not change direction (or refract) when they come through the slide and into the oil. This prevents the scattering of the most peripheral light rays and creates a more focused beam of light. This property effectively increases the numerical aperture and, when combined with the high magnification power of the oil immersion lens, greatly enhances resolution (figure 3.16).

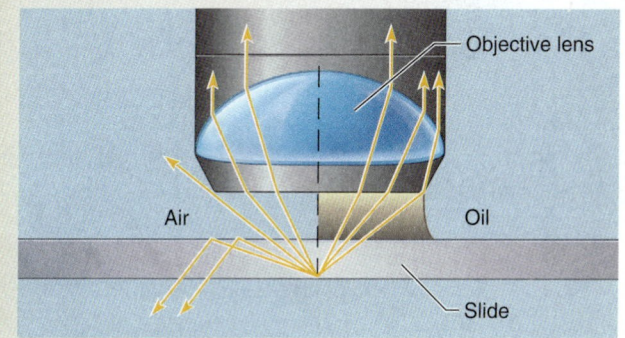

Figure 3.16 Workings of an oil immersion lens. Without oil, some of the peripheral light that passes through the specimen is scattered into the air or onto the glass slide; this scattering decreases resolution.

the ocular lens forms a second image, the **virtual image.** The virtual image is the one that will be received by the eye and converted to a retinal and visual image. The magnifying power of the objective alone usually ranges from 4× to 100×, and the power of the ocular alone ranges from 10× to 20×. The total power of magnification of the final image formed by the combined lenses is a product of the separate powers of the two lenses.

Power of Objective	×	Usual Power of Ocular	=	Total Magnification
10× low power objective	×	10×	=	100×
40× high dry objective	×	10×	=	400×
100× oil immersion objective	×	10×	=	1,000×

Resolution: Distinguishing Magnified Objects Clearly
As important as magnification is for visualizing tiny objects or cells, an additional optical property is essential for seeing clearly. That property is resolution, or **resolving power.** Resolution is the capacity of an optical system to distinguish or separate two adjacent objects or points from one another. For example, at a certain fixed distance, the lens in the human eye can resolve two small objects as separate points just as long as the two objects are no closer than 0.2 millimeters apart. The eye examination given by optometrists is in fact a test of the resolving power of the human eye for various-size letters read at a particular distance. Because microorganisms are extremely small and usually very close together, they will not be seen with clarity or any degree of detail unless the microscope's lenses can resolve them.

Resolving power is determined by a combination of characteristics of the objective lens and the wavelength of the light being used to illuminate the sample. The light source for optical microscopes consists of a band of colored wavelengths in the visible spectrum. The shortest visible wavelengths are in the violet-blue portion of the spectrum (400 nanometers), and the longest are in the red portion (750 nanometers) **(figure 3.17).** Because the wavelength must pass between the objects that are being resolved, shorter wavelengths (in the 400–500 nanometer range) will provide better resolution **(figure 3.18).** Some microscopes have a special blue filter placed over the lamp to limit the longer wavelengths of light from entering the specimen.

In practical terms, the oil immersion lens can resolve any cell or cell part as long as it is at least 0.2 micron in diameter, and it can resolve two adjacent objects as long as they are at least 0.2 micron apart **(figure 3.19).** In general, organisms that are 0.5 micron or more in diameter are readily seen. This includes fungi and protozoa, some of their internal structures, and most bacteria. However, a few bacteria and most viruses are far too small to be resolved by the optical microscope and require electron microscopy (discussed later in this chapter). In summary then, the factor

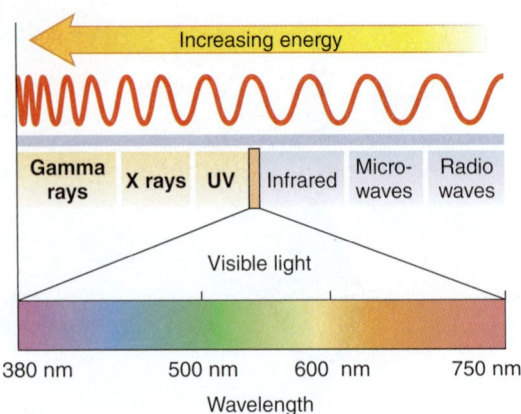

Figure 3.17 The electromagnetic spectrum.

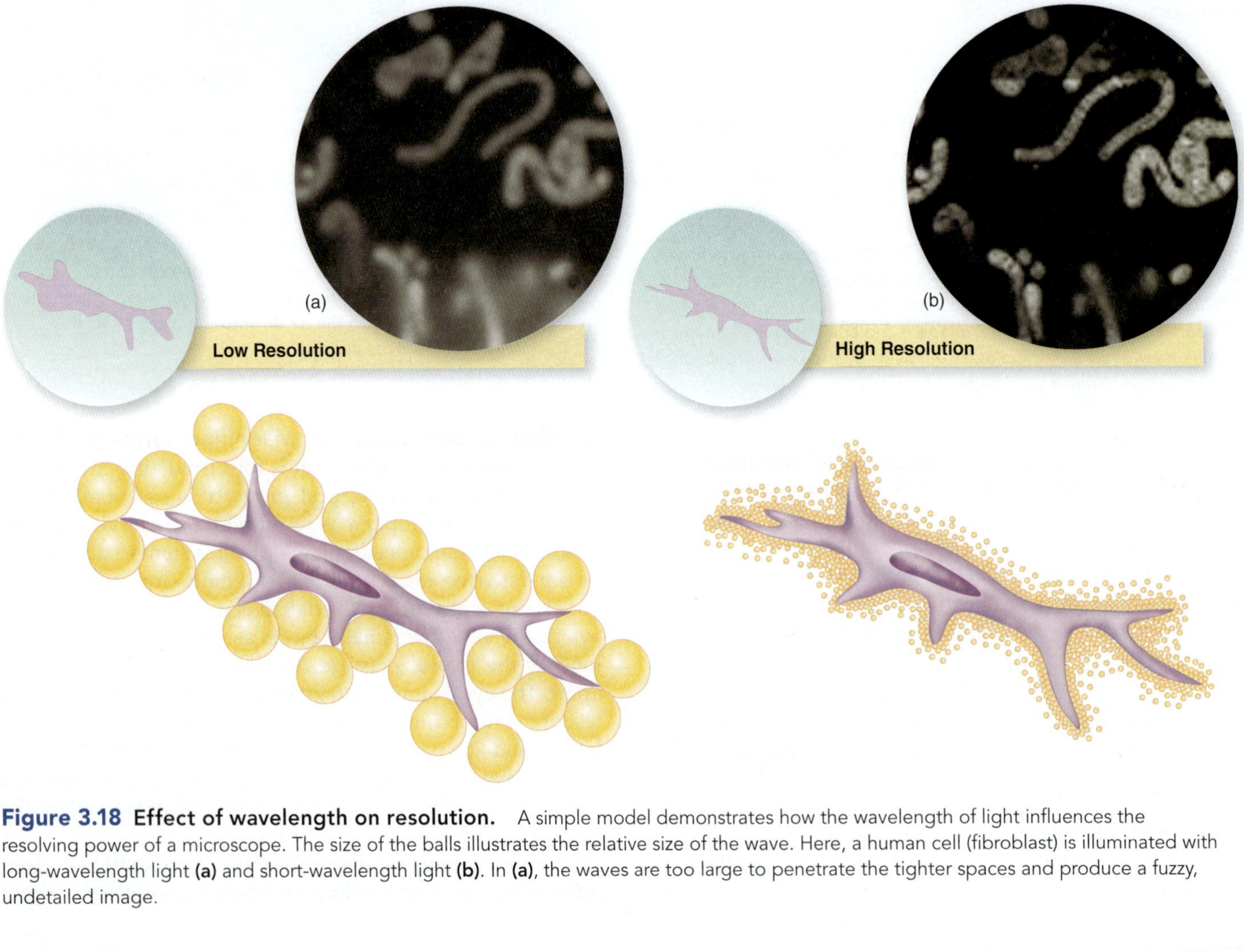

that most limits the clarity of a microscope's image is its resolving power. Even if a light microscope were designed to magnify several thousand times, its resolving power could not be increased and the image it produced would simply be enlarged and fuzzy.

(a)

Low Resolution

(b)

High Resolution

Figure 3.18 Effect of wavelength on resolution.   A simple model demonstrates how the wavelength of light influences the resolving power of a microscope. The size of the balls illustrates the relative size of the wave. Here, a human cell (fibroblast) is illuminated with long-wavelength light **(a)** and short-wavelength light **(b)**. In **(a)**, the waves are too large to penetrate the tighter spaces and produce a fuzzy, undetailed image.

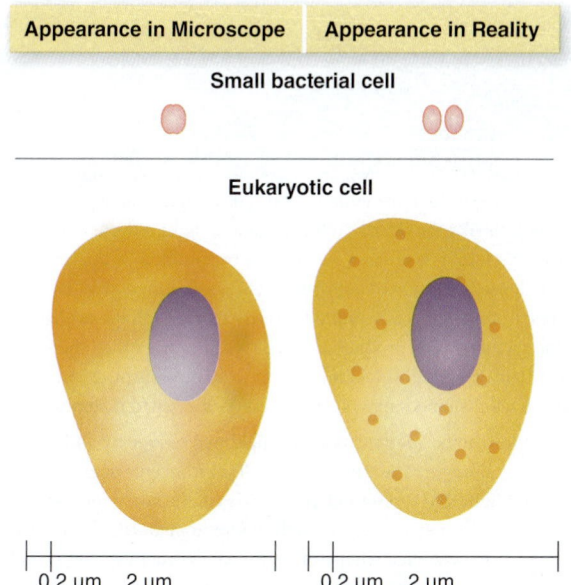

Appearance in Microscope	Appearance in Reality

Small bacterial cell

Eukaryotic cell

0.2 μm   2 μm   0.2 μm   2 μm

Figure 3.19 The Importance of resolution. If a microscope has a resolving power of 0.2 μm, then the bacterial cells will not be resolvable as two separate cells. Likewise, the small specks inside the eukaryotic cell will not be visible.

Contrast The third quality of a well-magnified image is its degree of contrast from its surroundings. The contrast is measured by a quality called the **refractive index.** Refractive index refers to the degree of bending that light undergoes as it passes from one medium (such as water or glass) to another medium, such as some bacterial cells. The higher the difference in refractive indexes (the more bending of light), the sharper the contrast that is registered by the microscope and the eye. Because too much light can reduce contrast and burn out the image, an adjustable iris diaphragm on most microscopes controls the amount of light entering the condenser. The lack of contrast in cell components is compensated for by using special lenses (the phase-contrast microscope) and by adding stains.

Variations on the Light Microscope

Optical microscopes that use visible light can be described by the nature of their field, meaning the circular area viewed through the ocular lens. There are four types of visible-light microscopes: bright-field, dark-field, phase-contrast, and interference. A fifth type of optical microscope, the fluorescence microscope, uses ultraviolet radiation as the illuminating source; another, the confocal microscope, uses a laser beam. Each of these microscopes is adapted for viewing specimens in a particular way, as described in **table 3.6.**

Preparing Specimens for Optical Microscopes

A specimen for optical microscopy is generally prepared by mounting a sample on a suitable glass slide that sits on the stage between the condenser and the objective lens.

The manner in which a slide specimen, or mount, is prepared depends upon (1) the condition of the specimen, either in a living or preserved state; (2) the aims of the examiner, whether to observe overall structure, identify the microorganisms, or see movement; and (3) the type of microscopy available, whether it is bright-field, dark-field, phase-contrast, or fluorescence.

Fresh, Living Preparations

Live samples of microorganisms are placed in wet mounts or in hanging drop mounts so that they can be observed as near to their natural state as possible. The cells are suspended in a suitable fluid (water, broth, saline) that temporarily maintains viability and provides space and a medium for locomotion. A wet mount consists of a drop or two of the culture placed on a slide and overlaid with a coverslip. Although this type of mount is quick and easy to prepare, it has certain disadvantages. The coverslip can damage larger cells, and the slide is very susceptible to drying and can contaminate the handler's fingers. A more satisfactory alternative is the hanging drop preparation made with a special concave (depression) slide, a Vaseline adhesive or sealant, and a coverslip from which a tiny drop of sample is suspended **(figure 3.20).** These types of short-term mounts provide a true assessment of the size, shape, arrangement, color, and motility of cells. Greater cellular detail can be observed with phase-contrast or interference microscopy.

Fixed, Stained Smears

A more permanent mount for long-term study can be obtained by preparing fixed, stained specimens. The smear technique, developed by Robert Koch more than 100 years ago, consists of spreading a thin film made from a liquid suspension of cells on a slide and air-drying it. Next, the air-dried smear is usually heated gently by a process called heat fixation that simultaneously kills the specimen and secures it to the slide. Another important action of fixation is to preserve various cellular components in a natural state with minimal distortion. Sometimes fixation of microbial cells is performed with chemicals such as alcohol and formalin.

Like images on undeveloped photographic film, the unstained cells of a fixed smear are quite indistinct, no matter how great the magnification or how fine the resolving power of the microscope. The process of "developing" a smear to create contrast and make inconspicuous features stand out requires staining techniques. Staining is any procedure that applies colored chemicals called dyes to specimens. Dyes impart a color to cells or cell parts by becoming

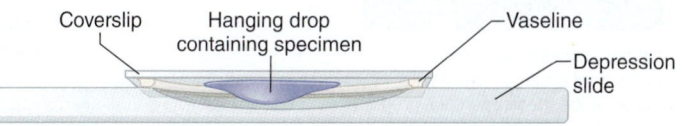

Coverslip   Hanging drop containing specimen   Vaseline   Depression slide

Figure 3.20 Hanging drop technique. Cross-section view of slide and coverslip. (Vaseline actually surrounds entire well of slide.)

Table 3.6 Comparison of Types of Microscopy

Microscope	Maximum Practical Magnification	Resolution	
Visible light as source of illumination			
Bright-field *Paramecium* (400×)	2,000×	0.2 µm (200 nm)	The bright-field microscope is the most widely used type of light microscope. Although we ordinarily view objects like the words on this page with light reflected off the surface, a bright-field microscope forms its image when light is transmitted through the specimen. The specimen, being denser and more opaque than its surroundings, absorbs some of this light, and the rest of the light is transmitted directly up through the ocular into the field. As a result, the specimen will produce an image that is darker than the surrounding brightly illuminated field. The bright-field microscope is a multipurpose instrument that can be used for both live, unstained material and preserved, stained material.
Dark-field *Paramecium* (400×)	2,000×	0.2 µm	A bright-field microscope can be adapted as a dark-field microscope by adding a special disc called a *stop* to the condenser. The stop blocks all light from entering the objective lens—except peripheral light that is reflected off the sides of the specimen itself. The resulting image is a particularly striking one: brightly illuminated specimens surrounded by a dark (black) field. The most effective use of dark-field microscopy is to visualize living cells that would be distorted by drying or heat or that cannot be stained with the usual methods.
Phase-contrast *Paramecium* (400×)	2,000×	0.2 µm	If similar objects made of clear glass, ice, cellophane, or plastic are immersed in the same container of water, an observer would have difficulty telling them apart because they have similar optical properties. Internal components of a live, unstained cell also lack contrast and can be difficult to distinguish. But cell structures do differ slightly in density, enough that they can alter the light that passes through them in subtle ways. The phase-contrast microscope has been constructed to take advantage of this characteristic. This microscope contains devices that transform the subtle changes in light waves passing through the specimen into differences in light intensity. For example, denser cell parts such as organelles alter the pathway of light more than less dense regions (the cytoplasm). Light patterns coming from these regions will vary in contrast. The amount of internal detail visible by this method is greater than by either bright-field or dark-field methods. The phase-contrast microscope is most useful for observing intracellular structures such as bacterial spores, granules, and organelles, as well as the locomotor structures of eukaryotic cells such as cilia. (This image has been colorized; the actual microscopic image is black and white.)
Differential interference *Amoeba proteus* (160×)	2,000×	0.2 µm	Like the phase-contrast microscope, the differential interference contrast (DIC) microscope provides a detailed view of unstained, live specimens by manipulating the light. But this microscope has additional refinements, including two prisms that add contrasting colors to the image and two beams of light rather than a single one. DIC microscopes produce extremely well-defined images that are vividly colored and appear three-dimensional.
Ultraviolet rays as source of illumination			
Fluorescent Cheek epithelial cells (the larger unfocused green or red cells). Bacteria are the filamentous green and red rods and the green diplococci (400×).	2,000×	0.2 µm	The fluorescent microscope is a specially modified compound microscope furnished with an ultraviolet (UV) radiation source and a filter that protects the viewer's eye from injury by these dangerous rays. The name of this type of microscopy originates from the use of certain dyes (acridine, fluorescein) and minerals that show **fluorescence.** The dyes emit visible light when bombarded by short ultraviolet rays. For an image to be formed, the specimen must first be coated or placed in contact with a source of fluorescence. Subsequent illumination by ultraviolet radiation causes the specimen to give off light that will form its own image, usually an intense color such as red against a black field. Fluorescence microscopy is most useful in diagnosing infections caused by specific bacteria, protozoans, and viruses.

Table 3.6 Comparison of Types of Microscopy, *continued*

Microscope	Maximum Practical Magnification	Resolution	
Confocal Myofibroblasts, cells involved in tissue repair (400×)	2,000×	0.2 µm	The scanning confocal microscope overcomes the problem of cells or structures being too thick, a problem resulting in other microscopes being unable to focus on all their levels. This microscope uses a laser beam of light to scan various depths in the specimen and deliver a sharp image focusing on just a single plane. It is thus able to capture a highly focused view at any level, ranging from the surface to the middle of the cell. It is most often used on fluorescently stained specimens but it can also be used to visualize live unstained cells and tissues.
Electron beam forms image of specimen.			
Transmission electron microscope (TEM) Coronavirus, causative agent of many respiratory infections (100,000×)	100,000,000×	0.5 nm	Transmission electron microscopes are the instruments of choice for viewing the detailed structure of cells and their organelles and viruses. This microscope produces its image by transmitting electrons through the specimen. Because electrons cannot readily penetrate thick preparations, the specimen must be sectioned into extremely thin slices (20–100 nm thick) and stained or coated with metals that will increase image contrast. The darkest areas of TEM micrographs represent the thicker (denser) parts, and the lighter areas indicate the more transparent and less dense parts.
Scanning electron microscope (SEM) An alga showing a cell wall made of calcium disks (10,000×)	100,000,000×	10 nm	The scanning electron microscope provides some of the most dramatic and realistic images. This instrument is designed to create an extremely detailed three-dimensional view of all kinds of objects—from plaque on teeth to tapeworm heads. To produce its images, the SEM does not transmit electrons; it bombards the surface of a whole metal-coated specimen with electrons while scanning back and forth over it. A shower of electrons deflected from the surface is picked up with great fidelity by a sophisticated detector, and the electron pattern is displayed as an image on a television screen. The contours of the specimen resolved with scanning electron microscopy are very revealing. Areas that look smooth and flat with the light microscope display intriguing surface features with the SEM. (This image has been colorized; the actual microscopic image is black and white.)
Atomically sharp tip probes surface of specimen.			
Atomic force microscope (AFM) Prion fibrils, which may only be 5 nm in diameter	100,000,000×	0.01 Angstroms	In atomic force microscopy, a diamond or metal tip with a radius of 1–50 nanometers scans a specimen and moves up and down with contour of the surface at the atomic level. The movement of the tip is measured with a laser and translated into an image.
Scanning tunneling microscope (STM) Strands of DNA (Inset is a magnified view of the bare gold surface, verifying that it is clean).	100,000,000×	0.01 Angstroms	In scanning tunneling microscopy, a tungsten tip hovers over specimen while electrical voltage is applied, generating a current that is dependent on the distance between the tip and surface. Image is produced from the electrical signal of the tip's pathway.

INSIGHT 3.2   Microscopy: Now on Your Smartphone

Viewing microbes—at least the larger ones—just got a little easier. Instead of dragging out a clunky microscope and fiddling with the oil immersion lens, now you can whip out your smartphone. Scientists at the VTT Technical Research Centre of Finland have developed an app that can convert your camera phone into a high-resolution microscope. Images are produced by the combined effect of an LED light and an optical lens when users attach a thin, magnetic microscope module in front of the camera lens. The field of view is 3×2 mm and has a resolution of 1 micron, and the LEDs allow objects to be viewed from different angles. The device has many different applications, from studying surface formations on printed materials, to studying trees, leaves, and insects, with potential use in the health care field.

Another product called SkyLight allows users to attach their smartphones to a traditional microscope with a simple plastic smartphone holder. This device aligns the smartphone's camera with the eyepiece of the microscope, allowing the user to take still images or video for viewing later. In addition to being an

excellent study aid for students, Andrew Miller and Tess Bakke, the developers of the SkyLight, hope to make an impact on global health by allowing clinicians to share images with doctors in remote areas. The device is compatible with virtually every model of microscope, even those dating back to the 1980s.

Now you can view microbes everywhere—with your smartphone.

affixed to them through a chemical reaction. Dyes can be classified as basic (cationic), which have a positive charge, or acidic (anionic) dyes, which have a negative charge. Because chemicals of opposite charge are attracted to each other, cell parts that are negatively charged will attract basic dyes and those that are positively charged will attract acidic dyes **(table 3.7).** Many cells, especially those of bacteria, have numerous negatively charged acidic substances on their surfaces and thus stain more readily with basic dyes. Acidic dyes, on the other hand, tend to be repelled by cells, so they are good for negative staining (discussed in the next section).

Negative Versus Positive Staining   Two basic types of staining technique are used, depending upon how a dye reacts with the specimen (summarized in table 3.7). Most procedures involve a **positive stain,** in which the dye actually sticks to the specimen and gives it color. A **negative stain,** on the other hand, is just the reverse (like a photographic negative). The dye does not stick to the specimen but settles around its outer boundary, forming a silhouette. In a sense, negative staining "stains" the glass slide to produce a dark background around the cells. Nigrosin (blue-black) and India ink (a black suspension of carbon particles) are the dyes most commonly used for negative staining. The cells themselves do not stain because these dyes are negatively charged and are repelled by the negatively charged surface of the cells. The value of negative staining is its relative simplicity and the reduced shrinkage or distortion of cells, as the smear is not heat fixed. A quick assessment can thus be made regarding cellular size, shape, and arrangement. Negative staining is also used to accentuate the capsule that surrounds certain bacteria and yeasts **(figure 3.21c).**

Table 3.7 Comparison of Positive and Negative Stains

Medium	Positive Staining	Negative Staining
Appearance of cell	Colored by dye	Clear and colorless
Background	Not stained (generally white)	Stained (dark gray or black)
Dyes employed	Basic dyes: Crystal violet, Methylene blue, Safranin, Malachite green	Acidic dyes: Nigrosin, India ink
Subtypes of stains	Several types: Simple stain, Differential stains — Gram stain, Acid-fast stain, Spore stain; Special stains — Capsule, Flagella, Spore, Granules, Nucleic acid	Few types: Capsule, Spore

Disease Connection

Sometimes, a capsule can make all the difference. *Streptococcus pneumoniae* is a very common and very dangerous pathogen that possesses a large capsule. It colonizes the upper respiratory tract (URT) and from there can cause pneumonia, meningitis, and bloodstream infections. Scientists have discovered that if the capsule is absent, the bacterium cannot attach to the URT and, therefore, causes no disease.

Simple Versus Differential Staining Positive staining methods are classified as simple, differential, or special **(figure 3.21).** Whereas **simple stains** require only a single dye and an uncomplicated procedure, **differential stains** use two differently colored dyes, called the primary dye and the counterstain, to distinguish between cell types or parts. These staining techniques tend to be more complex and sometimes require additional chemical reagents to produce the desired reaction. Special stains are those that were developed for a single purpose.

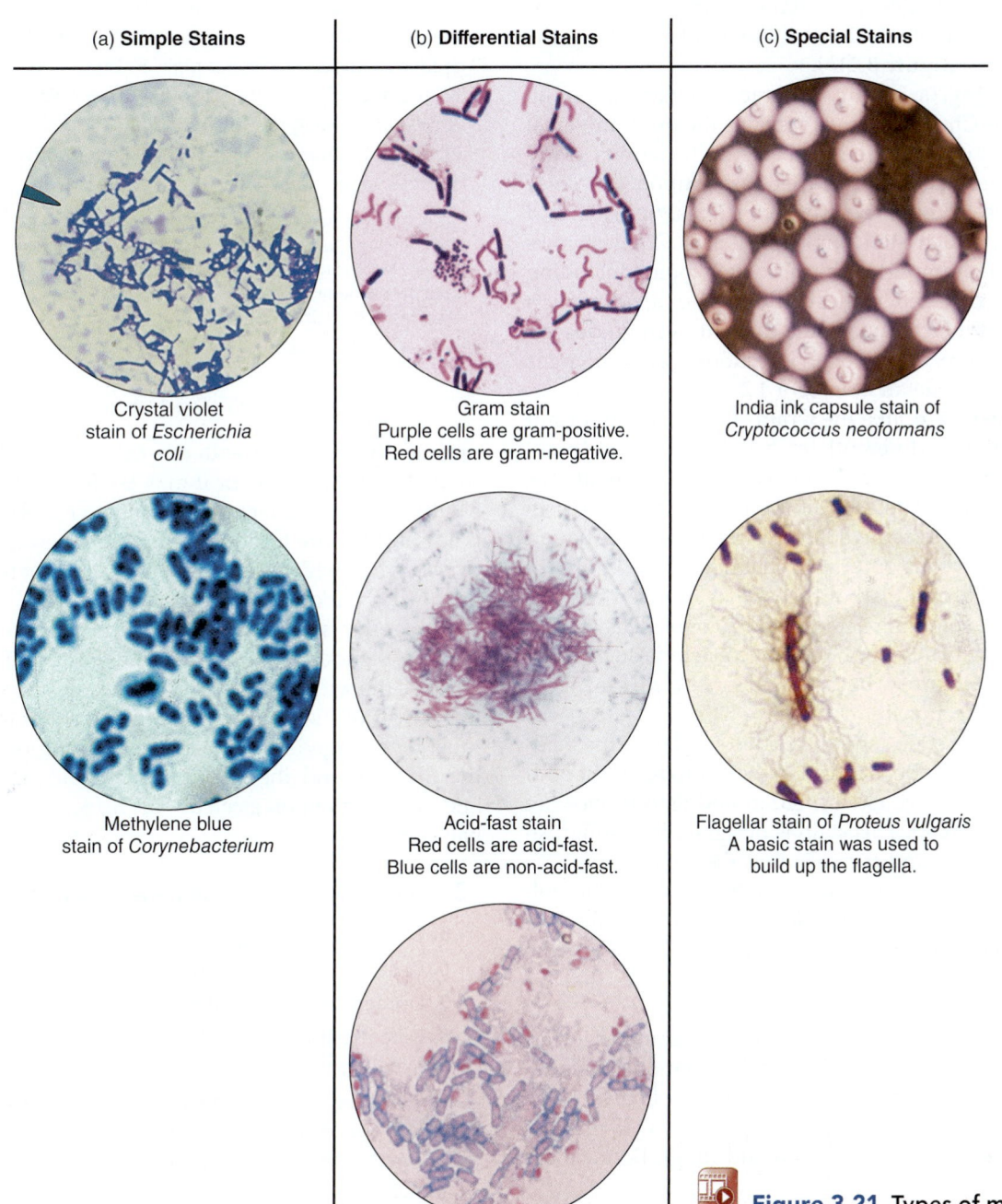

(a) Simple Stains

Crystal violet
stain of *Escherichia
coli*

Methylene blue
stain of *Corynebacterium*

(b) Differential Stains

Gram stain
Purple cells are gram-positive.
Red cells are gram-negative.

Acid-fast stain
Red cells are acid-fast.
Blue cells are non-acid-fast.

Endospore stain, showing
endospores (red)
and vegetative cells (blue)

(c) Special Stains

India ink capsule stain of
Cryptococcus neoformans

Flagellar stain of *Proteus vulgaris*
A basic stain was used to
build up the flagella.

Figure 3.21 Types of microbiological stains. **(a)** Simple stains. **(b)** Differential stains: Gram, acid-fast, and spore. **(c)** Special stains: capsule and flagellar.

Most simple staining techniques **(figure 3.21a)** take advantage of the ready binding of bacterial cells to dyes like malachite green, crystal violet, basic fuchsin, and safranin. Simple stains cause all cells in a smear to appear more or less the same color, regardless of type, but they can still reveal bacterial characteristics such as shape, size, and arrangement.

Types of Differential Stains   A satisfactory differential stain uses differently colored dyes to clearly contrast two cell types or cell parts. Common combinations are red and purple, red and green, or pink and blue. Differential stains can also pinpoint other characteristics, such as the size, shape, and arrangement of cells. Typical examples include Gram, acid-fast, and endospore stains. Some staining techniques (spore, capsule) which are differential are also in the "special" category **(figure 3.21b).**

Gram staining, a century-old method named for its developer, Hans Christian Gram, remains the most universal diagnostic staining technique for bacteria. It permits ready differentiation of major categories based upon the color reaction of the cells: **gram-positive,** which stain purple, and **gram-negative,** which stain pink (red). The Gram stain is the basis of several important bacteriologic topics, including bacterial taxonomy, cell wall structure, and identification and diagnosis of infection; in some cases, it even guides the selection of the correct drug for an infection. Gram staining is discussed in greater detail in Insight 4.2.

The **acid-fast stain,** like the Gram stain, is an important diagnostic stain that differentiates acid-fast bacteria (pink) from non-acid-fast bacteria (blue). This stain originated as a specific method to detect *Mycobacterium tuberculosis* in specimens. It was determined that these bacterial cells have a particularly impervious outer wall that holds fast (tightly or tenaciously) to the dye (carbol fuchsin) even when washed with a solution containing acid or acid alcohol. This stain is used for other medically important mycobacteria such as the Hansen's disease (leprosy) bacillus and for *Nocardia,* an agent of lung or skin infections.

The endospore stain (spore stain) is similar to the acid-fast method in that a dye is forced by heat into resistant bodies called endospores (their formation and significance are discussed in chapter 4). This stain is designed to distinguish between endospores and the cells that they come from (so-called **vegetative** cells). Of significance in medical microbiology are the gram-positive, endospore-forming members of the genus *Bacillus* (the cause of anthrax) and *Clostridium* (the cause of botulism and tetanus)—dramatic diseases that we consider in later chapters.

Special stains **(figure 3.21c)** are used to emphasize certain cell parts that are not revealed by conventional staining methods. **Capsule staining** is a method of observing the microbial capsule, an unstructured protective layer surrounding the cells of some bacteria and fungi. Because the

capsule does not react with most stains, it is often negatively stained with India ink, or it may be demonstrated by special positive stains. The fact that not all microbes exhibit capsules is a useful feature for identifying pathogens. One example is *Cryptococcus,* which causes a serious form of fungal meningitis in AIDS patients (see chapter 19).

Flagellar staining is a method of revealing flagella, the tiny, slender filaments used by bacteria for locomotion. Because the width of bacterial flagella lies beyond the resolving power of the light microscope, in order to be seen, they must be enlarged by depositing a coating on the outside of the filament and then staining it. Their presence, number, and arrangement on a cell are useful for identification of the bacteria.

3.2 Learning Outcomes—Assess Your Progress

5. Convert among different lengths within the metric system.
6. Describe the earliest microscopes.
7. List and describe the three elements of good microscopy.
8. Differentiate between the principles of light and electron microscopy.
9. Compare and contrast the three main categories of stains, and provide examples of each.

Case File 3   *Wrap-Up*

Selective and differential media, such as MacConkey agar, are most often used to detect the presence of *E. coli* and other enteric bacteria. However, due to its virulence and antibiotic resistance, *E. coli* O104:H4 is considered an extended-spectrum beta-lactamase-producing bacterium (ESBL) and requires specialized media for isolation. In this case, CHROMagar ESBL medium was selective for the pathogen because it contains ingredients that encourage the growth of and allow detection of ESBL bacteria, while inhibiting most other enteric bacteria. This medium was differential in that ESBL strains of *E. coli,* such as the O104:H4 strain in this epidemic, produce distinct pink- to burgundy-colored colonies on the agar. Epidemiologists combined isolation of the organism on CHROMagar ESBL medium with genomic techniques to identify the pathogenic strain causing the outbreak and eventually its source.

Source: Martina Bielaszewska, Alexander Mellmann, Wenlan Zhang, Robin Köck, Angelika Fruth, Andreas Bauwens, and others, "Characterisation of the *Escherichia coli* Strain Associated with an Outbreak of Haemolytic Uraemic Syndrome in Germany, 2011: A Microbiological Study," *The Lancet Infectious Diseases* 11, no. 9 (2011): 671–76.

Chapter Summary

3.1 Methods of Culturing Microorganisms: The Five I's (ASM Guidelines* 3.4, 8.2, 8.3)

- The Five I's—inoculation, incubation, isolation, inspection, and identification—summarize the kinds of laboratory procedures used in microbiology.
- Following *inoculation*, cultures are *incubated* at a specified temperature to encourage growth.
- Many microorganisms can be cultured on artificial media, but some can be cultured only in living tissue or in cells.
- Artificial media are classified by their *physical state* as either liquid, semisolid, liquefiable solid, or nonliquefiable solid.
- Artificial media are classified by their *chemical composition* as either *defined* or *complex*, depending on whether the exact chemical composition is known.
- Artificial media are classified by their *function* as either general-purpose media or media with one or more specific purposes. Enriched, selective, differential, transport, assay, and enumerating media are all examples of media designed for specific purposes.
- *Isolated colonies* that originate from single cells are composed of large numbers of cells piled up together.
- A culture may exist in one of the following forms: A pure culture contains only one species or type of microorganism. A mixed culture contains two or more known species. A contaminated culture contains both known and unknown (unwanted) microorganisms.
- During inspection, the cultures are examined and evaluated macroscopically and microscopically.
- Microorganisms are identified in terms of their macroscopic or immunologic morphology, their microscopic morphology, their biochemical reactions, and their genetic characteristics.

*ASM Curriculum Guidelines (American Society for Microbiology, 2012). Complete guidelines in appendix B of this book.

- Microbial cultures are usually disposed of in two ways: steam sterilization or incineration.

3.2 The Microscope: Window on an Invisible Realm (ASM Guidelines 2.4, 8.1)

- Magnification, resolving power, and contrast all influence the clarity of specimens viewed through the optical microscope.
- The maximum resolving power of the optical microscope is 200 nm, or 0.2 μm. This is sufficient to see the internal structures of eukaryotes and the morphology of most bacteria.
- There are at least six types of optical microscopes. Four types use visible light for illumination: bright-field, dark-field, phase-contrast, and interference microscopes. The fluorescence microscope uses UV light for illumination, but it has the same resolving power as the other optical microscopes. The confocal microscope can use UV light or visible light reflected from specimens.
- Electron microscopes (EM) use electrons, not light waves, as an illumination source to provide high magnification (5,000× to 1,000,000×) and high resolution (0.5 nm). Electron microscopes can visualize cell ultrastructure (transmission EM) and three-dimensional images of cell and virus surface features (scanning EM).
- The newest generation of microscope is called the scanning probe microscope and uses precision tips to image structures at the atomic level.
- Specimens viewed through optical microscopes can be either alive or dead, depending on the type of specimen preparation, but all EM specimens are dead because they must be viewed in a vacuum.
- Stains increase the contrast of specimens and they can be designed to differentiate cell shape, structure, and biochemical composition of the specimens being viewed.

Multiple-Choice and True-False Questions | Bloom's Levels 1 and 2: Remember and Understand

Multiple-Choice Questions. Select the correct answer from the options provided.

1. The term *culture* refers to the ____ growth of microorganisms in ____ .
   a. rapid, an incubator
   b. macroscopic, media
   c. microscopic, the body
   d. artificial, colonies

2. A mixed culture is
   a. the same as a contaminated culture.
   b. one that has been adequately stirred.
   c. one that contains two or more known species.
   d. a pond sample containing algae and protozoa.

3. Resolution is ____ with a longer wavelength of light.
   a. improved
   b. worsened
   c. not changed
   d. not possible

4. A real image is produced by the
   a. ocular.
   b. objective.
   c. condenser.
   d. eye.

5. A microscope that has a total magnification of 1,500× when using the oil immersion objective has an ocular of what power?
   a. 150×   c. 15×
   b. 1.5×   d. 30×

6. The specimen for an electron microscope is always
   a. stained with dyes.   c. killed.
   b. sliced into thin sections.   d. viewed directly.

7. Motility is best observed with a
   a. hanging drop preparation.
   b. negative stain.
   c. streak plate.
   d. flagellar stain.

8. Bacteria tend to stain more readily with cationic (positively charged) dyes because bacterial surfaces
   a. contain large amounts of alkaline substances.
   b. contain large amounts of acidic substances.
   c. are neutral.
   d. have thick cell walls.

9. **Multiple Matching.** For each type of medium, select all descriptions that fit. For media that fit more than one description, briefly explain why this is the case.
   _____ mannitol salt agar   a. selective medium
   _____ chocolate agar   b. differential medium
   _____ MacConkey agar   c. chemically defined
   _____ nutrient broth   (synthetic) medium

   _____ Sabouraud's agar   d. enriched medium
   _____ triple-sugar iron agar   e. general-purpose medium
   _____ nutrient agar   f. complex medium
   _____ SIM medium   g. transport medium

10. A fastidious organism must be grown on what type of medium?
   a. general-purpose medium
   b. differential medium
   c. defined medium
   d. enriched medium

True-False Questions. If the statement is true, leave as is. If it is false, correct it by rewriting the sentence.

11. Agar has the disadvantage of being easily decomposed by microorganisms.

12. A subculture is a culture made from an isolated colony.

13. The factor that most limits the clarity of an image in a microscope is the _magnification_.

14. Living specimens can be examined either by light microscopy or electron microscopy.

15. The best stain to use to visualize a microorganism with a large capsule is a simple stain.

Critical Thinking Questions | Bloom's Levels 3, 4, and 5: Apply, Analyze, and Evaluate

Critical thinking is the ability to reason and solve problems using facts and concepts. These questions can be approached from a number of angles and, in most cases, they do not have a single correct answer.

1. Provide a concise summary of the five basic techniques used to manipulate, grow, examine, and characterize microorganisms in a laboratory.

2. Define the term _sterile_, and explain why sterile instruments and media must be used for microbial sampling and inoculation.

3. Summarize three isolation techniques. Explain which method is utilized most often in a laboratory and how a colony produced by this method can be used to create a pure culture.

4. What is the functional type of mannitol salt agar (MSA)? Explain how it is utilized in the isolation and identification of _Staphylococcus_ species.

5. Create a short paragraph to differentiate among the following terms: _pure culture, subculture, mixed culture, contaminated culture,_ and _stock culture._

6. Briefly summarize the three qualities of an effective microscope and discuss how each of these contributes to the quality of a microscopic image.

7. For each of the following scenarios, explain which type of microscope(s) would provide the best image.
   a. when motility in a live specimen must be viewed
   b. when intracellular structures must be viewed
   c. when identification of a microbe based on surface structures must be determined
   d. when diagnosis of a prion disease must be determined

8. a. Create a paragraph to differentiate among the following terms: _positive stain, negative stain, simple stain,_ and _differential stain._
   b. For each of the following scenarios, explain which staining technique(s) should be used.
      • analyzing cell wall composition
      • observing structures for locomotion
      • identifying a structure enhancing pathogenicity

9. Provide an argument in support of or refuting the following statement: "Viruses are readily cultured in standard complex media and are easily identified using bright-field microscopy."

10. You are a scientist studying a marsh area contaminated with PCBs, toxic chemical compounds found commonly in industrial waste. Initial microscopic analysis of the soil reveals the presence of motile cells that measure in the micrometer range. You hypothesize that these microbes may be useful in bioremediation of the toxic waste. Thinking about microbial sampling and isolation, describe a method for culturing these microbes back in your laboratory.

Concept Connections | Bloom's Levels 4 and 6: Analyze and Create

This activity ties together multiple concepts in the chapter.

1. What is the purpose of solid microbial media?
2. What are the benefits of agar?
3. Give an example of each: defined and complex media.

4. What are the purposes of general-purpose, selective, differential, and enriched media? Give examples of each.
5. Give examples of media that are both selective and differential.

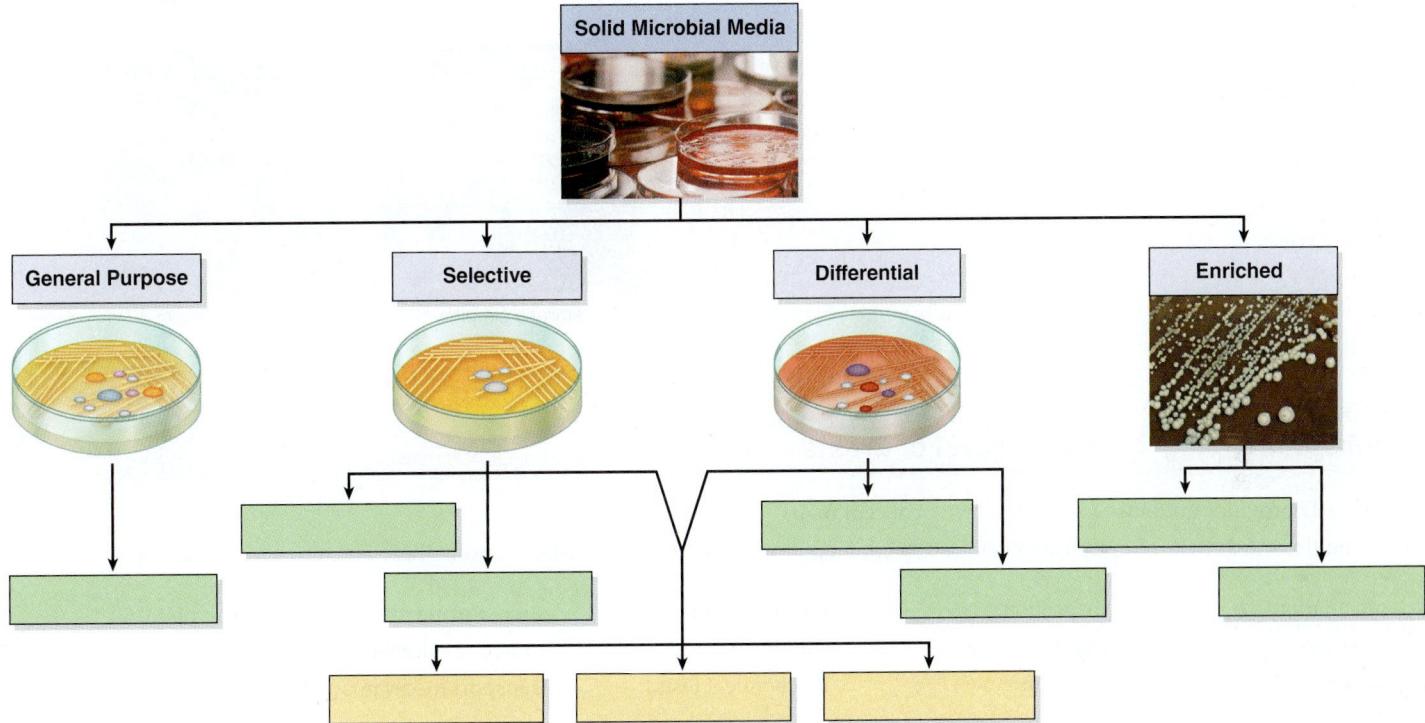

Visual Connections | Bloom's Level 5: Evaluate

These questions use visual images or previous content to make connections to this chapter's concepts.

1. **Process Figure 3.10.** If you were using the streak plate method to plate a very dilute broth culture (with many fewer bacteria than the broth used for 3*a*) would you expect to see single, isolated colonies in area 4 or area 3? Explain your answer.

(a) Steps in a streak plate

(b)

2. **From chapter 1, figure 1.6.** Which of these photos from chapter 1 is an SEM image? Which is a TEM image?

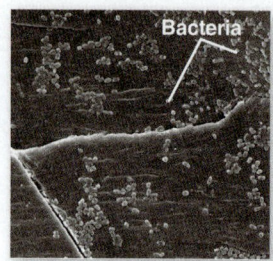

Bacterium: *E. coli*

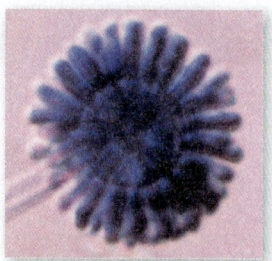

Fungus: *Syncephalastrum*

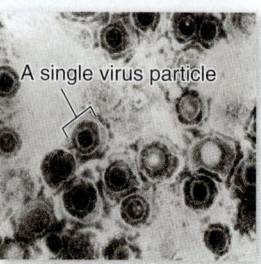

Virus: Herpes simplex

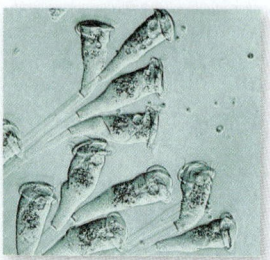

Protozoan: *Vorticella*

Concept Mapping | Bloom's Level 6: Create

Appendix D provides guidance for working with concept maps.

1. Using the words that follow, please create a concept map illustrating the relationships among these key terms from chapter 3.

inoculation	inspection	multiplication	subculturing	streak plate
isolation	identification	staining	source of microbes	
incubation	medium	biochemical tests	transport medium	

www.mcgrawhillconnect.com

Enhance your study of this chapter with study tools and practice tests. Also ask your instructor about the resources available through ConnectPlus, including the media-rich eBook, interactive learning tools, and animations.

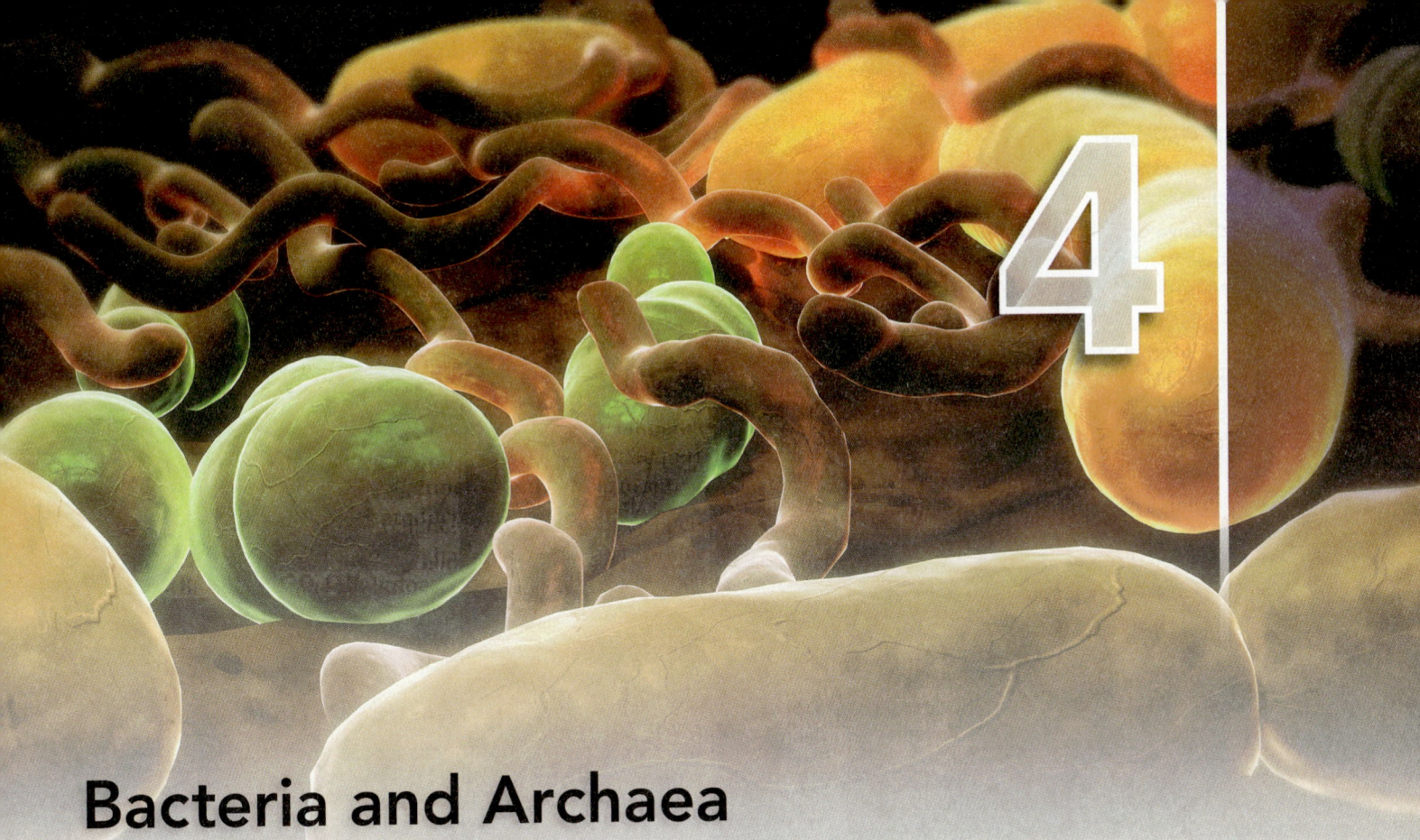

Bacteria and Archaea

Case File 4

You're Outnumbered

They're everywhere, and they outnumber you 10 to 1. They are your microbiota. The thought of bacteria living on and in your body may make you squeamish, but these compatriots are vital to your health. Take the microbiota in your gut, for example: There are approximately 10^{14} bacteria, archaea, and fungi living in your gut, accounting for about 1,100 different species, with about 10^{10} cells per gram of feces. There are also about a billion viruses per gram of feces. The gut microbiota is a complex ecosystem comprised of numerous genera including *Bacteroides*, *Clostridium*, *Streptococcus*, *Lactobacillus*, *Bifidobacterium*, and others. These organisms play an important role in your health: from helping to digest food, to producing vitamins and amino acids, to stimulating the immune system, to protecting you from harmful pathogens, to even having an influence on your behavior.

Some recent studies have shown that the gut microbiota can have an influence on obesity. Scientists have found that the gut microbiota in obese individuals can extract more energy from the diet that is stored as fat. Other studies have shown that obese individuals have higher levels of *Lactobacillus* species than nonobese populations. In studies with mice, scientists demonstrated that a change in diet from a low-fat, plant-based diet to a high-fat, high-sugar diet shifted the population of gut microbiota within a single day, favoring the microbiota that influence increased fat storage.

- What is the connection between gut microbiota and type II diabetes?
- What other human conditions can be influenced by gut microflora?

Continuing the Case appears on page 91

Outline and Learning Outcomes

4.1 Bacterial Form and Function

1. List the structures all bacteria possess.
2. Identify at least four structures that some, but not all, bacteria possess.
3. Describe the three major shapes of bacteria.
4. Describe other more unusual shapes of bacteria.
5. Provide at least four terms to describe bacterial arrangements.

In chapter 1, we described bacteria and archaea as being cells with no true nucleus. (Eukaryotes have a membrane around their DNA, and this structure is called the *nucleus*.) Let's look at bacteria and archaea as different from eukaryotes.

- *The way their DNA is packaged:* Bacteria and archaea have nuclear material that is free inside the cytoplasm (i.e., they do not have a nucleus). Eukaryotes have a membrane around their DNA (making up a nucleus). Eukaryotes wind their DNA around proteins called histones and archaea use similar proteins to do the same thing. Bacteria do not wind their DNA around proteins.
- *The makeup of their cell wall:* Bacteria and archaea generally have a wall structure that is unique compared to eukaryotes. Bacteria have sturdy walls made of a chemical called peptidoglycan. Archaeal walls are also tough and made of other chemicals, distinct from bacteria and distinct from eukaryotic cells.
- *Their internal structures:* Bacteria and archaea do not have complex, membrane-bounded organelles in their cytoplasm (eukaryotes do). A few bacteria and archaea have internal membranes, but they don't surround organelles.

Both non-eukaryotic and eukaryotic microbes are ubiquitous in the world today. Although both can cause infectious disease, drug targeting of bacterial and eukaryotic pathogens will be influenced by their unique cellular characteristics. In this chapter and coming chapters, you will discover why that is.

4.1 Bacterial Form and Function

The evolutionary history of non-eukaryotic cells extends back at least 2.9 billion years. The fact that these organisms have endured for so long in such a variety of habitats indicates a cellular structure and function that are amazingly versatile and adaptable.

The general cellular organization of a bacterial cell can be represented with this flowchart.

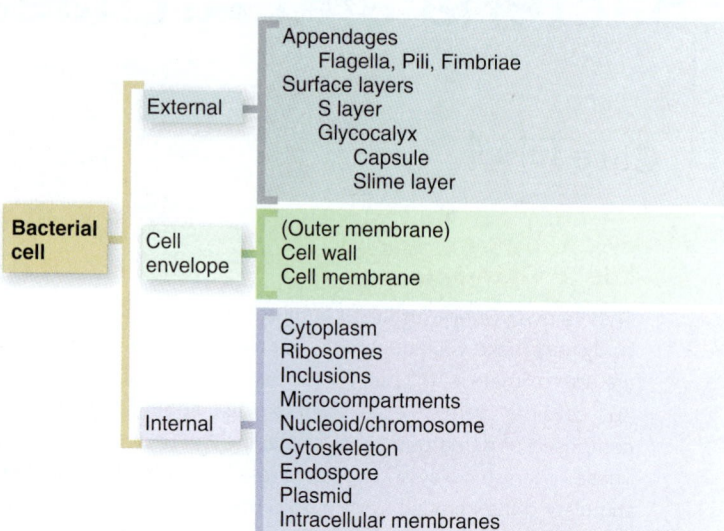

All bacterial cells invariably have a cell membrane, cytoplasm, ribosomes, and one (or a few) chromosome(s); the majority have a cell wall and some form of surface coating or glycocalyx. Specific structures that are found in some, but not all, bacteria are flagella, pili, fimbriae, an S layer, a cytoskeleton, inclusions, microcompartments, endospores, and intracellular membranes.

The Structure of a Generalized Bacterial Cell

Bacterial cells appear featureless and two-dimensional when viewed with an ordinary microscope. Not until they are subjected to the scrutiny of the electron microscope and biochemical studies does their intricate and functionally complex nature become evident. **Figure 4.1** presents a three-dimensional anatomical view of a generalized, rod-shaped, bacterial cell. As we survey the principal anatomical features

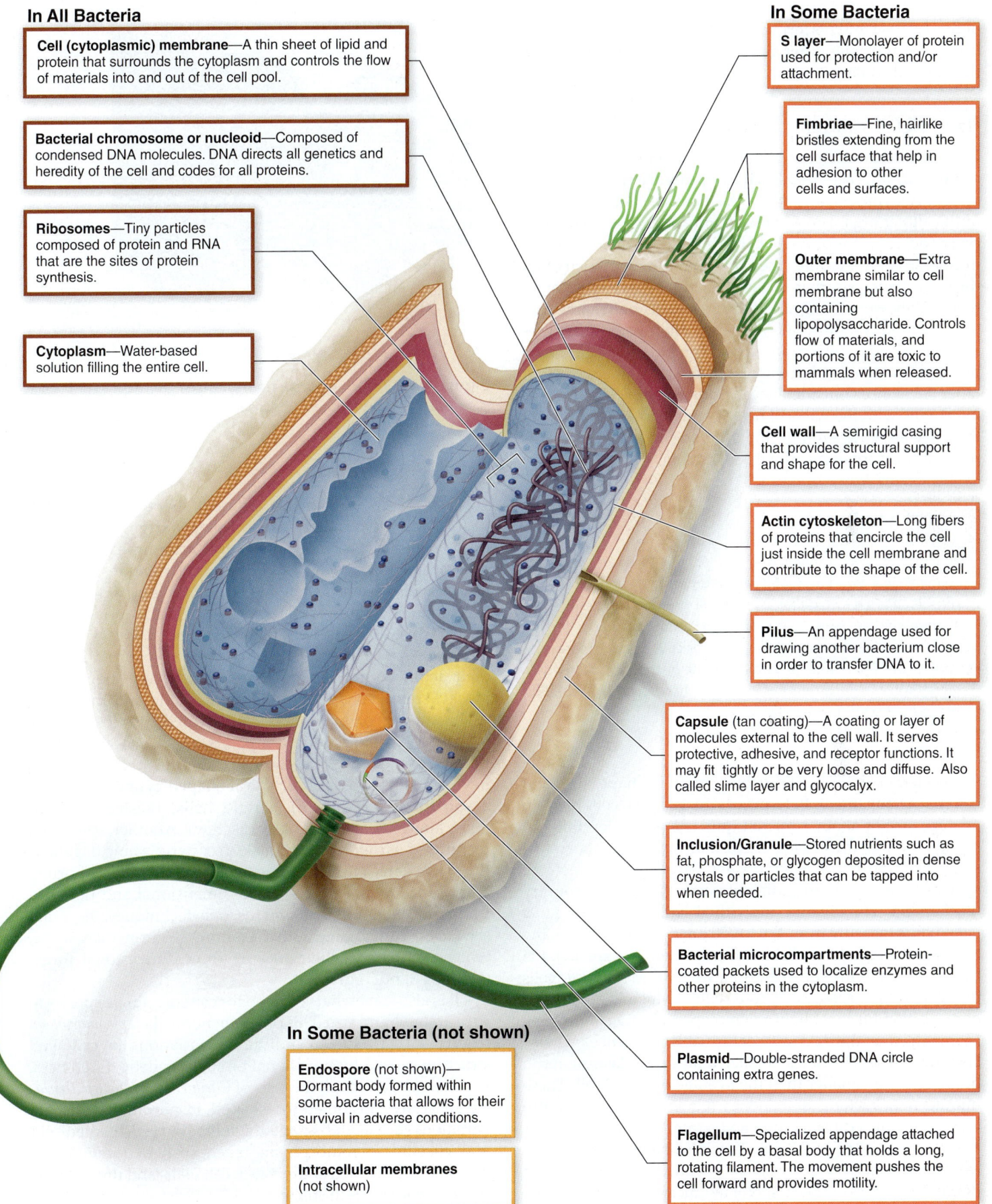

In All Bacteria

Cell (cytoplasmic) membrane—A thin sheet of lipid and protein that surrounds the cytoplasm and controls the flow of materials into and out of the cell pool.

Bacterial chromosome or nucleoid—Composed of condensed DNA molecules. DNA directs all genetics and heredity of the cell and codes for all proteins.

Ribosomes—Tiny particles composed of protein and RNA that are the sites of protein synthesis.

Cytoplasm—Water-based solution filling the entire cell.

In Some Bacteria

S layer—Monolayer of protein used for protection and/or attachment.

Fimbriae—Fine, hairlike bristles extending from the cell surface that help in adhesion to other cells and surfaces.

Outer membrane—Extra membrane similar to cell membrane but also containing lipopolysaccharide. Controls flow of materials, and portions of it are toxic to mammals when released.

Cell wall—A semirigid casing that provides structural support and shape for the cell.

Actin cytoskeleton—Long fibers of proteins that encircle the cell just inside the cell membrane and contribute to the shape of the cell.

Pilus—An appendage used for drawing another bacterium close in order to transfer DNA to it.

Capsule (tan coating)—A coating or layer of molecules external to the cell wall. It serves protective, adhesive, and receptor functions. It may fit tightly or be very loose and diffuse. Also called slime layer and glycocalyx.

Inclusion/Granule—Stored nutrients such as fat, phosphate, or glycogen deposited in dense crystals or particles that can be tapped into when needed.

Bacterial microcompartments—Protein-coated packets used to localize enzymes and other proteins in the cytoplasm.

Plasmid—Double-stranded DNA circle containing extra genes.

Flagellum—Specialized appendage attached to the cell by a basal body that holds a long, rotating filament. The movement pushes the cell forward and provides motility.

In Some Bacteria (not shown)

Endospore (not shown)—Dormant body formed within some bacteria that allows for their survival in adverse conditions.

Intracellular membranes (not shown)

Figure 4.1 Structure of a bacterial cell.   Cutaway view of a typical rod-shaped bacterium, showing major structural features.

of this cell, we will perform a microscopic dissection of sorts, beginning with the outer cell structures and proceeding to the internal contents.

Bacterial Arrangements and Sizes

Each individual bacterial cell is fully capable of carrying out all necessary life activities, such as reproduction, metabolism, and nutrient processing, unlike the more specialized cells of a multicellular organism. On the other hand, sometimes bacteria *can* act as a group. When bacteria are close to one another in colonies or in biofilms, they communicate with each other through chemicals that cause them to behave differently than if they were living singly. More surprisingly, some bacteria seem to communicate with each other using structures called *nanowires,* which are appendages that can be many micrometers long and are used for transferring electrons or other substances outside the cell onto metals. The wires also intertwine with the wires of neighboring bacteria. This is not the same as being a multicellular organism, but it represents new findings about microbial cooperation.

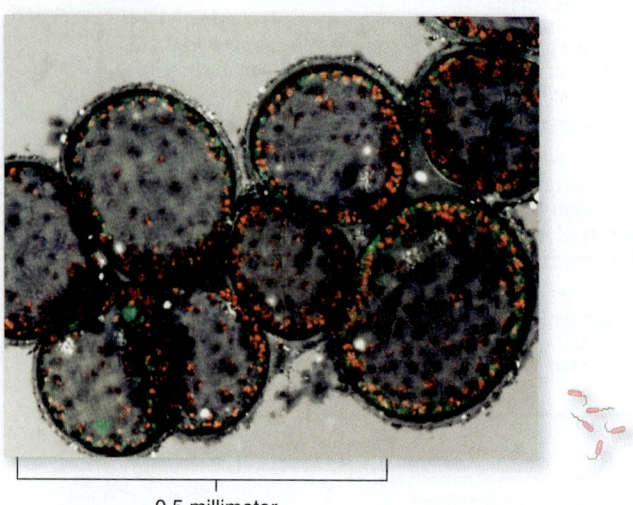

0.5 millimeter

Figure 4.2. *Thiomargarita namibiensis.*   A drawing of *E. coli* is included on right for size comparison.

Disease Connection

Biofilms can play a major role in infectious diseases. Scientists definitively showed in 2006 that children suffering from chronic ear infections had biofilms of bacteria growing on the mucosa of their middle ears. These biofilms were not eradicated by repeated courses of antibiotics. This discovery gave more support to the procedure of putting tubes in the ears of children with chronic or recurrent ear infections (to drain infected fluids) instead of treating with antibiotics.

Bacteria exhibit considerable variety in shape, size, and colonial arrangement. In terms of size, bacterial cells have an average size of about 1 micron (μm). As with everything in nature, though, there is a great deal of variation in microbial size. The largest non-eukaryote yet discovered is a bacterial species living in ocean sediments near the African country of Namibia. The gigantic individual cocci of *Thiomargarita namibiensis* measure from 100 to 750 μm (3/4 mm), and many are large enough to see with the naked eye **(figure 4.2).** On the other end of the size spectrum, we have *Mycoplasma* cells that generally measure 0.15 to 0.30 μm, which is at the limit of resolution for most light microscopes. A new controversy is brewing over the discovery of tiny cells that look like dwarf bacteria but are 10 times smaller than mycoplasmas and a hundred times smaller than the average bacterial cell. These minute nanobacteria or nanobes (Gr. *nanos,* one-billionth) were first isolated from blood and serum samples, and have a size range of 0.05 to 0.2 μm. They also have been found in sandstone rock deposits in the ocean and deeply embedded in billion-year-old minerals. Not all microbiologists are convinced that they are true microbes, but they expand our view of the size limitations that define life.

Bacteria come in many different shapes, but the vast majority are one of three general shapes **(figure 4.3).** If the cell is spherical or ball-shaped, the bacterium is described as a **coccus** (kok'-us). Cocci can be perfect spheres, but they also can exist as oval, bean-shaped, or even pointed variants. A cell that is cylindrical (longer than wide) is termed a *rod,* or **bacillus** (bah-sil'-lus). There is also a genus named *Bacillus.* As may be expected, rods are also quite varied in their actual form. Depending on the species, they can be blocky, spindle-shaped, round-ended, long and threadlike (filamentous), or even club-shaped or drumstick-shaped. When a rod is short and plump, it is called a **coccobacillus;** if it is gently curved, it is a **vibrio** (vib'-ree-oh). A bacterium having a slightly curled or spiral-shaped cylinder is called a **spirillum** (spy-ril'-em), a rigid helix, twisted twice or more along its axis (like a corkscrew). Another spiral cell mentioned earlier in the discussion of periplasmic flagella is the **spirochete,** a more flexible form that resembles a spring. Because bacterial cells look two-dimensional and flat with traditional staining and microscope techniques, they are seen to best advantage with a scanning electron microscope, which emphasizes their striking three-dimensional forms **(figure 4.3, micrographs).**

It is also somewhat common for cells of a single species to vary in shape and size. This phenomenon, called *pleomorphism* **(figure 4.4),** is due to individual variations in cell wall structure caused by nutritional or slight genetic differences. For example, although the cells of *Corynebacterium diphtheriae* are generally considered rod-shaped, in culture they display variations such as club-shaped, swollen, curved, filamentous, and coccoid. Pleomorphism reaches an extreme in the mycoplasmas, which entirely lack cell walls and thus display extreme variations in shape.

Bacterial cells can also be categorized according to arrangement, or style of grouping (see figure 4.3). The main

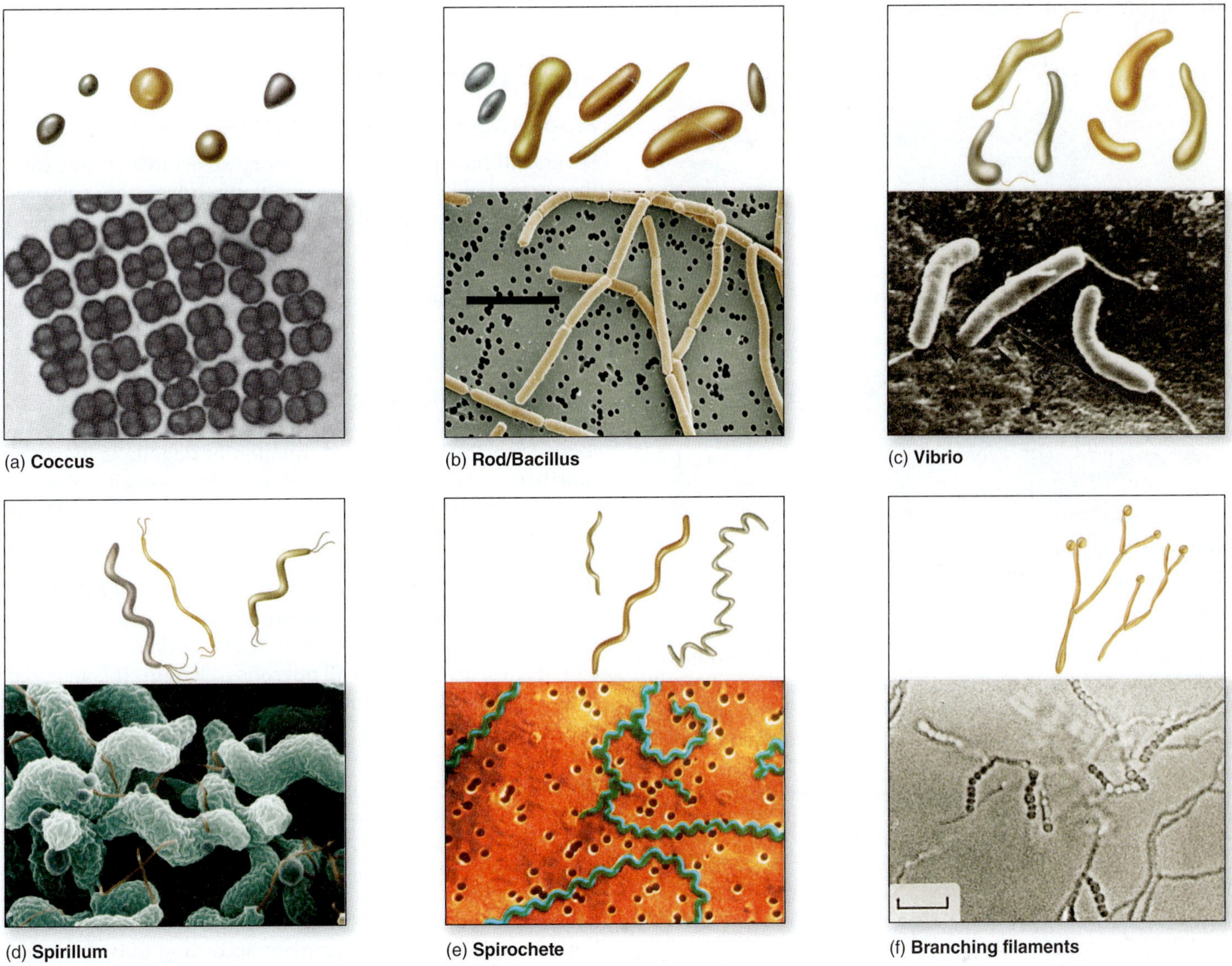

(a) **Coccus**

(b) **Rod/Bacillus**

(c) **Vibrio**

(d) **Spirillum**

(e) **Spirochete**

(f) **Branching filaments**

Key to Micrographs
(a) *Deinococcus* (2,000×) (b) *Lactobacillus bulgaricus* (5,000×) (c) *Vibrio cholerae* (13,000×) (d) *Aquaspirillum* (7,500×)
(e) Spirochetes on a filter (14,000×) (f) *Streptomyces* (1,500×)

Figure 4.3 Bacterial shapes and arrangements.   Drawings show examples of shape variations for cocci, rods, vibrios, spirilla, spirochetes, and branching filaments. Below each shape is a micrograph of a representative example.

factors influencing the arrangement of a particular cell type are its pattern of division and how the cells remain attached afterward. The greatest variety in arrangement occurs in cocci, which can be single, in pairs (diplococci), in **tetrads** (groups of four), in irregular clusters (as in staphylococci and micrococci), or in chains of a few to hundreds of cells (as in streptococci). An even more complex grouping is a cubical packet of 8, 16, or more cells called a **sarcina** (sar'-sih-nah). These different coccal groupings are the result of the division of a coccus in a single plane, in two perpendicular planes, or in several intersecting planes; after division, the resultant daughter cells remain attached.

Bacilli are less varied in arrangement because they divide only in the transverse plane (perpendicular to the axis). They occur either as single cells, as a pair of cells with their ends attached (diplobacilli), or as a chain of several cells (streptobacilli). A palisades arrangement is formed when

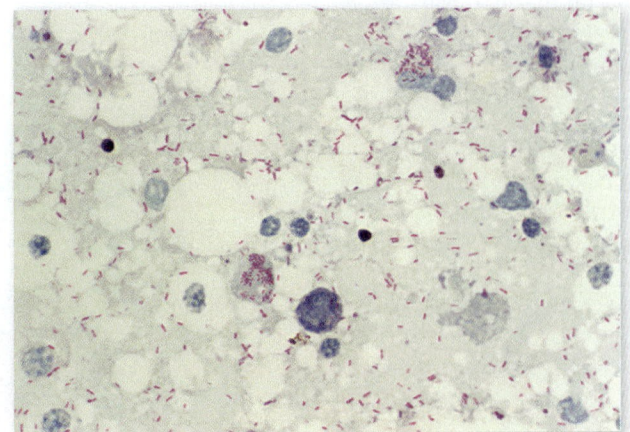

Figure 4.4 Pleomorphic bacteria.   If you look closely at this micrograph of stained *Rickettsia rickettsii* bacteria, you will see some coccoid cells, some rod-shaped cells, and some hybrid forms.

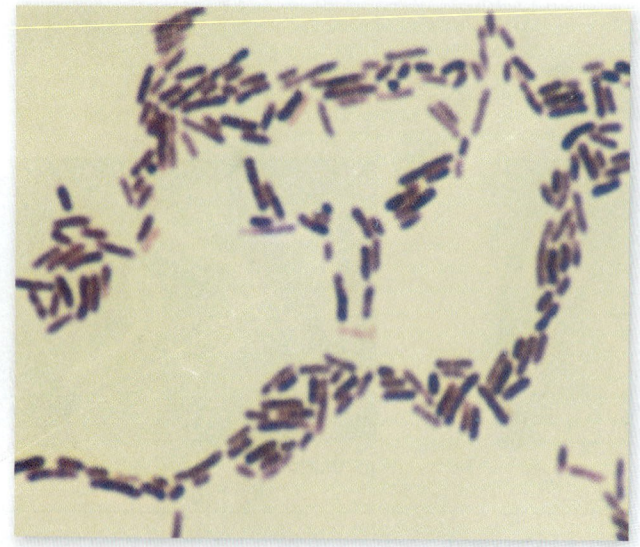

Figure 4.5. *Corynebacterium* **cells illustrating the palisades arrangement.**

cells of a chain remain partially attached at the ends; this hinge area can fold back creating a side-by-side row of cells **(figure 4.5).** Spirilla are occasionally found in short chains, but spirochetes rarely remain attached after division.

4.1 Learning Outcomes—Assess Your Progress

1. List the structures all bacteria possess.
2. Identify at least four structures that some, but not all, bacteria possess.
3. Describe the three major shapes of bacteria.
4. Describe other more unusual shapes of bacteria.
5. Provide at least four terms to describe bacterial arrangements.

4.2 External Structures

Appendages: Cell Extensions

Several different types of accessory structures sprout from the surface of bacteria. These long **appendages** are common but are not present on all species. Appendages can be divided into two major groups: those that provide motility (flagella and axial filaments) and those that provide attachment points or channels (fimbriae and pili).

Flagella—Little Propellers

The bacterial **flagellum** (flah-jel′-em), an appendage of truly amazing construction, is certainly unique in the biological world. The primary function of flagella is to confer **motility,** or self-propulsion—that is, the capacity of a cell to swim freely through an aqueous habitat. The extreme thinness of a bacterial flagellum necessitates high magnification to reveal its special architecture, which has three distinct parts: the filament, the hook (sheath), and the basal body **(figure 4.6).** The **filament,** a helical structure composed of proteins, is approximately 20 nanometers in diameter and varies from 1 to 70 microns in length. It is inserted into a curved, tubular hook. The hook is anchored to the cell by the basal body, a stack of rings firmly anchored through the cell wall to the cell membrane and the outer membrane. This arrangement permits the hook with its filament to rotate 360°, rather than undulating back and forth like a whip as was once thought.

One can generalize that all spirilla, about half of the bacilli, and a small number of cocci have flagella (these bacterial shapes are shown in **figure 4.7**). Flagella vary both in number and arrangement according to two general patterns:

1. In a **polar** arrangement, the flagella are attached at one or both ends of the cell. Three subtypes of this pattern are

Figure 4.6 **Details of the basal body of a flagellum in a gram-negative cell.** **(a)** The hook, rings, and rod function together as a tiny device that rotates the filament 360°. **(b)** An electron micrograph of the basal body of a bacterial flagellum.

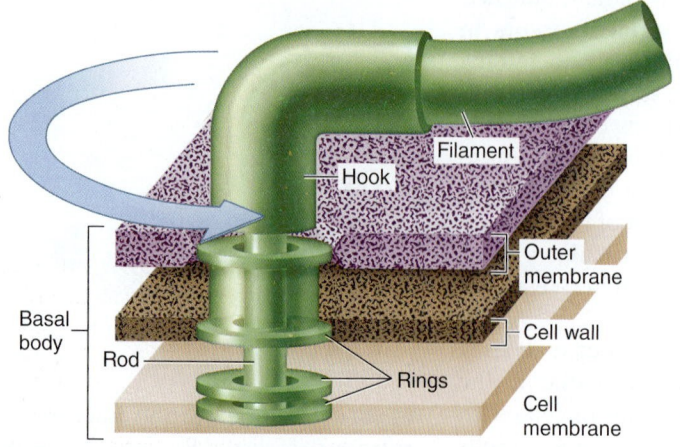

(a)

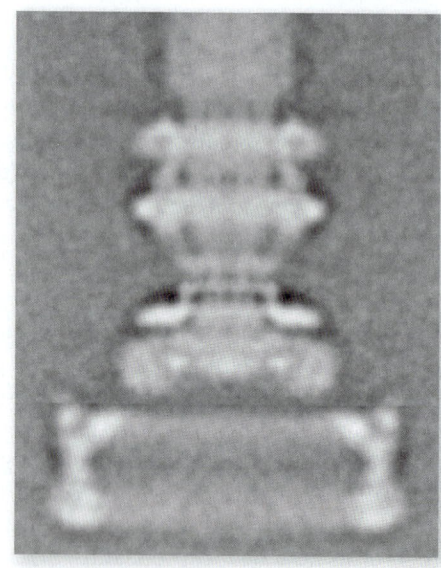

(b)

Figure 4.7 Electron micrographs depicting types of flagellar arrangements. **(a)** Monotrichous polar flagellum on the bacterium *Bdellovibrio*. **(b)** Lophotrichous polar flagella on *Plesiomonas*. **(c)** Amphitrichous polar flagella on *Campylobacter*. **(d)** Peritrichous flagella on *Escherichia coli*.

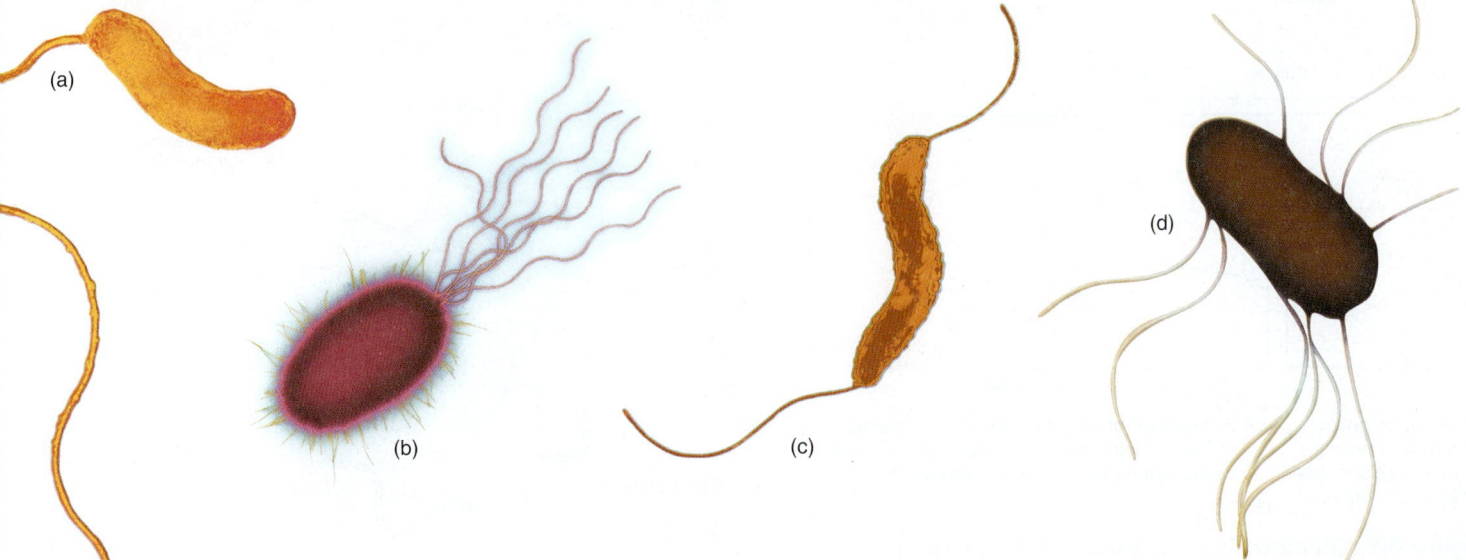

monotrichous (mah″-noh-trik′-us), with a single flagellum **(figure 4.7a); lophotrichous** (lo″-foh-), with small bunches or tufts of flagella emerging from the same site **(figure 4.7b);** and **amphitrichous** (am″-fee-), with flagella at both poles of the cell **(figure 4.7c).**

2. In a **peritrichous** (per″-ee-) arrangement, flagella are dispersed randomly over the surface of the cell **(figure 4.7d).**

The presence of motility is one piece of information used in the laboratory identification or diagnosis of pathogens. Flagella are hard to visualize in the laboratory, but often it is sufficient to know simply whether a bacterial species is motile. Motility can be assessed using semisolid media or through the hanging drop technique (see chapter 3).

Fine Points of Flagellar Function Flagellated bacteria can perform some rather sophisticated feats. They can detect and move in response to chemical signals—a type of behavior called **chemotaxis** (ke″-moh-tak′ -sis). Positive chemotaxis is movement of a cell in the direction of a favorable chemical stimulus (usually a nutrient); negative chemotaxis is movement away from a repellent (potentially harmful) compound.

The flagellum is effective in guiding bacteria through the environment primarily because the system for detecting chemicals is linked to the mechanisms that drive the flagellum. Located in the cell membrane are clusters of receptors[1] that bind specific molecules coming from the immediate environment. The attachment of sufficient numbers of these molecules transmits signals to the flagellum and sets it into rotary motion. The actual "fuel" for the flagellum to turn is a gradient of protons (hydrogen ions) that are generated by the metabolism of the bacterium and that bind to and detach from

parts of the flagellar motor within the cell membrane, causing the filament to rotate. If several flagella are present, they become aligned and rotate as a group **(figure 4.8).** As a flagellum rotates counterclockwise, the cell itself swims in a smooth linear direction toward the stimulus; this action is called a **run.** Runs are interrupted at various intervals by **tumbles,** during which the flagellum reverses direction and causes the cell to stop and change its course. It is believed that attractant molecules inhibit tumbles and permit progress toward the stimulus; these appear to play a major role in the process of quorum sensing. Repellents cause numerous tumbles, allowing the

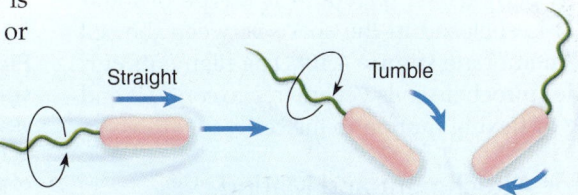

(a) General motility of a singular flagellum

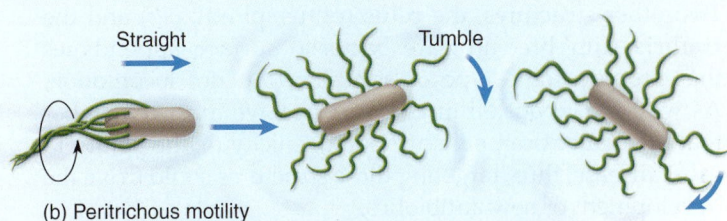

(b) Peritrichous motility

Figure 4.8 The operation of flagella and the mode of locomotion in bacteria with polar and peritrichous flagella. **(a)** In general, when a polar flagellum rotates in a counterclockwise direction, the cell swims forward. When the flagellum reverses direction and rotates clockwise, the cell stops and tumbles. **(b)** In peritrichous forms, all flagella sweep toward one end of the cell and rotate as a single group. During tumbles, the flagella lose coordination.

1. Cell surface molecules that bind specifically with other molecules.

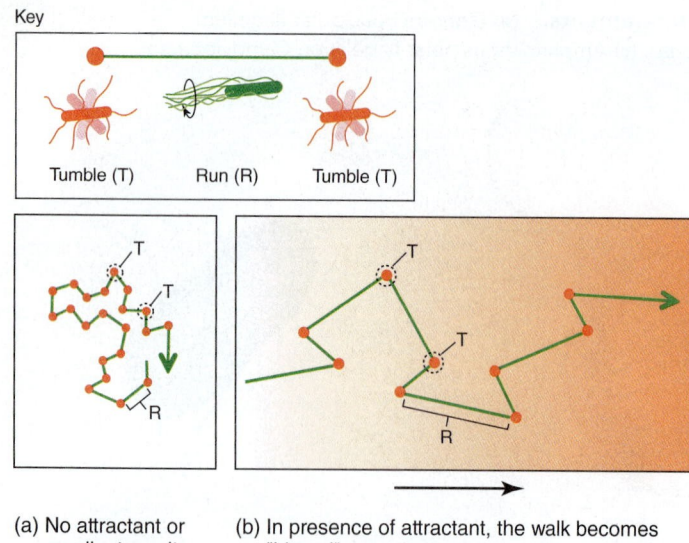

Key

Tumble (T)   Run (R)   Tumble (T)

(a) No attractant or
    repellent results
    in a "random walk"
    form of motion.

(b) In presence of attractant, the walk becomes
    "biased" toward more runs and fewer tumbles,
    directing the cell towards the attractant.

Figure 4.9 Chemotaxis in bacteria. **(a)** A bacterium moves via a random series of short runs and tumbles when there is no attractant or repellent. **(b)** The cell spends more time on runs as it gets closer to the attractant.

bacterium to redirect itself away from the stimulus **(figure 4.9).** Some photosynthetic bacteria exhibit **phototaxis**, movement in response to light rather than chemicals.

Periplasmic Flagella

Corkscrew-shaped bacteria called *spirochetes* (spy'-roh-keets) show an unusual, wriggly mode of locomotion caused by two or more long, coiled threads, the periplasmic flagella or **axial filaments**. A periplasmic flagellum is a type of internal flagellum that is enclosed in the space between the cell wall and the cell membrane **(figure 4.10).** The filaments curl closely around the spirochete coils yet are free to contract and impart a twisting or flexing motion to the cell.

Appendages for Attachment and Mating

Although their main function is motility, bacterial flagella can be used for attachment to surfaces in some species. Two other structures, the **pilus** (pil-us; plural, *pili*) and the **fimbria** (fim'-bree-ah), are bacterial surface appendages that provide some type of adhesion, but not locomotion. As we think about all three structures, we must remember that attachment can enhance pathogenicity or the ability to cause disease; thus, targeting these structures could drive the development of new antibiotics.

Fimbriae are small, bristlelike fibers sprouting off the surface of many bacterial cells **(figure 4.11).** Their exact composition varies, but most of them contain protein. Fimbriae have an inherent tendency to stick to each other and to surfaces. They may be responsible for the mutual clinging of cells that leads to biofilms and other thick aggregates of cells on the surface of liquids and for the microbial colonization of inanimate solids such as rocks and glass **(Insight 4.1).** Some

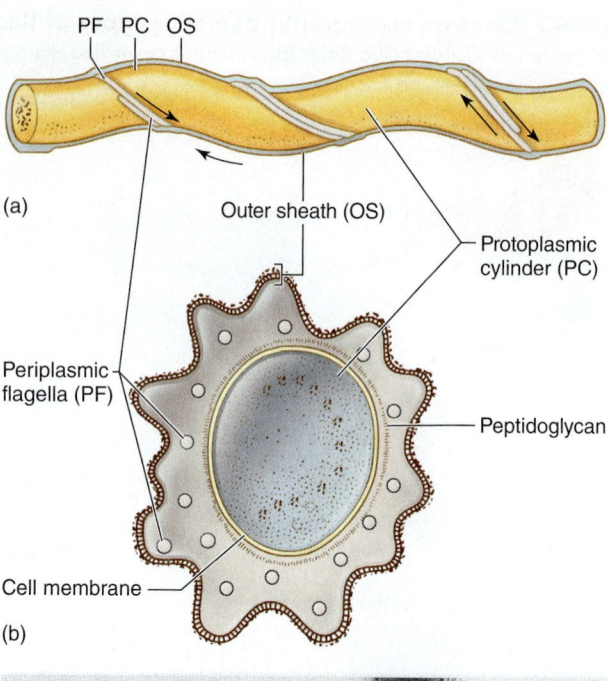

PF PC OS

(a)

Outer sheath (OS)

Protoplasmic cylinder (PC)

Periplasmic flagella (PF)

Peptidoglycan

Cell membrane

(b)

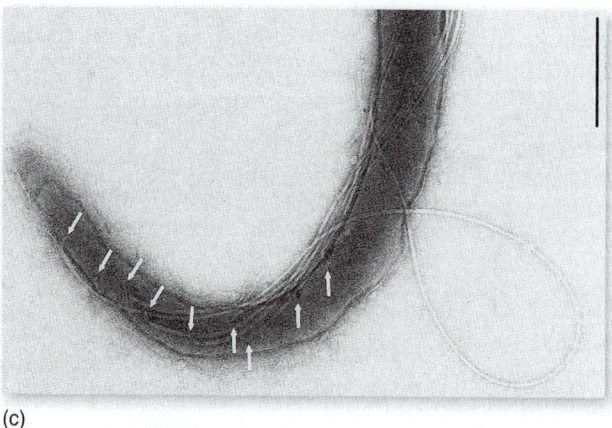

(c)

Figure 4.10 The orientation of periplasmic flagella on the spirochete cell. **(a)** Longitudinal section. **(b)** Cross section (end-on view). Contraction of the filaments imparts a spinning and undulating pattern of locomotion. **(c)** Electron micrograph captures the details of periplasmic flagella and their insertion points (arrows) in *Borrelia burgdorferi*, the cause of Lyme disease. One flagellum has escaped the outer sheath, probably during preparation for EM. (Bar = 0.2 microns)

pathogens, such as the gonococcus and *Escherichia coli*, can colonize and infect host tissues because of a tight adhesion between their fimbriae and epithelial cells **(figure 4.11b).** Mutant forms of these pathogens that lack a fimbriae, however, are unable to cause infections.

A pilus is a long, rigid tubular structure made of a special protein, **pilin.** So far, true pili have been found only on gram-negative bacteria, where they are utilized in a "mating" process between cells called **conjugation,**[2] which involves partial transfer of DNA from one cell to another **(figure 4.12).** A pilus from the donor cell unites with a recipient cell, bring-

---

2. Although the term *mating* is sometimes used for this process, it is not a form of sexual reproduction.

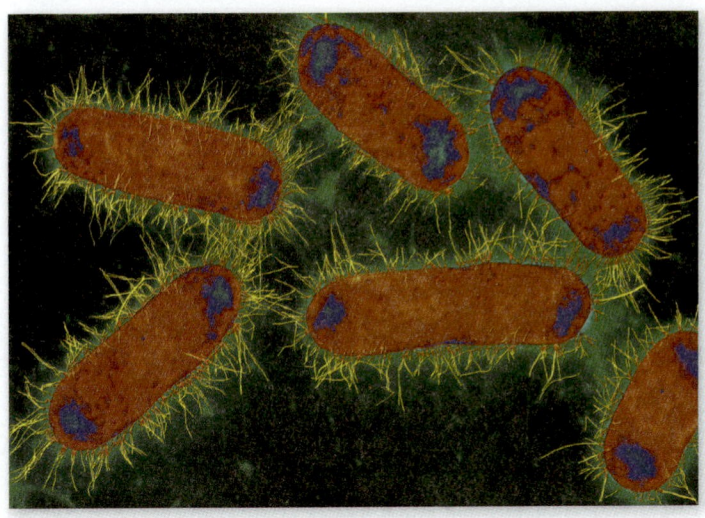

(a)

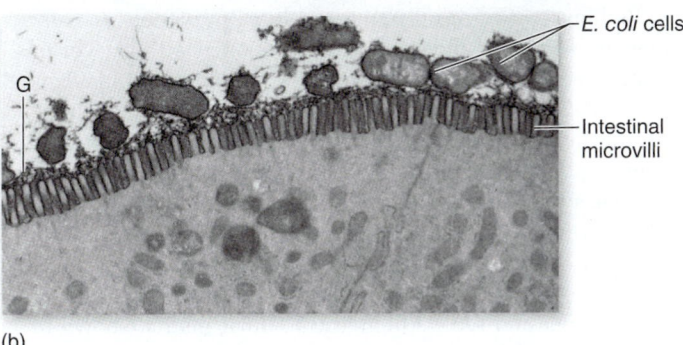

E. coli cells

G

Intestinal microvilli

(b)

Figure 4.11 Form and function of bacterial fimbriae.
(a) Several cells of pathogenic *Escherichia coli* covered with numerous stiff fibers called fimbriae (30,000×). Note also the dark-blue granules, which are the chromosomes. **(b)** A row of *E. coli* cells tightly adheres by their fimbriae to the surface of intestinal cells (12,000×). This is how the bacterium clings to the body during an infection. (G = Glycocalyx)

ing it close enough for DNA transfer. Production of pili is controlled genetically, and conjugation takes place only between compatible gram-negative cells. Conjugation in gram-positive bacteria does occur but involves aggregation proteins rather than pili. There is a special type of structure in some bacteria called a Type IV pilus. Like the pili described here, it can transfer genetic material. In addition, it can act like fimbriae and assist in attachment, and act like flagella and make a bacterium motile. The roles of pili and conjugation are further explored in chapter 9.

Surface Coatings: The S Layer and the Glycocalyx

The bacterial cell surface is frequently exposed to severe environmental conditions. Bacterial cells protect themselves with either an **S layer** or a **glycocalyx,** or both. S layers are single layers of thousands of copies of a single protein linked together like tiny chain mail. They are often called "the armor" of a bacterial cell. It took scientists a long time to discover them because bacteria only produce them when they are in a hostile environment. The nonthreatening conditions

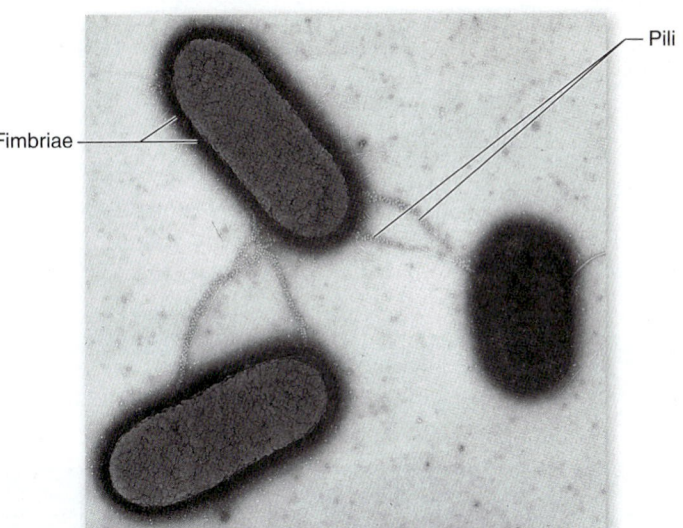

Pili

Fimbriae

Figure 4.12 Three bacteria in the process of conjugating.
Clearly evident are the sex pili forming mutual conjugation bridges between a donor (middle cell) and two recipients (cells on left side). Fimbriae can also be seen on the two left-hand cells.

of growing in a lab in a nutritious broth with no competitors around ensured that bacteria did not produce the layer. We now know that many different species have the ability to produce an S layer, including pathogens such as *Clostridium difficile* and *Bacillus anthracis*. Some bacteria use S layers to aid in attachment, as well.

The glycocalyx develops as a coating of repeating polysaccharide units that may or may not include protein. This protects the cell and, in some cases, helps it adhere to surfaces in its environment. Glycocalyces differ among bacteria in thickness, organization, and chemical composition. Some bacteria are covered with a loose shield called a **slime layer** that evidently protects them from loss of water and nutrients **(figure 4.13a)**. A glycocalyx is called a **capsule** when it is

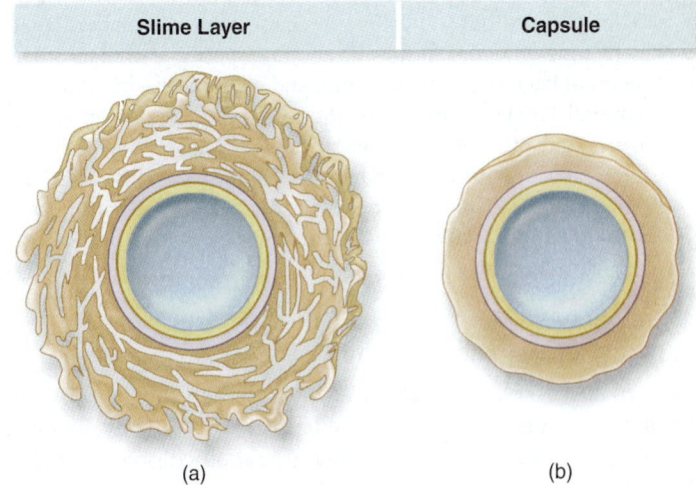

| Slime Layer | Capsule |

(a)   (b)

Figure 4.13 Drawings of sectioned bacterial cells to show the types of glycocalyces. **(a)** The slime layer is a loose structure that is easily washed off. **(b)** The capsule is a thick, structured layer that is not readily removed.

INSIGHT 4.1   Biofilms: Biological Glue

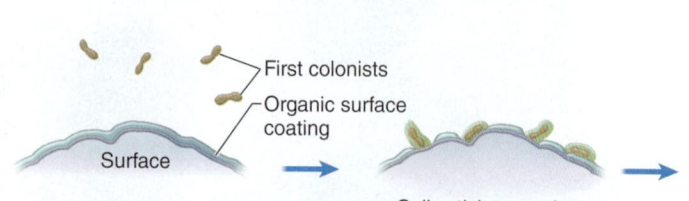
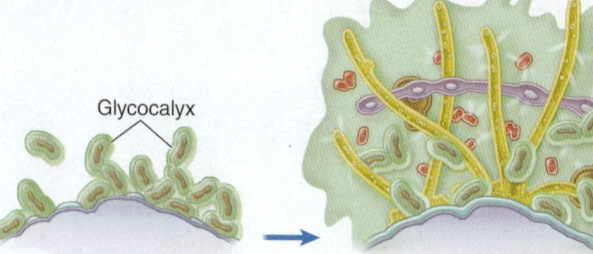

First colonists

Organic surface coating

Surface

Cells stick to coating.

Glycocalyx

As cells divide, they form a dense mat bound together by sticky extracellular deposits.

Additional microbes are attracted to developing film and create a mature community with complex function.

You've seen it before—the scum that builds up on the inside of your toilet, in your shower, or even on your teeth. This slimy gunk isn't merely evidence that you haven't cleaned in a while; it is a community of microbes called a **biofilm**. Scientists are discovering that bacteria often do not exist in a **planktonic** or single-cell form but rather live in cooperative associations that can include other organisms of the same species as well as other species of bacteria, archaea, fungi, and algae. These biofilms are microbial habitats with adequate access to food, water, atmosphere, and other environmental factors that are beneficial to each type of organism living there. Often, the biofilm is stratified, with the aerobic microbes near the surface where the oxygen levels are high and the anaerobic microbes near the bottom

where oxygen levels are low. Each member of the biofilm community finds its niche.

Biofilms can form on numerous inert substances, usually when the surface is moist and has developed a thin layer of organic material such as polysaccharides or glycoproteins. This slightly sticky texture attracts the first single-celled "colonists" that attach and begin to multiply on the surface. As the first colonizing organisms grow, they secrete substances such as cell signal receptors, fimbriae, slime layers, capsules, and even DNA molecules that attract other microbes to the surface as well. This cell-to-cell communication, including a process called *quorum sensing* (see chapter 7), allows for microbes of various species to grow together and secrete more extracellular matrix (shown in green in part *a* of the diagram). The biofilm can vary in thickness, depending on where it begins growing and how long it has been growing there (or how long it has been since you brushed and flossed your teeth).

Biofilms also have serious medical implications. Often, microbes will accumulate on damaged tissue such as heart valves or hard surfaces such as teeth. Bacteria also have an affinity for implanted medical devices such as IUDs, catheters, shunts, gastrostomy tubes, and urinary catheters, and readily form biofilms on these surfaces. Treating these types of infections is extremely difficult, and it has always been assumed that this was due to antibiotics being unable to penetrate the thick glycocalyx of the biofilm. Recent studies have shown that in biofilm form, microbes turn on different genes, allowing them to be impervious to antibiotic treatment. Finding novel ways to treat biofilm infections is an ongoing battle, and it is estimated that treating biofilm infections costs more than 1 billion dollars a year in the United States alone.

---

bound more tightly to the cell than a slime layer is and it is denser and thicker (**figure 4.13b**). Capsules can be viewed after a special staining technique (**figure 4.14a**). They are also often visible on agar because they give their colonies a sticky (mucoid) appearance (**figure 4.14b**).

Specialized Functions of the Glycocalyx

Capsules are formed by many pathogenic bacteria, such as *Streptococcus pneumoniae* (a cause of pneumonia, an infection

of the lung), *Haemophilus influenzae* (one cause of meningitis), and *Bacillus anthracis* (the cause of anthrax). Encapsulated bacterial cells generally have greater pathogenicity because capsules protect the bacteria against white blood cells called phagocytes. Phagocytes are a natural body defense that can engulf and destroy foreign cells through phagocytosis, thus preventing infection. A capsular coating blocks the mechanisms that phagocytes use to attach to and engulf bacteria. By escaping phagocytosis, the bacteria are free to multiply

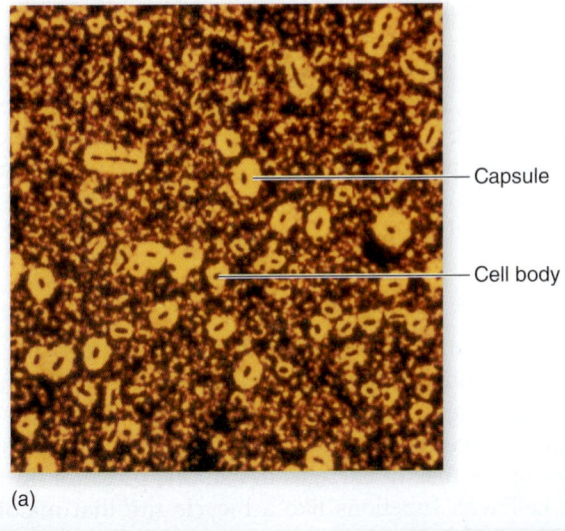

— Capsule

— Cell body

(a)

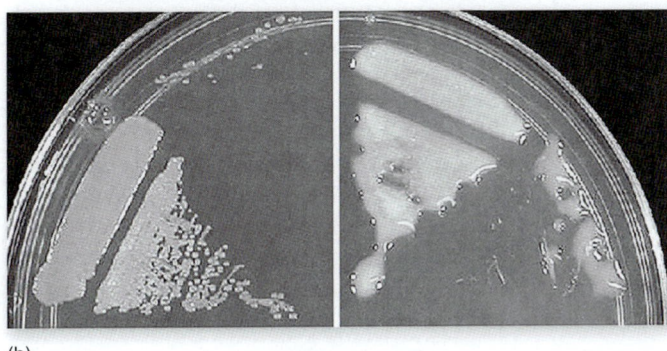

(b)

Figure 4.14 Encapsulated bacteria. **(a)** Negative staining reveals the microscopic appearance of a large, well-developed capsule. **(b)** Colony appearance of a nonencapsulated (left) and encapsulated (right) version of a soil bacterium called *Sinorhizobium*.

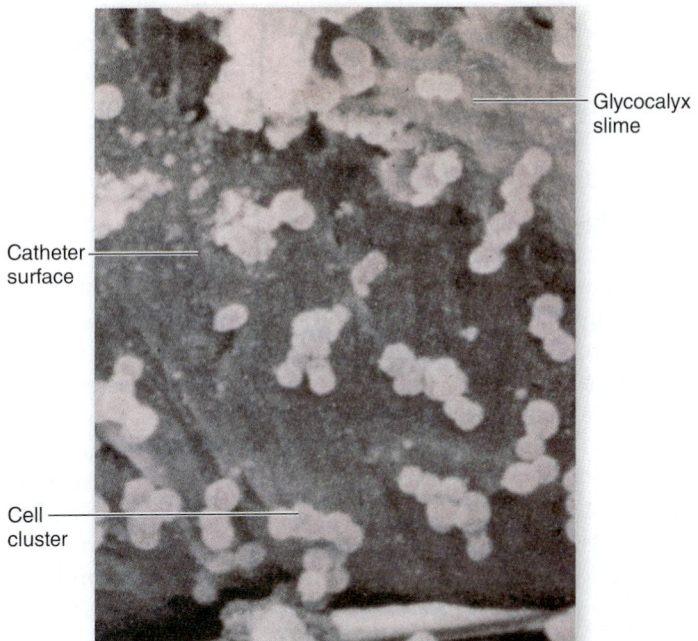

— Glycocalyx slime

Catheter surface —

Cell cluster —

 **Figure 4.15 Biofilm formation.** Scanning electron micrograph of *Staphylococcus aureus* cells attached to a catheter by a slime secretion.

and infect body tissues. Encapsulated bacteria that mutate to nonencapsulated forms usually lose their ability to cause disease.

Other types of glycocalyces can be important in the formation of biofilms. The thick, white plaque that forms on teeth comes in part from the surface slimes produced by certain streptococci in the oral cavity. This slime protects them from being dislodged from the teeth and provides a niche for other oral bacteria that, in time, can lead to dental disease. The glycocalyx of some bacteria is so highly adherent that it is responsible for persistent colonization of nonliving materials such as plastic catheters, intrauterine devices, and metal pacemakers that are in common medical use **(figure 4.15).**

4.2 Learning Outcomes—Assess Your Progress

6. Describe the structure and function of five different types of bacterial external structures.
7. Explain how a flagellum works in the presence of an attractant.

4.3 The Cell Envelope: The Boundary Layer of Bacteria

The majority of bacteria have a chemically complex external covering, termed the *cell envelope,* that lies outside of the cytoplasm. It is composed of two or three basic layers: the cell wall; the cell membrane; and, in some bacteria, the outer membrane. The layers of the envelope are stacked one upon

Case File 4   *Continuing the Case*

In a recent study, the fecal composition of normal, healthy adult males was compared with that of adult males with type II diabetes. Scientists used PCR to analyze the 16S rRNA of the gut microbiota, and found that patients with type II diabetes also had elevated levels of bacteria from the phyla *Bacteroidetes* and *Proteobacteria*. Bacteria from these phyla are gram-negative, and this study suggests that the lipopolysaccharide outer membrane of these organisms may induce inflammation in the gut that could play a role in the development of type II diabetes.

Yet another study suggests that multiple sclerosis (MS), an autoimmune disease that attacks the myelin sheath of nerve cells, may be triggered by gut microbiota. Scientists found that mice raised without gut microbiota and then colonized with certain intestinal bacteria developed MS-like symptoms. Researchers still need to determine if a faulty immune system overreacting to gut bacteria is the cause of MS or if a specific organism triggers the autoimmunity.

■ What happens when your gut microflora is disrupted?

another and are often tightly bonded together like the outer husk and casings of a coconut. Although each envelope layer performs a distinct function, together they act as a single protective unit.

Differences in Cell Envelope Structure

More than a hundred years ago, long before the detailed anatomy of bacteria was even remotely known, a Danish physician named Hans Christian Gram developed a staining technique, the **Gram stain,** that delineates two generally different groups of bacteria **(Insight 4.2).** The two major groups shown by this technique are the **gram-positive** bacteria and the **gram-negative** bacteria.

The structural differences denoted by the designations *gram-positive* and *gram-negative* lie in the cell envelope **(figure 4.16).** In gram-positive cells, a microscopic section reveals two layers: the thick cell wall, composed primarily of peptidoglycan (defined in the next section), and the

cytoplasmic membrane. A similar section of a gram-negative cell envelope shows three layers: an outer membrane, a thin cell wall, and the cytoplasmic membrane.

Moving from outside to in, the outer membrane (if present) lies just under the glycocalyx. Next comes the cell wall. Finally, the innermost layer is always the cytoplasmic membrane. Because only some bacteria have an outer membrane, we discuss the cell wall first.

Structure of the Cell Wall

The **cell wall** accounts for a number of important bacterial characteristics. In general, it helps determine the shape of a bacterium, and it also provides the kind of strong structural support necessary to keep a bacterium from bursting or collapsing because of changes in osmotic pressure. In this way, the cell wall functions like a bicycle tire that maintains the necessary shape and prevents the more delicate inner tube (the cytoplasmic membrane) from bursting when it is expanded.

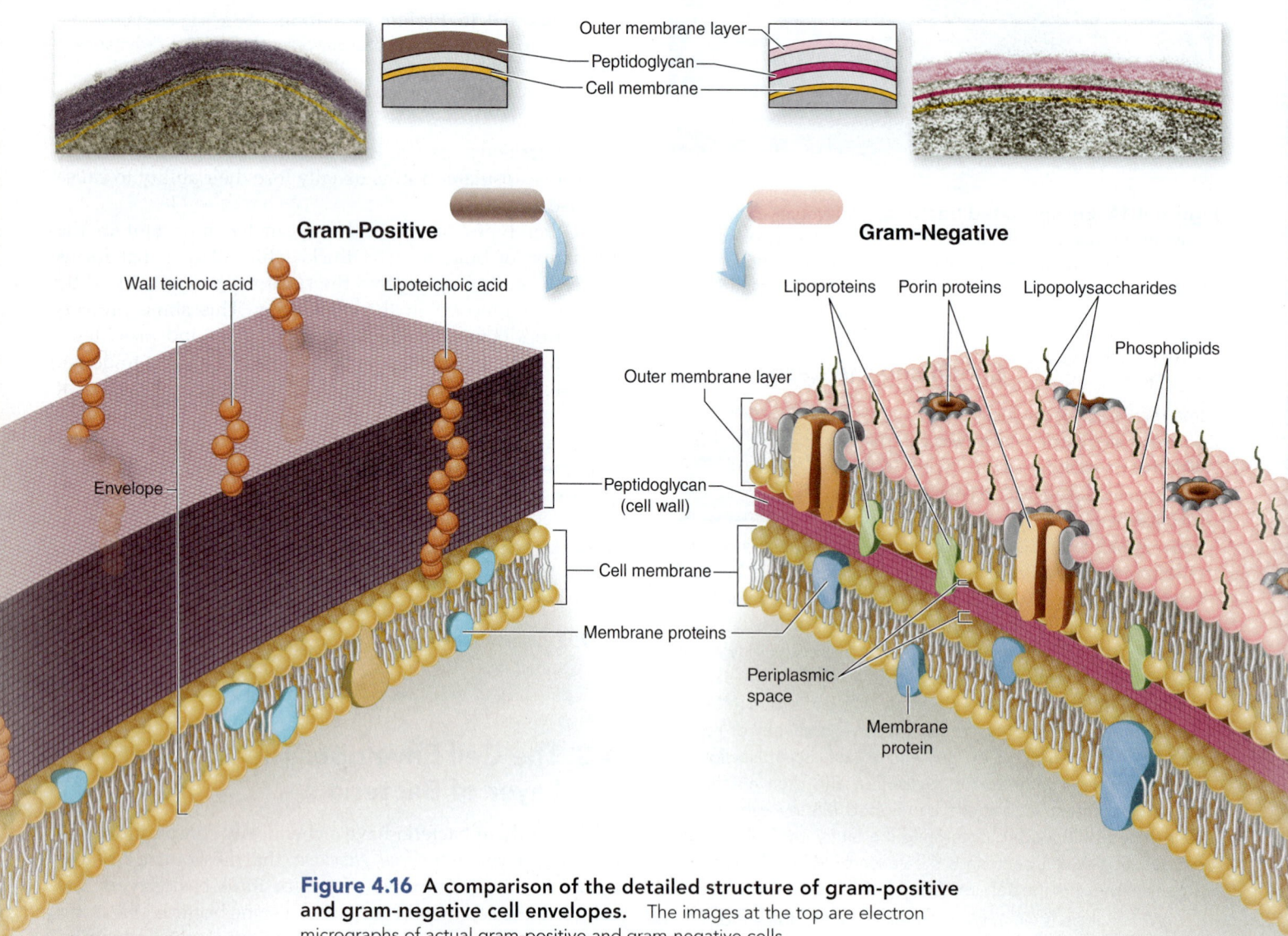

Figure 4.16 A comparison of the detailed structure of gram-positive and gram-negative cell envelopes.   The images at the top are electron micrographs of actual gram-positive and gram-negative cells.

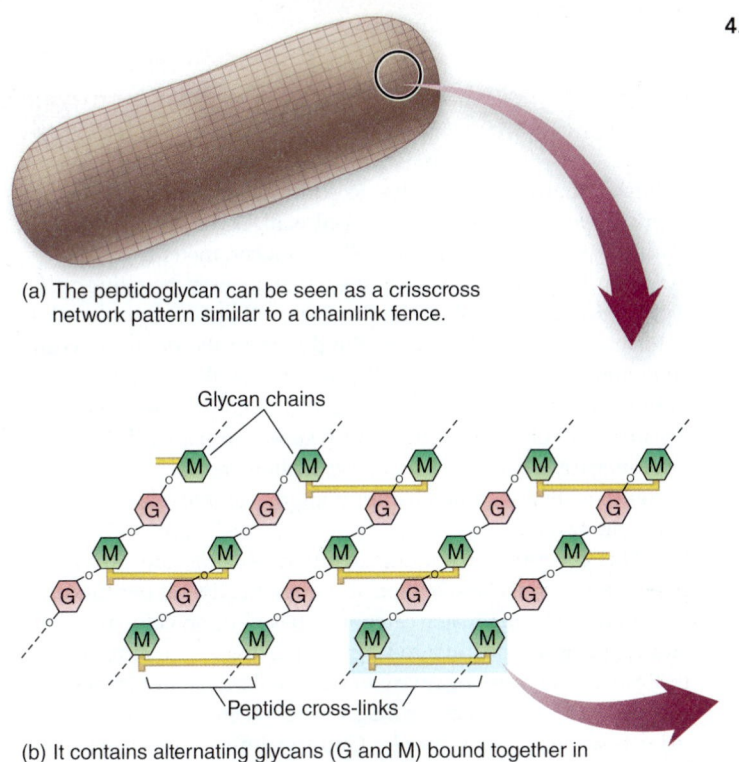

(a) The peptidoglycan can be seen as a crisscross network pattern similar to a chainlink fence.

(b) It contains alternating glycans (G and M) bound together in long strands. The G stands for *N*-acetyl glucosamine, and the M stands for *N*-acetyl muramic acid.

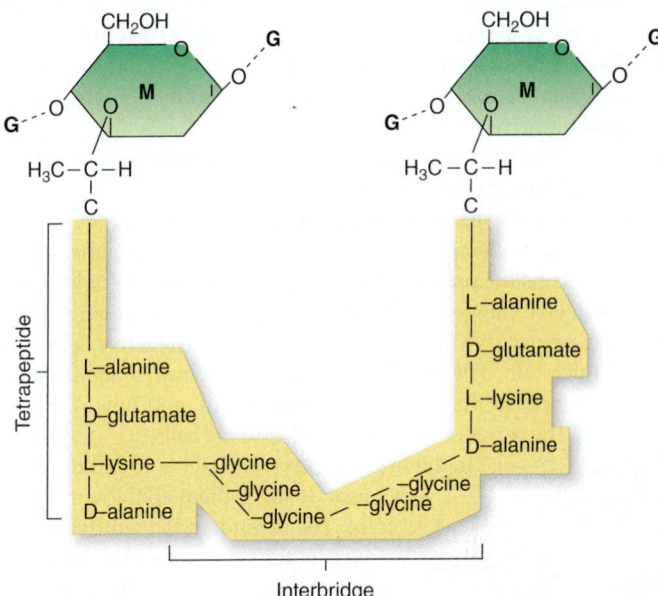

(c) A detailed view of the links between the muramic acids. Tetrapeptide chains branching off the muramic acids connect by interbridges also composed of amino acids. It is this linkage that provides rigid yet flexible support to the cell and that may be targeted by drugs like penicillin.

Figure 4.17 Structure of peptidoglycan in the cell wall.

The cell walls of most bacteria gain their relatively rigid quality from a unique macromolecule called **peptidoglycan** (PG). This compound is composed of a repeating framework of long **glycan** (sugar) chains cross-linked by short peptide (protein) fragments to provide a strong but flexible support framework **(figure 4.17)**. The amount and exact composition of peptidoglycan vary among the major bacterial groups.

Because many bacteria live in aqueous habitats with a low concentration of dissolved substances, they are constantly absorbing excess water by osmosis. Were it not for the strength and relative rigidity of the peptidoglycan in the cell wall, they would rupture from internal pressure. This function of the cell wall has been a tremendous boon to the drug industry. Several types of drugs used to treat infection (penicillin, cephalosporins) are effective because they target the peptide cross-links in the peptidoglycan, thereby disrupting its integrity. With their cell walls incomplete or missing, such cells have very little protection from **lysis** (ly′-sis), which is the disintegration or rupture of the cell. Lysozyme, an enzyme contained in tears and saliva, provides a natural defense against certain bacteria by hydrolyzing the bonds in the glycan chains and causing the wall to break down. (Chapter 11 discusses the actions of antimicrobial chemical agents.)

The Gram-Positive Cell Wall

The bulk of the gram-positive cell wall is a thick, homogeneous sheath of peptidoglycan ranging from 20 to 80 nm in thickness. It also contains tightly bound acidic polysaccharides, including **teichoic acid** and **lipoteichoic acid** (see figure 4.16). Teichoic acid is a polymer of ribitol or glycerol (alcohols) and phosphate that is embedded in the peptidoglycan sheath. Lipoteichoic acid is similar in structure but is attached to the lipids in the plasma membrane. These molecules probably function in cell wall maintenance and enlargement during cell division, and they also contribute to the acidic charge on the cell surface.

The Gram-Negative Cell Wall

The gram-negative wall is a single, thin (1–3 nm) sheet of peptidoglycan. Although it acts as a somewhat rigid protective structure as previously described, its thinness gives gram-negative bacteria a relatively greater flexibility and sensitivity to lysis.

Nontypical Cell Walls

Several bacterial groups lack the cell wall structure of gram-positive or gram-negative bacteria, and some bacteria have

INSIGHT 4.2   The Gram Stain: A Grand Stain

In 1884, Hans Christian Gram discovered a staining technique that could be used to make bacteria in infectious specimens more visible. His technique consisted of timed, sequential applications of crystal violet (the primary dye), Gram's iodine (the mordant), an alcohol rinse (decolorizer), and a contrasting counterstain. The initial counterstain used was yellow or brown and was later replaced by the red dye safranin. Bacteria that stain purple are called gram-positive, and those that stain red are called gramnegative.

Although these staining reactions involve an attraction of the cell to a charged dye (see chapter 3), it is important to note that the terms **gram-positive** and **gram-negative** are not used to indicate the electrical charge of cells or dyes but whether or not a cell retains the primary dye-iodine complex after decolorization. There is nothing specific in the reaction of gram-positive cells to the primary dye or in the reaction of gram-negative cells to the counterstain. The different results in the Gram stain are due to differences in the structure of the cell wall and how it reacts to the series of reagents applied to the cells.

In the first step, crystal violet is added to the cells in a smear. It stains them all the same purple color. The second and key differentiating step is the addition of the mordant— Gram's iodine. The mordant is a stabilizer that causes the dye to form large complexes in the peptidoglycan meshwork of the cell wall. Because the peptidoglycan layer in gram-positive cells is thicker, the entrapment of the dye is far more extensive in them than in gram-negative cells. Application of alcohol in the third step dissolves lipids in the outer membrane and removes the dye from the peptidoglycan layer and the gramnegative cells. By contrast, the crystals of dye tightly embedded in the peptidoglycan of gram- positive bacteria are relatively inaccessible and resistant to removal. Because gram-negative bacteria are colorless after decolorization, their presence is demonstrated by applying the counterstain safranin in the final step.

This century-old staining method remains the universal basis for bacterial classification and identification. It permits differentiation of four major categories based upon color reaction and shape: gram-positive rods, gram-positive cocci, gram-negative rods, and gram-negative cocci (see table 4.2). The Gram stain can also be a practical aid in diagnosing infection and in guiding drug treatment. For example, Gram staining a fresh urine or throat specimen can help pinpoint the possible cause of infection, and in some cases it is possible to begin drug therapy on the basis of this stain. Even in this day of elaborate and expensive medical technology, the Gram stain remains an important and unbeatable first tool in diagnosis.

Step	Microscopic Appearance of Cell		Chemical Reaction in Cell Wall (very magnified view)	
	Gram (+)	Gram (−)	Gram (+)	Gram (−)
1. Crystal violet First, crystal violet is added to the cells in a smear. It stains them all the same purple color.				Both cell walls affix the dye
2. Gram's iodine Then, the mordant, Gram's iodine, is added. This is a stabilizer that causes the dye to form large complexes in the peptidoglycan meshwork of the cell wall. The thicker gram-positive cell walls are able to more firmly trap the large complexes than those of the gram-negative cells.			Dye complex trapped in wall	No effect of iodine
3. Alcohol Application of alcohol dissolves lipids in the outer membrane and removes the dye from the peptidoglycan layer—only in the gram-negative cells.			Crystals remain in cell wall	Outer membrane weakened; wall loses dye
4. Safranin (red dye) Because gram-negative bacteria are colorless after decolorization, their presence is demonstrated by applying the counterstain safranin in the final step.			Red dye masked by violet	Red dye stains the colorless cell

no cell wall at all. Although these exceptional forms can stain positive or negative in the Gram stain, examination of their fine structure and chemistry shows that they do not really fit the descriptions for typical gram-negative or -positive cells. For example, the cells of *Mycobacterium* and *Nocardia* contain peptidoglycan and stain gram-positive, but the bulk of their cell wall is composed of unique types of lipids. One of these is a very-long-chain fatty acid called **mycolic acid**, or cord factor, that contributes to the pathogenicity of this group (see chapter 21). The thick, waxy nature imparted to the cell wall by these lipids is also responsible for a high degree of resistance to certain chemicals and dyes. Such resistance is the basis for the **acid-fast stain** used to diagnose tuberculosis and leprosy. In this stain, hot carbol fuchsin dye becomes tenaciously attached (is held fast) to these cells so that an acid-alcohol solution will not remove the dye (see chapter 3).

Mycoplasmas and Other Cell-Wall-Deficient Bacteria

Mycoplasmas are bacteria that naturally lack a cell wall. Although other bacteria require an intact cell wall to prevent the bursting of the cell, the mycoplasma cell membrane is stabilized by sterols and is resistant to lysis. These extremely tiny, **pleomorphic** cells are very small bacteria, ranging from 0.1 to 0.5 µm in size. They range in shape from filamentous to coccus or doughnut-shaped. They are *not* obligate parasites and can be grown on artificial media, although added sterols are required for the cell membranes of some species. Mycoplasmas are found in many habitats, including plants, soil, and animals. The most important medical species is *Mycoplasma pneumoniae* **(figure 4.18),** which adheres to the epithelial cells in the lung and causes an atypical form of pneumonia in humans (described in chapter 21).

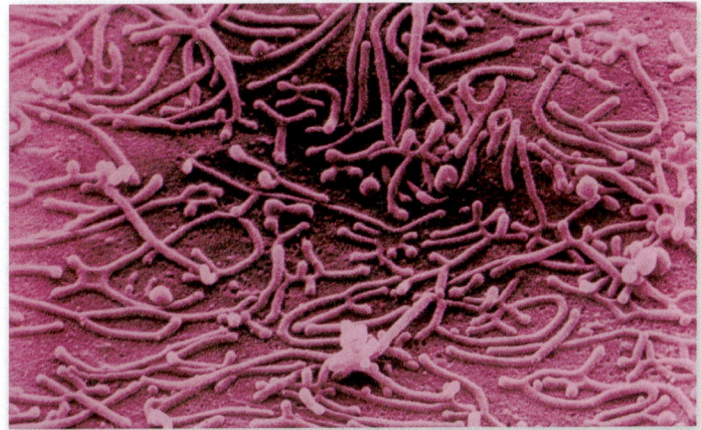

Figure 4.18 Scanning electron micrograph of *Mycoplasma pneumoniae* (magnified 62,000×). Cells like these that naturally lack a cell wall exhibit extreme variation in shape.

The Gram-Negative Outer Membrane

The **outer membrane** (OM) is somewhat similar in construction to the cell membrane, except that it contains specialized types of polysaccharides and proteins. The uppermost layer of the OM "sandwich" contains **lipopolysaccharide** (LPS). The polysaccharide chains extending off the surface function as cell markers and receptors. The lipid portion of LPS has been referred to as **endotoxin** because it stimulates fever and shock reactions in gram-negative infections such as meningitis and typhoid fever. The innermost layer of the OM is a phospholipid layer anchored by means of lipoproteins to the peptidoglycan layer below. The outer membrane serves as a partial chemical sieve by allowing only relatively small molecules to penetrate. Access is provided by special membrane channels formed by **porin proteins** that completely span the outer membrane. The size of these porins can be altered so as to block the entrance of harmful chemicals, making them one defense of gram-negative bacteria against certain antibiotics (see figure 4.16).

Cell Membrane Structure

Appearing just beneath the cell wall is the **cell membrane,** which is often called the **cytoplasmic membrane.** It is a very thin (5–10 nm), flexible sheet molded completely around the cytoplasm. Its general composition was described in chapter 2 as a lipid bilayer with proteins embedded to varying degrees (see Insight 2.3). Bacterial cell membranes have this typical structure, containing primarily phospholipids (making up about 30%–40% of the membrane mass) and proteins (contributing 60%–70%). Major exceptions to this description are the membranes of mycoplasmas, which contain high amounts of sterols—rigid lipids that stabilize and reinforce the membrane; and the membranes of archaea, which contain unique branched hydrocarbons rather than fatty acids.

Some environmental bacteria, including photosynthesizers and ammonia oxidizers, contain dense stacks of internal membranes that are studded with enzymes or photosynthetic pigments. The inner membranes allow a higher concentration of these enzymes and pigments and also accomplish a compartmentalization that allows for higher energy production.

Functions of the Cell Membrane

Because bacteria have none of the eukaryotic organelles, the cell membrane provides a site for functions such as energy reactions, nutrient processing, and synthesis. A major action of the cell membrane is to regulate **transport,** the passage of nutrients into the cell and the discharge of wastes. Although water and small uncharged molecules can diffuse across the membrane unaided, the membrane is a **selectively permeable** structure with special carrier mechanisms for passage of most molecules (see chapter 7). The cell membrane is also involved in **secretion,** or the discharge of a metabolic product into the extracellular environment.

The membranes of bacteria are an important site for a number of metabolic activities. Many enzymes of respiration and ATP synthesis reside in the cell membrane because these cells lack mitochondria (see chapter 8). Enzyme structures located in the cell membrane also help synthesize structural macromolecules to be incorporated into the cell envelope and appendages. Other products (enzymes and toxins) are secreted by the membrane into the extracellular environment.

Practical Considerations of Differences in Cell Envelope Structure

Variations in cell envelope anatomy contribute to several other differences between the two cell types. The outer membrane contributes an extra barrier in gram-negative bacteria that makes them impervious to some antimicrobial chemicals such as dyes and disinfectants, so they are generally more difficult to inhibit or kill than are gram-positive bacteria. One exception is alcohol-based compounds, which can dissolve the lipids in the outer membrane and disturb its integrity. Treating infections caused by gram-negative bacteria often requires different drugs from gram-positive infections, especially drugs that can cross the outer membrane.

The cell envelope or its parts can interact with human tissues and contribute to disease. Proteins attached to the outer portion of the cell wall of several gram-positive species, including *Corynebacterium diphtheriae* (the agent of diphtheria) and *Streptococcus pyogenes* (the cause of strep throat), also have toxic properties. The lipids in the cell walls of certain *Mycobacterium* species are harmful to human cells as well. Because most macromolecules in the cell walls are foreign to humans, they stimulate antibody production by the immune system (see chapter 15).

Looking at the unique structures within both gram-negative and gram-positive cell envelopes, we gain insight into the potential targets for new drug development by researchers today.

4.3 Learning Outcomes—Assess Your Progress

8. Differentiate between the two main types of bacterial envelope structure.
9. Discuss why gram-positive cell walls are stronger than gram-negative cell walls.
10. Name a substance in the envelope structure of some bacteria that can cause severe symptoms in humans.

4.4 Bacterial Internal Structure

Contents of the Cell Cytoplasm

Cytoplasm is a gelatinous solution encased by the cell membrane. It is another prominent site for many of the cell's biochemical and synthetic activities. Its major component is water (70%–80%), which serves as a solvent for the cell

pool, a complex mixture of nutrients including sugars, amino acids, and salts. The components of this pool serve as building blocks for cell synthesis or as sources of energy. The cytoplasm also contains larger, discrete cell masses such as the chromatin body, ribosomes, granules, and fibers resembling actin and tubulin strands that act as a cytoskeleton in bacteria that have them.

Bacterial Chromosomes and Plasmids: The Sources of Genetic Information

The hereditary material of most bacteria exists in the form of a single circular strand of DNA designated as the **bacterial chromosome.** Some bacteria have multiple chromosomes. By definition, bacteria do not have a nucleus; that is, their DNA is not enclosed by a nuclear membrane but instead is aggregated in a dense area of the cell called the **nucleoid (figure 4.19).** (Note that a very few species of bacteria have been found to have a nucleus-like structure, but these remain the exception.) The chromosome is actually an extremely long molecule of double-stranded DNA that is tightly coiled around special basic protein molecules so as to fit inside the cell compartment. Arranged along its length are genetic units (genes) that carry information required for bacterial maintenance and growth.

Although the chromosome is the minimal genetic requirement for bacterial survival, many bacteria contain other nonessential pieces of DNA called **plasmids** (refer to figure 4.1 for a representation of the nuclear material). Plasmids exist as separate double-stranded circles of DNA, although at times they can become integrated into the chromosome. During conjugation, they may be duplicated and passed on to related nearby bacteria. During bacterial reproduction, they are duplicated and passed on to offspring. They are not essential to bacterial growth and metabolism, but they often confer protective traits such as the ability to resist drugs and to produce toxins and enzymes (see chapter 9). Because they can be readily manipulated in the laboratory and transferred from one bacterial cell to another, plasmids are an important agent in genetic engineering techniques.

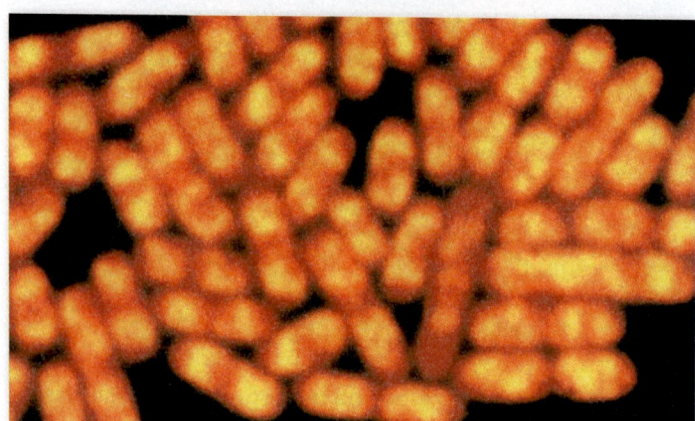

Figure 4.19 Chromosome structure. Fluorescent staining highlights the chromosomes of the bacterial pathogen *Salmonella enteritidis.* The cytoplasm is orange, and the chromosome(s) fluoresce(s) bright yellow.

Disease Connection

Typhoid fever is a major disease in the developing world. It is caused by *Salmonella enterica* serovar Typhi. Before 1995, these bacteria carried a variety of different plasmids carrying a variety of different genes. Ninety-eight percent of the isolates collected after 1995, however, carry a single type of plasmid, which confers resistance to multiple drugs and also allows the bacterium to survive in high-salt environments. Researchers speculate that these abilities give the bacterium a competitive advantage over bacteria that had the other plasmid types.

Ribosomes: Sites of Protein Synthesis

All cells contain thousands of tiny **ribosomes,** which are made of RNA and protein. When viewed even by very high magnification, ribosomes show up as fine, spherical specks dispersed throughout the cytoplasm and often occur in chains called polysomes. Many are also attached to the cell membrane. Chemically, a ribosome is a combination of a special type of RNA called ribosomal RNA, or rRNA (about 60%), and protein (40%). One method of characterizing ribosomes is by S, or Svedberg,[3] units, which rate the molecular sizes of various cell parts that have been spun down and separated by molecular weight and shape in a centrifuge. Heavier, more compact structures sediment faster and are assigned a higher S rating. Combining this method of analysis with high-resolution electron microscopy has revealed that the ribosome in bacteria, which has an overall rating of 70S, is actually composed of two smaller subunits **(figure 4.20).** They fit together to form a miniature platform upon which protein synthesis is performed. Note that eukaryotic ribosomes have a rating of 80S, making bacterial ribosomes a unique target for therapeutic drugs. We examine the more detailed functions of ribosomes in chapter 9 and drug targeting in chapter 12.

3. Named in honor of T. Svedberg, the Swedish chemist who developed the ultracentrifuge in 1926.

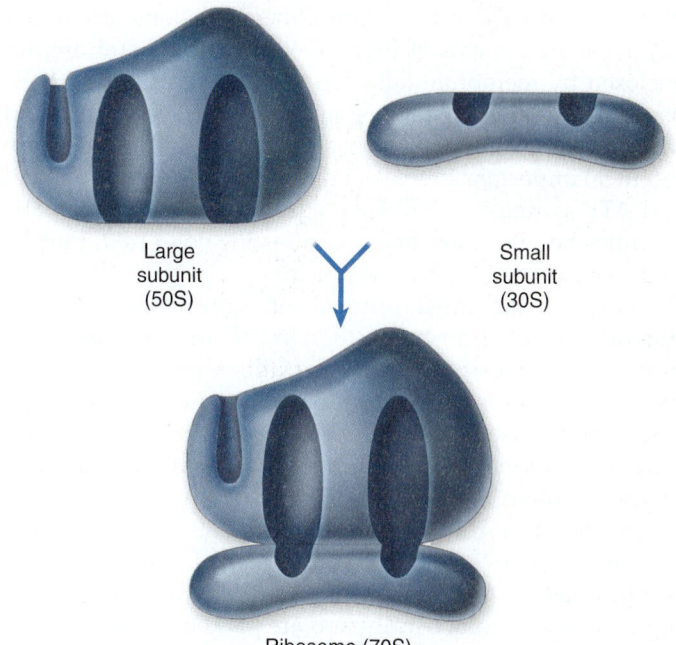

Large subunit (50S)

Small subunit (30S)

Ribosome (70S)

Figure 4.20 A model of a bacterial ribosome, showing the small (30S) and large (50S) subunits, both separate and joined.

Inclusion Bodies and Microcompartments

Most bacteria are exposed to severe shifts in the availability of food. During periods of nutrient abundance, some can compensate by laying down nutrients intracellularly in **inclusion bodies,** or **inclusions,** of varying size, number, and content. As the environmental source of these nutrients becomes depleted, the bacterial cell can mobilize its own storehouse as required. Some inclusion bodies carry condensed, energy-rich organic substances, such as glycogen and polyhydroxybutyrate (PHB), within special single-layered membranes **(figure 4.21).** A unique type of inclusion found in some aquatic bacteria is gas vesicles that

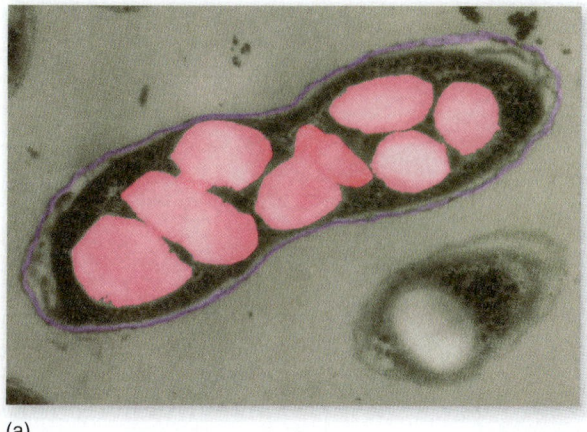

(a)

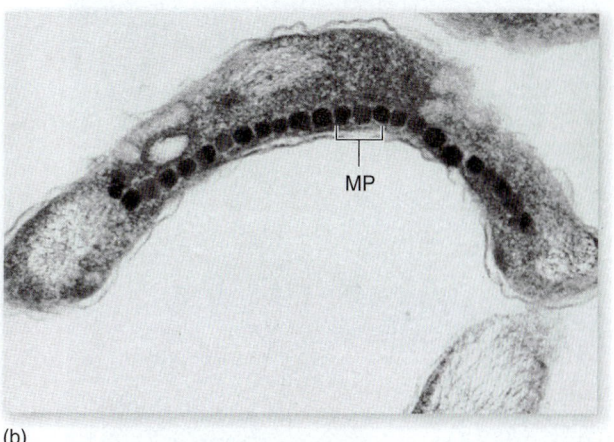

(b)

Figure 4.21 Bacterial inclusion bodies. **(a)** Large particles (pink) of polyhydroxybutyrate are deposited in a concentrated form that provides an ample long-term supply of that nutrient (32,500×). **(b)** A section through *Aquaspirillum* reveals a chain of tiny iron magnets, or magnetosomes (MP). These unusual bacteria use these inclusions to orient themselves within their habitat (123,000×).

provide buoyancy and flotation. Other inclusions, also called granules, are crystals of inorganic compounds and are not enclosed by membranes. Sulfur granules of photosynthetic bacteria and polyphosphate granules of *Corynebacterium* and *Mycobacterium*, described later, are of this type. The latter represent an important source of building blocks for nucleic acid and ATP synthesis. They have been termed **metachromatic granules** because they stain a contrasting color (red, purple) in the presence of methylene blue dye.

Perhaps the most unique cell granule is involved not in cell nutrition but rather in cell orientation. Magnetotactic bacteria contain crystalline particles of iron oxide (magnetosomes) that have magnetic properties. The bacteria use these granules to be pulled by the polar and gravitational fields into deeper habitats with a lower oxygen content.

In the early 2000s, new compartments inside bacterial cells were discovered. These were named bacterial microcompartments, or BMCs. Their outer shells are made of protein, arranged geometrically, and are packed full of enzymes that are designed to work together in pathways, thereby ensuring that they are in close proximity to one another.

The Cytoskeleton

Until very recently, scientists thought that the shape of all bacteria was completely determined by the peptidoglycan layer (cell wall). Although this is true of many bacteria, particularly the cocci, other bacteria produce long polymers of proteins that are very similar to eukaryotic **actin**. In bacteria, these are arranged in helical ribbons around the cell just under the cell membrane **(figure 4.22)**. Fibers contribute to cell shape, perhaps by influencing the way peptidoglycan is manufactured, and also function in cell division. The fibers have been found in rod-shaped and spiral bacteria. They are composed in part of proteins unique to bacterial cells, making them a potentially powerful target for future antibiotic development.

Bacterial Endospores: An Extremely Resistant Stage

Ample evidence indicates that the anatomy of bacteria helps them adjust rather well to adverse habitats. But of all microbial structures, nothing can compare to the bacterial **endospore** (or, simply, *spore*) for withstanding hostile conditions and facilitating survival.

Endospores are dormant bodies produced by bacteria of the genera *Bacillus*, *Clostridium*, and *Sporosarcina*. These bacteria have a two-phase life cycle—a vegetative cell and an endospore **(figure 4.23)**. The vegetative cell is a metabolically active and growing entity that can be induced by environmental conditions to undergo spore formation, or **sporulation.** Once formed, the spore exists in an inert, resting condition that shows up prominently in a spore or Gram stain (figure 4.23). Features of spores, including size, shape, and position in the vegetative cell, are somewhat useful in identifying some species. Both gram-positive and gram-negative bacteria can form endospores, but the medically relevant ones are all gram-positive. Most bacteria form only one endospore; therefore, this is not a reproductive function for them.

Bacterial endospores are the hardiest of all life forms, capable of withstanding extremes in heat, drying, freezing, radiation, and chemicals that would readily kill vegetative cells. Their survival under such harsh conditions is due to several factors. The heat resistance of spores has been linked to their high content of calcium and **dipicolinic acid.** We know, for instance, that heat destroys cells by inactivating proteins and DNA and that this process requires a certain amount of water in the protoplasm. Because the deposition of calcium dipicolinate in the endospore removes water and leaves the endospore very dehydrated, it is less vulnerable to the effects of heat. It is also metabolically inactive and highly resistant to damage from further drying. The thick, impervious cortex and spore coats also protect against radiation and chemicals. The longevity of bacterial spores verges on immortality. One record describes the isolation of

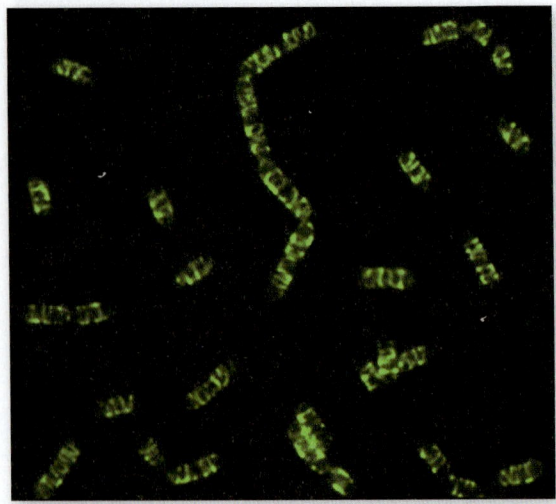

Figure 4.22 Bacterial cytoskeleton.   The fibers in these rod-shaped bacteria are fluorescently stained.

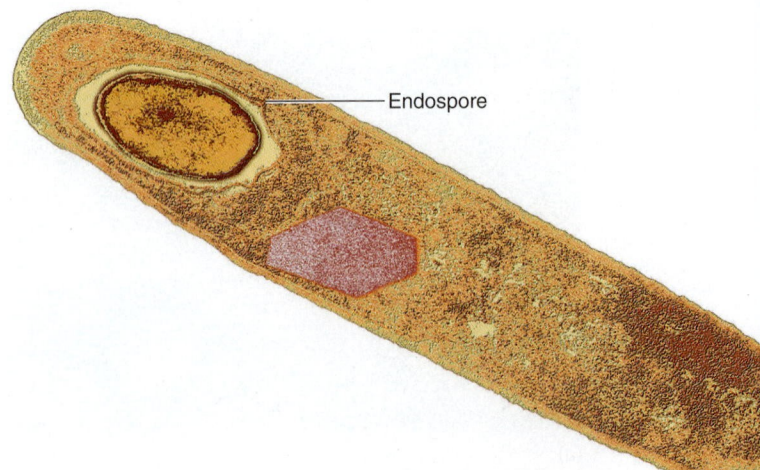

Endospore

Figure 4.23 Endospore inside *Bacillus thuringiensis.*
The genus *Bacillus* forms endospores. *B. thuringiensis* additionally forms crystalline bodies (pink) that are used as insecticides.

viable endospores from a fossilized bee that was 25 million years old. More recently, microbiologists unearthed a viable endospore from a 250-million-year-old salt crystal. Initial analysis of this ancient microbe indicates it is a species of *Bacillus* that is genetically different from known species.

Endospore Formation and Resistance

The depletion of nutrients, especially an adequate carbon or nitrogen source, is the stimulus for a vegetative cell to begin endospore formation. Once this stimulus has been received by the vegetative cell, it undergoes a conversion to become a sporulating cell called a **sporangium.** Complete transformation of a vegetative cell into a sporangium and then into an endospore requires 6 to 8 hours in most spore-forming species. **Process Figure 4.24** illustrates some major physical and chemical events in this process.

The Germination of Endospores

After lying in a state of inactivity for an indefinite time, endospores can be revitalized when favorable conditions arise. The breaking of dormancy, or germination, happens in the presence of water and a specific chemical or environmental stimulus (germination agent). Once initiated, it proceeds to completion quite rapidly (1½ hours). Although the specific germination agent varies among species, it is generally a small organic molecule such as an amino acid or an inorganic salt. This agent stimulates the formation of hydrolytic (digestive) enzymes by the endospore membranes. These enzymes digest the cortex and expose the core to water. As the core rehydrates and takes up nutrients, it begins to grow out of the endospore coats. In time, it reverts to a fully active vegetative cell, resuming the vegetative cycle.

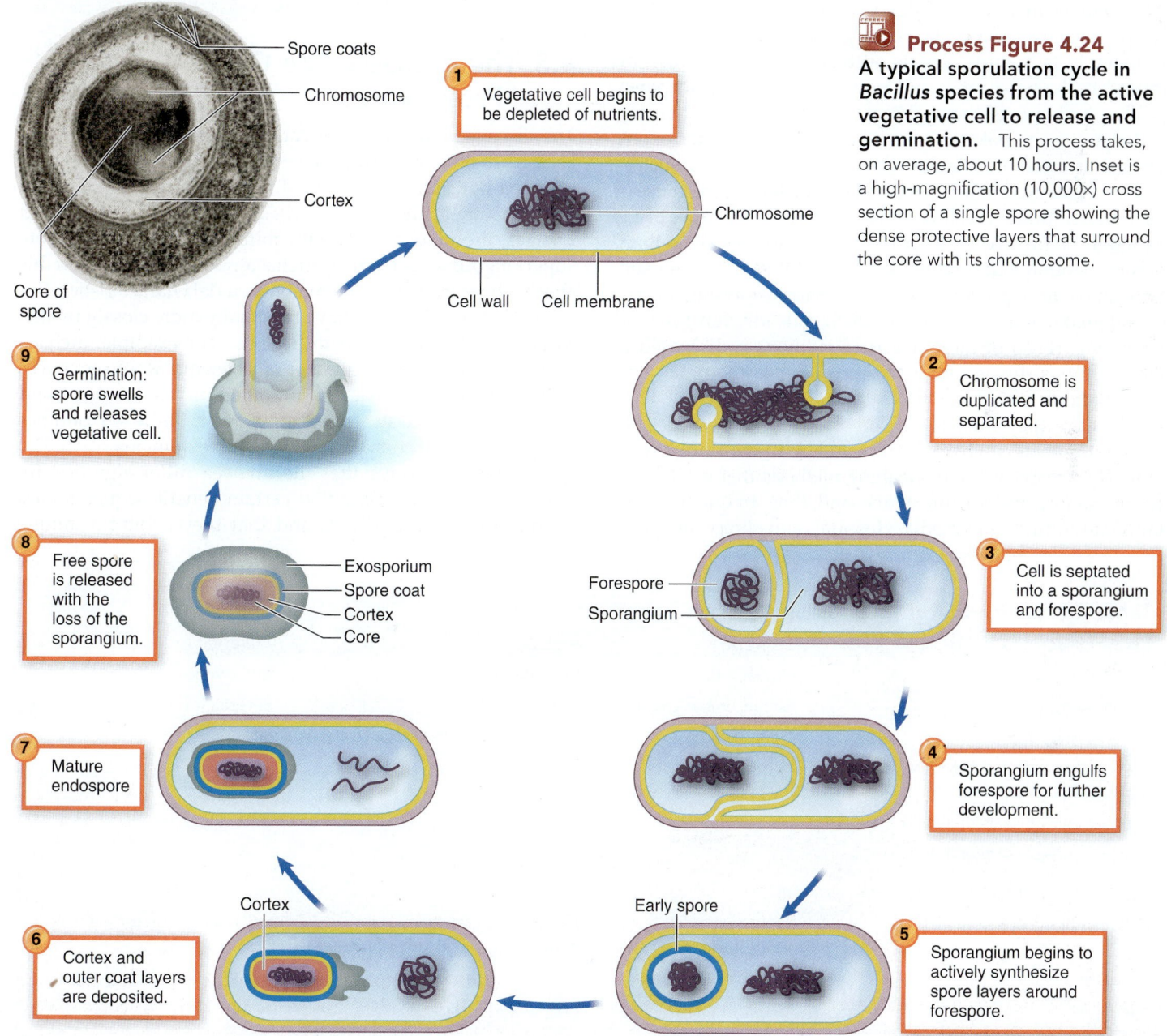

Process Figure 4.24
A typical sporulation cycle in *Bacillus* species from the active vegetative cell to release and germination. This process takes, on average, about 10 hours. Inset is a high-magnification (10,000×) cross section of a single spore showing the dense protective layers that surround the core with its chromosome.

Spore coats

Chromosome

Cortex

Core of spore

1 Vegetative cell begins to be depleted of nutrients.

Chromosome

Cell wall   Cell membrane

2 Chromosome is duplicated and separated.

3 Cell is septated into a sporangium and forespore.

Forespore

Sporangium

4 Sporangium engulfs forespore for further development.

5 Sporangium begins to actively synthesize spore layers around forespore.

Early spore

6 Cortex and outer coat layers are deposited.

Cortex

7 Mature endospore

8 Free spore is released with the loss of the sporangium.

Exosporium
Spore coat
Cortex
Core

9 Germination: spore swells and releases vegetative cell.

A Note on Terminology

The word *spore* can have more than one usage in microbiology. It is a generic term that refers to any tiny compact cell that is produced by vegetative or reproductive structures of microorganisms. Fungi have spores that serve as reproductive structures. The bacterial type discussed here is most accurately called an **endospore,** because it is produced inside a cell. They function in *survival,* not in reproduction, because no increase in cell numbers is involved in their formation. In contrast, the fungi produce many different types of spores for both survival and reproduction (see chapter 5).

Medical Significance of Bacterial Spores

Although the majority of endospore-forming bacteria are relatively harmless, several bacterial pathogens are endospore-formers. In fact, some aspects of the diseases they cause are related to the persistence and resistance of their endospores. *Bacillus anthracis* is the agent of anthrax; its persistence in endospore form makes it an ideal candidate for bioterrorism. The genus *Clostridium* includes even more pathogens, such as *C. tetani,* the cause of tetanus (lockjaw); *C. difficile,* the cause of pseudomembranous colitis; and *C. perfringens,* the cause of gas gangrene. When the endospores of these species are embedded in a wound that contains dead tissue, they can germinate, grow, and release potent toxins. Another toxin-forming species, *C. botulinum,* is the agent of botulism, a deadly form of food poisoning. (Each of these disease conditions is discussed in the infectious disease chapters, according to the organ systems it affects.)

Because they inhabit the soil and dust, endospores are constant intruders where sterility and cleanliness are important. They resist ordinary cleaning methods that use boiling water, soaps, and disinfectants; and they frequently contaminate cultures and media. Hospitals and clinics must take precautions to guard against the potential harmful effects of endospores in wounds, especially those of *Clostridium difficile,* the causative agent of a gastrointestinal disease commonly known as *C. diff.* Endospore destruction is a particular concern of the food-canning industry. Several endospore-forming species cause food spoilage or poisoning. Ordinary boiling (100°C) will usually not destroy such endospores, so canning is carried out in pressurized steam at 120°C for 20 to 30 minutes. Such rigorous conditions ensure that the food is sterile and free from viable bacteria.

4.4 Learning Outcomes—Assess Your Progress

11. Identify five structures that may be contained in bacterial cytoplasm.
12. Detail the causes and mechanisms of sporogenesis and germination.

4.5 The Archaea: The Other "Prokaryotes"

The discovery and characterization of novel cells resembling bacteria that have unusual anatomy, physiology, and genetics changed our views of microbial taxonomy and classification (see chapter 1). These single-celled, simple organisms, called **archaea,** are now considered a third cell type in a separate superkingdom (the Domain Archaea). We include them in this chapter because they share many bacterial characteristics. But it has become clear that they are actually more closely related to Domain Eukarya than to Bacteria. For example, archaea and eukaryotes share a number of ribosomal RNA sequences that are not found in bacteria, and their protein synthesis and ribosomal subunit structures are similar. **Table 4.1** outlines selected points of comparison of the three domains.

Among the ways that the archaea differ significantly from other cell types are that certain genetic sequences are found only in their rRNA, and that they exhibit a unique

Table 4.1 Comparison of Three Cellular Domains			
Characteristic	**Bacteria**	**Archaea**	**Eukarya**
Cell type	Prokaryotic	Prokaryotic	Eukaryotic
Chromosomes	Single, or few, circular	Single, circular	Several, linear
Types of ribosomes	70S	70S but structure is similar to 80S	80S
Contains unique ribosomal RNA signature sequences	+	+	+
Number of sequences shared with Eukarya	1	3	All
Protein synthesis similar to Eukarya	−	+	
Presence of peptidoglycan in cell wall	+	−	−
Cell membrane lipids	Fatty acids with ester linkages	Long-chain, branched hydrocarbons with ether linkages	Fatty acids with ester linkages
Sterols in membrane	− (Some exceptions)	−	+
Pili	N-linked glycans	O-linked glycans	None

method of DNA compaction. They also have unique membrane lipids, cell wall composition, and pilin proteins.

The archaea exhibit unusual and chemically distinct cell walls. In some, the walls are composed almost entirely of polysaccharides, and in others, the walls are pure protein; but as a group, they all lack the true peptidoglycan structure described previously. Because a few archaea lack a cell wall entirely, their cell membrane must serve the dual functions of support and transport.

It is clear that the archaea are the most primitive of all current life forms and are most closely related to the first cells that originated on the earth 4 billion years ago. The early earth is thought to have contained a hot, anaerobic "soup" with sulfuric gases and salts in abundance. The modern archaea still live in the remaining habitats on the earth that have these same ancient conditions—the most extreme habitats in nature. It is for this reason that they are often called extremophiles, meaning that they "love" extreme conditions in the environment.

Metabolically, the archaea exhibit incredible adaptations to what would be deadly conditions for other organisms. These hardy microbes have adapted to multiple combinations of heat, salt, acid, pH, pressure, and atmosphere. Included in this group are methane producers, hyperthermophiles, extreme halophiles, and sulfur reducers.

Members of the group called **methanogens** can convert CO_2 and H_2 into methane gas (CH_4) through unusual and complex pathways. These archaea are common inhabitants of anaerobic swamp mud, the bottom sediments of lakes and oceans, and even the digestive systems of animals. The gas they produce collects in swamps and may become a source of fuel. Methane may also contribute to the "greenhouse effect," which maintains the earth's temperature and can contribute to global warming (see chapter 24).

Other types of archaea—the extreme halophiles—require salt to grow, and some have such a high salt tolerance that they can multiply in sodium chloride solutions (36% NaCl) that would destroy most cells. They exist in the saltiest places

INSIGHT 4.3   CSI: Bacteria?

What if, instead of fingerprints, crime scene investigators looked for patterns of specific bacteria left behind by suspects? Noah Fierer at the University of Colorado, Boulder, studied the variability in bacterial communities on human fingertips and found that this scenario isn't as unlikely as you may think.

Fierer and his colleagues took samples from computer keyboards and computer mice, analyzed the bacterial DNA from the samples, and came up with a bacterial "fingerprint" that could be matched to the individual that had used the keyboard or mouse. His analysis showed that there is only about a 13% correspondence of bacterial species between any two individuals and that these communities of bacteria on the skin are stable over time, recovering themselves within a few hours after washing. Additionally, because skin bacteria are resistant to varying environmental conditions, these bacterial fingerprints can persist on surfaces for up to 2 weeks.

Current forensic analysis requires enough DNA from a crime scene (blood, semen, tissue, or saliva) to amplify, analyze, and match to a suspect. Bacterial cells are much more abundant on skin surfaces, and their DNA is much easier to recover and amplify. Further testing is needed to determine if these methods could actually be feasible for use in forensics, but Fierer brings up an excellent question: "Could our microbial fingerprint be more personally identifying than our human genome?"

Fierer's work is part of a much larger project designed to identify and study the microbes that live on and in our bodies: the Human Microbiome Project (HMP). Using metagenomic techniques developed for the Human Genome Project and J. Craig Venter's quest to discover novel microbes in the oceans, the HMP has been launched to study all of the microbes colonizing the human body. The goal of the HMP is to sequence as many as 3,000 genomes from cultured and uncultured

bacteria as well as other microbes such as archaea, fungi, and protozoa that make up the human microbiome. This preliminary genome collection will serve as a reference to determine whether there is a core microbiome common to all human beings or the degree to which there are commonalities. The main body sites that are being studied are the oral cavity, skin, vagina, gut, and nasal/lung cavities. The HMP will also study the relationship between diseases and changes in the human microbiome.

Could the next clues at a crime scene be bacterial?

Source: 2010. *Proc. Natl. Acad. Sci. U.S.A.* vol. 107, no. 14, p. 6477.

(a) (b)

Figure 4.25 Halophiles around the world. **(a)** Lake Natron in the Great Rift Valley on the border of Tanzania and Kenya. Halophilic algae give the lake its color. **(b)** A sample taken from a saltern in Australia viewed by fluorescent microscopy (1,000×). Note the range of cell shapes (cocci, rods, and squares) found in this community.

on the earth—inland seas, salt lakes, salt mines, and in salted fish. They are not particularly common in the ocean because the salt content is not high enough. Many of the "halobacteria" use a red pigment to synthesize ATP in the presence of light. These pigments are responsible for the color of the Red Sea, and the red color of salt ponds **(figure 4.25).**

Archaea that are adapted to growth at very low temperatures are called **psychrophilic** (loving cold temperatures); those growing at very high temperatures are **hyperthermophilic** (loving high temperatures). Hyperthermophiles flourish at temperatures between 80°C and 113°C and cannot grow at 50°C. They live in volcanic waters and soils and submarine vents and are also often salt- and acid-tolerant as well. One member, *Thermoplasma,* lives in hot, acidic habitats in the waste piles around coal mines that regularly sustain a pH of 1 and a temperature of nearly 60°C. Because many archaea are unculturable, the technology of rRNA sequencing has been invaluable in the identification of these microbes (see section 4.6 and chapter 17). Analysis of these unique sequences has advanced not only the process of identification but also our knowledge of transcription, translation, and cellular evolution.

Archaea are not just environmental microbes. They have been isolated from human tissues such as the colon, the mouth, and the vagina. Recently, an association was found between the degree of severity of periodontal disease and the presence of archaeal RNA sequences in the gingiva, suggesting—but not proving—that archaea may be capable of causing human disease.

4.5 Learning Outcomes—Assess Your Progress

13. List some differences between archaea and bacteria.

4.6 Classification Systems for Bacteria and Archaea

Classification systems serve both practical and academic purposes. They aid in differentiating and identifying unknown species in medical and applied microbiology. They are also useful in organizing microbes and as a means of studying their relationships and origins. Since classification was started around 200 years ago, several thousand species of bacteria and archaea have been identified, named, and cataloged.

For years, scientists have had intense interest in tracing the origins of and evolutionary relationships among bacteria and archaea, but doing so has not been an easy task. One of the questions that has plagued taxonomists is, What characteristics are the most indicative of closeness in ancestry? Early bacteriologists found it convenient to classify bacteria according to shape, variations in arrangement, growth characteristics, and habitat. However, as more species were discovered and as techniques for studying their biochemistry were developed, it soon became clear that similarities in cell shape, arrangement, and staining reactions do not automatically indicate relatedness. Even though the gram-negative rods look alike, there are hundreds of different species, with highly significant differences in biochemistry and genetics. If we attempted to classify them on the basis of Gram stain and shape alone, we could not assign them to a more specific level than class. Increasingly, classification schemes are turning to genetic and molecular traits that cannot be visualized under a microscope or in culture.

One of the most viable indicators of evolutionary relatedness and affiliation is comparison of the sequence of nitrogen bases in ribosomal RNA, a major component of ribosomes.

Ribosomes have the same function (protein synthesis) in all cells, and they tend to remain more or less stable in their nucleic acid content over long periods. Thus, any major differences in the sequence, or "signature," of the rRNA is likely to indicate some distance in ancestry. This technique is powerful at two levels: It is effective for differentiating general group differences, allowing for the creation of branching tree diagrams showing evolutionary relatedness among microbes (see figure 1.14); and it can be fine-tuned for bacterial identification at the species level (for example in *Mycobacterium* and *Legionella*). Elements of these and other identification methods are presented in more detail in chapter 17.

The definitive published source for bacterial and archaea classification, called *Bergey's Manual,* has been in print continuously since 1923. The basis for the early classification in *Bergey's* was the **phenotypic** traits of bacteria, such as their shape, cultural behavior, and biochemical reactions. These traits are still used extensively by clinical microbiologists or researchers who need to quickly identify unknown bacteria. As methods for RNA and DNA analysis became available, this information was used to supplement the phenotypic information. The current version of the publication, called *Bergey's Manual of Systematic Bacteriology,* presents a comprehensive view of bacterial and archaea relatedness, combining phenotypic information with rRNA sequencing information to classify them; it is a huge, five-volume set. (We need to remember that all classification systems are in a state of constant flux; no system is ever finished.)

With the explosion of information about evolutionary relatedness among bacteria, the need for a *Bergey's Manual* that contained easily accessible information for identifying unknown bacteria became apparent. Now there is a separate book, called *Bergey's Manual of Determinative Bacteriology,* based entirely on phenotypic characteristics. It is utilitarian in focus, categorizing bacteria by traits commonly assayed in clinical, teaching, and research labs. It is widely used by microbiologists who need to identify bacteria but need not know their evolutionary backgrounds. This phenotypic classification is more useful for students of medical microbiology, as well.

Taxonomic Scheme

Bergey's Manual of Determinative Bacteriology organizes the bacteria and archaea into four major divisions. These somewhat natural divisions are based on the nature of the cell wall. The **Gracilicutes** (gras"-ih-lik'-yoo-teez) have gram-negative cell walls and thus are thin-skinned; the **Firmicutes** have gram-positive cell walls that are thick and strong; the **Tenericutes** (ten"-er-ik'-yoo-teez) lack a cell wall and thus are soft; and the **Mendosicutes** (men-doh-sik'-yoo-teez) are the archaea. The first two divisions contain the greatest number of species. The 200 or so species that are so-far

known to cause human and animal diseases can be found in four classes: the Scotobacteria, Firmibacteria, Thallobacteria, and Mollicutes. The system used in *Bergey's Manual* organizes bacteria and archaea into subcategories such as classes, orders, and families, but these are not available for all groups.

Diagnostic Scheme

As mentioned earlier, many medical microbiologists prefer an informal working system that outlines the major families and genera. **Table 4.2** is an example of an adaptation of the phenotypic method of classification that may be used in clinical microbiology. This system is more applicable for diagnosis because it is restricted to bacterial disease agents, depends less on nomenclature, and is based on readily accessible morphological and physiological tests rather than on phylogenetic relationships. It also divides the bacteria into gram-positive, gram-negative, and those without cell walls and then subgroups them according to cell shape, arrangement, and certain physiological traits such as oxygen usage. **Aerobic** bacteria use oxygen in metabolism; **anaerobic** bacteria do not use oxygen in metabolism; and **facultative** bacteria may or may not use oxygen. Further tests not listed in the

Table 4.2 Medically Important Families and Genera of Bacteria, with Notes on Some Diseases*

I. Bacteria with Gram-Positive Cell Wall Structure

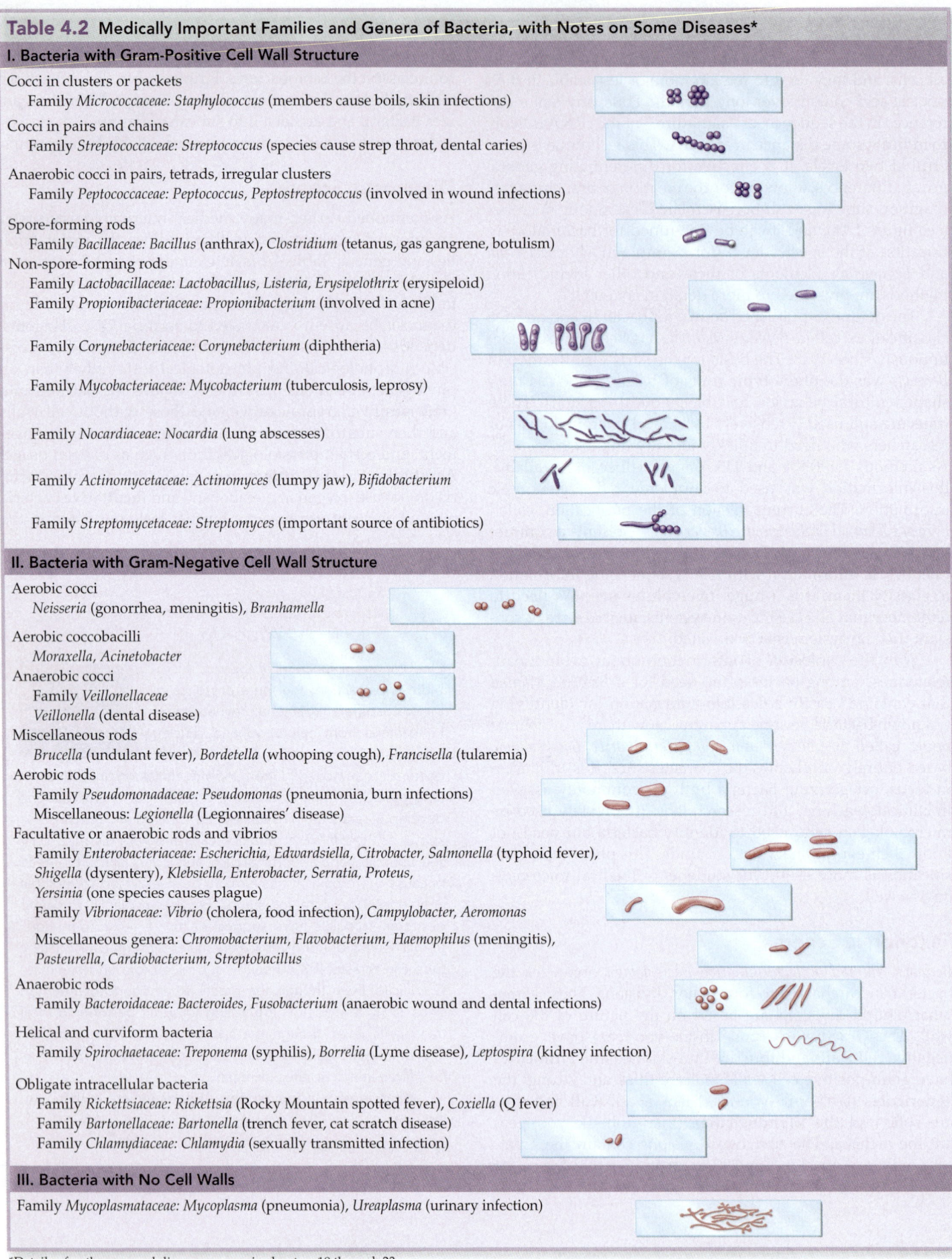

Cocci in clusters or packets
Family *Micrococcaceae*: *Staphylococcus* (members cause boils, skin infections)

Cocci in pairs and chains
Family *Streptococcaceae*: *Streptococcus* (species cause strep throat, dental caries)

Anaerobic cocci in pairs, tetrads, irregular clusters
Family *Peptococcaceae*: *Peptococcus, Peptostreptococcus* (involved in wound infections)

Spore-forming rods
Family *Bacillaceae*: *Bacillus* (anthrax), *Clostridium* (tetanus, gas gangrene, botulism)

Non-spore-forming rods
Family *Lactobacillaceae*: *Lactobacillus, Listeria, Erysipelothrix* (erysipeloid)
Family *Propionibacteriaceae*: *Propionibacterium* (involved in acne)

Family *Corynebacteriaceae*: *Corynebacterium* (diphtheria)

Family *Mycobacteriaceae*: *Mycobacterium* (tuberculosis, leprosy)

Family *Nocardiaceae*: *Nocardia* (lung abscesses)

Family *Actinomycetaceae*: *Actinomyces* (lumpy jaw), *Bifidobacterium*

Family *Streptomycetaceae*: *Streptomyces* (important source of antibiotics)

II. Bacteria with Gram-Negative Cell Wall Structure

Aerobic cocci
Neisseria (gonorrhea, meningitis), *Branhamella*

Aerobic coccobacilli
Moraxella, Acinetobacter

Anaerobic cocci
Family *Veillonellaceae*
Veillonella (dental disease)

Miscellaneous rods
Brucella (undulant fever), *Bordetella* (whooping cough), *Francisella* (tularemia)

Aerobic rods
Family *Pseudomonadaceae*: *Pseudomonas* (pneumonia, burn infections)
Miscellaneous: *Legionella* (Legionnaires' disease)

Facultative or anaerobic rods and vibrios
Family *Enterobacteriaceae*: *Escherichia, Edwardsiella, Citrobacter, Salmonella* (typhoid fever),
Shigella (dysentery), *Klebsiella, Enterobacter, Serratia, Proteus,*
Yersinia (one species causes plague)
Family *Vibrionaceae*: *Vibrio* (cholera, food infection), *Campylobacter, Aeromonas*

Miscellaneous genera: *Chromobacterium, Flavobacterium, Haemophilus* (meningitis),
Pasteurella, Cardiobacterium, Streptobacillus

Anaerobic rods
Family *Bacteroidaceae*: *Bacteroides, Fusobacterium* (anaerobic wound and dental infections)

Helical and curviform bacteria
Family *Spirochaetaceae*: *Treponema* (syphilis), *Borrelia* (Lyme disease), *Leptospira* (kidney infection)

Obligate intracellular bacteria
Family *Rickettsiaceae*: *Rickettsia* (Rocky Mountain spotted fever), *Coxiella* (Q fever)
Family *Bartonellaceae*: *Bartonella* (trench fever, cat scratch disease)
Family *Chlamydiaceae*: *Chlamydia* (sexually transmitted infection)

III. Bacteria with No Cell Walls

Family *Mycoplasmataceae*: *Mycoplasma* (pneumonia), *Ureaplasma* (urinary infection)

*Details of pathogens and diseases appear in chapters 18 through 23.

table would be required to separate closely related genera and species. Many of these are included in later chapters on specific bacterial groups.

Species and Subspecies in Bacteria and Archaea

Among most organisms, the species level is a distinct, readily defined, and natural taxonomic category. In animals, for instance, a species is a distinct type of organism that can produce viable offspring only when it mates with others of its own kind. This definition does not work for bacteria and archaea primarily because they do not exhibit a typical mode of sexual reproduction. They can accept genetic information from unrelated forms, and they can also alter their genetic makeup by a variety of mechanisms. Thus, it is necessary to hedge a bit when we define a bacterial species. Theoretically, it is a collection of bacterial cells, all of which share an overall similar pattern of traits, in contrast to other groups whose patterns differ significantly. Although the boundaries that separate two closely related species in a genus are in some cases arbitrary, this definition still serves as a method to separate the bacteria into various kinds that can be cultured and studied.

Individual members of a given species can show variations, as well. Therefore, more categories within species exist, but they are not well defined. Microbiologists use terms like **subspecies**, **strain**, or **type** to designate bacteria of the same species that have differing characteristics. **Serotype** refers to representatives of a species that stimulate a distinct pattern of antibody (serum) responses in their hosts because of distinct surface molecules.

4.6 Learning Outcomes—Assess Your Progress

14. Differentiate between *Bergey's Manual of Systematic Bacteriology* and *Bergey's Manual of Determinative Bacteriology*.

15. Name four divisions ending in –*cutes* and describe their characteristics.

16. Define a *species* in terms of bacteria.

Chapter Summary

4.1 Bacterial Form and Function (ASM Guideline* 2.4)

- Bacteria and archaea are ancient forms of cellular life. They are also the most widely dispersed, occupying every conceivable niche on the planet.
- Most bacteria and archaea have one of three general shapes: coccus (round), bacillus (rod), or spiral, based on the configuration of the cell wall. Two types of spiral cells are the spirochetes and the spirilla.
- Shape and arrangement of cells are key means of describing bacteria and archaea. Arrangements of cells are based on the number of planes in which a given species divides.
- Cocci can divide in many planes to form pairs, chains, packets, or clusters. Bacilli divide only in the transverse plane. If they remain attached, they form chains or palisades.

4.2 External Structures (ASM Guidelines 2.1, 2.2, 2.4)

- The external structures of bacteria include appendages (flagella, fimbriae, and pili) and surface coatings (the S layer and the glycocalyx).
- Flagella vary in number and arrangement as well as in the type and rate of motion they produce.

4.3 The Cell Envelope: The Boundary Layer of Bacteria (ASM Guidelines 2.1, 2.4, 3.4, 8.1)

- The cell envelope is the complex boundary structure surrounding a bacterial cell. In gram-negative bacteria, 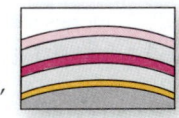 the envelope consists of an outer membrane, the cell wall, and the cell membrane. Gram-positive bacteria have only the cell wall and cell membrane.
- In a Gram stain, gram-positive bacteria retain the crystal violet and stain purple. Gram-negative bacteria lose the crystal violet and stain red from the safranin counterstain.
- Gram-positive bacteria have thick cell walls of peptidoglycan and acidic polysaccharides such as teichoic acid. The cell walls of gram-negative bacteria are thinner and have a wide periplasmic space.
- The outer membrane of gram-negative cells contains lipopolysaccharide (LPS). LPS is toxic to mammalian hosts.
- The bacterial cell membrane is typically composed of phospholipids and proteins, and it performs many metabolic functions as well as transport activities.

4.4 Bacterial Internal Structure (ASM Guidelines 1.1, 2.1, 2.2, 2.4, 3.4, 4.2, 5.4)

- The cytoplasm of bacterial cells serves as a solvent for materials used in all cell functions.
- The genetic material of bacteria is DNA. Genes are arranged on large, circular chromosomes. Additional genes are carried on plasmids.
- Bacterial ribosomes are dispersed in the cytoplasm in chains (polysomes) and are also embedded in the cell membrane.
- Bacteria may store nutrients in their cytoplasm in structures called *inclusions*. Inclusions vary in structure and the materials that are stored.
- Packets in the cytoplasm called *bacterial microcompartments* are shells of protein packed with enzymes.

ASM Curriculum Guidelines (American Society for Microbiology, 2012). Complete guidelines in appendix B of this book.

- Some bacteria manufacture long actin- and tubulin-like filaments that help determine their cellular shape.
- A few families of bacteria produce dormant bodies called *endospores,* which are the hardiest of all life forms, surviving for hundreds or thousands of years.
- The genera *Bacillus* and *Clostridium* are endospore formers, and both contain deadly pathogens.

4.5 The Archaea: The Other "Prokaryotes" (ASM Guidelines 1.1, 1.4, 1.5, 4.2)

- Archaea constitute the third domain of life. They exhibit unusual biochemistry and genetics that make them different from bacteria. Many members are adapted to extreme habitats with low or high temperature, salt, pressure, or acid.

4.6 Classification Systems for Bacteria and Archaea (ASM Guidelines 1.1, 1.4, 1.5)

- Bacteria and archaea are formally classified by phylogenetic relationships and phenotypic characteristics.

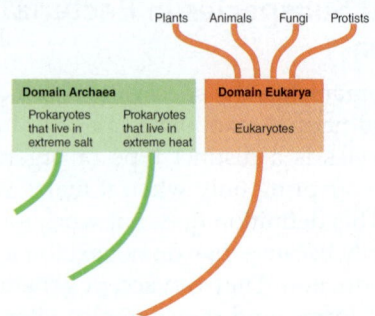

- Medical identification of pathogens uses an informal system of classification based on Gram stain, morphology, biochemical reactions, and metabolic requirements.
- A *bacterial species* is loosely defined as a collection of bacterial cells that share an overall similar pattern of traits different from other groups of bacteria.
- Variant forms within a species (subspecies) include strains and types.

Multiple-Choice and True-False Questions | Bloom's Levels 1 and 2: Remember and Understand

Multiple-Choice Questions. Select the correct answer from the options provided.

1. Which of the following is not found in all bacterial cells?
   a. cell membrane
   b. a nucleoid
   c. ribosomes
   d. actinlike cytoskeleton

2. Pili are tubular shafts in _____ bacteria that serve as a means of _____.
   a. gram-positive, genetic exchange
   b. gram-positive, attachment
   c. gram-negative, genetic exchange
   d. gram-negative, protection

3. An example of a glycocalyx is
   a. a capsule.
   b. a pilus.
   c. an outer membrane.
   d. a cell wall.

4. Which of the following is a primary bacterial cell wall function?
   a. transport
   b. motility
   c. support
   d. adhesion

5. Which of the following is present in both gram-positive and gram-negative cell walls?
   a. an outer membrane
   b. peptidoglycan
   c. teichoic acid
   d. lipopolysaccharides

6. Darkly stained granules are concentrated crystals of _____ that are found in _____.
   a. fat, *Mycobacterium*
   b. dipicolinic acid, *Bacillus*
   c. sulfur, *Thiobacillus*
   d. PO_4, *Corynebacterium*

7. Bacterial endospores usually function in
   a. reproduction.
   b. survival.
   c. protein synthesis.
   d. storage.

8. A bacterial arrangement in packets of eight cells is described as a _____.
   a. micrococcus
   b. diplococcus
   c. tetrad
   d. sarcina

9. To which division of bacteria does *E. coli* belong?
   a. Tenericutes
   b. Gracilicutes
   c. Firmicutes
   d. Mendosicutes

10. Which stain is used to distinguish differences between the cell walls of medically important bacteria?
    a. simple stain
    b. acridine orange stain
    c. Gram stain
    d. negative stain

True-False Questions. If the statement is true, leave as is. If it is false, correct it by rewriting the sentence.

11. One major difference in the envelope structure between gram-positive bacteria and gram-negative bacteria is the presence or absence of a cytoplasmic membrane.

12. A research microbiologist looking at evolutionary relatedness between two bacterial species is more likely to use *Bergey's Manual of Determinative Bacteriology* than *Bergey's Manual of Systematic Bacteriology.*

13. Nanobes may or may not actually be bacteria.

14. Both bacteria and archaea used to be known as prokaryotes.

15. A collection of bacteria that share an overall similar pattern of traits is called a *species.*

Critical Thinking Questions | Bloom's Levels 3, 4, and 5: Apply, Analyze, and Evaluate

Critical thinking is the ability to reason and solve problems using facts and concepts. These questions can be approached from a number of angles and, in most cases, they do not have a single correct answer.

1. Define the term *ubiquitous* and explain whether this term can be used appropriately to describe bacteria and archaea.

2. Draw a picture to illustrate how bacteria and archaea can be distinguished from eukaryotic cells and to provide details on the general characteristics of these cell types.

3. Quorum sensing is a process used by many bacteria for communication. It involves the production of molecules called *autoinducers,* which act as bacterial chemoattractants. Describe how a motile bacterium would use its flagellum to respond to such a stimulus in its environment.

4. Provide examples of at least three different structures used for bacterial attachment and explain which are useful in the formation of biofilms.

5. You are a scientist viewing a Gram-stained slide of a bacterial culture obtained from a patient. You note purple spherical cells that are attached to one another in groups of four. Make a sketch of this bacterium; using proper terminology from this chapter, describe the morphology and arrangement of this bacterium as well as its cell wall composition.

6. Based upon your knowledge of cell wall structure, explain how the microbes causing meningitis and typhoid fever can induce fever and systemic shock in an infected patient.

7. Provide evidence in support of or refuting the following statement: The cell, or cytoplasmic, membrane is a nonessential structure in bacteria because its function is replaced by the cell wall in these microbes.

8. a. Describe the characteristics of an endospore-producing bacterium that make it an ideal candidate for bioterrorism but an undesirable intruder in a hospital setting.

   b. Explain why the production of endospores is not considered a method of reproduction in most bacterial species.

9. Compare and contrast main characteristics of the three domains of life: Bacteria, Archaea, and Eukarya.

   a. Create a branched-tree diagram showing the evolutionary relationship among the three domains of life, and list the characteristics that have led microbiologists to believe the archaea are more closely related to eukaryotes than to bacteria.

   b. Which archaeal adaptations make these microbes most suited to extreme habitats?

10. A microbe has been found in the boiling hot waters of a deep ocean hydrothermal vent. It cannot be readily stained or cultured in the laboratory, but its rRNA has just been sequenced and analyzed.

    a. Make a hypothesis regarding the domain of life to which this microbe most likely belongs.

    b. Explain which edition of *Bergey's Manual* should be used in this case to determine the identity of this microbe.

    c. Explain whether the rRNA sequence information is enough to define the species of a microbe.

Concept Connections | Bloom's Levels 4 and 6: Analyze and Create

This activity ties together multiple concepts in the chapter.

1. Name at least five structures that are not found in all bacteria but are important in some.

2. What characteristics are unique to gram-positive bacteria? What characteristics are unique to gram-negative bacteria? What characteristics are common to both?

3. What is the clinical importance of identifying the Gram reaction of an organism cultured from a patient sample?

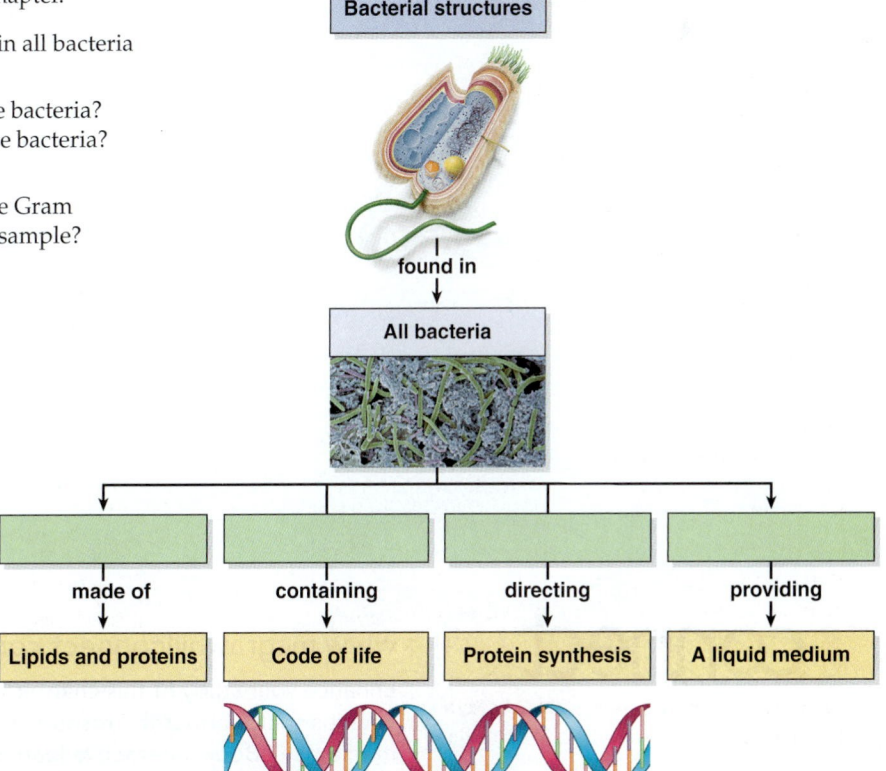

Bacterial structures

found in

All bacteria

made of	containing	directing	providing
Lipids and proteins	Code of life	Protein synthesis	A liquid medium

Visual Connections | Bloom's Level 5: Evaluate

These questions use visual images or previous content to make connections to this chapter's concepts.

1. **From chapter 3, figure 3.6*b*.** Do you believe that the bacteria spelling *"Klebsiella"* or the bacteria spelling *"S. aureus"* possess the larger capsule? Defend your answer.

2. **From chapter 1, figure 1.14.** Study this figure. How would it be drawn differently if the archaea were more closely related to bacteria than to eukaryotes?

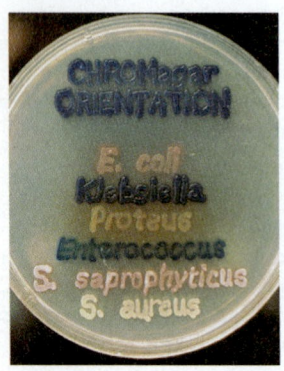

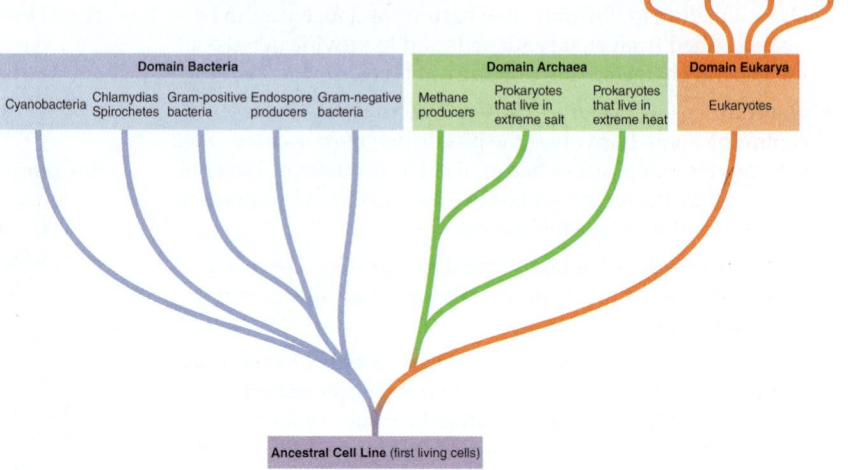

Concept Mapping | Bloom's Level 6: Create

Appendix D provides guidance for working with concept maps.

1. Using the words that follow, please create a concept map illustrating the relationships among these key terms from chapter 4.

genus	serotype	*Borrelia burgdorferi*
species	domain	spirochete

www.mcgrawhillconnect.com

Enhance your study of this chapter with study tools and practice tests. Also ask your instructor about the resources available through ConnectPlus, including the media-rich eBook, interactive learning tools, and animations.

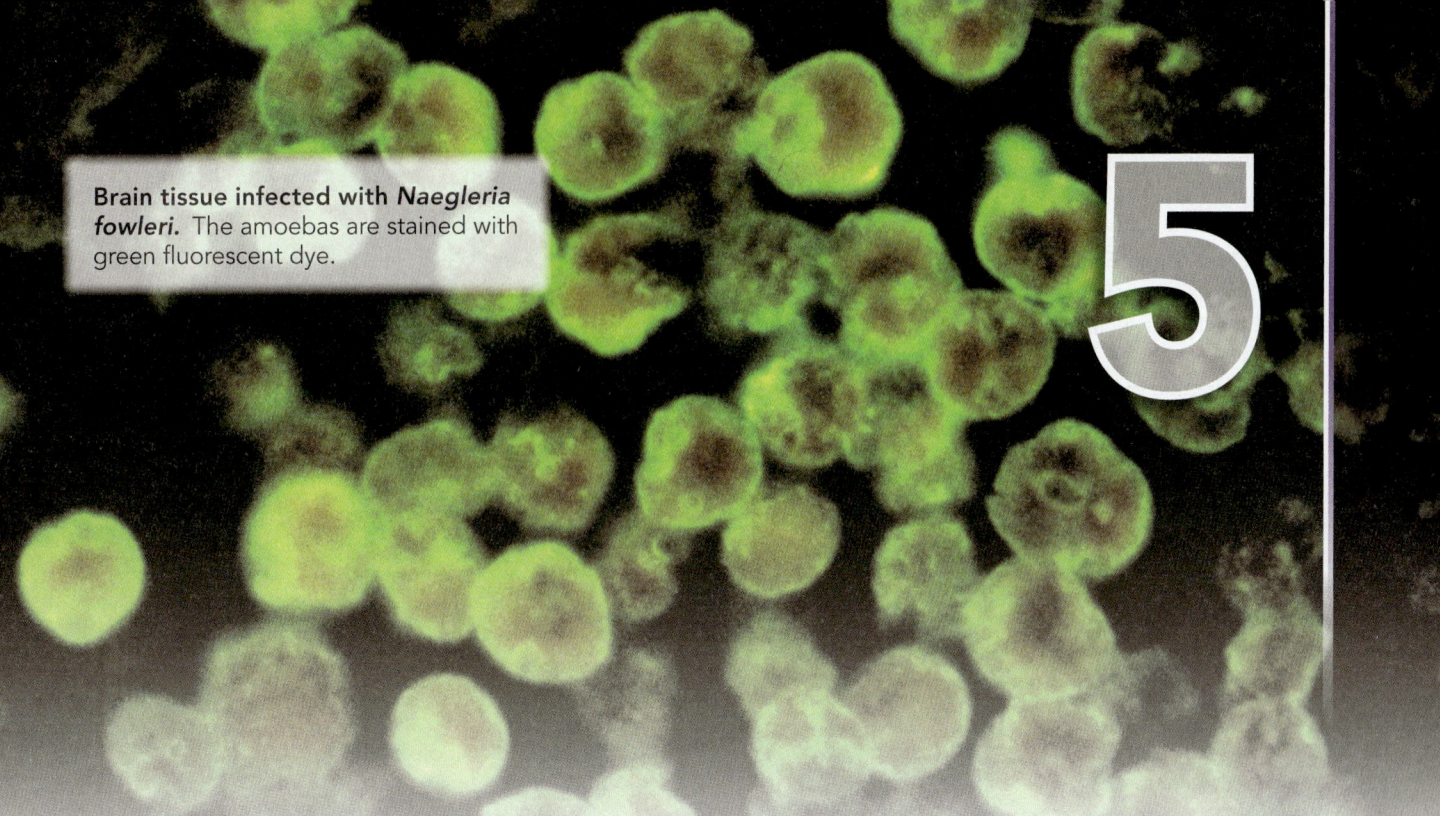

Brain tissue infected with *Naegleria fowleri.* The amoebas are stained with green fluorescent dye.

Eukaryotic Cells and Microorganisms

Case File 5

Swim at Your Own Risk

Ah, the dog days of summer. Nothing is more refreshing than diving into the local fishing hole to cool off, but swim at your own risk. Warm weather and rising water temperatures may bring an unwelcome swimming partner: *Naegleria fowleri.* This **eukaryotic** amoeba is found in freshwater lakes, swimming pools without adequate chlorination, geothermal pools, and even tap water. Usually, the organism lives in the sediment at the bottom of lakes and in the soil. During the colder times of the year, the organism exists in a cyst form, in which the organism is dormant and covered in a durable coating. In the warmer months of the year, *Naegleria* enters a **trophozoite**, or amoebic, phase in which it actively divides and can be infectious.

In August 2010, Kyle Lewis and his family were camping in Glen Rose, Texas, where he swam in several warm lakes and rivers. When he got home, he developed a severe headache, then fever, nausea, and vomiting. His parents didn't know the cause of his symptoms, which continued to worsen. Over the next few days, he developed symptoms of meningitis, began having hallucinations and fevers, and was hospitalized. His condition got progressively worse. Kyle later died.

- How is *Naegleria fowleri* transmitted?
- How can infections be prevented?

Continuing the Case appears on page 129.

Outline and Learning Outcomes

5.1 The History of Eukaryotes
  1. Relate bacterial, archaeal, and eukaryotic cells to the Last Common Ancestor.
  2. List the types of eukaryotic microorganisms and denote which are unicellular and which are multicellular.

5.2 Form and Function of the Eukaryotic Cell: External Structures and Boundary Structures
  3. Differentiate between cilia and flagella in eukaryotes, and differentiate flagellar structure among bacteria, archaea, and eukaryotes.

4. Describe the important characteristics of a glycocalyx in eukaryotes.

5. List which eukaryotic microorganisms have a cell wall.

6. List similarities and differences between eukaryotic and bacterial cytoplasmic membranes.

5.3 Form and Function of the Eukaryotic Cell: Internal Structures

7. Describe the main structural components of a nucleus.

8. Diagram how the nucleus, endoplasmic reticulum, and Golgi apparatus act together with vesicles during the transport process.

9. Explain the function of the mitochondrion.

10. Discuss the function of chloroplasts, explaining which cells contain them and how they arose.

11. Explain the importance of ribosomes and differentiate between eukaryotic and bacterial types.

12. List and describe the three main fibers of the cytoskeleton.

5.4 The Fungi

13. List three general features of fungal anatomy.

14. Differentiate among the terms *heterotroph*, *saprobe*, and *parasite*.

15. Explain the relationship between fungal hyphae and the production of a mycelium.

16. Describe two ways in which fungal spores arise.

17. List two detrimental and two beneficial activities of fungi (from the viewpoint of humans).

5.5 The Protists

18. Note the protozoan characteristics that illustrate why they are informally placed into a single group.

19. List three means of locomotion exhibited by protozoa.

20. Explain why a cyst stage may be useful in a protozoan.

21. Give an example of a human disease caused by each of the four types of protozoa.

5.6 The Helminths

22. List the two major groups of helminths and provide examples representing each body type.

23. Summarize the stages of a typical helminth life cycle.

5.1 The History of Eukaryotes

Evidence from paleontology indicates that the first eukaryotic cells appeared on the earth approximately 2 billion to 3 billion years ago. Some fossilized cells that look remarkably like modern-day algae or protozoa appear in shale sediments from China, Russia, and Australia that date from 850 million to 950 million years ago **(figure 5.1).** While it used to be thought that eukaryotic cells evolved directly from ancient "prokaryotic" cells, we now believe that bacteria, archaea, and eukaryotes evolved from a different kind of cell, a precursor to both prokaryotes and eukaryotes that biologists call the Last Common Ancestor. This ancestor was neither prokaryotic nor eukaryotic but gave rise to bacteria, archaea, and eukarya separately. It now seems clear that some of the **organelles** inside eukaryotic cells originated from more primitive cells that became trapped in them **(Insight 5.1).** The structure of these first eukaryotic cells was so versatile that eukaryotic microorganisms soon spread out into available habitats and adopted greatly diverse styles of living.

The first primitive eukaryotes were probably single-celled and independent, but, over time, some forms began to aggregate, forming colonies. With further evolution, some of the cells within colonies became *specialized,* or adapted to perform

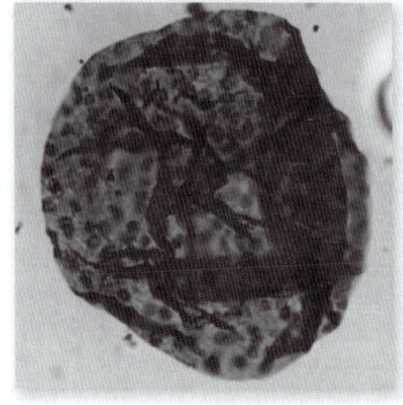

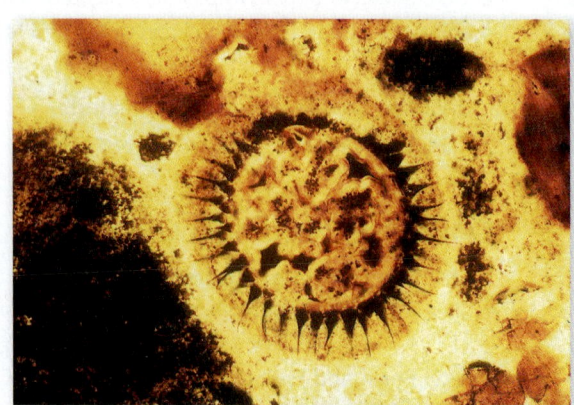

Figure 5.1 Ancient eukaryotic protists caught up in fossilized rocks. **(a)** An alga-like cell found in Siberian shale deposits and dated from 850 million to 950 million years ago. **(b)** A large, disclike cell bearing a crown of spines is from Chinese rock dated 590 million to 610 million years ago.

(a)   (b)

For years, biologists have grappled with the problem of how a cell as complex as the modern eukaryotic cell originated. The explanation seems to be **endosymbiosis,** which suggests that over 2 billion years ago eukaryotic cells arose when a very large precursor cell engulfed small bacterial cells that began to live and reproduce inside the large cell rather than being destroyed. As the smaller cells took up permanent residence, they came to perform specialized functions for the larger cell, from (perhaps) serving as a nucleus to performing functions such as food synthesis and oxygen utilization. Many of these endosymbionts enhanced the cell's versatility and survival. Over time, the engulfed bacterium gave up its ability to live independently and transferred some of its genes to the host cell.

The biologist responsible for early consideration of the theory of endosymbiosis is Dr. Lynn Margulis. Using molecular techniques, she accumulated convincing evidence of the relationships between the organelles of modern eukaryotic cells and the structure of bacteria. In many ways, the mitochondrion of eukaryotic cells is something like a tiny cell within a cell. It is capable of independent division, contains a circular chromosome that has bacterial DNA sequences, and has ribosomes that are clearly bacterial. Mitochondria also have bacterial membranes and can be inhibited by drugs that affect only bacteria.

Chloroplasts likely arose when endosymbiotic cyanobacteria provided their host cells with a built-in feeding mechanism. Margulis also found convincing evidence that eukaryotic cilia are the consequence of endosymbiosis between spiral bacteria and the cell membrane of early eukaryotic cells.

As molecular techniques improve, more evidence accumulates for the endosymbiont theory, which is now widely accepted among evolutionary scientists. Recently, scientists have even discovered a model organism in which they can study the earliest stages of chloroplast development. *Paulinella chromatophora* is an amoeba that contains photosynthetic organelles called chromatophores that evolved fairly recently in time—60 million years ago. Researchers have been able to observe the processes of how genes were incorporated and how proteins were transported into the newly derived chromatophores in real time. This has provided new insight into the genesis of organelles and additional molecular evidence supporting the endosymbiotic theory.

Larger Precursor Cell	Smaller Prokaryotic Cell

Cell would have flexible membrane.

Early nucleus

Larger cell engulfs smaller one; smaller one survives and begins an endosymbiotic association.

Smaller bacterium becomes established in its host's cytoplasm and multiplies; it can utilize aerobic metabolism and increase energy availability for the host.

Early endoplasmic reticulum

Nuclear envelope

Early mitochondria

Ancestral eukaryotic cell develops extensive membrane pouches that become the endoplasmic reticulum and nuclear envelope.

Photosynthetic bacteria (cyanobacteria) are also engulfed; they develop into chloroplasts.

Ancestral cell

Chloroplast

(a) Dr. Lynn Margulis

(b) Protozoa, fungi, animals

Algae, higher plants

Table 5.1	Eukaryotic Organisms Studied in Microbiology	
Always Unicellular	**May Be Unicellular or Multicellular**	**Always Multicellular**
Protozoa	Fungi	Helminths (have unicellular egg or larval forms)
	Algae	

a particular function advantageous to the whole colony, such as locomotion, feeding, or reproduction. Complex multicellular organisms evolved as individual cells in the organism lost the ability to survive apart from the intact colony. Although a multicellular organism is composed of many cells, it is more than just a disorganized assemblage of cells like a colony. Rather, it is composed of distinct groups of cells that cannot exist independently of the rest of the body. The cell groupings of multicellular organisms that have a specific function are termed *tissues*, and groups of tissues make up *organs*.

Looking at modern eukaryotic organisms, we find examples of many levels of cellular complexity **(table 5.1)**. All protozoa, as well as numerous algae and fungi, are unicellular. Truly multicellular organisms are found only among plants and animals and some of the fungi (mushrooms) and algae (seaweeds). Only certain eukaryotes are traditionally studied by microbiologists—primarily the protozoa, the microscopic algae and fungi, and animal parasites, or helminths.

5.1 Learning Outcomes—Assess Your Progress

1. Relate bacterial, archaeal, and eukaryotic cells to the Last Common Ancestor.
2. List the types of eukaryotic microorganisms and denote which are unicellular and which are multicellular.

5.2 Form and Function of the Eukaryotic Cell: External Structures and Boundary Structures

The cells of eukaryotic organisms are so varied that no one member can serve as a typical example. **Figure 5.2** presents the generalized structure of typical algal, fungal, and protozoan cells. The flowchart on this page shows the organization of a eukaryotic cell and compares it to the organization for bacterial cells that you already saw in chapter 4.

In general, eukaryotic microbial cells have a cytoplasmic membrane, nucleus, mitochondria, endoplasmic reticulum, Golgi apparatus, vacuoles, cytoskeleton, and glycocalyx. A cell wall, locomotor appendages, and chloroplasts are found only in some groups. In the following sections, we cover the microscopic structure and functions of the eukaryotic cell. As with the bacteria, we begin on the outside and proceed inward through the cell.

Locomotor Appendages: Cilia and Flagella

Motility allows a microorganism to locate nutrients and to migrate toward positive stimuli such as sunlight. It also permits microorganisms to avoid harmful substances and stimuli.

Although they share the same name, eukaryotic flagella are much different from those of bacteria and archaea. The eukaryotic flagellum is thicker (by a factor of 10), structurally more complex, and covered by an extension of the cell membrane. A single flagellum is a long, sheathed cylinder containing regularly spaced hollow tubules—microtubules— that extend along its entire length **(figure 5.3a)**. A cross section reveals nine pairs of closely attached microtubules surrounding a single central pair. This scheme, called the 9 + 2 arrangement, is the pattern of eukaryotic flagella and cilia **(figure 5.3b)**. During locomotion, the adjacent microtubules slide past each other, whipping the flagellum back and forth. Although details of this process are too complex to discuss here, it involves expenditure of energy and a coordinating mechanism in the cell membrane. The placement and number of flagella can be useful in identifying flagellated protozoa and certain algae.

Disease Connection

As far as we know, only one ciliated protozoan causes disease in humans. *Balantidium coli* makes its home in the intestines of pigs and other mammals. If it gets transmitted to the human digestive tract, it can cause a disease called balantidiasis, which is a diarrheal disease.

Cilia are very similar in overall architecture to flagella, but they are shorter and more numerous (some cells have several thousand). They are found only on a single group of

Structure Flowchart

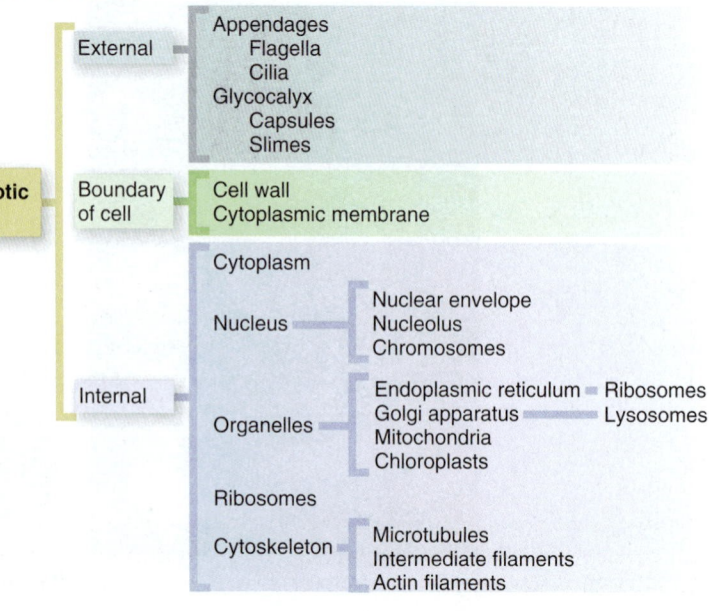

In All Eukaryotes

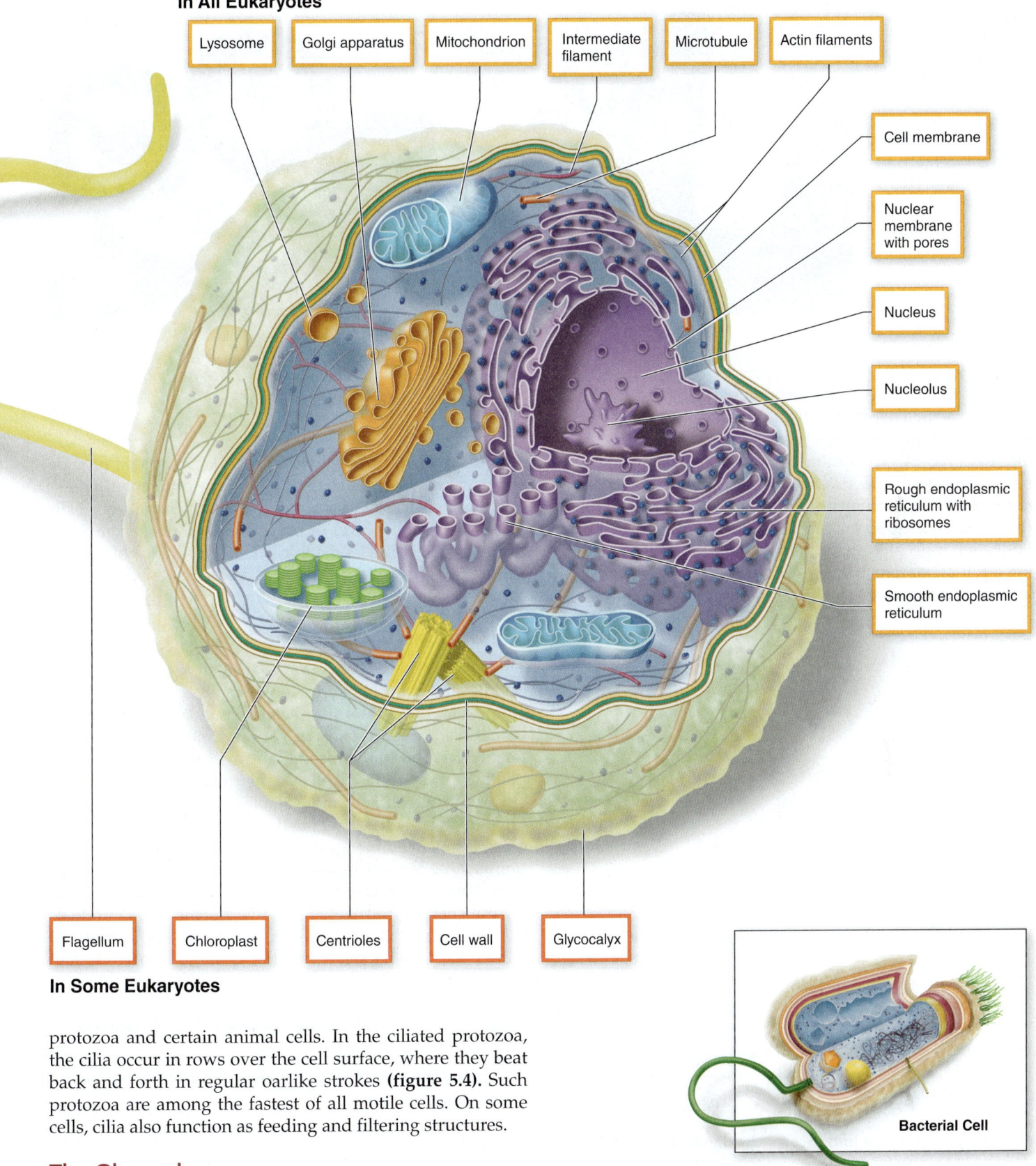

| Lysosome | Golgi apparatus | Mitochondrion | Intermediate filament | Microtubule | Actin filaments |

Cell membrane

Nuclear membrane with pores

Nucleus

Nucleolus

Rough endoplasmic reticulum with ribosomes

Smooth endoplasmic reticulum

| Flagellum | Chloroplast | Centrioles | Cell wall | Glycocalyx |

In Some Eukaryotes

protozoa and certain animal cells. In the ciliated protozoa, the cilia occur in rows over the cell surface, where they beat back and forth in regular oarlike strokes **(figure 5.4).** Such protozoa are among the fastest of all motile cells. On some cells, cilia also function as feeding and filtering structures.

The Glycocalyx

Most eukaryotic cells have a **glycocalyx,** an outermost boundary that comes into direct contact with the environment (see figure 5.2). This structure, which is sometimes called an

Bacterial Cell

Figure 5.2 Structure of a eukaryotic cell. The figure of a bacterial cell from chapter 4 is included here for comparison.

Figure 5.3
Microtubules in flagella.
(a) Longitudinal section through a flagellum, showing microtubules.
(b) A cross section that reveals the typical 9 + 2 arrangement found in both flagella and cilia.

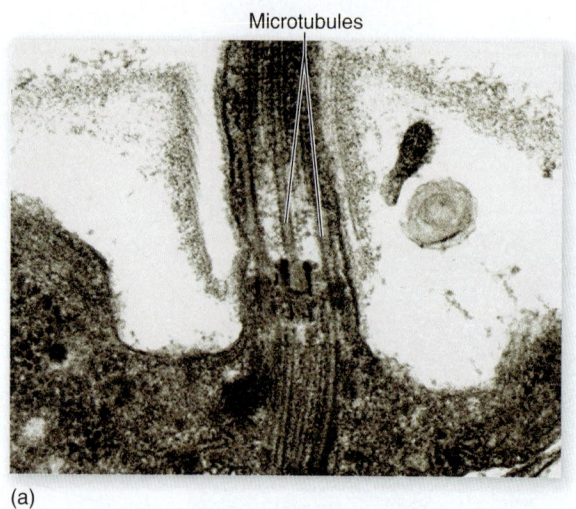

Microtubules

(a)

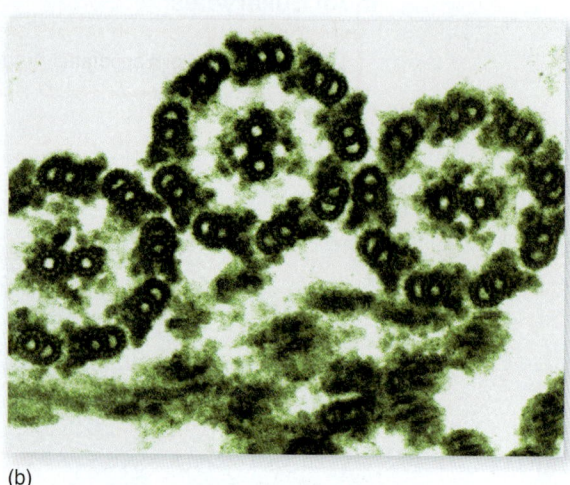

(b)

extracellular matrix, is usually composed of polysaccharides and appears as a network of fibers, a slime layer, or a capsule much like the glycocalyx of bacteria. Because of its positioning, the glycocalyx contributes to protection, adherence of cells to surfaces, and reception of signals from other cells and from the environment. The nature of the layer beneath the glycocalyx varies among the several eukaryotic groups. Fungi and most algae have a thick, rigid cell wall surrounding a cell membrane, whereas protozoa, a few algae, and all animal cells lack a cell wall and have only a cell membrane.

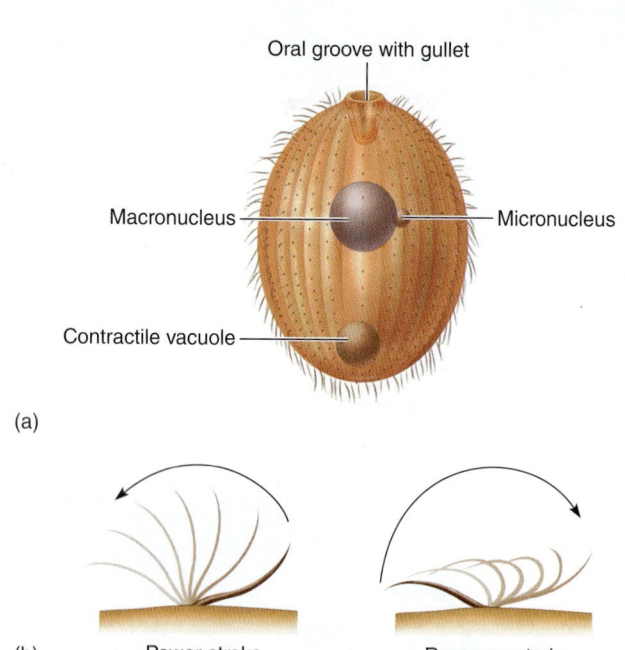

Oral groove with gullet

Macronucleus — — Micronucleus

Contractile vacuole

(a)

(b)   Power stroke   Recovery stroke

Figure 5.4 Structure and locomotion in ciliates. **(a)** The structure of a simple representative, *Holophrya*, with a regular pattern of cilia in rows. **(b)** Cilia beat in coordinated waves, driving the cell forward and backward. View of a single cilium shows that it has a pattern of movement like a swimmer, with a power forward stroke and a repositioning stroke.

Boundary Structures

The Cell Wall

Fungi and algae have cell walls. They are rigid and provide structural support and shape, but they are different in chemical composition from bacterial cell walls. Fungal cell walls have a thick, inner layer of polysaccharide fibers composed of chitin or cellulose and a thin outer layer of mixed glycans **(figure 5.5)**. The cell walls of algae are quite varied in chemical composition.

The Cytoplasmic Membrane

The cytoplasmic (cell) membrane of eukaryotic cells is a typical bilayer of phospholipids in which protein molecules are embedded. In addition to phospholipids, eukaryotic membranes also contain *sterols* of various kinds. Sterols are different from phospholipids in both structure and behavior, as you may recall from chapter 2. Their relative rigidity makes eukaryotic membranes more stable. This strengthening feature is extremely important in those cells that lack a cell wall. Cytoplasmic membranes of eukaryotes have the same function as those of bacteria, serving as selectively permeable barriers. Membranes have extremely sophisticated mechanisms for transporting nutrients *in* and waste and other products *out*. You'll read about these transport systems in bacterial membranes in chapter 7, but the systems in bacteria and eukaryotes are very similar.

5.2 Learning Outcomes—Assess Your Progress

3. Differentiate between cilia and flagella in eukaryotes, and differentiate flagellar structure between bacteria and eukaryotes.

4. Describe the important characteristics of a glycocalyx in eukaryotes.

5. List which eukaryotic microorganisms have a cell wall.

6. List similarities and differences between eukaryotic and bacterial cytoplasmic membranes.

Cell Wall

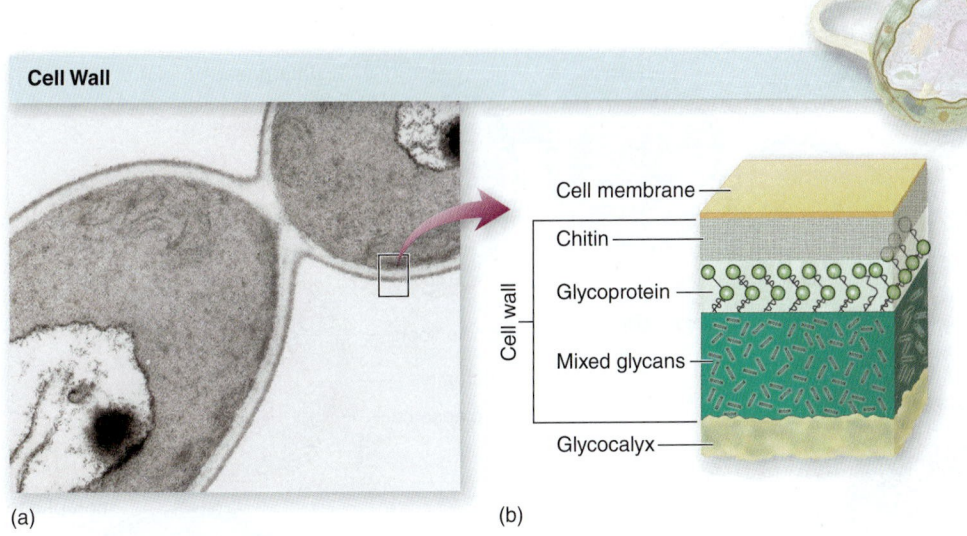

Cell membrane
Chitin
Glycoprotein
Mixed glycans
Glycocalyx

Cell wall

(a)   (b)

Figure 5.5 Cross-sectional views of fungal cell walls. (a) An electron micrograph of two fungal cells. (b) A drawing of the section of the wall inside the square in part (a).

5.3 Form and Function of the Eukaryotic Cell: Internal Structures

Unlike bacteria and archaea, eukaryotic cells contain a number of individual membrane-bound organelles that are extensive enough to account for 60% to 80% of their volume.

The Nucleus: The Control Center

The nucleus is a compact sphere that is the most prominent organelle of eukaryotic cells. It is separated from the cell cytoplasm by an external boundary called a *nuclear envelope.* The envelope has a unique architecture. It is composed of two parallel membranes separated by a narrow space, and it is perforated with small, regularly spaced openings, or pores, formed at sites where the two membranes unite **(figure 5.6).** The nuclear pores are passageways through which macromolecules migrate from the nucleus to the cytoplasm, and

vice versa. The nucleus contains an inner substance called the *nucleoplasm* and a granular mass, the **nucleolus,** that stains more intensely than the immediate surroundings because of its RNA content. The nucleolus is the site for ribosomal RNA synthesis and a collection area for ribosomal subunits. The subunits are transported through the nuclear pores into the cytoplasm for final assembly into ribosomes.

A prominent feature of the nucleoplasm in stained preparations is a network of dark fibers known as **chromatin.** Analysis has shown that chromatin makes up the eukaryotic **chromosomes,** large units of genetic information in the cell. The chromosomes in the nucleus of most cells are not readily visible because they are long, linear DNA molecules bound in varying degrees to **histone** proteins, and they are far too fine to be resolved as distinct structures without extremely high magnification. During **mitosis,** however, when the duplicated chromosomes are separated equally into daughter cells, the chromosomes themselves become readily visible as discrete

Nucleus

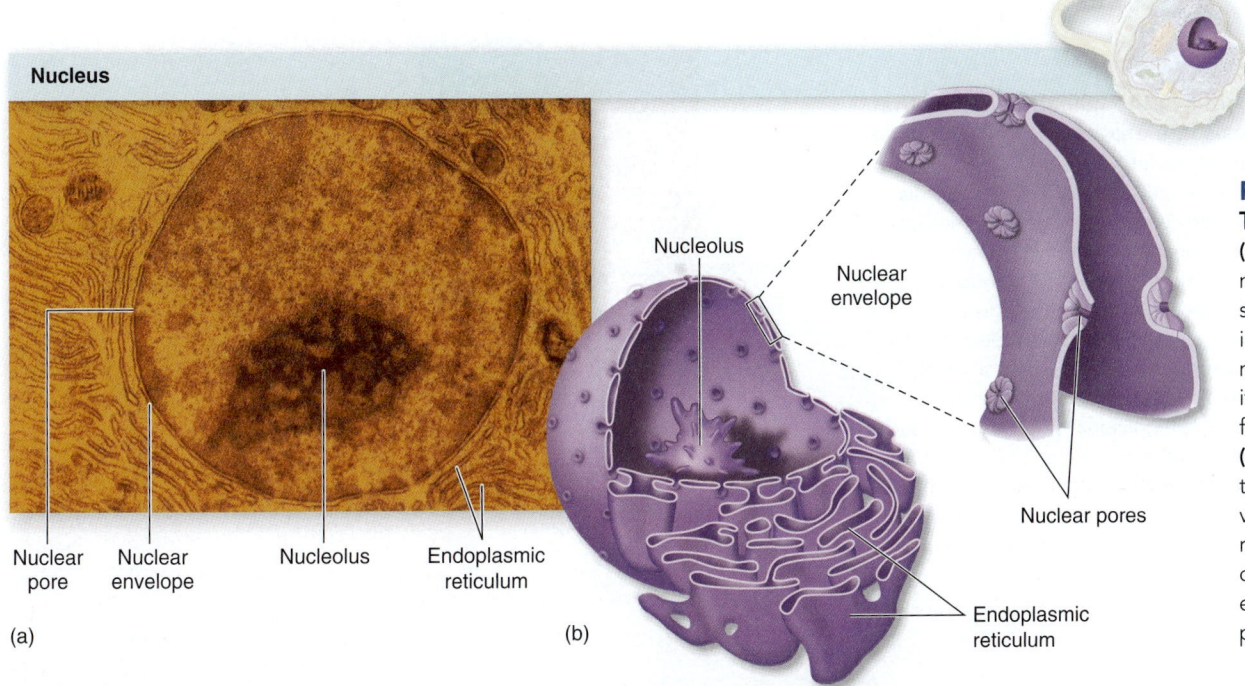

Nucleolus
Nuclear envelope
Nuclear pores
Endoplasmic reticulum

Nuclear pore   Nuclear envelope   Nucleolus   Endoplasmic reticulum

(a)   (b)

Figure 5.6 The nucleus. (a) Electron micrograph section of an interphase nucleus, showing its most prominent features. (b) Cutaway three-dimensional view of the relationships of the nuclear envelope and pores.

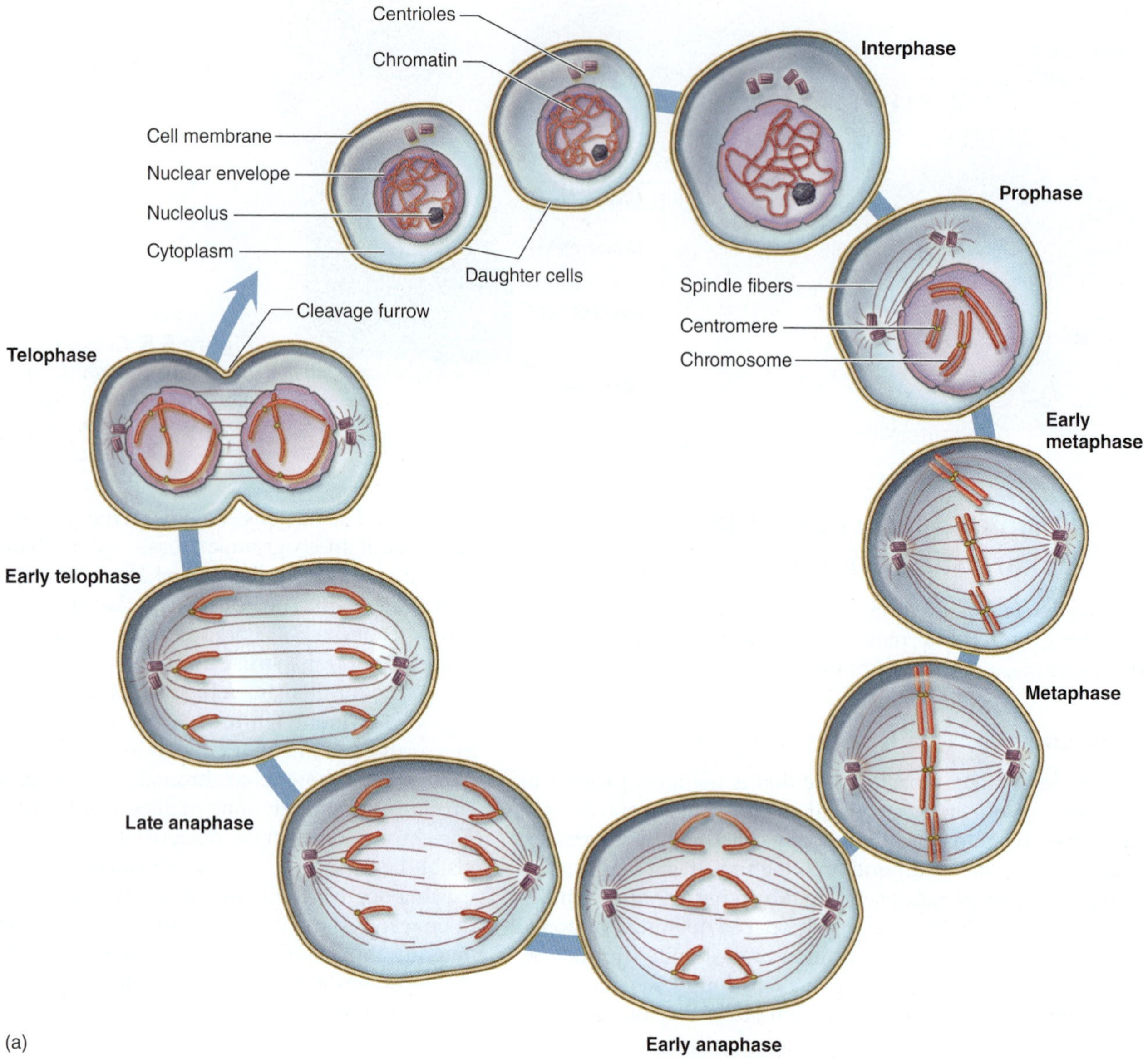

Centrioles

Chromatin

Interphase

Cell membrane

Nuclear envelope

Nucleolus

Cytoplasm

Daughter cells

Prophase

Spindle fibers

Centromere

Chromosome

Cleavage furrow

Telophase

Early metaphase

Early telophase

Metaphase

Late anaphase

Early anaphase

(a)

Figure 5.7 Changes in the cell and nucleus that accompany mitosis in a eukaryotic cell such as a yeast. **(a)** Before mitosis (at a period called *interphase*), chromosomes are visible only as chromatin. As mitosis proceeds (known as *early prophase*), chromosomes take on a fine, threadlike appearance as they condense, and the nuclear membrane and nucleolus are temporarily disrupted. **(b)** By metaphase, the chromosomes are fully visible as X-shaped structures. The shape is due to duplicated chromosomes attached at a central point, the centromere. Spindle fibers attach to these and facilitate the separation of individual chromosomes during metaphase. Later phases serve in the completion of chromosomal separation and division of the cell proper into daughter cells.

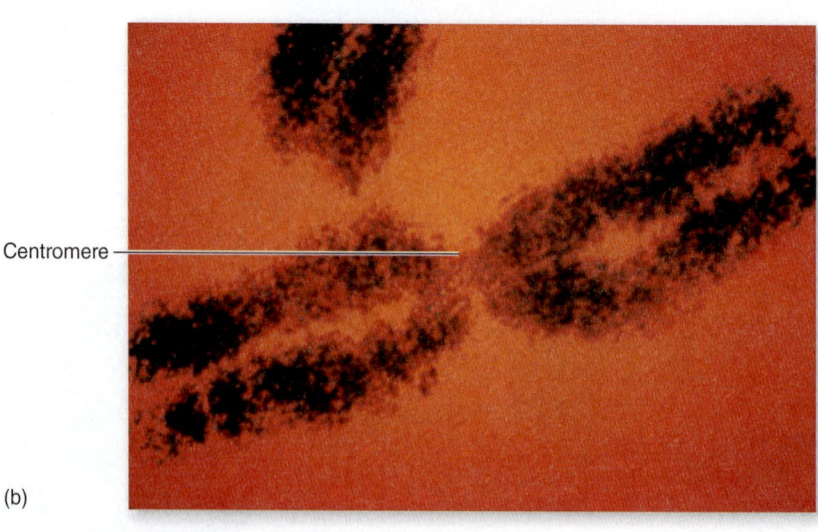

Centromere

(b)

bodies **(figure 5.7)**. This happens when the DNA becomes highly condensed by forming coils and supercoils around the histones to prevent the chromosomes from tangling as they are separated into new cells. This process is described in more detail in chapter 9.

The nucleus, as you've just seen, contains instructions in the form of DNA. Elaborate processes have evolved for transcription and duplication of this genetic material. In addition to mitosis, some cells also undergo **meiosis,** the process by which sex cells are created. Much of the protein synthesis and other work of the cell takes place outside the nucleus in the cell's other organelles.

Endoplasmic Reticulum: A Passageway in the Cell

The **endoplasmic reticulum (ER)** is a microscopic series of tunnels used in transport and storage. Two kinds of endoplasmic reticulum are the **rough endoplasmic reticulum (RER) (figure 5.8)** and the **smooth endoplasmic reticulum (SER).** The RER originates from the outer membrane of the nuclear envelope and extends in a continuous network through the cytoplasm, even all the way out to the cell membrane. This architecture permits the spaces in the RER, called *cisternae* (singular, *cistern*), to transport materials from the nucleus to the cytoplasm and ultimately to the cell's exterior. The RER appears rough because of large numbers of ribosomes partly attached to its membrane surface. Proteins synthesized on the ribosomes are shunted into the inside space (the lumen)

of the RER and held there for later packaging and transport. In contrast to the RER, the SER is a closed tubular network without ribosomes that functions in nutrient processing and in synthesis and storage of nonprotein macromolecules such as lipids.

Golgi Apparatus: A Packaging Machine

The **Golgi**[1] **apparatus,** also called the *Golgi complex* or *Golgi body,* is the site in the cell in which proteins are modified and then sent to their final destinations. It is a discrete organelle consisting of a stack of several flattened, disc-shaped sacs, or cisternae. These sacs have outer limiting membranes and cavities like those of the endoplasmic reticulum, but they do not form a continuous network **(figure 5.9)**. This organelle is always closely associated with the endoplasmic reticulum both in its location and function. At a site where it meets the Golgi apparatus, the endoplasmic reticulum buds off tiny membrane-bound packets of protein called *transitional vesicles* that are picked up by the forming face of the Golgi apparatus. Once in the complex itself, the proteins are often modified by the addition of polysaccharides and lipids. The final action of this apparatus is to pinch off finished *condensing vesicles* that will be conveyed to organelles such as lysosomes or transported outside the cell as secretory vesicles **(figure 5.10)**.

---

1. Named for C. Golgi, an Italian histologist who first described the apparatus in 1898.

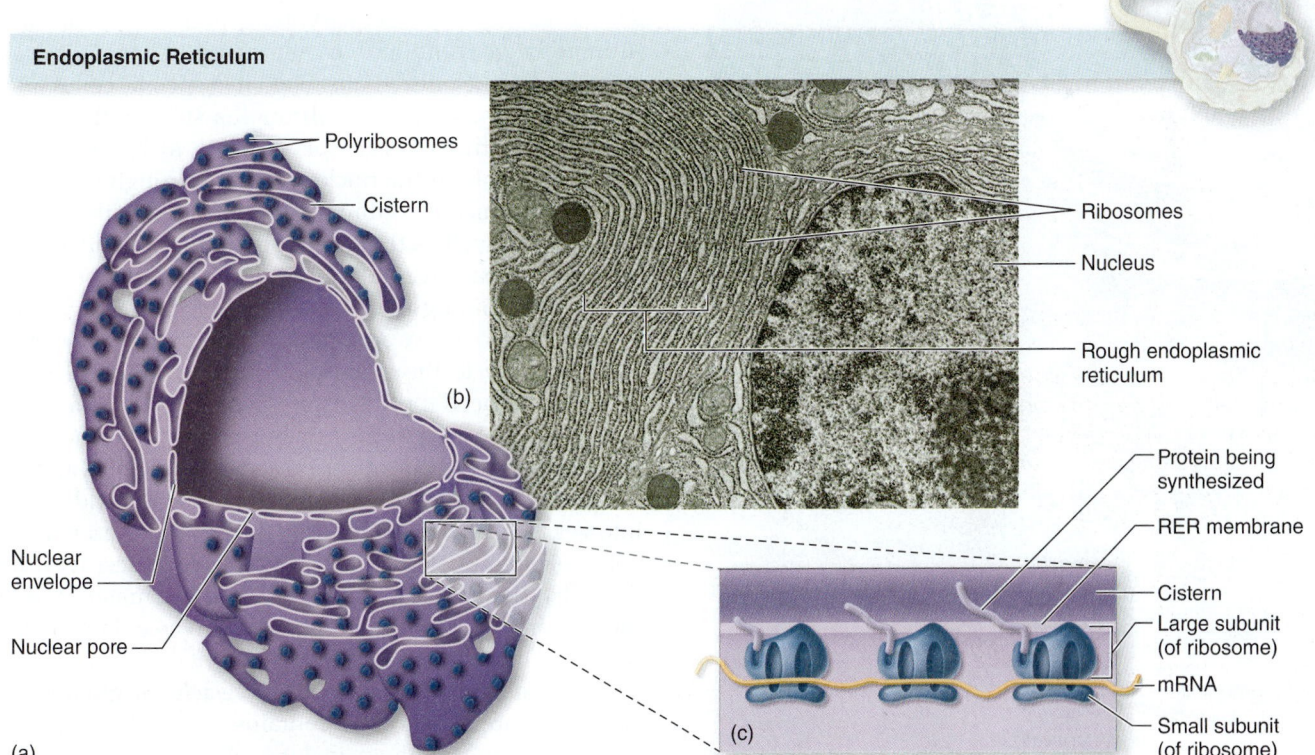

Endoplasmic Reticulum

Polyribosomes

Cistern

Ribosomes

Nucleus

Rough endoplasmic reticulum

(b)

Nuclear envelope

Nuclear pore

(a)

Protein being synthesized

RER membrane

Cistern

Large subunit (of ribosome)

mRNA

Small subunit (of ribosome)

(c)

Figure 5.8 The origin and structure of the rough endoplasmic reticulum (RER). **(a)** Schematic view of the origin of the RER from the outer membrane of the nuclear envelope. **(b)** Three-dimensional projection of the RER. **(c)** Detail of the orientation of a ribosome on the RER membrane.

Golgi Apparatus

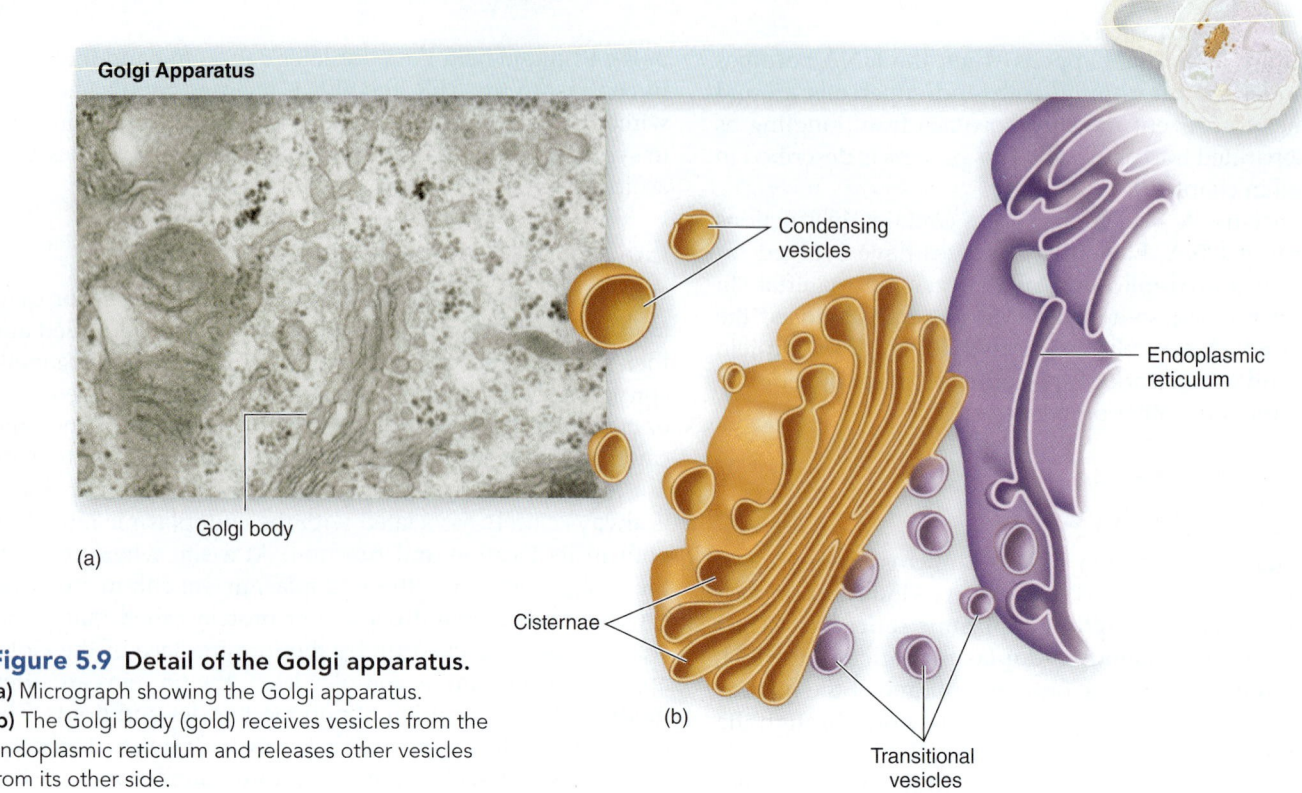

Figure 5.9 Detail of the Golgi apparatus.
(a) Micrograph showing the Golgi apparatus.
(b) The Golgi body (gold) receives vesicles from the endoplasmic reticulum and releases other vesicles from its other side.

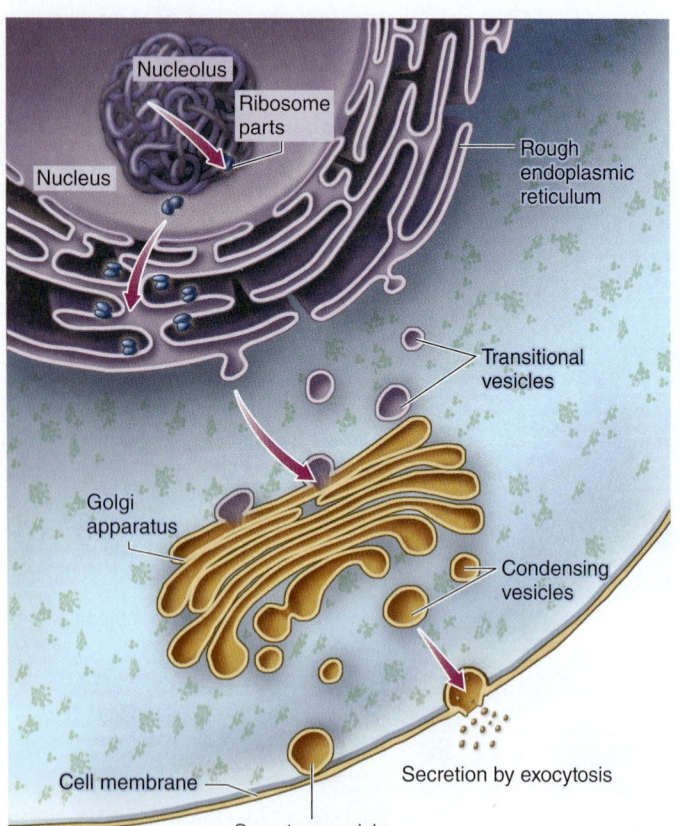

Figure 5.10 The transport process. The cooperation of organelles in protein synthesis and transport: nucleus ⟶ RER ⟶ Golgi apparatus ⟶ vesicles ⟶ secretion.

Nucleus, Endoplasmic Reticulum, and Golgi Apparatus: Nature's Assembly Line

As the keeper of the eukaryotic genetic code, the nucleus ultimately governs and regulates all cell activities. But, because the nucleus remains fixed in a specific cellular site, it must direct these activities through a structural and chemical network (figure 5.10). This network includes ribosomes, which originate in the nucleus, and the rough endoplasmic reticulum, which is continuously connected with the nuclear envelope, as well as the smooth endoplasmic reticulum and the Golgi apparatus. Initially, a segment of the genetic code of DNA containing the instructions for producing a protein is copied into RNA and passed out through the nuclear pores directly to the ribosomes on the endoplasmic reticulum. Here, specific proteins are synthesized from the RNA code and deposited in the lumen (space) of the endoplasmic reticulum. After being transported to the Golgi apparatus, the protein products are chemically modified and packaged into vesicles that can be used by the cell in a variety of ways. Some of the vesicles contain enzymes to digest food inside the cell; other vesicles are secreted to digest materials outside the cell, and yet others are important in the enlargement and repair of the cell wall and membrane.

A **lysosome** is one type of vesicle originating from the Golgi apparatus that contains a variety of enzymes. Lysosomes are involved in intracellular digestion of food particles and in protection against invading microorganisms. They also participate in digestion and removal of cell debris in damaged tissue. The formation of lysosomes involves the so-called GERL, or Golgi-endoplasmic reticulum-lysosomal, complex. Lysosomal enzymes are synthesized by the rough endoplasmic reticulum and are then translocated through the smooth endoplasmic reticulum. Transitional vesicles then transport the enzymes to the Golgi, where unique chemical tags on these proteins localize them to the condensing vesicles that will become the primary lysosomes.

Other types of vesicles include **vacuoles** (vak′-yoo-ohlz), which are membrane-bound sacs containing fluids or solid particles to be digested, excreted, or stored. They are formed in phagocytic cells (certain white blood cells and protozoa) in response to food and other substances that have been engulfed. The contents of a food vacuole are digested through the merger of the vacuole with a lysosome. This merged structure is called a *phagosome* **(figure 5.11).** Other types of vacuoles are used in storing reserve food such as fats and glycogen. Protozoa living in freshwater habitats regulate osmotic pressure by means of contractile vacuoles, which regularly expel excess water that has diffused into the cell (described later).

Mitochondria: Energy Generators of the Cell

Although the nucleus is the cell's control center, none of the cellular activities it commands could proceed without a constant supply of energy, the bulk of which is generated in most eukaryotes by **mitochondria** (my″-toh-kon′-dree-uh). When viewed with light microscopy, mitochondria appear as round or elongated particles scattered throughout the cytoplasm. A single mitochondrion consists of a smooth, continuous outer membrane that forms the external contour, and an inner, folded membrane nestled neatly within the outer membrane **(figure 5.12).** The folds on the inner membrane, called **cristae** (kris′-te), may be tubular, like fingers, or folded into shelflike bands.

The cristae membranes hold the enzymes and electron carriers of aerobic respiration. This is an oxygen-using process that extracts chemical energy contained in nutrient molecules and stores it in the form of high-energy molecules, or ATP. More detailed functions of mitochondria are covered in chapter 8. The spaces around the cristae are filled with a chemically complex fluid called the **matrix,** which holds ribosomes, DNA, and the pool of enzymes and other compounds involved in the metabolic cycle. Mitochondria (along with chloroplasts) are unique among organelles in that they divide independently of the cell, contain circular strands of DNA, and have bacteria-size 70S ribosomes. These findings provide evidence that the mitochondria were bacterial cells engulfed by other cells, which were then destined to become these organelles.

Chloroplasts: Photosynthesis Machines

Chloroplasts are remarkable organelles found in algae and plant cells that are capable of converting the energy of sunlight into chemical energy through photosynthesis. The photosynthetic role of chloroplasts makes them the primary producers of organic nutrients upon which all

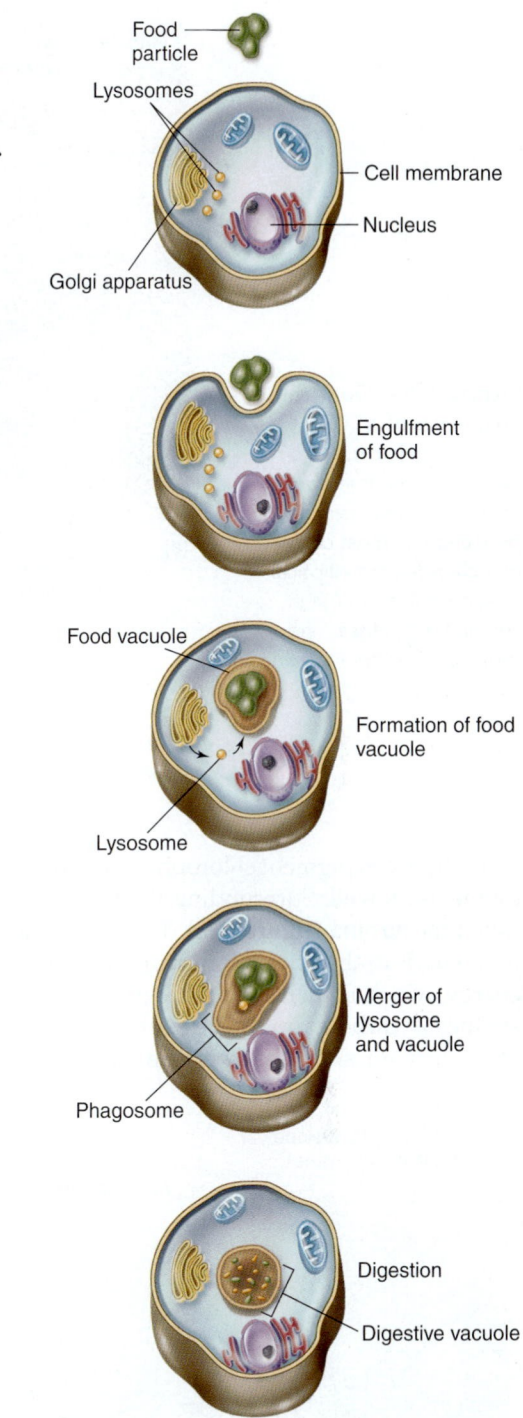

Figure 5.11
The origin and action of lysosomes in phagocytosis.

other organisms (except certain bacteria) ultimately depend. Another important photosynthetic product of chloroplasts is oxygen gas. Although chloroplasts resemble mitochondria, chloroplasts are larger, contain special pigments, and are much more varied in shape.

There are differences among various algal chloroplasts, but most are generally composed of two membranes, one enclosing the other. There is a smooth, outer membrane in addition to an inner membrane. Inside the chloroplast is a third membrane folded into small, disclike sacs called **thylakoids** that are stacked upon one another into **grana.** These structures

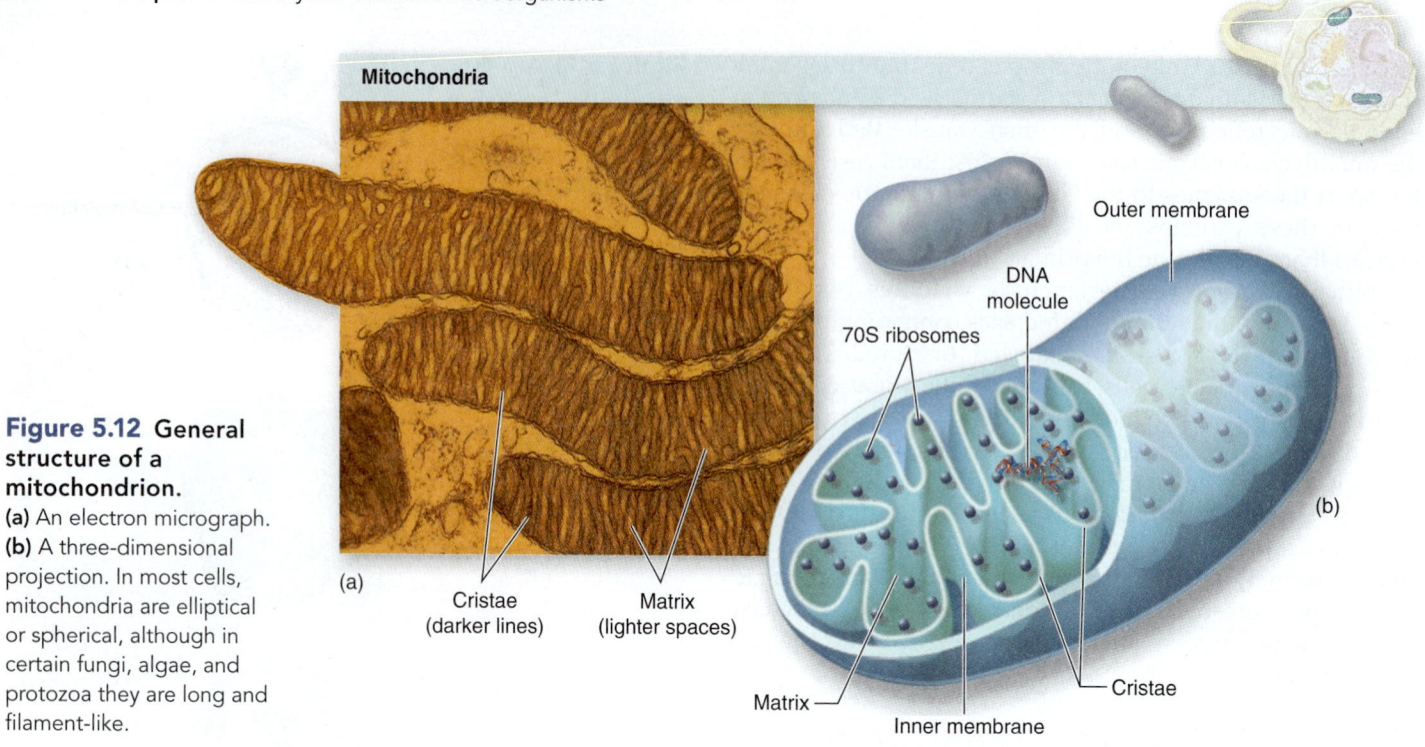

Mitochondria

(a)

Cristae (darker lines)

Matrix (lighter spaces)

70S ribosomes

DNA molecule

Outer membrane

(b)

Cristae

Matrix

Inner membrane

Figure 5.12 **General structure of a mitochondrion.** **(a)** An electron micrograph. **(b)** A three-dimensional projection. In most cells, mitochondria are elliptical or spherical, although in certain fungi, algae, and protozoa they are long and filament-like.

carry the green pigment chlorophyll and sometimes additional pigments as well. Surrounding the thylakoids is a substance called the **stroma (figure 5.13).** The role of the photosynthetic pigments is to absorb and transform solar energy into chemical energy, which is then used during reactions in the stroma to synthesize carbohydrates. We explore more important aspects of photosynthesis in chapters 8 and 24.

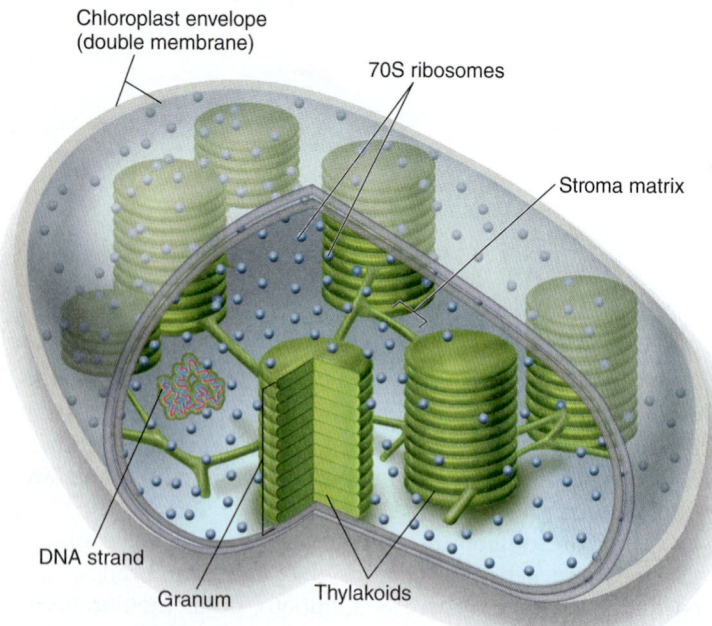

Chloroplast envelope (double membrane)

70S ribosomes

Stroma matrix

DNA strand

Granum

Thylakoids

Figure 5.13 **Detail of an algal chloroplast.**

Ribosomes: Protein Synthesizers

In an electron micrograph of a eukaryotic cell, ribosomes are numerous, tiny particles that give a dotted appearance to the cytoplasm. Ribosomes are distributed throughout the cell: Some are scattered freely in the cytoplasm and cytoskeleton; others are attached to the rough endoplasmic reticulum as previously described. Still others appear inside the mitochondria and in chloroplasts. Multiple ribosomes are often found arranged in short chains called *polyribosomes* (polysomes). The basic structure of eukaryotic ribosomes is similar to that of bacterial ribosomes, described in chapter 4. Both are composed of large and small subunits of ribonucleo-protein (see figure 5.8). By contrast, however, the eukaryotic ribosome (except in the mitochondrion) is the larger 80S variety that is a combination of 60S and 40S subunits. As in the bacteria, eukaryotic ribosomes are the staging areas for protein synthesis.

Disease Connection

The difference in bacterial and eukaryotic ribosome structure has important implications for our ability to fight infections with antibiotics. Because the goal of antimicrobial treatment is to harm the microbe without harming the host (made of eukaryotic cells), drugs that target parts of the ribosome that are unique to the bacterial variety are effective antibiotics that cause minimal harm by way of side effects to the host.

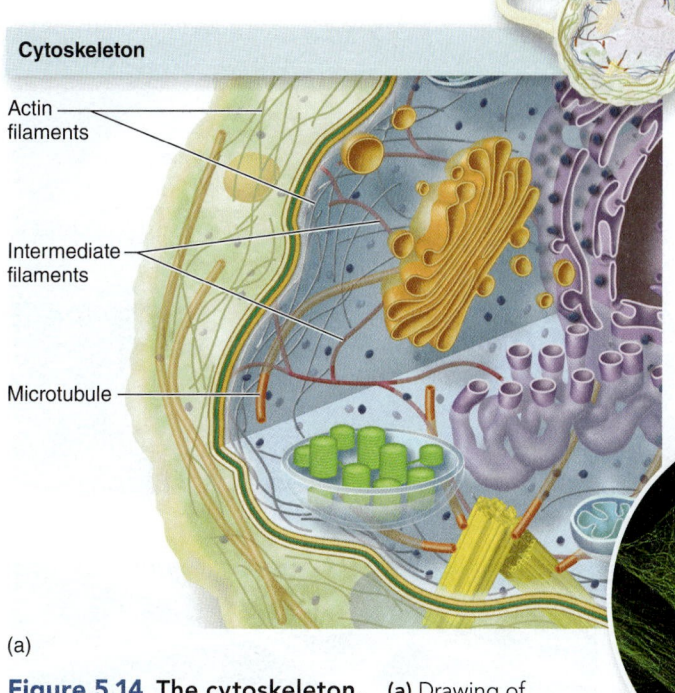

Cytoskeleton

Actin filaments

Intermediate filaments

Microtubule

(a)

Figure 5.14 The cytoskeleton. (a) Drawing of microtubules, actin filaments, and intermediate filaments. (b) Microtubules are dyed fluorescent green in this micrograph.

(b)

The Cytoskeleton: A Support Network

The cytoplasm of a eukaryotic cell is crisscrossed by a flexible framework of molecules called the cytoskeleton **(figure 5.14).** This framework appears to have several functions, such as anchoring organelles, moving RNA and vesicles, and permitting shape changes and movement in some cells. The three main types of cytoskeletal elements are *actin filaments, intermediate filaments,* and *microtubules.* **Actin filaments** are long, thin, protein strands about 7 nanometers in diameter. They are found throughout the cell but are most highly concentrated just inside the cell membrane. Actin filaments are responsible for cellular movements such as contraction, crawling, pinching during cell division, and formation of cellular extensions. **Microtubules** are long, hollow tubes that maintain the shape of eukaryotic cells when they don't have walls and transport substances from one part of a cell to another. The spindle fibers that play an essential role in mitosis are actually microtubules that attach to chromosomes and separate them into daughter cells. As indicated earlier,

Table 5.2 The Major Elements of Life and Their Primary Characteristics

Function or Structure	Characteristic*	Bacterial/ Archaeal Cells	Eukaryotic Cells	Viruses**
Genetics	Nucleic acids	+	+	+
	Chromosomes	+	+	−
	True nucleus	−	+	−
	Nuclear envelope	−	+	−
Reproduction	Mitosis	−	+	−
	Production of sex cells	+/−	+	−
	Binary fission	+	+	−
Biosynthesis	Independent	+	+	−
	Golgi apparatus	−	+	−
	Endoplasmic reticulum	−	+	−
	Ribosomes	+***	+	−
Respiration	Mitochondria	−	+	
Photosynthesis	Pigments	+/−	+/−	−
	Chloroplasts	−	+/−	−
Motility/locomotor structures	Flagella	+/−***	+/−	−
	Cilia	−	+/−	−
Shape/protection	Membrane	+	+	+/−
	Cell wall	+***	+/−	− (have capsids instead)
	Capsule	+/−	+/−	−
Complexity of function		+	+	+/−
Size (in general)		0.5–3 μm****	2–100 μm	< 0.2 μm

*+ means most members of the group exhibit this characteristic; − means most lack it; +/− means some members have it and some do not.

**Viruses cannot participate in metabolic or genetic activity outside their host cells.

***The bacterial/archaeal type is functionally similar to the eukaryotic type, but it is structurally unique.

****Much smaller and much larger bacteria do exist, but they are not common.

microtubules are also responsible for the movement of cilia and flagella. **Intermediate filaments** are ropelike structures that are about 10 nanometers in diameter. (Their name comes from their intermediate size, between actin filaments and microtubules.) Their main role is in structural reinforcement of the cell and of organelles. For example, they support the structure of the nuclear envelope.

Table 5.2 summarizes the differences between eukaryotic and bacterial cells. Viruses (discussed in chapter 6) are included as well.

Survey of Eukaryotic Microorganisms

With the general structure of the eukaryotic cell in mind, let us next examine the amazingly wide range of adaptations that this cell type has undergone. The following sections contain a general survey of the principal eukaryotic microorganisms—fungi, algae, protozoa, and parasitic worms—while also introducing elements of their structure, life history, classification, identification, and importance.

5.3 Learning Outcomes—Assess Your Progress

7. Describe the main structural components of a nucleus.
8. Diagram how the nucleus, endoplasmic reticulum, and Golgi apparatus act together with vesicles during the transport process.
9. Explain the function of the mitochondrion.
10. Discuss the function of chloroplasts, explaining which cells contain them and how they arose.
11. Explain the importance of ribosomes and differentiate between eukaryotic and bacterial types.
12. List and describe the three main fibers of the cytoskeleton.

5.4 The Fungi

The Kingdom Fungi, or Myceteae, is large and filled with forms of great variety and complexity. For practical purposes, the approximately 100,000 species of fungi can be divided into two groups: the *macroscopic fungi* (mushrooms, puffballs, gill fungi) and the *microscopic fungi* (molds, yeasts). Although the majority of fungi are either unicellular or colonial, a few complex forms such as mushrooms and puffballs are considered multicellular. Cells of the microscopic fungi exist in two basic morphological types: yeasts and hyphae. A yeast cell has a round to oval shape and uses asexual reproduction. It grows swellings on its surface called *buds*, which then become separate cells. **Hyphae** (hy′-fee) are long, threadlike cells found in the bodies of filamentous fungi, or molds **(figure 5.15).** Some species form a **pseudohypha,** a chain of yeasts formed when buds remain attached in a row **(figure 5.16).** Because of its manner of formation, it is not a true hypha like that of molds. While some fungal cells exist only in a yeast form and others occur primarily as hyphae, a few, called **dimorphic,** can take either form, depending on

(a)

Septum

(b)

Septa

Septate hyphae   Nonseptate hyphae

as in *Penicillium*   as in *Rhizopus*

(c)

Figure 5.15 *Diplodia maydis*, **a pathogenic fungus of corn plants.** **(a)** Scanning electron micrograph of a single colony showing its filamentous texture (24×). **(b)** Close-up of hyphal structure (1,200×). **(c)** Basic structural types of hyphae.

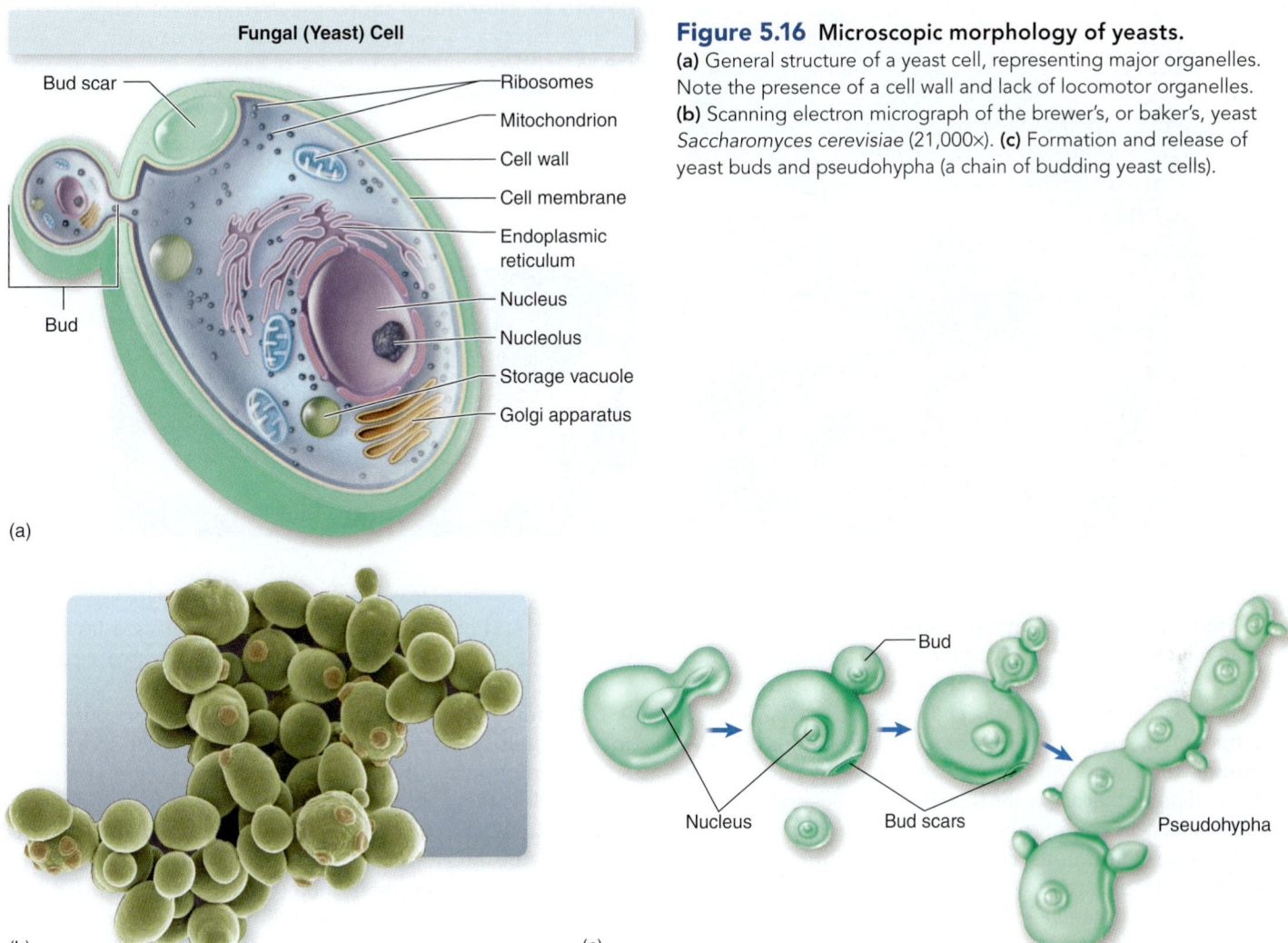

Figure 5.16 Microscopic morphology of yeasts.
(a) General structure of a yeast cell, representing major organelles. Note the presence of a cell wall and lack of locomotor organelles. (b) Scanning electron micrograph of the brewer's, or baker's, yeast *Saccharomyces cerevisiae* (21,000×). (c) Formation and release of yeast buds and pseudohypha (a chain of budding yeast cells).

growth conditions such as changing temperature. This variability in growth form is particularly characteristic of some pathogenic molds.

Fungal Nutrition

All fungi are **heterotrophic.** This means that they acquire nutrients from a wide variety of organic materials called **substrates (figure 5.17).** Most fungi are **saprobes,** meaning that they obtain these substrates from the remnants of dead plants and animals in soil or aquatic habitats. Fungi can also be **parasites,** meaning they live on the bodies of living animals or plants, although very few fungi absolutely require a living host. In general, the fungus penetrates the substrate and secretes enzymes that reduce it to small molecules that can be absorbed by the cells. Fungi have enzymes for digesting an incredible array of substances, including feathers, hair, cellulose, petroleum products, wood, and rubber. It has been said that every naturally occurring organic material on the earth can be attacked by some type of fungus. Fungi are often found in nutritionally poor or adverse environments. Various fungi thrive in substrates with high salt or sugar content, at relatively high temperatures, and even in snow and glaciers. Their medical and agricultural impact is extensive. A number of species cause **mycoses** (fungal infections) in animals, and thousands of species are important plant pathogens. Fungal toxins may cause disease in humans, and airborne fungi are a frequent cause of allergies and other medical conditions.

Organization of Microscopic Fungi

The cells of most microscopic fungi grow in loose associations or colonies. The colonies of yeasts are much like those of bacteria in that they have a soft, uniform texture and appearance. The colonies of filamentous fungi are noted for the striking cottony, hairy, or velvety textures that arise from their microscopic organization and morphology. The woven, intertwining mass of hyphae that makes up the body or colony of a mold is called a **mycelium.**

(a)

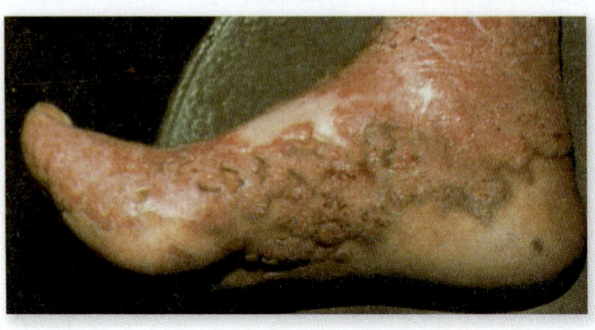

(b)

Figure 5.17 **Nutritional sources (substrates) for fungi.** **(a)** A fungal mycelium growing on raspberries. The fine hyphal filaments and black sporangia are typical of *Rhizopus*. **(b)** The skin of the foot infected by a soil fungus, *Fonsecaea pedrosoi.*

Although hyphae contain the usual eukaryotic organelles, they also have some unique organizational features. In most fungi, the hyphae are divided into segments by cross walls, or **septa,** a condition called *septate* (see figure 5.15*c*). The structure of the septa varies from solid partitions with no communication between the compartments to partial walls with small pores that allow the flow of organelles and nutrients between adjacent compartments. Nonseptate hyphae consist of one long, continuous cell *not* divided into individual compartments by cross walls. With this construction, the cytoplasm and organelles move freely from one region to another, and each hyphal element can have several nuclei.

Hyphae can also be classified according to their particular function. Vegetative hyphae (mycelia) are responsible for the visible mass of growth that appears on the surface of a substrate and penetrates it to digest and absorb nutrients. During the development of a fungal colony, the vegetative hyphae give rise to structures called reproductive, or fertile, hyphae, which branch off a vegetative mycelium. These hyphae are responsible for the production of fungal reproductive bodies called **spores.** A variety of hyphae are illustrated in **figure 5.18.**

Reproductive Strategies and Spore Formation

Fungi have many complex and successful reproductive strategies. Most can propagate by the simple outward growth of existing hyphae or by fragmentation, in which a separated piece of mycelium can generate a whole new colony. But the primary reproductive mode of fungi involves the production of various types of spores. (Do not confuse fungal spores with the more resistant, nonreproductive bacterial spores.) Fungal spores are responsible not only for multiplication but also for survival, producing genetic variation, and dissemination. Because of their compactness and relatively light weight, spores are dispersed widely through the environment by

air, water, and living things. Upon encountering a favorable substrate, a spore will germinate and produce a new fungus colony in a very short time (see figure 5.18).

The fungi contain such a marked diversity of spores that they are largely classified and identified by their spores and spore-forming structures. There are elaborate systems for naming and classifying spores, but we won't cover them. The most general subdivision is based on the way the spores arise. Asexual spores are the products of mitotic division of a single parent cell, and sexual spores are formed through a process involving the fusing of two parental nuclei followed by meiosis.

Asexual Spore Formation

There are two subtypes of asexual spores, **sporangiospores** and **conidiospores,** also called *conidia* **(figure 5.19):**

1. Sporangiospores **(figure 5.19*a*)** are formed by successive cleavages within a saclike head called a **sporangium,** which is attached to a stalk, the sporangiophore. These spores are initially enclosed but are released when the sporangium ruptures.
2. Conidiospores, or **conidia,** are free spores not enclosed by a spore-bearing sac. They develop either by the pinching off of the tip of a special fertile hypha or by the segmentation of a preexisting vegetative hypha. There are many different forms of conidia, illustrated in **figure 5.19*b*.**

Sexual Spore Formation

Fungi can propagate themselves successfully with their millions of asexual spores. That being the case, what is the function of their sexual spores? The answer lies in important variations that occur when fungi of different genetic makeup combine their genetic material. Just as in plants and animals, this linking of genes from two parents creates offspring with combinations of genes different from that of either parent.

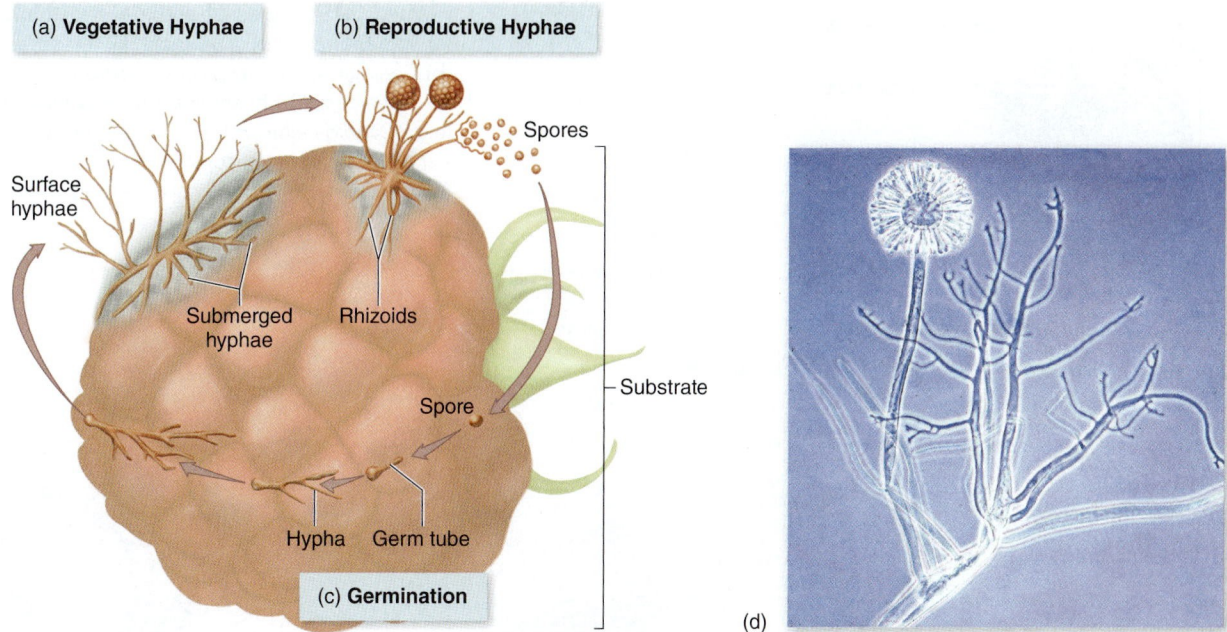

Figure 5.18 Functional types of hyphae using the mold *Rhizopus* as an example. **(a)** Vegetative hyphae are those surface and submerged filaments that digest, absorb, and distribute nutrients from the substrate. This species also has special anchoring structures called *rhizoids*. **(b)** Later, as the mold matures, it sprouts reproductive hyphae that produce asexual spores. **(c)** During the asexual life cycle, the free mold spores settle on a substrate and send out germ tubes that elongate into hyphae. Through continued growth and branching, an extensive mycelium is produced. So prolific are the fungi that a single colony of mold can easily contain 5,000 spore-bearing structures. If each of these released 2,000 single spores and if every spore were able to germinate, we would soon find ourselves in a sea of mycelia. Most spores do not germinate, but enough are successful to keep the numbers of fungi and their spores very high in most habitats. **(d)** *Syncephalastrum* demonstrates all major stages in the life cycle of a zygomycota.

INSIGHT 5.2 **The Zombie Ant Apocalypse**

When you think of a fungus, you usually think of a delicious pizza topping or an annoying growth on your bread. Little did you know that a species of *Ophiocordyceps* is bringing on the zombie apocalypse—in carpenter ants. Scientists studied the nervous system of ants in Thailand and found that infection with the fungus causes them to behave in a manner less like a carpenter ant and more like a fungus-spreading zombie. As the fungus grows inside the ant, it begins to behave erratically, wandering aimlessly, and even suffering convulsions. After a few days, the fungus causes the ant to attach to a leaf with its jaws, preventing it from letting go. Then the fungus grows through the top of the ant's head and releases spores to the environment. Scientists studying the zombie ants have found that the fungus synchronizes the timing of the ant clamping onto the leaf with high noon, when the sun is strongest. Then the fungus can burst out of the ant during the cooler temperatures of the night, assuring more effective fungal growth.

Source: K. Than, "'Zombie' Ants Bite at High Noon, Then Die," *National Geographic Daily News* (May 11, 2011). Retrieved from http://news.nationalgeographic.com/news/2011/05/110511-zombies-ants-fungus-infection-spores-bite-noon-animals-science/.

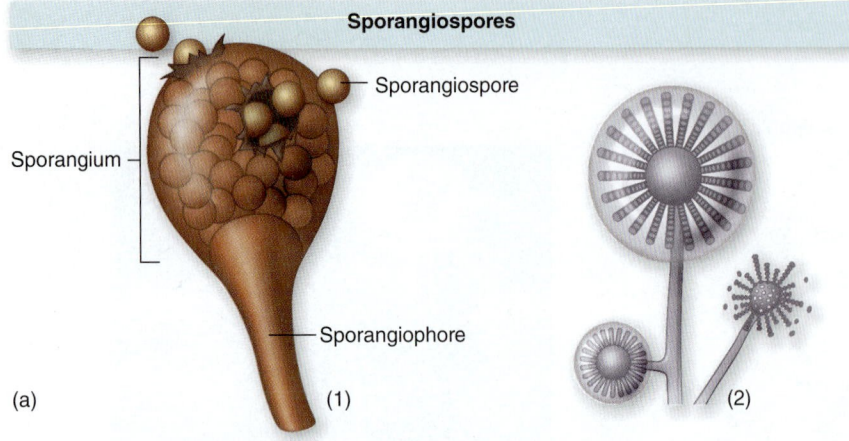

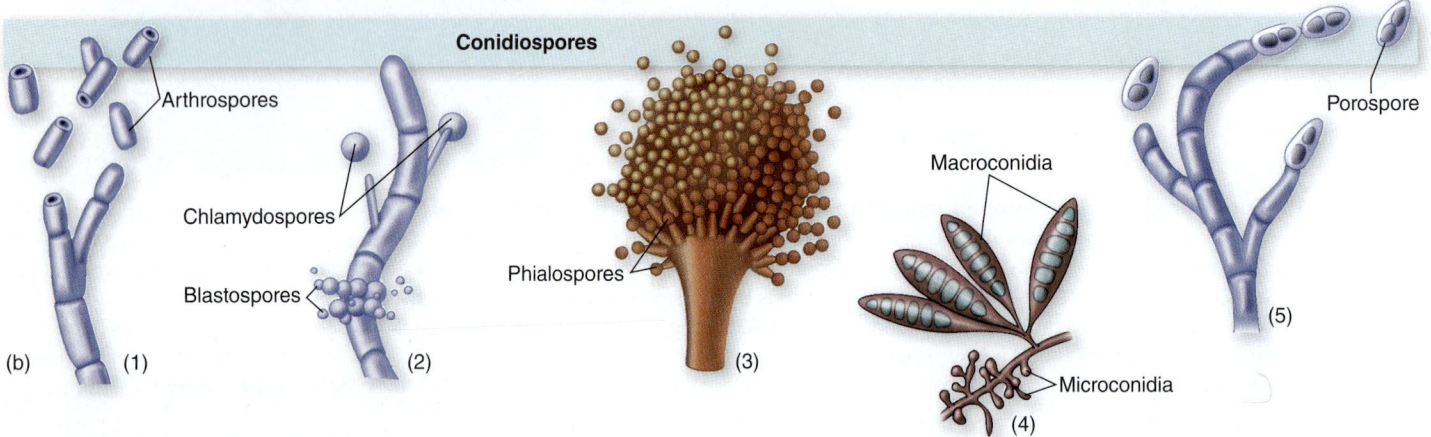

Figure 5.19 Types of asexual mold spores.
(a) Sporangiospores: (1) *Absidia*, (2) *Syncephalastrum*.
(b) Conidial variations: (1) arthrospores (e.g.,
Coccidioides), (2) chlamydospores and blastospores
(e.g., *Candida albicans*), (3) phialospores (e.g.,
Aspergillus), (4) macroconidia and microconidia (e.g.,
Microsporum), and (5) porospores (e.g., *Alternaria*).

The offspring from such a union can have slight variations in form and function that are potentially advantageous in the adaptation and survival of their species.

The majority of fungi produce sexual spores at some point. The nature of this process varies from the simple fusion of fertile hyphae of two different strains to a complex union of differentiated male and female structures and the development of special fruiting structures. It may be a surprise to discover that the fleshy part of a mushroom is actually a fruiting body designed to protect and help disseminate its sexual spores.

Fungal Identification and Cultivation

Fungi are identified in medical specimens by first being isolated on special types of media and then being observed macroscopically and microscopically. Because the fungi are classified into general groups by the presence and type of sexual spores, it would seem logical to identify them in the same way, but sexual spores are difficult to detect in the laboratory setting. As a result, the asexual spore-forming structures and spores are usually used to identify organisms to the level of genus and species. Other characteristics that contribute to identification are hyphal type, colony texture and pigmentation, physiological characteristics, and genetic makeup. Even as bacterial and viral identification relies increasingly on molecular techniques, fungi are some of the most strikingly beautiful life forms, and their appearance under the microscope is still heavily relied on to identify them **(figure 5.20a,b).**

The Roles of Fungi in Nature and Industry

Nearly all fungi are free-living and do not require a host to complete their life cycles. Even among those fungi that are pathogenic, most human infection occurs through accidental contact with an environmental source such as soil, water, or dust. Humans are generally quite resistant to fungal infection, except for two main types of fungal pathogens: the primary pathogens, which can sicken even healthy persons, and the opportunistic pathogens, which attack persons who are already weakened in some way. So far, about 270 species of fungi have been found to be able to cause human disease.

Mycoses (fungal infections) vary in the way the agent enters the body and the degree of tissue involvement **(table 5.3).** The list of opportunistic fungal pathogens has been increasing in the past few years because of newer medical techniques that keep immunocompromised patients alive. Even so-called harmless species found in the air and dust around us may be able to cause opportunistic infections in patients who already have AIDS, cancer, or diabetes.

Fungi are involved in other medical conditions besides infections. Fungal cell walls give off chemical substances that can cause allergies. The toxins produced by poisonous mushrooms can induce neurological disturbances and even death. The mold *Aspergillus flavus* synthesizes a potentially lethal poison called aflatoxin, which is the cause of a disease in domestic animals that have eaten grain contaminated with the mold and is also a cause of liver cancer in humans.

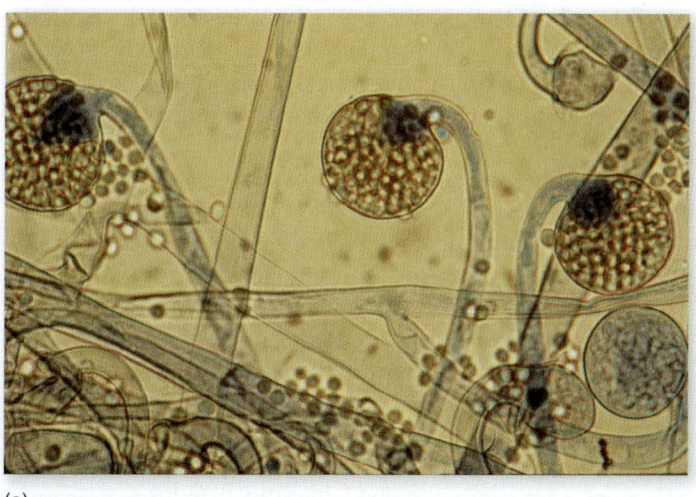

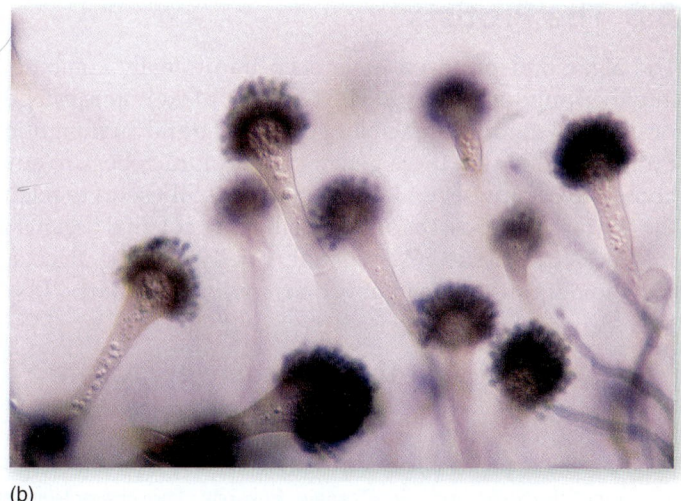

(a)   (b)

Figure 5.20 Representative fungi.   (a) *Circinella*, a fungus associated with soil and decaying nuts. (b) *Aspergillus*, a ubiquitous environmental fungus that can be associated with human disease.

Disease Connection

Cryptococcus neoformans is an example of a primary fungal pathogen. This fungus can be transmitted to humans from bird droppings and causes lung infections and central nervous system (brain) inflammation. *Candida albicans* commonly causes opportunistic disease. This is the organism responsible for "yeast infections" and thrush. It lives as normal biota on mucous membranes; if the host becomes immunocompromised or if the host's normal biota is disturbed through antibiotic treatment, *Candida* will overgrow and cause symptoms.

Fungi pose an ever-present economic hindrance to the agricultural industry. A number of species are pathogenic to field plants such as corn and grain, and fungi also rot fresh produce during shipping and storage. It has been estimated that as much as 40% of the yearly fruit crop is consumed not by humans but by fungi. On the beneficial side, however, fungi play an essential role in decomposing organic matter and returning essential minerals to the soil.

They form stable associations with plant roots that increase the ability of the roots to absorb water and nutrients. Industry has tapped the biochemical potential of fungi to produce large quantities of antibiotics, alcohol, organic acids, and vitamins. Some fungi are eaten or used to impart flavorings to food. The yeast *Saccharomyces* produces the alcohol in beer and wine and the gas that causes bread to rise. Blue cheese, soy sauce, and cured meats derive their unique flavors from the actions of fungi (see chapter 25).

5.4 Learning Outcomes—Assess Your Progress

13. List three general features of fungal anatomy.
14. Differentiate among the terms *heterotroph*, *saprobe*, and *parasite*.
15. Explain the relationship between fungal hyphae and the production of a mycelium.
16. Describe two ways in which fungal spores arise.
17. List two detrimental and two beneficial activities of fungi (from the viewpoint of humans).

Table 5.3 Major Fungal Infections of Humans

Degree of Tissue Involvement and Area Affected	Name of Infection	Name of Causative Fungus
Superficial (Not Deeply Invasive)		
Outer epidermis	Tinea versicolor	*Malassezia furfur*
Epidermis, hair, and dermis can be attacked.	Dermatophytosis, also called tinea or ringworm of the scalp, body, feet (athlete's foot), toenails	*Microsporum, Trichophyton,* and *Epidermophyton*
Mucous membranes, skin, nails	Candidiasis, or yeast infection	*Candida albicans*
Systemic (Deep; Organism Enters Lungs; Can Invade Other Organs)		
Lung	Coccidioidomycosis (San Joaquin Valley fever)	*Coccidioides immitis dermatitidis*
	North American blastomycosis (Chicago disease)	*Blastomyces*
	Histoplasmosis (Ohio Valley fever)	*Histoplasma capsulatum*
	Cryptococcosis (torulosis)	*Cryptococcus neoformans*
Lung, skin	Paracoccidioidomycosis (South American blastomycosis)	*Paracoccidioides brasiliensis*

5.5 The Protists

The algae and protozoa have been traditionally combined into the Kingdom Protista. The two major taxonomic categories of this kingdom are Subkingdom Algae and Subkingdom Protozoa. Although these general types of microbes are now known to occupy several kingdoms, it is still useful to retain the concept of a *protist* as any unicellular or colonial organism that lacks true tissues. We will only briefly mention algae, as they do not cause human infections for the most part.

A Note About the Taxonomy of Protists

Exploring the origins of eukaryotic cells with molecular techniques has significantly clarified our understanding of relationships among the organisms in Domain Eukarya. The characteristics traditionally used for placing plants, animals, and fungi into separate kingdoms are general cell type, level of organization (body plan), and nutritional type. While it now appears that these criteria often do reflect accurate differences among these organisms and give rise to the same classifications as molecular techniques, in many cases the molecular data point to new and different classifications.

Because our understanding of the phylogenetic relationships is still in development, there is not yet a single official system of taxonomy for presenting all of the eukaryotes. This is especially true of the protists (which contain algae and protozoa). Genetic analysis has determined that this group, generally classified at the kingdom level, is far more diverse than previously appreciated and probably should instead be divided into several different kingdoms. Some organisms we call *protists* are more related to fungi than they are to other protists, for instance. For that reason, most scientists believe that the labels "protist" and "protozoa" are meaningless, taxonomically.

For the purposes of this book and your class, we will use the term *protozoa* to refer to eukaryotic organisms that are not animals, plants, or fungi. But be aware that the science is still developing.

The Algae: Photosynthetic Protists

The **algae** are a group of photosynthetic organisms usually recognized by their larger members, such as seaweeds and kelps. In addition to being beautifully colored and diverse in appearance, they vary in length from a few micrometers to 100 meters. Algae occur in unicellular, colonial, and filamentous forms; the larger forms can possess tissues and simple organs. **Figure 5.21** depicts various types of algae. Algal cells as a group exhibit all of the eukaryotic organelles. The most noticeable of these are the chloroplasts, which contain, in addition to the green pigment chlorophyll, a number of other pigments that create the yellow, red, and brown coloration of some groups.

Algae are widespread inhabitants of fresh and marine waters. They are one of the main components of the large floating community of microscopic organisms called **plankton.** In this capacity, they play an essential role in the aquatic food web and produce most of the earth's oxygen. Other algal habitats include the surface of soil, rocks, and plants; several species are even hardy enough to live in hot springs or snowbanks.

Animal tissues are rather inhospitable to algae, so algae are rarely infectious. One exception is *Prototheca*, an unusual nonphotosynthetic alga, which has been associated with skin and subcutaneous infections in humans and animals.

The primary medical threat from algae is due to a type of food poisoning caused by the toxins of certain marine algae. During particular seasons of the year, the overgrowth of these motile algae imparts a brilliant red color to the water, which is referred to as a *red tide*. When intertidal animals feed, their bodies accumulate toxins given off by the algae that can persist for several months. Paralytic shellfish poisoning is caused by eating exposed clams or other invertebrates. It is marked by severe neurological symptoms and can be fatal. Ciguatera is another serious intoxication caused by algal toxins that have accumulated in fish such as bass and mackerel. Cooking does not destroy the toxin, and there is no antidote.

Several episodes of a severe infection caused by *Pfiesteria piscicida*, a toxic algal form, have been reported over the past several years in the United States. The disease was first reported in fish and was later transmitted to humans. This newly identified species occurs in at least 20 forms, including spores, cysts, and amoebas **(figure 5.21c)**, that can release potent toxins. Both fish and humans develop neurological symptoms and bloody skin lesions. The cause of the epidemic has been traced to nutrient-rich agricultural runoff water that promoted the sudden "bloom" of *Pfiesteria*. These microbes first attacked and killed millions of fish and later people whose occupations exposed them to fish and contaminated water.

Biology of the Protozoa

If a poll were taken to choose the most engrossing and vivid group of microorganisms, many biologists would choose the protozoa. Although their name comes from the Greek for "first animals," they are far from being simple, primitive organisms. The protozoa constitute a very large group (about 65,000 species) of creatures that, although single-celled, have startling properties when it comes to movement, feeding, and behavior. Although most members of this group are harmless, free-living inhabitants of water and soil, a few species are parasites collectively responsible for hundreds of millions of infections of humans each year. Before we consider a few examples of important pathogens, let us examine some general aspects of protozoan biology, remembering that the term *protozoan* is more of a convenience than an accurate taxonomic designation. As we describe them in the coming paragraphs, you will see why they are categorized together. It is because of their similar physical characteristics rather than their genetic relatedness.

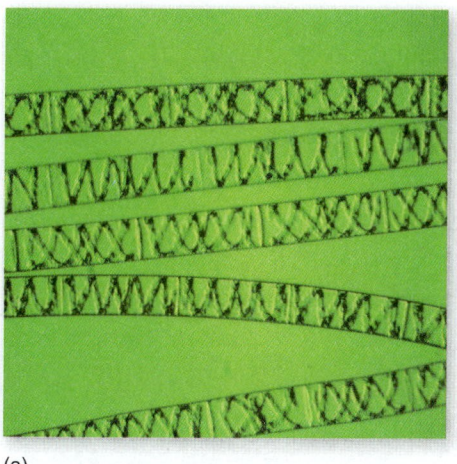
(a)

(b)

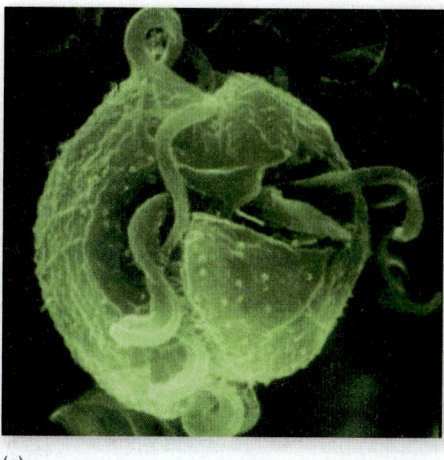
(c)

Figure 5.21 Representative microscopic algae. **(a)** *Spirogyra*, a colonial filamentous form with spiral chloroplasts. **(b)** A collection of beautiful algae called diatoms shows the intricate and varied structure of their silica cell walls. **(c)** *Pfiesteria piscicida*. Although it is free-living, it is known to parasitize fish and release potent toxins that kill fish and sicken humans.

Case File 5   *Continuing the Case*

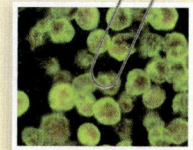

N. fowleri gains entry to the body through the nose and mouth, and in Kyle's case, it likely was due to inhalation of water while he was swimming. The organism invades the soft tissues of the olfactory mucosa and penetrates the cribiform plate in the skull. From there, the amoeba is able to access the nerve fibers through the cribriform plate and enter the brain, where it begins to infect brain cells. This causes primary amoebic meningoencephalitis (PAM). *N. fowleri* is susceptible to the drug amphotericin B *in vitro* (in the test tube), but there is little evidence that the drug is effective *in vivo* (in a living organism). Unfortunately, survival rates for PAM are less than 1%.

The Centers for Disease Control and Prevention recommends avoiding swimming in or participating in athletic activities in warm freshwater. If you do choose to swim in warm water, hold your nose or use nose clips, and avoid disturbing the sediment in shallow water. Only 42 cases have been documented in the United States from 2001 to 2010, making it an uncommon phenomenon. However, this "brain-eating" amoeba has caught the attention of the media and scientists alike.

■ In what other ways can *Naegleria fowleri* be spread?

Protozoan Form and Function

Most protozoan cells are single cells containing all the major eukaryotic organelles except chloroplasts. Their organelles can be highly specialized and are essentially analogous to mouths, digestive systems, reproductive tracts, and "legs"—or means of locomotion. The cytoplasm is usually divided into a clear outer layer called the **ectoplasm** and a granular inner region called the **endoplasm.** Ectoplasm is involved in locomotion, feeding, and protection. Endoplasm houses the nucleus, mitochondria,

and food and contractile vacuoles. Some ciliates and flagellates even have organelles that work somewhat like a primitive nervous system to coordinate movement. Because protozoa lack a cell wall, they have a certain amount of flexibility. Their outer boundary is a cell membrane that regulates the movement of food, wastes, and secretions. Cell shape can remain constant (as in most ciliates) or can change constantly (as in amoebas). Certain amoebas (foraminiferans) encase themselves in hard shells made of calcium carbonate. The size of most protozoan cells falls within the range of 3 to 300 μm. Some notable exceptions are giant amoebas and ciliates that are large enough (3 to 4 mm in length) to be seen swimming in pond water.

Nutritional and Habitat Range   Protozoa are heterotrophic and usually require their food in a complex organic form. Free-living species scavenge dead plant or animal debris and even graze on live cells of bacteria and algae. Some protozoa absorb food directly through the cell membrane. Parasitic species live on the fluids of their host, such as plasma and digestive juices, or they can actively feed on tissues.

Although protozoa have adapted to a wide range of habitats, their main limiting factor is the availability of moisture. Their predominant habitats are fresh and marine water, soil, plants, and animals. Even extremes in temperature and pH are not a barrier to their existence; hardy species are found in hot springs, ice, and habitats with low or high pH. Many protozoa can convert to a resistant, dormant stage called a *cyst.*

Styles of Locomotion   Except for one group (the Apicomplexa), protozoa can move through fluids by means of **pseudopods** ("false feet"), **flagella,** or **cilia.** A few species have both pseudopods (also called *pseudopodia*) and flagella. Some unusual protozoa move by a gliding or twisting movement that does not appear to involve any of these locomotor structures. Pseudopods are blunt, branched, or long and pointed, depending on the particular species. The flowing action of

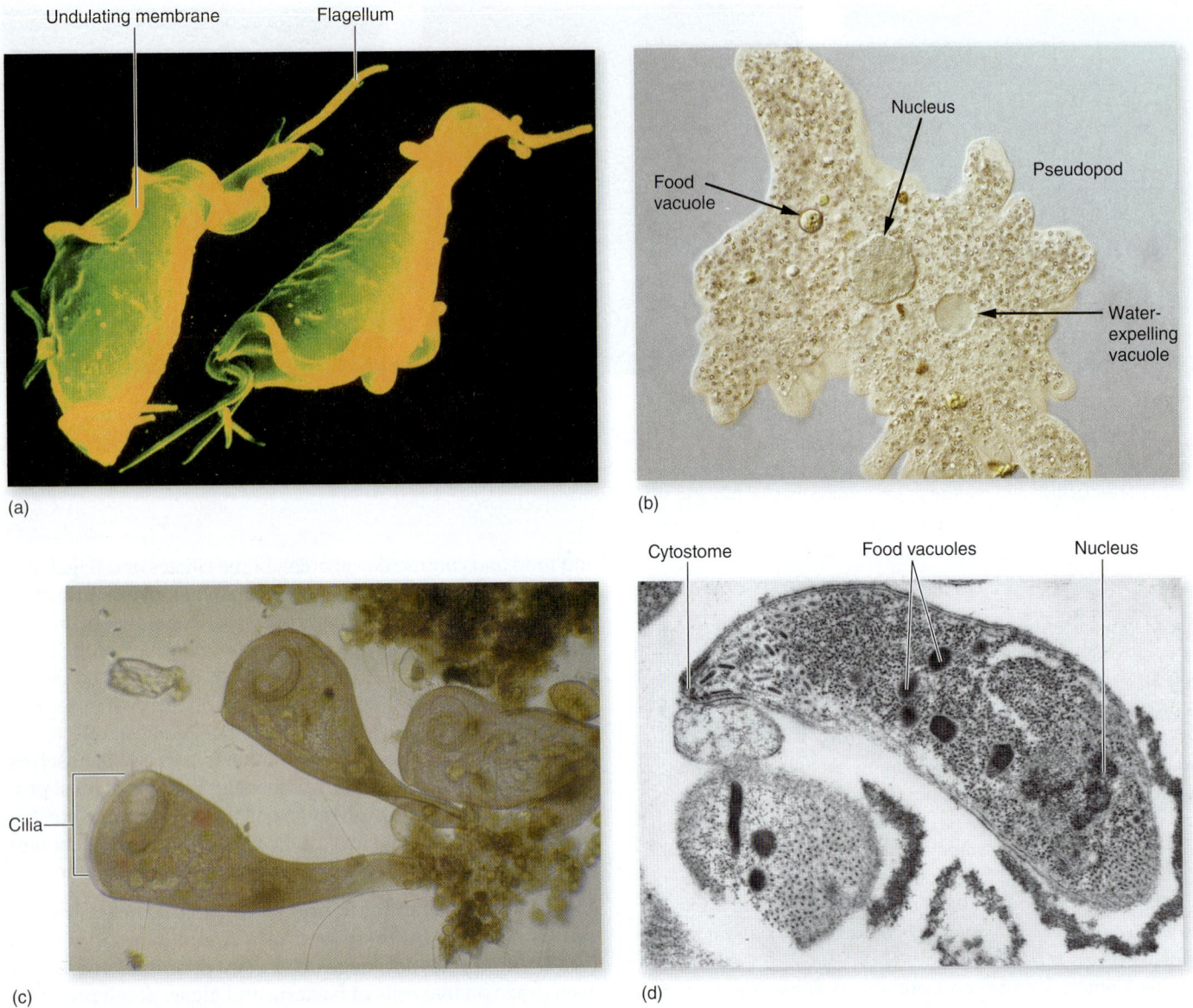

Figure 5.22 Examples of the four types of locomotion in protozoa. **(a)** Using flagella: *Trichomonas vaginalis*, displaying flagella. **(b)** Using amoeboid motion: *Amoeba*, with pseudopods. **(c)** Using cilia: *Stentor*, displaying cilia. **(d)** Sporozoan: *Cryptosporidium*. Sporozoa have no specialized locomotion organelles.

the pseudopods results in amoeboid motion, and pseudopods also serve as feeding structures in many amoebas. (The structure and behavior of flagella and cilia were discussed in the first section of this chapter.) Flagella vary in number from one to several, and in certain species they are attached along the length of the cell by an extension of the cytoplasmic membrane called the *undulating membrane* **(figure 5.22a)**. In most ciliates, the cilia are distributed over the entire surface of the cell in characteristic patterns. Because of the tremendous variety in ciliary arrangements and functions, ciliates are among the most diverse and awesome cells in the biological world. In certain protozoa, cilia line the oral groove and function in feeding; in others, they fuse together to form stiff props that serve as primitive rows of walking legs.

Life Cycles and Reproduction Most protozoa can be recognized in their motile feeding stage called the **trophozoite.** This is a stage that requires ample food and moisture to remain active. A large number of species are also capable of entering into a dormant, resting stage called a **cyst** when conditions in the environment become unfavorable for growth and feeding. During *encystment*, the trophozoite cell rounds up into a sphere, and its ectoplasm secretes a tough, thick cuticle around the cell membrane **(process figure 5.23).** Because cysts are more resistant than ordinary cells to heat, drying, and chemicals, they can survive adverse periods. They can be dispersed by air currents and may even be an important factor in the spread of diseases such as amoebic dysentery. If provided with moisture and

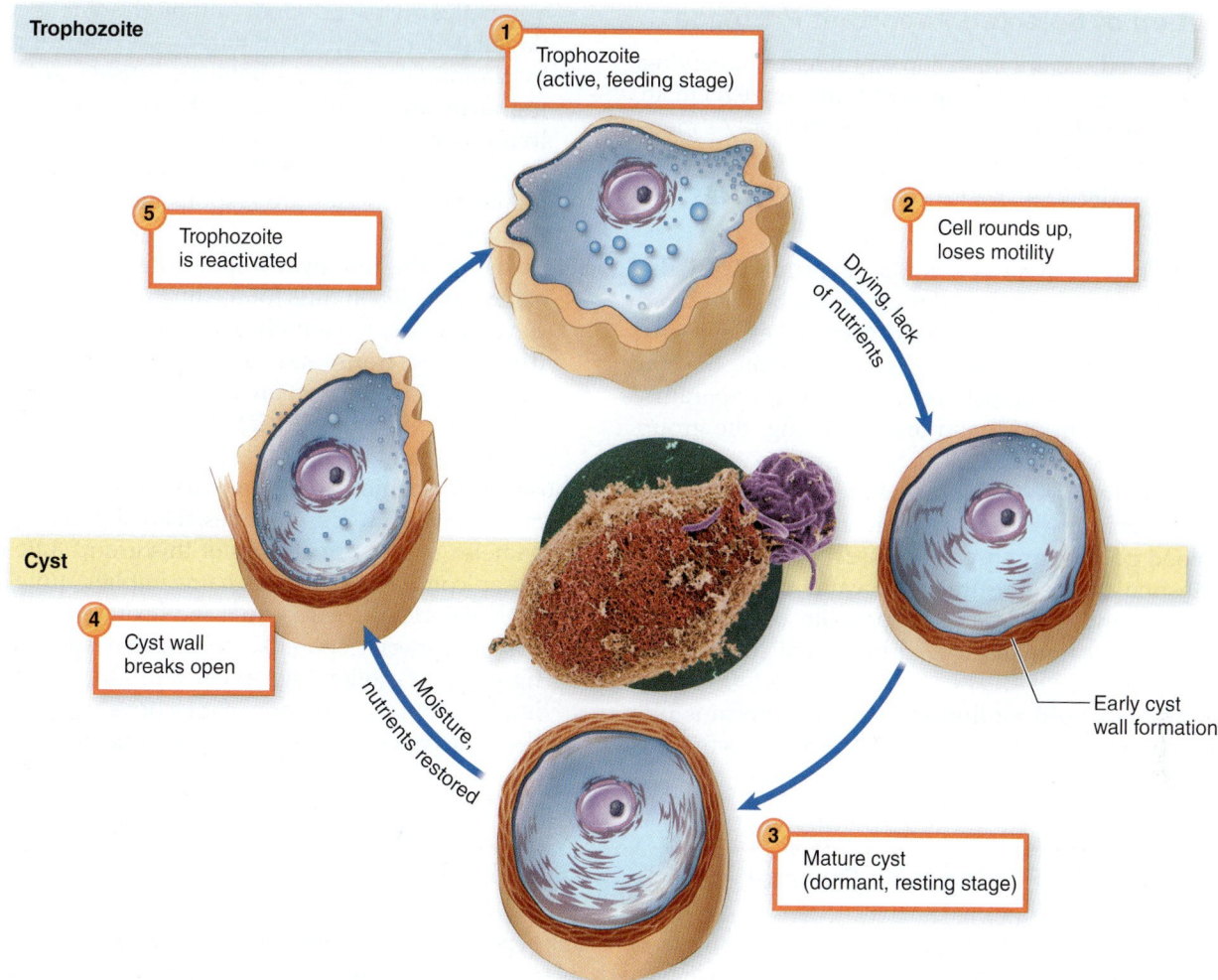

Trophozoite

1 Trophozoite (active, feeding stage)

5 Trophozoite is reactivated

2 Cell rounds up, loses motility

Drying, lack of nutrients

Cyst

4 Cyst wall breaks open

Early cyst wall formation

Moisture, nutrients restored

3 Mature cyst (dormant, resting stage)

Process Figure 5.23 The general life cycle exhibited by many protozoa. All protozoa have a trophozoite form, but not all produce cysts. The photo in the center shows a *Giardia* trophozoite (purple) emerging from its cyst form (orange).

nutrients, a cyst breaks open and releases the active trophozoite. Both the cyst and trophozoite forms of protozoan pathogens can be identified through O & P (ova and parasite) testing of patient stool samples. This method, combined with immunology-based tests, is currently used for disease diagnosis in cases of giardiasis and cryptosporidiosis (see chapters 17 and 22).

The life cycles of protozoans vary from simple to complex. Several protozoan groups exist only in the trophozoite state. Many alternate between a trophozoite and a cyst stage, depending on the conditions of the habitat. The life cycle of a parasitic protozoan dictates its mode of transmission to other hosts. For example, the flagellate *Trichomonas vaginalis* causes a common sexually transmitted disease. Because it does not form cysts, it is more delicate and must be transmitted by intimate contact between sexual partners. In contrast, intestinal pathogens such as *Cryptosporidium* and *Giardia lamblia* (process figure 5.23) form cysts and are readily transmitted in contaminated water and foods.

All protozoa reproduce by relatively simple, asexual methods, usually mitotic cell division. Several parasitic species, including the agents of malaria and toxoplasmosis, reproduce asexually by multiple fission inside a host cell. Sexual reproduction also occurs during the life cycle of most protozoa. Ciliates participate in **conjugation,** a form of genetic exchange in which two cells fuse temporarily and exchange micronuclei. This process of sexual recombination yields new and different genetic combinations that can be advantageous in evolution.

Many protozoa engulf toxic bacteria and maintain them in their cytoplasm. This, in turn, can make them toxic.

Classification of Selected Important Protozoa

As has been stated, taxonomists have problems classifying protozoa. We will use a functional way to categorize them, in a way that will be most useful in a clinical situation. As mentioned above, protozoa can be placed in four groups based on how they move. These categories are summarized here and in figure 5.22.

Those using flagella to move. Motility is primarily by flagella alone or by both flagellar and amoeboid motion. Single nucleus. Sexual reproduction, when present, by syngamy; division by longitudinal fission. Several parasitic forms lack mitochondria and Golgi apparatus. Most species form cysts and are free-living; the group also includes several parasites. Some species are found in loose aggregates or colonies, but most are solitary. Members include *Trypanosoma* and *Leishmania*, important blood pathogens spread by insect vectors; *Giardia*, an intestinal parasite spread in water contaminated with feces; and *Trichomonas*, a parasite of the reproductive tract of humans spread by sexual contact (see figure 5.22*a*).

Those using amoeboid motion to move. Cell form is primarily an amoeba (see figure 5.22*b*). Major locomotor organelles are pseudopods, although some species have flagellated reproductive states. Asexual reproduction by fission. Two groups have an external shell; mostly uninucleate; usually encyst. Most amoebas are free-living and not infectious; *Entamoeba* is a pathogen or parasite of humans; shelled amoebas called foraminifera and radiolarians are responsible for chalk deposits in the ocean.

Those using cilia to move. Trophozoites are motile by cilia; some have cilia in tufts for feeding and attachment; most develop cysts; have both macronuclei and micronuclei; division by transverse fission; most have a definite mouth and feeding organelle; show relatively advanced behavior (see figure 5.22*c*). The majority of ciliates are free-living and harmless.

Those with no motility (Sporozoa). Although motility is absent in most representatives, it is exhibited by the male gametes of many members of this group. Life cycles of the apicomplexa are, as the name implies, quite complex, with well-developed asexual and sexual stages. Sporozoa produce special sporelike cells called **sporozoites** (see figure 5.22*d*) following sexual reproduction, which are important in transmission of infections and have recently been discovered to exhibit a unique form of gliding

---

INSIGHT 5.3   Flirting with Disaster

Toxoplasma gondii is a eukaryotic protozoan excreted in the feces of cats. Although this is a parasite to the cat, it can have a hidden benefit: It will attract an easy meal. Studies have shown that rats infected with *Toxoplasma* show a reduced fear response to cats. In fact, the neural response in the rat is more closely related to a sexual response. Instead of a fight-or-flight response, the rat cozies up to the cat and unwittingly becomes its next meal. Robert Sapolsky, a professor of biology at Stanford, has shown that *Toxoplasma* has a preference for the amygdala, the emotional center of the brain, where it confuses the fear and attraction responses. Nearly one-third of the world's human population is infected with *Toxoplasma* as well. The organism is capable of crossing the placenta in pregnant women, causing complications, and can be fatal in individuals with compromised immune systems. In individuals with normal immune systems, the effects of infection are subtler: Studies in the United Kingdom have shown that *Toxoplasma gondii* affects dopamine levels in the brain. Dr. Glenn McConkey at the University of Leeds has found a link between *Toxoplasma* infection and schizophrenia. Although there are numerous genetic triggers for schizophrenia, Dr. McConkey's work shows that infection with *Toxoplasma* can have an impact on the development of schizophrenia in some individuals. Other studies in Europe have indicated that individuals infected with the parasite may be at greater risk for suicide.

Scientists have also noted that infection with *Toxoplasma* can lead to thrill-seeking behaviors and slower reaction times. In 2009, a study in Czechoslovakia found that individuals who were infected with *Toxoplasma* and had an Rh-negative blood type were 2.5 times more likely to be in a car accident than drivers who were free of infection. The study extrapolated that 400,000 to 1 million car crash deaths worldwide could be linked to *Toxoplasma* infection. Apparently, a protozoan infection can leave you flirting with disaster.

Source: Kathleen McAuliffe, "How Your Cat Is Making You Crazy," *The Atlantic* (March 5, 2012). Retrieved from www.theatlantic.com/magazine/archive/2012/03/how-your-cat-is-making-you-crazy/308873/#.

motility. Most sporozoa form thick-walled zygotes called oocysts, and this entire group of organisms is parasitic. *Plasmodium*, the most prevalent protozoan parasite, causes 100 million to 300 million cases of malaria each year worldwide. It is an intracellular parasite with a complex cycle alternating between humans and mosquitoes. *Toxoplasma gondii* causes infection (toxoplasmosis) in humans, which is acquired from cats and other animals.

Just as with bacteria and other eukaryotes, protozoans that cause disease produce symptoms in different organ systems. These diseases are covered in chapters 18 through 23.

Protozoan Identification and Cultivation

The unique appearance of most protozoa makes it possible for a knowledgeable person to identify them to the level of genus—and often species—by microscopic morphology alone. Characteristics to consider in identification include the shape and size of the cell; the type, number, and distribution of locomotor structures; the presence of special organelles or cysts; and the number of nuclei. Medical specimens taken from blood, sputum, cerebrospinal fluid, feces, or the vagina are smeared directly onto a slide and observed with or without special stains. Occasionally, protozoa are cultivated on artificial media or in laboratory animals for further identification or study.

Important Protozoan Pathogens

Although protozoan infections are very common, they are actually caused by only a small number of species often restricted geographically to the tropics and subtropics **(table 5.4).** In this section, we look at an example of a very common protozoan disease that illustrates some of the main features of protozoan diseases.

The study of protozoa and helminths is sometimes called *parasitology*. Although, technically, a parasite can be any organism that obtains food and other requirements at the expense of a host, the term *parasite* is most often used to denote protozoan and helminth pathogens.

Pathogenic Flagellates: Trypanosomes

Trypanosomes are protozoa belonging to the genus *Trypanosoma* (try-pan"-oh-soh'-mah). The two most important representatives are *T. brucei* and *T. cruzi*, species that are closely related but geographically distant. *Trypanosoma brucei* occurs in Africa, where it causes approximately 5,000 new cases of sleeping sickness each year (see chapter 19). *Trypanosoma cruzi*, the cause of Chagas disease, is endemic to South and Central America, where it infects several million people a year. Both species have long, crescent-shaped cells with a single flagellum that is sometimes attached to the cell body by an undulating membrane. Both are found in the blood during infection and are transmitted by blood-sucking vectors. We use *T. cruzi* to illustrate the phases of a trypanosomal life cycle and to demonstrate the complexity of parasitic relationships.

The trypanosome of Chagas disease relies on the close relationship of a warm-blooded mammal and an insect that feeds on mammalian blood. There are many different mammalian hosts, including dogs, cats, opossums, armadillos, and foxes. The insect vector is a group of bugs called the triatomines. These bugs are sometimes called "kissing bugs" because of their habit of biting their hosts at the corner of the mouth. Transmission occurs from bug to mammal and from

Table 5.4 Major Pathogenic Protozoa

Protozoan	Disease	Reservoir/Source
Amoeboid Protozoa		
Entamoeba histolytica	Amoebiasis (intestinal and other symptoms)	Humans, water and food
Naegleria, Acanthamoeba	Brain infection	Free-living in water
Ciliated Protozoa		
Balantidium coli	Balantidiosis (intestinal and other symptoms)	Pigs, cattle
Flagellated Protozoa		
Giardia lamblia	Giardiasis (intestinal distress)	Animals, water and food
Trichomonas vaginalis	Trichomoniasis (vaginal symptoms)	Human
Trypanosoma brucei, T. cruzi	Trypanosomiasis (intestinal distress and widespread organ damage)	Animals, vector-borne
Leishmania donovani, L. tropica, L. brasiliensis	Leishmaniasis (either skin lesions or widespread involvement of internal organs)	Animals, vector-borne
Nonmotile Protozoa		
Plasmodium vivax, P. falciparum, P. malariae	Malaria (cardiovascular and other symptoms)	Human, vector-borne
Toxoplasma gondii	Toxoplasmosis (flulike illness)	Animals, vector-borne
Cryptosporidium	Cryptosporidiosis (intestinal and other symptoms)	Free-living, water, food
Cyclospora cayetanensis	Cyclosporiasis (intestinal and other symptoms)	Water, fresh produce

mammal to bug but usually not from mammal to mammal, except across the placenta during pregnancy. The general phases of this cycle are presented in **figure 5.24.**

The trypanosome trophozoite multiplies in the intestinal tract of the reduviid bug and is harbored in the feces. The bug seeks a host and bites the mucous membranes, usually of the eye, nose, or lips. As it fills with blood, the bug defecates on the bite site and contaminates it with feces containing the trypanosome. Ironically, the victims themselves inadvertently contribute to the entry of the microbe by scratching the bite wound. The trypanosomes ultimately become established and multiply in muscle and white blood cells. Periodically, these parasitized cells rupture, releasing large numbers of new trophozoites into the blood. Eventually, the trypanosome can spread to many systems, including the lymphoid organs, heart, liver, and

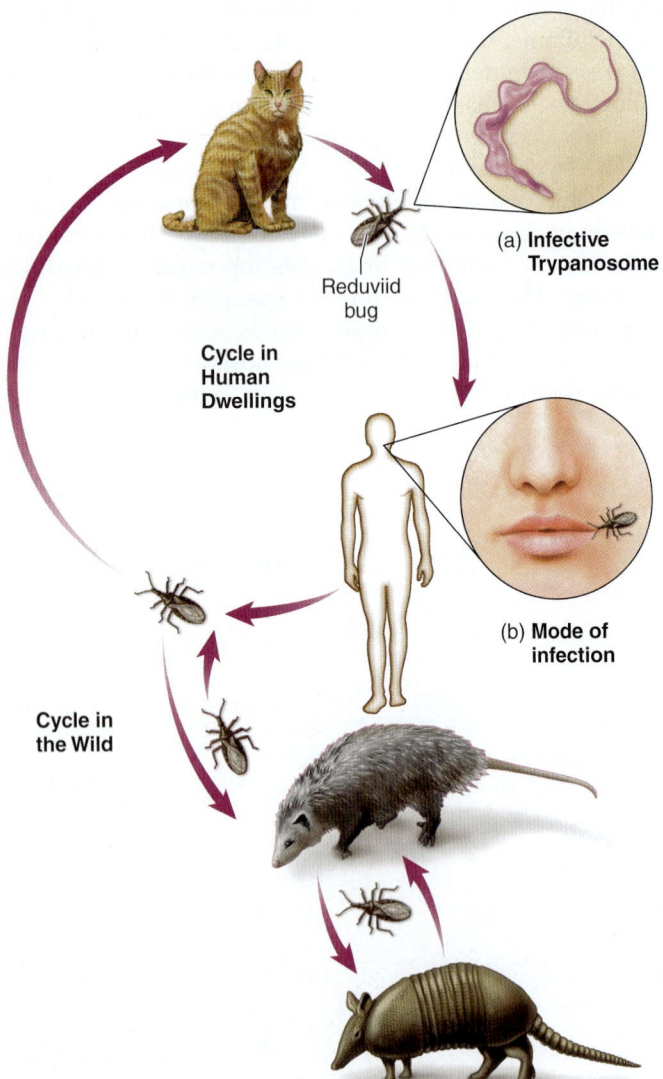

Figure 5.24 Cycle of transmission in Chagas disease.
Trypanosomes (inset *a*) are transmitted among mammalian hosts and human hosts by means of a bite from the kissing bug (inset *b*).

brain. Manifestations of the resultant disease range from mild to very severe and include fever, inflammation, and heart and brain damage. In many cases, the disease continues for an extended time and can cause death.

Infective Amoebas: *Entamoeba*

Several species of amoebas cause disease in humans (see the Case File in this chapter), but probably the most common disease is amoebiasis, or amoebic dysentery, caused by *Entamoeba histolytica* (see chapter 22). This microbe is widely distributed in the world, from northern zones to the tropics, and is nearly always associated with humans. Amoebic dysentery—or severe diarrhea—is the fourth most common protozoan infection in the world. This microbe has a life cycle quite different from the trypanosomes in that it does not involve multiple hosts and a blood-sucking vector. It lives part of its cycle as a trophozoite and part as a cyst. Because the cyst is the more resistant form and can survive in water and soil for several weeks, it is the more important stage for transmission. The primary way that people become infected is by ingesting food or water contaminated with human feces.

5.5 Learning Outcomes—Assess Your Progress

18. Note the protozoan characteristics that illustrate why they are informally placed into a single group.
19. List three means of locomotion exhibited by protozoa.
20. Explain why a cyst stage may be useful in a protozoan.
21. Give an example of a human disease caused by each of the four types of protozoa.

5.6 The Helminths

Tapeworms, flukes, and roundworms are collectively called *helminths*, from the Greek word meaning "worm." Adult animals are usually large enough to be seen with the naked eye, and they range from the longest tapeworms, measuring up to about 25 m in length, to roundworms less than 1 mm in length. Nevertheless, they are included among microorganisms because of their infective abilities and because the microscope is necessary to identify their eggs and larvae.

On the basis of body type, the two major groups of parasitic helminths are the flatworms (Phylum Platyhelminthes) and the roundworms (Phylum Aschelminthes, also called **nematodes**). Flatworms have a very thin, often segmented body plan **(figure 5.25),** and roundworms have a long, cylindrical, unsegmented body **(figure 5.26).** The flatworm group is subdivided into the **cestodes,** or tapeworms, named for their long, ribbonlike arrangement, and the **trematodes,** or flukes, characterized by flat, ovoid bodies. Not all flatworms and roundworms are parasites by nature; many live free in soil and water. Because most disease-causing helminths spend part of their lives in the gastrointestinal tract, they are discussed in chapter 22.

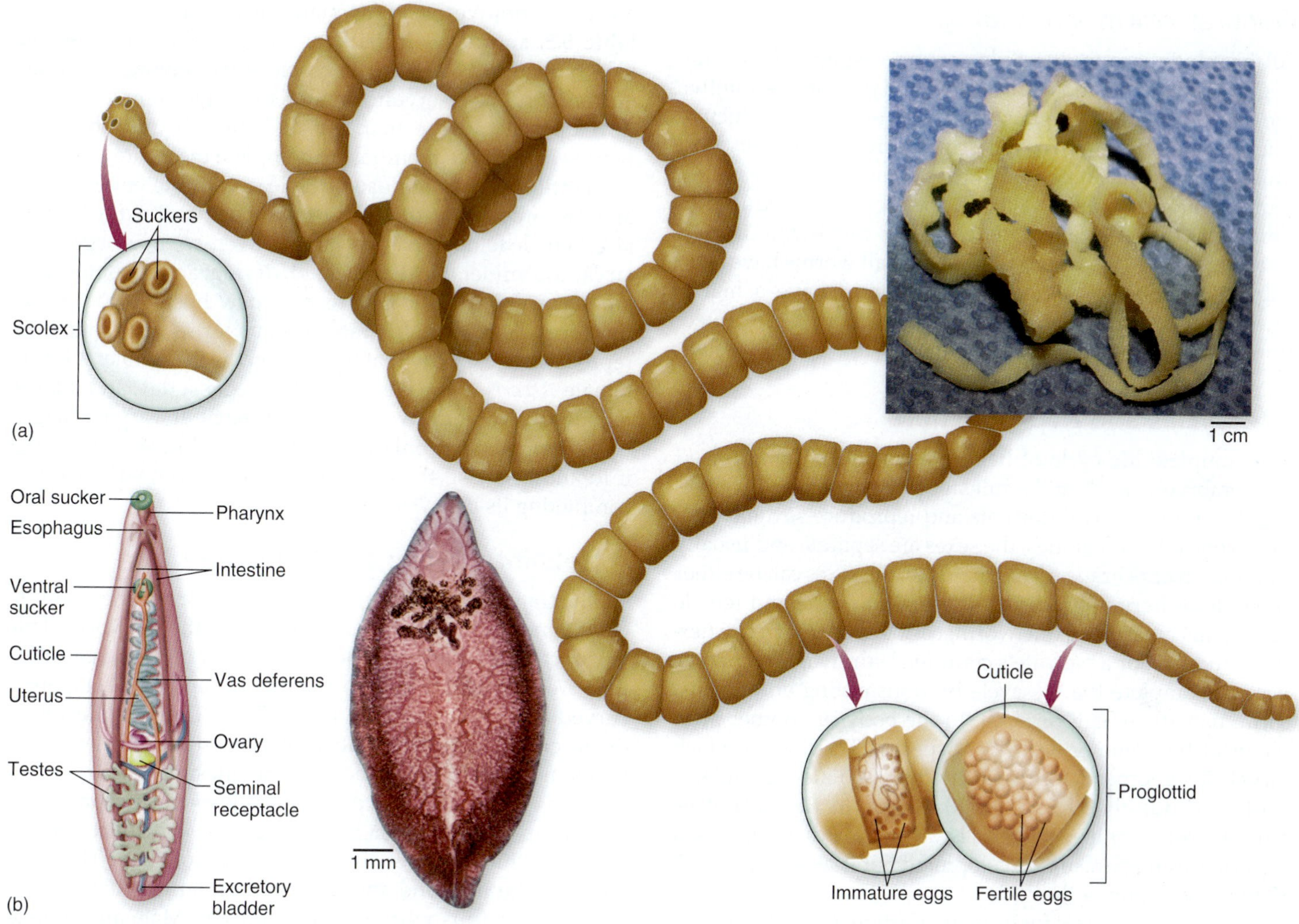

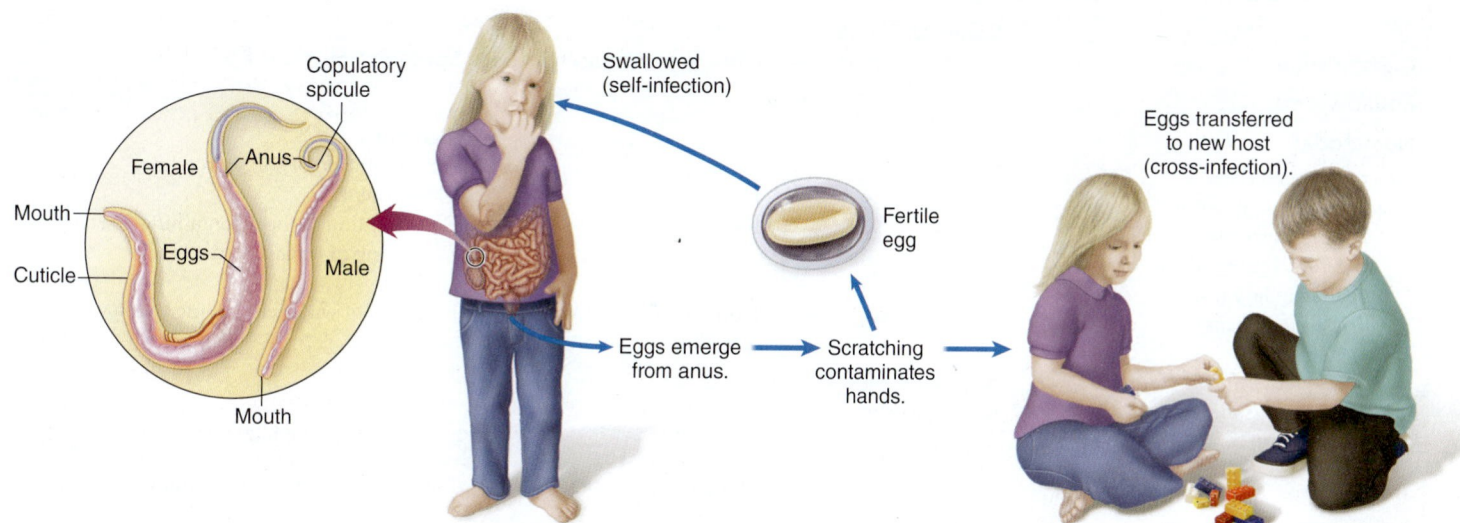

Figure 5.25 Parasitic flatworms. **(a)** A cestode (tapeworm), showing the scolex; long, tapelike body; and magnified views of immature and mature proglottids (body segments). The photo shows an actual tapeworm. **(b)** The structure of a trematode (liver fluke). Note the suckers that attach to host tissue and the dominance of reproductive and digestive organs. The photo shows an actual liver fluke.

Figure 5.26 The life cycle of the pinworm, a roundworm. Eggs are the infective stage and are transmitted by unclean hands. Children frequently reinfect themselves and also pass the parasite on to others.

General Worm Morphology

All helminths are multicellular animals equipped to some degree with organs and organ systems. In parasitic helminths, the most developed organs are those of the reproductive tract, with more primitive digestive, excretory, nervous, and muscular systems. In particular groups, such as the cestodes, reproduction is so dominant that the worms are reduced to little more than a series of flattened sacs filled with ovaries, testes, and eggs (see figure 5.25a,b). Not all worms have such extreme adaptations as cestodes, but most have a highly developed reproductive potential, thick cuticles for protection, and mouth glands for breaking down the host's tissue.

Life Cycles and Reproduction

The complete life cycle of helminths includes the fertilized egg (embryo), larval, and adult stages. In the majority of helminths, adults derive nutrients and reproduce sexually in a host's body. In nematodes, the sexes are separate and usually different in appearance; in trematodes, the sexes can be either separate or **hermaphroditic,** meaning that male and female sex organs are in the same worm; cestodes are generally hermaphroditic. For a parasite's continued survival as a species, it must complete the life cycle by transmitting an infective form, usually an egg or larva, to the body of another host, either of the same or a different species. The host in which larval development occurs is the intermediate (secondary) host, and adulthood and mating occur in the **definitive (final) host.** A transport host is an intermediate host that experiences no parasitic development but is an essential link in the completion of the cycle.

In general, sources for human infection are contaminated food, soil, and water or infected animals; routes of infection are by oral intake or penetration of unbroken skin. Humans are the definitive hosts for many of the parasites listed in **table 5.5,** and in about half the diseases, they are also the sole biological reservoir. In other cases, animals or insect vectors serve as reservoirs or are required to complete worm development. In the majority of helminth infections, the worms must leave their host to complete the entire life cycle.

Fertilized eggs are usually released to the environment and are provided with a protective shell and extra food to aid their development into larvae. Even so, most eggs and larvae are vulnerable to heat, cold, drying, and predators and are destroyed or unable to reach a new host. To counteract this formidable mortality rate, certain worms have adapted a reproductive capacity that borders on the incredible: A single female *Ascaris* worm can lay 200,000 eggs a day, and a large female can contain over 25 million eggs at varying stages of development! If only a tiny number of these eggs make it to another host, the parasite will have been successful in completing its life cycle.

A Helminth Cycle: The Pinworm

To illustrate a helminth cycle in humans, we use the example of a roundworm, *Enterobius vermicularis,* the pinworm. This worm causes a very common infestation of the large intestine. Worms range from 2 to 12 mm long and have a tapered, curved cylinder shape (see figure 5.26). The condition they cause, enterobiasis, is usually a simple, uncomplicated infection that does not spread beyond the intestine.

A cycle starts when a person swallows microscopic eggs picked up from another infected person by direct contact or by touching articles that person has touched. The eggs hatch in the intestine and then release larvae that mature into adult worms within about 1 month. Male and female worms mate, and the female migrates out to the anus to deposit eggs, which cause intense itchiness that is relieved

Table 5.5 Examples of Helminths and Their Modes of Transmission

Classification	Common Name of Disease or Worm	Life Cycle Requirement	Spread to Humans By
Roundworms			
Nematodes			Ingestion
Intestinal Nematodes		Humans	Fecal pollution of soil with eggs
Infective in egg (embryo) stage		Humans	Close contact
Ascaris lumbricoides	Ascariasis	Pigs, wild mammals	Consumption of meat containing larvae
Enterobius vermicularis	Pinworm		Burrowing of larva into tissue
Infective in larval stage		Humans, black flies	Fly bite
Trichinella spiralis	Trichina worm	Humans and *Cyclops* (an aquatic invertebrate)	Ingestion of water containing *Cyclops*
Tissue Nematodes			
Onchocerca volvulus	River blindness		
Dracunculus medinensis	Guinea worm		
Flatworms			
Trematodes		Humans and snails	Ingestion of fresh water containing larval stage
Schistosoma japonicum	Blood fluke		
Cestodes		Humans, swine	Consumption of undercooked or raw pork
Taenia solium	Pork tapeworm	Humans, fish	Consumption of undercooked or raw fish
Diphyllobothrium latum	Fish tapeworm		

Table 5.6 Variation Among Eukaryotes

	Protozoa	Fungi	Algae	Helminths	Humans
Level of complexity	Always unicellular	Unicellular/Multicellular	Unicellular/Multicellular	Multicellular (adults) Unicellular (ova, larva)	Multicellular
Cell wall	None	Chitin or cellulose	Cellulose	None	None
Cytoplasm	Divided (endoplasm/ectoplasm)	Not divided	Not divided	Not divided	Not divided
Nutritional type	Heterotrophic/Autotrophic	Heterotrophic	Heterotrophic/Autotrophic	Heterotrophic	Heterotrophic
Motility	Flagella, cilia, pseudopodia, or none	Flagella (gametes)	Flagella (gametes)	Flagella (gametes)	Limbs
Important structures for identification	Cysts	Hyphae/spores	Chloroplasts	Ova	None

by scratching. Herein lies a significant means of dispersal: Scratching contaminates the fingers, which, in turn, transfer eggs to bedclothes and other inanimate objects. This person becomes a host and a source of eggs and can spread them to others in addition to reinfesting himself or herself. Enterobiasis occurs most often among families and in other close living situations. Its distribution is worldwide among all socioeconomic groups, but it seems to attack children more frequently than adults.

Helminth Classification and Identification

The helminths are classified according to their shape; their size; the degree of development of various organs; the presence of hooks, suckers, or other special structures; the mode of reproduction; the kinds of hosts; and the appearance of eggs and larvae. They are identified in the laboratory by microscopic detection of the adult worm or its larvae and eggs, which often have distinctive shapes or external and internal structures. Occasionally, they are cultured in order to verify all of the life stages.

Distribution and Importance of Parasitic Worms

About 50 species of helminths parasitize humans. They are distributed in all areas of the world that support human life. Some worms are restricted to a given geographic region, and many have a higher incidence in tropical areas. This knowledge must be tempered with the realization that jet-age travel, along with human migration, is gradually changing the patterns of worm infections, especially of those species that do not require alternate hosts or special climatic conditions for development. The yearly estimate of worldwide cases numbers in the billions, and these are not confined to developing countries. A conservative estimate places 50 million helminth infections in North America alone.

It is important to realize that humans evolved on this planet in the constant presence of helminths—and indeed *all* types of microbes. Only very recently, in evolutionary

terms, have some pockets of humankind been relatively free of helminth colonization. The absence of worm infections may in fact be leading to some of the "newer" conditions we encounter, such as autoimmunity and allergy. We'll talk more about this later.

You have now learned about the variety of organisms that microbiologists study and classify. This chapter contained a very short description of the extremely complex variety of eukaryotic organisms. **Table 5.6** will help you differentiate among these and compare them to a familiar eukaryotic organism, the human. In chapter 6, you will continue the exploration of potential pathogens as you investigate the "not-quite-organisms," namely, viruses.

Case File 5   *Wrap-Up*

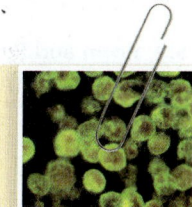

Sinus sufferers often use a sinus-irrigation device called a neti pot. Warm water mixed with a saline solution is flushed through the sinuses, drawing out mucus and alleviating sinus pressure. This technique originated in India and has been used for centuries to relieve sinus symptoms caused by the common cold, allergies, or infection. Unfortunately, the neti pot is also an effective way to introduce *Naegleria fowleri* into the nasal cavity. In 2011, two patients died from PAM in Louisiana after using a neti pot. Although tap water is safe for drinking, it may still harbor the amoeba. State epidemiologists recommend using distilled, sterile, or previously boiled water in neti pots.

Petrochko, C. 2011. http://www.medpagetoday.com/InfectiousDisease/InfectionControl/30283

5.6 Learning Outcomes—Assess Your Progress

22. List the two major groups of helminths and provide examples representing each body type.

23. Summarize the stages of a typical helminth life cycle.

Chapter Summary

5.1 The History of Eukaryotes (ASM Guideline* 1.1)

- Eukaryotes are cells with a nucleus and organelles compartmentalized by membranes. They, like bacteria, originated from a primitive cell referred to as the Last Common Ancestor. Eukaryotic cell structure enabled eukaryotes to diversify from single cells into a huge variety of complex multicellular forms.

- The cell structures common to most eukaryotes are the cell membrane, nucleus, vacuoles, mitochondria, endoplasmic reticulum, Golgi apparatus, and a cytoskeleton. Cell walls, chloroplasts, and locomotor organs are present in some eukaryote groups.

5.2 Form and Function of the Eukaryotic Cell: External Structures and Boundary Structures (ASM Guideline 2.4)

- Microscopic eukaryotes use locomotor organs such as flagella or cilia for moving themselves or their food.

- The glycocalyx is the outermost boundary of most eukaryotic cells. Its functions are protection, adherence, and reception of chemical signals from the environment or from other organisms. The glycocalyx is supported by either a cell wall or a cell membrane.

- The cytoplasmic (cell) membrane of eukaryotes is similar in function to that of bacteria, but it differs in composition, possessing sterols as additional stabilizing agents.

5.3 Form and Function of the Eukaryotic Cell: Internal Structures (ASM Guideline 2.4)

- The genome of eukaryotes is located in the nucleus, a spherical structure surrounded by a double membrane. The nucleus contains the nucleolus, the site of ribosome synthesis. DNA is organized into chromosomes in the nucleus.

- The endoplasmic reticulum (ER) is an internal network of membranous passageways extending throughout the cell.

- The Golgi apparatus is a packaging center that receives materials from the ER and then forms vesicles around them for storage or for transport to the cell membrane for secretion.

- The mitochondria generate energy in the form of ATP to be used in numerous cellular activities.

- Chloroplasts, membranous packets found in plants and algae, are used in photosynthesis.

- Ribosomes are the sites for protein synthesis present in both eukaryotes and bacteria.

- The cytoskeleton maintains the shape of cells and produces movement of cytoplasm within the cell; movement of chromosomes at cell division; and, in some groups, movement of the cell as a unit.

5.4 The Fungi (ASM Guidelines 2.4, 5.4)

- The fungi are nonphotosynthetic species with cell walls. They are either saprobes or parasites and may be unicellular, colonial, or multicellular.

- All fungi are heterotrophic.

- Fungi have many reproductive strategies, including both asexual and sexual.

- Fungi have asexual spores called sporangiospores and conidiospores.

- Fungal sexual spores enable the organisms to acquire variation in their form and function.

- Fungi are often identified on the basis of their microscopic appearance.

- There are two categories of fungi that cause human disease: the primary pathogens, which infect healthy persons, and the opportunistic pathogens, which cause disease only in compromised hosts.

5.5 The Protists (ASM Guidelines 2.4, 5.4)

- The protists are mostly unicellular or colonial eukaryotes that lack specialized tissues. There are two major organism types: the algae and the protozoa.

- Algae are photosynthetic organisms that contain chloroplasts with chlorophyll and other pigments.

- Protozoa are heterotrophs that are categorized based on how they move. Most are single-celled trophozoites, and many produce a resistant stage, or cyst.

5.6 The Helminths (ASM Guidelines 2.4, 5.4)

- The Kingdom Animalia has only one group that contains members that are studied in microbiology. These are the helminths or worms. Parasitic members include flatworms and roundworms that are able to invade and reproduce in human tissues.

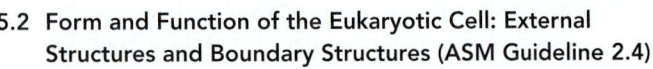
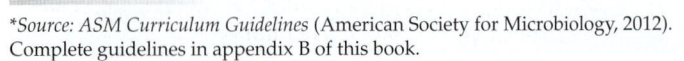

*Source: ASM Curriculum Guidelines (American Society for Microbiology, 2012). Complete guidelines in appendix B of this book.

Multiple-Choice and True-False Questions | Bloom's Levels 1 and 2: Remember and Understand

Multiple-Choice Questions. Select the correct answer from the options provided.

1. Both flagella and cilia are found primarily in
   a. algae.   c. fungi.
   b. protozoa.   d. both b and c.

2. Features of the nuclear envelope include
   a. ribosomes.
   b. a double-membrane structure.
   c. pores that allow communication with the cytoplasm.
   d. b and c.
   e. all of these.

3. The cell wall is found in which eukaryotes?
   a. fungi   c. protozoa
   b. algae   d. a and b

4. Yeasts are _____ fungi, and molds are _____ fungi.
   a. macroscopic, microscopic
   b. unicellular, filamentous
   c. motile, nonmotile
   d. water, terrestrial

5. Algae generally contain some type of
   a. spore.   c. locomotor organelle.
   b. chlorophyll.   d. toxin.

6. Almost all protozoa have a
   a. locomotor organelle. c. pellicle.
   b. cyst stage.   d. trophozoite stage.

7. All mature sporozoa are
   a. parasitic.   c. carried by vectors.
   b. nonmotile.   d. both a and b.

8. Parasitic helminths reproduce with
   a. spores.   d. cysts.
   b. eggs and sperm.   e. all of these.
   c. mitosis.

9. Mitochondria likely originated from
   a. archaea.
   b. invaginations of the cell membrane.
   c. bacteria.
   d. chloroplasts.

10. Most helminth infections
    a. are localized to one site in the body.
    b. spread through major systems of the body.
    c. develop within the spleen.
    d. develop within the liver.

True-False Questions. If the statement is true, leave as is. If it is false, correct it by rewriting the sentence.

11. Bacteria and eukaryotes arose from the same kind of primordial cell.

12. Hyphae that are divided into compartments by cross walls are called septate hyphae.

13. The infective stage of a protozoan is the trophozoite.

14. In humans, fungi can only infect the skin.

15. Fungi generally derive nutrients through photosynthesis.

Critical Thinking Questions | Bloom's Levels 3, 4, and 5: Apply, Analyze, and Evaluate

Critical thinking is the ability to reason and solve problems using facts and concepts. These questions can be approached from a number of angles and, in most cases, they do not have a single correct answer.

1. Summarize the endosymbiotic theory and explain how it accounts for major structural similarities and differences between bacterial and eukaryotic cells.

2. Using the analogy that a cell is like a city, describe the anatomy and functions of each of the major eukaryotic organelles in terms of the role it would play within that fictional city.

3. Compare and contrast the structure and function of the following among bacteria and eukaryotes:
   a. ribosome
   b. flagellum
   c. glycocalyx

4. Write a paragraph illustrating the life of a protein, from DNA to mature polypeptide, and the course of its travels within a cell throughout its synthesis.

5. a. Describe the process of spore formation and explain how a single fungal spore can lead to the development of a mature mycelium.
   b. *Candida albicans* is a *dimorphic fungus*. Explain what is meant by this term and discuss which form of growth allows this organism to establish infections within human tissues.

6. Provide at least four examples illustrating both beneficial and detrimental aspects of fungi in the modern world today.

7. a. Describe three means of motility exhibited by protozoans.
   b. Briefly outline the characteristics of the four protozoan groups, listing at least one important pathogen in each group.

8. Summarize the general life cycle of a protozoan, explaining the importance of the various stages in disease transmission and species identification.

9. a. List the three stages of a complete helminth life cycle and define the role of a definitive host and an intermediate host within this process.
   b. Discuss at least three major structural adaptations exhibited by the helminths that provide them the ability to withstand a host environment and multiply to great numbers.

10. Sam has been suffering from diarrhea and abdominal bloating for nearly 2 weeks. Based upon the results of his stool analysis in the laboratory, answer the following:
    a. Single-celled organisms were found in the stool sample. Explain whether this evidence indicates an infection caused by a fungus, a protozoan, or a helminth.

b. The single-celled organisms in the stool sample were flagellated. Explain whether this evidence indicates an infection caused by a fungus, a protozoan, or a helminth.

c. Small structures called cysts were found in the stool sample as well as in the well water in the home of the patient. Explain whether this evidence indicates an infection caused by a fungus, a protozoan, or a helminth.

Concept Connections | Bloom's Levels 4 and 6: Analyze and Create

This activity ties together multiple concepts in the chapter.

1. What is the endosymbiotic theory?
2. What eukaryotic organelles most closely resemble bacteria?
3. What other characteristics of these organelles suggest they evolved from bacteria?

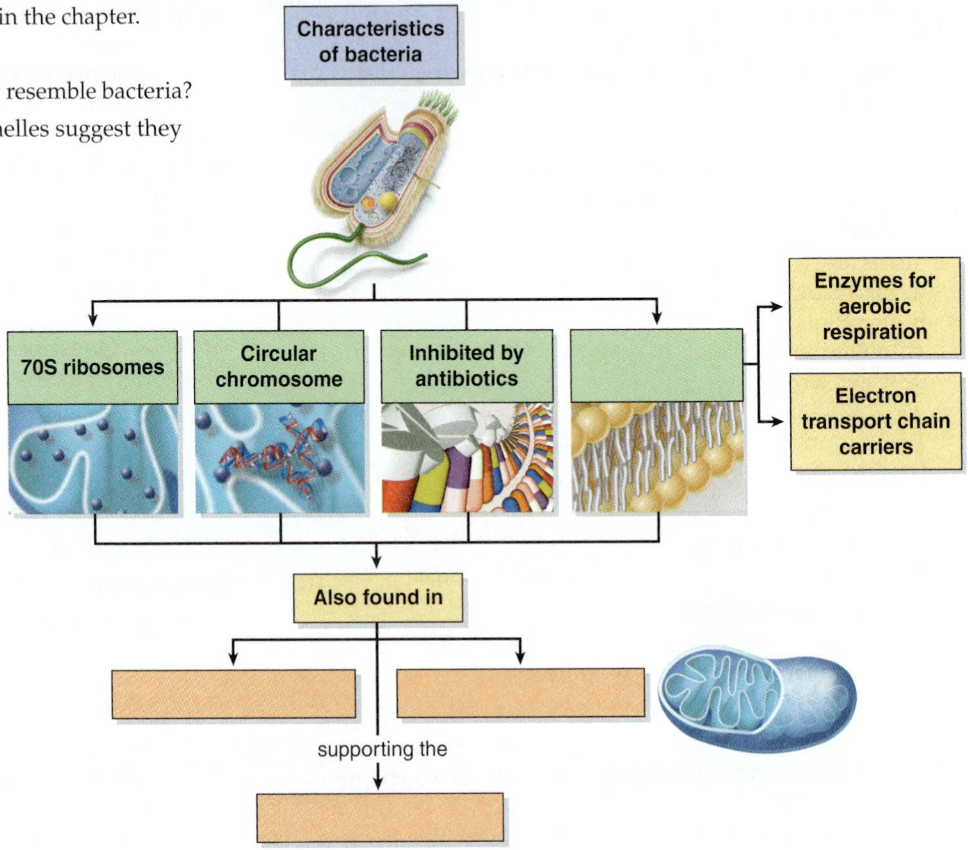

Characteristics of bacteria

70S ribosomes

Circular chromosome

Inhibited by antibiotics

Enzymes for aerobic respiration

Electron transport chain carriers

Also found in

supporting the

Visual Connections | Bloom's Level 5: Evaluate

These questions use visual images or previous content to make connections to this chapter's concepts.

1. **From chapter 4, figure 4.25a.** You may have seen similar sites in the environment. Can you think of two locations you have encountered that showed colorful evidence of microbial growth?

2. **From chapter 1, figure 1.14.** Which of the groups of organisms from this figure will contain a nucleus? Why?

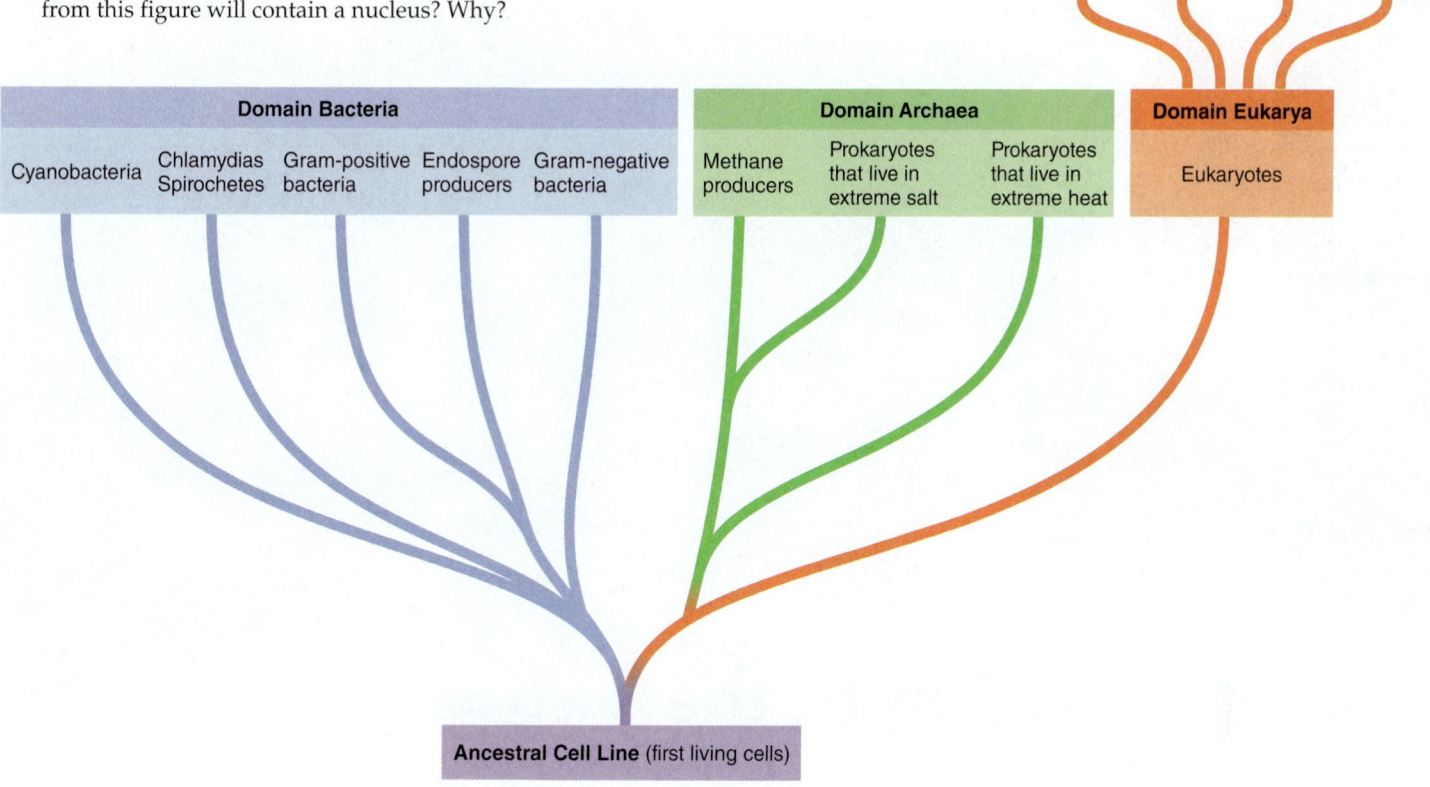

Concept Mapping | Bloom's Level 6: Create

Appendix D provides guidance for working with concept maps.

1. Using the words that follow, please create a concept map illustrating the relationships among these key terms from chapter 5.

Golgi apparatus	cytoplasm	ribosomes	nucleolus
chloroplasts	endospore	flagella	

www.mcgrawhillconnect.com

Enhance your study of this chapter with study tools and practice tests. Also ask your instructor about the resources available through ConnectPlus, including the media-rich eBook, interactive learning tools, and animations.

6

An Introduction to the Viruses

Case File 6

Flu: The Real Deal

Influenza: The word strikes fear in the hearts of many. These tiny viral particles, composed of no more than a bit of genetic material and a protein coat, can hijack your cells, produce more viruses, and spread to other cells and other people, often with disastrous consequences. Not only do people dread the symptoms of fever, cough, sore throat, stuffy nose, head and body aches, and fatigue, but pandemics of influenza have swept across the globe in waves, killing millions. Since 1918, there have been four major pandemics: "The Spanish flu" occurred in 1918 and sickened 20%–40% of the world's population, killing nearly 50 million people. In 1957, "the Asian flu" killed 1–2 million people. Nearly a decade later, "the Hong Kong flu" spread throughout the world, with deaths estimated around 1 million, with the highest death rates among individuals 65 and over. It wasn't until 2009 that another flu pandemic began in Mexico, spreading to the United States and affecting nearly 74 countries—it was dubbed "the swine flu." Despite 80 million people vaccinated against the disease, the CDC estimates that 43–89 million people became ill with the virus and it caused tens of thousands of deaths.

Each of these pandemics was caused by a different strain of influenza. Strains of influenza virus are identified by characteristics of the hemagglutinin (H) and neuraminidase (N) **capsid** proteins. The influenza virus, with its segmented RNA genome, is able to mutate rapidly. Mutations can result in changes to the capsid proteins that allow the virus to evade the immune system. Sixteen different types of H antigens and nine different types of N antigens have been identified; each viral strain is identified by the numbered type of each H and N antigen found on its capsid. For example, the Spanish flu pandemic was caused by the H1N1 strain, the Asian flu pandemic was caused by the H2N2 strain, and the Hong Kong flu was caused by the H3N2 strain. The 2009 swine flu pandemic was caused by a novel strain of H1N1.

- What other animals can develop influenza infections?
- What are the possible consequences of influenza virus transmission between animals and humans?

Continuing the Case appears on page 161.

Outline and Learning Outcomes

6.1 The Search for the Elusive Viruses
1. Describe the significance of viruses being recognized as "filterable."

6.2 The Position of Viruses in the Biological Spectrum
2. Summarize arguments on both sides of the debate regarding the classification of viruses as living organisms.
3. Identify effective terms to describe the behavior of viruses.

6.3 The General Structure of Viruses
4. Discuss the size of viruses relative to other microorganisms.
5. Describe the function and structure(s) of viral capsids.
6. Distinguish between enveloped and naked viruses.
7. Explain the importance of viral surface proteins, or spikes.
8. Compare and contrast the composition of a viral genome to that of a cellular organism's genome.
9. Diagram the possible nucleic acid configurations exhibited by viruses.

6.4 How Viruses Are Classified and Named
10. Develop two arguments against assigning species names to viruses.
11. Demonstrate how family and genus names in viruses are written.

6.5 Modes of Viral Multiplication
12. Diagram the six-step life cycle of animal viruses.
13. Define the term *cytopathic effect* and provide one example.
14. Provide examples of persistent and transforming infections, describing their effects on the host.
15. Provide a thorough description of lysogenic and lytic bacteriophage infections.

6.6 Techniques in Cultivating and Identifying Animal Viruses
16. List the three principal purposes for cultivating viruses.
17. Describe three ways in which viruses are cultivated.

6.7 Other Noncellular Infectious Agents
18. List three noncellular infectious agents besides viruses.

6.8 Viruses and Human Health
19. Analyze the relative importance of viruses in human infection and disease.
20. Discuss the primary reason that antiviral drugs are more difficult to design than antibacterial drugs.

6.1 The Search for the Elusive Viruses

For many years, the cause of viral infections such as smallpox and polio was unknown, even though it was clear that the diseases were transmitted from person to person. The French scientist Louis Pasteur was certainly on the right track when he postulated that rabies was caused by a "living thing" smaller than bacteria, and in 1884 he was able to develop the first vaccine for rabies. Pasteur also proposed the term **virus** (Latin, "poison") to denote this special group of infectious agents.

The first substantial revelations about the unique characteristics of viruses occurred in the 1890s. First, D. Ivanovski and M. Beijerinck showed that a disease in tobacco was caused by a virus (tobacco mosaic virus). Then, Friedrich Loeffler and Paul Frosch discovered an animal virus that causes foot-and-mouth disease in cattle. These early researchers found that when infectious fluids from host organisms were passed through porcelain filters designed to trap bacteria, the filtrate remained infectious. This result proved that an infection could be caused by a cell-free fluid containing agents smaller than bacteria and thus first introduced the concept of a *filterable virus.*

Over the succeeding decades, a remarkable picture of the physical, chemical, and biological nature of viruses began to take form. Viruses continue to fascinate us. In addition to being the causes of a wide variety of diseases (ones you would expect and ones you would not), they seem to have actually determined how biological organisms, including humans, have turned out **(Insight 6.1).** Viruses continue to surprise us. In recent years we have discovered that not all viruses are tiny. Some can be just as big as bacteria.

6.1 Learning Outcomes—Assess Your Progress

1. Describe the significance of viruses being recognized as "filterable."

6.2 The Position of Viruses in the Biological Spectrum

Viruses are a unique group of biological entities known to infect every type of cell, including bacteria, algae, fungi, protozoa, plants, and animals, and are extremely abundant on our planet. Norwegian ocean waters have been found to contain 60,000 viruses in a single milliliter (less than a thimbleful) of water. Lake water contains many more—as many as 250 million viruses per milliliter. We are just beginning to understand the impact of these huge numbers of viruses in our environment. The exceptional and curious nature of viruses prompts numerous questions, including the following:

1. Are they organisms, that is, are they alive?
2. What role did viruses play in the evolution of life?
3. What are their distinctive biological characteristics?
4. How can particles so small, simple, and seemingly insignificant be capable of causing disease and death?
5. What is the connection between viruses and cancer?
6. What role did they play in the development of all other organisms?

In this chapter, we address these questions and many others.

The unusual structure and behavior of viruses have led to debates about their connection to the rest of the microbial world. One viewpoint holds that because viruses are unable to multiply independently from the host cell, they are not living things but are more akin to infectious molecules. Another viewpoint proposes that even though viruses do not exhibit most of the life processes of cells, they can direct them and thus are certainly more than inert and lifeless molecules. This view is the predominant one among scientists today. This debate has greater philosophical than practical importance when discussing disease because viruses are agents of disease and must be dealt with through control, therapy, and prevention, whether we regard them as living or not. In keeping with their special position in the biological spectrum, it is best to describe viruses as *infectious particles* (rather than organisms) and as either *active* or *inactive* (rather than alive or dead).

Viruses are not just agents of disease. They have many positive uses (as shown in Insight 6.1). By infecting other cells, and sometimes influencing their genetic makeup, they have shaped the way cells, tissues, bacteria, plants, and animals have evolved to their present forms. For example, scientists think that the human genome contains up to 80,000 viral genes, sequences that come from viruses that have incorporated their genetic material permanently into human DNA. Bacterial DNA contains 10% to 20% viral sequences. As you learn more about how viruses work, you will see how bacterial DNA can end up in viral genomes.

Viruses are different from their host cells in size, structure, behavior, and physiology. They are a type of *obligate intracellular parasite* that cannot multiply unless they invade a specific host cell and instruct its genetic and

Table 6.1 **Properties of Viruses**

- Are not cells
- Are obligate intracellular parasites of bacteria, protozoa, fungi, algae, plants, and animals
- Do not independently fulfill the characteristics of life
- Are inactive macromolecules outside the host cell and active only inside host cells
- Have basic structure of protein shell (capsid) surrounding nucleic acid core
- Are ubiquitous in nature and have had major impact on development of biological life
- Are ultramicroscopic in size, ranging from 20 nm to 450 nm (diameter)
- Can have either DNA or RNA but not both
- Can have double-stranded DNA, single-stranded DNA, single-stranded RNA, or double-stranded RNA
- Carry molecules on surface that determine specificity for attachment to host cell
- Multiply by taking control of host cell's genetic material and regulating the synthesis and assembly of new viruses
- Lack enzymes for most metabolic processes
- Lack machinery for synthesizing proteins

metabolic machinery to make and release quantities of new viruses. Other unique properties of viruses are summarized in **table 6.1.**

6.2 Learning Outcomes—Assess Your Progress

2. Summarize arguments on both sides of the debate regarding the classification of viruses as living organisms.
3. Identify effective terms to describe the behavior of viruses.

6.3 The General Structure of Viruses

Size Range

Many viruses are much smaller than the average bacterium. These viruses cannot be seen with a light microscope; an electron microscope is necessary to detect them or examine their fine structure. More than 2,000 bacterial viruses could fit into an average bacterial cell, and more than 50 million polioviruses could be accommodated by an average human cell. Animal viruses range in size from the small parvoviruses[1] (around 20 nm [0.02 μm] in diameter) to mimiviruses that are larger than small bacteria (up to 450 nm [0.4 μm] in length) **(figure 6.1).** Some cylindrical viruses are relatively long (800 nm [0.8 μm] in length) but so narrow in diameter (15 nm [0.015 μm]) that their visibility is still

---

1. DNA viruses that cause respiratory infections in humans.

limited without the high magnification and resolution of an electron microscope. As you can see in figure 6.1, the mimivirus, discovered in 2003, is bigger than some bacteria. The pandoravirus, discovered in 2013, is as big as the *Streptococcus* bacterium (purple in figure 6.1). Figure 6.1 compares the sizes of several viruses with bacteria and eukaryotic cells and molecules.

Viral architecture is most readily observed through special stains in combination with electron microscopy (figure 6.2).

(a)

(b)

(c)

Figure 6.2 Methods of viewing viruses.
(a) Negative staining of an orf virus (a type of poxvirus), revealing details of its outer coat. (b) Positive stain of the Ebola virus, a type of filovirus, so named because of its tendency to form long strands. Note the textured capsid. (c) Shadowcasting image of a vaccinia virus.

Figure 6.1 Size comparison of viruses with a eukaryotic cell (yeast) and bacteria. A molecule of protein is included to indicate proportion of macromolecules.

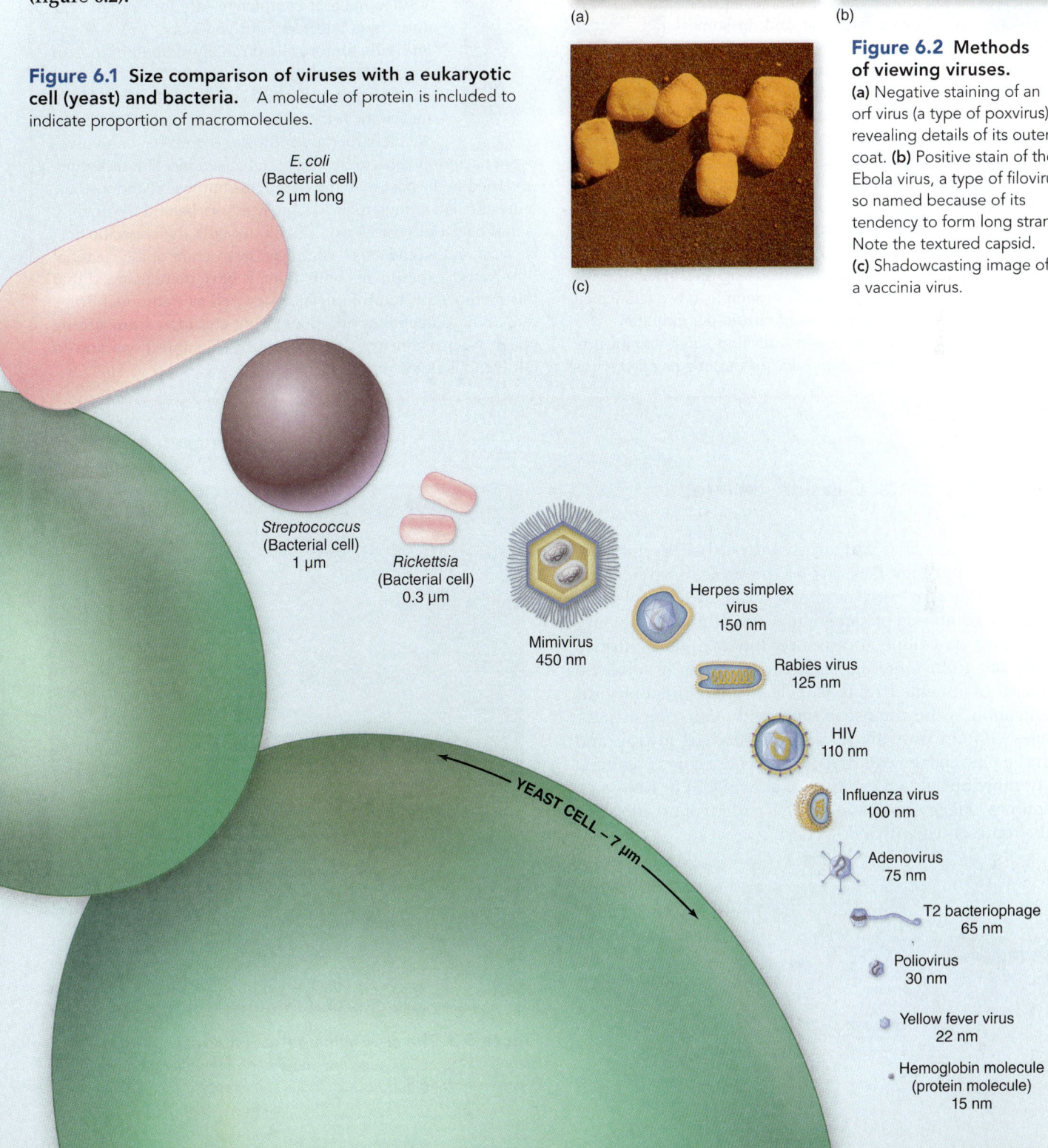

E. coli
(Bacterial cell)
2 μm long

Streptococcus
(Bacterial cell)
1 μm

Rickettsia
(Bacterial cell)
0.3 μm

Mimivirus
450 nm

Herpes simplex virus
150 nm

Rabies virus
125 nm

HIV
110 nm

Influenza virus
100 nm

Adenovirus
75 nm

T2 bacteriophage
65 nm

Poliovirus
30 nm

Yellow fever virus
22 nm

Hemoglobin molecule
(protein molecule)
15 nm

YEAST CELL – 7 μm

Looking at this beautiful tulip, one would never guess that it derives its pleasing appearance from a viral infection. It contains tulip mosaic virus, which alters the development of the plant cells and causes complex patterns of colors in the petals. Aside from this, the virus does not cause severe harm to the plants. Despite the reputation of viruses as cell killers, there is another side of viruses—that of being harmless and, in some cases, even beneficial.

Although there is no agreement on the origins of viruses, it is very likely that they have been in existence for billions of years. Virologists are convinced that viruses have been an important force in the evolution of living things. This is based on the fact that they interact with the genetic material of their host cells and that they carry genes from one host to another (transduction). It is easy to imagine that viruses arose early in the history of cells as loose pieces of genetic material that became dependent nomads, moving from cell to cell. Viruses are also a significant factor in the functioning of many ecosystems. For example, it is documented that seawater can contain millions of viruses per milliliter.

Experts have used viruses to develop safer vaccines by combining a less harmful virus such as vaccinia or adenovirus with some genetic material from a pathogen such as herpes simplex. This technique creates a vaccine that provides immunity but does not expose the person to the intact pathogen.

Scientists have recently had important successes using a virus called vesicular stomatitis virus (VSV) to cure cancer. They alter a gene in VSV to make it completely safe for normal cells and then inject it intravenously. VSV targets and kills tumor cells (in many different kinds of cancers, including brain, prostate, and ovarian cancers) and has even been shown to track down metastatic tumor cells in distant parts of the body.

An older therapy getting a second chance involves use of bacteriophages to treat bacterial infections. This technique was tried in the past with mixed success but was abandoned for more efficient antimicrobial drugs. The basis behind the therapy is that bacterial viruses would seek out only their specific host bacteria and would cause complete destruction of the bacterial cell. Newer experiments with animals have demonstrated that this method can control infections as well as traditional drugs can. Some potential applications being considered are adding phage suspension to grafts to control skin infections and to intravenous fluids for blood infections.

Viral Components: Capsids, Envelopes, and Nucleic Acids

It is important to realize that viruses bear no real resemblance to cells and that they lack any of the protein-synthesizing machinery found in even the simplest cells. Their molecular structure is composed of regular, repeating subunits that give rise to their crystalline appearance. Indeed, many purified viruses can form large aggregates or crystals if subjected to special treatments **(figure 6.3)**. The general plan of virus organization is the utmost in simplicity and compactness. Viruses contain only those parts needed to invade and control a host cell: an external coating and a core containing one or more nucleic acid strands of either DNA or RNA and, sometimes, one or two enzymes. This pattern of organization can be represented with a flowchart:

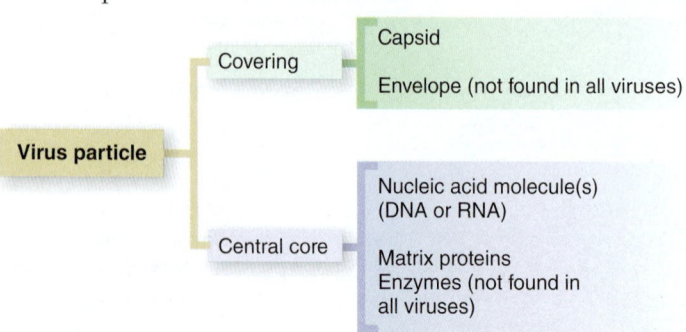

Figure 6.3 The crystalline nature of viruses. Highly magnified (150,000×) electron micrograph of purified poliovirus crystals, showing hundreds of individual viruses.

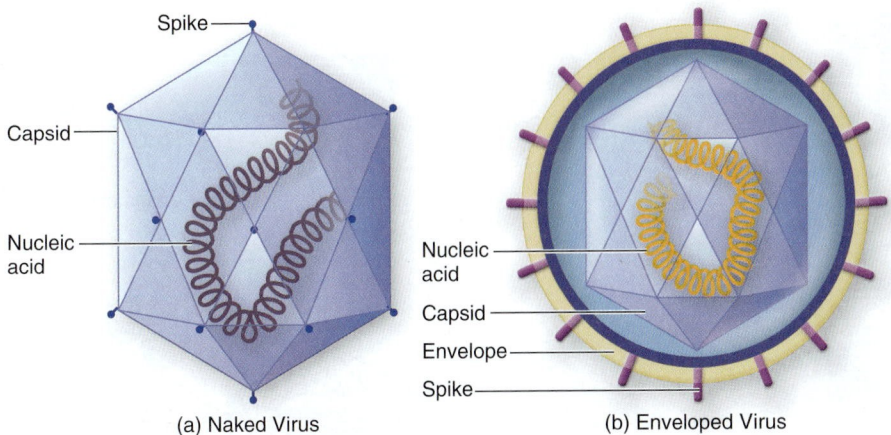

Figure 6.4 Generalized structure of viruses. **(a)** The simplest virus is a naked virus (nucleocapsid) consisting of a geometric capsid assembled around a nucleic acid strand or strands. **(b)** An enveloped virus is composed of a nucleocapsid surrounded by a flexible membrane called an envelope.

(a) Naked Virus

(b) Enveloped Virus

All viruses have a protein **capsid,** or shell, that surrounds the nucleic acid in the central core. Together, the capsid and the nucleic acid are referred to as the **nucleocapsid (figure 6.4).** Members of many families of animal viruses possess an additional covering external to the capsid called an envelope, which is usually a modified piece of the host's cell membrane **(figure 6.4b).** Viruses that consist of only a nucleocapsid are considered *naked viruses* **(figure 6.4a).** Both naked and enveloped viruses possess proteins on their outer surfaces that project from either the nucleocapsid or the envelope. They are the molecules that allow viruses to dock with their host cells. As we shall see later, the enveloped viruses differ from the naked viruses in the way that they enter and leave a host cell. A fully formed virus that is able to establish an infection in a host cell is often called a **virion.**

The Viral Capsid: The Protective Outer Shell

When a virus particle is magnified several hundred thousand times, the capsid appears as the most prominent geometric feature. In general, each capsid is constructed from identical subunits called **capsomers** that are constructed from protein molecules. The capsomers spontaneously self-assemble into the finished capsid. Depending on how the capsomers are shaped and arranged, this assembly results in two different types for animal viruses: helical and icosahedral.

The simpler **helical** capsids have rod-shaped capsomers that bond together to form a series of hollow discs resembling a bracelet. During the formation of the nucleocapsid, these discs link with other discs to form a continuous helix into which the nucleic acid strand is coiled **(figure 6.5).** In electron micrographs, the appearance of a helical capsid varies with the type of virus. The nucleocapsids of naked helical viruses are very rigid and tightly wound into a cylinder-shaped package **(figure 6.6a,b).** An example is the tobacco mosaic virus, which attacks tobacco leaves. Enveloped helical nucleocapsids are more flexible and tend to be arranged as a looser helix within the envelope **(figure 6.6c, d).** This type of

morphology is found in several enveloped human viruses, including influenza, measles, and rabies.

Disease Connection

You will see in chapter 19 that the dreaded rabies virus, which has a near 100% fatality rate in humans, has a very distinctive shape. Its characteristic bullet shape is a product of a matrix protein that lies between the helical capsid and the envelope.

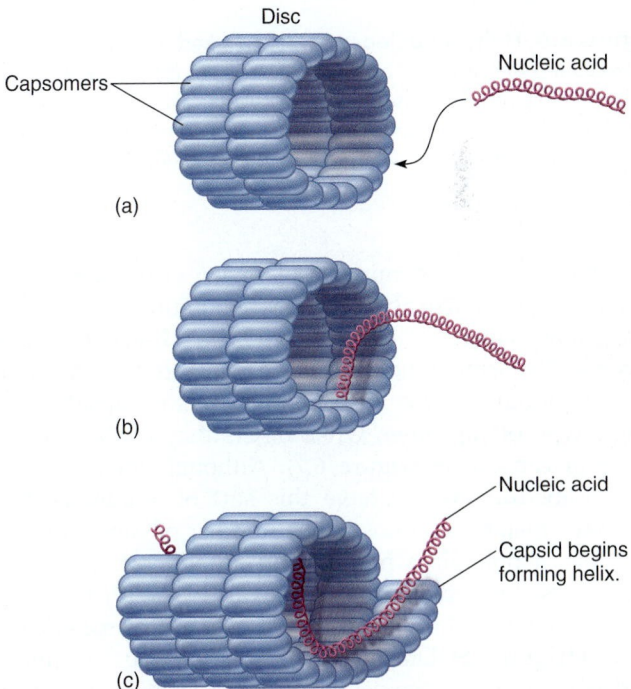

Figure 6.5 Assembly of helical nucleocapsids.
(a) Capsomers assemble into hollow discs. **(b)** The nucleic acid is inserted into the center of the disc. **(c)** Elongation of the nucleocapsid progresses from one or both ends, as the nucleic acid is wound "within" the lengthening helix.

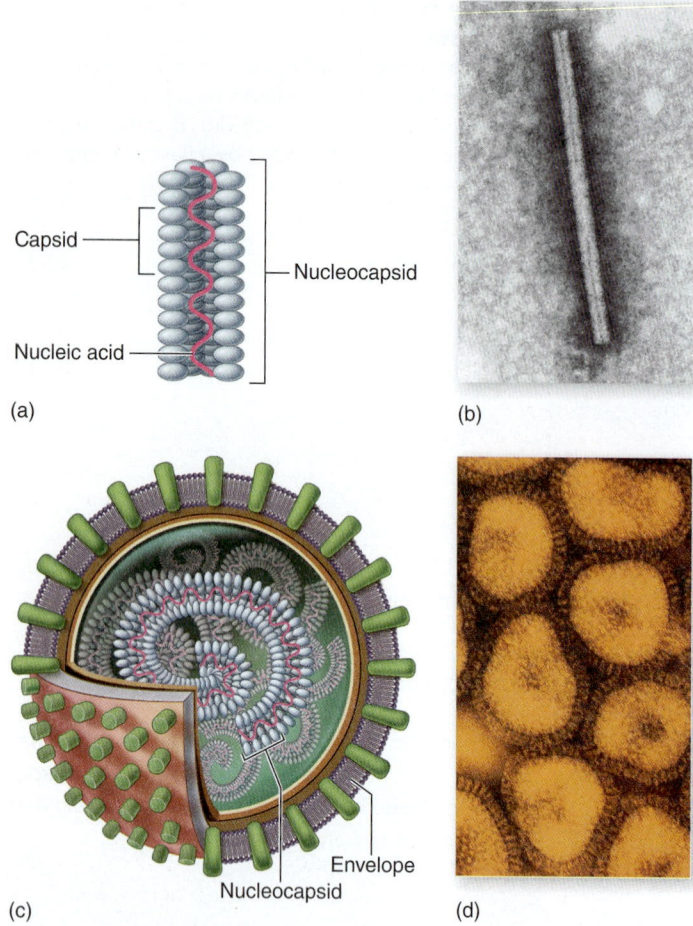

Capsid

Nucleocapsid

Nucleic acid

(a)

(b)

Envelope

Nucleocapsid

(c)

(d)

Figure 6.6 Helical nucleocapsids.   **Naked helical virus (tobacco mosaic virus): (a)** a schematic view and **(b)** a greatly magnified micrograph. Note the overall cylindrical morphology. **Enveloped helical virus (influenza virus): (c)** a schematic view and **(d)** an electron micrograph of the same virus (350,000×).

The capsids of a number of major virus families are arranged in an **icosahedron** (eye-koh-suh-hee'-drun)—a three-dimensional, 20-sided figure with 12 evenly spaced corners. The arrangements of the capsomers vary from one virus to another. Some viruses construct the capsid from a single type of capsomer, while others may contain several types of capsomers **(figure 6.7).** Although the capsids of all icosahedral viruses have this sort of symmetry, they can have major variations in the number of capsomers; for example, a poliovirus has 32, and an adenovirus has 252 capsomers. Individual capsomers can look either ring- or dome-shaped, and the capsid itself can appear spherical or cubical **(figure 6.8).** During assembly of the virus, the nucleic acid is packed into the center of this icosahedron, forming a nucleocapsid. While most viruses have capsids that are either icosahedral or helical, there is another category of capsid that is simply called *complex*. Complex capsids, found in the viruses that infect bacteria, may have multiple types of proteins and take shapes that are not symmetrical. An example of a complex virus is shown in **figure 6.9.** Another

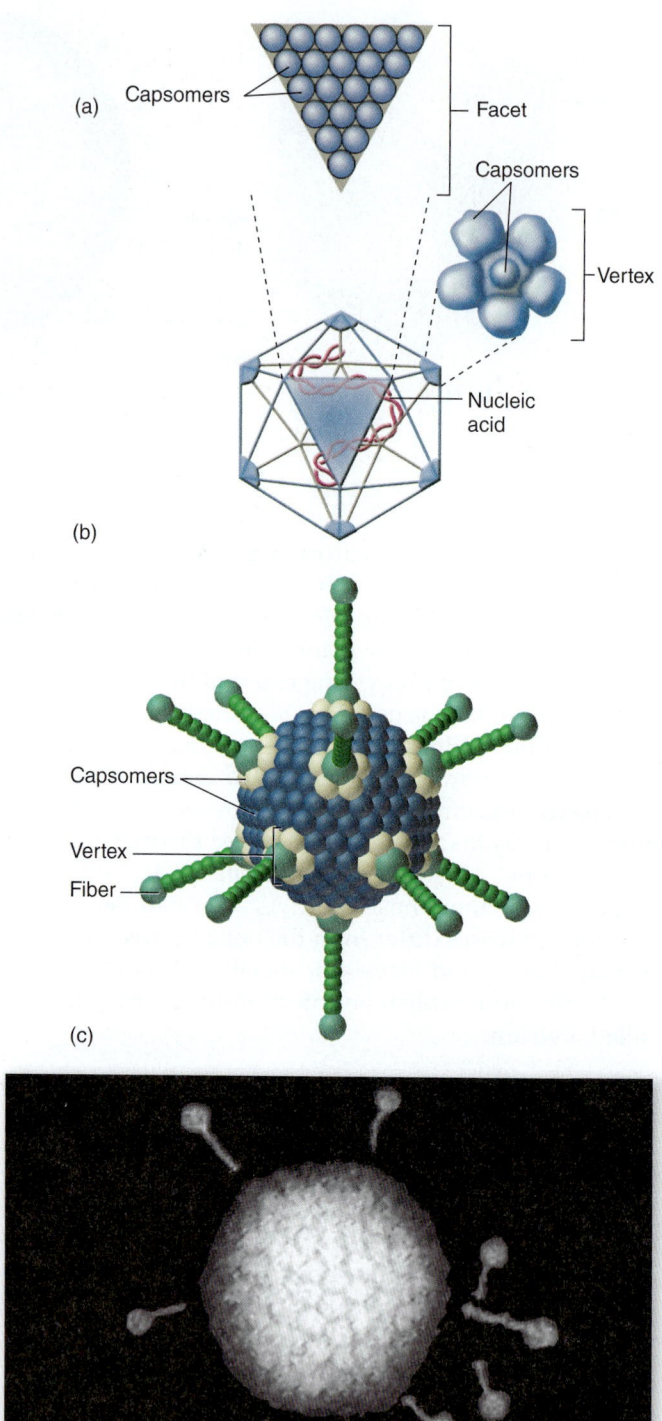

(a)   Capsomers

Facet

Capsomers

Vertex

Nucleic acid

(b)

Capsomers

Vertex

Fiber

(c)

(d)

Figure 6.7 How icosahedral capsids are formed. Adenovirus is the model. **(a)** A facet or "face" of the capsid is composed of 21 identical capsomers arranged in a triangular shape. A vertex or "point" consists of a different type of capsomer with a single penton in the center. Other viruses can vary in the number, types, and arrangement of capsomers. **(b)** An assembled virus shows how the facets and vertices come together to form a shell around the nucleic acid. **(c)** A three-dimensional model of this virus shows fibers (spikes) attached to the pentons. **(d)** A negative stain of this virus (640,000×) highlights its texture and fibers that have fallen off.

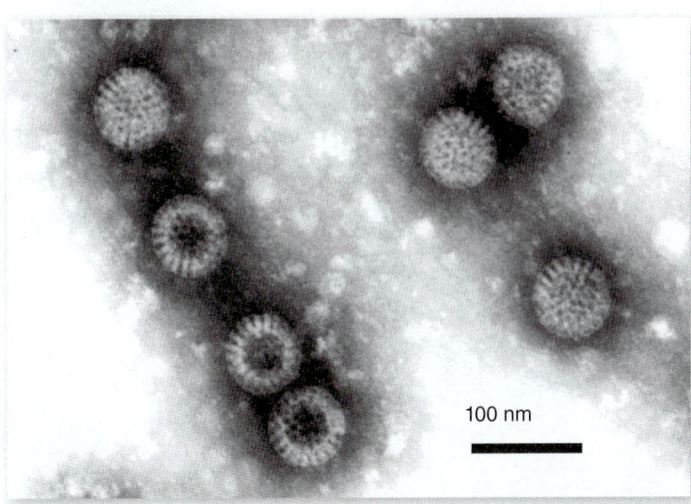

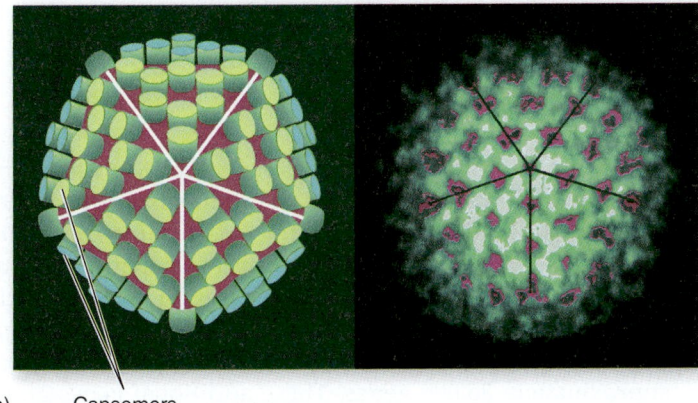

(a)   Capsomers

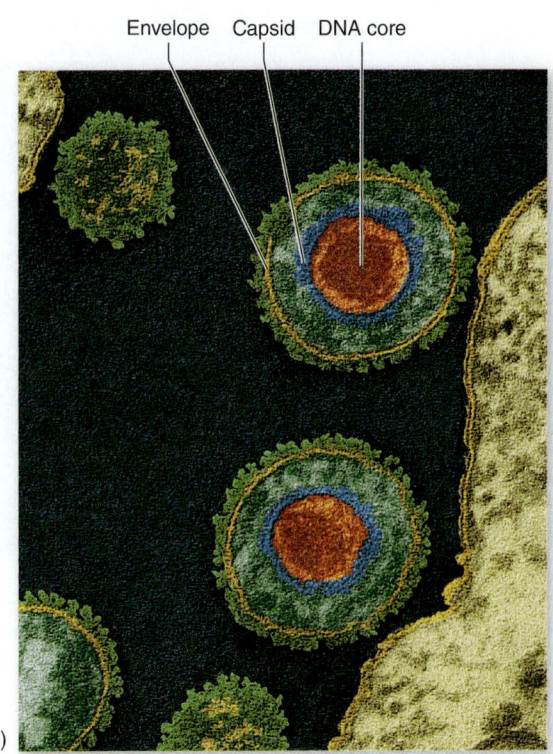

Envelope   Capsid   DNA core

(b)

Figure 6.8 Icosahedral capsids. **(a)** Upper view: a negative stain of rotaviruses with capsomers that look like spokes on a wheel; lower view is a three-dimensional model of this virus. **(b)** Electron micrograph of herpes simplex virus, an enveloped icosahedral virus (300,000×).

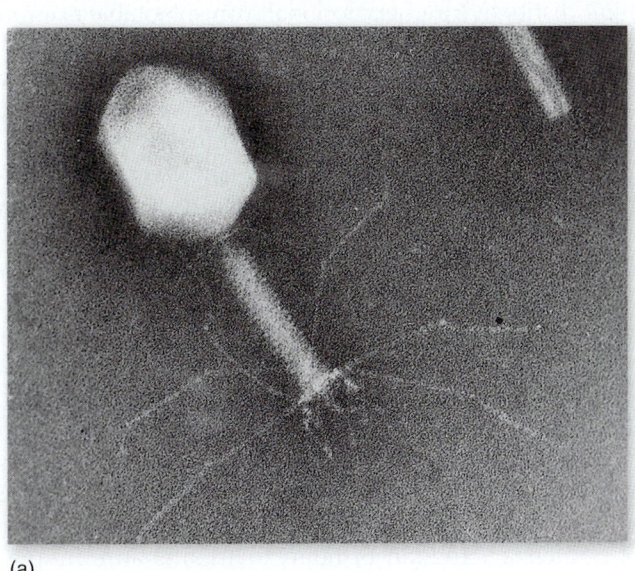

(a)

Figure 6.9 Structure of complex viruses. **(a)** Photomicrograph and **(b)** diagram of a T4 bacteriophage, a virus that infects bacteria.

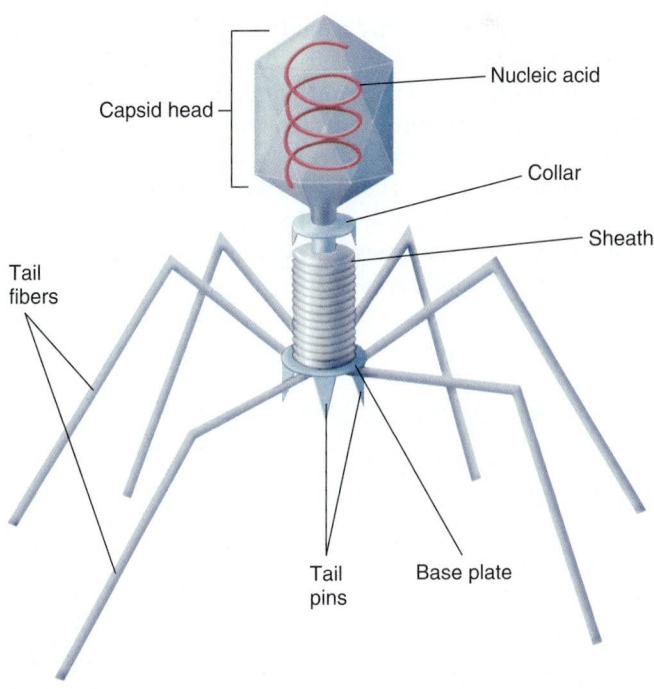

Capsid head

Nucleic acid

Collar

Sheath

Tail fibers

Tail pins

Base plate

(b)

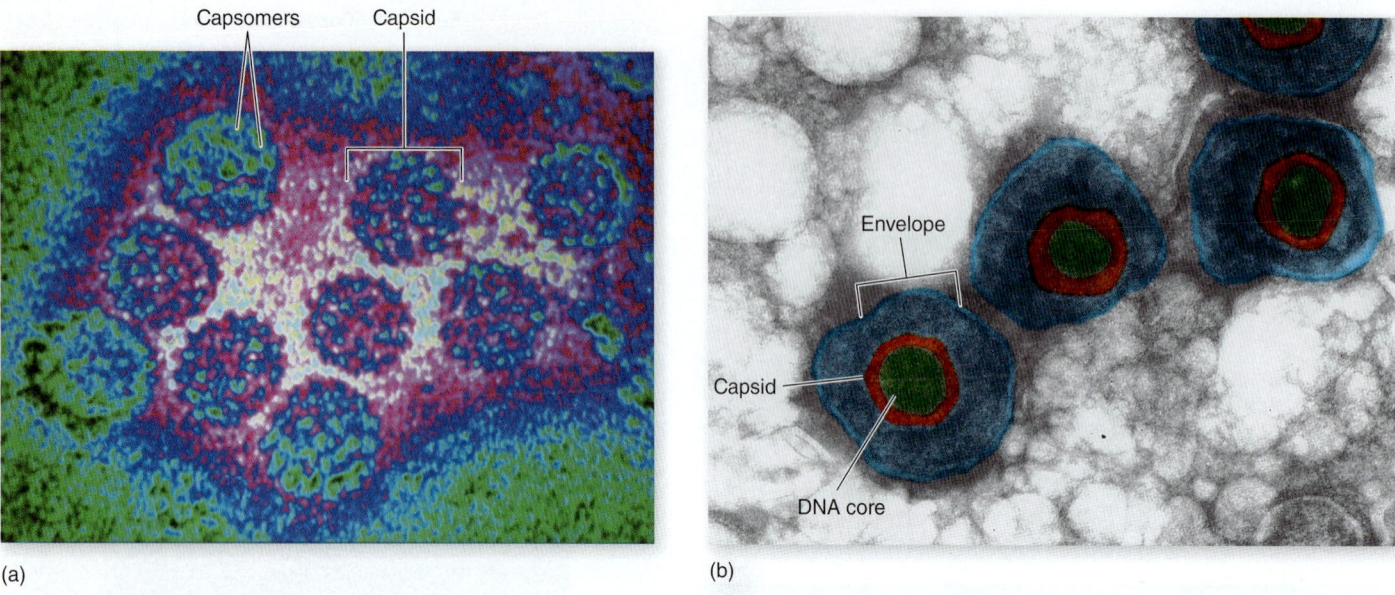

Figure 6.10 Nonenveloped and enveloped viruses. **(a)** Micrograph of nonenveloped papillomaviruses with unusual, ring-shaped capsomers. **(b)** Herpesvirus, an enveloped icosahedron (300,000×). Both micrographs have been colorized.

factor that alters the appearance of icosahedral viruses is whether or not they have an outer envelope; contrast a papillomavirus (causes warts) and its naked nucleocapsid with herpes simplex (causes cold sores) and its enveloped nucleocapsid **(figure 6.10).**

The Viral Envelope

When enveloped viruses are released from the host cell, they take with them a bit of its membrane in the form of an envelope, as described later. Some viruses bud off the cell membrane; others leave via the nuclear envelope or the endoplasmic reticulum. Whichever avenue of escape, the viral envelope differs significantly from the host's membranes. In the envelope, some or all of the regular membrane proteins are replaced with special viral proteins. Some of the envelope proteins attach to the capsid of the virus, and glycoproteins (proteins bound to a carbohydrate) remain exposed on the outside of the envelope. These protruding molecules, called *spikes* when they are on enveloped viruses, are essential for the attachment of viruses to the next host cell. Because the envelope is more supple than the capsid, the surface appearance of enveloped viruses is pleomorphic, and these viruses range from spherical to filamentous in shape.

Nucleic Acids: At the Core of a Virus

The sum total of the genetic information carried by any organism is known as its **genome.** So far, one biological constant is that the genetic information of living cells is carried by nucleic acids (DNA, RNA). Viruses, although not technically alive and definitely not cells, are no exception to this rule, but there is a significant difference. Unlike cells, which contain both DNA and RNA, viruses contain either DNA or RNA *but not both*. Because viruses must pack into a tiny space all of the genes necessary to instruct the host cell to make new viruses, the number of viral genes is often—but not always—quite small compared with that of a cell. It varies from four genes in hepatitis B virus to hundreds of genes in herpesviruses to over 2,500 in pandoraviruses. By comparison, the bacterium *Escherichia coli* has approximately 4,000 genes, and a human cell has approximately 23,000 genes. These additional genes allow cells to carry out the complex metabolic activity necessary for independent life.

In chapter 2, you learned that DNA usually exists as a double-stranded molecule and that RNA is single-stranded. Although most viruses follow this same pattern, a few exhibit distinctive and exceptional forms. Notable examples are the parvoviruses, which contain single-stranded DNA, and reoviruses (a cause of respiratory and intestinal tract infections), which contain double-stranded RNA. In fact, viruses exhibit wide variety in how their RNA or DNA is configured. DNA viruses can have single-stranded (ss) or double-stranded (ds) DNA; the dsDNA can be arranged linearly or in ds circles. RNA viruses can be double-stranded but are more often single-stranded. You will learn in chapter 9 that all proteins are made by "translating" the nucleic acid code on a single strand of RNA into an amino acid sequence. Single-stranded RNA genomes that are ready for immediate translation into proteins are called **positive-sense RNA.** Other RNA genomes have to be converted into the proper form to be made into proteins, and these are called **negative-sense RNA.** RNA genomes may also be *segmented,* meaning that the individual genes exist on separate pieces of RNA. The influenza virus (an orthomyxovirus) is an example of this. A special type of RNA virus is called a *retrovirus.* We'll discuss it later. **Table 6.2** provides the structures of some medically relevant DNA and RNA viruses.

Table 6.2 Viral Nucleic Acid

	Diagram	Virus Name	Disease It Causes
DNA Viruses			
Double-stranded DNA		Variola virus	Smallpox
		Herpes simplex 2	Genital herpes
Single-stranded DNA		Parvovirus	Erythema infectiosum (skin condition)
RNA Viruses			
Single-stranded (+) polarity		Poliovirus	Poliomyelitis
Single-stranded (−) polarity		Influenza virus	Influenza
Double-stranded RNA		Rotavirus	Gastroenteritis
Single-stranded RNA reverse transcriptase		HIV	AIDS

In all cases, these tiny strands of genetic material carry the blueprint for viral structure and functions. In a very real sense, viruses are genetic parasites because they cannot multiply until their nucleic acid has reached the internal habitat of the host cell. At the minimum, they must carry genes for synthesizing the viral capsid and genetic material, for regulating the actions of the host, and for packaging the mature virus.

Other Substances in the Virus Particle

In addition to the protein of the capsid, the proteins and lipids of envelopes, and the nucleic acid of the core, viruses can contain enzymes for specific operations within their host cell. They may come with preformed enzymes that are required for viral replication. Examples include **polymerases** (pol-im'-ur-ace-uz) that synthesize DNA and RNA, and replicases that copy RNA. HIV comes equipped with

reverse transcriptase (RT) for synthesizing DNA from RNA. However, viruses completely lack the genes for synthesis of metabolic enzymes. As we shall see, this deficiency has little consequence, because viruses have adapted to completely take over their hosts' metabolic resources. Some viruses can actually carry away substances from their host cell. For instance, arenaviruses pack along host ribosomes, and retroviruses "borrow" the host's tRNA molecules.

6.3 Learning Outcomes—Assess Your Progress

4. Discuss the size of viruses relative to other microorganisms.
5. Describe the function and structure(s) of viral capsids.
6. Distinguish between enveloped and naked viruses.
7. Explain the importance of viral surface proteins, or spikes.
8. Compare and contrast the composition of a viral genome to that of a cellular organism's genome.
9. Diagram the possible nucleic acid configurations exhibited by viruses.

6.4 How Viruses Are Classified and Named

Although viruses are not classified as members of the domains or kingdoms discussed in chapter 1, they are diverse enough to require their own classification scheme to help with their study and identification. In an informal way, we have already begun classifying viruses—as animal, plant, or bacterial viruses; enveloped or naked viruses; DNA or RNA viruses; and helical or icosahedral viruses. These introductory categories are certainly useful in organization and description, but the study of specific viruses requires a more standardized method of nomenclature. For many years, the animal viruses were classified mainly on the basis of their hosts and the kind of diseases they caused. Newer systems for naming viruses also take into account the actual nature of the virus particles themselves, with only partial emphasis on host and disease. The main criteria presently used to group viruses are structure, chemical composition, and similarities in genetic makeup.

In 2011, the International Committee on the Taxonomy of Viruses issued its latest report on the classification of viruses. The committee listed six orders, 94 families, and 395 genera of viruses. Previous to 2000, there had been only a single recognized order of viruses. Examples of each of the six orders of viruses are presented in **table 6.3.** (An additional order is being considered for viruses that infect archaea.) Note the naming conventions: Virus families are written with -*viridae* on the end of the name, and genera end with -*virus*.

In the last 10 years, scientists have discovered a group of viruses that they call *nucleocytoplasmic large DNA viruses (NCLDVs)*. The mimivirus pictured in figure 6.1 is one of these and the pandoravirus is another. Some of these scientists argue that these organisms are closely enough related to each other, and distant enough from other viruses,

Table 6.3 Examples from the Six Orders of Viruses

Order	Family	Genus	Species	Host
Caudovirales	*Myoviridae*	*SPO1-like virus*	Bacillus phage	Bacterium
Herpesvirales	*Herpesviridae* *Alloherpesviridae*	*Simplexvirus* *Salmonivirus*	Human herpesvirus 2 Salmonid herpesvirus 3	Animal Animal
Mononegavirales	*Paramyxoviridae* *Filoviridae*	*Morbillivirus* *Ebola virus*	Measles virus Ebola virus	Animal Animal
Nidovirales	*Togaviridae* *Luteoviridae*	*Rubivirus* *Tobamovirus*	Rubella virus Tobacco mosaic virus	Animal Plant
Picornavirales	*Iflaviridae* *Picornaviridae*	*Iflavirus* *Enterovirus*	Sacbrood virus Human enterovirus A	Animal Animal
Tymovirales	*Betaflexiviridae* *Alphaflexiviridae*	*Citrivirus* *Lolavirus*	Citrus leaf blotch virus Lolium latent virus	Plant Plant

that they constitute a fourth domain of life, or at the very least, a new order.

Historically, some virologists had created an informal **species** naming system that mirrors the species names in higher organisms, using genus and species epithets such as *Measles morbillivirus.* This has not been an official designation, however. Because the use of standardized species names has not been widely accepted, the genus or common English vernacular names (for example, poliovirus and rabies virus) will be used in discussions of specific viruses in this text. **Table 6.4** illustrates the naming system for important viruses and the diseases they cause.

6.4 Learning Outcomes—Assess Your Progress

10. Develop two arguments against assigning species names to viruses.
11. Demonstrate how family and genus names in viruses are written.

6.5 Modes of Viral Multiplication

The process of viral multiplication is an extraordinary biological phenomenon. Viruses are minute parasites that seize control of the synthetic and genetic machinery of cells. The nature of this cycle dictates the way the virus is transmitted and what it does to its host, the responses of the immune defenses, and human attempts to control viral infections. From these perspectives, we cannot over-emphasize the importance of a working knowledge of the relationship between viruses and their host cells.

Multiplication Cycles in Animal Viruses

The general phases in the life cycle of animal viruses are

- **adsorption,**
- **penetration,**
- **uncoating,**
- **synthesis,**

- **assembly,** and
- **release** from the host cell.

The length of the entire multiplication cycle varies from 8 hours in polioviruses to 36 hours in some herpesviruses. See **process figures 6.11** and **6.14** for the major phases of the viral life cycle.

Adsorption and Host Range

Invasion begins when the virus encounters a susceptible host cell and adsorbs specifically to receptor sites on the cell membrane. (Note the difference between the words *absorb,* as in what paper towels do, and *adsorb,* which means "attach.") The membrane receptors that viruses attach to are usually glycoproteins the cell requires for its normal function. For example, the rabies virus binds to the acetylcholine receptor of nerve cells, and the human immunodeficiency virus (HIV) attaches to the CD4 protein on certain white blood cells. The mode of attachment varies between the two general types of viruses. In enveloped forms such as influenza virus and HIV, glycoprotein spikes bind to the cell membrane receptors **(figure 6.12).** Viruses with naked nucleocapsids (adenovirus, for example) use molecules on their capsids that adhere to cell membrane receptors. Because a virus can invade its host cell only through making an exact fit with a specific host molecule, the range of hosts it can infect in a natural setting is limited. This limitation, known as the **host range,** may be restricted as in the case of hepatitis B, which infects only liver cells of humans; intermediate like the poliovirus, which infects intestinal and nerve cells of primates (humans, apes, and monkeys); or as broad as the rabies virus, which can infect various cells of all mammals. Cells that lack compatible virus receptors are resistant to adsorption and invasion by that virus. This explains why, for example, human liver cells are not infected by the canine hepatitis virus and dog liver cells cannot host the human hepatitis A virus. It also explains why viruses usually have tissue specificities called *tropisms* (troh'-pizmz) for certain cells in the body. The hepatitis B virus targets the liver, and the mumps virus targets salivary glands.

Table 6.4 Important Human Virus Families, Genera, Common Names, and Types of Diseases

	Family	Genus of Virus	Common Name of Genus Members	Name of Disease
DNA Viruses				
Double-stranded	Poxviridae	Orthopoxvirus	Variola and vaccinia	Smallpox, cowpox
	Herpesviridae	Simplexvirus	Herpes simplex (HSV) 1 virus	Fever blister, cold sores
			Herpes simplex (HSV) 2 virus	Genital herpes
		Varicellovirus	Varicella zoster virus (VZV)	Chickenpox, shingles
	Papillomaviridae	Papillomavirus	Human papillomavirus (HPV)	Several types of warts
	Hepadnaviridae	Hepadnavirus	Hepatitis B virus (HBV or Dane particle)	Serum hepatitis
Single-stranded	Parvoviridae	Erythrovirus	Parvovirus B19	Erythema infectiosum
RNA Viruses				
+ polarity	Picornaviridae	Enterovirus	Poliovirus	Poliomyelitis
			Coxsackievirus	Hand-foot-mouth disease
		Hepatovirus	Hepatitis A virus (HAV)	Short-term hepatitis
		Rhinovirus	Human rhinovirus	Common cold, bronchitis
	Caliciviridae	Calicivirus	Norwalk virus	Viral diarrhea, Norwalk virus syndrome
	Togaviridae	Alphavirus	Eastern equine encephalitis virus	Eastern equine encephalitis (EEE)
		Rubivirus	Rubella virus	Rubella (German measles)
	Flaviviridae	Flavivirus	Dengue fever virus	Dengue fever
			West Nile fever virus	West Nile fever
	Coronaviridae	Coronavirus	Infectious bronchitis virus (IBV)	Bronchitis
			SARS virus	Severe acute respiratory syndrome
– polarity	Filoviridae	Filovirus	Ebola, Marburg virus	Ebola fever
	Orthomyxoviridae	Influenza virus	Influenza virus, type A (Asian, Hong Kong viruses)	Influenza or "flu" and swine influenza
	Paramyxoviridae	Paramyxovirus	Parainfluenza virus, types 1–5	Parainfluenza
			Mumps virus	Mumps
	Rhabdoviridae	Lyssavirus	Rabies virus	Rabies (hydrophobia)
	Bunyaviridae	Bunyavirus	Bunyamwera viruses	California encephalitis
		Hantavirus	Sin Nombre virus	Respiratory distress syndrome
dsRNA	Reoviridae	Rotavirus	Human rotavirus	Rotavirus gastroenteritis
		Morbillivirus	Measles virus	Measles (red)
Special RNA	Retroviridae	Oncornavirus	Human T-cell leukemia virus (HTLV)	T-cell leukemia
		Lentivirus	HIV (human immunodeficiency viruses 1 and 2)	Acquired immunodeficiency syndrome (AIDS)

Disease Connection

The fact that the poliovirus has tropisms for both neural and intestinal cells explains how it wreaks havoc on humans. Most people know that it causes paralysis; this is because it affects the neurons that make muscles work. But most people have no idea how you "catch" it. You catch it by ingesting water or food that is contaminated with the virus because it attaches to intestinal cells, and from there invades the nervous system. Polio is gone in the Western Hemisphere but still hangs on in three developing countries (as of 2013), despite the world health community's best efforts.

Penetration/Uncoating of Animal Viruses

Animal viruses exhibit some impressive mechanisms for entering a host cell. The flexible cell membrane of the host is penetrated either by the whole virus or by its nucleic acid **(figure 6.13)**. In penetration by **endocytosis (figure 6.13a)**, the entire virus is engulfed by the cell and enclosed in a vacuole or vesicle. When enzymes in the vacuole dissolve the envelope and capsid, the virus is said to be uncoated, a process that releases the viral nucleic acid into the cytoplasm. The exact manner of uncoating varies, but in most cases, the virus fuses with the wall of the vesicle. Another means of entry involves direct fusion of the viral

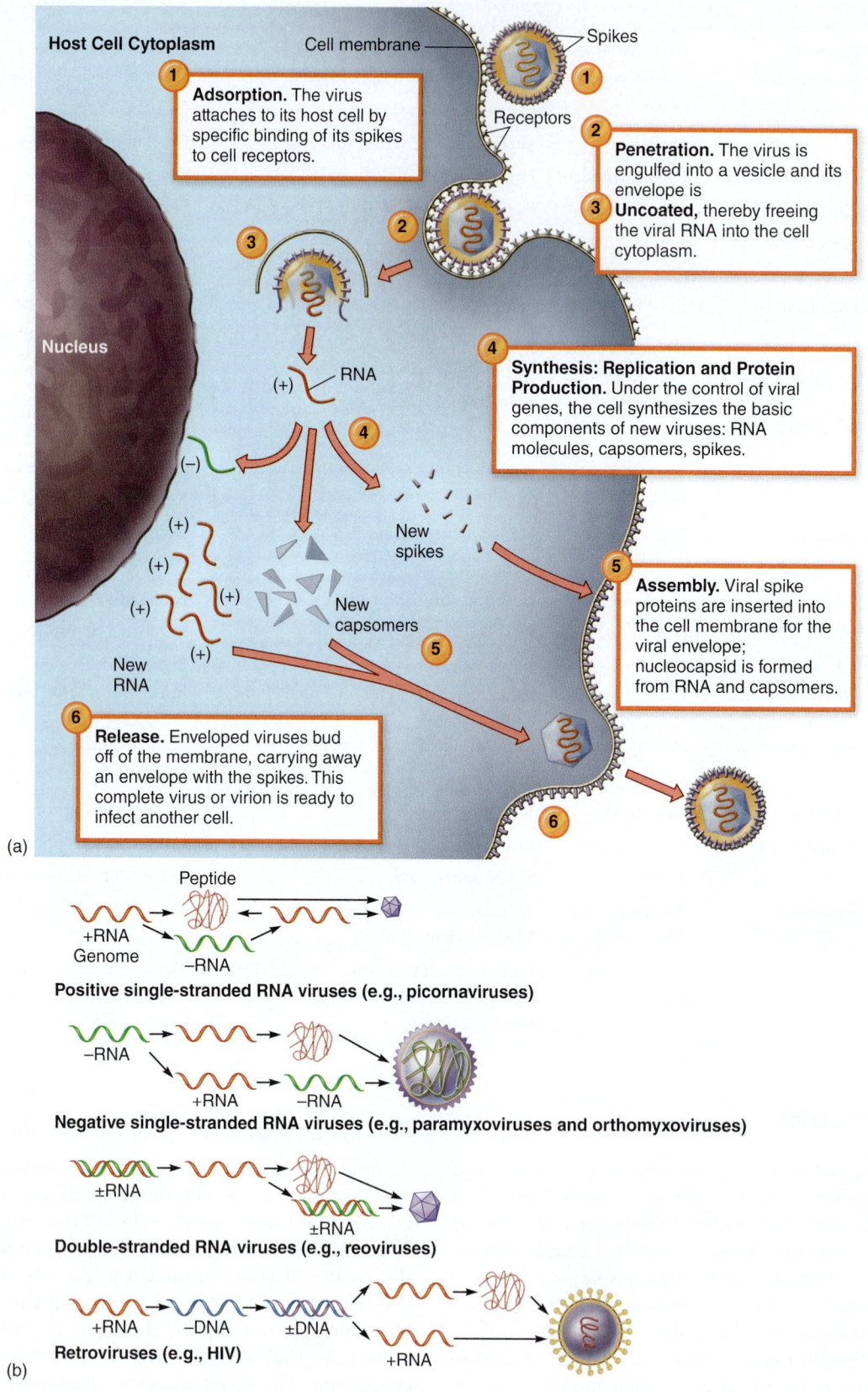

Host Cell Cytoplasm

1 Adsorption. The virus attaches to its host cell by specific binding of its spikes to cell receptors.

Cell membrane

Spikes

Receptors

2 Penetration. The virus is engulfed into a vesicle and its envelope is **3 Uncoated,** thereby freeing the viral RNA into the cell cytoplasm.

Nucleus

(+) RNA

(−)

(+)

(+)

(+) (+)

(+)

New RNA

New spikes

New capsomers

4 Synthesis: Replication and Protein Production. Under the control of viral genes, the cell synthesizes the basic components of new viruses: RNA molecules, capsomers, spikes.

5 Assembly. Viral spike proteins are inserted into the cell membrane for the viral envelope; nucleocapsid is formed from RNA and capsomers.

6 Release. Enveloped viruses bud off of the membrane, carrying away an envelope with the spikes. This complete virus or virion is ready to infect another cell.

(a)

Peptide

+RNA Genome

−RNA

Positive single-stranded RNA viruses (e.g., picornaviruses)

−RNA

+RNA −RNA

Negative single-stranded RNA viruses (e.g., paramyxoviruses and orthomyxoviruses)

±RNA

±RNA

Double-stranded RNA viruses (e.g., reoviruses)

+RNA −DNA ±DNA

+RNA

Retroviruses (e.g., HIV)

(b)

Process Figure 6.11 **General features in the multiplication cycle of RNA animal viruses.** **(a)** The major events in the life cycle of an enveloped + strand RNA virus. **(b)** Summary of synthesis in RNA viruses.

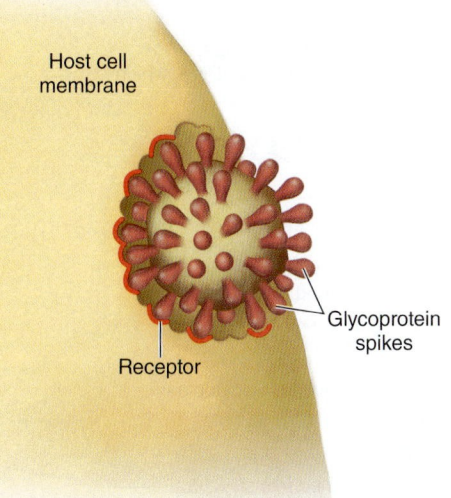

Host cell membrane

Glycoprotein spikes

Receptor

Figure 6.12 The mode by which animal viruses adsorb to the host cell membrane. An enveloped coronavirus with prominent spikes. The configuration of the spike has a complementary fit for cell receptors. The process in which the virus lands on the cell and plugs into receptors is termed *docking*.

there. With few exceptions (such as retroviruses), RNA viruses are replicated and assembled in the cytoplasm.

In process figure 6.11, we provide an overview of the process, using a + strand RNA virus as a model. Rubella viruses are an example of this type of virus. Almost immediately upon entry, the viral nucleic acid begins to synthesize the building blocks for new viruses. First, the +ssRNA, which can serve immediately upon entry as mRNA, starts being translated into viral proteins, especially those needed for further viral replication (illustrated in process figure 6.11a). The + strand is then replicated by host machinery into −ssRNA. This RNA becomes the template for the creation of many new +ssRNAs, which are used as the viral genomes for new viruses. Additional +ssRNAs are synthesized and used for late-stage mRNAs. Some viruses come equipped with the necessary enzymes for synthesis of viral components; others utilize those of the host. Proteins for the capsid, spikes, and viral enzymes are synthesized on the host's ribosomes using host amino acids. Process figure 6.11b shows how the synthesis of new genomes and mRNAs for translation differs among the various types of RNA viruses. Note that the retroviruses turn their RNA genomes into DNA. This step is accomplished by a viral enzyme called reverse transcriptase and has important implications in infections with these viruses, one of which is HIV. The retroviral cycle is explained in more detail in chapter 20.

DNA viruses generally follow the same steps of adsorption, penetration, uncoating, synthesis, assembly,

envelope with the host cell membrane (as in influenza and mumps viruses) **(figure 6.13b).** In this form of penetration, the envelope merges directly with the cell membrane, thereby liberating the nucleocapsid into the cell's interior.

Synthesis: Replication and Protein Production

The synthetic and replicative phases of animal viruses are highly regulated and extremely complex at the molecular level. The viral nucleic acid takes control over the host's synthetic and metabolic machinery. How this control proceeds will vary, depending on whether the virus is a DNA or an RNA virus. In general, the DNA viruses (except poxviruses) enter the host cell's nucleus and are replicated and assembled

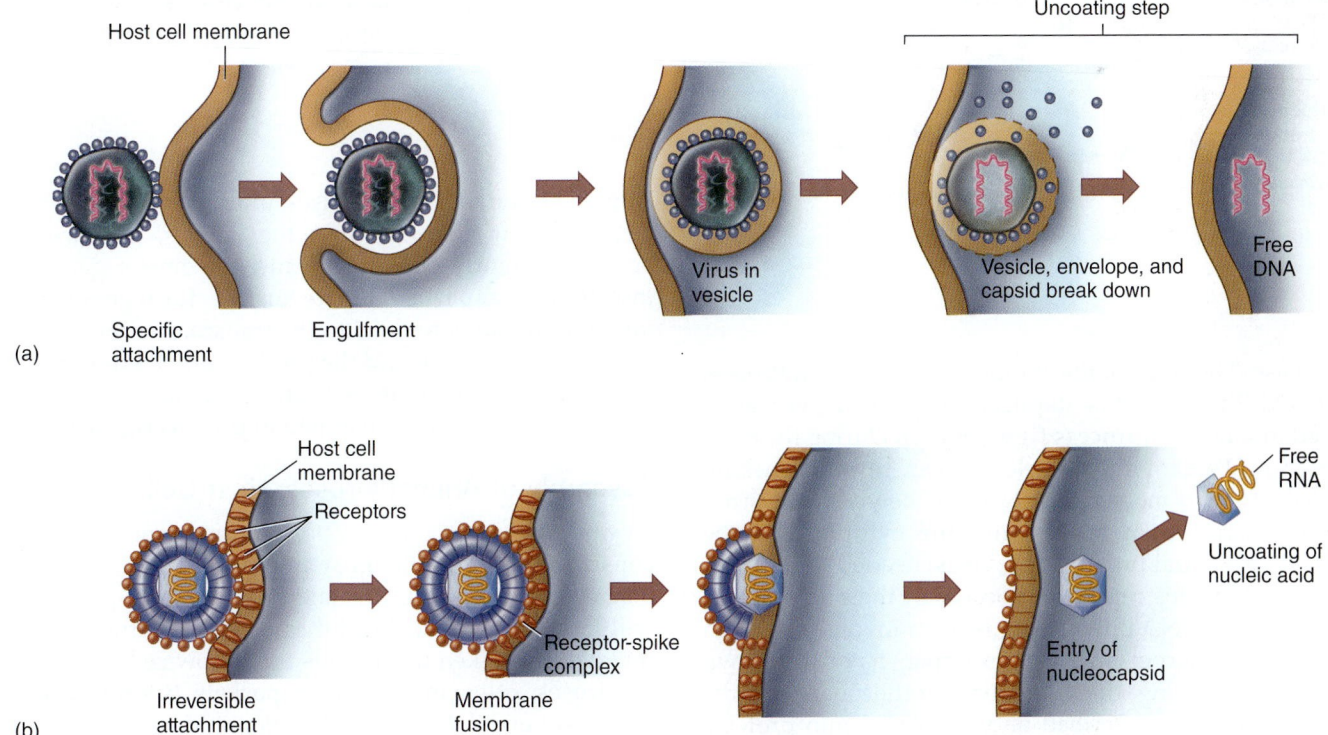

Uncoating step

Host cell membrane

Virus in vesicle

Vesicle, envelope, and capsid break down

Free DNA

(a)   Specific attachment   Engulfment

Host cell membrane

Receptors

Free RNA

Receptor-spike complex

Entry of nucleocapsid

Uncoating of nucleic acid

(b)   Irreversible attachment   Membrane fusion

Figure 6.13 Two principal means by which animal viruses penetrate.   **(a)** Endocytosis (engulfment) and uncoating of a herpesvirus. **(b)** Fusion of the cell membrane with the viral envelope (mumps virus).

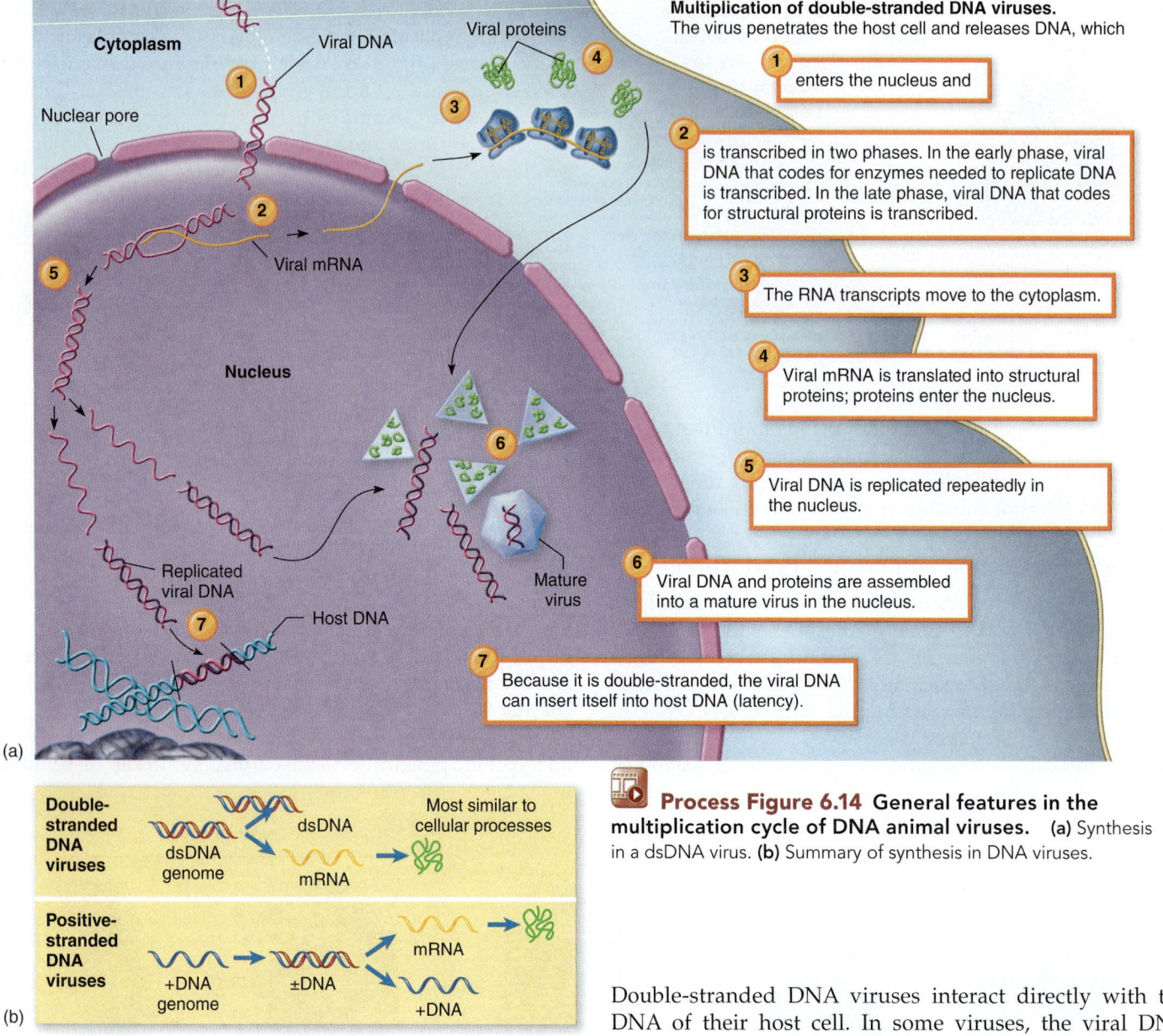

Multiplication of double-stranded DNA viruses.
The virus penetrates the host cell and releases DNA, which

1. enters the nucleus and

2. is transcribed in two phases. In the early phase, viral DNA that codes for enzymes needed to replicate DNA is transcribed. In the late phase, viral DNA that codes for structural proteins is transcribed.

3. The RNA transcripts move to the cytoplasm.

4. Viral mRNA is translated into structural proteins; proteins enter the nucleus.

5. Viral DNA is replicated repeatedly in the nucleus.

6. Viral DNA and proteins are assembled into a mature virus in the nucleus.

7. Because it is double-stranded, the viral DNA can insert itself into host DNA (latency).

Process Figure 6.14 **General features in the multiplication cycle of DNA animal viruses.** **(a)** Synthesis in a dsDNA virus. **(b)** Summary of synthesis in DNA viruses.

Double-stranded DNA viruses interact directly with the DNA of their host cell. In some viruses, the viral DNA becomes silently integrated into the host's genome by insertion at a particular site on the host genome. This integration may later lead to the transformation of the host cell into a cancer cell and the production of a tumor.

A slightly different replication mechanism is used by ssDNA viruses; this is illustrated in **process figure 6.14b**.

Assembly of Animal Viruses: Host Cell as Factory

Toward the end of the cycle, mature virus particles are constructed from the growing pool of parts. In most instances, the capsid is first laid down as an empty shell that will serve as a receptacle for the nucleic acid strand. Electron micrographs taken during this time show cells with masses of viruses, often in crystalline packets **(figure 6.15)**. One important event leading to the release of enveloped viruses is the insertion of viral spikes into the host's cell membrane so they can be picked up as the virus buds off with its envelope, as discussed earlier.

and release. The steps of the synthesis process vary. Process figure 6.14 illustrates this. Replication of dsDNA viruses is divided into phases **(process figure 6.14a)**. During the early phase, viral DNA enters the nucleus, where several genes are transcribed into a messenger RNA. The newly synthesized RNA transcript then moves into the cytoplasm to be translated into viral proteins (enzymes) needed to replicate the viral DNA; this replication occurs in the nucleus. The host cell's own DNA polymerase is often involved, though some viruses (herpesvirus, for example) have their own polymerase. During the late phase, other parts of the viral genome are transcribed and translated into proteins required to form the capsid and other structures. The new viral genomes and capsids are assembled, and the mature viruses are released by budding or cell disintegration.

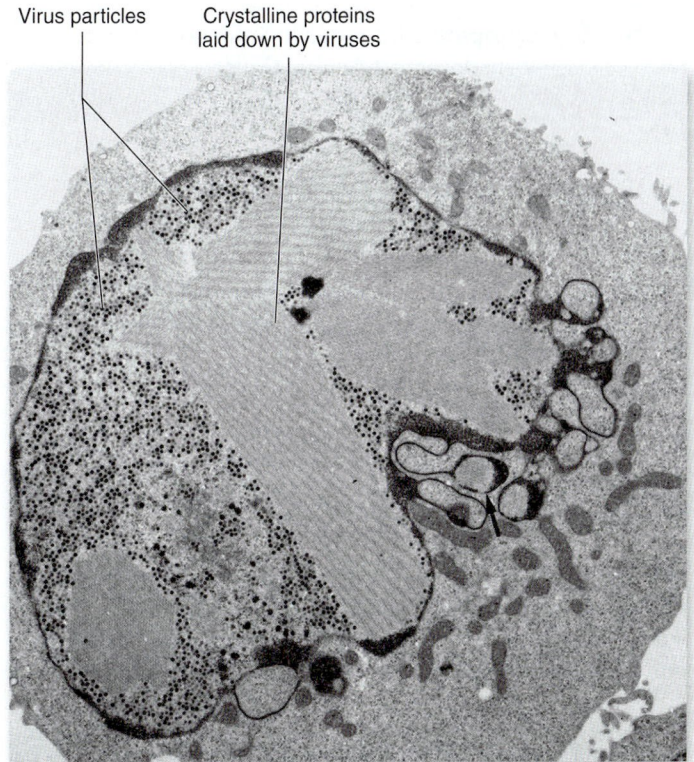

Virus particles

Crystalline proteins laid down by viruses

Figure 6.15 Nucleus of a eukaryotic cell, containing hundreds of adenovirus virions.

Release of Mature Viruses

To complete the cycle, assembled viruses leave their host in one of two ways. Nonenveloped and complex viruses that reach maturation in the cell nucleus or cytoplasm are released when the cell lyses or ruptures. Enveloped viruses are liberated by **budding** or **exocytosis**[2] from the membranes of the cytoplasm, nucleus, endoplasmic reticulum, or vesicles. During this process, the nucleocapsid binds to the membrane, which curves completely around it and forms a small pouch. Pinching off the pouch releases the virus with its envelope **(figure 6.16)**. Budding of enveloped viruses causes them to be shed gradually, without the sudden destruction of the cell. But regardless of how the virus leaves, most active viral infections are ultimately lethal to the cell because of accumulated damage. The number of viruses released by infected cells is variable, controlled by factors such as the size of the virus and the health of the host cell. About 3,000 to 4,000 virions are released from a single cell infected with poxviruses, whereas a poliovirus-infected cell can release over 100,000 virions. If even a small number of these virions happen to meet another susceptible cell and infect it, the potential for rapid viral proliferation is immense.

2. For enveloped viruses, the terms *budding* and *exocytosis* are interchangeable. They mean the release of a virus from an animal cell by enclosing it in a portion of membrane derived from the cell.

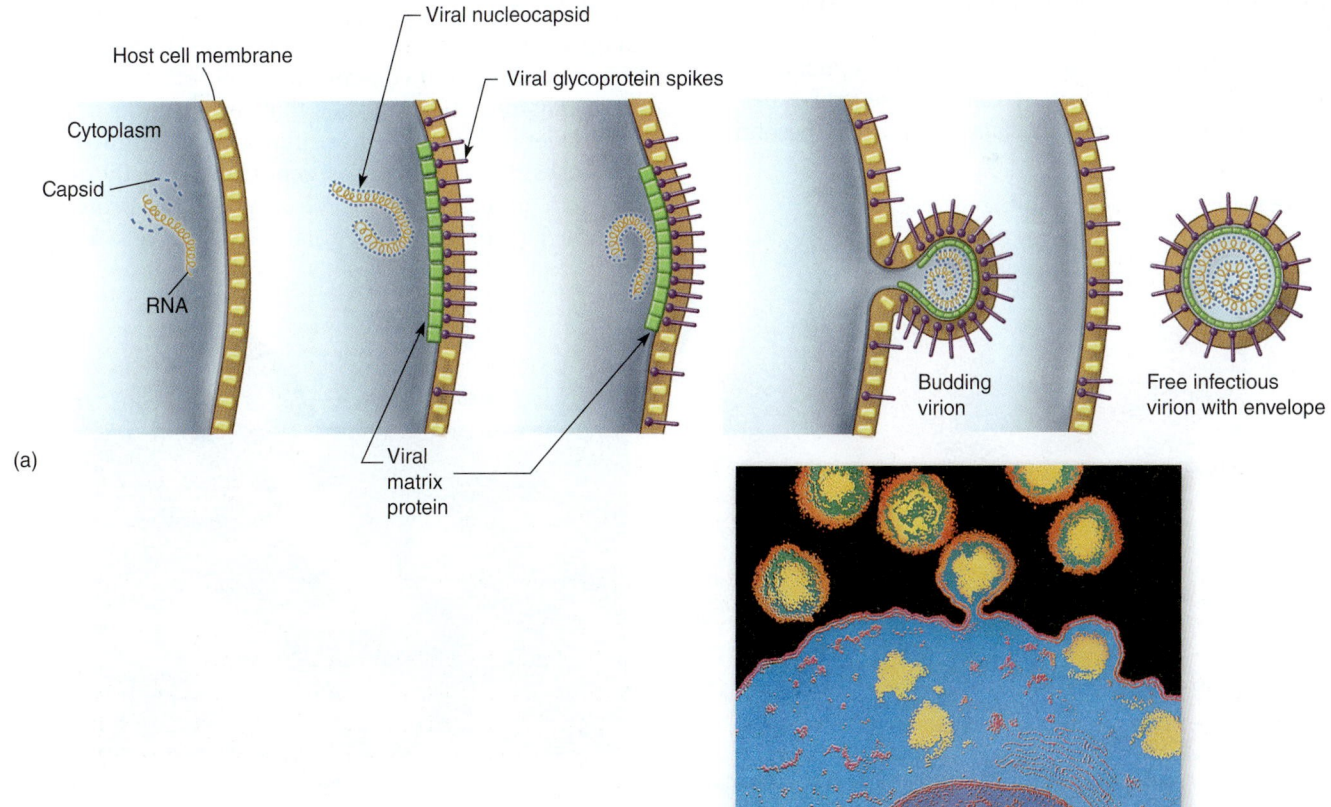

Host cell membrane

Viral nucleocapsid

Cytoplasm

Viral glycoprotein spikes

Capsid

RNA

Viral matrix protein

Budding virion

Free infectious virion with envelope

(a)

(b)

Figure 6.16 Maturation and release of enveloped viruses. **(a)** As parainfluenza virus is budded off the membrane, it simultaneously picks up an envelope and spikes. **(b)** HIV virions leave their host T cell by budding off its surface.

Damage to the Host Cell and Persistent Infections

The short- and long-term effects of viral infections on animal cells are well documented. Virus-induced damage to the cell that alters its microscopic appearance is termed **cytopathic effects (CPEs).** Individual cells can become disoriented, undergo major changes in shape or size, or develop intracellular changes **(figure 6.17a).** It is common to find **inclusion bodies,** or compacted masses of viruses or damaged cell organelles, in the nucleus and cytoplasm **(figure 6.17b).** Examination of cells and tissues for cytopathic effects is an important part of the diagnosis of viral infections. **Table 6.5** summarizes some prominent cytopathic effects associated with specific viruses. One very common CPE is the fusion of multiple host cells into single large cells containing multiple nuclei. These **syncytia** (singular, *syncytium*) are a result of some viruses' ability to fuse membranes. One virus (respiratory syncytial virus) is even named for this effect.

Although accumulated damage from a virus infection kills most host cells, some cells maintain a carrier relationship, in which the cell harbors the virus and is not immediately lysed. These so-called *persistent infections* can last from a few weeks to the remainder of the host's life. Viruses can remain latent in the cytoplasm of a host cell, or can incorporate into the DNA of the host. When viral DNA is incorporated into the DNA of the host, it is called a **provirus.** The virus that causes roseola has been found to be passed down from parent to infant in the provirus state, the first such instance of this form of transmission that can result in disease symptoms. One of the more serious complications occurs with the measles virus. It may remain hidden in brain cells for many years, causing progressive damage and loss of function. Several types of viruses remain in a *chronic latent state,*[3] periodically becoming

---

3. Meaning that they exist in an inactive state over long periods.

Table 6.5 Cytopathic Changes in Selected Virus-Infected Animal Cells

Virus	Response in Animal Cell
Smallpox virus	Cells round up; inclusions appear in cytoplasm
Herpes simplex	Cells fuse to form multinucleated syncytia; nuclear inclusions (see figure 6.17)
Adenovirus	Clumping of cells; nuclear inclusions
Poliovirus	Cell lysis; no inclusions
Reovirus	Cell enlargement; vacuoles and inclusions in cytoplasm
Influenza virus	Cells round up; no inclusions
Rabies virus	No change in cell shape; cytoplasmic inclusions (Negri bodies)
Measles virus	Syncytia form (multinucleate)

reactivated. Examples of this are herpes simplex viruses (cold sores and genital herpes) and herpes zoster virus (chickenpox and shingles). Both viruses can go into latency in nerve cells and later emerge under the influence of various stimuli to cause recurrent symptoms. Specific damage that occurs in viral diseases is covered more completely in chapters 18 through 23.

Viruses and Cancer

Some animal viruses enter their host cell and permanently alter its genetic material, leading to cancer. Experts estimate that up to 20% of human cancers are caused by viruses. These viruses are termed **oncogenic,** and their effect on the cell is called **transformation.** Viruses that cause cancer in animals act in several different ways, illustrated in **figure 6.18.** In some cases, the virus carries genes that directly cause the cancer. In other cases, the virus produces proteins that induce a loss of growth regulation in the cell, leading to

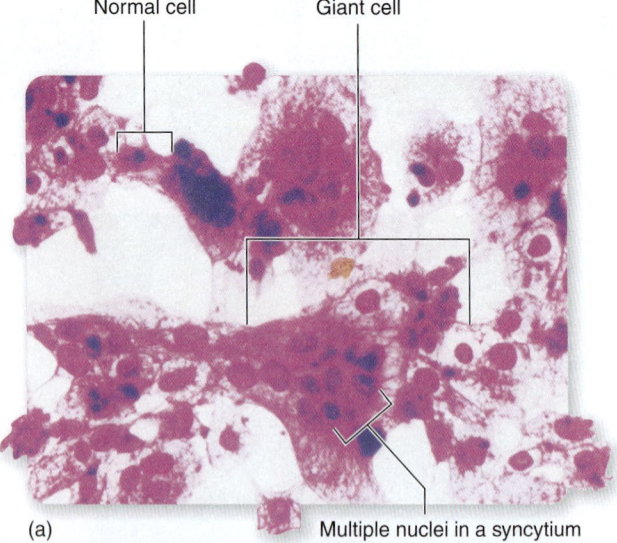

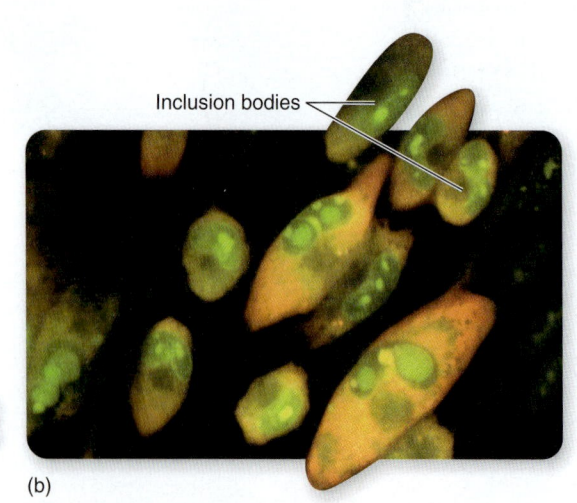

Normal cell   Giant cell

(a)   Multiple nuclei in a syncytium   (b)

Inclusion bodies

Figure 6.17 Cytopathic changes in cells and cell cultures infected by viruses.   **(a)** Human epithelial cells infected by herpes simplex virus demonstrate giant cells with multiple nuclei. **(b)** Fluorescent-stained human cells infected with cytomegalovirus. Note the inclusion bodies (labeled). Note also that both viruses disrupt the cohesive junctions between cells, which would ordinarily be arranged side by side in neat patterns.

INSIGHT 6.2   Coral Decline Linked to Herpesvirus?

Coral reefs are often referred to as the "rainforests of the seas." Thriving coral reefs harbor an abundance of plants, animals, and even microbes. Built from tiny coral polyps, these immense structures teeming with life have been built over millennia, and are now facing massive destruction due to global warming, pollution, and destructive fishing practices. Scientists estimate that coral reefs have declined 80% in the last 30 to 40 years and are threatened with extinction.

Coral reef ecologists have linked bleaching and destruction of coral to human excrement due to improper waste management practices in the Caribbean, and most studies have linked coral disease to bacterial causes. But more recent studies have found that viruses may play a role in coral disease and decline as well. Rebecca Vega-Thurber, assistant professor of Microbiology at Oregon State University, studies metagenomics in corals, analyzing the genomes found in these complex systems. She has found evidence that the predominant types of viruses in coral reefs are herpesviruses. It's not a shocking finding; herpesviruses are ancient and infect many different types of animals. But it is significant because it is the first association of this virus with the corals. Vega-Thurber noted that after episodes of acute stress (reef disturbance by boats or storms, for example), there were higher levels of herpesvirus-like genetic sequences in the coral. Because corals represent some of the oldest life forms on the earth, Thurber postulates that the coral and virus may have evolved together. Her studies also show that warm water, physical handling of coral, and nutrient increases in water due to pollution increase virus levels as well.

Declining coral levels throughout the earth's oceans are indicative of larger ecological issues stemming from increased

Healthy coral

Diseased coral

pollution and global warming. Studies such as these linking coral decline to bacterial and viral infection give scientists greater perspective on how to prevent further infection and decline of these "rainforests of the seas."

Source: 2012. *Science Daily.* http://www.sciencedaily.com/releases/2012/03/120328090941.htm. March 28.

cancer. Transformed cells have an increased rate of growth; alterations in chromosomes; changes in the cell's surface molecules; and the capacity to divide for an indefinite period, unlike normal animal cells. Mammalian viruses capable of initiating tumors are called **oncoviruses.** Some of these are DNA viruses such as papillomavirus (genital warts are associated with cervical cancer), herpesviruses (Epstein-Barr

virus causes Burkitt's lymphoma), and hepatitis B virus (liver cancer). A virus related to HIV—HTLV-I[4]—is also involved in human cancers. These findings have spurred a great deal of speculation on the possible involvement of viruses in cancers whose causes are still unknown.

4. Human T-cell lymphotropic virus, type I: causes type of leukemia.

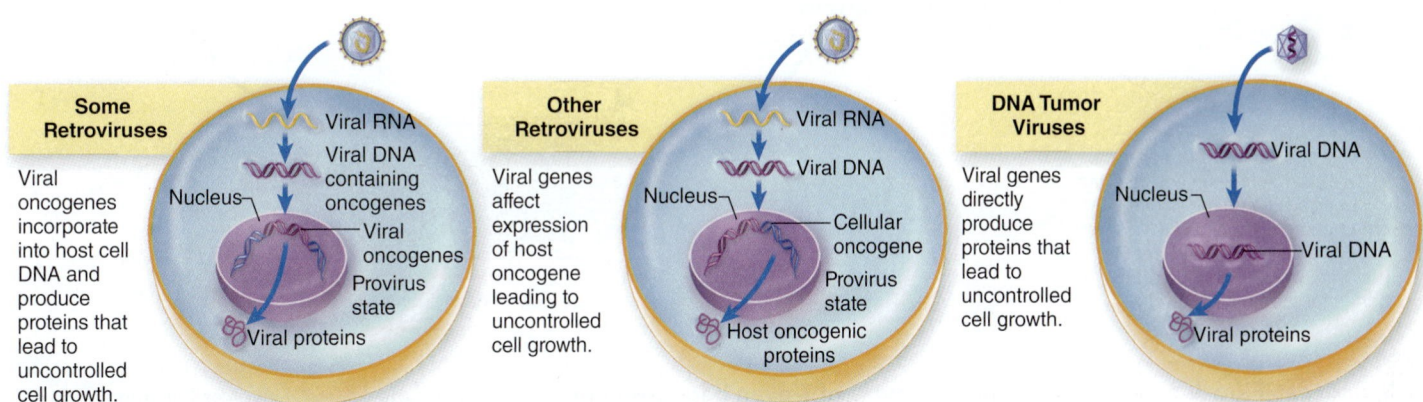

Figure 6.18 Three mechanisms for viral induction of cancer.

Viruses That Infect Bacteria

We now turn to the life cycle of another type of virus called bacteriophage. When Frederick Twort and Felix d'Herelle discovered bacterial viruses in 1915, it first appeared that the bacterial host cells were being eaten by some unseen parasite; hence, the name *bacteriophage* was used (*phage* coming from the Greek word for "eating"). Most bacteriophages (often shortened to **phage**) contain double-stranded DNA, although single-stranded DNA and RNA types exist as well. So far as is known, every bacterial species is parasitized by various specific bacteriophages. Bacteriophages are of great interest to medical microbiologists because they often make the bacteria they infect more pathogenic for humans (more about this later). Probably the most widely studied bacteriophages

are those of the intestinal bacterium *Escherichia coli*—especially the ones known as the "T-even" phages such as T2 and T4. They have an icosahedral capsid head containing DNA, a central tube (surrounded by a sheath), collar, base plate, tail pins, and fibers, which in combination make an efficient package for infecting a bacterial cell (see figure 6.9). Momentarily setting aside a strictly scientific and objective tone, it is tempting to think of these extraordinary viruses as minute spacecrafts docking on an alien planet, ready to unload their genetic cargo.

T-even bacteriophages go through similar stages as the animal viruses described earlier **(process figure 6.19)**. They *adsorb* to host bacteria using specific receptors on the bacterial surface. Although the entire phage does not enter the host

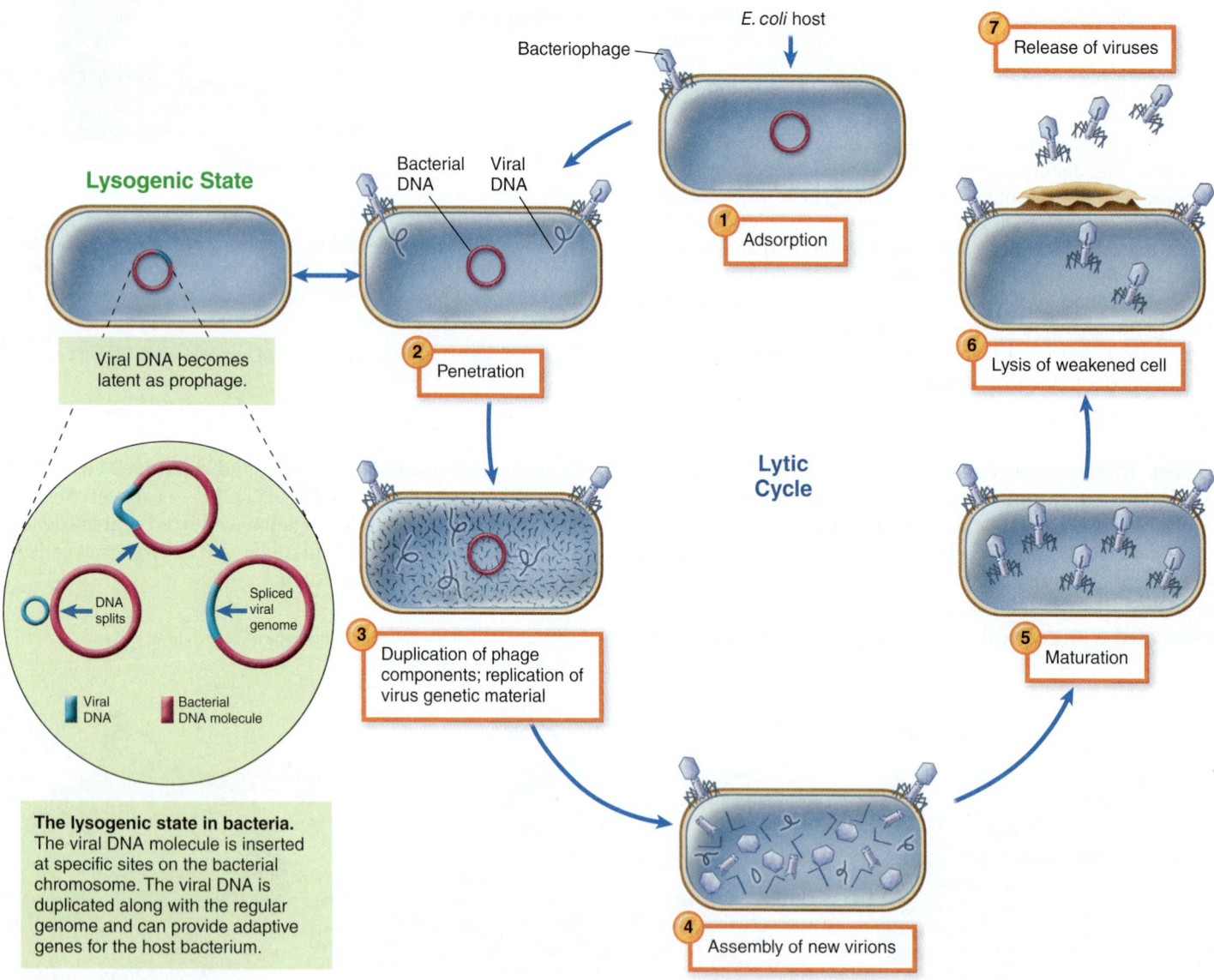

Lysogenic State

Viral DNA becomes latent as prophage.

DNA splits

Spliced viral genome

Viral DNA

Bacterial DNA molecule

The lysogenic state in bacteria. The viral DNA molecule is inserted at specific sites on the bacterial chromosome. The viral DNA is duplicated along with the regular genome and can provide adaptive genes for the host bacterium.

E. coli host

Bacteriophage

Bacterial DNA   Viral DNA

1 Adsorption

2 Penetration

3 Duplication of phage components; replication of virus genetic material

4 Assembly of new virions

5 Maturation

6 Lysis of weakened cell

7 Release of viruses

Lytic Cycle

Process Figure 6.19 Events in the lytic cycle of T-even bacteriophages. The lytic cycle (**1**–**7**) involves full completion of viral infection through lysis and release of virions. Occasionally, the virus enters a reversible state of lysogeny (left) and its genetic material is incorporated into the host's genetic material.

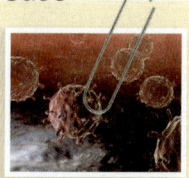

Strains of influenza can infect birds, pigs, horses, and dogs, as well as humans. Although transmission of influenza strains from animals to humans is rare, scientists have determined that it does in fact occur. Genomic analysis of past pandemic strains has revealed the presence of avian (bird) influenza genes in these human influenza viruses. Swine influenza virus is not usually transmitted to humans, but certain variant strains have been found to infect humans that live or work in close contact with pigs. Because pigs can contract both human and avian influenza, there is a higher probability of several viruses circulating in the same organism at one time. This can result in the production of a novel variant strain, as occurred during the 2009 swine flu pandemic. Analysis of this strain revealed the presence of two types of swine influenza genes in addition to avian and human influenza viral sequences, and it became known to scientists as a "quadruple reassortant" virus.

Today, two main types of avian influenza are infecting humans: low pathogenic avian influenza A (LPAI) viruses and highly pathogenic influenza A (HPAI) viruses. The Centers for Disease Control and Prevention (CDC) reports that there have been approximately 600 cases of HPAI in humans throughout the world since 2003. These illnesses are due to infection with the H5N1 strain of influenza, which exhibits a nearly 60% mortality rate in affected individuals. The CDC and the World Health Organization (WHO) have been carefully monitoring this strain of influenza because it has the potential to cause another pandemic.

■   What is being done to prevent a pandemic caused by H5N1 influenza?

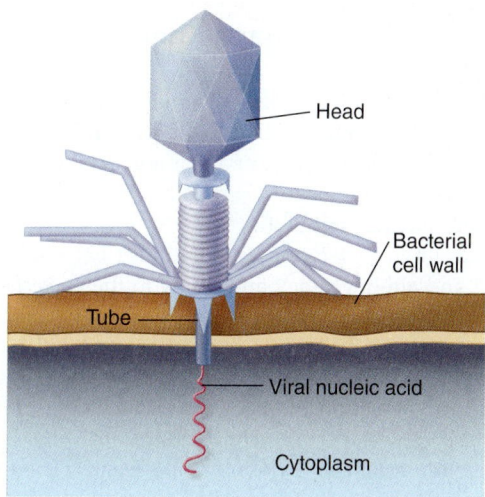

Figure 6.20 Penetration of a bacterial cell by a T-even bacteriophage.   After adsorption, the phage plate becomes embedded in the cell wall and the sheath contracts, pushing the tube through the cell wall and releasing the nucleic acid into the interior of the cell.

Lysogeny: The Silent Virus Infection

Special DNA phages, called **temperate phages,** while they can participate in a lytic phase, also have the ability to undergo adsorption and penetration into the bacterial host and not undergo replication or release immediately. Instead, the viral DNA enters an inactive **prophage** state, during which it is inserted into the bacterial chromosome. This viral DNA will be retained by the bacterial cell and copied during its normal cell division so that the cell's progeny will also have the temperate phage DNA (see green part of figure 6.19). This condition, in which the host chromosome carries bacteriophage DNA, is termed **lysogeny** (ly-soj'-uhn-ee). Because viral particles are not produced, the bacterial cells carrying temperate phages do not lyse, and they appear

cell, the nucleic acid *penetrates* the host after being injected through a rigid tube the phage inserts through the bacterial membrane and wall **(figure 6.20).** This eliminates the need for *uncoating.* Entry of the nucleic acid brings host cell DNA replication and protein synthesis to a halt. Soon the host cell machinery is used for viral *replication* and synthesis of viral proteins. As the host cell produces new phage parts, the parts spontaneously *assemble* into bacteriophages.

An average-size *E. coli* cell can contain up to 200 new phage units at the end of this period. Eventually, the host cell becomes so packed with viruses that it **lyses**—splits open— thereby releasing the mature virions **(figure 6.21).** This process is hastened by viral enzymes produced late in the infection cycle that digest the cell envelope, thereby weakening it. Upon release, the virulent phages can spread to other susceptible bacterial cells and begin a new cycle of infection.

Bacteriophage infection may result in lysis of the cell, as just described. When this happens, the phage is said to have been in the *lytic phase* or *cycle.* Alternatively, phages can be less obviously damaging, in a cycle called the *lysogenic cycle.*

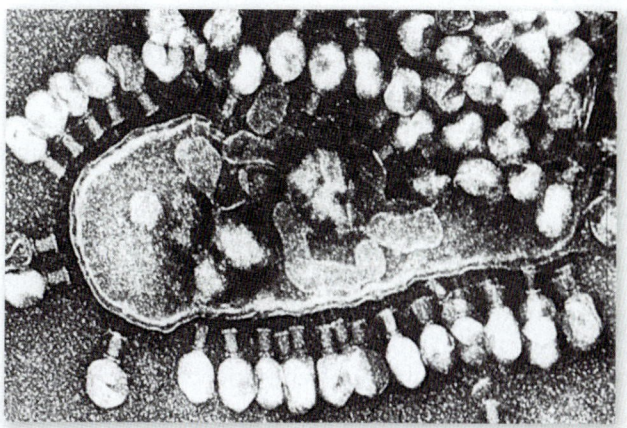

Figure 6.21 A weakened bacterial cell, crowded with viruses.   The cell has ruptured and released numerous virions that can then attack nearby susceptible host cells. Note the empty heads of "spent" phages lined up around the ruptured wall.

INSIGHT 6.3　Phage Therapy

Cystic fibrosis (CF) is a genetic disorder that causes buildup of mucus in the lungs and other organs. The mucus-filled lungs harbor *Pseudomonas aeruginosa,* an organism that normally lives in soil and water and is not usually dangerous to humans. In patients with CF, however, lung infections caused by *P. aeruginosa* are a persistent problem requiring nebulized (inhaled) antibiotics.

Bacteriophages are viruses that infect bacterial cells. Their name means "bacteria eaters," and although they are deadly to the bacteria they infect, they have no affinity for human cells. Bacteriophages can be more specific than antibiotics, targeting only the organisms they infect, while antibiotics often kill off helpful bacteria. Treating bacterial infections with bacteriophages, or "phage therapy," has been utilized, especially in Eastern Bloc countries, since the early 1900s, with little acceptance in westernized countries.

Recently, scientists working at the Teagasc Food Research Centre in Cork, Ireland, have found that certain bacteriophages can eliminate *P. aeruginosa* in mouse models of CF. The researchers found that a combination of two bacteriophages reduced the number of bacteria from 10 million to a thousand within 6 hours. This novel treatment could someday eliminate these infections in humans, improving and prolonging the lives of CF patients.

Source: 2012. mBio. DOI:10.1128/mBio.00029-12.

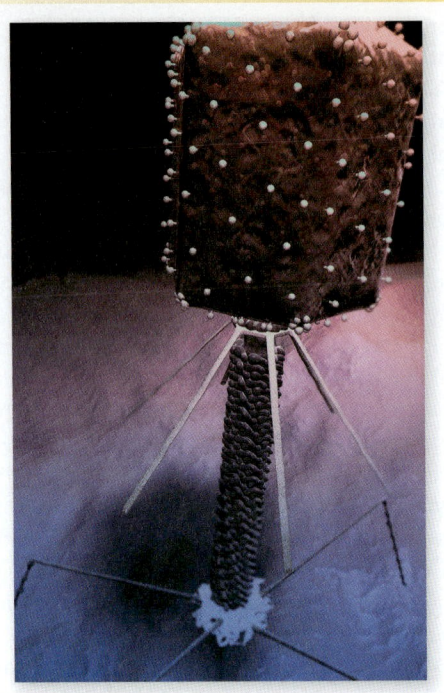

entirely normal. (This will remind you, appropriately, of the provirus state of animal viruses.) On occasion, in a process called **induction,** the prophage in a lysogenic cell will be activated and progress directly into viral replication and the lytic cycle. Lysogeny is a less deadly form of parasitism than the full lytic cycle and is thought to be an advancement that allows the virus to spread without killing the host.

Bacteriophages are just now receiving their due as important shapers of biological life. Scientists believe that there are more bacteriophages than all other forms of life combined in the biosphere. As we mentioned in the opening paragraphs of this chapter, viral genes linger in human, animal, plant, and bacterial genomes in huge numbers. It is estimated that 10% to 20% of DNA in bacteria is actually prophage DNA. As such, viruses can contribute what are essentially permanent traits to the bacteria, so much so that it could be said that all bacteria—indeed, all organisms—are really hybrids of themselves and the viruses that infect them.

In 2008, a discovery was made that set the world of virology on end. The identification of Sputnik, a subcellular particle now termed a *virophage,* revealed the existence of viruses that parasitize other viruses. When a virophage enters a host cell such as an amoeba that is already infected with a larger virus, the virophage is able to utilize genes from the other virus for its own replication and production. A second—called Maverick virus, or Mavirus—was discovered in 2011 and has provided more evidence that these may truly represent a unique form of virus.

The Danger of Lysogeny in Human Disease

Many bacteria that infect humans are lysogenized by phages. Sometimes, that is very bad news for the human; occasionally, phage genes in the bacterial chromosome cause the production of toxins or enzymes that cause pathology in the human. When a bacterium acquires a new trait from its temperate phage, it is called **lysogenic conversion.** The phenomenon was first discovered in the 1950s in *Corynebacterium diphtheriae,* the bacteria that cause diphtheria. The diphtheria toxin responsible for the deadly nature of the disease is a bacteriophage product. *C. diphtheriae* without the phage are harmless. Other bacteria that are made virulent by their prophages are *Vibrio cholerae,* the agent of cholera, and *Clostridium botulinum,* the cause of botulism.

Disease Connection

Streptococcus pyogenes hardly needs any help to cause disease. This pathogen causes "strep throat," flesh-eating disease, and serious bloodstream infections. But when it carries genes from its bacteriophage, it can cause even greater damage. Lysogenized *S. pyogenes* carries genes coding for something called erythrogenic toxin. These strains lead to postinfection problems such as scarlet fever, which results in a skin rash and a high fever, all courtesy of the bacteriophage.

Table 6.6 Comparison of Bacteriophage and Animal Virus Multiplication

	Bacteriophage	Animal Virus
Adsorption	Precise attachment of special tail fibers to cell wall	Attachment of capsid or envelope to cell surface receptors
Penetration	Injection of nucleic acid through cell wall; no uncoating of nucleic acid	Whole virus is engulfed and uncoated, or virus surface fuses with cell membrane; nucleic acid is released
Synthesis and assembly	Occurs in cytoplasm Cessation of host synthesis Viral DNA or RNA is replicated and begins to function Viral components synthesized	Occurs in cytoplasm and nucleus Cessation of host synthesis Viral DNA or RNA is replicated and begins to function Viral components synthesized
Viral persistence	Lysogeny	Latency, chronic infection, cancer
Release from host cell	Cell lyses when viral enzymes weaken it	Some cells lyse; enveloped viruses bud off host cell membrane
Cell destruction	Immediate or delayed	Immediate or delayed

The life cycles of animal and bacterial viruses (see process figure 6.11, process figure 6.14, and process figure 6.19) illustrate general features of viral multiplication in a very concrete way. The two cycles are compared in **table 6.6.** Because of the intimate association between the genetic material of the virus and that of the host, phages occasionally serve as transporters of bacterial genes from one bacterium to another and consequently can play a profound role in bacterial genetics. This phenomenon, called *transduction,* is one way that genes for toxin production and drug resistance are transferred between bacteria (see chapters 9 and 12).

6.5 Learning Outcomes—Assess Your Progress

12. Diagram the six-step life cycle of animal viruses.

13. Define the term *cytopathic effect* and provide one example.

14. Provide examples of persistent and transforming infections, describing their effects on the host.

15. Provide a thorough description of lysogenic and lytic bacteriophage infections.

6.6 Techniques in Cultivating and Identifying Animal Viruses

One problem hampering earlier animal virologists was their inability to grow specific viruses routinely in pure culture—and in sufficient quantities for their studies. Early on, viruses could only be grown in an organism that was the usual host for the virus. But this method had its limitations. How could researchers have ever traced the stages of viral multiplication if they had been restricted to the natural host, especially in the case of human viruses? Fortunately, systems of cultivation with broader applications were developed, called *in vivo* (in vee'-voh) and *in vitro* (in vee'-troh) methods. *In vivo* means that incubations are carried out in lab animals or embryonic bird tissues (whole organisms). *In vitro* refers to the use of cell (or tissue) culture methods. Such use of substitute host systems permits greater control, uniformity, and wide-scale harvesting of viruses.

The primary purposes of viral cultivation are

1. to isolate and identify viruses in clinical specimens;
2. to prepare viruses for vaccines; and
3. to do detailed research on viral structure, multiplication cycles, genetics, and effects on host cells.

Using Live Animal Inoculation

Specially bred strains of white mice, rats, hamsters, guinea pigs, and rabbits are the usual choices for animal cultivation of viruses. Invertebrates (insects) or nonhuman primates are occasionally used as well. Because viruses can exhibit some host specificity, certain animals can propagate a given virus more readily than others. The animal is exposed to the virus by injection of a viral preparation or specimen into the brain, blood, muscle, body cavity, skin, or footpads.

Using Bird Embryos

An embryo is an early developmental stage of animals marked by rapid differentiation of cells. Birds undergo their embryonic period within the closed protective case of an egg, which makes an incubating bird egg a nearly perfect system for viral propagation. It is an intact and self-supporting unit, complete with its own sterile environment and nourishment. Furthermore, it furnishes several embryonic tissues that readily support viral multiplication.

Chicken, duck, and turkey eggs are the most common choices for inoculation. The egg must be injected through the shell, usually by drilling a hole or making a small window. Rigorous sterile techniques must be used to prevent contamination by bacteria and fungi from the air and the outer surface of the shell. The exact part of the egg that is

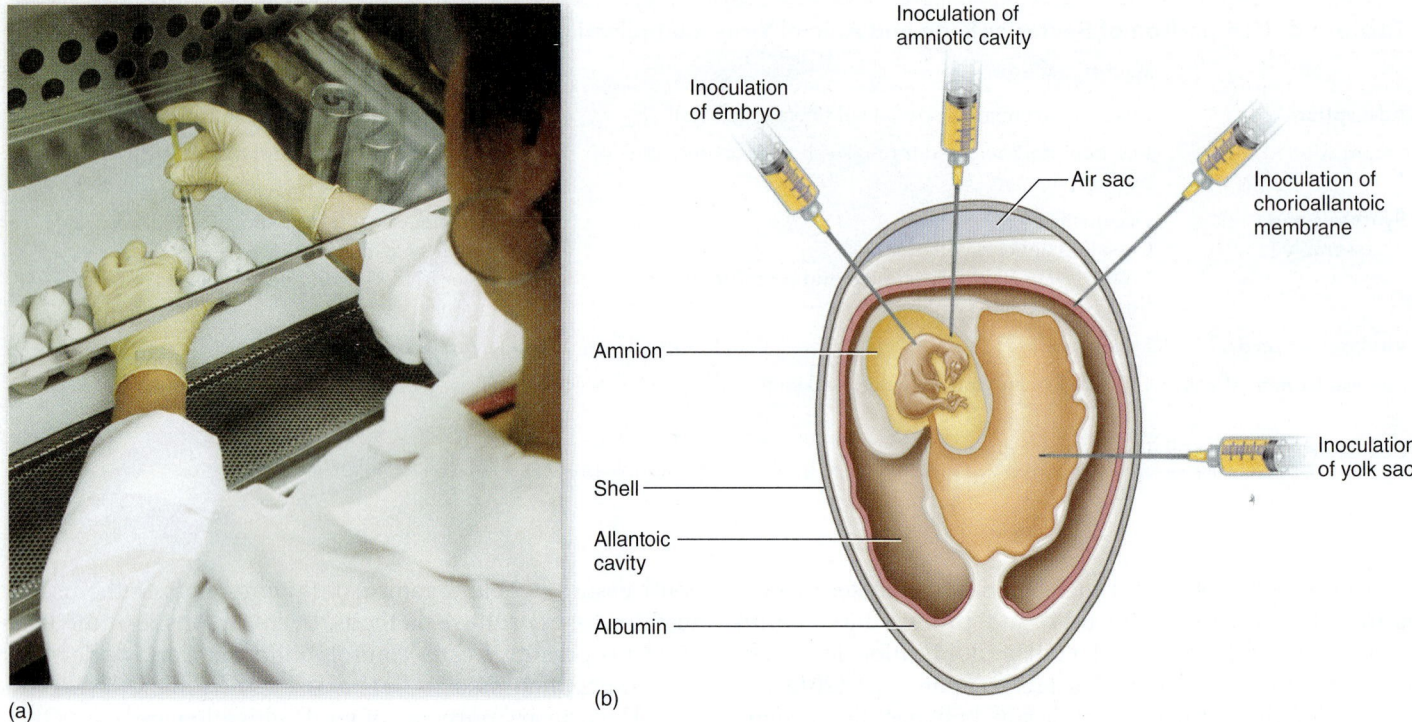

(a)

(b)

Figure 6.22 Cultivating animal viruses in a developing bird embryo. (a) A technician inoculates fertilized chicken eggs with viruses in the first stage of preparing vaccines. Current influenza vaccine is prepared this way. (b) The shell is perforated using sterile techniques, and a virus preparation is injected into a site selected to grow the viruses. Targets include the allantoic cavity, a fluid-filled sac that functions in embryonic waste removal; the amniotic cavity, a sac that cushions and protects the embryo itself; the chorioallantoic membrane, which functions in embryonic gas exchange; the yolk sac, a membrane that mobilizes yolk for the nourishment of the embryo; and the embryo itself.

inoculated is guided by the type of virus being cultivated and the goals of the experiment **(figure 6.22).**

Viruses multiplying in embryos may or may not cause effects visible to the naked eye. The signs of viral growth include death of the embryo; defects in embryonic development; and localized areas of damage in the membranes, resulting in discrete, opaque spots called *pocks*. If a virus does not produce overt changes in the developing embryonic tissue, virologists have other methods of detection. Embryonic fluids and tissues can be prepared for direct examination with an electron microscope. Certain viruses can also be detected by their ability to agglutinate red blood cells (form big clumps) or by their reaction with an antibody of known specificity that will affix to its corresponding virus, if it is present.

Using Cell (Tissue) Culture Techniques

The most important early discovery that led to easier cultivation of viruses in the laboratory was the development of a simple and effective way to grow populations of isolated animal cells in culture dishes. These types of *in vitro* cultivation systems are termed *cell culture* or *tissue culture*. (Although these terms are used interchangeably, *cell culture* is probably a more accurate description.) So prominent is this method that most viruses are propagated in some sort of cell culture, and much of the virologist's work involves developing and maintaining

these cultures. Animal cell cultures are grown in sterile dishes or bottles with special media that contain the correct nutrients required by animal cells to survive. The cultured cells grow in the form of a *monolayer,* a single, confluent sheet of cells that supports viral multiplication and permits close inspection of the culture for signs of infection **(figure 6.23).**

Cultures of animal cells usually exist in the primary or continuous form. *Primary cell cultures* are prepared by placing freshly isolated animal tissue in a growth medium. The cells undergo a series of mitotic divisions to produce a monolayer on the surface of the dish. Embryonic, fetal, adult, and even cancerous tissues have served as sources of primary cultures. Primary cultures retain several characteristics of the original tissues from which they were derived, but they generally have a limited existence. Eventually, they will either die out or mutate into a line of cells that can grow continuously. Continuous cell lines tend to have altered chromosome numbers, grow rapidly, and show changes in morphology; they can be continuously subcultured, provided they are routinely transferred to fresh nutrient medium. One very clear advantage of cell culture is that a specific cell line can be available for viruses with a very narrow host range.

Ongoing worries about influenza pandemics and the need for vaccines have prompted scientists to look for faster and more efficient ways to grow the vaccine strains of influenza virus, which has been grown in chicken eggs since the 1950s. Scientists have succeeded in propagating the

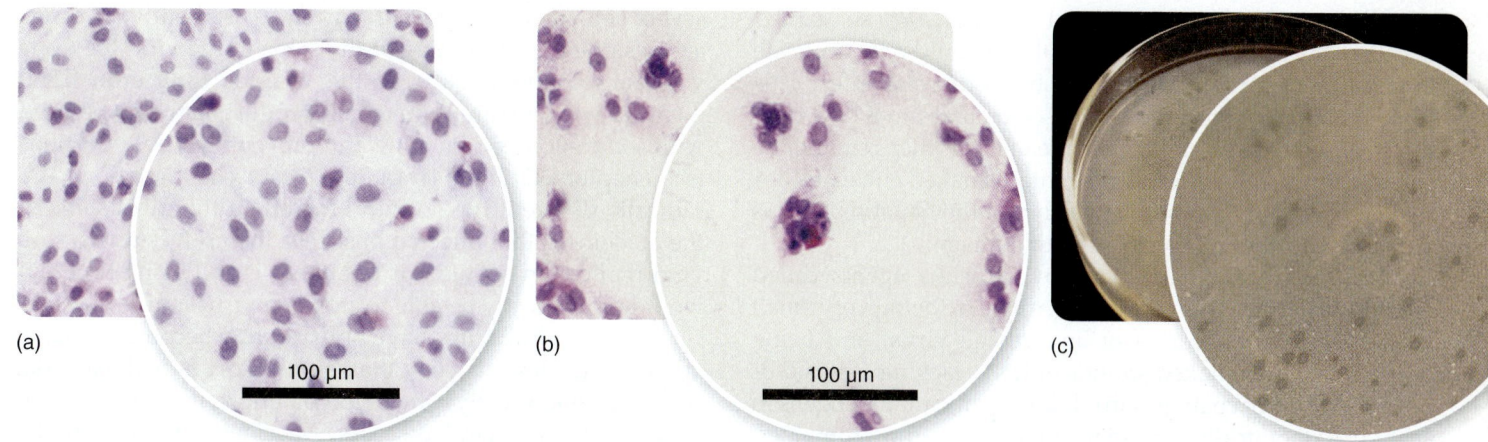

Figure 6.23 Appearance of normal and infected cell cultures. **(a)** Microscopic view of an undisturbed layer of animal cells. **(b)** Plaques in the animal cell layer. These are open spaces where cells have been disrupted by viral infection. **(c)** Macroscopic view of a lawn of *E. coli* on agar. The clear, round spaces are plaques, points of infection by bacteriophages.

viruses in a continuous cell line derived from animal kidney cells. Since 2009, several clinical trials of vaccine strains grown in culture were found to be as effective as those grown in chicken eggs, but these new vaccines are not likely to be available for at least a few years.

One way to detect the growth of a virus in culture is to observe degeneration and lysis of infected cells in the monolayer of cells. The areas where virus-infected cells have been destroyed show up as clear, well-defined patches in the cell sheet called **plaques** (see figure 6.23). Plaques are essentially the macroscopic manifestation of cytopathic effects (CPEs), discussed earlier. This same technique is used to detect and count bacteriophages, because they also produce plaques when grown in soft agar cultures of their host cells (bacteria). A plaque develops when the viruses released by an infected host cell radiate out to adjacent host cells (see figure 6.23). As new cells become infected, they die and release more viruses, and so on. As this process continues, the infection spreads gradually and symmetrically from the original point of infection, causing the macroscopic appearance of round, clear spaces that correspond to areas of dead cells. Such technology advances the study of viruses, but also creates critical ethical issues that will continually be reviewed by both federal and international organizations.

6.6 Learning Outcomes—Assess Your Progress

16. List the three principal purposes for cultivating viruses.

17. Describe three ways in which viruses are cultivated.

6.7 Other Noncellular Infectious Agents

Not all noncellular infectious agents have typical viral morphology. One group of unusual forms, even smaller and simpler than viruses, is implicated in chronic, persistent diseases in humans and animals. These diseases are called spongiform encephalopathies because the brain tissue

removed from affected animals resembles a sponge. The infection has a long period of latency (usually several years) before the first clinical signs appear. Signs range from mental derangement to loss of muscle control. The diseases are progressive and universally fatal.

A common feature of these conditions is the deposition of distinct protein fibrils in the brain tissue. Researchers have hypothesized that these fibrils are the agents of the disease and have named them **prions** (pree'-onz).

Creutzfeldt-Jakob disease (CJD) afflicts the central nervous system of humans and causes gradual degeneration and death. It is transmissible—but by an unknown mechanism. Several animals (sheep, mink, elk) are victims of similar transmissible diseases. Bovine spongiform encephalopathy (BSE), or "mad cow disease," was recently the subject of fears and a crisis in Europe when researchers found evidence that the disease could be acquired by humans who consumed contaminated beef. This was the first incidence of prion disease transmission from animals to humans. Several hundred Europeans developed symptoms of a variant form of Creutzfeldt-Jakob disease, leading to strict governmental controls on exporting cattle and beef products. In 2003, isolated cows with BSE were found in Canada and in the United States. Extreme precautionary measures have been taken to protect North American consumers. As of 2012, only a few BSE-positive cows have been found in the United States, compared to over 184,000 in the United Kingdom. (This disease is described in more detail in chapter 19.)

The exact mode of prion infection is currently being analyzed. The fact that prions are composed primarily of protein (no nucleic acid) has certainly revolutionized our ideas of what can constitute an infectious agent. One of the most compelling questions is just how a prion could be replicated, because all other infectious agents require some nucleic acid.

Other fascinating viruslike agents in human disease are defective forms called satellite viruses that are actually dependent on other viruses for replication. One remarkable example is the adeno-associated virus (AAV), so named because it was originally thought that it could only replicate in cells infected with adenovirus; but it can also infect cells that are infected with other viruses or that have had their DNA disrupted through other means. Another remarkable satellite virus, called the delta agent, is a naked circle of RNA that is expressed only in the presence of the hepatitis B virus and can worsen the severity of liver damage.

Plants are also parasitized by viruslike agents called **viroids** that differ from ordinary viruses by being very small (about one-tenth the size of an average virus) and being composed of only naked strands of RNA, lacking a capsid or any other type of coating. Viroids are significant pathogens in several economically important plants, including tomatoes, potatoes, cucumbers, citrus trees, and chrysanthemums.

6.7 Learning Outcomes—Assess Your Progress

18. List three noncellular infectious agents besides viruses.

6.8 Viruses and Human Health

The number of viral infections that occur on a worldwide basis is nearly impossible to measure accurately. Certainly, viruses are the most common cause of acute infections that do not result in hospitalization, especially when one considers widespread diseases such as colds, chickenpox, influenza, herpes, and warts. If one also takes into account prominent viral infections found only in certain regions of the world, such as Dengue fever, Rift Valley fever, and yellow fever, the total could easily exceed several billion cases each year. Although most viral infections do not result in death, some, such as rabies or Ebola, have very high mortality rates, and others can lead to long-term debility (polio, neonatal rubella). Current research is focused on the possible connection of viruses to chronic afflictions of unknown cause, such as type I diabetes, multiple sclerosis, various cancers, and even obesity. Additionally, as mentioned earlier, several cancers have their origins in viral infection.

The rate at which many pathogenic viruses mutate is quite rapid (see Case File). In order for scientists to protect the public health, swift identification of new viral strains is a must. One potential way of doing this is to identify changes in the host response to the novel pathogen at the cellular level. Recently, it has been determined that virally infected cells display a unique protein "bar code." The collection of protein types with elevated expression in the patient during a viral infection can be scanned through a database, and results can quickly alert scientists to a new and potentially deadly strain of virus circulating within a population.

The nature of viruses makes it difficult to design effective therapies against them. Because viruses are not bacteria, antibiotics aimed at disrupting bacterial cells do not work on them. Out of necessity, many antiviral drugs block virus replication by targeting the function of host cells and can cause

severe side effects. Almost all currently used antiviral drugs are designed to target one of the steps in the viral life cycle you learned about earlier in this chapter. **Interferon (IFN)** a naturally occurring human cell product, can also be used with some success in treating and preventing viral infections (see chapters 12 and 14). In 2011, researchers announced a radically different type of viral treatment that uses some of the machinery that infected host cells themselves use to stop viral replication. While this treatment is still being tested, it would theoretically be able to stop any virus infection.

Because viral drugs have only recently become available and are often less effective than antibiotics are with bacteria, scientists historically put a lot of effort into developing vaccines against viral diseases. Vaccines can prevent the diseases before they start (see chapter 15).

We have completed our survey of bacteria, archaea, eukaryotes, and viruses and have described characteristics of different representatives of these four groups. Chapters 7 and 8 explore how microorganisms maintain themselves, beginning with nutrition (chapter 7) and then looking into microbial metabolism (chapter 8).

6.8 Learning Outcomes—Assess Your Progress

19. Analyze the relative importance of viruses in human infection and disease.

20. Discuss the primary reason that antiviral drugs are more difficult to design than antibacterial drugs.

Case File 6 *Wrap-Up*

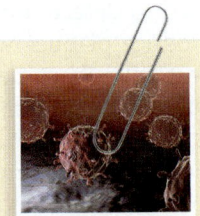

Because of the grave threat of a pandemic caused by the H5N1 influenza virus, scientists have begun studying it for possible mutations that would allow easier passage from birds to humans. Yoshihiro Kawaoka, a scientist at the University of Wisconsin-Madison, and an international team of researchers identified an initial set of mutations that enhanced virus transmission in ferrets, an experimental mammalian model. This research has sparked a global debate among the scientific community, and many have called for censorship of this research, insisting that this information in the wrong hands could lead to the development of a lethal biological weapon. In December 2011, the National Science Advisory Board for Biosecurity requested that the scientific journals *Science* and *Nature* not publish details of the studies. The scientific teams agreed to hold their research for 60 days until the WHO could hold a meeting to discuss the implications of publishing this type of research. The study was ultimately published in the May 3, 2012, issue of *Nature*.

Source: T. H. Saey, "Controversial Flu Research Published," *Science News Online* (May 2, 2012). Retrieved from www.sciencenews.org/view/generic/id/340373/title/Controversial_flu_research_published.

Chapter Summary

6.1 The Search for the Elusive Viruses (ASM Guideline* 2.2)

- Viruses are noncellular entities whose properties have been identified through microscopy, tissue culture, and molecular biology.

6.2 The Position of Viruses in the Biological Spectrum (ASM Guidelines 1.5, 3.3, 4.4, 5.4)

- Viruses are infectious particles that invade every known type of cell. They are not alive, yet they are able to redirect the metabolism of living cells to reproduce virus particles.
- Viruses have a profound influence on the genetic makeup of the biosphere.
- Viral replication inside a cell usually causes death or loss of function of that cell.

6.3 The General Structure of Viruses (ASM Guidelines 2.3, 2.4, 4.4)

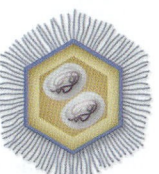

Mimivirus
450 nm

- Virus size range is from 20 nm to 1000 nm (diameter). Viruses are composed of an outer protein capsid containing either DNA or RNA plus a variety of enzymes. Some viruses also possess an envelope around the capsid.
- Spikes on the surface of the virus capsid or envelope are critical for their attachment to host cells.

6.4 How Viruses Are Classified and Named (ASM Guideline 1.4)

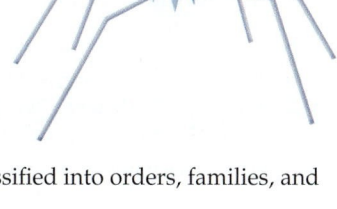

- Viruses are grouped in various ways. This textbook uses their structure, genetic composition, and host range to categorize them.
- The International Committee on the Taxonomy of Viruses oversees naming and classification of viruses. Viruses are classified into orders, families, and genera.

6.5 Modes of Viral Multiplication (ASM Guidelines 2.3, 4.4, 5.4)

- Viruses go through a multiplication cycle that generally involves adsorption, penetration (sometimes followed by uncoating), viral synthesis and assembly, and viral release by lysis or budding.
- These events turn the host cell into a factory solely for making and shedding new viruses. This results in the ultimate destruction of the cell.

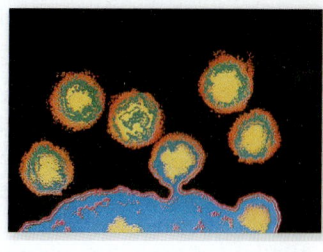

- Animal viruses can cause acute infections or can persist in host tissues as chronic latent infections that can reactivate periodically throughout the host's life. Some persistent animal viruses are oncogenic.
- Bacteriophages vary significantly from animal viruses in their methods of adsorption, penetration, site of replication, and method of exit from host cells.
- Lysogeny is a condition in which viral DNA is inserted into the bacterial chromosome and remains inactive for an extended period. It is replicated right along with the chromosome every time the bacterium divides.
- Some bacteria express virulence traits that are coded for by the bacteriophage DNA in their chromosomes. This phenomenon is called *lysogenic conversion.*

6.6 Techniques in Cultivating and Identifying Animal Viruses (ASM Guidelines 8.2, 8.3, 8.6)

- Animal viruses must be studied in some type of host cell environment such as laboratory animals, bird embryos, or tissue cultures.
- Cell and tissue cultures are cultures of host cells grown in special sterile chambers containing correct types and proportions of growth factors using aseptic techniques to exclude unwanted microorganisms.
- Virus growth in cell culture and bacteriophage growth on bacterial lawns are detected by the appearance of plaques.

6.7 Other Noncellular Infectious Agents (ASM Guidelines 5.4, 6.4)

- Other noncellular agents of disease are the prions, which are not viruses at all but protein fibers; viroids, extremely small lengths of protein-coated nucleic acid; and satellite viruses, which require the presence of larger viruses to cause disease.

6.8 Viruses and Human Health (ASM Guideline 5.4)

- Viruses are easily responsible for several billion infections each year. It is conceivable that many chronic diseases of unknown cause will eventually be connected to viral agents.
- Viral infections are difficult to treat because the drugs that attack viral replication also cause side effects in the host.

ASM Curriculum Guidelines (American Society for Microbiology, 2012). Complete guidelines in appendix B of this book.

Multiple-Choice and True-False Questions | Bloom's Levels 1 and 2: Remember and Understand

Multiple-Choice Questions. Select the correct answer from the options provided.

1. A virus is a tiny infectious
   a. cell.   c. particle.
   b. living thing.   d. nucleic acid.

2. Viruses are known to infect
   a. plants.   c. fungi.
   b. bacteria.   d. all organisms.

3. The nucleic acid of a virus is
   a. DNA only.   c. both DNA and RNA.
   b. RNA only.   d. either DNA or RNA.

4. The general steps in a viral multiplication cycle are
   a. adsorption, penetration, synthesis, assembly, and release.
   b. endocytosis, uncoating, replication, assembly, and budding.
   c. adsorption, uncoating, duplication, assembly, and lysis.
   d. endocytosis, penetration, replication, maturation, and exocytosis.

5. A prophage is a stage in the development of a/an
   a. bacterial virus.   c. lytic virus.
   b. poxvirus.   d. enveloped virus.

6. In general, RNA viruses multiply in the cell _____, and DNA viruses multiply in the cell _____.
   a. nucleus, cytoplasm   c. vesicles, ribosomes
   b. cytoplasm, nucleus   d. endoplasmic reticulum, nucleolus

7. Viruses cannot be cultivated in/on
   a. tissue culture.   c. live mammals.
   b. bird embryos.   d. blood agar.

8. Clear patches in cell cultures that indicate sites of virus infection are called
   a. plaques.   c. colonies.
   b. pocks.   d. prions.

9. Label the parts of this virus. Identify the capsid, nucleic acid, and other features of this virus.

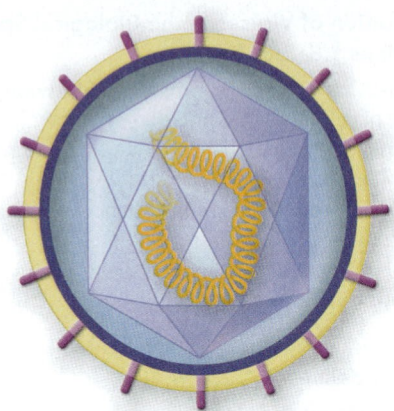

10. Circle the viral infections from this list: cholera, rabies, plague, cold sores, whooping cough, tetanus, genital warts, gonorrhea, mumps, Rocky Mountain spotted fever, syphilis, rubella.

True-False Questions. If the statement is true, leave as is. If it is false, correct it by rewriting the sentence.

11. In lysogeny, viral DNA is inserted into the host chromosome.

12. A viral capsid is composed of subunits called virions.

13. The envelope of an animal virus is derived from the peptidoglycan of its host cell.

14. The nucleic acid of animal viruses enters the cell through a process called translocation.

15. Viruses that persist in the (host) cell and cause recurrent disease are called latent.

Critical Thinking Questions | Bloom's Levels 3, 4, and 5: Apply, Analyze, and Evaluate

Critical thinking is the ability to reason and solve problems using facts and concepts. These questions can be approached from a number of angles and, in most cases, they do not have a single correct answer.

1. Provide evidence in support of or refuting the following statement: Viruses are simple cellular agents of disease.

2. Summarize the unique properties of viruses and explain which of these characteristics allow them to function as "parasites."

3. a. Sketch the basic structure of both a nonenveloped and an enveloped virus, labeling all parts.
   b. Discuss the validity of the following statement: The viral capsid and envelope only provide functions that enhance the pathogenicity of a virus.

4. a. You identify a novel microbe in your laboratory and find that it possesses two types of nucleic acid. Explain why you immediately rule out the fact that this microbe is a virus.
   b. Describe the nucleic acid configuration of a positive-sense RNA virus and explain why its multiplication cycle is less complex than that of a retrovirus.

5. Define the term *tropism,* and provide at least one example illustrating how viral structure determines this property of a virus.

6. a. Provide one example of an oncogenic virus and explain the unique properties of its multiplication cycle that allow it to trigger the development of cancer.
   b. Compare and contrast the processes of latency and lysogeny, providing examples of latent viruses and lysogenic viruses.

7. Summarize the method used by most companies to manufacture influenza vaccine today, providing one clear advantage and one disadvantage of this process.

8. a. Provide at least three examples of noncellular infectious agents that do not display typical viral morphology.

   b. Of these examples, explain which are associated with the development of human disease.

9. a. Why are viral diseases more difficult to treat than bacterial diseases?

   b. You are a researcher aiming to develop a drug to reduce the multiplication of chickenpox (varicella zoster) virus. Thinking about the steps of the viral multiplication cycle,

describe at least three potential targets that would reduce the amount of virus produced by infected cells.

10. It is estimated that nearly 10% of the human genome is comprised of viral sequences. Conduct additional research and explain the following:

    a. What is a HERV?

    b. What is the potential role of HERVs in disease development?

Concept Connections | Bloom's Levels 4 and 6: Analyze and Create

This activity ties together multiple concepts in the chapter.

1. How does capsid and/or envelope structure determine the type of cells a virus infects?

2. Describe the composition of the viral envelope.

3. How are enveloped viruses different from nonenveloped viruses?

4. Provide examples of enveloped and nonenveloped viruses in each category as well as examples of complex viruses to complete the flowchart.

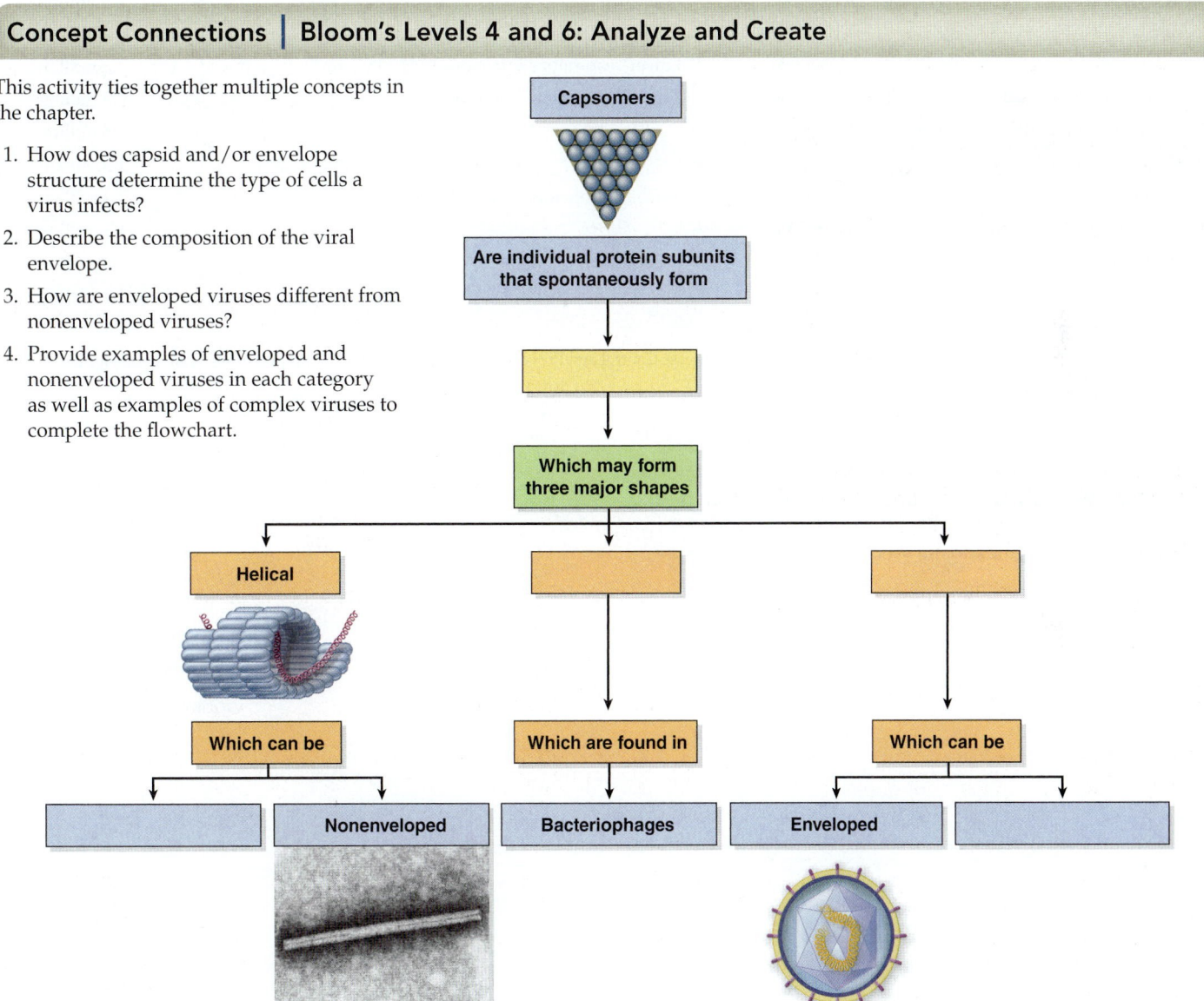

Visual Connections | Bloom's Level 5: Evaluate

This question uses visual images or previous content to make connections to this chapter's concepts.

1. **From chapter 1, table 1.2.** This table from chapter 1 identifies diseases most clearly caused by microorganisms. Considering what you have learned in this chapter, are there more deaths caused by microorganisms than may be accounted for by the red-labeled diseases? Can you make a rough guess of how many total deaths might be caused by viruses?

Table 1.2 Top Causes of Death—All Diseases

United States	No. of Deaths	Worldwide	No. of Deaths
1. Heart disease	617,000	1. Heart disease	7.3 million
2. Cancer	565,000	2. Stroke	6.2 million
3. Chronic lower-respiratory disease	141,000	3. Lower-respiratory infections (influenza and pneumonia)*	3.5 million
4. Cerebrovascular disease	134,000	4. Chronic obstructive pulmonary disease	3.3 million
5. Accidents (unintentional injuries)	122,000	5. Diarrheal diseases	2.5 million
6. Alzheimer's disease	82,000	6. HIV/AIDS	1.8 million
7. Diabetes	71,000	7. Trachea, bronchus, lung cancers	1.4 million
8. Influenza and pneumonia	56,000	8. Tuberculosis	1.3 million
9. Kidney disease	48,000	9. Diabetes	1.3 million
10. Suicide	36,000	10. Road traffic accidents	1.2 million

*Diseases in red are those most clearly caused by microorganisms.

Source: Data from the World Health Organization and the Centers for Disease Control and Prevention. Data published in 2011 representing final figures for the year 2008.

Concept Mapping | Bloom's Level 6: Create

Appendix D provides guidance for working with concept maps.

1. Using the words that follow, please create a concept map illustrating the relationships among these key terms from chapter 6.

adsorption	exocytosis	endocytosis	synthesis
tropism	penetration	nucleic acids	assembly
release	uncoating	integration	virions

www.mcgrawhillconnect.com

Enhance your study of this chapter with study tools and practice tests. Also ask your instructor about the resources available through ConnectPlus, including the media-rich eBook, interactive learning tools, and animations.

Microbial Nutrition, Ecology, and Growth

Case File 7

The Haitian Earthquake: Unexpected Devastation

On January 12, 2010, a 7.0 magnitude earthquake struck Haiti. The already poor and struggling nation was quickly reduced to rubble: 250,000 were killed, 300,000 were injured, and more than 1.3 million people were left homeless. As foreign nations rushed to aid the decimated country, displaced-persons camps cropped up all over the countryside, as Haiti attempted to rebuild. Ten months later, the devastated country was still rebuilding when reports of diarrheal disease began coming in from far-flung displaced-persons camps in the countryside. Stool samples were sent to hospitals in Port-au-Prince, and fears were confirmed: cholera—one of the most dreaded infections known to man. Infections with *Vibrio cholerae* spread quickly through the displaced-persons camps where clean water and sanitation were scarce. Between October and November 2010, a total of 11,125 cases and 724 deaths from cholera were confirmed, making it one of the worst cholera outbreaks in recorded history. Hospitals and clinics were already stretched thin with the wounded from the earthquake, and it was difficult to transport needed treatment and supplies through the earthquake-ravaged countryside. In a country where only 27% of the population had access to basic sewage treatment and over 70% had either rudimentary or no toilets in their households *before* the earthquake, the need for clean water became dire after the disaster struck.

- What are the physiological characteristics of *V. cholerae*?
- How does *V. cholerae* cause such severe diarrhea?

Continuing the Case appears on page 184.

Outline and Learning Outcomes

7.1 Microbial Nutrition

1. List the essential nutrients of a bacterial cell.
2. Differentiate between *macronutrients* and *micronutrients.*
3. List and define four different terms that describe an organism's sources of carbon and energy.
4. Define *saprobe* and *parasite,* and provide microbial examples of each.
5. Compare and contrast the processes of diffusion and osmosis.

6. Identify the effects of isotonic, hypotonic, and hypertonic conditions on a cell.
7. Name two types of passive transport and three types of active transport.

7.2 Environmental Factors That Influence Microbes

8. List and define five terms used to express a microbe's optimal growth temperature.
9. Summarize three ways in which microorganisms function in the presence of oxygen.
10. Identify three other physical factors that microbes must contend with in the environment.
11. List and describe the five major types of microbial association.
12. Discuss characteristics of biofilms that differentiate them from planktonic bacteria and their infections.

7.3 The Study of Microbial Growth

13. Summarize the steps of cell division used by most bacteria; describe another method used by fewer bacterial species.
14. Define *doubling time*, and describe how it leads to exponential growth.
15. Compare and contrast the four phases of growth in a bacterial growth curve.
16. Identify one quantitative and one qualitative method used for analyzing bacterial growth.

7.1 Microbial Nutrition

Nutrition is a process by which chemical substances called **nutrients** are acquired from the environment and used in cellular activities such as metabolism and growth. With respect to nutrition, microbes are not really so different from humans. Bacteria living in mud on a diet of inorganic sulfur or protozoa digesting wood in a termite's intestine seem to live radical lifestyles, but even these organisms require a constant influx of certain substances from their habitat. (There are even bacteria that live on caffeine—just as many students claim *they* do.) No matter how strange their diet seems, all living things require a source of elements such as carbon, hydrogen, oxygen, phosphorus, potassium, nitrogen, sulfur, calcium, iron, sodium, chlorine, magnesium, and certain other elements. But the ultimate source of a particular element, its chemical form, and how much of it the microbe needs are all points of variation between different types of organisms. Any substance, whether in elemental or molecular form, that must be provided to an organism is called an **essential nutrient**. For microbes, the essential nutrients are carbon, hydrogen, oxygen, nitrogen, phosphorus (phosphate), and sulfur—often referred to by the acronym CHONPS. Once absorbed, nutrients are processed and transformed into the chemicals of the cell.

Two categories of essential nutrients are **macronutrients** and **micronutrients**. Macronutrients are required in relatively large quantities and play principal roles in cell structure and metabolism. Examples of macronutrients are carbon, hydrogen, and oxygen. Micronutrients, or **trace elements**, such as manganese, zinc, and nickel, are present in much smaller amounts and are involved in enzyme function and maintenance of protein structure. What constitutes a micronutrient can vary from one microbe to another.

Another way to categorize nutrients is according to their carbon content. An inorganic nutrient is an atom or simple molecule that contains a combination of atoms other than carbon and hydrogen. The natural reservoirs of inorganic compounds are mineral deposits in the crust of the earth, bodies of water, and the atmosphere. Examples

include metals and their salts (magnesium sulfate, ferric nitrate, sodium phosphate), gases (oxygen, carbon dioxide), and water **(table 7.1)**. In contrast, the molecules of organic nutrients contain carbon and hydrogen atoms and are usually the products of living things. They range from the simplest organic molecule, methane (CH_4), to large polymers (carbohydrates, lipids, proteins, and nucleic acids). The source of nutrients is extremely varied: Some microbes obtain their nutrients entirely from inorganic sources, and others require a combination of organic and inorganic sources. Parasites capable of invading and living on the human body derive all essential nutrients from host tissues, tissue fluids, secretions, and wastes.

Table 7.1 Principal Inorganic Reservoirs of Elements

Element	Inorganic Environmental Reservoir
Carbon	CO_2 in air; CO_3^{2-} in rocks and sediments
Oxygen	O_2 in air, certain oxides, water
Nitrogen	N_2 in air; NO_3^-, NO_2^-, NH_4^+ in soil and water
Hydrogen	Water, H_2 gas, mineral deposits
Phosphorus	Mineral deposits (PO_4^{3-}, H_3PO_4)
Sulfur	Mineral deposits, volcanic sediments (SO_4^{2-}, H_2S)
Potassium	Mineral deposits, the ocean (KCl, K_3PO_4)
Sodium	Mineral deposits, the ocean ($NaCl$, $NaSi$)
Calcium	Mineral deposits, the ocean ($CaCO_3$, $CaCl_2$)
Magnesium	Mineral deposits, geologic sediments ($MgSO_4$)
Chloride	The ocean ($NaCl$, NH_4Cl)
Iron	Mineral deposits, geologic sediments ($FeSO_4$)
Manganese, molybdenum, cobalt, nickel, zinc, copper, other micronutrients	Various geologic sediments

Chemical Analysis of Microbial Cytoplasm

Examining the chemical composition of a bacterial cell can indicate its nutritional requirements. **Table 7.2** lists the major contents of the intestinal bacterium *Escherichia coli*. Some of these components are absorbed in a ready-to-use form, and others must be synthesized by the cell from simple nutrients. Several important features of cell composition can be summarized as follows.

- Water is the most abundant of all the components (70%).
- Proteins are the next most prevalent chemical.
- About 97% of the dry cell weight is composed of organic compounds.
- About 96% of the dry cell weight is composed of six elements (represented by CHONPS).
- Chemical elements are needed in the overall scheme of cell growth, but most of them are available to the cell as compounds and not as pure elements (see table 7.2).
- A cell as "simple" as *E. coli* contains on the order of 5,000 different compounds, yet it needs to absorb only a few types of nutrients to synthesize this great diversity. These include $(NH_4)_2SO_4$, $FeCl_2$, $NaCl$, trace elements, glucose, KH_2PO_4, $MgSO_4$, $CaHPO_4$, and water.

Sources of Essential Nutrients

In their most basic form, elements that make up nutrients exist in environmental inorganic reservoirs. These reservoirs not only serve as a permanent, long-term source of these elements but also can be replenished by the activities of organisms. In fact, as we shall see in chapter 24, the ability of microbes to keep elements cycling is essential to all life on the earth.

For convenience, this section on nutrients is organized by element. You will no doubt notice that some categories overlap and that many of the compounds furnish more than one element.

Carbon Sources

It seems worthwhile to emphasize a point about the *extracellular source* of carbon as opposed to the *intracellular function* of carbon compounds. Although a distinction is made between the type of carbon compound cells absorb as nutrients (inorganic or organic), the majority of carbon compounds involved in the normal structure and metabolism of all cells are organic.

A **heterotroph** is an organism that must obtain its carbon in an organic form. Because organic carbon originates from the bodies of other organisms, heterotrophs are dependent on other life forms (*hetero-* is a Greek prefix meaning "other"). Among the common organic molecules that can satisfy this requirement are proteins, carbohydrates, lipids, and nucleic acids. In most cases, these nutrients provide several other elements as well. Some organic nutrients available to heterotrophs already exist in a form that is simple enough for absorption (e.g., monosaccharides and amino acids), but many larger molecules must be digested by the cell before absorption. Moreover, heterotrophs vary in their capacities to use different organic carbon sources. Some are restricted to a few substrates, whereas others (certain *Pseudomonas* species, for example) are so versatile that they can metabolize more than 100 different substrates.

Disease Connection

Pseudomonas aeruginosa is a gram-negative rod that is ubiquitous in the environment, due to its ability to live on so many different chemicals. It does not cause disease in healthy humans, but in immunocompromised patients it is a major pathogen. It is involved in 90% of deaths of cystic fibrosis patients.

An **autotroph** ("self-feeder") is an organism that uses inorganic CO_2 as its carbon source. Because autotrophs have the special capacity to convert CO_2 into organic compounds, they are not nutritionally dependent on other living things.

Nitrogen Sources

The main reservoir of nitrogen is nitrogen gas (N_2), which makes up 79% of the earth's atmosphere. This element is indispensable to the structure of proteins, DNA, RNA, and ATP. Such nitrogenous compounds are the primary nitrogen source for heterotrophs, but to be useful, they must first be degraded into their basic building blocks (proteins into amino acids; nucleic acids into nucleotides). Some bacteria and algae utilize inorganic nitrogenous nutrients (NO_3^-, NO_2^-, or NH_3). A small number of bacteria can transform N_2 into compounds usable by other organisms through the process of nitrogen fixation (see chapter 24). Regardless of the initial form in which the inorganic nitrogen enters the cell, it must first be converted to NH_3, the only form that can

Table 7.2 Analysis of the Chemical Composition of an *Escherichia coli* Cell

	% Total Weight	% Dry Weight		% Dry Weight
Organic Compounds			**Elements**	
Proteins	15	50	Carbon (C)	50
Nucleic acids			Oxygen (O)	20
RNA	6	20	Nitrogen (N)	14
DNA	1	3	Hydrogen (H)	8
Carbohydrates	3	10	Phosphorus (P)	3
Lipids	2	10	Sulfur (S)	1
Miscellaneous	2	4	Potassium (K)	1
			Sodium (Na)	1
Inorganic Compounds			Calcium (Ca)	0.5
Water	70		Magnesium (Mg)	0.5
All others	1	3	Chlorine (Cl)	0.5
			Iron (Fe)	0.2
			Trace metals	0.3

be directly combined with carbon to synthesize amino acids and other compounds.

Oxygen Sources

Because oxygen is a major component of organic compounds such as carbohydrates, lipids, nucleic acids, and proteins, it plays an important role in the structural and enzymatic functions of the cell. Oxygen is likewise a common component of inorganic salts such as sulfates, phosphates, nitrates, and water. Free gaseous oxygen (O_2) makes up 20% of the atmosphere. It is absolutely essential to the metabolism of many organisms, as we shall see later in this chapter and in chapter 8.

Hydrogen Sources

Hydrogen is a major element in all organic and several inorganic compounds, including water (H_2O), salts ($Ca[OH]_2$), and certain naturally occurring gases (H_2S, CH_4, and H_2). These gases are both used and produced by microbes. Hydrogen performs these overlapping roles in the biochemistry of cells by

1. maintaining **pH,**
2. forming **hydrogen bonds** between molecules, and
3. serving as the source of **free energy** in oxidation-reduction reactions of respiration (see chapter 8).

Phosphorus (Phosphate) Sources

The main inorganic source of phosphorus is phosphate (PO_4^{3-}), derived from phosphoric acid (H_3PO_4) and found in rocks and oceanic mineral deposits. Phosphate is a key component of nucleic acids and is therefore essential to the genetics of cells and viruses. It is also found in ATP, an important energy molecule in cells. Other phosphate-containing compounds are phospholipids in cell membranes and coenzymes such as NAD^+ (see chapter 8). Certain environments have very little available phosphate for use by organisms and therefore limit the ability of these organisms to grow. However, *Corynebacterium* is able to concentrate and store phosphate in metachromatic granules in the cytoplasm.

Sulfur Sources

Sulfur is widely distributed throughout the environment in mineral form. Rocks and sediments (such as gypsum) can contain sulfate (SO_4^{2-}), sulfides (FeS), hydrogen sulfide gas (H_2S), and elemental sulfur (S). Sulfur is an essential component of some vitamins (vitamin B_1) and the amino acids methionine and cysteine; the latter help determine shape and structural stability of proteins by forming unique linkages called disulfide bonds (described in chapter 2).

Other Nutrients Important in Microbial Metabolism

Other important elements in microbial metabolism include mineral ions. Potassium is essential to protein synthesis and membrane function. Sodium is important for certain types of cell transport. Calcium is a stabilizer of the cell wall and endospores of bacteria. Magnesium is a component of chlorophyll and a stabilizer of membranes and ribosomes. Iron is an important component of the cytochrome proteins of cell respiration. Zinc is an essential regulatory element for eukaryotic genetics. It is a major component of "zinc fingers"—binding factors that help enzymes adhere to specific sites on DNA. Copper, cobalt, nickel, molybdenum, manganese, silicon, iodine, and boron are needed in small amounts by some microbes but not others. On the other hand, in chapter 11 you will see that metals can also be very toxic to microbes. The concentration of metal ions can even influence the diseases microbes cause. For example, the bacteria that cause gonorrhea and meningitis grow more rapidly in the presence of iron ions.

Growth Factors: Essential Organic Nutrients

Few microbes are as versatile as *Escherichia coli* in assembling molecules from scratch. Many fastidious bacteria lack the genetic and metabolic mechanisms to synthesize every organic compound they need for survival. An organic compound such as an amino acid, nitrogenous base, or vitamin that cannot be synthesized by an organism and must be provided as a nutrient is a **growth factor.** For example, although all cells require 20 different amino acids for proper assembly of proteins, many cells cannot synthesize all of them. Those that must be obtained from food are called essential amino acids.

How Microbes Feed: Nutritional Types

The earth's limitless habitats and microbial adaptations are matched by an elaborate menu of microbial nutritional schemes. Fortunately, most organisms show consistent trends and can be described by a few general categories **(table 7.3)** and a few selected terms (see "A Note About Terminology" on page 177). The main determinants of a microbe's nutritional type are its sources of carbon and energy. In a previous section, microbes were defined according to their carbon sources as autotrophs or heterotrophs. Now we will subdivide all bacteria according to their energy source as **phototrophs** or **chemotrophs.** Microbes that photosynthesize are phototrophs and those that gain energy from chemical compounds are chemotrophs. The terms for carbon and energy source are often merged into a single word for convenience (see table 7.3). The categories described here are meant to describe only the major nutritional groups and do not include unusual exceptions.

Autotrophs and Their Energy Sources

Autotrophs derive energy from one of two possible nonliving sources: sunlight (photoautotrophs) and chemical reactions involving simple chemicals (chemoautotrophs). **Photoautotrophs** are photosynthetic—that is, they capture the energy of light rays and transform it into chemical energy that can be used in cell metabolism **(figure 7.1).** Because photosynthetic organisms (algae, plants, some bacteria) produce organic molecules that can be used by themselves and heterotrophs, they form the basis for most food webs. Their role as primary producers of organic matter is discussed in chapter 24.

Table 7.3 Nutritional Categories of Microbes by Energy and Carbon Source

Category/Carbon Source	Energy Source	Example
Autotroph/CO₂	**Nonliving Environment**	
Photoautotroph	Sunlight	Photosynthetic organisms, such as algae, plants, cyanobacteria
Chemoautotroph	Simple inorganic chemicals	Only certain bacteria or archaea, such as methanogens, deep-sea vent bacteria
Heterotroph/Organic	**Other Organisms or Sunlight**	
Photoheterotroph	Sunlight	Purple and green photosynthetic bacteria
Chemoheterotroph	Metabolizing the organic matter of dead organisms	Fungi, bacteria (decomposers)
Parasite	Utilizing the tissues, fluids of a live host	Various parasites and pathogens; can be bacteria, fungi, protozoa, animals
Saprobe	Metabolizing the organic matter of dead organisms	Fungi, bacteria

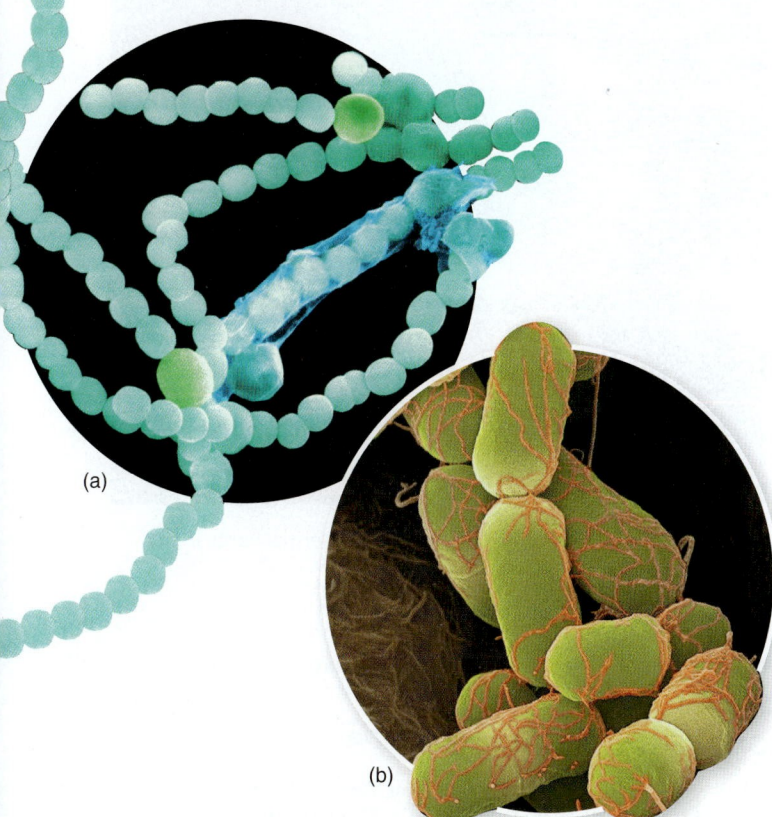

Figure 7.1 A photoautotroph and a chemoheterotroph.
(a) *Cyanobacterium*, a photosynthetic autotroph. **(b)** *Escherichia coli*, a chemoheterotroph.

Chemoautotrophs are of two types: One of these is the group called chemoorganic autotrophs. These use organic compounds for energy and inorganic compounds as a carbon source. The second type of chemoautotroph is a group called **lithoautotrophs,** which require neither sunlight nor organic nutrients, relying totally on inorganic minerals. These bacteria derive energy in diverse and rather amazing ways. In very simple terms, they remove electrons from inorganic substrates such as hydrogen gas, hydrogen sulfide, sulfur, or iron and combine them with carbon dioxide and hydrogen. This reaction provides simple organic molecules and a modest amount of energy to drive the synthetic processes of the cell. Lithoautotrophic bacteria play an important part in recycling inorganic nutrients. An interesting group of chemoautotrophs is **methanogens** (meth-an′-oh-genz), which produce methane (CH₄) from hydrogen gas and carbon dioxide:

$$4H_2 + CO_2 \longrightarrow CH_4 + 2H_2O$$

Methane, sometimes called "swamp gas" or "natural gas," is formed in anaerobic, hydrogen-containing microenvironments of soil, swamps, mud, and even in the intestines of some animals. Methanogens are archaea, some of which live in extreme habitats such as ocean vents and hot springs, where temperatures reach up to 400°C **(Insight 7.1).** Methane, which is used as a fuel in some homes, can also be produced in limited quantities using a type of generator primed with a mixed population of microbes (including methanogens) and fueled with various waste materials that can supply enough methane to drive a steam generator. Methane also plays a role as one of the greenhouse gases that is currently an environmental concern (see chapter 24).

Heterotrophs and Their Energy Sources

The majority of heterotrophic microorganisms are **chemoheterotrophs** that derive both carbon and energy from organic compounds. Processing these organic molecules by respiration or fermentation releases energy in the form of ATP. An example of chemoheterotrophy is **aerobic respiration,** the principal energy-yielding pathway in animals, most protozoa and fungi, and aerobic bacteria. It can be simply represented by the equation

$$\text{Glucose } [(CH_2O)_n] + O_2 \longrightarrow CO_2 + H_2O + \text{Energy (ATP)}$$

INSIGHT 7.1   Life in the Extremes

Any extreme habitat—whether hot, cold, salty, acidic, alkaline, high pressure, arid, oxygen-free, or toxic—is likely to harbor microorganisms that have made special adaptations to their conditions. Although in most instances the inhabitants are archaea and bacteria, certain fungi, protozoa, and algae are also capable of living in harsh habitats. Microbiologists have termed such remarkable organisms **extremophiles.**

Hot and Cold

Some of the most extreme habitats are hot springs, geysers, volcanoes, and ocean vents, all of which support flourishing microbial populations. Temperatures in these regions range from 50°C to well above the boiling point of water, with some ocean vents even approaching 350°C. Many heat-adapted microbes are archaea whose genetics and metabolism are extremely modified for this mode of existence. A unique ecosystem based on hydrogen-sulfide-oxidizing bacteria exists in the hydrothermal vents lying along deep oceanic ridges. It is hypothesized that the bacteria survive through photosynthesis by scavenging the few photons of light that come from infrared radiation given off by the hydrothermal vent water.

A particular species of **hyperthermophile**, *Pyrococcus furiosis,* was discovered near a hydrothermal vent near Vulcano Island, Italy, and has proven to be very useful in industrial processes. Its enzymes contain an unusual element—tungsten, which is not found naturally in any other organisms. Because it prefers scalding temperatures, the metabolic enzymes of *P. furiosis* are extremely heat stable, and its DNA polymerase is useful in the polymerase chain reaction (PCR), which requires high temperatures to denature the DNA double helix. Genetic engineering of *P. furiosis* has also made this organism useful in biotechnology and manufacturing for products such as biofuels.

A large part of the earth exists at cold temperatures. Microbes settle and grow throughout the Arctic and Antarctic, and in the deepest parts of the ocean, in temperatures that hover near the freezing point of water. Although the sea ice of Antarctica appears to be completely solid, it is honeycombed by various-size pores and tunnels filled with liquid water. These frigid microhabitats harbor a microcosm of planktonic life, including predators (fish and shrimp) that live on these algae and bacteria. Scientists are particularly interested in bacteria that can live at extremely cold temperatures. Finding bacteria on this planet that can thrive at those temperatures suggests that life may exist on other planets exhibiting cold environments.

Salt, Acidity, Alkalinity

The growth of most microbial cells is inhibited by high amounts of salt; for this reason, salt is a common food preservative. Yet whole communities of salt-dependent bacteria and algae occupy habitats in oceans, salt lakes, and inland seas, some of which are saturated with salt (30%—which is almost 10 times as salty as a normal ocean). Most of these microbes have demonstrable metabolic requirements for high levels of

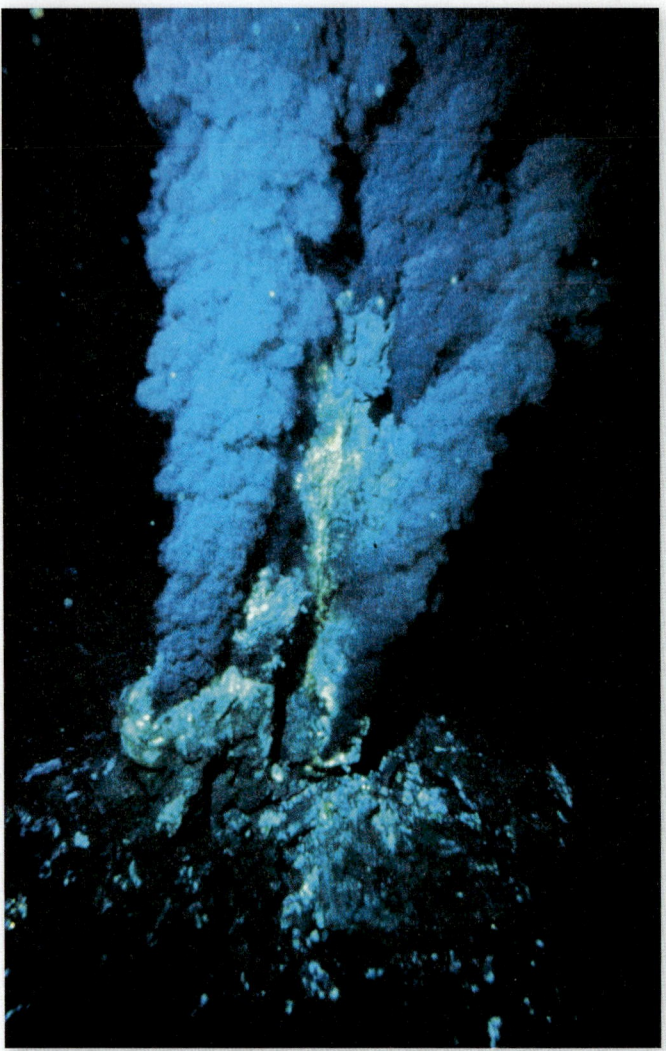

Hyperthermophilic bacteria live near hydrothermal vents in the ocean floor.

minerals such as sodium, potassium, magnesium, chlorides, or iodides.

Other Frontiers to Conquer

It was once thought that the region far beneath the soil and upper crust of the earth's surface was sterile. However, work with deep core samples (from 330 meters down) indicates a vast microbial population in these zones. Many biologists believe these are very similar to the first ancient microbes to have existed on earth. Numerous species have carved a niche for themselves in the depths of mud, swamps, and oceans, where oxygen gas and sunlight cannot penetrate. The predominant living things in the deepest part of the oceans (10,000 meters or below) are pressure- and cold-loving microorganisms. Thriving bacterial populations can also be found in petroleum, coal, and mineral deposits containing copper, zinc, gold, and uranium.

A Note About Terminology

Much of the vocabulary for describing microbial adaptations is based on some common root words. These are combined in various ways that assist in discussing the types of nutritional or ecological adaptations, as shown in this partial list.

Root	Meaning	Example of Use
troph-	Food	Trophozoite—the feeding stage of protozoa
-phile	To love	Extremophile—an organism that has adapted to ("loves") extreme environments
-obe	To live	Microbe—to live "small"
hetero-	Other	Heterotroph—an organism that requires nutrients from other organisms
auto-	Self	Autotroph—an organism that does not need other organisms for food (obtains nutrients from a nonliving source)
photo-	Light	Phototroph—an organism that uses light as an energy source
chemo-	Chemical	Chemotroph—an organism that uses chemicals for energy, rather than light
sapro-	Rotten	Saprobe—an organism that lives on dead organic matter
halo-	Salt	Halophile—an organism that can grow in high-salt environments
thermo-	Heat	Thermophile—an organism that grows best at high temperatures
psychro-	Cold	Psychrophile—an organism that grows best at cold temperatures
aero-	Air (O_2)	Aerobe—an organism that uses oxygen in metabolism

Modifier terms are also used to specify the nature of an organism's adaptations. *Obligate* or *strict* refers to being restricted to a narrow niche or habitat, such as an obligate thermophile that requires high temperatures to grow. By contrast, *facultative* means not being so restricted but being able to adapt to a wider range of metabolic conditions and habitats. A facultative halophile can grow with or without high salt concentration.

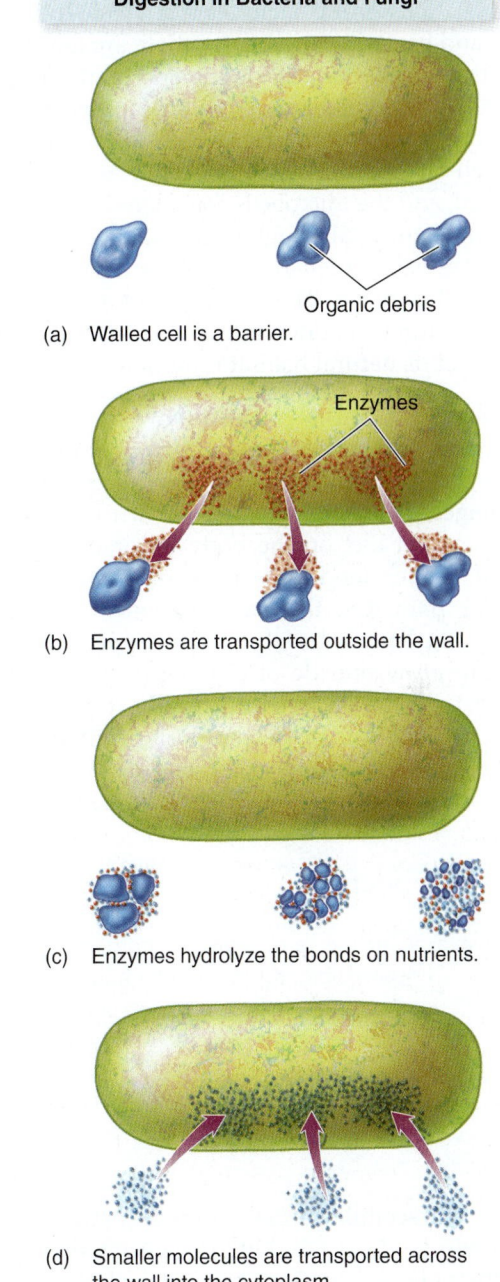

Digestion in Bacteria and Fungi

(a) Walled cell is a barrier.

Enzymes

(b) Enzymes are transported outside the wall.

(c) Enzymes hydrolyze the bonds on nutrients.

(d) Smaller molecules are transported across the wall into the cytoplasm.

Figure 7.2 Extracellular digestion in a saprobe with a cell wall (bacterium or fungus).   **(a)** A walled cell is inflexible and cannot engulf large pieces of organic debris. **(b)** In response to a usable substrate, the cell synthesizes enzymes that are transported across the wall into the extracellular environment. **(c)** The enzymes hydrolyze the bonds in the debris molecules. **(d)** Digestion produces molecules small enough to be transported into the cytoplasm.

This reaction is complementary to photosynthesis. Here, glucose and oxygen are reactants, and carbon dioxide is given off. Indeed, the earth's balance of both energy and metabolic gases is greatly dependent on this relationship. Chemoheterotrophic microorganisms belong to one of two main categories that differ in how they obtain their organic nutrients: **Saprobes** are free-living microorganisms that feed primarily on organic detritus from dead organisms, and **parasites** ordinarily derive nutrients from the cells or tissues of a living host.

Saprobic Microorganisms Saprobes occupy a niche as decomposers of plant litter, animal matter, and dead microbes. If not for the work of decomposers, the earth would

gradually fill up with organic material, and the nutrients it contains would not be recycled. Most saprobes, notably bacteria and fungi, have a rigid cell wall and cannot engulf large particles of food. To compensate, they release enzymes to the extracellular environment and digest the food particles into smaller molecules that can be transported into the cell **(figure 7.2)**. *Obligate saprobes* exist strictly on dead organic matter in soil and water and are unable to adapt to the body of a live host. This group includes many free-living protozoa, fungi, and bacteria. Apparently, there are fewer of these strict species than was once thought, and many supposedly nonpathogenic saprobes can infect a susceptible host. When a saprobe does infect a host, it is considered a *facultative parasite*. Such an infection usually occurs when the host is compromised, and the microbe is considered an *opportunistic pathogen*. For example, although its natural habitat is soil and water, *Pseudomonas aeruginosa* frequently causes infections in hospitalized patients. The yeast *Cryptococcus neoformans* causes a severe lung and brain infection in AIDS patients (see chapter 19), yet its natural habitat is the soil.

Parasitic Microorganisms   Parasites live in or on the body of a host, which they harm to some degree. Because parasites cause damage to tissues (disease) or even death, they are also called **pathogens.** Parasites range from viruses to helminths (worms); they can live on the body (ectoparasites), in the organs and tissues (endoparasites), or even within cells (intracellular parasites, the most extreme type). *Obligate parasites* (e.g., the leprosy bacillus and the syphilis spirochete) are unable to grow outside of a living host. Parasites that are less strict can be cultured artificially if provided with the correct nutrients and environmental conditions. Bacteria such as *Streptococcus pyogenes* (the cause of strep throat) and *Staphylococcus aureus* can grow on artificial media.

Disease Connection

You may have heard the term *leprosy* before. It is even referenced in the Christian Bible. It is an ancient human disease that causes skin lesions and deformities and, if left unchecked, death. It is caused by *Mycobacterium leprae*. People suffering from leprosy have been stigmatized in society for a long time. For that reason, most professionals now refer to it as *Hansen's disease*, after its Norwegian discoverer.

Obligate intracellular parasitism is an extreme but relatively common mode of life. Microorganisms that spend all or part of their life cycle inside a host cell include the viruses, a few bacteria (rickettsias, chlamydias), and certain protozoa (apicomplexans). Contrary to what one may think, the cell interior is not completely without hazards and microbes must overcome some difficult challenges. They must find a way into the cell, keep from being destroyed, not destroy the host cell too soon, multiply, and find a way to infect other cells. Intracellular parasites obtain different substances from the host cell, depending on the group. Viruses are extreme, parasitizing the host's genetic and metabolic machinery. Rickettsias are primarily energy parasites, and the malaria protozoan is a hemoglobin parasite.

How Microbes Feed: Nutrient Absorption

A microorganism's habitat provides necessary nutrients—some abundant, others scarce—that must still be taken into the cell. Survival also requires that cells transport waste materials out of the cell (and into the environment). Whatever the direction, transport occurs across the cell membrane, the structure specialized for this role. This is true even in organisms with cell walls (bacteria, algae, and fungi), because the cell wall is usually too nonselective to screen the entrance or exit of molecules. Before we talk about transport mechanisms, let's examine the basic principles of diffusion.

The Movement of Molecules: Diffusion and Transport

The driving force of transport is atomic and molecular movement—the natural tendency of atoms and molecules to be in constant random motion. The existence of this motion is evident in Brownian movement of particles suspended in liquid. It can also be demonstrated by a variety of simple observations. A drop of perfume released into one part of a room is soon smelled in another part, or a lump of sugar in a cup of tea spreads through the whole cup without stirring. This phenomenon of molecular movement, in which atoms or molecules move in a gradient from an area of higher density or concentration to an area of lower density or concentration, is **diffusion (figure 7.3).**

Figure 7.3 Diffusion of molecules in aqueous solutions. A high concentration of sugar exists in the cube at the bottom of the liquid. An imaginary molecular view of this area shows that sugar molecules are in a constant state of motion. Those at the edge of the cube diffuse from the concentrated area into more dilute regions. As diffusion continues, the sugar will spread randomly throughout the aqueous phase, and eventually there will be no gradient. At that point, the system is said to be in equilibrium.

Diffusion

All molecules, regardless of being in a solid, liquid, or gas, are in continuous movement, and as the temperature increases, the molecular movement becomes faster. This is called "thermal" movement. In any solution, including cytoplasm, these moving molecules cannot travel very far without having collisions with other molecules and, therefore, will bounce off each other like millions of pool balls every second. As a result of each collision, the directions of the colliding molecules are altered and the direction of any one molecule is unpredictable and is therefore "random." If we start with a solution in which the solute, or dissolved substance, is more concentrated in one area than another, then the random thermal movement of molecules in this solution will eventually distribute the molecules from the area of higher concentration to the area of lower concentration, thus evenly distributing the molecules. This net movement of molecules down their concentration gradient by random thermal motion is known as *diffusion.* Diffusion of molecules across the cell membrane is largely determined by the concentration gradient and permeability of the substance. But before we talk about movement of nutrients (molecules, solutes) in and out of cells, we will address the movement of water, or osmosis. You may want to take a moment to review solutes and solvents on page 37 in chapter 2.

The Movement of Water: Osmosis

Diffusion of water through a selectively permeable membrane, a process called **osmosis,** is also a physical phenomenon that is easily demonstrated in the laboratory with nonliving materials. It provides a model of how cells deal with various solute concentrations in aqueous solutions **(process figure 7.4).** In an osmotic system, the membrane is *selectively,* or *differentially, permeable,* having passageways that allow free diffusion of water but can block certain other dissolved molecules. When this membrane is placed between solutions of differing concentrations and the solute is not diffusible (protein, for example), then under the laws of diffusion, water will diffuse at a faster rate from the side that has more water to the side that has less water. As long as the concentrations of the solutions differ, one side will experience a net loss of water and the other a net gain of water, until equilibrium is reached and the rate of diffusion is equalized.

Osmosis in living systems is similar to the model shown in process figure 7.4. Living membranes generally block the entrance and exit of larger molecules and permit free diffusion of water. Because most cells are surrounded by some free water, the amount of water entering or leaving has a far-reaching impact on cellular activities and survival. This osmotic relationship between cells and their environment is determined by the relative concentrations of the solutions

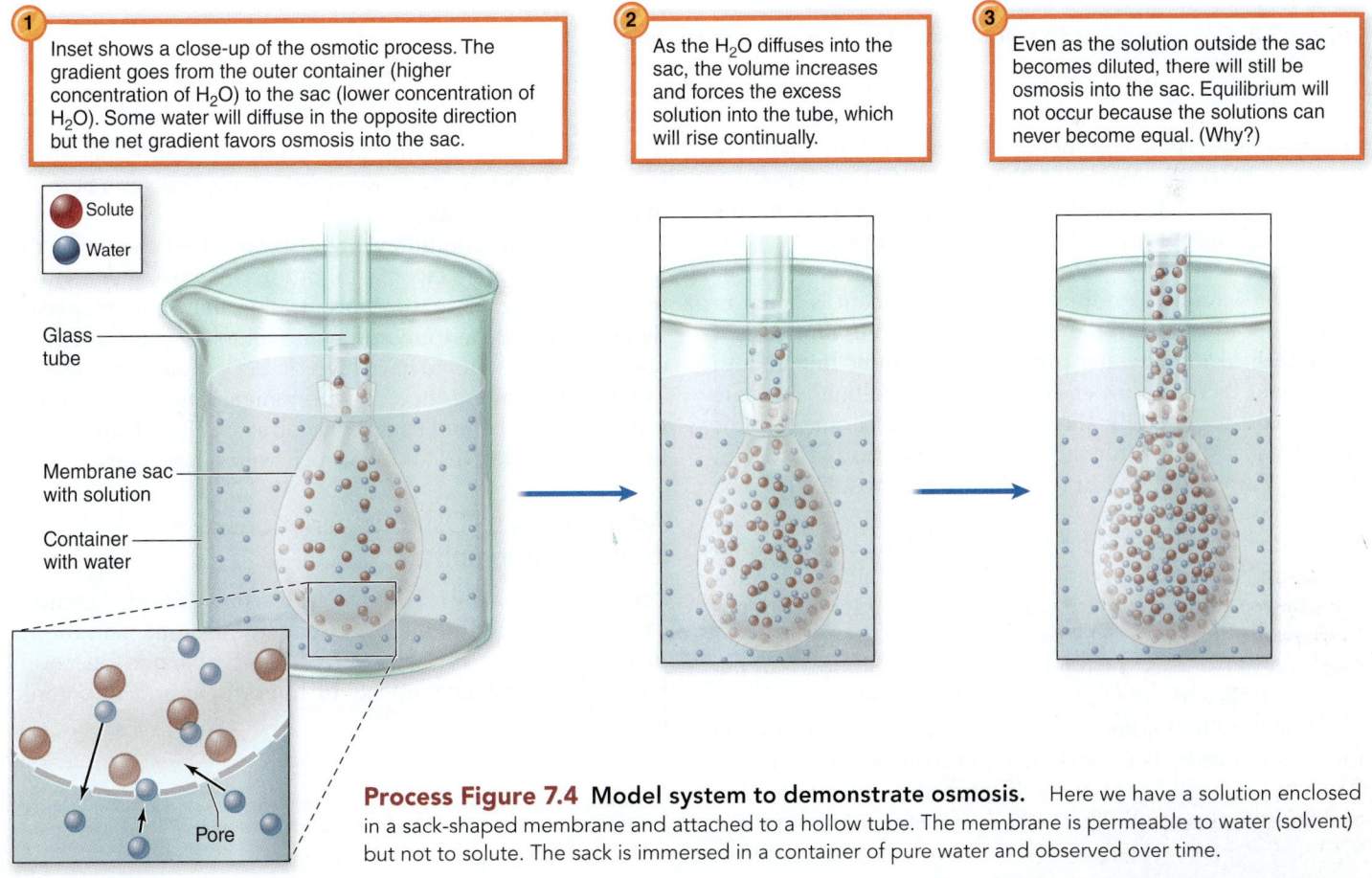

1 Inset shows a close-up of the osmotic process. The gradient goes from the outer container (higher concentration of H_2O) to the sac (lower concentration of H_2O). Some water will diffuse in the opposite direction but the net gradient favors osmosis into the sac.

2 As the H_2O diffuses into the sac, the volume increases and forces the excess solution into the tube, which will rise continually.

3 Even as the solution outside the sac becomes diluted, there will still be osmosis into the sac. Equilibrium will not occur because the solutions can never become equal. (Why?)

- Solute
- Water

Glass tube

Membrane sac with solution

Container with water

Pore

Process Figure 7.4 Model system to demonstrate osmosis. Here we have a solution enclosed in a sack-shaped membrane and attached to a hollow tube. The membrane is permeable to water (solvent) but not to solute. The sack is immersed in a container of pure water and observed over time.

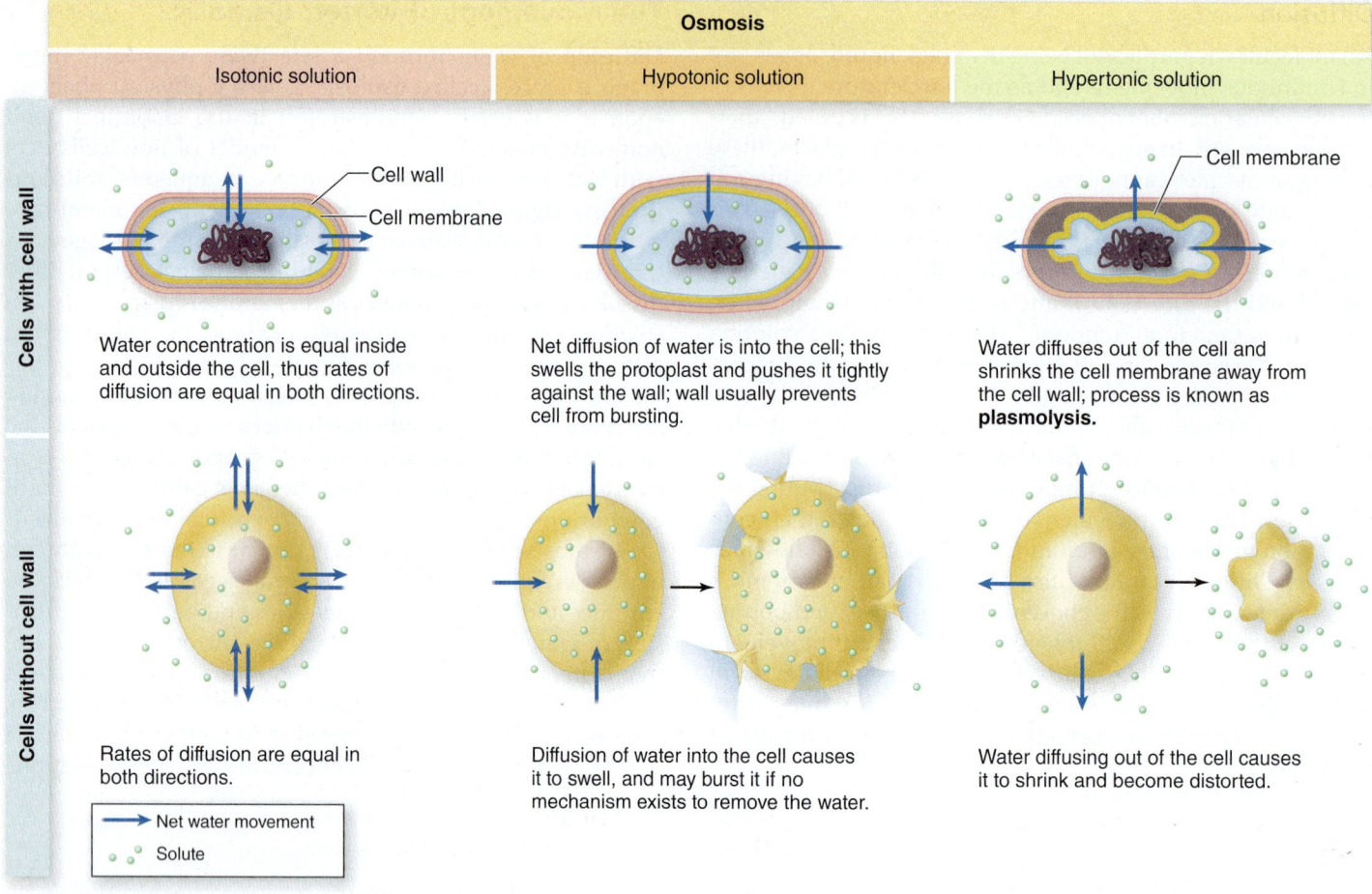

Figure 7.5 Cell responses to solutions of differing osmotic content.

on either side of the cell membrane **(figure 7.5).** Such systems can be compared using the terms *isotonic, hypotonic,* and *hypertonic.* (The root *-tonic* means "tension"; *iso-* means "the same," *hypo-* means "less," and *hyper-* means "over" or "more.")

Under **isotonic** conditions, the environment is equal in solute concentration to the cell's internal environment, and because diffusion of water proceeds at the same rate in both directions, there is no net change in cell volume. Isotonic solutions are generally the most stable environments for cells, because they are already in an osmotic steady state with the cell. Parasites living in host tissues are most likely to be living in isotonic habitats.

Under **hypotonic** conditions, the solute concentration of the external environment is lower than that of the cell's internal environment. Pure water provides the most hypotonic environment for cells because it has no solute. The net direction of osmosis is from the hypotonic solution into the cell, and cells without walls swell and can burst.

A slightly hypotonic environment can be quite favorable for bacterial cells. The constant slight tendency for water to flow into the cell keeps the cell membrane fully extended and the cytoplasm full. This is the optimum condition for the many processes occurring in and on the membrane. Slight hypotonicity is tolerated quite well by most bacteria because of their rigid cell walls.

Hypertonic[1] conditions are also out of balance with the tonicity of the cell's cytoplasm, but in this case, the environment has a higher solute concentration than the cytoplasm. Because a hypertonic environment will force water to diffuse out of a cell, it is said to have high *osmotic pressure* or potential. The growth-limiting effect of hypertonic solutions on microbes is the principle behind using concentrated salt and sugar solutions as preservatives for food, such as in salted hams.

Adaptations to Osmotic Variations in the Environment

Let us now see how specific microbes have adapted osmotically to their environments. In general, isotonic conditions pose little stress on cells, so survival depends on counteracting the adverse effects of hypertonic and hypotonic environments.

---

1. It will help you to recall these osmotic conditions if you remember that the prefixes iso-, hypo-, and hyper- refer to the environment *outside* of the cell.

A bacterium and an amoeba living in fresh pond water are examples of cells that live in constantly hypotonic conditions. The rate of water diffusing across the cell membrane into the cytoplasm is rapid and constant, and the cells would die without a way to adapt. As just mentioned, the majority of bacterial cells compensate by having a cell wall that protects them from bursting even as the cytoplasmic membrane becomes *turgid* (ter'-jid) from pressure. The amoeba's (without a cell wall) adaptation is an anatomical and physiological one that requires the constant expenditure of energy. It has a water, or contractile, vacuole that moves excess water back out into the habitat like a tiny pump.

A microbe living in a high-salt environment (hypertonic) has the opposite problem and must either restrict its loss of water to the environment or increase the salinity of its internal environment. Halobacteria living in the Great Salt Lake and the Dead Sea actually absorb salt to make their cells isotonic with the environment; thus, they have a physiological need for a high-salt concentration in their habitats (see discussion of halophiles on page 187).

So far, the discussion of passive or simple diffusion has not included the added complexity of membranes or cell walls, which hinder simple diffusion by adding a physical barrier. Therefore, simple diffusion is limited to small nonpolar molecules like oxygen or lipid-soluble molecules that may pass through the membranes. It is imperative that a cell be able to move polar molecules and ions across the plasma membrane; given the greatly decreased permeability of these chemicals, simple diffusion will not allow this movement. The multiple ways that substances move in and out of cells are summarized in **table 7.4,** and described here. First, there is **facilitated diffusion.** This type of mediated transport mechanism utilizes a carrier protein that will bind a specific substance. This binding changes the conformation of the carrier proteins so that the substance is moved across the membrane. Once the substance is transported, the carrier protein resumes its original shape and is ready to transport again. These carrier proteins exhibit **specificity,** which means that they bind and transport only one or a few types of molecules. For example, a carrier protein that transports sodium will not bind glucose.

A second characteristic exhibited by facilitated diffusion is **saturation.** The rate of transport of a substance is limited by the number of binding sites on the transport proteins. As the substance's concentration increases, so does the rate of transport until the concentration of the transported substance is such that all of the transporters' binding sites are occupied. Then the rate of transport reaches a steady state and cannot move faster despite further increases in the substance's concentration. A third characteristic of these carrier proteins is that they exhibit competition. This is when two molecules of similar shape can bind to the same binding site on a carrier protein. The chemical with the higher binding affinity, or the chemical in the higher concentration, will be transported at a greater rate.

Neither simple diffusion nor facilitated diffusion requires energy, because molecules are moving down a concentration gradient, from an area of higher concentration to an area of lower concentration.

Active Transport: Bringing in Molecules Against a Gradient

Free-living microbes exist under relatively nutrient-starved conditions and cannot rely completely on slow and rather inefficient passive transport mechanisms. To ensure a constant supply of nutrients and other required substances, microbes must capture those that are in extremely short supply and actively transport them into the cell. Features inherent in **active transport** systems are

1. the transport of nutrients against the diffusion gradient or in the same direction as the natural gradient but at a rate faster than by diffusion alone;
2. the presence of specific membrane proteins (permeases and pumps); and
3. the expenditure of energy. Examples of substances transported actively are monosaccharides, amino acids, organic acids, phosphates, and metal ions. Some freshwater algae have such efficient active transport systems that an essential nutrient can be found in intracellular concentrations 200 times that of the habitat.

An important type of active transport involves specialized pumps, which can rapidly carry ions such as K^+, Na^+, and H^+ across the membrane. This behavior is particularly important in membrane ATP formation and protein synthesis, as described in chapter 8. Another type of active transport, **group translocation,** couples the transport of a nutrient with its conversion to a substance that is immediately useful inside the cell. This method is used by certain bacteria to transport sugars (glucose, fructose), while simultaneously adding molecules such as phosphate that prepare them for the next stage in metabolism.

Endocytosis: Eating and Drinking by Cells

Some eukaryotic cells transport large molecules, particles, liquids, or even other cells across the cell membrane. Because the cell usually expends energy to carry out this transport, it is also a form of active transport. The substances transported do not pass physically through the membrane but are carried into the cell by **endocytosis.** First the cell encloses the substance in its membrane, simultaneously forming a vacuole and engulfing it. Amoebas and certain white blood cells ingest whole cells or large solid matter by a type of endocytosis called **phagocytosis.** Liquids, such as oils or molecules in solution, enter the cell through **pinocytosis.**

Table 7.4 **Transport Processes in Cells**

Examples	Description	Energy Requirements	
Passive			
Simple diffusion	A fundamental property of atoms and molecules that exist in a state of random motion	None. Substances move on a gradient from higher concentration to lower concentration.	
Facilitated diffusion	Molecule binds to a specific receptor in membrane and is carried to other side. Molecule-specific. Goes both directions. Rate of transport is limited by the number of binding sites on transport proteins.	None. Substances move on a gradient from higher concentration to lower concentration.	
Active			
Carrier-mediated active transport	Atoms or molecules are pumped into or out of the cell by specialized receptors.	Driven by ATP or the proton motive force	
Group translocation	Molecule is moved across membrane and simultaneously converted to a metabolically useful substance	ATP	
Bulk transport	Mass transport of large particles, cells, and liquids by engulfment and vesicle formation. Processes generally called endocytosis. Phagocytosis moves solids into cell; pinocytosis moves liquids into cell.	ATP	

7.1 Learning Outcomes—Assess Your Progress

1. List the essential nutrients of a bacterial cell.
2. Differentiate between *macronutrients* and *micronutrients*.
3. List and define four different terms that describe an organism's sources of carbon and energy.
4. Define *saprobe* and *parasite*, and provide microbial examples of each.
5. Compare and contrast the processes of diffusion and osmosis.
6. Identify the effects of isotonic, hypotonic, and hypertonic conditions on a cell.
7. Name two types of passive transport and three types of active transport.

7.2 Environmental Factors That Influence Microbes

Microbes are exposed to a wide variety of environmental factors in addition to nutrients. Microbial ecology focuses on ways that microorganisms deal with or adapt to such factors as heat, cold, gases, acid, radiation, osmotic and hydrostatic pressures, and even other microbes. Adaptation is a complex adjustment in biochemistry or genetics that enables long-term survival and growth. For most microbes, environmental factors fundamentally affect the function of metabolic enzymes. Thus, survival in a changing environment is largely a matter of whether the enzyme systems of microorganisms can adapt to alterations in their habitat. Incidentally, one must be careful to differentiate between *growth* in a given condition and *tolerance*, which implies survival without growth.

Temperature

Microbial cells are unable to control their temperature and therefore assume the ambient temperature of their natural habitats. Their survival is dependent on adapting to whatever temperature variations are encountered in that habitat. The range of temperatures for the growth of a given microbial species can be expressed as three *cardinal temperatures*. The **minimum temperature** is the lowest temperature that permits a microbe's continued growth and metabolism; below this temperature, its activities are inhibited. The **maximum temperature** is the highest temperature at which growth and metabolism can proceed. If the temperature rises slightly above maximum, growth will stop, but if it continues to rise beyond that point, the enzymes and nucleic acids will eventually become permanently inactivated (otherwise known as *denaturation*) and the cell will die. This is why heat works so well as an agent in microbial control. The **optimum temperature** covers a small range, intermediate between the minimum and maximum, which

promotes the fastest rate of growth and metabolism (rarely is the optimum a single point). Small chemical differences in bacterial membranes, which affect their fluidity, also allow them to thrive at different temperatures.

Depending on their natural habitats, some microbes have a narrow cardinal range, others a broad one. Some strict parasites will not grow if the temperature varies more than a few degrees below or above the host's body temperature. For instance, the typhus rickettsia multiplies only in the range of 32°C to 38°C, and rhinoviruses (one cause of the common cold) multiply most successfully in tissues that are slightly below normal body temperature (33°C to 35°C). Other organisms are not so limited. Strains of *Staphylococcus aureus* grow within the range of 6°C to 46°C, and the intestinal bacterium *Enterococcus faecalis* grows within the range of 0°C to 44°C.

Another way to express temperature adaptation is to describe whether an organism grows optimally in a cold, moderate, or hot temperature range. The terms used for these ecological groups are *psychrophile, mesophile,* and *thermophile,* respectively **(figure 7.6).**

A **psychrophile** (sy′-kroh-fyl) is a microorganism that has an optimum temperature below 15°C and is capable of growth at 0°C. It is obligate with respect to cold and generally cannot grow above 20°C. Laboratory work with true psychrophiles can be a real challenge. Inoculations have to be done in a cold room because room temperature can be lethal to the organisms. Unlike most laboratory cultures, storage in the refrigerator incubates, rather than inhibits, them. As one may predict, the habitats of psychrophilic bacteria, fungi, and algae are lakes and rivers, snowfields **(figure 7.7),** polar ice, and the deep ocean. Rarely, if ever, are they pathogenic. True psychrophiles must be distinguished from *psychrotrophs* or *facultative psychrophiles* that grow slowly in cold but have an optimum temperature between 15°C and 30°C. Bacteria

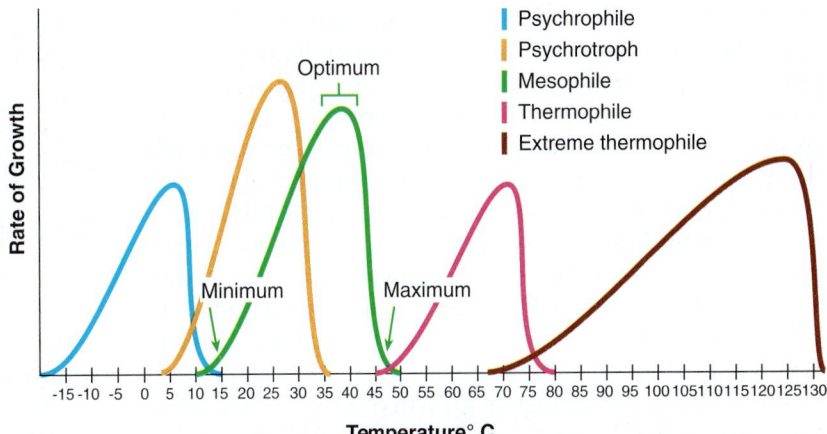

Figure 7.6 Ecological groups by temperature of adaptation.
Psychrophiles can grow at or near 0°C and have an optimum below 15°C. Psychrotrophs have an optimum of from 15°C to 30°C. As a group, mesophiles can grow between 10°C and 50°C, but their optima usually fall between 20°C and 40°C. Generally speaking, thermophiles require temperatures above 45°C and grow optimally between this temperature and 80°C. Extreme thermophiles have optima above 80°C. Note that the ranges can overlap to an extent.

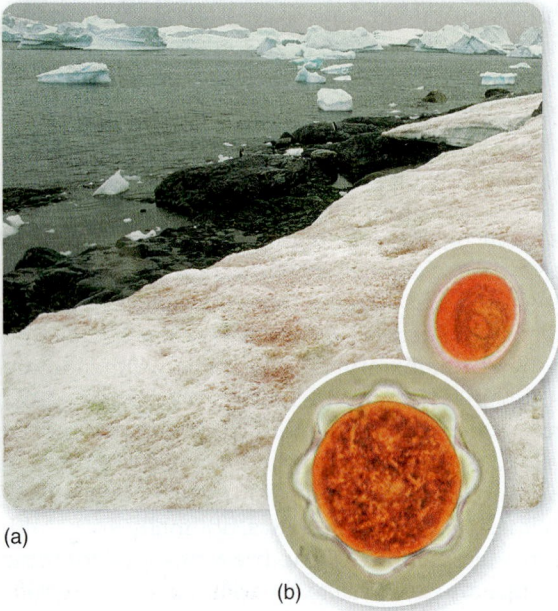

(a)

(b)

Figure 7.7 Red snow. **(a)** An early summer snowbank provides a perfect habitat for psychrophilic photosynthetic organisms like *Chlamydomonas nivalis.* **(b)** Microscopic view of this snow alga (actually classified as a "green" alga although a red pigment dominates at this stage of its life cycle).

such as *Staphylococcus aureus* and *Listeria monocytogenes* are a concern because they can grow in refrigerated food and cause food-borne illness.

The majority of medically significant microorganisms are **mesophiles** (mez'-oh-fylz), organisms that grow at intermediate temperatures. Although an individual species can grow at the extremes of 10°C or 50°C, the optimum growth temperatures (optima) of most mesophiles fall into the range of 20°C to 40°C. Organisms in this group inhabit animals and plants as well as soil and water in temperate, subtropical, and tropical regions. Most human pathogens have optima somewhere between 30°C and 40°C (human body temperature is 37°C). *Thermoduric* microbes, which can survive short exposure to high temperatures but are normally mesophiles, are common contaminants of heated or pasteurized foods (see chapter 11). Examples include heat-resistant cysts such as *Giardia* or sporeformers such as *Bacillus* and *Clostridium.*

A **thermophile** (thur'-moh-fyl) is a microbe that grows optimally at temperatures greater than 45°C. Such heat-loving microbes live in soil and water associated with volcanic activity, in compost piles, and in habitats directly exposed to the sun. Thermophiles vary in heat requirements, with a general range of growth of 45°C to 80°C. Most eukaryotic forms cannot survive above 60°C, but a few thermophilic bacteria, called extreme thermophiles, grow between 80°C and 121°C (currently thought to be the temperature limit established by enzymes and cell structures). Strict thermophiles are so heat tolerant that researchers may use an autoclave to isolate them in culture. Currently, there is intense interest in thermal microorganisms on the part of biotechnology companies (see Insight 7.1).

Gases

The atmospheric gases that most influence microbial growth are O_2 and CO_2. Of these, oxygen gas has the greatest impact on microbial growth. Not only is it an important respiratory gas, but it is also a powerful oxidizing agent that exists in many toxic forms. In general, microbes fall into one of three categories:

- those that use oxygen and can detoxify it,
- those that can neither use oxygen nor detoxify it, and
- those that do not use oxygen but can detoxify it.

How Microbes Process Oxygen

As oxygen enters into cellular reactions, it is transformed into several toxic products. Singlet oxygen written either as (O) or as $\dot{O}_2$ is an extremely reactive molecule produced by both living and nonliving processes. Notably, it is one of the substances produced by phagocytes to kill invading bacteria (see chapter 14). The buildup of singlet oxygen and the oxidation of membrane lipids and other molecules can damage and destroy a cell. The highly reactive superoxide ion (O_2^-), hydrogen peroxide (H_2O_2), and hydroxyl radicals (OH^-) are other destructive metabolic by-products of oxygen. To protect themselves against damage, most cells have developed enzymes that go about the business of scavenging and neutralizing these chemicals. The complete conversion of superoxide ion into harmless oxygen requires a two-step process and at least two enzymes:

Step 1. $2O_2^- + 2H^+ \xrightarrow{\text{Superoxide dismutase}} H_2O_2 \text{ (hydrogen peroxide)} + O_2$

Step 2. $2H_2O_2 \xrightarrow{\text{Catalase}} 2H_2O + O_2$

In this series of reactions (essential for aerobic organisms), the superoxide ion is first converted to hydrogen peroxide and normal oxygen by the action of an enzyme called superoxide dismutase. Because hydrogen peroxide is also toxic to cells (it is used as a disinfectant and antiseptic), it must be degraded by the enzyme catalase into water and oxygen. If a microbe is not capable of dealing with toxic oxygen by these or similar mechanisms, it is forced to live in habitats free of oxygen.

With respect to oxygen requirements, several general categories are recognized. An **aerobe** (air'-ohb) (aerobic organism) can use gaseous oxygen in its metabolism and possesses the enzymes needed to process toxic oxygen products. An organism that cannot grow without oxygen is an **obligate** aerobe. Most fungi and protozoa, as well as many bacteria, have to have oxygen in their metabolism.

A **facultative anaerobe** is an aerobe that does not require oxygen for its metabolism and is capable of growth in the absence of it. This type of organism metabolizes by aerobic respiration when oxygen is present; but in its absence, it adopts an anaerobic mode of metabolism such as fermentation. Facultative anaerobes usually possess catalase and superoxide dismutase. A large number of bacterial pathogens fall into this group (e.g., gram-negative intestinal bacteria and staphylococci). A **microaerophile** (myk"-roh-air'-oh-fyl) does not grow at normal atmospheric concentrations of oxygen but requires a small amount of it in metabolism. Most organisms in this category live in a habitat (soil, water, or the human body) that provides small amounts of oxygen but is not directly exposed to the atmosphere.

An **anaerobe** (anaerobic microorganism) lacks the metabolic enzyme systems for using oxygen in respiration. Because strict, or obligate, anaerobes also lack the enzymes for processing toxic oxygen, they cannot tolerate any free oxygen in the immediate environment and will die if exposed to it. Strict anaerobes live in highly reduced habitats, such as deep muds, lakes, oceans, and soil. Even though human cells use oxygen and oxygen is found in the blood and tissues, some body sites present anaerobic pockets or microhabitats where colonization or infection can occur. One region that is an important site for anaerobic infections is the oral cavity. Dental caries are partly due to the complex actions of aerobic and anaerobic bacteria, and most gingival infections consist of similar mixtures of oral bacteria that have invaded damaged gum tissues (see chapter 22). Another common site for anaerobic infections is the large intestine, a relatively oxygen-free habitat that harbors a rich assortment of strictly anaerobic bacteria. Anaerobic infections can occur following abdominal surgery and traumatic injuries (gas gangrene and tetanus). Growing anaerobic bacteria usually requires special media, methods of incubation, and handling chambers that exclude oxygen **(figure 7.8a).**

INSIGHT 7.2   Ancient Aerobes and the Great Oxidation Event

There has been a lot of debate about when oxygen-producing microbes appeared on the earth. Until recently, scientists have estimated that only anaerobic cells existed until between 2 and 3 billion years ago. It was not until what is called the *Great Oxidation Event* (GOE) occurred that life, as we know it, was possible on the earth. What is the Great Oxidation Event, and why is it significant? Until that time, atmospheric oxygen in the form of O_2 was very low—probably less than 1/10,000 of present levels. Previous to the GOE, the atmosphere contained significant amounts of methane, a greenhouse gas, which kept the earth's temperatures at high levels. The production of oxygen was a significant event in the evolution of life on earth. Until O_2 first accumulated in the atmosphere, the earth was too hot, too full of methane, and too acidic for life to exist.

Scientists led by geomicrobiologist Kirt Konhauser at the University of Alberta have studied oxidation of iron pyrite (FeS_2, also known as fool's gold) into iron oxide, which released acid that dissolved rocks into chromium and other metals in ancient sea beds. Their data show that chromium levels in the sea beds increased significantly about 2.48 billion years ago, about 100 million years earlier than was previously thought. These **acidophilic** microbes thrived in the extremely low pH of

the ocean waters of ancient earth. Konhauser has found evidence of these same bacterial life forms living off of pyrite in the highly acidic wastewater of mining sites (pictured). These organisms are the most ancient oxygen producers known and may be the driving force behind the atmosphere in which we live today.

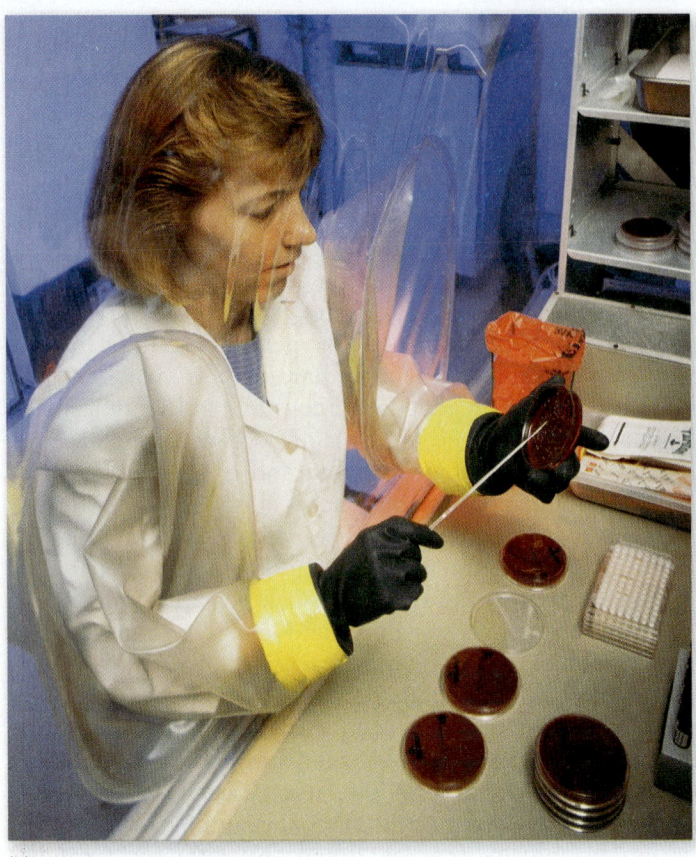

(a)

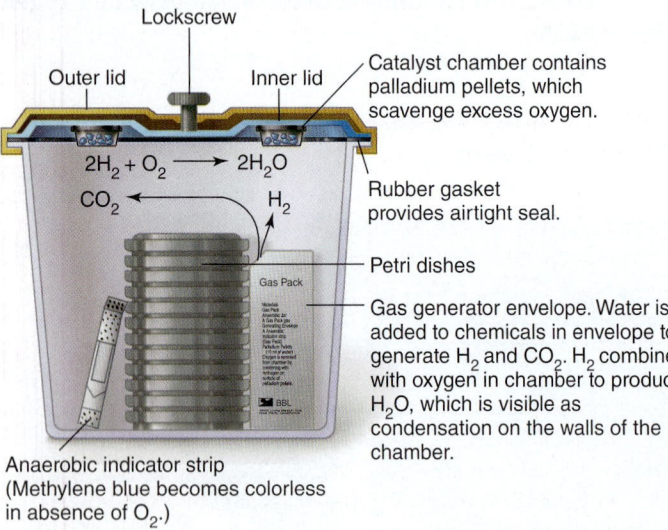

Lockscrew

Outer lid   Inner lid

Catalyst chamber contains palladium pellets, which scavenge excess oxygen.

$2H_2 + O_2 \rightarrow 2H_2O$

CO_2 H_2

Rubber gasket provides airtight seal.

Petri dishes

Gas Pack

Gas generator envelope. Water is added to chemicals in envelope to generate H_2 and CO_2. H_2 combines with oxygen in chamber to produce H_2O, which is visible as condensation on the walls of the chamber.

Anaerobic indicator strip (Methylene blue becomes colorless in absence of O_2.)

(b)

Figure 7.8 Culturing techniques for anaerobes.
(a) A special anaerobic environmental chamber makes it possible to handle strict anaerobes without exposing them to air. It has provisions for incubation and inspection in a completely O_2-free system. **(b)** A simpler anaerobic, or CO_2, incubator system. To create an anaerobic environment, a packet is activated to produce hydrogen gas and the chamber is sealed tightly. The gas reacts with available oxygen to produce water. Carbon dioxide can also be added to the system for growth of organisms needing high concentrations of it.

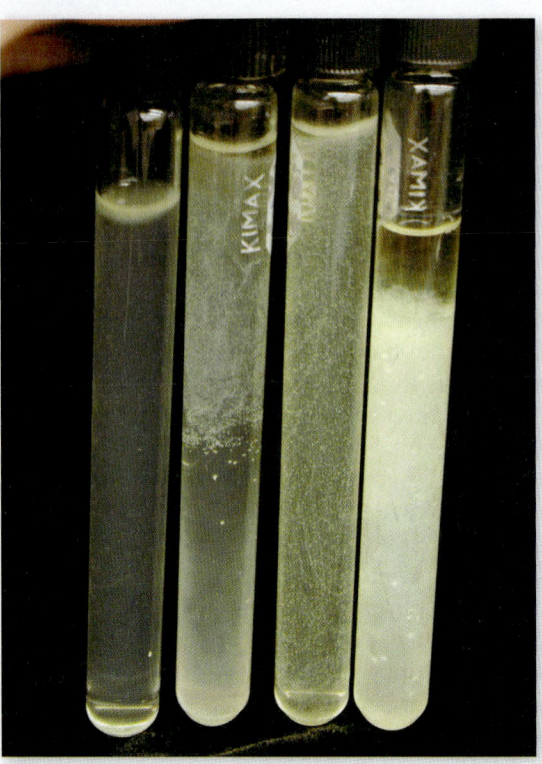

Figure 7.9 Examples of three of the gas categories.
Tube 1 (on the left): aerobic (*Pseudomonas aeruginosa*); Tubes 2 and 3: facultative (*Staphylococcus aureus* on left and *Escherichia coli* on right); Tube 4: obligate anaerobe (*Clostridium butyricum*).

Disease Connection

Tetanus, also known as lockjaw, is caused by an anaerobic bacterium, *Clostridium tetani*. You are aware that you may be at risk for this infection through puncture wounds such as stepping on a nail. This reflects the oxygen requirements of the bacterium. Deep puncture wounds contain limited oxygen at the "bottom" of the wound and often are not perfused with a lot of blood (another source of oxygen). In that situation, *Clostridium* endospores can germinate and multiply.

Aerotolerant anaerobes do not utilize oxygen but can survive and grow to a limited extent in its presence. These anaerobes are not harmed by oxygen, mainly because they possess alternate mechanisms for breaking down peroxides and superoxide. Certain lactobacilli and streptococci use manganese ions or peroxidases to perform this task.

Determining the oxygen requirements of a microbe from a biochemical standpoint can be a very time-consuming process. Often it is illuminating to perform culture tests with reducing media (those that contain an oxygen-absorbing chemical). One such technique demonstrates oxygen requirements by the location of growth in a tube of fluid thioglycollate **(figure 7.9).**

Although all microbes require some carbon dioxide in their metabolism, *capnophiles* grow best at a higher CO_2

tension than is normally present in the atmosphere. This becomes important in the initial isolation of some pathogens from clinical specimens, notably *Neisseria* (gonorrhea, meningitis), *Brucella* (undulant fever), and *Streptococcus pneumoniae.* Incubation is carried out in a CO_2 incubator that provides 3% to 10% CO_2 **(figure 7.8b).**

pH

Microbial growth and survival are also influenced by the pH of the habitat. The term *pH* was defined in chapter 2 as the degree of acidity or alkalinity (basicity) of a solution. It is expressed by the pH scale, a series of numbers ranging from 0 to 14. The pH of pure water (7.0) is neutral, neither acidic nor basic. As the pH value decreases toward 0, the acidity increases; and as the pH increases toward 14, the alkalinity increases. The majority of organisms live or grow in habitats between pH 6 and 8 because strong acids and bases can be highly damaging to enzymes and other cellular substances.

A few microorganisms live at pH extremes. Obligate *acidophiles* include *Euglena mutabilis,* an alga that grows in acid pools between 0 and 1.0 pH, and *Thermoplasma,* an archaea that lacks a cell wall, lives in hot coal piles at a pH of 1 to 2, and will lyse if exposed to pH 7. *Picrophilus* thrives at a pH of 0.7, and can grow at a pH of 0. Because many molds and yeasts tolerate moderate acid, they are the most common spoilage agents of pickled foods. Alkalinophiles, such as *Natronomonas* species, live in hot pools and soils that contain high levels of basic minerals (up to pH 12.0). Bacteria that decompose urine create alkaline conditions, because ammonium (NH_4^+) can be produced when urea (a component of urine) is digested. Metabolism of urea is one way that *Proteus* spp. can neutralize the acidity of the urine to colonize and infect the urinary system.

Osmotic Pressure

Although most microbes exist under hypotonic or isotonic conditions, a few, called **osmophiles,** live in habitats with a high solute concentration. One common type of osmophile prefers high concentrations of salt; these organisms are called **halophiles** (hay'-loh-fylz). Obligate halophiles such as *Halobacterium* and *Halococcus* inhabit salt lakes, ponds, and other hypersaline habitats **(Insight 7.3).** They grow optimally in solutions of 25% NaCl but require at least 9% NaCl (combined with other salts) for growth. These archaea have significant modifications in their cell walls and membranes and will lyse in hypotonic habitats. *Facultative halophiles* are remarkably resistant to salt, even though they do not normally reside in high-salt environments. For example, *Staphylococcus aureus* can grow on NaCl media ranging from 0.1% up to 20%. Although it is common to use high concentrations of salt and sugar to preserve food (jellies, syrups, and brines), many bacteria and fungi actually thrive under these conditions and are common spoilage agents.

Radiation and Hydrostatic Atmospheric Pressure

Various forms of electromagnetic radiation (ultraviolet, infrared, visible light) stream constantly onto the earth from the

---

INSIGHT 7.3   The BP Deepwater Horizon Spill: Oil Isn't the Only Environmental Disaster

The British Petroleum (BP) Deepwater Horizon oil spill in April 2010 dumped 4.9 million barrels of oil into the Gulf of Mexico, and has been called the worst environmental disaster in history. In spite of a massive cleanup by BP and government agencies, tar balls continued to wash up on the once pristine beaches of Alabama, Louisiana, Mississippi, and Florida for months

afterward. However, toxic crude oil wasn't the only dangerous substance in the tar balls; scientists found high levels of *Vibrio vulnificus* growing in them as well.

Vibrio vulnificus is a cousin to the deadly *Vibrio cholera* and causes 95% of all seafood-related deaths. It thrives in marine habitats, particularly in the sediments, where it is taken up by filter feeders such as clams, scallops, mussels, and other shellfish. *V. vulnificus* is considered a moderate **halophile,** meaning that it requires a higher concentration of salt than other organisms to grow. If ingested in seafood, it causes nausea, abdominal pain, and vomiting. If it enters through a skin wound, it can cause skin lesions that can lead to septicemia. *V. vulnificus* is particularly dangerous to individuals who are immunocompromised.

Recent research has shown that tar balls from the Deepwater Horizon spill harbor levels of *V. vulnificus* that are 10 to 100 times higher than that found in seawater, sand, and sediment. Scientists were surprised to find such high levels of a salt-loving organism in the tar balls, but they surmise that *V. vulnificus* is living off of by-products of other bacteria metabolizing the oil in the tar balls. Further research is needed to determine if these high concentrations pose a threat to humans. Beachgoers are advised to avoid picking up tar balls and, if they do come into contact with tar balls, to wash their skin thoroughly.

sun. Some microbes (phototrophs) can use visible light rays as an energy source, but nonphotosynthetic microbes tend to be damaged by the toxic oxygen products produced by contact with light. Some microbial species produce yellow carotenoid pigments to protect against the damaging effects of light by absorbing and dismantling toxic oxygen. Other types of radiation that can damage microbes are ultraviolet and ionizing rays (X rays and cosmic rays). In chapter 11, you will see just how these types of energy are applied in microbial control.

Descent into the ocean depths subjects organisms to increasing hydrostatic pressure. Deep-sea microbes called **barophiles** exist under pressures that range from a few times to over 1,000 times the pressure of the atmosphere. These bacteria are so strictly adapted to high pressures that they will rupture when exposed to normal atmospheric pressure.

Because of the high water content of cytoplasm, all cells require water from their environment to sustain growth and metabolism. Water is the solvent for cell chemicals, and it is needed for enzyme function and digestion of macromolecules. A certain amount of water on the external surface of the cell is required for the diffusion of nutrients and wastes. Even in apparently dry habitats, such as sand or dry soil, the particles retain a thin layer of water usable by microorganisms. Only dormant, dehydrated cell stages (e.g., spores and cysts) tolerate extreme drying because of the inactivity of their enzymes.

Other Organisms

Up to now, we have considered the importance of nonliving environmental influences on the growth of microorganisms. Another profound influence comes from other organisms that share (or sometimes are) their habitats. In all but the rarest instances, microbes live in shared habitats, which give rise to complex and fascinating associations. Some associations are between similar or dissimilar types of microbes; others involve multicellular organisms such as animals or plants. Interactions can have beneficial, harmful, or no particular effects on the organisms involved; they can be obligatory or nonobligatory to the members; and they often involve nutritional interactions. This outline provides an overview of the major types of microbial associations:

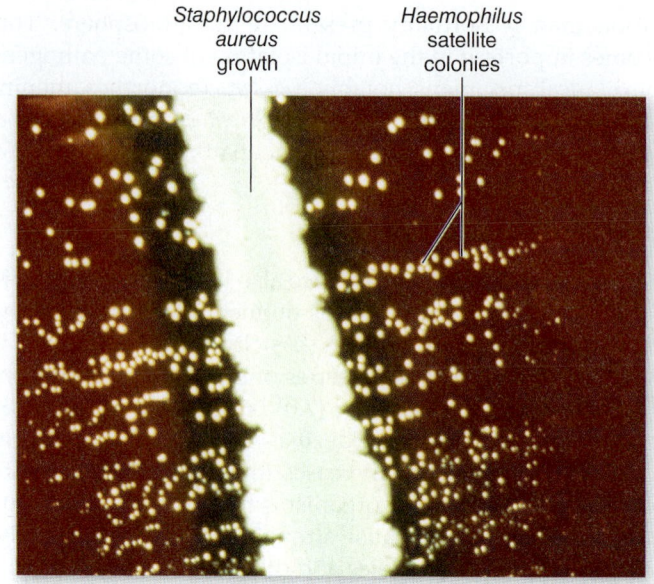

Staphylococcus aureus growth   Haemophilus satellite colonies

Figure 7.10 Satellitism, a type of commensalism between two microbes.   In this example, *Staphylococcus aureus* provides growth factors to *Haemophilus influenzae*, which grows as tiny satellite colonies near the streak of *Staphylococcus*. By itself, *Haemophilus* could not grow on blood agar. The *Staphylococcus* gives off several nutrients such as vitamins and amino acids that diffuse out to the *Haemophilus*, thereby promoting its growth.

A general term used to denote a situation in which two organisms live together in a close partnership is **symbiosis**,[2] and the members are termed *symbionts*. Three main types of symbiosis occur. **Mutualism** exists when organisms live in an obligatory but mutually beneficial relationship. This association is rather common in nature because of the survival value it has for the members involved. In other symbiotic relationships, the relationship tends to be unequal, meaning it benefits one member and not the other, and it can be obligatory.

In a relationship known as **commensalism,** the member called the commensal receives benefits, while its coinhabitant is neither harmed nor benefited. A classic commensal interaction between microorganisms called **satellitism** arises when one member provides nutritional or protective factors needed by the other **(figure 7.10).** Some microbes can break down a substance that would be toxic or inhibitory to another microbe. Relationships between humans and resident commensals that derive nutrients from the body are discussed in a later section.

In an earlier section, we introduced the concept of **parasitism** as a relationship in which the host organism provides the parasitic microbe with nutrients and a habitat. Multiplication of the parasite usually harms the host to some extent. As this relationship evolves, the host may even develop tolerance for or dependence on a parasite,

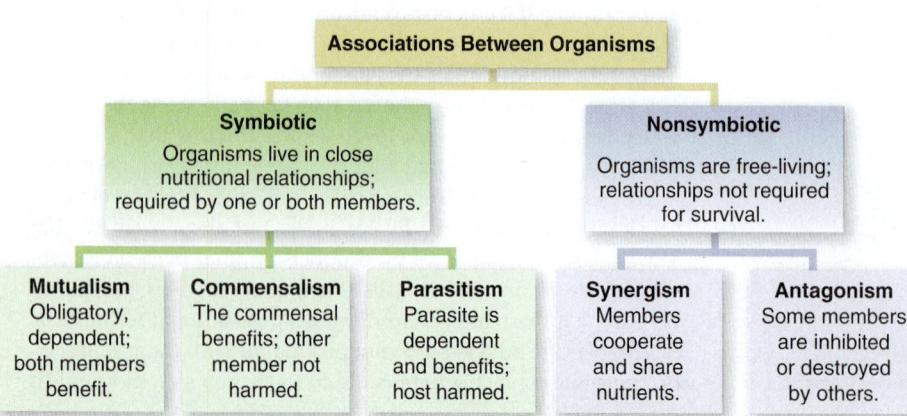

Associations Between Organisms

Symbiotic	Nonsymbiotic
Organisms live in close nutritional relationships; required by one or both members.	Organisms are free-living; relationships not required for survival.

Mutualism	Commensalism	Parasitism	Synergism	Antagonism
Obligatory, dependent; both members benefit.	The commensal benefits; other member not harmed.	Parasite is dependent and benefits; host harmed.	Members cooperate and share nutrients.	Some members are inhibited or destroyed by others.

---

2. Note that *symbiosis* is a neutral term and does not by itself imply benefit or detriment.

at which point we call the relationship commensalism or mutualism.

Antagonism is an association between free-living species that arises when members of a community compete. In this interaction, one microbe secretes chemical substances into the surrounding environment that inhibit or destroy another microbe in the same habitat. The first microbe may gain a competitive advantage by increasing the space and nutrients available to it. Interactions of this type are common in the soil, where mixed communities often compete for space and food. *Antibiosis*—the production of inhibitory compounds such as antibiotics—is actually a form of antagonism. Hundreds of naturally occurring antibiotics have been isolated from bacteria and fungi and used as drugs to control diseases (see chapter 12).

Synergism is an interrelationship between two or more free-living organisms that benefits them but is not necessary for their survival. Together, the participants cooperate to produce a result that none of them could do alone. Biofilms are the best examples of synergism.

In synergistic infections, a combination of organisms can produce tissue damage that a single organism would not cause alone. Gum disease, dental caries, and gas gangrene involve mixed infections by bacteria interacting synergistically.

Biofilms: The Epitome of Synergy

You have already heard about the importance of biofilms, both in chapter 1 and in chapter 4 (Insight 4.1). The National Institutes of Health estimate that 80% of chronic infections are caused by biofilms **(process figure 7.11).** These include chronic ear infections, prostate infections, lung infections in cystic fibrosis patients, and wound infections, as well as many others. Ordinary antibiotic treatment does not work against most biofilms (which is why the infections remain chronic). We'll learn more about that aspect of biofilms in chapter 12.

A Note About Coevolution

Organisms that have close, ongoing relationships with each other participate in **coevolution,** the process whereby a change in one of the partners leads to a change in the other partner, which may in turn lead to another change in the first partner, and so on. This is another example of the interconnectedness of biological entities on this planet. There are many well-documented examples of the relationships between plants and insects. One of the earliest is the discovery by Charles Darwin of a plant that had a nectar tube that was 10 inches long. Knowing that the plant depended on insects for pollination, Darwin predicted the existence of an insect with a 10-inch tongue—and 41 years later one was discovered.

The plant and the insect had influenced each other's evolution over time. Commensal gut bacteria are considered to have coevolved with their mammalian hosts, with the hosts evolving mechanisms to prevent the disease effects of their bacterial passengers, and the bacteria evolving mechanisms to not only be less pathogenic to their hosts, but also to provide important benefits to them.

Many, if not all, biofilms are mixed communities of different kinds of bacteria and other microbes. Usually there is a "pioneer" colonizer, a bacterium that initially attaches to a surface, such as a tooth or the lung tissue. Other microbes then attach either to those bacteria or to the polymeric sugar and protein substance that inevitably is secreted by microbial colonizers of surfaces. In many cases, once the cells are attached, they are stimulated to release chemicals that accumulate as the cell population grows. By this means, they can monitor the size of their own population. This is a process called **quorum sensing.** Bacteria can use quorum sensing to interact with other members of the same species as well as members of other species that are close by (in a biofilm,

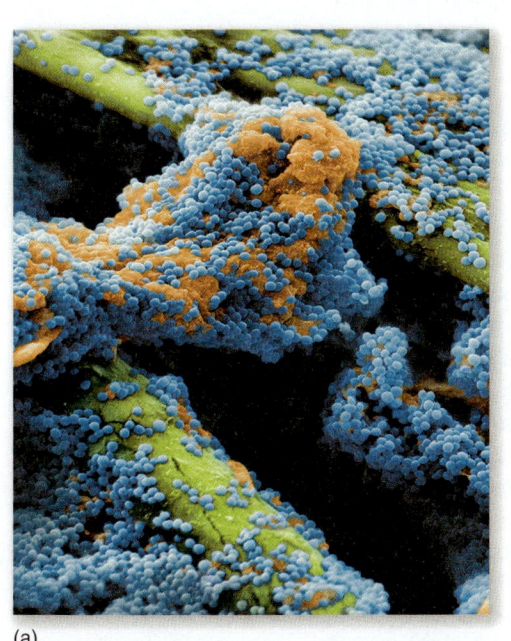

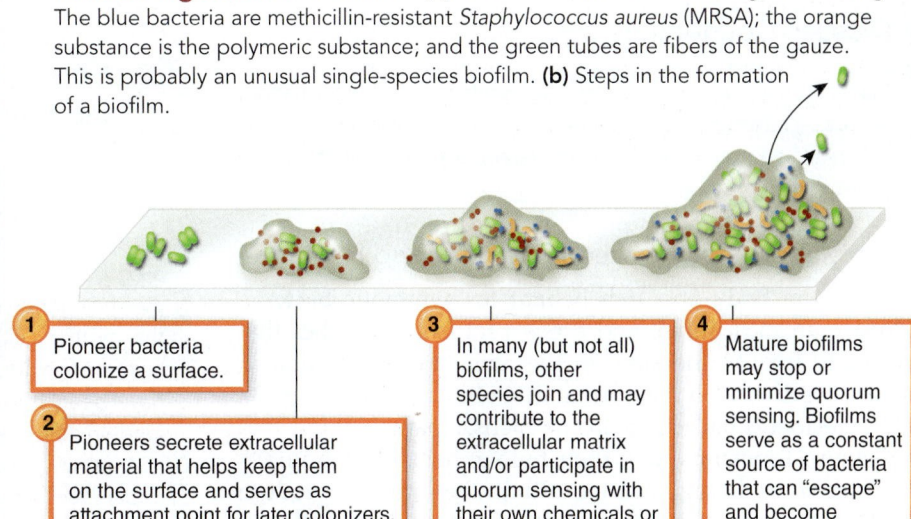

Process Figure 7.11 Biofilms. **(a)** SEM of a biofilm formed on a gauze bandage. The blue bacteria are methicillin-resistant *Staphylococcus aureus* (MRSA); the orange substance is the polymeric substance; and the green tubes are fibers of the gauze. This is probably an unusual single-species biofilm. **(b)** Steps in the formation of a biofilm.

1 Pioneer bacteria colonize a surface.

2 Pioneers secrete extracellular material that helps keep them on the surface and serves as attachment point for later colonizers. Quorum sensing chemicals (red dots) are released by bacteria.

3 In many (but not all) biofilms, other species join and may contribute to the extracellular matrix and/or participate in quorum sensing with their own chemicals or the ones released by other species.

4 Mature biofilms may stop or minimize quorum sensing. Biofilms serve as a constant source of bacteria that can "escape" and become planktonic again.

(a)

(b)

for example). Eventually, large complex communities are formed, in which physical and biological characteristics vary in different locations of the community. The very bottom of a biofilm may have very different pH and oxygen conditions than the surface of a biofilm, for example. It is now clearly established that microbes in a biofilm, as opposed to those in a planktonic (free-floating) state, behave and respond very differently to their environments. Different genes are even utilized in the two situations. The same chemicals cells secrete during quorum sensing are responsible for some of this change in gene expression. At any rate, a single biofilm is usually a partnership among multiple microbial inhabitants and thus cannot be eradicated by traditional methods targeting individual infections. This kind of synergism has led to the necessity of rethinking how microbes progress from colonization to the development of true disease (see chapter 13).

Biofilms are so prevalent that they dominate the structure of most natural environments on earth. This tendency of microbes to form biofilm communities is an ancient and effective adaptive strategy. Not only do biofilms favor microbial persistence in habitats, but they also offer greater access to life-sustaining conditions for the microbes.

For many years, biologists regarded most single-celled microbes as simple individuals that did not work together other than to cling together in colonies as they multiplied. But these assumptions have turned out to be incorrect. It is now evident that microbes show a well-developed capacity to communicate and cooperate in the formation and function of biofilms. This is especially true of bacteria, although fungi and other microorganisms can participate in these activities.

Biofilms are known to be a rich ground for genetic transfers among neighboring cells (see chapter 9). As our knowledge of biofilm formation and quorum sensing grows, it will likely lead to greater understanding of their involvement in infections and their contributions to disinfectant and drug resistance. It may also be the key to new drugs that successfully target these biological formations.

Disease Connection

Many human diseases can be the result of biofilm formation. Anthrax (caused by *Bacillus anthracis*), typhoid fever (*Salmonella enterica*), and cholera (*Vibrio cholerae*) are some of the more serious ones.

Interrelationships Between Microbes and Humans

The human body is a rich habitat for symbiotic bacteria, viruses, archaea, fungi, and a few protozoa. Microbes that normally live on the skin, in the alimentary tract, and in other sites are called the **normal microbiota** (see chapter 13). These residents participate in commensal, parasitic, and synergistic relationships with their human hosts. For example, *Escherichia coli* living symbiotically in the intestine produce vitamin K, and species of symbiotic *Lactobacillus* residing in the vagina help maintain an acidic environment that protects against infection by other microorganisms. Hundreds

of commensal species "make a living" on the body without either harming or benefiting it. For example, many bacteria and yeasts reside in the outer dead regions of the skin; oral microbes feed on the constant flow of nutrients in the mouth; and billions of bacteria live on the wastes in the large intestine. Because the normal microbiota and the body are in a constant state of change, these relationships are not absolute, and a commensal can convert to a parasite by invading body tissues and causing disease.

7.2 Learning Outcomes—Assess Your Progress

8. List and define five terms used to express a microbe's optimal growth temperature.
9. Summarize three ways in which microorganisms function in the presence of oxygen.
10. Identify three other physical factors that microbes must contend with in the environment.
11. List and describe the five major types of microbial association.
12. Discuss characteristics of biofilms that differentiate them from planktonic bacteria and their infections.

7.3 The Study of Microbial Growth

When microbes are provided with nutrients and the required environmental factors, they become metabolically active and grow. Growth takes place on two levels. On one level, a cell synthesizes new cell components and increases its size; on the other level, the number of cells in the population increases. This capacity for multiplication, increasing the size of the population by cell division, has tremendous importance in microbial control, infectious disease, and biotechnology. In the following sections, we will focus primarily on the characteristics of bacterial growth that are generally representative of single-celled microorganisms.

A Note About Bacterial Reproduction— and the "Culture Bias"

By far most of the bacteria that have ever been studied reproduce via binary fission, as described in this chapter. But there are important exceptions. In recent years, researchers have discovered bacteria that produce multiple offspring within their cytoplasm and then split open to release multiple new bacteria (killing the mother cell). One example is *Epulopiscium*, a symbiont of surgeon fish. Most of these bacteria have never been cultured but have been studied by dissecting the animals they colonize. The long-standing belief that bacteria always multiply by binary fission is another by-product of the "culture bias"—meaning that we understand most about the bacteria that we were able to cultivate in the lab, even though there are many more bacteria that exist in the biosphere that have not yet been cultivated.

The Basis of Population Growth: Binary Fission

The division of a bacterial cell occurs mainly through **binary fission;** *binary* means that one cell becomes two. During binary fission, the parent cell enlarges, duplicates its chromosome, and then starts to pull its cell envelope together in the center of the cell using a band of protein that is made up of proteins that resemble actin and tubulin—the protein components of microtubules in eukaryotic cells. The cell wall eventually forms a complete central septum. This process divides the cell into two daughter cells. This process is repeated at intervals by each new daughter cell in turn, and with each successive round of division, the population increases. The stages in this continuous process are shown in greater detail in **process figure 7.12** and **figure 7.13.**

The Rate of Population Growth

The time required for a complete fission cycle—from parent cell to two new daughter cells—is called the **generation,** or **doubling, time.** The term *generation* has a similar meaning as it does in humans—the period between an individual's birth and the time of producing offspring. In bacteria, each new fission cycle or generation increases the population by a

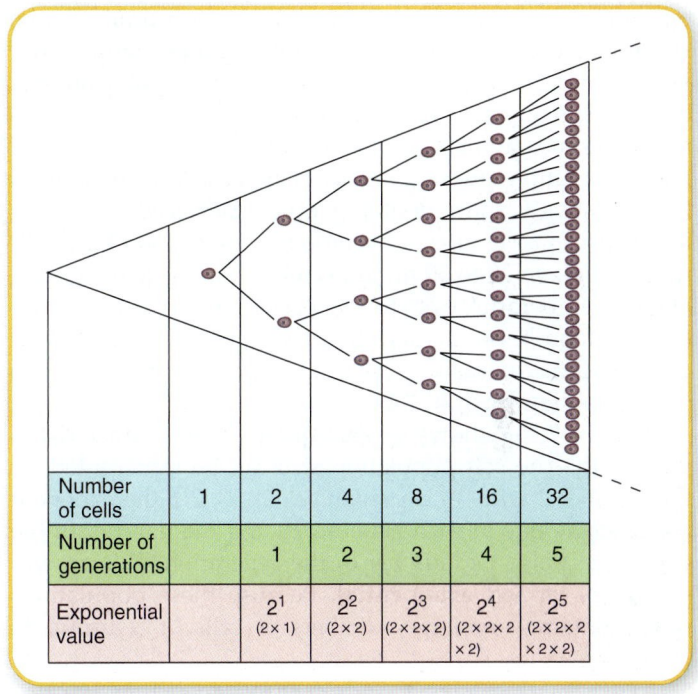

(a)

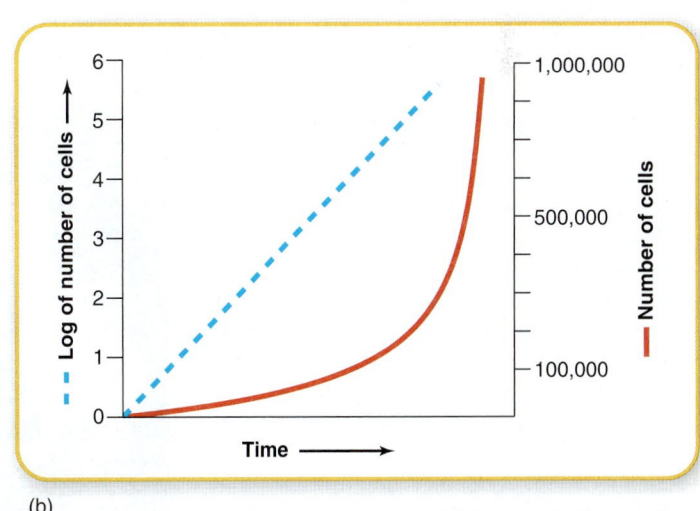

(b)

Figure 7.13 The mathematics of population growth.
(a) Starting with a single cell, if each product of reproduction goes on to divide by binary fission, the population doubles with each new cell division or generation. This process can be represented by logarithms (2 raised to an exponent) or by simple numbers. **(b)** Plotting the logarithm of the cells produces a straight line indicative of exponential growth, whereas plotting the cell numbers arithmetically gives a curved slope.

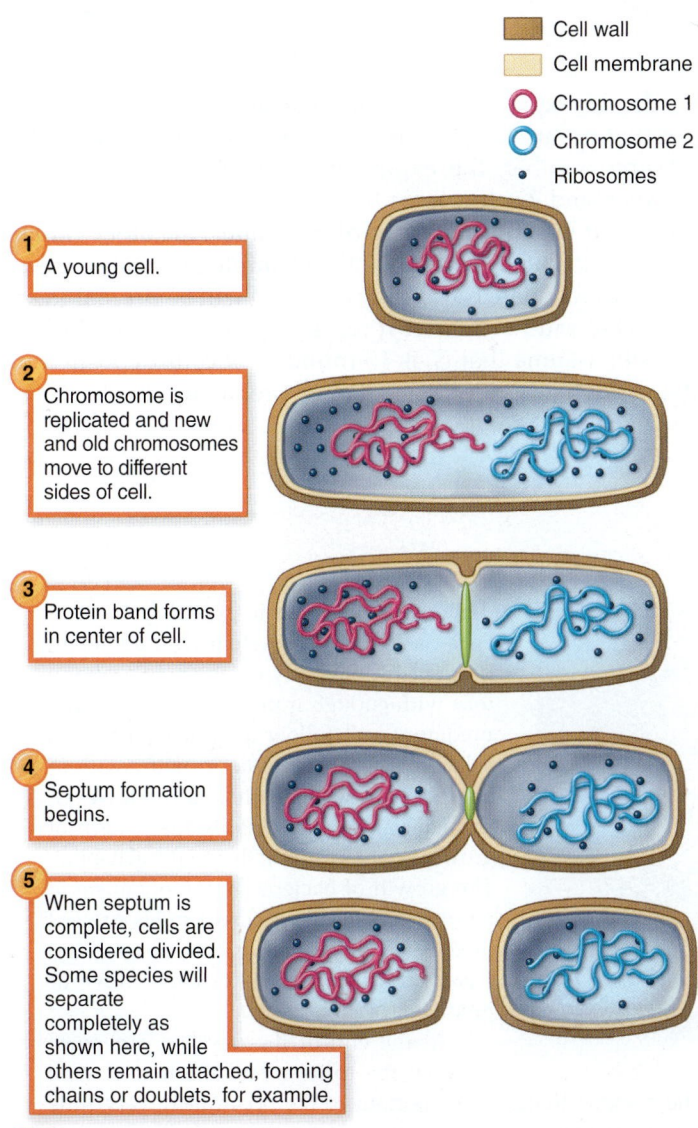

- Cell wall
- Cell membrane
- Chromosome 1
- Chromosome 2
- Ribosomes

1. A young cell.
2. Chromosome is replicated and new and old chromosomes move to different sides of cell.
3. Protein band forms in center of cell.
4. Septum formation begins.
5. When septum is complete, cells are considered divided. Some species will separate completely as shown here, while others remain attached, forming chains or doublets, for example.

Process Figure 7.12 Steps in binary fission of a rod-shaped bacterium.   Note that even though the two chromosomes are colored differently, the new one is an exact copy of the old one (with some mistakes incorporated that you will learn about later).

factor of 2, or doubles it. Thus, the initial parent stage consists of 1 cell, the first generation consists of 2 cells, the second 4, the third 8, then 16, 32, 64, and so on. As long as the environment remains favorable, this doubling effect can continue at a constant rate. With the passing of each generation, the population will double, over and over again.

The length of the generation time is a measure of the growth rate of an organism. Compared with the growth rates of most other living things, bacteria are notoriously rapid. The average generation time is 30 to 60 minutes under optimum conditions. The shortest generation times can be 10 to 12 minutes, and longer generation times require days. For example, *Mycobacterium leprae,* the cause of Hansen's disease, has a generation time of 10 to 30 days—as long as that of some animals. Environmental bacteria commonly have generation times measured in months. Most pathogens have relatively short doubling times. *Salmonella enteritidis* and *Staphylococcus aureus,* bacteria that cause food-borne illness, double in 20 to 30 minutes, which is why leaving food at room temperature even for a short period has caused many cases of food-borne disease. In a few hours, a population of these bacteria can easily grow from a small number of cells to several million **(Insight 7.4).**

Figure 7.13a shows several quantitative characteristics of growth: The cell population size can be represented by the number 2 with an exponent (2^1, 2^2, 2^3, 2^4); the exponent increases by one in each generation; and the number of the exponent is also the number of the generation. This growth pattern is termed **exponential.** Because these populations often contain very large numbers of cells, it is useful to express them by means of exponents or logarithms (see appendix A). The data from a growing bacterial population are graphed by plotting the number of cells as a function of time **(figure 7.13b).** The cell number can be represented logarithmically or arithmetically. Plotting the logarithm number over time provides a straight line indicative of exponential growth. Plotting the data arithmetically gives a constantly curved slope. In general, logarithmic graphs are preferred because an accurate cell number is easier to read, especially during early growth phases.

Predicting the number of cells that will arise during a long growth period (yielding millions of cells) is based on a relatively simple concept. One could use the method of addition $2 + 2 = 4$, $4 + 4 = 8$, $8 + 8 = 16$, $16 + 16 = 32$, and so on; or a method of multiplication (for example, $2^5 = 2 \times 2 \times 2 \times 2 \times 2$); but it is easy to see that for 20 or 30 generations, this calculation could be very tedious. An easier way to calculate the size of a population over time is to use an equation such as

$$N_f = (N_i)2^n$$

In this equation, N_f is the total number of cells in the population at some point in the growth phase, N_i is the starting number, the exponent n denotes the generation number, and 2^n represents the number of cells in that generation. If we know any two of the values, the other values can be calculated. Let us use the example of *Staphylococcus aureus* to calculate how many cells (N_f) will be present in an egg salad sandwich after it sits in a warm car for 4 hours. We will assume that N_i is 10 (number of cells deposited in the sandwich while it was being prepared). To derive n,

---

INSIGHT 7.4   The Tortoise and the Hare

Scientists have recently discovered the slowest-growing bacteria on the planet. Analyzing the amino acids deposited in the sediment in the seabed, microbiologists at Aarhus University in Denmark have found bacteria with a generation time of 1,000 to 3,000 *years.* These organisms live under extreme pressures—several hundred times normal atmospheric pressure—in total darkness, with very few nutrients. Despite their extremely slow rate of reproduction, the organisms play an important role in the global carbon cycle, recycling nutrients that fall to the ocean depths.

In contrast, *Escherichia coli* exhibit a positively breakneck pace of reproduction, doubling itself every 20 minutes. *Bacillus subtilis* is a close second with generation times measured at around 30 minutes. What is the difference between these microbial tortoises and hares? A lot of it has to do with the availability of nutrients. Figure 7.15 shows the bacterial growth curve measured in the laboratory, which demonstrates the basic growth pattern of bacteria in a closed system with abundant nutrients. Almost any organism in a laboratory with enough nutrients and no natural predators will follow a similar pattern of a lag phase, logarithmic growth, stationary phase, and a death phase. However, this isn't always necessarily the pattern of growth of organisms in their natural habitat. The growth of bacteria or any organism in nature is drastically different and is affected by the availability of nutrients, oxygen, and water and the presence of competitive or predatory organisms.

At the end of the day, the difference between the tortoise and the hare is fuel: The bacteria living at the bottom of the ocean have very little compared to the ones in your gut. The more nutrients an organism has available to it, the closer it will come to its maximum growth rate.

Source: 2012. *ScienceDaily.* March 19.

we need to divide 4 hours (240 minutes) by the generation time (we will use 20 minutes). This calculation comes out to 12, so 2^n is equal to 2^{12}. Using a calculator, we find that 2^{12} is 4,096.

Final number $(N_f) = 10 \times 4,096$
$\qquad\qquad\qquad = 40,960$ bacterial cells in the sandwich

This same equation, with modifications, is used to determine the generation time, a more complex calculation that requires knowing the number of cells at the beginning and end of a growth period. Such data are obtained through actual testing by a method discussed in the following section.

The Population Growth Curve

In reality, a population of bacteria does not maintain its potential growth rate and does not double endlessly, because in most systems numerous factors prevent the cells from continuously dividing at their maximum rate. A population typically displays a predictable pattern, or **growth curve,** over time.

The method traditionally used to observe the population growth pattern is a viable count technique, in which the total number of live cells is counted over a given time period.

This is a fundamental method of laboratory microbiology. A growing population is established by inoculating a flask containing a known quantity of sterile liquid medium with a few cells of a pure culture. The flask is incubated at that bacterium's optimum temperature and timed. The population size at any point in the growth cycle is quantified by removing a tiny measured sample of the culture from the growth chamber and plating it out on a solid medium to develop isolated colonies. This procedure is repeated at evenly spaced intervals (i.e., every hour for 24 hours).

Evaluating the samples involves a common and important principle in microbiology: One colony on the plate represents one cell or colony-forming unit (CFU) from the original sample. Because the CFU of some bacteria is actually composed of several cells (consider the clustered arrangement of *Staphylococcus,* for instance), using a colony count can underestimate the exact population size to an extent. This is not a serious problem because, in such bacteria, the CFU is the smallest unit of colony formation and dispersal. Multiplication of the number of colonies in a single sample by the container's volume gives a fair estimate of the total population size (number of cells) at any given point. The growth curve is determined by graphing the number for each sample in sequence for the whole incubation period **(figure 7.14).**

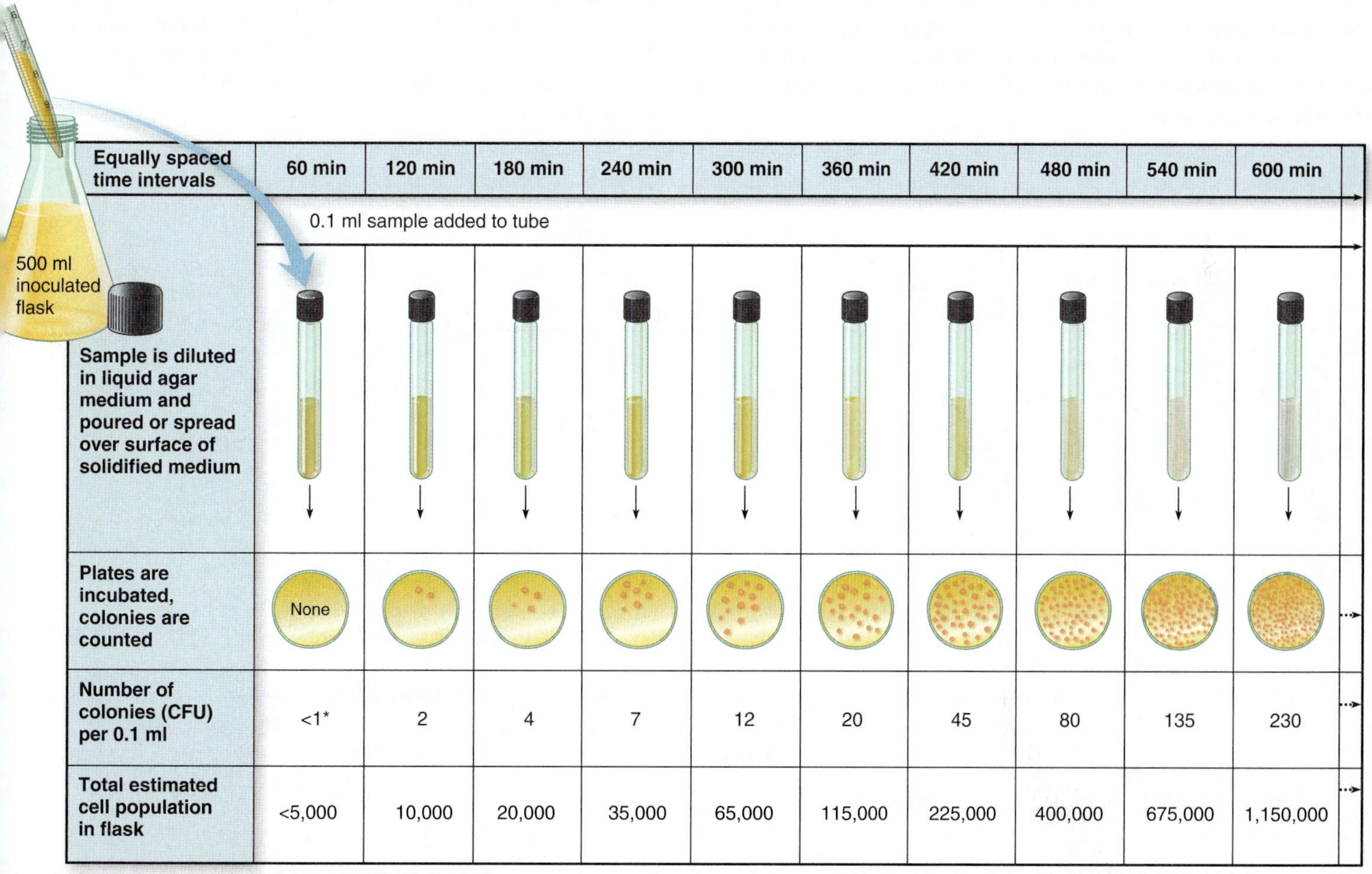

Equally spaced time intervals	60 min	120 min	180 min	240 min	300 min	360 min	420 min	480 min	540 min	600 min
Number of colonies (CFU) per 0.1 ml	<1*	2	4	7	12	20	45	80	135	230
Total estimated cell population in flask	<5,000	10,000	20,000	35,000	65,000	115,000	225,000	400,000	675,000	1,150,000

*Only means that too few cells are present to be assayed.

Figure 7.14 Steps in a viable plate count: batch culture method.

Stages in the Normal Growth Curve

The system of culturing described in figure 7.14 is *closed*, meaning that nutrients and space are finite and there is no mechanism for the removal of waste products. Data from an entire growth period of 3 to 4 days typically produce a curve with a series of phases termed the *lag phase*, the *exponential growth (log) phase*, the *stationary phase*, and the *death phase* **(figure 7.15).**

The **lag phase** is a relatively "flat" period on the graph when the population appears not to be growing or is growing at less than the exponential rate. Growth lags primarily because

1. the newly inoculated cells require a period of adjustment, enlargement, and synthesis;
2. the cells are not yet multiplying at their maximum rate; and
3. the population of cells is so sparse or dilute that the sampling misses them.

The length of the lag period varies somewhat from one population to another. It is important to note that even though the population of cells is not increasing (growing), individual cells are metabolically active as they increase their contents and prepare to divide.

The cells reach the maximum rate of cell division during the **exponential growth** (logarithmic or **log**) **phase,** a period during which the curve increases geometrically. This phase will continue as long as cells have adequate nutrients and the environment is favorable.

At the **stationary growth phase,** the population enters a survival mode in which cells stop growing or grow slowly. The curve levels off because the rate of cell inhibition or death balances out the rate of multiplication. The decline in the growth rate is caused by depleted nutrients and oxygen plus excretion of organic acids and other biochemical pollutants into the growth medium, due to the increased density of cells.

As the limiting factors intensify, cells begin to die at an exponential rate (literally perishing in their own wastes), and they are unable to multiply. The curve now dips downward as the **death phase** begins. The speed with which death occurs depends on the relative resistance of the species and how toxic the conditions are, but it is usually slower than the exponential growth phase. Viable cells often remain many weeks and months after this phase has begun. In the laboratory, refrigeration is used to slow the progression of the death phase so that cultures will remain viable as long as possible.

Practical Importance of the Growth Curve

The tendency for populations to exhibit phases of rapid growth, slow growth, and death has important implications in microbial control, infection, food microbiology, and culture technology. Antimicrobial agents such as heat and disinfectants rapidly accelerate the death phase in all populations, but microbes in the exponential growth phase are more vulnerable to these agents than are those that have entered the stationary phase. In general, actively growing cells are more vulnerable to conditions that disrupt cell metabolism and binary fission.

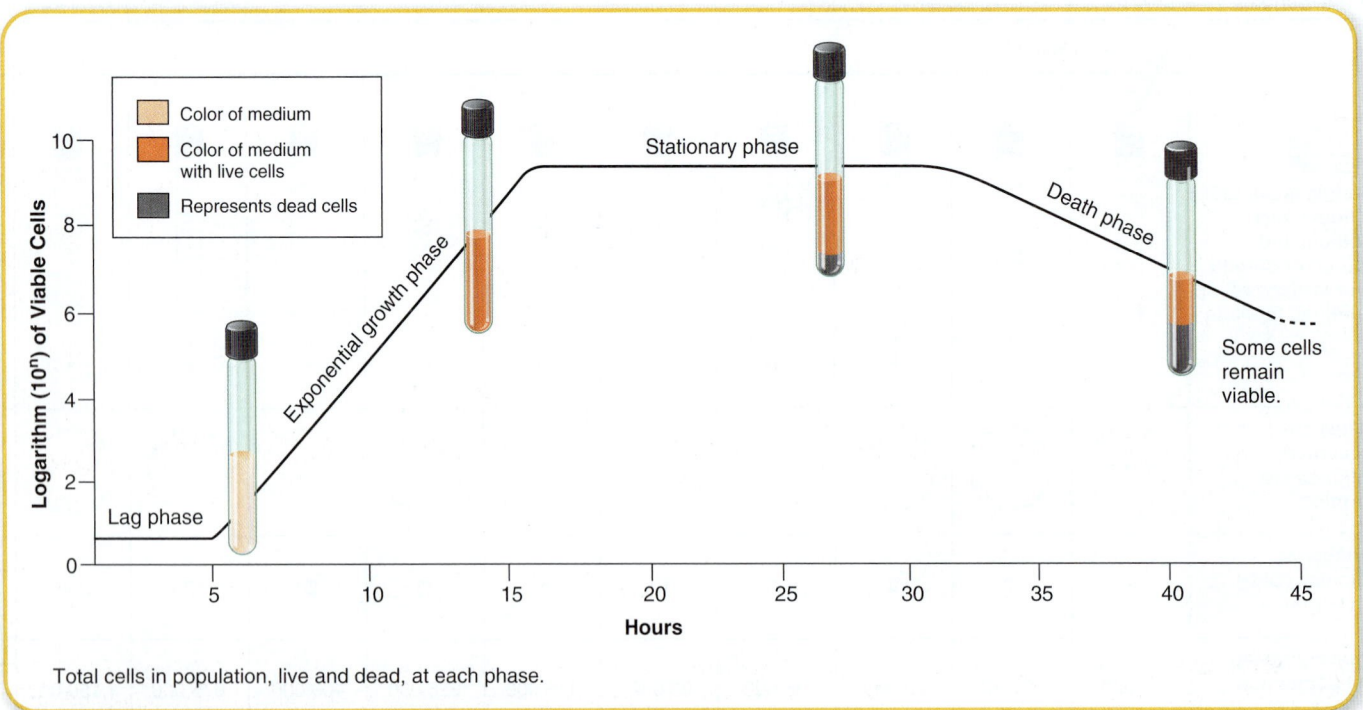

Figure 7.15 The growth curve in a bacterial culture.   On this graph, the number of viable cells expressed as a logarithm (log) is plotted against time. See text for discussion of the various phases. Note that with a generation time of 30 minutes, the population has risen from 10 (10^1) cells to 1,000,000,000 (10^9) cells in only 16 hours.

Growth patterns in microorganisms can account for the stages of infection (see chapter 13). A person shedding bacteria in the early and middle stages of an infection is more likely to spread it to others than is a person in the late stages. The course of an infection is also influenced by the relatively faster rate of multiplication of the microbe, which can overwhelm the slower growth rate of the host's own cellular defenses.

Understanding the stages of cell growth is crucial for work with cultures. Sometimes a culture that has reached the stationary phase is incubated under the mistaken impression that enough nutrients are present for the culture to multiply. In most cases, it is unwise to continue incubating a culture beyond the stationary phase, because doing so will reduce the number of viable cells and the culture could die out completely. It is also preferable to use young cultures to do stains (an exception is the spore stain) and motility tests, because the cells will show their natural size and correct reaction and motile cells will have functioning flagella.

For certain research or industrial applications, closed batch culturing with its four phases is inefficient. The alternative is an automatic growth chamber called the **chemostat,** or continuous culture system. This device can admit a steady stream of new nutrients and siphon off used media and old bacterial cells, thereby stabilizing the growth rate and cell number. It has the advantage of maintaining the culture in a biochemically active state and preventing it from entering the death phase.

Other Methods of Analyzing Population Growth

Turbidometry

Microbiologists have developed several alternative ways of analyzing bacterial growth qualitatively and quantitatively. One of the simplest methods for estimating the size of a population is through turbidometry. This technique relies on the simple observation that a tube of clear nutrient solution becomes cloudy, or **turbid,** as microbes grow in it. In general, the greater the turbidity, the larger the population size, which can be measured by means of sensitive instruments **(figure 7.16).**

Enumeration of Bacteria

Turbidity readings are useful for evaluating relative amounts of growth, but if a more quantitative evaluation is required, the viable colony count described in this chapter or some other enumeration (counting) procedure is necessary. The **direct (total) cell count** involves counting the number of cells in a sample microscopically **(figure 7.17).** This technique, very similar to that used in blood cell counts, employs a special microscope slide (cytometer) calibrated to accept a tiny sample that is spread over a premeasured grid. The cell count from a cytometer can be used to estimate the total number of cells in a larger sample (i.e., of milk or water). One inherent inaccuracy in this method as well as in spectrophotometry is

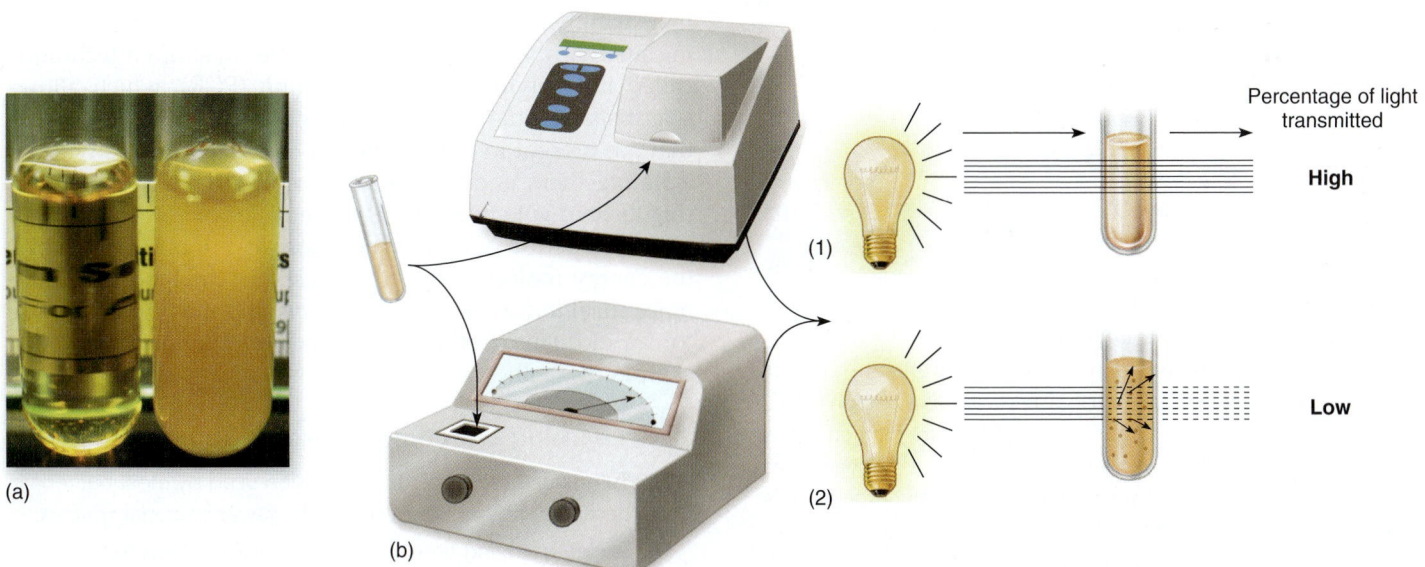

Figure 7.16 Turbidity measurements as indicators of growth. **(a)** Holding a broth to the light is one method of checking for gross differences in cloudiness (turbidity). The broth on the left is transparent, indicating little or no growth; the broth on the right is cloudy and opaque, indicating heavy growth. **(b)** The eye is not sensitive enough to pick up fine degrees in turbidity; more sensitive measurements can be made with a spectrophotometer. (1) A tube with no growth will allow light to easily pass. Therefore, more light will reach the photodetector and give a higher transmittance value. (2) In a tube with growth, the cells scatter the light, resulting in less light reaching the photodetector, which, therefore, gives a lower transmittance value. Both a newer digital and an older analog spectrophotometer are pictured.

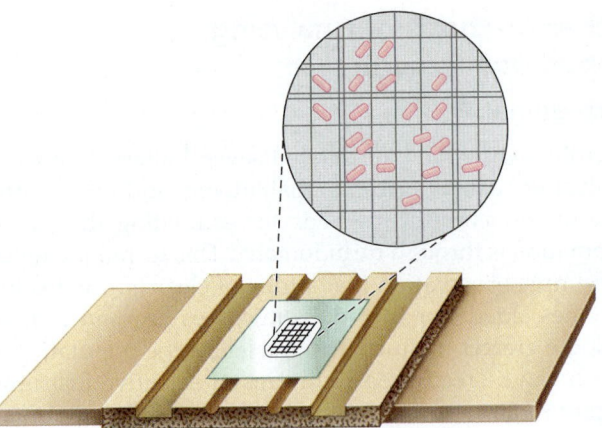

Figure 7.17 Direct microscopic count of bacteria.   A small sample is placed on the grid under a cover glass. Individual cells, both living and dead, are counted. This number can be used to calculate the total count of a sample.

that no distinction can be made between dead and live cells, both of which are included in the count.

Counting can be automated by sensitive devices such as the *Coulter counter,* which electronically scans a culture as it passes through a tiny pipette. As each cell flows by, it is detected and registered on an electronic sensor **(figure 7.18).** A *flow cytometer* works on a similar principle, but in addition to counting, it can measure cell size and even differentiate between live and dead cells. When used in conjunction with fluorescent dyes and antibodies to tag cells, it has been used to differentiate between gram-positive and gram-negative bacteria. It has been adapted for use as a rapid method to identify pathogens in patient specimens and to differentiate blood cells. More sophisticated forms of the flow cytometer can actually sort cells of different types into separate compartments of a collecting device.

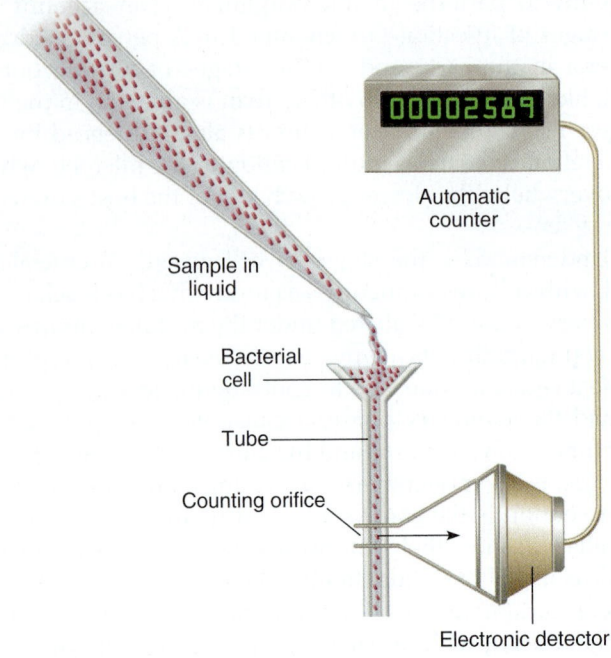

Figure 7.18 Coulter counter.   As cells pass through this device, they trigger an electronic sensor that tallies their numbers.

Although flow cytometry can be used to count bacteria in natural samples without the need for culturing them, it requires fluorescent labeling of the cells you are interested in detecting, which is not always possible.

New Methods

Increasingly, nonculture methods are being used for counting microbes. In chapter 9, you will learn about a technique called the polymerase chain reaction (PCR), which allows scientists to quantify bacteria and other microorganisms that are present in environmental or tissue samples without isolating them and without culturing them. A variation of this technique is proving very useful for counting microbes in a short period of time. In addition, tests that measure ATP, the energy molecule, have been used in the food and pharmaceutical industries for some time, and may hold promise for rapid quantification of microbes in other environmental samples as well.

Case File 7   *Wrap-Up*

Oral rehydration therapy (ORT) is the standard treatment for cholera. It replaces fluids and electrolytes and is vital in the early stages of infection to prevent death through dehydration. Patients can also be given intravenous rehydration if they cannot tolerate the oral therapy. The Centers for Disease Control and Prevention (CDC) and the World Health Organization (WHO) recommend that the rehydration fluids be of low osmolarity to effectively treat malnourished children. Antibiotic treatment with tetracycline is also recommended and should accompany ORT or IV rehydration. A cholera vaccine is available, but it is in short supply and is relatively expensive. Researchers recommend a comprehensive global plan for vaccinating areas of the world such as Haiti that are at risk for cholera to prevent a devastating outbreak from accompanying a natural disaster in the future.

Source: 2011. *New England Journal of Medicine.* Volume 364, p. 3.

7.3 Learning Outcomes—Assess Your Progress

13. Summarize the steps of cell division used by most bacteria; describe another method used by fewer bacterial species.

14. Define *doubling time* and describe how it leads to exponential growth.

15. Compare and contrast the four phases of growth in a bacterial growth curve.

16. Identify one quantitative and one qualitative method used for analyzing bacterial growth.

Chapter Summary

7.1 Microbial Nutrition (ASM Guidelines* 3.1, 5.1, 6.3, 6.4)

- Nutrition is a process by which all living organisms obtain substances from their environment to convert to metabolic uses.
- Although the chemical form of nutrients varies widely, all organisms require six elements—carbon, hydrogen, oxygen, nitrogen, phosphorus, and sulfur—to survive, grow, and reproduce.
- Nutrients are categorized by the amount required (macronutrients or micronutrients), by chemical structure (organic or inorganic), and by their importance to the organism's survival (essential or nonessential).
- Microorganisms are classified both by the chemical form of their nutrients and the energy sources they utilize.
- Nutrients are transported into microorganisms by two kinds of processes: active transport that expends energy and passive transport that occurs independently of energy input.

7.2 Environmental Factors That Influence Microbes (ASM Guidelines 3.1, 3.2, 3.3, 5.1, 5.2)

- The environmental factors that control microbial growth are temperature, pH, moisture, radiation, gases, and the presence of other microorganisms.
- Environmental factors control microbial growth mainly by their influence on microbial enzymes.
- Each microbe has three "cardinal" temperatures described for its growth: its minimum temperature, its maximum temperature, and its optimum temperature at which it grows best.

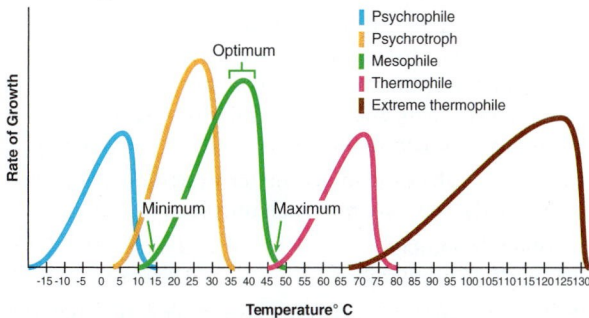

Source: ASM Curriculum Guidelines (American Society for Microbiology, 2012). Complete guidelines in appendix B of this book.

- Microorganisms are classified by their temperature requirements as psychrophiles, mesophiles, or thermophiles. Organisms that can withstand very harsh environments are termed *extremophiles*.
- Most eukaryotic microorganisms are aerobic, while bacteria vary widely in their oxygen requirements from obligately aerobic to anaerobic.
- Microorganisms live in associations with other species that range from mutually beneficial symbiosis to parasitism and antagonism.
- Biofilms are examples of complex synergistic communities of microbes that behave differently than planktonic microorganisms.

7.3 The Study of Microbial Growth (ASM Guidelines 3.3, 3.4, 5.3, 8.4)

- The splitting of a parent bacterial cell to form a pair of similar-size daughter cells is known as binary fission.
- Microbial growth refers both to increase in cell size and increase in number of cells in a population.
- The generation time is a measure of the growth rate of a microbial population. It varies in length according to environmental conditions.
- Microbial cultures in a nutrient-limited batch environment exhibit four distinct stages of growth: the lag phase, the exponential growth (log) phase, the stationary phase, and the death phase.
- Microbial cell populations in a natural environment show distinct phases of growth in response to changing nutrient and waste conditions.
- Population growth can be quantified by measuring turbidity, colony counts, and direct cell counts. Other techniques can be used to count bacteria without growing them.

Multiple-Choice and True-False Questions | Bloom's Levels 1 and 2: Remember and Understand

Multiple-Choice Questions. Select the correct answer from the options provided.

1. The source of the necessary elements of life is
   a. an inorganic environmental reservoir.
   b. the sun.
   c. rocks.
   d. the air.

2. An organism that can synthesize all its required organic components from CO_2 using energy from the sun is a
   a. photoautotroph.   c. chemoautotroph.
   b. photoheterotroph.   d. chemoheterotroph.

3. Chemoautotrophs can survive on _____ alone.
   a. minerals
   b. CO_2
   c. minerals and CO_2
   d. methane

4. Which of the following statements is true for *all* organisms?
   a. They require organic nutrients.
   b. They require inorganic nutrients.
   c. They require growth factors.
   d. They require oxygen gas.

5. A pathogen would most accurately be described as a
   a. parasite.
   b. commensal.
   c. saprobe.
   d. symbiont.

6. Which of the following is true of passive transport?
   a. It requires a gradient.
   b. It uses the cell wall.
   c. It includes endocytosis.
   d. It only moves water.

7. A cell exposed to a hypertonic environment will _____ by osmosis.
   a. gain water
   b. lose water
   c. neither gain nor lose water
   d. burst

8. Psychrophiles would be expected to grow
   a. in hot springs.
   b. on the human body.
   c. at refrigeration temperatures.
   d. at low pH.

9. Superoxide ion is toxic to strict anaerobes because they lack
   a. catalase.
   b. peroxidase.
   c. dismutase.
   d. oxidase.

10. In a viable plate count, each _____ represents a _____ from the sample population.
    a. cell, colony
    b. colony, cell
    c. hour, generation
    d. cell, generation

True-False Questions. If the statement is true, leave as is. If it is false, correct it by rewriting the sentence.

11. Active transport of a substance across a membrane requires a concentration gradient.

12. An organic nutrient essential to an organism's metabolism that it cannot synthesize is called a growth factor.

13. Biofilms often consist of multiple species of bacteria.

14. An obligate halophile is an organism that requires high osmotic pressure.

15. An anaerobe can grow with or without oxygen.

Critical Thinking Questions | Bloom's Levels 3, 4, and 5: Apply, Analyze, and Evaluate

Critical thinking is the ability to reason and solve problems using facts and concepts. These questions can be approached from a number of angles and, in most cases, they do not have a single correct answer.

1. Provide an example of an organism within each of the four main nutritional categories of microbes and describe how each obtains its essential nutrients.

2. Provide evidence in support of or refuting the following statement: Microbial life can exist in the complete absence of sunlight or organic nutrients.

3. a. Compare and contrast passive and active forms of transport, providing an example of a molecule transported by each and the basic requirements of each process.
   b. Describe the process used by a saprobic microbe to transport food particles. How does this compare to how an amoeba feeds?

4. You are working in a laboratory and are told to prepare a blood sample for microscopic analysis. You prepare a small amount of concentrated blood cells and then suspend the cells in sterile water. When you view the slide, you see nothing but what appears to be fragments of cell membranes.
   a. Using principles learned in this chapter, explain why you were not able to see actual red blood cells in your sample.
   b. Discuss whether or not you would have been able to visualize bacterial cells prepared in this manner.

5. a. Define the three cardinal temperatures for a given microbial species.
   b. Explain how these temperatures play a role in pathogenicity, or the ability of a microbe to cause disease.
   c. Illustrate how it is possible for food to spoil even while properly stored at refrigeration temperatures.

6. a. Define the term *osmophile* and provide examples of osmophilic microbes living in the environment and on the human body.
   b. Opening the last jar of strawberry jam that he processed at home last June, Sam finds a lawn of fuzzy growth on the surface of the jam. Explain what type of microbe he is most likely seeing and how it could have survived the high sugar environment within the strawberry preserves.

7. Define each of the following terms and describe where in or on the body you may find such microbes:

   obligate aerobes   anaerobes
   facultative anaerobes   aerotolerant anaerobes

8. Explain what is happening to the bacterial population in the diagram at the top of page 199. Discuss at which point on the graph it would be best to test the effectiveness of a new antibiotic drug.

9. While preparing food for the class picnic, Morgan introduces 20 bacterial cells into the pasta salad.
   a. During the 3 hours prior to the picnic, the salad sits at room temperature in the classroom. How many bacterial cells are now present, assuming that the generation time is 20 minutes?
   b. Using principles learned in this and previous chapters, explain how the microbial contamination of the salad could have been prevented or reduced.

10. What type of ecological association do biofilm communities exhibit? Explain why biofilm infections are so difficult to treat and discuss one potential method for targeting these microbial communities based upon new biofilm research.

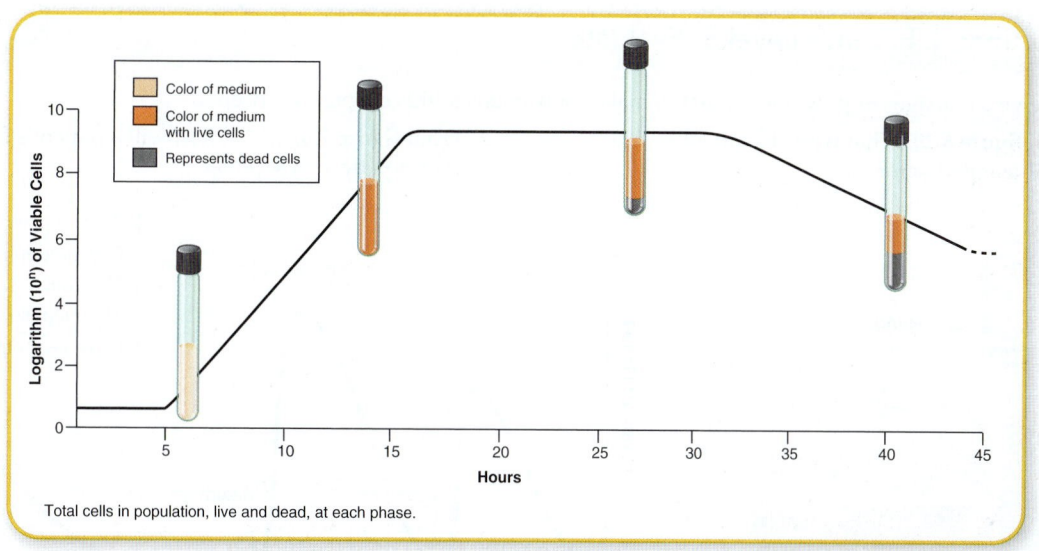

Total cells in population, live and dead, at each phase.

Concept Connections | Bloom's Levels 4 and 6: Analyze and Create

This activity ties together multiple concepts in the chapter.

1. Discuss how oxygen can damage cells.
2. Describe how organisms can overcome the toxic effects of oxygen.

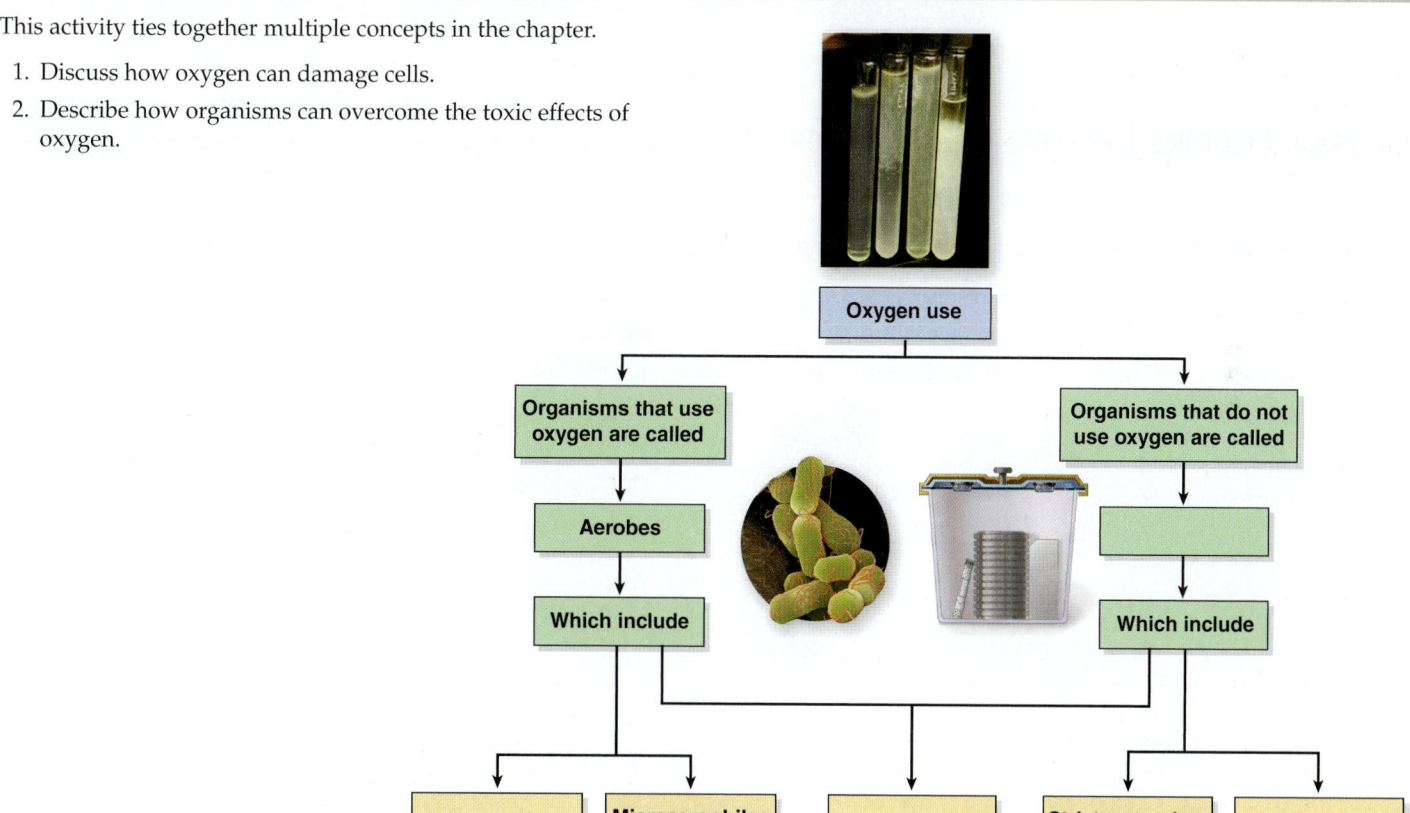

Visual Connections | Bloom's Level 5: Evaluate

These questions use visual images or previous content to make connections to this chapter's concepts.

1. **From chapter 6, figure 6.20.** What type of symbiotic relationship is illustrated here?

2. **From figure 7.6.** What effect will a patient's fever have on infection by a mesophile?

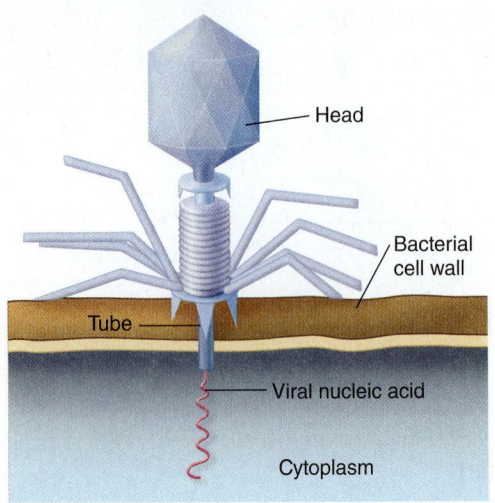

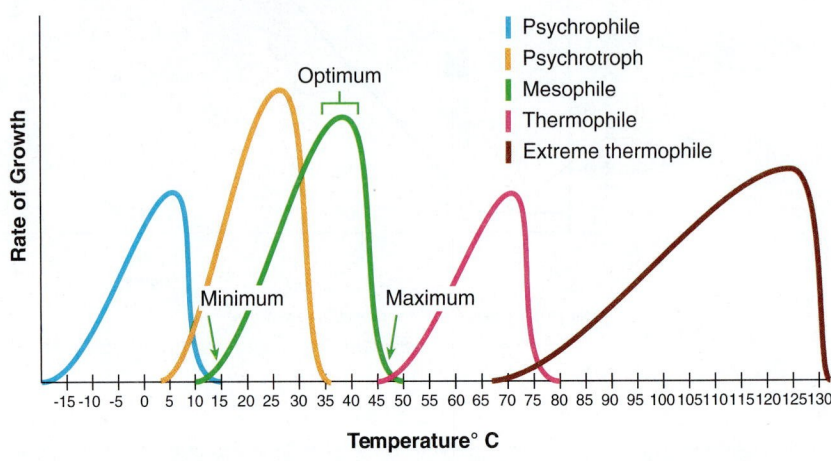

Concept Mapping | Bloom's Level 6: Create

Appendix D provides guidance for working with concept maps.

1. Using the words that follow, please create a concept map illustrating the relationships among these key terms from chapter 7.

symbiosis	parasitism	disease
protection	commensalism	pathogens
mutualism	nonsymbiosis	normal biota

www.mcgrawhillconnect.com

Enhance your study of this chapter with study tools and practice tests. Also ask your instructor about the resources available through ConnectPlus, including the media-rich eBook, interactive learning tools, and animations.

Microbial Metabolism
The Chemical Crossroads of Life

Case File 8

Bacteria and Biofuels

Rising oil prices and rising global temperatures have spurred entrepreneurs and scientists alike to find renewable sources of fuel that have less environmental impact. The most common types of biofuels are ethanol and biodiesel. Biodiesel can be used to power engines directly, and ethanol can be combined with gasoline to reduce carbon emissions. Ethanol is produced in the same way as the ethanol in alcoholic beverages—through fermentation of sugars and distillation. Unfortunately, the traditional methods for mass-producing ethanol from corn or other products through these methods are very expensive, giving biofuels very little advantage over crude oil in terms of cost. Another concern with using this method for biofuel production is that the corn, sugar beets, wheat, and other foods could be used for food for humans or livestock.

Scientists are now looking toward other ways to produce fuel, including using algae, and also solid biomass—the solid portion of plant products made of cellulose—that can be converted into ethanol as the source of a cheaper biofuel alternative. Solid biomass, such as wood chips, grass trimmings, or sawdust, is a waste product and is not considered competition for food production. The problem with using solid biomass is that cellulose is not easily broken down into its component sugars for fermentation. Lignin, the component of woody plants that makes them sturdy and rigid, is particularly difficult to digest.

■ How is glucose extracted from cellulose in plants for use in biofuel production?

■ What organisms are involved in this process?

Continuing the Case appears on page 218.

Outline and Learning Outcomes

8.1 The Metabolism of Microbes

   1. Describe the relationship among metabolism, catabolism, and anabolism.
   2. Fully discuss the structure and function of enzymes.
   3. Differentiate between an apoenzyme and a holoenzyme.

4. Differentiate between an endoenzyme and exoenzyme, and between constitutive and regulated enzymes.

5. Diagram the four major patterns of metabolism.

6. Describe how enzymes are controlled.

8.2 The Pursuit and Utilization of Energy

7. Name the chemical in which energy is stored in cells.

8. Create a general diagram of a redox reaction.

9. Identify electron carriers used by cells.

8.3 Catabolism: Getting Materials and Energy

10. List three basic catabolic pathways and the estimated ATP yield for each.

11. Construct a paragraph summarizing glycolysis.

12. Describe the Krebs cycle and compare the process between bacteria and eukaryotes.

13. Discuss the significance of the electron transport system.

14. State two ways in which anaerobic respiration differs from aerobic respiration.

15. Summarize the steps of microbial fermentation and list three useful products it can create.

16. Describe how noncarbohydrate compounds are catabolized.

8.4 Biosynthesis and the Crossing Pathways of Metabolism

17. Provide an overview of the anabolic stages of metabolism.

18. Define *amphibolism*.

8.5 Photosynthesis: It All Starts with Light

19. Summarize the overall process of photosynthesis in a single sentence.

20. Discuss the relationship between light-dependent and light-independent reactions.

21. Explain the role of the Calvin cycle in the process of photosynthesis.

8.1 The Metabolism of Microbes

Metabolism, from the Greek term *metaballein,* meaning "change," pertains to all chemical reactions and physical workings of the cell. Although metabolism entails thousands of different reactions, most of them fall into one of two general categories. The first, **anabolism,** sometimes also called *biosynthesis,* is any process that results in synthesis of cell molecules and structures. It is a building and bond-making process that forms larger macromolecules from smaller ones, and it usually requires the input of energy. The second, **catabolism,** is the opposite of anabolism. Catabolic reactions break the bonds of larger molecules into smaller molecules and often release energy. In a cell, linking anabolism to catabolism ensures the efficient completion of many thousands of processes.

In summary, metabolism accomplishes the following **(figure 8.1):**

1. It assembles smaller molecules into larger macromolecules needed for the cell; in this process, ATP (energy) is utilized to form bonds (anabolism).

2. It degrades macromolecules into smaller molecules, a process that yields energy (catabolism).

3. It conserves energy in the form of ATP (adenosine triphosphate) or heat.

Enzymes: Catalyzing the Chemical Reactions of Life

A microbial cell could be viewed as a microscopic factory, complete with basic building materials, a source of energy, and a "blueprint" for running its extensive network of metabolic reactions. But the chemical reactions of life, even when highly organized and complex, cannot proceed without a special class of macromolecules called **enzymes.** Enzymes are a remarkable example of **catalysts,** chemicals that increase the rate of a chemical reaction without becoming part of the products or being consumed in the reaction. It is easy to think that an enzyme creates a reaction, but that is not true. Chemical reactions could occur spontaneously at some point even without an enzyme—but at a very slow rate. A study of the enzyme urease shows that it increases the rate of the breakdown of urea by a factor of 100 trillion as compared to an uncatalyzed reaction. Uncatalyzed reactions do not generally occur fast enough for cellular processes. Therefore, enzymes, which speed up the rate of reactions, are indispensable to life. Other major characteristics of enzymes are summarized in **table 8.1.**

How Do Enzymes Work?

An enzyme speeds up the rate of a metabolic reaction, but just how does it do this? During a chemical reaction, reactants

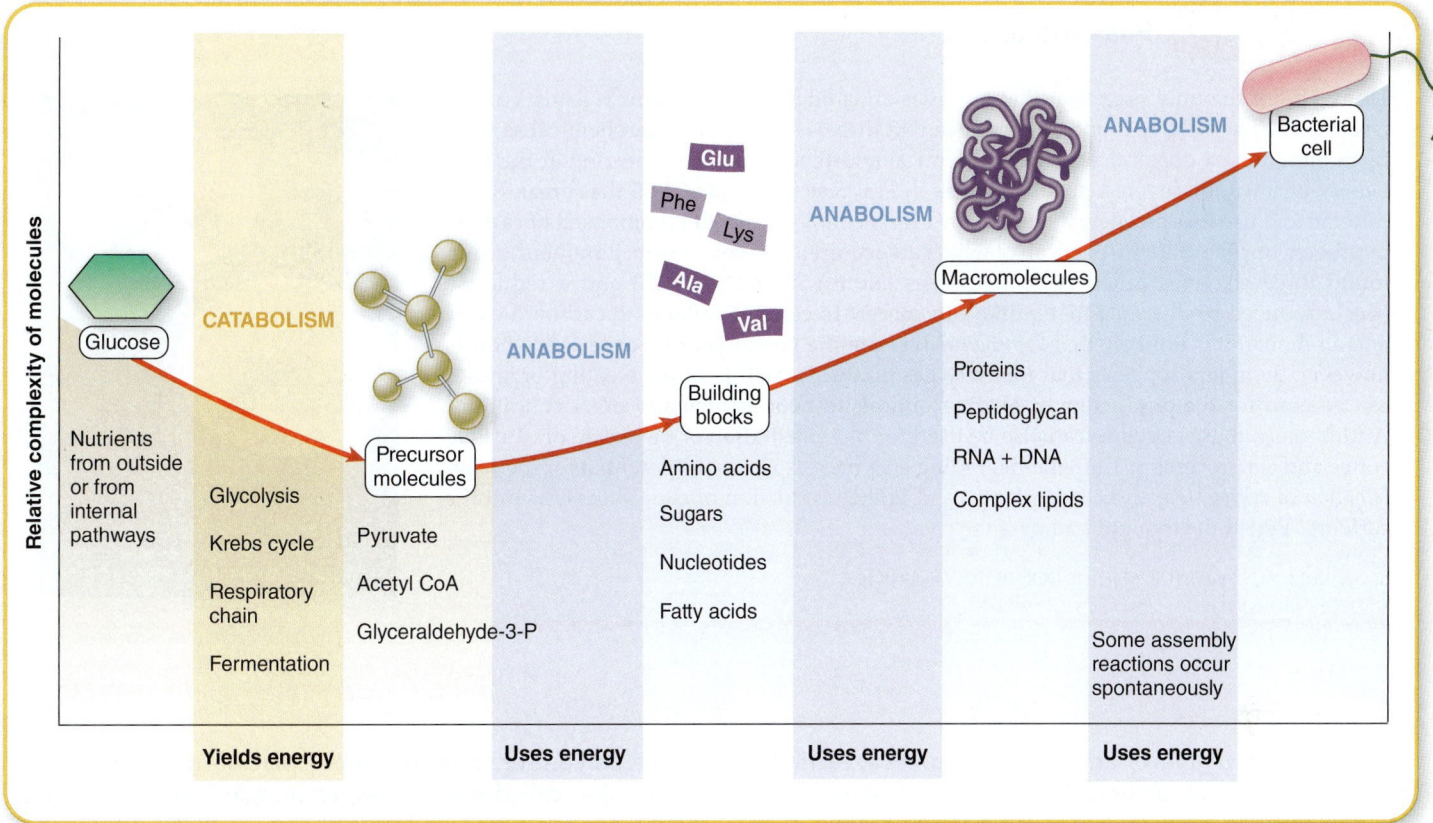

Figure 8.1 Simplified model of metabolism. Cellular reactions fall into two major categories. Catabolism (yellow) involves the breakdown of complex organic molecules to extract energy and form simpler end products. Anabolism (blue) uses the energy to synthesize necessary macromolecules and cell structures from precursors.

are converted to products by bond formation or breakage. A certain amount of energy is required to initiate every such reaction, which limits its rate. This resistance to a reaction, which must be overcome for a reaction to proceed, is measurable and is called the **energy of activation** or activation energy. In the laboratory, overcoming this initial resistance can be achieved by

1. increasing thermal energy (heating) to increase the velocity of molecules,
2. increasing the concentration of reactants to increase the rate of molecular collisions, or
3. adding a catalyst.

In most living systems, the first two alternatives are not feasible, because elevating the temperature is potentially harmful and higher concentrations of reactants are not practical. This leaves only the action of catalysts, and enzymes fill this need efficiently and potently. Enzymatic catalysts effectively lower the energy of activation, allowing a reaction to progress at a faster pace and with reduced energy input **(Insight 8.1).**

Table 8.1 Checklist of Enzyme Characteristics

- Most composed of protein; may require cofactors
- Act as organic catalysts to speed up the rate of cellular reactions
- Lower the activation energy required for a chemical reaction to proceed
- Have unique characteristics such as shape, specificity, and function
- Enable metabolic reactions to proceed at a speed compatible with life
- Have an active site for target molecules called substrates
- Are much larger in size than their substrates
- Associate closely with substrates but do not become integrated into the reaction products
- Are not used up or permanently changed by the reaction
- Can be recycled, thus function in extremely low concentrations
- Are greatly affected by temperature and pH
- Can be regulated by feedback and genetic mechanisms

At the molecular level, an enzyme promotes a reaction by serving as a physical site upon which the reactant molecules, called **substrates,** can be positioned for various interactions. The enzyme is much larger in size than its substrate, and it presents a unique active site that fits only that particular substrate. Although an enzyme binds to the substrate and participates directly in changes to the substrate, it does not become a part of the products, is not used up by the reaction, and can function over and over again. *Enzyme speed,* defined as the number of substrate molecules converted per enzyme per second, is well documented. Speeds range from several million for catalase to a thousand for lactate dehydrogenase.

Enzyme Structure

Most enzymes are proteins, and they can be classified as simple or conjugated. Simple enzymes consist of protein alone, whereas conjugated enzymes **(figure 8.2)** contain protein and some other nonprotein molecule or molecules. A conjugated enzyme, sometimes referred to as a **holoenzyme,** is a combination of the protein, called the **apoenzyme** in these

cases, and one or more **cofactors.** Cofactors are either organic molecules, called **coenzymes,** or inorganic elements (metal ions). For example, catalase, an enzyme that we learned in chapter 7 breaks down hydrogen peroxide, requires iron as a metallic cofactor.

In the early 1980s, a special class of enzymes was identified and found to be made of RNA. Named **ribozymes,** these molecules are remarkable because they are RNA molecules that catalyze reactions on other RNA. Ribozymes are thought to be remnants of the earliest molecules on earth that could have served as both catalysts and genetic material. Their discovery has lent support for what is known today as the "RNA hypothesis," which states that RNA was in fact the first genetic material within ancient cells. In natural systems, ribozymes are involved in self-splicing or cutting of RNA molecules during final processing of the genetic code (see chapter 9). Ribozymes can inhibit gene expression, and the development of ribozyme-based therapies for the treatment of cancer and HIV infection is ongoing.

Apoenzymes: Specificity and the Active Site

Apoenzymes range in size from small polypeptides with about 100 amino acids and a molecular weight of 12,000 to large polypeptide conglomerates with thousands of amino acids and a molecular weight of over 1 million. Like all proteins, an apoenzyme exhibits levels of molecular complexity called the primary, secondary, tertiary, and—in larger enzymes—quaternary organization **(figure 8.3).** As we saw in chapter 2, the first three levels of structure arise when a single polypeptide chain undergoes an automatic folding process and achieves stability by forming disulfide and other types of bonds. The actual site where the substrate binds is a crevice or groove called the **active site,** or **catalytic site,**

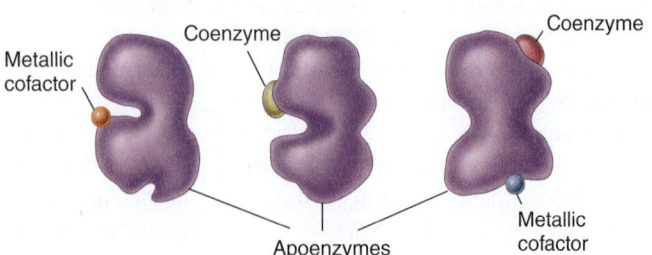

Figure 8.2 Conjugated enzyme structure. All have an apoenzyme (polypeptide or protein) component and one or more cofactors.

Levels of Structure

Primary	Secondary	Tertiary

(a) As the polypeptide forms intrachain bonds, it folds into a three-dimensional (tertiary) state. Active sites (AS) are created by the 3D shape.

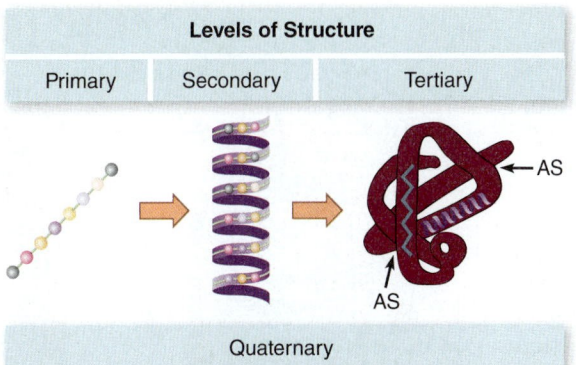

Quaternary

(b) More complex enzymes have a quaternary structure consisting of several polypeptides bound by weak forces. Often the active site is formed by the junction of two polypeptides.

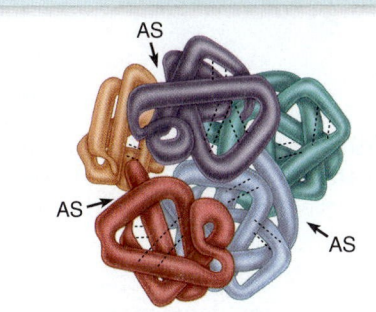

Figure 8.3 How the active site and specificity of the apoenzyme arise. The active site is always formed by the three-dimensional structure of the tertiary or quaternary folding, which means that amino acids that may be distant from one another in the primary structure can be adjacent in the active site.

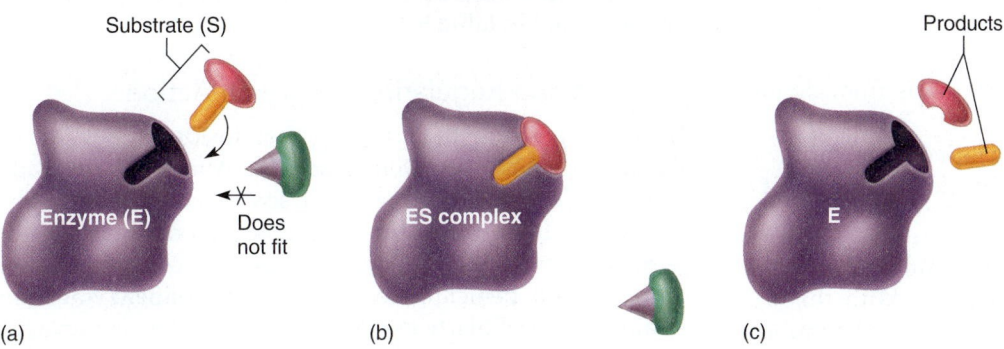

Figure 8.4 Enzyme-substrate reactions. (a) When the enzyme and substrate come together, the substrate (S) must show the correct fit and position with respect to the enzyme (E). (b) When the ES complex is formed, it enters a transition state. During this temporary but tight interlocking union, the enzyme participates directly in breaking or making bonds. (c) Once the reaction is complete, the enzyme releases the products.

and there can be from one to several such sites (as shown in figure 8.3). The three-dimensional shape of each site is formed by the way the amino acid chain or chains are folded. Each type of enzyme has a different primary structure (type and sequence of amino acids), variations in folding, and unique active sites.

Enzyme-Substrate Interactions

For a reaction to take place, the substrate has to nestle into the active site **(figure 8.4).** The fit is so specific that it is often described as a "lock-and-key" fit in which the substrate is inserted into the active site's pocket.

The bonds formed between the substrate and enzyme are weak, but there are many of them. Once the enzyme-substrate complex has formed, appropriate reactions occur on the substrate, often with the aid of a cofactor, and a product is formed and released. The enzyme can then attach to another substrate molecule and repeat this action. Although enzymes can potentially catalyze reactions in both directions, most examples in this chapter depict them working in one direction only.

Cofactors: Supporting the Work of Enzymes

In chapter 7, you learned that microorganisms require specific metal ions called trace elements and certain organic growth factors. In many cases, the need for these substances arises from their roles as cofactors for enzymes. The metallic cofactors, including iron, copper, magnesium, manganese, zinc, cobalt, selenium, and many others, assist with precise functions between the enzyme and its substrate. In general, metals activate enzymes, help bring the active site and substrate close together, and participate directly in chemical reactions with the enzyme-substrate complex.

Coenzymes are a type of cofactor. They are organic compounds that work together with an apoenzyme to perform the necessary alteration of a substrate. The general function of a coenzyme is to remove a chemical group from one substrate molecule and add it to another substrate, thereby serving as a transient carrier of this group. In a later section of this chapter, we shall see that coenzymes carry and transfer hydrogen atoms, electrons, carbon dioxide, and amino groups. One of the most important components of coenzymes is **vitamins,** which explains why vitamins are important to nutrition and may be required as growth factors for living things. Vitamin deficiencies prevent the complete holoenzyme from forming. Consequently, both the chemical reaction and the structure or function dependent upon that reaction are compromised.

Classification of Enzyme Functions

Most metabolic reactions require separate and unique enzymes. A standardized system of nomenclature and classification was developed to keep things clear. In general, an enzyme name is composed of two parts: (1) a prefix or stem word derived from a certain characteristic—usually the substrate acted upon or the type of reaction catalyzed, or both—followed by (2) the ending *-ase.*

The system classifies the enzyme in one of these six classes, on the basis of its general biochemical action:

1. *Oxidoreductases* transfer electrons from one substrate to another, and *dehydrogenases* transfer a hydrogen from one compound to another.
2. *Transferases* transfer functional groups from one substrate to another.
3. *Hydrolases* cleave bonds on molecules with the addition of water.
4. *Lyases* add groups to or remove groups from double-bonded substrates.
5. *Isomerases* change a substrate into its isomeric[1] form.
6. *Ligases* catalyze the formation of bonds with the input of ATP and the removal of water.

Each enzyme is also assigned a common name that indicates the specific reaction it catalyzes. With this system, an enzyme that digests a carbohydrate substrate is a *carbohydrase;* a specific carbohydrase, *amylase,* acts on starch (amylose is a major component of starch). An enzyme that hydrolyzes peptide bonds of a protein is a *proteinase, protease,* or *peptidase,* depending on the size of the protein substrate. Some fats and other lipids are digested by *lipases.* DNA is hydrolyzed by *deoxyribonuclease,* generally shortened to *DNase.* A *synthetase* or *polymerase* bonds together many small

---

1. An isomer is a compound that has the same molecular formula as another compound but differs in arrangement of the atoms.

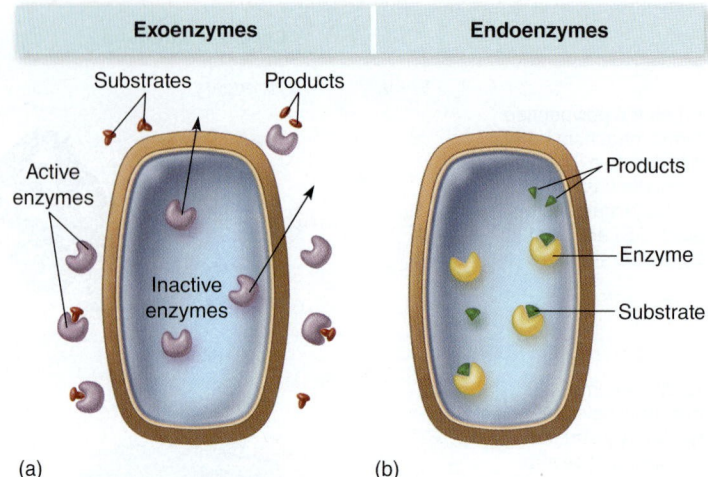

Figure 8.5 Types of enzymes, as described by their location of action. **(a)** Exoenzymes are released outside the cell to function. **(b)** Endoenzymes remain in the cell and function there.

molecules into large molecules. Other examples of enzymes are presented in **table 8.2**.

Location and Regularity of Enzyme Action

Enzymes perform their tasks either inside or outside of the cell in which they were produced. After they are made inside the cell, **exoenzymes** are transported extracellularly, where they break down (hydrolyze) large food molecules or harmful chemicals. Examples of exoenzymes are cellulase, amylase, and penicillinase. By contrast, **endoenzymes** are retained intracellularly and function there. Most enzymes of the metabolic pathways are endoenzymes **(figure 8.5)**.

Enzymes are not all produced in equal amounts or at equal rates. Some, called **constitutive enzymes (figure 8.6a)**, are always present and in relatively constant amounts, regardless of the cellular environment. The enzymes involved in utilizing glucose, for example, are very important in metabolism and thus are constitutive. Other enzymes are **regulated enzymes (figure 8.6b)**, the production of which is either turned on (induced) or turned off (repressed) in response to changes in concentration of the substrate.

Table 8.2 A Sampling of Enzymes, Their Substrates, and Their Reactions

Common Name	Systematic Name	Enzyme Class	Substrates	Action
Lactase	β-D-galactosidase	Hydrolase	Lactose	Breaks lactose down into glucose and galactose
Penicillinase	Beta-lactamase	Hydrolase	Penicillin	Hydrolyzes beta-lactam ring
DNA polymerase	DNA nucleotidyl-transferase	Transferase	DNA nucleosides	Synthesizes a strand of DNA using the complementary strand as a model
Lactate dehydrogenase	Same as common name	Oxidoreductase	Pyruvic acid	Catalyzes the conversion of pyruvic acid to lactic acid
Oxidase	Cytochrome oxidase	Oxidoreductase	Molecular oxygen	Catalyzes the reduction of O_2 (addition of electrons and hydrogen)

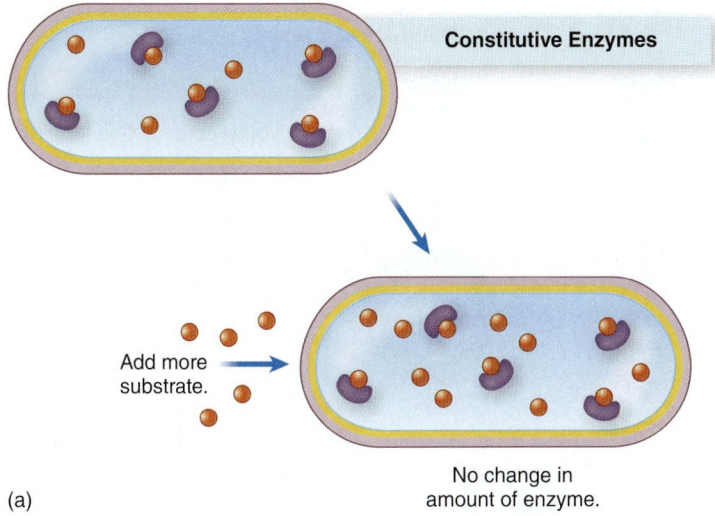

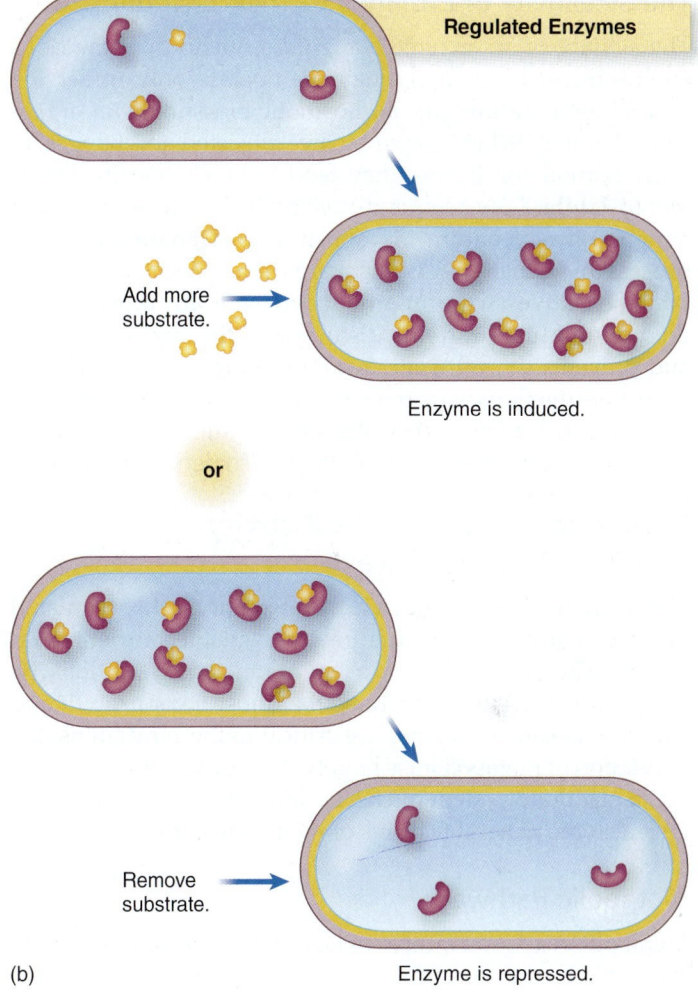

Figure 8.6 Constitutive and regulated enzymes.
(a) Constitutive enzymes are present in constant amounts in a cell. The addition of more substrate does not increase the numbers of these enzymes. **(b)** The concentration of regulated enzymes in a cell increases or decreases in response to substrate levels.

Transfer Reactions by Enzymes Other enzyme-driven processes that involve the simple addition or removal of a functional group are important to the overall economy of the cell. Oxidation-reduction and other transfer activities are examples of these types of reactions.

Some atoms and compounds readily give or receive electrons and participate in oxidation (the loss of electrons) or reduction (the gain of electrons). The compound that loses the electrons is **oxidized,** and the compound that receives the electrons is **reduced.** Such oxidation-reduction (redox) reactions are common in the cell and indispensable to the energy transformations discussed later in this chapter (see also Insight 2.2). Important components of cellular redox reactions are oxidoreductases, which remove electrons from one substrate and add them to another. Their coenzyme carriers are nicotinamide adenine dinucleotide (NAD) and flavin adenine dinucleotide (FAD). Oxidation and reduction are covered in more detail later in this chapter.

The Role of Microbial Enzymes in Disease Many pathogens secrete unique exoenzymes that help them avoid host defenses or promote their multiplication in tissues. Because these enzymes contribute to pathogenicity, they are referred to as virulence factors, or toxins in some cases. *Streptococcus pyogenes* (a cause of throat and skin infections) produces a streptokinase that digests blood clots and apparently assists in invasion of wounds. Another exoenzyme from this bacterium is called streptolysin. In mammalian hosts, streptolysin damages blood cells and tissues. It is also responsible for lysing red blood cells used in blood agar dishes, and this trait is used for identifying the bacteria growing in culture (see chapter 21). *Pseudomonas aeruginosa,* a respiratory and skin pathogen, produces elastase and collagenase, which digest elastin and collagen, two proteins found in connective tissue. These increase the severity of certain lung diseases and burn infections. *Clostridium perfringens,* an agent of gas gangrene, synthesizes lecithinase C, a lipase that profoundly damages cell membranes and accounts for the tissue death associated with this disease. Not all microbial enzymes digest tissues; some, such as penicillinase, inactivate penicillin and thereby protect a microbe from its effects.

Disease Connection

Infections with penicillinase-producing bacteria, which were resistant to the antibiotic penicillin, first caught the public attention in the 1980s. However, its presence in bacteria was first established in 1940, before penicillin had been introduced to the public. This should not be a surprise, though. Penicillin is a natural product produced by the fungus *Penicillium*. Bacteria coexisting with the fungus in the environment would be expected to develop a strategy against the substance that is toxic to them.

The Sensitivity of Enzymes to Their Environment

The activity of an enzyme is highly influenced by the cell's environment. In general, enzymes operate only under the natural temperature, pH, and osmotic pressure of an organism's habitat. When enzymes are subjected to changes in these normal conditions, they tend to be chemically unstable, or **labile.** Low temperatures inhibit catalysis, and high temperatures denature the apoenzyme. **Denaturation** is a process by which the weak bonds that collectively maintain the native shape of the apoenzyme are broken. This disruption causes extreme distortion of the enzyme's shape and prevents the substrate from attaching to the active site. Such nonfunctional enzymes block metabolic reactions and thereby can lead to cell death. Low or high pH or certain chemicals (heavy metals, alcohol) are also denaturing agents.

Regulation of Enzymatic Activity and Metabolic Pathways

Metabolic reactions proceed in a systematic, highly regulated manner that maximizes the use of available nutrients and energy. The cell responds to environmental conditions by using those metabolic reactions that most favor growth and survival. Because enzymes are critical to these reactions, the regulation of metabolism is largely the regulation of enzymes by an elaborate system of checks and balances. Let us take a look at some general features of metabolic pathways.

Metabolic Pathways

Metabolic reactions rarely consist of a single action or step. More often, they occur in a multistep series or pathway, with each step catalyzed by an enzyme. An individual reaction is shown in various ways, depending on the purpose at hand **(figure 8.7).** The product of one reaction is often the reactant (substrate) for the next, forming a linear chain of reactions. Many pathways have branches that provide alternate methods for nutrient processing. Others take a cyclic form, in which the starting molecule is regenerated to initiate another turn of the cycle. On top of that, pathways generally do not stand alone; they are interconnected and merge at many sites.

Direct Controls on the Action of Enzymes

The bacterial cell has many ways of directly influencing the activity of its enzymes. It can inhibit enzyme activity by supplying a molecule that resembles the enzyme's normal substrate. The "mimic" can then occupy the enzyme's active site, preventing the actual substrate from binding there. Because the mimic cannot actually be acted on by the enzyme or function in the way the product would have, the enzyme is effectively shut down. This form of inhibition is called **competitive inhibition,** because the mimic is competing with the substrate for the binding site **(figure 8.8).** (In chapter 12, you will see that some antibiotics use the same strategy of competing with enzymatic active sites to shut down metabolic processes.)

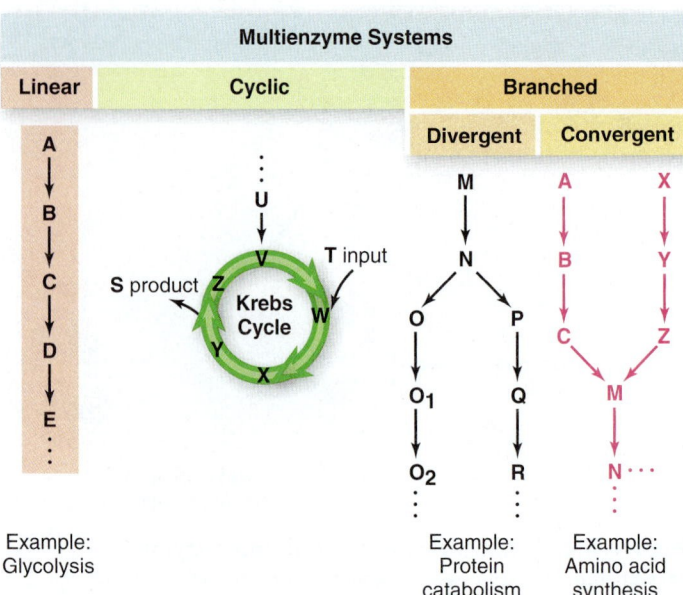

Figure 8.7 Patterns of metabolism. In general, metabolic pathways consist of a linked series of individual chemical reactions that produce intermediary metabolites and lead to a final product. These pathways occur in several patterns, including linear, cyclic, and branched. Anabolic pathways involved in biosynthesis result in a more complex molecule, each step adding on a functional group, whereas catabolic pathways involve the dismantling of molecules and can generate energy. Virtually every reaction in a series—represented by an arrow—involves a specific enzyme.

Another form of inhibition can occur with special types of enzymes that have two binding sites—the active site and another area called the regulatory site (as shown in figure 8.8). These enzymes are regulated by the binding of molecules other than the substrate in their regulatory sites. Often, the regulatory molecule is the product of the enzymatic reaction itself. This provides a negative feedback mechanism that can slow down enzymatic activity once a certain concentration of product is produced. This is **noncompetitive inhibition,** because the regulator molecule does not bind in the same site as the substrate.

Controls on Enzyme Synthesis

Controlling enzymes by controlling their synthesis is another effective mechanism, because enzymes do not last indefinitely. Some wear out, some are deliberately degraded, and others are diluted with each cell division. For catalysis to continue, enzymes eventually must be replaced. This cycle works into the scheme of the cell, where replacement of enzymes can be regulated according to cell demand.

Enzyme repression is a means to stop further synthesis of an enzyme somewhere along its pathway. As the level of the end product from a given enzymatic reaction has built to excess, the genetic apparatus responsible for replacing these enzymes is automatically suppressed **(process figure 8.9).** The response time is longer than for feedback inhibition, but its effects are more enduring.

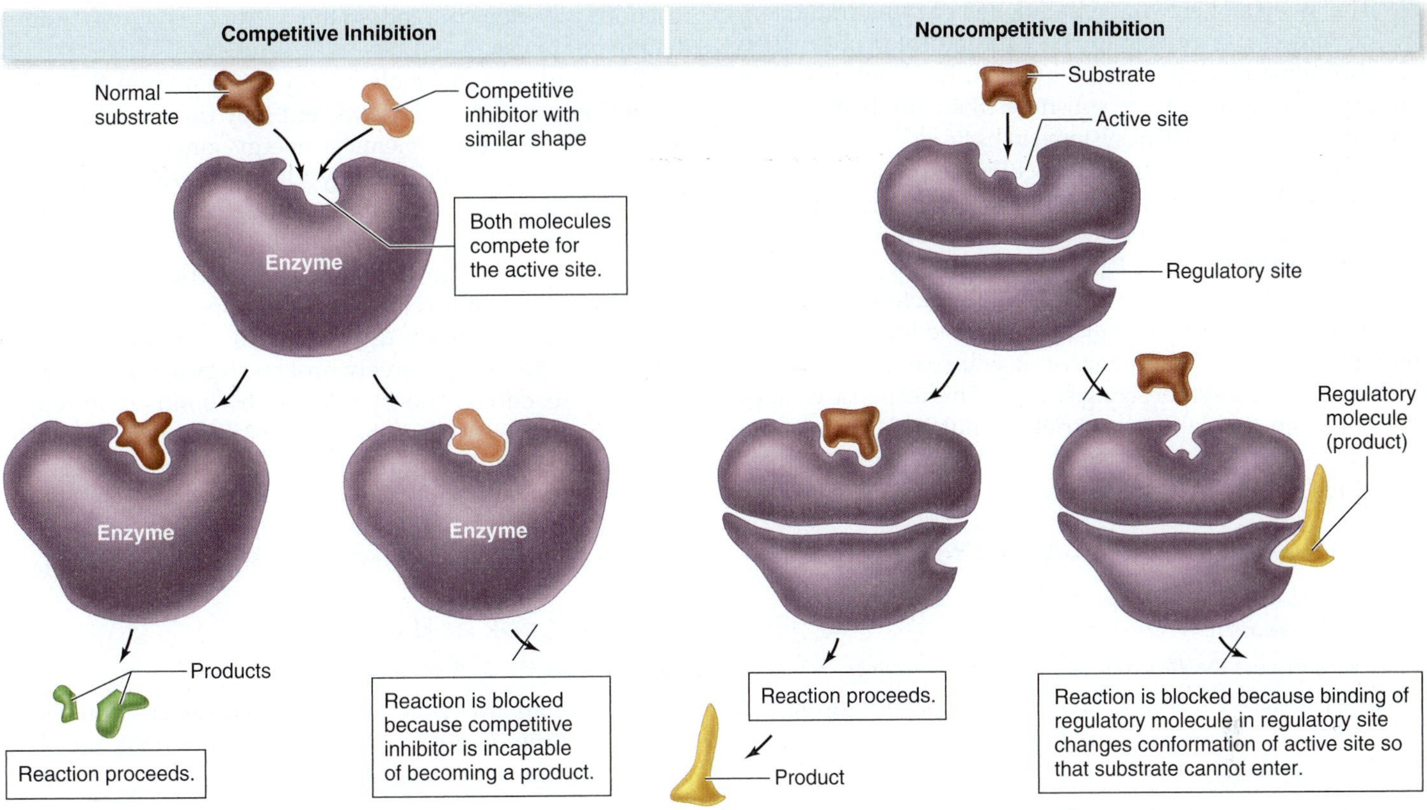

Competitive Inhibition | **Noncompetitive Inhibition**

Normal substrate

Competitive inhibitor with similar shape

Both molecules compete for the active site.

Enzyme

Substrate

Active site

Enzyme

Regulatory site

Enzyme

Enzyme

Products

Reaction proceeds.

Reaction is blocked because competitive inhibitor is incapable of becoming a product.

Reaction proceeds.

Regulatory molecule (product)

Product

Reaction is blocked because binding of regulatory molecule in regulatory site changes conformation of active site so that substrate cannot enter.

Figure 8.8 Examples of two common control mechanisms for enzymes.

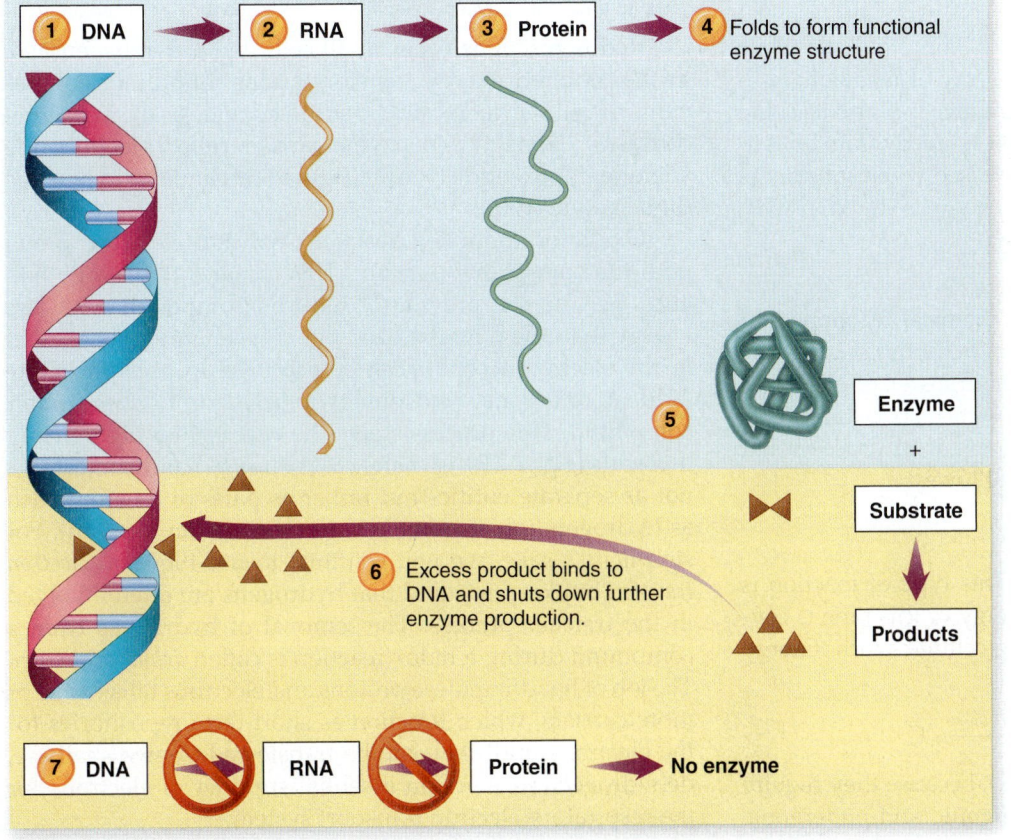

1 DNA → 2 RNA → 3 Protein → 4 Folds to form functional enzyme structure

5 Enzyme
+
Substrate

6 Excess product binds to DNA and shuts down further enzyme production.

Products

7 DNA ⊘ RNA ⊘ Protein → No enzyme

Process Figure 8.9 One type of genetic control of enzyme synthesis: enzyme repression. 1–5: The enzyme is synthesized continuously via uninhibited transcription and translation (chapter 9) until enough product has been made. 6, 7: Excess product reacts with a site on DNA that regulates the enzyme's synthesis, thereby inhibiting further enzyme production.

The inverse of enzyme repression is **enzyme induction.** In this process, enzymes appear (are induced) only when suitable substrates are present—that is, the synthesis of an enzyme is induced by its substrate. Both mechanisms are important genetic control systems in bacteria.

A classic model of enzyme induction occurs in the response of *Escherichia coli* to certain sugars. For example, if a particular strain of *E. coli* is inoculated into a medium whose principal carbon source is lactose, it will produce the enzyme lactase to hydrolyze it into glucose and galactose. If the bacterium is subsequently inoculated into a medium containing only sucrose as a carbon source, it will cease synthesizing lactase and begin synthesizing sucrase. This response enables the organism to adapt to a variety of nutrients, and it also prevents a microbe from wasting energy by making enzymes for which no substrates are present.

8.1 Learning Outcomes—Assess Your Progress

1. Describe the relationship among metabolism, catabolism, and anabolism.
2. Fully discuss the structure and function of enzymes.
3. Differentiate between an apoenzyme and a holoenzyme.
4. Differentiate between an endoenzyme and an exoenzyme, and between constitutive and regulated enzymes.
5. Diagram the four major patterns of metabolism.
6. Describe how enzymes are controlled.

8.2 The Pursuit and Utilization of Energy

In order to carry out the work of an array of metabolic processes, cells require constant input and expenditure of some form of usable energy. The energy comes directly from light or is contained in chemical bonds and released when substances are catabolized, or broken down. The energy is stored in ATP.

Energy in Cells

Cells manage energy in the form of chemical reactions that change molecules. This often involves activities such as the making or breaking of bonds and the transfer of electrons. Not all cellular reactions are equal with respect to energy. Some release energy, and others require it to proceed. For example, a reaction that proceeds as follows:

$$X + Y \xrightarrow{\text{Enzyme}} Z + \text{Energy}$$

releases energy as it goes forward. This type of reaction is termed **exergonic** (ex-er-gon´-ik). Energy of this type is considered free—it is available for doing cellular work. Energy transactions such as the following:

$$\text{Energy} + A + B \xrightarrow{\text{Enzyme}} C$$

are called **endergonic** (en-der-gon´-ik), because they require the addition of energy. In cells, exergonic and endergonic

reactions are often coupled, so that released energy is immediately put to use.

Summaries of metabolism may make it seem that cells "create" energy from nutrients, but they do not. What they actually do is extract chemical energy already present in nutrients and apply that energy toward useful work in the cell, much like a gasoline engine releases energy as it burns fuel. The engine does not actually produce energy, but it converts some of the potential energy to do work.

At the simplest level, cells possess specialized enzyme systems that trap the energy present in the bonds of nutrients as they are progressively broken **(figure 8.10).** During exergonic reactions, energy released by bonds is stored in certain high-energy phosphate bonds, such as in ATP. The ability of ATP to temporarily store and release the energy of chemical bonds fuels endergonic cell reactions. Before discussing ATP, we examine redox reactions, which provide the electrons that are critical to energy production.

A Closer Look at Biological Oxidation and Reduction

Redox reactions always occur in pairs, with an electron donor and an electron acceptor, which constitute a *redox pair.* The reaction can be represented as follows:

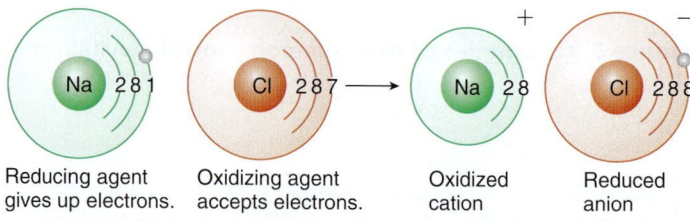

| Reducing agent gives up electrons. | Oxidizing agent accepts electrons. | Oxidized cation | Reduced anion |

Redox reactions occur in all cells and are indispensable to the required energy transformations. Important components of cellular redox reactions are enzymes called oxidoreductases. They have coenzyme carriers called nicotinamide adenine dinucleotide (NAD) **(figure 8.11)** and flavin adenine dinucleotide (FAD).

Oxidation-reduction reactions salvage electrons along with the energy they contain. This changes the energy balance, leaving the previously reduced compound with less energy than the now oxidized one. The energy now present in the electron acceptor can be captured to **phosphorylate** (add an inorganic phosphate) to ADP or to some other compound. This process stores the energy in a high-energy molecule (e.g., ATP). In many cases, the cell handles electrons not as separate entities but rather as parts of an atom such as hydrogen (which contains a proton and an electron). For simplicity's sake, we will continue to use the term *electron transfer,* but keep in mind that hydrogens are often involved in the transfer process. The removal of hydrogens from a compound during a redox reaction is called *dehydrogenation.* The job of handling these protons and electrons falls to one or more carriers, which function as short-term repositories for the electrons until they can be transferred. As we shall see, dehydrogenations are an essential supplier of electrons for the respiratory electron transport system.

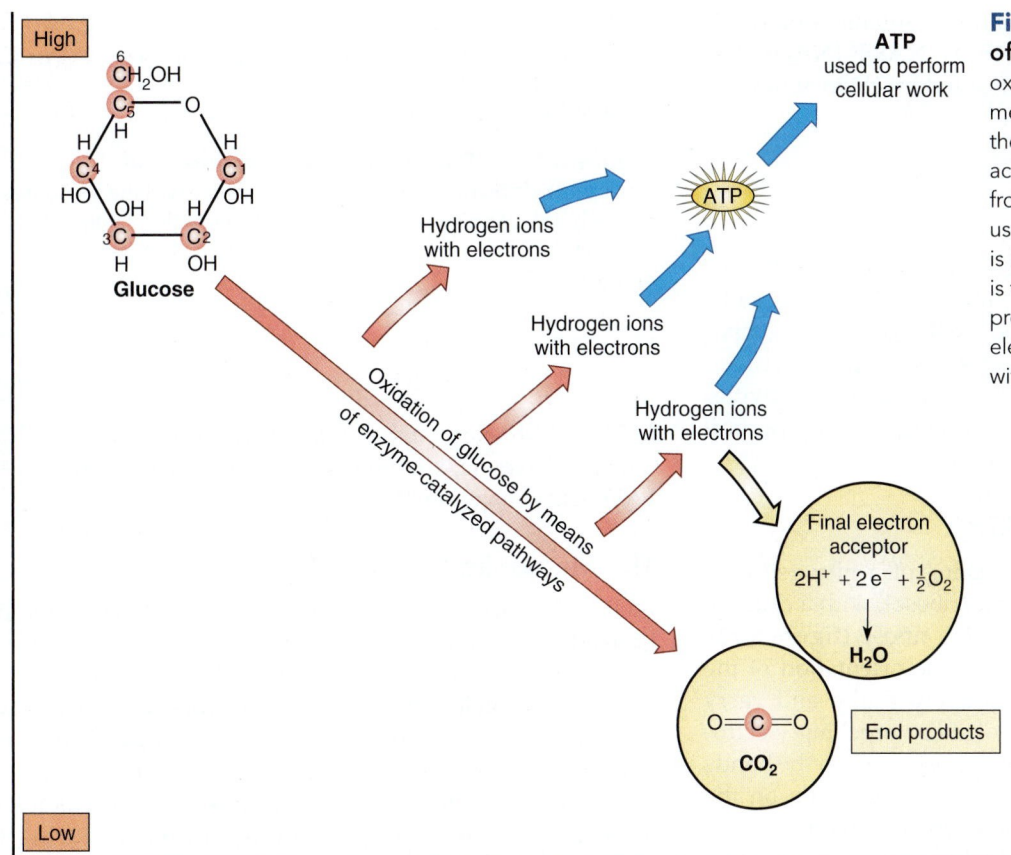

Figure 8.10 **A simplified model of energy production.** Glucose is oxidized as it passes through sequential metabolic pathways, resulting in the removal of hydrogens and their accompanying electrons. The energy from the hydrogens and electrons is used to generate ATP. Eventually, all that is left of the carbon skeleton of glucose is the end product CO_2. Another by-product of aerobic metabolism (due to electrons and hydrogen ions combining with oxygen) is H_2O.

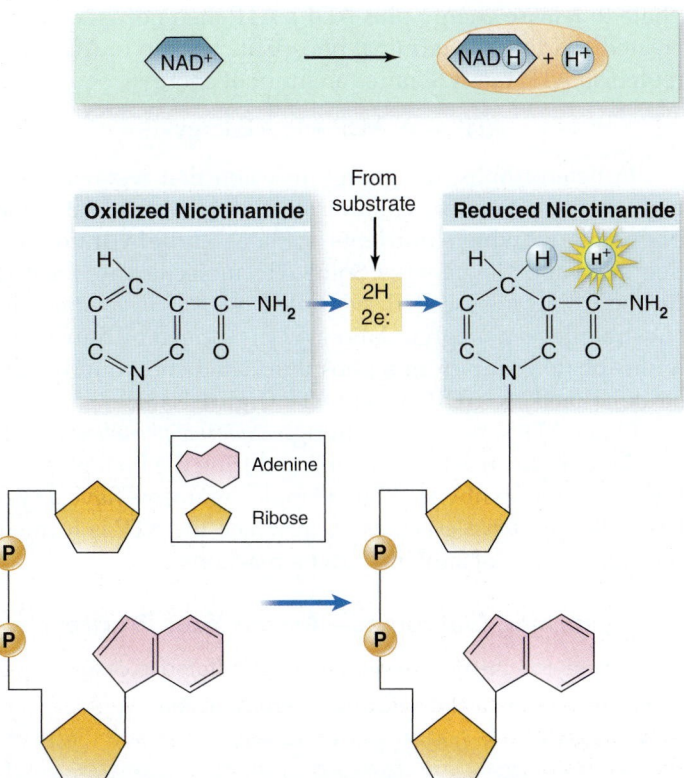

Figure 8.11 **Details of NAD reduction.** The coenzyme NAD contains the vitamin nicotinamide (niacin) and the purine adenine attached to two ribose phosphate molecules. The principal site of action is on the nicotinamide (boxed areas). Hydrogens and electrons donated by a substrate interact with a carbon on the top of the ring. One hydrogen bonds there, carrying two electrons (H:), and the other hydrogen is carried in solution as H^+ (a proton).

Electron Carriers: Molecular Shuttles

Electron carriers resemble shuttles that are alternately loaded and unloaded, repeatedly accepting and releasing electrons and hydrogens to facilitate the transfer of redox energy. Most carriers are coenzymes that transfer both electrons and hydrogens, but some transfer electrons only. The most common carrier is NAD, which carries hydrogens (and a pair of electrons) from dehydrogenation reactions (as shown in figure 8.11). Reduced NAD can be represented in various ways. Because 2 hydrogens are added, the actual carrier state is $NADH + H^1$, but this is cumbersome, so we will write it as "NADH." In catabolic pathways, electrons are extracted and carried through a series of redox reactions until the final electron acceptor at the end of a particular pathway is reached (see figure 8.10). In aerobic metabolism, this acceptor is molecular oxygen; in anaerobic metabolism, it is some other inorganic or organic compound. Other common redox carriers are FAD, NADP (NAD phosphate), and the compounds of the respiratory chain, which are fixed into membranes.

Adenosine Triphosphate: Metabolic Money

Let's look more closely at the energy molecule ATP. ATP has been described as "metabolic money" because it can be earned, banked, saved, spent, and exchanged. As a temporary energy repository, ATP provides a connection between energy-yielding catabolism and the other cellular activities that require energy. Some clues to its energy-storing properties lie in its unique molecular structure.

The Molecular Structure of ATP

ATP is a three-part molecule consisting of a nitrogen base (adenine) linked to a 5-carbon sugar (ribose), with a chain of three phosphate groups bonded to the ribose **(figure 8.12)**. The high energy of ATP comes from the orientation of the phosphate groups, which are relatively bulky and carry negative charges. The proximity of these repelling electrostatic charges imposes a strain that is most acute on the bonds between the last two phosphate groups. The strain on the phosphate bonds accounts for the energetic quality of ATP because removal of the terminal phosphates releases free energy.

Breaking the bonds between the two outermost phosphates of ATP yields adenosine diphosphate (ADP), which is then converted to adenosine monophosphate

Figure 8.13 ATP formation by substrate-level phosphorylation.   The inorganic phosphate and the substrates form a bond with high potential energy. In a reaction catalyzed enzymatically, the phosphate is transferred to ADP, thereby producing ATP.

(AMP). AMP derivatives help form the backbone of RNA and are also a major component of certain coenzymes (NAD, FAD, and coenzyme A).

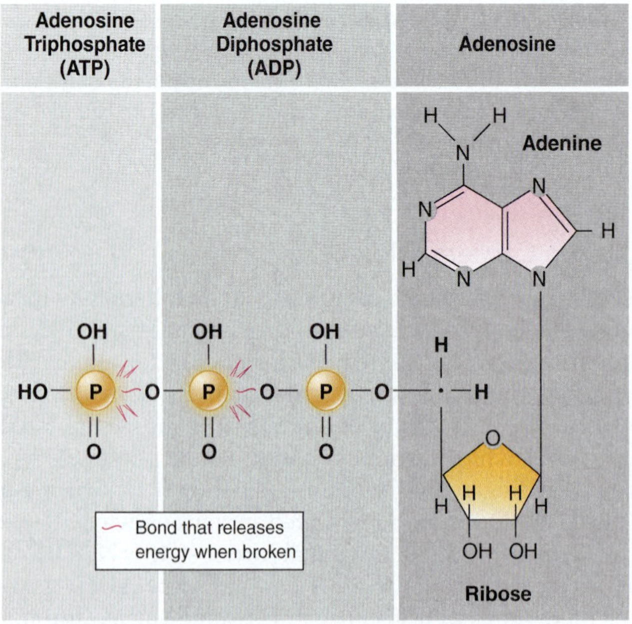

The Metabolic Role of ATP

ATP is the primary energy currency of the cell; when it is used in a chemical reaction, it must then be replaced. Therefore, ATP utilization and replenishment make up an ongoing cycle. Often, the energy released during ATP hydrolysis drives biosynthesis by activating individual substrates before they are enzymatically linked together. ATP is also used to prepare molecules for catabolism such as when a 6-carbon sugar is phosphorylated during the early stages of glycolysis.

$$\text{Glucose} \xrightarrow{\quad \overset{\text{ATP} \quad \text{ADP}}{\curvearrowright} \quad} \text{Glucose-6-phosphate}$$

When ATP is utilized (the removal of the terminal phosphate to release energy plus ADP), ATP then needs to be re-created. Adding the terminal phosphate back in to ADP will replenish ATP, but it requires an input of energy:

$$\text{ATP} \rightleftharpoons \text{ADP} + P_i + \text{Energy}$$

In heterotrophs, the energy infusion that regenerates a high-energy phosphate comes from certain steps of catabolic pathways in which nutrients such as carbohydrates are degraded and yield energy. Some ATP molecules are formed through a process called *substrate-level phosphorylation*. In substrate-level phosphorylation, ATP is formed by transfer of a phosphate group from a phosphorylated compound (substrate) directly to ADP to yield ATP **(figure 8.13)**.

Other ATPs are formed through *oxidative phosphorylation*, a series of redox reactions occurring during the final phase of the respiratory pathway. Phototrophic organisms have a system called *photophosphorylation*, in which the ATP is formed through a series of sunlight-driven reactions.

8.2 Learning Outcomes—Assess Your Progress

7. Name the chemical in which energy is stored in cells.

8. Create a general diagram of a redox reaction.

9. Identify electron carriers used by cells.

Adenosine Triphosphate (ATP)	Adenosine Diphosphate (ADP)	Adenosine

Figure 8.12 The structure of adenosine triphosphate (ATP).   Removing the left-most phosphate group yields ADP; removing the next one yields AMP.

8.3 Catabolism: Getting Materials and Energy

Now you have an understanding of all the tools a cell needs to *metabolize*. Metabolism uses *enzymes* to catalyze reactions that break down (*catabolize*) organic molecules to materials (*precursor molecules*) that cells can then use to build (*anabolize*) larger, more complex molecules that are particularly suited to them. This process is presented symbolically in figure 8.1, which is repeated as an icon in this section to guide the discussion. Another very important point about metabolism is that *reducing power* (the electrons available in NADH and FADH$_2$) and *energy* (stored in the bonds of ATP) are needed in large quantities for the anabolic parts of metabolism (the blue bars in our figure). They are produced during the catabolic part of metabolism (the yellow bar).

Metabolism starts with "nutrients" from the environment, usually discarded molecules from other organisms. Cells have to get the nutrients inside; to do this, they use the mechanisms discussed in chapter 7. Some of these require energy, which is available from catabolism already occurring in the cell. In the next step, intracellular nutrients have to be broken down to the appropriate precursor molecules. These catabolic pathways are discussed next.

Overview of Catabolism

Nutrient processing is extremely varied, especially in bacteria, yet in most cases it is based on three basic catabolic pathways. Frequently, the nutrient is glucose. There are several pathways that can be used to break down glucose, but the most common one is **glycolysis** (gly-kol'-ih-sis). In previous discussions, microorganisms were categorized according to their requirement for oxygen gas, which is related directly to their mechanisms of energy release. **Figure 8.14** provides an overview of the three major pathways for producing the needed precursors and energy (i.e., catabolism).

Aerobic respiration is a series of reactions (glycolysis, the Krebs[2] cycle, and the respiratory chain) that converts glucose to CO_2 and allows the cell to recover significant amounts of energy (review figure 8.10). Aerobic respiration relies on

---

2. "Krebs" is in honor of Sir Hans Krebs who, with F. A. Lipmann, delineated this pathway, an achievement for which they won the Nobel Prize in Physiology or Medicine in 1953.

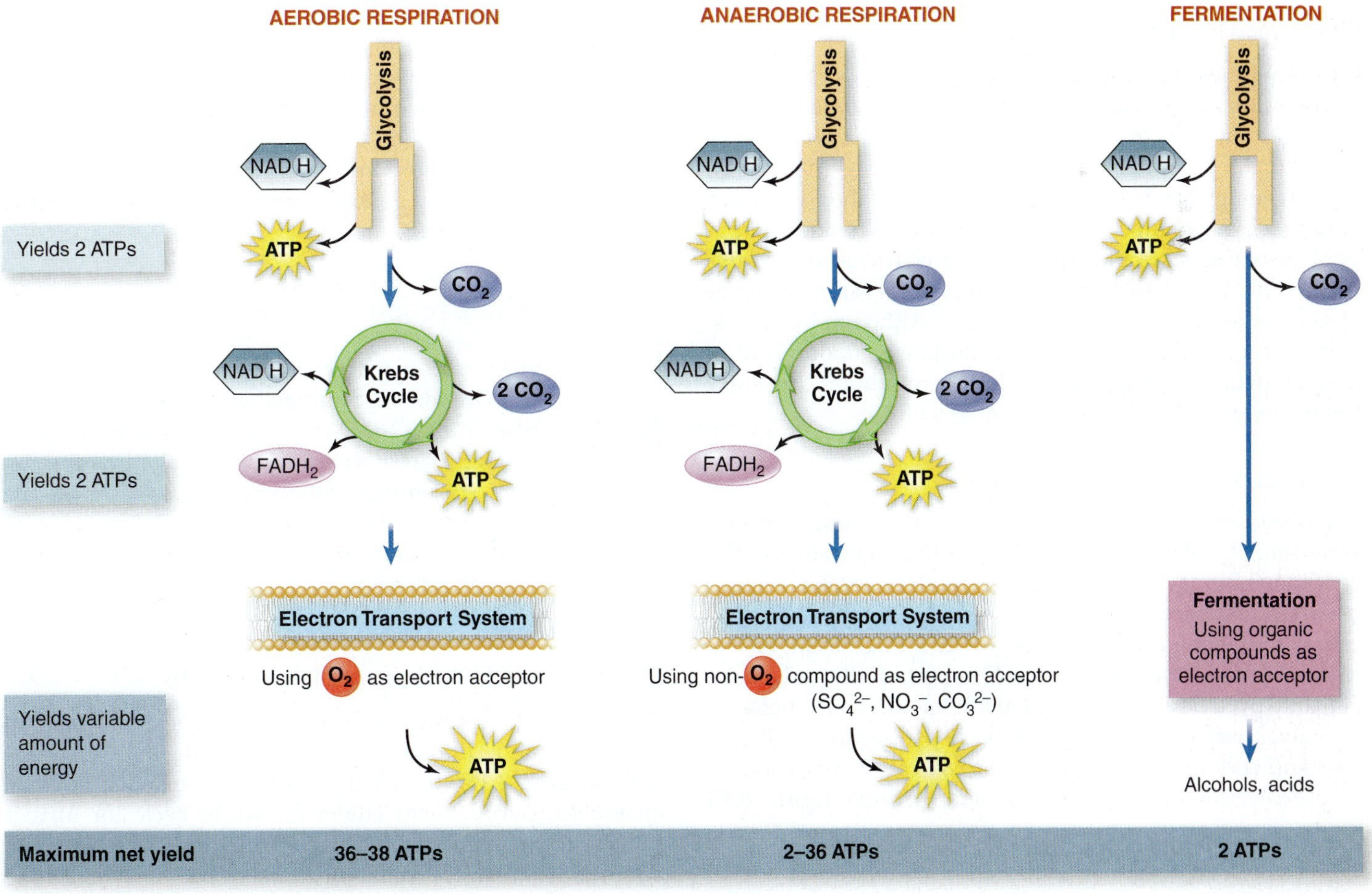

	AEROBIC RESPIRATION	**ANAEROBIC RESPIRATION**	**FERMENTATION**
Yields 2 ATPs	Glycolysis → NADH, ATP, CO_2	Glycolysis → NADH, ATP, CO_2	Glycolysis → NADH, ATP, CO_2
Yields 2 ATPs	NADH, Krebs Cycle, 2 CO_2, FADH$_2$, ATP	NADH, Krebs Cycle, 2 CO_2, FADH$_2$, ATP	
Yields variable amount of energy	Electron Transport System / Using O_2 as electron acceptor → ATP	Electron Transport System / Using non-O_2 compound as electron acceptor (SO_4^{2-}, NO_3^-, CO_3^{2-}) → ATP	Fermentation / Using organic compounds as electron acceptor → Alcohols, acids
Maximum net yield	**36–38 ATPs**	**2–36 ATPs**	**2 ATPs**

Figure 8.14 Overview of the three main pathways of catabolism.

free oxygen as the final acceptor for electrons and hydrogens and produces a relatively large amount of ATP. Aerobic respiration is characteristic of many bacteria, fungi, protozoa, and animals. Facultative and aerotolerant anaerobes may use only the glycolysis scheme to incompletely oxidize (ferment) glucose. In this case, oxygen is not required, organic compounds are the final electron acceptors, and a relatively small amount of ATP is produced. While the growth of aerobic bacteria is usually limited by the availability of substrates, the growth of anaerobes is likely to be stopped when final electron acceptors run out. Some strictly anaerobic microorganisms metabolize by means of **anaerobic respiration.** This system involves the same three pathways as aerobic respiration, but it does not use molecular oxygen as the final electron acceptor; instead, NO_3^-, SO_4^{2-}, CO_3^{2-}, and other oxidized compounds are utilized. Aspects of fermentation and anaerobic respiration are covered in subsequent sections of this chapter.

Aerobic Respiration

Aerobic respiration is a series of enzyme-catalyzed reactions in which electrons are transferred from fuel molecules such as glucose to oxygen as a final electron acceptor. This pathway is the principal energy-yielding scheme for aerobic heterotrophs, and it provides both ATP and metabolic intermediates for many other pathways in the cell, including those of protein, lipid, and carbohydrate synthesis.

Glucose: The Starting Compound

Carbohydrates such as glucose are good fuels because these compounds are readily oxidized; that is, they are excellent hydrogen and electron donors. The enzymatic withdrawal of hydrogen from them also removes electrons that can be used in energy transfers. The end products of the conversion of these carbon compounds are energy-rich ATP and energy-poor carbon dioxide and water. Polysaccharides (starch, glycogen) and disaccharides (maltose, lactose) are stored sources of glucose for the respiratory pathways. Although in our discussion we use glucose as the main starting compound, other hexoses (fructose, galactose) and fatty acid subunits can enter the pathways of aerobic respiration as well.

Glycolysis: The Starting Lineup

Glycolysis enzymatically converts glucose through several steps into pyruvic acid. Depending on the organism and the conditions, it may be only the first phase of aerobic respiration, or it may serve as the primary metabolic pathway (in the case of fermentation). Glycolysis provides a significant means to synthesize a small amount of ATP anaerobically and also to generate pyruvic acid, an essential intermediary metabolite.

Glycolysis proceeds along nine steps, starting with glucose and ending with pyruvic acid (pyruvate[3]). An overview of glycolysis will be presented here; **process figure 8.15**

---

3. In biochemistry, the terms used for organic acids appear as either the acid form (*pyruvic acid*) or its salt (*pyruvate*).

contains the chemical structures and a visual representation of the reactions. Each of the nine reactions is catalyzed by a specific enzyme with a specific name (not mentioned here).

First, glucose is activated by adding a phosphate to it, resulting in glucose-6-phosphate. It is then converted (another reaction, another enzyme) to fructose-6-phosphate, and another phosphate is added. The resulting molecule—fructose diphosphate—is more symmetrical and can be split into two 3-carbon molecules **(process figure 8.15, ④).** At this point, no oxidation-reduction has occurred; in fact, 2 ATPs have been used. The next step involves converting 3C molecule (DHAP) into the other (glyceraldehyde-3-P; G-3-P), resulting in two G-3-Ps.

From here to the end, everything that happens in glycolysis happens twice—once to each of the 3-carbon molecules. First, the G-3-Ps each receive another phosphate. At the same time, 2 NADs in the vicinity are reduced to NADHs. These NADHs will be used in the last step of catabolism (the electron transport system) to produce ATP.

In the last four steps of glcolysis **(process figure 8.15, ⑥–⑨),** the 3-carbon molecule is manipulated enzymatically to donate both of its phosphates to ADPs via substrate-level phosphorylation. This results in four new ATPs and two 3-carbon molecules with no phosphates, called pyruvic acid. However, because two ATPs were already spent in the early steps of glycolysis, the net yield (of ATP) from glycolysis of one glucose molecule is 2 ATPs.

Although glycolysis is the main route to pyruvate production for most organisms, some microbes lack the enzymes needed for this pathway to function. There are alternate biochemical reactions such as the Entner-Doudoroff pathway (by *Pseudomonas* and *Enterococcus* species) and the pentose phosphate pathway (by some photosynthetic microbes). Our aim here is to focus on general principles, so we will restrict ourselves to glycolysis.

Disease Connection

Enterococci are bacteria that can live peacefully in the intestinal tract or the female reproductive tract, but they can occasionally cause disease in those areas or in surgical wounds. In recent years, *Enterococci* that are resistant to the antibiotic vancomycin have become problematic, particularly in hospitals. They are termed *VREs* (vancomycin-resistant enterococci). They have an extremely flexible metabolism and can use a huge variety of substrates to produce energy.

Pyruvic Acid: A Central Metabolite

Pyruvic acid occupies an important position in several pathways, and different organisms handle it in different ways **(figure 8.16).** In strictly aerobic organisms and some anaerobes, pyruvic acid enters the Krebs cycle for further processing and energy release. Facultative anaerobes can adopt a fermentative metabolism, in which pyruvic acid is re-reduced into acids or other products.

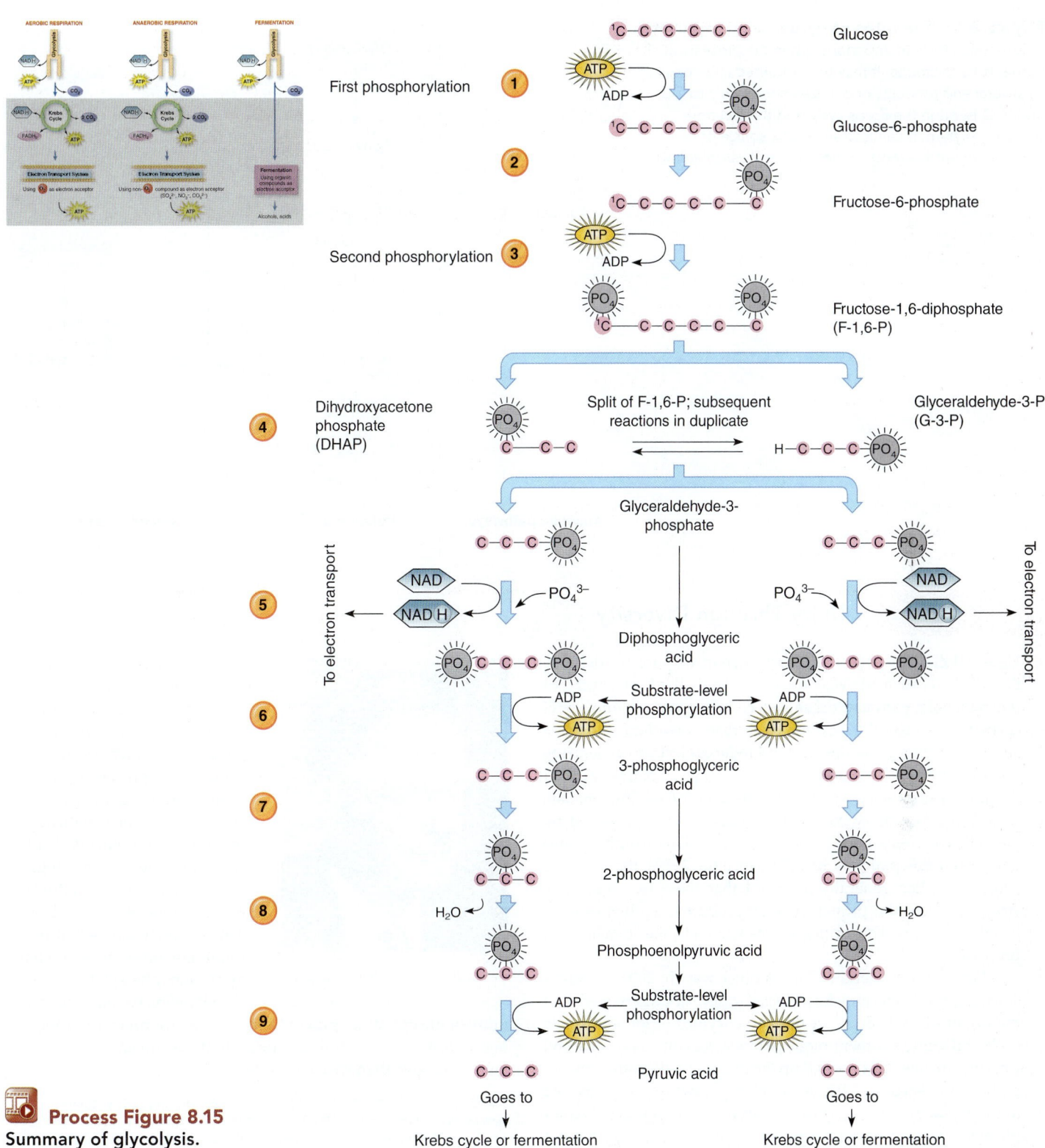

Process Figure 8.15
Summary of glycolysis.

The Krebs Cycle: A Carbon and Energy Wheel

In glycolysis, the oxidation of glucose yields a comparatively small amount of energy and gives off pyruvic acid. Pyruvic acid is still energy-rich, containing a number of extractable hydrogens and electrons to power ATP synthesis, but this can be achieved only through the work of the second and third phases of respiration, in which pyruvic acid's hydrogens are transferred to oxygen, producing CO_2 and H_2O. In the following section, we examine the next phase of this process, the Krebs cycle. This set of reactions takes place in the cytosol of bacteria and is catalyzed by a group of enzymes (some of which are associated with the plasma membrane).

Figure 8.16 The fates of pyruvic acid (pyruvate).
This metabolite is an important hub in the processing of nutrients by microbes. It may be fermented anaerobically to several end products or oxidized completely to CO_2 and H_2O through the Krebs cycle and the electron transport system. It can also serve as a source of raw material for synthesizing amino acids and carbohydrates.

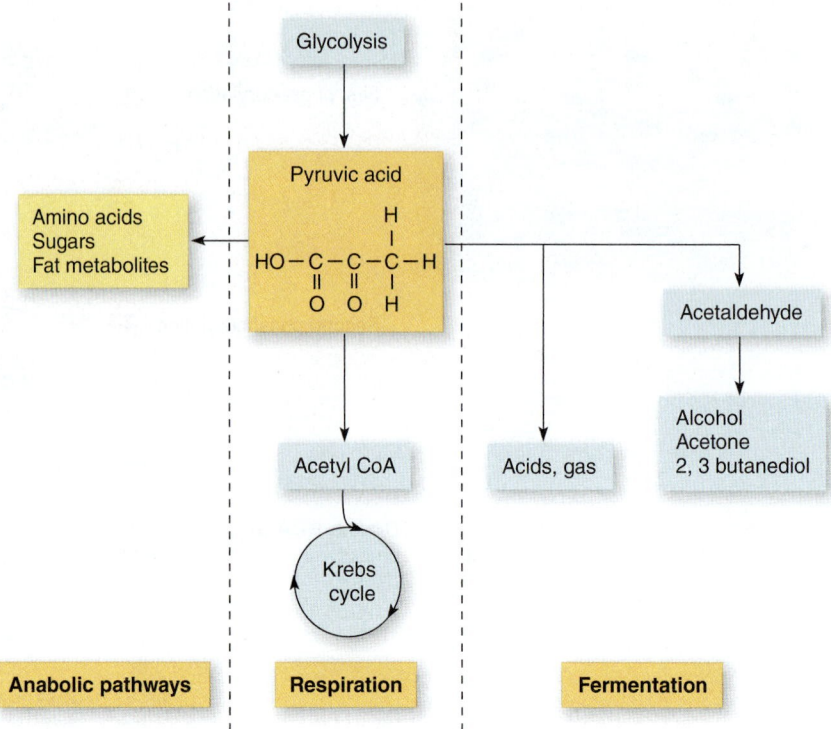

INSIGHT 8.2   Unity Through Diversity

Hans Adolf Krebs won the Nobel Prize in 1953 for his description of the citric acid cycle, often known as the Krebs cycle. In this metabolic pathway, organisms convert pyruvate into acetyl coenzyme A (acetyl CoA), which is then converted into citric acid, which then generates ATP and reduces electron carriers (see process figure 8.17). Some organisms use another pathway called the glyoxylate cycle, which converts acetyl CoA into malate, which can then be converted into other molecules needed by the cell. Dr. Krebs and his colleague Dr. Hans Kornberg also discovered this pathway in 1959. Since the 1950s, the scientific community has generally accepted that these two metabolic strategies were the primary energy-producing pathways used by all organisms. This principle was known as the "biochemical unity of life."

Leave it to bacteria to find another way. In 2007, scientists studying the bacterium *Rhodobacter sphaeroides* discovered that the organism lacked one of the key enzymes of the glyoxylate cycle and found another unique pathway for conversion of acetyl CoA into malate. This opened up the possibility that other unique pathways existed. In 2011, yet another pathway was discovered in the archaeon *Halobacterium marismortui*. This pathway is complicated and requires nine extra steps catalyzed by enzymes not

Source: 2011. Science. Jan. 21. p. 294.

found in the glyoxylate cycle. When these enzymes were studied, it appeared that *H. marismortui* borrowed genes from other organisms rather than producing their own unique enzymes. In addition, this new pathway enhances the survival of this microbe in the high salt content of its natural environment—the Dead Sea. This research shows that the relatively ubiquitous Krebs cycle and glyoxylate cycles do not define the biochemical unity of life but rather provide a framework for diversity among many forms of life—and it appears that this diversity is much more extensive than we ever thought.

Michael Marshall, "Need a New Metabolic Pathway? Steal a Few Genes," *New Scientist* (January 20, 2011). Retrieved from www.newscientist.com/article/dn20009-need-a-new-metabolic-pathway-steal-a-few-genes.html.

In eukaryotic cells, this process takes place in the mitochondrial matrix.

To connect the glycolysis pathway to the Krebs cycle, for either aerobic or anaerobic respiration, the pyruvic acid is first converted to a starting compound for that cycle **(process figure 8.17).** This step starts with an oxidation-reduction reaction, which also releases the first carbon dioxide molecule. It involves a cluster of enzymes and coenzyme A that participate in the dehydrogenation (oxidation) of pyruvic acid, the reduction of NAD to NADH, and the decarboxylation of pyruvic acid to a 2-carbon acetyl group. The acetyl group remains attached to coenzyme A, forming acetyl coenzyme A (acetyl CoA) that feeds into the Krebs cycle **(Insight 8.2).**

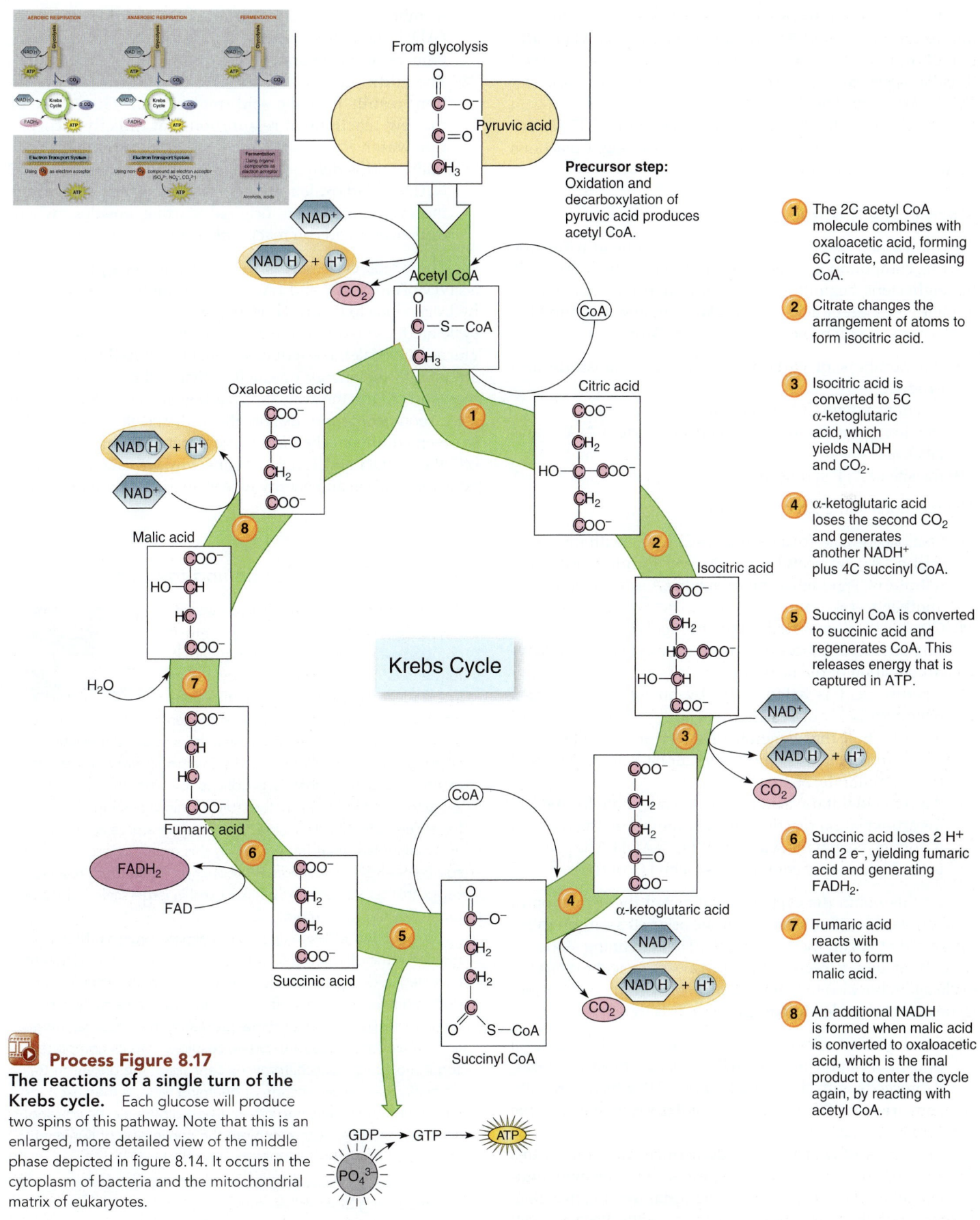

Process Figure 8.17

The reactions of a single turn of the Krebs cycle. Each glucose will produce two spins of this pathway. Note that this is an enlarged, more detailed view of the middle phase depicted in figure 8.14. It occurs in the cytoplasm of bacteria and the mitochondrial matrix of eukaryotes.

1. The 2C acetyl CoA molecule combines with oxaloacetic acid, forming 6C citrate, and releasing CoA.

2. Citrate changes the arrangement of atoms to form isocitric acid.

3. Isocitric acid is converted to 5C α-ketoglutaric acid, which yields NADH and CO_2.

4. α-ketoglutaric acid loses the second CO_2 and generates another $NADH^+$ plus 4C succinyl CoA.

5. Succinyl CoA is converted to succinic acid and regenerates CoA. This releases energy that is captured in ATP.

6. Succinic acid loses 2 H^+ and 2 e^-, yielding fumaric acid and generating $FADH_2$.

7. Fumaric acid reacts with water to form malic acid.

8. An additional NADH is formed when malic acid is converted to oxaloacetic acid, which is the final product to enter the cycle again, by reacting with acetyl CoA.

Precursor step: Oxidation and decarboxylation of pyruvic acid produces acetyl CoA.

Krebs Cycle

The NADH formed during this reaction will be shuttled into electron transport and used to generate ATP via oxidative phosphorylation. *Keep in mind that all reactions described actually happen twice for each glucose because of the two pyruvates that are formed during glycolysis.*

The Krebs cycle as depicted in process figure 8.17 always looks intimidating. Think of it as a series of eight reactions catalyzed by eight different enzymes.

Steps in the Krebs Cycle

As you learned earlier, a cyclic pathway is one in which the starting compound is regenerated at the end. The Krebs cycle has eight steps, beginning with citric acid formation and ending with oxaloacetic acid. As we take a single spin around the Krebs cycle, it will be helpful to keep track of

- the numbers of carbons (#C) of each substrate and product,
- reactions where CO_2 is generated,
- the involvement of the electron carriers NAD and FAD, and
- the site of ATP synthesis.

The reactions in the Krebs cycle follow.

1. Oxaloacetic acid (oxaloacetate; 4C) reacts with the acetyl group (2C) on acetyl CoA, thereby forming citric acid (citrate; 6C) and releasing coenzyme A so it can join with another acetyl group.
2. Citric acid is converted to isocitric acid (isocitrate; 6C) to prepare this substrate for the decarboxylation and dehydrogenation of the next step.
3. Isocitric acid is acted upon by an enzyme complex including NAD or NADP (depending on the organism) in a reaction that generates NADH or NADPH, splits off a carbon dioxide, and leaves alpha-ketoglutaric acid (α-ketoglutarate; 5C).
4. Alpha-ketoglutaric acid serves as a substrate for the last decarboxylation reaction and yet another redox reaction involving coenzyme A and yielding NADH. The product is the high-energy compound succinyl CoA (4C).

At this point, the cycle has completed the formation of 3 CO_2 molecules that balance out the original 3-carbon pyruvic acid that began the Krebs cycle. The remaining steps are needed not only to regenerate the oxaloacetic acid to start the cycle again but also to extract more energy from the intermediate compounds leading to oxaloacetic acid.

5. Succinyl CoA is the source of the one substrate-level phosphorylation in the Krebs cycle. In most microbes, it proceeds with the formation of GTP, which is readily converted to ATP. The product of this reaction is succinic acid (succinate; 4C).
6. Succinic acid next becomes dehydrogenated, but in this case, the electron and H+ acceptor is flavin adenine dinucleotide (FAD). The enzyme that catalyzes this reaction, succinyl dehydrogenase, is found in the bacterial cell

membrane and mitochondrial cristae of eukaryotic cells. $FADH_2$ then directly enters the electron transport system. Fumaric acid (fumarate; 4C) is the product of this reaction.

7. The addition of water to fumaric acid (called hydration) results in malic acid (malate; 4C). This is one of the few reactions in respiration that directly incorporate water.
8. Malic acid is dehydrogenated (with formation of a final NADH), and oxaloacetic acid is formed. This step brings the cycle back to its original starting position, where oxaloacetic acid can react with acetyl coenzyme A.

The Krebs cycle serves to transfer the energy stored in acetyl CoA to NAD+ and FAD by reducing them (transferring hydrogen ions to them). Thus, the main products of the Krebs cycle are these reduced molecules (as well as 2 ATPs for each glucose molecule). The reduced coenzymes NADH and $FADH_2$ are vital to the energy production that will occur in electron transport. Along the way, the 2-carbon acetyl CoA joins with a 4-carbon compound, oxaloacetic acid, and then participates in seven additional chemical transformations while "spinning off" the NADH and $FADH_2$. That's why we called the Krebs cycle the "carbon and energy wheel" in a preceding heading.

Case File 8   *Continuing the Case*

As with any catabolic process, an enzyme is required to break lignin down into its component sugars, which then can be fermented. Several types of enzymes are involved in this process. First, the large, woody molecule lignin must be broken down into smaller pieces of cellulose. Recently, researchers at the Universities of Warwick and British Columbia discovered an enzyme capable of lignin digestion that is produced by *Rhodococcus jostii*, a bacterium that lives in the soil. Lignin-digesting enzymes have been identified in fungi, but bacteria are easier to grow and manipulate than fungi, making this organism an excellent source of enzymes for biofuel production. This recent research suggests that this enzyme can be used in large-scale industrial processes to produce biofuels.

Production of biofuels is not a simple process. Manufacture of ethanol from solid biomass requires many different enzymes and organisms working in concert in order to produce ethanol in a cost-effective and environmentally sound manner. *Cellulase*, an enzyme produced by bacteria, fungi, and protozoans, is used to break cellulose into its component sugar molecules. *Saccharomyces cerevisiae*, the organism traditionally used in production of beer and wine, is then used for fermenting the sugars into ethanol-based biofuels. As you see, no single organism can complete the conversion of lignin into ethanol on its own—it is truly a team effort.

■ What are the advantages and disadvantages of producing biofuels?

The Respiratory Chain: Electron Transport and Oxidative Phosphorylation

We now come to the energy chain, which is the final "processing mill" for electrons and hydrogen ions and the major generator of ATP. Overall, the electron transport system (ETS) consists of a chain of special redox carriers (proteins) that receives electrons from reduced carriers (NADH, FADH$_2$) generated by glycolysis and the Krebs cycle and passes them in a sequential and orderly fashion from one redox molecule to the next. The flow of electrons down this chain is highly energetic and allows the active transport of hydrogen ions to the outside of the membrane where the respiratory chain is located. The step that finalizes the transport process is the acceptance of electrons and hydrogen by oxygen, producing water. Obviously, this process consumes oxygen (see p. 218, Continuing the Case). Some variability exists from one organism to another, but the principal compounds that carry out these complex reactions are NADH dehydrogenase, flavoproteins, coenzyme Q (ubiquinone), and **cytochromes**

(sy'-toh-krohmz). The cytochromes contain a tightly bound metal atom at their center that is actively involved in accepting electrons and donating them to the next carrier in the series. The highly compartmentalized structure of the respiratory chain is an important factor in its function. Note in **figure 8.18** that the electron transport carriers and enzymes are embedded in the cell membrane in bacteria. The equivalent structure for housing them in eukaryotes is the inner mitochondrial membranes pictured in **figure 8.19**. We will describe the electron transport system in both bacteria and eukaryotes.

Elements of Electron Transport: The Energy Cascade

The principal questions about the electron transport system are: How are the electrons passed from one carrier to another in the series? How does this progression result in ATP synthesis? Where and how is oxygen (or another electron acceptor) utilized? Although the biochemical details of this process are rather complicated, the basic reactions consist of a number

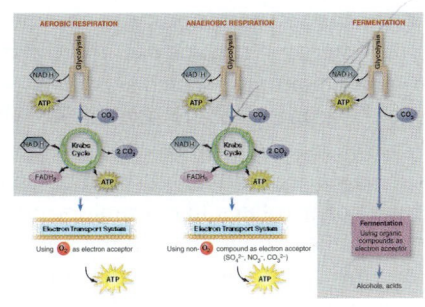

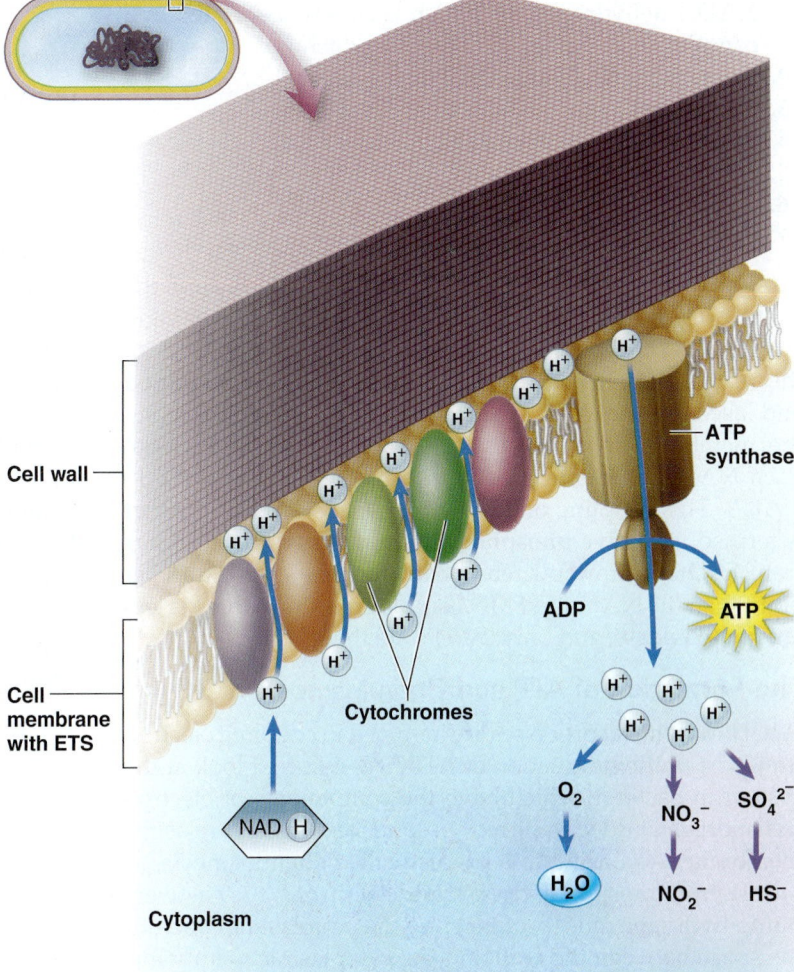

Figure 8.18 The electron transport system and oxidative phosphorylation in bacterial membranes. Starting at NADH dehydrogenase, electrons brought in from the Krebs cycle by NADH are passed along the chain of electron transport carriers. Each adjacent pair of transport molecules undergoes a redox reaction. Coupled to the transport of electrons is the simultaneous active transport of H$^+$ into the periplasm by specific carriers. These processes set the scene for ATP synthesis and final H$^+$ and e$^-$ acceptance. Note the differences in final electron acceptors in aerobic versus anaerobic respirers.

Figure 8.19 The electron transport system on the inner membrane of the mitochondrial cristae. As the carriers in the mitochondrial cristae transport electrons, they also actively pump H$^+$ ions (protons) to the intermembrane space, producing a chemical and charge gradient between the outer and inner mitochondrial compartments.

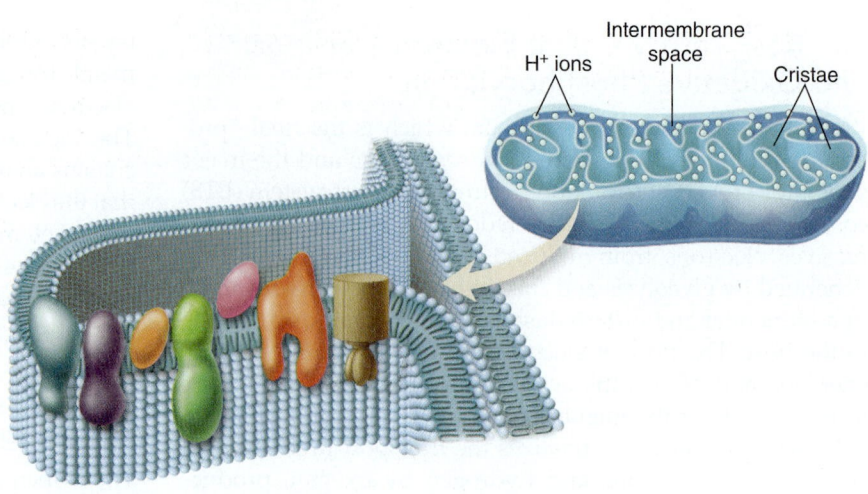

of redox reactions now familiar to us. In general, the carrier compounds and their enzymes are arranged in linear sequence and are reduced and oxidized in turn.

The sequence of electron carriers in the respiratory chain of most aerobic organisms is

1. NADH dehydrogenase, which is closely associated in a complex with the adjacent carrier, which is
2. flavin mononucleotide (FMN);
3. coenzyme Q;
4. cytochrome b;
5. cytochrome c_1;
6. cytochrome c; and
7. cytochromes a and a_3, which are complexed together.

Conveyance of the NADHs from glycolysis and the Krebs cycle to the first carrier sets in motion the remaining six steps. With each redox exchange, the energy level of the reactants is lessened. The released energy is captured and used by the **ATP synthase** complex, stationed along the membrane in close association with the ETS carriers. Each NADH that enters electron transport can give rise to 3 ATPs. This coupling of ATP synthesis to electron transport is termed **oxidative phosphorylation.** Because the electrons from FADH$_2$ from the Krebs cycle enter the cycle at a later point than the NAD and FMN complex reactions, there is less energy to release, and only 2 ATPs are the result.

The Formation of ATP and Chemiosmosis

What biochemical processes are involved in coupling electron transport to the production of ATP? We will first look at the system in bacteria, which have the components of electron transport embedded in a precise sequence on the cytoplasmic membrane. According to a process called **chemiosmosis,** as the electron transport carriers shuttle electrons, they actively pump hydrogen ions (protons) into the periplasmic space, or the space between the wall and the cytoplasmic membrane, depending on whether the bacterium is gram-positive or gram-negative. This process sets up a concentration gradient of hydrogen ions called the *proton motive force* (PMF). The

PMF consists of a difference in charge between the outside of the membrane (+) and the inside of the membrane (−) (see figure 8.18).

Separating the charge has the effect of a battery, which can temporarily store potential energy. This charge will be maintained by the impermeability of the membrane to H$^+$. The only site where H$^+$ can diffuse into the cytoplasm is at the ATP synthase complex, which sets the stage for the final processing of H$^+$ leading to ATP synthesis.

ATP synthase is a complex enzyme composed of two large units (see figures 8.18 and 8.19). It is embedded in the membrane, but part of it rotates like a motor and traps chemical energy. As the H$^+$ ions flow through the center of the enzyme by diffusion, the other compartments pull in ADP and P$_i$. Rotation causes a three-dimensional change in the enzyme that bonds these two molecules, thereby releasing ATP into the cytoplasm (see figure 8.18). The enzyme is then rotated back to the start position and will continue the process.

Eukaryotic ATP synthesis occurs by means of the same overall process. However, eukaryotes have the ETS stationed in mitochondrial membranes, between the inner mitochondrial matrix and the outer intermembrane space (see figure 8.19). This difference will affect the amount of ATP produced (discussed in the next section).

Potential Yield of ATPs from Oxidative Phosphorylation

The total of five NADHs (four from the Krebs cycle and one from glycolysis) can be used to synthesize

15 ATPs for ETS (5 × 3 per electron pair)

and

15 × 2 × 30 ATPs per glucose

The single FADH produced during the Krebs cycle results in

2 ATPs per electron pair

and

2 × 2 × 4 ATPs per glucose

Figure 8.20 summarizes the total of ATP and other products for the entire aerobic pathway. These totals are the potential yields possible but may not be fulfilled by many organisms.

Summary of Aerobic Respiration

Originally, we presented a summary equation for respiration. We are now in a position to tabulate the input and output of this equation at various points in the pathways and sum up the final ATP. Close examination of figure 8.20 will reveal several important facets of aerobic respiration:

1. The total possible production of ATP is 40: 4 from glycolysis, 2 from the Krebs cycle, and 34 from electron transport. However, because 2 ATPs were expended in early glycolysis, this leaves a maximum of **38 ATPs.**

The actual totals may be lower in certain eukaryotic cells because energy is expended in transporting the NADH produced during glycolysis across the mitochondrial membrane. Certain aerobic bacteria come closest to achieving the full total of 38 because they lack mitochondria and thus do not have to use ATP in the transport of NADH across the outer mitochondrial membrane.

2. Six carbon dioxide molecules are generated during the Krebs cycle.
3. Six oxygen molecules are consumed during electron transport.
4. Six water molecules are produced in electron transport and 2 in glycolysis, but because 2 are used in the Krebs cycle, this leaves a net number of 6.

The Terminal Step

The terminal step, during which oxygen accepts the electrons, is catalyzed by cytochrome aa_3, also called cytochrome oxidase. This large enzyme complex is specifically adapted to receive electrons from cytochrome c, pick up hydrogens from the solution, and react with oxygen to form a molecule of water. This reaction, though in actuality more complex, is summarized as follows:

$$2H^+ + 2e^- + \tfrac{1}{2}O_2 \rightarrow H_2O$$

Most eukaryotic aerobes have a fully functioning cytochrome system, but bacteria exhibit wide-ranging variations in this part of the system. Some species lack one or more of the redox steps; others have several alternative electron transport schemes. Because many bacteria lack

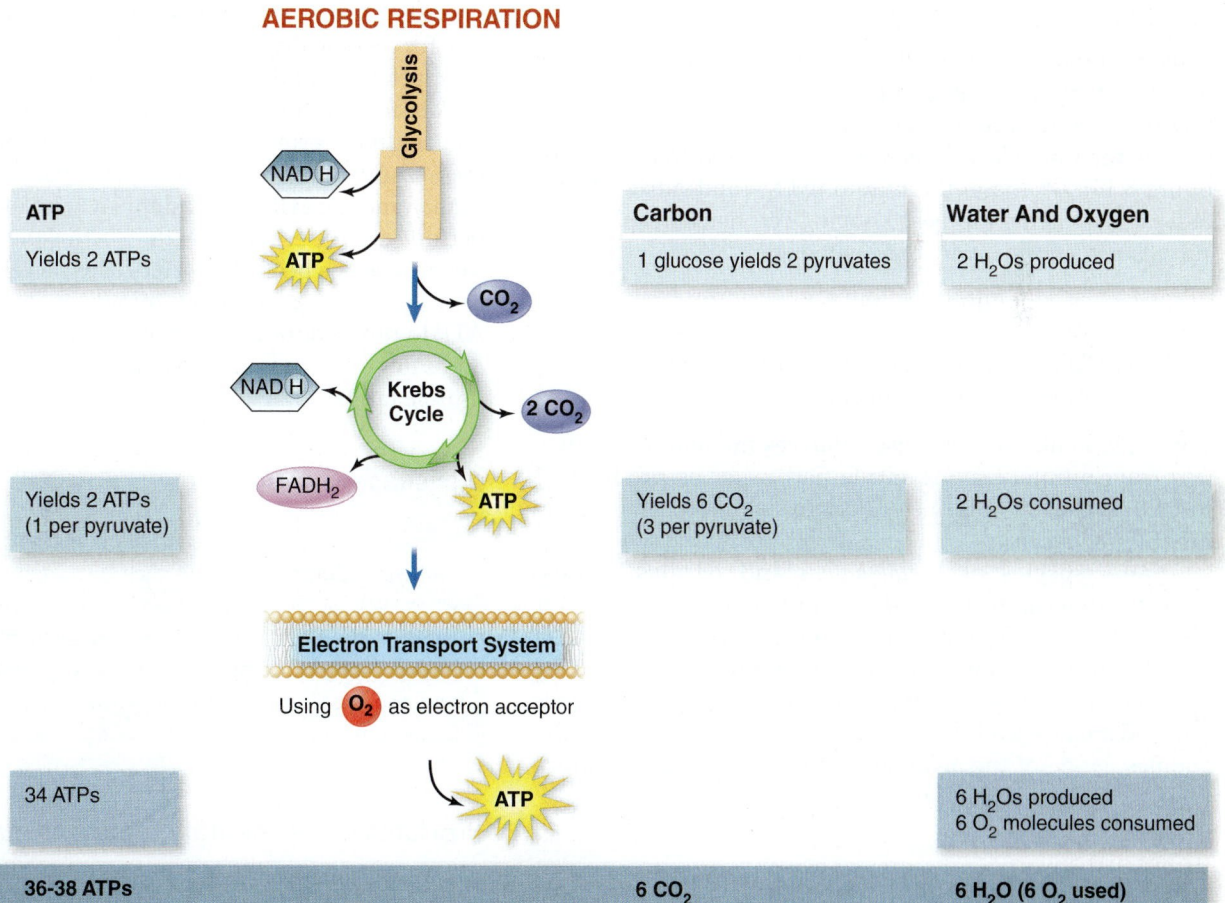

Figure 8.20 Theoretic ATP yield from aerobic respiration.   To attain the theoretic maximum yield of ATP, we assume a ratio of 3 for the oxidation of NADH to 2 for $FADH_2$. The actual yield is generally lower and varies between eukaryotes and bacteria and among bacterial species.

cytochrome *c* oxidase, this variation can be used to differentiate among certain genera of bacteria. An oxidase detection test can be used to help identify members of the genera *Neisseria* and *Pseudomonas* and some species of *Bacillus*. Another variation in the cytochrome system is evident in certain bacteria (*Klebsiella, Enterobacter*) that can grow even in the presence of cyanide because they lack cytochrome oxidase. Cyanide will cause rapid death in humans and other eukaryotes because it blocks cytochrome oxidase, thereby completely terminating aerobic respiration, but it is harmless to these bacteria.

A potential side reaction of the respiratory chain in aerobic organisms is the incomplete reduction of oxygen to superoxide ion (O_2^-) and hydrogen peroxide (H_2O_2). As mentioned in chapter 7, these toxic oxygen products can be very damaging to cells. Aerobes have neutralizing enzymes to deal with these products, including *superoxide dismutase* and *catalase*. One exception is the genus *Streptococcus*, which can grow well in oxygen yet lacks both cytochromes and catalase. The tolerance of these organisms to oxygen can be explained by the neutralizing effects of a special peroxidase. The lack of cytochromes, catalase, and peroxidases in anaerobes as a rule limits their ability to process free oxygen and contributes to its toxic effects on them.

Anaerobic Respiration

Some bacteria have evolved an anaerobic respiratory system that functions like the aerobic cytochrome system except that it utilizes oxygen-containing ions, rather than free oxygen, as the final electron acceptor in electron transport (see figure 8.14). Of these, the nitrate (NO_3^-) and nitrite (NO_2^-) reduction systems are best known. The reaction in species such as *Escherichia coli* is represented as

$$\text{Nitrate reductase}$$
$$\downarrow$$
$$\underset{\text{nitrate}}{NO_3^-} + NADH \rightarrow \underset{\text{nitrite}}{NO_2^-} + H_2O + NAD^+$$

The enzyme nitrate reductase catalyzes the removal of oxygen from nitrate, leaving nitrite and water as products. A test for this reaction is one of the physiological tests used in identifying bacteria.

Some species of *Pseudomonas* and *Bacillus* possess enzymes that can further reduce nitrite to nitric oxide (NO), nitrous oxide (N_2O), and even nitrogen gas (N_2). This process, called **denitrification,** is a very important step in recycling nitrogen in the biosphere. Other oxygen-containing nutrients reduced anaerobically by various bacteria are carbonates and sulfates. None of the anaerobic pathways produce as much ATP as aerobic respiration.

Fermentation

The definition of **fermentation** is *the incomplete oxidation of glucose or other carbohydrates in the absence of oxygen.* This process uses organic compounds as the terminal electron acceptors and yields a small amount of ATP (see figure 8.14).

Over time, the term *fermentation* has acquired several looser connotations. Originally, Pasteur called the microbial action of yeast during wine production *ferments,* and to this day, biochemists use the term in reference to the production of ethyl alcohol by yeasts acting on glucose and other carbohydrates. Fermentation is also what bacteriologists call the formation of acid, gas, and other products by the action of various bacteria on pyruvic acid. The process is a common metabolic strategy among bacteria. Industrial processes that produce chemicals on a massive scale through the actions of microbes are also called fermentations (see chapter 25). Each of these usages is acceptable for one application or another.

Without the use of an electron transport chain, it may seem that fermentation would yield only meager amounts of energy (2 ATPs maximum per glucose) and that would slow down growth. What actually happens, however, is that many bacteria can grow as fast as they would in the presence of oxygen. This rapid growth is made possible by an increase in the rate of glycolysis. From another standpoint, fermentation permits independence from molecular oxygen and allows colonization of anaerobic environments. It also enables microorganisms with a versatile metabolism to adapt to variations in the availability of oxygen. For them, fermentation provides a means to grow even when oxygen levels are too low for aerobic respiration.

Bacteria that digest cellulose in the rumens of cattle are largely fermentative. After initially hydrolyzing cellulose to glucose, they ferment the glucose to organic acids, which are then absorbed as the bovine's principal energy source. Even human muscle cells can undergo a form of fermentation that permits short periods of activity after the oxygen supply in the muscle has been exhausted. Muscle cells convert pyruvic acid into lactic acid, which allows anaerobic production of ATP to proceed for a time. But this cannot go on indefinitely, and after a few minutes, the accumulated lactic acid causes muscle fatigue.

Disease Connection

Bacteria that cause disease in humans also can use fermentative pathways. Surprisingly, one such bacterium colonizes the lungs, where you would think oxygen would be abundant. However, in some circumstances (particularly in cystic fibrosis patients), the bacterium *Pseudomonas aeruginosa* creates biofilms in the lungs and the biofilms become anaerobic. In these conditions, the bacterium ferments pyruvate to acetate and lactic acid to survive.

Products of Fermentation in Microorganisms

Alcoholic beverages (wine, beer, whiskey) are perhaps the most prominent among fermentation products; others are solvents (acetone, butanol), organic acids (lactic, acetic), dairy products, and many other foods. Derivatives of proteins, nucleic acids, and other organic compounds are fermented to produce vitamins, antibiotics, and even hormones such as hydrocortisone.

Fermentation products can be grouped into two general categories: alcoholic fermentation products and acidic fermentation products **(figure 8.21)**. **Alcoholic fermentation** occurs in yeast or bacterial species that have metabolic pathways for converting pyruvic acid to ethanol. This process involves a decarboxylation of pyruvic acid to acetaldehyde, followed by a reduction of the acetaldehyde to ethanol. In oxidizing the NADH formed during glycolysis, NAD is regenerated, thereby allowing the glycolytic pathway to continue. These processes are crucial in the production of beer and wine, though the actual techniques for arriving at the desired amount of ethanol and the prevention of unwanted side reactions are important tricks of the brewer's trade. Note that the products of alcoholic fermentation are not only ethanol but also CO_2, a gas that accounts for the bubbles in champagne and beer (see chapter 25).

The pathways of **acidic fermentation** are extremely varied. Lactic acid bacteria ferment pyruvate in the same way

that humans do—by reducing it to lactic acid. If the product of this fermentation is mainly lactic acid, as in certain species of *Streptococcus* and *Lactobacillus*, it is termed *homolactic*. The souring of milk is due largely to the production of this acid by bacteria. When glucose is fermented to a mixture of lactic acid, acetic acid, and carbon dioxide, as is the case with *Leuconostoc* and other species of *Lactobacillus*, the process is termed *heterolactic fermentation* **(Insight 8.3)**.

Many members of the family *Enterobacteriaceae* (*Escherichia, Shigella,* and *Salmonella*) possess enzyme systems for converting pyruvic acid to several acids simultaneously. **Mixed acid fermentation** produces a combination of acetic, lactic, succinic, and formic acids, and it lowers the pH of a medium to about 4.0. *Propionibacterium* produces primarily propionic acid, which gives the characteristic flavor to Swiss cheese while fermentation gas (CO_2) produces the holes. Some members also further decompose formic acid completely to carbon dioxide and hydrogen gases. Because enteric bacteria commonly occupy the intestine, this fermentative activity accounts for the accumulation of some types of gas in the intestine. Some bacteria reduce the organic acids and produce the neutral end product 2,3-butanediol and other solvents.

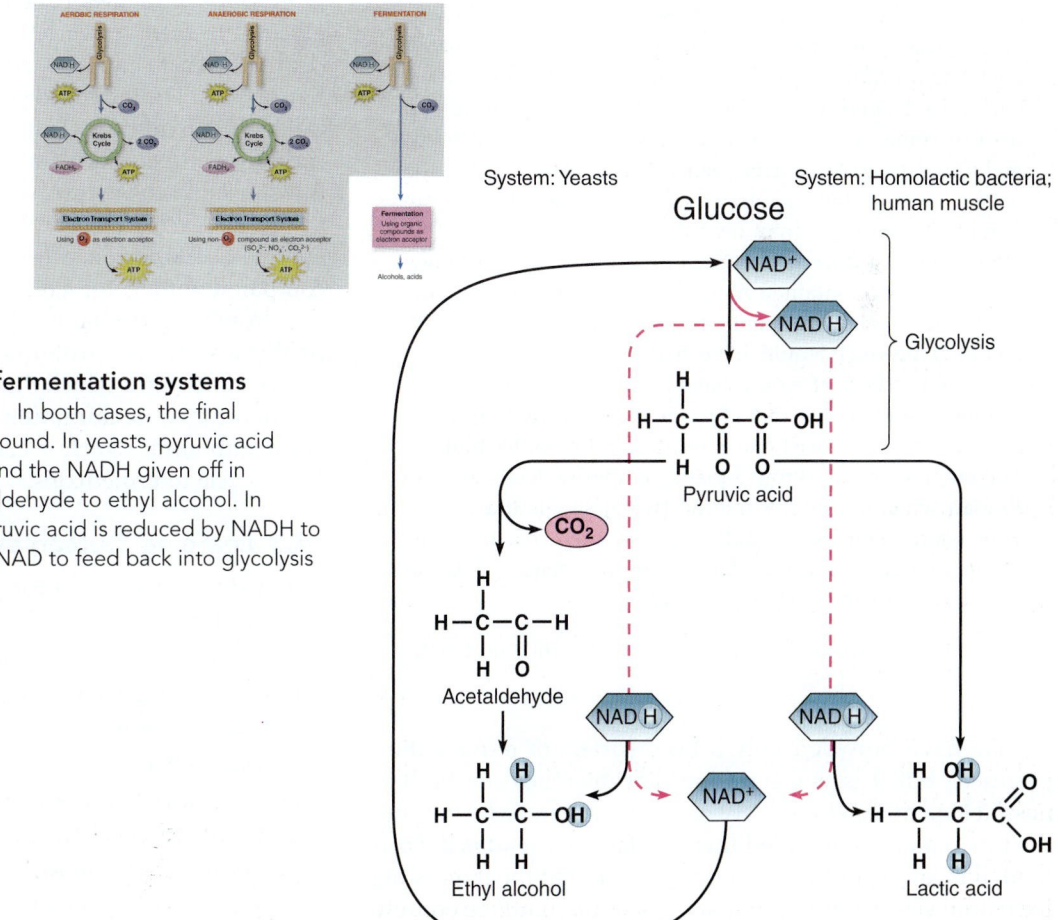

Figure 8.21 The chemistry of fermentation systems that produce acid and alcohol. In both cases, the final electron acceptor is an organic compound. In yeasts, pyruvic acid is decarboxylated to acetaldehyde, and the NADH given off in the glycolytic pathway reduces acetaldehyde to ethyl alcohol. In homolactic fermentative bacteria, pyruvic acid is reduced by NADH to lactic acid. Both systems regenerate NAD to feed back into glycolysis or other cycles.

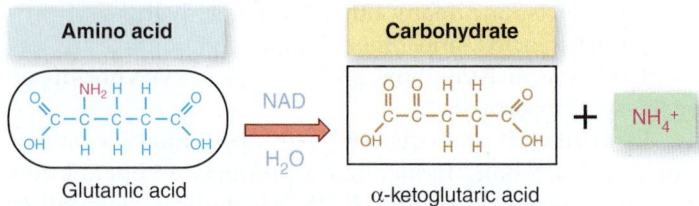

Figure 8.22 Deamination. Removal of an amino group converts an amino acid to an intermediate of carbohydrate metabolism. Ammonium is a by-product.

Catabolism of Noncarbohydrate Compounds

We have given you one version of events for catabolism, using glucose, a carbohydrate, as our example. Other compounds serve as fuel, as well. The more complex polysaccharides are easily broken down into their component sugars, which can enter glycolysis at various points. Microbes also break down other molecules for their own use, of course. Two other major sources of energy and building blocks for microbes are lipids (fats) and proteins. Both of these must be broken down to their component parts to produce precursor metabolites and energy.

Recall from chapter 2 that fats are fatty acids joined to glycerol. Enzymes called **lipases** break these apart. The glycerol is then converted to dihydroxyacetone phosphate (DHAP), which can enter step ④ of glycolysis (see process figure 8.15). The fatty acid component goes through a process called **beta oxidation.** Fatty acids have a variable number of carbons; in beta oxidation, 2-carbon units are successively transferred to coenzyme A, creating acetyl CoA, which enters the Krebs cycle. This process can yield a large amount of energy. Oxidation of a 6-carbon fatty acid yields 50 ATPs, compared with 38 for a 6-carbon sugar.

Proteins are chains of amino acids. Enzymes called **proteases** break proteins down to their amino acid components, after which the amino groups are removed by a reaction called **deamination (figure 8.22).** This leaves a carbon compound, which is easily converted to one of several Krebs cycle intermediates.

8.3 Learning Outcomes—Assess Your Progress

10. List three basic catabolic pathways and the estimated ATP yield for each.
11. Construct a paragraph summarizing glycolysis.
12. Describe the Krebs cycle and compare the process between bacteria and eukaryotes.
13. Discuss the significance of the electron transport system.
14. State two ways in which anaerobic respiration differs from aerobic respiration.
15. Summarize the steps of microbial fermentation and list three useful products it can create.
16. Describe how noncarbohydrate compounds are catabolized.

We have provided only a brief survey of fermentation products, but it is worth noting that microbes can be harnessed to synthesize a variety of other substances by varying the raw materials provided them. In fact, so broad is the colloquial meaning of the word *fermentation* that the large-scale industrial syntheses by microorganisms often utilize entirely different mechanisms from those described here, and they even occur aerobically, particularly in antibiotic, hormone, vitamin, and amino acid production (see chapter 25).

8.4   Biosynthesis and the Crossing Pathways of Metabolism

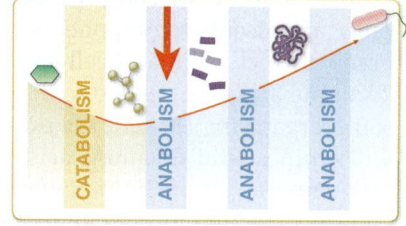

Our discussion now turns from catabolism and energy extraction to anabolic functions and biosynthesis. In this section, we present aspects of intermediary metabolism, including amphibolic pathways, the synthesis of simple molecules, and the synthesis of macromolecules.

The Frugality of the Cell: Waste Not, Want Not

It must be obvious by now that cells have mechanisms for careful management of carbon compounds. Rather than being dead ends, most catabolic pathways contain strategic molecular intermediates (metabolites) that can be diverted into anabolic pathways. In this way, a given molecule can serve multiple purposes, and the maximum benefit can be derived from all nutrients and metabolites of the cell pool. The property of a system to integrate catabolic and anabolic pathways to improve cell efficiency is termed **amphibolism** (am-fee-bol'-izm).

At this point in the chapter, you can appreciate a more complex view of metabolism than that presented at the beginning in figure 8.1. **figure 8.23** demonstrates the amphibolic nature of intermediary metabolism. The pathways of glucose catabolism are an especially rich "metabolic marketplace." The principal moments of amphibolic interaction occur during glycolysis (glyceraldehyde-3-phosphate and pyruvic acid) and the Krebs cycle (acetyl coenzyme A and various organic acids).

Amphibolic Sources of Cellular Building Blocks

Glyceraldehyde-3-phosphate can be diverted away from glycolysis and converted into precursors for amino acid, carbohydrate, and triglyceride (fat) synthesis. (A *precursor molecule* is a compound that is the source of another compound.) Earlier, we noted the numerous directions that pyruvic acid catabolism can take. In terms of synthesis, pyruvate also plays a pivotal role in providing intermediates for amino acids. In the

event of an inadequate glucose supply, pyruvate serves as the starting point in glucose synthesis from various metabolic intermediates, a process called **gluconeogenesis** (gloo'-koh-nee'-oh-gen'-uh-sis).

The acetyl group that starts the Krebs cycle is another extremely versatile metabolite that can be fed into a number of synthetic pathways. This 2-carbon fragment can be converted as a single unit into one of several amino acids, or a number of these fragments can be condensed into hydrocarbon chains that are important building blocks for fatty acid and lipid synthesis. Note that the reverse is also true—fats can be degraded to acetyl through beta oxidation and thereby enter the Krebs cycle as acetyl coenzyme A.

Pathways that synthesize the nitrogen bases (purines, pyrimidines), which are components of DNA and RNA, originate in amino acids and so can be dependent on intermediates from the Krebs cycle as well. Because the coenzymes NAD, NADP, FAD, and others contain purines and pyrimidines similar to the nucleic acids, their synthetic

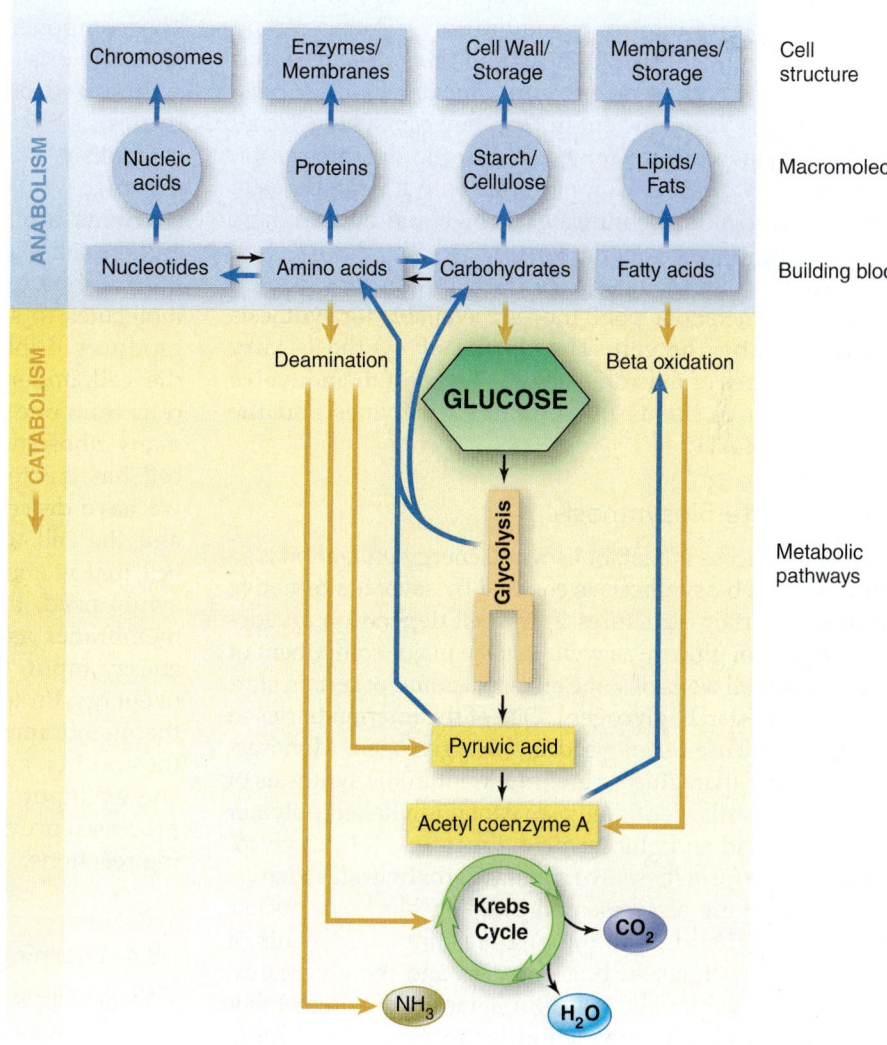

Figure 8.23 An amphibolic view of metabolism. A simple compound such as glucose is both broken down into smaller parts (catabolism) and modified and used as a building block in anabolism.

pathways are also dependent on amino acids. During times of carbohydrate deprivation, organisms can likewise convert amino acids to intermediates of the Krebs cycle by deamination and thereby derive energy from proteins (see figure 8.22).

Anabolism: Formation of Macromolecules

Monosaccharides, amino acids, fatty acids, nitrogen bases, and vitamins—the building blocks that make up the various macromolecules and organelles of the cell—

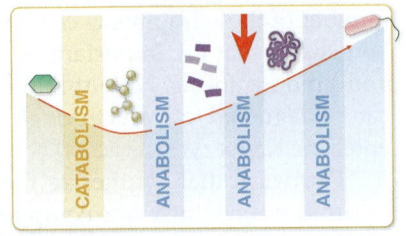

come from two possible sources. They can enter the cell from the outside "ready to use," or they can be synthesized through various cellular pathways. The degree to which an organism can synthesize its own building blocks (simple molecules) is determined by its genetic makeup, a factor that varies tremendously from group to group. In chapter 7, you learned that autotrophs require only CO_2 as a carbon source, a few minerals to synthesize all cell substances, and no organic nutrients. Some heterotrophic organisms (*E. coli*, yeasts) are also very efficient in that they can synthesize all cellular substances from minerals and one organic carbon source such as glucose. Compare this with a strict parasite that has few synthetic abilities of its own and derives most precursor molecules from the host.

Whatever their source, once these building blocks are added to the metabolic pool, they are available for synthesis of polymers by the cell. The details of synthesis vary among the types of macromolecules, but all of them involve the formation of bonds by specialized enzymes and the expenditure of ATP.

Carbohydrate Biosynthesis

The role of glucose in metabolism and energy utilization is so crucial that its biosynthesis is ensured by several alternative pathways. Certain structures in the cell depend on an adequate supply of glucose as well. It is the major component of the cellulose cell walls of some eukaryotes and of certain storage granules (starch, glycogen). One of the intermediaries in glycolysis, glucose-6-P, is used to form glycogen. Monosaccharides other than glucose are important in the synthesis of bacterial cell walls. Peptidoglycan contains a linked polymer of muramic acid and glucosamine. Fructose-6-P from glycolysis is used to form these two sugars. Carbohydrates (deoxyribose, ribose) are also essential building blocks in nucleic acids. Polysaccharides are the predominant components of cell surface structures such as capsules and the glycocalyx, and they are commonly found in slime layers. Remember that most polymerization reactions occur via loss of a water molecule and the input of energy.

Amino Acids, Protein Synthesis, and Nucleic Acid Synthesis

Proteins account for a large proportion of a cell's constituents. They are essential components of enzymes, the cell membrane, the cell wall, and cell appendages. As a general rule, 20 amino acids are needed to make these proteins. Although some organisms (*E. coli*, for example) have pathways that will synthesize all 20 amino acids, others, including animals, lack some or all of the pathways for amino acid synthesis and must acquire the essential ones from their diets. Protein synthesis itself is a complex process that requires a genetic blueprint and the operation of intricate cellular machinery, as you will see in chapter 9.

DNA and RNA are responsible for the hereditary continuity of cells and the overall direction of protein synthesis. Because nucleic acid synthesis is a major topic of genetics and is closely allied to protein synthesis, it will likewise be covered in chapter 9.

Assembly of the Cell

The component parts of a bacteria cell are synthesized on a continuous basis, and catabolism is also taking place as long as nutrients are present and the cell is in a non-

dormant state. When anabolism produces enough macromolecules to serve two cells, and when DNA replication produces duplicate copies of the cell's genetic material, the cell undergoes binary fission, which results in two cells from one parent cell. The two cells will need twice as many ribosomes, twice as many enzymes, and so on. The cell has created these during the initial anabolic phases we have described. Before cell division, the membrane(s) and the cell wall will have increased in size to create a cell that is almost twice as big as a "newborn" cell. Once synthesized, the phospholipid bilayer components of the membranes assemble themselves spontaneously with no energy input. Other assembly reactions require the input of energy. Proteins and other components must be added to the membranes. Growth of the cell wall, accomplished by the addition and coupling of sugars and peptides, requires energy input. The energy accumulated during catabolic processes provides all the energy for these complex building reactions.

8.4 Learning Outcomes—Assess Your Progress

17. Provide an overview of the anabolic stages of metabolism.
18. Define *amphibolism*.

8.5 Photosynthesis: It All Starts with Light

As mentioned earlier, the ultimate source of most of the chemical energy in cells comes from the sun. Most organisms depend either directly or indirectly on the sunlight's energy, which is converted into chemical energy through photosynthesis. (Some chemoautotrophs derive their energy and nutrients solely from inorganic substrates.) The other major products of photosynthesis are organic carbon compounds, which are produced from carbon dioxide through a process called carbon fixation.

With few exceptions, the energy that drives all life processes comes from the sun, but this source is directly available only to the cells of photosynthesizers. In the terrestrial biosphere, green plants are the primary photosynthesizers; in aquatic ecosystems, algae, green and purple bacteria, and cyanobacteria fill this role. It was also recently discovered that bacteriophages that infect marine cyanobacteria provide some of the genes allowing these organisms to carry out photosynthesis.

Photosynthetic organisms use light energy to produce high-energy glucose from low-energy CO_2 and water. They do this through a series of reactions involving light, pigment, CO_2, and water, which is used as a source for electrons.

Photosynthesis proceeds in two phases: the **light-dependent reactions,** which proceed only in the presence of sunlight, and the **light-independent reactions,** which proceed regardless of the lighting conditions (light or dark).

Solar energy is delivered in discrete energy packets called **photons** (also called *quanta*) that travel as waves. The wavelengths of light operating in photosynthesis occur in the visible spectrum between 400 (violet) and 700 nanometers (red). As this light strikes photosynthetic pigments, some wavelengths are absorbed, some pass through, and some are reflected. The activity that has greatest impact on photosynthesis is the absorbance of light by photosynthetic pigments. These include the **chlorophylls,** which are green; **carotenoids,** which are yellow, orange, or red; and **phycobilins,** which are red or blue-green.[4] The most important of these pigments are the bacterial chlorophylls, which contain a *photocenter* that consists of a magnesium atom held in the center of a complex, ringed molecule called a *porphyrin.* As we will see, the chlorophyll molecule harvests the energy of photons and converts it to chemical energy. Accessory photosynthetic pigments such as carotenes trap light energy and shuttle it to chlorophyll, thereby functioning like antennae. These light-dependent reactions are catabolic (energy-producing) reactions, which pave the way for the next set of reactions, the light-independent reactions, which use the produced energy for synthesis (anabolism). During this phase, carbon atoms from CO_2 are added to the carbon backbones of organic molecules.

The detailed biochemistry of photosynthesis is beyond the scope of this text, but we will provide an overview of the general process as it occurs in green plants, algae, and cyanobacteria **(figure 8.24).** Many of the basic activities (electron transport and phosphorylation) are biochemically similar to certain pathways of respiration.

Light-Dependent Reactions

The same systems that carry the photosynthetic pigments are also the sites for the light reactions. They occur in the **thylakoid** membranes of compartments called grana (singular, *granum*) in chloroplasts **(process figure 8.25)** and in specialized parts of the cell membranes in bacteria (see chapter 4). These systems exist as two separate complexes called *photosystem I* (P700) and *photosystem II* (P680).[5] Both systems contain chlorophyll and they are simultaneously activated by light, but the reactions in photosystem II help

5. The numbers refer to the wavelength of light to which each system is most sensitive.

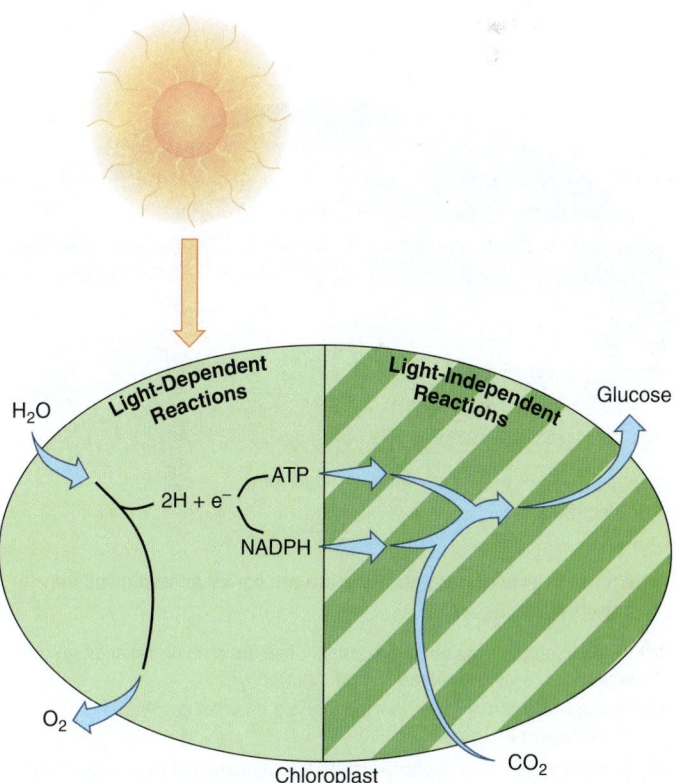

Figure 8.24 Overview of photosynthesis. The general reactions of photosynthesis, divided into two phases called light-dependent reactions and light-independent reactions. The dependent reactions require light to activate chlorophyll pigment and use the energy given off during activation to split an H_2O molecule into oxygen and hydrogen, producing ATP and NADPH. The independent reactions, which occur either with or without light, utilize ATP and NADPH produced during the light reactions to fix CO_2 into organic compounds such as glucose.

4. The color of the pigment corresponds to the wavelength of light it reflects.

drive photosystem I. Together the systems are activated by light, transport electrons, pump hydrogen ions, and form ATP and NADPH.

When photons enter the photocenter of the P680 system (PS II), the magnesium atom in chlorophyll becomes excited

and releases 2 electrons. The loss of electrons from the photocenter has two major effects:

1. It creates a vacancy in the chlorophyll molecule forceful enough to split an H_2O molecule into hydrogen (H^+) (electrons and hydrogen ions) and oxygen (O_2). This splitting of water, termed **photolysis,** is the ultimate source of the O_2 gas that is an important product of photosynthesis. The electrons released from the lysed water regenerate photosystem II for its next reaction with light.

2. Electrons generated by the first photoevent are immediately shunted through a series of carriers (cytochromes) to the P700 system. At this same time, hydrogen ions accumulate in the internal space of the thylakoid complex, thereby producing an electrochemical gradient.

The P700 system (PS I) has been activated by light so that it is ready to accept electrons generated by the PS II. The electrons it receives are passed along a second transport chain to a complex that uses electrons and hydrogen ions to reduce NADP to NADPH. (Recall that reduction in this sense entails the addition of electrons and hydrogens to a substrate.)

A second energy reaction involves synthesis of ATP by a chemiosmotic mechanism similar to that shown in figures 8.18 and 8.19. Channels in the thylakoids of the granum actively pump H^+ into the inner chamber, producing a charge gradient. ATP synthase located in this same thylakoid uses the energy from H^+ transport to phosphorylate ADP to ATP. Because it occurs in light, this process is termed **photophosphorylation.** Both NADPH and ATP are released into the stroma of the chloroplast, where they drive the reactions of the **Calvin cycle.**

Light-Independent Reactions

The subsequent photosynthetic reactions that do not require light occur in the chloroplast stroma or the cytoplasm of cyanobacteria. These reactions use energy produced by the light phase to synthesize glucose by means of the Calvin cycle **(figure 8.26).**

The cycle begins at the point where CO_2 is combined with a doubly phosphorylated 5-carbon acceptor molecule called ribulose-1,5-bisphosphate (RuBP). This process, called **carbon fixation,** generates a 6-carbon intermediate compound that immediately splits into two 3-carbon molecules of 3-phosphoglyceric acid (PGA). The subsequent steps use the ATP and NADPH generated by the photosystems to form high-energy intermediates. First, ATP adds a second phosphate to 3-PGA and produces 1,3-bisphosphoglyceric acid (BPG). Then, during the same step, NADPH contributes its hydrogen to BPG, and one high-energy phosphate is removed. These events give rise to glyceraldehyde-3-phosphate (PGAL). This molecule and its isomer dihydroxyacetone phosphate (DHAP) are key molecules in hexose synthesis leading to fructose and glucose. You may notice that this pathway is very similar to glycolysis, except that it runs in reverse (see process figure 8.15). Bringing the cycle

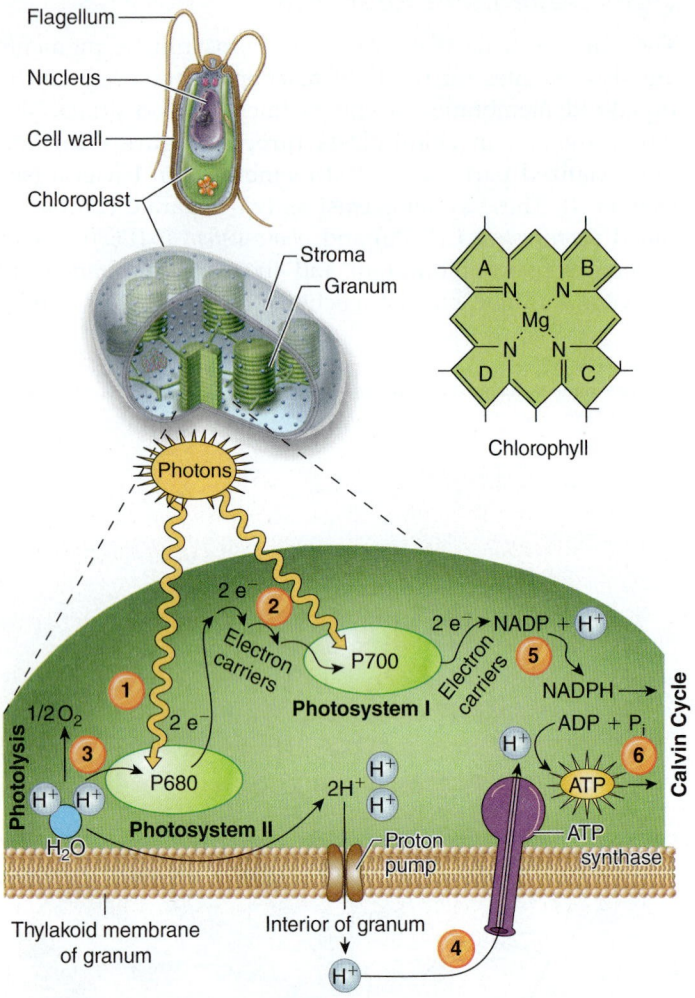

The main events of the light reaction shown as an exploded view in one granum.

1. When light activates photosystem II, it sets up a chain reaction, in which electrons are released from chlorophyll.

2. These electrons are transported along a chain of carriers to photosystem I.

3. The empty position in photosystem II is replenished by photolysis of H_2O. Other products of photolysis are O_2 and H^+.

4. Pumping of H^+ into the interior of the granum produces conditions for ATP to be synthesized.

5. The final electron and H^+ acceptor is NADP, which receives these from photosystem I.

6. Both NADPH and ATP are fed into the stroma for the Calvin cycle.

Process Figure 8.25 The reactions of photosynthesis.
A cell of the eukaryotic motile alga *Chlamydomonas*, with a single large chloroplast (magnified cutaway view). The chloroplast contains membranous compartments called grana where chlorophyll molecules and the photosystems for the light molecules are located.

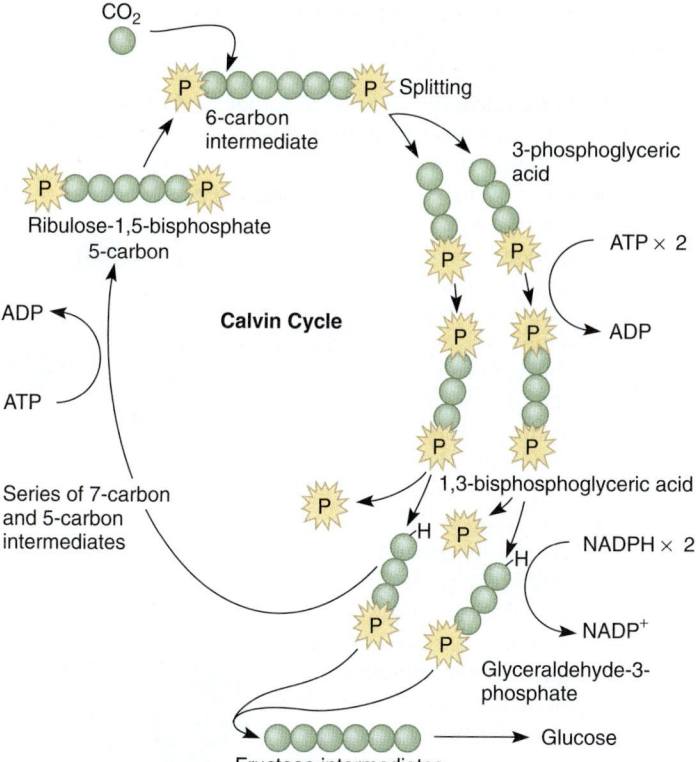

Figure 8.26 The Calvin cycle. The main events of the reactions in photosynthesis that do not require light. It is during this cycle that carbon is fixed into organic form using the energy (ATP and NADPH) released by the light reactions. The end product, glucose, can be stored as complex carbohydrates, or it can be used in various amphibolic pathways to produce other carbohydrate intermediates or amino acids.

back to regenerate RuBP requires PGAL and several steps not depicted in figure 8.26.

Other Mechanisms of Photosynthesis

The **oxygenic,** or oxygen-releasing, photosynthesis that occurs in plants, algae, and cyanobacteria is the dominant type on the earth. Other photosynthesizers such as green and purple bacteria possess bacteriochlorophyll, which is more versatile in capturing light. They have only a cyclic photosystem I, which routes the electrons from the photocenter to the electron carriers and back to the photosystem again. This pathway generates a relatively small amount of ATP, and it may not produce NADPH. As photolithotrophs, these bacteria use H_2, H_2S, or elemental sulfur rather than H_2O as a source of electrons and reducing power. As a consequence, they are **anoxygenic** (non-oxygen-producing), and many are strict anaerobes.

While most of the mechanisms just described involve chlorophyll or bacteriochlorophyll as the light-absorbing pigment, archaea use a pigment called bacteriorhodopsin. You may recognize the root *rhodopsin*, which is a pigment present in vertebrate eyes (in the rods and cones). This type of photosynthesis does not involve electron transport but instead uses a light-driven proton pump. Through chemiosmosis, it generates ATP. Until recently, it was thought that only archaea can photosynthesize in this manner, but in 2000 evidence became available that many "common" bacteria in the ocean—as many as 13% of surface marine bacteria—use a rhodopsin-like molecule, called proteorhodopsin, to photosynthesize.

Case File 8 — *Wrap-Up*

According to recent reports, production of biofuels has increased 17% worldwide. Brazil and the United States are the largest producers of biofuels, with the majority of biofuels produced from corn and sugarcane. The increase in biofuel production is largely motivated by rising oil prices and increased mandates in the United States and worldwide for greater climate protection and reduction of greenhouse gas emissions.

It seems like an excellent idea: Biofuels will reduce dependence on fossil fuels, will reduce carbon emissions, and are sustainable sources of energy. Unfortunately, biofuel production is not as simple as it seems. Fossil fuels are often heavily employed during the production and transportation of biofuels. The nitrogen fertilizer for biofuel crops is produced using natural gas. Increased production of plants for biofuel use can lead to increased water use and increased pollution. Some researchers have shown that biofuel production is less effective than implementing policies aimed at reducing use of fossil fuels and increasing energy efficiency.

Source: W. H. van Zyl, A. F. A. Chimphango, R. den Haan, J. F. Gorgens, and P. W. C. Chirwa, "Next-Generation Cellulosic Ethanol Technologies and Their Contribution to a Sustainable Africa," *Interface Focus* 1, no. 2 (2011): 196–211.

8.5 Learning Outcomes—Assess Your Progress

19. Summarize the overall process of photosynthesis in a single sentence.

20. Discuss the relationship between light-dependent and light-independent reactions.

21. Explain the role of the Calvin cycle in the process of photosynthesis.

Chapter Summary

8.1 The Metabolism of Microbes (ASM Guidelines* 3.1, 3.3, 3.4, 4.3, 5.4)

- Metabolism is the sum of cellular chemical and physical activities. It consists of anabolism, energy-requiring reactions that convert small molecules into large molecules, and catabolism, in which large molecules are degraded and energy is produced.
- Metabolism is made possible by organic catalysts called enzymes that speed up reactions by lowering the energy of activation.
- Enzymes are not consumed and can be reused. Each enzyme acts specifically upon its matching molecule or substrate.
- Substrates attach to enzymes in a special pocket called the active, or catalytic, site.
- Many pathogens secrete enzymes or toxins, referred to as virulence factors, that enable them to avoid host defenses.

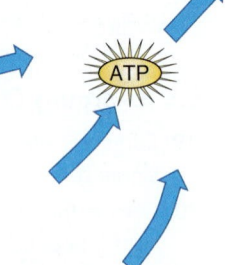

- Enzymes are labile (unstable) and function only within narrow operating ranges of temperature, osmotic pressure, and pH, and they are especially vulnerable to denaturation.
- Enzymes are frequently the targets for physical and chemical agents used in control of microbes.
- Regulatory controls can act on enzymes directly or on the process that gives rise to the enzymes.

8.2 The Pursuit and Utilization of Energy (ASM Guideline 3.1)

- Energy is the capacity of a system to perform work. It is consumed in endergonic reactions and is released in exergonic reactions.
- Extracting energy requires a series of electron carriers arrayed in a redox chain between electron donors and electron acceptors.

8.3 Catabolism: Getting Materials and Energy (ASM Guidelines 1.1, 3.1, 4.3, 6.1, 6.3)

- Carbohydrates, such as glucose, are energy-rich because when catabolized they can yield a large number of electrons per molecule.
- Glycolysis is a pathway that degrades glucose to pyruvic acid without requiring oxygen.

- Pyruvic acid is processed in aerobic and anaerobic respiration via the Krebs cycle and its associated electron transport chain.
- Acetyl coenzyme A is the product of pyruvic acid processing that undergoes further oxidation and decarboxylation in the Krebs cycle, which generates ATP, CO_2, and H_2O.

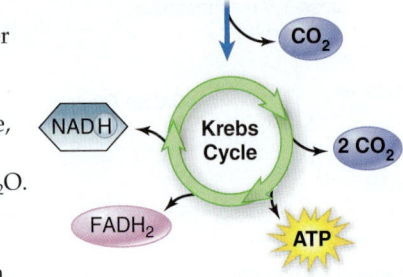

- The respiratory chain completes energy extraction.
- The final electron acceptor in aerobic respiration is oxygen. In anaerobic respiration, compounds such as sulfate, nitrate, or nitrite serve this function.
- Bacteria serve as important agents in the nitrogen cycle, and in other processes such as nitrogen fixation.
- Fermentation is an anaerobic process in which both the electron donor and final electron acceptors are organic compounds.
- Production of alcohol, vinegar, and certain industrial solvents relies upon fermentation.
- Glycolysis and the Krebs cycle are central pathways that link catabolic and anabolic pathways, allowing cells to break down different classes of molecules in order to synthesize compounds required by the cell.
- Intermediates such as pyruvic acid can be made into amino acids through amination.
- Amino acids can be deaminated and used as precursors to glucose and other carbohydrates (gluconeogenesis).
- Two-carbon acetyl molecules from pyruvate can be used in fatty acid synthesis.

8.4 Biosynthesis and the Crossing Pathways of Metabolism (ASM Guidelines 3.1, 4.3)

- The ability of a cell or system to integrate catabolic and anabolic pathways to improve efficiency is called amphibolism.
- Macromolecules, such as proteins, carbohydrates, and nucleic acids, are made of building blocks from two possible sources: from outside the cell (preformed) or via synthesis in one of the anabolic pathways.

8.5 Photosynthesis: It All Starts with Light (ASM Guidelines 1.1, 3.1, 6.1)

- Photosynthesis converts the sun's energy into chemical energy and organic carbon compounds, which are produced from carbon dioxide.

*Source: ASM Curriculum Guidelines (American Society for Microbiology, 2012). Complete guidelines in appendix B of this book.

Multiple-Choice and True-False Questions | Bloom's Levels 1 and 2: Remember and Understand

Multiple-Choice Questions. Select the correct answer from the options provided.

1. Catabolism is a form of metabolism in which ____ molecules are converted into ____ molecules.
   a. large, small   c. amino acid, protein
   b. small, large   d. food, storage

2. An enzyme
   a. becomes part of the final products.
   b. is nonspecific for substrate.
   c. is consumed by the reaction.
   d. is heat and pH labile.

3. An apoenzyme is where the ____ is located.
   a. cofactor   c. redox reaction
   b. coenzyme   d. active site

4. Many coenzymes are
   a. metals.   c. proteins.
   b. vitamins.   d. substrates.

5. To digest cellulose in its environment, a fungus produces a/an
   a. endoenzyme.   c. catalase.
   b. exoenzyme.   d. polymerase.

6. Energy is carried from catabolic to anabolic reactions in the form of
   a. ADP.
   b. high-energy ATP bonds.
   c. coenzymes.
   d. inorganic phosphate.

7. A product or products of glycolysis is/are
   a. ATP.   c. CO_2.
   b. H_2O.   d. both a and b.

8. Fermentation of a glucose molecule has the potential to produce a net number of ____ ATPs.
   a. 4   c. 40
   b. 2   d. 0

9. Complete oxidation of glucose in aerobic respiration can yield a net output of ____ ATPs.
   a. 40   c. 38
   b. 6   d. 2

10. ATP synthase complexes can generate ____ ATPs for each NADH that enters electron transport.
    a. 1   c. 3
    b. 2   d. 4

True-False Questions. If the statement is true, leave as is. If it is false, correct it by rewriting the sentence.

11. All photosynthesis begins with light.

12. An enzyme lowers the activation energy required for a chemical reaction.

13. One cycle of fermentation yields more energy than one cycle of aerobic respiration.

14. Energy in biological systems is primarily chemical.

15. Exoenzymes are produced outside the cell.

Critical Thinking Questions | Bloom's Levels 3, 4, and 5: Apply, Analyze, and Evaluate

Critical thinking is the ability to reason and solve problems using facts and concepts. These questions can be approached from a number of angles and, in most cases, they do not have a single correct answer.

1. a. List four characteristics of an enzyme and explain the role each plays in an enzyme's catalytic activity.
   b. Diagram an enzyme-substrate reaction involving an apoenzyme, making sure to label the active site, the substrate molecule, and the products formed in the reaction.

2. a. As you learned in previous chapters, fungi are decomposers. Discuss how they might use exoenzymes or endoenzymes to feed.
   b. Provide brief definitions of *constitutive enzyme* and *regulated enzyme,* explaining one use for each enzyme within a biological cell.

3. Polymerase chain reaction (PCR) is a technology that requires high temperatures to reproduce DNA fragments. Explain why the discovery of thermophilic archaea and their associated DNA polymerases was critical to the success of this technique.

4. a. Describe the chemical properties of ATP that make it suitable as "metabolic money" within a biological cell.
   b. Compare and contrast the processes of substrate-level phosphorylation, oxidative phosphorylation, and photophosphorylation. Provide one example of how each is used in a biological cell.

5. a. List two common electron carriers used in biological cells and summarize the role they play in the production of ATP.
   b. Based upon the information in table 8.2, explain which type or types of enzyme supply electrons to these carriers.

6. a. Compare the main reactions of aerobic respiration, anaerobic respiration, and fermentation in a diagram.
   b. What are the final electron acceptors in aerobic, anaerobic, and fermentative metabolism? How many ATPs are produced in each process, explaining why this value varies among these biochemical reactions.

7. a. Draw a bacterial cell and a eukaryotic cell side by side. Label where each of the following steps would take place:
      glycolysis
      Krebs cycle
      electron transport
   b. Describe the fate of pyruvic acid in strict aerobes, anaerobes, and facultative anaerobes.

8. Summarize how the electron transport chain functions to produce ATP in a bacterial cell versus a eukaryotic cell. Explain whether or not the term *chemiosmosis* accurately illustrates this process.

9. *Saccharomyces cerevisiae* is used in the production of wine and beer, while *Lactobacillus acidophilus* is used in the making of yogurt. Explain why each is biochemically suited to the manufacturing of each of these products.

10. Provide evidence in support of or refuting the following statement: The evolution of aerobic respiration was driven by the success of photosynthetic microbes.

Concept Connections | Bloom's Levels 4 and 6: Analyze and Create

This activity ties together multiple concepts in the chapter.

1. What are the three types of fermentation discussed in this chapter?

2. What are the by-products of each type of fermentative activity?

3. What types of organisms accomplish each type of fermentation? Give the genus name or general type of organism.

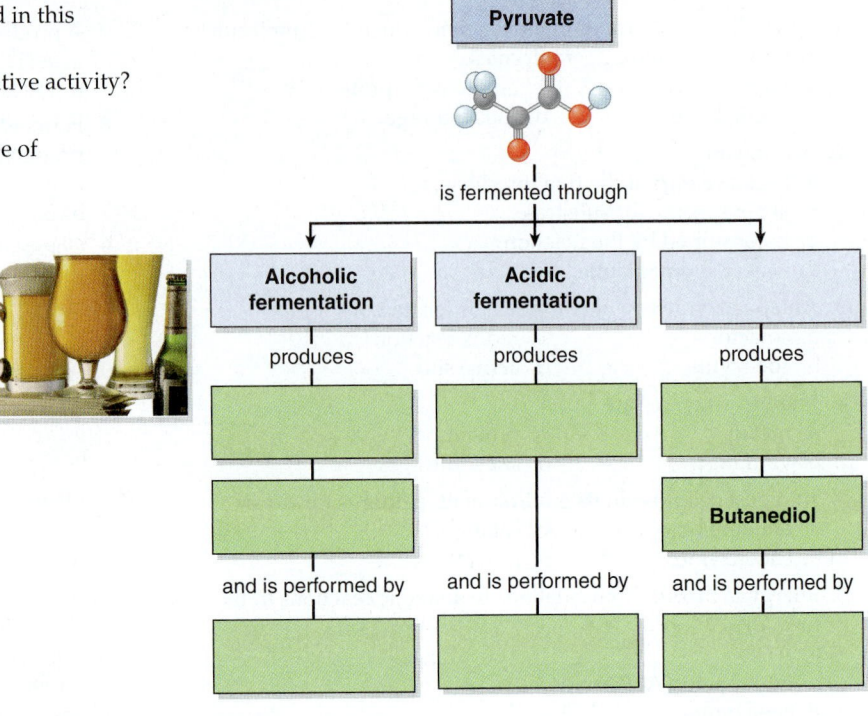

Visual Connections | Bloom's Level 5: Evaluate

These questions use visual images or previous content to make connections to this chapter's concepts.

1. **From chapter 4, figure 4.16.** On the enlarged sections of both **(a)** and **(b)**, draw protons in the proper compartment in such a way that it illustrates the creation of a proton motive force.

Concept Mapping | Bloom's Level 6: Create

Appendix D provides guidance for working with concept maps.

1. Using the words that follow, please create a concept map illustrating the relationships among these key terms from chapter 8.

metabolism	products	activation energy	substrates
anabolism	pH	catalysts	
catabolism	temperature	enzymes	

www.mcgrawhillconnect.com

Enhance your study of this chapter with study tools and practice tests. Also ask your instructor about the resources available through ConnectPlus, including the media-rich eBook, interactive learning tools, and animations.

Microbial Genetics

Case File 9

Bioluminescence: The Oceans Are Glowing

Maybe you've been walking along a lovely beach at night when you noticed flashes of light coming from the waves. Or you might have noticed it in the wake of a ship in the night—iridescent blue water rippling around the hull. The glow has even been seen from space: In 1995, satellites detected luminescent water off the coast of Somalia for three consecutive nights. Sailors have described these "milky seas" for hundreds of years—there was even mention of a "milk sea" in Jules Verne's classic *20,000 Leagues Under the Sea*. What causes these shimmering displays? Dinoflagellates, a group of eukaryotic marine plankton, cause some of the iridescent surf, but in many cases, bacteria are the source of light in the oceans. The phenomenon of living organisms giving off light is called **bioluminescence.**

Many species of animals, fish, insects, and bacteria are bioluminescent. Scientists have studied these organisms and have discovered that certain species of bioluminescent bacteria live symbiotically with organisms that use the light they emit to attract prey or to scare off predators. *Vibrio, Photorhabdus,* and *Photobacterium* are some of the genera that live symbiotically with certain crustaceans and fish. Some of these organisms are free-living as well and cause the beautiful oceanic light displays we have discussed.

But how do these bacteria emit light? An **operon**—a genetic operational unit that regulates a genetic product by controlling mRNA production—is responsible. In addition to containing more than one structural gene, operons have a regulatory gene, control sites, and an operator region that acts as an on/off switch. In bioluminescent bacteria, this is known as the *lux* operon, which controls production of **luciferase,** an enzyme that leads to the emission of light.

- Why would arranging genes in an operon be advantageous to a bacterium?
- What stimulates a bacterium to produce light?

Continuing the Case appears on page 251.

Outline and Learning Outcomes

9.1 Introduction to Genetics and Genes: Unlocking the Secrets of Heredity

1. Define the terms *genome* and *gene*.
2. Differentiate between genotype and phenotype.

3. Diagram a segment of DNA, labeling all important chemical groups within the molecule.
4. Summarize the steps of bacterial DNA replication and the enzymes used in this process.
5. Compare and contrast the synthesis of leading and lagging strands during DNA replication.

9.2 Applications of the DNA Code: Transcription and Translation

6. Explain how the classical view of the "central dogma" has been changed by modern science.
7. Identify important structural and functional differences between RNA and DNA.
8. Illustrate the steps of transcription, noting the key elements and the direction of mRNA synthesis.
9. List the three types of RNA directly involved in translation.
10. Define the terms *codon* and *anticodon*, and list the four known start and stop codons.
11. Identify the locations of the promoter, the start codon, and the A and P sites during translation.
12. Indicate how eukaryotic transcription and translation differ from these processes in bacteria and archaea.
13. Explain the relationship between genomics and proteomics.

9.3 Genetic Regulation of Protein Synthesis

14. Define the term *operon* and explain one advantage it provides to a bacterial cell.
15. Differentiate between repressible and inducible operons and provide an example of each.
16. List several antibiotic drugs and their targets within the transcription and translation machinery.

9.4 DNA Recombination Events

17. Explain the defining characteristics of a recombinant organism.
18. Describe three forms of horizontal gene transfer used in bacteria.

9.5 Mutations: Changes in the Genetic Code

19. Define the term *mutation* and discuss one positive and one negative example of it in microorganisms.
20. Differentiate among frameshift, nonsense, silent, and missense mutations.

9.1 Introduction to Genetics and Genes: Unlocking the Secrets of Heredity

Genetics is the study of the inheritance, or **heredity,** of living things. It is a wide-ranging science that explores

1. the transmission of biological properties (traits) from parent to offspring,
2. the expression and variation of those traits,
3. the structure and function of the genetic material, and
4. how this material changes.

This chapter will explore DNA, which is the genetic material, and the proteins and other products that it produces in a cell. Coming out of chapter 8, we should point out that the production of new DNA, RNA, and proteins is an example of an anabolic process.

The study of genetics takes place on several levels **(figure 9.1).** *Organismal genetics* observes the heredity of the whole organism or cell; *chromosomal genetics* examines the characteristics and actions of chromosomes; and *molecular genetics* deals with the biochemistry of the genes. All of these levels are useful areas of exploration, but in order to understand the expressions of microbial structure, physiology, mutations, and pathogenicity, we need to examine the operation of genes at the cellular and molecular levels. The study of *microbial genetics* provides a greater understanding of human genetics and an increased appreciation for the role microbes play in human ecology. It also serves as the basis for

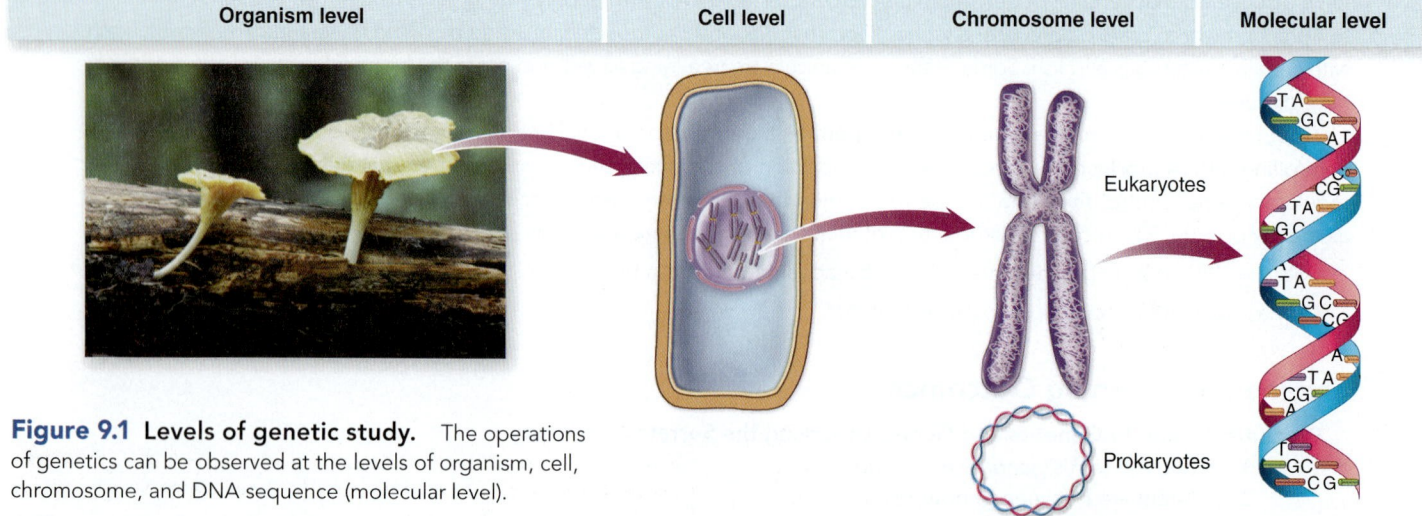

| Organism level | Cell level | Chromosome level | Molecular level |

Eukaryotes

Prokaryotes

Figure 9.1 Levels of genetic study. The operations of genetics can be observed at the levels of organism, cell, chromosome, and DNA sequence (molecular level).

the astounding advances in genetic engineering we are currently witnessing, which we will learn about in chapter 10.

The Nature of the Genetic Material

For a species to survive, it must have the capacity of self-replication. In single-celled microorganisms, reproduction usually involves the division of the cell by means of binary fission or budding and the accurate duplication and separation of genetic material into each daughter cell. Before we look at how DNA is copied, let us explore the organization of this genetic material, proceeding from the general to the specific.

The Levels of Structure and Function of the Genome

The **genome** is the sum total of genetic material of an organism. Although most of the genome exists in the form of chromosomes, genetic material can appear in nonchromosomal sites as well **(figure 9.2)**. For example, bacteria and some fungi contain tiny extra pieces of DNA (plasmids), and certain organelles of eukaryotes (the mitochondria and chloroplasts) are equipped with their own DNA. Genomes of cells are composed exclusively of DNA, but viruses contain either DNA or RNA as the principal genetic material. Although the specific genome of an individual organism is unique, the general pattern of nucleic acid structure and function is similar among all organisms. **Genomics** is the study of an organism's entire genome; research in this area has grown exponentially over the past 10 years, leading to new discoveries about microbial evolution, ecology, and pathogenesis.

In general, a **chromosome** is a discrete cellular structure composed of a neatly packaged DNA molecule. The chromosomes of eukaryotes and bacterial cells differ in several respects. The structure of eukaryotic chromosomes consists of a DNA molecule tightly wound around histone proteins, whereas a bacterial chromosome is condensed and secured into a packet by means of histonelike proteins. Eukaryotic chromosomes are located in the nucleus; they vary in number from a few to hundreds; they can occur in pairs (diploid) or singles (haploid); and they have a linear appearance. In contrast, most bacteria have a single, circular (double-stranded) chromosome, although many bacteria have multiple circular chromosomes and some have linear chromosomes.

The chromosomes of all cells are subdivided into basic informational packets called genes. A **gene** can be defined from more than one perspective. In classical genetics, the term *gene* refers to the fundamental unit of heredity responsible for a given trait in an organism. In the molecular and biochemical sense, it is a site on the chromosome that provides information for a certain cell function. With new findings in the area of gene expression, we now prefer to speak of a gene as a segment of DNA that contains the necessary code to make a **protein** or an RNA.

Genes fall into three basic categories: (1) structural genes that code for proteins, (2) genes that code for the RNA machinery used in protein production, and (3) regulatory genes that control gene expression. The sum of all of these types of genes constitutes an organism's distinctive genetic makeup, or **genotype** (jee″-noh-tīp). The expression of the genotype creates traits (certain structures or functions) referred to as the **phenotype** (fee″-noh-tīp). Just as a person inherits a

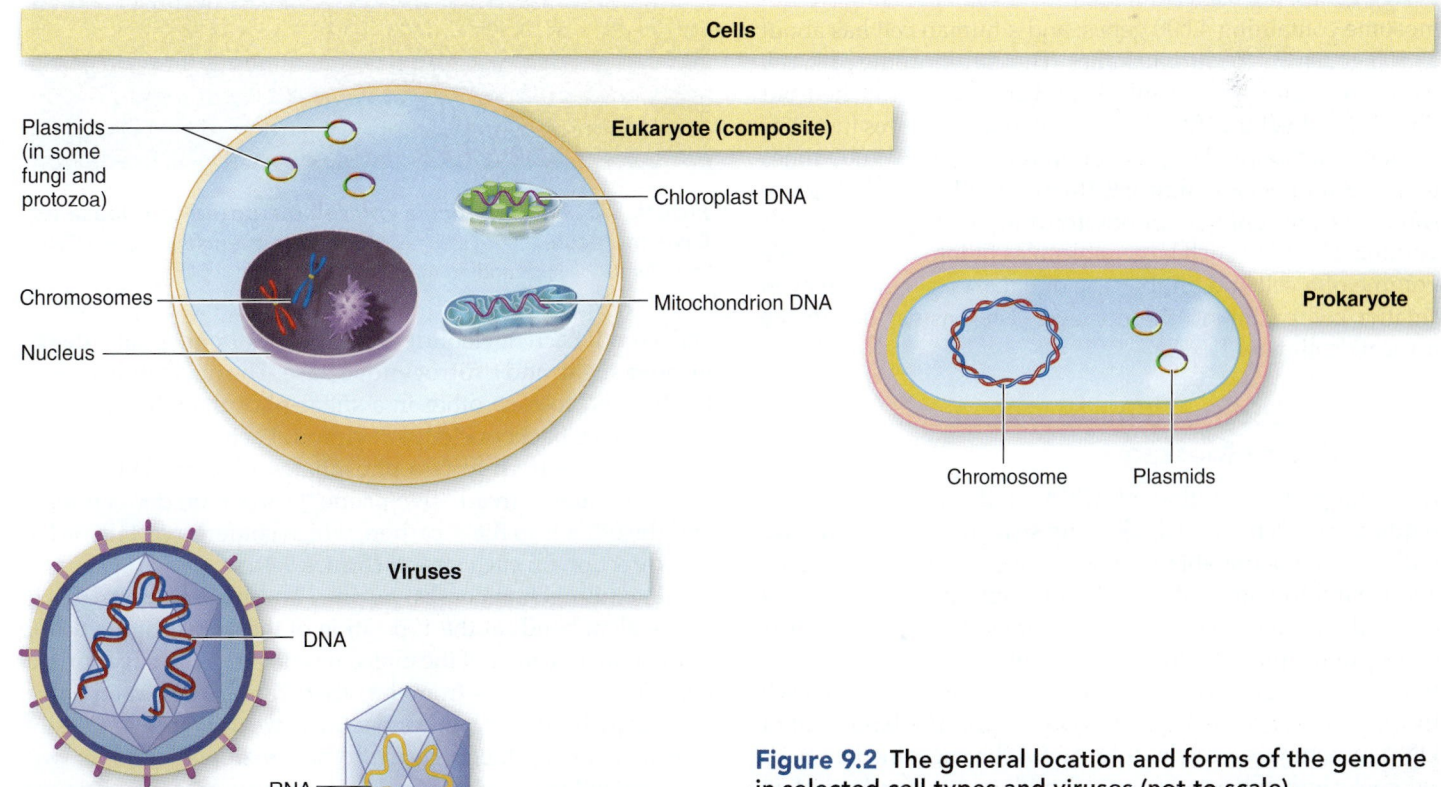

Figure 9.2 The general location and forms of the genome in selected cell types and viruses (not to scale).

INSIGHT 9.1   How Much DNA Does One Bacterium Need?

Every living thing contains thousands, if not millions, of bases of DNA. The human genome contains approximately 250 million base pairs. The smallest living organism contains about 600,000 base pairs. How many of those base pairs are *really* necessary for the function and survival of an organism? Recently, a team at the Stanford University School of Medicine has determined—to the base pair—exactly how many bases of DNA one bacterium needs to survive. The team found that only 12% of the genome of *Caulobacter crescentus* is required for growth under laboratory conditions. The other 88% can be mutated or discarded without preventing the organism from growing and reproducing. The entire genome of *C. crescentus* was sequenced in 2001, but that map of the genetic code did not show what genes were necessary for the survival of the organism. This research shows that the genome of *C. crescentus* contains 1,012 essential genes, 402 regulatory sequences, and 130 noncoding sequences, including 90 segments of unknown function. This will allow scientists to determine what essential genetic elements are conserved through evolution as well as to identify genes that make bacteria infectious, leading to development of new antibiotics.

Source: 2011. *Mol. Systems. Bio.* vol. 7, p. 528.

gene combination (genotype) that gives a certain eye color or height (phenotype), a bacterium inherits genes that direct the formation of a flagellum, and a virus contains genes for its capsid structure. All organisms contain more genes in their genotypes than are manifested as a phenotype at any given time **(Insight 9.1)**. In other words, the phenotype can change depending on which genes are "turned on" (expressed).

The Size and Packaging of Genomes

Genomes vary greatly in size. The smallest viruses have four or five genes; the bacterium *Escherichia coli* has a single chromosome containing 4,000 genes; and a human cell has about 23,000 genes on 46 chromosomes. The chromosome of *E. coli* would measure about 1 mm if unwound and stretched out linearly, and yet this fits within a cell that measures just over 1 micron across, making the stretched-out DNA 1,000 times longer than the cell **(figure 9.3)**. Still, the bacterial chromosome takes up only about one-third to one-half of the cell's volume. How can such large molecules fit into the minuscule volume of a cell and, in the case of eukaryotes, into an even smaller compartment, the nucleus? The answer lies in the intricate coiling of DNA.

The DNA Code: A Simple yet Profound Message

Examining the function of DNA at the molecular level requires an even closer look at its structure. To do this, we will imagine being able to magnify a small piece of a gene about 5 million times. What such fine scrutiny will disclose is one of the great marvels of biology. James Watson and Francis Crick put the pieces of the puzzle together in 1953 to discover that DNA is a gigantic molecule, a type of nucleic acid, with two strands combined into a double helix. The basic unit of DNA structure is a **nucleotide,** and a chromosome in a typical bacterium consists of several million nucleotides linked

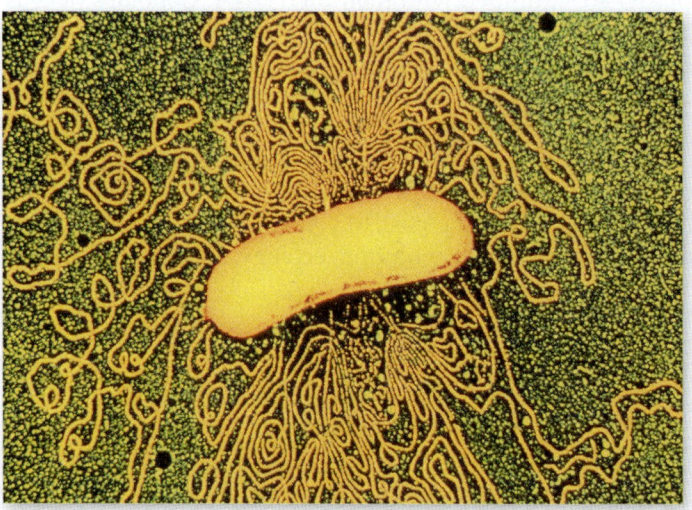

Figure 9.3 An *Escherichia coli* cell disrupted to release its DNA molecule.   The cell has spewed out its single, uncoiled DNA strand into the surrounding medium.

end to end. Each nucleotide is composed of **phosphate, deoxyribose** sugar, and a **nitrogenous base.** The nucleotides covalently bond to each other in a sugar-phosphate linkage that becomes the backbone of each strand. Each sugar attaches in a repetitive pattern to two phosphates. One of the bonds is to the number 5' (read "five prime") carbon on deoxyribose, and the other is to the 3' carbon, which confers a certain order and direction on each strand **(figure 9.4)**.

The nitrogenous bases, **purines** and **pyrimidines,** attach by covalent bonds at the 1' position of the sugar **(figure 9.4*a*)**. They span the center of the molecule and pair with appropriate complementary bases from the other strand. The paired bases are joined by hydrogen bonds. Such weak bonds are easily broken, allowing the molecule to be "unzipped" into its two complementary strands. This feature is of great importance

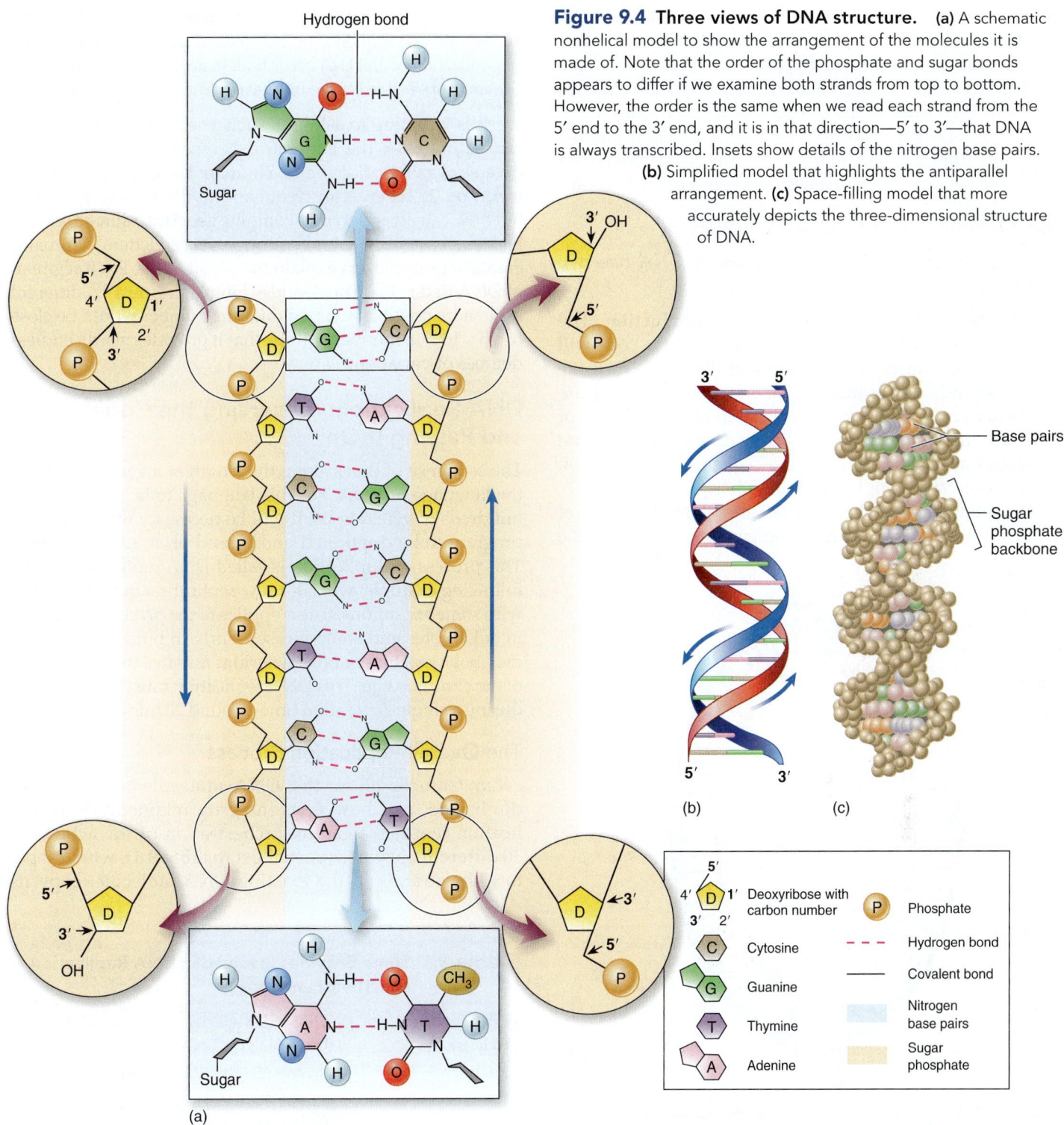

Figure 9.4 Three views of DNA structure. (a) A schematic nonhelical model to show the arrangement of the molecules it is made of. Note that the order of the phosphate and sugar bonds appears to differ if we examine both strands from top to bottom. However, the order is the same when we read each strand from the 5′ end to the 3′ end, and it is in that direction—5′ to 3′—that DNA is always transcribed. Insets show details of the nitrogen base pairs. (b) Simplified model that highlights the antiparallel arrangement. (c) Space-filling model that more accurately depicts the three-dimensional structure of DNA.

in gaining access to the information encoded in the nitrogenous base sequence. Pairing of purines and pyrimidines is not random; it is dictated by the formation of hydrogen bonds between certain bases. Thus, in DNA, the purine **adenine (A)** always pairs with the pyrimidine **thymine (T)**, and the purine **guanine (G)** always pairs with the pyrimidine **cytosine (C)**. The bases are attracted to each other in this pattern because

each has a complementary three-dimensional shape that matches its pair. Although the base-pairing partners generally do not vary, the sequence of base pairs along the DNA molecule can assume any order, resulting in an infinite number of possible nucleotide sequences.

Other important considerations of DNA structure concern the nature of the double helix itself. The halves are not oriented in the same direction. One side of the helix runs in the opposite direction of the other, in what is called an **antiparallel** arrangement **(figure 9.4b)**. The order of the bond between the carbon on deoxyribose and the phosphates is used to keep track of the direction of the two sides of the helix. Thus, one helix runs from the 5′ to 3′ direction, and the other runs from the 3′ to 5′ direction. This characteristic is a significant factor in DNA synthesis and protein production.

The Significance of DNA Structure

The arrangement of nitrogenous bases in DNA has two essential effects:

1. **Maintenance of the code during reproduction.** The constancy of base-pairing guarantees that the code will be retained during cell growth and division. When the two strands are separated, each one provides a template (pattern or model) for the replication (exact copying) of a new molecule **(figure 9.5)**. Because the sequence of one strand automatically gives the sequence of its partner, the code can be duplicated with fidelity.
2. **Providing variety.** The order of bases along the length of the DNA strand constitutes the genetic program, or the

language, of the DNA code. The message present in a gene is a precise sequence of these bases, and the genome is the collection of all DNA bases that, in an ordered combination, are responsible for the unique qualities of each organism.

It is tempting to ask how such a seemingly simple code can account for the extreme differences among forms as diverse as a virus, *E. coli,* and a human. The English language, based on 26 letters, can create an infinite variety of words, but how can an apparently complex genetic language such as DNA be based on just four nitrogen base "letters"? A mathematical example can explain the possibilities. For a segment of DNA that is 1,000 nucleotides long, there are $4^{1,000}$ different sequences possible. Carried out, this number would be close to 1.5×10^{602}, a number so huge that it provides nearly endless degrees of variation.

DNA Replication: Preserving the Code and Passing It On

The sequence of bases along the length of a gene constitutes the language of DNA. For this language to be preserved for hundreds of generations, it will be necessary for the genetic program to be duplicated and passed on to each offspring. This process of duplication is called DNA replication. In the following example, we will show replication in bacteria; but, with some exceptions, it also applies to the process as it works in eukaryotes and some viruses. Early in binary fission, the metabolic machinery of a bacterium initiates the duplication of the chromosome. This DNA replication must be completed during a single generation time (around 20 minutes in *E. coli*).

The Overall Replication Process

What features allow the DNA molecule to be exactly duplicated, and how is its integrity retained? DNA replication requires a careful orchestration of the actions of 30 different enzymes (partial list in **table 9.1**), which separate the strands of the existing DNA molecule, copy its

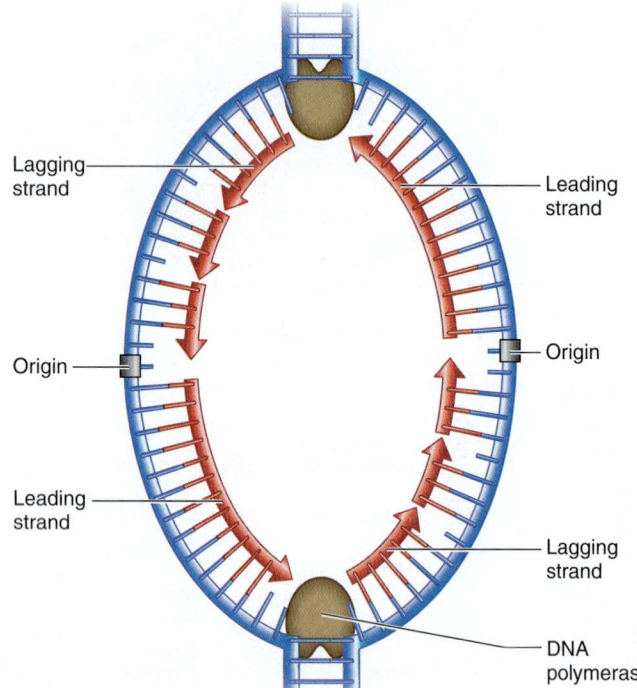

Figure 9.5 Simplified steps to show the semiconservative replication of DNA.   Circular DNA has a special origin site where replication originates. When strands are separated, two replication forks form, and a DNA polymerase III complex enters at each fork. The DNA polymerases proceed in both directions along the DNA molecule, attaching the correct nucleotides according to the pattern of the template. An A on the template will pair with a T on the new molecule, and a C will pair with a G. The resultant new DNA molecules contain one strand of the newly synthesized DNA and the original template strand. The integrity of the code is kept intact because the linear arrangement of the bases is maintained during this process. Note that the actual details of the process are presented in process figure 9.6.

Table 9.1	Some Enzymes Involved in DNA Replication and Their Functions
Enzyme	**Function**
Helicase	Unzipping the DNA helix
Primase	Synthesizing an RNA primer
DNA polymerase III	Adding bases to the new DNA chain; proofreading the chain for mistakes
DNA polymerase I	Removing primer, closing gaps, repairing mismatches
Ligase	Final binding of nicks in DNA during synthesis and repair
Topoisomerase I	Making single-stranded DNA breaks to relieve supercoiling at origin
Topoisomerase II (DNA gyrase) and IV	Making double-stranded DNA breaks to remove supercoiling ahead of origin and separate replicated daughter DNA molecules

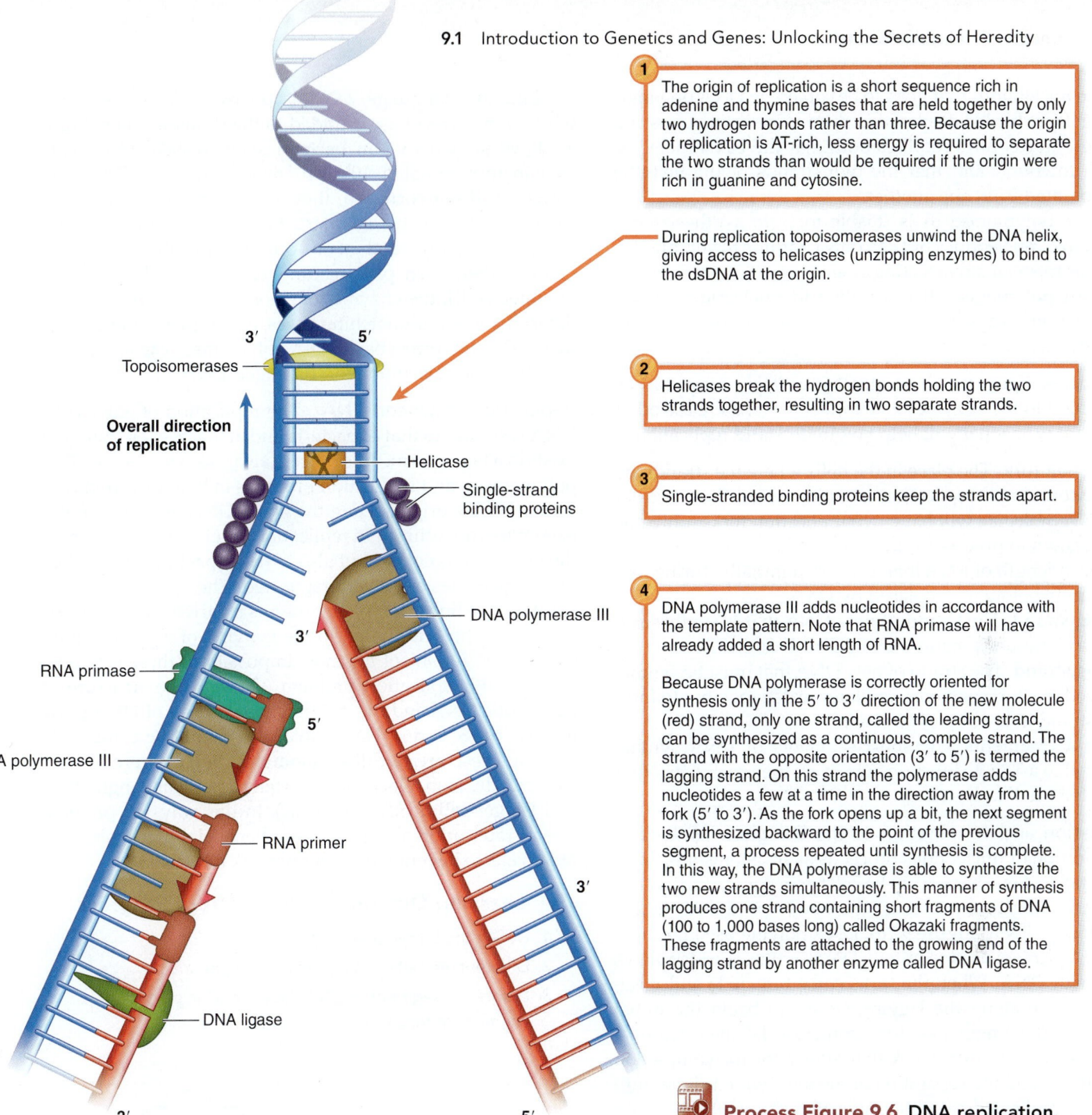

1 The origin of replication is a short sequence rich in adenine and thymine bases that are held together by only two hydrogen bonds rather than three. Because the origin of replication is AT-rich, less energy is required to separate the two strands than would be required if the origin were rich in guanine and cytosine.

During replication topoisomerases unwind the DNA helix, giving access to helicases (unzipping enzymes) to bind to the dsDNA at the origin.

2 Helicases break the hydrogen bonds holding the two strands together, resulting in two separate strands.

3 Single-stranded binding proteins keep the strands apart.

4 DNA polymerase III adds nucleotides in accordance with the template pattern. Note that RNA primase will have already added a short length of RNA.

Because DNA polymerase is correctly oriented for synthesis only in the 5′ to 3′ direction of the new molecule (red) strand, only one strand, called the leading strand, can be synthesized as a continuous, complete strand. The strand with the opposite orientation (3′ to 5′) is termed the lagging strand. On this strand the polymerase adds nucleotides a few at a time in the direction away from the fork (5′ to 3′). As the fork opens up a bit, the next segment is synthesized backward to the point of the previous segment, a process repeated until synthesis is complete. In this way, the DNA polymerase is able to synthesize the two new strands simultaneously. This manner of synthesis produces one strand containing short fragments of DNA (100 to 1,000 bases long) called Okazaki fragments. These fragments are attached to the growing end of the lagging strand by another enzyme called DNA ligase.

Process Figure 9.6 DNA replication.

template, and produce two complete daughter molecules. A simplified version of replication is shown in figure 9.6 and includes the following:

1. uncoiling the parent DNA molecule;
2. unzipping the hydrogen bonds between the base pairs, thus separating the two strands and exposing the nucleotide sequence of each strand (which is normally buried in the center of the helix) to serve as templates; and
3. synthesizing two new strands by attachment of the correct complementary nucleotides to each single-stranded template.

A critical feature of DNA replication is that each daughter molecule will be identical to the parent in composition, but

neither one is completely new; the strand that serves as a template is an original parental DNA strand. The preservation of the parent molecule in this way, termed **semiconservative replication**—*semi-* meaning "half," as in *semicircle*—helps explain the reliability and fidelity of replication.

Refinements and Details of Replication

The process of synthesizing a new daughter strand of DNA using the parental strand as a template is carried out by the enzyme DNA polymerase III. The entire process of replication does, however, depend on several enzymes and can be most easily understood by keeping in mind a few points concerning both the structure of the DNA molecule and the limitations of DNA polymerase III.

1. The nucleotides that need to be copied by DNA polymerase III are buried deep within the double helix. Accessing these nucleotides requires both that the DNA molecule be unwound and that the two strands of the helix be separated from one another.
2. DNA polymerase III is unable to *begin* synthesizing a chain of nucleotides but can only continue to add nucleotides to an already existing chain.
3. DNA polymerase III can only add nucleotides in one direction, so a new strand is always synthesized in a 5′ to 3′ direction.

Process figure 9.6 illustrates the details of replication. In addition to the enzymes listed in table 9.1, there are other important terms that will help you understand replication:

replication fork The place in the helix where the strands are unwound and replication is taking place. Each circular DNA molecule will have two replication forks (only one is shown in process figure 9.6).
primer A length of RNA that is inserted initially during replication before it is replaced by DNA.
leading strand The strand of new DNA that is synthesized in a continuous manner in the 5′ to 3′ direction.
lagging strand The strand of new DNA that must be synthesized in short segments and later sealed together to form a strand in the 3′ to 5′ direction.
Okazaki fragments The short segments of DNA synthesized in a 5′ to 3′ direction, which are then sealed together to form a 3′ to 5′ strand.

Elongation and Termination of the Daughter Molecules

The addition of nucleotides proceeds at an astonishing pace, estimated in some bacteria to be 750 bases per second at each fork. As replication proceeds, the newly produced double strand loops away **(figure 9.7a)**. DNA polymerase I removes the RNA primers used to initiate DNA synthesis and replaces them with DNA. When the forks come full circle and meet, ligases move along the lagging strand to begin the initial linking of the fragments. Topoisomerase IV then causes a double-stranded DNA break that allows for the completion of synthesis and the separation of the intertwined circles into two fully replicated daughter molecules **(figure 9.7b)**.

Like any language, DNA is occasionally "misspelled" when an incorrect base is added to the growing chain. Studies have shown that in bacteria such mistakes are made once in approximately 10^8 to 10^9 bases, but most of these are corrected. If not corrected, they are referred to as *mutations* (covered later in this chapter). Because continued cellular integrity is very dependent on accurate replication, cells have evolved their own proofreading function for DNA. DNA polymerase III, the enzyme that elongates the molecule, can detect incorrect, unmatching bases; excise them; and replace them with the correct base. DNA polymerase I can also proofread the molecule and repair damaged DNA.

Replication of Linear DNA The replication of eukaryotic DNA is similar to that of bacteria and archaea, even though it exists in a linear form. This process also uses a variety of DNA polymerases, and replication proceeds in both directions but from multiple origins along the linear DNA molecule. Topoisomerases are utilized in replication to relieve the tension on the DNA as it is copied but also to recompact the DNA when the molecule is completely replicated. The synthesis of new DNA from a linear template presents a variety of challenges, however, when compared to the copying of a circular molecule of DNA. One of the most important of these is known as the "end replication problem." Due to the structure of eukaryotic DNA and the unidirectional action of DNA polymerase, the 3′ end of DNA molecules cannot be completely copied. These areas, called **telomeres,** begin to erode with each cell division. Once they shorten to a certain length, they will trigger cell death (apoptosis). In this way, the problem of end replication also provides a beneficial mechanism for older cells to be removed in higher eukaryotes.

9.1 Learning Outcomes—Assess Your Progress

1. Define the terms *genome* and *gene*.
2. Differentiate between genotype and phenotype.
3. Diagram a segment of DNA, labeling all important chemical groups within the molecule.

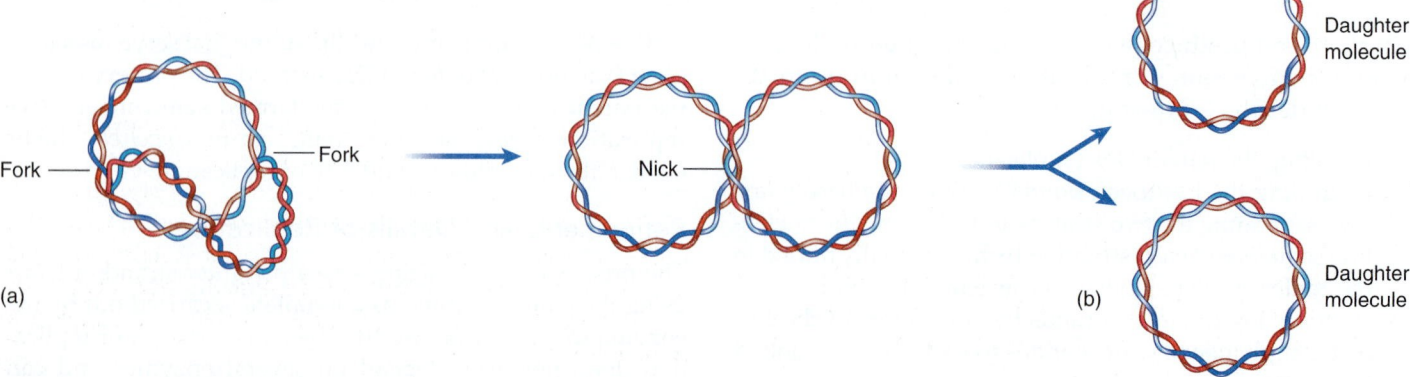

(a)

Fork

Fork

Nick

Daughter molecule

Daughter molecule

(b)

Figure 9.7 Completion of chromosome replication in bacteria. **(a)** As replication proceeds, one double strand loops down. **(b)** Final separation is achieved through action of topoisomerase IV and the final release of two completed molecules. The daughter cells receive these during binary fission.

4. Summarize the steps of bacterial DNA replication and the enzymes used in this process.

5. Compare and contrast the synthesis of leading and lagging strands during DNA replication.

9.2 Applications of the DNA Code: Transcription and Translation

We have explored how the genetic message in the DNA molecule is conserved through replication. Now we must consider the precise role of DNA in the cell. Given that the sequence of bases in DNA is a genetic code, just what is the nature of this code and how is it utilized by the cell? Although the genome is full of critical information, the molecule itself does not perform cell processes directly. Its stored information is conveyed to RNA molecules, which carry out instructions. The concept that genetic information flows from DNA to RNA to protein is a central theme of molecular biology **(figure 9.8a).** More precisely, it states that the master code of DNA is first used to synthesize an RNA molecule via a process called **transcription,** and the information contained in the RNA is then used to produce proteins in a process known as **translation.** The only exceptions to this pattern are found in RNA viruses, which convert RNA to other RNA, and in retroviruses, which convert RNA to DNA.

Disease Connection

The most well-known retrovirus is HIV, which causes AIDS.

This "central dogma," which outlined the primary understanding of genetics during the first half century of the genetic revolution (since the 1950s), has recently been shown to be incomplete. While it is true that proteins are made in accordance with this central dogma, there is more to the story **(figure 9.8b).** In addition to the RNA that is used to produce proteins, a wide variety of RNAs are used to regulate gene function. Many of the genetic malfunctions that cause human disease are in fact found in these regulatory RNA segments—and not in genes for proteins as was once thought. The DNA that codes for these very crucial RNA molecules was called "junk" DNA until very recently **(Insight 9.2).**

The Gene-Protein Connection

Several questions invariably arise concerning the relationship between genes and cell function. For instance, how does gene structure lead to the expression of traits in the individual, and what features of gene expression cause one organism to be so distinctly different from another? For answers, we turn to the correlation between gene and protein structure. We know that each structural gene is a linear sequence of nucleotides that codes for a protein. Because each protein is different, each gene must also differ somehow in its composition. In fact, the language of DNA exists in the order of groups of three

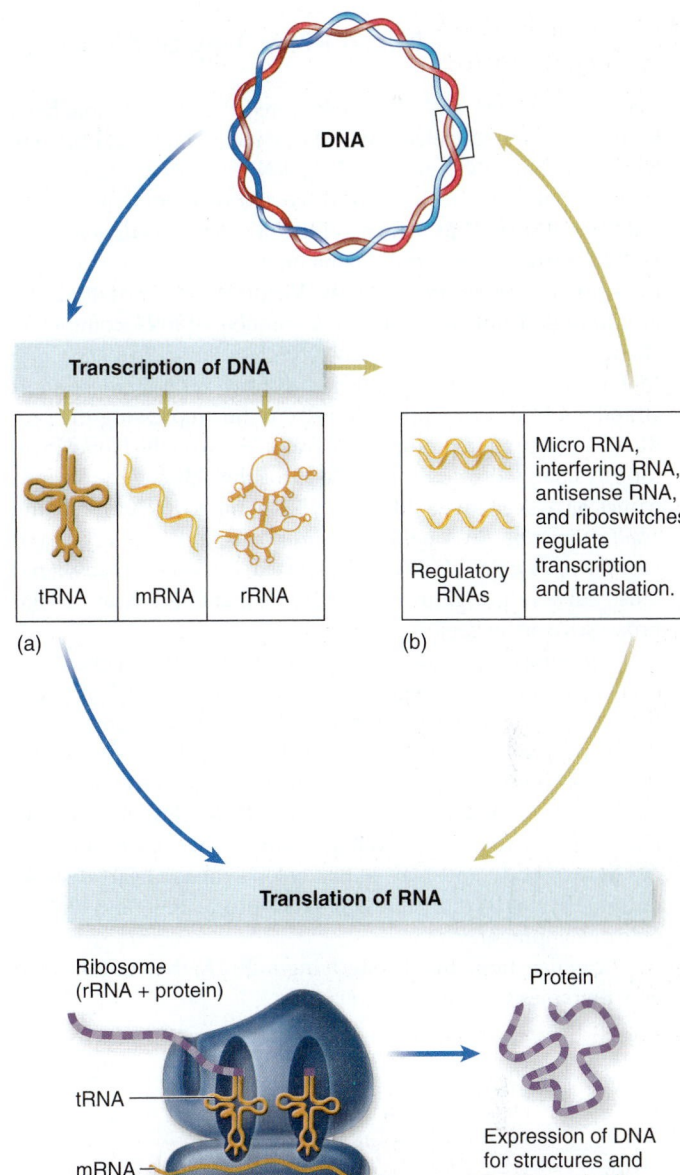

Figure 9.8 Summary of the flow of genetic information in microbes. DNA is the ultimate storehouse and distributor of genetic information. **(a)** DNA must be deciphered into a usable cell language. It does this by transcribing its code into RNA helper molecules that translate that code into protein. **(b)** Other sections of the DNA produce very important RNA molecules that regulate genes and their products.

consecutive bases called *triplets* on one DNA strand **(figure 9.9).** Thus, one gene differs from another in its composition of triplets. An equally important part of this concept is that each triplet represents a code for a particular amino acid. When the triplet code is transcribed and translated, it dictates the type and order of amino acids in a polypeptide (protein) chain.

The final key points that connect DNA and an organism's traits follow:

1. A protein's primary structure—the order and type of amino acids in the chain—determines its characteristic shape and function (see figures 2.21 and 2.22).

INSIGHT 9.2   Micro RNA: Tiny but Mighty

Ever since Watson and Crick discovered the DNA double helix in the 1950s, the so-called "central dogma" of molecular biology has been that DNA makes RNA, RNA makes proteins, and proteins make us. In 1993, the discovery of short sequences of mRNA that were 21 to 35 nucleotides long, termed *micro RNA* or *miRNA,* turned the central dogma on its ear. First discovered in the worm *Caenorhabditis elegans* by Victor Ambros, Rosalind Lee, and Rhonda Feinbaum, these tiny snippets of RNA completely changed the way scientists view DNA regulation. In 1998, Craig Mello and Andrew Fire published research on how these short strands of RNA can bind with mRNA, thus repressing the production of a protein product. Mello and Fire won the Nobel Prize for Physiology or Medicine in 2006 for their discovery. Various names have been given different types of micro RNAs, including *small interfering RNA (siRNA), riboswitches, antisense RNA,* and *piwi-interacting RNA (piRNA).* The system is found in many, if not most, eukaryotic organisms and viruses that infect them. Similar processes exist in bacteria, as well.

Since the original discoveries, hundreds of RNA genes have been discovered that regulate as many as one-third of genes that code for protein products in the human genome alone. For many years, scientists believed that long stretches of seemingly noncoding DNA were just evolutionary junk. Since the discovery of miRNA, it is now believed that this "junk DNA" plays a major role in controlling cellular processes throughout the human body and may play a role in cancer, heart disease, immune system dysfunction, memory production, and embryonic development.

Scientists hope that by studying miRNA, they can develop new therapies to treat certain diseases.

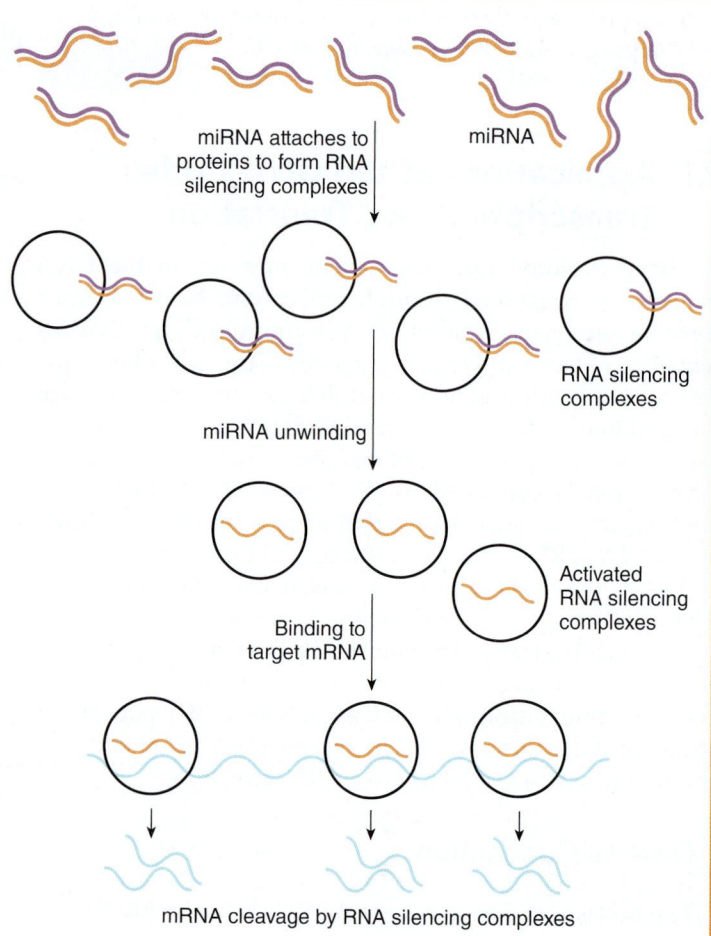

Figure labels: miRNA attaches to proteins to form RNA silencing complexes; miRNA; RNA silencing complexes; miRNA unwinding; Activated RNA silencing complexes; Binding to target mRNA; mRNA cleavage by RNA silencing complexes

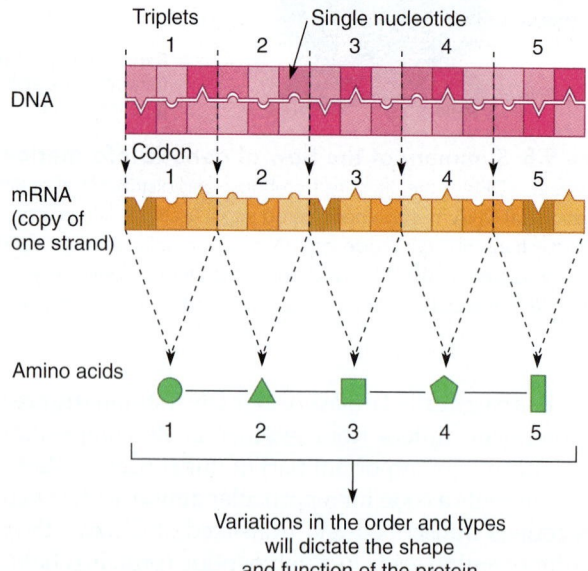

Figure 9.9 Simplified view of the DNA-protein relationship.   The DNA molecule is a continuous chain of base pairs, but the sequence must be interpreted in groups of three base pairs (a triplet). Each triplet as copied into mRNA codons will translate into one amino acid; consequently, the ratio of base pairs to amino acids is 3:1.

Figure labels: Triplets; Single nucleotide; DNA; Codon; mRNA (copy of one strand); Amino acids; Variations in the order and types will dictate the shape and function of the protein.

2. Proteins ultimately determine phenotype, the expression of all aspects of cell function and structure. Put more simply, living things are what their proteins make them. Regulatory RNAs help determine which proteins are made. **Proteomics** is the study of an organism's complete set of expressed proteins.

3. DNA is mainly a blueprint that tells the cell which kinds of proteins and RNAs to make and how to make them.

The Major Participants in Transcription and Translation

Transcription, the formation of RNA using DNA as a template, and translation, the synthesis of proteins using RNA as a template, are highly complex. A number of components participate: most prominently, messenger RNA, transfer RNA, regulatory RNAs, ribosomes, several types of enzymes, and a storehouse of raw materials. After first examining each of these components, we shall see how they come together in the assembly line of the cell.

RNAs: Tools in the Cell's Assembly Line

Ribonucleic acid is similar to DNA, but its general structure (see figure 2.23) is different in several ways:

1. It is a single-stranded molecule that exists in helical form. This single strand can assume secondary and tertiary levels of complexity due to bonds within the molecule, leading to specialized forms of RNA (tRNA and rRNA—see figure 9.8a).

2. RNA contains **uracil (U),** instead of thymine, as the complementary base-pairing mate for adenine. This does not change the inherent DNA code in any way because the uracil still follows the pairing rules.

3. Although RNA, like DNA, contains a backbone that consists of alternating sugar and phosphate molecules, the sugar in RNA is **ribose** rather than deoxyribose.

The many functional types of RNA range from small regulatory pieces to large structural ones **(table 9.2** and Insight 9.2). All types of RNA are formed through transcription of a DNA gene, but only mRNA is further translated into another type of molecule (protein).

Messenger RNA: Carrying DNA's Message

Messenger RNA (mRNA) is a transcript (copy) of a structural gene or genes in the DNA. It is synthesized by a process similar to synthesis of the leading strand during DNA replication, and the complementary base-pairing rules ensure that the code will be faithfully copied in the mRNA transcript. The message of this transcribed strand is later read as a series of triplets called **codons (figure 9.10),** and the length of the mRNA molecule varies from about 100 nucleotides to several thousand. The details of transcription and the function of mRNA in translation will be covered shortly.

Table 9.2 Types of Ribonucleic Acid

RNA Type	Contains Codes For	Function in Cell	Translated
Messenger (mRNA)	Sequence of amino acids in protein	Transports the DNA master code to the ribosome	Yes
Transfer (tRNA)	A cloverleaf tRNA to carry amino acids	Brings amino acids to ribosome during translation	No
Ribosomal (rRNA)	Several large structural rRNA molecules	Forms the major part of a ribosome and participates in protein synthesis	No
Micro (miRNA), antisense, riboswitch, and small interfering (siRNA)	Regulatory RNAs	Regulation of gene expression and coiling of chromatin	No
Primer	An RNA that can begin DNA replication	Primes DNA	No
Ribozymes and spliceosomes (snRNA)	RNA enzymes, parts of splicer enzymes	Remove introns from other RNAs in eukaryotes	No

Figure 9.10 Characteristics of transfer and messenger RNA.

(a) **Transfer RNA (tRNA)**
Left: The tRNA strand loops back on itself to form intrachain hydrogen bonds. The result is a cloverleaf structure, shown here in simplified form. At its bottom is an anticodon that specifies the attachment of a particular amino acid at the 3′ end. Right: A three-dimensional view of tRNA structure.

(b) **Messenger RNA (mRNA)**
A short piece of messenger RNA (mRNA) illustrates the general structure of RNA: single strandedness, repeating phosphate-ribose sugar backbone attached to single nitrogen bases; use of uracil instead of thymine.

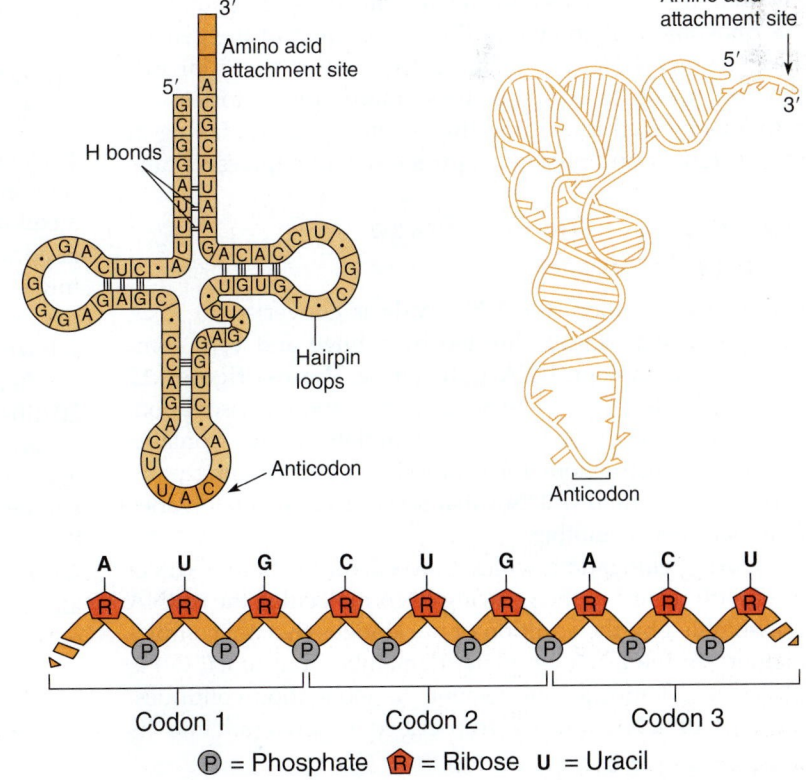

Transfer RNA: The Key to Translation

Transfer RNA (tRNA) is also a copy of a specific region of DNA; however, it differs from mRNA. It is uniform in length, 75 to 95 nucleotides long, and it contains sequences of bases that form hydrogen bonds with complementary sections of the same tRNA strand. At these points, the molecule bends back upon itself into several *hairpin loops*, giving the molecule a secondary *cloverleaf* structure that folds even further into a complex, three-dimensional helix (as shown in figure 9.10). This compact molecule is an adaptor that converts RNA language into protein language. At the bottom loop of the cloverleaf, there is an exposed triplet, the **anticodon,** that both designates the specificity of the tRNA and complements mRNA's codons. At the opposite end of the molecule is a binding site for the amino acid that is specific for that tRNA's anticodon. For each of the 20 amino acids, there is at least one specialized type of tRNA to carry it. Binding of an amino acid to its specific tRNA, a process known as "charging" the tRNA, takes place in two enzyme-driven steps: First, an ATP activates the amino acid; then, this group binds to the acceptor end of the tRNA. Because tRNA is the molecule that will convert the master code on mRNA into a protein, the accuracy of this step is crucial.

The Ribosome: A Mobile Molecular Factory for Translation

The bacterial (70S) ribosome is a particle composed of tightly packaged **ribosomal RNA (rRNA)** and protein. The rRNA component of the ribosome is also a long polynucleotide molecule. It forms complex three-dimensional figures that contribute to the structure and function of ribosomes. The interactions of proteins and rRNA create the two subunits of the ribosome that engage in final translation of the genetic code **(figure 9.11).** A metabolically active bacterial cell can contain up to 20,000 of these minuscule factories—all actively engaged in reading the genetic program, taking in raw materials, and producing proteins at an impressive rate.

Transcription: The First Stage of Gene Expression

During transcription, the DNA code is converted to RNA through several stages, directed by a huge and very complex enzyme system, **RNA polymerase. Process figure 9.12** supplies the details you will need to know about transcription. Only one strand of the DNA—the **template strand**—contains meaningful instructions for synthesis of a functioning polypeptide. The strand of DNA that serves as a template varies from one gene to another.

During elongation, which proceeds in the 5′ to 3′ direction (with regard to the growing RNA molecule), the mRNA is assembled by the addition of nucleotides that are complementary to the DNA template. Remember that uracil (U) is placed as adenine's complement. As elongation continues, the part of DNA already transcribed is rewound into its

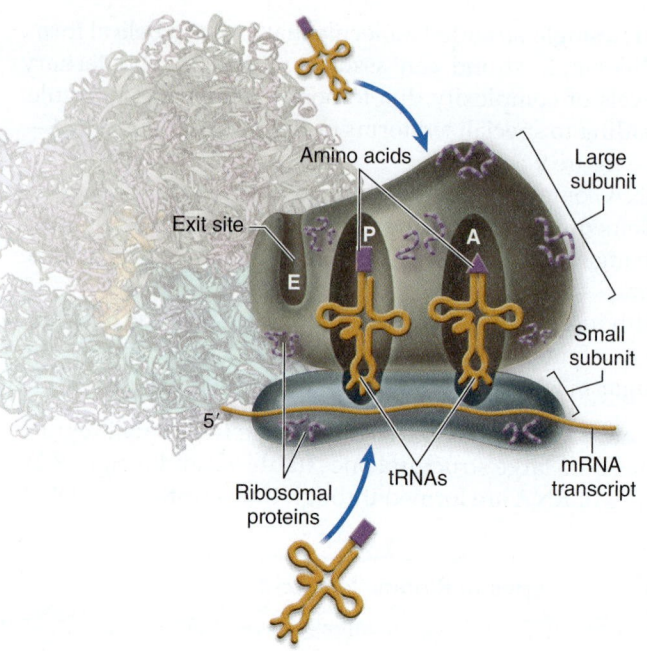

Figure 9.11 The "players" in translation.   A ribosome serves as the stage for protein synthesis. Assembly of the small and large subunits results in specific sites for holding the mRNA and two tRNAs with their amino acids. This depiction of the ribosome matches the shaded depiction of the molecular view (seen in background).

original helical form. At termination, the polymerases recognize another code that signals the separation and release of the mRNA strand, also called the **transcript.** How long is the mRNA? The smallest mRNA may consist of 100 bases; an average-size mRNA may consist of 1,200 bases; and a large one may consist of several thousand.

The Master Genetic Code: The Message in Messenger RNA

Translation relies on a central principle: The mRNA nucleotides are read in groups of three. Three nucleotides are called a **codon,** and it is the codon that dictates which amino acid is added to the growing peptide chain. In **figure 9.13,** the mRNA codons and their corresponding amino acid specificities are given. Except in a very few cases, this code is universal, whether for bacteria, archaea, eukaryotes, or viruses.

Because there are 64 different triplet codes[1] and only 20 different amino acids, it is not surprising that some amino acids are represented by several codons. For example, leucine and serine can each be represented by any of six different triplets, and only tryptophan and methionine are represented by a single codon. This property—of an amino acid being represented by several codons—is called **redundancy** and allows for the insertion of correct amino acids (sometimes) even when mistakes occur in the DNA sequence, as they do

---

1. $164 5 4^3$ (the four different codons in all possible combinations of three).

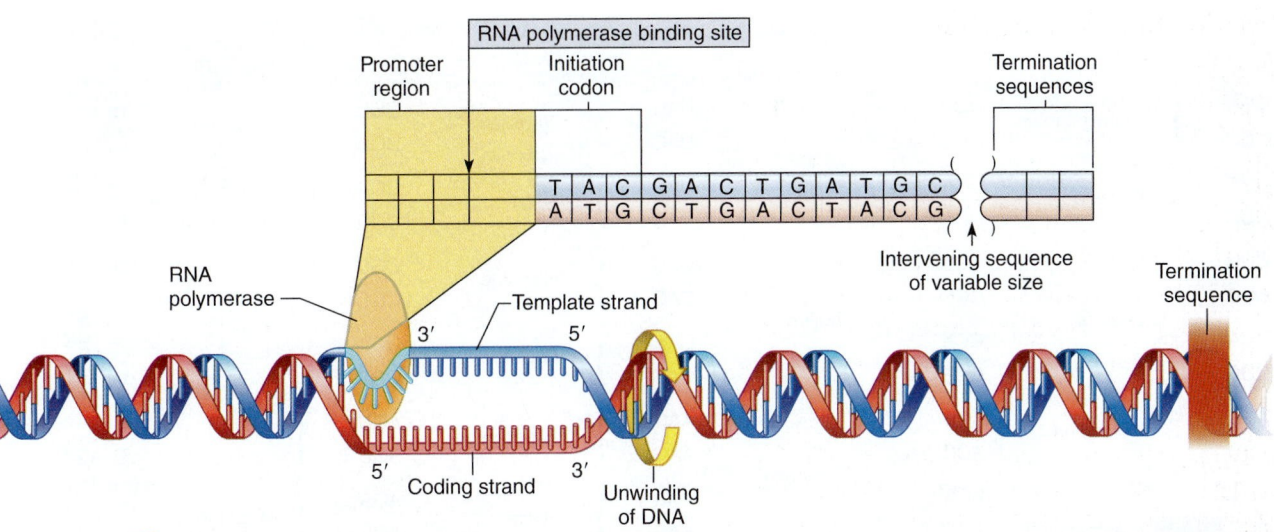

1 Initiation. Transcription is initiated when RNA polymerase recognizes a segment of the DNA called the **promoter** region. This region consists of two sequences of DNA just prior to the beginning of the gene to be transcribed. These promoter sequences provide the signal for RNA polymerase to bind to the DNA. Then there is a special codon called the initiation codon, which is where the RNA polymerase begins its transcription. As the DNA helix unwinds, the polymerase first pulls the early parts of the DNA into itself, a process called "DNA scrunching," and then, having acquired energy from the scrunching process, begins to advance down the DNA strand to continue synthesizing an RNA molecule complementary to the template strand of DNA. The nucleotide sequence of promoters differs only slightly from gene to gene, with all promoters being rich in adenine and thymine.

Only one strand of DNA, called the **template** strand, is copied by RNA polymerase.

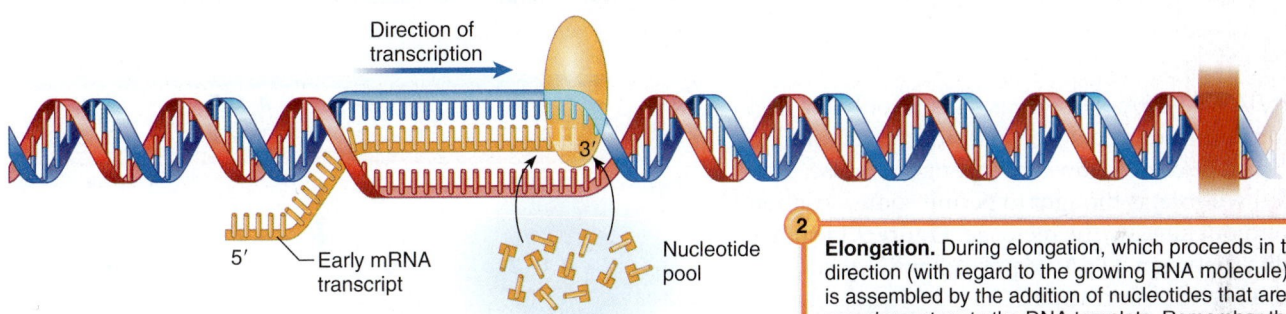

2 Elongation. During elongation, which proceeds in the 5′ to 3′ direction (with regard to the growing RNA molecule), the mRNA is assembled by the addition of nucleotides that are complementary to the DNA template. Remember that uracil (U) is placed as adenine's complement. As elongation continues, the part of DNA already transcribed is rewound into its original helical form.

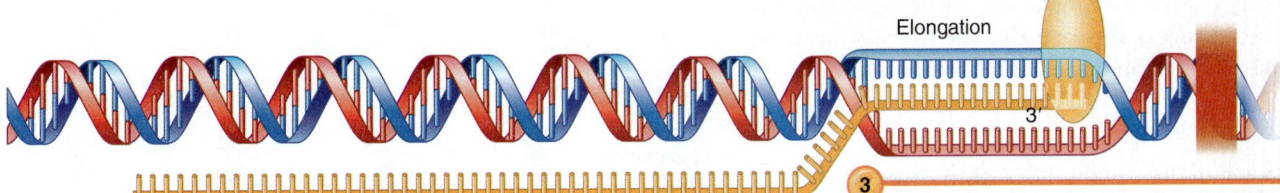

3 Termination. At termination, the polymerases recognize another code that signals the separation and release of the mRNA strand, or transcript. The smallest mRNA might consist of 100 bases; an average-size mRNA might consist of 1,200 bases; and a large one might consist of several thousand.

 Process Figure 9.12 Transcription.

First Base Position	Second Base Position: U		Second Base Position: C		Second Base Position: A		Second Base Position: G		Third Base Position
U	UUU	Phenylalanine	UCU	Serine	UAU	Tyrosine	UGU	Cysteine	U
U	UUC	Phenylalanine	UCC	Serine	UAC	Tyrosine	UGC	Cysteine	C
U	UUA	Leucine	UCA	Serine	UAA	STOP**	UGA	STOP**	A
U	UUG	Leucine	UCG	Serine	UAG	STOP**	UGG	Tryptophan	G
C	CUU	Leucine	CCU	Proline	CAU	Histidine	CGU	Arginine	U
C	CUC	Leucine	CCC	Proline	CAC	Histidine	CGC	Arginine	C
C	CUA	Leucine	CCA	Proline	CAA	Glutamine	CGA	Arginine	A
C	CUG	Leucine	CCG	Proline	CAG	Glutamine	CGG	Arginine	G
A	AUU	Isoleucine	ACU	Threonine	AAU	Asparagine	AGU	Serine	U
A	AUC	Isoleucine	ACC	Threonine	AAC	Asparagine	AGC	Serine	C
A	AUA	Isoleucine	ACA	Threonine	AAA	Lysine	AGA	Arginine	A
A	AUG	START f-Methionine*	ACG	Threonine	AAG	Lysine	AGG	Arginine	G
G	GUU	Valine	GCU	Alanine	GAU	Aspartic acid	GGU	Glycine	U
G	GUC	Valine	GCC	Alanine	GAC	Aspartic acid	GGC	Glycine	C
G	GUA	Valine	GCA	Alanine	GAA	Glutamic acid	GGA	Glycine	A
G	GUG	Valine	GCG	Alanine	GAG	Glutamic acid	GGG	Glycine	G

* This codon initiates translation.

**For these codons, which give the orders to stop translation, there are no corresponding tRNAs and no amino acids.

Figure 9.13 The genetic code: codons of mRNA that specify a given amino acid. The master code for translation is found in the mRNA codons.

with regularity. Also, in codons such as leucine, only the first two nucleotides are required to encode the correct amino acid, and the third nucleotide does not change its sense. This property, called **wobble,** is thought to permit some variation or mutation without altering the message. **Figure 9.14** shows the relationship between DNA sequence, mRNA codons, tRNA anticodons, and amino acids.

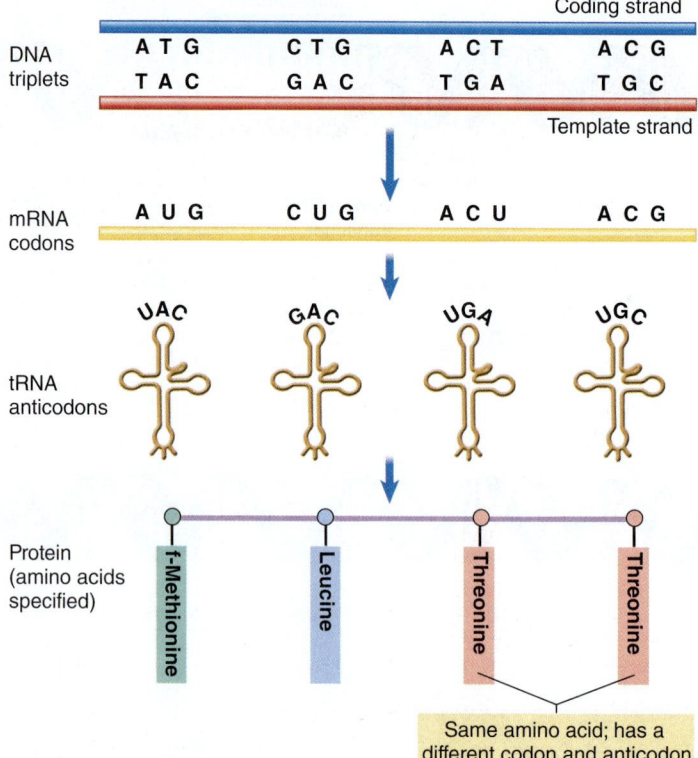

Figure 9.14 Interpreting the DNA code. If the DNA sequence is known, the mRNA codon can be surmised. If a codon is known, the anticodon and, finally, the amino acid sequence can be determined. The reverse is not as straightforward (determining the exact codon or anticodon from amino acid sequence) due to the redundancy of the code.

Translation: The Second Stage of Gene Expression

In translation, all of the elements needed to synthesize a protein, from the mRNA to the amino acids, are brought together on the ribosomes **(process figure 9.15).** The process has three main stages: initiation, elongation, and termination.

Initiation of Translation

The mRNA molecule leaves the DNA transcription site and is transported to ribosomes also in the cytoplasm. Ribosomal subunits function by coming together and forming sites to hold the mRNA and tRNAs. The ribosome thus recognizes these molecules and stabilizes reactions between them. The small subunit binds to the 5′ end of the mRNA, and the large subunit supplies enzymes for making peptide bonds on the protein.

The Terms of Protein Synthesis

With mRNA serving as the guide, the stage is finally set for actual protein assembly. Process figure 9.15 lays out the

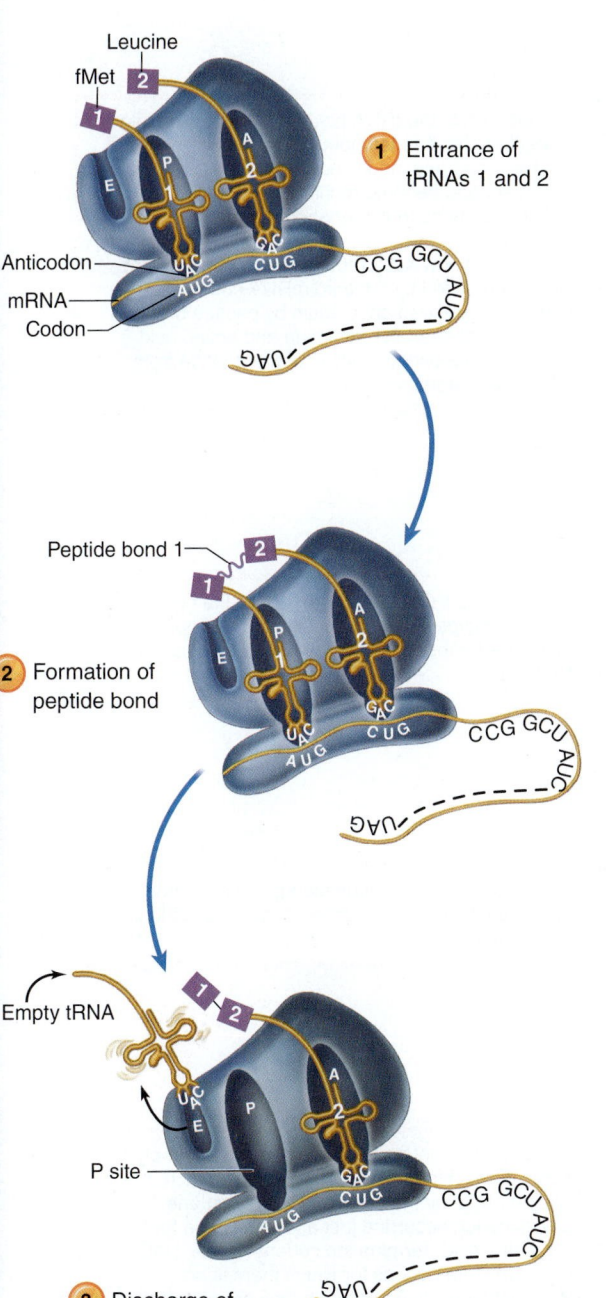

1 Entrance of tRNAs 1 and 2

Leucine
fMet
Anticodon
mRNA
Codon

2 Formation of peptide bond

Peptide bond 1

3 Discharge of tRNA 1 at E site

Empty tRNA
P site

1

The mRNA molecule leaves the DNA transcription site and is transported to ribosomes in the cytoplasm. Ribosomal subunits come together and form sites to hold the mRNA and tRNAs. The ribosome begins to scan the mRNA by moving in the 5′ to 3′ direction along the mRNA. The first codon it encounters is called the START codon, which is almost always AUG (and, rarely, GUG). With the mRNA message in place on the assembled ribosome, the next step in translation involves entrance of tRNAs with their amino acids. The pool of cytoplasm contains a complete array of tRNAs, previously charged by having the correct amino acid attached. The step in which the complementary tRNA meets with the mRNA code is guided by the two sites on the large subunit of the ribosome called the P site (left) and the A site (right). The ribosome also has an exit or E site where used tRNAs are released. (P stands for peptide site; A stands for aminoacyl [amino acid] site; E stands for exit site.)

2

Rules of pairing dictate that the anticodon of this tRNA must be complementary to the mRNA codon AUG; thus, the tRNA with anticodon UAC will first occupy site P. It happens that the amino acid carried by the initiator tRNA in bacteria is **formyl methionine.** The formyl group provides a special signal that this amino acid is not part of the translated protein because usually fMet does not remain a permanent part of the finished protein but instead is cleaved from the finished peptide. The ribosome shifts its "reading frame" to the right along the mRNA from one codon to the next. This brings the next codon into place on the ribosome and makes a space for the next tRNA to enter the A position. A peptide bond is formed between the amino acids on the adjacent tRNAs, and the polypeptide grows in length.

Elongation begins with the filling of the A site by a second tRNA. The identity of this tRNA and its amino acid is dictated by the second mRNA codon.

3

The entry of tRNA 2 into the A site brings the two adjacent tRNAs in favorable proximity for a peptide bond to form between the amino acids (aa) they carry. The fMet is transferred from the first tRNA to aa 2, resulting in two coupled amino acids called a dipeptide.

For the next step to proceed, some room must be made on the ribosome, and the next codon in sequence must be brought into position for reading. This process is accomplished by translocation, the enzyme-directed shifting of the ribosome to the right along the mRNA strand, which causes the blank tRNA 1 to be discharged from the ribosome at the E site.

Process Figure 9.15 Translation. (*continues on following page*)

details of translation. Several terms are important for understanding the process.

start codon The first three RNA nucleotides that signal the beginning of the message. The start codon is always AUG.

stop codon One of three codons—UAA, UAG, or UGA—that has no corresponding tRNA and therefore causes translation to be terminated; also called **nonsense codon.**

translocation The process of shifting the ribosome down the mRNA strand to read new codons.

Completion of Protein Synthesis

Before newly made proteins can carry out their structural or enzymatic roles, they often require finishing touches. Even before the peptide chain is released from the ribosome, it begins folding upon itself to achieve its biologically active tertiary conformation. Other alterations, called **posttranslational** modifications, may be necessary. Some proteins must have the starting amino acid (formyl methionine) clipped off; proteins destined to become complex

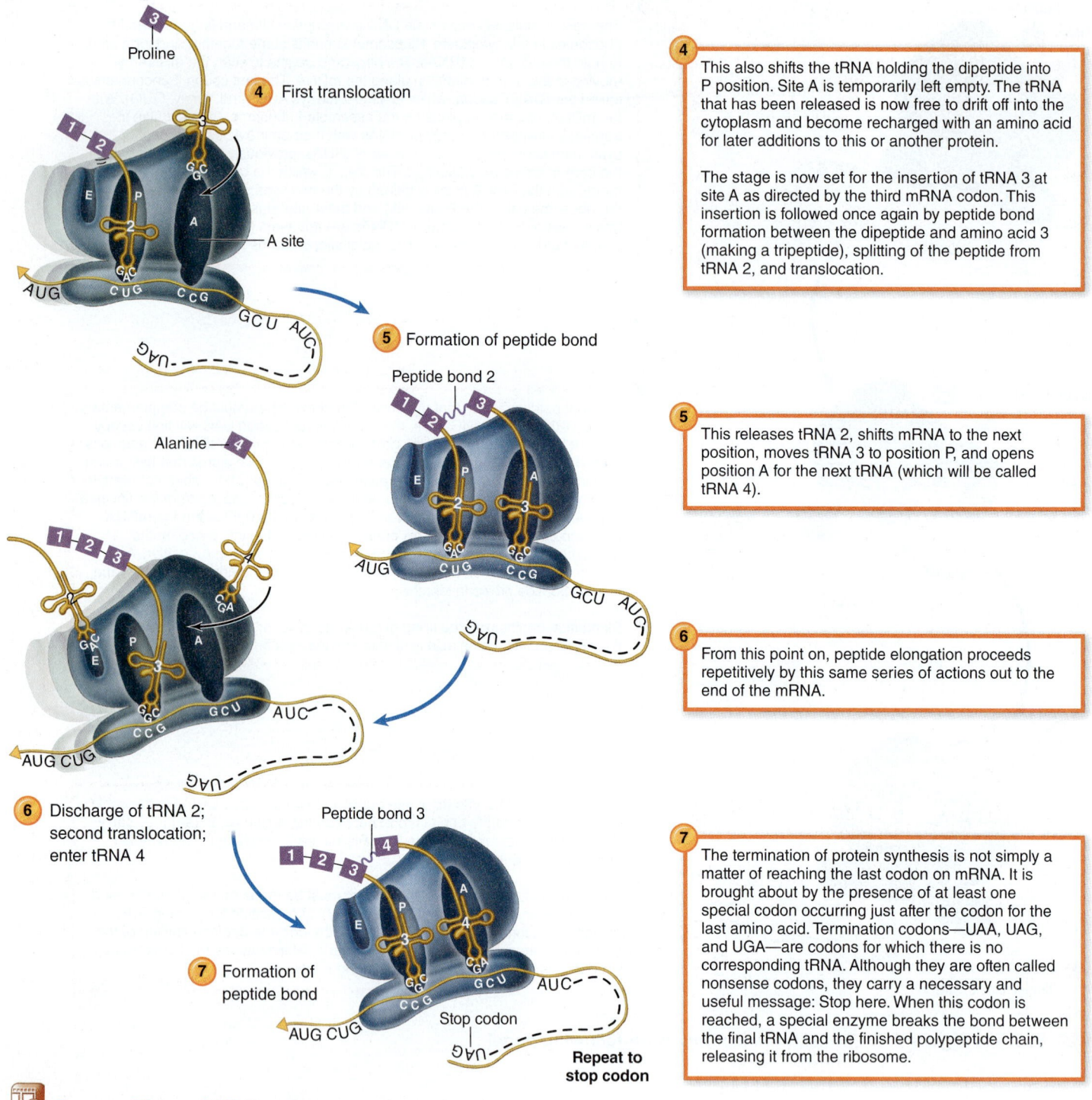

4 First translocation

Proline

A site

4

This also shifts the tRNA holding the dipeptide into P position. Site A is temporarily left empty. The tRNA that has been released is now free to drift off into the cytoplasm and become recharged with an amino acid for later additions to this or another protein.

The stage is now set for the insertion of tRNA 3 at site A as directed by the third mRNA codon. This insertion is followed once again by peptide bond formation between the dipeptide and amino acid 3 (making a tripeptide), splitting of the peptide from tRNA 2, and translocation.

5 Formation of peptide bond

Peptide bond 2

Alanine

5

This releases tRNA 2, shifts mRNA to the next position, moves tRNA 3 to position P, and opens position A for the next tRNA (which will be called tRNA 4).

6 Discharge of tRNA 2; second translocation; enter tRNA 4

6

From this point on, peptide elongation proceeds repetitively by this same series of actions out to the end of the mRNA.

Peptide bond 3

7 Formation of peptide bond

Stop codon

Repeat to stop codon

7

The termination of protein synthesis is not simply a matter of reaching the last codon on mRNA. It is brought about by the presence of at least one special codon occurring just after the codon for the last amino acid. Termination codons—UAA, UAG, and UGA—are codons for which there is no corresponding tRNA. Although they are often called nonsense codons, they carry a necessary and useful message: Stop here. When this codon is reached, a special enzyme breaks the bond between the final tRNA and the finished polypeptide chain, releasing it from the ribosome.

Process Figure 9.15 Translation. (*continued*)

enzymes have cofactors added; and some join with other completed proteins to form quaternary levels of structure.

The operation of transcription and translation is machine-like in its precision. Protein synthesis in bacteria is both efficient and rapid, as the translation of mRNA starts while transcription is still occurring **(figure 9.16)**, a process called cotranscriptional translation. A single mRNA is long enough to be fed through more than one ribosome simultaneously. This permits the synthesis of hundreds of protein molecules from the same mRNA transcript arrayed along a chain of

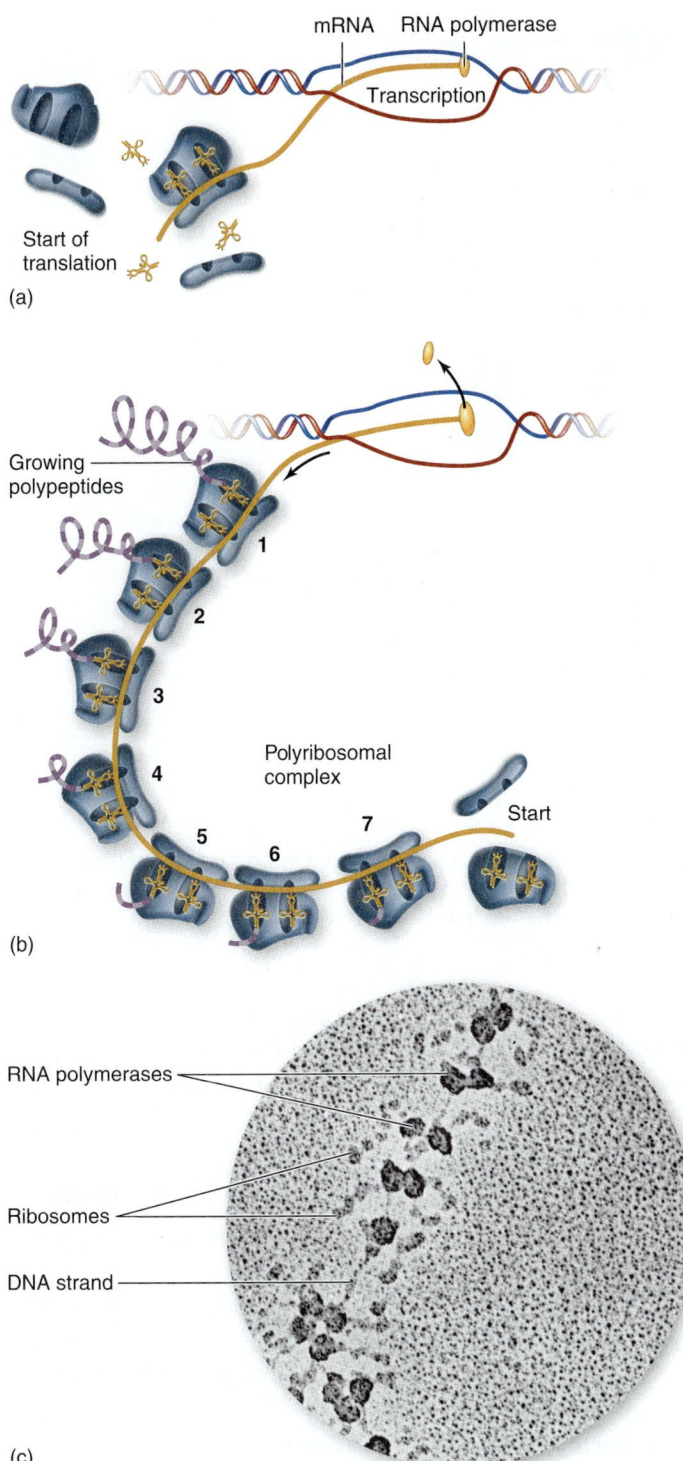

Figure 9.16 Speeding up the protein assembly line in bacteria. **(a)** The mRNA transcript encounters ribosomal parts immediately as it leaves the DNA. **(b)** The ribosomal factories assemble along the mRNA in a chain, each ribosome reading the message and translating it into protein. Many products will thus be well along the synthetic pathway before transcription has even terminated. **(c)** Photomicrograph of a polyribosomal complex in action. Note that the protein "tails" vary in length, depending on the stage of translation.

ribosomes. This **polyribosomal complex** is indeed an assembly line for mass production of proteins. Cotranscriptional translation only occurs in bacteria and archaea, because there is no nucleus and transcription and translation both occur in the cytoplasm. (In eukaryotes, transcription occurs in the nucleus.) Remember that all of the processes involved in gene expression are anabolic processes; nearly 1,200 ATPs are required just for synthesis of an average-size protein.

Eukaryotic Transcription and Translation: Similar yet Different

Although they share many similarities in protein synthesis, there are important differences in this process among eukaryotes and the noneukaryotes. Only bacteria and archaea exhibit cotranscriptional translation. Although the start codon in eukaryotes is also AUG, it codes for a different form of methionine. Another difference is that eukaryotic mRNAs code for just one protein, unlike bacterial mRNAs, which often contain information from several genes in series.

We have given the simplified definition of *gene*, which works well for bacteria, but most eukaryotic genes do *not* exist as an uninterrupted series of triplets coding for a protein. A eukaryotic gene contains the code for a protein, but located along the gene are one to several intervening sequences of bases, called **introns,** that do not code for protein. Introns are interspersed between coding regions, called **exons,** that will be translated into protein **(figure 9.17).** We can use words as examples. A short section of colinear bacterial gene might read TOM SAW OUR DOG DIG OUT; a eukaryotic gene that codes for the same portion would read TOM SAW XZKP FPL OUR DOG QZWVP DIG OUT. The recognizable words are the exons, and the nonsense letters represent the introns.

This unusual genetic architecture, sometimes called a split gene, requires further processing before translation. Transcription of the entire gene with both exons and introns occurs first, producing a pre-mRNA. A series of adenosines is added to the mRNA molecule. This protects the molecule and eventually directs it out of the nucleus for translation. Next, a type of RNA and protein called a **spliceosome** recognizes the exon-intron junctions and enzymatically cuts through them. The action of this splicer enzyme loops the introns into lariat-shaped pieces, excises them, and joins the exons end to end. By this means, a strand of mRNA with no intron material is produced. This completed mRNA strand can then proceed to the cytoplasm to be translated. In some eukaryotes, however, the mRNA may be alternatively spliced into multiple different mRNAs or its sequence may be edited to produce a variety of different proteins from one single gene.

The Genetics of Animal Viruses

The genetic nature of viruses was described in chapter 6. Viruses essentially consist of one or more pieces of DNA or RNA enclosed in a protective coating. Above all, they are

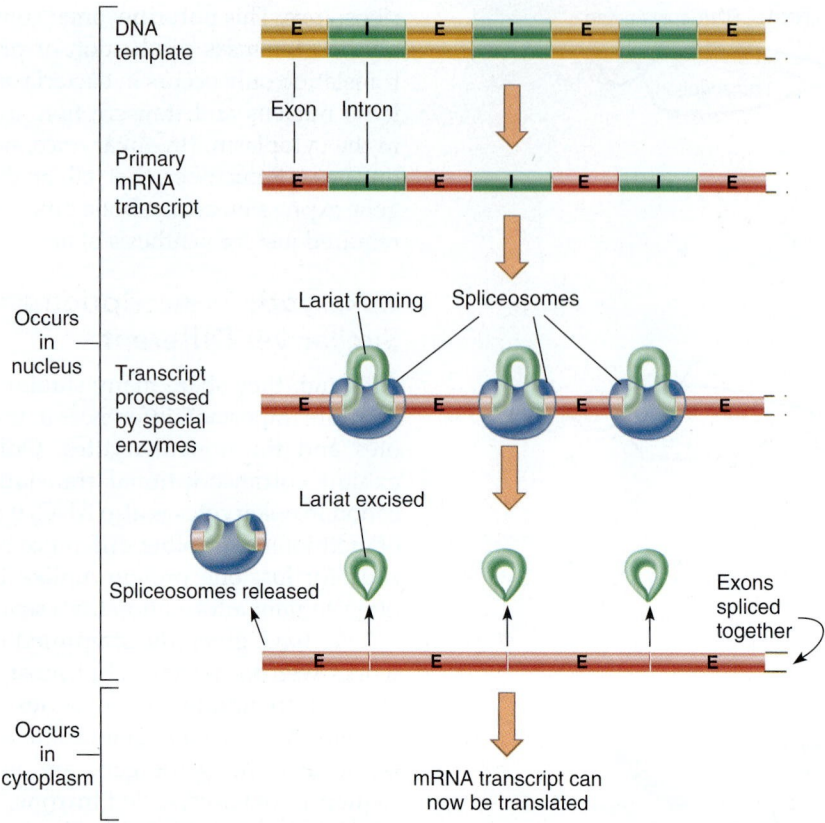

DNA template

Exon   Intron

Primary mRNA transcript

Occurs in nucleus

Lariat forming   Spliceosomes

Transcript processed by special enzymes

Lariat excised

Spliceosomes released

Exons spliced together

Occurs in cytoplasm

mRNA transcript can now be translated

Figure 9.17 The split gene of eukaryotes. Eukaryotic genes have an additional complicating factor in their translation. Their coding sequences, or exons (E), are interrupted at intervals by segments called introns (I) that are not part of that protein's code. Introns are transcribed but not translated, which necessitates their removal by RNA splicing enzymes before translation.

genetic parasites that require access to their host cell's genetic and metabolic machinery to be replicated, transcribed, and translated; they also have the potential for genetically changing the cells. Because they contain only those genes needed for the production of new viruses, the genomes of viruses tend to be very compact and economical. In fact, this very simplicity makes them excellent subjects for the study of gene function.

The genetics of viruses is quite diverse. In many viruses, the nucleic acid is linear in form; in others, it is circular. The genome of most viruses exists in a single molecule, though in a few it is segmented into several smaller molecules. Most viruses contain normal double-stranded (ds) DNA or single-stranded (ss) RNA, but other patterns exist. There are ssDNA viruses, dsRNA viruses, and retroviruses, which work backward by making dsDNA from ssRNA.

A few generalities can be stated about viral genetics. In all cases, the viral nucleic acid penetrates the cell and is introduced into the host's gene-processing machinery at some point. In successful infection, an invading virus instructs the host's machinery to synthesize large numbers of new virus particles by a mechanism specific to a particular group. With few exceptions, replication of the DNA molecule of DNA animal viruses occurs in the nucleus, where the cell's DNA

replication machinery lies and the genome of RNA viruses is replicated in the cytoplasm. In all viruses, viral mRNA is translated into viral proteins on host cell ribosomes using host tRNA.

9.2 Learning Outcomes—Assess Your Progress

6. Explain how the classical view of the "central dogma" has been changed by modern science.

7. Identify important structural and functional differences between RNA and DNA.

8. Illustrate the steps of transcription, noting the key elements and the direction of mRNA synthesis.

9. List the three types of RNA directly involved in translation.

10. Define the terms *codon* and *anticodon*, and list the four known start and stop codons.

11. Identify the locations of the promoter, the start codon, and the A and P sites during translation.

12. Indicate how eukaryotic transcription and translation differ from these processes in bacteria and archaea.

13. Explain the relationship between genomics and proteomics.

9.3 Genetic Regulation of Protein Synthesis

In chapter 8, we surveyed the metabolic reactions in cells and the enzymes involved in those reactions. At that time, we mentioned that some enzymes are expressed constitutively in cell, whereas others are tightly regulated. Such regulation can occur at the genetic level, and this control mechanism ensures that genes are active only when their products are required. This prevents the waste of energy and materials in dead-end protein synthesis. Antisense RNAs, micro RNAs, and riboswitches (see Insight 9.2) provide regulation in many kinds of cells. But bacteria and archaea have an additional strategy: They organize collections of genes into **operons.** Operons consist of a coordinated set of genes, all of which are regulated as a single unit. Operons are described as either inducible or repressible. The category each operon falls into is determined by how transcription is affected by the environment surrounding the cell. Many catabolic operons, or operons encoding enzymes that act in catabolism, are inducible, meaning that the operon is turned on (induced) by the substrate of the enzyme(s) for which the structural genes code. In this way, the enzymes needed to metabolize a nutrient (lactose, for example) are only produced when that nutrient is present in the environment. Repressible operons often contain genes coding for anabolic enzymes, such as those used to synthesize amino acids. In the case of these operons, several genes in series are turned off (repressed) by the product synthesized by the enzyme.

The Lactose Operon: A Model for Inducible Gene Regulation in Bacteria

The best understood cell system for explaining control through genetic induction is the **lactose (*lac*) operon.** This system, first described in 1961 by François Jacob and Jacques Monod, accounts for the regulation of lactose metabolism in *Escherichia coli.* Many other operons with similar modes of action have since been identified, and together they show us that the environment of a cell can have great impact on gene expression.

The lactose operon has three important features **(process figure 9.18):**

1. the **regulator,** composed of the gene that codes for a protein capable of repressing the operon (a **repressor**);
2. the *control locus,* composed of two areas, the **promoter** (recognized by RNA polymerase) and the **operator,** a sequence that acts as an on/off switch for transcription; and
3. the *structural locus,* made up of three genes, each coding for a different enzyme needed to catabolize lactose.

One of the enzymes, β-galactosidase, hydrolyzes the lactose into its monosaccharides; another, permease, brings lactose across the cell membrane. The third enzyme is an acetyltransferase that helps in the metabolism of lactose.

The operon provides an efficient strategy that permits genes for a particular metabolic pathway to be induced or repressed in unison by a single regulatory element. The promoter, operator, and structural components usually lie adjacent to one another, but the regulator can be at a distant site.

In inducible systems like the *lac* operon, the operon is normally in an *off* mode and does not initiate transcription when the appropriate substrate is absent **(process figure 9.18).** How is the operon maintained in this mode? The key is in the repressor protein that is coded by the regulatory gene. This relatively large molecule is **allosteric,** meaning it has two binding sites, one for the operator sequence on the DNA and another for lactose. In the absence of lactose, this repressor binds to the operator locus, thereby blocking the transcription of the structural genes lying downstream. Think of the repressor as a lock on the operator, and if the operator is locked, the structural genes cannot be transcribed.

Case File 9   *Continuing the Case*

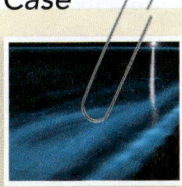

The *lux* operon in *Vibrio fischeri* and *Vibrio harveyi* has been carefully studied. *Vibrio fischeri* is an organism that lives either on its own or symbiotically with the bobtail squid; *Vibrio harveyi* lives on shrimp and other marine species. The *lux* operon is composed of five different genes that control bioluminescence. Protein products of this operon combine to form luciferase and bioluminesce by emitting photons of light.

Because transcription and translation of the genes in the *lux* operon require energy, the cell only translates and transcribes these genes at certain times. Having the genes adjacent to one another and under the control of a signal mechanism is very efficient for the bacterium. The bacteria continually produce a signaling molecule called an **autoinducer,** which is not unlike a hormone. When the bacteria reach a certain concentration, the autoinducer reaches a concentration that is detectable by all of the bacteria in the immediate area. The higher concentration of the autoinducer signals the bacteria that they are at the right concentration to produce luciferase and bioluminesce. This phenomenon is known as **quorum sensing.** The bacteria do not produce light until they have reached a specific concentration. At low concentrations of bacteria, not enough autoinducer is produced, the signaling pathway is shut down, and the *lux* operon is turned off. When bacteria reach concentrations of up to 10^{12} cells/ml, the autoinducer initiates a signaling pathway that turns on the *lux* operon.

- How does understanding bioluminescent bacteria and the *lux* operon help us to understand other bacterial processes?

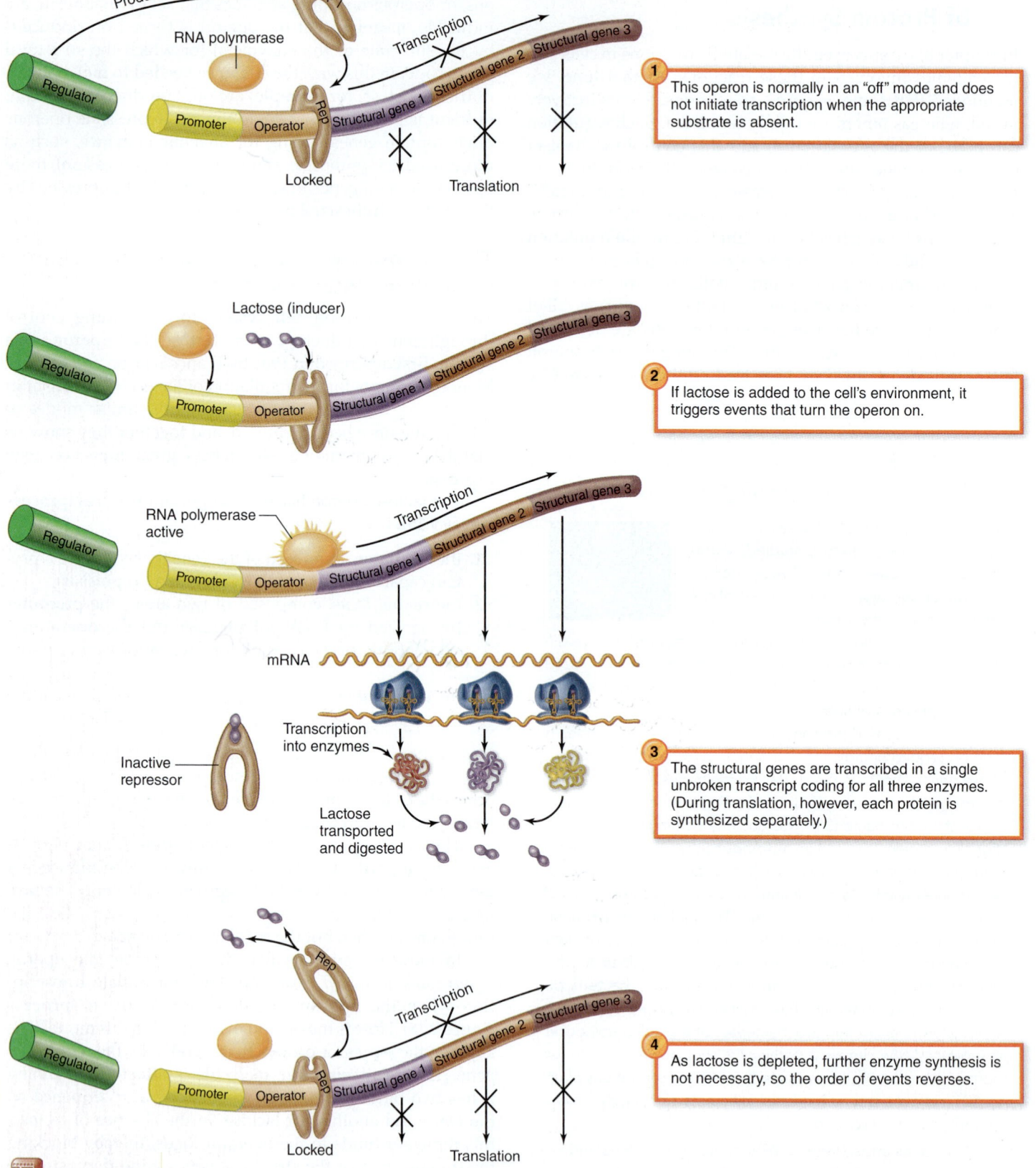

1 This operon is normally in an "off" mode and does not initiate transcription when the appropriate substrate is absent.

2 If lactose is added to the cell's environment, it triggers events that turn the operon on.

3 The structural genes are transcribed in a single unbroken transcript coding for all three enzymes. (During translation, however, each protein is synthesized separately.)

4 As lactose is depleted, further enzyme synthesis is not necessary, so the order of events reverses.

Process Figure 9.18 The lactose *(lac)* operon in bacteria: how inducible genes are controlled by substrate.

Importantly, the regulator gene lies upstream (to the left) of the operator region and is transcribed constitutively because it is not controlled in tandem with the operon.

If lactose is added to the cell's environment, it triggers several events that turn the operon *on*. The binding of lactose to the repressor protein causes a conformational change in the repressor that dislodges it from the operator segment of the DNA (process figure 9.18). With the operator opened up, RNA polymerase can now bind to the promoter and proceed. The structural genes are transcribed in a single unbroken transcript coding for all three enzymes. (During translation, however, each protein is synthesized separately.) Because lactose is ultimately responsible for stimulating protein synthesis, it is called the **inducer.**

As lactose is depleted, further enzyme synthesis is not necessary, so the order of events reverses. At this point, there is no longer sufficient lactose to inhibit the repressor; hence, the repressor is again free to attach to the operator. The operator is locked, and transcription of the structural genes and enzyme synthesis related to lactose both stop.

A fine but important point about the *lac* operon is that it functions only in the absence of glucose or if the cell's energy needs are not being met by the available glucose. Glucose is the preferred carbon source because it can be used immediately in growth and does not require induction of an operon. When glucose is present, a second regulatory system ensures that the *lac* operon is inactive, regardless of lactose levels in the environment.

Disease Connection

A negative connection has been found between the *lac* operon and virulence, at least in *Salmonella*. In 2011, researchers found that *Salmonella* species that contained a *lac* operon were less virulent than those who had lost their *lac* operons.

A Repressible Operon

Bacterial systems for synthesis of amino acids, purines and pyrimidines, and many other processes work on a slightly different principle—that of repression. Similar factors such as repressor proteins, operators, and a series of structural genes exist for this operon but with some important differences. Unlike the *lac* operon, this operon is normally in the *on* mode and will be turned *off* only when this nutrient is no longer required. The excess nutrient serves as a **corepressor** needed to block the action of the operon.

A growing cell that needs the amino acid arginine (*arg*) effectively illustrates the operation of a repressible operon. Under these conditions, the *arg* operon is set to *on* and arginine is being actively synthesized through the action of the operon's enzymatic products **(figure 9.19a).** In an active cell, the arginine will be used immediately, and the repressor will remain inactive (unable to bind the operator) because there is too little free arginine to activate it. As the cell's metabolism begins to slow down, however, the

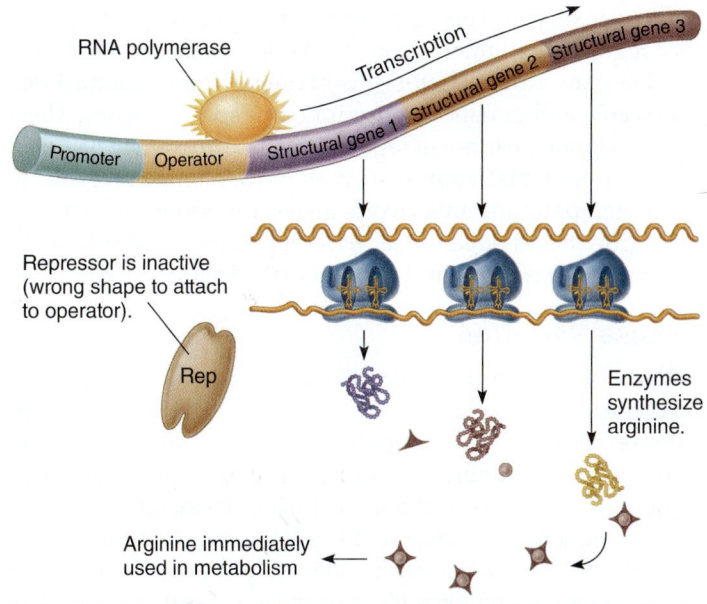

(a) **Operon On.** A repressible operon remains on when its nutrient products (here, arginine) are in great demand by the cell because the repressor is unable to bind to the operator at low nutrient levels.

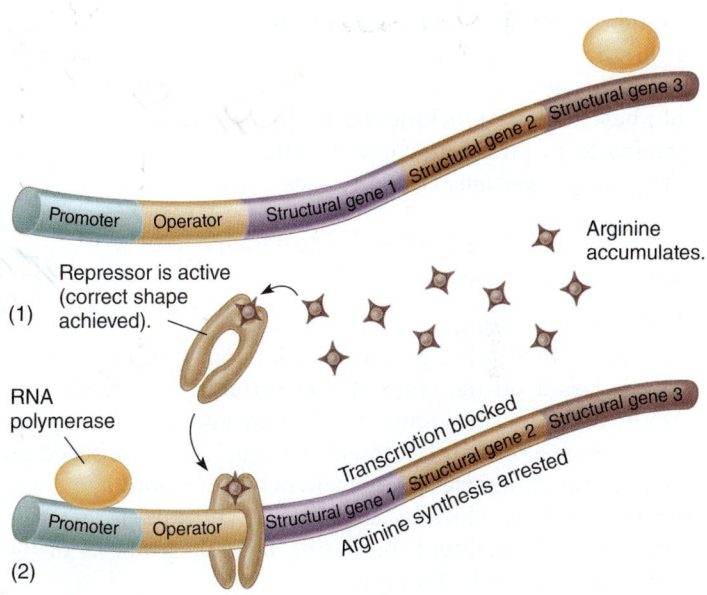

(b) **Operon Off.** The operon is repressed when (1) arginine builds up and, serving as a corepressor, activates the repressor. (2) The repressor complex affixes to the operator and blocks the RNA polymerase and further transcription of genes for arginine synthesis.

Figure 9.19 Repressible operon: control of a gene through excess nutrient.

synthesized arginine will no longer be used up and will accumulate. The free arginine is then available to act as a corepressor by attaching to the repressor. This reaction changes the shape of the repressor, making it capable of binding to the operator. Transcription stops; arginine is no longer synthesized **(figure 9.19b).**

In eukaryotic cells, gene function can be altered by intrinsic regulatory segments similar to operons. Some molecules, called transcription factors, insert on the grooves of the DNA molecule and enhance transcription of specific genes. These transcription factors can regulate gene expression in response to environmental stimuli such as nutrients, toxin levels, or even temperature. Eukaryotic genes are also regulated during growth and development, leading to the hundreds of different tissue types found in higher multicellular organisms.

Phase Variation

When bacteria turn on or off a complement of genes that leads to obvious phenotypic changes, it is sometimes called **phase variation.** Phase variation is a type of phenotypic variation that is heritable—meaning it is passed down to subsequent generations—but may be further changed as it passes to subsequent generations. This process involves the turning on of genes mediated by regulatory proteins, as described with operons. The term *phase variation* is most often applied to traits affecting the bacterial cell surface and was originally coined to describe the ability of bacteria to change components of their surface that marked them for targeting by the host's immune system. Because these surface molecules also influenced the bacterium's ability to attach to surfaces, the ability to undergo phase variation allowed the microbes to adapt to—and stick in—different environments. Examples of phase variation include the ability of *Neisseria gonorrhoeae* strains to produce attachment fimbriae, and the ability of *Streptococcus pneumoniae* to produce a capsule.

Antibiotics That Affect Transcription and Translation

Naturally occurring cell nutrients are not the only agents capable of modifying gene expression. Some infection therapy is based on the concept that certain drugs react with DNA, RNA, or ribosomes and thereby alter genetic expression (see chapter 12). Treatment with such drugs is based on an important premise: that growth of the infectious agent will be inhibited by blocking its protein-synthesizing machinery selectively, without disrupting the cell synthesis of the patient receiving the therapy.

Drugs that inhibit protein synthesis exert their influence on transcription or translation. For example, the rifamycins used in therapy for tuberculosis bind to RNA polymerase, blocking the initiation step of transcription, and are selectively more active against bacterial RNA polymerase than the corresponding eukaryotic enzyme. Actinomycin D binds to bacterial DNA and halts mRNA chain elongation, but it also binds to human DNA. For this reason, it is very toxic and never used to treat bacterial infections, though it can be applied in tumor treatment.

The ribosome is a frequent target of antibiotics that inhibit ribosomal function and ultimately protein synthesis. The value and safety of these antibiotics again depend upon the differential susceptibility of bacterial and eukaryotic ribosomes. One problem with drugs that selectively disrupt bacterial ribosomes is that the mitochondria of humans contain a bacterial type of ribosome, and these drugs may inhibit the function of the host's mitochondria. One group of antibiotics (including erythromycin and spectinomycin) prevents translation by interfering with the attachment of mRNA to ribosomes. Chloramphenicol, lincomycin, and tetracycline bind to the ribosome in a way that blocks the elongation of the polypeptide, and aminoglycosides (such as streptomycin) inhibit peptide initiation and elongation. It is interesting to note that these drugs have served as important tools to explore genetic events because they can arrest specific stages in these processes.

9.3 Learning Outcomes—Assess Your Progress

14. Define the term *operon* and explain one advantage it provides to a bacterial cell.
15. Differentiate between repressible and inducible operons and provide an example of each.
16. List several antibiotic drugs and their targets within the transcription and translation machinery.

9.4 DNA Recombination Events

Genetic recombination through sexual reproduction is an important means of genetic variation in eukaryotes. Although bacteria have no exact equivalent to sexual reproduction, they exhibit a primitive means for sharing or recombining parts of their genome. An event in which one bacterium donates DNA to another bacterium is a type of genetic transfer termed **recombination,** the end result of which is a new strain different from both the donor and the original recipient strain. Recombination in bacteria depends in part on the fact that bacteria contain extrachromosomal DNA—that is, plasmids—and are adept at moving between cells. Genetic exchanges have tremendous effects on the genetic diversity of bacteria. They provide additional genes for resistance to drugs and metabolic poisons, new nutritional and metabolic capabilities, and increased virulence and adaptation to the environment.

In general, any organism that contains (and expresses) genes that originated in another organism is called a **recombinant.**

Horizontal Gene Transfer in Bacteria

Any transfer of DNA that results in organisms acquiring new genes that did not come directly from parent organisms is called **horizontal gene transfer.** (Acquiring genes from parent organisms during reproduction is vertical gene transfer.) For decades, it has been known that bacteria engage in horizontal gene transfer. It is now becoming clear that eukaryotic organisms—including humans—also engage in horizontal gene transfer, often aided and abetted by microbes such as viruses. This revelation has upended traditional views about eukaryotic evolution, taxonomy, and even "human-ness." Remember in chapter 1 the assertion that 40% to 50% of DNA in humans comes from nonhuman species, transferred to us via viruses? Here, we will study the mechanisms used by bacteria to acquire genes horizontally.

DNA transfer between bacterial cells typically involves small pieces of DNA in the form of plasmids or chromosomal fragments. Plasmids are small, circular pieces of DNA that contain their own origin of replication and therefore can replicate independently of the bacterial chromosome. Plasmids are found in many bacteria (as well as some fungi) and typically contain, at most, only a few dozen genes. Although plasmids are not necessary for bacterial survival, they often carry useful traits, such as antibiotic resistance. Chromosomal fragments that have escaped from a lysed bacterial cell are also commonly involved in the transfer of genetic information between cells. An important difference between plasmids and fragments is that while a plasmid has its own origin of replication and is stably replicated and inherited, chromosomal fragments must integrate themselves into the bacterial chromosome in order to be replicated and eventually passed to progeny cells. While the process of genetic recombination is relatively rare in nature, its frequency can be increased in the laboratory, where the ability to shuffle genes between organisms is highly prized.

Depending on the mode of transmission, the means of genetic recombination in bacteria is called conjugation, transformation, or transduction. **Conjugation** requires the attachment of two related species and the formation of a bridge that can transport DNA. **Transformation** entails the transfer of naked DNA and requires no special vehicle. **Transduction** is DNA transfer mediated through the action of a bacterial virus **(table 9.3)**.

Conjugation: Bacterial "Sex"

Conjugation is a mode of genetic exchange in which a plasmid or other genetic material is transferred by a donor to a recipient cell via a direct connection **(figure 9.20)**. Both gram-negative and gram-positive cells can conjugate. In gram-negative cells, the donor has a plasmid known as the **fertility (F) factor** that allows the synthesis of a conjugative **pilus**. The recipient cell has a recognition site on its surface. A cell's role in conjugation is denoted by F^1 for the cell that has the F plasmid and by F^2 for the cell that lacks it. Contact is made when a pilus grows out from the F^1 cell, attaches to the surface of the F^2 cell, contracts, and draws the two cells together (as shown in figure 9.20; see also figure 4.12). In both gram-positive and gram-negative cells, an opening is created between the connected cells, and the replicated DNA passes across from one cell to the other. The DNA probably does not pass through the pilus—that structure is used to bring the cells in contact. The actual transfer takes place through membrane secretion systems. Conjugation is a conservative process, in that the donor bacterium generally retains ("conserves") a copy of the genetic material being transferred.

There are hundreds of conjugative plasmids with some variations in their properties. One of the best understood plasmids is the F factor in *E. coli*, which exhibits these patterns of transfer:

1. The donor (F^1) cell makes a copy of its F factor and transmits this to a recipient (F^2) cell. The F^2 cell is thereby changed into an F^1 cell capable of producing a pilus and conjugating with other cells. No additional donor genes are transferred at this time.
2. In high-frequency recombination (Hfr) donors, the plasmid becomes integrated into the F^1 donor chromosome.

The term *high-frequency recombination* was adopted to denote a cell with an integrated F factor that transmits its chromosomal genes. These genes become integrated into recipient chromosomes at a very high frequency.

The F factor can direct a more comprehensive transfer of part of the donor chromosome to a recipient cell. This transfer occurs through duplication of the DNA, after which one strand of DNA is retained by the donor, and the other

Table 9.3 Types of Horizontal Gene Transfer in Bacteria

Examples of Mode	Factors Involved	Direct or Indirect*	Examples of Products of Transferred Genes
Conjugation	Donor cell with pilus Fertility plasmid in donor Both donor and recipient alive Bridge forms between cells to transfer DNA.	Direct	Drug resistance; resistance to metals; toxin production; enzymes; adherence molecules; degradation of toxic substances; uptake of iron
Transformation	Free donor DNA (fragment) Live, competent recipient cell	Indirect	Polysaccharide capsule; unlimited with cloning techniques
Transduction	Donor is lysed bacterial cell. Defective bacteriophage is carrier of donor DNA. Live recipient cell of same species as donor	Indirect	Toxins; enzymes for sugar fermentation; drug resistance

Direct means the donor and recipient are in contact during exchange; *indirect* means they are not.

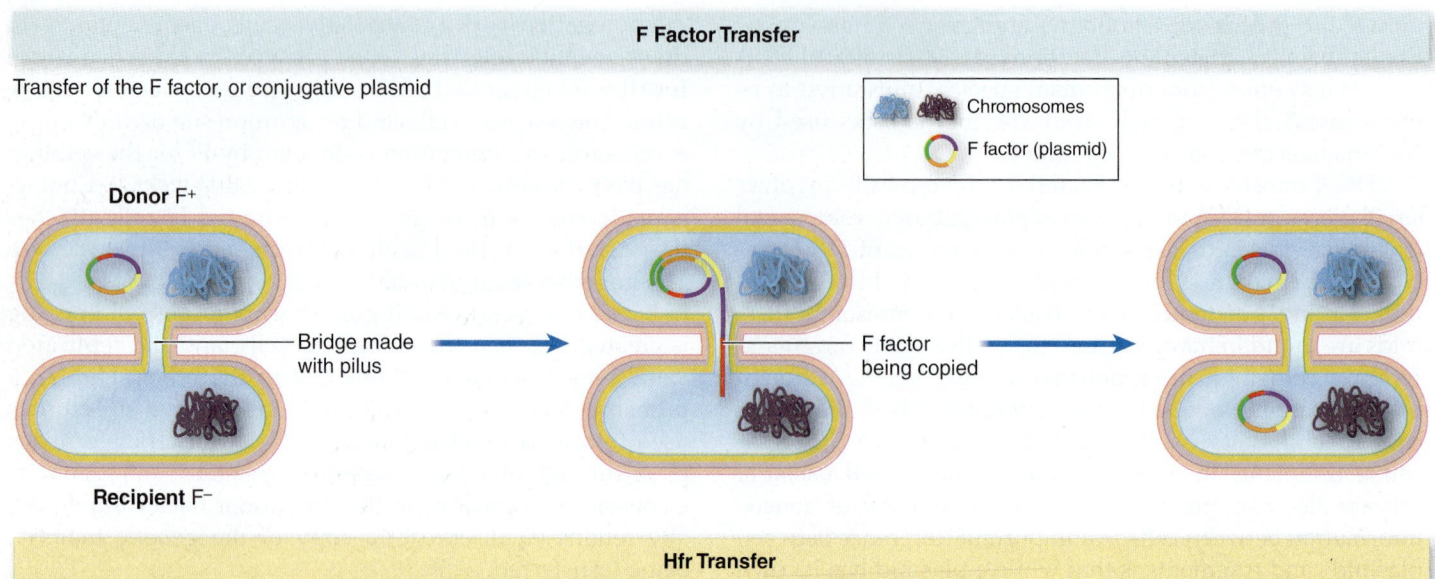

F Factor Transfer

Transfer of the F factor, or conjugative plasmid

Chromosomes
F factor (plasmid)

Donor F⁺

Bridge made with pilus

F factor being copied

Recipient F⁻

Hfr Transfer

High-frequency (Hfr) transfer involves transmission of chromosomal genes from a donor cell to a recipient cell. The plasmid jumps into the chromosome, and when the chromosome is duplicated the plasmid and part of the chromosome are transmitted to a new cell through conjugation. This plasmid/chromosome hybrid then incorporates into the recipient chromosome.

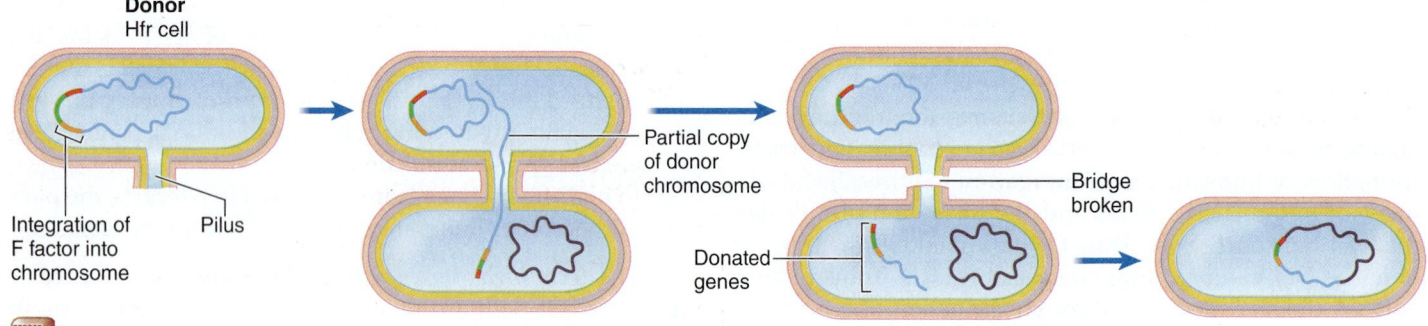

Donor Hfr cell

Partial copy of donor chromosome

Integration of F factor into chromosome

Pilus

Bridge broken

Donated genes

Figure 9.20 Conjugation: genetic transmission through direct contact between two cells.

strand is transported across to the recipient cell. The F factor may or may not be transferred during this process. The transfer of an entire chromosome takes about 100 minutes, but the connection between cells is ordinarily broken before this time, and rarely is the entire genome of the donor cell transferred.

Conjugation has great biomedical importance. Special **resistance (R) plasmids,** or **factors,** that bear genes for resisting antibiotics and other drugs are commonly shared among bacteria through conjugation. Transfer of R factors can confer multiple resistance to antibiotics such as tetracycline, chloramphenicol, streptomycin, sulfonamides, and penicillin. This phenomenon is discussed further in chapter 12. Other types of R factors carry genetic codes for resistance to heavy metals (nickel and mercury) or for synthesizing virulence factors (toxins, enzymes, and adhesion molecules) that increase the pathogenicity of the bacterial strain. Conjugation studies have also provided an excellent way to map the bacterial chromosome.

Transformation: Capturing DNA from Solution

The acceptance by a bacterial cell of small fragments of soluble DNA from the surrounding environment is termed **transformation (figure 9.21).** Cells that are capable of

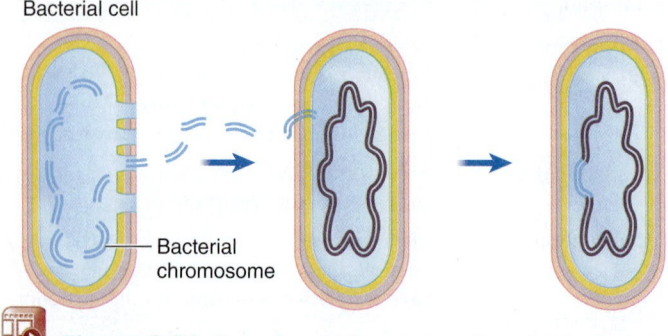

Bacterial cell

Bacterial chromosome

Figure 9.21 Transformation. DNA from dead cells is released into the environment and taken up into living cells. Some portion of the DNA may recombine into the genome of the recipient cell.

accepting genetic material through this means are termed **competent.** The new DNA traverses the outer membrane (in gram-negatives) using special protein channels, and moves through the cell wall via DNA-binding proteins. The DNA is then processed by the cell membrane and transported into the cytoplasm, where some of it is inserted into the bacterial chromosome. Transformation is a natural event found in several groups of gram-positive and gram-negative bacterial species.

The phenomenon was discovered in a famous experiment that elegantly illustrates both how transformation works and how the exchange of DNA between bacteria in a single host can have real effects on the host. The experiment was conducted in the late 1920s by the English biochemist Frederick Griffith working with *Streptococcus pneumoniae* and laboratory mice. This bacterium can be found in two different forms based on the presence of a capsule—the presence or absence of which affects colony morphology and pathogenicity. Encapsulated strains have a smooth (S) colony appearance and are virulent; strains lacking a capsule have a rough (R)

appearance and are nonvirulent. (Recall from chapter 4 that the capsule protects a bacterium from the phagocytic host defenses.) To set the groundwork, Griffith showed that when mice were injected with a live, virulent (S) strain, they soon died **(figure 9.22a).** Mice injected with a live, nonvirulent (R) strain remained alive and healthy **(figure 9.22b).** Next, he tried a variation on this theme. First, he heat-killed an S strain and injected it into mice, which remained healthy **(figure 9.22c).** Then came the ultimate test: Griffith injected both dead S cells and live R cells into mice, with the result that the mice died from pneumococcal blood infection **(figure 9.22d).** If killed bacterial cells do not come back to life and the nonvirulent live strain was harmless, why did the mice die? Although he did not know it at the time, Griffith had demonstrated that dead S cells, while passing through the body of the mouse, broke open and released some of their DNA (by chance, that part containing the genes for making a capsule). A few of the live R cells subsequently picked up this loose DNA and were transformed by it into virulent, capsule-forming strains. Later studies supported the concept that a chromosome released by

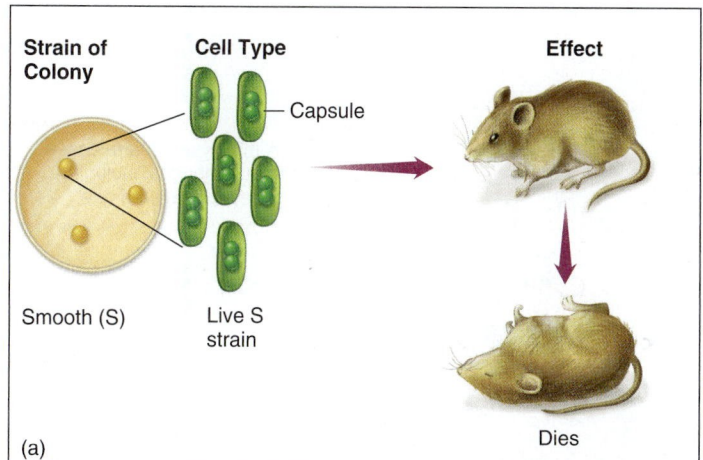

(a)

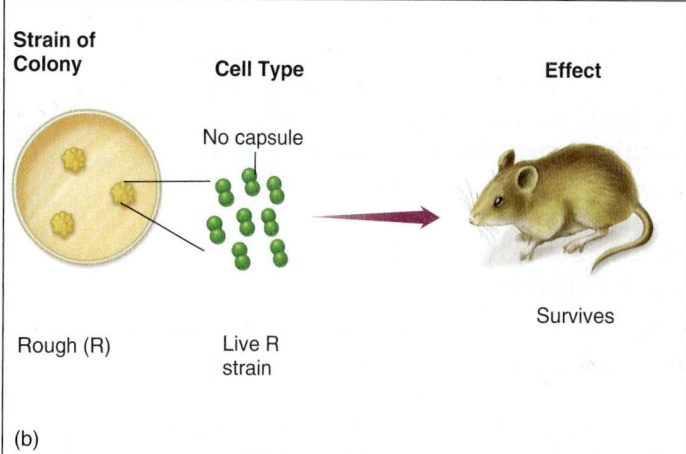

(b)

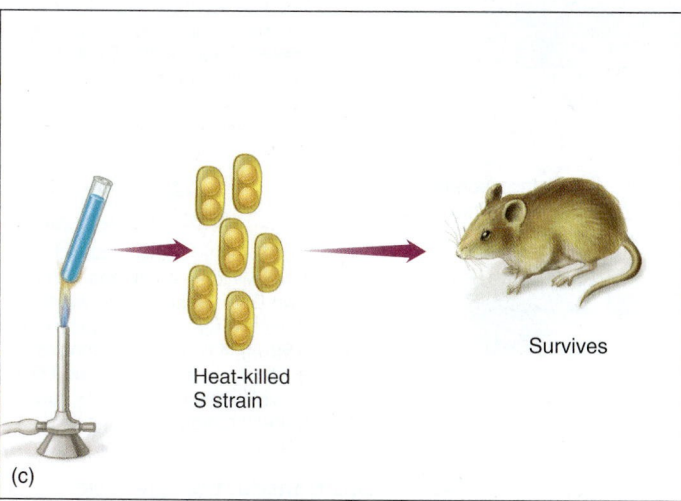

(c)

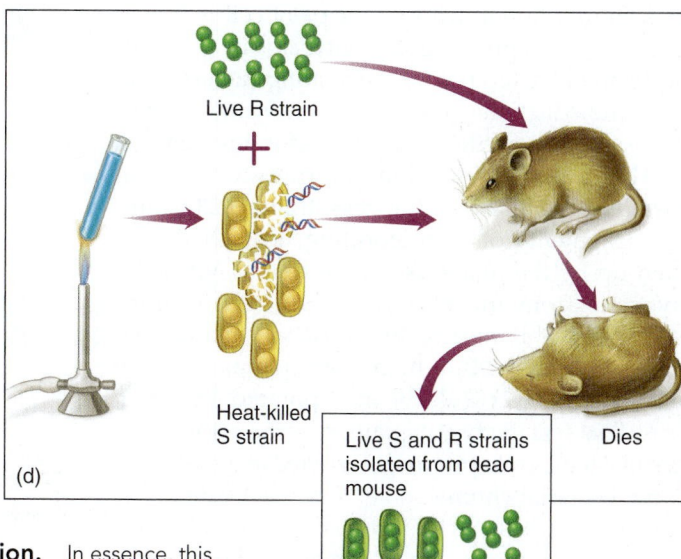

(d)

Figure 9.22 Griffith's classic experiment in transformation.   In essence, this experiment proved that DNA released from a killed cell can be acquired by a live cell. The cell receiving this new DNA is genetically transformed—in this case, from a nonvirulent strain to a virulent one.

a lysed cell breaks into fragments small enough to be accepted by a recipient cell and that DNA, even from a dead cell, retains its genetic code.

Disease Connection

Streptococcus pneumoniae is a versatile human pathogen. It causes pneumonia, ear infections, meningitis, and other conditions. Its capsule is vital for its disease-causing capacity.

Because transformation requires no special appendages and the donor and recipient cells do not have to be in direct contact, the process is useful for certain types of recombinant DNA technology. With this technique, foreign genes from a completely unrelated organism are inserted into a plasmid, which is then introduced into a competent bacterial cell through transformation. These recombinations can be carried out in a test tube, and human genes can be experimented upon and even expressed outside the human body by placing them in a microbial cell. This same phenomenon in eukaryotic cells, termed **transfection,** is an essential aspect of genetically engineered yeasts, plants, and mice, and it has been proposed as a future technique for curing genetic diseases in humans. These topics are covered in more detail in chapter 10.

Transduction: The Case of the Piggyback DNA

Bacteriophages (bacterial viruses) have been previously described as destructive bacterial parasites. Viruses can in fact serve as genetic vectors (an entity that can bring foreign DNA into a cell). The process by which a bacteriophage serves as the carrier of DNA from a donor cell to a recipient cell is transduction. Although it occurs naturally in a broad spectrum of bacteria, the participating bacteria in a single transduction event must be the same species because of the specificity of viruses for host cells.

There are two versions of transduction. In *generalized transduction* **(process figure 9.23),** random fragments of disintegrating host DNA are taken up by the phage during assembly. Virtually any gene from the bacterium can be transmitted through this means. In *specialized transduction* **(process figure 9.24),** a highly specific part of the host genome is regularly incorporated into the virus. This specificity is explained by the prior existence of a temperate prophage inserted in a fixed site on the bacterial chromosome. When activated, the prophage DNA separates from the bacterial chromosome, carrying a small segment of host genes with it. During a lytic cycle, these specific viral-host

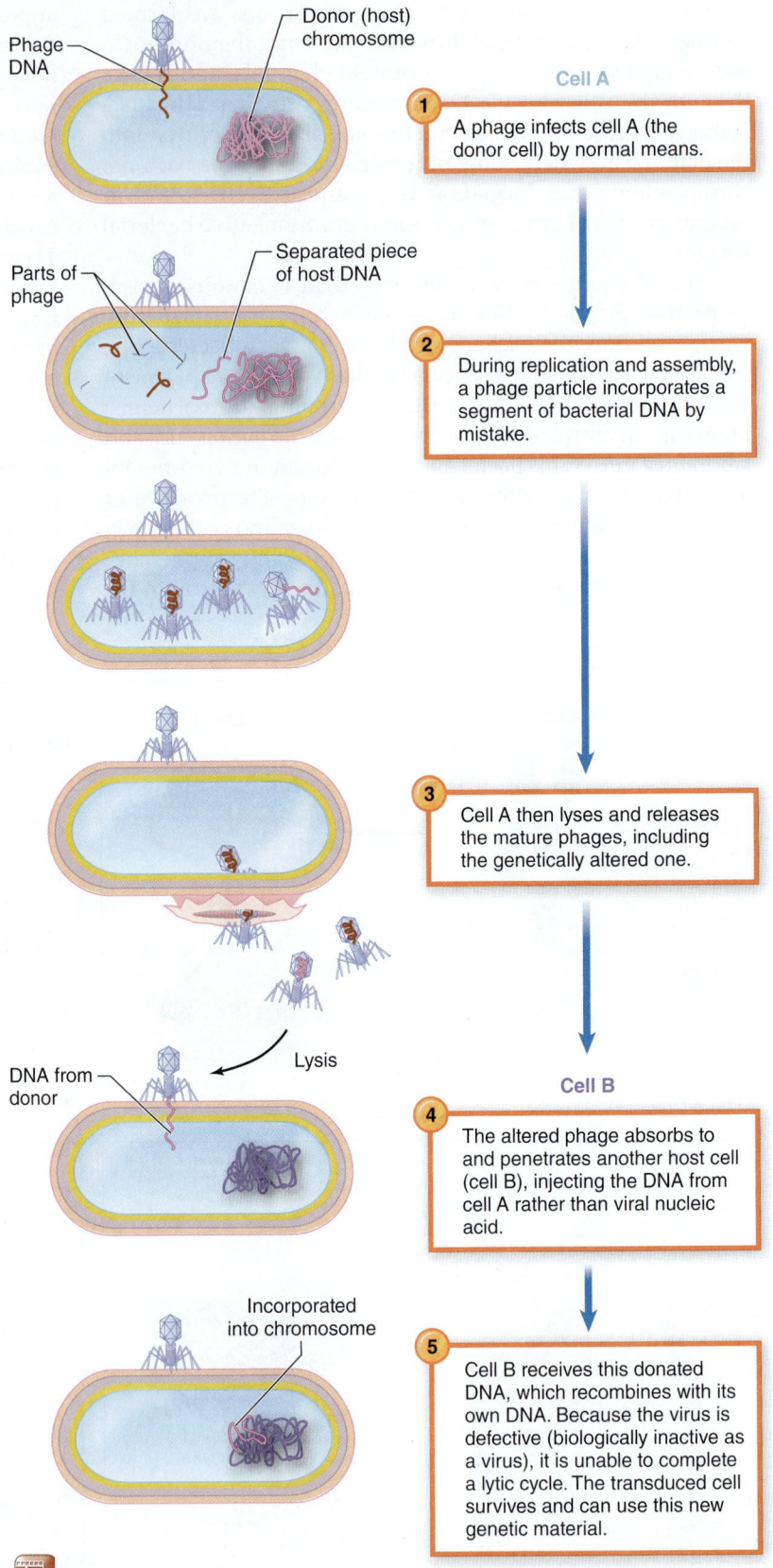

Process Figure 9.23 Generalized transduction: genetic transfer by means of a virus carrier.

1. A phage infects cell A (the donor cell) by normal means.

2. During replication and assembly, a phage particle incorporates a segment of bacterial DNA by mistake.

3. Cell A then lyses and releases the mature phages, including the genetically altered one.

4. The altered phage absorbs to and penetrates another host cell (cell B), injecting the DNA from cell A rather than viral nucleic acid.

5. Cell B receives this donated DNA, which recombines with its own DNA. Because the virus is defective (biologically inactive as a virus), it is unable to complete a lytic cycle. The transduced cell survives and can use this new genetic material.

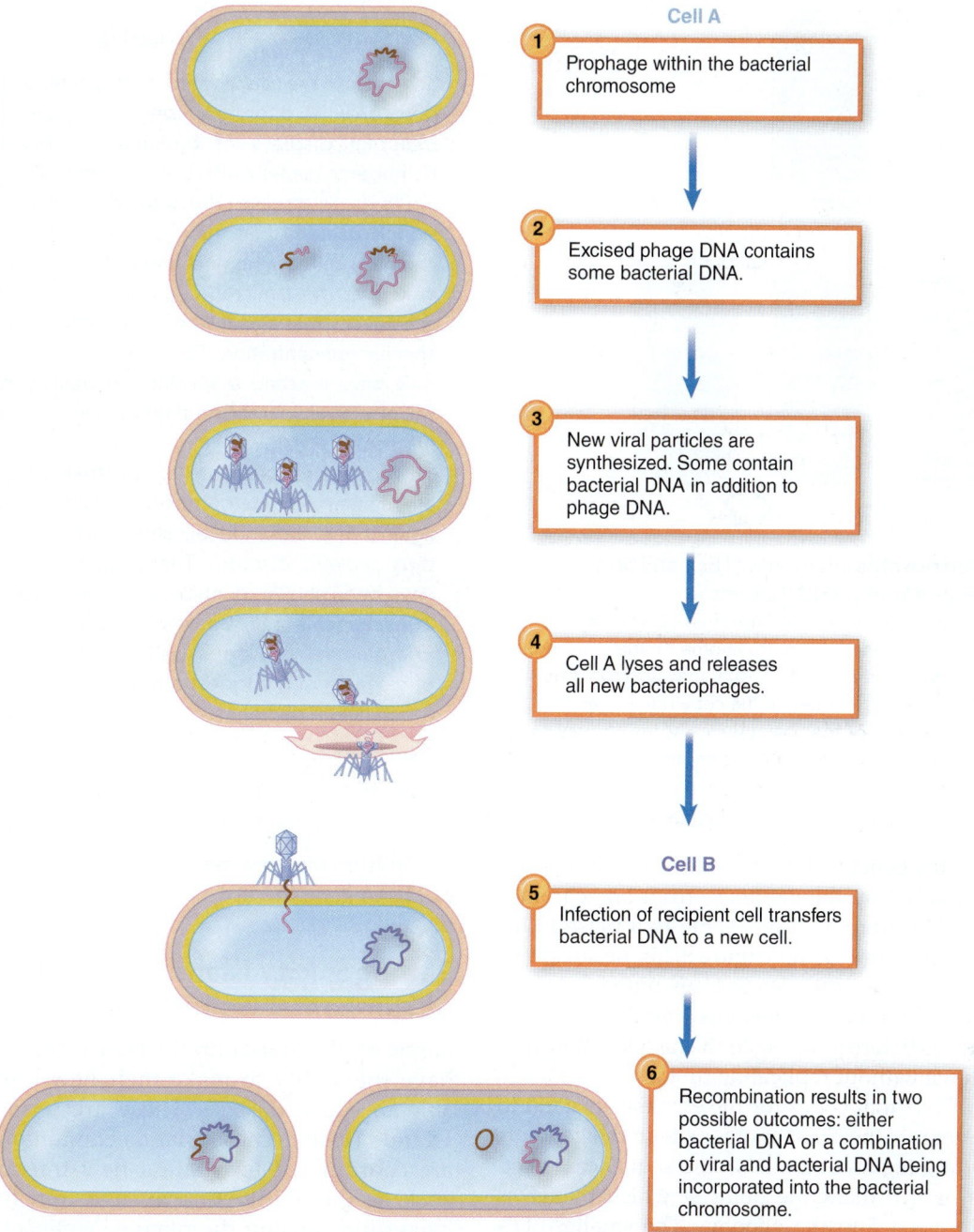

Cell A

1 Prophage within the bacterial chromosome

2 Excised phage DNA contains some bacterial DNA.

3 New viral particles are synthesized. Some contain bacterial DNA in addition to phage DNA.

4 Cell A lyses and releases all new bacteriophages.

Cell B

5 Infection of recipient cell transfers bacterial DNA to a new cell.

6 Recombination results in two possible outcomes: either bacterial DNA or a combination of viral and bacterial DNA being incorporated into the bacterial chromosome.

Process Figure 9.24 **Specialized transduction: transfer of specific genetic material by means of a virus carrier.**

gene combinations are incorporated into the viral particles and carried to another bacterial cell.

Several cases of specialized transduction have medical importance. The virulent strains of bacteria such as *Corynebacterium diphtheriae, Clostridium* spp., and *Streptococcus pyogenes* all produce toxins with profound physiological effects, whereas nonvirulent strains do not produce toxins. It turns out that the toxins are produced by bacteriophage genes that have been introduced by transduction. Only those bacteria infected with a temperate phage are toxin formers. (Details of toxin action are discussed in the organ-system-specific disease chapters.)

Transposable Elements

Another type of genetic transfer of great interest involves **transposons,** or transposable elements (TEs). TEs have the distinction of shifting from one part of the genome to another and so are termed "jumping genes." When the idea of their existence in corn plants was first proposed by geneticist Barbara McClintock in 1951, it was greeted with nearly universal skepticism because it had long been believed that the location of a given gene was set and that genes did not or could not move around. Now it is evident that jumping genes are widespread among cells and viruses.

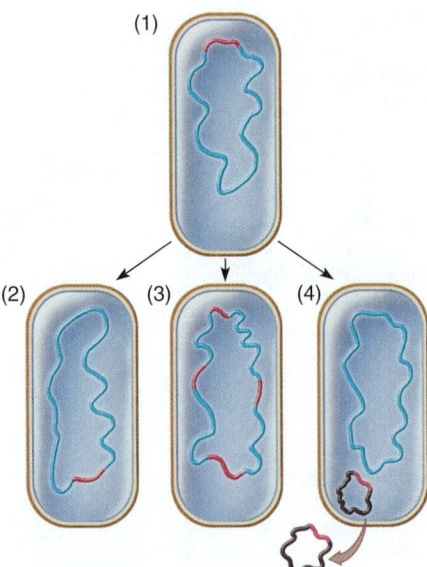

Figure 9.25 Transposable elements (TEs): shifting segments of the genome. **(1)** A TE exists as a small piece of DNA integrated into the host cell chromosome. **(2)** The TE may excise itself and move from one location to another in the genome, maintaining itself at a single copy per cell. **(3)** It may also replicate prior to moving, leading to an increase in the copy number and a greater effect on the genome of the host. **(4)** Finally, the TE may jump to a plasmid, which can then be transferred to another bacterial cell.

Case File 9   Wrap-Up

Scientists have discovered that bacteria don't only use quorum sensing to produce beautiful displays of bioluminescence. Pathogenic bacteria also use quorum sensing to cause disease. The autoinducer that signals bacteria that they have reached a quorum is a small chemical called a homoserine lactone. Scientists have discovered that *most* microbes produce this autoinducer or other substances that signal when a bacterial species has reached a specific concentration. Bacteria will only cause disease when they have reached a specific cell density, or a quorum, that will effectively overcome the host's immune system and cause disease.

Recent research has shown that certain chemicals that are similar to the homoserine lactone structure can act against the autoinducer, shut down quorum sensing, and thus *prevent* disease. These quorum-sensing antagonists have been shown to inhibit operons similar to the *lux* operon in *Escherichia coli, Chromobacterium violaceum, Burkholderia glumae,* and other pathogens. This exciting new field of research may suggest a novel way to overcome antibiotic resistance in bacteria.

Source: 2011. Mol. Cell vol. 42, no. 2, pp. 199–209.

All TEs share the general characteristic of traveling from one location to another on the genome—from one chromosomal site to another, from a chromosome to a plasmid, or from a plasmid to a chromosome (**figure 9.25**). Because TEs can occur in plasmids, they can also be transmitted from one cell to another in bacteria and a few eukaryotes. Some TEs replicate themselves before jumping to the next location, and others simply move without replicating first.

TEs contain DNA that codes for the enzymes needed to remove and reintegrate the TE at another site in the genome. Flanking the coding region of the DNA are sequences called tandem repeats, which mark the point at which the TE is removed or reinserted into the genome. The smallest TEs consist of only these two genetic sequences and are often referred to as **insertion elements.** A type of TE called a **retrotransposon** can transcribe DNA into RNA and then back into DNA for insertion in a new location. Other TEs contain additional genes that provide traits such as antibiotic resistance or toxin production.

The overall effect of TEs—to scramble the genetic language—can be beneficial or adverse to its host, depending upon such variables as where insertion occurs in a chromosome, what kinds of genes are relocated, and the type of cell involved. In bacteria, TEs are known to be involved in

1. changes in traits such as colony morphology, pigmentation, and antigenic characteristics;
2. replacement of damaged DNA; and
3. the intermicrobial transfer of drug resistance (in bacteria).

In humans, new research has pointed to a beneficial use of TEs in gene therapy (see chapter 10).

Pathogenicity Islands: Special "Gifts" of Horizontal Gene Transfer?

Some of the horizontally transferred genes in bacteria have the ability to make their new hosts pathogenic, or able to cause disease. These are termed **pathogenicity islands (PAIs).** These islands contain multiple genes that are coordinated to create a new trait on the bacterium, such as the ability to scavenge iron (important for the bacterium causing the plague, *Yersinia pestis*) or the ability to produce exotoxins (seen in *Staphylococcus aureus*). The islands are usually flanked by sequences that look like genes for TE enzymes. The field of bioinformatics has allowed for the identification of several PAIs in bacteria. Data derived from these studies have shown that organisms "share" their genes, sometimes in great chunks, with one another, essentially leap-frogging the evolution process by shuffling genes in this manner.

9.4 Learning Outcomes—Assess Your Progress

17. Explain the defining characteristics of a recombinant organism.
18. Describe three forms of horizontal gene transfer used in bacteria.

9.5 Mutations: Changes in the Genetic Code

As precise and predictable as the rules of genetic expression seem, permanent changes do occur in the genetic code. Indeed, genetic change is the driving force of evolution. In microorganisms, such changes may become evident in altered gene expression, such as the appearance or disappearance of anatomical or physiological traits. For example, a pigmented bacterium can lose its ability to form pigment, or a strain of the malarial parasite can develop resistance to a drug. Any change to the nucleotide sequence in the genome is called a **mutation.** Mutations are most noticeable when the genotypic change leads to a change in phenotype. Mutations can involve the loss of base pairs, the addition of base pairs, or a rearrangement in the order of base pairs. Do not confuse this with genetic recombination, in which microbes transfer whole segments of genetic information among themselves.

A microorganism that exhibits a natural, nonmutated characteristic is known as a **wild type,** or wild strain with respect to that trait. You may ask, In a constantly changing population of microbes, what is the natural, nonmutated state? For that reason, most scientists prefer to define *wild type* as the trait present in the highest numbers in a population. If a microorganism bears a mutation, it is called a **mutant strain.** Mutant strains can show variance in morphology, nutritional characteristics, genetic control mechanisms, resistance to chemicals, temperature preference, or nearly any type of enzymatic function. Mutant strains are very useful for tracking genetic events, unraveling genetic organization, and pinpointing genetic markers. A classic method of detecting mutant strains involves addition of various nutrients to a culture to screen for its use of that nutrient. For example, in a culture of a wild-type bacterium that is lactose-positive (meaning it has the necessary enzymes for fermenting this sugar), a small number of mutant cells have become lactose-negative, having lost the capacity to ferment this sugar. If the culture is plated on a medium containing indicators for fermentation, each colony can be observed for its fermentation reaction and the negative strain isolated.

Causes of Mutations

Mutations can be spontaneous or induced, depending upon their origin. A **spontaneous mutation** is a random change in the DNA arising from errors in replication that occur randomly. The frequency of spontaneous mutations has been measured for a number of organisms. Mutation rates vary tremendously, from one mutation in 10^5 replications (a high rate) to one mutation in 10^{10} replications (a low rate). The rapid rate of bacterial reproduction allows these mutations to be observed more readily in bacteria than in most eukaryotes.

Induced mutations result from exposure to known **mutagens,** which are primarily physical or chemical agents that interact with DNA in a disruptive manner **(table 9.4).** Chemical mutagenic agents act in a variety of ways to change the DNA. Some agents insert completely

Table 9.4 Selected Mutagenic Agents and Their Effects

Agent	Effect
Chemical	
Nitrous acid, bisulfite	Remove an amino group from some bases
Ethidium bromide	Inserts between the paired bases
Acridine dyes	Cause frameshifts due to insertion between base pairs
Nitrogen base analogs	Compete with natural bases for sites on replicating DNA
Radiation	
Ionizing (gamma rays, X rays)	Form free radicals that cause single or double breaks in DNA
Ultraviolet	Causes cross-links between adjacent pyrimidines

across the DNA helices between adjacent bases to produce mutations that distort the helix. Analogs of the nitrogen bases (5-bromodeoxyuridine and 2-aminopurine, for example) are chemical mimics of natural bases that are incorporated into DNA during replication. Addition of these abnormal bases leads to mistakes in base-pairing. Many chemical mutagens also act as carcinogens, or cancer-causing agents, when vertebrates are exposed to them (see "The Ames Test" later in this chapter).

Physical agents can also alter DNA, especially radiation. High-energy gamma rays and X rays introduce major physical changes into DNA, accumulating breaks that may not be repairable. Ultraviolet (UV) radiation induces abnormal bonds between adjacent pyrimidines that prevent normal replication. Exposure to large doses of radiation can be fatal, which is why radiation is so effective in microbial control; it can also be carcinogenic in animals. (The intentional use of UV to control microorganisms is described further in chapter 11.)

Categories of Mutations

Mutations range from large mutations, in which large genetic sequences are gained or lost, to small ones that affect only a single base on a gene. These latter mutations, which involve addition, deletion, or substitution of single bases, are called **point mutations.**

To understand how a change in DNA influences the cell, remember that the DNA code appears in a particular order of triplets (three bases) that is transcribed into mRNA codons, each of which specifies an amino acid. A permanent alteration in the DNA that is copied faithfully into mRNA and translated can change the structure of the protein. A change in a protein can likewise change the morphology and physiology of a cell. Some mutations have a harmful effect on the cell, leading to cell dysfunction or death; these are called *lethal mutations. Neutral mutations* produce neither adverse nor helpful changes. Of course, mutations can also be beneficial if they provide the cell with a useful change in structure or physiology.

Any change in the code that leads to placement of a different amino acid is called a **missense mutation.** A missense mutation can

1. create a faulty, nonfunctional (or less functional) protein;
2. produce a protein that functions in a different manner; or
3. cause no significant alteration in protein function (**table 9.5b** shows how missense mutations look).

A **nonsense mutation,** on the other hand, changes a normal codon into a stop codon that does not code for an amino acid and stops the production of the protein wherever it occurs. A nonsense mutation almost always results in a nonfunctional protein. (**Table 9.5d** shows a nonsense mutation resulting from a frameshift, which is described in the next paragraph.) A **silent mutation (table 9.5c)** alters a base but does not change the amino acid and thus has no effect. For example, because of the redundancy of the code, ACU, ACC, ACG, and ACA all code for threonine, so a mutation that changes only the last base will not alter the sense of the message in any way. A **back-mutation** occurs when a gene that has undergone mutation reverses (mutates back) to its original base composition.

Mutations also occur when one or more bases are inserted into or deleted from a newly synthesized DNA strand. This type of mutation, known as a **frameshift (table 9.5d,e),** is so named because the reading frame of the mRNA has been changed. Frameshift mutations nearly always result in a nonfunctional protein because every amino acid after the mutation is different from what was coded for in the original DNA. Also note that insertion or deletion of bases in multiples of three (3, 6, 9, etc.) results in the addition or deletion of amino acids but does not disturb the reading frame. The effects of all of these types of mutations can be seen in the table.

Repair of Mutations

Earlier, we indicated that DNA has a proofreading mechanism to repair mistakes in replication that may otherwise become permanent (see page 241). Because mutations are potentially life-threatening, the cell has additional systems for finding and repairing DNA that has been damaged by various mutagenic agents and processes. Most ordinary DNA damage is resolved by enzymatic systems specialized for finding and fixing such defects.

DNA that has been damaged by ultraviolet radiation can be restored by photoactivation, or light repair. This repair mechanism requires visible light and a light-sensitive enzyme, DNA photolyase, which can detect and attach to the damaged areas (sites of abnormal pyrimidine binding). Ultraviolet repair mechanisms are successful only for a relatively small number of UV mutations. Cells cannot repair severe, widespread damage and will die. In humans, the genetic disease *xeroderma pigmentosa* is due to nonfunctioning genes for enzymes responsible for excising pyrimidine dimers caused by UV light. Persons suffering from this rare disorder develop severe skin cancers.

Mutations can be excised by a series of enzymes that remove the incorrect bases and add the correct ones. This process is known as excision repair. First, enzymes break the bonds between the bases and the sugar-phosphate strand at the site of the error. A different enzyme subsequently removes the defective bases one at a time, leaving a gap that will be filled in by DNA polymerase I and ligase (**figure 9.26**). A repair system can also locate mismatched bases that were missed during proofreading, for example, C mistakenly paired with A, or G with T. The base must be replaced soon after the mismatch is made, or it will not be recognized by the repair enzymes.

The Ames Test

New agricultural, industrial, and medicinal chemicals are constantly being added to the environment, and exposure to them is widespread. The discovery that many such

Table 9.5 Categories of Point Mutations and Their Effects

(a)	DNA RNA Protein	TAC AUG Met	TGG ACC Thr	CTG GAC Asp	CTC GAG Glu	TAC AUG Met	TTT... AAA... Lys...	Normal gene
(b)	DNA RNA Protein	TAC AUG Met	TGG ACC Thr	CTT GAA Glu	CTC GAG Glu	TAC AUG Met	TTT... AAA... Lys...	Missense mutation: leading to amino acid switch (may or may not function well)
(c)	DNA RNA Protein	TAC AUG Met	TGG ACC Thr	CTA GAU Asp	CTC GAG Glu	TAC AUG Met	TTT... AAA... Lys...	Base substitution: silent (no change in function)

→ G

(d)	DNA RNA Protein	TAC AUG Met	TGC ACG Thr	TGC ACG Thr	TCT AGA Arg	ACT UGA STOP	TT AAA...	

Frameshift and premature stop

(e)	DNA RNA Protein	TAC AUG Met	TGG ACC Thr	GCT CGA Arg	GCT CGA Arg	CTA GAU Asp	CTT... GAA... Glu...	

Frameshift

Frameshift mutation

Deletion mutation (d)
↓
Both lead to frameshifts and can lead to premature stop codons and/or poorly functioning protein
↑
Insertion mutation (e)

compounds are mutagenic and that many of these mutagens are linked to cancer is significant. Although animal testing has been a standard method of detecting chemicals with carcinogenic potential, a more rapid screening system called the **Ames test** is also commonly used. In this ingenious test, the experimental subjects are bacteria whose gene expression and mutation rate can be readily observed and monitored. The premise is that any chemical capable of mutating bacterial DNA could similarly mutate mammalian (and thus human) DNA and is therefore potentially hazardous.

One indicator organism in the Ames test is a mutant strain of *Salmonella typhimurium* that has lost the ability to synthesize the amino acid histidine, a defect highly susceptible to back-mutation because the strain also lacks DNA repair mechanisms. Mutations that cause reversion to the wild strain, which is capable of synthesizing histidine, occur spontaneously at a very low rate. A test agent is considered a mutagen if it enhances the rate of back-mutation beyond levels that would occur spontaneously. One version of this testing procedure is outlined in **figure 9.27**. The Ames test has proved invaluable for screening an assortment of environmental and dietary chemicals for mutagenicity and carcinogenicity without resorting to animal studies. Because many mutagens affect bacteria differently than humans, the Ames test is considered a first step that must be followed by more rigorous testing.

Positive and Negative Effects of Mutations

Many mutations are not repaired. How the cell copes with them depends on the nature of the mutation and the strategies available to that organism. Mutations are permanent and heritable and will be passed on to the offspring of organisms and new viruses and become a long-term part of the gene pool.

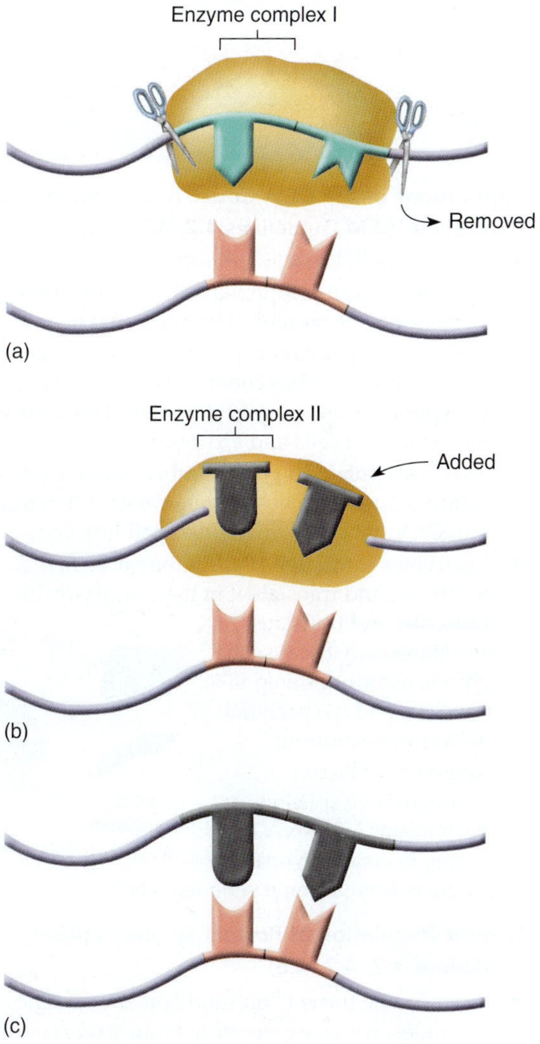

Figure 9.26 Excision repair of mutation by enzymes.
(a) The first enzyme complex recognizes one or several incorrect bases and removes them. (b) The second complex (DNA polymerase I and ligase) places correct bases and seals the gaps. (c) Repaired DNA.

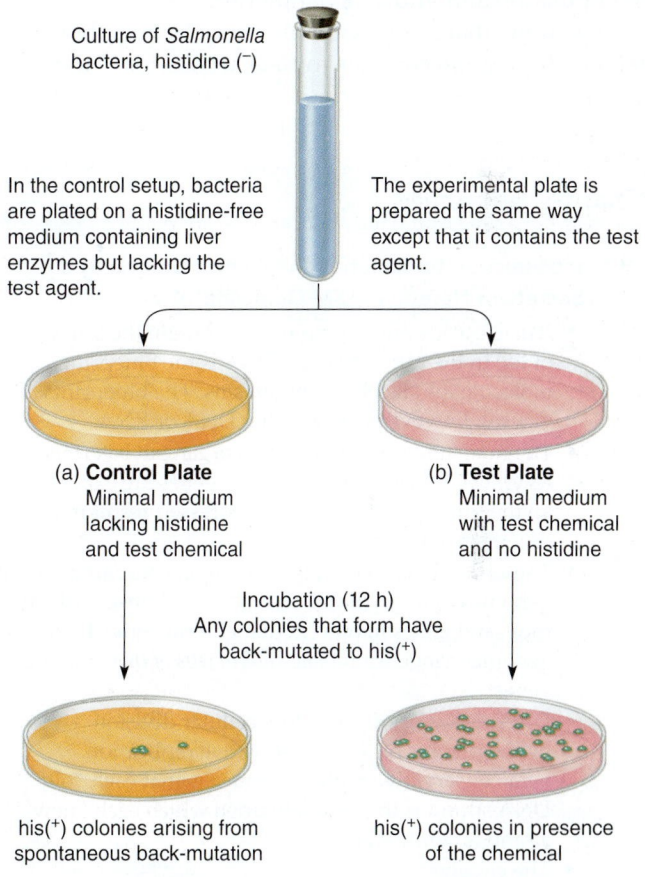

(a) **Control Plate**
Minimal medium lacking histidine and test chemical

(b) **Test Plate**
Minimal medium with test chemical and no histidine

Incubation (12 h)
Any colonies that form have back-mutated to his(+)

his(+) colonies arising from spontaneous back-mutation

his(+) colonies in presence of the chemical

(c) The degree of mutagenicity of the chemical agent can be calculated by comparing the number of colonies growing on the control plate with the number on the test plate. Chemicals that induce an increased incidence of back-mutation (right side) are considered carcinogens.

Figure 9.27 The Ames test.   This test is based on a strain of *Salmonella typhimurium* that cannot synthesize histidine [his(−)]. It lacks the enzymes to repair DNA so that mutations show up readily, and it has leaky cell walls that permit the ready entrance of chemicals. Many potential carcinogens (benzanthracene and aflatoxin, for example) are mutagenic agents only after being acted on by mammalian liver enzymes, so an extract of these enzymes is added to the test medium.

If a mutation leading to a nonfunctional protein occurs in a gene for which there is only a single copy, as in haploid or simple organisms, the cell will probably die. This happens when certain mutant strains of *E. coli* acquire mutations in the genes needed to repair damage by UV radiation. Mutations of the human genome affecting the action of a single protein (mostly enzymes) are responsible for more than 3,500 diseases.

Although most spontaneous mutations are not beneficial, a small number contribute to the success of the individual and the population by creating variant strains with alternate ways of expressing a trait. Microbes are not "aware" of this advantage and do not direct these changes; they simply respond to the environment they encounter. Those organisms with beneficial mutations can more readily adapt, survive, and reproduce. In the long-range view, mutations and the variations they produce are the raw materials for change in the population and, thus, for evolution.

Mutations that create variants occur frequently enough that any population contains mutant strains for a number of characteristics, but as long as the environment is stable, these mutants will usually never comprise more than a tiny percentage of the population. When the environment changes, however, it can become hostile for the survival of certain individuals, and only those microbes bearing protective mutations will be equipped to survive in the new environment. In this way, the environment naturally selects certain mutant strains that will reproduce, give rise to subsequent generations and, in time, be the dominant strain in the population. Through these means, any change that confers an advantage during selection pressure will be retained by the population. One of the clearest models for this sort of selection and adaptation is acquired drug resistance in bacteria (see chapter 12).

9.5 Learning Outcomes—Assess Your Progress

19. Define the term *mutation* and discuss one positive and one negative example of it in microorganisms.

20. Differentiate among frameshift, nonsense, silent, and missense mutations.

Chapter Summary

9.1 Introduction to Genetics and Genes: Unlocking the Secrets of Heredity (ASM Guideline* 4.2)

- Nucleic acids are molecules that contain the blueprints of life in the form of genes. DNA is the blueprint molecule for all cellular organisms. The blueprints of viruses, however, can be either DNA or RNA.

- The total amount of DNA in an organism is termed its *genome* (also *genotype*). Not all genes are expressed all the time; the ones that are expressed result in an organism's phenotype.

- The genome of bacteria is quite small compared with the genome of eukaryotes. Bacterial DNA consists of a few thousand genes in one circular chromosome. Eukaryotic genomes range from *thousands to tens of thousands* of genes.

- DNA copies itself just before cellular division by the process of semiconservative replication. Semiconservative replication means that each "old" DNA strand is the template upon which each "new" strand is synthesized.

- The circular bacterial chromosome is replicated at two forks as directed by DNA polymerase III. At each fork, two new strands are synthesized—one continuously and one in short fragments—and mistakes are proofread and removed.

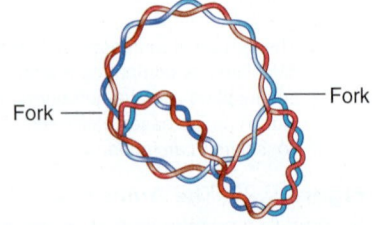

9.2 Applications of the DNA Code: Transcription and Translation (ASM Guidelines 4.2, 4.3, 4.4)

- Information in DNA is converted to proteins by the processes of transcription and translation. These proteins may be structural or functional in nature.

- The DNA code occurs in groups of three bases; this code is copied onto RNA as codons; the message determines the types of amino acids in a protein. This code is universal in all cells and viruses.

- DNA also contains a great number of non-protein-coding sequences. These sequences are often transcribed into RNA that serves to regulate cell function.

- Eukaryotes transcribe DNA in the nucleus, remove its introns, and translate it in the cytoplasm. Bacteria transcribe and translate simultaneously because the DNA is not sequestered in a nucleus and the bacterial DNA is free of introns.

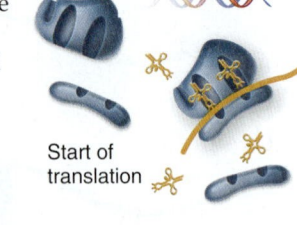

Start of translation

- Eukaryotic cells can use alternative splicing mechanisms and RNA editing to create diverse products from a single gene sequence.

9.3 Genetic Regulation of Protein Synthesis (ASM Guidelines 4.2, 4.3, 4.5)

- Genes can be turned "on" and "off" by specific molecules, which expose or hide their nucleotide codes for transcribing proteins.

- Operons are collections of genes in bacteria that code for products with a coordinated function.

- Nutrients can combine with regulator gene products to turn a set of structural genes on (inducible genes)

ASM Curriculum Guidelines (American Society for Microbiology, 2012). Complete guidelines in appendix B of this book.

or off (repressible genes). The *lac* (lactose) operon is an example of an inducible operon. The *arg* (arginine) operon is an example of a repressible operon.

- The rifamycins, tetracyclines, and aminoglycosides are classes of antibiotics that interfere with transcription and translation processes in microorganisms.

9.4 DNA Recombination Events (ASM Guidelines 4.1, 4.4, 4.5)

- Genetic recombination occurs in eukaryotes through sexual reproduction and through horizontal gene transfer.
- In bacteria, recombination occurs only through horizontal gene transfer.
- The three main types of horizontal gene transfer in bacteria are transformation, conjugation, and transduction.
- Transposable elements (TEs) are genes that can relocate from one part of the genome to another, causing rearrangement of genetic material.
- Some of the horizontally transferred genes in bacteria are pathogenicity islands, which confer the ability to cause disease on their hosts.

9.5 Mutations: Changes in the Genetic Code (ASM Guidelines 1.2, 4.1, 4.5)

- Changes in the genetic code can occur by two means: mutation and recombination. Mutation means a change in the nucleotide sequence of the organism's genome.
- Mutations can be either spontaneous or induced by exposure to some external mutagenic agent.
- All cells have enzymes that repair damaged DNA. When the degree of damage exceeds the ability of the enzymes to make repairs, mutations occur.
- Mutation-induced changes in DNA nucleotide sequences range from a single nucleotide to addition or deletion of large sections of genetic material.

Multiple-Choice and True-False Questions | Bloom's Levels 1 and 2: Remember and Understand

Multiple-Choice Questions. Select the correct answer from the options provided.

1. What is the smallest unit of heredity?
   a. chromosome
   b. gene
   c. codon
   d. nucleotide

2. The nitrogen bases in DNA are bonded to the
   a. phosphate.
   b. deoxyribose.
   c. ribose.
   d. hydrogen.

3. DNA replication is semiconservative because the _____ strand will become half of the _____ molecule.
   a. RNA, DNA
   b. template, finished
   c. sense, mRNA
   d. codon, anticodon

4. In DNA, adenine is the complementary base for _____, and cytosine is the complement for _____.
   a. guanine, thymine
   b. uracil, guanine
   c. thymine, guanine
   d. thymine, uracil

5. Transfer RNA is the molecule that
   a. contributes to the structure of ribosomes.
   b. adapts the genetic code to protein structure.
   c. transfers the DNA code to mRNA.
   d. provides the master code for amino acids.

6. As a general rule, the template strand on DNA will always begin with
   a. TAC.
   b. AUG.
   c. ATG.
   d. UAC.

7. The *lac* operon is usually in the _____ position and is activated by a/an _____ molecule.
   a. on, repressor
   b. off, inducer
   c. on, inducer
   d. off, repressor

8. Which genes can be transferred by all three methods of horizontal gene transfer?
   a. capsule production
   b. toxin production
   c. F factor
   d. drug resistance

9. Which of the following would occur through specialized transduction?
   a. acquisition of Hfr plasmid
   b. transfer of genes for toxin production
   c. transfer of genes for capsule formation
   d. transfer of a plasmid with genes for degrading pesticides

10. When genes are turned on differently under different environmental conditions, this represents a change in
    a. species.
    b. genotype.
    c. phenotype.
    d. growth rate.

True-False Questions. If the statement is true, leave as is. If it is false, correct it by rewriting the sentence.

11. The DNA pairs are held together primarily by covalent bonds.

12. Mutation usually has a negative outcome.

13. The lagging strand of DNA is replicated in short pieces because DNA polymerase can synthesize in only one direction.

14. Messenger RNA is formed by translation of a gene on the DNA template strand.

15. A nucleotide is composed of a 5-carbon sugar, a phosphate group, and a nitrogenous base.

Critical Thinking Questions | Bloom's Levels 3, 4, and 5: Apply, Analyze, and Evaluate

Critical thinking is the ability to reason and solve problems using facts and concepts. These questions can be approached from a number of angles and, in most cases, they do not have a single correct answer.

1. Explain the relationship among the following terms: *genomics, proteomics, gene, protein, genotype,* and *phenotype.*

2. Sketch a molecule of double-stranded DNA, and explain each of the following:
   a. proper nitrogenous base-pairing between the strands
   b. proper orientation of the strands
   c. proper positioning of the sugar-phosphate backbone

3. On paper, replicate the following segment of DNA:
   5′ A T C G G C T A C G T T C A C 3′
   3′ T A G C C G A T G C A A G T G 5′
   a. Show the direction of replication of the new strands and indicate the location of the lagging and leading strands.
   b. Explain one challenge in the replication of circular DNA and in the replication of linear DNA and how each is resolved in a cell.

4. Provide evidence in support of or refuting the following statement: The life cycle of a retrovirus follows the classical view of the central dogma of biology.

5. Compare the structure and functions of DNA and RNA.

6. The following sequence represents triplets on DNA:
   3′ TAC CAG ATA CAC TCC CCT GCG ACT 5′
   Draw a diagram, and describe the following:
   a. the mRNA codons that correspond to this DNA sequence
   b. the location of ribosome binding within the first two mRNA codons
   c. the tRNA anticodons that correspond with the entire mRNA sequence
   d. the sequence of amino acids in the polypeptide coded by the mRNA sequence

7. Define the terms *redundancy* and *wobble* in terms of gene expression; using your knowledge of these processes, provide another mRNA sequence that could be used to synthesize the same protein as that in question 6.

8. Using the DNA sequence in question 6, illustrate and explain the following mutations:
   a. a deletion
   b. an insertion
   c. a substitution
   d. a nonsense mutation
   e. a frameshift mutation

9. a. Summarize how bacterial and eukaryotic cells differ in gene structure as well as the processes involved in gene expression.
   b. A new organism was identified by members of your lab. DNA analysis revealed the presence of introns and what appear to be split genes. Explain whether you have identified a bacterium or a eukaryotic organism.

10. Use your knowledge of DNA recombination events to complete the following:
   a. Propose two ways in which antibiotic resistance may develop in a bacterium.
   b. Explain how transposable elements may be used to treat humans with mutations in insulin-producing genes.
   c. Describe how bacterial cells acquire the ability to produce toxins.

Concept Connections | Bloom's Levels 4 and 6: Analyze and Create

This activity ties together multiple concepts in the chapter.

1. What are the four bases that make up RNA?

2. What are the functions of mRNA, tRNA, and rRNA? What structures does each form?

3. What are the functions of antisense RNA, riboswitches, and small interfering RNA?

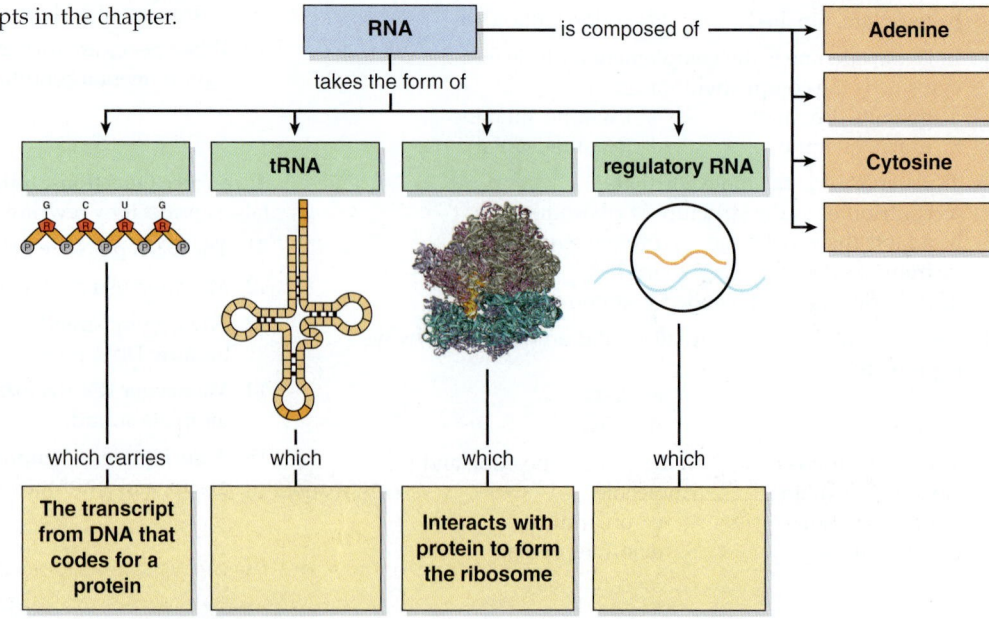

Visual Connections | Bloom's Level 5: Evaluate

This question uses visual images or previous content to make connections to this chapter's concepts.

1. **Process figure 9.15, step ①**. Label each of the parts of the illustration.

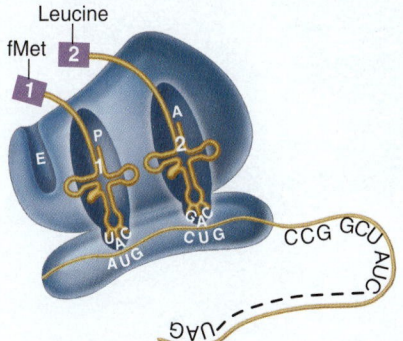

2. **From chapter 4, figure 4.11a.** Speculate on why these cells contain two chromosomes (shown in blue).

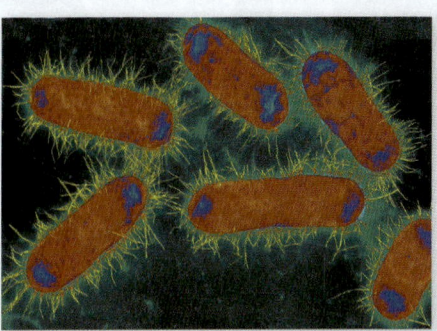

Concept Mapping | Bloom's Level 6: Create

Appendix D provides guidance for working with concept maps.

1. Using the words that follow, please create a concept map illustrating the relationships among these key terms from chapter 9.

ribozyme	mRNA	transcription
primer	tRNA	translation
riboswitch	rRNA	DNA

www.mcgrawhillconnect.com

Enhance your study of this chapter with study tools and practice tests. Also ask your instructor about the resources available through ConnectPlus, including the media-rich eBook, interactive learning tools, and animations.

10

Genetic Engineering and Recombinant DNA

Case File 10

Identifying the Victims of 9/11 Over a Decade Later

It has been over a decade since the terrorist attacks of 9/11/2001, and still the remains of many of the victims of the World Trade Center attacks have not been identified, leaving families and loved ones with unresolved grief. As of September 2011, 1,632 victims had been identified and 1,121 had not. The positive identifications were achieved mainly through DNA analysis of 21,800 body parts recovered from the site. Fire from the explosions, steel, concrete, glass, and water used to douse the flames all contributed to the degradation of the remains, making DNA extraction extremely difficult. Many of the remains were so badly damaged that there was little to no genetic material for analysis.

Relatives of the victims soon began to provide DNA samples to the newly developed Mass Disaster Kinship Analysis Program (MDKAP). These donated samples were used to create a database that would facilitate the identification of lost family members. Families gave hair, toothbrushes, razors, or other personal effects of the victims that contained their DNA to the MDKAP. Using the DNA extracted from these samples, forensic scientists were able to identify some of the unknown remains by scanning the MDKAP database for genetic similarities. The quest to identify the victims of the WTC attacks has spurred innovations in DNA analysis that have been helpful in identifying other individuals such as missing persons or war casualties.

■ How is DNA from the victims of the WTC attacks analyzed?

■ What improvements to this technology have been made in the past decade to accurately identify the victims?

Source: L. Geddes, "Massive DNA Effort to Name 1121 Unknown Dead of 9/11," *New Scientist* 2829 (2011). Retrieved from www.newscientist.com/article/dn20878-massive-dna-effort-to-name-1121-unknown-dead-of-911.html.
2011. *New Scientist*. vol. 2829, p. 7.

Continuing the Case appears on page 284.

Outline and Learning Outcomes

10.1 Introduction to Genetic Engineering

1. Provide examples of practical applications of modern genetic technologies.

10.1 Introduction to Genetic Engineering

In chapter 9, we looked at the ways in which microorganisms duplicate, exchange, and use their genetic information. In scientific vocabulary, this is called *basic science* because no product or application is directly derived from it. Humans, however, being who they are, are quick to derive applications from basic research findings. As an example, basic science regarding the workings of the electron has led to the development of television, computers, and cell phones. None of these staples of modern life were envisioned when early physicists were deciphering the nature of subatomic particles, but without the knowledge of how electrons worked, our ability to harness them for our own uses never would have materialized.

The same scenario has played out with regard to genetics. The knowledge of how DNA was manipulated within the cell to carry out the goals of a microbe allowed scientists to use these processes to accomplish goals important to human beings. Examples of human goals that have been more efficiently attained through the use of modern and not-so-modern genetic technologies can be seen in each of these scenarios:

1. A farmer mates his two largest pigs in the hopes of producing larger offspring. Unfortunately, he quite often ends up with small or unhealthy animals due to other genes that are transferred during mating. Genetic manipulation allows for the transfer of specific genes so that only advantageous traits are selected.

2. Courts have, for thousands of years, relied on a description of a person's phenotype (eye color, hair color, etc.) as a means of identification. By remembering that a phenotype is the product of a particular sequence of DNA, you can quickly see how looking at someone's DNA (perhaps from a drop of blood) gives a better clue as to his or her identification.

3. We have understood for a long time that many diseases are the result of a missing or dysfunctional protein, and we have generally treated the diseases by replacing the protein as best we can, usually resulting in only temporary relief and limited success. Examples include insulin-dependent diabetes, adenosine deaminase deficiency, and blood-clotting disorders. Genetic engineering offers the promise that fixing the underlying mutation responsible for the lack of a particular protein can treat these diseases far more successfully than we've been able to do in the past.

4. New results from whole-organism sequencing show us that RNA regulatory molecules may be even more useful in permanently "fixing" many diseases.

Information on genetic engineering and its biotechnological applications is growing at such an accelerating rate that some new discovery or product is disclosed on an almost daily basis. To keep this subject somewhat manageable, we present essential concepts and applications, organized under the following five topics:

- Tools and Techniques of Genetic Engineering
- Products of Recombinant DNA Technology
- Genetic Treatments
- Genome Analysis
- Proteome Analysis

10.1 Learning Outcomes—Assess Your Progress

   1. Provide examples of practical applications of modern genetic technologies.

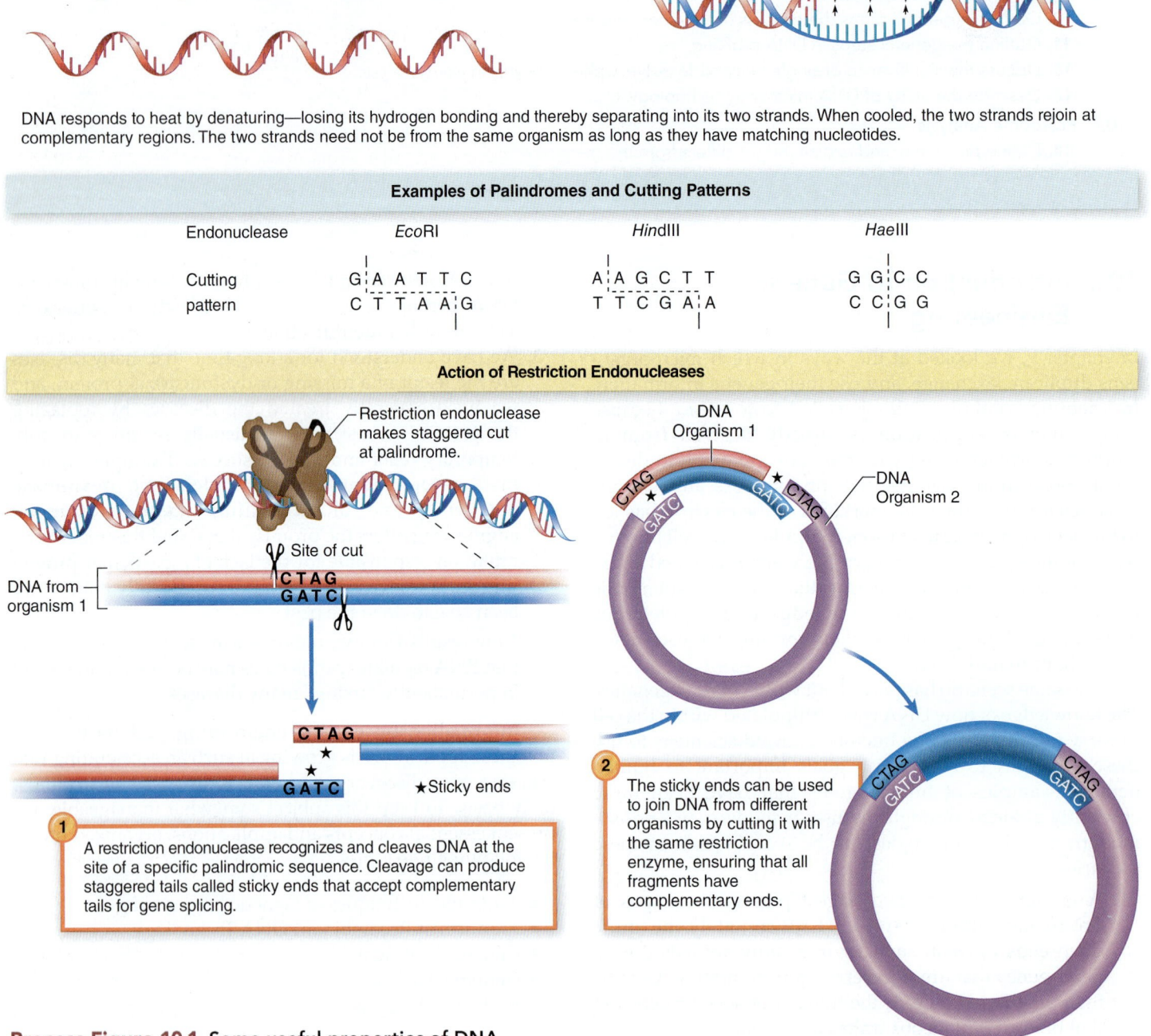

Process Figure 10.1 **Some useful properties of DNA.**

10.2 Tools and Techniques of Genetic Engineering

DNA: The Raw Material

All of the intrinsic properties of DNA are true whether the DNA is in a bacterium or a test tube. For example, the enzyme helicase is able to unwind the two strands of the double helix just as easily in the lab as it does in a bacterial cell. But in the laboratory we can take advantage of our knowledge of DNA chemistry to make helicase unnecessary. It turns out that when DNA is heated to just below boiling (90°C to 95°C), the two strands separate, revealing the information contained in their bases. With the nucleotides exposed, DNA can be more easily identified, replicated, or transcribed. If heat-denatured DNA is then slowly cooled, complementary nucleotides will hydrogen bond with one another and the strands will renature, or regain their familiar double-stranded form (**process figure 10.1**). As we shall see, this process is a necessary feature of the polymerase chain reaction and in the application of nucleic acid probes described later.

Enzymes for Dicing, Splicing, and Reversing Nucleic Acids

The polynucleotide strands of DNA can also be clipped crosswise at selected positions by means of enzymes called **restriction endonucleases.**[1] These enzymes come from bacterial and archaeal cells, and their discovery in 1971 has made almost everything discussed in this section possible. They recognize foreign DNA and are capable of breaking the phosphodiester bonds between adjacent nucleotides on both strands of DNA, leading to a break in the DNA strand. In bacteria and archaea in nature, this ability protects against the incompatible DNA of bacteriophages or plasmids. In the biotechnologist's lab, the enzymes can be used to cleave DNA at desired sites and are necessary for the techniques of recombinant DNA technology.

So far, about 3,000 restriction endonucleases have been discovered in bacteria and archaea. Each type has a known sequence of 4 to 10 base pairs as its target, so sites of cutting can be finely controlled. These enzymes have the unique property of recognizing and clipping at base sequences called **palindromes** (see process figure 10.1). Palindromes are sequences of DNA that are identical when read from the 5' to 3' direction on one strand and the 5' to 3' direction on the other strand.

Endonucleases are usually named by combining the first letter of the bacterial genus, the first two letters of the species, and the endonuclease number. Thus, *Eco*RI is the first endonuclease found in *Escherichia coli* (in the R strain), and *Hind*III is the third endonuclease discovered in *Haemophilus influenzae* type d (see process figure 10.1).

Endonucleases are used in the laboratory to cut DNA into smaller pieces for further study as well as to remove and insert sequences during recombinant DNA techniques, described in a subsequent section. Endonucleases such as

*Hae*III make straight, blunt cuts on DNA. But more often, the enzymes make staggered symmetrical cuts that leave short tails called "sticky ends." The enzymes cut four to five bases on the 3' strand, and four to five bases on the 5' strand, leaving overhangs on each end. Such adhesive tails will base-pair with complementary tails on other DNA fragments or plasmids (see process figure 10.1). This effect makes it possible to splice genes into specific sites.

The pieces of DNA produced by restriction endonucleases are termed **restriction fragments.** Because DNA sequences vary, even among members of the same species, differences in the cutting pattern of specific restriction endonucleases give rise to restriction fragments of differing lengths, known as **restriction fragment length polymorphisms (RFLPs).** RFLPs allow the direct comparison of the DNA of two different organisms at a specific site, which, as we will see, has many uses.

Another enzyme, called a **ligase,** is necessary to seal the sticky ends together by rejoining the phosphate-sugar bonds cut by endonucleases. Its main application is in final splicing of genes into plasmids and chromosomes.

An enzyme called **reverse transcriptase** is best known for its role in the replication of HIV and other retroviruses. It also provides geneticists with a valuable tool for converting RNA into DNA. Copies called **complementary DNA (cDNA)** can be made from messenger, transfer, ribosomal, and other forms of RNA. The technique provides a valuable means of synthesizing eukaryotic genes from mRNA transcripts (**figure 10.2**). The advantage is that the synthesized gene

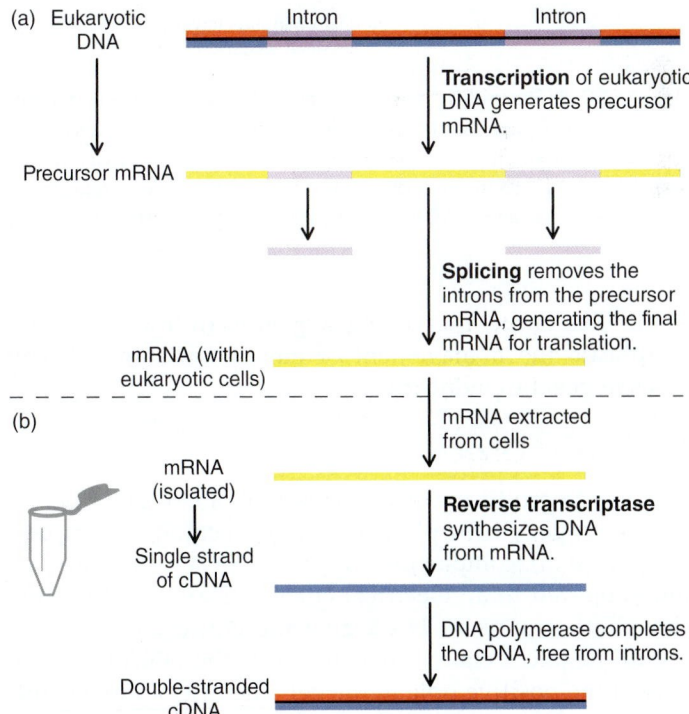

Figure 10.2 Making cDNA from eukaryotic mRNA. In order for eukaryotic genes to be expressed by a bacterial cell, a copy of DNA without introns must be cloned. The cDNA encodes the same protein as the original DNA but lacks introns.

---

1. The meaning of *restriction* is that the enzymes do not act upon the bacterium's own DNA; an *endo*nuclease nicks DNA internally, not at the ends.

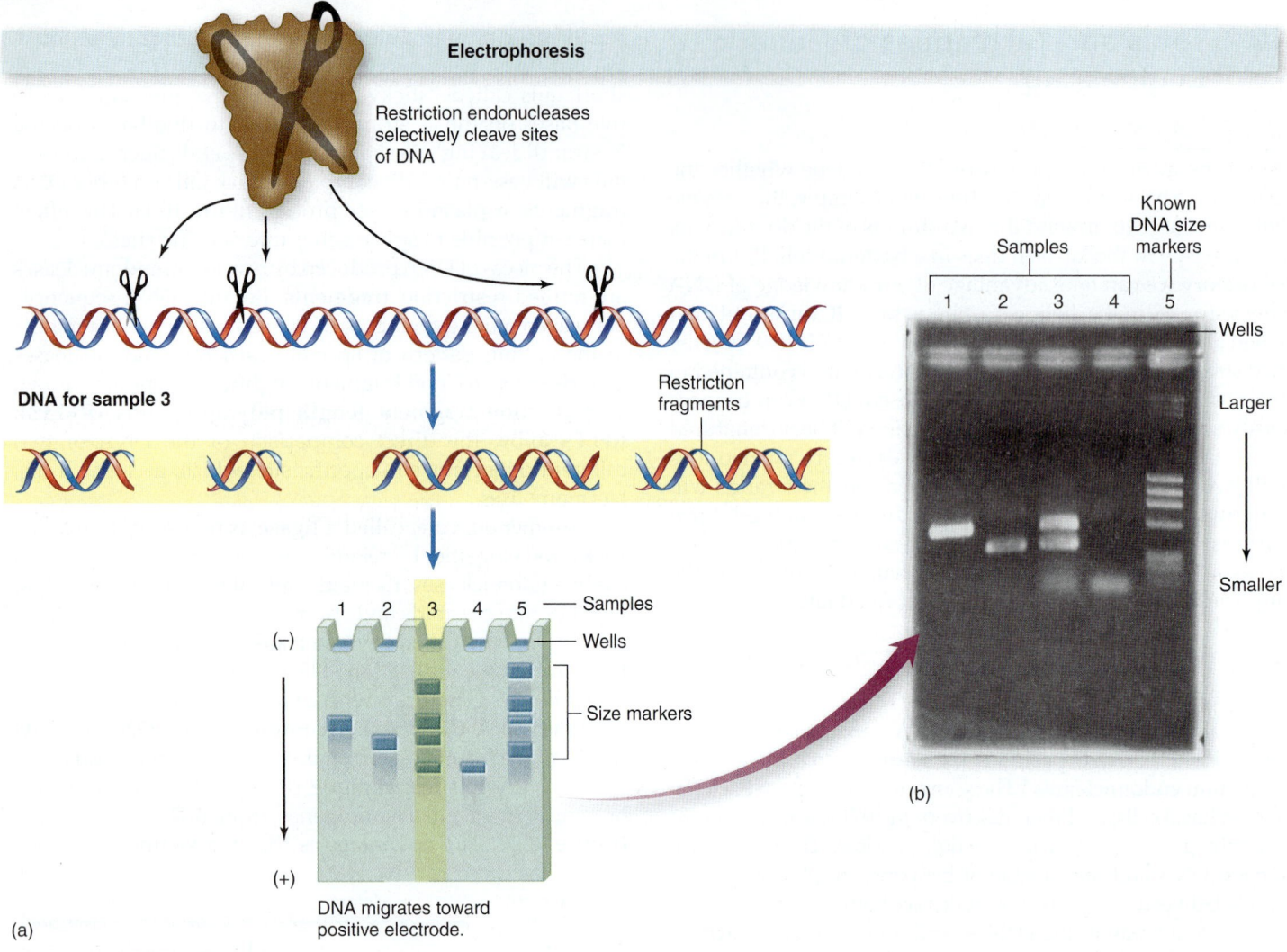

Figure 10.3 Revealing the patterns of DNA with electrophoresis. **(a)** After cleavage into fragments, DNA is loaded into wells on one end of an agarose gel. When an electrical current is passed through the gel (from the negative pole to the positive pole), the DNA, being negatively charged, migrates toward the positive pole. The larger fragments, measured in numbers of base pairs, migrate more slowly and remain nearer the wells than the smaller (shorter) fragments. **(b)** An actual stained gel reveals a separation pattern of the fragments of DNA. The size of a given DNA band can be determined by comparing the distance it traveled to the distance traveled by a set of DNA fragments of known size (lane 5).

will be free of the intervening sequences (introns) that can complicate the management of eukaryotic and archaeal genes in genetic engineering.

Analysis of DNA

One way to produce a readable pattern of DNA fragments is through **gel electrophoresis.** In this technique, samples are placed in compartments (wells) in a soft agar gel and subjected to an electrical current **(figure 10.3a).** The phosphate groups in DNA give the entire molecule an overall negative charge, which causes the DNA to move toward the positive pole in the gel. The rate of movement is based primarily on the size of the fragments. The larger fragments move more slowly and remain nearer the top of the gel, whereas the smaller fragments migrate faster and are positioned farther from the wells. The positions

of DNA fragments are determined by staining the DNA fragments in the gel **(figure 10.3b).** Electrophoresis patterns can be quite distinctive and are very useful in characterizing DNA fragments and comparing the degree of genetic similarities among samples as in a genetic fingerprint (discussed later).

Disease Connection

One very useful application of gel electrophoresis is in the investigation of food-borne disease outbreaks. The CDC and public health departments use a variation of electrophoresis called pulse-field gel electrophoresis to determine if microbes isolated from food samples or ill patients are exactly the same strain and, thus, part of a common outbreak.

Nucleic Acid Hybridization and Probes

Two different nucleic acids can **hybridize** by uniting at their complementary regions. All different combinations are possible: Single-stranded DNA can unite with other single-stranded DNA or RNA, and RNA can hybridize with other RNA. This property has allowed for the development of specially formulated tracers called **gene probes.** These probes consist of a short stretch of DNA of a known sequence that will base-pair with a stretch of DNA with a complementary sequence, if one exists in the test sample. So that areas of hybridization can be visualized, the probes carry reporter molecules such as fluorescent dyes, which are visible under UV light, or luminescent labels, which give off visible light. Enzyme-linked probes are detected when nonpigmented substrates are turned into colored molecules by the enzyme's action.

Probes are commonly used for diagnosing the cause of an infection from a patient's specimen and identifying a culture of an unknown bacterium or virus. A simple and rapid method called a *hybridization test* does not require electrophoresis. DNA from a test sample is isolated, denatured, placed on an absorbent filter, and combined with a microbe-specific probe **(process figure 10.4).** The blot is then developed and observed for areas of hybridization. Commercially available diagnostic kits are used to identify intestinal pathogens such as *Salmonella, Campylobacter, Shigella, Clostridium difficile,* rotaviruses, and adenoviruses. Other bacterial probes exist for *Mycobacterium, Legionella, Mycoplasma,* and *Chlamydia;* viral probes are available for herpes simplex and zoster, papilloma (genital warts), hepatitis A and B, and HIV. DNA probes have also been developed for human genetic markers and some types of cancer.

With another method, called **fluorescent *in situ* hybridization (FISH),** probes are applied to intact cells and observed microscopically for the presence and location of specific genetic marker sequences on genes. It is a very effective way to locate genes on chromosomes. *In situ* techniques can also be used to identify unknown bacteria living in natural habitats without having to culture them, and they can be used to detect RNA in cells and tissues (see chapter 17).

The Size of DNA

The relative sizes of nucleic acids are usually denoted by the number of base pairs (bp) or nucleotides they contain. For example, the palindromic sequences recognized by endonucleases are usually 4 to 10 bp in length; an average gene in *E. coli* is approximately 1,300 bp, or 1.3 kilobases (kb); and its entire genome is approximately 4,700,000 bp, 4,700 kb, or 4.7 megabases (Mb). The DNA of the human mitochondrion contains 16 kb, and the Epstein-Barr virus (a cause of infectious mononucleosis) has 172 kb. Humans have approximately 3.1 billion base pairs arrayed along 46 chromosomes.

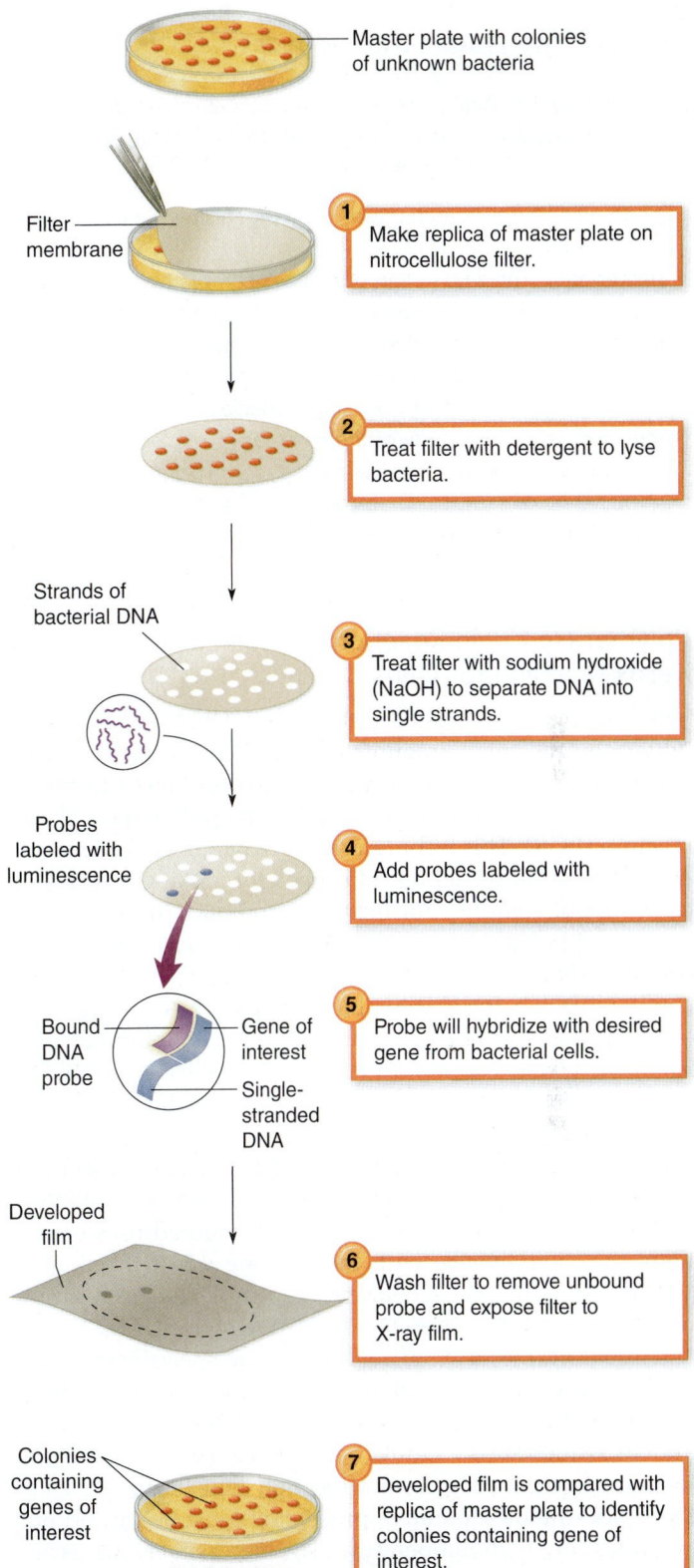

Master plate with colonies of unknown bacteria

Filter membrane

1 Make replica of master plate on nitrocellulose filter.

2 Treat filter with detergent to lyse bacteria.

Strands of bacterial DNA

3 Treat filter with sodium hydroxide (NaOH) to separate DNA into single strands.

Probes labeled with luminescence

4 Add probes labeled with luminescence.

Bound DNA probe — Gene of interest — Single-stranded DNA

5 Probe will hybridize with desired gene from bacterial cells.

Developed film

6 Wash filter to remove unbound probe and expose filter to X-ray film.

Colonies containing genes of interest

Replica plate

7 Developed film is compared with replica of master plate to identify colonies containing gene of interest.

Process Figure 10.4 A hybridization test relies on the action of microbe-specific probes to identify an unknown bacterium or virus.

A Note About the "-omics"

The ability to obtain the entire sequences of organisms has spawned new vocabulary that refers to the "total picture" of some aspect of a cell or organism.

genomics   The systematic study of an organism's genes and their functions.

proteomics   The study of an organism's complement of proteins (its "proteome") and functions mediated by the proteins.

metagenomics (also called "community genomics")   The study of all the genomes in a particular ecological niche, as opposed to individual genomes from single species.

metabolomics   The study of the complete complement of small chemicals present in a cell at any given time. Provides a snapshot of the physiological state of the cell and the end products of its metabolism.

Polymerase Chain Reaction: A Molecular Xerox Machine for DNA

Some of the techniques used to analyze DNA and RNA are limited by the small amounts of test nucleic acid available. This problem was largely solved by the invention of a simple, versatile way to amplify DNA called the **polymerase chain reaction (PCR).** This technique rapidly increases the amount of DNA in a sample without the need for making cultures or carrying out complex purification techniques. It is so sensitive that it holds the potential to detect cancer from a single cell or to diagnose an infection from a single gene copy. It is comparable to being able to pluck a single DNA "needle" out of a "haystack" of other molecules and make unlimited copies of the DNA. The rapid rate of PCR makes it possible to replicate a target DNA from a few copies to billions of copies in a few hours.

To understand the idea behind PCR, it will be helpful to review process figure 9.6, which describes synthesis of DNA as it occurs naturally in cells. The PCR method uses essentially the same events, with the opening up of the double strand, using the exposed strands as templates, the addition of primers, and the action of a DNA polymerase.

Initiating the reaction requires a few specialized ingredients **(process figure 10.5).** As we saw earlier, **primers** are synthetic oligonucleotides (short DNA strands) of a known sequence of 15 to 30 bases that serve as landmarks to indicate where DNA amplification will begin. To keep the DNA strands separated, processing must be carried out at a relatively high temperature. This necessitates the use of special **DNA polymerases** isolated from thermophilic bacteria. Examples of these unique enzymes are Taq polymerase obtained from *Thermus aquaticus* and Vent polymerase from *Thermococcus litoralis.* Enzymes isolated from these thermophilic organisms remain active at the elevated temperatures used in PCR (see Insight 7.1). Another useful component of

PCR is a machine called a thermal cycler that automatically performs the cyclic temperature changes.

The PCR technique operates by repetitive cycling of three basic steps: denaturation, priming, and extension. The process is fully described in process figure 10.5.

It is through cyclic repetition of these steps that DNA becomes amplified. When the DNAs formed in the first cycle are denatured, they become amplicons to be primed and extended in the second cycle. Each subsequent cycle converts the new DNAs to amplicons and doubles the number of copies. The number of cycles required to produce a million molecules is 20, but the process is usually carried out to 30 or 40 cycles. One significant advantage of this technique has been its natural adaptability to automation. A PCR machine can perform 20 cycles on nearly 100 samples in 2 or 3 hours.

Once the PCR is complete, the amplified DNA can be analyzed by any of the techniques discussed in this chapter. In addition, a newer technique, called real-time PCR, can detect products during the reaction instead of at the end. PCR can also be adapted to analyze RNA by initially converting an RNA sample to DNA with reverse transcriptase. This cDNA can then be amplified by PCR in the usual manner. It is by such means that ribosomal RNA and messenger RNA are readied for sequencing. The polymerase chain reaction has found prominence as a powerful workhorse of molecular biology, medicine, and biotechnology. It often plays an essential role in gene mapping, the study of genetic defects and cancer, forensics, infectious disease diagnosis **(figure 10.6),** and taxonomy studies.

Methods in Recombinant DNA Technology: How to Imitate Nature

The primary intent of **recombinant DNA technology** is to deliberately remove genetic material from one organism and combine it with that of a different organism. An important objective of this technique is to form genetic **clones.** Cloning

A Note About Clones

Like so many words in biology, the word *clone* has two different, although related, meanings. In this chapter, we will discuss genetic clones created within microorganisms. What we are cloning is *genes.* We use microorganisms to allow us to manipulate and replicate genes outside of the original host of that gene. You are much more likely to be familiar with the other type of cloning— which we will call *whole-organism cloning.* It is also known as *reproductive cloning.* This is the process of creating an identical organism using the DNA from an original. Dolly the sheep was the first cloned whole organism, and many others followed in her wake. These processes are beyond the scope of this book.

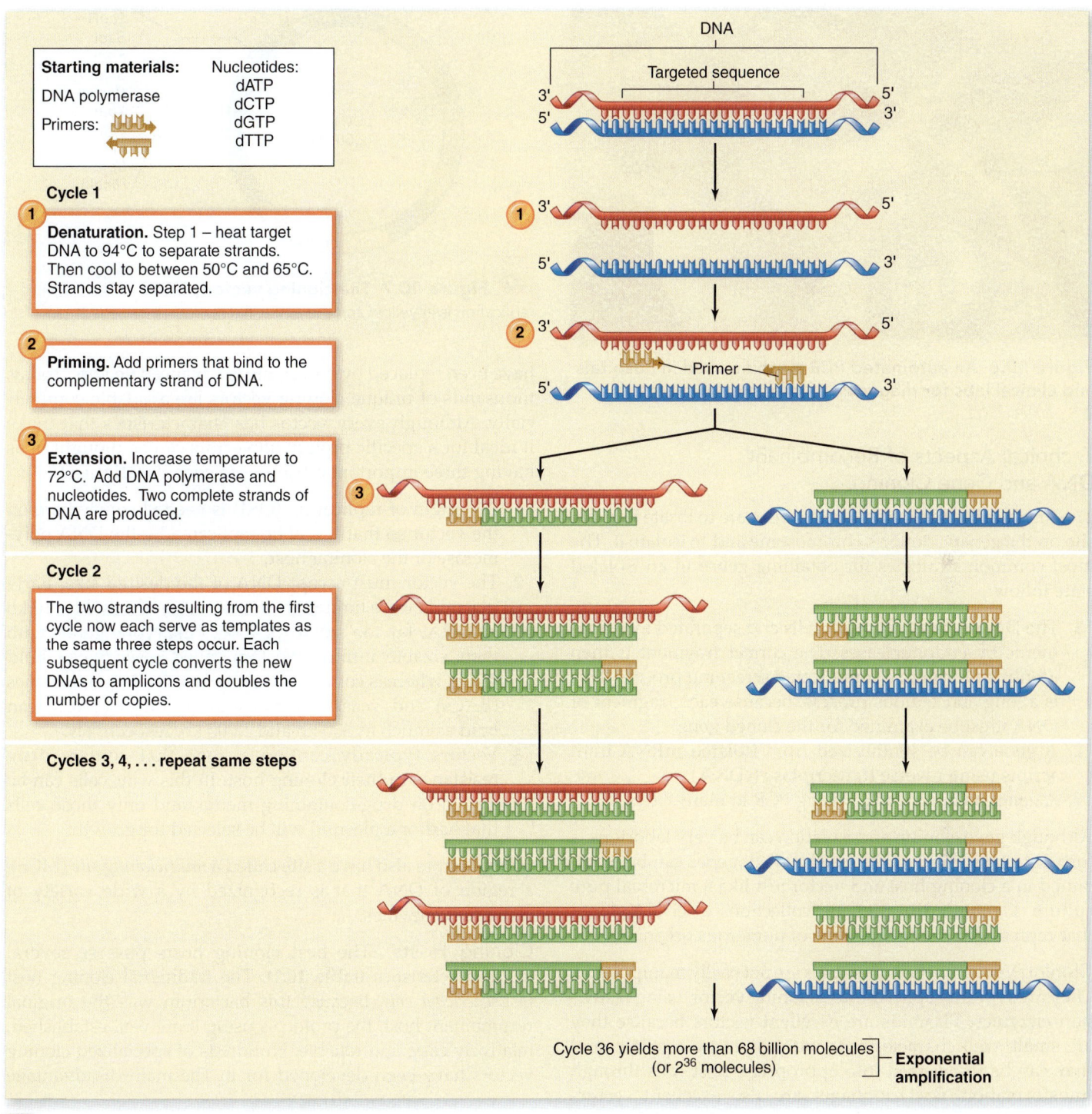

Starting materials: Nucleotides:
DNA polymerase   dATP
                  dCTP
Primers: dGTP
                  dTTP

Cycle 1

1 Denaturation. Step 1 – heat target DNA to 94°C to separate strands. Then cool to between 50°C and 65°C. Strands stay separated.

2 Priming. Add primers that bind to the complementary strand of DNA.

3 Extension. Increase temperature to 72°C. Add DNA polymerase and nucleotides. Two complete strands of DNA are produced.

Cycle 2

The two strands resulting from the first cycle now each serve as templates as the same three steps occur. Each subsequent cycle converts the new DNAs to amplicons and doubles the number of copies.

Cycles 3, 4, . . . repeat same steps

DNA
Targeted sequence

Primer

Cycle 36 yields more than 68 billion molecules (or 2^{36} molecules) — **Exponential amplification**

Process Figure 10.5 The polymerase chain reaction.

involves the removal of a selected gene from an animal, plant, or microorganism (the genetic donor) followed by its propagation in a different host organism. Cloning requires that the desired donor gene first be selected, excised by restriction endonucleases, and isolated. The gene is next inserted into a **vector** (usually a plasmid or a virus) that will insert the DNA

into a **cloning host.** The cloning host is usually a bacterium or a yeast that can replicate the gene and translate it into the protein product for which it codes. In the next section, we examine the elements of gene isolation, vectors, and cloning hosts and show how they participate in a complete recombinant DNA procedure.

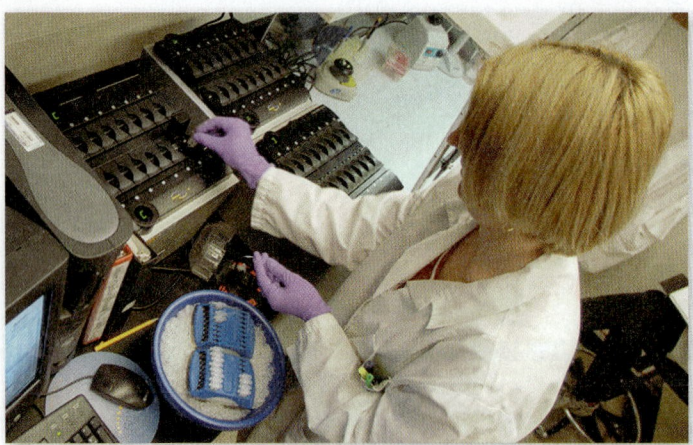

Figure 10.6 An automated PCR machine used in hospitals and clinical labs for diagnosis of infectious diseases.

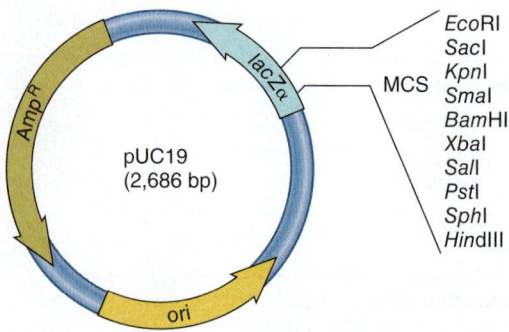

Figure 10.7 The cloning vector pUC19. The origin of replication is in yellow and the ampicillin-resistance gene is in tan.

Technical Aspects of Recombinant DNA and Gene Cloning

The first hurdles in cloning a target gene are to locate its exact site on the genetic donor's chromosome and to isolate it. The most common strategies for obtaining genes in an isolated state follow:

1. The DNA is removed from cells and separated into fragments by endonucleases. The correct fragment is then identified through a complicated screening process. This is a long and tedious process, because each fragment of DNA must be examined for the cloned gene.
2. A gene can be synthesized from isolated mRNA transcripts using reverse transcriptase (cDNA).
3. A gene can be amplified using PCR in many cases.

Although gene cloning and isolation can be very laborious, a fortunate outcome is that, once isolated, genes can be maintained in a cloning host and vector just like a microbial pure culture. **Genomic libraries** are collections of cDNA clones that represent the entire genome of numerous organisms.

Cloning Vectors Isolated genes are not easily manipulated. They are typically spliced into a cloning vector, using restriction enzymes. Plasmids are excellent vectors because they are small, well characterized, and easy to manipulate; and they can be transferred into appropriate host cells through transformation. Bacteriophages are also excellent vectors because they have the natural ability to inject DNA into bacterial hosts through transduction. A common vector in early work was an *E. coli* plasmid that carried genetic markers for resistance to antibiotics, although it was restricted by the relatively small amount of foreign DNA it could accept. A modified phage vector, the *Charon*[2] phage, is missing large sections of its genome, so it can carry a fairly large segment of foreign DNA. The simple plasmids and bacteriophages that were a staple of early recombinant DNA methodologies

have been replaced by newer, more advanced vectors. Today, thousands of unique cloning vectors are available commercially. Although every vector has characteristics that make it ideal for a specific project, all vectors can be thought of as having three important attributes to consider **(figure 10.7):**

1. An origin of replication (ORI) is needed somewhere on the vector so that it will be replicated by the DNA polymerase of the cloning host.
2. The vector must accept DNA of the desired size. Early plasmids were limited to an insert size of less than 10 kb of DNA, far too small for most eukaryotic genes with their sizable introns. Vectors called *cosmids* can hold 45 kb, whereas complex bacterial artificial chromosomes (BACs) and yeast artificial chromosomes (YACs) can hold as much as 300 kb and 1,000 kb, respectively.
3. Vectors typically contain a gene that confers drug resistance to their cloning host. In this way, cells can be grown on drug-containing media, and only those cells that harbor a plasmid will be selected for growth.

Many vectors also have a site called a *multicloning site* (MCS), a region of DNA that is recognized by a wide variety of restriction enzymes.

Cloning Hosts The best cloning hosts possess several key characteristics **(table 10.1).** The traditional cloning host is *Escherichia coli.* Because this bacterium was the original recombinant host, the protocols using it are well established, relatively easy, and reliable. Hundreds of specialized cloning vectors have been developed for it. The main disadvantage

---

Table 10.1 Desirable Features in a Microbial Cloning Host

Rapid turnover; fast growth rate
Can be grown in large quantities using ordinary culture methods
Nonpathogenic
Genome that is well delineated (mapped)
Capable of accepting plasmid or bacteriophage vectors
Maintains foreign genes through multiple generations
Will secrete a high yield of proteins from expressed foreign genes

---

2. Named for the mythical boatman, Charon, in Hades who carried souls across the River Styx.

with this species is that the splicing of mRNA and the modification of proteins that would normally occur in the eukaryotic endoplasmic reticulum and Golgi apparatus are unavailable in this bacterial cloning host. One alternative host for certain industrial processes and research is the yeast *Saccharomyces cerevisiae,* which, being eukaryotic, already possesses mechanisms for processing and modifying eukaryotic gene products. Certain techniques may also employ different bacteria (*Bacillus subtilis*), animal cell cultures, and even live animals and plants to serve as cloning hosts. In our coverage, we present the recombinant process as it is performed in bacteria and yeasts.

Construction of a Recombinant, Insertion into a Cloning Host, and Genetic Expression

This section illustrates one example of recombinant DNA technology, in this case, to produce a drug called alpha-2a interferon (Roferon-A). This form of interferon is used to treat chronic hepatitis C (described in chapter 22) and cancers such as hairy cell leukemia and Kaposi's sarcoma in AIDS patients (described in chapter 20). The human alpha interferon gene is a DNA molecule of approximately 500 bp that codes for a polypeptide of 166 amino acids. It was originally isolated and identified from human blood cells and prepared from processed mRNA transcripts that are free of introns. This step is necessary because the bacterial cloning host has none of the machinery needed to excise this nontranslated part of a gene. The rest of the process is outlined in **process figure 10.8.**

The bacterial cells' ability to express the eukaryotic gene is ensured, because the plasmid has been modified with the necessary transcription and translation recognition sequences. As the *E. coli* culture grows, it transcribes and translates the interferon gene, synthesizes the peptide, and secretes it into the growth medium. At the end of the process, the cloning cells and other chemical and microbial impurities are removed from the medium. Final processing to excise a terminal amino acid from the peptide yields the interferon product in a relatively pure form. The scale of this procedure can range from test tube size to gigantic industrial vats that can manufacture thousands of gallons of product (see chapter 25).

Although the process we have presented here produces interferon, some variation of it can be used to mass produce a variety of hormones, enzymes, and agricultural products such as pesticides. Recent advances even allow scientists to produce functions that weren't originally present in the biological world. **Insight 10.1** describes attempts by do-it-yourselfers to conduct these types of experiments at home.

INSIGHT 10.1   Biohackers and DIYbio: Genetics in Your Garage

It is a rapidly growing community: scientists working in their basements, kitchens, or garages, searching for cures for genetic diseases, tinkering with bacterial genomes, or creating artificial life forms. Not unlike Gregor Mendel, who cross-bred pea plants in his garden, these biology hobbyists are working outside the mainstream of university centers funded by government grants or corporations influenced by profit. Often, they have no formal training and are self-taught, using equipment from their kitchen or purchased cheaply on eBay. They collaborate through websites such as DIYBio.org, and many share lab equipment and bench space at places like Genspace, a community laboratory in downtown Brooklyn.

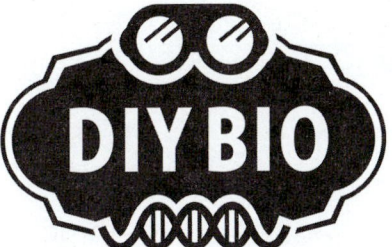

They call themselves biohackers: a community of amateurs and professionals with a passion for science and who work outside the traditional confines of a lab. Many of these at-home scientists have a personal motivation. Kay Aull, a graduate of MIT, set up a lab in her bedroom closet. For her first project, she genetically engineered *E. coli* to mimic a computer process by pulsing blue light, mimicking the 1's and 0's of binary code. She then turned her attention to developing a genetic test for hemochromatosis, a genetic disorder that her father and grandfather had, which was most likely passed on to her. Other biohackers have similar personal motivations. Hugh Reinhoff has studied his daughter Beatrice's genetic code to find a rare genetic disorder that has eluded diagnosis by her doctors. By taking samples of his daughter's blood, extracting the DNA, and sequencing it at home with his own PCR machine, and after hours of tedious analysis, Reinhoff was able to find a very rare genetic mutation.

He is now in the process of confirming his data in preparation for publication.

Some scientists and policy makers worry about the growing DIYbio movement. In a preemptive move, Ellen Jorgensen, founder of Genspace, reached out to the local FBI when she created the lab. The FBI and local law enforcement support the efforts of the Brooklyn-based lab because of the community outreach efforts of Genspace. Members give free classes to schoolchildren and the public, raising awareness about what may constitute a biological threat. Genspace, DIYbio.org, and others in the biohacker community have developed a set of safety standards to minimize risks and foster innovation.

Source: 2012. *The New York Times.* January 17, p. D4. Retrieved from www.nytimes.com/2012/01/17/science/for-bio-hackers-lab-work-often-begins-at-home.html.

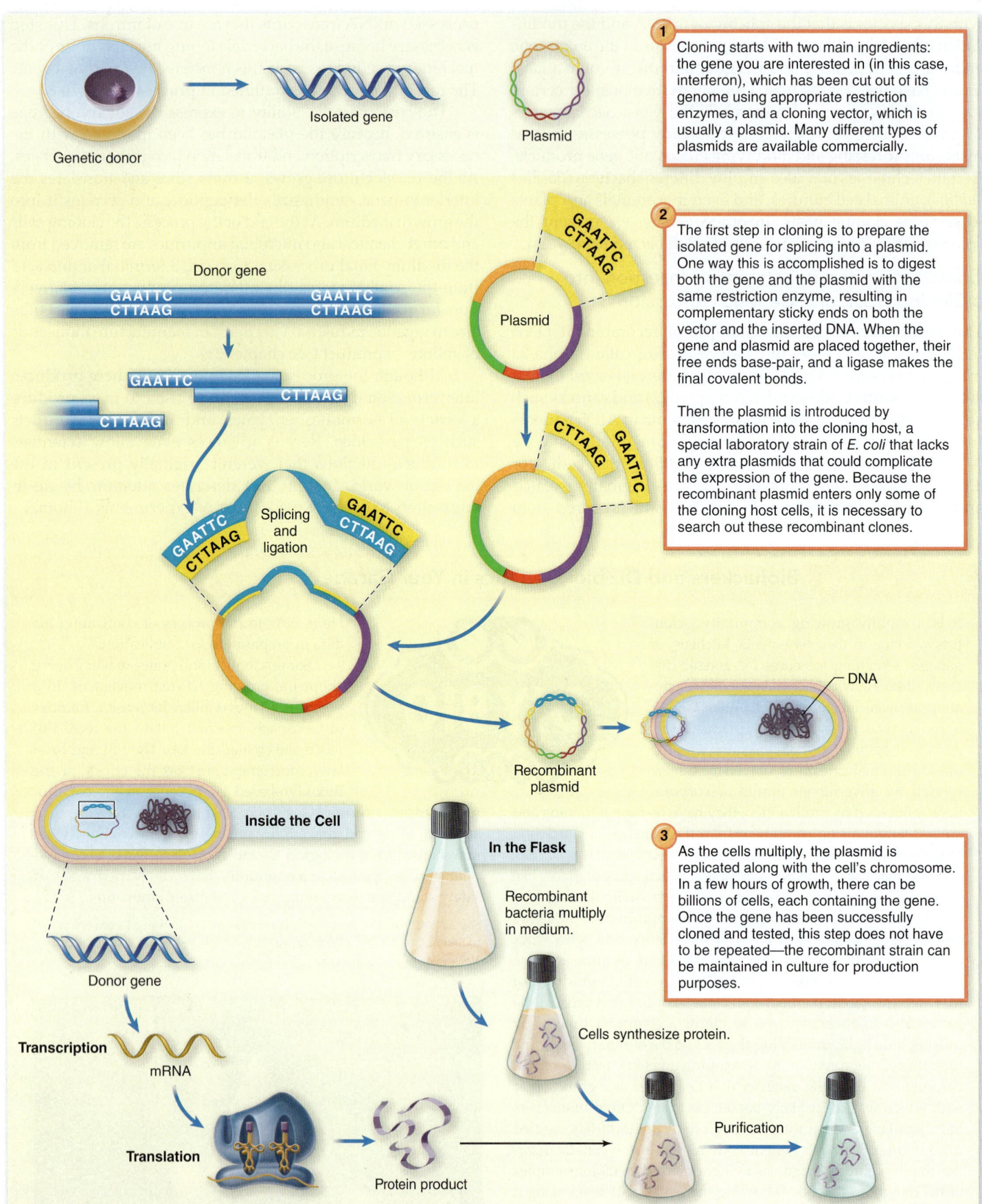

1 Cloning starts with two main ingredients: the gene you are interested in (in this case, interferon), which has been cut out of its genome using appropriate restriction enzymes, and a cloning vector, which is usually a plasmid. Many different types of plasmids are available commercially.

Genetic donor

Isolated gene

Plasmid

Donor gene

GAATTC
CTTAAG

GAATTC
CTTAAG

GAATTC
CTTAAG

CTTAAG

2 The first step in cloning is to prepare the isolated gene for splicing into a plasmid. One way this is accomplished is to digest both the gene and the plasmid with the same restriction enzyme, resulting in complementary sticky ends on both the vector and the inserted DNA. When the gene and plasmid are placed together, their free ends base-pair, and a ligase makes the final covalent bonds.

Plasmid

GAATTC
CTTAAG

CTTAAG   GAATTC

Splicing and ligation

GAATTC
CTTAAG

GAATTC
CTTAAG

Then the plasmid is introduced by transformation into the cloning host, a special laboratory strain of *E. coli* that lacks any extra plasmids that could complicate the expression of the gene. Because the recombinant plasmid enters only some of the cloning host cells, it is necessary to search out these recombinant clones.

DNA

Recombinant plasmid

Inside the Cell

In the Flask

3 As the cells multiply, the plasmid is replicated along with the cell's chromosome. In a few hours of growth, there can be billions of cells, each containing the gene. Once the gene has been successfully cloned and tested, this step does not have to be repeated—the recombinant strain can be maintained in culture for production purposes.

Recombinant bacteria multiply in medium.

Donor gene

Transcription

mRNA

Cells synthesize protein.

Translation

Protein product

Purification

Process Figure 10.8 Using recombinant DNA for gene cloning.

10.2 Learning Outcomes—Assess Your Progress

2. Explain the role of restriction endonucleases in the process of genetic engineering.
3. Describe how gel electrophoresis is used to analyze DNA.
4. List the steps in the polymerase chain reaction; discuss one disadvantage to this technique.
5. Describe how recombinant DNA is created; discuss its role in gene cloning.

10.3 Products of Recombinant DNA Technology

Recombinant DNA technology can be used to produce recombinant organisms, but it can also be used to create abundant sources of protein products or nucleotide sequences. Recombinant technology is used by pharmaceutical companies to manufacture medications that cannot be manufactured by any other means.

Recombinant DNA technology changed the outcome of diseases such as diabetes, dwarfism, and many other conditions by enabling large-scale manufacture of lifesaving hormones and enzymes of human origin. Recombinant human insulin can now be prescribed for diabetics, and recombinant HGH can now be administered to children with dwarfism and other conditions. HGH is also used to prevent the wasting syndrome that occurs in AIDS and cancer patients. In all of these applications, recombinant DNA technology has led to both a safer product and one that can be manufactured in quantities previously unfathomable. Other protein-based hormones, enzymes, and vaccines produced through recombinant DNA technology are summarized in **table 10.2**.

Genetically Modified Organisms

Recombinant organisms produced through the introduction of foreign genes are called **transgenic** or genetically modified organisms (GMOs). Foreign genes have been inserted into a variety of microbes, plants, and animals through recombinant DNA techniques developed especially for them. Transgenic "designer" organisms are available for a variety of biotechnological applications.

Recombinant Microbes: Modified Bacteria and Viruses

One of the first practical applications of recombinant DNA in agriculture was to create a genetically altered strain of the bacterium *Pseudomonas syringae*. The wild strain ordinarily contains a gene that promotes ice or frost formation on moist plant surfaces. Genetic alteration of the frost gene using recombinant plasmids created a different strain that could prevent ice crystals from forming. A strain of *Pseudomonas fluorescens* has been engineered with the gene from a bacterium (*Bacillus thuringiensis*) that codes for an insecticide. These recombinant bacteria are released to colonize plant roots and help destroy invading insects. This bacterial gene

Table 10.2 Examples of Current Protein Products from Recombinant DNA Technology

Immune Treatments
Interferons—peptides used to treat some types of cancer, multiple sclerosis, and viral infections such as hepatitis and genital warts
Interleukins—types of cytokines that regulate the immune function of white blood cells; used in cancer treatment
Tumor necrosis factor (TNF)—used to treat cancer
Hormones
Erythropoietin (EPO)—a peptide that stimulates bone marrow used to treat some forms of anemia
Human growth hormone (HGH)—stimulates growth in children with dwarfism; prevents wasting syndrome
Enzymes
rhDNase (Pulmozyme)—a treatment that can break down the thick lung secretions of cystic fibrosis
Tissue plasminogen activating factor (tPA)—can dissolve potentially dangerous blood clots
PEG-SOD—a form of superoxide dismutase that minimizes damage to brain and other tissues after surgery or severe trauma
Vaccines
Vaccines for hepatitis B, human papillomavirus, and *Haemophilus influenzae* type b meningitis
Miscellaneous
Factor VIII—needed as replacement blood-clotting factor in type A hemophilia
Bovine growth hormone or bovine somatotropin (BST)—given to cows to increase milk production

has also been added to transgenic corn and potato crops to help make them more resistant to insect pests. All releases of recombinant microbes must be approved by the Environmental Protection Agency (EPA) and are closely monitored.

Another significant bioengineering interest has been to create microbes to bioremediate disturbed environments. Biotechnologists have already developed and tested several types of bacteria that clean up oil spills and degrade pesticides and toxic substances (see chapter 25). Recently, scientists created a recombinant strain of bacterium that could incorporate artificial amino acids into newly synthesized proteins, making them capable of producing biofuels, new drugs with novel properties, or industrial chemicals in a more environmentally friendly manner.

The movement toward release of engineered plants into the environment has led to some controversy. Many plant geneticists and ecologists are seriously concerned that transgenic plants will share their genes for herbicide, pesticide, and virus resistance with natural plants, leading to "superweeds" that could flourish and become indestructible. There is promising news, however. In 2008, a study found that rice that was manipulated through genetic engineering had fewer unintentional genetic changes than did crops bred through conventional methods. The U.S. Department of Agriculture is carefully regulating all releases of transgenic plants. Still, screening the food supply for genetically modified plants is now standard in some countries around the world today **(Insight 10.2).**

Transgenic Animals: Engineering Embryos

Animals, too, can be genetically engineered. In fact, animals are so amenable to gene transfer that several hundred strains of transgenic animals have been introduced by research and industry. One reason for this movement toward animals is that, unlike bacteria and yeasts, they can express human genes in organs and organ systems that are very similar to those of humans. This advantage has led to the design of animal models to study human genetic diseases and then to use these natural systems to test new genetic therapies before they are used in humans. Animals such as sheep or goats can also be engineered to become "factories" capable of manufacturing proteins useful to humans and excreting them in their milk or semen, a process often referred to (when done to produce medically useful proteins) as "pharming."

Synthetic Biology

In recent years, researchers have staked out entirely new territory in genetic manipulation: They are trying to create new organisms from scratch. This field is called *synthetic biology.* In Insight 6.4, you read about the creation of an "artificial" poliovirus and influenza virus. Viruses are simpler than cells, as you know, and now researchers are trying to create cells. One pioneer in the field is one of the same men who sequenced the human genome, Craig Venter. In 2010, he successfully created a self-replicating bacterial cell from four bottles of chemicals: the four nucleotides of DNA. This was a breakthrough of major proportions as it was the first time a living, replicating cell had been synthesized from chemicals. In an interesting side note, the researchers added some "letters" to the DNA code in the bacterium (inventing their own code) that spelled out their names, the name of the laboratory, and an e-mail address that you could write to if you managed to decode the genome. They also added some relevant quotations in the DNA, such as one from the Irish writer James Joyce: "To live, to err, to fall, to triumph, to recreate life out of life."

Synthetic biology has yielded some surprising results. We have become accustomed to thinking that only the four nucleotides adenine, guanine, thymine, and cytosine were capable of carrying and copying the genetic code. But in 2012, synthetic biologists created what they call xeno-nucleic acids, or XNAs (*xeno* meaning "other" or "strange"). The molecules contained different sugars than RNA and DNA, and in some cases no sugars at all. The scientists found enzymes that were able to copy the XNAs, resulting in the ability to store information and pass it on to descendant cells. One of the

---

INSIGHT 10.2   Detecting Genetically Modified Food Using Bioluminescence

Genetically modified (GM) food has been the center of an ongoing debate. On one side, there is a dire need for crops that are pest-, drought-, and cold-resistant in a world where 25,000 people die per day of starvation and over 8 million are malnourished. On the other side, governments, scientists, and many individuals are concerned that these modified plants could cause unintended harm to the environment, transfer their genes to other plants or organisms, or pose risks to human health. Since 2004, the European Union (EU) has required the clear labeling of foods that have GMO (genetically modified organism) components. Because genetically engineered sequences can unintentionally be incorporated into crops, the EU has restricted the proportion of GM food to no more than 0.9% in a non-GM (non-labeled) product. This restriction requires careful monitoring of all imports of GM plants to the EU. Polymerase chain reaction (PCR) has been the traditional method of testing for GM plants, but this process is expensive and can be time-consuming.

Recently, a procedure was developed by researchers from Lumora, Ltd., in the United Kingdom that can identify as little as 0.1% GM content in corn. The process is called "loop mediated isothermal amplification" (LAMP) coupled with "bioluminescent output produced in real-time" (BART) or LAMP-BART. Unlike PCR, LAMP amplifies short sequences of DNA at a single temperature, rather than heating the DNA strand to high temperatures. BART produces a light signal when specific DNA sequences are detected using luciferase, the same enzyme pro-

duced by bioluminescent bacteria (and, as it happens, fireflies). Utilization of LAMP and BART together allows scientists to test for GM crops easily and simply in the field. The debate over GM food will continue to rage, but this method makes GM crops easier to detect and monitor.

Source: www.sciencedaily.com/releases/2012/04/120429234637.htm.

implications of this finding is that life on other planets does/did not require DNA or RNA to exist.

This kind of research activity raises some obvious ethical questions. For many, it raises fears that whatever we create can get away from our control and have unintended consequences, as with Frankenstein in the famous Mary Shelley novel. Others argue that the potential benefits of being able to design cells that create biofuels or medical or agricultural products outweigh such hypothetical concerns.

10.3 Learning Outcomes—Assess Your Progress

6. Provide several examples of recombinant products that have contributed to human health.

7. List examples of genetically modified bacteria, plants, and animals and a purpose for each.

10.4 Genetic Treatments: Introducing DNA into the Body

Gene Therapy

We have known for decades that for certain diseases, the disease phenotype is due to the lack of a single specific protein. For example, type I diabetes is caused by a lack of insulin, leaving those with the disease unable to properly regulate their blood sugar. The initial treatment for this disease was simple: Provide diabetics with insulin isolated from a different source, in most cases the pancreas of pigs or cows. Whereas this treatment was adequate for most diabetics, it presented problems, including the large number of animals required to meet demand and the occurrence of sensitivities or allergies to animal proteins. We've already discussed the first way in which genetic engineering has been used in the treatment of disease, namely, producing recombinant proteins in bacteria or yeast rather than isolating the protein from animals or humans. In fact, recombinant human insulin was the first genetically engineered drug to be approved for use in humans. The next logical step is to see if we can correct or repair a faulty gene in humans suffering from a fatal or debilitating disease, a process known as **gene therapy.**

The inherent benefit of this therapy is to permanently cure the physiological dysfunction by repairing the genetic defect. There are various strategies for this therapy. In general, the normal gene is cloned in vectors such as retroviruses (mouse leukemia virus) or adenoviruses that are infectious but relatively harmless. In one technique, tissues can be removed from the patient and incubated with these genetically modified viruses to transfect them with the normal gene. The transfected cells are then reintroduced into the patient's body by transfusion **(process figure 10.9).** Alternatively, naked DNA or a virus vector is directly introduced into the patient's tissues. This is the basis of a successful immunotherapy treatment for melanoma today.

Experimentation with various types of gene therapy, or clinical testing, is performed on human volunteers with the particular genetic condition. Thousands of these trials have been and are being carried out in the United States and other countries. Most trials target cancer, single-gene defects, and infections; and most gene deliveries are carried out by virus vectors. Early therapeutic trials were hampered by several difficulties relating to effectiveness and safety. Some of the safety issues were related to the use of (seemingly safe) viruses as delivery vehicles, which then ended up causing malignancies.

In 2009, scientists successfully treated humans with a particular hereditary eye disease using genes surgically delivered to the eye using a virus vector. They later confirmed that the virus did not leave the eye and therefore concluded that viral vectors may indeed sometimes be appropriate. More recently, viral vectors carrying genes for healthy proteins have been injected intramuscularly into humans. The muscles produce the appropriate protein, fixing defects such

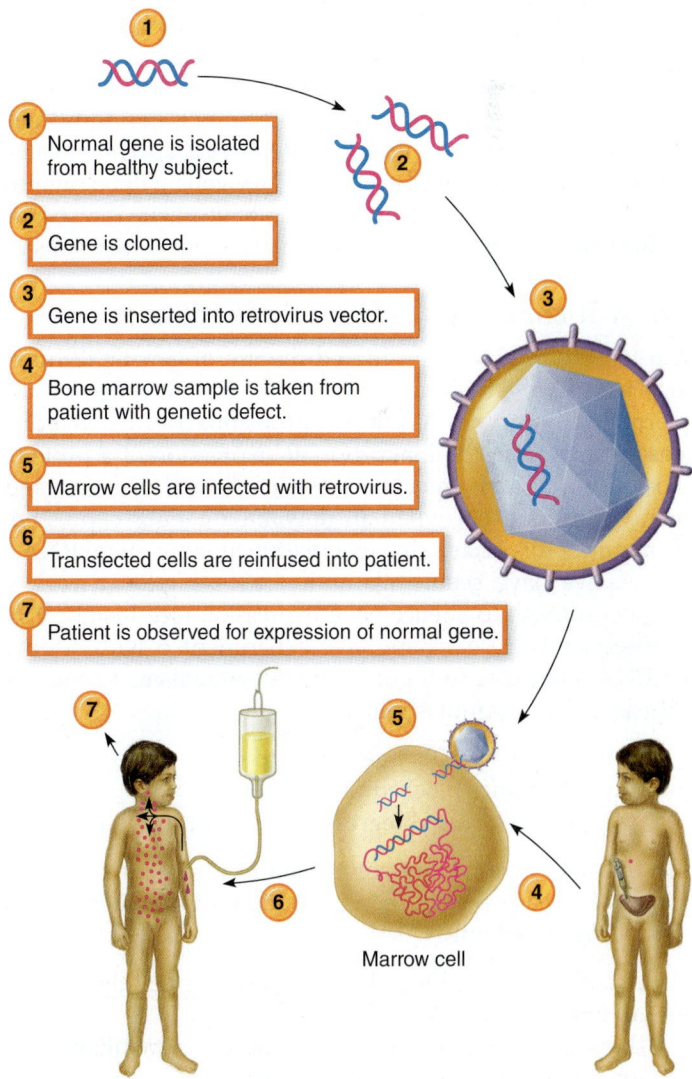

1. Normal gene is isolated from healthy subject.
2. Gene is cloned.
3. Gene is inserted into retrovirus vector.
4. Bone marrow sample is taken from patient with genetic defect.
5. Marrow cells are infected with retrovirus.
6. Transfected cells are reinfused into patient.
7. Patient is observed for expression of normal gene.

Marrow cell

Process Figure 10.9 Protocol for the *ex vivo* type of gene therapy in humans.

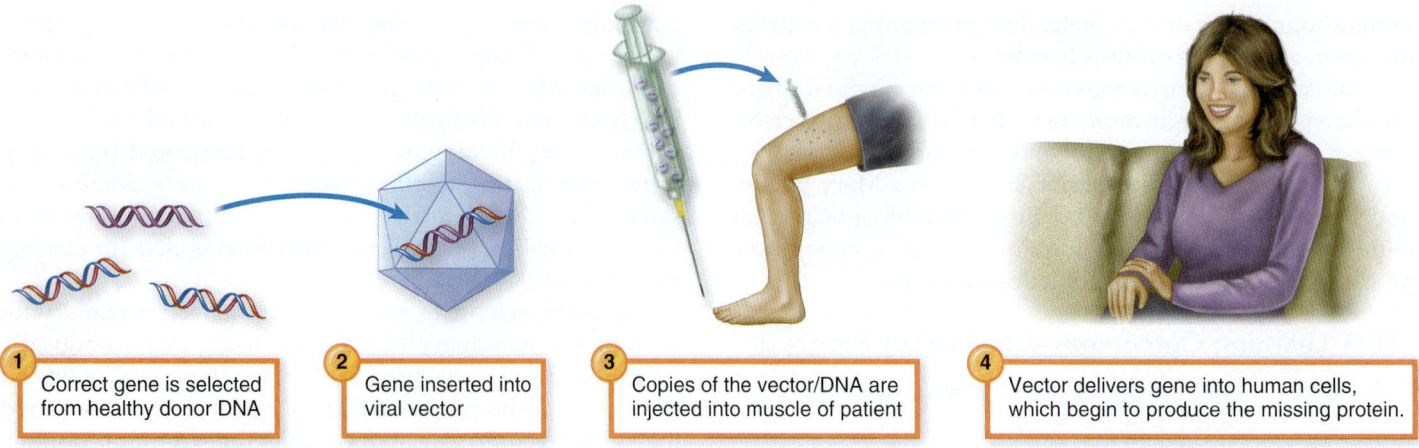

1 Correct gene is selected from healthy donor DNA

2 Gene inserted into viral vector

3 Copies of the vector/DNA are injected into muscle of patient

4 Vector delivers gene into human cells, which begin to produce the missing protein.

Process Figure 10.10 **Intramuscular gene therapy.**

as those seen in muscular dystrophy and possibly conditions such as diabetes and cancer **(process figure 10.10)**.

The strategies described so far are called *somatic cell gene therapy*. This means that the changes are permanent in the individual who is treated, but are not passed on to their offspring. The ultimate sort of gene therapy is germline therapy, in which genes are inserted into an egg, sperm, or early embryo. In this type of therapy, the new gene will be present in all cells of the individual. The therapeutic gene is also heritable (that is, can be passed on to subsequent generations).

DNA Technology as Genetic Medicine

Up to now, we have considered the use of genetic technology to replace a missing or faulty protein that is needed for normal cell function. A different problem arises when a disease results from the inappropriate expression of a protein. For example, Alzheimer's disease, most viral diseases, and many cancers occur when an unwanted gene is expressed. Sometimes this occurs due to a malfunctioning micro RNA.

As we have seen, cells and viruses employ a variety of micro RNAs to silence gene expression (see Insight 9.3). Consequently, researchers have experimented with adding miRNA molecules to tissues or whole organisms to counter diseases. Two examples follow:

- Introducing specific miRNAs into small-cell lung cancer tumors in mice blocked tumor growth.
- Micro RNAs caused regression of liver cancers in mice.

On the other hand, out-of-control miRNAs sometimes are the cause of disease. Scientists have found multiple successes in inhibiting errant miRNAs:

- Inhibiting a specific miRNA in mice stopped the growth of both breast and ovarian cancers.
- Inhibiting a micro RNA in African green monkeys lowered their plasma cholesterol levels.
- Multiplication of hepatitis C virus in chimpanzees was inhibited by shutting down one class of miRNAs (this strategy is now being tested in humans).

The first successful human application of interfering RNAs was conducted at the University of Tennessee in 2010. There, healthy human volunteers were given a nasal spray containing miRNA designed to silence a gene from respiratory syncytial virus (RSV). They were infected with RSV and then used the spray for 5 days. Only 44% of those who had received the interfering RNA became infected, compared with 71% of control subjects. Among those who were infected, symptoms were significantly reduced.

Disease Connection

Respiratory syncytial virus causes a relatively mild disease in children and adults, but it can be a killer in newborn babies, especially premature infants. (Note that the study described here was performed in adults, for ethical reasons.)

In summary, the discovery of miRNAs and their role in disease has opened a vast new area of research and hope for treatment, cure, and even prevention of disease.

10.4 Learning Outcomes—Assess Your Progress

8. Differentiate between somatic and germline gene therapy.
9. Describe miRNAs and ways in which their discovery can impact human disease.

10.5 Genome Analysis: Maps and Profiles

As was mentioned earlier, DNA technology has allowed us to accomplish many age-old goals by new and improved means. Analysis of DNA provides a better, more accurate, mechanism of differentiating among organisms than simply looking at their phenotype can. Additionally, DNA can be used to identify an organism no longer present, as when a criminal is identified by DNA extracted from a strand of hair left behind at a crime scene. Finally, possession of a particular sequence of DNA may indicate an increased risk of a genetic disease. Detection of this piece of DNA (known as a marker) can identify a person as being at increased risk for cancer or Alzheimer's disease long before symptoms arise. The ability to detect diseases before symptoms arise is especially important for diseases such as cancer, for which early treatment is sometimes the difference between life and death. With examples like this in mind, let's look at several ways in which new DNA technology is allowing us to accomplish goals in ways that were only dreamed of a few years ago.

Genome Mapping and Screening: An Atlas of the Genome

By far the most detailed maps of a genome are **sequence maps,** which give an exact order of bases in a plasmid, chromosome, or entire genome. Genome sequencing projects have been highly successful. As of 2012, the genome sequences of more than 2,800 viruses, close to 2,000 bacteria and archaea, and about 160 eukaryotes had been published. The eukaryotic genomes include human, mouse, *Caenorhabditis elegans* (a nematode worm), yeast, fruit fly, *Arabidopsis* (a small flowering plant), and rice. One of the remarkable discoveries in this huge enterprise has been how similar the genomes of relatively unrelated organisms are. Humans share approximately 80% of their DNA codes with mice, about 60% with rice, and even 30% with the worm *C. elegans.*

So how is this sequencing performed? The approach most commonly used is called whole-genome shotgun sequencing and is illustrated in **process figure 10.11.** The process can be broken down into seven steps:

1. First, the whole genome of an organism is broken down into smaller, manageable fragments.
2. The fragments are separated through gel electrophoresis.
3. Each fragment is inserted into a plasmid and is cloned into an *E. coli* cell. This produces a complete **library** of fragments. The library exists in the bacterial cultures, which can be preserved indefinitely and sampled repeatedly.
4. The plasmids are purified and the DNA fragments are sequenced by automated sequencers, machines that add labeled primers to each fragment. The primers usually recognize the plasmid sequences that flank the genome insert, so that the machine can tell where the fragment begins and ends. Each section of the genome ends up being sequenced multiple times in an overlapping fashion.
5. A computer program takes all the sequence data and is able to find where the sequence overlaps. This automated process results in a larger, contiguous set of nucleotide sequences called **contigs.**
6. The contigs are put in the proper order to determine the entire sequence. This step is tricky; there are often gaps between the contigs, but there are a variety of methods to resolve this issue.
7. An important last step is editing. A human examines the sequence, looking for irregularities, frameshifts, and ambiguities.

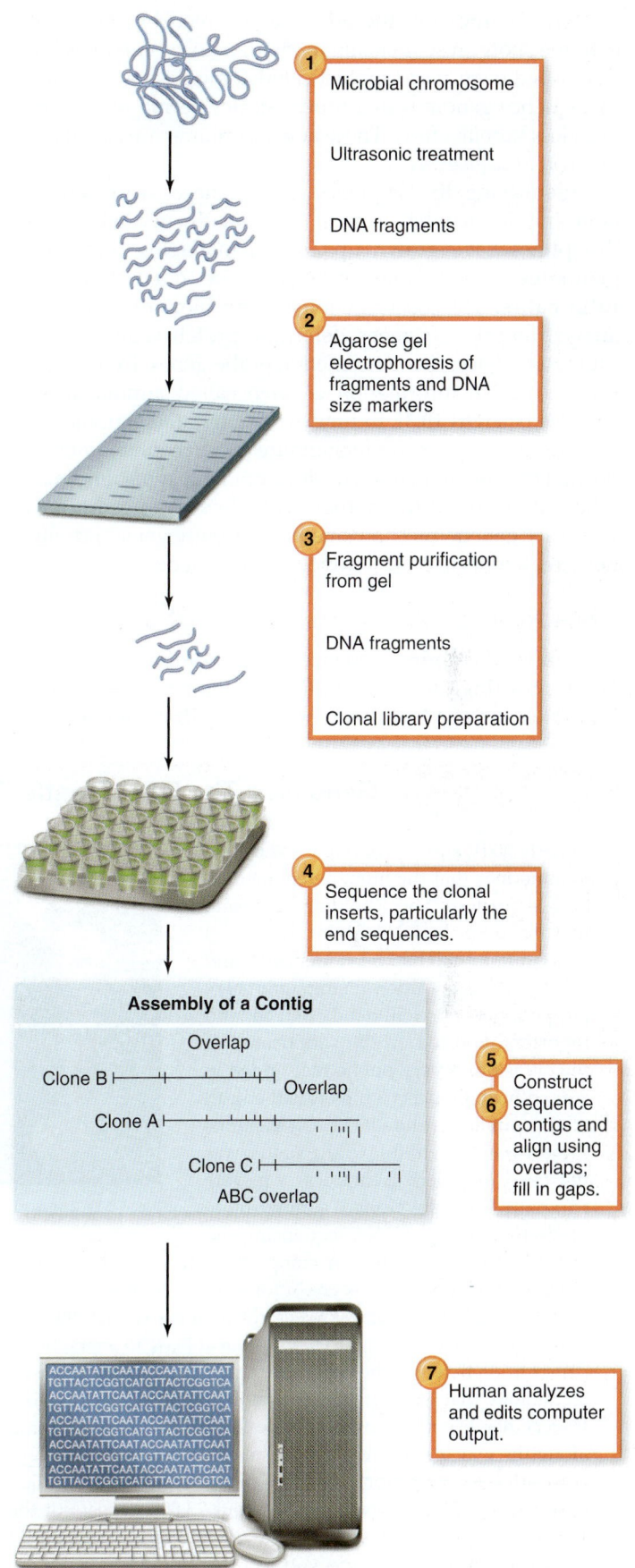

1. Microbial chromosome

   Ultrasonic treatment

   DNA fragments

2. Agarose gel electrophoresis of fragments and DNA size markers

3. Fragment purification from gel

   DNA fragments

   Clonal library preparation

4. Sequence the clonal inserts, particularly the end sequences.

   Assembly of a Contig

   Overlap
   Clone B
   Overlap
   Clone A
   Clone C
   ABC overlap

5. / 6. Construct sequence contigs and align using overlaps; fill in gaps.

7. Human analyzes and edits computer output.

Process Figure 10.11 Whole-genome shotgun sequencing.   See text for explanation of numbers.

Since witnessing the advances enabled by the whole-genome shotgun sequencing method, many companies have developed even more sophisticated computers and systems to sequence genomes in a much shorter period of time and with less human effort. These new techniques are called *high-throughput sequencing.*

Identifying the sequence doesn't necessarily tell you anything about what it does. As a result, two whole new disciplines have grown up around managing these data: **genomics** (see "A Note About the New -omics") and **bioinformatics.** The job of genomics and bioinformatics is to analyze and classify genes, determine protein sequences, and ultimately determine the function of the genes. Determining this functional information is often called **annotating** the genome **(Insight 10.3).** In time, well-annotated genomes will provide a complete understanding of such phenomena as normal cell function, disease, development, aging, and many other issues. In addition, they will allow us to characterize the exact genetic mechanisms behind pathogens and allow new treatments to be developed against them.

DNA Profiles: A Unique Picture of a Genome

Sometimes the entire sequence of a genome is not needed. **DNA profiling** (also called DNA typing or fingerprinting) is best known as a tool of forensic science first devised in the

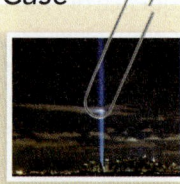

Case File 10   *Continuing the Case*

When a piece of DNA is analyzed through the process called polymerase chain reaction (PCR), primers—short strands of DNA—attach to the target DNA in order to make multiple copies for analysis. If the DNA is badly damaged, as is the case with the WTC remains, the large DNA primers do not stick to the DNA strand and it cannot be amplified. This problem spurred scientists to look for smaller pieces of DNA called short tandem repeats (STRs) that could be more easily amplified. These STRs are patterns of two to five nucleotides repeated directly adjacent to each other on the DNA strand in a pattern such as this, "GATAGATAGATAGATA." These repeated sequences are unique to each individual and when these regions are compared against a database of STRs (such as in the MDKAP), a match can be made. The push to identify the degraded DNA from the WTC site spurred scientists to find more effective ways to analyze STRs. The result was a product called the MiniFiler kit that can amplify even very small pieces of DNA for analysis. This product is now used worldwide in identification of disaster victims in addition to the WTC victim samples.

- What other methods have been recently discovered to improve the identification of WTC victims?

INSIGHT 10.3   Genomics: The Final Frontier?

In early 2001, a press conference was held to announce that the human genome had been sequenced. The 3.1 billion base pairs that make up the DNA found in (nearly) every human cell had been identified and put in the proper order. Champagne corks popped, balloons fell, bands played, and reporters reported on the significance of the occasion. A very public race ended in a dead heat. Francis Collins was the head of the publicly funded Human Genome Project (HGP), while J. Craig Venter was the head of Celera Genomics, a private company that developed a new, more powerful method of DNA sequencing and competed with the HGP. A compromise was finally reached whereby both groups took credit for sequencing the genome.

Craig Venter (left) and Francis Collins (right)

The Human Genome Project got under way in 1990 and, although it was supposed to have taken at least 15 years, fierce competition between the two teams spurred major advances in technology that led to its being completed in just over 10. The Project met its goal of describing the entire human genome and, along the way, produced startling revelations about our DNA: Only 1.5% of our genome encodes protein; and the 3.1 billion base pairs of DNA code for only about 23,000 genes, not the 100,000 or so that was once thought. In comparison, a fruit fly has about 15,000 genes—do you feel that you are much more complex than an insect? On the basis of genes, you are not!

Since the human genome was sequenced, a huge variety of other genomes have also been sequenced. The information from all of these organisms revealed that the most important sequences may well be the ones in between the protein-coding genes. These regions regulate the genes. Seemingly "junk" DNA plays a major role in the modification of gene expression providing further insight into the genetic complexity of disease development.

Moreover, the advances made in the field of genomics have led to the ability to screen the human body and other ecological settings for microbial life. The field of metagenomics, defined as the study of all genomes in a given habitat, has exploded with information as of late. For example, we now know the environment hosts a much bigger array of microbes than previously imagined. J. Craig Venter, the same person referenced earlier, sampled the world's seas from 2002 to 2010 and discovered millions of previously unknown genes and thousands of new proteins in oceans, indicating forms of life that are new to us. The Human Microbiome Project, which conducts metagenomics studies of the microbes present in and on healthy and diseased persons, has likewise discovered many new microbes and redefined what "normal" and "disease" states are in humans.

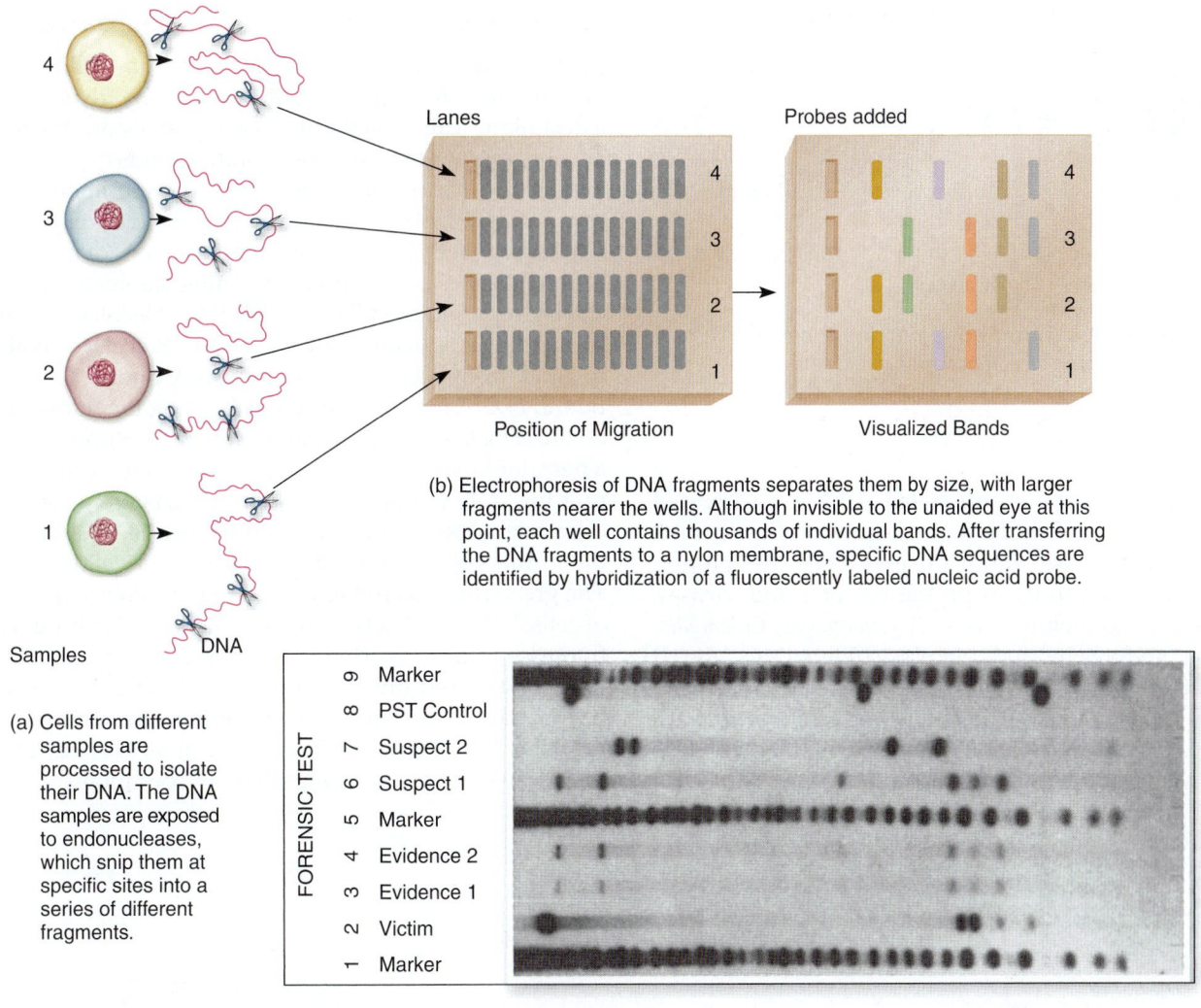

(b) Electrophoresis of DNA fragments separates them by size, with larger fragments nearer the wells. Although invisible to the unaided eye at this point, each well contains thousands of individual bands. After transferring the DNA fragments to a nylon membrane, specific DNA sequences are identified by hybridization of a fluorescently labeled nucleic acid probe.

(a) Cells from different samples are processed to isolate their DNA. The DNA samples are exposed to endonucleases, which snip them at specific sites into a series of different fragments.

(c) An actual DNA profile used in a rape trial. Control lanes with known markers are in lanes 1, 5, 8, and 9. The second lane contains a sample of DNA from the victim's blood. Evidence samples 1 and 2 (lanes 3 and 4) contain semen samples taken from the victim. Suspects 1 and 2 (lanes 6 and 7) were tested. Can you tell by comparing evidence and suspect lanes which individual committed the rape?

Figure 10.12 DNA profiles: the bar codes of life.

mid-1980s by Alex Jeffreys of Great Britain (figure 10.12). It is based on the principle that although DNA is made of the same four nucleotides, the exact way these nucleotides are combined is unique for each organism. It is typically used in more practical settings, such as in public health labs to differentiate between microbes that might be causing an outbreak, or in criminal investigations trying to match a suspect to the DNA left at a crime scene.

One type of analysis depends on the ability of a restriction enzyme to cut DNA at a specific recognition site. If a given strand of DNA possesses the recognition site for a particular restriction enzyme, the DNA strand is cut, resulting in two smaller pieces of DNA. If the same strand from a different person does not contain the recognition site (perhaps due to a mutation many generations ago), it is not cut by the restriction enzyme and remains as a single large piece of DNA. When each of these DNA samples is digested with

restriction enzymes and separated on an electrophoretic gel, the first displays two small bands, while the second displays one larger band. This is an example of a restriction fragment length polymorphism (RFLP), which was discussed earlier. All methods of DNA fingerprinting depend on some variation of this strategy to ferret out differences in DNA sequence at the same location in the genome.

One type of polymorphism recently found to be important in individual traits is called **single nucleotide polymorphism (SNP)** because only a single nucleotide is altered. Tens of thousands of these differences at a single locus (when two different individuals are compared) are known to exist throughout the genome. The human genome contains 10 million SNPs. These variations are currently a hot area of research and also commerce. Scientists are mapping SNPs and determining which ones put humans at greater risk of cancers and other diseases. Many companies have sprung up that will

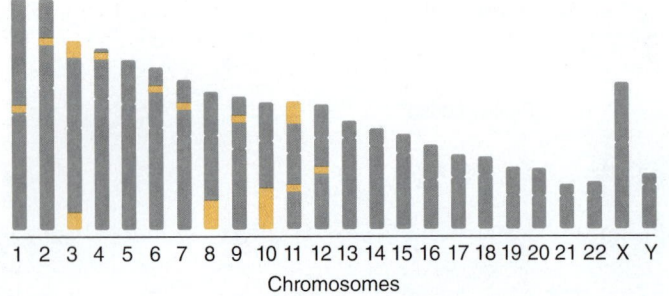

Figure 10.13 The location of one pertinent set of SNPs on human chromosomes. Genetic variations at the sites indicated can be used to estimate the risk of type II diabetes.

map SNPs and give you an estimated risk for a variety of diseases. An example of an SNP map is shown in **figure 10.13.**

Infectious disease laboratories have developed numerous test procedures to profile bacteria and viruses. Profiling is used to identify *Neisseria gonorrhoeae, Chlamydia,* the syphilis spirochete, and *Mycobacterium tuberculosis.* It was instrumental in identifying the anthrax strain used in the mail attacks of 2001. It is also an essential tool for determining genetic relationships between microbes, such as those involved in food-borne disease outbreaks.

Measuring Gene Expression: Microarrays

Twin advances in biology and electronics have allowed biologists to view the expression of genes in any given cell using a technique called DNA **microarray** analysis. Microarrays are able to track the expression of thousands of genes at once and are able to do so in a single efficient experiment. Microarrays consist of a "chip" made of glass, silicon, or nylon, onto which have been bound sequences from tens of thousands of different genes. A solution containing fluorescently labeled cDNA, representing all of the mRNA molecules in a cell at a given time, is added to the chip. The labeled cDNA is allowed to bind to any complementary DNA bound to the chip. Bound cDNA is then detected by exciting the fluorescent tag on the cDNA with a laser and recording the fluorescence with a detector linked to a computer. The computer can then interpret this data to determine what mRNAs are present in the cell under a variety of conditions **(figure 10.14).** In the example in this figure, you see green, red, and yellow reactions. The green fluorescent tag was added to a control population of cells, let's say a bacterium growing in fluid culture. The red fluorescent tag was added to bacteria growing in a biofilm. Genes expressed only in fluid culture appear green on the microarray; those expressed in biofilm growth are colored red; and the yellow reactions were dual-labeled, meaning those genes are expressed under both conditions.

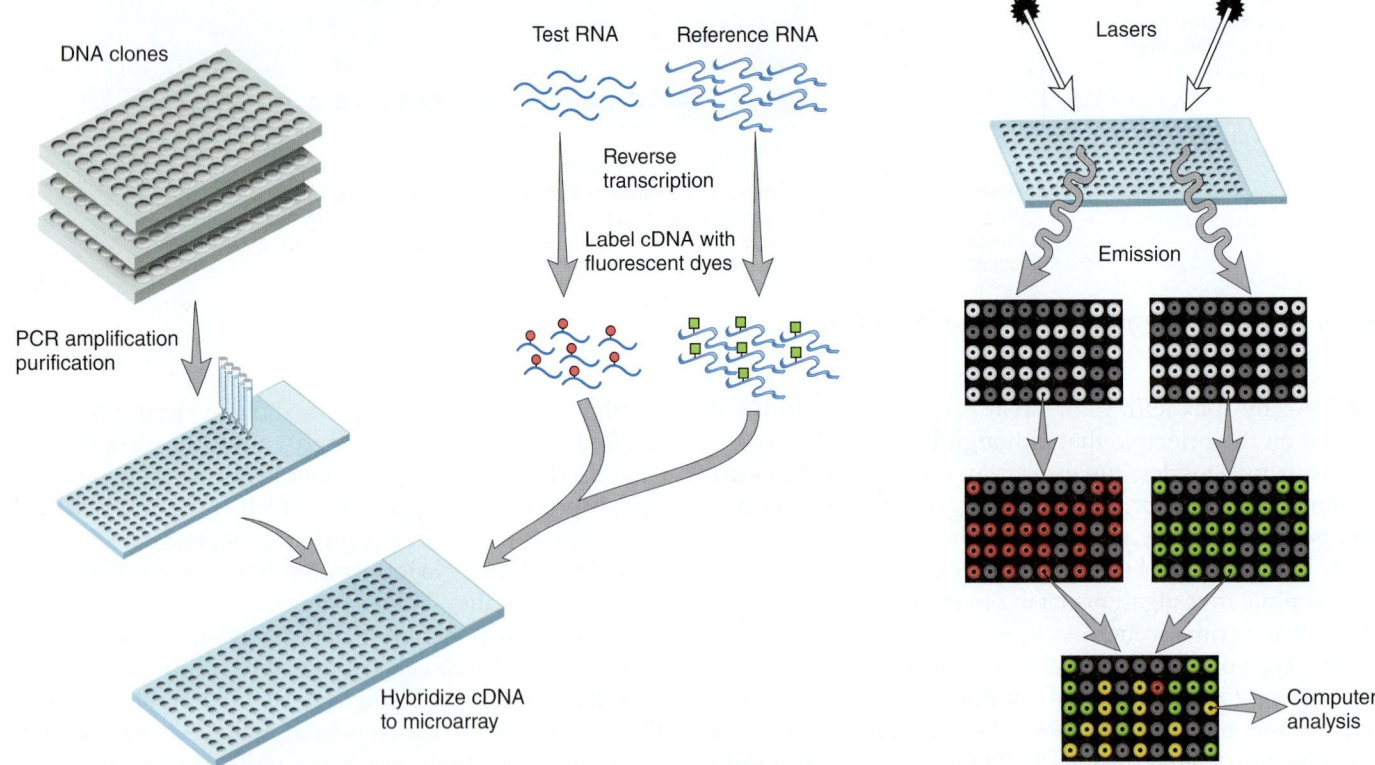

Figure 10.14 Gene expression analysis using microarrays. Cloned genes from an organism are amplified by PCR; after purification, samples are applied to a chip to generate a spotted microarray. mRNA from test and reference cultures are converted to cDNA by reverse transcription and labeled with two different fluorescent dyes. The labeled mixture is hybridized to the microarray and scanned. Gray reactions represent no hybridization having occurred. Green and red reactions indicate that expression was significantly higher in either the test or reference cultures. Yellow reactions represent genes expressed equally in both reference and test RNA samples.

Possible uses of microarrays include developing extraordinarily sensitive diagnostic tests that search for a specific pattern of gene expression. As an example, being able to identify a patient's cancer as one of many subtypes (rather than just, for instance, breast cancer) will allow pharmacologists and doctors to treat each cancer with the drug that will be most effective. Again, we see that genetic technology can be a very effective way to reach long-held goals.

10.5 Learning Outcomes—Assess Your Progress

10. Outline in general terms the process of DNA sequencing.

11. Outline the general steps in DNA profiling.

12. Discuss the significance of single nucleotide polymorphisms (SNPs) in DNA analysis.

13. Describe the utility of DNA microarray technology.

10.6 Proteome Analysis

As you saw in the previous section, microarray analysis can tell us what genes are being transcribed. That does not mean they are all being translated, however. A very important way to study the phenotypes of cells and organisms is by looking at exactly what proteins they are producing. This, in keeping with the *-omics* nomenclature of genomics, is called **proteomics.** The term refers to the entire collection of proteins being produced at a defined point in time.

Though scientists have been studying individual proteins for a long time, the field of proteomics is young (the name was coined in 1997) and it is rapidly evolving. It sprung from a realization that the proteome is an extremely dynamic entity; the production of proteins in a cell changes rapidly. Also, the fact that many cells exert control on translation meant that understanding gene expression (which resulted in mRNA transcripts) was not sufficient. Early attempts to characterize the proteome involved adaptations of gel electrophoresis. Increasingly, methods utilizing advanced instruments (i.e., mass spectrometers) and procedures such as X-ray diffraction are becoming dominant.

10.6 Learning Outcomes—Assess Your Progress

14. Define *proteome*, and explain how it differs from the *genome*.

Case File 10   *Wrap-Up*

Another innovation that has been developed as a result of the quest to identify WTC victims is improved extraction of DNA from bone samples. Often, when a body is completely decomposed, all that remains are teeth and bone. The matrix of the bone is impervious to most types of decomposition, either through chemicals or bacteria in the environment. The mineral matrix of the bone that protects it from decomposition is exactly what prevents scientists from extracting DNA for analysis. In many cases of samples from the WTC attacks, only a small fragment of bone has been found. Traditional methods of identification through dental records and structural analysis of a skeleton are impossible, leaving only DNA analysis as a viable option. Previous methods extracted only a small amount of DNA and wasted a great deal of the mineralized portion of the bone that still contained DNA. New methods of DNA extraction from bone have been developed using new enzymes that break down the crystallized protein in bone, allowing the DNA to be amplified through PCR and analyzed. These newly discovered enzymes are able to more efficiently demineralize bone to extract the DNA and perform the STR analysis discussed previously.

Recently discovered improvements in the technology of DNA extraction and analysis not only have brought peace and closure to the families of the WTC attacks but have provided new technologies for identification of victims of other large-scale disasters as well as the remains of soldiers from Vietnam and even World War II.

Source: 2010. For. Sci. Int: Gen. vol. 4, pp. 275–80.

Chapter Summary

10.1   Introduction to Genetic Engineering (ASM Guidelines* 4.5, 6.2, 6.3)

- The genetic revolution has produced a wide variety of technologies that allow humans to radically alter the blueprints of life.

10.2   Tools and Techniques of Genetic Engineering (ASM Guidelines 4.5, 6.2, 6.3)

- Genetic engineering utilizes a wide range of methods that physically manipulate DNA for purposes of visualizing, sequencing, hybridizing, and identifying specific sequences.
- The tools of genetic engineering include restriction endonucleases, gel electrophoresis, and gene probes.
- The polymerase chain reaction (PCR) technique amplifies small amounts of DNA into much larger quantities for further analysis.
- Recombinant DNA techniques combine DNA from different sources to produce microorganism "factories" that produce hormones, enzymes, and vaccines on an industrial scale.
- Cloning is the process by which genes are removed from the original host and duplicated for transfer into a cloning host by means of cloning vectors.

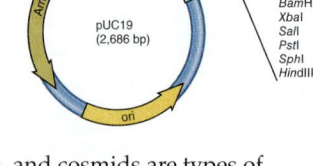

- Plasmids, bacteriophages, and cosmids are types of cloning vectors used to transfer recombinant DNA into a cloning host.

10.3   Products of Recombinant DNA Technology (ASM Guidelines 4.5, 6.2, 6.3)

- Bioengineered hormones, enzymes, and vaccines are often safer and more effective than similar substances isolated directly from animals.
- Recombinant microorganisms are genetically designed for medical treatments and immunizations, crop improvement, pest reduction, and bioremediation.

*ASM Curriculum Guidelines (American Society for Microbiology, 2012). Complete guidelines in appendix B of this book.

- Synthetic biology is the attempt to create novel organisms in the laboratory that can perform valuable functions and produce useful products.

10.4   Genetic Treatments: Introducing DNA into the Body (ASM Guidelines 4.5, 6.2, 6.3)

- Gene therapy is the replacement of faulty host genes with functional genes using delivery vehicles such as viruses or polymers. This treatment can be used to correct genetic disorders and acquired diseases.
- Micro RNAs are used to block expression of undesirable host genes, to silence deleterious miRNAs in host cells, and to assist in defense against microbial attack.
- DNA technology has advanced understanding of basic genetic principles that have significant applications in a wide range of disciplines, particularly medicine, evolution, forensics, and anthropology.

10.5   Genome Analysis: Maps and Profiles (ASM Guidelines 4.5, 6.2, 6.3)

- The Human Genome Project and other genome sequencing projects have revolutionized our understanding of organisms and led to two new biological disciplines, genomics and bioinformatics.
- Whole-genome shotgun sequencing is the method used to determine the nucleotide sequences of humans, other eukaryotes, bacteria and archaea, and viruses.

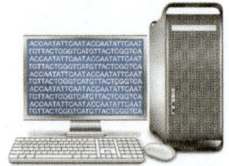

- DNA profiling is a technique by which organisms are identified for purposes of medical diagnosis, genetic ancestry, and forensics.
- Microarray analysis can determine what genes are transcribed in a given tissue. It is used to identify and devise treatments for diseases based on the phenotypic profile of the disease.

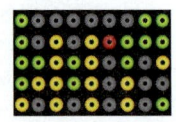

10.6   Proteome Analysis (ASM Guidelines 4.3, 4.5)

- To determine the on-the-spot phenotype of a given cell, the proteome needs to be analyzed. The proteome is the entire collection of proteins present in a cell.

Multiple-Choice and True-False Questions | Bloom's Levels 1 and 2: Remember and Understand

Multiple-Choice Questions. Select the correct answer from the options provided.

1. Which of the following is/are *not* essential to carry out the polymerase chain reaction?
   a. primers   c. gel electrophoresis
   b. DNA polymerase   d. high temperature

2. Which of the following is the closest synonym to *contig*?
   a. nucleotide sequence   c. library
   b. restriction enzyme   d. primer

3. The function of ligase is to
   a. rejoin segments of DNA.
   b. make longitudinal cuts in DNA.
   c. synthesize cDNA.
   d. break down ligaments.

4. The creation of customized organisms with customized properties is called
   a. recombination
   b. bioremediation
   c. sequencing
   d. synthetic biology
   e. artificial biology

5. Which of the following sequences, when combined with its complement, could be clipped by an endonuclease?
   a. ATCGATCGTAGCTAGC
   b. AAGCTTTTCGAA
   c. ACCATTGGTA

6. A region of DNA in a plasmid that is recognized by a wide variety of restriction enzymes is called the
   a. origin.
   b. regulator.
   c. multicloning site.
   d. vector.

7. Short sequences of RNA that are used in a wide variety of cells to regulate gene expression are called
   a. ribosomal RNAs.
   b. ribosomes.
   c. micro RNAs.
   d. messenger RNAs.

8. Which of the following is a primary participant in cloning an isolated gene?
   a. restriction endonuclease
   b. vector

   c. host organism
   d. all of these

9. Single nucleotide polymorphisms are found in
   a. DNA.
   b. RNA.
   c. plasmids.
   d. siRNA.

10. Microarrays are used to monitor
   a. the rate of DNA replication.
   b. the presence of particular genes in DNA.
   c. antisense DNA.
   d. which genes are being expressed.

True-False Questions. If the statement is true, leave as is. If it is false, correct it by rewriting the sentence.

11. The synthetic unit of the polymerase chain reaction is the replica.

12. A nucleic acid probe can be used to identify unknown bacteria or viruses in clinical samples.

13. A DNA fragment with 450 bp will be closer to the top (negative pole) of an electrophoresis gel than one with 2,500 bp.

14. In order to detect recombinant cells, plasmids contain antibiotic resistance genes.

15. Plasmids are the only vectors currently available for use in recombinant procedures.

Critical Thinking Questions | Bloom's Levels 3, 4, and 5: Apply, Analyze, and Evaluate

Critical thinking is the ability to reason and solve problems using facts and concepts. These questions can be approached from a number of angles and, in most cases, they do not have a single correct answer.

1. You are a public health official trying to determine the identity of the pathogen circulating within your city. Explain which genetic technologies would be most useful in this process.

2. a. Construct a strand of complementary DNA (cDNA) from the following mRNA transcript:
   3'-UAUGAACCCCGCUUU-5'
   b. What enzyme is used to copy DNA from an mRNA transcript, and why is it necessary to utilize this process to synthesize eukaryotic genes for use in bacterial cells?

3. Based on the following image:

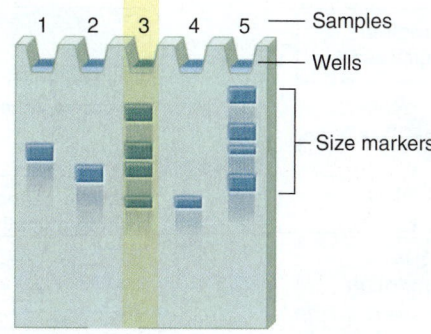

DNA migrates toward positive electrode.

   a. Explain where the charges should be applied to this gel in order for the DNA samples to migrate properly.
   b. Identify the largest DNA fragment in sample 5. Explain your selection.

4. Summarize the goal of the Human Genome Project and discuss three new fields of science that have developed from this research.

5. a. Explain whether or not DNA polymerase from a mesophilic bacterium could be used successfully in a PCR reaction.
   b. If starting with a single double-stranded DNA molecule, how many copies of the DNA would be synthesized after 25 PCR cycles?
   c. Provide two applications of PCR technology today.

6. a. Outline the main steps in cloning a gene, explaining how horizontal gene transfer is used in this process.
   b. Describe one method used to determine whether a bacterial culture has received a recombinant plasmid.

7. a. Define the term *RFLP*. Explain how RFLPs are created and why they are useful in DNA analysis.
   b. In the following DNA profile, identify the pathogen that is making the two patients ill and explain your answer.

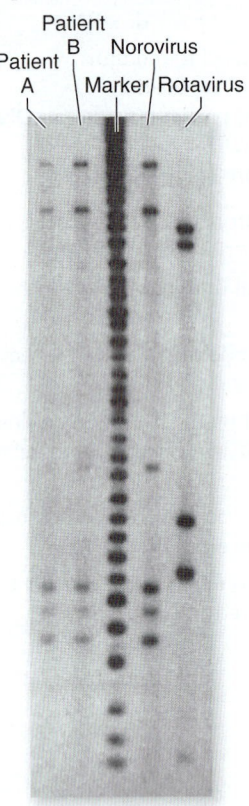

Patient A
Patient B
Marker
Norovirus
Rotavirus

8. Conduct additional research and provide evidence in support of or refuting the following statement: All forms of gene therapy used today result in heritable changes.

9. In the field of pharmacogenetics, scientists attempt to develop drugs or drug treatment regimens for patients based upon their genetic profiles. Provide an example of how DNA microarray analysis could assist in the treatment of a cancer patient using pharmacogenomic methods.

10. If you were given free access to testing using gene probes, profiling, and mapping, would you wish to use this technology to find out if you or your children are at risk for genetic disease? Before answering, conduct additional research at the National Human Genome Research Institute's Ethical, Legal and Social Implications (ELSI) Research website to obtain information. Then develop an informed opinion on this subject, recognizing all benefits and possible consequences to your decision.

Concept Connections | Bloom's Levels 4 and 6: Analyze and Create

This activity ties together multiple concepts in the chapter.

1. What are restriction endonucleases? How are they used by bacteria and archaea?

2. What is a "sticky end"?

3. How are restriction endonucleases used in genetic engineering?

4. Give at least one example of an application for each of the methods listed.

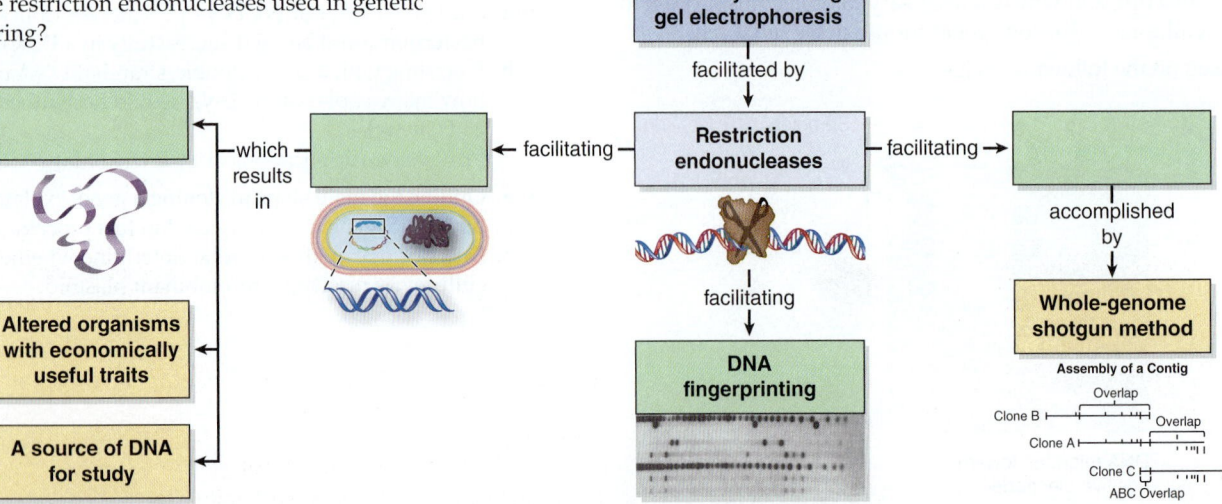

Visual Connections │ Bloom's Level 5: Evaluate

These questions use visual images or previous content to make connections to this chapter's concepts.

1. **From chapter 6, figure 6.22.** What has happened to the bacterial DNA in this illustration? What effect can this have on a bacterium? Is this temporary or permanent?

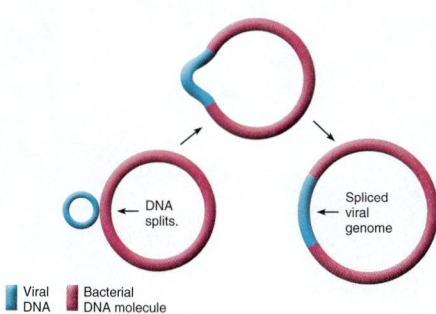

2. **From chapter 9, process figure 9.23.** Study the series of events in this illustration. What do cell A (step ①) and cell B (step ⑤) have in common?

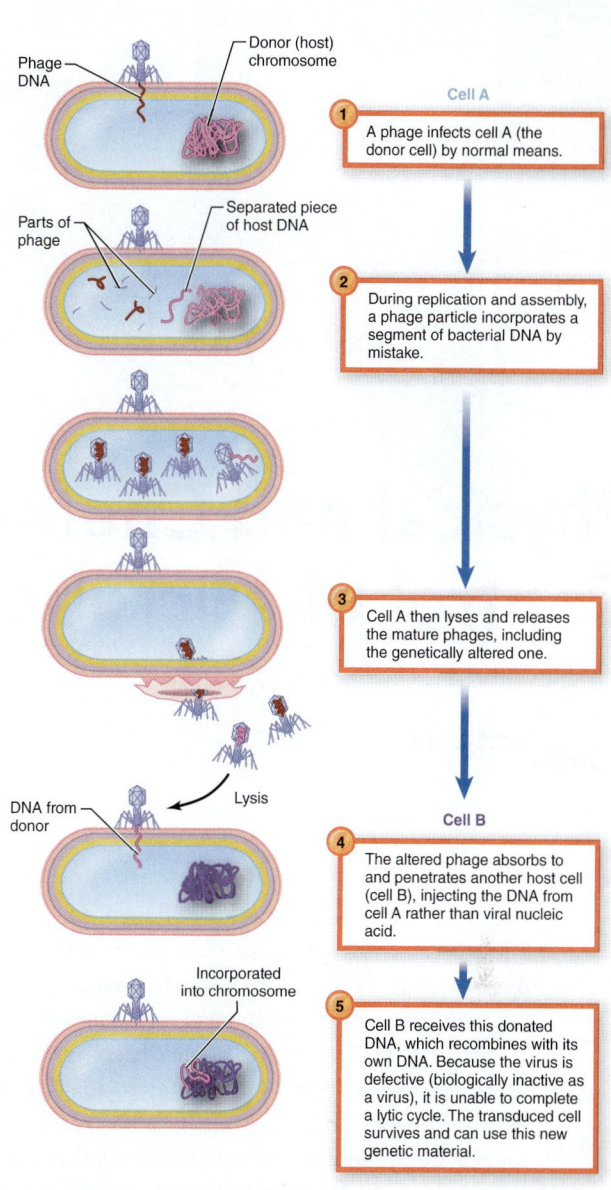

Concept Mapping │ Bloom's Level 6: Create

Appendix D provides guidance for working with concept maps.

1. Using the words that follow, please create a concept map illustrating the relationships among these key terms from chapter 10.

DNA	palindrome	plasmid	origin of replication
restriction endonuclease	ligase	vector	recombinant

www.mcgrawhillconnect.com

Enhance your study of this chapter with study tools and practice tests. Also ask your instructor about the resources available through ConnectPlus, including the media-rich eBook, interactive learning tools, and animations.

Physical and Chemical Control of Microbes

Case File 11

Contaminated Antiseptic

Microbial control is extremely important in a hospital setting. Hospitalized patients have numerous open portals where bacteria can enter and cause infection, from simple skin breaks due to injections or intravenous lines to surgical sites and open wounds. Extreme care needs to be taken when preparing skin for injections, IVs, and surgery. Sterile equipment and fluids must be used to ensure that no bacteria are introduced into these open portals.

In October 2010, a child at the Children's Hospital at Aurora, Colorado, was diagnosed with leukemia, and a vascular access device, or "port," was inserted to administer chemotherapy drugs. Twenty-four hours after the port was inserted, the child developed clinical **sepsis,** or bacteria in the blood, as well as **cellulitis** at the insertion site. The wound required surgical **debridement,** and extensive wound care post-surgery. The child also required antibiotics and outpatient treatment. Cultures were taken of the child's blood and tissues, and *Bacillus cereus* grew from the samples.

A month later at the same hospital, an infant with congenital heart disease received an internal jugular line. Four days after the line was placed, the infant became clinically septic and developed a fever and respiratory distress. The jugular line had to be removed, and the infant was treated with intravenous antibiotics over an extended period before being discharged from the hospital. Blood and tissue samples from the infant were also positive for *B. cereus*.

- What procedures are used to prepare skin for injections, IVs, and surgical procedures?
- How could *B. cereus* have been introduced into these patients?

Continuing the Case appears on page 312.

Outline and Learning Outcomes

11.1 Controlling Microorganisms

1. Distinguish among the terms *sterilization, disinfection, antisepsis,* and *decontamination*.
2. Identify the types of microorganisms that are most resistant and least resistant to control measures.

3. Define *-static* and *-cidal*. Compare the action of microbicidal and microbistatic agents, providing an example of each.
4. Name four categories of cellular targets for physical and chemical agents.

11.2 Methods of Physical Control
5. Name six methods of physical control of microorganisms.
6. Compare and contrast moist and dry heat methods of control, and identify multiple examples of each.
7. Define *thermal death time* and *thermal death point*, and describe their role in proper sterilization.
8. Explain four different methods of moist heat control.
9. Explain two methods of dry heat control.
10. Identify advantages and disadvantages of cold treatment and desiccation.
11. Differentiate between the two types of radiation control methods, providing an application of each.
12. Outline the process of filtration and describe its two advantages in microbial control. Explain how filtration functions as a control method.
13. Identify some common uses of osmotic pressure as a control method.

11.3 Chemical Agents in Microbial Control
14. Name the desirable characteristics of chemical control agents.
15. Discuss several different halogen agents and their uses in microbial control.
16. List advantages and disadvantages to the use of phenolic compounds as control agents.
17. Explain the mode of action of alcohols and their limitations as effective antimicrobials.
18. Pinpoint the most appropriate applications of hydrogen peroxide agents.
19. Define the term *surfactant*, and explain this antimicrobial's mode of action.
20. Identify examples of some heavy metal control agents and their most common applications.
21. Discuss the advantages and disadvantages of aldehyde agents in microbial control.
22. Identify applications for ethylene oxide sterilization.

11.1 Controlling Microorganisms

Much of the time in our daily existence, we take for granted tap water that is drinkable, food that is not spoiled, shelves full of products to eradicate "germs," and drugs to treat infections. Controlling our degree of exposure to potentially harmful microbes is a monumental concern in our lives, and it has a long and eventful history. Salting, smoking, pickling, and drying foods and exposing food, clothing, and bedding to sunlight were prevalent practices among early civilizations. The Greeks and Romans burned clothing and corpses during epidemics, and they stored water in copper and silver containers. During the great plague pandemic of the Middle Ages, it was commonplace to bury corpses in mass graves, burn the clothing of plague victims, and ignite aromatic woods in the houses of the sick in the belief that fumes would combat the disease. These attempts may sound foolish and antiquated, but the release of formaldehyde from burning wood may have in fact acted as a disinfectant. Each of these early methods, although somewhat crude, laid the foundations for microbial control methods that are still in use today.

General Considerations in Microbial Control

The methods of microbial control used outside of the body are designed to result in four possible outcomes: sterilization, disinfection, antisepsis, or decontamination.

Sterilization is the destruction of all microbial life.

Disinfection destroys most microbial life, reducing contamination on inanimate surfaces.

Antisepsis (also called **degermation**) is the same as disinfection except a living surface is involved.

Decontamination (also called **sanitization**) is the mechanical removal of most microbes from an animate or inanimate surface. The **figure 11.1** flowchart summarizes the major applications and aims in microbial control.

Relative Resistance of Microbial Forms

The primary targets of microbial control are microorganisms capable of causing infection or spoilage, which are constantly present in the external environment and on the human body. This targeted population is rarely simple or uniform; in fact, it often contains mixtures of microbes with extreme differences in resistance and harmfulness. Contaminants that can have far-reaching effects if not adequately controlled include bacterial vegetative cells and endospores, fungal hyphae and spores, yeasts, protozoan trophozoites and cysts, worms, viruses, and prions. **Figure 11.2** compares the general resistance these forms have to physical and chemical methods of control.

Actual comparative figures on the requirements for destroying various groups of microorganisms are shown in **table 11.1.** Bacterial endospores have traditionally been considered the most resistant microbial entities, being as much as

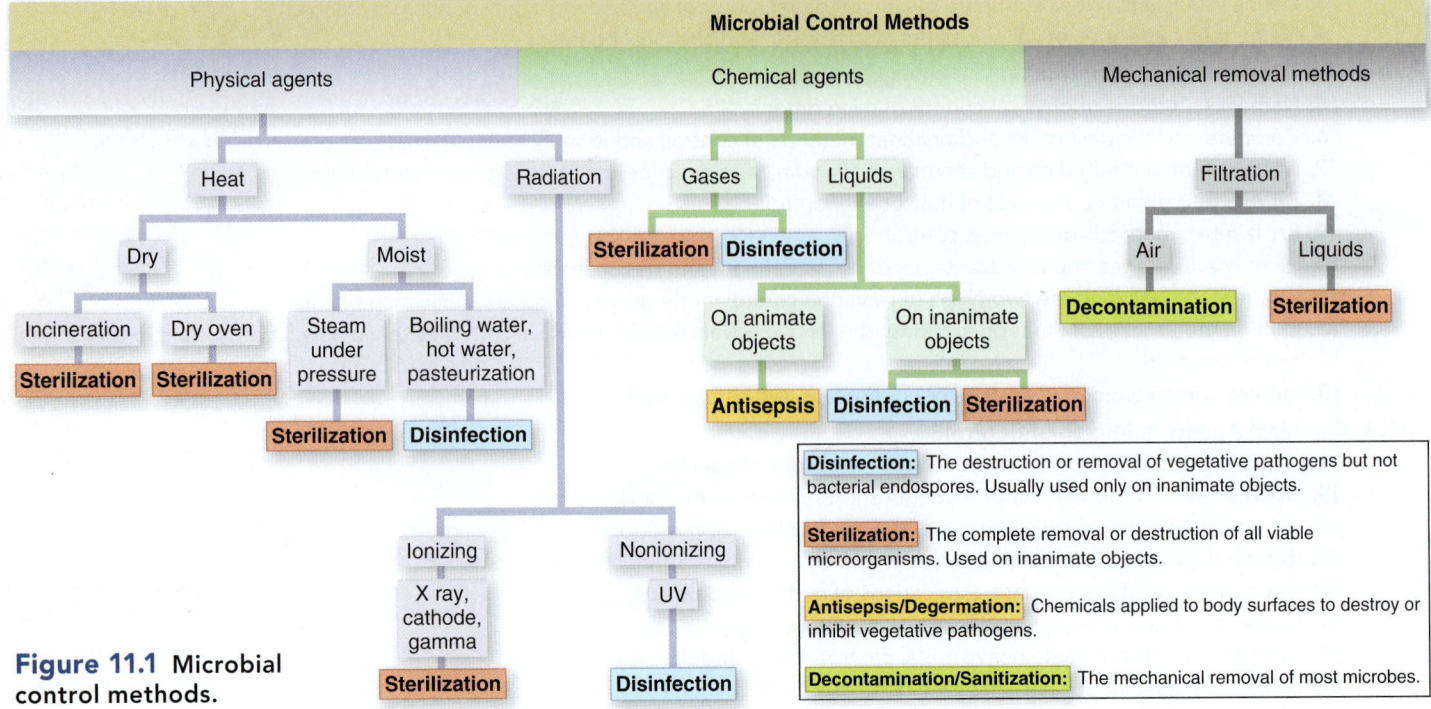

Figure 11.1 Microbial control methods.

18 times harder to destroy than their counterpart vegetative cells. Because of their resistance to microbial control, their destruction is the goal of *sterilization* because any process that kills endospores will invariably kill all less resistant microbial forms. Other methods of control (disinfection, antisepsis) act primarily upon microbes that are less hardy than endospores.

Methods of Microbial Control

Through the years, a growing terminology has emerged for describing and defining measures that control microbes. To complicate matters, the everyday use of some of these terms can at times be vague and inexact. For example, occasionally

one may be directed to *sterilize* or *disinfect* a patient's skin, even though this usage does not fit the technical definition of either term. To lay the groundwork for the concepts in microbial control to follow, we present here a series of concepts, definitions, and usages in antimicrobial control.

Terminology

Sterilization is a process that destroys or removes all viable microorganisms, including viruses. Any material that has been subjected to this process is said to be **sterile.** These terms should be used only in the strictest sense for methods that have been proved to sterilize. An object cannot be

Figure 11.2 Relative resistance of different microbial types to microbial control agents. This is a very general hierarchy; different control agents are more or less effective against the various microbes.

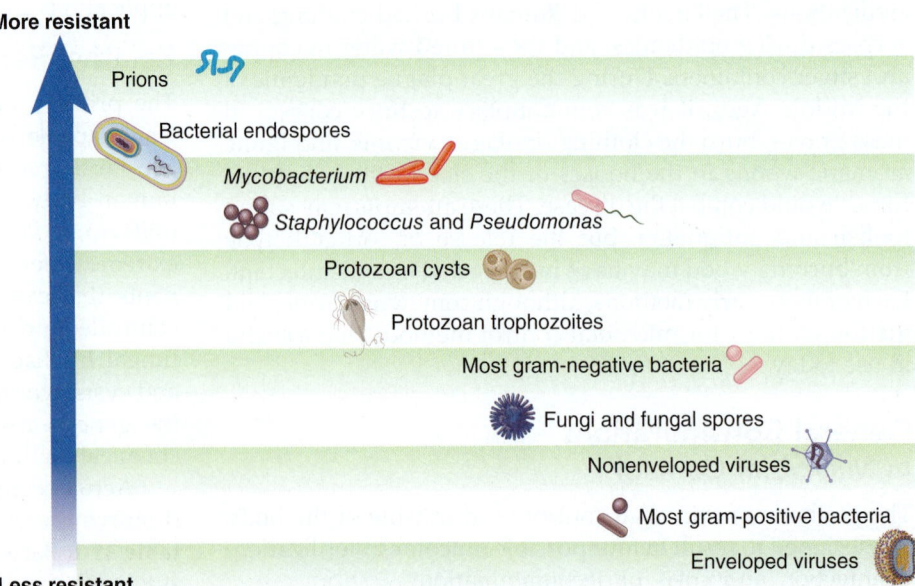

A Note About Prions

Prions are in a class of their own when it comes to "sterilization" procedures. This chapter defines *sterile* as the absence of all viable microbial life—but none of the procedures described in this chapter are necessarily sufficient to destroy prions. Prions are extraordinarily resistant to heat and chemicals. If instruments or other objects become contaminated with these unique agents, they must either be discarded as biohazards or, if this is not possible, enhanced sterilization procedures must be applied in accordance with CDC guidelines. The guidelines themselves are constantly evolving as new information becomes available. In the meantime, this chapter discusses sterilization using bacterial endospores as the toughest form of microbial life. When tissues, fluids, or instruments are suspected of containing prions, consultation with infection control experts and/or the CDC is recommended when determining effective sterilization conditions. Chapter 19 describes prions in detail.

Table 11.1 Comparative Resistance of Bacterial Endospores and Vegetative Cells to Control Agents

Method	Endospores*	Vegetative Forms*	Endospores Are ___ More Resistant**
Heat (moist)	120°C	80°C	1.5×
Radiation (X-ray) dosage	4,000 Grays	1,000 Grays	4×
Sterilizing gas (ethylene oxide)	1,200 mg/L	700 mg/L	1.7×
Sporicidal liquid (2% glutaraldehyde)	3 h	10 min	18×

*Values are based on methods (concentration, exposure time, intensity) that are required to destroy the most resistant pathogens in each group.
**The greater resistance of spores versus vegetative cells given as an average figure.

slightly sterile or almost sterile—it is either sterile or not sterile. Control methods that sterilize are generally reserved for inanimate objects, because sterilizing parts of the human body would call for such harsh treatment that it would be highly dangerous and impractical.

Sterilized products—surgical instruments, syringes, and commercially packaged foods, just to name a few—are essential to human well-being. Although most sterilization is performed with a physical agent such as heat, a few chemicals can be classified as sterilizing agents because of their ability to destroy endospores.

In many situations, sterilization is neither practical nor necessary, and only certain groups of microbes need to be controlled. Some antimicrobial agents eliminate only the susceptible vegetative states of microorganisms but do not destroy the more resistant endospore and cyst stages. Keep in mind that the destruction of endospores is not always a necessity, because most of the infectious diseases of humans and animals are caused by non-endospore-forming microbes.

Disinfection refers to the use of a physical process or a chemical agent (a disinfectant) to destroy vegetative pathogens but not bacterial endospores. It is important to note that disinfectants are normally used only on inanimate objects because, in the concentrations required to be effective, they can be toxic to human and other animal tissue. Disinfection processes also remove the harmful products of microorganisms (toxins) from materials. Examples of disinfection include applying a solution of 5% bleach to an examining table, boiling food utensils used by a sick person, and immersing thermometers in an iodine solution between uses.

In modern usage, *sepsis* is defined as the growth of microorganisms in the blood and other tissues. The term **asepsis** refers to any practice that prevents the entry of infectious agents into sterile tissues and thus prevents infection. Aseptic techniques commonly practiced in health care range from sterile methods that exclude all microbes to *antisepsis*. In antisepsis, chemical agents called **antiseptics** are applied directly to exposed body surfaces (skin and mucous membranes), wounds, and surgical incisions to destroy or inhibit vegetative pathogens. Examples of antisepsis include preparing the skin before surgical incisions with iodine compounds, swabbing an open root canal with hydrogen peroxide, and ordinary hand washing with a germicidal soap.

The Agents Versus the Processes

The terms *sterilization, disinfection,* and so on refer to processes. You will encounter other terms that describe the agents used in the process. Two examples of these are the terms *bactericidal* and *bacteristatic*. The root *-cide*, meaning "to kill," can be combined with other terms to define an antimicrobial agent aimed at destroying a certain group of microorganisms. For example, a **bactericide** is a chemical that destroys bacteria except for those in the endospore stage. It may or may not be effective on other microbial groups. A *fungicide* is a chemical that can kill fungal spores, hyphae, and yeasts. A *virucide* is any chemical known to inactivate viruses, especially on living tissue. A *sporicide* is an agent capable of destroying bacterial endospores. A sporicidal agent can also be a sterilant because it can destroy the most resistant of all microbes. **Germicide** and **microbicide** are additional terms for chemical agents that kill microorganisms.

The Greek words *stasis* and *static* mean "to stand still." They can be used in combination with various prefixes to denote a condition in which microbes are temporarily prevented from multiplying but are not killed outright. Although killing or permanently inactivating microorganisms is the usual goal of microbial control, microbistasis does have meaningful applications. **Bacteristatic** agents prevent the growth of bacteria on tissues or on objects in the environment, and *fungistatic* chemicals inhibit fungal growth. Chemicals used to control microorganisms in the body (antiseptics and drugs) often have **microbistatic** effects because many microbicidal compounds can be highly toxic to human cells. Note that a *-cidal* agent does not necessarily result in sterilization.

Decontamination

Several applications in commerce and medicine do not require actual sterilization, disinfection, or antisepsis but are based on reducing the levels of microorganisms (the microbial load) so that the possibility of infection or spoilage is greatly decreased. Restaurants, dairies, breweries, and other food industries consistently handle large numbers of soiled utensils that could readily become sources of infection and spoilage. These industries must keep microbial levels to a minimum during preparation and processing. **Sanitization** is any cleansing technique that mechanically removes microorganisms as well as other debris to reduce contamination to safe levels. A sanitizer is a compound such as soap or detergent used to perform this task.

Cooking utensils, dishes, bottles, cans, and clothing that have been washed and dried may not be completely free of microbes, but they are considered safe for normal use (sanitary). Air sanitization with ultraviolet lamps reduces airborne microbes in hospital rooms, veterinary clinics, and laboratory installations. Note that some sanitizing processes (such as dishwashing machines) may be rigorous enough to sterilize objects, but this is not true of all sanitization methods. Also note that sanitization is often preferable to sterilization. In a restaurant, for example, you could be given a sterile fork with someone else's old food on it and a sterile glass with lipstick on the rim, but wouldn't you rather have a sanitized glass with no remnants of the previous guest? On top of this, realize that the costs associated with sterilization would lead to the advent of the $50 fast-food meal. Thus, the usefulness of a technique depends on the context.

It is often necessary to reduce the numbers of microbes on the human skin through antisepsis (degermation). This process usually involves scrubbing the skin or immersing it in chemicals, or both. It also emulsifies oils that lie on the outer cutaneous layer and mechanically removes potential pathogens on the outer layers of the skin. Examples of degerming procedures are the surgical hand scrub, the application of alcohol wipes to the skin, and the cleansing of a wound with germicidal soap and water.

Practical Concerns in Microbial Control

Numerous considerations govern the selection of a workable method of microbial control. These are among the most pressing concerns:

1. Does the item in question require sterilization, or is disinfection adequate? In other words, must spores be destroyed or is it necessary to destroy only vegetative pathogens?
2. Is the item to be reused or permanently discarded? If it will be discarded, then the quickest and least expensive method should be chosen.
3. If it will be reused, can the item withstand heat, pressure, radiation, or chemicals?
4. Is the control method suitable for a given application? (For example, ultraviolet radiation is a possible sporicidal agent, but it will not penetrate solid materials.) Or, in the case of a chemical, will it leave an undesirable residue?
5. Will the agent penetrate to the necessary extent?
6. Is the method cost- and labor-efficient, and is it safe?

One useful framework for determining how devices that come in contact with patients should be handled is whether they are considered *critical, semicritical,* or *noncritical.* Critical medical devices are those that are expected to come into contact with sterile tissues. A good example of this would be a syringe needle or an artificial hip. These must be sterilized before use. Semicritical devices are those that come into contact with mucosal membranes. An endoscopy tube is an example. These must receive at least high-level disinfection and, preferably, should be sterilized. Noncritical items are those that do not touch the patient or are only expected to touch intact skin, such as blood pressure cuffs or crutches. They require only low-level disinfection unless they become contaminated with blood or body fluids.

A remarkable variety of substances can require sterilization. They range from durable solids such as rubber to sensitive liquids such as serum, and even to entire office buildings, as seen in 2001 when the Hart Senate Office Building was contaminated with *Bacillus anthracis* endospores **(Insight 11.1).** Hundreds of situations requiring sterilization confront the network of persons involved in health care, whether technician, nurse, doctor, or manufacturer, and no universal method works well in every case.

Considerations such as cost, effectiveness, and method of disposal are all important. For example, disposable plastic items such as catheters and syringes that are used in invasive medical procedures have the potential for infecting the tissues. These must be sterilized during manufacture by a nonheating method (gas or radiation), because heat can damage plastics. After these items have been used, it is often necessary to destroy or decontaminate them before they are discarded because of the potential risk to the handler (from needlesticks). Steam sterilization, which is quick and sure, is a sensible choice at this point, because it does not matter if the plastic is destroyed.

What Is Microbial Death?

Death is a phenomenon that involves the permanent termination of an organism's vital processes. Signs of life in complex organisms such as animals are self-evident, and death is made clear by loss of nervous function, respiration, or heartbeat. In contrast, death in microscopic organisms that are composed of just one or a few cells is often hard to detect, because they reveal no conspicuous vital signs to begin with. Lethal agents (such as radiation and chemicals) do not necessarily alter the overt appearance of microbial cells. Even the loss of movement in a motile microbe cannot

INSIGHT 11.1   "Suspicious Powder"

In late February 2012, popular talk show hosts Jon Stewart and Stephen Colbert received letters containing a "suspicious powdery substance." The letters were postmarked from Oregon; similar letters had been sent to House of Representatives Speaker John Boehner, Senate Sergeant at Arms Terrance Gainer, and other government officials, as well as other media outlets including the *Washington Post, USA Today, NPR,* and *Fox News.* After analyzing the white powder, the FBI concluded that none of the letters contained harmful material.

This recent anthrax scare calls to mind the anthrax attacks on Congress following the 9/11 attacks in 2001. In that case, letters containing *Bacillus anthracis* endospores were mailed to members of Congress and several news agencies. The FBI called this the worst case of bioterrorism in U.S. history, in which five people were killed and 17 sickened by anthrax. The FBI was able to trace the attacks to Dr. Bruce Ivins, a senior biodefense researcher at the U.S. Army Medical Research Institute of Infectious Disease (USAMRIID), who took his own life before formal charges could be filed.

In addition to the investigation into the source of the anthrax-laced mailings, the immediate task was to decontaminate the Hart Office Building where the letters were opened. The Environmental Protection Agency (EPA) was charged with the cleanup of the building, a daunting task for several reasons: the large square footage of the office building populated by thousands of individuals; the complexity of decontaminating heating and air conditioning vents and a variety of different surfaces such as carpeting, furniture, office equipment, and sensitive papers; and the fact that *Bacillus anthracis* produces endospores that are difficult to eradicate.

The first step of this process was to take swab samples from various offices of Congress to determine the extent of the contamination. Endospores were found in many offices in the Hart Office Building, and Senator Tom Daschle's offices were the most highly contaminated. When the extent of the contamination was determined, the EPA devised a strategy to decontaminate the offices, keeping in mind that they contained delicate, expensive, and irreplaceable paperwork, office equipment, and personal items left by evacuated workers. Most areas were cleaned by a combination of HEPA vacuuming followed by treatment with liquid chlorine dioxide or Sandia decontamination foam, an antibacterial foam that combines surfactants with oxidizing agents. Finally, the buildings were fumigated with chlorine dioxide gas, a sterilant that has been used since 1988 to treat medical waste and other contaminated substances.

Workers in protective garments prepare the Hart Office Building for decontamination.

In order to determine the success of the treatments, the EPA used 3,000 test strips impregnated with *Bacillus stearothermophilus,* which were distributed throughout the offices. *B. stearothermophilus* is a nonpathogenic endospore-forming relative of *B. anthracis,* generally considered to be more difficult to kill than *B. anthracis.* If, after fumigation, endospores of *B. stearothermophilus* remained on the test strips, then further decontamination would be necessary. Before fumigation began, humidity in the buildings was increased to 75% to improve the adherence of the chlorine dioxide to any lingering spores, and office machines were turned on in order for their fans to effectively disperse the gas. Fumigation lasted 20 hours, after which the gas was neutralized and ventilated. Areas that showed trace contamination due to positive growth of *B. stearothermophilus* were again cleaned with liquid chlorine dioxide.

The EPA has used this experience to develop protocols and emergency preparedness measures to respond to other potential bioterrorism attacks such as these. The science and technology research developed as a result has culminated in the Bio-response Operational Testing and Evaluation Program co-led by the EPA and the Department of Homeland Security. Fortunately, the offices of Jon Stewart, Stephen Colbert, congressmen, and news outlets did not test positive for *B. anthracis* in the February 2012 mailings, but the EPA is prepared in the event of other attacks.

Source: 2012. www.reuters.com. February 22.

---

be used to indicate death. This fact has made it necessary to develop special qualifications that define and delineate microbial death.

The destructive effects of chemical or physical agents occur at the level of a single cell. As the cell is continuously exposed to an agent such as intense heat or toxic chemicals, various cell structures become dysfunctional. The entire cell can sustain irreversible damage in the process. At present, the

most practical way to detect this damage is to determine if a microbial cell can still reproduce when exposed to a suitable environment. If the microbe has sustained metabolic or structural damage to such an extent that it can no longer reproduce, even under ideal environmental conditions, then it is no longer viable. The permanent loss of reproductive capability, even under optimum growth conditions, has become the accepted microbiological definition of death.

Factors That Affect Death Rate

The cells of a culture can show significant variation in susceptibility to a given microbicidal agent. Death of the whole population is not instantaneous but begins when a certain threshold of microbicidal agent (some combination of time and concentration) is met. Death continues in a logarithmic manner as the time or concentration of the agent is increased **(figure 11.3a).** Because many microbicidal agents target the cell's metabolic processes, active cells (younger, rapidly dividing) tend to die more quickly than those that are less metabolically active (older, inactive). Eventually, a point is reached at which survival of any cells is highly unlikely; this point is equivalent to sterilization.

The effectiveness of a particular agent is governed by several factors besides time. These additional factors influence the action of antimicrobial agents:

1. The number of microorganisms **(figure 11.3b).** A higher load of contaminants requires more time to destroy.
2. The nature of the microorganisms in the population **(figure 11.3c).** In most actual circumstances of disinfection

and sterilization, the target population is not a single species of microbe but a mixture of bacteria, fungi, spores, and viruses, presenting a broad spectrum of microbial resistance.

3. The type of microbial growth. Planktonic bacterial populations grow freely within fluid environments and do not attach to surfaces. In general, they are more susceptible to control agents as compared to the microbes within well-developed biofilms adhering to surfaces such as medical devices and human tissues.
4. The temperature and pH of the environment.
5. The concentration (dosage, intensity) of the agent. For example, UV radiation is most effective at 260 nm, and most disinfectants are more active at higher concentrations.
6. The mode of action of the agent **(figure 11.3d).** How does it kill or inhibit the microorganism?
7. The presence of solvents, interfering organic matter, and inhibitors. Saliva, blood, and feces can inhibit the actions of disinfectants and even of heat.

The influence of these factors is discussed in greater detail in subsequent sections.

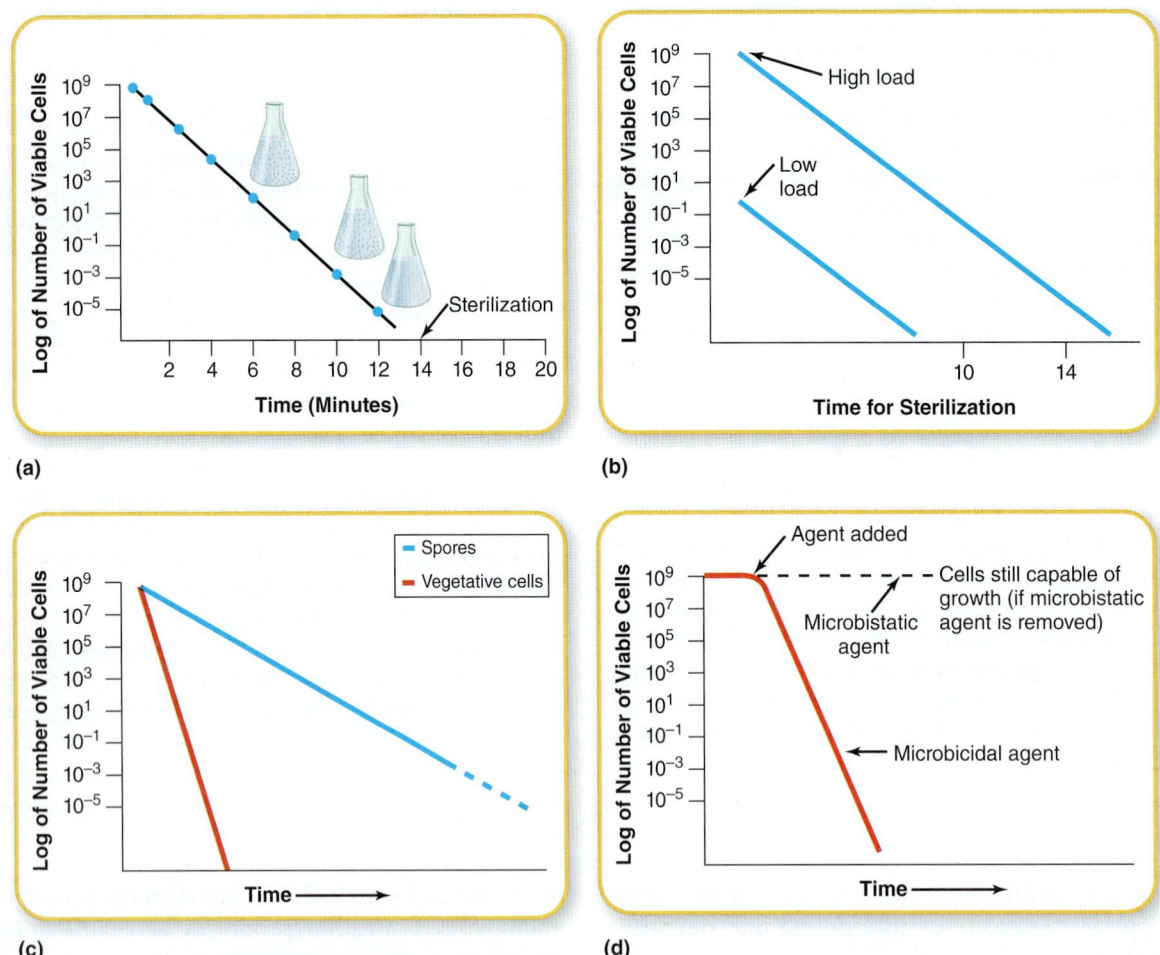

(a)

(b)

(c)

(d)

Figure 11.3 Factors that influence the rate at which microbes are killed by antimicrobial agents.   (a) Length of exposure to the agent. During exposure to a chemical or physical agent, all cells of a microbial population, even a pure culture, do not die simultaneously. Over time, the number of viable organisms remaining in the population decreases logarithmically, giving a straight-line relationship on a graph. The point at which the number of survivors is infinitesimally small is considered sterilization. **(b)** Effect of the microbial load. **(c)** Relative resistance of spores versus vegetative forms. **(d)** Action of the agent, whether microbicidal or microbistatic.

How Antimicrobial Agents Work: Their Modes of Action

An antimicrobial agent's adverse effect on cells is known as its *mode* (or *mechanism*) *of action*. Agents affect one or more cellular targets, inflicting damage progressively until the cell is no longer able to survive. Antimicrobials have a range of cellular targets, with the agents that are least selective in their targeting tending to be effective against the widest range of microbes (examples include heat and radiation). More selective agents (drugs, for example) tend to target only a single cellular component and are much more restricted as to the microbes they are effective against.

The cellular targets of physical and chemical agents fall into four general categories:

1. the cell wall,
2. the cell membrane,
3. cellular synthetic processes (DNA, RNA), and
4. proteins.

The Effects of Agents on the Cell Wall

The cell wall maintains the structural integrity of bacterial and fungal cells. Several types of chemical agents damage the cell wall by blocking its synthesis, digesting it, or breaking down its surface. A cell deprived of a functioning cell wall becomes fragile and is lysed very easily. Detergents and alcohol can also disrupt cell walls, especially in gram-negative bacteria.

How Agents Affect the Cell Membrane

All microorganisms have a cell membrane composed of lipids and proteins, and many viruses have an outer membranous envelope. As we learned in previous chapters, a cell's membrane provides a two-way system of transport. If this membrane is disrupted, a cell loses its selective permeability and can neither prevent the loss of vital molecules nor bar the entry of damaging chemicals. Loss of those abilities leads to cell death. Detergents called **surfactants** (sir-fak´-tunts) work as microbicidal agents **(Insight 11.2).** Surfactants are polar molecules with hydrophilic and hydrophobic regions that can physically bind to the lipid layer and penetrate the internal hydrophobic region of membranes. In effect, this process "opens up" the once tight interface, leaving leaky spots that allow injurious chemicals to seep into the cell and important ions to seep out **(figure 11.4).**

Agents That Affect Protein and Nucleic Acid Synthesis

Microbial life depends upon an orderly and continuous supply of proteins to function as enzymes and structural molecules. As we saw in chapter 9, these proteins are synthesized via the ribosomes through a complex process called translation. The antibiotic chloramphenicol binds to the ribosomes of bacteria in a way that stops peptide bonds from forming. In its presence, many bacterial cells are inhibited

---

INSIGHT 11.2 — A Green Clean

Detergent, soap, shampoo, household cleaners—we use them every day. These surface-acting agents, or **surfactants**, have a hydrophilic head and a hydrophobic tail, not unlike a phospholipid. Surfactant molecules surround oils and other molecules, creating small droplets called **micelles.** They loosen surface tension, breaking up or emulsifying fats. These unique molecules create the foaming action of soaps and are indispensable cleaners of every type. Unfortunately, many surfactants are manufactured from fossil fuels, which are in limited supply and not biodegradable.

Researchers at the Fraunhofer Institute for Interfacial Engineering and Biotechnology IGB are searching for environmentally friendly **biosurfactants** derived from fungi and bacteria. Suzanne Zibek and her team in Stuttgart are testing a type of fungus that can infect corn plants. Her group has found that this organism produces lipids that also have antibacterial properties. "We produce biosurfactants microbially, based on sustainable resources such as sugar and plant oil," she states. Zibek's group has noted that these molecules have increased molecular diversity, giving them a broad range of applications, and are also biodegradable. Unfortunately, these unique "green" surfactants are only used in a limited number of products such as cosmetics and household cleaners because they are difficult to produce. Zibek and her fellow researchers are currently looking for ways to increase production of these biosurfactants in bioreactors. If these methods are successful, you could be washing your hair or cleaning your toilet with an environmentally friendly product derived from bacteria or fungi.

Source: www.sciencedaily.com/releases/2012/03/120309104841.htm.

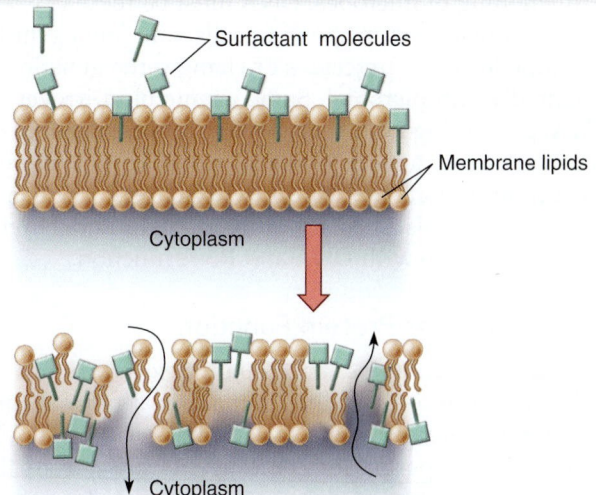

Figure 11.4 Mode of action of surfactants on the cell membrane. Surfactants inserting in the lipid bilayer disrupt it and create abnormal channels that alter permeability and cause leakage both into and out of the cell.

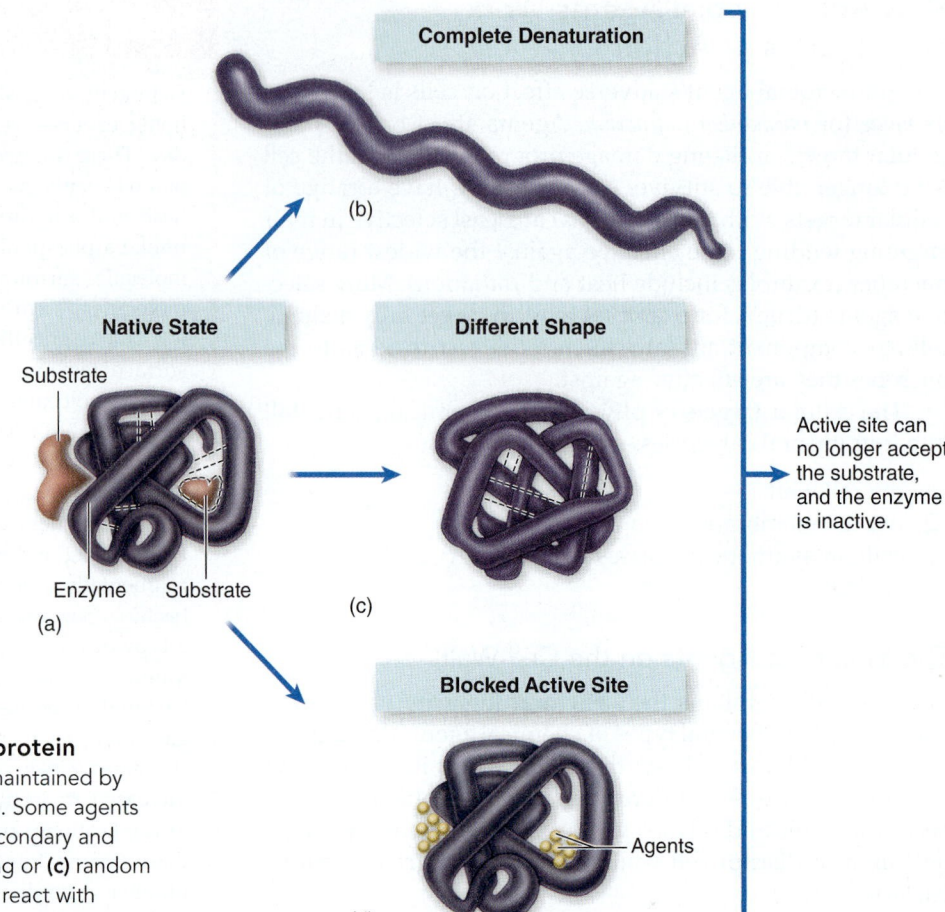

Figure 11.5 Modes of action affecting protein function. **(a)** The native (functional) state is maintained by bonds that create active sites to fit the substrate. Some agents denature the protein by breaking all or some secondary and tertiary bonds. Results are **(b)** complete unfolding or **(c)** random bonding and incorrect folding. **(d)** Some agents react with functional groups on the active site and interfere with bonding.

from forming proteins required in growth and metabolism and are thus inhibited from multiplying.

The nucleic acids are likewise necessary for the continued functioning of microbes. DNA must be regularly replicated and transcribed in growing cells, and any agent that either impedes these processes or changes the genetic code is potentially antimicrobial. Some agents bind irreversibly to DNA, preventing both transcription and translation; others are mutagenic agents. Gamma, ultraviolet, or X radiation cause mutations that result in permanent inactivation of DNA. Chemicals such as formaldehyde and ethylene oxide also interfere with DNA and RNA function.

Agents That Alter Protein Function

A microbial cell contains large quantities of proteins that function properly only if they remain in a normal three-dimensional configuration called the *native state*. The antimicrobial properties of some agents arise from their capacity to disrupt, or **denature,** proteins. In general, denaturation occurs when the bonds that maintain the secondary and tertiary structure of the protein are broken. Breaking these bonds will cause the protein to unfold or create random, irregular loops and coils **(figure 11.5).** One

way that proteins can be denatured is through coagulation by moist heat (the same reaction seen in the irreversible solidification of the white of an egg when boiled). Chemicals such as strong organic solvents (alcohols, acids) and phenolics also coagulate proteins. Other antimicrobial agents, such as metallic ions, attach to the active site of the protein and prevent it from interacting with its correct substrate. Regardless of the exact mechanism, such losses in normal protein function can promptly arrest metabolism. Most antimicrobials of this type are nonselective as to the microbes they affect.

11.1 Learning Outcomes—Assess Your Progress

1. Distinguish among the terms *sterilization, disinfection, antisepsis,* and *decontamination.*
2. Identify the types of microorganisms that are most resistant and least resistant to control measures.
3. Define *-static* and *-cidal.* Compare the action of microbicidal and microbistatic agents, providing an example of each.
4. Name four categories of cellular targets for physical and chemical agents.

11.2 Methods of Physical Control

We can divide our methods of controlling microorganisms into two broad categories: physical and chemical. We will start with physical methods. Microorganisms have adapted to the tremendous diversity of habitats the earth provides, even severe conditions of temperature, moisture, pressure, and light. For microbes that normally withstand such extreme physical conditions, our attempts at control would probably have little effect. Fortunately for us, we are most interested in controlling microbes that flourish in the same environment in which humans live. The vast majority of these microbes are readily controlled by abrupt changes in their environment. Most prominent among antimicrobial physical agents is heat. Other less widely used agents include radiation, filtration, ultrasonic waves, and even cold. The following sections examine some of these methods and explore their practical applications in medicine, commerce, and the home.

Heat as an Agent of Microbial Control

A sudden departure from a microbe's temperature of adaptation is likely to have a detrimental effect on it. As a rule, elevated temperatures (exceeding the maximum growth temperature) are microbicidal, whereas lower temperatures (below the minimum growth temperature) are microbistatic. Heat can be applied in either moist or dry forms. *Moist heat* occurs in the form of hot water, boiling water, or steam (vaporized water). In practice, the temperature of moist heat usually ranges from 60°C to 135°C. As we shall see, the temperature of steam can be regulated by adjusting its pressure in a closed container. The expression *dry heat* denotes air with a low moisture content that has been heated by a flame or electric heating coil. In practice, the temperature of dry heat ranges from 160°C to several thousand degrees Celsius.

Mode of Action and Relative Effectiveness of Heat

Moist heat and dry heat differ in their modes of action as well as in their efficiency. Moist heat operates at lower temperatures and shorter exposure times to achieve the same effectiveness as dry heat **(table 11.2).** Although many cellular

Table 11.2 Comparison of Times and Temperatures to Achieve Sterilization with Moist and Dry Heat

	Temperature (°C)	Time to Sterilize (Min)
Moist heat	121	15
	125	10
	134	3
Dry heat	121	600
	140	180
	160	120
	170	60

structures are damaged by moist heat, its most microbicidal effect is the coagulation and denaturation of proteins, which quickly and permanently halts cellular metabolism.

Dry heat dehydrates the cell, removing the water necessary for metabolic reactions, and it also denatures proteins. However, the lack of water actually increases the stability of some protein conformations, necessitating the use of higher temperatures when dry heat is employed as a method of microbial control. At very high temperatures, dry heat oxidizes cells, burning them to ashes. This method is the one used in the laboratory when a loop is flamed or in industry when medical waste is incinerated.

Heat Resistance and Thermal Death: Spores and Vegetative Cells

Bacterial endospores exhibit the greatest resistance, and vegetative states of bacteria and fungi are the least resistant to both moist and dry heat. Destruction of endospores usually requires temperatures above boiling, although resistance varies widely.

Vegetative cells also vary in their sensitivity to heat. Among bacteria, the death times with moist heat range from 50°C for 3 minutes (*Neisseria gonorrhoeae*) to 60°C for 60 minutes (*Staphylococcus aureus*). It is worth noting that vegetative cells of sporeformers are just as susceptible as vegetative cells of non-sporeformers and that pathogens are neither more nor less susceptible than nonpathogens. Other microbes, including fungi, protozoa, and worms, are rather similar in their sensitivity to heat. Viruses are surprisingly resistant to heat, with a tolerance range extending from 55°C for 2 to 5 minutes (adenoviruses) to 60°C for 600 minutes (hepatitis A virus). For practical purposes, all non-heat-resistant forms of bacteria, yeasts, molds, protozoa, worms, and viruses are destroyed by exposure to 80°C for 20 minutes.

Susceptibility of Microbes to Heat: Thermal Death Measurements

Adequate sterilization requires that both temperature and length of exposure be considered. As we have seen, higher temperatures allow shorter exposure times, and lower temperatures require longer exposure times. A combination of these two variables constitutes the **thermal death time** (TDT), defined as the shortest length of time required to kill all test microbes at a specified temperature. The TDT has been experimentally determined for the microbial species that are common or important contaminants in various heat-treated materials. Another way to compare the susceptibility of microbes to heat is the **thermal death point** (TDP), defined as the lowest temperature required to kill all microbes in a sample in 10 minutes.

Many perishable substances are processed with moist heat. Some of these products are intended to remain on the shelf at room temperature for several months or even years. The chosen heat treatment must render the product free of agents of spoilage or disease. At the same time, the quality of the product and the speed and cost of processing must be considered. For example, in the commercial preparation of canned green beans, one of the manufacturer's greatest concerns is to prevent growth of the agent of botulism. From several possible TDTs (that is, combinations of time and temperature) for *Clostridium botulinum* endospores, the manufacturer must choose one that kills all endospores but does not turn the beans to mush. Out of these many considerations emerges an optimal TDT for a given processing method. Commercial canneries heat low-acid foods at 121°C for 30 minutes, a treatment that sterilizes these foods. Because of such strict controls in canneries, cases of botulism due to commercially canned foods are rare.

Common Methods of Moist Heat Control

The four ways that moist heat is employed to control microbes are

1. boiling water,
2. pasteurization,
3. nonpressurized steam, and
4. steam under pressure.

These methods are described in detail in **table 11.3.**

Disease Connection

A disease known as CJD (Creutzfeldt-Jakob disease), is caused by the mysterious infectious protein known as a prion (see A Note About Prions earlier in this chapter). There are different forms of the disease. Some cases occur spontaneously, some are due to an inherited genetic marker, and some are transmitted from animals. In the 1980s, an outbreak of this disease occurred in cows in Great Britain and elsewhere. The condition in cattle is called *mad cow disease*. Before the existence of prions was appreciated, the disease was occasionally transmitted to patients during surgical procedures, even though the instruments being used had been sterilized after being used on the previous patient. That is because methods of sterilization used in those settings, including dry heat and autoclaving, do not destroy the prion. Current medical instrument sterilization guidelines account for the enhanced techniques required to sterilize devices that may have come in contact with prion-infected tissue. The methods include autoclaving at a higher temperature or pretreating devices with chemical sterilants before autoclaving.

Dry Heat: Hot Air and Incineration

Dry heat is not as versatile or as widely used as moist heat, but it has several important sterilization applications. The temperatures and times employed in dry heat vary according to the particular method, but in general, they are greater than with moist heat. **Table 11.4** describes two dry heat sterilization methods.

The Effects of Cold and Desiccation

The principal benefit of cold treatment is slow growth of cultures and microbes in food during processing and storage. *It must be emphasized that cold merely retards the activities of most microbes.* Although it is true that some microbes are killed by cold temperatures, most are not adversely affected by gradual cooling, long-term refrigeration, or deep-freezing. In fact, freezing temperatures, ranging from −70°C to −135°C, are often used in research labs to preserve cultures of bacteria, viruses, and fungi for long periods. Some psychrophiles grow very slowly even at freezing temperatures and can continue to secrete toxic products. Ignorance of these facts is probably responsible for numerous cases of food poisoning from frozen foods that have been defrosted at room temperature and then inadequately cooked. Pathogens able to survive several months in the refrigerator are *Staphylococcus aureus; Clostridium* species (sporeformers); *Streptococcus* species; and several types of yeasts, molds, and viruses. Outbreaks of *Salmonella* food infection traced back to refrigerated foods such as ice cream, eggs, and tiramisu are testimony to the inability of freezing temperatures to reliably kill pathogens.

Vegetative cells directly exposed to normal room air gradually become dehydrated, or **desiccated.** Delicate pathogens such as *Streptococcus pneumoniae*, the spirochete of syphilis, and *Neisseria gonorrhoeae* can die after a few hours of air drying, but many others are not killed and some are even preserved. Endospores of *Bacillus* and *Clostridium* are viable for millions of years under extremely arid conditions. Staphylococci and streptococci in dried secretions and the tubercle bacillus surrounded by sputum can remain viable in air and dust for lengthy periods. Many viruses (especially nonenveloped) and fungal spores can also withstand long periods of desiccation. Desiccation can be a valuable way to preserve foods because it greatly reduces the amount of water available to support microbial growth.

It is interesting to note that a combination of freezing and drying—**lyophilization** (ly-off″-il-ih-za′-shun)—is a common method of preserving microorganisms and other cells in a viable state for many years. Pure cultures are frozen instantaneously and exposed to a vacuum that rapidly removes the water (it goes right from the frozen state into the vapor state). This method avoids the formation of ice crystals that would damage the cells. Although not all cells survive this process, enough of them do to permit future reconstitution of that culture.

As a general rule, chilling, freezing, and desiccation should not be construed as methods of disinfection or sterilization because their antimicrobial effects are erratic and uncertain, and one cannot be sure that pathogens subjected to them have been killed.

Table 11.3   Moist Heat Methods

Techniques and chemicals that are capable of sterilizing are highlighted with a pink background.

Method	Applications
 **Boiling Water: Disinfection** A simple boiling water bath or chamber can quickly decontaminate items in the clinic and home. Because a single processing at 100°C will not kill all resistant cells, this method can be relied on only for disinfection and not for sterilization. Exposing materials to boiling water for 30 minutes will kill most nonspore-forming pathogens, including resistant species such as the tubercle bacillus and staphylococci. Probably the greatest disadvantage with this method is that the items can be easily recontaminated when removed from the water.	Useful in the home for disinfection of water, materials for babies, food and utensils, bedding, and clothing from the sickroom

 **Pasteurization: Disinfection of Beverages**   Fresh beverages such as milk, fruit juices, beer, and wine are easily contaminated during collection and processing. Because microbes have the potential for spoiling these foods or causing illness, heat is frequently used to reduce the microbial load or destroy pathogens. **Pasteurization** is a technique in which heat is applied to liquids to kill potential agents of infection and spoilage, while at the same time retaining the liquid's flavor and food value.

Milk, wine, beer, other beverages

Ordinary pasteurization techniques require special heat exchangers that expose the liquid to 71.6°C for 15 seconds (flash method) or to 63°C to 66°C for 30 minutes (batch method). The first method is preferable because it is less likely to change flavor and nutrient content, and it is more effective against certain resistant pathogens such as *Coxiella* and *Mycobacterium*. Although these treatments inactivate most viruses and destroy the vegetative stages of 97% to 99% of bacteria and fungi, they do not kill endospores or particularly heat-resistant microbes (mostly nonpathogenic lactobacilli, micrococci, and yeasts). Milk is not sterile after regular pasteurization. In fact, it can contain 20,000 microbes per milliliter or more, which explains why even an unopened carton of milk will eventually spoil. (Newer techniques can also produce *sterile milk* that has a storage life of 3 months. This milk is processed with ultrahigh temperature [UHT]—134°C—for 1 to 2 seconds.) This is not generally considered pasteurization, so we don't consider pasteurization a sterilization method.

| Monday / Tuesday / Wednesday / Thursday | **Nonpressurized Steam**   Selected substances that cannot withstand the high temperature of the autoclave can be subjected to *intermittent sterilization*, also called **tyndallization.** This technique requires a chamber to hold the materials and a reservoir for boiling water. Items in the chamber are exposed to free-flowing steam for 30 to 60 minutes. This temperature is not sufficient to reliably kill endospores, so a single exposure will not suffice. On the assumption that surviving endospores will germinate into less resistant vegetative cells, the items are incubated at appropriate temperatures for 23 to 24 hours, and then again subjected to steam treatment. This cycle is repeated for 3 days in a row. Because the temperature never gets above 100°C, highly resistant endospores that do not germinate may survive even after 3 days of this treatment. | Heat-sensitive culture media, such as those containing sera, egg, or carbohydrates (which can break down at higher temperatures) and some canned foods. Probably not effective in sterilizing items such as instruments and dressings that provide no environment for endospore germination, but it certainly can disinfect them. |

Even though this is sometimes called "intermittent sterilization," sterilization is not guaranteed so we don't consider it a reliable sterilization method.

Table 11.3 Moist Heat Methods (continued)

Method	Applications
Steam Under Pressure: Sterilization At sea level, normal atmospheric pressure is 15 pounds per square inch (psi), or 1 atmosphere. At this pressure, water will boil (change from a liquid to a gas) at 100°C, and the resultant steam will remain at exactly that temperature, which is unfortunately too low to reliably kill all microbes. In order to raise the temperature of steam, the pressure at which it is generated must be increased. As the pressure is increased, the temperature at which water boils and the temperature of the steam produced both rise. For example, at a pressure of 20 psi (5 psi above normal), the temperature of steam is 109°C. As the pressure is increased to 10 psi above normal, the steam's temperature rises to 115°C, and at 15 psi above normal (a total of 2 atmospheres), it will be 121°C. It is not the pressure by itself that is killing microbes but the increased temperature it produces. Such pressure-temperature combinations can be achieved only with a special device that can subject pure steam to pressures greater than 1 atmosphere. Health and commercial industries use an **autoclave** for this purpose, and a comparable home appliance is the pressure cooker. The most efficient pressure-temperature combination for achieving sterilization is 15 psi, which yields 121°C. It is important to avoid overpacking or haphazardly loading the chamber, which prevents steam from circulating freely around the contents and impedes the full contact that is necessary. The duration of the process is adjusted according to the bulkiness of the items in the load (thick bundles of material or large flasks of liquid) and how full the chamber is. The range of holding times varies from 10 minutes for light loads to 40 minutes for heavy or bulky ones; the average time is 20 minutes.	Heat-resistant materials such as glassware, cloth (surgical dressings), metallic instruments, liquids, paper, some media, and some heat-resistant plastics. If items are heat-sensitive (plastic Petri dishes) but will be discarded, the autoclave is still a good choice. However, it is ineffective for sterilizing substances that repel moisture (oils, waxes), or for those that are harmed by it (powders).

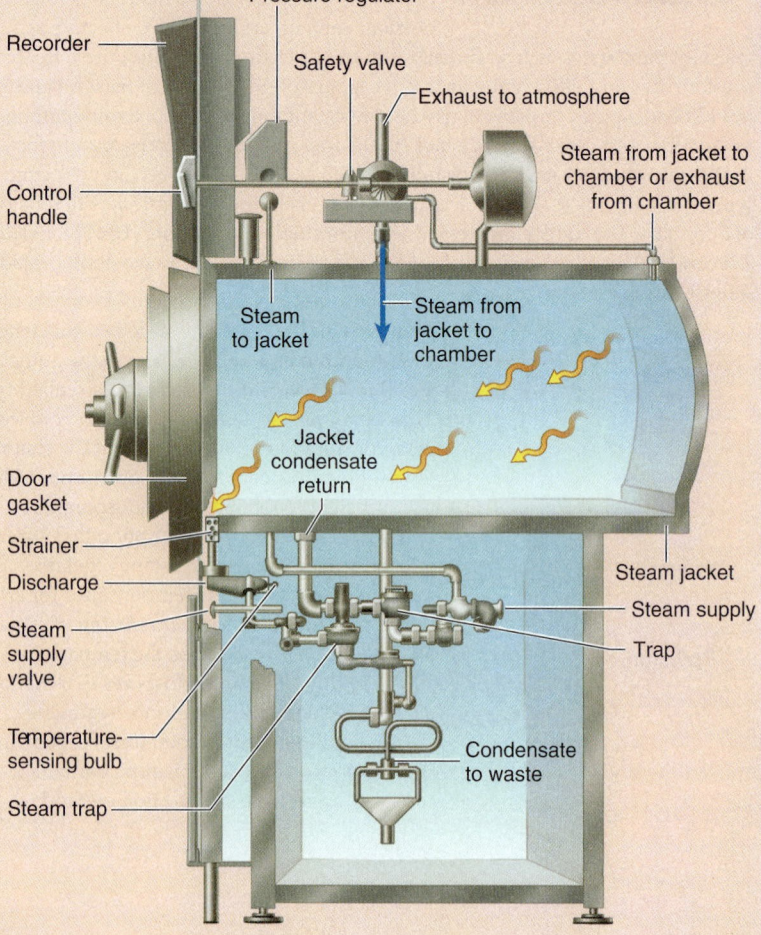

Figure 8.2 on p. 170 from Perkins, Principles and Methods of Sterilization in Health Science, 2nd ed. Courtesy of Charles C. Thomas Publisher, Ltd., Springfield, Illinois.

Table 11.4   Dry Heat Methods
Techniques and chemicals that are capable of sterilizing are highlighted with a pink background.

Method		Applications
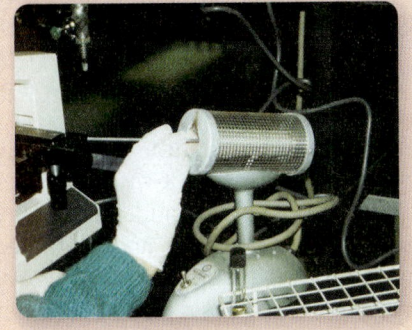	**Incineration** in a flame is perhaps the most rigorous of all heat treatments. The flame of a Bunsen burner reaches 1,870°C at its hottest point, and furnaces/incinerators operate at temperatures of 800°C to 6,500°C. Direct exposure to such intense heat ignites and reduces microbes and other substances to ashes and gas. Incineration of microbial samples on inoculating loops and needles using a Bunsen burner is a very common practice in the microbiology laboratory. This method is fast and effective, but it is also limited to metals and heat-resistant glass materials. This method also presents hazards to the operator (an open flame)	Bunsen burners/small incinerators: laboratory instruments such as inoculating loops. Large incinerators: syringes, needles, culture materials, dressings, bandages, bedding, animal carcasses, and pathology samples.

and to the environment (contaminants on needle or loop often spatter when placed in flame). Tabletop infrared incinerators have replaced Bunsen burners in many labs for these reasons. Large incinerators are regularly employed in hospitals and research labs for complete destruction of infectious materials.

	**The hot-air oven** provides another means of dry-heat sterilization. The so-called *dry oven* is usually electric (occasionally gas) and has coils that radiate heat within an enclosed compartment. Heated, circulated air transfers its heat to the materials in the oven. Sterilization requires exposure to 150°C to 180°C for 2 to 4 hours, which ensures thorough heating of the objects and destruction of endospores.	Glassware, metallic instruments, powders, and oils that steam does not penetrate well. Not suitable for plastics, cotton, and paper, which may burn at the high temperatures, or for liquids, which will evaporate.

Radiation as a Microbial Control Agent

Another way in which energy can serve as an antimicrobial agent is through the use of radiation. **Radiation** is defined as energy emitted from atomic activities and dispersed at high velocity through matter or space. **Figure 11.6** illustrates the different wavelengths of radiation. Although radiation exists in many states and can be described and characterized in various ways, in this discussion we consider only those types suitable for microbial control: gamma rays, X rays, and ultraviolet radiation.

Modes of Action of Ionizing Versus Nonionizing Radiation

The actual physical effects of radiation on microbes can be understood by visualizing the process of **irradiation,** or bombardment with radiation, at the cellular level **(figure 11.7).** When a cell is bombarded by certain waves or particles, its molecules absorb some of the available energy, leading to one of two consequences: (1) If the radiation ejects orbital electrons from an atom, it causes ions to form; this type of radiation is termed **ionizing radiation.** It was previously believed that the most sensitive target for ionizing radiation is DNA, which undergoes mutations on a broad scale, but studies conducted in 2007 suggest that protein damage is the culprit. If proteins are not destroyed, apparently they can almost always repair the DNA. Secondary lethal effects appear to be chemical changes in organelles and the production of toxic substances. Gamma rays, X rays, and high-speed

Figure 11.6 The electromagnetic spectrum, showing different types of radiation.

electrons are all ionizing in their effects. (2) **Nonionizing radiation,** best exemplified by ultraviolet (UV), excites atoms by raising them to a higher energy state, but it does not ionize them. This atomic excitation, in turn, leads to the formation of abnormal bonds within molecules such as DNA and is thus a source of mutations.

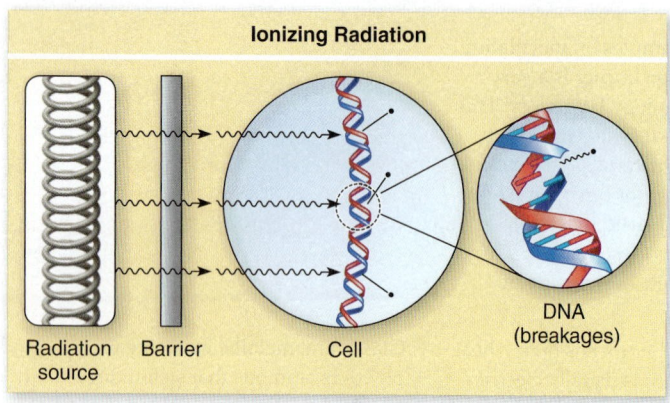

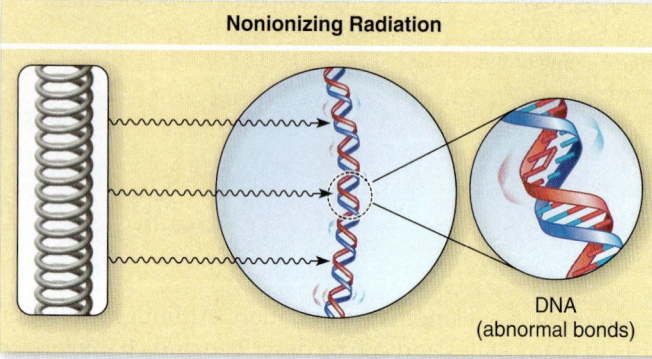

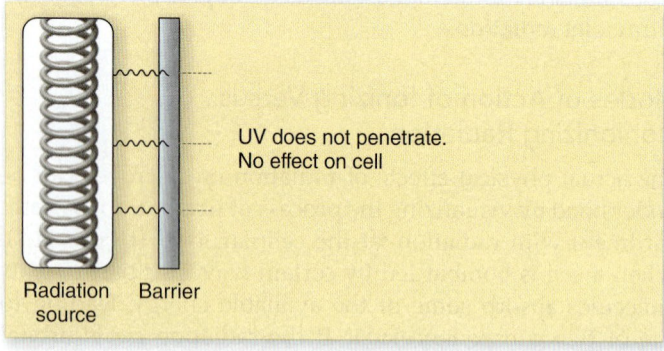

Figure 11.7 Cellular effects of irradiation. **(a)** Ionizing radiation can penetrate a solid barrier, bombard a cell, enter it, and dislodge electrons from molecules. Breakage of DNA creates massive mutations, and damage to proteins prevents them from repairing it. **(b)** Nonionizing radiation enters a cell, strikes molecules, and excites them. The effect on DNA is mutation by formation of abnormal bonds. **(c)** A solid barrier cannot be penetrated by nonionizing radiation.

Ionizing Radiation: Gamma Rays, X Rays, and Cathode Rays

Over the past several years, ionizing radiation has become safer and more economical to use, and its applications have mushroomed. It is a highly effective alternative for sterilizing materials that are sensitive to heat or chemicals **(figure 11.8).** Because it sterilizes without heating, irradiation is a type of **cold** (or low-temperature) **sterilization (table 11.5).**

Devices that emit ionizing rays include gamma-ray machines containing radioactive cobalt, X-ray machines similar to those used in medical diagnosis, and cathode-ray machines. Items are placed in these machines and irradiated for a short time with a carefully chosen dosage. The dosage of radiation is measured in *Grays* (which has replaced the older term, *rads*). Depending on the application, exposure ranges from 5 to 50 kiloGrays (a kiloGray is equal to 1,000 Grays). Although all ionizing radiation can penetrate liquids and most solid materials, gamma rays are most penetrating, X rays are intermediate, and cathode rays are least penetrating.

Applications of Ionizing Radiation

Foods have been subject to irradiation in limited circumstances for more than 50 years. From flour to pork and ground beef to fruits and vegetables, radiation is used to kill not only bacterial pathogens but also insects and worms and even to inhibit the sprouting of white potatoes (see figure 11.8). As soon as radiation is mentioned, however, consumer concern arises that food may be made less nutritious, unpalatable, or even unsafe by its having been subjected to ionizing radiation. But irradiated food has been extensively studied, and each of these concerns has been addressed.

Irradiation may lead to a small decrease in the amount of thiamine (vitamin B_1) in food, but this change is small enough to be inconsequential. The irradiation process does produce short-lived free radical oxidants, which disappear almost immediately (this same type of chemical intermediate is produced through cooking as well). Certain foods do

Figure 11.8 Foods commonly irradiated. Regulations dictate that the universal symbol for irradiation must be affixed to all irradiated materials.

Table 11.5 Radiation Methods

Techniques and chemicals that are capable of sterilizing are highlighted with a pink background.

Ionizing Radiation: Gamma Rays and X Rays Ionizing radiation is a highly effective alternative for sterilizing materials that are sensitive to heat or chemicals. Devices that emit ionizing rays include gamma-ray machines containing radioactive cobalt.

Nonionizing Radiation: Ultraviolet Rays Ultraviolet (UV) radiation ranges in wavelength from approximately 100 to 400 nm. It is most lethal from 240 to 280 nm (with a peak at 260 nm).

not irradiate well and are not good candidates for this type of antimicrobial control. The whites of eggs become milky and liquid, grapefruits get mushy, and alfalfa seeds do not germinate properly. Lastly, it is important to remember that food is not made radioactive by the irradiation process; many studies, in both animals and humans, have concluded that there are no ill effects from eating irradiated food. In fact, NASA relies on irradiated meat for its astronauts.

Nonionizing Radiation: Ultraviolet Rays

Ultraviolet (UV) radiation ranges in wavelength from approximately 100 nm to 400 nm. It is most lethal from 240 nm to 280 nm (with a peak at 260 nm). In everyday practice, the source of UV radiation is the germicidal lamp, which generates radiation at 254 nm. Owing to its lower energy state, UV radiation is not as penetrating as ionizing radiation. Because UV radiation passes readily through air, slightly

through liquids, and only poorly through solids, the object to be disinfected must be directly exposed to it for full effect.

As UV radiation passes through a cell, it is initially absorbed by DNA. Specific molecular damage occurs on the pyrimidine bases (thymine and cytosine), which form abnormal linkages with each other called **pyrimidine dimers (figure 11.9).** These bonds occur between adjacent bases on the same DNA strand and interfere with normal DNA replication and transcription. The results are inhibition of growth and cellular death. In addition to altering DNA directly, UV radiation also disrupts cells by generating toxic photochemical products called free radicals. These highly reactive molecules interfere with essential cell processes by binding to DNA, RNA, and proteins. Ultraviolet rays are a powerful tool for destroying fungal cells and spores, bacterial vegetative cells, protozoa, and viruses. Bacterial endospores are about 10 times more resistant to radiation than are vegetative cells, but they can be killed by increasing the time of exposure.

Ultraviolet treatment has proved effective in freeing vaccines and plasma from contaminants. The surfaces of solid, nonporous materials such as walls and floors, as well as meat, nuts, tissues for grafting, and drugs, have been successfully disinfected with UV.

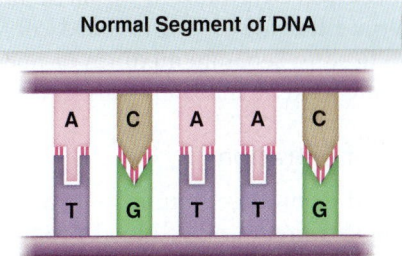

Figure 11.9 Formation of pyrimidine dimers by the action of ultraviolet (UV) radiation. This shows what occurs when two adjacent thymine bases on one strand of DNA are induced by UV rays to bond laterally with each other. The result is a thymine dimer (shown in greater detail). Dimers can also occur between adjacent cytosines and thymine and cytosine bases. If they are not repaired, dimers can prevent that segment of DNA from being correctly replicated or transcribed. Massive dimerization is lethal to cells.

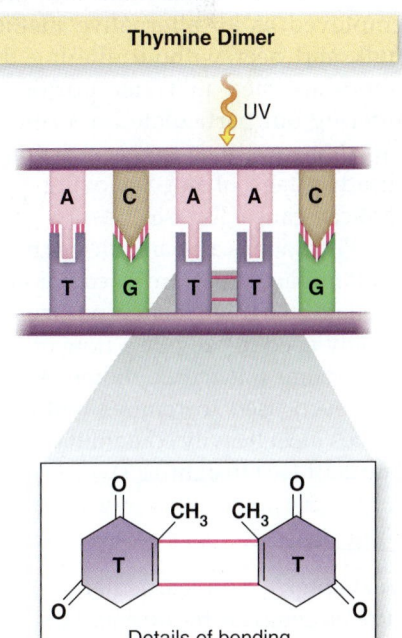

One major disadvantage of UV is its poor powers of penetration through solid materials such as glass, metal, cloth, plastic, and even paper. Another drawback to UV is the damaging effect of overexposure on human tissues, including sunburn, retinal damage, cancer, and skin wrinkling.

Decontamination by Filtration: Techniques for Removing Microbes

Filtration is an effective method to remove microbes from air and liquids. In practice, a fluid is strained through a filter with openings large enough for the fluid to pass through but too small for microorganisms to pass through **(figure 11.10*a*).**

Most modern microbiological filters are thin membranes of cellulose acetate, polycarbonate, and a variety of plastic materials (Teflon, nylon) whose pore size can be carefully controlled and standardized. Ordinary substances such as charcoal, diatomaceous earth, or unglazed porcelain are also used in some applications. Viewed microscopically, most filters are perforated by very precise, uniform pores **(figure 11.10*b*).** The pore diameters vary from coarse (8 microns) to ultrafine (0.02 micron), permitting selection of the minimum particle size to be trapped. Those with even smaller pore diameters permit true sterilization by removing viruses, and some will even remove large proteins. A sterile liquid filtrate is typically produced by suctioning the liquid through a sterile filter into a presterilized container. These filters are also used to separate mixtures of microorganisms and to enumerate bacteria in water analysis (see chapter 25).

Applications of Filtration

Filtration is used to prepare liquids that cannot withstand heat, including serum and other blood products, vaccines, drugs, IV fluids, enzymes, and media. Filtration has been employed as an alternative method for decontaminating milk and beer without altering their flavor. It is also an important step in water purification. Its use extends to filtering out particulate impurities (crystals, fibers, and so on) that can cause severe reactions in the body. It has the disadvantage of not removing soluble molecules (toxins) that can cause disease.

Filtration is also an efficient means of removing airborne contaminants that are a common source of infection and spoilage. High-efficiency particulate air (HEPA) filters are widely used to provide a flow of decontaminated air to hospital rooms and sterile rooms. A vacuum with a HEPA filter was even used to remove anthrax spores from the Senate offices most heavily contaminated after the terrorist attack in late 2001 (see Insight 11.1).

Osmotic Pressure

In chapter 7, you learned about the effects of osmotic pressure on cells (see figure 7.5). This fact has long been exploited as a means of preserving food. Adding large amounts of salt or sugar to foods creates a hypertonic environment for

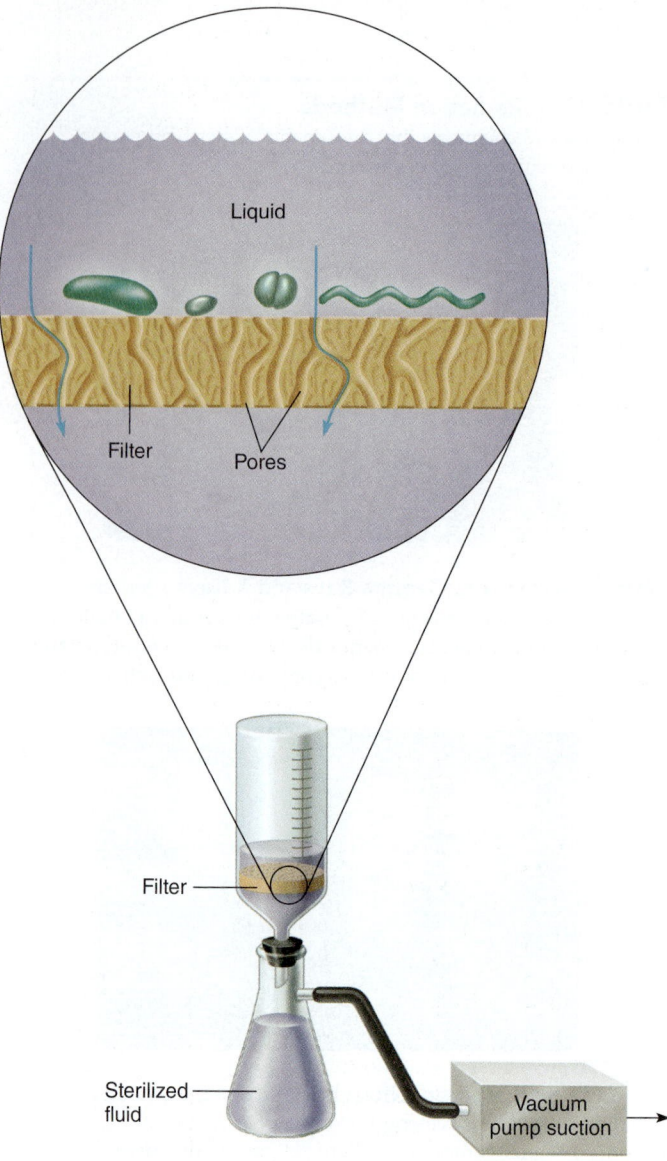

(a)

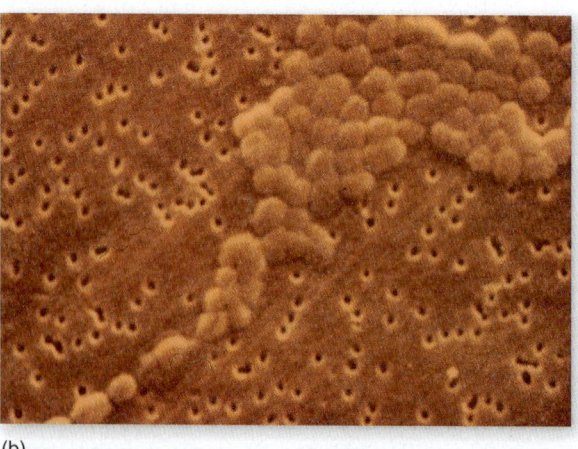

(b)

Figure 11.10 Membrane filtration. **(a)** Vacuum assembly for achieving filtration of liquids through suction. Inset shows filter as seen in cross section, with tiny passageways (pores) too small for the microbial cells to enter but large enough for liquid to pass through. **(b)** Scanning electron micrograph of filter, showing relative size of pores and bacteria trapped on its surface (5,900×).

bacteria in the foods, causing plasmolysis and making it impossible for the bacteria to multiply. People knew that these techniques worked long before the discovery of bacteria. This is why meats are "cured," or treated with high salt concentrations so they can be kept for long periods without refrigeration. High sugar concentrations in foods like jellies have the same effect.

11.2 Learning Outcomes—Assess Your Progress

5. Name six methods of physical control of microorganisms.
6. Compare and contrast moist and dry heat methods of control, and identify multiple examples of each.
7. Define *thermal death time* and *thermal death point*, and describe their role in proper sterilization.
8. Explain four different methods of moist heat control.
9. Explain two methods of dry heat control.
10. Identify advantages and disadvantages of cold treatment and desiccation.
11. Differentiate between the two types of radiation control methods, providing an application of each.
12. Outline the process of filtration and describe its two advantages in microbial control. Explain how filtration functions as a control method.
13. Identify some common uses of osmotic pressure as a control method.

11.3 Chemical Agents in Microbial Control

Chemical control of microbes probably emerged as a serious science in the 1800s, when physicians used chloride of lime and iodine solutions to treat wounds and to wash their hands before surgery. At the present time, more than 10,000 different antimicrobial chemical agents are manufactured; probably 1,000 of them are used routinely in the health care arena and the home. A genuine need exists to avoid infection and spoilage, but the abundance of products available to "kill germs," "disinfect," "antisepticize," "clean and sanitize," "deodorize," "fight plaque," and "purify the air" indicates a preoccupation with eliminating microbes from the environment that, at times, seems excessive.

Antimicrobial chemicals occur in the liquid, gaseous, or even solid state, and they range from disinfectants and antiseptics to sterilants and preservatives (chemicals that inhibit the deterioration of substances). For the sake of convenience (and sometimes safety), many solid or gaseous antimicrobial chemicals are dissolved in water, alcohol, or a mixture of the two to produce a liquid solution. Solutions containing pure water as the solvent are termed **aqueous**, whereas those dissolved in pure alcohol or water-alcohol mixtures are termed **tinctures.**

Table 11.6 provides an overview of chemicals that are routinely used in health care.

Table 11.6 Qualities of Chemical Agents Used in Health Care

Agent	Target Microbes	Level of Activity	Toxicity	Comments
Chlorine	Sporicidal (slowly)	Intermediate	Gas is highly toxic; solution irritates skin	Inactivated by organics; unstable in sunlight
Phenolics	Some bacteria, viruses, fungi	Low to intermediate	Can be absorbed by skin; can cause CNS damage	Poor solubility; expensive
Chlorhexidine*	Most bacteria, some viruses, fungi	Low to intermediate	Low toxicity	Fast-acting, mild, has residual effects
Alcohols	Most bacteria, viruses, fungi	Intermediate	Toxic if ingested; a mild irritant; dries skin	Flammable, fast-acting
Hydrogen peroxide,* stabilized	Sporicidal	High	Toxic to eyes; toxic if ingested	Improved stability; works well in organic matter
Quaternary ammonium compounds	Some bactericidal, virucidal, fungicidal activity	Low	Irritating to mucous membranes; poisonous if taken internally	Weak solutions can support microbial growth; easily inactivated
Soaps	Certain very sensitive species	Very low	Nontoxic; few if any toxic effects	Used for removing soil, oils, debris, and reducing load
Silver nitrate	Bactericidal	Low	Toxic, irritating	Discolors skin
Glutaraldehyde*	Sporicidal	High	Can irritate skin; toxic if absorbed	Not inactivated by organic matter; unstable
Ethylene oxide gas*	Sporicidal	High	Very dangerous to eyes, lungs; carcinogenic	Explosive in pure state; good penetration; materials must be aerated

*These chemicals approach the ideal by having many of the following characteristics: broad-spectrum, low toxicity, fast action, penetrating abilities, residual effects, stability, potency in organic matter, and solubility.

Selecting a Microbicidal Chemical

The choice and appropriate use of antimicrobial chemical agents are of constant concern in medicine and dentistry. Although actual clinical practices of chemical decontamination vary widely, some desirable qualities in a germicide have been identified, including

1. rapid action even in low concentrations,
2. solubility in water or alcohol and long-term stability,
3. broad-spectrum microbicidal action without toxicity to human and animal tissues,
4. penetration of inanimate surfaces to sustain a cumulative or persistent action,
5. resistance to becoming inactivated by organic matter,
6. noncorrosive or nonstaining properties,
7. sanitizing and deodorizing properties, and
8. affordability and ready availability.

As yet, no chemical can completely fulfill all of those requirements, but glutaraldehyde and hydrogen peroxide approach this ideal. At the same time, we should question the rather overinflated claims made about certain commercial agents such as mouthwashes and disinfectant air sprays. In 2012, researchers identified a chemical, polyhexamethylene-guanidine hydrochloride (PHMGH), that seems to satisfy most of these requirements and destroys spores. After further testing, it may become an important part of the chemical arsenal.

Germicides are evaluated in terms of their effectiveness in destroying microbes in medical and dental settings. Echoing the language we used earlier in the chapter, referring to medical devices as critical, semicritical, or noncritical, three levels of chemical decontamination procedures exist. These are *high, intermediate,* and *low.* High-level germicides kill endospores and, if properly used, are sterilants. Critical items, for example, are catheters, heart-lung equipment, and implants. These are not heat-sterilizable and are intended to enter body tissues during medical procedures. Intermediate-level germicides kill fungal (but not bacterial) spores, resistant pathogens such as the tubercle bacillus, and viruses. They are used to disinfect semicritical items (respiratory equipment, thermometers). Low levels of disinfection eliminate only vegetative bacteria, vegetative fungal cells, and some viruses. They are used to clean noncritical materials such as electrodes, straps, and pieces of furniture that touch the skin surfaces but not the mucous membranes.

Factors Affecting the Microbicidal Activity of Chemicals

Factors that control the effect of a germicide include the nature of the microorganisms being treated, the nature of the material being treated, the degree of contamination, the time of exposure, and the strength and chemical action of the germicide. The strength of a germicide also plays a role in its ability to act upon a microbial population. The modes of action of most germicides are to attack the cellular targets discussed earlier: proteins, nucleic acids, the cell wall, and the cell membrane.

A chemical's strength or concentration is expressed in various ways, depending on convention and the method of preparation. The content of many chemical agents can be expressed by more than one notation. In dilutions, a small volume of the liquid chemical (solute) is diluted in a larger volume of solvent to achieve a certain ratio. For example, a common laboratory phenolic disinfectant such as Lysol is usually diluted 1:200; that is, one part of chemical has been added to 200 parts of water by volume. Solutions such as chlorine that are effective in very diluted concentrations are expressed in parts per million (ppm). In percentage solutions, the solute is added to water by weight or volume to achieve a certain percentage in the solution. Alcohol, for instance, is used in percentages ranging from 50% to 95%. In general, solutions of low dilution or high percentage have more of the active chemical (are more concentrated) and tend to be more germicidal, but expense and potential toxicity can necessitate using the minimum strength that is effective.

As previously discussed, most compounds require adequate contact time to allow the chemical to penetrate and to act on the microbes present. The composition of the material being treated must also be considered as it may greatly impact the effectiveness of these germicidal agents. Smooth, solid objects are more reliably disinfected than are those with pores or pockets that can trap soil. An item contaminated with common biological matter such as serum, blood, saliva, pus, fecal material, or urine presents a problem in disinfection. Large amounts of organic material can hinder the penetration of a disinfectant and, in some cases, can form bonds that possibly reduce its activity. Adequate cleaning of instruments and other reusable materials ensures that the germicidal agent will be able to do its job.

Germicidal Categories According to Chemical Group

Several general groups of chemical compounds are widely used for antimicrobial purposes in medicine and commerce. Prominent agents include halogens, heavy metals, alcohols, phenolic compounds, oxidizers, aldehydes, detergents, and gases. These groups are surveyed in the following section from the standpoint of each agent's specific forms, modes of action, indications for use, and limitations.

The Halogen Antimicrobial Chemicals

The **halogens** are fluorine, bromine, chlorine, and iodine, a group of nonmetallic elements, all of which are found in group VII of the periodic table. These elements are highly effective components of disinfectants and antiseptics because they are microbicidal and not just microbistatic, and they are sporicidal with longer exposure. For these reasons, halogens are the active ingredients in nearly one-third of all antimicrobial chemicals currently marketed.

Chlorine and Its Compounds   Chlorine has been used for disinfection and antisepsis for approximately 200 years. The major forms used in microbial control are liquid and gaseous chlorine (Cl_2), hypochlorites (ClO^1), and chloramines (NH_2Cl). In solution, these compounds combine with water and release hypochlorous acid (HOCl), which oxidizes the sulfhydryl (S—H) group on the amino acid cysteine and interferes with disulfide (S—S) bridges on numerous enzymes. The resulting denaturation of the enzymes is permanent and suspends metabolic reactions. Chlorine kills not only bacteria and endospores but also fungi and viruses. Chlorine compounds are less effective if exposed to light, alkaline pH, and excess organic matter.

Chlorine Compounds in Disinfection and Antisepsis
Gaseous and liquid chlorine are used almost exclusively for large-scale disinfection of drinking water, sewage, and wastewater from such sources as agriculture and industry. Chlorination to a concentration of 0.6 to 1.0 parts of chlorine per million parts of water will usually ensure that water is safe to drink. This treatment rids the water of most pathogenic vegetative microorganisms without unduly affecting its taste (some people may debate this). In chapter 22, however, you will learn about pathogenic organisms that can survive water chlorination.

Hypochlorites are perhaps the most extensively used of all chlorine compounds. The scope of applications is broad, including sanitization and disinfection of food equipment in dairies, restaurants, and canneries and treatment of swimming pools, spas, drinking water, and even fresh foods. Hypochlorites are used in allied health areas to treat wounds and to disinfect equipment, bedding, and instruments. Common household bleach is a weak solution (5%) of sodium hypochlorite that serves as an all-around disinfectant, deodorizer, and stain remover.

Chloramines (dichloramine, halazone) are being employed more frequently as alternatives to pure chlorine in treating water supplies. Because standard chlorination of water is now believed to produce unsafe levels of cancer-causing substances such as trihalomethanes, some water districts have been directed by federal agencies to adopt chloramine treatment of water supplies. Chloramines also serve as sanitizers and disinfectants, and for treating wounds and skin surfaces.

Iodine and Its Compounds   Iodine is a pungent chemical that forms brown-colored solutions when dissolved in water or alcohol. The two primary iodine preparations are *free iodine* in solution (I_2) and *iodophors*. Iodine rapidly penetrates the cells of microorganisms, where it apparently disturbs a variety of metabolic functions by interfering with the hydrogen and disulfide bonding of proteins (a mode of action similar to chlorine). All classes of microorganisms are killed by iodine if proper concentrations and exposure times are used. Iodine activity is not as adversely affected by organic matter and pH as chlorine is.

Applications of Iodine Solutions   Aqueous iodine contains 2% iodine and 2.4% sodium iodide; it is used as a topical antiseptic before surgery and occasionally as a treatment for burned and infected skin. A stronger iodine solution (5% iodine and 10% potassium iodide) is used primarily as a disinfectant for plastic items, rubber instruments, cutting blades, thermometers, and other inanimate items. Iodine tincture is a 2% solution of iodine and sodium iodide in 70% alcohol that can be used in skin antisepsis. Because iodine can be extremely irritating to the skin and toxic when absorbed, strong aqueous solutions and tinctures (5% to 7%) are no longer considered safe for routine antisepsis. Iodine tablets are available for disinfecting water during emergencies or destroying pathogens in impure water supplies.

Iodophors are complexes of iodine and alcohol. This formulation allows the slow release of free iodine and increases its degree of penetration. These compounds have largely replaced free iodine solutions in medical antisepsis because they are less prone to staining or irritating tissues. Common iodophor products marketed as Betadine, Povidone (PVP), and Isodine contain 2% to 10% of available iodine. They are used to prepare skin and mucous membranes for surgery and injections, in surgical hand scrubs, to treat burns, and to disinfect equipment and surfaces. Although pure iodine is toxic to the eye, studies show that Betadine solution is an effective means of preventing eye infections in newborn infants, and it may replace antibiotics and silver nitrate as the method of choice.

Phenol and Its Derivatives

Phenol (carbolic acid) is a poisonous compound derived from the distillation of coal tar. First adopted by Joseph Lister in 1867 as a surgical germicide, phenol was the major antimicrobial chemical until other phenolics with fewer toxic and irritating effects were developed. Solutions of phenol are now used only in certain limited cases, but phenol remains one standard against which other phenolic disinfectants are rated. The *phenol coefficient* quantitatively compares a chemical's antimicrobial properties to those of phenol. Substances chemically related to phenol are often referred to as phenolics. Hundreds of these chemicals are now available.

Phenolics consist of one or more aromatic carbon rings with added functional groups **(figure 11.11)**. Among the most important are alkylated phenols (cresols), chlorinated phenols, and bisphenols. In high concentrations, they are cellular poisons, rapidly disrupting cell walls and membranes and precipitating proteins; in lower concentrations, they inactivate certain critical enzyme systems. The phenolics are strongly microbicidal and will destroy vegetative bacteria (including the tuberculosis bacterium), fungi, and most viruses (not hepatitis B), but they are not reliably sporicidal. Their ability to act in the presence of organic matter and their detergent actions contribute to their usefulness. Unfortunately, the toxicity of many of the phenolics makes them too dangerous to use as antiseptics.

Applications of Phenolics   Phenol itself is still used for general disinfection of drains, cesspools, and animal quarters, but it is seldom applied as a medical germicide. The

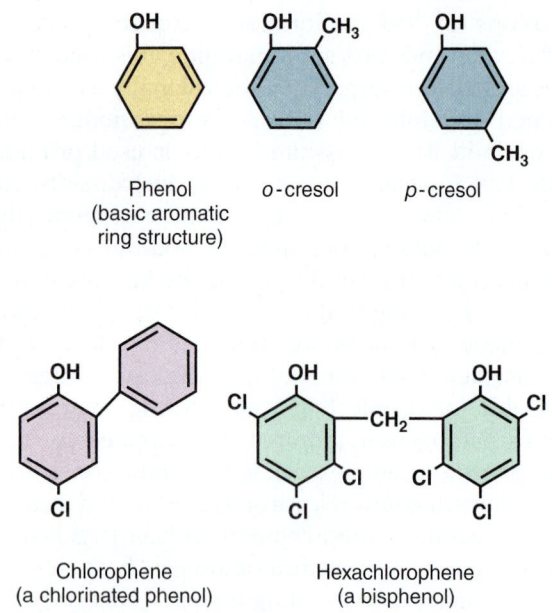

Figure 11.11 Some phenolics. All contain a basic aromatic ring, but they differ in the types of additional compounds such as Cl and CH₃.

cresols are simple phenolic derivatives that are combined with soap for intermediate or low levels of disinfection in the hospital. Lysol and creolin, in a 1% to 3% emulsion, are common household versions of this type.

The bisphenols are also widely employed in commerce, clinics, and the home. One type, orthophenyl phenol, is the major ingredient in disinfectant aerosol sprays. This same phenolic is also found in some proprietary compounds (Lysol) often used in hospital and laboratory disinfection. One particular bisphenol, hexachlorophene, was once a common additive of cleansing soaps (pHisoHex) used in the hospital and home. When hexachlorophene was found to be absorbed through the skin and a cause of neurological damage, it was no longer available without a prescription. It is occasionally used to control outbreaks of skin infections.

Perhaps the most widely used phenolic is *triclosan,* chemically known as dichlorophenoxyphenol. It is the antibacterial compound added to dozens of products, from soaps to kitty litter. It acts as both disinfectant and antiseptic and is broad-spectrum in its effects. Unfortunately, new research has indicated that widespread use of triclosan can lead to the development of resistance—both to it and to antibiotics—in microbes.

Chlorhexidine

The compound chlorhexidine (Hibiclens, Hibitane, Peridex) is a complex organic base containing chlorine and two phenolic rings. Its mode of action targets both cell membranes (lowering surface tension until selective permeability is lost) and protein structure (causing denaturation). At moderate to high concentrations, it is bactericidal for both gram-positive and gram-negative bacteria but inactive against endospores. Its effects on viruses and fungi vary. It possesses distinct advantages over many other antiseptics because of its mildness, low

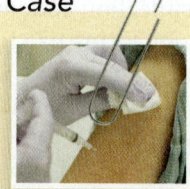
toxicity, and rapid action; it is not absorbed into deeper tissues to any extent. Alcoholic or aqueous solutions of chlorhexidine are now commonly used for hand scrubbing, preparing skin sites for surgical incisions and injections, and whole-body washing. Chlorhexidine solution also serves as an obstetric antiseptic, a neonatal wash, a wound degermer, a mucous membrane irrigant, and a preservative for eye solutions. It is sold in many over-the-counter mouthwashes as well.

Alcohols as Antimicrobial Agents

Alcohols are colorless hydrocarbons with one or more —OH functional groups. Of several alcohols available, only ethyl and isopropyl are suitable for microbial control. Methyl alcohol is not particularly microbicidal, and more complex alcohols are either poorly soluble in water or too expensive for routine use. Alcohols are employed alone in aqueous solutions or as solvents for tinctures (iodine, for example).

Alcohol's mechanism of action depends in part upon its concentration. Concentrations of 50% and higher dissolve membrane lipids, disrupt cell surface tension, and compromise membrane integrity. Alcohol that has entered the cytoplasm denatures proteins through coagulation but only in alcohol-water solutions of 50% to 95%. Alcohol is the exception to the rule that higher concentrations of an antimicrobial chemical have greater microbicidal activity. Because water is needed for proteins to coagulate, alcohol shows a greater microbicidal activity at 70% concentration (that is, 30% water) than at 100% (0% water). Absolute alcohol (100%) dehydrates cells and inhibits their growth but is generally not a protein coagulant.

Although useful in intermediate- to low-level germicidal applications, alcohol does not destroy bacterial endospores at room temperature. Alcohol can, however, destroy resistant vegetative forms, including tuberculosis bacteria and fungal spores, provided the time of exposure is adequate. Alcohol is generally more effective in inactivating enveloped viruses than nonenveloped viruses such as poliovirus and hepatitis A virus.

Applications of Alcohols

Ethyl alcohol, also called ethanol or grain alcohol, is known for being germicidal, nonirritating, and inexpensive. Solutions of 70% to 95% are routinely used as skin degerming agents because the surfactant action removes skin oil, soil, and some microbes sheltered in deeper skin layers. One limitation to its effectiveness is the rate at which it evaporates. Ethyl alcohol is occasionally used to disinfect electrodes, face masks, and thermometers, which are first cleaned and then soaked in alcohol for 15 to 20 minutes. Most alcohol-based hand sanitizers and foams contain 60% to 70% ethyl alcohol and are more resistant to evaporation, but to be microbicidal they must still be exposed to skin for 30 seconds. Isopropyl alcohol, sold as rubbing alcohol, is even more microbicidal and less expensive than ethanol, but these benefits must be weighed against its toxicity. It must be used with caution in disinfection or skin cleansing, because inhalation of its vapors can adversely affect the nervous system.

Hydrogen Peroxide and Related Germicides

Hydrogen peroxide (H_2O_2) is a colorless, caustic liquid that decomposes in the presence of light, metals, or catalase into water and oxygen gas. The germicidal effects of hydrogen peroxide are due to the direct and indirect actions of oxygen. Oxygen forms hydroxyl free radicals ($\cdot OH$), which, like the superoxide radical (see chapter 7), are highly toxic and reactive to cells. Although most microbial cells produce catalase to inactivate the metabolic hydrogen peroxide, it cannot neutralize the amount of hydrogen peroxide entering the cell during disinfection and antisepsis. Hydrogen peroxide is bactericidal, virucidal, fungicidal, and, in higher concentrations, sporicidal.

Applications of Hydrogen Peroxide

As an antiseptic, 3% hydrogen peroxide serves a variety of needs, including skin and wound cleansing, bedsore care, and mouthwashing. It is especially useful in treating infections by anaerobic bacteria because of the lethal effects of oxygen on these forms. Hydrogen peroxide is also a versatile disinfectant for soft contact lenses, surgical implants, plastic equipment, utensils, bedding, and room interiors.

A number of clinical procedures involve delicate reusable instruments such as endoscopes and dental handpieces. Because these devices can become heavily contaminated by tissues and fluids, they need to undergo sterilization, not just disinfection, between patients to prevent transmission of infections such as hepatitis, tuberculosis, and genital warts. These very effective and costly diagnostic tools (a colonoscope may cost up to $30,000) have created another dilemma. They may trap infectious agents where they cannot be easily removed, and they are delicate, complex, and difficult to clean. Traditional methods are either too harsh (heat) to protect the instruments from damage or too slow (ethylene oxide) to sterilize them in a timely fashion between patients. The need for effective rapid sterilization has led to the development of low-temperature sterilizing cabinets that contain liquid chemical sterilants **(figure 11.12)**. The major types of chemical sterilants used in these machines are powerful oxidizing agents such as hydrogen peroxide (35%) and peracetic acid (35%) that penetrate into delicate machinery, kill the most resistant microbes, and do not corrode or damage the working parts.

Vaporized hydrogen peroxide can also be used as a sterilant in enclosed areas. Hydrogen peroxide plasma sterilizers exist for those applications involving small industrial or medical items. For larger enclosed spaces, such as isolators and pass-through rooms, peroxide generators can be used to fill a room with hydrogen peroxide vapors at concentrations high enough to be sporicidal.

Another compound with effects similar to those of hydrogen peroxide is ozone (O_3), used to disinfect air, water, and industrial air conditioners and cooling towers.

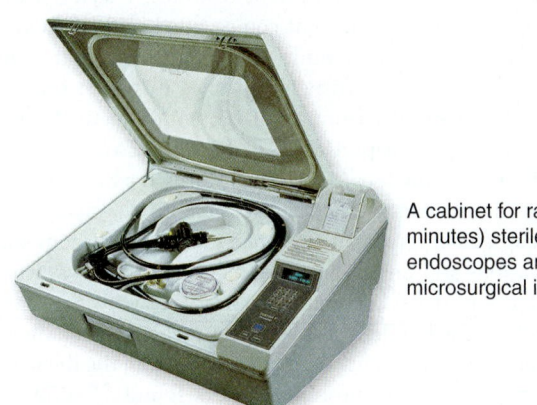

A cabinet for rapid (within 30 minutes) sterile processing of endoscopes and other microsurgical instruments

Figure 11.12 Sterile processing of invasive equipment protects patients.

Chemicals with Surface Action: Detergents

Detergents are polar molecules that act as **surfactants.** Most anionic detergents have limited microbicidal power. This includes most soaps. Much more effective are positively charged (cationic) detergents, particularly the quaternary ammonium compounds (usually shortened to *quats*).

The activity of cationic detergents arises from the amphipathic (two-headed) nature of the molecule. The positively charged end binds well with the predominantly negatively charged bacterial surface proteins, while the long, uncharged hydrocarbon chain allows the detergent to disrupt the cell membrane **(figure 11.13).** Eventually, the cell membrane loses selective permeability, leading to the death of the cell. Several other effects are seen, but the loss of integrity of the cell membrane is most important.

The effects of detergents are varied. When used at high enough concentrations, quaternary ammonium compounds are effective against some gram-positive bacteria, viruses, fungi, and algae. In low concentrations, they exhibit only microbistatic effects. Drawbacks to the quats include their ineffectiveness against the tuberculosis bacterium, hepatitis virus, *Pseudomonas*, and endospores at any concentration. Furthermore, their activity is greatly reduced in the presence of organic matter and they function best in alkaline solutions. As a result of these limitations, quats are rated only for low-level disinfection in the clinical setting.

Applications of Detergents and Soaps   Quaternary ammonium compounds **(quats)** include benzalkonium chloride, Zephiran, and cetylpyridinium chloride (Ceepryn). In dilutions ranging from 1:100 to 1:1,000, quats are mixed with cleaning agents to simultaneously disinfect and sanitize floors, furniture, equipment surfaces, and restrooms. They are used to

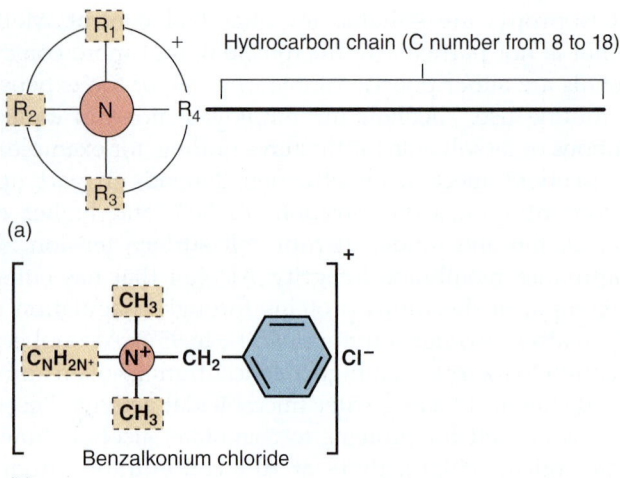

(a)

(b)

Figure 11.13 The structure of detergents. **(a)** In general, detergents are polar molecules with a positively charged head and at least one long, uncharged hydrocarbon chain. The head contains a central nitrogen nucleus with various alkyl (R) groups attached. **(b)** A common quaternary ammonium detergent, benzalkonium chloride.

clean restaurant eating utensils, food-processing equipment, dairy equipment, and clothing. They are common preservatives for ophthalmic solutions and cosmetics. Their level of disinfection is far too low for disinfecting medical instruments.

Soaps are alkaline compounds made by combining the fatty acids in oils with sodium or potassium salts. In usual practice, soaps are only weak microbicides, and they destroy only highly sensitive forms such as the agents of gonorrhea, meningitis, and syphilis. The common hospital pathogen *Pseudomonas* is so resistant to soap that various species grow abundantly in soap dishes **(Insight 11.3).**

INSIGHT 11.3   Dispensing More Than Just Soap

What do you do after you use the restroom? The answer should be obvious: Wash your hands. Hand washing is essential for reducing bacterial counts on your hands and preventing disease transmission, especially in a health care setting. But the next time you wash your hands in a public restroom, check out the soap dispenser. Is it the kind that comes filled with individually packaged refills, or is it a bulk-refillable dispenser? It turns out that the bulk-refillable dispensers (ones in which soap is poured into the dispenser from a bulk container) actually dispense more than just soap: They can dispense *Escherichia coli* and other fecal bacteria, too. In a study published by the journal *Applied and Environmental Microbiology*, scientists tested bulk-refillable soap dispensers in an elementary school bathroom and in other locations and found that all of the soap dispensers were contaminated with fecal bacteria. After washing with the soap, researchers found a 26-fold *increase* in gram-negative bacteria on the hands of the students and staff using soap from the dispensers. In another study, Japanese researchers found 17 different species of bacteria on soaps from public-use restrooms, including *Klebsiella pneumoniae, Serratia marcescens, Enterobacter* species, and

Pseudomonas species. In both studies, bacterial loads were measured at one million cells per milliliter of soap.

A number of disease outbreaks have been linked to contaminated soap dispensers. As a result, the Centers for Disease Control and Prevention (CDC) has issued guidelines for hand soaps used in health care settings. Along with guidelines for the types of hand-washing products to be used, the CDC states, "Do not add soap to a partially empty soap dispenser. This practice of 'topping off' dispensers can lead to bacterial contamination of soap." In the elementary school study noted, when bulk-refillable dispensers were replaced with single-use sealed-soap refills, none of them were found to be contaminated after a year of use.

Bulk-refillable dispenser

Source: 2011. *Appl. Env. Microb.* 77, pp. 2898–2904.

Soaps function primarily as cleansing agents and sanitizers in industry and the home. The superior sudsing and wetting properties of soaps help to mechanically remove large amounts of surface soil, greases, and other debris that contains microorganisms. Soaps gain greater germicidal value when mixed with agents such as chlorhexidine or iodine. They can be used for cleaning instruments before heat sterilization, degerming patients' skin, routine hand washing by medical and dental personnel, and preoperative hand scrubbing. Vigorously brushing the hands with germicidal soap over a 15-second period is an effective way to remove dirt, oil, and surface contaminants as well as some resident microbes, but it will never sterilize the skin (**figure 11.14**).

Heavy Metal Compounds

Various forms of the metallic elements mercury, silver, gold, copper, arsenic, and zinc have been applied in microbial control over several centuries. These are often referred to as *heavy metals* because of their relatively high atomic weight. However, from this list, only preparations containing mercury and silver still have any significance as germicides. Although some metals (zinc, iron) are actually needed in small concentrations as cofactors on enzymes, the higher molecular weight metals (mercury, silver, gold) can be very toxic, even in minute quantities (parts per million). This property of having antimicrobial effects in exceedingly small amounts is called an **oligodynamic** (ol'-ih-goh-dy-nam'-ik) **action (figure 11.15)**. Heavy metal germicides contain either an inorganic or an organic metallic salt, and

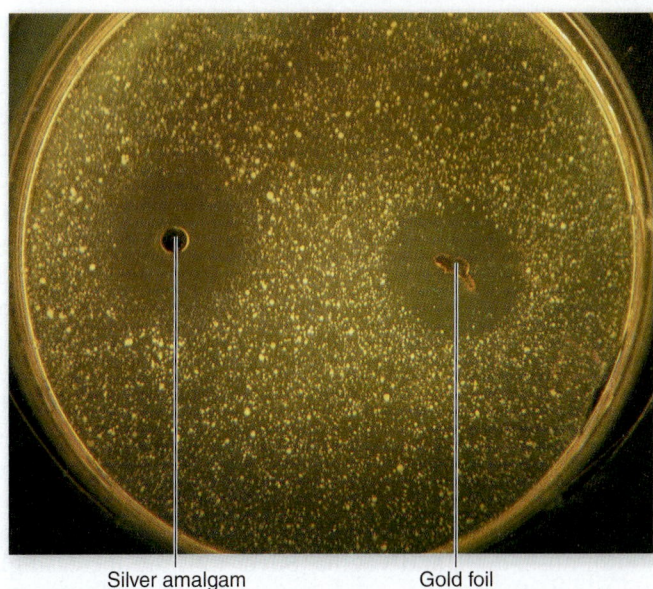

Silver amalgam   Gold foil

Figure 11.15 Demonstration of the oligodynamic action of heavy metals. A pour plate inoculated with saliva has small fragments of heavy metals pressed lightly into it. During incubation, clear zones indicating growth inhibition develop around both fragments. The slightly larger zone surrounding the amalgam (used in tooth fillings) probably reflects the synergistic effect of the silver and mercury it contains.

they come in the form of aqueous solutions, tinctures, ointments, or soaps.

Mercury, silver, and most other metals exert microbicidal effects by binding onto functional groups of proteins and inactivating them, rapidly bringing metabolism to a standstill (see figure 11.5d). This mode of action can destroy many types of microbes, including vegetative bacteria, fungal cells and spores, algae, protozoa, and viruses (but not endospores).

Unfortunately, there are several drawbacks to using metals in microbial control:

1. Metals can be very toxic to humans if ingested, inhaled, or absorbed through the skin, even in small quantities, for the same reasons that they are toxic to microbial cells.
2. They often cause allergic reactions.
3. Large quantities of biological fluids and wastes neutralize their actions.
4. Microbes can develop resistance to metals.

Health and environmental considerations have dramatically reduced the use of metallic antimicrobial compounds in medicine, dentistry, commerce, and agriculture.

Applications of Heavy Metals Weak (0.001% to 0.2%) organic mercury tinctures such as thimerosal (Merthiolate) and nitro-mersol (Metaphen) are fairly effective antiseptics and infection preventives, but they should never be used on broken skin because they are harmful and can delay healing. The organic mercurials also serve as preservatives in cosmetics and ophthalmic solutions. Mercurochrome, that old staple

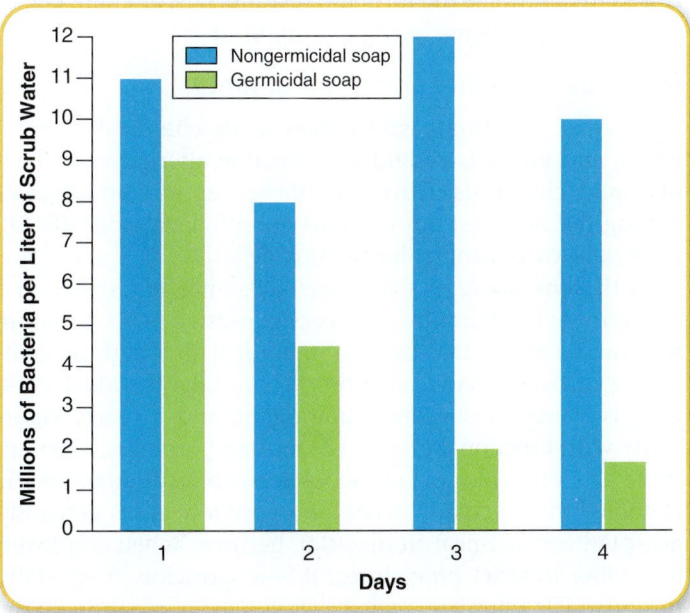

Figure 11.14 Graph showing effects of hand scrubbing. Comparison of scrubbing over several days with a nongermicidal soap versus a germicidal soap. The vertical axis shows number of live bacteria in the collected scrub water. Germicidal soap has persistent effects on skin over time, keeping the microbial count low. Without germicide, soap does not show this sustained effect.
Source: Nolte, et al. *Oral Microbiology*, 4/e, (c) 1982, Mosby.

of the medicine cabinet, is now considered among the poorest of antiseptics.

A silver compound with several applications is silver nitrate ($AgNO_3$) solution. German professor of obstetrics Carl Siegmund Franz Credé introduced it in the late-19th century for preventing gonococcal infections in the eyes of newborn infants who had been exposed to an infected birth canal (described in chapter 23). This preparation is not used as often now because many pathogens are resistant to it. It has been replaced by antibiotics in most instances. Solutions of silver nitrate (1% to 2%) can also be used as topical germicides on mouth ulcers and occasionally root canals. Silver sulfadiazine ointment, when added to dressings, effectively prevents infection in second- and third-degree burn patients, and pure silver is now incorporated into catheters to prevent urinary tract infections in the hospital. Colloidal silver preparations are mild germicidal ointments or rinses for the mouth, nose, eyes, and vagina. Silver ions are increasingly incorporated into many hard surfaces, such as plastics and steel, as a way to control microbial growth on items such as toilet seats, stethoscopes, and even refrigerator doors. Companies have even found ways to impregnate textiles with silver and quaternary ammonium compounds to produce antimicrobial fabrics that stay stain and odor-free over long periods of use.

Aldehydes as Germicides

Organic substances bearing a —CHO functional group (a strong reducing group) on the terminal carbon are called aldehydes. Several common substances such as sugars and some fats are technically aldehydes. The two aldehydes used most often in microbial control are *glutaraldehyde* and *formaldehyde.*

Glutaraldehyde is a yellow liquid with a mild odor. The mechanism of activity involves cross-linking protein molecules on the cell surface. In this process, amino acids are alkylated, meaning that a hydrogen atom on an amino acid is replaced by the glutaraldehyde molecule itself. It can also irreversibly disrupt the activity of enzymes within the cell. Glutaraldehyde is rapid and broad-spectrum and is one of the few chemicals officially accepted as a sterilant and high-level disinfectant. It destroys endospores in 3 hours and fungi and vegetative bacteria (even *Mycobacterium* and *Pseudomonas*) in a few minutes. Viruses, including the most resistant forms, appear to be inactivated after relatively short exposure times. Glutaraldehyde retains its potency even in the presence of organic matter, is noncorrosive, does not damage plastics, and is less toxic or irritating than formaldehyde. Its principal disadvantage is that it is somewhat unstable, especially with increased pH and temperature.

Formaldehyde is a sharp, irritating gas that readily dissolves in water to form an aqueous solution called **formalin.** Pure formalin is a 37% solution of formaldehyde gas dissolved in water. The chemical is microbicidal through its attachment to nucleic acids and functional groups of amino acids. Formalin is an intermediate- to high-level disinfectant,

although it acts more slowly than glutaraldehyde. Formaldehyde's extreme toxicity (it is classified as a carcinogen) and irritating effects on the skin and mucous membranes greatly limit its clinical usefulness.

A third aldehyde, ortho-phthalaldehyde (OPA), is another high-level disinfectant. OPA is a pale blue liquid with a barely detectable odor and can be most directly compared to glutaraldehyde. It has a mechanism of action similar to glutaraldehyde, is stable, is nonirritating to the eyes and nasal passages, and, for most uses, is much faster acting than glutaraldehyde. It is effective against vegetative bacteria, including *Mycobacterium* and *Pseudomonas,* fungi, and viruses. Chief among its disadvantages are an inability to reliably destroy endospores and, on a more practical note, its tendency to stain proteins, including those in human skin.

Applications of the Aldehydes   Glutaraldehyde is a milder chemical for sterilizing materials that are damaged by heat. Commercial products (Cidex, Sporicidin) diluted to 2% are used to sterilize respiratory therapy equipment, hemostats, fiberoptic endoscopes (laparoscopes, arthroscopes), and kidney dialysis equipment. Glutaraldehyde is employed on dental instruments (usually in combination with autoclaving) to inactivate hepatitis B and other blood-borne viruses. It also serves to preserve vaccines, sanitize poultry carcasses, and degerm cows' teats.

Formalin tincture (8%) has limited use as a disinfectant for surgical instruments, and formalin solutions have applications in aquaculture to kill fish parasites and control growth of algae and fungi. Any object that is intended to come into intimate contact with the body must be thoroughly rinsed to neutralize the formalin residue. It is, after all, one of the active ingredients in embalming fluid.

Gaseous Sterilants and Disinfectants

Processing inanimate substances with chemical vapors, gases, and aerosols provides a versatile alternative to heat or liquid chemicals. Currently, those vapors and aerosols having the broadest applications are ethylene oxide (ETO), propylene oxide, and chlorine dioxide.

Ethylene oxide is a colorless substance that exists as a gas at room temperature. It is very explosive in air, a feature that can be eliminated by combining it with a high percentage of carbon dioxide or fluorocarbon. Like the aldehydes, ETO is a very strong alkylating agent, and it reacts vigorously with functional groups of DNA and proteins. Through these actions, it blocks both DNA replication and enzymatic actions. Ethylene oxide is one of a very few gases generally accepted for chemical sterilization because, when employed according to strict procedures, it is a sporicide. A specially designed ETO sterilizer called a *chemiclave,* a variation on the autoclave, is equipped with a chamber; gas ports; and temperature, pressure, and humidity controls. Ethylene oxide is rather penetrating but relatively slow-acting, requiring from 90 minutes to 3 hours. Some items absorb ETO residues and must be aerated with sterile air for several hours after exposure to ensure dissipation of as much residual gas as possible.

For all of its effectiveness, ETO has some unfortunate features. Its explosiveness makes it dangerous to handle; it can damage the lungs, eyes, and mucous membranes if contacted directly; and it is rated as a carcinogen by the government.

Chlorine dioxide is another gas that has of late been used as a sterilant. Despite the name, chlorine dioxide works in a completely different way from the chlorine compounds discussed earlier in the chapter. It is a strong alkylating agent, which disrupts proteins and is effective against vegetative bacteria, fungi, viruses, and endospores. Although chlorine dioxide is used for the treatment of drinking water, wastewater, food processing equipment, and medical waste, its most well-known use was in the decontamination of the Senate offices after the anthrax attack of 2001 (see Insight 11.1).

Applications of Gases and Aerosols   Ethylene oxide (carboxide, cryoxide) is an effective way to sterilize and disinfect plastic materials and delicate instruments in hospitals and industries. It can sterilize prepackaged medical devices, surgical supplies, syringes, and disposable Petri dishes. Ethylene oxide has been used extensively to disinfect sugar, spices, dried foods, and drugs.

Propylene oxide is a close relative of ETO, with similar physical properties and mode of action, although it is less toxic. Because it breaks down into a relatively harmless substance, it is safer than ETO for sterilization of foods (nuts, powders, starches, spices).

Dyes as Antimicrobial Agents

Dyes are important in staining techniques and as selective and differential agents in media; they are also a primary source of certain drugs used in chemotherapy. Because aniline dyes such as crystal violet and malachite green are very active against gram-positive species of bacteria and various fungi, they are incorporated into solutions and ointments to treat skin infections (ringworm, for example). The yellow acridine dyes, acriflavine and proflavine, are sometimes utilized for antisepsis and wound treatment in medical and veterinary clinics. For the most part, dyes will continue to have limited applications because they stain and have a narrow spectrum of activity.

Acids and Alkalis

Conditions of very low or high pH can destroy or inhibit microbial cells; but they are limited in applications due to their corrosive, caustic, and hazardous nature. Aqueous solutions of ammonium hydroxide remain a common component of detergents, cleansers, and deodorizers. Organic acids are widely used in food preservation because they prevent endospore germination and bacterial and fungal growth and because they are generally regarded as safe to eat. Acetic acid (in the form of vinegar) is a pickling agent that inhibits bacterial growth; propionic acid is commonly incorporated into breads and cakes to retard molds; lactic acid is added to sauerkraut and olives to prevent growth of anaerobic bacteria (especially the clostridia); and benzoic and sorbic acids are added to beverages, syrups, and margarine to inhibit yeasts.

For a look at the antimicrobial chemicals found in some common household products, see **table 11.7**. Although the development of advanced microbial control methods has benefited humans greatly, our society is at the point where negative effects may have caught up with the positive ones.

The most disturbing trend comes from the widespread use of triclosan in numerous products today. According to the

Table 11.7 Active Ingredients of Various Commercial Antimicrobial Products

Product	Specific Chemical Agent	Antimicrobial Category
Lysol Sanitizing Wipes	Dimethyl benzyl ammonium chloride	Detergent (quat)
Clorox Disinfecting Wipes	Dimethyl benzyl ammonium chloride	Detergent (quat)
Tilex Mildew Remover	Sodium hypochlorites	Halogen
Lysol Mildew Remover	Sodium hypochlorites	Halogen
Ajax Antibacterial Hand Soap	Triclosan	Phenolic
Dawn Antibacterial Hand Soap	Triclosan	Phenolic
Dial Antibacterial Hand Soap	Triclosan	Phenolic
Lysol Disinfecting Spray	Alkyl dimethyl benzyl ammonium saccharinate/ethanol	Detergent (quats)/alcohol
ReNu Contact Lens Solution	Polyaminopropyl biguanide	Chlorhexidine
Wet Ones Antibacterial Moist Towelettes	Benzethonium chloride	Detergents (quat)
Noxzema Triple Clean	Triclosan	Phenolic
Scope Mouthwash	Ethanol	Alcohol
Purell Instant Hand Sanitizer	Ethanol	Alcohol
Pine-Sol	Phenolics and surfactant	Mixed
Allergan Eye Drops	Sodium chlorite	Halogen
Colgate Total Toothpaste	Triclosan	Phenolic

Centers for Disease Control, triclosan is excreted in the urine of 75% of Americans. The rinsing and flushing of this chemical down drains adds detectable levels of triclosan in various groundwater sources (rivers, wells, tap water). This antimicrobial degrades into a dioxin-like compound when exposed to sunlight, posing risks to the environment and the plants and animals in the surrounding area. Moreover, researchers around the world have identified triclosan resistance in various microorganisms, including both environmental microbes and human pathogens. The use of triclosan-containing water for crop irrigation has the potential for triggering the evolution of novel resistant microbes that may wind up in the food supply. More troublesome is the evidence showing that triclosan resistance contributes to the development of additional drug resistance in clinically relevant microorganisms such as *E. coli*. Compounded by the data that the addition of triclosan does not in fact enhance the antimicrobial effectiveness of most products, many companies have begun to remove the chemical from their manufacturing processes.

All of this information has led many scientists down the path to investigate alternative sources of microbial control chemicals today. Natural compounds like chitosan, derived from chitin in crustaceans, have been found to exhibit antimicrobial properties making them medically useful in the manufacturing of gauzes, sutures, and wound dressings. Nanotechnology presents another avenue by which researchers may also develop new methods for microbial control.

11.3 Learning Outcomes—Assess Your Progress

14. Name the desirable characteristics of chemical control agents.

15. Discuss several different halogen agents and their uses in microbial control.

16. List advantages and disadvantages to the use of phenolic compounds as control agents.

17. Explain the mode of action of alcohols and their limitations as effective antimicrobials.

18. Pinpoint the most appropriate applications of hydrogen peroxide agents.

19. Define the term *surfactant*, and explain this antimicrobial's mode of action.

20. Identify examples of some heavy metal control agents and their most common applications.

21. Discuss the advantages and disadvantages of aldehyde agents in microbial control.

22. Identify applications for ethylene oxide sterilization.

Case File 11   *Wrap-Up*

Bacillus cereus is a gram-positive endospore-forming bacillus usually found in the soil and is most often known for causing food poisoning, causing 25% of food-borne intoxications annually. *B. cereus* is also an **opportunistic** pathogen, causing infections in immunocompromised individuals. The children in this case were especially susceptible to infection with *B. cereus* because of their underlying conditions of leukemia and congenital heart disease. Additionally, they were further compromised by the insertion of an open portal—the vascular access device in the patient with leukemia—and the internal jugular line in the patient with congenital heart disease. These portals exit through the skin and provide an opening for bacteria to enter.

As an endospore-forming bacterium, *B. cereus* is especially resistant to the action of disinfectants and antiseptics. The

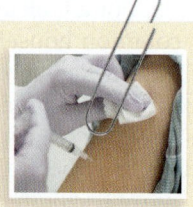

endospore coat is made up of proteins and small amounts of carbohydrates and lipids that are not affected by the membrane-dissolving action of isopropyl alcohol. It is likely that endospores of *B. cereus* contaminated the inner and outer packaging of the alcohol prep pads and germinated once they were introduced onto the skin.

The alcohol prep pads used by the hospital came in two forms: sterile and nonsterile. The packaging of the alcohol prep pads used in these cases did not state whether the pads were sterile or not. It is extremely important, especially with immunocompromised patients, that sterile procedures and equipment be used, and that all packaging is clearly labeled as being sterile.

Source: www.cdc.gov/mmwr/preview/mmwrhtml/mm6011a5.htm?s_cid=mm6011 a5_w. 2011. Morbidity and Mortality Weekly Report 60, no. 11, p. 347.

Chapter Summary

11.1 Controlling Microorganisms (ASM Guidelines* 2.1, 3.3, 3.4, 5.2)

- Microbial control methods involve the use of physical and chemical agents to eliminate or reduce the numbers of microorganisms from a specific environment to prevent the spread of infectious agents, retard spoilage, and keep commercial products safe.
- The population of microbes that cause spoilage or infection varies widely, so microbial control methods must be adjusted to fit individual situations.
- The type of microbial control is indicated by the terminology used. *Sterilization*

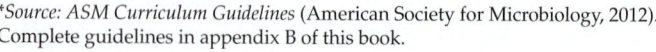

Prions

Bacterial endospores

Source: ASM Curriculum Guidelines (American Society for Microbiology, 2012). Complete guidelines in appendix B of this book.

agents destroy all viable organisms, including viruses. *Antisepsis, disinfection,* and *decontamination* reduce the numbers of viable microbes to a specified level.

- Antimicrobial agents are described according to their ability to destroy or inhibit microbial growth. Microbicidal agents cause microbial death. They are described by what they are -*cidal* for: sporocides, bactericides, fungicides, viricides.
- An antiseptic agent is applied to living tissue to destroy or inhibit microbial growth.
- A disinfectant agent is used on inanimate objects to destroy vegetative pathogens but not bacterial endospores.
- Sanitization reduces microbial numbers on inanimate objects to safe levels by physical or chemical means.
- Degermation refers to the process of mechanically removing microbes from the skin.
- Microbial death is defined as the permanent loss of reproductive capability in microorganisms.
- Antimicrobial agents attack specific cell sites to cause microbial death or damage. The four major cell targets are the cell wall, the cell membrane, biosynthesis pathways for DNA or RNA, or protein (enzyme) function.

11.2 Methods of Physical Control (ASM Guidelines 1.2, 1.3, 2.1, 2.2, 3.3, 3.4, 4.1, 5.2)

- Physical methods of microbial control include heat, cold, radiation, drying, filtration, and osmotic pressure.
- Heat is the most widely used method of microbial control. It is used in combination with water (moist heat) or as dry heat (oven, flames).
- The thermal death time (TDT) is the shortest length of time required to kill all microbes at a specific temperature.
- The thermal death point (TDP) is the lowest temperature at which all microbes are killed in a specified length of time (10 minutes).
- Autoclaving, or steam sterilization, is the process by which steam is heated under pressure to sterilize a wide range of materials in a comparatively short time (minutes to hours).
- Boiling water and pasteurization of beverages disinfect but do not sterilize materials.
- Dry heat is microbicidal under specified times and temperatures. Flame heat, or incineration, is microbicidal.
- Chilling, freezing, and desiccation are microbistatic but not microbicidal. They are not considered true methods of disinfection because they are not consistent in their effectiveness.

- Ionizing radiation (cold sterilization) by gamma rays and X rays is used to sterilize medical products, meats, and spices. It damages DNA and cell organelles by producing disruptive ions.
- Ultraviolet light, or nonionizing radiation, has limited penetrating ability. It is therefore usually restricted to disinfecting air and certain liquids.
- Decontamination by filtration removes microbes from heat-sensitive liquids and circulating air. The pore size of the filter determines what kinds of microbes are removed.
- The addition of high amounts of salt or sugar to food results in preservation through osmotic pressure.

11.3 Chemical Agents in Microbial Control (ASM Guidelines 1.2, 1.3, 2.1, 2.2, 3.3, 3.4, 4.1, 5.2)

- Chemical agents of microbial control are classified by their physical state and chemical nature.
- Chemical agents can be either microbicidal or microbistatic and can be classified as high-, medium-, or low-level germicides.
- Halogens are effective chemical agents at both microbicidal and microbistatic levels. Chlorine, iodine, and iodophors are examples.
- Phenols are strong microbicidal agents used in general disinfection. Milder phenol compounds, the bisphenols, are also used as antiseptics.
- Alcohols dissolve membrane lipids and destroy cell proteins. Their action depends upon their concentration, but they are generally only microbistatic.
- Hydrogen peroxide is a versatile microbicide that can be used as an antiseptic for wounds and a disinfectant for utensils. A high concentration is an effective sporicide.
- Surfactants are of two types: detergents and soaps. They reduce cell membrane surface tension, causing membrane rupture. Cationic detergents, or quats, are low-level germicides limited by the amount of organic matter present and the microbial load.
- Aldehydes are potent sterilizing agents and high-level disinfectants that irreversibly disrupt microbial enzymes.
- Ethylene oxide and chlorine dioxide are gaseous sterilants that work by alkylating protein and DNA.

Multiple-Choice and True-False Questions | Bloom's Levels 1 and 2: Remember and Understand

Multiple-Choice Questions. Select the correct answer from the options provided.

1. Microbial control methods that kill/destroy _____ are able to sterilize.
   a. viruses
   b. the tubercle bacillus
   c. endospores
   d. cysts

2. Sanitization is a process by which
   a. the microbial load on objects is reduced.
   b. objects are made sterile with chemicals.
   c. utensils are scrubbed.
   d. skin is debrided.

3. An example of an agent that lowers the surface tension of cells is
   a. phenol.   c. alcohol.
   b. chlorine.   d. formalin.

4. High temperatures _____ and low temperatures _____.
   a. sterilize, disinfect
   b. kill cells, inhibit cell growth
   c. denature proteins, burst cells
   d. speed up metabolism, slow down metabolism

5. Microbe(s) that is/are the target(s) of pasteurization include
   a. *Clostridium botulinum.*   c. *Salmonella* species.
   b. *Mycobacterium* species.   d. both *b* and *c.*

6. The primary mode of action of nonionizing radiation is to
   a. produce superoxide ions.
   b. make pyrimidine dimers.
   c. denature proteins.
   d. break disulfide bonds.

7. The most practical method of disinfecting municipal water supplies would be
   a. UV radiation.   c. beta propiolactone.
   b. exposure to ozone.   d. filtration.

8. A chemical with sporicidal properties is
   a. phenol.
   b. alcohol.
   c. quaternary ammonium compound.
   d. glutaraldehyde.

9. Silver nitrate is used
   a. in antisepsis of burns.
   b. as a mouthwash.
   c. to treat genital gonorrhea.
   d. to disinfect water.

10. Detergents are
    a. high-level germicides.
    b. low-level germicides.
    c. excellent antiseptics.
    d. used in disinfecting surgical instruments.

True-False Questions. If the statement is true, leave as is. If it is false, correct it by rewriting the sentence.

11. The process of destroying non-spore-forming organisms on inanimate objects fits within the definition of disinfection.

12. The acceptable temperature-pressure combination for an autoclave is 131°C and 9 psi.

13. Ionizing radiation dislodges protons from atoms.

14. A microbicide is an agent that destroys microorganisms.

15. Prions are easily denatured by heat.

Critical Thinking Questions | Bloom's Levels 3, 4, and 5: Apply, Analyze, and Evaluate

Critical thinking is the ability to reason and solve problems using facts and concepts. These questions can be approached from a number of angles and, in most cases, they do not have a single correct answer.

1. You work in a laboratory that regularly analyzes patient specimens for the presence of *Clostridium difficile.*
   a. Knowing that many specimens test positive, explain which type of antimicrobial chemical agent you should choose to effectively clean your laboratory each day.
   b. Provide an explanation in support of or refuting the following statement: "The goal of disinfection is the destruction of bacterial endospores." Briefly explain how the type of microorganisms present will influence the effectiveness of an antimicrobial agent.

2. a. Discuss the factors that influence an antimicrobial agent's ability to completely remove microbial contaminants from dentures used on a daily basis.
   b. Improper cleaning of endoscopes has led to the transmission of hepatitis C infection among patients given colonoscopies at the same facility. Develop one hypothesis to explain how the virus may have persisted and remained infectious.

3. a. Precisely what is microbial death?
   b. Why does a population of microbes not die instantaneously when exposed to an antimicrobial agent?

4. Compare and contrast the terms *bactericidal* and *bacteristatic* in relation to microbial control agents. Develop a simple experimental method that could be used to test whether a control agent exhibits bactericidal or bacteristatic effects.

5. a. Provide a valid reason for whether or not heat is a selective microbial control agent.
   b. Explain the concepts of TDT and TDP, and discuss one example of how these measurements can be modified to produce food for human consumption that is free of microbial contaminants yet still tasty, using examples. What are the minimum TDTs for vegetative cells and endospores?

6. a. Explain why desiccation and cold are not reliable methods of disinfection. Discuss how food poisoning can occur through the consumption of frozen food that has been properly stored prior to thawing and cooking.

b. Lack of refrigeration on the battlefield makes milk hard to come by. Describe the microbial control method that can allow soldiers to have cream in their coffee and milk on their cereal each morning.

7. Explain what is wrong with this statement: "Prior to vaccination, the patient's skin was sterilized with alcohol." What would be a more correct wording?

8. For each item on the following list, select the method that will provide proper sterilization of the item without destroying the material or product. You cannot use the same method more than three times. After considering a workable method, think of one method that would *not* work. *Note:* Where an object containing something is given, you must sterilize everything (e.g., both the jar and the Vaseline in it). Some examples of methods are autoclave, ethylene oxide gas, dry oven, and ionizing radiation.

room air

a vial of influenza vaccine

a pot of soil

plastic Petri dishes

cloth dressings

soiled bedding and clothing from an infected patient

a cheese sandwich

green beans for consumption

carcasses of cows infected with "mad cow" disease

inside of a refrigerator

apple juice

a jar of Vaseline

fruit in plastic bags

human hair (for wigs)

a flask of nutrient agar

an entire room (walls, floor, etc.)

rubber gloves

disposable syringes

metal surgical instruments

mail contaminated with anthrax spores

9. Tissue transplantation is a procedure that saves lives every day, but pathogen transmission can occur from donor to recipient. Discuss which microbial control methods can be used to reduce disease transmission during the transplantation process.

10. Conduct additional research on the use of triclosan and other chemical agents in antimicrobial products today. Develop an opinion on whether this process should continue, providing evidence to support your stance.

Concept Connections | Bloom's Levels 4 and 6: Analyze and Create

This activity ties together multiple concepts in the chapter.

1. What physical control methods are used to sterilize? What materials require sterilization?

2. What physical control methods disinfect? What types of materials require disinfection but not sterilization?

3. Describe the differences between dry heat and moist heat sterilization.

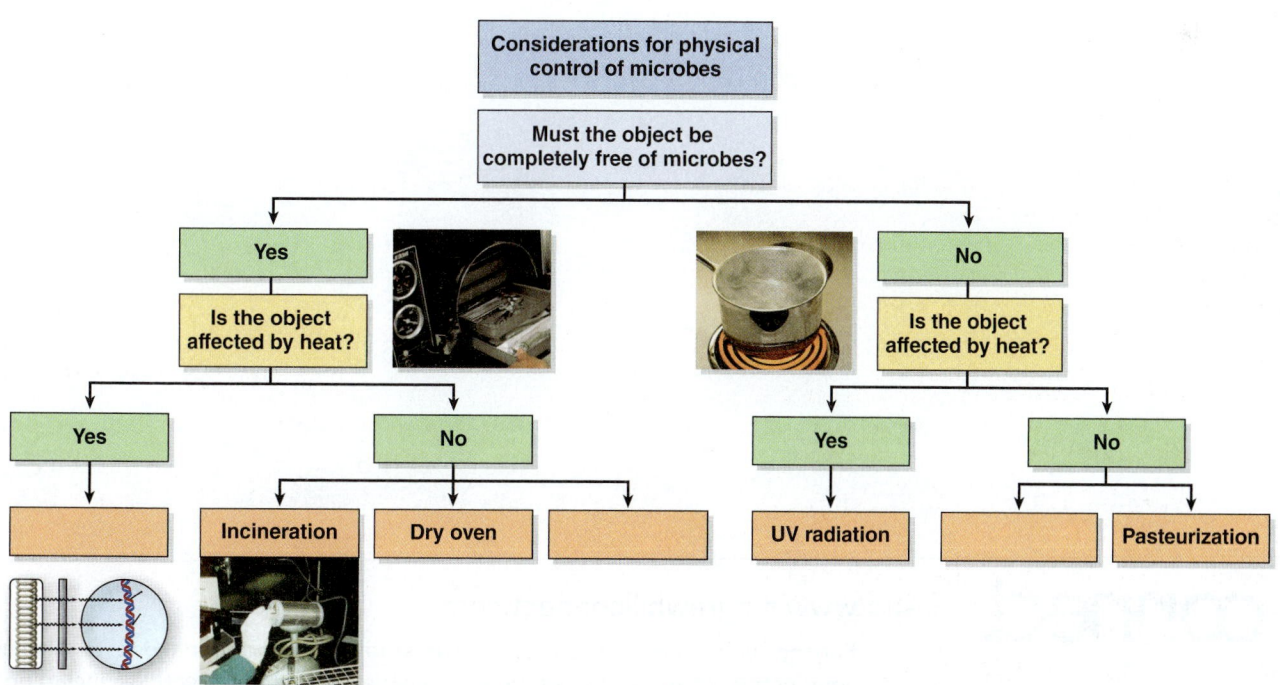

Visual Connections | Bloom's Level 5: Evaluate

These questions use visual images or previous content to make connections to this chapter's concepts.

1. **From chapter 2, figure 2.20.** Study this illustration of a cell membrane. In what ways could alcohol damage this membrane? How would that harm the cell? (*Hint:* Alcohol is a solvent.)

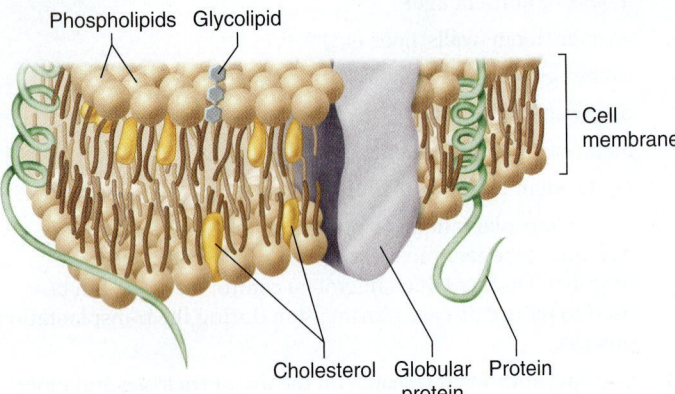

2. **From chapter 4, figure 4.23.** Why would many chemical control agents be ineffective in controlling this organism?

Concept Mapping | Bloom's Level 6: Create

Appendix D provides guidance for working with concept maps.

1. Using the words that follow, please create a concept map illustrating the relationships among these key terms from chapter 11.

halogens	alcohol	chemical
oligodynamic	phenolics	physical
surfactants	sporicidal	silver

www.mcgrawhillconnect.com

Enhance your study of this chapter with study tools and practice tests. Also ask your instructor about the resources available through ConnectPlus, including the media-rich eBook, interactive learning tools, and animations.

Drugs, Microbes, Host—
The Elements of Chemotherapy

Case File 12

World Travel, Antibiotic Resistance, and Political Discord

In May 2011, a woman from Rhode Island visited her native Cambodia. While she was there, she was treated for spinal cord compression in Ho Chi Minh City, Vietnam. During her stay in the hospital, she developed a urinary tract infection due to a long-term indwelling catheter and was prescribed antibiotics for treatment. Upon returning home to Rhode Island in January of 2012, she was immediately hospitalized and diagnosed with lymphoma. Her cancer treatment required her to again be catheterized, and subsequently she developed another urinary tract infection. When the infection did not respond to the normal course of antibiotic treatments, culturing of urine specimens revealed the growth of a unique drug-resistant strain of *Klebsiella pneumoniae*.

Given the woman's recent travel history, the isolate was sent to the Centers for Disease Control and Prevention (CDC) for further testing. The CDC confirmed that the patient was harboring a strain of bacterium known as carbapenem-resistant *Enterobacteriaceae* (CRE). This particular strain of *K. pneumoniae*, however, contained a gene (bla_{NDM-1}) encoding an enzyme called New Dehli metallo-beta-lactamase 1 (NDM-1), which confers resistance to 24 different antibiotics. Because the organism was highly resistant to antibiotics, the patient was placed on isolation precautions during her stay in the hospital; she was discharged on March 26, 2012. On March 30, a rectal swab taken from one of seven individuals treated in the same unit was positive for *K. pneumoniae*. Further testing revealed that it was genetically identical to the CRE NDM-1 strain found initially in the discharged patient, indicating that this bacterium had somehow spread within the unit. Hospital records showed that although no physicians or nurse practitioners were in contact with both patients during their stay, there were 23 nurses that had cared for both patients.

- What is New Delhi metallo-beta-lactamase 1 (NDM-1)?
- How is the gene for this enzyme spread among bacteria?

Continuing the Case appears on page 335.

Outline and Learning Outcomes

12.1 Principles of Antimicrobial Therapy

A hundred years ago in the United States, one out of three children was expected to die of an infectious disease before the age of 5. Early death or severe lifelong debilitation from scarlet fever, diphtheria, tuberculosis, meningitis, and many other bacterial diseases was a fearsome yet undeniable fact of life to most of the world's population. The introduction of modern drugs to control infections in the 1930s was a medical revolution that has added significantly to the life span and health of humans. It is no wonder that for many years, antibiotics in particular were regarded as miracle drugs. Although antimicrobial drugs have greatly reduced the incidence of certain infections, they have definitely not eradicated infectious disease and probably never will. In fact, many doctors are now warning that we are dangerously close to a post-antibiotic era, wherein the drugs we have are no longer effective. Part of the history of humans' struggle to chemically control disease is outlined in **Insight 12.1**.

The goal of antimicrobial chemotherapy is deceptively simple: Administer a drug to an infected person, which destroys the infective agent without harming the host's cells. In actuality, this goal is rather difficult to achieve, because many (often contradictory) factors must be taken into account. The ideal drug should be easy to administer yet be able to reach the infectious agent anywhere in the body, be toxic to the infectious agent (or cripple its ability to multiply) while simultaneously being nontoxic to the host, and remain active in the body as long as needed yet be safely and easily broken down and excreted. Additionally, microbes in biofilms often require different drugs than when they are not in biofilms. In short, the perfect drug does not exist; but by balancing drug characteristics against one another, we can usually achieve a satisfactory compromise **(table 12.1)**.

The Origins of Antimicrobial Drugs

Chemotherapeutic agents are described with regard to their origin, range of effectiveness, and whether they are naturally produced or chemically synthesized. A few of the more important terms you will encounter are found in **table 12.2**.

Nature is a prolific producer of antimicrobial drugs. Antibiotics, after all, are common metabolic products of bacteria and fungi. By inhibiting the growth of other microorganisms in the same habitat (antagonism), antibiotic producers presumably enjoy less competition for nutrients and

INSIGHT 12.1 Penicillin and Winning World War II

The discovery of penicillin by Alexander Fleming was a landmark in modern medicine. In 1928, seemingly by accident, Fleming found that colonies of *Penicillium* species inhibited growth of *Staphylococcus aureus* growing on discarded Petri dishes. Fleming thought the species was *Penicillium chrysogenum,* but scientists have since determined that it is a different species of *Penicillium* from Fleming's lab. Fleming extracted a chemical that he called "penicillin" from the mold and showed that it was responsible for inhibiting the growth of bacteria.

Microbiology instructor Andrea Rediske visiting the grave of Alexander Fleming in St. Paul's Cathedral in London.

Ten years later, Howard Florey and Ernst Chain at Oxford University developed a method to produce 500 liters of penicillin extract a week. They tested the drug on Albert Alexander, a 43-year-old policeman who had scratched himself on a rose bush while doing yard work and developed a serious infection on his face. They injected him with penicillin, and the infection began to clear. Supplies of the experimental penicillin were extremely limited at the time, and doctors resorted to extracting penicillin excreted in his urine to continue the treatments. Unfortunately, supplies of penicillin ran out, and Alexander died a few days later. But, Florey and Chain, along with biochemists Norman Heatley and Edward Abraham, were determined to develop a process to bring penicillin to British soldiers on the battlefield during World War II.

Florey published his landmark paper describing their work with penicillin and the preliminary tests with the new drug the day before Germany began the London Blitz. Despite the constant bombing in London, Florey and Chain continued to manufacture and test penicillin; but eventually it became too difficult for any British manufacturing company to continue research and clinical trials. Florey, Chain, Abraham, and Heatley then turned to the Rockefeller Foundation in the United States for help in finding a U.S. drug company who could scale up manufacture of the drug. Florey and Heatley traveled to the United States under utmost secrecy—the miracle of penicillin had to be kept from the Germans, and traveling over the Atlantic during the war was extremely difficult. Even their partner Chain was unaware of their travel plans. Florey and Heatley flew from Bristol to Lisbon, and Florey kept his briefcase—filled with samples of penicillin, mold samples, and his notes and papers—on his lap for the entire trip to ensure their safety. Throughout their harrowing journey, Florey kept close tabs on his briefcase, making sure it was a locked in a safe every night.

Many pharmaceutical companies and chemical companies worked together to efficiently produce large quantities of penicillin, and on March 1, 1944, Pfizer opened the first large-scale production plant for penicillin. The production of penicillin greatly contributed to the war effort, and the first limited supplies were given to the military to treat soldiers. Penicillin produced by the Pfizer plant was used by the Allies for the rest of World War II and helped to save countless soldiers' lives. Eventually, the miracle drug became available to civilians as well. Fleming, Florey, and Chain won the Nobel Prize in Physiology or Medicine for their discovery that many believe helped to win World War II.

space. This selective advantage has allowed the genes for antibiotic production to be preserved in evolution. The greatest numbers of current antibiotics are derived from bacteria in the genera *Streptomyces* and *Bacillus* and from molds in the genera *Penicillium* and *Cephalosporium.* Whatever benefit the microbes derive, these compounds have been extremely profitable for humans. Scientists have become "bioprospectors" as they continue to mine various environments for new metabolic compounds with antimicrobial effects. With the explosion of information coming from the field of metagenomics, computational biology has come to play a vital role in drug discovery today. Because 99% of microbial life is unculturable, researchers use computer-based screening tools to analyze metagenomic, proteomic, and metabolomic data collected from various sources in order to direct them to potential candidate drugs.

Chemistry has been responsible for the other major source of antimicrobial compounds available today. Chemists are able to create new drugs by altering the structure of naturally occurring antibiotics. Called "click chemistry," this method takes a natural microbial product and joins it with various preselected molecules in order to create **semisynthetic** drugs that offer advantages over other, related drugs. Paul Ehrlich was one of the first researchers to systematically alter a parent molecule to create a new *derivative*, eventually arriving at a compound

Table 12.1 Characteristics of the Ideal Antimicrobial Drug

- Selectively toxic to the microbe but nontoxic to the host
- Microbicidal rather than microbistatic
- Relatively soluble; functions even when highly diluted in body fluids
- Remains potent long enough to act and is not broken down or excreted prematurely
- Doesn't lead to the development of antimicrobial resistance
- Complements or assists the activities of the host's defenses
- Remains active in tissues and body fluids
- Readily delivered to the site of infection
- Reasonably priced
- Does not disrupt the host's health by causing allergies or predisposing the host to other infections

Table 12.2 Terminology of Chemotherapy

Chemotherapeutic drug	Any chemical used in the treatment, relief, or prophylaxis of a disease
Prophylaxis	Use of a drug to prevent imminent infection of a person at risk
Antimicrobial chemotherapy	The use of chemotherapeutic drugs to control infection
Antimicrobials	All-inclusive term for any antimicrobial drug, regardless of what type of microorganism it targets.
Antibiotics	Substances produced by the natural metabolic processes of some microorganisms—or created by scientists—that can inhibit or destroy microorganisms
Semisynthetic drugs	Drugs that are chemically modified in the laboratory after being isolated from natural sources
Synthetic drugs	Drugs produced entirely by chemical reactions within a laboratory setting
Narrow-spectrum (limited spectrum)	Antimicrobials effective against a limited array of microbial types—for example, a drug effective mainly on gram-positive bacteria
Broad-spectrum (extended spectrum)	Antimicrobials effective against a wide variety of microbial types—for example, a drug effective against both gram-positive and gram-negative bacteria

Disease Connection

Salvarsan was a derivative of the poisonous chemical arsenic. It was toxic to patients, and the injections were painful. But the disease it treated, syphilis, was potentially fatal; since salvarsan did result in a significant cure rate, it was hailed as a medical breakthrough and called the magic bullet. Today syphilis is often treated with penicillin G, which, when discovered, became a true magic bullet: harming the pathogen without harming the host.

a gonorrhea infection reveals the characteristic appearance: bean-shaped gram-negative diplococci. This appearance is unique enough (when combined with the symptoms) that the physician can assume it is *Neisseria gonorrhoeae* and begin treatment accordingly. In most cases, further testing is needed to identify the offending microbe. These methods may involve culturing the microbe or may instead rely on a variety of nonculture methods. All of these diagnostic methods are discussed in chapter 17. Sometimes identification efforts are inconclusive. In these cases, epidemiological statistics may be required to predict the most likely agent in a given infection. For example, *Streptococcus pneumoniae* accounts for the majority of cases of meningitis in children, followed by *Neisseria meningitidis* (discussed in detail in chapter 19).

Testing for the Drug Susceptibility of Microorganisms

Determining which antimicrobial agent or agents are most effective against isolated pathogens is essential in those groups of bacteria commonly showing resistance, such as *Staphylococcus* species, *Neisseria gonorrhoeae*, *Streptococcus pneumoniae*, *Enterococcus faecalis*, and the aerobic gram-negative enteric bacilli. However, not all infectious agents require antimicrobial sensitivity testing. For example, drug testing in fungal or protozoan infections is difficult and is often unnecessary, because the antimicrobial agents generally target all representatives of these groups. Lastly, when certain groups, such as group A streptococci, are known to be uniformly susceptible to penicillin G, testing may not be necessary unless the patient is allergic to penicillin.

Selection of a proper antimicrobial agent begins by demonstrating the *in vitro* activity of several drugs against the infectious agent by means of standardized methods. In general, these tests involve exposing a pure culture of the bacterium to several different drugs and observing the effects of the drugs on growth.

The *Kirby-Bauer* technique is an agar diffusion test that provides useful data on antimicrobial susceptibility. In this test, the surface of a plate of special medium is spread with the test bacterium, and small discs containing a premeasured amount of antimicrobial are dispensed onto the bacterial lawn. After incubation, the zone of inhibition surrounding the discs is measured and compared with a standard for each drug (**table 12.3** and **figure 12.1**). This profile of

he called salvarsan. Some natural compounds, however, cannot be obtained in limitless supply without the destruction of a habitat or an organismal population. In this case, chemists have created **synthetic** drugs in the laboratory that mimic the action of these compounds. The potential for using bioengineering techniques to design such drugs seems almost limitless, and, indeed, several drugs are produced by manipulating the genes of antibiotic producers.

Starting Treatment

Before antimicrobial therapy using any type of drug can begin, it is important that at least three factors be known:

1. the identity of the microorganism causing the infection,
2. the degree of the microorganism's susceptibility (also called sensitivity) to various drugs, and
3. the overall medical condition of the patient.

Identifying the Agent

Identification of infectious agents from body specimens should be attempted as soon as possible. It is especially important that such specimens be taken before any antimicrobial drug is given—in case the drug eliminates the infectious agent. Direct examination of body fluids, sputum, or stool is a rapid initial method for detecting and perhaps even identifying bacteria or fungi. A doctor often begins the therapy on the basis of such immediate findings or even on the basis of an informed best guess. For instance, a gram stain of pus produced from

Table 12.3 Results of a Sample Kirby-Bauer Test

| Drug | Zone Sizes (mm) Required For: | | Example Results (mm) for *Staphylococcus aureus* | Evaluation* |
	Susceptibility (S)	Resistance (R)		
Bacitracin	>13	<8	15	S
Chloramphenicol	>18	<12	20	S
Erythromycin	>18	<13	15	I
Gentamicin	>13	<12	16	S
Kanamycin	>18	<13	20	S
Neomycin	>17	<12	12	R
Penicillin G	>29	<20	10	R
Polymyxin B	>12	<8	10	R
Streptomycin	>15	<11	11	R
Tetracycline	>19	<14	25	S
Vancomycin	>12	<9	15	S

*R = resistant, I = intermediate, S = sensitive

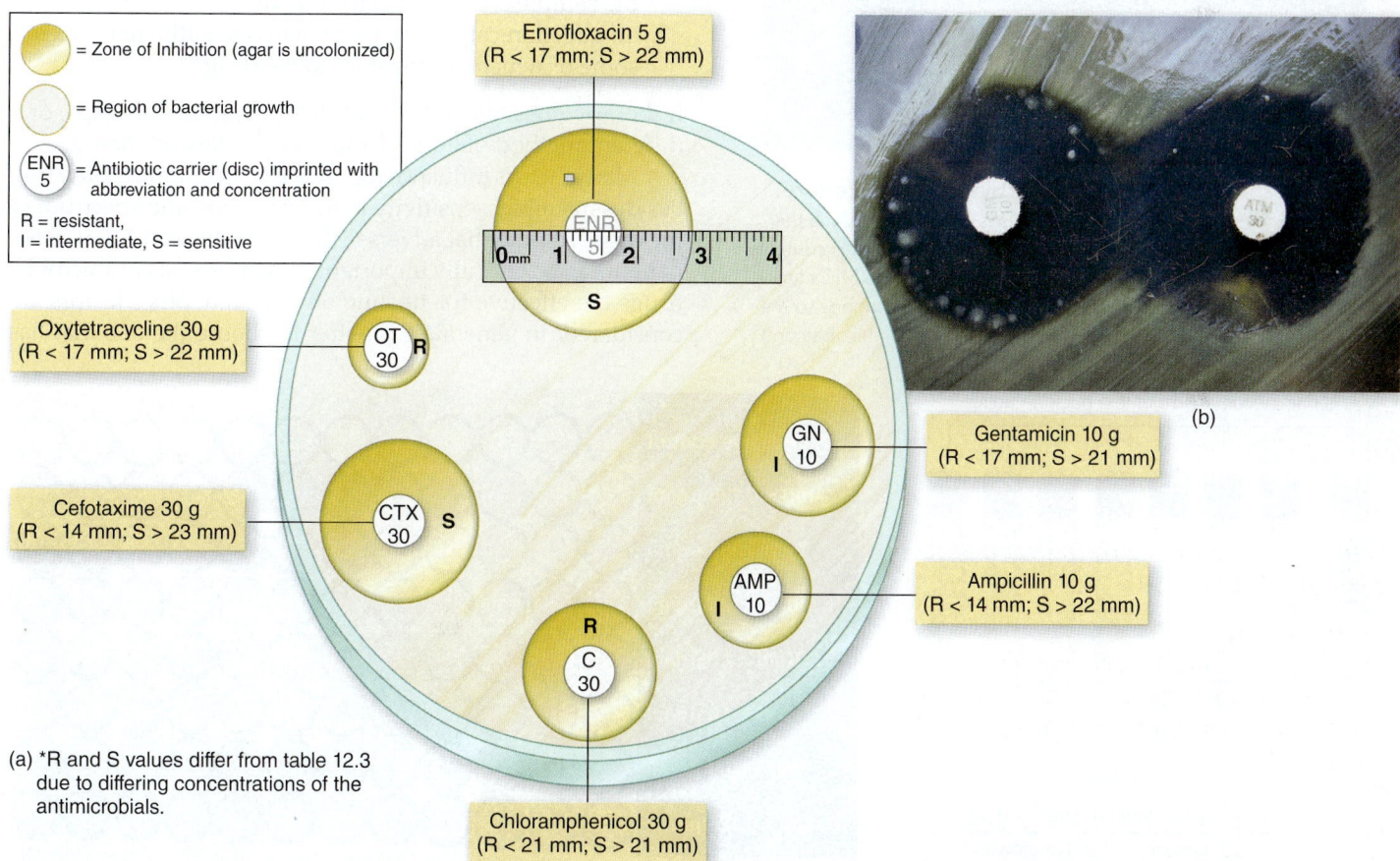

= Zone of Inhibition (agar is uncolonized)

= Region of bacterial growth

ENR 5 = Antibiotic carrier (disc) imprinted with abbreviation and concentration

R = resistant,
I = intermediate, S = sensitive

Enrofloxacin 5 g
(R < 17 mm; S > 22 mm)

Oxytetracycline 30 g
(R < 17 mm; S > 22 mm)

Cefotaxime 30 g
(R < 14 mm; S > 23 mm)

Chloramphenicol 30 g
(R < 21 mm; S > 21 mm)

Gentamicin 10 g
(R < 17 mm; S > 21 mm)

Ampicillin 10 g
(R < 14 mm; S > 22 mm)

(b)

(a) *R and S values differ from table 12.3 due to differing concentrations of the antimicrobials.

Figure 12.1 Technique for preparation and interpretation of disc diffusion tests. **(a)** Standardized methods are used to seed a lawn of bacteria over the medium. A dispenser delivers several drug-impregnated discs onto a plate, followed by incubation. Interpretation of results: During incubation, antimicrobials become increasingly diluted as they diffuse out of the discs into the medium. If the test bacterium is sensitive to a drug, a zone of inhibition develops around its disc. Roughly speaking, the larger the size of this zone, the greater is the bacterium's sensitivity to the drug. The diameter of each zone is measured in millimeters and evaluated for susceptibility or resistance by means of a comparative standard (see table 12.3). **(b)** Results of test with *Escherichia hermannii* indicate a synergistic effect between ticarcillin (TIC) and AMC (note the expanded zone between these two drugs).

antimicrobial sensitivity is called an *antibiogram*. The Kirby-Bauer procedure is less effective for bacteria that are anaerobic, highly fastidious, or slow-growing (*Mycobacterium*). An alternative diffusion system that provides additional information on drug effectiveness is the Etest **(figure 12.2)**.

More sensitive and quantitative results can be obtained with tube dilution tests. First, the antimicrobial is diluted serially in tubes of broth; then each tube is inoculated with a small uniform sample of pure culture, incubated, and examined for growth (turbidity). The smallest concentration (highest dilution) of drug that visibly inhibits growth is called the **minimum inhibitory concentration (MIC)**. The MIC is useful in determining the smallest effective dosage of a drug and in providing a comparative index against other antimicrobials **(figure 12.3)**. In many clinical laboratories, these antimicrobial testing procedures are performed in automated machines that can test dozens of drugs simultaneously.

The MIC and Therapeutic Index

The results of antimicrobial sensitivity tests guide the physician's choice of a suitable drug. If therapy has already begun, it is imperative to determine if the tests bear out the use of that particular drug. Once therapy has begun, it is important to observe the patient's clinical response, because the *in vitro* activity of the drug is not always correlated with its *in vivo* effect. When antimicrobial treatment fails, the failure is due to

1. the inability of the drug to diffuse into that body compartment (the brain, joints, skin);
2. resistant microbes in the infection that didn't make it into the sample collected for testing; or
3. an infection caused by more than one pathogen (mixed), some of which are resistant to the drug.

If therapy does fail, a different drug, combined therapy, or a different method of administration must be considered.

Many factors influence the choice of an antimicrobial drug besides microbial sensitivity to it. The nature and spectrum of the drug, its potential adverse effects, and the condition of the patient can be critically important. When several antimicrobial drugs are effective for treating an infection, other factors are considered. In general, it is better to choose the one that has

Figure 12.2 Alternative to the Kirby-Bauer procedure. Another diffusion test is the Etest, which uses a strip to produce the zone of inhibition. The advantage of the Etest is that the strip contains a gradient of drug calibrated in micrograms. This way, the MIC can be measured by observing the mark on the strip that corresponds to the edge of the zone of inhibition. (IP = imipenem and TZ = tazobactam)

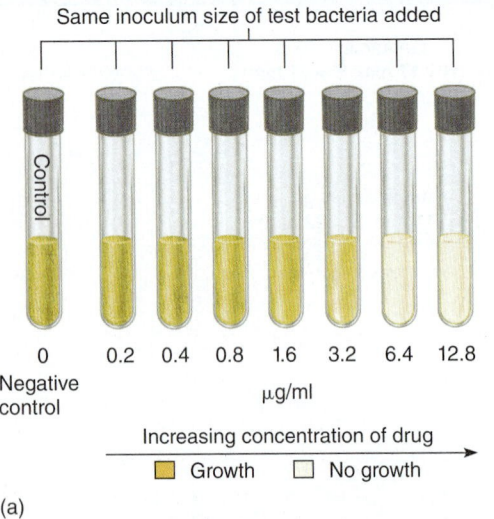

(a)

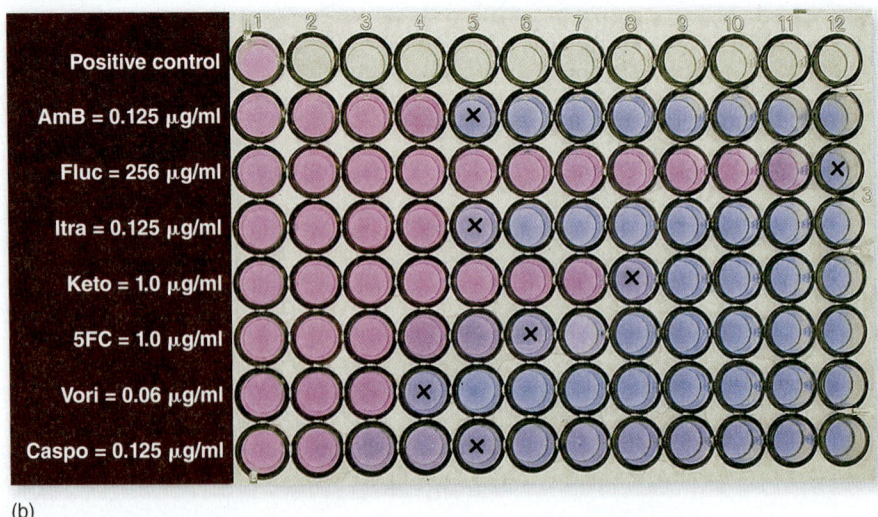

(b)

Figure 12.3 Tube dilution test for determining the minimum inhibitory concentration (MIC).   (a) The antibiotic is diluted serially through tubes of liquid nutrient from right to left. All tubes are inoculated with an identical amount of a test bacterium and then incubated. The first tube on the left is a control that lacks the drug and shows maximum growth. The first tube in the series that shows no growth (no turbidity) contains the concentration of antibiotic that is the MIC. (b) Microbroth dilution in a multiwell plate. Here, amphotericin B, flucytosine, and several azole drugs are tested on a pathogenic yeast. Pink indicates growth, and blue indicates no growth. Numbers indicate the dilution of the MIC, and the X in each row shows the first well without growth.

fewest effects on microbes other than the one being targeted. This decreases the potential of a variety of adverse reactions.

Because drug toxicity is of concern, it is best to choose the one with high selective toxicity for the infectious agent and low human toxicity. The **therapeutic index (TI)** is defined as the ratio of the dose of the drug that is toxic to humans to its minimum effective (therapeutic) dose. The closer these two figures are (the smaller the ratio), the greater the potential is for toxic drug reactions. For example, a drug that has a therapeutic index of

$$\frac{10 \; \mu g/ml: \text{toxic dose}}{9 \; \mu g/ml \; (\text{MIC})} \quad TI = 1.1$$

is a riskier choice than one with a therapeutic index of

$$\frac{10 \; \mu g/ml}{1 \; \mu g/ml} \quad TI = 10$$

When a series of drugs being considered for therapy have similar MICs, the drug with the highest therapeutic index usually has the widest margin of safety.

The physician must also take a careful history of the patient to discover any preexisting medical conditions that will influence the activity of the drug or the response of the patient. A history of allergy to a certain class of drugs precludes the use of that drug and any drugs related to it. Underlying liver or kidney disease will ordinarily require changing the drug therapy, because these organs play such an important part in metabolizing or excreting the drug. Infants, the elderly, and pregnant women require special precautions. For example, age can diminish gastrointestinal absorption and organ function, and most antimicrobial drugs cross the placenta and could affect fetal development.

Patients must be asked about other drugs they are taking, because incompatibilities can result in increased toxicity or failure of one or more of the drugs. For example, the combination of aminoglycosides and cephalosporins increases toxic effects on the kidney; antacids reduce the absorption of isoniazid; and the interaction of tetracycline or rifampin with oral contraceptives can abolish the contraceptive's effect. Some drug combinations (penicillin with certain aminoglycosides, or amphotericin B with flucytosine) act synergistically, so that reduced doses of each can be used in combined therapy. Other concerns in choosing drugs include any genetic or metabolic abnormalities in the patient, the site of infection, the route of administration, and the cost of the drug.

The Art and Science of Choosing an Antimicrobial Drug

Even when all the information is in, the final choice of a drug is not always easy or straightforward. Consider the hypothetical case of an elderly alcoholic patient with pneumonia caused by *Klebsiella* and complicated by diminished liver and kidney function. All drugs must be given by injection because of prior damage to the gastrointestinal lining and poor absorption. Drug tests show that the infectious agent

is sensitive to third-generation cephalosporins, gentamicin, imipenem, and azlocillin. The patient's history shows previous allergy to the penicillins, so these would be ruled out. Drug interactions occur between alcohol and the cephalosporins, which are also associated with serious bleeding in elderly patients, so this may not be a good choice. Aminoglycosides such as gentamicin are toxic to kidneys and poorly cleared by damaged kidneys. Imipenem causes intestinal discomfort, but it has less toxicity and would be a viable choice.

In the case of a cancer patient with severe systemic *Candida* infection, there will be fewer criteria to weigh. Intravenous amphotericin B or fluconazole are the only possible choices, despite drug toxicity and other possible adverse side effects. In a life-threatening situation in which a dangerous chemotherapy is perhaps the only chance for survival, the choices are reduced and the priorities are different.

Whereas choosing the right drug is an art and science requiring the consideration of many different things, the process has been made simpler—or at least more portable—with the advent of smartphones and relevant applications ("apps"). Most doctors now have the information literally at their fingertips when they pull their smartphones out of their pockets.

In the following section, various types of antibiotic drugs, their mechanism of action, and the types of microbes on which they are effective are described. The organ system chapters 18 through 23 list specific disease agents and the drugs used to treat them.

It is interesting to note that although the golden age of antibiotics ushered in a drastic reduction in infectious disease mortality in the 20th century, this rate is actually on the rise today. Is this all due to drug resistance? What can be done to alleviate the situation? We will explore these questions in subsequent sections.

12.1 Learning Outcomes—Assess Your Progress

1. State the main goal of antimicrobial treatment.
2. Identify sources of the most commonly used antimicrobial drugs.
3. Summarize two methods for testing antimicrobial susceptibility.
4. Define *therapeutic index*, and identify whether a high or a low index is preferable in a drug.

12.2 Interactions Between Drug and Microbe

The goal of antimicrobial drugs is either to disrupt the cell processes or structures of pathogens or to inhibit their replication. Most of the drugs used in chemotherapy interfere with the function of enzymes required to synthesize or assemble macromolecules, or they destroy structures already formed in the cell. Above all, drugs should be **selectively toxic,** which means they should kill or inhibit microbial cells without simultaneously damaging host tissues. This concept of selective toxicity is central to antibiotic treatment, and the best drugs in current use are those that block the actions or synthesis of molecules in microorganisms but not in vertebrate cells. (We will see later that the drugs of the future may indeed inhibit microbial growth by attacking some aspect of the host cell they utilize.) Examples of drugs with excellent selective toxicity are those that block the synthesis of the cell wall in bacteria (penicillins). They have low toxicity and few direct effects on human cells because human cells lack the chemical peptidoglycan and are thus unaffected by this action of the antibiotic. Among the most toxic to human cells are drugs that act upon a structure common to both the infective agent and the host cell, such as the cell membrane (e.g., amphotericin B, used to treat fungal infections). As the characteristics of the infectious agent become more and more similar to those of the host cell, selective toxicity becomes more difficult to achieve and undesirable side effects are more likely to occur. We examine the subject in more detail in a later section.

Mechanisms of Drug Action

If the goal of chemotherapy is to disrupt the structure or function of an organism to the point where it can no longer survive, then the first step toward this goal is to identify the structural and metabolic needs of a living cell. Once the requirements of a living cell have been determined, methods of removing, disrupting, or interfering with these requirements can be used as potential chemotherapeutic strategies.

A Note About Chemotherapy

The word *chemotherapy* is commonly associated with the treatment of cancer. As you see in table 12.2, its official meaning is broader than that and can also be applied to antimicrobial treatment.

The metabolism of an actively dividing cell is marked by the production of new cell wall components (in most cells), DNA, RNA, proteins, and cell membrane. Consequently, current antimicrobial drugs are divided into categories based on which of these metabolic targets they affect. These categories are outlined in **figure 12.4** and include

1. inhibition of cell wall synthesis,
2. inhibition of nucleic acid (RNA and DNA) structure and function,
3. inhibition of protein synthesis,
4. interference with cell membrane structure or function, and
5. inhibition of folic acid synthesis.

As you will see, these categories are not completely discrete, and some effects can overlap. **Table 12.4** describes these categories, as well as common drugs within each.

Figure 12.4 Primary sites of action of antimicrobial drugs on bacterial cells.

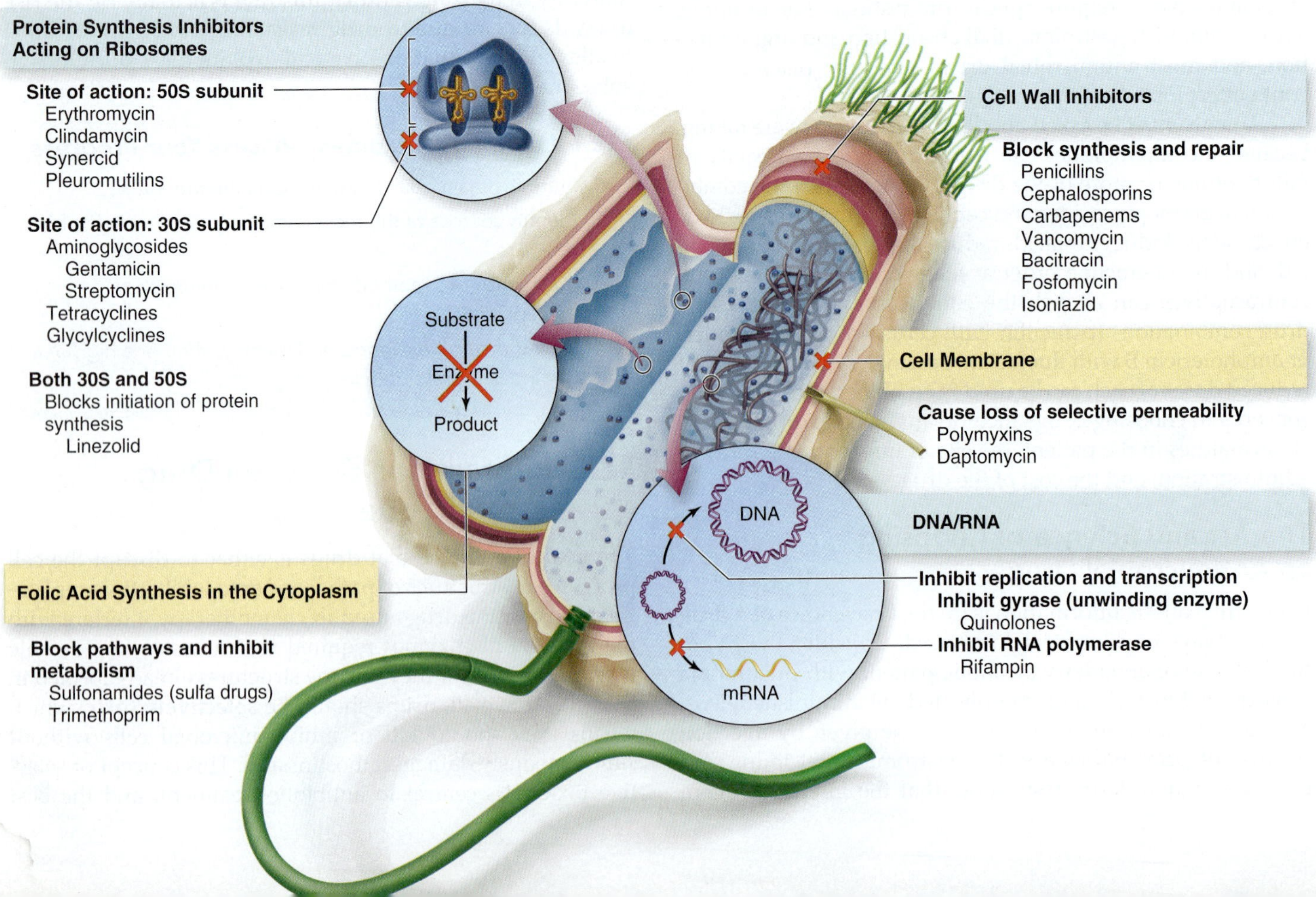

Protein Synthesis Inhibitors Acting on Ribosomes

Site of action: 50S subunit
Erythromycin
Clindamycin
Synercid
Pleuromutilins

Site of action: 30S subunit
Aminoglycosides
Gentamicin
Streptomycin
Tetracyclines
Glycylcyclines

Both 30S and 50S
Blocks initiation of protein synthesis
Linezolid

Substrate
Enzyme
Product

Folic Acid Synthesis in the Cytoplasm

Block pathways and inhibit metabolism
Sulfonamides (sulfa drugs)
Trimethoprim

Cell Wall Inhibitors

Block synthesis and repair
Penicillins
Cephalosporins
Carbapenems
Vancomycin
Bacitracin
Fosfomycin
Isoniazid

Cell Membrane

Cause loss of selective permeability
Polymyxins
Daptomycin

DNA

DNA/RNA

Inhibit replication and transcription
Inhibit gyrase (unwinding enzyme)
Quinolones
Inhibit RNA polymerase
Rifampin

mRNA

Table 12.4 Specific Antibacterial Drugs and Their Metabolic Targets

Drug Class/Mechanism of Action	Subgroups	Uses/Characteristics
Drugs That Target the Cell Wall		
Penicillins	Penicillins G and V	Most important natural forms used to treat gram-positive cocci, some gram-negative bacteria
	Ampicillin, carbenicillin, amoxicillin	Have a broad spectra of action, are semisynthetic; used against gram-negative enteric rods
	Methicillin, nafcillin, cloxacillin	Useful in treating infections caused by some penicillinase-producing bacteria (enzymes capable of destroying the beta-lactam ring of penicillin, which makes some bacteria resistant to penicillin)
	Mezlocillin, azlocillin	Extended spectrum; can be substituted for combinations of antibiotics
	Clavulanic acid	Inhibits beta-lactamase enzymes; added to penicillins to increase their effectiveness in the presence of penicillinase-producing bacteria
Cephalosporins	Cephalothin, cefazolin	First generation*; most effective against gram-positive cocci, few gram-negative bacteria
	Cefaclor, cefonicid	Second generation; more effective than first generation against gram-negative bacteria such as *Enterobacter, Proteus,* and *Haemophilus*
	Cephalexin, cefotaxime	Third generation; broad-spectrum, particularly against enteric bacteria that produce beta-lactamases
	Ceftriaxone	Third generation; semisynthetic broad-spectrum drug that treats wide variety of urinary, skin, respiratory, and nervous system infections
	Cefpirome, cefepime	Fourth generation
	Ceftobiprole	Fifth generation; used against methicillin-resistant *Staphylococcus aureus* (MRSA) and also against penicillin-resistant gram-positive and gram-negative bacteria
Carbapenems	Doripenem, imipenem	Powerful but potentially toxic; reserved for use when other drugs are not effective
	Aztreonam	Narrow-spectrum; used to treat gram-negative aerobic bacilli causing pneumonia, septicemia, and urinary tract infections; effective for those who are allergic to penicillin
Miscellaneous Drugs That Target the Cell Wall	Bacitracin	Narrow-spectrum; used to combat superficial skin infections caused by streptococci and staphylococci; main ingredient in Neosporin
	Isoniazid	Used to treat *Mycobacterium tuberculosis*, but only against growing cells; used in combination with other drugs in active tuberculosis
	Vancomycin	Narrow spectrum of action; used to treat staphylococcal infections in cases of penicillin and methicillin resistance or in patients with an allergy to penicillin
	Fosfomycin tromethamine	Phosphoric acid agent; effective as an alternative treatment for urinary tract infection caused by enteric bacteria
Drugs That Target Protein Synthesis		
Aminoglycosides Insert on sites on the 30S subunit and cause the misreading of the mRNA, leading to abnormal proteins	Streptomycin	Broad-spectrum; used to treat infections caused by gram-negative rods, certain gram-positive bacteria; used to treat bubonic plague, tularemia, and tuberculosis; vancomycin also targets protein synthesis as well as cell walls
Tetracyclines Block the attachment of tRNA on the A acceptor site and stop further protein synthesis	Tetracycline, oxytetracycline (Terramycin)	Effective against gram-positive and gram-negative rods and cocci, aerobic and anaerobic bacteria, mycoplasmas, rickettsias, and spirochetes
Glycylcyclines	Tigecycline	Derivative of tetracycline; effective against bacteria that have become resistant to tetracyclines

Cell lyses.

10,000×

*New improved versions of drugs are referred to as new "generations."

(continued)

Table 12.4 Specific Antibacterial Drugs and Their Metabolic Targets (*continued*)

Drug Class/Mechanism of Action	Subgroups	Uses/Characteristics
Drugs That Target Protein Synthesis (*continued*)		
Macrolides Inhibit translocation of the subunit during translation (erythromycin)	Erythromycin, clarithromycin, azithromycin	Relatively broad-spectrum, semisynthetic; used in treating ear, respiratory, and skin infections, as well as *Mycobacterium* infections in AIDS patients
Miscellaneous Drugs That Target Protein Synthesis	Clindamycin	Broad-spectrum antibiotic used to treat penicillin-resistant staphylococci, serious anaerobic infections of the stomach and intestines unresponsive to other antibiotics
	Quinupristin and dalfopristin (Synercid)	A combined antibiotic from the streptogramin group of drugs; effective against *Staphylococcus* and *Enterococcus* species causing endocarditis and surgical infections, including resistant strains
	Linezolid	Synthetic drug from the oxazolidinones; inhibits the initiation of protein synthesis; used to treat antibiotic-resistant organisms such as MRSA and VRE
Drugs That Target Folic Acid Synthesis		
Sulfonamides Interfere with folate metabolism by blocking enzymes required for the synthesis of tetrahydrofolate, which is needed by the cells for folic acid synthesis and eventual production of DNA, RNA, and amino acids	Sulfasoxazole	Used to treat shigellosis, acute urinary tract infections, certain protozoan infections
	Silver sulfadiazine	Used to treat burns, eye infections (in ointment and solution forms)
	Trimethoprim	Inhibits the enzymatic step immediately preceding the step inhibited by sulfonamides; trimethoprim often given in conjunction with sulfamethoxazole because of this synergistic effect; used to treat *Pneumocystis jiroveci* in AIDS patients
Drugs That Target DNA or RNA		
Fluoroquinolones Inhibit DNA unwinding enzymes or helicases, thereby stopping DNA transcription	Nalidixic acid	First generation; rarely used anymore
	Ciprofloxacin, ofloxacin	Second generation
	Levofloxacin	Third generation; used against gram-positive organisms, including some that are resistant to other drugs
	Trovafloxacin	Fourth generation; effective against anaerobic organisms
Miscellaneous Drugs That Target DNA or RNA	Rifamycin (altered chemically into rifampin)	Limited in spectrum because it cannot pass through the cell envelope of many gram-negative bacilli; mainly used to treat infections caused by gram-positive rods and cocci and a few gram-negative bacteria; used to treat leprosy and tuberculosis
Drugs That Target Cell Membranes		
Polymyxins Interact with membrane phospholipids; distort the cell surface and cause leakage of protein and nitrogen bases, particularly in gram-negative bacteria	Polymyxin B and E	Used to treat drug-resistant *Pseudomonas aeruginosa* and severe urinary tract infections caused by gram-negative rods
	Daptomycin	Most active against gram-positive bacteria

Sulfa drug (inhibitor)

Higher levels of sulfa drug more likely to bind to enzyme

Enzyme

Helicase

12.2 Learning Outcomes—Assess Your Progress

5. Explain the concept of selective toxicity.
6. Describe the five major targets of antimicrobial agents, and list major drugs associated with each.
7. Identify which categories of drugs are most selectively toxic, and explain why they exhibit this effect.

12.3 Survey of Major Antimicrobial Drug Groups

Scores of antimicrobial drugs are marketed in the United States. Although the medical and pharmaceutical literature contains a wide array of names for antimicrobials, most of them are variants of a small number of drug families. One of the most useful ways of categorizing antimicrobials is to designate them as either **broad-spectrum** or **narrow-spectrum.** Broad-spectrum drugs are effective against more than one group of bacteria, while narrow-spectrum drugs generally target a specific group. **Table 12.5** demonstrates that tetracyclines are broad-spectrum, while polymyxin and even penicillins are narrow-spectrum agents.

The rest of this section provides details about drugs based on the five major mechanisms they target. There will also be a discussion of the special—and important—case of treating biofilm infections with antibiotics.

Antibacterial Drugs Targeting the Cell Wall

Penicillin and Its Relatives

The **penicillin** group of antibiotics, named for the parent compound, is a large, diverse group of compounds, most of which end in the suffix *-cillin.* Penicillins can be either completely synthesized in the laboratory from simple raw materials or obtained naturally through microbial fermentation. The natural product can then be used either in unmodified form or to make semisynthetic derivatives. For 80 years, it was thought that *Penicillium chrysogenum* was the source of the drug, but it was recently discovered that the original penicillin-producing fungus was a different *Penicillium* species that is yet unnamed. All penicillins consist of three parts: a thiazolidine ring, a *beta-lactam* (bey'-tuh-lak'-tam) ring, and a variable side chain that dictates its microbicidal activity **(figure 12.5).**

Subgroups and Uses of Penicillins The characteristics of certain penicillin drugs are shown in **table 12.6.** Penicillins G and V are the most important natural forms. Penicillin is considered the drug of choice for infections by known sensitive, gram-positive cocci (some streptococci) and some gram-negative bacteria (e.g., the syphilis spirochete).

Certain semisynthetic penicillins such as ampicillin, carbenicillin, and amoxicillin have broader spectra and thus can be used to treat infections by gram-negative enteric rods. Many bacteria produce enzymes that are capable of destroying the beta-lactam ring of penicillin. The enzymes are referred to as **penicillinases** or *beta-lactamases,* and they make

Table 12.5 Spectrum of Activity for Antibiotics

Bacteria	Mycobacteria	Gram-Negative Bacteria	Gram-Positive Bacteria	Chlamydias	Rickettsias
Examples of diseases	Tuberculosis	Salmonellosis, plague, gonorrhea	Strep throat, staph infections*	Chlamydia, trachoma	Rocky Mountain spotted fever
Spectrum of activity of various antibiotics	Isoniazid — Streptomycin — Tobramycin — Polymyxin — Carbapenems — Tetracyclines — Sulfonamides Cephalosporins — Penicillins				
Are there normal biota in this group?	Yes	Yes	Yes	Probably	None known

*Note that some members of a bacterial group may not be affected by the antibiotics indicated, due to acquired or natural resistance. In other words, exceptions do exist.

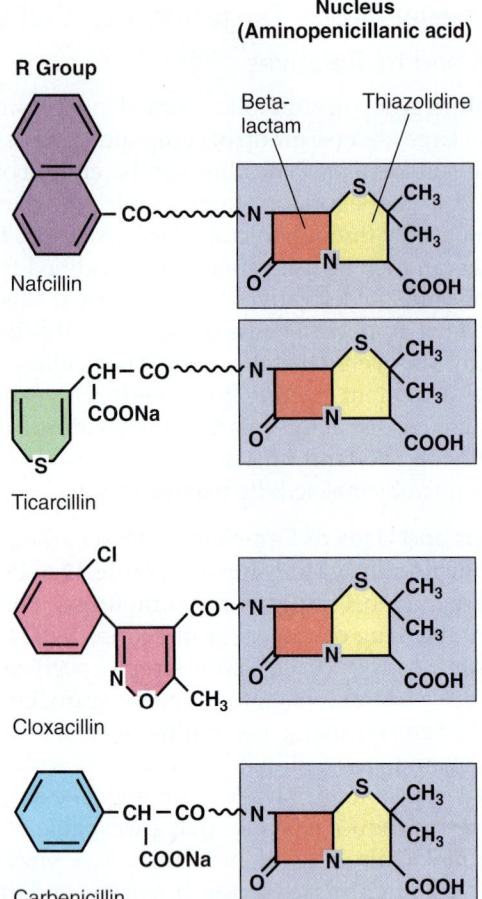

Figure 12.5 Chemical structure of penicillins. All penicillins contain a thiazolidine ring (yellow) and a beta-lactam ring (red), but each differs in the nature of the side chain (R group), which is also responsible for differences in biological activity.

the bacteria that possess them resistant to many penicillins. Researchers have counted as many as 532 different beta-lactamases in bacteria. This points to how versatile bacteria can be in resisting our attacks on them. In response, scientists have created penicillinase-resistant penicillins such as methicillin, nafcillin, and cloxacillin, some of which are useful in treating infections caused by some penicillinase-producing bacteria. All of the "-cillin" drugs are relatively mild and well tolerated because of their specific mode of action on cell walls (which humans lack). The primary problems in therapy include allergy, which is altogether different from toxicity, and resistant strains of pathogens. Clavulanic acid is a chemical that inhibits beta-lactamase enzymes, thereby increasing the effectiveness of beta-lactam antibiotics in the presence of penicillinase-producing bacteria. For this reason, clavulanic acid is often added to semisynthetic penicillins to augment their effectiveness. For example, a combination of amoxicillin and clavulanate is marketed under the trade name Augmentin. Zosyn is a similar combination of tazobactam, a beta-lactamase inhibitor, and piperacillin that is used for a wide variety of systemic infections.

The Cephalosporin Group of Drugs

The **cephalosporins** are a group of antibiotics that were originally isolated in the late 1940s from the mold *Cephalosporium acremonium*. Cephalosporins are similar to penicillins; they have a beta-lactam structure that can be synthetically altered **(figure 12.6)** and have a similar mode of action. The generic names of these compounds are often recognized by the presence of the root *cef, ceph,* or *kef* in their names.

Subgroups and Uses of Cephalosporins The cephalosporins are versatile. They are relatively broad-spectrum, resistant to most penicillinases, and cause fewer allergic reactions than penicillins. Although some cephalosporins are given orally, many are poorly absorbed from the intestine and must be administered **parenterally** (par-ehn'-tur-ah-lee) by injection into a muscle or a vein.

Table 12.6 Characteristics of Selected Penicillin Drugs

Name	Spectrum of Action	Uses, Advantages	Disadvantages
Penicillin G	Narrow	Best drug of choice when bacteria are sensitive; low cost; low toxicity	Can be hydrolyzed by penicillinase; allergies occur; requires injection
Penicillin V	Narrow	Good absorption from intestine; otherwise, similar to penicillin G	Hydrolysis by penicillinase; allergies
Oxacillin, dicloxacillin	Narrow	Not susceptible to penicillinase; good absorption	Allergies; expensive
Methicillin, nafcillin	Narrow	Not usually susceptible to penicillinase	Poor absorption; allergies; growing resistance
Ampicillin	Broad	Works on some gram-negative bacteria	Can be hydrolyzed by penicillinase; allergies; only fair absorption
Amoxicillin	Broad	Gram-negative infections; good absorption	Hydrolysis by penicillinase; allergies
Carbenicillin	Broad	Works on many gram-negative bacteria	Poor absorption; used only parenterally
Azlocillin, mezlocillin, ticarcillin	Very broad	Effective against *Pseudomonas* species; low toxicity compared with aminoglycosides	Allergies, susceptible to many beta-lactamases

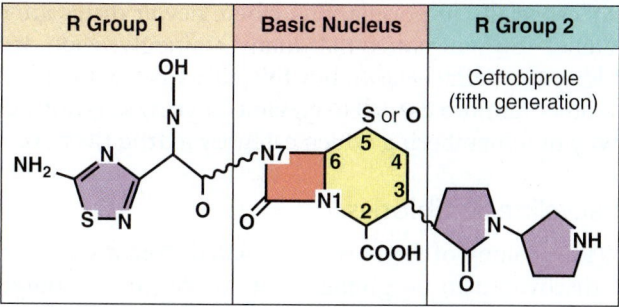

R Group 1	Basic Nucleus	R Group 2

Figure 12.6 The structure of cephalosporins. Like penicillin, cephalosporins have a beta-lactam ring (red), but they have a different main ring (yellow). However, unlike penicillins, they have two sites for placement of R groups (at positions 3 and 7). This makes possible greater versatility in function and complexity in structure.

Five generations of cephalosporins exist, and display different levels of antibacterial activity. First-generation cephalosporins such as cephalothin and cefazolin are most effective against gram-positive cocci and a few gram-negative bacteria. Second-generation forms include cefaclor and cefonicid, which are more effective than the first-generation forms in treating infections by gram-negative bacteria such as *Enterobacter, Proteus,* and *Haemophilus.* Third-generation cephalosporins, such as cephalexin (Keflex) and cefotaxime, are broad-spectrum with especially well-developed activity against enteric bacteria that produce beta-lactamases. Ceftriaxone (Rocephin) is a semisynthetic broad-spectrum drug for treating a wide variety of respiratory, skin, urinary, and nervous system infections. The fourth-generation cephalosporins include cefpirome and cefepime. The fifth-generation drug ceftobiprole exhibits activity against methicillin-resistant *Staphylococcus aureus* and also against penicillin-resistant gram-positive and gram-negative bacteria.

Other Beta-Lactam Antibiotics

Newer antibiotics such as doripenem and imipenem belong to a new class of cell wall antibiotics called carbapenems. They are powerful but potentially dangerous and reserved for use in hospitals when other drugs aren't working. Aztreonam, isolated from the bacterium *Chromobacterium violaceum,* is a narrow-spectrum drug for treating pneumonia, septicemia, and urinary tract infections by gram-negative aerobic bacilli. Aztreonam is especially useful when treating persons who are allergic to penicillin. Because of similarities in their chemical structure, allergies to penicillin often are accompanied by allergies to cephalosporins and carbapenems. The structure of aztreonam is chemically distinct so that persons with allergies to penicillin are not usually adversely affected by treatment with aztreonam.

Recently, the appearance of a gene coding for the NDM enzyme (see Case File) in gram-negative bacteria has caused great concern because it confers resistance to carbapenems and is highly transmissible from bacterium to bacterium. This development has caused some scientists to declare that the end of the antibiotic era is nearing.

Other Drugs Targeting the Cell Wall

Bacitracin is a narrow-spectrum antibiotic produced by a strain of the bacterium *Bacillus subtilis.* Since it was first isolated, its greatest claim to fame has been as a major ingredient in a common drugstore antibiotic ointment (Neosporin) for combating superficial skin infections by streptococci and staphylococci. For this purpose, it is usually combined with neomycin (an aminoglycoside) and polymyxin.

Isoniazid (INH) is bactericidal to *Mycobacterium tuberculosis* but only against growing cells. It is generally used in combination with two or three additional drugs in active tuberculosis cases. **Vancomycin** is a narrow-spectrum antibiotic most effective in treating staphylococcal infections in cases of penicillin and methicillin resistance or in patients with an allergy to penicillins. Vancomycin belongs to the first generation of glycopeptide antibiotics, initially

Disease Connection

Tuberculosis, or the "White Death," has been a widespread and deadly disease throughout recorded history. The discovery in the 1950s that isoniazid was effective against the disease gave the world the first real hope that the disease could be cured. Isoniazid is still the first-line drug for tuberculosis, but many strains of the bacterium are resistant to it, and it is always administered with at least one other antibiotic—and sometimes three or four.

used in the 1960s; it is currently used more widely because gram-positive bacteria have become resistant to methicillin, consequently plaguing hospitals and the community at large. Second-generation glycopeptides include telavancin and oritavancin. Fosfomycin is a phosphoric acid agent effective as alternate treatment for urinary tract infections caused by enteric bacteria. It works by inhibiting an enzyme necessary for cell wall synthesis.

Antibacterial Drugs Targeting Protein Synthesis

The Aminoglycoside Drugs

Antibiotics composed of one or more amino sugars and an aminocyclitol (6-carbon) ring are referred to as **aminoglycosides (figure 12.7a)**. These complex compounds are exclusively the products of various species of soil **actinomycetes** in the genera *Streptomyces* **(figure 12.7b)** and *Micromonospora*.

Subgroups and Uses of Aminoglycosides The aminoglycosides have a relatively broad antimicrobial spectrum because they inhibit protein synthesis **(figure 12.8)**. They are especially useful in treating infections caused by aerobic gram-negative rods and certain gram-positive bacteria. Streptomycin is among the oldest of the drugs and has gradually been replaced by newer forms with less mammalian toxicity. It is still the antibiotic of choice for treating bubonic plague and tularemia and is considered a good antituberculosis

agent, especially in populations where newer drugs are not available. You will notice that many aminoglycoside drugs end with the suffix *-mycin*, but this suffix is used for drugs from other families as well (e.g., vancomycin) so is not a useful way of remembering which category a drug fits into.

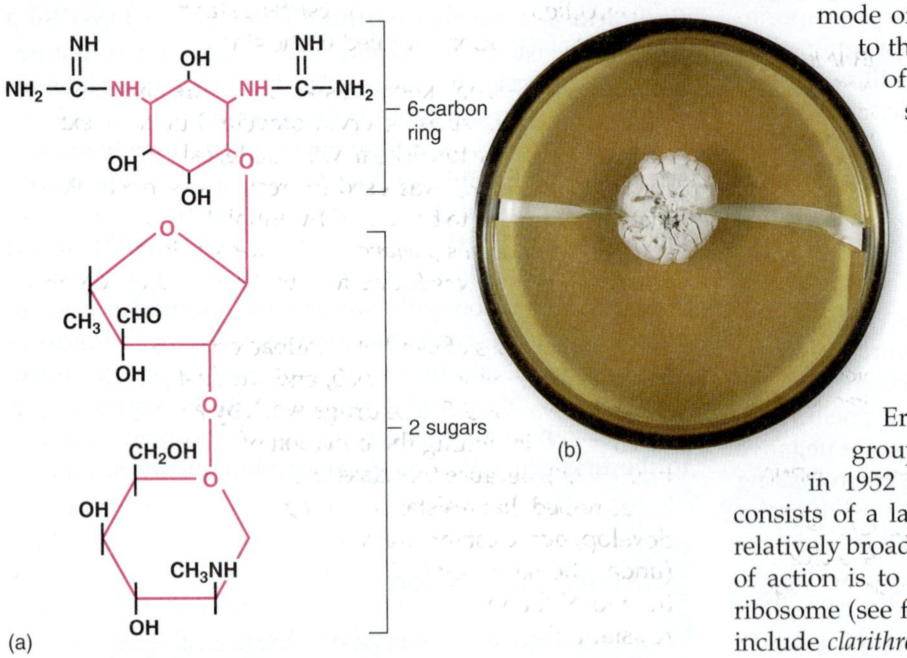

Tetracycline Antibiotics

In 1948, a colony of *Streptomyces* isolated from a soil sample was discovered to be giving off a substance, aureomycin, with strong antimicrobial properties. This antibiotic was used to synthesize its relatives terramycin and tetracycline. These natural parent compounds and semisynthetic derivatives are known as the **tetracyclines (figure 12.9a)**. Their action of binding to ribosomes and blocking protein synthesis accounts for the broad-spectrum effects in the group (see figure 12.8).

Subgroups and Uses of Tetracyclines The scope of microorganisms inhibited by tetracyclines includes gram-positive and gram-negative rods and cocci, aerobic and anaerobic bacteria, mycoplasmas, rickettsias, and spirochetes. Although generic tetracycline is low in cost and easy to administer, its side effects—namely, gastrointestinal disruption due to changes in the normal biota of the gastrointestinal tract and deposition in hard tissues—can limit its use (see table 12.9).

Glycylcyclines

Glycylcyclines are newer derivatives of tetracyclines. Their mode of action, like that of tetracyclines, is to bind to the 30S ribosomal subunit and block the entry of the tRNA bearing an amino acid into the A site of the ribosome (see figure 12.8). However, differences in the structure of the drug make it effective against bacteria that have become resistant to the tetracyclines. The first antibiotic licensed in this group is tigecycline, marketed as Tygacil.

Erythromycin, Clindamycin, and Telithromycin

Erythromycin is an antibiotic with a chemical group called a macrolide ring; it was first isolated in 1952 from a strain of *Streptomyces*. Its structure consists of a large ring with sugars attached. This drug is relatively broad-spectrum and of fairly low toxicity. Its mode of action is to block protein synthesis by attaching to the ribosome (see figure 12.8). Newer semisynthetic macrolides include *clarithromycin* and *azithromycin*. Both drugs are useful for middle ear, respiratory, and skin infections and have also been approved for *Mycobacterium* infections in AIDS patients. Clarithromycin has additional applications in controlling infectious stomach ulcers (see chapter 22).

Clindamycin is a broad-spectrum antibiotic derived from the less effective lincomycin. The tendency of clindamycin to

Figure 12.7 Streptomycin. **(a)** Chemical structure of the antibiotic. Colored portions of the molecule show the general arrangement of an aminoglycoside. **(b)** A colony of *Streptomyces*, one of nature's most prolific antibiotic producers. (*Note:* The white streaks are places where the agar has cracked, which often happens as it dries out.)

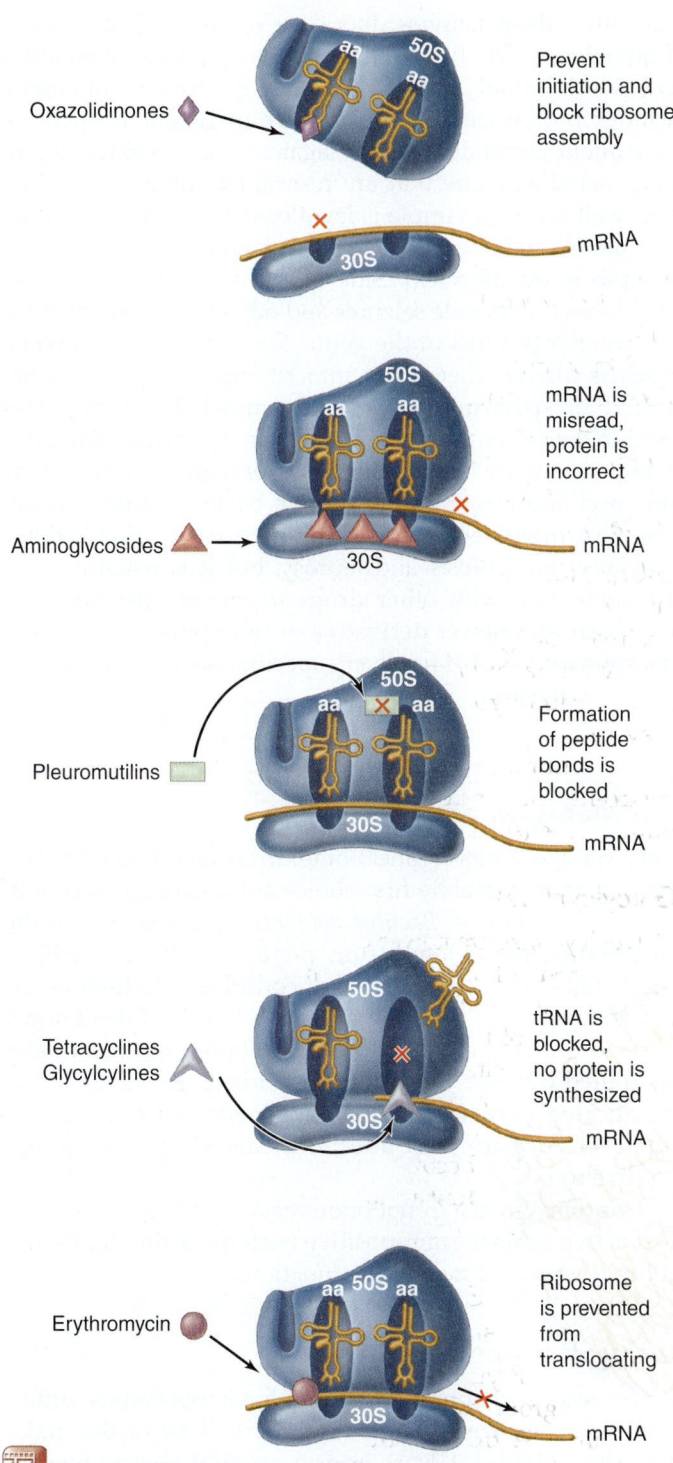

Figure 12.8 Sites of inhibition on the bacterial ribosome and major antibiotics that act on these sites. All have the general effect of blocking protein synthesis. Blockage actions are indicated by a red X.

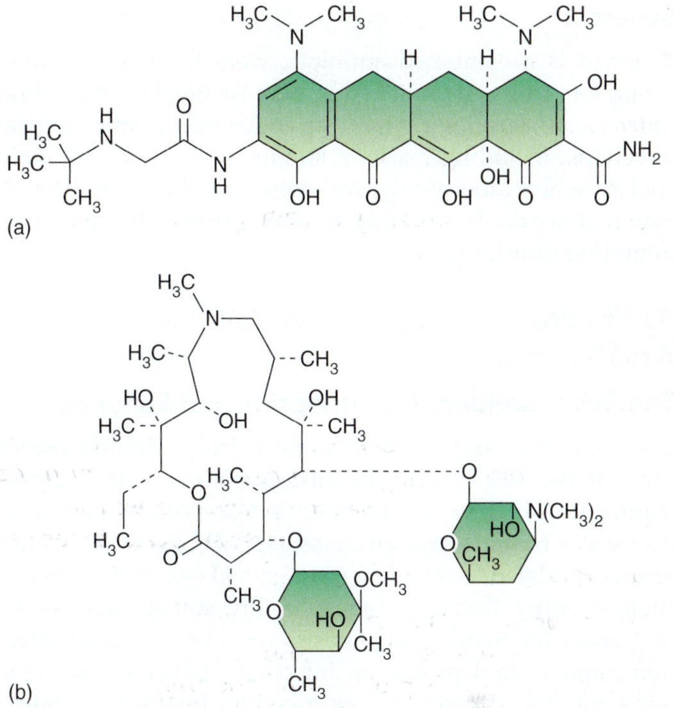

cause adverse reactions in the gastrointestinal tract limits its applications to

1. serious infections in the large intestine and abdomen due to anaerobic bacteria (*Bacteroides* and *Clostridium*) that are unresponsive to other antibiotics,

Figure 12.9 Structures of two broad-spectrum antibiotics. **(a)** Tigecycline, a glycylcycline, a newer tetracycline. This is named for its regular group of four rings. The several types vary in structure and activity by substitution at the four R groups. **(b)** Azithromycin, an example of a macrolide drug. Its central feature is a large lactone ring to which two hexose sugars are attached.

2. infections with penicillin-resistant staphylococci, and
3. acne medications applied to the skin.

A class of drugs known as ketolides are similar to macrolide antibiotics such as erythromycin, but they exhibit a different ring structure.The new ketolide, called telithromycin (trade name Ketek), was used for respiratory tract infections that are suspected to be caused by antibiotic-resistant bacteria such as *Streptococcus pneumoniae*. However, its usefulness has been limited as it was found to cause serious liver damage in some patients.

Another class of synthetic antibacterial drugs, oxazolidinones, was developed in 2000, and the first member of that class was linezolid. These drugs work by a completely novel mechanism, inhibiting the initiation of protein synthesis (see figure 12.8). Because this class of drug is not found in nature, it was hoped that resistance among bacteria would be slow to develop, but resistant strains have been identified. Linezolid (under the name Zyvox) is used to treat infections caused by two of the most difficult clinical pathogens, methicillin-resistant *Staphylococcus aureus* (MRSA) and vancomycin-resistant *Enterococcus* (VRE).

Drugs called pleuromutilins block the action of peptidyl transferase (see figure 12.8). The first representative of the pleuromutilins is retapamulin (Altabax) and is approved only for external application for skin infections such as impetigo (see chapter 18).

Synercid

Synercid is a combined antibiotic from the streptogramin group of drugs. It is effective against *Staphylococcus* and *Enterococcus* species that cause endocarditis and surgical infections, including resistant strains. It is one of the main choices when other drugs are ineffective due to resistance. Synercid works by binding to sites on the 50S ribosome, inhibiting translation.

Antibacterial Drugs Targeting Folic Acid Synthesis

The Sulfonamides, Trimethoprim, and Sulfones

The very first modern antimicrobial drugs, finding use as early as the 1930s, were the **sulfonamides,** or sulfa drugs **(figure 12.10).** They are synthetic and do not originate from bacteria or fungi. Although thousands of sulfonamides have been formulated, only a few have gained any importance in chemotherapy. Because of its solubility, sulfisoxazole is the best agent for treating shigellosis, acute urinary tract infections, and certain protozoan infections. Silver sulfadiazine ointment and solution are prescribed for treatment of burns and eye infections. Another drug, trimethoprim (Septra, Bactrim), inhibits the enzymatic step immediately following the step inhibited by sulfonamides in the synthesis of folic acid. Because of this, trimethoprim is often given in combination with sulfamethoxazole to take advantage of the synergistic effect of the two drugs. This combination is one of the primary treatments for *Pneumocystis jiroveci* pneumonia (PCP) in AIDS patients.

Antibacterial Drugs Targeting DNA or RNA

Even though nucleic acids in bacteria and humans are chemically similar, DNA and RNA have proven to be useful targets for antimicrobials. **Fluoroquinolones** exhibit several ideal traits, including high potency and broad-spectrum. Even in minimal concentrations, quinolones inhibit a wide variety of gram-positive and gram-negative bacterial species. In addition, they are readily absorbed from the intestine. Just as

with other drug families, there are multiple "generations" of quinolones. The first generation was typified by nalidixic acid, which is rarely used now. Second-generation quinolones include ciprofloxacin and ofloxacin. Third-generation quinolones exhibit expanded activity against gram-positive organisms, including some that are resistant to other drugs. The most well-known example is levofloxacin. Fourth-generation quinolones are effective against anaerobic organisms; an example is trovafloxacin. Side effects that limit the use of quinolones can include seizures and other brain disturbances.

Another product of the genus *Streptomyces* is rifamycin, which is altered chemically into rifampin. It is somewhat limited in spectrum because the molecule cannot pass through the cell envelope of many gram-negative bacilli. It is mainly used to treat infections by several gram-positive rods and cocci and a few gram-negative bacteria. Rifampin figures most prominently in treating mycobacterial infections, especially tuberculosis and leprosy, but it is usually given in combination with other drugs to prevent development of resistance. A newer derivative of rifampin called Xifaxan was approved in 2004 for a very specific use: the treatment of traveler's diarrhea.

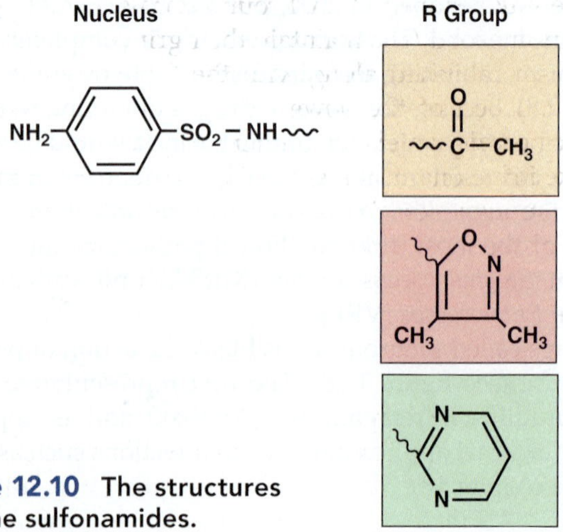

Antibacterial Drugs Targeting Cell Membranes

Every cell has a membrane. Some drugs target membranes, but they are not usually first-choice antimicrobials except in a few circumstances. *Bacillus polymyxa* is the source of the **polymyxins,** narrow-spectrum peptide antibiotics with a unique fatty acid component that contributes to their detergent activity. Only two polymyxins—B and E (also known as colistin)—have any routine applications, and even these are limited by their toxicity to the kidney. Either drug can be indicated to treat drug-resistant *Pseudomonas aeruginosa* and severe urinary tract infections caused by other gram-negative rods.

Daptomycin is a lipopeptide made by *Streptomyces*. It is most active against gram-positive bacteria, acting to disrupt multiple aspects of membrane function.

Antibiotics and Biofilms

As you read in chapter 7, biofilm inhabitants behave differently than their free-living counterparts. One of the major ways they differ—at least from a medical perspective—is that they are as much as 1,000 times less sensitive to the same antimicrobials that work against them when they are free-living. When this was first recognized, it was assumed that it was a problem of penetration, that the (often ionically charged) antimicrobial drugs could not penetrate the sticky extracellular material surrounding biofilm organisms. While that is a factor, there is something more important contributing to biofilm resistance: the different phenotype expressed by biofilm bacteria. When secured to surfaces, they express different genes and therefore have different antibiotic susceptibility profiles.

Figure 12.10 The structures of some sulfonamides.

Years of research have so far not yielded an obvious solution to this problem, though there are several partially successful strategies. One of these involves interrupting the quorum-sensing pathways that mediate communication between cells and may change phenotypic expression. Daptomycin, a lipopeptide that is effective in deep-tissue infections with resistant bacteria, has also shown some success in biofilm infection treatment. Also, some researchers have found that adding DNase to their antibiotics can help with penetration of the antibiotic through the extracellular debris—apparently some of which is DNA from lysed cells. Recent studies have shown that pretreatment of biofilms with newly found, natural antimicrobial compounds before exposure to common antibiotics increases the effectiveness of treatment against even multidrug-resistant strains of bacteria.

Many biofilm infections can be found on biomaterials inserted in the body, such as cardiac or urinary catheters. These can be impregnated with antibiotics prior to insertion to prevent colonization. This, of course, cannot be done with biofilm infections of natural tissues, such as the prostate or middle ear.

Interestingly, it appears that chemotherapy with some antibiotics—notably, aminoglycosides—can cause bacteria to form biofilms at a higher rate than they otherwise would. Obviously, there is much more to come in understanding biofilms and their control.

Agents to Treat Fungal Infections

Because the cells of fungi are eukaryotic, they present special problems in chemotherapy. For one, the great majority of chemotherapeutic drugs are designed to act on bacteria and are generally ineffective in combating fungal infections. For another, the similarities between fungal and human cells often mean that drugs toxic to fungal cells are also capable of harming human tissues. A few agents with special antifungal properties have been developed for treating systemic and superficial fungal infections. Four main drug groups currently in use are the polyene antibiotics, the azoles, the echinocandins, and the allylamines, in addition to a few miscellaneous drugs **(table 12.7).**

Antiprotozoal and Antihelminthic Chemotherapy

The enormous diversity among protozoan and helminth parasites and their corresponding therapies reach far beyond the scope of this textbook; however, a few of the more common drugs are surveyed here and described again for particular diseases in the organ systems chapters.

Antimalarial Drugs: Quinine and Its Relatives

Quinine, extracted from the bark of the cinchona tree, was the principal treatment for malaria for hundreds of years, but it has been replaced by the synthesized quinolones, mainly chloroquine and primaquine, which have less toxicity to humans. Because there are several species of *Plasmodium* (the malaria parasite) and many stages in its life cycle, no single drug is universally effective for every species and stage, and each drug is restricted in application. For instance, primaquine eliminates the liver phase of infection, and chloroquine suppresses acute attacks associated with infection of red blood cells. Chloroquine is taken alone for prophylaxis and in combination with doxycycline or other antibiotics for the suppression of acute forms of malaria. Primaquine is administered to patients with relapsing cases of malaria. Artemisinin combination therapy (ACT) is recommended for the treatment of certain types of malaria today and uses artemisinin (a compound from wormwood) with quinine derivatives or other drugs.

Chemotherapy for Other Protozoan Infections

A widely used amoebicide, metronidazole (Flagyl), is effective in treating mild and severe intestinal infections and hepatic disease caused by *Entamoeba histolytica.* Given orally, it also has applications for infections by *Giardia lamblia* and *Trichomonas vaginalis* (described in chapters 22 and 23, respectively). Other drugs with antiprotozoan activities are quinacrine (a quinine-based drug), sulfonamides, and tetracyclines.

Antihelminthic Drug Therapy

Treating helminthic infections has been one of the most difficult and challenging of all chemotherapeutic tasks. Flukes, tapeworms, and roundworms are much larger parasites than other microorganisms and, being animals, have greater similarities to human physiology. Also, the usual strategy of using drugs to block their reproduction is usually not successful in eradicating the adult worms. The most effective drugs immobilize, disintegrate, or inhibit the metabolism of all stages of the life cycle but especially nondividing helminths.

Mebendazole and albendazole are broad-spectrum antiparasitic drugs used in several roundworm intestinal infestations. These drugs work locally in the intestine to inhibit the function of the microtubules of worms, eggs, and larvae. This means the parasites can no longer utilize glucose, which leads to their demise. The compound pyrantel paralyzes the muscles of intestinal roundworms. Consequently, the worms are unable to maintain their grip on the intestinal wall and are expelled along with the feces by the normal peristaltic action of the bowel. Two newer antihelminthic drugs are praziquantel, a treatment for various tapeworm and fluke infections, and ivermectin, a veterinary drug now used for strongyloidiasis and oncocercosis in humans. Helminthic diseases are described in chapter 22 because these organisms spend at least some part of their life cycles in the digestive tract.

Antiviral Chemotherapeutic Agents

The chemotherapeutic treatment of viral infections presents unique problems. With a virus, we are dealing with an infectious agent that relies upon the host cell for the vast majority

Table 12.7 Agents Used to Treat Fungal Infections

Drug Group	Drug	Action
Macrolide polyenes	Amphotericin (shown) Natamycin Nystatin	• Bind to fungal membranes, causing loss of selective permeability; extremely versatile • Can be used to treat skin, mucous membrane lesions caused by *Candida albicans*; injectable form can be used to treat histoplasmosis and *Cryptococcus* meningitis • Natamycin—used for treatment of fungal keratitis • Nystatin—used in treatment of *Candida* and *Cryptococcus* infection
Azoles	Imidazoles (ketoconazole, clotrimazole (shown), miconazole) Triazoles (fluconazole, itraconazole, voriconazole) Thiazoles (abafungin) 	• Interfere with sterol synthesis in fungi • Ketoconazole—cutaneous mycoses, vaginal and oral candidiasis, systemic mycoses • Fluconazole—AIDS-related mycoses (aspergillosis, *Cryptococcus* meningitis) • Clotrimazole and miconazole—used to treat infections in the skin, mouth, and vagina • Abafungin—synthetic antifungal drug that targets both dividing and undividing cells
Echinocandins	Micafungin, caspofungin (shown) Anidulafungin 	• Inhibit fungal cell wall synthesis • Useful in immunocompromised patients • Used against *Candida* strains and aspergillosis
Allylamines	Terbinafine (Lamisil)	• Block sterol synthesis in fungi • Highly lipophilic making them useful in the treatment of dermatophytic infections
Miscellaneous	Griseofulvin	• Inhibits microtubule function in fungi • Use in oral form for dermatophytic infections
	Flucytosine 	• Rapidly absorbed orally, readily dissolves in the blood and CSF (cerebrospinal fluid) • Used to treat cutaneous mycoses • Usually combined with amphotericin B to treat systemic mycoses because many fungi are resistant to this drug

of its metabolic functions. In currently used drugs, disrupting viral metabolism requires that we disrupt the metabolism of the host cell. Put another way, selective toxicity with regard to viral infection is difficult to achieve because a single metabolic system is responsible for the well-being of both virus and host. Although viral diseases such as measles, mumps, and hepatitis are routinely prevented by the use of effective vaccinations, epidemics of AIDS, influenza, and even the common cold attest to the need for more effective medications for the treatment of viral pathogens.

The currently used antiviral drugs were developed to target specific points in the infectious cycle of viruses. Three major modes of action are

1. barring penetration of the virus into the host cell,
2. blocking the transcription and translation of viral molecules, and
3. preventing the maturation of viral particles.

Table 12.8 presents a comprehensive overview of the most widely used antiviral drugs. The following examples provide some additional detail about the principles in table 12.8. Although antiviral drugs protect uninfected cells by keeping viruses from being synthesized and released, most are unable to destroy extracellular viruses or those in a latent state.

Fuzeon (generic name enfuvirtide), an anti-HIV drug approved in 2003, keeps the virus from attaching to its cellular receptor and thereby prevents the initial fusion of HIV to the host cell. Relenza and Tamiflu medications can be effective treatments for influenza A and B and useful prophylactics as well. Because one action of these drugs is to inhibit the fusion and uncoating of the virus, they must be given rather early in an infection. Also, viruses can quickly become resistant to antivirals. The dominant flu virus circulating in 2009–2010 was mostly resistant to Tamiflu, for example.

Several antiviral agents mimic the structure of nucleotides and compete for sites on replicating DNA. The incorporation of these synthetic nucleotides inhibits further DNA synthesis. **Acyclovir** (Zovirax) and its relatives are synthetic purine compounds that block DNA synthesis in a small group of viruses, particularly the herpesviruses (see chapters 18 and 23). Some derivatives of this drug are valganciclovir, famciclovir, and penciclovir.

HIV is classified as a retrovirus, meaning it carries its genetic information in the form of RNA rather than DNA (HIV and AIDS are discussed in chapter 20). Upon infection, the RNA genome is used as a template by the enzyme **reverse transcriptase** to produce a DNA copy of the virus' genetic information. Because this particular reaction is not seen outside of the retroviruses, it offers two ideal targets for chemotherapy. The first is interfering with the synthesis of the new DNA strand, which is accomplished using *nucleoside reverse transcriptase inhibitors* (nucleotide analogs), while the second involves interfering with the action of the enzyme responsible for the synthesis, which is accomplished using *nonnucleoside reverse transcriptase inhibitors*.

Azidothymidine (AZT or zidovudine) is a thymine analog that exerts its effect by incorporating itself into the growing DNA chain of HIV and terminating synthesis, in a manner analogous to that seen with acyclovir. AZT is used at all stages of HIV infection, including prophylactically with people accidentally exposed to blood or other body fluids.

Nonnucleoside reverse transcriptase inhibitors (such as nevirapine) accomplish the same goal (preventing reverse transcription of the HIV genome) by binding to the reverse transcriptase enzyme itself, inhibiting its ability to synthesize DNA.

Assembly and release of mature viral particles are also targeted in HIV through the use of protease inhibitors. These drugs (indinavir, saquinavir), usually used in combination with nucleotide analogs and reverse transcriptase inhibitors, have been shown to reduce the HIV load to undetectable levels by preventing the maturation of virus particles in the cell. Refer to table 12.8 for a summary of HIV drug mechanisms and see chapter 20 for further coverage of this topic.

A sensible alternative to artificial drugs has been a human-based substance, **interferon (IFN).** Interferon is a glycoprotein produced primarily by fibroblasts and leukocytes in response to various immune stimuli. It has numerous biological activities, including antiviral and anticancer properties. Studies have shown that it is a versatile part of animal host defenses, having a major role in natural immunities. (Its mechanism is discussed in chapter 14.)

Many scientists believe that a breakthrough made in 2011 has the potential to revolutionize antiviral treatment. Researchers at the Massachusetts Institute of Technology created a drug that could be used to treat a wide variety of different viral infections, unlike current antivirals that tend to be effective against one or, at most, a few specific viruses. The new drug, called DRACO, detects the presence of long dsRNA strands in host cells. While host cells make short dsRNA (siRNAs), they are generally never longer than 30 nucleotides. If viruses are multiplying inside a cell, DRACO detects the long dsRNA and causes the cell to undergo self-destruction. Neighboring healthy cells are untouched. Studies have shown DRACO to be successful against a variety of viruses, because it targets a process that is common to almost all virally infected cells.

12.3 Learning Outcomes—Assess Your Progress

8. Distinguish between broad-spectrum and narrow-spectrum antimicrobials, and explain the significance of the distinction.
9. Trace the development of penicillin antimicrobials, and identify which microbes they are effective against.
10. Describe the action of beta-lactamases, and explain their importance in drug resistance.
11. List examples of other beta-lactam antibiotics.
12. Describe common cell wall antibiotics that are not in the beta-lactam class of drugs.
13. Identify the ribosomal targets of several antibiotics that inhibit protein synthesis.

Table 12.8 Actions of Antiviral Drugs

Mode of Action	Examples	Effects of Drug	
Inhibition of Virus Entry: Receptor/ fusion/uncoating inhibitors	Enfuvirtide (Fuzeon)	Blocks *HIV* infection by preventing the binding of viral GP-41 receptors to cell receptor (1), thereby preventing fusion of virus with cell	
	Amantadine and its relatives, zanamivir (Relenza), oseltamivir (Tamiflu)	Block entry of *influenza virus* by interfering with fusion of virus with cell membrane (also release); stop the action of influenza neuraminidase, required for entry of virus into cell (also assembly) (2) (3)	
Inhibition of Nucleic Acid Synthesis	Acyclovir (Zovirax), other "-cyclovirs," vidarabine	Purine analogs that terminate DNA replication in *herpesviruses* (4)	
	Ribavirin	Purine analog, used for *respiratory syncytial virus (RSV)* and some *hemorrhagic fever viruses*	
	Zidovudine (AZT), lamivudine (3T3), didanosine (ddI), zalcitabine (ddC), and stavudine (d4T)	Nucleotide analog reverse transcriptase (RT) inhibitors; stop the action of RT in *HIV*, blocking viral DNA production (5)	
	Nevirapine, efavirenz, delavirdine	Nonnucleotide analog reverse transcriptase inhibitors; attach to *HIV* RT binding site, stopping its action (6)	
Inhibition of Viral Assembly/ Release	Indinavir, saquinavir	Protease inhibitors; insert into *HIV* protease, stopping its action and resulting in inactive noninfectious viruses (7)	

14. Identify the cellular target of quinolones, and provide two examples of these drugs.

15. Name two drugs that target the cellular membrane.

16. Describe the unique methods used to treat biofilm infections.

17. Name the four main categories of antifungal agents, and provide one example of each.

18. List four antiprotozoal drugs and three antihelminthic drugs used today.

19. Describe two major modes of action of antiviral drugs.

12.4 Antimicrobial Resistance

Interactions Between Microbes and Drugs: The Acquisition of Drug Resistance

One unfortunate outcome of the use of antimicrobials is the development of microbial **drug resistance,** an adaptive response in which microorganisms begin to tolerate an amount of drug that would ordinarily be inhibitory. The ability to circumvent or inactivate antimicrobial drugs is due largely to the genetic versatility and adaptability of microbial populations. The property of drug resistance can be intrinsic as well as acquired. Intrinsic drug resistance can best be exemplified by the fact that bacteria must, of course, be resistant to any antibiotic that they themselves produce. Of much greater importance is the acquisition of resistance to a drug by a microbe that was previously sensitive to the drug **(Insight 12.2).** In our context, the term *drug resistance* will refer to this last type of acquired resistance.

How Does Drug Resistance Develop?

Contrary to popular belief, antibiotic resistance is not a recent phenomenon. In 2012, 93 bacterial species were discovered in a cave in New Mexico, which had been cut off from the surface for millions of years. Most of these species were found to have resistance to multiple antibiotics—antibiotics naturally produced by other microbes. Because most of our oldest therapeutically used antibiotics are natural products from fungi and bacteria, resistance to them has been a survival strategy for *other* microbes for as long as the microbes have been around. The scope of the problem in terms of using the antibiotics as treatments for humans became apparent in the 1980s and 1990s, when scientists and physicians observed treatment failures on a large scale.

Microbes become newly resistant to a drug after one of the following two events occurs: (1) spontaneous mutations in critical chromosomal genes, or (2) acquisition of entire new genes or sets of genes via horizontal transfer from another species. Chromosomal drug resistance usually results from spontaneous random mutations in bacterial populations. The chance that such a mutation will be advantageous is minimal, and the chance that it will confer resistance to a specific drug is lower still. Nevertheless, given the huge numbers of microorganisms in any population and

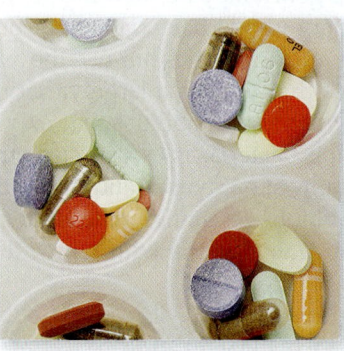
the constant rate of mutation, such mutations do occur. The end result varies from slight changes in microbial sensitivity, which can be overcome by larger doses of the drug, to complete loss of sensitivity.

Then we have the occurrence of resistance originating through horizontal transfer from plasmids called **resistance (R) factors** that are transferred through conjugation, transformation, or transduction. Studies have shown that plasmids encoded with drug resistance are naturally present in microorganisms before they have been exposed to the drug. Such traits are "lying in wait" for an opportunity to be expressed and to confer adaptability on the species. Many bacteria also maintain transposable drug resistance sequences (transposons) that are duplicated and inserted from one plasmid to another or from a plasmid to the chromosome. Chromosomal genes and plasmids containing codes for drug resistance are faithfully replicated and inherited by all subsequent progeny. This sharing of resistance genes accounts for the rapid proliferation of drug-resistant species. As you have read in earlier chapters, gene transfers are extremely frequent in nature, with genes coming from totally unrelated bacteria, viruses, and other organisms living in the body's normal biota and the environment.

We also have a new appreciation for where the reservoirs of antibiotic-resistance genes may be. Recently, it was discovered that a wide variety of soil bacteria cannot only survive in the presence of many antibiotics but use the antibiotics as fuel. This indicates that there is a large variety of natural environmental bacteria with capabilities that may be transferred to disease-causing bacteria. It is also clear that non-disease-causing inhabitants of our bodies and the bodies of our pets harbor many antibiotic-resistance genes that can and do easily jump to pathogenic bacteria with which they share space.

Specific Mechanisms of Drug Resistance

The two events that precipitate microbes becoming resistant to a drug (described earlier) have as their net effect one of the following actions, which actually cause the bacterium to be resistant (illustrated in **figure 12.11**):

1. New enzymes are synthesized; these inactivate the drug (only occurs when new genes are acquired).
2. Permeability or uptake of drug into bacterium is decreased (usually occurs via mutation).
3. Drug is immediately eliminated (usually occurs via acquisition of new genes).
4. Binding sites for drug are decreased in number or affinity (can occur via mutation or acquisition of new genes).
5. An affected metabolic pathway is shut down or an alternative pathway is used (occurs due to mutation of original enzyme or enzymes).

Some bacteria can become resistant indirectly by lapsing into dormancy by slowing down metabolism or, in the case of penicillin, by converting to a cell-wall-deficient form (L form) that penicillin cannot affect.

Drug Inactivation Mechanisms   Microbes inactivate drugs by producing enzymes that permanently alter drug structure. One example, bacterial enzymes called **beta-lactamases,** hydrolyze the beta-lactam ring (a critical structure) of penicillins and cephalosporins, rendering the drugs inactive. Two beta-lactamases—penicillinase and cephalosporinase—disrupt the structure of certain penicillin or cephalosporin molecules so their activity is lost. So many strains of *Staphylococcus aureus* produce penicillinase that regular penicillin is rarely a possible therapeutic choice. Different forms

of beta-lactamases are spreading among other pathogenic human bacteria as well (see Case File).

Decreased Drug Permeability or Increased Drug Elimination   The resistance of some bacteria can be due to a mechanism that prevents the drug from entering the cell and acting on its target. For example, the outer membrane of the cell wall of certain gram-negative bacteria is a natural blockade for some of the penicillin drugs. In addition, resistance to the tetracyclines can arise from plasmid-encoded proteins that pump the drug out of the cell.

Many bacteria possess multidrug-resistant (MDR) pumps that actively transport drugs and other chemicals out of cells. These pumps are proteins encoded by plasmids or chromosomes. They are stationed in the cell membrane and expel molecules by a proton-motive force similar to ATP synthesis (see figure 12.11). They confer drug resistance on many gram-positive pathogens (*Staphylococcus*, *Streptococcus*) and gram-negative pathogens (*Pseudomonas*, *E. coli*). Because they lack selectivity, one type of pump can expel a broad array of antimicrobial drugs, detergents, and other toxic substances.

Change of Drug Receptors   Because most drugs act on a specific target such as protein, RNA, DNA, or membrane structure, microbes can get around the effects of drugs by altering the nature of this target. Bacteria can become resistant to aminoglycosides when point mutations in ribosomal proteins arise (see figure 12.11). Erythromycin and clindamycin resistance is associated with an alteration on the 50S ribosomal binding site. Penicillin resistance in *Streptococcus pneumoniae* and methicillin resistance in *Staphylococcus aureus* are related to an alteration in the binding proteins in the cell wall. Enterococci have acquired resistance to vancomycin through a similar alteration of cell wall proteins. Fungi can become resistant by decreasing their synthesis of ergosterol, the principal receptor for certain antifungal drugs.

Changes in Metabolic Patterns   The action of antimetabolites can be avoided by a microbe if it develops an alternative metabolic pathway or enzyme. For example, sulfonamide and trimethoprim resistance develop when microbes deviate from the usual patterns of folic acid synthesis. Fungi can acquire resistance to flucytosine by completely shutting off certain metabolic activities.

Natural Selection and Drug Resistance

So far, we have been considering drug resistance at the cellular and molecular levels, but its full impact is felt only if this resistance occurs throughout the cell population. Let's examine how this may happen and its long-term therapeutic consequences.

Any large population of microbes is likely to contain a few individual cells that are already drug resistant because of prior mutations or transfer of plasmids **(figure 12.12a).** As long as the drug is not present in the habitat, the numbers of these resistant forms will remain low because they have no particular growth advantage (and may be disadvantaged

Mechanism

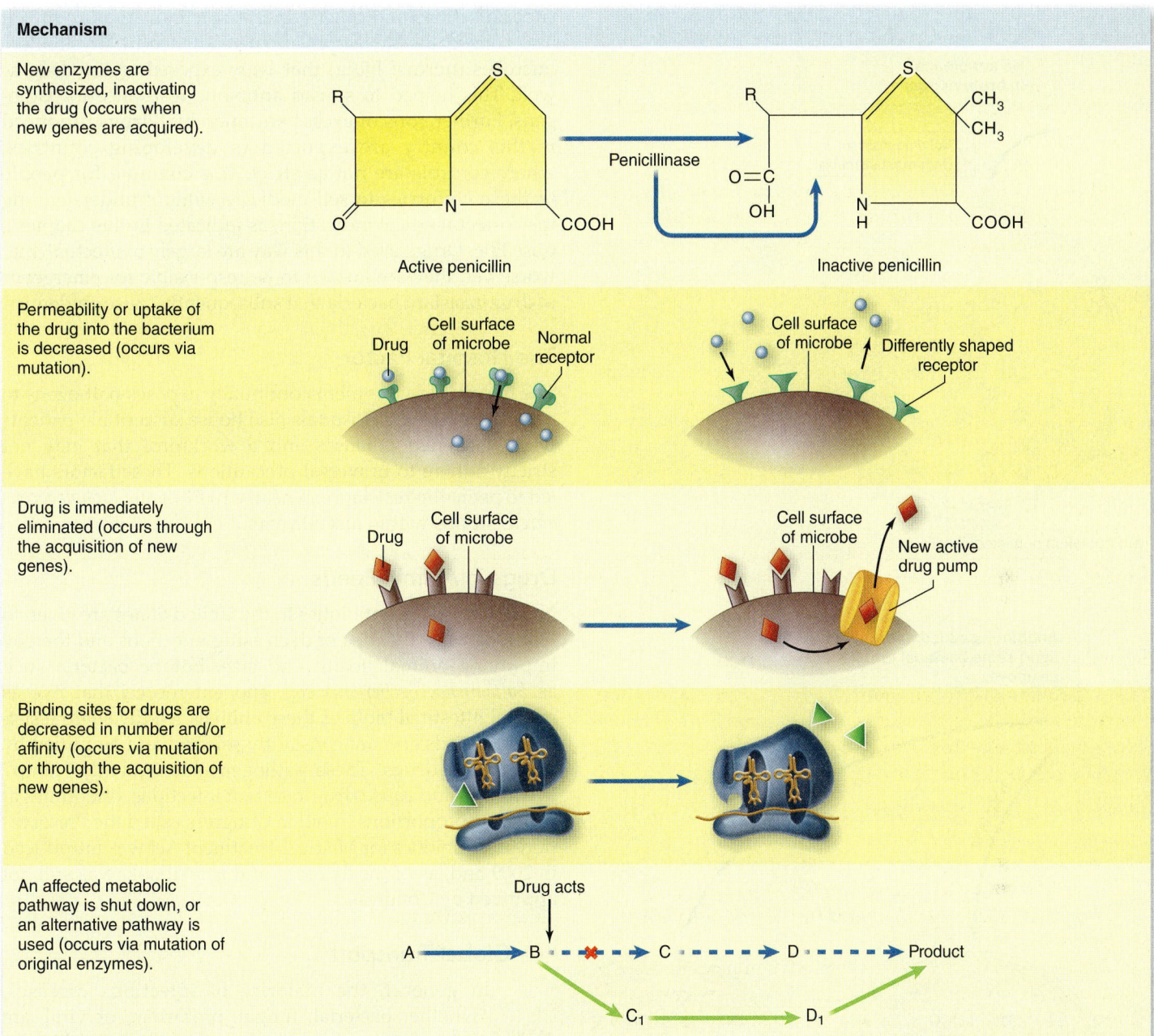

New enzymes are synthesized, inactivating the drug (occurs when new genes are acquired).

Active penicillin

Penicillinase

Inactive penicillin

Permeability or uptake of the drug into the bacterium is decreased (occurs via mutation).

Drug | Cell surface of microbe | Normal receptor

Cell surface of microbe | Differently shaped receptor

Drug is immediately eliminated (occurs through the acquisition of new genes).

Drug | Cell surface of microbe

Cell surface of microbe | New active drug pump

Binding sites for drugs are decreased in number and/or affinity (occurs via mutation or through the acquisition of new genes).

An affected metabolic pathway is shut down, or an alternative pathway is used (occurs via mutation of original enzymes).

Drug acts

A → B ··✗··➤ C ····➤ D ····➤ Product

C_1 → D_1

Figure 12.11 Mechanisms of drug resistance.

relative to their nonmutated counterparts). But if the population is subsequently exposed to this drug **(figure 12.12b),** sensitive individuals are inhibited or destroyed and resistant forms survive and proliferate. During subsequent population growth, offspring of these resistant microbes will inherit this drug resistance. In time, the replacement population will have a preponderance of the drug-resistant forms and can eventually become completely resistant **(figure 12.12c).** In ecological terms, the environmental factor (in this case, the drug) has put selection pressure on the population, allowing the more "fit" microbe (the drug-resistant one) to survive, and the population has evolved to a condition of drug resistance.

The Human Role in Antimicrobial Resistance

Many individuals are completely unaware that everyday human activities greatly facilitate the development of drug resistance in environmental microbes and human pathogens as well. A recent study found that 75% of antimicrobial prescriptions are for pharyngeal, sinus, lung, and upper respiratory infections. A fairly high percentage of these are viral in origin and will have little or no benefit from antibacterial drugs. In the past, many physicians tended to use a "shotgun" antimicrobial therapy for minor infections, which involves administering a broad-spectrum drug instead of a more specific narrow-spectrum one. This practice led to

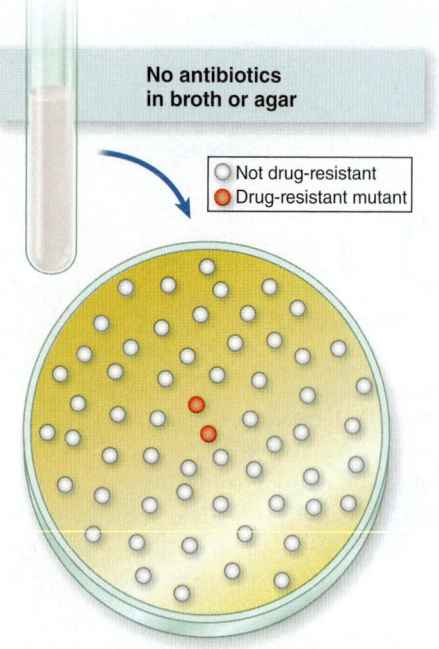

No antibiotics in broth or agar

○ Not drug-resistant
● Drug-resistant mutant

(a) Population of microbial cells

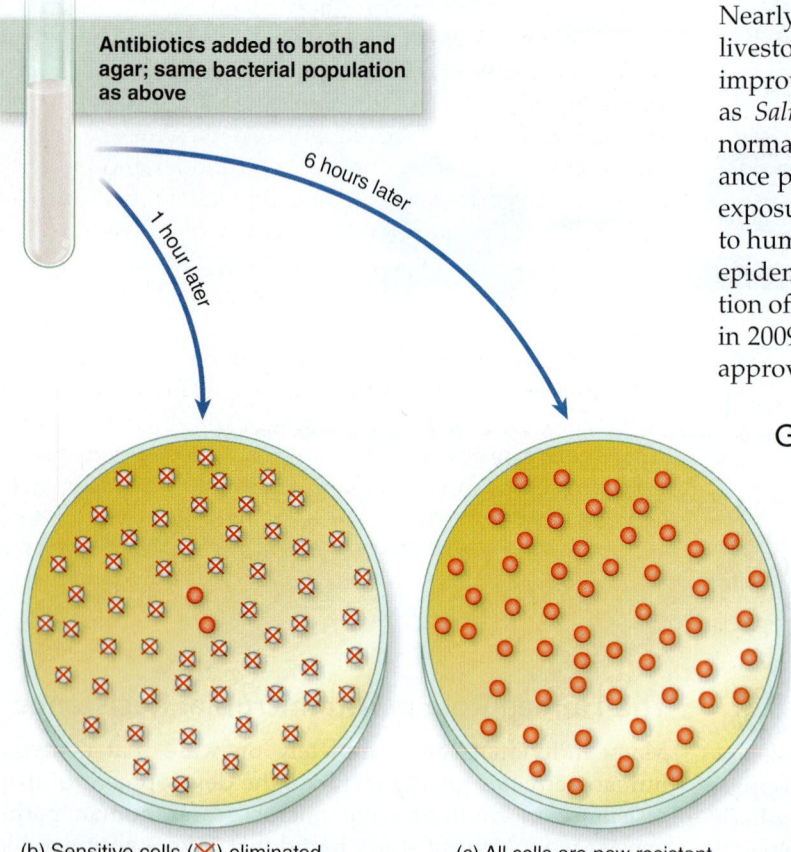

Antibiotics added to broth and agar; same bacterial population as above

1 hour later

6 hours later

(b) Sensitive cells (⊠) eliminated by drug; resistant mutants survive

(c) All cells are now resistant

Figure 12.12 The events in natural selection for drug resistance. (a) Populations of microbes can harbor some members with a mutation that confers drug resistance. (b) Environmental pressure (here, the presence of the drug) selects for survival of these mutants. (c) They eventually become the dominant members of the population.

superinfections and other adverse reactions. Importantly, it also caused the development of resistance in "bystander" microbes (normal biota) that were exposed to the drug as well. This helped to spread antibiotic resistance to pathogens. Further, tons of excess antimicrobial drugs produced in this country are exported to developing countries, where controls are not as strict. It is common for people in these countries to self-medicate without understanding the correct medical indication, as indicated in this chapter's Case File. Drugs used in this way are largely ineffectual, but, worse yet, they are known to be responsible for emergence of drug-resistant bacteria that subsequently cause epidemics.

The Hospital Factor

The hospital environment continually exposes pathogens to a variety of drugs. Hospitals also house susceptible patients with weakened defenses and a workforce that may not strictly adhere to universal precautions. These factors have led to penicillin resistance in nearly 100% of all *Staphylococcus aureus* strains within just 30 years.

Drugs in Animal Feeds

Nearly 80% of all antibiotics in the United States are given to livestock, with the idea of decreasing infections and thereby improving animal health and size. Enteric bacteria such as *Salmonella, Escherichia coli,* and enterococci that live as normal intestinal biota of these animals readily share resistance plasmids and are constantly selected and amplified by exposure to drugs. These pathogens subsequently "jump" to humans and cause drug-resistant infections, oftentimes at epidemic proportions. A bill in Congress called the Preservation of Antibiotics for Medical Treatment Act was introduced in 2009 and, as of the printing of this textbook, was still not approved by Congress.

Global Transport

In general, the majority of infectious diseases, whether bacterial, fungal, protozoan, or viral, are showing increases in drug resistance in all areas of the world. To add to the problem, global travel and globalization of food products mean that drug resistance can be rapidly exported. It is clear that we are in a race with microbes and we are falling behind. If the trend is not contained, the world may return to a time when there are few effective drugs left. We simply cannot develop them as rapidly as microbes can develop resistance. In this light, it is essential to fight the battle on more than one front as we will see next.

Strategies to Limit Drug Resistance

Armed with a vast amount of information, scientists must now take this knowledge and create working policies that will give humans an edge in controlling drug resistance. In terms of drug usage, physicians must be sure to accurately prescribe antibiotics—and only if test results indicate they are needed. Health care professionals also hold responsibility in educating patients of the risks of noncompliance, or the failure to correctly complete their course of drug therapy. In cases where microbes are suspected of being—or becoming—resistant, treatment with two antibiotics simultaneously is indicated to avoid the development of drug resistance.

More importantly, we must also look at implementing long-term strategies in the control of drug resistance. Proposals have been made to restrict the use of first-line antibiotics, or those that are most powerful in the treatment of the most resistant microbes. The most valuable long-term changes that are currently in practice are enhanced surveillance mechanisms. Government agencies like the Centers for Disease Control and Prevention and the World Health Organization have created databases (PulseNet, WHONET) of information from the tracking of food-borne pathogens and other pathogenic microbial populations. Such monitoring is also being extended into the environment so that scientists will have an up-to-date view of the changes in resistance genes occurring within many microbes. Knowing the enemy's strategy is half the battle.

New Approaches to Antimicrobial Therapy

Often, the quest for new antimicrobial strategies focuses on finding new targets in the bacterial cell and custom-designing drugs that aim for them. However, very recently a new approach has been in the spotlight: disabling host molecules that the invaders use to enhance their position. One promising development is to target a host cell protein that bacteria use to move from one cell compartment to another. This would be most effective for bacteria that live inside host cells. In the domain of direct damage to the invading bacteria, there are many interesting new strategies as well.

The real good news is that there is now an uptick in research into novel antimicrobial strategies. When antibiotic resistance first became widely problematic, the specter of "no new drugs" was very real, because drug makers had turned their efforts to looking for drugs for chronic diseases, such as heart disease and diabetes, that a patient would take for life, rather than for a week. They cannot really be blamed; there was a time from the 1950s to the 1980s when the medical world assumed that eventually all infections could be wiped out with antibiotics. That proved to be a premature assessment, as we have seen.

Sometimes the low-tech solution can be the best one. Eastern European countries have gained a reputation for using mixtures of bacteriophages as medicines for bacterial infections. There is little argument about the effectiveness of these treatments, though they have never been approved for use in the West. One recent human trial used a mixture of bacteriophages specific for *Pseudomonas aeruginosa* to treat ear infections caused by the bacterium. These infections are found in the form of biofilms and have been extremely difficult to treat. The phage preparation called Biophage-PA successfully treated patients who had experienced long-term

antibiotic-resistant infections. Other researchers are incorporating phages into wound dressings. One clear advantage to bacteriophage treatments is the extreme specificity of the phages— only one species of bacterium is affected, leaving the normal inhabitants of the body alone.

Helping Nature Along

Other novel approaches to controlling infections include the use of **probiotics** and **prebiotics.** Probiotics are preparations of live microorganisms that are fed to animals and humans to improve the intestinal biota. This can serve to replace microbes lost during antimicrobial therapy or simply to augment the biota that is already there. This is a slightly more sophisticated application of methods that have long been used in an empiric fashion, for instance, by people who consume yogurt because of the beneficial microbes it contains. Recent years have seen a huge increase in the numbers of probiotic products sold in ordinary grocery stores **(figure 12.13).** Experts generally find these products safe, and in some cases they can be effective. Probiotics are thought to be useful for the management of food allergies; their role in the stimulation of mucosal immunity is also being investigated.

Prebiotics are nutrients that encourage the growth of beneficial microbes in the intestine. For instance, certain sugars such as fructans are thought to encourage the growth of the beneficial *Bifidobacterium* in the large intestine and to discourage the growth of potential pathogens. You can be sure that you will hear more about prebiotics and probiotics as the concepts become increasingly well studied by scientists. Clearly, the use of these agents is a different type of antimicrobial strategy than we are used to, but it may have its place in a future in which traditional antibiotics are more problematic.

A technique that is being employed in some medical communities is the use of fecal transplants in the treatment of recurrent *Clostridium difficile* infection and ulcerative colitis. This procedure involves the transfer of feces from healthy patients, containing beneficial normal biota, to affected patients via colonoscopy. This is, in fact, just an adaptation of probiotics. Instead of a few beneficial bacteria being given

Figure 12.13 Examples of probiotic grocery items.

INSIGHT 12.3 A New Twist on Antibiotic Therapy: Vampire Bacteria

Since the discovery of penicillin, we have been fighting off bacterial infections with chemicals that kill bacteria by inhibiting cell processes or structures. Recently, scientists discovered a bacterium that could change the course of the battle: "vampire bacteria" that prey on other bacteria. Researchers at the University of Virginia found that *Micavibrio aeruginosavorus* is an organism that inhabits wastewater, attaches itself to the cell walls of its victims, and sucks out nutrients in order to survive. The organism has not been well studied because it is difficult to grow in the laboratory, but the researchers sequenced its genome and found that although *M. aeruginosavorus* has

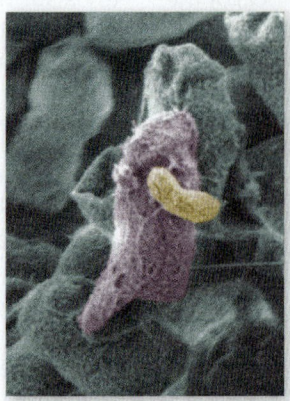

genes for all major metabolic pathways, it cannot make or import essential amino acids from the environment. This requires the organism to attack other organisms that have these amino acids. They found that *M. aeruginosavorus* is able to prey on bacteria that are pathogenic to humans, such as *Pseudomonas aeruginosa*, *Burkholderia cepacia*, *Klebsiella pneumoniae*, and *Escherichia coli*.

According to these studies, *M. aeruginosavorus* has two life phases, the first of which is a free-living attack phase during which it produces flagella and is able to swim through water and even viscous mucus or biofilms looking for prey. When it finds its prey, the second phase begins. The bacterium uses its pili to dock to the victim bacterium. In this way it accesses the victim's cytoplasm and steals some of its essential amino acids to use for its own binary fission. The exact mechanism by which *M. aeruginosavorus* obtains nutrients is unknown, but it is hypothesized that it creates a cytoplasmic bridge with its prey to import the needed amino acids.

This research suggests that these vampire-like bacteria could be used as antibiotic therapy. *M. aeruginosavorus* attacks only specific pathogens and has the ability to swim through layers of mucus. This is especially promising for infections of *P. aeruginosa* in patients with cystic fibrosis who produce thick mucous layers in the lungs. The vampire bacteria also may be able to reduce biofilms in pipes or on medical devices.

Source: 2011. BMC Genomics 12. doi: 10.1186/1471-2164-12-453

---

orally, with the hope that they will establish themselves in the intestines, a rich microbiota is administered directly to the site it must colonize—the large intestine. This method has had documented success in farm animals, and studies have shown a therapeutic effect in humans as well. However, this procedure is not widely accepted nor is it or its research currently regulated by U.S. government agencies.

12.4 Learning Outcomes—Assess Your Progress

20. Discuss two possible ways that microbes acquire antimicrobial resistance.
21. List five cellular or structural mechanisms that microbes use to resist antimicrobials.
22. Discuss at least three novel antimicrobial strategies that are under investigation.

12.5 Interactions Between Drug and Host

Although selective antimicrobial toxicity is the ideal constantly being sought, chemotherapy by its very nature involves contact with foreign chemicals that can harm human tissues. In fact, estimates indicate that at least 5% of all persons taking an antimicrobial drug experience some type of serious adverse reaction to it. The major side effects of drugs fall into one of three categories: direct damage to tissues through toxicity, allergic reactions, and disruption in the balance of normal microbial biota. The damage incurred by antimicrobial drugs can be short term and reversible or permanent, and it ranges in severity from cosmetic to lethal. In the field of antimicrobial therapy, one must always remember the ancient axiom: *Graviora quaedum sunt remedia periculus* ("Some remedies are worse than the disease").

Toxicity to Organs

Drugs most often adversely affect the following organs: the liver (hepatotoxic), kidneys (nephrotoxic), gastrointestinal tract, cardiovascular system and blood-forming tissue (hemotoxic), nervous system (neurotoxic), respiratory tract, skin, bones, and teeth. The potential toxic effects of drugs on the body, along with the responsible drugs, are detailed in **table 12.9.**

The skin is a frequent target of drug-induced side effects. The skin response can be a symptom of drug allergy or a direct toxic effect. Some drugs interact with sunlight to cause photodermatitis, a skin inflammation. Tetracyclines are contraindicated (not advisable) for children from birth to 8 years of age because they bind to the enamel of the teeth, creating a permanent gray to brown discoloration **(figure 12.14).** Pregnant women should avoid tetracyclines because they can cause liver damage. They also cross the placenta and can be deposited in the developing fetal bones and teeth. However, the most common complaint associated with oral antimicrobial therapy is diarrhea, which can progress to severe intestinal irritation or colitis. Although some drugs directly irritate the intestinal lining, the usual gastrointestinal complaints are caused by disruption of the intestinal microbiota.

Table 12.9 Major Adverse Toxic Reactions to Common Drug Groups

Antimicrobial Drug	Primary Damage or Abnormality Produced
Antibacterials	
Penicillin G	Skin abnormalities
Cephalosporins	Inhibition of platelet function Decreased circulation of white blood cells Nephritis
Tetracyclines	Diarrhea and enterocolitis Discoloration of tooth enamel Reactions to sunlight (photosensitization)
Aminoglycosides (streptomycin, gentamicin, amikacin)	Diarrhea and enterocolitis Malabsorption Loss of hearing, dizziness, kidney damage
Isoniazid	Hepatitis Seizures Dermatitis
Sulfonamides	Formation of crystals in kidney; blockage of urine flow Hemolysis Reduction in number of red blood cells
Polymyxin	Kidney damage Weakened muscular responses
Quinolones (ciprofloxacin, norfloxacin)	Headache, dizziness, tremors, GI distress
Antifungals	
Amphotericin B	Disruption of kidney function
Flucytosine	Decreased number of white blood cells
Antiprotozoan Drugs	
Metronidazole	Nausea, vomiting
Chloroquine	Vomiting Headache Itching
Antihelminthics	
Niclosamide	Nausea, abdominal pain
Pyrantel	Irritation Headache, dizziness
Antivirals	
Amantadine	Nervousness, light-headedness Nausea
AZT	Immunosuppression, anemia

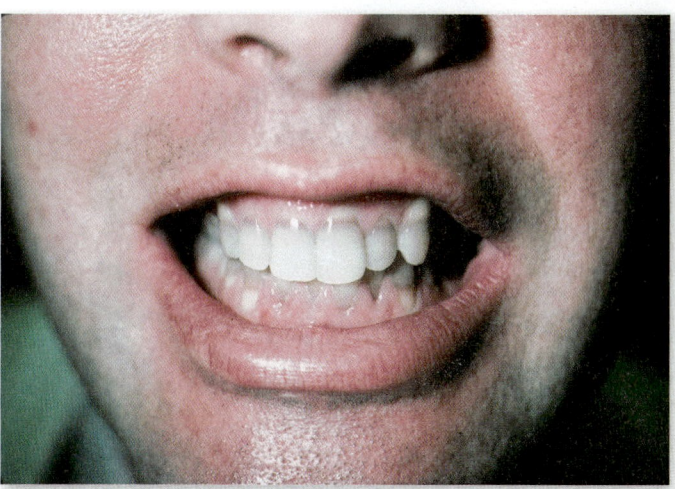

Figure 12.14 Drug-induced side effect. An adverse effect of tetracycline given to young children is the permanent discoloration of tooth enamel.

and stimulates an allergic response. This response can be provoked by the intact drug molecule or by substances that develop from the body's metabolic alteration of the drug. In the case of penicillin, for instance, it is not the penicillin molecule itself that causes the allergic response but a product, *benzylpenicilloyl*. Allergic reactions have been reported for every major type of antimicrobial drug, but the penicillins account for the greatest number of antimicrobial allergies, followed by the sulfonamides.

People who are allergic to a drug become sensitized to it during the first contact, usually without symptoms. Once the immune system is sensitized, a second exposure to the drug can lead to a reaction such as a skin rash (hives); respiratory inflammation; and, rarely, anaphylaxis, an acute, overwhelming allergic response that develops rapidly and can be fatal. (This topic is discussed in greater detail in chapter 16.)

Suppression and Alteration of the Microbiota by Antimicrobials

Most body surfaces, such as the skin, large intestine, and outer openings of the urogenital tract and oral cavity, provide numerous habitats for a virtual "garden" of microorganisms. These normal colonists or residents, called the **biota** or microbiota, consist mostly of harmless or beneficial bacteria, but a small number can potentially be pathogens. Although we defer a more detailed discussion of this topic to chapter 13 and later chapters, here we focus on the general effects of drugs on this population.

If a broad-spectrum antimicrobial is introduced into a host to treat infection, it will destroy microbes regardless of their roles as normal biota, affecting not only the targeted infectious agent but also many others in sites far removed from the original infection **(figure 12.15).** When this therapy destroys beneficial resident species, other microbes that were once in small numbers can begin to overgrow and cause disease. This complication is called a **superinfection.**

Allergic Responses to Drugs

One of the most frequent drug reactions is **allergy.** This reaction occurs because the drug acts as an antigen (a foreign material capable of stimulating the immune system)

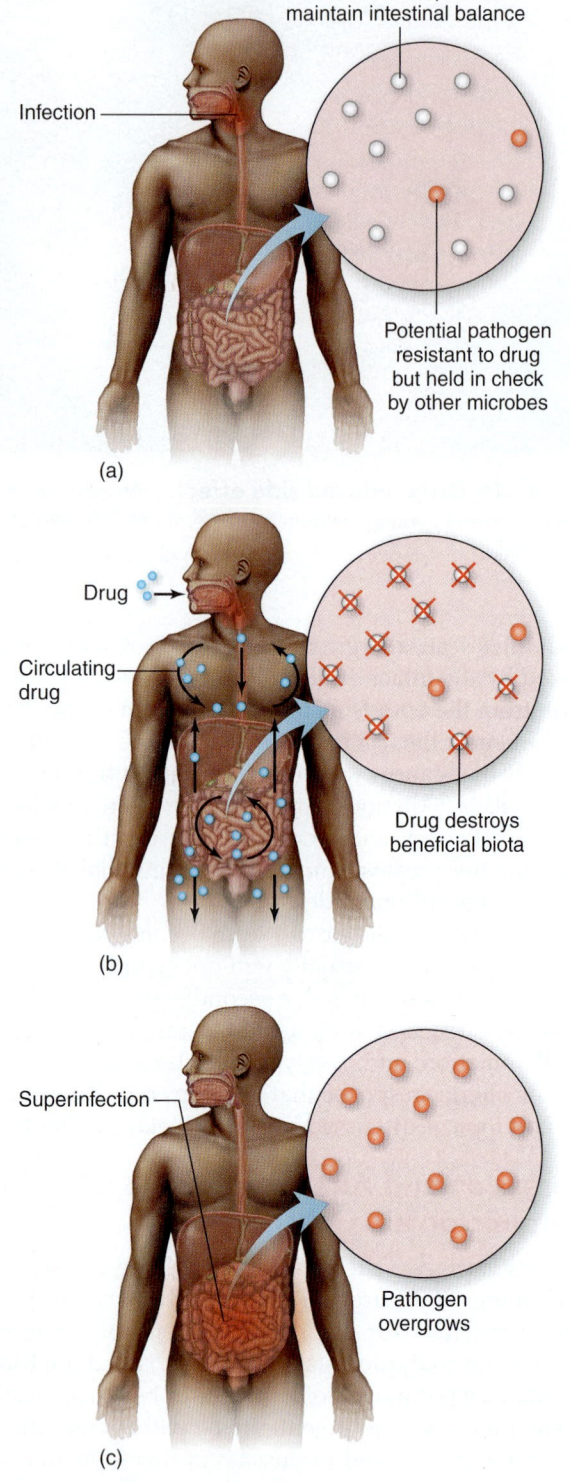

(a)

(b)

(c)

Figure 12.15 The role of antimicrobials in disrupting microbial biota and causing superinfections. **(a)** A primary infection in the throat is treated with an oral antibiotic. **(b)** The drug is carried to the intestine and is absorbed into the circulation. **(c)** The primary infection is cured, but drug-resistant pathogens have survived and create an intestinal superinfection.

Some common examples demonstrate how a disturbance in microbial biota leads to replacement biota and superinfection. A broad-spectrum cephalosporin used to treat a urinary tract infection by *Escherichia coli* will cure the infection, but it will also destroy the lactobacilli in the vagina that normally maintain a protective acidic environment there. The drug has no effect, however, on *Candida albicans*, a yeast that also resides in normal vaginas. Released from the inhibitory environment provided by lactobacilli, the yeasts proliferate and cause symptoms. *Candida* can cause similar superinfections of the oropharynx (thrush) and the large intestine.

Oral therapy with tetracyclines, clindamycin, and broad-spectrum penicillins and cephalosporins is associated with a serious and potentially fatal condition known as *antibiotic-associated colitis* (pseudomembranous colitis). This condition is due to the overgrowth in the bowel of *Clostridium difficile* ("*C. diff*"), an endospore-forming bacterium that is resistant to many antibiotics. It invades the intestinal lining and releases toxins that induce diarrhea, fever, and abdominal pain. It is very difficult to eradicate, which is why experimental treatments such as fecal transplants are sometimes considered. (You'll learn more about infectious diseases of the gastrointestinal tract, including *C. difficile,* in chapter 22.)

12.5 Learning Outcomes—Assess Your Progress

23. Distinguish between drug toxicity and allergic reactions to drugs.

24. Define the term *superinfection*, and summarize how it develops in a patient.

Case File 12   *Wrap-Up*

It is not an uncommon practice for scientists to name beta-lactamase genes after the country in which they were first identified. The Verona integron-encoded metallo-beta-lactamase, or VIM, was discovered in the Verona University hospital in Italy; a similar gene named GIM was first isolated in Germany. NDM-1 was first detected in hospitals in and around New Delhi, India.

In India, a prescription is not required to purchase antibiotics. Dr. V. M. Katoch, director general of the Indian Council of Medical Research stated, "Several parts of rural India do not even have doctors to prescribe an antibiotic. At present, people go to shops and purchase the antibiotics needed to cure their illnesses." This unchecked dissemination of antibiotics has led to the spread of resistant organisms in this area. A study recently published in *The Lancet* showed the *bla*$_{NDM-1}$ gene in two of 50 drinking water samples and 51 of 171 water seepage samples from New Delhi. The study isolated 20 strains of bacteria from these samples, 12 of which carried *bla*$_{NDM-1}$.

Sources: 2012. CDC *MMWR* 61. June 22.; 2011. *The Lancet* 11. p. 355.

Chapter Summary

12.1 Principles of Antimicrobial Therapy (ASM Guidelines* 2.1, 3.4, 4.1, 6.3, 8.3)

- Antimicrobial chemotherapy involves the use of drugs to control infection on or in the body.
- Antimicrobial drugs are produced either synthetically or from natural sources. They inhibit or destroy microbial growth in the infected host.
- Antimicrobial drugs are classified by their range of effectiveness. Broad-spectrum antimicrobials are effective against many types of microbes. Narrow-spectrum antimicrobials are effective against a limited group of microbes.
- Bacteria and fungi are the primary sources of most currently used antibiotics. The molecular structures of these compounds can be chemically altered or mimicked in the laboratory.
- The three major considerations necessary to choose an effective antimicrobial are the identity of the infecting microbe, the microbe's sensitivity to available drugs, and the overall medical status of the infected host.
- The Kirby-Bauer test identifies antimicrobials that are effective against a specific infectious bacterial isolate.
- The MIC (minimum inhibitory concentration) identifies the smallest effective dose of an antimicrobial toxic to the infecting microbe.
- The therapeutic index is a ratio of the amount of drug toxic to the infected host and the MIC. The smaller the ratio, the greater the potential for toxic host-drug reactions.

12.2 Interactions Between Drug and Microbe (ASM Guidelines 2.1, 3.4, 4.1, 6.3)

- Ideally, an antibiotic should be selectively toxic, meaning that it harms the microbe but not the host.
- Antimicrobials are classified into major drug families based on chemical composition, source or origin, and their site of action.

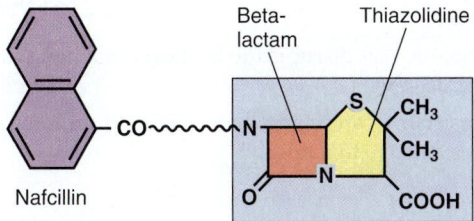

- The majority of antimicrobials are effective against bacteria, and a more limited number are effective against protozoa, helminths, fungi, and viruses.

- There are five main cellular targets for antibiotics in microbes: cell wall synthesis, nucleic acid structure and function, protein synthesis, cell membranes, and folic acid synthesis.

12.3 Survey of Major Antimicrobial Drug Groups (ASM Guidelines 2.1, 3.4, 6.3)

- Narrow-spectrum antibiotics target one or a few types of bacteria. Broad-spectrum antimicrobials affect multiple types.
- Penicillins, cephalosporins, carbapenems, and vancomycin block cell wall synthesis.
- Aminoglycosides, tetracyclines, erythromycin, and ketolides block protein synthesis in bacteria.
- Synercid, oxazolidinone, and pleuromutilins are newer antimicrobials that affect protein synthesis.
- Sulfonamides and trimethoprim block enzymatic steps in the synthesis of folic acid by bacteria.
- The fluoroquinolones are synthetic antimicrobials effective against a broad range of microorganisms. They block steps in the synthesis of nucleic acids.
- Polymyxins and daptomycin are the major drugs that disrupt cell membranes.
- Bacteria in biofilms respond differently to antibiotics than when they are free-floating. It is therefore difficult to eradicate biofilms in the human body.
- Fungal antimicrobials, such as macrolide polyenes, azoles, echinocandins, and allylamines, must be monitored carefully because of the potential toxicity to the infected host.
- There are fewer antiparasitic drugs than antibacterial drugs because parasites are eukaryotes like their human hosts and they have several life stages, some of which can be resistant to the drug.
- Antihelminthic drugs immobilize or disintegrate infesting helminths or inhibit their metabolism.
- Antiviral drugs interfere with viral replication by blocking viral entry into cells, blocking the replication process, or preventing the assembly of viral subunits into complete virions.
- Many antiviral agents are analogs of nucleotides. They inactivate the replication process when incorporated into viral nucleic acids. HIV antivirals interfere with reverse transcriptase or proteases to prevent the maturation of viral particles.
- Commercial interferon has some use against viral infections and cancer.

12.4 Antimicrobial Resistance (ASM Guidelines 1.2, 2.1, 4.1)

- Microorganisms are termed *drug resistant* when they are no longer inhibited by an antimicrobial to which they were previously sensitive.
- Most drug resistance is genetic; microbes acquire genes that code for methods of inactivating or escaping the antimicrobial, or acquire mutations that affect the drug's impact.
- Varieties of microbial drug resistance include drug inactivation, decreased drug uptake, decreased drug

Source: ASM Curriculum Guidelines (American Society for Microbiology, 2012). Complete guidelines in appendix B of this book.

receptor sites, and modification of metabolic pathways formerly attacked by the drug.

- Widespread indiscriminate use of antimicrobials has resulted in an explosion of microorganisms resistant to all common drugs.
- Research strategies for new types of antimicrobials include targeting host cell proteins that are important for microbial multiplication, the use of RNA interference, mimicking natural defense peptides, and the use of bacteriophages.
- Pro- and prebiotics are methods of crowding out pathogenic bacteria and providing a favorable environment for the growth of beneficial bacteria.

12.5 Interactions Between Drug and Host (ASM Guidelines 3.4, 5.4)

- The three major side effects of antimicrobials are toxicity to organs, allergic reactions, and problems resulting from alteration of normal biota.
- Antimicrobials that destroy most but not all normal biota can allow the unaffected normal biota to overgrow, causing a superinfection.

Multiple-Choice and True-False Questions | Bloom's Levels 1 and 2: Remember and Understand

Multiple-Choice Questions. Select the correct answer from the options provided.

1. A compound synthesized by bacteria or fungi that destroys or inhibits the growth of other microbes is a/an
   a. synthetic drug.
   b. antibiotic.
   c. interferon.
   d. competitive inhibitor.

2. Which statement is *not* an aim in the use of drugs in antimicrobial chemotherapy? The drug should
   a. have selective toxicity.
   b. be active even in high dilutions.
   c. be broken down and excreted rapidly.
   d. be microbicidal.

3. Drugs that prevent the formation of the bacterial cell wall are
   a. quinolones.
   b. beta-lactams.
   c. tetracyclines.
   d. aminoglycosides.

4. Microbial resistance to drugs is acquired through
   a. conjugation.
   b. transformation.
   c. transduction.
   d. all of these.

5. R factors are _____ that contain a code for _____ .
   a. genes, replication
   b. plasmids, drug resistance
   c. transposons, interferon
   d. plasmids, conjugation

6. Phage therapy is a technique that uses
   a. chemicals to destroy phages infecting human cells.
   b. chemicals to foster the growth of beneficial phages in the body.
   c. phages to foster the growth of normal biota.
   d. phages to target pathogenic bacteria in the body.

7. Most antihelminthic drugs function by
   a. weakening the worms so they can be flushed out by the intestine.
   b. inhibiting worm metabolism.
   c. blocking the absorption of nutrients.
   d. inhibiting egg production.

8. Which of the following modes of action would be most selectively toxic?
   a. interrupting ribosomal function
   b. dissolving the cell membrane
   c. preventing cell wall synthesis
   d. inhibiting DNA replication

9. The MIC is the _____ of a drug that is required to inhibit growth of a microbe.
   a. largest concentration
   b. standard dose
   c. smallest concentration
   d. lowest dilution

10. An antimicrobial drug with a ____ therapeutic index is a better choice than one with a _____ therapeutic index.
    a. low, high
    b. high, low

True-False Questions. If the statement is true, leave as is. If it is false, correct it by rewriting the sentence.

11. Most antiviral agents work by destroying active viruses.

12. Sulfonamide drugs work by disrupting protein synthesis.

13. Biofilms are difficult to treat and don't always respond to antibiotics.

14. An antibiotic that disrupts the host's normal biota can cause a superinfection.

15. Drug resistance can occur when a patient's immune system becomes reactive to a drug.

Critical Thinking Questions | Bloom's Levels 3, 4, and 5: Apply, Analyze, and Evaluate

Critical thinking is the ability to reason and solve problems using facts and concepts. These questions can be approached from a number of angles and, in most cases, they do not have a single correct answer.

1. Construct a paragraph describing the interrelationship among the microbial pathogen, the affected host, and potential antimicrobial drugs in the development of an appropriate chemotherapeutic treatment.

2. a. Is it more beneficial for a drug to be microbistatic or microbicidal? Defend your answer.

   b. Draw a microbial growth curve, and indicate during which phase an antibiotic should be added to a bacterial culture in order to accurately assess the drug's microbicidal activity.

3. A critically ill patient enters your emergency room, exhibiting signs and symptoms of severe septic shock. In this case, should you immediately begin treatment with a broad-spectrum drug or a narrow-spectrum drug? Explain your answer and discuss any possible consequences of using either drug in the patient.

4. Explain how protein-inhibiting drugs are able to exhibit selective toxicity, providing examples of these drugs and their cellular targets.

5. Amphotericin B is often referred to as "Amphi-Terrible" in medical settings due to its effects in treated patients. Describe when this drug should be prescribed, and provide a biological reason for its damaging activity on host cells.

6. HAART, or highly active antiretroviral therapy, is currently recommended for the treatment of HIV and involves the administration of three or more medicines at one time. Explain why this combined therapy approach is more effective than single drug treatment in the management of HIV infection, providing current evidence to support your answer.

7. Antibiotic-resistance genes, as well as other virulence factor genes, are easily passed between bacterial cells through horizontal gene transfer.

   a. Conduct additional research and summarize the unique pathogenic characteristics of *Escherichia coli* O157:H7; describe how it acquired these traits over time.

   b. Conduct additional research on New Delhi metallo-beta-lactamase 1 strains of bacteria, and explain why medical tourism poses a serious threat to the spread of this organism. Provide evidence to support your explanation.

8. A friend was recently diagnosed with strep throat. One week after his treatment, he redeveloped the infection. In conversation, your friend tells you, "I must have become immune to the drug the doctor gave me!"

   a. Discuss the validity of your friend's statement, providing evidence in support of or refuting his claim.

   b. After further conversation, your friend tells you that he stopped taking his initial antibiotics after 2 days because he "felt 100% better." Explain how this action might have played a role in the redevelopment of his infection.

   c. When he returned to his physician, she ordered a test to determine which antibiotic should be prescribed to treat his reinfection. Summarize a test that could be used to obtain this information.

9. You have been directed to aseptically obtain a sample from a growth-free portion of the zone of inhibition in a completed Kirby-Bauer test and inoculate it onto a plate of nonselective medium.

   a. After incubation at 37°C for 48 hours, you observe growth on the plate. Explain this result and how it impacts the analysis of the Kirby-Bauer test.

   b. The drug tested was amoxicillin. Based upon your results, explain whether or not this is an effective drug choice for the treatment of an infection caused by this bacterium.

10. a. Referring to figure 12.1a, determine whether the cultured bacterium is sensitive, intermediate, or resistant to each of the antibiotic drugs tested and provide an explanation in each case.

    b. Referring to figure 12.3a, determine the minimum inhibitory concentration (MIC) of the drugs being tested, providing an explanation for your answer.

Concept Connections | Bloom's Levels 4 and 6: Analyze and Create

This activity ties together multiple concepts in the chapter.

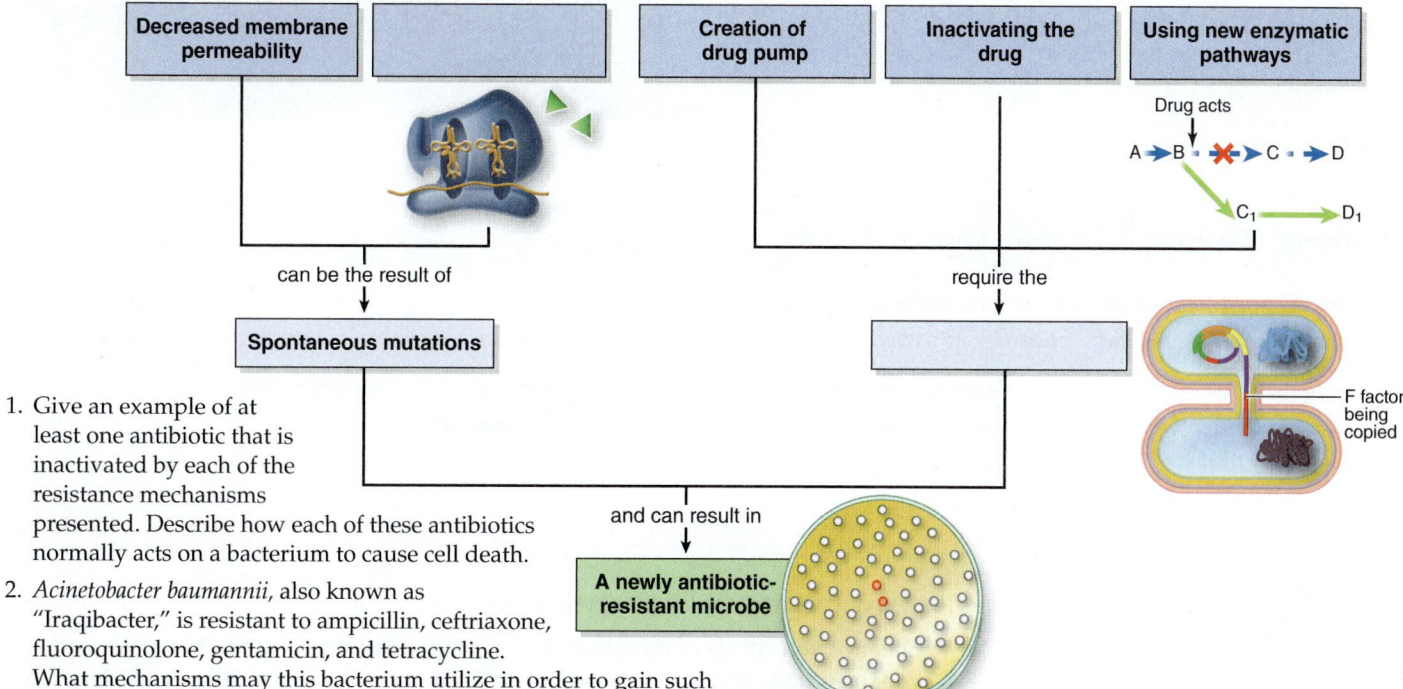

1. Give an example of at least one antibiotic that is inactivated by each of the resistance mechanisms presented. Describe how each of these antibiotics normally acts on a bacterium to cause cell death.

2. *Acinetobacter baumannii*, also known as "Iraqibacter," is resistant to ampicillin, ceftriaxone, fluoroquinolone, gentamicin, and tetracycline. What mechanisms may this bacterium utilize in order to gain such multidrug resistance?

Visual Connections | Bloom's Level 5: Evaluate

These questions use visual images or previous content to make connections to this chapter's concepts.

1. **Figure 12.5.** Where could penicillinase affect each of these antibiotics?

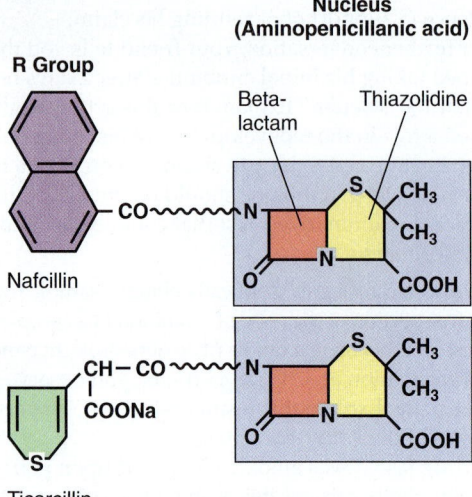

2. **From chapter 6, process figure 6.14a.** Describe as many ways as possible for an antiviral drug to interfere with the activity illustrated in the figure. How is each effective in controlling the viral cycle?

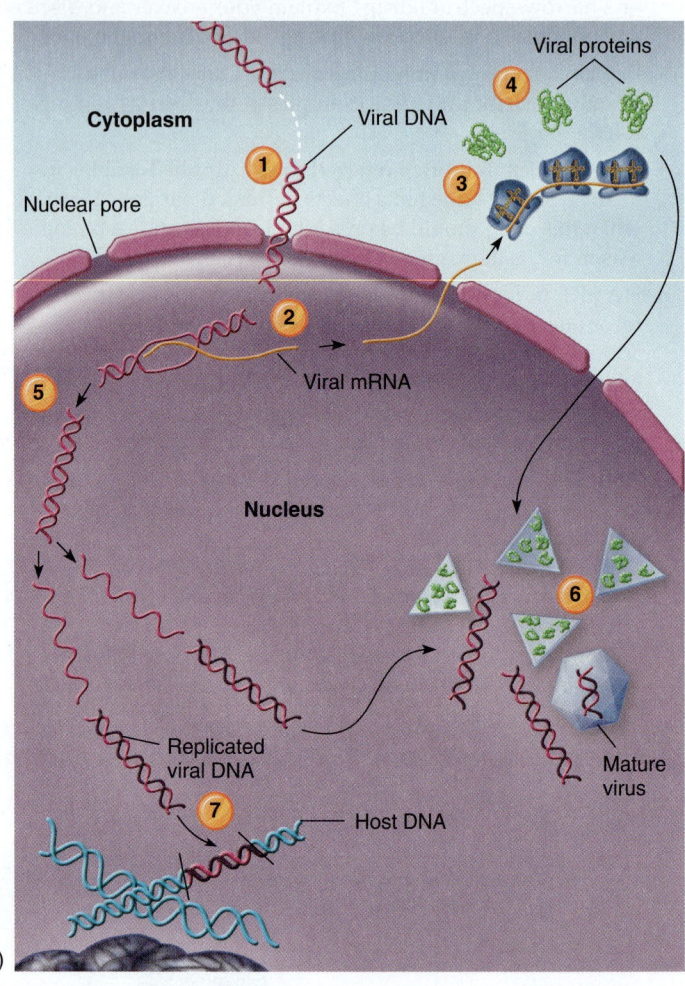

(a)

Concept Mapping | Bloom's Level 6: Create

Appendix D provides guidance for working with concept maps.

1. Using the words that follow, please create a concept map illustrating the relationships among these key terms from chapter 12.

chemotherapy   narrow-spectrum   minimum inhibitory concentration (MIC)
selectively toxic   superinfection   therapeutic index (TI)
broad-spectrum

www.mcgrawhillconnect.com

Enhance your study of this chapter with study tools and practice tests. Also ask your instructor about the resources available through ConnectPlus, including the media-rich eBook, interactive learning tools, and animations.

Microbe-Human Interactions
Infection and Disease

Case File 13

Hospital Workers: What's Coming Home on Your Clothing?

It's not an uncommon sight in the health care field: tired doctors, nurses, and technicians leaving the hospital at the end of their shift, heading to their cars and home to their loved ones. What do they all have in common, in addition to being exhausted after a long shift? They are all wearing hospital scrubs. To a member of the general public, this may seem innocuous and does nothing more than identify someone as a health care worker by his or her attire. As a microbiology student, this should raise some red flags: These workers, whether they are doctors, nurses, respiratory technicians, X-ray technicians, phlebotomists, or even custodial staff, have been exposed to microbes in the hospital and may be carrying them on their clothing.

Why is the clothing of health care workers any different than clothing that you're wearing? Everyone carries **resident biota,** or bacteria on the skin that is rubbed off into clothing, and everyone comes into contact with environmental surfaces, other people, and animals harboring microbes that may be transferred onto their clothing or skin. The difference is the hospital setting. **Healthcare-associated** or **nosocomial infections** are not uncommon, and many of these infections are passed from patient to patient via hospital staff. Health care workers are exposed to potential pathogens through aerosolized body fluids or through direct contact with patients and their body fluids. Even without visible stains from blood, feces, urine, mucus, or other emissions, microscopic particles can become embedded in the clothing. Because the hospital can be the source of multidrug-resistant bacteria and other potential pathogens, the contamination of hospital clothing can increase the risk for transmission of these microbes to the cars, homes, and loved ones of health care workers.

- How common is contamination of hospital scrubs among health care workers?
- What microbes most commonly are found on the clothing of health care workers?
- Is laundering at home sufficient to remove microbial contamination?

Continuing the Case appears on page 380.

355

Outline and Learning Outcomes

13.1 The Human Host

The human body exists in a state of dynamic equilibrium with microorganisms. In the healthy individual, this balance is maintained as a peaceful coexistence and lack of disease. In this chapter, we explore each component of the host-parasite relationship, beginning with the nature and function of normal biota, moving to the stages of infection and disease, and closing with a study of epidemiology and the patterns of disease in populations. These topics will set the scene for chapters 14 and 15, which deal with the ways the host defends itself against assault by microorganisms, and also for chapters 18 through 23, which examine diseases affecting different organ systems.

Colonization, Infection, Disease— A Continuum

The human body exists in a cloud of microbes. Our own bodies consist of 10 times as many microbes as our own human cells. Some microbes are colonists (normal biota), some are rapidly lost (transients), and others invade the tissues. Such intimate contact with microbes often leads to **infection,** a condition in which pathogenic microorganisms penetrate the host defenses, enter the tissues, and multiply. When the cumulative effects of the infection damage or disrupt tissues and organs, the **pathologic** state that results

is known as a **disease.** A disease is defined as any deviation from health. There are hundreds of different diseases caused by such factors as infections, diet, genetics, and aging. In this chapter, however, we discuss only **infectious disease**—the disruption of a tissue or organ caused by microbes or their products.

The pattern of the host-parasite relationship can be viewed as a series of stages that begins with contact, progresses to infection, and ends in disease. Because of numerous factors relating to host resistance and degree of pathogenicity, not all contacts lead to colonization, not all colonizations lead to infection, and not all infections lead to disease. In fact, contamination without colonization and colonization without disease are the rule.

Resident Biota: The Human as a Habitat

With its constant source of nourishment and moisture, relatively stable pH and temperature, and extensive surfaces upon which to settle, the human body provides a favorable habitat for an abundance of microorganisms. The large and diverse collection of microbes living on and in the body has been variously called the **normal biota,** resident or indigenous biota, though some microbiologists prefer to use the term *normal flora.* The normal residents include an array of bacteria, fungi, protozoa, and viruses. These organisms have a profound effect on human biology.

The Human Microbiome Project

In a nearly unprecedented cooperative effort, the **Human Microbiome Project (HMP)** is being conducted at laboratories all over the country and funded by the National Institutes of Health. The aim is to collect genetic sequences in the gut, respiratory tract, skin, and other body sites to determine which microbes are there, even when they cannot be grown in the laboratory. A second aim is to determine what role these normal biota play in health and disease. While results have trickled out since 2009, in June 2012, members of the HMP consortium published several papers containing results about resident microbes they found in their studies of 242 healthy U.S. volunteers. Samples were collected from 15 body sites in men and 18 body sites in women. In all, they analyzed 5,000 samples. The sites included the mouth, nose, skin, lower intestine (in the form of stool) and three vaginal sites. This kind of analysis has just become possible due to the knowledge gained from the Human Genome Project and the decrease in cost and time involved in doing whole-genome sequencing. Several surprising results were announced:

- Human cells contain about 23,000 protein-encoding genes. The microbes that inhabit us contain about 8 million, meaning their genetic contribution is about 360 times greater than our own. Most of the proteins produced by these genes are enzymes that help us digest our food and metabolize all kinds of substances that we could not otherwise use.
- We have a lot of microbes in places we used to think were sterile. The most striking example is the lungs; they were previously thought to be sterile but actually contain 2,000 microbes per square centimeter.
- All healthy people seem to also harbor potentially dangerous pathogens, in low numbers. This is a tribute to the normal biota that rarely cause disease; their presence keeps the pathogens in check.

Disease Connection

One reminder of the fact that even healthy people harbor small numbers of pathogenic microbes is the growing problem of "C. diff" infections. *Clostridium difficile* is an endospore-forming gram-positive bacterium that lives in our guts in very small quantities, kept in check by the microbial antagonism a diverse microbiota provides. When the healthy microbiota is damaged, especially through broad-spectrum or long-term antibiotic usage, *C. difficile* can flourish, leading to severe and long-lasting intestinal disease.

It seems clear that humans are who and what they are as a result of the microbes with whom they evolved. Continuing research with the HMP is looking at the different microbiomes that may be associated with deviations from health. For example, ongoing studies are finding differences in the microbiomes in healthy subjects and those with autism, and also the microbiomes of people with conditions as diverse as

Crohn's disease and psoriasis. Although the results are just beginning to emerge, the HMP is going to have a profound effect on our understanding of disease and the role our microbiome plays in it.

Acquiring Resident Biota

The human body offers a seemingly endless variety of environmental niches, with wide variations in temperature, pH, nutrients, and oxygen tension occurring from one area to another. Because the body provides such a range of habitats, it should not be surprising that the body supports a wide range of microbes. **Tables 13.1** and **13.2** outline our current understanding of where microbes reside on our bodies and where they do not. This information is in flux, of course, as we are in the midst of a revolution in understanding what microbes are in and on our bodies, due to the Human Microbiome Project (HMP).

Table 13.1 Sites That Harbor a Known Normal Biota

- Skin and its adjacent mucous membranes
- Respiratory tract and lungs
- Gastrointestinal tract (various parts)
- Outer opening of urethra
- External genitalia
- Vagina
- External ear canal
- External eye (lids, conjunctiva)

Table 13.2 Anatomical Sites and Fluids Thought to Be Sterile

All internal tissues and organs
Heart and circulatory system*
Liver*
Kidneys and bladder
Brain and spinal cord
Muscles
Bones
Ovaries/testes
Glands (pancreas, salivary, thyroid)
Sinuses**
Middle and inner ear
Fluids within an organ or tissue
Blood**
Urine in kidneys, ureters, bladder
Cerebrospinal fluid
Saliva prior to entering the oral cavity
Semen prior to entering the urethra
Amniotic fluid surrounding the embryo and fetus**

*Data analysis pending, Human Microbiome Project.
**Preliminary Human Microbiome Project data show the presence of microbial species.

The vast majority of microbes that come in contact with the body are removed or destroyed by the host's defenses long before they are able to colonize a particular area. Of those microbes able to establish an ongoing presence, an even smaller number are able to remain without attracting the unwanted attention of the body's defenses. This last group of organisms has evolved, along with its human hosts, to produce a complex relationship in which the effects of normal biota are generally not deleterious to the host. Recall from chapter 7 that microbes exist in different kinds of relationships with their hosts. Normal biota are generally either in a commensal or a mutualistic association with their hosts.

A Note About Viruses as Normal Biota

Microbiologists have never known how to characterize the non-disease-causing viruses and viral sequences we know are present in mammalian organisms. As you read in previous chapters, DNA sequences that we know come from viruses account for a large percentage of human DNA. Some of those are just genes carried to you by viruses, but some of those are also entire viruses. In 2006, researchers at Texas A&M showed that viruses called endogenous retroviruses (ERVs) are present in all mammals, and, in fact, are vital to healthy development of placentas in the sheep they studied.

They seem to have originated from infections of mammals from thousands of years ago and have remained in mammals because the proteins they produce provide some real benefit to their hosts. In this case, the ERVs studied in sheep seem to be vital to healthy development of placentas and embryos. When researchers blocked the action of the ERVs, sheep miscarried at a high rate. Apparently, over time, the sheep and the viruses have coevolved to their mutual benefit.

We now know that bacterial biota benefit the human host in many ways. The very development of our organs is influenced by the presence of resident biota. They also prevent the overgrowth of harmful microorganisms. A common example is the fermentation of glycogen by lactobacilli, which keep the pH in the vagina quite acidic and prevent the overgrowth of the yeast *Candida albicans*.

The generally antagonistic effect "good" microbes have against intruder microorganisms is called **microbial antagonism.** Normal biota exist in a steady, established relationship with the host and are unlikely to be displaced by incoming microbes. This antagonism is also enabled by the chemical or physiological environment created by the resident biota, which are hostile to other microbes. There are often members of the "normal" biota that would be pathogenic if they were allowed to multiply to larger numbers. Microbial antagonism is also responsible for keeping them in check.

The Importance of the Gut Biota

In 2012 and 2013, it has become clear that the makeup of your intestinal biota can influence many facets of your overall health. In the most obvious examples, data show that people who have chronic intestinal conditions such as Crohn's disease and other inflammatory bowel conditions possess a different bowel microbiota than healthy people. Patients who are chronically ill with *Clostridium difficile* ("*C. diff*") infection, which is often acquired after intense antibiotic use, can go months without relief, no matter the therapy. Some of these patients have elected to receive a fecal transplant, in which they receive an infusion of a healthy patient's fecal microbiota; often, this is the only thing that cures them.

Differences in the gut microbiome have preliminarily been associated with differences in the risk for heart disease, asthma, autism, rheumatoid arthritis, and even one's thoughts and moods and propensity for mental illness. Clearly, we are on the verge of learning a lot about human health and variability and how they are affected by those organisms under whose influence we evolved. It does seem clear that normal intestinal development and normal immune system development are absolutely reliant on our normal intestinal microbiota.

It is also the case that hosts with compromised immune systems can very easily experience disease caused by their (previously normal) biota **(table 13.3).** This outcome is seen when AIDS patients become sick with pneumococcal pneumonia, the causative agent of which (*Streptococcus pneumoniae*) is often carried as normal biota in the nasopharynx. **Endogenous** infections (those caused by biota already present in the body) can also occur when normal biota is introduced to a site that was previously sterile, as when *Escherichia coli* enters the bladder, resulting in a urinary tract infection.

Initial Colonization of the Newborn

The uterus and its contents used to be considered sterile during embryonic and fetal development. A growing number of doctors and scientists believe that fetuses are seeded with normal microbiota in utero, and that these microbes are important for healthy full-term pregnancies, and healthy newborns. At any rate, we know that exposure occurs during the birth process itself, when the baby becomes colonized with the mother's vaginal biota **(figure 13.1).** (Babies born by cesarean section typically are colonized by adult skin biota.) Within 8 to 12 hours after delivery, the vaginally delivered newborn typically has been colonized by bacteria

Table 13.3 Factors That Weaken Host Defenses and Increase Susceptibility to Infection*
• Age: the very young and the very old
• Genetic defects in immunity, and acquired defects in immunity (AIDS)
• Surgery and organ transplants
• Underlying disease: cancer, liver malfunction, diabetes
• Chemotherapy/immunosuppressive drugs
• Physical and mental stress
• Other infections

*These conditions compromise defense barriers or immune responses.

INSIGHT 13.1 "Normal" Microbiota Influenced Human Evolution

Ancient human ancestors were nearing extinction 100,000 years ago, their numbers reduced to 5,000 to 10,000 individuals on the African continent. The cause? Bacterial infection. Recent research shows that organisms such as *Escherichia coli* and group B *Streptococcus* (GBS) may have been the cause of high rates of infant mortality that decimated species at the time. Yet, the fossil record shows that the precursors to modern humans eventually emerged from this group, branching off from the Neanderthals and the Denisovans. How did they survive the bacterial attack?

Researchers at the University of California San Diego studied two genes present in chimpanzees that are absent or mutated in humans: *Siglec-13* and *Siglec-17*. (Chimpanzees are evolutionary cousins to *Homo sapiens*, sharing more than 98% of our genome.) These genes are significant because they help regulate the immune response. The researchers showed that certain microbes, such as *E. coli* and GBS, use the products of these human genes as receptors to more effectively cause infections such as sepsis and meningitis in humans. When researchers inserted these missing genes into modern human cells, they found that the bacteria attached themselves more effectively

to human cells and elicited a much weaker immune response.

The presence of these genes in the primitive human genome may have led to massive bacterial epidemics and near extinction of the species. Although *E. coli* and GBS are now considered normal microbiota, the organisms can be dangerous to newborn infants. With a weakened immune response due to the presence of *Siglec-13* and *Siglec-17*, infant mortality would have been high. The only way for the population to survive was to get rid of these two genes, which may have occurred as recently as 46,000 years ago. This genetic data suggest that the genes may have been turned off as early as 440,000 to 270,000 years ago, but it would have taken thousands of years for the effect to spread through the entire population. Those without the mutation or loss of *Siglec-13* and *Siglec-17* would not have survived the bacterial onslaught. Although many factors come into play in the evolution of a species, bacterial infections of organisms that are now considered normal microbiota certainly played a role in the evolution of the human species.

Source: 2012. PNAS Jun. 4. p. 9935.

such as *Lactobacillus*, *Prevotella*, and *Sneathia*, acquired primarily from the birth canal. Data from the Human Microbiome Project revealed that the microbial composition of the vagina changes significantly in pregnant women. Early on, a

Figure 13.1 The origins of microbiota in newborns. A vaginal birth exposes babies to the biota of the mother's reproductive tract. Babies delivered via cesarean section become colonized with maternal skin biota. The second major influence on the infant's microbiome is its early diet.

Lactobacillus species that digests milk begins to populate the vagina. Immediately prior to delivery, additional bacterial species colonize the birth canal. Scientists suggest that the lactobacilli provide the newborn baby with the enzymes necessary to digest milk and that the later colonizers are better equipped to protect a newborn baby from skin disorders and other conditions. After the baby is born, the mother's vaginal microbiota returns to its former state.

The baby continues to acquire resident microbiota from the environment, notably from its diet. Throughout most of evolutionary history, of course, that means human breast milk. Scientists have found that human milk contains around 600 species of bacteria and a lot of sugars that babies cannot digest. The sugars *are* used by healthy gut bacteria, suggesting a role for breast milk in maintaining a healthy gut microbiome in the baby.

Indigenous Biota of Specific Regions

The Human Microbiome Project has shown that among healthy adults, the normal microbiota varies significantly. For instance, the microbiota on a person's right hand was found to be significantly different than that on the left hand. What seemed to be more important than the exact microbial profile of any given body site was the profile of proteins, especially the enzymatic capabilities. That profile remained stable across subjects, though the microbes that were supplying those enzymes could differ broadly. With that caveat, we present in **table 13.4** a summary of the types of normal, indigenous biota present in specific body sites. This table represents the results of the Human Microbiome Project with respect to bacteria, as well as information we have long had about the presence of fungi and other microbes in some sites. Scientists are in the process of cataloging other microorganisms via metagenomics and are just beginning to appreciate their numbers in the human microbiome. For example, we now know there are at least 100 types of fungi in the intestine and as many as a billion viruses per gram of feces.

13.1 Learning Outcomes—Assess Your Progress

1. Differentiate among the terms *colonization*, *infection*, and *disease*.
2. Enumerate the sites where normal biota is found in humans.
3. Discuss how the Human Microbiome Project has changed our understanding of normal biota.

13.2 The Progress of an Infection

A microbe whose relationship with its host is parasitic and results in infection and disease is termed a **pathogen.** The type and severity of infection depend both on the pathogenicity of the organism and the condition of the host **(figure 13.2). Pathogenicity,** you will recall, is a broad concept that describes an organism's potential to cause infection or disease, and is used to divide pathogenic microbes into one of two groups. **True pathogens** (also termed *primary pathogens*) are capable of causing disease in healthy persons with normal immune defenses. They are often associated with a specific, recognizable disease, which may vary in severity from mild (colds) to severe (malaria) to fatal (rabies). Examples of true pathogens include the influenza virus, plague bacillus, and malarial protozoan.

Opportunistic pathogens cause disease when the host's defenses are compromised[1] or when they become established

1. People with weakened immunity are often termed *immunocompromised*.

Table 13.4 Life on Humans: Sites Containing Well-Established Biota and Representative Examples*

Anatomical Sites	Common Genera	Remarks
Skin	**Gram-positive bacteria:** *Staphylococcus* (including *S. aureus*), *Propionibacterium*, *Streptococcus*, *Corynebacterium*, *Lactobacillus* **Gram-negative bacteria:** *Bacteroides*, *Prevotella*, *Haemophilus* **Fungi:** *Candida*	Skin biota varies with body location and with age; in different individuals, different genera predominate; approx. 4% of subjects carry *Staphylococcus aureus* on their skin.
Gastrointestinal Tract Oral cavity	**Gram-positive bacteria:** *Streptococcus* predominates; *Actinomyces*, *Corynebacterium* **Gram-negative bacteria:** *Haemophilus*, *Prevotella*, *Veillonella*, *Bacteroides*, *Moraxella* **Fungi:** *Candida* **Protozoa:** *Entamoeba*	More than a dozen species of *Streptococcus*; microbes colonize the epidermal layer of cheeks, gingiva, pharynx; surface of teeth; found in saliva in huge numbers
Intestinal tract	**Gram-negative bacteria:** *Bacteroides*, *Prevotella* **Fewer gram-positives:** *Streptococcus*, *Lactobacillus* **Fungi:** *Candida*	Fecal biota consists predominantly of anaerobes; other microbes are aerotolerant or facultative. *E. coli* present in majority of subjects, but in relatively low abundance.
Respiratory Tract Nose	**Gram-positive bacteria:** *Propionibacterium*, *Corynebacterium*, *Staphylococcus* **Gram-negative bacteria:** *Moraxella*, *Prevotella*	Approx. 30% of subjects carry *Staphylococcus aureus* in nose.
Throat	**Gram-positive bacteria:** *Streptococcus*, *Corynebacterium* **Gram-negative bacteria:** *Haemophilus*, *Prevotella*, *Veillonella*, *Moraxella*	Biota similar to oral cavity
Lungs	**Gram-negative bacteria:** *Prevotella*, *Veillonella*	Previously thought to be sterile; asthmatic and COPD lungs colonized by different species than healthy
Vagina	**Gram-positive bacteria:** *Lactobacillus* predominates; *Streptococcus* **Gram-negative bacteria:** *Prevotella* **Fungi:** *Candida*	Biota responds to hormonal changes during life, with significant changes in preparation for birth, and with more variety of species after menopause
Urinary Tract	**Gram-positive bacteria:** *Lactobacillus* (predominant) **Gram-negative bacteria:** *Prevotella*, *Gardnerella*	In females, biota exists only in the first portion of the urethral mucosa; the remainder of the tract is sterile. In males, the entire reproductive and urinary tract is thought to be sterile except for a short portion of the anterior urethra.

*Information in this table subject to significant change as results of Human Microbiome Project become available.

Figure 13.2 Will disease result from an encounter between a (human) host and a microorganism?   In most cases, all of the slider bars must be in the correct ranges and the microbe's on-off switch in the "on" position with the host's on-off switch in the "off" position in order for disease to occur. These are just a few examples and not the only options.

Microbe X			Host			Outcomes
Virulence	Percentage of optimal infectious dose	Correct portal of entry	Genetic profile that resists Microbe X (nonspecific defenses)	Previous exposure to Microbe X (specific immunity)	General level of health	
Low	~30	Off	mid	On	mid	Microbe passes through unnoticed.
Low	~70	On	mid	On	high	Microbe passes through unnoticed. *or* Microbe becomes established without disease (colonization or infection).
mid-Low	0	Off	high	Off	high	Microbe passes through unnoticed. *or* Microbe becomes established without disease (colonization or infection).
High	~50	On	mid-Low	Off	high	Microbe causes disease.

in a part of the body that is not natural to them. Opportunists are not considered pathogenic to a normal healthy person and, unlike primary pathogens, do not generally possess well-developed disease properties. Examples of opportunistic pathogens include *Pseudomonas* species and the yeast *Candida albicans.* Factors that greatly predispose a person to infections, both primary and opportunistic, are shown in table 13.3.

The relative severity of the disease caused by a particular microorganism depends on the **virulence** of the microbe. Although the terms *pathogenicity* and *virulence* are often used interchangeably, *virulence* is the accurate term for describing the *degree* of pathogenicity. The virulence of a microbe is determined by its ability to (1) establish itself in the host and (2) cause damage.

A Note About Pathogens

So far, science has documented a total of 1,407 microbes that cause disease in humans. Of these, 538 are bacteria, 317 are fungi, 287 are helminths, 208 are viruses, and 57 are protozoa. Of course, we don't know how many pathogens we *don't* know about. There are plenty of conditions and diseases that have no known cause as of yet.

Much is involved in both of these steps. To establish themselves in a host, microbes must enter the host, attach firmly to host tissues, and survive the host defenses. To cause damage, microbes produce toxins or induce a host response that is actually injurious to the host. Any characteristic or structure of the microbe that contributes to the preceding activities is called a **virulence factor.** Virulence can be due to single or multiple factors. In some microbes, the causes of virulence are clearly established; in others, they are not. In the following section, we examine the effects of virulence factors while simultaneously outlining the stages in the progress of an infection.

Note that different healthy individuals have widely varying responses to the same microorganism. This is determined in part by genetic variation in the specific components of their defense systems. That is why the same infectious agent can cause severe disease in one individual and mild or no disease in another.

Why is there variation? In chapter 7, we described *coevolution* as changes in genetic composition by one species in response to changes in another. Infectious agents evolve in response to their interaction with a host (as in the case of antibiotic resistance). Hosts evolve, too; although their pace of change is much slower than that of a microbe, changes eventually show up in human populations due to their past experiences with pathogens. One striking example is sickle-cell disease. Persons who are carriers of a mutation in their hemoglobin gene (i.e., who inherited one mutated hemoglobin gene and one normal) have few or no sickle-cell disease symptoms but are more resistant to malaria than people who have no mutations in their hemoglobin genes. When a person inherits two alleles for the mutation (from both parents), that person enjoys some protection from malaria but will suffer from sickle-cell disease.

People of West African descent are much more likely to have one or two sickle-cell alleles. Malaria is endemic in West Africa. It seems the hemoglobin mutation is an adaptation of the human host to its long-standing relationship with the malaria protozoan.

In another example, researchers have found a gene that correlates with how people react to infection with the swine flu virus. The gene codes for a protein that blocks viral entry into cells. People who experienced only mild flu symptoms during the 2009 swine flu epidemic were found to have the intact gene, whereas those who became most gravely ill or who died were much more likely to have a mutated variant of this gene.

These are examples of the variability represented by the slider bars in the column marked "genetic profile" in figure 13.2. Scientists have also found that bacteria in the human body respond to human stress hormones, such as norepinephrine. For example, when nerve cells in the gut produced this hormone, *E. coli* were found to increase their numbers up to 10,000-fold. Other studies have found that some stress hormones can induce bacteria to adhere to hard surfaces, raising the possibility of biofilm formation, and that bacteria increase the expression of pathogenic genes under the influence of these hormones. These phenomena suggest an intriguing new area of research into the prevention of microbial disease.

Even though human factors are important, the Centers for Disease Control and Prevention has adopted a system of biosafety categories for pathogens based on their general degree of pathogenicity and the relative danger in handling them. **Table 13.5** summarizes the four major biosafety levels.

Becoming Established: Step One— Portals of Entry

To initiate an infection, a microbe enters the tissues of the body by a characteristic route, the **portal of entry,** usually through the skin or a mucous membrane. The source of the infectious agent can be **exogenous,** originating from a source outside the body (the environment or another person or animal), or endogenous, already existing on or in the body (normal biota or a previously silent infection).

For the most part, the portals of entry are the same anatomical regions that also support normal biota: the skin, gastrointestinal tract, respiratory tract, and urogenital tract. The majority of pathogens have adapted to a specific portal of entry, one that provides a habitat for further growth and spread. This adaptation can be so restrictive that if certain pathogens enter the "wrong" portal, they will not be infectious. For instance, inoculation of the nasal mucosa with the influenza virus invariably gives rise to the flu, but if this virus contacts only the skin, no infection will result. Likewise, contact with athlete's foot fungi in small cracks in the toe webs can induce an infection, but inhaling the fungus spores will not infect a healthy individual. Occasionally, an infective agent can enter by more than one portal. For instance, *Mycobacterium tuberculosis* enters through both the respiratory and gastrointestinal tracts, and pathogens in the genera *Streptococcus* and *Staphylococcus* have adapted to invasion through several portals of entry such as the skin, urogenital tract, and respiratory tract.

Infectious Agents That Enter the Skin

The skin is a very common portal of entry. The actual sites of entry are usually nicks, abrasions, and punctures (many of which are tiny and inapparent) rather than unbroken skin. Intact skin is a very tough barrier that few microbes can penetrate. *Staphylococcus aureus* (the cause of boils), *Streptococcus pyogenes* (an agent of impetigo), the fungal dermatophytes, and agents of gangrene and tetanus gain access through damaged skin. The viral agent of cold sores (herpes simplex, type I) enters through the mucous membranes near the lips.

Table 13.5 Primary Biosafety Levels and Agents of Disease

Biosafety Level	Facilities and Practices	Risk of Infection and Class of Pathogens
1	Standard, open bench, no special facilities needed; typical of most microbiology teaching labs; access may be restricted.	Low infection hazard; microbes not generally considered pathogens and will not colonize the bodies of healthy persons; *Micrococcus luteus, Bacillus megaterium, Lactobacillus, Saccharomyces.*
2	At least level 1 facilities and practices; plus personnel must be trained in handling pathogens; lab coats and gloves required; safety cabinets may be needed; biohazard signs posted; access restricted.	Agents with moderate potential to infect; class 2 pathogens can cause disease in healthy people but can be contained with proper facilities; most pathogens belong to class 2; includes *Staphylococcus aureus, Escherichia coli, Salmonella* spp., *Corynebacterium diphtheriae;* pathogenic helminths; hepatitis A, B, and rabies viruses; *Cryptococcus* and *Blastomyces.*
3	Minimum of level 2 facilities and practices; plus all manipulation performed in safety cabinets; lab designed with special containment features; only personnel with special clothing can enter; no unsterilized materials can leave the lab; personnel warned, monitored, and vaccinated against infection dangers.	Agents can cause severe or lethal disease especially when inhaled; class 3 microbes include *Mycobacterium tuberculosis, Francisella tularensis, Yersinia pestis, Brucella* spp., *Coxiella burnetii, Coccidioides immitis,* and yellow fever, WEE, and HIV.
4	Minimum of level 3 facilities and practices; plus facilities must be isolated with very controlled access; clothing changes and showers required for all people entering and leaving; materials must be autoclaved or fumigated prior to entering and leaving lab.	Agents are highly virulent microbes that pose extreme risk for morbidity and mortality when inhaled in droplet or aerosol form; most are exotic flaviviruses; arenaviruses, including Lassa fever virus; or filoviruses, including Ebola and Marburg viruses.

Some infectious agents create their own passageways into the skin using digestive enzymes. For example, certain helminth worms burrow through the skin directly to gain access to the tissues. Other infectious agents enter through bites. The bites of insects, ticks, and other animals offer an avenue to a variety of viruses, rickettsias, and protozoa. An artificial means for breaching the skin barrier is contaminated hypodermic needles by intravenous drug abusers. Users who inject drugs are predisposed to a disturbing list of well-known diseases: hepatitis, AIDS, tetanus, tuberculosis, osteomyelitis, and malaria. Contaminated needles often contain bacteria from the skin or environment that induce heart disease (endocarditis), lung abscesses, and chronic infections at the injection site.

Although the conjunctiva, the outer protective covering of the eye, is ordinarily a relatively good barrier to infection, bacteria such as *Haemophilus aegyptius* (pinkeye), *Chlamydia trachomatis* (trachoma), and *Neisseria gonorrhoeae* have a special affinity for this membrane.

The Gastrointestinal Tract as Portal

The gastrointestinal tract is the portal of entry for pathogens contained in food, drink, and other ingested substances. They are adapted to survive digestive enzymes and abrupt pH changes. The best-known enteric agents of disease are gram-negative rods in the genera *Salmonella, Shigella, Vibrio,* and certain strains of *Escherichia coli.* Viruses that enter through the gut are poliovirus, hepatitis A virus, echovirus, and rotavirus. Important enteric protozoans are *Entamoeba histolytica* (amoebiasis) and *Giardia lamblia* (giardiasis). See chapter 22 for details of these diseases.

INSIGHT 13.2   The Microscopic Elephant in the Room

When you enter an empty room, are you *really* alone? A recent study at Yale University says no. A single person's presence in a room adds 37 million bacteria to the air in the room *per hour.* Jordan Peccia, associate professor of environmental engineering at Yale, along with his fellow researchers, analyzed the air in a ground-level university classroom over 8 days. The study included 4 days when the room was occupied for instructional use and 4 days when the room was empty; the doors and windows were kept closed, and the HVAC system was intact. Peccia's group found that the majority of bacteria in the air were resuspended from the floor and had been left behind by previous occupants. Also, 18% of all bacteria in a room, either fresh or previously deposited bacteria, came from humans rather than from animals or plants.

Source: www.sciencedaily.com/releases/2012/03/120328172255.htm

The Respiratory Portal of Entry

The oral and nasal cavities are also the gateways to the respiratory tract, the portal of entry for the greatest number of pathogens. Because there is a continuous mucous membrane surface covering the upper respiratory tract, the sinuses, and the auditory tubes, microbes are often transferred from one site to another. The extent to which an agent is carried into the respiratory tree is based primarily on its size. In general, small cells and particles are inhaled more deeply than larger ones. Infectious agents with this portal of entry include the bacteria of streptococcal sore throat, meningitis, diphtheria, and whooping cough and the viruses of influenza, measles, mumps, rubella, chickenpox, and the common cold. Pathogens that are inhaled into the lower regions of the respiratory tract (bronchioles and lungs) can cause **pneumonia,** an inflammatory condition of the lung. Bacteria (*Streptococcus pneumoniae, Klebsiella, Mycoplasma*) and fungi (*Cryptococcus* and *Pneumocystis*) are a few of the agents involved in pneumonias. Other agents causing unique recognizable lung diseases are *Mycobacterium tuberculosis* and fungal pathogens such as *Histoplasma.* Chapter 21 describes infections of the respiratory system.

Urogenital Portals of Entry

The urogenital tract is the portal of entry for many pathogens that are contracted by sexual means (intercourse or intimate direct contact). **Sexually transmitted infections (STIs)** account for an estimated 4% of infections worldwide, with approximately 13 million new cases occurring in the United States each year. The most recent available statistics for the estimated incidence of common STIs are provided in **table 13.6.**

The microbes causing STIs enter the skin or mucosa of the penis, external genitalia, vagina, cervix, and urethra. Some can penetrate an unbroken surface; others require a cut or abrasion. The once predominant sexual diseases syphilis and gonorrhea have been supplanted by a large and growing list of STIs led by genital warts, chlamydia, and herpes. Evolving sexual practices have increased the incidence of STIs that were once uncommon, and diseases that were

not originally considered STIs are now so classified.[2] Other common sexually transmitted agents are HIV, *Trichomonas* (a protozoan), *Candida albicans* (a yeast), and hepatitis B virus. STIs are described in detail in chapter 23, with the exception of HIV (see chapter 20) and hepatitis B (see chapter 22).

Disease Connection

Despite its name, *Trichomonas vaginalis* infects both men and women. Previously thought to be a relatively mild infection, it is now known to enhance your risk of contracting other STIs, including HIV. A strong association has been found between *Trichomonas* infection and aggressive forms of prostate cancer. *Trichomonas* is discussed in chapter 23.

Not all urogenital infections are STIs. Some of these infections are caused by displaced organisms (as when normal biota from the gastrointestinal tract cause urinary tract infections) or by opportunistic overgrowth of normal biota ("yeast infections").

Pathogens That Infect During Pregnancy and Birth

The placenta is an exchange organ—formed by maternal and fetal tissues—that separates the blood of the developing fetus from that of the mother yet permits diffusion of dissolved nutrients and gases to the fetus. Whether or not normal biota colonize the fetus, we know that some pathogens such as the syphilis spirochete can cross the placenta, enter the umbilical vein, and spread by the fetal circulation into the fetal tissues **(figure 13.3).**

Other infections, such as herpes simplex, can occur perinatally when the child is contaminated by the birth canal. The common infections of fetus and neonate are grouped together in a unified cluster known by the acronym **TORCH,** which medical personnel must monitor. *TORCH* stands for **t**oxoplasmosis, **o**ther diseases (syphilis, coxsackievirus, varicella-zoster virus, AIDS, and chlamydia), **r**ubella, **c**ytomegalovirus, and **h**erpes simplex virus. The most serious complications of TORCH infections are spontaneous abortion, congenital abnormalities, brain damage, prematurity, and stillbirths.

The Size of the Inoculum

Another factor crucial to the course of an infection is the quantity of microbes in the inoculating dose. For most agents, infection will proceed only if a minimum number, called the *infectious dose* (ID), is present. This number has been determined experimentally for many microbes. In general, microorganisms with smaller infectious doses have greater virulence. On the low end of the scale, the ID for rickettsia, the causative agent of Q fever, is only a single cell; it is only about 10 cells in tuberculosis, giardiasis, and

Table 13.6 Incidence of Common Sexually Transmitted Infections	
STI	**Estimated Number of New Cases per Year in United States**
Trichomoniasis	7,400,000
Human papillomavirus	6,200,000
Chlamydiosis	2,800,000
Herpes simplex	1,600,000
Gonorrhea	309,000
Hepatitis B	77,000
HIV	60,000
Syphilis	41,000

---

2. Amoebic dysentery, scabies, salmonellosis, and *Strongyloides* worms are examples.

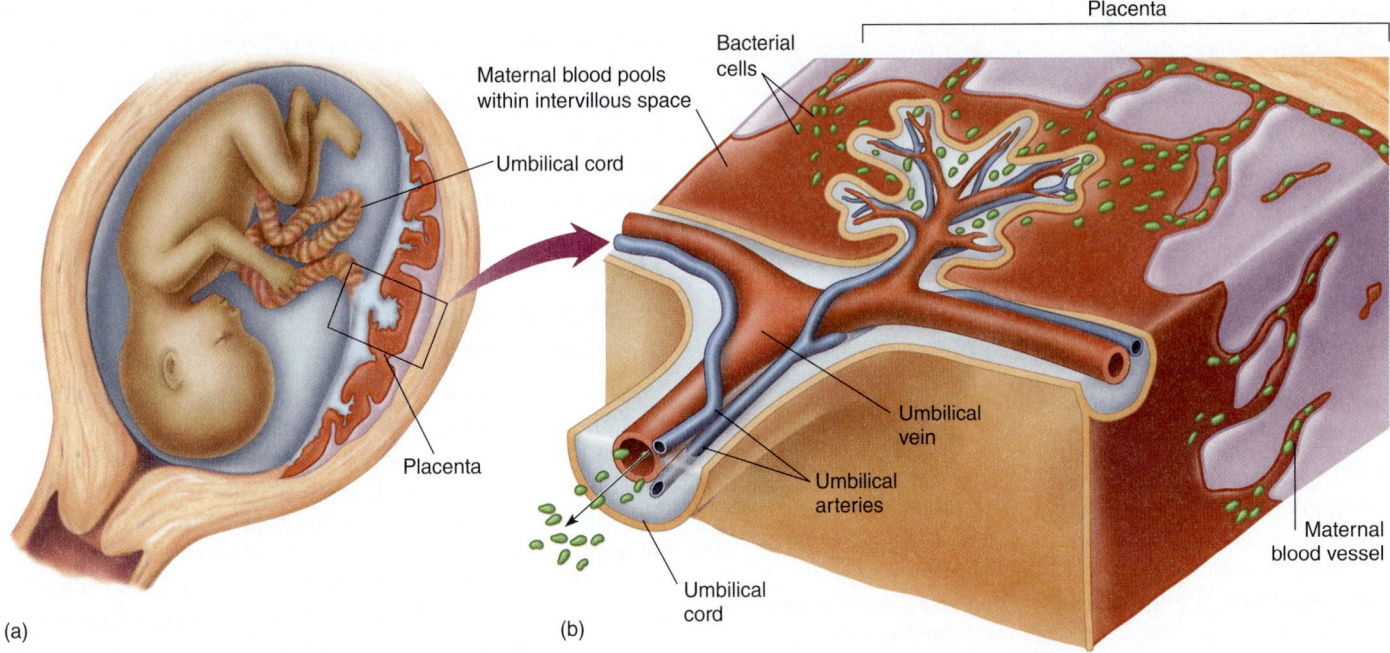

Figure 13.3 Transplacental infection of the fetus. (a) Fetus in the womb. (b) In a closer view, microbes are shown penetrating the maternal blood vessels and entering the blood pool of the placenta. They then invade the fetal circulation by way of the umbilical vein.

coccidioidomycosis. The ID is 1,000 bacteria for gonorrhea and 10,000 bacteria for typhoid fever, in contrast to 1,000,000,000 bacteria in cholera. Numbers below an infectious dose will generally not result in an infection. However, if the quantity is far in excess of the ID, the onset of disease can be extremely rapid.

Recent research has shown that communication among microbes is critical to the establishment of infection. If quorum-sensing chemicals are blocked, the bacteria are not able to sense the presence of enough cells to mount an effective attack against the host. This forces them to sit "silently" waiting for enough members to arrive at the site of colonization, making them vulnerable to the host immune system in the meantime. Development of drugs that block this critical communication process may be on the horizon for the treatment of many microbial infections.

Becoming Established: Step Two—Attaching to the Host

Adhesion is a process by which microbes gain a more stable foothold on host tissues. Because adhesion is dependent on binding between specific molecules on both the host and pathogen, a particular pathogen is limited to only those cells (and organisms) to which it can bind. Once attached, the pathogen is poised advantageously to invade the body compartments. Bacterial, fungal, and protozoal pathogens attach most often using fimbriae (pili), surface proteins, and adhesive slimes or capsules; viruses attach by means of specialized spikes, or glycoproteins, on their surfaces **(figure 13.4)**. In addition, parasitic helminths are mechanically fastened to the portal of entry by suckers, hooks, and barbs. Adhesion

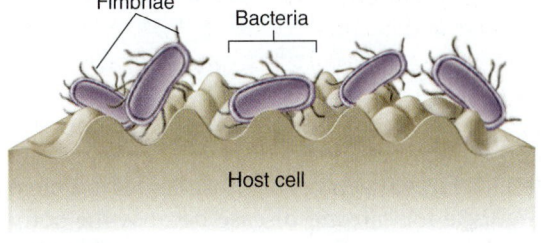

(a) **Fimbriae**

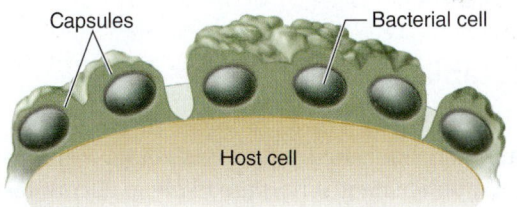

(b) **Capsules**

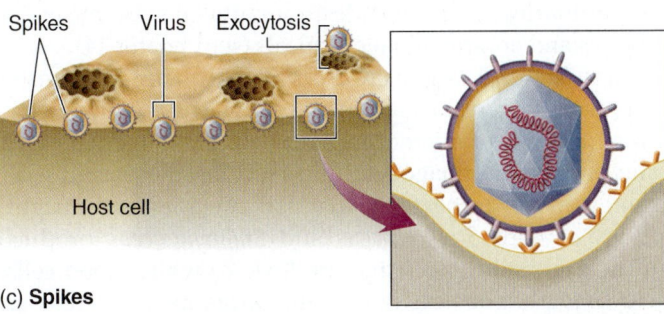

(c) **Spikes**

Figure 13.4 Mechanisms of adhesion by pathogens.
(a) Fimbriae, minute bristlelike appendages. (b) Adherent extracellular capsules made of slime or other sticky substances. (c) Viral envelope spikes.

Table 13.7 Adhesive Properties of Microbes

Microbe	Disease	Adhesion Mechanism
Neisseria gonorrhoeae	Gonorrhea	Fimbriae attach to genital epithelial cells.
Escherichia coli	Diarrhea	Fimbrial adhesin
Shigella	Dysentery	Fimbriae attach to intestinal epithelium.
Mycoplasma	Pneumonia	Specialized tip at ends of bacteria fuse tightly to lung epithelium.
Pseudomonas aeruginosa	Burn, lung infections	Fimbriae and slime layer
Streptococcus pyogenes	Pharyngitis, impetigo	Lipoteichoic acid and capsule anchor cocci to epithelium.
Streptococcus mutans, S. sobrinus	Dental caries	Dextran slime layer glues cocci to tooth surface after initial attachment.
Influenza virus	Influenza	Viral spikes attach to receptor on cell surface.
Poliovirus	Polio	Capsid proteins attach to receptors on susceptible cells.
HIV	AIDS	Viral spikes adhere to white blood cell receptors.
Giardia lamblia (protozoan)	Giardiasis	Small suction disc on underside attaches to intestinal surface.

methods of various microbes and the diseases they lead to are shown in **table 13.7.** Firm attachment to host tissues is almost always a prerequisite for causing disease because the body has so many mechanisms for flushing microbes and foreign materials from its tissues.

Becoming Established: Step Three—Surviving Host Defenses

Microbes that are not established in a normal biota relationship in a particular body site in a host are likely to encounter resistance from host defenses when first entering, especially from certain white blood cells called **phagocytes.** These cells ordinarily engulf and destroy pathogens by means of enzymes and antimicrobial chemicals (see chapter 14).

Antiphagocytic factors are a type of virulence factor used by some pathogens to avoid phagocytes. The antiphagocytic factors of resistant microorganisms help them to circumvent some part of the phagocytic process. The most aggressive strategy involves bacteria that kill phagocytes outright. Species of both *Streptococcus* and *Staphylococcus* produce **leukocidins,** substances that are toxic to white blood cells, including phagocytes. Some microorganisms secrete an extracellular surface layer (slime or capsule) that makes it physically difficult for the phagocyte to engulf them. *Streptococcus pneumoniae, Salmonella typhi, Neisseria meningitidis,* and *Cryptococcus neoformans* are notable examples. Some bacteria

are well adapted to survive inside phagocytes after ingestion. For instance, pathogenic species of *Legionella, Mycobacterium,* and many rickettsias are readily engulfed but are capable of avoiding further destruction. The ability to survive intracellularly in phagocytes has special significance because it provides a place for the microbes to hide, grow, and be spread throughout the body.

Step Four—Causing Disease
How Virulence Factors Contribute to Tissue Damage

Virulence factors are structures, products, or capabilities that allow a pathogen to cause infection in a host. From a microbe's perspective, they are simply adaptations it uses to invade and establish itself in the host. (You will remember from chapter 9 that many virulence factors can be found on pathogenicity islands, genetic regions that have been passed horizontally from other microbes.) These same factors determine the degree of tissue damage that occurs. The effects of a pathogen's virulence factors on tissues vary greatly. Cold viruses, for example, invade and multiply but cause relatively little damage to their host. At the other end of the spectrum, pathogens such as *Clostridium tetani* or HIV severely damage or kill their host. Microorganisms either inflict direct damage on hosts through the use of enzymes or toxins **(figure 13.5a,b),** or they cause damage indirectly when their presence causes an excessive or inappropriate host response **(figure 13.5c).** For convenience, we divide the "directly damaging" virulence factors into exoenzymes and toxins. Although this distinction is useful, there is often a very fine line between enzymes and toxins because many substances called toxins actually function as enzymes.

Microbial virulence factors are often responsible for inducing the host to cause damage, as well. The capsule of *Streptococcus pneumoniae* is a good example. Its presence prevents the bacterium from being cleared from the lungs by phagocytic cells, leading to a continuous influx of fluids into the lung spaces and the condition we know as pneumonia.

Extracellular Enzymes Many pathogenic bacteria, fungi, protozoa, and worms secrete **exoenzymes** that break down and inflict damage on tissues. Other enzymes dissolve the host's defense barriers and promote the spread of microbes to deeper tissues.

Examples of enzymes are

1. mucinase, which digests the protective coating on mucous membranes and is a factor in amoebic dysentery;
2. keratinase, which digests the principal component of skin and hair, and is secreted by fungi that cause ringworm;
3. collagenase, which digests the principal fiber of connective tissue and is an invasive factor of *Clostridium* species and certain worms; and
4. hyaluronidase, which digests hyaluronic acid, the ground substance that cements animal cells together. This enzyme is an important virulence factor in staphylococci, clostridia, streptococci, and pneumococci.

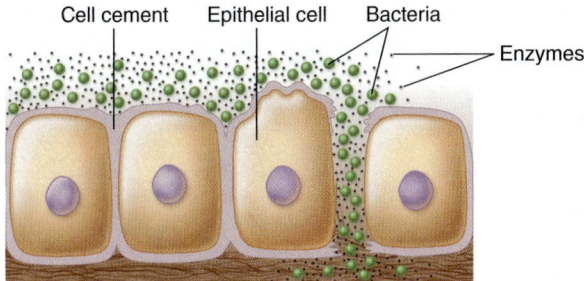

(a) **Exoenzymes**

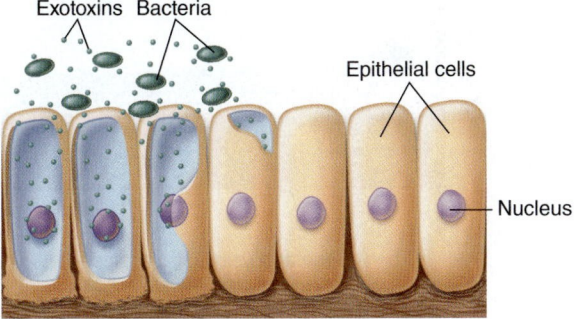

(b) **Toxins**

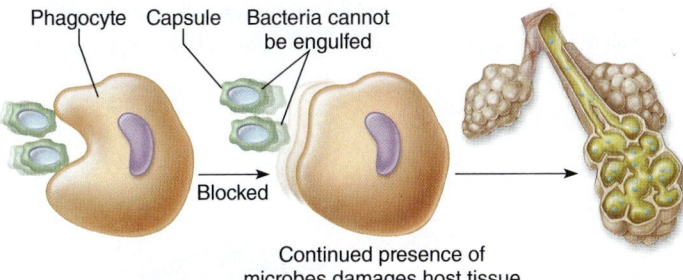

Continued presence of
microbes damages host tissue

(c) **Blocked phagocytic response**

Figure 13.5 Three ways microbes damage the host.
(a) Exoenzymes. Bacteria produce extracellular enzymes that dissolve
intracellular connections and penetrate through or between cells
to underlying tissues. **(b)** Toxins. Exotoxins (pictured) are secreted
by bacteria and damage target cells; endotoxin is released from
disintegrating gram-negative bacteria, which activates the immune
system inappropriately. **(c)** Bacterium has a property that enables it
to escape phagocytosis and remain as an "irritant" to host defenses,
which are deployed excessively.

Some enzymes react with components of the blood.
Coagulase, an enzyme produced by pathogenic staphylococci, causes clotting of blood or plasma. By contrast, the
bacterial kinases (streptokinase, staphylokinase) do just the
opposite, dissolving fibrin clots and expediting the invasion of damaged tissues. In fact, one form of streptokinase
(Streptase) is marketed as a therapy to dissolve blood clots in
patients with problems with thrombi and emboli.[3]

---

3. These conditions are intravascular blood clots that can cause circulatory
   obstructions.

Bacterial Toxins: A Potent Source of Cellular Damage

A **toxin** is a specific chemical product of microbes, plants,
and some animals that is poisonous to other organisms.
Toxigenicity, the power to produce toxins, is a genetically
controlled characteristic of many species and is responsible
for the adverse effects of a variety of diseases generally called
toxinoses. Toxinoses in which the toxin is spread by the
blood from the site of infection are called **toxemias** (tetanus and diphtheria, for example), whereas those caused by
ingestion of toxins are **intoxications** (botulism). A toxin is
named according to its specific target of action: Neurotoxins
act on the nervous system; enterotoxins act on the intestine;
hemotoxins lyse red blood cells; and nephrotoxins damage
the kidneys.

Another useful scheme classifies toxins according to
their origins **(figure 13.6).** A toxin molecule secreted by a
living bacterial cell into the infected tissues is an **exotoxin.**
There are many different types of exotoxins. A toxin that is
not actively secreted but is shed from the outer membrane is
an **endotoxin.** There is only one endotoxin, which is found
on all gram-negative bacteria. Other important differences
between the two groups are summarized in **table 13.8.**

Exotoxins are proteins with a strong specificity for a target cell and extremely powerful, sometimes deadly, effects.
They generally affect cells by damaging the cell membrane
and initiating lysis or by disrupting intracellular function.

Table 13.8	Differential Characteristics of Bacterial Exotoxins and Endotoxin	
Characteristic	**Exotoxins**	**Endotoxin**
Toxicity	Toxic in minute amounts	Toxic in high doses
Effects on the body	Specific to a cell type (blood, liver, nerve); induces TNF production resulting in fever	Systemic: fever, inflammation
Chemical composition	Small proteins	Lipopolysaccharide of cell wall
Heat denaturation at 60°C	Unstable	Stable
Toxoid formation	Can be converted to toxoid*	Cannot be converted to toxoid
Immune response	Stimulate antitoxins**	Does not stimulate antitoxins
Fever stimulation	Usually not	Yes
Manner of release	Secreted from live cell	Released by cell via shedding or during lysis
Typical sources	A few gram-positive and gram-negative	All gram-negative bacteria

*A toxoid is an inactivated toxin used in vaccines.
**An antitoxin is an antibody that reacts specifically with a toxin.

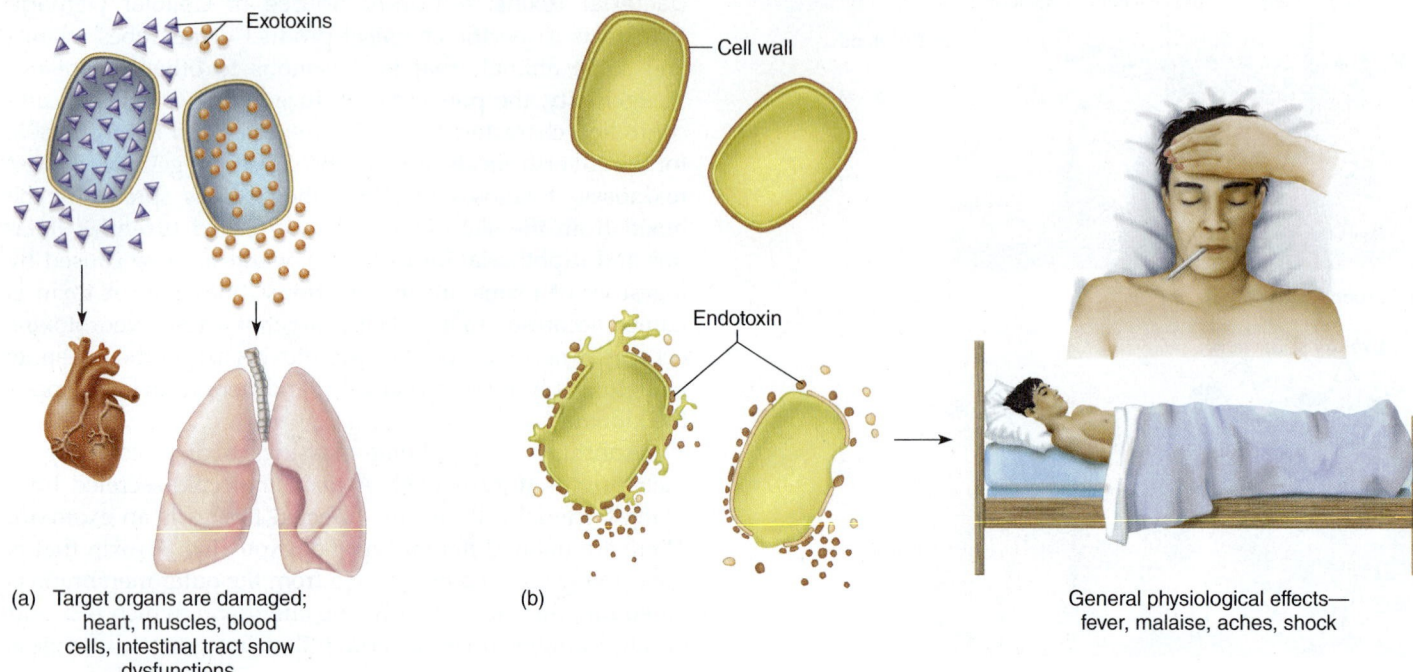

(a) Target organs are damaged; heart, muscles, blood cells, intestinal tract show dysfunctions.

(b)

General physiological effects— fever, malaise, aches, shock

Figure 13.6 The origins and effects of circulating exotoxins and endotoxin.   **(a)** Exotoxins, given off by live cells, have highly specific targets and physiological effects. **(b)** Endotoxin, given off when the cell wall of gram-negative bacteria disintegrates, has more generalized physiological effects.

Hemolysins (hee-mahl'-uh-sinz) are a class of bacterial exotoxin that disrupts the cell membrane of red blood cells (and some other cells, too). This damage causes the red blood cells to **hemolyze**—to burst and release hemoglobin pigment. Hemolysins that increase pathogenicity include the streptolysins of *Streptococcus pyogenes* and the alpha (α) and beta (β) toxins of *Staphylococcus aureus* **(figure 13.7).** When colonies of bacteria growing on blood agar produce hemolysin, distinct zones appear around the colony. The pattern of hemolysis is often used to identify bacteria and determine their degree of pathogenicity.

The exotoxins of diphtheria, tetanus, and botulism, among others, attach to a particular target cell, become internalized, and interrupt an essential cell pathway. The consequences of cell disruption depend upon the target. One toxin of *Clostridium tetani* blocks the action of certain spinal neurons; the toxin of *Clostridium botulinum* prevents the transmission of nerve-muscle stimuli; pertussis toxin inactivates the respiratory cilia; and cholera toxin provokes profuse salt and water loss from intestinal cells. More details of the pathology of exotoxins are found in later chapters on specific diseases.

In contrast to the category of *exotoxins*, which contains many specific examples, the word *endotoxin* refers to a single substance. Endotoxin is actually a chemical called lipopolysaccharide (LPS), which is part of the outer membrane of gram-negative cell walls. Gram-negative bacteria shed these LPS molecules into tissues or into the circulation. Endotoxin differs from exotoxins in having a variety of systemic effects on tissues and organs. Depending upon the amounts present, endotoxin

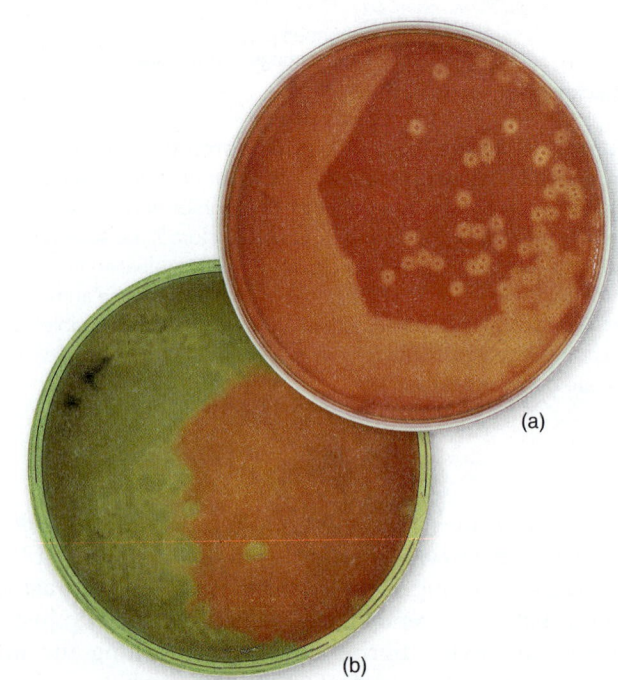

(a)

(b)

Figure 13.7 Beta-hemolysis and alpha-hemolysis by different bacteria on blood agar.   **(a)** Beta-hemolysis results in complete lysis of the red blood cells incorporated in the agar, leaving an area of complete clearing around the bacterial colony. **(b)** Alpha-hemolysis incompletely hemolyzes the red blood cells, leading to a green tint in the areas of agar where the exotoxin has acted.

can cause fever, inflammation, hemorrhage, and diarrhea. Blood infection by gram-negative bacteria such as *Salmonella, Shigella, Neisseria meningitidis,* and *Escherichia coli* are particularly dangerous because it can lead to fatal endotoxic shock.

Inducing an Injurious Host Response   Despite the extensive discussion on direct virulence factors, such as enzymes and toxins, it is probably the case that just as many microbial diseases are the result of indirect damage, or the host's excessive or inappropriate response to a microorganism. This is an extremely important point because it means that pathogenicity is not a trait solely determined by microorganisms but is really a consequence of the interplay between microbe and host.

Of course, it is easier to study and characterize the microbes that cause direct damage through toxins or enzymes. For this reason, these true pathogens were the first to be fully understood as the science of microbiology progressed. But in the last 15 to 20 years, microbiologists have come to appreciate exactly how important the relationship between microbe and host is, and this has greatly expanded our understanding of infectious diseases.

The Process of Infection and Disease

Aided by virulence factors, microbes eventually settle in a particular target organ and cause damage at the site. The type and scope of injuries inflicted during this process account for the typical stages of an infection, the patterns of the infectious disease, and its manifestations in the body.

In addition to the adverse effects of enzymes, toxins, and other factors, the mere multiplication of a pathogen can weaken host tissues. Pathogens can obstruct tubular structures such as blood vessels, lymphatic channels, fallopian tubes, and bile ducts. Accumulated damage can lead to cell and tissue death, a condition called **necrosis.** Although viruses do not produce toxins or destructive enzymes, they destroy cells by multiplying in and lysing them. Many of the cytopathic effects of viral infection arise from the impaired metabolism and death of cells (see chapter 6).

Patterns of Infection   Patterns of infection are many and varied. In the simplest situation, a **localized infection,** the microbe enters the body and remains confined to a specific tissue **(figure 13.8a).** Examples of localized infections are boils, fungal skin infections, and warts.

Many infectious agents do not remain localized but spread from the initial site of entry to other tissues. In fact, spreading is necessary for pathogens such as rabies and hepatitis A virus, whose target tissue is some distance from the site of entry. The rabies virus travels from a bite wound along nerve tracts to its target in the brain, and the hepatitis A virus moves from the intestine to the liver via the circulatory system. When an infection spreads to several sites and tissue fluids, usually in the bloodstream, it is called a **systemic infection (figure 13.8b).** Examples of systemic infections are viral diseases (measles, rubella, chickenpox, and AIDS); bacterial diseases (brucellosis, anthrax, typhoid fever, and

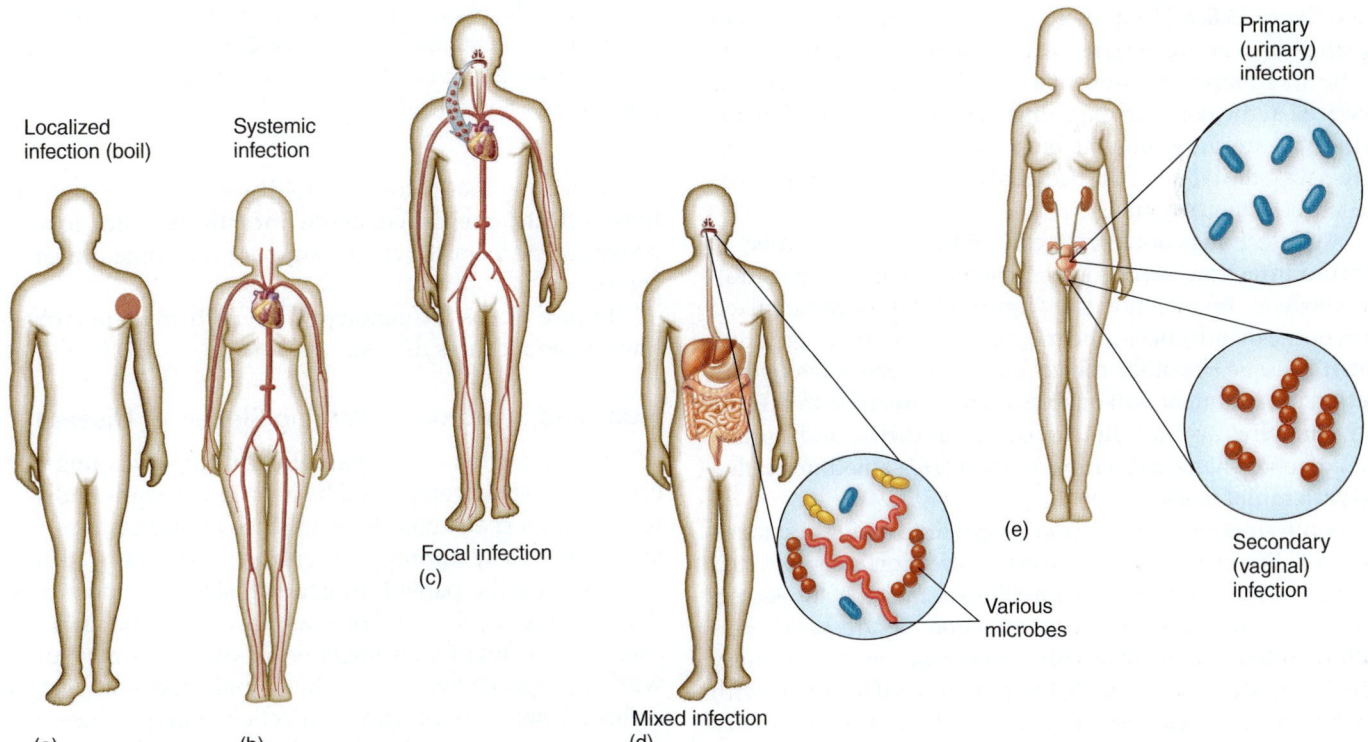

Figure 13.8 The occurrence of infections with regard to location, type of microbe, and order of infection.   (a) A localized infection, in which the pathogen is restricted to one specific site. (b) Systemic infection, in which the pathogen spreads through circulation to many sites. (c) A focal infection occurs initially as a local infection, but circumstances cause the microbe to be carried to other sites systemically. (d) A mixed infection, in which the same site is infected with several microbes at the same time. (e) In a primary-secondary infection, an initial infection is complicated by a second one in the same or a different location and caused by a different microbe.

A Note About Terminology

Words in medicine have great power and economy. A single technical term can often replace a whole phrase or sentence, thereby saving time and space in patient charting. The beginning student may feel overwhelmed by what seems like a mountain of new words. However, having a grasp of a few root words and a fair amount of anatomy can help you learn many of these words and even deduce the meaning of unfamiliar ones. Some examples of medical shorthand follow:

- The suffix *-itis* means "an inflammation" and, when affixed to the end of an anatomical term, indicates an inflammatory condition in that location. Thus, meningitis is an inflammation of the meninges surrounding the brain; encephalitis is an inflammation of the brain itself; hepatitis involves the liver; vaginitis, the vagina; gastroenteritis, the intestine; and otitis media, the middle ear. Although not all inflammatory conditions are caused by infections, many are.

- The suffix *-emia* is derived from the Greek word *haeima*, meaning "blood." When added to a word, it means "associated with the blood." Thus, septicemia means sepsis (infection) of the blood; bacteremia, bacteria in the blood; viremia, viruses in the blood; and fungemia, fungi in the blood. It is also applicable to specific conditions such as toxemia, gonococcemia, and spirochetemia.

- The suffix *-osis* means "a disease or morbid process." It is frequently added to the names of pathogens to indicate the disease they cause: for example, listeriosis, histoplasmosis, toxoplasmosis, shigellosis, salmonellosis, and borreliosis. A variation of this suffix is *-iasis*, as in trichomoniasis and candidiasis.

- The suffix *-oma* comes from the Greek word *onkomas* (swelling) and means "tumor." Although it is often used to describe cancers (sarcoma, melanoma), it is also applied in some infectious diseases that cause masses or swellings (tuberculoma, leproma).

syphilis); and fungal diseases (histoplasmosis and cryptococcosis). Infectious agents can also travel to their targets by means of nerves (as in rabies) or cerebrospinal fluid (as in meningitis).

A **focal infection** is said to exist when the infectious agent breaks loose from a local infection and is carried into other tissues **(figure 13.8c)**. This pattern is exhibited by tuberculosis or by streptococcal pharyngitis, which gives rise to scarlet fever. In the condition called toxemia,[4] the infection itself remains localized at the portal of entry, but the toxins produced by the pathogens are carried by the blood to the actual target tissue. In this way, the target of the bacterial cells can be different from the target of their toxin.

An infection is not always caused by a single microbe. In a **mixed infection,** several agents establish themselves simultaneously at the infection site **(figure 13.8d)**. In some mixed or synergistic infections, the microbes cooperate in breaking down a tissue. In other mixed infections, one microbe creates an environment that enables another microbe to invade. Gas gangrene, wound infections, dental caries, and human bite infections tend to be mixed. These are sometimes called **polymicrobial** diseases.

Some diseases are described according to a sequence of infection. When an initial, or **primary, infection** is complicated by another infection caused by a different microbe, the second infection is termed a **secondary infection (figure 13.8e)**. This pattern often occurs in a child with chickenpox (primary infection) who may scratch his pox and infect them with *Staphylococcus aureus* (secondary infection). The secondary infection need not be in the same site as the primary infection, and it usually indicates altered host defenses.

Disease Connection

Perhaps the most well-tracked secondary infection is pneumonia when it occurs during the course of an influenza infection. Pneumonia is so often the cause of death in people who are infected with the influenza virus that one very prominent method of tracking influenza outbreaks used by the Centers for Disease Control and Prevention tracks the two causes of death using a single metric (see figure 13.20b).

Infections that come on rapidly, with severe but short-lived effects, are called **acute infections.** Infections that progress and persist over a long period of time are **chronic infections.**

Figure 13.9 is a summary of the pathway a microbe follows when it causes disease.

Signs and Symptoms: Warning Signals of Disease

When an infection causes pathologic changes leading to disease, it is often accompanied by a variety of signs and symptoms. A **sign** is any objective evidence of disease as noted by an observer; a **symptom** is the subjective evidence of disease as sensed by the patient. In general, signs are more precise than symptoms, though both can have the same underlying cause. For example, an infection of the brain may present with the sign of bacteria in the spinal fluid and symptom of headache; a streptococcal infection may produce a sore throat (symptom) and inflamed pharynx (sign). A disease indicator that can be sensed and observed can qualify as either a sign or a symptom. When a disease can be identified or defined by a certain complex of signs and symptoms, it is termed a **syndrome.** Signs and symptoms with considerable

---

4. This is not to be confused with toxemia of pregnancy, which is a metabolic disturbance and not an infection.

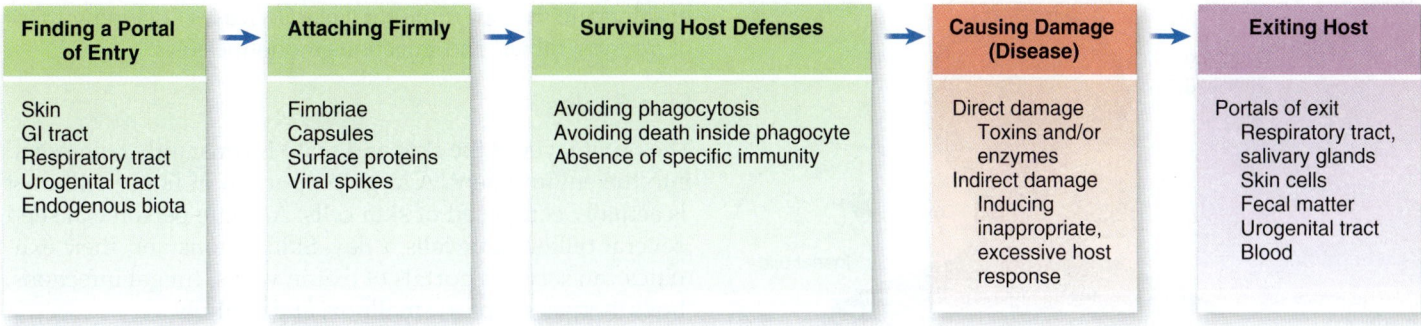

Figure 13.9 The steps involved when a microbe causes disease in a host.

importance in diagnosing infectious diseases are shown in **table 13.9.** Specific signs and symptoms for particular infectious diseases are covered in chapters 18 through 23.

Signs and Symptoms of Inflammation The earliest symptoms of disease result from the activation of the body defense process called **inflammation.** The inflammatory response includes cells and chemicals that respond nonspecifically to disruptions in the tissue. This subject is discussed in greater detail in chapter 14, but as noted earlier, many signs and symptoms of infection are caused by the mobilization of this system. Some common symptoms of inflammation include fever, pain, soreness, and swelling. Signs of inflammation include **edema,** the accumulation of fluid in an afflicted tissue; **granulomas** and **abscesses,** walled-off collections of inflammatory cells and microbes in the tissues; and **lymphadenitis,** swollen lymph nodes.

Rashes and other skin eruptions are common symptoms and signs in many diseases; because they tend to mimic each other, it can be difficult to differentiate among diseases on this basis alone. The general term for the site of infection or disease is **lesion.** Skin lesions can be restricted to the epidermis and its glands and follicles, or they can extend into the dermis and subcutaneous regions. The lesions of some infections undergo characteristic changes in appearance during the course of disease and thus fit more than one category.

Signs of Infection in the Blood Changes in the number of circulating white blood cells, as determined by special counts, are considered to be signs of possible infection. **Leukocytosis** (loo'-koh-sy-toh'-sis) is an increase in the level of white blood cells, whereas **leukopenia** (loo'-koh-pee'-nee-uh) is a decrease. Other signs of infection revolve around the occurrence of a microbe or its products in the blood. The clinical term for blood infection, **septicemia,** refers to a general state in which microorganisms are multiplying in the blood and are present in large numbers. When small numbers of bacteria are present in the blood but not necessarily multiplying, the correct term is **bacteremia. Viremia** is the term used to describe the presence of viruses in the blood, whether or not they are actively multiplying.

During infection, a normal host will show signs of an immune response in the form of antibodies in the serum or some type of sensitivity to the microbe. This fact is the basis for several serological tests used in diagnosing many infectious diseases. Such specific immune reactions indicate the body's attempt to develop specific immunities against pathogens. We concentrate on this role of the host defenses in chapters 14 and 15.

Infections That Go Unnoticed In more cases than you might think, true infections go unnoticed. In other words, although infected, the host does not manifest the disease. Infections of this nature are known as **asymptomatic, subclinical,** or **inapparent** because the patient experiences no symptoms or disease and does not seek medical attention. However, it is important to note that most infections are attended by some sort of sign. In the section on epidemiology, we further address the significance of subclinical infections in the transmission of infectious agents.

Step Five—Vacating the Host: Portals of Exit

Earlier, we introduced the idea that a parasite is considered *unsuccessful* if it does not have a provision for leaving its host and moving to other susceptible hosts. With few exceptions,

Table 13.9	Common Signs and Symptoms of Infectious Diseases
Signs	**Symptoms**
Fever	Chills
Septicemia	Pain, ache, soreness, irritation
Microbes in tissue fluids	Malaise
Chest sounds	Fatigue
Skin eruptions	Chest tightness
Leukocytosis	Itching
Leukopenia	Headache
Swollen lymph nodes	Nausea
Abscesses	Abdominal cramps
Tachycardia (increased heart rate)	Anorexia (lack of appetite)
Antibodies in serum	Sore throat

of saliva are the exit route for several viruses, including those of mumps, rabies, and infectious mononucleosis.

Skin Scales

The outer layer of the skin and scalp is constantly being shed into the environment. A large proportion of household dust is actually composed of skin cells. A single person can shed several billion skin cells a day. Skin lesions and their exudates can serve as portals of exit in warts, fungal infections, boils, herpes simplex, smallpox, and syphilis.

Fecal Exit

Feces are a very common portal of exit. Some intestinal pathogens grow in the intestinal mucosa and create an inflammation that increases the motility of the bowel. This increased motility speeds up peristalsis, resulting in diarrhea, and the more fluid stool provides a rapid exit for the pathogen. A number of helminth worms release cysts and eggs through the feces (see chapter 22). Feces containing pathogens are a public health problem when allowed to contaminate drinking water or when used to fertilize crops.

Urogenital Tract

A number of agents involved in sexually transmitted infections leave the host in vaginal discharge or semen. This is also the source of neonatal infections such as herpes simplex, *Chlamydia,* and *Candida albicans,* which infect the infant as it passes through the birth canal. Certain pathogens that infect the kidney are discharged in the urine: for instance, the agents of leptospirosis, typhoid fever, tuberculosis, and schistosomiasis.

Removal of Blood or Bleeding

Although the blood does not have a direct route to the outside, it can serve as a portal of exit when it is removed or released through a vascular puncture made by natural or artificial means. Blood-feeding animals such as ticks and fleas are common transmitters of pathogens. The AIDS and hepatitis viruses are transmitted by shared needles or through small breaks in a mucous membrane caused by sexual intercourse. Blood donation is also a means for certain microbes to leave the host, though this means of exit is now unusual because of close monitoring of the donor population and blood used for transfusions.

The Persistence of Microbes and Pathologic Conditions

The apparent recovery of the host does not always mean that the microbe has been completely removed or destroyed by the host defenses. After the initial symptoms in certain chronic infectious diseases, the infectious agent retreats into a dormant state called **latency.** Throughout this latent state, the microbe can periodically become active and produce a recurrent disease. The viral agents of herpes simplex, herpes zoster, hepatitis B, AIDS, and Epstein-Barr can persist in the host for long

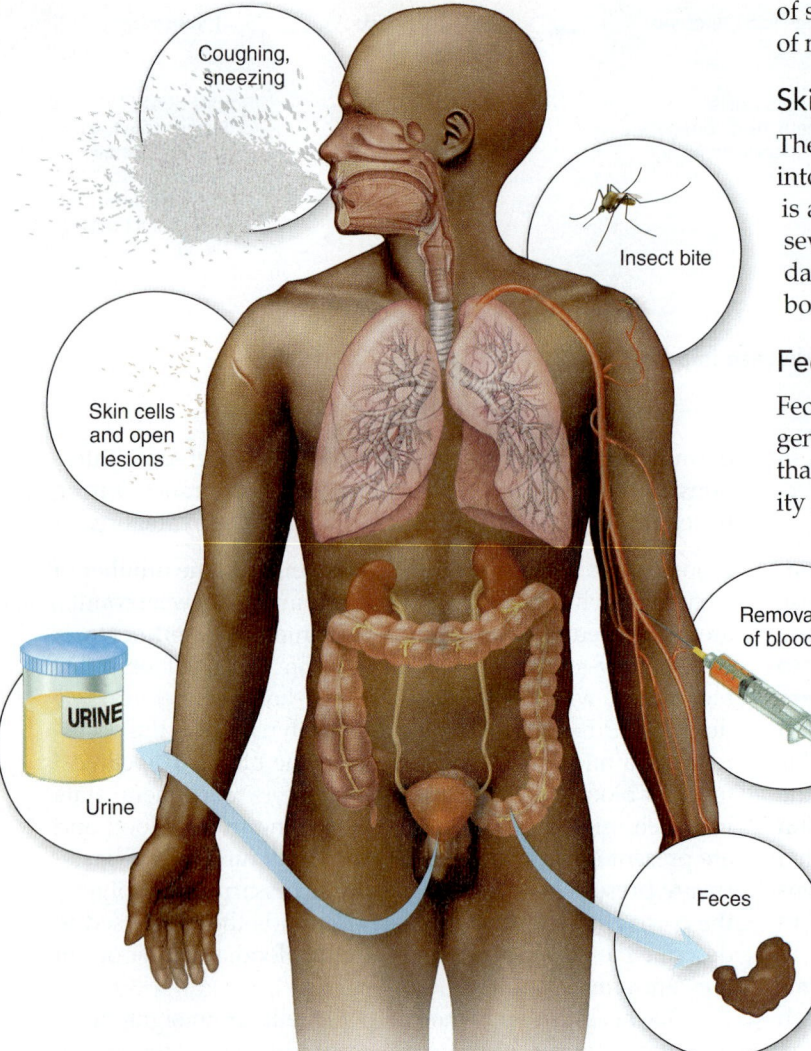

Figure 13.10 Major portals of exit of infectious diseases.

pathogens depart by a specific avenue called the **portal of exit** (figure 13.9 and **figure 13.10).** In most cases, the pathogen is shed or released from the body through secretion, excretion, discharge, or sloughed tissue. The usually very high number of infectious agents in these materials increases the likelihood that the pathogen will reach other hosts. In many cases, the portal of exit is the same as the portal of entry, but some pathogens use a different route.

Respiratory and Salivary Portals

Mucus, sputum, nasal drainage, and other moist secretions are the media of escape for the pathogens that infect the lower or upper respiratory tract. The most effective means of releasing these secretions are coughing and sneezing (see figure 13.15), although they can also be released during talking and laughing. Tiny particles of liquid released into the air form aerosols or droplets that can spread the infectious agent to other people. The agents of tuberculosis, influenza, measles, and chickenpox most often leave the host through airborne droplets. Droplets

periods. The agents of syphilis, typhoid fever, tuberculosis, and malaria also enter into latent stages. The person harboring a persistent infectious agent may or may not shed it during the latent stage. If it is shed, such persons are chronic carriers who serve as sources of infection for the rest of the population.

Some diseases leave **sequelae** in the form of long-term or permanent damage to tissues or organs. For example, meningitis can result in deafness, strep throat can lead to rheumatic heart disease, Lyme disease can cause arthritis, and polio can produce paralysis.

What Happens in Your Body

There are four distinct phases of infection and disease: the incubation period, the prodrome, the period of invasion, and the convalescent period **(figure 13.11).** The **incubation period** is the time from initial contact with the infectious agent (at the portal of entry) to the appearance of the first symptoms. During the incubation period, the agent is multiplying at the portal of entry but has not yet caused enough damage to elicit symptoms. Although this period is relatively well defined and predictable for each microorganism, it does vary according to host resistance, degree of virulence, and distance between the target organ and the portal of entry (the farther apart, the longer the incubation period). Overall, an incubation period can range from several hours in pneumonic plague to several years in leprosy. The majority of infections, however, have incubation periods ranging between 2 and 30 days.

The earliest notable symptoms of infection usually appear as a vague feeling of discomfort, such as head and muscle aches, fatigue, upset stomach, and general malaise. This short period (1 to 2 days) is known as the **prodromal stage.** Some diseases have very specific prodromal symptoms. Next, the infectious agent enters a **period of invasion,** during which it multiplies at high levels, exhibits its greatest virulence, and becomes well established in its target tissue. This period is often marked by fever and other prominent and more specific signs and symptoms, which can include cough, rashes, diarrhea, loss of muscle control, swelling, jaundice, discharge of exudates, or severe pain, depending on the particular infection. The length of this period is extremely variable.

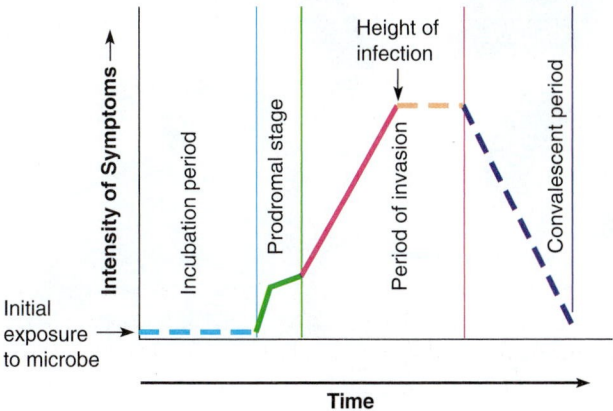

Figure 13.11 Stages in the course of infection and disease. Dashed lines represent periods with a variable length.

INSIGHT 13.3 — Viable but Nonculturable: I'm Not Dead Yet!

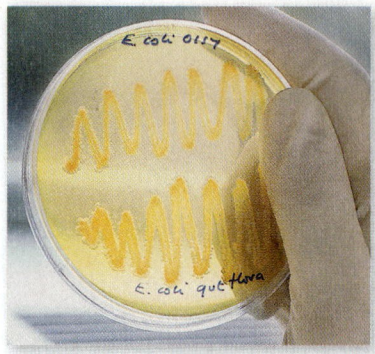

Bacteria have a number of virulence factors, such as toxins, enzymes, capsules, pili, and the ability to evade phagocytosis. These allow organisms to survive in the host or overcome the host immune system. Recently, scientists have detected another mechanism that allows bacteria to remain in the host: playing dead. When bacteria like *Escherichia coli* are exposed to stressful environments such as low temperatures or lack of nutrients, a set of genes is induced that allows them to remain alive but in a dormant state. Scientists have referred to bacteria that play dead in this way as "viable but nonculturable," or VBNC.

The VBNC state poses a problem in screening foods for bacteria, because the bacteria may be present on foods such as lettuce or other vegetables but undetectable through traditional culturing methods. Cells in the VBNC state will not grow readily when exposed to culture media or rapid screening tests. This was possibly the case in the 2011 outbreak of *E. coli* O104:H4 in Germany, in which thousands of people were sickened by consumption of fresh sprouts. Researchers at the Robert Koch Institute in Wernigerode, Germany, exposed *E. coli* O104:H4 to copper and cold tap water, both of which induced stress in the organism and caused it to go into the VBNC state. When copper was removed, the bacteria were resuscitated and still able to produce the virulence factors that caused bloody diarrhea and kidney failure in the 2011 outbreak. Potentially pathogenic *E. coli* could be hiding in your food, undetectable by common food safety screening methods.

Source: 2011 *Environ. Microbiol.* DOI: 10.1111

Figure 13.12 Carrier states.

Carrier State	Explanation	Example
Asymptomatic carriers	Infected but show no symptoms of disease	Gonorrhea, genital herpes with no lesions
Incubating carriers	Infected but show no symptoms of disease	Infectious mono-nucleosis
Convalescent carriers	Recuperating patients without symptoms; they continue to shed viable microbes and convey the infection to others	Hepatitis A
Chronic carriers	Individuals who shelter the infectious agent for a long period after recovery because of the latency of the infectious agent	Tuberculosis, typhoid fever
Passive carriers	Medical and dental personnel who must constantly handle patient materials that are heavily contaminated with patient secretions and blood risk picking up pathogens mechanically and accidentally transferring them to other patients	Various healthcare-associated infections

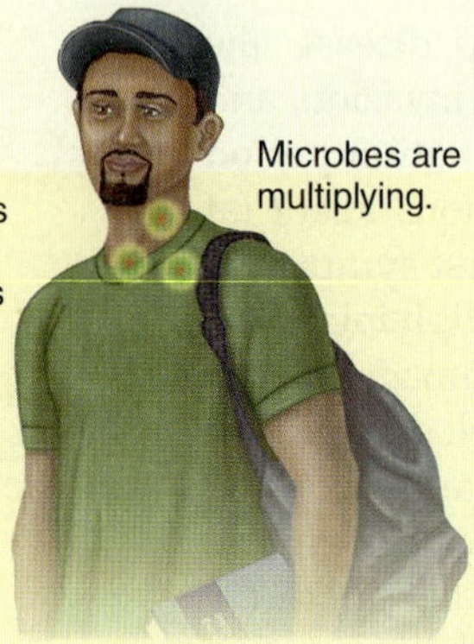

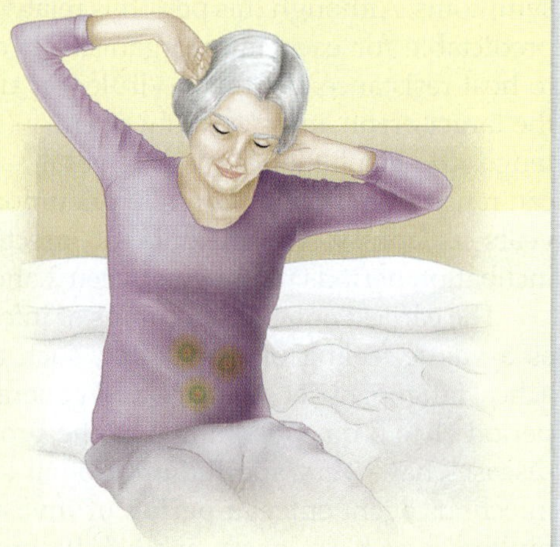

Microbes are multiplying.

Asymptomatic STI

Incubation

Convalescent

Chronic

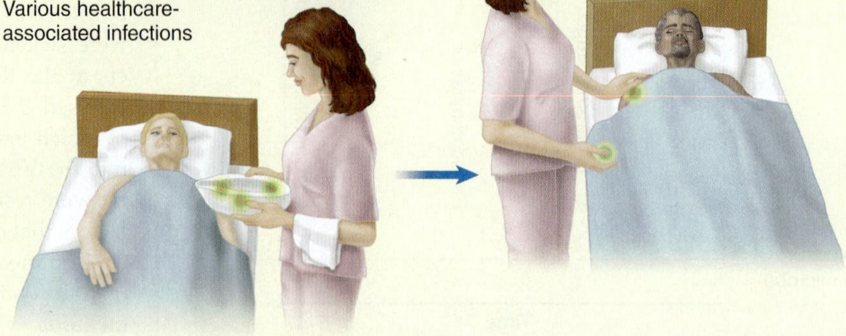

As the patient begins to respond to the infection, the symptoms decline—sometimes dramatically, other times slowly. During the recovery that follows, called the **convalescent period,** the patient's strength and health gradually return owing to the healing nature of the immune response. During this period, many patients stop taking their antibiotics, even though there are still pathogens in their system. This noncompliance means that bacteria with higher resistance are left behind to repopulate, putting the patient at risk for redevelopment of the infection that will not be treatable with the previously used antibiotic. The transmissibility of the microbe during these four stages must be considered on an individual basis. A few agents are released mostly during incubation (measles, for example); many are released during the invasive period (*Shigella*); and others can be transmitted during all of these periods (hepatitis B).

Reservoirs: Where Pathogens Persist

In order for an infectious agent to continue to exist and be spread, it must have a permanent place to reside. The **reservoir** is the primary habitat in the natural world from which a pathogen originates. Often it is a human or animal carrier, although soil, water, and plants are also reservoirs. The reservoir can be distinguished from the infection **source,** which is the individual or object from which an infection is actually acquired. In diseases such as syphilis, the reservoir and the source are the same (the human body). In the case of hepatitis A, the reservoir (a human carrier) is usually different from the source of infection (contaminated food).

Living Reservoirs

Persons or animals with obvious symptomatic infection are obvious sources of infection, but a **carrier** is, by definition, an individual who *inconspicuously* shelters a pathogen and can spread it to others without knowing. Although human carriers are occasionally detected through routine screening (blood tests, cultures) and other epidemiological devices, they are unfortunately very difficult to discover and control. As long as a pathogenic reservoir is maintained by the carrier state, the disease will continue to exist in that population and the potential for epidemics will be a constant threat. The duration of the carrier state can be short or long term, and it is important to remember that the carrier may or may not have experienced disease due to the microbe.

Several situations can produce the carrier state **(figure 13.12). Asymptomatic** (apparently healthy) **carriers** are infected but they show no symptoms. A few asymptomatic infections (gonorrhea and human papillomavirus, for instance) may carry out their entire course without overt manifestations. **Incubating carriers** spread the infectious agent during the incubation period. For example, AIDS patients can harbor and spread the virus for months and years before their first overt symptoms appear. Recuperating patients without symptoms are considered **convalescent carriers** when they continue to shed viable microbes and convey the infection to others. Diphtheria patients, for example, spread the microbe for up to 30 days after the disease has subsided.

An individual who shelters the infectious agent for a long period after recovery because of the latency of the infectious agent is a **chronic carrier.** Patients who have recovered from tuberculosis or hepatitis infections frequently carry the agent chronically. About one in 20 victims of typhoid fever continues to harbor *Salmonella typhi* in the gallbladder for several years, and sometimes for life. The most infamous of these was "Typhoid Mary," a cook who spread the infection to hundreds of victims in the early 1900s. (*Salmonella* infection is described in chapter 22.)

The **passive carrier** state is of great concern during patient care (see a later section on healthcare-associated infections). Medical and dental personnel who must constantly handle materials that are heavily contaminated with patient secretions and blood risk picking up pathogens mechanically and accidently transferring them to other patients. Proper hand washing, handling of contaminated materials, and aseptic techniques greatly reduce this likelihood.

Animals as Reservoirs and Sources   Up to now, we have lumped animals with humans in discussing living reservoirs or carriers, but animals deserve special consideration as vectors of infections. The word **vector** is used by epidemiologists to indicate a live animal that transmits an infectious agent from one host to another. (The term is sometimes misused to include any object that spreads disease.) The majority of vectors are arthropods such as fleas, mosquitoes, flies, and ticks, although larger animals can also spread infection—for example, mammals (rabies), birds (psittacosis), or lizards (salmonellosis).

By tradition, vectors are placed into one of two categories, depending on the animal's relationship with the microbe **(figure 13.13).** A **biological vector** actively participates in a pathogen's life cycle, serving as a site in which it can multiply or complete its life cycle. A biological vector communicates the infectious agent to the human host by biting, aerosol formation, or touch. In the case of biting vectors, the animal can

1. inject infected saliva into the blood (the mosquito, **figure 13.13***a*),
2. defecate around the bite wound (the flea), or
3. regurgitate blood into the wound (the tsetse fly).

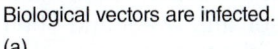

Biological vectors are infected.
(a)

Mechanical vectors are not infected.
(b)

Figure 13.13 Two types of vectors.   **(a)** Biological vectors serve as hosts during pathogen development. Examples are mosquitoes, which carry malaria; bats, which carry rabies and other viral diseases; and chickens, which can transmit their flu viruses to humans. **(b)** Mechanical vectors such as the housefly and the cockroach transport pathogens on their feet and mouthparts.

Mechanical vectors are not necessary to the life cycle of an infectious agent and merely transport it without being infected. The external body parts of these animals become contaminated when they come into physical contact with a source of pathogens. The agent is subsequently transferred to humans indirectly by an intermediate such as food or, occasionally, by direct contact (as in certain eye infections). Houseflies **(figure 13.13b)** are noxious mechanical vectors. They feed on decaying garbage and feces, and while they are feeding, their feet and mouthparts easily become contaminated. They also regurgitate juices onto food to soften and digest it. Flies spread more than 20 bacterial, viral, protozoan, and helminth infections. Various flies transmit tropical ulcers, yaws, and trachoma. Cockroaches, which have similar unsavory habits, play a role in the mechanical transmission of fecal pathogens.

Many vectors and animal reservoirs spread their own infections to humans. An infection indigenous to animals but naturally transmissible to humans is a **zoonosis** (zoh'-uh-noh'-sis). In these types of infections, the human is essentially a dead-end host and does not contribute to the natural persistence of the microbe. Some zoonotic infections (rabies, for instance) can have multihost involvement, and others can have very complex cycles in the wild (see discussion of plague in chapter 20). Zoonotic spread of disease is promoted by close associations of humans with animals, and people in animal-oriented or outdoor professions are at greatest risk. At least 150 zoonoses exist worldwide; the most common ones are listed in **table 13.10.** Zoonoses make up a full 70% of all new emerging diseases worldwide. It is worth noting that zoonotic infections are impossible to completely eradicate without also eradicating the animal reservoirs. Attempts have been made to eradicate mosquitoes and certain rodents, and in 2004 China slaughtered tens of thousands of civet cats who were thought (incorrectly) to be a source of the respiratory disease SARS.

A 2005 U.N. study warned that one of the most troublesome trends is the increase in infectious diseases due to environmental destruction. Deforestation and urban sprawl cause animals to find new habitats, often leading to new patterns of disease transmission. For example, the fatal Nipah virus seems to have begun to infect humans although it previously only infected Asian fruit bats. The bats were pushed out of their forest habitats by the creation of palm plantations. They encountered domesticated pigs, passing the virus to them, and the pigs in turn transmitted it to their human handlers.

Nonliving Reservoirs

Clearly, microorganisms have adapted to nearly every habitat in the biosphere. They thrive in soil and water and often find their way into the air. Although most of these microbes are saprobic and cause little harm and considerable benefit to humans, some are opportunists and a few are regular pathogens. Because human hosts are in regular contact with these environmental sources, acquisition of pathogens from natural habitats is of diagnostic and epidemiological importance.

Table 13.10 Common Zoonotic Infections

Disease	Primary Animal Reservoirs
Viruses	
Rabies	All mammals
Yellow fever	Wild birds, mammals, mosquitoes
Viral fevers	Wild mammals
Hantavirus	Rodents
Influenza	Chickens, birds, swine
West Nile virus	Wild birds, mosquitoes
Bacteria	
Rocky Mountain spotted fever	Dogs, ticks
Psittacosis	Birds
Leptospirosis	Domestic animals
Anthrax	Domestic animals
Brucellosis	Cattle, sheep, pigs
Plague	Rodents, fleas
Salmonellosis	Variety of mammals, birds, and rodents
Tularemia	Rodents, birds, arthropods
Miscellaneous	
Ringworm	Domestic mammals
Toxoplasmosis	Cats, rodents, birds
Trypanosomiasis	Domestic and wild mammals
Trichinosis	Swine, bears
Tapeworm	Cattle, swine, fish

Soil harbors the vegetative forms of bacteria, protozoa, helminths, and fungi, as well as their resistant or developmental stages such as endospores, cysts, ova, and larvae. Bacterial pathogens include the anthrax bacillus and species of *Clostridium* that are responsible for gas gangrene, botulism, and tetanus. Pathogenic fungi in the genera *Coccidioides* and *Blastomyces* are spread by spores in the soil and dust. The invasive stages of the hookworm *Necator* occur in the soil. Natural bodies of water carry fewer nutrients than soil does but still support pathogenic species such as *Legionella, Cryptosporidium,* and *Giardia.*

As we saw in Insight 13.2, the built environment—the buildings where we live, work, and spend leisure time—can also serve as nonliving reservoirs of infection.

The Acquisition and Transmission of Infectious Agents

Infectious diseases can be categorized on the basis of how they are acquired. A disease is **communicable** when an infected host can transmit the infectious agent to another host and establish infection in that host. (Although this terminology is standard, one must realize that it is not the disease that is communicated but the microbe. Also be aware that the word *infectious* is sometimes used interchangeably with the word *communicable,* but this is not precise usage.) The transmission of the agent can be direct or indirect, and the ease with which the disease is transmitted varies considerably from one agent to another. If the agent is highly communicable, especially through direct contact, the disease is **contagious.** Influenza and measles move readily from host to host and thus are contagious, whereas Hansen's disease (leprosy) is only weakly communicable. Because they can be spread through the population, communicable diseases are our main focus in the following sections.

In contrast, a **noncommunicable** infectious disease does *not* arise through transmission of the infectious agent from host to host. The infection and disease are acquired through some other special circumstance. Noncommunicable infections occur primarily when a compromised person is invaded by his or her own microbiota (as with certain pneumonias, for example) or when an individual has accidental contact with a microbe that exists in a nonliving reservoir such as soil. Some examples are certain mycoses, acquired through inhalation of fungal spores, and tetanus, in which *Clostridium tetani* endospores from a soiled object enter a cut or wound. Persons thus infected do not become a source of disease to others.

Patterns of Transmission in Communicable Diseases

The routes or patterns of disease transmission are many and varied. The spread of diseases is by direct or indirect contact with animate or inanimate objects and can be horizontal or vertical. The term *horizontal* means the disease is spread through a population from one infected individual to another; *vertical* signifies transmission from parent to offspring via the ovum, sperm, placenta, or milk. The extreme complexity of transmission patterns among microorganisms makes it very difficult to generalize. However, for easier organization, we will divide microorganisms into two major groups, as shown in **figure 13.14**. Horizontal transmission occurs by some form of direct contact or transmission by indirect routes, in which some vehicle is involved.

Modes of Direct Transmission In order for microbes to be directly transferred, some type of contact must occur between the skin or mucous membranes of the infected person and that of the new infectee. It may help to think of this route as the portal of exit meeting the portal of entry without the involvement of an intermediate object, substance, or space. Most sexually transmitted infections are spread

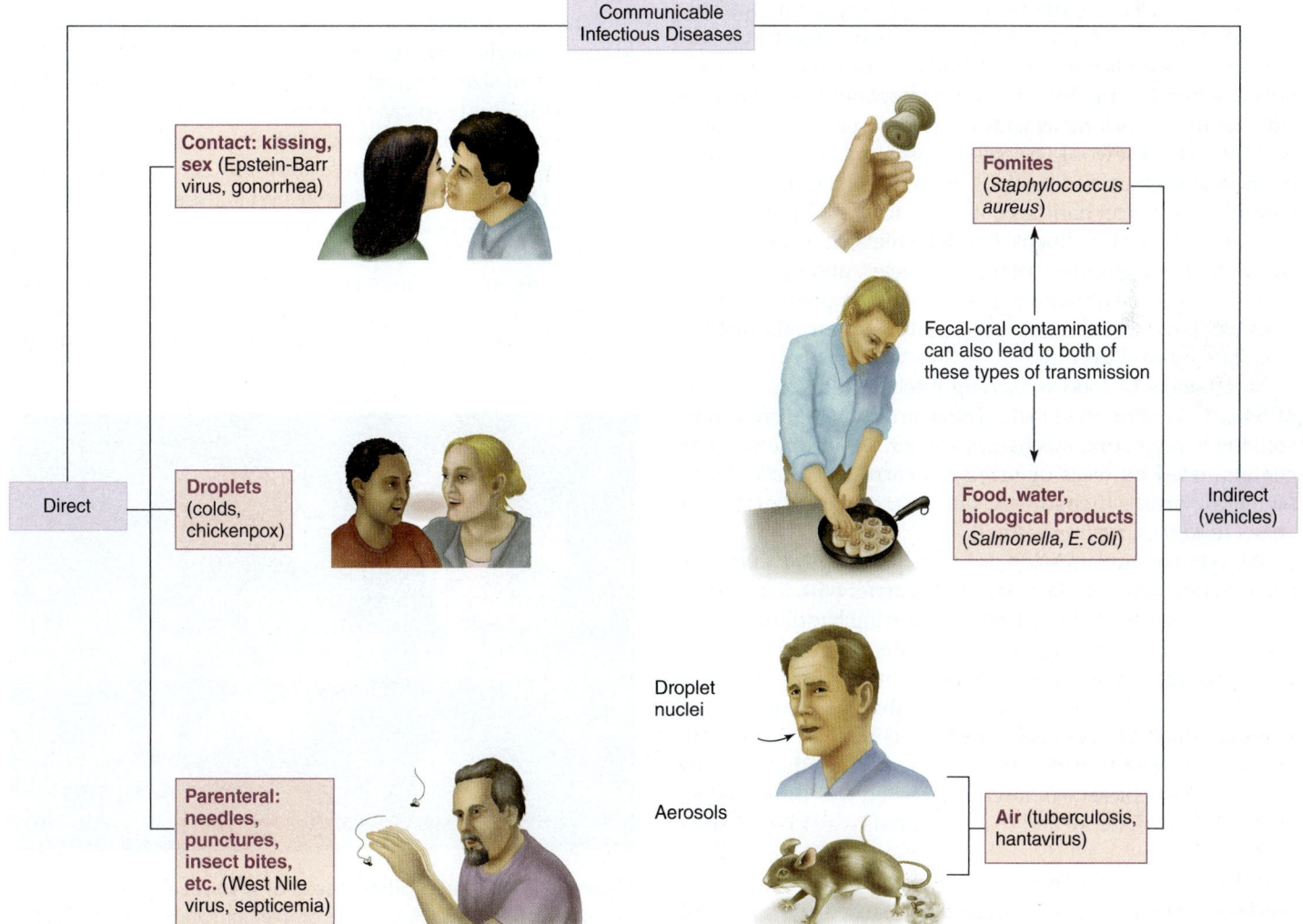

Figure 13.14 **Summary of how communicable infectious diseases are transmitted horizontally.**

directly. In addition, infections that result from kissing or bites by biological vectors are direct. Most obligate parasites are far too sensitive to survive for long outside the host and can be transmitted only through direct contact. The trickiest type of "contact" transmission is droplet contact, in which fine droplets are sprayed directly upon a person during sneezing or coughing (as distinguished from droplet nuclei that are transmitted over a meter or more by air). While there is some space between the infecter and the infectee, it is still considered a form of contact because the two people have to be in each other's presence, as opposed to indirect forms of contact.

Routes of Indirect Transmission

For microbes to be indirectly transmitted, the infectious agent must pass from an infected host to an intermediate conveyor (a vehicle) and from there to another host. The transmitter of the infectious agent can be either openly infected or a carrier.

Indirect Spread by Vehicles: Fomites

The term **vehicle** specifies any inanimate material commonly used by humans that can transmit infectious agents. A *common vehicle* is a single material that serves as the source of infection for many individuals. Some specific types of vehicles are food, water, various biological products (such as blood, serum, and tissue), and fomites. A **fomite** is an inanimate object that harbors and transmits pathogens. Unlike a reservoir, however, a fomite is not a continuous source of infection. The list of possible fomites is as long as your imagination allows. Probably highest on the list would be objects commonly in contact with the public such as doorknobs, telephones, handheld remote controls, and faucet handles that are readily contaminated by touching. Shared bed linens, handkerchiefs, toilet seats, toys, eating utensils, clothing, personal articles, and syringes are other examples. Although paper money is impregnated with a disinfectant to inhibit microbes, pathogens are still isolated from bills as well as coins.

Outbreaks of food poisoning often result from the role of food as a common vehicle. The source of the agent can be soil, the handler, or a mechanical vector. Water that has been contaminated by feces or urine can carry *Salmonella, Vibrio* (cholera), viruses (hepatitis A, polio), and pathogenic protozoans (*Giardia, Cryptosporidium*).

A type of transmission termed the *oral-fecal route* can occur in two ways. In the first, a fecal carrier with inadequate personal hygiene contaminates food during handling and an unsuspecting person ingests it. Hepatitis A, amoebic dysentery, shigellosis, and typhoid fever are often transmitted this way. Oral-fecal transmission can also involve contaminated materials such as toys and diapers. It is really a special category of indirect transmission, which specifies that the way in which the vehicle became contaminated was through contact with fecal material and that it found its way to someone's mouth.

A recent investigation of a small outbreak of diarrhea among a traveling girls' soccer team caused by norovirus showed that the first person with the infection had probably contaminated some reusable grocery bags that were stored in the hotel bathroom. Fine aerosols of her feces settled on the bags, which were then handled by other girls on the team, leading to eight of them contracting the infection. Here, the bags were the fomites, and the girls handling the bags apparently transferred the virus to their mouths.

Indirect Spread by Vehicles: Water, Soil, and Air as Vehicles

As discussed in the section on reservoirs, soil and water harbor a variety of microbes that can sicken humans. Also, they can become temporarily contaminated with pathogens that come from humans, as in the case of water becoming contaminated during a cholera outbreak. Unlike soil and water, however, outdoor air cannot provide nutritional support for microbial growth and seldom transmits airborne pathogens. On the other hand, indoor air (especially in a closed space) can serve as an important medium for the suspension and dispersal of certain respiratory pathogens via droplet nuclei and aerosols. **Droplet nuclei** are dried microscopic residues created when microscopic pellets of mucus and saliva are ejected from the mouth and nose. They are generated forcefully in a sneeze or cough **(figure 13.15)** or mildly during talking or singing. The larger beads of moisture settle rapidly. If these settle in or on another person, it is considered droplet contact, as described earlier; but the smaller particles evaporate and remain suspended for longer periods. After evaporation, microscopic pellets 1 to 4 microns in size are created. Their small size enhances their pathogenic ability to ease their passage into the lungs. They can be encountered by a new host who is geographically or chronologically distant; thus, they are considered indirect contact. Droplet nuclei are implicated in the spread of hardier pathogens such as the tubercle bacillus and the influenza virus. **Aerosols** are suspensions of fine dust or moisture particles in the air that contain live pathogens. Q fever is spread by dust from animal quarters, and psittacosis is spread by aerosols from infected birds. An unusual outbreak

Figure 13.15 The explosiveness of a sneeze. Special photography dramatically captures droplet formation in an unstifled sneeze. When such droplets dry and remain suspended in air, they become droplet nuclei.

of coccidioidomycosis (a lung infection) occurred during the 1994 Southern California earthquake. Epidemiologists speculate that disturbed hillsides and soil gave off clouds of dust containing the spores of *Coccidioides.*

In the disease chapters of this book (chapters 18 through 23), the modes of transmission appearing in the pink boxes in figure 13.14 will be used to describe the diseases.

Healthcare-Associated Infections: The Hospital as a Source of Disease

Infectious diseases that are acquired or develop during a hospital stay (or a stay in another health care facility, such as a rehabilitation hospital) are known as **healthcare-associated infections.** This concept seems incongruous at first thought, because a hospital is regarded as a place to get treatment for a disease, not a place to acquire a disease. Yet it is not uncommon for a surgical patient's incision to become infected or a burn patient to develop a case of pneumonia in the clinical setting. The rate of healthcare-associated infections (HAIs) can be as low as 0.1% or as high as 20% of all admitted patients depending on the clinical setting, with an average of about 5%. In light of the number of admissions, this adds up to 2 to 4 million cases a year, which result in nearly 90,000 deaths. HAIs cost time and money as well as suffering. By one estimate, they amount to 8 million additional days of hospitalization a year and an increased cost of $5 to $10 billion.

So many factors unique to the hospital environment are tied to HAIs that a certain number of infections are virtually unavoidable. After all, the hospital both attracts and creates compromised patients, and it serves as a collection point for pathogens. Some patients become infected when surgical procedures or lowered defenses permit resident biota to invade their bodies. Other patients acquire infections directly or indirectly from fomites, medical equipment, other patients, medical personnel, visitors, air, and water. It is often difficult to determine if healthcare-associated infections are endogenous or exogenous in nature. The health care process itself increases the likelihood that infectious agents will be transferred from one patient to another. Treatments using reusable instruments such as respirators and thermometers constitute a possible source of infectious agents. Indwelling devices such as catheters, prosthetic heart valves, grafts, drainage tubes, and tracheostomy tubes form ready portals of entry and habitats for infectious agents. Because such a high proportion of the hospital population receives antimicrobial drugs during its stay, drug-resistant microbes are selected for at a much greater rate than is the case outside the hospital.

The most common healthcare-associated infections involve the urinary tract, the respiratory tract, and surgical incisions **(figure 13.16).** Gram-negative intestinal biota (*Escherichia coli, Klebsiella, Pseudomonas*) are cultured in more than half of patients with HAIs. Gram-positive bacteria (staphylococci and streptococci) and yeasts make up most of the remainder. True pathogens such as *Mycobacterium tuberculosis, Salmonella,* hepatitis B, and influenza virus can be transmitted in the clinical setting as well.

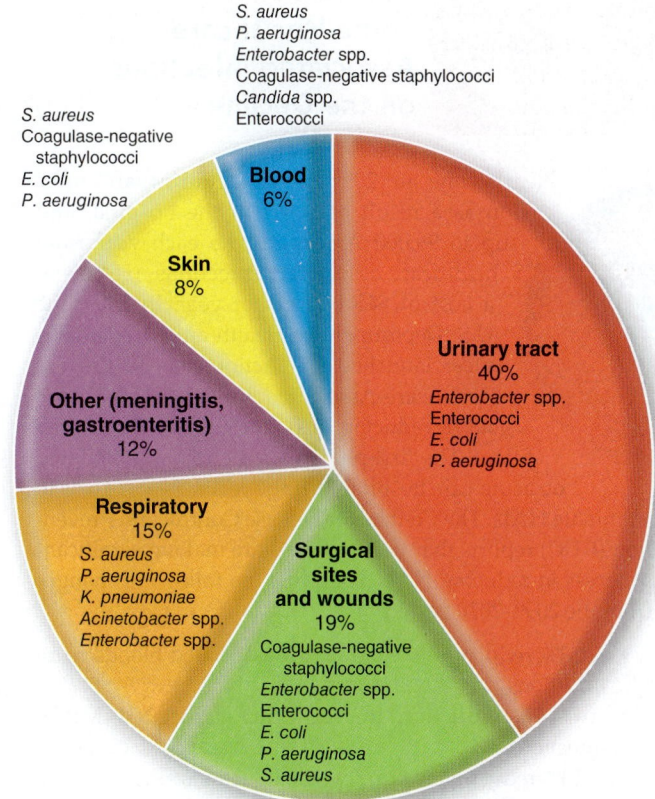

Figure 13.16 Most common healthcare-associated infections. Relative frequency by target area.

The federal government has taken steps to incentivize hospitals to control HAI transmission. In the fall of 2008, the Medicare and Medicaid programs announced they would not reimburse hospitals for catheter-associated urinary tract infections, vascular catheter-associated bloodstream infections, and surgical site infections acquired during hospital care. As can be seen in **Insight 13.4,** the measures seem to have helped reduce the rates, although these gains are not universal and many hospitals still struggle. Also, troublesome outbreaks still occur in facilities with otherwise good records.

Medical asepsis includes practices that lower the microbial load in patients, caregivers, and the hospital environment. These practices include proper hand washing, disinfection, and sanitization, as well as patient isolation. The goal of these procedures is to limit the spread of infectious agents from person to person. An even higher level of stringency is seen with *surgical asepsis,* which involves all of the strategies listed previously plus ensuring that all surgical procedures are conducted under sterile conditions. This includes sterilization of surgical instruments, dressings, sponges, and the like, as well as clothing personnel in sterile garments and scrupulously disinfecting the room surfaces and air (see chapter 11).

Hospitals generally employ an *infection-control officer* who not only implements proper practices and procedures throughout the hospital but also is charged with tracking potential outbreaks, identifying breaches in asepsis, and training other health care workers in aseptic technique. Among

INSIGHT 13.4   Some Healthcare-Associated Infections on the Decline

Nearly one in 20 patients acquires an infection as a result of being in the hospital, and up to 98,000 Americans die each year from healthcare-associated infections (HAIs) at a cost of $45 billion per year. In 2009, the U.S. Department of Health and Human Services launched an "Action Plan to Prevent Healthcare-Associated Infections" with the goal of reducing HAIs throughout the United States. The first phase of this effort was centered on acute-care hospitals, using specific 5-year goals to eliminate HAIs. The Centers for Disease Control and Prevention (CDC) reports that U.S. hospitals are making significant progress in reducing HAIs, including the following changes from the 2009–2010 period:

- A 33% reduction in central line–associated bloodstream infections
- An 18% reduction in healthcare–associated MRSA infections
- A 10% reduction in surgical site infections
- A 7% reduction in catheter–associated urinary tract infections

Although significant progress is being made in reducing HAIs, CDC Director Thomas R. Frieden, MD, MPH, emphasizes, "Hospitals and state health departments need to translate this progress to other areas of health care delivery and health care infections, such as dialysis and ambulatory surgery centers, and diarrheal infections such as *Clostridium difficile*."

Source: 2011. http://www.cdc.gov/media/releases/2011/p1019_healthcare_infections.html

Case File 13   *Continuing the Case*

Researchers in Israel collected swab samples from the uniforms of registered nurses (RNs) and medical doctors (MDs), testing the ends of sleeves, the abdominal region, and pockets. The samples were pressed onto blood agar plates to detect microbial growth. Microbial pathogens were isolated from 63% of those sampled; multidrug-resistant *Staphylococcus aureus* (MRSA) was isolated from 21 cultures of RN uniforms and 6 cultures of MD uniforms.

In another study researchers sampled hospital surfaces and clothing that were routinely touched by health care workers. The results showed that 75% of white lab coats were contaminated with *Pseudomonas oryzihabitans*, MRSA, and *Moraxella*, among other pathogens. Doctors' ties were also sampled, and 66% were contaminated with MRSA and other pathogens.

Many hospitals allow hospital workers to launder their own uniforms as a cost-saving measure; often, scrubs and white lab coats are taken home and washed together with personal clothing items at varying temperatures and with different detergents. The main difference between a hospital laundry and a home laundry is the water temperature—hospital laundry water can reach 160°F and home laundry can reach only 130°F on the hottest setting, which may not destroy all pathogens. A recent study in the journal *Infection Control and Hospital Epidemiology* showed that hot water with detergent killed MRSA but was only moderately successful against *Acinetobacter*, an organism known to have resistance to multiple antibiotics. If clothes are washed with warm water but not detergent, MRSA is removed, but the clothing becomes contaminated by *Enterobacter, Serratia*, and *Klebsiella*, all gram-negative organisms found in fecal material. Another study showed that home-laundered scrubs showed significantly higher levels of coliform bacteria than hospital-laundered scrubs.

- What are the guidelines for the laundering of potentially contaminated hospital clothing?
- What measures are being taken to reduce the levels of bacteria on hospital clothing?

Sources: 2011. *Infect. Contr. Hosp. Epidem.* vol. 32, no. 11, p. 1103. 2011. *Am. J. Infect. Contr.* vol. 39, no. 7, p. 555.

---

those most in need of this training are nurses and other caregivers whose work, by its very nature, exposes them to needlesticks, infectious secretions, blood, and physical contact with the patient. The same practices that interrupt the routes of infection in the patient can also protect the health care worker. It is for this reason that most hospitals have adopted universal precautions that recognize that all secretions from all persons in the clinical setting are potentially infectious and that transmission can occur in either direction.

Universal Blood and Body Fluid Precautions

Medical and dental settings require stringent measures to prevent the spread of HAIs from patient to patient, from patient to worker, and from worker to patient. Previously, control guidelines were disease-specific and clearly identified infections were managed with particular restrictions and techniques. With this arrangement, personnel tended to handle materials labeled *infectious* with much greater care than those that were not so labeled. The AIDS epidemic spurred a reexamination

of that policy. Because of the potential for increased numbers of undiagnosed HIV-infected patients, the Centers for Disease Control and Prevention laid down more stringent guidelines for handling patients and body substances. These guidelines have been termed **universal precautions (UPs),** because they are based on the assumption that all patient specimens could harbor infectious agents and so must be treated with the same degree of care. They also include body substance isolation (BSI) techniques to be used in known cases of infection.

It is worth mentioning that these precautions are designed to protect all individuals in the clinical setting—patients, workers, and the public alike. In general, they include techniques designed to prevent contact with pathogens and contamination and, if prevention is not possible, to take purposeful measures to decontaminate potentially infectious materials.

The universal precautions recommended for all health care settings follow:

1. Barrier precautions, including masks and gloves, should be taken to prevent contact of skin and mucous membranes with patients' blood or other body fluids. Because gloves can develop small invisible tears, double gloving decreases the risk further. For protection during surgery, venipuncture, or emergency procedures, gowns, aprons, and other body coverings should be worn. Dental workers should wear eyewear and face shields to protect against splattered blood and saliva.

2. More than 10% of health care personnel are pierced each year by sharp (and usually contaminated) instruments. These accidents carry risks not only for HIV infection but also for hepatitis B, hepatitis C, and other diseases. Preventing inoculation infection requires vigilant observance of proper techniques. All disposable needles, scalpels, or sharp devices from invasive procedures must immediately be placed in puncture-proof containers for sterilization and final discard. Under no circumstances should a worker attempt to recap a syringe, remove a needle from a syringe, or leave unprotected used syringes where they pose a risk to others. Reusable needles or other sharp devices must be heat-sterilized in a puncture-proof holder before they are handled. If a needlestick or other injury occurs, immediate attention to the wound, such as thorough degermation and application of strong antiseptics, can prevent infection.

3. Dental handpieces should be sterilized between patients, but if this is not possible, they should be thoroughly disinfected with a high-level disinfectant (peroxide, hypochlorite). Blood and saliva should be removed completely from all contaminated dental instruments and intraoral devices prior to sterilization.

4. Hands and other skin surfaces that have been accidently contaminated with blood or other fluids should be scrubbed immediately with a germicidal soap. Hands should likewise be washed after removing rubber gloves, masks, or other barrier devices.

5. Because saliva can be a source of some types of infections, barriers should be used in all mouth-to-mouth resuscitations.

6. Health care workers with active, draining skin or mucous membrane lesions must refrain from handling patients or equipment that will come into contact with other patients. Pregnant health care workers risk infecting their fetuses and must pay special attention to these guidelines. Personnel should be protected by vaccination whenever possible.

Isolation procedures for known or suspected infections should still be instituted on a case-by-case basis.

Which Agent Is the Cause? Using Koch's Postulates to Determine Etiology

An essential aim in the study of infection and disease is determining the precise **etiologic,** or causative, **agent** of a newly recognized condition. More than a century ago, Robert Koch realized that in order to prove the germ theory of disease he would have to develop a standard for determining causation that would stand the test of scientific scrutiny. Out of his experimental observations on the transmission of anthrax in cows came a series of proofs, called **Koch's postulates,** that established the principal criteria for etiologic studies **(process figure 13.17).** These postulates direct an investigator to

1. find evidence of a particular microbe in every case of a disease,
2. isolate that microbe from an infected subject and cultivate it in pure culture in the laboratory,

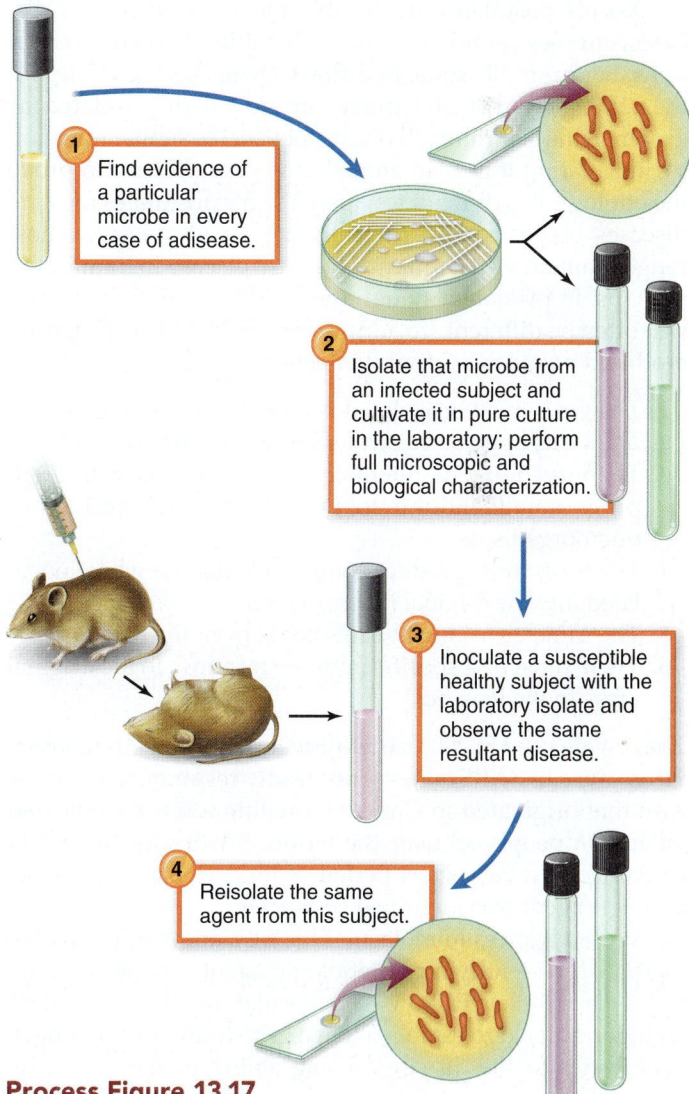

Process Figure 13.17
Koch's postulates: Is this the etiologic agent?
The microbe in the initial and second isolations and the disease in the patient and experimental animal must be identical for the postulates to be satisfied.

③ inoculate a susceptible healthy subject with the laboratory isolate and observe the same resultant disease, and
④ reisolate the agent from this subject.

Valid application of Koch's postulates requires attention to several critical details. Each isolated culture must be pure, observed microscopically, and identified by means of characteristic tests; the first and second isolates must be identical; and the pathologic effects, signs, and symptoms of the disease in the first and second subjects must be the same. Once established, these postulates were rapidly put to the test, and within a short time, they had helped determine the causative agents of tuberculosis, diphtheria, and plague. Today, most known infectious diseases have been directly linked to a known infectious agent.

Koch's postulates continue to play an essential role in modern epidemiology. Every decade, new diseases challenge the scientific community and require application of the postulates.

Koch's postulates are reliable for many infectious diseases, but they cannot be completely fulfilled in certain situations. For example, some infectious agents, such as *M. leprae,* the etiologic agent of leprosy, are not readily isolated or grown in the laboratory. If one cannot elicit a similar infection by inoculating it into an animal, it is very difficult to prove the etiology. It is difficult to satisfy Koch's postulates for viral diseases because viruses usually have a very narrow host range. Human viruses may only cause disease in humans or, perhaps, in primates, though the disease symptoms in apes will often be different. To address this, T. M. Rivers proposed modified postulates for viral infections:

1. The virus must be isolated from each diseased host.
2. The virus must be cultivated in cell culture.
3. The virus must be filterable, that is, must pass through pores small enough to impede bacteria and other microorganisms.
4. The virus must produce comparable disease when inoculated into the original host species or a related one.
5. The same virus must be reisolated from the new host.
6. There must be a specific immune response to the original virus in the new host.

These were used in 2003 to definitively determine the coronavirus cause of SARS, a new and deadly respiratory tract disease that originated in China and within a few months had killed 916 people all over the world. SARS was effectively contained in a very short period of time, in part because its etiologic agent was identified so quickly.

With advances in molecular biology, another alternative method for identifying an etiologic agent has been developed. Dr. Stanley Falkow's "molecular Koch's postulates" were formulated to establish that a gene found in a pathogen contributes to the disease-causing ability of the organism. Today, this set of postulates is often used to investigate candidate virulence genes.

It is also usually not possible to use Koch's postulates to determine causation in polymicrobial diseases. Diseases such as periodontitis and soft tissue abscesses are caused by complex mixtures of microbes. While it is theoretically possible to isolate each member and to re-create the exact proportions of individual cultures for step ③, it is not attempted in practice.

13.2 Learning Outcomes—Assess Your Progress

4. Differentiate between a microbe's pathogenicity and its virulence.
5. Define *opportunism*, and list examples of common opportunistic pathogens.
6. List the steps a microbe has to take to get to the point where it can cause disease.
7. List several portals of entry and exit.
8. Define *infectious dose*, and explain its role in establishing infection.
9. Describe three ways microbes cause tissue damage.
10. Compare and contrast major characteristics of exotoxins and endotoxin.
11. List several virulence factors, and summarize their actions within a host.
12. Draw and label a curve representing the course of clinical infection.
13. Differentiate among the various types of reservoirs, providing examples of each.
14. List six different modes of horizontal transmission, providing an infectious disease spread by each.
15. Define *healthcare-associated infection*, listing common types and their causative agents.
16. List Koch's postulates, and explain alternative methods for identifying an etiologic agent.

13.3 Epidemiology: The Study of Disease in Populations

So far, our discussion has revolved primarily around the impact of an infectious disease in a single individual. Let us now turn our attention to the effects of diseases on the community—the realm of **epidemiology**. By definition, this term involves the study of the frequency and distribution of disease and other health-related factors in defined populations. It involves many disciplines—not only microbiology but also anatomy, physiology, immunology, medicine, psychology, sociology, ecology, and statistics—and it considers all forms of disease, including heart disease, cancer, drug addiction, and mental illness.

A groundbreaking British nurse named Florence Nightingale helped to lay the foundations of modern epidemiology. In the mid-1850s, she arrived in the Crimean war zone in Turkey, where the British were fighting and dying at an astonishing rate. Estimates suggest that 20% of the soldiers there died (by contrast, 2.6% of U.S. soldiers in the Vietnam war died). Even though this was some years before the discovery of the germ theory, Nightingale understood that filth contributed to disease and instituted methods that had never been seen in military field hospitals. She insisted that separate linens and towels be used for each patient, and that the floors be cleaned and the pipes of sewage unclogged. She kept meticulous notes of what was killing the patients and was able to demonstrate that many more men died of disease than of their traumatic injuries. She used statistical analysis as well to convince government officials that these trends were real. This was indeed one of the earliest forays into epidemiology—trying to understand how diseases were being transmitted and using statistics to do so.

The techniques of epidemiology are also used to track behaviors, such as exercise or smoking. The epidemiologist is a medical sleuth who collects clues on the causative agent, pathology, sources, and modes of transmission and tracks the numbers and distribution of cases of disease in the community. In fulfilling these demands, the epidemiologist asks who, when, where, how, why, and what about diseases. The outcome of these studies helps public health departments develop prevention and treatment programs and establish a basis for predictions.

Tracking Disease in a Population

Epidemiologists are concerned with all of the factors covered earlier in this chapter: virulence, portals of entry and exit, and the course of disease. But they are also interested in surveillance—that is, collecting, analyzing, and reporting data on the rates of occurrence, mortality, morbidity, and transmission of infections. Surveillance involves keeping data for a large number of diseases seen by the medical community and reported to public health authorities. By law, certain **reportable,** or notifiable, **diseases** must be reported to authorities; others are reported on a voluntary basis.

A well-developed network of individuals and agencies at the local, district, state, national, and international levels keeps track of infectious diseases. Physicians and hospitals report all notifiable diseases that are brought to their attention. These reports are either made about individuals or in the aggregate, depending on the disease.

Traditionally, local public health agencies first receive the case data and determine how they will be handled. In most cases, health officers investigate the history and movements of patients to trace their prior contacts and to control the further spread of the infection as soon as possible through drug therapy, immunization, and education. In notifiable sexually transmitted infections, patients are asked to name their partners so that these persons can be notified, examined, and treated. It is very important to maintain the confidentiality of the persons in these reports. The principal government agency responsible for keeping track of infectious diseases nationwide is the Centers for Disease Control and Prevention (CDC) in Atlanta, Georgia; the CDC is a

part of the U.S. Public Health Service. The CDC publishes a weekly notice of diseases (the *Morbidity and Mortality Report*) that provides weekly and cumulative summaries of the case rates and deaths for about 50 notifiable diseases, highlights important and unusual diseases, and presents data concerning disease occurrence in the major regions of the United States. It is available to anyone at www.cdc.gov/mmwr/. Ultimately, the CDC shares its statistics on disease with the World Health Organization (WHO) for worldwide tabulation and control.

Access to the Internet has revolutionized disease tracking. For example, in 2008, Google launched a service called Google Flu Trends. This application compiles aggregated data from key word searches for terms such as *thermometer, chest congestion, muscle aches*, or *flu symptoms*. The company publishes the data on a website, which serves as an early warning system for the locations of new flu activity. Analysis of the Google Flu Trends data from the H1N1 outbreak that began in Mexico shows that its data predicted the epidemic about a week before CDC data did. Twitter is also being used as an early monitoring method for flu epidemics and even dengue fever in South America.

Epidemiological Statistics: Frequency of Cases

The **prevalence** of a disease is the total number of existing cases with respect to the entire population. It is often thought of as a snapshot and is usually reported as the percentage of the population having a particular disease at any given time. Disease **incidence** measures the number of new cases over a certain time period. This statistic, also called the *case,* or *morbidity, rate,* indicates both the rate and the risk of infection. The equations used to figure these rates follow:

$$\text{Prevalence} = \frac{\text{Total number of cases in population}}{\text{Total number of persons in population}} \times 100 = \%$$

$$\text{Incidence} = \frac{\text{Number of new cases}}{\text{Total number of susceptible persons}} \quad \begin{array}{l}\text{(Usually reported per 100,000 persons per unit of time)}\end{array}$$

Disease Connection

Prevalence is affected by three factors: A new (incident) case adds to prevalence, and death or recovery is the only occurrence that decreases prevalence. Therefore, in the United States, HIV prevalence has *increased* even though there is much more awareness of how it is transmitted. Why? Because of our ability to treat it and keep it from killing its hosts, death removes cases from the prevalence calculation less frequently now.

The changes in incidence and prevalence are usually followed over a seasonal, yearly, and long-term basis and are helpful in predicting trends **(figure 13.18).** Statistics of concern to the epidemiologist are the rates of disease with regard to sex, race, or geographic region. Also of importance is the **mortality rate,** which measures the number of deaths in a population due to a certain disease. Over the past century, the overall death rate from infectious diseases in the developed world has dropped, although the number of persons afflicted with infectious diseases (the **morbidity rate**) has remained relatively high.

When there is an increase in disease in a particular geographic area, it can be helpful to examine the epidemic curve (incidence over time) to determine if the infection is a **point-source, common-source,** or **propagated epidemic.**

A *point-source epidemic*, illustrated in **figure 13.19a,** is one in which the infectious agent came from a single source and all of its "victims" were exposed to it from that source. The classic example of this is food illnesses brought on by exposure to a contaminated food item at a potluck dinner or restaurant. *Common-source epidemics* or outbreaks result from common exposure to a single source of infection that can occur over a period of time **(figure 13.19b).** Think of a contaminated water plant that infects multiple people over the course of a week, or even of a single restaurant worker who is a carrier of hepatitis A and does not practice good hygiene. Lastly, a *propagated epidemic* **(figure 13.19c)** results from an infectious agent that is communicable from person to person and therefore is sustained—propagated—over time in a population.

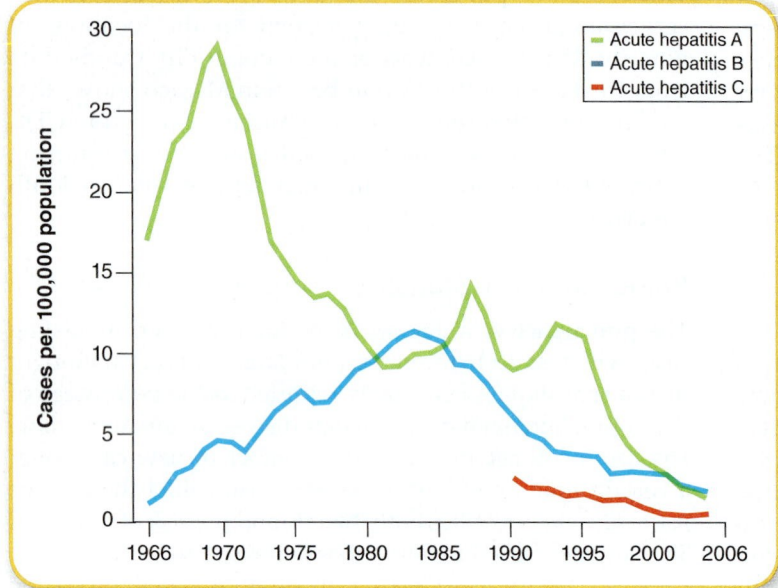

(a) Hepatitis incidence: cases per year, United States, 1966–2006.

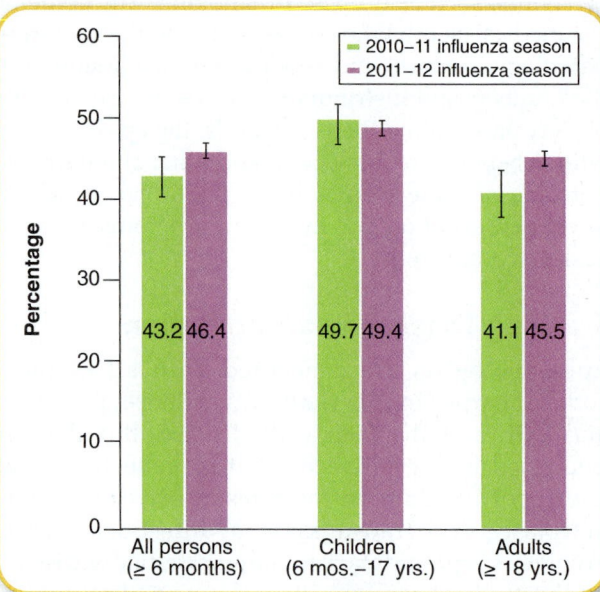

(b) Percentage of people in the United States who received the influenza vaccine in two recent years.

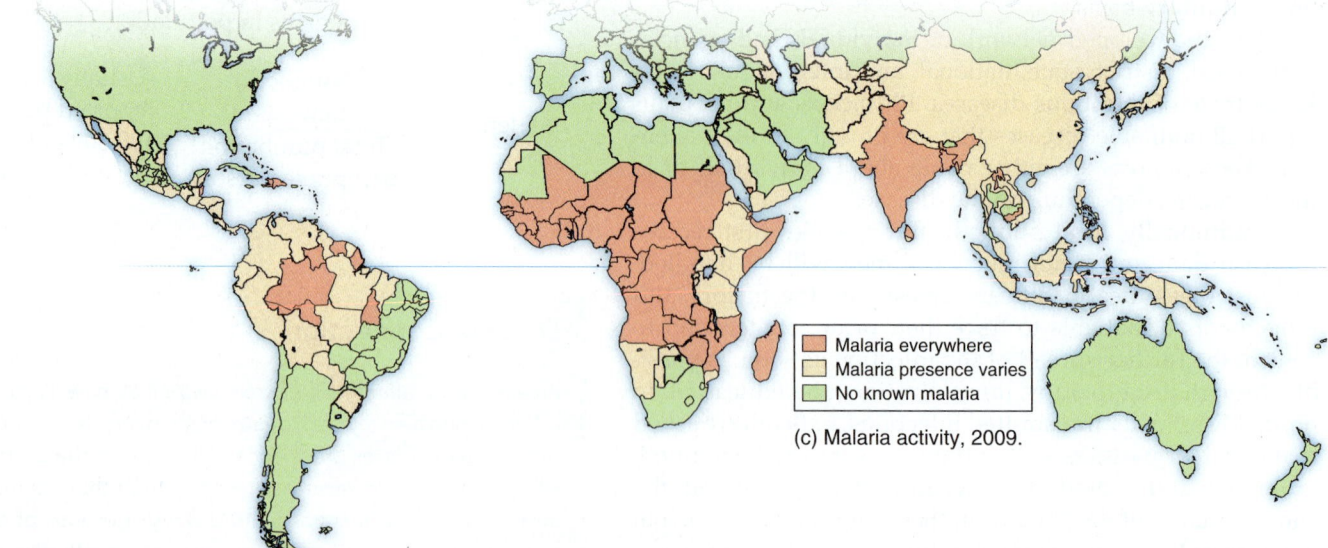

(c) Malaria activity, 2009.

Figure 13.18 Graphical representation of epidemiological data.   The Centers for Disease Control and Prevention collect epidemiological data that are analyzed with regard to **(a)** time, **(b)** age and other characteristics, and **(c)** geographic region.

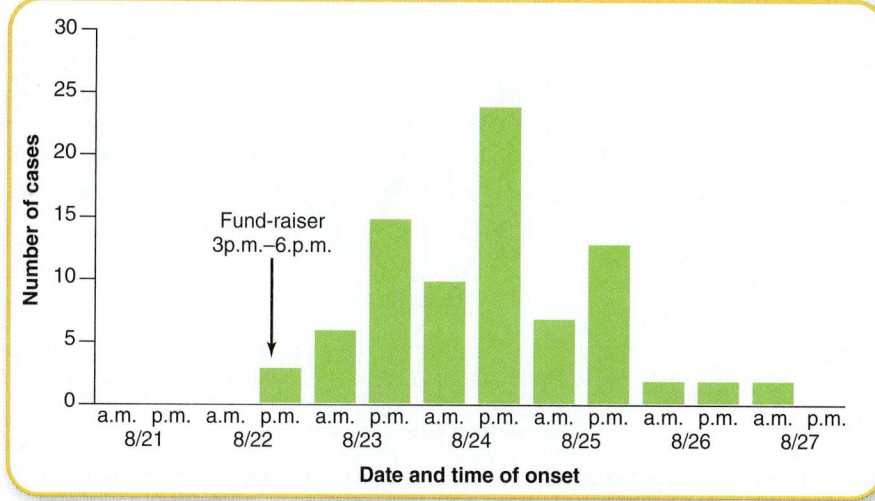

(a) Point-source epidemic traced to crab cakes at a fund-raiser in Maryland in 2003.

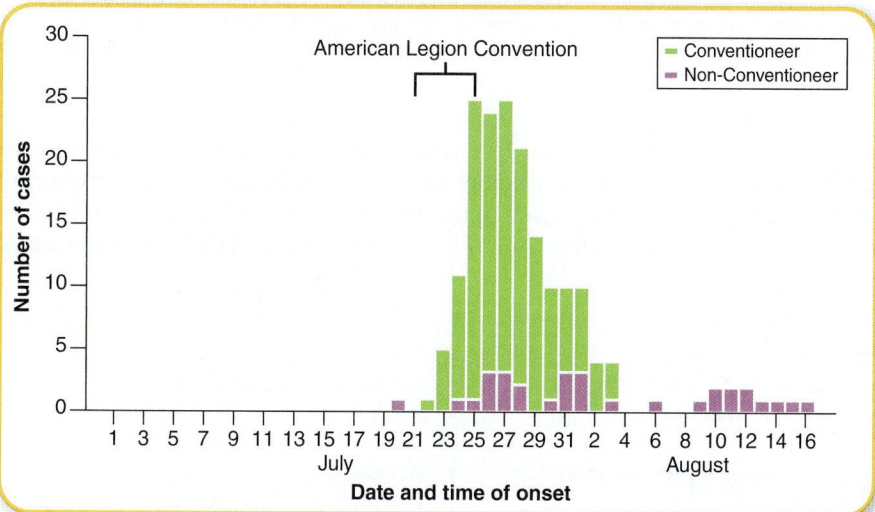

(b) Common-source epidemic graph illustrating the first outbreak of Legionnaires' disease at the American Legion Convention in 1976 in Philadelphia.

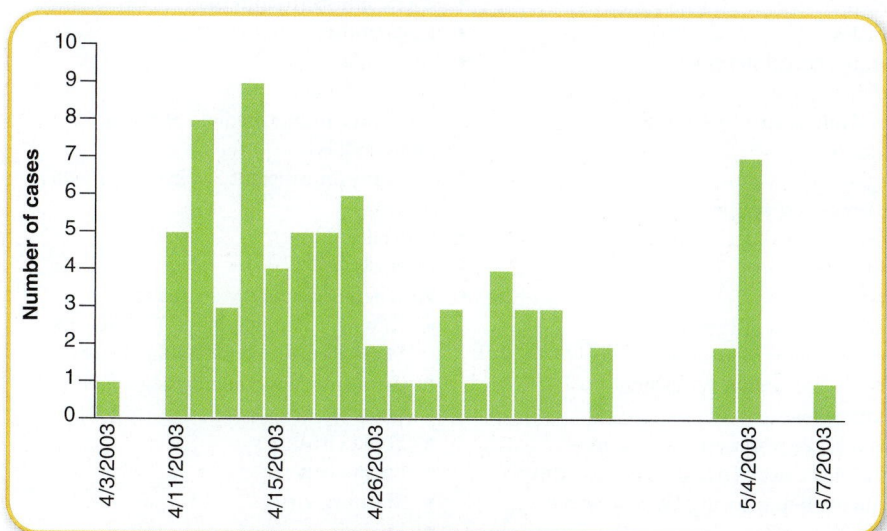

(c) "Curve" representing the propagated epidemic of the SARS virus in 2003. This is only one of several transmission chains.

Figure 13.19 Different outbreak or epidemic "curves."   **(a)** Point-source epidemic; **(b)** common-source epidemic; **(c)** propagated epidemic.

Influenza is the classic example of this. The point is that each of these types of spread become apparent from the shape of the outbreak or epidemic "curves."

An additional term, the **index case,** refers to the first patient found in an epidemiological investigation. How the cases unfurl from this case helps explain the type of epidemic it is. The index case may not turn out to be the first case—as the investigation continues, earlier cases may be found—but the index case is the case that brought the epidemic to the attention of officials. Monitoring statistics also makes it possible to define the frequency of a disease in the population. An infectious disease that exhibits a relatively steady frequency over a long time period in a particular geographic locale is **endemic (figure 13.20*a*).** For example, Lyme disease is endemic to certain areas of the United States where the tick vector is found. A certain number of new cases are expected in these areas every year. Of course, in order to know whether the incidence is remaining the same or close to the same year after year, you have to plot the incidence over time **(figure 13.20*b*).** When a disease is **sporadic,** occasional cases are reported at irregular intervals in random locales. A single disease can be endemic in certain areas and sporadic in others. For example, in figure 13.20*a*, the occurrence of Lyme disease in New Mexico, Utah, and Idaho can be called sporadic. Some diseases, such as tetanus and diphtheria, are reported sporadically across the United States (fewer than 50 cases a year).

When statistics indicate that the prevalence of an endemic or sporadic disease is increasing beyond what is expected for that population, the pattern is described as an **epidemic.** To see this, the incidence must be visualized over time (figure 13.20*b*). The time period over which this change occurs differs for each disease. It can range from hours in food poisoning to years in syphilis. Also, the exact percentage of increase needed before an outbreak can qualify as an epidemic is specific for each disease. Figure 13.20*b* shows the expected percentage of deaths from influenza across seasons for several years. When the actual rate significantly exceeds the "normal" or baseline rate, indicated by the top black line, it indicates an epidemic. The spread of an epidemic across continents is a **pandemic,** as exemplified by HIV and influenza.

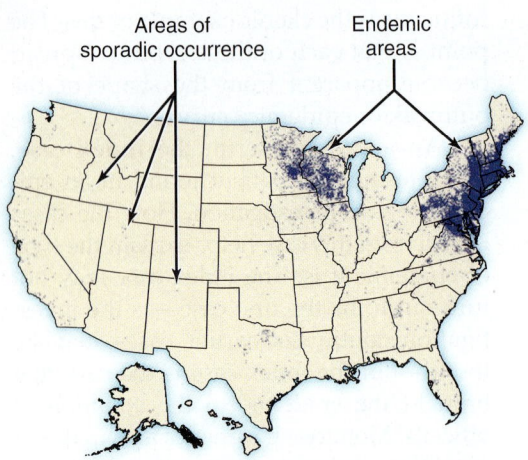

(a) Lyme disease incidence in 2010. One dot is placed randomly within county of residence for each confirmed case.

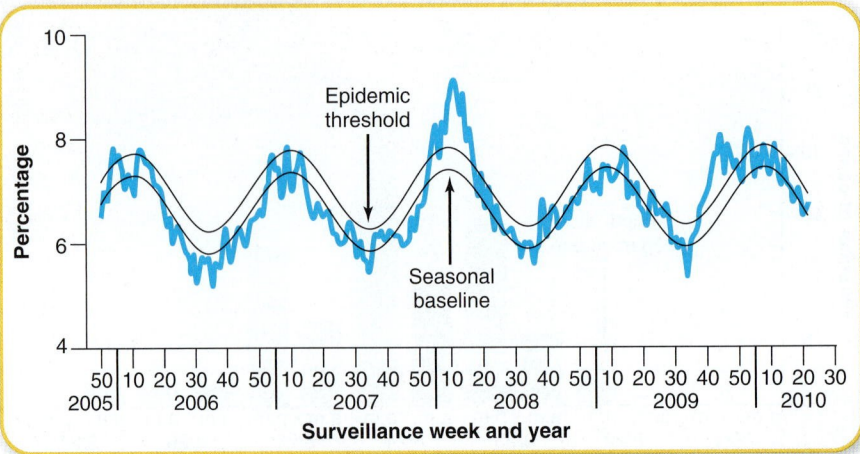

(b) The blue line indicates the percentage of all U.S. deaths that were caused by influenza and pneumonia over a 5-year period.

Figure 13.20 Patterns of infectious disease occurrence. **(a)** In an endemic occurrence, cases are concentrated in one area at a relatively stable rate. In a sporadic occurrence, a few cases occur randomly over a wide area. **(b)** An epidemic is an increased number of cases over the customary rate.

One important epidemiological truism may be called the "iceberg effect," which refers to the fact that only a small portion of an iceberg is visible above the surface of the ocean, with a much more massive part lingering unseen below the surface. Regardless of case reporting and public health screening, a large number of cases of infection in the community go undiagnosed and unreported. (For a list of reportable diseases in the United States, see **table 13.11**.) In the instance of

---

Table 13.11 Reportable Diseases in the United States, 2012*

- Anthrax
- Arboviral neuroinvasive and non-neuroinvasive diseases
  - California serogroup virus disease
  - Eastern equine encephalitis virus disease
  - Powassan virus disease
  - St. Louis encephalitis virus disease
  - West Nile virus disease
  - Western equine encephalitis disease
- Babesiosis
- Botulism
- Brucellosis
- Chancroid
- *Chlamydia trachomatis* genital infections
- Cholera
- Coccidioidomycosis
- Cryptosporidiosis
- Cyclosporiasis
- Dengue (including Dengue fever, hemorrhagic fever, and shock syndrome)
- Diphtheria
- Ehrlichiosis/Anaplasmosis
- Giardiasis
- Gonorrhea
- *Haemophilus influenzae*, invasive disease
- Hansen disease (leprosy)
- Hantavirus pulmonary syndrome

- Hemolytic uremic syndrome, post-diarrheal hepatitis A, B, C
- HIV infection
- Influenza-associated pediatric mortality
- Legionellosis
- Listeriosis
- Lyme disease
- Malaria
- Measles
- Meningococcal disease
- Mumps
- Novel influenza A infections
- Pertussis
- Plague
- Poliomyelitis, paralytic
- Poliovirus infection, nonparalytic
- Psittacosis
- Q fever
- Rabies, animal or human
- Rubella (German measles)
- Rubella, congenital syndrome
- Salmonellosis
- Severe acute respiratory syndrome–associated coronavirus (SARS-CoV) disease
- Shiga toxin–producing *Escherichia coli* (STEC)
- Shigellosis
- Smallpox

- Spotted Fever Rickettsiosis
- Streptococcal toxic-shock syndrome
- *Streptococcus pneumoniae*, invasive disease
- Syphilis
- Tetanus
- Toxic shock syndrome (other than streptococcal)
- Trichinellosis (trichinosis)
- Tuberculosis
- Tularemia
- Typhoid fever
- Vancomycin-intermediate *Staphylococcus aureus* (VISA)
- Vancomycin-resistant *Staphylococcus aureus* (VRSA)
- Varicella
- Vibriosis
- Viral hemorrhagic fevers due to:
  - New world arenaviruses (Guanarito, Machupo, Junin, and Sabia viruses)
  - Crimean-Congo hemorrhagic fever virus
  - Ebola virus
  - Lassa virus
  - Marburg virus
  - Lujo virus
- Yellow fever

*Reportable to the CDC; other diseases may be reportable to state departments of health.
Source: Centers for Disease Control and Prevention, 2012.

salmonellosis, approximately 40,000 cases are reported each year. Epidemiologists estimate that the actual number is more likely somewhere between 400,000 and 4,000,000. The iceberg effect can be even more lopsided for sexually transmitted infections or for infections that are not brought to the attention of reporting agencies. As you will see in chapter 23, it is a major uphill battle to get vulnerable populations tested for STIs.

Global Issues in Epidemiology

In the early 1900s, it was assumed by many that antibiotics would be the "magic bullet" that would eradicate all infectious disease from the human population. Although the mortality rates from such diseases declined dramatically after the advent of antibiotic drugs, an alarming trend was noted in the early 1980s: The incidence of infectious diseases began to increase—and it increased quite dramatically. It rose due to the appearance of a newly identified virus, HIV, but more importantly continues to grow even today due to *emerging* and *reemerging diseases*. Emerging diseases are caused by newly identified microbes, such as the SARS virus and novel strains of human influenza virus. Reemerging diseases are those that have affected the human population in the past but are now becoming more prevalent due to travel, habitat invasion, or the development of drug resistance. Dengue fever, tuberculosis, and yellow fever are just a few examples of diseases once thought to be conquered by modern medicine, but they have come back with a vengeance in many areas around the world today **(figure 13.21).**

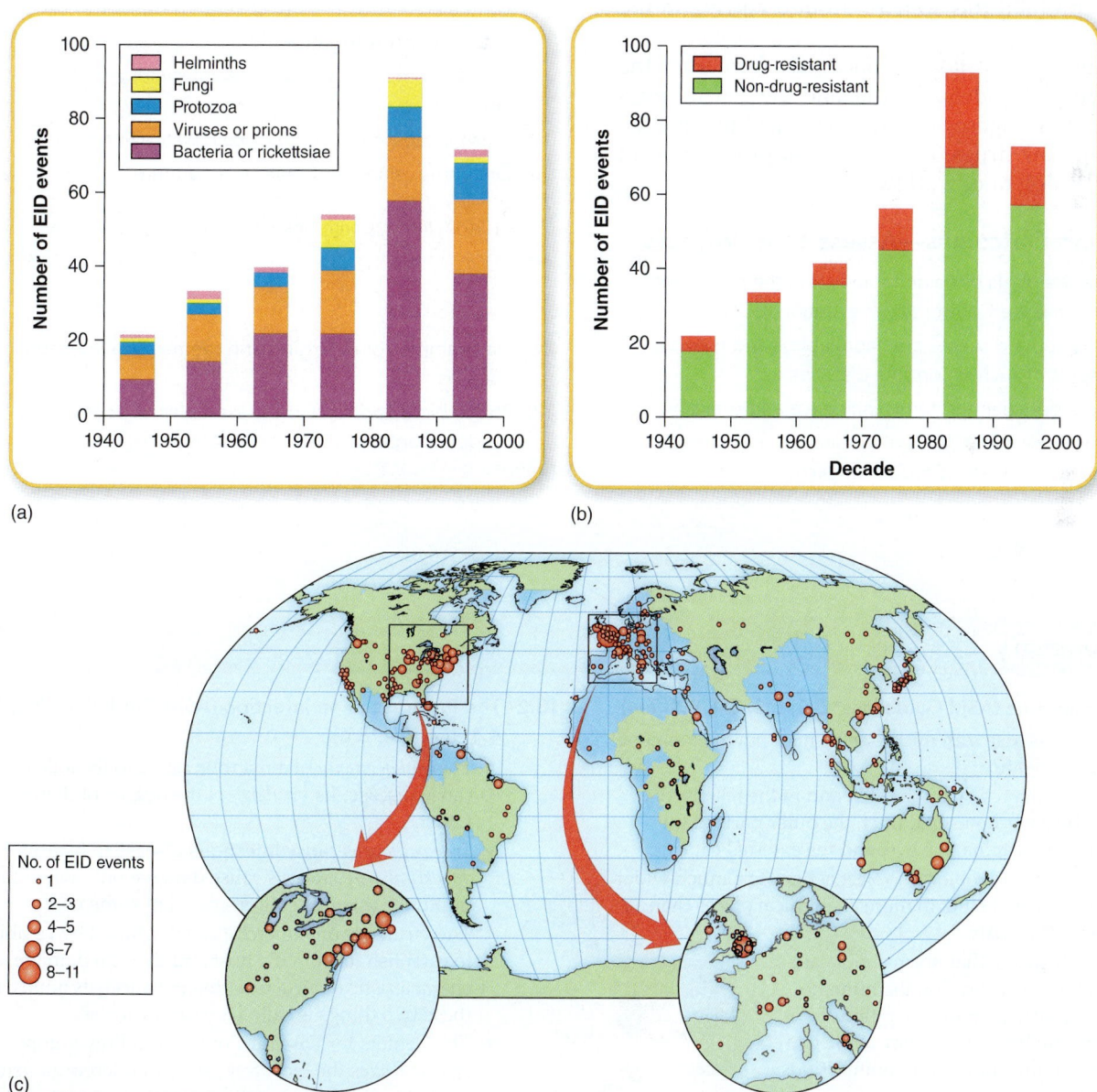

Figure 13.21 Analysis of emerging infectious diseases (EIDs). Data represent the number of EIDs per decade from 1940 to 2000 by **(a)** pathogen group or **(b)** drug resistance. **(c)** Map of geographic origins of EIDs from 1940 to 2004. Sizes of red circles are proportional to the number of events.

Adding to this issue is the threat of *bioterrorism*, the intentional or threatened use of microorganisms or toxins from living organisms to cause death or disease in humans, livestock, or plants. Although use of microbes to inflict damage on human populations or to cause political discord is not new, the stakes have become much higher with the advancement of biotechnology. The spread of anthrax in the United States in 2001 was a fairly controlled event in hindsight; had the microbe been genetically altered in a laboratory to enhance its pathogenicity, it would have created a much more devastating scenario. Beyond the targeting of humans, *agroterrorism* involves the use of microorganisms to decimate the agricultural industry. Rather than making humans ill, agroterrorists target the food supply to exert their damage. Although no documented cases have occurred, many government agencies are conducting surveillance and developing policies to prohibit this scenario from occurring in the world today.

In the future, it will take a concerted effort from the medical community, epidemiologists, and the general public to keep infectious agents in check. This will involve the development of new drugs, new education programs, and increased vaccination rates worldwide.

13.3 Learning Outcomes—Assess Your Progress

17. Summarize the goals of epidemiology and the role of the Centers for Disease Control and Prevention (CDC).
18. Identify why some diseases are "notifiable," and provide four examples of such reportable diseases.
19. Differentiate between the terms *incidence* and *prevalence*.
20. Discuss the three major types of epidemics, and identify the epidemic curves associated with each.

21. List examples of emerging and reemerging infectious diseases.
22. Define *bioterrorism*, and list examples of possible biological agents.

Case File 13   *Wrap-Up*

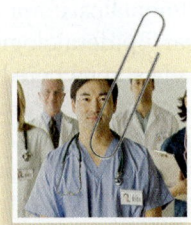

Each hospital, doctor's office, and clinic has its own standards and guidelines for the laundering of hospital attire, but the Association of periOperative Registered Nurses (AORN) is recognized as an authority for recommending standards and practices in health care settings. The AORN strongly recommends *against* home laundering of surgical scrubs and recommends outsourcing to an accredited laundry facility to process contaminated scrubs. If this is not available and workers launder their hospital attire at home, AORN recommends the following:

- Contain and confine scrubs.
- Remove and gather soiled scrubs in a single location and container.
- Wash all scrubs together.
- Refrain from mixing in soiled scrubs with personal clothing.
- Wash scrubs in hot water or cold water with bleach.
- Dry items using a hot dryer temperature.

Source: 2012. Infect. Contr. Hosp. Epi. vol. 33, no. 3, p. 268.

Chapter Summary

13.1 The Human Host (ASM Guidelines* 5.1, 5.3, 5.4, 6.1, 6.4)

- Humans coexist with microorganisms from the moment of birth onward.
- Normal biota reside on the skin and in the respiratory tract, the gastrointestinal tract, the outer parts of the urethra, the vagina, the eye, and the external ear canal.
- The Human Microbiome Project is finding a much wider array of normal biota in more anatomical places than known previously.
- Research shows that normal biota are needed to fully develop the human immune system, and disruptions to the normal biota likely play a role in a variety of infectious and noninfectious conditions.

13.2 The Progress of an Infection (ASM Guidelines 2.2, 3.2, 4.1, 5.4, 6.4, 8.6)

- The *pathogenicity* of a microbe refers to its ability to cause disease. Its v*irulence* is the degree of damage it can inflict.
- *True pathogens* cause infectious disease in healthy hosts; *opportunistic pathogens* cause damage only when the host immune system is compromised in some way.
- The virulence of a microbe is determined by its ability to establish itself in the host and then do damage. Any characteristic or structure that enhances its ability to do these two things is called a *virulence factor.*
- The Centers for Disease Control and Prevention characterizes the biosafety levels of microorganisms to protect individuals handling infectious agents in the laboratory.

*Source: ASM Curriculum Guidelines (American Society for Microbiology, 2012). Complete guidelines in appendix B of this book.

- To cause disease, microbes must enter the host, attach to host tissue, avoid host defenses, and then result in damage.

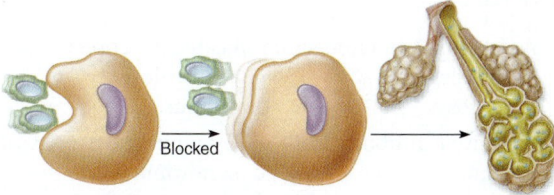

Blocked

- The site at which a microorganism first contacts host tissue is called the *portal of entry*. Most pathogens have one preferred portal of entry, although some have more than one.
- The respiratory system is the portal of entry for the greatest number of pathogens.
- The *infectious dose,* or ID, refers to the minimum number of microbial cells required to initiate infection in the host. The ID is often influenced by quorum-sensing chemicals.
- Fimbriae and adhesive capsules allow pathogens to physically attach to host tissues.
- Antiphagocytic factors produced by microorganisms include leukocidins, capsules, and factors that resist digestion by white blood cells.
- Secreted enzymes, secreted toxins, and the ability to induce injurious host responses are the three main types of *virulence factors* pathogens use to combat host defenses and damage host tissue.
- Exotoxins and endotoxin differ in their chemical composition and tissue specificity.
- Inappropriate or extreme host responses are a major factor in most infectious diseases.
- Patterns of infection vary with the pathogen or pathogens involved. Examples are local, focal, and systemic.
- Mixed infections are more common than previously appreciated.
- Infections can be characterized by their sequence as primary or secondary and by their duration as either acute or chronic.
- The portal of exit by which a pathogen leaves its host is often but not always the same as the portal of entry.
- The portals of exit and entry determine how pathogens spread in a population.
- Some pathogens persist in the body in a latent state.
- There are four distinct phases of infection and disease: the incubation period, the prodrome, the period of invasion, and the convalescent period.
- The primary habitat of a pathogen is called its reservoir. A human reservoir is also called a carrier.

- Animals can be either reservoirs or vectors of pathogens. An infected animal is a biological vector. Uninfected animals, especially insects, that transmit pathogens mechanically are called mechanical vectors.
- Soil and water are nonliving reservoirs for pathogenic bacteria, protozoa, fungi, and worms.
- A communicable disease can be transmitted from an infected host to others, but not all infectious diseases are communicable.
- The spread of infectious disease from person to person is called horizontal transmission. The spread from parent to offspring is called vertical transmission.
- Infectious diseases are spread horizontally by direct or indirect routes of transmission. Vehicles of indirect transmission include soil, water, food, air, the built environment, and fomites (inanimate objects).
- Healthcare-associated infections are acquired in a hospital from surgical procedures, equipment, personnel, and exposure to drug-resistant microorganisms.
- Causative agents of infectious disease may be identified according to Koch's postulates. Alternative methods for the identification of etiologic agents have been developed for use when suitable animal models of disease are not present or if the agent is unculturable.

13.3 Epidemiology: The Study of Disease in Populations (ASM Guidelines 5.4, 6.4)

- Epidemiology is the study of the determinants and distribution of infectious and noninfectious diseases in populations.
- Data on specific, reportable diseases are collected by local, national, and worldwide agencies.
- The *prevalence* of a disease is the percentage of existing cases in a given population. The disease *incidence,* or *morbidity rate,* is the number of newly infected members in a population during a specified time period.
- Outbreaks and epidemics are described as point-source, common-source, or propagated based on the source of the pathogen.

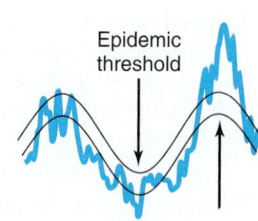

Epidemic threshold

- Disease frequency is described as sporadic, epidemic, pandemic, or endemic.
- Emerging diseases are caused by microbes that have never been seen before, while reemerging diseases are due to microorganisms that have become more prevalent in a population often due to increased virulence, travel, or lack of vaccination.
- Bioterrorism has been in use for hundreds of years, but it still remains a viable threat to global populations today.

Multiple-Choice and True-False Questions | Bloom's Levels 1 and 2: Remember and Understand

Multiple-Choice Questions. Select the correct answer from the options provided.

1. The best descriptive term for the resident biota is
   a. commensals.
   b. parasites.
   c. pathogens.
   d. mutualists.

2. Resident biota are absent from the
   a. pharynx.
   b. heart.
   c. intestine.
   d. hair follicles.

3. Virulence factors include
   a. toxins.
   b. enzymes.
   c. capsules.
   d. all of these.

4. The specific action of hemolysins is to
   a. damage white blood cells.
   b. cause fever.
   c. damage red blood cells.
   d. cause leukocytosis.

5. The ____ is the time that lapses between encounter with a pathogen and the first symptoms.
   a. prodrome
   b. period of invasion
   c. period of convalescence
   d. period of incubation

6. A short period early in a disease that may manifest with general malaise and achiness is the
   a. period of incubation.
   b. prodrome.
   c. sequela.
   d. period of invasion.

7. A/an ____ is a passive animal transporter of pathogens.
   a. zoonosis
   b. biological vector
   c. mechanical vector
   d. asymptomatic carrier

8. An example of a noncommunicable infection is
   a. measles.
   b. leprosy.
   c. tuberculosis.
   d. tetanus.

9. A positive antibody test for HIV would be a ____ of infection.
   a. sign
   b. symptom
   c. syndrome
   d. sequela

10. An outbreak caused by a batch of bad potato salad at a picnic is a _____ outbreak.
    a. point-source
    b. common-source
    c. propagated
    d. all of the above

True-False Questions. If the statement is true, leave as is. If it is false, correct it by rewriting the sentence.

11. The presence of a few bacteria in the blood is called septicemia.

12. Resident microbiota are commonly found in the kidney.

13. A subclinical infection is one that is acquired in a hospital or medical facility.

14. The general term that describes an increase in the number of white blood cells is *leukopenia*.

15. The index case is the first case found in an epidemiological investigation.

Critical Thinking Questions | Bloom's Levels 3, 4, and 5: Apply, Analyze, and Evaluate

Critical thinking is the ability to reason and solve problems using facts and concepts. These questions can be approached from a number of angles and, in most cases, they do not have a single correct answer.

1. Based upon data from the Human Microbiome Project (HMP):
   a. Provide evidence in support of or refuting the following statement: Babies are born sterile.
   b. Define *microbial antagonism,* and discuss how the various microbial populations keep each other "in check," with consequences for human health.

2. Summarize one example illustrating how microbes benefit human health. Can the removal of microbes actually lead to disease development?

3. Trace the path taken by pathogens from various portals of entry, through the establishment of disease, and ending with their exit from various portals in the host.

4. Conduct additional research, then complete the following.
   a. Describe a historical example illustrating how quarantine was used to stop the spread of disease within a population.
   b. Discuss what policies are in place at local, state, and federal government levels to prohibit the spread of disease in the case of a pandemic.

5. There are some who believe that HIV does not cause AIDS. Have all of Koch's postulates been met for HIV as the causative agent that leads to the development of AIDS? Cite evidence to explain your answer.

6 Describe each type of infection in the following list and include the mode of transmission in each scenario. Use terms such as *primary, secondary, healthcare-associated, STI, mixed, latent, toxemia, chronic, zoonotic, asymptomatic, local, systemic,* and so on to describe the types of infections (more than one term may apply).

   hepatitis B infection caused by a needlestick during a dental procedure

   the development of *Pneumocystis* pneumonia in an AIDS patient

   bubonic plague acquired through the bite of a rat flea

   hantavirus pulmonary syndrome infection acquired while vacationing in a log cabin

   salmonellosis

   undiagnosed chlamydiosis

   mononucleosis transmitted via a shared drinking glass

   neonatal gonorrhea

7. Drugs that block quorum sensing are being tested for the potential treatment of bacterial infections. Provide an explanation for how such drugs may prohibit the establishment of an infection in a host.

8. Discuss the factors that enable the development of healthcare-associated infections. Explain which infection-control measures have been implemented successfully to decrease the spread of these infections.

9. Dr. John Snow is famous for bringing a cholera epidemic to an end in London by identifying the source of infection as a contaminated water pump used by many in the city.
    a. Based upon this information, explain what type of epidemic occurred and draw a curve exhibiting the incidence over time in this type of disease spread.
    b. Explain whether or not identifying an index case would have helped Dr. Snow in this situation.

10. Conduct additional research, then complete the following.
    a. List historical examples of bioterrorism worldwide.
    b. Discuss the goal of agroterrorism and potential agents that could be used by an agroterrorist.

Concept Connections | Bloom's Levels 4 and 6: Analyze and Create

This activity ties together multiple concepts in the chapter.

1. Fill in the blanks regarding the types of reservoirs.

2. How are incubating carriers different from asymptomatic carriers?

3. What type of carrier was Typhoid Mary?

4. Describe the difference between biological and mechanical vectors.

5. Name three common zoonoses.

6. Of all the potential reservoirs, living and nonliving, which has the greatest potential to spread infection to all age groups in the human population?

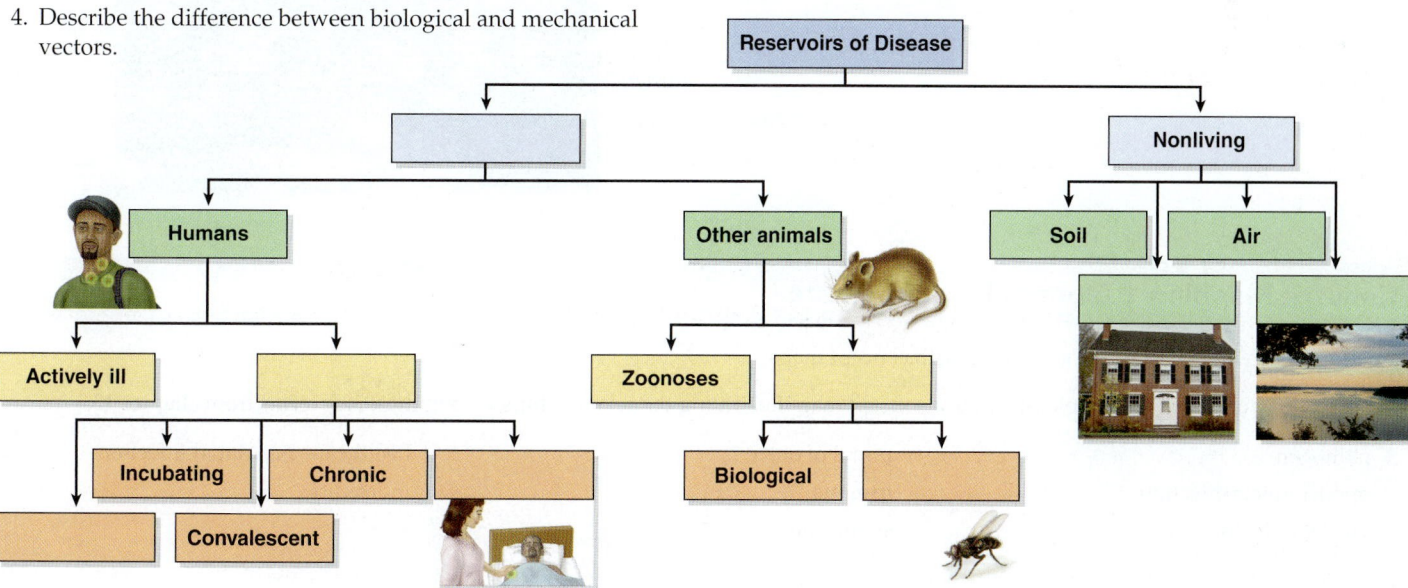

Visual Connections | Bloom's Level 5: Evaluate

These questions use visual images or previous content to make connections to this chapter's concepts.

1. **From chapter 3, figure 3.3a.** What chemical is the organism in this illustration producing? How does this add to an organism's pathogenicity?

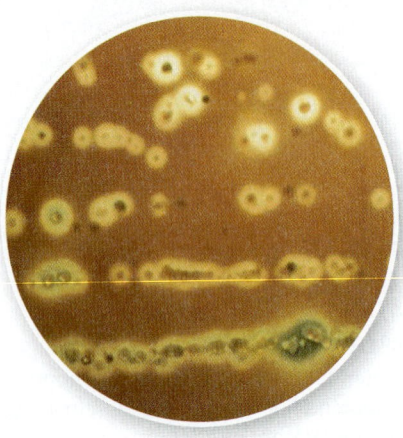

2. **From chapter 4, figure 4.15.** In what setting was this infection most likely acquired? What is this type of infection called? What could be the source of the pathogen?

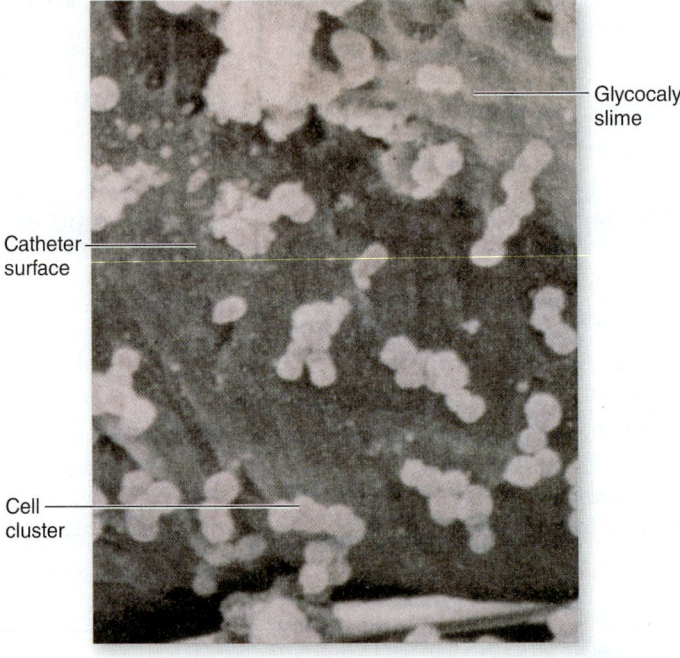

Glycocalyx slime

Catheter surface

Cell cluster

Concept Mapping | Bloom's Level 6: Create

Appendix D provides guidance for working with concept maps.

1. Using the words that follow, please create a concept map illustrating the relationships among these key terms from chapter 13.

pathogen

opportunistic infection

virulence factor

portal of entry

toxemia

syndrome

healthcare-associated infection

morbidity

www.mcgrawhillconnect.com

Enhance your study of this chapter with study tools and practice tests. Also ask your instructor about the resources available through ConnectPlus, including the media-rich eBook, interactive learning tools, and animations.

Host Defenses I
Overview and Nonspecific Defenses

Case File 14

Donations Accepted

According to 2012 statistics from the American Red Cross, every 2 seconds someone needs donated blood, and more than 44,000 units of blood are needed every day. On average, a patient needing a blood donation requires 3 pints of blood, and a car accident victim can require up to 100 pints of blood. The average adult has about 10 pints of blood in his or her body. On average, 1 pint of blood is given per donation, so up to 100 donors would be needed to save the life of a car accident victim. Patients with cancer, organ transplant recipients, and those with other diseases also require blood transfusions.

Four types of products are used from a blood donation: red blood cells; platelets; plasma; and cryoprecipitate, a frozen blood product prepared from plasma. Donors can give whole blood or specific blood components, but any and all donations are desperately needed. Healthy donors, especially those with rare or especially needed blood types, can donate whole blood every 56 days. Red blood cells can be donated every 16 weeks. The American Association of Blood Banks (AABB) estimates that 38% of the U.S. population is eligible to donate blood, but less than 10% donate. To be eligible to donate blood, a donor must be at least 16 years of age and at least 110 pounds. Supplies of blood fluctuate throughout the year, and blood supplies are especially low during the height of winter months. Regardless of the time of year, the need for donated blood is always critical.

- What types of tests are performed on donated blood?
- Who should not or cannot donate blood?

Continuing the Case appears on page 404.

Outline and Learning Outcomes

14.1 Defense Mechanisms of the Host in Perspective

The survival of the host depends upon an elaborate network of defenses that keep harmful microbes and other foreign materials from penetrating the body. Should they penetrate, additional host defenses are summoned to prevent them from becoming established in tissues. This chapter introduces the main lines of defense intrinsic to all humans. Topics included in this overview are the anatomical and physiological systems that detect, recognize, and destroy foreign substances and the general adaptive responses that account for an individual's long-term immunity or resistance to infection and disease.

In chapter 13, we explored the host-parasite relationship, with emphasis on the role of microorganisms in disease. In this chapter, we examine the other side of the relationship—that of the host defending itself against microorganisms. As previously stated, whether an encounter between a human and a microbe results in disease is dependent on many factors (see figure 13.2). The encounters occur constantly. In the battle against all sorts of invaders, microbial and otherwise, the body erects a series of barriers, sends in an army of cells, and emits a flood of chemicals to protect tissues from harm.

The host defenses are a multilevel network of innate, nonspecific protections and specific **immunities** referred to as the first, second, and third lines of defense **(figure 14.1)**. The interaction and cooperation of these three levels of defense normally provide complete protection against infection. The *first line of defense* includes any barrier that blocks invasion at the portal of entry. This mostly nonspecific line of

defense limits access to the internal tissues of the body. However, it is not considered a true immune response because it does not involve recognition of a specific foreign substance but is very general in action. The *second line of defense*, also nonspecific, is a more internalized system of protective cells and fluids that includes inflammation and phagocytosis. It acts rapidly at both the local and systemic levels once the first line of defense has been circumvented. The highly specific *third line of defense* is acquired on an individual basis as each foreign substance is encountered by white blood cells called **lymphocytes.** The reaction with each different microbe produces unique protective substances and cells that can come into play if that microbe is encountered again. The third line of defense provides long-term immunity, which is discussed in detail in chapter 15. This chapter focuses on the first and second lines of defense.

Humans are armed with various levels of defense that do not operate in a completely separate fashion; most defenses overlap and are even redundant in some of their effects. This literally bombards microbial invaders with an entire assault force, making their survival unlikely. Because of the interwoven nature of host defenses, we will introduce basic concepts of structure and function that will prepare you for later information on specific reactions of the immune system.

Barriers: A First Line of Defense

A number of defenses are a normal part of the body's anatomy and physiology. These inborn, nonspecific defenses can be divided into physical, chemical, and genetic barriers that impede the entry of not only microbes but also any foreign agent, whether living or not **(figure 14.2).**

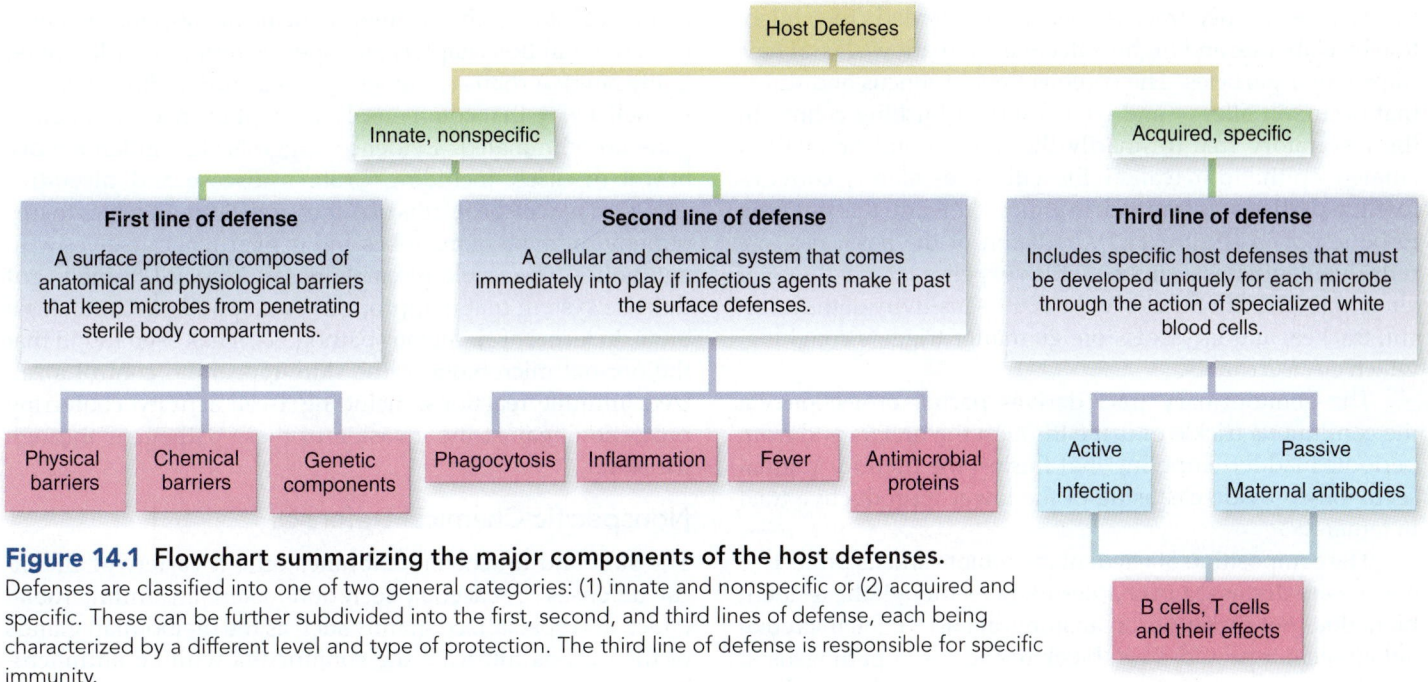

Figure 14.1 **Flowchart summarizing the major components of the host defenses.**
Defenses are classified into one of two general categories: (1) innate and nonspecific or (2) acquired and specific. These can be further subdivided into the first, second, and third lines of defense, each being characterized by a different level and type of protection. The third line of defense is responsible for specific immunity.

Physical or Anatomical Barriers at the Body's Surface

The skin and mucous membranes of the respiratory and digestive tracts have several built-in defenses. The outermost layer (stratum corneum) of the skin is composed of epithelial cells that have become compacted, cemented together, and impregnated with an insoluble protein—keratin. The result is a thick, tough layer that is highly impervious and waterproof. Few pathogens can penetrate this unbroken barrier, especially in regions such as the soles of the feet or the palms of the hands, where the stratum corneum is much thicker than on other parts of the body. It is so obvious as to be overlooked: The skin separates our inner bodies from the microbial assaults of the environment. It is a surprisingly tough and sophisticated barrier. The keratin in the top layer of cells is a protective and waterproofing protein. In addition, outer layers of skin are constantly sloughing off, taking associated microbes with them. Other cutaneous barriers include hair follicles and skin glands. The hair shaft is periodically extruded, and the follicle cells are **desquamated** (des'-kwuh-mayt-ud). The flushing effect of sweat glands also helps remove microbes.

The mucous membranes of the digestive, urinary, and respiratory tracts and of the eye are moist and permeable. They do provide barrier protection but without a keratinized layer. The mucous coat on the free surface of some membranes impedes the entry and attachment of bacteria. Blinking and tear production (lacrimation) flush the eye's surface with tears and rid it of irritants. The constant flow of saliva helps carry microbes into the harsh conditions of the stomach. Vomiting and defecation also evacuate noxious substances or microorganisms from the body.

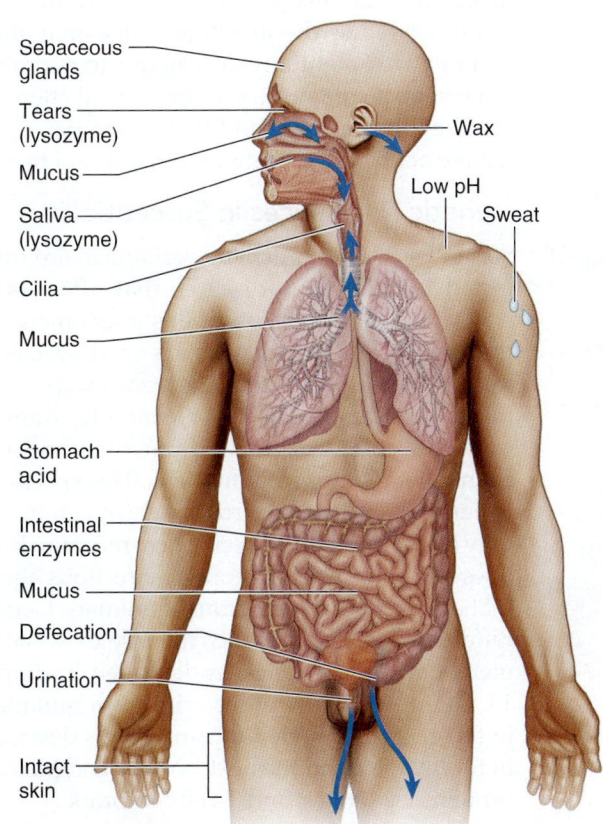

Figure 14.2 **The primary physical and chemical defense barriers.**

The respiratory tract is constantly guarded from infection by elaborate and highly effective adaptations. Nasal hair traps larger particles. The copious flow of mucus and fluids that occurs in allergy and colds exerts a flushing action. In the respiratory tree (primarily the trachea and bronchi), a ciliated epithelium (called the ciliary escalator) conveys foreign particles entrapped in mucus toward the pharynx to be removed (figure 14.3). Irritation of the nasal passage reflexively initiates a sneeze, which expels a large volume of air at high velocity. Similarly, the acute sensitivity of the bronchi, trachea, and larynx to foreign matter triggers coughing, which ejects irritants.

The genitourinary tract derives partial protection via the continuous trickle of urine through the ureters and from periodic bladder emptying that flushes the urethra. Vaginal secretions provide cleansing of the lower reproductive tract in females.

The composition of resident microbiota and its protective effect were discussed in chapter 13. Even though the resident biota does not constitute an anatomical barrier, the microbial antagonism it provides can block the access of pathogens to epithelial surfaces and can create an unfavorable environment for pathogens by competing for limited nutrients or by altering the local pH.

New research stemming from the Human Microbiome Project (HMP) has continued to highlight the importance of microbiota on the development of nonspecific defenses (described in this chapter) and specific immunity (described in the next). A robust commensal biota "trains" host defenses in such a way that commensals are kept in check and pathogens are eliminated. Evidence suggests that inflammatory bowel diseases, including Crohn's disease and ulcerative colitis, may well be results of our overzealous attempts to free our environment of microbes and to overtreat ourselves with antibiotics. The result, researchers say, is an "ill-trained" gut defense system that inappropriately responds to commensal biota. In terms of cutaneous pathogens, it has been found that the normal microbiota of the skin appears to control localized immune reactions, including T-cell activity, conferring protection against the invasion of these pathogens through the skin.

Nonspecific Chemical Defenses

The skin and mucous membranes offer a variety of chemical defenses. Sebaceous secretions exert an antimicrobial effect, and specialized glands such as the meibomian glands of the eyelids lubricate the conjunctiva with an antimicrobial secretion. An additional defense in tears and saliva is lysozyme, an enzyme that hydrolyzes the peptidoglycan in the cell wall of bacteria. The high lactic acid and electrolyte concentrations of sweat and the skin's acidic pH and fatty acid content are also inhibitory to many microbes. Likewise, the hydrochloric acid in the stomach renders protection against many pathogens that are swallowed, and the intestine's digestive juices and bile are potentially destructive to microbes. Even semen contains an antimicrobial chemical that inhibits bacteria, and the vagina has a protective acidic pH maintained by normal biota.

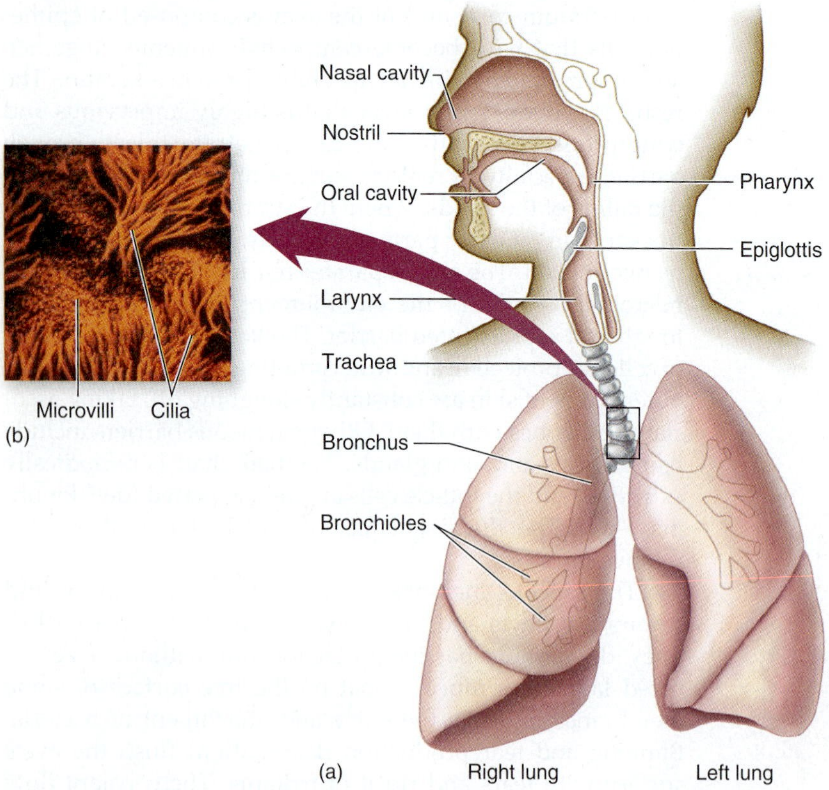

Genetic Differences in Susceptibility

Some hosts, due to genetic variations, are unaffected by infectious diseases that affect other hosts. One manifestation of this phenomenon is that some pathogens have such great specificity for one host species that they are incapable of infecting other species. For example, humans cannot acquire distemper from cats, and cats cannot get mumps from humans. This specificity is particularly true of viruses, which can invade only by attaching to a specific host receptor. But it does not hold true for zoonotic infectious agents that attack a broad spectrum of animals. Genetic differences in susceptibility can also exist within members of one species, as described in chapter 13. Often these differences arise from mutations in the genes that code for components described in this chapter and the next, such as complement proteins, cytokines, and T-cell receptors.

The vital contribution of barriers is clearly demonstrated in people who have lost them or never had them. Patients with severe skin

Nasal cavity
Nostril
Oral cavity
Pharynx
Epiglottis
Larynx
Trachea
Bronchus
Bronchioles
Microvilli   Cilia
(b)
(a)   Right lung   Left lung

Figure 14.3 The ciliary defense of the respiratory tree. (a) The epithelial lining of the airways contains a brush border of cilia to entrap and propel particles upward toward the pharynx. (b) Tracheal mucosa (5,000×).

damage due to burns are extremely susceptible to infections; those with blockages in the salivary glands, tear ducts, intestine, and urinary tract are also at greater risk for infection. But as important as it is, the first line of defense alone is not sufficient to protect against infection. Because many pathogens find a way to circumvent the barriers by using their virulence factors (discussed in chapter 13), a whole new set of defenses—inflammation, phagocytosis, specific immune responses—are brought into play.

14.1 Learning Outcomes—Assess Your Progress

1. Summarize the three lines of host defenses.
2. Identify three components of the first line of defense.
3. Discuss the role of normal biota as a first-line defense mechanism.

14.2 The Second and Third Lines of Defense: An Overview

Immunology encompasses the study of all features of the body's second and third lines of defense. Although this chapter is concerned, not surprisingly, with infectious microbial agents, be aware that immunology is central to the study of fields as diverse as cancer and allergy.

In the body, the mandate of the immune system can be easily stated. A healthy functioning immune system is responsible for

1. surveillance of the body,
2. recognition of foreign material, and
3. destruction of entities deemed to be foreign **(process figure 14.4).**

Because infectious agents could potentially enter through any number of portals, the cells of the immune system constantly move about the body, searching for potential pathogens. This process is carried out primarily by white blood cells, which have been trained to recognize body cells —so-called **self**—and differentiate them from any foreign material in the body, such as an invading bacterial cell— **nonself.** The ability to evaluate cells and macromolecules as either self or nonself is central to the functioning of the immune system. While foreign substances must be recognized as a potential threat and dealt with appropriately, self cells and chemicals must not come under attack by the immune defenses.

The immune system evaluates cells by examining certain molecules on their surfaces called **markers.**[1] These markers, which generally consist of proteins and/or sugars, can be thought of as the cellular equivalent of facial characteristics in humans and allow the cells of the immune system to identify whether or not a newly discovered cell poses a threat.

---

1. The term *marker* is also employed in genetics in a different sense—that is, to denote a detectable characteristic of a particular genetic mutant.

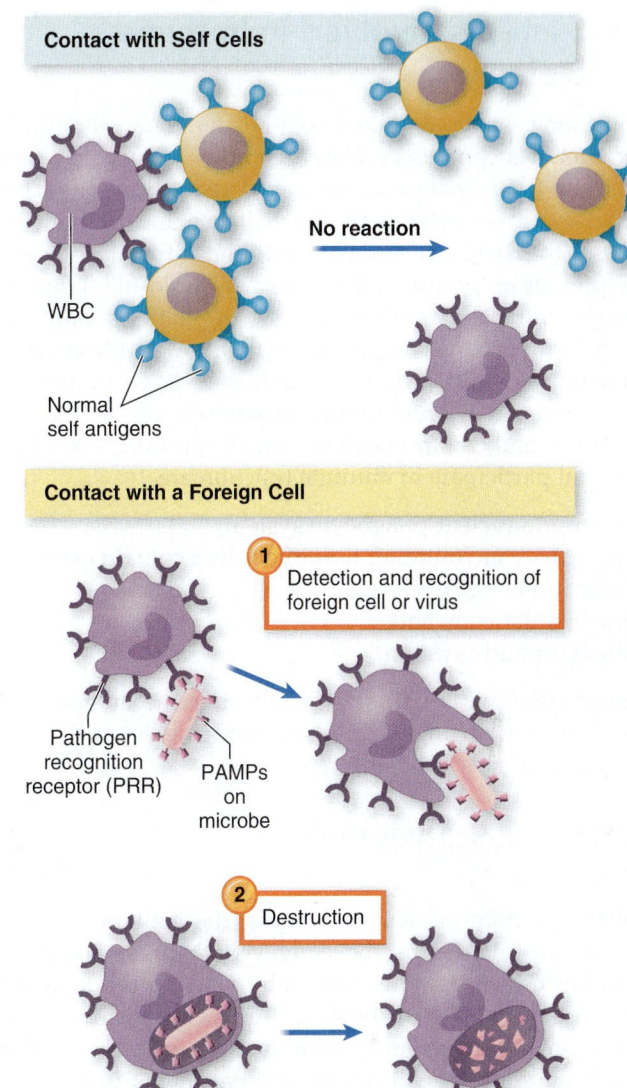

Process Figure 14.4 *Search, recognize,* and *destroy* is the mandate of the immune system.   White blood cells are equipped with a very sensitive sense of "touch." As they sort through the tissues, they feel surface markers that help them determine what is self and what is not. When self markers are recognized, no response occurs. However, when nonself is detected, a reaction to destroy it is mounted.

While cells deemed to be self are left alone, cells and other objects designated as foreign are marked for destruction by a number of methods, the most common of which is phagocytosis. There is a middle ground as well. Nonself proteins that are not harmful—such as those found in food we ingest and on commensal microorganisms—are generally recognized as such and the immune system is signaled to not react.

14.2 Learning Outcomes—Assess Your Progress

4. Define *marker,* and discuss its importance in the second and third lines of defense.

14.3 Systems Involved in Immune Defenses

The immune system does not exist in a single, well-defined site; rather, it encompasses a large, complex, diffuse network of cells and fluids that permeate every organ and tissue. This very arrangement enables the surveillance and recognition processes that help screen the body for harmful substances.

The body is partitioned into several fluid-filled spaces called the intracellular, extracellular, lymphatic, cerebrospinal, and circulatory compartments. Although these compartments are physically separated, they have numerous connections. Their structure and position permit extensive interchange and communication. Among the body compartments that participate in immune function are

1. the **mononuclear phagocyte system,**
2. the spaces surrounding tissue cells that contain extracellular fluid (ECF),
3. the bloodstream, and
4. the **lymphatic system.**

In the following section, we consider the anatomy of these main compartments and how they interact in the second and third lines of defense.

The Communicating Body Compartments

For effective immune responsiveness, the activities in one fluid compartment must be conveyed to other compartments. Let us see how this occurs by viewing tissue at the microscopic level (**figure 14.5**). At this level, clusters of tissue cells are in direct contact with the reticular cells and the extracellular fluid (ECF). Other compartments (vessels) present at this level are blood and lymphatic capillaries. This close association allows cells and chemicals that originate in the reticular system and ECF to diffuse or migrate into the blood and lymphatics; any products of a lymphatic reaction can be transmitted directly into the blood through the connection between these two systems; and certain cells and chemicals originating in the blood can move through the vessel walls into the extracellular spaces and migrate into the lymphatic system.

The flow of events among these systems depends on where an infectious agent or foreign substance first intrudes. A typical progression may begin in the extracellular spaces and reticular tissue, move to the lymphatic circulation, and ultimately end up in the bloodstream. Regardless of which compartment is first exposed, an immune reaction in any one of them will eventually be communicated to the others at the microscopic level. An obvious benefit of such an integrated system is that no cell of the body is far removed from competent protection, no matter how isolated. Let us take a closer look at each of these compartments.

Immune Functions of the Mononuclear Phagocyte System

The tissues of the body are permeated by a support network of connective tissue fibers, the reticular system, which originates in the cellular basal lamina, interconnects nearby cells, and meshes with the massive connective tissue network surrounding all organs. This network, called the **mononuclear phagocyte system** (MPS; **figure 14.6**), is intrinsic to the immune function because it provides a passageway within and between tissues and organs. The MPS consists of the thymus, where important white blood cells mature; lymph nodes; tonsils; spleen; and lymphoid tissue in the mucosa of the gut and respiratory tract, where most of the MPS "action" takes place. The lymphoid tissue in the gut is sometimes called **gut-associated lymphoid tissue (GALT);** more generally, lymphoid tissue associated with the mucosal surfaces anywhere is called **mucosa-associated lymphoid tissue (MALT).** The MPS is a source of white blood cells called macrophages waiting to attack passing foreign intruders as they arrive in the skin, lungs, liver, lymph nodes, spleen, and bone marrow.

Components and Functions of the Lymphatic System

The lymphatic system is a compartmentalized network of vessels, cells, and specialized accessory organs (**figure 14.7**). It begins in the farthest reaches of the tissues as tiny capillaries

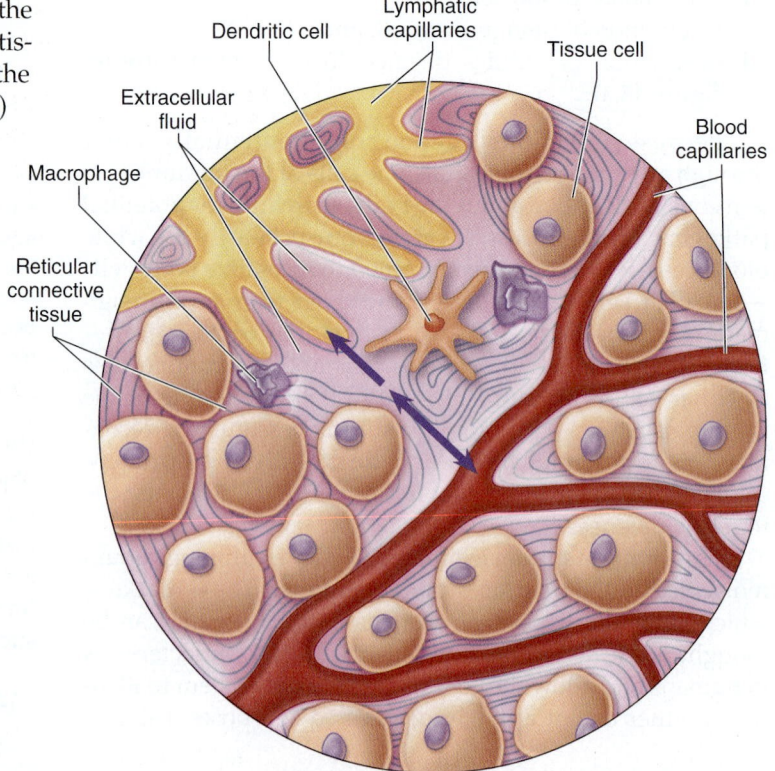

Figure 14.5 Connections between the body compartments.
The meeting of the major fluid compartments at the microscopic level.

Lymphatic Fluid Lymph is a plasmalike liquid carried by the lymphatic circulation. It is formed when certain blood components move out of the blood vessels into the extracellular spaces and diffuse or migrate into the lymphatic capillaries. Lymph is made up of water, dissolved salts, and 2% to 5% protein (especially antibodies and albumin). Like blood, it transports numerous white blood cells (especially lymphocytes) and miscellaneous materials such as fats, cellular debris, and infectious agents that have gained access to the tissue spaces.

Lymphatic Vessels The system of vessels that transports lymph is constructed along the lines of blood vessels. Because the lymph is never subjected to high pressure, the lymphatic vessels appear most similar to thin-walled veins rather than thicker-walled arteries. The tiniest vessels, lymphatic capillaries, accompany the blood capillaries and permeate all parts of the body except the central nervous system and certain organs such as bone, placenta, and thymus. Their thin walls are easily permeated by extracellular fluid that has escaped from the circulatory system. Lymphatic vessels are found in particularly high numbers in the hands, feet, and breast.

In the next section, you will read about the bloodstream and blood vessels. Two overriding differences between the bloodstream and the lymphatic system should be mentioned. First, because one of the main functions of the lymphatic system is returning lymph to the circulation, the flow of lymph is in one direction only, with lymph moving from the extremities toward the heart. Eventually, lymph will be returned to the bloodstream through the thoracic duct or the right lymphatic duct to the subclavian vein near the heart. The second difference concerns how lymph travels through the vessels of the lymphatic system. While blood is transported through the body by means of a dedicated pump (the heart), lymph is moved only through the contraction of the skeletal muscles through which the lymphatic ducts wend their way. This dependence on muscle movement helps to explain the swelling of the hands and feet that sometimes occurs during the night (when muscles are inactive) yet dissipates soon after waking.

Lymphoid Organs and Tissues Other organs and tissues that perform lymphoid functions are the thymus, lymph nodes (glands), spleen, and clusters of tissues that appear in mucosal surfaces (MALT). A trait common to these organs is a loose connective tissue framework that houses aggregations of lymphocytes, the important class of white blood cells mentioned previously.

The Thymus: Site of T-Cell Maturation The **thymus** originates in the embryo as two lobes in the pharyngeal region that fuse into a triangular structure. The size of the thymus is greatest proportionately at birth **(figure 14.8),** and it continues to exhibit high rates of activity and growth until puberty, after which it shrinks gradually through

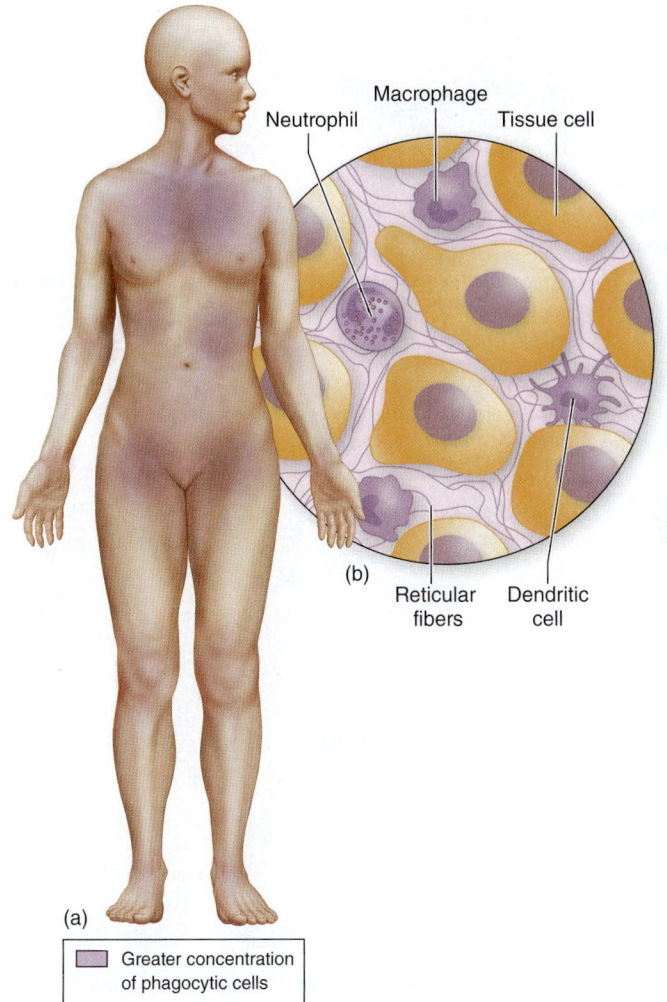

(a)

Neutrophil
Macrophage
Tissue cell
(b)
Reticular fibers
Dendritic cell

■ Greater concentration of phagocytic cells

Figure 14.6 The mononuclear phagocyte system is a pervasive, continuous connective tissue framework throughout the body. **(a)** The degrees of shading in the body indicate variations in phagocyte concentration (darker = greater). **(b)** This system begins at the microscopic level with a fibrous support network (reticular fibers) enmeshing each cell. This web connects one cell to another within a tissue or organ and provides a niche for phagocytic white blood cells, which can crawl within and between tissues.

that transport a special fluid (lymph) through an increasingly larger tributary system of vessels and filters (lymph nodes), and it leads to major vessels that drain back into the regular circulatory system. Some major functions of the lymphatic system are

1. to provide an auxiliary route for the return of extracellular fluid to the circulatory system proper;
2. to act as a "drain-off" system for the inflammatory response; and
3. to render surveillance, recognition, and protection against foreign materials through a system of lymphocytes, phagocytes, and antibodies.

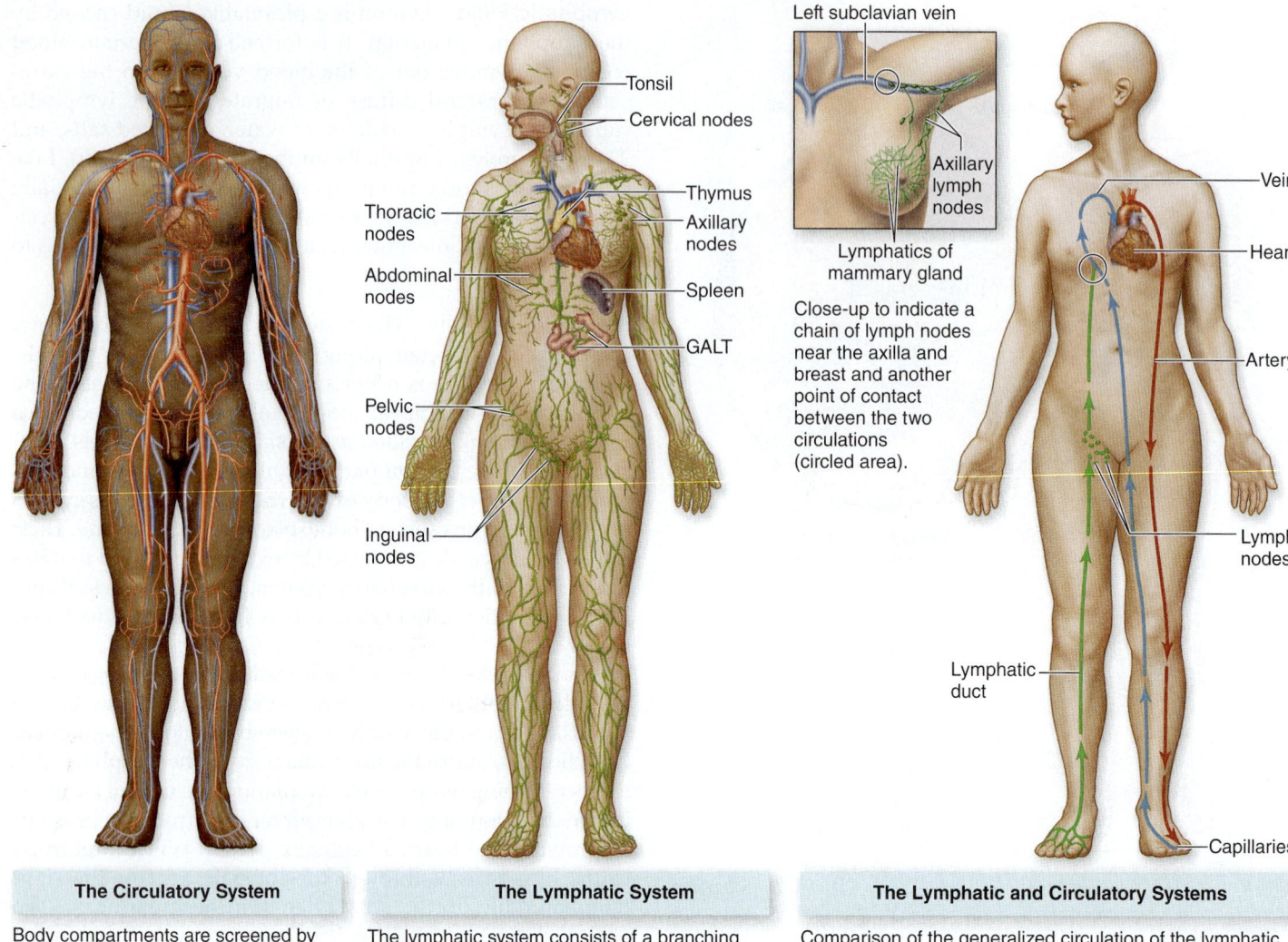

The Circulatory System

Body compartments are screened by circulating WBCs in the cardiovascular system.

The Lymphatic System

The lymphatic system consists of a branching network of vessels that extend into most body areas. Note the higher density of lymphatic vessels in the "dead-end" areas of the hands, feet, and breast, which are frequent contact points for infections. Other lymphatic organs include the lymph nodes, spleen, gut-associated lymphoid tissue (GALT), the thymus, and the tonsils.

The Lymphatic and Circulatory Systems

Comparison of the generalized circulation of the lymphatic system and the blood. Although the lymphatic vessels parallel the regular circulation, they transport in only one direction unlike the cyclic pattern of blood. Direct connection between the two circulations occurs at points near the heart where large lymph ducts empty their fluid into veins (circled area).

Figure 14.7 The circulatory and lymphatic systems.

adulthood. Under the influence of thymic hormones, thymocytes develop specificity and are released into the circulation as mature T cells. The T cells subsequently migrate to and settle in other lymphoid organs (e.g., the lymph nodes and spleen), where they occupy the specific sites described previously.

Children born without a thymus (DiGeorge syndrome, see chapter 16) or who have had their thymus surgically removed are severely immunodeficient and fail to thrive. Adults have developed enough mature T cells that removal of the thymus or reduction in its function has milder effects. Do not confuse the thymus with the thyroid gland, which is located nearby but has an entirely different function.

Lymph Nodes   Lymph nodes are small, encapsulated, bean-shaped organs stationed, usually in clusters, along lymphatic channels and large blood vessels of the thoracic and abdominal cavities (see figure 14.7). Major aggregations of nodes occur in the loose connective tissue of the armpit (axillary nodes), groin (inguinal nodes), and neck (cervical nodes). Both the location and architecture of these nodes make them ideal for filtering out materials that have entered the lymph and providing appropriate cells and niches for immune reactions.

The Spleen   The spleen is a lymphoid organ in the upper left portion of the abdominal cavity. It is somewhat similar to a lymph node except that it serves as a filter for blood

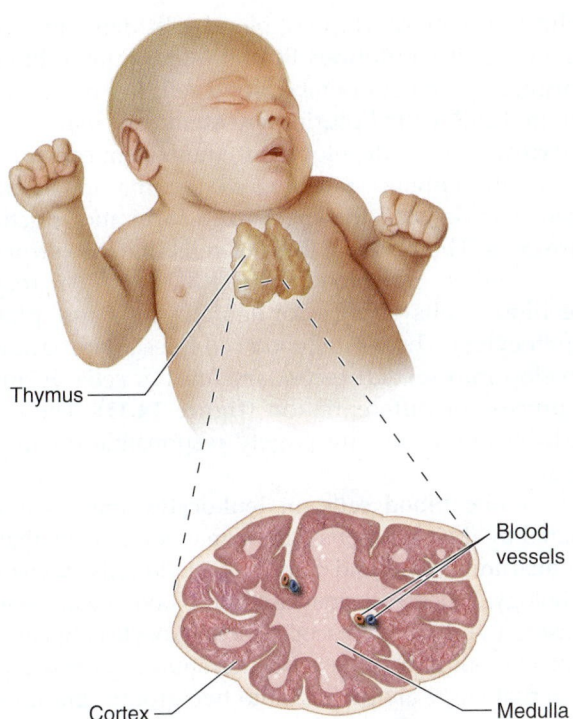

Thymus

Blood vessels

Cortex

Medulla

Figure 14.8 The thymus.   Immediately after birth, the thymus is a large organ that nearly fills the region over the midline of the upper thoracic region. In the adult, however, it is proportionately smaller (to compare, see figure 14.7). The section shows the main anatomical regions of the thymus. Immature T cells enter through the cortex and migrate into the medulla as they mature.

instead of lymph. While the spleen's primary function is to remove worn-out red blood cells from circulation, its most important immunologic function centers on the filtering of pathogens from the blood and their subsequent phagocytosis by resident macrophages. Although adults whose spleens have been surgically removed can live a relatively normal life, asplenic children are severely immunocompromised.

Miscellaneous Lymphoid Tissue   At many sites on or just beneath the mucosa of the gastrointestinal and respiratory tracts lie discrete bundles of lymphocytes. The positioning of this diffuse system provides an effective first-strike potential against the constant influx of microbes and other foreign materials in food and air. In the pharynx, a ring of tissues called the tonsils provides an active source of lymphocytes. The breasts of pregnant and lactating women also become temporary sites of antibody-producing lymphoid tissues. The intestinal tract houses GALT, the best-developed collection of lymphoid tissue. Examples of GALT include the appendix; the lacteals (special lymphatic vessels stationed in each intestinal villus); and **Peyer's patches,** compact aggregations of lymphocytes in the ileum of the small intestine. GALT provides immune functions against intestinal pathogens and is a significant source of some types of antibodies. Other, less well-organized collections of secondary

lymphoid tissue include the mucosal-associated lymphoid tissue (MALT), skin-associated lymphoid tissue (SALT), and bronchial-associated lymphoid tissue (BALT).

Disease Connection

New therapies for HIV may be on the horizon in light of new research showing that the diarrhea and malnutrition accompanying the infection appear to be due to viral disruption of GALT activity.

Origin, Composition, and Functions of the Blood

The circulatory system consists of the circulatory system proper, which includes the heart, arteries, veins, and capillaries that circulate the blood, and the lymphatic system, which includes lymphatic vessels and lymphatic organs (lymph nodes) that circulate lymph. These two circulations parallel, interconnect with, and complement one another.

The substance that courses through the arteries, veins, and capillaries is **whole blood,** a liquid consisting of **blood cells** (formed elements) suspended in **plasma.** One can visualize these two components with the naked eye when a tube of unclotted blood is allowed to sit or is spun in a centrifuge. The cells' density causes them to settle into an opaque layer at the bottom of the tube, leaving the plasma, a clear, yellowish fluid, on top **(figure 14.9).** In chapter 15, we introduce the concept of **serum.** This substance is essentially the same as plasma, and contains no clotting factors. Serum is often used in immune testing and therapy.

Fundamental Characteristics of Plasma   Plasma contains hundreds of different chemicals produced by the liver, white blood cells, endocrine glands, and nervous system and

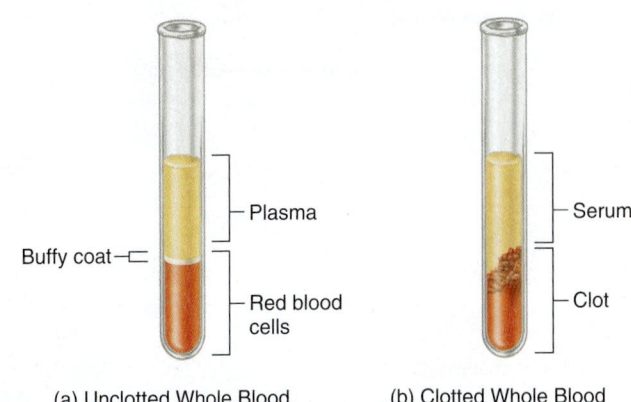

Plasma

Buffy coat

Red blood cells

Serum

Clot

(a) Unclotted Whole Blood

(b) Clotted Whole Blood

Figure 14.9 The macroscopic composition of whole blood.   **(a)** When blood containing anticoagulants is allowed to sit for a period, it stratifies into a clear layer of plasma; a thin layer of off-white material called the buffy coat (which contains the white blood cells); and a layer of red blood cells in the bottom, thicker layer. **(b)** Serum is the clear fluid that separates from clotted blood.

absorbed from the digestive tract. The main component of this fluid is water (92%), and the remainder consists of proteins such as albumin and globulins (including antibodies); other immunochemicals; fibrinogen and other clotting factors; hormones; nutrients (glucose, amino acids, fatty acids); ions (sodium, potassium, calcium, magnesium, chloride, phosphate, bicarbonate); dissolved gases (O_2 and CO_2); and waste products (urea). These substances support the normal physiological functions of nutrition, development, protection, homeostasis, and immunity. We return to the subject of plasma and its function in immune interactions later in this chapter and in chapter 15.

A Survey of Blood Cells   The production of blood cells, or **hematopoiesis** (hee"-mat-o-poy-ee'-sis), begins early in embryonic development in the yolk sac (an embryonic membrane). Later, most of it is taken over by the liver and lymphatic organs and is finally assumed permanently by the red bone marrow **(figure 14.10).** Although much of a newborn's red marrow is devoted to hematopoietic function, the active marrow sites gradually recede; by the age of 4 years, only the ribs, sternum, pelvic girdle, flat bones of the skull and spinal column, and proximal portions of the humerus and femur are devoted to blood cell production.

The relatively short life of blood cells demands a rapid turnover that is continuous throughout a human life span. The primary precursor of new blood cells is a pool of undifferentiated cells called pluripotent **stem cells** maintained in the marrow. During development, these stem cells proliferate and differentiate—meaning that immature or unspecialized cells develop the specialized form and function of mature cells. The primary lines of cells that arise from this process produce red blood cells (RBCs, or erythrocytes), white blood cells (WBCs, or leukocytes), and platelets (thrombocytes). The white blood cell lines are programmed to develop into several secondary lines of cells during the final process of differentiation **(figure 14.11).** These committed lines of WBCs are largely responsible for immune function.

The **white blood cells,** or **leukocytes,** are traditionally evaluated by their reactions with hematologic stains that contain a mixture of dyes and can differentiate cells by color and morphology. When this stain used on blood smears is evaluated using the light microscope, the leukocytes appear either with or without noticeable colored granules in the cytoplasm and, on that basis, are divided into two groups: **granulocytes** and **agranulocytes.** Greater magnification reveals that even the agranulocytes have tiny granules in their cytoplasm, so some

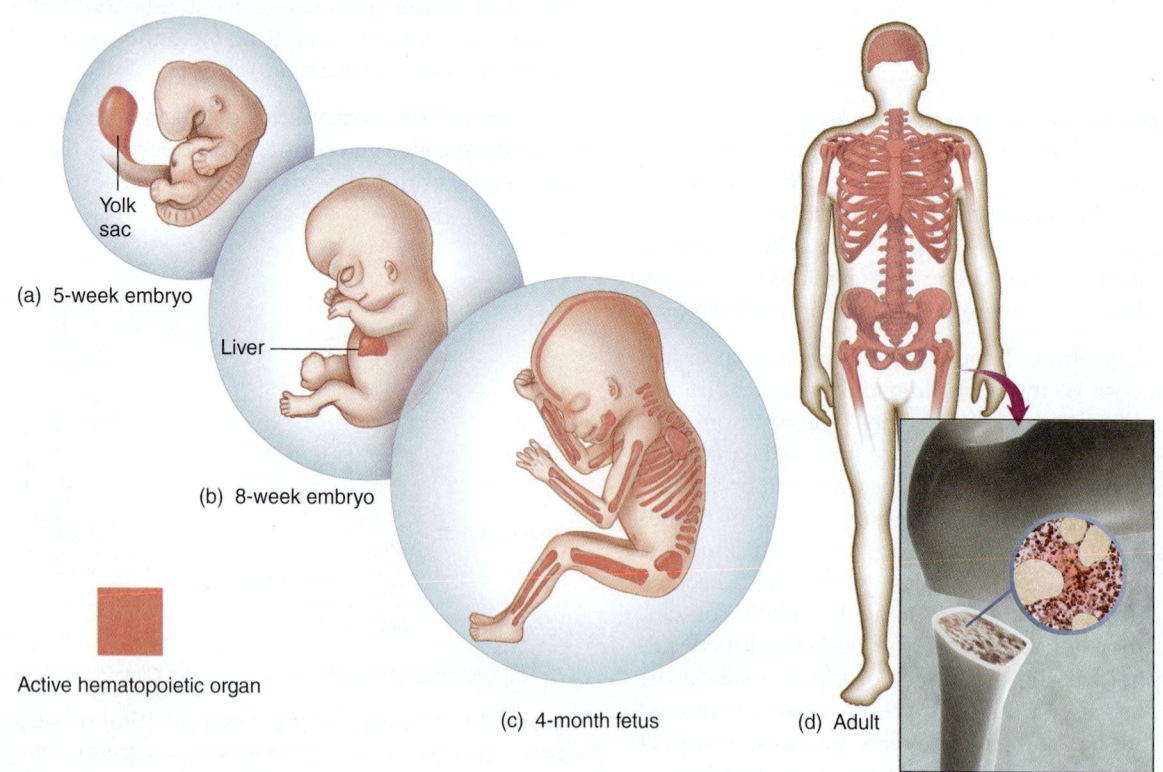

Yolk sac

(a) 5-week embryo

Liver

(b) 8-week embryo

Active hematopoietic organ

(c) 4-month fetus

(d) Adult

Figure 14.10 Stages in hematopoiesis.   The sites of blood cell production change as development progresses from **(a, b)** yolk sac and liver in the embryo to **(c)** extensive bone marrow sites in the fetus and **(d)** selected bone marrow sites in the child and adult. **(Inset)** Red marrow occupies the spongy bone (circle) in these areas.

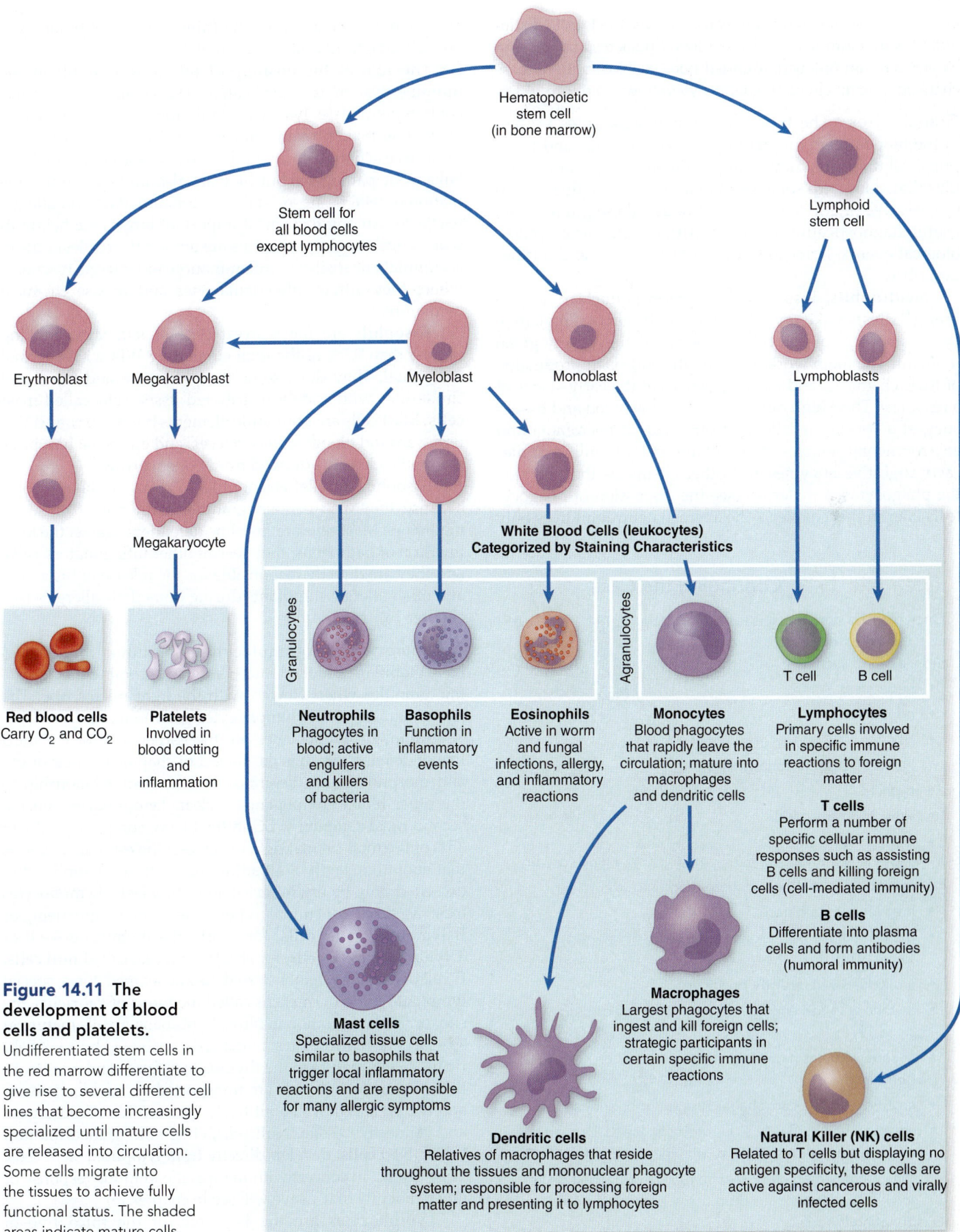

White Blood Cells (leukocytes) Categorized by Staining Characteristics

Granulocytes

Agranulocytes

T cell   B cell

Red blood cells
Carry O$_2$ and CO$_2$

Platelets
Involved in blood clotting and inflammation

Neutrophils
Phagocytes in blood; active engulfers and killers of bacteria

Basophils
Function in inflammatory events

Eosinophils
Active in worm and fungal infections, allergy, and inflammatory reactions

Monocytes
Blood phagocytes that rapidly leave the circulation; mature into macrophages and dendritic cells

Lymphocytes
Primary cells involved in specific immune reactions to foreign matter

T cells
Perform a number of specific cellular immune responses such as assisting B cells and killing foreign cells (cell-mediated immunity)

B cells
Differentiate into plasma cells and form antibodies (humoral immunity)

Mast cells
Specialized tissue cells similar to basophils that trigger local inflammatory reactions and are responsible for many allergic symptoms

Macrophages
Largest phagocytes that ingest and kill foreign cells; strategic participants in certain specific immune reactions

Dendritic cells
Relatives of macrophages that reside throughout the tissues and mononuclear phagocyte system; responsible for processing foreign matter and presenting it to lymphocytes

Natural Killer (NK) cells
Related to T cells but displaying no antigen specificity, these cells are active against cancerous and virally infected cells

Figure 14.11 The development of blood cells and platelets.
Undifferentiated stem cells in the red marrow differentiate to give rise to several different cell lines that become increasingly specialized until mature cells are released into circulation. Some cells migrate into the tissues to achieve fully functional status. The shaded areas indicate mature cells.

hematologists also use the appearance of the nucleus to distinguish them. Granulocytes have a lobed nucleus, and agranulocytes have an unlobed, rounded nucleus. Note both of these characteristics in circulating leukocytes shown in figure 14.11.

Granulocytes　The types of granular leukocytes present in the bloodstream are neutrophils, eosinophils, and basophils. All three are known for prominent cytoplasmic granules that stain with some combination of acidic dye (eosin) or basic dye (methylene blue). Although these granules are useful diagnostically, they also function in numerous physiological events. Refer to figure 14.11 to view the cell types described.

Neutrophils, also called polymorphonuclear neutrophils (PMNs), make up 55% to 90% of the circulating leukocytes—about 25 billion cells in the circulation at any given moment. The main work of the neutrophils is in production of toxic chemicals and in phagocytosis at the early stages of a response. Their high numbers in both the blood and tissues suggest a constant challenge from resident microbiota and environmental sources. Most of the cytoplasmic granules carry digestive enzymes and other chemicals that degrade the phagocytosed materials (see the discussion of phagocytosis later in this chapter). The average neutrophil lives only about 8 days, spending much of this time in the tissues and only about 6 to 12 hours in circulation.

The role of the **eosinophil** (ee"-oh-sin'-oh-fil) in the immune system is complicated. The eosinophil granules contain peroxidase, lysozyme, and other digestive enzymes, as well as toxic proteins and inflammatory chemicals. The protective action of eosinophils is to attack and destroy large eukaryotic pathogens, but they are also involved in the formation of fetal tissue as well as in inflammation and allergic reactions. Among their most important targets are helminth worms and fungi. Eosinophils are among the earliest cells to accumulate near sites of inflammation and allergic reactions, where they attract other leukocytes and release chemical mediators.

Basophils are the scarcest type of leukocyte, making up less than 0.5% of the total circulating WBCs in a normal individual. They share some morphological and functional similarities with widely distributed tissue cells called **mast cells.** Mast cells are nonmotile elements bound to connective tissue around blood vessels, nerves, and epithelia; basophils are motile elements derived from bone marrow.

Basophils parallel eosinophils in many of their actions, because they also contain granules with potent chemical mediators. Mast cells are first-line defenders against the local invasion of pathogens; they recruit other inflammatory cells; and they are directly responsible for the release of histamine and other allergic stimulants during immediate allergies (see chapter 16).

Agranulocytes　Agranulocytes (agranular leukocytes) have globular, nonlobed nuclei and lack prominent cytoplasmic granules when viewed with the light microscope. The two general types are lymphocytes and monocytes.

Although lymphocytes are the cornerstone of the third line of defense, which is the subject of chapter 15, their origin and morphology are described here so their relationship to the other blood components is clear. **Lymphocytes** are the second most common WBC in the blood, comprising 20% to 35% of the total circulating leukocytes. One estimate suggests that about one-tenth of all adult body cells are lymphocytes, exceeded only by erythrocytes and fibroblasts. Lymphocytes exist as three functional types—the bursal-equivalent, or **B lymphocytes** (**B cells,** for short); the thymus-derived, or **T lymphocytes** (**T cells,** for short); and a set called **null cells.** B cells were first demonstrated in and named for a special lymphatic gland of chickens called the **bursa of Fabricius,** the site for their maturation in birds. In humans, B cells mature in special bone marrow sites; humans do not have a bursa of Fabricius. T cells mature in the thymus in all birds and mammals. Both populations of cells are transported by the bloodstream and lymph and move about freely between lymphoid organs and connective tissue. Null cells, chief among them **natural killer (NK) cells,** develop directly from lymphoid stem cells. They can act in concert with the specific immune response or independently of it, as we will see in chapter 15.

Lymphocytes are the key cells of the third line of defense—the specific immune response. When stimulated by

Case File 14　*Continuing the Case*

Once blood is donated, it is tested for specific infectious diseases that are transmitted via blood such as

- hepatitis B,
- hepatitis C,
- HIV,
- human T-cell leukemia viruses,
- syphilis,
- West Nile virus, and
- Trypanosoma cruzi.

According to the American Association of Blood Banks (AABB), individuals who should not donate blood include

- intravenous drug users;
- men who have had sexual contact with other men since 1977;
- individuals who are HIV positive;
- anyone who has engaged in sex for money or drugs since 1977;
- anyone who has had hepatitis since his or her 11th birthday;
- anyone who has had Chagas disease or babesiosis; and
- anyone who has Creutzfeldt-Jakob disease (CJD), who has a family member with CJD, or who may be at risk for CJD.

■ Why are gay men or those who may have had exposure to HIV (i.e., intravenous drug users) prevented from donating blood?

foreign substances (antigens), lymphocytes are transformed into activated cells that neutralize and destroy those foreign substances. The contribution of B cells is mainly in **antibody-mediated immunity,** defined as protective molecules carried in the fluids of the body; for this reason, it used to be called "humoral immunity," as in "in the humors." When activated B cells divide, they form specialized **plasma cells,** which produce **antibodies,** large protein molecules that interlock with an antigen and participate in its destruction. Activated T cells engage in a spectrum of immune functions characterized as **cell-mediated immunity** in which T cells modulate immune functions and kill foreign cells. The action of both classes of lymphocytes accounts for the recognition and memory typical of immunity.

Monocytes are generally the largest of all white blood cells and the third most common in the circulation (3% to 7%). Their cytoplasm holds many fine vacuoles containing digestive enzymes. Monocytes are discharged by the bone marrow into the bloodstream, where they live as phagocytes for a few days. Later, they leave the circulation to undergo final differentiation into **macrophages.** Unlike many other WBCs, the monocyte series is relatively long-lived and retains an ability to multiply. Macrophages are among the most versatile and important of cells. In general, they are responsible for

1. many types of specific and nonspecific phagocytic and killing functions (they assume the job of cellular housekeepers, "mopping up the messes" created by infection and inflammation);
2. processing foreign molecules and presenting them to lymphocytes; and
3. secreting biologically active compounds that assist, mediate, attract, and inhibit immune cells and reactions.

We touch upon these functions in several ensuing sections.

Another product of the monocyte cell line is the **dendritic cell (figure 14.12),** named for its long, thin cell processes.

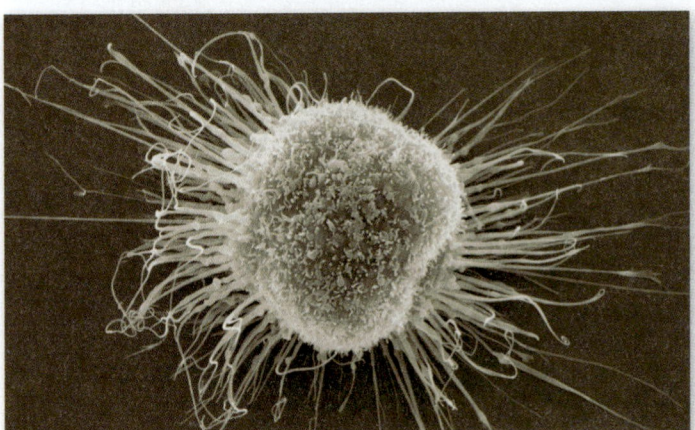

Figure 14.12 A dendritic cell. Dendritic cells reside in most tissue where they survey their local environments for pathogens and altered host cells (infected and cancerous cells).

Immature dendritic cells move from the blood to the MPS and lymphatic tissues, where they trap pathogens. Ingestion of bacteria and viruses stimulates dendritic cells to migrate to lymph nodes and the spleen. Here, they mature into highly effective processors and presenters of foreign proteins (see chapter 15).

Erythrocyte and Platelet Lines   These elements stay in the circulatory system proper. Their development is also shown in figure 14.11. **Erythrocytes** develop from stem cells in the bone marrow and lose their nucleus just prior to entering the circulation. The resultant red blood cells are simple, biconcave sacs of hemoglobin that transport oxygen and carbon dioxide to and from the tissues. These are the most numerous of circulating blood cells, appearing in stains as small pink circles. Red blood cells do not ordinarily have immune functions, though they can be the target of immune reactions (see chapter 16).

Platelets are sticky cell fragments in circulating blood that are *not* whole cells. In stains, platelets are blue-gray with fine red granules and are readily distinguished from cells by their small size. Until recently, it was believed that platelets' main function was in hemostasis (plugging broken blood vessels to stop bleeding) and in releasing chemicals that act in blood clotting and inflammation. These are important functions of platelets, but new studies show that platelets also play a role in immunity. These sticky fragments attach to bloodborne bacteria, which tags them for transport to the spleen, where they sensitize cells that trigger a specific immune response. Platelets "gone rogue" have also been implicated in exacerbating inflammation in autoimmune diseases and tumor spreading in cancer.

14.3 Learning Outcomes—Assess Your Progress

5. Name four body compartments that participate in immunity.
6. List the components of the mononuclear phagocyte system.
7. Fully describe the structure and function of the lymphatic system.
8. Differentiate between whole blood, plasma, and serum.
9. Name six types of blood cells that function in nonspecific immunity, and specify the most important function of each.
10. Describe the major characteristics of the two major types of lymphocytes involved in specific immunity.

14.4 The Second Line of Defense

Now that we have introduced the principal anatomical and physiological framework of the immune system, we address some mechanisms that play important roles in host defenses: (1) phagocytosis, (2) inflammation, (3) fever, and (4) antimicrobial proteins. Because of the generalized nature of these defenses, they are primarily nonspecific in their effects, but they also support and interact with the specific immune responses described in chapter 15.

Phagocytosis: Cornerstone of Inflammation and Specific Immunity

By any standard, a phagocyte represents an impressive piece of living machinery, meandering through the tissues to seek, capture, and destroy a target. The general activities of phagocytes are

1. to survey the tissue compartments and discover microbes, particulate matter (dust, carbon particles, antigen-antibody complexes), and injured or dead cells;
2. to ingest and eliminate these materials; and
3. to recognize immunogenic information (antigens) in foreign matter.

It is generally accepted that all cells have some capacity to engulf materials, but professional **phagocytes** do it for a living. The three main types of phagocytes are neutrophils, monocytes, and macrophages.

Neutrophils and Eosinophils

As previously stated, neutrophils are general-purpose phagocytes that react early in the inflammatory response to bacteria, other foreign materials, and damaged tissue. A common sign of bacterial infection is a high neutrophil count in the blood (neutrophilia), and neutrophils are also a primary component of pus. Eosinophils are attracted to sites of parasitic infections and antigen-antibody reactions, though they play only a minor phagocytic role.

Monocytes and Macrophages: Kings of the Phagocytes

After emigrating out of the bloodstream into the tissues due to chemical stimuli, monocytes are transformed by various inflammatory mediators into macrophages **(figure 14.13).** This process is marked by an increase in size and by enhanced development of lysosomes and other organelles **(figure 14.14).** At one time, macrophages were classified as either *fixed* (adherent to tissue) or *wandering*, but this terminology can be misleading. All macrophages retain the capacity to move about. Whether they reside in a specific organ or wander depends upon their stage of development and the immune stimuli they receive. Specialized macrophages called **histiocytes** (*histio* = "tissue") migrate to a certain tissue and remain there during their life span. Examples are alveolar (lung) macrophages; the Kupffer cells in the liver; dendritic cells in the skin (see figure 14.12); and macrophages in the spleen, lymph nodes, bone marrow, kidney, bone, and brain. Other macrophages do not reside permanently in a particular tissue and drift nomadically throughout the MPS. Not only are macrophages dynamic scavengers **(Insight 14.1),** but they also process foreign substances and prepare them for reactions with B and T lymphocytes.

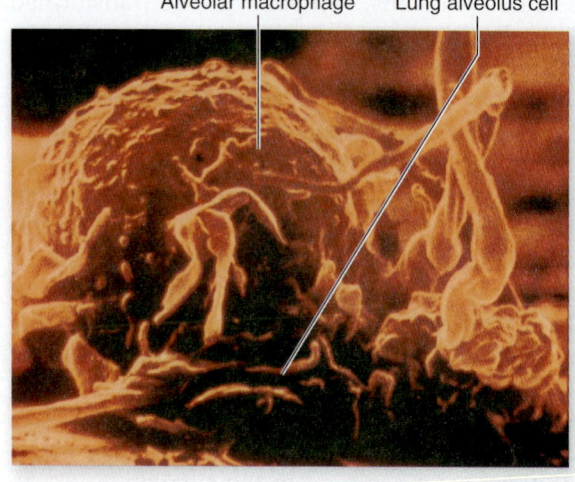
Alveolar macrophage   Lung alveolus cell

(a)

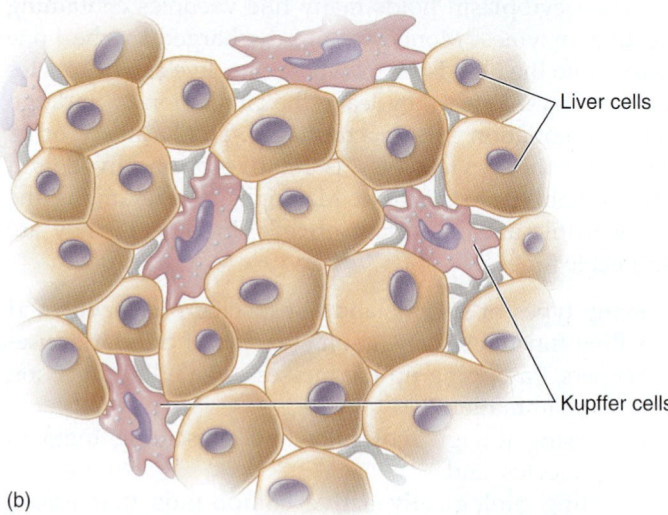
Liver cells

Kupffer cells

(b)

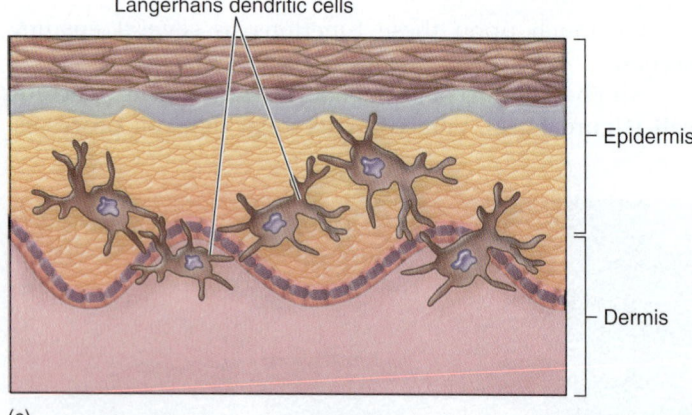
Langerhans dendritic cells

Epidermis

Dermis

(c)

Figure 14.13 Sites containing macrophages. **(a)** Scanning electron micrograph of a lung with an alveolar macrophage. **(b)** Liver tissue with Kupffer cells. **(c)** Langerhans cells deep in the epidermis.

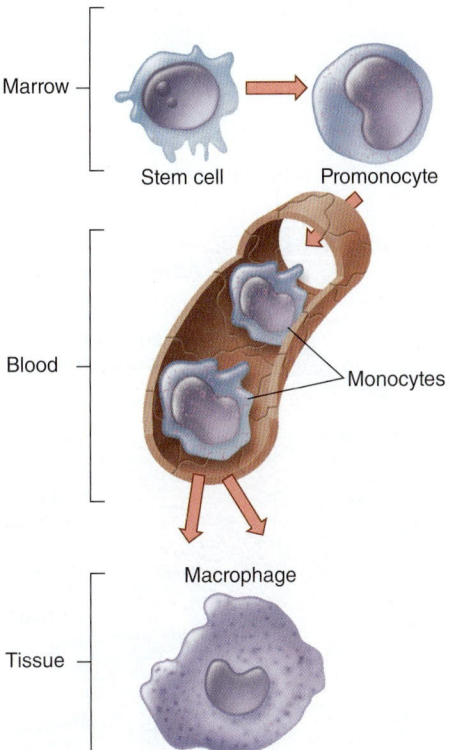

Marrow

Stem cell   Promonocyte

Blood

Monocytes

Macrophage

Tissue

Figure 14.14 The developmental stages of monocytes and macrophages. The cells progress through maturational stages in the bone marrow and peripheral blood. Once in the tissues, a macrophage can remain nomadic or take up residence in a specific organ.

Mechanisms of Phagocytic Recognition, Engulfment, and Killing

The term phagocyte literally means "eating cell." But *phagocytosis* (the term for what phagocytes do) is more than just the physical process of engulfment, because phagocytes also actively attack and dismantle foreign cells with a wide array of antimicrobial substances. Phagocytosis can occur as an isolated event performed by a lone phagocytic cell responding to a minor irritant in its area or as part of the orchestrated events of inflammation described in the next section. The events in phagocytosis include chemotaxis, ingestion, phagolysosome formation, destruction, and excretion (**process figure 14.15**).

Chemotaxis and Ingestion   Phagocytes and other defensive cells are able to recognize many microorganisms as foreign because of various signal molecules that the microbes have on their surfaces. Some important examples of these are called **pathogen-associated molecular patterns (PAMPs).** They are molecules shared by many microorganisms—but not present in mammals—and therefore serve as "red flags" for phagocytes and other cells of innate immunity.

Bacterial PAMPs include peptidoglycan and lipopolysaccharide. Double-stranded RNA, which is found only in some viruses, is also a PAMP. On the host side, phagocytes, dendritic cells, endothelial cells, and even lymphocytes possess molecules on their surfaces—called **pattern recognition receptors (PRRs)**—that recognize and bind PAMPs. The cells possess these PRRs all the time, whether or not they have

---

INSIGHT 14.1   Don't Drink and Phagocytose

We all know that drinking can impair your decision-making and driving skills, but scientists have recently discovered another downside to alcohol consumption: It can slow down your immune response. Gyongyi Szabo and her colleagues at the University of Massachusetts Medical School studied the effects of alcohol on monocytes (pictured), the precursors to macrophages and dendritic cells. They compared the function of monocytes exposed to alcohol levels equivalent to four or five drinks a day to those without any alcohol. The group found that the boozing monocytes reacted much more slowly than teetotaling monocytes and produced only half the amount of *interferon-1,* a chemical key to fighting viral infection. The drunk monocytes also overproduced the inflammatory chemical tumor necrosis factor-*alpha,* which can damage tissue. Szabo's group found evidence from medical records indicating that heavy drinkers die from HIV and hepatitis C sooner than nondrinkers.

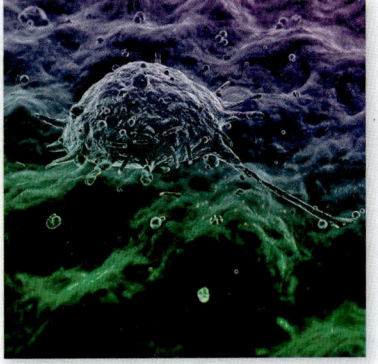

Another study showed that cells responsible for activating T cells were also impaired by alcohol. **Dendritic cells** are key components of the specific immune response (discussed in chapter 15) in activating T cells to fight specific pathogens. In a study performed by Jack R. Wands and his colleagues at Brown University, the activity of dendritic cells in alcoholic versus nonalcoholic mice was compared. In the alcoholic mouse model, dendritic cells showed lowered T-cell activation and lowered cytokine production. Both of these studies illustrate why alcoholics have a greater susceptibility to bacterial and viral infections and do not respond well to vaccines. When you get drunk, so does your immune system.

Sources: 2011. *Clin. and Vacc. Immun.* vol. 18, no. 7, p. 1157.
2011. BMC *Immun.* vol. 12, p. 55.

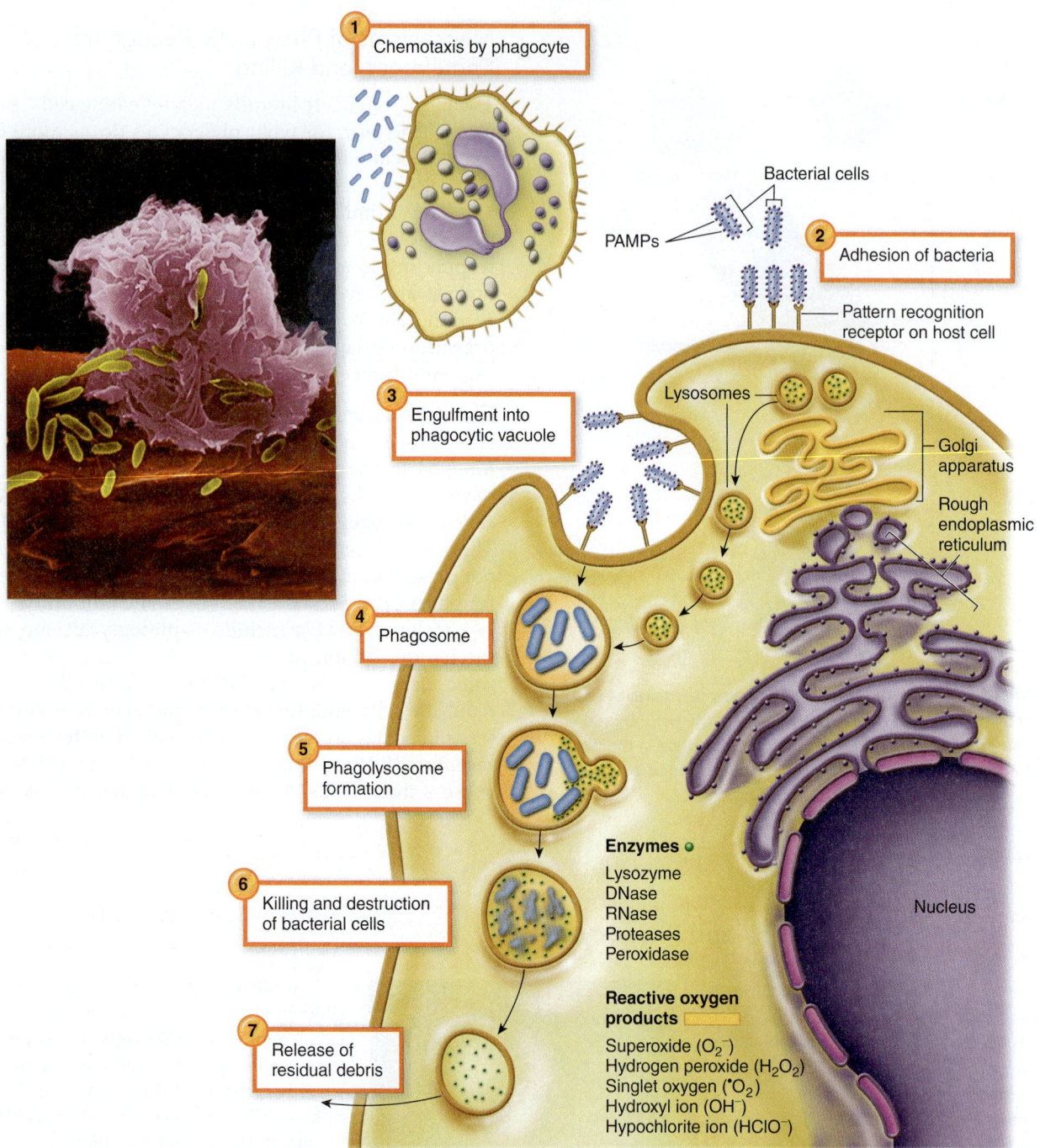

1 Chemotaxis by phagocyte

Bacterial cells

2 Adhesion of bacteria

PAMPs

Pattern recognition receptor on host cell

Lysosomes

3 Engulfment into phagocytic vacuole

Golgi apparatus

Rough endoplasmic reticulum

4 Phagosome

5 Phagolysosome formation

Enzymes ●
Lysozyme
DNase
RNase
Proteases
Peroxidase

Nucleus

6 Killing and destruction of bacterial cells

Reactive oxygen products ▭

Superoxide (O_2^-)
Hydrogen peroxide (H_2O_2)
Singlet oxygen ($^•O_2$)
Hydroxyl ion (OH^-)
Hypochlorite ion ($HClO^-$)

7 Release of residual debris

Process Figure 14.15 The phases of phagocytosis. **1** Phagocyte is attracted to bacteria. **2** Close-up view of process showing bacteria adhering to phagocyte PRRs by their PAMPs. **3** Vacuole is formed around bacteria during engulfment. **4** Phagosome, a digestive vacuole results. **5** Lysosomes fuse with phagosome, forming a phagolysosome. **6** Enzymes and toxic oxygen products kill and digest bacteria. **7** Undigested particles are released. **(Inset)** Scanning electron micrograph of a macrophage actively engaged in devouring bacteria (10,000×).

encountered PAMPs before. This is different than the situation with specific immunity. One category of PRRs is the **toll-like receptors (TLRs; figure 14.16)** (called "toll-like" because similar proteins called "toll" were originally discovered in fruit flies). The receptors not only recognize PAMPs but, upon binding, set in motion a cascade of events inside the host cell that amplifies and orchestrates a defensive response to the pathogen. This may include the formation of what is known as inflammasome, a protein complex that promotes a fully developed inflammatory response; or the response may be the initiation of a specific immune response. (There are a lot of acronyms in immunology and especially in this last paragraph. Don't let them get away from you; keep up with them. If you know what all the acronyms stand for and what they do, you are a good deal of the way there in understanding host defenses!)

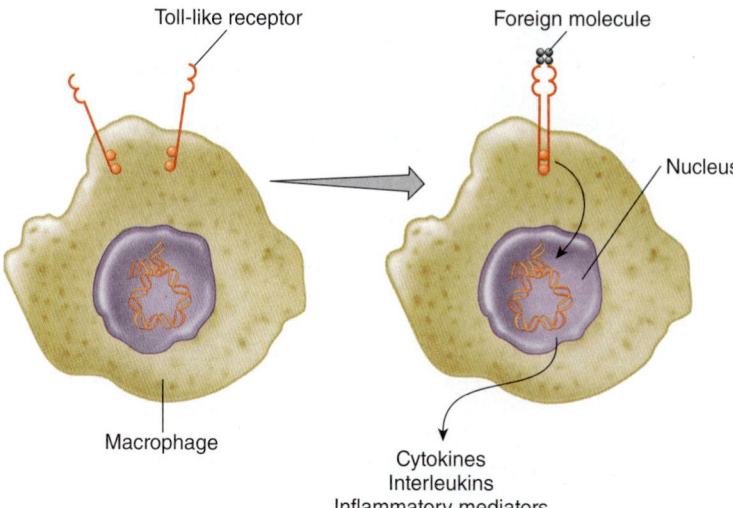

Toll-like receptor

Foreign molecule

Nucleus

Macrophage

Cytokines
Interleukins
Inflammatory mediators

Figure 14.16 Phagocyte detection and signaling with pattern recognition receptors. The example here shows the actions of a toll-like receptor that spans the membrane of many cells of the immune system. When a molecule specific to a particular class of pathogen is recognized by a receptor, the toll-like receptors merge and bind the foreign molecule. This induces production of chemicals that stimulate an immune response.

There is a class of PRRs that are not part of the cell membrane of phagocytes. **Collectins** are soluble molecules that roam through blood and tissues, bind to microbial PAMPs, and mark them for phagocytic destruction.

On the scene of an inflammatory reaction, phagocytes often trap cells or debris against the fibrous network of connective tissue or the wall of blood and lymphatic vessels. Once the phagocyte has made contact with its prey, it extends pseudopods that enclose the cells or particles in a pocket and internalize them in a vacuole called a **phagosome.** It also secretes more cytokines to further amplify the innate response.

Phagolysosome Formation and Killing In a short time, **lysosomes** migrate to the scene of the phagosome and fuse with it to form a **phagolysosome.** Other granules containing antimicrobial chemicals are released into the phagolysosome, forming a potent brew designed to poison and then dismantle the ingested material. The destructiveness of phagocytosis is evident by the death of bacteria within 30 minutes after contacting this battery of antimicrobial substances.

Disease Connection

Many pathogenic bacteria have developed mechanisms to resist phagocytic digestion. The bacteria that cause tuberculosis, listeriosis, plague, and many other infections survive inside the phagocyte. This can provide them with protection from the rest of the host's defenses and allows them to be transported throughout the body.

Destruction and Elimination Systems Two separate systems of destructive chemicals await the microbes in the phagolysosome. The oxygen-dependent system (known as the respiratory burst, or oxidative burst) involves several substances that were described in chapters 7 and 11. Myeloperoxidase, an enzyme found in granulocytes, forms halogen ions (OCl^-) that are strong oxidizing agents. Other products of oxygen metabolism such as hydrogen peroxide, the superoxide anion (O_2^-), activated or so-called singlet oxygen ($^\bullet O_2$), and the hydroxyl free radical ($OH^\bullet$) separately and together have formidable killing power. Other mechanisms that come into play are the liberation of lactic acid, lysozyme, and nitric oxide (NO), a powerful mediator that kills bacteria and inhibits viral replication. Cationic proteins that injure bacterial cell membranes and a number of proteolytic and other hydrolytic enzymes complete the job. The small bits of undigestible debris are released from the macrophage by exocytosis.

Interestingly, recent studies suggest that in the human gut, only the epithelial cells deep in intestinal crypts express large numbers of the toll-like receptors, which function to recognize microbes. The commensal bacteria occupy the "top" of the epithelium, not the crypts. This phenomenon may help explain why commensals are tolerated and pathogens (which are likely to colonize the crypts) are not.

Inflammation: A Complex Concert of Reactions to Injury

At its most general level, the inflammatory response is a reaction to any traumatic event in the tissues. It is so commonplace that all of us manifest inflammation in some way every day. It appears in the nasty flare of a cat scratch, the blistering of a burn, the painful lesion of an infection, and the symptoms of allergy. When close to our external surfaces, it is readily identifiable by a classic series of signs and symptoms characterized succinctly by four Latin terms: *rubor, calor, tumor,* and *dolor. Rubor* ("redness") is caused by increased circulation and vasodilation in the injured tissues; *calor* ("warmth") is the heat given off by the increased flow of blood; *tumor* ("swelling") is caused by increased fluid escaping into the tissues; and *dolor* ("pain") is caused by the stimulation of nerve endings **(figure 14.17).** A fifth symptom,

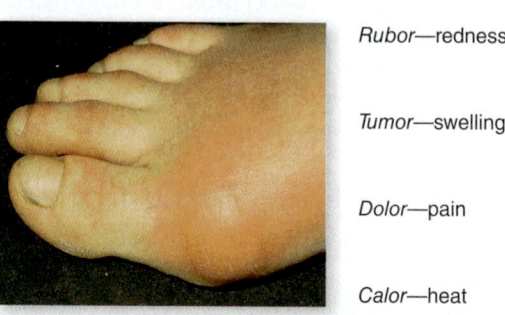

Rubor—redness

Tumor—swelling

Dolor—pain

Calor—heat

Figure 14.17 The response to injury. This classic checklist encapsulates the reactions of the tissues to an assault. Each of the events is an indicator of one of the mechanisms of inflammation described in this chapter.

loss of function, has been added to give a complete picture of the effects of inflammation. Although these manifestations can be unpleasant, they serve an important warning that injury has taken place and set in motion responses that save the body from further injury.

It is becoming increasingly clear that some chronic diseases, such as cardiovascular disease, can be caused by chronic inflammation. While we speak of inflammation at a local site (such as a finger), inflammation can affect an entire system—such as blood vessels, lungs, skin, the joints, and so on. Some researchers believe that the very act of aging is a consequence of increasing inflammation in multiple body systems.

Factors that can elicit inflammation include trauma from infection (the primary emphasis here), tissue injury or necrosis due to physical or chemical agents, and specific immune reactions. Although the details of inflammation are very complex, its chief functions are

1. to mobilize and attract immune components to the site of the injury,
2. to set in motion mechanisms to repair tissue damage and localize and clear away harmful substances, and
3. to destroy microbes and block their further invasion **(process figure 14.18).**

The inflammatory response is a powerful defensive reaction, a means for the body to maintain stability and restore itself after an injury. But when it is chronic, it has the potential to actually *cause* tissue injury, destruction, and disease. Some brain infections, including those seen in cases of trypanosomiasis and cryptococcosis, lead to devastating and permanent damage to the nervous system due to the buildup of inflammatory products. Inflammation may actually trap pathogens within a localized area, leading to the formation of an abscess over time. Granuloma formation is characteristic of many chronic infectious diseases, including tuberculosis and leprosy, and is due to incomplete immune activity against pathogens in localized tissue areas.

The Stages of Inflammation

The process leading to inflammation is a dynamic, predictable sequence of events that can be acute, lasting from a few minutes or hours, to chronic, lasting for days, weeks, or years. Once the initial injury has occurred, a chain reaction takes place at the site of damaged tissue, summoning beneficial cells and fluids into the injured area. As an example, we will look at an injury at the microscopic level and observe the flow of major events (as shown in process figure 14.18).

Vascular Changes: Early Inflammatory Events

Following an injury, some of the earliest changes occur in the vasculature (arterioles, capillaries, venules) in the vicinity of the damaged tissue. These changes are controlled by nervous stimulation, **chemical mediators,** and **cytokines** released by blood cells, tissue cells, and platelets in the injured area. Some mediators are **vasoactive**—that is, they affect the endothelial

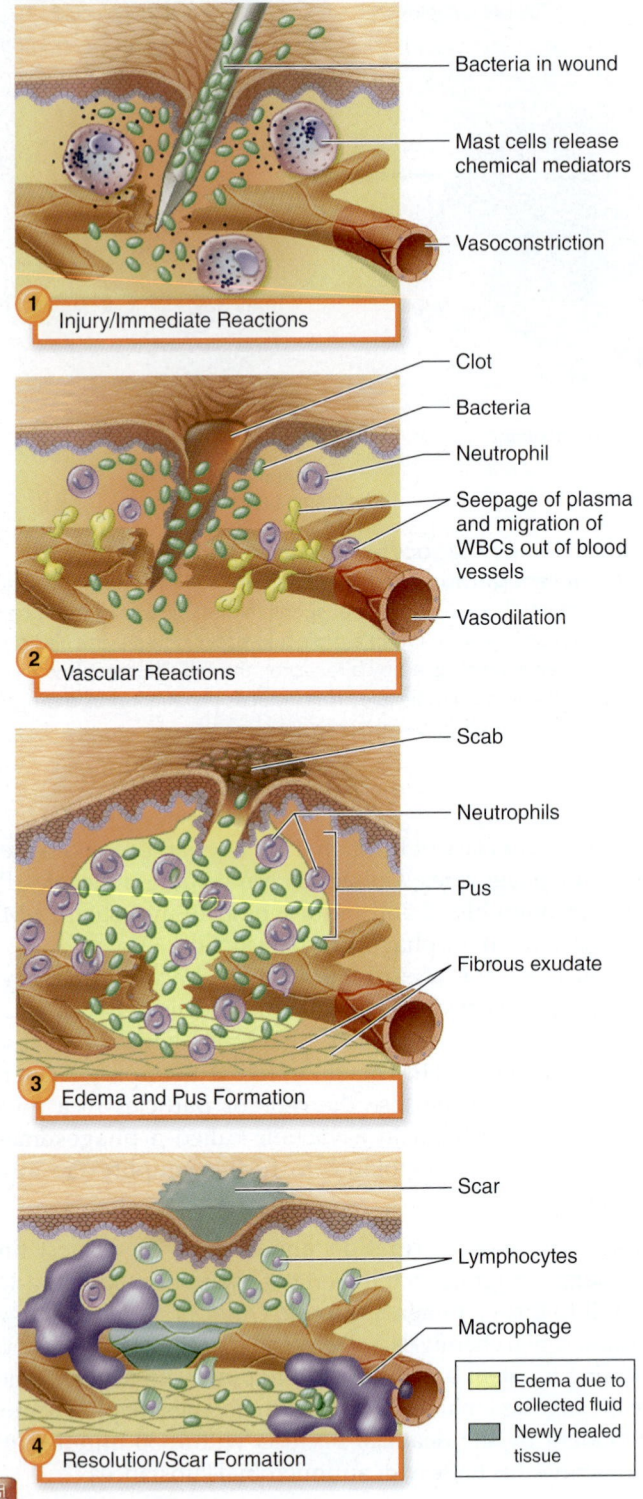

Process Figure 14.18 The major events in inflammation. ① Injury → Reflex narrowing of the blood vessels (vasoconstriction) lasting for a short time → Release of chemical mediators into area. ② Increased diameter of blood vessels (vasodilation) → Increased blood flow → Increased vascular permeability → Leakage of fluid (plasma) from blood vessels into tissues (exudate formation). ③ Edema → Infiltration of site by neutrophils and accumulation of pus. ④ Macrophages and lymphocytes → Repair, either by complete resolution and return of tissue to normal state or by formation of scar tissue.

cells and smooth muscle cells of blood vessels; others are **chemotactic factors,** also called **chemokines,** that affect white blood cells. Inflammatory mediators cause fever, stimulate lymphocytes, prevent virus spread, and cause allergic symptoms **(table 14.1).** Although the constriction of arterioles is stimulated first, it lasts for only a few seconds or minutes and is followed in quick succession by the opposite reaction, vasodilation. The overall effect of vasodilation is to increase the flow of blood into the area, which facilitates the influx of immune components and also causes redness and warmth.

Edema: Leakage of Vascular Fluid into Tissues

Some vasoactive substances cause the endothelial cells in the walls of postcapillary venules to contract and form gaps through which blood-borne components exude into the extracellular spaces. The fluid part that escapes is called the **exudate.** Accumulation of this fluid in the tissues gives rise to local swelling and firmness called **edema.** The edematous exudate contains varying amounts of plasma proteins, such as globulins, albumin, the clotting protein fibrinogen, blood cells, and cellular debris. Depending on its content, the exudate may be clear—called **serous,** or it may contain red blood cells or pus. Pus is composed mainly of white blood cells and the debris generated by phagocytosis. In some types of edema, the fibrinogen is converted to fibrin threads that enmesh the injury site. Within an hour, multitudes of neutrophils responding chemotactically to special signaling molecules converge on the injured site (see process figure 14.18, step **3**).

Unique Dynamic Characteristics of White Blood Cells

In order for WBCs to leave the blood vessels and enter the tissues, they adhere to the inner walls of the smaller blood vessels. From this position, they are poised to migrate out of the blood into the tissue spaces by a process called **diapedesis** (dye″-ah-puh-dee′-sis).

Diapedesis, also known as *transmigration,* is aided by several related characteristics of WBCs. For example, they are actively motile and readily change shape. Diapedesis is also assisted by the nature of the endothelial cells lining the venules. They contain complex adhesive receptors that capture the WBCs and participate in their transport from the venules into the extracellular spaces **(figure 14.19).**

Another factor in the migratory habits of these WBCs is **chemotaxis,** defined as the tendency of cells to migrate in response to a specific chemical stimulus given off at a site of injury or infection (see inflammation and phagocytosis later in this chapter). Through this means, cells swarm from many compartments to the site of infection and remain there to perform general and specific immune functions. These basic properties are absolutely essential for the sort of intercommunication and deployment of cells required for most immune reactions (see process figure 14.18).

Phagocytes migrate into a region of inflammation with a deliberate sense of direction, attracted by a gradient of stimulant products from the parasite and host tissue at the site of injury. The endpoint function of the white blood cells is the phagocytosis of microbes or other invading substances.

Table 14.1 Inflammatory Mediators and Other Cytokines

Mediators of Inflammation and Immunity	**Tumor necrosis factor (TNF),** a substance from macrophages, lymphocytes, and other cells that increases chemotaxis and phagocytosis and stimulates other cells to secrete inflammatory cytokines. It also serves as an endogenous pyrogen that induces fever, increases blood coagulation, suppresses bone marrow, and suppresses appetite.
	Interferons (IFN), produced by leukocytes, fibroblasts, and other cells, inhibit virus replication and cell division and increase the action of certain lymphocytes that kill other cells.
	Interleukin (IL) 1, a product of macrophages and dendritic cells that has many of the same biological activities as TNF, such as inducing fever and activating certain white blood cells.
	Interleukin-6, secreted by macrophages and T cells. Its primary effects are to stimulate the growth of B cells and to increase the synthesis of liver proteins.
	Prostaglandins, produced by most body cells; complex chemical mediators that can have opposing effects (e.g., dilation or constriction of blood vessels) and are powerful stimulants of inflammation and pain.
	Platelet-activating factor, a substance released from basophils, causes the aggregation of platelets and the release of other chemical mediators during immediate allergic reactions.
Vasoactive Mediators	**Histamine,** a vasoactive mediator produced by mast cells and basophils, causes vasodilation, increased vascular permeability, and mucus production. It functions primarily in inflammation and allergy.
	Serotonin, a mediator produced by platelets and intestinal cells, causes smooth muscle contraction, inhibits gastric secretion, and acts as a neurotransmitter.
	Bradykinin, a vasoactive amine from the blood or tissues, stimulates smooth muscle contraction and increases vascular permeability, mucus production, and pain. It is particularly active in allergic reactions.
Cytokines That Regulate Lymphocyte Growth and Activation	**Interleukin-2,** the primary growth factor from T cells. Interestingly, it acts on the same cells that secrete it. It stimulates mitosis and secretion of other cytokines. In B cells, it is a growth factor and stimulus for antibody synthesis.
	Macrophage colony-stimulating factor (M-CSF), produced by a variety of cells. M-CSF promotes the growth and development of macrophages from undifferentiated precursor cells.

Figure 14.19 Diapedesis and chemotaxis of leukocytes. **(a)** View of a venule depicts white blood cells squeezing themselves between spaces in the blood vessel wall via diapedesis. This process, shown in cross section, indicates how the pool of leukocytes adheres to the endothelial wall. From this site, they are poised to migrate out of the vessel into the tissue space. **(b)** This photograph captures neutrophils in the process of diapedesis.

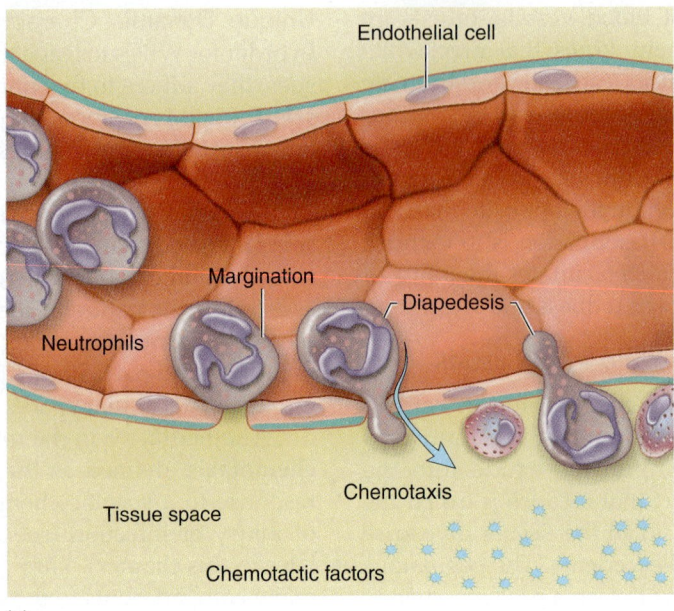
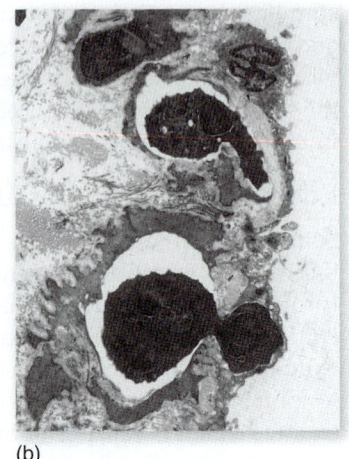

(a) (b)

The Benefits of Edema and Chemotaxis Both the formation of edematous exudate and the infiltration of neutrophils are physiologically beneficial activities. The influx of fluid dilutes toxic substances, and the fibrin clot can effectively trap microbes and prevent their further spread. The neutrophils that aggregate in the inflamed site are immediately involved in phagocytosing and destroying bacteria, dead tissues, and particulate matter (by mechanisms discussed in a later section on phagocytosis). In some types of inflammation, accumulated phagocytes contribute to **pus,** a whitish mass of cells, liquefied cellular debris, and bacteria.

Disease Connection

Certain bacteria (streptococci, staphylococci, gonococci, and meningococci) are especially powerful attractants for neutrophils and are thus termed **pyogenic,** or pus-forming, bacteria.

Late Reactions of Inflammation Sometimes a mild inflammation can be resolved by edema and phagocytosis. Inflammatory reactions that are more long-lived attract a collection of monocytes, lymphocytes, and macrophages to the reaction site. Clearance of pus, cellular debris, dead neutrophils, and damaged tissue is performed by macrophages, the only cells that can engulf and dispose of such large masses. At the same time, B lymphocytes react with foreign molecules and cells by producing specific antimicrobial proteins (antibodies), and T lymphocytes kill intruders directly. Late in the process, the tissue is completely repaired, if possible, or replaced by connective tissue in the form of a scar (see process figure 14.18, step ④). If the inflammation cannot be relieved or resolved in this way, it can become chronic and create a long-term pathologic condition.

Fever: An Adjunct to Inflammation

An important systemic component of inflammation—and innate immunity in general—is *fever,* defined as an abnormally elevated body temperature. Although fever is a nearly universal symptom of infection, it is also associated with certain allergies, cancers, and other organic illnesses.

The body temperature is normally maintained by a control center in the hypothalamus region of the brain. This thermostat regulates the body's heat production and heat loss and sets the core temperature at around 37°C (98.6°F) with slight fluctuations (1°F) during a daily cycle. Fever is initiated when circulating substances called pyrogens (py'-roh-jenz) reset the hypothalamic thermostat to a higher setting. This change signals the musculature to increase heat production and peripheral arterioles to decrease heat loss through vasoconstriction. Fevers range in severity from low-grade (37.7°C to 38.3°C, or 100°F to 101°F) to moderate (38.8°C to 39.4°C, or 102°F to 103°F) to high (40.0°C to 41.1°C, or 104°F to 106°F). Pyrogens are described as **exogenous** (coming from outside the body) or **endogenous** (originating internally). Exogenous pyrogens are products of infectious agents such as viruses, bacteria, protozoans, and fungi. One well-characterized exogenous pyrogen is endotoxin, the lipopolysaccharide found in the cell walls of gram-negative bacteria. Blood, blood products, vaccines, or injectable solutions can also contain exogenous pyrogens. Endogenous pyrogens are released by monocytes, neutrophils, and macrophages during the process of phagocytosis and appear to be a natural part of the immune response. Two potent pyrogens released by macrophages are interleukin-1 (IL-1) and tumor necrosis factor (TNF).

Benefits of Fever

The association of fever with infection strongly suggests that it serves a beneficial role, a view still being debated but gaining acceptance. Aside from its practical and medical importance as a sign of a physiological disruption, increased body temperature has additional benefits:

- Fever inhibits multiplication of temperature-sensitive microorganisms such as the poliovirus, cold viruses, herpes zoster virus, systemic and subcutaneous fungal pathogens, *Mycobacterium* species, and the syphilis spirochete.
- Fever impedes the nutrition of bacteria by reducing the availability of iron. It has been demonstrated that during fever, the macrophages stop releasing their iron stores, which could retard several enzymatic reactions needed for bacterial growth.
- Fever increases metabolism and stimulates immune reactions and naturally protective physiological processes. It speeds up hematopoiesis, phagocytosis, and specific immune reactions. It increases the ability of specific lymphocytes to home in on sites of infection.

Treatment of Fever

With this revised perspective on fever, whether to suppress it or not can be a difficult decision **(Insight 14.2).** Some advocates feel that a slight to moderate fever in an otherwise healthy person should be allowed to run its course, in light of its potential benefits and minimal side effects. All medical experts do agree that high and prolonged fevers or fevers in patients with cardiovascular disease, head trauma, seizures, or respiratory ailments are risky and must be treated immediately with fever-reducing drugs. The classic therapy for fever is aspirin or acetaminophen (Tylenol). These drugs lower the setting of the hypothalamic center and restore normal temperature. Any physical technique that stimulates heat loss (tepid baths, for example) can also help reduce the core temperature.

Antimicrobial Proteins: (1) Interferon

Interferon (IFN) was described in chapter 12 as a small protein produced naturally by certain white blood and tissue cells; it is used in therapy against certain viral infections and cancer. Although the interferon system was originally thought to be directed exclusively against viruses, it is now known to be involved also in defenses against other microbes and in immune regulation and intercommunication. Three major types are *interferons alpha* and *beta*, products of many cells, including lymphocytes, fibroblasts, and macrophages; and *interferon gamma*, a product of T cells.

All three classes of interferon are produced in response to viruses, RNA, immune products, and various antigens. Their biological activities are extensive. In all cases, they bind to cell

---

INSIGHT 14.2   Autism Risk Doubled by Fever During Pregnancy

Autism is a disease that is in many ways a mystery to parents, doctors, and scientists. A fraudulent study published in 1998 by Andrew Wakefield erroneously linked the MMR vaccine to autism. Wakefield was later found to be guilty of medical, ethical, and scientific misconduct; and his paper was called the worst medical hoax in 100 years. Although *The Lancet*, where it had been originally published, retracted Wakefield's paper, the damage was done and millions of parents stopped vaccinating their children. Yet, rates of autism continue to rise with few clues as to its cause. Autism is generally considered to be a genetic disorder, but there is evidence that it is a complex disease with many contributing factors.

It should be stated that there are many solid studies suggesting causal factors for autism, one of which is this well-researched possibility. A large study entitled, "Childhood Autism Risk from Genetics and the Environment (CHARGE)," conducted at the University of California Davis, found that women who had a fever during pregnancy were more than twice as likely to have a child with autism or who exhibited developmental delays compared to women who did not have a fever or who used fever-reducing medications. Another researcher within the CHARGE study found a link between children

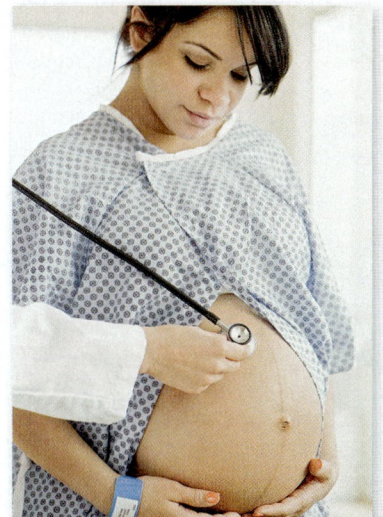

with autism and mothers who were obese or who had diabetes. Infections with bacteria or viruses, as well as the conditions diabetes and obesity, result in an inflammatory response that includes production of cytokines, specifically IL-1, IL-6, tumor necrosis factor-alpha, and interferon gamma. These cytokines are responsible for the onset of fever as a response to infection in addition to the inflammatory response. The researchers in the CHARGE study note that cytokines are able to cross the placenta and can have an effect on neurotransmitters in the developing fetal brain. Even though this study shows correlation (and not causation), researchers suggest that women who experience fever during pregnancy for any reason should take anti-fever medications to reduce the cytokine response and thus reduce the chance for development of autism or developmental delays later in life. It should be emphasized that no definitive cause for autism has been found. But health care practitioners should stay informed about solid scientific research to help questioning patients weed through the information and the misinformation.

Source: 2012. *J. Autism Dev. Disord.* vol. 43, no. 1, p. 25.

surfaces and induce changes in genetic expression, but the exact results vary. In addition to antiviral effects discussed in the next section, all three IFNs can inhibit the expression of cancer genes and have tumor suppressor effects. IFN alpha and beta stimulate phagocytes, and IFN gamma is an immune regulator of macrophages and T and B cells.

Characteristics of Antiviral Interferon

When a virus binds to the receptors on a host cell, a signal is sent to the nucleus that directs the cell to synthesize interferon. After transcribing and translating the interferon gene, newly synthesized interferon molecules are rapidly secreted by the cell into the extracellular space, where they bind to other host cells. The binding of interferon to a second cell induces the production of proteins in that cell that inhibit viral multiplication either by degrading the viral RNA or by preventing the translation of viral proteins **(figure 14.20).** Interferon is not virus-specific, so its synthesis in response to one type of virus will also protect against other types. Because this protein is an inhibitor of viruses, it has been produced industrially and used as a treatment for a number of viral infections.

Other Roles of Interferon

Interferons are also important immune regulatory cytokines that activate or instruct the development of white blood cells. For example, interferon alpha produced by T lymphocytes activates a subset of cells called natural killer (NK) cells. In addition, one type of interferon beta plays a role in the maturation of B and T lymphocytes and in inflammation. Interferon gamma inhibits cancer cells, stimulates B lymphocytes, activates macrophages, and enhances the effectiveness of phagocytosis. It was recently discovered that interferon also reduces the amount of cholesterol in the body. Because cholesterol is used by bacteria and viruses as a nutrient, this provides another source of innate protection.

Antimicrobial Proteins: (2) Complement

Among its many overlapping functions, the immune system has another complex and multiple-duty system called **complement** that, like inflammation and phagocytosis, is brought into play at several levels. The complement system, named for its property of "complementing" immune reactions, consists of over 30 blood proteins that work in concert to destroy bacteria and certain viruses. Some knowledge of this important system will help in your understanding of topics in chapter 15.

The concept of a cascade reaction is helpful in understanding how complement functions. A *cascade reaction* is a sequential physiological response like that of blood clotting, in which the first substance in a chemical series activates the next substance, which activates the next, and so on, until a desired end product is reached. There are three different complement pathways, distinguished by how they become activated. The final stages of the three pathways are the same and yield a similar end result. For our discussion, we will focus on shared characteristics. **Process figure 14.21** illustrates the classical pathway, which is kicked off when antibody (part of the specific immune response) binds to microbial cells. **Table 14.2** points out, though, that there are two other ways complement gets activated that do not

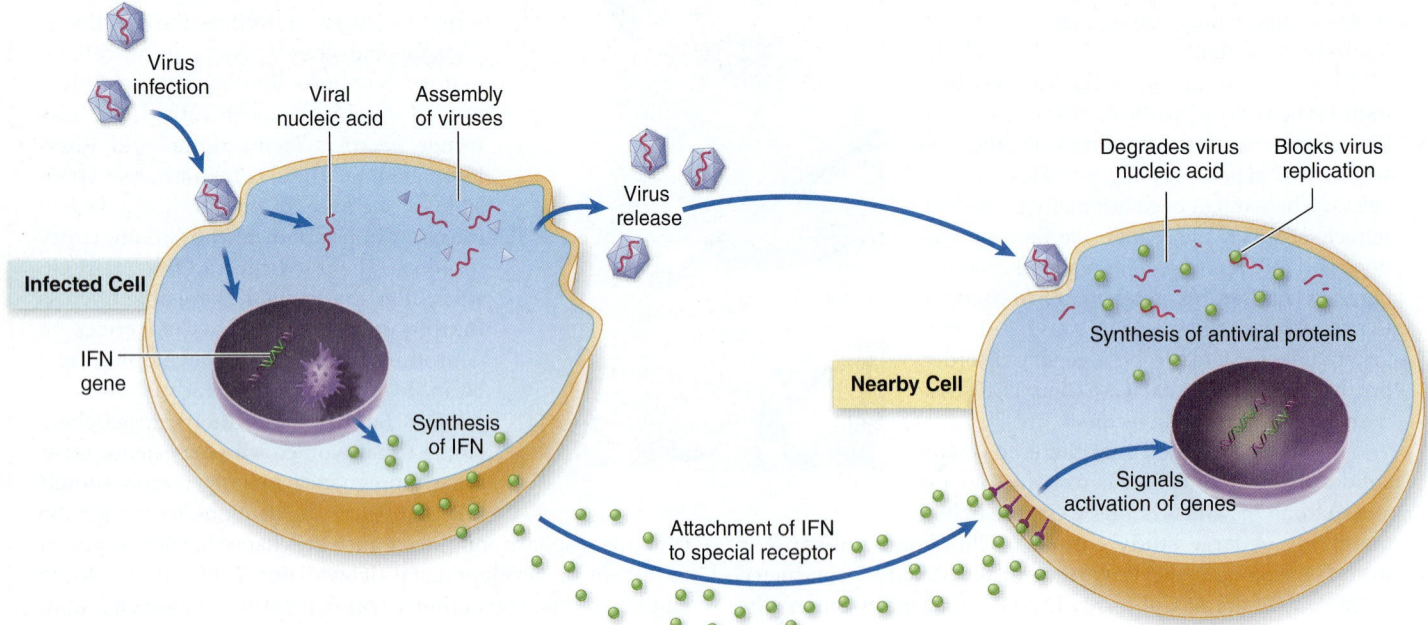

Figure 14.20 The antiviral activity of interferon. When a cell is infected by a virus, its nucleus is triggered to transcribe and translate the interferon (IFN) gene. Interferon diffuses out of the cell and binds to IFN receptors on nearby uninfected cells, where it induces production of proteins that eliminate viral genes and block viral replication. Note that the original cell is not protected by IFN and that IFN does not prevent viruses from invading the protected cells.

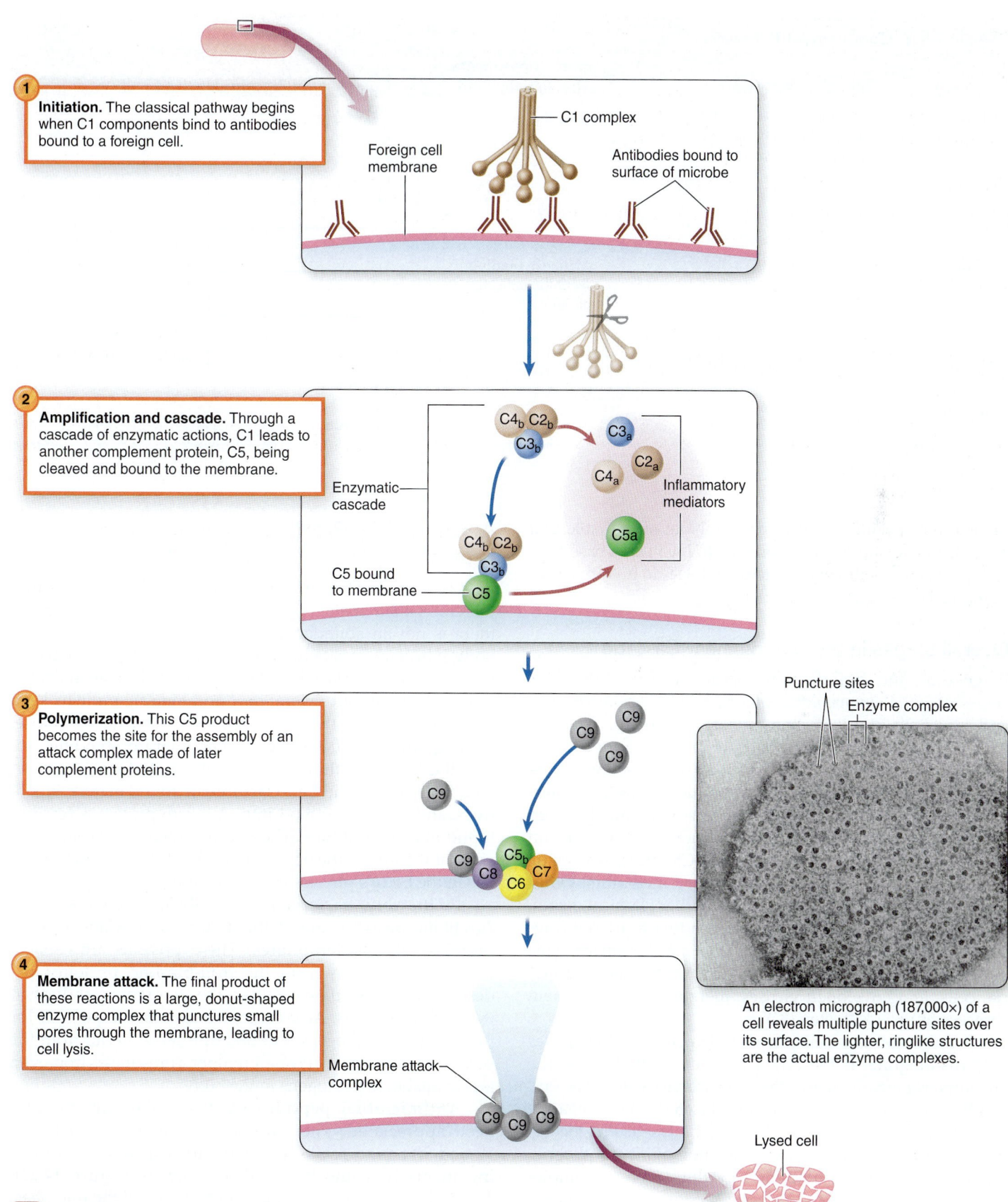

1 Initiation. The classical pathway begins when C1 components bind to antibodies bound to a foreign cell.

C1 complex

Foreign cell membrane

Antibodies bound to surface of microbe

2 Amplification and cascade. Through a cascade of enzymatic actions, C1 leads to another complement protein, C5, being cleaved and bound to the membrane.

Enzymatic cascade

C4$_b$ C2$_b$
C3$_b$

C3$_a$
C2$_a$
C4$_a$ Inflammatory mediators

C4$_b$ C2$_b$
C3$_b$

C5a

C5 bound to membrane

C5

3 Polymerization. This C5 product becomes the site for the assembly of an attack complex made of later complement proteins.

C9
C9
C9
C9
C9
C8   C5$_b$
C6   C7

Puncture sites
Enzyme complex

4 Membrane attack. The final product of these reactions is a large, donut-shaped enzyme complex that punctures small pores through the membrane, leading to cell lysis.

Membrane attack complex

C9 C9 C9

An electron micrograph (187,000×) of a cell reveals multiple puncture sites over its surface. The lighter, ringlike structures are the actual enzyme complexes.

Lysed cell

Process Figure 14.21 Steps in the classical complement pathway. All complement pathways function in a similar way, but the details differ.

Photo: © 1966 Rockefeller University Press. Originally published in *The Journal of Experimental Medicine*, June 1966, Vol. 123, pp. 969–84.

Table 14.2 Complement Pathways

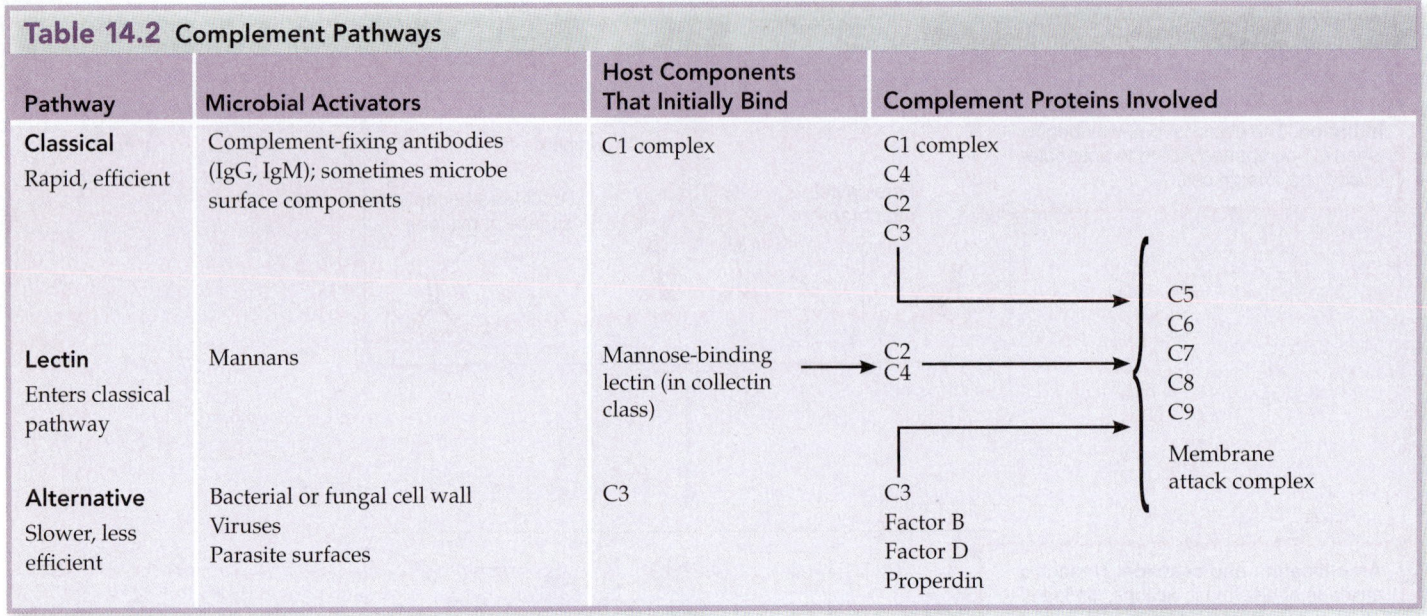

Pathway	Microbial Activators	Host Components That Initially Bind	Complement Proteins Involved
Classical Rapid, efficient	Complement-fixing antibodies (IgG, IgM); sometimes microbe surface components	C1 complex	C1 complex C4 C2 C3
Lectin Enters classical pathway	Mannans	Mannose-binding lectin (in collectin class)	C2 C4
Alternative Slower, less efficient	Bacterial or fungal cell wall Viruses Parasite surfaces	C3	C3 Factor B Factor D Properdin

C5, C6, C7, C8, C9 — Membrane attack complex

involve the specific immune system. Note in the table that because the complement numbers (C1–C9) are based on the order of their discovery, factors C1–C4 do not appear in numerical order during activation.

Overall Stages in the Complement Cascade

In general, the complement cascade includes the four stages of *initiation, amplification and cascade, polymerization,* and *membrane attack.* At the outset, an initiator (such as microbes, or antibodies, see table 14.2) reacts with the first complement chemical, which propels the reaction on its course. There is a recognition site on the surface of the target cell where the initial C components will bind. Through a stepwise series, each component reacts with another on or near the recognition site. In the C2–C5 series, enzymatic cleavage produces several inflammatory mediators. Other details of the pathways differ, but whether classical, lectin, or alternative, the functioning end product is a large ring-shaped protein termed the *membrane attack complex.* This complex can digest holes in the cell membranes of bacteria, cells, and enveloped viruses, thereby destroying them (process figure 14.21, step ④).

If the target is a cell, the complement reaction causes it to disintegrate. If the target is an enveloped virus, the envelope is perforated and the virus is inactivated. The end result of complement action is multifaceted. Gram-negative bacteria and infected host cells may be lysed. Phagocytes will be attracted to the site in greater numbers. Overall inflammation will be amplified by the action of complement. In recent years, the excessive actions of complement have been implicated as aggravators of several autoimmune diseases, such as lupus, rheumatoid arthritis, and myasthenia gravis.

Antimicrobial Proteins: (3) Iron-Binding Proteins and (4) Antimicrobial Peptides

Both humans and bacteria require iron for their enzymes and, therefore, their metabolisms to function properly. Because there is not an abundance of available iron in the human body, it becomes a rate-limiting factor in the growth of bacteria that have invaded a host. There is generally a great deal of iron in the body, but several host iron-binding proteins keep it bound tightly so that it is not available for microbial use. **Hemoglobin** is probably the most familiar iron-binding protein. It is located within red blood cells. **Transferrin,** found in blood and tissue fluids, and **lactoferrin,** found in milk, blood, tears, and saliva, also bind iron. A fourth protein is **ferritin,** found in every cell type. These four proteins make sure that most of the iron in the body is bound tightly, sequestered for use in the body's physiological reactions.

Predictably, bacteria have evolved mechanisms for "grabbing away" some of this bound iron. Many possess proteins called **siderophores.** These proteins are capable of scavenging iron from the iron-binding proteins just described because they can bind the iron more tightly than the human proteins. As you see, it becomes a tug-of-war for iron between pathogens and body proteins. Recall from the section on fever that elevated temperatures make iron even less available to pathogens in the body.

Antimicrobial peptides (AMPs) have only recently been appreciated. They are short proteins—of between 12 and 50 amino acids—that have the capability of inserting themselves into bacterial membranes **(figure 14.22).** Through this mechanism and others, they kill the microbes. They have names like bacteriocins, defensin, magainins, and protegrins. They are part of the innate immune system and also have an effect on other actions of nonspecific and

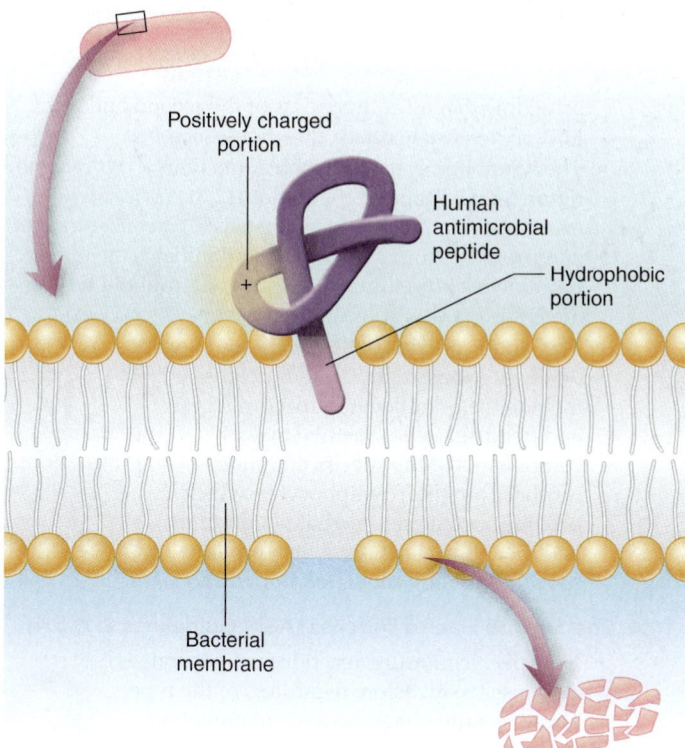

Positively charged portion

Human antimicrobial peptide

Hydrophobic portion

+

Bacterial membrane

Figure 14.22 Antimicrobial peptides. These peptides have various mechanisms, but a very common one is for the peptide to insert itself into pathogen membranes using a positive charge plus a hydrophobic tail.

specific immunity. Many researchers are trying to turn these antimicrobial peptides into practical use as therapeutic drugs. Their ability to modulate immune responses would distinguish them from other antibiotics on the market and may represent a new weapon in the war against microbial drug resistance. Although their clinical use has not fully developed, researchers are discovering ways to utilize computer programs to design the most effective antimicrobial peptides in the laboratory.

Case File 14 *Wrap-Up*

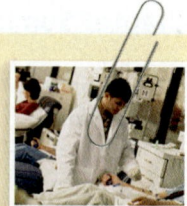

Gay men, or men who have sex with men (MSMs, as they are called by the public health community) are prevented from donating blood. The ban on MSMs donating blood can be traced back to the early days of the AIDS epidemic in the 1980s. Before doctors and scientists knew the cause of AIDS, there were no limitations on blood donation. However, patients with hemophilia were dying of these types of infections, including children who did not engage in practices that would otherwise expose them to the causative agent of AIDS. At that time, the causative agent of AIDS and its mechanism of transmission were unknown, but it was hypothesized that it was a blood-borne disease. Between 1981 and 1984, 50% of the U.S. hemophiliac population became infected and acquired AIDS before a ban on MSM blood donation was put in place.

Today, there is greater understanding of the cause of AIDS and the transmission of HIV. Blood and blood products are carefully screened to prevent transmission of HIV and other blood-borne diseases. HIV transmission through blood donation has been curbed. For that reason, efforts are underway to allow MSMs to donate blood in order to increase the supply of a desperately needed resource.

Source: 2011. www.sciencedaily.com/releases/2011/09/110908191048.htm

14.4 Learning Outcomes—Assess Your Progress

11. List the four major categories of nonspecific immunity.

12. Summarize the steps in phagocytosis, and describe the roles of PAMPs in this process.

13. Outline the steps in inflammation.

14. Discuss the mechanism of fever and its role in nonspecific immunity.

15. Compare and contrast the three different complement pathways.

16. Name four types of antimicrobial proteins.

Chapter Summary

14.1 Defense Mechanisms of the Host in Perspective (ASM Guidelines* 2.1, 5.4)

- The interconnecting network of host protection against microbial invasion is organized into three lines of defense.
- The first line of defense consists of physical and chemical barriers associated with the skin and mucous membranes. Normal biota, though not considered a physical barrier, have been found to contribute to this and other lines of defense as well.
- The second line encompasses all the nonspecific cells and chemicals found in the tissues and blood.
- The third line, the specific immune response, is customized to react to specific antigens of a microbial invader.

14.2 The Second and Third Lines of Defense: An Overview (ASM Guidelines 2.1, 5.4)

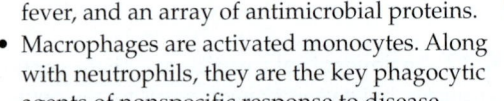

- The immune system operates as a surveillance system that discriminates between the host's self identity markers and the nonself identity markers of foreign cells.

14.3 Systems Involved in Immune Defenses (ASM Guidelines 2.1, 5.4)

- The immune system is a complex collection of fluids and cells that penetrate every organ, tissue space, fluid compartment, and vascular network of the body. The four major subdivisions of this system are the mononuclear phagocyte system (MPS), the extracellular fluid (ECF), the lymphatic system, and the blood vascular system.
- The MPS is a network of connective tissue fibers inhabited by macrophages ready to attack and ingest microbes that have managed to bypass the first line of defense.
- The ECF compartment surrounds all tissue cells and is penetrated by both blood and lymph vessels, which

bring together all components of the second and third lines of defense to attack infectious microbes.

- The lymphatic system has three functions: (1) It returns tissue fluid to general circulation; (2) it carries away excess fluid in inflamed tissues; and (3) it concentrates and processes foreign invaders and initiates the specific immune response. Important sites of lymphoid tissues are lymph nodes, spleen, thymus, tonsils, and GALT.
- The blood contains both specific and nonspecific defenses. Nonspecific cellular defenses include the granulocytes, macrophages, and dendritic cells. The two major components of the specific immune response are the T lymphocytes, which provide specific cell-mediated immunity, and the B lymphocytes, which produce specific antibody-mediated immunity.

Dendritic cells

14.4 The Second Line of Defense (ASM Guidelines 2.1, 5.4)

- Nonspecific immune reactions are generalized responses to invasion, regardless of the type. These include phagocytosis, inflammation, fever, and an array of antimicrobial proteins.
- Macrophages are activated monocytes. Along with neutrophils, they are the key phagocytic agents of nonspecific response to disease.
- The four symptoms of inflammation are rubor (redness), calor (heat), tumor (edema), and dolor (pain). Loss of function often accompanies these.
- Fever is another component of nonspecific immunity. It is caused by both endogenous and exogenous pyrogens. Fever increases the rapidity of the host immune responses and reduces the viability of many microbial invaders.

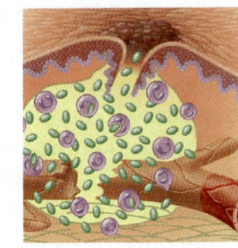

- There are four main types of antimicrobial proteins: the complement system, interferons, iron-binding proteins, and antimicrobial peptides.

Source: ASM Curriculum Guidelines (American Society for Microbiology, 2012). Complete guidelines in appendix B of this book.

Multiple-Choice and True-False Questions | Bloom's Levels 1 and 2: Remember and Understand

Multiple-Choice Questions. Select the correct answer from the options provided.

1. An example of a nonspecific chemical barrier to infection is/are
   a. unbroken skin.
   b. lysozyme in saliva.
   c. cilia in respiratory tract.
   d. all of these.

2. Which nonspecific host defense is associated with the trachea?
   a. lacrimation
   b. ciliary lining
   c. desquamation
   d. lactic acid

3. Which of the following blood cells function primarily as phagocytes?
   a. eosinophils
   b. basophils
   c. lymphocytes
   d. neutrophils

4. Which of the following is *not* a lymphoid tissue?
   a. spleen
   b. thyroid gland
   c. lymph node
   d. GALT

5. What is included in GALT?
   - a. thymus
   - b. Peyer's patches
   - c. tonsils
   - d. breast lymph nodes

6. Monocytes are _____ leukocytes that develop into _____.
   - a. granular, phagocytes
   - b. agranular, mast cells
   - c. agranular, macrophages
   - d. granular, T cells

7. An example of an exogenous pyrogen is
   - a. interleukin-1.
   - b. complement.
   - c. interferon.
   - d. endotoxin.

8. Which of the following is an antimicrobial protein that has a much greater role in the third line of defense than in the second line of defense?
   - a. antibody
   - b. complement
   - c. protegrin
   - d. interferon

9. Which of the following substances is/are not produced by phagocytes to destroy engulfed microorganisms?
   - a. hydroxyl radicals
   - b. superoxide anion
   - c. hydrogen peroxide
   - d. bradykinin

10. Which of the following is the end product of the complement system?
    - a. properdin
    - b. cascade reaction
    - c. membrane attack complex
    - d. complement factor C9

True-False Questions. If the statement is true, leave as is. If it is false, correct it by rewriting the sentence.

11. The liquid component of clotted blood is called plasma.

12. Pyogenic bacteria are commonly associated with fever.

13. Communication between cells of the immune system is accomplished using chemical signals.

14. Lysozyme is an enzyme found in tears and saliva that hydrolyzes peptidoglycan in bacterial cell walls.

15. The immune system uses DNA content to distinguish self from nonself.

Critical Thinking Questions | Bloom's Levels 3, 4, and 5: Apply, Analyze, and Evaluate

Critical thinking is the ability to reason and solve problems using facts and concepts. These questions can be approached from a number of angles and, in most cases, they do not have a single correct answer.

1. Individuals who smoke have much higher rates of lung infection. Explain which first-line defense mechanisms may be impaired by smoking, allowing pathogens to more readily enter the lower respiratory tract.

2. Conduct additional research and summarize one example of a genetic variation that in humans prevents the entry of HIV into host cells.

3. a. Describe how the immune system distinguishes foreign cells from cells deemed as self.
   b. Type I diabetes may be triggered by immune cells attacking one's own insulin-secreting pancreatic cells. Research shows that this may occur in an individual after a viral infection. Develop a hypothesis to explain this situation, explaining how PAMPs may play a role in this process.

4. a. Viewing a prepared slide of a blood smear with a brightfield microscope, you visualize relatively few large stained cells that contain a rounded nucleus. Upon closer inspection, you note the lack of granules in the cytoplasm. What type of blood cell are you most likely viewing based upon this description?
   b. You would look for which type(s) of immune cell in a blood sample for the (1) diagnosis of a parasitic helminth infection; and (2) diagnosis of an allergic reaction?

5. Myasthenia gravis is an autoimmune disease caused by T cells attacking healthy tissue. When diagnosed in early childhood, the treatment often involves the removal of the thymus. Discuss why this treatment is therapeutic for this condition based upon your knowledge of basic immunology.

6. a. Explain why white cells but not red cells are normally found in lymph.
   b. Explain why swollen lymph nodes are often indicative of a microbial infection.

7. Summarize the action of phagocytes in reaction to an invading microbe. Can a pathogen ever evade phagocytosis? Why or why not?

8. a. Inflammation is characterized by heat, pain, redness, and swelling. Discuss the vascular changes that lead to the development of these signs and symptoms.
   b. Construct a paragraph explaining how immune cells migrate to the site of injury. Use the following terms: *vasodilation, margination, diapedesis, chemotaxis,* and *chemoattractant.*

9. The diagnosis of tuberculosis involves the observation of lung structures called tubercles on an X ray. What immunologic process leads to the formation of tubercles, and what type of immune cells may comprise these lesions?

10. HIV predominantly infects T helper cells, cells that are responsible for coordinating B- and T-cell activity. Based upon this information, explain why HIV-infected individuals are at a very high risk for developing microbial infections.

Concept Connections | Bloom's Levels 4 and 6: Analyze and Create

This activity ties together multiple concepts in the chapter.

Scenario: It has been a busy day and you haven't had time to have lunch. While you're getting gas in your car between your classes and work, you decide to run into the convenience store and grab an egg salad sandwich and a soda. Later that day, after you get home from work, you feel chilled and nauseated and have abdominal pain. When you take your temperature with an oral thermometer, you start to gag and vomit. While you're recovering on the bathroom floor, you notice that the thermometer says your fever is 101°F. A few hours later, while you're lying on the couch watching reruns on TV, you feel impending diarrhea and race back to the bathroom. For the rest of the night, you alternate between vomiting and diarrhea. When you call your mom to tell her of

your woes, she says, "It might be *Salmonella* food poisoning. What did you eat today?" It's only then that you remember the egg salad sandwich from the convenience store.

1. Which of the symptoms that you experienced can be attributed to the first line of defense of the immune system?

2. Which of the symptoms that you experienced can be attributed to the second line of defense of the immune system?

3. What other aspects of the first and second lines of defense are working to defend your body against *Salmonella*?

4. Fill in the boxes on the chart to match symptoms with lines of defense.

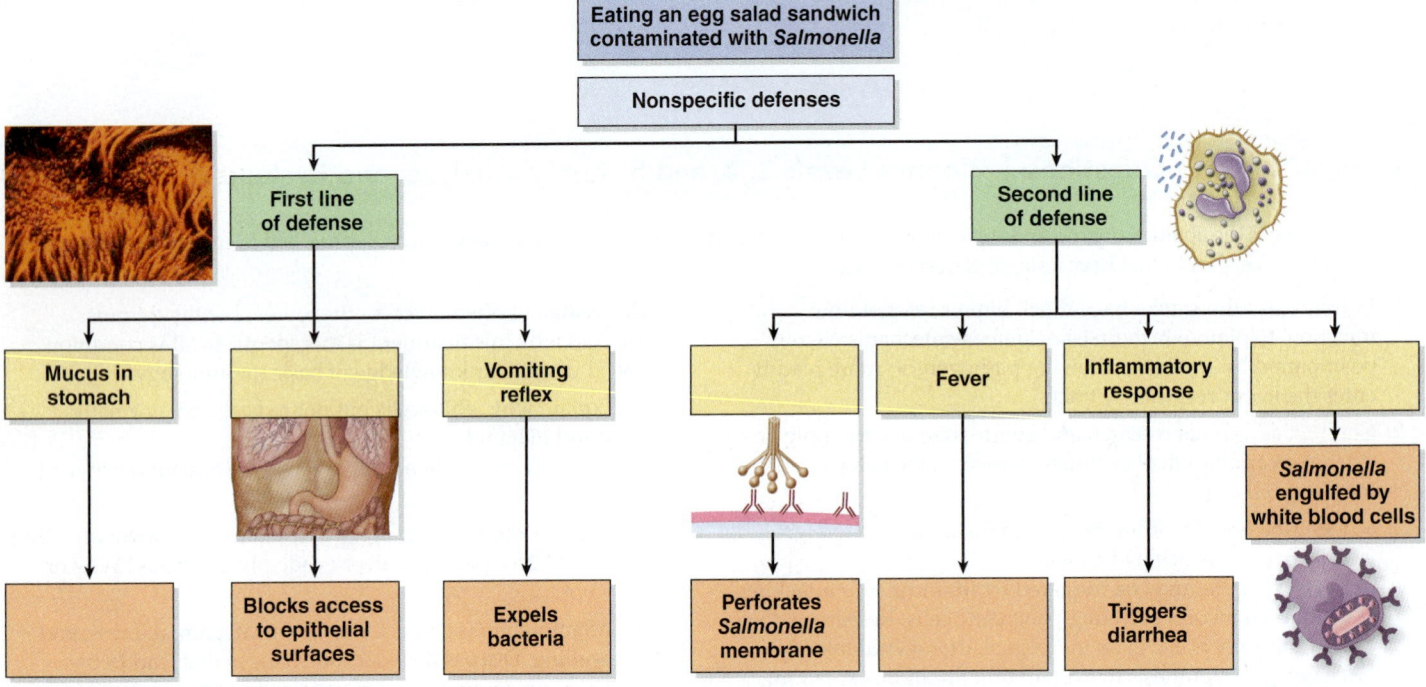

Visual Connections | Bloom's Level 5: Evaluate

This question uses visual images or previous content to make connections to this chapter's concepts.

1. **From chapter 4, figure 4.16.**
   a. In both cell types shown, sketch where the membrane attack complex (MAC) would form.

b. Speculate on whether gram-positive or gram-negative bacterial cells are more resistant to the formation of a membrane attack complex.

Outer membrane layer
Peptidoglycan
Cell membrane

Gram-Positive

Wall teichoic acid
Lipoteichoic acid

Envelope

Peptidoglycan

Cell membrane

Membrane proteins

Gram-Negative

Lipoproteins
Porin proteins
Lipopolysaccharides
Phospholipids

Outer membrane layer

Periplasmic space

Membrane protein

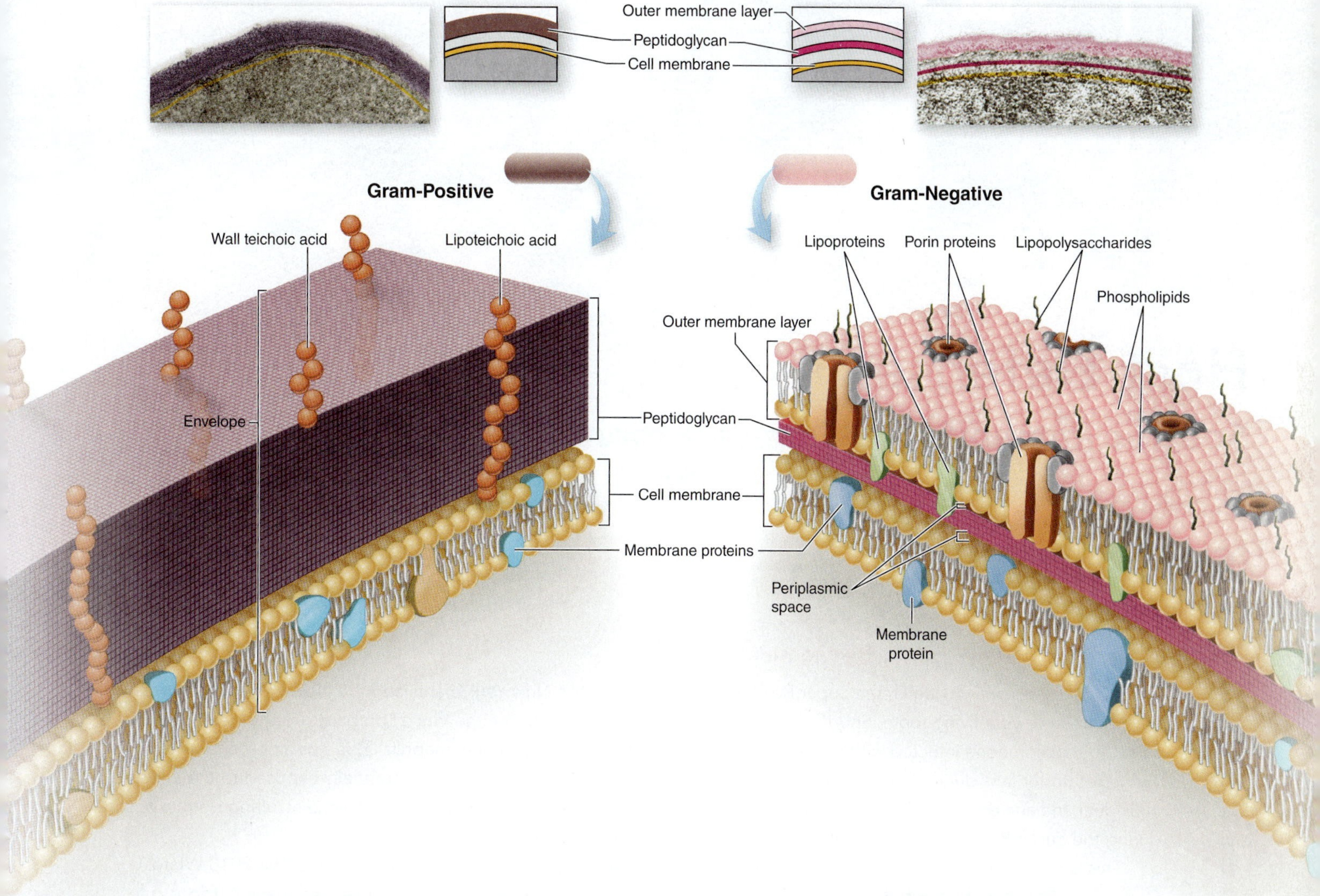

Concept Mapping | Bloom's Level 6: Create

Appendix D provides guidance for working with concept maps.

1. Using the words that follow, please create a concept map illustrating the relationships among these key terms from chapter 14.

defenses	monocytes	antibodies
leukocytes	macrophages	neutrophils
lymphocytes	inflammation	fever

15

Host Defenses II
Specific Immunity and Immunization

Case File 15

Super Bowl Fever? It May Be the Measles

In 2012, the safest way to watch the Super Bowl matchup between the New York Giants and the New England Patriots may have been from your living room. Shortly after the New York Giants took home the Vince Lombardi Trophy, news reports surfaced that someone infected with measles visited the Super Bowl Village in Indianapolis. This was an entertainment venue set up for fans during Super Bowl week, and an estimated 200,000 visitors were potentially exposed to the infected individual. Indiana officials later reported a total of 14 confirmed cases of measles in central Indiana resulting from exposure during the Super Bowl celebrations, all among individuals who were not vaccinated.

This isn't the first incident in which a measles outbreak has accompanied a major sporting event. The 2010 Winter Olympics in Vancouver attracted over 60,000 visitors to the opening ceremonies alone and spawned many outdoor celebrations following Canada's record of 14 gold medals at the games. Following the Olympics, 85 individuals in the province of British Columbia became infected with measles, the majority of whom were individuals who also had not been vaccinated.

■ What type of infectious agent is responsible for causing the measles?

■ How is measles spread?

■ Why would individuals attending sporting events be more susceptible to exposure to the measles?

Continuing the Case appears on page 444.

Outline and Learning Outcomes

15.1 Specific Immunity: The Third Line of Defense
1. Describe how the third line of defense is different from the other host defense mechanisms.
2. List the four stages of a specific immune response.
3. Discuss four major functions of immune system markers.
4. Define the role of the major histocompatibility complex (MHC), and list the three classes of MHC genes.
5. Compare and contrast the process of antigen recognition in T cells and B cells.

15.1 Specific Immunity: The Third Line of Defense

In chapter 14, we described the capacity of the immune system to survey, recognize, and react to foreign cells and molecules; and we overviewed the characteristics of nonspecific host defenses, blood cells, phagocytosis, inflammation, and complement. In addition, we introduced the concepts of acquired immunity and specificity. In this chapter, we take a closer look at those topics.

When host barriers and nonspecific defenses fail to control an infectious agent, a person with a normally functioning immune system has a mechanism to resist the pathogen—the third, specific line of immunity. Immunity is the resistance developed after contracting ailments such as chickenpox or measles, and it provides long-term protection against future attacks. This sort of immunity is not innate but adaptive; it is acquired only after an immunizing event such as an infection. The absolute need for acquired or adaptive immunity is impressively documented in children who have genetic defects in this system or in AIDS patients who have lost it. Even with heroic measures to isolate the patient, combat infection, or restore lymphoid tissue, the victim is constantly vulnerable to life-threatening infections.

Acquired specific immunity is the product of a dual system that we have previously mentioned—the B and T lymphocytes. During development, these lymphocytes undergo a selective process that specializes them for reacting only to one specific antigen or immunogen. During this time, **immunocompetence,**

the ability of the body to react with countless foreign substances, develops. An infant is born with the theoretical potential to react to an extraordinary array of different substances.

Antigens or immunogens figure very prominently in specific immunity. They are defined as molecules that stimulate a response by T and B cells. They are usually protein or polysaccharide molecules on or inside all cells and viruses, including our own. (Environmental chemicals can also be antigens. These are covered in chapter 16.) In fact, any exposed or released protein or polysaccharide is potentially an antigen, even those on our own cells. For reasons we discuss later, our own antigens do not usually evoke a response from our own immune systems.

In chapter 14, we discussed pathogen-associated molecular patterns (PAMPs) that stimulate responses by phagocytic cells during an innate defense response. While PAMPs are molecules shared by many types of microbes that stimulate a nonspecific response, antigens are highly individual and stimulate specific immunity. The two types of molecules do share two characteristics: (1) They are "parts" of foreign cells (microbes or other foreign materials), and (2) they provoke a defensive reaction from the host.

Two features that most characterize this third line of defense are **specificity** and **memory.** Unlike mechanisms such as anatomical barriers or phagocytosis, acquired immunity is highly specific. For example, the antibodies produced during an infection against the chickenpox virus will function against that virus and not against the measles virus. The property of memory refers to the rapid mobilization of lymphocytes that have been programmed to "recall" their first engagement with the invader and rush to the attack once again.

The elegance and complexity of immune function are largely due to lymphocytes working closely together with phagocytes. To simplify and clarify the network of immunologic development and interaction, we present it here as a series of stages, with each stage covered in a separate section **(figure 15.1)**. The principal stages are

I. lymphocyte development and differentiation;
II. the presentation of antigens;
III. the challenge of B and T lymphocytes by antigens; and
IV. B-lymphocyte response (the production and activities of antibodies) and T-lymphocyte response (cell-mediated immunity).

This sequence is illustrated here and in figure 15.1. We will give an overview here and spend the rest of the chapter filling in the details.

A Brief Overview of the Immune Response

Lymphocyte Development

Lymphocytes are central to immune responsiveness. They undergo development that begins in the embryonic yolk sac and shifts to the liver and bone marrow. Although all lymphocytes arise from the same basic stem cell type, at some point in development they diverge into two distinct types.

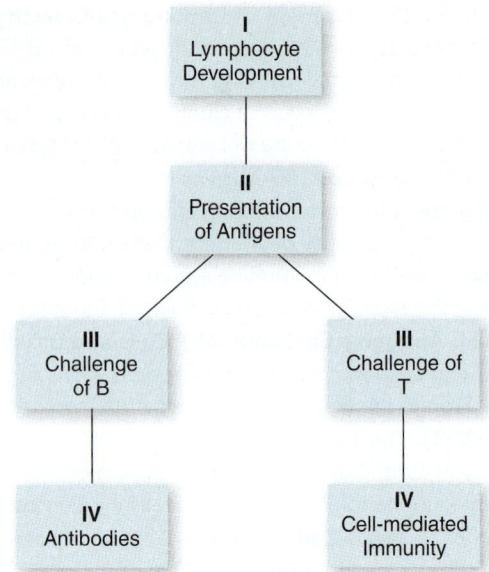

Final maturation of B cells occurs in specialized bone marrow sites, and that of T cells occurs in the thymus. Both cell types subsequently migrate to separate areas in the lymphoid organs (e.g., nodes and spleen). B and T cells constantly recirculate through the circulatory system and lymphatics, migrating into and out of the lymphoid organs.

Microbiological concepts are more than simply facts about disease. For example: You meet a guy at a party. Your eyes lock. You fall madly in love. Was it his suave manner, his witty conversation, or because he looked like George Clooney? Possibly none of these. Research suggests that what you're most attracted to may be his immune system. Fhionna Moore of Abetay University in the United Kingdom recruited 74 Latvian men in their early 20s and sampled their blood before and after they were given a hepatitis B vaccine. Then, she recruited 94 Latvian women to study photographs of these men and rate them on a 10-point scale of attractiveness. Moore and her fellow researchers found that the men found most attractive by the women were those who had the highest levels of testosterone and the highest levels of antibodies generated in response to the vaccine. The most attractive men also had the lowest levels of the stress hormone cortisol, which suggests that stress and accompanying high levels of cortisol may take a toll on the immune system and thus attractiveness.

Or—maybe it was his cologne or his body odor. Claus Wedekind at the University of Lausanne, Switzerland, found that the particular scent of one's body odor is tied to his or her unique major histocompatibility complex (MHC) molecules. In Wedekind's study, the MHC molecules of male and female students were typed and male students were asked to wear a T-shirt

to bed for two consecutive nights. Female students were then asked to sniff the T-shirts and rate them on how pleasant the odors were. The T-shirts that were rated most pleasant by the female students were from male students who had the greatest difference in the MHC complex from the female. Wedekind has developed several hypotheses to explain just how the MHC is involved in attractiveness. "In our evolutionary past, humans would have lived in small groups where the risk of inbreeding was high, so a method of distinguishing the most dissimilar mates would have been useful," he stated. Another possibility is that the skin's MHC may have an influence on the bacteria that thrive there.

Sources: 2012. *Nature Comm.* vol. 3. doi:10.1038/ncomms1696; 1995. *Proc. Royal Soc.* vol. 260, no. 1359, p. 245.

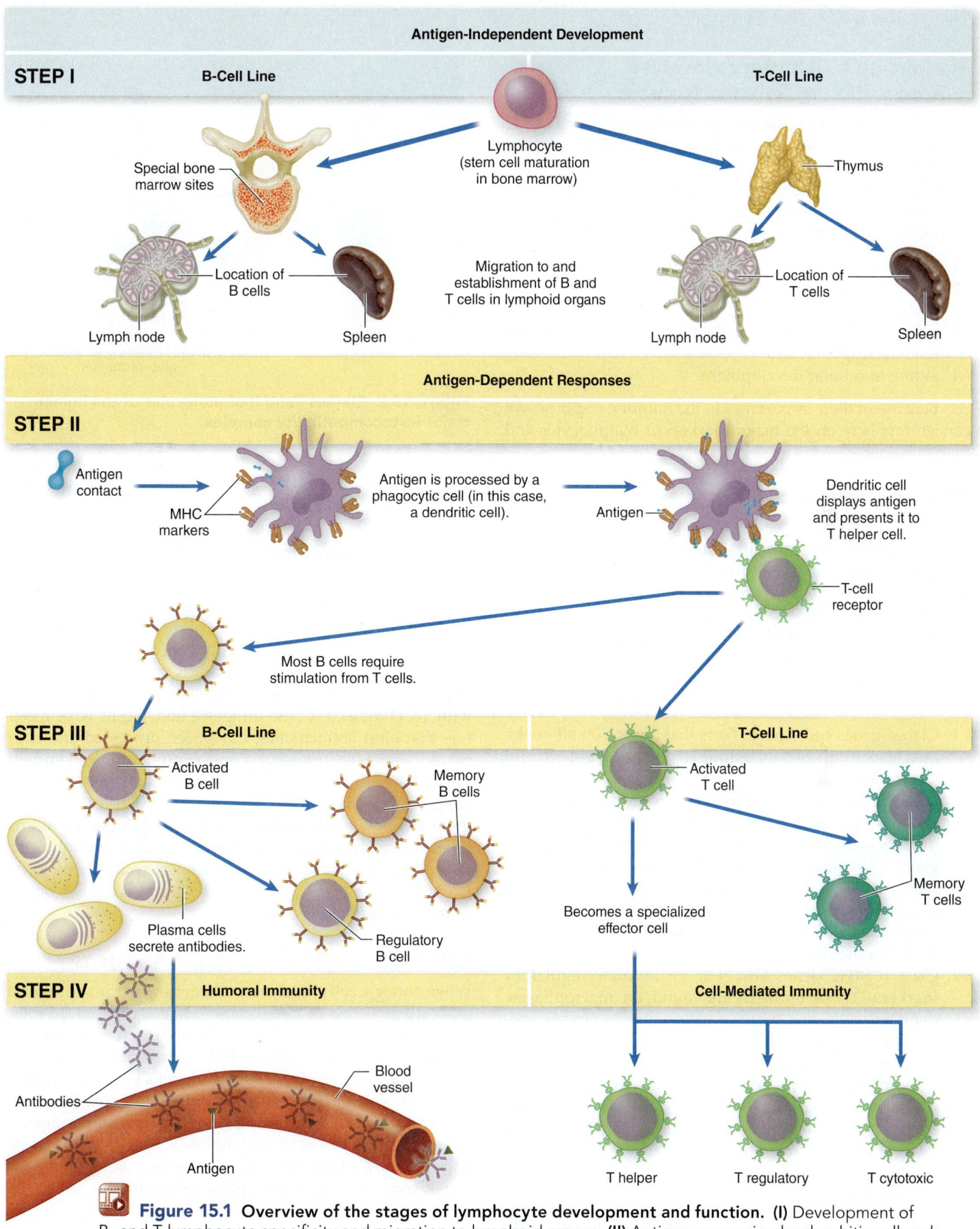

Figure 15.1 Overview of the stages of lymphocyte development and function. **(I)** Development of B- and T-lymphocyte specificity and migration to lymphoid organs. **(II)** Antigen processing by dendritic cell and presentation to lymphocytes; assistance to B cells by T cells. **(III)** Lymphocyte activation, clonal expansion, and formation of memory B and T cells. **(IV)** End result of lymphocyte activation. *Left-hand side:* antibody release; *right-hand side:* cell-mediated immunity. Details of these processes are covered in each corresponding section heading.

Markers on Cell Surfaces Involved in Recognition of Self and Nonself

Chapter 14 touched on the fundamental idea that cell markers (sometimes called receptors) confer specificity and identity. A given cell can express several different markers, each type playing a distinct and significant role in detection, recognition, and cell communication. Major functions of immune system markers are

1. attachment to nonself or foreign antigens;
2. binding to cell surface receptors that indicate self, such as MHC molecules (discussed next);
3. receiving and transmitting chemical messages to coordinate the response; and
4. aiding in cellular development.

Because of their importance in the immune response, we concentrate here on the major markers of lymphocytes and macrophages.

Major Histocompatibility Complex

One set of genes that codes for human cell receptors is the **major histocompatibility complex (MHC).** This gene complex gives rise to a series of glycoproteins (called MHC molecules) found on all cells except red blood cells. The MHC is also known as the human leukocyte antigen (HLA) system. This marker complex plays a vital role in recognition of self by the immune system and in rejection of foreign tissue.

Three classes of MHC genes have been identified:

1. Class I genes code for markers that appear on all nucleated cells. They display unique characteristics of self and allow for the recognition of self molecules and the regulation of immune reactions. The system is rather complicated in its details, but in general, each human being inherits a particular combination of class I MHC (HLA) genes in a relatively predictable fashion. Although millions of different combinations and variations of these genes are possible among humans, the closer the blood relationship, the greater the probability for similarity in MHC profile (see chapter 16).
2. Class II MHC genes also code for immune regulatory markers. These markers are found on macrophages, dendritic cells, and B cells and are involved in presenting antigens to T cells during cooperative immune reactions.
3. Class III MHC genes encode proteins involved with the complement system among others. We'll focus on classes I and II in this chapter. See **figure 15.2** for depictions of the first two MHC classes.

CD Molecules

Another set of markers that are important in immunity are the CD molecules. (CD stands for "cluster of differentiation.") CDs are molecules on the membranes of a variety of different cells involved in the immune response. Over

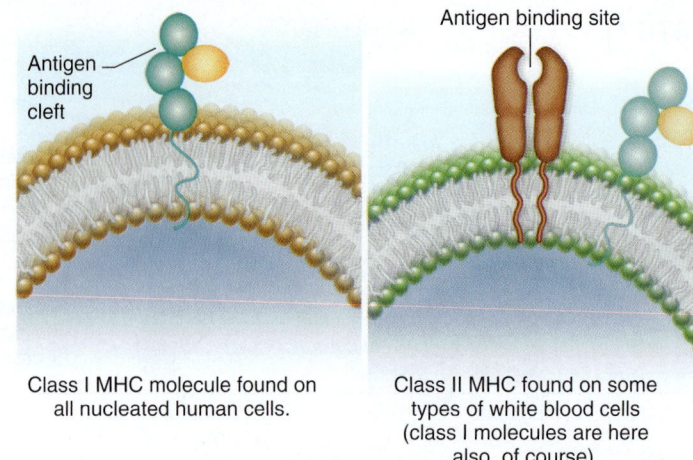

Figure 15.2 Classes I and II of molecules of the human major histocompatibility complex.

Antigen binding site

Antigen binding cleft

Class I MHC molecule found on all nucleated human cells.

Class II MHC found on some types of white blood cells (class I molecules are here also, of course).

300 have been described. The major ones will be described in the following sections.

Lymphocyte Receptors and Specificity to Antigen

The part lymphocytes play in immune surveillance and recognition emphasizes the essential role of their markers. These markers are even more frequently called receptors, a name that emphasizes that their major role is to "accept" or "grasp" antigens in some form. B cells have receptors that, together with a CD molecule, bind antigens, and T cells have receptors that bind antigens that have been processed and complexed with MHC molecules on the presenting cell surface. **Figure 15.3** illustrates the surfaces of B and T cells, and their antigen receptors. Antigen molecules exist in great diversity; there are potentially millions and even billions of unique types. The many sources of antigens include microorganisms as well as an endless array of chemical compounds in the environment. We will soon see how T and B cells recognize so many different antigens.

Entrance and Presentation of Antigens

When foreign cells, such as pathogens (carrying antigens), cross the first line of defense and enter the tissue, resident phagocytes migrate to the site. Tissue macrophages ingest the pathogens and induce an inflammatory response in the tissue if appropriate. Tissue dendritic cells ingest the antigen and migrate to the nearest lymphoid organ (often the draining lymph nodes). Here they process and present antigen to T lymphocytes. Pieces of the pathogens also drain into these lymph nodes. Along with dendritic cells, macrophages and B cells serve as antigen-presenting cells (APCs).

Antigen Challenge and Clonal Selection

When challenged by antigen, both B cells and T cells further proliferate and differentiate. The multiplication of a

Figure 15.3 The surfaces of T cells and B cells.

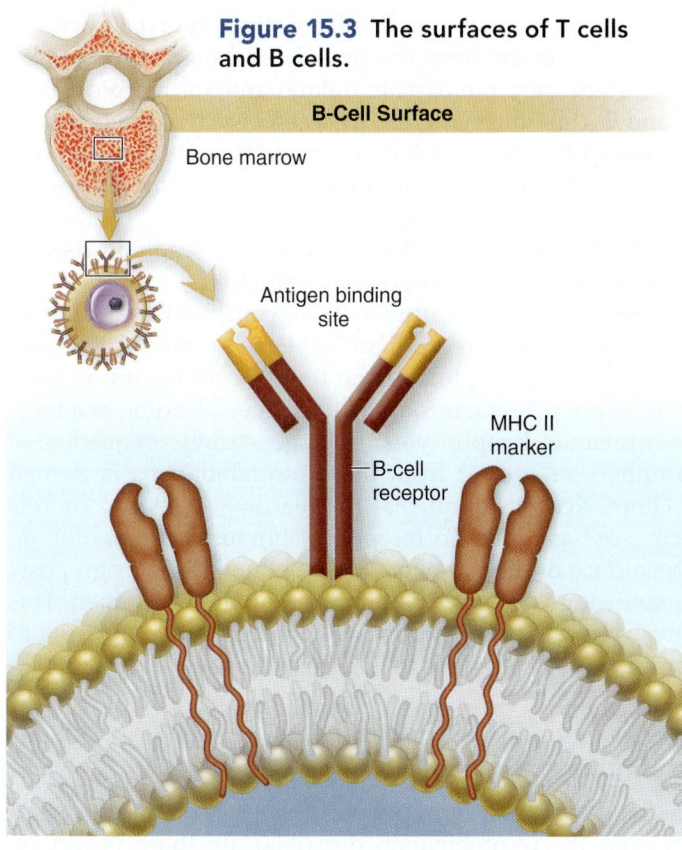

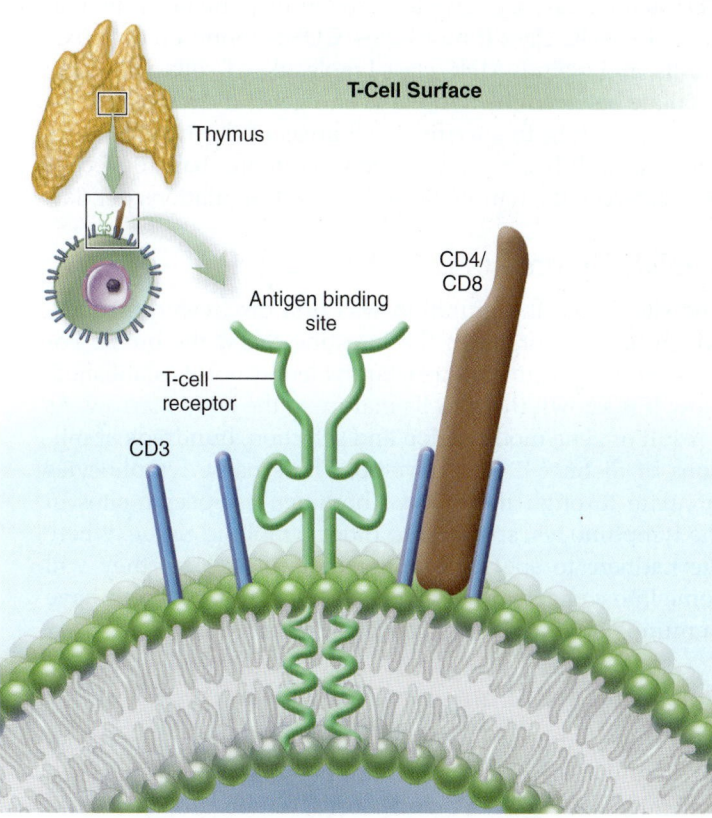

particular lymphocyte creates a **clone**, or group of genetically identical cells, some of which are memory cells that will ensure future reactiveness against that antigen. Because the B-cell and T-cell responses differ significantly from this point on in the sequence, they are summarized separately.

How T Cells Respond to Antigen: Cell-Mediated Immunity (CMI)

T-cell types and responses are extremely varied. When activated (sensitized) by antigen, a T cell gives rise to one of three different types of progeny, each involved in a cell-mediated immune function. The three main functional types of T cells are:

1. helper T cells that activate macrophages, assist B-cell processes, and help activate cytotoxic T cells;
2. regulatory T cells that control the T-cell response; and
3. cytotoxic T cells that lead to the destruction of infected host cells and other "foreign" cells.

Although T cells secrete cytokines that help destroy pathogens and regulate immune responses, they do not produce antibodies.

How B Cells Respond to Antigen: Antibody Release

When activated by antigen, a B cell divides, giving rise to plasma cells, each with the same reactive profile. Plasma cells

release antibodies into the tissue and blood. When these antibodies attach to the antigen for which they are specific, the antigen is marked for destruction or neutralization.

15.1 Learning Outcomes—Assess Your Progress

1. Describe how the third line of defense is different from the other host defense mechanisms.
2. List the four stages of a specific immune response.
3. Discuss four major functions of immune system markers.
4. Define the role of the major histocompatibility complex (MHC), and list the three classes of MHC genes.
5. Compare and contrast the process of antigen recognition in T cells and B cells.

15.2 Step I: The Development of Lymphocyte Diversity

Specific Events in T-Cell Maturation

The maturation of most T cells and the development of their specific receptors are directed by the thymus and its hormones. Other T cells reach full maturity in the gastrointestinal tract. In addition to the antigen-specific T-cell receptor, all mature T lymphocytes express CD3 markers. CD3 molecules surround the T-cell receptor and assist in binding. T cells also express either a CD4 or a CD8 coreceptor (see figure 15.3).

CD4 is an accessory receptor protein on T helper cells that binds to MHC class II molecules. CD8 is found on cytotoxic T cells, and it binds MHC class I molecules. T cells constantly circulate between the lymphatic and general circulatory systems, migrating to specific T-cell areas of the lymph nodes and spleen. It has been estimated that more than 10^9 T cells pass between the lymphatic and general circulations per day.

Specific Events in B-Cell Maturation

The site of B-cell maturation was first discovered in birds, which have an organ in the intestine called the bursa. For some time, the human bursal equivalent was not established. Now it is known that B cells mature in the bone marrow. As a result of gene modification and selection, hundreds of millions of distinct B cells develop. These naive lymphocytes circulate through the blood, "homing" to specific sites in the lymph nodes, spleen, and other lymphoid tissue, where they adhere to specific binding molecules. Here they will come into contact with antigens throughout life. B cells have immunoglobulins as surface receptors **(table 15.1)**.

The Origin of Immunologic Diversity

Each naive lymphocyte bears an antigen receptor that recognizes a unique antigen. How is this possible? The mechanism, generally true for both B and T cells, can be summarized as follows: In the bone marrow, stem cells can become granulocytes, monocytes, or lymphocytes. The lymphocytes then become either T cells or B cells. Cells destined to become B cells stay in the bone marrow; T cells migrate to the thymus. Here they build their unique antigen receptor. Both B and T cells then migrate to secondary lymphoid tissues **(figure 15.4)**. The secondary lymphoid tissues are resupplied with B and T cells because some self-destruct if they are not used and others become activated and leave.

By the time T and B cells reach the lymphoid tissues, each one is already equipped to respond to a single unique antigen. This amazing diversity is generated by extensive rearrangements of more than 500 gene segments that code for the antigen receptors on the T and B cells **(figure 15.5)**. In time, every possible recombination occurs, leading to a huge assortment of lymphocytes.[1] Each genetically unique line of lymphocytes arising from these recombinations is termed a **clone**. Keep in mind that the rearranged genetic code is expressed as a protein receptor of unique configuration on the surface of the lymphocyte, something like a "sign post" announcing its specificity and reactivity for an antigen. This *proliferative* stage of lymphocyte development occurs prior to lymphocytes' contact with foreign antigens.

The Specific B-Cell Receptor: An Immunoglobulin Molecule

In the case of B lymphocytes, the receptor genes that undergo the recombination described are those governing **immunoglobulin** (im"-yoo-noh-glahb'-yoo-lin) **(Ig)** synthesis.

1. Estimates of the theoretical number of possible variations that may be created vary from 10^{14} to 10^{18} different specificities.

Table 15.1 Contrasting Properties of B Cells and T Cells		
	B Cells	**T Cells**
Site of Maturation	Bone marrow	Thymus
Specific Surface Markers	Immunoglobulin; Distinct CD molecules	T-cell receptor; Distinct CD molecules
Circulation in Blood	Low numbers	High numbers
Receptors for Antigen	B-cell receptor (immunoglobulin)	T-cell receptor
Distribution in Lymphatic Organs	Cortex (in follicles)	Paracortical sites (interior to the follicles)
Require Antigen Presented with MHC	No	Yes
Product of Antigenic Stimulation	Plasma cells and memory cells	Several types of sensitized T cells and memory cells
General Functions	Production of antibodies to inactivate, neutralize, target antigens	Cells function in helping other immune cells, suppressing, killing abnormal cells; hypersensitivity; synthesize cytokines

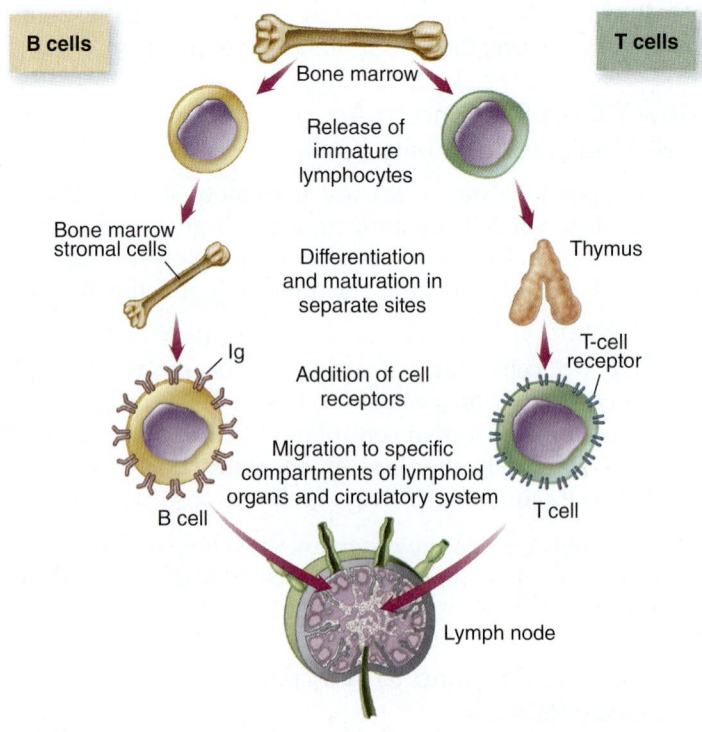

Figure 15.4 Major stages in the development of B and T cells.

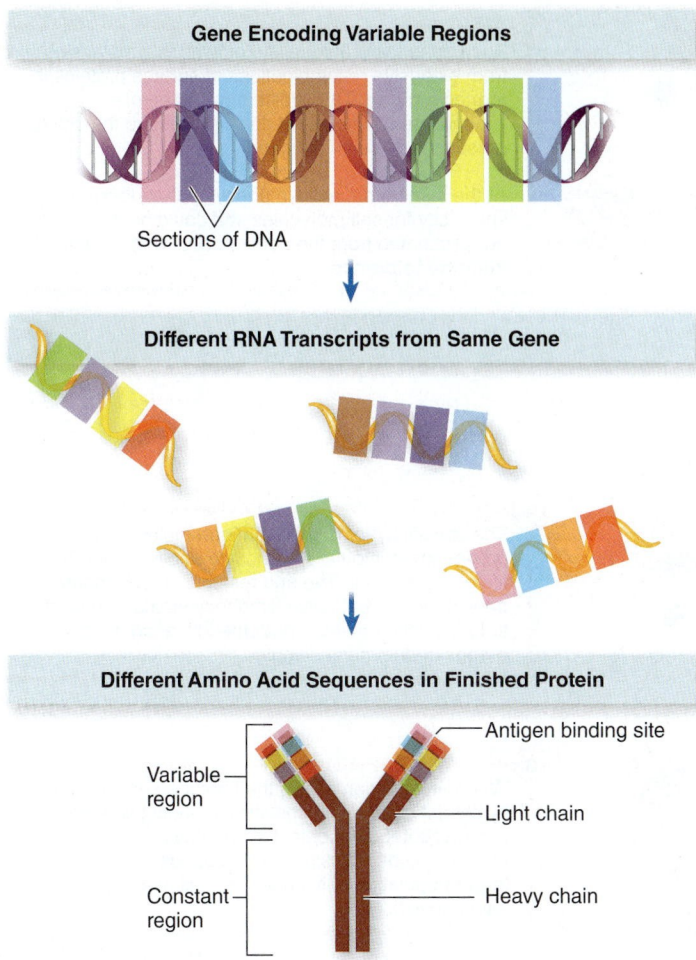

Gene Encoding Variable Regions

Sections of DNA

Different RNA Transcripts from Same Gene

Different Amino Acid Sequences in Finished Protein

Variable region

Constant region

Antigen binding site

Light chain

Heavy chain

Figure 15.5 The mechanism behind antibody variability.
The genes coding for the variable regions of antibody molecules have multiple different sections along their lengths. As a result of alternative splicing, very different RNA transcripts are created from the same original gene. When those transcripts are translated, the resulting protein will have extremely variable amino acid sequences—and, therefore, extremely variable shapes.

Immunoglobulins are large glycoprotein molecules that serve as the antigen receptors of B cells and, when secreted, as antibodies. The basic immunoglobulin molecule is a composite of four polypeptide chains: a pair of identical heavy (H) chains and a pair of identical light (L) chains (see figure 15.5). One light chain is bonded to one heavy chain, and the two heavy chains are bonded to one another with disulfide bonds, creating a symmetrical, Y-shaped arrangement.

The ends of the forks formed by the light and heavy chains contain pockets called the **antigen binding sites.** These sites can be highly variable in shape to fit a wide range of antigens. This extreme versatility is due to **variable regions (V)** in antigen binding sites, where amino acid composition is highly varied from one clone of B lymphocytes to another, a result of the genetic reassortment we discussed earlier. The remainder of the light chains and heavy chains consist of constant (C) regions whose amino acid content does not vary greatly from one antibody to another.

T-Cell Receptors

The T-cell receptor for antigen belongs to the same protein family as the B-cell receptor. It is similar to the B-cell receptor in being formed by genetic modification, having variable and constant regions, being inserted into the membrane, and having an antigen binding site formed from two parallel polypeptide chains (see figure 15.3). Unlike the immunoglobulins, the T-cell receptor is relatively small and is never secreted.

Clonal Selection

The second stage of development—**clonal selection** and expansion—does require stimulation by an antigen such as a microbe. When this antigen enters the immune surveillance system, it encounters specific lymphocytes ready to recognize it. Such contact stimulates that clone to undergo mitotic divisions and expands it into a larger population of lymphocytes all bearing the same specificity **(process figure 15.6).** This increases the capacity of the immune response to that antigen. Two important generalities one can derive from the phenomenon of clonal selection are that (1) lymphocyte specificity is preprogrammed, existing in the genetic makeup before an antigen has ever entered the tissues; and (2) each genetically distinct lymphocyte expresses only a single specificity and can react to only one type of antigen. Other important features of the lymphocyte response system are expanded in later sections.

One potentially problematic outcome of random genetic assortment is the development of clones of lymphocytes able to react to *self*. This outcome could lead to severe damage when the immune system actually perceives self molecules as foreign and mounts a harmful response against the host's tissues. Any such clones are destroyed during development through clonal *deletion*. The removal of such potentially harmful clones is the basis of **immune tolerance** or tolerance to self. Because humans are exposed to many new antigenic substances during their lifetimes, such as animal and plant cells that we consume as food, T cells and B cells in the periphery of the body have mechanisms for *not* reacting to innocuous antigens. Some diseases (i.e., autoimmune diseases) are thought to be caused by the loss of immune tolerance through the survival of certain "forbidden clones" or failure of these other systems (see chapter 16).

15.2 Learning Outcomes—Assess Your Progress

6. Summarize the maturation process of both B cells and T cells.
7. Draw a diagram illustrating how lymphocytes are capable of responding to nearly any antigen imaginable.
8. Outline the processes of clonal selection and expansion.
9. Describe the structure of both a B-cell receptor and a T-cell receptor.

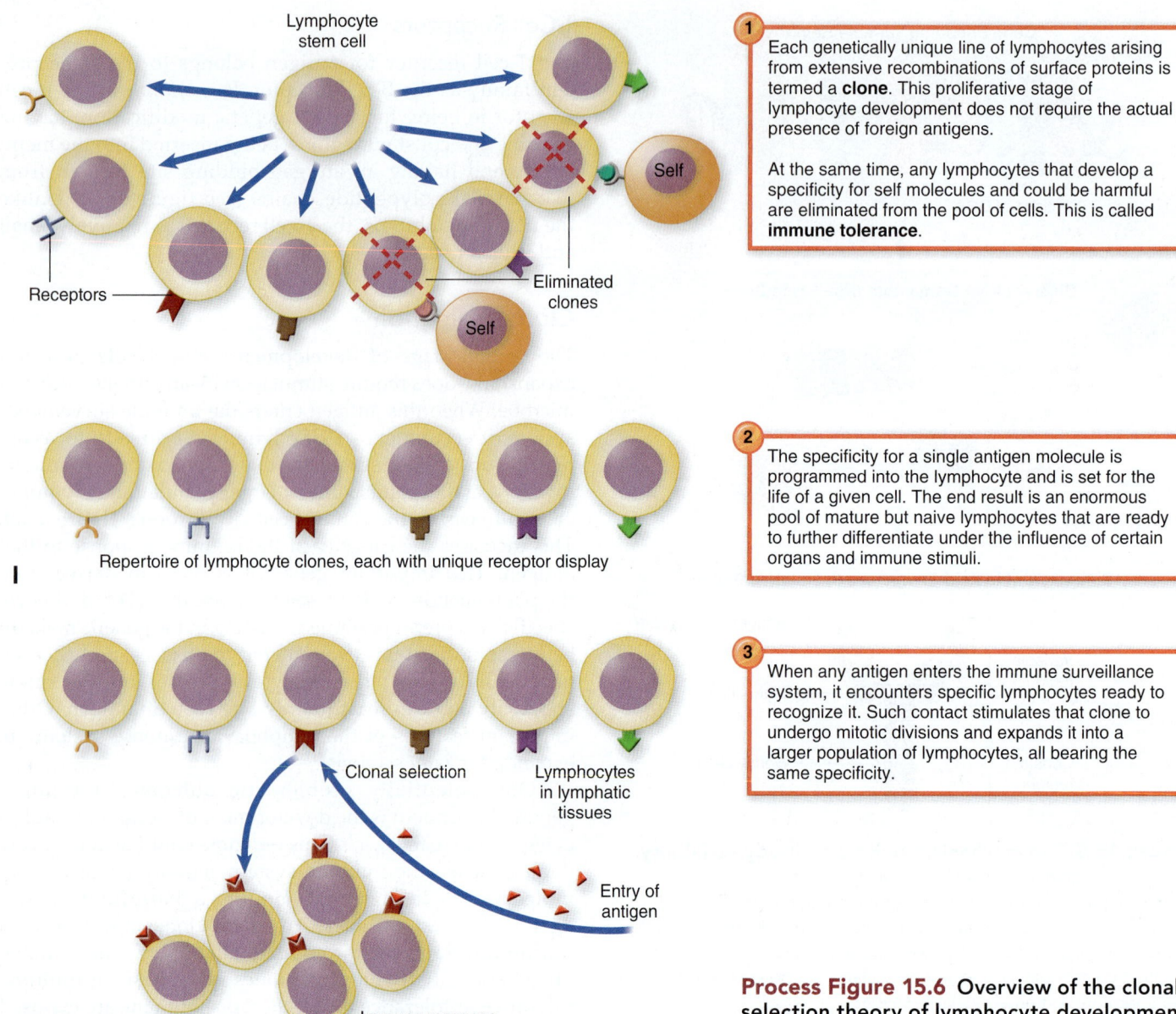

1. Each genetically unique line of lymphocytes arising from extensive recombinations of surface proteins is termed a **clone**. This proliferative stage of lymphocyte development does not require the actual presence of foreign antigens.

   At the same time, any lymphocytes that develop a specificity for self molecules and could be harmful are eliminated from the pool of cells. This is called **immune tolerance**.

2. The specificity for a single antigen molecule is programmed into the lymphocyte and is set for the life of a given cell. The end result is an enormous pool of mature but naive lymphocytes that are ready to further differentiate under the influence of certain organs and immune stimuli.

3. When any antigen enters the immune surveillance system, it encounters specific lymphocytes ready to recognize it. Such contact stimulates that clone to undergo mitotic divisions and expands it into a larger population of lymphocytes, all bearing the same specificity.

Process Figure 15.6 Overview of the clonal selection theory of lymphocyte development and diversity.

15.3 Step II: Presentation of Antigens

Having reviewed the characteristics of lymphocytes, let us more deeply examine the properties of antigens, the substances that cause them to react. As discussed earlier, an **antigen (Ag)** is a substance that provokes an immune response in specific lymphocytes. The property of behaving as an antigen is called **antigenicity**. To be perceived as an antigen or immunogen, a substance must meet certain requirements in foreignness, shape, size, and accessibility.

Characteristics of Antigens

One important characteristic of an antigen is that it is perceived as foreign, meaning that it is not a normal constituent of the body. Whole microbes or their parts, cells, or substances that arise from other humans, animals, plants, and various molecules all possess this quality of foreignness and thus are potentially antigenic to the immune system of an individual **(figure 15.7)**. Molecules of complex composition such as proteins and protein-containing compounds prove to be more immunogenic than repetitious polymers composed of a single type of unit. Most materials that serve as antigens fall into these chemical categories:

- proteins and polypeptides (enzymes, cell surface structures, hormones, exotoxins);
- lipoproteins (cell membranes);
- glycoproteins (blood cell markers);
- nucleoproteins (DNA complexed to proteins but not pure DNA); and
- polysaccharides (certain bacterial capsules) and lipopolysaccharides.

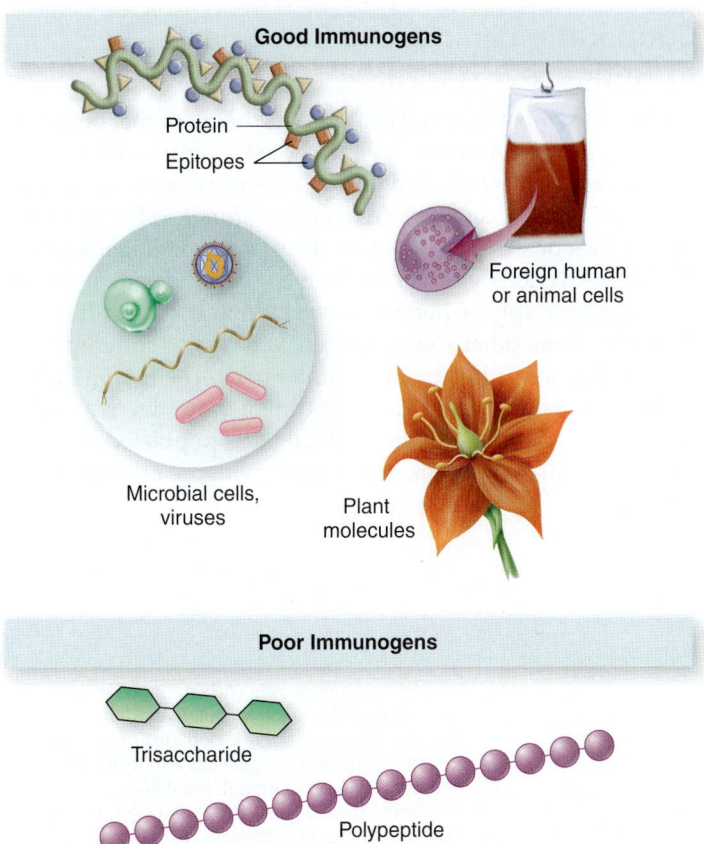

Good Immunogens

Protein

Epitopes

Foreign human
or animal cells

Microbial cells,
viruses

Plant
molecules

Poor Immunogens

Trisaccharide

Polypeptide

Figure 15.7   A comparison of good immunogens and poor immunogens.   *Top:* Good immunogens are large and complex. *Bottom:* Small molecules and linear molecules are less likely to be good immunogens.

Effects of Molecular Shape and Size

To initiate an immune response, a substance must also be large enough to "catch the attention" of the surveillance cells. Molecules with a molecular weight (MW) of less than 1,000 are seldom complete antigens, and those between 1,000 MW and 10,000 MW are typically weak antigens. Complex macromolecules approaching 100,000 MW are the most immunogenic. Note that large size alone is not sufficient for antigenicity; glycogen, a polymer of glucose with a highly repetitious structure, has a molecular weight over 100,000 and is not normally antigenic, whereas insulin, a protein with a molecular weight of 6,000, can be antigenic.

A lymphocyte's capacity to discriminate differences in molecular shape is so fine that it recognizes and responds to only a portion of the antigen molecule. This molecular fragment, called the **epitope** (shown in figure 15.7), is the primary signal that the molecule is foreign.

A Note About Epitopes and Antigens

While up to now we have been calling the immunogenic substance "the antigen," it is more precisely termed the *epitope*. You could say, for instance, "The antigenic portion of the protein on a microbe is the epitope." You will also note that in practice, clinicians, and even other parts of this book, use the word *antigen* when the precise term is *epitope*. You will know, however, that the part of the molecule that is actually recognized by the immune system is the epitope. This means that every epitope can be recognized by B- and T-cell receptors that were formed during genetic reassortment. The particular tertiary structure and shape of this determinant must conform like a key to the receptor "lock" of the lymphocyte, which then responds to it.

Small foreign molecules that consist only of a determinant group and are too small by themselves to elicit an immune response are termed **haptens.** However, if such an incomplete antigen is linked to a larger carrier molecule, the combined molecule develops immunogenicity **(figure 15.8).** The carrier group contributes to the size of the complex and enhances the proper spatial orientation of the determinative group, while the hapten serves as the epitope. Haptens include molecules such as drugs; metals; and ordinarily innocuous household, industrial, and environmental chemicals. Many haptens inappropriately develop antigenicity in the body by combining with large carrier molecules such as serum proteins.

Because each human being is genetically and biochemically unique (except for identical twins), the proteins and

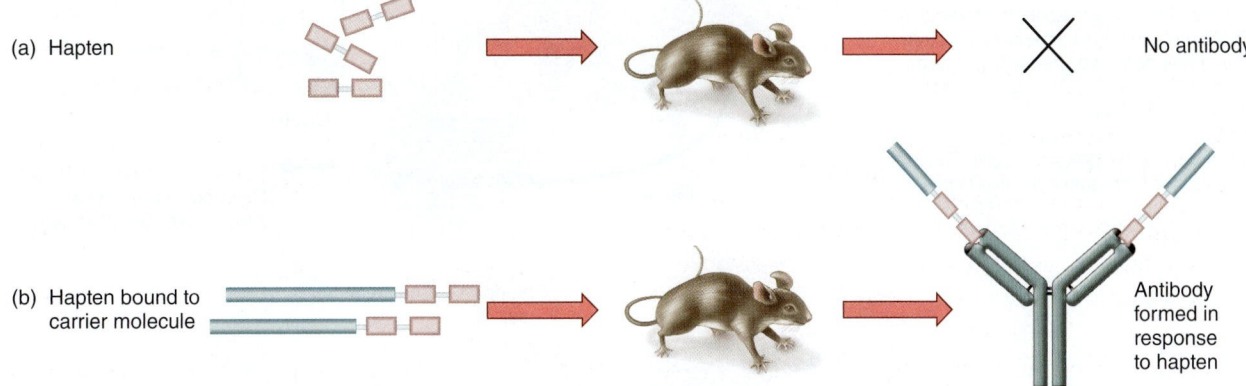

(a) Hapten

No antibody

(b) Hapten bound to
carrier molecule

Antibody
formed in
response
to hapten

Figure 15.8   The hapten-carrier phenomenon.   **(a)** Haptens are too small to be discovered by an animal's immune system; no response. **(b)** A hapten bound to a large molecule will serve as an epitope and stimulate a response and an antibody that is specific for it.

other molecules of one person can be antigenic to another. **Alloantigens** are cell surface markers and molecules that occur in some members of the same species but not in others. Alloantigens are the basis for an individual's blood group and major histocompatibility profile, and they are responsible for incompatibilities that can occur in blood transfusion or organ grafting.

Some bacterial toxins, which belong to a group of immunogens called **superantigens,** are potent stimuli for T cells. Their presence in an infection activates T cells at a rate 100 times greater than ordinary antigens. The result can be an overwhelming release of cytokines and cell death. Such diseases as toxic shock syndrome and certain autoimmune diseases are associated with this class of antigens.

Antigens that evoke allergic reactions, called **allergens,** are characterized in detail in chapter 16.

Cooperation in Immune Reactions to Antigens

The basis for most immune responses is the encounter between antigens and white blood cells. Microbes and other foreign substances enter most often through the respiratory or gastrointestinal mucosa and less frequently through other mucous membranes or the skin or across the placenta. Antigens introduced intravenously travel through the bloodstream and end up in the liver, spleen, bone marrow, kidney, and lung. If introduced by some other route, antigens are carried in lymphatic fluid and concentrated by the lymph nodes. The lymph nodes and spleen are important in concentrating the antigens and circulating them thoroughly through all areas populated by lymphocytes so that they come into contact with the proper clone.

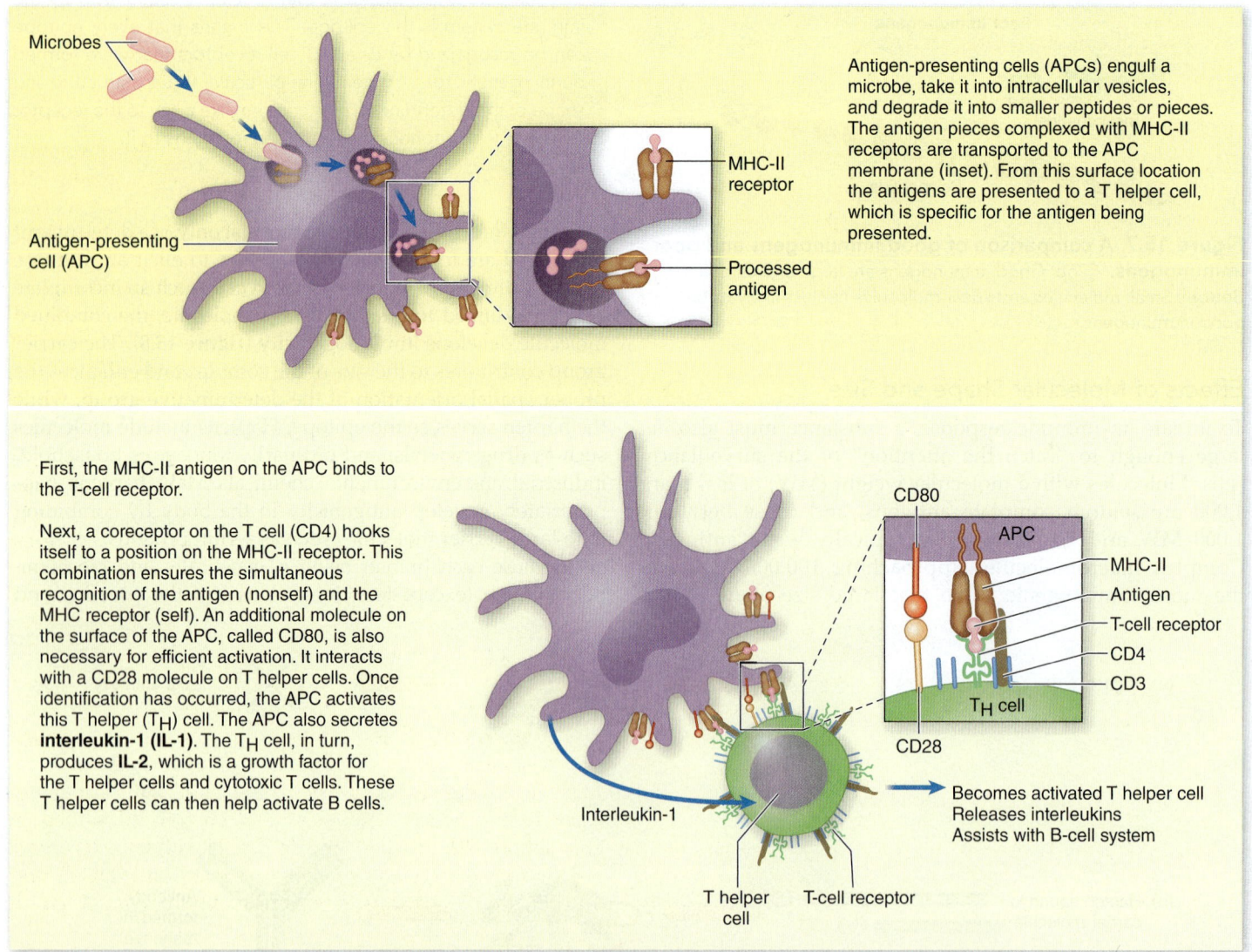

Process Figure 15.9 Interactions between antigen-presenting cells (APCs) and T helper (CD4) cells required for T-cell activation.   For T cells to recognize foreign antigens, they must have the antigen processed and presented by a professional APC such as a dendritic cell.

The Role of Antigen Processing and Presentation

In most immune reactions, the antigen must be further acted upon and formally presented to lymphocytes by cells called antigen-presenting cells (APCs). Three different cells can serve as APCs: macrophages, B cells, and dendritic (den'-drih-tik) cells. **Process figure 15.9** illustrates how this process works. **Figure 15.10** illustrates the activity of one particular APC, a dendritic cell. All three types of APCs can engulf antigens and process them intracellularly. After processing is complete, the antigen is inserted into a cleft on the MHC receptor and the complex is moved to the surface of the APC so that it will be readily accessible to T lymphocytes during presentation.

Disease Connection

Microbial pathogens sometimes exploit components of our defense system for their own purposes. For example, when HIV is taken up by dendritic cells in the normal course of the immune response, HIV gets transmitted to T cells, which are the main target of HIV infection. So we make it easy for the virus! There is also evidence that HIV changes the dendritic cell so that the virus can maintain itself inside them and provide a constant source of virus in the body.

15.3 Learning Outcomes—Assess Your Progress

10. Compare the terms *antigen* and *epitope*.

11. List characteristics of antigens that optimize their immunogenicity.

12. Describe how the immune system responds to alloantigens, superantigens, and allergens.

13. List the types of cells that can act as antigen-presenting cells.

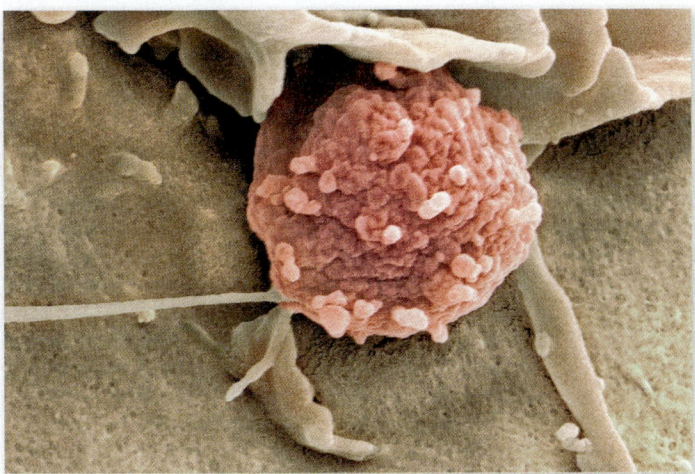

Figure 15.10 Dendritic cell.   This is a close-up view of a dendritic cell (brown) beginning to engulf a spore of the fungus *Aspergillus*.

15.4 Step III: Antigenic Challenge of T Cells and B Cells

The Activation of T Cells and Their Differentiation into Subsets

Now that the antigen is presented on the surface of the APC, these cells are ready to activate T cells bearing CD4 markers. This class of T cells is called the T helper class; it bears an antigen-specific T-cell receptor that binds to the antigen (epitope) held by the MHC molecule. At the same time, the T helper cell's CD4 marker also binds to the MHC molecule. Once identification has occurred, the APC activates this T helper (T_H) cell. The T_H cell, in turn, produces a cytokine, **interleukin-2 (IL-2),** which is a growth factor for the T helper cells and cytotoxic T cells.

A stimulated T cell multiplies through successive mitotic divisions and produces a large population of genetically identical daughter cells in the process of clonal expansion **(figure 15.11).** Some cells that are activated stop short of becoming fully differentiated, such as the memory cells. However, the expansion of this cell type's clone size accounts for the increased speed and intensity of the memory response.

A T helper cell activated in this way can then help activate B cells. The manner in which B and T cells subsequently become activated by the APC–T helper cell complex and their individual responses to antigen are addressed in later sections.

Not all antigens require T helper cell intervention to activate B cells. A few antigens can trigger a response from B lymphocytes without the cooperation of APCs or T helper cells. These are called T-cell-independent antigens, and are usually simple molecules such as carbohydrates with many repeating and invariable determinant groups. Because so few antigens are of this type, most B-cell reactions require T helper cells. We call these T-cell-dependent antigens.

As you see at the bottom of figure 15.11, the activation of another type of T cell, which bears CD8—not CD4—markers, is very similar. CD8-bearing T cells are called T cytotoxic cells. They are activated by antigen on the surface of APCs that has been complexed with MHC-I (not MHC-II) molecules. They are then ready to do their job, which is very different than that of T helper cells, as we will see later.

In summary, mature T cells in lymphoid organs are primed to react with antigens that have been processed and presented to them by dendritic cells, B cells, or macrophages. They recognize an antigen only when it is presented in association with an MHC carrier. T cells with CD4 receptors recognize endocytosed peptides presented on MHC-II, and T cells with CD8 receptors recognize peptides presented on MHC-I.

Activated T cells transform in preparation for mitotic divisions, and they differentiate into one of the subsets of effector cells and memory cells that can respond quickly to MHC plus antigen on APCs upon subsequent contact

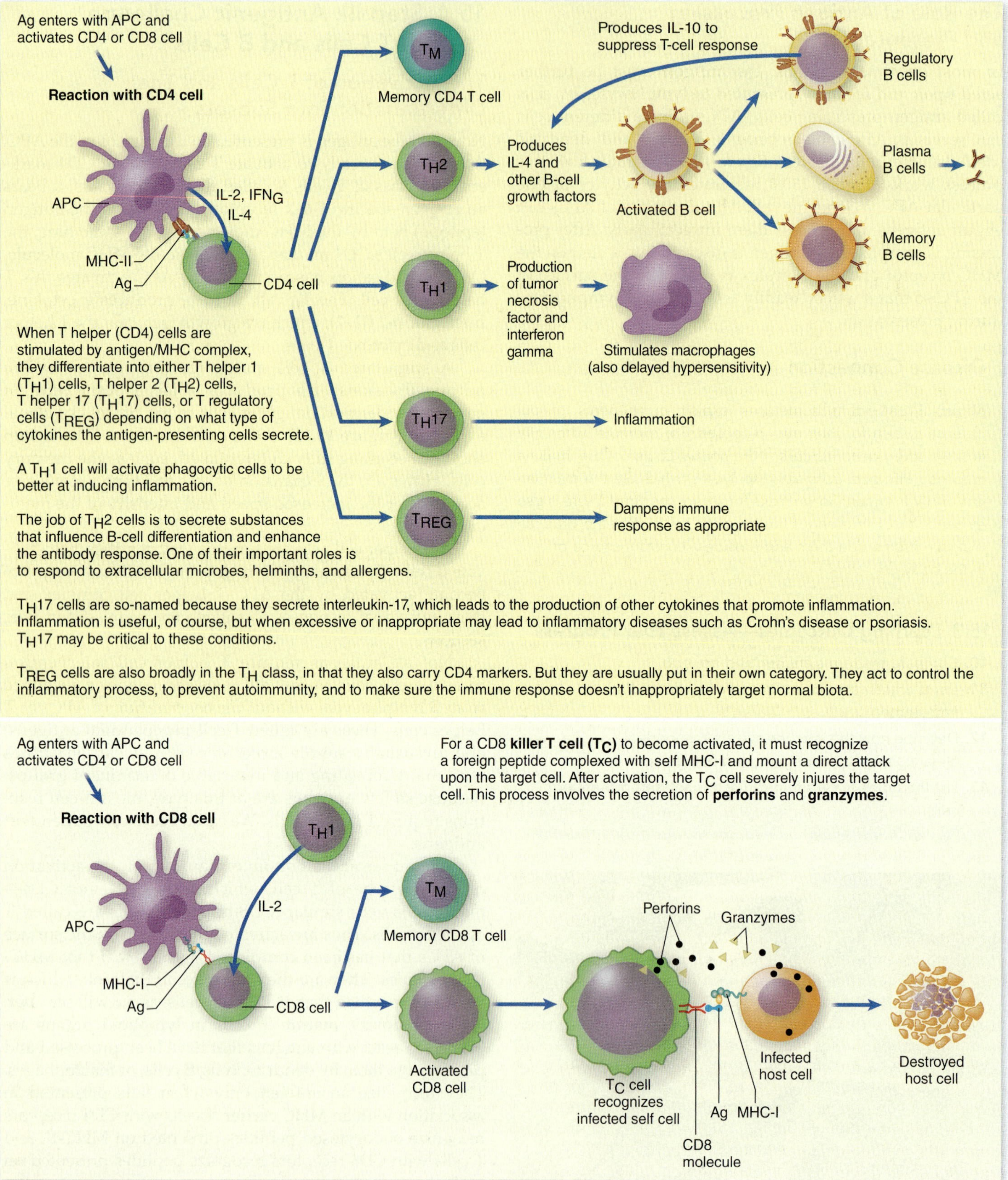

Ag enters with APC and
activates CD4 or CD8 cell

Reaction with CD4 cell

APC

IL-2, IFN$_G$
IL-4

MHC-II

Ag ——— CD4 cell

T$_M$

Memory CD4 T cell

T$_H$2 → Produces
IL-4 and
other B-cell
growth factors

T$_H$1 → Production
of tumor
necrosis
factor and
interferon
gamma

Stimulates macrophages
(also delayed hypersensitivity)

Produces IL-10 to
suppress T-cell response

Regulatory
B cells

Activated B cell

Plasma
B cells

Memory
B cells

When T helper (CD4) cells are
stimulated by antigen/MHC complex,
they differentiate into either T helper 1
(T$_H$1) cells, T helper 2 (T$_H$2) cells,
T helper 17 (T$_H$17) cells, or T regulatory
cells (T$_{REG}$) depending on what type of
cytokines the antigen-presenting cells secrete.

A T$_H$1 cell will activate phagocytic cells to be
better at inducing inflammation.

The job of T$_H$2 cells is to secrete substances
that influence B-cell differentiation and enhance
the antibody response. One of their important roles is
to respond to extracellular microbes, helminths, and allergens.

T$_H$17 → Inflammation

T$_{REG}$ → Dampens immune
response as appropriate

T$_H$17 cells are so-named because they secrete interleukin-17, which leads to the production of other cytokines that promote inflammation.
Inflammation is useful, of course, but when excessive or inappropriate may lead to inflammatory diseases such as Crohn's disease or psoriasis.
T$_H$17 may be critical to these conditions.

T$_{REG}$ cells are also broadly in the T$_H$ class, in that they also carry CD4 markers. But they are usually put in their own category. They act to control the
inflammatory process, to prevent autoimmunity, and to make sure the immune response doesn't inappropriately target normal biota.

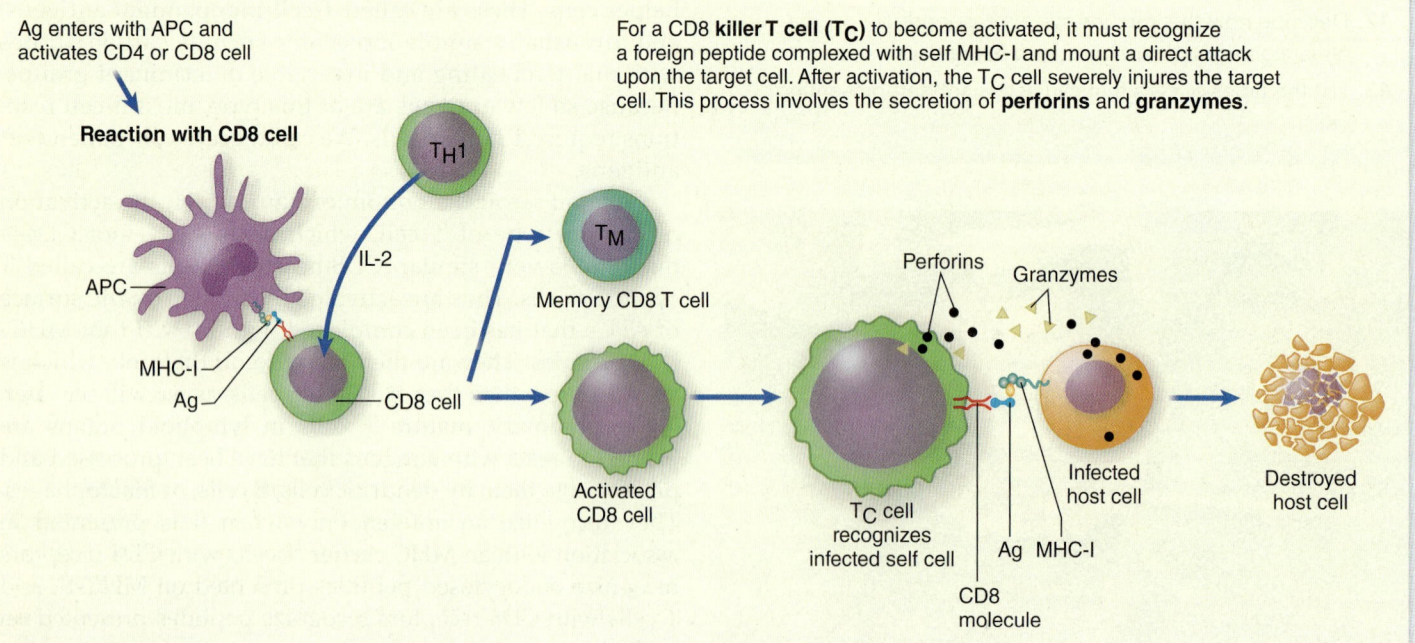

Ag enters with APC and
activates CD4 or CD8 cell

Reaction with CD8 cell

APC

T$_H$1

IL-2

MHC-I

Ag ——— CD8 cell

T$_M$

Memory CD8 T cell

Activated
CD8 cell

For a CD8 **killer T cell (T$_C$)** to become activated, it must recognize
a foreign peptide complexed with self MHC-I and mount a direct attack
upon the target cell. After activation, the T$_C$ cell severely injures the target
cell. This process involves the secretion of **perforins** and **granzymes**.

Perforins   Granzymes

T$_C$ cell
recognizes
infected self cell

CD8
molecule

Ag MHC-I

Infected
host cell

Destroyed
host cell

    Figure 15.11 Events in T-cell activation.

Table 15.2 Characteristics of Subsets of T-Cell Types

Types	Primary Marker on T Cell	Functions/Important Features
T helper cell 1 (T_H1)	CD4 (requires MHC-II for activation)	Activates the cell-mediated immunity pathway; secretes tumor necrosis factor and interferon gamma; also responsible for delayed hypersensitivity (allergy occurring several hours or days after contact)
T helper cell 2 (T_H2)	CD4	Drives B-cell proliferation; secretes IL-4, IL-5, IL-13
T helper cell 17 (T_H17)	CD4	Promotes inflammation; secretes IL-17; important in lung immunity
T regulatory cell (T_R)	CD4	Controls specific immune response; prevents autoimmunity
T cytotoxic cell (T_C)	CD8 (requires MHC-I for activation)	Destroys a target foreign cell by lysis; important in destruction of complex microbes, cancer cells, virus-infected cells; graft rejection; requires MHC-I for activation
Memory T cells	CD4 or CD8	Differentiate after activation of T helper or T cytotoxic cell; do not participate in initial response but seed immune tissues to be activated in future responses; always bear receptors for the specific antigen that originally activated T_H or T_C cell from which they are derived

(table 15.2). Memory T cells are some of the longest-lived blood cells known (70 years in one well-documented case).

The Activation of B Cells: Clonal Expansion and Antibody Production

The activation of B cells by most antigens (T-dependent antigens) involves a series of events **(process figure 15.12):**

1. **Binding of antigen.**
2. **Antigen processing and presentation.**
3. **B cell/T_H cell cooperation and recognition.**
4. **B-cell activation.**
5. **Differentiation.**
6. **Clonal expansion.** The primary action of plasma cells is to secrete into the surrounding tissues copious amounts of antibodies with the same specificity as the original receptor. Although an individual plasma cell can produce around 2,000 antibodies per second, production does not continue indefinitely. The plasma cells do not survive for long and deteriorate after they have synthesized antibodies.

As mentioned before, some antigens are able to stimulate a strong B-cell response without the involvement of T cells. These antigens are often very large polymers of repeating units. Examples include lipopolysaccharide from the cell wall of *Escherichia coli,* polysaccharide from the capsule of *Streptococcus pneumoniae,* and molecules from rabies and Epstein-Barr virus. They are capable of activating naive B cells simply by binding to their antigen receptors directly (left-hand side of process figure 15.12).

15.4 Learning Outcomes—Assess Your Progress

14. Summarize the process of T-cell activation, and list major types of T cells produced in this process.
15. Diagram the steps of B-cell activation, and list the types of B cells produced in this process.

15.5 Step IV (1): The T-Cell Response

The responses of T cells are **cell-mediated immunities,** which require the direct involvement of T lymphocytes throughout the course of the reaction. These reactions are among the most complex and diverse in the immune system and involve several subsets of T cells whose particular actions are dictated by the APCs that activate them. T cells require some type of MHC (self) recognition before they can be activated, and all produce cytokines with a spectrum of biological effects.

Rather than making antibodies to control foreign antigens, T cells stimulate other T cells, B cells, and phagocytes. This activity requires the cooperation of a variety of cell types (see table 15.2).

T Helper (T_H) Cells

T helper cells play a central role in regulating immune reactions to antigens, including those of B cells and other T cells. They are also involved in activating macrophages. They do this directly by receptor contact and indirectly by releasing cytokines like interferon gamma (IFNγ). T helper cells secrete interleukin-2, which stimulates the primary growth and activation of many types of T cells, including cytotoxic T cells. Some T helper cells secrete interleukins-4, -5, and -6, which stimulate various activities of B cells. T helper cells are the most prevalent type of T cell in the blood and lymphoid organs, making up about 65% of this population. The severe depression of this class of T cells (with CD4 receptors) by HIV contributes greatly to the immunopathology of AIDS.

When T helper (CD4) cells are stimulated by an antigen/MHC complex, they differentiate into either T helper 1 (T_H1) cells, T helper 2 (T_H2) cells, or T helper 17 (T_H17) cells, depending on what type of cytokines the antigen-presenting cells secrete. A T helper 1 cell will activate phagocytic cells to be better at inducing inflammation, resulting in a delayed hypersensitivity reaction. If the APC secretes another set of cytokines, the T cell will differentiate into a T_H2 cell. These cells have the functions of (1) secreting substances that

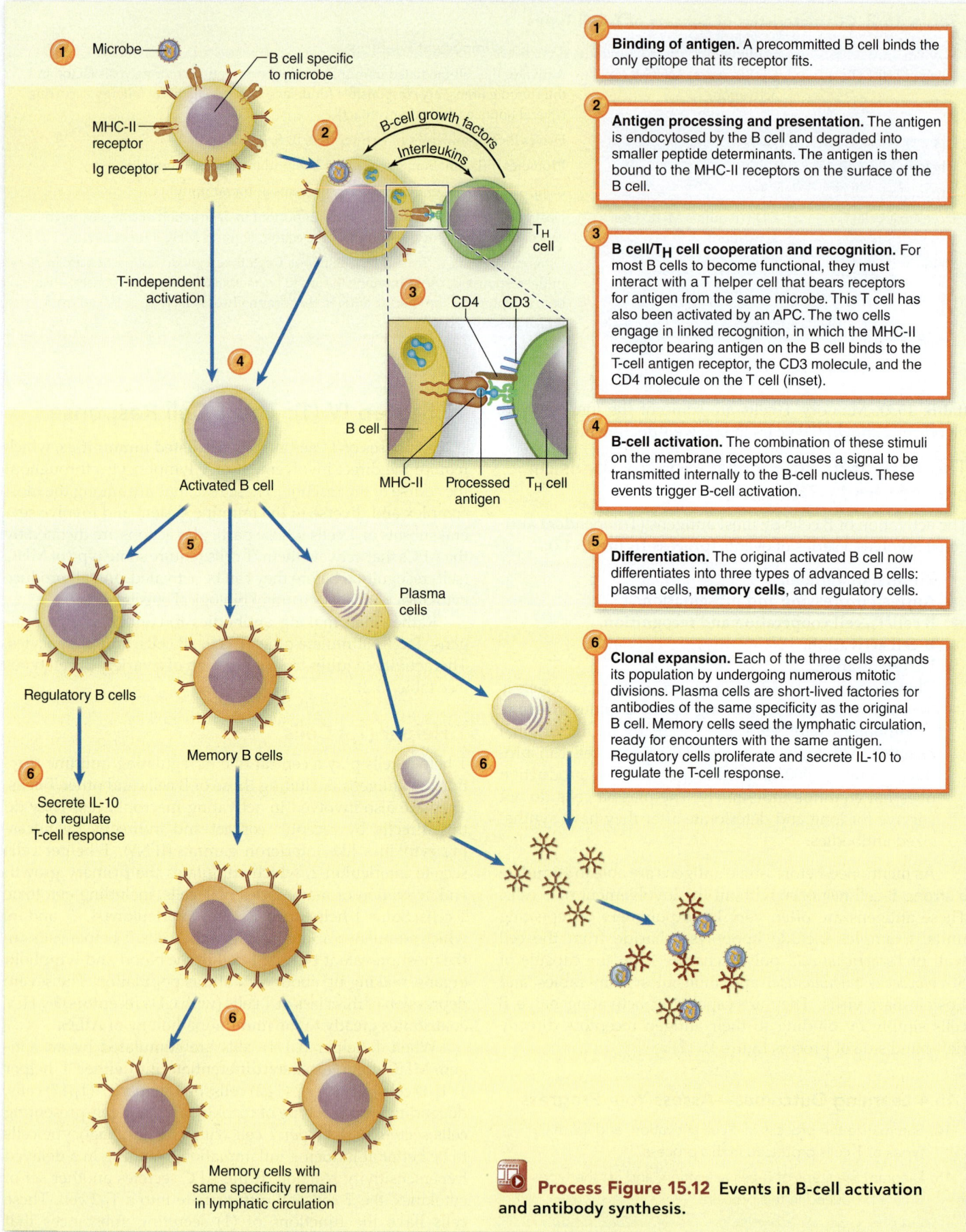

1   **Binding of antigen.** A precommitted B cell binds the only epitope that its receptor fits.

2   **Antigen processing and presentation.** The antigen is endocytosed by the B cell and degraded into smaller peptide determinants. The antigen is then bound to the MHC-II receptors on the surface of the B cell.

3   **B cell/T_H cell cooperation and recognition.** For most B cells to become functional, they must interact with a T helper cell that bears receptors for antigen from the same microbe. This T cell has also been activated by an APC. The two cells engage in linked recognition, in which the MHC-II receptor bearing antigen on the B cell binds to the T-cell antigen receptor, the CD3 molecule, and the CD4 molecule on the T cell (inset).

4   **B-cell activation.** The combination of these stimuli on the membrane receptors causes a signal to be transmitted internally to the B-cell nucleus. These events trigger B-cell activation.

5   **Differentiation.** The original activated B cell now differentiates into three types of advanced B cells: plasma cells, **memory cells**, and regulatory cells.

6   **Clonal expansion.** Each of the three cells expands its population by undergoing numerous mitotic divisions. Plasma cells are short-lived factories for antibodies of the same specificity as the original B cell. Memory cells seed the lymphatic circulation, ready for encounters with the same antigen. Regulatory cells proliferate and secrete IL-10 to regulate the T-cell response.

Process Figure 15.12 Events in B-cell activation and antibody synthesis.

influence B-cell differentiation and (2) enhancing the antibody response. One of their important roles is to respond to extracellular microbes, helminths, and allergens. T helper 17 cells are so-named because they secrete interleukin-17, which leads to the production of other cytokines that promote inflammation. Inflammation is useful, of course, but when excessive or inappropriate may lead to inflammatory diseases such as Crohn's disease or psoriasis; T_H17 cells may be critical to these conditions.

Regulatory T (T_R) Cells: Cells That Maintain the Happy Medium

Regulatory T cells are also broadly in the T_H class, in that they also carry CD4 markers. But they are not "helpers" in the sense that they encourage immune activity. They act to control the inflammatory process, to prevent autoimmunity, and to make sure the immune response does not inappropriately target normal biota. Regulatory B cells also regulate the degree of response from T cells. So B cells are involved in two ways in the T-cell response: (1) They can become activated to become plasma cells by cytokines from activated T cells, and (2) already activated regulatory B cells can secrete their own cytokines to dampen the T-cell response.

Cytotoxic T (T_C) Cells: Cells That Kill Other Cells

Cytotoxicity is the capacity of certain T cells to kill a specific target cell. It is a fascinating and powerful property that accounts for much of our immunity to foreign cells and cancer; yet, under some circumstances, it can lead to disease. For a CD8 **killer T cell** to become activated, it must recognize a foreign peptide complexed with self MHC-I presented to it and mount a direct attack upon the target cell. After activation, the T_C cell severely injures the target cell (figure 15.11). This process involves the secretion of **perforins**[2] and **granzymes.** Perforins are proteins that can punch holes in the membranes of target cells. Granzymes are enzymes that attack proteins of target cells. The action of the perforins causes ions to leak out of target cells and creates a passageway for granzymes to enter. These events are usually followed by targeted cell death through a process called *apoptosis.*

Target cells that T_C cells can destroy include the following:

- *Virally infected cells.* Cytotoxic T cells recognize these because of telltale virus peptides expressed on their surface. Cytotoxic defenses are an essential protection against viruses.
- *Cancer cells.* T cells constantly survey the tissues and immediately attack any abnormal cells they encounter **(figure 15.13).** The importance of this function is clearly demonstrated in the susceptibility of T-cell-deficient people to cancer (chapter 16).
- *Cells from other animals and humans.* Cytotoxic cell-mediated immunity is the most important factor in graft rejection. In this instance, the T_C cells attack the foreign tissues that have been implanted into a recipient's body.

---

2. From the term *perforate,* meaning "to penetrate with holes."

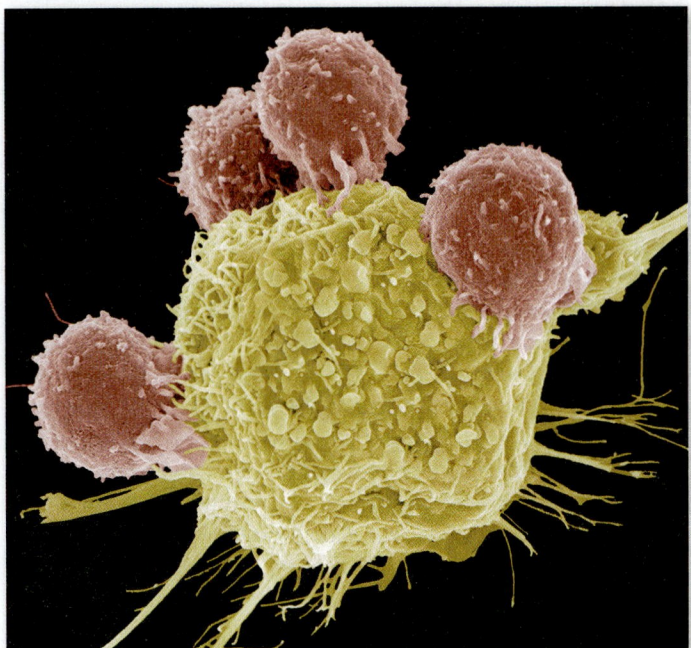

Figure 15.13 Cytotoxic T cells (pink cells) mount an attack on a tumor cell (large yellow cell). These small killer cells perforate their cellular targets with holes that lead to lysis and death.

Additional Cells with Orders to Kill

Natural killer (NK) cells are a type of lymphocyte related to T cells that lack specificity for antigens. They circulate through the spleen, blood, and lungs and are probably the first killer cells to attack cancer cells and virus-infected cells. They destroy those cells by similar mechanisms as T_C cells. They are generally not considered part of specific cell-mediated immunity because they themselves do not possess antigen receptors.

Natural killer T cells (NKT cells) were more recently discovered and appear to be a hybrid type of cell sharing properties of both T cells and NK cells. They express both T-cell receptors and NK-cell markers and are stimulated by glycolipids on foreign cells. They exhibit the ability to rapidly produce cytokines as well as granzymes and perforins and, in turn, can trigger self-destruction in target cells.

As you can see, the T-cell system is very complex. In summary, T cells differentiate into many different types of cells (including memory cells), each of which contributes to the orchestrated immune response under the influence of a multitude of cytokines. The T-cell system is summarized in figure 15.11; compare it with the B-cell system summary depicted in process figure 15.12 for further study.

15.5 Learning Outcomes—Assess Your Progress

16. Describe the main functions of the major T-cell types and their subsets.

17. Explain the role of cytotoxic T cells in apoptosis, and list the potential targets of this process.

15.6 Step IV (2): The B-Cell Response

The Structure of Immunoglobulins

In section 15.4, you saw an overview of how B cells are activated. The end result of their activation is the secretion of highly specific antibodies, also known as immunoglobulins. Earlier we saw that a basic immunoglobulin (Ig) molecule contains four polypeptide chains connected by disulfide bonds. Let us view this structure once again using an IgG molecule as a model **(figure 15.14)**. Two functionally distinct segments called *fragments* can be differentiated. The two "arms" that bind antigen are the antigen binding fragments (abbreviated "Fabs"), and the rest of the molecule is the crystallizable fragment (Fc), so called because it was the first to be crystallized in pure form. The amino-terminal end of each Fab fragment (consisting of the variable regions of the heavy and light chains) folds into a groove that will accommodate one epitope. The presence of a special region at the site of attachment between the Fab and Fc fragments allows swiveling of the Fab fragments. The Fc fragment serves as an anchor, involved in binding to various cells and molecules of the immune system itself.

Antibody-Antigen Interactions and the Function of the Fab

The site on the antibody where the epitope binds is composed of a *hypervariable region* whose amino acid content can be extremely varied. Antibodies differ somewhat in the exactness of this groove for antigen, but a minimal complementary fit is necessary for the antigen to be held effectively

(figure 15.15). The specificity of antigen binding sites for antigens is very similar to enzymes and substrates. Because the specificity of the two Fab sites is identical, an Ig molecule can bind epitope on the same cell or on two separate cells and thereby link them.

The principal activity of an antibody is to unite with, immobilize, call attention to, or neutralize the antigen for which it was formed **(figure 15.16)**. Antibodies called *opsonins* stimulate **opsonization** (ahp"-son-uh-zaz'-shun), a process in which microorganisms or other particles are coated with specific antibodies so that they will be more readily recognized by phagocytes, which dispose of them. Opsonization has been likened to putting handles on a slippery object to provide phagocytes a better grip. The capacity for antibodies to aggregate, or **agglutinate**, antigens is the consequence of their cross-linking cells or particles into large clumps. Agglutination renders microbes immobile and enhances their phagocytosis. (This is also a principle behind certain immunologic tests discussed in chapter 17.) The interaction of an antibody with complement can result in the specific rupturing of cells and some viruses. In **neutralization** reactions, antibodies fill the surface receptors on a virus or the active site on a microbial enzyme to prevent it from attaching normally. An **antitoxin** is a special type of antibody that neutralizes bacterial exotoxins.[3] Antibodies may also function to kill targets by inducing production of H_2O_2 and ozone.

---

3. There are other uses for the term *antitoxin*, notably, substances used to counteract snake bites and so on. But in immunology, an antitoxin is an antibody that binds to microbial toxins.

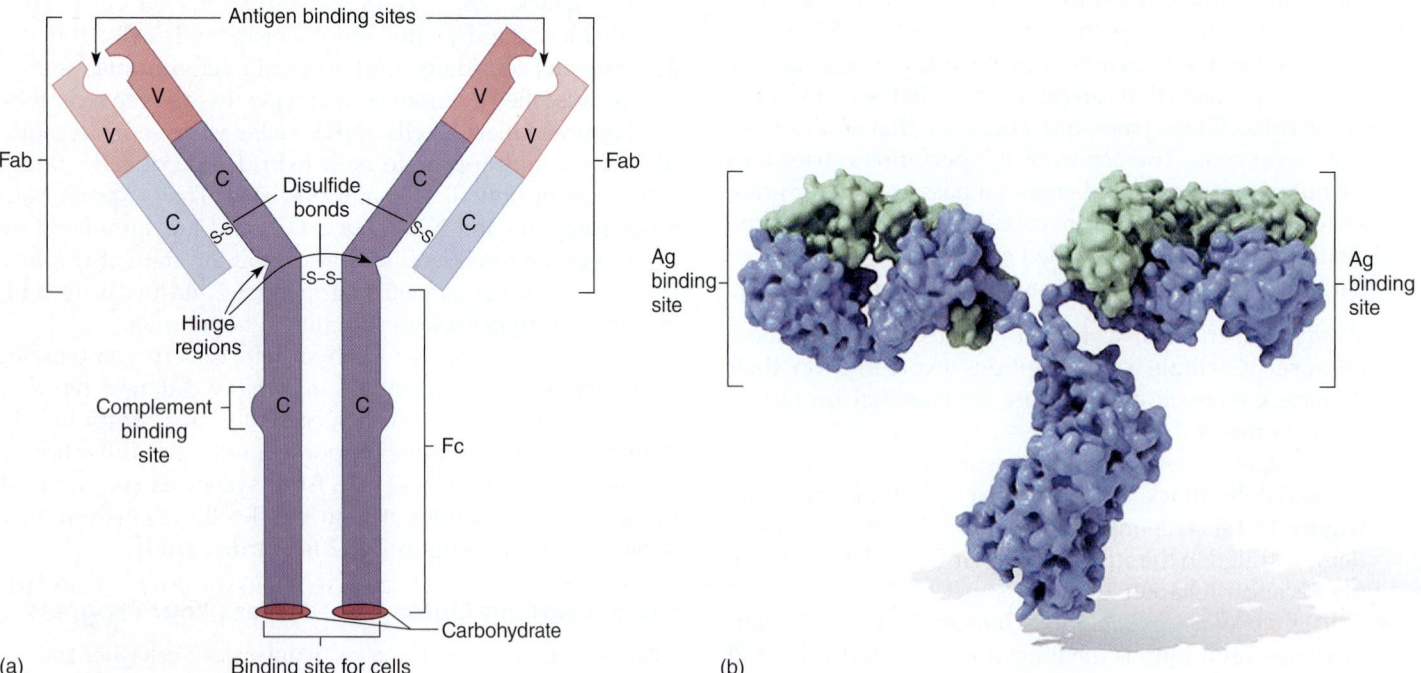

(a)   (b)

Figure 15.14 Working models of antibody structure.   **(a)** Diagrammatic view of IgG depicts the principal functional areas (Fabs and Fc) of the molecule. **(b)** Realistic model of immunoglobulin shows the tertiary and quaternary structure achieved by additional intrachain and interchain bonds.

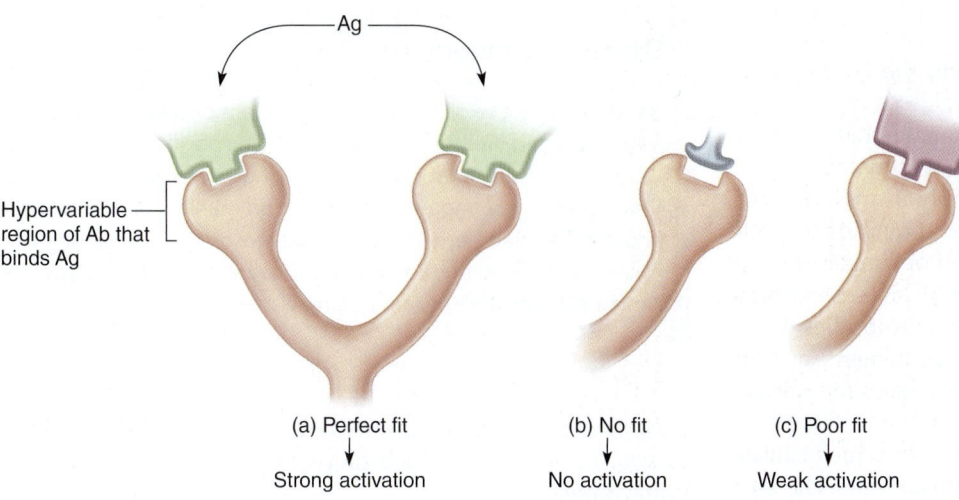

Ag

Hypervariable region of Ab that binds Ag

(a) Perfect fit
↓
Strong activation

(b) No fit
↓
No activation

(c) Poor fit
↓
Weak activation

Figure 15.15 Antigen-antibody binding. The union of antibody (Ab) and antigen (Ag) is characterized by a certain degree of fit and is supported by a multitude of weak linkages, especially hydrogen bonds and electrostatic attraction. The better the fit—that is, antigen in **(a)** versus antigen in **(c)**—the stronger the stimulation of the lymphocyte during the activation stage.

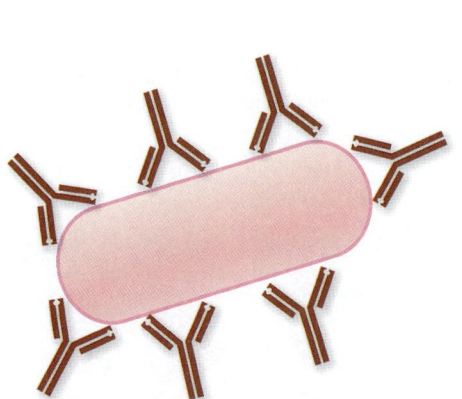

Antibodies coat the surface of a bacterium, preventing its normal function and reproduction in various ways.

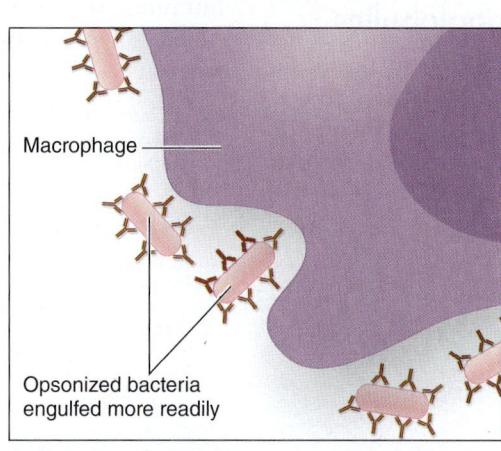

Macrophage

Opsonized bacteria engulfed more readily

When antibodies bind to microbes, they encourage the uptake of the microbe by phagocytes. This process is called **opsonization**. Opsonization has been likened to putting handles on a slippery object to provide phagocytes a better grip.

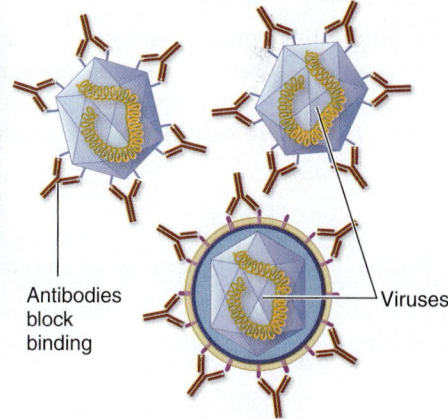

Antibodies block binding

Viruses

In **neutralization** reactions, antibodies fill the surface receptors on a virus or the active site on a microbial enzyme to prevent it from attaching normally.

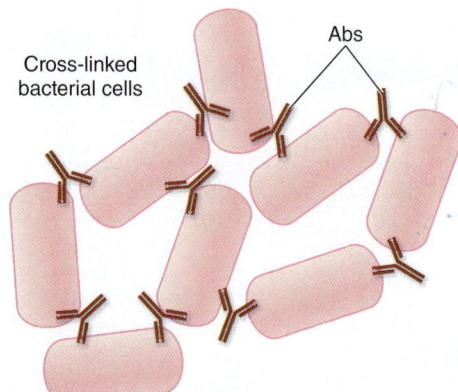

Abs

Cross-linked bacterial cells

The capacity for antibodies to aggregate, or **agglutinate**, antigens is the consequence of their cross-linking cells or particles into large clumps. Agglutination renders microbes immobile and enhances their phagocytosis. This is a principle behind certain immune tests discussed in chapter 17.

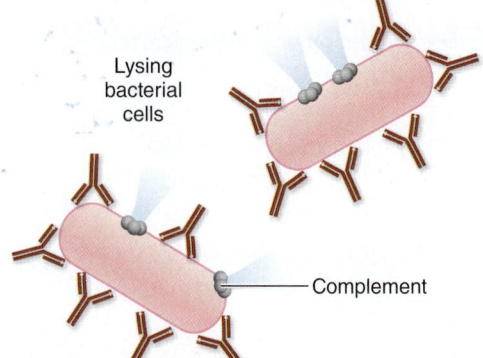

Lysing bacterial cells

Complement

The interaction of an antibody with complement can result in the specific rupturing of cells and some viruses.

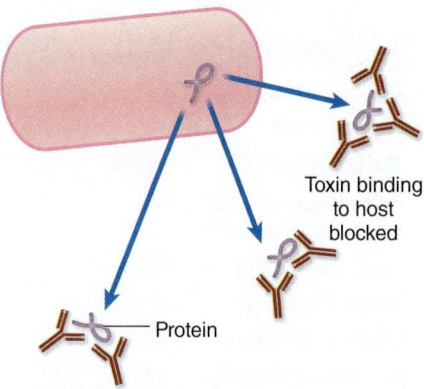

Toxin binding to host blocked

Protein

An **antitoxin** is a special type of antibody that neutralizes bacterial exotoxins.

Figure 15.16 Summary of antibody functions. Complement fixation, agglutination, and precipitation are covered further in chapter 17.

Functions of the Fc Fragment

Although the Fab fragments bind antigen, the Fc fragment has a different binding function. In most classes of immunoglobulin, the Fc end contains an effector portion that can bind to receptors on the membranes of cells, such as macrophages, neutrophils, eosinophils, mast cells, basophils, and lymphocytes. The effect of an antibody's Fc fragment binding to a cell depends upon that cell's role. In the case of opsonization, the attachment of antibody to foreign cells and viruses exposes the Fc fragments to phagocytes. Certain antibodies have regions on the Fc portion for binding complement; and in some immune reactions, the binding of Fc causes the release of cytokines. For example, the Fc end of the antibody of allergy (IgE) binds to basophils and mast cells, which causes the release of allergic mediators such as histamine. The size and amino acid composition of Fc also determine an antibody's permeability, its distribution in the body, and its class.

Accessory Molecules on Immunoglobulins

All antibodies contain molecules in addition to the basic polypeptide structure. Varying amounts of carbohydrates are affixed to the constant regions in most instances **(table 15.3).** Two additional accessory molecules are the

Disease Connection

Researchers are exploiting the fact that antibodies have two Fab fragments and creating what they call "bispecific antibodies" to treat cancer. In the laboratory, they engineer a single antibody that has one Fab fragment that will bind markers on cancer cells and one Fab fragment that binds markers on the host's own cytotoxic T cells, bringing the cytotoxic T cell in close proximity to the cancer cell, to which it can then react.

J chain, which joins the monomers[4] of IgA and IgM, and the *secretory component,* which helps move IgA across mucous membranes.

The Classes of Immunoglobulins

Immunoglobulins exist as structural and functional classes called *isotypes* (compared and contrasted in table 15.3). The

---

4. *Monomer* means "one unit" or "one part." Accordingly, *dimer* means "two units," *pentamer* means "five units," and *polymer* means "many units."

Table 15.3 Characteristics of the Immunoglobulin (Ig) Classes

	IgG	IgA (dimer shown)	IgM	IgD	IgE
	Monomer	Dimer, Monomer	Pentamer	Monomer	Monomer
Number of Antigen Binding Sites	2	4, 2	10	2	2
Molecular Weight	150,000	170,000-385,000	900,000	180,000	200,000
Percentage of Total Antibody in serum	80%	13%	6%	1%	0.002%
Average Half-Life in Serum (Days)	23	6	5	3	2.5
Crosses Placenta?	Yes	No	No	No	No
Fixes Complement?	Yes	No	Yes	No	No
Fc Binds To	Phagocytes				Mast cells and basophils
Biological Function	Long-term immunity; memory antibodies; neutralizes toxins, opsonizes, fixes complement	Secretory antibody; on mucous membranes	Produced at first response to antigen; can serve as B-cell receptor	Receptor on B cells	Antibody of allergy; worm infections

differences in these classes are due primarily to variations in the Fc fragment. The classes are differentiated with shorthand names (Ig, followed by a letter: IgG, IgA, IgM, IgD, IgE).

- The structure of IgG has already been presented. It is a monomer produced by plasma cells in a primary response and by memory cells responding the second time to a given antigenic stimulus. It is by far the most prevalent antibody circulating throughout the tissue fluids and blood. It has numerous functions: It neutralizes toxins, opsonizes, and fixes (binds) complement; and it is the only antibody capable of crossing the placenta.

- The two forms of IgA are (1) a monomer that circulates in small amounts in the blood; and (2) a dimer that is a significant component of the mucous and serous secretions of the salivary glands, intestine, nasal membrane, breast, lung, and genitourinary tract. The dimer, called secretory IgA, is formed by two monomers held together by a J chain. To facilitate the transport of IgA across membranes, a secretory piece is later added. IgA coats the surface of these membranes and is also found in saliva, tears, colostrum, and mucus. It provides the most important specific local immunity to enteric, respiratory, and genitourinary pathogens. During lactation, the breast becomes a site for the proliferation of lymphocytes that produce IgA. The very earliest secretion of the breast—a thin, yellow milk called **colostrum**—is very high in IgA. These antibodies form a protective coating in the gastrointestinal tract of a nursing infant that guards against infection by a number of enteric pathogens (*Escherichia coli*, *Salmonella*, poliovirus, rotavirus). Protection at this level is especially critical because an infant's own IgA and natural intestinal barriers are not yet developed. As with immunity *in utero*, the necessary antibodies will be donated only if the mother herself has active immunity to the microbe through a prior infection or vaccination.

- IgM is a huge molecule composed of five monomers (making it a pentamer) attached by the Fc portions to a central J chain. With its 10 binding sites, this molecule has tremendous capacity for binding antigen. IgM is the first class synthesized following the host's first encounter with antigen. Its complement-fixing qualities make it an important antibody in many immune reactions. It circulates mainly in the blood and does not cross the placental barrier.

- IgD is a monomer found in minuscule amounts in the serum, and it does not fix complement, opsonize, or cross the placenta. Its main function is that it is the receptor for antigen on B cells, usually along with IgM. It seems to be the triggering molecule for B-cell activation.

- IgE is also an uncommon blood component unless one is allergic or has a parasitic worm infection. Its Fc region interacts with receptors on mast cells and basophils. Its biological role is to stimulate an inflammatory response through the release of potent physiological substances by the basophils and mast cells. Because inflammation enlists blood cells such as eosinophils and lymphocytes to the site of infection, it is an important defense against parasites. Unfortunately, IgE has another, more insidious effect—that of mediating anaphylaxis, asthma, and certain other allergies (explained in chapter 16).

Monitoring Antibody Production over Time: Primary and Secondary Responses to Antigens

We can learn a great deal about how the immune system reacts to an antigen by studying the levels of antibodies in serum over time (**process figure 15.17**). This level is expressed quantitatively as the **titer** (ty'-tur), or concentration of antibodies. Upon the first exposure to an antigen, the system undergoes a **primary response.** The earliest part of this response, the *latent period,* is marked by a lack of antibodies for that antigen, but much activity is occurring. During this time, the antigen is being concentrated in lymphoid tissue and is being processed by the correct clones of B lymphocytes. As plasma cells synthesize antibodies, the serum titer increases to a certain plateau and then tapers off to a low level over a few weeks or months. It turns out that, early in the primary response, most of the antibodies are the IgM type, which is the first class to be secreted by plasma cells. Later, the class of the antibodies is switched to IgG or some other class (IgA or IgE). The specificity of the antibodies does not change, only the class (IgM vs. IgG or something else).

When the immune system is exposed again to the same immunogen within weeks, months, or even years, a **secondary response** occurs. The rate of antibody synthesis, the peak titer, and the length of antibody persistence are greatly increased over the primary response. The speed and intensity seen in this response are attributable to the memory B cells that were formed during the primary response. The secondary response is also called the **anamnestic response** (from the Greek word for "memory"). The advantage of this response is evident: It provides a quick and potent strike against subsequent exposures to infectious agents. This memory effect is the fundamental basis for vaccination, which we discuss later.

15.6 Learning Outcomes—Assess Your Progress

18. Diagram an antibody binding antigen, and list the possible end results of this process.

19. List the five types of antibodies and important characteristics of each.

20. Draw and label a graph illustrating the development of a secondary immune response.

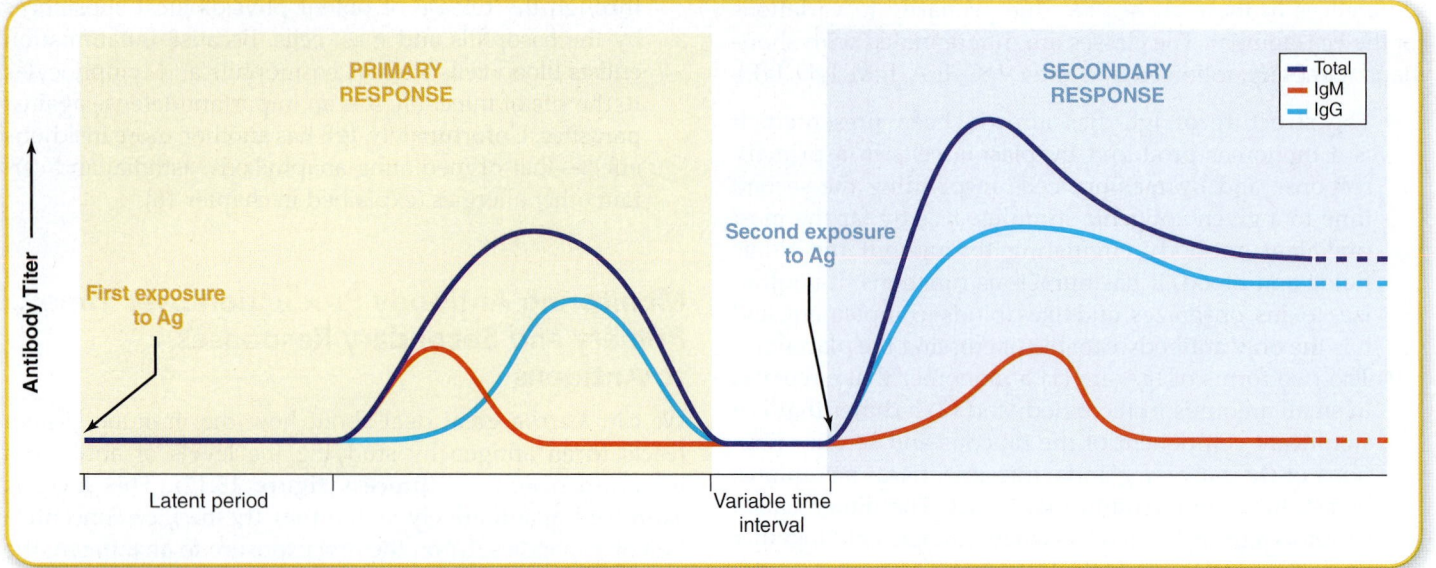

Process Figure 15.17 Primary and secondary responses to antigens.

Upon the first exposure to an antigen, the system undergoes a **primary response**. The earliest part of this response, the *latent period*, is marked by a lack of antibodies for that antigen, but much activity is occurring. During this time, the antigen is being concentrated in lymphoid tissue and is being processed by the correct clones of B lymphocytes. As plasma cells synthesize antibodies, the serum titer increases to a certain plateau and then tapers off to a low level over a few weeks or months. Early in the primary response, most of the antibodies are the IgM type, which is the first class to be secreted by plasma cells. Later, the class of the antibodies (but not their specificity) is switched to IgG or some other class (IgA or IgE).

After the initial response, there is no activity, but memory cells of the same specificity are seeded throughout the lymphatic system.

When the immune system is exposed again to the same immunogen within weeks, months, or even years, a **secondary response** occurs. The rate of antibody synthesis, the peak titer, and the length of antibody persistence are greatly increased over the primary response. The speed and intensity seen in this response are attributable to the memory B cells that were formed during the primary response. The secondary response is also called the **anamnestic response**. The advantage of this response is evident: It provides a quick and potent strike against subsequent exposures to infectious agents.

15.7 Specific Immunity and Vaccination

Specific immunity in humans and other mammals is categorized using two different sets of criteria that, when combined, result in four specific descriptors of the immune state. Immunity can either be active or passive. Also, it can be either natural or artificial.

- **Active immunity** occurs when an individual receives an immune stimulus (antigen) that activates the B and T cells, causing the body to produce immune substances such as antibodies. Active immunity is marked by several characteristics: (1) It creates a memory that renders the person ready for quick action upon reexposure to that same antigen; (2) it requires several days to develop; and (3) it lasts for a relatively long time, sometimes for life. Active immunity can be stimulated by natural or artificial means.
- **Passive immunity** occurs when an individual receives immune substances (usually antibodies) that were produced actively in the body of another human or animal donor. The recipient is protected for a short time even though he or she has not had prior exposure to the antigen. It is characterized by (1) lack of memory for the original antigen; (2) lack of production of new antibodies against that disease; (3) immediate protection; and (4) short-term effectiveness, because antibodies have a

limited period of function and, ultimately, the recipient's body disposes of them. Passive immunity can also be natural or artificial in origin.

- **Natural immunity** encompasses any immunity that is acquired during the normal biological experiences of an individual rather than through medical intervention.
- **Artificial immunity** is protection from infection obtained through medical procedures. This type of immunity is induced by immunization with vaccines or the administration of immune serum.

Table 15.4 illustrates the various possible combinations of acquired immunities. An outline summarizing the system of host defenses covered in chapters 14 and 15 was presented in figure 14.1. You may want to use this resource to review major aspects of immunity.

Artificial Passive Immunization: Immunotherapy

The first attempts at passive immunization involved the transfusion of horse serum containing antitoxins to prevent tetanus and to treat patients exposed to diphtheria. Since then, antisera from animals have been replaced with products of human origin that function with various degrees of specificity. Immune serum globulin (ISG), sometimes called *gamma globulin*, contains immunoglobulin extracted from

Table 15.4 The Four Types of Acquired Immunity

Natural Immunity is acquired through the normal life experiences of a human and is not induced through medical means.

Active

After recovering from infectious disease, a person will generally be actively resistant to reinfection for a period that varies according to the disease. In the case of childhood viral infections such as measles, mumps, and rubella, this natural active stimulus provides nearly lifelong immunity. Other diseases result in a less extended immunity of a few months to years (such as pneumococcal pneumonia and shigellosis), and reinfection is possible. Even a subclinical infection can stimulate natural active immunity. This probably accounts for the fact that some people are immune to an infectious agent without ever having been noticeably infected with or vaccinated for it.

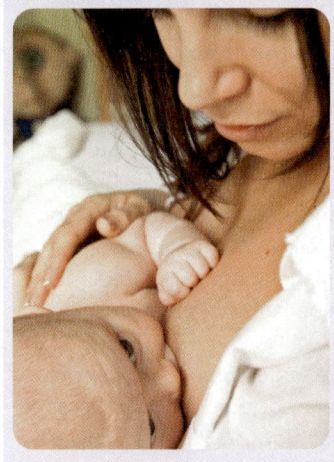

Passive

Natural, passively acquired immunity occurs only as a result of the prenatal and postnatal mother-child relationship. During fetal life, IgG antibodies circulating in the maternal bloodstream are small enough to pass or be actively transported across the placenta. This natural mechanism provides an infant with a mixture of many maternal antibodies that can protect it for the first few critical months outside the womb, while its own immune system is gradually developing active immunity. Depending on the microbe, passive protection lasts anywhere from a few months to a year.

Another source of natural passive immunity comes to the baby by way of the mother's milk. Although the human infant acquires 99% of natural passive immunity *in utero* and only about 1% through nursing, the milk-borne antibodies provide a special type of intestinal protection that is not available from transplacental antibodies.

Artificial Immunity is that produced purposefully through medical procedures.

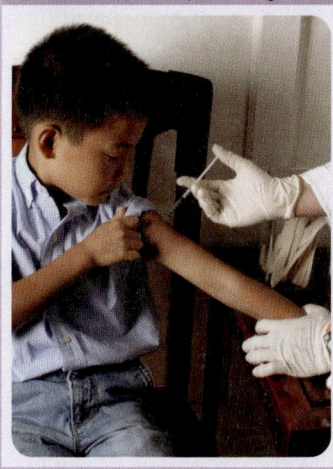

Active

Vaccination exposes a person to a specially prepared microbial (antigenic) stimulus, which then triggers the immune system to produce antibodies and lymphocytes to protect the person upon future exposure to that microbe. As with natural active immunity, the degree and length of protection vary.

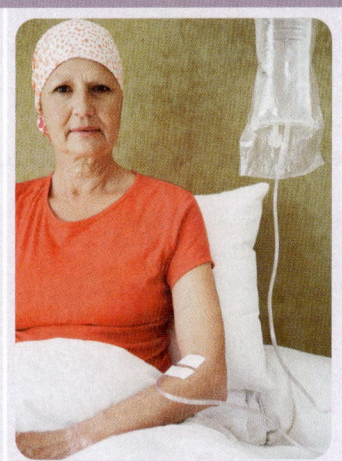

Passive

Passive immunotherapy involves a preparation that contains specific antibodies against a particular infectious agent. Pooled human serum from donor blood (gamma globulin) and immune serum globulins containing high quantities of antibodies are frequently used.

---

the pooled blood of human donors. The method of processing ISG concentrates the antibodies to increase potency and eliminates potential pathogens (such as the hepatitis B and HIV viruses). It is a treatment of choice in preventing measles and hepatitis A and in replacing antibodies in immunodeficient patients. Most forms of ISG are injected intramuscularly to minimize adverse reactions, and the protection it provides lasts 2 to 3 months.

A preparation called specific immune globulin (SIG) is derived from a more defined group of donors. Companies that prepare SIG obtain serum from patients who are convalescing and in a hyperimmune state after such infections as pertussis, tetanus, chickenpox, and hepatitis B. These globulins are preferable to ISG because they contain higher titers of specific antibodies obtained from a smaller pool of patients. Although useful for prophylaxis in persons who have been exposed or may be exposed to infectious agents, these sera are often limited in availability.

Disease Connection

The famous Iditarod sled dog race from Anchorage to Nome is run as a commemoration of a heroic trek made in 1925. At that time, twenty different mushers and 100 dogs ran in relay fashion to deliver desperately needed antiserum to children in Nome who were suffering from a diphtheria outbreak.

Case File 15   *Continuing the Case*

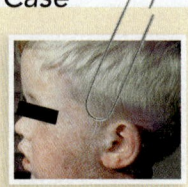

Measles infection is caused by an enveloped, single-stranded, negative-sense RNA *Morbillivirus* and is spread primarily through the respiratory route. In most cases, it causes a mild, maculopapular rash, a fever of 101°F or higher, cough, sore throat, and conjunctivitis. Most people recover with few side effects. However, in a small number of cases, patients can develop pneumonia; and in 1 in 100 cases, encephalitis can occur, resulting in permanent brain damage or epilepsy. The incubation period for measles is 7 to 18 days, so an individual may be harboring the virus and unknowingly spreading it to others.

Sporting events usually involve large numbers of people often crowded into a stadium, arena, or outdoor field. Over 68,000 people attended the Super Bowl in 2012, and over 200,000 visited the Super Bowl Village in the days before the big game. Close proximity to others, lots of screaming and yelling for favorite teams, and an airborne virus all combined in a perfect storm for the transmission of the measles virus.

- Why should health officials be concerned about measles outbreaks in the United States?
- Why have vaccination rates declined?
- What are physicians doing to counter the decline in vaccination against the measles virus?

When a human immune globulin is not available, antisera and antitoxins of animal origin can be used. Sera produced in horses are available for diphtheria, botulism, and spider and snake bites. Unfortunately, the presence of horse antigens can stimulate allergies such as serum sickness or anaphylaxis (see chapter 16). Although donated immunities only last a relatively short time, they act immediately and can protect patients for whom no other useful medication or vaccine exists.

Artificial Active Immunity: Vaccination

Active immunity can be conferred artificially by **vaccination**—exposing a person to material that is antigenic but not pathogenic. The discovery of vaccination was one of the farthest reaching and most important developments in medical science **(Insight 15.2).** The basic principle behind vaccination is to stimulate a primary response that primes the immune system for future exposure to a virulent pathogen. If the actual pathogen later enters the body, the immune response—because it will be a secondary response—will be immediate, powerful, and sustained.

Vaccines have profoundly reduced the prevalence and impact of many infectious diseases that were once common and often deadly. For decades, the emphasis was on immunizing babies and children against formerly common childhood diseases like measles, mumps, and rubella. Recent years have seen a new emphasis on also immunizing adolescents and adults against conditions such as human papillomavirus (HPV), *Streptococcus pneumoniae,* and shingles.

INSIGHT 15.2   The Lively History of Vaccination

Smallpox is one of the deadliest diseases known to humans, causing horrible pustules all over the body and massive scarring, with about 30% of the cases ending in death. It is also the only disease to have been eradicated from the earth through vaccination. The earliest recorded evidence of vaccination was 1000 BC in China when attempts were made to prevent this terrible scourge. Smallpox scabs were collected, ground into a fine powder, and blown into the nostrils of susceptible individuals. In 1661, Emperor K'ang survived an epidemic of smallpox through this method of vaccination and recommended this method to his family and subjects.

In the early 1700s, another method, called **variolation** (named after the smallpox virus, variola), in which smallpox scabs were ground up and inoculated into the arm of a susceptible individual, was used in Africa and the Middle East. An Englishwoman named Lady Mary Montagu utilized this method to protect her son from smallpox when she lived in Turkey in 1718 and brought the method back to England. Although the principles of the technique had merit, many recipients of variolation died because the material was still pathogenic. This resulted in a ban on this procedure in England.

In the late 1700s, Edward Jenner, a British doctor, noted that dairymaids had scars from cowpox on their hands but were immune to smallpox. Cowpox is related to smallpox and causes a similar illness in cattle but a much milder condition in humans. He reasoned that exposing humans to the cowpox virus would provide "cross-protection" against the very similar smallpox virus. He decided to test his hypothesis on an 8-year-old boy named James Phipps. He took material from a cowpox scab on the hand of Sarah Nelmes, a dairymaid, and inoculated it into young Phipps. The boy had a local reaction and was sick for a few days but eventually recovered. Later, he exposed Phipps to material from a fresh human smallpox sore, and he remained healthy. Jenner's method was termed *vaccination* (from Latin for *vacca*, meaning "cow"), and it was first met with fear and skepticism but soon became the standard for preventing smallpox. After a massive vaccination effort, the World Health Organization declared smallpox eradicated in 1979. However, recent threats of bioterrorism have revived the need for vigilance against this disease.

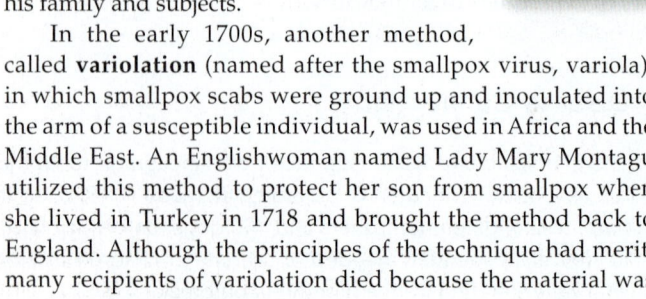

Principles of Vaccine Preparation

A vaccine must be considered from the standpoints of antigen selection, effectiveness, ease in administration, safety, and cost. In natural immunity, an infectious agent stimulates a relatively long-term protective response. In artificial active immunity, the objective is to obtain this same response with a modified version of the microbe or its components. Qualities of an effective vaccine are listed in **table 15.5**. Vaccine preparations can be broadly categorized as either whole-organism or part-of-organism preparations. These categories also have subcategories:

1. Whole cells or viruses
   a. Live, attenuated cells or viruses
   b. Killed cells or inactivated viruses
2. Part-of-organism preparations: antigenic molecules derived from bacterial cells or viruses (subunits)
   a. Subunits derived from cultures of cells or viruses
   b. Subunits chemically synthesized to mimic molecules found on pathogens
   c. Subunits manufactured via genetic engineering
   d. Subunits conjugated with proteins (often from other microbes) to make them more immunogenic—called **conjugated vaccines**

These categories are also shown in **table 15.6**.

As you may know, some childhood vaccines are given as complexes—such as the MMR vaccine, used for measles, mumps, and rubella. This trend has increased in recent years, and a wide variety of vaccine combinations are available in a single administration. These are listed in **table 15.7**.

Development of New Vaccines

Despite considerable successes, dozens of bacterial, viral, protozoan, and fungal diseases still remain without a functional vaccine. At the present time, no reliable vaccines are available for malaria, HIV/AIDS, various diarrheal diseases, respiratory diseases, and worm infections that affect over 200 million people per year worldwide. Worse than that, most existing vaccines are out of reach for much of the world's population.

New vaccine development is a very active area of research. Some of the newer strategies are presented here. Almost all of the new strategies involve genetic engineering techniques. These capabilities have quickly revitalized the quest for improved vaccination.

Genetic technology provides a means of isolating the genes that encode various microbial antigens, inserting them into plasmid vectors, and cloning them in appropriate hosts. The outcome of recombination can be as varied as desired. For instance, the cloning host can be stimulated to synthesize and secrete a protein product (antigen), which is then harvested and purified. A vaccine for hepatitis B is prepared in this way.

Another ingenious technique using genetic recombination has been nicknamed the *Trojan horse* vaccine. The term derives from an ancient legend in which the Greeks sneaked soldiers into the fortress of their Trojan enemies by hiding them inside a large, mobile wooden horse. In the microbial equivalent, genetic material from a selected infectious agent is inserted into a live carrier microbe that is nonpathogenic. Ideally, the recombinant microbe will multiply and express the foreign genes, and the vaccine recipient will be immunized against the microbial antigens. Vaccinia, the virus originally used to vaccinate for smallpox, and adenoviruses are frequently used as the carrier viruses for this technique. Vaccinia is being used as the carrier in experimental vaccines for HIV, herpes simplex 2, leprosy, and tuberculosis. A vaccine against the skin cancer melanoma was also created using this method, and its development is promising for the field of cancer immunotherapy.

DNA vaccines are one of the newer approaches to immunization. The technique in these formulations is very similar to gene therapy as described in chapter 10, except in this case, microbial (not human) DNA is inserted into a plasmid vector and inoculated into a recipient **(process figure 15.18)**. The expectation is that the human cells will take up some of the plasmids and express the microbial DNA in the form of proteins. Because these proteins are foreign, they will be recognized during immune surveillance and cause B and T cells to be sensitized and form memory cells.

In the past 20 years, more than 30 DNA vaccines have been developed and tested, but so far none have proved effective or safe enough to be licensed for humans. A few DNA vaccines are being used in animals. One of these is a DNA vaccine against West Nile virus in horses.

Another very active area of research is the development of vaccines for threats to human health that do not involve microbes at all **(Insight 15.3)**.

Route of Administration and Side Effects of Vaccines

Most vaccines are injected by subcutaneous, intramuscular, or intradermal routes. One form of the influenza vaccine comes in the form of a nasal spray. Oral (or nasal) vaccines are available for only a few diseases, but they have some distinct

Table 15.5	**Checklist of Requirements for an Effective Vaccine**

- It should have a low level of adverse side effects or toxicity and not cause serious harm.
- It should protect against exposure to natural, wild forms of pathogen.
- It should stimulate both antibody (B-cell) response and cell-mediated (T-cell) response.
- It should have long-term, lasting effects (produce memory).
- It should not require numerous doses or boosters.
- It should be inexpensive, have a relatively long shelf life, and be easy to administer.

Table 15.6 Types of Vaccines

Whole Cell Vaccines

Whole cells or viruses are very effective immunogens, since they are so large and complex. Depending on the vaccine, these are either killed or attenuated.

Killed vaccines (viruses are termed "inactivated" instead of "killed") are prepared by cultivating the desired strain or strains of a bacterium or virus and treating them with chemicals, radiation, heat, or some other agent that does not destroy antigenicity. The hepatitis A vaccine and three forms of the influenza vaccine contain inactivated viruses. Because the microbe does not multiply, killed vaccines often require a larger dose and more boosters to be effective.

Live attenuated vaccines contain live microbes whose virulence has been attenuated, or lessened/eliminated. This is usually achieved by modifying the growth conditions or manipulating microbial genes in a way that eliminates virulence factors. Vaccines for measles, mumps, polio (Sabin), and rubella contain live, nonvirulent viruses.

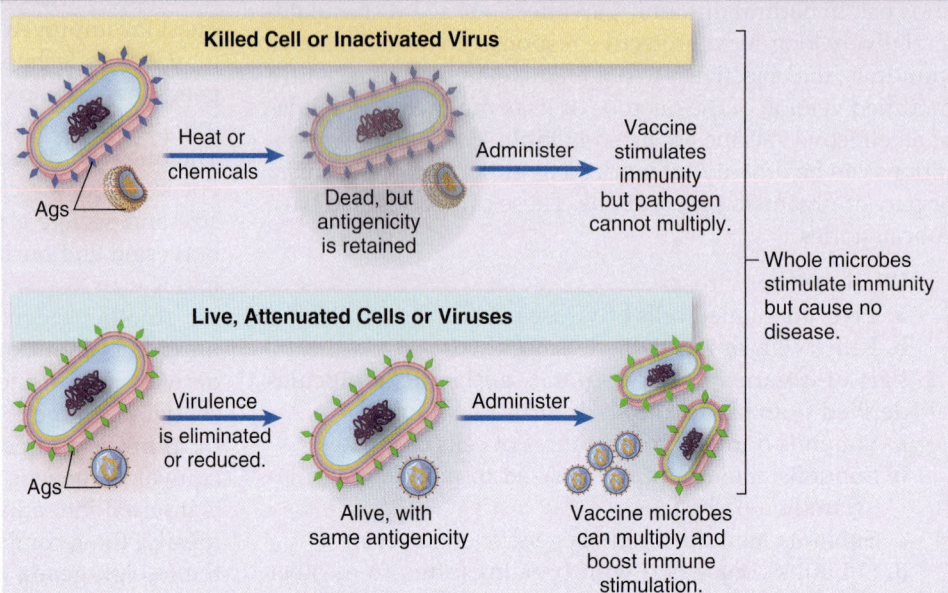

The advantages of live preparations are as follows:

1. Viable microorganisms can multiply and produce infection (but not disease) like the natural organism.
2. They confer long-lasting protection.
3. They usually require fewer doses and boosters than other types of vaccines.
4. They are particularly effective at inducing cell-mediated immunity.

Disadvantages of using live microbes in vaccines are that they require special storage facilities, can possibly be transmitted to other people, and can conceivably mutate back to become virulent again.

Subunit Vaccines (Parts of Organisms)

If the exact epitopes that stimulate immunity are known, it is possible to produce a vaccine based on a selected component of a microorganism. These vaccines for bacteria are called **subunit vaccines**. The antigens used in these vaccines may be taken from cultures of the microbes, produced by genetic engineering or synthesized chemically.

Examples of component antigens currently in use are the capsules of the pneumococcus and meningococcus, the protein surface antigen of anthrax, and the surface proteins of hepatitis B virus. A special type of vaccine is the **toxoid**, which consists of a purified bacterial exotoxin that has been chemically denatured. By eliciting the production of antitoxins that can neutralize the natural toxin, toxoid vaccines provide protection against diseases such as diphtheria, tetanus, and pertussis.

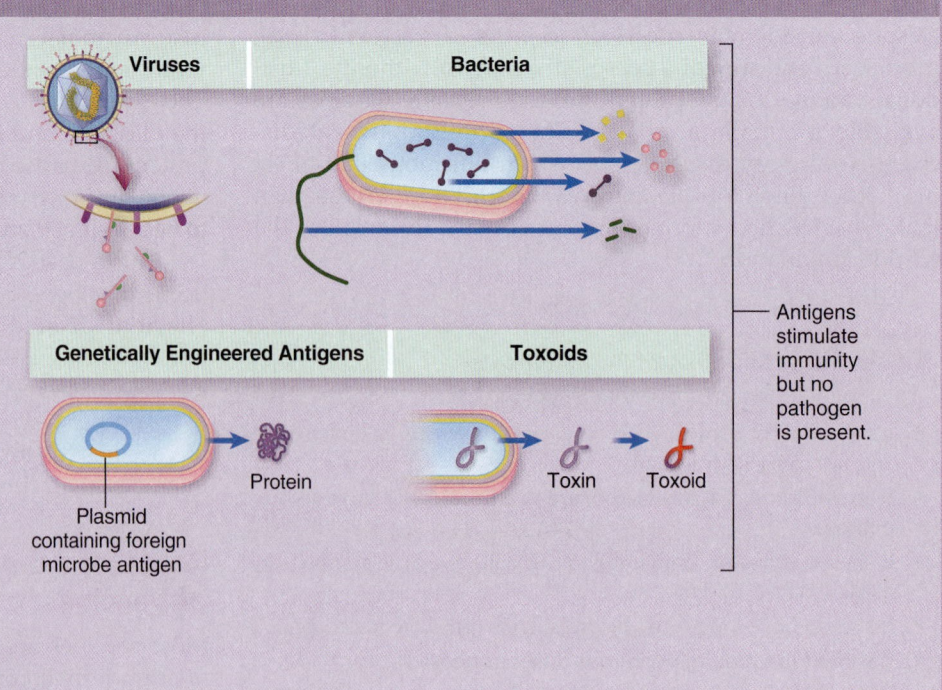

Table 15.7 Multiple-Disease Vaccines

Diseases	Vaccines
Diphtheria, tetanus	Decavac
	DT (generics)
Diphtheria, pertussis, tetanus	DTaP (Daptacel, Infanrix, Tripedia)
	Tdap (Boostrix, Adacel)
Diphtheria, pertussis, tetanus, *Haemophilus influenzae* b	TriHIBit
Diphtheria, pertussis, tetanus, hepatitis B, polio	Pediarix
Hepatitis A, hepatitis B	Twinrix
Hepatitis B, *Haemophilus influenzae* b	Comvax
Measles, mumps, rubella	MMR II
Measles, mumps, rubella, chickenpox	ProQuad

advantages. An oral or nasal dose of a vaccine can stimulate protection (IgA) on the mucous membrane of the portal of entry. Oral and nasal vaccines are also easier to give than are injections, are more readily accepted, and are well tolerated.

Some vaccines require the addition of a special binding substance, or **adjuvant** (ad'-joo-vunt). An adjuvant is any compound that enhances immunogenicity and prolongs antigen retention at the injection site. The adjuvant precipitates the antigen and holds it in the tissues so that it will be released gradually. Its gradual release presumably facilitates contact with antigen-presenting cells and lymphocytes. The most common adjuvant is alum (aluminum hydroxide salts).

Vaccines must go through many years of trials in experimental animals and human volunteers before they are licensed for general use. Even after they have been approved, like all therapeutic products, they are not without complications. The most common of these are local reactions at the injection site, fever, and allergies. Relatively rare reactions (about 1 case out of 220,000 vaccinations) are panencephalitis (from measles vaccine), back-mutation to a virulent strain (from polio vaccine), disease due to contamination with dangerous viruses or chemicals, and neurological effects of unknown cause (from pertussis and swine flu vaccines). Some patients experience allergic reactions to the medium (eggs or tissue culture) rather than to vaccine antigens. When known or suspected adverse effects have been detected, vaccines are altered or withdrawn. Several years ago, the whole cell pertussis vaccine was replaced by the acellular capsule (aP) form when it was associated with adverse neurological effects. The first oral rotavirus vaccine had to be withdrawn when children experienced intestinal blockage. An improved version was licensed in 2006. Polio vaccine was switched from live oral vaccine to inactivated preparations when occasional cases of paralytic disease occurred from back-mutated vaccine stocks. Vaccine companies have also phased out certain preservatives, such as thimerosal, that are thought to cause allergies and other potential side effects.

In the recent past, some people have attempted to link childhood vaccinations to the development of diabetes, asthma, and autism. These have fueled a very public debate about the safety of vaccines. It can be difficult for parents and consumers to discriminate between good and bad information. Scientists are trying to address parent fears and provide reliable information.

In 2011, the Institute of Medicine, an independent, nonprofit agency of the widely respected National Academies of Science, published the results of its comprehensive examination of childhood vaccines and stated unequivocally that the MMR vaccine does not cause autism. At the same time, the price of not being vaccinated has become painfully clear. Outbreaks of measles, mumps, diphtheria, polio, typhoid fever, and whooping cough have popped up all over the United States in college dormitories; in antivaccination religious

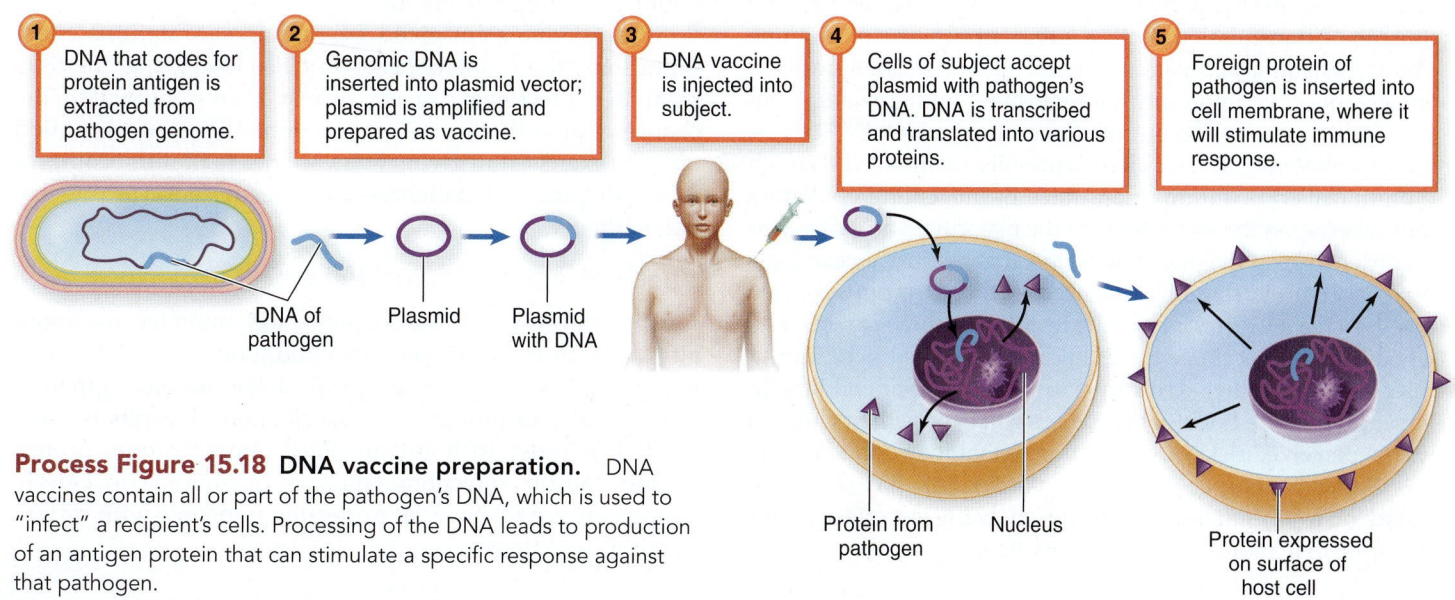

1 DNA that codes for protein antigen is extracted from pathogen genome.

2 Genomic DNA is inserted into plasmid vector; plasmid is amplified and prepared as vaccine.

3 DNA vaccine is injected into subject.

4 Cells of subject accept plasmid with pathogen's DNA. DNA is transcribed and translated into various proteins.

5 Foreign protein of pathogen is inserted into cell membrane, where it will stimulate immune response.

DNA of pathogen   Plasmid   Plasmid with DNA

Protein from pathogen   Nucleus

Protein expressed on surface of host cell

Process Figure 15.18 DNA vaccine preparation. DNA vaccines contain all or part of the pathogen's DNA, which is used to "infect" a recipient's cells. Processing of the DNA leads to production of an antigen protein that can stimulate a specific response against that pathogen.

INSIGHT 15.3   There's a Vaccine for That. . . .

You're familiar with standard vaccines against microbial pathogens such as measles, mumps, rubella, diphtheria, tetanus, pertussis, influenza, hepatitis, pneumonia, and other bacteria and viruses. Each of these targets an infectious agent and prevents a specific disease by priming the immune system for future attack. Two recently developed vaccines against human papillomavirus (HPV) and hepatitis B virus (HBV) prevent infections by these viruses, but the intent of their design is to prevent cancer. HPV is the leading cause of cervical cancer in women, and HBV is one of the leading causes of liver cancer. For that reason, these vaccines are sometimes called *cancer vaccines.*

A number of other vaccines in development are aimed at preventing diseases or conditions not associated with microbial infection. For example, a vaccine is currently being tested that would help in treating metastatic breast and ovarian cancer. A vaccine targeting prostate cancer has recently been approved by the Food and Drug Administration (FDA). These vaccines stimulate the immune system to activate T cells and antibody-producing B cells to attack cancer cells rather than microbes, stopping cancer growth or shrinking tumors. According to the National Cancer Institute, clinical trials of vaccines for lung cancer, brain tumors, pancreatic cancer, leukemia, and many others are currently underway.

Vaccine innovations don't stop there. Neurologists in Sweden have recently completed successful trials of a vaccine

against Alzheimer's disease. Researchers at Weill Cornell Medical College in New York are currently testing a vaccine to treat nicotine and cocaine addiction. These uses of vaccination are not preventive but therapeutic—a completely new way to think about vaccines. Soon, you may be diagnosed with a disease, only to have your doctor say, "But wait, there's a vaccine for that. . . ."

Sources: www.sciencedaily.com/releases/2012/06/120607092616.htm; www.livescience.com/21132-cocaine-vaccine-cure-addiction.html; www.sciencedaily.com/releases/2012/06/120627142419.htm.

communities; in airplanes; and, as you saw in the Case File at the opening of the chapter, at the Super Bowl **(Insight 15.4).**

These outbreaks are often attributed to a decrease in the level of **herd immunity.** We first encountered this concept in chapter 13. To recap, each microorganism requires a certain density of susceptible individuals in a population (herd) so that the chain of transmission will continue. Individuals who are immune to that particular microbe are a dead end for the microbe's transmission. With a sufficient number of immune individuals in a population, the microbe does not spread. In effect, collective immunity through mass immunization confers indirect protection on the nonimmune members (such as nonvaccinated children). Herd immunity maintained through immunization is an important force in preventing epidemics but relies upon the willingness of the majority to be vaccinated in order to keep the population safe. It is not just Vulcan logic that has us understanding "the needs of the many outweigh the needs of the few." It is interesting to note that some parents who choose not to vaccinate their children have cited herd immunity as a reason they don't have to vaccinate their children. In essence, they are hopeful that others will expose their children to the perceived risks of vaccination so that they don't have to.

Some have speculated that vaccination has done too good of a job—at least in terms of being so effective for so long that many young parents have no memory of the prevaccination era and don't appreciate the much greater risk of not vaccinating compared to vaccinating. In the decade before measles vaccination began, 3 to 4 million cases occurred each year in the United States. Typically, 300 to 400 children died annually and 1,000 more were chronically disabled due to measles encephalitis. Put simply, childhood vaccines save the lives of 2.5 million children a year (worldwide), according to UNICEF.

Professionals involved in giving vaccinations must understand their inherent risks but also realize that the risks from the infectious disease almost always outweigh the chance of an adverse vaccine reaction. The greatest caution must be exercised in giving live vaccines to immunocompromised or pregnant patients—the latter because of possible risk to the fetus.

Vaccinating: Who and When?

As you read earlier in the chapter, vaccination has traditionally been most prominent in childhood. With advanced understanding of disease control, it has become apparent to public health officials that vaccination of adults is often needed in order to boost an older immunization, protect against "adult" infections (such as pneumonia in elderly people), or provide special protection in people with certain medical conditions.

INSIGHT 15.4   How Anti-Vaxxers Were Misled

Actress and model Jenny McCarthy blamed her child's autism on the measles, mumps, rubella (MMR) vaccine, while promoting her book on the Oprah Winfrey show. Oprah lauded her for her bravery, and although Oprah read a short statement from the Centers for Disease Control and Prevention (CDC) stating that there was no link between autism and the MMR, Jenny responded, "My science is named Evan, and he's at home. That's my science."

In 1998, a British physician named Andrew Wakefield published a paper that seemed to establish a link between vaccination with the MMR vaccine and the occurrence of bowel disease associated with autism. The effect was immediate: MMR vaccination rates in the United Kingdom fell from 92% to 80% and were as low as 50% in some parts of London. Consequently, measles infections have increased. In 1998, there were 56 measles cases in the United Kingdom; in 2008, there were 1,348. In 2011, there were 222 cases of measles in the United States, the largest

number since 1957; and there was a total of 1,086 confirmed cases in the United Kingdom in 2011.

Let's examine the data: Wakefield's 1998 study followed just 12 children. In 2004, 10 of the coauthors on the paper asked that the paper be withdrawn, saying that the data were insufficient. And in 2009, two disturbing facts came to light. *The Times* of London found that the data in the paper had been altered to support Wakefield's hypothesis. In most of the cases, the problems Wakefield cited as a result of the vaccination were present *before* the vaccination. Hospital pathologists testified that the majority of the cases were normal and that the data were altered for publication. Second, Wakefield was receiving money from lawyers who represented families suing vaccine manufacturers for alleged harm done them by the vaccines. Finally, recent investigations have shown that Wakefield was pursuing business options to develop a "replacement" vaccine for MMR, predicting he could make up to £28 million ($43 million) on a new vaccine. The medical journal that published the original paper completely retracted the paper in February 2010. In May 2010, the General Medical Council of Britain found Wakefield guilty of serious ethical violations and removed his medical license.

Soon after the 1998 Wakefield study came out, dozens of well-controlled studies on the link between autism and the MMR vaccine and the thimerosal preservative used in it refuted Wakefield's bogus claims. Unfortunately, more people chose to listen to celebrities over science and not vaccinate their children. The results have been devastating. Numerous reports of outbreaks of measles in the United States, the United Kingdom, and around the globe have resulted in a loss of "herd immunity" that occurs when a majority of a population is vaccinated, protecting the few that are not. Even one person on an airplane, in a crowded stadium, or in a college dormitory can be the source of numerous cases of measles if enough people are not protected by the vaccine.

In a recent study by Lenisa Chang of the University of Cincinnati, a direct correlation was found between Wakefield's fraudulent study in 1998 and a decrease in parents obtaining the MMR vaccine for their children. Surprisingly, Chang found that college-educated mothers were *less* likely to obtain the MMR vaccine for their children than non-college-educated mothers. She also found that the controversy surrounding the MMR vaccine spilled over into other vaccines, with a corresponding drop in polio; diphtheria, tetanus, and pertussis (DTP); and other vaccinations. The time has come to stop getting medical information from celebrities and listen to your doctor: Vaccinate.

Sources: http://phys.org/news/2011-01-autism-fraud-vaccine-booster.html; www.sciencedaily.com/releases/2012/06/120604142726.htm

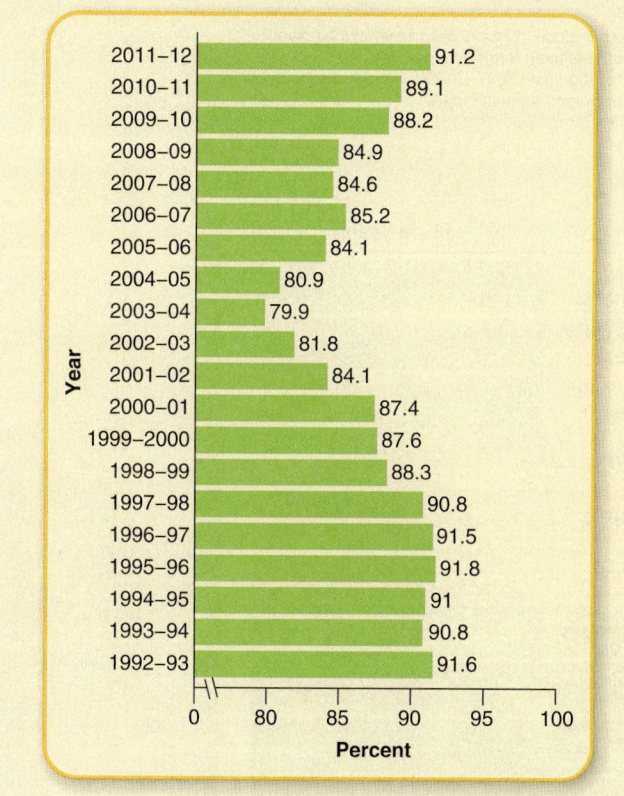

Table 15.A Uptake of the First Dose of MMR Vaccine from 1992–93 to 2011–12

Year	Percent
2011–12	91.2
2010–11	89.1
2009–10	88.2
2008–09	84.9
2007–08	84.6
2006–07	85.2
2005–06	84.1
2004–05	80.9
2003–04	79.9
2002–03	81.8
2001–02	84.1
2000–01	87.4
1999–2000	87.6
1998–99	88.3
1997–98	90.8
1996–97	91.5
1995–96	91.8
1994–95	91
1993–94	90.8
1992–93	91.6

Table 15.8 provides the current recommended schedule for childhood and adolescent immunizations. As you have seen, some vaccines are mixtures of antigens from several pathogens, notably Pediarix (DTaP, IPV, and HB).

Table 15.9 contains the recommended adult immunization schedule.

Table 15.8 Recommended Childhood and Adolescent Immunization Schedules, United States, 2012

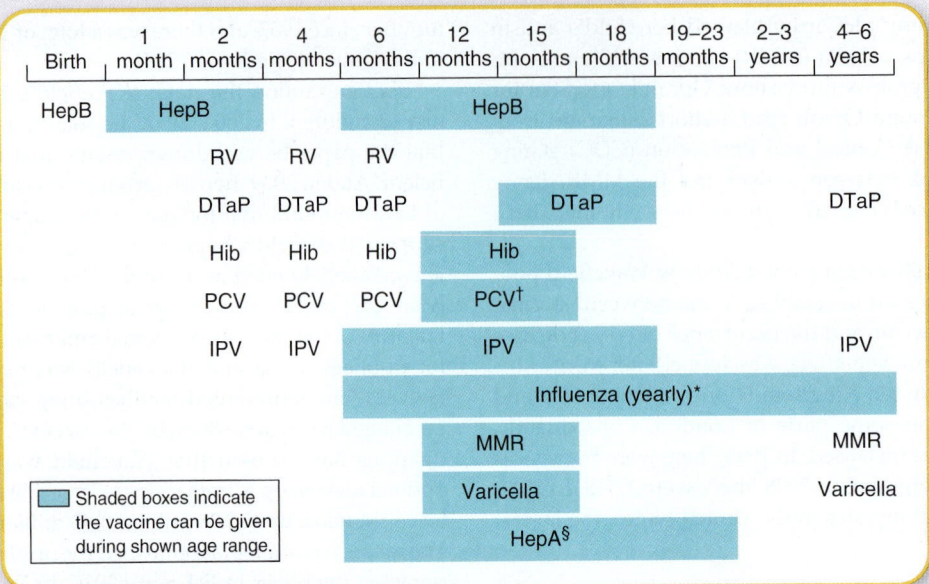

† Children 2 years old and older with certain medical conditions may need a dose of pneumococcal vaccine (PPSV) and meningococcal vaccine (MCV4). See vaccine-specific recommendations at http://www.cdc.gov/vaccines/pubs/ACIP-list.htm.

* Two doses given at least four weeks apart are recommended for children aged 6 months through 8 years of age who are getting a flu vaccine for the first time.

§ Two doses of HepA vaccine are needed for lasting protection. The first dose of HepA vaccine should be given between 12 months and 23 months of age. The second dose should be given 6 to 18 months later. HepA vaccination may be given to any child 12 months and older to protect against HepA. Children and adolescents who did not receive the HepA vaccine and are at high risk should be vaccinated against HepA.

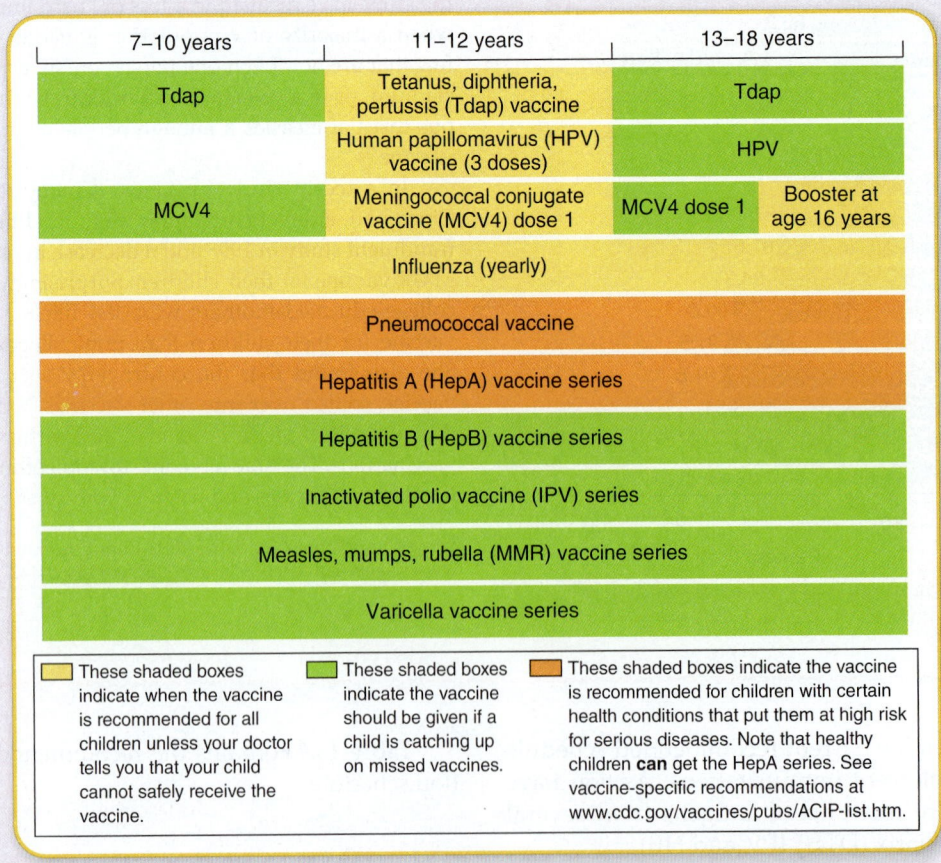

Table 15.9   Recommended Adult Immunization Schedule, United States, 2012

Then you should get these vaccines:	If you are this age,					
	19–21 years	22–26 years	27–49 years	50–59 years	60–64 years	65+ years
Influenza (flu)*	Get a flu vaccine every year					
Tetanus, diptheria, pertusis (Td/Tdap)	Get a Tdap vaccine once, then a Td booster vaccine every 10 years					
Varicella (chickenpox)	2 doses					
HPV vaccine for women	3 doses					
HPV vaccine for men	3 doses	3 doses				
Zoster (shingles)					1 dose	
Measles, mumps, rubella (MMR)	1 or 2 doses			1 or 2 doses		
Pneumococcal vaccine (pneumonia)	1–3 doses					1 dose
Meningococcal	1 or more doses					
Hepatitis A	2 doses					
Hepatitis B	3 doses					

Boxes this color show that the vaccine is recommended for all adults unless your doctor or nurse tells you that you cannot safely receive the vaccine.

Boxes this color show when the vaccine is recommended for adults with certain risks related to their health, job, or lifestyle that put them at higher risk for serious diseases. Talk to your doctor or nurse to see if you are at higher risk.

No recommendation

*There are four different flu vaccines available—talk to your doctor or nurse about which flu vaccine is right for you.

Case File 15   *Wrap-Up*

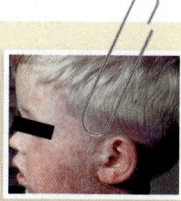

Fourteen cases of measles after the Super Bowl may not seem like a large outbreak, but in the United States, this infectious disease *had* been effectively "eliminated" due to nearly 90% vaccination coverage since 2000. (The MMR vaccine is used to provide protection against measles, mumps, and rubella.) However, rates of vaccination have been in decline in the United States and abroad, largely due to a fraudulent study published by Dr. Andrew Wakefield linking autism to the MMR vaccine. Although Wakefield's study was debunked by the scientific community, the paper was retracted, and he was stripped of his medical license, millions of parents worldwide stopped vaccinating their children (see Insight 15.4). The World Health Organization reported that in 2010 there were 139,300 measles deaths

globally, the majority among unvaccinated children in developing countries with poor health infrastructure.

In 2011, there were 222 cases of measles reported in the United States, the highest since measles elimination was achieved in 2000. Ninety percent of the cases in the United States were "imported" from countries where vaccination rates against measles are low. The CDC states that the increased numbers of imported cases of measles underscores the risk for unvaccinated individuals within the United States and the importance of vaccination against measles and the establishment of herd immunity. Once exposed, an unvaccinated individual has a 90% chance of developing measles, which makes even one person in a crowded Super Bowl Village the source of a potential outbreak that could spread to those who are not protected by the vaccine.

15.7 Learning Outcomes—Assess Your Progress

21. List the four categories of acquired immunity, and provide examples of each.

22. Discuss the qualities of an effective vaccine.

23. List several types of vaccines, and discuss how they are utilized today.

24. Explain the principle of herd immunity and the risks that unfold when it is not maintained.

Chapter Summary

15.1 Specific Immunity: The Third Line of Defense (ASM Guidelines* 3.4, 5.4)

- Acquired specific immunity is an elegant but complex matrix of interrelationships between lymphocytes and antigen-presenting cells consisting of several stages.
- Step I. Lymphocytes originate in hematopoietic tissue but go on to diverge into two distinct types: B cells, which produce antibody, and T cells, which destroy cells and produce cytokines that mediate and coordinate the entire immune response.
- Step II. Antigen-presenting cells detect invading pathogens and present these antigens to lymphocytes, which recognize the antigen and initiate the specific immune response.
- Step III. Lymphocytes proliferate, producing clones of progeny that include groups of responder cells, regulator cells, and memory cells.
- Step IV. Activated T lymphocytes (one of three subtypes) regulate and participate directly in the specific immune responses. Activated B lymphocytes become plasma cells that produce and secrete large quantities of antibodies. Regulatory B cells are also produced.

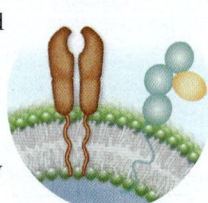

15.2 Step I: The Development of Lymphocyte Diversity (ASM Guidelines 3.4, 5.4)

- During development, both B and T cells develop millions of genetically different clones. Together, these clones possess enough genetic variability to respond to many millions of different antigens. Each clone, however, can respond to only one specific antigen.
- Binding of antigen to a particular clone is called clonal selection. That clone is exclusively amplified in a process called clonal expansion, which leads to an army of cells with that individual specificity.

15.3 Step II: Presentation of Antigens (ASM Guidelines 3.4, 5.4)

- Immature lymphocytes released from hematopoietic tissue migrate (home) to one of two sites for further development. T cells mature in the thymus. B cells mature in the stromal cells of the bone marrow.
- Antigens or immunogens are proteins or other complex molecules of high molecular weight that trigger the immune response in the host.

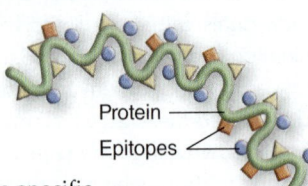

Protein

Epitopes

- Lymphocytes respond to a specific portion of an antigen called the epitope. A

given microorganism has many such epitopes, all of which stimulate individual specific immune responses.
- Antigen must be formally presented to lymphocytes by antigen-presenting cells (APCs) in most immune reactions. Macrophages, B cells, and dendritic cells can serve as APCs.

15.4 Step III: Antigenic Challenge of T Cells and B Cells (ASM Guidelines 3.4, 5.4)

- Physical contact between the APC, T cells, and B cells activates these lymphocytes to proceed with their respective immune responses.
- T cells do not produce antibodies. Instead, they produce different cytokines that play diverse roles in the immune response.
- The main classes of T cells are T helper cells, T regulatory cells, T cytotoxic cells, and NKT cells.
- B-cell activation produces memory B cells, regulatory B cells, and plasma cells. Most B-cell reactions require T helper cells to develop.

15.5 Step IV (1): The T-Cell Response (ASM Guidelines 3.4, 5.4)

- Each subset of T cell produces a distinct set of cytokines that stimulate lymphocytes or destroy foreign cells. T helper cells release cytokines that stimulate macrophages and B cells, among other functions. Regulatory T cells guard against excessive or inappropriate inflammation and immunity. Cytotoxic T cells can kill targeted cells directly.
- Cytotoxic T cells induce apoptosis in target cells through the action of perforins and granzymes.

15.6 Step IV (2): The B-Cell Response (ASM Guidelines 3.4, 5.4)

- B-cell activation leads to the generation of plasma cells that secrete antibodies, regulatory B cells that modulate the autoimmune response; and long-term memory B cells.
- B cells produce five classes of antibody: IgM, IgG, IgA, IgD, and IgE. IgM and IgG predominate in plasma. IgA predominates in body secretions. IgD is expressed on B cells as an antigen receptor. IgE binds to mast cells and basophils in tissues, promoting inflammation.
- Antibodies bind physically to the specific antigen that stimulates their production, thereby immobilizing the antigen and enabling it to be destroyed by other components of the immune system.
- The memory response means that the second exposure to antigen calls forth a much faster and more vigorous response than the first.

*Source: ASM Curriculum Guidelines (American Society for Microbiology, 2012). Complete guidelines in appendix B of this book.

15.7 Specific Immunity and Vaccination (ASM Guidelines 3.4, 4.5, 5.4)

- Active immunity means that your body produces antibodies to a disease agent. If you contract the disease, you can develop natural active immunity. If you are vaccinated, your body will produce artificial active immunity.
- In passive immunity, you receive antibodies from another person. Natural passive immunity comes from the mother. Artificial passive immunity is administered medically.
- Artificial passive immunity usually involves administration of antiserum. This means that antibodies collected from donors (human or otherwise) are injected into people who need protection immediately.
- Vaccines are artificial active agents that provoke a protective immune response in the recipient but do not cause the actual disease. Vaccination is the process of challenging the immune system with a specially selected antigen. Vaccines currently in use consist either of whole cells or viruses or of subunits from them that are immunogenic.

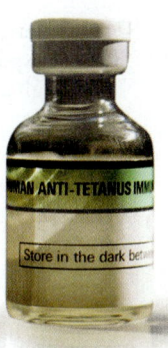

Multiple-Choice and True-False Questions | Bloom's Levels 1 and 2: Remember and Understand

Multiple-Choice Questions. Select the correct answer from the options provided.

1. The primary B-cell receptor is
   a. IgD.
   c. IgE.
   b. IgA.
   d. IgG.

2. In humans, B cells mature in the _____, and T cells mature in the _____.
   a. GALT, liver
   c. bone marrow, thymus
   b. bursa, thymus
   d. lymph nodes, spleen

3. Small, simple molecules are _____ antigens.
   a. poor
   c. good
   b. never
   d. heterophilic

4. The cross-linkage of antigens by antibodies is known as
   a. opsonization.
   c. agglutination.
   b. a cross-reaction.
   d. complement fixation.

5. T cells assist in the functions of certain B cells and other _____ T cells.
   a. sensitized
   c. helper
   b. cytotoxic
   d. natural killer

6. T_C cells are important in controlling
   a. virus infections.
   c. autoimmunity.
   b. allergy.
   d. all of these.

7. Which of the following can serve as antigen-presenting cells (APCs)?
   a. T cells
   d. dendritic cells
   b. B cells
   e. b, c, and d
   c. macrophages

8. A vaccine that contains parts of viruses is called
   a. acellular.
   c. a subunit.
   b. a recombinant.
   d. attenuated.

9. Conjugated vaccines combine antigens and
   a. antibodies.
   c. epitopes.
   b. adjuvants.
   d. foreign proteins.

10. Widespread immunity that protects the population from the spread of disease is called
    a. seropositivity.
    c. epidemic prophylaxis.
    b. cross-reactivity.
    d. herd immunity.

True-False Questions. If the statement is true, leave as is. If it is false, correct it by rewriting the sentence.

11. Cell surface markers are also often called receptors.

12. Antibodies are secreted by monocytes.

13. Vaccination could be described as artificial passive immunity.

14. IgE antibodies are found in body secretions.

15. The process of reducing the virulence of microbes so that they can be used in vaccines is called denaturation.

Critical Thinking Questions | Bloom's Levels 3, 4, and 5: Apply, Analyze, and Evaluate

Critical thinking is the ability to reason and solve problems using facts and concepts. These questions can be approached from a number of angles and, in most cases, they do not have a single correct answer.

1. Explain the two main features that characterize the third line of host defense mechanisms.

2. Chronic lymphocytic leukemia (CLL) leads to the production of cancerous B cells, and treatment often involves bone marrow transplantation. Based upon your knowledge of lymphocyte development, explain how this procedure can lead to therapeutic effects in some patients.

3. Describe three types of cell markers that allow for the specificity of B- and T-cell action and how they function in immune reactions.

4. Provide an explanation in support of or refuting the following statement: Humans would never develop natural immunity to a novel biological agent created in a laboratory.

5. How many types of B cells are produced in the process of clonal expansion? Summarize why each is needed for a thorough and long-lasting humoral immune response to an antigen.

6. Explain how superantigens, such as toxic shock syndrome toxin produced by *Staphylococcus aureus,* often lead to the development of life-threatening symptoms in an infected individual.

7. a. Summarize how T cells are stimulated by antigen, and compare this process with how B cells are activated.
   b. Define the term *tolerance,* and explain whether it is necessary for effective humoral and cell-mediated immunity.

8. Describe the structure of a B-cell receptor. Which portion binds to antigen? Which portion keeps the receptor tethered to the B cell?

9. a. Explain how the anamnestic response is triggered by vaccination.
   b. Conduct additional research and discuss one current example illustrating how lack of herd immunity within a population has led to localized disease outbreaks in the United States.

10. Conduct additional research and discuss whether or not macrophages play a role in the development of inflammation and chronic disease.

Concept Connections | Bloom's Levels 4 and 6: Analyze and Create

This activity ties together multiple concepts in the chapter.

Scenario: It has been a week since your bout of *Salmonella* food poisoning, and you may never look at an egg salad sandwich the same way again. Your fever has gone down, you're starting to feel better, and you have started eating normal foods again. In your microbiology class, you are studying the specific immune response; and you start to ponder how your immune system has reacted to the infection. You have learned through your sad experience that *Salmonella* infects the gut mucosa and may gain entry to the bloodstream. Fill in the following flowchart to outline the components of your immune system that have been activated as a result of its encounter with *Salmonella.*

1. Which cells produce antibodies?
2. Which cells activate antibody production?
3. What antibody type is produced first?
4. What antibody type found in the blood is produced later in the infection?
5. What antibody type would be found in the gut mucosa to fight *Salmonella*?
6. How will your immune system remember your encounter with *Salmonella*?

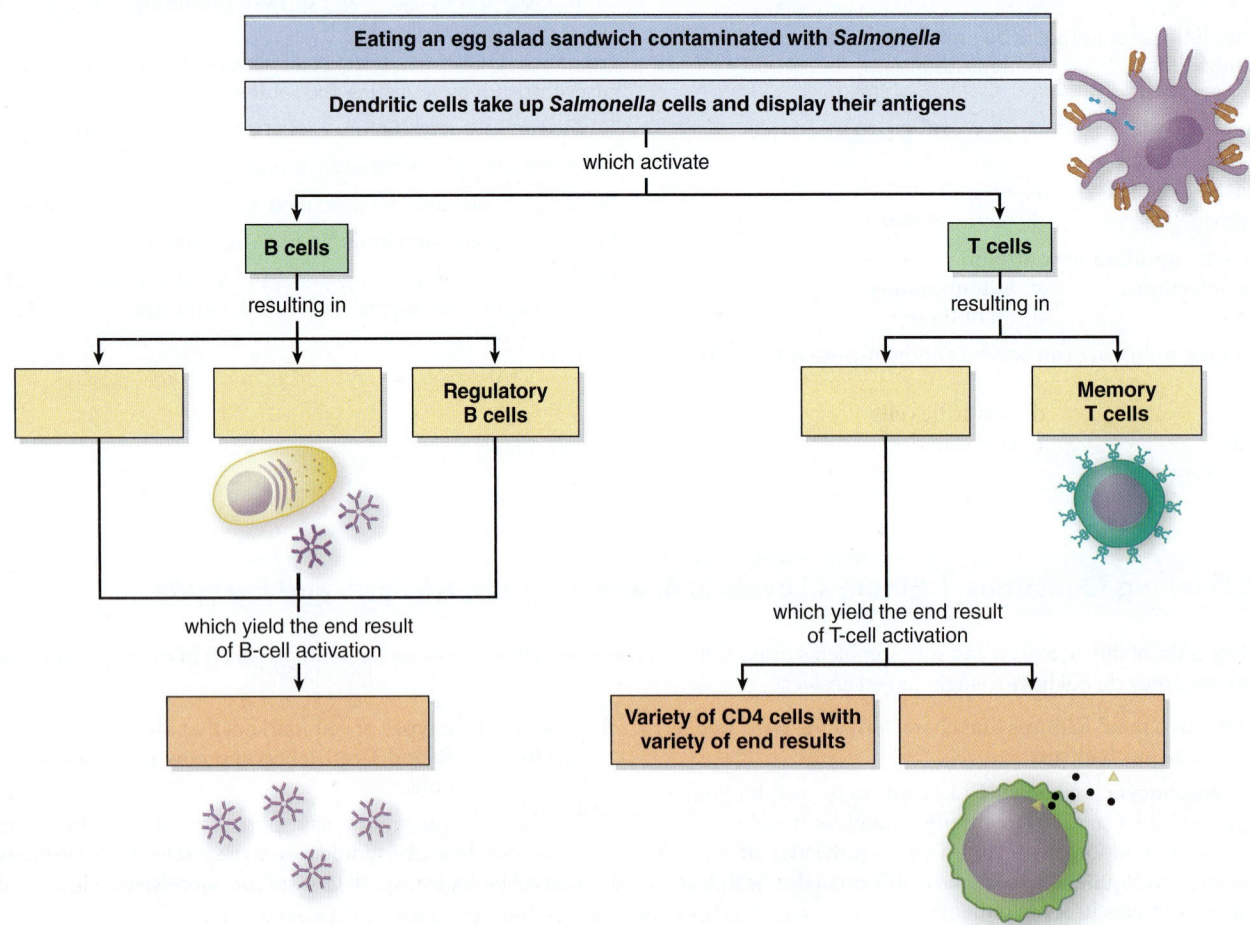

Visual Connections | Bloom's Level 5: Evaluate

This question uses visual images or previous content to make connections to this chapter's concepts.

1. **From this chapter, process figure 15.17.** In this figure describing primary and secondary responses to antigen, indicate where a vaccination may be most effective, and also indicate where natural infection would play a role.

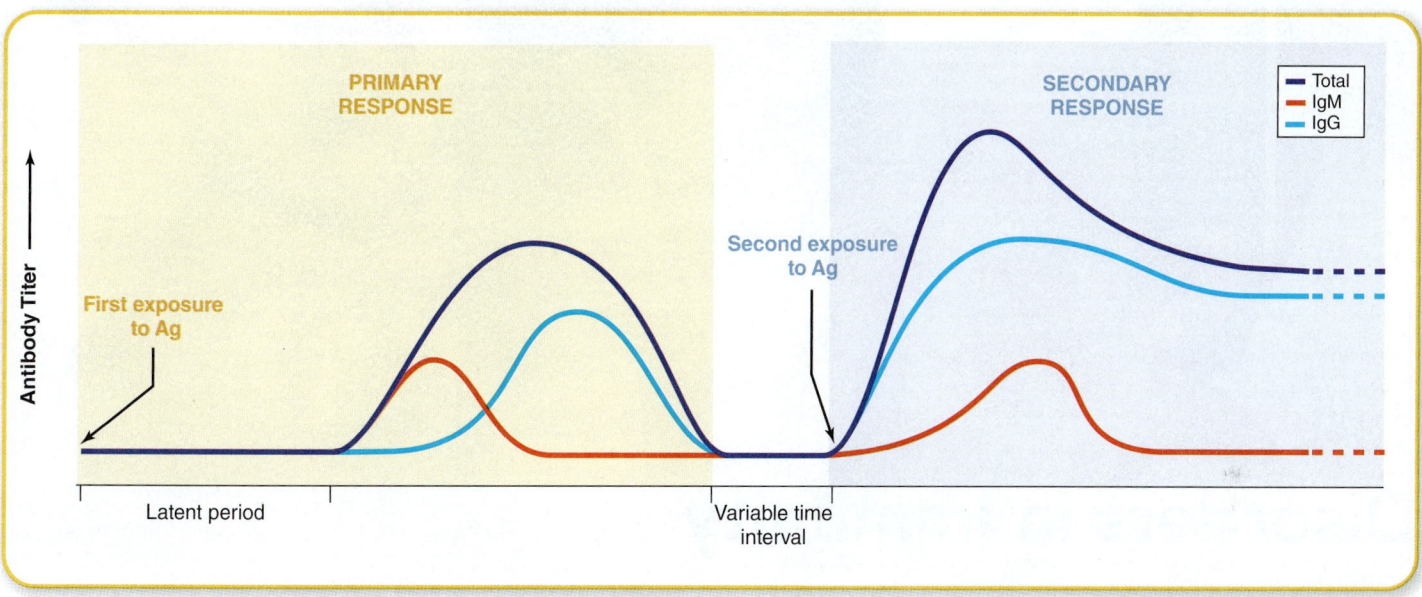

Concept Mapping | Bloom's Level 6: Create

Appendix D provides guidance for working with concept maps.

1. Using the words that follow, please create a concept map illustrating the relationships among these key terms from chapter 15.

active immunity	antibody secretion	inflammation	innate immunity
vaccines	activated T_C and T_H cells	artificial immunity	
passive immunity	natural immunity	memory	

Enhance your study of this chapter with study tools and practice tests. Also ask your instructor about the resources available through ConnectPlus, including the media-rich eBook, interactive learning tools, and animations.

16

Disorders in Immunity

Case File 16

Let Them Eat Dirt

Your child is crawling around on the floor and happens upon a remnant of a cookie. Your first reaction may be to swipe that dirty cookie out of his little hands, but you may be doing the child a favor by letting him gnaw on that dusty crumb. Doctors and scientists agree—let kids get dirty. In 1989, Dr. David P. Strachan investigated the "hygiene hypothesis," which had previously been cited to explain the increased rates of allergies, eczema, and asthma seen in children living in urban areas. Other studies have found that children growing up in rural farm areas with many animals have lower incidences of asthma and allergies. This is most likely due to the fact that children growing up on a farm come from larger families, they have pets and farm animals, and their houses are heated with wood and coal, among other factors—all of which expose them to microbes and other substances. Children who grow up in relatively affluent urban areas have less exposure to animals, dust, dirt, and microbes and have higher rates of **atopy,** or chronic local allergy, than their rural counterparts. Dr. Strachan's research showed that hay fever, asthma, and eczema were also less common in children growing up in larger families than families with only one child.

Hay fever, asthma, hives, and other atopic allergies are all mediated by IgE antibodies and are among the most common chronic illnesses seen in children in the industrialized world. As you have learned, exposure to microbes often occurs without infection, and the constant but subtle exposure to microbes and their products is vital to the proper development of a child's immune system. If there is little exposure to materials that stimulate the developing immune system, the more likely it is that there will be an overreaction to innocuous substances such as dust, pollen, or pet dander.

- How does exposure to microbes early in life affect the immune system?
- What cells are involved in the atopic response?

Continuing the Case appears on page 469.

Outline and Learning Outcomes

16.1 The Immune Response: A Two-Sided Coin

Humans possess a powerful and intricate system of defense, which by its very nature also carries the potential to cause injury and disease. Evolutionarily, these reactions may have developed to fend off infection caused by multicellular organisms. However, in some individuals the quantity and quality of these reactions are excessive and uncontrolled. In most instances, a defect in immune function is expressed in commonplace but miserable symptoms such as those of hay fever and dermatitis. Abnormal or undesirable immune functions are also actively involved in debilitating or life-threatening diseases such as asthma, anaphylaxis, diabetes, rheumatoid arthritis, and graft rejection.

With few exceptions, our previous discussions of the immune response have centered on its numerous beneficial effects. The precisely coordinated system that seeks out, recognizes, and destroys an unending array of foreign materials is clearly protective, but it also presents another side—a side that promotes rather than prevents disease. In this chapter, we survey **immunopathology,** the study of disease states associated with overreactivity or underreactivity of the immune response **(figure 16.1).** Overreactivity takes the forms of allergy, hypersensitivity, and autoimmunity. In the cases of allergies and autoimmunity, the tissues are innocent bystanders attacked by immunologic functions that can't distinguish one's own tissues from those expressing foreign material. In immunodeficiency or **hyposensitivity diseases,** immune function is incompletely developed, suppressed, or destroyed. Cancer falls into a special category because it is both a cause and an effect of immune dysfunction.

Hypersensitivity: Four Types

The most widely accepted classification of the four types of allergy and hypersensitivity includes four major categories: type I ("common" allergy and anaphylaxis), type II (IgG- and IgM-mediated cell damage), type III (immune complex), and type IV (delayed hypersensitivity) **(table 16.1).** In general, types I, II, and III involve a B-cell–immunoglobulin response, and type IV involves a T-cell response (see figure 16.1). The antigens that elicit these reactions can be exogenous, originating from outside the body (microbes, pollen grains, and foreign

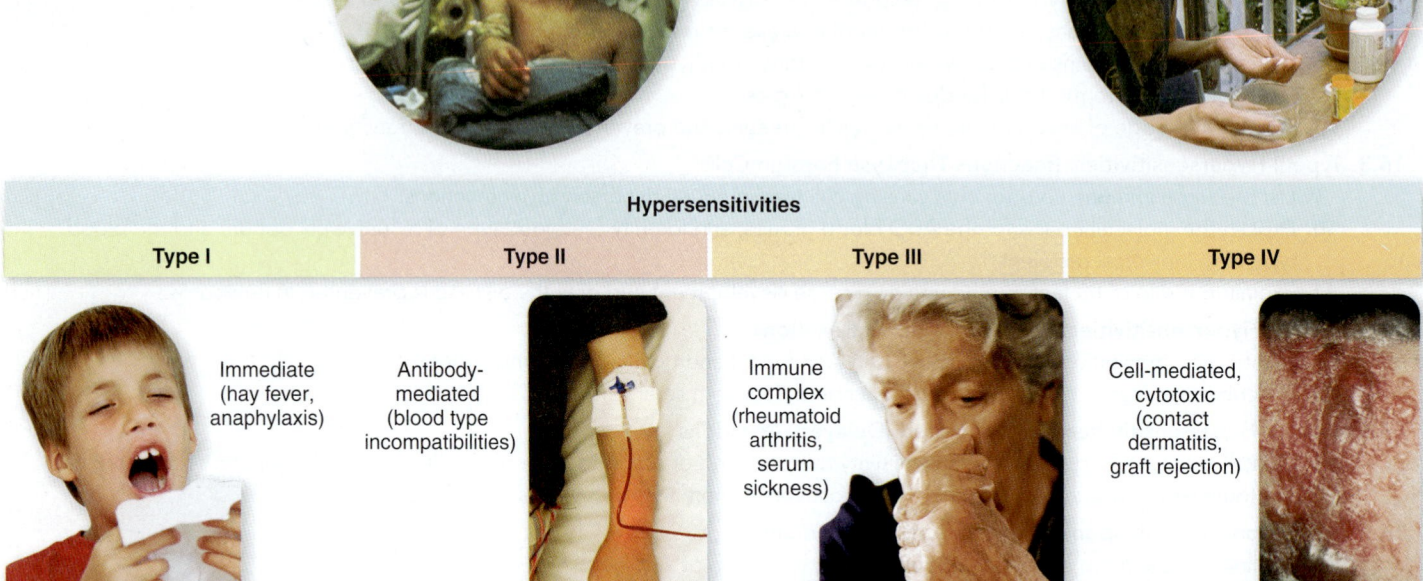

Figure 16.1 Overview of disorders of the immune system.   Just as the system of T cells and B cells provides necessary protection against infection and disease, the same system can cause serious and debilitating conditions by overreacting or underreacting to antigens.

cells and proteins), or endogenous, arising from self tissue (autoimmunities).

One of the reasons allergies are easily mistaken for infections is that both involve damage to the tissues and thus trigger the inflammatory response, as described in chapter 14. Many symptoms and signs of inflammation (redness, heat, skin eruptions, edema, and granuloma) are prominent features of allergies.

16.1 Learning Outcomes—Assess Your Progress

1. Define *immunopathology*, and describe the two major categories of immune dysfunction.

2. Identify the four major categories of hypersensitivities, or overreaction to antigens.

Table 16.1 Hypersensitivity States			
Type		**Systems and Mechanisms Involved**	**Examples**
I	Immediate hypersensitivity	IgE-mediated; involves mast cells, basophils, and allergic mediators	Anaphylaxis, allergies such as hay fever, asthma
II	Antibody-mediated	IgG, IgM antibodies act upon cells with complement and cause cell lysis; includes some autoimmune diseases	Blood group incompatibility; pernicious anemia; myasthenia gravis
III	Immune complex–mediated	Antibody-mediated inflammation; circulating IgG complexes deposited in basement membranes of target organs; includes some autoimmune diseases	Systemic lupus erythematosus; rheumatoid arthritis; serum sickness; rheumatic fever
IV	T-cell-mediated	Delayed hypersensitivity and cytotoxic reactions in tissues; includes some autoimmune diseases	Infection reactions; contact dermatitis; graft rejection

16.2 Type I Allergic Reactions: Atopy and Anaphylaxis

The term **allergy** refers to an exaggerated immune response that is manifested by inflammation. Although it is sometimes used interchangeably with hypersensitivity, some experts refer to immediate reactions such as hay fever as allergies and to delayed reactions as hypersensitivities. Allergic individuals are acutely sensitive to repeated contact with antigens, called **allergens,** that do not noticeably affect nonallergic individuals. All type I allergies share a similar physiological mechanism, are immediate in onset, and are associated with exposure to specific antigens. However, there are two levels of severity: **Atopy** is any chronic local allergy such as hay fever or asthma; **anaphylaxis** (an"-uh-fuh-lak'-sis) is a systemic—sometimes fatal—reaction that involves airway obstruction and circulatory collapse.

Although the general effects of hypersensitivity are detrimental, we must be aware that it involves the very same types of immune reactions as those at work in protective immunities. These include humoral and cell-mediated actions, the inflammatory response, phagocytosis, and complement. Such an association means that all humans have the potential to develop hypersensitivity under particular circumstances. In the following sections, we consider the epidemiology of type I allergies, allergens and routes of inoculation, mechanisms of disease, and specific syndromes.

Who Is Affected, and How?

Allergies exert profound medical and economic impact. Allergists have observed a steady increase in the prevalence of allergic disease in the world today. In fact, it is estimated that 40% of the world's population suffers from one or more allergic conditions. This rate is expanding most rapidly in young children. Experts believe that morbidity and mortality from allergic diseases will continue to rise as air quality diminishes around the world. This can be compounded by increasing global temperatures, which could lead to increased pollen production, biting insect populations, and mold spore levels. Although these statistics are alarming, they may actually be an underrepresentation of the total number of affected individuals worldwide as self-treatment with over-the-counter medicines accounts for significant underreporting of cases, especially in developed countries.

In the United States, nearly half of the population is affected by airborne allergens, such as dust, pollen, and mold. Treatment of asthma, hay fever, and eczema associated with these allergens results in a price tag of about $21 million annually, making it the sixth most costly condition in the United States. Monetary loss due to employee debilitation and absenteeism is immeasurable, as is the loss of school and playtime for affected children. The majority of type I allergies are relatively mild, but certain forms such as asthma and anaphylaxis may require hospitalization and can cause death, especially in the youngest patients. Some individuals have atopic allergies for a lifetime; others "outgrow" them, and still others suddenly develop them later in life.

The predisposition for type I allergies has a strong familial association. Be aware that what is hereditary is a generalized *susceptibility*, not the allergy to a specific substance. For example, a parent who is allergic to ragweed pollen can have a child who is allergic to cat hair. The prospect of a child's developing atopic allergy is at least 25% if one parent exhibits symptoms and increases to nearly 50% if grandparents or siblings are also afflicted. The actual basis for atopy appears to be a genetic program that favors allergic antibody (IgE) production, increased reactivity of mast cells, and increased susceptibility of target tissue to allergic mediators.

The "hygiene hypothesis" (introduced in the opening Case File) provides one explanation for the occurrence of allergies. Also, researchers have found that the combination of being delivered by cesarean section *and* a maternal history of allergy elevates the risk that a child will be allergic to foods by a factor of eight. Scientists suggest that delivery by cesarean section keeps the baby from being exposed to vaginal and stool bacteria. Additional work shows that babies need to be exposed to commensal bacteria in order for the IgA system to develop normally. As one author notes: "We must keep in mind that the current epidemic of allergy in industrialized countries is a small price to pay for the remarkable reduction of infant mortality provided by the elimination of pathogens through improved hygiene. Having too few microbes in our immediate environment seems to be problematic, but having many pathogens is far, far worse."[1]

Newborn babies that are breast-fed exclusively for the first 4 months of life have a lower risk of asthma and eczema, especially if they have a family history of allergy. This is thought to come from the presence of cytokines and growth factors in human milk that act on the baby's gut mucosa to induce tolerance, rather than reactivity, to allergens. New information from the Human Microbiome Project reveals that nearly 600 species of bacteria can be transferred to infants through breast milk. Combined with data from other studies showing that a disruption of microbial populations in the gut may influence the development of asthma (see Continuing the Case), it is clear that these organisms play an important role in the development of tolerance to foreign antigens.

Is there any value to allergies? Most allergy sufferers would answer with a resounding "No!" Why would humans and other mammals evolve an allergic response that is capable of causing suffering, tissue damage, and even death? The answer seems to be that the components involved in an allergic response exist to defend against helminthic worms and other multicellular human parasites. It is only relatively recently in evolutionary history that highly developed countries have dramatically fewer infections with these parasites. An arm of the immune system that has serious benefits in a worm-laden population is left idle in a population that has recently been "scrubbed" of these parasites—and goes awry.

---

1. P. Brandtzaeg, "Why We Develop Food Allergies," *American Scientist* 95, no. 1 (2007): 28–35.

The Nature of Allergens and Their Portals of Entry

As with other antigens, allergens have certain immunogenic characteristics. Proteins are more allergenic than carbohydrates, fats, or nucleic acids, mainly due to the fact that their structure tends to be unique within a particular human individual or a species. Some allergens are haptens, nonproteinaceous substances with a molecular weight of less than 1,000, which can form complexes with carrier molecules in the body (shown in figure 15.8). Organic and inorganic chemicals found in industrial and household products, cosmetics, food, and drugs are commonly of this type. **Table 16.2** lists a number of common allergenic substances.

Allergens typically enter through epithelial portals in the respiratory tract, gastrointestinal tract, and skin **(figure 16.2)**. The mucosal surfaces of the gut and respiratory system present a thin, moist surface that is normally quite penetrable. The dry, tough keratin coating of skin is less permeable, but access still occurs through tiny breaks, glands, and hair follicles. It is worth noting that the organ where an allergy is expressed may or may not be the same as the portal of entry.

Airborne environmental allergens such as pollen, house dust, dander (shed skin scales), or fungal spores are termed *inhalants* **(figure 16.2a)**. Each geographic region harbors a particular combination of airborne substances that varies

with the season and humidity. Pollen is given off seasonally by trees and other flowering plants, while mold spores are released throughout the year. Airborne animal hair and dander, feathers, and the saliva of dogs and cats are common sources of allergens. The component of house dust that appears to account for most dust allergies is not soil or other debris but the decomposed bodies and feces of tiny mites that commonly live in this dust. Some people are allergic to their work, in the sense that they are exposed to allergens on the job. Examples include florists, woodworkers, farmers, drug processors, welders, and plastics manufacturers whose work can aggravate inhalant and contact allergies.

Table 16.2 Common Allergens, Classified by Portal of Entry

Inhalants	Ingestants	Injectants	Contactants
Pollen	Food (milk,	Hymenopteran	Drugs
Dust	peanuts,	venom (bee,	Cosmetics
Mold spores	wheat,	wasp)	Heavy metals
Dander	shellfish,	Drugs	Detergents
Animal hair	soybeans, nuts,	Vaccines	Formalin
Insect parts	eggs, fruits)	Serum	Latex
Formalin	Food additives	Enzymes	Glue
Drugs	Drugs (aspirin,	Hormones	Solvents
	penicillin)		Dyes

Figure 16.2 Common allergens, classified by portal of entry.
(a) Common inhalants, or airborne environmental allergens, include pollen and insect parts. **(b)** Common ingestants, allergens that enter by mouth. **(c)** Common injectants, allergens that enter via the parenteral route. **(d)** Common contactants, allergens that enter through the skin.

Dust mites

Pollen

(a) Inhalants

Type I—Immediate

Bees

Penicillin

(c) Injectants

Red dye

Peanuts

Strawberries

Shrimp

(b) Ingestants

Detergent

Latex glove

Lotion

(d) Contactants

Allergens that enter by mouth, called *ingestants,* often cause food allergies **(figure 16.2b).** *Injectant* allergies are triggered by drugs, vaccines, or hymenopteran (bee) venom **(figure 16.2c).** *Contactants* are allergens that enter through the skin **(figure 16.2d).** Many contact allergies are of the type IV (delayed) variety, discussed later in this chapter. It is also possible to be exposed to certain allergens—penicillin among them—during sexual intercourse due to the presence of allergens in the semen.

Mechanisms of Type I Allergy: Sensitization and Provocation

What causes some people to sneeze and wheeze every time they step out into the spring air, while others suffer no ill effects? In order to answer this question, we must examine what occurs in the tissues of the allergic individual that does not occur in the nonallergenic person. In general, type I allergies develop in stages **(process figure 16.3).** This figure tells the whole story, beginning with the initial encounter with an allergen that sets up the conditions for the allergy to manifest on subsequent encounters.

The Physiology of IgE-Mediated Allergies

During primary contact and sensitization, the allergen penetrates the portal of entry **(process figure 16.3a).** When large particles such as pollen grains, hair, and spores encounter a moist membrane, they release molecules of allergen that pass into the tissue fluids and lymphatics. The lymphatics then carry the allergen to the lymph nodes, where specific clones of B cells recognize it, are activated, and proliferate into plasma cells. These plasma cells produce immunoglobulin E (IgE), the antibody of allergy. IgE is different from other immunoglobulins in having an Fc region with great affinity for mast cells and basophils. The binding of IgE to these cells in the tissues sets the scene for the reactions that occur upon repeated exposure to the same allergen **(process figure 16.3b).**

The Role of Mast Cells and Basophils

Mast cells and basophils play an important role in allergy due to the following:

1. Their ubiquitous location in tissues. Mast cells are located in the connective tissue of virtually all organs, but particularly high concentrations exist in the lungs, skin, gastrointestinal tract, and genitourinary tract. Basophils circulate in the blood but migrate readily into tissues.
2. Their capacity to bind IgE during sensitization (see process figure 16.3) and *degranulate.* Each cell carries 30,000 to 100,000 cell receptors, which trigger the release of inflammatory cytokines from cytoplasmic granules (secretory vesicles) when bound by allergen-associated IgE.

Let us now see what occurs when sensitized cells are challenged with the allergen a second time.

The Second Contact with an Allergen

After sensitization, the IgE-primed mast cells can remain in the tissues for years. Even after long periods without contact, a person can retain the capacity to react immediately upon reexposure. The next time allergen molecules contact these sensitized cells, they bind across adjacent receptors and stimulate degranulation. As chemical mediators are released, they diffuse into the tissues and bloodstream. Cytokines give rise to numerous local and systemic reactions, many of which appear quite rapidly (see process figure 16.3b). The symptoms of allergy are not caused by the direct action of allergens on tissues but by the physiological effects of mast cell mediators on target organs.

Cytokines, Target Organs, and Allergic Symptoms

Numerous substances involved in mediating allergy (and inflammation) have been identified. The principal chemical mediators produced by mast cells and basophils are histamine, serotonin, leukotriene, platelet-activating factor, prostaglandins, and bradykinin **(figure 16.4).** These chemicals, acting alone or in combination, account for the tremendous scope of allergic symptoms. Targets of these mediators include the skin, upper respiratory tract, gastrointestinal tract, and conjunctiva. The general responses of these organs include rashes, itching, redness, rhinitis, sneezing, diarrhea, and shedding of tears. Systemic targets include smooth muscle, mucous glands, and nervous tissue. Because smooth muscle is responsible for regulating the size of blood vessels and respiratory passageways, changes in its activity can profoundly alter blood flow, blood pressure, and respiration. Pain, anxiety, agitation, and lethargy are also attributable to the effects of mediators on the nervous system.

Histamine is the most profuse and fastest-acting allergic mediator. It is a potent stimulator of smooth muscle, glands, and eosinophils. Histamine's actions on smooth muscle vary with location. It *constricts* the smooth muscle layers of the small bronchi and intestine, thereby causing labored breathing and increased intestinal motility. In contrast, histamine *relaxes* vascular smooth muscle and dilates arterioles and venules. It is responsible for the *wheal-and-flare* reaction in the skin (see figure 16.6), pruritus (itching), and headache. More severe reactions such as anaphylaxis can be accompanied by edema and vascular dilation, which lead to hypotension, tachycardia, circulatory failure, and, frequently, shock. Salivary, lacrimal, mucous, and gastric glands are also histamine targets. Histamine can also stimulate eosinophils to release inflammatory cytokines, escalating the aforementioned symptoms.

Although the role of **serotonin** in human allergy is uncertain, its effects appear to complement those of histamine. In experimental animals, serotonin increases vascular permeability, capillary dilation, smooth muscle contraction, intestinal peristalsis, and respiratory rate; but it diminishes central nervous system activity.

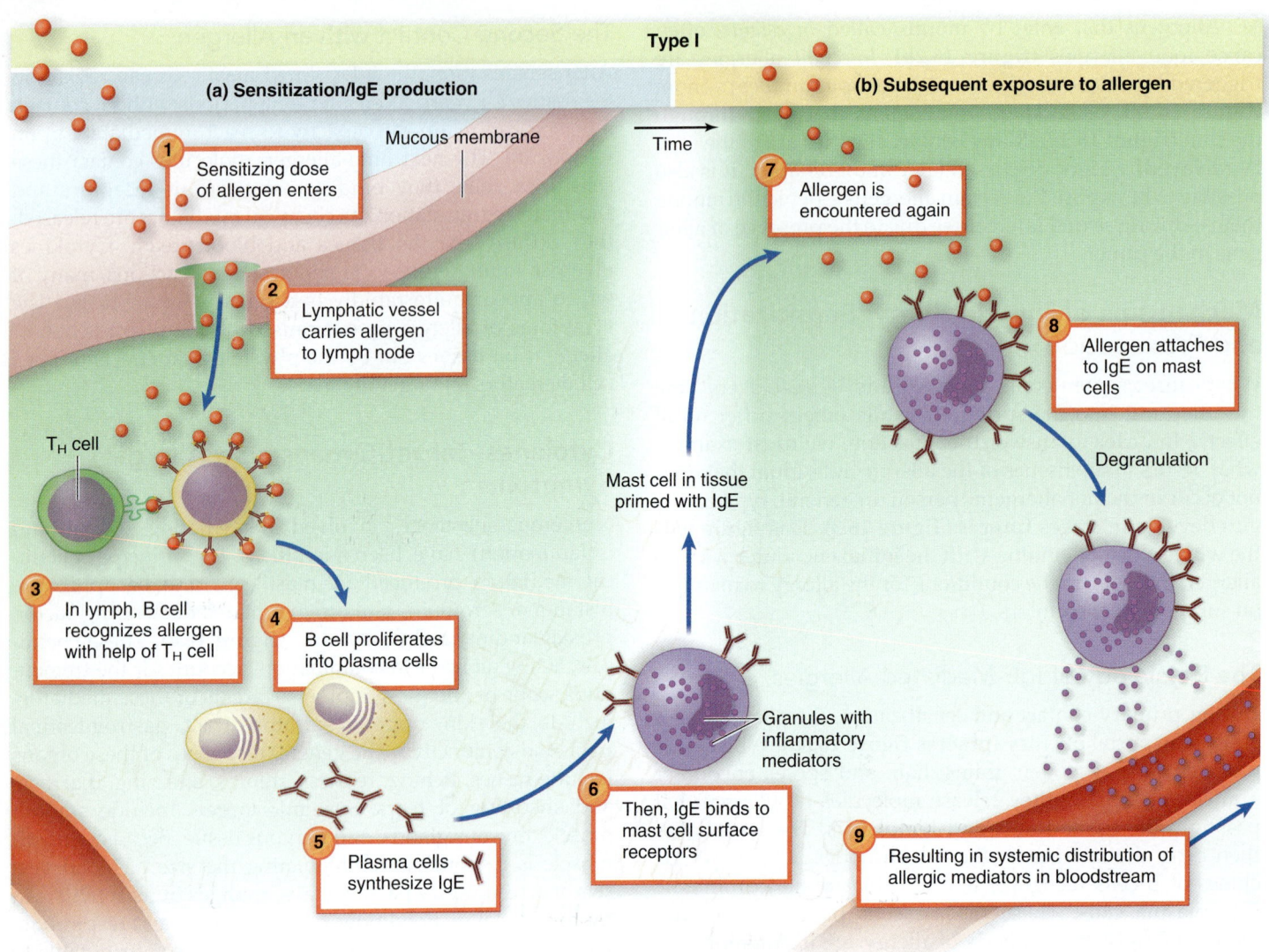

Process Figure 16.3 A schematic view of cellular reactions during the type I allergic response. **(a)** Sensitization (initial contact with sensitizing dose), ❶–❺. **(b)** Provocation (later contacts with provocative dose), ❻–❾.

Before the specific types were identified, **leukotriene** (loo'-koh-try"-een) was known as the "slow-reacting substance of anaphylaxis" for its property of inducing gradual contraction of smooth muscle. This type of leukotriene is responsible for the prolonged bronchospasm, vascular permeability, and mucus secretion of the asthmatic individual. Other leukotrienes stimulate the activities of polymorphonuclear leukocytes, or granulocytes, which play a role in various immune functions (see chapter 14).

Platelet-activating factor is a lipid released by basophils, neutrophils, monocytes, and macrophages. The physiological response to stimulation by this factor is similar to that of histamine, including increased vascular permeability, pulmonary smooth muscle contraction, pulmonary edema, hypotension, and a wheal-and-flare response in the skin.

Prostaglandins are a group of powerful inflammatory agents. Normally, these substances regulate smooth muscle contraction (i.e., they stimulate uterine contractions during delivery). In allergic reactions, they are responsible for vasodilation, increased vascular permeability, increased sensitivity to pain, and bronchoconstriction. Nonsteroidal anti-inflammatory drugs (NSAIDs), such as aspirin and ibuprofen, work by preventing the actions of prostaglandins. **Bradykinin** is related to a group of plasma and tissue peptides known as kinins that participate in blood clotting and chemotaxis. In allergic reactions, it causes prolonged smooth muscle contraction of the bronchioles, dilatation of peripheral arterioles, increased capillary permeability, and increased mucus secretion.

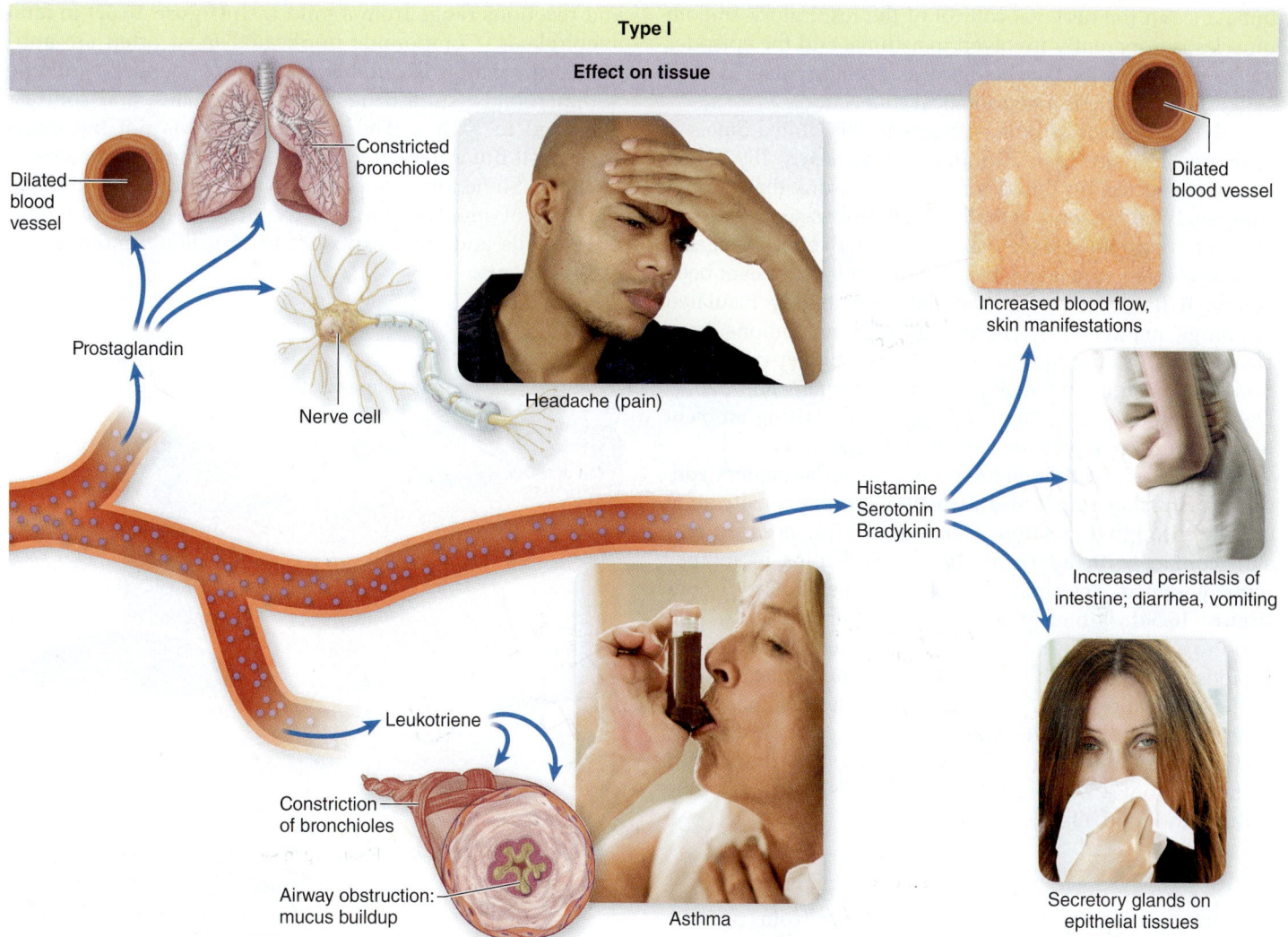

Figure 16.4 The spectrum of reactions to inflammatory cytokines released by mast cells and the common symptoms they elicit in target tissues and organs. Note the extensive overlapping effects.

IgE- and Mast-Cell-Mediated Allergic Conditions

The mechanisms just described are basic to hay fever, allergic asthma, food allergy, drug allergy, eczema, and anaphylaxis. In this section, we cover the main characteristics of these conditions, followed by methods of detection and treatment.

Atopic Diseases

Hay fever is a generic term for **allergic rhinitis,** a seasonal reaction to inhaled plant pollen or molds, or a chronic, year-round reaction to a wide spectrum of airborne allergens or inhalants (see table 16.2). The targets are typically respiratory membranes, and the symptoms include nasal congestion; sneezing; coughing; profuse mucus secretion; itchy, red, and teary eyes; and mild bronchoconstriction.

Asthma is a respiratory disease characterized by episodes of impaired breathing due to severe bronchoconstriction. The airways of asthmatic people are exquisitely responsive to minute amounts of inhalant allergens, food, or other stimuli, such as infectious agents. The symptoms of asthma range from occasional, annoying bouts of difficult breathing to fatal suffocation. Labored breathing, shortness of breath, wheezing, cough, and ventilatory **rales** are present to one degree or another. The respiratory tract of an asthmatic person is chronically inflamed and severely overreactive to allergy chemicals, especially leukotrienes and serotonin from pulmonary mast cells. Upon activation of the allergic response, natural killer T (NKT) cells are recruited and activated, adding to the cytokine storm brewing in the lungs. Other pathologic components are thick mucous plugs in the air sacs and lung damage that can result in long-term respiratory compromise. An imbalance in the nervous control of the respiratory smooth muscles is apparently involved in asthma, and the episodes are influenced by the psychological state of the person, which suggests that there is a neurological connection.

The number of asthma sufferers in the United States is estimated at more than 25 million, with nearly 10% of all children affected by this disorder. For reasons that are not completely understood, asthma is on the increase, and deaths from it have doubled since 1982, even though effective agents to control it are more available now than they have ever been before. It has been suggested that more highly insulated buildings, mandated by energy efficiency regulations, have created indoor air conditions that harbor higher concentrations of contaminants, including insect remains and ozone. Decreasing air quality due to pollution and rising ambient temperatures may play an influential role as well.

Atopic dermatitis is an intensely itchy inflammatory condition of the skin, sometimes also called **eczema.** Sensitization occurs through ingestion, inhalation, and, occasionally, skin contact with allergens. It usually begins in infancy with reddened, vesicular, weeping, encrusted skin lesions **(figure 16.5a).** It then progresses in childhood and adulthood to a dry, scaly, thickened skin condition **(figure 16.5b).** Lesions can occur on the face, scalp, neck, and inner surfaces of the limbs and trunk. The itchy, painful lesions cause considerable discomfort, and they are often predisposed to secondary bacterial infections. Recent studies show that infants suffering from eczema exhibit a higher risk of developing asthma and food allergies as they age.

Food Allergy

The most common food allergens come from peanuts **(Insight 16.1),** fish, cow's milk, eggs, shellfish, and soybeans. Although the mode of entry is intestinal, food allergies can also affect the skin and respiratory tract. Gastrointestinal symptoms include vomiting, diarrhea, and abdominal pain. Other manifestations of food allergies include eczema, hives, rhinitis, asthma, and, occasionally, anaphylaxis. Classic food hypersensitivity involves IgE and degranulation of mast cells, but not all reactions involve this mechanism. (Do not confuse food allergy with food intolerance. Many people are lactose intolerant, for example, due to a deficiency in the enzyme that degrades the milk sugar.) Food (egg) allergies must be considered when vaccinating individuals, due to the presence of egg protein in many vaccine preparations.

Drug Allergy

Modern chemotherapy has been responsible for many medical advances. Unfortunately, drugs are foreign compounds capable of stimulating allergic reactions in some people. In fact, allergy to drugs is one of the most common side effects of treatment (present in 5% to 10% of hospitalized patients). Depending on the allergen, route of entry, and individual sensitivities, virtually any tissue of the body can be affected,

and reactions range from a mild rash **(figure 16.5c)** to fatal anaphylaxis. Compounds implicated most often are antibiotics (penicillin is number one in prevalence), synthetic antimicrobials (sulfa drugs), aspirin, opiates, and contrast dye used in X rays. The actual allergen is not the intact drug itself but a hapten given off when the liver processes the drug. Some forms of penicillin sensitivity are due to the presence of small amounts of the drug in meat, milk, and other foods and from exposure to *Penicillium* mold in the environment.

Type I—Immediate

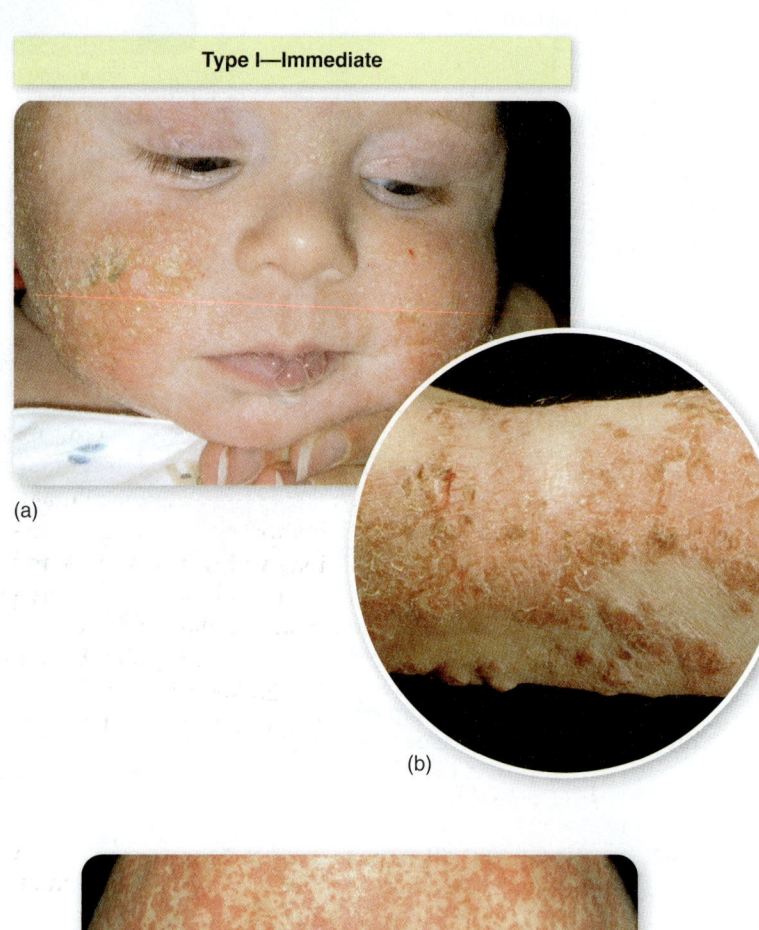

(a)

(b)

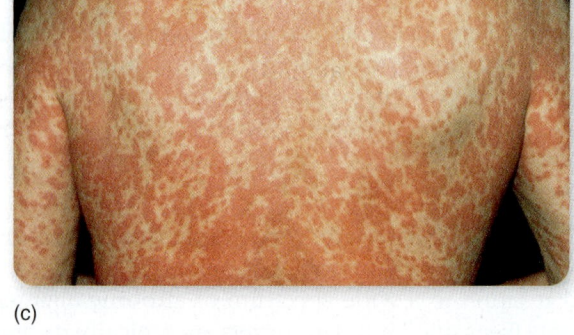

(c)

Figure 16.5 Skin manifestations in atopic and drug allergies. **(a)** Vesicular, weepy, encrusted lesions are typical in afflicted infants. **(b)** In adulthood, lesions are more likely to be dry, scaly, and thickened. **(c)** A typical rash that develops in an allergic reaction to an antibiotic.

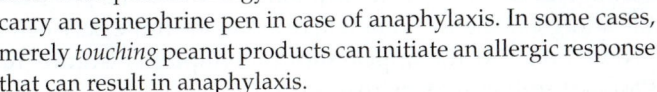

INSIGHT 16.1   Treatment for Deadly Peanut Allergy

Allergies to peanuts and other nuts can be deadly, causing anything from hives to eczema, asthma, vomiting, or anaphylactic shock in extremely sensitive individuals. Peanut allergy is listed as one of the most common causes of food-related death by the Asthma and Allergy Foundation of America. Individuals with peanut allergy are advised to avoid peanuts or traces of peanuts in foods. Parents of children with peanut allergy are advised to carry an epinephrine pen in case of anaphylaxis. In some cases, merely *touching* peanut products can initiate an allergic response that can result in anaphylaxis.

Using a mouse model that mimics peanut allergy, researchers at Northwestern University Feinberg School of Medicine have found a way to overcome this peanut allergy. Peanut proteins were attached to the membranes of white blood cells extracted from susceptible mice and then re-infused into their bloodstream. After two treatments, mice were fed a peanut extract that should have induced deadly anaphylaxis, but none of the mice exhibited any allergic reaction to peanuts. Researchers hope to use this treatment to help peanut allergy sufferers as well as treat other types of food allergies.

Source: www.sciencedaily.com/releases/2011/10/111011112801.htm

Anaphylaxis: An Overpowering Systemic Reaction

The term **anaphylaxis,** or anaphylactic shock, was first used to denote a reaction of animals injected with a foreign protein. Although the animals showed no response during the first contact, upon reinoculation with the same protein at a later time, they exhibited acute symptoms—itching, sneezing, difficult breathing, prostration, and convulsions—and many died in a few minutes. Systemic anaphylaxis is characterized by sudden respiratory and circulatory disruption that can be fatal in a few minutes. In humans, the allergen and route of entry are variable, though bee stings and injections of antibiotics or serum are implicated most often. Bee venom is a complex material containing several allergens and enzymes that can create a sensitivity that can last for decades after exposure.

The underlying physiological events in anaphylaxis parallel those of atopy, but the concentration of chemical mediators and the strength of the response are greatly amplified. The immune system of a sensitized person exposed to a provocative dose of an allergen or allergens responds with a sudden, massive release of chemicals into the tissues and blood, which act rapidly on the target organs. Anaphylactic persons have been known to die in 15 minutes from complete airway blockage.

Diagnosis of Allergy

Because allergy mimics infection and other conditions, it is important to determine if a person is actually allergic and to identify the specific allergen or allergens involved. Allergy diagnosis involves several levels of tests, including nonspecific, specific, *in vitro,* and *in vivo* methods.

The most widely used blood test is a radioallergosorbent test (RAST), which measures levels of IgE to specific allergens. It cannot be used for all allergens, but recently more tests have become available for food allergens. Another test that can distinguish whether a patient has experienced an allergic attack measures elevated blood levels of tryptase, an enzyme released by mast cells that increases during an allergic response. Several types of specific *in vitro* tests can determine the allergic potential of a patient's blood sample. A differential blood cell count can indicate the levels of basophils and eosinophils, indicating allergy; the leukocyte histamine-release test measures the amount of histamine released from the patient's basophils when exposed to a specific allergen.

Skin Testing

A tried and true *in vivo* method to detect precise atopic or anaphylactic sensitivities is skin testing. With this technique, a patient's skin is injected, scratched, or pricked with a small amount of a pure allergen extract. There are hundreds of these allergen extracts containing common airborne allergens (plant and mold pollen) and more unusual allergens (mule dander, theater dust, bird feathers). Unfortunately, skin tests for food allergies using food extracts are unreliable in most cases. In patients with numerous allergies, the allergist maps the skin on the inner aspect of the forearms or back and injects the allergens intradermally according to this predetermined pattern **(figure 16.6a)**. Approximately 20 minutes after antigenic challenge, each site is appraised for a wheal response indicative of histamine release. The diameter of the wheal is measured and rated on a scale of 0 (no reaction) to 4 (greater than 15 mm) **(figure 16.6b)**.

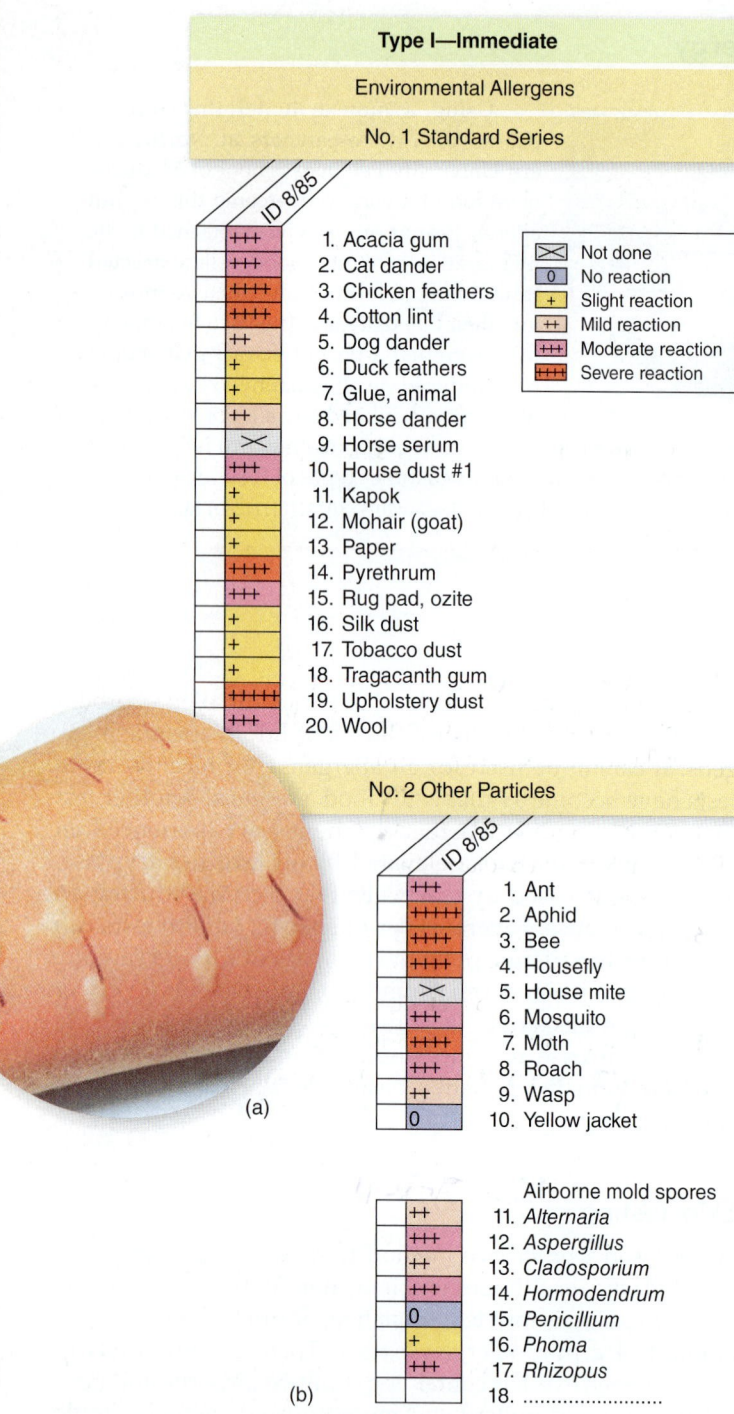

Type I—Immediate	
Environmental Allergens	
No. 1 Standard Series	

ID 8/85

+++	1. Acacia gum
+++	2. Cat dander
++++	3. Chicken feathers
++++	4. Cotton lint
++	5. Dog dander
+	6. Duck feathers
+	7. Glue, animal
++	8. Horse dander
✕	9. Horse serum
+++	10. House dust #1
+	11. Kapok
+	12. Mohair (goat)
+	13. Paper
++++	14. Pyrethrum
+++	15. Rug pad, ozite
+	16. Silk dust
+	17. Tobacco dust
+	18. Tragacanth gum
+++++	19. Upholstery dust
+++	20. Wool

✕	Not done
0	No reaction
+	Slight reaction
++	Mild reaction
+++	Moderate reaction
++++	Severe reaction

No. 2 Other Particles

ID 8/85

+++	1. Ant
+++++	2. Aphid
++++	3. Bee
++++	4. Housefly
✕	5. House mite
+++	6. Mosquito
++++	7. Moth
+++	8. Roach
++	9. Wasp
0	10. Yellow jacket

Airborne mold spores

++	11. *Alternaria*
+++	12. *Aspergillus*
++	13. *Cladosporium*
+++	14. *Hormodendrum*
0	15. *Penicillium*
+	16. *Phoma*
+++	17. *Rhizopus*
	18. ....

(a)

(b)

Figure 16.6 A method for conducting an allergy skin test. The forearm (or back) is mapped and then injected with a selection of allergen extracts. The allergist must be very aware of potential anaphylaxis attacks triggered by these injections. **(a)** Close-up of skin wheals showing a number of positive reactions (dark lines are measurer's marks). **(b)** An actual skin test record for some common environmental allergens [not related to **(a)**].

Treatment and Prevention of Allergy

In general, the methods of treating and preventing type I allergy involve

1. avoiding the allergen, although this may be very difficult in many instances;
2. taking drugs that block the action of lymphocytes, mast cells, or chemical mediators; and
3. using injections of the allergen in such a way that the allergic reaction is short-circuited ("allergy shots").

Taking Drugs to Block Allergy

The aim of antiallergy medication is to block the progress of the allergic response somewhere along the route between IgE production and the appearance of symptoms **(figure 16.7).** Oral anti-inflammatory drugs such as corticosteroids inhibit the activity of lymphocytes and thereby reduce the production of IgE, but they also have dangerous side effects and should not be taken for prolonged periods. Some drugs block the degranulation of mast cells and reduce the levels of inflammatory cytokines. The most effective of these are diethylcarbamazine and cromolyn. Asthma and rhinitis sufferers can find relief with a drug that blocks synthesis of leukotriene and a monoclonal antibody that inactivates IgE (omalizumab [Xolair]).

Widely used medications for preventing symptoms of atopic allergy are **antihistamines,** the active ingredients in most over-the-counter allergy-control drugs. Antihistamines interfere with histamine activity by binding to histamine receptors on target organs. Most of them have side effects, however, such as drowsiness. Newer antihistamines lack this side effect because they do not cross the blood-brain barrier. Other drugs that relieve inflammatory symptoms are aspirin and acetaminophen, which reduce pain by interfering with prostaglandin; and theophylline, a bronchodilator that reverses spasms in the respiratory smooth muscles. Persons who are prone to anaphylactic attacks are urged to carry injectable epinephrine (adrenaline) and an identification tag indicating their sensitivity. An aerosol inhaler containing epinephrine can also provide rapid relief. Epinephrine reverses constriction of the airways and slows the release of allergic mediators. Although epinephrine works quickly and well, it has a very short half-life. It is very common to require more than one dose in anaphylactic reactions. Injectable epinephrine buys the individual time to get to a hospital for continuing treatment.

Allergy "Vaccines"

Approximately 70% of allergic patients benefit from controlled injections of specific allergens as determined by skin tests. This technique, called **desensitization** or **hyposensitization,** is a therapeutic way to prevent reactions between allergen, IgE, and mast cells. The allergen preparations contain pure suspensions of plant antigens, venoms, dust mites, dander, and molds. The immunologic basis of

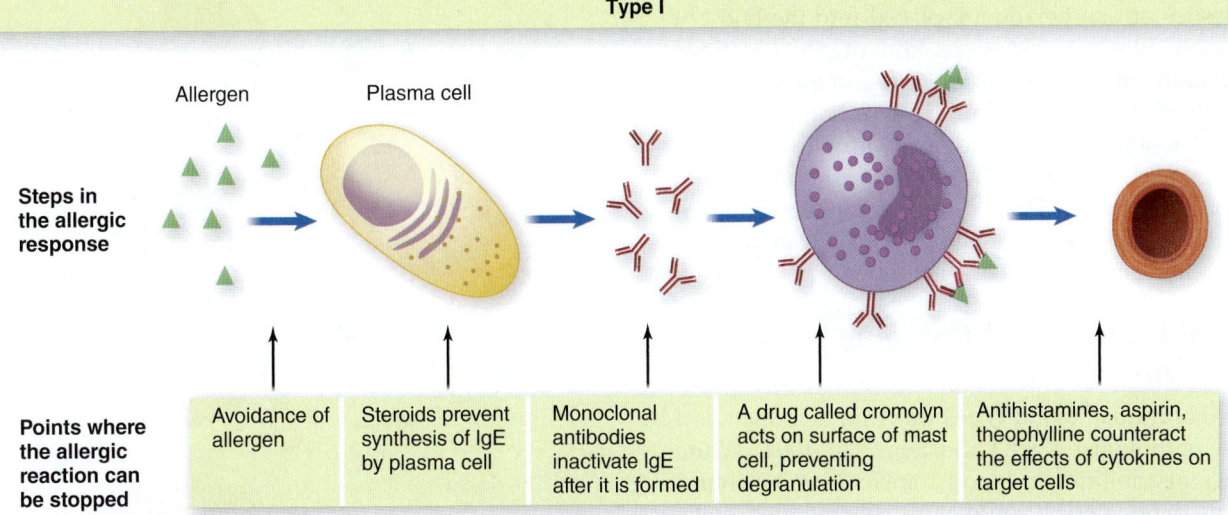

Figure 16.7 Strategies for circumventing allergy attacks.

this treatment is open to differences in interpretation. One hypothesis suggests that injected allergens stimulate the formation of high levels of allergen-specific **IgG "blocking antibodies"** that can remove allergen from the system before it can bind to IgE and trigger mast cell degranulation **(figure 16.8)**. It is also possible that allergen delivered in this fashion combines with the IgE itself and takes it from circulation before it can react with the mast cells.

Newer experimental therapy is focused on the development of allergy treatments that do not contain the allergen itself. These new injections instead contain a "decoy," an innocuous molecule that merely resembles an allergen. The particles are made using genetic engineering technology and appear to effectively increase IgG levels and block the formation of inflammation-inducing IgE-allergen complexes. Ultimately, allergy sufferers theoretically can be stimulated to develop natural immunity to any allergen, provided it can be synthesized in a laboratory.

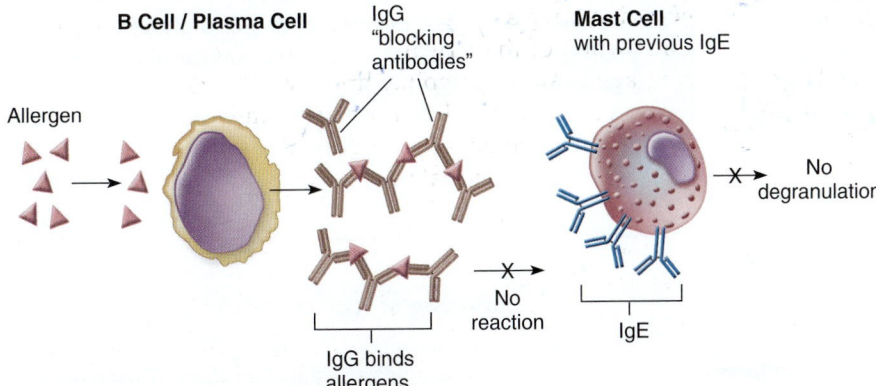

Figure 16.8 Blocking antibodies allow for allergic desensitization.
An injection of allergen causes IgG antibodies to be formed instead of IgE; these blocking antibodies cross-link and effectively remove the allergen before it can react with the IgE in the mast cell.

16.2 Learning Outcomes—Assess Your Progress

3. Summarize genetic and environmental factors that influence allergy development.
4. Outline the steps of a type I allergic response, and discuss the effects on target organs and tissue.
5. Identify three conditions caused by IgE-mediated allergic reactions.
6. Describe the symptoms of anaphylaxis and link these to physiological events.
7. Briefly describe two methods for diagnosing allergies.
8. Explain the mode of action of two strategies for treating and preventing type I allergic reactions.

16.3 Type II Hypersensitivities: Reactions That Lyse Foreign Cells

The diseases termed *type II hypersensitivities* are a complex group of syndromes that involve complement-assisted destruction (lysis) of cells by antibodies (IgG and IgM) directed against those cells' surface antigens. This category includes transfusion reactions and some types of autoimmunities (discussed in a later section). The cells targeted for destruction are often red blood cells, but other cells can be involved.

Chapters 14 and 15 described the functions of unique surface markers on cell membranes. Ordinarily, these molecules play essential roles in transport, recognition, and development, but they become medically important when the tissues of one person are placed into the body of another person. Blood transfusions and organ donations introduce alloantigens (molecules that differ in the same species) on donor cells that are recognized by the lymphocytes of the recipient. These reactions are not really immune dysfunctions as allergy and autoimmunity are. The immune system is in fact working normally, but it is not equipped to distinguish between the desirable foreign cells of a transplanted tissue and the undesirable ones of a microbe.

The Basis of Human ABO Antigens and Blood Types

The existence of human blood types was first demonstrated by Austrian pathologist Karl Landsteiner in 1904. While studying incompatibilities in blood transfusions, he found that the serum of one person could clump the red blood cells of another. Landsteiner identified four distinct types, subsequently called the **ABO blood groups.**

Like the MHC antigens on white blood cells, the ABO antigen markers on red blood cells are genetically determined and composed of glycoproteins. These ABO antigens are inherited as two (one from each parent) of three alternative **alleles:** A, B, or O. As **table 16.3** indicates, this mode of inheritance gives rise to four blood types (phenotypes), depending on the particular combination of genes. Thus, a person with an *AA* or *AO* genotype has type A blood; genotype *BB* or *BO* gives type B; genotype *AB* produces type AB; and genotype *OO* produces type O. Some important points about the blood types follow:

1. They are named for the dominant antigen(s).
2. The RBCs of type O persons have antigens but not A and B antigens.
3. Tissues other than RBCs carry A and B antigens.

A diagram of the AB antigens and blood types is shown in **figure 16.9.** Each of the A and B genes codes for an enzyme that adds a terminal carbohydrate to RBC surface molecules during maturation. RBCs of type A contain an enzyme that adds *N*-acetylgalactosamine to the molecule; RBCs of type B have an enzyme that adds D-galactose; RBCs of type AB contain both enzymes that add both carbohydrates; and RBCs of type O lack the genes and enzymes to add a terminal molecule.

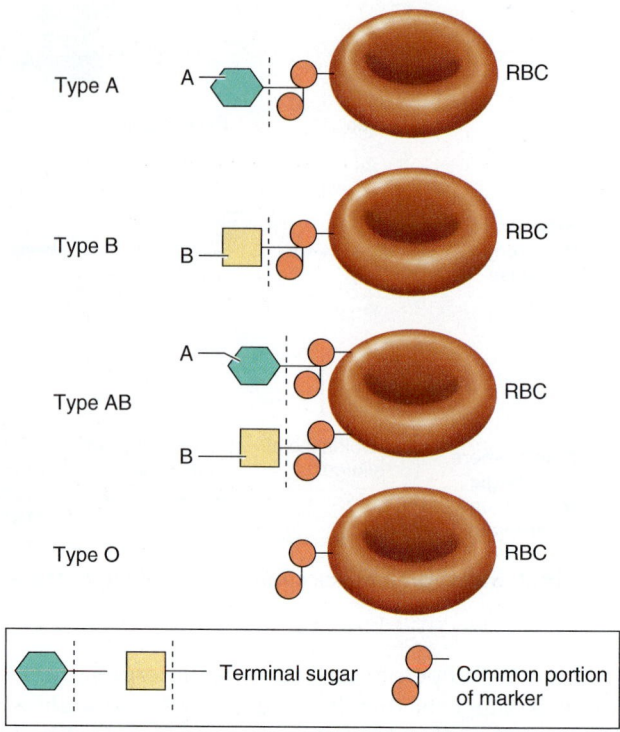

Figure 16.9 The genetic/molecular basis for the A and B antigens (receptors) on red blood cells.   In general, persons with blood types A, B, and AB inherit a gene for the enzyme that adds a certain terminal sugar to the basic RBC receptor. Type O persons do not have such an enzyme and lack the terminal sugar.

Disease Connection

Since the 1980s, we have known that the bacterium *Helicobacter pylori* is a major cause of gastritis and stomach ulcers. Since that time, we have also discovered that people with the O blood type seem to have a higher susceptibility to the condition, perhaps because the bacterium utilizes blood group glycoproteins exposed on cells in the stomach for purposes of attachment. Alternatively, antibodies directed to A group glycoproteins and B group glycoproteins may bind to the surface of *H. pylori* and induce enhanced inflammation.

Table 16.3 Characteristics of ABO Blood Groups

Genotype	Blood Type	Antigen Present on Erythrocyte Membranes	Antibody in Plasma	Incidence of Type in United States		
				Among Whites (%)	Among Asians (%)	Among Those of African and Caribbean Descent (%)
AA, AO	A	A	Anti-B	41	28	27
BB, BO	B	B	Anti-A	10	27	20
AB	AB	A and B	Neither anti-A nor anti-B	4	5	7
OO	O	Neither A nor B	Anti-A and anti-B	45	40	46

Recently, researchers at Brigham and Women's Hospital compared the immune systems of "germ-free" mice—mice that had never been exposed to microbes and lived in completely sterile conditions with their normal, "dirty" counterparts. The sterile, germ-free mice had an exaggerated inflammatory response in their lungs resembling asthma and in their colon resembling irritable bowel syndrome (IBS). This out-of-control inflammatory response was caused by a class of T cells called invariant natural killer T cells (INKT cells). INKT cells release several interleukins that initiate the inflammatory response in the airways, leading to symptoms of hay fever and asthma. When germ-free mice were exposed to microbes early in life, they developed healthy, normal immune systems; but when adult germ-free mice were exposed, they developed the hyperactive inflammatory response typical of asthma and IBS. These data parallel the results of another study showing that young children who frequently take antibiotics appear to develop higher rates of asthma and allergies. Although further study is needed, this research demonstrates the importance of what scientists call "proper immune conditioning" early in life and the fact that microbes may play an important role in this process.

- What is autoimmunity?
- Why is autoimmunity more common in women than men?

Source: www.sciencedaily.com/releases/2012/03/120322142157.htm

Antibodies Against A and B Antigens

Although an individual does not normally produce antibodies in response to his or her own RBC antigens, the serum can contain antibodies that react with blood of another antigenic type even though contact with this other blood type has *never* occurred. These preformed antibodies account for the immediate and intense quality of transfusion reactions. As a rule, type A blood contains antibodies (anti-B) that react against the B antigens on types B and AB red blood cells. Type B blood contains antibodies (anti-A) that react with A antigen on types A and AB red blood cells. Type O blood contains antibodies against both A and B antigens. Type AB blood does not contain antibodies against either A or B antigens[2] (see table 16.3). What is the source of these anti-A and anti-B antibodies? It appears that they develop in early infancy because of exposure to certain antigens that are widely distributed in nature. These antigens are surface molecules on bacteria and plant cells that mimic the structure of A and B antigens. Exposure to these sources stimulates the production of corresponding antibodies.

2. Why would this be true? The answer lies in the first sentence of the paragraph.

Clinical Concerns in Transfusions

The presence of ABO antigens and A and B antibodies can present problems in giving blood transfusions. First, the individual blood types of donor and recipient must be determined. By use of a standard technique, drops of blood are mixed with antisera that contain antibodies against the A and B antigens and are then observed for the evidence of agglutination (**figure 16.10**).

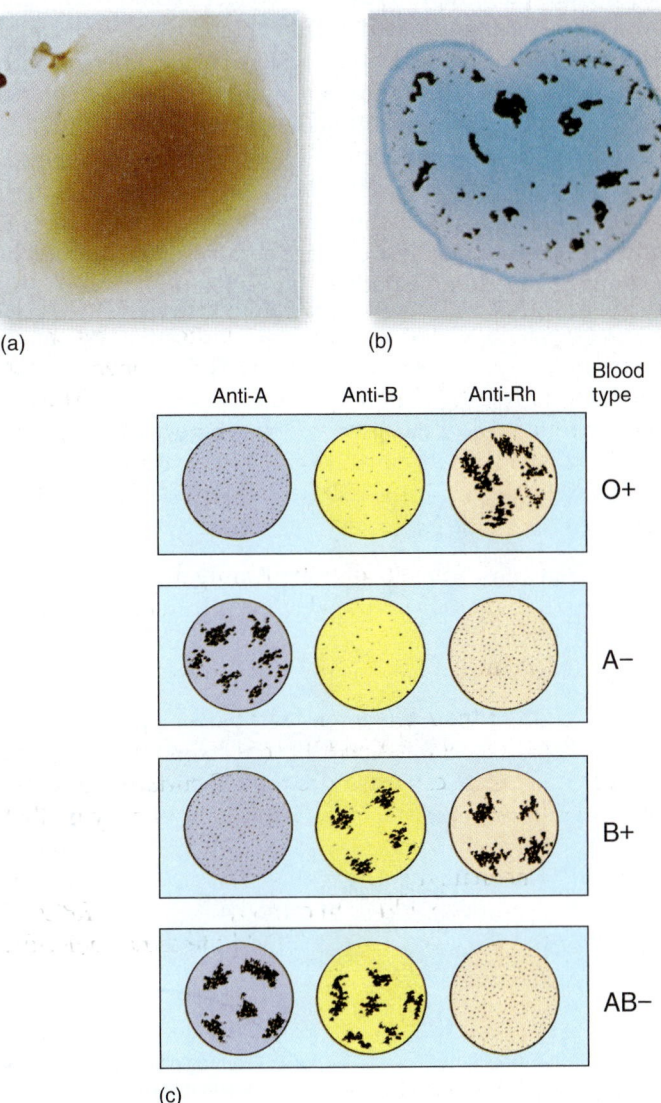

Figure 16.10 Interpretation of blood typing. In this test, a drop of blood is mixed with a specially prepared antiserum known to contain antibodies against the A, B, or Rh antigens. **(a)** If that particular antigen is not present, the red blood cells in that droplet do not agglutinate and form an even suspension. **(b)** If that antigen is present, agglutination occurs and the RBCs form visible clumps. **(c)** Several patterns and their interpretations. *Anti-A, anti-B,* and *anti-Rh* are shorthand for the antisera applied to the drops. (In general, O+ is the most common blood type, and AB− is the rarest.)

Knowing the blood types involved makes it possible to determine which transfusions are safe to do. The general rule of compatibility is that the RBC antigens of the donor must not be agglutinated by antibodies in the recipient's blood **(figure 16.11)**. The ideal practice is to transfuse blood that is a perfect match (A to A, B to B). But even in this event, blood samples must be cross-matched before the transfusion because other blood group incompatibilities can exist. This test involves mixing the blood of the donor with the serum of the recipient to check for agglutination.

Under certain circumstances (emergencies, the battle-field), the concept of universal transfusions can be used. To appreciate how this works, we must apply the rule stated in the previous paragraph. Type O blood lacks A and B antigens and will not be agglutinated by other blood types, so it could theoretically be used in any transfusion. Hence, a person with this blood type is called a **universal donor.** Because type AB blood lacks agglutinating antibodies, an individual with this blood could conceivably receive any type of blood. Type AB persons are consequently called *universal recipients.* Although both types of transfusions involve antigen-antibody incompatibilities, these are of less concern because of the dilution of the donor's blood in the body of the recipient.

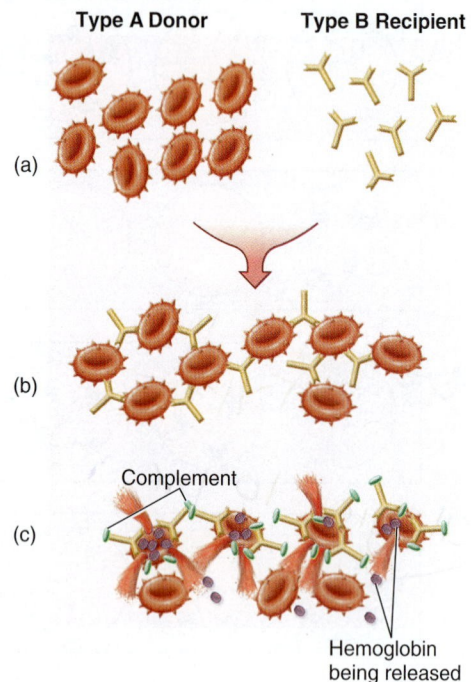

Figure 16.11 Microscopic view of a transfusion reaction. **(a)** Incompatible blood. The red blood cells of the type A donor have antigen A, while the serum of the type B recipient contains anti-A antibodies that can agglutinate donor cells. **(b)** Agglutination complexes can block the circulation in vital organs. **(c)** Activation of complement by antibody on the RBCs can cause hemolysis and anemia. This sort of incorrect transfusion is very rare because of the great care taken by blood banks to ensure a correct match.

Transfusion of the wrong blood type causes differing degrees of adverse reaction. The most severe reaction is massive hemolysis when the donated red blood cells react with recipient antibody and trigger the complement cascade (see figure 16.11). The resultant destruction of red cells leads to systemic shock and kidney failure brought on by the blockage of glomeruli (blood-filtering apparatuses) by cell debris. Death is a common outcome. Other reactions caused by RBC destruction are fever, anemia, and jaundice. A transfusion reaction is managed by immediately halting the transfusion, administering drugs to remove hemoglobin from the blood, and beginning another transfusion with red blood cells of the correct type. The development of synthetic blood is important to establishing a safe, plentiful blood supply in many areas of the world, including war zones. "Pharmed" blood sources created from stem cell populations are showing great progress and contain cells that are of the O-negative blood phenotype.

The Rh Factor and Its Clinical Importance

Another RBC antigen of major clinical concern is the **Rh factor** (or D antigen). This factor was first discovered in experiments exploring the genetic relationships among animals. Rabbits inoculated with the RBCs of rhesus monkeys produced an antibody that also reacted with human RBCs. Further tests showed that this monkey antigen (termed *Rh* for "rhesus") was present in about 85% of humans and absent in the other 15%. The details of Rh inheritance are more complicated than those of ABO, but in simplest terms, a person's Rh type results from a combination of two possible alleles—a dominant one that codes for the factor and a recessive one that does not. A person inheriting at least one Rh gene will be Rh-positive (Rh+); only those persons inheriting two recessive genes are Rh-negative (Rh–). The "+" or "–" appearing after a blood type (i.e., O+) reflects the Rh status of the person (see figure 16.10*c*). Unlike the case with ABO antigens, the only way one can develop antibodies against this antigen is through transfusion or being sensitized in the womb. Although the Rh factor should be matched for a transfusion to avoid this situation, it is acceptable to transfuse Rh– blood if the Rh type is not known.

Hemolytic Disease of the Newborn and Rh Incompatibility

The potential for placental sensitization occurs when a mother is Rh– and her unborn child is Rh+. It is possible for fetal RBCs to leak into the mother's circulation during childbirth, when the detachment of the placenta creates avenues for fetal blood to enter the maternal circulation. The mother's immune system detects the foreign Rh factors on the fetal RBCs and is sensitized to them by producing antibodies and memory B cells. The first Rh+ child is usually not affected because the process begins so late in pregnancy that the child is born before maternal sensitization is completed. However, the mother's immune system has been strongly primed for

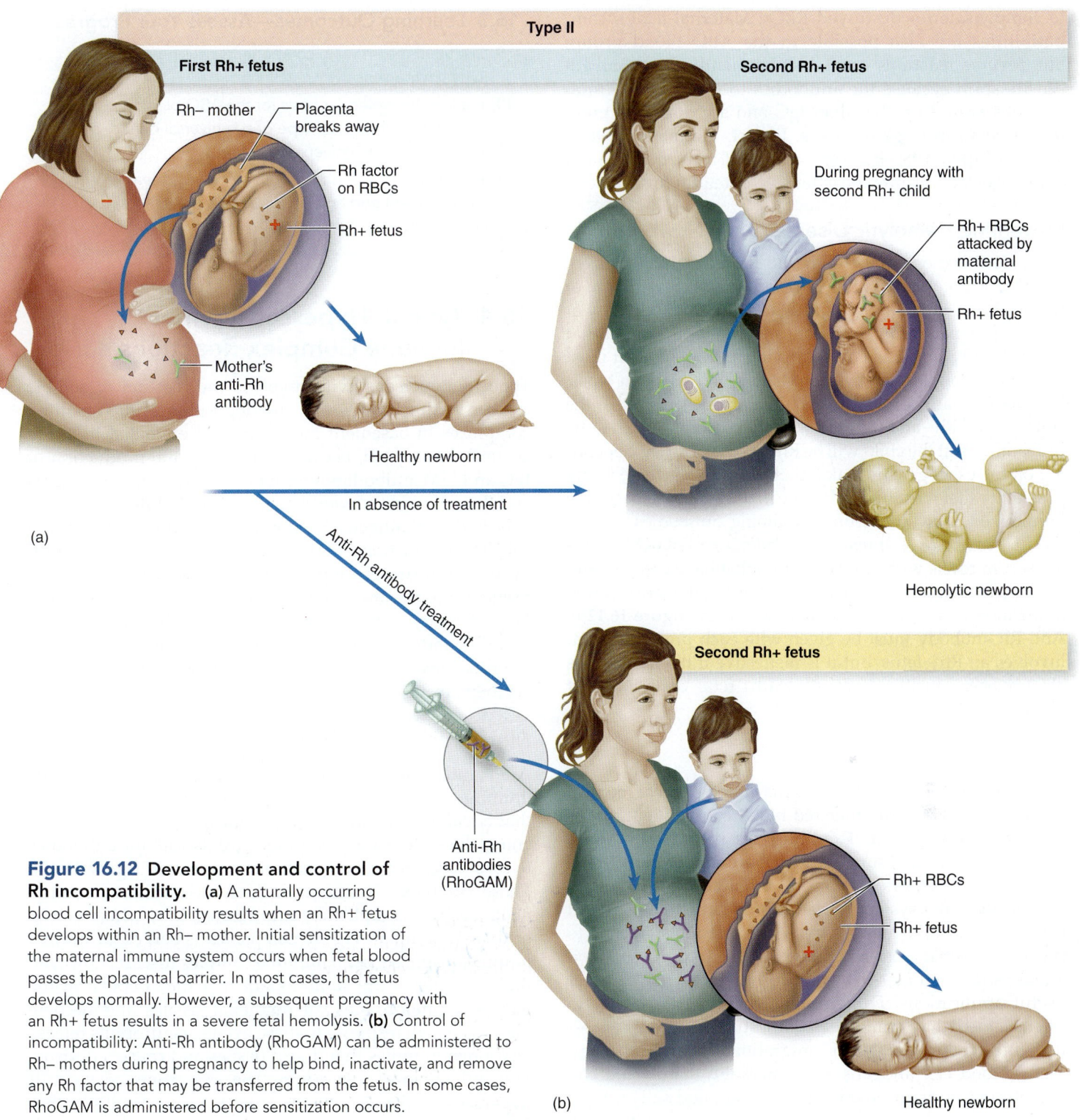

Figure 16.12 Development and control of Rh incompatibility. **(a)** A naturally occurring blood cell incompatibility results when an Rh+ fetus develops within an Rh– mother. Initial sensitization of the maternal immune system occurs when fetal blood passes the placental barrier. In most cases, the fetus develops normally. However, a subsequent pregnancy with an Rh+ fetus results in a severe fetal hemolysis. **(b)** Control of incompatibility: Anti-Rh antibody (RhoGAM) can be administered to Rh– mothers during pregnancy to help bind, inactivate, and remove any Rh factor that may be transferred from the fetus. In some cases, RhoGAM is administered before sensitization occurs.

a second contact with this factor in a subsequent pregnancy **(figure 16.12a).**

In the next pregnancy with an Rh+ fetus, fetal blood cells escape into the maternal circulation late in pregnancy and elicit a memory response. Maternal anti-Rh antibodies then cross the placenta into the fetal circulation, where they affix to fetal RBCs and cause complement-mediated lysis. The outcome is a potentially fatal **hemolytic disease of the newborn (HDN),** characterized by severe anemia and jaundice. It is also called *erythroblastosis fetalis* (eh-rith"-roh-blas-toh'-sis fee-tal'-is), reflecting the release of immature nucleated RBCs (erythroblasts) into the blood to compensate for destroyed RBCs. Maternal-fetal incompatibilities are also possible in the ABO blood group, but adverse reactions occur less frequently than with Rh sensitization because the antibodies to these blood group antigens are IgM rather than IgG and are unable to cross the placenta in large numbers. In fact, the maternal-fetal relationship is a fascinating instance of foreign tissue not being rejected, despite the extensive potential for contact.

Preventing Hemolytic Disease of the Newborn

Once sensitization of the mother to Rh factor has occurred, all other Rh+ fetuses will be at risk for hemolytic disease of the newborn. Prevention requires a careful family history of an Rh– pregnant woman. It can predict the likelihood that she is already sensitized or is carrying an Rh+ fetus. It must take into account other children she has had, their Rh types, and the Rh status of the father. If the father is also Rh–, the child will be Rh– and free of risk; but if the father is Rh+, the probability that the child will be Rh+ is 50% or 100%, depending on the exact genetic makeup of the father. If there is any possibility that the fetus is Rh+, the mother must be passively immunized with antiserum containing antibodies against the Rh factor (Rh_0 [D] *immune globulin*, or RhoGAM).[3] This antiserum reacts with any fetal RBCs that have escaped into the maternal circulation, thereby preventing the sensitization of the mother's immune system to Rh factor **(figure 16.12b)**. Anti-Rh antibody must be given with each pregnancy that involves an Rh+ fetus, but it is ineffective if the mother has already been sensitized by a prior Rh+ fetus or an incorrect blood transfusion.

Other RBC Antigens

Although the ABO and Rh systems are of greatest medical significance, about 20 other red blood cell antigen groups have been discovered. Examples are the MN, Ss, Kell, and P blood groups, some of which are unique membrane proteins while others are simply carbohydrate antigens as seen in the ABO system. Because of incompatibilities that these blood groups present, transfused blood is screened to prevent possible cross-reactions. The study of these blood antigens (as well as ABO and Rh) has given rise to other useful applications. For example, they can be useful in forensic medicine (crime detection), studying ethnic ancestry, and tracing prehistoric migrations in anthropology. Many blood cell antigens are remarkably hardy and can be detected in dried blood stains, semen, and saliva. Even the 3,300-year-old mummy of King Tutankhamun has been typed A_2MN!

In section 16.6, "An Inappropriate Response Against Self: Autoimmunity," you will read about special cases of type II hypersensitivity in which it is directed against self.

16.3 Learning Outcomes—Assess Your Progress

9. List the three immune components causing cell lysis in type II hypersensitivity reactions.
10. Explain the molecular basis for the ABO blood groups, and identify the blood type of a "universal donor" and the blood type of a "universal recipient."
11. Explain the role of Rh factor in hemolytic disease development and how the disease is prevented in newborns.

16.4 Type III Hypersensitivities: Immune Complex Reactions

Type III hypersensitivity involves the reaction of soluble antigen with antibody and the deposition of the resulting complexes in basement membranes of epithelial tissue. It is similar to type II, because it involves the production of IgG and IgM antibodies after repeated exposure to antigens and the activation of complement. Type III differs from type II because its antigens are not attached to the surface of a cell. The interaction of these antigens with antibodies produces free-floating complexes that can be deposited in the tissues, causing an **immune complex reaction** or disease. This category includes therapy-related disorders (serum sickness and the Arthus reaction) and a number of autoimmune diseases (such as glomerulonephritis and lupus erythematosus).

Mechanisms of Immune Complex Disease

After initial exposure to a profuse amount of antigen, the immune system produces large quantities of antibodies that circulate in the fluid compartments. When this antigen enters the system a second time, it reacts with the antibodies to form antigen-antibody complexes **(process figure 16.13)**. These complexes summon various inflammatory components such as complement and neutrophils, which would ordinarily eliminate Ag-Ab complexes as part of the normal immune response. In an immune complex disease, however, these complexes are so abundant that they deposit in the **basement membranes**[4] of epithelial tissues and become inaccessible. In response to these events, neutrophils release lysosomal granules that digest tissues and cause a destructive inflammatory condition. The symptoms of type III hypersensitivities are due in great measure to this pathologic state.

Types of Immune Complex Disease

During the early tests of immunotherapy using animals, hypersensitivity reactions to serum and vaccines were

---

3. RhoGAM: Immunoglobulin fraction of human anti-Rh serum, prepared from pooled human sera.

4. Basement membranes are the bottom layers of epithelia that normally filter out circulating antigen-antibody complexes.

Process Figure 16.13 Pathogenesis of immune complex disease.

Ab
Ag

Immune complexes

Lodging of complexes in basement membrane

Neutrophils — Ag-Ab complexes

Basement membrane

Epithelial tissue

Blood vessels Heart/Lungs Joints Skin Kidney

Major organs that can be targets of immune complex deposition

Steps:

1 Antibody combines with excess soluble antigen, forming large quantities of Ag-Ab complexes.

2 Circulating immune complexes become lodged in the basement membrane of epithelia in sites such as kidney, lungs, joints, skin.

3 Fragments of complement cause release of histamine and other mediator substances.

4 Neutrophils migrate to the site of immune complex deposition and release enzymes that cause severe damage in the tissues and organs involved.

The Arthus Reaction

The Arthus reaction is usually an acute response to a second injection of vaccines (boosters) or drugs at the same site as the first injection. In a few hours, the area becomes red, hot to the touch, swollen, and very painful **(figure 16.14a).** These symptoms are mainly due to the destruction of tissues in and around the blood vessels and the release of histamine from mast cells and basophils. Although the reaction is usually self-limiting and rapidly cleared, intravascular blood clotting can occasionally cause necrosis and loss of tissue.

Serum Sickness

Serum sickness was named for a condition that appeared in soldiers after repeated injections of horse serum to treat tetanus. It can also be caused by injections of animal hormones

common. In addition to anaphylaxis, two syndromes, the **Arthus reaction**[5] and **serum sickness,** were identified. These syndromes are associated with certain types of passive immunization (especially with animal serum).

Serum sickness and the Arthus reaction are like anaphylaxis in that all of them require sensitization and preformed antibodies. However, serum sickness and Arthus are set apart from anaphylaxis because

1. they depend on IgG, IgM, or IgA (precipitating antibodies) rather than IgE;
2. they require large doses of antigen (not a minuscule dose as in anaphylaxis); and
3. their symptoms are delayed (a few hours to days).

The Arthus reaction and serum sickness differ from each other in some important ways. The Arthus reaction is a *localized* dermal injury due to inflamed blood vessels in the vicinity of any injected antigen. Serum sickness is a *systemic* injury initiated by antigen-antibody complexes that circulate in the blood and settle into membranes at various sites.

5. Named after Maurice Arthus, the physiologist who first identified this localized inflammatory response.

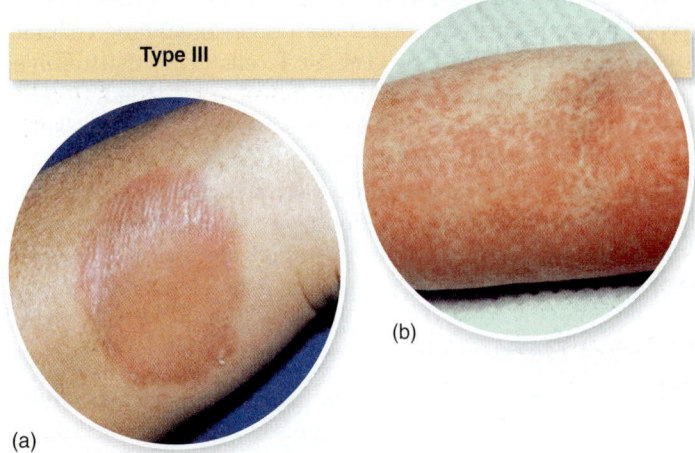

Type III

(a)   (b)

Figure 16.14 Typical presentations of (a) the Arthus reaction and (b) serum sickness, two immune complex diseases.

and drugs. The immune complexes enter the circulation; are carried throughout the body; and are eventually deposited in blood vessels of the kidney, heart, skin, and joints (see process figure 16.13). The condition can become chronic, causing symptoms such as enlarged lymph nodes, rashes **(figure 16.14b)**, painful joints, swelling, fever, and renal dysfunction.

16.4 Learning Outcomes—Assess Your Progress

12. Identify commonalities and differences between type II and type III hypersensitivities.

13. Describe the ways in which the Arthus reaction differs from serum sickness.

16.5 Type IV Hypersensitivities: Cell-Mediated (Delayed) Reactions

The adverse immune responses we have covered so far are explained primarily by B-cell involvement and antibodies. Type IV hypersensitivity is different; it involves primarily the T-cell branch of the immune system. In general, type IV diseases result when T cells respond to antigens displayed on self tissues or transplanted foreign cells. Type IV immune dysfunction has traditionally been known as delayed hypersensitivity because the symptoms arise one to several days following the second contact with an antigen. Examples of type IV hypersensitivity include delayed allergic reactions to infectious agents, contact dermatitis, and graft rejection.

Delayed-Type Hypersensitivity

Infectious Allergy

A classic example of delayed-type hypersensitivity occurs when a person sensitized by tuberculosis infection is injected with an extract (tuberculin) of the bacterium *Mycobacterium tuberculosis*. The so-called tuberculin reaction is an acute skin inflammation at the injection site appearing within 24 to 48 hours. Other infections that use similar skin testing are leprosy, syphilis, histoplasmosis, toxoplasmosis, and candidiasis. Type IV hypersensitivity arises from time-consuming cellular events involving a specific class of T cells (T_H1) that receive the processed allergens from dendritic cells. Activated T_H cells release cytokines that recruit various inflammatory cells such as macrophages, neutrophils, and eosinophils. The buildup of fluid and cells at the site gives rise to a red papule **(figure 16.15).** Delayed hypersensitivity reactions can play a role in a chronic infection (tertiary syphilis, for example), leading to extensive damage to organs through granuloma formation.

Contact Dermatitis

The most common delayed allergic reaction, contact dermatitis, is caused by exposure to resins in poison ivy or poison oak, to simple haptens in household and personal articles (jewelry, cosmetics, elasticized undergarments), or to certain drugs **(Insight 16.2).** Like immediate atopic dermatitis, the reaction

Figure 16.15 Positive tuberculin test.   Intradermal injection of tuberculin extract in a person sensitized to tuberculosis yields a slightly raised red bump greater than 10 mm in diameter.

INSIGHT 16.2	Is Rock and Roll Hazardous to Your Health?

Stories of rock musicians dying of drug overdoses, choking on their vomit after a night of binge drinking, or being electrocuted by a microphone on stage are familiar to fans. But there is another danger lurking in the strings of your favorite axe: contact dermatitis. Metals such as chromium, cobalt, gold, nickel, palladium, and silver found in brass instruments and the strings of instruments such as guitars, violins, violas, and cellos can cause the skin to become red, scaly, and inflamed. Other ingredients such as exotic woods, staining agents, glues, and rosin in string instruments can also exacerbate skin conditions. Researchers recommend that all afflicted rock stars see a dermatologist to be evaluated and treated, take a break from shredding, and change to strings made out of different metals. Once your fingers have healed up, rock on!

Source: www.sciencedaily.com/releases/2012/03/120316101149.htm

to these allergens requires a sensitizing and a provocative dose. The allergen first penetrates the outer skin layers, is processed by Langerhans cells (skin dendritic cells), and is presented to T cells. When subsequent exposures attract lymphocytes and macrophages to this area, these cells give off enzymes and inflammatory cytokines that severely damage the epidermis in the immediate vicinity **(process figure 16.16a)**. This response accounts for the intensely itchy papules and blisters that are the early symptoms **(process figure 16.16b)**. As healing progresses, the epidermis is replaced by a thick, keratinized layer. Depending on the dose and the sensitivity of the individual, the time from initial contact to healing can be a week to 10 days.

T Cells and Their Role in Organ Transplantation

Transplantation or grafting of organs and tissues is a common medical procedure. Although it is life-giving, this technique is plagued by the natural tendency of lymphocytes to seek out foreign antigens and mount a campaign to destroy them. The bulk of the damage that occurs in graft rejections can be attributed to cytotoxic T-cell action. This section covers the mechanisms involved in graft rejection, tests for transplant compatibility, reactions against grafts, prevention of graft rejection, and types of grafts.

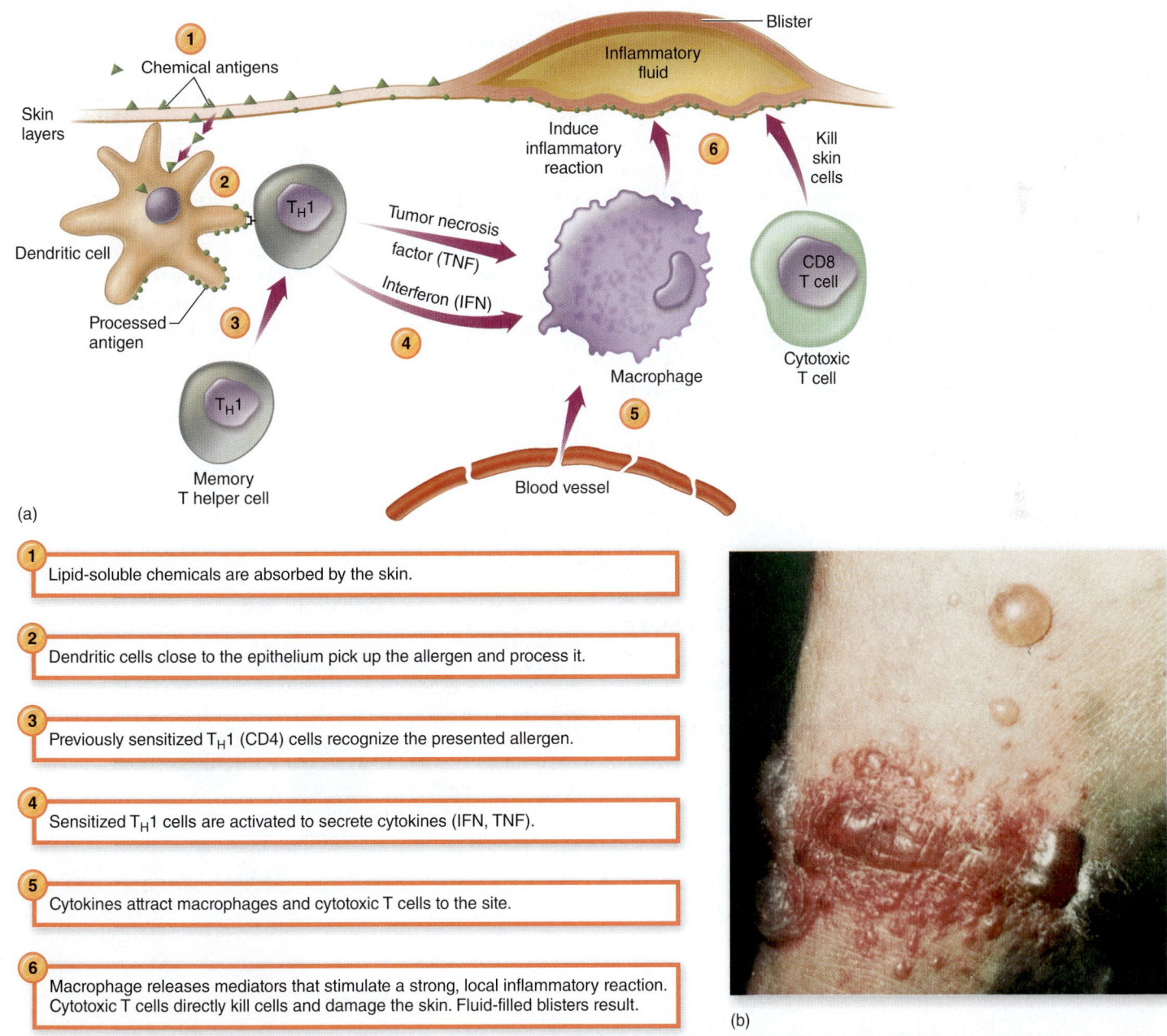

(a)

1. Lipid-soluble chemicals are absorbed by the skin.

2. Dendritic cells close to the epithelium pick up the allergen and process it.

3. Previously sensitized T_H1 (CD4) cells recognize the presented allergen.

4. Sensitized T_H1 cells are activated to secrete cytokines (IFN, TNF).

5. Cytokines attract macrophages and cytotoxic T cells to the site.

6. Macrophage releases mediators that stimulate a strong, local inflammatory reaction. Cytotoxic T cells directly kill cells and damage the skin. Fluid-filled blisters result.

(b)

Process Figure 16.16 Contact dermatitis. (a) Genesis of contact dermatitis. (b) Contact dermatitis from poison oak, showing various stages of involvement: blister, scales, and thickened patches.

The Genetic and Biochemical Basis for Graft Rejection

In chapter 15, we learned that the genes and markers in major histocompatibility (MHC or HLA) classes I and II are extremely important in recognizing self and in regulating the immune response. These molecules also set the events of graft rejection in motion. Although the cells of each person can exhibit variability in the pattern of these cell surface molecules, the pattern is identical in different cells of the same person. Similarity is seen among related siblings and parents, but the more distant the relationship, the less likely that the MHC genes and markers will be similar. When donor tissue (a graft) displays surface molecules of a different MHC class, the T cells of the recipient (called the host) will recognize its foreignness and react against it.

T-Cell-Mediated Recognition of Foreign MHC Receptors

Host Rejection of Graft   When the cytotoxic T cells of a host recognize foreign class I MHC markers on the surface of grafted cells, they release interleukin-2 as part of a general immune mobilization **(figure 16.17).** Antigen-specific helper and cytotoxic T cells bind to the grafted tissue and secrete lymphokines that begin the rejection process within 2 weeks of transplantation. Late in this process, antibodies formed against the graft tissue contribute to immune damage resulting in the destruction of the vascular supply and death of the graft.

Graft Rejection of Host   In certain severe immunodeficiencies, the host cannot or does not reject a graft. But this failure may not protect the host from serious damage because graft incompatibility is a two-way phenomenon. Some grafted tissues (especially bone marrow) contain an indigenous population called passenger lymphocytes (as shown in figure 16.17). This makes it quite possible for the graft to reject the host, causing **graft versus host disease (GVHD).** Because any host tissue bearing MHC markers foreign to the graft can be attacked, the effects of GVHD are widely systemic and toxic. A papular, peeling skin rash is the most common symptom. Other organs affected are the liver, intestine, muscles, and mucous membranes. GVHD typically occurs within 100 to 300 days of the graft; overall, such reactions are declining due to better screening and more sophisticated means of selecting tissues.

Classes of Grafts

Grafts are generally classified according to the genetic relationship between the donor and the recipient. Tissue transplanted from one site on an individual's body to another site on his or her body is known as an **autograft.** Typical examples are skin replacement in burn repair and the use of a vein to fashion a coronary artery bypass. In an **isograft,** tissue from an identical twin is used. Because isografts do not contain foreign antigens, they are not rejected. **Allografts,** the most common type of grafts, are exchanges between genetically different individuals belonging to the same species (two humans). A close genetic correlation is sought for most allograft transplants (see next section). A **xenograft** is a tissue exchange between individuals of different species.

Types of Transplants

Over 28,000 people receive transplants each year in the United States, which reflects the beneficial nature of this medical procedure today. Transplantation involving every

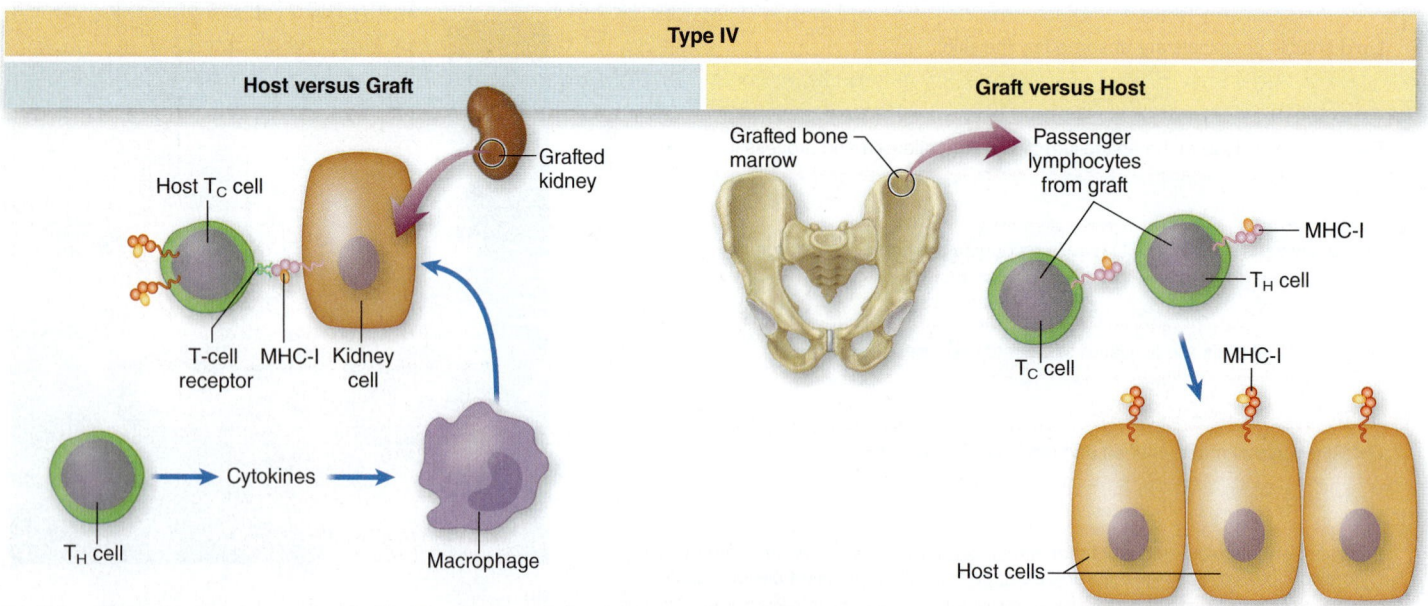

Type IV	
Host versus Graft	**Graft versus Host**

Host T$_C$ cells (and macrophages recruited by T$_H$ cells to assist) attack grafted cells with foreign MHC-I markers.

Passenger lymphoctes from grafted tissue have donor MHC-I markers; attack recipient cells with different MHC-I specificity.

Figure 16.17 Development of type IV incompatible tissue graft reactions.

major organ, including parts of the brain, has been performed, but most often involves skin, liver, heart, kidney, coronary artery, cornea, and bone marrow. The sources of organs and tissues are live donors (kidney, skin, bone marrow, liver), cadavers (heart, kidney, cornea), and fetal tissues.

In the past decade, advancements in transplantation science have expanded the possibilities for treatment and survival. Fetal tissues have been used in the treatment of diabetes and Parkinson disease, while parents have successfully donated portions of their organs to help save their children suffering from the effects of cystic fibrosis or liver disease. Recent advances in stem cell technology have made it possible to isolate stem cells more efficiently from blood donors, and the use of umbilical cord blood cells has furthered progress in this area of science. Though many hurdles still exist, scientists are using genetic engineering technology to develop an ample supply of immunologically compatible, safe tissues for xenotransplantation.

Bone marrow transplantation is a rapidly growing medical procedure for patients with immune deficiencies, aplastic anemia, leukemia and other cancers, and radiation damage. Before closely matched bone marrow can be infused, the patient is pretreated with chemotherapy and whole-body irradiation, a procedure designed to destroy the person's own blood stem cells and thus prevent rejection of the new marrow cells. While the donor is sedated, a bone marrow sample is aspirated by inserting a special needle into an accessible marrow cavity. The most favorable sites are the crest and spine of the ilium (major bone of the pelvis). During this procedure, which lasts 1 to 2 hours, 3% to 5% of the donor's marrow is withdrawn in 20 to 30 separate extractions. Between 500 and 800 milliliters of marrow are removed. The donor may experience some pain and soreness, but serious complications are rare. In a few weeks, the depleted marrow will naturally replace itself. Implanting the harvested bone marrow is rather convenient, because it is not necessary to place it directly into the marrow cavities of the recipient. Instead, it is dripped intravenously into the circulation, and the new marrow cells automatically settle in the appropriate bone marrow regions. Within 2 weeks to a month after infusion, the grafted cells are established in the host. Because donor lymphoid cells can still cause GVHD, antirejection drugs may be necessary. Interestingly, after bone marrow transplantation a recipient's blood type may change to the blood type of the donor!

Autoimmune diseases in which type IV hypersensitivities play a role are rheumatoid arthritis, type I diabetes, and multiple sclerosis (see next section and table 16.4). As you read in chapter 15, certain B cells called regulatory B cells have a role in controlling the T-cell response. It is thought that when the B_{Reg} cells malfunction, T cells are able to respond inappropriately as discussed in this section and the next.

16.5 Learning Outcomes—Assess Your Progress

14. Identify one type IV delayed hypersensitivity reaction, and describe the role of T cells in the pathogenesis of this condition.

15. List four classes of grafts, and explain how host versus graft and graft versus host diseases develop.

16.6 An Inappropriate Response Against Self: Autoimmunity

The immune diseases we have covered so far are all caused by foreign antigens. In the case of autoimmunity, an individual actually develops hypersensitivity to him- or herself. This pathologic process accounts for **autoimmune diseases,** in which **autoantibodies,** T cells; and, in some cases, both mount an abnormal attack against self antigens. The scope of autoimmune diseases is extremely varied. In general, they are either *systemic,* involving several major organs, or *organ-specific,* involving only one organ or tissue. There are more than 80 recognized autoimmune diseases affecting roughly 5% to 8% of the U.S. population. Some major diseases, their targets, and basic pathology are presented in **table 16.4.** (For a reminder of hypersensitivity types, refer to table 16.1.)

Table 16.4 Selected Autoimmune Diseases

Disease	Target	Type of Hypersensitivity	Characteristics
Systemic lupus erythematosus (SLE)	Systemic	III	Inflammation of many organs; antibodies against red and white blood cells, platelets, clotting factors, nucleus DNA
Rheumatoid arthritis and ankylosing spondylitis	Systemic	II, III, and IV	Vasculitis; frequent target is joint lining; antibodies against other antibodies (rheumatoid factor), T-cell cytokine damage
Graves' disease	Thyroid	III	Antibodies against thyroid-stimulating hormone receptors
Myasthenia gravis	Muscle	III	Antibodies against the acetylcholine receptors on the nerve-muscle junction alter function.
Type I diabetes	Pancreas	IV	T cells attack insulin-producing cells.
Multiple sclerosis	Myelin	II and IV	T cells and antibodies sensitized to myelin sheath destroy neurons.

Genetic and Gender Correlation in Autoimmune Disease

In most cases, the precipitating cause of autoimmune disease remains obscure, but we do know that susceptibility is determined by genetics and influenced by gender. Cases cluster in families, and even unaffected members tend to develop the autoantibodies for that disease. More direct evidence comes from studies of the major histocompatibility gene complex. Particular genes in the class I and II major histocompatibility complex coincide with certain autoimmune diseases. For example, autoimmune joint diseases such as rheumatoid arthritis and ankylosing spondylitis are more common in persons with the B-27 HLA type; systemic lupus erythematosus, Graves' disease, and myasthenia gravis are associated with the B-8 HLA antigen. With the expansion of genomic technology and the screening of whole genomes, many novel genes have recently been found to play a role in the pathway to autoimmunity. Sequencing of genomes may represent a new avenue for clinical diagnosis or treatment of disease, and studies have suggested that seemingly unrelated disorders, such as autism, may share a common genetic basis with autoimmune disease.

Women account for nearly 75% of all cases of diagnosed autoimmune disease, but the biological basis for this fact largely remains a mystery. A number of autoimmunities have been linked to genes on the X chromosome, and one hypothesis centers on the role of X-chromosome inactivation in the development of these diseases. Although research is still ongoing, it appears that serum from women with autoimmune diseases is often reactive with Barr bodies, a remnant of X-chromosome inactivation in nuclei.

The Origins of Autoimmune Disease

Otherwise healthy individuals produce autoantibodies, albeit at very low levels, indicating that a moderate, regulated amount of autoimmunity is probably required to dispose of old cells and cellular debris. Disease apparently arises when this regulatory or recognition apparatus goes awry. Sometimes the processes go awry due to genetic irregularities or inherent errors in the host's immunologic processes. This is a very active area of research. Here are some of the possibilities that are currently being investigated:

- *Sequestered antigens* trigger the development of autoimmune reactions. During embryonic growth, some tissues are immunologically privileged; that is, they are sequestered behind anatomical barriers and cannot be scanned by the immune system. Eventually, some of these antigens become exposed by means of infection, trauma, or deterioration. When they finally encounter immune cells, they are recognized as a foreign substance, triggering a reaction to self antigen.
- *Forbidden clones* erroneously target self tissues leading to autoimmunity. According to the **clonal selection theory** (see chapter 15), the immune system of a fetus develops tolerance by eradicating all self-reacting lymphocyte clones, called *forbidden clones*, while retaining only those clones that react to foreign antigens. Some of these forbidden clones may survive; and because they have not been subjected to this tolerance process, these autoreactive B or T cells can inappropriately attack tissues with self antigens.
- The *bystander effect* may lead to the development of autoreactive immune cell populations. T-cell activation may incorrectly "turn on" B cells that can react with self antigens, or chemical compounds such as heavy metals may stimulate autoreactive T-cell populations within the immune system. Both circumstances may lead to the development of autoimmune disease.
- *Molecular mimicry* leads the immune system to misidentify self antigens. In some cases, microbial antigens display molecular determinants similar to normal human cells. An infection could cause formation of antibodies that can cross-react with tissues. Mimicry of bacterial antigens has been linked to the development of autoimmune diseases such as rheumatic fever and psoriasis. Type I diabetes and multiple sclerosis are disorders possibly triggered by viral infection. Viruses can display epitopes similar to self antigens, which may induce autoimmunity; but more importantly they can noticeably alter normal cell receptors, thereby causing immune cells to attack virus-infected tissues.

Examples of Autoimmune Disease

Systemic Autoimmunities

One of the most severe chronic autoimmune diseases is systemic lupus erythematosus (SLE), or lupus. This name originated from the characteristic butterfly-shaped rash that drapes across the nose and cheeks **(figure 16.18a),** as ancient physicians thought the rash resembled a wolf bite on the face (*lupus* is Latin for "wolf"). Although the manifestations of the disease vary considerably, all SLE patients produce autoantibodies against a great variety of organs and tissues or intracellular materials, such as the nucleoprotein of the nucleus and mitochondria.

In SLE, autoantibody-autoantigen complexes appear to be deposited in the basement membranes of various organs. Kidney failure, blood abnormalities, lung inflammation, myocarditis, and skin lesions are the predominant symptoms. One form of chronic lupus (called discoid) is influenced by exposure to the sun and primarily afflicts the skin. The etiology of lupus is still a puzzle, and it is not exactly clear how such a generalized loss of self-tolerance arises. Viral infection and the loss of normal suppression of immune response are suspected. Another possible cause is the inefficient clearing of normal cellular debris. Genomics research has also led to the identification of several genetic variations linked to SLE susceptibility and will most likely lead to the development of new targeted therapies for this disease. The diagnosis of SLE can usually be made with blood tests. Antibodies against the

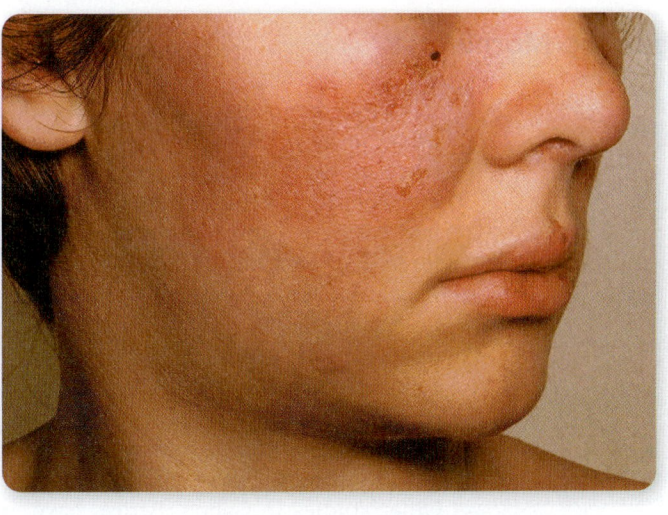

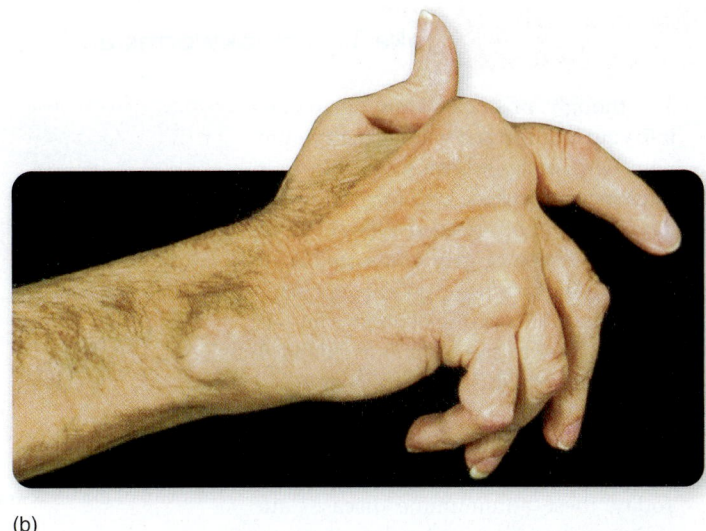

(a)

(b)

Figure 16.18 Common autoimmune diseases. (a) Systemic lupus erythematosus. One symptom is a prominent rash across the bridge of the nose and on the cheeks. These papules and blotches can also occur on the chest and limbs. (b) Rheumatoid arthritis commonly targets the synovial membrane of joints. Over time, chronic inflammation causes thickening of this membrane, erosion of the articular cartilage, and fusion of the joint. These effects severely limit motion and can eventually swell and distort the joints.

nucleus and various tissues are common, and a positive test for the lupus factor (an antinuclear factor) is also very indicative of the disease.

Rheumatoid arthritis, another systemic autoimmune disease, incurs progressive, debilitating damage to the joints. In some patients, the lungs, eyes, skin, and nervous system are also involved. In the joint form of the disease, autoantibodies form immune complexes that bind to the synovial membrane of the joints and activate phagocytes and stimulate release of cytokines. Chronic inflammation leads to scar tissue and joint destruction. The joints in the hands and feet are affected first, followed by the knee and hip joints **(figure 16.18b).** These cytokines (such as tumor necrosis factor [TNF]) can then trigger additional type IV delayed hypersensitivity responses. Epstein-Barr virus has been implicated as a precipitating cause; and the presence of an IgM antibody, called rheumatoid factor (RF), can be used in diagnosis of this disease. Treatment has traditionally involved the targeting of TNF or TNF-mediated pathways, but new drugs that target other immune system components are appearing frequently.

Autoimmunities of the Endocrine Glands

The underlying cause of **Graves' disease** is the attachment of autoantibodies to receptors on the thyroxin-secreting follicle cells of the thyroid gland. The abnormal stimulation of these cells causes the overproduction of this hormone and the symptoms of hyperthyroidism, which affect nearly every body system.

Type I diabetes is another condition that may be a result of autoimmunity. Insulin, secreted by the beta cells in the pancreas, regulates and is essential to the utilization of glucose by cells. Molecular mimicry has been implicated in the sensitization of cytotoxic T cells in type I diabetes. The

inappropriate immune response then leads to the lysis of important pancreatic cells.

Neuromuscular Autoimmunities

Myasthenia gravis is a syndrome caused by autoantibodies binding to the receptors for acetylcholine, a chemical required to transmit a nerve impulse across the synaptic junction to a muscle. The immune attack so severely damages the muscle cell membrane that transmission is blocked and paralysis ensues. The first effects are usually felt in the muscles of the eyes and throat but eventually progress to complete loss of skeletal muscle function and death. Current treatment usually includes immunosuppressive drugs and therapy to remove the autoantibodies from the circulation. Experimental therapy using immunotoxins to destroy lymphocytes that produce autoantibodies shows some promise in affected patients as does the use of complement-inhibiting drugs.

Multiple sclerosis (MS) is a paralyzing neuromuscular disease associated with lesions in the insulating myelin sheath that surrounds neurons of the central nervous system. T-cell-induced and autoantibody-induced damage severely compromises the capacity of neurons to send impulses, resulting in muscular weakness and tremors, difficulties in speech and vision, and some degree of paralysis. Most MS patients first experience symptoms as young adults, and they tend to experience remissions (periods of relief) alternating with recurrences of disease throughout their lives. Data suggest a possible association between infection with human herpesvirus 6 and the onset of disease. Immunosuppressants like cortisone and interferon beta alleviate symptoms, and the disease can be treated passively with antibody therapy targeted to specific T-cell antigens. Alternative medical

INSIGHT 16.3   Take Two Hookworms and Call Me in the Morning

The thought of hookworms burrowing into your bare feet and making their way to your lungs or tapeworms lodging in your intestines may cause you to shudder. In developed countries, these types of parasites have been largely eliminated with safe farming practices, good water sanitation, and the wearing of shoes. But our immune system needs time to adapt to these changes. Diseases such as multiple sclerosis, Crohn's disease, asthma, and other atopic and autoimmune diseases were unknown in the pre-sanitation era. Today, these autoimmune diseases are almost unheard of in developing countries where over 1 billion people are still afflicted by parasites.

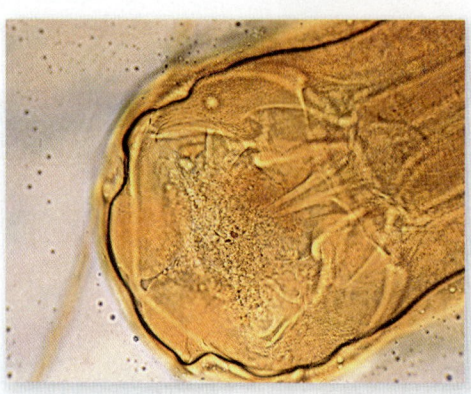

The head of the hookworm *Necator americanus*.

The working hypothesis is that important portions of the mammalian immune system have evolved over millennia to manage the colonization by parasites. Suddenly (in terms of evolutionary time) removing the helminth from the mammalian ecosystem leaves the immune system primed to act but with no appropriate target. Thus, inappropriate targets (self antigens) act as stimuli.

Researchers have noted that in urban areas of Ethiopia, children are twice as likely to develop asthma as are their rural counterparts living in less sanitary conditions. Physicians Jorge Correale and Mauricio Farez in Argentina studied patients with multiple sclerosis who had recently been infected with a parasitic worm and compared them to a similar group of patients without a parasite. The patients with parasites had no relapses of MS during the course of the study. Other studies in Gabon, Africa, have shown similar results: Patients with parasitic worm infections have lower incidences of asthma, allergies, and autoimmune diseases. Controlled clinical trials are needed to confirm these anecdotal observations, but soon, doctors may be prescribing a roundworm along with traditional treatments.

Source: 2011. *Sci. News*, vol. 179, no. 3, p. 26.

---

approaches have also been investigated, including the use of helminths to reduce autoimmunity **(Insight 16.3)**.

16.6 Learning Outcomes—Assess Your Progress

16. Outline at least three different explanations for the origin of autoimmunity.

17. List three autoimmune diseases, and describe immunologic features common to all.

16.7 Immunodeficiency Diseases: Hyposensitivity of the Immune System

Occasionally, errors occur in the development of the immune system and a person is born with or develops weakened immune responses called immunodeficiencies. The predominant consequences of immunodeficiencies are recurrent, overwhelming infections, often with opportunistic microbes. Immunodeficiencies fall into two general categories: *primary diseases*, present at birth (congenital) and usually stemming from genetic errors, and *secondary diseases*, acquired after birth and caused by natural or artificial agents.

Primary Immunodeficiency Diseases

Primary deficiencies affect both specific immunities such as antibody production and less-specific ones such as phagocytosis. Consult **figure 16.19** to survey the places in the normal

sequential development of lymphocytes where defects can occur and the possible consequences. In many cases, the deficiency is due to an inherited abnormality, though the exact nature of the abnormality is not known for a number of diseases. In some deficiencies, the lymphocyte in question is completely absent or is present at very low levels; in others, lymphocytes are present but do not function normally. Because the development of B cells and T cells diverges at some point, an individual can lack one or both cell lines. It must be emphasized, however, that some deficiencies affect the function of other cells in the immune system as well.

Clinical Deficiencies in B-Cell Development or Expression

Genetic deficiencies in B cells usually appear as an abnormality in immunoglobulin expression. In some instances, only certain immunoglobulin classes are absent; in others, the levels of all types of immunoglobulins (Ig) are reduced. A significant number of B-cell deficiencies are X-linked (also called sex-linked) recessive traits, meaning that the gene occurs on the X chromosome and the disease appears primarily in male children.

The term **agammaglobulinemia** literally means the absence of gamma globulin, the fraction of serum that contains immunoglobulins. Because it is very rare for Ig to be completely absent, some physicians prefer the term **hypogammaglobulinemia.** T-cell function in these patients is usually normal. The symptoms of recurrent, serious bacterial infections usually appear about 6 months after birth. The bacteria most often implicated are pyogenic cocci,

Primary Immunodeficiency

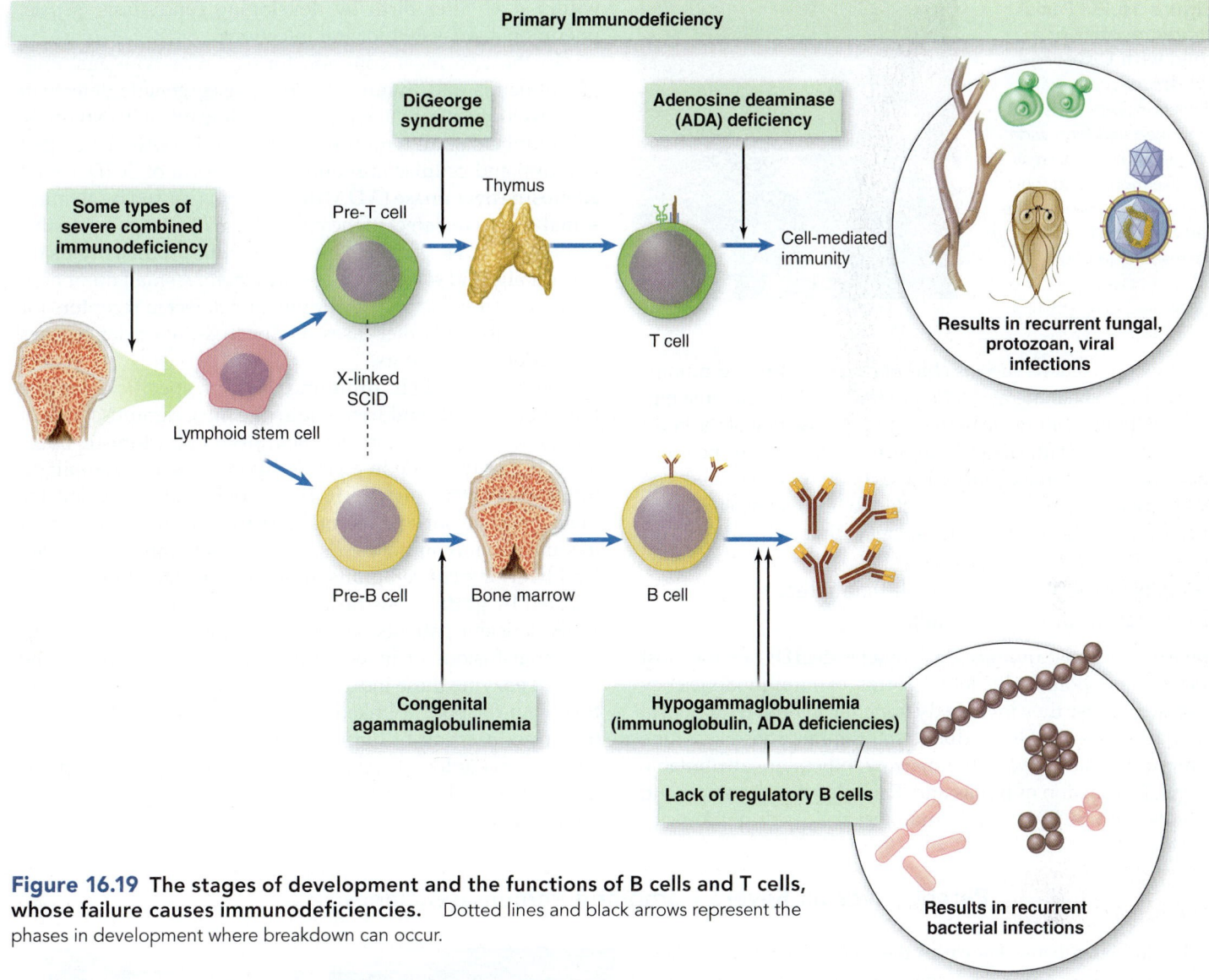

Figure 16.19 The stages of development and the functions of B cells and T cells, whose failure causes immunodeficiencies.   Dotted lines and black arrows represent the phases in development where breakdown can occur.

Pseudomonas, and *Haemophilus influenzae.* The most common infection sites are the lungs, sinuses, meninges, and blood. Many Ig-deficient patients can have recurrent infections with viruses and protozoa, as well. Patients often manifest a wasting syndrome and have a reduced life span, but modern therapy has improved their prognosis. The current treatment for this condition is passive immunotherapy with immune serum globulin and continuous antibiotic therapy.

The lack of a particular class of immunoglobulin is a relatively common condition, though its underlying genetic mechanisms are not clear. IgA deficiency is the most prevalent form, in which patients have normal quantities of B cells and other immunoglobulins but they are unable to synthesize IgA. Consequently, they lack protection against local microbial invasion of the mucous membranes and suffer frequent respiratory and gastrointestinal infections. There is no existing treatment for IgA deficiency, because conventional preparations of immune serum globulin are high in IgG, not IgA.

Clinical Deficiencies in T-Cell Development or Expression

Due to the critical role of T cells in immune defenses, a genetic defect in T-cell development results in a broad spectrum of diseases, including severe opportunistic infections, wasting, and cancer. In fact, a defective T-cell line is usually more devastating than a defective B-cell line because T helper cells are required to assist in most specific immune reactions. The deficiency can occur anywhere along the developmental spectrum, from thymus to mature, circulating T cells.

Abnormal Development of the Thymus   The most severe of the T-cell deficiencies involve the congenital absence or immaturity of the thymus. Thymic aplasia, or **DiGeorge syndrome,** comes about due to errors during embryogenesis or deletions in chromosome 22. It is characterized by a lack of cell-mediated immunity, making children highly susceptible to persistent infections by fungi, protozoa, and viruses. Vaccinations using live, attenuated microbes pose a danger, and common

Figure 16.20 Facial characteristics of a child with DiGeorge syndrome.   Typical defects include low-set, deformed earlobes; wide-set, slanted eyes; a small, bowlike mouth; and the absence of a philtrum (the vertical furrow between the nose and upper lip).

childhood infections such as chickenpox, measles, and mumps can be overwhelming and fatal in these children. Other symptoms of thymic failure are reduced growth, wasting of the body, unusual facial characteristics **(figure 16.20),** and an increased incidence of lymphatic cancer. These children can have reduced antibody levels, and they are unable to reject transplants. The major therapy for them is a transplant of thymus tissue.

Severe Combined Immunodeficiencies: Dysfunction in B and T Cells

Severe combined immunodeficiencies (SCIDs) are the most serious and potentially lethal forms of immunodeficiency disease because they involve dysfunction in both lymphocyte systems. Some SCIDs are due to the complete absence of the lymphocyte stem cell in the marrow; others are attributable to the dysfunction of B cells and T cells later in development. Infants with SCIDs usually manifest the T-cell deficiencies

within days after birth by developing candidiasis, sepsis, pneumonia, or systemic viral infections.

In the two most common forms, Swiss-type agamma-globulinemia and thymic alymphoplasia, genetic defects in the development of the lymphoid cell line result in extremely low numbers of all lymphocyte types and poorly developed humoral and cellular immunity. A rare form of SCID, called **adenosine deaminase (ADA) deficiency,** is caused by an autosomal recessive defect in the metabolism of adenosine. In this case, lymphocytes develop but a metabolic product builds up abnormally and selectively destroys them. A small number of SCID cases are due to a developmental defect in receptors for B and T cells, and other cases are due to X-linked deficiencies in interleukin receptors.

Because of their profound lack of specific adaptive immunities, SCID children require the most rigorous kinds of aseptic techniques to protect them from opportunistic infections. Aside from life in a sterile plastic bubble, exemplified by David Vetter **(Insight 16.4),** the only serious option for their longtime survival is total replacement or correction of dysfunctional lymphoid cells. Some infants can benefit from fetal liver or stem cell grafts, though transplantation is complicated by graft versus host disease. The condition of some ADA-deficient patients has been partially corrected by periodic transfusions of blood containing large amounts of the normal enzyme these individuals are missing. A more lasting treatment for both X-linked and ADA types of SCID is gene therapy—insertion of normal genes to replace the defective genes (see chapter 10). Although there have been some problems with gene therapy trials, to date the most successful gene therapy treatments have been used for SCID.

INSIGHT 16.4   Perspectives on Severe Combined Immunodeficiency

David Vetter, better known as the "bubble boy," would have been 41 years old on September 21, 2012. Born with severe combined immunodeficiency (SCID), he spent his life in sterile enclosures designed by NASA to protect him from germs. David died at age 12, succumbing to lymphoma resulting from a bone marrow transplant that infected him with the Epstein-Barr virus. Since his death in the early 1980s, great strides have been made in treating those living with SCID and other primary immune deficiencies. Immune deficiencies are inherited and are very rare—only 1 in 200,000 babies is born with SCID. Children with SCID today are successfully treated with intravenous immune globulin (IVIG) and adenosine deaminase (ADA) enzymes. Recent research has indicated that gene therapy and bone marrow transplants can be effective treatments for SCID.

Early diagnosis is key in preserving the lives of SCID patients. Laboratory clinicians are able to rapidly identify the absence of B and T cells in a newborn's blood, preventing infection and beginning IVIG and ADA therapy. Many states have added screening for SCID and other primary immunodeficiency diseases (PIDs) to standard newborn screening, and SCID screening has been recently approved by the Secretary of Health and Human

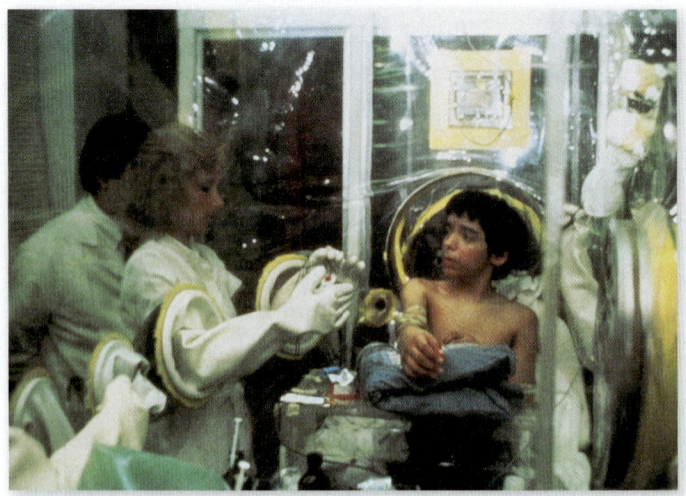

Services to be implemented nationwide. Thanks to David and the doctors who worked with him, children with SCID can live normal, healthy lives.

Source: 2012. J Allergy Clin. Immun. vol. 129, no. 3, p. 604.

Secondary Immunodeficiency Diseases

Secondary acquired deficiencies in B cells and T cells are caused by one of four general agents:

1. infection,
2. noninfectious metabolic disease,
3. chemotherapy, or
4. radiation.

The most recognized infection-induced immunodeficiency is **AIDS** (see chapter 20). This syndrome is caused when several types of immune cells, including T helper cells, monocytes, macrophages, and antigen-presenting cells, are infected by the human immunodeficiency virus (HIV). It is generally thought that the depletion of T helper cells and functional impairment of immune responses ultimately account for the cancers and opportunistic protozoan, fungal, and viral infections associated with this disease. Other infections that can deplete immunities are measles, leprosy, and malaria.

Cancers that target the bone marrow or lymphoid organs can be responsible for extreme malfunction of both humoral and cellular immunity. In leukemia, a massive number of cancer cells compete for space and literally displace the normal cells of the bone marrow and blood. Plasma cell tumors produce large amounts of nonfunctional antibodies, and thymus tumors cause severe T-cell deficiencies.

An ironic outcome of lifesaving medical procedures is the possible suppression of a patient's immune system. Drugs that prevent graft rejection or decrease the symptoms of rheumatoid arthritis can likewise suppress beneficial immune responses, while radiation and anticancer drugs are extremely damaging to the bone marrow and other body cells.

16.7 Learning Outcomes—Assess Your Progress

18. Distinguish between primary and secondary immunodeficiencies, explaining how each develops.
19. Define *severe combined immunodeficiency*, and discuss current therapeutic approaches to this type of disease.
20. List three conditions that can lead to the development of secondary immunodeficiency diseases.

Case File 16 *Wrap-Up*

A recent study by the National Institutes of Health (NIH) shows that as many as 32 million people in the United States have antibodies directed at their own tissues, or autoantibodies. Lupus, rheumatoid arthritis, and multiple sclerosis affect women up to 10 times more than men. Women usually begin showing signs of autoantibodies around the ages of 40 to 49, usually associated with the onset of menopause, suggesting that declining estrogen and progesterone may play a role in the development of autoimmune disease.

However, recent research suggests that hormones may not be entirely to blame for autoimmune disease. Scientists have discovered a new subset of B cells that make autoantibodies that express the surface protein called integrin. They named these cells *age-associated B cells* or ABCs. They found higher levels of ABCs in elderly female mice and older mice prone to autoimmune disease, even before they developed any disease. The researchers also found ABCs in humans with autoimmune diseases. When ABC levels were depleted in mice, levels of autoantibodies dropped as well. It turns out that activation of these cells is governed by a gene on the X chromosome. Since females have two X chromosomes, the likelihood is greater that ABCs will be activated.

Back to the dirt: It turns out that there may be another reason why females have higher rates of asthma and autoimmune diseases. Little girls are often "cleaner" than little boys. A study published in the journal *Social Science and Medicine* by Sharyn Clough, a philosopher of science at Oregon State University, showed that little girls are more often dressed in "outfits" and expected to stay clean, while little boys are often free to play in the dirt. This lack of dirt in little girls' lives may be part of the reason for development of autoimmune diseases later in life.

Sources: 2012. *Science* vol. 336, no. 6080, p. 489.
2011. *Soc. Sci. Med.* vol. 72, no. 4, p. 486.

Chapter Summary

16.1 The Immune Response: A Two-Sided Coin (ASM Guidelines* 2.1, 3.4, 5.4)

- Immunopathology is the study of diseases associated with excesses and deficiencies of the immune response. Such diseases include allergies, autoimmunity, grafts, transfusions, immunodeficiency disease, and cancer.
- There are four categories of hypersensitivity reactions: type I (allergy and anaphylaxis), type II (IgG and IgM tissue destruction), type III (immune complex reactions), and type IV (delayed hypersensitivity reactions).

16.2 Type I Allergic Reactions: Atopy and Anaphylaxis (ASM Guidelines 2.1, 3.4, 5.4)

- Antigens that trigger hypersensitivity reactions are allergens. They can be either exogenous (originate outside the host) or endogenous (caused by the host's own tissue).

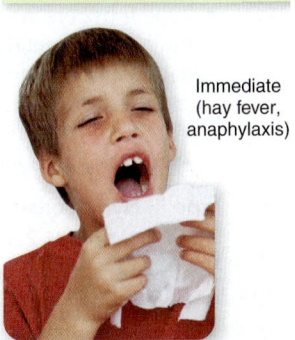

Immediate (hay fever, anaphylaxis)

Source: ASM Curriculum Guidelines (American Society for Microbiology, 2012). Complete guidelines in appendix B of this book.

- Type I hypersensitivity reactions result from excessive IgE production in response to an exogenous antigen.
- Atopy is a chronic, local allergy, whereas anaphylaxis is a systemic, potentially fatal allergic response. Both result from excessive IgE production in response to exogenous antigens.
- The predisposition to type I hypersensitivities is inherited, but age, geographic locale, and infection also influence allergic response.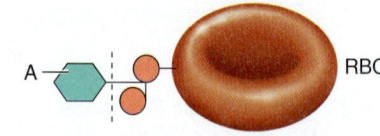
- Type I allergens include inhalants, ingestants, injectants, and contactants; potential portals of entry are the skin, respiratory tract, gastrointestinal tract, and genitourinary tract.
- Type I hypersensitivities are set up by a sensitizing dose of allergen and expressed when a second provocative dose triggers the allergic response.
- The primary participants in type I hypersensitivities are IgE, basophils, mast cells, and agents of the inflammatory response.
- Allergies are diagnosed by a variety of *in vitro* and *in vivo* tests that assay specific cells, IgE, and local reactions.
- Allergies are treated by medications that interrupt the allergic response at certain points. Allergic reactions can also be prevented by injection of allergens in controlled amounts.

16.3 Type II Hypersensitivities: Reactions That Lyse Foreign Cells (ASM Guidelines 2.1, 3.4, 5.4)

- Type II hypersensitivity reactions are complement-assisted reactions that occur when preformed antibodies (IgG or IgM) react with foreign cell-bound antigens, leading to membrane attack complex formation and lysis.
- ABO blood groups are based upon antigens present on red blood cells. These markers are genetically determined and composed of glycoproteins.
- Cross-matching donor and recipient blood is necessary to determine which transfusions are safe to perform. The most common type II reactions occur when transfused blood is mismatched to the recipient's ABO type. The concepts of universal donor (type O) and universal recipient (type AB) apply only under emergency circumstances.
- Type II hypersensitivity disease can develop when Rh− mothers are sensitized to Rh+ RBCs of their unborn babies and the mother's anti-Rh antibodies cross the placenta, causing hemolysis of the newborn's RBCs. This is called hemolytic disease of the newborn, or erythroblastosis fetalis.

16.4 Type III Hypersensitivities: Immune Complex Reactions (ASM Guidelines 2.1, 3.4, 5.4)

- Type III hypersensitivity reactions occur when large quantities of antigen react with host antibody to form immune complexes that settle in tissue cell membranes,

causing chronic destructive inflammation. The reactions appear hours or days after the antigen challenge.
- Like type II reactions, type III hypersensitivities involve the production of IgG and IgM and the activation of complement; they differ in that the antigen recognized in type III reactions is soluble.
- The mediators of type III hypersensitivity reactions include soluble IgA, IgG, or IgM and agents of the inflammatory response.
- Localized (Arthus) and systemic (serum sickness) reactions are two forms of type III hypersensitivities.

16.5 Type IV Hypersensitivities: Cell-Mediated (Delayed) Reactions (ASM Guidelines 2.1, 3.4, 5.4)

- Type IV or delayed hypersensitivity reactions, like the tuberculin reaction and transplant reactions (host rejection and GVHD), occur when cytotoxic T cells attack either self tissue or transplanted foreign cells. Target cells must display both a foreign MHC and a nonself receptor site.
- The four classes of transplants or grafts are determined by the degree of MHC similarity between graft and host. From most to least similar, these are autografts, isografts, allografts, and xenografts.
- Graft rejection can be minimized by tissue matching procedures, immunosuppressive drugs, and use of tissues that do not provoke a type IV response.

16.6 An Inappropriate Response Against Self: Autoimmunity (ASM Guidelines 2.1, 3.4, 5.4)

- Autoimmune reactions occur when autoantibodies or host T cells mount an abnormal attack against self antigens.
- Susceptibility to autoimmune disease appears to be influenced by gender and by genes in the MHC complex.
- Autoimmune disease may be an excessive response of a normal immune function, the appearance of sequestered antigens, "forbidden" clones of lymphocytes that react to self antigens, or the result of alterations in the immune response caused by infectious agents, particularly viruses.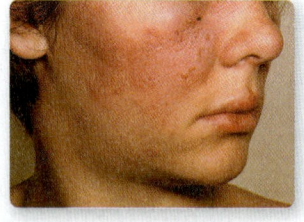
- Examples of autoimmune diseases include systemic lupus erythematosus, rheumatoid arthritis, diabetes mellitus, myasthenia gravis, and multiple sclerosis.

16.7 Immunodeficiency Diseases: Hyposensitivity of the Immune System (ASM Guidelines 2.1, 3.4, 5.4)

- Immunodeficiency diseases occur when the immune response is reduced or absent.
- Primary immune diseases are genetically induced deficiencies of B cells, T cells, the thymus, or combinations of these. SCIDs are the most severe forms of these diseases due to the loss of both humoral and cell-mediated immunity.
- Secondary immune diseases are caused by infection, organic disease, chemotherapy, or radiation. The best-known infection-induced immunodeficiency is AIDS.

Multiple-Choice and True-False Questions │ Bloom's Levels 1 and 2: Remember and Understand

Multiple-Choice Questions. Select the correct answer from the options provided.

1. Pollen is which type of allergen?
   a. contactant   c. injectant
   b. ingestant   d. inhalant

2. B cells are responsible for which allergies?
   a. asthma   c. tuberculin reactions
   b. anaphylaxis   d. both a and b

3. The contact with allergen that results in symptoms is called the
   a. sensitizing dose.   c. provocative dose.
   b. degranulation dose.   d. desensitizing dose.

4. The direct, immediate cause of allergic symptoms is the action of
   a. the allergen directly on smooth muscle.
   b. the allergen on B lymphocytes.
   c. allergic mediators released from mast cells and basophils.
   d. IgE on smooth muscle.

5. Theoretically, type _____ blood can be donated to all persons because it lacks _____.
   a. AB, antibodies   c. AB, antigens
   b. O, antigens   d. O, antibodies

6. An example of a type III immune complex disease is
   a. serum sickness.   c. graft rejection.
   b. contact dermatitis. d. atopy.

7. Type II hypersensitivities are due to
   a. IgE reacting with mast cells.
   b. activation of cytotoxic T cells.
   c. IgG-allergen complexes that clog epithelial tissues.
   d. complement-induced lysis of cells in the presence of antibodies.

8. Production of autoantibodies may be due to
   a. emergence of forbidden clones of B cells.
   b. production of antibodies against sequestered tissues.
   c. infection-induced change in receptors.
   d. possibly all of these.

9. Rheumatoid arthritis is an _____ that affects the _____.
   a. immunodeficiency disease, muscles
   b. autoimmune disease, nerves
   c. allergy, cartilage
   d. autoimmune disease, joints

10. Which disease would be most similar to AIDS in its pathology?
    a. X-linked agammaglobulinemia
    b. SCID
    c. ADA deficiency
    d. DiGeorge syndrome

True-False Questions. If the statement is true, leave as is. If it is false, correct it by rewriting the sentence.

11. T cells are associated with type IV hypersensitivities.

12. A positive tuberculin skin test is an example of antibody-mediated inflammation.

13. Contact dermatitis can be caused by proteins found in foods.

14. Antibody-mediated degranulation of mast cells is involved in anaphylaxis.

15. Rejection of transplanted tissue is dependent on MHC/HLA markers.

Critical Thinking Questions │ Bloom's Levels 3, 4, and 5: Apply, Analyze, and Evaluate

Critical thinking is the ability to reason and solve problems using facts and concepts. These questions can be approached from a number of angles and, in most cases, they do not have a single correct answer.

1. Conduct additional research and discuss examples that illustrate how cancer can be both a cause of immune dysfunction and an effect of this process.

2. Summarize the roles of normal biota and genetics in the development of type I allergic reactions. Discuss how probiotics or gene therapy could be used to alter an individual's allergic response to antigen.

3. a. Outline the steps of cellular reactions in response to an antigen.
   b. You have developed a new treatment for type I allergic responses in your laboratory. This new therapeutic approach completely blocks the degranulation of mast cells. Discuss whether or not this would be an effective treatment for allergy sufferers.

4. Describe a test that could be used to distinguish true food allergy from food intolerance.

5. a. Draw a diagram illustrating whether or not each of the following transfusions would be immunologically compatible.
      Type A donor into a type B recipient

   Type B donor into a type AB recipient
   Type O– donor into a type O+ recipient
   b. Explain how xenotransplantation can be successful in light of the immune system's robust ability to recognize foreign antigen.

6. Summarize the role of the immune system in the development of type I diabetes. Propose a strategy that could be used to protect young children from developing an autoimmune reaction, and subsequently type I diabetes, after a viral infection.

7. a. Explain which have a greater impact on immune function and patient health: T-cell deficiencies or B-cell deficiencies.
   b. A patient in your unit exhibits frequent bouts of microbial infections and is found to produce extremely low levels of IgG and IgM antibodies. Your colleague suggests that the patient receive numerous vaccinations against a broad spectrum of common pathogens; you disagree. Why? Explain another treatment that may be beneficial to this patient.

8. A 31-year-old male develops a severe, full-body rash after receiving penicillin. He is certain that he has never received

this antibiotic in the past. Summarize the immunologic reaction that is occurring in this patient, and discuss one hypothesis that would explain why the symptoms occurred when he had never been exposed to the drug before.

9. Explain why it is advisable for an Rh– woman who has had an abortion, miscarriage, or an ectopic pregnancy to be immunized against the Rh factor.

10. A young couple brings their 7-month-old daughter to the pediatrician. The doctor identifies that the baby is suffering from *Haemophilus influenzae* and *Pseudomonas* infection and, additionally, has oral thrush (candidiasis). Blood tests reveal that the infant suffers from agammaglobulinemia. Explain which type of lymphocyte is malfunctioning in this baby, and discuss why her symptoms took so long after birth to develop.

Concept Connections | Bloom's Levels 4 and 6: Analyze and Create

This activity ties together multiple concepts in the chapter.

1. From what blood types can a person with type A– blood receive donations?

2. From what blood types can a person with type B+ blood receive donations?

3. What are the consequences if an incompatible blood type is given as a donation?

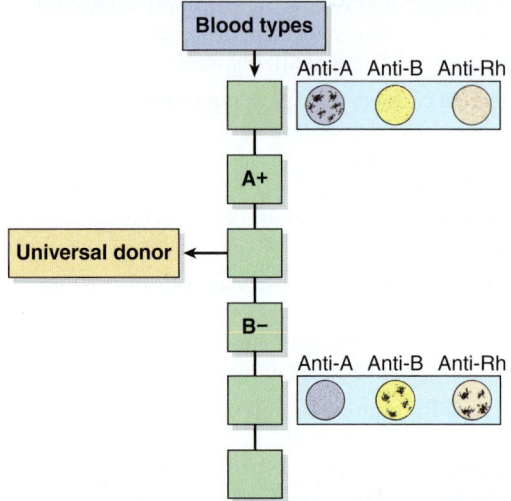

Visual Connections | Bloom's Level 5: Evaluate

These questions use visual images or previous content to make connections to this chapter's concepts.

1. **From chapter 15, figure 15.1.** How would a person's immunity be affected if he or she had a deficiency in cytotoxic T cells? Would a deficiency in T helper cells have a greater or lesser effect? Explain your answer.

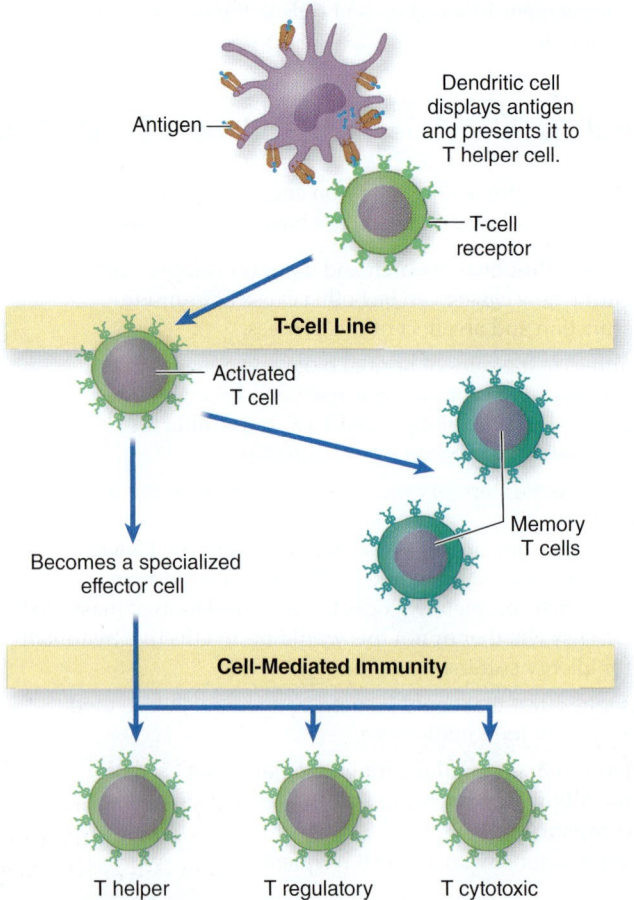

2. **Figure 16.10c.** Draw the expected agglutination pattern for a universal donor blood type.

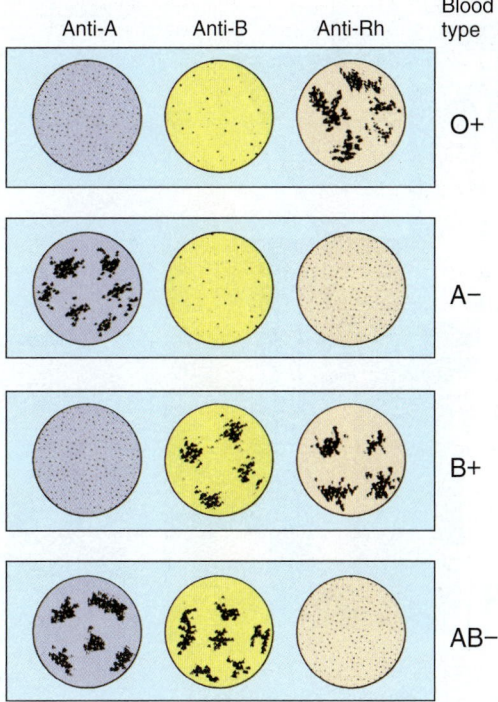

Concept Mapping | Bloom's Level 6: Create

Appendix D provides guidance for working with concept maps.

1. Using the words that follow, please create a concept map illustrating the relationships among these key terms from chapter 16.

lysed cells	immune complexes	cell-bound antibody
degranulation (release of mediators)	damage by T cells	processed antigen
	allergens	soluble antigen

www.mcgrawhillconnect.com

Enhance your study of this chapter with study tools and practice tests. Also ask your instructor about the resources available through ConnectPlus, including the media-rich eBook, interactive learning tools, and animations.

17

Diagnosing Infections

Case File 17

Whooping Cough in Washington

In 2012, Washington State experienced an epidemic of pertussis, more commonly known as whooping cough. Pertussis is a bacterial disease caused by *Bordetella pertussis*, an aerobic gram-negative coccobacillus that is highly transmissible via respiratory droplets. Whooping cough is a name derived from a characteristic symptom exhibited by many affected by the disease: a paroxysmal (violent) cough followed by a sharp intake of breath that sounds like a "whoop." In the first 6 months of 2012, there were 2,520 cases of pertussis in Washington State, compared to 180 cases in the same period of the previous year. Incidence of the disease was highest among children less than a year old and children between the ages of 10 and 14, and among the Hispanic population as well. The statewide incidence of pertussis was 37.5 cases per 100,000 compared to the rate of 4.2 cases per 100,000 nationwide.

Reported cases of pertussis were confirmed by clinical and state health laboratories through culturing of patient specimens. The samples were also subjected to polymerase chain reaction (PCR) analysis for molecular typing of the microbe. This nucleic test led to a rapid diagnosis of disease in patients because the tests detect the presence of microorganisms in a sample without culturing.

- What are the limitations of culture methods and PCR analysis in the diagnosis of pertussis?
- What other diagnostic methods were used in diagnosing pertussis in this outbreak?

Continuing the Case appears on page 497.

Outline and Learning Outcomes

17.1 Identifying the Infectious Agent

1. List the three major categories of microbial identification techniques.
2. Summarize factors that may affect the identification of an infectious agent from a patient sample.

3. Compare microbial identification tests performed on microbial isolates with those performed on patients themselves.
4. Differentiate between presumptive and confirmatory data in the process of specimen analysis.

17.2 Phenotypic Methods
5. Summarize the most common methods used for direct examination of a specimen.
6. Describe how the correct isolation medium is chosen when cultivating specimens, and provide examples.
7. Explain the main principle behind biochemical testing, and identify examples of such tests.

17.3 Genotypic Methods
8. Explain how PCR is used in microbial identification tests.
9. Summarize two genotypic methods that involve DNA analysis.
10. Describe how RNA analysis has influenced the process of infectious disease diagnosis.

17.4 Immunologic Methods
11. Define the term *serology*, and explain the immunologic principle behind serological tests.
12. Differentiate between sensitivity and specificity.
13. Explain how fluorescent antibodies can be used in the diagnosis of microbial disease.
14. Compare and contrast agglutination and precipitation reactions, and provide one example of how each is used to diagnose infectious disease.
15. Summarize the process of Western blotting, and explain how it can be used in microbial identification.
16. Describe two rapid immunologic methods for the detection of either microbial antigens or microbe-specific antibodies within a specimen.
17. Explain the difference between a direct and an indirect ELISA, and provide a clinical application for each.

17.5 Breakthrough Methodologies
18. Explain why isolating a pathogen through standard culture methods may become an outdated diagnosis strategy.
19. List the major advantages of microarray methods of diagnosis.
20. Summarize the benefits of using whole-genome sequencing of patient samples for disease diagnosis.
21. Describe how a pathogen's protein fingerprint can be used to determine its identity and disease diagnosis.

17.1 Identifying the Infectious Agent

In chapters 18 through 23, the most clinically significant bacterial, fungal, parasitic, and viral diseases are covered. The chapters survey the most prevalent infectious conditions and the organisms that cause them. This chapter gets us started with an introduction to the how-to of diagnosing the infections.

For many students (and professionals), the most pressing topic in microbiology is *how to identify unknown bacteria* in patient specimens or in samples from nature. Methods microbiologists use to identify bacteria to the level of genus and species fall into three main categories: **phenotypic,** which includes a consideration of morphology (microscopic and macroscopic) as well as bacterial physiology or biochemistry; **immunologic,** which entails serological analysis; and **genotypic** (or genetic) techniques.

We are on the verge of a revolution in infectious disease diagnosis. Most experts predict that within 5 years technology will allow large-scale and affordable adoption of genetic and other diagnostic methods that will replace many of the phenotypic and immunologic methods. We will summarize these new technologies at the end of this chapter.

Specimen Collection

Regardless of the method of diagnosis, specimen collection is the common point that guides the health care decisions of every member of a clinical team. Indeed, the success of identification and treatment depends on how specimens are collected, handled, stored, and cultured. Specimens can be taken by a clinical laboratory scientist or medical technologist, nurse, physician, or even the patient. However, it is imperative that general aseptic procedures be used, including sterile sample containers and other tools to prevent contamination from the environment or the patient. **Figure 17.1** delineates the most common sampling sites and procedures.

In sites that normally contain resident microbiota, care should be taken to sample only the infected site and not surrounding areas. Saliva is an especially undesirable contaminant because it contains millions of bacteria per milliliter, most of which are normal biota. Sputum, the mucus secretion that coats the lower respiratory surfaces, especially the lungs, is discharged by coughing or taken by a thin tube called a catheter to avoid contamination with saliva. In addition, throat and nasopharyngeal swabs should not touch the

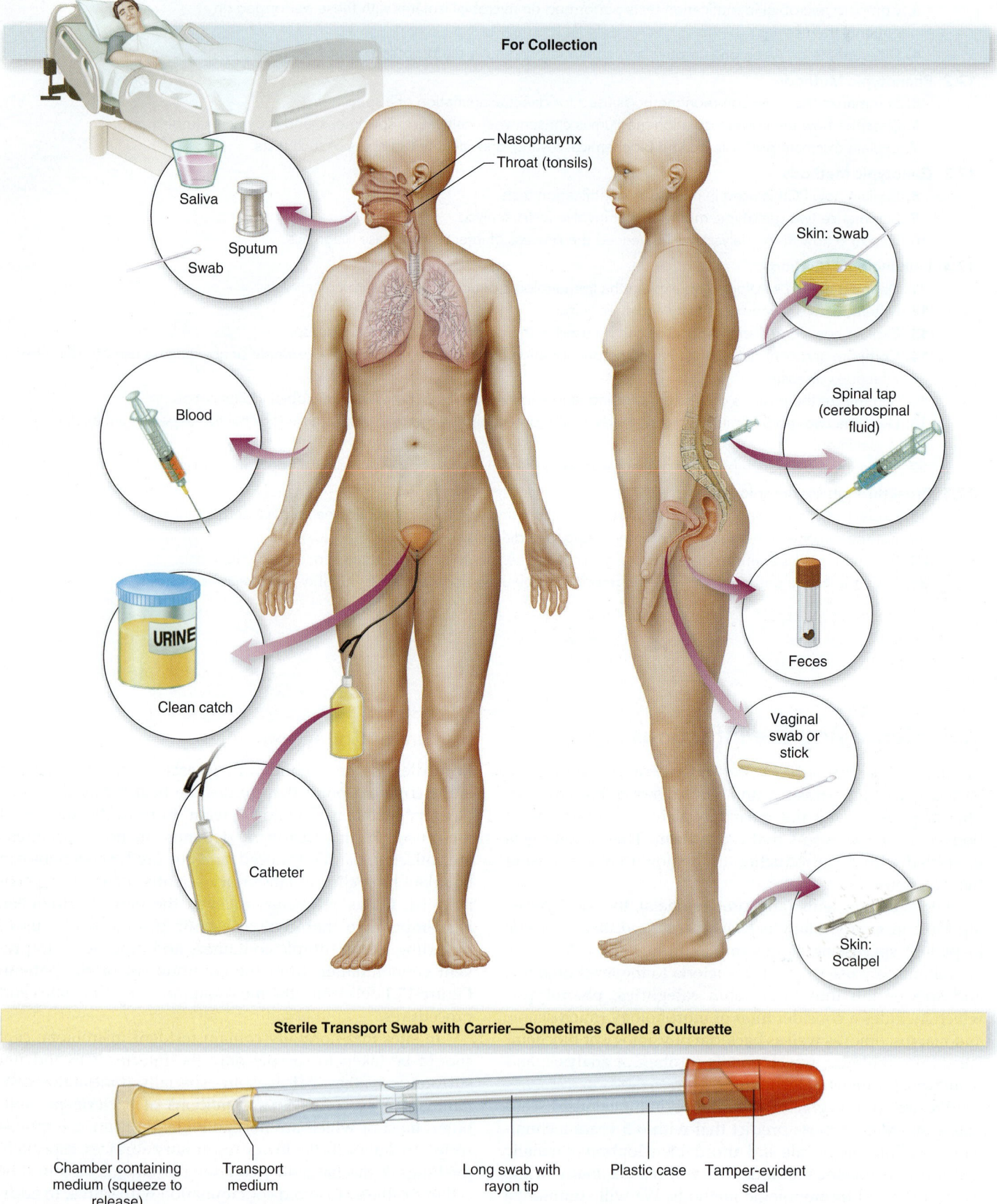

For Collection

Saliva

Sputum

Swab

Nasopharynx

Throat (tonsils)

Skin: Swab

Blood

Spinal tap
(cerebrospinal
fluid)

URINE

Clean catch

Feces

Catheter

Vaginal
swab or
stick

Skin:
Scalpel

Sterile Transport Swab with Carrier—Sometimes Called a Culturette

Chamber containing
medium (squeeze to
release)

Transport
medium

Long swab with
rayon tip

Plastic case

Tamper-evident
seal

Figure 17.1 **Sampling sites and methods of collection for clinical laboratories.**

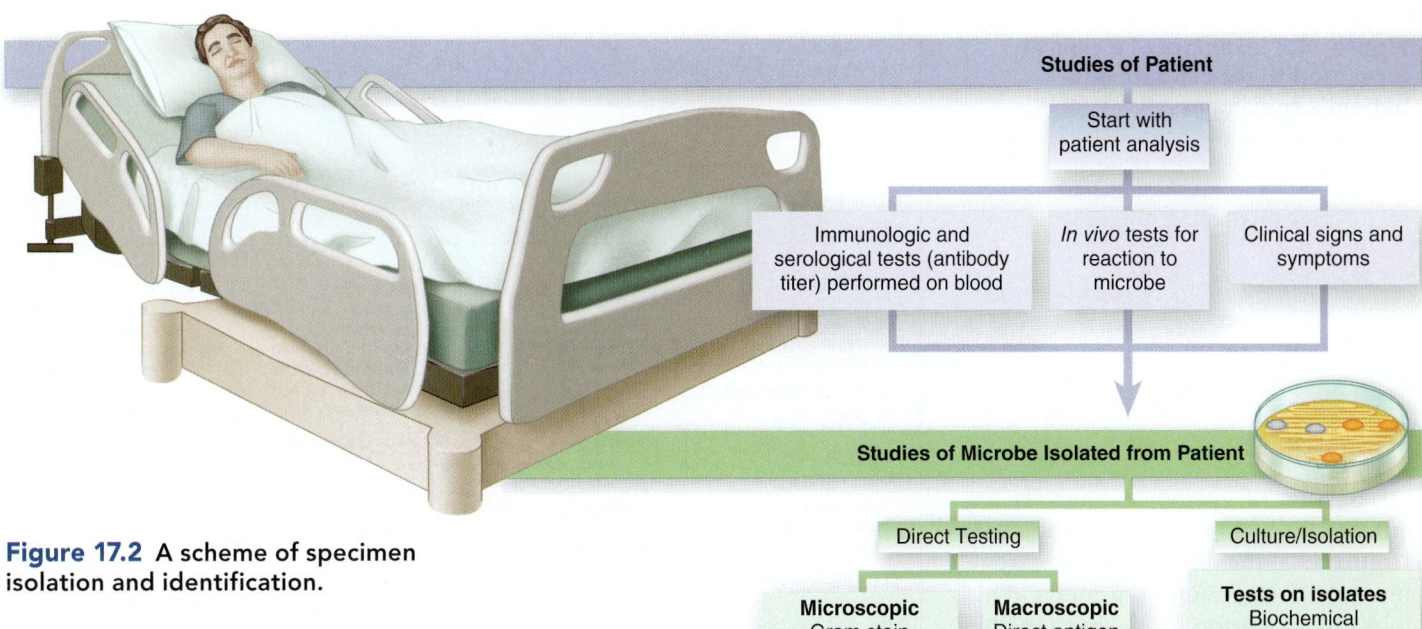

Figure 17.2 A scheme of specimen isolation and identification.

tongue, cheeks, or saliva. On occasion, saliva samples are needed for dental diagnosis and are obtained by having the patient expectorate into a container.

Urine is taken aseptically from the bladder with a catheter designed for that site. Another method, called a "clean catch," is taken by washing the external urethra and collecting the urine midstream. The latter method inevitably incorporates a few normal biota into the sample, but these can usually be differentiated from pathogens in an actual infection. Sometimes diagnostic techniques require first-voided "dirty catch" urine. The mucous lining of the urethra, vagina, or cervix can be sampled with a swab or applicator stick. Depending on the nature of a skin lesion, skin can be swabbed or scraped with a scalpel to expose deeper layers. Wounds are cleansed prior to swabbing for culture to avoid collecting the many normal microbiota of the skin. Sterile materials such as blood, cerebrospinal fluid, and tissue fluids must be taken by sterile needle aspiration. Antisepsis of the puncture site is extremely important in these cases. Additional sources of specimens are the eye, ear canal, synovial fluid, nasal cavity (all by swab), and diseased tissue that has been surgically removed (biopsied).

After proper collection, the specimen is promptly transported to a lab and stored appropriately (usually refrigerated) if it must be held for a time. Nonsterile samples in particular, such as urine, feces, and sputum, are especially prone to deterioration at room temperature. Special swab and transport systems are designed to collect the specimen and maintain it in stable condition for several hours. These devices contain nonnutritive maintenance media (so the microbes survive but do not grow), a buffering system, and an anaerobic environment to prevent possible destruction of oxygen-sensitive bacteria.

Overview of Laboratory Techniques

Analyzing the patient for signs of microbial infection (i.e., fever, wound exudate, mucus production, abnormal lesion) comes first; after that, specimens are collected and analyzed. This involves (1) direct tests using microscopic, immunologic, or genetic methods that provide immediate clues as to the identity of the microbe or microbes in the sample; and (2) cultivation, isolation, and identification of pathogens using a wide variety of general and specific tests **(figure 17.2).** Most test results fall into two categories: presumptive data, which place the isolated microbe (isolate) in a preliminary category such as a genus; and more specific, confirmatory data, which can pinpoint the microbe's specific identity. The total time required for analysis ranges from a few minutes for a streptococcal sore throat to several weeks in the diagnosis of tuberculosis infection.

Figure 17.3 Example of a clinical form used to report data on a patient's specimens.

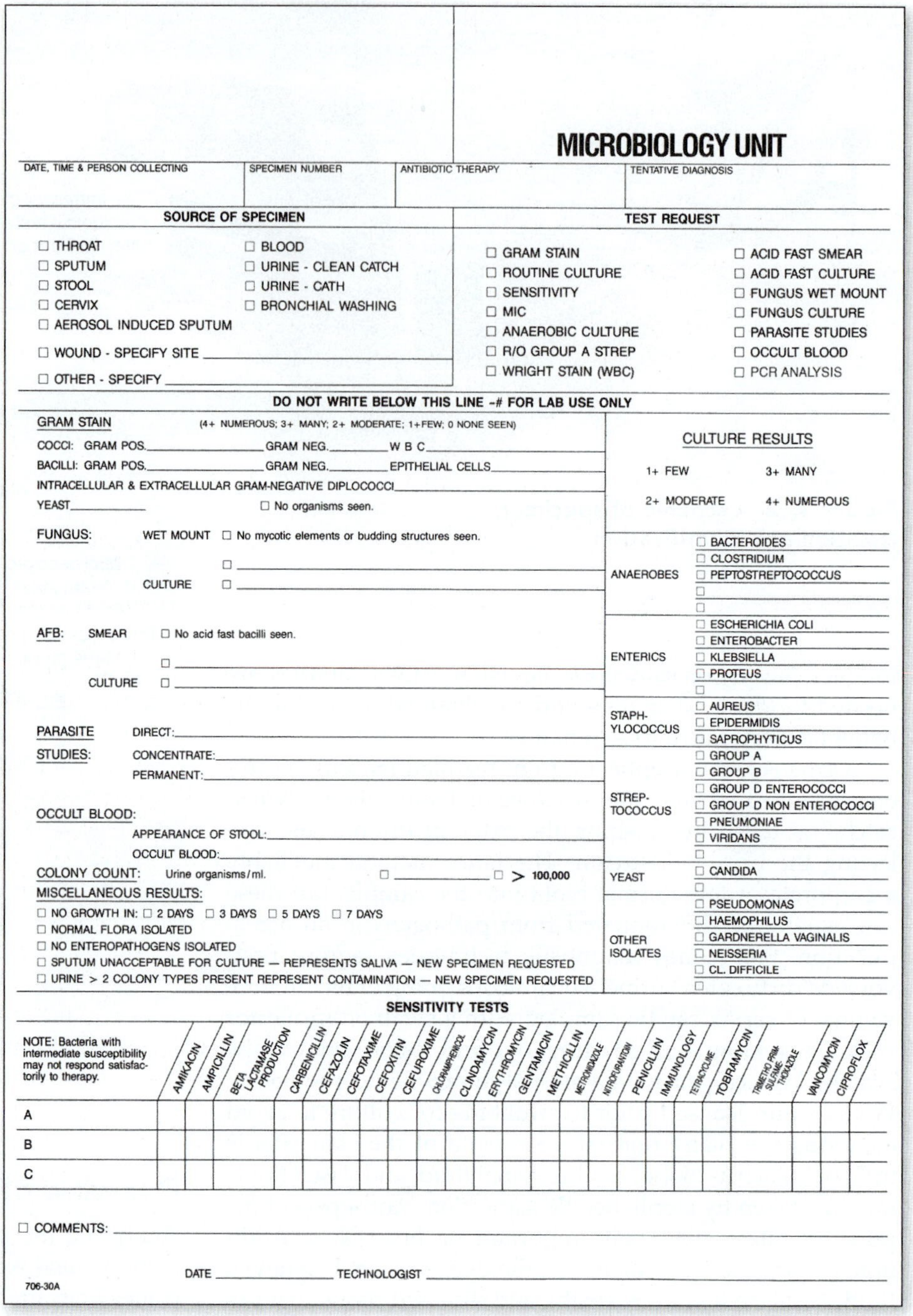

MICROBIOLOGY UNIT

| DATE, TIME & PERSON COLLECTING | SPECIMEN NUMBER | ANTIBIOTIC THERAPY | TENTATIVE DIAGNOSIS |

SOURCE OF SPECIMEN

- ☐ THROAT
- ☐ SPUTUM
- ☐ STOOL
- ☐ CERVIX
- ☐ AEROSOL INDUCED SPUTUM
- ☐ WOUND - SPECIFY SITE _____
- ☐ OTHER - SPECIFY _____
- ☐ BLOOD
- ☐ URINE - CLEAN CATCH
- ☐ URINE - CATH
- ☐ BRONCHIAL WASHING

TEST REQUEST

- ☐ GRAM STAIN
- ☐ ROUTINE CULTURE
- ☐ SENSITIVITY
- ☐ MIC
- ☐ ANAEROBIC CULTURE
- ☐ R/O GROUP A STREP
- ☐ WRIGHT STAIN (WBC)
- ☐ ACID FAST SMEAR
- ☐ ACID FAST CULTURE
- ☐ FUNGUS WET MOUNT
- ☐ FUNGUS CULTURE
- ☐ PARASITE STUDIES
- ☐ OCCULT BLOOD
- ☐ PCR ANALYSIS

DO NOT WRITE BELOW THIS LINE -# FOR LAB USE ONLY

GRAM STAIN   (4+ NUMEROUS; 3+ MANY; 2+ MODERATE; 1+FEW; 0 NONE SEEN)

COCCI: GRAM POS._____ GRAM NEG._____ W B C_____
BACILLI: GRAM POS._____ GRAM NEG._____ EPITHELIAL CELLS_____
INTRACELLULAR & EXTRACELLULAR GRAM-NEGATIVE DIPLOCOCCI_____
YEAST_____   ☐ No organisms seen.

FUNGUS: WET MOUNT ☐ No mycotic elements or budding structures seen.
                 ☐ _____
           CULTURE   ☐ _____

AFB: SMEAR   ☐ No acid fast bacilli seen.
                 ☐ _____
        CULTURE ☐ _____

PARASITE   DIRECT:_____
STUDIES: CONCENTRATE:_____
            PERMANENT:_____

OCCULT BLOOD:
            APPEARANCE OF STOOL:_____
            OCCULT BLOOD:_____

COLONY COUNT: Urine organisms / ml.   ☐ _____ ☐ > 100,000
MISCELLANEOUS RESULTS:
- ☐ NO GROWTH IN: ☐ 2 DAYS ☐ 3 DAYS ☐ 5 DAYS ☐ 7 DAYS
- ☐ NORMAL FLORA ISOLATED
- ☐ NO ENTEROPATHOGENS ISOLATED
- ☐ SPUTUM UNACCEPTABLE FOR CULTURE — REPRESENTS SALIVA — NEW SPECIMEN REQUESTED
- ☐ URINE > 2 COLONY TYPES PRESENT REPRESENT CONTAMINATION — NEW SPECIMEN REQUESTED

CULTURE RESULTS

| 1+ FEW | 3+ MANY |
| 2+ MODERATE | 4+ NUMEROUS |

ANAEROBES	☐ BACTEROIDES ☐ CLOSTRIDIUM ☐ PEPTOSTREPTOCOCCUS ☐ ☐
ENTERICS	☐ ESCHERICHIA COLI ☐ ENTEROBACTER ☐ KLEBSIELLA ☐ PROTEUS
STAPH-YLOCOCCUS	☐ AUREUS ☐ EPIDERMIDIS ☐ SAPROPHYTICUS
STREP-TOCOCCUS	☐ GROUP A ☐ GROUP B ☐ GROUP D ENTEROCOCCI ☐ GROUP D NON ENTEROCOCCI ☐ PNEUMONIAE ☐ VIRIDANS ☐
YEAST	☐ CANDIDA ☐
OTHER ISOLATES	☐ PSEUDOMONAS ☐ HAEMOPHILUS ☐ GARDNERELLA VAGINALIS ☐ NEISSERIA ☐ CL. DIFFICILE ☐

SENSITIVITY TESTS

NOTE: Bacteria with intermediate susceptibility may not respond satisfactorily to therapy.

Columns: AMIKACIN, AMPICILLIN, BETA LACTAMASE PRODUCTION, CARBENICILLIN, CEFAZOLIN, CEFOTAXIME, CEFOXITIN, CEFUROXIME, CHLORAMPHENICOL, CLINDAMYCIN, ERYTHROMYCIN, GENTAMICIN, METHICILLIN, METRONIDAZOLE, NITROFURANTOIN, PENICILLIN, IMMUNOLOGY, TETRACYCLINE, TOBRAMYCIN, TRIMETHOPRIM SULFAMETHOXAZOLE, VANCOMYCIN, CIPROFLOX

Rows: A, B, C

☐ COMMENTS: _____

706-30A   DATE _____   TECHNOLOGIST _____

Results of specimen analysis are entered in a summary patient chart **(figure 17.3)** that can be used in assessment and treatment regimens. This looks like a boring form—but take the time to read it! You will notice that it compiles information on tests you are already familiar with, such as the Gram stain.

Some diseases are diagnosed without analyzing actual microbes within specimens. Serological tests on patient sera provide indirect evidence for specific pathogens through analysis of the antibody response. Measurement of microbe-specific antibody levels over time can differentiate current from prior infection, while skin testing can identify those in the general population who have had past exposures to infectious agents such as rubella or tuberculosis. Additionally, some pathogens are identified almost solely on patient signs and symptoms. AIDS, for example, is diagnosed by serological tests and a complex of signs and symptoms without ever isolating the virus. Some diseases, such as athlete's foot, are diagnosed purely on the typical presenting symptoms and may require no lab tests at all.

17.1 Learning Outcomes—Assess Your Progress

1. List the three major categories of microbial identification techniques.
2. Summarize factors that may affect the identification of an infectious agent from a patient sample.
3. Compare microbial identification tests performed on microbial isolates with those performed on patients themselves.
4. Differentiate between presumptive and confirmatory data in the process of specimen analysis.

17.2 Phenotypic Methods

Immediate Direct Examination of Specimen

Direct microscopic observation of a fresh or stained specimen is one of the most rapid methods of determining presumptive and sometimes confirmatory characteristics. The Gram stain (see Insight 4.2) and the acid-fast stain (see figure 3.21b) are most often used for bacterial identification. As useful as these stains are, they can identify only a few organisms on their own and are typically conducted in combination with a variety of other techniques to further this process. We will start by reviewing additional phenotypic methods.

Cultivation of Specimen

Isolation Media

Such a wide variety of media exist for microbial isolation that a certain amount of preselection must occur, based on the nature of the specimen. In cases in which the suspected pathogen is present in small numbers or is easily overgrown, the specimen can be initially enriched with specialized media. Nonsterile specimens containing a diversity of bacterial species, such as urine and feces, are cultured on selective media to encourage the growth of only the suspected pathogen. For example, approximately 80% of urinary tract infections are known to be caused by *Escherichia coli,* so selective media that will be sure to allow for the growth of this common pathogen are chosen. Specimens are also inoculated into differential media to identify definitive characteristics such as reactions in blood (blood agar) and fermentation patterns (mannitol salt and MacConkey agar). A patient's blood is usually cultured in a special bottle of broth that can be periodically sampled for analysis using a number of isolation, differential, and biochemical media (see chapter 3). Working with a mixed or contaminated culture causes misleading and ambiguous results. So that subsequent steps in identification will be as accurate as possible, pure cultures of the microbe must be obtained from culturing on isolation media. Clinical microbiologists can then observe the suspected pathogen's microscopic morphology and staining reactions, cultural appearance, motility, and oxygen requirements, in addition to biochemical analysis and antibiotic-sensitivity tests.

Biochemical Testing

The physiological reactions of bacteria to nutrients and other substrates provide excellent indirect evidence of the types of enzyme systems present in a particular species. Many of these tests are based on enzyme-mediated metabolic reactions that are visualized by a color change. These types of reactions are particularly meaningful in bacteria, which are haploid organisms that generally express their genes for utilizing a given nutrient.

The microbe is cultured in a medium with a special substrate and then tested for a particular end product. Microbial expression of the enzyme is made visible by the colored dye; no coloration means it lacks the enzyme for utilizing the substrate in that particular way. Although routinely performed

Disease Connection

One disease that is often diagnosed based solely on a Gram stain is gonorrhea, particularly in symptomatic men. The condition is caused by *Neisseria gonorrhoeae.* When discharge from this condition is Gram-stained, distinctively shaped gram-negative diplococci, frequently seen inside phagocytes, are diagnostic.

on cultured isolates, direct biochemical testing of patient samples can be performed today producing results within hours versus days.

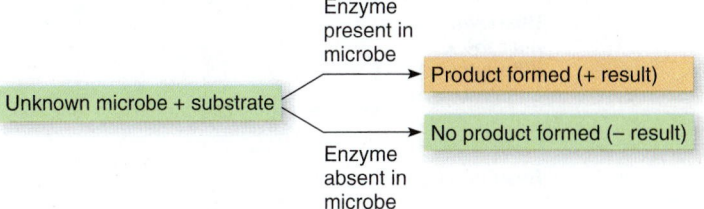

Among the prominent biochemical tests are carbohydrate fermentation (production of acid and/or gas); hydrolysis of gelatin, starch, and other polymers; the actions of enzymes such as catalase, oxidase, and coagulase; and various by-products of metabolism. Many are presently performed with rapid, miniaturized systems that can simultaneously determine up to 23 characteristics in small individual cups or spaces **(figure 17.4).** An important plus, given the complexity of biochemical profiles, is that such systems are readily adapted to computerized analysis.

Common schemes exist for identifying bacteria. These are based on easily recognizable characteristics such as motility, oxygen requirements, Gram-stain reactions, shape, spore formation, and various biochemical reactions. **Dichotomous keys** are flowcharts **(figure 17.5)** used to trace a route of identification by offering pairs of opposing characteristics (positive versus negative, for example) with two choices from which to select at each level. Eventually, an endpoint is reached, and the name of a genus or species that fits that particular combination of characteristics appears. Although useful in student and research laboratories, diagnostic tables providing more complete microbial information are preferred over dichotomous keys in most clinical laboratories.

Miscellaneous Tests

When morphological and biochemical tests are insufficient to complete identification, other tests come into play. Phage typing involves the use of bacteriophages, viruses that attack bacteria in very species-specific and strain-specific ways. Phage typing is useful in identifying some bacteria, primarily *Salmonella,* and is often used for tracing bacterial strains in epidemics. The technique involves inoculating a lawn of cells onto a Petri dish, mapping it off into blocks, and applying a different phage to each sectioned area of growth **(figure 17.6).** Cleared areas corresponding to lysed cells indicate sensitivity to that phage, and bacterial identification may be determined based upon this unique pattern.

Antimicrobial sensitivity tests are not only important in determining the drugs to be used in treatment (see figure 12.1), but the patterns of sensitivity can also be used in presumptive identification of some species of *Streptococcus, Pseudomonas,* and *Clostridium.* Antimicrobials are also used as selective agents in many media.

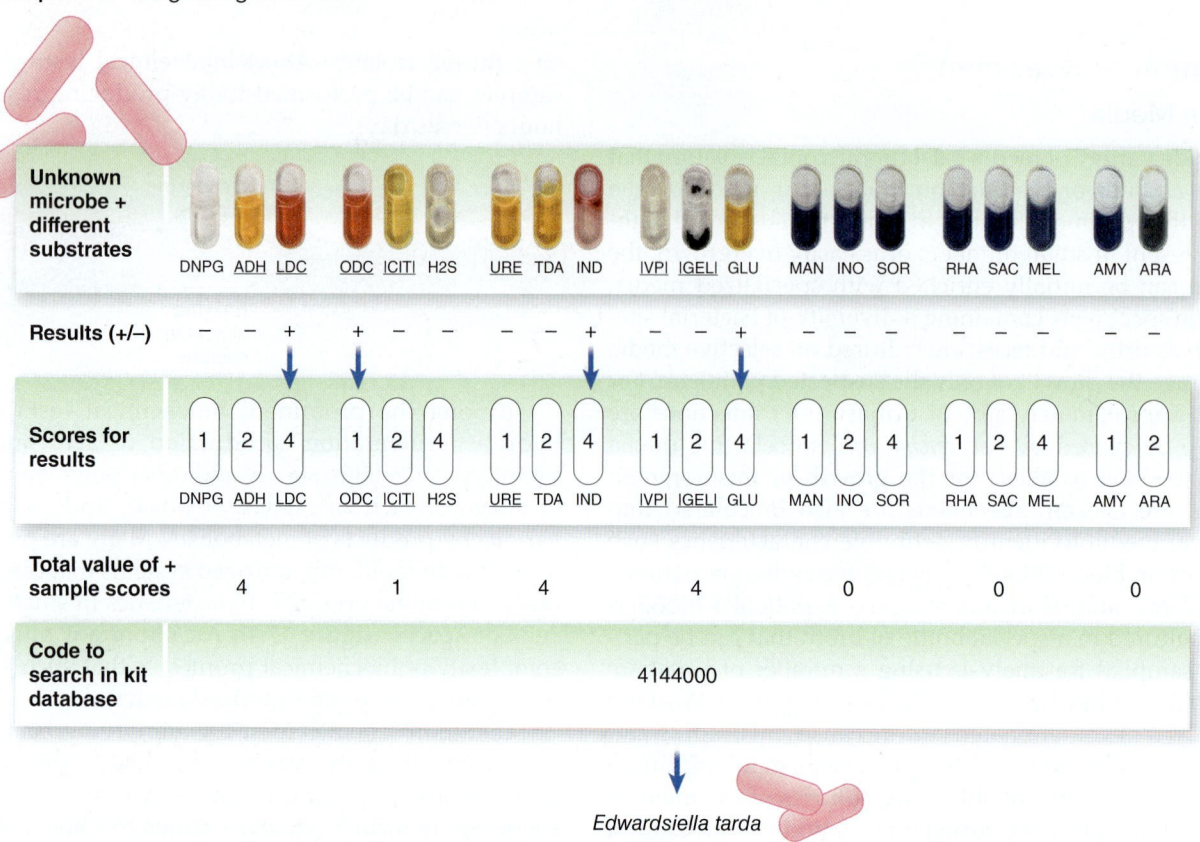

Figure 17.4 Rapid tests—a biochemical system for microbial identification. Samples of a single bacterial isolate (an unknown gram-negative microbe) were placed in the different cups, which contain growth media and chemicals designed to test for a particular enzyme. The organism produced four positive results, as indicated. Scoring of the results is completed by adding the designated values of each positive result per set of cups indicated. This string of seven numbers creates a code that can be referenced in a database to determine the identity of the unknown microbe.

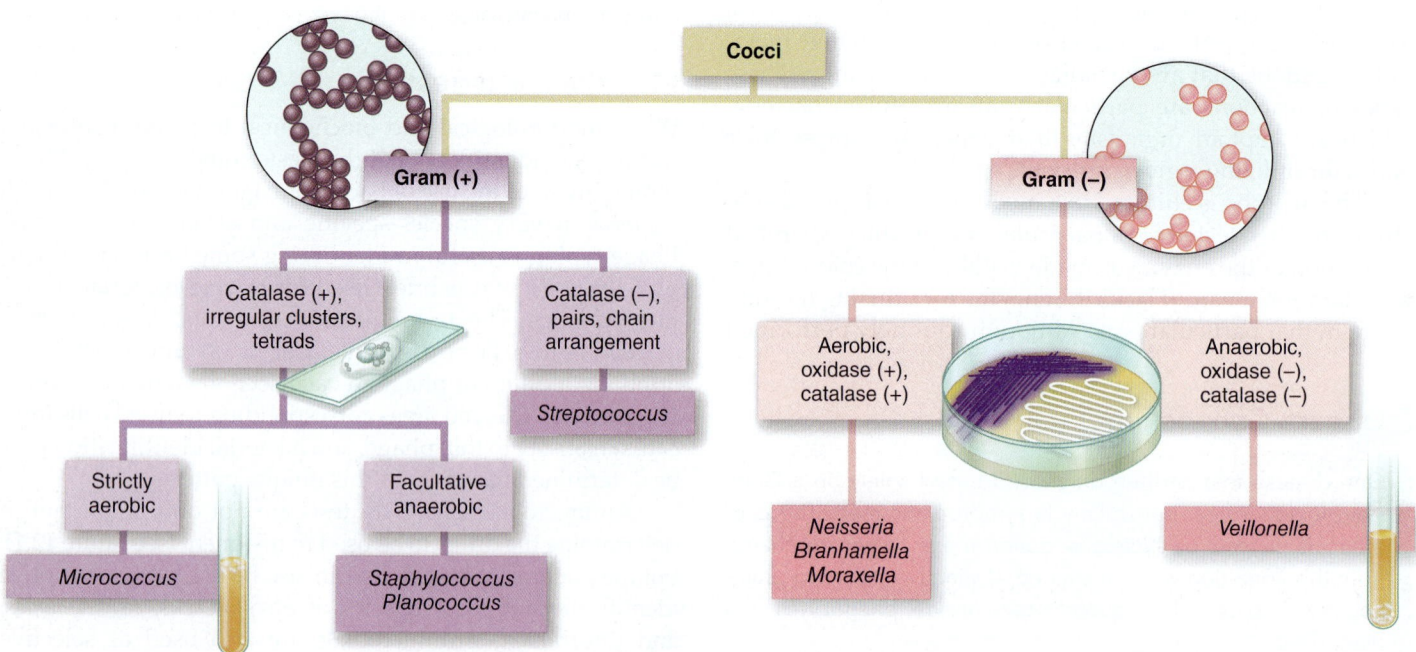

Figure 17.5 Flowchart to separate primary genera of gram-positive and gram-negative bacteria. Identification scheme for cocci commonly involved in human diseases.

17.3 Genotypic Methods

The sequence of nitrogenous bases within DNA or RNA is unique to every microorganism. Due to this fact and the numerous advances that have been made in genomic technology, nucleic acid tests have become a mainstay of microbial identification today.

Polymerase Chain Reaction: Amplifying the Information

Many nucleic acid tests employ the use of the **polymerase chain reaction (PCR).** PCR results in the production of numerous identical copies of DNA or RNA molecules within hours (see process figure 10.5). This method can amplify even minute quantities of nucleic acids present in a sample, which greatly improves the sensitivity of these tests. Diagnosis with PCR can be performed on genetic material from a wide variety of bacteria, viruses, protozoa, and fungi. Metagenomic analysis of the human body and the environment also depends on the ability of PCR analysis to amplify the amount of microbial information available for nucleic acid testing from these sites. In some cases, where the microbial populations are relatively unknown, a form of PCR called **random amplified polymorphic DNA (RAPD)** may be used because it employs primers of random sequence in an attempt to pick a microbial needle out of a haystack.

Hybridization: Probing for Identity

Hybridization is a technique that makes it possible to identify a microbe by analyzing segments of its genetic material. This requires small fragments of single-stranded DNA (or RNA) called **probes** that are known to be complementary to the specific sequences of nucleic acid isolated from a particular microbe. Base-pairing of the known probe to the nucleic acid can be observed providing evidence of the microbe's identity. Although hybridization techniques are quite specific, control probes must be used in order to rule out cross-reactivity. These tests have become more convenient and portable over the years. Probes are typically fluorescently labeled or attached to an enzyme that triggers a colorimetric change when hybridization occurs. The property of dyes such as fluorescein and rhodamine to emit visible light in response to ultraviolet radiation was discussed in chapter 3. This property of fluorescence has found numerous applications in diagnostic testing.

Several variations on the principle of hybridization are used in infectious disease diagnosis today. In the oldest method, unknown test DNA is extracted from cells in specimens or cultures and is bound to special blotter paper. After several different probes have been added to the blotter, it is observed for visible signs that the probes have become fixed (hybridized) to the test DNA. Such rapid hybridization test cards are still used routinely in the diagnosis of infections such as vaginitis.

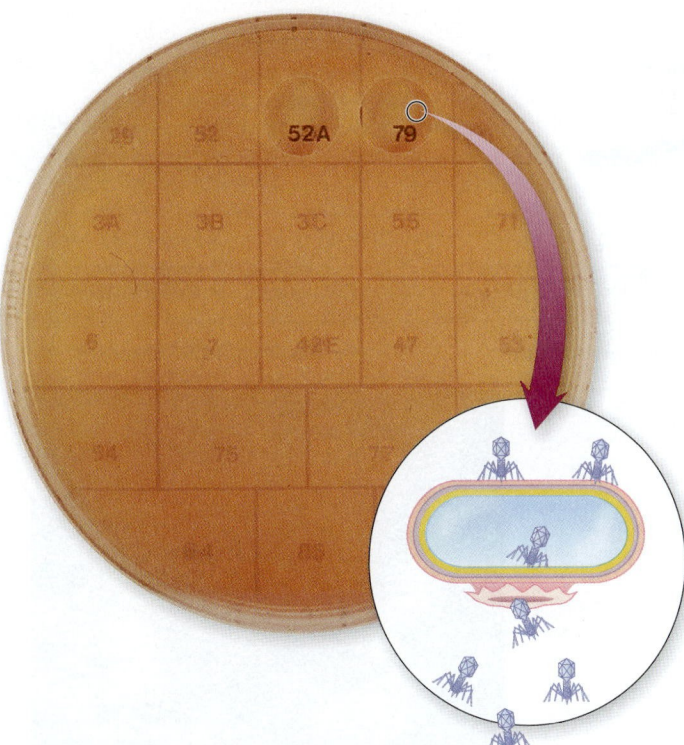

Figure 17.6 Phage typing of an unknown bacterium. A cleared area within a square of the bacterial lawn forms due to phage-induced lysis of cells, indicating sensitivity of the bacterium to the corresponding phage.

Determining Clinical Significance of Cultures

It is important to rapidly determine if an isolate from a specimen is clinically important or if it is merely a contaminant or normal biota. Although answering these questions may prove difficult, one can first focus on the number of microbes in a specimen. For example, a few colonies of *Escherichia coli* in a urine sample can simply indicate normal biota, whereas several hundred can mean active infection. In contrast, the presence of a single colony of a true pathogen, such as *Mycobacterium tuberculosis* in a sputum culture or an opportunist in sterile sites such as cerebrospinal fluid or blood, is highly suggestive of its role in disease.

17.2 Learning Outcomes—Assess Your Progress

5. Summarize the most common methods used for direct examination of a specimen.
6. Describe how the correct isolation medium is chosen when cultivating specimens, and provide examples.
7. Explain the main principle behind biochemical testing, and identify examples of such tests.

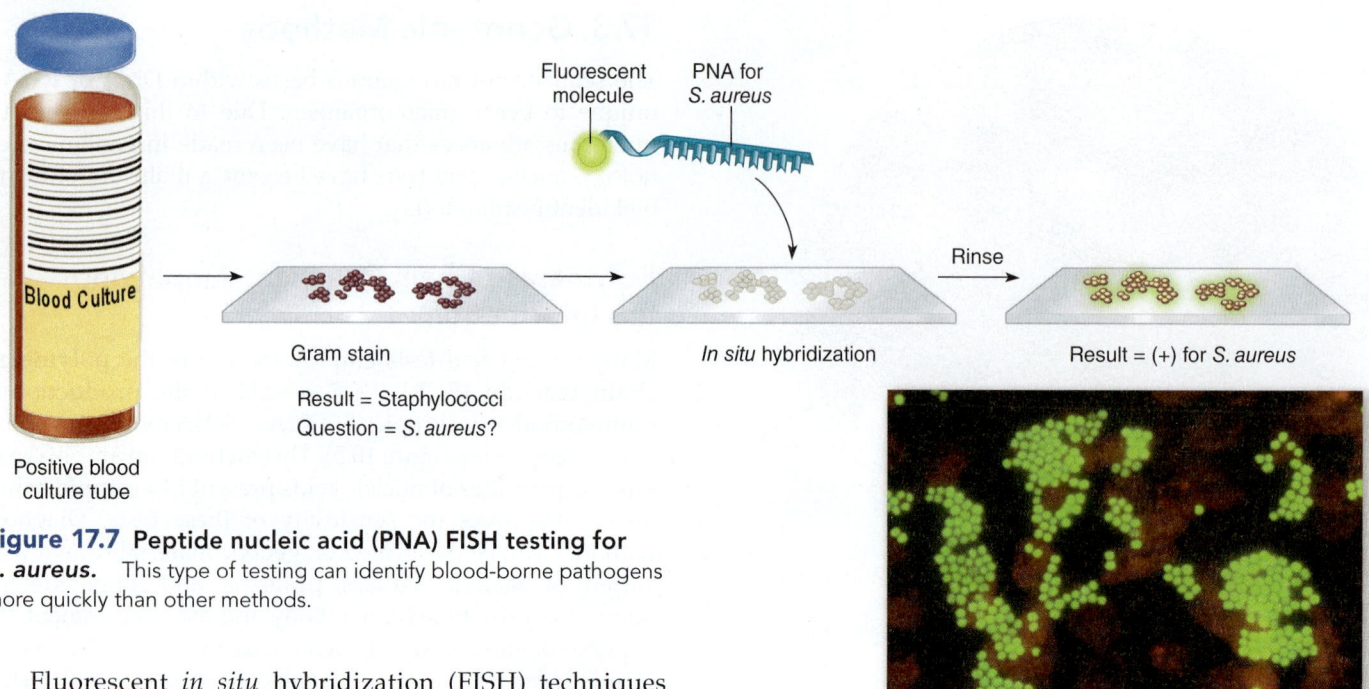

Figure 17.7 Peptide nucleic acid (PNA) FISH testing for S. aureus. This type of testing can identify blood-borne pathogens more quickly than other methods.

Fluorescent *in situ* hybridization (FISH) techniques involve the application of fluorescently labeled probes to intact cells within a patient specimen or an environmental sample **(figure 17.7)**. Microscopic analysis is used to locate "glowing" cells and determine the identity of a specific microbe. FISH is often used to confirm a diagnosis or to identify the microbial components within a biofilm.

Pulse-Field Gel Electrophoresis: Microbial Fingerprints

In chapter 10, DNA fingerprinting was described as a method for analyzing short segments of DNA within a sample. Pulse-field gel electrophoresis (PFGE) is a similar technique, but it involves the separation of DNA fragments that are too large for conventional gel electrophoresis methods **(process figure 17.8)**. This separation is accomplished by slowly applying alternating voltage levels to the gel from three different directions, allowing even similarly sized DNA fragments to fully separate. Although developed in 1984, this technique first came to prominence in 1993 when it was used by the CDC to determine that *E. coli* O157:H7 was the source of a food-borne outbreak in the United States. Since then, it has become an important method in epidemiological studies due to its accuracy in assessing microbial subtype and identification from patient samples. PulseNet is a program established by the CDC to assist in the investigation of possible infectious-disease outbreaks, especially those caused by food-borne pathogens. Scientists from public health facilities across the country are able to rapidly communicate and compare PFGE data from patient specimens and other samples, allowing identification of outbreaks to occur within hours versus days or even weeks.

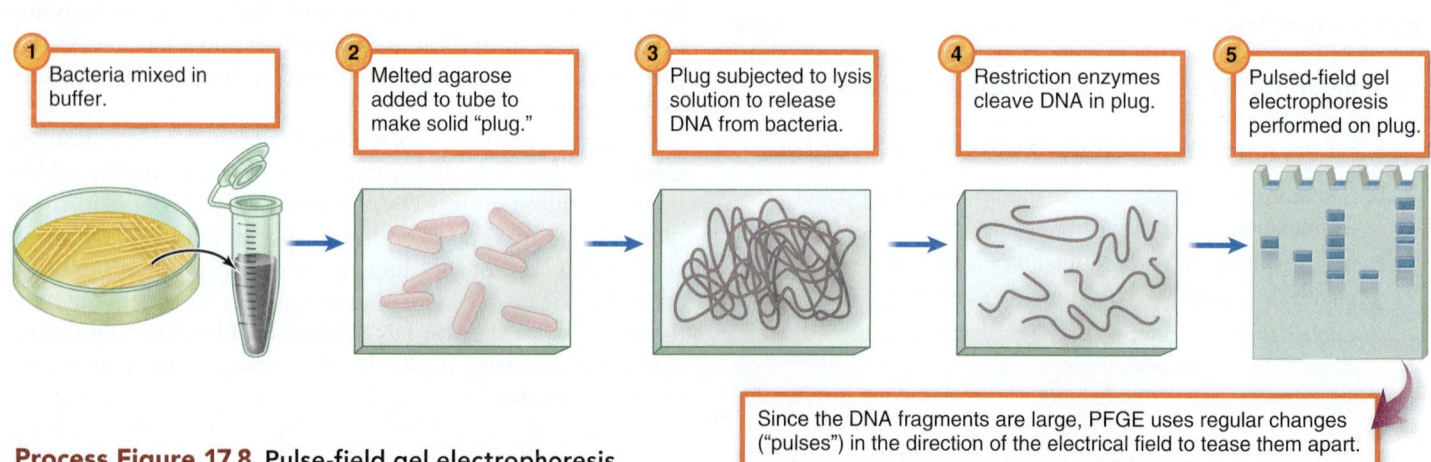

Since the DNA fragments are large, PFGE uses regular changes ("pulses") in the direction of the electrical field to tease them apart.

Process Figure 17.8 Pulse-field gel electrophoresis.

Many methods of detection and diagnosis require time to culture and isolate a pathogen, followed by additional inspection and microbial testing. Genetic analysis from patient samples through probes, FISH, and PCR are more rapid but are also costly and can be time-consuming as well. Recently, a method has been developed that rapidly identifies RNA produced when pathogenic microbes switch on certain genes. Researchers at Boston-area universities, research laboratories, and hospitals have developed RNA probes that glow when they are activated. Amy Barczak and her team of researchers also developed RNA probes that can distinguish between antibiotic-susceptible and -resistant microbes. The rapid detection of antibiotic-resistant microbes is especially valuable in treating cases of tuberculosis, for which culturing and antibiotic susceptibility testing can take weeks. Using this method, fast-growing

pathogens have been identified in as little as 30 minutes, and slow-growing microbes such as *Mycobacterium tuberculosis* were identified in around 3 hours. Although further testing is needed to determine sensitivity and specificity, researchers have been able to analyze blood and urine samples from patients and have identified antibiotic-resistant strains of *M. tuberculosis*, *Escherichia coli*, and *Pseudomonas aeruginosa*. Researchers were also able to successfully identify *Candida albicans*, the fungus implicated in thrush and vaginal yeast infections, *Plasmodium falciparum*, the malaria parasite, and viral pathogens such as HIV, influenza, and herpes simplex 1. A simple test kit that can be used by doctors and hospitals is yet to be developed, but this research suggests that glowing RNA probes may soon be used to rapidly and accurately diagnose infections.

Colonies of *Mycobacterium tuberculosis*.
Using glowing RNA probes, diagnosis of antibiotic-resistant tuberculosis can now take hours rather than days to weeks.

Source: 2012. *Proc. Nat. Acad. Sci.* vol. 109, no. 16, p. 6217.

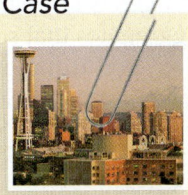

Although bacterial culture of nose and throat swabs is the standard protocol for diagnosing pertussis, *B. pertussis* is often difficult to culture under laboratory conditions since it is a fastidious organism. Obtaining quality specimens from patients is also a challenge, especially if they have already received antibiotic. Culture methods are also time-consuming and can delay treatment if pertussis is suspected. PCR is sometimes used because a positive diagnosis can be made rapidly. However, sometimes false-positive or false-negative results can occur, and PCR is only optimally sensitive during the first 3 weeks of the pertussis cough.

Pulse-field gel electrophoresis (PFGE), a type of DNA fingerprinting, was performed by the Centers for Disease Control and Prevention (CDC) in the Washington State outbreak. The DNA profiles generated from the samples taken from patients were compared to a national database of PFGE profiles of *B. pertussis*. According to the CDC, 54% of the isolates represented the 4 most common isolates of *B. pertussis*, 20 isolates represented the 7 less common isolates, and five samples had PFGE profiles that had not been seen in the CDC database previously.

■ Pertussis is part of the standard DTaP (diphtheria, tetanus, and pertussis) vaccination series. What was the reason for such a large increase in cases of pertussis from 2011 to 2012?

Ribotyping: rRNA Analysis

You may remember from chapter 1 that one of the most viable indicators of evolutionary relatedness and affiliation is comparison of the sequence of nitrogen bases in 16s ribosomal RNA (rRNA). The 16s ribosomal RNA is a component of the 30s subunit of bacterial and archaeal ribosomes. Other rRNA sequences vary between species, whereas 16s rRNA is highly conserved across species and evolutionary time. This makes such RNA analysis, or ribotyping, perfectly suited for bacterial identification and subsequent diagnosis of infection. Ribosomal RNA is isolated, sequenced, and analyzed from cultured cells obtained from a patient site or environmental sample to obtain this information.

The FISH method discussed earlier is also being used for rRNA analysis. It can rapidly identify 16s rRNA sequences without first culturing the organism. The turnaround time for identifying microorganisms present in blood cultures has been reduced from 24 hours to 90 minutes using this technique (see figure 17.7).

In this section, we have discussed widely used genomic diagnostic methods. A wide array of newer genomic methods have been developed, many of which will become standard tools for infectious disease diagnosis in the near future. These are discussed in section 17.5, "Breakthrough Methodologies."

17.3 Learning Outcomes—Assess Your Progress

8. Explain how PCR is used in microbial identification tests.

9. Summarize two genotypic methods that involve DNA analysis.

10. Describe how RNA analysis has influenced the process of infectious disease diagnosis.

17.4 Immunologic Methods

The antibodies formed during an immune reaction are important in combating infection, but they hold additional practical value. Characteristics of antibodies (such as their quantity or specificity) can reveal the history of a patient's contact with microorganisms or other antigens. This is the underlying basis of serological testing. **Serology** is the term used for *in vitro* (meaning "taking place outside of the body") diagnostic testing of serum. Serological testing is based on the familiar concept that antibodies have extreme specificity for antigens, so when a particular antigen is exposed to its specific antibody, it will fit like a hand in a glove. The ability to visualize this interaction provides a powerful tool for detecting, identifying, and quantifying antibodies—or for that matter, antigens. The scheme works both ways, depending on the situation. One can detect or identify an unknown antibody using a known antigen, or one can use an antibody of known specificity to help detect or identify an unknown antigen **(figure 17.9)**. Modern serological methods have evolved beyond the ability to test sera; urine, cerebrospinal fluid, whole tissues, and saliva can also be analyzed for the presence of specific antibodies. These and other immune tests help to determine the immunologic status of patients, confirm a suspected diagnosis, or screen individuals for disease.

General Features of Immune Testing

The most effective serological tests have a high degree of specificity and sensitivity **(figure 17.10)**. *Specificity* is the property of a test to focus on only a certain antibody or antigen and not to react with unrelated or distantly related ones. Better said, specificity is the degree to which a test does not (falsely) detect people who do not have a condition. *Sensitivity* refers to the detection of even minute quantities of antibodies or antigens in a specimen and reflects the degree to which a test will detect every positive person. Since no procedure will be 100% accurate all the time, tests are chosen for higher sensitivity or higher specificity depending on whether it is more important to avoid false-positive results or avoid false-negative results.

Visualizing Antigen-Antibody Interactions

To be useful in a clinical setting, antigen-antibody binding must be visible to the naked eye or be evident in a light microscope. If whole-human or microbial cells are subjected to antibody of the correct specificity, large clumps or aggregates form and can easily be seen. Since the formation of smaller antigen-antibody complexes may not be readily observed, tests requiring special indicators are used that employ dyes or fluorescent reagents to visualize the endpoint of reactions.

Agglutination and Precipitation Reactions

The essential differences between agglutination and precipitation, as seen in **table 17.1,** are in size, solubility, and location of the antigen. In agglutination, the antigens are whole cells such as red blood cells or microbes such as bacteria or viruses displaying surface antigens; in precipitation, the antigen examined is a soluble molecule. In both reactions, when antigen and antibody concentrations are optimal, one antigen is interlinked by several antibodies to form insoluble aggregates that settle out in solution.

Agglutination is easily seen because it consists of visible clumps of cells **(figure 17.11b)**. Agglutination tests are

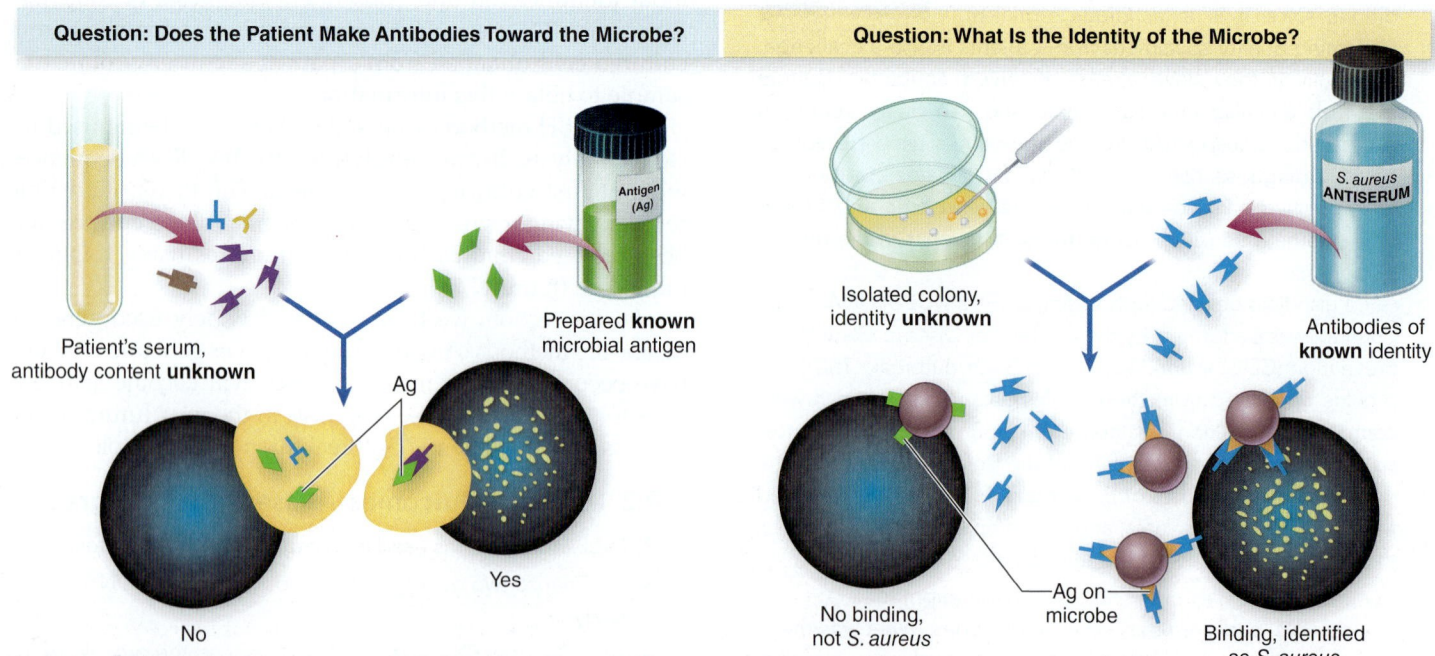

Figure 17.9 Basic principles of serological testing using antibodies and antigens.

Question: Does the Patient Make Antibodies Toward the Microbe?

Patient's serum, antibody content **unknown**

Prepared **known** microbial antigen

Ag

No

Yes

Question: What Is the Identity of the Microbe?

Isolated colony, identity **unknown**

S. aureus ANTISERUM

Antibodies of **known** identity

No binding, not *S. aureus*

Ag on microbe

Binding, identified as *S. aureus*

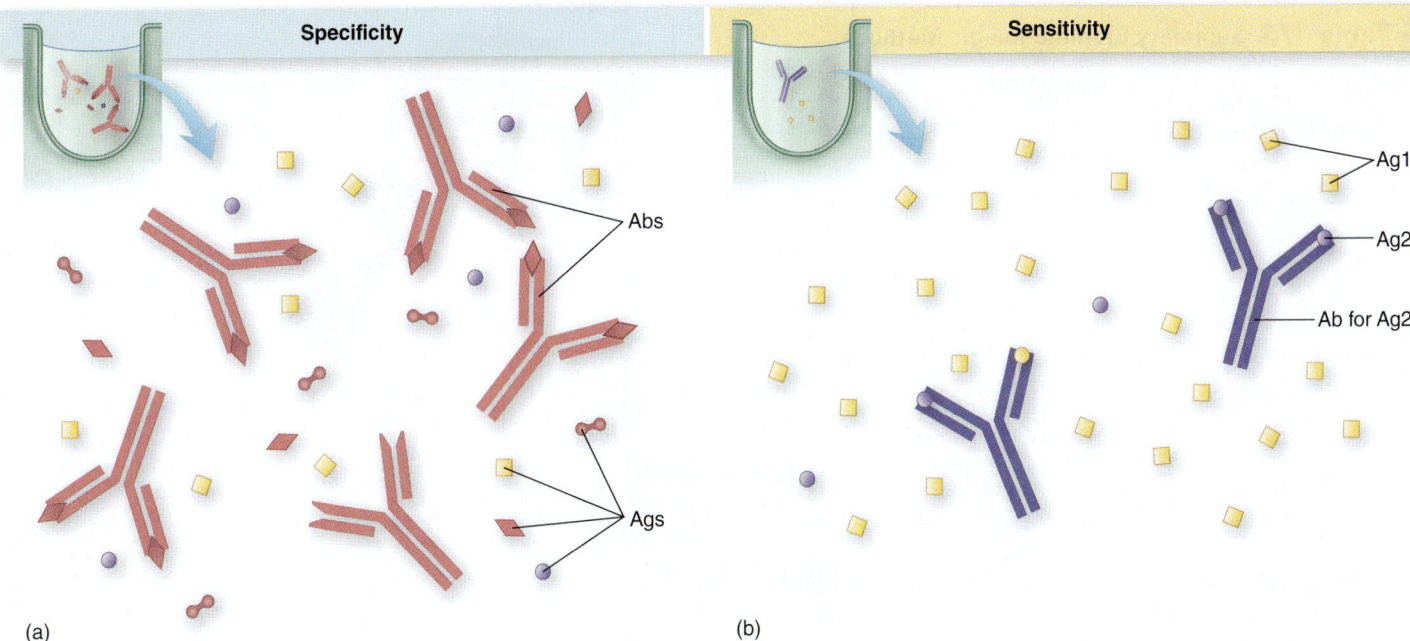

Figure 17.10 Specificity and sensitivity in immune testing. **(a)** This test shows specificity in which an antibody (Ab) attaches with great exactness with only one type of antigen (Ag). Often, high-specificity antibodies sacrifice a degree of sensitivity, meaning that they will miss some of the antigen of the correct identity. **(b)** Sensitivity is demonstrated by the fact that Ab can pick up antigens even when the antigen is greatly diluted. Often, high-sensitivity antibodies sacrifice some specificity, meaning that occasionally they will bind an incorrect antigen.

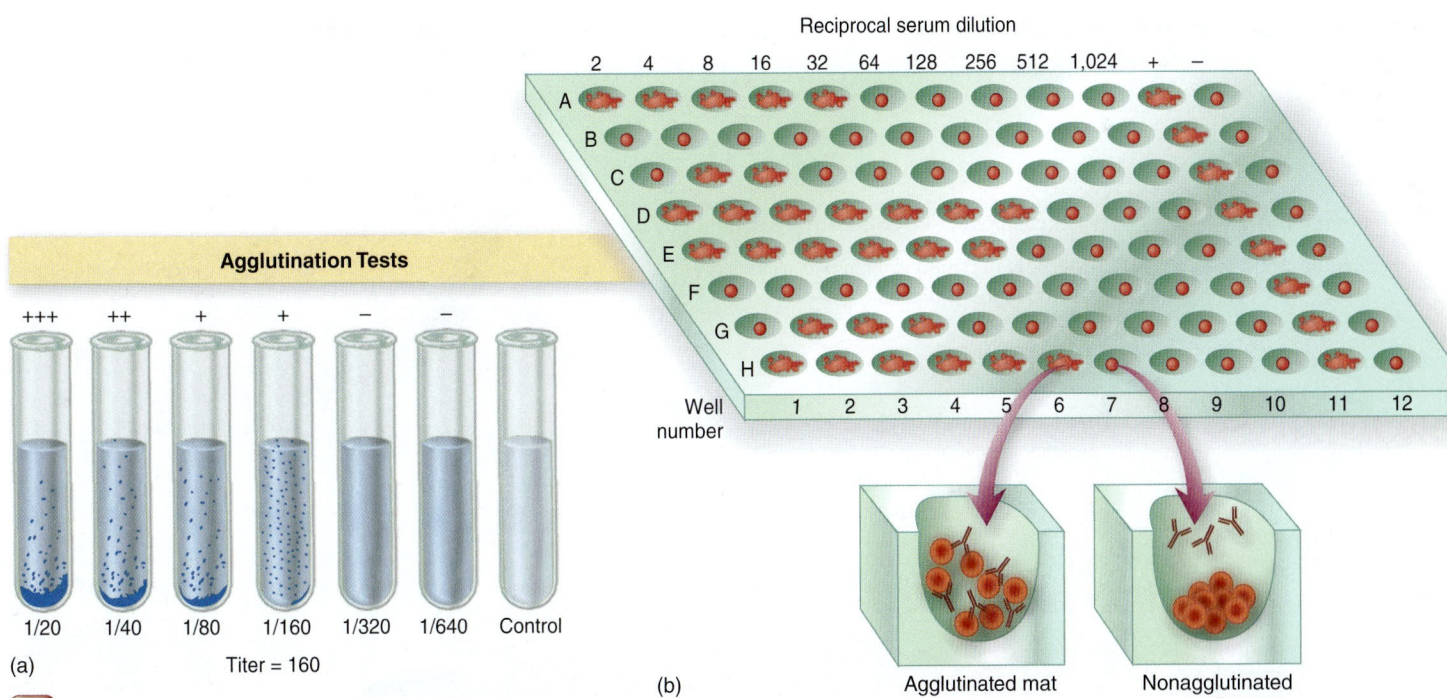

Figure 17.11 Agglutination tests. **(a)** Tube agglutination test for determining antibody titer. The same number of cells (antigen) are added to each tube, and the patient's serum is diluted in series. The titer in this example is 160 because there is no agglutination in the next tube in the dilution series (1/320). **(b)** A microtiter plate illustrating hemagglutination. The antibody is placed in the wells of all rows in columns 1 through 10. Positive controls (row 11) and negative controls (row 12) are included for comparison. Red blood cells are added to each well. If sufficient antibody is present to agglutinate the cells, they sink as a diffuse mat to the bottom of the well. If insufficient antibody is present, they form a tight pellet at the bottom.

Table 17.1 Summary of Immunologic Methods

Test	Description	Example
Agglutination	Involves antibody-mediated clumping of whole cells	Blood typing, Ab titering
Precipitation	Produces antibody-antigen complexes in a cell-free system	RPR test, serotyping
Western blot	Separation of proteins followed by antibody-mediated detection	HIV verification test
Immunofluorescence	Antibodies labeled with fluorescent dyes (Fabs)	
Direct	Unknown specimen is exposed to a known Fab solution	Meningitis, plague tests
Indirect	Fc region of patient's antibody binds with known Fab	FTA-ABS test for syphilis
Immunoassay	Sensitive, rapid tests for trace levels of antibody or antigen	
Radioimmunoassay	Measure of bound radioactively labeled antibodies or antigens	RAST or RIST for allergies
Immunochromatographic testing	Direct antigen and "dipstick" tests	HIV, *Leishmania*
ELISA	Colorimetric test to detect unknown antigen or antibody	*Helicobacter*, HIV screening
In vivo methods	Antigen or antibody introduced into patient to elicit a reaction	Tuberculin test

Microscopic view

performed routinely by blood banks to determine ABO and Rh (rhesus) blood types in preparation for transfusions. In this type of test, antisera containing antibodies against the blood group antigens on red blood cells are mixed with a small sample of blood and read for the presence or absence of clumping. Agglutination is also made possible by the use of latex beads coated with antibody. Latex agglutination tests are used to diagnose a variety of infectious diseases rapidly, including strep throat. In these tests, growth from microbial isolates or patient samples are applied directly to a preparation of the beads coated in microbe-specific antibody. If the microbe is present in the sample, antigen-antibody complexes form and agglutination of the beads can be easily visualized.

An antigen-antibody reaction that takes place in liquid is read as a **titer**, or the concentration of antibodies in a sample. Titer is determined by serially diluting patient serum into test tubes or wells of a microtiter plate, all containing equal amounts of cells (antigen) (**figure 17.11***a*). *Titer* is defined as the highest dilution of serum that still produces agglutination. In general, the more a serum sample can be diluted and still react with antigen, the greater the concentration of antibodies and thus its titer. Antibody titers are often used to diagnose autoimmune disorders such as rheumatoid arthritis and lupus, as well as to determine past exposure to certain diseases such as rubella.

In **precipitation** reactions, the soluble antigen is precipitated (made insoluble) by an antibody. This reaction is observable in a test tube in which antiserum has been carefully laid over an antigen solution. At the boundary of the two solutions, a cloudy or opaque zone forms. New immunoprecipitation methods for the detection of prions are also being developed today. Although precipitation is a useful detection tool, the precipitates are so easily disrupted in liquid media that most precipitation reactions are now conducted with antigen or antibody that is anchored to a large insoluble particle so that the reactions are visible without magnification (see table 17.1).

Serotyping is an antigen-antibody technique for identifying, classifying, and subgrouping certain bacteria into categories called serotypes (see table 17.1). This method employs antisera against cell antigens such as the capsule, flagellum, and cell wall. Serotyping is widely used in identifying *Salmonella* species and strains and is the basis for differentiating the numerous pneumococcal and streptococcal serotypes.

Disease Connection

One example of the precipitation technique is the Venereal Disease Research Laboratory (VDRL) test that detects antibodies produced when a person is infected with syphilis. There is an easier form of the test, however. The rapid plasma reagin (RPR) test for syphilis is basically the VDRL test made visible to the naked eye by attaching the antigen to carbon particles.

The Western Blot for Detecting Proteins

Western blotting involves the electrophoretic separation of proteins, followed by antibody-mediated detection of these proteins (**figure 17.12**). First, a sample of proteins from a bacterial cell or virus is separated via electrical charge within a gel. The proteins distributed throughout the gel are then transferred and immobilized to a special filter. The filter is incubated with a patient's serum (containing antibody). If the serum contains antibodies to the microbe, they will bind to the antigens on the filter paper. At this point the binding is not visible, since the patient's antibodies are not "labeled" in any way. To see if there are antibodies bound to the antigens, a second antibody—one that is designed to see the Fc portion of human antibody as the antigen—is applied to the filter paper. This secondary antibody has been engineered to contain a fluorescent or luminescent molecule or an enzyme that will turn a colorless substrate into a colored product. After incubation, sites of specific antigen-antibody binding will appear as a pattern of bands that can be compared with known positive and negative controls. It is a highly specific and sensitive way to identify or verify the presence of microbial-specific antigens or antibodies in a patient sample. Western blotting is the second (verification) test for preliminary antibody-positive HIV screening tests and has significant applications for detecting microbes and their antigens in specimens.

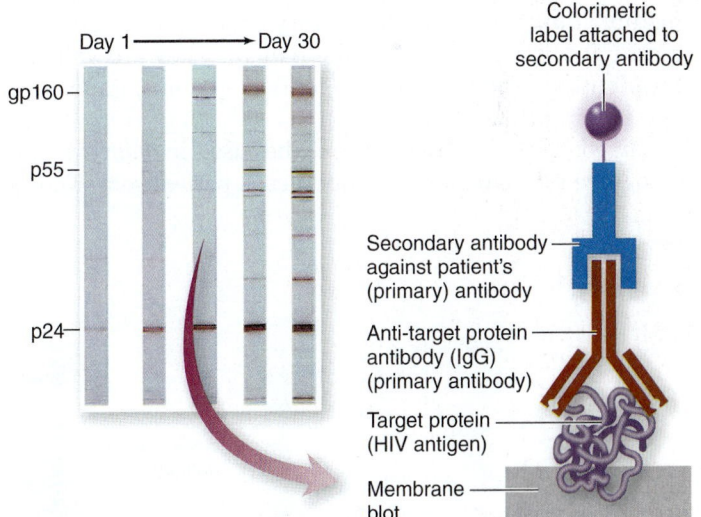

Figure 17.12 The Western blot procedure.   Major HIV surface and core antigens were separated via electrophoresis and transferred to a filter. The filter was incubated with the patient's sera collected on 5 separate days over the course of a month. Sera contain primary antibodies. Secondary antibodies (blue in the drawing) and a colorimetric label were added to visualize the patient's bound antibody (maroon in the drawing). HIV antigen-antibody complexes are visualized as bands (labels correspond with glycoproteins [gp] or proteins [p] that are part of HIV-1 antigen structure).

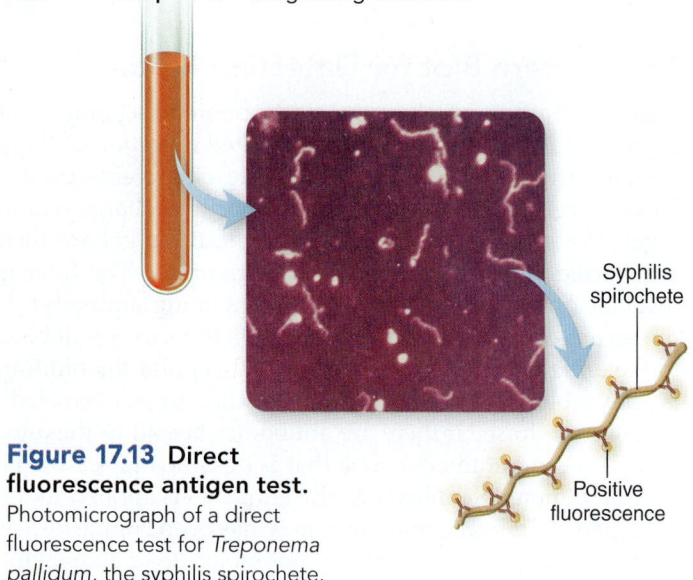

Figure 17.13 Direct fluorescence antigen test. Photomicrograph of a direct fluorescence test for *Treponema pallidum*, the syphilis spirochete.

Syphilis spirochete

Positive fluorescence

Immunofluorescence Testing

The fundamental tool in immunofluorescence testing is a fluorescent antibody—an antibody labeled by a fluorescent dye. Fluorescent antibodies can be used for diagnosis in two ways. In *direct testing*, an unknown test specimen or antigen is fixed to a slide and exposed to a Fab solution of known composition. If antigen-antibody complexes form, they will remain bound to the sample and can be visualized by fluorescence microscopy, thus indicating a positive result **(figure 17.13)**. Direct fluorescence antibody tests are particularly useful in the analysis of organisms that are not readily cultivated in the laboratory or if rapid diagnosis is essential for the survival of the patient. They are valuable for identifying the causative agents of syphilis, gonorrhea, and meningitis, among others.

In contrast, fluorescent antibodies used in *indirect testing* recognize the Fc region of antibodies in patient sera. Known antigen (i.e., bacterial cells) is added to the test serum of unknown antibody content. Binding of the fluorescent-tagged antibody is visualized through fluorescence microscopy. Fluorescing aggregates or cells indicate that the Fabs have complexed with the microbe-specific antibodies in the test serum.

Radioimmunoassay (RIA)

Antibodies or antigens labeled with a radioactive isotope can be used to pinpoint minute amounts of a corresponding antigen or antibody (see table 17.1). Although very complex in practice, these assays compare the amount of label present in a sample before and after incubation with a known, labeled antigen or antibody. The amount of radioactivity is measured with an isotope counter **(figure 17.14)**. Radioimmunoassays are often used to detect hormone levels in samples and to diagnose allergies in patients. In the latter case, the procedure is often called the radioallergosorbent test (RAST).

Immunochromatographic Testing

Another way in which samples can be analyzed is through *direct antigen testing*, a technique similar to direct fluorescence in that known antibodies are used to identify antigens on the surface of bacterial isolates or patient specimens. In direct antigen testing, however, the reactions can be seen with the naked eye. Lateral flow immunochromatographic tests were first developed for home testing for pregnancy but are now commonplace in point-of-care testing in health care clinics. Such quick tests greatly speed clinical diagnosis and are available for diagnosing infections caused by *Staphylococcus aureus, Streptococcus pyogenes, Neisseria gonorrhoeae, Haemophilus influenzae,* and *Neisseria meningitidis,* among others. However, when the microbes are very sparse in the specimen, direct testing is like looking for a needle in a haystack and more sensitive methods are necessary.

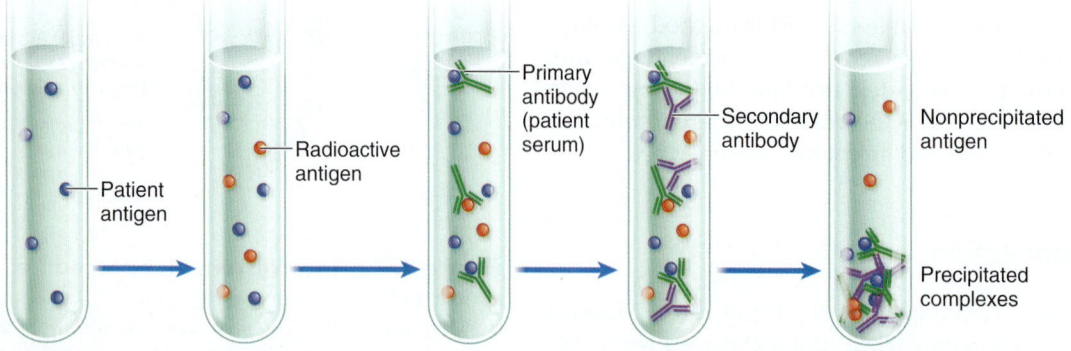

Patient antigen

Radioactive antigen

Primary antibody (patient serum)

Secondary antibody

Nonprecipitated antigen

Precipitated complexes

Figure 17.14 Radioimmunoassay. The patient's serum, containing unlabeled antigen, is added to buffer in a tube. Next, a known amount of antibody and radioactive antigen is added. The extent to which the patient's antigen displaces the labeled antigen indicates the concentration of the antigen in the patient's serum. The secondary antibody, which binds to the Fc portion of the primary antibody, acts to precipitate the antigen-antibody complexes. The concentration of "free" antigen in the soluble phase can then be measured.

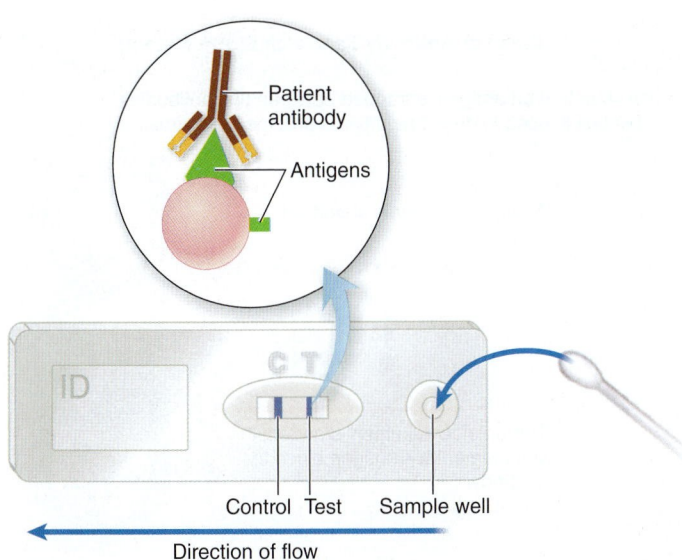

Figure 17.15 An immunochromatographic test relying on antigen-antibody binding. This is a positive test for *Neisseria gonorrhoeae*.

Immunochromatographic "dipstick" tests are also available to analyze specimens for the presence of antigen-specific antibody. In this case, known antigen is immobilized within the cartridge and the sample (i.e., saliva, serum) is collected on its filter paper tip. As the sample migrates along the length of the cartridge, it may interact with the bound antigen producing a color change in positive samples. Such tests are used for rapid HIV screening **(figure 17.15)** and for identifying *Leishmania* and *Trypanosoma* infections in developing countries.

Enzyme-Linked Immunosorbent Assay (ELISA)

The **ELISA** test, also known as enzyme immunoassay (EIA), uses an enzyme-linked indicator antibody to visualize antigen-antibody reactions. This technique also relies on a solid support such as a plastic microtiter plate that can *adsorb* (bind to its surface) the reactants **(figure 17.16)**.

The *indirect ELISA* test detects microbe-specific antibodies in patient sera. A known antigen is adsorbed to the surface of a well and mixed with the patient's serum **(figure 17.16a)**. If an antigen-antibody complex forms, the patient's antibody will remain in the well even after being rinsed with solution. A secondary antibody (with enzyme) is then added. It will bind to the Fc portion of the patient's antibody. Then, when the correct (colorless) substrate for the enzyme is added, the substrate is acted on by the enzyme. The product is a colored substance, so the color change in the well indicates a positive result. If the patient's serum did not contain the correct

antibody, then the secondary antibody will just be rinsed out of the well; and when the colorless substrate is added, there will be no enzyme to produce a colored product. This is the common test used for antibody screening for HIV, various rickettsial species, hepatitis A and C, and *Helicobacter*. Because false positives can occur, a verification test (such as a Western blot) may be necessary.

In *direct ELISA* (or sandwich) tests, a known antibody is adsorbed to the bottom of a well and incubated with an unknown antigen **(figure 17.16b)**. If an antigen-antibody complex forms, it will attract the indicator antibody, and color will develop in these wells, indicating a positive result. The direct technique is used to detect antibodies to hantavirus, rubella virus, and *Toxoplasma* in patient sera.

Newer versions of this technique involve an enzyme called alkaline phosphatase, which produces visible light instead of a color change. The light is detected or quantified by machines and photographic films.

In Vivo Testing

In practice, *in vivo* (meaning "taking place in or on the body") tests employ principles similar to serological tests, except in this case an antigen or antibody is introduced into a patient to elicit some sort of visible reaction. The **tuberculin test,** in which a small amount of purified protein derivative (PPD) from *Mycobacterium tuberculosis* is injected into the skin, is a classic example (see table 17.1). The appearance of a red, raised, thickened lesion in 48 to 72 hours can indicate previous exposure to tuberculosis (shown in figure 16.15). Similar diagnostic skin tests are useful for evaluating infections due to fungi (coccidioidin and histoplasmin tests, for example) or allergens.

17.4 Learning Outcomes—Assess Your Progress

11. Define the term *serology,* and explain the immunologic principle behind serological tests.
12. Differentiate between sensitivity and specificity.
13. Explain how fluorescent antibodies can be used in the diagnosis of microbial disease.
14. Compare and contrast agglutination and precipitation reactions, and provide one example of how each is used to diagnose infectious disease.
15. Summarize the process of Western blotting, and explain how it can be used in microbial identification.
16. Describe two rapid immunologic methods for the detection of either microbial antigens or microbe-specific antibodies within a specimen.
17. Explain the difference between a direct and an indirect ELISA, and provide a clinical application for each.

| **Indirect ELISA** | **Direct or Antibody Sandwich ELISA Method** |

(a) Comparing a positive versus negative reaction. This is the basis for HIV screening tests.

(b) Note that an antigen is trapped between two antibodies. This test is used to detect hantavirus and measles virus.

Well A   Well B

Known antigen is adsorbed to well.

Serum samples with unknown antibodies applied.

Well is rinsed to remove unbound (nonbinding) antibodies.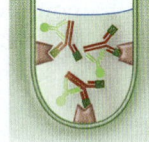

Indicator antibody outfitted with an enzyme attaches to any bound antibody.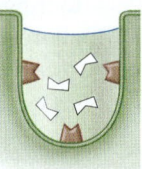

Wells are rinsed to remove unbound indicator antibody. A colorless substrate for enzyme is added.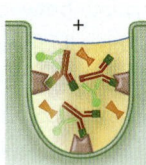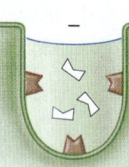

Enzymes linked to indicator Ab hydrolyze the substrate, which releases a dye. Wells that develop color are positive for the antibody; colorless wells are negative.

\+   –

Applied antibody is absorbed to well.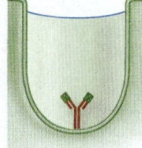

The unknown antigen is added; if complementary, antigen binds to antibody.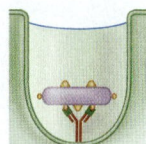

Enzyme-linked antibody specific for test antigen then binds to antigen, forming a sandwich.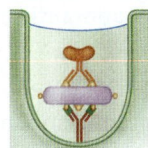

Enzyme's substrate (□) is added, and reaction produces a visible color change (■).

Microtiter ELISA Plate with 96 Tests for HIV Antibodies

Colored wells indicate a positive reaction.

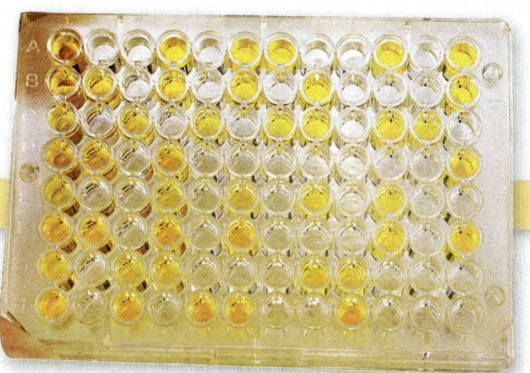

 Figure 17.16 Methods of ELISA testing. **(a)** Indirect ELISAs detect patient antibody. Here, both positive and negative tests are illustrated. **(b)** Direct ELISAs detect microbial antigens. Here, only a positive reaction is illustrated.

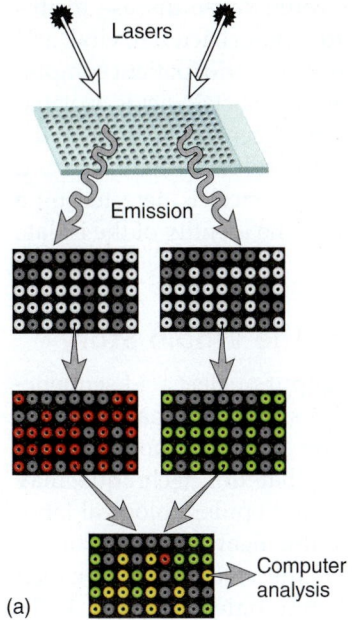

Lasers

Emission

Computer analysis

(a)

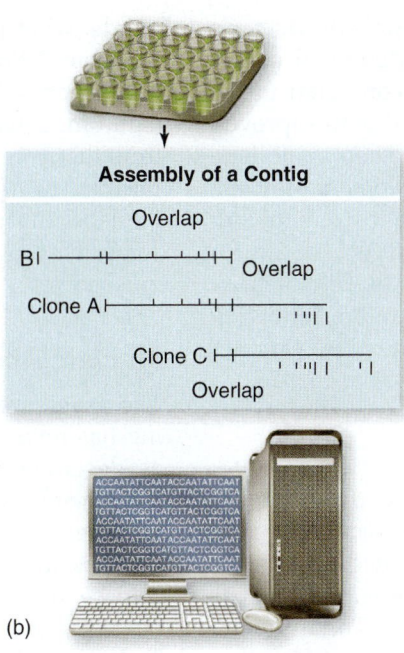

Assembly of a Contig

Overlap

B⊢

Overlap

Clone A⊢

Clone C⊢

Overlap

(b)

Figure 17.17 Overview of emerging diagnostic technologies. **(a)** Microarrays. **(b)** Nucleic acid sequencing. **(c)** Mass spectrometry. **(d)** Lab-on-a-chip.

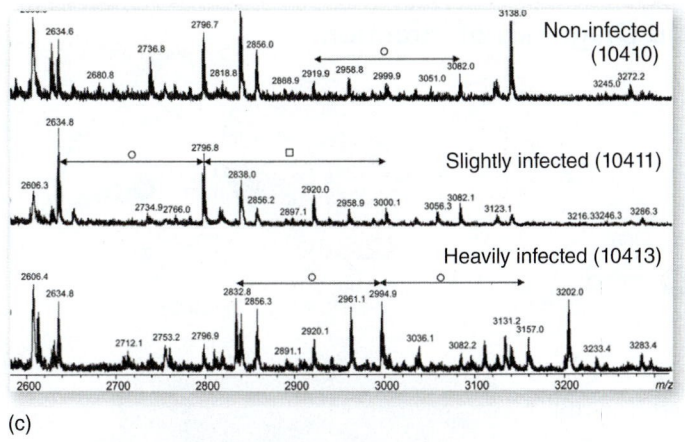

Non-infected (10410)

Slightly infected (10411)

Heavily infected (10413)

(c)

(d)

17.5 Breakthrough Methodologies

Diagnostic microbiology has entered a new era. A vast array of new technologies in the area of genetics, physics, and information science (in the form of massive databases) has led the medical profession to begin to adopt radically new diagnostic techniques **(figure 17.17).** While the gold standard of diagnosis has always been growing an isolated culture of the offending microbe (and then identifying it), that technique has its downfalls. What if the microbe you isolate and grow is not the cause of the disease but just a bystander? Also, culturing takes time—18 hours minimum—and many organisms require much longer incubation times. In addition, more and more, we realize that many infections are polymicrobial. Our single-minded efforts to isolate one disease-causing organism can result in serious misdiagnoses.

Take the very common example of infection of the bloodstream (septicemia). This is a condition that can kill very quickly. The critical time frame for appropriate management is estimated to be less than 6 hours. However, traditionally, the course of events is as follows:

- A blood sample is drawn and inoculated into a blood culture tube.
- Broad-spectrum antimicrobials are begun.
- After 18 to 24 hours, identification of bacteria in blood culture is attempted and antimicrobial sensitivities determined.
- More appropriate antimicrobial therapy is instituted.

Studies show that during the 18- to 24-hour incubation period, if the patient is improving based on the broad-spectrum antimicrobial he or she is receiving, many physicians do not rewrite the antibiotic order for a more appropriate drug. In the cases in which the patients do not improve, the delay in changing the antimicrobial is frequently fatal.

For these reasons, there is a call for more sophisticated diagnostic techniques that can occur immediately and, often, at the point-of-care: at the patient's bedside or in the doctor's office. Although many of the techniques described earlier in this chapter are relatively inexpensive compared to the newer methods, the consensus seems to be that improved patient outcomes with the new tests will soon drive hospitals and clinics to adopt the new technologies. After all, infectious disease specialists point out that even though diagnostic tests influence approximately 70% of health care treatment decisions, currently only 2% of U.S. health care costs are expended on them. Therefore, following is an overview of the most likely new diagnostic techniques to be seen in the next 5 years in U.S. health care.

Microarrays

Microarrays designed for infectious disease diagnosis are "chips" (absorbent plates) that contain gene sequences from potentially thousands of different possible infectious agents, selected based on the syndrome being investigated (such as respiratory infection or meningitis symptoms). The arrays are selected based on a very large differential diagnosis; in other words, what possible microbes could cause disease in this syndrome? Arrays can be made to contain bacterial, viral, and fungal genes in a single test. In this scenario, patient samples (sputum, cerebrospinal fluid) or the nucleic acids isolated from them are incubated with labeled gene sequences on the microarray. Matching sequences hybridize to the chip, and the label (in most cases, it is fluorescence) is detected by a computer program, which provides the identity of the isolate or isolates.

Nucleic Acid Sequencing: The Whole Story

The development of high-throughput nucleic acid sequencing has revolutionized the analysis of the human genome, as described in chapter 10. The cost of whole-genome sequencing is becoming so low that this technique may become commonplace in clinical and epidemiological laboratories around the world in the near future. Nucleic acid sequencing has also led to the creation of so-called next-generation sequencing technologies **(Insight 17.2)**. Techniques like *random amplified polymorphic DNA analysis*

INSIGHT 17.2   The Human Microbiome Project and Diagnosis of Infection

For decades, scientists have realized that culturing microbes for identification is not only time-consuming and inefficient but also incomplete. Many microbes are fastidious, requiring highly specialized growth conditions, and many others are considered "viable but nonculturable" (VBNC). Some estimate that 99.9% of sequences found in the human body fall under the heading of VBNC, and little is understood about the role that these elusive microbes play in the human microbiome and in disease.

Using techniques developed by J. Craig Venter in mapping the genome of microbes in the oceans, scientists have been able to map the genomes of VBNC microorganisms in the Human Microbiome Project (HMP). Two important technologies have been developed to amplify and analyze the DNA from these microbes. First, scientists at the Venter Institute developed multiple displacement amplification (MDA) technology. MDA is able to copy fragments of DNA copies from a single cell until they reach the equivalent of the billions required for analysis. MDA is able to capture 90% of the genes from a single cell.

Using MDA, scientists are able to obtain information about susceptibility to antibiotics, signaling proteins used, and even how a VBNC microbe lives and moves. All of this information is vital to the HMP. As you read in chapter 13, in the HMP, researchers sampled microbes from the mouth, nose, skin, intestine, and genitals of 242 volunteers (129 male and 113 female), resulting in over 5,000 samples from body sites. Using MDA, they collected about 3.5 terabases* of DNA sequences that have been entered into a gigantic comparative analysis system called the Integrated

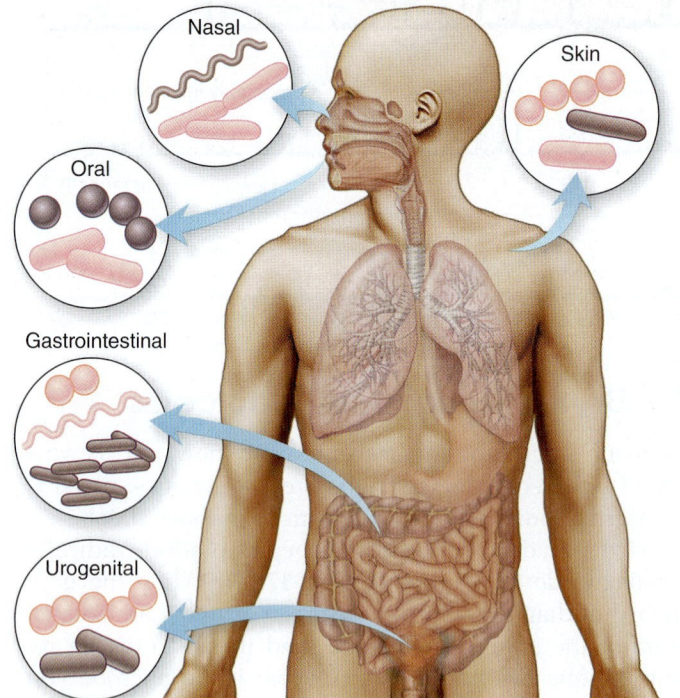

Microbial Genomes and Metagenomes for the Human Microbiome Project (IMG/M HMP).

This massive accumulation of genetic information has given researchers insight into how microbes play a role in our heath and in microbial disease. This research represents a new frontier of science—breaking through the barrier of traditional culture techniques to genetic analysis of the full human microbiome.

*A *terabase* is a genetic sequence of 10^{12} nucleic acid bases.

(described in the section on genetic identification) uses random primers to allow for the identification of novel nucleic acid sequences. More importantly, a single genome can be scanned and analyzed multiple times in a process called *deep sequencing,* which minimizes errors. Some scientists suspect that these types of sequencing will become so cheap and so routine that we will soon just "sequence everything" from a patient sample to find the one or more microbes causing symptoms. Others believe that microarrays or other technologies will take over in the race for better diagnostics.

Another potential next-generation sequencing technology is the analysis of the **transcriptome,** or all RNA molecules formed from the genetic information within a microbe. Deep sequencing of these areas of the genome will continue to provide vital information regarding microbial evolution, pathogenesis, and drug treatment.

Mass Spectrometry

Mass spectrometry has been used for years to determine the structure and composition of various chemical compounds and biological molecules. Analysis of samples by mass spectrometry is poised to become the new cutting-edge technology for providing rapid and highly accurate microbial identification within just minutes. This technique, which is often called MALDI-TOF,[1] can be used to analyze a protein fingerprint from pure culture isolates or directly from patient specimens. It works by adding the patient sample to a metal plate and then striking it with a laser. This causes the sample to become ionized. The ions from the sample are guided into a machine that separates them and identifies them according to their mass-to-charge ratio. This technology has been applied to the identification of bacteria, viruses, and fungi so far; it may become commonplace in many clinical and research laboratories due to its ability to produce rapid, precise, and cost-effective results compared to conventional phenotypic, genotypic, and immunologic methods.

Both the microarray and mass spectrometry technologies can also be used to simultaneously detect antibiotic susceptibilities—obviously, an important advantage.

Lab-on-a-Chip

The newest cutting-edge technology employs detection systems that contain computer chips to measure minute changes in electrical current that occur when antigen-antibody complexes are formed. The potential for sensitivity here is extreme. Tiny amounts of fluids are required, and it is thought that as few as 12 molecules can be detected in a sample. These technologies are being called "Lab-on-a-Chip." There is great hope that this technology can be commercialized and used especially in developing countries.

Imaging

An old way of diagnosing infections, which found use in only occasional infections, involves various imaging techniques. Infections associated with hip implants, for example, may be difficult to access through blood samples. The bacteria may be growing in biofilms on the implanted materials, or they may be growing in an abscess deep in the hip joint. Magnetic resonance imaging, computerized tomography (CT) scans, and positron emission tomography (PET) scans have been increasingly employed to find areas of localized infection in deep tissue, which can later be biopsied to aspirate samples for culture. In the event no infection is found on the image, the patient has been spared an invasive procedure. This seems very new and "high tech," but imaging in the form of X rays has been used for centuries in the diagnosis of tuberculosis.

17.5 Learning Outcomes—Assess Your Progress

18. Explain why isolating a pathogen through standard culture methods may become an outdated diagnosis strategy.

19. List the major advantages of microarray methods of diagnosis.

20. Summarize the benefits of using whole-genome sequencing of patient samples for disease diagnosis.

21. Describe how a pathogen's protein fingerprint can be used to determine its identity and disease diagnosis.

Case File 17   *Wrap-Up*

According to the CDC, the number of cases of pertussis in the United States in 2012 surpassed the total number of cases in the previous 5 years. In the Washington State outbreak, there were multiple strains of *B. pertussis* causing infections, even in individuals who had received all of the standard vaccines. Children who are vaccinated can still develop pertussis, but they experience milder symptoms, are less likely to transmit disease, and have a shorter illness. Children who are not vaccinated are eight times as likely to be infected with pertussis. In 1997, there was a switch in the DTaP vaccine from a whole-cell vaccine to an acellular vaccine with only toxoids. Researchers suggest that the acellular vaccine may not last as long as the whole-cell vaccine, which is why there was a greater incidence among 13- to 14-year-old patients. Additionally, according to a federal study, Washington State had the highest percentage of parents who opted out of vaccinations for their children. The combination of waning vaccine efficacy and an under immunized population may be the cause for the epidemic of whooping cough in Washington State.

Source: 2012. CDC *Morbidity and Mortality Weekly Report* vol. 61, no. 28, p. 517.

---

1. The full name for this mass spectrometry technique is "matrix-assisted laser desorption/ionization time of flight," which is why it is always abbreviated.

Chapter Summary

17.1 Identifying the Infectious Agent (ASM Guidelines* 2.1, 5.4, 6.3, 7.1, 8.3, 8.5, 8.6)

- Microbiologists use three categories of techniques to diagnose infections: *phenotypic, genotypic,* and *immunologic.*
- The first step in clinical diagnosis (after observing the patient) is obtaining a sample. If this step is not performed correctly, specimen analysis will not be accurate no matter how good the test.

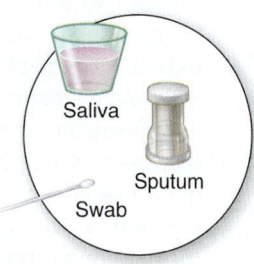

Saliva

Sputum

Swab

17.2 Phenotypic Methods (ASM Guidelines 2.1, 5.4, 6.3, 7.1, 8.3, 8.5, 8.6)

- The main phenotypic methods include the direct examination of specimens, observing the growth of specimen cultures on special media, and biochemical testing of pure cultures.

17.3 Genotypic Methods (ASM Guidelines 2.1, 5.4, 6.3, 7.1, 8.3, 8.5, 8.6)

- The use of genotypic methods in microbial identification has grown exponentially.
- Polymerase chain reaction has quickly become a standard diagnostic technique.
- Hybridization techniques exploit the base-pairing characteristics of nucleic acids.
- Pulse-field gel electrophoresis and ribotyping have specific applications in diagnosis.

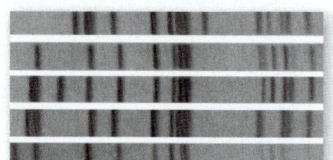

17.4 Immunologic Methods (ASM Guidelines 2.1, 5.4, 6.3, 7.1, 8.3, 8.5, 8.6)

- Serological tests can be performed on a variety of body fluids or tissues and are based on the principle that antibodies have extreme specificity for antigens.
- Testing for microbial-specific antigens or antibodies is typically performed *in vitro*, and antigen-antibody interactions are made macroscopically or microscopically visible.
- Agglutination reactions occur between antibody and antigens bound to cells or latex beads, resulting in visible clumping, which is the basis of determining titer, or antibody concentration, in patient sera.

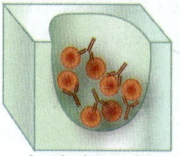 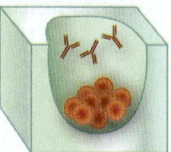

Agglutinated mat   Nonagglutinated pellet

Enlarged side view of wells

- Precipitation reactions also occur between antibody and antigen and produce insoluble, visible precipitates, but they are typically made visible by adding radioactive or enzyme markers.
- In immunoelectrophoretic techniques such as the Western blot, proteins that have been separated by electrical current are identified by labeled antibodies.
- Direct fluorescence antibody tests indicate presence of microbial antigens; indirect fluorescence tests indicate the presence of microbe-specific antibodies.
- Radioimmunoassays can detect very small quantities of antigen, antibody, or other substances and use dyes or radioactive isotopes to visualize antigen-antibody complexes.
- Immunochromatographic tests are used in rapid point-of-care testing for microbial antigens or microbe-specific antibodies in clinical specimens.
- The ELISA test is widely used to detect microbial antigens (direct methods) or microbe-specific antibodies (indirect method) in patient samples.
- *In vivo* serological testing, such as the tuberculin test, involves injection of antigen to elicit a visible immune response in the host.

17.5 Breakthrough Methodologies (ASM Guidelines 2.1, 5.4, 6.3, 7.1, 8.3, 8.5, 8.6)

- The next 5 years are likely to bring many new technologies to the widespread diagnosis of infectious diseases.
- Whole-genome sequencing relies on DNA sequencing of microbes.
- Mass spectrometry detects microbes via their protein fingerprints.

Source: ASM Curriculum Guidelines (American Society for Microbiology, 2012). Complete guidelines in appendix B of this book.

Multiple-Choice and True-False Questions | Bloom's Levels 1 and 2: Remember and Understand

Multiple-Choice Questions. Select the correct answer from the options provided.

1. The most likely interpretation of the isolation of two colonies of *E. coli* on a plate streaked from a urine sample is
   a. probable infection.
   b. normal biota.
   c. contamination.

2. The most likely interpretation of the isolation of 50 colonies of *Streptococcus pneumoniae* is
   a. probable infection.
   b. normal biota.
   c. contamination.

3. The most likely interpretation of the isolation of 80 colonies of various streptococci on a culture from a throat swab is
   a. probable infection.
   b. normal biota.
   c. contamination.

4. The most likely interpretation of colonies of black bread mold on selective media used to isolate bacteria from stool is
   a. probable infection.
   b. normal biota.
   c. contamination.

5. Which of the following methods can identify different strains of a microbe?
   a. microscopic examination
   b. radioimmunoassay
   c. DNA typing
   d. agglutination test

6. In agglutination reactions, the antigen is a
   _____; in precipitation reactions, it is a
   _____.
   a. soluble molecule, whole cell
   b. whole cell, soluble molecule
   c. bacterium, virus
   d. protein, carbohydrate

7. A patient with a _____ titer of antibodies to an infectious agent generally has greater protection than a patient with a _____ titer.
   a. high, low   c. negative, positive
   b. low, high   d. old, new

8. Direct immunofluorescence tests use a labeled antibody to identify
   a. an unknown microbe.   c. fixed complement.
   b. an unknown antibody.   d. agglutinated antigens.

9. The Western blot test can be used to identify
   a. unknown antibodies.
   b. unknown antigens.
   c. specific DNA.
   d. both a and b.

10. Which of the following methods looks for protein signatures?
    a. Nucleic acid sequencing
    b. Serological methods
    c. Mass spectrometry
    d. Ribotyping

True-False Questions. If the statement is true, leave as is. If it is false, correct it by rewriting the sentence.

11. A PNA FISH test utilizes both fluorescence and nucleic acids.

12. DNA probes are used to search for complementary segments of DNA.

13. Biochemical identification methods are based on a microbe's utilization of nutrients.

14. All microorganisms that grow from a clinical sample should be considered significant.

15. The differential diagnosis drives the selection of genes placed on a microarray chip.

Critical Thinking Questions | Bloom's Levels 3, 4, and 5: Apply, Analyze, and Evaluate

Critical thinking is the ability to reason and solve problems using facts and concepts. These questions can be approached from a number of angles and, in most cases, they do not have a single correct answer.

1. Explain why specimens should be taken aseptically, even when nonsterile sites are being sampled and selective media are to be used, and why speed is important in the clinical testing process.

2. Summarize in order the general steps one would take to identify the exact strain of the bacterium causing a patient's infection, using one particular methodology as your example.

3. Differentiate between the serological tests used to identify isolated cultures of pathogens and those used to diagnose disease from patient sera.

4. Which category of identification methods does lipid analysis belong to, and how it could be used to differentiate respiratory disease caused by *Streptococcus pneumoniae* infection from *Mycobacterium tuberculosis* infection?

5. a. Define the term *seropositivity*, and explain whether or not it develops at the same rate in all patients affected by the same microbe.
   b. Would a high rate of false-positive reactions result from a decrease in the sensitivity or specificity of the test?

6. a. Explain which type of ELISA can be used to determine an individual's past exposure to a pathogen.
   b. Discuss whether immunochromatographic assays are indirect or direct methods of testing.

7. Discuss how new information on the human microbiome may lead to the identification of specific microbes associated with diseases currently thought to be noninfectious in origin.

8. You are working at a health clinic, and a woman enters suspecting that she has been exposed to HIV two nights ago.
   a. Discuss whether or not she can be tested for HIV infection at this point.
   b. Summarize how you would respond to this patient, providing her with appropriate information regarding testing for HIV infection.

9. a. Compare and contrast the process of restriction analysis used in traditional DNA fingerprinting with the procedure used in pulse-field gel electrophoresis.
   b. What is the "basic local alignment search tool" (BLAST) maintained by the National Institutes of Health? How can this database be used in genotypical analysis for microbial identification?

10. Conduct additional research and summarize two examples illustrating how the transcriptome is being used today in disease diagnosis.

Concept Connections | Bloom's Levels 4 and 6: Analyze and Create

This activity ties together multiple concepts in the chapter.

Scenario: You're still not quite back to normal after your bout with *Salmonella*, and your mother convinces you to go to the doctor. After taking your history and doing an examination, the doctor fills out paperwork for you to have some lab work done, concerned that you might have been infected with a particularly virulent strain of *Salmonella*. When you look at the paperwork, you notice that you will have to give a stool sample and have some blood drawn. As a microbiology student studying chapter 17 in this textbook, you're interested in what tests will be done. When you go into the lab clinic to have the tests performed, you ask the technician about the tests they will perform on your blood and stool.

1. What direct tests will be performed on the stool sample?
2. What culture and isolation tests will be performed on the stool sample?
3. What other tests may be performed after isolating *Salmonella* from your stool sample?
4. The lab technician tells you that in addition to a complete blood count (CBC), the blood sample will be used to test for levels of IgM antibody against *Salmonella*. What test will be performed to detect antibodies in your blood?

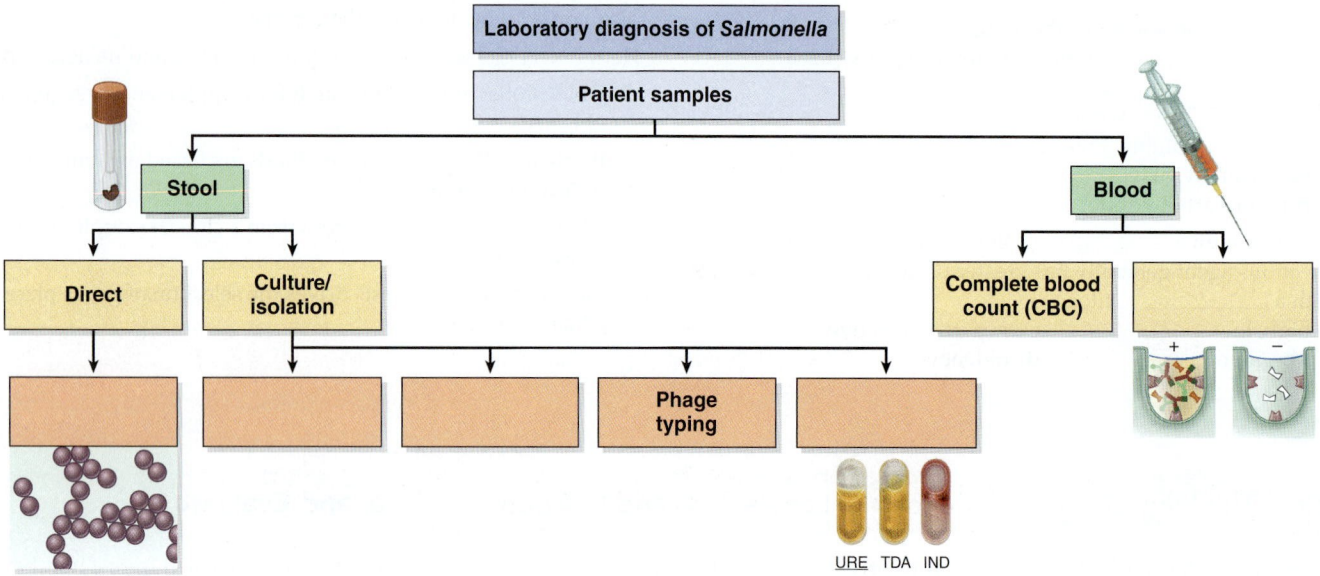

Visual Connections | Bloom's Level 5: Evaluate

These questions use visual images or previous content to make connections to this chapter's concepts.

1. **From chapter 3, figure 3.5b.** What biochemical characteristic does this figure illustrate? How could this characteristic be used to begin the identification of these organisms? Explain your answer.

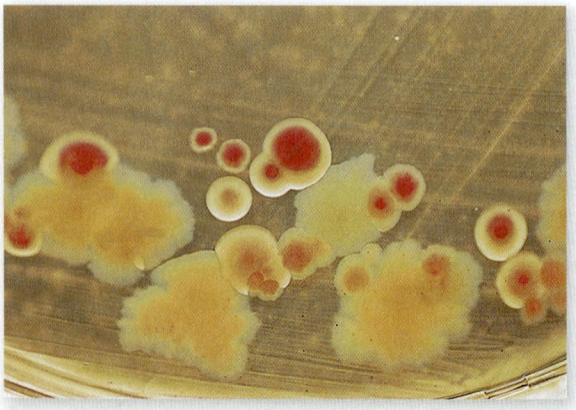

2. **From chapter 16, process figure 16.16b.** Imagine that this patient is being seen by his or her physician for this unknown rash. What rapid phenotypic test could suggest that this condition is caused by a bacterium? Could a rapid immunoassay or fluorescence procedure be used to identify a specific viral cause? Explain your answer.

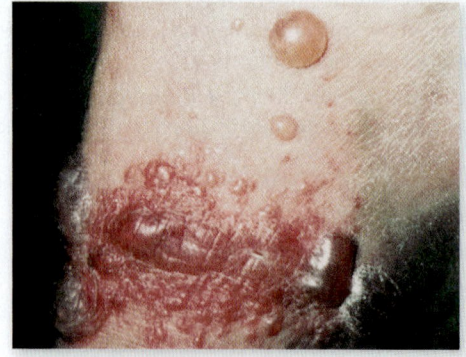

Concept Mapping | Bloom's Level 6: Create

Appendix D provides guidance for working with concept maps.

1. Using the words that follow, please create a concept map illustrating the relationships among these key terms from chapter 17.

direct testing methods

indirect testing methods

phenotypic methods

genotypic methods

immunologic methods

presumptive data

confirmatory data

microscopic, macroscopic, and
biochemical analysis

PCR, nucleic acid sequencing, and
rRNA analysis

ELISA and Western blotting

MALDI-TOF

www.mcgrawhillconnect.com

Enhance your study of this chapter with study tools and practice tests. Also ask
your instructor about the resources available through ConnectPlus, including the
media-rich eBook, interactive learning tools, and animations.

The New York City skyline. Between 1997 and 2006, CA-MRSA infections nearly tripled in New York City, especially among children, individuals with HIV or diabetes, and people who were homeless.

Infectious Diseases Affecting the Skin and Eyes

Case File 18

Community-Acquired MRSA in the Big Apple

For decades, methicillin-resistant *Staphylococcus aureus* (MRSA) has been associated with hospitals. In the hospital setting, there is a higher concentration of infected and compromised patients that require antibiotics and a greater likelihood that microbes exposed to antibiotics will develop resistance through mutation or horizontal gene transfer. Until fairly recently, individuals outside the hospital setting did not have to worry about MRSA. However, in the past several years, reports have emerged of MRSA in the community. Young, healthy athletes, among others, have developed deadly cases of MRSA. Clinicians and researchers have now begun distinguishing between hospital-acquired MRSA (HA-MRSA) and community-acquired MRSA (CA-MRSA).

CA-MRSA usually begins as an innocuous skin infection such as an insect bite, scrape, cut, or pimple and may develop into a more serious deep tissue or bloodstream infection. In a recent study published in the July 2012 issue of *Infection Control and Hospital Epidemiology*, the number of people admitted to the hospital with CA-MRSA skyrocketed in New York City from 1997 to 2006. In 1997, there were approximately 1.47 admissions to hospitals for CA-MRSA per 100,000 people; in 2006, there were 10.65 admissions per 100,000 people.

- ■ What was the cause of the increase in CA-MRSA hospital admissions?
- ■ What groups are most susceptible to CA-MRSA?

Continuing the Case appears on page 521.

Outline and Learning Outcomes

18.1 The Skin and Its Defenses
    1. Describe the important anatomical features of the skin.
    2. List the natural defenses present in the skin.

18.2 Normal Biota of the Skin
    3. List the types of normal biota presently known to occupy the skin.

18.3 Skin Diseases Caused by Microorganisms

4. List the possible causative agents for each of the infectious skin conditions: acne, impetigo, cellulitis, staphylococcal scalded skin syndrome, gas gangrene, vesicular/pustular rash diseases, maculopapular rash diseases, wart-like eruptions, large pustular skin lesions, and cutaneous mycoses.
5. Identify which of these conditions are transmitted to the respiratory tract through droplet contact.
6. List the skin conditions for which vaccination is recommended.
7. Summarize methods used to distinguish infections caused by *Staphylococcus aureus* and *Streptococcus pyogenes*, and discuss the spectrum of skin and tissue diseases caused by each.
8. Provide an update of the status of MRSA infections in the United States.
9. Discuss the relative dangers of rubella and rubeola viruses in different populations.

18.4 The Surface of the Eye and Its Defenses

10. Describe the important anatomical features of the eye.
11. List the natural defenses present in the eye.

18.5 Normal Biota of the Eye

12. List the types of normal biota presently known to occupy the eye.

18.6 Eye Diseases Caused by Microorganisms

13. List the possible causative agents for each of the infectious eye diseases: conjunctivitis, trachoma, keratitis, and river blindness.
14. Discuss why there are distinct differential diagnoses for neonatal and non-neonatal conjunctivitis.

A Note About the Chapter Organization

Beginning in this chapter, we discuss disease conditions caused by microbial infection. The chapter organization mirrors the clinical experience. Patients present themselves to health care practitioners with a set of symptoms, and the health care team makes an "anatomical" diagnosis—such as *a generalized vesicular rash.* The anatomical diagnosis allows practitioners to narrow down the list of possible causes to microorganisms that are known to be capable of creating such a condition. Then the proper tests can be performed to arrive at an etiologic diagnosis (that is, to determine the exact microbial cause). So the order of events is (1) anatomical diagnosis based on signs and symptoms; (2) consideration of a number of agents that are known to cause disease in that anatomical location (often called the differential diagnosis); followed by (3) the etiologic diagnosis. In practice, this process may be shortened. For instance, if a patient has a disease such as mumps, the distinctive signs and symptoms of that disease may allow the practitioner to make the anatomical and the etiologic diagnosis at the same time, followed by confirmation of the etiology through laboratory methods, if necessary. In other cases, such as the common cold, the physician may consider only the anatomical diagnosis and never advance to the etiologic diagnosis because a cold is a mild self-limiting disease.

The chapters are organized by anatomical diagnosis (for example, "Infectious Diseases Affecting the Skin and Eyes"). Specific diseases and the microorganisms that cause them are then detailed. Some diseases are the result of infection by a single type of microorganism. Scalded skin syndrome is an example of this. In other cases, single diseases or conditions can be caused by many different microorganisms, including bacteria, viruses, and so on. The classic examples are pneumonia in the respiratory tract, meningitis in the central nervous system, and diarrhea in the gastrointestinal system. In this chapter, for instance, a maculopapular rash may be caused by the measles virus, the rubella virus, or a parvovirus. A Disease Table at the end of each disease/condition makes it clear whether a single microorganism or multiple agents are to be considered in the diagnosis.

18.1 The Skin and Its Defenses

The skin makes contact directly with the environment—not only with solid objects but also with water and other fluids—and with the atmosphere. What's more, many infectious diseases include skin eruptions or lesions as part of the course of illness and often as a major symptom, even if the infective agent does not enter via the skin.

The eye surface, like the skin, is also exposed constantly to the environment. For this reason, we include diseases of both organ systems in this chapter. The organs under consideration in this chapter form the boundary between the organism and the environment. The skin, together with the hair, nails, and sweat and oil glands, forms the **integument.** The skin has a total surface area of 1.5 to 2 square meters. Its thickness varies from 1.5 millimeters at places such as the eyelids to 4 millimeters on the soles of the feet. Several distinct layers can be found in this thickness, and we summarize them here. Follow **figure 18.1** as you read. This figure also depicts where certain diseases have their effects.

The outermost portion of the skin is the epidermis, which is further subdivided into four or five distinct layers. On top is a thick layer of epithelial cells called the stratum corneum, about 25 cells thick. The cells in this layer are dead and have migrated from the deeper layers during the normal course of cell division. They are packed with a protein called **keratin,** which the cells have been producing ever since they arose from the deepest level of the epidermis. Because this process is continuous, the entire epidermis is replaced every 25 to 45 days. Keratin gives the cells their ability to withstand damage and abrasion; the surface of the skin is termed **keratinized** for this reason. The spaces between individual cells of the stratum corneum are packed with a special kind of lipid that has super water-repellent properties. Below the stratum corneum are three or four more layers of epithelial cells. The lowest layer, the stratum basale, or basal layer, is attached to the underlying dermis and is the source for all of the cells that make up the epidermis.

The dermis, underneath the epidermis, is composed of connective tissue instead of epithelium. This means that it is a rich matrix of fibroblast cells and fibers such as collagen, and it contains macrophages and mast cells. The dermis also harbors a dense network of nerves, blood vessels, and lymphatic vessels. Damage to the epidermis generally does not result in bleeding, whereas damage deep enough to penetrate the dermis results in broken blood vessels. Blister formation, the result of friction trauma or burns, causes a separation between the dermis and epidermis.

The "roots" of hairs, called follicles, are in the dermis. **Sebaceous** (oil) **glands** and scent glands are associated with the hair follicle. Separate sweat glands are also found in this tissue. All of these glands have openings on the surface of the skin, so they pass through the epidermis as well.

It could be said that the skin is its own defense—in other words, the very nature of its keratinized surface prevents most microorganisms from penetrating into sensitive deeper tissues. Millions of cells from the stratum corneum slough off every day, and attached microorganisms slough off with them. The skin is also brimming with antimicrobial substances. Perhaps the most effective skin defense against infection is the one most recently discovered. In the past 20 years,

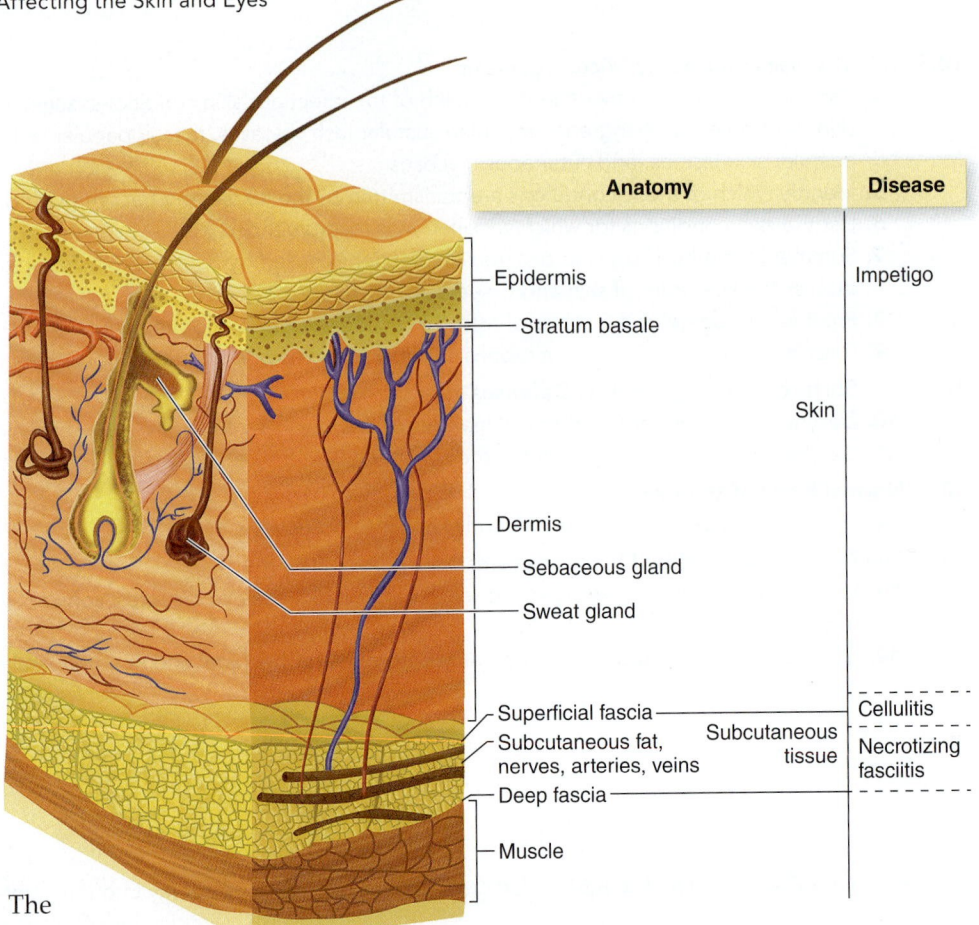

Figure 18.1 A cross section of skin and the levels at which some disease processes take place.

small molecules called **antimicrobial peptides** have been identified in epithelial cells. These are positively charged chemicals that act by disrupting (negatively charged) membranes of bacteria. There are many different types of these peptides, and they seem to be chiefly responsible for keeping the microbial count on skin relatively low.

The sebaceous glands' secretion, called **sebum,** has a low pH, which makes the skin inhospitable to most microorganisms. Sebum is oily due to its high concentration of lipids. The lipids can serve as nutrients for normal microbiota, but breakdown of the fatty acids contained in lipids leads to toxic by-products that inhibit the growth of microorganisms not adapted to the skin environment. This mechanism helps control the growth of potentially pathogenic bacteria. Sweat is also inhibitory to microorganisms because of both its low pH and its high salt concentration. **Lysozyme** is an enzyme found in sweat (and tears and saliva) that specifically breaks down peptidoglycan, which you learned in chapter 4 is a unique component of bacterial cell walls.

18.1 Learning Outcomes—Assess Your Progress

1. Describe the important anatomical features of the skin.
2. List the natural defenses present in the skin.

18.2 Normal Biota of the Skin

Overall, the skin presents an environment that is inhospitable to the growth of many microorganisms. The vast dry, salty surfaces, many of which are coated in toxic antimicrobial lipids, were once thought to be sparsely covered in biota. Early studies showed more dense microbial populations in moist areas and skin folds, such as the underarm and groin areas, and in the protected environment of the hair follicles and glandular ducts.

The first published data from the Human Microbiome Project (HMP) have brought some surprises, however. Some general observations about the skin microbiota follow:

- Hundreds of species of microbes were found distributed over many different areas of the body.
- Although five major taxa were represented in the microbiota, the predominance of these groups varied in the different regions of the body sampled.
- There are large differences among people with respect to the types of microbes found on various skin sites.
- An individual's own skin microbiota seems to be relatively stable over time.

Staphylococcus epidermidis and *Propionibacterium acnes* have long been thought to be the most numerous normal biota on skin due to their tolerance of high salt conditions. The HMP does indeed find these two species in large numbers, along with many other bacteria. Also, approximately 4% of the population carry *Staphylococcus aureus* on their skin. Continuing studies look to establish differences among healthy microbiomes and those of patients affected by skin disorders and how a person's living environment may dictate the composition and distribution of microbes on the skin.

Many microbes were also identified living under the skin. The vast majority of microbes resided within the uppermost layers of the epidermis, and nearly 25% of all bacteria were localized to hair follicles. It will be important to determine the role of these microbes in dermatologic disease. The HMP stresses the importance of developing new antiseptics that efficiently remove microbial contaminants from even the most protected areas of the skin and will continue to expand our view of the skin microbiome, revealing "micro habitats" and more diversity than we ever imagined in this environment.

Defenses and Normal Biota of the Skin		
	Defenses	**Normal Biota**
Skin	Keratinized surface, sloughing, low pH, high salt, lysozyme	Bacteria such as *Staphylococcus epidermidis, Propionibacterium, Corynebacterium, Lactobacillus, Bacteroides, Prevotella, Haemophilus*; yeasts such as *Malassezia, Candida*

18.2 Learning Outcomes—Assess Your Progress

3. List the types of normal biota presently known to occupy the skin.

18.3 Skin Diseases Caused by Microorganisms

Acne

The term *acne* encompasses all follicle-associated lesions, from the isolated pimple to severe widespread acne. Normally, the sebaceous glands associated with hair follicles (see figure 18.1) are a self-contained system for protecting, softening, and lubricating the skin. As hair and skin grow, dead epidermal cells and sebum work their way upward and are discharged from the pore to the skin surface.

Skin prone to pimples and acne has a structure that traps the mass of sebum and dead cells, clogging the pores. An exaggerated process of keratinization occurs in skin cells in and around the follicle, which also helps to block the pore. An added factor is overproduction of sebum when the sebaceous gland is stimulated by hormones (especially male hormones). *Propionibacterium acnes* present in the follicle releases lipases to digest this surplus of oil. The combination of digestive products (fatty acids) and bacterial antigens stimulates an intense local inflammation that eventually can burst the follicle. In time, the lesion can erupt on the surface.

Different types of lesions are associated with this process. When the skin initially swells over the pore leading out of a hair follicle, it is called a **comedo.** If the pore is closed, this comedo is commonly called a whitehead. If the pore remains open to the surface but is blocked with a dark plug of sebum, it appears as a blackhead. When the lesion erupts on the surface, it is called a pustule or papule. At this point, the lesion contains sebum and pus, a collection of bacteria, dead skin cells, and white blood cells from the inflammatory reaction. Pustules that come to involve deeper layers of skin are called cysts, and they can be quite painful. Widespread lesions of this type are called cystic acne.

▶ Causative Agent

Propionibacterium acnes is the bacterium associated with acne, but the "cause" of acne is multifactorial, requiring other conditions to be just right before the presence of this otherwise benign bacterium results in acne.

The bacterium is an anaerobic or aerotolerant gram-positive rod arranged in short chains or clumps. It releases a variety of enzymes that contribute to its virulence. The most important of these appears to be lipase, although it also releases proteases, neuraminidase, and a hyaluronidase. In addition, it secretes a low-molecular-weight protein that is a strong attractant for white blood cells that contribute to inflammation.

▶ Transmission and Epidemiology

Because *P. acnes* is normal biota on human skin, it is not considered a transmissible infection. When speaking of the epidemiology of acne, we are really considering what groups have the combination of factors that can result in acne. About 85% of adolescents and young adults experience acne of some degree at some time in their lives, and it appears linked to the amount of sebaceous fluid produced in the skin. More severe forms of adolescent acne are more common in males than females, probably because male hormones, or androgens, aggravate the condition. A genetic link to the development of acne is also apparent.

▶ Prevention and Treatment

There is no effective prevention of acne; it is not the result of poor hygiene or even of eating the wrong foods. For many years, the only treatment options were (1) topical agents that enhanced the sloughing of skin cells, which could help to prevent comedo formation or to keep comedos from becoming pustules or papules; and (2) either topical or oral antibiotics, such as erythromycin or tetracycline. It has become apparent that such long-term use of antibiotics causes a high rate of antibiotic resistance in skin bacteria (in the case of topical application) or in whole-body normal biota. This result should have been predicted because oral antibiotics are typically given for long periods of time in low, sublethal doses—the perfect set of conditions for creating antibiotic resistance in bacteria. It has even been shown that live-in family members of people taking antibiotics for their acne also eventually harbor skin bacteria resistant to the same antibiotic the acne patient is taking. In this way, resistant bacteria can spread beyond the original host.

For patients with severe acne and for whom other treatment options have failed, isotretinoin may be prescribed. This drug can have severe side effects, the most severe of which affect fetuses. These are called **teratogenic** (ter-at'-oh-jen"-ik) effects. Because of this, women taking the drugs should also be using two different methods of birth control. Other side effects include psychological depression, and patients must be closely monitored. Benzoyl peroxide is often used topically, as well. It speeds up skin turnover and has direct antimicrobial effects.

Scientists are currently working on a way to use bacteriophages specific for *P. acnes* as an alternative to antibiotic or isotretinoin treatment.

Disease Table 18.1	Acne
Causative Organism(s)	*Propionibacterium acnes*
Most Common Mode(s) of Transmission	Endogenous
Virulence Factors	Lipase, inflammatory mediator, other enzymes
Culture/Diagnosis	Based on clinical picture
Prevention	None
Treatment	Antibiotics (topical or oral), isotretinoin, benzoyl peroxide
Epidemiological Features	Highest rates in Western industrialized countries; 40–50 million annual incidence in the United States

Impetigo

Impetigo is a superficial bacterial infection that causes the skin to flake or peel off **(figure 18.2).** It is not a serious disease but is highly contagious, and children are the primary victims. Impetigo can be caused by either *Staphylococcus aureus* or *Streptococcus pyogenes*, and a mixture of the two probably causes some cases. As you may know, these two bacteria cause a wide variety of skin conditions; these are summarized in **Insight 18.1**. It has been suggested that *S. pyogenes*

A Note About Statistics in the Disease Tables

Each condition we study is summarized in a disease table. The last row of each table contains information about the epidemiology of the disease. The type of epidemiological information that is most relevant to a particular disease can vary. For example, it is vitally important to know the numbers of new cases **(incidence)** of some diseases. This is the case for measles in the United States (see Disease Table 18.7), as we track the resurgence of a disease once controlled by vaccination. For other conditions, such as fifth disease in the same table, it is more useful to know how many people have been affected by the time they reach a certain age. **Prevalence,** or the current number of people affected by the condition, is another common measure. And for some diseases the most informative statistic is how deadly they are **(mortality** rate). These tables contain the most useful information about each condition.

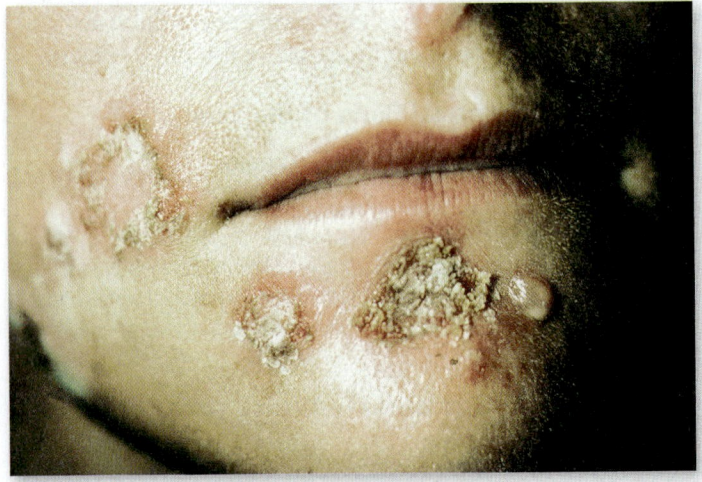

Figure 18.2 Impetigo lesions on the face.

INSIGHT 18.1   Skin, Staph, and Strep

The skin is the largest organ of the body and has a unique landscape with folds, ridges, regions that are warm and moist, and regions that are cool and dry. It is also host to hundreds of different types of bacteria, some of which cannot be cultured in the laboratory and have only been discovered through 16S rRNA sampling. Most bacteria on the skin are either harmless or beneficial and prevent pathogenic microbes from establishing themselves. However, there are a few pathogens in the mix, specifically *Staphylococcus* and *Streptococcus*, and these are the cause of the majority of bacterial skin infections. Some skin infections range from relatively mild skin eruptions such as folliculitis to deadly necrotizing fasciitis, caused by "flesh-eating bacteria." What makes the difference in these types of infections is the microbe causing them and the toxins it produces.

Skin Infections Caused by *Staphylococcus aureus*

Folliculitis, carbuncles, and furuncles (boils) are all infections of the hair follicle and are mainly caused by *Staphylococcus aureus*, although other microbes may be involved. **Folliculitis** is a relatively superficial infection of the hair follicle characterized by a rash or a pimple. **Furuncles,** also known as boils, are skin infections involving an entire hair follicle and the surrounding skin tissue. Furuncles may begin as a pea-sized lesion and then spread, oozing pus that may require drainage. **Carbuncles** involve a group of hair follicles and have similar symptoms as furuncles but are clustered and form a lump that occurs deep in the skin. Good skin hygiene is important in preventing and treating all of these conditions; if not treated properly, they can spread to other areas of the body and can result in more serious infections.

Skin Infections Caused by *Streptococcus pyogenes*

Streptococcus pyogenes is the main cause of cellulitis, impetigo, and erysipelas. In **erysipelas,** the bacteria enter the skin through a small cut or break in the skin and cause blisters or swollen lesions accompanied by fever, shaking, and chills. Twenty percent of erysipelas cases are seen on the face with 80% affecting the legs. Similar to skin infections caused by *S. aureus*, erysipelas can result in more serious infections such as bacteremia and septic shock if it is not treated correctly or promptly. Cellulitis and impetigo will be covered in more detail later in the text.

Necrotizing Fasciitis

Necrotizing fasciitis or "flesh-eating bacteria" has gained wide notoriety due to the speed with which it infects and its lethality. Like the skin infections already mentioned, necrotizing fasciitis can be caused by numerous bacteria such as *Clostridium perfringens* and *Vibrio vulnificus*, but the main causes are *S. aureus* and *S. pyogenes*. Symptoms begin in a similar fashion as simple folliculitis or erysipelas—a minor infection at a hair follicle or break in the skin. However, the similarity ends there. Infection proceeds rapidly as toxins produced by the bacteria break down tissues, the tissues die, and the bacteria spread through the bloodstream. Additionally, these bacteria can produce **superantigens,** which produce powerful stimulation of T cells and a massive release of cytokines that can lead to shock and death. Necrotizing fasciitis is fatal in about 25% of those infected without treatment. Treatment is with powerful broad-spectrum antibiotics and surgery to remove dead and damaged tissue and bone. Amputation may be necessary in severe cases.

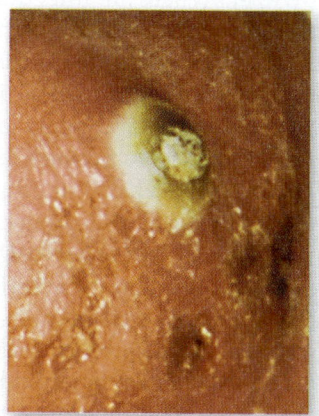 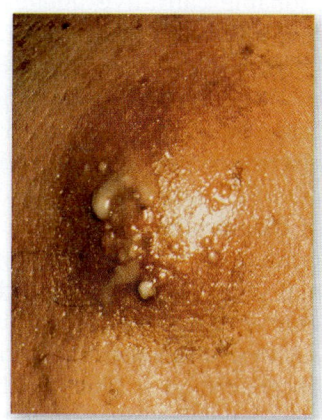 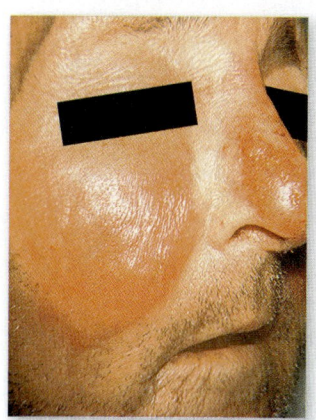 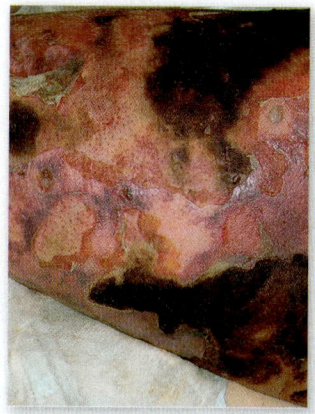

A furuncle or boil caused by *Staphylococcus aureus*.

A carbuncle caused by *Staphylococcus aureus*.

Facial erysipelas caused by *Streptococcus pyogenes*.

Necrotizing fasciitis, or "flesh-eating disease."

causes the initial infection of this disease, but in some cases *S. aureus* later takes over and becomes the predominant bacterium cultured from lesions. Because *S. aureus* produces a bacteriocin (toxin) that can destroy *S. pyogenes*, it is possible that *S. pyogenes* is often missed in culture-based diagnosis of impetigo.

▶ Signs and Symptoms

The "lesion" of impetigo looks variously like peeling skin, crusty and flaky scabs, or honey-colored crusts. Lesions are most often found around the mouth, face, and extremities, though they can occur anywhere on the skin. It is very superficial and it itches. The symptomatology does not indicate whether the infection is caused by *Staphylococcus* or *Streptococcus*.

Impetigo Caused by *Staphylococcus aureus*

Staphylococcus aureus is one of the most exquisitely tuned microorganisms for causing disease in humans. It is responsible for a long list of different diseases in addition to the ones highlighted in this chapter. *S. aureus* can cause pneumonia, food poisoning, serious bloodstream infections, bone infections, toxic shock syndrome, and meningitis. It is a gram-positive coccus that grows in clusters resembling a bunch of grapes **(figure 18.3a)**. It is a nonmotile organism, and much of its destructiveness is due to its array of virulence factors as well as the fact that it acts as a superantigen, meaning that it stimulates (inappropriately) T cells that are not supposed to be triggered by *S. aureus*. These T cells then produce cytokines that increase tissue damage and symptomology.

S. *aureus* in culture produces large, round, opaque colonies at an optimum temperature of 37°C, although it can grow anywhere between 10°C and 46°C. The species is a facultative anaerobe whose growth is enhanced in the presence of O_2 and CO_2. Its nutrient requirements can be satisfied by routine laboratory media, and most strains are metabolically versatile—that is, they can digest proteins and lipids and ferment a variety of sugars. This species is considered the sturdiest of all non-endospore-forming pathogens, with well-developed capacities to withstand high salt (7.5% to 10%), extremes in pH, and high temperatures (up to 60°C for 60 minutes). *S. aureus* also remains viable after months of air drying and resists the effect of many disinfectants and antibiotics. These properties contribute to the fact that *S. aureus* is a troublesome hospital pathogen.

▶ Pathogenesis and Virulence Factors

The most important virulence factors relevant to *S. aureus* impetigo are exotoxins called exfoliative toxins A and B, which are coded for by a phage that infects some *S. aureus* strains. At least one of the toxins attacks a protein that is very important for epithelial cell-to-cell binding in the outermost layer of the skin. Breaking up this protein leads to the characteristic blistering seen in the condition. The breakdown of skin architecture also facilitates the spread of the bacterium. All pathogenic *S. aureus* strains typically produce **coagulase,** an enzyme that coagulates plasma and blood. It is thought that this enzyme causes fibrin to be deposited around the bacteria, concentrating the exotoxins in an area of local damage. Ninety-seven percent of all human isolates of *S. aureus* produce this enzyme, and it is considered the most diagnostic biochemical species characteristic.

Other enzymes expressed by *S. aureus* include hyaluronidase, which digests the intercellular "glue" (hyaluronic acid) that binds connective tissue in host tissues; staphylokinase, which digests blood clots; a nuclease that digests DNA (DNase); and lipases that help the bacteria colonize oily skin surfaces.

▶ Culture and/or Diagnosis

Doctors usually diagnose impetigo by visual inspection and typically treat it with antibiotics that target both probable causative agents. However, when an infection requires identification (for instance, if initial treatment fails), well-established methods exist to establish *S. aureus* as the etiologic agent. Primary isolation of *S. aureus* is achieved by inoculation on sheep or rabbit blood agar; for heavily contaminated specimens, selective media such as mannitol salt agar are used **(figure 18.3b)**. Gram staining of the specimen shows characteristic, irregular clusters of gram-positive cocci. However, it is not possible to use colonial or microscopic characteristics to differentiate among gram-positive cocci, including *Staphylococcus epidermidis,* so additional biochemical testing is required. The production of catalase, an enzyme that breaks down hydrogen peroxide accumulated during oxidative metabolism, can be used to differentiate the staphylococci, which produce it, from the streptococci, which do not. The ability of *Staphylococcus* to grow anaerobically and to ferment sugars separates it from *Micrococcus,* a nonpathogenic genus that is a common specimen contaminant.

One key technique for identifying *S. aureus* from other species of *Staphylococcus* is the coagulase test **(figure 18.4).** By definition, any isolate that coagulates plasma is *S. aureus;* all others are coagulase negative. Rapid multitest systems are also available today and used routinely in clinical settings.

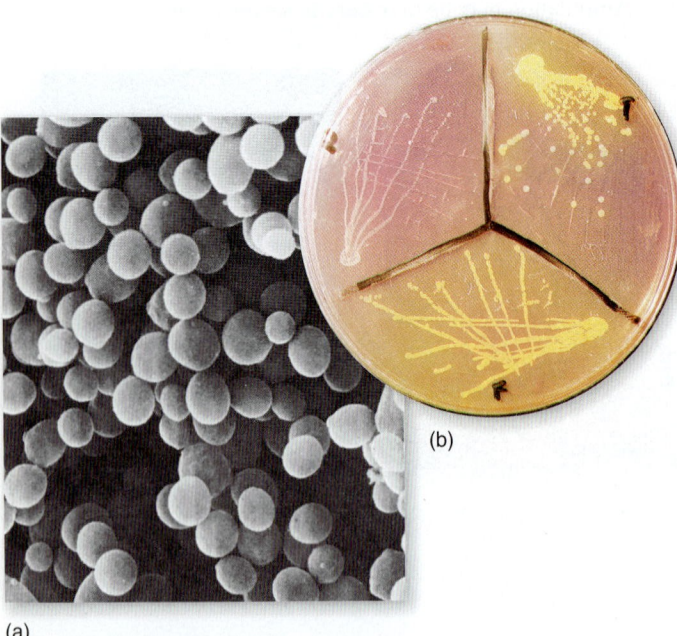

(b)

(a)

Figure 18.3 *Staphylococcus aureus.* (a) Scanning electron micrograph of *S. aureus*. (b) Mannitol salt agar is both differential and selective for the growth of *S. aureus*. Unlike many organisms, *S. aureus* can withstand the relatively high levels of salt in this medium. In addition, it can ferment the mannitol found in the medium, leading to a yellow color change over time.

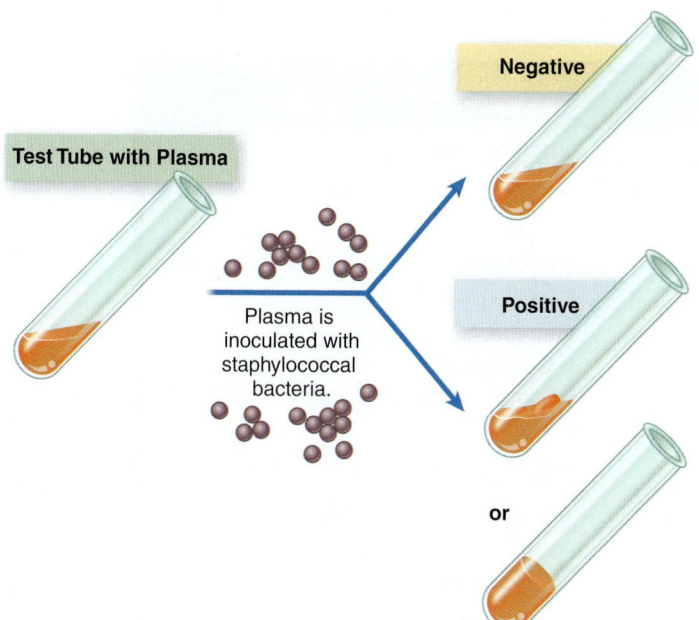

Figure 18.4 The coagulase test. Staphylococcal coagulase is an enzyme that reacts with factors in plasma to initiate clot formation. In the coagulase test, a tube of plasma is inoculated with the bacterium. If it remains liquid, the test is negative. If the plasma develops a lump or becomes completely clotted, the test is positive.

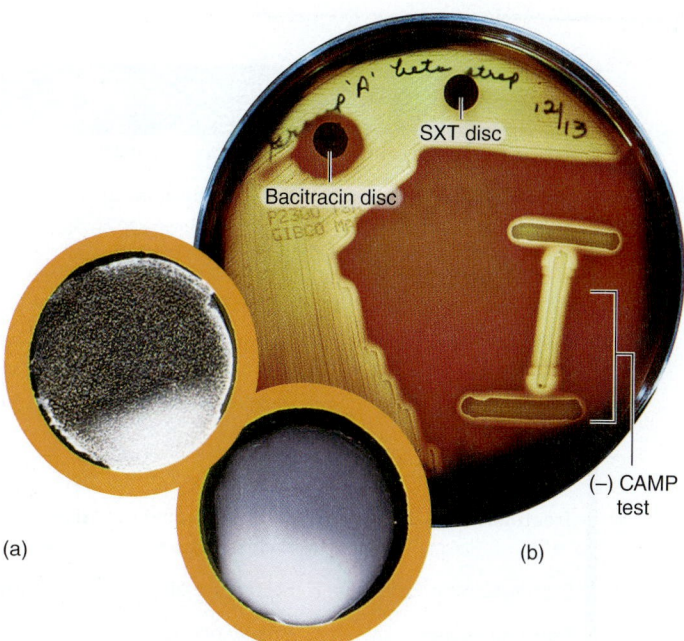

Figure 18.5 Identification tests for *Streptococcus pyogenes*. (a) A rapid, direct test kit for diagnosis of group A infections. With this method, a patient's throat swab is introduced into a system composed of latex beads and monoclonal antibodies. (*Left*) In a positive reaction, the C carbohydrate on group A streptococci produces visible clumps. (*Right*) A smooth, milky reaction is negative. (b) Bacitracin disc test. With very few exceptions, only *Streptococcus pyogenes* is sensitive to a minute concentration (0.02 µg) of bacitracin. Any zone of inhibition around the B disc is interpreted as a presumptive indication of this species. (*Note:* Group A streptococci are negative for sulfamethoxazole-trimethoprim [SXT] sensitivity and the CAMP test.)

Impetigo Caused by *Streptococcus pyogenes*

Streptococcus pyogenes is thoroughly described in chapter 21 in the section on pharyngitis. The important features are briefly summarized here and the features pertinent to impetigo are listed in **Disease Table 18.2.**

   S. pyogenes is a gram-positive coccus in Lancefield group A and is beta-hemolytic on blood agar. In addition to impetigo, it causes streptococcal pharyngitis (strep throat), scarlet fever, pneumonia, puerperal fever, necrotizing fasciitis, serious bloodstream infections, and poststreptococcal conditions such as rheumatic fever.

   If the precise etiologic agent must be identified, there are well-established methods for identifying group A streptococci. **Figure 18.5a** illustrates a rapid direct test, while **figure 18.5b** depicts the beta-hemolysis the colonies display on blood agar, as well as their sensitivity to bacitracin.

▸ ## Pathogenesis and Virulence Factors

The symptoms of *S. pyogenes* impetigo are indistinguishable from those caused by *S. aureus*. Like *S. aureus*, this bacterium possesses a huge arsenal of enzymes and toxins. As mentioned earlier, it anchors itself to surfaces (including skin) using a variety of adhesive elements on its surface (LTA, M protein and other proteins, and a hyaluronic acid capsule). M protein also protects it from phagocytosis. Like *S. aureus*, it possesses hyaluronidase.

   Rarely, impetigo caused by *S. pyogenes* can be followed by acute poststreptococcal glomerulonephritis (see chapter 21).

▸ ## Transmission and Epidemiology of Impetigo

Impetigo, whether it is caused by *S. pyogenes*, *S. aureus*, or both, is highly contagious and transmitted through direct contact but also via fomites and mechanical vectors. It affects mostly preschool children, but individuals of all ages can acquire the disease. The peak incidence is in the summer and fall. *S. pyogenes* is more often the cause of impetigo in newborns, and *S. aureus* is more often the cause of impetigo in older children, but both microbes can cause infection in either age group.

▸ ## Prevention

The only current prevention for impetigo is good hygiene.

▸ ## Treatment

Impetigo is sometimes treated with a drug that will target either bacterium, *S. pyogenes* or *S. aureus*, eliminating the need to determine the exact etiologic agent. The drug of choice is topical mupirocin (brand name Bactroban), a protein synthesis inhibitor, although resistance to this drug is increasing in *S. aureus*. Retapamulin is also used topically. Often, cases caused by *S. aureus* require oral antibiotics as well. Dicloxacillin or cephalexin are used for sensitive strains. Trimethoprim-sulfamethoxazole (TMP-SMZ) is the

Disease Table 18.2	Impetigo	
Causative Organism(s)	*Staphylococcus aureus*	*Streptococcus pyogenes*
Most Common Modes of Transmission	Direct contact, indirect contact	Direct contact, indirect contact
Virulence Factors	Exfoliative toxin A, coagulase, other enzymes	Streptokinase, plasminogen-binding ability, hyaluronidase, M protein
Culture/Diagnosis	Routinely based on clinical signs, when necessary, culture and Gram stain, coagulase and catalase tests, multitest systems, PCR	Routinely based on clinical signs, when necessary, culture and Gram stain, coagulase and catalase tests, multitest systems, PCR
Prevention	Hygiene practices	Hygiene practices
Treatment	Topical mupirocin or retapamulin, oral dicloxacillin, cephalexin, or TMP-SMZ	Topical mupirocin or retapamulin
Distinguishing Features	Seen more often in older children, adults	Seen more often in newborns
Epidemiological Features	Prevalence approximately 1% of children in North America	

first alternative for methicillin-resistant *S. aureus* (MRSA), but these strains are often resistant to multiple drugs (**Disease Table 18.2**).

Cellulitis

Cellulitis is a condition caused by a fast-spreading infection in the dermis and in the subcutaneous tissues below. It causes pain, tenderness, swelling, and warmth. Fever and swelling of the lymph nodes draining the area may also occur. Frequently, red lines leading away from the area are visible (a phenomenon called *lymphangitis*); this symptom is the result of microbes and inflammatory products being carried by the lymphatic system. Although septicemia can develop, most cases of the disease are uncomplicated and patients have a good prognosis.

Cellulitis generally follows the introduction of bacteria or fungi into the dermis, either through trauma or by subtle means (with no obvious break in the skin). Cellulitis is very common on the lower leg, and it is thought the bacteria can enter through breaks in the skin between the toes caused by fungal infection (athlete's foot). Symptoms take several days to develop. The most common causes of the condition in healthy people are *Staphylococcus aureus* and *Streptococcus pyogenes*, although almost any bacterium and some fungi can cause this condition in an immunocompromised patient. In infants, group B streptococci are a frequent cause of this infection (see chapter 23).

People who are immunocompromised or who have cardiac insufficiency are at higher risk for this condition compared to healthy individuals. They also risk complications, such as spread to the bloodstream, rapid spreading through

adjacent tissues, and, especially in children, meningitis. Occasionally, cellulitis is a complication of varicella (chickenpox) infections.

Mild cellulitis responds well to oral antibiotics chosen to be effective against both *S. aureus* and *S. pyogenes*. More involved infections and infections in the immunocompromised require intravenous antibiotics. If there are extensive areas of tissue damage, surgical debridement (duh-breed'-munt) may be warranted (**Disease Table 18.3**).

A Note About MRSA

As documented in the Case File in this chapter, during the period between 2004 and 2006, a disturbing *S. aureus* trend that had been bubbling under the surface emerged into the public eye. While the health care establishment had grown accustomed to (though not complacent about) hospital-acquired MRSA (HA-MRSA), people who had no recent history in hospitals or health care facilities started turning up with MRSA infections. These infections are classified as *community-acquired* and given the acronym CA-MRSA. They are growing at a fast rate, at the same time that HA-MRSA rates are falling.

Perhaps more disturbing, there have been 13 cases of VRSA (vancomycin-resistant *S. aureus*) in the United States since 2002. In each of these cases, the pathogen acquired vancomycin resistance on its own, showing that this process can indeed occur in a natural environment. The VRSA strains all shared some genetic characteristics that made them capable of coexisting with bacteria that could transfer antibiotic-resistance genes to them.

Disease Table 18.3	Cellulitis		
Causative Organism(s)	*Staphylococcus aureus*	*Streptococcus pyogenes*	Other bacteria or fungi
Most Common Modes of Transmission	Parenteral implantation	Parenteral implantation	Parenteral implantation
Virulence Factors	Exfoliative toxin A, coagulase, other enzymes	Streptokinase, plasminogen-binding ability, hyaluronidase, M protein	–
Culture/Diagnosis	Based on clinical signs	Based on clinical signs	Based on clinical signs
Prevention	–	–	–
Treatment	Oral or IV antibiotic (dicloxacillin if sensitive, vancomycin if MRSA); surgery sometimes necessary	Oral or IV antibiotic (penicillin); surgery sometimes necessary	Aggressive treatment with oral or IV antibiotic; surgery sometimes necessary
Distinguishing Features	–	–	More common in immunocompromised
Epidemiological Features	Incidence highest among males 45–64		

Staphylococcal Scalded Skin Syndrome (SSSS)

This syndrome is another **dermolytic** condition caused by *Staphylococcus aureus*. Although children and adults can be affected, SSSS develops predominantly in newborns and babies. Newborns are susceptible when sharing a nursery with another newborn who is colonized with *S. aureus*. Transmission may occur when caregivers carry the bacterium from one baby to another. Adults in the nursery can also directly transfer *S. aureus* because approximately 30% of adults are asymptomatic carriers. Carriers can harbor the bacterium in the nasopharynx, axilla, perineum, and even the vagina. (Fortunately, only about 5% of *S. aureus* strains are lysogenized by the type of phage that codes for the toxins responsible for the pathogenesis of this disease.)

SSSS can be thought of as a systemic form of impetigo. Like impetigo, it is an exotoxin-mediated disease. The phage-encoded exfoliative toxins A and B are responsible for the damage. Unlike impetigo, the toxins enter the bloodstream from the site of initial infection (the throat, the eye, or sometimes an impetigo infection) and then travel throughout the body interacting with the skin at many different sites. The A

Case File 18 *Continuing the Case*

The authors of the study speculated that there were several factors that contributed to the increase in CA-MRSA cases in New York City. They found that cases of CA-MRSA were higher among patients with diabetes and HIV. Both of these groups tend to have a greater likelihood for open sores or skin lesions, which are the portals of entry for *S. aureus*. Higher rates of CA-MRSA were seen in the Bronx where residents have more limited access to health care than in the other boroughs of New York. Additionally, higher rates of CA-MRSA were seen among men and children who play contact sports where the possibility of transmission through skin-to-skin contact is higher. Children are also often in daycare settings where there is a high likelihood of disease transmission. According to the study, men, children, individuals with HIV and diabetes, and people who were homeless had the highest rates of CA-MRSA as compared to the general population.

■ What steps could be taken in New York City and elsewhere to prevent further spread of CA-MRSA?

and B toxins cause **bullous lesions,** which often appear first around the umbilical cord (in neonates) or in the diaper or axilla area. The lesions begin as red areas, take on the appearance of wrinkled tissue paper, and then form very large blisters. Fever may precede the skin manifestations. Eventually, the top layers of epidermis peel off completely. The split occurs in the epidermal tissue layers just above the stratum basale (see figure 18.1). Widespread **desquamation** of the skin follows, leading to the burned appearance referred to in the name of this condition **(figure 18.6).**

At this point, the protective keratinized layer of the skin is gone, and the patient is vulnerable to secondary infections, cellulitis, and bacteremia. In the absence of these complications, young patients nearly always recover if treated promptly. Adult patients have a higher mortality rate—as high as 50%. Once a tentative diagnosis of SSSS is made, immediate antibiotic therapy should be instituted.

It is important, however, to differentiate this disease from a similar skin condition called *toxic epidermal necrolysis (TEN)*, which is caused by a reaction to antibiotics, barbiturates, or other drugs. TEN has a significant mortality rate. The treatments for the two diseases are very different, so it is important to distinguish between them before instituting therapy. In TEN, the split in skin tissue occurs *between* the dermis and the epidermis, not within the epidermis as is the case with SSSS. Histological examination of tissue from a lesion is usually a better way to diagnose the disease than reliance on culture. Because SSSS is caused by the dissemination of exotoxin, *S. aureus* may not be found in lesions. Nevertheless, culture should be attempted so that antibiotic sensitivities can be established.

Disease Table 18.4	Scalded Skin Syndrome
Causative Organism(s)	*Staphylococcus aureus*
Most Common Modes of Transmission	Direct contact, droplet contact
Virulence Factors	Exfoliative toxins A and B
Culture/Diagnosis	Histological sections; culture performed but false negatives common because toxins alone are sufficient for disease
Prevention	Eliminate carriers in contact with neonates
Treatment	Immediate systemic antibiotics (nafcillin if sensitive; vancomycin if MRSA)
Distinguishing Features	Split in skin occurs *within* epidermis
Epidemiological Features	Mortality 1%–5% in children, 50%–60% in adults

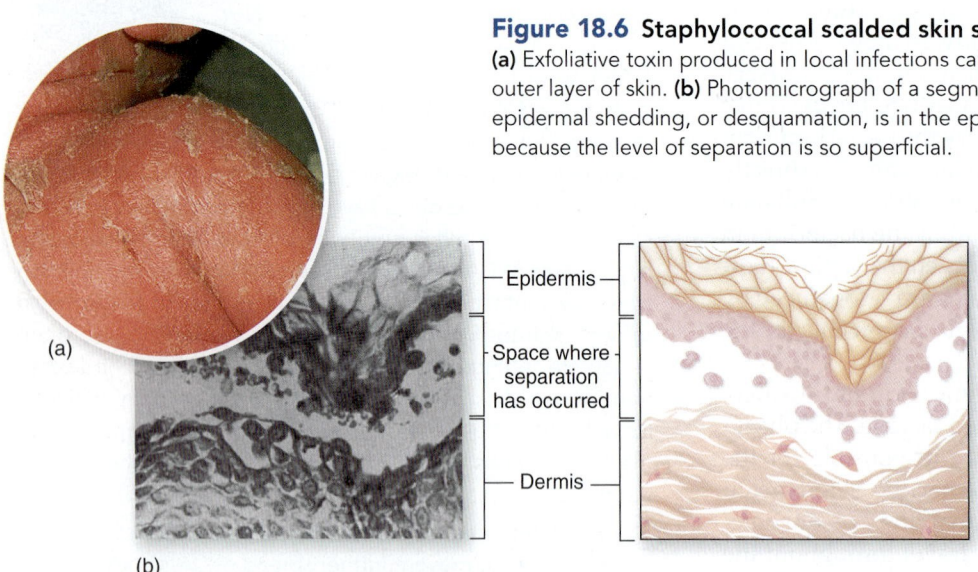

Figure 18.6 Staphylococcal scalded skin syndrome (SSSS) in a newborn child. **(a)** Exfoliative toxin produced in local infections causes blistering and peeling away of the outer layer of skin. **(b)** Photomicrograph of a segment of skin affected with SSSS. The point of epidermal shedding, or desquamation, is in the epidermis. The lesions will usually heal well because the level of separation is so superficial.

(a)

Epidermis

Space where separation has occurred

Dermis

(b)

Gas Gangrene

Clostridium perfringens, a gram-positive endospore-forming bacterium, causes a serious condition called **gas gangrene**, or clostridial **myonecrosis** (my"-oh-neh-kro'-sis). The endospores of this species can be found in soil, on human skin, and in the human intestine and vagina. This bacterium is anaerobic and requires anaerobic conditions to manufacture and release the exotoxins that mediate the damage in this disease.

▶ Signs and Symptoms

Two forms of gas gangrene have been identified. In anaerobic cellulitis, the bacteria spread within damaged necrotic muscle tissue, producing toxins and gas as the infection proceeds. However, the infection remains localized and does not spread into healthy tissue. The pathology of true myonecrosis is more destructive. Toxins produced in large muscles, such as the thigh, shoulder, and buttocks, diffuse into nearby healthy tissue and cause local necrosis at these sites. This damaged tissue then serves as a focus for continued bacterial growth, toxin formation, and gas production. The disease can quickly progress through an entire limb or body area, destroying tissues as it goes **(figure 18.7).** Initial symptoms of pain, edema, and a bloody exudate in the lesion are followed by fever, tachycardia, and blackened necrotic tissue filled with bubbles of gas. Gangrenous infections of the uterus caused by septic abortions and clostridial septicemia are particularly serious complications that can arise. If treatment is not begun early, the disease is invariably fatal.

▶ Pathogenesis and Virulence Factors

Because clostridia are not highly invasive, infection requires damaged or dead tissue, which supplies growth factors, and an anaerobic environment. The low-oxygen environment results from an interrupted blood supply and the presence of additional aerobic bacteria, which deplete oxygen. Such conditions stimulate endospore germination, rapid vegetative growth in the dead tissue, and release of exotoxins. *C. perfringens* produces several active exotoxins; the most potent one, *alpha toxin,* causes red blood cell rupture, edema, and tissue destruction **(figure 18.8).** Additional virulence factors that enhance tissue destruction are collagenase, hyaluronidase, and DNase. The gas formed in tissues, resulting from fermentation of muscle carbohydrates, can also destroy muscle structure. Histology or MRI can visualize these disruptions.

▶ Transmission and Epidemiology

The conditions that may predispose a person to gangrene are surgical incisions, compound fractures, diabetic ulcers, septic abortions, puncture and gunshot wounds, and crushing injuries contaminated by endospores from the body or the environment.

▶ Prevention and Treatment

One of the most effective ways to prevent clostridial wound infections is immediate and rigorous cleansing and surgical repair of deep wounds, decubitus ulcers (bedsores), compound fractures, and infected incisions. Debridement of diseased tissue eliminates the conditions that promote the spread of gangrenous infection. This procedure is most difficult in the intestine or body cavity, where only limited amounts of tissue can be removed. Surgery is supplemented by large doses of antibiotics to control infection. Hyperbaric oxygen therapy, in which the affected part is exposed to an increased oxygen mix in a pressurized chamber, can also lessen the severity of infection by inhibiting the growth of anaerobic bacteria.

Extensive myonecrosis of a limb may call for amputation. Because there are so many different antigenic subtypes in this bacterial group, active immunization is not possible.

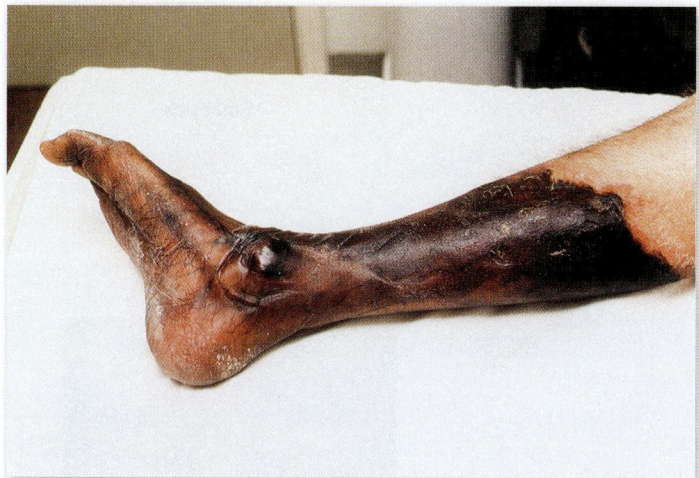

Figure 18.7 The clinical appearance of myonecrosis in a compound fracture of the leg.

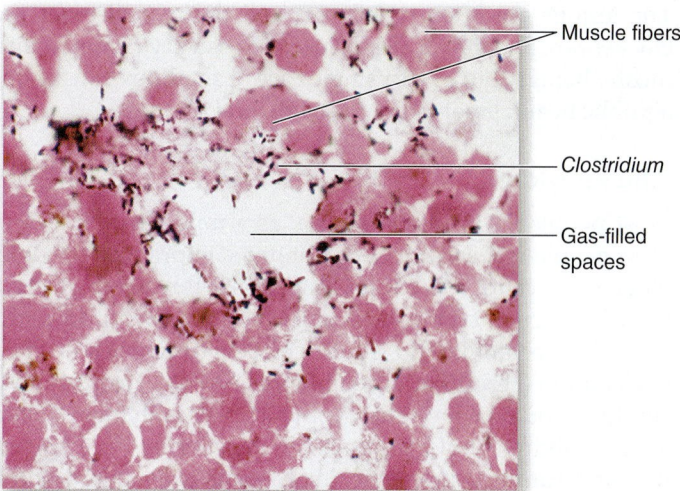

Figure 18.8 Growth of *Clostridium perfringens* (plump rods), causing gas formation and separation of the fibers. A microscopic analysis of clostridial myonecrosis, showing a histological section of gangrenous skeletal muscle.

Disease Table 18.5	Gas Gangrene
Causative Organism(s)	*Clostridium perfringens*, other species
Most Common Modes of Transmission	Vehicle (soil), endogenous transfer from skin, GI tract, reproductive tract
Virulence Factors	Alpha toxin, other exotoxins, enzymes, gas formation
Culture/Diagnosis	Gram stain, CT scans (abdominal infections), X ray, clinical picture
Prevention	Clean wounds, debride dead tissue
Treatment	Surgical removal, clindamycin + penicillin, oxygen therapy
Epidemiological Features	U.S. incidence: 900–1,000 annually; mortality 25% but approaches 100% when treatment is delayed

Vesicular or Pustular Rash Diseases

There are two diseases that present as generalized "rashes" over the body in which the individual lesions contain fluid. The lesions are often called *pox*, and the two diseases are chickenpox and smallpox. Chickenpox is very common and mostly benign, but even a single case of smallpox constitutes a public health emergency. Both are viral diseases.

Chickenpox

Most people think of chickenpox as a mild disease, and in most people it is. However, in immunocompromised people, older adolescents, and adults, it can be life-threatening. Before the introduction of the vaccine in 1995, it was not unheard of for some families to hold "chickenpox parties." When one child in a group of acquaintances had chickenpox, other children would be brought together to play with them so that all the children could contract the infection at once so that it would run its course in a timely manner. Parents wanted to ensure that their children got the disease while they were young because they knew that getting the disease at an older age could lead to more serious disease. However, purposeful infection of a child with a potentially damaging pathogen is not warranted, when a safe, effective vaccine exists.

▶ **Signs and Symptoms**

After an incubation period of 10 to 20 days, the first symptoms to appear are fever and an abundant rash that begins on the scalp, face, and trunk and radiates in sparse crops to the extremities. Skin lesions progress quickly from macules and papules to itchy vesicles filled with a clear fluid. In several days, they encrust and drop off, usually healing completely but sometimes leaving a tiny pit or scar. Lesions number from a few to hundreds and are more abundant in adolescents and adults than in young children. **Figure 18.9** contains images of chickenpox lesions. The lesion distribution is *centripetal*, meaning that there are more in the center of the body and fewer on the extremities, in contrast to the centrifugal distribution seen with smallpox. The illness usually lasts 4 to 7 days; new lesions stop appearing after about 5 days. Patients are considered contagious until all of the lesions have crusted over.

Most cases resolve without event within 2 to 3 weeks of onset. Some patients may experience secondary infections of the lesions caused by group A streptococci or staphylococci, and these require antibiotic therapy. Immunocompromised patients, as well as some adults and adolescents, may experience pneumonia as a result of chickenpox. The immunocompromised may

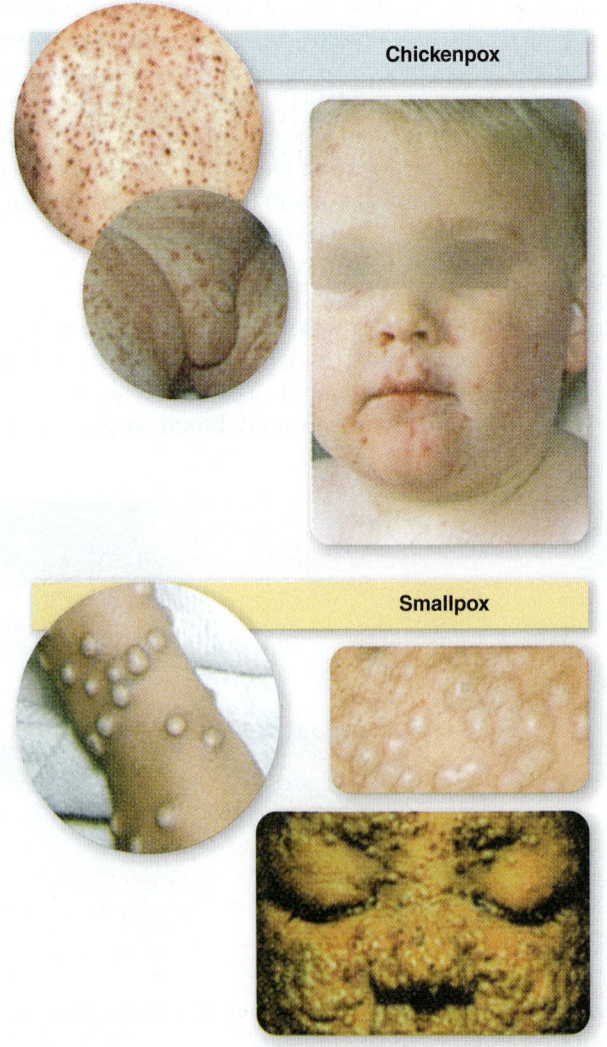

Figure 18.9 Images of chickenpox and smallpox.

also experience infection of the heart, liver, and kidney, resulting in a 20% mortality rate for this population.

Approximately 0.1% of chickenpox cases are followed by encephalopathy, or inflammation of the brain caused by the virus. It can be fatal, but in most cases recovery is complete.

Women who become infected with chickenpox during the early months of pregnancy are at risk for fetal infection. This virus can be teratogenic, and affected babies may be born with serious birth defects such as cataracts and missing limbs. Also, women who develop chickenpox just before or after giving birth may have passed the infection to the baby just before birth, resulting in serious infection in the newborn infant.

Shingles

After recuperation from chickenpox, the virus enters into the sensory endings that innervate dermatomes, regions of the skin supplied by the cutaneous branches of nerves, especially the thoracic **(figure 18.10a)** and trigeminal nerves. From there, it goes into latency within the ganglia; the virus may reemerge from these cells resulting in **shingles** (also known as **herpes zoster**). Shingles presents with a characteristic asymmetrical distribution of lesions on the skin of the trunk or head **(figure 18.10b)**.

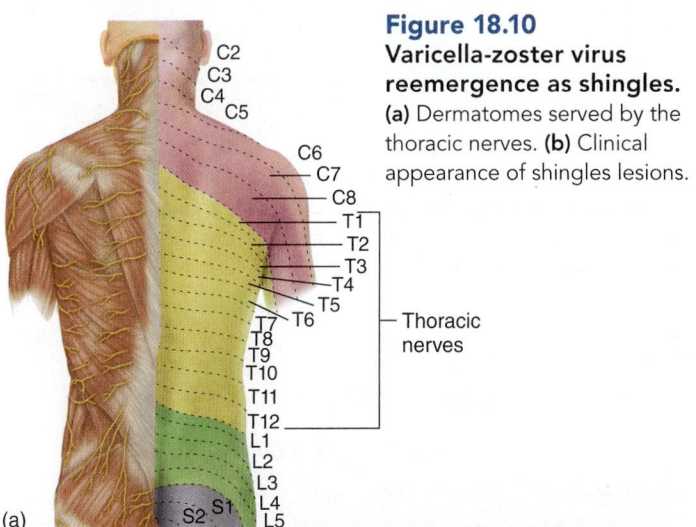

Figure 18.10
Varicella-zoster virus reemergence as shingles.
(a) Dermatomes served by the thoracic nerves. **(b)** Clinical appearance of shingles lesions.

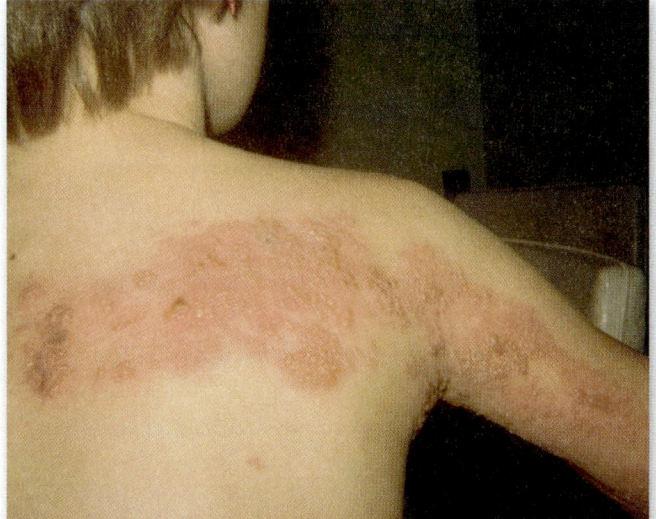

Shingles develops abruptly after the virus is reactivated by such stimuli as psychological stress, X-ray treatments, immunosuppressive and other drug therapy, surgery, or a developing malignancy. The virus is believed to migrate down the infected ganglia to the skin, where multiplication resumes and produces crops of tender, persistent vesicles. Inflammation of the ganglia and the associated pathways of nerves can cause pain and tenderness that can last for several months. Involvement of cranial nerves can lead to eye inflammation and ocular and facial paralysis.

▶ Causative Agent

Human herpesvirus 3 (HHV-3, also called **varicella** (var"-ih-sel'-ah) virus, causes chickenpox, as well as the condition called herpes zoster or shingles. The virus is sometimes referred to as the varicella-zoster virus (VZV). Like other herpesviruses, it is an enveloped DNA virus.

▶ Pathogenesis and Virulence Factors

HHV-3 enters the respiratory tract, attaches to respiratory mucosa, and then enters the bloodstream. The viremia disseminates the virus to the skin, where the virus causes adjacent cells to fuse and eventually lyse, resulting in the characteristic lesions of this disease. The virus enters sensory nerves at this site, traveling to the ganglia.

The ability of HHV-3 to remain latent in ganglia is an important virulence factor, because resting in this site protects it from attack by the immune system and provides a reservoir of virus for the reactivation condition of shingles.

▶ Transmission and Epidemiology

Humans are the only natural hosts for HHV-3. The virus is harbored in the respiratory tract but is communicable via both respiratory droplets and the fluid of active skin lesions. People can acquire a chickenpox infection by being exposed to the fluid of shingles lesions. (It is not possible to "get" shingles from someone with shingles. If you are not immune to HHV-3, you can acquire HHV-3, which will manifest as chickenpox or, very occasionally, as an asymptomatic infection. Once you have the virus, whether you experience shingles or not is dependent on your own host factors.)

Infected persons are most infectious a day or two prior to the development of the rash. Only in rare instances will a person acquire chickenpox more than once. Chickenpox is so contagious that if you are exposed to the virus and you do not have established immunity, you almost always become infected. Some people experience subclinical cases of the disease, meaning that their lesions never erupt. They will still develop lifelong immunity and will likely harbor the virus in their ganglia, making them subject to shingles in the future. When people think they have never had chickenpox, yet they don't seem to get it when exposed to infected persons, it is likely that they have had a subclinical case at some time in their lives.

Epidemics of the disease used to occur in winter and early spring. The introduction of the varicella vaccine in 1995 reduced the occurrence of the disease, so now cases develop sporadically.

▶ Prevention

A live attenuated vaccine was licensed in 1995. Varivax consists of a weakened form of the Oka strain of the HHV-3 virus, which was isolated from a Japanese boy named Oka. It is recommended that infants receive a first dose of the vaccine between the ages of 12 and 15 months, and an additional dose between the ages of 4 and 6 years. ProQuad is a multi-pathogen vaccine and can be used to immunize individuals against varicella in addition to measles, mumps, and rubella. In 2006, the CDC recommended that young adults receive an additional varicella vaccine booster to ensure effective protection.

In 2006, the FDA approved a unique vaccine called Zostavax. It is intended for adults ages 60 and over and is for the prevention of shingles due to varicella reactivation.

▶ Treatment

Uncomplicated varicella is self-limiting and requires no therapy aside from alleviation of discomfort. Oral acyclovir or related antivirals should be administered within 24 hours of onset of the rash to people considered to be at risk for serious complications. The acyclovir may diminish viral load and prevent complications. Non-immune people over the age of 20 and the immunocompromised who are exposed to the virus are often given passive varicella immunoglobulin. If they develop the disease, they are then given acyclovir.

Smallpox

Largely through the World Health Organization's comprehensive global efforts, naturally occurring smallpox is now a disease of the past. However, after the terrorist attacks in the United States on September 11, 2001, and the anthrax bioterrorism event that followed shortly thereafter, the U.S. government began taking the threat of smallpox bioterrorism very seriously. Vaccination, which had been discontinued, was once again offered to certain U.S. populations.

▶ Signs and Symptoms

Infection begins with fever and malaise; later, a rash begins in the pharynx, spreads to the face, and progresses to the extremities. Initially the rash is *macular,* evolving in turn to *papular, vesicular,* and *pustular* appearance before eventually crusting over, leaving behind nonpigmented sites pitted with scar tissue (see **Insight 18.2** for a description of these terms).

INSIGHT 18.2 | Skin Deep?

When studying the skin, there are so many conditions, lesions, and irregularities that it is difficult to keep track of them all. When describing a skin condition in a medical setting, it is important to use the proper terminology in order to avoid confusion. For example, simply using the word *rash* may not be adequate in conveying the symptom seen on the skin. None of the terms used in the accompanying table point to an etiologic cause, but each is precise enough in its description to narrow down the possibilities when making a diagnosis.

Terms Used in Describing Skin Conditions and Infections

Descriptive Name	Appearance	Examples
Bulla	Large (wide) vesicle	Blister, gas blisters in gangrene
Cyst	Raised, encapsulated lesion, usually solid or semisolid when palpated	Severe acne
Macule	Flat, well-demarcated lesion characterized mainly by color change	Freckle, tinea versicolor (fungus infection)
Maculopapular rash	Flat to slightly raised colored bump	Measles, rubella, fifth disease, roseola
Papule	Small, elevated, solid bump	Warts, cutaneous leishmaniasis
Petechiae	Small purpura (see below)	Meningococcal bloodstream infection
Plaque	Elevated, flat-topped lesion larger than 1 cm (i.e., a wider papule)	Psoriasis
Purpura	Reddish-purple discoloration due to blood in small areas of tissue; does not blanch when pressed	Meningococcal bloodstream infection
Pustule	Small, elevated lesion filled with purulent fluid (i.e., pus)	Acne, smallpox, mucocutaneous leishmaniasis, cutaneous anthrax
Scale	Flaky portions of skin separated from deeper skin layers	Ringworm of body and scalp, athlete's foot
Vesicle	Elevated lesion with clear fluid	Chickenpox

There are two principal forms of smallpox: variola minor and variola major. Variola major is a highly virulent form that causes toxemia, shock, and intravascular coagulation. People who have survived any form of smallpox nearly always develop lifelong immunity.

It is vitally important for health care workers to be able to recognize the early signs of smallpox. The diagnosis of even a single suspected case must be treated as a health and law enforcement emergency. The symptoms of variola major progress as follows: After the prodrome period of high fever and malaise, a rash emerges; the rash typically develops first in the mouth. Severe abdominal and back pain can accompany this phase of the disease. As lesions develop, they break open and spread virus into the mouth and throat, making the patient highly contagious. A rash then appears on the skin and spreads throughout the body within 24 hours.

By the third or fourth day of the rash, the bumps become larger and fill with a thick opaque fluid. A major distinguishing feature of this disease is that the pustules are indented in the middle (see Disease Table 18.6). Also, patients report that the lesions feel as if they contain a BB pellet. Within a few days, these pustules begin to scab over. After 2 weeks, most of the lesions will have crusted over; the patient remains contagious until the last scabs fall off because the crusts contain the virus. During the entire rash phase, the patient is very ill.

A patient with variola minor has a rash that is less dense and generally experiences weaker symptoms than someone affected by variola major.

▶ Causative Agent

The causative agent of smallpox, the variola virus, is an orthopoxvirus, an enveloped DNA virus. Variola is shaped like a brick and is 200 nanometers in diameter. Other members of this group are the monkeypox virus and the vaccinia virus from which smallpox vaccine is made. Variola is a hardy virus, surviving outside the host longer than most viruses.

Disease Table 18.6   Vesicular/Pustular Rash Diseases

Disease	Chickenpox	Smallpox
Causative Organism(s)	Human herpesvirus 3 (varicella-zoster virus)	Variola virus
Most Common Modes of Transmission	Droplet contact, inhalation of aerosolized lesion fluid	Droplet contact, indirect contact
Virulence Factors	Ability to fuse cells, ability to remain latent in ganglia	Ability to dampen, avoid immune response
Culture/Diagnosis	Based largely on clinical appearance	Based largely on clinical appearance
Prevention	Live attenuated vaccine; there is also vaccine to prevent reactivation of latent virus (shingles)	Live virus vaccine (vaccinia virus)
Treatment	None in uncomplicated cases; acyclovir for high risk	Cidofovir, immune globulin
Distinguishing Features	No fever prodrome; lesions are superficial; in centripetal distribution (more in center of body)	Fever precedes rash, lesions are deep and in centrifugal distribution (more on extremities)
Epidemiological Features	Chickenpox: vaccine decreased hospital visits by 88%, ambulatory visits by 59%; shingles: 1 million cases annually	Last natural case worldwide was in 1977
Appearance of Lesions		

▶ Pathogenesis and Virulence Factors

The infection begins by implantation of the virus in the nasopharynx. The virus invades the mucosa and multiplies in the regional lymph nodes, leading to viremia. Variola multiplies within white blood cells and then travels to the small blood vessels in the dermis. The lesions occur at the dermal level, which is the reason that scars remain after the lesions are healed.

▶ Transmission and Epidemiology

Before the eradication of smallpox, almost everyone contracted the disease over the course of his or her lifetime, either surviving with lifelong immunity or dying. It is spread primarily through droplet transmission, although fomites such as contaminated bedding and clothing can also spread the disease. Traditionally, the incidence of smallpox was highest in the winter and early spring.

In the early 1970s, smallpox was endemic in 31 countries. Every year, 10 to 15 million people contracted the disease, and approximately 2 million people died from it. By 1977, after 11 years of intensive effort by the world health community, the last natural case occurred in Somalia.

▶ Prevention

In chapter 15, you read about Edward Jenner and his development of vaccinia virus to inoculate against smallpox (see Insight 15.2). To this day, the vaccination for smallpox is based on the vaccinia virus. The United States stopped using the vaccine in 1972 after a massive effort to eradicate the virus worldwide. In 1980, the WHO declared that the war against smallpox was "won" and recommended that all laboratories destroy their stocks of the virus. However, two laboratories maintained stocks of the virus: One stock was housed at the Centers for Disease Control and Prevention (CDC) in the United States, and the other was held at the Institute of Virus Preparation in Moscow. Despite WHO recommendations, scientific and governmental organizations balked at destroying a potentially useful research subject. Many had reservations about *intentionally* causing the extinction of another species.

Concerns have been expressed about the existence of smallpox stocks in other regions of the world, with allegations that Russian scientists shared the virus with other countries. Iraqi prisoners captured in the 1991 Gulf War were reported to have high titers of antibodies to smallpox, which suggested they had been immunized. In 2002, the *Washington Post* reported that the CIA identified four countries with clandestine supplies of smallpox, including North Korea and Iraq. Since the terrorist events of 2001, the U.S. government has taken the possibility of smallpox bioterrorism very seriously. Currently, the United States has a large enough stockpile of smallpox vaccine to vaccinate the entire U.S. population in the event of an emergency. Today, most military branches are requiring that their personnel receive smallpox vaccination—among many others—before deploying to certain parts of the world.

In 2007, a new vaccine called ACAM 2000 was approved by the Food and Drug Administration and is used only for military personnel. Because of the potential side effects of the vaccine, ACAM 2000 is not approved for use in the general public, and a safer smallpox vaccine is still needed to protect infants, the elderly, and those who are immunocompromised.

▶ Treatment

There is no treatment for smallpox. Some advocate the use of cidofovir, which is labeled for use in cytomegalovirus infection. Immune globulin may also be useful. If lesions become infected secondarily with bacteria, antibiotics can be used for treatment of that complication **(Disease Table 18.6).**

Maculopapular Rash Diseases

Insight 18.2 contains a description of the different infectious conditions that can result in a rash of some sort on the skin. The infectious conditions described in this section are those with their major manifestations on the skin. (Meningococcal meningitis, for instance, can result in a diffuse rash on the skin, but its major manifestations are in the central nervous system, so it is discussed in chapter 19.) In this section, we examine measles, rubella, "fifth disease," and roseola. They all cause skin eruptions classified as maculopapular.

Measles

Most of us living in the United States don't think twice about measles. It is just another disease we receive vaccination against when we are children. But every year, hundreds of thousands of children in the developing world die from this disease (at last count 430 a day), even though an extremely effective vaccine has been available since 1963. Health campaigns all over the world seek to make measles vaccine available to all, and have been very effective in doing so. Roughly 85% of children throughout the world received a single dose of measles vaccine in 2010, an increase of over 10% since 2000. Consequently, since 2002, worldwide deaths from measles have dropped 74%. Ironically, it seems that more work and education needs to be done in developed countries now. Many parents are opting not to have their children vaccinated, due to unfounded fears about the link between the vaccine and autism (see Insight 15.4). We would do well to remember that before the vaccine was introduced, measles killed 6 million people each year worldwide.

Measles is also known as **rubeola.** Be very careful not to confuse it with the next maculopapular rash disease, rubella.

▶ Signs and Symptoms

The initial symptoms of measles are sore throat, dry cough, headache, conjunctivitis, lymphadenitis, and fever. In a short time, unusual oral lesions called *Koplik's spots* appear as a prelude to the characteristic red maculopapular **exanthem** (eg-zan'-thum) that erupts on the head and then progresses to the trunk and extremities until most of the body is covered **(figure 18.11).** The rash gradually coalesces into red patches that fade to brown.

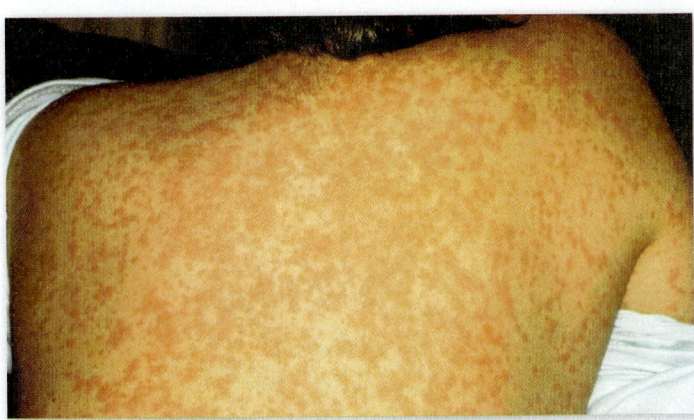

Figure 18.11 The rash of measles.

In a small number of cases, children develop laryngitis, bronchopneumonia, and bacterial secondary infections such as ear and sinus infections. Children afflicted with leukemia or thymic deficiency are especially predisposed to pneumonia because of their lack of a natural T-cell defense.

In about 6% of cases, the virus can cause pneumonia. Affected patients are very ill and often have a characteristic dusky skin color from lack of oxygen. On occasion (1 in 100 cases), measles progresses to encephalitis, resulting in various CNS changes ranging from disorientation to coma. Permanent brain damage or epilepsy can result.

A large number of measles patients experience secondary bacterial infections with *Haemophilus influenzae, Streptococcus pneumoniae,* or other streptococci or staphylococci. These can also lead to pneumonia or upper respiratory tract complications.

The most serious complication is **subacute sclerosing panencephalitis (SSPE)**, a progressive neurological degeneration of the cerebral cortex, white matter, and brain stem. Its incidence is approximately one case in a million measles infections, and it afflicts primarily male children and adolescents. The pathogenesis of SSPE appears to involve a defective virus, one that has lost its ability to form a capsid and be released from an infected cell. Instead, it spreads unchecked through the brain by cell fusion, gradually destroying neurons and accessory cells and breaking down myelin. The disease can cause profound intellectual and neurological impairment. The course of the disease invariably leads to coma and death in a matter of months or years.

Measles during pregnancy has been associated with spontaneous miscarriage and low-birthweight babies, but severe birth defects have not been reported.

▶ Causative Agent

The measles virus is a member of the *Morbillivirus* genus. It is a single-stranded enveloped RNA virus in the Paramyxovirus family.

▶ Pathogenesis and Virulence Factors

The virus implants in the respiratory mucosa and infects the tracheal and bronchial cells. From there it travels to the lymphatic system, where it multiplies and then enters the bloodstream. Viremia carries the virus to the skin and to various organs.

The measles virus induces the cell membranes of adjacent host cells to fuse into large **syncytia** (sin-sish'-uh), giant cells with many nuclei. These cells no longer perform their proper function. The virus seems proficient at disabling many aspects of the host immune response, especially cell-mediated immunity and delayed-type hypersensitivity. The host may be left vulnerable for many weeks after infection; this immune response disruption is one of the reasons that secondary bacterial infections are so common.

▶ Transmission and Epidemiology

Measles is one of the most contagious infectious diseases, transmitted principally by respiratory droplets. Epidemic spread is favored by crowding, low levels of herd immunity, malnutrition, and inadequate medical care. There is no reservoir other than humans, and a person is contagious during the periods of incubation, prodrome phase, and the skin rash phase but usually not during convalescence. Only relatively large, dense populations of susceptible individuals can sustain the continuous chain necessary for transmission. However, this is occurring in many areas around the globe due to families opting against vaccination against measles. This has led to outbreaks of disease among populations in the United States and abroad. In 2011, 222 individuals in the United States developed measles; of these cases, 200 were linked to people who had traveled abroad and brought the virus home with them.

2011 was deemed the worst measles year in the United States in 15 years. One-third of the 2011 cases were hospitalized, although none of them died.

▶ Culture and Diagnosis

The disease can be diagnosed on clinical presentation alone; but if further identification is required, an ELISA test is available that tests for patient IgM to measles antigen, indicating a current infection. For best results, blood should be drawn on the third day of onset or later, because before that time titers of IgM may not be high enough to be detected by the test. Also, the method of comparing acute and convalescent sera may be used to confirm a measles infection after the fact. As you may recall from chapter 17, much higher IgG titers 14 days after onset when compared to titers at day 1 or 2 are a clear indication of current or recent infection. This knowledge allows health care providers to be on the lookout for complications and to be ahead of the game if a person who has had contact with the patient presents with similar symptoms.

▶ Prevention

The MMR vaccine (for measles, mumps, and rubella) contains live attenuated measles virus, which confers protection for about 20 years. Measles immunization is recommended for all healthy children at the age of 12 to 15 months, with a booster before the child enters kindergarten. Failing that, the preadolescent health check serves as a good time to get the second dose of measles vaccine. There is also an MMRV vaccine, which includes the varicella vaccine strain in addition to the measles, mumps, and rubella strains.

▶ Treatment

Treatment relies on reducing fever, suppressing cough, and replacing lost fluid. Complications require additional remedies to relieve neurological and respiratory symptoms and to sustain nutrient, electrolyte, and fluid levels. Therapy includes antibiotics for bacterial complications and doses of immune globulin. Vitamin A supplements are recommended by some physicians; they have been found effective in reducing the symptoms and decreasing the rate of complications.

Rubella

This disease is also known as German measles. Rubella is derived from the Latin for "little red," and that's a good way to remember it because it causes a relatively minor rash disease with few complications. Sometimes it is called the 3-day measles. The only exception to this mild course of events is when a fetus is exposed to the virus while in its mother's womb (*in utero*). Serious damage can occur, and for that reason women of childbearing years must be sure to have been vaccinated well before they plan to conceive.

▶ Signs and Symptoms

The two clinical forms of rubella are referred to as postnatal infection, which develops in children or adults, and **congenital** (prenatal) infection of the fetus, expressed in the newborn as various types of birth defects.

Postnatal Rubella   During an incubation period of 2 to 3 weeks, the rubella virus multiplies in the respiratory epithelium, infiltrates local lymphoid tissue, and enters the bloodstream. Early symptoms include malaise, mild fever, sore throat, and lymphadenopathy. The rash of pink macules and papules first appears on the face and progresses down the trunk and toward the extremities, advancing and resolving in about 3 days. The rash is milder looking than the measles rash (see Disease Table 18.7). Adult rubella is often accompanied by joint inflammation and pain rather than a rash. Very occasionally, complications such as arthralgia/arthritis, or even encephalitis, can occur but more often in adults than in children.

Congenital Rubella   Rubella is a strongly teratogenic virus. Transmission of the rubella virus to a fetus *in utero* can result in a serious complication called **congenital rubella (figure 18.12)**. The mother is able to transmit the virus even if she is asymptomatic. Fetal injury varies according to the time of infection. Infection in the first trimester is most likely to induce miscarriage or multiple permanent defects in the newborn. The most common of these is deafness and may be the only defect seen in some babies.

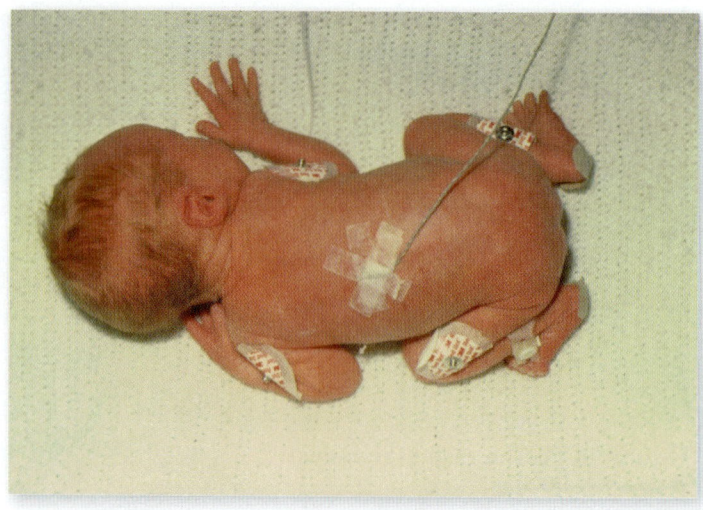

Figure 18.12 An infant born with congenital rubella can manifest a papular pink or purple rash.

Other babies may experience cardiac abnormalities, ocular lesions, deafness, and mental and physical retardation in varying combinations. Less drastic sequelae that usually resolve in time are anemia, hepatitis, pneumonia, carditis, and bone infection.

▶ Causative Agent

The rubella virus is a *Rubivirus*, in the family Togavirus. It is a nonsegmented single-stranded RNA virus with a loose lipid envelope. There is only one known serotype of the virus, and humans are the only natural host. Its envelope contains two different viral proteins.

▶ Pathogenesis and Virulence Factors

The course of disease in postnatal rubella is mostly unremarkable; but when exposed to a fetus, the virus creates havoc. It has the ability to stop mitosis, which is an important process in a rapidly developing embryo and fetus. It also induces apoptosis of normal tissue cells. This inappropriate cell death can do irreversible harm to organs it affects. Last, the virus damages vascular endothelium, leading to poor development of many organs.

▶ Transmission and Epidemiology

Rubella is an endemic disease with worldwide distribution. Infection is initiated through contact with respiratory secretions and occasionally urine. The virus is shed during the prodromal phase and up to a week after the rash appears. Congenitally infected infants are contagious for a much longer period of time. Because the virus is only moderately communicable, close living conditions are required for its spread. Although this disease is well controlled overall in the United States, recent outbreaks have led to an unprecedented increase in infections over the past few years. Normally, most sporadic cases are reported among adolescents and young adults in military training camps, colleges, and summer camps. Note that it is always a concern that nonimmune women of childbearing age may be caught up in this cycle, raising the prospect of congenital rubella.

▶ **Culture and Diagnosis**

Diagnosing rubella relies on the same twin techniques discussed earlier for measles. Because it mimics other diseases, rubella should not be diagnosed on clinical grounds alone. IgM antibody to rubella virus can be detected early using an ELISA technique or a latex-agglutination card. Other conditions and infections can lead to false positives, however, and the IgM test should be augmented by an acute and convalescent measurement of IgG antibody. It is important to know whether the infection is indeed rubella, especially in women, because if so, they will be immune to reinfection.

▶ **Prevention**

The attenuated rubella virus vaccine is usually given to children in the combined form (MMR or MMRV vaccination) at 12 to 15 months and a booster at 4 or 6 years of age.

▶ **Treatment**

Postnatal rubella is generally benign and requires only symptomatic treatment. No specific treatment is available for the congenital manifestations.

Fifth Disease

This disease, more precisely called *erythema infectiosum,* is so named because about 100 years ago it was the fifth of the diseases recognized by doctors to cause rashes in children. The first four were scarlet fever, measles, rubella, and another rash called "fourth disease" that was thought to be a distinct illness but was later found to be misdiagnosed rubella or scarlet fever. The name "fifth disease" has stuck for this viral condition. It is a very mild disease that often results in a characteristic "slapped-cheek" appearance because of a confluent reddish rash that begins on the face. Within 2 days, the rash spreads on the body but is most prominent on the arms, legs, and trunk. The rash is maculopapular in appearance, and the blotches tend to run together rather than to appear as distinct bumps. The illness is rather mild, featuring low-grade fever and malaise and lasting 5 to 10 days. The rash may persist for days to weeks, and it tends to recur under stress or with exposure to sunlight. As with almost any infectious agent, it can cause more serious disease in people with underlying immune disease.

The causative agent is parvovirus B19. You may have heard of "parvo" as a disease of dogs, but parvovirus B19 does not cause disease in dogs, nor does dog "parvo" cause disease in humans. Fifth disease is usually diagnosed by the clinical presentation, but sometimes it is helpful to rule out rubella by testing for IgM against rubella. Specific serological tests for fifth disease are available if they are considered necessary.

This infection is very contagious. It is transmitted through respiratory droplets or even direct contact. It can be transmitted through the placenta, with a range of possible effects, from no symptoms to stillbirth. There is no vaccine and no treatment for this usually mild disease.

Roseola

This disease is common in young children and babies and is sometimes known as "sixth disease." It can result in a maculopapular rash, but a high percentage (up to 70%) of cases proceeds without development of the rash stage. Children sick with this disease exhibit a high fever (up to 41°C, or 105°F) that comes on quickly and lasts for up to 3 days. Seizures may occur during this period, but other than that patients remain alert and do not act terribly ill. On the fourth day, the fever disappears, and it is at this point that a rash can appear, first on the chest and trunk and less prominently on the face and limbs. By the time the rash appears, the disease is almost over.

Roseola is caused by a human herpesvirus called HHV-6. Like all herpesviruses, it can remain latent in its host indefinitely after the disease has cleared. Very occasionally, the virus reactivates in childhood or adulthood, leading to mononucleosis-like or hepatitis-like symptoms. Immunocompetent hosts generally do not experience reactivation. It is thought that 100% of the U.S. population becomes infected with this virus by adulthood, though some cases are subclinical or asymptomatic. HHV-6 can cause severe disseminated disease in AIDS patients and other people with compromised immunity.

There is no vaccine. For uncomplicated roseola, no treatment is recommended. Severe manifestations in immunocompromised patients can be treated with ganciclovir.

Scarlet Fever

To complete our survey of infections that can cause maculopapular rashes, we include a disease that has primary symptoms elsewhere but can produce a distinctive red rash on the skin as well. Scarlet fever is most often the result of a respiratory infection with *Streptococcus pyogenes* (most often, pharyngitis). Occasionally, scarlet fever will follow a streptococcal skin infection, such as impetigo or cellulitis. If the *S. pyogenes* strain contains a bacteriophage carrying a gene for an exotoxin called erythrogenic toxin, scarlet fever can result. More details on scarlet fever are given in chapter 21; it is included here mainly for purposes of differentiating the rash from the others in this group **(Disease Table 18.7).**

Disease Table 18.7   Maculopapular Rash Diseases

Disease	Measles (Rubeola)	Rubella	Fifth Disease
Causative Organism(s)	Measles virus	Rubella virus	Parvovirus B19
Most Common Modes of Transmission	Droplet contact	Droplet contact	Droplet contact, direct contact
Virulence Factors	Syncytium formation, ability to suppress CMI	In fetuses: inhibition of mitosis, induction of apoptosis, and damage to vascular endothelium	–
Culture/Diagnosis	ELISA for IgM, acute/convalescent IgG	Acute IgM, acute/convalescent IgG	Usually diagnosed clinically
Prevention	Live attenuated vaccine (MMR or MMRV)	Live attenuated vaccine (MMR or MMRV)	–
Treatment	No antivirals; vitamin A, antibiotics for secondary bacterial infections	–	–
Distinguishing Features of the Rashes	Starts on head, spreads to whole body, lasts over a week	Milder red rash, lasts approximately 3 days	"Slapped-face" rash first, spreads to limbs and trunk, tends to be confluent rather than distinct bumps
Epidemiological Features	Incidence increasing in North America; in developing countries incidence is 30 million cases/yr and 1 million deaths	3 cases reported in United States in 2009; worldwide: 100,000 infants/yr born with congenital rubella syndrome	60% of population seropositive by age 20
Appearance of Lesions			

Wart-like Eruptions

All types of warts are caused by viruses. Most common warts you have seen on yourself and others are probably caused by one of more than 100 human papillomaviruses, or HPVs. HPVs are also the cause of genital warts, described in chapter 23. Another virus causes a condition called **molluscum contagiosum,** which causes bumps that may look like warts.

Warts

Warts, also known as **papillomas,** can develop in nearly all individuals. Children seem to get them more frequently than

adults, and there is speculation that people gradually build up immunity to the various HPVs that they encounter over time, as is the case with the viruses that cause the common cold.

The warts caused by HPV are benign, squamous epithelial growths. Some HPVs can infect mucous membranes; others invade skin. The appearance and seriousness of the infection vary somewhat from one anatomical region to another. Painless, elevated, rough growths on the fingers and occasionally on other body parts are called common, or seed, warts **(Disease Table 18.8).** These growths commonly occur in children and young adults. Just as certain types of HPVs are associated with particular outcomes in the genital area, common warts are most often caused by HPV-2, -4, -27, and -29. **Plantar warts** are often caused by HPV-1. They are

Roseola	Scarlet Fever
Human herpesvirus 6	*Streptococcus pyogenes* (lysogenized)
Unknown	Droplet or direct contact
Ability to remain latent	Erythrogenic toxin
Usually diagnosed clinically	Examination of skin lesions, throat culture (beta-hemolytic on blood agar, sensitive to bacitracin, rapid antigen tests)
–	Hygiene practices
–	Penicillin, clindamycin
High fever precedes rash stage; rash not always present	Sandpaper feel to affected skin; severe sore throat
> 90% seropositive; 90% of disease cases occur before age of 2	In United States, 10% of those with strep throat progress to scarlet fever

The warts caused by papillomaviruses are usually distinctive enough to permit reliable clinical diagnosis without much difficulty. However, a biopsy and histological examination can help clarify ambiguous cases. Warts disappear on their own 60% to 70% of the time, usually over the course of 2 to 3 years. Physicians do approve of home remedies for resolving warts, including nonprescription salicylic acid preparations. Physicians have other techniques for removing warts, including a number of drugs and/or cryosurgery. No treatment guarantees that the viruses are eliminated because the virus can integrate into the DNA of the host; therefore, warts can always grow back.

Molluscum Contagiosum

This disease is distributed throughout the world, with highest incidence occurring on certain regions of the Pacific Islands, although its incidence in North America has been increasing since the 1980s. Skin lesions take the form of smooth, waxy nodules on the face, trunk, and limbs. The firm nodules may be indented in the middle (see Disease Table 18.8), and they contain a milky fluid containing epidermal cells filled with virus particles in intracytoplasmic inclusion bodies. This condition is common in children, where it most often causes nodules on the face, arms, legs, and trunk. In adults, it appears mostly in the genital areas. In immunocompromised patients, the lesions can be more disfiguring and more widespread on the body. It is particularly common in AIDS patients and often presents as facial lesions.

The molluscum contagiosum virus is a poxvirus, containing double-stranded DNA and possessing an envelope. It is spread via direct contact and also through fomites. Adults who acquire this infection usually acquire it through sexual contact. Autoinoculation can spread the virus from existing lesions to new places on the body, resulting in new nodules.

The condition may be diagnosed on clinical appearance alone, or a skin biopsy may be performed and histological analysis undertaken. A clinician can perform a more simple "squash procedure," in which fluid from the lesion is extracted onto a microscope slide, squashed by another microscope slide, stained, and examined for the presence of the characteristic inclusion bodies in the epithelial cells. PCR can also be used to detect the virus in skin lesions. In most cases, no treatment is indicated, although a physician may remove the lesions or treat them with a topical chemical. Treatment of lesions, however, does not ensure elimination of the virus (Disease Table 18.8).

deep, painful papillomas on the soles of the feet. Flat warts (HPV types 3, 10, 28, and 49) are smooth, skin-colored lesions that develop on the face, trunk, elbows, and knees.

The warts contain variable amounts of virus. Transmission occurs through direct contact, and often warts are transmitted from one part of the body to another by autoinoculation. Because the viruses are fairly stable in the environment, they can also be transmitted indirectly from towels, shower stalls, or pedicure equipment, where they persist inside the protective covering of sloughed-off keratinized skin cells. The incubation period can be from 1 to 8 months. Almost all nongenital warts are harmless, and they tend to resolve themselves over time. Rarely, a wart can become malignant when caused by specific strains of HPV.

Disease Table 18.8   Warts and Wart-like Eruptions

Causative Organism(s)	Human papillomaviruses	Molluscum contagiosum viruses
Most Common Modes of Transmission	Direct contact, autoinoculation, indirect contact	Direct contact, including sexual contact, autoinoculation
Virulence Factors	–	–
Culture/Diagnosis	Clinical diagnosis, also histology, microscopy, PCR	Clinical diagnosis, also histology, microscopy, PCR
Prevention	Avoid contact	Avoid contact
Treatment	Home treatments, cryosurgery (virus not eliminated)	Usually none, although mechanical removal can be performed (virus not eliminated)
Epidemiological Features	United States: 6 million new cases/yr, prevalence is 13%; worldwide HPV prevalence: 12%	Worldwide incidence is 2%–3% with greater distribution in tropical areas and in overcrowded communities where there is poor hygiene
Appearance of Lesions		

Large Pustular Skin Lesions

Leishmaniasis

Two infections that result in large lesions (greater than a few millimeters across) deserve mention in this chapter on skin infections. The first is leishmaniasis, a zoonosis transmitted among various mammalian hosts by female sand flies. This infection can express itself in several different forms depending on which species of the protozoan *Leishmania* is involved. Cutaneous leishmaniasis is a localized infection of the capillaries of the skin caused by *L. tropica,* found in Mediterranean, African, and Indian regions. A form of mucocutaneous leishmaniasis called espundia is caused by *L. braziliensis,* endemic to parts of Central and South America. It affects both the skin and mucous membranes. Another form of this infection is systemic leishmaniasis.

Leishmania is transmitted to the mammalian host by the sand fly when it ingests the host's blood. The disease is endemic to equatorial regions that provide favorable conditions for the sand fly. Numerous wild and domesticated animals, especially dogs, serve as reservoirs for the protozoan. Although humans are usually accidental hosts, the flies freely feed on them. At particular risk are travelers or immigrants who have never had contact with the protozoan and lack specific immunity. However, recent outbreaks of leishmaniasis have occurred in the Dallas–Fort Worth region of Texas, indicating that a species of this protozoan may be spreading northward from Mexico and becoming endemic to southern areas of the United States.

In cutaneous leishmaniasis, a small red papule occurs at the site of the bite and spreads laterally into a large ulcer **(Disease Table 18.9).** The edges of the ulcer are raised and the base is moist. It can be filled with a serous/purulent exudate or covered with a crust. Satellite lesions may occur. Mucocutaneous leishmaniasis usually begins with a skin lesion on the head or face and then progresses to single or multiple lesions, usually in the mouth and nose. Lesions can be quite extensive, eventually involving and disfiguring the hard palate, the nasal septum, and the lips.

Disease Table 18.9 Large Pustular Skin Lesions

Disease	Leishmaniasis	Cutaneous Anthrax
Causative Organism(s)	*Leishmania* spp.	*Bacillus anthracis*
Most Common Modes of Transmission	Biological vector	Direct contact with endospores
Virulence Factors	Multiplication within macrophages	Endospore formation; capsule, lethal factor, edema factor (see chapter 20)
Culture/Diagnosis	Culture of protozoa, microscopic visualization	Culture on blood agar; serology, PCR performed by CDC
Prevention	Avoiding sand fly	Avoid contact; vaccine available but not widely used
Treatment	Sodium stibogluconate, pentamidine	Ciprofloxacin, doxycycline, levofloxacin
Distinguishing Features	Mucocutaneous and systemic forms	Can be fatal
Epidemiological Features	Untreated visceral leishmaniasis mortality rate is 100%; 10% for cutaneous leishmaniasis	Untreated cutaneous anthrax mortality rate: 20%; treated mortality rate less than 1%
Appearance of Lesions		

Cutaneous Anthrax

This form of anthrax is the most common and least dangerous version of infection with *Bacillus anthracis*. (The spectrum of anthrax disease is discussed fully in chapter 20.) It is caused by endospores entering the skin through small cuts or abrasions. Germination and growth of the pathogen in the skin are marked by the production of a papule that becomes increasingly necrotic and later ruptures to form a painless, black **eschar** (ess'-kar) (see Disease Table 18.9). In the fall of 2001, 11 cases of cutaneous anthrax occurred in the United States as a result of bioterrorism (along with 11 cases of inhalational anthrax). Mail workers and others contracted the infection when endospores were sent through the mail. The infection can be naturally transmitted by contact with hides of infected animals (especially goats).

Left untreated, even the cutaneous form of anthrax is fatal approximately 20% of the time. A vaccine exists but is recommended only for high-risk persons and the military. Upon suspicion of cutaneous anthrax, ciprofloxacin, levofloxacin, and/or doxycycline should be used initially. If the isolate is found to be sensitive to penicillin, patients can be switched to that drug **(Disease Table 18.9).**

Ringworm (Cutaneous Mycoses)

A group of fungi that is collectively termed **dermatophytes** causes a constellation of integument conditions. These mycoses are strictly confined to the nonliving epidermal tissues (stratum corneum) and their derivatives (hair and nails). All these conditions have different names that begin with the word **tinea** (tin'-ee-ah), which derives from the erroneous belief that they were caused by worms. That misconception is also the reason these diseases are often called *ringworm*—ringworm of the scalp (tinea capitis), beard (tinea barbae), body (tinea corporis), groin (tinea cruris), foot (tinea pedis), and hand (tinea manuum). (Don't confuse these "tinea" terms with genus and species names. It is simply an old practice for naming conditions.) Most of these conditions are caused by one of three different dermatophytes. The signs and symptoms of ringworm infections are summarized in **table 18.1.**

▶ Causative Agents

There are about 39 species in the genera *Trichophyton*, *Microsporum*, and *Epidermophyton* that can cause the tinea conditions. The causative agent of a given type of ringworm

Table 18.1 Signs and Symptoms of Cutaneous Mycoses

Ringworm of the Scalp (Tinea Capitis)	This mycosis results from the fungal invasion of the scalp and the hair of the head, eyebrows, and eyelashes.	
Ringworm of the Beard (Tinea Barbae)	This tinea, also called *barber's itch*, affects the chin and beard of adult males. Although once a common aftereffect of unhygienic barbering, it is now contracted mainly from animals.	
Ringworm of the Body (Tinea Corporis)	This extremely prevalent infection of humans can appear nearly anywhere on the body's glabrous (smooth and bare) skin.	
Ringworm of the Groin (Tinea Cruris)	Sometimes known as *jock itch*, crural ringworm occurs mainly in males on the groin, perianal skin, scrotum, and, occasionally, the penis. The fungus thrives under conditions of moisture and humidity created by sweating.	
Ringworm of the Foot (Tinea Pedis)	Tinea pedis has more colorful names as well, including athlete's foot and jungle rot. Infections begin with blisters between the toes that burst, crust over, and can spread to the rest of the foot and nails.	
Ringworm of the Nail (Tinea Unguium)	Fingernails and toenails, being masses of keratin, are often sites for persistent fungus colonization. The first symptoms are usually superficial white patches in the nail bed. A more invasive form causes thickening, distortion, and darkening of the nail.	

varies from one geographic location to another and is not restricted to a particular genus and species. These fungi are so closely related and morphologically similar that they can be difficult to differentiate. Various species exhibit unique macroconidia, microconidia, and unusual types of hyphae. In general, *Trichophyton* produces thin-walled, smooth macroconidia and numerous microconidia (**figure 18.13a**);

Microsporum produces thick-walled, rough macroconidia and sparser microconidia (**figure 18.13b**); and *Epidermophyton* has ovoid, smooth, clustered macroconidia and no microconidia (**figure 18.13c**).

The presenting symptoms of a cutaneous mycosis occasionally are so dramatic and suggestive of these genera that no further testing is necessary. In most cases, however,

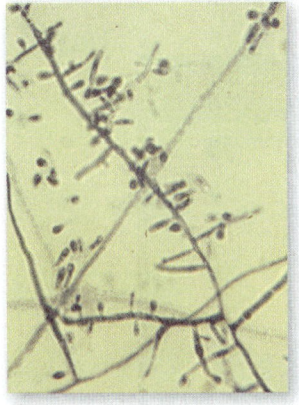

(a)

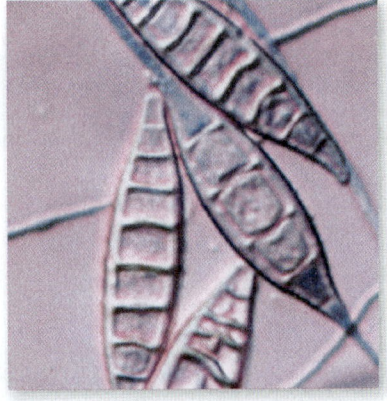

(b)

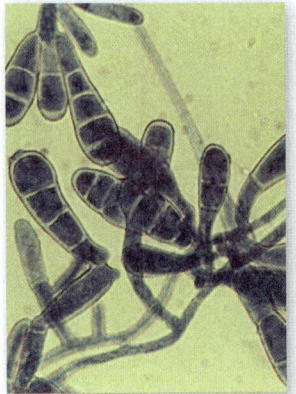

(c)

Figure 18.13 Examples of dermatophyte spores.
(a) Regular, numerous microconidia of *Trichophyton*. (b) Macroconidia of *Microsporum canis*, a cause of ringworm in cats, dogs, and humans. (c) Smooth-surfaced macroconidia in clusters characteristic of *Epidermophyton*.

direct microscopic examination and culturing are required. Diagnosis of tinea of the scalp caused by some species is aided by use of a long-wave ultraviolet lamp that causes infected hairs to fluoresce. Samples of hair, skin scrapings, and nail debris treated with heated potassium hydroxide (KOH) show a thin, branching fungal mycelium if infection is present.

▶ **Pathogenesis and Virulence Factors**

The dermatophytes have the ability to invade and digest keratin, which is naturally abundant in the cells of the stratum corneum. The fungi do not invade deeper epidermal layers. Important factors that promote infection are the hardiness of the dermatophyte spores (they can last for years on fomites), presence of abraded skin, and intimate contact. Most infections exhibit a long incubation period (months), followed by localized inflammation and allergic reactions to fungal proteins. As a general rule, infections acquired from animals and soil cause more severe reactions than do infections acquired from other humans, and infections eliciting stronger immune reactions are resolved faster.

▶ **Transmission and Epidemiology**

Transmission of the fungi that cause these diseases is direct and indirect contact with other humans or with infected animals. Some of these fungi can be acquired from the soil.

▶ **Prevention and Treatment**

The only way to prevent these infections is to avoid contact with the dermatophytes, which is impractical. Keeping susceptible skin areas dry is helpful. Treatment of ringworm is based on the knowledge that the dermatophyte is feeding on dead epidermal tissues. These regions undergo constant replacement from living cells deep in the epidermis, so if multiplication of the fungus can be blocked, the fungus will eventually be sloughed off along with the skin or nail. Unfortunately, this takes time. Most infections are treated with topical antifungal agents. Ointments containing tolnaftate, miconazole, itraconazole, terbinafine, or thiabendazole are applied regularly for several weeks. Some drugs work by speeding up loss of the outer skin layer. Often tinea capitis is treated with oral terbinafine.

Superficial Mycoses

Agents of **superficial mycoses** involve the outer epidermal surface and are ordinarily innocuous infections with cosmetic rather than inflammatory effects. Tinea versicolor is caused by the yeast genus *Malassezia*, a genus that has at least 10 species living on human skin. The yeast feeds on the high oil content of the skin glands. Even though these yeasts are very common normal biota (carried by nearly 100% of humans tested), in some people their growth elicits mild, chronic scaling and interferes with production of pigment by melanocytes. The trunk, face, and limbs may take on a mottled appearance **(figure 18.14).** The disease is most pronounced in young people who are frequently exposed to the sun, because the area affected doesn't tan well. Other superficial skin conditions in which *Malassezia* is implicated are folliculitis, psoriasis, and seborrheic dermatitis (dandruff). It is also occasionally associated with systemic infections and catheter-associated sepsis in compromised patients **(Disease Table 18.10).**

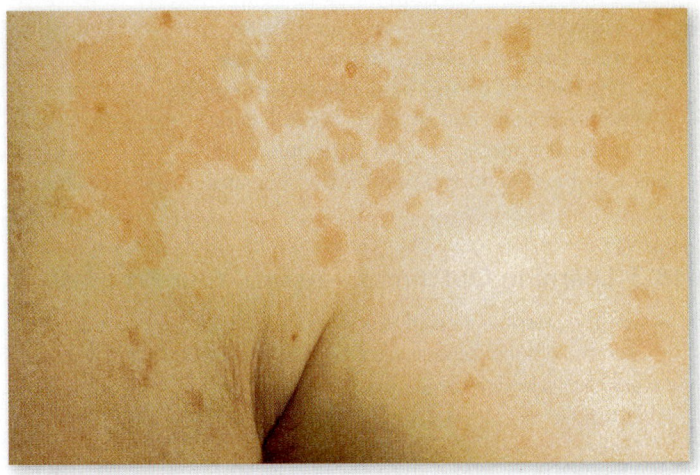

Figure 18.14 Tinea versicolor. Mottled, discolored skin pigmentation is characteristic of superficial skin infection by *Malassezia furfur.*

Disease Table 18.10 Cutaneous and Superficial Mycoses

Disease	Cutaneous Infections	Superficial Infections (Tinea Versicolor)
Causative Organism(s)	*Trichophyton, Microsporum, Epidermophyton*	*Malassezia* species
Most Common Modes of Transmission	Direct and indirect contact, vehicle (soil)	Endogenous "normal biota"
Virulence Factors	Ability to degrade keratin, invoke hypersensitivity	–
Culture/Diagnosis	Microscopic examination, KOH staining, culture	Usually clinical, KOH can be used
Prevention	Avoid contact	None
Treatment	Topical tolnaftate, itraconazole, terbinafine, miconazole, thiabendazole, oral terbinafine	Topical antifungals
Epidemiological Features	Among schoolchildren, 0%–19% prevalence, in humid climates up to 30%	Highest incidence among adolescents

INSIGHT 18.3   Do Mosquitoes Love You?

Has this ever happened to you? You spend the weekend camping or you go for a hike, and you come back covered in mosquito bites, but your friends remain relatively unscathed? Did you forget to pack your insect repellant, or do the bugs just love to feast on you? According to research conducted at Wageningen University in the Netherlands, some people may be more attractive to mosquitoes than others. Nels Verhulst and his team of researchers studied the mosquito that is one of the main vectors for malaria, called *Anopheles gambiae*. These mosquitoes are able to locate their host through physical cues such as heat and moisture and chemical cues given off as scent from the skin.

Human sweat is odorless, so why do we stink when we sweat? Blame the bacteria. The bacteria on our skin metabolize our sweaty secretions and give off volatile compounds that cause body odor. Each individual has a distinctive odor based on the type and number of bacteria living on their skin. Verhulst and his group studied the skin bacteria of 48 men and were able to correlate the microbial community to their attractiveness to mosquitoes. They found that individuals with a diverse population of microbes on their skin were *less* attractive to mosquitoes than those that had fewer species of bacteria in their skin biota. Specifically, they found that individuals that were populated with *Pseudomonas* and *Variovorax* were less attractive to mosquitoes. Individuals that harbored *Leptotrichia, Delftia*, and *Actinobacteria* were more likely to get bitten. These findings are important in understanding the host range of mosquitoes and the transmission of malaria. So, the next time you go camping, make sure to pack DEET and maybe throw in a little *Pseudomonas* and *Variovorax* to avoid the bite.

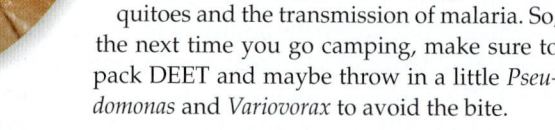

Your skin biota may determine your attractiveness to mosquitoes.

Source: 2011. PLoS ONE, vol. 6, no. 12, e28991.

18.3 Learning Outcomes—Assess Your Progress

4. List the possible causative agents for each of the infectious skin conditions: acne, impetigo, cellulitis, staphylococcal scalded skin syndrome, gas gangrene, vesicular/pustular rash diseases, maculopapular rash diseases, wart-like eruptions, large pustular skin lesions, and cutaneous mycoses.

5. Identify which of these conditions are transmitted to the respiratory tract through droplet contact.

6. List the skin conditions for which vaccination is recommended.

7. Summarize methods used to distinguish infections caused by *Staphylococcus aureus* and *Streptococcus pyogenes*, and discuss the spectrum of skin and tissue diseases caused by each.

8. Provide an update of the status of MRSA infections in the United States.

9. Discuss the relative dangers of rubella and rubeola viruses in different populations.

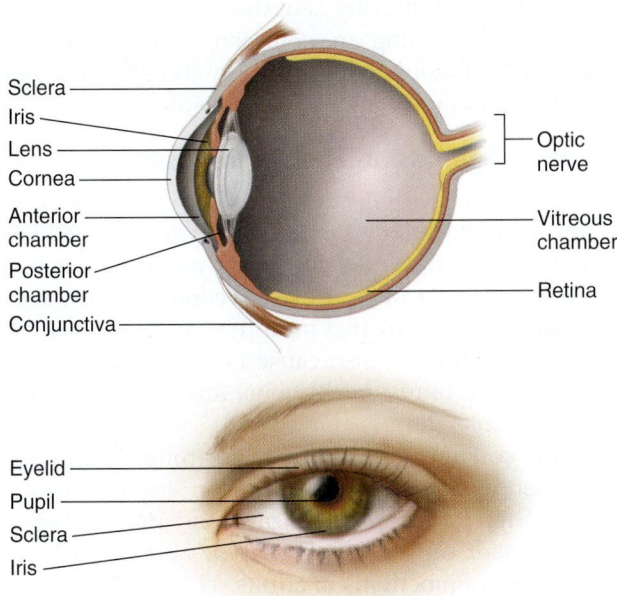

Figure 18.15 The anatomy of the eye.

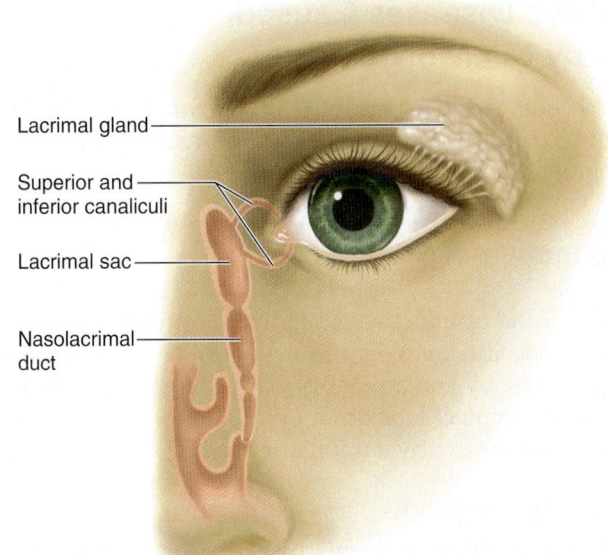

Figure 18.16 The lacrimal apparatus of the eye.

18.4 The Surface of the Eye and Its Defenses

The eye is a complex organ with many different tissue types, but for the purposes of this chapter we consider only its exposed surfaces, the *conjunctiva* and the *cornea* (figure 18.15). The **conjunctiva** is a very thin membranelike tissue that covers the eye (except for the cornea) and lines the eyelids. It secretes an oil- and mucus-containing fluid that lubricates and protects the eye surface. The **cornea** is the dome-shaped central portion of the eye lying over the iris (the colored part of the eye). It has five to six layers of epithelial cells that can regenerate quickly if they are superficially damaged. It has been called "the windshield of the eye."

The eye's best defense is the film of tears, which consists of an aqueous fluid, oil, and mucus. The tears are formed in the lacrimal gland at the outer and upper corner of each eye (figure 18.16), and they drain into the lacrimal duct at the inner corner. The aqueous portion of tears contains sugars, lysozyme, and lactoferrin. These last two substances have antimicrobial properties. The mucous layer contains proteins and sugars and plays a protective role. And, of course, the flow of the tear film prevents the attachment of microorganisms to the eye surface.

Because the eye's primary function is vision, anything that hinders vision would be counterproductive. For that reason, inflammation does not occur in the eye as readily as it does elsewhere in the body. Flooding the eye with fluid containing a large number of light-diffracting objects such as lymphocytes and phagocytes in response to every irritant would mean almost constantly blurred vision. So even though the eyes are relatively vulnerable to infection (not being covered by keratinized epithelium), the evolution of the vertebrate eye has of necessity favored reduced innate immunity. This characteristic is sometimes known as **immune privilege.**

The specific immune response, involving B and T cells, is also somewhat restricted in the eye. The anterior chamber (see figure 18.15) is largely cut off from the blood supply. Lymphocytes that do gain access to this area are generally less active than lymphocytes elsewhere in the body.

18.4 Learning Outcomes—Assess Your Progress

10. Describe the important anatomical features of the eye.
11. List the natural defenses present in the eye.

18.5 Normal Biota of the Eye

The normal biota of the eye—so far as is currently known—is generally sparse. When people are tested, up to 20% have no recoverable (i.e., culturable) bacteria in their eyes. The few bacteria that are found resemble the normal biota of the skin—namely, diphtheroids, coagulase-negative staphylococci, *Micrococcus*, nonhemolytic streptococci, and some yeast. *Neisseria* species can also live on the surface of the eye.

Defenses and Normal Biota of the Eyes		
	Defenses	**Normal Biota**
Eyes	Mucus in conjunctiva and in tears; lysozyme and lactoferrin in tears	Sparsely populated with *Staphylococcus epidermidis*, *Micrococcus*, *Streptococcus*, and *Corynebacterium* species

18.5 Learning Outcomes—Assess Your Progress

12. List the types of normal biota presently known to occupy the eye.

18.6 Eye Diseases Caused by Microorganisms

In this section, we cover the infectious agents that cause diseases of the surface structures of the eye—namely, the cornea and conjunctiva.

Conjunctivitis

Infection of the conjunctiva is relatively common. It can be caused by specific microorganisms that have a predilection for eye tissues, by contaminants that proliferate due to the presence of a contact lens or an eye injury, or by accidental inoculation of the eye by a traumatic event.

▶ Signs and Symptoms

Just as there are many different causes of conjunctivitis, there are many different clinical presentations. Inflammation of this tissue almost always causes a discharge of some sort. Most bacterial infections produce a milky discharge, whereas viral infections tend to produce a clear exudate. It is typical for a patient to wake up in the morning with an eye "glued" shut by secretions that have accumulated and solidified through the night. Some conjunctivitis cases are caused by an allergic response, and these often produce copious amounts of clear fluid as well. The pain generally is mild, although often patients report a gritty sensation in their eye(s). Redness and eyelid swelling are common, and in some cases patients report photophobia (sensitivity to light). The informal name for common conjunctivitis is pinkeye.

▶ Causative Agents and Their Transmission

Cases of neonatal eye infection caused by *Neisseria gonorrhoeae* or *Chlamydia trachomatis* are usually transmitted vertically from a genital tract infection in the mother (discussed in chapter 23). Either one of these eye infections can lead to serious eye damage if not treated promptly **(figure 18.17).** Note that herpes simplex can also cause neonatal conjunctivitis, but it is usually accompanied by generalized herpes infection (covered in chapter 23).

Bacterial conjunctivitis in other age groups is most commonly caused by *Staphylococcus aureus* or *Streptococcus pneumoniae*, although *Haemophilus influenzae* and *Moraxella* species are also frequent causes. *N. gonorrhoeae* and *C. trachomatis* can also cause conjunctivitis in adults. These infections may result from autoinoculation from a genital infection or from sexual activity, although *N. gonorrhoeae* can be part of the normal biota in the respiratory tract. A wide variety of bacteria, fungi, and protozoa can contaminate contact lenses and lens cases and then be transferred to the eye, resulting in disease that may be very serious. This means of infection is considered vehicle transmission, with the lens or the solution being the vehicle.

Disease Table 18.11	Conjunctivitis		
Disease	**Neonatal Conjunctivitis**	**Bacterial Conjunctivitis**	**Viral Conjunctivitis**
Causative Organism(s)	*Chlamydia trachomatis* or *Neisseria gonorrhoeae*	*Streptococcus pneumoniae, Staphylococcus aureus, Haemophilus influenzae, Moraxella,* and also *Neisseria gonorrhoeae, Chlamydia trachomatis*	Adenoviruses and others
Most Common Modes of Transmission	Vertical	Direct, indirect contact	Direct, indirect contact
Virulence Factors	–	–	–
Culture/Diagnosis	Gram stain and culture	Clinical diagnosis	Clinical diagnosis
Prevention	Screen mothers, apply antibiotic or silver nitrate to newborn eyes	Hygiene	Hygiene
Treatment	Topical and oral antibiotics	Gatifloxacin or levofloxacin ophthalmic solution	None, although antibiotics often given because type of infection not distinguished
Distinguishing Features	In babies <28 days old	Mucopurulent discharge	Serous (clear) discharge
Epidemiological Features	Less than 0.5% in developed world; higher incidence in developing world	More common in children	More common in adults

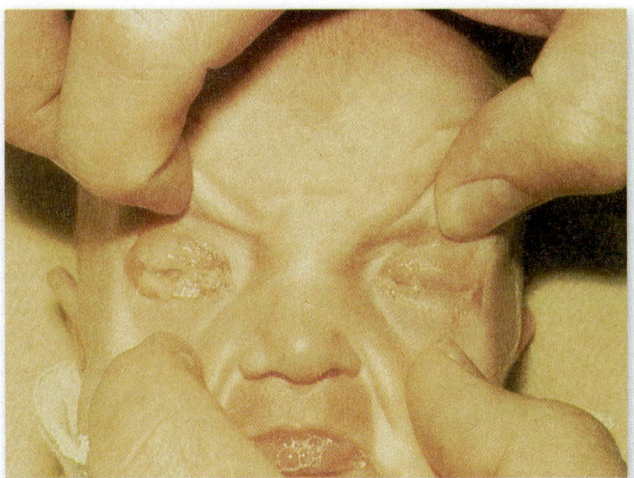

Figure 18.17 Neonatal conjunctivitis.

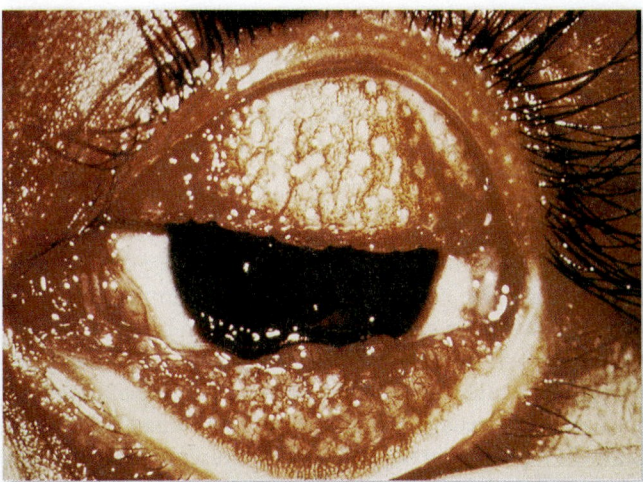

Figure 18.18 Ocular trachoma caused by *C. trachomatis*.

Viral conjunctivitis is commonly caused by adenoviruses, although other viruses may be responsible. (Herpesvirus infection of the eye is discussed later on.) Both bacterial and viral conjunctivitis are transmissible by direct and even indirect contact and are usually highly contagious.

▶ Prevention and Treatment

Good hygiene is the only way to prevent conjunctivitis in adults and children other than neonates. Newborn children in the United States are administered antimicrobials in their eyes after delivery to prevent neonatal conjunctivitis from either *N. gonorrhoeae* or *C. trachomatis*. Treatment of those infections, if they are suspected, is started before lab results are available and usually is accomplished with erythromycin, both topical and oral. If *N. gonorrhoeae* is confirmed, oral therapy is usually switched to ceftriaxone. If antibacterial therapy is prescribed for other conjunctivitis cases, it should cover all possible bacterial pathogens. Ciprofloxacin eye drops are a common choice. Erythromycin or gentamicin are also often used. Because conjunctivitis is usually diagnosed based on clinical signs, a physician may prescribe prophylactic antibiotics even if a viral cause is suspected. If symptoms don't begin improving within 48 hours, more extensive diagnosis may be performed. **Disease Table 18.11** lists the most common causes of conjunctivitis; keep in mind that other microorganisms can also cause conjunctival infections **(Disease Table 18.13).**

Trachoma

Ocular trachoma is a chronic *Chlamydia trachomatis* infection of the epithelial cells of the eye. It is an ancient disease and a major cause of blindness in certain parts of the world. Although a few cases occur annually in the United States, several million cases occur endemically in parts of Africa, Asia, the Middle East, Latin America, and the Pacific Islands. Transmission is favored by contaminated fingers, fomites, fleas, and a hot, dry climate. It is caused by a different *C. trachomatis* strain than the one that causes simple conjunctivitis. Ongoing infection or

many recurrent infections with this strain eventually lead to chronic inflammatory damage and scarring.

The first signs of infection are a mild conjunctival discharge and slight inflammation of the conjunctiva. These symptoms are followed by marked infiltration of lymphocytes and macrophages into the infected area. As these cells build up, they impart a pebbled (rough) appearance to the inner aspect of the upper eyelid **(figure 18.18)**. In time, a vascular pseudomembrane of exudates and inflammatory leukocytes forms over the cornea, a condition called *pannus,* which lasts a few weeks. Chronic and secondary infections can lead to corneal damage and impaired vision. Early

Disease Table 18.12	Trachoma
Causative Organism(s)	*C. trachomatis* serovars A–C
Most Common Modes of Transmission	Indirect contact, mechanical vector
Virulence Factors	Intracellular growth
Culture/Diagnosis	Detection of inclusion bodies in stained preparations
Prevention	Hygiene, vector control, prompt treatment of initial infection
Treatment	Azithromycin or topical erythromycin
Epidemiological Features	Highest prevalence among children between ages 3 and 5. Prevalence as high as 60% in endemic areas

treatment of this disease with azithromycin is highly effective and prevents all of the complications. It is a tragedy that in this day of sophisticated preventive medicine, millions of children worldwide will develop blindness for lack of a few dollars' worth of antibiotics **(Disease Table 18.12).**

Keratitis

Keratitis is a more serious eye infection than conjunctivitis. Invasion of deeper eye tissues occurs and can lead to complete corneal destruction. Any microorganism can cause this condition, especially after trauma to the eye. In developed countries, herpes simplex virus is the most common cause. It can cause keratitis in the absence of predisposing trauma. In developing countries, bacterial and fungal causes are more common.

The usual cause of herpetic keratitis is a "misdirected" reactivation of (oral) herpes simplex virus type 1 (HSV-1). The virus, upon reactivation, travels into the ophthalmic rather than the mandibular branch of the trigeminal nerve. Infections with HSV-2 can also occur as a result of virus exposure during sexual activity, via transfer of the virus from the genital to eye area or through autoinoculation from a recurrent HSV-2 oral infection. Preliminary symptoms are a gritty feeling in the eye, conjunctivitis, sharp pain, and sensitivity to light. Some patients develop characteristic branched or opaque corneal lesions as well. In 25% to 50% of cases, this keratitis is recurrent and chronic and can interfere with vision. Blindness due to herpes is the leading infectious cause of blindness in the United States. The viral condition is treated with topical trifluridine, sometimes supplemented with oral acyclovir.

In the last few years, another form of keratitis has been increasing in incidence. An amoeba called *Acanthamoeba* has been causing serious keratitis cases, especially in people who wear contact lenses. This free-living amoeba is everywhere—it lives in tap water, freshwater lakes, and the like. The infections are usually associated with less-than-rigorous contact lens hygiene, or previous trauma to the eye **(figure 18.19) (Disease Table 18.13).**

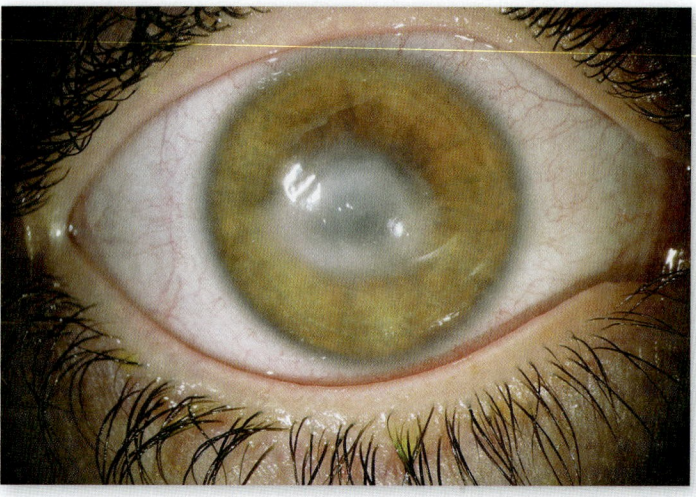

Figure 18.19 *Acanthamoeba* infection of the eye.

River Blindness

River blindness is a chronic parasitic (helminthic) infection. It is endemic in dozens of countries in Latin America, Africa, Asia, and the Middle East. At any given time, approximately 37 million people are infected with the worm called *Onchocerca volvulus* (ong"-koh-ser'-kah volv'-yoo-lus). This organism is a filarial (threadlike) helminthic worm transmitted by small biting vectors called black flies. These voracious flies often attack in large numbers, and it is not uncommon in endemic areas to be bitten several hundred times a day. The disease gets its name from the habitat where these flies are most often found, rural settlements along rivers bordered with overhanging vegetation.

The *Onchocerca* larvae are deposited into a bite wound and develop into adults in the immediate subcutaneous tissues, where disfiguring nodules form within 1 to 2 years after initial contact. Microfilariae (immature worm forms) given off by the adult female migrate via the bloodstream to many

Disease Table 18.13	Keratitis	
Causative Organism(s)	Herpes simplex virus	Miscellaneous microorganisms
Most Common Modes of Transmission	Reactivation of latent virus, although primary infections can occur in the eye	Often traumatic introduction (parenteral)
Virulence Factors	Latency	Various
Culture/Diagnosis	Usually clinical diagnosis; viral culture or PCR if needed	Various
Prevention	–	–
Treatment	Topical trifluridine +/− oral acyclovir	Specific antimicrobials
Epidemiological Features	One-third worldwide population infected; in United States, annual incidence of 500,000	

locations but especially to the eyes. While the worms are in the blood, they can be transmitted to other feeding black flies.

Some cases of onchocerciasis result in a severe itchy rash that can last for years. It was previously thought that the condition was caused by degeneration of the worms and the inflammation and granulomatous lesion formation that result from the release of their antigens. It is in fact the case that the worms eventually invade the entire eye, producing much inflammation and permanent damage to the retina and optic nerve. In fact, half a million people are blind due to this infection worldwide today. In 1999, researchers first discovered large colonies of bacteria called *Wolbachia* living *inside* the *Onchocerca* worms. There is convincing evidence that the damage caused to human tissues is induced by the bacteria rather than by the worms. Of course, the worms serve as the delivery system to the human as it does not appear that the bacteria can infect humans on their own. These bacteria enjoy a mutualistic relationship with their hosts; they are essential for normal *Onchocerca* development.

In regions of high prevalence, it is not unusual for an ophthalmologist to see microfilariae wiggling in the anterior chamber during a routine eye checkup. Microfilariae die in several months, but adults can exist for up to 15 years in skin nodules.

River blindness has been a serious problem in many areas of Africa. In some villages, nearly half of the residents are affected by the disease. A campaign to eradicate onchocerciasis is currently underway, supported by the Carter Center, an organization run by former U.S. President Jimmy Carter. The approach is to treat people with *ivermectin*, a potent antifilarial drug, and to use insecticides to control the black flies. The drug company that manufactures ivermectin has promised to provide the drug for free for as long as the need for it exists. Combined with work by the World Health Organization and the African Programme for Onchocerciasis Control (APOC), there is hope that this disease will be eradicated from these populations in the near future.

Case File 18   *Wrap-Up*

Education is key. This study urges the New York Department of Health to begin education programs in homeless shelters and areas where access to health care is more limited, in recognizing and limiting exposure to CA-MRSA. Efforts should also be made to educate high-risk groups such as individuals with HIV and diabetes in recognizing symptoms and seeking proper treatment.

Of course these infections are increasing all over the country. But some locales have successfully addressed the problem. During nearly the same time period as this study, northern Saskatchewan, Canada, experienced a nearly 20-fold increase in CA-MRSA cases from 8.2 per 10,000 population to 142.6 per 10,000 population in 8 years. In this region, two educational strategies were employed. The first was an educational program for health care providers and an active surveillance program that monitored CA-MRSA in the area. The second approach was community outreach through posters, radio interviews, and slide shows that educated the public about CA-MRSA and soft tissue infections. Radio broadcasts were transmitted in English as well as local aboriginal languages in order to have the greatest impact on the public. A "Germs Away" program was also implemented in the school system to educate children on proper hygiene and prevention of disease transmission as part of the community outreach program. After these educational efforts were implemented, researchers found a two-fold decrease of CA-MRSA infections from 243 to 129 infections per 10,000 population from 2006 to 2008.

Source: 2012. Infect. Cont. Hosp. Epidem. vol. 33, no. 7, p. 725.

Disease Table 18.14	River Blindness
Causative Organism(s)	*Wolbachia* plus *Onchocerca volvulus*
Most Common Modes of Transmission	Biological vector
Virulence Factors	Induction of inflammatory response
Culture/Diagnosis	"Skin snips": small piece of skin in NaCl solution examined under microscope and microfilariae counted
Prevention	Avoiding black fly
Treatment	Ivermectin
Distinguishing Features	Worms often visible in eye
Epidemiological Features	18–40 million afflicted worldwide; 99% of cases found in 30 African countries; also in Central and South America

18.6 Learning Outcomes—Assess Your Progress

13. List the possible causative agents for each of the infectious eye diseases: conjunctivitis, trachoma, keratitis, and river blindness.

14. Discuss why there are distinct differential diagnoses for neonatal and non-neonatal conjunctivitis.

▶ **Summing Up**

Taxonomic Organization Microorganisms Causing Diseases of the Skin and Eyes

Microorganism	Disease	Location of Disease Table
Gram-positive bacteria		
Propionibacterium acnes	Acne	Acne, p. 516
Staphylococcus aureus	Impetigo, cellulitis, scalded skin syndrome, folliculitis, abscesses (furuncles and carbuncles), necrotizing fasciitis	Impetigo, p. 520 Cellulitis, p. 521 Scalded skin syndrome, p. 522; Insight 18.1, p. 517
Streptococcus pyogenes	Impetigo, cellulitis, erysipelas, necrotizing fasciitis, scarlet fever	Cellulitis, p. 521; Insight 18.1, p. 517 scarlet fever, p. 532
Clostridium perfringens	Gas gangrene	Gas gangrene, p. 524
Bacillus anthracis	Cutaneous anthrax	Large pustular skin lesions, p. 535
Gram-negative bacteria		
Neisseria gonorrhoeae	Neonatal conjunctivitis	Conjunctivitis, p. 540
Chlamydia trachomatis	Neonatal conjunctivitis, trachoma	Conjunctivitis, p. 540 Trachoma, p. 541
Wolbachia (in combination with *Onchocerca*)	River blindness	River blindness, p. 543
DNA viruses		
Human herpesvirus 3 (varicella) virus	Chickenpox	Vesicular or pustular rash diseases, p. 527
Variola virus	Smallpox	Vesicular or pustular rash diseases, p. 527
Parvovirus B19	Fifth disease	Maculopapular rash diseases, p. 532
Human herpesvirus 6	Roseola	Maculopapular rash diseases, p. 532
Human papillomavirus	Warts	Warts and wart-like eruptions, p. 534
Molluscum contagiosum virus	Molluscum contagiosum	Warts and wart-like eruptions, p. 534
Herpes simplex virus	Keratitis	Keratitis, p. 542
RNA viruses		
Measles virus	Measles	Maculopapular rash diseases, p. 532
Rubella virus	Rubella	Maculopapular rash diseases, p. 532
Fungi		
Trichophyton	Ringworm	Ringworm, p. 538
Microsporum	Ringworm	Ringworm, p. 538
Epidermophyton	Ringworm	Ringworm, p. 538
Malassezia species	Superficial mycoses	Superficial mycoses, p. 538
Protozoa		
Leishmania spp.	Leishmaniasis	Large pustular skin lesions, p. 535
Acanthamoeba	Keratitis	Keratitis, p. 542
Helminths		
Onchocerca volvulus (in combination with *Wolbachia*)	River blindness	River blindness, p. 543

INFECTIOUS DISEASES AFFECTING
The Skin and Eyes

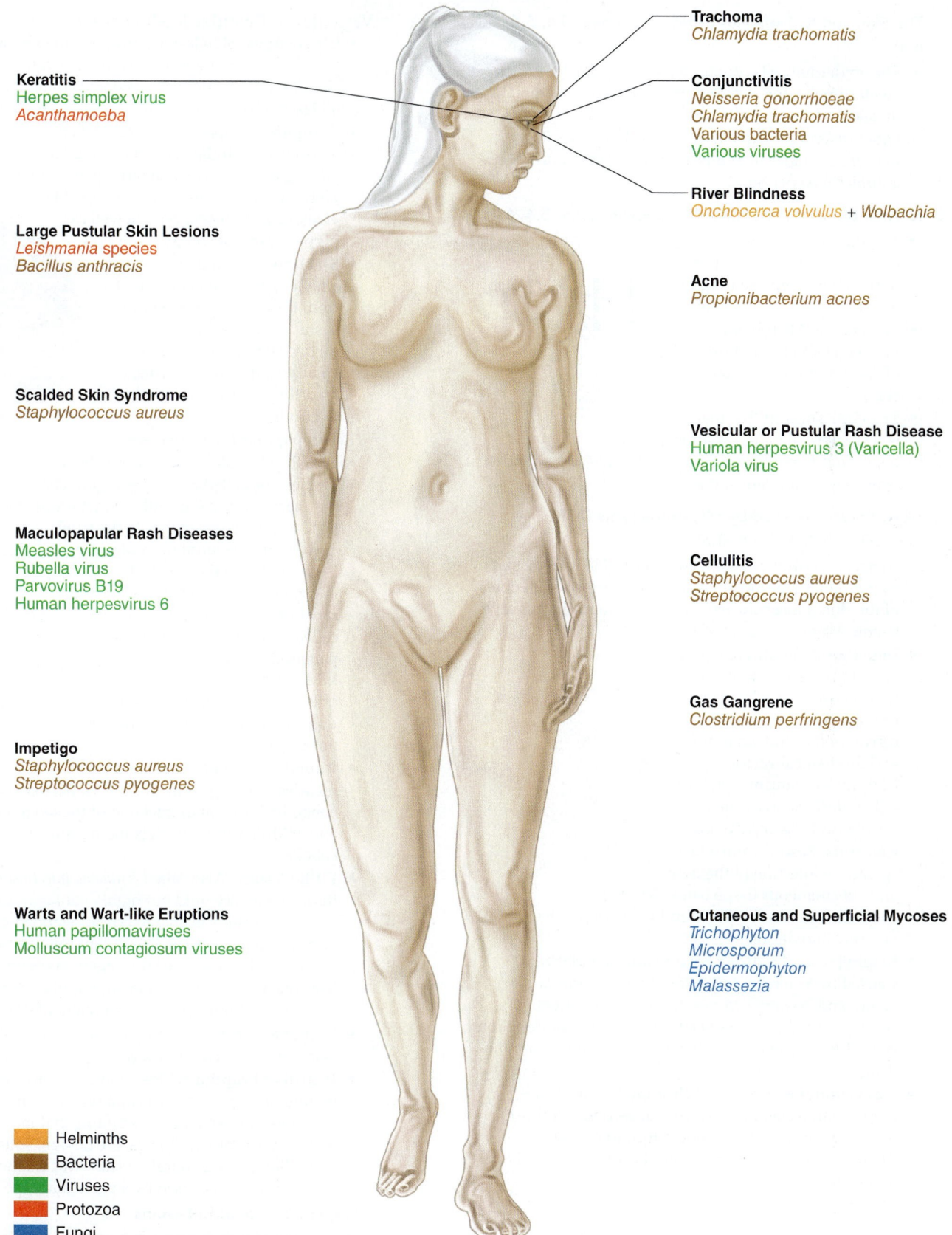

Keratitis
Herpes simplex virus
Acanthamoeba

Large Pustular Skin Lesions
Leishmania species
Bacillus anthracis

Scalded Skin Syndrome
Staphylococcus aureus

Maculopapular Rash Diseases
Measles virus
Rubella virus
Parvovirus B19
Human herpesvirus 6

Impetigo
Staphylococcus aureus
Streptococcus pyogenes

Warts and Wart-like Eruptions
Human papillomaviruses
Molluscum contagiosum viruses

Trachoma
Chlamydia trachomatis

Conjunctivitis
Neisseria gonorrhoeae
Chlamydia trachomatis
Various bacteria
Various viruses

River Blindness
Onchocerca volvulus + *Wolbachia*

Acne
Propionibacterium acnes

Vesicular or Pustular Rash Disease
Human herpesvirus 3 (Varicella)
Variola virus

Cellulitis
Staphylococcus aureus
Streptococcus pyogenes

Gas Gangrene
Clostridium perfringens

Cutaneous and Superficial Mycoses
Trichophyton
Microsporum
Epidermophyton
Malassezia

Helminths
Bacteria
Viruses
Protozoa
Fungi

System Summary Figure 18.20

Chapter Summary

18.1 The Skin and Its Defenses (ASM Guidelines* 3.4, 5.4, 6.4)

- The epidermal cells contain the protein keratin, which "waterproofs" the skin and protects it from microbial invasion.
- Other defenses include antimicrobial peptides, low pH sebum, high salt and lysozyme in sweat, and antimicrobial peptides.

18.2 Normal Biota of the Skin (ASM Guidelines 3.4, 5.4, 6.4)

- The skin has a diverse array of microbes that make up its normal biota, representing five major taxa.
- The Human Microbiome Project (HMP) found hundreds of species of microbes on the skin.
- The HMP showed that there is wide normal biota variation between different people and that there is less variation over time on the same person.

18.3 Skin Diseases Caused by Microorganisms (ASM Guidelines 5.3, 5.4, 6.4, 8.3)

- **Acne:** A syndrome of follicle-associated lesions caused by microbial digestion of excess sebum trapped in pores of the skin. *Propionibacterium acnes* is the main causative agent.
- **Impetigo:** A highly contagious superficial bacterial infection that can cause skin to peel or flake off; transmitted by direct contact and via fomites and mechanical vectors. Causative organisms can be either *Staphylococcus aureus* or *Streptococcus pyogenes* or both.
- **Cellulitis:** Results from a fast-spreading infection of the dermis and subcutaneous tissue below. Most commonly caused by the introduction of *S. aureus* or *S. pyogenes* into dermis.
- **Staphylococcal Scalded Skin Syndrome (SSSS):** Caused by *S. aureus*. Affects mostly newborns and babies and is similar to a systemic form of impetigo. *Toxic epidermal necrolysis* (*TEN*) is a similar manifestation caused by a reaction to antibiotics, barbiturates, or other drugs.
- **Gas Gangrene:** Also called clostridial myonecrosis, can be manifested in two forms: anaerobic cellulitis or myonecrosis. The endospore-forming anaerobe, *Clostridium perfringens*, is the most common causative organism.

- **Vesicular or Pustular Rash Diseases**
  - **Chickenpox:** Skin lesions progress quickly from macules and papules to itchy vesicles filled with clear fluid. Patients are considered contagious until all lesions have crusted over.
  - **Shingles:** Recuperation from chickenpox is associated with the virus becoming latent in the ganglia and possibly reemerging as shingles. Human herpesvirus 3, an enveloped DNA virus, causes chickenpox, as well as herpes zoster or shingles.
  - **Smallpox:** Naturally occurring smallpox has been eradicated from the world, but the virus is still considered a bioterror threat. The causative agent of smallpox, the variola virus, is an orthopoxvirus, an enveloped DNA virus.
- **Maculopapular Rash Diseases**
  - **Measles:** Measles or *rubeola* results in oral lesions called *Koplik's spots* and characteristic red maculopapular exanthem that erupts on the head and then progresses to the trunk and extremities until most of body is covered. The disease itself is serious, and can lead to a rare but serious complication called subacute sclerosing panencephalitis (SSPE). The MMR and MMRV vaccines (measles, mumps, and rubella +/− varicella) contains attenuated measles virus.
  - **Rubella:** Also known as German measles, can appear in two forms: postnatal and congenital (prenatal) infection of the fetus. The MMR and MMRV vaccinations contain protection from rubella.
  - **Fifth Disease:** Also called *erythema infectiosum*, fifth disease is a very mild but highly contagious disease that often results in characteristic "slapped-cheek" appearance because of a confluent reddish rash that begins on the face. Causative agent is parvovirus B19.
  - **Roseola:** Can result in a maculopapular rash; is caused by a human herpesvirus called HHV-6.
  - **Scarlet Fever:** May accompany infection of throat or skin with *Streptococcus pyogenes*.
  - **Wart-like Eruptions:** Most common warts are caused by human papillomavirus or a poxvirus, molluscum contagiosum, which causes bumps that may look like warts. Warts, or papillomas, are benign, squamous epithelial growths. Rarely, a skin wart can become malignant when caused by a particular type of HPV.
- **Larger Pustular Skin Lesions**
  - **Leishmaniasis:** A zoonosis transmitted by the female sand fly when it ingests host's blood. A protozoan causes this equatorial disease, and the infection can

Source: ASM Curriculum Guidelines (American Society for Microbiology, 2012). Complete guidelines in appendix B of this book.

be localized in the skin or mucous membranes, or be systemic.

- **Cutaneous Anthrax:** Most common and least dangerous version of infection with *Bacillus anthracis*. The skin shows a papule that becomes necrotic and later ruptures to form a painless, black eschar.
- **Ringworm (Cutaneous Mycoses):** A group of fungi that are collectively termed dermatophytes cause mycoses to the nonliving epidermal tissues, hair, and nails. Diseases are often called "ringworm"—ringworm of the scalp (tinea capitis), beard (tinea barbae), body (tinea corporis), groin (tinea cruris), foot (tinea pedis), and hand (tinea manuum). Species in the genera *Trichophyton*, *Microsporum*, and *Epidermophyton* cause the cutaneous mycoses.

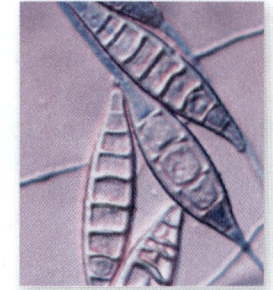

- **Superficial Mycoses:** Agents of superficial mycoses involve the outer epidermis. Tinea versicolor is caused by the yeast genus *Malassezia*, a normal inhabitant of human skin that feeds on the high oil content of the skin glands.

18.4 The Surface of the Eye and Its Defenses (ASM Guidelines 3.4, 5.4, 6.4)

- The flushing action of the tears, which contain lysozyme and lactoferrin, is the major protective feature of the eye.

18.5 Normal Biota of the Eye (ASM Guidelines 3.4, 5.4, 6.4)

- The eye has similar microbes as the skin but in lower numbers.

18.6 Eye Diseases Caused by Microorganisms (ASM Guidelines 5.3, 5.4, 6.4, 8.3)

- **Conjunctivitis:** Infection of the conjunctiva (commonly called pinkeye) has many different clinical presentations. Neonatal eye infection is usually associated with *Neisseria gonorrhoeae* or *Chlamydia trachomatis*; they are transmitted vertically via a genital tract infection in the mother. Bacterial conjunctivitis in other age groups is most commonly caused by *Staphylococcus aureus*, *Streptococcus pneumoniae*, *Haemophilus influenzae*, or *Moraxella* species. Viral conjunctivitis is commonly caused by adenoviruses. Both bacterial and viral conjunctivitis are highly contagious.
- **Trachoma:** Ocular trachoma is a chronic *Chlamydia trachomatis* infection of the epithelial cells of the eye and a major cause of blindness in certain parts of the world. Trachoma and simple conjunctivitis are caused by different strains of *C. trachomatis*.
- **Keratitis:** A more serious eye infection than conjunctivitis. Herpes simplex viruses (HSV-1 and HSV-2) and *Acanthamoeba* cause two different forms of the disease.
- **River Blindness:** A chronic parasitic helminth infection endemic in dozens of countries in Latin America, Africa, Asia, and the Middle East. The condition is caused by a symbiotic pair, the bacterium *Wolbachia* living inside the helminth *Onchocerca*. The worm is transmitted to humans by small biting black flies.

Multiple-Choice and True-False Questions | Bloom's Levels 1 and 2: Remember and Understand

Multiple-Choice Questions. Select the correct answer from the options provided.

1. An effective treatment for a cutaneous mycosis like tinea pedis would be
   a. penicillin.
   b. miconazole
   c. griseofulvin.
   d. doxycycline.

2. What is the antimicrobial enzyme found in sweat, tears, and saliva that can specifically break down peptidoglycan?
   a. lysozyme
   b. beta-lactamase
   c. catalase
   d. coagulase

3. Which of the following is probably the most important defense factor for skin?
   a. phagocytes
   b. sebum
   c. dryness
   d. antimicrobial peptides

4. Name the organism(s) most commonly associated with cellulitis.
   a. *Staphylococcus aureus*
   b. *Propionibacterium acnes*
   c. *Streptococcus pyogenes*
   d. both a and b
   e. both a and c

5. Due to a highly successful vaccination program, the WHO has managed the worldwide eradication of the naturally occurring disease
   a. chickenpox.
   b. anthrax.
   c. smallpox.
   d. German measles.

6. Warts are caused by
   a. human herpesvirus 3.
   b. papillomavirus.
   c. herpes simplex virus.
   d. measles virus.

7. Herpesviruses can cause all of the following diseases, except
   a. chickenpox.
   b. shingles.
   c. keratitis.
   d. smallpox.
   e. roseola.

8. Which disease is incorrectly matched with the causative agent?
   a. viral conjunctivitis—adenovirus
   b. river blindness—*Onchocerca volvulus*
   c. smallpox—variola virus
   d. gas gangrene—*Staphylococcus aureus*

9. Dermatophytes are fungi that infect the epidermal tissue by invading and attacking
   a. collagen.
   b. keratin.
   c. fibroblasts.
   d. sebaceous glands.

10. Poor contact lens hygiene is likely to get you a case of
    a. herpetic keratitis.
    b. *Wolbachia* infection.
    c. *Acanthamoeba* keratitis.
    d. ophthalmic gonorrhea.

True-False Questions. If the statement is true, leave as is. If it is false, correct it by rewriting the sentence.

11. The enzyme catalase is associated with pathogenic strains of *Staphylococcus aureus*.

12. Fifth disease can be treated with acyclovir and prevented by immunization.

13. Measles can potentially be eradicated because humans are the only reservoir.

14. The blistering and peeling of the skin in scalded skin syndrome are due to the ability of *Staphylococcus aureus* to produce catalase.

15. The normal skin biota is similar among different people.

Critical Thinking Questions | Bloom's Levels 3, 4, and 5: Apply, Analyze, and Evaluate

Critical thinking is the ability to reason and solve problems using facts and concepts. These questions can be approached from a number of angles and, in most cases, they do not have a single correct answer.

1. A microscopic cluster of *Staphylococcus aureus* is transferred to the surface of your skin.
   a. Describe any defense mechanisms these cells will have to avoid at this portal of entry.
   b. The microbial cells have managed to survive these attacks and now enter into a hair follicle. Discuss what skin disease(s) may arise due to the entry of this pathogen.

2. Conduct additional research, and discuss examples of how normal biota of the skin contributes to protection from pathogens at this portal of entry.

3. A young boy was at the playground when he felt a sharp pain on his leg. Upon inspection, his mother realized he was stung by a bee. They went home and she carefully removed the stinger and washed the site well. Within a week, the site became swollen and painful; a red line appeared at the site, trailing up his leg.
   a. Explain what condition the young boy appears to suffering from and the most likely causative agent involved.
   b Discuss how the microbe may have gained access to the portal of entry.

4. A farmer working on a piece of machinery gets his shirtsleeve caught in a moving piece of the equipment. His shirt is sliced, and a sharp blade covered in mud slices through his upper arm. He attempts to control the bleeding and immediately seeks medical attention. After 3 days, he develops a fever and his arm becomes extremely swollen and painful. Pulling back the bandages, he finds that the wound has become blackened and is leaking a bloody fluid. Microscopic analysis of the fluid reveals the presence of gram-positive bacilli.
   a. Discuss what condition the patient is suffering from and the likely causative agent of this infection.
   b. Explain how the patient contracted this pathogenic microbe, and what virulence factors contributed to the pathogenesis seen at the wound site.
   c. In addition to antibiotics, the physician prescribes hyperbaric therapy. Describe what this treatment involves and how it could be therapeutic to this patient.

5. a. Conduct additional research, and discuss whether "pox parties" represent a safe method of developing immunity to varicella zoster virus.
   b. Provide evidence in support of or refuting the following statement: Shingles develops when you are reinfected with varicella zoster virus later on in life.

6. a. Name the three genera of fungi associated with tinea.
   b. Individuals being treated for fungal infections must often undergo routine blood testing for analysis of liver and kidney function. Explain why antifungal drugs have such toxic side effects in humans.

7. a. List the pathogens targeted by the MMR vaccine, and summarize how the vaccine provides immunologic protection.
   b. Numerous cases of measles have surfaced recently in the United States. Explain how such extensive outbreaks can occur when the MMR vaccine program aims to establish herd immunity within a population.

8. A 17-year-old male goes to his physician with an uncomfortable condition: a painful bump on the sole of his foot. As the physician examines the lesion, the young man states that he is an active member of the school swim team.
   a. What type of skin condition do you feel the patient is suffering from, based upon this information?
   b. What causative agent is most likely involved, and how might the patient have acquired the pathogen?
   c. After receiving his diagnosis, the young man is very disturbed. He says that it is impossible because he has never been sexually active. Form an explanation that would clarify this misconception for the patient.

9. Smallpox has a rich history—from prompting the first vaccine to potential use as a bioterrorism agent. Given what you know about the etiology of the disease and the current state of the world's immunity to smallpox, discuss how effective (or ineffective) a smallpox biological weapon could be against a human population.

10. A 26-year-old female develops what she thinks is pinkeye: reddened eyes producing a clear discharge. She notices that she cannot tolerate bright sunlight and that she constantly feels as if she has sand in her eyes.
    a. Based upon this evidence, what condition is the patient suffering from? Explain whether or not the patient will be prescribed an antibiotic to treat her condition.

    b. The physician asks whether or not she has recently had a sexual encounter with an individual who had a sexually transmitted infection. She is shocked by the question. Formulate an explanation that will allow her to see the connection between her diagnosis and this very personal question.

Concept Connections | Bloom's Levels 4 and 6: Analyze and Create

This activity ties together multiple concepts in the chapter.

1. *Staphylococcus aureus* and *Streptococcus pyogenes* are implicated as the causative agents of least three skin diseases. What are they?

2. How are *S. aureus* and *S. pyogenes* phenotypically similar? How are they different?

3. What diagnostic tests are used to distinguish between *S. aureus* and *S. pyogenes*? How does each organism react to each test?

4. What would be the outcome of a catalase test if it was performed on *S. aureus*? On *S. pyogenes*?

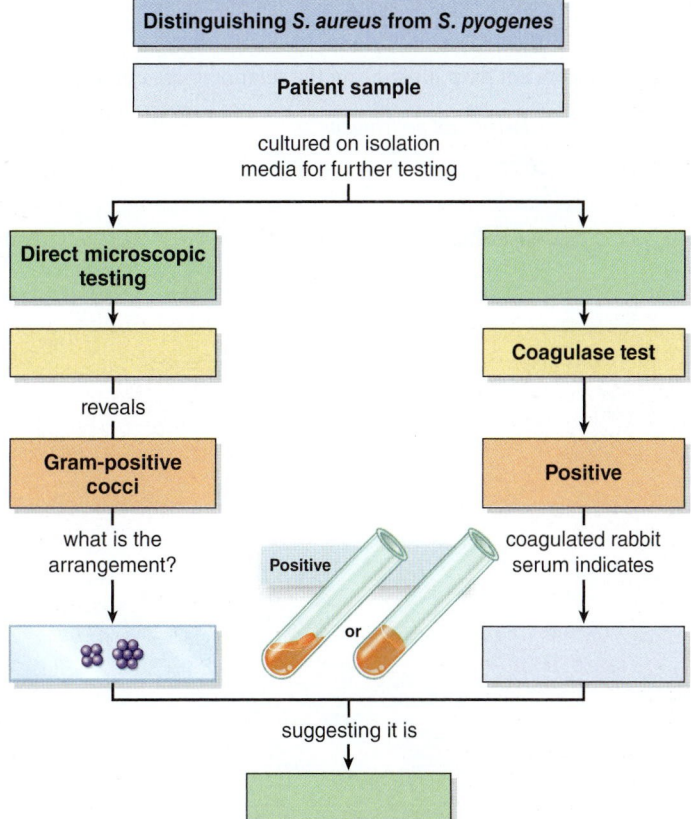

Visual Connections | Bloom's Level 5: Evaluate

These questions use visual images or previous content to make connections to this chapter's concepts.

1. **From chapter 13, figure 13.5a.** Discuss whether this figure illustrates the pathogenesis of impetigo caused by *Staphylococcus aureus* or *Streptococcus pyogenes*.

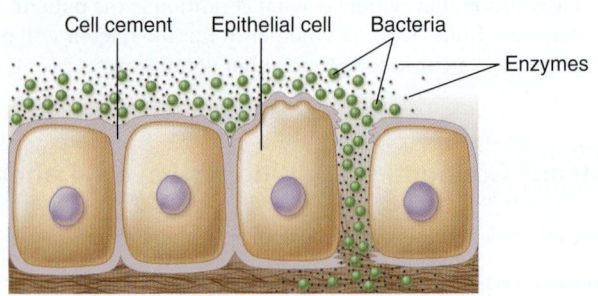

Concept Mapping | Bloom's Level 6: Create

Appendix D provides guidance for working with concept maps.

1. Using the words that follow, please create a concept map illustrating the relationships among these key terms from chapter 18.

Streptococcus pyogenes	*Clostridium perfringens*	*Staphylococcus aureus*
exfoliative toxins A and B	SSSS	gas gangrene
alpha toxin	scarlet fever	erythrogenic toxin
sandpaper-like rash	blistering of epidermis	myonecrosis

www.mcgrawhillconnect.com

Enhance your study of this chapter with study tools and practice tests. Also ask your instructor about the resources available through ConnectPlus, including the media-rich eBook, interactive learning tools, and animations.

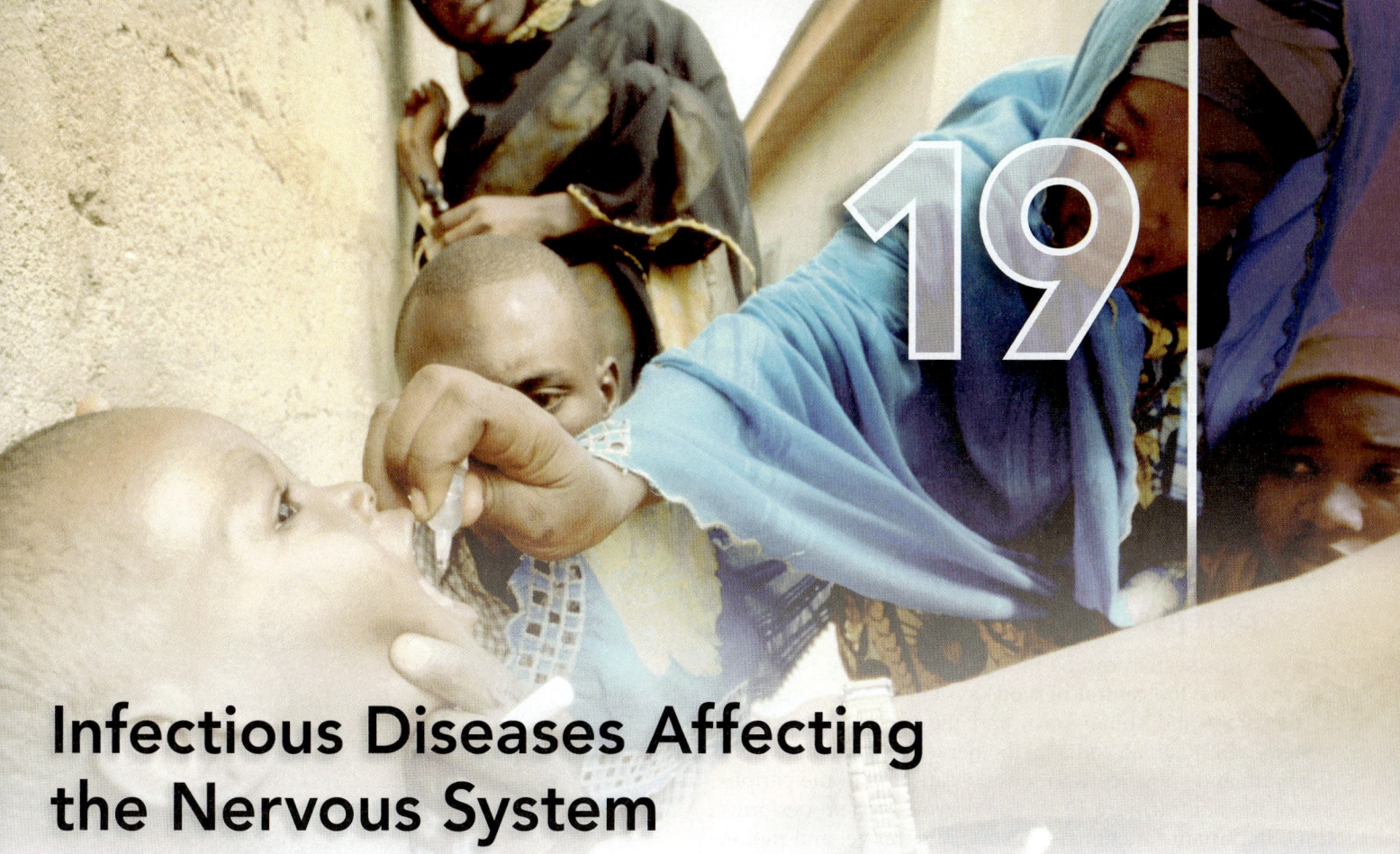

Infectious Diseases Affecting the Nervous System

Case File 19

Ridding the Planet of Polio

Polio has been nearly eradicated in most of the industrialized world, so why do concerns over this disease still exist worldwide? Because polio has not been eradicated globally; millions of children are still at risk of death or paralysis due to poliovirus infection. The Global Polio Eradication Initiative (GPEI) was launched in 1988 by the World Health Organization (WHO) with the intent of vaccinating every child. The GPEI was largely successful in interrupting endemic polio in India, which has been considered polio-free since February 2012. However, as of mid-2013 there are three countries in which polio is still endemic: Afghanistan, Nigeria, and Pakistan. Three countries that were previously deemed polio-free have recently seen a resurgence in cases of the disease: Angola, Chad, and the Democratic Republic of the Congo (DRC). The most recent outbreak of polio began in November 2010 in the DRC; a total of 554 cases of acute flaccid paralysis occurred and the outbreak was not brought under control until March 2011 through the immunization of all citizens. Travelers to and from the country were also advised to receive booster vaccinations to ensure that the disease would not spread beyond the country's borders. This is just one example of a number of polio outbreaks that have occurred worldwide in recent years. The WHO declared eradication of polio a "programmatic emergency for global public health" in January 2012, and it is hoped that its vaccination efforts will reduce the suffering of many individuals around the world today.

- Why does polio continue to be an "emergency for global public health"?
- What are the difficulties in vaccinating children around the world and eradicating polio?

Source: www.polioeradication.org/infectedcountries.aspx

Continuing the Case appears on page 576.

Outline and Learning Outcomes

19.1 The Nervous System and Its Defenses
1. Describe the important anatomical features of the nervous system.
2. List the natural defenses present in the nervous system.

19.1 The Nervous System and Its Defenses

The nervous system can be thought of as having two component parts: the central nervous system (CNS), consisting of the brain and spinal cord, and the peripheral nervous system (PNS), which contains the nerves that emanate from the brain and spinal cord to sense organs and to the periphery of the body **(figure 19.1)**. The nervous system performs three important functions—sensory, integrative, and motor. The sensory function is fulfilled by sensory receptors at the ends of peripheral nerves. They generate nerve impulses that are transmitted to the central nervous system. There, the impulses are translated, or integrated, into sensation or thought, which in turn drives the motor function. The motor function necessarily involves structures outside of the nervous system, such as muscles and glands.

The brain and the spinal cord are dense structures made up of cells called **neurons.** They are both surrounded by bone. The brain is situated inside the skull, and the spinal cord lies within the spinal column **(figure 19.2),** which is composed of a stack of interconnected bones called vertebrae. The soft tissue of the brain and spinal cord is encased within a tough casing of three membranes called the **meninges.** The layers of membranes, from outermost to innermost position, are the dura mater, the arachnoid mater, and the pia mater. Between the arachnoid mater and pia mater is the subarachnoid space (that is, the space under the arachnoid mater). The subarachnoid space is filled with a clear, serumlike fluid called cerebrospinal fluid (CSF). The CSF provides nutrition to the CNS, while also providing a liquid cushion for the sensitive brain and spinal cord. The meninges are a common site of infection, and microorganisms can often be found in the CSF when meningeal infection **(meningitis)** occurs.

The PNS consists of nerves and ganglia (see figure 19.1). A ganglion is a swelling in the nerve where the cell bodies of the neurons congregate. Nerves are bundles of neuronal axons that receive and transmit nerve signals. The axons and dendrites of adjacent neurons communicate with each other over a very small space, called a synapse. Chemicals called neurotransmitters are released from one cell and act on the next cell in the synapse.

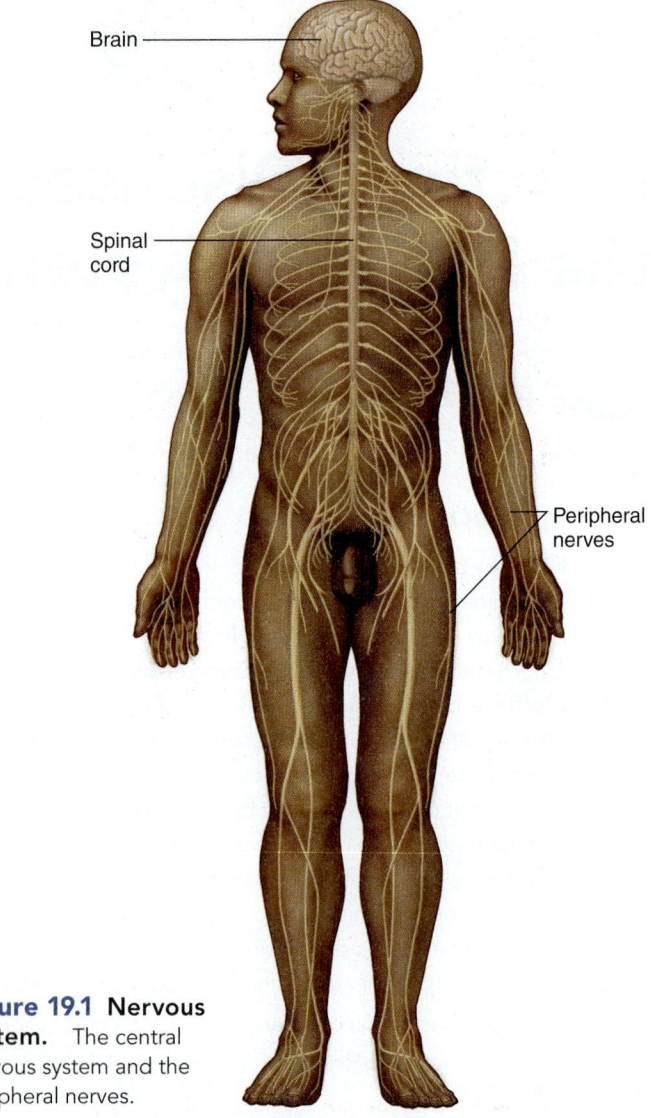

Figure 19.1 Nervous system. The central nervous system and the peripheral nerves.

The defenses of the nervous system are mainly structural. The bony casings of the brain and spinal cord protect them from traumatic injury. The surrounding CSF also serves as a cushion against impact. The entire nervous system is served by the vascular system, but the interface

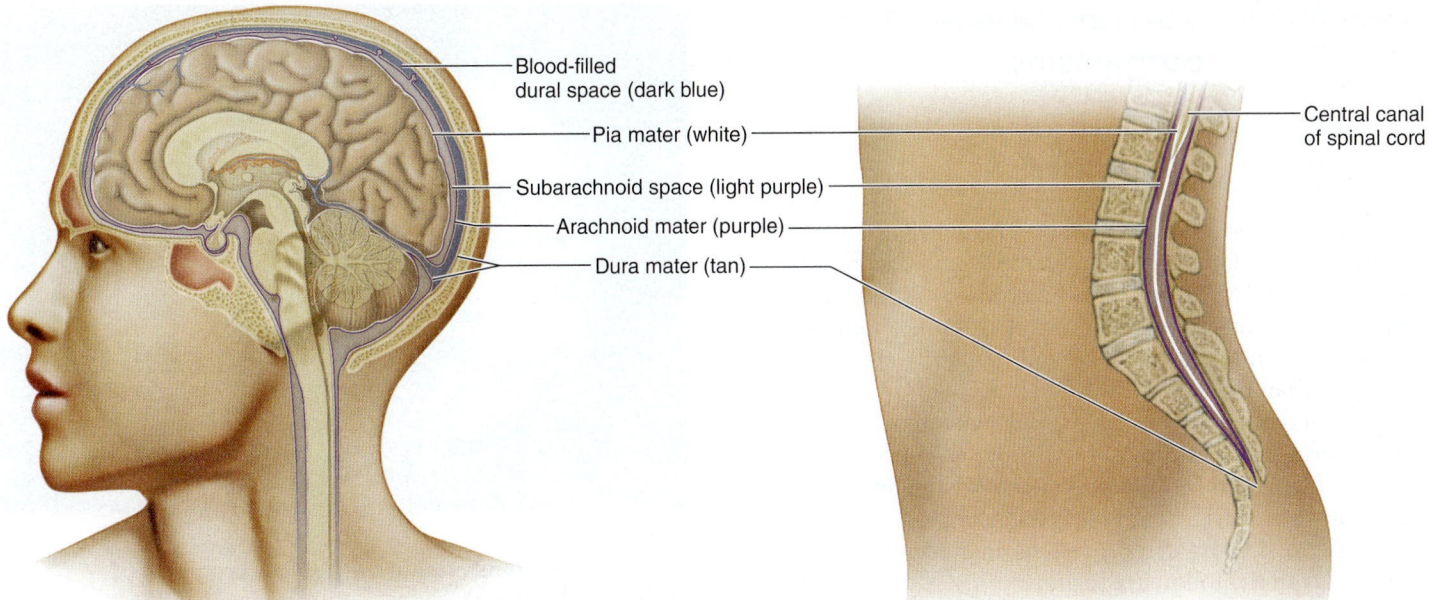

Figure 19.2 Detailed anatomy of the brain and spinal cord.

between the blood vessels serving the brain and the brain itself is different from that of other areas of the body and provides a third structural protection. The cells that make up the walls of the blood vessels allow very few molecules to pass through. In other parts of the body, there is freer passage of ions, sugars, and other metabolites through the walls of blood vessels. The restricted permeability of blood vessels in the brain is called the **blood-brain barrier,** and it prohibits most microorganisms from passing into the central nervous system. The drawback of this phenomenon is that drugs and antibiotics are difficult to introduce into the CNS when needed.

The CNS is considered an "immunologically privileged" site. These sites are able to mount only a partial, or at least a different, immune response when exposed to immunologic challenge. The functions of the CNS are so vital for the life of an organism that even temporary damage that could potentially result from "normal" immune responses would be very detrimental. The uterus and parts of the eye are other examples of immunologically privileged sites. Cells in the CNS express lower levels of MHC antigens. Complement proteins are also in much lower quantities in the CNS. Researchers have established that MHC markers and complement proteins play a role in the development, regulation, and repair of neurons and nervous tissues through their signaling mechanisms. They speculate that malfunctions with or absence of these molecules in the CNS may be responsible for a variety of conditions such as schizophrenia or autism. Other specialized cells in the central nervous system perform defensive functions. Microglial cells exhibit phagocytic activity, which is beneficial in terms of both immunity and brain development. Brain macrophages also exist in the CNS, although the activity of both of these types of cells is thought to be less than that of phagocytic cells elsewhere in the body.

19.1 Learning Outcomes—Assess Your Progress

1. Describe the important anatomical features of the nervous system.
2. List the natural defenses present in the nervous system.

19.2 Normal Biota of the Nervous System

It is still believed that the CNS and PNS both lack normal biota of any kind and that finding microorganisms of any type in these tissues represents a deviation from the healthy state. Viruses such as herpes simplex live in a dormant state in the nervous system between episodes of acute disease, but they are not considered normal biota. Although the Human Microbiome Project is not sampling this system at the present time, information from this project is revealing a potential link between the gut microbiome and the nervous system. Gut microbiota may actually induce central nervous system autoimmunity and appear to cause changes in brain chemistry and behavior. This phenomenon is known as the gut-brain axis.

19.2 Learning Outcomes—Assess Your Progress

3. Discuss the current state of knowledge regarding the normal biota of the nervous system.

Nervous System Defenses and Normal Biota		
	Defenses	**Normal Biota**
Nervous System	Bony structures, blood-brain barrier, microglial cells, and macrophages	None

19.3 Nervous System Diseases Caused by Microorganisms

Meningitis

Meningitis, an inflammation of the meninges, is an excellent example of an anatomical syndrome. Many different microorganisms can cause an infection of the meninges, and they produce a similar constellation of symptoms. Noninfectious causes of meningitis exist as well, but they are much less common than the infections listed here.

The more serious forms of acute meningitis are caused by bacteria, but it is thought that their entrance to the CNS is often facilitated by coinfection or previous infection with respiratory viruses. Meningitis in neonates is most often caused by different microorganisms; therefore, it is described separately in the following section.

Whenever meningitis is suspected, a lumbar puncture (spinal tap) is performed to obtain CSF, which is then examined by Gram stain and/or culture. Most physicians will begin treatment with a broad-spectrum antibiotic immediately and shift treatment if necessary after a diagnosis has been confirmed.

▶ Signs and Symptoms

No matter the cause, meningitis results in these typical symptoms: photophobia (sensitivity to light), headache, painful or stiff neck, fever, and usually an increased number of white blood cells in the CSF. Many patients have described the headache associated with this disease as the "worst headache I have ever had." Specific microorganisms may cause additional, and sometimes characteristic, symptoms, which are described in the individual sections that follow.

Like many other infectious diseases, meningitis can manifest as acute or chronic disease. Some microorganisms are more likely to cause acute meningitis, and others are more likely to cause chronic disease.

In a normal healthy person, it is very difficult for microorganisms to gain access to the nervous system. Those that are successful usually have specific virulence factors.

Neisseria meningitidis

Neisseria meningitidis appears as gram-negative diplococci lined up side by side in microscopic analysis **(figure 19.3)** and is commonly known as the meningococcus. It is often associated with epidemic forms of meningitis. This organism causes the most serious form of acute meningitis, and it is responsible for about 25% of all meningitis cases. Most cases occur in young children, since vaccination of otherwise healthy children against this disease is not recommended until age 11. Although 12 different strains of capsular antigens exist, serotypes A, B, C, Y, and W135 are responsible for most cases of the disease.

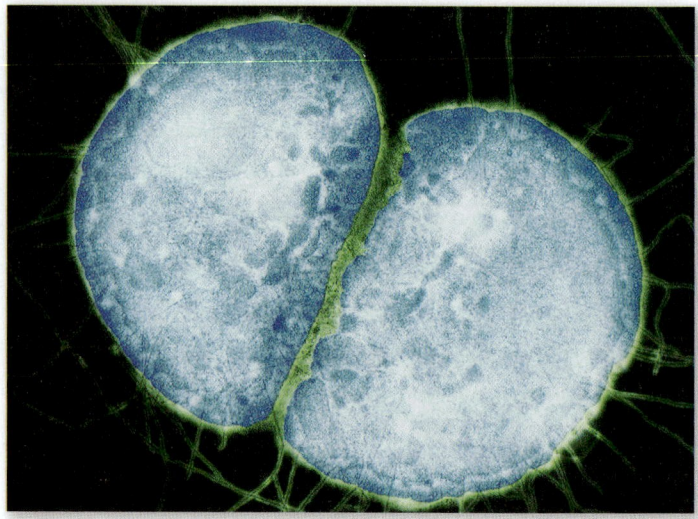

Figure 19.3 Transmission electron micrograph of *Neisseria* (52,000x).

▶ Pathogenesis and Virulence Factors

The portal of entry for this pathogen is the upper respiratory tract. The bacterium passes into surrounding blood vessels **(figure 19.4)**, rapidly penetrating the meninges and producing symptoms of meningitis. Meningitis is marked by fever, sore throat, headache, stiff neck, convulsions, and vomiting. The most serious complications of meningococcal infection are due to meningococcemia, which can accompany meningitis but can also occur on its own. The pathogen releases endotoxin within the bloodstream, which acts as a potent

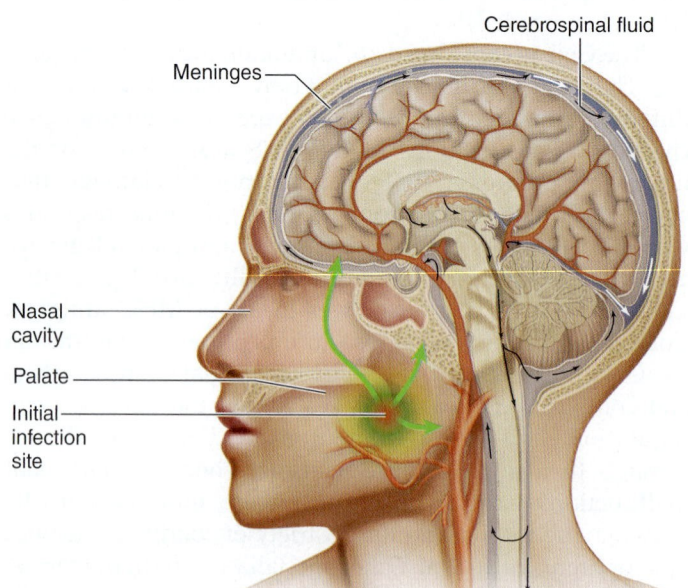

Figure 19.4 Dissemination of the meningococcus from a nasopharyngeal infection. Bacteria spread to the roof of the nasal cavity, which borders a highly vascular area at the base of the brain. From this location, they can enter the blood and escape into the cerebrospinal fluid, leading to infection of the meninges.

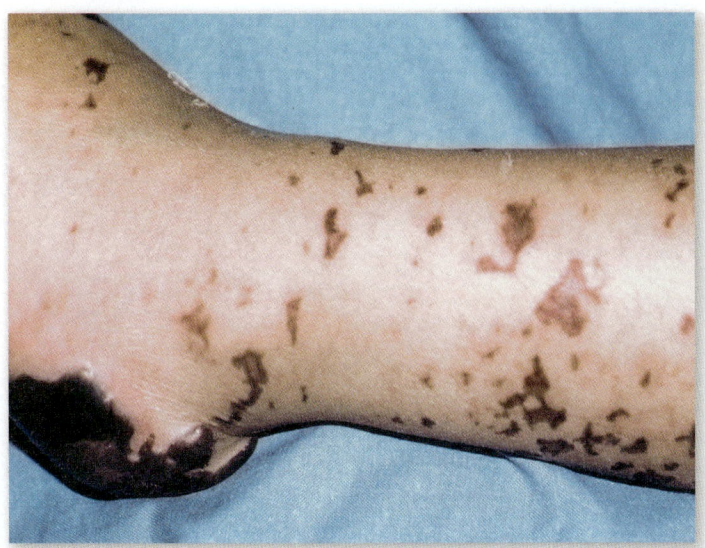

Figure 19.5 Petechiae and ecchymosis associated with meningococcal meningitis.

white blood cell stimulator. Damage to blood vessels caused by cytokines released by the white blood cells leads to vascular collapse, hemorrhage, and crops of red or purple lesions called **petechiae** (pee-tee'-kee-ay) on the trunk and appendages **(figure 19.5).**

In a small number of cases, meningococcemia becomes an overwhelming disease with a high mortality rate. The disease has a sudden onset, marked by fever higher than 40°C (104°F), chills, delirium, severe widespread ecchymosis (ek"-ih-moh'seez) (areas of bleeding under the skin larger than petechiae), shock, and coma. Generalized intravascular clotting, cardiac failure, damage to the adrenal glands, and death can occur within a few hours. Recent evidence suggests a genetic role in this form of the disease, as a significant number of patients contain changes in the genes that encode toll-like receptors (see chapter 14). These variations reduce the host's ability to initiate an early defensive response to the bacterium. The pathogen has a natural ability to avoid destruction through its production of IgA protease and the presence of a capsule.

▶ Transmission and Epidemiology

Because meningococci do not survive long in the environment, these bacteria are usually acquired through close contact with secretions or droplets. Upon reaching their portal of entry in the nasopharynx, the meningococci use attachment pili to adhere to mucosal membranes. In many people, this can result in simple asymptomatic colonization. In the more vulnerable individual, however, the meningococci are engulfed by epithelial cells of the mucosa and penetrate into the nearby blood vessels. Damage to the epithelium causes pharyngitis as the pathogen continues on its way to the meninges.

Meningococcal meningitis has a sporadic or epidemic incidence in late winter or early spring. The continuing reservoir of infection is humans who harbor the pathogen in the nasopharynx. The scene is set for transmission when carriers live in close quarters with nonimmune individuals,

as may be expected in families, day care facilities, college dormitories, and military barracks. The carriage state, which can last from a few days to several months, exists in roughly 10% of the adult population. This rate can exceed 50% in institutional settings or endemic regions, and it appears to be influenced by the duration of time living in close quarters. The highest carriage of meningococci is seen in young adults (15 to 24 years old) with decreased rates occurring in young children (less than 4 years old) and individuals over the age of 50.

Every year, in what is called "the meningitis belt" in sub-Saharan Africa, a meningococcal epidemic sweeps through, coinciding with the dry season, which runs from approximately December to May **(Insight 19.1).** In 2009, a particularly large outbreak killed more than 2,100 people in Niger and Nigeria and infected tens of thousands. Many more would have been affected except for a massive mobilization of vaccine. In the space of 4 months, 7.5 million people were vaccinated.

▶ Culture and Diagnosis

Suspicion of bacterial meningitis constitutes a medical emergency, and differential diagnosis must be done with great haste and accuracy. It is most important to confirm (or rule out) meningococcal meningitis, because it can be rapidly fatal. Treatment is usually begun with this bacterium in mind until it can be ruled out. Cerebrospinal fluid, blood, or nasopharyngeal samples are stained and observed directly for the characteristic gram-negative diplococci. Cultivation may be necessary to differentiate the bacterium from other species. Specific rapid tests are also available for detecting the capsular polysaccharide or the cells directly from specimens without culturing.

It is usually necessary to differentiate this species from normal *Neisseria* that also live in the human body and can be present in infectious fluids. Immediately after collection, specimens are streaked on Modified Thayer-Martin (MTM) medium or chocolate agar and incubated in a high CO_2 atmosphere. Presumptive identification of the genus is obtained by a Gram stain and oxidase testing on isolated colonies **(figure 19.6).** Further testing may be necessary to differentiate *N. meningitidis* and *N. gonorrhoeae* from one another, from other oxidase-positive species, and from normal biota of the oropharynx that may be mistaken for these pathogens. If no samples were obtained prior to antibiotic treatment, a PCR test is the best bet for identifying the pathogen. Susceptibility testing is also warranted to ensure that proper treatment protocols are developed.

▶ Prevention and Treatment

The infection rate in most populations is about 1%, so well-developed natural immunity to the meningococcus appears to be the rule. In the United States, immunization begins at the age of 11 with the conjugated MCV4 vaccine (Menveo or Menactra) that is effective against groups A, C, Y, and W135 but not B (see chapter 15). Although it was thought to provide protection for 10 years, it is now recognized that a booster dose is needed to ensure protection through adolescence. This vaccine can also be used in young children who are at high risk for infection. Adults 55 or older should

INSIGHT 19.1   The African Meningitis Belt

Twenty-one sub-Saharan countries from Senegal to Ethiopia make up the "African Meningitis Belt" where epidemics of meningitis occur during the dry summer months. The World Health Organization (WHO) reports that in the past 15 years, there were 800,000 cases with a 10% case fatality rate. Most of the cases are caused by *Neisseria meningitidis* serogroups A and W135. The Meningitis Vaccine Project (MVP) was implemented by the WHO along with the Program for Appropriate Technology in Health (PATH) to distribute meningococcal vaccines throughout the meningitis belt with the hope of eventually eradicating the disease. Recently, an effective, long-lasting, and affordable vaccine (less than $0.50 per dose) was developed through a grant from the Bill and Melinda Gates Foundation. The vaccine, known as MenAfriVac, was launched in 2010 in Burkina Faso, Mali, and Niger in preparation for more widespread use in Africa. Individuals between the ages of 1 and 29 were vaccinated because they represent the portion of the population most susceptible to meningitis.

Researchers who worked on the vaccine are optimistic about its efficacy. During the vaccination campaign, 12 million people were vaccinated in Burkina Faso. Researchers quickly saw a decline in the incidence of meningococcal meningitis in the area: from 6,145 cases in 2010 to 2,825 cases in 2011. Similar results were seen in Mali and Niger. The vaccine also appears to be particularly safe and effective in children under the age

of 2 and confers long-term protection as compared to traditional polysaccharide vaccines. It is hoped that by 2016, all 26 nations within the African Meningitis Belt will employ MenAfriVac for immunization in order to eliminate meningococcal A disease epidemics from this region.

Source: www.who.int/csr/don/2012_05_24/en/

---

receive either of the conjugated vaccines or a polysaccharide vaccine (Menomune) if their health or environmental conditions place them at risk for disease development. In European countries, serogroup B is more common than A, C, Y, and W135. In 2012, the European Union approved a new vaccine for serogroup B prevention.

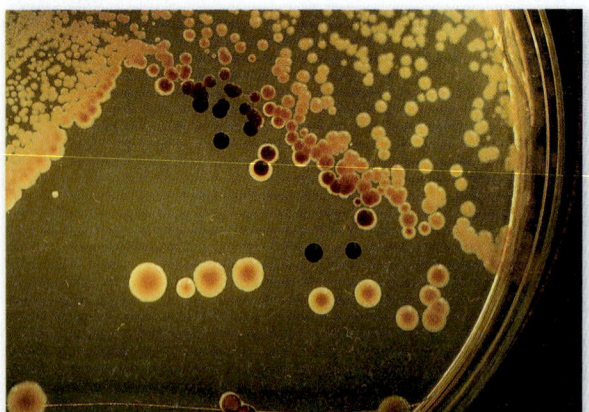

Figure 19.6 The oxidase test.   A drop of oxidase reagent is placed on a suspected *Neisseria* or *Branhamella* colony. If the colony reacts with the chemical to produce a purple to black color, it is oxidase-positive; those that remain white to tan are oxidase-negative. Because several species of gram-negative rods are also oxidase-positive, this test is presumptive for these two genera only if a Gram stain has verified the presence of gram-negative cocci.

Because even treated meningococcemial disease has a mortality rate of up to 15%, it is vital that chemotherapy begin as soon as possible with one or more drugs. Ceftriaxone is the first-line antibiotic for this condition; aztreonam or chloramphenicol may also be used. Patients may also require treatment for shock and intravascular clotting in addition to antibiotic therapy. When family members, medical personnel, or children in day care or school have come in close contact with infected people, preventive therapy with ciprofloxacin, rifampin, or ceftriaxone may be warranted.

Streptococcus pneumoniae

You will see in chapter 21 that *Streptococcus pneumoniae* causes the majority of bacterial pneumonias. (It is also referred to as the **pneumococcus.**) Pneumococcal meningitis is also caused by this bacterium; indeed, this pathogen is the most frequent cause of community-acquired meningitis and often causes a severe form of the disease. It does not cause the petechiae associated with meningococcal meningitis, and that difference is useful diagnostically. Pneumococcal meningitis is most likely to occur in patients with underlying susceptibility, such as patients with alcoholism, patients with sickle-cell disease, or those with absent or defective spleen function. Up to 25% of pneumococcal meningitis patients will also develop pneumococcal pneumonia. Pneumococcal infections occur worldwide and today are most prevalent in developing countries. The pneumococcus is considered one of the three main causative agents of bacterial community-acquired pneumonias.

Although carriage rates exceed 30% in some populations (the highest rate in adolescents), *S. pneumoniae* obviously exhibits the potential to be highly pathogenic. It can penetrate the respiratory mucosa; gain access to the bloodstream; and then, under certain conditions, enter the meninges.

The bacterium is a small, gram-positive, flattened coccus that appears in end-to-end pairs. Its appearance is distinctive in a Gram stain of cerebrospinal fluid; testing of nasopharynx specimens is not useful because of its role as normal biota in many individuals. Like the meningococcus, this bacterium has a polysaccharide capsule that protects it against phagocytosis. Over 90 serotypes with varying capsular antigenicity have been identified so far. *S. pneumoniae* produces an alpha-hemolysin (observable on blood agar) and hydrogen peroxide, both of which have been shown to induce damage in the CNS, as well as inducing brain cell apoptosis.

Treatment requires a drug to which the bacterium is not resistant; resistance to penicillin, cephalosporins and macrolide antibiotics is a problem worldwide. Suspected cases of pneumococcal meningitis should initially be treated with vancomycin in combination with ceftriaxone until the resistance pattern of the organism is determined. If it is sensitive to penicillin G, the therapy should be shifted to that drug. It is recommended that a steroid be administered 20 minutes prior to antibiotic administration to dampen the inflammatory response to cell wall components that are released by antibiotic treatment of the gram-positive bacterium and increase the efficacy of the antibiotic.

As mentioned in chapter 21, three vaccines are now available in the United States for protection against *S. pneumoniae* infection. The 7-valent conjugated vaccine (Prevnar) has been a part of the childhood immunization schedule (see chapter 15) since 2000 but is now being replaced by the 13-valent conjugated vaccine (Prevnar 13). A 23-valent polysaccharide vaccine (Pneumovax) is available for vaccination of adults aged 65 and older as well as at-risk patients. In many cases older adults are also offered the 13-valent vaccine.

Haemophilus influenzae

Haemophilus influenzae is a gram-negative coccobacillus that causes one of the most severe forms of meningitis in humans. Humans are the only known reservoir, and the portal of entry for this bacterium is the nasopharynx; asymptomatic carriage rates vary worldwide but have been greatly reduced in the United States due to successful vaccination. Disease caused by *H. influenzae* is often called "Hib" because it is due primarily to infection with the B serotype, though capsular serotypes A through F exist in addition to other pathogenic but untypable strains. Routine vaccination with one of two subunit vaccines (both contain capsular polysaccharide conjugated to a protein) is recommended for all children, beginning at age 2 months, with the recommendation of a follow-up booster dose. Combination vaccines containing the Hib conjugate vaccine are also available for use in the current U.S. vaccine schedule (see chapter 15). Before the first vaccine was introduced in 1985, *H. influenzae* was a very common cause of severe meningitis and death. Invasive Hib disease has been virtually eliminated in the United States today, a clear victory for successful vaccination programs. Physicians recognize, however, that this situation can rapidly change if vaccine coverage falls and herd immunity is compromised. Today, *H. influenzae* is recognized as a key causative agent of bacterial community-acquired pneumonia in immunocompromised patients. Resistance to beta-lactam drugs is on the rise.

Listeria monocytogenes

Listeria monocytogenes is a gram-positive bacterium that ranges in morphology from coccobacilli to long filaments in palisade formation **(figure 19.7).** Cells do not produce capsules or endospores and have from one to four flagella. *Listeria* is not fastidious and is resistant to cold, heat, salt, pH extremes, and bile. It grows inside host cells and can move directly from an infected host cell to an adjacent healthy cell.

Listeriosis in healthy adults is often a mild or subclinical infection with nonspecific symptoms of fever, diarrhea, and sore throat. However, listeriosis in elderly or immunocompromised patients, fetuses, and neonates (described later) usually affects the brain and meninges and results in septicemia. The death rate is around 20%. Pregnant women are especially susceptible to infection, which can be transmitted to the infant prenatally when the microbe crosses the placenta or postnatally through the birth canal. Intrauterine infections are widely systemic and usually result in premature abortion and fetal death.

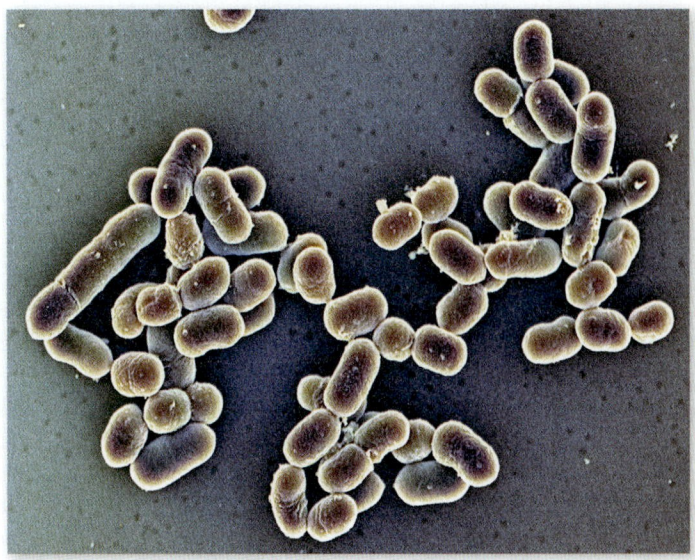

Figure 19.7   *Listeria monocytogenes.* The bacterium is generally rod shaped. In Gram stains, individual cells tend to stack up in structures called palisades.

The distribution of *L. monocytogenes* is so broad that its reservoir has been difficult to determine. It has been isolated all over the world from water, soil, plant materials, and the intestines of healthy mammals (including humans), birds, fish, and invertebrates. Apparently, the primary reservoir is soil and water; animals, plants, and food are secondary sources of infection. Most cases of listeriosis are associated with ingesting contaminated dairy products, poultry, and meat. Recent epidemics have spurred an in-depth investigation into the prevalence of *L. monocytogenes* in these sources. Reports have established that consumers are routinely exposed to low to moderate levels of *L. monocytogenes*, as the pathogen has been isolated in 10% to 15% of ground beef and in 25% to 30% of chicken and turkey carcasses and is also present in 5% to 10% of luncheon meats, hot dogs, and cheeses.

In 2011, produce was implicated in the third deadliest food-borne disease outbreak in U.S. history. This occurred due to *Listeria* contamination of cantaloupe from a farm in Colorado. Nearly 40 people died, and more than 110 other Americans across 28 states were sickened by listeriosis transmitted by the contaminated fruit. The outbreak served as a wake-up call to the U.S. government regarding the need for stricter enforcement of existing policies regarding food safety, as well as possible new regulations.

Diagnosing listeriosis is hampered by the difficulty in isolating the microbe. The chances of isolation, however, can be improved by using a procedure called *cold enrichment*, in which the specimen is held at 4°C and periodically plated onto media, but this procedure can take 4 weeks. Rapid diagnostic kits using ELISA, immunofluorescence, and nucleic acid sequencing technologies are now available for direct testing of dairy products and cultures. Antibiotic therapy should be started as soon as listeriosis is suspected. Ampicillin and trimethoprim-sulfamethoxazole are the first choices, followed by meropenem. Prevention can be improved by adequate pasteurization temperatures and by proper washing, refrigeration, and cooking of foods that are suspected of being contaminated with animal manure or sewage. Pregnant women are cautioned by the U.S. Food and Drug Administration not to eat soft, unpasteurized cheeses or deli meats unless they are properly reheated.

Cryptococcus neoformans

The fungus *Cryptococcus neoformans* causes a more chronic form of meningitis with a more gradual onset of symptoms, although in AIDS patients the onset may be fast and the course of the disease more acute. It is sometimes classified as a meningoencephalitis (inflammation of both brain and meninges). Headache is the most common symptom, but nausea and neck stiffness are very common. This fungus is a widespread resident of human habitats. It has a spherical to ovoid shape, with small, constricted buds and a large capsule that is important in its pathogenesis **(figure 19.8)**.

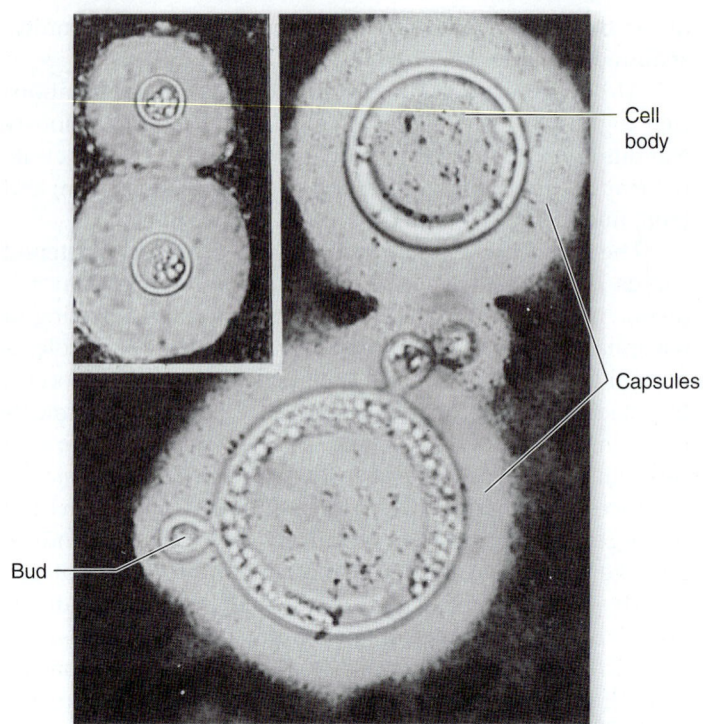

Figure 19.8 *Cryptococcus neoformans* **from infected spinal fluid stained negatively with India ink.** Halos around the large spherical yeast cells are thick capsules. Also note the buds forming on one cell. Encapsulation is a useful diagnostic sign for cryptococcosis, although the capsule is fragile and may not show up in some preparations (150×).

▶ **Transmission and Epidemiology**

The primary ecological niche of *C. neoformans* is the bird population. It is prevalent in urban areas where pigeons congregate, and it proliferates in the high-nitrogen environment of droppings that accumulate on pigeon roosts. Masses of dried yeast cells are readily scattered into the air and dust. Its role as an opportunist is supported by evidence that healthy humans have strong resistance to it and that clinically obvious infection occurs primarily in debilitated patients. Most cryptococcal infections cause symptoms in the respiratory and central nervous systems.

By far the highest rates of cryptococcal meningitis occur among patients with AIDS. This meningitis is frequently fatal. Other conditions that predispose individuals to infection are steroid treatment, diabetes, and cancer. It is not considered communicable among humans.

The primary portal of entry for *C. neoformans* is the respiratory tract, but most lung infections are subclinical and rapidly resolved.

▶ **Pathogenesis and Virulence Factors**

The escape of this pathogen from the respiratory system into the blood is intensified by weakened host defenses and results in severe complications. *Cryptococcus* shows an extreme affinity for the meninges and brain. The tumorlike masses formed in these locations can cause headache, mental changes, coma, paralysis, eye disturbances, and seizures. In some cases, the infection disseminates into the skin, bones, and viscera.

▶ Culture and Diagnosis

Cryptococcosis can be diagnosed via negative staining of specimens to detect encapsulated budding yeast cells that do not occur as pseudohyphae. Rapid tests, such as the cryptococcal antigen test, have largely replaced this method of disease diagnosis in many labs today. However, a fungal culture is required to differentiate between the various *Cryptococcus* species. Isolated colonies can be used to perform screening tests that presumptively differentiate *C. neoformans* from other important cryptococcal species like *C. gattii*, a recognized emerging infectious agent in the United States. Confirmatory results include a negative nitrate assimilation, pigmentation on birdseed agar, fluorescent antibody testing, and nucleic acid analysis.

▶ Prevention and Treatment

Systemic cryptococcosis requires immediate treatment with amphotericin B and fluconazole over a period of weeks or months. There is no prevention.

Coccidioides immitis

The morphology of *Coccidioides immitis* is very distinctive. At 25°C, it forms a moist white to brown colony with abundant, branching, septate hyphae. These hyphae fragment into thick-walled, blocklike **arthroconidia** (arthrospores) at maturity **(figure 19.9a)**. On special media incubated at 37°C to 40°C, an arthrospore germinates into the parasitic phase, a small, spherical cell called a spherule **(figure 19.9b)** that can be found in infected tissues as well. This structure swells into a giant sporangium that cleaves internally to form numerous endospores that look like bacterial endospores but lack their resistance traits.

▶ Pathogenesis and Virulence Factors

This is a true systemic fungal infection of high virulence, as opposed to an opportunistic infection. It usually begins with pulmonary infection but can disseminate quickly throughout the body. Coccidioidomycosis of the meninges is the most serious manifestation. All persons inhaling the arthrospores probably develop some degree of infection, but certain groups have a genetic susceptibility that gives rise to more serious disease. After the arthrospores are inhaled, they develop into spherules in the lungs. These spherules release scores of endospores into the lungs. At this point the patient either experiences mild respiratory symptoms, which resolve themselves, or the endospores can lead to the development of disseminated disease. Disseminated disease can include meningitis, osteomyelitis, and skin granulomas.

▶ Transmission and Epidemiology

C. immitis occurs endemically in various natural reservoirs and casually in areas where it has been carried by wind and animals. Conditions favoring its settlement include high carbon and salt content and a semiarid, relatively hot climate. The fungus has been isolated from soils, plants, and a large number of vertebrates. The natural history of *C. immitis* follows a cyclic pattern—a period of dormancy in winter and spring, followed by growth in summer and fall. Growth and spread are greatly increased by cycles of drought and heavy rains.

Skin testing has disclosed that the highest incidence of coccidioidomycosis, known commonly as Valley Fever, occurs in the southwestern United States **(figure 19.10)**, although it also occurs in Mexico and parts of Central and South America. Especially concentrated reservoirs exist in the San Joaquin Valley of California and in southern Arizona. Outbreaks are usually associated with farming activity, archaeological digs, construction, and mining. A highly unusual outbreak of coccidioidomycosis was traced to the Northridge, California, earthquake in 1994. Clouds of dust bearing loosened spores were given off by landslides, and local winds then carried the dust into the outlying residential

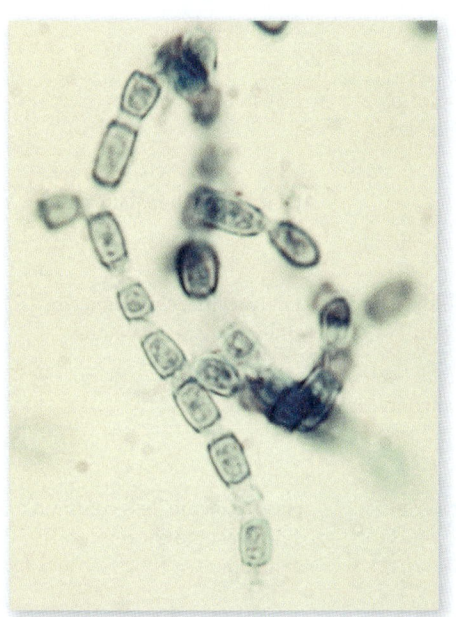

(a) Arthrospores

(b) Spherules containing endospores

Figure 19.9 Two phases of *Coccidioides* infection.   **(a)** Arthrospores are present in the environment and are inhaled. **(b)** In the lungs, the brain, or other tissues, arthrospores develop into spherules that are filled with endospores. Endospores are released and induce damage.

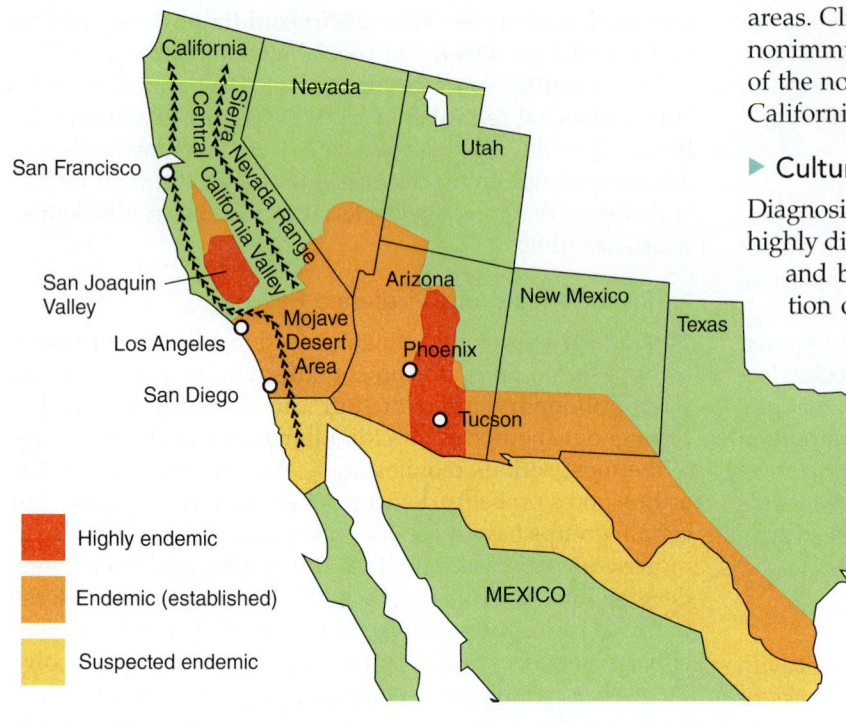

Figure 19.10 Areas in the United States endemic for *Coccidioides immitis*.

Highly endemic

Endemic (established)

Suspected endemic

areas. Climate change, construction, and the immigration of nonimmune individuals are being hypothesized as causes of the now epidemic levels of disease seen in many areas of California today.

▶ **Culture and Diagnosis**

Diagnosis of coccidioidomycosis is straightforward when the highly distinctive spherules are found in sputum, spinal fluid, and biopsies. This finding is further supported by isolation of typical mycelia and arthrospores on Sabouraud's agar. Newer specific antigen tests have been effective tools to identify and differentiate *Coccidioides* from other fungi. All cultures must be grown in closed tubes or bottles and opened in a biological containment hood to prevent laboratory infections.

▶ **Prevention and Treatment**

The majority of patients do not require treatment. In people with disseminated disease, however, oral fluconazole is used. Minimizing contact with the fungus in its natural habitat has been of some value. For example, oiling dirt roads and planting vegetation help reduce spore aerosols,

Disease Table 19.1	Meningitis		
Causative Organism(s)	*Neisseria meningitidis*	*Streptococcus pneumoniae*	*Haemophilus influenzae*
Most Common Modes of Transmission	Droplet contact	Droplet contact	Droplet contact
Virulence Factors	Capsule, endotoxin, IgA protease	Capsule, induction of apoptosis, hemolysin and hydrogen peroxide production	Capsule
Culture/Diagnosis	Gram stain/culture of CSF, blood, rapid antigenic tests, oxidase test	Gram stain/culture of CSF	Culture on chocolate agar
Prevention	Conjugated vaccine; ciprofloxacin, rifampin, or ceftriaxone used to protect contacts	Two vaccines: Prevnar (children and adults), and Pneumovax (adults)	Hib vaccine, ciprofloxacin, rifampin, or ceftriaxone
Treatment	Ceftriaxone, aztreonam, chloramphenicol	Penicillin G if sensitive; vancomycin + ceftriaxone if resistant	Ceftriaxone
Distinctive Features	Petechiae, meningococcemia rapid decline	Serious, acute, most common meningitis in adults	Serious, acute, less common since vaccine became available
Epidemiological Features	United States: 0.9–1.5 cases per 100,000 annually; meningitis belt: 1,000 cases per 100,000 annually	U.S. incidence before Prevnar: 7.7 hospitalizations per 100,000. After Prevnar: 2.6 per 100,000	Before vaccine, 300,000–400,000 deaths worldwide per year

and using dust masks while excavating soil prevents workers from inhaling as many spores. Many Californians are pushing for the development of a protective vaccine, though sufficient funds are not being allocated to this research at the moment.

Viruses

A wide variety of viruses can cause meningitis. Because no bacteria or fungi are found in the CSF in viral meningitis, the condition is often called *aseptic meningitis*. Aseptic meningitis may also have noninfectious causes.

By far the majority of cases of viral meningitis occur in children, and 90% are caused by enteroviruses. But many other viruses also gain access to the central nervous system on occasion. An initial infection with herpes simplex type 2 is sometimes known to cause meningitis; other herpesviruses, such as HHV-6, HHV-7, and HHV-3 (the chickenpox virus), can infect the meninges as well. New research shows that herpesviruses can actually take over neuronal cells and use them as highways to travel throughout the central nervous system. Arboviruses, arenaviruses, and adenoviruses have also been identified as causative agents of meningitis; and it is recognized that HIV infection may manifest as meningitis even when the virus is controlled in the rest of the body.

Viral meningitis is generally milder than bacterial or fungal meningitis, and it is usually resolved within 2 weeks. The mortality rate is less than 1%. Diagnosis begins with the failure to find bacteria, fungi, or protozoa in CSF and can be confirmed, depending on the virus, by viral culture or specific antigen tests. In most cases, no treatment is indicated. Acyclovir can be used when the causative agent is a herpesvirus; of course, if HIV is involved, the entire HIV antiviral regimen is merited (HIV is discussed in chapter 20). **Disease Table 19.1** summarizes the agents causing meningitis.

Neonatal and Infant Meningitis

Meningitis in neonates is usually a result of infection transmitted by the mother, either *in utero* or (more frequently) during passage through the birth canal. (Infections caused by *Cronobacter* are the exception, as discussed subsequently.) As more premature babies survive, the rates of neonatal meningitis increase, because the condition is favored in patients with immature immune systems. Although morbidity has increased, mortality rates have significantly declined. In the

Listeria monocytogenes	*Cryptococcus neoformans*	*Coccidioides immitis*	Viruses
Vehicle (food)	Vehicle (air, dust)	Vehicle (air, dust, soil)	Droplet contact
Intracellular growth	Capsule, melanin production	Granuloma (spherule) formation	Lytic infection of host cells
Cold enrichment, rapid methods	Negative staining, biochemical tests, DNA probes, cryptococcal antigen test	Identification of spherules, cultivation on Sabouraud's agar	Initially, absence of bacteria/fungi/protozoa, followed by viral culture or antigen tests
Cooking food, avoiding unpasteurized dairy products	–	Avoiding airborne endospores	–
Ampicillin, trimethoprim-sulfamethoxazole	Amphotericin B and fluconazole	Amphotericin B or oral or IV itraconazole	Usually none (unless specific virus identified and specific antiviral exists)
Asymptomatic in healthy adults; meningitis in neonates, elderly, and immunocompromised	Acute or chronic, most common in AIDS patients	Almost exclusively in endemic regions	Generally milder than bacterial or fungal
Mortality can be as much as 33%	Incidence before AIDS: >1 case per million per year; 66 cases per year in pre-HAART era; worldwide: 1 million new cases per year	Incidence in endemic areas: 200–300 annually	In United States, 26,000–42,000 hospitalizations/year

United States, the two most common causes are *Streptococcus agalactiae* and *Escherichia coli*. *Listeria monocytogenes* is also found frequently in neonates. It has already been covered here but is included in **Disease Table 19.2** as a reminder that it can cause neonatal cases as well. In the developing world, neonatal meningitis is more commonly caused by other organisms.

Streptococcus agalactiae

This species of *Streptococcus* belongs to the Lancefield group B of the streptococci. It colonizes 10% to 30% of female genital tracts and is the most frequent cause of neonatal meningitis (for details about this condition in women, see chapter 23). The treatment for neonatal disease is intravenous penicillin G, sometimes supplemented with an aminoglycoside.

Escherichia coli

The K1 strain of *Escherichia coli* is the second most common cause of neonatal meningitis. Most babies who suffer from this infection are premature, and their prognosis is poor.

Twenty percent of them die, even with aggressive antibiotic treatment, and those who survive often have brain damage.

The bacterium is usually transmitted from the mother's birth canal. It causes no disease in the mothers but can infect the vulnerable tissues of a neonate. It seems to have a predilection for the tissues of the central nervous system. Ceftazidime or cefepime +/− gentamicin is usually administered intravenously **(Disease Table 19.2).**

Cronobacter sakazakii

Cronobacter, formerly known as *Enterobacter sakazakii*, is a gram-negative bacillus found mainly in the environment and can survive very dry conditions. It has been implicated in outbreaks of neonatal and infant meningitis transmitted through contaminated powdered infant formula. Although cases of *Cronobacter* meningitis are rare, mortality rates can reach 40%. The FDA and the CDC advise hospitals to use ready-to-feed or concentrated liquid formulas and that home caregivers make fresh formula for each feeding and discard

Disease Table 19.2 Neonatal and Infant Meningitis

Causative Organism(s)	*Streptococcus agalactiae*	*Escherichia coli*, strain K1	*Listeria monocytogenes*	*Cronobacter sakazakii*
Most Common Modes of Transmission	Vertical (during birth)	Vertical (during birth)	Vertical	Vehicle (baby formula)
Virulence Factors	Capsule	–	Intracellular growth	Ability to survive dry conditions
Culture/Diagnosis	Culture mother's genital tract on blood agar; CSF culture of neonate	CSF Gram stain/culture	Cold enrichment, rapid methods	Chromogenic differential agar, or rapid detection kits
Prevention	Culture and treatment of mother	–	Cooking food, avoiding unpasteurized dairy products	Safe preparation and use of, or avoidance of, powdered formula
Treatment	Penicillin G plus aminoglycosides	Ceftazidime or cefepime +/− gentamicin	Ampicillin, trimethoprim-sulfamethoxazole	Begin with broad-spectrum drugs until susceptibilities determined
Distinctive Features	Most common; positive culture of mother confirms diagnosis	Suspected if infant is premature		–
Epidemiological Features	Before intrapartum antibiotics in 1996: 1.8 cases per 1,000 live births After intrapartum antibiotics: 0.32 cases per 1,000 live births	Estimated at 0.2–5 per 1,000 live births; 20% of pregnant women colonized	Mortality can be as high as 33%	12 cases in United States in 2011

any leftover formula. Care should be taken to wash hands and use clean feeding equipment when preparing formula to avoid infections with *Cronobacter*.

Meningoencephalitis

Up to this point, we have described microorganisms causing meningitis (inflammation of the meninges). Next we discuss microorganisms that cause **encephalitis,** inflammation of the brain. Because the brain and the spinal cord (and the meninges) are so closely connected, infections of one of these structures may also involve the other.

Two microorganisms cause a distinct disease called *meningoencephalitis,* and they are both amoebas. *Naegleria fowleri* and *Acanthamoeba* are protozoans considered to be accidental parasites that invade the body only under unusual circumstances.

Naegleria fowleri

The trophozoite of *Naegleria* is a small, flask-shaped amoeba that moves by means of a single, broad pseudopod. It forms a rounded, thick-walled, uninucleate cyst that is resistant to temperature extremes and mild chlorination.

Most cases of *Naegleria* infection reported worldwide occur in people who have been swimming in warm, natural bodies of freshwater. Infection can begin when amoebas are forced into human nasal passages as a result of swimming, diving, or other aquatic activities. The amoeba infects the olfactory epithelium and utilizes the olfactory nerve to travel to the brain. Infection then spreads as the pathogen enters into the fluid-filled subarachnoid space, making diagnosis from CSF possible. The result is primary amoebic meningoencephalitis (PAM), a rapid, massive destruction of brain and spinal tissue that causes hemorrhage and coma and invariably ends in death within a week of onset **(figure 19.11)**.

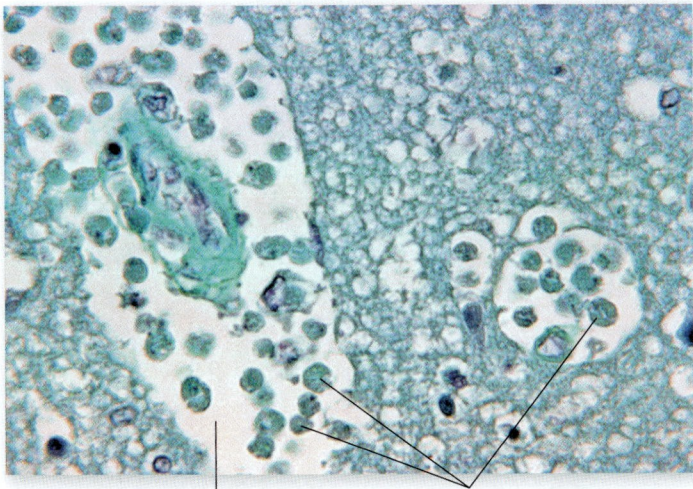

Pathologic changes in brain   *Naegleria*

Figure 19.11 *Naegleria fowleri* **in the brain.** The trophozoite form invades brain tissue, destroying it.

Although cases of disease are extremely rare, *Naegleria* meningoencephalitis advances so rapidly that treatment usually proves futile. This is illustrated by the fact that only one individual out of 128 total documented cases of infection in the United States (as of late 2013) has ever survived. Studies have indicated that early therapy with amphotericin B, sulfadiazine, or tetracycline in some combination can be of some benefit. Because of the wide distribution of the amoeba in nature and its hardiness, no general means of control exists. Public swimming pools and baths must be adequately chlorinated and checked periodically for the amoeba. Recent cases involving individuals unknowingly using infected tap water in a neti pot for nasal cleansing have prompted the CDC to advise against the use of tap water for this purpose.

Acanthamoeba

The protozoan *Acanthamoeba* is characterized by a large, amoeboid trophozoite with spiny pseudopods and a double-walled cyst. It differs from *Naegleria* in its portal of entry; it invades broken skin, the conjunctiva, and occasionally the lungs and urogenital epithelia. Although it causes a meningoencephalitis somewhat similar to that of *Naegleria*, the course of infection is lengthier. The disease is called granulomatous amoebic meningoencephalitis (GAM) and has only a 2% to 3% survival rate. We discussed ocular infections caused by this pathogen in chapter 18. Cutaneous and CNS infections with this organism are occasional complications in AIDS patients **(Disease Table 19.3)**.

Acute Encephalitis

Encephalitis can present as acute or **subacute.** It is always a serious condition, as the tissues of the brain are extremely sensitive to damage by inflammatory processes. Acute encephalitis is almost always caused by viral infection. One category of viral encephalitis is caused by viruses borne by insects (arboviruses), including West Nile virus. Alternatively, other viruses, such as members of the herpes family, are causative agents. Bacteria such as those covered under meningitis can also cause encephalitis, but the symptoms are almost always more pronounced in the meninges than in the brain.

The signs and symptoms of encephalitis vary, but they may include behavior changes or confusion because of inflammation. Decreased consciousness and seizures frequently occur. Symptoms of meningitis are often also present. Few of these agents have specific treatments, but because swift initiation of acyclovir therapy can save the life of a patient suffering from herpesvirus encephalitis, most physicians will begin empiric therapy with acyclovir in all seriously ill neonates and most other patients showing evidence of encephalitis. Treatment will, in any case, do no harm in patients who are infected with other agents.

Disease Table 19.3 Meningoencephalitis

	Primary Amoebic Meningoencephalitis	Granulomatous Amoebic Meningoencephalitis
Causative Organism(s)	*Naegleria fowleri*	*Acanthamoeba*
Most Common Modes of Transmission	Vehicle (exposure while swimming in water)	Direct contact
Virulence Factors	Invasiveness	Invasiveness
Culture/Diagnosis	Examination of CSF; brain imaging, biopsy	Examination of CSF; brain imaging, biopsy
Prevention	Limit warm fresh water or untreated tap water entering nasal passages	–
Treatment	Amphotericin B; mostly ineffective	Surgical excision of granulomas; ketoconazole may help
Epidemiological Features	United States: 37 infections in 10-year period	Predominantly occurs in immunocompromised patients

Arboviruses

Wherever there are arthropods, there are also arboviruses, so collectively their distribution is worldwide. The vectors and viruses tend to be clustered in the tropics and subtropics, but many temperate zones report periodic epidemics. A given arbovirus type may have very restricted distribution, even to a single isolated region, but some types range over several continents, and others can spread along with their vectors **(figure 19.12).**

Most arthropods that serve as infectious disease vectors feed on the blood of hosts, a process that infects them for varying time periods. Peak incidence of infection typically occurs when the arthropod is actively feeding and reproducing, usually from late spring through early fall. Warm-blooded vertebrates also maintain the virus during the cold and dry seasons. Humans can serve as dead-end, accidental hosts, as in equine encephalitis, or they can be a maintenance reservoir, as in yellow fever (discussed in chapter 20).

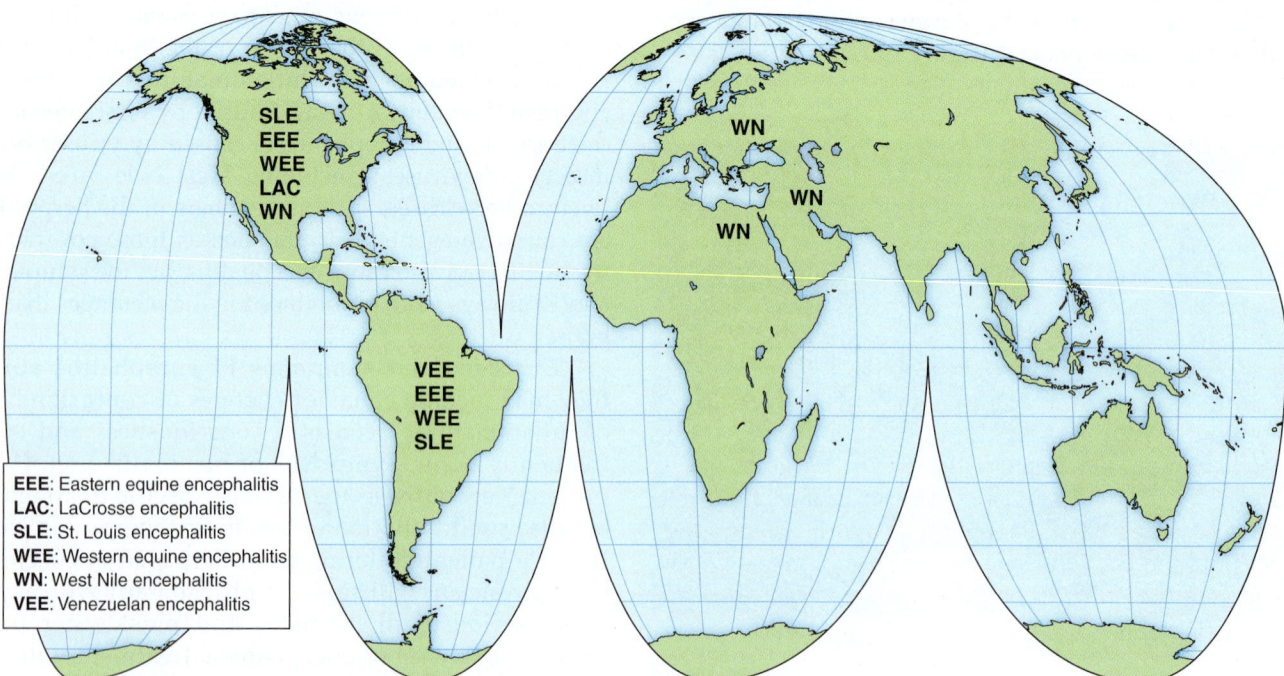

EEE: Eastern equine encephalitis
LAC: LaCrosse encephalitis
SLE: St. Louis encephalitis
WEE: Western equine encephalitis
WN: West Nile encephalitis
VEE: Venezuelan encephalitis

Figure 19.12 Worldwide distribution of major arboviral encephalitides.

Arboviral diseases have a great impact on humans. Although exact statistics are unavailable, it is believed that millions of people acquire infections each year and thousands of them die. One common outcome of arboviral infection is an acute fever, often accompanied by rash. Viruses that primarily cause these symptoms are covered in chapter 20.

The arboviruses discussed in this chapter can cause encephalitis, and we consider them as a group because the symptoms and management are similar (see Disease Table 19.4). The transmission and epidemiology of individual viruses are different, however, and are discussed for each virus. **Insight 19.2** discusses West Nile virus, an arbovirus that has spread across North America in recent years.

▶ Pathogenesis and Virulence Factors

Arboviral encephalitis begins with an arthropod bite, releasing the virus into the bloodstream where it will travel to nearby lymphatic tissues for replication. Prolonged viremia establishes viral infection in the brain, leading to inflammation-induced swelling and damage to the brain, nerves, and meninges. Symptoms are extremely variable and can include coma, convulsions, paralysis, tremor, loss of coordination, memory deficits, changes in speech and personality, and heart disorders. In some cases, survivors experience some degree of permanent brain damage. Young children and the elderly are most sensitive to injury by arboviral encephalitis.

The virulence of these viruses is not well understood, but much research has focused on proteins that the virus uses to attach to host tissues or to induce fusion with host cell membranes. Both of these functions facilitate invasion of the virus.

▶ Culture and Diagnosis

Except during epidemics, detecting arboviral infections can be difficult. The patient's history of travel to endemic areas or contact with vectors, along with serum analysis, helps with the diagnosis. Rapid serological tests are available for some of the viruses as are nucleic acid amplification tests.

▶ Prevention and Treatment

No satisfactory treatment exists for any of the arboviral encephalitides (plural of *encephalitis*). As mentioned earlier, empiric acyclovir treatment may be begun in case the infection is actually caused by either herpes simplex virus or varicella zoster. Treatment of the other infections relies entirely on support measures to control fever, convulsions, dehydration, shock, and edema.

Most of the control safeguards for arbovirus disease are aimed at the arthropod vectors. Mosquito abatement by eliminating breeding sites and by broad use of insecticides has been highly effective in restricted urban settings. Birds play a role as reservoirs of the virus, but direct transmission between birds and humans does not appear to occur.

At the present time, no commercial vaccines for these diseases are available in the United States for human use.

INSIGHT 19.2 The West Nile Virus . . . in Texas?

In 2012, there was a large outbreak of West Nile virus (WNV)—but not along the banks of the river from which the virus gets its name. The Centers for Disease Control and Prevention (CDC) reported that from January through August 2012 1,118 cases and 41 deaths due to WNV were reported in the United States. WNV cases were found in Mississippi, Louisiana, South Dakota, and Oklahoma; but more than half of the cases were centered in Texas. On average, there have been about 300 cases reported in the same time period in previous years, so the 2012 outbreak was over three times the yearly average. The summer of 2012 was one of the hottest summers on record, according to the National Oceanic and Atmospheric Administration (NOAA) and may have fostered breeding of more mosquitoes and greater spread of the disease. Recent studies also suggest that elevated environmental temperatures increase the rate of virus replication inside the insect vector.

West Nile virus was first reported in New York City in 1999 and has gradually spread over the United States, carried by birds that harbor the disease and transmitted by mosquitoes. All 50 states have reported cases since 1999. According to the CDC, symptoms of WNV occur in only 20% of those infected and include fever, headache, body aches, nausea, vomiting, and swollen lymph nodes. Sometimes patients experience a skin rash on the chest, stomach, and back. Individuals who are 50 and older are at higher risk of developing a severe illness. The CDC recommends wearing an insect repellant outdoors, especially at dusk and dawn, putting screens in windows, and removing sources of standing water that can be mosquito-breeding sites. There is no specific treatment for WNV infections, and usually the illness resolves on its own. CDC officials urge individuals suffering from severe headaches or confusion to seek medical attention, especially after being bitten by mosquitoes.

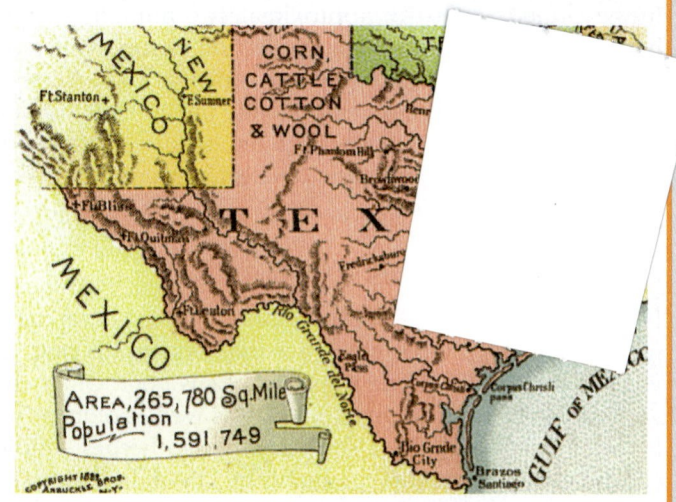

Source: http://www.cdc.gov/westnile/index.html

Western Equine Encephalitis (WEE) This disease occurs sporadically in the western United States and Canada, appearing first in horses and later in humans. The mosquito that carries the virus emerges in the early summer when irrigation begins in rural areas and breeding sites are abundant. The disease is extremely dangerous to infants and small children, with a case fatality rate of 3% to 7%.

Eastern Equine Encephalitis (EEE) EEE is endemic to an area along the eastern coast of North America and Canada. The usual pattern is sporadic, but occasional epidemics can occur in humans and horses. High periods of rainfall favor mosquito populations, leading to increased incidence of disease. Cases of disease usually appear first in horses and caged birds; a vaccine exists for horses, and its use is strongly urged to eliminate the virus from this reservoir. In humans, the case fatality rate can be very high (70%).

California Encephalitis This condition is most often caused by two viral strains belonging to what are known as the California serotype viruses. The California strain occurs occasionally in the western United States and has little impact on humans. The LaCrosse strain is widely distributed in the eastern United States and Canada and is a prevalent cause of viral encephalitis in North America. Children living in rural areas are the primary target group; most of them exhibit mild, transient symptoms. Fatalities are rare.

St. Louis Encephalitis St. Louis encephalitis is a very common viral encephalitis. Cases appear throughout North and South America, but epidemics in the United States occur most often in the Midwest and South. Asymptomatic infection is very common, and the actual number of cases is probably thousands of times greater than the mere 50 to 100 reported each year. The seasons of peak activity are spring and summer, depending on the region and species of mosquito.

West Nile Encephalitis The West Nile virus is a close relative of the SLE virus. It emerged in the United States in 1999, and by 2008 the CDC was reporting that 1% of people in the United States—or approximately 3 million people—had evidence of past or present infection. Seroprevalence in Egypt where the virus originated is nearly 40%. All signs point to the virus becoming endemic in the United States as well. Insight 19.2 summarizes the history of West Nile virus in the United States and details a current outbreak of disease.

Herpes Simplex Virus

Herpes simplex type I and II viruses can cause encephalitis in newborns born to HSV-positive mothers. In this case, the virus is disseminated and the prognosis is poor. Older children and young adults (ages 5 to 30), as well as older adults (over 50 years old), are also susceptible to herpes simplex encephalitis caused most commonly by HSV-I. In these cases, the HSV encephalitis usually represents a reactivation of dormant HSV from the trigeminal ganglion.

It should be noted the varicella-zoster virus (see chapter 18) can also reactivate from the dormant state, and it is responsible for rare cases of encephalitis.

JC Virus

The **JC virus (JCV)** gets its name from the initials of the patient in whom it was first diagnosed as the cause of illness. Seroprevalence of this polyomavirus nears 80% in many parts of the United States and Europe, though most infections are asymptomatic. In patients with immune dysfunction, especially in those with AIDS, this pathogen can cause a condition called **progressive multifocal leukoencephalopathy** (loo"-koh-en-sef"uh-lop'-uh-thee) **(PML).** This uncommon but generally fatal infection is a result of JC virus attack of accessory brain cells. The infection demyelinizes certain parts of the cerebrum. This virus should be considered when encephalitis symptoms are observed in AIDS patients. Administration of high doses of zidovudine has shown clinical benefits in recent cases of disease.

Other Virus-Associated Encephalitides

Infection with measles and other childhood rash diseases can result 1 to 2 weeks later in an inappropriate immune response with consequences in the CNS. The condition is called postinfectious encephalitis (PIE), and it is thought to be a result of immune system action and not of direct viral invasion of neural tissue. Very rarely, PIE can occur after immunization with vaccines comprised of live attenuated virus. Note that PIE is distinct from another possible sequela of measles virus infection called SSPE (discussed later in this chapter) **(Disease Table 19.4).**

Subacute Encephalitis

When encephalitis symptoms take longer to show up and when the symptoms are less striking, the condition is known as **subacute encephalitis.** The most common cause of subacute encephalitis is the protozoan *Toxoplasma*. Another form of subacute encephalitis can be caused by persistent measles virus as many as 7 to 15 years after the initial infection. Finally, a class of infectious agents known as prions can cause a condition called spongiform encephalopathy.

Toxoplasma gondii

Toxoplasma gondii is a flagellated parasite with such extensive cosmopolitan distribution that some experts estimate it affects the majority of the world's population at some time in their lives. Infection in the fetus and in immunodeficient people, especially those with AIDS, is severe and often fatal. Although infection in otherwise healthy people is generally unnoticed, recent data tell us it can have profound effects on their brain and the responses it controls. It seems that people with a history of *Toxoplasma* infection are more likely to

Disease Table 19.4   Acute Encephalitis

Causative Organism(s)	Arboviruses (viruses causing WEE, EEE, California encephalitis, SLE, West Nile encephalitis)	Herpes simplex 1 or 2	JC virus	Immunologic reaction to other viral infections
Most Common Modes of Transmission	Vector (arthropod bites)	Vertical or reactivation of latent infection	? Ubiquitous	Sequelae of measles, other viral infections, and occasionally, vaccination
Virulence Factors	Attachment, fusion, invasion capabilities	–	–	–
Culture/ Diagnosis	History, rapid serological tests, nucleic acid amplification tests	Clinical presentation, PCR, Ab tests, growth of virus in cell culture	PCR of cerebrospinal fluid	History of viral infection or vaccination
Prevention	Insect control; vaccines for WEE and EEE available	Maternal screening for HSV	None	–
Treatment	None	Acyclovir	Zidovudine or other antivirals	Steroids, anti-inflammatory agents
Distinctive Features	History of exposure to insect important	In infants, disseminated disease present; rare between 30 and 50 years	In severely immunocompromised, especially AIDS	History of virus/ vaccine exposure critical
Epidemiological Features	For West Nile virus: In Egypt, up to half infected during childhood and do not experience neurological disease

In United States approx. 5,000 cases (half of them neuroinvasive) since appearance in 1999 | HSV-1 most with common cause of encephalitis 2 cases per million per year | Affects 5% of adults with untreated AIDS | Rare in United States due to vaccination; more common in developing countries, more common in boys than girls |

display thrill-seeking behaviors. Also, people with infection histories seem to have slower reaction times.

T. gondii is a very successful parasite with so little host specificity that it can attack at least 200 species of birds and mammals. However, its primary reservoir and hosts are members of the feline family, both domestic and wild.

▶ Signs and Symptoms

As just mentioned, most cases of toxoplasmosis are asymptomatic or marked by mild symptoms such as sore throat, lymph node enlargement, and low-grade fever. In patients whose immunity is suppressed by infection, cancer, or drugs, the outlook may be grim. The infection causes a more chronic or subacute form of encephalitis than do most viruses, often producing extensive brain lesions and fatal disruptions of the heart and lungs. A pregnant woman with toxoplasmosis has a 33% chance of transmitting the infection to her fetus. Congenital infection occurring in the first or second trimester is associated with stillbirth and severe abnormalities such as liver and spleen enlargement, liver failure, hydrocephalus, convulsions, and damage to the retina that can result in blindness.

▶ Pathogenesis and Virulence Factors

Toxoplasma is an obligate intracellular parasite, making its ability to invade host cells an important factor for virulence.

▶ Transmission and Epidemiology

To follow the transmission of toxoplasmosis, we must first look at the general stages of the *Toxoplasma* life cycle in the cat **(figure 19.13a).** The parasite undergoes a sexual phase in the intestine and is then released in feces, where it becomes an infective *oocyst* that survives in moist soil for several months. Ingested oocysts release an invasive asexual tissue phase called a *tachyzoite* that infects many different tissues and often causes disease in the cat. These forms eventually enter an asexual cyst state in tissues, called a *pseudocyst.* Most of the time, the parasite does not cycle in cats alone and is spread by oocysts to intermediate hosts, usually rodents and birds. The cycle returns to cats when they eat these infected prey animals.

In 2007, scientists at Stanford University found that the protozoan crowds into a part of the rat brain that usually directs the rat to avoid the smell of cat urine (a natural defense against a domestic rat's major predator). When *Toxoplasma* infects rat brains, the rats lose their fear of cats. Infected rats are then easily eaten by cats, ensuring the continuing *Toxoplasma* life cycle. All other neurological functions in the rat are left intact.

Other vertebrates become a part of this transmission cycle **(figure 19.13b).** Herbivorous animals such as cattle and sheep ingest oocysts that persist in the soil of grazing areas and then develop pseudocysts in their muscles and other organs. Carnivores such as canines are infected by eating pseudocysts in the tissues of carrier animals.

Humans appear to be constantly exposed to the pathogen. The rate of prior infections, as detected through serological tests, can be as high as 90% in some populations. Many cases are caused by ingesting pseudocysts in contaminated meats. A common source is raw or undercooked meat. The grooming habits of cats spread fecal oocysts on their body surfaces, and unhygienic handling of them presents an opportunity to ingest oocysts. Infection can also occur when oocysts are inhaled in air or dust contaminated with cat droppings and when tachyzoites cross the placenta to the fetus.

▶ Culture and Diagnosis

This infection can be differentiated from viral encephalitides by means of serological tests that detect antitoxoplasma antibodies, especially those for IgM, which appears early in infection. Disease can also be diagnosed by culture or histological analysis for the presence of cysts or tachyzoites.

▶ Prevention and Treatment

The most effective drugs are pyrimethamine and leucovorin and sulfadiazine alone or in combination. Because these drugs do not destroy the cyst stage, they must be given for long periods to prevent recurrent infection.

In view of the fact that the oocysts are so widespread and resistant, hygiene is of paramount importance in controlling toxoplasmosis. Adequate cooking or freezing below −20°C destroys both oocysts and tissue cysts. Oocysts can also be avoided by washing the hands after handling cats or soil possibly contaminated with cat feces, especially sandboxes

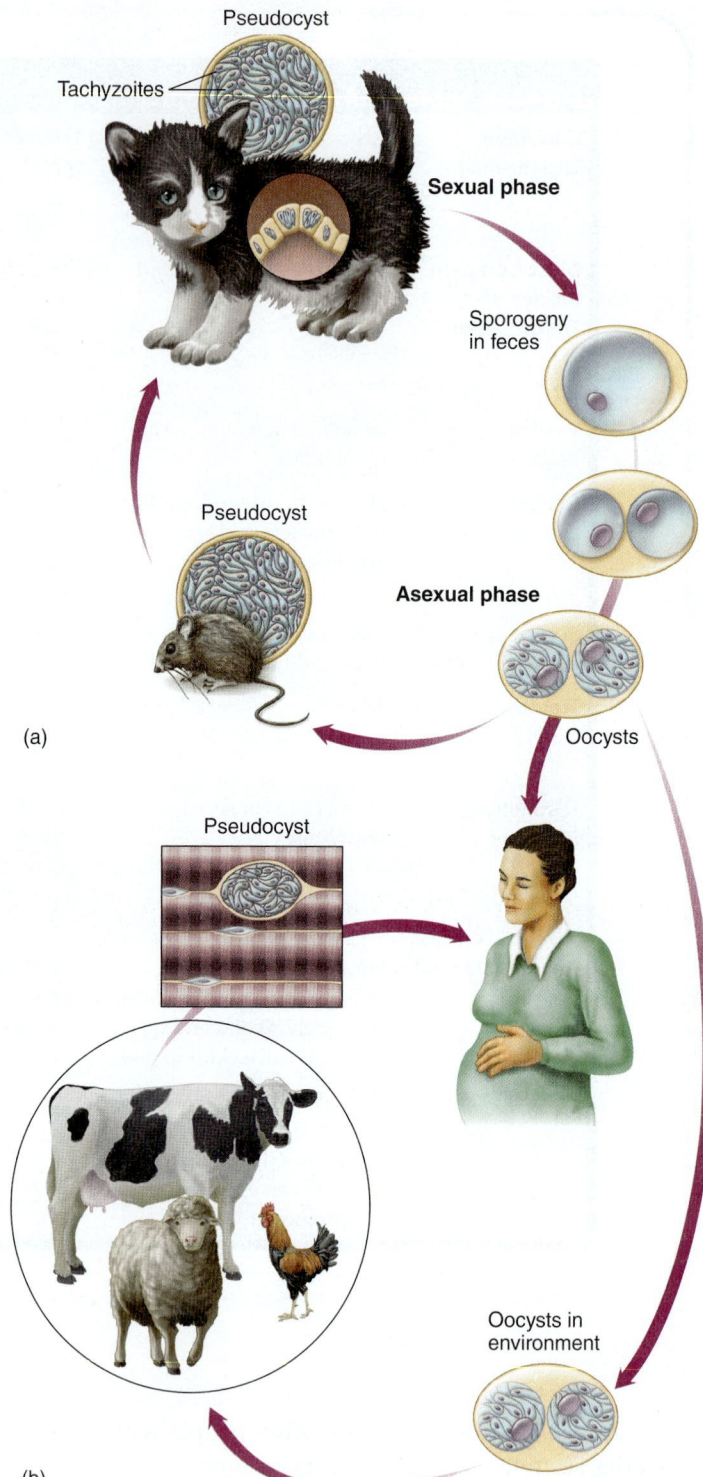

(a)

(b)

Figure 19.13 The life cycle and morphological forms of *Toxoplasma gondii.* **(a)** The cycle in cats and their prey. **(b)** The cycle in other animal hosts. The zoonosis has a large animal reservoir (domestic and wild) that becomes infected through contact with oocysts in the soil. Humans can be infected through contact with cats or ingestion of pseudocysts in animal flesh. Infection in pregnant women is a serious complication because of the potential damage to the fetus.

and litter boxes. A new blood test was recently developed to test women early in pregnancy for asymptomatic infection as a means of sparing the developing fetus permanent damage or death.

Measles Virus: Subacute Sclerosing Panencephalitis

Subacute sclerosing panencephalitis (SSPE) is sometimes called a "slow virus infection." The symptoms appear years after an initial measles episode and are different from those of immune-mediated postinfectious encephalitis, described earlier. SSPE seems to be caused by direct viral invasion of neural tissue. It is not clear what factors lead to persistence of the virus in some people. SSPE's important features are listed in **Disease Table 19.5.**

Prions

As you read in chapter 6, **prions** are *proteinaceous infectious particles* containing, apparently, no genetic material. They are known to cause diseases called **transmissible spongiform encephalopathies (TSEs)**, neurodegenerative diseases with long incubation periods but rapid progressions

once they begin. The human TSEs include **Creutzfeldt-Jakob disease (CJD),** Gerstmann-Sträussler-Scheinker disease, kuru, and fatal familial insomnia. TSEs are also found in animals and include a disease called scrapie in sheep and goats, transmissible mink encephalopathy, and bovine spongiform encephalopathy (BSE). This last disease is commonly known as "mad cow disease" and was in the headlines in the 1990s due to its apparent link to a variant form of CJD in humans in Great Britain.

▶ **Signs and Symptoms of CJD**

Symptoms of all forms of CJD include altered behavior, dementia, memory loss, impaired senses, delirium, and premature senility. Uncontrollable muscle contractions continue until death, which usually occurs within 1 year of diagnosis.

▶ **Causative Agent of CJD**

It is thought that 10% to 15% of CJD cases are due to an inherited mutation within a single gene. These are termed familial or hereditary CJD. In the 1980s, Stanley Prusiner identified a protein called PrP^C that spontaneously transforms into a nonfunctional form in CJD. This altered protein (PrP^{SC}), which he termed a *prion*, triggers damage in the brain and other areas of the central nervous system. The PrP^{SC} actually becomes

Disease Table 19.5 **Subacute Encephalitis**			
Causative Organism(s)	*Toxoplasma gondii*	Subacute sclerosing panencephalitis	Prions
Most Common Modes of Transmission	Vehicle (meat) or fecal-oral	Persistence of measles virus	CJD = direct/parenteral contact with infected tissue, or inherited vCJD = vehicle (meat, parenteral)
Virulence Factors	Intracellular growth	Cell fusion, evasion of immune system	Avoidance of host immune response
Culture/Diagnosis	Serological detection of IgM, culture, histology	EEGs, MRI, serology (Ab versus measles virus)	Biopsy, image of brain
Prevention	Personal hygiene, food hygiene	None	Avoiding tissue
Treatment	Pyrimethamine and/or leucovorin and/or sulfadiazine	None	None
Distinctive Features	Subacute, slower development of disease	History of measles	Long incubation period; fast progression once it begins
Epidemiological Features	15%–29% of U.S. population is seropositive; internationally, seroprevalence is to 90%; disease occurs in 3%–15% of AIDS patients	United States: fewer than 10 cases/year; incidence has declined 90% in countries who vaccinate against measles	CJD: 1 case per year per million worldwide; seen in older adults vCJD: 98% cases originated in United Kingdom

catalytic and able to spontaneously convert other normal human PrP^C proteins into the abnormal form. This becomes a self-propagating chain reaction that creates a massive accumulation of PrP^{SC}, leading to plaques, spongiform damage (that is, holes in the brain), and severe loss of brain function.

Further studies showed that prions could cause disease when transferred to a new host, confirming for the first time that a prion could function as an infectious transmissible agent. This led to the recognition of cases called iatrogenic CJD, in which a patient acquired the disease through contaminated equipment during a medical procedure. In the late 1990s, it became apparent that humans were contracting a variant form of CJD (vCJD) after ingesting meat from cattle that had been afflicted by a related disease called bovine spongiform encephalopathy. Presumably, meat products had been contaminated with fluid or tissues infected with the prion. Cases of this disease were concentrated in Great Britain, where many cows were found to have BSE. As of late 2012, a total of 224 people worldwide had developed the disease, 176 of them in the United Kingdom. There have been three cases in the United States, two of which are thought to have acquired the infection during travel to the United Kingdom. A 2012 examination of 13,878 appendixes in the United Kingdom revealed that four of them were positive for vCJD, which suggests that the positivity rate there is 288 per million.

Since Prusiner first described prions, much has been learned about them. The nonpathogenic forms of them are vital for normal brain development and seem to be very important for memory and other vital functions in the nervous system.

▶ **Pathogenesis and Virulence Factors**

Autopsies of the brain of all CJD patients reveal spongiform lesions as well as tangled protein fibers (neurofibrillary tangles) and enlarged astroglial cells (**figure 19.14**). These changes affect the gray matter of the CNS and seem to be caused by the massive accumulation of altered PrP, which may be toxic to neurons. The altered PrPs apparently stimulate no host immune response. Prions are also incredibly hardy "pathogens." They are highly resistant to chemicals, radiation, and heat and can even withstand prolonged autoclaving.

▶ **Transmission and Epidemiology**

Hereditary CJD and sporadic CJD are most common in elderly people. The median age at death of patients with vCJD is 28 years. In contrast, the median age at death of patients with the classic forms (sporadic, hereditary) of CJD is 68 years.

Aside from genetic transmission, prions can be spread through direct or indirect contact with infected brain tissue or cerebrospinal fluid. Ingestion of contaminated tissue has been documented to cause disease, and it is suggested that aerosols may represent another possible mode of transmission.

Health care professionals should be aware of the possibility of CJD in patients, especially when surgical procedures are performed, since iatrogenic cases have been reported due to transmission due to CJD via contaminated

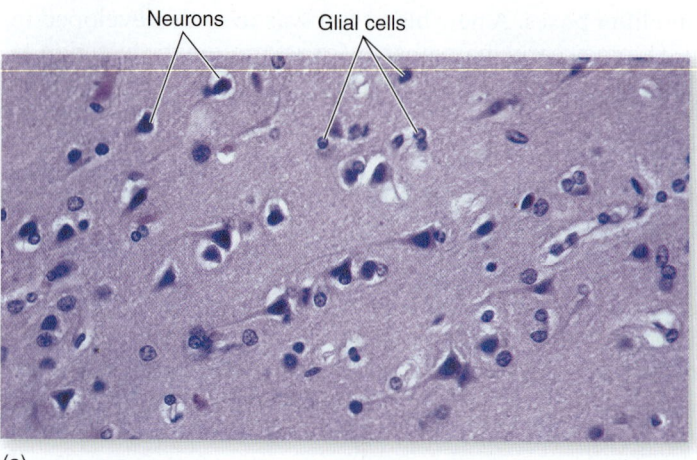

(a)

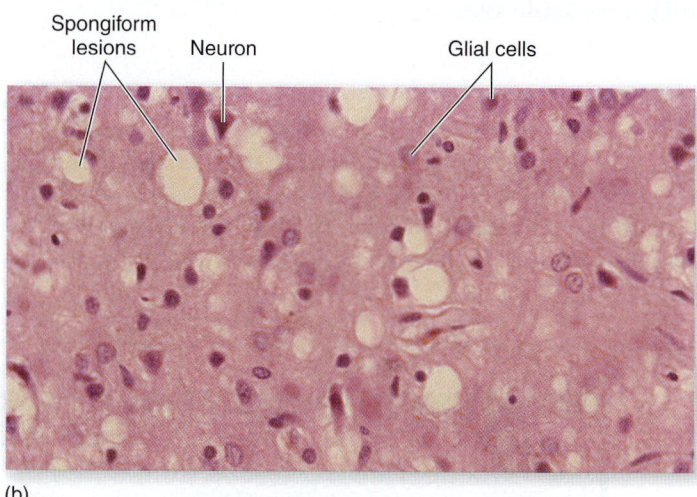

(b)

Figure 19.14 The microscopic effects of spongiform encephalopathy. (a) Normal cerebral cortex section, showing neurons and glial cells. (b) Sectioned cortex in CJD patient shows numerous round holes, producing a "spongy" appearance. This destroys brain architecture and causes massive loss of neurons and glial cells.

surgical instruments. Due to the heat and chemical resistance of prions, normal disinfection and sterilization procedures are usually not sufficient to eliminate them from instruments and surfaces. The latest CDC guidelines for handling of CJD patients in a health care environment should be consulted. CJD has also been transmitted through corneal grafts and administration of contaminated human growth hormone. In 2003, a British patient died of CJD after receiving a blood transfusion in 1996 from a donor who had CJD. Experiments suggest that vCJD seems to be more transmissible through blood than classic CJD. For that reason, blood donation programs screen for possible exposure to BSE by asking about travel and residence history.

▶ **Culture and Diagnosis**

It is very difficult to diagnose CJD. Definitive diagnosis requires examination of biopsied brain or nervous tissue, and this procedure is usually considered too risky because of both the trauma induced in the patient and the undesirability of contaminating surgical instruments and operating rooms. Electroencephalograms and magnetic resonance imaging can provide important clues. New tests are being developed to identify prions in cerebrospinal fluid samples, making diagnosis possible before the patient's death.

▶ Prevention and Treatment

Prevention of this disease relies on avoiding infected tissues. Avoiding vCJD entails not ingesting tainted meats. No known treatment currently exists for any form of CJD; patients inevitably die. There is active research in treatments for prion diseases **(Insight 19.3)**. Medical intervention focuses on easing symptoms and making the patient as comfortable as possible (see Disease Table 19.5). New studies involving the role of microglia and other immune system components may however show new promise for a potential treatment in the future.

INSIGHT 19.3 Treatment for Mad Cow Disease?

Prions are unlike any other infectious agent—they are composed entirely of protein and withstand nearly all forms of disinfection, including radiation, incineration, and the strongest disinfectants and detergents. If a patient is affected by a prion disease, there are few options for treatment. However, recently, researchers at the University of Alberta, Canada, have discovered chemicals that can remove prions from infected brain cells in mice. Dr. Frederick West and his team are still in the initial testing phases, and the compounds are still too large to be injected into cows, but they show promise in treating diseases like vCJD and Alzheimer's. Another study at the NYU School of Medicine has found that two drugs currently in use—trimipramine, an antidepressant, and fluphenazine, an antipsychotic—also act against prions. These drugs are already approved for clinical use, so they can be tested in patients with CJD and vCJD. A team of researchers at NYU, led by Dr. Thomas Wisniewski, has only tested these treatments against prion diseases in animals but are hopeful that they will offer some hope for treatment of human prion diseases.

Source: 2011. *PLoS One*, vol. 6, no. 9, article 24844.

Rabies

Rabies is a slow, progressive zoonotic disease characterized by fatal encephalitis. It is so distinctive in its pathogenesis and its symptoms that we discuss it separately from the other encephalitides. It is distributed nearly worldwide, except for perhaps two dozen countries that have remained rabies-free by practicing rigorous animal control.

▶ Signs and Symptoms

The average incubation period of rabies is 1 to 2 months or more, depending on the wound site, its severity, and the inoculation dose. The incubation period is shorter in facial, scalp, or neck wounds because of closer proximity to the brain. The prodromal phase begins with fever, nausea, vomiting, headache, fatigue, and other nonspecific symptoms.

In the form of rabies termed "furious," the first acute signs of neurological involvement are periods of agitation, disorientation, seizures, and twitching. Spasms in the neck and pharyngeal muscles lead to severe pain upon swallowing, leading to a symptom known as **hydrophobia** (fear of water). Throughout this phase, the patient is fully coherent and alert. With the "dumb" form of rabies, a patient is not hyperactive but is paralyzed, disoriented, and stuporous. Ultimately, both forms progress to the coma phase, resulting in death from cardiac or respiratory arrest. Until recently, humans were never known to survive rabies. But a handful of patients have recovered in recent years after receiving intensive, long-term treatment.

▶ Causative Agent

The rabies virus is in the family *Rhabdoviridae,* genus *Lyssavirus.* This virus has a distinctive bulletlike appearance, round on one end and flat on the other. Additional features are a helical nucleocapsid and spikes that protrude through the envelope **(figure 19.15).** The family contains about 60 different viruses, but only the rabies *Lyssavirus* infects humans.

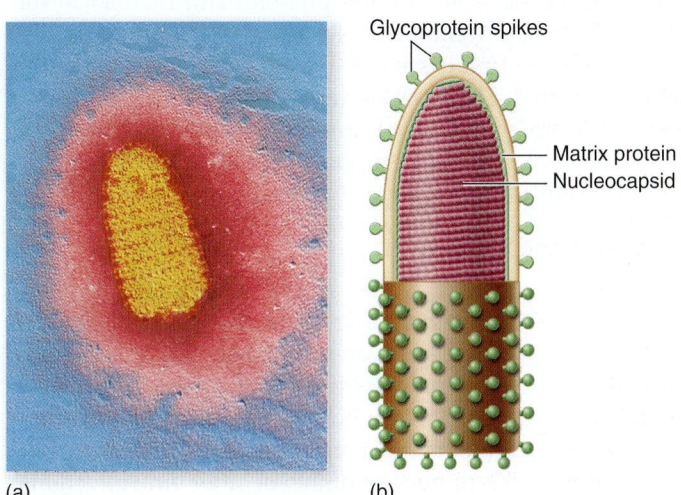

(a)   (b)

Figure 19.15 The structure of the rabies virus. **(a)** Color-enhanced virion shows internal serrations, which represent the tightly coiled nucleocapsid. **(b)** A schematic model of the virus, showing its major features.

▶ Pathogenesis and Virulence Factors

Infection with rabies virus typically begins when an infected animal's saliva enters a puncture site. The virus occasionally is inhaled or inoculated through the membranes of the eye. The rabies virus remains up to a week at the trauma site, where it multiplies. The virus then gradually enters nerve endings and advances toward the ganglia, spinal cord, and brain. Viral multiplication throughout the brain is eventually followed by migration to such diverse sites as the eye, heart, skin, and oral cavity. The infection cycle is completed when the virus replicates in the salivary gland and is shed into the saliva. Clinical rabies proceeds through several distinct stages that almost inevitably end in death, unless vaccination is performed before symptoms begin.

Scientists have discovered that virulence is associated with an envelope glycoprotein that seems to give the virus its ability to spread in the CNS and to invade certain types of neural cells.

▶ Transmission and Epidemiology

The primary reservoirs of the virus are wild mammals such as canines, skunks, raccoons, badgers, cats, and bats that can spread the infection to domestic dogs and cats. Both wild and domestic mammals can spread the disease to humans through bites, scratches, and inhalation of droplets **(Insight 19.4)**. The annual worldwide total for human rabies is estimated to be from 35,000 to 50,000 cases, but only a tiny number of these cases occur in the United States. Most U.S. cases of rabies occur in wild animals (about 6,000 to 7,000 cases per year), while dog rabies has declined **(figure 19.16)**.

The epidemiology of animal rabies in the United States varies. The most common wild animal reservoir host has changed from foxes to skunks to raccoons. Regional differences in the dominant reservoir also occur. Rats, skunks, and bobcats are the most common carriers of rabies in California; raccoons are the predominant carriers in the East; and coyotes dominate in Texas.

In 2004, the first cases of rabies in recipients of donated organs occurred. The lungs, kidneys, and liver of a man were donated to four patients; three of them died of rabies (the fourth died of surgical complications). The virus has also been transmitted through cornea transplants.

▶ Culture and Diagnosis

When symptoms appear after an attack by a rabid animal, the disease is readily diagnosed. But the diagnosis can be obscured when contact with an infected animal is not clearly defined or when symptoms are absent or delayed. Anxiety, agitation, and depression can pose as a psychoneurosis; muscle spasms resemble tetanus; and encephalitis with convulsions and paralysis mimics a number of other viral infections. Often the disease is diagnosed only at autopsy. The direct

---

INSIGHT 19.4   Bats on a Plane?

I hope they checked for "Eye of newt, and toe of frog / Wool of bat, and tongue of dog" as well. Never mind the risks associated with driving to the airport... No, it's not the title of the latest Samuel L. Jackson thriller—the story is all too real. According to CDC reports, on August 5, 2011, shortly after a commercial airline's takeoff from Madison, Wisconsin, a bat flew through the cabin several times before being trapped in the lavatory. The plane, whose destination was Atlanta, Georgia, returned to the airport and the passengers disembarked while maintenance crew workers attempted to remove the bat. The high-flying mammal proved too elusive for the ground crew, and it made its way through the airport terminal and was seen exiting through the automatic doors. After the plane was searched for other nocturnal stowaways, the passengers re-boarded the plane and continued on their journey. What was the risk to the passengers—aside from a good scare—on an early morning commuter flight? Rabies. Physical contact with the bat or its saliva could potentially expose passengers and crew members to the virus if the bat was infected. Because the animal was not captured, the rabies status of the animal was unknown. After the incident, the CDC tracked down 45 passengers from 11 different states along with the pilots, flight attendants, and ground crew to determine if they had been in contact with the bat or its saliva and if they required postexposure prophylaxis against the rabies virus. Because bats had previously been seen at the airport, the Wisconsin Department of Natural Resources, Public Health Madison & Dane County, worked with airport officials to determine if any other bats had taken up residence at the airport. Airport employees were also trained on correct procedures for capturing bats.

Source: http://www.npr.org/blogs/health/2012/04/12/150496100/bat-on-a-plane-triggers-rabies-hunt

---

fluorescent antibody test is the standard for postmortem identification.

Diagnosis before death requires multiple tests. Reverse transcription PCR is used with saliva samples but must be accompanied by detection of antibodies to the virus in serum or spinal fluid. Skin biopsies are also used.

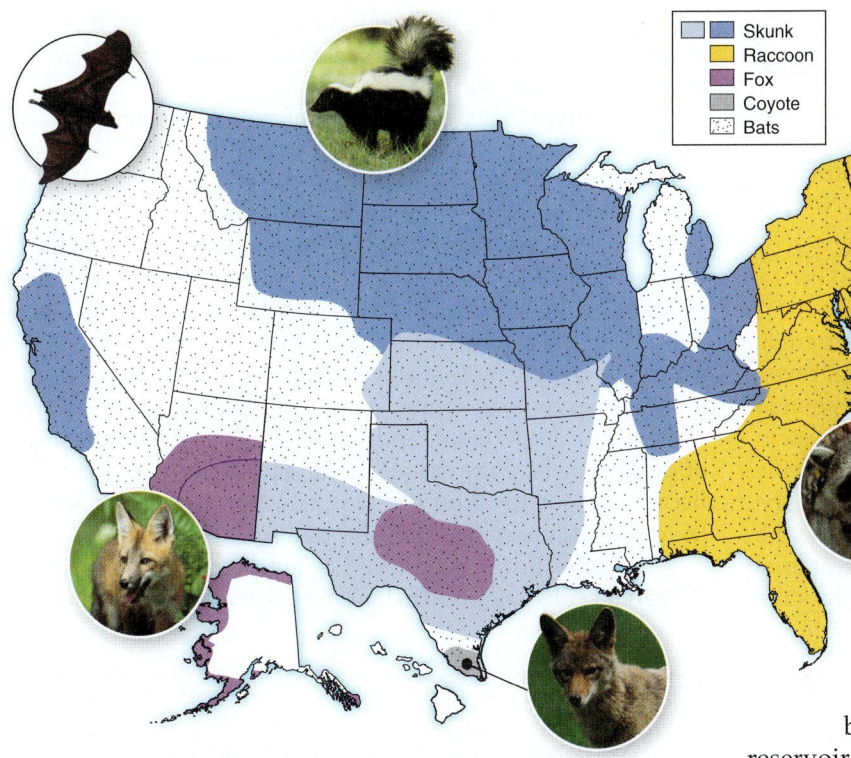

Figure 19.16 Distribution of rabies in the United States.
Rabies is found in 10 distinct geographic areas. In each area, a particular animal is the dominant reservoir as illustrated by four different colors. In addition to the cases shown, insectivorous bats are responsible for sporadic cases of rabies in wild animals throughout the country.

▶ Prevention and Treatment

A bite from a wild or stray animal demands assessment of the animal, meticulous care of the wound, and a specific treatment regimen. A wild mammal—especially a skunk, raccoon, fox, or coyote—that bites without provocation is presumed to be rabid, and therapy is immediately begun. If the animal is captured, brain samples and other tissue are examined for verification of rabies. Healthy domestic animals are observed closely for signs of disease and sometimes quarantined. Preventive therapy is initiated if any signs of rabies appear.

Rabies is one of the few infectious diseases for which a combination of passive and active postexposure immunization is indicated (and successful). Initially the wound is infused with human rabies immune globulin (HRIG) to impede the spread of the virus, and globulin is also injected intramuscularly to provide immediate systemic protection. A full course of vaccination is started simultaneously. The routine postexposure vaccination entails intramuscular or intradermal injection on the 1st, 3rd, 7th, and 14th day. Sometimes putting a patient in whom disease has already manifested in a drug-induced coma and on

ventilator support can save his or her life. High-risk groups such as veterinarians, animal handlers, laboratory personnel, and travelers should receive three doses to protect against possible exposure. A DNA vaccine for rabies is in development.

Control measures such as vaccination of domestic animals, elimination of strays, and strict quarantine practices have helped reduce the virus reservoir. However, rabid dogs remain a dangerous source of infection in developing countries around the world where over 55,000 people still die of this preventable disease each year. In recent years, the United States and other countries have utilized a live oral vaccine made with a vaccinia virus that carries the gene for the rabies virus surface antigen for mass immunization of wild animals. The vaccine has been incorporated into bait (sometimes peanut butter sandwiches!) placed in the habitats of wild reservoir species such as skunks and raccoons. See **Disease Table 19.6** for a summary of this infectious disease.

Disease Table 19.6 Rabies

Causative Organism(s)	Rabies virus
Most Common Modes of Transmission	Parenteral (bite trauma), droplet contact
Virulence Factors	Envelope glycoprotein
Culture/Diagnosis	RT-PCR of saliva; Ab detection of serum or CSF; skin biopsy
Prevention	Inactivated vaccine
Treatment	Postexposure passive and active immunization; induced coma and ventilator support
Epidemiological Features	United States: 1–5 cases per year Worldwide: 35,000–55,000 cases annually

Poliomyelitis

Poliomyelitis (poh"-lee-oh-my"-eh-ly'tis) (polio) is an acute enteroviral infection of the spinal cord that can cause neuromuscular paralysis. Because it often affects small children, in the past it was called infantile paralysis **(Insight 19.5).** No civilization or culture has escaped the devastation of polio.

▶ Signs and Symptoms

Most infections are contained as short-term, mild viremia. Some persons develop mild nonspecific symptoms of fever, headache, nausea, sore throat, and myalgia. If the viremia persists, viruses can be carried to the central nervous system through its blood supply. The virus then spreads along specific pathways in the spinal cord and brain. Being **neurotropic,** the virus infiltrates the motor neurons of the anterior horn of the spinal cord, although it can also attack spinal ganglia, cranial nerves, and motor nuclei. Nonparalytic disease involves the invasion but not the destruction of nervous tissue. It gives rise to muscle pain and spasm, meningeal inflammation, and vague hypersensitivity.

In paralytic disease, invasion of motor neurons causes various degrees of flaccid paralysis over a period of a few hours to several days. Depending on the level of damage to motor neurons, paralysis of the muscles of the legs, abdomen, back, intercostals, diaphragm, pectoral girdle, and bladder can result. In rare cases of **bulbar poliomyelitis,** the brain stem, medulla, or even cranial nerves are affected. This situation leads to loss of control of cardiorespiratory regulatory centers, requiring mechanical respirators. In time, the unused muscles begin to atrophy, growth is slowed, and severe deformities of the trunk and limbs develop. Common sites of deformities are the spine, shoulder, hips, knees, and feet. Because motor function but not sensation is compromised, the crippled limbs are often very painful.

In recent times, a condition called post-polio syndrome (PPS) has been diagnosed in long-term survivors of childhood

---

INSIGHT 19.5   Polio

In 1926, when Louis Ellis (pictured) was 8 years old, he was walking home from school in Manola, Alberta, Canada. It was winter, and his parents' farmhouse was several miles from the school. All of a sudden, he felt weak and feverish and lay down in the snow to cool himself off. After several days of fighting the fever, his parents took him to a hospital in Edmonton, where doctors diagnosed him with polio and told his parents he would never walk again. Not to be deterred, his mother took him home, laid him on the kitchen table in her little farmhouse, and massaged his muscles three times a day with olive oil to keep them from atrophying. He eventually learned to walk. Although one leg was shorter than the other, and he had to wear a built-up shoe, he was able to live a normal life. In 1955, when the Salk polio vaccine became available, Louis marched his two young children to the doctor's office and made sure they were vaccinated.

Polio was one of the most feared childhood diseases in the 20th century. Many families have someone in their history like Louis Ellis. Small, localized outbreaks of polio occurred throughout the late 1800s in the United States, Canada, and Europe, but the first major epidemic was reported in Brooklyn, New York, in 1916; 27,000 people were infected, and there were 6,000 deaths. Panic swept through the city; public gatherings were canceled, movie theaters were closed, and children were told to avoid drinking from water fountains and swimming in the ocean. From that time forward through the 1960s, there was a polio epidemic every summer. Children could be seen wearing splints and braces, walking with crutches, or living in iron lungs. When the Salk polio vaccine was made available in 1955, parents flocked to have their children vaccinated, and rates of polio dropped. In 1962, the Sabin vaccine was approved for use, further lowering the number of cases. The last reported outbreak of polio in the United States was in 1979 among individuals in an Amish community.

It was a presidential hopeful who was instrumental in spurring development of the polio vaccine. Franklin D. Roosevelt

Louis Ellis (standing), before polio.

became paralyzed in 1921 with what was believed to be polio. After trying various unsuccessful therapies, he refused to allow the disease to limit his political aspirations. In 1938, he helped found the National Foundation for Infantile Paralysis, known today as the March of Dimes. The foundation altered the method of funding vaccine research from soliciting donations from wealthy philanthropists to collecting small amounts—pennies and dimes—from millions of people, and eventually raising $25.5 million for polio vaccine research. Roosevelt became a champion of the March of Dimes, hosting "Birthday Balls" every year on his birthday to raise money for polio research. Roosevelt is quoted as saying, "Once you've spent two years trying to wiggle one toe, everything is in proportion."

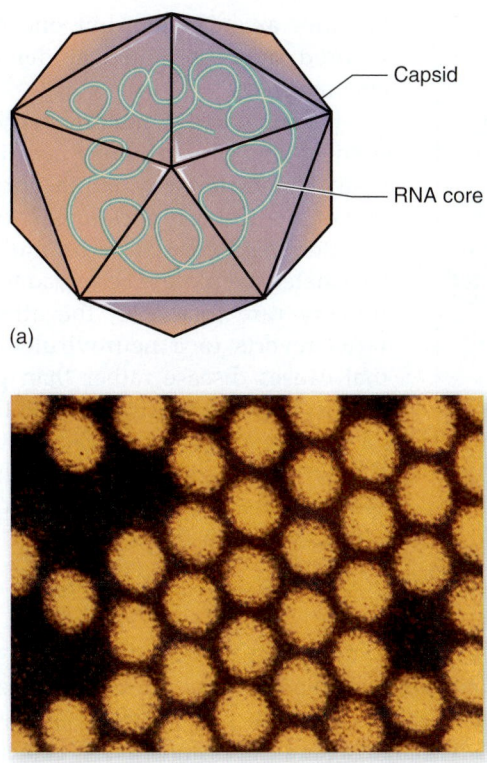

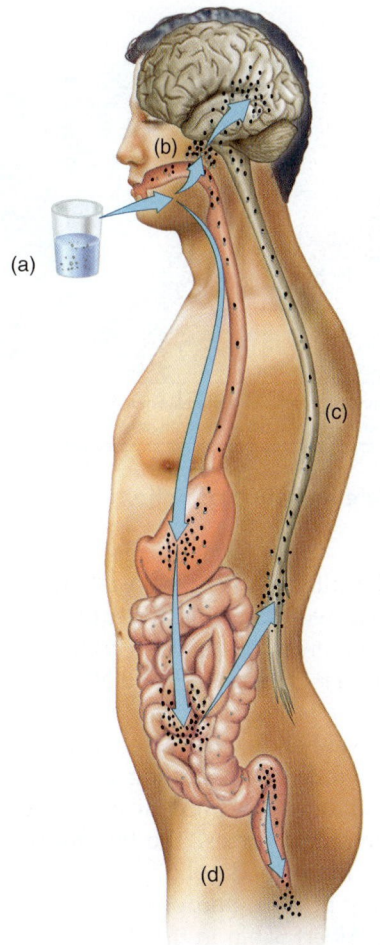

Figure 19.18 The stages of infection and pathogenesis of poliomyelitis. (a) First, the virus is ingested and carried to the throat and intestinal mucosa. (b) The virus then multiplies in the tonsils. Small numbers of viruses escape to the regional lymph nodes and blood. (c) The viruses are further amplified and cross into certain nerve cells of the spinal column and central nervous system. (d) Last, the intestine actively sheds viruses.

Figure 19.17 Typical structure of a picornavirus.
(a) A poliovirus, a type of picornavirus that is one of the simplest and smallest viruses (30 nm). It consists of an icosahedral capsid shell around a molecule of RNA. (b) A crystalline mass of stacked poliovirus particles in an infected host cell (300,000×).

infection. PPS manifests as a progressive muscle deterioration that develops in about 25% to 50% of patients several decades after their original polio attack.

▶ **Causative Agent**

The poliovirus is in the family *Picornaviridae*, genus *Enterovirus*—named for its small size and its RNA core **(figure 19.17)**. It is nonenveloped and nonsegmented. There are three strains, types I, II, and III. The naked capsid of the virus confers chemical stability and resistance to acid, bile, and detergents. By this means, the virus survives the gastric environment and other harsh conditions, which contributes to its ease of transmission.

▶ **Pathogenesis and Virulence Factors**

After being ingested, polioviruses adsorb to receptors of mucosal cells in the oropharynx and intestine **(figure 19.18)**. There, they multiply in the mucosal epithelia and lymphoid tissue. Multiplication results in large numbers of viruses being shed into the throat and feces, and some of them leak into the blood. Depending on the number of viruses in the blood and their duration of stay there, an individual may exhibit no symptoms, mild nonspecific symptoms such as fever or short-term muscle pain, or devastating paralysis. Scientists studying poliovirus virulence focus on components of the virus that allow attachment and penetration of host cells.

▶ **Transmission and Epidemiology**

Sporadic cases of polio can break out at any time of the year, but its incidence is more pronounced during the summer and fall. Humans are the only known reservoir, and the virus is passed within the population through food, water, hands, objects contaminated with feces, and mechanical vectors. Although the 20th century saw a very large rise in paralytic polio cases, it was also the century during which effective vaccines were developed. The infection was eliminated from the Western Hemisphere in the late 20th century **(figure 19.19)**. Sadly, it is proving extremely difficult to eradicate from the developing world (see Continuing the Case).

▶ **Culture and Diagnosis**

Poliovirus can usually be isolated by inoculating cell cultures with stool or throat washings in the early part of the disease. Viruses are sometimes then subjected to DNA fingerprinting or whole-genome sequencing to determine if they are wild strains or vaccine strains. The stage of the patient's infection can also be demonstrated by testing serum samples for the type and amount of antibody.

▶ **Prevention and Treatment**

Treatment of polio rests largely on alleviating pain and suffering. During the acute phase, muscle spasm, headache, and

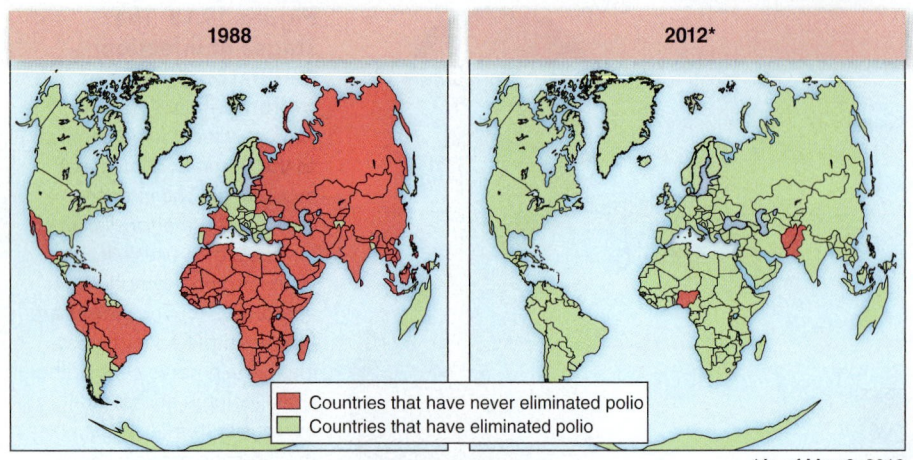

| 1988 | 2012* |

Countries that have never eliminated polio
Countries that have eliminated polio

*As of May 3, 2012

Figure 19.19 Progress in the elimination of polio.

associated discomfort can be alleviated by pain-relieving drugs. Respiratory failure may require artificial ventilation maintenance. Prompt physical therapy to diminish crippling deformities and to retrain muscles is recommended after the acute febrile phase subsides.

The mainstay of polio prevention is vaccination as early in life as possible, usually in four doses starting at about 2 months of age. Adult candidates for immunization are travelers and members of the armed forces. The two forms of vaccine currently in use are inactivated poliovirus vaccine (IPV), developed by Jonas Salk in 1954, and oral poliovirus vaccine (OPV), developed by Albert Sabin in the 1960s. Both are prepared from animal cell cultures and are trivalent (combinations of the three serotypes of the poliovirus). Both

vaccines are effective, but one may be favored over the other under certain circumstances.

For many years, the oral vaccine was used in the United States because it is easily administered by mouth, but it is not free of medical complications. It contains an attenuated virus that can multiply in vaccinated people and be spread to others. In very rare instances, the attenuated virus reverts to a neurovirulent strain that causes disease rather than protects against it. For this reason, IPV is the only vaccine used in the United States (see chapter 15 for current vaccine schedule). See **Disease Table 19.7** for additional information on this infectious disease.

Tetanus

Tetanus is a neuromuscular disease whose alternate name, lockjaw, refers to an early effect of the disease on the jaw muscle. The etiologic agent, *Clostridium tetani*, is a common resident of soil and the gastrointestinal tracts of animals. It is a gram-positive, endospore-forming rod. The endospores it produces often swell the vegetative cell **(figure 19.20)** but are only produced under anaerobic conditions.

▶ Signs and Symptoms

C. tetani releases a powerful neurotoxin, **tetanospasmin,** that binds to target sites on peripheral motor neurons, on

---

Case File 19   *Continuing the Case*

Massive efforts have been undertaken by the GPEI, UNICEF, the WHO, the CDC, and Rotary International to vaccinate children in countries where polio is still endemic or where there have been outbreaks. A trivalent (three-part) vaccine is available, but studies have shown that this vaccine is less effective than newer versions of monovalent oral vaccines consisting of only type I or type III poliovirus or a bivalent vaccine consisting of type I and type III. In 2003, with new cases of disease dropping rapidly, it appeared that the effort had paid off—it was estimated that disease eradication would occur by 2005. However, within the year, new outbreaks began to develop in Nigeria leading to nearly 800 new cases in this country alone and many as far away as Saudi Arabia, where travelers to Mecca furthered its spread. This was disheartening news to health care workers around the world.

What prompted this reemergence of polio in countries that were on the brink of eradication? Geography is one factor, as there is great difficulty in reaching children in remote areas. The countryside of many affected countries, such as Pakistan and

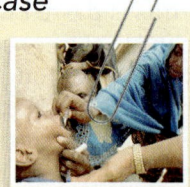

Afghanistan, is mountainous, rocky, and treacherous, making it difficult to transport the vaccine to children in need. Also, the vaccine itself was found to contribute to the new cases. Oral polio vaccine (OPV) is used for mass vaccination of individuals in developing countries, because it is effective and much cheaper than the alternate inactivated polio vaccine (IPV) used in other areas of the world. The trade-off comes in the potential for the OPV to cause actual polio disease. It is made up of live attenuated virus, which in rare cases reverts to a virulent form causing disease rather than protecting against it. This contributed to a small portion of new disease cases. This would not have happened if the population had been at the 100% level of vaccination, and it illustrates the importance of achieving vaccination in all countries around the world.

■ What difficulties are encountered today in the continued global effort to eradicate polio?

■ Besides vaccination, what other measures can be taken to interrupt the polio transmission cycle?

Source: 2012. CDC *MMWR,* vol. 61, no. 19, p. 353.

Disease Table 19.7 Poliomyelitis	
Causative Organism(s)	Poliovirus
Most Common Modes of Transmission	Fecal-oral, vehicle
Virulence Factors	Attachment mechanisms
Culture/Diagnosis	Viral culture, serology
Prevention	Live attenuated (developing world) or inactivated vaccine (developed world)
Treatment	None, palliative, supportive
Epidemiological Features	223 cases reported worldwide in 2012; still endemic in Nigeria, Pakistan, and Afghanistan

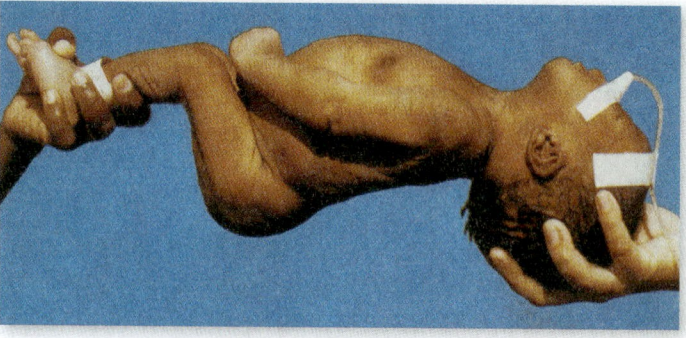

Figure 19.21 Neonatal tetanus. Baby with neonatal tetanus, showing spastic paralysis of the paravertebral muscles, which locks the back into a rigid, arched position. Also note the abnormal flexion of the arms and legs.

the spinal cord and brain, and in the sympathetic nervous system. The toxin acts by blocking the inhibition of muscle contraction. Without inhibition of contraction, the muscles contract uncontrollably, resulting in spastic paralysis. The first symptoms are clenching of the jaw, followed in succession by extreme arching of the back, flexion of the arms, and extension of the legs **(figure 19.21)**. Lockjaw confers the bizarre appearance of *risus sardonicus* (sardonic grin), which looks eerily as though the person is smiling **(figure 19.22c)**. Death most often occurs due to paralysis of the respiratory muscles and respiratory arrest.

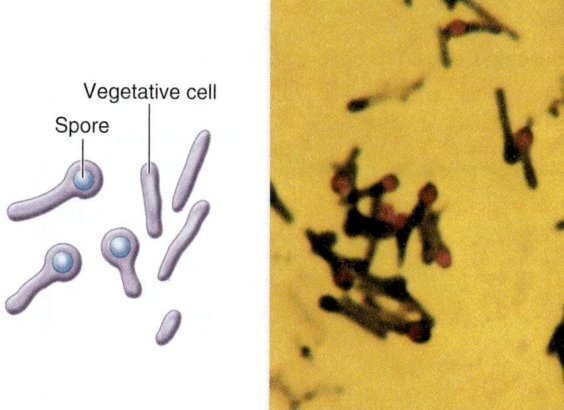

Figure 19.20 *Clostridium tetani.* Its typical tennis racket morphology is created by terminal endospores that swell the end of the cell (170×).

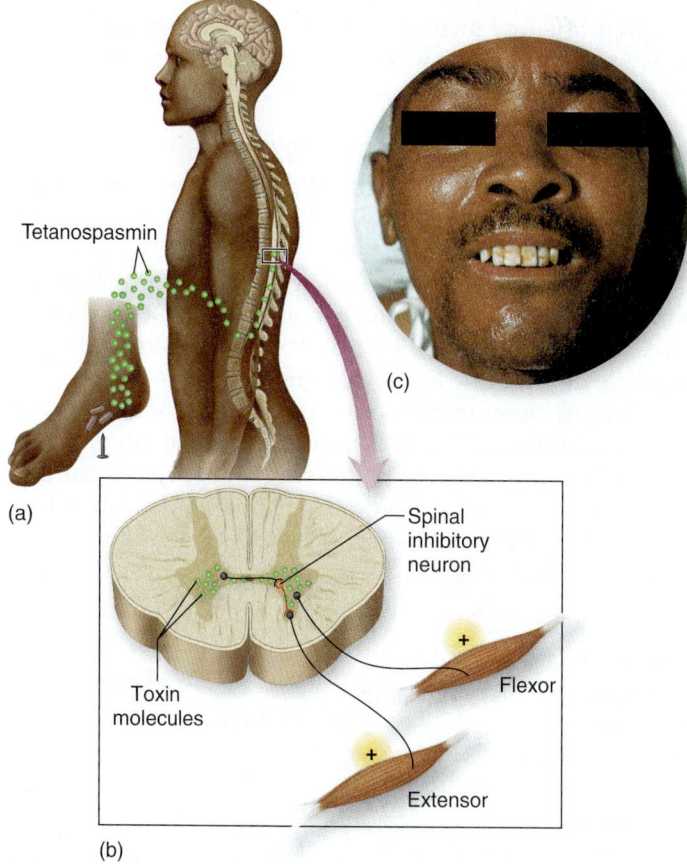

Figure 19.22 The events in tetanus. **(a)** After traumatic injury, bacteria infecting the local tissues secrete tetanospasmin, which is absorbed by the peripheral axons and is carried to the target neurons in the spinal column. **(b)** In the spinal cord, the toxin attaches to the junctions of regulatory neurons that inhibit inappropriate contraction. Released from inhibition, the muscles, even opposing members of a muscle group, receive constant stimuli and contract uncontrollably. **(c)** Muscles contract spasmodically, without regard to regulatory mechanisms or conscious control. Note the clenched jaw typical of *risus sardonicus*.

▶ Pathogenesis and Virulence Factors

The mere presence of endospores in a wound is not sufficient to initiate infection because the bacterium is unable to invade damaged tissues readily. It is also a strict anaerobe, and the endospores cannot become established unless tissues at the site of the wound are necrotic and poorly supplied with blood, conditions that favor germination.

As the vegetative cells grow, various metabolic products are released into the infection site, including the tetanospasmin toxin. The toxin spreads to nearby motor nerve endings in the injured tissue, binds to them, and travels via axons to the ventral horns of the spinal cord **(figure 19.22b).** The toxin blocks the release of neurotransmitter, and only a small amount is required to initiate the symptoms. The incubation period varies from 4 to 10 days, and shorter incubation periods signify a more serious condition.

The muscle contractions are intermittent and extremely painful, and they may be forceful enough to break bones, especially the vertebrae. The fatality rate, ranging from 10% to 70%, is highest in cases involving delayed medical attention, a short incubation time, or head wounds. Full recovery requires a few weeks, and no permanent damage to the muscles usually remains.

▶ Transmission and Epidemiology

Endospores usually enter the body through accidental puncture wounds, burns, umbilical stumps, frostbite, and crushed body parts. The incidence of tetanus is low in North America. Most cases occur among geriatric patients, people who are intravenous drug abusers, and people who are unvaccinated. In developing countries, however, new mothers and neonates are at high risk for developing disease. A majority of infections in these countries are a direct result of unhygienic practices during childbirth, including the use of dung, ashes, or mud to arrest bleeding or for religious purposes during this process. Through the promotion of more hygienic delivery practices and vaccination, the WHO has reduced the incidence of maternal and neonatal tetanus by over 90% and is close to eliminating this disease in many countries today **(figure 19.23).**

▶ Prevention and Treatment

Tetanus treatment is aimed at deterring the degree of toxemia and infection and maintaining patient homeostasis. A patient with a clinical appearance suggestive of tetanus should immediately receive antitoxin therapy with human tetanus immune globulin (TIG). Penicillin G is also administered. Although the antitoxin inactivates circulating toxin, it will not counteract the effect of toxin already bound to neurons. Other methods include thoroughly cleansing and removing the afflicted tissue, controlling infection with penicillin or tetracycline, and administering muscle relaxants. The patient may require the assistance of a respirator, and a **tracheotomy**[1] is sometimes performed to prevent respiratory complications such as aspiration pneumonia or lung collapse.

Tetanus is one of the world's most preventable diseases, chiefly because four effective vaccines containing tetanus toxoid exist today. These are combination vaccines that provide protection against tetanus and additional infectious diseases (diphtheria, pertussis). DTaP and DT preparations are used to immunize children under the age of 7 years, while Tdap and Td are used for vaccinating older adolescents and adults (see chapter 15).

Immunized children are considered to be protected for 10 years. At that point, and every 10 years thereafter, they should receive a dose of Td, tetanus-diphtheria vaccine. Additional protection against neonatal tetanus may be achieved by vaccinating pregnant women, whose antibodies will be passed to the fetus. Toxoid should also be given to injured persons who have never been immunized, have not completed the series, or whose last booster was received more than 10 years previously. The vaccine can be given simultaneously with passive TIG immunization to achieve immediate as well as long-term protection. See **Disease Table 19.8** for a summary of this infectious disease.

---

1. The surgical formation of an air passage by perforation of the trachea.

Figure 19.23 The effect of vaccination on global tetanus incidence.

Source: Reproduced, with the permission of the publisher, from http://www.who.int/immunization_monitoring/diseases/tetanus/en/index.html, by permission of World Health Organization.

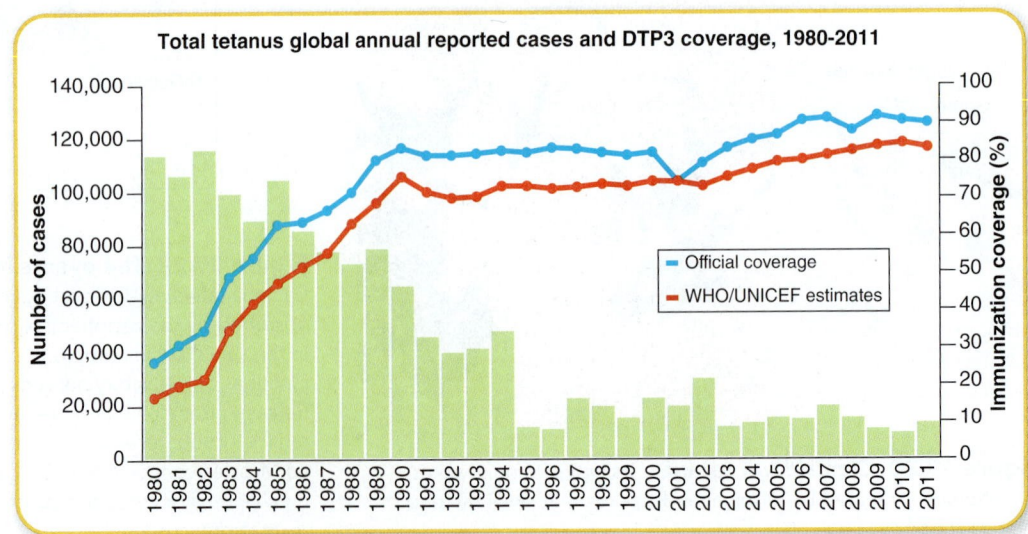

Total tetanus global annual reported cases and DTP3 coverage, 1980-2011

Legend: Official coverage; WHO/UNICEF estimates

Disease Table 19.8	Tetanus
Causative Organism(s)	*Clostridium tetani*
Most Common Modes of Transmission	Parenteral, direct contact
Virulence Factors	Tetanospasm exotoxin
Culture/Diagnosis	Symptomatic
Prevention	Tetanus toxoid immunization
Treatment	Combination of passive antitoxin and tetanus toxoid active immunization, penicillin G, muscle relaxants
Epidemiological Features	United States: Approximately 30 cases/year; worldwide estimated 1 million cases annually, 50% in newborns

Botulism

Botulism is an **intoxication** (that is, caused by an exotoxin) associated with eating poorly preserved foods, although it can also occur as a true infection. Until recent times, it was relatively common and frequently fatal, but modern techniques of food preservation and medical treatment have reduced both its incidence and its fatality rate. However, botulism is a common cause of death in livestock that have grazed on contaminated food and in aquatic birds that have eaten decayed vegetation. In the United States, there are between 10 and 30 outbreaks of human botulism a year.

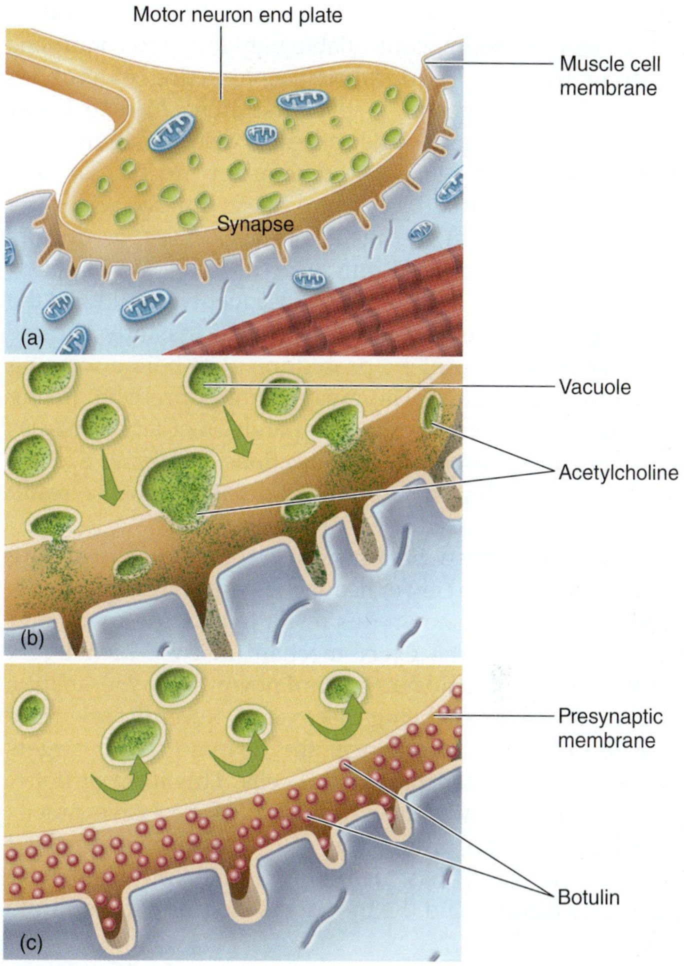

▶ ## Signs and Symptoms

There are three major forms of botulism, distinguished by their means of transmission and the population they affect. These are *food-borne botulism* (in children and adults), *infant botulism,* and *wound botulism.* Food-borne botulism in children and adults is an intoxication resulting from the ingestion of preformed toxin; the other two types of botulism are infections that are followed by the entrance of an exotoxin called **botulinum toxin** into the bloodstream (that is, toxemia). The symptoms are largely the same in all three forms, however. From the circulatory system, the toxin travels to its principal site of action, the neuromuscular junctions of skeletal muscles **(figure 19.24).** The effect of botulinum is to

Figure 19.24 The physiological effects of botulism toxin (botulinum). **(a)** The relationship between the motor neuron and the muscle at the neuromuscular junction. **(b)** In the normal state, acetylcholine released at the synapse crosses to the muscle and creates an impulse that stimulates muscle contraction. **(c)** In botulism, the toxin enters the motor end plate and attaches to the presynaptic membrane, where it blocks release of the chemical. This prevents impulse transmission and keeps the muscle from contracting. This causes flaccid paralysis.

prevent the release of the acetylcholine, the neurotransmitter that initiates the signal for muscle contraction. The usual time before onset of symptoms is 12 to 72 hours, depending on the size of the dose. Neuromuscular symptoms first affect the muscles of the head and include double vision, difficulty in swallowing, and dizziness, but there is no sensory or mental lapse. Although nausea and vomiting can occur at an early stage, they are not common. Later symptoms are descending muscular paralysis and respiratory compromise. In the past, death resulted from respiratory arrest, but mechanical respirators have reduced the fatality rate to about 10%.

Surprisingly, doctors and scientists have been able to utilize the deadly effects of the botulinum toxin for effective medical treatments. In 1989, Botox was first approved to treat cross-eyes and uncontrollable blinking, two conditions resulting from the inappropriate contracting of muscles around the eye. Success in this first arena lead to Botox treatment for a variety of neurological disorders that cause painful contraction of neck and shoulder muscles, as well muscle spasms caused by cerebral palsy. A much wider use of Botox occurred in so-called "off-label" uses, as doctors found that injecting facial muscles with the toxin inhibited contraction of these muscles and consequent wrinkling of the overlying skin. The "lunch-hour facelift" instantly became the most popular cosmetic procedure, even before winning FDA approval in 2002. The most common problem arising from Botox treatment is excessive paralysis of facial muscles resulting from poorly targeted injections. Drooping eyelids, facial paralysis, slurred speech, and drooling can also result from improperly placed injections.

In a surprise twist, patients undergoing Botox treatment for wrinkles reported fewer headaches, especially migraines. Clinical trials have shown this result to be widespread and reproducible, but the exact mechanism by which Botox works to prevent headaches is unknown. New therapeutic uses for the toxin are being discovered. Recently, scientists have found a way to engineer the toxin so that it targets epithelial cells rather than neural cells. This form of Botox is able to shut down the production of excessive mucus in asthma and the production of cytokines that can lead to cancer. Research has also shown that Botox can treat overactive bladder and the tremors that occur with multiple sclerosis.

► Causative Agent

Clostridium botulinum, like *Clostridium tetani,* is an endospore-forming anaerobe that does its damage through the release of an exotoxin. *C. botulinum* commonly inhabits soil and water and occasionally the intestinal tract of animals. It is distributed worldwide but occurs most often in the Northern Hemisphere. The species has eight distinctly different types (designated A, B, C_a, C_b, D, E, F, and G) that vary in distribution among animals, regions of the world, and types of exotoxin. Human disease is usually associated with types A, B, E, and F; and animal disease is associated with types A, B, C, D, and E.

Both *C. tetani* and *C. botulinum* produce neurotoxins; but tetanospasmin, the toxin made by *C. tetani,* results in spastic paralysis (uncontrolled muscle contraction). In contrast, botulinum, the *C. botulinum* neurotoxin, results in flaccid paralysis, a loss of ability to contract the muscles.

► Pathogenesis and Virulence Factors

As just described, the symptoms are caused entirely by the exotoxin botulinum. Its action is very potent in an affected individual, as the td50 (or median toxic dose) for botulinum toxin is only 0.03 µg per kilogram of body weight!

► Transmission and Epidemiology of Food-Borne Botulism in Children and Adults

In the United States, the disease is often associated with low-acid vegetables (green beans, corn), fruits, and occasionally meats, fish, and dairy products. Many botulism outbreaks occur in home-processed foods, including canned vegetables, smoked meats, and cheese spreads.

Several factors in food processing can lead to botulism. Endospores are present on the vegetables or meat at the time of gathering and are difficult to remove completely. When contaminated food is put in jars and steamed in a pressure cooker that does not reach reliable pressure and temperature, some endospores survive. At the same time, the pressure is sufficient to evacuate the air and create anaerobic conditions. Storage of the jars at room temperature favors endospore germination and vegetative growth, and one of the products of the cell's metabolism is botulinum, the most potent microbial toxin known.

Bacterial growth may not be evident in the appearance of the jar or can or in the food's taste or texture, and only minute amounts of toxin may be present. Botulism is never transmitted from person to person.

► Transmission and Epidemiology of Infant Botulism

Infant botulism was first described in the late 1970s in children between the ages of 2 weeks and 6 months who had ingested endospores. It is currently the most common type of botulism in the United States, with approximately 75 cases reported annually. The exact food source is not always known, although raw honey has been implicated in some cases. Apparently, the immature state of the neonatal intestine and microbial biota allows the endospores to gain a foothold, germinate, and give off neurotoxin. As in adults, babies exhibit flaccid paralysis, usually manifested as a weak sucking response, generalized loss of tone (the "floppy baby syndrome"), and respiratory complications. Although adults can also ingest botulinum endospores in contaminated vegetables and other foods, the adult intestinal tract normally inhibits this sort of infection.

► Transmission and Epidemiology of Wound Botulism

Perhaps three or four cases of wound botulism occur each year in the United States. In this form of the disease, endospores enter a wound or puncture, much as in tetanus, but the symptoms are similar to those of food-borne botulism. Occasionally, this form of botulism is reported in people who are intravenous drug users as a result of needle puncture.

► Culture and Diagnosis

Diagnostic standards are slightly different for the three different presentations of botulism. In food-borne botulism, some laboratories attempt to identify the toxin in the offending food. Alternatively, if multiple patients present with the same symptoms after ingesting the same food, a presumptive diagnosis can be made. The cultivation of *C. botulinum* in feces is considered confirmation of the diagnosis since the carrier rate is very low.

In infant botulism, finding the toxin or the organism in the feces confirms the diagnosis. In wound botulism, the toxin should be demonstrated in the serum, or the organism should be grown from the wound. Because minute amounts of the toxin are highly dangerous, laboratory testing should only be performed by experienced personnel. A suspected case of botulism should trigger a phone call to the state health department or the CDC before proceeding with diagnosis or treatment.

▶ Prevention and Treatment

The CDC maintains a supply of type A, B, and E trivalent horse antitoxin, which, when administered early, can prevent the worst outcomes of the disease. Wound botulism is also treated with penicillin G. Patients are also managed with respiratory and cardiac support systems. In all cases, hospitalization is required and recovery takes weeks. There is an overall 5% mortality rate.

Disease Table 19.9	Botulism
Causative Organism(s)	*Clostridium botulinum*
Most Common Modes of Transmission	Vehicle (food-borne toxin, airborne organism); direct contact (wound); parenteral (injection)
Virulence Factors	Botulinum exotoxin
Culture/Diagnosis	Culture of organism; demonstration of toxin
Prevention	Food hygiene; toxoid immunization available for laboratory professionals
Treatment	Antitoxin, penicillin G for wound botulism, supportive care
Epidemiological Features	United States: 75% of botulism is infant botulism; approximately 100–150 cases annually

African Sleeping Sickness

This condition is caused by *Trypanosoma brucei,* a member of the protozoan group known as hemoflagellates because of their propensity to live in the blood and tissues of the human host. The disease, also called trypanosomiasis, has greatly affected the living conditions of Africans since ancient times. Millions of individuals residing in 36 countries are at risk for disease; although the total number of new cases has dropped significantly over the past 10 years, many more cases are thought to go unreported in these areas.

▶ Signs and Symptoms

Trypanosomiasis affects the lymphatics and areas surrounding blood vessels. Usually a long asymptomatic period precedes onset of symptoms. Symptoms include intermittent fever, enlarged spleen, swollen lymph nodes, and joint pain. There are two variants of the disease, caused by two different subspecies of the protozoan. In both forms, the central nervous system is affected, the initial signs being personality and behavioral changes that progress to lassitude and sleep disturbances. The disease is commonly called *sleeping sickness;* but, in fact, uncontrollable sleepiness occurs primarily in the day and is followed by sleeplessness at night. Signs of advancing neurological deterioration are muscular tremors, shuffling gait, slurred speech, seizures, and local paralysis. Death results from coma, secondary infections, or heart damage.

▶ Causative Agent

Trypanosoma brucei is a flagellated protozoan, an obligate parasite that is spread by a blood-sucking insect called the tsetse fly, which serves as its intermediate host. It shares a complicated life cycle with other hemoflagellates. *T. brucei gambiense* is found in west and central Africa, is associated with chronic disease, and accounts for over 95% of total reported cases; *T. brucei rhodesiense* is found in eastern and southern Africa. It is much less common (<5% of all cases) and causes acute infection that leads to rapid disease development **(figure 19.25a).** Note that in chapter 5, we first described the trypanosome life cycle using the example of *T. cruzi,* the agent that causes Chagas disease.

▶ Transmission and Epidemiology

The cycle begins when a tsetse fly becomes infected after feeding on an infected reservoir host, such as a wild animal (antelope, pig, lion, hyena), domestic animal (cow, goat), or human **(figure 19.25b).** In the fly's gut, the trypanosome multiplies, migrates to the salivary glands, and develops into the infectious stage. When the fly bites a new host, it releases the large, fully formed stage of the parasite into the wound. At this site, the trypanosome multiplies and produces a sore called the *primary chancre.* From there, the pathogen moves into the lymphatics and the blood (as shown in the figure). The trypanosome can also cross the placenta and damage a developing fetus.

(a) The distribution of African trypanosomiasis.

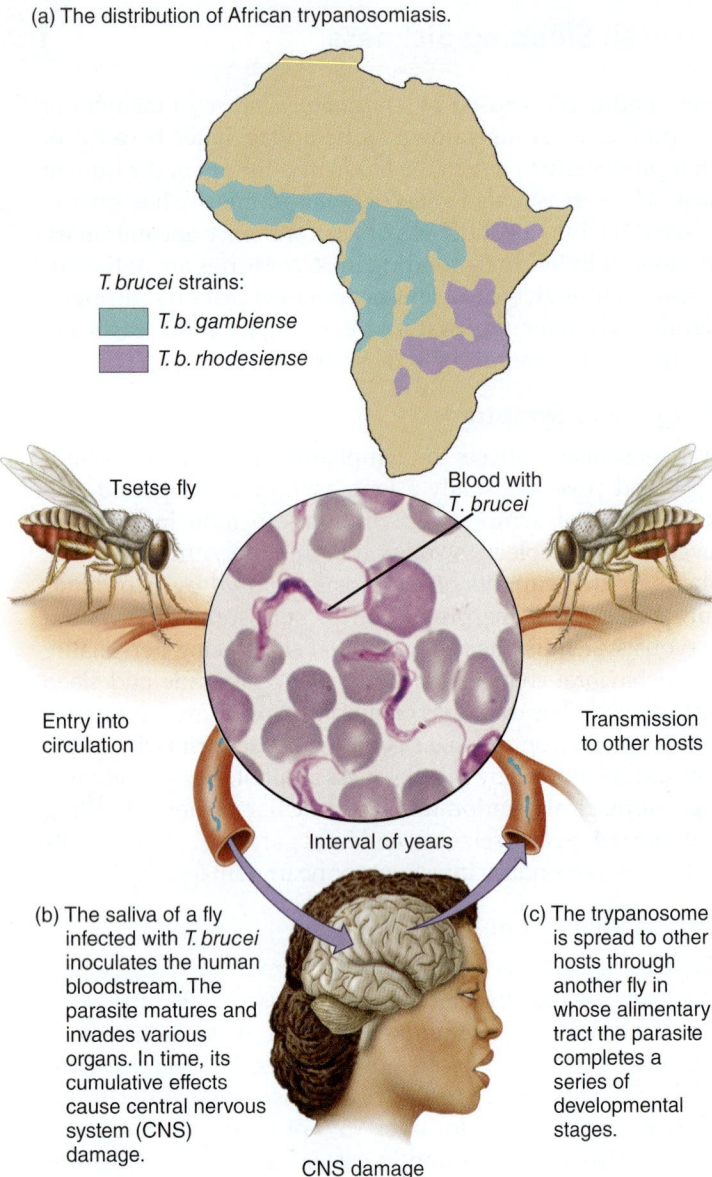

T. brucei strains:

- T. b. gambiense
- T. b. rhodesiense

Tsetse fly

Blood with T. brucei

Entry into circulation

Transmission to other hosts

Interval of years

(b) The saliva of a fly infected with T. brucei inoculates the human bloodstream. The parasite matures and invades various organs. In time, its cumulative effects cause central nervous system (CNS) damage.

(c) The trypanosome is spread to other hosts through another fly in whose alimentary tract the parasite completes a series of developmental stages.

CNS damage

Figure 19.25 The generalized cycle between humans and the tsetse fly vector.

African sleeping sickness occurs only in sub-Saharan Africa. Tsetse flies exist elsewhere, and it is not known why they do not support *Trypanosoma* in other regions. Cases seen in the United States are only those that were acquired by travelers or emigrants from Africa.

► Pathogenesis and Virulence Factors

The protozoan manages to flourish in the blood even though it stimulates a strong immune response. The immune response is counteracted by an unusual adaptation of the trypanosome. As soon as the host begins manufacturing IgM antibodies to the trypanosome, surviving organisms change the structure of their surface glycoprotein antigens. This change in specificity (sometimes referred to as an antigenic shift) renders the existing IgM ineffective, so that the parasite

eludes control and multiplies in the blood. The host responds by producing IgM of a new specificity, but the protozoan changes its antigens again. The host eventually becomes exhausted and overwhelmed by repeated efforts to catch up with this trypanosome masquerade.

► Culture and Diagnosis

Trypanosomes are readily demonstrated in blood smears, as well as in spinal fluid or lymph nodes. Serological tests are available for diagnosis of *T. brucei gambiense* infection in endemic areas of Africa.

► Prevention and Treatment

Control of trypanosomiasis in western Africa, where humans are the main reservoir hosts, involves eliminating tsetse flies by applying insecticides, trapping flies, or destroying the shelter and breeding sites. In eastern regions, where cattle herds and large wildlife populations are reservoir hosts, control is less practical because large mammals are the hosts and flies are less concentrated in specific sites. The antigenic shifting exhibited by the trypanosome makes the development of a vaccine very difficult.

Chemotherapy is most successful if administered prior to nervous system involvement. Two drugs are available for the early stages of the disease. Suramin works against *T. brucei rhodesiense*, and pentamidine is used for *T. brucei gambiense*. Brain infection must be treated with drugs that can cross the

Disease Table 19.10	African Sleeping Sickness
Causative Organism(s)	*Trypanosoma brucei* subspecies *gambiense* or *rhodesiense*
Most Common Modes of Transmission	Vector, vertical
Virulence Factors	Immune evasion by antigen shifting
Culture/Diagnosis	Microscopic examination of blood, CSF
Prevention	Vector control
Treatment	Suramin or pentamidine (early), eflornithine or melarsoprol (late)
Epidemiological Features	*T. brucei gambiense*: 7,000–10,000 cases reported annually; actual occurrence estimated at 600,000; *T. brucei rhodesiense*: estimated 30,000 cases occur annually

blood-brain barrier. One of these is a highly toxic arsenic-based drug called melarsoprol; another is called eflornithine.

19.3 Learning Outcomes—Assess Your Progress

4. List the possible causative agents for meningitis and neonatal/infant meningitis.

5. Identify which of the agents causing meningitis is the most common and which is the most deadly.

6. Discuss important features of meningoencephalitis, encephalitis, and subacute encephalitis.

7. Identify which encephalitis-causing viruses you should be aware of in your geographic area.

8. List the possible causative agents for each of the following conditions: rabies, poliomyelitis, tetanus, botulism, and African sleeping sickness.

9. Identify the conditions for which vaccination is available.

10. Explain the difference between the oral polio vaccine and the inactivated polio vaccine, and identify under which circumstances each is appropriate.

Case File 19   *Wrap-Up*

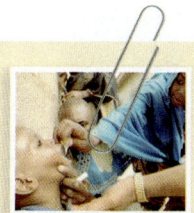

Often there are cultural and even military barriers to vaccinating children against polio. Workers attempting to administer the polio vaccine have been murdered by militants in Pakistan, who claim that the vaccinators are actually working for the U.S. government by locating targets for drone strikes and using the injections to sterilize Muslims. These fears were abetted by the U.S. Central Intelligence Agency's use of polio vaccination as a "cover" to find Osama Bin Laden in Pakistan in 2011. Uneducated people in rural areas of Afghanistan and Pakistan refuse vaccines for their children based on the admonishments of their cultural leaders, not only putting their children at risk, but hampering the effort to eradicate polio in their country and around the world.

The Bill & Melinda Gates Foundation is actively involved in philanthropic efforts to eradicate polio, donating nearly a billion dollars to the polio vaccine effort. He has met with doctors, aid workers, and even the Sultan of Sokoto, the Muslim leader of Nigeria, to encourage understanding and smooth the way for vaccine efforts. Polio is a problem in countries like Nigeria, Afghanistan, and Pakistan, but so are HIV, tuberculosis, hepatitis, malaria, and other infectious diseases. In addition to better communication and understanding about the need for polio vaccination and eradication, many countries need better health infrastructure, water sanitation, and nutrition for their citizens. The WHO, UNICEF, Rotary International, and the Gates Foundation are working in concert on a "horizontal" approach, addressing the need for better hygiene and sanitation in addition to vaccination in countries where polio is still endemic. Although Rotary International proclaims that we are "this close" to completely silencing a disease that is currently 99% eradicated, these efforts can easily be derailed by lack of funding, political unrest, and nature itself.

Source: 2012. *Morbidity and Mortality Weekly Report* vol. 61, no. 19, p. 353.

▶ Summing Up

Taxonomic Organization Microorganisms Causing Disease in the Nervous System

Microorganism	Disease	Location of Disease Table
Gram-positive endospore-forming bacteria		
Clostridium botulinum	Botulism	Botulism, p. 581
Clostridium tetani	Tetanus	Tetanus, p. 579
Gram-positive bacteria		
Streptococcus agalactiae	Neonatal meningitis	Neonatal meningitis, p. 562
Streptococcus pneumoniae	Meningitis	Meningitis, p. 560
Listeria monocytogenes	Meningitis, neonatal meningitis	Meningitis, p. 560
		Neonatal meningitis, p. 562
Gram-negative bacteria		
Cronobacter sakazakii	Neonatal and infant meningitis	Neonatal and infant
Escherichia coli	Neonatal meningitis	meningitis, p. 562
Haemophilus influenzae	Meningitis	Meningitis, p. 560
Neisseria meningitidis	Meningococcal meningitis	Meningitis, p. 560
DNA viruses		
Herpes simplex virus 1 and 2	Encephalitis	Encephalitis, p. 567
JC virus	Progressive multifocal leukoencephalopathy	Encephalitis, p. 567
RNA viruses		
Arboviruses	Encephalitis	Encephalitis, p. 567
Western equine encephalitis virus, Eastern equine encephalitis virus, California encephalitis virus (California and LaCrosse strains), St. Louis encephalitis virus, West Nile virus		
Measles virus	Subacute sclerosing panencephalitis	Subacute encephalitis, p. 569
Poliovirus	Poliomyelitis	Poliomyelitis, p. 577
Rabies virus	Rabies	Rabies, p. 573
Fungi		
Cryptococcus neoformans	Meningitis	Meningitis, p. 560
Coccidioides immitis	Meningitis	Meningitis, p. 560
Prions		
Creutzfeldt-Jakob prion	Creutzfeldt-Jakob disease	Subacute encephalitis, p. 569
Protozoa		
Acanthamoeba	Meningoencephalitis	Meningoencephalitis, p. 564
Naegleria fowleri	Meningoencephalitis	Meningoencephalitis, p. 564
Toxoplasma gondii	Subacute encephalitis	Subacute encephalitis, p. 569
Trypanosoma brucei subspecies *gambiense* and *rhodesiense*	African sleeping sickness	African sleeping sickness, p. 582

INFECTIOUS DISEASES AFFECTING
The Nervous System

Encephalitis
Arboviruses
Herpes simplex virus 1 or 2
JC virus

Subacute Encephalitis
Toxoplasma gondii
Measles virus
Prions

Rabies
Rabies virus

Tetanus
Clostridium tetani

African Sleeping Sickness
Trypanosoma brucei

Creutzfeldt-Jakob Disease
Prion

Meningoencephalitis
Naegleria fowleri
Acanthamoeba

Meningitis
Neisseria meningitidis
Streptococcus pneumoniae
Haemophilus influenzae
Listeria monocytogenes
Cryptococcus neoformans
Coccidioides immitis
Various viruses

Neonatal and Infant Meningitis
Streptococcus agalactiae
Escherichia coli
Listeria monocytogenes
Cronobacter sakazakii

Polio
Poliovirus

Botulism
Clostridium botulinum

■ Bacteria
■ Viruses
■ Protozoa
■ Fungi
■ Prions

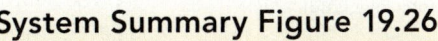

System Summary Figure 19.26

Chapter Summary

19.1 The Nervous System and Its Defenses (ASM Guidelines* 5.4)

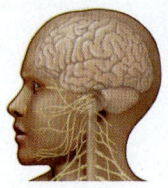

- The nervous system has two parts: the central nervous system (the brain and spinal cord), and the peripheral nervous system (nerves and ganglia).
- The soft tissue of the brain and spinal cord is encased within the tough casing of three membranes called the *meninges.* The subarachnoid space is filled with a clear, serumlike fluid called cerebrospinal fluid (CSF).
- The nervous system is protected by the *blood-brain barrier,* which limits the passage of substances from the bloodstream to the brain and spinal cord.

19.2 Normal Biota of the Nervous System (ASM Guidelines 5.4, 6.4)

- At the present time, we believe there is no normal biota in either the central nervous system (CNS) or the peripheral nervous system (PNS). However, normal biota of the gut may indirectly influence functioning of the nervous system.

19.3 Nervous System Diseases Caused by Microorganisms (ASM Guidelines 5.3, 5.4, 6.4, 8.3)

- **Meningitis:** Inflammation of the meninges. The most serious forms of this disease are caused by bacteria, often facilitated by coinfection or previous infection with respiratory viruses.
  - *Neisseria meningitidis:* Gram-negative diplococcus, commonly known as the meningococcus; causes most serious form of acute meningitis.
  - *Streptococcus pneumoniae:* Gram-positive coccus, commonly known as the pneumococcus; most frequent cause of community-acquired pneumococcal meningitis.
  - *Haemophilus influenzae:* Cases have declined sharply due to vaccination.
  - *Listeria monocytogenes:* Most cases are associated with ingesting contaminated dairy products, poultry, and meat.
  - *Cryptococcus neoformans:* Fungus; causes more chronic form of meningitis with more gradual onset of symptoms. Closely related to emerging infectious agent, *C. gattii.*
  - *Coccidioides immitis:* True systemic fungal infection; begins in lungs but can disseminate quickly throughout the body; highest incidence occurs in southwestern United States, Mexico, and parts of Central and South America.
  - Viruses: Viral meningitis is very common, particularly in children; 90% of cases are caused by enteroviruses.
- **Neonatal and Infant Meningitis:** Usually transmitted vertically. Primary causes in this country are *Streptococcus agalactiae, Escherichia coli,* and *Listeria monocytogenes. Cronobacter* is a rare but deadly cause.

- **Meningoencephalitis:** Caused mainly by two amoebas, *Naegleria fowleri* and *Acanthamoeba.*
- **Acute Encephalitis:** Usually caused by viral infection. Arboviruses carried by arthropods are often responsible.
  - Western Equine Encephalitis (WEE): Occurs sporadically in western United States and Canada.
  - Eastern Equine Encephalitis (EEE): Endemic to eastern coast of North America and Canada.
  - California Encephalitis: Caused by two different viral strains, the California strain and the LaCrosse strain.
  - St. Louis Encephalitis (SLE): May be most common of American viral encephalitides. Appears throughout North, South America; epidemics occur most often in Midwest and South.
  - West Nile Encephalitis: West Nile virus is close relative of SLE virus. Emerged in United States in 1999.
  - Herpes simplex virus: Herpes simplex types 1 and 2 cause encephalitis in newborns born to HSV-positive mothers, older children and young adults (ages 5 to 30), and older adults (over 50 years old).
  - JC Virus: Can cause progressive multifocal leukoencephalopathy (PML), particularly in immunocompromised individuals. Fatal infection.
- **Subacute Encephalitis:** Symptoms take longer to manifest.

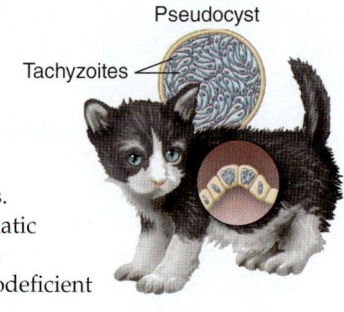

Pseudocyst

Tachyzoites

  - *Toxoplasma gondii:* Protozoan; causes toxoplasmosis, most common form of subacute encephalitis. Relatively asymptomatic in the healthy; can be severe in the immunodeficient and fetuses.
  - Measles Virus: Can produce subacute sclerosing panencephalitis (SSPE) years after initial measles infection.
  - Prions: Proteinaceous infectious particles containing no genetic material. Cause transmissible spongiform encephalopathies (TSEs), neurodegenerative diseases with long incubation periods but rapid progressions once they begin. Human TSEs include **Creutzfeldt-Jakob disease (CJD)** and kuru.

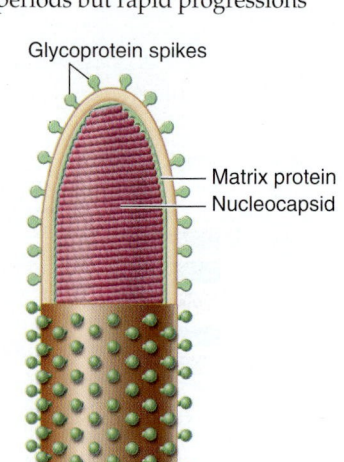

Glycoprotein spikes

Matrix protein
Nucleocapsid

- **Rabies:** Slow, progressive zoonotic disease characterized by fatal encephalitis. Rabies virus is in the family *Rhabdoviridae.*
- **Poliomyelitis:** Acute enterovirus infection of spinal cord; can cause neuromuscular paralysis. Two effective

Source: ASM Curriculum Guidelines (American Society for Microbiology, 2012). Complete guidelines in appendix B of this book.

vaccines exist: Inactivated Salk poliovirus vaccine (IPV) is the only one used now in the United States; attenuated oral Sabin poliovirus vaccine (OPV) is still being used in the developing world.

- **Tetanus:** Neuromuscular disease, also called lockjaw; caused by *Clostridium tetani* neurotoxin, tetanospasmin, which binds target sites on spinal neurons, blocks inhibition of muscle contraction.
- **Botulism:** Caused by exotoxin of *C. botulinum*; associated with eating poorly preserved foods; can also occur as true infection. Three major forms of botulism:

food-borne botulism (in children and adults), infant botulism, and wound botulism.

Tsetse fly

- **African Sleeping Sickness:** Caused primarily by two subspecies of the protozoan, *Trypanosoma brucei*. Affects central nervous system, leading to neurological deterioration: muscular tremors, shuffling gait, slurred speech, seizures, and local paralysis.

Multiple-Choice and True-False Questions | Bloom's Levels 1 and 2: Remember and Understand

Multiple-Choice Questions. Select the correct answer from the options provided.

1. Which of the following organisms does *not* cause meningitis?
   a. *Haemophilus influenzae*
   b. *Streptococcus pneumoniae*
   c. *Neisseria meningitidis*
   d. *Clostridium tetani*

2. The first-choice antibiotic for bacterial meningitis is
   a. ceftriaxone.   c. ampicillin.
   b. penicillin.   d. vancomycin.

3. Meningococcal meningitis is caused by
   a. *Haemophilus influenzae*.
   b. *Streptococcus pneumoniae*.
   c. *Neisseria meningitidis*.
   d. *Listeria monocytogenes*.

4. Which of the following neurological diseases is not caused by a prion?
   a. Creutzfeldt-Jakob disease
   b. scrapie
   c. mad cow disease
   d. St. Louis encephalitis

5. *Cryptococcus neoformans* is primarily transmitted by
   a. direct contact.   c. fomites.
   b. bird droppings.   d. sexual activity.

6. Which of the following is *not* caused by an arbovirus?
   a. St. Louis encephalitis
   b. Eastern equine encephalitis
   c. West Nile encephalitis
   d. PAM

7. CJD is caused by a(n)
   a. arbovirus.   c. protozoan.
   b. prion.   d. bacterium.

8. What food should you avoid feeding a child under 1 year old because of potential botulism?
   a. honey   c. apple juice
   b. milk   d. applesauce

9. *Naegleria fowleri* meningoencephalitis is commonly acquired via
   a. bird droppings.
   b. swimming in ponds and streams.
   c. mosquito bites.
   d. chickens.

10. Which organism is responsible for progressive multifocal leukoencephalopathy?
    a. JC virus   c. *E. coli*
    b. herpesvirus   d. *Haemophilus influenzae*

True-False Questions. If the statement is true, leave as is. If it is false, correct it by rewriting the sentence.

11. *Toxoplasma gondii* is a bacterium.

12. Penicillin G is the first line of treatment for coccidioidomycosis.

13. A diagnosis of bacterial meningitis can be made by analyzing cerebral spinal fluid (CSF).

14. In the United States, dogs are a common reservoir for rabies.

15. The protein PrP is beneficial before it is transformed into an abnormal protein.

Critical Thinking Questions | Bloom's Levels 3, 4, and 5: Apply, Analyze, and Evaluate

Critical thinking is the ability to reason and solve problems using facts and concepts. These questions can be approached from a number of angles and, in most cases, they do not have a single correct answer.

1. a. Explain why the nervous system is described as "immunologically privileged," and discuss whether this provides a beneficial or disadvantageous effect in this system.
   b. Discuss the defenses a pathogen encounters as it attempts to gain entry into the nervous system.

2. a. Describe the defining symptoms of meningitis.
   b. Pathogens causing meningitis enter the body through which portal(s) of entry?

3. a. Summarize the classes of pathogens that can cause meningitis, noting which are most pathogenic to the human host.
   b. What tests are used to identify these causative agents?
   c. Conduct additional research and summarize the causative agent and mode of transmission behind the multistate meningitis outbreak linked to steroid use that occurred in 2012. Thinking back to chapter 11, how did improper physical and chemical control methods play a major role in this outbreak?

4. In 2011, two cases of meningoencephalitis occurred in adults living in the southern United States. Both patients were found to have used untreated tap water in a neti pot to flush their sinuses.
   a. Based upon the information provided, which microbe was most likely the causative agent of disease in these cases?
   b. Explain why the disease was rapidly fatal in both patients.

5. a. Compare and contrast common characteristics of meningitis and encephalitis.
   b. Discuss how biological vectors and travel greatly impact the range of arboviral encephalitides today, and explain whether vaccination is a viable option to impede the spread of these diseases.

6. a. A patient undergoing a medical procedure was unaware that he had a prion disease. Discuss the risks of using nondisposable medical equipment in the treatment of this patient.
   b. Chronic wasting disease (CWD) is a transmissible spongiform encephalopathy affecting white-tailed deer and elk. In 2005, venison from a deer testing positive for CWD was served during an annual wild-game dinner in the town of Verona in upstate New York. Based upon your knowledge of vCJD and related spongiform encephalopathies and any additional research you conduct, discuss whether any of the attendees are at risk for developing disease in the future.

7. An HIV-positive individual is prone to developing a variety of nervous system infections. Provide two examples, noting the causative agent involved and summarizing the pathogenesis of disease in the two examples.

8. As you learned in chapter 8, many types of fruits and vegetables can be fermented into alcoholic beverages. Such was the case recently when prisoners in Utah attempted to make an illegal beverage called "pruno"; however, someone added a weeks-old baked potato to the mix, letting a microbe into the party who was clearly uninvited. Consumers of the pruno began to develop difficulty swallowing, vomiting, double vision, and muscle weakness; and three required ventilation therapy. No deaths were attributed to the contaminated beverage.
   a. What disease were the prisoners suffering from, and what was the causative agent involved?
   b. Based upon your knowledge of this disease, what form of treatment was used to successfully avoid the worst outcomes of the disease in these patients?

9. a. Summarize the most common reservoirs of rabies virus today.
   b. In August of 2011, a soldier from Fort Drum in Watertown, New York, tested positive for rabies; he died less than 3 weeks later. Further investigation revealed that he actually became infected when he was bitten by a dog in January of the same year while stationed in Afghanistan. Discuss any risks the soldier posed to his platoon, explaining whether or not this fatal outcome could have been avoided.

10. The Sabin oral polio vaccine (OPV) is currently used in India and other developing countries around the world. Conduct additional research and discuss whether this method of immunization has been safe and effective.

Concept Connections | Bloom's Levels 4 and 6: Analyze and Create

This activity ties together multiple concepts in the chapter.

1. What are the signs and symptoms common to all forms of meningitis?

2. How is bacterial meningitis different from viral meningitis?

3. What are the treatment options for meningitis caused by bacteria, viruses, or fungi?

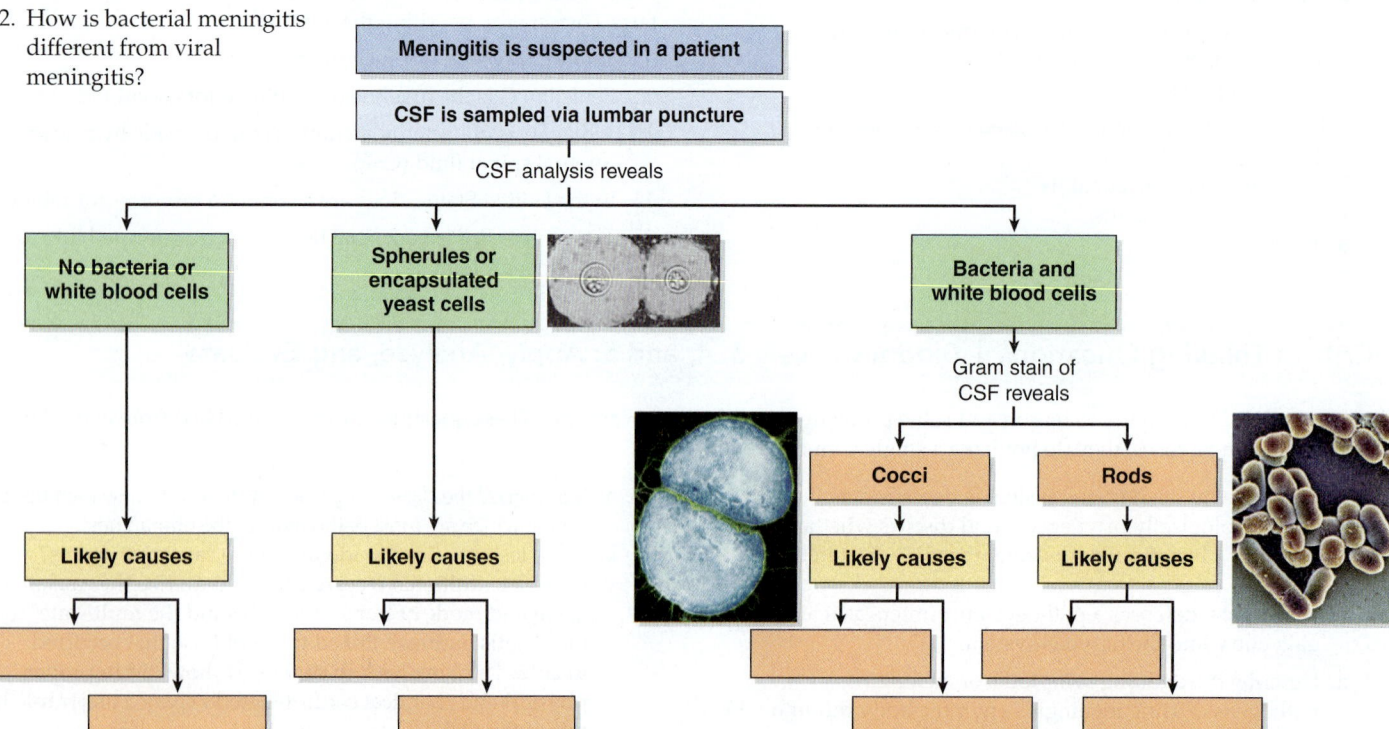

Visual Connections | Bloom's Level 5: Evaluate

These questions use visual images or previous content to make connections to this chapter's concepts.

1. **From chapter 3, figure 3.21.** Without looking back to the figure in chapter 3, speculate on which meningitis-causing organism you are seeing here. How could your presumptive diagnosis be confirmed?

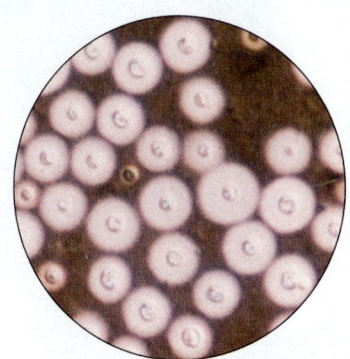

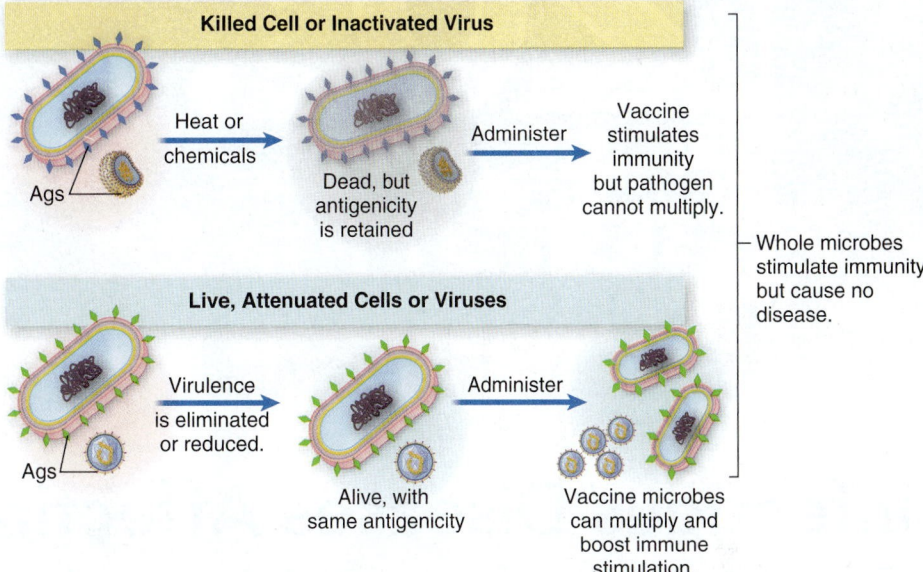

Killed Cell or Inactivated Virus

Ags → Heat or chemicals → Dead, but antigenicity is retained → Administer → Vaccine stimulates immunity but pathogen cannot multiply.

Live, Attenuated Cells or Viruses

Ags → Virulence is eliminated or reduced. → Alive, with same antigenicity → Administer → Vaccine microbes can multiply and boost immune stimulation.

Whole microbes stimulate immunity but cause no disease.

2. **From chapter 15, table 15.6.** A vaccine used to immunize individuals against meningococcal meningitis is described as containing "meningococcal capsular polysaccharide antigens." Which of the vaccine production strategies shown in this illustration could be used to produce this vaccine? Explain your answer.

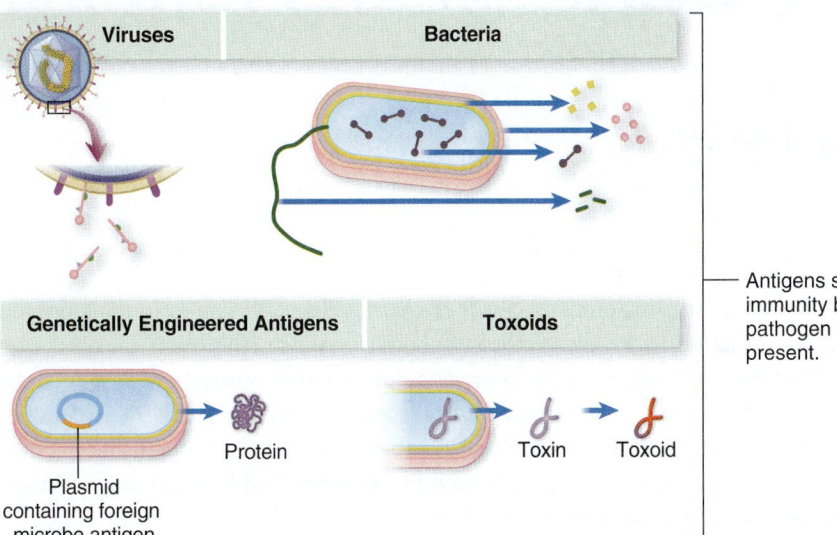

Viruses   **Bacteria**

Genetically Engineered Antigens   **Toxoids**

Plasmid containing foreign microbe antigen → Protein

Toxin → Toxoid

Antigens stimulate immunity but no pathogen is present.

Concept Mapping | Bloom's Level 6: Create

Appendix D provides guidance for working with concept maps.

1. Using the words that follow, please create a concept map illustrating the relationships among these key terms from chapter 19.

bacteria	droplets	meningitis	vaccination
viruses	vehicles	colonization	
fungi	vaccines	transmission	

www.mcgrawhillconnect.com

Enhance your study of this chapter with study tools and practice tests. Also ask your instructor about the resources available through ConnectPlus, including the media-rich eBook, interactive learning tools, and animations.

20

The entrance of the Kikwit General Hospital, photographed during the Ebola outbreak in what is now the Democratic Republic of the Congo (formerly Zaire).

Infectious Diseases Affecting the Cardiovascular and Lymphatic Systems

Case File 20

Ebola in the Congo

In 1994, Richard Preston's *The Hot Zone* electrified readers with gruesome descriptions of an exotic virus that caused its victims to bleed out of every orifice in a horrifying death. This book was not a work of fiction but was a factual account of outbreaks of the Ebola virus in Kenya and Sudan and its presence at an animal storage facility in Reston, Virginia. Although Ebola is a fairly rare disease, case fatality rates range from 25% to 90% and outbreaks often strike in impoverished nations with limited medical resources. Usually outbreaks are small and self-limiting, but they are deadly and quickly capture the attention of public health agencies such as the World Health Organization (WHO) and the Centers for Disease Control and Prevention (CDC) and news outlets alike.

Ebola first grabbed the attention of the world in 1976 with simultaneous outbreaks in Sudan and Zaire (currently the Democratic Republic of the Congo, or D.R.C.). Since then, there have been sporadic outbreaks in Sudan, Zaire/D.R.C., Gabon, Ivory Coast, Uganda, South Africa, and the Philippines. In the summer of 2012, an outbreak occurred in the D.R.C. with 15 cases and 10 deaths and in Uganda with 60 cases and 16 deaths. One case was confirmed in a man who stole a cell phone from a patient in a hospital ward where Ebola patients were quarantined. He later started showing symptoms of the disease and checked himself into the same hospital where he stole the phone.

- What makes Ebola such a deadly virus?
- What is the reservoir of the virus?
- How is the virus transmitted?

Continuing the Case appears on page 607.

590

Outline and Learning Outcomes

20.1   The Cardiovascular and Lymphatic Systems and Their Defenses

    1. Describe the important anatomical features of the cardiovascular system.

    2. List the natural defenses present in the cardiovascular and lymphatic systems.

20.2   Normal Biota of the Cardiovascular and Lymphatic Systems

    3. Discuss the current state of knowledge of the normal biota of the cardiovascular and lymphatic systems.

20.3   Cardiovascular and Lymphatic System Diseases Caused by Microorganisms

    4. List the possible causative agents for each of the following infectious cardiovascular conditions: acute and subacute endocarditis, plague, tularemia, Lyme disease, infectious mononucleosis, anthrax, Chagas disease, and malaria.

    5. Discuss what series of events may lead to septicemia and how it should be prevented and treated.

    6. Describe what makes anthrax a good agent for bioterrorism, and list the important presenting signs to look for in patients.

    7. Discuss the difference between hemorrhagic and nonhemorrhagic fever diseases.

    8. List the possible causative agents and modes of transmission for hemorrhagic fever diseases.

    9. List the possible causative agents and modes of transmission for nonhemorrhagic fever diseases.

    10. Discuss all aspects of malaria, with special emphasis on epidemiology.

    11. Describe important events in the course of an HIV infection in the absence of treatment.

    12. Explain the rationale behind the recommended treatment for HIV infection.

    13. Discuss the epidemiology of HIV infection in the developed and the developing world.

20.1   The Cardiovascular and Lymphatic Systems and Their Defenses

The Cardiovascular System

The cardiovascular system is the pipeline of the body. It is composed of the blood vessels, which carry blood to and from all regions of the body, and the heart, which pumps the blood. This system moves the blood in a closed circuit, and it is therefore known as the *circulatory system*. The cardiovascular system provides tissues with oxygen and nutrients and carries away carbon dioxide and waste products, delivering them to the appropriate organs for removal. A closely related but largely separate system, the **lymphatic system** is a major source of immune cells and fluids, and it serves as a one-way passage, returning fluid from the tissues to the cardiovascular system.

The heart is a fist-size muscular organ that pumps blood through the body. It is divided into two halves, each of which is divided into an upper and lower chamber **(figure 20.1)**. The upper chambers are called atria (singular, *atrium*), and the lower are ventricles. The entire organ is encased in a fibrous covering, the pericardium, which is an occasional site of infection. The actual wall of the heart has three layers: from outer to inner, they are the epicardium, the myocardium, and the endocardium. The endocardium also covers the valves of the heart, and it is a relatively common target of microbial infection.

The atria receive blood coming from the body. This blood, which is low in oxygen and high in carbon dioxide, enters the right atrium and passes through to the right ventricle. From there it is pumped to the pulmonary arteries in the lung, where it becomes oxygenated and reenters the heart through the left atrium. Finally, the blood moves into the left

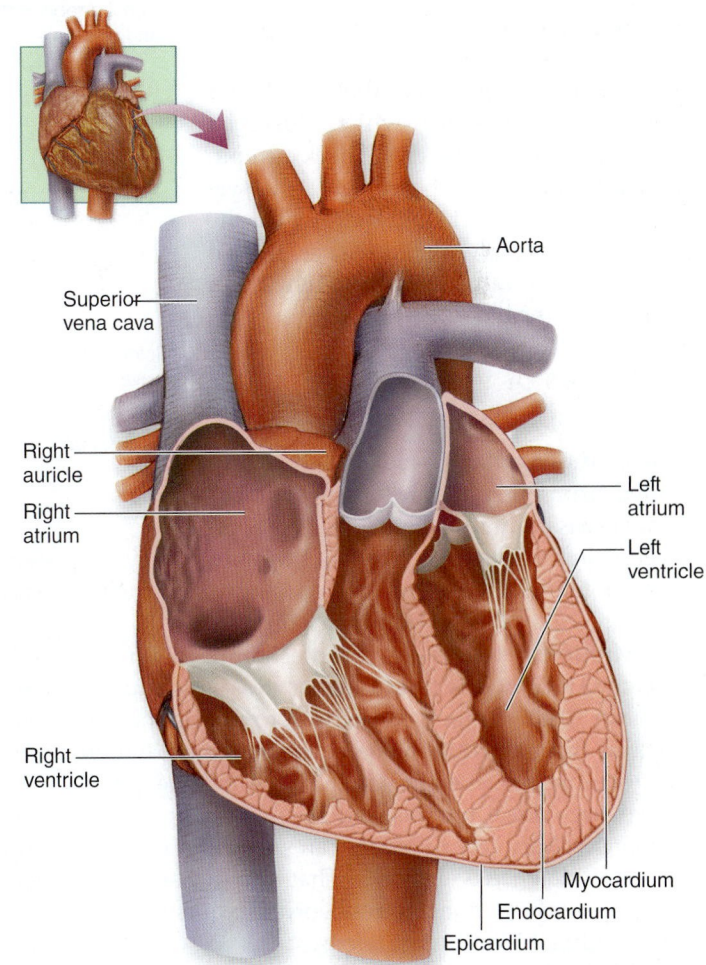

Figure 20.1   The heart.

ventricle and is pumped into the aorta and the rest of the body. Valves control the movement of blood into and out of the chambers of the heart.

The blood vessels consist of *arteries, veins,* and *capillaries.* Arteries carry oxygenated blood away from the heart under relatively high pressure. They branch into smaller vessels called arterioles. Veins actually begin as smaller venules in the periphery of the body and coalesce into veins. The smallest blood vessels, the capillaries, connect arterioles to venules. Both arteries and veins have walls made of three layers of tissue. The innermost layer is composed of a smooth epithelium called endothelium. Its smooth surface encourages the smooth flow of cells and platelets through the system. The next layer is composed of connective tissue and muscle fibers. The outside layer is a thin layer of connective tissue. Capillaries, the smallest vessels, have walls made of only one layer of endothelium. **Figure 20.2** illustrates the complete cardiovascular system.

The Lymphatic System

Chapter 14 provided a detailed description of the lymphatic system; you may wish to review page 399 and figure 14.7 before continuing. In short, the lymphatic system consists mainly of the lymph vessels, which roughly parallel the blood vessels; lymph nodes, which cluster at body sites such as the groin, neck, armpit, and intestines; and the spleen. This system serves to collect fluid that has left the blood vessels and entered tissues, filter it of impurities and infectious agents, and return it to the blood.

Defenses of the Cardiovascular and Lymphatic Systems

The cardiovascular system is highly protected from microbial infection. Microbes that successfully invade the system, however, gain access to every part of the body, and every system may potentially be affected. For this reason, bloodstream infections are called **systemic infections.**

Multiple defenses against infection reside in the bloodstream. The blood is full of leukocytes, with approximately 5,000 to 10,000 white blood cells per milliliter of blood. The various types of white blood cells include the lymphocytes, responsible for specific immunity, and the phagocytes, which are critical to nonspecific as well as specific immune responses. Very few microbes can survive in the blood with so many defensive elements. That said, a handful of infectious agents have nonetheless evolved exquisite mechanisms for avoiding blood-borne defenses.

Medical conditions involving the blood often have the suffix -*emia.* For instance, viruses that cause meningitis can

travel to the nervous system via the bloodstream. Their presence in the blood is called **viremia.** When fungi are in the blood, the condition is termed **fungemia;** bacterial presence in the blood is called **bacteremia,** a general term denoting only their *presence.* Although the blood contains no normal

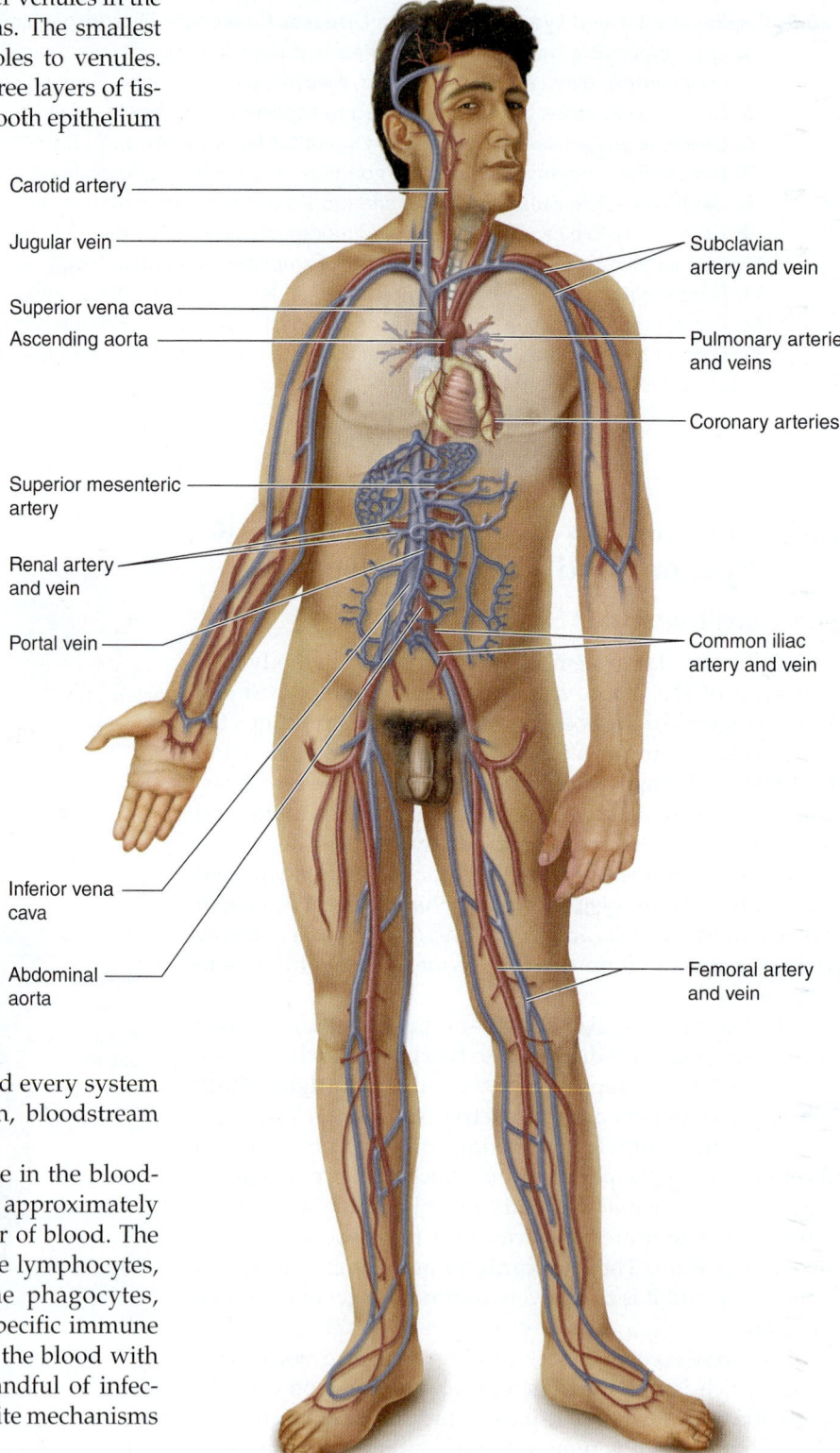

Carotid artery

Jugular vein

Superior vena cava

Ascending aorta

Superior mesenteric artery

Renal artery and vein

Portal vein

Inferior vena cava

Abdominal aorta

Subclavian artery and vein

Pulmonary arteries and veins

Coronary arteries

Common iliac artery and vein

Femoral artery and vein

Figure 20.2 The anatomy of the cardiovascular system.

biota (see next section), bacteria frequently are introduced into the bloodstream during the course of daily living. Brushing your teeth or tearing a hangnail can introduce bacteria from the mouth or skin into the bloodstream. This situation is usually temporary, but there is mounting evidence of the development of more long-term effects as oral microbes have been localized in arterial plaques of heart disease patients. When bacteria flourish and grow in the bloodstream, the condition is termed **septicemia.** Septicemia can very quickly lead to cascading immune responses, resulting in decreased systemic blood pressure, which can lead to **septic shock,** a life-threatening condition.

20.1 Learning Outcomes—Assess Your Progress

1. Describe the important anatomical features of the cardiovascular system.
2. List the natural defenses present in the cardiovascular and lymphatic systems.

20.2 Normal Biota of the Cardiovascular and Lymphatic Systems

Like the nervous system, the cardiovascular and lymphatic systems are "closed" systems with no normal access to the external environment. Therefore, current science believes they possess no normal biota. In the absence of disease, microorganisms may be transiently present in either system as just described. The lymphatic system serves to filter microbes and their products out of tissues. Thus, in the healthy state, no microorganisms *colonize* either the lymphatic or cardiovascular systems. Of course, this is biology, and it is never quite that simple. Recent data from the Human Microbiome Project suggest that the bloodstream is not completely sterile, even during periods of apparent health. It is tempting to speculate that these low-level microbial "infections" may contribute to diseases for which no infectious cause has previously been found.

Cardiovascular and Lymphatic Systems Defenses and Normal Biota

	Defenses	Normal Biota
Cardiovascular System	Blood-borne components of nonspecific and specific immunity—including phagocytosis, specific immunity	None
Lymphatic System	Numerous immune defenses reside here.	None

20.2 Learning Outcomes—Assess Your Progress

3. Discuss the current state of knowledge of the normal biota of the cardiovascular and lymphatic systems.

20.3 Cardiovascular and Lymphatic System Diseases Caused by Microorganisms

Categorizing cardiovascular and lymphatic infections according to clinical presentation is somewhat difficult because most of these conditions are systemic, with effects on multiple organ systems. We begin with infections involving the heart and the blood in general and then discuss conditions with more specific causes.

Endocarditis

Endocarditis is an inflammation of the endocardium, or inner lining of the heart. Most of the time, endocarditis refers to an infection of the valves of the heart, often the mitral or aortic valves **(figure 20.3).** Two variations of infectious endocarditis have been described: acute and subacute. Each has distinct groups of possible causative agents, most of which are bacterial organisms. Fungi have been documented to cause rare cases of endocarditis, while the role of viruses in this disease is still under investigation. Rarely, endocarditis can also be caused by vascular trauma or by circulating immune complexes in the absence of infectious agents.

The surgical innovation of prosthetic valves presents a new hazard for development of endocarditis. Patients with prosthetic valves can acquire acute endocarditis if bacteria are introduced during the surgical procedure, and such infections result in high rates of morbidity and mortality today.

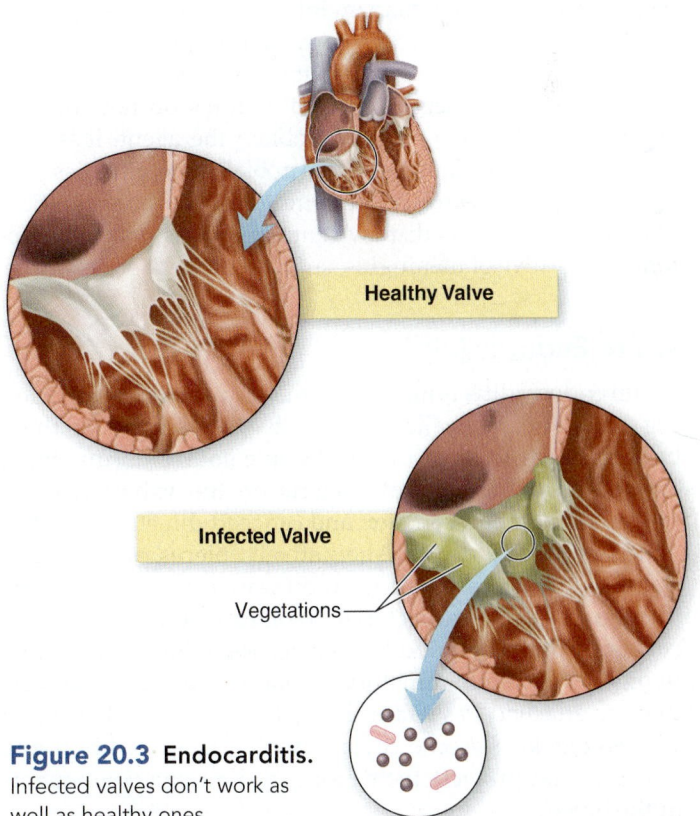

Healthy Valve

Infected Valve

Vegetations

Figure 20.3 Endocarditis. Infected valves don't work as well as healthy ones.

Alternatively, the prosthetic valves can serve as infection sites for the subacute form of endocarditis long after the surgical procedure. Because the symptoms and the diagnostic procedures are similar for both forms of endocarditis, they are discussed first; then the specific aspects of acute and subacute endocarditis are addressed.

▶ Signs and Symptoms

The signs and symptoms are similar for both types of endocarditis, except that in the subacute condition they develop more slowly and are less pronounced than with the acute disease. Symptoms include fever, anemia, abnormal heartbeat, and sometimes symptoms similar to myocardial infarction (heart attack). Shortness of breath is a common symptom; additionally, chills may develop.

Abdominal or side pain is sometimes reported. The patient may look very ill and may have petechiae (small red-to-purple spots) over the upper half of the body and under the fingernails. Red, painless skin spots on the palms and the soles (Janeway lesions) and small painful nodes on the pads of fingers and toes (Osler's nodes) may also be apparent on examination. In subacute cases, an enlarged spleen may have developed over time; cases of extremely long duration can lead to clubbed fingers and toes due to lack of oxygen in the blood.

▶ Culture and Diagnosis

The diagnostic procedures for the two forms of endocarditis are essentially the same. One of the most important diagnostic tools is a high index of suspicion. A history of risk factors, or behaviors, such as abnormal valves, intravenous drug use, recent surgery, or bloodstream infections, should lead one to consider endocarditis when the symptoms just described are observed. Blood cultures, if positive, are the gold standard for diagnosis, but negative blood cultures do not rule out endocarditis. If it is possible to obtain the agent, it is very important to determine its antimicrobial susceptibilities.

In acute endocarditis, the symptoms may be magnified. The patient may also display central nervous system symptoms suggestive of meningitis, such as stiff neck or headache.

Acute Endocarditis

Acute endocarditis is most often the result of an overwhelming bloodstream challenge with bacteria. Some of these bacteria seem to have the ability to colonize normal heart valves. Accumulations of bacteria on the valves (vegetations) hamper their function and can lead directly to cardiac malfunction and death. Alternatively, pieces of the bacterial vegetation can break off and create emboli (blockages) in vital organs. The bacterial colonies can also provide a constant source of blood-borne bacteria, with the accompanying systemic inflammatory response and shock. Bacteria that are attached to surfaces bathed by blood (such as heart valves) quickly become covered with a mesh of fibrin and platelets that protects them from the immune components in the blood.

▶ Causative Agents

The acute form of endocarditis is most often caused by *Staphylococcus aureus*. Other agents that cause it are *Streptococcus pyogenes, Streptococcus pneumoniae,* and *Neisseria gonorrhoeae,* as well as a host of other bacteria. Each of these bacteria is described elsewhere in this book; all are pathogenic.

▶ Transmission and Epidemiology

The most common route of transmission for acute endocarditis is parenteral—that is, via direct entry into the body. Intravenous or subcutaneous drug users are a significant risk group for the condition. Traumatic injuries and surgical procedures can also introduce the large number of bacteria required for the acute form of endocarditis.

▶ Prevention and Treatment

Prevention is based on avoiding the introduction of bacteria into the bloodstream during surgical procedures or injections. Untreated, this condition is invariably fatal. Recommended antibiotics are nafcillin or oxacillin with or without gentamicin. Alternatively, vancomycin in combination with gentamicin can be used. High, continuous blood levels of antibiotics are required to resolve the infection because the bacteria exist in biofilm vegetations. In addition to the decreased access of antibiotics to bacteria deep in the biofilm, these bacteria often express a phenotype of lower susceptibility to antibiotics. Surgical debridement of the valves, accompanied by antibiotic therapy, is sometimes required **(Disease Table 20.1).**

Subacute Endocarditis

Subacute forms of this condition are almost always preceded by some form of damage to the heart valves or by congenital malformation. Irregularities in the valves encourage the attachment of bacteria, which then form biofilms and impede normal function, as well as provide an ongoing source of bacteria to the bloodstream. People who have suffered rheumatic fever and the accompanying damage to heart valves are particularly susceptible to this condition (see chapter 21 for a complete discussion of rheumatic fever).

▶ Causative Agents

Most commonly, subacute endocarditis is caused by bacteria of low pathogenicity, often originating in the oral cavity. Alpha-hemolytic streptococci, such as *Streptococcus sanguinis, S. oralis,* and *S. mutans,* are most often responsible, although normal biota from the skin and other bacteria can also colonize abnormal valves and lead to this condition.

▶ Transmission and Epidemiology

Minor disruptions in the skin or mucous membranes, such as those induced by vigorous tooth brushing, dental procedures, or relatively minor cuts and lacerations, can introduce bacteria into the bloodstream and lead to valve colonization. The bacteria are not, therefore, transmitted from other people or from the environment. The average age of onset for subacute endocarditis has increased in recent decades from

Disease Table 20.1 Endocarditis

Disease	Acute Endocarditis	Subacute Endocarditis
Causative Organism(s)	*Staphylococcus aureus, Streptococcus pyogenes, S. pneumoniae, Neisseria gonorrhoeae,* others	Alpha-hemolytic streptococci, others
Most Common Modes of Transmission	Parenteral	Endogenous transfer of normal biota to bloodstream
Culture/Diagnosis	Blood culture	Blood culture
Prevention	Aseptic surgery, injections	Prophylactic antibiotics before invasive procedures
Treatment	Nafcillin or oxacillin +/− gentamicin or tobramycin OR vancomycin + gentamicin; surgery may be necessary	Surgery may be necessary
Distinctive Features	Acute onset, high fatality rate	Slower onset
Epidemiological Features	Three times more common in males than females	–

the mid-20s to the mid-50s. Males are slightly more likely to experience it than females.

▶ **Prevention and Treatment**

The practice of prophylactic antibiotic therapy in advance of surgical and dental procedures on patients with underlying valve irregularities has decreased the incidence of this infection. When it occurs, treatment is similar to treatment for the acute form of the disease, described earlier (see Disease Table 20.1).

Septicemia

Septicemia occurs when organisms are actively multiplying in the blood. Many different bacteria (and a few fungi) can cause this condition. Patients suffering from these infections are sometimes described as "septic." One infection that should be considered in cases of aggressive septicemia, especially if respiratory symptoms are also present, is anthrax.

▶ **Signs and Symptoms**

Fever is a prominent feature of septicemia. The patient appears very ill and may have an altered mental state, shaking chills, and gastrointestinal symptoms. Often an increased breathing rate is exhibited, accompanied by respiratory alkalosis (increased tissue pH due to breathing disorder). Low blood pressure is a hallmark of this condition and is caused by the inflammatory response to infectious agents in the bloodstream, which leads to a loss of fluid from the vasculature. This condition is the most dangerous feature of the disease, often culminating in death.

▶ **Causative Agents**

Bacteria cause the vast majority of septicemias and are evenly divided between gram-positive and gram-negative organisms. Perhaps 10% are caused by fungal infections. Polymicrobial bloodstream infections involving more than one microorganism are increasingly being identified today.

▶ **Pathogenesis and Virulence Factors**

Gram-negative bacteria multiplying in the blood release large amounts of endotoxin into the bloodstream, stimulating a massive inflammatory response mediated by a host of cytokines. This response invariably leads to a drastic drop in blood pressure, a condition called **endotoxic shock.** Gram-positive bacteria can instigate a similar cascade of events when fragments of their cell walls are released into the blood.

▶ **Transmission and Epidemiology**

In many cases, septicemias can be traced to parenteral introduction of the microorganisms via intravenous lines or surgical procedures. Other infections may arise from serious urinary tract infections or from renal, prostatic, pancreatic, or gallbladder abscesses. Patients with underlying spleen malfunction may be predisposed to multiplication of microbes in the bloodstream. Meningeal infections, bone infections (osteomyelitis), or pneumonia can all occasionally lead to sepsis. It should be noted that hospitalization for sepsis has more than doubled in recent years. More alarming is the fact there is a 20% to 50% mortality rate associated with septic infections, reflecting the high risks of this disease even when treatment is available.

▶ Culture and Diagnosis

Because the infection is in the bloodstream, a blood culture is the obvious route to diagnosis. A full regimen of media should be inoculated to ensure isolation of the causative microorganism. Antibiotic susceptibilities should be assessed. Empiric therapy should be started immediately before culture and susceptibility results are available. The choice of antimicrobial agent should be informed by knowledge of any suspected source of the infection, such as an intravenous catheter (in which case, skin biota should be considered), urinary tract infections (in which case, gram-negative microbes and *Streptococci* should be considered), and so forth.

▶ Prevention and Treatment

Empiric therapy, which is begun immediately after blood cultures are taken, often begins with a broad-spectrum antibiotic. Once the organism is identified and its antibiotic susceptibility is known, treatment can be adjusted accordingly. Statistics show that successful patient outcomes depend on rapid diagnosis and treatment. The recent approval of new rapid nucleic acid tests will provide physicians with vital information on the causative agent and its drug susceptibility within hours rather than days **(Disease Table 20.2)**.

Disease Table 20.2	Septicemia
Causative Organism(s)	Bacteria or fungi
Most Common Modes of Transmission	Parenteral, endogenous transfer
Virulence Factors	Cell wall or membrane components
Culture/Diagnosis	Blood culture
Prevention	–
Treatment	Broad-spectrum antibiotic until identification and susceptibilities tested
Epidemiological Features	In United States: 200,000 cases and 100,000 deaths per year

Plague

Although pandemics of plague have probably occurred since antiquity, the first one that was reliably chronicled killed an estimated 100 million people in the sixth century AD. The last great pandemic occurred in the late 1800s and was transmitted around the world, primarily by rat-infested ships. The disease was brought to the United States through the port of San Francisco around 1906. Infected rats eventually mingled with native populations of rodents and gradually spread the disease throughout the West and Southwest, where it is endemic today.

▶ Signs and Symptoms

Three possible manifestations of infection occur with the bacterium causing plague. **Pneumonic plague** is a respiratory disease, described in chapter 21. In **bubonic plague,** the bacterium, which is injected by the bite of a flea, enters the lymph and is filtered by a local lymph node. Infection causes inflammation and necrosis of the node, resulting in a swollen lesion called a **bubo,** usually in the groin or axilla **(figure 20.4a)**. The incubation period lasts 2 to 8 days, ending abruptly with the onset of fever, chills, headache, nausea, weakness, and tenderness of the bubo. Mortality rates, even with treatment, can reach up to 15%.

These cases often progress to massive bacterial growth in the blood termed **septicemic plague.** The presence of the bacteria in the blood results in disseminated intravascular coagulation, subcutaneous hemorrhage, and purpura that may degenerate into necrosis and gangrene. Mortality rates, once the disease has progressed to this point, are 30% to 50% with treatment and 100% without treatment. Because of the visible darkening of the skin, the plague has often been called the "Black Death."

▶ Causative Agent

The cause of this dreadful disease is a tiny, harmless-looking, gram-negative rod, *Yersinia pestis,* a member of the family *Enterobacteriaceae. Y. pestis* displays unusual bipolar staining that makes it look like a safety pin **(figure 20.4b)**.

▶ Pathogenesis and Virulence Factors

The number of bacteria required to initiate a plague infection is small—perhaps only 3 to 50 cells for bubonic or septicemic cases. Scientists have discovered that *Y. pestis* carries additional genes that help it to cause disease in mice and to survive in the flea vector. These genes include a gene for capsule formation and a gene for plasminogen activation (similar to the streptokinase expressed by *S. pyogenes;* see chapter 18). Plasminogen activation leads to clotting, which helps the microbe resist phagocytosis.

▶ Transmission and Epidemiology

The principal agents in the transmission of the plague bacterium are fleas. These tiny, bloodsucking insects have a special relationship with the bacterium. After a flea ingests a blood meal from an infected animal, the bacteria multiply in its gut. In fleas that effectively transmit the bacterium, the esophagus becomes blocked due to coagulation factors produced by the pathogen. Being unable to feed properly, the ravenous flea jumps from animal to animal in a futile attempt to get

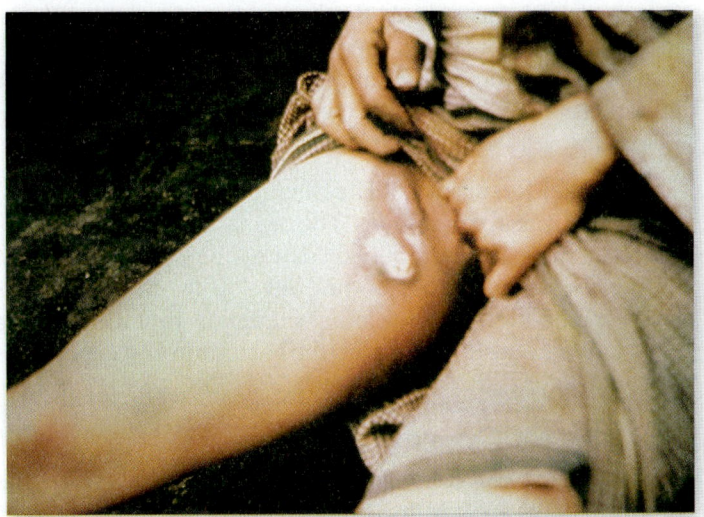

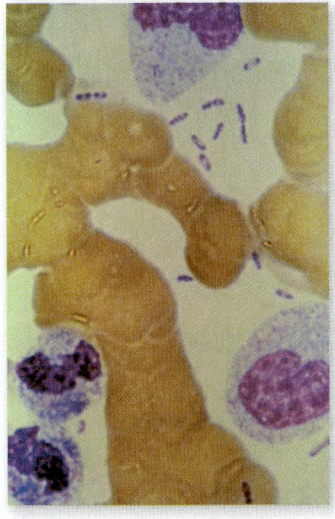

(a)

(b)

Figure 20.4 Bubonic plague.
(a) A classic inguinal bubo of bubonic plague. This hard nodule is very painful and can rupture onto the surface. **(b)** *Yersinia pestis.* These bacteria are said to have a "safety pin" appearance.

INSIGHT 20.1   The Cause of Black Plague: An Alternate Hypothesis?

"Bring out your dead!" The scene from *Monty Python and the Holy Grail* is a cult classic where impoverished serfs in medieval England load corpses onto a cart, victims of the Black Plague. While the scene is hilarious, the Black Plague, also known as the Black Death, was no laughing matter. This plague epidemic, from 1347 to 1351, originated in China and spread through Europe, killing one-third of the world's population at the time. There were periodic epidemics throughout the Dark Ages and even into modern history, and each brought fear and dread as the disease rapidly spread throughout the population leaving death in its wake. The plague is caused by *Yersinia pestis,* a gram-negative bacillus that is spread by fleas. However, in a paper published in the *Postgraduate Medical Journal* in 2005, it was postulated that the Black Death in the 1300s was caused not by a bacterium but rather by a viral hemorrhagic fever. The paper makes several assertions about the nature of the disease. For example, the authors note that the infection spread more rapidly than rats could travel from village to village, and that the incubation period in the historical record was 32 days, while the incubation period of modern *Y. pestis* is about 16 days.

However, there are a few holes in this hypothesis. DNA extracted from the dental pulp from the ancient remains of plague victims in mass graves shows evidence of *Y. pestis,* while corpses from the same time period that were not killed by the plague show no evidence of *Y. pestis.* In fact, the DNA evidence extracted from medieval plague victims shows that the ancient scourge is genetically very similar to the modern-day *Y. pestis.*

This episode illustrates that science is dynamic. Questions are raised all the time, and answers are settled upon using the

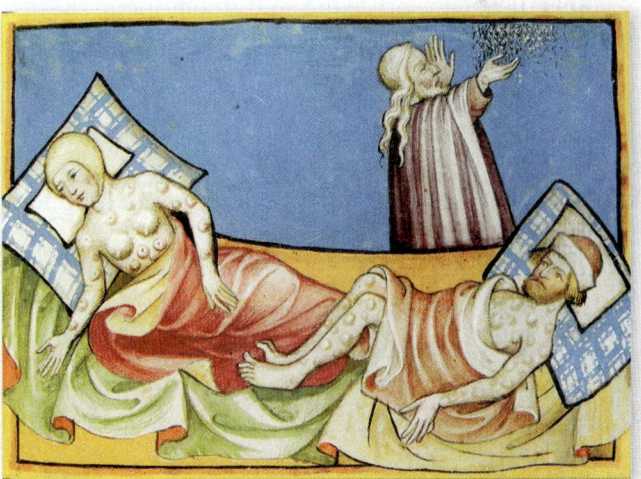

The Black Death killed nearly one in three people in Europe during the 1347 pandemic. This is an illustration of the Black Death from a Bible printed in 1411.

best data and technology available. In the end, the consensus is that the Black Death in the 14th century was indeed caused by *Y. pestis.* But questions such as the ones raised here have led investigators to study what caused the differences in spread and virulence of the disease between the 1300s and now. And science marches on, questioning always.

Source: 2005. *Post. Med. J.* vol. 81, no. 955, p. 315.

nourishment. During this process, regurgitated infectious material is inoculated into the bite wound.

The plague bacterium exists naturally in many animal hosts. Plague still exists endemically in large areas of Africa, South America, the Mideast, Asia, and the former Soviet Union; upwards of 2,000 cases of disease are reported in these areas each year. In the United States, sporadic cases (usually less than 10 per year) occur as a result of contact with wild

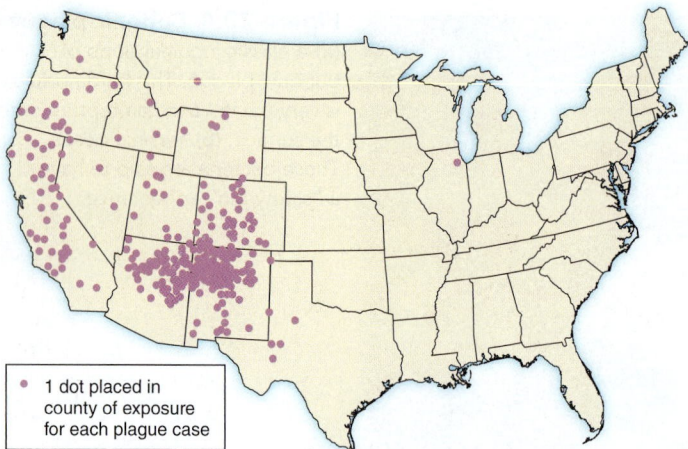

Figure 20.5 **Reported cases of human plague—United States, 1970–2010.**

and domestic animals **(figure 20.5)**. In 2012, a young girl in Colorado was diagnosed with plague after attempting to bury a dead squirrel while camping. Her sweatshirt was on the ground during the procedure and at some point she tied it around her waist, where later several insect bites were found. In the United States, the disease is considered endemic in western and southwestern states. Persons most at risk for developing plague are veterinarians and people living and working near woodlands and forests. Dogs and cats can be infected with the plague, often from contact with infected wild animals such as prairie dogs.

The epidemiology of plague is among the most complex of all diseases. It involves several different types of vertebrate hosts and flea vectors, and its exact cycle varies from one region to another. A general scheme of the cycle is presented in **figure 20.6**. Humans can develop plague through contact with the fleas of wild or domestic or semidomestic animals.

Contact with infected body fluids can also spread the disease. If a person has breaks in the skin on his or her hands, handling infected animals or animal skins is a possible means of transmission. (Persons with the pneumonic form of the disease can spread *Y. pestis* through respiratory droplets.)

The Animal Reservoirs The plague bacillus occurs in 200 different species of mammals. The primary long-term *endemic reservoirs* are various rodents, such as mice and voles, that harbor the organism but do not develop the disease. These hosts spread the disease to other mammals called *amplifying hosts* that become infected with the bacterium and experience massive die-offs during epidemics.

▶ **Culture and Diagnosis**

Because death can occur as quickly as 2 to 4 days after the appearance of symptoms, prompt diagnosis and treatment of plague are imperative. Culture of the organism is the definitive method of diagnosis, although a Gram stain of aspirate from buboes often reveals the presence of the safety-pin-shaped bacteria. Today, rapid genomic and immunochromatographic tests have been developed to quickly diagnose infection. These are critical tools in light of the potential use of *Y. pestis* as a biological weapon.

▶ **Prevention and Treatment**

Plague is one of a handful of internationally quarantinable diseases (other examples are cholera and yellow fever). A vaccine for plague is no longer in the United States, though a subunit vaccine for pneumonic plague may be ready for use in 2015. This would be an important step in protecting the human population from potential bioterrorism attacks involving plague. Streptomycin or gentamicin can be used in the treatment of disease in most cases, though the looming threat of drug resistance in *Y. pestis* could make these drugs ineffective in the future **(Disease Table 20.3)**.

Figure 20.6. **The infection cycle of *Yersinia pestis*.**

Chipmunk   Flea

Flea

Mouse   Ground squirrel   Flea

Infective fluids/ breaks in human's skin

Bubonic plague   Pneumonic plague

Endemic reservoir hosts   Amplifying hosts   Human (accidental) hosts

Disease Table 20.3 Plague

Causative Organism(s)	*Yersinia pestis*
Most Common Modes of Transmission	Vector, biological; also droplet contact (pneumonic) and direct contact with body fluids
Virulence Factors	Capsule, plasminogen activator
Culture/Diagnosis	Rapid genomic methods
Prevention	Flea and or animal control; vaccine available for high-risk individuals
Treatment	Streptomycin or gentamicin
Epidemiological Features	United States: endemic in all western and southwestern states; internationally, 95% of human cases occur in Africa, including Madagascar

Tularemia

▶ Signs and Symptoms

Tularemia is a zoonotic disease that is endemic throughout the Northern Hemisphere. After an incubation period ranging from a few days to 3 weeks, acute symptoms of headache, backache, fever, chills, malaise, and weakness appear. Further clinical manifestations are tied to the portal of entry. They include ulcerative skin lesions, swollen lymph glands, conjunctival inflammation, sore throat, intestinal disruption, and pulmonary involvement. The death rate in the most serious forms of disease is 30%, but proper treatment reduces mortality to almost zero.

▶ Causative Agent

The causative agent of tularemia is a facultative intracellular gram-negative bacterium called *Francisella tularensis*. It has several characteristics in common with *Yersinia pestis*, and the two species were previously included in a single genus called *Pasteurella*. It is a zoonotic disease of assorted mammals endemic to the Northern Hemisphere. Because it has been associated with outbreaks of disease in wild rabbits, it is sometimes called "rabbit fever." It is currently listed as a pathogen of concern on the lists of Category A bioterrorism agents that have the highest priority in research and funding. This list also includes pulmonary anthrax, pneumonic plague, botulism, smallpox, and viral hemorrhagic fevers.

▶ Transmission and Epidemiology

Tularemia is abundantly distributed through numerous animal reservoirs and vectors in northern Europe, Asia, and North America but not in the tropics. This disease is noteworthy for its complex epidemiology and spectrum of symptoms. Although rabbits and rodents (muskrats and ground squirrels) are the chief reservoirs, other wild animals (skunks, beavers, foxes, opossums), and some domestic animals are implicated as well. The chief route of transmission in the past had been through the activity of skinning rabbits, but with the decline of rabbit hunting, transmission via tick bites is more common. Ticks are the most frequent arthropod vector, followed by biting flies, mites, and mosquitoes. Due to this shift in transmission, disease cases now occur more often in summer months rather than the winter as was previously seen.

Tularemia is strikingly varied in its portals of entry and disease manifestations. Although bites by a vector are the most common source of infection, in many cases infection results when the skin or eye is inoculated through contact with infected animals, animal products, contaminated water, and dust. Pulmonary forms of the infection can result from aerosolized soils or animal fluids and also from spread of the bacterium in the bloodstream. The disease is not communicated from human to human. With disease developing after exposure to just 10 to 50 organisms, *F. tularensis* is often considered one of the most infectious of all bacterial pathogens. The name "lawnmower" tularemia refers to tularemia acquired when people have accidentally run over dead rabbits while lawn mowing, presumably from inhaling aerosolized bacteria. In 2009, two different people in Alaska acquired tularemia after wresting infected rabbits from their dogs' mouths. In that same year, two people in Oregon became infected after being bitten by cats and a third after removing an infected squirrel from her cat's clenched teeth.

▶ Prevention and Treatment

Treatment typically involves the use of gentamicin or streptomycin. Because the intracellular persistence of *F. tularensis* can lead to relapses, antimicrobial therapy must not be discontinued prematurely. Protection is available in the form of a live attenuated vaccine, but research in this area is still ongoing. Laboratory workers and other occupationally exposed personnel must wear gloves, masks, and eyewear **(Disease Table 20.4)**.

Disease Table 20.4	Tularemia
Causative Organism(s)	*Francisella tularensis*
Most Common Modes of Transmission	Vector, biological; also direct contact with body fluids from infected animal; airborne
Virulence Factors	Intracellular growth
Culture/Diagnosis	Culture dangerous to lab workers and not reliable; serology most often used
Prevention	Live attenuated vaccine for high-risk individuals
Treatment	Gentamicin or streptomycin
Epidemiological Features	United States: several hundred cases per year; internationally, 500,000 cases per year

Lyme Disease

In the 1970s, an enigmatic cluster of arthritis cases appeared in the town of Old Lyme, Connecticut. The phenomenon caught the attention of nonprofessionals and professionals alike, whose persistence and detective work ultimately disclosed the unusual nature and epidemiology of Lyme disease. The process of discovery began in the home of Polly Murray, who, along with her family, was beset for years by recurrent bouts of stiff neck, swollen joints, malaise, and fatigue that seemed vaguely to follow a rash from tick bites. When Mrs. Murray's son was diagnosed as having juvenile rheumatoid arthritis, she became skeptical. Conducting her own literature research, she began to discover inconsistencies. Rheumatoid arthritis was described as a rare, noninfectious disease, yet over an 8-year period, she found that 30 of her neighbors had experienced similar illnesses. Ultimately, this cluster of cases and others were reported to state health authorities. Eventually Lyme disease was shown to be caused by *Borrelia burgdorferi*. It is now recognized that Lyme disease has been around for centuries.

▶ Signs and Symptoms

Lyme disease is slow-acting, but it often evolves into a progressive syndrome that mimics neuromuscular and rheumatoid conditions. An early symptom in 70% of cases is a rash at the site of a tick bite. The lesion, called *erythema migrans*, looks something like a bull's-eye, with a raised erythematous (reddish) ring that gradually spreads outward and a pale central region **(figure 20.7).** Other early symptoms are fever, headache, stiff neck, and dizziness. If not treated or if treated too late, the disease can advance to the second stage, during which cardiac and neurological symptoms, such as facial palsy, can develop. After several weeks or months, a crippling polyarthritis can attack joints. Some people acquire chronic neurological complications that are severely disabling.

▶ Causative Agent

Borrelia burgdorferi was discovered in 1981 by Dr. Willy Burgdorfer, although he did not realize at that time its connection with disease. Although it is a spirochete bacterium, it is morphologically distinct from other pathogenic spirochetes. *Borrelia burgdorferi* is comparatively larger, ranging from 0.2 to 0.5 micrometer in width and from 10 to 20 micrometers in length, and exhibits 3 to 10 irregularly spaced and loose coils **(figure 20.8).** The nutritional requirements of *Borrelia* are complex, and culturing the bacterium in artificial media is difficult at best.

▶ Pathogenesis and Virulence Factors

The bacterium is a master of immune evasion. It changes its surface antigens while it is in the tick and again after it has been transmitted to a mammalian host. It provokes a strong humoral and cellular immune response, but this response is mainly ineffective, perhaps because of the bacterium's ability to switch its antigens. Indeed, it is possible that the immune response contributes to the pathology of the infection; research on the outer-membrane proteins of this pathogen is ongoing.

B. burgdorferi also has multiple proteins for attachment to host cells; these are considered virulence factors as well.

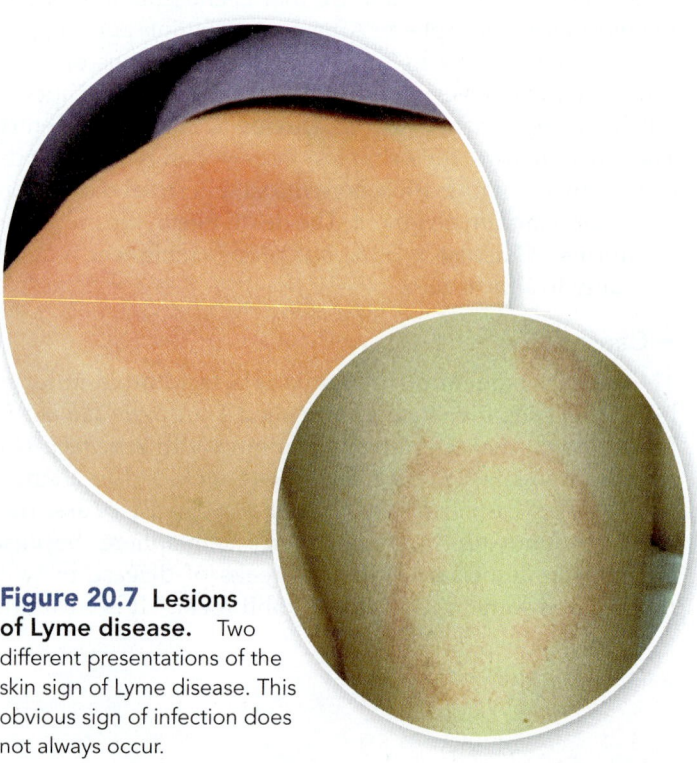

Figure 20.7 Lesions of Lyme disease. Two different presentations of the skin sign of Lyme disease. This obvious sign of infection does not always occur.

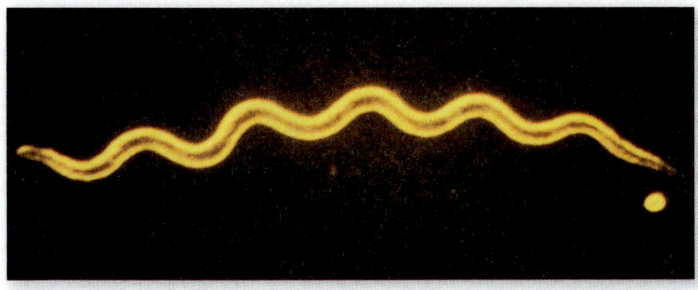

Figure 20.8 *Borrelia burgdorferi.* This spirochete has 3–10 loose, irregular coils.

▶ Transmission and Epidemiology

B. burgdorferi is transmitted primarily by hard ticks of the genus *Ixodes.* In the northeastern part of the United States, *Ixodes scapularis* (the black-legged deer tick) passes through a complex 2-year cycle that involves two principal hosts (**process figure 20.9**). As a larva or nymph, it feeds on either the white-footed mouse or birds or raccoons, where it picks up the infectious agent. The nymph is relatively nonspecific and will try to feed on nearly any type of vertebrate—thus, it is the form most likely to bite humans. The adult tick reproductive phase of the cycle is completed on deer. In California, the transmission cycle involves *Ixodes pacificus,*

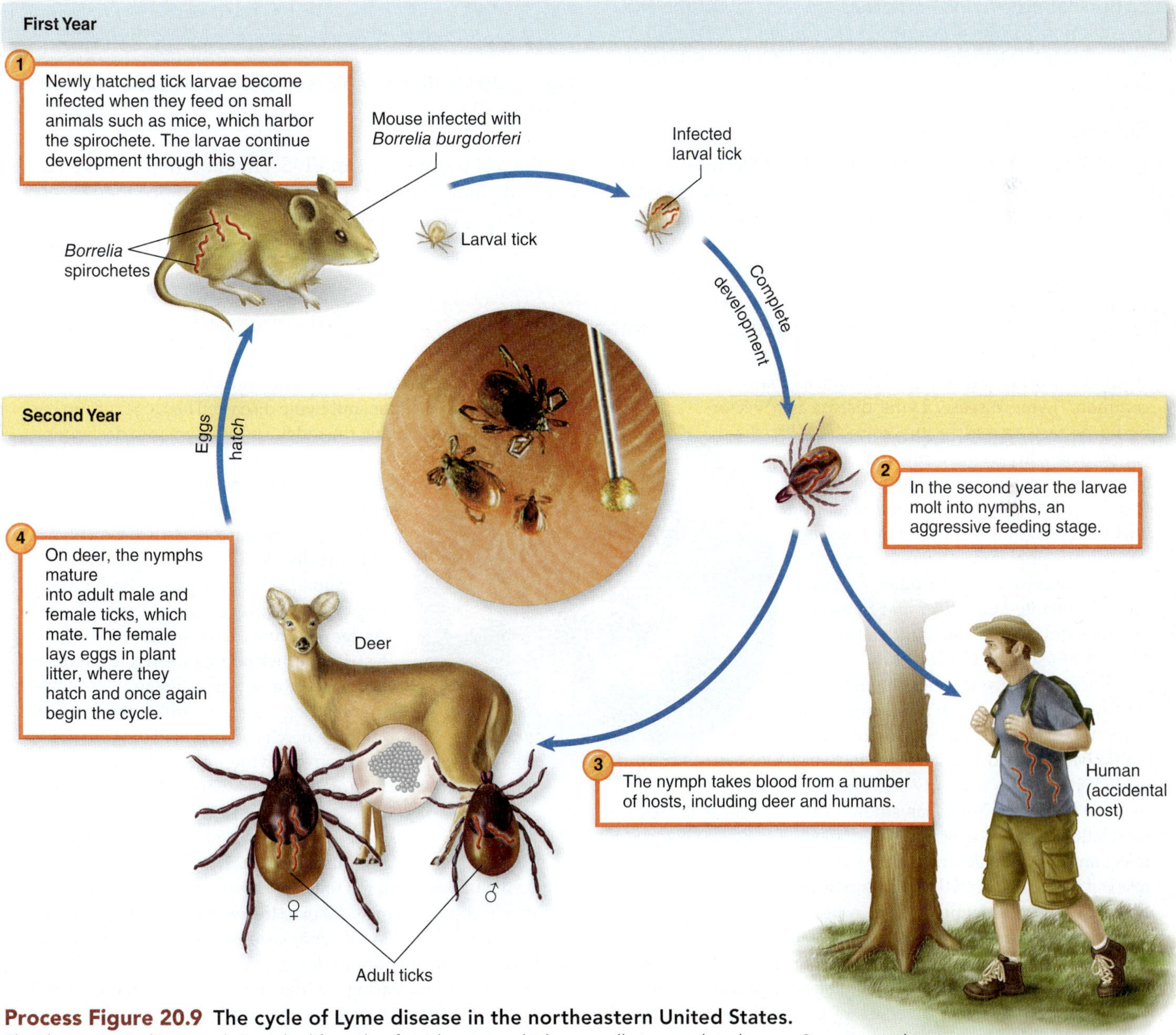

First Year

1 Newly hatched tick larvae become infected when they feed on small animals such as mice, which harbor the spirochete. The larvae continue development through this year.

Mouse infected with *Borrelia burgdorferi*

Infected larval tick

Borrelia spirochetes

Larval tick

Complete development

Second Year

Eggs hatch

4 On deer, the nymphs mature into adult male and female ticks, which mate. The female lays eggs in plant litter, where they hatch and once again begin the cycle.

2 In the second year the larvae molt into nymphs, an aggressive feeding stage.

Deer

3 The nymph takes blood from a number of hosts, including deer and humans.

Human (accidental host)

♀   ♂

Adult ticks

Process Figure 20.9 The cycle of Lyme disease in the northeastern United States.
The disease is tied intimately into the life cycle of a tick vector, which generally is completed over a 2-year period.
The exact hosts and species of tick vary from region to region but still display this basic pattern. The photograph gives an idea of the actual size of the nymph and adult black-legged deer ticks displayed on a human finger. Many people may not realize how very small and difficult to detect the feeding nymph can be.

another black-legged tick, and the dusky-footed woodrat as a reservoir.

The incidence of Lyme disease showed a gradual upward trend from about 10,000 cases per year in 1991 to 27,000 in 2007. This increase may be partly due to improved diagnosis, but it also reflects changes in the numbers of hosts and vectors, as described in **Insight 20.2**. The greatest concentrations of Lyme disease are found in areas having high deer populations **(figure 20.10)**. Most of the cases have occurred in New York, Pennsylvania, Connecticut, New Jersey, Rhode Island, and Maryland; but the numbers in the Midwest and West are growing. Highest-risk groups include hikers, backpackers, and people living in newly developed communities near woodlands and forests. It should be noted that over 60,000 cases of tick-borne Lyme borreliosis are seen in Europe each year. Infections in these countries are caused by a variety of *Borrelia* species, each causing a distinct set of symptoms in patients.

▶ Culture and Diagnosis

Diagnosis of Lyme disease can be difficult because of the range of symptoms it presents. Most suggestive are the ring-shaped lesions, isolation of spirochetes from the patient, and

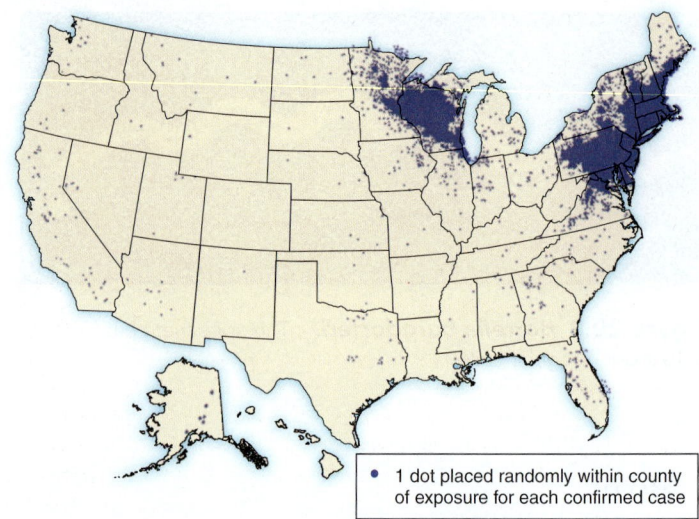

> • 1 dot placed randomly within county of exposure for each confirmed case

Figure 20.10   **Reported cases of Lyme disease—United States, 2011.**

serological testing with an ELISA method that tracks a rising antibody titer. Tests for spirochete DNA in specimens is especially helpful for late-stage diagnosis.

INSIGHT 20.2   Acorns, Red Foxes, and Climate Change

What do acorns, red foxes, and climate change have in common? Lyme disease. Lyme disease is a vector-borne disease caused by the spirochete *Borrelia burgdorferi*. In its complicated life cycle described in process figure 20.9, larval *Ixodes* (black-legged) ticks take a blood meal from a white-footed mouse infected with *B. burgdorferi,* develop into a nymph tick, take another blood meal from other animals including white-tailed deer and humans, and then develop into an adult tick where they take their third blood meal.

Scientists have developed methods for tracking and predicting surges in Lyme disease based on various models. Dr. Richard Ostfeld, a disease ecologist at the Cary Institute of Ecosystem Studies in Milbrook, New York, has studied the correlation between acorns, white-footed mice, and black-legged ticks. He and his research team have found that 2 years after a boom in acorn crops there is a boom in Lyme disease. A heavy acorn crop provides more food for the white-footed mouse, whose population spikes, which in turn increases the larval tick population, leading to a spike in cases of Lyme disease in humans the following year.

Researchers at the University of California at Santa Cruz have also shown that declining populations of the red fox can also lead to an increase in Lyme infections. The red fox is the main predator of small mammals, such as the white-footed

The white-footed mouse, known to carry *Borrelia burgdorferi.*

mouse, the Eastern chipmunk, and short-tailed shrews, all of which transmit Lyme disease. The UCSC group has found that the red fox population is in decline and coyotes are becoming the top predator in the eastern states, preying on the red fox. As the red fox population declines, small mammal populations increase, as does the incidence of Lyme disease.

Climate change has also had a major impact on the prevalence of Lyme disease. As temperatures have increased over the last decade, tick populations have migrated north to Canada, where before 1990, Lyme-transmitting ticks were unknown. According to research conducted at the University of Montreal, Lyme-transmitting ticks have migrated to areas where 18% of the Canadian population now lives and are predicted to impact 80% of the Canadian population by 2020.

Lyme disease isn't the only tick-borne disease that is on the rise due to climate change. All over the world, tick-borne diseases such as human granulocytic anaplasmosis, tick-borne encephalitis, Rocky Mountain spotted fever, and Crimean-Congo hemorrhagic fever are increasing. As the climate warms, more animal carriers of tick-borne disease become available to ticks and transmission of disease increases.

Source: 2012.www.sciencedaily.com/releases/2012/06/120618153714.htm

▶ Prevention and Treatment

A vaccine for Lyme disease was available for a brief period of time, but it was withdrawn from the market in early 2002. Other vaccines are in development. Because dogs can also acquire the disease, there is a vaccine for them. Anyone involved in outdoor activities should wear protective clothing, boots, leggings, and insect repellant containing DEET.[1] Individuals exposed to heavy infestation should routinely inspect their bodies for ticks and remove ticks gently without crushing, preferably with forceps or fingers protected with gloves, because it is possible to become infected by tick feces or body fluids.

Early, prolonged (3 to 4 weeks) treatment with doxycycline and amoxicillin is effective, and other antibiotics such as ceftriaxone and penicillin are used in late Lyme disease therapy. Roughly 10% to 20% of treated patients, however, go on to develop posttreatment Lyme disease syndrome or chronic Lyme disease **(Disease Table 20.5)**.

Disease Table 20.5	Lyme Disease
Causative Organism(s)	*Borrelia burgdorferi*
Most Common Modes of Transmission	Vector, biological
Virulence Factors	Antigenic shifting, adhesins
Culture/Diagnosis	ELISA for Ab, PCR
Prevention	Tick avoidance
Treatment	Doxycycline and/or amoxicillin (3–4 weeks), also cephalosporins and penicillin
Epidemiological Features	Endemic in North America, Europe, and Asia

Infectious Mononucleosis

This lymphatic system disease, which is often simply called "mono" or the "kissing disease," can be caused by a number

1. *N,N*-Diethyl-*m*-toluamide—the active ingredient in OFF! and Cutter brand insect repellants.

of bacteria or viruses, but the vast majority of cases are caused by the **Epstein-Barr virus (EBV),** a member of the family *Herpesviridae.*

▶ Signs and Symptoms

The symptoms of mononucleosis are sore throat, high fever, and cervical lymphadenopathy, which develop after a long incubation period (30 to 50 days). Many patients also have a gray-white exudate in the throat, a skin rash, and enlarged spleen and liver. A notable sign of mononucleosis is sudden leukocytosis, consisting initially of infected B cells and later T cells. Fatigue is a hallmark of the disease. Patients remain fatigued for a period of weeks. During that time, they are advised to not engage in strenuous activity due to the possibility of injuring their enlarged spleen (or liver).

Eventually, the strong, cell-mediated immune response is decisive in controlling the infection and preventing complications. But after recovery, people usually remain chronically infected with EBV.

Epstein-Barr Virus

The Epstein-Barr virus shares morphological and antigenic features with other herpesviruses; in addition, it contains a circular form of DNA that is readily spliced into the host cell DNA.

▶ Pathogenesis and Virulence Factors

The latency of the virus and its ability to splice its DNA into host cell DNA make it an extremely versatile virus that can avoid the host's immune response.

▶ Transmission and Epidemiology

More than 90% of the world's population has been infected with EBV. In general, the virus causes no noticeable symptoms, but the time of life when the virus is first encountered seems to matter. In the case of EBV, infection during the teen years seems to result in disease, whereas infection before or after this period is usually asymptomatic.

Direct oral contact and contamination with saliva are the principal modes of transmission, although transfer through blood transfusions, sexual contact, and organ transplants is possible. True outbreaks of this disease rarely occur.

▶ Culture and Diagnosis

A differential blood count that shows excess lymphocytes, reduced neutrophils, and large, atypical lymphocytes with lobulated nuclei and vacuolated cytoplasm is suggestive of

Figure 20.11 Histology of lymphocytes infected with Epstein-Barr virus (1,000×). **(a)** Blood smear of a normal lymphocyte, with round nucleus. **(b)** Blood smear of patient with infectious mononucleosis. Note a large atypical lymphocyte with irregular nucleus and indented border (arrows).

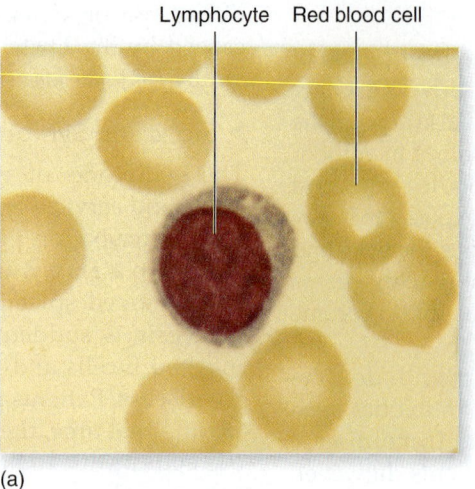
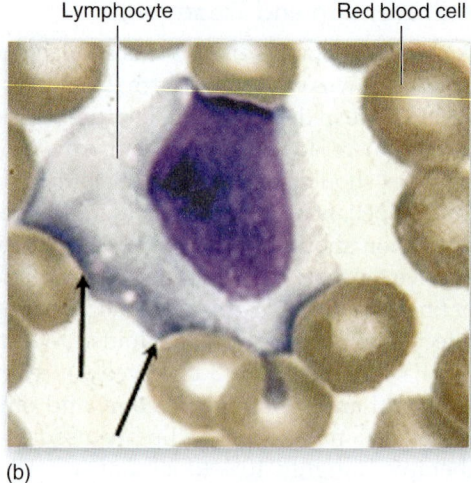

(a) (b)

EBV infection **(figure 20.11)**. A test called the "Monospot test" detects *heterophile antibodies*—which are antibodies that are not directed against EBV but are seen when a person has an EBV infection. This test is not reliable in children younger than age 4, in which case a specific EBV antigen/antibody test is conducted.

▶ **Prevention and Treatment**

The usual treatments for infectious mononucleosis are directed at symptomatic relief of fever and sore throat. Hospitalization is rarely needed. Occasionally, rupture of the spleen necessitates immediate surgery to remove it **(Disease Table 20.6)**.

Disease Table 20.6	Infectious Mononucleosis
Causative Organism(s)	Epstein-Barr virus (EBV)
Most Common Modes of Transmission	Direct, indirect contact; parenteral
Virulence Factors	Latency, ability to incorporate into host DNA
Culture/Diagnosis	Differential blood count, Monospot test for heterophile antibody, specific ELISA
Prevention	–
Treatment	Supportive
Distinctive Features	Most common in teens
Epidemiological Features	United States: 500 cases per 100,000 per year

Anthrax

Anthrax is discussed in other chapters as well as this one (see a discussion of cutaneous anthrax on page 535 in chapter 18). We discuss anthrax in this chapter because it multiplies in large numbers in the blood and because septicemic anthrax is a possible outcome of all forms of anthrax.

For centuries, anthrax has been known as a zoonotic disease of herbivorous livestock (sheep, cattle, and goats). It has an important place in the history of medical microbiology because Robert Koch used anthrax as a model for developing his postulates in 1877 and, later, Louis Pasteur used the disease to prove the usefulness of vaccination.

▶ **Signs and Symptoms**

As anthrax infection can exhibit its primary symptoms in various locations of the body: on the skin (cutaneous anthrax), in the lungs (pulmonary anthrax), in the gastrointestinal tract (acquired through ingestion of contaminated foods), and in the central nervous system (anthrax meningitis). The cutaneous and pulmonary forms of the disease are the most common. In all of these forms, the anthrax bacterium gains access to the bloodstream, and death, if it occurs, is usually a result of an overwhelming septicemia. Pulmonary anthrax—and the accompanying pulmonary edema and hemorrhagic lung symptoms—can sometimes be the primary cause of death, although it is difficult to separate the effects of septicemia from the effects of pulmonary infection.

In addition to symptoms specific to the site of infection, septicemic anthrax results in headache, fever, and malaise. Bleeding in the intestine and from mucous membranes and orifices may occur in late stages of septicemia.

▶ **Causative Agent**

Bacillus anthracis is a gram-positive endospore-forming rod that is among the largest of all bacterial pathogens. It is composed of block-shaped, angular rods 3 to 5 micrometers long and 1 to 1.2 micrometers wide. Central endospores develop under all growth conditions except in the living body of the

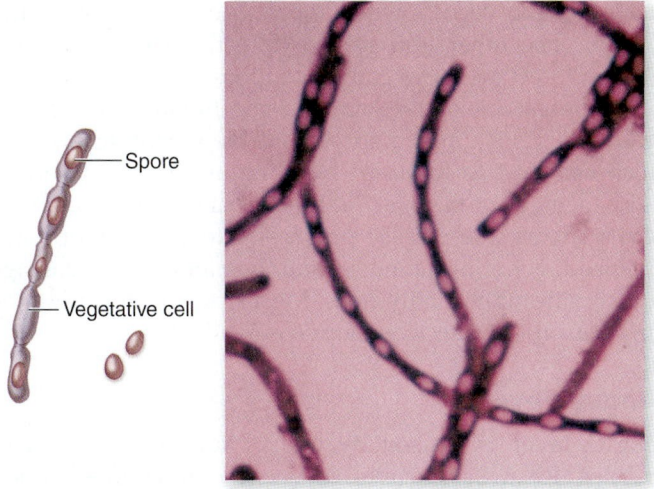

Figure 20.12 *Bacillus anthracis.* Note the centrally placed endospores and streptobacillus arrangement (600×).

host **(figure 20.12).** The genus *Bacillus* is aerobic and catalase-positive, and none of the species are fastidious. *Bacillus* as a genus is noted for its versatility in degrading complex macromolecules, and it is also a common source of antibiotics. Because the primary habitat of many species, including *B. anthracis,* is the soil, endospores are continuously dispersed by means of dust into water and onto the bodies of plants and animals.

▶ **Pathogenesis and Virulence Factors**

The main virulence factor of *B. anthracis* is what is referred to as a "tripartite" toxin—an exotoxin complex composed of three separate proteins. One of the proteins is called *edema factor,* which increases cellular cyclic AMP levels leading to disruption of water balance and ultimately edema. Another part of the toxin is *protective antigen,* so named because it is a good target for vaccination, not because it protects the bacterium or the host directly. It helps the edema factor get to its target site. The third exotoxin is called *lethal factor.* It combines with edema factor to form lethal toxin, which appears to target alveolar epithelium triggering massive inflammation and initiation of shock—especially in cases of pulmonary anthrax.

The *B. anthracis* exotoxin complex is of the "A-B toxin" type. The bacteria that cause cholera, shigellosis, pertussis, and diphtheria all use A-B exotoxins. Most A-B toxins have two components: a "B" component that binds to host cells and an "A," or active, component that enters the cell and exerts the toxic effect. *B. anthracis* is a bit different; its protective antigen is the B component, and both the lethal factor and edema factor are A components.

▶ **Transmission and Epidemiology**

The anthrax bacillus is a facultative parasite that undergoes its cycle of vegetative growth and sporulation in the soil. Animals become infected while grazing on grass contaminated with endospores. When the pathogen is returned to the soil in animal excrement or carcasses, it can sporulate and become

a long-term reservoir of infection for the animal population. The majority of natural anthrax cases are reported in livestock from Africa, Asia, and the Middle East. Most recent (natural) cases in the United States have occurred in textile workers handling imported animal hair, hide, or products made from them. Because of effective control procedures, the number of cases in the United States is extremely low (fewer than 10 per year).

After the terrorist attacks of 2001, anthrax dominated the public consciousness as never before. The anthrax attack aimed at two senators and several media outlets focused a great deal of attention on the threat of bioterrorism. During that attack, 22 people acquired anthrax and 5 people died. The exact infectious dose for each form of anthrax in humans is not known, though studies using data from the 2001 attacks suggest that pulmonary anthrax may develop from exposure to as few as 2 to 9 endospores.

▶ **Culture and Diagnosis**

Diagnosis requires a high index of suspicion. This means that anthrax must be present as a possibility in the clinician's mind or it is likely not to be diagnosed, because it is such a rare disease in the developed world and because, in all of its manifestations, it can mimic other infections that are not so rare. (A very astute public health clinician in Florida first suspected anthrax in the attacks of 2001 and called for the proper tests.) First-level (presumptive) diagnosis begins with culturing the bacterium on blood agar and performing a Gram stain. Further tests can be performed to provide evidence regarding presence of *B. anthracis* as opposed to other *Bacillus* species. These tests include motility (*B. anthracis* is nonmotile) and a lack of hemolysis on blood agar. Ultimately, samples should be handled by the Centers for Disease Control and Prevention, which will perform confirmatory tests, usually involving direct fluorescent antibody testing and phage lysis tests.

▶ **Prevention and Treatment**

A vaccine containing live endospores and a toxoid prepared from a special strain of *B. anthracis* are used to protect livestock in areas of the world where anthrax is endemic. Humans should be vaccinated with the purified toxoid—AVA (BioThrax), currently the only licensed human vaccine—if they have occupational contact with livestock or products such as hides and bone or if they are members of the military. Effective vaccination requires six inoculations given over 1.5 years, with yearly boosters. The cumbersome nature of vaccination has spurred research into more manageable vaccines but has also led to changes in protocols for safety evaluation due to the ethical nature of testing such a vaccine.

Carcasses of animals that have died from anthrax must be burned or chemically decontaminated before burial to prevent establishing the microbe in the soil. Imported items containing animal hides, hair, and bone should be gas sterilized.

The recommended treatment for anthrax has traditionally been penicillin, doxycycline, or ciprofloxacin. As of 2012, there is still debate about the best way to treat anthrax exposure, as antibiotic usage can potentially worsen the symptoms by releasing large amounts of toxin into the blood-stream. Alternatives being debated include administering passive antibody to the toxins and also administering the vaccine postexposure. At any rate, treatment of human cases of the disease will be conducted in consultation with the CDC **(Disease Table 20.7).**

Disease Table 20.7 Anthrax	
Causative Organism(s)	*Bacillus anthracis*
Most Common Modes of Transmission	Vehicle (air, soil), indirect contact (animal hides), vehicle (food)
Virulence Factors	Triple exotoxin
Culture/Diagnosis	Culture, direct fluorescent antibody tests
Prevention	Vaccine for high-risk population
Treatment	In consultation with the CDC
Epidemiological Features	Internationally, 2,000–20,000 cases annually, most cutaneous

Hemorrhagic Fever Diseases

A number of agents that infect the blood and lymphatics cause extreme fevers, some of which are accompanied by internal hemorrhaging. The hemorrhagic fever diseases described here are caused by viruses in one of three families: *Arenaviridae, Filoviridae,* and *Flaviviridae.* All of these viruses are RNA enveloped viruses and are classified as biosafety level 4 pathogens. Most of these viruses are zoonotic and their geographic pattern of distribution is determined by the presence of their natural hosts. *Bunyaviridae* is a fourth family with members that cause hemorrhagic fevers, such as Rift Valley fever, which is endemic to Africa. Although we do not discuss examples of such diseases here, it is important to note that the prevalence of many of these diseases fluctuates today due to global warming patterns.

Yellow Fever

This hemorrhagic fever disease is caused by an arbovirus, a single-stranded RNA flavivirus that is generally called the yellow fever virus. It currently occurs only in parts of Africa and South America. Two patterns of transmission are seen in nature. One is an urban cycle between humans and the mosquito *Aedes aegypti,* which reproduces in standing water in cities. The other is a sylvan (forest) cycle, maintained between forest monkeys and mosquitoes.

The presence of the virus in the bloodstream causes capillary fragility and disrupts the blood-clotting system, which can lead to localized bleeding and shock. Infection begins acutely with fever, headache, and muscle pain. In some patients, the disease progresses to oral hemorrhage, nosebleed, vomiting, jaundice, and liver and kidney damage with significant mortality rates. Most cases occur during the rainy season. A protective vaccine is available.

Dengue Fever

Dengue fever is caused by one of four related single-stranded RNA flaviviruses carried by *Aedes* mosquitoes. Dengue fever is also called "breakbone fever" because of the severe pain it induces in muscles and joints (it does not actually cause fractures). The illness is endemic to Southeast Asia and India, and several epidemics have occurred in South America and Central America, the Caribbean, and Mexico. Although dengue fever typically presents as a mild infection, a new form of disease called dengue hemorrhagic fever (DHF) has emerged that causes high rates of morbidity and mortality in endemic areas. Dengue shock syndrome (DSS) can develop in DHF patients exhibiting life-threatening hypotension. Both forms of disease represent a major public health concern, as global warming has increased the geographic distribution of both the viruses and their vector to put nearly 3 billion people at risk for transmission of infection. DengueNet has been established by the WHO to enhance global surveillance of disease.

A low-tech approach has led to big successes in Vietnam. There, health officials urged local citizens to round up tiny crustaceans that are common in natural water sources and to put them in water tanks and wells. The crustaceans, which are not harmful to humans, eat the mosquitoes that carry dengue. Officials reported a complete elimination of the disease in communities where the strategy was used.

Chikungunya

The Chikungunya virus was discovered in 1955 and has caused sporadic outbreaks of disease since then. The name comes from an African phrase meaning, "that which bends up," a reference to the arthritic stance people infected with this virus often assume. It is an alphavirus that is transmitted by *Aedes albopictus* mosquitoes. Symptoms are similar to dengue fever with the additional complication of severe joint pain, sometimes lasting for years, and occasional neurological impairment. There is growing concern about this virus, since mosquitos carrying it showed up in Western Europe (in 2007) and in New York City soon after. Research in 2012 showed that there was one of these mosquitoes per five people in New York City. When the ratio changes to five mosquitoes per person, scientists predict that Chikungunya could become endemic in the city (and elsewhere in the United States).

Ebola and Marburg

Unlike the viruses causing yellow fever and dengue fever, the Ebola and Marburg viruses are filoviruses (family *Filoviridae*). The two viruses are related and cause similar symptoms, although Ebola has received the greatest share of media attention. Its gruesome symptoms are extreme manifestations of the same kind of hemorrhagic events described for yellow fever and dengue fever. The virus in the bloodstream leads to extensive capillary fragility and disruption of clotting. Patients bleed from their orifices, even from their mucous membranes, and experience massive internal and external hemorrhage. Very often, they manifest a rash on their trunk in early stages of the disease. The mortality rate for Marburg infection is 25%, while it is a staggering 97% in cases of Ebola infection. There is currently no effective treatment for either disease, though new research focuses on the development of monoclonal antibodies for future therapy.

It is thought that bats are the natural reservoir of these viruses. Direct contact with an infected person or with the person's body fluids will also transmit the virus. Hospital workers caring for Ebola patients are at high risk of becoming infected. The highly infectious nature of this virus was illustrated in 2012 when a man developed Ebola after stealing a cell phone from an infected patient.

Outbreaks with Marburg virus are also rare, but individuals have been infected sporadically since it was first recognized in 1967. In 2005, the largest Marburg outbreak in history occurred in and around a hospital in Angola. Sixty-three people died during the 5-month outbreak. Symptoms are similar to Ebola virus infection. In 2012, Uganda had outbreaks of both Ebola and Marburg.

Lassa Fever

The Lassa fever virus is an arenavirus. Several related arenaviruses cause the diseases Argentine hemorrhagic fever, Bolivian hemorrhagic fever, and lymphocytic choriomeningitis (an infection of the brain and meninges). Lassa fever virus is found primarily in West Africa, but imported cases of disease have been identified in the United Kingdom. This means that although they became ill while in the United Kingdom, the patients acquired their actual infection while in Africa. In most cases, infection with this virus is asymptomatic, but in 20% of the cases a severe hemorrhagic syndrome develops. The syndrome includes chest pain, hemorrhaging, sore throat, back pain, vomiting, diarrhea, and sometimes encephalitis. Patients who recover suffer from deafness at a significant rate. An outbreak in Nigeria in 2012 resulted in over 900 cases of infection and over 100 deaths.

Case File 20 *Continuing the Case*

Ebola is an RNA virus in the *Filoviridae* family and is most easily recognized in electron micrographs by its long, thin, spaghetti-like shape. The virus infects the endothelial cells lining blood vessels and interferes with coagulation of the blood. The combination of damaged blood vessels and lack of coagulation leads to massive internal bleeding and shock, usually resulting in death. What makes Ebola so insidious is that its early symptoms mimic so many other tropical diseases: Fatigue, headache, muscle pain, fever, vomiting, and diarrhea are often mistaken for malaria, typhoid fever, cholera, relapsing fever, and other diseases commonly seen in this area of the world.

The reservoir of the virus is still unknown, but massive efforts have been undertaken to collect samples from live animals and corpses of infected animals during an Ebola outbreak to detect antibodies against the virus. Fruit bats are generally considered to be the host of the virus, but to date, only 3% of all fruit bat species have been sampled, making it difficult to determine the exact reservoir.

The virus is mainly transmitted through direct contact with bodily fluids or contaminated fomites such as bed linens or clothing from infected individuals. There are three main pathways for person-to-person transmission of the virus during an outbreak: between family members and caregivers of sick individuals in a home setting, through contact with dead bodies during funeral preparations, and between infected patients and the medical staff in health care settings. Of utmost importance is prevention of transmission of Ebola in the hospital setting. There have been numerous cases of health care workers contracting and dying from the virus. During an outbreak in the Democratic Republic of the Congo in 1994, the only physician at the Watsa district hospital in D.R.C. died after treating a patient. This meant that there was no physician available at that hospital from 1994 to 1996. In 2000, the medical director and 11 staff members died in Gulu, Uganda, during an outbreak.

■ What are some of the difficulties in controlling and treating Ebola outbreaks?

■ In terms of "viral success," or the ability of the virus to transmit itself, Ebola is not a "successful" virus because it kills its victims so quickly. Why then is Ebola such a great concern globally?

Source: 2012. www.cdc.gov/ncidod/dvrd/spb/mnpages/dispages/ebola/ebolatable.htm

The reservoir of the virus is a rodent found in sub-Saharan Africa called the multimammate rat. The virus is spread to humans through aerosolization of rat droppings, urine, hair, and so forth. Eating food contaminated by rat excretions also transmits the virus. Infected persons can spread it to other people through their own secretions.

Vertical transmission also occurs, and the disease leads to spontaneous abortions in 95% of infected pregnant women.

This hemorrhagic fever has been shown to respond to the antiviral agent ribavirin, especially if administered in the early stages of infection. There is no vaccine (**Disease Table 20.8**).

Nonhemorrhagic Fever Diseases

In this section, we examine some infectious diseases that result in a syndrome characterized by high fever but without the capillary fragility that leads to hemorrhagic symptoms. All of the diseases in this section are caused by bacteria.

Brucellosis

This disease goes by several different names (besides brucellosis): Malta fever, undulant fever, and Bang's disease.[2]

▶ Signs and Symptoms

The *Brucella* species responsible for this disease live in phagocytic cells. These cells carry the bacteria into the bloodstream, creating focal lesions in the liver, spleen, bone marrow, and kidney. The cardinal manifestation of human brucellosis is a fluctuating pattern of fever, which is the origin of the common name *undulant fever* (**figure 20.13**). It is also

2. After B. L. Bang, a Danish physician.

Disease Table 20.8 Hemorrhagic Fevers					
Disease	**Yellow Fever**	**Dengue Fever**	**Chikungunya**	**Ebola and/or Marburg**	**Lassa Fever**
Causative Organism(s)	Yellow fever virus	Dengue fever virus	Chikungunya virus	Ebola virus, Marburg virus	Lassa fever virus
Most Common Modes of Transmission	Biological vector	Biological vector	Biological vector	Direct contact, body fluids	Droplet contact (aerosolized rodent excretions), direct contact with infected fluids
Virulence Factors	Disruption of clotting factors	Disruption of clotting factors	Disruption of clotting factors	Disruption of clotting factors	Disruption of clotting factors
Culture/Diagnosis	ELISA, PCR	Rise in IgM titers	PCR	PCR, viral culture (conducted at CDC)	ELISA
Prevention	Live attenuated vaccine available	–	–	–	Avoiding rats, safe food storage
Treatment	Supportive	Supportive	Supportive	Supportive	Ribavirin
Distinctive Features	Accompanied by jaundice	"Breakbone fever"—so named due to severe pain	Arthritic symptoms	Massive hemorrhage; rash sometimes present	Chest pain, deafness as long-term sequelae
Epidemiological Features	United States: only sporadic cases in travelers; internationally, 200,000 cases annually, 30,000 deaths; 90% of cases in Africa	United States: only sporadic cases in travelers; internationally, 50–100 million people infected every year and 22,000 deaths occur, mostly among children	United States: no reported cases; internationally, periodic epidemics	United States: infections only associated with facilities handling imported monkeys; internationally, sporadic outbreaks in Africa	United States: no reported cases; internationally, occasional outbreaks in West Africa

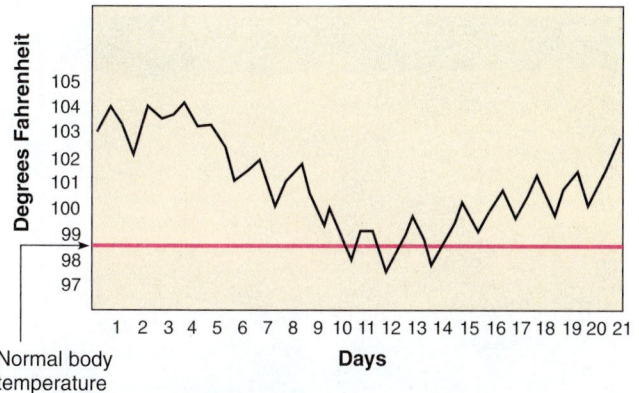

Figure 20.13 The temperature cycle in classic brucellosis. Body temperature undulates between day and night and between elevated, normal, and subnormal.

Source: Alice Lorraine Smith, *Principles of Microbiology*, 10th ed., 1985. Times Mirror/Mosby College Pub.

accompanied by chills, profuse sweating, headache, muscle pain and weakness, and weight loss. Fatalities are not common, although the syndrome can last for a few weeks to a year, even with treatment.

▶ Causative Agent

The bacterial genus *Brucella* contains tiny, aerobic gram-negative coccobacilli. At least six species are known to cause disease in humans: *B. abortus* (cattle), *B. melitensis* (goats, sheep), *B. canis* (canines), *B. suis* (pigs), and at least two species living in marine mammals. In humans, infection with *B. melitensis* is most common. Even though a principal manifestation of the disease in animals is an infection of the placenta and fetus, human placentas do not become infected. The CDC lists *Brucella* species as possible bioterror agents, though they are not designated as being "of highest concern."

▶ Pathogenesis and Virulence Factors

Brucella enters through damaged skin or via mucous membranes of the digestive tract, conjunctiva, and respiratory tract. From there, it is taken up by phagocytic cells. Because it is able to avoid destruction in the phagocytes, the bacterium is transported easily through the bloodstream and to various organs, such as the liver, kidney, breast tissue, or joints. Scientists suspect that the up-and-down nature of the fever is related to unusual properties of the bacterial lipopolysaccharide.

▶ Transmission and Epidemiology

Brucellosis is one of the most common zoonotic diseases, as more than 500,000 human cases are reported worldwide each year in areas of Europe, Africa, India, and Latin America. It is associated predominantly with occupational contact in slaughterhouses, livestock handling, and the veterinary profession. Infection takes place through contact with blood, urine, and placentas and through consumption of raw milk and cheese. Human-to-human transmission is rare, but brucellosis is considered the most common laboratory-acquired infection. In 2007, a researcher in a university lab that studied possible bioweapons agents contracted brucellosis while cleaning a chamber used to infect mice.

Brucellosis is also a common disease of wild herds of bison and elk. Cattle that share grazing land with these wild herds often suffer severe outbreaks of the placental infections (called Bang's disease). Along with the toll on human health, the worldwide economic impact from animal loss due to disease is immense.

▶ Culture and Diagnosis

The patient's history can be very helpful in diagnosis, as are serological tests of the patient's blood and blood culture of the pathogen. In areas where *Brucella* is endemic, serology is of limited use because significant proportions of the population already display antibodies to the bacterium. Blood culture is positive in less than 40% of cases and may take up to 4 weeks to perform; Gram staining of biopsy material from lymph nodes or bone marrow (from the sternum) is considered more reliable. PCR-based testing is currently in development for more rapid and accurate patient diagnosis.

▶ Prevention and Treatment

Prevention is effectively achieved by testing and elimination of infected animals, quarantine of imported animals, and pasteurization of milk. Although several types of animal vaccines are available, those developed so far for humans are ineffective or unsafe. The status of this pathogen as a potential germ warfare agent makes a reliable vaccine even more urgent. A reemergence of brucellosis has occurred recently in countries that have been free of disease for over 50 years. This underscores the need for continued vaccination of animals and enhanced surveillance for disease in not only animal herds but also the human population as well.

A combination of doxycycline and gentamicin or rifampin taken for 3 to 6 weeks is usually effective in controlling infection.

Q Fever

The name of this disease arose from the frustration created by not being able to identify its cause. The Q stands for "query." Its cause, a bacterium called *Coxiella burnetii,* was finally identified in the mid-1900s. The clinical manifestations of acute Q fever are abrupt onset of fever, chills, head and muscle ache, and, occasionally, a rash. The disease is sometimes complicated by pneumonitis (30% of cases), hepatitis, and endocarditis. About a quarter of the cases are chronic rather than acute and result in vascular damage and endocarditis-like symptoms.

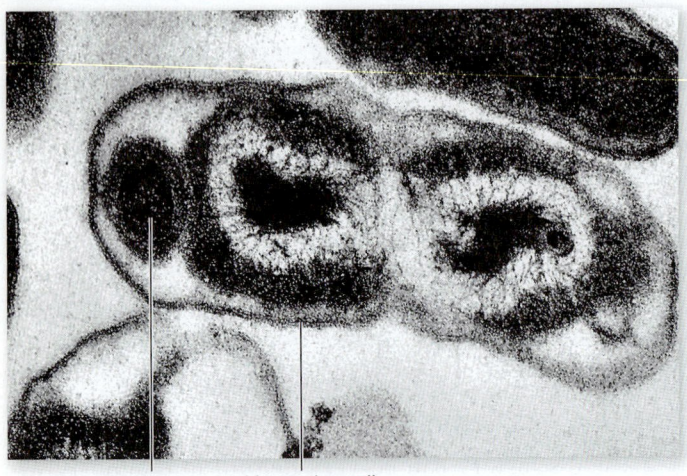

Endospore   Vegetative cell

Figure 20.14 The agent of Q fever.   The vegetative cells of *Coxiella burnetii* produce unique endospore-like structures that are released when the cell disintegrates. Free endospores survive outside the host and are important in transmission.

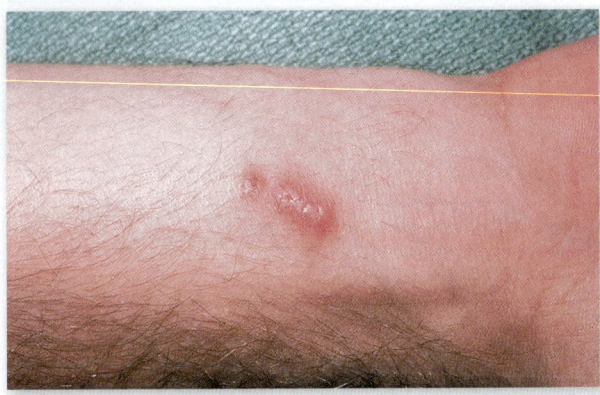

Figure 20.15 Cat-scratch disease.   A primary nodule appears at the site of the scratch in about 21 days. In time, large quantities of pus collect and the regional lymph nodes swell.

C. burnetii is a very small pleomorphic gram-negative bacterium, and for a time it was considered a rickettsia. It is an intracellular parasite, and it produces an unusual type of endospore-like structure **(figure 20.14).** *C. burnetii* is apparently harbored by a wide assortment of vertebrates and arthropods, especially ticks, which play an essential role in transmission between wild and domestic animals. Ticks do not transmit the disease to humans, however. Humans acquire infection largely by means of environmental contamination and airborne spread. Birth products such as placentas of infected domestic animals contain large numbers of bacteria. Other sources of infectious material include urine, feces, milk, and airborne particles from infected animals. The primary portals of entry are the lungs, skin, conjunctiva, and gastrointestinal tract.

C. burnetii has been isolated from most regions of the world. California and Texas have the highest numbers of cases in the United States, although most cases probably go undetected. People at highest risk are farm workers, meat cutters, veterinarians, laboratory technicians, and consumers of raw milk products. In 1984, 18 individuals working with sheep at an Idaho research station were infected with the bacterium; milk-producing goats in the same state tested positive for the disease in 2012.

Mild or subclinical cases resolve spontaneously, and more severe cases respond to doxycycline therapy. A vaccine is available in many parts of the world but not in the United States. Q fever is of potential concern as a bioterror agent because it is very resistant to heat and drying, it can be inhaled, and even a single bacterium is enough to cause disease. It is an organism that the U.S. military worked with during the period when potential biowarfare agents were being developed in this country (the 1950s and 1960s).

Cat-Scratch Disease

This disease is one of a group of diseases caused by different species of the small gram-negative rod *Bartonella. Bartonella* species are considered to be emerging zoonotic pathogens. They are fastidious but not obligate intracellular parasites, so they will grow on blood agar. In addition to cat-scratch disease and trench fever, discussed next, a new species of *Bartonella* (*B. rochalimae*) that causes high fever and life-threatening anemia was identified in 2007.

Bartonella henselae is the agent of cat-scratch disease (CSD), an infection connected with being clawed or bitten by a cat. The pathogen is present in over 40% of cats, especially kittens. There are approximately 25,000 cases per year in the United States, 80% of them in children 2 to 14 years old. The symptoms start after 1 to 2 weeks, with a cluster of small papules at the site of inoculation **(figure 20.15).** In a few weeks, the lymph nodes along the lymphatic drainage swell and can become pus-filled. Only about one-third of patients experience high fever. It is a particular problem in AIDS patients. Most infections remain localized and resolve in a few weeks, but drugs such as azithromycin, erythromycin, and rifampin can be effective therapies. The disease can be prevented by thorough antiseptic cleansing of a cat bite or scratch.

Trench Fever

This disease has a long history. Trench fever was once a common condition of soldiers in battle. The causative agent, *Bartonella quintana,* is carried by lice. The feces of the lice contain the bacterium, and transmission usually occurs when the feces enter the bite wound. Most cases occur in endemic regions of Europe, Africa, and Asia, although the disease is reemerging in poverty-stricken areas of large cities in the developed world. This version of the disease is called "urban trench fever," and recent studies documented that 33% of homeless individuals in San Francisco carried body lice harboring the bacterium. Highly variable symptoms can include a 5- to 6-day fever (the species epithet, *quintana,* refers to a 5-day fever). Symptoms also include leg pains, especially in the tibial region (the disease is sometimes called "shinbone fever"); headache; chills; and muscle aches. A macular rash can also occur. (See Insight 18.2 for definitions of skin lesions.) Endocarditis can develop, especially in the urban version of the disease. The microbe can persist in the blood long after convalescence and is responsible for later relapses.

Trench fever may be treated with doxycycline or erythromycin.

Ehrlichioses: HGA and HME

There are two similar tick-borne, fever-producing diseases caused by members of the genera *Ehrlichia* and *Anaplasma*. The causative organisms for the two diseases were thought to be contained within the genus *Ehrlichia* until 2005, when a reclassification distinguished them as members of two different genera.

Members of the two genera are small intracellular bacteria, and they share many characteristics with rickettsia, including a strict parasitic existence and association with ticks. *Ehrlichia chaffeensis* causes human monocytic ehrlichiosis (HME). *Anaplasma phagocytophilum* causes human granulocytic anaplasmosis (HGA). Another species, *Ehrlichia ewingii*, can cause either syndrome. The diseases are sometimes referred to as "spotless" Rocky Mountain spotted fever.

Both bacteria spend part of their life cycle in ticks in the genus *Ixodes*. The species of tick varies with the various regions of the United States and Europe. Serology samples taken from residents of endemic areas suggest that between 15% and 36% of them have been infected with *A. phagocytophilum*, mostly without symptoms. Both HME and HGA are showing increased incidence, probably due to improved diagnosis.

The signs and symptoms of HGA and HME are similar: an acute febrile state manifesting headache, muscle pain, and rigors. Most patients recover rapidly with no lasting effects, but around 5% of older, chronically ill patients die from disseminated infection. Rapid diagnosis is enabled by PCR tests and indirect fluorescent antibody tests. It can be critical to differentiate or detect coinfection with Lyme disease *Borrelia*, which is carried by the same tick. Doxycycline will clear up most infections within 7 to 10 days.

Rocky Mountain Spotted Fever (RMSF)

This disease is named for the region in which it was first detected in the United States—the Rocky Mountains of Montana and Idaho. In spite of its name, the disease occurs infrequently in the western United States. The majority of cases are concentrated in the Southeast and eastern seaboard regions. It also occurs in Canada and Central and South America. Infections occur most frequently in the spring and summer, when the tick vector is most active. The yearly rate of RMSF is 20 to 40 cases per 10,000 population, with fluctuations coinciding with weather and tick infestations.

RMSF is caused by a bacterium called *Rickettsia rickettsii*, which is transmitted by hard ticks such as the wood tick (*Dermacentor andersoni*), the American dog tick (*D. variabilis*, among others), and the Lone Star tick (*Amblyomma americanum*). The dog tick is probably most responsible for transmission to humans because it is the major vector in the southeastern United States.

After 2 to 4 days of incubation, the first symptoms are sustained fever, chills, headache, and muscular pain. A distinctive spotted rash usually comes on within 2 to 4 days after the prodrome **(figure 20.16)** and develops first on the wrists, forearms, and ankles before spreading to other areas of the body. Early lesions are slightly mottled like measles, but later ones are macular, maculopapular, and even petechial. In the most severe untreated cases, the enlarged lesions merge and can become necrotic, predisposing to gangrene of the toes or fingertips.

Although the spots are the most obvious symptom of the disease, the most grave manifestations are cardiovascular disruption, including hypotension, thrombosis, and hemorrhage. Conditions of restlessness, delirium, convulsions, tremor, and coma are signs of the often overwhelming effects on the central nervous system. Fatalities occur in an average of 20% of untreated cases and 5% to 10% of treated cases.

Suspected cases of RMSF require immediate treatment even before laboratory confirmation. A recent aid to early diagnosis is a method for staining rickettsias directly in a tissue biopsy using fluorescent antibodies. Isolating rickettsias from the patient's blood or tissues is desirable, but it is expensive and requires specially qualified lab personnel and lab facilities. Specimens taken from the rash lesions are suitable for PCR assay, which is very specific and sensitive and can circumvent the need for culture.

Figure 20.16 The rash in RMSF. This case occurred in a child several days after the onset of fever. Also pictured is an example of the hard ticks that transmit the infection.

The drug of choice for suspected and known cases is doxycycline administered for 1 week. Preventive measures parallel those for Lyme disease: wearing protective clothing, using insect sprays, and fastidiously removing ticks **(Disease Table 20.9)**.

Chagas Disease

This disease was described in chapter 5, as an illustration of conditions caused by flagellated protozoa. Chagas disease is

Disease Table 20.9 Nonhemorrhagic Fever Diseases

Disease	Brucellosis	Q Fever	Cat-Scratch Disease	Trench Fever	Ehrlichioses	Rocky Mountain Spotted Fever
Causative Organism(s)	*Brucella abortus* or *B. suis*	*Coxiella burnetii*	*Bartonella henselae*	*Bartonella quintana*	*Ehrlichia* and *Anaplasma* species	*Rickettsia rickettsii*
Most Common Modes of Transmission	Direct contact, airborne, parenteral (needlesticks)	Airborne, direct contact, food-borne	Parenteral (cat scratch or bite)	Biological vector (lice)	Biological vector (tick)	Biological vector (tick)
Virulence Factors	Intracellular growth; avoidance of destruction by phagocytes	Endospore-like structure	Endotoxin	Endotoxin	–	Induces apoptosis in cells lining blood vessels
Culture/ Diagnosis	Gram stain of biopsy material; PCR	Serological tests for antibody; PCR	Biopsy of lymph nodes plus Gram staining; ELISA (performed by CDC)	ELISA (performed by CDC)	PCR, indirect antibody test	Fluorescent antibody, PCR
Prevention	Animal control, pasteurization of milk	Vaccine for high-risk population	Clean wound sites	Avoid lice	Avoid ticks	Avoid ticks
Treatment	Doxycycline plus gentamicin or rifampin	Doxycycline	Azithromycin or doxycycline	Azithromycin or doxycycline	Doxycycline	Doxycycline
Distinctive Features	Undulating fever, muscle aches	Airborne route of transmission, variable disease presentation	History of cat bite or scratch; fever not always present	Endocarditis common, 5-day fever	Seasonal occurrence (April–Oct.)	Most common in east and southeast United States
Epidemiological Features	United States: fewer than 100 cases per year; internationally, 500,000 cases per year	United States: number of cases increased from 21 in 1999 to 169 in 2006	United States: estimated incidence is 9.3 cases per 100,000; internationally, seroprevalence from 0.6%–37% depending on cat population	Most infections asymptomatic; found on every continent except Antarctica	Slightly higher incidence rate in males than females	Only in Americas

sometimes called "the American trypanosomiasis." The causative agent is *Trypanosoma cruzi*. A different trypanosome, *T. brucei,* causes sleeping sickness on the African continent.

▶ Signs and Symptoms

Once the trypanosomes are transmitted by a group of insects called the triatomines **(figure 20.17)**, they multiply in muscle and blood cells. From time to time, the blood cells rupture and large numbers of trypanosomes are released into the bloodstream. The disease manifestations are divided into acute and chronic phases. Soon after infection, the acute phase begins; symptoms are relatively nondescript and range from mild to severe fever, nausea, and fatigue. A swelling called a "chagoma" at the site of the bug bite may be present. If the bug bite was close to the eyes, a distinct condition called Romana's sign, swelling of the eyelids, may appear. The acute phase lasts for weeks or months after which the condition becomes chronic and virtually asymptomatic for a period of years or indefinitely. Eventually the trypanosomes are found in numerous sites around the body, which in later years may lead to inflammation and disruption of function in organs such as the heart, the brain, and the intestinal tract.

▶ Causative Agent

T. cruzi is a flagellated protozoan.

▶ Pathogenesis and Virulence Factors

T. cruzi is equipped with special antioxidant enzymes that act to neutralize the lysosomal attack of cells they infect. This allows it to live inside host cells without being killed by them. In addition, it can cloak itself in host proteins, disguising itself from the immune system. It can also induce autoimmunity, so that the same immune cells trained to recognize it begin to react with host tissues, causing the symptoms of late-stage Chagas disease.

▶ Transmission and Epidemiology

This disease is endemic in Central and South America but not in North America, even though the insects that transmit it are found here. It was recently called "the new AIDS of the Americas" as it has a long incubation time and is difficult to cure. Estimates put the prevalence of this disease at 8 million people, 300,000 of whom live in the United States. Most U.S. cases were acquired in an endemic area by travelers or people who have since immigrated to this country.

Triatomine bugs are often called "kissing bugs" because of their tendency to bite humans on the face. The bugs become infected by biting an animal or human that has *T. cruzi* in its blood. A wide range of animals can be infected, including raccoons, armadillos, and rodents. Domesticated animals like dogs and guinea pigs can also be infected. The disease multiplies in the intestinal tract of the bugs. After the insect finishes a blood meal, it defecates. Its feces contain the trypanosome, which can gain access to the bloodstream via the bite puncture. This process is often facilitated by the subject's scratching of the bite site.

The trypanosome can also be transmitted vertically, since it crosses the placenta, and via blood transfusion with infected blood. Recently, the United States began screening all donated blood for this disease.

▶ Culture and Diagnosis

During the acute phase of the disease, there are large numbers of trypanosomes in the blood so a peripheral blood smear will detect the organism **(figure 20.18)**. In later stages of the disease, it can be diagnosed with serological methods.

▶ Prevention

There is no vaccine for Chagas disease. In endemic areas, pesticides and improved building materials in houses are used to minimize the presence of the bug.

Figure 20.17 **A representative triatomine bug, the carrier of *T. cruzi*.**

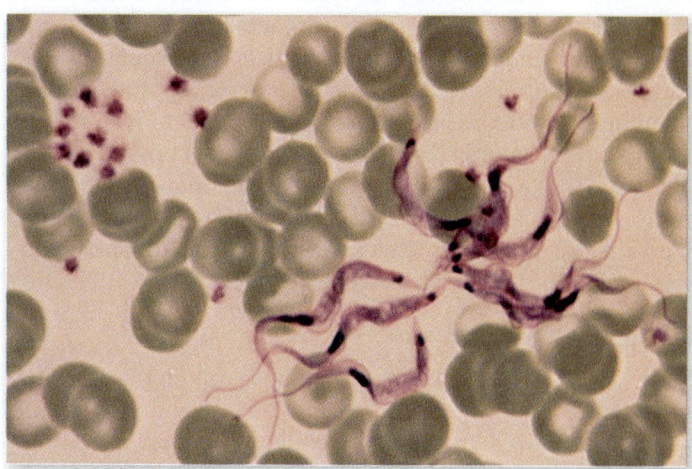

Figure 20.18 **_T. cruzi_ seen in a blood smear.** The circular objects are red blood cells.

▶ Treatment

Treatment is most successful if begun during the acute phase. However, it is often not accomplished because the acute phase of the disease is not necessarily suggestive of Chagas. Drugs for treatment are only available through the CDC. During the chronic phase of the disease, symptomatic treatment of cardiac and other problems may also be indicated.

Disease Table 20.10	Chagas Disease
Causative Organism	*Trypanosoma cruzi*
Most Common Modes of Transmission	Biological vector, vertical
Virulence Factors	Antioxidant enzymes, co-opting host antigens; induces autoimmunity
Culture/Diagnosis	Blood smear in acute phase; serological methods in later stages
Prevention	Insect control
Treatment	Consult CDC
Epidemiological Features	Endemic in Central and South America; 300,000 cases present in United States

Malaria

Throughout history, including prehistoric times, malaria has been one of humankind's greatest afflictions, in the same rank as bubonic plague, influenza, and tuberculosis. Even now, as the dominant protozoan disease, it threatens 40% of the world's population every year. The origin of the name is from the Italian words *mal*, "bad," and *aria*, "air." The superstitions of the Middle Ages alleged that evil spirits or mists and vapors arising from swamps caused malaria, because many victims came down with the disease after this sort of exposure. We now know that a swamp was mainly involved as a habitat for the mosquito vector.

▶ Signs and Symptoms

After a 10- to 16-day incubation period, the first symptoms are malaise, fatigue, vague aches, and nausea with or without diarrhea, followed by bouts of chills, fever, and sweating. These symptoms occur at 48- or 72-hour intervals, as a result of the synchronous rupturing of red blood cells. The interval, length, and regularity of symptoms reflect the type of malaria

(described next). Patients with falciparum malaria, the most virulent type, often manifest persistent fever, cough, and weakness for weeks without relief. Complications of malaria are hemolytic anemia from lysed blood cells and organ enlargement and rupture due to cellular debris that accumulates in the spleen, liver, and kidneys. One of the most serious complications of falciparum malaria is termed *cerebral malaria*. In this condition, small blood vessels in the brain become obstructed due to the increased ability of red blood cells (RBCs) to adhere to vessel walls (a condition called *cytoadherence* induced by the infecting protozoan). The resulting decrease in oxygen in brain tissue can result in coma and death. In general, malaria has the highest death rate in the acute phase, especially in children. Certain kinds of malaria (those caused by *Plasmodium vivax* and *P. ovale*) are subject to relapses because some infected liver cells harbor dormant protozoans for up to 5 years.

▶ Causative Agent

Plasmodium species are protozoans in the sporozoan group. They are **apicomplexans,** which live in animal hosts and lack locomotor organelles in the mature state (chapter 5 describes protozoan classification). Apicomplexans alternate between sexual and asexual phases, often in different animal hosts. The genus *Plasmodium* contains five species causing disease in humans: *P. malariae, P. vivax, P. falciparum, P. ovale,* and *P. knowlesi.* Humans are the primary vertebrate hosts for most of the species. The five species show variations in the pattern and severity of disease.

Development of the malarial parasite is divided into two distinct phases: the asexual phase, carried out in the human **(figure 20.19),** and the sexual phase, carried out in the mosquito. **Process figure 20.20** depicts the entire infection cycle.

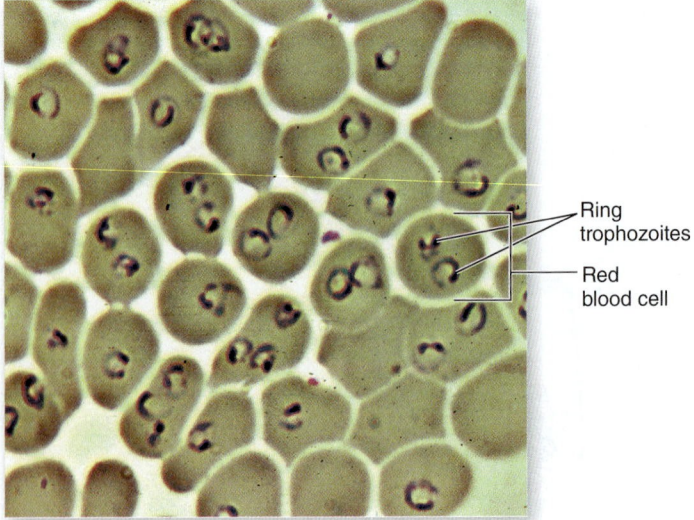

Figure 20.19 **The ring trophozoite stage in a *Plasmodium falciparum* infection.** A smear of peripheral blood shows ring forms in red blood cells. Some RBCs have multiple trophozoites.

Ring trophozoites

Red blood cell

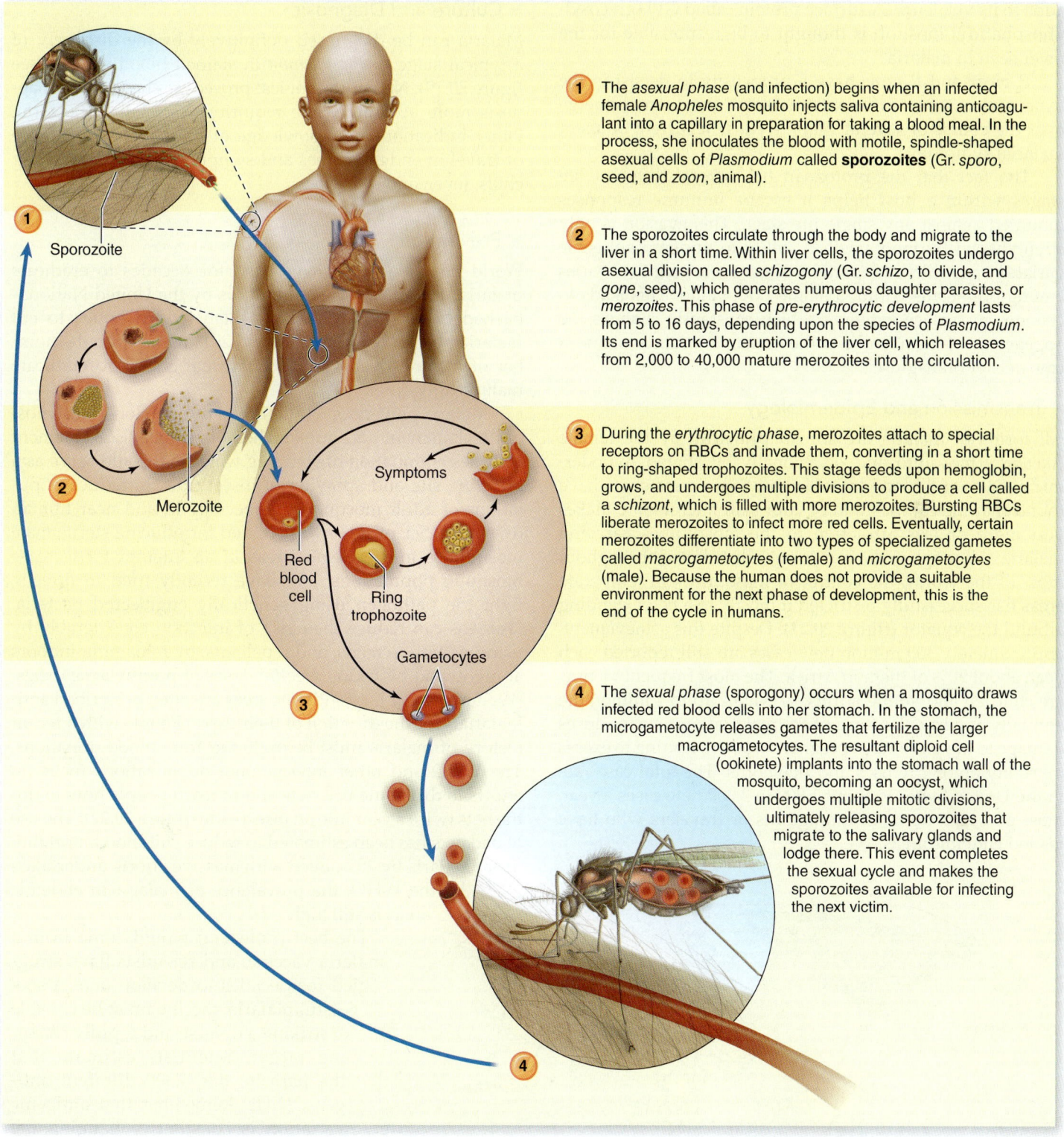

1. The *asexual phase* (and infection) begins when an infected female *Anopheles* mosquito injects saliva containing anticoagulant into a capillary in preparation for taking a blood meal. In the process, she inoculates the blood with motile, spindle-shaped asexual cells of *Plasmodium* called **sporozoites** (Gr. *sporo*, seed, and *zoon*, animal).

2. The sporozoites circulate through the body and migrate to the liver in a short time. Within liver cells, the sporozoites undergo asexual division called *schizogony* (Gr. *schizo*, to divide, and *gone*, seed), which generates numerous daughter parasites, or *merozoites*. This phase of *pre-erythrocytic development* lasts from 5 to 16 days, depending upon the species of *Plasmodium*. Its end is marked by eruption of the liver cell, which releases from 2,000 to 40,000 mature merozoites into the circulation.

3. During the *erythrocytic phase*, merozoites attach to special receptors on RBCs and invade them, converting in a short time to ring-shaped trophozoites. This stage feeds upon hemoglobin, grows, and undergoes multiple divisions to produce a cell called a *schizont*, which is filled with more merozoites. Bursting RBCs liberate merozoites to infect more red cells. Eventually, certain merozoites differentiate into two types of specialized gametes called *macrogametocytes* (female) and *microgametocytes* (male). Because the human does not provide a suitable environment for the next phase of development, this is the end of the cycle in humans.

4. The *sexual phase* (sporogony) occurs when a mosquito draws infected red blood cells into her stomach. In the stomach, the microgametocyte releases gametes that fertilize the larger macrogametocytes. The resultant diploid cell (ookinete) implants into the stomach wall of the mosquito, becoming an oocyst, which undergoes multiple mitotic divisions, ultimately releasing sporozoites that migrate to the salivary glands and lodge there. This event completes the sexual cycle and makes the sporozoites available for infecting the next victim.

Process Figure 20.20 The life and transmission cycle of *Plasmodium*, the cause of malaria.

▶ Pathogenesis and Virulence Factors

The invasion of the merozoites into RBCs leads to the release of fever-inducing chemicals into the bloodstream. Chills and fevers often occur in a cyclic pattern. *Plasmodium* also metabolizes glucose at a very high rate, leading to hypoglycemia in the human host. The damage to RBCs results in anemia. The accumulation of malarial products in the liver and the immune stimulation in the spleen can lead to enlargement of these organs. The inducement of RBC adhesion to blood vessels in the brain (cytoadherence) adds to its virulence. A surface protein called GPI (glycosyl-phosphatidyl inositol) is thought to be responsible for the fever seen in malaria.

P. vivax and P. ovale have a propensity to persist in the liver; and without sufficient treatment, they can reemerge over the course of several years to cause recurrent bouts of malarial symptoms.

The fact that the protozoan has several different life stages within a host helps it escape immune responses mounted against any single life stage. This evasion is only strengthened by the parasite's ability to undergo antigenic variation, in which the pattern of gene transcription varies among organisms within a single population. This leads to changes in surface antigens, which constantly changes the appearance of the microbe to the human immune system—a true master of disguise!

▶ Transmission and Epidemiology

All forms of malaria are spread primarily by the female *Anopheles* mosquito and occasionally by shared hypodermic needles and blood transfusions. Some researchers have found that the composition of your skin microbiome makes you more or less attractive to the mosquitoes carrying malaria. Although malaria was once distributed throughout most of the world, the control of mosquitoes in temperate areas has successfully restricted it mostly to a belt extending around the equator **(figure 20.21)**. Despite this achievement, approximately 300 million new cases are still reported each year, about 90% of them in Africa. The most frequent victims are children and young adults, of whom up to 1 million die annually. A particular form of the malarial protozoan causes damage to the placenta in pregnant women, leading to excess mortality among fetuses and newborns. The total case rate in the United States is about 1,000 to 2,000 new cases a year, most of which occur in immigrants or travelers who have visited endemic areas.

▶ Culture and Diagnosis

Malaria can be diagnosed definitively by the discovery of a typical stage of *Plasmodium* in stained blood smears (see figure 20.19). Newer serological procedures have made diagnosis more accurate while requiring less skill to perform. Other indications are knowledge of the patient's residence or travel in endemic areas and symptoms such as recurring chills, fever, and sweating.

▶ Prevention

World health officials have tried for decades to eradicate malaria. The most recent attempt is by the United Nations–backed Global Malaria Action Plan, which hopes to cut malaria by 75% between 2000 and 2015 and reduce the number of malaria deaths to zero. Its final goal is to eradicate malaria altogether.

Malaria prevention is attempted through long-term mosquito abatement and human chemoprophylaxis. Abatement includes elimination of standing water that could serve as a breeding site and spraying of insecticides to reduce populations of adult mosquitoes, especially in and near human dwellings. Scientists have also tried introducing sterile male mosquitoes into endemic areas in an attempt to decrease mosquito populations, and have recently tried to directly fight the pathogen using genetically engineered bacteria. Humans can reduce their risk of infection considerably by using netting, screens, and repellants; by remaining indoors at night; and by taking weekly doses of prophylactic drugs. (Western travelers to endemic areas are often prescribed antimalarials for the duration of their trips.) People with a recent history of malaria must be excluded from blood donations. The WHO and other international organizations focus on efforts to distribute bed nets and to teach people how to dip the nets twice a year into an insecticide **(figure 20.22)**. The use of bed nets has been estimated to reduce childhood mortality from malaria by 20%. Even with massive efforts undertaken by the WHO, the prevalence of malaria in endemic areas is still high.

The best protection would come from a malaria vaccine, and scientists have struggled for decades to develop one. A successful malaria vaccine must be capable of striking a diverse and rapidly changing target. Scientists estimate that the parasite has 5,300 different antigens. (*Note:* Remember that antigenic variation is constantly changing the expression of these antigens on the cell surface.) Despite these odds, one vaccine has reached human trials. It is called RTS,S and contains a molecule expressed by the parasite as it enters the human in mosquito saliva. Data published in 2013 showed only a 17% protection rate, however.

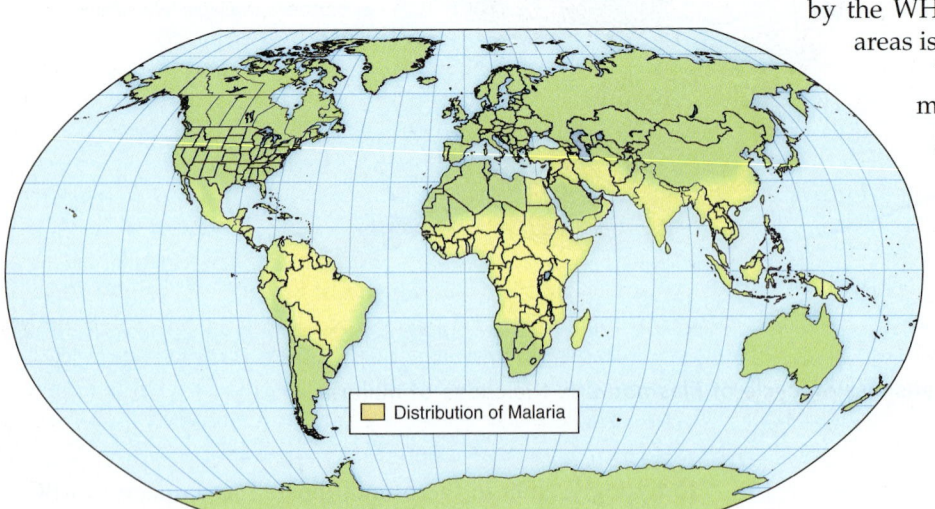

Figure 20.21   The malaria belt.   Yellow zones outline the major regions that harbor malaria. The malaria belt corresponds to a band around the equator.

Distribution of Malaria

Figure 20.22 An African family sits under a treated mosquito net acquired through the UNICEF mosquito nets program.

Disease Table 20.11 Malaria	
Causative Organism(s)	*Plasmodium falciparum, P. vivax, P. ovale, P. malariae, P. knowlesi*
Most Common Modes of Transmission	Biological vector (mosquito), vertical
Virulence Factors	Multiple life stages; multiple antigenic types, ability to scavenge glucose, GPI, cytoadherence
Culture/Diagnosis	Blood smear; serological methods
Prevention	Mosquito control; use of bed nets; no vaccine yet available; prophylactic antiprotozoal agents
Treatment	Chloroquine, mefloquine, artemisinin, pyrimethamine plus sulfadoxine (Fansidar), quinine, or proguanil; consult WHO
Epidemiological Features	United States: cases are generally in travelers or immigrants; internationally, 300 million cases in "malaria belt"; 3 million deaths per year; more deadly in children

▶ Treatment

Chloroquine, a drug that had been effective for decades, became ineffective due to resistance to it starting in the 1950s and culminating in the early 2000s when it stopped being recommended. It has not been widely used since 2003. Tests in 2012 showed that, due to it not being used, *Plasmodium* species in many areas were again sensitive to it. It is much cheaper than other options and, unlike other drugs, can be given to pregnant women. In areas of the world where resistant strains of *P. falciparum* and *P. vivax* predominate (particularly in Southeast Asia), a course of mefloquine or pyrimethamine plus sulfadoxine (Fansidar) may be indicated. In recent years, artemisinin, another plant compound, has been most effective. The World Health Organization now recommends only administering artemisinin in combination with other antimalarials, in order to prevent resistance development. News out of Southeast Asia in 2012, however, revealed the development of artemisinin combination therapy (ACT) resistant strains **(Disease Table 20.11).** Clinicians in the United States should be aware of where the infection might have been acquired and then consult the WHO's online map depicting antimalarial resistance.

A growing problem contributing to the development of antimalarial resistance, particularly in Southeast Asia, seems to be the sale of counterfeit antimalarial drugs, many of which contain only a fraction of the appropriate dosage.

HIV Infection and AIDS

The sudden emergence of AIDS in the early 1980s focused an enormous amount of public attention, research studies, and financial resources on the virus and its disease. Physicians in Los Angeles, San Francisco, and New York City saw the first cases of AIDS. They observed clusters of young male patients with one or more of a complex of symptoms: severe pneumonia caused by *Pneumocystis jirovecii* (ordinarily a harmless fungus), a rare vascular cancer called Kaposi's sarcoma, sudden weight loss, swollen lymph nodes, and general loss of immune function. Eventually, virologists at the Pasteur Institute in France isolated a novel retrovirus, later named the **human immunodeficiency virus (HIV).** This cluster of symptoms was therefore clearly a communicable infectious disease, and the medical community termed it **acquired immunodeficiency syndrome (AIDS).**

▶ Signs and Symptoms

A spectrum of clinical disease is associated with HIV infection. To understand the progression, follow **figure 20.23** closely. Symptoms in HIV infection are directly tied to two things: the level of virus in the blood and the level of T cells in the blood. This figure shows two different lines that correspond to virus and T-cell levels in the blood, in addition to another line depicting the amount of circulating antibody against the virus. Note that the figure depicts the course of HIV infection in the absence of medical intervention or chemotherapy.

Initial symptoms may be fatigue, diarrhea, weight loss, and neurological changes, but most patients first notice this phase of infection because of one or more opportunistic infections or neoplasms. These are detailed in **Insight 20.3**. Other disease-related symptoms appear to accompany severe immune deregulation, hormone imbalances, and metabolic disturbances. Pronounced wasting of body mass is a consequence of weight loss, diarrhea, and poor nutrient absorption. Protracted fever, fatigue, sore throat, and night sweats are significant and debilitating. Both a rash and generalized lymphadenopathy in several chains of lymph nodes are the presenting symptoms in many AIDS patients.

Some of the most virulent complications are neurological. Lesions occur in the brain, meninges, spinal column, and peripheral nerves. Patients with nervous system involvement show social withdrawal, persistent memory loss, spasticity, sensory loss, and progressive AIDS dementia.

▶ Causative Agent

HIV is a retrovirus, in the genus *Lentivirus*. Many retroviruses have the potential to cause cancer and produce dire, often fatal diseases and are capable of altering the host's DNA in profound ways. They are named "retroviruses" because they reverse the usual order of transcription. They contain an unusual enzyme called **reverse transcriptase (RT)** that catalyzes the replication of double-stranded DNA from single-stranded RNA. The association of retroviruses with their hosts can be so intimate that viral genes are permanently integrated into the host genome. In fact, as you have read in earlier chapters, it has become increasingly evident that retroviral sequences are integral parts of host chromosomes. Not only can this retroviral DNA be incorporated into the host genome as a provirus that can be passed on to progeny cells, but also some retroviruses also transform cells and regulate certain host genes.

There are two major types of HIV, namely HIV-1, which is the dominant form in most of the world, and HIV-2. Genetic sequencing of HIV-1 shows that it is most related to simian immunodeficiency viruses in chimpanzees, while HIV-2 evolved from related viruses in sooty mangabeys. Both highlight the evidence that HIV in humans was derived from a zoonotic primate virus. Subtypes of both HIV-1 and HIV-2 have been identified, and each displays a distinct geographic distribution in the world today in addition to particular trends in disease pathogenesis. HIV and other retroviruses display structural features typical

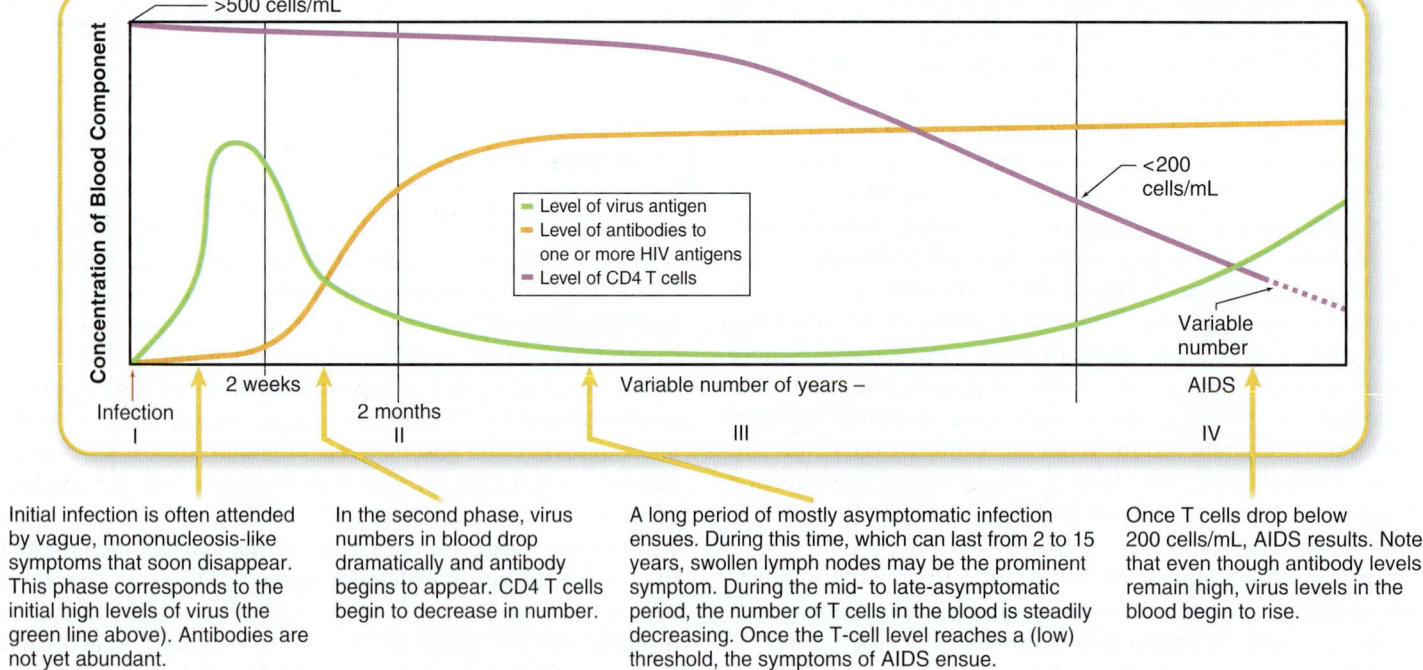

Figure 20.23 **Dynamics of virus antigen, antibody, and T cells in circulation.**

of enveloped RNA viruses **(figure 20.24a).** The outermost component is a lipid envelope with transmembrane glyco-protein spikes and knobs that mediate viral adsorption to the host cell. HIV can only infect host cells that present the required receptors, which is a combination receptor consisting of the CD4 marker plus a coreceptor called CCR-5. The virus uses these receptors to gain entrance to several types of leukocytes and tissue cells **(figure 20.24b).**

▶ Pathogenesis and Virulence Factors

As summarized in **process figure 20.25,** HIV enters a mucous membrane or the skin and travels to dendritic cells, a type of phagocyte living beneath the epithelium. In the dendritic cell, the virus grows and is shed from the cell without killing it. The virus is amplified by macrophages in the skin, lymph organs, bone marrow, and blood. One of the great ironies of HIV is that it infects and destroys many of the very cells needed to combat it, including the helper (T4 or CD4) class of lymphocytes, monocytes, macrophages, and even B lympho-cytes. The virus is adapted to docking onto its host cell's surface receptors (see figure 20.24). It then induces viral fusion with the cell membrane and creates syncytia.

Once the virus is inside the cell, its reverse transcriptase converts its RNA into DNA. Although initially it can produce a lytic infection, in many cells it enters a latent period in the nucleus of the host cell and integrates its DNA into host DNA (as shown in process figure 20.25). This latency accounts for the lengthy course of the disease. Despite being described as a "latent" stage, research suggests that new viruses are

constantly being produced and new T cells are constantly being manufactured, in an ongoing race that ultimately the host cells lose (in the absence of treatment).

The primary effects of HIV infection—those directly due to viral action—are harm to T cells and the central nervous system. The death of T cells and other white blood cells results in extreme **leukopenia** and loss of essential T4 memory clones and stem cells. The viruses also cause formation of giant T cells and other syncytia, which allow the spread of viruses directly from cell to cell, followed by mass destruction of the syncytia. The destruction of T4 lymphocytes paves the way for invasion by opportunistic agents and malignant cells. The central nervous system is affected when infected macrophages cross the blood-brain barrier and release viruses, which then invade nervous tissue. Studies have indicated that some of the viral envelope proteins can have a direct toxic effect on the brain's glial cells and other cells.

The secondary effects of HIV infection are the opportunistic infections and malignancies that occur as the immune system becomes progressively crippled by viral attack. These are summarized in Insight 20.3.

▶ Transmission

In general, HIV is spread only by direct and rather specific routes. Because the blood of HIV-infected individuals harbors high levels of free virus in both very early and very late stages of infection and high levels of infected leukocytes throughout infection, any form of intimate contact involving transfer of

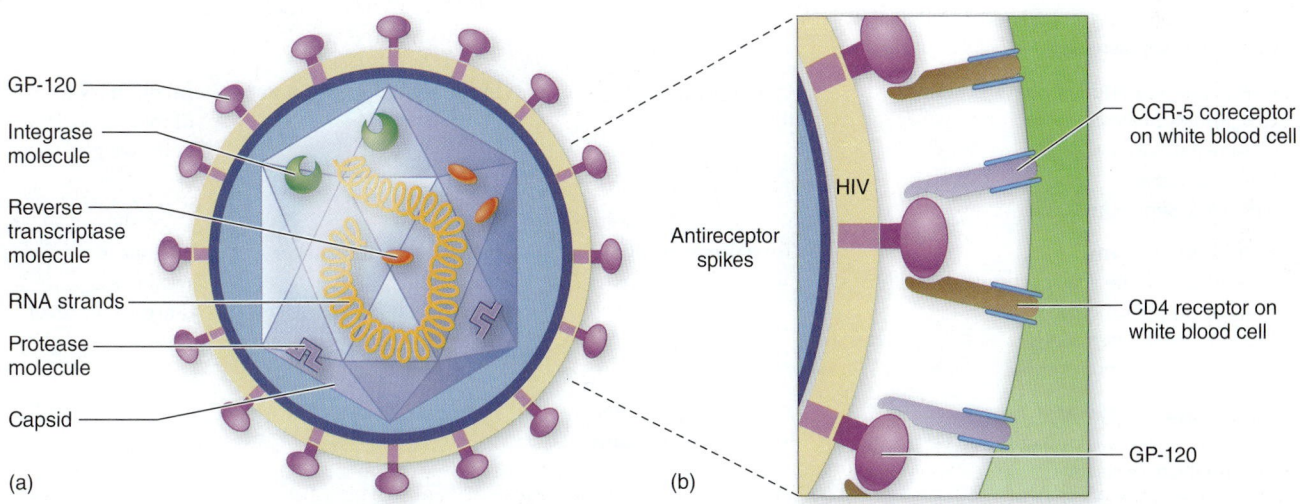

Figure 20.24 The general structure of HIV. **(a)** The envelope contains two types of glycoprotein (GP) spikes, two identical RNA strands, and several molecules of reverse transcriptase, protease, and integrase encased in a protein capsid. **(b)** The snug attachment of HIV glycoprotein molecules (GP-41 and GP-120) to their specific receptors on a human cell membrane. These receptors are CD4 and a coreceptor called CCR-5 (fusin) that permit docking with the host cell and fusion with the cell membrane.

INSIGHT 20.3   AIDS-Defining Illnesses (ADIs)

If an individual is infected with HIV, does he or she have AIDS? When in the course of an HIV infection does an individual develop AIDS? In order to answer these questions and standardize statistical surveillance data, the Centers for Disease Control and Prevention (CDC) has set a standard for defining when an HIV-infected individual can be classified as having AIDS. Since the HIV epidemic began in the 1980s, the standards have been revised as new virus-detection methods have been developed. The most recent revisions are as follows. A diagnosis of AIDS is warranted when a person has a positive HIV test and one of the following:

- CD4+ T-lymphocyte count of less than 200 cells/mL or a CD4+ T-lymphocyte percentage of less than 14% of total lymphocytes, or
- Documentation of an AIDS-defining illness (ADI)

There are a number of conditions that can be considered AIDS-defining illnesses, and they fall under four broad categories listed in the table that follows. Each of these illnesses reflects the impact that HIV has on the human immune system. Seemingly innocuous infections that cause little to no harm in individuals with a healthy immune system can be devastating to a patient with AIDS. Some of the most common ADIs are *Pneumocystitis jirovecii* pneumonia, *Mycobacterium avium* complex, Kaposi's sarcoma, cerebral toxoplasmosis, chronic candidiasis, and cytomegalovirus (CMV) infections. Many of the cancers listed are associated with viral infections and could also be considered opportunistic microbial infections.

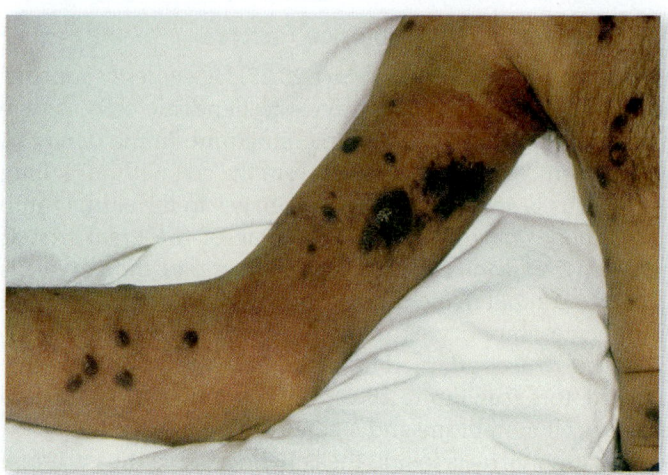

Kaposi's sarcoma in an AIDS patient.

Opportunistic Infection	Diseases of the Nervous System	Cancer	Wasting Syndrome
Skin and mucous membranes Cytomegalovirus retinitis (loss of vision) Herpes simplex chronic ulcers	AIDS dementia complex	Kaposi's sarcoma	Inadequate nutrition
Nervous system HIV encephalopathy Progressive multifocal leukoencephalopathy Cerebral toxoplasmosis	Peripheral neuropathy	Burkitt's lymphoma	Metabolic disturbances
Cardiovascular and lymphatic systems Coccidioidomycosis Cytomegalovirus Histoplasmosis *Salmonella* septicemia	Lymphoma, primarily in the brain	Immunoblastic lymphoma	Involuntary loss of more than 10% body weight
Respiratory system Candidiasis of the trachea, bronchi, and lungs *P. jirovicecii* pneumonia *Mycobacterium tuberculosis* *Mycobacterium avium* complex	Toxoplasmosis of the brain	Invasive cervical cancer carcinoma	30 days of weakness, diarrhea, or fever
Gastrointestinal system Candidiasis of the esophagus and GI tract Intestinal isosporiasis Chronic cryptosporidiosis CMV colitis	Progressive, multifocal leukoencephalopathy	Lymphoma	
Genitourinary tract and reproductive tract Herpes simplex chronic ulcers Vaginal candidiasis Genital warts		Non-Hodgkin's lymphoma	

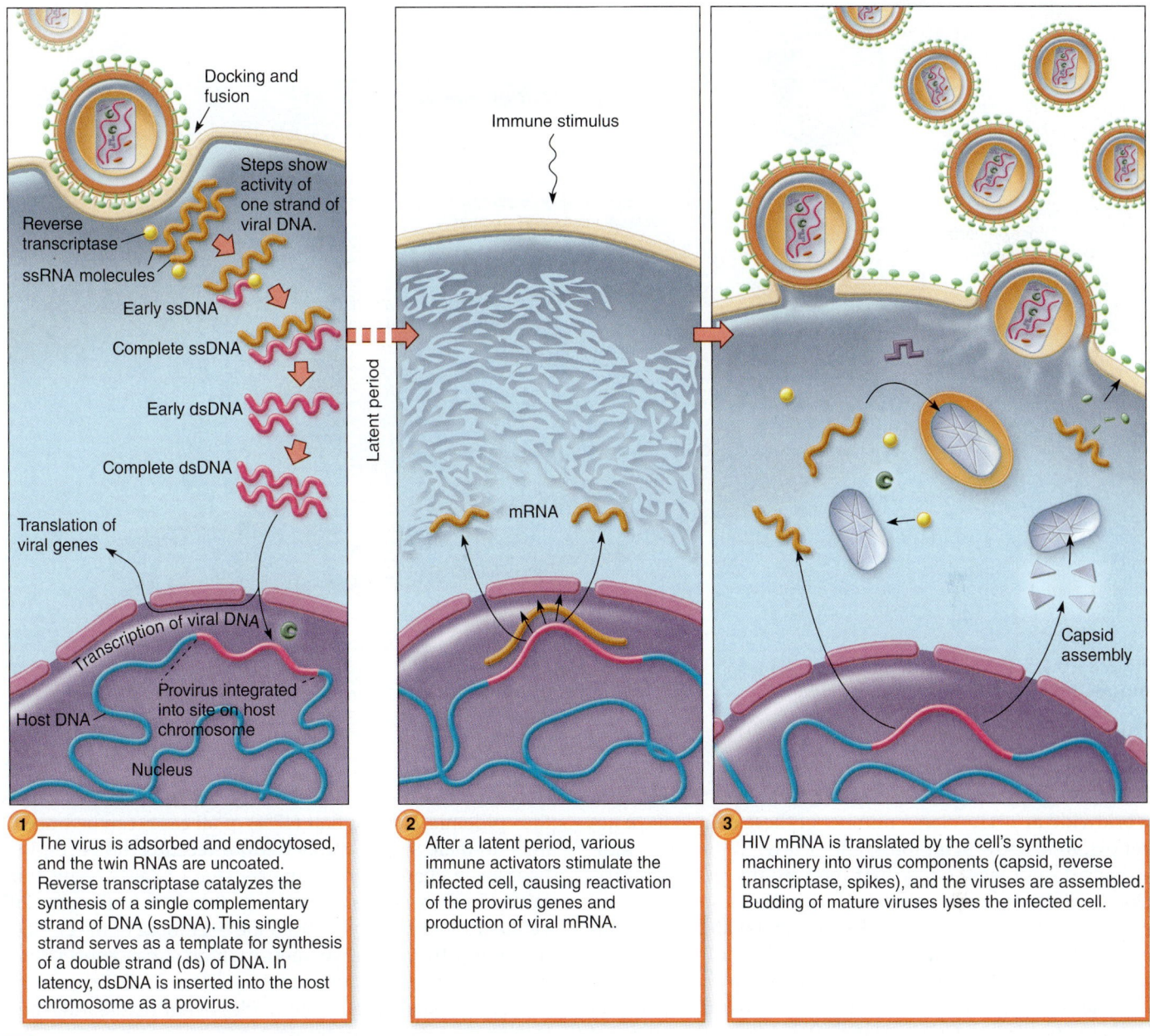

1 The virus is adsorbed and endocytosed, and the twin RNAs are uncoated. Reverse transcriptase catalyzes the synthesis of a single complementary strand of DNA (ssDNA). This single strand serves as a template for synthesis of a double strand (ds) of DNA. In latency, dsDNA is inserted into the host chromosome as a provirus.

2 After a latent period, various immune activators stimulate the infected cell, causing reactivation of the provirus genes and production of viral mRNA.

3 HIV mRNA is translated by the cell's synthetic machinery into virus components (capsid, reverse transcriptase, spikes), and the viruses are assembled. Budding of mature viruses lyses the infected cell.

Process Figure 20.25 The general multiplication cycle of HIV.

blood (trauma, needle sharing) can be a potential source of infection **(figure 20.26)**. Semen and vaginal secretions also harbor free virus and infected white blood cells; thus, they are significant factors in sexual transmission. The virus can be isolated from urine, tears, sweat, and saliva but in such small numbers that these fluids are not considered sources of infection. Because breast milk contains significant numbers of leukocytes, neonates who have escaped infection prior to and during birth can still become infected through nursing.

▶ Epidemiology

Since the beginning of the HIV epidemic in the early 1980s, over 60 million people have become infected with HIV and more than 30 million have died of HIV-related causes worldwide. The best global estimate of the number of individuals currently infected with HIV (in 2011) is 34 million, with approximately 1.1 million currently in the United States. The WHO estimates that 2.5 million new infections occurred

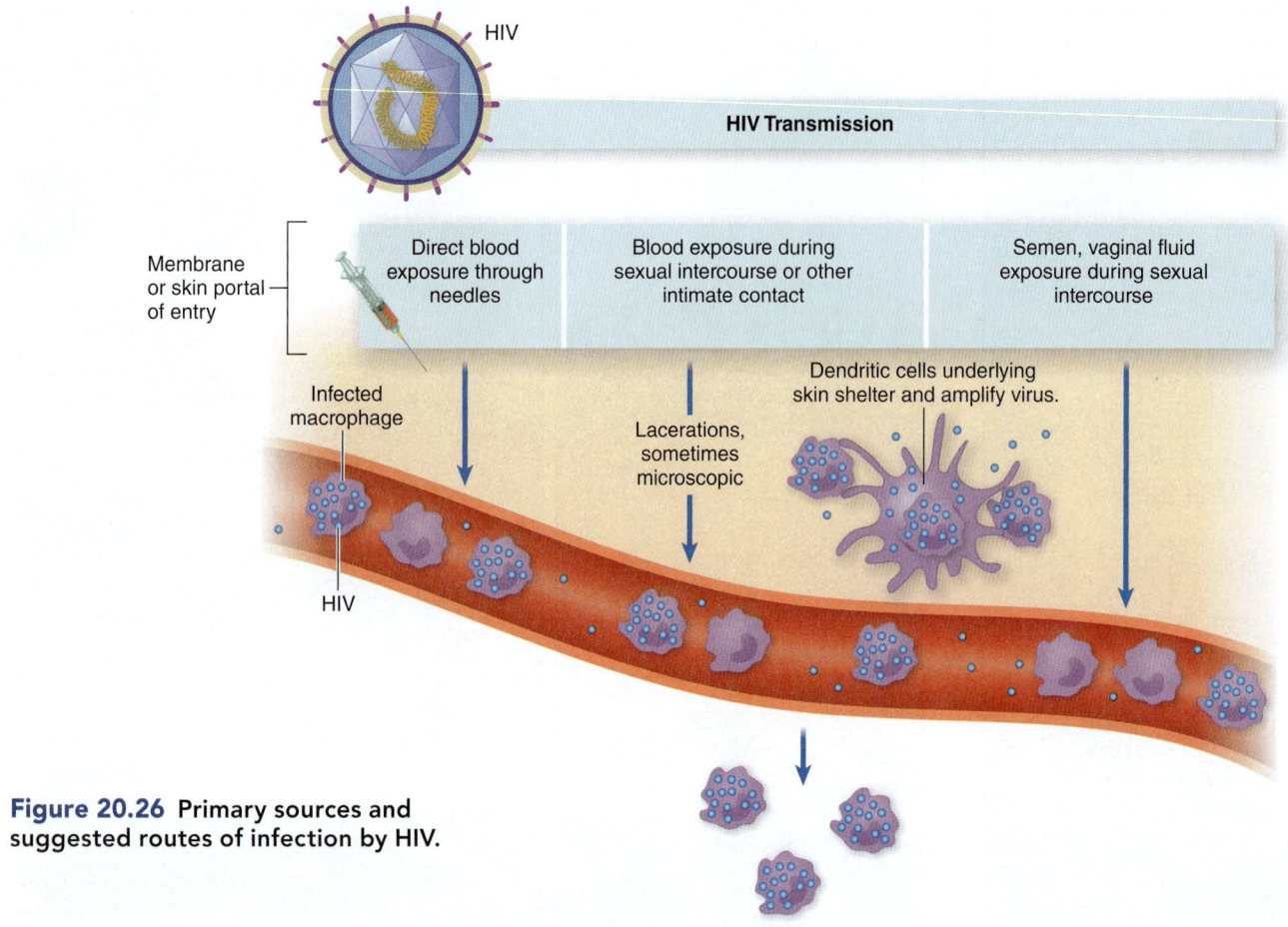

Figure 20.26 Primary sources and suggested routes of infection by HIV.

worldwide in 2011. A large number of these people have not yet begun to show symptoms, and nearly 60% of these individuals do not even know that they are infected with HIV. Due to efforts of many global AIDS initiatives, many more people in the developing world are receiving lifesaving treatments. In spite of these efforts, over 7 million infected people still have no access to any form of HIV therapy.

Table 20.1 gives a snapshot of behaviors that result in HIV infection in the United States. Throughout the course of the epidemic, close to half (47%) of all cases can be traced to male-to-male sexual contact. More recently, however, there have been changes in the percentage of cases being transmitted by heterosexual contact (nearly 30% in the year 2010 versus 18% cumulatively from the beginning of the epidemic through 2010). In large metropolitan areas especially, as many as 60% of intravenous drug users (IDUs) can be HIV carriers. In most parts of the world, heterosexual intercourse is the primary mode of transmission.

Table 20.1 HIV Infection in the United States by Exposure Category

Transmission Category	Estimated Number of Diagnoses of HIV infection, 2010		
	Adult and Adolescent Males	Adult and Adolescent Females	Total
Male-to-male sexual contact	28,782	–	28,782
Injection drug use	2,373	1,393	3,766
Male-to-male sexual contact and injection drug use	1,443	–	1,443
Hetrosexual Contact[a]	4,416	8,459	12,875
Other[b]	31	16	47

[a] Heterosexual contact with a person known to have, or to be at high risk for, HIV infection.

[b] Includes hemophilia, blood transfusion, perinatal exposure, and risk not reported or not identified.

Source: Data from the Centers for Disease Control and Prevention, 2012.

Now that donated blood is routinely tested for antibodies to HIV, transfusions are no longer considered a serious risk. Rarely, organ transplants can carry HIV, so they too should be tested. Other blood products (serum, coagulation factors) were once implicated in HIV transmission. Thousands of hemophiliacs died from the disease in the 1980s and 1990s. It is now standard practice to heat-treat any therapeutic blood products to destroy all viruses.

We should note that not everyone who becomes infected or is antibody-positive develops AIDS. About 1% of people who are antibody-positive remain free of disease, indicating that functioning immunity to the virus can develop. Any person who remains healthy despite HIV infection is termed a *nonprogressor*. These people are the object of intense scientific study. Some have been found to lack the cytokine receptors that HIV requires for entry. Others are infected by a strain of virus weakened by genetic mutation.

Treatment of HIV-infected mothers with a simple anti-HIV drug has dramatically decreased the rate of maternal-to-infant transmission of HIV during pregnancy. Current treatment regimens result in a transmission rate of approximately 11%, with some studies of multidrug regimens claiming rates as low as 5%. Evidence suggests that giving mothers protease inhibitors can reduce the transmission rate to around 1%. (Untreated mothers pass the virus to their babies at the rate of 33%.) Combining such drug treatment with cesarean delivery, vertical transmission of HIV from infected mothers to newborn infants has been reduced to 0% in some major U.S. cities.

Medical and dental personnel are not considered a high-risk group, although several hundred medical and dental workers are known to have acquired HIV or become antibody-positive as a result of clinical accidents. A health care worker involved in an accident in which gross inoculation with contaminated blood occurs (as in the case of a needlestick) has a less than 1 in 1,000 chance of becoming infected. We should emphasize that transmission of HIV will not occur through casual contact or routine patient care procedures and that universal precautions for infection control (see chapter 13) were designed to give full protection for both worker and patient.

▶ Culture and Diagnosis

First, let's define some terms. A person is diagnosed as having HIV infection if he or she has tested positive for exposure to the human immunodeficiency virus. This diagnosis is not the same as having AIDS. In 2012, the U.S. Preventive Services Task Force recommended that all people between the ages of 15 and 64 be tested for HIV. People outside of that age group who are at high risk, as well as pregnant women, should also be tested. Most viral testing is based on detection of antibodies specific to the virus in serum or other fluids, which allows for the rapid, inexpensive screening of large numbers of samples. Testing usually proceeds at two levels. The initial screening tests include the older ELISA and newer latex agglutination and rapid antibody tests.

In 2012, the FDA approved an over-the-counter testing method called OraQuick. It is available at drugstores and uses a mouth swab to detect antibodies to the virus in 20 to 40 minutes. There is some controversy over the easy accessibility to the test (without counseling) because users may not understand that their—or their partners'—results may not be accurate if they are inside the period before antibodies develop. However, public health officials believe that wider access to testing will help decrease the spread of the virus by those who don't know they have it. Positive tests always require follow-up with a more specific test—usually a *Western blot* analysis. This test detects several different anti-HIV antibodies and can usually rule out false-positive results.

Blood and blood products are sometimes tested for HIV antigens (rather than for HIV antibodies) to close the window of time between infection and detectable levels of antibodies during which infection could be missed by antibody tests.

In the United States, people are diagnosed with AIDS if they meet the following criteria: (1) They are positive for the virus, *and* (2) they fulfill one of these additional criteria:

- They have a CD4 (helper T cell) count of fewer than 200 cells per microliter of blood.
- Their CD4 cells account for fewer than 14% of all lymphocytes.
- They experience one or more of a CDC-provided list of AIDS-defining illnesses (ADIs).

The list of ADIs is long and includes opportunistic infections such as *Pneumocystis jirovecii* pneumonia and *Cryptosporidium* diarrhea; neoplasms such as Kaposi's sarcoma and invasive cervical cancer; and other conditions such as wasting syndrome (see Insight 20.3).

▶ Prevention

Avoidance of sexual contact with infected persons is a cornerstone of HIV prevention. Monogamous or not, a sexually active person should consider every partner to be infected unless proven otherwise. This may sound harsh, but it is the only sure way to avoid infection during sexual encounters. Barrier protection (condoms) should be used when having sex with anyone whose HIV status is not known with certainty to be negative. Although avoiding intravenous drugs is an obvious deterrent, many drug addicts do not, or cannot, choose this option. In such cases, risk can be decreased by not sharing syringes or needles or by cleaning needles with bleach and then rinsing before another use.

New research has shown that treating newly infected people with antiretrovirals can prevent the progression to AIDS. The WHO recently recommended that HIV-positive people with uninfected partners should immediately begin a regimen of antiretroviral drugs. The risk of transmission to the uninfected partner can be cut by 96%. There is also new research that shows that putting uninfected heterosexuals on low doses of antiretrovirals—at the cost of 25 cents per day—can prevent HIV from beginning an infection if it is contracted. A newer

HIV drug (Truvada—which actually contains two HIV drugs) was recently recommended for use by HIV-negative people in high-risk groups. This category would include partners of HIV-positive people, homosexuals with multiple partners, injecting drug users, and sex workers. The risk groups would also include the female partners of nonmonogamous men.

From the very first years of the AIDS epidemic, the potential for creating a vaccine has been regarded as slim, because the virus presents many seemingly insurmountable problems. Among them, HIV becomes latent in cells; its cell surface antigens mutate rapidly; and although it does elicit immune responses, it is apparently not completely controlled by them. In view of the great need for a vaccine, however, none of those facts has stopped the medical community from moving ahead. Currently, more than 30 potential HIV vaccines are in clinical trials.

▶ Treatment

Clear-cut guidelines exist for treating people who test HIV-positive. These guidelines are updated regularly. The most recent update involves beginning treatment much earlier than previously. Until now, recommendations called for

beginning aggressive antiviral chemotherapy only after AIDS manifested itself. The newer recommendations call for treatment to begin soon after HIV diagnosis. In addition to antiviral chemotherapy, HIV-positive persons should receive a wide array of drugs to prevent or treat a variety of opportunistic infections and other ADIs such as wasting disease. These treatment regimens vary according to each patient's profile and needs.

Figure 20.27 illustrates the various strategies anti-HIV drugs utilize. The first effective drugs developed were the synthetic nucleoside analogs (reverse transcriptase inhibitors), azidothymidine (AZT), didanosine (ddI), lamivudine (Epivir; 3TC), and stavudine (d4T) **(figure 20.27a)**. They interrupt the HIV multiplication cycle by mimicking the structure of actual nucleosides and being added to viral DNA by reverse transcriptase. Because these drugs lack all of the correct binding sites for further DNA synthesis, viral replication and the viral cycle are terminated. Another important class of drugs that can be used is the protease inhibitors **(figure 20.27b)**, which block the action of the HIV enzyme (protease) involved in the final assembly and maturation of the virus. Another class of drugs

Location of reaction

☐ External to cell

☐ Cytoplasm

☐ Nucleus

Figure 20.27 Mechanisms of action of anti-HIV drugs.

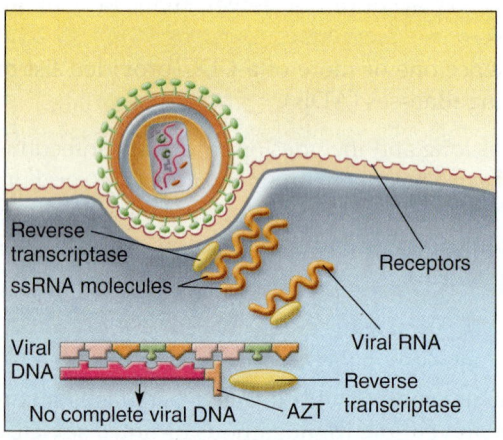

(a) A prominent group of drugs (AZT, ddI, 3TC) are **nucleoside analogs that inhibit reverse transcriptase**. They are inserted in place of the natural nucleotide by reverse transcriptase but block further action of the enzyme and synthesis of viral DNA. Non-nucleoside RT inhibitors are also in use.

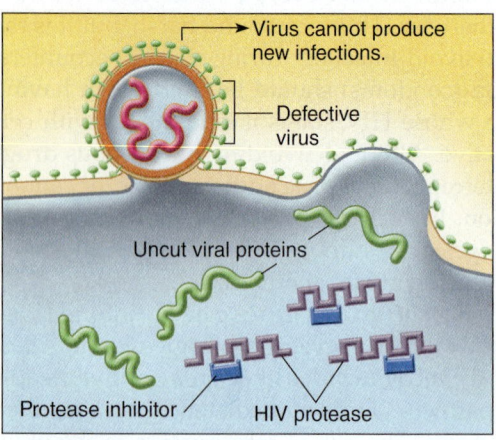

(b) **Protease inhibitors** plug into the active sites on HIV protease. This enzyme is necessary to cut elongated HIV protein strands and produce functioning smaller protein units. Because the enzyme is blocked, the proteins remain uncut, and abnormal defective viruses are formed.

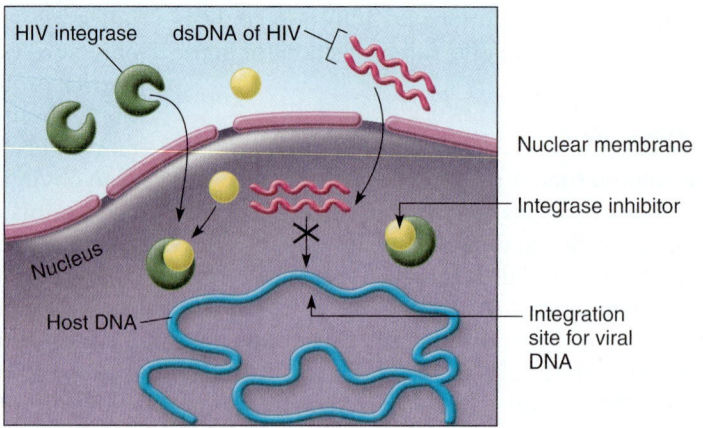

(c) **Integrase inhibitors** are a class of experimental drugs that attach to the enzyme required to splice the dsDNA from HIV into the host genome. This will prevent formation of the provirus and block future virus multiplication in that cell.

called integrase inhibitors targets the enzyme that allows viral DNA to splice into human DNA. Dolutegravir, tradename Tivicay, is an integrase inhibitor approved in 2013 **(figure 20.27c)**.

In 2012, the FDA approved a four-drug combination named Stribild, sometimes called the "quadpill." By combining two reverse transcriptase inhibitors and one integrase inhibitor in a "cocktail," the virus is interrupted in two different phases of its cycle. The fourth drug has no anti-HIV activity but inhibits the liver enzymes that break down the other drugs, enabling higher effective concentrations of these drugs to remain in the body. This therapy has been successful in reducing viral load to undetectable levels and facilitating the improvement of immune function. It has also reduced the incidence of viral drug resistance, because the virus would have to undergo three separate mutations simultaneously, at nearly impossible odds.

In 2007, an HIV-positive man received a bone marrow transplant from a person who was known to possess two copies of a gene that prevents HIV from invading lymphocytes. The gene from the donor continued a mutation that eliminated the T-cell co-receptor for HIV on T cells. Since then, at least two more patients have been "cured" by this treatment. That strategy is not likely to be an answer for worldwide HIV treatment, as bone marrow transplants would be too drastic to treat the millions of HIV-positive people in the world and the donor genotype (two copies of the relevant gene) is very rare **(Disease Table 20.12)**.

Disease Table 20.12	HIV Infection and AIDS
Causative Organism(s)	Human immunodeficiency virus 1 or 2
Most Common Modes of Transmission	Direct contact (sexual), parenteral (blood-borne), vertical (perinatal and via breast milk)
Virulence Factors	Attachment, syncytia formation, reverse transcriptase, high mutation rate
Culture/Diagnosis	Initial screening for antibody followed by Western blot confirmation of antibody
Prevention	Avoidance of contact with infected sex partner, contaminated blood, breast milk; antiretrovirals for high-risk individuals
Treatment	Stribild (two reverse transcriptase inhibitors, one integrase inhibitor, and one "booster" drug)
Epidemiological Features	United States: HIV infection is 1.1 million; internationally, HIV infection is 34 million

20.3 Learning Outcomes—Assess Your Progress

4. List the possible causative agents for each of the following infectious cardiovascular conditions: acute and subacute endocarditis, plague, tularemia, Lyme disease, infectious mononucleosis, anthrax, Chagas disease, and malaria.

5. Discuss what series of events may lead to septicemia and how it should be prevented and treated.

6. Describe what makes anthrax a good agent for bioterrorism, and list the important presenting signs to look for in patients.

7. Discuss the difference between hemorrhagic and nonhemorrhagic fever diseases.

8. List the possible causative agents and modes of transmission for hemorrhagic fever diseases.

9. List the possible causative agents and modes of transmission for nonhemorrhagic fever diseases.

10. Discuss all aspects of malaria, with special emphasis on epidemiology.

11. Describe important events in the course of an HIV infection in the absence of treatment.

12. Explain the rationale behind the recommended treatment for HIV infection.

13. Discuss the epidemiology of HIV infection in the developed and the developing world.

Case File 20   *Wrap-Up*

The majority of Ebola outbreaks have been in tropical African countries with limited medical resources. Ebola is classified as a biosafety level 4 (BSL-4) pathogen, one of the most deadly pathogens known, requiring specialized containment for patients as well as health care workers. Working in a BSL-4 laboratory requires positive-pressure, air-supplied, full-body suits to be worn, as well as specialized air supply and ventilation of the laboratory to prevent airborne release of the pathogen and other safety measures. Health care workers in hospitals should use effective barrier precautions including masks, gloves, and goggles to prevent contact with patients' blood or other secretions. Patients must be quarantined to prevent transmission of the virus to other patients in the hospital. Unfortunately, remote hospitals in rural areas of Africa lack the protective equipment required to prevent transmission, which is why the disease is often transmitted to hospital workers and between patients by hospital workers.

Sadly, in addition to the epidemiological impact that Ebola has on a population, it has social and psychological effects as well. "Alarm and near panic" were reported in the Maridi hospital in Sudan in 1976, where 154 of the nursing staff were infected and 33 died. During the 1995 outbreak in Zaire/D.R.C., 25% of the 315 Ebola cases were in health care workers, and many of the hospital staff quit out of fear of being infected. Those who chose to stay were stigmatized as likely carriers of the disease.

In comparison to other tropical diseases like malaria, Ebola is extremely rare. Only about 2,300 cases total have been reported since its discovery in 1976. One of the reasons it is a concern is because of its case fatality rate of 25% to 90%. Another is that there is no vaccine or treatment for the disease. Finally, many are concerned that Ebola will be used as a bioterrorism agent.

▶ Summing Up

Taxonomic Organization Microorganisms Causing Diseases in the Cardiovascular and Lymphatic System

Microorganism	Disease	Location of Disease Table
Gram-positive endospore-forming bacteria		
Bacillus anthracis	Anthrax	Anthrax, p. 606
Gram-positive bacteria		
Staphylococcus aureus	Acute endocarditis	Endocarditis, p. 595
Streptococcus pyogenes	Acute endocarditis	Endocarditis, p. 595
Streptococcus pneumoniae	Acute endocarditis	Endocarditis, p. 595
Gram-negative bacteria		
Yersinia pestis	Plague	Plague, p. 599
Francisella tularensis	Tularemia	Tularemia, p. 600
Borrelia burgdorferi	Lyme disease	Lyme disease, p. 603
Brucella abortus, B. suis	Brucellosis	Nonhemorrhagic fever diseases, p. 612
Coxiella burnetii	Q fever	Nonhemorrhagic fever diseases, p. 612
Bartonella henselae	Cat-scratch disease	Nonhemorrhagic fever diseases, p. 612
Bartonella quintana	Trench fever	Nonhemorrhagic fever diseases, p. 612
Ehrlichia chaffeensis, E. phagocytophila, E. ewingii	Ehrlichiosis	Nonhemorrhagic fever diseases, p. 612
Neisseria gonorrhoeae	Acute endocarditis	Endocarditis, p. 595
Rickettsia rickettsii	Rocky Mountain spotted fever	Nonhemorrhagic fever diseases, p. 612
DNA viruses		
Epstein-Barr virus	Infectious mononucleosis	Infectious mononucleosis, p. 604
RNA viruses		
Yellow fever viruses	Yellow fever	Hemorrhagic fevers, p. 608
Dengue fever viruses	Dengue fever	Hemorrhagic fevers, p. 608
Ebola and Marburg viruses	Ebola and Marburg hemorrhagic fevers	Hemorrhagic fevers, p. 608
Lassa fever virus	Lassa fever	Hemorrhagic fevers, p. 608
Chikungunya virus	Hemorrhagic fever	Hemorrhagic fevers, p. 608
Retroviruses		
Human immunodeficiency virus 1 and 2	HIV infection and AIDS	HIV infection and AIDS, p. 625
Protozoa		
Plasmodium falciparum, P. vivax, P. ovale, P. malariae	Malaria	Malaria, p. 617
Trypanosoma cruzi	Chagas disease	Chagas disease, p. 614

INFECTIOUS DISEASES AFFECTING
The Cardiovascular and Lymphatic Systems

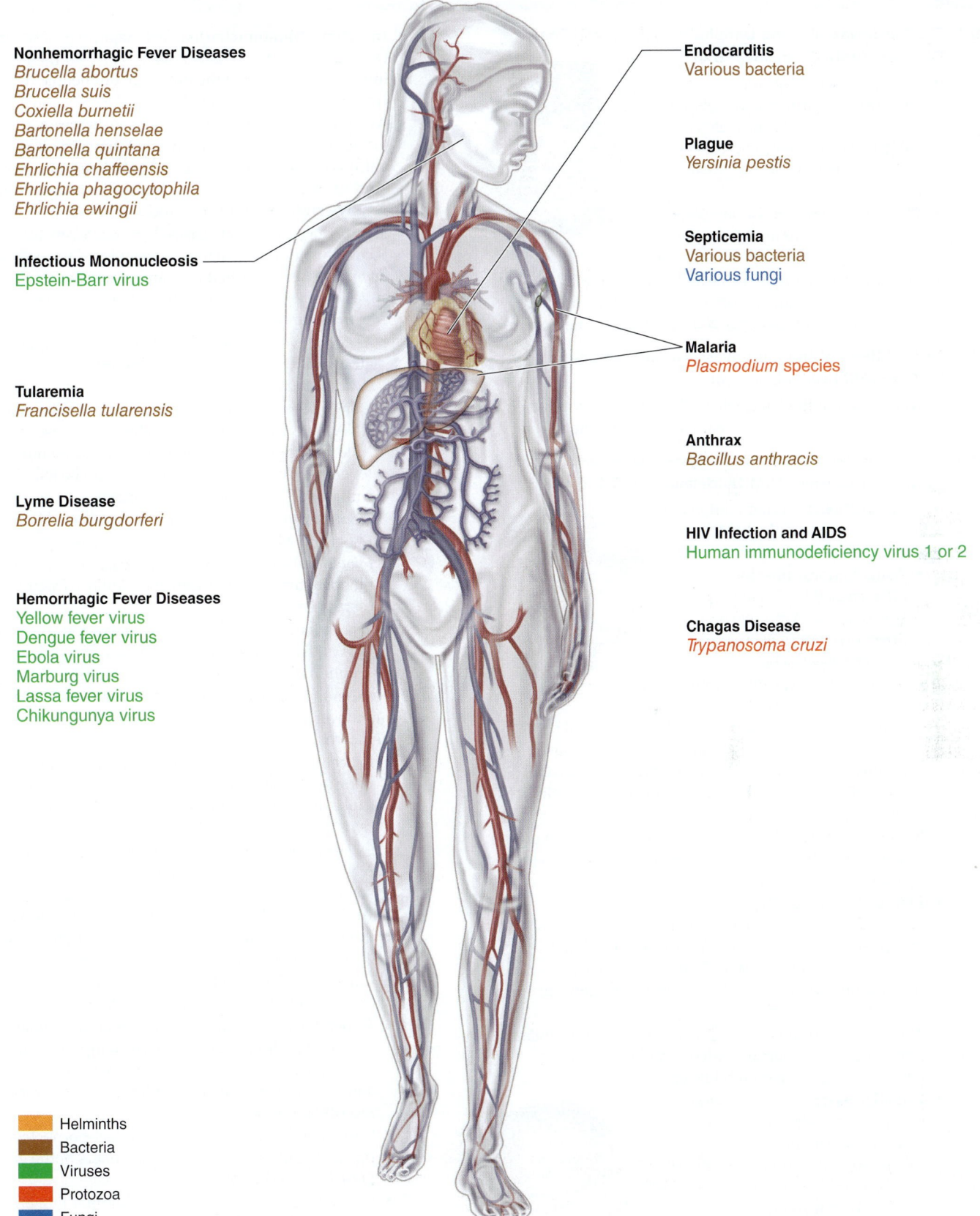

Nonhemorrhagic Fever Diseases
Brucella abortus
Brucella suis
Coxiella burnetii
Bartonella henselae
Bartonella quintana
Ehrlichia chaffeensis
Ehrlichia phagocytophila
Ehrlichia ewingii

Infectious Mononucleosis
Epstein-Barr virus

Tularemia
Francisella tularensis

Lyme Disease
Borrelia burgdorferi

Hemorrhagic Fever Diseases
Yellow fever virus
Dengue fever virus
Ebola virus
Marburg virus
Lassa fever virus
Chikungunya virus

Endocarditis
Various bacteria

Plague
Yersinia pestis

Septicemia
Various bacteria
Various fungi

Malaria
Plasmodium species

Anthrax
Bacillus anthracis

HIV Infection and AIDS
Human immunodeficiency virus 1 or 2

Chagas Disease
Trypanosoma cruzi

☐ Helminths
☐ Bacteria
☐ Viruses
☐ Protozoa
☐ Fungi

System Summary Figure 20.28

Chapter Summary

20.1 The Cardiovascular and Lymphatic Systems and Their Defenses (ASM Guidelines* 5.4)

- *The cardiovascular system is* composed of the blood vessels and the heart. It provides tissues with oxygen and nutrients and carries away carbon dioxide and waste products.
- *The lymphatic system is* a one-way passage, returning fluid from the tissues to the cardiovascular system.
- The systems are highly protected from microbial infection, as they are not open body systems and they contain many components of the host's immune system.

20.2 Normal Biota of the Cardiovascular and Lymphatic Systems (ASM Guidelines 5.4)

- At the present time, we believe that the cardiovascular and lymphatic systems contain no normal biota.

20.3 Cardiovascular and Lymphatic System Diseases Caused by Microorganisms (ASM Guidelines 5.3, 5.4, 6.4, 8.3)

- **Endocarditis:** An inflammation of the endocardium, usually due to an infection of the valves of the heart.
  - Acute Endocarditis: Most often caused by *Staphylococcus aureus,* group A streptococci, *Streptococcus pneumoniae,* and *Neisseria gonorrhoeae.*
  - Subacute Endocarditis: Almost always preceded by some form of damage to the heart valves or by congenital malformation. Alpha-hemolytic streptococci, such as *Streptococcus sanguis, S. oralis,* and *S. mutans,* are most often responsible; normal biota can also colonize abnormal valves.
- **Septicemia:** Caused by organisms actively multiplying in the blood. Most caused by bacteria, to a lesser extent by fungi.
- **Plague:** *Pneumonic plague* is a respiratory disease; *bubonic plague* causes inflammation and necrosis of the lymph nodes; *septicemic plague* is the result of multiplication of bacteria in the blood. *Yersinia pestis* is the causative organism. Fleas are principal agents in transmission of the bacterium.
- **Tularemia:** Causative agent is a facultative intracellular gram-negative bacterium called *Francisella tularensis.* Disease is often called rabbit fever.
- **Lyme Disease:** Caused by *Borrelia burgdorferi.* Syndrome mimics neuromuscular and rheumatoid conditions. *B. burgdorferi* is a unique spirochete transmitted primarily by *Ixodes* ticks.

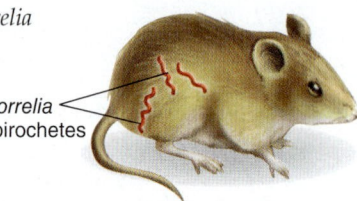

Borrelia spirochetes

- **Infectious Mononucleosis:** Vast majority of cases are caused by the *Epstein-Barr virus (EBV).* Cell-mediated immunity can control the infection, but people usually remain chronically infected.
- **Anthrax:** Exhibits primary symptoms in various locations: skin (cutaneous anthrax), lungs (pulmonary anthrax), gastrointestinal tract, central nervous system (anthrax meningitis). Caused by *Bacillus anthracis,* gram-positive endospore-forming rod found in soil.
- **Hemorrhagic Fever Diseases:** Extreme fevers often accompanied by internal hemorrhaging. Hemorrhagic fever diseases described here are caused by RNA enveloped viruses in one of three families: *Arenaviridae, Filoviridae,* and *Flaviviridae.*
  - Yellow Fever: Caused by an arbovirus, a single-stranded RNA flavivirus transmitted by the mosquito *Aedes aegypti.*
  - Dengue Fever: Caused by one of four single-stranded RNA flaviviruses, also carried by *Aedes* mosquitoes. Mild infection is most common; dengue hemorrhagic fever and dengue shock syndrome can be lethal.
  - Chikungunya: Caused by an alphavirus transmitted by *Aedes* mosquitoes, first hemorrhagic virus in Europe.
  - Ebola and Marburg viruses are filoviruses (family *Filoviridae*) endemic to Central Africa. Virus in the bloodstream leads to extensive capillary fragility and disruption of clotting.
  - The Lassa fever virus is an arenavirus found primarily in West Africa. Reservoir of the virus is a rodent found in Africa called the multimammate rat.
- **Nonhemorrhagic Fever Diseases:** Characterized by high fever without the capillary fragility that leads to hemorrhagic symptoms.
  - Brucellosis: Also called Malta fever, undulant fever, Bang's disease. Multiple species of the genus *Brucella* cause disease in humans, including *B. melitensis, B. canis, B. abortis,* and *B. suis.*
  - Q Fever: Caused by *Coxiella burnetii,* a very small pleomorphic gram-negative bacterium and intracellular parasite. *C. burnetii* is harbored by a wide assortment of vertebrates and arthropods, especially ticks. However, humans acquire infection mainly by environmental contamination and airborne transmission.
  - Cat-Scratch Disease: Infection by *Bartonella henselae,* connected with being clawed or bitten by a cat.
  - Trench Fever: Causative agent, *Bartonella quintana,* is carried by lice. Highly variable symptoms can include a 5- to 6-day fever, leg pains, headache, chills, and muscle aches. Related pathogen, *B. rochalimae,* recently identified.
  - Ehrlichioses: There are two major tick-borne, fever-producing diseases caused by members of the genus *Ehrlichia* and *Anaplasma.*
  - Rocky Mountain Spotted Fever: Another tick-borne disease; causes a distinctive rash. Caused by *Rickettsia rickettsii.*

Source: ASM Curriculum Guidelines (American Society for Microbiology, 2012).
Complete guidelines in appendix B of this book.

- **Chagas Disease:** *Trypanosoma cruzi* transmitted by insects; disease has acute and chronic phases. Endemic in South and Central America.
- **Malaria:** Symptoms are malaise, fatigue, vague aches, and nausea, followed by bouts of chills, fever, and sweating. Symptoms occur at 48- or 72-hour intervals, as a result of synchronous rupturing of red blood cells. Causative organisms are *Plasmodium* species: *P. malariae, P. vivax, P. falciparum, P. ovale,* and *P. knowlesi.* Carried by *Anopheles* mosquito.
- **HIV Infection and AIDS:** Symptoms directly tied to the level of virus in the blood versus the level of T cells in the blood.

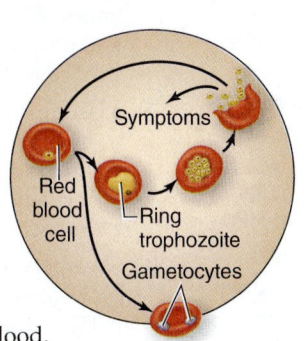

- HIV is a retrovirus (genus *Lentivirus*). Contains *reverse transcriptase,* which catalyzes the replication of double-stranded DNA from single-stranded RNA. Retroviral DNA incorporated into the host genome as provirus that can be passed on to progeny cells in latent state.
- Destruction of T4 lymphocytes paves way for invasion by opportunistic agents and malignant cells.
- HIV transmission occurs mainly through sexual intercourse and transfer of blood or blood products.

Multiple-Choice and True-False Questions | Bloom's Levels 1 and 2: Remember and Understand

Multiple-Choice Questions. Select the correct answer from the options provided.

1. When bacteria flourish and grow in the bloodstream, this is referred to as
   a. viremia.
   b. bacteremia.
   c. septicemia.
   d. fungemia.

2. Which of the following diseases is caused by a retrovirus?
   a. Lassa fever
   b. cat-scratch disease
   c. anthrax
   d. HIV

3. The plague bacterium, *Yersinia pestis,* is transmitted mainly by
   a. mosquitoes.
   b. fleas.
   c. dogs.
   d. birds.

4. Rabbit fever is caused by
   a. *Yersinia pestis.*
   b. *Francisella tularensis.*
   c. *Borrelia burgdorferi.*
   d. *Chlamydia bunnyensis.*

5. A distinctive bull's-eye rash results from a tick bite transmitting
   a. Lyme disease.
   b. tularemia.
   c. Q fever.
   d. Rocky Mountain spotted fever.

6. Cat-scratch disease is caused by
   a. *Coxiella burnetii.*
   b. *Bartonella henselae.*
   c. *Bartonella quintana.*
   d. *Brucella abortus.*

7. The bite of the tick *Ixodes scapularis* can cause
   a. ehrlichioses.
   b. Lyme disease.
   c. trench fever.
   d. both a and b.
   e. both b and c.

8. Cat-scratch disease is effectively treated with
   a. rifampin.
   b. penicillin.
   c. amoxicillin.
   d. azithromycin.

9. Wool-sorter's disease is caused by
   a. *Brucella abortus.*
   b. *Bacillus anthracis.*
   c. *Coxiella burnetii.*
   d. rabies virus.

10. Which of the following is *not* a hemorrhagic fever?
    a. Lassa fever
    b. Marburg fever
    c. Ebola fever
    d. trench fever

True-False Questions. If the statement is true, leave as is. If it is false, correct it by rewriting the sentence.

11. Brucellosis can be transmitted to humans by drinking contaminated milk.

12. Respiratory tract infection with *Bartonella henselae* is considered an AIDS-defining condition.

13. Lyme disease is caused by *Rickettsia rickettsii.*

14. Yellow fever is caused by a protozoan transmitted by fleas.

15. HIV in the United States is mainly transmitted via male homosexual sex.

Critical Thinking Questions | Bloom's Levels 3, 4, and 5: Apply, Analyze, and Evaluate

Critical thinking is the ability to reason and solve problems using facts and concepts. These questions can be approached from a number of angles and, in most cases, they do not have a single correct answer.

1. What is endotoxic shock? Explain why the pharmaceutical industry must be mindful of this condition when creating drugs using recombinant DNA technology.

2. Explain why cases of dengue fever have been observed beyond endemic regions of the world today. Discuss whether or not completely eradicating mosquito (vector) populations from disease-ridden areas is advisable.

3. Summarize the life cycle of the malarial parasite, including the significant events of sexual and asexual reproduction.

4. Explain why a highly effective vaccine does not yet exist for malaria. What progress is currently being made today?

5. a. Discuss whether or not genetics plays a role in HIV infection, providing at least one example to illustrate your position.
   b. Provide evidence in support of or refuting the following statement: An HIV-positive individual will always harbor the virus even if no viral load is detectable by PCR or other methods.

6. a. Explain why over the years the incidence of HIV infection has declined in the United States while the prevalence of AIDS has increased.
   b. Summarize the four drugs used in HIV therapy today, and describe the mechanism of action of each drug against HIV.

7. Some South African sex workers regularly exposed to HIV have been known to resist infection for years. Also, the Sydney Blood Bank Cohort represents a group of individuals termed "long-term nonprogressors," who have been infected with HIV for over 30 years. Conduct additional research and see what scientists have learned from these two groups and how this information may shape the development of protective treatments or vaccines against HIV.

8. Today, few cases of natural anthrax are seen in the United States. What group of individuals is primarily affected today, and how do they become infected? Which form of anthrax do they commonly develop?

9. a. Compare and contrast various characteristics of hemorrhagic and nonhemorrhagic fever diseases.
   b. Provide an explanation for the observed increase in incidence of these zoonotic infections around the world today.

10. Several pathogens in this chapter are listed as potential bioweapons by the Centers for Disease Control and Prevention (CDC). Discuss two examples and why they are considered effective for use in bioterrorism.

Concept Connections | Bloom's Levels 4 and 6: Analyze and Create

This activity ties together multiple concepts in the chapter.

1. What cardiovascular and lymphatic system diseases are transmitted by arthropod vectors?

2. Which of these vector-borne diseases are bacterial? Which are protozoan? Which are viral?

3. What are the difficulties in preventing and/or controlling vector-borne diseases?

4. What might be the ecological consequences of eliminating a particular vector in an effort to control infection?

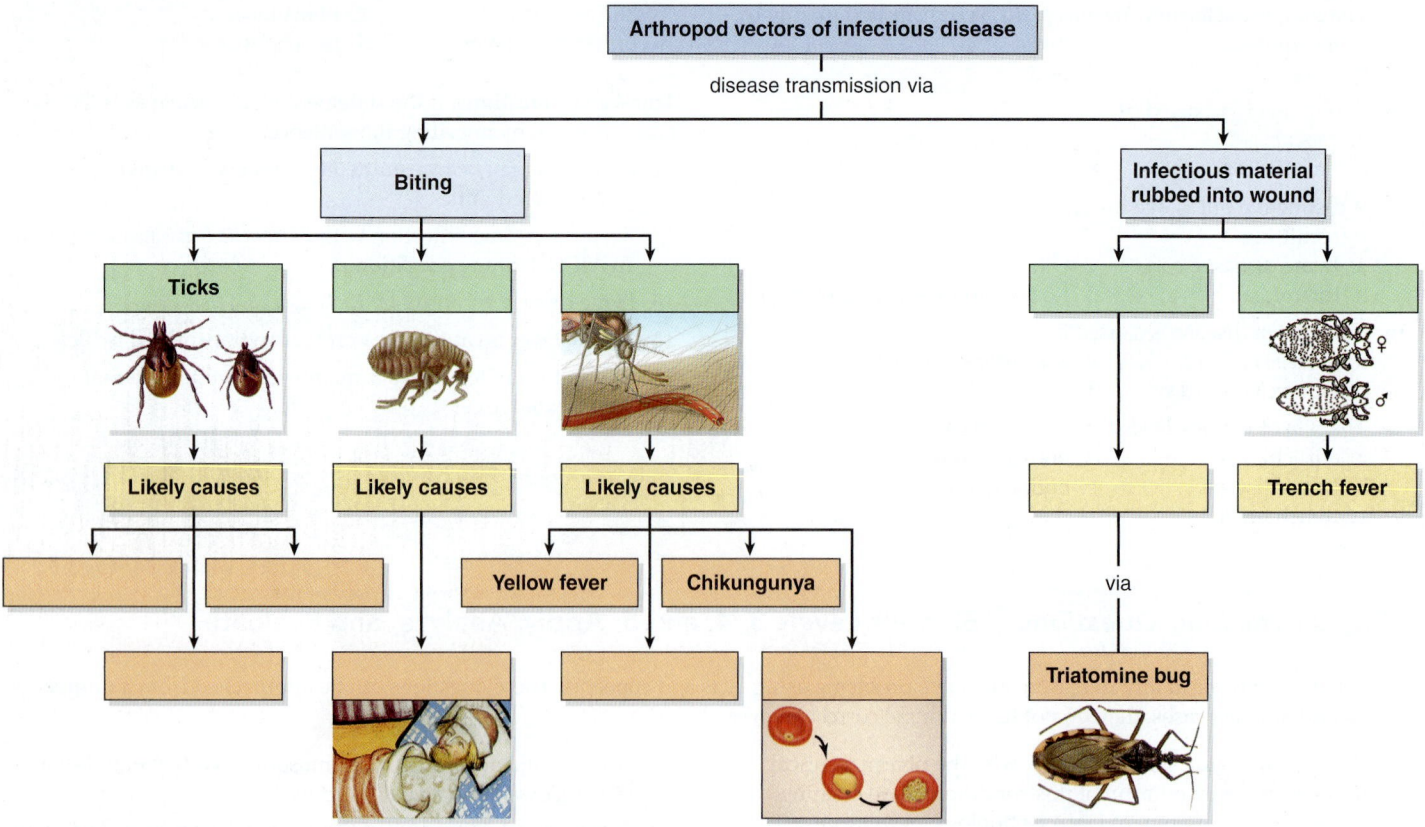

Visual Connections | Bloom's Level 5: Evaluate

This question uses visual images or previous content to make connections to this chapter's concepts.

1. a. **From chapter 14, figure 14.19.** Imagine that the WBCs shown in this illustration are unable to control the microorganisms. Could the change that has occurred in the vessel wall help the organism spread to other locations? If so, how?

   b. If the organisms are able to survive phagocytosis, how could that impact the progress of this disease? Explain your answer.

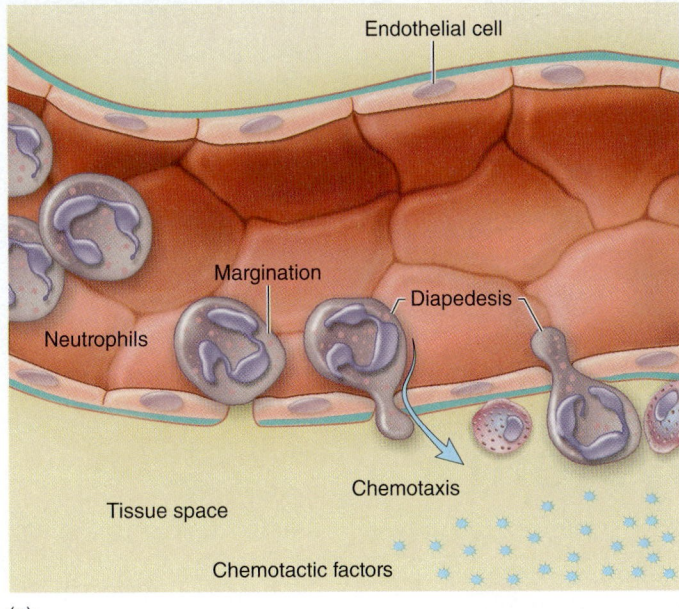

(a)

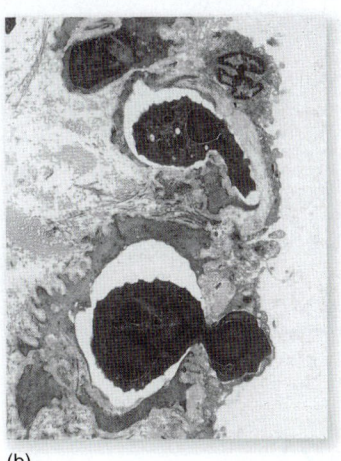

(b)

Concept Mapping | Bloom's Level 6: Create

Appendix D provides guidance for working with concept maps.

1. Using the words that follow, please create a concept map illustrating the relationships among these key terms from chapter 20.

HIV	AIDS	CD8 lymphocytes	specific immunity
reverse transcriptase	CD4 lymphocytes	macrophages	opportunistic infections
latency	B lymphocytes	leukopenia	

 www.mcgrawhillconnect.com

Enhance your study of this chapter with study tools and practice tests. Also ask your instructor about the resources available through ConnectPlus, including the media-rich eBook, interactive learning tools, and animations.

Infectious Diseases Affecting the Respiratory System

Case File 21

Tuberculosis Outbreak Among the Homeless in Florida

During the period 2004 to 2012, Florida experienced the worst outbreak of tuberculosis (TB) seen by the Centers for Disease Control and Prevention (CDC) in 20 years. The outbreak included at least 99 cases and 13 deaths, mainly among homeless people living in Duval County. According to reports, the outbreak was traced in part to a homeless man with schizophrenia who complained of a cough in 2008. He visited numerous facilities seeking treatment and was housed in homeless shelters, mental hospitals, and prisons before he was diagnosed with tuberculosis. The patient infected at least 17 other individuals during the intervening 4 years.

When the CDC became involved in tracking the TB outbreak in Florida, it found that all of the cases could be traced to homeless shelters. TB has a relatively low prevalence in the U.S. population, but there are still significant numbers of cases among homeless people. CDC officials found that this single individual had infected 88% of his contacts in the facilities in which he was housed.

Interestingly, at the same time this outbreak investigation was taking place, the Florida legislature voted to shut down the Holley State Tuberculosis Hospital in a cost-saving move. Patients were moved to Miami's Jackson Memorial Hospital, but 18 Holley patients were released to the community to be cared for by their own doctors.

- How is tuberculosis transmitted in a population?
- Why are homeless people at risk for contracting tuberculosis?

Continuing the Case appears on page 653.

Outline and Learning Outcomes

21.1 The Respiratory Tract and Its Defenses
  1. Draw or describe the anatomical features of the respiratory tract.
  2. List the natural defenses present in the respiratory tract.

21.2 Normal Biota of the Respiratory Tract
  3. List the main genera of normal biota presently known to occupy the respiratory tract.

21.3 Upper Respiratory Tract Diseases Caused by Microorganisms
  4. List the possible causative agents, modes of transmission, virulence factors, diagnostic techniques, and prevention and treatment for each of the diseases of the upper respiratory tract: the common cold, sinusitis, otitis media, pharyngitis, and diphtheria.
  5. Identify which disease is often caused by a mixture of microorganisms.
  6. Identify two bacteria that can cause dangerous pharyngitis cases.

21.4 Diseases Caused by Microorganisms Affecting Both the Upper and Lower Respiratory Tracts
  7. List the possible causative agents for each of the infectious conditions affecting both the upper and lower respiratory tract: whooping cough, RSV disease, and influenza.
  8. Describe the symptoms appearing in each stage of whooping cough.
  9. Discuss reasons for the increase of pertussis cases over the past three decades.
  10. Identify the age group at most risk for serious disease from RSV.
  11. Plan a response for a situation in which a patient declines influenza vaccination because he or she believes that the warnings about pandemics are always exaggerated.
  12. Compare and contrast *antigenic drift* and *antigenic shift* in influenza viruses.

21.5 Lower Respiratory Tract Diseases Caused by Microorganisms
  13. List the possible causative agents for each of the diseases affecting the lower respiratory tract: tuberculosis, community-acquired pneumonia, healthcare-associated pneumonia, and hantavirus pulmonary syndrome.
  14. Discuss the problems associated with MDR-TB and XDR-TB.
  15. Demonstrate an in-depth understanding of the current epidemiology of tuberculosis infection.
  16. Explain why so many diverse microorganisms can cause the condition of pneumonia.
  17. Identify the top three causes of community-acquired pneumonia.
  18. List the distinguishing characteristics of healthcare-associated pneumonia compared to community-acquired pneumonia.
  19. Outline the chain of transmission of the causative agent of hantavirus pulmonary syndrome.

21.1 The Respiratory Tract and Its Defenses

The respiratory tract is the most common place for infectious agents to gain access to the body. Obviously, we breathe 24 hours a day and anything in the air we breathe passes at least temporarily into this organ system.

The structure of the system is illustrated in **figure 21.1a.** Most clinicians divide the system into two parts, the *upper* and *lower respiratory tracts.* The upper respiratory tract includes the mouth, the nose, the nasal cavity and sinuses above it, the throat or pharynx, and the epiglottis and larynx. The lower respiratory tract begins with the trachea, which feeds into the bronchi and bronchioles in the lungs. Attached to the bronchioles are small balloonlike structures called alveoli, which inflate and deflate with inhalation and exhalation. These are the sites of oxygen exchange in the lungs.

Several anatomical features of the respiratory system protect it from infection. As described in chapter 14, nasal hair serves to trap particles. Cilia **(figure 21.1b)** on the epithelium of the trachea and bronchi (the ciliary escalator) propel particles upward and out of the respiratory tract. Mucus on the surface of the mucous membranes lining the respiratory tract is a natural trap for invading microorganisms. Once the microorganisms are trapped, involuntary responses such as coughing, sneezing, and swallowing can move them out of sensitive areas. These are first-line defenses.

The second and third lines of defense also help protect the respiratory tract. Complement action in the lungs helps to protect against invading pathogens, and increased levels of cytokines and antimicrobial peptides further reduce the ability of microbes to cause disease. Macrophages inhabit the alveoli of the lungs and the clusters of lymphoid tissue (tonsils) in the throat. Secretory IgA that targets specific pathogens can be found in the mucus secretions as well.

21.1 Learning Outcomes—Assess Your Progress

  1. Draw or describe the anatomical features of the respiratory tract.
  2. List the natural defenses present in the respiratory tract.

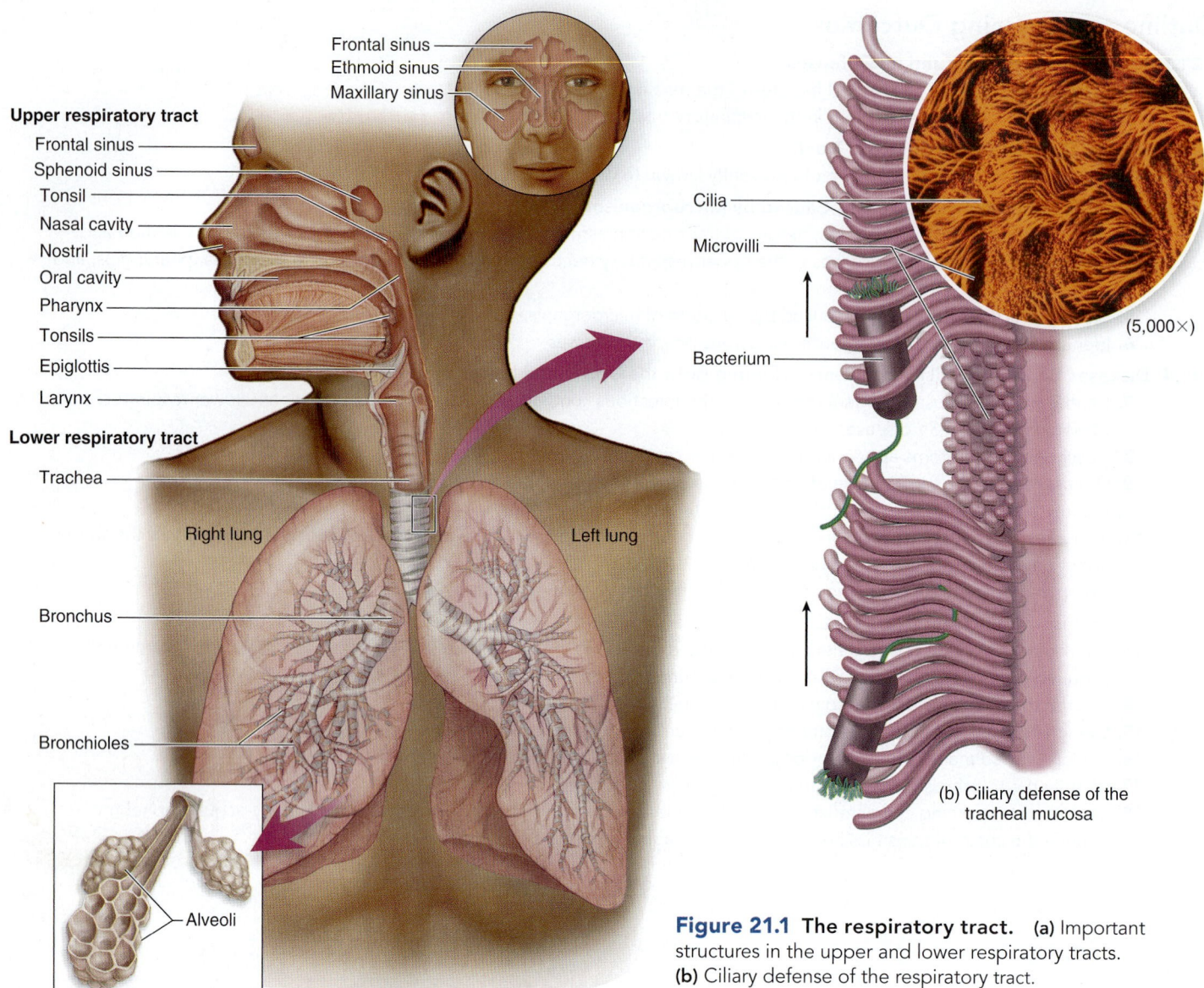

Upper respiratory tract
- Frontal sinus
- Sphenoid sinus
- Tonsil
- Nasal cavity
- Nostril
- Oral cavity
- Pharynx
- Tonsils
- Epiglottis
- Larynx

Frontal sinus
Ethmoid sinus
Maxillary sinus

Lower respiratory tract
- Trachea

Right lung   Left lung

Bronchus

Bronchioles

Alveoli

(a) Anatomy of the respiratory system

Cilia
Microvilli

Bacterium

(5,000×)

(b) Ciliary defense of the tracheal mucosa

Figure 21.1 The respiratory tract. (a) Important structures in the upper and lower respiratory tracts. **(b)** Ciliary defense of the respiratory tract.

21.2 Normal Biota of the Respiratory Tract

Because of its constant contact with the external environment, the respiratory system harbors a large number of commensal microorganisms. It was previously thought that such microbes were only localized to the upper respiratory tract. Recent metagenomic analysis of sputum samples, however, has shown that healthy lungs are a virtual tapestry of microorganisms. The bronchial tree in these studies harbored an average of 2,000 bacterial genomes per sample analyzed. A significant portion of the normal biota belongs to nine major bacterial genera: *Prevotella, Sphingomonas, Pseudomonas, Acinetobacter, Fusobacterium, Megasphaera, Veillonella, Staphylococcus,* and *Streptococcus.* Yeasts, especially *Candida albicans,* also colonize the mucosal surfaces of the mouth in the upper respiratory tract.

Note that some bacteria considered "normal biota" in the respiratory tract can cause serious disease especially in immunocompromised individuals; these include *Streptococcus pyogenes, Haemophilus influenzae, Streptococcus pneumoniae, Neisseria meningitidis,* and *Staphylococcus aureus.* In addition, researchers have discovered that the overall composition of the lung microbiome is altered in patients suffering from lung disorders such as chronic obstructive pulmonary disease (COPD), asthma, and cystic fibrosis. Pockets of distinct microbial composition even appear to develop within the lungs of these individuals suggesting that pathogenesis may be due in large part to the action of microorganisms. Further studies may reveal new treatments that will involve reestablishing the normal biota composition within the lungs of these patients.

In the respiratory system, as in some other organ systems, the normal biota performs the important function of microbial antagonism (see chapter 13). This reduces the chances of pathogens establishing themselves in the same area by competing with them for resources and space. To illustrate this point, *Lactobacillus sakei,* a known member of the sinus microbiome, can suppress the pathogenic potential

Respiratory System Defenses and Normal Biota		
	Defenses	**Normal Biota**
Respiratory System	Nasal hair, ciliary escalator, mucus, involuntary responses such as coughing and sneezing, secretory IgA, alveolar macrophages, cytokines and complement	Large number of genera. Most abundant: *Streptococcus, Prevotella, Sphingomonas, Pseudomonas, Acinetobacter, Fusobacterium, Megasphaera, Veillonella,* and *Staphylococcus* Note: *Streptococcus pyogenes, Streptococcus pneumoniae, Haemophilus influenzae, Neisseria meningitidis,* and *Staphylococcus aureus* often present as "normal" biota.

of another normal biota organism, *Corynebacterium tuberculo-stearicum*, reducing the incidence of sinus infection. It may be that in the future, a dose of probiotics may be the most effective way to treat a variety of disorders.

21.2 Learning Outcomes—Assess Your Progress

3. List the main genera of normal biota presently known to occupy the respiratory tract.

21.3 Upper Respiratory Tract Diseases Caused by Microorganisms

The Common Cold

In the course of a year, people in the United States suffer from about 67 million colds caused by viruses. The common cold is often called *rhinitis,* from the Latin word *rhin,* meaning "nose," and the suffix *-itis,* meaning "inflammation." Many people have several episodes a year, and economists estimate that this fairly innocuous infection costs the United States $40 billion a year in trips to the doctors, medications, and 22 million missed days of work.

▶ Signs and Symptoms

Everyone is familiar with the symptoms of a cold: sneezing, scratchy throat, and runny nose (rhinorrhea), which usually begin 2 or 3 days after infection. An uncomplicated cold generally is not accompanied by fever, although children can experience low fevers (less than 102°F). Note that people with asthma and other underlying respiratory conditions (i.e., chronic obstructive pulmonary disease, or COPD) often suffer more severe symptoms triggered by the common cold.

▶ Causative Agents

The common cold is caused by one of over 200 different kinds of viruses. The particular virus is almost never identified, and the symptoms and handling of the infection are the same no matter which of the viruses is responsible.

The most common type of virus leading to the common cold is the group called rhinoviruses, of which there are 99 serotypes. Coronaviruses and adenoviruses are also major causes. Most viruses causing the common cold never lead to any serious consequences, but some of them can be serious for some patients. The respiratory syncytial virus (RSV) causes colds in most people, but in some, especially infants,

infection with this virus can lead to more serious respiratory tract symptoms (discussed later in the chapter). In this section, we consider all cold-causing viruses together as a group because they are treated similarly.

Viral infection of the upper respiratory tract can predispose a patient to secondary infections by other microorganisms, such as bacteria. Secondary infections may explain why some people report that their colds improved when they were given antibiotics. A virus originally caused the cold, but a bacterial infection might have followed.

▶ Pathogenesis and Virulence Factors

Viruses that induce the common cold do not have many virulence mechanisms. They must penetrate the mucus that coats the respiratory tract and then find firm attachment points. Once they are attached, they use host cells to produce more copies of themselves (see chapter 6). The symptoms we experience as the common cold are mainly the result of our body fighting back against the viral invaders. Virus-infected cells in the upper respiratory tract release chemicals that attract certain types of white blood cells to the site, and these cells release cytokines and other inflammatory mediators, as described earlier in chapters 14 and 16. These mediators generate a localized inflammatory reaction, characterized by swelling and inflammation of the nasal mucosa, leakage of fluid from capillaries and lymph vessels, and increased production of mucus.

▶ Transmission and Epidemiology

Cold viruses are transmitted by droplet contact, but indirect transmission may be more common, such as when a healthy person touches a fomite and then touches one of his or her own vulnerable surfaces, such as the mouth, nose, or an eye. In some cases, the viruses can remain airborne in droplet nuclei and aerosols and can be transmitted via the respiratory route.

The epidemiology of the common cold is fairly simple: Practically everybody gets colds—and fairly frequently. Children have more frequent infections than adults, probably because nearly every virus they encounter is a new one and they have no secondary immunity to it. People can acquire some degree of immunity to a cold virus that they have encountered before, but because there are more than 200 viruses, this immunity doesn't provide much overall protection.

▶ Prevention

There is no vaccine for the common cold. A traditional vaccine would need to contain antigens from about 200 viruses to provide complete protection. Researchers are studying novel types of immunization strategies, however. Because most of the viruses causing the common cold use only a few different chemicals on host epithelium for their attachment site, some scientists have proposed developing a vaccine that would stimulate antibody to the docking site on the host. Other approaches include inducing antibody to the sites of action for the inflammatory mediators. But for now, the best prevention is to stop the transmission between hosts. The best way to prevent transmission is frequent hand washing, followed closely by stopping droplets from traveling away from the mouth and nose by covering them when sneezing or coughing. It is better to do this by covering the face with the crook of the arm rather than the hand, because subsequent contact with surfaces is less likely.

▶ Treatment

No chemotherapeutic agents cure the common cold. A wide variety of over-the-counter agents, such as antihistamines and decongestants, improve symptoms by blocking inflammatory mediators and their action. The use of these agents may also cut down on transmission to new hosts, because fewer virus-loaded secretions are produced. Zinc appears to block the replication of rhinovirus; however, it appears to only reduce the duration of the common cold and not prevent the disease. Scientists are currently working on a unique therapeutic treatment called DRACO (discussed in chapter 12), which may protect against virtually any viral infectious agent, including those causing the common cold **(Disease Table 21.1).**

Disease Table 21.1 The Common Cold

Causative Organism(s)	Approximately 200 viruses
Most Common Modes of Transmission	Indirect contact, droplet contact
Virulence Factors	Attachment proteins; most symptoms induced by host response
Culture/Diagnosis	Not necessary
Prevention	Hygiene practices
Treatment	For symptoms only
Epidemiological Features	Highest incidence among preschool and elementary schoolchildren with average of three to eight colds per year; adults and adolescents: two to four colds per year

Sinusitis

Commonly called a *sinus infection*, this inflammatory condition of any of the four pairs of sinuses in the skull (see figure 21.1) can actually be caused by allergy (most common), or infections. The infectious agents that may be responsible for the condition include a variety of viruses or bacteria and, less commonly, fungi. Infections of the sinuses often follow a bout with the common cold. The inflammatory symptoms of a cold produce a large amount of fluid and mucus and when trapped in the sinuses, and these secretions provide an excellent growth medium for bacteria or fungi. This is why it is common for patients suffering from the common cold to then develop sinusitis caused by bacteria or fungi.

▶ Signs and Symptoms

A person suffering from any form of sinusitis typically experiences nasal congestion, pressure above the nose or in the forehead, and sometimes the feeling of a headache or a toothache. Facial swelling and tenderness are common. Discharge from the nose and mouth appears opaque and may have a green or yellow color in the case of bacterial infections. Viral infections are less likely to produce colored discharge. Discharge caused by an allergy is usually clear, and the symptoms may be accompanied by itchy, watery eyes.

▶ Causative Agents

Viruses   Viral infection is probably the most common cause of mild sinusitis. The viruses involved are the same as with the common cold.

Bacteria   Any number of bacteria that are normal biota in the upper respiratory tract may cause sinus infections. Many cases are caused by *Streptococcus pneumoniae, Streptococcus pyogenes, Staphylococcus aureus, Corynebacterium tuberculostearicum,* and *Haemophilus influenzae.* The causative organism is usually not identified, but treatment is begun empirically, based on the symptoms.

The bacteria that cause these infections are most often normal biota in the host and don't have an arsenal of virulence factors that lead to their ability to cause disease. The pathogenesis of this condition is brought about by the confluence of several factors: predisposition to infection because of underlying infection; buildup of fluids, providing a rich environment for bacterial multiplication; and sometimes the anatomy of the sinuses, which can contribute to entrapment of mucus and bacterial growth.

Bacterial sinusitis is not a communicable disease. Of course, the virus causing a preceding cold is transmissible, but the host takes it from there by creating the conditions favorable for respiratory tract microorganisms to multiply in the sinus spaces.

Sinusitis is extremely common and results in approximately 11.5 million office visits a year in the United States. A large proportion of these cases are allergic sinusitis episodes, but approximately 30% of them are caused by bacterial overgrowth in the sinuses. Women and residents of the southern United States have slightly higher rates. As with many upper respiratory tract infections, smokers have higher rates of infection than nonsmokers. Children who are exposed to large amounts of secondhand smoke are also more susceptible. Infections lasting for longer than 12 weeks are termed *chronic sinusitis*, and such cases are often difficult to treat due to loss of mucociliary defenses and bacterial biofilm formation in the sinuses.

Broad-spectrum antibiotics may be prescribed when the physician feels that the sinusitis is bacterial in origin. However, such antibiotic treatment may not actually increase the patient's recovery rate or symptomology. Most uncomplicated cases may be best treated by having the patient "wait it out" while his or her own immune system clears the infection.

Fungi Fungal sinusitis is rare, but it is often recognized when antibacterial drugs fail to alleviate symptoms. Simple fungal infections may normally be found in the maxillary sinuses and are noninvasive in nature. These colonies are generally not treated with antifungal agents but instead are simply mechanically removed by a physician. *Aspergillus fumigatus* is a common fungus involved in this type of infection, but *Bipolaris* species are an emerging cause of fungal sinusitis today. The growth of fungi in this type of sinusitis may be encouraged by trauma to the area.

More serious invasive fungal infections of the sinuses may be found in severely immunocompromised patients. Fungi such as *Aspergillus* and *Mucor* species may invade the bony structures in the sinuses and even travel to the brain or eye. These infections are treated aggressively with a combination of surgical removal of the fungus and intravenous antifungal therapy **(Disease Table 21.2).**

Acute Otitis Media (Ear Infection)

This condition is another common sequela of the common cold—for reasons similar to the ones described for sinusitis. Viral infections of the upper respiratory tract lead to inflammation of the eustachian tubes and the buildup of fluid in the middle ear, which can lead to bacterial multiplication in those fluids. Bacteria can migrate along the eustachian tube from the upper respiratory tract **(figure 21.2).** When bacteria encounter mucus and fluid buildup in the middle ear, they multiply rapidly. Their presence increases the inflammatory response, leading to pus production and continued fluid secretion. This fluid is referred to as *effusion*.

Another condition, known as chronic otitis media, occurs when fluid remains in the middle ear for indefinite periods of time. Until recently, physicians considered it to be the result of a noninfectious immune reaction because they could not culture bacteria from the site and because antibiotics were not effective. New data suggest that this form of otitis media is caused by a mixed biofilm of bacteria that are attached to the membrane of the inner ear. Biofilm bacteria generally are less susceptible to antibiotics (as discussed in chapter 4), and their presence in biofilm form would explain the inability to culture them from ear fluids.

▶ **Signs and Symptoms**

Otitis media may be accompanied by a sensation of fullness or pain in the ear and loss of hearing. Younger children may exhibit irritability, fussiness, and difficulty in sleeping,

Disease Table 21.2	Sinusitis		
Causative Organism(s)	Viruses	Various bacteria, often mixed infection	Various fungi
Most Common Modes of Transmission	Direct contact, indirect contact	Endogenous (opportunism)	Introduction by trauma or opportunistic overgrowth
Virulence Factors	–	–	–
Culture/Diagnosis	Culture not usually performed; diagnosis based on clinical presentation.	Culture not usually performed; diagnosis based on clinical presentation, occasionally X rays or other imaging technique used	Same
Prevention	Hygiene	–	–
Treatment	None	Broad-spectrum antibiotics or none	Physical removal of fungus; in severe cases, antifungals used
Distinctive Features	Viral and bacterial much more common than fungal	Viral and bacterial much more common than fungal	Suspect in immunocompromised patients
Epidemiological Features	–	United States: affects 1 of 7 adults; between 12 and 30 million diagnoses per year	Fungal sinusitis varies with geography; in United States, more common in SE and SW; internationally: more common in India, North Africa, Middle East

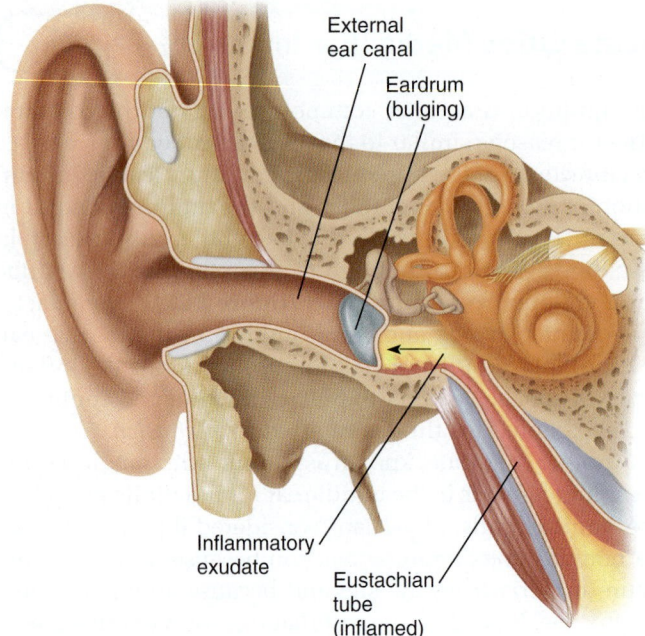

External
ear canal

Eardrum
(bulging)

Inflammatory
exudate

Eustachian
tube
(inflamed)

Figure 21.2 An infected middle ear.

eating, or hearing. Severe or untreated infections can lead to rupture of the eardrum because of pressure of pus buildup, or to internal breakthrough of these infected fluids, which can lead to more serious conditions such as mastoiditis, meningitis, or intracranial abscess.

▶ Causative Agents

Many different viruses and bacteria can cause acute otitis media, but the most common cause is *Streptococcus pneumoniae* (also discussed in the section on pneumonia later in this chapter). *Streptococcus pneumoniae* appears as pairs of elongated, gram-positive cocci joined end to end. It is often called by the familiar name *pneumococcus*, and diseases caused by it are termed *pneumococcal*.

Haemophilus influenzae is another common cause of this condition; however, the incidence of all types of infections with this bacterium was significantly reduced with the introduction of a childhood vaccine against it in the 1980s. Scientists now believe that the majority of acute and chronic otitis media cases are mixed infections with viruses and bacteria acting together to cause disease.

▶ Transmission and Epidemiology

Otitis media is a sequela of upper respiratory tract infection and is not communicable, although the upper respiratory infection preceding it is. Children are particularly susceptible, and boys have a slightly higher incidence than do girls.

▶ Prevention

A vaccine against *S. pneumoniae* has been a part of the recommended childhood vaccination schedule since 2000. The vaccine (Prevnar) is a seven-valent conjugated vaccine (see chapter 15). It contains polysaccharide capsular material from seven different strains of the bacterium complexed with a chemical that makes

Disease Table 21.3	Otitis Media		
Causative Organism(s)	*Streptococcus pneumoniae*	*Haemophilus influenzae*	Other bacteria/viruses
Most Common Modes of Transmission	Endogenous (may follow upper respiratory tract infection by *S. pneumoniae* or other microorganisms)	Endogenous (follows upper respiratory tract infection)	Endogenous
Virulence Factors	Capsule, hemolysin	Capsule, fimbriae	–
Culture/Diagnosis	Usually relies on clinical symptoms and failure to resolve within 72 hours	Same	Same
Prevention	Pneumococcal conjugate vaccine (heptavalent)	Hib vaccine	None
Treatment	Wait for resolution; if needed, amoxicillin (are high rates of resistance) or amoxicillin + clavulanate or cefuroxime	Same as for *S. pneumoniae*	Wait for resolution; if needed, a broad-spectrum antibiotic (azithromycin) might be used in absence of etiologic diagnosis
Distinctive Features	–	–	Suspect if fully vaccinated against other two
Epidemiological Features	United States: 70% of children experience at least one case before age 2; in developing world: chronic otitis media results in significant hearing loss in 100s of millions and death in approx. 30,000 per year (in absence of treatment)		

it more antigenic. In 2012, the FDA approved a new Prevnar vaccine that contains antigens from 13 strains of this pathogenic bacterium. There is an older vaccine—called Pneumovax—for the same bacterium, which is primarily targeted to the older population to prevent pneumococcal pneumonia.

▶ Treatment

Until the late 1990s, broad-spectrum antibiotics were routinely prescribed for otitis media. When it became clear that frequently treating children with these drugs was producing a bacterial biota with high rates of antibiotic resistance, the treatment regimen was reexamined.

The current recommendation for uncomplicated acute otitis media with a fever below 104°F is "watchful waiting" for 72 hours to allow the body to clear the infection, avoiding the use of antibiotics. When antibiotics are used, antibiotic resistance must be considered. Children who experience frequent recurrences of ear infections sometimes have small tubes placed through the tympanic membranes into their middle ears to provide a means of keeping fluid out of the site when inflammation occurs. Scientists believe that normal biota of the upper respiratory tract may play a role in suppressing the action of pathogens associated with acute otitis media, which may lead to probiotic-based treatments for this disease in the future **(Disease Table 21.3).**

Pharyngitis

▶ Signs and Symptoms

The name says it all—this is an inflammation of the throat, which the host experiences as pain and swelling. The severity of pain can range from moderate to severe, depending on the causative agent. Viral sore throats are generally mild and sometimes lead to hoarseness. Sore throats caused by bacteria are generally more painful than those caused by viruses, and they are more likely to be accompanied by fever, headache, and nausea.

Clinical signs of a sore throat are reddened mucosa, swollen tonsils, and sometimes white packets of inflammatory products visible on the walls of the throat, especially in streptococcal disease **(figure 21.3).** The mucous membranes may be swollen, affecting speech and swallowing. Often pharyngitis results in foul-smelling breath. The incubation period for most sore throats is generally 2 to 5 days.

▶ Causative Agents

The same viruses causing the common cold most commonly cause a sore throat. It can also accompany other diseases, such as infectious mononucleosis (described in chapter 20). Pharyngitis may simply be the result of mechanical irritation from prolonged shouting or from drainage of an infected sinus cavity. The bacteria that cause the most serious cases of pharyngitis are *Streptococcus pyogenes* and *Fusobacterium necrophorum.*

Streptococcus pyogenes

S. pyogenes is a gram-positive coccus that grows in chains. It does not form endospores, is nonmotile, and forms capsules and slime layers. *S. pyogenes* is a facultative anaer⦁ ferments a variety of sugars. It does not produce catal⦁ it does have a peroxidase system for inactivating hyⅾrogen peroxide, which allows its survival in the presence of oxygen.

▶ Pathogenesis

Untreated streptococcal throat infections occasionally can result in serious complications, either right away or days to weeks after the throat symptoms subside. These complications include scarlet fever, rheumatic fever, and glomerulonephritis. More rarely, invasive and deadly conditions—such as necrotizing fasciitis, which is described in chapter 18—can result from infection by *S. pyogenes.* The apparent mechanism behind some of these complications is autoimmunity. The bacterium has antigens on its surface that resemble heart, joint, and brain proteins. This initially allows the bacterium to evade immune detection. Eventually though, the immune system learns to respond to these antigens and then also may attack the human analogs.

There is an as-yet unproved link between "strep throat" and the sudden onset in children of a form of obsessive-compulsive disorder. The hypothesis is that the immune system attacks brain cells leading to abnormal behaviors. This syndrome is called PANDAS (pediatric autoimmune neuropsychiatric disorder associated with *Streptococcus*) and has in the past been treated with antibiotics. Studies are still in progress trying to establish definite causality.

Scarlet Fever   Scarlet fever is the result of infection with an *S. pyogenes* strain that is itself infected with a bacteriophage. This lysogenic virus confers on the streptococcus the ability to produce erythrogenic toxin, described in the section on virulence. Scarlet fever is characterized by a sandpaper-like rash, most often on the neck, chest, elbows, and inner surfaces of the thighs. High fever accompanies the rash. It most often affects school-age children and was a source of great suffering in the United States in the early part of the 20th century. In epidemic form, the disease can have a fatality rate of up to 95%. Most cases seen today are mild. They are easily

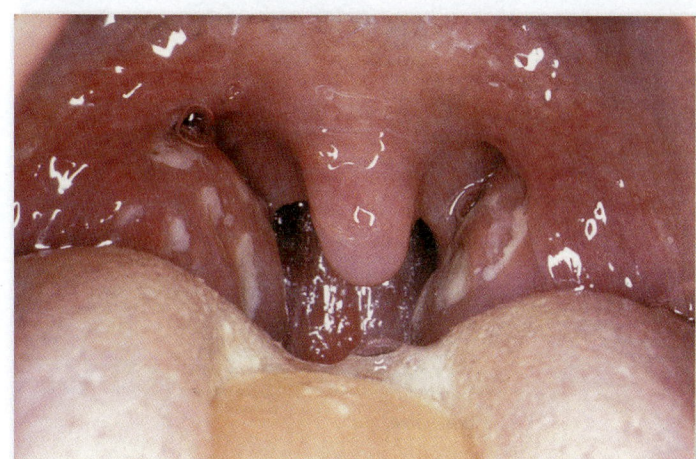

Figure 21.3 The appearance of the throat in pharyngitis and tonsillitis.   The pharynx and tonsils become bright red and suppurative. Whitish pus nodules may also appear on the tonsils.

recognizable and amenable to antibiotic therapy. Because of the fear elicited by the name "scarlet fever," the disease is often called "scarlatina" in North America.

Rheumatic Fever   Rheumatic fever is thought to be due to an immunologic cross-reaction between the streptococcal M protein and heart muscle. It tends to occur approximately 3 weeks after pharyngitis has subsided. It can result in permanent damage to heart valves **(figure 21.4).** Other symptoms include arthritis in multiple joints and the appearance of nodules over bony surfaces just under the skin.

Glomerulonephritis   Glomerulonephritis is thought to be the result of streptococcal proteins participating in the formation of antigen-antibody complexes, which then are deposited in the basement membrane of the glomeruli of the kidney. It is characterized by **nephritis** (appearing as swelling in the hands and feet and low urine output), blood in the urine, increased blood pressure, and occasionally heart failure. It can result in permanent kidney damage. The incidence of poststreptococcal glomerulonephritis has been declining in the United States, but it is still common in Africa, the Caribbean, and South America.

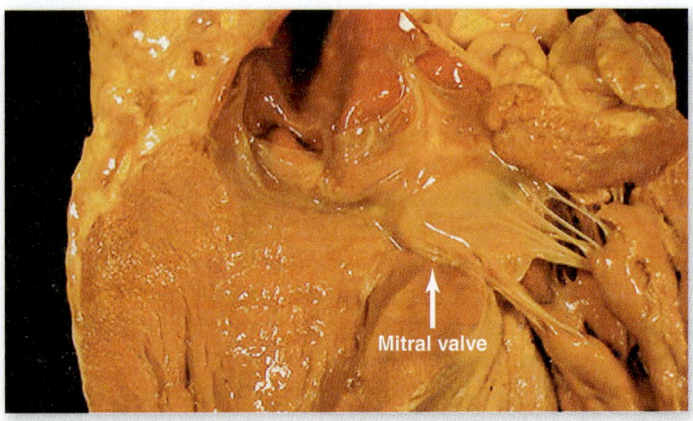

(a)

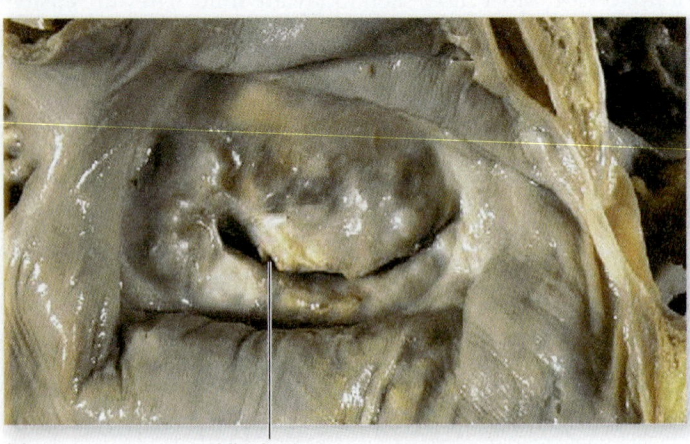

(b)   Mitral valve

Figure 21.4 The cardiac complications of rheumatic fever.
Scarring and deformation change the capacity of the valves to close and shunt the blood properly. **(a)** A normal valve, viewed from above. **(b)** A scarred mitral valve. The color difference in the two views is artificial.

▶ **Virulence Factors**

As already noted, surface antigens of *S. pyogenes* mimic host proteins, leading to collateral damage by the immune system. Its virulence is also enhanced by the substantial array of surface antigens, toxins, and enzymes it can generate.

Streptococci display numerous surface antigens **(figure 21.5).** Specialized polysaccharides on the surface of the cell wall help to protect the bacterium from being dissolved by the lysozyme of the host. Lipoteichoic acid (LTA) contributes to the adherence of *S. pyogenes* to epithelial cells in the pharynx. A spiky surface projection called *M protein* contributes to virulence by resisting phagocytosis and possibly by contributing to adherence. A capsule made of *hyaluronic* acid (HA) is formed by most *S. pyogenes* strains. It probably contributes to the bacterium's adhesiveness.

Extracellular Toxins   Group A streptococci owe some of their virulence to the effects of hemolysins called streptolysins. The two types are streptolysin O (SLO) and streptolysin S (SLS).[1] Both types cause beta-hemolysis of sheep blood agar (see "Culture and Diagnosis"). Both hemolysins rapidly injure many cells and tissues, including leukocytes and liver and heart muscle.

A key toxin in the development of scarlet fever is **erythrogenic** (eh-rith"-roh-jen'-ik) **toxin.** This toxin is responsible for the bright red rash typical of this disease **(figure 21.6a),** and it also induces fever by acting upon the temperature regulatory center in the brain. Only lysogenic strains of *S. pyogenes* that contain genes from a temperate bacteriophage can synthesize this toxin. (For a review of the concept of lysogeny, see chapter 6.)

Some of the streptococcal toxins (erythrogenic toxin and streptolysin O) contribute to increased tissue injury by acting as *superantigens.* These toxins elicit excessively strong reactions from monocytes and T lymphocytes. When activated, these cells proliferate and produce *tumor necrosis factor (TNF),* which leads to a cascade of immune responses resulting in vascular injury. This is the likely mechanism for the severe pathology of toxic shock syndrome and necrotizing fasciitis.

---

1. In SLO, O stands for oxygen because the substance is inactivated by oxygen. In SLS, *S* stands for serum because the substance has an affinity for serum proteins. SLS is oxygen-stable.

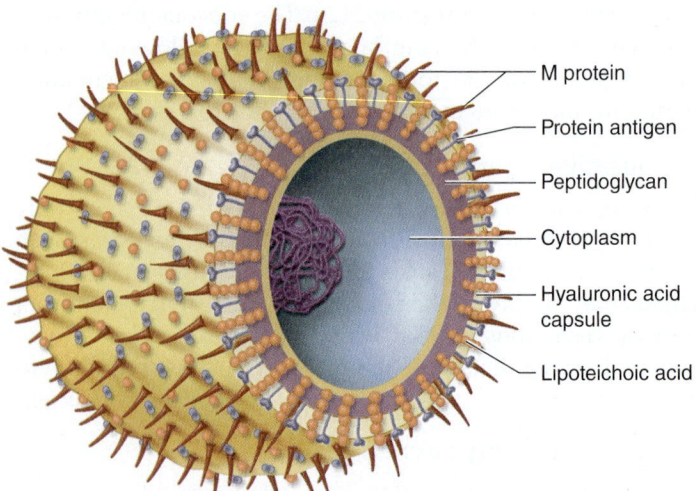

- M protein
- Protein antigen
- Peptidoglycan
- Cytoplasm
- Hyaluronic acid capsule
- Lipoteichoic acid

Figure 21.5 Cutaway view of group A streptococcus.

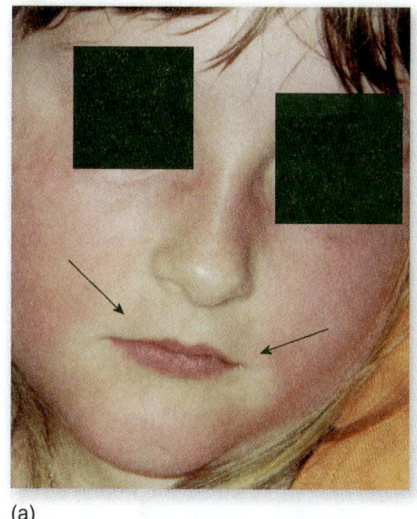

(a)

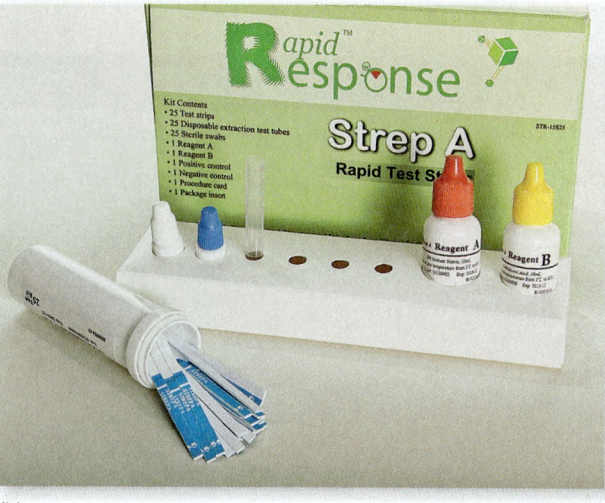

(b)

Figure 21.6 **Streptococcal infections.** **(a)** The bright red rash characteristic of scarlet fever. The area around the mouth remains unaffected. **(b)** A rapid immunologic test for diagnosis of group A infections. With this method, a patient's throat swab is introduced into an immunochromatographic or lateral flow system (see chapter 17).

▶ Transmission and Epidemiology

Physicians estimate that 30% of sore throats may be caused by *S. pyogenes*, adding up to several million cases each year. Most transmission of *S. pyogenes* is via respiratory droplets or direct contact with mucus secretions. This bacterium is carried as "normal" biota by 15% of the population, but transmission from this reservoir is less likely than from a person who is experiencing active disease from the infection because of the higher number of bacteria present in the disease condition. It is less common but possible to transmit this infection via fomites. Humans are the only significant reservoir of *S. pyogenes*.

More than 80 serotypes of *S. pyogenes* exist, meaning that people can experience multiple infections throughout their lives because immunity is serotype-specific. Even so, only a minority of encounters with the bacterium result in disease.

▶ Culture and Diagnosis

The failure to recognize group A streptococcal infections can have devastating effects. Rapid cultivation and diagnostic techniques to ensure proper treatment and prevention measures are essential. Several different rapid diagnostic test kits are used in clinics and doctors' offices to detect group A streptococci from pharyngeal swab samples. These tests are based on antibodies that react with the outer carbohydrates of group A streptococci (**figure 21.6b**). Because the rapid tests have a significant possibility of returning a false-negative result, guidelines call for confirming the negative finding with a culture, which can be read the following day.

A culture is generally taken at the same time as the rapid swab and is plated on sheep blood agar. *S. pyogenes* displays a beta-hemolytic pattern due to its streptolysins (and hemolysins) (see figure 18.5). Group A streptococci are by far the most common beta-hemolytic isolates in human diseases, but lately an increased number of infections by group B streptococci (also beta-hemolytic), as well as the existence of beta-hemolytic enterococci, have made it important to use differentiation tests. A positive bacitracin disc test provides additional evidence for group A organisms.

▶ Prevention

No vaccine exists for group A streptococci, although many researchers are working on the problem. A vaccine against this bacterium would also be a vaccine against rheumatic fever, and thus it is in great demand. In the meantime, infection can be prevented by good hand washing, especially after coughing and sneezing and before preparing foods or eating.

▶ Treatment

The antibiotic of choice for *S. pyogenes* is penicillin; many group A streptococci have become resistant to erythromycin, a macrolide antibiotic. In patients with penicillin allergies, a first-generation cephalosporin, such as cephalexin, is prescribed.

Fusobacterium necrophorum

In the last several years, it has been recognized that a bacterium called *Fusobacterium necrophorum* is causing potentially serious pharyngitis cases in adolescents and young adults. Some studies suggest it is as common as *S. pyogenes* in this age group. It can invade the bloodstream and cause serious infections of the bloodstream and other organs, a condition called Lemierre's syndrome. Doctors speculate that this disease was previously rarely seen since most sore throats were empirically treated with broad-spectrum antibiotics, a treatment that generally kills *F. necrophorum*. Now that physicians are being much more judicious with antibiotic treatment and generally not treating at all if strep tests are negative, this bacterium has a doorway to cause disease. This bacterium is sensitive to penicillin and related drugs, which are the first-line drugs for *S. pyogenes* as well. It does make the use of second-line drugs for strep throats less desirable as some of them, such as azithromycin, have no effect on this bacterium. If left untreated, pharyngitis caused by this pathogen can lead to the development of sequelae including otitis media, sinusitis, and meningitis and has a significant mortality rate. It should be suspected if the pharyngitis worsens for several days and if there is marked (usually unilateral) neck swelling.

Disease Table 21.4	Pharyngitis		
Causative Organism(s)	*Streptococcus pyogenes*	*Fusobacterium necrophorum*	Viruses
Most Common Modes of Transmission	Droplet or direct contact	Opportunistic	All forms of contact
Virulence Factors	LTA, M protein, hyaluronic acid capsule, SLS and SLO, superantigens, induction of autoimmunity	Invasiveness, endotoxin, leukotoxin	–
Culture/Diagnosis	Beta-hemolytic on blood agar, sensitive to bacitracin, rapid antigen tests	Growth on anaerobic agar	Goal is to rule out *S. pyogenes*, further diagnosis usually not performed
Prevention	Hygiene practices	Hygiene practices	Hygiene practices
Treatment	Penicillin, cephalexin in penicillin-allergic	Penicillin, cefuroxime	Symptom relief only
Distinctive Features	Generally more severe than viral pharyngitis	Common in adolescents and young adults, neck swelling common; infections spread to cardiovascular system or deeper tissues	Hoarseness frequently accompanies viral pharyngitis
Epidemiological Features	United States: 20%–30% of all cases of pharyngitis	United States: 10%–30% of cases	Ubiquitous; responsible for 40%–60% of all pharyngitis

F. necrophorum is currently estimated to cause from 10% to 30% of all pharyngitis cases today in adolescents and young adults. The exact mode of transmission for this pathogen is currently unknown but may involve animal reservoirs.

There are currently no rapid diagnostic tests for *F. necrophorum*. Identification of this pathogen is possible through Gram staining of specimens, as the trained eye of a laboratory clinician should pick up on the unique pleomorphic morphology of *F. necrophorum* versus the distinct streptococcal appearance seen in *S. pyogenes* infections.

A negative rapid test for group A streptococcal infection typically results in no drug treatment for the patient. If a patient with pharyngitis is negative for a rapid strep test and displays neck swelling and rapidly worsening conditions, it is recommended to treat with penicillin or cefuroxime **(Disease Table 21.4).**

Diphtheria

For hundreds of years, diphtheria was a significant cause of morbidity and mortality, but in the last 50 years, both the number of cases and the fatality rate have steadily declined throughout the world. In the United States, no cases have been reported since 2003; but when healthy people are screened for the presence of the bacterium, it is found in a significant percentage of them, indicating that the lack of cases is due to the protection afforded by immunization with the diphtheria toxoid, which is part of the childhood immunization series.

Indeed, during the 1990s, a diphtheria epidemic occurred in the former Soviet Union in which 157,000 people became ill

with diphtheria and 5,000 people died. This upsurge of cases was attributed to a breakdown in immunization practices and production of vaccine, which followed the breakup of the Soviet Union. These examples and other smaller current outbreaks of disease today emphasize the importance of maintaining vaccination, even for diseases that have long been kept under control.

▶ Signs, Symptoms, and Causative Organism

The disease is caused by *Corynebacterium diphtheriae*, a non-spore-forming, gram-positive club-shaped bacterium **(figure 21.7).** The symptoms of diphtheria are experienced initially in the upper respiratory tract. At first the patient experiences a sore throat, lack of appetite, and low-grade fever. The most striking symptom of this disease is the characteristic membrane, usually referred to as a pseudomembrane, which forms on the tonsils or pharynx **(figure 21.8).** The membrane is formed by the bacteria and consists of bacterial cells, fibrin, lymphocytes, and dead tissue cells and may be quite extensive. It adheres to tissues and cannot easily be removed. It may eventually completely block respiration. The patient may or may not recover after this crisis. Exotoxin manufactured by the bacterium may penetrate the bloodstream and travel throughout the body.

▶ Prevention and Treatment

Diphtheria can easily be prevented by a series of vaccinations with toxoid, usually given as part of a mixed vaccine against tetanus and pertussis as well, called the *DTaP* (for diphtheria, tetanus, and acellular pertussis), in the routine childhood vaccination program. A single dose of Tdap is recommended as a

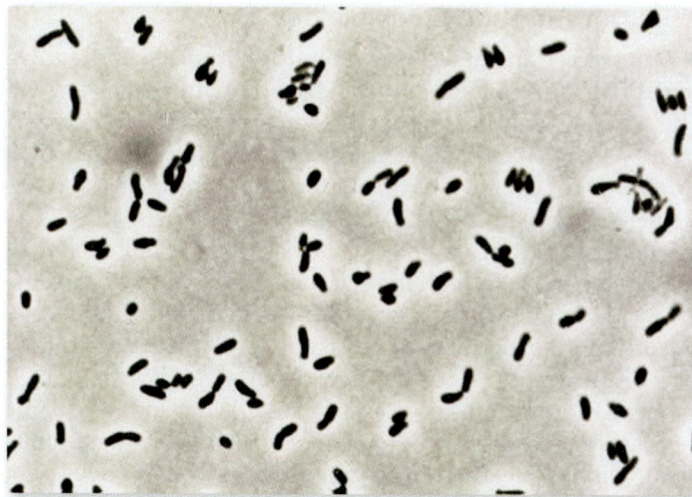

Figure 21.7 *Corynebacterium diphtheriae.*

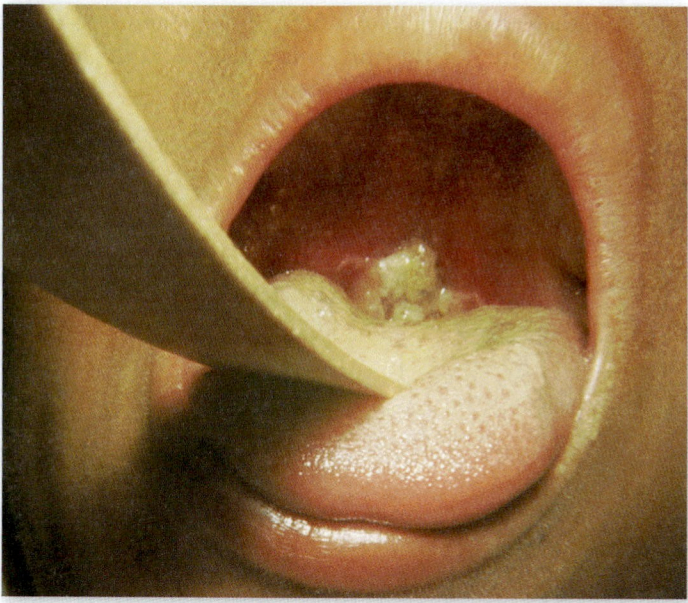

Figure 21.8 Diphtheria. The clinical appearance in diphtheria infection includes gross inflammation of the pharynx and tonsils marked by grayish patches (a pseudomembrane) and swelling over the entire area.

Disease Table 21.5 Diphtheria	
Causative Organism(s)	*Corynebacterium diphtheriae*
Most Common Modes of Transmission	Droplet contact, direct contact or indirect contact with contaminated fomites
Virulence Factors	Exotoxin: diphtheria toxin
Culture/Diagnosis	Tellurite medium—gray/black colonies, club-shaped morphology on Gram stain; treatment begun before definitive identification
Prevention	Diphtheria toxoid vaccine (part of DTaP, Tdap, and Td)
Treatment	Antitoxin plus penicillin or erythromycin
Epidemiological Features	United States: no cases since 2003; internationally: +/− 5,000 cases per year, even though there is 83% vaccine coverage

booster for individuals aged 11 to 64 years in order to maintain immunity to the pathogen today **(Disease Table 21.5).**

21.3 Learning Outcomes—Assess Your Progress

4. List the possible causative agents, modes of transmission, virulence factors, diagnostic techniques, and prevention and treatment for each of the diseases of the upper respiratory tract: the common cold, sinusitis, otitis media, pharyngitis, and diphtheria.

5. Identify which disease is often caused by a mixture of microorganisms.

6. Identify two bacteria that can cause dangerous pharyngitis cases.

21.4 Diseases Caused by Microorganisms Affecting Both the Upper and Lower Respiratory Tracts

A number of infectious agents affect both the upper and lower respiratory tract regions. We discuss the more common diseases in this section; specifically, they are whooping cough, respiratory syncytial virus (RSV), and influenza.

Whooping Cough

Whooping cough is also known as *pertussis* (the suffix *-tussis* is Latin for "cough"). A vaccine for this potentially serious infection has been available since 1926. The disease is still troubling to the public health community because its incidence is increasing in the United States, despite high vaccine coverage among children. In addition, some parents have recently become concerned about the safety of the vaccine. It is vitally important for health care professionals to convey accurate information about this disease and the safety of the vaccine.

▶ Signs and Symptoms

The disease has two distinct symptom phases called the catarrhal and paroxysmal stages, which are followed by a long recovery (or convalescent) phase, during which a patient is particularly susceptible to other respiratory infections. After an incubation period of from 3 to 21 days, the **catarrhal** stage begins when bacteria present in the respiratory tract cause what appear to be cold symptoms, most notably a runny nose. This stage lasts 1 to 2 weeks. The disease worsens in the second **(paroxysmal)** stage, which is characterized by severe and uncontrollable coughing (a *paroxysm* is a convulsive attack). The common name for the disease comes from the whooping sound a patient makes as he or she tries to grab a breath between uncontrollable bouts of coughing. The violent coughing spasms can result in burst blood vessels in the eyes or even vomiting. In the worst cases, seizures result from small hemorrhages in the brain.

As in any disease, the **convalescent phase** is the time when numbers of bacteria are decreasing and no longer cause ongoing symptoms. But the active stages of the disease damage the cilia on respiratory tract epithelial cells, and complete recovery of these surfaces requires weeks or even months. During this time, other microorganisms can more easily colonize and cause secondary infection.

▶ Causative Agent

Bordetella pertussis is a very small gram-negative rod. Sometimes it looks like a coccobacillus. It is strictly aerobic and fastidious, having specific nutritional requirements for successful culture.

▶ Pathogenesis and Virulence Factors

The progress of this disease can be clearly traced to the virulence mechanisms of the bacterium. It is absolutely essential for the bacterium to attach firmly to the epithelial cells of the mouth and throat, and it does so using specific adhesive molecular structures on its surface. One of these structures is called *filamentous hemagglutinin* (*FHA*). It is a fibrous structure that surrounds the bacterium like a capsule and is also secreted in soluble form. In that form, it can act as a bridge between the bacterium and the epithelial cell.

Once the bacteria are attached in large numbers, production of mucus increases and localized inflammation ensues, resulting in the early stages of the disease. Then the real damage begins: The bacteria release multiple exotoxins that damage ciliated respiratory epithelial cells and cripple other components of the host defense, including phagocytic cells.

The two most important exotoxins are *pertussis toxin* and *tracheal cytotoxin*. Pertussis toxin is a toxin that triggers excessive amounts of cyclic AMP to accumulate in affected cells. This results in copious production of mucus and a variety of other effects in the respiratory tract and the immune system.

Tracheal cytotoxin results in more direct destruction of ciliated cells. The cells are no longer capable of clearing mucus and secretions, leading to the extraordinary coughing required to get relief. Another important contributor to the pathology of the disease is *B. pertussis* endotoxin. As always with endotoxins, its release leads to the production of a host of cytokines that have direct and indirect effects on physiological processes and on the host response.

▶ Transmission and Epidemiology

B. pertussis is transmitted via respiratory droplets. It is highly contagious during both the catarrhal and paroxysmal stages. The disease manifestations are most serious in infants. Twenty-five percent of infections occur in older children and adults, who generally have milder symptoms. The disease results in approximately 200,000 deaths annually worldwide.

Pertussis outbreaks continue to occur in the United States and elsewhere. Although high vaccination coverage has kept the incidence of pertussis low in the United States, the number of reported pertussis cases has steadily increased since 1980 **(figure 21.9)**. With the 2011 level at 18,719, the CDC reported more than 40,000 cases in 2012. What could explain the startling increases in light of good vaccine coverage? First of all, we have learned over time that the vaccine does not provide lifelong protection. Immunity begins to wane a few years after the childhood series of vaccinations is completed. That has resulted in adults becoming infected with the bacterium, often with no or mild symptoms. They then pass it on to infants who are not yet fully immunized, and the infants

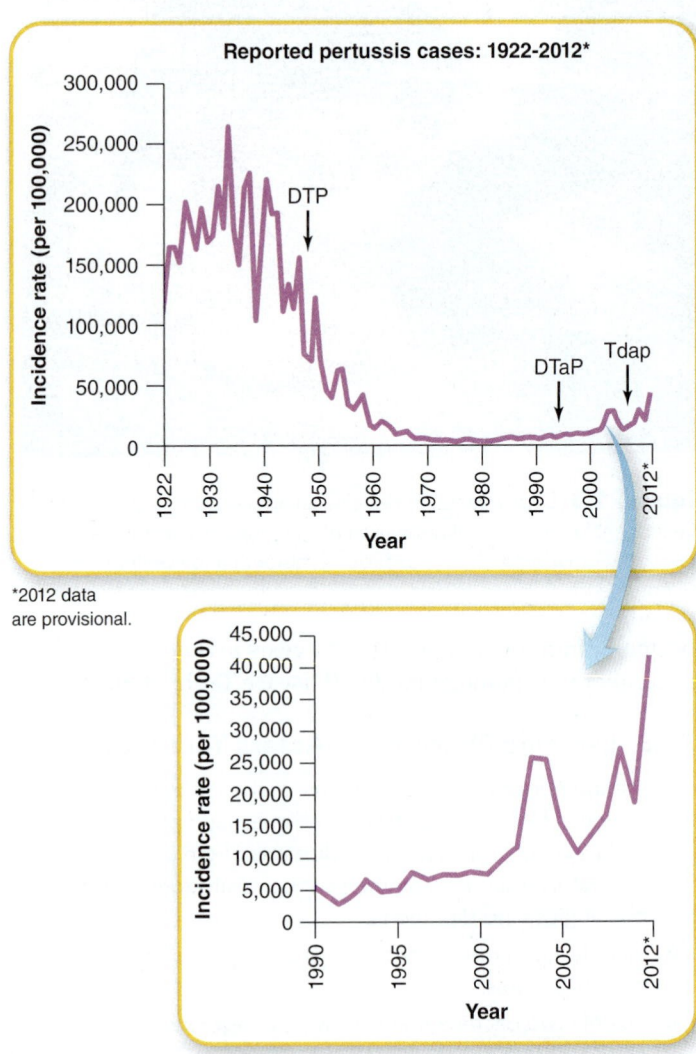

*2012 data are provisional.

Figure 21.9 Reported pertussis incidence over time.

experience serious illness. Secondly, it appears that *B. pertussis* is evolving over time and the current vaccine may not be providing the best protection against infection today. Lastly, there has been a rise in the percentage of unvaccinated individuals. In many cases, the highest rate of nonvaccination (in children or in adults) occurs in states that have shown the highest increases in pertussis cases.

▶ Culture and Diagnosis

This disease is often diagnosed based solely on its symptoms because they are so distinctive. When culture confirmation is desired, nasopharyngeal swabs can be inoculated on specific media—Bordet-Gengou (B-G) medium, charcoal agar, or potato-glycerol agar.

▶ Prevention and Treatment

The current vaccine for pertussis is an acellular formulation of important *B. pertussis* antigens. It is generally given in the form of the DTaP vaccine. A single booster with Tdap after the age of 11 is especially important to maintain immunity against this disease. A second prevention strategy is the administration of antibiotics to contacts of people who have been diagnosed with the disease to prevent disease in those who may have been infected.

To treat an active case of the disease, azithromycin is the drug of choice **(Disease Table 21.6).**

Disease Table 21.6	Pertussis (Whooping Cough)
Causative Organism(s)	*Bordetella pertussis*
Most Common Modes of Transmission	Droplet contact
Virulence Factors	FHA (adhesion), pertussis toxin and tracheal cytotoxin, endotoxin
Culture/Diagnosis	PCR or growth on B-G, charcoal, or potato-glycerol agar; diagnosis can be made on symptoms
Prevention	Acellular vaccine (DTaP), azithromycin for contacts
Treatment	Azithromycin
Epidemiological Features	United States: great increase in cases—more than 40,000 in 2012; internationally: 140,000 cases in 2012

Respiratory Syncytial Virus Infection

As its name indicates, respiratory syncytial virus (RSV) infects the respiratory tract and produces giant multinucleated cells (syncytia). It is a member of the paramyxovirus family and contains single-stranded, negative-sense RNA. It is an enveloped virus. Outbreaks of droplet-spread RSV disease occur regularly throughout the world, with peak incidence in the winter and early spring. Children 6 months of age or younger, as well as premature babies, are especially susceptible to serious disease caused by this virus. RSV is the most prevalent cause of respiratory infection in the newborn age group, and nearly all children have experienced it by age 2. An estimated 100,000 children are hospitalized with RSV disease each year in the United States. The mortality rate is highest for children with complications such as prematurity, congenital disease, and immunodeficiency. Infection in older children and adults usually manifests as a cold.

The first symptoms are fever that lasts for approximately 3 days, rhinitis, pharyngitis, and otitis. More serious infections progress to the bronchial tree and lung parenchyma, giving rise to symptoms of croup, which include acute bouts of coughing, wheezing, difficulty in breathing (called **dyspnea**), and abnormal breathing sounds (called rales). This condition is often called "croup" and also bronchiolitis; be aware that both of these terms are clinical descriptions of diseases caused by a variety of viruses (in addition to RSV) and sometimes bacteria.

The virus is highly contagious and is transmitted through droplet contact but also through fomite contamination. Diagnosis of RSV infection is more critical in babies than in older children or adults. The afflicted child is conspicuously ill, with signs typical of pneumonia and bronchitis. The best diagnostic procedures are those that demonstrate the viral antigen directly from specimens (direct and indirect fluorescent staining, ELISA, and DNA probes).

There is no RSV vaccine available yet, but an effective passive antibody preparation (Synagis) is used as prevention in high-risk children and babies born prematurely. It is very expensive (about $6,000 for a five-dose treatment); therefore, insurance companies will only reimburse for it when children meet stringent criteria. Ribavirin, an antiviral drug, can be administered as an inhaled aerosol to very sick children, although the clinical benefit is uncertain **(Disease Table 21.7).**

Influenza

The "flu" is a very important disease to study for several reasons. First of all, the familiar annual "flu seasons" have the potential of turning deadly for very many people very quickly. Second, many conditions are erroneously termed the "flu," while in fact only diseases caused by influenza viruses are actually the flu. Third, the way that influenza viruses behave provides an excellent illustration of the way other viruses can, and do, change to cause more serious diseases than they did previously.

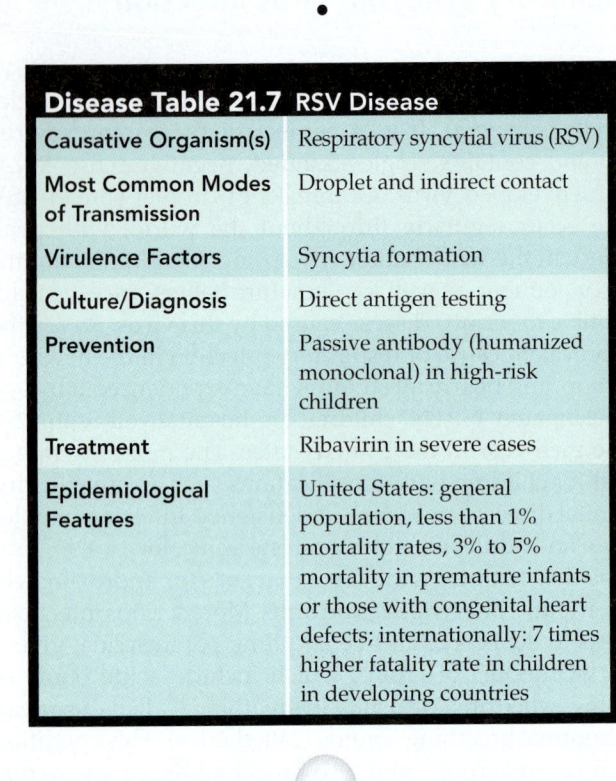

Disease Table 21.7	RSV Disease
Causative Organism(s)	Respiratory syncytial virus (RSV)
Most Common Modes of Transmission	Droplet and indirect contact
Virulence Factors	Syncytia formation
Culture/Diagnosis	Direct antigen testing
Prevention	Passive antibody (humanized monoclonal) in high-risk children
Treatment	Ribavirin in severe cases
Epidemiological Features	United States: general population, less than 1% mortality rates, 3% to 5% mortality in premature infants or those with congenital heart defects; internationally: 7 times higher fatality rate in children in developing countries

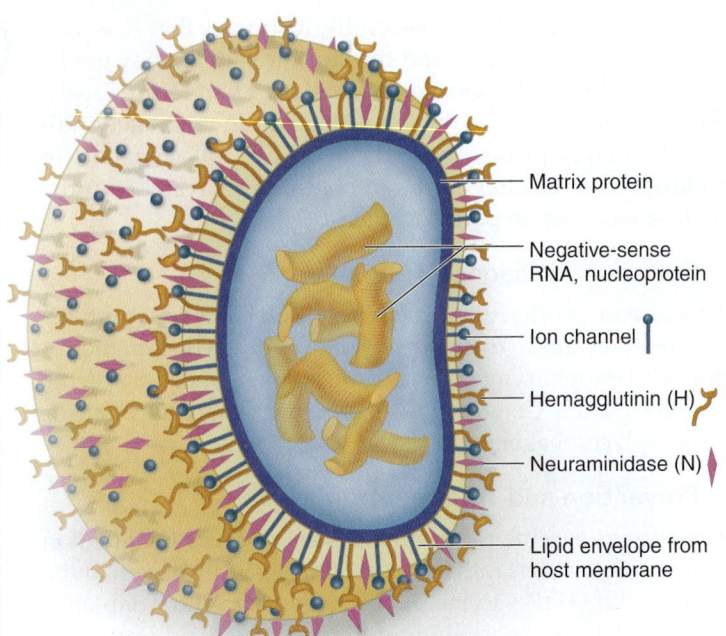

Figure 21.10 Schematic drawing of influenza virus.

Influenza viruses that circulate every year are called "seasonal" flus. Often these are the only influenza viruses that circulate each year. Occasionally, another influenza strain appears, one that is new and may cause worldwide pandemics. In some years, such as in 2009, both types of influenza infection are problematic. They are characterized by different symptoms, affect different age groups, and require separate vaccine protocols.

▶ Signs and Symptoms

Influenza begins in the upper respiratory tract but in serious cases may also affect the lower respiratory tract. There is a 1- to 4-day incubation period, after which symptoms begin very quickly. These include headache, chills, dry cough, body aches, fever, stuffy nose, and sore throat. Even the sum of all these symptoms can't describe how a person actually feels: lousy. The flu is known to "knock you off your feet." Extreme fatigue can last for a few days or even a few weeks. An infection with influenza can leave patients vulnerable to secondary infections, often bacterial. Influenza infection often leads to a pneumonia that can cause rapid death, even in young healthy adults.

Patients with emphysema or cardiopulmonary disease, along with very young, elderly, or pregnant patients, are more susceptible to serious complications.

The latest pandemic virus, H1N1, or the swine flu of 2009, caused typical flulike symptoms but with a couple of

differences. Not all patients had a fever (very unusual for influenza), and many patients had gastrointestinal distress or developed multiorgan system failure.

▶ Causative Agent

All cases of influenza are caused by one of three influenza viruses: A, B, or C. They belong to the family *Orthomyxoviridae*. They are spherical particles with an average diameter of 80 to 120 nanometers. Each virion is covered with a lipoprotein envelope that is studded with glycoprotein spikes acquired during viral maturation **(figure 21.10).** Also note that the envelope contains proteins that form a channel for ion transport into the virus. The two glycoproteins that make up the spikes of the envelope and contribute to virulence are called hemagglutinin (H) and neuraminidase (N). The name hemagglutinin is derived from this glycoprotein's agglutinating action on red blood cells, which is the basis for viral assays used to identify the viruses. Hemagglutinin contributes to infectivity by binding to host cell receptors of the respiratory mucosa, a process that facilitates viral penetration. Neuraminidase breaks down the protective mucus coating of the respiratory tract, assists in viral budding and release, keeps viruses from sticking together, and participates in host cell fusion.

The ssRNA genome of the influenza virus is known for its extreme variability. It is subject to constant genetic changes that alter the structure of its envelope glycoproteins. Research has shown that genetic changes are very frequent in the area of the glycoproteins recognized by the host immune response but very rare in the areas of the glycoproteins used for attachment to the host cell **(figure 21.11).** In this way, the virus can continue to attach to host cells while managing to decrease the effectiveness of the host response to its presence. This constant mutation of the glycoproteins is called

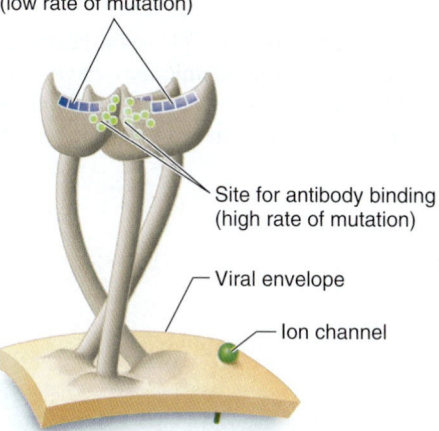

Binding sites used to anchor virus to host cell receptors (low rate of mutation)

Site for antibody binding (high rate of mutation)

Viral envelope

Ion channel

Figure 21.11 Schematic drawing of hemagglutinin (H) of influenza virus. Blue boxes depict site used to attach virus to host cells; green circles depict sites for anti-influenza antibody binding.

antigenic drift—the antigens gradually change their amino acid composition, resulting in decreased ability of host memory cells to recognize them.

An even more serious phenomenon is known as **antigenic shift.** The genome of the virus consists of eight separate RNA strands, except for influenza C, which has seven. Antigenic shift is the swapping out of one of those strands with a gene or strand from a different influenza virus. Some explanation is in order. First, we know that certain influenza viruses infect both humans and swine (pigs). Other influenza viruses infect birds (or ducks) and swine. All of these viruses have genes coding for the same important influenza proteins (including H and N)—but the actual sequence of the genes is different in the different types of viruses. Second, when the two viruses just described infect a single swine host, with both virus types infecting the same host cell, the viral packaging step can accidentally produce a human influenza virus that contains seven human influenza virus RNA strands plus a single duck influenza virus RNA strand **(figure 21.12).** When that virus infects a human, no immunologic recognition of the protein that came from the duck virus occurs. Experts have traced the flu pandemics of 1918, 1957, 1968, 1977, and 2009 to strains of a virus that came from pigs (swine flu). In fact, sequencing of the 2009 strain showed it to contain genes from swine, avian (bird), and human viruses. Influenza A viruses are named according to the different types of H and N spikes they display on their surfaces. For instance, in 2011, the most common circulating subtype of influenza virus was H3N2. H2 is a subtype that has not been seen in humans since 1968 and has many in the health field concerned about reemergence in a vulnerable human population in the future. Influenza B viruses are not divided into subtypes because they are thought only to undergo antigenic drift and not antigenic shift. Influenza C viruses are thought to cause only minor respiratory disease and are probably not involved in epidemics.

Insight 21.1 gives a breakdown of some of the important developments in the history of influenza.

▶ **Pathogenesis and Virulence Factors**

The influenza virus binds primarily to ciliated cells of the respiratory mucosa. Infection causes the rapid shedding of these cells along with a load of viruses. Stripping the respiratory epithelium to the basal layer eliminates protective ciliary clearance. Combine that with what is often called a "cytokine storm" caused by the viral stimulus and the lungs experience severe inflammation and irritation. During the 2009 swine flu pandemic, young,

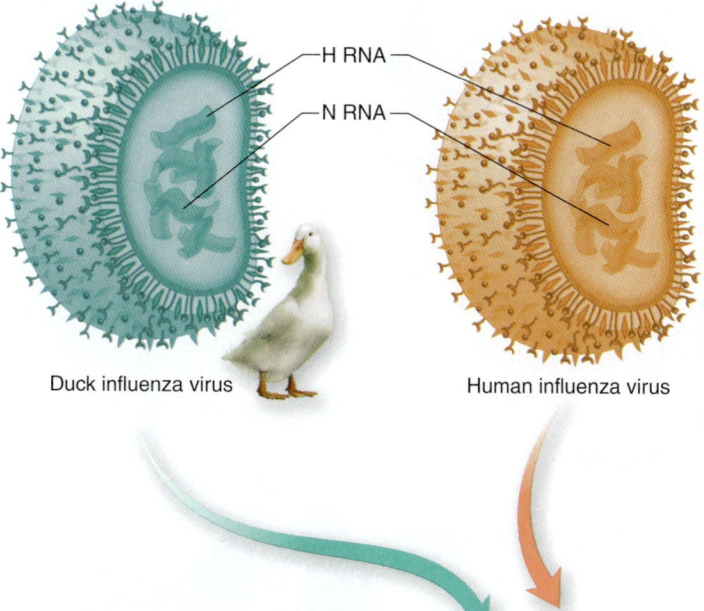

H RNA

N RNA

Duck influenza virus

Human influenza virus

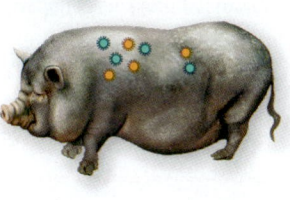

Figure 21.12 Antigenic shift event. Where ducks, swine, and humans live close together, the swine can serve as a melting pot for creating "hybrid" influenza viruses that are not recognized by the human immune system.

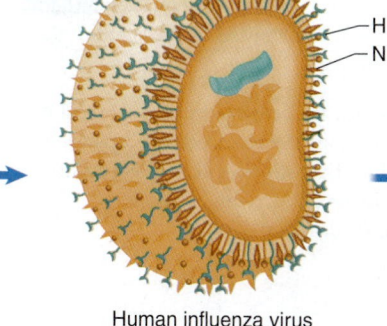

H

N

Human influenza virus with duck H spike

INSIGHT 21.1   Influenza: A Time Line

There have been a number of influenza pandemics throughout the course of history that have had a major impact on human history and on the human population. Influenza was first described in 412 in early Greek writings, and the first recorded influenza epidemic may have occurred from 1173 to 1174. Historical records are scarce, but an influenza pandemic may have originated in Asia and spread through Europe in 1510. In the following table are some of the more notable influenza events in recent history.

Date	Influenza Event	Historical Event
1889	Called the "Russian" pandemic, circled the globe in 4 months due to improved railroads. Possibly caused by the H3N8 strain. Over a million deaths were recorded.	Johnstown flood
1918	Called the "Spanish flu." The H1N1 virus evolved from bird flu to a human virus. Fifty million worldwide died—more than the number killed in World War I. The war contributed to the spread of the virus.	World War I
1957	Called the "Asian flu." The H2N2 virus replaced H1N1 and killed 1.5 million. People born after this date will have less immunity to H1N1.	*American Bandstand*'s television premier.
1968	Called the "Hong Kong flu." Caused by the H3N2 virus. Killed 1 million, with the greatest deaths among those over 65.	Martin Luther King Jr. assassinated
1976	H1N1 virus infected four soldiers on a U.S. Army base in New Jersey; one died. Forty-eight million people were vaccinated against this new virus, leading to 532 people acquiring Guillain-Barré syndrome, and no pandemic.	Jimmy Carter elected president
1998	H1N1 reemerged in livestock among U.S. pigs. It was now a combination of human/bird/swine flu.	War raged in Kosovo
2004–2006	H5N1 bird flu sickened 47 people in Thailand and Vietnam; 34 died. Transmitted by birds to humans, but not between humans.	Tsunami in the Indian Ocean
2009	Called the "swine flu," caused by a new H1N1 strain, spread quickly from Mexico. Declared a pandemic by the WHO. The CDC estimates that 43–89 million people were infected with H1N1 with between 8,800 and 18,300 deaths.	Financial markets collapse

healthy adults were severely affected. Although this was in seeming contradiction to the typical profile expected, in which the most vulnerable patients are the very young and very old, it makes sense that those with the strongest immune system had the most severe cases of inflammation-mediated side effects from the disease. This paralleled what scientists now believe occurred during the influenza pandemic of 1918. Scientists also found that the disease was worse in people who had previously experienced a seasonal flu and therefore had antibodies to other strains. In those cases, the "old" antibodies bound to the virus but not strongly enough to initiate immunity. The antibody-virus complexes congregated in the lungs and kidneys, activating complement and worsening the symptoms.

As just noted, the glycoproteins and their structure are important virulence determinants because of their ability to change. But they also mediate the adhesion of the virus to host cells. One feature of the 2009 H1N1 virus is that it bound to cells lower in the respiratory tract—and at a much higher rate, leading to massive damage, and often death, in the worst-affected patients. There was a total of around 12,000 deaths worldwide in the 2009 pandemic, either from primary influenza pneumonia or secondary bacterial infection that resulted from opportunistic microbes infecting already compromised lungs.

▶ Transmission and Epidemiology

Inhalation of virus-laden aerosols and droplets constitutes the major route of influenza infection, although fomites can play a secondary role. Transmission is greatly facilitated by crowding and poor ventilation in classrooms, barracks, nursing homes, dormitories, and military installations in the late fall and winter. The drier air of winter facilitates the spread of the virus, as the moist particles expelled by sneezes and coughs become dry very quickly, helping the virus remain airborne for longer periods of time. In addition, the dry cold air makes respiratory tract mucous membranes more brittle, with microscopic cracks that facilitate invasion by viruses. Influenza is highly contagious and affects people of all ages. Annually, from 17,000 to 52,000 deaths occur in the United States from seasonal influenza and its complications, mainly among the very young and the very old.

▶ Culture and Diagnosis

A wide variety of culture-based and non-culture-based methods are used to diagnose the infection. Rapid influenza tests (such as PCR, ELISA-type assays, or immunofluorescence) provide results within 24 hours; viral culture provides results in 3 to 10 days. Cultures are not typically performed at the point of care; they must be sent to diagnostic laboratories. Despite these disadvantages, culture can be useful for identifying the particular subtype of influenza involved. Identifying the particular subtype is critical for public health authorities to know. In 2009, officials did not often test for H1N1 but tested for influenza A or B, assuming if it was A

that it was H1N1, since the circulating seasonal virus was influenza B. When specimens were tested, 100% of the influenza A isolates were found to be the H1N1 strain. As the epidemic progressed, the vast majority of flu cases that were identified were influenza A, indicating that it had replaced the already established seasonal virus.

▶ Prevention

Preventing influenza infections and epidemics is one of the top priorities for public health officials. The standard vaccine for seasonal flu contains inactivated viruses that were grown in embryonated eggs. It has an overall effectiveness of 70% to 90%. The vaccine consists of three different influenza viruses (usually two influenza A strains and one influenza B strain) that have been judged to most resemble the virus variants likely to cause infections in the coming flu season. Because of the changing nature of the antigens on the viral surface, annual vaccination is considered the best way to avoid infection. The CDC recommends that anyone over the age of 6 months receive the vaccine, and it is particularly recommended for anyone in a high-risk group or for people who have a high degree of contact with the public.

A vaccine called FluMist made available in 2003 is a nasal mist vaccine consisting of the three strains of influenza virus in live attenuated form. In 2012, a quadrivalent version of FluMist was approved by the FDA. It contains two influenza A strains and two influenza B strains. Because FluMist contains active virus, it is not advised for immunocompromised individuals or asthma sufferers, and it is significantly more expensive than the injected vaccine. In 2013 the first vaccine not manufactured in chicken eggs was approved. Flublok is a recombinant vaccine approved for use in people ages 18-49.

During the 2009 H1N1 pandemic, seasonal vaccine had already been distributed when officials realized a different pandemic strain was causing disease. A new vaccine containing the pandemic strain was quickly prepared. The existence of two vaccines confused the public and made flu vaccination more cumbersome.

When a new "reassortment virus" evolves, humans have greatly reduced immunologic protection against the new pathogen. If the new virus possesses the optimal genetic combination for pathogenesis and transmission in humans, it can rapidly trigger a new pandemic. Particularly worrisome are "bird flu" strains that have been circulating around the world for years; they could cause major devastation if they change to be transmissible human to human. Current research is being performed on strains of avian influenza to discover how to develop new vaccines and stay ahead of these novel viruses. In 2012, scientists "created" a strain of bird flu that was transmissible among animal models. The science world was launched into a heated debate about whether those experiments should have been done in the first place, and whether the methods should be published in the scientific literature. The fears arose due to the potential of the engineered virus to actually trigger an outbreak if it was accidentally released and from the possibility that rogue scientists could construct their own strains using the published reports. On the other side, many argued for the long-honored scientific tradition of publishing to share information with peers and the public. The first results were in fact published in August of 2013.

▶ Treatment

Antiviral treatment for influenza must be taken early in the infection, preferably by the second day. This requirement is an inherent difficulty because most people do not realize until later that they may have the flu. Currently, the recommended drug is Tamiflu (oseltamivir). It can also be used for prevention in epidemic situations. Since 2009, there has been controversy about Tamiflu's effectiveness, but as of 2012 it was still the drug of choice. An alternative is Relenza (zanamivir) **(Disease Table 21.8).**

Disease Table 21.8 Influenza

Causative Organism(s)	Influenza A, B, and C viruses
Most Common Modes of Transmission	Droplet contact, direct contact or indirect contact
Virulence Factors	Glycoprotein spikes, overall ability to change genetically, ability to slow down immune system
Culture/Diagnosis	Viral culture (3–10 days) or rapid antigen-based or PCR tests
Prevention	Inactivated injected vaccine, inhaled live attenuated vaccine, or new recombinant vaccine—taken annually
Treatment	Oseltamivir or zanamivir
Epidemiological Features	For seasonal flu, deaths vary from year to year. United States: range from 17,000–52,000; internationally: range from 250,000–500,000

21.4 Learning Outcomes—Assess Your Progress

7. List the possible causative agents for each of the infectious conditions affecting both the upper and lower respiratory tract: whooping cough, RSV disease, and influenza.

8. Describe the symptoms appearing in each stage of whooping cough.

9. Discuss reasons for the increase of pertussis cases over the past three decades.

10. Identify the age group at most risk for serious disease from RSV.

11. Plan a response for a situation in which a patient declines influenza vaccination because he or she believes that the warnings about pandemics are always exaggerated.

12. Compare and contrast *antigenic drift* and *antigenic shift* in influenza viruses.

21.5 Lower Respiratory Tract Diseases Caused by Microorganisms

In this section, we consider a microbial disease that affects the lower respiratory tract primarily—namely, the bronchi, bronchioles, and lungs—with minimal involvement of the upper respiratory tract. Our discussion focuses on tuberculosis and pneumonia.

Tuberculosis

Mummies from the Stone Age, ancient Egypt, and Peru provide unmistakable evidence that tuberculosis (TB) is an ancient human disease. In fact, historically it has been such a prevalent cause of death that it was called "Captain of the Men of Death" and "White Plague." After the discovery of streptomycin in 1943, the rates of tuberculosis in the developed world declined rapidly, although that did not happen in the developing world. Tuberculosis reemerged in the United States and other developed countries in the mid-1980s, fueled by the HIV epidemic and its resistance to multiple antibiotics. That resurgence was eventually quelled again in the developed world. However, in Southeast Asia, the Western Pacific region, and Africa, a multidrug-resistant TB epidemic is raging. Worldwide, 2 billion people—nearly one-third of the world's population—are currently infected. The cause of tuberculosis is primarily the bacterial species *Mycobacterium tuberculosis,* informally called the tubercle bacillus.

▶ Signs and Symptoms

A clear-cut distinction can be made between infection with the TB bacterium and the disease it causes. In general, humans are rather easily infected with the bacterium but are resistant to the disease. Estimates project that only about 5% of infected people actually develop a clinical case of tuberculosis. The majority of TB cases are contained in the lungs, even though disseminated TB bacteria can give rise to tuberculosis in any organ of the body. Clinical tuberculosis is divided into primary tuberculosis, secondary (reactivation or reinfection) tuberculosis, and disseminated tuberculosis.

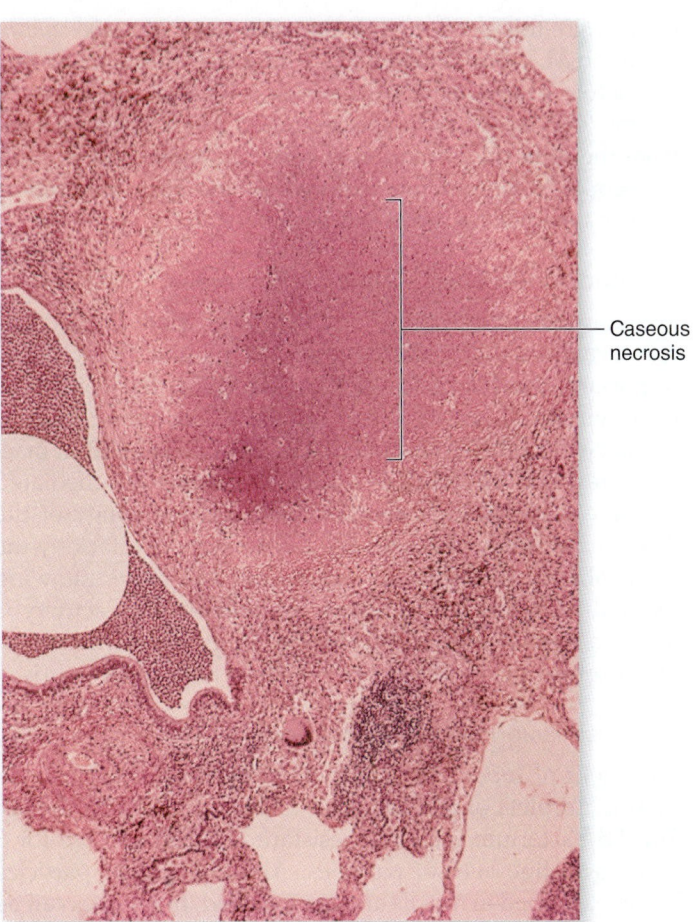

Figure 21.13 **Tubercle formation.** Photomicrograph of a tubercle (16×). The core of this tubercle is a caseous (cheesy) material containing the bacilli.

Caseous necrosis

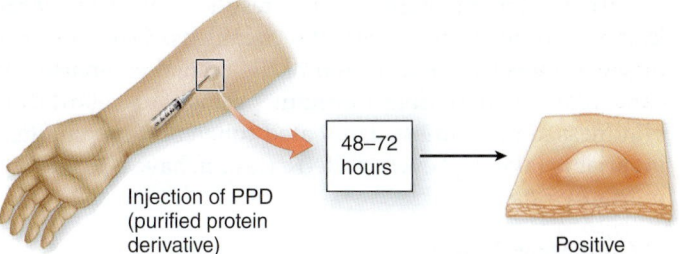

Injection of PPD (purified protein derivative)

48–72 hours

Positive

Figure 21.14 Skin testing for tuberculosis. The Mantoux test. Tuberculin is injected into the dermis. A small bleb from the injected fluid develops but will be absorbed in a short time. After 48 to 72 hours, the skin reaction is rated by the degree (or size) of the raised area. The surrounding red area is not counted in the measurement.

skin test called the tuberculin reaction, a valuable diagnostic and epidemiological tool **(figure 21.14).**

Secondary (Reactivation) Tuberculosis   Although the majority of adequately treated TB patients recover more or less completely from the primary episode of infection, live bacteria can remain dormant and become reactivated weeks, months, or years later, especially in people with weakened immunity. In chronic tuberculosis, tubercles filled with masses of bacteria expand, cause cavities in the lungs, and drain into the bronchial tubes and upper respiratory tract. The patient gradually experiences more severe symptoms, including violent coughing, greenish or bloody sputum, low-grade fever, anorexia, weight loss, extreme fatigue, night sweats, and chest pain. It is the gradual wasting of the body that accounts for an older name for tuberculosis—*consumption.* Untreated secondary disease has nearly a 60% mortality rate.

Extrapulmonary Tuberculosis   TB infection outside of the lungs is more common in immunosuppressed patients and young children. Organs most commonly involved in **extrapulmonary TB** are the regional lymph nodes, intestines, kidneys, long bones, genital tract, brain, and meninges. Because of the debilitation of the patient and the high load of TB bacteria, these complications are usually grave. Renal tuberculosis results in necrosis and scarring of the kidney and the pelvis, ureters, and bladder. This damage is accompanied by painful urination, fever, and the presence of blood and the TB bacterium in urine. Genital tuberculosis in males damages the prostate gland, epididymis, seminal vesicle, and testes; and in females, the fallopian tubes, ovaries, and uterus. Tuberculosis of the bones and joints is a common complication. The spine is a frequent site of infection, although the hip, knee, wrist, and elbow can also be involved. Advanced infiltration of the vertebral column produces degenerative changes that collapse the vertebrae, resulting in abnormal curvature of the thoracic region (humpback) or of the lumbar region (swayback). Neurological damage stemming from compression on nerves can cause extensive paralysis and sensory loss.

Primary Tuberculosis   The minimum infectious dose for lung infection in primary tuberculosis is around 10 bacterial cells. Alveolar macrophages phagocytose these cells, but they are not killed and continue to multiply inside the macrophages. This period of hidden infection is asymptomatic or is accompanied by mild fever. Some bacteria escape from the lungs into the blood and lymphatics. After 3 to 4 weeks, the immune system mounts a complex, cell-mediated assault against the bacteria. The large influx of mononuclear cells into the lungs plays a part in the formation of specific infection sites called tubercles. Tubercles are granulomas that consist of a central core containing TB bacteria in enlarged macrophages and an outer wall made of fibroblasts, lymphocytes, and macrophages **(figure 21.13).** Although this response further checks spread of infection and helps prevent the disease, it also carries a potential for damage. Frequently, as neutrophils come on the scene and release their enzymes, the centers of tubercles break down into necrotic caseous (kay′-see-us) lesions that gradually heal by calcification—normal lung tissue is replaced by calcium deposits. The response of T cells to *M. tuberculosis* proteins also causes a cell-mediated immune response evident in the

Tubercular meningitis is the result of an active brain lesion seeding bacteria into the meninges. Over a period of several weeks, the infection of the cranial compartments can create mental deterioration, permanent retardation, blindness, and deafness. Untreated tubercular meningitis is invariably fatal, and even treated cases can have a 30% to 50% mortality rate.

▶ Causative Agents

Mycobacterium tuberculosis is the cause of tuberculosis in most patients. It is a long and thin acid-fast rod. It is a strict aerobe and is not referred to as gram-positive or gram-negative because its acid-fast nature is much more relevant in a clinical setting. It grows very slowly; with a generation time of 15 to 20 hours, a period of up to 6 weeks is required for colonies to appear in culture. (*Note:* The prefix *Myco-* may make you think of fungi, but this is a bacterium. The prefix in the name came from the mistaken impression that colonies growing on agar [**figure 21.15**] resembled fungal colonies. And be sure to differentiate this bacterium from *Mycoplasma*—they are unrelated.)

Robert Koch identified that *M. tuberculosis* often forms serpentine cords while growing, and he called the unknown substance causing this style of growth "cord factor." Cord factor appears to be associated with virulent strains, and it is a lipid component of the mycobacterial cell wall. All mycobacterial species have walls that have a very high content of complex lipids, including mycolic acid and waxes. This chemical characteristic makes them relatively impermeable to stains and difficult to decolorize (acid-fast) once they are stained. The lipid wall of the bacterium also influences its virulence and makes it resistant to drying and disinfectants.

In recent decades, tuberculosis-like conditions caused by *Mycobacterium avium*, and related mycobacterial species (sometimes referred to as the *M. avium* complex, or MAC) have been found in AIDS patients and other immunocompromised people. In this section, we consider only *M. tuberculosis*, although *M. avium* is discussed briefly near the conclusion.

Before routine pasteurization of milk, humans acquired bovine TB, caused by a species called *Mycobacterium bovis*, from the milk they drank. It is very rare today, but in 2004, six people in a nightclub acquired bovine TB from a fellow reveler. One person died from her infection

▶ Pathogenesis and Virulence Factors

The course of the infection—and all of its possible variations—was previously described under "Signs and Symptoms." Important characteristics of the bacterium that contribute to its virulence are its waxy surface (contributing both to its survival in the environment and its survival within macrophages) and its ability to stimulate a strong cell-mediated immune response that contributes to the pathology of the disease. The tubercle bacilli are able to survive attack by the macrophages that tried to engulf and kill them, allowing the pathogen to avoid immune recognition. This activity is due to a special enzyme that blocks the proper formation of phagosomes, acidic compartments used to degrade debris.

▶ Transmission and Epidemiology

The agent of tuberculosis is transmitted almost exclusively by fine droplets of respiratory mucus suspended in the air. The TB bacterium is highly resistant and can survive for 8 months in fine aerosol particles. Although larger particles become trapped in mucus and are expelled, tinier ones can be inhaled into the bronchioles and alveoli. This effect is especially pronounced among people sharing small closed rooms with limited access to sunlight and fresh air.

The epidemiological patterns of *M. tuberculosis* infection vary with the living conditions in a community or an area of the world. Factors that significantly affect people's susceptibility to tuberculosis are inadequate nutrition, debilitation of the immune system, poor access to medical care, lung damage, and their own genetics. Put simply, TB is an infection of poverty. People in developing countries are often infected as infants and harbor the microbe for many years until the disease is reactivated in young adulthood. In 2010, 1.4 million people died from TB, a reduction from the 1.8 million deaths seen in 2008.

Case rates have begun to drop in the United States, from a high in 2004. However, it is still a threat. In 2011, an outbreak of TB occurred after an infected teacher spread the pathogen in a Texas high school. The opening case study for this chapter describes an outbreak in Florida. About 60% of cases in the United States are in foreign-born persons.

▶ Culture and Diagnosis

The CDC recommends one of two tests for the initial diagnosis of TB in the United States: skin testing or a blood test called the interferon-gamma release assay (IGRA).

Skin Testing Because infection with the TB bacillus can lead to delayed hypersensitivity to tuberculoproteins, testing for hypersensitivity has been an important way to screen

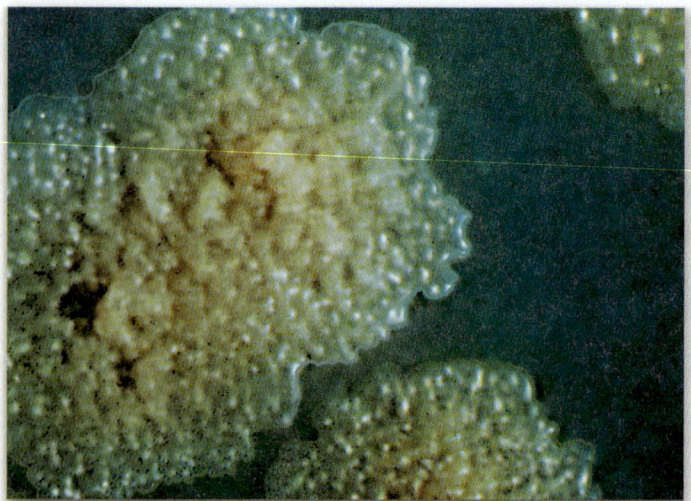

Figure 21.15 Cultural appearance of *Mycobacterium tuberculosis*. Colonies with a typical granular, waxy pattern of growth.

Case File 21   *Continuing the Case*

Tuberculosis is caused by *Mycobacterium tuberculosis*, an aerobic, acid-fast bacillus that is very slow growing. It can take up to 6 weeks for bacterial colonies to form when specimens are cultured on artificial laboratory media. Transmission of tuberculosis occurs through droplets of respiratory mucus expelled through coughing. Particles can remain viable for up to 8 months on surfaces or in aerosolized dust.

Homeless people are at risk for contracting TB for a number of reasons. First, they have limited access to appropriate health care. A persistent cough may be attributed to a cold or pneumonia, and many doctors have never seen a case of TB and may misdiagnose it as something less serious. Second, homeless people visit overcrowded homeless shelters or mental health facilities where many individuals sleep in a large, open areas, increasing their chances of becoming infected through aerosolized droplets. Finally, treatment of TB requires several months of antibiotic therapy that must be taken on a strict schedule. Because of their transient lifestyle, homeless people may not complete their treatment schedule or may take antibiotics sporadically. Such noncompliance increases the chances of promoting the growth of multidrug-resistant strains of the pathogen. Homeless people rarely have the means to pay for antibiotic treatment. Treating a nonresistant strain of *M. tuberculosis* costs about $500, but the cost of treating a drug-resistant strain is approximately $275,000 per patient.

The Florida outbreak was not the first TB outbreak among homeless people. In Kane County, Illinois, there was an outbreak of 25 cases between 2007 and 2011, all tied to a single homeless shelter. Other outbreaks in New York City; King County, Washington; and Portland, Maine, have been reported in the last decade.

- How are strains of resistant or nonresistant *M. tuberculosis* identified?
- What precautions should be taken in homeless shelters, mental health facilities, and jails to prevent transmission of TB?

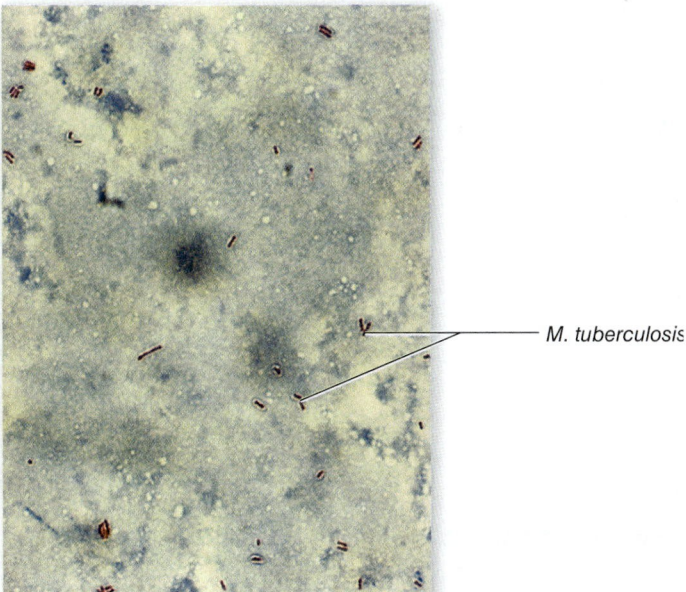

— *M. tuberculosis*

Figure 21.16 Ziehl-Neelsen staining of *Mycobacterium tuberculosis* in sputum.

negative, meaning that the person is infected but is not yet reactive. One cause of a false-negative test may be that it is administered too early in the infection, requiring retesting at a later time. Subgroups with severely compromised immune systems, such as those with AIDS, advanced age, or chronic disease, may be unable to mount a reaction even though they are infected. Skin testing may not be a reliable diagnostic indicator in these populations. False-positives may result when a person born in another country is tested if that person received a common TB vaccine called BCG (discussed in the "Prevention" section). This vaccine is not used in the United States.

IGRA   In the IGRA test, a patient's blood is drawn and incubated in test kits that detect the presence of T cells that react with *M. tuberculosis* antigens. If they have been so sensitized, they will release interferon-gamma (IFN-γ) after binding the antigens. High levels of IFN-γ trigger a positive response. The advantage of these tests is that no return visit is required.

Other Tests   In developing countries, the most common means of diagnosis is the sputum smear sample. The sample is then processed and stained in an acid-fast procedure. The most common one, the Ziehl-Neelsen stain, produces bright red acid-fast bacilli (AFB) against a blue background **(figure 21.16)**. A positive diagnosis has typically required serial sputum samples on two or three subsequent days. Because many people find it difficult to return to the clinic for a variety of reasons, often they fail to complete the series, therefore not receiving the diagnosis or treatment. The WHO has found that a single Ziehl-Neelsen smear, if positive, is as accurate as the series, and the case definition may soon be updated accordingly.

There is a PCR method available to simultaneously detect *M. tuberculosis* and determine its rifampin sensitivity within 100 minutes. The WHO started encouraging its use in 2010; as

populations for tuberculosis infection and disease. Although there are newer methods available, the most widely used test is still the tuberculin skin test, called the **Mantoux test.** It involves local injection of purified protein derivative (PPD), a standardized solution taken from culture fluids of *M. tuberculosis*. The injection is done intradermally into the forearm to produce an immediate small bleb. After 48 hours, the site is observed for a red wheal called an **induration,** which is measured and interpreted as positive or negative according to size (see figure 21.14). The disadvantage to skin testing is that a second visit to the health care provider is required.

A negative skin test usually indicates that ongoing TB infection is not present. In some cases, it may be a false

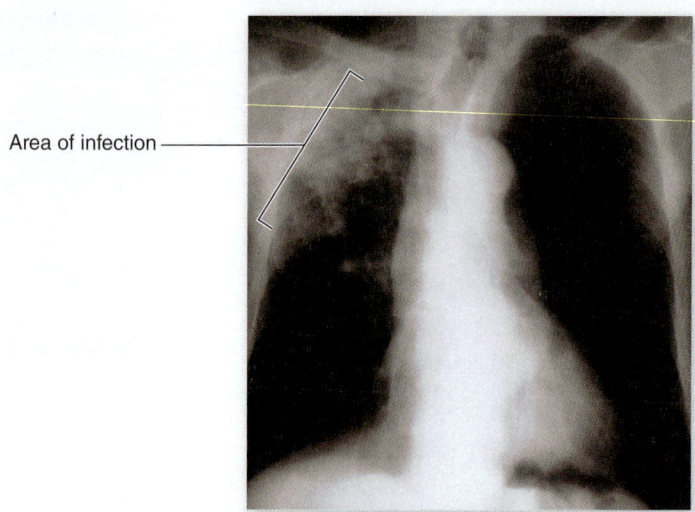

Area of infection

Figure 21.17 Chest X ray showing primary tuberculosis.

of 2012, 73 low- and middle-income countries had purchased the kits. If one of these initial diagnostic tests yields a positive result, sputum samples are cultured to grow the bacterium, even though treatment is usually begun immediately. Even though newer cultivation schemes exist that shorten the incubation period from 6 weeks to several days, this delay is unacceptable for beginning treatment or isolation precautions. However, culture still must be performed because growing colonies are required to determine antibiotic sensitivities.

Because the specimens are often contaminated with rapid-growing bacteria that will interfere with the isolation of *M. tuberculosis,* they are pretreated with chemicals to remove contaminants and are plated onto selective medium (such as Lowenstein-Jensen medium). *M. tuberculosis* colonies are depicted in figure 21.15.

Chest X rays can help verify TB when other tests have given indeterminate results. X rays are generally used for further verification after a positive test. X-ray films reveal abnormal radiopaque patches, the appearance and location of which can be very indicative. Primary tubercular infection presents the appearance of fine areas of infection **(figure 21.17)** and enlarged lymph nodes in the lower and central areas of the lungs. Secondary tuberculosis films show more extensive infiltration in the upper lungs and bronchi and marked tubercles. Scars from older infections often show up on X rays and can furnish a basis for comparison when trying to identify newly active disease.

▶ Prevention

Preventing TB in the United States is accomplished by limiting exposure to infectious airborne particles. Extensive precautions, such as isolation in negative-pressure rooms, are used in health care settings when a person with active TB is identified. Vaccine is not used in the United States, although an attenuated vaccine, called BCG, is used in many countries. *BCG* stands for "Bacille Calmette-Guerin," named for two French scientists who created the vaccine in the early 1900s. It is a live strain of a bovine tuberculosis bacterium that has been made avirulent by long passage through artificial media.

Prevention in the context of tuberculosis may also refer to preventing a person with latent TB from experiencing reactivation. This strategy is more accurately referred to as "treatment of latent infection" and is considered in the next section.

▶ Treatment

Treatment of latent TB infection is effective in preventing full-blown disease in persons who have positive skin tests and who are at risk for reactivated TB. Treatment of latent TB is with three drugs: isoniazid, rifampin, and rifapentine. The rifampin and rifapentine are taken for 4 and 3 months, respectively, and the isoniazid is continued for 9 months.

Treatment of active TB infection occurs in two phases. In the first, four drugs—rifampin, isoniazid, ethambutol, and pyrazinamide—are used for 2 months. The second phase uses only rifampin and isoniazid and lasts either 4 or 7 months, decided on a case-by-case basis.

One of the biggest problems with TB therapy is noncompliance on the part of the patient. It is very difficult, even under the best of circumstances, to keep to a regimen of multiple antibiotics daily for months—and most TB patients are not living under the best of circumstances. Failure to adhere to the antibiotic regimen leads to antibiotic resistance in the slow-growing microorganism; in fact, many *M. tuberculosis* isolates are now found to be **MDR-TB,** or multidrug-resistant TB. **Figure 21.18** demonstrates how treatment failure leads to MDR-TB. The threat to public health is so great when patients do not adhere to treatment regimens that the United States and other countries have occasionally incarcerated people—and isolated them—for not following their treatment schedules.

In 2006, a new strain of *M. tuberculosis* was identified in Africa. It is particularly lethal for HIV-infected people and has been named **XDR-TB** (extensively drug-resistant TB). XDR-TB is defined as resistance to isoniazid and rifampin plus resistance to any fluoroquinolone and at least one of three injectable second-line anti-TB drugs. Since 2006, XDR-TB has spread to at least 77 countries around the world; the CDC estimates that 500,000 new cases are seen every year. In the United States, a handful of cases of XDR-TB occur each year.

Mycobacterium avium Complex (MAC)

Before the introduction of effective HIV treatments, described in chapter 20, disseminated tuberculosis infection with MAC was one of the biggest killers of AIDS patients. It mainly affects patients with CD4 counts below 50 cells per milliliter of blood.

In 2009, scientists discovered that *M. avium* is a frequent inhabitant of showerheads that are served by city water systems, which can be an important source of infection for people with a variety of underlying respiratory conditions. *M. intracellulare* is recognized as a contributing member to MAC today. It affects the lungs of patients more severely

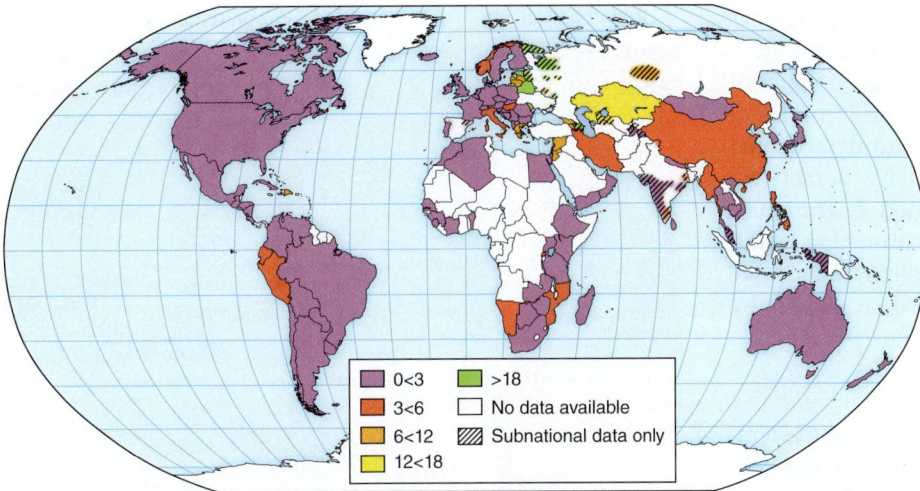

Figure 21.18 The effect of incomplete treatment on TB strains. Percentage of MDR among new TB cases, 1994–2010.

	0<3		>18
	3<6		No data available
	6<12		Subnational data only
	12<18		

Disease Table 21.9	Tuberculosis	
Causative Organism(s)	*Mycobacterium tuberculosis*	*Mycobacterium avium, M. intracellulare* complex
Most Common Modes of Transmission	Vehicle (airborne)	Vehicle (airborne)
Virulence Factors	Lipids in wall, ability to stimulate strong cell-mediated immunity (CMI)	–
Culture/Diagnosis	United States: skin test or IGRA; developing world: PCR test or sputum smear test	Positive blood culture and chest X ray
Prevention	Avoiding airborne *M. tuberculosis*, BCG vaccine in other countries	Rifabutin or azithromycin given to AIDS patients at risk
Treatment	Latent TB: rifampin, isoniazid, rifapentine Active TB: rifampin, isoniazid, ethambutol, pyrazinamide followed by rifampin and isoniazid only	Azithromycin or clarithromycin plus one additional antibiotic
Distinctive Features	Responsible for nearly all TB except for some HIV-positive patients	Suspect this in HIV-positive patients
Epidemiological Features	United States: 16% among non-Hispanic whites, 84% among ethnic minorities (29% Hispanics, 30% Asians, 23% African Americans); internationally: 1.4 million deaths in 2010	Occurs in 22% of AIDS patients

than *M. avium*, making it important to distinguish the two causative agents for effective patient treatment **(Disease Table 21.9)**.

Pneumonia

Pneumonia is a classic example of an *anatomical diagnosis*. It is defined as an inflammatory condition of the lung in which fluid fills the alveoli. The set of symptoms that we call pneumonia can be caused by a wide variety of different microorganisms. In a sense, the microorganisms need only to have appropriate characteristics to allow them to circumvent the

host's defenses and to penetrate and survive in the lower respiratory tract. In particular, the microorganisms must avoid being phagocytosed by alveolar macrophages, or at least avoid being killed once inside the macrophage. Bacteria and a wide variety of viruses can cause pneumonias. Viral pneumonias are usually—but not always—milder than those caused by bacteria. At the same time, some bacterial pneumonias are very serious and others are not. In addition, fungi such as *Histoplasma* can also cause pneumonia. Overall, U.S. residents experience 2 to 3 million cases of pneumonia and more than 50,000 deaths due to this condition every year. It is much more common in the winter. Globally, more than 1.5 million children younger than 5 years of age die from pneumonia every year. In short, pneumonia kills more children than any other infectious disease in the world today.

Physicians distinguish between various forms of pneumonia today, each characterized by different modes of transmission and pathogenic agents. Community-acquired pneumonias (CAP) are those experienced by persons in the general population. Healthcare-associated pneumonias (HCAP) develop in individuals receiving treatment at health care facilities, including hospitals. All pneumonias have similar symptoms, which we describe next, followed by separate sections for each cause of pneumonia.

▶ Signs and Symptoms

Pneumonias of all types usually begin with upper respiratory tract symptoms, including runny nose and congestion. Headache is common. Fever is often present, and the onset of lung symptoms follows. These symptoms are chest pain, fever, cough, and the production of discolored sputum. Because of the pain and difficulty of breathing, the patient appears pale and presents an overall sickly appearance. The severity and speed of onset of the symptoms vary according to the etiologic agent.

Community-Acquired Pneumonia

▶ Causative Agents

Streptococcus pneumoniae accounts for about 40% of community-acquired pneumonia cases. Respiratory tract viruses account for an additional 30%, and *Mycoplasma* causes 20%. (This leaves 10% of the cases caused by all the rest of the organisms in this section.) *Legionella* is an uncommon but serious cause of the disease. *Haemophilus influenzae* had been a major cause of community-acquired pneumonia, but the introduction of the Hib vaccine in 1988 has reduced its incidence. A number of bacteria cause a milder form of pneumonia that is often referred to as "walking pneumonia." By far the most common of these is *Mycoplasma pneumoniae*. *Histoplasma capsulatum* is a fungus that infects many people but causes a pneumonia-like disease in relatively few. Hantavirus, which emerged in 1993 in the United States, causes a type of pneumonia that can be very serious. Pneumonia can also be a secondary effect of influenza disease.

The rest of this section covers pneumonias caused by *S. pneumoniae*, *Legionella*, *Mycoplasma*, the hantavirus, and the fungi *Histoplasma* and *Pneumocystis* in more detail.

Streptococcus pneumoniae

This bacterium, which is often simply called the pneumococcus, is a small, gram-positive flattened coccus that often appears in pairs, lined up end to end (**figure 21.19a**). It is alpha-hemolytic on blood agar (**figure 21.19b**). *S. pneumoniae* is part of the normal biota in the upper respiratory tract of up to 50% of healthy people. Infection can occur when the bacterium is inhaled into deep areas of the lung or by transfer of the bacterium between two people via respiratory droplets. *S. pneumoniae* is very delicate and does not survive long out of its habitat. Factors that favor the ability of the pneumococcus

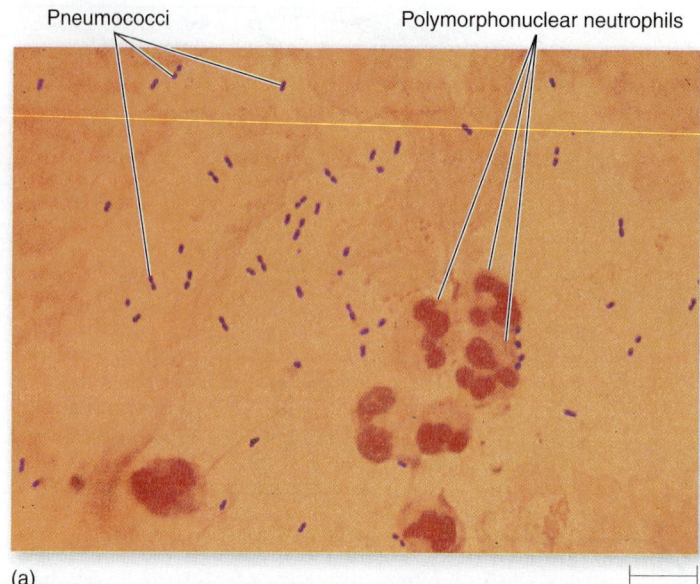

Pneumococci   Polymorphonuclear neutrophils

(a)   10 μm

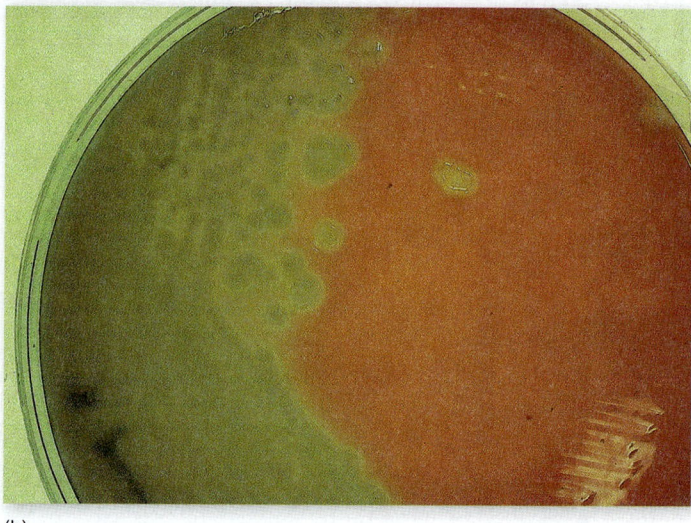

(b)

Figure 21.19 *Streptococcus pneumoniae.* **(a)** Gram stain of sputum. **(b)** Alpha-hemolysis of *S. pneumoniae* on blood agar.

to cause disease are old age, the season (rate of infection is highest in the winter), underlying viral respiratory disease, diabetes, and chronic abuse of alcohol or narcotics. Healthy people commonly inhale this and other microorganisms into the respiratory tract without serious consequences because of the host defenses present there.

This pneumonia is likely to occur when mucus containing a load of bacterial cells passes into the bronchi and alveoli. The pneumococci multiply and induce an overwhelming inflammatory response. The polysaccharide capsule of the bacterium prevents efficient phagocytosis, apparently by blocking the attachment of complement, with the result that the fluids of inflammation are continuously released into the lungs. As the infection and inflammation spread rapidly through the lung, the patient can actually "drown" in his or her own secretions. If this mixture of exudates, cells, and bacteria

A Note About SARS

In 2003, a virus from a family previously known only to cause coldlike symptoms burst onto the world stage as it started to cause pneumonias and death in Hong Kong. The SARS epidemic ended nearly as quickly as it started; since 2004, new cases of SARS have not been detected anywhere on the planet.

Severe Acute Respiratory Syndrome–Associated Coronavirus
In the winter of 2002, reports of an acute respiratory illness, originally termed an *atypical pneumonia*, began to filter in from Asia. In March of 2003, the World Health Organization issued a global health alert about the new illness. By mid-April, scientists had sequenced the entire genome of the causative virus, making the creation of diagnostic tests possible and paving the way for intensive research on the virus. The epidemic was contained by the end of July 2003, but in less than a year it had sickened more than 8,000 people. About 9% of those died. The disease was given the name **SARS,** for **severe acute respiratory syndrome.** It was concentrated in China and Southeast Asia, although several dozen countries, from Australia and Canada to the United States, reported cases. Most of the cases seem to have originated in people who had traveled to Asia or who had close contact with people from that region. Close contact (direct or droplet) seems to be required for its transmission. The virus was a previously unknown strain of coronavirus (family *Coronaviridae*). It seems

to have been a virus of bats, which mutated to be able to infect humans and be transmitted between them.

Symptoms begin with a fever of above 38°C (100.4°F) and progress to body aches and an overall feeling of malaise. Early in the infection, there seems to be little virus in the patient and a low probability of transmission. Within a week, viral numbers surge and transmissibility is very high. After 3 weeks, if the patient survives, viral levels decrease significantly and symptoms subside. Patients may or may not experience classic respiratory symptoms. They may develop breathing problems. Severe cases of the illness can result in respiratory distress and death.

In 2012, whole-genome sequencing resulted in the rapid identification of a new SARS-like virus causing deaths in the Middle East. This virus is not identical to the one circulating in 2002 and 2003 but is a coronavirus with increased virulence, just like the first one. The symptoms of the new virus also include severe respiratory distress. The cases in the Middle East were found shortly before the Hajj, the annual pilgrimage to Mecca in Saudi Arabia by millions of Muslims from all over the world. The prospect of these pilgrims taking a pandemic virus back home to all areas of the globe was a nightmare to epidemiologists. That did not happen, but by September 2013 114 laboratory-confirmed cases of disease from this virus had been recorded.

Source: World Health Organization, 2013

---

solidifies in the air spaces, a condition known as *consolidation,* occurs **(figure 21.20).** Systemic complications of pneumonia are pleuritis and endocarditis, but pneumococcal bacteremia and meningitis are the greatest danger to the patient.

Because the pneumococcus is such a frequent cause of pneumonia in older adults, this population is encouraged to seek immunization with the older pneumococcal polysaccharide vaccine, which stimulates immunity to the capsular

polysaccharides of 23 different strains of the bacterium. Active disease is treated with antibiotics, but the choice of antibiotic is often difficult. Many isolates of *S. pneumoniae* are resistant to penicillin and its derivatives, as well as to the macrolides, tetracyclines, and fluoroquinolones; therefore, broad-spectrum cephalosporins are now typically prescribed for drug therapy with or without vancomycin. Treatment also varies based on whether the patient receives outpatient or inpatient care. This

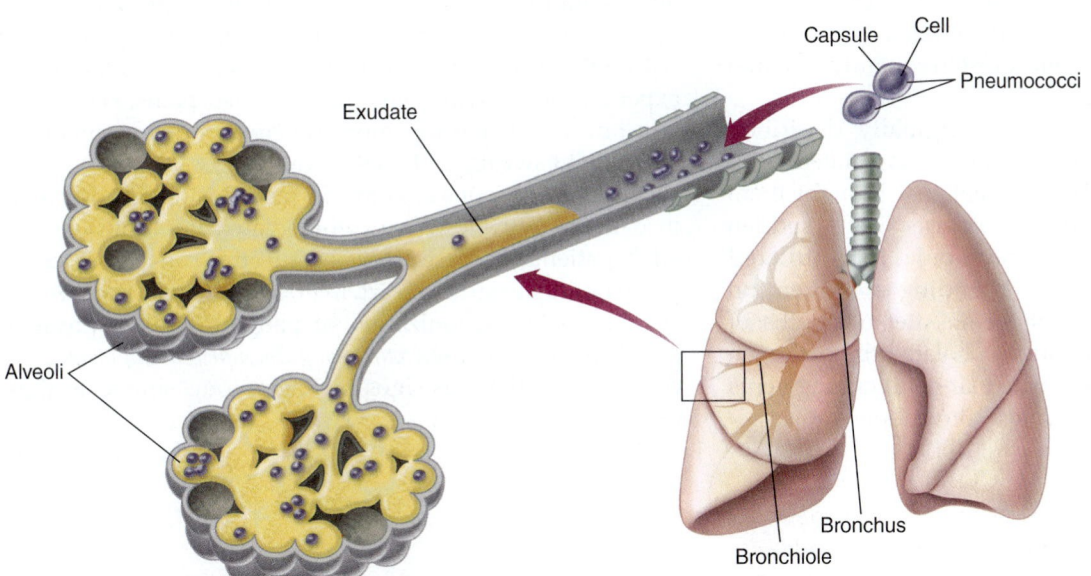

Figure 21.20 The course of bacterial pneumonia. As the pneumococcus traces a pathway down the respiratory tree, it provokes intense inflammation and exudate formation. The blocking of the bronchioles and alveoli by consolidation of inflammatory cells and products is evident.

bacterium is clearly capable of rapid development of resistance, and effective treatment requires that the practitioner be familiar with local resistance trends.

Mycoplasma pneumoniae

Pneumonias caused by a handful of bacteria, most often *Mycoplasma,* are often called atypical pneumonia—atypical in the sense that the symptoms do not resemble those of pneumococcal or other severe pneumonia. They are more like a bad case of the common cold, except that there is mucus accumulation in the lungs.

Mycoplasmas, as you learned in chapter 4, are among the smallest known self-replicating microorganisms. They naturally lack a cell wall and are therefore irregularly shaped. They may resemble cocci, filaments, doughnuts, clubs, or helices. They are free-living but fastidious, requiring complex medium to grow in the lab. (This genus should not be confused with *Mycobacterium.*)

Mycoplasma pneumonia is transmitted by aerosol droplets among people confined in close living quarters, especially families, students, and the military. It accounts for nearly 20% to 40% of all cases of community-acquired pneumonia today. Lack of acute illness in most patients has given rise to the name "walking pneumonia." A cyclical incidence of *Mycoplasma* pneumonia seems to occur every 3 to 6 years in the United States, indicating that an individual's immunity to the pathogen is short-lived or that the bacterium is experiencing antigenic changes.

Diagnosis of *Mycoplasma* may begin with ruling out other bacteria or viral agents. Serological or PCR tests confirm the diagnosis. These bacteria do not stain with Gram's stain and are not visible in direct smears of sputum. More advanced genotypic testing is performed today to monitor outbreaks and the development of macrolide resistance in this pathogen.

Legionella pneumophila

Legionella is a weakly gram-negative bacterium that displays a range of shapes, from coccus to filaments. Several species or subtypes have been characterized, but *L. pneumophila* (lung-loving) is the one most frequently isolated from infections.

Although the organisms were originally described in the late 1940s, they were not clearly associated with human disease until 1976. The incident that brought them to the attention of medical microbiologists was a sudden and mysterious epidemic of pneumonia that afflicted 200 American Legion members attending a convention in Philadelphia and killed 29 of them. After 6 months of painstaking analysis, epidemiologists isolated the pathogen and traced its source to contaminated air-conditioning vents in the hotel hosting the Legionnaires' convention.

Legionella is widely distributed in aqueous habitats as diverse as tap water, cooling towers, spas, ponds, and other freshwaters. It is resistant to chlorine. It is released during aerosol formation and can be carried for long distances. Cases have been traced to supermarket vegetable sprayers, hotel fountains, and even the fallout from the Mount St. Helens volcano eruption in 1980. Cases of this disease tripled in the United States in the 2000–2009 period, possibly due to the aging population, since it more often causes disease in the elderly. In 2012, an outbreak occurred in Quebec in Canada. It was traced to an office cooling tower. At least 180 people became ill and 13 died.

Although this bacterium can cause another disease called Pontiac fever, pneumonia is the more serious disease, with a fatality rate of 3% to 30%. *Legionella* pneumonia is thought of as an opportunistic disease, usually affecting elderly people and rarely being seen in children and healthy adults. It is difficult to diagnose, even with specific antibody tests. It is not transmitted person to person.

Histoplasma capsulatum

Pulmonary infections with this dimorphic fungus have probably afflicted humans since antiquity, but it was not described until 1905 by Dr. Samuel Darling. Through the years, it has been known by various names: Darling's disease, Ohio Valley fever, and spelunker's disease. Certain aspects of its current distribution and epidemiology suggest that it has been an important disease for as long as humans have practiced agriculture.

▶ Pathogenesis and Virulence Factors

Histoplasmosis presents a formidable array of manifestations. It can be benign or severe, acute or chronic; and it can show pulmonary, systemic, or cutaneous lesions. Inhaling a small dose of microconidia into the deep recesses of the lung establishes a primary pulmonary infection that is usually asymptomatic. Its primary location of growth is in the cytoplasm of phagocytes such as macrophages. It flourishes within these cells and is carried to other sites. Some people experience mild symptoms such as aches, pains, and coughing; but a few develop more severe symptoms, including fever, night sweats, and weight loss.

The most serious systemic forms of histoplasmosis occur in patients with defective cell-mediated immunity such as AIDS patients. In these cases, the infection can lead to lesions in the brain, intestines, heart, liver, spleen, bone marrow, and skin. Persistent colonization of patients with emphysema and bronchitis causes *chronic pulmonary histoplasmosis,* a complication that has signs and symptoms similar to those of tuberculosis.

After the terrorist attacks of September 11, 2001, and the anthrax attacks via the U.S. Postal Service that occurred later that fall, the U.S. government renewed its interest in preparing for bioterrorism or biowarfare attacks of all kinds. The U.S. Public Health Service designated six infectious diseases as "Category A," meaning they have the highest priority in research and funding. Category A agents have the following characteristics:

Image from 2001 anthrax attack in the United States.

1. They can be easily disseminated or transmitted from person to person.
2. They result in high mortality rates and have the potential for major public health impact.
3. They have the ability to cause public panic and social disruption.
4. They require special action for public health preparedness.

Of the six diseases, three of them have their primary effects on the respiratory tract: pulmonary anthrax, pneumonic plague, and tularemia. The other three diseases on the A list are botulism, smallpox, and viral hemorrhagic fevers.

One of the most important components of a successful bioterrorism prevention strategy is early detection of infected persons. Because most of the conditions on the A list are rarely seen in the United States, clinicians' index of suspicion may be low. Following are the symptoms of the three agents that cause overt respiratory symptoms.

Pulmonary Anthrax (or Inhalation Anthrax)

This disease is the result of a lung infection with *Bacillus anthracis* (see chapter 20). Pulmonary anthrax should be considered when there is lung congestion accompanied by fever, malaise, and headache. Chest X rays are very useful because a widened mediastinum (the interpleural space that appears as the dark divider in the center of most chest X rays) is a **pathognomic** (path-oh-nōm-ik) for this disease. Typical bronchopneumonia does not occur. In about half of patients, a hemorrhagic meningitis accompanies the pneumonitis. It is not transmitted from person to person, but because the bacterium forms endospores, these are easily disseminated through a variety of methods.

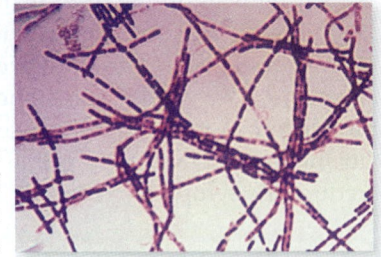

Bacillus anthracis

The most useful test for pulmonary anthrax is blood culture, because the organism is abundant in blood, as discussed in chapter 20. Treatment should be coordinated with the CDC. A vaccine for anthrax is currently administered only to military personnel and to some with occupational exposure to livestock.

Pneumonic Plague

This pneumonia illness is caused by *Yersinia pestis*, the same agent responsible for bubonic plague (see chapter 20). The first signs of the pneumonic form are fever, headache, weakness, and rapidly developing pneumonia. Sometimes sputum is bloody or watery. Within 2 to 4 days, respiratory failure and shock can ensue. The incidence of plague in the United States is low and generally of the bubonic type, which is transmitted by fleas from a small mammal host. *Y. pestis* used as a bioterrorism agent would likely be disseminated as an aerosol, leading to large numbers of pneumonic cases. Gram staining of sputum, blood, or lymph node aspirates would reveal gram-negative rods, and additional staining with Wright or Giemsa stain would result in rods with characteristic bipolar staining.

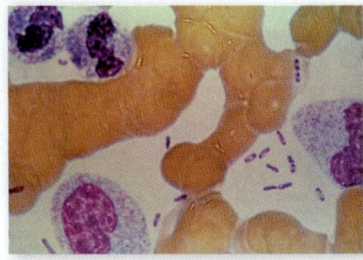

Image of Wright-Giemsa stain of *Y. pestis* in a blood smear.

Without treatment, patients die within 2 to 6 days; but swift antibiotic therapy with streptomycin or gentamicin can save lives. A vaccine exists, but it does not protect against the pneumonic form of the disease and is no longer available in the United States.

Tularemia

This infection, caused by *Francisella tularensis*, is not widely known in the United States (see chapter 20). It can cause skin and bloodstream infections, lung disease, and severe ocular lesions. The infectious dose is extremely low; as few as 10 bacteria can initiate serious disease. As a bioterrorism weapon, it would most probably be disseminated via the aerosol route, and most of the infections would no doubt be of the respiratory variety. The abrupt appearance of large numbers of people with acute pneumonitis that progresses rapidly to sepsis would be the first sign that a tularemia bioterrorism incident has occurred. Because *F. tularensis* does not seem to be transmitted person to person, it would be unusual to find large numbers of infected people over a short period of time, which would raise the possibility that there was an intentional release.

Tularemia is difficult to diagnose, and the first steps in a suspected bioterrorism incident would be to rule out plague or anthrax pneumonic disease. The bacterium is extremely dangerous to laboratory workers, so caution must be used if *Francisella* is suspected. Antibiotics such as streptomycin and gentamicin can prevent death in most cases.

As you can see, one of the greatest difficulties associated with managing a bioterrorism incident is that initial symptoms in patients are nonspecific. The time it takes for public health officials to begin to suspect one of these unusual etiologic agents (as opposed to common community-acquired or hospital-acquired respiratory infections) may make the difference between life and death for large numbers of people. We already have one advantage, however. Since the fall of 2001, U.S. health practitioners are much more alert to the possibility of intentional dissemination of infectious agents.

Figure 21.21 Sign in wooded area in Kentucky. The sign is covered in bird droppings. Up to 90% of the population in the Ohio Valley show evidence of past infection with *Histoplasma*.

▶ Transmission and Epidemiology

The organism is endemically distributed on all continents except Australia. Its highest rates of incidence occur in the eastern and central regions of the United States, especially in the Ohio Valley. This fungus appears to grow most abundantly in moist soils high in nitrogen content, especially those supplemented by bird and bat droppings **(figure 21.21).**

In high-prevalence areas such as southern Ohio, Illinois, Missouri, Kentucky, Tennessee, Michigan, Georgia, and Arkansas, 80% to 90% of the population shows signs of prior infection. Histoplasmosis incidence in the United States is estimated at about 250,000 cases per year, with several thousand of them requiring hospitalization and a small number resulting in death. In the summer of 2012, a small outbreak at a day camp in Nebraska sickened at least 32 people. The largest outbreak was in Indianapolis in 1978 when over 100,000 people became ill.

People of both sexes and all ages incur infection, but adult males experience the majority of symptomatic cases. The oldest and youngest members of a population are most likely to develop serious disease.

▶ Culture and Diagnosis

Discovering *Histoplasma* in clinical specimens is a substantial diagnostic indicator. Usually it appears as spherical, "fish-eye" yeasts intracellularly in macrophages and occasionally as free yeasts in samples of sputum and cerebrospinal fluid. There are rapid antigen tests available as well.

▶ Prevention and Treatment

Avoiding the fungus is the only way to prevent this infection, and in many parts of the country this is impossible. Luckily, undetected or mild cases of histoplasmosis resolve without medical management. Chronic or disseminated disease calls for systemic antifungal chemotherapy. Itraconazole is the drug of choice. Surgery to remove affected masses in the lungs or other organs is sometimes also useful.

Pneumocystis jiroveci

Although the fungus *Pneumocystis jiroveci* (formerly called *P. carinii*) was discovered in 1909, it remained relatively obscure until it was suddenly propelled into clinical prominence as the agent of *Pneumocystis* pneumonia (called PCP because of the old name of the fungus). PCP is the most frequent opportunistic infection in AIDS patients, most of whom will develop one or more episodes during their lifetimes.

▶ Symptoms, Pathogenesis, and Virulence Factors

In people with intact immune defenses, *P. jiroveci* is usually held in check by lung phagocytes and lymphocytes; but in those with deficient immune systems, it multiplies intracellularly and extracellularly. The massive numbers of fungi adhere tenaciously to the lung pneumocytes and cause an inflammatory condition. The lung epithelial cells slough off, and a foamy exudate builds up. Symptoms are nonspecific and include cough, fever, shallow respiration, and cyanosis (sī-ǝh-nō-sis).

▶ Transmission and Epidemiology

There is some debate about how *Pneumocystis* is acquired. Inhalation of spores is probably common. Healthy people may even harbor it as normal biota in their lungs. Contact with the agent is so widespread that in some populations, a majority of people show serological evidence of infection by the age of 3 or 4. Until the AIDS epidemic, symptomatic infections by this organism were very rare, occurring only among elderly people, premature infants, or patients that were severely debilitated or malnourished.

▶ Culture and Diagnosis

A sputum sample should be obtained and subjected to a commercially available PCR test.

▶ Prevention and Treatment

Traditional antifungal drugs are ineffective against *Pneumocystis* pneumonia because the chemical makeup of the organism's cell wall differs from that of most fungi. The primary treatment is trimethoprim-sulfamethoxazole. This combination should be administered even if disease appears mild or is only suspected. It is sometimes given to patients with low T-cell counts to prevent the disease. The airways of patients in the active stage of infection often must be suctioned to reduce the symptoms **(Disease Table 21.10).**

Healthcare-Associated Pneumonia

▶ Causative Agents

About 1% of hospitalized or institutionalized people experience the complication of pneumonia. It is most commonly associated with mechanical ventilation, via an endotracheal or tracheostomy tube. This is sometimes labeled "ventilator-associated pneumonia," or VAP. The mortality rate is quite high—between 30% and 50%. The most frequent causes of

Disease Table 21.10 Community-Acquired Pneumonia

	Streptococcus pneumoniae	Respiratory viruses	*Mycoplasma pneumoniae*	*Legionella* species	*Histoplasma capsulatum*	*Pneumocystis jiroveci*
Causative Organism(s)						
Most Common Modes of Transmission	Droplet contact or endogenous transfer	Droplet contact or endogenous transfer	Droplet contact	Vehicle (water droplets)	Vehicle—inhalation of fungal spores in contaminated soil	Vehicle—inhalation of fungal spores
Virulence Factors	Capsule		Adhesins	–	Survival in phagocytes	–
Culture/ Diagnosis	Gram stain often diagnostic, alpha-hemolytic on blood agar	Failure to find bacteria or fungi	Rule out other etiologic agents; serology; PCR	Requires selective charcoal yeast extract agar; serology unreliable	Rapid antigen tests, microscopy	PCR
Prevention	Pneumococcal polysaccharide vaccine (23-valent)	Hygiene	No vaccine, no permanent immunity	–	Avoid contaminated soil/bat, bird droppings	Antibiotics given to AIDS patients to prevent this
Treatment	Cefotaxime, ceftriaxone, with or without vancomycin; much resistance	None	Doxycycline	Fluoroquinolone, azithromycin, clarithromycin	Itraconazole	Trimethoprim-sulfamethoxazole
Distinctive Features	Patient usually severely ill	Usually mild	Usually mild; "walking pneumonia"	Mild pneumonias in healthy people; can be severe in elderly or immunocompromised	Many infections asymptomatic	Vast majority occur in AIDS patients
Epidemiological Features	40% of CAP cases; in 2009 H1N1 epidemic, 29% of fatalities were co-infected with this bacterium	30% of CAP cases	20%–40% of CAP cases	United States: 8,000–18,000 cases per year; internationally: 2 million cases per year	In United States, 250,000 infected per year; 5%–10% have symptoms	80% of untreated AIDS patients are infected

all forms of HCAP today are *Pseudomonas aeruginosa* and *Acinetobacter baumannii*, although infections with *Streptococcus pneumoniae* and *Klebsiella pneumoniae* are common as well. Pneumonia due to *S. aureus* infection arises frequently in HCAP and is frequently caused by MRSA strains of the bacterium. Further complicating matters, many cases of HCAP appear to be polymicrobial in origin—meaning that there are multiple microorganisms multiplying in the alveolar spaces.

▶ Prevention and Treatment

Because microorganisms aspirated from the upper respiratory tract cause many healthcare-associated pneumonias, measures that discourage the transfer of microbes into the lungs are very useful for preventing the condition. Elevating patients' heads to a 45-degree angle helps reduce aspiration of secretions. Good preoperative education of patients about the importance of deep breathing and frequent coughing can reduce postoperative infection rates. Proper care of mechanical ventilation and respiratory therapy equipment is essential as well.

Studies have shown that delaying antibiotic treatment of suspected healthcare-associated pneumonia leads to a greater likelihood of death. Even in this era of conservative antibiotic use, empiric therapy should be started as soon as healthcare-associated pneumonia is suspected, using multiple antibiotics that cover both gram-negative and gram-positive organisms **(Disease Table 21.11)**.

Disease Table 21.11	Healthcare-Associated Pneumonia
Causative Organism(s)	Gram-negative and gram-positive bacteria from upper respiratory tract or stomach; environmental contamination of ventilator
Most Common Modes of Transmission	Endogenous (aspiration)
Virulence Factors	–
Culture/Diagnosis	Culture of lung fluids
Prevention	Elevating patient's head, preoperative education, care of respiratory equipment
Treatment	Broad-spectrum antibiotics
Epidemiological Features	United States: 300,000 cases per year; occurs in 0.5%–1.0% of admitted patients; mortality rate in United States and internationally is 20%–50%

Hantavirus Pulmonary Syndrome

In 1993, hantavirus suddenly burst into the American consciousness. A cluster of unusual cases of severe lung edema among healthy young adults arose in the Four Corners area of New Mexico. Most of the patients died within a few days. They were later found to have been infected with hantavirus, an agent that had previously only been known to cause severe kidney disease and hemorrhagic fevers in other parts of the world. The new condition was named hantavirus pulmonary syndrome (HPS). Since 1993, the disease has occurred sporadically, but it has a mortality rate of at least 33%. It is considered an emerging disease.

▶ Symptoms, Pathogenesis, and Virulence Factors

Common features of the prodromal phase of this infection include fever, chills, myalgias (muscle aches), headache, nausea, vomiting, and diarrhea or a combination of these symptoms. A cough is common but is not a prominent early feature. Initial symptoms resemble those of other common viral infections. Soon a severe pulmonary edema occurs and causes acute respiratory distress (acute respiratory distress syndrome [ARDS] has many microbial and nonmicrobial causes; this is but one of them).

The acute lung symptoms appear to be due to the presence of large amounts of hantavirus antigen, which becomes disseminated throughout the bloodstream (including the capillaries surrounding the alveoli of the lung). Massive amounts of fluid leave the blood vessels and flood the alveolar spaces in response to the inflammatory stimulus, causing severe breathing difficulties and a drop in blood pressure. The propensity to cause a massive inflammatory response could be considered a virulence factor for this organism.

▶ Transmission and Epidemiology

Hantavirus is transmitted via airborne dust contaminated with the urine, feces, or saliva of infected rodents. Deer mice and other rodents can harbor one of the multiple strains of hantavirus identified throughout the world today, exhibiting few apparent symptoms. Small outbreaks of the disease are usually correlated with increases in the local rodent population **(Insight 21.3)**. Epidemiologists suspect that rodents have been infected with this pathogen for centuries. It has no doubt been the cause of sporadic cases of unexplained acute respiratory distress and disease in humans for decades, but the incidence seems to be increasing, especially in areas of the United States west of the Mississippi River **(figure 21.22)**.

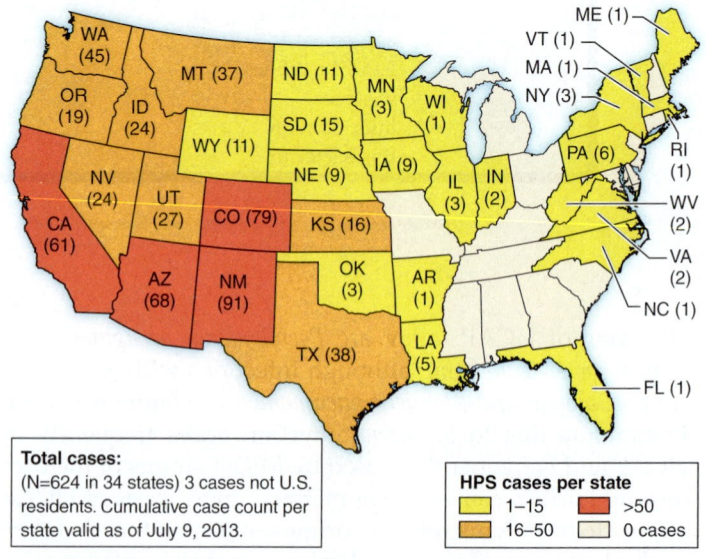

Total cases:
(N=624 in 34 states) 3 cases not U.S. residents. Cumulative case count per state valid as of July 9, 2013.

HPS cases per state
- 1–15
- 16–50
- >50
- 0 cases

Figure 21.22 Distribution of hantavirus pulmonary syndrome (HPS) cases in the United States.

Disease Table 21.12	Hantavirus Pulmonary Syndrome
Causative Organism(s)	Hantavirus
Most Common Modes of Transmission	Vehicle—airborne virus emitted from rodents
Virulence Factors	Ability to induce inflammatory response
Culture/Diagnosis	Serology (IgM), PCR identification of antigen in tissue
Prevention	Avoid mouse habitats and droppings
Treatment	Supportive
Epidemiological Features	United States: 10–25 cases per year; similar rates internationally. Previously thought to be universally lethal, but now known fatality rate is 25%–50%

▶ Treatment and Prevention

The diagnosis is established by detection of IgM to hantavirus in the patient's blood or by using PCR techniques to find hantavirus genetic material in clinical specimens. Treatment consists mainly of supportive care. Mechanical ventilation is often required.

There is no specific treatment other than supportive care. **(Disease Table 21.12).**

21.5 Learning Outcomes—Assess Your Progress

13. List the possible causative agents for each of the diseases affecting the lower respiratory tract: tuberculosis, community-acquired pneumonia, healthcare-associated pneumonia, and hantavirus pulmonary syndrome.

14. Discuss the problems associated with MDR-TB and XDR-TB.

15. Demonstrate an in-depth understanding of the current epidemiology of tuberculosis infection.

16. Explain why so many diverse microorganisms can cause the condition of pneumonia.

17. Identify the top three causes of community-acquired pneumonia.

18. List the distinguishing characteristics of healthcare-associated pneumonia compared to community-acquired pneumonia.

19. Outline the chain of transmission of the causative agent of hantavirus pulmonary syndrome.

INSIGHT 21.3   Hantavirus in Yosemite

In 1993, a mysterious illness appeared in the "Four Corners" region (Arizona, New Mexico, Colorado, and Utah) in the United States. A young, healthy Navajo man succumbed to a rapidly advancing pneumonia. Within a few weeks, several other cases occurred with patients whose symptoms matched the first one, all in young, healthy adults among the Navajo population. The CDC launched an investigation to determine the cause of the outbreak. After weeks of tense investigations, scientists tracked the source to a virus, which they named the "Sin Nombre Virus." It was later identified as a type of hantavirus, which was first discovered in South Korea. Scientists discovered that the virus was transmitted by the aerosolized urine of the deer mouse and several other species of mice and rats native to the area. The previous spring, there had been heavy snowfall and rains, which had increased the number of seed-producing plants in the area. The abundance of food led to a boom in the mouse population, which increased the likelihood that humans would come into contact with mice and their excreta. Since then, other strains of hantavirus have been discovered and cases of hantavirus infections have been documented in other areas of the United States, as well as Argentina, Brazil, Canada, Chile, Paraguay, and Uruguay.

Recently, there was an outbreak of hantavirus among campers at Yosemite National Park. Nine campers who vacationed in the "Signature" tent cabins in the park contracted hantavirus, and three adult males died of the infection. The tent cabins were insulated and had double walls where deer mice burrowed and made their nests, feasting on the crumbs and leftovers of the residents. One additional case was found in someone who had hiked in the park. Officials from Yosemite National Park notified over 10,000 campers who used the insulated tents between June 10 and August 24, 2012, that they could have been potentially exposed to the virus. Maintenance crews quickly moved in to clean the tents and replace insulation and canvas coverings to avoid further spread of the virus. Officials urged campers to be aware of the symptoms of hantavirus pulmonary syndrome, such as fever, chills, muscle aches, and gastrointestinal symptoms. There is no specific treatment or vaccine for hantavirus infection, but if symptoms are recognized early, patients can receive respiratory assistance in an intensive care unit.

Source: 2012. http://www.who.int/csr/don/2012_09_04/en/

Case File 21   *Wrap-Up*

Strains of *M. tuberculosis* are identified by genotyping through PCR analysis. The strain most commonly identified in the outbreak in Florida was the PCR00160 strain, which is relatively rare in the United States. Only 225 cases in the United States were found to be of this strain. Eighty-eight were reported in Florida, 70 of these in the Duval County outbreak.

The CDC investigation found that each site had different policies and procedures with respect to administrative, respiratory, and engineering controls for TB, and TB screenings were voluntary for residents. The CDC recommended that a formal infection control program with standard procedures for screening, treating, and tracking individuals with TB should be implemented in Duval County. Health care workers, volunteers in homeless shelters, and correctional facility staff should be made aware of the risk of TB transmission among the homeless population. According to CDC recommendations, individuals staying at homeless shelters and mental health facilities or who are incarcerated should be screened for TB and records should be shared electronically to track those infected with TB. The CDC also recommended that respiratory protection measures, including cough monitoring and hygiene awareness, should be implemented in sites housing the homeless. Engineering controls such as proper ventilation and air filtration were also recommended for these facilities.

Source: 2012. *MMWR* vol. 61, no. 28, p. 539.

▶ Summing Up

Taxonomic Organization Microorganisms Causing Disease in the Respiratory Tract

Microorganism	Disease	Location of Disease Table
Gram-positive bacteria		
Streptococcus pneumoniae	Otitis media, pneumonia	Otitis media, p. 638 Pneumonia, p. 661
Streptococcus pyogenes	Pharyngitis	Pharyngitis, p. 642
Corynebacterium diphtheriae	Diphtheria	Diphtheria, p. 643
Gram-negative bacteria		
Haemophilus influenzae	Otitis media	Otitis media, p. 638
Fusobacterium necrophorum	Pharyngitis	Pharyngitis, p. 642
Bordetella pertussis	Whooping cough	Whooping cough, p. 645
Mycobacterium tuberculosis,* M. avium, M. intracellulare	Tuberculosis	Tuberculosis, p. 655
Legionella spp.	Pneumonia	Pneumonia, p. 661
Other bacteria		
Mycoplasma pneumoniae	Pneumonia	Pneumonia, p. 661
RNA viruses		
Respiratory syncytial virus	RSV disease	RSV disease, p. 646
Influenza virus A, B, and C	Influenza	Influenza, p. 650
Hantavirus	Hantavirus pulmonary syndrome	Hantavirus pulmonary syndrome, p. 663
Fungi		
Pneumocystis jiroveci	*Pneumocystis* pneumonia	Pneumonia, p. 661
Histoplasma capsulatum	Histoplasmosis	Pneumonia, p. 661

*There is some debate about the Gram status of the genus *Mycobacterium*; it is generally not considered gram-positive or gram-negative.

INFECTIOUS DISEASES AFFECTING
The Respiratory System

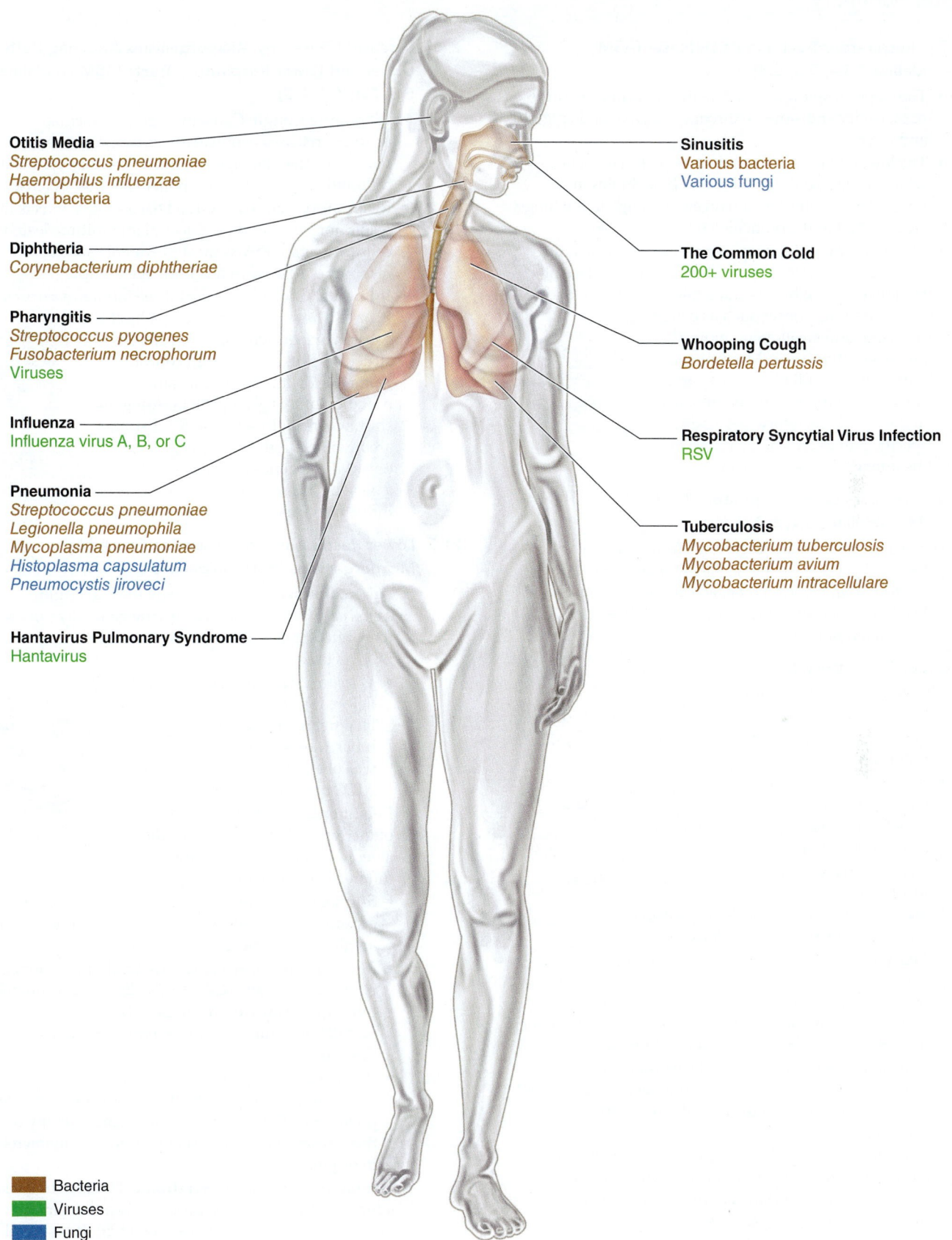

Otitis Media
Streptococcus pneumoniae
Haemophilus influenzae
Other bacteria

Diphtheria
Corynebacterium diphtheriae

Pharyngitis
Streptococcus pyogenes
Fusobacterium necrophorum
Viruses

Influenza
Influenza virus A, B, or C

Pneumonia
Streptococcus pneumoniae
Legionella pneumophila
Mycoplasma pneumoniae
Histoplasma capsulatum
Pneumocystis jiroveci

Hantavirus Pulmonary Syndrome
Hantavirus

Sinusitis
Various bacteria
Various fungi

The Common Cold
200+ viruses

Whooping Cough
Bordetella pertussis

Respiratory Syncytial Virus Infection
RSV

Tuberculosis
Mycobacterium tuberculosis
Mycobacterium avium
Mycobacterium intracellulare

Bacteria
Viruses
Fungi

System Summary Figure 21.23

Chapter Summary

21.1 The Respiratory Tract and Its Defenses (ASM Guidelines* 3.4, 5.4, 6.4)

- The upper respiratory tract includes the mouth, nose, nasal cavity and sinuses, throat (pharynx), and epiglottis and larynx.
- The lower respiratory tract begins with the trachea, which feeds into the bronchi and bronchioles in the lungs. Alveoli, the site of oxygen exchange in the lungs, are attached to the bronchioles.
- The ciliary escalator propels particles upward and out of the respiratory tract. Mucus on the surface of the mucous membranes traps microorganisms, and involuntary responses such as coughing, sneezing, and swallowing move them out of sensitive areas. Macrophages inhabit the alveoli of the lungs and clusters of lymphoid tissue (tonsils) in the throat. Secretory IgA against specific pathogens can be found in the mucus secretions as well.

21.2 Normal Biota of the Respiratory Tract (ASM Guidelines 3.4, 5.4, 6.4)

- Normal biota include organisms from nine major bacterial genera: *Prevotella, Sphingomonas, Pseudomonas, Acinetobacter, Fusobacterium, Megasphaera, Veillonella, Staphylococcus,* and *Streptococcus. Candida* is also considered normal biota.

21.3 Upper Respiratory Tract Diseases Caused by Microorganisms (ASM Guidelines 5.3, 5.4, 6.4, 8.3)

- **The Common Cold:** Caused by one of over 200 different kinds of viruses, most commonly the rhinoviruses, followed by the coronaviruses.
- **Sinusitis:** Inflammatory condition most commonly caused by allergy or by a variety of viruses or bacteria and, less commonly, fungi.
- **Acute Otitis Media (Ear Infection):** Most common cause is *Streptococcus pneumoniae,* though multiple organisms are usually present in infections.
- **Pharyngitis:** The same viruses causing the common cold commonly cause inflammation of the throat. However, two potentially serious causes of pharyngitis are *Streptococcus pyogenes* and *Fusobacterium necrophorum.* Untreated streptococcal throat infections can result in scarlet fever, rheumatic fever, glomerulonephritis, and necrotizing fasciitis. Untreated *F. necrophorum* infections can lead to Lemierre's syndrome.
- **Diphtheria:** Caused by *Corynebacterium diphtheriae,* a non-spore-forming, gram-positive club-shaped bacterium. An important exotoxin is encoded by a bacteriophage of *C. diphtheriae.*

21.4 Diseases Caused by Microorganisms Affecting Both the Upper and Lower Respiratory Tracts (ASM Guidelines 5.3, 5.4, 6.4, 8.3)

- **Whooping Cough:** Causative agent, *Bordetella pertussis,* releases exotoxins—*pertussis toxin* and *tracheal cytotoxin*—that damage ciliated respiratory epithelial cells and cripple other components of host defense.
- **Respiratory Syncytial Virus Disease:** RSV infects the respiratory tract and produces giant multinucleated cells (syncytia). RSV is most prevalent cause of respiratory disease in newborn age group.
- **Influenza:** Caused by one of three influenza viruses: A, B, or C. The ssRNA genome is subject to constant genetic changes that alter the structure of its envelope glycoprotein. Antigenic drift refers to constant mutation of this glycoprotein. Antigenic shift, where eight separate RNA strands are involved in the swapping out of one of those genes or strands with a gene or strand from a different influenza virus, is even more serious.

21.5 Lower Respiratory Tract Diseases Caused by Microorganisms (ASM Guidelines 5.3, 5.4, 6.4, 8.3)

- **Tuberculosis:** Cause is primarily the bacterium *Mycobacterium tuberculosis.* Vaccine generally not used in the United States, although an attenuated vaccine, called BCG, is used in many countries.
  - *Mycobacterium avium* Complex (MAC): Before introduction of effective HIV treatments, disseminated tuberculosis infection with MAC was one of the biggest killers of AIDS patients.
- **Pneumonia:** Inflammatory condition of the lung in which fluid fills the alveoli. Caused by wide variety of different microorganisms.
  - *Streptococcus pneumoniae:* Main agent for community-acquired pneumonia (CAP) cases. Respiratory viruses are the second-most common cause.
  - *Mycoplasma pneumoniae* causes a mild pneumonia, referred to as "atypical." *Legionella* is a less common but serious cause of the disease. *Histoplasma capsulatum,* a fungus, causes a pneumonia-like disease.
  - *Pseudomonas aeruginosa* and *Klebsiella pneumoniae* are commonly responsible for healthcare-associated pneumonia (HCAP) cases. Furthermore, many of these cases of pneumonia appear to be polymicrobial in origin.
- **Hantavirus Pulmonary Syndrome:** Hantavirus causes a form of severe acute respiratory distress named hantavirus pulmonary syndrome (HPS). It is spread in mouse excretions.

*Source: ASM Curriculum Guidelines (American Society for Microbiology, 2012). Complete guidelines in appendix B of this book.

Multiple-Choice and True-False Questions | Bloom's Levels 1 and 2: Remember and Understand

Multiple-Choice Questions. Select the correct answer from the options provided.

1. The two most common groups of virus associated with the common cold are
   a. rhinoviruses.   d. both a and b.
   b. coronaviruses.   e. both a and c.
   c. influenza viruses.

2. Which of the following conditions is/are associated with *Streptococcus pyogenes*?
   a. pharyngitis   c. rheumatic fever
   b. scarlet fever   d. all of the above

3. Which is not a characteristic of *Streptococcus pyogenes*?
   a. group A streptococcus c. sensitive to bacitracin
   b. alpha-hemolytic   d. gram-positive

4. The common stain used to identify *Mycobacterium* species is
   a. Gram stain.   c. negative stain.
   b. acid-fast stain.   d. spore stain.

5. Which of the following techniques can be used to diagnose tuberculosis?
   a. tuberculin skin testing
   b. IGRA
   c. cultural isolation and antimicrobial testing
   d. all of the above

6. The DTaP vaccine provides protection against the following diseases, *except*
   a. diphtheria.   c. pneumonia.
   b. pertussis.   d. tetanus.

7. Which of the following infections often has/have a polymicrobial cause?
   a. otitis media   c. sinusitis
   b. hospital-acquired pneumonia   d. all of the above

8. Which of the following organisms causes the vast majority of pneumonias that occur in AIDS patients?
   a. hantavirus   c. *Pneumocystis jiroveci*
   b. *Histoplasma capsulatum* d. *Mycoplasma pneumoniae*

9. The beta-hemolysis of blood agar observed with *Streptococcus pyogenes* is due to the presence of
   a. streptolysin.   c. hyaluronic acid.
   b. M protein.   d. catalase.

10. An estimated ____ of the world population is infected with *Mycobacterium tuberculosis*.
    a. one-half   c. one-third
    b. one-fourth   d. three-fourths

True-False Questions. If the statement is true, leave as is. If it is false, correct it by rewriting the sentence.

11. *Bordetella pertussis* is the causative agent for whooping cough.

12. *Mycoplasma pneumoniae* causes "atypical" pneumonia and can be diagnosed by serology.

13. BCG vaccine is used in other countries to prevent Legionnaires' disease.

14. Respiratory syncytial virus (RSV) is a respiratory infection associated with elderly people.

15. The "flu shot" can cause the flu in immunocompromised people.

Critical Thinking Questions | Bloom's Levels 3, 4, and 5: Apply, Analyze, and Evaluate

Critical thinking is the ability to reason and solve problems using facts and concepts. These questions can be approached from a number of angles and, in most cases, they do not have a single correct answer.

1. a. Smokers tend to suffer from higher rates of lower respiratory tract infections. Based upon your knowledge of respiratory tract defenses, provide at least one explanation for this situation.
   b. Conduct additional research and discuss how the microbiome of smokers differs from that of healthy, nonsmoking individuals and how that change may impact their health.

2. a. What is DRACO? Describe how it could possibly be used to treat a variety of viral infections including the common cold.
   b. Provide evidence in support of or refuting the following statement: Antibiotic therapy should be used in the treatment of the common cold.

3. a. Explain why the current recommendation for uncomplicated acute otitis media is "watchful waiting."
   b. Compare and contrast the two vaccines currently available for the prevention of *S. pneumoniae* infection.

4. A 20-year-old male came to the ER with symptoms of a nearly closed-off airway and systemic shock. He was treated immediately with steroids to bring down the swelling in his neck and treat the shock. Culturing of blood specimens revealed the presence of gram-negative pleomorphic rods. Five days previously, he had gone to the doctor with a severe sore throat. His strep test was negative so the doctor prescribed no antibiotics. What disease is the patient suffering from?

5. Explain why individuals suffering from pertussis often develop secondary infections during the convalescent phase of the disease, and discuss aspects of today's protective vaccine against *B. pertussis*.

6. Construct a paragraph explaining the process of antigenic shift in the evolution of the H1N1 swine flu pandemic strain of influenza virus seen in 2009.

7. In an episode of the television show *House*, Dr. House pulled the patient's tracheostomy tube and viewed into the pharynx

to find evidence of a pseudomembrane. He then contacted the CDC to obtain the patient's needed treatment.

a. What did he think the patient was suffering from? Is this a notifiable disease?

b. What did he obtain from the CDC, and how will it be able to effectively treat the patient?

8. a. A graduate student from Ireland tests positive in the tuberculin skin test. Upon reading the patient history, the doctor determines that the test is a false positive and does not pursue further treatment. What is the possible explanation for the false-positive skin test?

9. Explain why patients on ventilators are at high risk for ventilator-associated pneumonia (VAP).

10. Conduct additional research and explain where cases of XDR-TB are most prevalent today and the potential risks for global spread of this disease.

Concept Connections | Bloom's Levels 4 and 6: Analyze and Create

This activity ties together multiple concepts in the chapter.

1. Why is pneumonia considered an anatomical diagnosis?

2. What are the signs and symptoms of pneumonia?

3. How is bacterial pneumonia different from viral pneumonia?

4. Which form of pneumonia is an AIDS-defining illness?

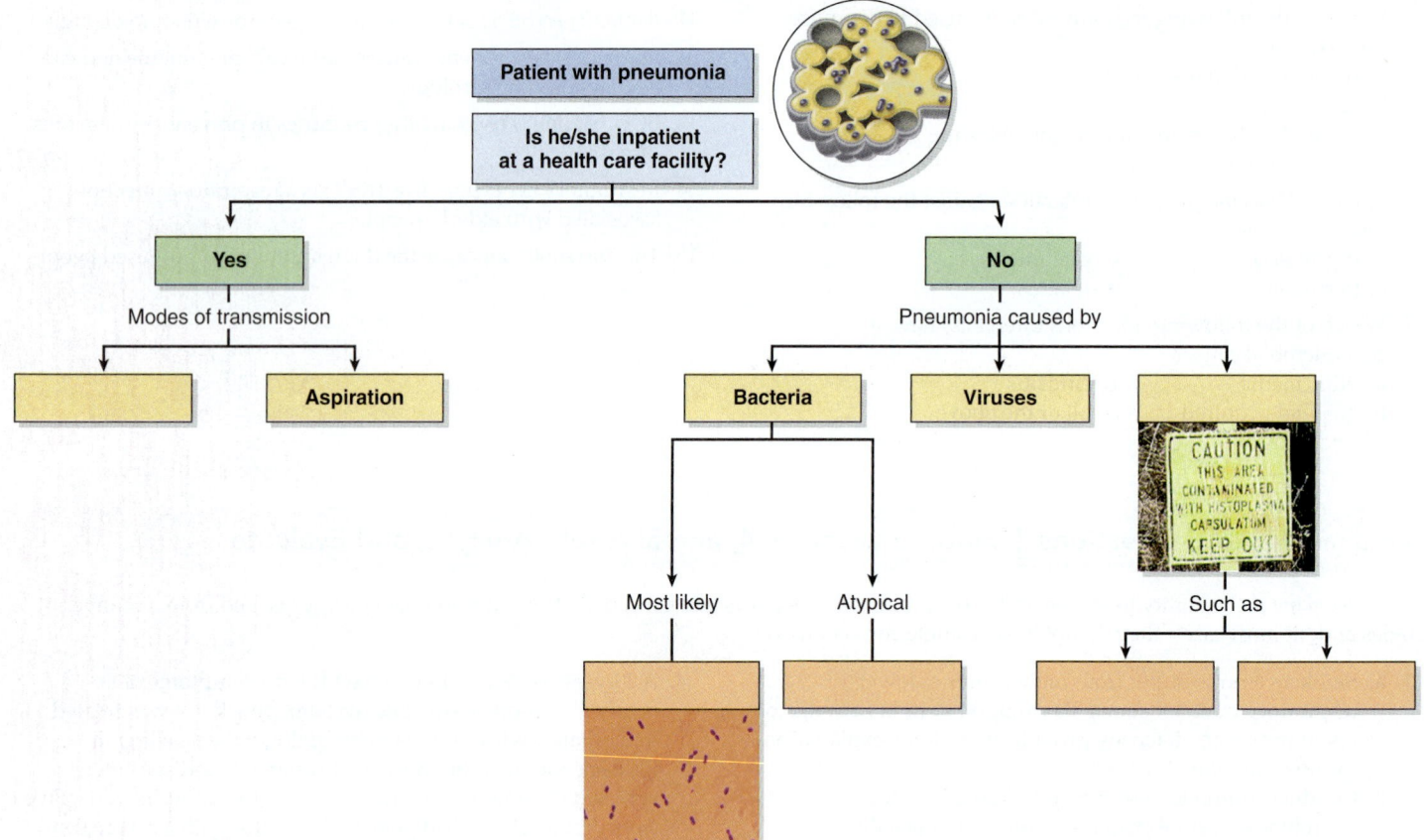

Visual Connections | Bloom's Level 5: Evaluate

These questions use visual images or previous content to make connections to this chapter's concepts.

1. **Figure 21.2.** Some doctors suggest that gently forcing one's ears to "pop" is an effective way to prevent ear infection. Use the following illustration to explain how this could work.

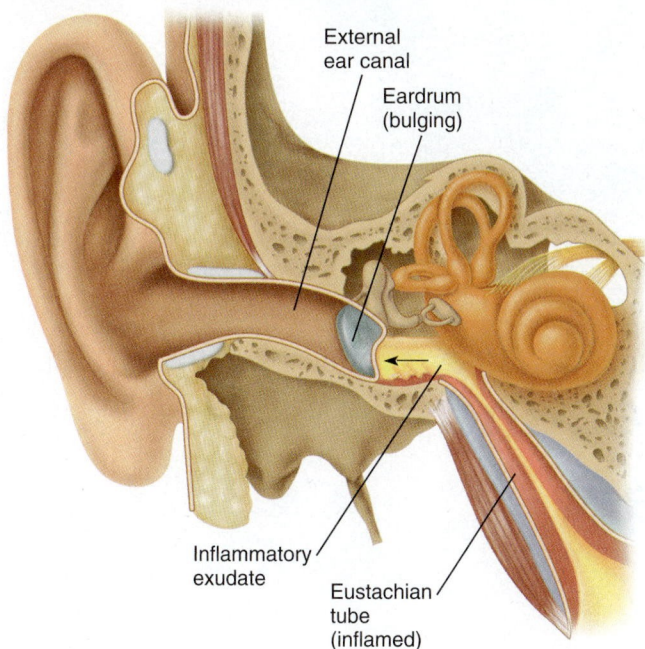

External ear canal

Eardrum (bulging)

Inflammatory exudate

Eustachian tube (inflamed)

2. **From chapter 3, figure 3.21.** Although there are many different organisms present in the respiratory tract, an acid-fast stain of sputum, like the one shown here, along with patient symptoms can establish a presumptive diagnosis of tuberculosis. Explain why.

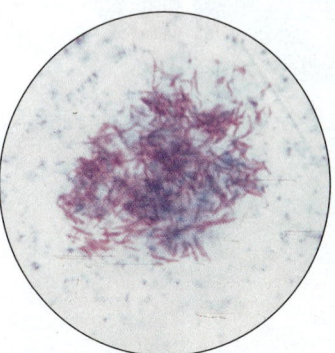

Acid-fast stain
Red cells are acid-fast.
Blue cells are non-acid-fast.

Concept Mapping | Bloom's Level 6: Create

Appendix D provides guidance for working with concept maps.

1. Using the words that follow, please create a concept map illustrating the relationships among these key terms from chapter 21.

FHA	pertussis toxin	cilia	*Bordetella pertussis*
coughing	tracheal cytotoxin	mucus	
multiplication	endotoxin	secondary infection	

www.mcgrawhillconnect.com

Enhance your study of this chapter with study tools and practice tests. Also ask your instructor about the resources available through ConnectPlus, including the media-rich eBook, interactive learning tools, and animations.

22

Infectious Diseases Affecting the Gastrointestinal Tract

> Cantaloupes from Jensen Farms were the source of a multistate outbreak of listeriosis that sickened 145 and killed 33.

Case File 22

Multistate *Listeria* Outbreak Tied to Cantaloupe

In September 2011, reports began to emerge of a food-borne outbreak of listeriosis. The initial CDC investigation found a number of cases in Colorado, Nebraska, Oklahoma, and Texas. By the time the outbreak was over in December 2011, it had stretched to 147 people in 28 states with 33 deaths and 1 miscarriage all tied to a single source: cantaloupe. Colorado was found to have the highest incidence of disease, with 40 cases and 9 deaths documented. Epidemiologists discovered why: Colorado was the home of Jensen Farms, the farm supplying the contaminated cantaloupes.

Shortly after the outbreak began, Jensen Farms voluntarily recalled shipments of cantaloupe. The FDA issued a press release warning consumers not to consume cantaloupes with the "Rocky Ford" label that Jensen Farms had sent to 17 states. After the FDA warning, several other companies issued recalls of products containing fresh-cut cantaloupes originating from Jensen Farms.

- What are the signs and symptoms of listeriosis?
- Who is at greatest risk of infection with *Listeria*?
- Why was this an unusual outbreak of listeriosis?

Continuing the Case appears on page 692.

Outline and Learning Outcomes

22.1 The Gastrointestinal Tract and Its Defenses
1. Draw or describe the anatomical features of the gastrointestinal tract.
2. List the natural defenses present in the gastrointestinal tract.

22.2 Normal Biota of the Gastrointestinal Tract
3. List the types of normal biota presently known to occupy the various regions of the gastrointestinal tract.
4. Summarize the known functions of the gastrointestinal microbiota and the role they may play in disease development.

22.1 The Gastrointestinal Tract and Its Defenses

The gastrointestinal (GI) tract can be thought of as a long tube, extending from mouth to anus. It is a very sophisticated delivery system for nutrients, composed of *eight* main sections and augmented by *four* accessory organs. The eight sections are the mouth, pharynx, esophagus, stomach, small intestine, large intestine, rectum, and anus. Along the way, the salivary glands, liver, gallbladder, and pancreas add digestive fluids and enzymes to assist in digesting and processing the food we take in **(figure 22.1).** The GI tract is often called the *digestive tract* or the *enteric tract.*

The GI tract has a very heavy load of microorganisms, and it encounters millions of new ones every day. Because of this, defenses against infection are extremely important. All intestinal surfaces are coated with a layer of mucus, which confers mechanical protection. Secretory IgA can also be found on most intestinal surfaces. The muscular walls of the GI tract keep food (and microorganisms) moving through the system through the action of peristalsis. Various fluids in the GI tract have antimicrobial properties. Saliva contains the antimicrobial proteins lysozyme and lactoferrin. The stomach fluid is antimicrobial by virtue of its extremely low pH. Bile is also antimicrobial.

The entire system is outfitted with cells of the immune system, collectively called gut-associated lymphoid tissue (GALT). The tonsils and adenoids in the oral cavity and pharynx, small areas of lymphoid tissue in the esophagus, Peyer's patches in the small intestine, and the appendix are all packets of lymphoid tissue consisting of T and B cells as

well as cells of nonspecific immunity. One of their jobs is to produce IgA, but they perform a variety of other immune functions. It is this vast assortment of immune players in the intestines, however, that puts some individuals at risk for

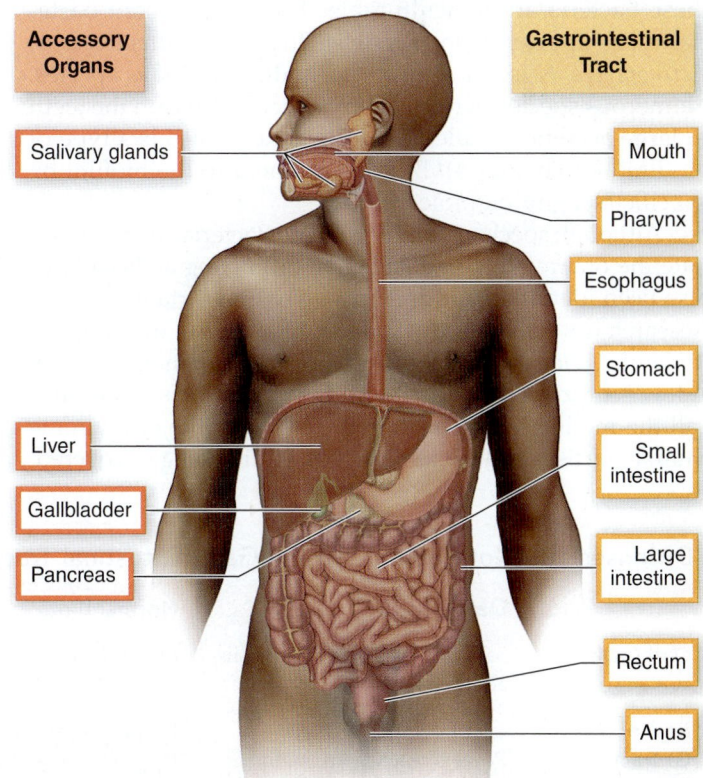

Figure 22.1 Major organs of the digestive system.

developing inflammatory bowel disease (IBD), as we will see in the next section.

22.1 Learning Outcomes—Assess Your Progress

1. Draw or describe the anatomical features of the gastrointestinal tract.
2. List the natural defenses present in the gastrointestinal tract.

22.2 Normal Biota of the Gastrointestinal Tract

The GI tract is home to a very large variety of normal biota. Every portion of it has a distinct microbial population. Antonie van Leeuwenhoek was one of the first to observe and describe oral microbes, through the observation of tooth scrapings using his newly developed lenses. Since then, the oral cavity has been found to harbor a vast number and variety of microorganisms, which is not surprising as it is in constant contact with the external environment. Breast-fed babies obtain at least part of their oral microbiome from breast milk, which contains abundant normal biota from the mother. Other babies pick up their oral biota from contact with their caregivers. The Human Microbiome Project (HMP) showed that microbial composition varies between individuals, which may have a great impact on the diagnosis and treatment of oral diseases in the future. Even though 300 bacterial species have been cultured and identified, the HMP has located over 600 species to date in the oral cavity, indicating that many of these bacteria are nonculturable. The predominant bacterial types appear to be *Prevotella, Treponema, Streptococcus, Actinomyces, Neisseria, Veillonella,* and *Lactobacillus* species. Numerous species of normal biota bacteria live on the teeth in large accretions called dental plaque, which is a type of biofilm (see chapter 4). Bacteria are held in the biofilm by specific recognition molecules.

Although species from the domain Bacteria clearly dominate the oral microbiome, methane-producing archaea have also been identified and may play a role in certain diseases. Research has also identified 85 fungal genera in the oral cavity, with *Candida albicans* being the most common member. A few protozoa (*Trichomonas tenax, Entamoeba gingivalis*) also call the mouth "home," while the extent of the oral virome (human viruses and bacteriophages) is still being uncovered.

Both the esophagus and stomach were thought to be sterile portions of the GI tract, due to the presence of physical and chemical barriers as well as immune defenses. Recent genetic analysis of these regions, however, has revealed the presence of nearly 200 different species of microorganisms. The most common types belong to the Firmicutes (*Streptococcus, Staphylococcus, Clostridium,* and *Bacillus* species). Though some of these are likely to be "just passing through," many of the microbes are true colonizers.

The large intestine has always been known to be a haven for billions of microorganisms (10^{11} per gram of contents), including the bacteria *Bacteroides, Fusobacterium, Bifidobacterium, Clostridium, Streptococcus, Peptostreptococcus, Lactobacillus, Escherichia,* and *Enterobacter;* the fungus *Candida;* and several protozoa as well. Researchers have also found archaeal species there. Recent studies have identified distinct overall gut microbiota profiles, or "enterotypes" in humans. This new evidence may pave the way for the development of personalized digestive medicine in terms of prevention, diagnosis, and patient treatment.

Currently, the accessory organs of the GI tract (salivary glands, gallbladder, liver, and pancreas) are considered to be free of resident microorganisms, just as all internal organs are.

The normal gut biota provide a protective function and can "teach" our immune system to react appropriately to microbial antigens. They also perform other jobs as well, such as aiding in digestion or providing nutrients that we can't produce ourselves. *E. coli,* for instance, synthesizes vitamin K. Its mere presence in the large intestine seems to be important for the proper formation of epithelial cell structure. One thing is becoming clear: A diverse gut microbiome is associated with health. When the gut microbiome loses its diversity, deviations from gastrointestinal—and systemic—health can occur. Disruptions may come from antibiotic treatment, illness, pregnancy, or dietary changes. This can result in such diverse conditions as obesity, diabetes, or other seemingly noninfectious disorders. For example, studies have linked high levels of intestinal proteobacteria to the onset of Crohn's disease, while ulcerative colitis seems to be influenced by the action of *Enterobacteriaceae.* Although the immune system is ultimately responsible for the injury to the GI tract in these cases of inflammatory bowel disease (IBD), it appears that an imbalance in the numbers of the normal biota may be the actual trigger of the overreaction. In chapter 19, we introduced the concept of the gut-brain axis, in which the composition of the GI tract microbiome can affect the central nervous system.

Defenses and Normal Biota of the Gastrointestinal Tract

	Defenses	Normal Biota
Oral Cavity	Saliva, sIgA, lysozyme, tonsils, adenoids	*Prevotella, Treponema, Streptococcus, Actinomyces, Neisseria, Veillonella, Lactobacillus*
Rest of GI Tract	GALT, lymphoid tissue, Peyer's patches, appendix, sIgA, rich normal biota	Esophagus, stomach: *Streptococcus, Staphylococcus, Clostridium, Bacillus* Large intestine: *Bacteroides, Fusobacterium, Bifidobacterium, Clostridium, Streptococcus, Peptostreptococcus, Lactobacillus, Escherichia,* and *Enterobacter, Candida* and protozoa

22.2 Learning Outcomes—Assess Your Progress

3. List the types of normal biota presently known to occupy the various regions of the gastrointestinal tract.

4. Summarize the known functions of the gastrointestinal microbiota and the role they may play in disease development.

22.3 Gastrointestinal Tract Diseases Caused by Microorganisms (Nonhelminthic)

Tooth and Gum Infections

It is difficult to pinpoint exactly when the "normal biota biofilm" described for the oral environment becomes a "pathogenic biofilm." If left undisturbed, the biofilm structure eventually contains anaerobic bacteria that can damage the soft tissues and bones (referred to as the periodontium) surrounding the teeth. Also, the introduction of carbohydrates to the oral cavity can result in breakdown of hard tooth structure (the dentition) due to the production of acid by certain oral streptococci in the biofilm. These two separate circumstances are discussed in the following sections.

Dental Caries (Tooth Decay)

Dental caries or tooth decay is the most common infectious disease of human beings. The process of decay involves the dissolution of solid tooth surface due to the metabolic action of bacteria. (**Figure 22.2** depicts the structure of a tooth.) The symptoms are often not noticeable but range from minor disruption in the outer (enamel) surface of the tooth to complete destruction of the enamel and then destruction of deeper layers (**process figure 22.3**). Deeper lesions can result in infection to the soft tissue inside the tooth, called the pulp, which contains blood vessels and nerves. These deeper infections lead to pain, referred to as a "toothache."

▶ Causative Agents

Two representatives of oral alpha-hemolytic streptococci, *Streptococcus mutans* and *Streptococcus sobrinus*, seem to be the main causes of dental caries, although a mixed species consortium, consisting of other *Streptococcus* species and some lactobacilli, is probably the best route to caries. A specific condition called *early childhood caries* may also be caused by a newly identified species, *Scardovia wiggsiae*. Note that in the absence of dietary carbohydrates bacteria do not cause decay.

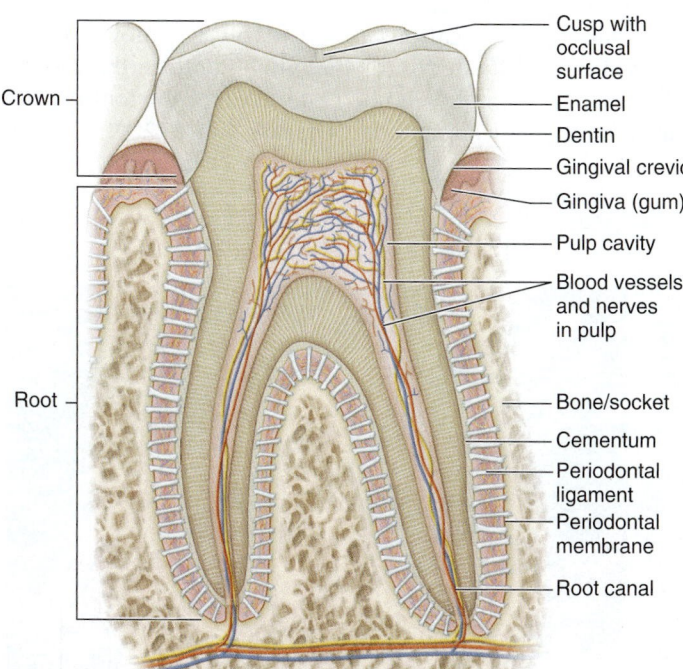

Figure 22.2 The anatomy of a tooth.

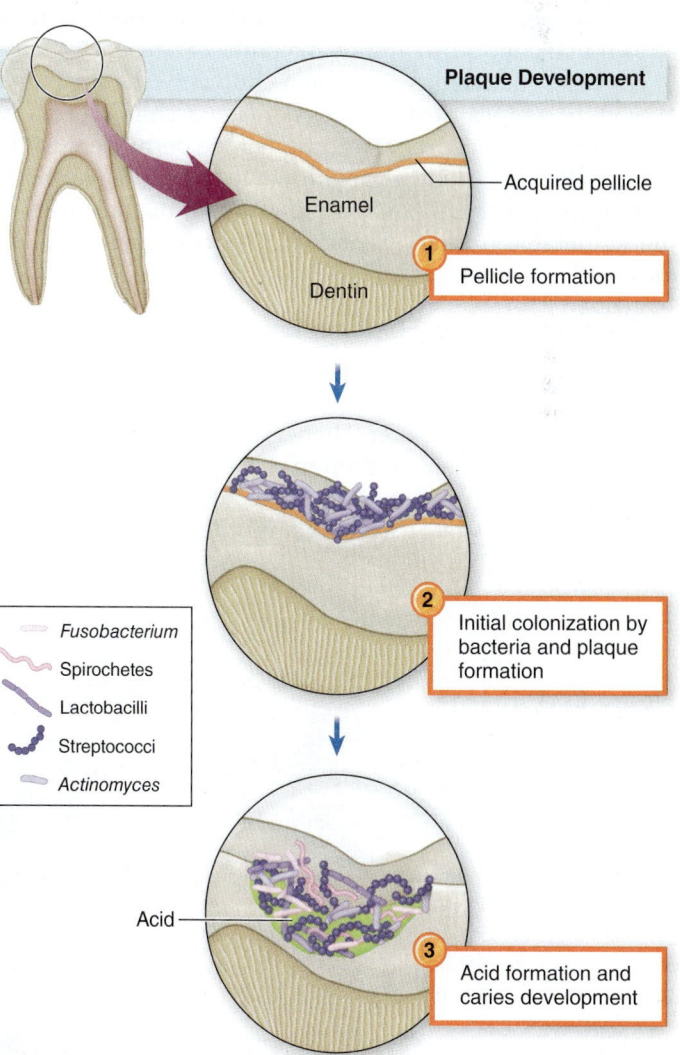

Process Figure 22.3 Stages in plaque development and cariogenesis.

▶ **Pathogenesis and Virulence Factors**

In the presence of sucrose and, to a lesser extent, other carbohydrates, *S. mutans* and other streptococci produce sticky polymers of glucose called fructans and glucans. These adhesives help bind them to the smooth enamel surfaces and contribute to the sticky bulk of the plaque biofilm **(figure 22.4)**. If mature plaque is not removed from sites that readily trap food, it can result in a carious lesion. This is due to the action of the streptococci and other bacteria that produce acid as they ferment the carbohydrates. If the acid is immediately flushed from the plaque and diluted in the mouth, it has little effect. However, in the denser regions of plaque, the acid can accumulate in direct contact with the enamel surface and lower the pH to below 5, which is acidic enough to begin to dissolve (decalcify) the calcium phosphate of the enamel in that spot. This initial lesion can remain localized in the enamel and can be repaired with various inert materials (fillings). Once the deterioration has reached the level of the dentin, tooth destruction speeds up and the tooth can be rapidly destroyed. Exposure of the pulp leads to severe tenderness and toothache, and the chance of saving the tooth is diminished.

Teeth become vulnerable to caries as soon as they appear in the mouth at around 6 months of age. Early childhood caries, defined as caries in a child between birth and 6 years of age, can extensively damage a child's primary teeth and affect the proper eruption of the permanent teeth. The practice of putting a baby down to nap with a bottle of fruit juice or formula can lead to rampant dental caries in the vulnerable primary dentition. This condition is called *nursing bottle caries*.

▶ **Transmission and Epidemiology**

The bacteria that cause dental caries are transmitted to babies and children by their close contacts, especially the mother or closest caregiver. There is evidence for transfer of oral bacteria between children in day care centers, as well. Although it was previously believed that humans don't acquire *S. mutans* or *S. sobrinus* until the eruption of teeth in the mouth, it now seems likely that both of these species may survive in the infant's oral cavity prior to appearance of the first teeth.

Dental caries has a worldwide distribution. Its incidence varies according to many factors, including amount of carbohydrate consumption, hygiene practices, and host genetic factors. Susceptibility to caries generally decreases with age, possibly due to the fact that grooves and fissures—common sites of dental caries—tend to become more shallow as teeth are worn down. As the population ages and natural teeth are retained for longer periods, the caries rate may well increase in the elderly, because receding gums expose the more susceptible root surfaces.

▶ **Culture and Diagnosis**

Dental professionals diagnose caries based on the tooth condition. Culture of the lesion is not routinely performed.

▶ **Prevention and Treatment**

The best way to prevent dental caries is through dietary restriction of sucrose and other refined carbohydrates. Regular brushing and flossing to remove plaque are also important. Most municipal communities in the United States add trace amounts of fluoride to their drinking water, because fluoride, when incorporated into the tooth structure, can increase tooth (as well as bone) hardness. Fluoride can also encourage the remineralization of teeth that have begun the demineralization process. These and other proposed actions of fluoride could make teeth less susceptible to decay. Fluoride is also added to toothpastes and mouth rinses and can be applied in gel form. Many European countries do not fluoridate their water due to concerns over additives in drinking water, and the same controversy exists in parts of the United States today.

Treatment of a carious lesion involves removal of the affected part of the tooth (or the whole tooth in the case of advanced caries), followed by restoration of the tooth structure with an artificial material **(Disease Table 22.1)**.

Figure 22.4 The macroscopic and microscopic appearance of plaque. **(a)** Disclosing tablets containing vegetable dye stain heavy plaque accumulations at the junction of the tooth and gingiva. **(b)** Scanning electron micrograph of the plaque biofilm with long filamentous forms and "corn cobs" that are mixed bacterial aggregates.

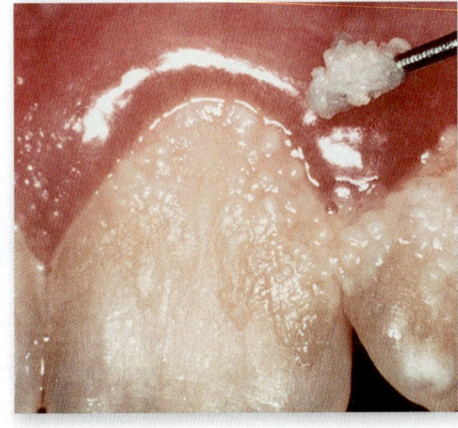

(a)

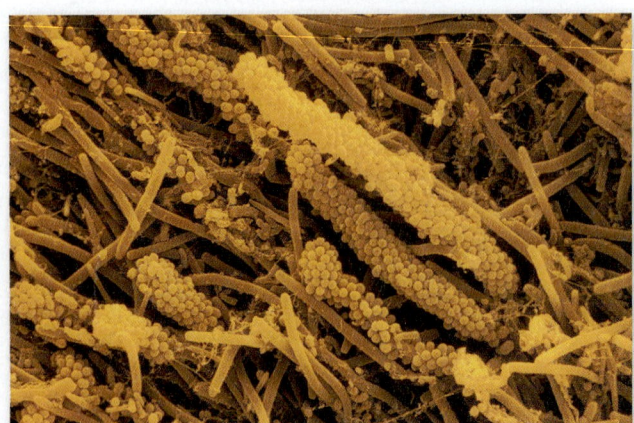

(b)

Disease Table 22.1	Dental Caries
Causative Organism(s)	*Streptococcus mutans, Streptococcus sobrinus,* others
Most Common Modes of Transmission	Direct contact
Virulence Factors	Adhesion, acid production
Culture/Diagnosis	–
Prevention	Oral hygiene, fluoride supplementation
Treatment	Removal of diseased tooth material
Epidemiological Features	Globally, 60%–90% prevalence in school-age children

Periodontal Disease

Periodontal disease is so common that 97% to 100% of the population have some manifestation of it by age 45. Most kinds are due to bacterial colonization and varying degrees of inflammation that occur in response to gingival damage. Microbes from pets have also been implicated in periodontal disease. Recent research has shown that people who have close contact with their dogs harbor some of the same periodontitis-causing bacteria as their pets have in their oral cavity.

Periodontitis

▶ Signs and Symptoms

The initial stage of periodontal disease is **gingivitis,** the signs of which are swelling, loss of normal contour, patches of redness, and increased bleeding of the gingiva. Spaces or pockets of varying depth also develop between the tooth and the gingiva. If this condition persists, a more serious disease called periodontitis results. This is the natural extension of the disease into the periodontal membrane and cementum. The deeper involvement increases the size of the pockets and can cause bone resorption severe enough to loosen the tooth in its socket. If the condition is allowed to progress, the tooth can be lost **(process figure 22.5).** Humans can sometimes help this process, as described in **Insight 22.1.**

▶ Causative Agent

Dental scientists stop short of stating that particular bacteria cause periodontal disease, because not all of the criteria for establishing causation have been satisfied. In fact, dental diseases (in particular, periodontal disease) provide an excellent model of disease mediated by communities of microorganisms rather than single organisms. When the polymicrobial biofilms consist of the right combination of bacteria, such as the anaerobes *Tannerella forsythia* (formerly *Bacteroides forsythus*), *Aggregatibacter actinomycetemcomitans*, *Porphyromonas gingivalis*, and perhaps *Fusobacterium* and spirochete species, the periodontal destruction process begins. Data collected through the Human Microbiome Project reveal that an individual's risk for dental caries or periodontitis is directly related to the composition of their personal oral microbiome. The presence of *Methanobrevibacter oralis* in the gingival crevice seems to be an important contributor to periodontal disease, which marks the first association between archaeal

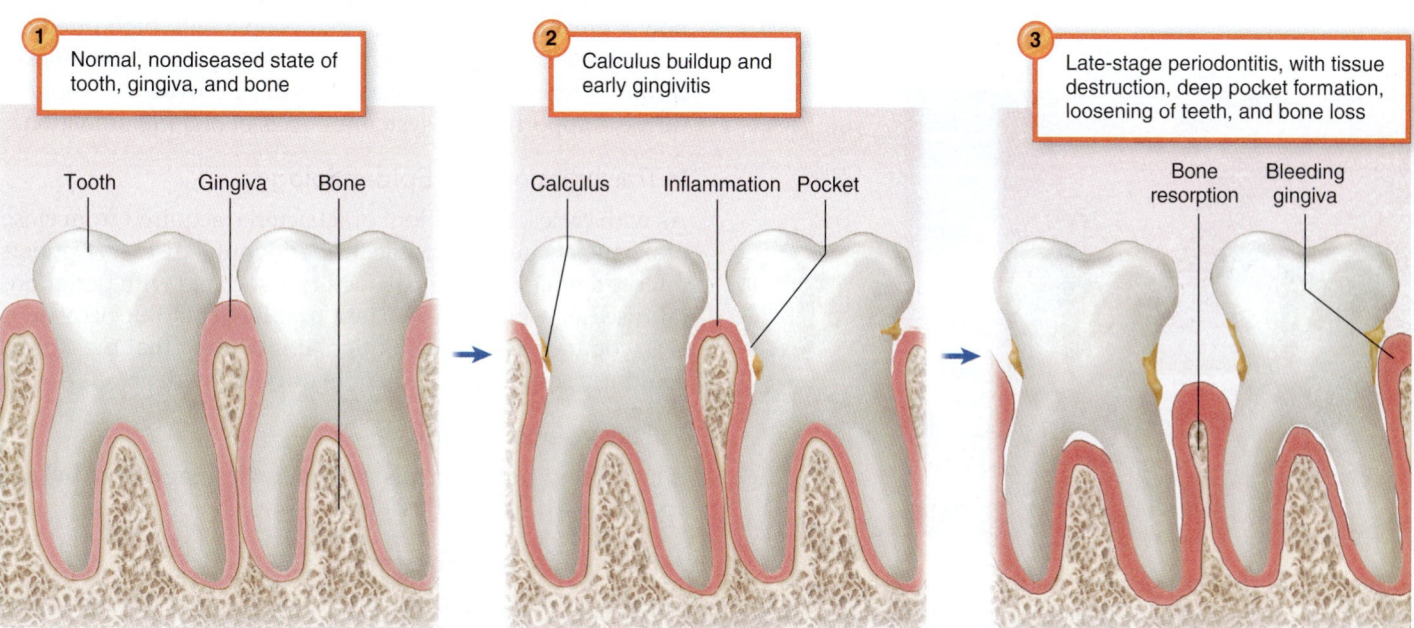

Process Figure 22.5 Stages in soft-tissue infection, gingivitis, and periodontitis.

INSIGHT 22.1 Metal in Your Mouth, Biofilms on the Barbell

Body art such as tattoos and piercings are becoming increasingly popular. People have found many creative ways to express themselves, including piercing parts of the body that your grandmother never would have considered. One of the most popular piercing sites—after the earlobes—is the tongue, with a metal barbell inserted through the muscle tissue of the tongue. Because of the increasing popularity of this type of piercing, dentists have seen a surge in dental and oral complications in individuals with tongue piercings including infections, damage to the enamel, chipped teeth, and gum and tongue recession. A recent study shed light on the bacterial populations that colonize tongue piercings. Researchers asked volunteers to remove their tongue stud and replace it with stainless steel, titanium, polypropylene (a type of plastic), or polytetrafluoroethylene (Teflon). Then, after 2 weeks, the studs were removed and samples were taken of the piercing canal and the stud itself. The study showed that biofilms formed more frequently on the stainless steel and titanium barbells than on the plastic or Teflon studs. The biofilms included 67 of 80 species of bacteria linked to oral infections. In rare cases, tongue piercings have been linked to infections of the oral cavity, head, and neck; systemic bacterial infection; toxic shock syndrome; endocarditis; and brain abscesses. In all of these cases, oral bacteria were the source of the infection. Hepatitis B, C, and D have also been linked to tongue piercing. Body piercing remains largely unregulated and may be performed by unlicensed individuals in unregistered body art parlors. The American Dental Association opposes the practice of intraoral piercings and supports legislation that requires a parent's consent for minors wanting these types of piercings.

species and the development of human disease. Other factors are also important in the development of periodontal disease, such as behavioral and genetic influences, as well as tooth position. The most common predisposing condition occurs when the plaque becomes mineralized (calcified) with calcium and phosphate crystals. This process produces a hard, porous substance called **calculus** above and below the gingival margin (edge) that can induce varying degrees of periodontal damage **(figure 22.6)**.

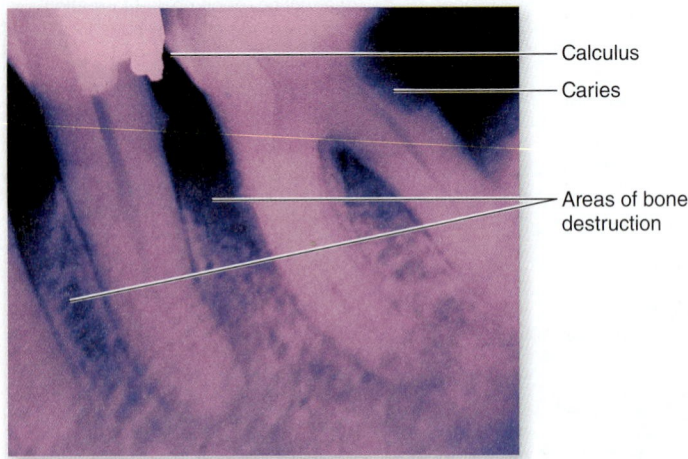
Calculus
Caries
Areas of bone destruction

Figure 22.6 The nature of calculus. Radiograph of lower premolar and molar teeth, showing calculus on the top and a caries lesion on the right. Bony defects caused by periodontitis affect both teeth.

▶ Pathogenesis and Virulence Factors

Calculus and plaque accumulating in the gingival sulcus cause abrasions in the delicate gingival membrane, and the chronic trauma causes a pronounced inflammatory reaction. The damaged tissues become a portal of entry for a variety of bacterial residents. The bacteria have an arsenal of enzymes, such as proteases, that destroy soft oral tissues. In response to the mixed infection, the damaged area becomes infiltrated by neutrophils and macrophages and, later, by lymphocytes, which cause additional inflammation and tissue damage. There is now a great deal of evidence that people with high numbers of the bacteria associated with periodontitis also have thicker carotid arteries and increased rates of cardiovascular disease, further supporting the systemic effects of oral inflammation.

▶ Transmission and Epidemiology

As with caries, the resident oral bacteria, acquired from close oral contact, are responsible for periodontal disease. Dentists refer to a wide range of risk factors associated with periodontal disease, especially deficient oral hygiene. But because it is so common in the population, it is evident that most of us could use some improvement in our oral hygiene.

▶ Culture and Diagnosis

Like caries, periodontitis is generally diagnosed by the appearance of the oral tissues.

▶ Prevention and Treatment

Regular brushing and flossing to remove plaque automatically reduce both caries and calculus production. Once calculus has formed on teeth, it cannot be removed by brushing but can be dislodged only by special mechanical procedures (scaling) in the dental office. Because much of the pathology results from inflammation, some scientists are testing the use of new anti-inflammatory peptides to control disease progression. As already noted, the identification of high-risk microbiome profiles may also lead to the development of early preventive measures in patients.

Most periodontal disease is treated by removal of calculus and plaque and maintenance of good oral hygiene. Often, surgery to reduce the depth of periodontal pockets is required. Antibiotic therapy, either systemic or applied in periodontal packings, may also be utilized. Steroid use may also benefit the patient by reducing inflammation.

Oral microbes causing dental caries and periodontitis can enter the bloodstream at the site of periodontal infection, spreading to distant sites within the body. In 2009, a wide variety of oral bacteria were detected via PCR in plaques in coronary arteries. Live cells of *Porphyromonas gingivalis* and *Aggregatibacter actinomycetemcomitans,* two periodontal pathogens, have also been localized to atherosclerotic tissue. A newly discovered oral bacterium, *Streptococcus tigurinus,* can escape into the bloodstream, increasing the risk for endocarditis and even meningitis. Evidence seems to be increasing that allowing our oral health to slide can have deeper consequences than once thought.

Necrotizing Ulcerative Gingivitis and Periodontitis

The most destructive periodontal diseases are necrotizing ulcerative gingivitis (NUG) and necrotizing ulcerative periodontitis (NUP). The two diseases were formerly lumped under one name, acute necrotizing ulcerative gingivitis (ANUG). It was commonly referred to as "trench mouth," reflecting the poor dental health of soldiers in the battlefield trenches of World War I. These diseases are synergistic infections involving *Treponema vincentii, Prevotella intermedia,* and *Fusobacterium* species. These pathogens together produce several invasive factors that cause rapid advancement into the periodontal tissues. The condition is associated with severe pain, bleeding, pseudomembrane formation, and necrosis. Scientists believe that NUP may be an extension of NUG, but the conditions can be distinguished by the advanced bone destruction that results from NUP. Both diseases seem to result from poor oral hygiene, altered host defenses, or prior gum disease rather than being communicable. The diseases are common in AIDS patients and other immunocompromised populations. Diabetes and cigarette smoking can predispose people to these conditions. NUG and NUP usually respond well to targeted antibiotics after debridement of damaged periodontal tissue **(Disease Table 22.2)**.

Mumps

The word *mumps* is Old English for "lump" or "bump." The symptoms of this viral disease are so distinctive that Hippocrates clearly characterized it in the fifth century BC as a self-limited, mildly epidemic illness associated with painful swelling at the angle of the jaw **(figure 22.7)**.

▶ Signs and Symptoms

After an average incubation period of 2 to 3 weeks, symptoms of fever, nasal discharge, muscle pain, and malaise develop. These may be followed by inflammation of the salivary glands (especially the parotids), producing the classic gopherlike swelling of the cheeks on one or both sides (as shown in figure 22.7). Swelling of the gland is called parotitis, and it can cause considerable discomfort. Viral multiplication

Disease Table 22.2 Periodontal Diseases		
Disease	**Periodontitis**	**Necrotizing Ulcerative Gingivitis and Periodontitis**
Causative Organism(s)	Polymicrobial community including some or all of: *Tannerella forsythia, Aggregatibacter actinomycetemcomitans, Porphyromonas gingivalis,* others	Polymicrobial community (*Treponema vincentii, Prevotella intermedia, Fusobacterium* species)
Most Common Modes of Transmission	–	–
Virulence Factors	Induction of inflammation, enzymatic destruction of tissues	Inflammation, invasiveness
Culture/Diagnosis	–	–
Prevention	Oral hygiene	Oral hygiene
Treatment	Removal of plaque and calculus, gum reconstruction, possibly anti-inflammatory treatments	Debridement of damaged tissue, possibly antibiotics
Epidemiological Features	United States: smokers = 11%, nonsmokers = 2%; internationally: 10%–15% of adults	–

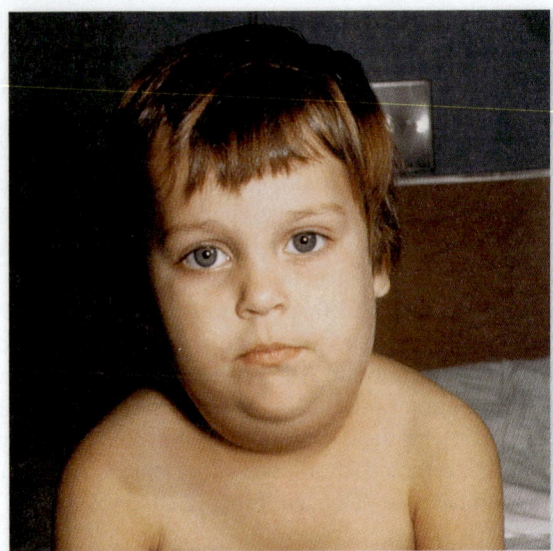

Figure 22.7 **The external appearance of swollen parotid glands in mumps (parotitis).**

in salivary glands is followed by invasion of other organs, especially the testes, ovaries, thyroid gland, pancreas, meninges, heart, and kidney. Despite the invasion of multiple organs, the prognosis of most infections is complete, uncomplicated recovery with permanent immunity.

Complications in Mumps In 20% to 30% of young adult males, mumps infection localizes in the epididymis and testis, usually on one side only. The resultant syndrome of orchitis and epididymitis may be rather painful, but no permanent damage usually occurs. The popular belief that mumps readily causes sterilization of adult males is still held, despite medical evidence to the contrary. Perhaps this notion has been reinforced by the tenderness that continues long after infection and by the partial atrophy of one testis that occurs in about half the cases. Permanent sterility due to mumps is very rare.

In mumps pancreatitis, the virus replicates in beta cells and pancreatic epithelial cells. Viral meningitis, characterized by fever, headache, and stiff neck, appears 2 to 10 days after the onset of parotitis, lasts for 3 to 5 days, and then dissipates, leaving few or no adverse side effects. Another rare event is infection of the inner ear that can lead to deafness.

▶ **Causative Agent**

Mumps is caused by an enveloped single-stranded RNA virus (mumps virus) from the genus *Paramyxovirus*, which is part of the family *Paramyxoviridae*. Other members of this family that infect humans are *Morbillivirus* (measles virus) and the respiratory syncytial virus. The envelopes of paramyxoviruses possess spikes that have specific functions.

▶ **Pathogenesis and Virulence Factors**

A virus-infected cell is modified by the insertion of proteins called HN spikes into its cell membrane. The HN spikes immediately bind an uninfected neighboring cell, and in the presence of another type of spike called F spikes, the two cells

permanently fuse. A chain reaction of multiple cell fusions then produces a *syncytium* (sin-sish'-yum) with cytoplasmic inclusion bodies, which is a diagnostically useful cytopathic effect **(figure 22.8).** The ability to induce the formation of syncytia is characteristic of the family *Paramyxoviridae*.

▶ **Transmission and Epidemiology**

Humans are the exclusive natural hosts for the mumps virus. It is communicated primarily through salivary and respiratory secretions. Transmission occurs readily among populations living in close proximity, at home or in dormitories, and the virus has a greater chance of spreading the longer one is in contact with an infected individual. Infection occurs worldwide, with increases in the late winter and early spring in temperate climates.

High rates of infection arise among crowded populations or in communities with poor herd immunity. Most cases occur in children under the age of 15, and as many as 40% are subclinical. Because lasting immunity follows any form of mumps infection, no long-term carrier reservoir exists in the population. Before 2006, the average incidence of mumps had been reduced in the United States to around 200 cases per year, and up to 90% were imported cases of disease—meaning the infection was acquired outside of the United States. The incidence has become more unpredictable since 2006, though. In that year, there were about 2,600 cases. The next 3 years saw cases in the low hundreds again, but then in 2010 there were more than 1,500 cases. This outbreak was directly linked to an 11-year-old boy who became infected after traveling to the United Kingdom; air travel has also been implicated in the 2006 outbreak. The recommendation is to be sure to get two doses of MMR vaccine.

▶ **Culture and Diagnosis**

Diagnosis is usually based on the clinical sign of swollen parotid glands and known exposure 2 or 3 weeks previously. Because parotitis is not always present, and the incubation period can range from 7 to 23 days, a practical diagnostic alternative is to perform a direct fluorescent antibody test for viral antigen or an ELISA test on a patient's serum.

▶ **Prevention and Treatment**

The general pathology of mumps is mild enough that symptomatic treatment to relieve fever, dehydration, and pain is usually adequate. The new vaccine recommendations call for a dose of MMR at 12 to 15 months and a second dose at 4 to 6 years. Health care workers and college students who haven't already had both doses are advised to do so. Even though the vaccine provides 80% to 90% protection against disease, this still leaves a susceptible population that may become infected with mumps virus even if adequately vaccinated. This is exactly what happened in the propagation of the 2010 outbreak, Researchers have determined that this resurgence of mumps infections is not due to the evolution of novel mutant strains of the virus but is in fact due to a reduced secondary immune response even after vaccination. This hypothesis is supported by the fact that administration of a third dose of vaccine appeared to help bring the 2010 outbreak to a halt **(Disease Table 22.3).**

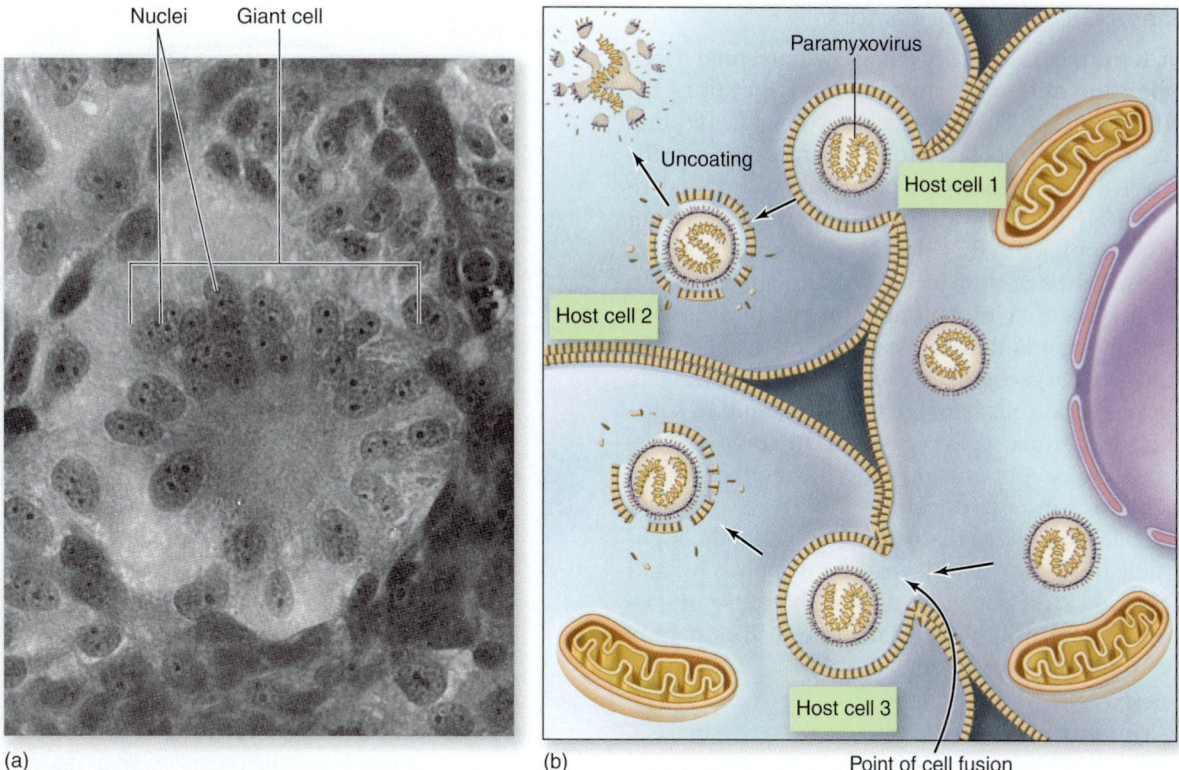

Figure 22.8 The effects of paramyxoviruses. **(a)** When they infect a host cell, paramyxoviruses induce the cell membranes of adjacent cells to fuse into large multinucleate giant cells, or syncytia. **(b)** This fusion allows direct passage of viruses from an infected cell to uninfected cells by communicating membranes. Through this means, the virus evades antibodies.

Disease Table 22.3	Mumps
Causative Organism(s)	Mumps virus (genus *Paramyxovirus*)
Most Common Modes of Transmission	Droplet contact
Virulence Factors	Spike-induced syncytium formation
Culture/Diagnosis	Clinical, fluorescent Ag tests, ELISA for Ab
Prevention	MMR live attenuated vaccine
Treatment	Supportive
Epidemiological Features	United States: fluctuates between a few hundred cases a year and a few thousand; internationally: are epidemic peaks every 2–5 years

Gastritis and Gastric Ulcers

The curved cells of *Helicobacter* were first detected by J. Robin Warren in 1979 in stomach biopsies from ulcer patients. He and an assistant, Barry J. Marshall, isolated the microbe in culture and even served as guinea pigs by swallowing a large inoculum to prove that it would cause gastric ulcers. Warren and Marshall won the Nobel Prize in Medicine in 2005 for their discovery.

▶ Signs and Symptoms

Gastritis is experienced as sharp or burning pain emanating from the abdomen. Gastric or peptic ulcers are actual lesions in either the mucosa of the stomach (gastric ulcers) or in the uppermost portion of the small intestine (duodenal ulcers). Severe ulcers can be accompanied by bloody stools, vomiting, or both. The symptoms are often worse at night, after eating, or under conditions of psychological stress.

The second most common cancer in the world is stomach cancer (although it has been declining in the United States), and ample evidence suggests that long-term infection with *H. pylori* is a major contributing factor.

▶ Causative Agent

Helicobacter pylori is a curved gram-negative rod, closely related to *Campylobacter*, which we study later in this chapter.

▶ Pathogenesis and Virulence Factors

Once the bacterium passes into the gastrointestinal tract, it bores through the outermost mucous layer that lines the stomach epithelial tissue. Then it attaches to specific binding sites on the cells and entrenches itself. One receptor specific for *Helicobacter* is the same molecule on human cells that confers the O blood type. This finding accounts for the higher rate of ulcers in people with this blood type. Another protective adaptation of the bacterium is the formation of urease, an enzyme that converts urea into ammonium and bicarbonate, both alkaline compounds that can neutralize stomach acid. As the immune system recognizes and attacks the pathogen, infiltrating white blood cells damage the epithelium to some degree, leading to chronic active gastritis. In some people, these lesions lead to deeper erosions and ulcers that can lay the groundwork for cancer to develop.

Before the bacterium was discovered, spicy foods, high-sugar diets (which increase acid levels in the stomach), and psychological stress were considered to be the cause of gastritis and ulcers. Now it appears that these factors merely aggravate the underlying infection.

▶ Transmission and Epidemiology

The mode of transmission of this bacterium remains a mystery. Studies have revealed that the pathogen is present in a large proportion of the human population. It occurs in the stomachs of 25% of healthy middle-age adults and in more than 60% of adults over 60 years of age. *H. pylori* is probably transmitted from person to person by the oral-oral or fecal-oral route. It seems to be acquired early in life mainly through what is called "familial transfer"—the microbe is acquired from family members, especially from infected mothers to their children. The pathogen is carried asymptomatically until its activities begin to damage the digestive mucosa. Because other animals are also susceptible to *H. pylori* and even develop chronic gastritis, it has been proposed that the disease is a zoonosis transmitted from an animal reservoir. The bacterium has also been found in water sources indicating that proper sanitation could reduce the transmission of the infections in developing nations.

Approximately one-half of the world's population is infected with *H. pylori*. It is not known what causes some people to experience symptoms, although it is most likely that those with the right combination of aggravating factors are those who experience disease.

▶ Culture and Diagnosis

Diagnosis has typically been accomplished with endoscopy, a procedure in which a long flexible tube **(figure 22.9)** is inserted through the throat into the stomach to visualize any lesions there. During this procedure, biopsies of the tissue can be obtained for histological examination, microbial culture, and rapid urea testing. The urea breath test is a noninvasive method that is sometimes used. In this test, patients ingest urea that has a radioactive tag on its carbon molecule. If *Helicobacter* is present in a patient's stomach, the bacterium's urease breaks down the urea and the patient exhales

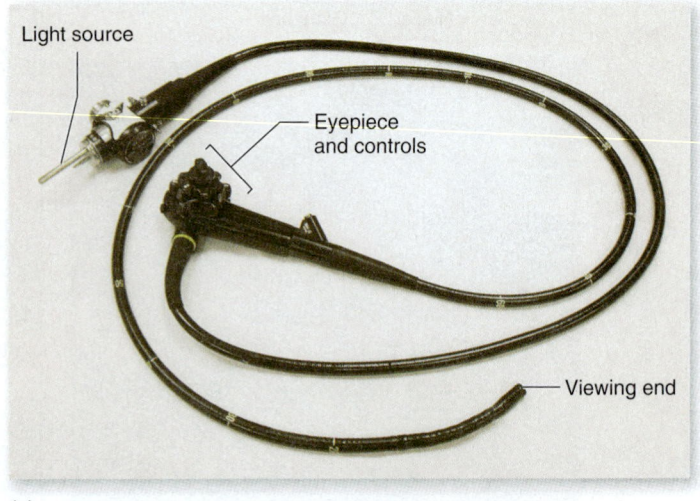

(a)

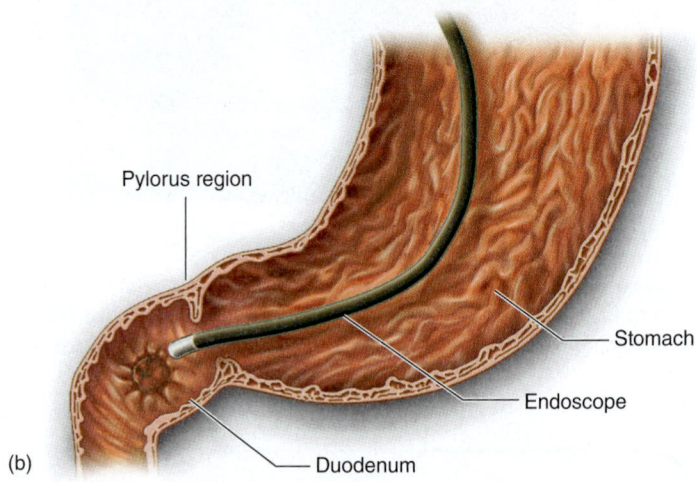

(b)

Figure 22.9 Endoscopy. **(a)** Flexible endoscope. **(b)** The flexible endoscope is able to provide images of many parts of the stomach, and even the duodenum.

radioactively labeled carbon dioxide. In the absence of urease, the intact urea molecule passes through the digestive system. Patients whose breath is positive for the radioactive carbon are considered positive for *Helicobacter*.

A stool test is also available. The HpSA (*H. pylori* stool antigen) test is an ELISA format test. Other immunologic tests, including Western blotting, can also be used to diagnose infection.

▶ Prevention and Treatment

The only preventive approaches available currently are those that diminish some of the aggravating factors just mentioned. Many over-the-counter remedies offer symptom relief by acting to neutralize stomach acid. The best treatment is a sequential course of antibiotics, usually amoxicillin followed by clarithromycin + tinidazole. A new candidate vaccine called HelicoVax has been identified that prohibits the colonization of *H. pylori* and may offer the promise of infection protection in the future. Some scientists question, however, if complete eradication of the bacterium is wise, as studies have revealed a potential protective effect of *H. pylori* against the development of asthma **(Disease Table 22.4).**

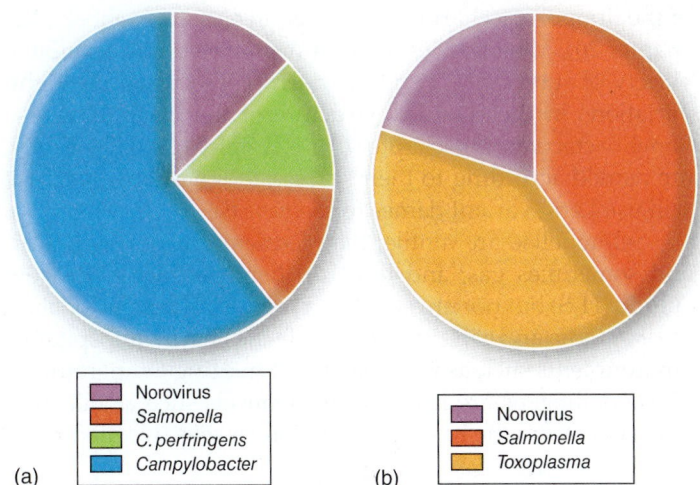

Disease Table 22.4	Gastritis and Gastric Ulcers
Causative Organism(s)	*Helicobacter pylori*
Most Common Modes of Transmission	?
Virulence Factors	Adhesins, urease
Culture/Diagnosis	Endoscopy, urea breath test, stool antigen test
Prevention	None
Treatment	Amoxicillin followed by clarithromycin + tinidazole
Epidemiological Features	United States: infection (not disease) rates at 35% of adults; internationally: infection rates at 50%

Figure 22.10 Profiles of food-borne illness in the United States. **(a)** The most common causes of food-borne illness. **(b)** The most common causes of death from food-borne illness.

Legend (a): Norovirus, *Salmonella*, *C. perfringens*, *Campylobacter*

Legend (b): Norovirus, *Salmonella*, *Toxoplasma*

Acute Diarrhea (With or Without Vomiting)

Diarrhea—usually defined as three or more loose stools in a 24-hour period—needs little explanation. In recent years, on average, citizens of the United States experienced 1.2 to 1.9 cases of diarrhea per person per year; among children, that number is twice as high. The incidence of diarrhea is even higher among children attending day care centers. In tropical countries, children may experience more than 10 episodes of diarrhea a year. In fact, more than 3 million children a year—mostly in developing countries—die from a diarrheal disease (see Insight 22.2). In developing countries, the high mortality rate is not the only issue. Children who survive dozens of bouts with diarrhea during their developmental years are likely to have permanent physical and cognitive effects. The effect on the overall well-being of these children is hard to estimate, but it is very significant.

In the United States, up to a third of all acute diarrhea is transmitted by contaminated food. In recent years, consumers have become much more aware of the possibility of contaminated hamburgers or *Salmonella*-contaminated ice cream. New food safety measures are being implemented all the time, including the development of more rapid testing methods. The use of ultraviolet light and food irradiation, in addition to more novel methods such as the application of bacteriophage, are also helping to create a safer food supply in the United States. Even with all of these measures in place,

it is still necessary for the consumer to be aware of and to practice good food-handling techniques. **Figure 22.10** shows the proportions of food-borne illnesses due to various pathogens considered here.

Although most diarrhea episodes are self-limiting and therefore do not require treatment, others (such as *E. coli* O157:H7) can have devastating effects. In most diarrheal illnesses, antimicrobial treatment is contraindicated (inadvisable), but some, such as shigellosis, call for quick treatment with antibiotics. For public health reasons, it is important to know which agents are causing diarrhea in the community, but in many cases identification of the agent is not performed.

In this section, we describe acute diarrhea having infectious agents as the cause. In the sections following this one, we discuss acute diarrhea and vomiting caused by toxins, commonly known as food poisoning, and chronic diarrhea and its causes.

Salmonella

A decade ago, one of every three chickens destined for human consumption was contaminated with *Salmonella*, but the rate has fallen to about 10%. Other poultry (e.g., ducks and turkeys) is also affected. Eggs may harbor the pathogen on their shells, but bacteria may actually be incorporated into the egg, while the shell is being formed within the chicken. In 2007 and again in 2012, raw peanuts and peanut butter were found to be the sources of *Salmonella* outbreaks in the United States. *Salmonella* is a very large genus of bacteria, but only one species is of interest to us: *S. enterica* is divided into many serotypes, based on variation in the major surface antigens.

As mentioned in chapter 4, serotype or variant analysis aids in bacterial identification. Many gram-negative enteric bacteria are named and designated according to the following antigens: H, the flagellar antigen; K, the capsular antigen; and O, the cell wall antigen. Not all enteric bacteria carry the H and K antigens, but all have O, the polysaccharide portion of the lipopolysaccharide implicated in endotoxic shock (see chapter 20). Most species of gram-negative enterics exhibit a variety of subspecies, variant, or serotypes caused by slight variations in the chemical structure of the HKO antigens. Some bacteria in this chapter (for example, *E. coli* O157:H7) are named according to their surface antigens; however, we will use Latin variant names for *Salmonella*.

Salmonellae are motile; they ferment glucose with acid and sometimes gas; and most of them produce hydrogen sulfide (H_2S) but not urease. They grow readily on most laboratory media and can survive outside the host in inhospitable environments such as freshwater and freezing temperatures. These pathogens are resistant to chemicals such as bile and dyes, which are the bases for isolation on selective media.

▶ Signs and Symptoms

The genus *Salmonella* causes a variety of illnesses in the GI tract and beyond. Roughly 1.2 million cases of illness are reported each year in the United States, with nearly 500 deaths attributed to *Salmonella* infection. Until the mid-1900s, its most severe manifestation was typhoid fever, which is discussed shortly. Since that time, a milder disease usually called salmonellosis has been much more common **(figure 22.11)**. Sometimes the condition is also called enteric fever or gastroenteritis. Whereas typhoid fever is caused by *Salmonella enterica* serotype Typhi, gastroenteritises in the United States are generally caused by the serotypes known as Typhimurium, Enteritidis, Heidelberg, Newport, and Javiana. *Salmonella* bacteria are normal intestinal biota in cattle, poultry, rodents, and reptiles, and each (including domesticated pets) has been documented as a source of infection in humans.

Salmonellosis can be relatively severe, with an elevated body temperature and septicemia as more prominent features than GI tract disturbance. But it can also be fairly mild, with gastroenteritis—vomiting, diarrhea, and mucosal irritation—as its

major feature. Blood can appear in the stool. In otherwise healthy adults, symptoms spontaneously subside after 2 to 5 days; death is infrequent except in debilitated persons.

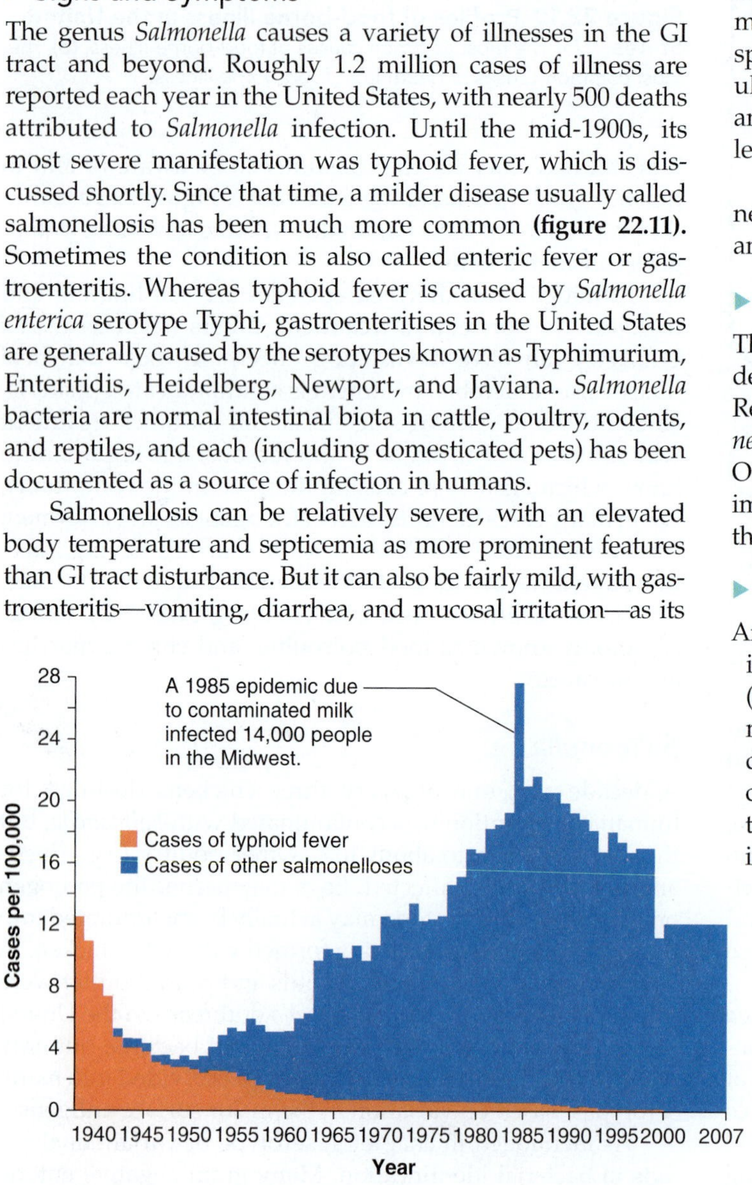

Typhoid fever is so named because it bears a superficial resemblance to typhus, a rickettsial disease, even though the two diseases are otherwise very different. In the United States, the incidence of typhoid fever has remained at a steady rate for the last 30 years, appearing sporadically (as shown in figure 22.11). Of the 50 to 100 cases reported annually, roughly half are imported from endemic regions. In other parts of the world, typhoid fever is still a serious health problem, responsible for 25,000 deaths each year and probably millions of cases.

Typhoid fever, caused by *Salmonella enterica* serotype Typhi, is characterized by a progressive, invasive infection that leads eventually to septicemia. Symptoms are fever, diarrhea, and abdominal pain. The bacterium infiltrates the mesenteric lymph nodes and the phagocytes of the liver and spleen. In some people, the small intestine develops areas of ulceration that are vulnerable to hemorrhage, perforation, and peritonitis. Its presence in the circulatory system may lead to nodules or abscesses in the liver or urinary tract.

Because it is so rare compared with the less severe salmonellosis, the rest of this section refers mainly to salmonellosis and not to typhoid fever.

▶ Pathogenesis and Virulence Factors

The ability of *Salmonella* to cause disease seems to be highly dependent on its ability to adhere effectively to the gut mucosa. Recent research has uncovered an "island" of genes in *Salmonella* that seems to confer enhanced attachment capabilities. Other pathogenicity islands encoding proteins allowing for immune evasion have also been identified. It is also believed that endotoxin is an important virulence factor for *Salmonella*.

▶ Transmission and Epidemiology

An important factor to consider in all diarrheal pathogens is how many organisms must be ingested to cause disease (their ID_{50}). It varies widely. *Salmonella* has a high ID_{50}, meaning many organisms have to be ingested in order for disease to result. Animal products such as meat and milk can be readily contaminated with *Salmonella* during slaughter, collection, and processing. Inherent risks are involved in eating poorly cooked chicken or unpasteurized fresh or dried milk, ice cream, and cheese.

Most cases are traceable to a common food source such as milk or eggs. Some cases may be due to poor sanitation. In one outbreak, about 60 people became infected after visiting the Komodo dragon exhibit at the Denver zoo. They picked up the infection by handling the rails and fence of the dragon's cage. In 2002, two people apparently acquired salmonellosis from a blood transfusion, and one of them died. The blood donor, who had an asymptomatic infection with *Salmonella*, had contracted the infection from his pet snake. People have become infected through contacting pet turtles and even contaminated pet food, further stressing the need for proper hand washing, especially in young children.

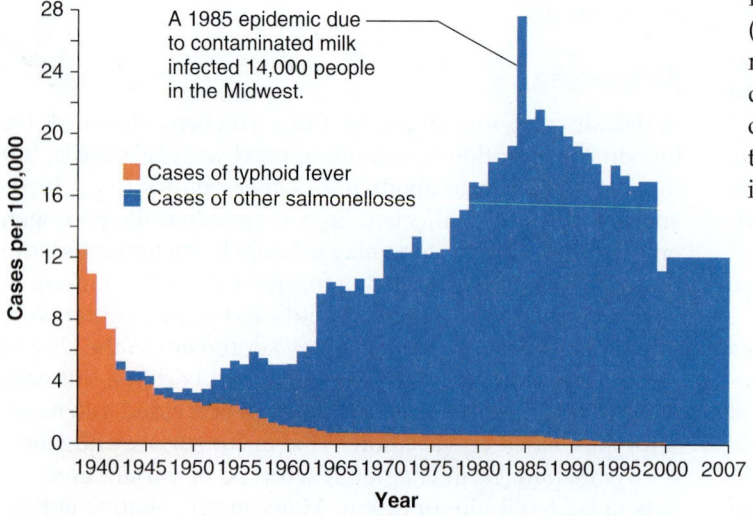

Figure 22.11 Data on the prevalence of typhoid fever and other salmonelloses from 1940 to 2007 in the United States. Nontyphoidal salmonelloses did occur before 1940, but the statistics are not available.

Chart legend and labels:
- A 1985 epidemic due to contaminated milk infected 14,000 people in the Midwest.
- Cases of typhoid fever
- Cases of other salmonelloses
- Y-axis: Cases per 100,000 (0, 4, 8, 12, 16, 20, 24, 28)
- X-axis: Year (1940, 1945, 1950, 1955, 1960, 1965, 1970, 1975, 1980, 1985, 1990, 1995, 2000, 2007)

▶ Prevention and Treatment

The only prevention for salmonellosis is avoiding contact with the bacterium. In 1998, a vaccine was approved for use in poultry, making it the first "food safety" vaccine. A vaccine for humans is undergoing testing as well.

Uncomplicated cases of salmonellosis are treated with fluid and electrolyte replacement; if the patient has underlying immunocompromise or if the disease is severe, antibiotics are recommended. However, multidrug-resistant *Salmonella* strains have evolved due in large part to the prophylactic use of antibiotics in animal herds.

Typhoid fever, by contrast, is always treated with antibiotics, in part to clear the patient of the bacterium, which has a tendency to be shed for weeks after recovery. A small number of people chronically carry the bacterium for longer periods in the gallbladder; from this site, the bacteria are constantly released into the intestine and feces. Such asymptomatic carriers can easily spread disease. In some people, gallbladder removal is necessary to stop the shedding. Two vaccines are available for protection against serotype Typhi and are recommended for people traveling to endemic areas.

Shigella

The *Shigella* bacteria are gram-negative straight rods, nonmotile and non-endospore-forming. They do not produce urease or hydrogen sulfide, traits that help in their identification. They are primarily human parasites, though they can infect apes. All produce a similar disease that can vary in intensity. These bacteria resemble some types of pathogenic *E. coli* very closely. Diagnosis is complicated by the fact that several alternative candidates can cause bloody diarrhea, such as *E. coli* and others. Isolation and identification follow the usual protocols for enterics. Stool culture is still the gold standard for identification in the case of *Shigella* infections.

Although *Shigella dysenteriae* causes the most severe form of dysentery, it is uncommon in the United States and occurs primarily in the Eastern Hemisphere. In the past decade, the prevalent agents in the United States have been *Shigella sonnei* and *Shigella flexneri*, which cause approximately 20,000 to 25,000 cases each year, half of them in children.

▶ Signs and Symptoms

The symptoms of shigellosis include frequent, watery stools, as well as fever, and often intense abdominal pain. Nausea and vomiting are common. Stools often contain obvious blood and even more often are found to have occult (not visible to the naked eye) blood. Diarrhea containing blood is also called **dysentery**. Mucus from the GI tract will also be present in the stools.

▶ Pathogenesis and Virulence Factors

Shigellosis is different from many GI tract infections in that *Shigella* invades the villus cells of the large intestine rather than the small intestine. In addition, it is not as invasive as *Salmonella* and does not perforate the intestine or invade the blood. It enters the intestinal mucosa by means of lymphoid cells in Peyer's patches. Once in the mucosa, *Shigella* instigates an inflammatory response that causes extensive tissue destruction. The release of endotoxin causes fever. **Enterotoxin**, an exotoxin that affects the enteric (or GI) tract, damages the mucosa and villi. Local areas of erosion give rise to bleeding and heavy secretion of mucus **(figure 22.12)**. *Shigella dysenteriae*

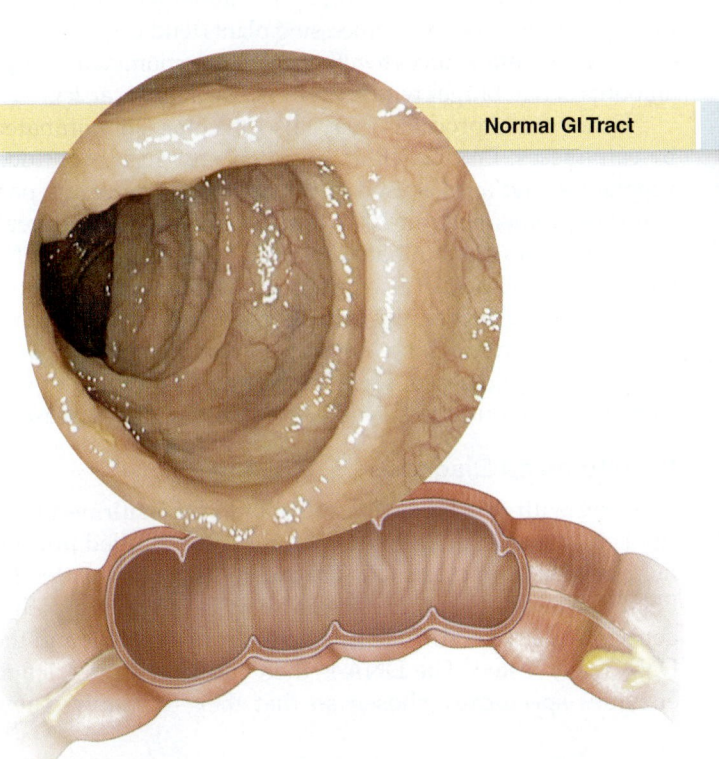
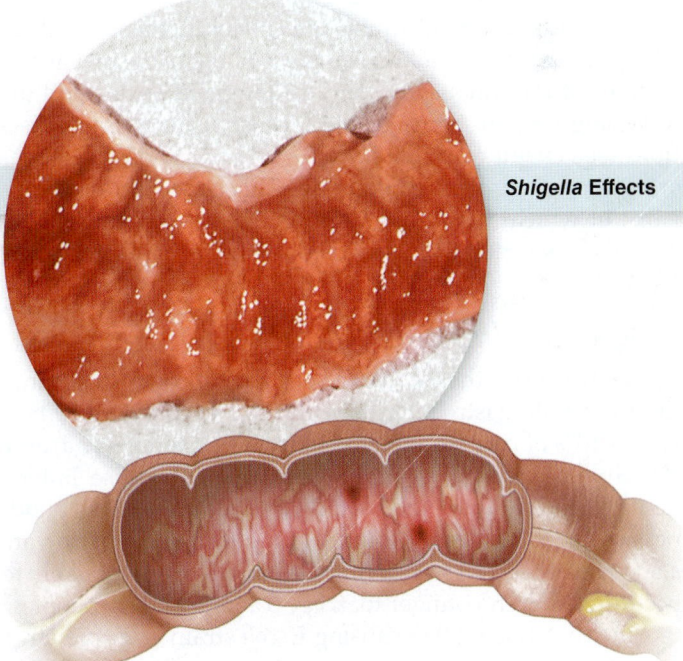

Normal GI Tract

Shigella **Effects**

Figure 22.12 The appearance of the large intestinal mucosa in *Shigella* dysentery. Note the patches of blood and mucus, the erosion of the lining, and the absence of perforation.

(and perhaps some of the other species) produces a heat-labile exotoxin called **shiga toxin**, which seems to be responsible for the more serious damage to the intestine as well as any systemic effects, including injury to nerve cells. You'll encounter shiga toxin again when we discuss *E. coli* O157:H7.

▶ Transmission and Epidemiology

In addition to the usual oral route, shigellosis is also acquired through direct person-to-person contact, largely because of the small infectious dose required (from 10 to 200 bacteria). The disease is mostly associated with lax sanitation, malnutrition, and crowding; it is spread epidemically in day care centers, prisons, mental institutions, nursing homes, and military camps. As in other enteric infections, *Shigella* can establish a chronic carrier condition in some people that lasts several months.

▶ Prevention and Treatment

The only prevention of this and most other diarrheal diseases is good hygiene and avoiding contact with infected persons. Most physicians recommend prompt treatment of shigellosis with ciprofloxacin.

Shiga-Toxin-Producing *E. coli* (STEC)

In January of 1993, a new *E. coli* strain burst into the public's consciousness when three children died after eating undercooked hamburgers at a fast-food restaurant in Washington State. The cause of their illness was determined to be *E. coli* O157:H7, which had actually been recognized since the 1980s. Since then, it has led to approximately 95,000 illnesses and about 50 deaths each year in the United States. It is considered an emerging pathogen.

Dozens of different strains of *E. coli* exist, many of which cause no disease at all. A handful of them cause various degrees of intestinal symptoms as described in this and the following section. Some of them cause urinary tract infections (see chapter 23). *E. coli* O157:H7 and its close relatives are the most virulent of them all. This collection of organisms, of which this *E. coli* strain is the most famous representative, is generally referred to as **shiga-toxin-producing *E. coli* (STEC).**

▶ Signs and Symptoms

E. coli O157:H7 is the agent of a spectrum of conditions, ranging from mild gastroenteritis with fever to bloody diarrhea. About 10% of patients develop **hemolytic uremic syndrome (HUS),** a severe hemolytic anemia that can cause kidney damage and failure. Neurological symptoms such as blindness, seizure, and stroke (and long-term debilitation) are also possible. These serious manifestations are most likely to occur in children younger than age 5 and in elderly people.

In 2011, a new HUS-causing *E. coli* strain caused a large and deadly outbreak in Germany. Contaminated fenugreek seeds were ultimately identified as the source of the pathogen. It was named *E. coli* O104:H4 and was identified as a STEC strain. A total of six additional STEC strains have been

identified, and the U.S. Department of Agriculture (USDA) started testing ground beef for all of these strains in 2012.

▶ Pathogenesis and Virulence Factors

These *E. coli* owe much of their virulence to shiga toxins (so named because they are identical to the shiga exotoxin secreted by virulent *Shigella* species). The shiga toxin genes are present on bacteriophage in *E. coli* but are on the chromosome of *Shigella dysenteriae,* suggesting that STEC strains of *E. coli* acquired the virulence factor through phage-mediated transfer. As described earlier for *Shigella,* the shiga toxin interrupts protein synthesis in its target cells. It seems to be responsible especially for the systemic effects of this infection.

Another important virulence determinant for STEC is the ability to efface (rub out or destroy) enterocytes, which are gut epithelial cells. The net effect is a lesion in the gut (effacement), usually in the large intestine. The microvilli are lost from the gut epithelium, and the lesions produce bloody diarrhea.

▶ Transmission and Epidemiology

The most common mode of transmission for STEC is the ingestion of contaminated and undercooked beef, although other foods and beverages can be contaminated as well. Ground beef is more dangerous than steaks or other cuts of meat, for several reasons. Consider the way that the beef becomes contaminated in the first place. The bacterium is a natural inhabitant of the GI tracts of cattle. Contamination occurs when intestinal contents contact the animal carcass, so bacteria are confined to the surface of meats. Because high heat destroys this bacterium, even a brief trip under the broiler is usually sufficient to kill *E. coli* on the surface of steaks or roasts. But in ground beef, the "surface" of meat is mixed and ground up throughout a batch, meaning any bacteria are mixed in also. This mixing explains why hamburgers should be cooked all the way through. Hamburger is also a common vehicle because meat processing plants tend to grind meats from several cattle sources together, thereby contaminating large amounts of hamburger with meat from one animal carrier.

Other farm products may also become contaminated by cattle feces. Products that are eaten raw, such as lettuce, vegetables, and apples used in unpasteurized cider, are particularly problematic. In 2006, a major nationwide outbreak stemming from contaminated spinach held the headlines for weeks. The disease can also be spread via the fecal-oral route of transmission, especially among young children in group situations. Even touching surfaces contaminated with cattle feces can cause disease, since ingesting as few as 10 organisms has been found to be sufficient to initiate this disease.

▶ Culture and Diagnosis

Infection with this type of *E. coli* should be confirmed with stool culture or with ELISA, PCR, or a process called pulsed-field gel electrophoresis (PFGE). As you saw in chapter 17, PFGE is a technique for restriction analysis in which a pathogen such as *E. coli* O157:H7 is isolated from a patient and the DNA is harvested. The DNA is then cut up with restriction enzymes specifically chosen so that they find only a few places to cut into the organism's genome. The lengths of the fragments and thus the pattern revealed by each microbe will be different—even for different strains of the same microbial species—because the enzymes cut in different places in the genome where small DNA changes exist, corresponding to different strain types.

In 1993, the CDC used PFGE for the first time to trace the outbreak of *E. coli* O157:H7 in undercooked hamburgers in Washington State. They determined that the strain of *E. coli* O157:H7 found in the patients had the same PFGE pattern as the strain found in the suspected hamburger patties that had been served at the fast-food restaurant. The use of the technique led to the creation of PulseNet, which contains PFGE patterns of common food-borne pathogens that have been implicated in outbreaks. Participating PulseNet laboratories all around the country can compare PFGE patterns they obtain from patients or suspected foods to patterns in the centralized database. This way, outbreaks that are geographically dispersed (for instance, those caused by contaminated meat or fruit that may have been distributed nationally) can be identified quickly. When new patterns come in, they are archived so that other laboratories submitting the same patterns will quickly realize that the cases are related.

▶ **Prevention and Treatment**

The best prevention for this disease is to never eat raw or even rare hamburger and to wash raw vegetables well. The shiga toxin is heat-labile and the *E. coli* is killed by heat as well. If you are thinking "I used to be able to eat rare hamburgers," you are correct, but things have changed. The emergence of this pathogen in the early 1990s, probably resulting from a regular *E. coli* picking up the shiga toxin from *Shigella*, has changed the rules for proper food handling.

No human vaccine exists for *E. coli* O157:H7 or other STEC strains. Some countries vaccinate cattle against *E. coli* O157:H7 as a means to protect human populations.

Antibiotics are contraindicated for this infection. Even with severe disease manifestations, antibiotics may increase the pathology by releasing more toxin, leading to HUS. Supportive therapy, including plasma transfusion to dilute toxin in the blood, is the only option.

Other *E. coli*

At least five other categories of *E. coli* can cause diarrheal diseases. In clinical practice, most physicians are interested in differentiating shiga-toxin-producing *E. coli* (STEC) from all the others. Each of the five other categories is considered separately and briefly here; in Disease Table 22.5, the non-shiga-toxin-producing *E. coli* are grouped together in one column.

- Enterotoxigenic *E. coli* (ETEC). The presentation varies depending on which type of *E. coli* is causing the disease. Traveler's diarrhea, characterized by watery diarrhea, low-grade fever, nausea, and vomiting, is usually caused by enterotoxigenic *E. coli* (ETEC). These strains also cause a great deal of illness in infants in developing countries.

Most infections with ETEC are self-limiting, however miserable they make you feel. They are treated only with fluid replacement, often due to the high rate of drug resistance. In infants, ETEC can be life-threatening, and fluid replacement is vital to survival.

- Enteroinvasive *E. coli* (EIEC). These strains cause bacillary dysentery, which is often mistaken for *Shigella* dysentery. The bacteria invade gut mucosa and cause widespread destruction. Blood and mucus will be found in the stool. Significant fever is often present. EIEC does not produce the heat-labile or heat-stable exotoxins just described and does not have a shiga toxin, despite the clinical similarity to *Shigella* disease. EIEC does seem to have a protein that is expressed inside host cells, which leads to its destruction.

Disease caused by this bacterium is more common in developing countries. It is transmitted primarily through contaminated food and water. Treatment is supportive (including rehydration).

- Enteropathogenic *E. coli* (EPEC). These strains result in a profuse, watery diarrhea. Fever and vomiting are also common. The EPEC bacteria are very similar to the shiga-toxin-producing *E. coli* (STEC) described earlier—they produce effacement of gut surfaces. The important difference between EPEC and STEC is that EPEC does not produce a shiga toxin and, therefore, does not produce the systemic symptoms characteristic of those bacteria.

Most disease is self-limiting. As with any other diarrhea, however, it can be life-threatening in young babies. Rehydration is the main treatment.

- Enteroaggregative *E. coli* (EAEC). These bacteria are most notable for their ability to cause chronic diarrhea in young children and in AIDS patients. EAEC is considered in the section on chronic diarrhea.
- Diffusely adherent *E. coli* (DAEC). These bacteria are identified based on virulence factors used to attach to host cells. They are typically associated with the development of urinary tract infections in addition to acute diarrhea in the developing world. DAEC strains are implicated in proinflammatory reactions and may play a role in the development of inflammatory bowel disease (IBD).

Campylobacter

Although you may never have heard of *Campylobacter*, it is considered to be the most common bacterial cause of diarrhea in the United States. It probably causes more diarrhea than *Salmonella* and *Shigella* combined, with 2.4 million cases (or >1% of the U.S. population!) of diarrhea credited to it per year.

The symptoms of campylobacteriosis are frequent watery stools, fever, vomiting, headaches, and severe abdominal pain. The symptoms may last longer than most acute diarrheal episodes, sometimes extending beyond 2 weeks. They may subside and then recur over a period of weeks.

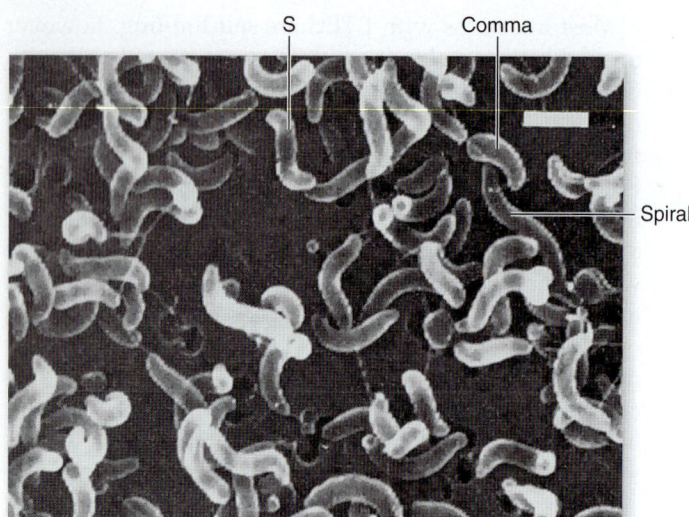

Figure 22.13 Scanning micrograph of *Campylobacter jejuni*, showing comma, S, and spiral forms.

Campylobacter jejuni is the most common cause, although there are other pathogenic *Campylobacter* species. Campylobacters are slender, curved or spiral, gram-negative bacteria propelled by polar flagella at one or both poles, often appearing in S-shaped or gull-winged pairs **(figure 22.13)**. These bacteria tend to be microaerophilic inhabitants of the intestinal tract, genitourinary tract, and oral cavity of humans and animals. A close relative, *Helicobacter pylori,* is the causative agent of most stomach ulcers (described earlier). Transmission of this pathogen takes place via the ingestion of contaminated beverages and food, especially water, milk, meat, and chicken. Recent studies suggest that *C. jejuni* is much more resistant to heating temperatures during the cooking process than was previously estimated, increasing the need for proper food handling.

Once ingested, *C. jejuni* cells reach the mucosa at the last segment of the small intestine (ileum) near its junction with the colon; they adhere, burrow through the mucus, and multiply. Symptoms commence after an incubation period of 1 to 7 days. The mechanisms of pathology appear to involve a heat-labile enterotoxin that stimulates a secretory diarrhea like that of cholera. In a small number of cases, infection with this bacterium can lead to a serious neuromuscular paralysis called Guillain-Barré syndrome.

Guillain-Barré syndrome (GBS) (pronounced gee"-luhn-buh-ray') is the leading cause of acute paralysis in the United States since the eradication of polio there. The good news is that many patients recover completely from this paralysis. The condition is still mysterious in many ways, but it seems to be an autoimmune reaction that can be brought on by infection with viruses and bacteria, by vaccination in rare cases, and even by surgery. The single most common precipitating event for the onset of GBS is *Campylobacter* infection. Twenty to forty percent of GBS cases are preceded by infection with *Campylobacter.* The reasons for this are not clear. (Note that even though 20% to 40% of GBS

cases are preceded by *Campylobacter* infection, only about 1 in 1,000 cases of *Campylobacter* infection results in GBS.)

Resolution of infection occurs in most instances with simple, nonspecific rehydration and electrolyte-balance therapy. In more severely affected patients, it may be necessary to administer azithromycin. Antibiotic resistance is growing in these bacteria, in large part due to the use of fluoroquinolones in the treatment of poultry destined for human consumption. Prevention depends on rigid sanitary control of water and milk supplies and care in food preparation. Vaccine development for use in poultry is ongoing.

Clostridium difficile

Clostridium difficile is a gram-positive, endospore-forming rod found as normal biota in the intestine. It was once considered relatively harmless but now is known to cause a condition called pseudomembranous colitis, also known as antibiotic-associated colitis. In most cases, this infection seems to be precipitated by therapy with broad-spectrum antibiotics. It is a major cause of diarrhea in hospitals, although community-acquired infections have been on the rise in the last few years. Also, new studies suggest that the use of gastric acid inhibitors for the treatment of heartburn can predispose patients to this infection. Although *C. difficile* is relatively noninvasive, it is able to superinfect the large intestine when drugs have disrupted the normal biota. It produces two enterotoxins, toxins A and B, that cause areas of necrosis in the wall of the intestine. The predominant symptom is diarrhea commencing late in therapy or even after therapy has stopped. More severe cases exhibit abdominal cramps, fever, and leukocytosis. The colon is inflamed and gradually sloughs off loose, membranelike patches called pseudomembranes consisting of fibrin and cells **(figure 22.14)**. If the condition is not stopped, perforation of the cecum and death can result.

Mild, uncomplicated cases respond to withdrawal of antibiotics and replacement therapy for lost fluids and electrolytes. Some physicians prescribe metronidazole. More severe infections are treated with oral vancomycin or a new drug approved in 2001 called fidaxomicin (Dificid) for several weeks until the intestinal biota returns to normal. Because infected persons often shed large numbers of endospores in their stools, increased precautions are necessary to prevent spread of the agent to other patients who may be on antimicrobial therapy. Nearly 15,000 people in the United States die each year from *C. difficile* infection, and a 400% increase in disease incidence has been observed over a period of just 7 years. Drug resistance is rapidly evolving as well, and such strains have caused global epidemics within the past decade. These facts highlight the immediate need for more effective prevention and treatment methods.

Some patients have tried the technique of fecal transplant. This is a revival of a very old-fashioned method of obtaining feces from a healthy person and instilling them in the colon of the patient. Many have found relief from this method, presumably because a diverse microbiome with "healthy" species replaces the now-depleted microbiome of the *C. diff* patient.

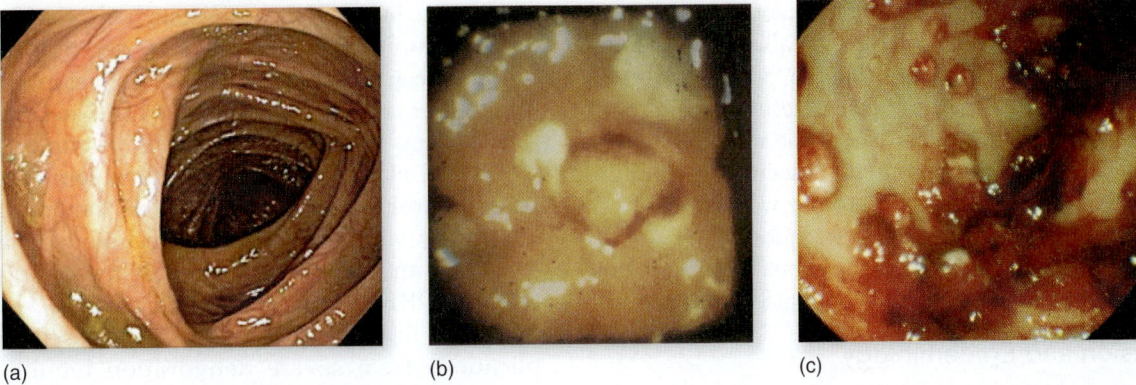

Figure 22.14 Antibiotic-associated colitis. (a) Normal colon. (b) A mild form of colitis with diffuse, inflammatory patches. (c) Heavy yellow plaques, or pseudomembranes, typical of more severe cases. Photographs were made by a sigmoidoscope, an instrument capable of photographing the interior of the colon.

Vibrio cholerae

Cholera has been a devastating disease for centuries. It is not an exaggeration to say that the disease has shaped a good deal of human history in Asia and Latin America, where it has been endemic. These days we have come to expect outbreaks of cholera to occur after natural disasters, war, or large refugee movements, especially in underdeveloped parts of the world.

Vibrios are comma-shaped rods with a single polar flagellum. They belong to the family *Vibrionaceae*. A freshly isolated specimen of *Vibrio cholerae* contains quick, darting cells that slightly resemble a comma **(figure 22.15)**. *Vibrio* shares many cultural and physiological characteristics with members of the *Enterobacteriaceae*, a closely related family. Vibrios are fermentative and grow on ordinary or selective media containing bile at 37°C. They possess unique O and H antigens and membrane receptor antigens that provide some basis for classifying members of the family. There are two major biotypes, called classic and *El Tor*.

▶ Signs and Symptoms

After an incubation period of a few hours to a few days, symptoms begin abruptly with vomiting, followed by copious watery feces called secretory diarrhea. The intestinal contents are lost very quickly, leaving only secreted fluids. This voided fluid contains flecks of mucus—hence, the description "rice-water stool." Fluid losses of nearly 1 liter per hour have been reported in severe cases, and an untreated patient can lose up to 50% of body weight during the course of this disease. The diarrhea causes loss of blood volume, acidosis from bicarbonate loss, and potassium depletion, which manifest in muscle cramps, severe thirst, flaccid skin, sunken eyes, and—in young children—coma and convulsions. Secondary circulatory consequences can include hypotension, tachycardia, cyanosis, and collapse from shock within 18 to 24 hours. If cholera is left untreated, death can occur in less than 48 hours; the mortality rate is between 55% and 70%.

▶ Pathogenesis and Virulence Factors

After being ingested with food or water, *V. cholerae* travels through the stomach to the small intestine. At the junction of the duodenum and jejunum, the vibrios penetrate the mucous barrier using their flagella, adhere to the microvilli of the epithelial cells, and multiply there. The bacteria never enter the

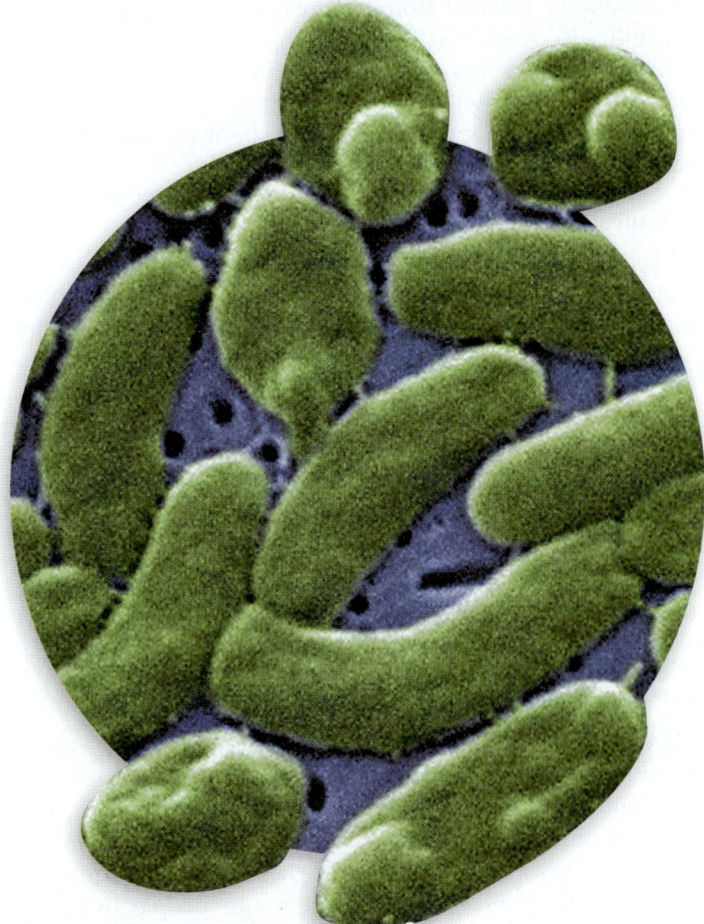

Figure 22.15 *Vibrio cholerae*. Note the characteristic curved shape of this bacterium.

host cells or invade the mucosa. The virulence of *V. cholerae* lies mainly in the action of an enterotoxin called cholera toxin (CT), which disrupts the normal physiology of intestinal cells. When this toxin binds to specific intestinal receptors, a secondary signaling system is activated. Under the influence of this system, the cells shed large amounts of electrolytes into the intestine, an event accompanied by profuse water loss. It was recently discovered that *V. cholerae* uses quorum sensing to regulate the precise expression of its virulence factors, making proteins used in this process potential targets for drug therapy.

▶ Transmission and Epidemiology

The pattern of cholera transmission and the onset of epidemics are greatly influenced by the season of the year and the climate. Cold, acidic, dry environments inhibit the migration and survival of *Vibrio,* whereas warm, monsoon, alkaline, and saline conditions favor them. The bacteria survive in water sources for long periods of time. Recent outbreaks in several parts of the world have been traced to giant cargo ships that pick up ballast water in one port and empty it in another elsewhere in the world.

In nonendemic areas such as the United States, the microbe is spread by water and food contaminated by asymptomatic carriers, but it is relatively uncommon. Sporadic outbreaks occur along the Gulf of Mexico, and *V. cholerae* is sometimes isolated from shellfish in that region. Due to its ability to produce chitinase, this pathogen can often live within marine copepods. Since various aspects of global warming can impact the growth of these plankton populations, many scientists are concerned about the risk of more frequent cholera epidemics in the future.

▶ Culture and Diagnosis

V. cholerae can be readily isolated and identified in the laboratory from stool samples. Direct dark-field microscopic observation reveals characteristic curved cells with brisk, darting motility as confirmatory evidence. Immobilization or fluorescent staining of feces with group-specific antisera is supportive as well. Difficult cases can be traced by detecting a rising antitoxin titer in the serum. In order to determine the initial source of infection, it is often necessary to determine the exact genetic nature of the strain causing an epidemic. This was the case in the massive outbreak of disease that occurred in Haiti in 2010. Real-time PCR methods are now being employed to rapidly test food and water supplies for the presence of *V. cholerae* and related pathogenic species (*V. fulnificus* and *V. parahaemolyticus*).

▶ Prevention and Treatment

Effective prevention is contingent on proper sewage treatment and water purification. Detecting and treating carriers with mild or asymptomatic cholera are serious goals, but they are difficult to accomplish because of inadequate medical provisions in those countries where cholera is endemic. Vaccines are available for travelers and people living in endemic regions. One vaccine contains killed *V. cholerae* but protects for only 6 months or less. An oral vaccine containing live, attenuated bacteria was developed to be a more effective

alternative, but evidence suggests it also confers only short-term immunity. It is not routinely used in the United States but has been used successfully to develop protective herd immunity in endemic nations like Haiti and Africa.

The key to cholera therapy is prompt replacement of water and electrolytes, because their loss accounts for the severe morbidity and mortality. This therapy can be accomplished by various rehydration techniques that replace the lost fluid and electrolytes. One of these, oral rehydration therapy (ORT), is described in **Insight 22.2.**

Cases in which the patient is unconscious or has complications from severe dehydration require intravenous replenishment as well. Oral antibiotics such as doxycycline are given as an adjunct to rehydration. They also diminish the period of vibrio excretion.

Cryptosporidium

Cryptosporidium is an intestinal protozoan of the apicomplexan type (see chapter 5) that infects a variety of mammals, birds, and reptiles. For many years, cryptosporidiosis was considered an intestinal ailment exclusive to calves, pigs, chickens, and other poultry, but it is clearly a zoonosis as well. The organism's life cycle includes a hardy intestinal oocyst as well as a tissue phase. Humans accidentally ingest the oocysts with water or food that has been contaminated by feces from infected animals. The oocyst "excysts" once it reaches the intestines and releases sporozoites that attach to the epithelium of the small intestine **(figure 22.16).**

The organism penetrates the intestinal cells and lives intracellularly within them. It undergoes asexual and sexual reproduction in these cells and produces more oocysts, which are released into the gut lumen, excreted from the host, and after

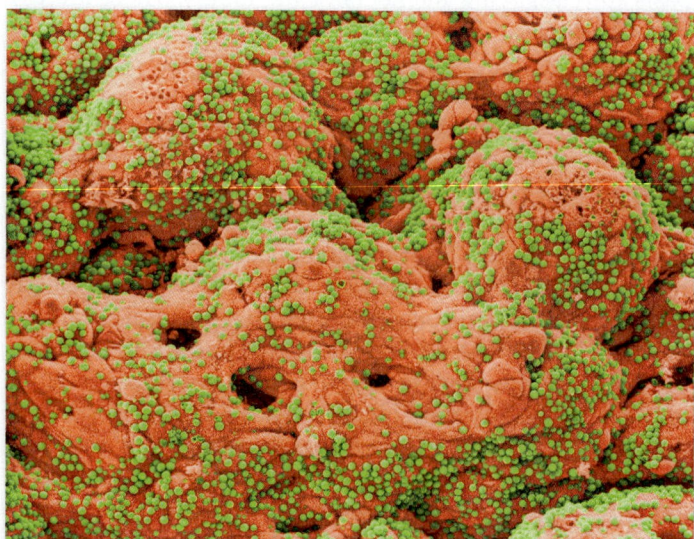

Figure 22.16 Scanning electron micrograph of *Cryptosporidium* (green) attached to the intestinal epithelium.

INSIGHT 22.2   "The Most Important Medical Advance This Century"—*The Lancet,* August 5, 1978

In 1970, the first clinical trials began on a simple treatment for a life-threatening problem. A solution of electrolytes—sodium chloride, sodium bicarbonate, potassium chloride, and glucose or sucrose—dissolved in water was administered to patients dying of diarrheal disease. With essential electrolytes and fluid restored, patients were brought back from the brink of death. This simple solution called oral rehydration therapy (ORT) or oral rehydration solution (ORS) became the treatment of choice for cholera and other diarrheal diseases in developing countries. The British medical journal *The Lancet* declared that this simple and inexpensive treatment was "the most important medical advance this century." Later studies showed that adding zinc further improved the outcome in patients with diarrhea in addition to decreasing susceptibility to lower respiratory infections.

Today, diarrheal disease is still the second leading cause of death among children under age 5, with an estimated 840,000 deaths annually, representing 10% of child deaths every year. Even though ORT and zinc supplements are inexpensive and available, according to the WHO, only 39% of children under age 5 have access to ORT in developing countries. In a recent study, it was found that caregivers recognized serious diarrhea in children only 55% of the time and, when seeking therapy, didn't utilize ORT or didn't recognize it as a viable treatment. Zinc treatments were either underutilized or unavailable in local treatment centers in this study.

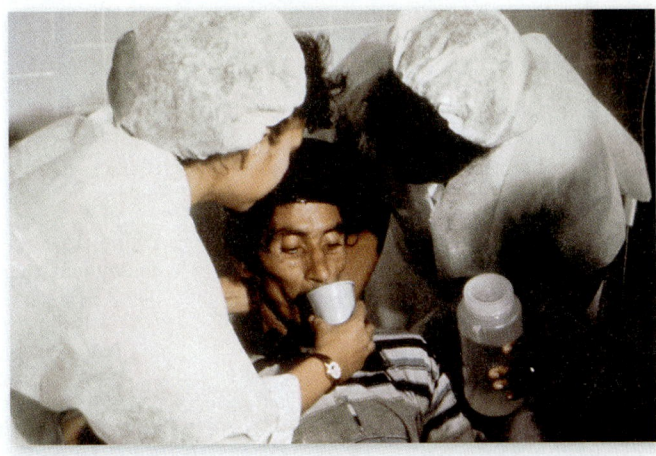

This cholera patient is drinking oral rehydration solution.

Even when ORT is available, it is relatively difficult to administer to a young child. One of the reasons for this is that the treatment has an unpleasant taste and the proper dose of electrolytes is usually dissolved in 1-liter volumes, which makes it unpopular with children. Anyone who has tried to get a child to eat something they don't want to will understand the difficulty in trying to administer even this lifesaving treatment. Efforts are being made by WHO and UNICEF to provide smaller packaging with flavors that would be more suitable for small children.

a short time become infective again. The oocysts are highly infectious and extremely resistant to treatment with chlorine and other disinfectants. The prominent symptoms mimic other types of gastroenteritis, with headache, sweating, vomiting, severe abdominal cramps, and diarrhea. AIDS patients may experience chronic persistent cryptosporidial diarrhea that can be used as a criterion to help diagnose AIDS. The agent can be detected in fecal samples or in biopsies **(figure 22.17)** using ELISA or acid-fast staining. Stool cultures should be performed to rule out other (bacterial) causes of infection.

Half of the outbreaks of diarrhea associated with swimming pools are caused by *Cryptosporidium*. Because chlorination is not entirely successful in eradicating the cysts, most treatment plants and recreational water parks utilize a combination of ultraviolet light treatment and filtration, but even this method is not foolproof.

Treatment is not usually required for otherwise healthy patients. Antidiarrheal agents (antimotility drugs) may be used. Although no curative antimicrobial agent exists for *Cryptosporidium*, physicians will often try a drug called nitazoxanide in immunocompetent patients. Immunocompromised patients should not receive specific treatment for this condition.

Rotavirus

Rotavirus is a member of the *Reovirus* group, which consists of an unusual double-stranded RNA genome with both an inner and an outer capsid. Globally, rotavirus is the primary viral cause of morbidity and mortality resulting from diarrhea, accounting for nearly 50% of all cases. It is estimated that there are 2 to 3 million cases of rotavirus infection in the United States every year, leading to 70,000 hospitalizations.

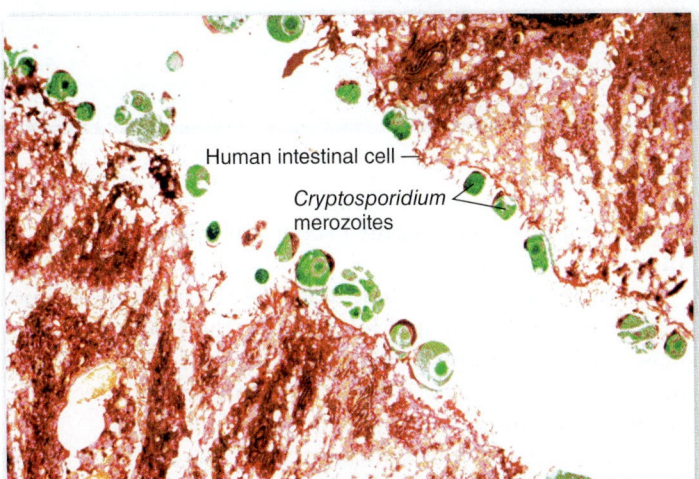

Human intestinal cell →

Cryptosporidium merozoites

Figure 22.17 A micrograph of *Cryptosporidium* merozoites in the act of penetrating the intestinal mucosa.

Disease Table 22.5   Acute Diarrhea (With or Without Vomiting)

Bacterial Causes

	Salmonella	*Shigella*	Shiga-toxin-producing *E. coli* (O157:H7 and others)	Other *E. coli* (non-shiga-toxin producing)
Causative Organism(s)				
Most Common Modes of Transmission	Vehicle (food, beverage), fecal-oral	Fecal-oral, direct contact	Vehicle (food, beverage), fecal-oral	Vehicle, fecal-oral
Virulence Factors	Adhesins, endotoxin	Endotoxin, enterotoxin, shiga toxins in some strains	Shiga toxins; proteins for attachment, secretion, effacement	Various: proteins for attachment, secretion, effacement; heat-labile and/or heat-stable exotoxins; invasiveness
Culture/Diagnosis	Stool culture, not usually necessary	Stool culture; antigen testing for shiga toxin	Stool culture, antigen testing for shiga toxin	Stool culture not usually necessary in absence of blood, fever
Prevention	Food hygiene and personal hygiene	Food hygiene and personal hygiene	Avoid live *E. coli* (cook meat and clean vegetables)	Food and personal hygiene
Treatment	Rehydration; no antibiotic for uncomplicated disease	Ciprofloxacin	Antibiotics contraindicated, supportive measures	Rehydration, antimotility agent
Fever Present	Usually	Often	Often	Sometimes
Blood in Stool	Sometimes	Often	Usually	Sometimes
Distinctive Features	Often associated with chickens, reptiles	Very low ID_{50}	Hemolytic uremic syndrome (HUS)	ETEC, EIEC, EPEC, DAEC, EAEC
Epidemiological Features	United States: 20% of all cases require hospitalization; death rate of 0.6%	United States: estimated 450,000 cases per year; internationally: 165 million cases per year	Internationally: causes HUS in 10% of patients; 25% of HUS patients suffer neurological complications, 50% have chronic renal sequelae	–

Peak occurrences of this infection are seasonal; in the Southwest, the peak is often in the late fall; in the Northeast, the peak comes in the spring.

Diagnosis of rotavirus infections is not always performed, as it is treated symptomatically. Nevertheless, studies are often conducted so that public health officials can maintain surveillance of how prevalent the infection is. Stool samples from infected persons contain large amounts of virus, which is readily visible using electron microscopy **(figure 22.18).** The virus gets its name from its physical appearance, which is said to resemble a "spoked wheel." A rapid antigen test for stool specimens is commonly used in clinical settings, and an ELISA test is also available.

The virus is transmitted by the fecal-oral route, including through contaminated food, water, and fomites. For this reason, disease is most prevalent in areas of the world with poor sanitation. In the United States, rotavirus infection is relatively common, but its course is generally mild.

The effects of infection vary with the age, nutritional state, general health, and living conditions of the patient. Babies from 6 to 24 months of age lacking maternal antibodies have the greatest risk for fatal disease. These children

			Nonbacterial Causes		
Campylobacter	*Clostridium difficile*	*Vibrio cholerae*	*Cryptosporidium*	Rotavirus	Norovirus
Vehicle (food, water), fecal-oral	Endogenous (normal biota)	Vehicle (water and some foods), fecal-oral	Vehicle (water, food), fecal-oral	Fecal-oral, vehicle, fomite	Indirect, vehicle (food), direct contact
Adhesins, exotoxin, induction of autoimmunity	Enterotoxins A and B	Cholera toxin (CT)	Intracellular growth	–	Limited immunity to reinfection
Stool culture not usually necessary; dark-field microscopy	Stool culture, PCR, ELISA demonstration of toxins in stool	Clinical diagnosis, microscopic techniques, serological detection of antitoxin	Acid-fast staining, ruling out bacteria	Rapid antigen test	Rapid antigen test
Food and personal hygiene	–	Water and food hygiene	Water treatment, proper food handling	Oral live-virus vaccine	Hygiene
Rehydration; azithromycin in severe cases (antibiotic resistance rising)	Metronidazole in mild cases; vancomycin or fidaxomicin for severe; fecal transplants	Rehydration and possibly doxycycline	None or nitazoxanide in immunocompetent patients	Rehydration	Rehydration
Usually	Sometimes	No	Often	Often	Sometimes
No	Not usually; mucus prominent	No	Not usually	No	No
Guillain-Barré syndrome	Associated with disruption of normal biota	Rice-water stools	Resistant to chlorine disinfection	Severe in infants	Resistant to disinfection
United States: 2.4 million cases per year; internationally: 400 million cases per per year	United States: 3 million cases per year	Global estimate: 100,000–130,000 deaths annually	United States: estimated 748,000 cases per year; 30% seropositive	United States: 2–3 million cases per year; internationally: 125 million cases of infantile diarrhea annually	United States: most common cause of diarrhea in <18-year-olds

present symptoms of watery diarrhea, fever, vomiting, dehydration, and shock. The intestinal mucosa can be damaged in a way that chronically compromises nutrition, and long-term or repeated infections can retard growth. Newborns seem to be protected by maternal antibodies. Adults can also acquire this infection, but it is generally mild and self-limiting.

Children are treated with oral replacement fluid and electrolytes. A new oral live virus vaccine called Rotarix was introduced in 2006, and hospital admissions due to rotavirus infection have decreased by nearly 90% since that time. There is now a second oral live-virus vaccine called RotaTeq.

Norovirus

A bewildering array of viruses can cause gastroenteritis, including adenoviruses, astroviruses, and noroviruses (sometimes known as Norwalk viruses). Norovirus is the most common of these and, indeed, the most common cause of food-borne illness in the United States (see figure 22.10).

Transmission is fecal-oral or via contamination of food and water. Viruses generally cause a profuse, watery diarrhea of 3 to 5 days' duration. Severe vomiting is a feature of the disease, especially in the early phases. Mild fever is often

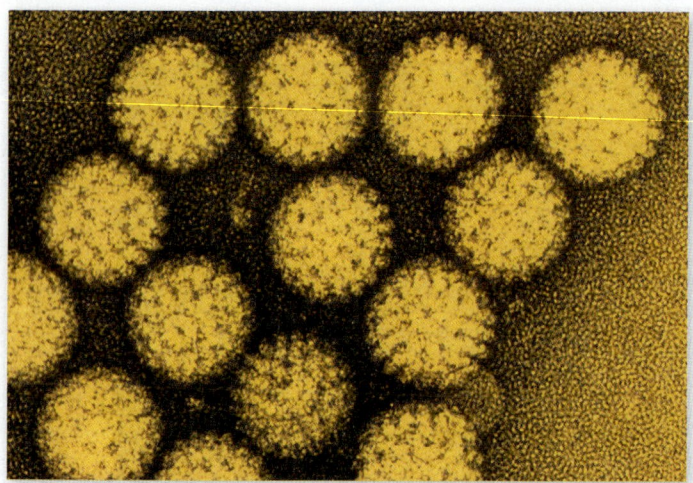

Figure 22.18 Rotavirus visible in a sample of feces from a child with gastroenteritis. Note the unique "spoked-wheel" morphology of the virus.

seen. Scientists consider this an exquisitely tuned pathogen because it has a very low infectious dose (1 to 20 viruses), causes the host to expel enormous amounts of the virus during illness, and survives for days on countertops and in the air. In a study from 2010, scientists found 21 different types of norovirus on a single hospital countertop.

In the years since 2002, a series of gastroenteritis outbreaks have occurred on cruise ships. Most of them have been attributed to noroviruses. This has led to the CDC's development of the Vessel Sanitation Program, aimed at protecting passengers and populations living in ports of call.

Treatment of these infections always focuses on rehydration **(Disease Table 22.5).**

Acute Diarrhea with Vomiting Caused by Exotoxins (Food Poisoning)

If a patient presents with severe nausea and frequent vomiting accompanied by diarrhea and reports that companions with whom he or she shared a recent meal (within the last 1 to 6 hours) are suffering the same fate, food poisoning should be suspected. **Food poisoning** refers to symptoms in the gut that are caused by a preformed toxin of some sort. In many cases, the toxin comes from *Staphylococcus aureus*. In others, the source of the toxin is *Bacillus cereus* or *Clostridium perfringens*. The toxin occasionally comes from nonmicrobial sources such as fish, shellfish, or mushrooms. In any case, if the symptoms are violent and the incubation period is very short, *intoxication* (the effects of a toxin) rather than *infection* should be considered.

Staphylococcus aureus Exotoxin

This illness is associated with eating foods such as custards, sauces, cream pastries, processed meats, chicken salad, or ham that have been contaminated by handling and then left unrefrigerated for a few hours. Because of the high salt tolerance of *S. aureus*, even foods containing salt as a preservative

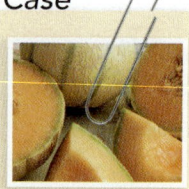

Listeria monocytogenes is a gram-positive rod and is primarily spread through contaminated food. Symptoms of listeriosis are generally muscle aches, fever, diarrhea, and other gastrointestinal symptoms. Depending on the individual and the invasiveness of the organism, symptoms can include headache, stiff neck, loss of balance, confusion, and convulsions. Pregnant women can experience mild flulike symptoms, but because the organism can cross the placenta, an infection during pregnancy can cause miscarriage, premature delivery, life-threatening infection of the newborn, or stillbirth. Those at greatest risk for serious illness with listeriosis are pregnant women; newborns; people with cancer, diabetes, liver disease, kidney disease, or alcoholism; older adults; and people with AIDS or others who have weakened immune systems due to organ transplants and certain diseases.

Usually, outbreaks of *Listeria* infection are associated with processed dairy or deli meats such as soft cheeses, hot dogs, and smoked seafood. Any unpasteurized milk or cheese is likely to be contaminated with *Listeria*. The bacterium can grow at refrigerator temperatures, which is why cheeses and processed meats can be the source of the pathogen. There have been numerous outbreaks associated with cheeses produced in the home such as "queso fresco" or locally produced cheeses sold in farmers' markets. This outbreak was the first outbreak in history associated with cantaloupe. Listeriosis is interesting because it is an unusual food-borne pathogen that causes its main symptoms outside of the gastrointestinal system. For that reason, the disease is covered in chapter 19, "Infectious Diseases Affecting the Nervous System." The case in this chapter will help you remember that listeriosis is a food-borne disease.

■ How did the CDC trace so many widespread cases to Jensen Farms?

■ What could have been done to prevent the outbreak?

are not exempt. The toxins produced by the multiplying bacteria do not noticeably alter the food's taste or smell. The exotoxin (which is an enterotoxin) is heat-stable; inactivation requires 100°C for at least 30 minutes. Thus, heating the food after toxin production may not prevent disease.

Over 20 enterotoxin-producing strains of *S. aureus* have been identified. The illness produced by these strains is caused by the toxin itself and does not require *S. aureus* to be present or alive in the contaminated food. The ingested toxin acts upon the gastrointestinal epithelium and stimulates nerves, with acute symptoms of cramping, nausea, vomiting, and diarrhea. Recovery is also rapid, usually within 24 hours. The disease is not transmissible person to person. Often, a single source will contaminate several people, leading to a small point-source outbreak.

As you learned earlier, many diarrheal diseases have symptoms caused by bacterial exotoxins. In most cases, the bacteria take up temporary residence in the gut and then start producing exotoxin, so the incubation period is longer than the 1 to 6 hours seen with the onset of *S. aureus* food poisoning. Proper food handling, preparation, and storage are required to prevent this form of food poisoning. This condition is almost always self-limiting, and antibiotics are usually not warranted.

Bacillus cereus Exotoxin

Bacillus cereus is a sporulating gram-positive bacterium that is naturally present in soil. As a result, it is a common resident on vegetables and other products in close contact with soil. It produces two exotoxins, one of which causes a diarrheal-type disease, the other of which causes an **emetic** (ee-met'-ik) or vomiting disease. The type of disease that takes place is influenced by the type of food that is contaminated by the bacterium. The emetic form is most frequently linked to fried rice, especially when it has been cooked and kept warm for long periods of time. These conditions are apparently ideal for the expression of the low-molecular-weight, heat-stable exotoxin having an emetic effect. The diarrheal form of the disease is usually associated with cooked meats or vegetables that are held at a warm temperature for long periods of time. These conditions apparently favor the production of the high-molecular-weight, heat-labile exotoxin. The symptom in these cases is a watery, profuse diarrhea that lasts only for about 24 hours.

Diagnosis of the emetic form of the disease is accomplished by finding the bacterium in the implicated food source. Microscopic examination of stool samples is used to diagnose the diarrheal form of the disease. Of course, in everyday practice, neither diagnosis nor treatment is performed because of the short duration of the disease.

For many years, *B. cereus* has been regarded as a relatively harmless bacterium in light of its ability to cause limiting disease. However, the rise of *B. cereus* as a formidable foe in immunocompromised individuals has recently made microbiologists take a closer look at its pathogenic capabilities.

Clostridium perfringens Exotoxin

Another sporulating gram-positive bacterium that causes intestinal symptoms is *Clostridium perfringens*. You first read about this bacterium as the causative agent of gas gangrene in chapter 18. Endospores from *C. perfringens* can also contaminate many kinds of foods. Those most frequently implicated in disease are animal flesh (meat, fish) and vegetables such as beans that have not been cooked thoroughly enough to destroy endospores. When these foods are cooled, endospores germinate and the germinated cells multiply, especially if the food is left unrefrigerated. If the food is eaten without adequate reheating, live *C. perfringens* cells enter the small intestine and release exotoxin. The toxin, acting upon epithelial cells, initiates acute abdominal pain, diarrhea, and nausea in 8 to 16 hours. Recovery is rapid, and deaths are rare. A recent outbreak at a psychiatric facility, however, led to a nearly 6% fatality rate leading clinicians to recognize the increased risk of *Clostridium* food poisoning in patients receiving psychiatric medication. This is due to the fact that these drugs slow down the functioning of the GI tract, enhancing the pathogen's ability to cause disease.

C. perfringens also causes an enterocolitis infection similar to that caused by *C. difficile*. This infectious type of diarrhea is acquired from contaminated food, or it may be transmissible by inanimate objects **(Disease Table 22.6).**

Disease Table 22.6	Acute Diarrhea with Vomiting Caused by Exotoxins (Food Poisoning)		
Causative Organism(s)	*Staphylococcus aureus* exotoxin	*Bacillus cereus*	*Clostridium perfringens*
Most Common Modes of Transmission	Vehicle (food)	Vehicle (food)	Vehicle (food)
Virulence Factors	Heat-stable exotoxin	Heat-stable toxin, heat-labile toxin	Heat-labile toxin
Culture/Diagnosis	Usually based on epidemiological evidence	Microscopic analysis of food or stool	Detection of toxin in stool
Prevention	Proper food handling	Proper food handling	Proper food handling
Treatment	Supportive	Supportive	Supportive
Fever Present	Not usually	Not usually	Not usually
Blood in Stool	No	No	No
Distinctive Features	Suspect in foods with high salt or sugar content	Two forms: emetic and diarrheal	Acute abdominal pain
Epidemiological Features	United States: estimated 240,000 cases per year	United States: estimated 63,000 cases per year	United States: estimated 966,000 cases per year

Chronic Diarrhea

Chronic diarrhea is defined as lasting longer than 14 days. It can have infectious causes or can reflect noninfectious conditions. Most of us are familiar with diseases that present a constellation of bowel syndromes, such as irritable bowel syndrome and ulcerative colitis. As previously discussed, these conditions may indeed represent an overreaction to the presence of an infectious agent or another irritant, but the host response seems to be responsible for the pathology. When the presence of an infectious agent is ruled out by a negative stool culture or other tests, these conditions are suspected.

People suffering from AIDS almost universally suffer from chronic diarrhea. Most of the patients who are not taking antiretroviral drugs have diarrhea caused by a variety of opportunistic microorganisms, including *Cryptosporidium, Mycobacterium avium,* and so forth. A patient's HIV status should be considered if he or she presents with chronic diarrhea.

Next we examine a few of the microbes that can be responsible for chronic diarrhea in otherwise healthy people. Keep in mind that practically any disease of the intestinal tract has a sexual mode of transmission in addition to the ones that are commonly stated. For example, any kind of oral-anal sexual contact efficiently transfers pathogens to the "oral" partner. This mode is more commonly seen in cases of chronic illness than it is in patients experiencing acute diarrhea, for obvious reasons.

Enteroaggregative *E. coli* (EAEC)

In the section on acute diarrhea, you read about the various categories of *E. coli* that can cause disease in the gut. One type, the enteroaggregative *E. coli* (EAEC), is particularly associated with chronic disease, especially in

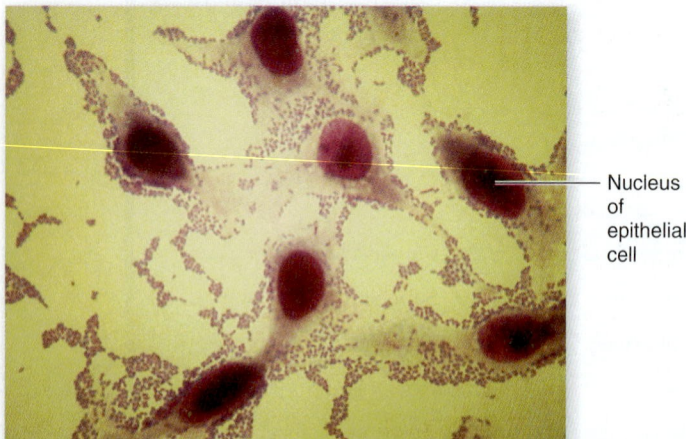

— Nucleus of epithelial cell

Figure 22.19 Enteroaggregative *E. coli* adhering to epithelial cells.

children. This bacterium was first recognized in 1987. It secretes neither the heat-stable nor heat-labile exotoxins previously described for enterotoxigenic *E. coli* (ETEC). It is distinguished by its ability to adhere to human cells in aggregates rather than as single cells **(figure 22.19).** Its presence appears to stimulate secretion of large amounts of mucus in the gut, which may be part of its role in causing chronic diarrhea. The bacterium also seems capable of exerting toxic effects on the gut epithelium, although the mechanisms are not well understood.

Transmission of the bacterium is through contaminated food and water. It is difficult to diagnose in a clinical lab because EAEC is not easy to distinguish from other *E. coli,* including normal biota. Genotypic methods such as PCR are needed for accurate identification during outbreaks. The designation EAEC is not actually a serotype but is functionally defined as an *E. coli* that adheres in an aggregative pattern.

This bacterium seems to be associated with chronic diarrhea in people who are malnourished. It is not exactly clear whether the malnutrition predisposes patients to this infection or whether this infection contributes to malnutrition. Probably both possibilities are operating in patients, who are usually children in developing countries. More recently, the bacterium has been associated with acute diarrhea in industrialized countries, perhaps providing a clue to this question. It may be that in well-nourished hosts, the bacterium produces acute, self-limiting disease.

Cyclospora

Cyclospora cayetanensis is an emerging protozoan pathogen. Since the first occurrence in 1979, hundreds of outbreaks of cylcosporiasis have been reported in the United States and Canada. Its mode of transmission is fecal-oral—though not through the ingestion of cysts themselves. This differentiates this pathogen from its relative, *Cryptosporidium.* Infection occurs only after the oocyst sporulates, a process that begins the formation of the infectious form of the pathogen. In most cases, sporulated oocysts are ingested through the consumption of fresh produce and water presumably contaminated with feces. This disease occurs worldwide and, although primarily of human origin, is not spread directly from person to person. Outbreaks have been traced to imported raspberries, salad made with fresh greens, and drinking water. A major outbreak of this organism occurred on a cruise ship in April of 2009, where 135 of 1,318 passengers and 25 crew members became ill with *Cyclospora.*

The organism is 8 to 10 micrometers in diameter and stains variably in an acid-fast stain. Diagnosis can be complicated by the lack of recognizable oocysts in the feces. Techniques that improve identification of the parasite are examination of fresh preparations under a fluorescent microscope and an acid-fast stain of a processed stool specimen **(figure 22.20).** Autofluorescence of the oocysts can be

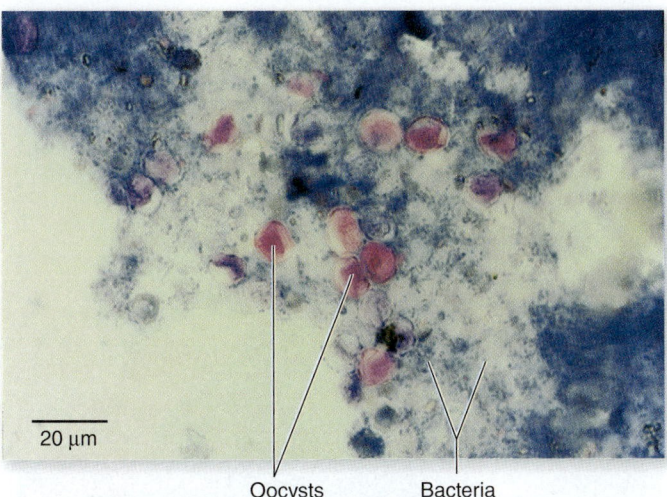

Figure 22.20 An acid-fast stain of *Cyclospora* in a human fecal sample. The large (8–10 μm) cysts stain pink to red and have a wrinkled outer wall. Bacteria stain blue.

visualized when specimens are exposed to ultraviolet (UV) light. A PCR-based test can also be used to identify *Cyclospora* and differentiate it from other parasites. This form of analysis is more sensitive and can detect protozoan genetic material even in the absence of actual cysts.

After an incubation period of about 1 week, symptoms of watery diarrhea, stomach cramps, bloating, fever, and muscle aches appear. Patients with prolonged diarrheal illness experience anorexia and weight loss.

Most cases of infection have been effectively controlled with trimethoprim-sulfamethoxazole lasting 1 week. Traditional antiprotozoan drugs are not effective. Some cases of disease may be prevented by cooking or freezing food to kill the oocysts.

Giardia

Giardia lamblia (also known as *Giardia intestinalis*) is a pathogenic flagellated protozoan first observed by Antonie van Leeuwenhoek in his own feces. For 200 years, it was considered a harmless or weak intestinal pathogen; and only since the 1950s has its prominence as a cause of diarrhea been recognized. In fact, it is the most common flagellate isolated in clinical specimens. Observed straight on, the trophozoite has a unique symmetrical heart shape with organelles positioned in such a way that it resembles a face **(figure 22.21)**. Four pairs of flagella emerge from the ventral surface, which is concave and acts like a suction cup

for attachment to a substrate. *Giardia* cysts are small, compact, and contain four nuclei.

▶ Signs and Symptoms

Typical symptoms include diarrhea of long duration, abdominal pain, and flatulence. Stools have a greasy, malodorous quality to them. Fever is usually not present.

▶ Pathogenesis and Virulence Factors

Ingested *Giardia* cysts enter the duodenum, germinate, and travel to the jejunum to feed and multiply. Some trophozoites remain on the surface, while others invade the deeper crypts to varying degrees. Superficial invasion by trophozoites causes damage to the epithelial cells, edema, and infiltration by white blood cells, but these effects are reversible. The presence of the protozoan leads to malabsorption (especially of fat) in the digestive tract and can cause significant weight loss.

▶ Transmission and Epidemiology of Giardiasis

Giardiasis has a complex epidemiological pattern. The protozoan has been isolated from the intestines of beavers, cattle, coyotes, cats, and human carriers, but the precise reservoir is unclear at this time. Although both trophozoites and cysts escape in the stool, the cysts play a greater role in transmission. Unlike other pathogenic flagellates, *Giardia* cysts can survive for 2 months in the environment. Cysts are usually ingested with water and food or swallowed after close

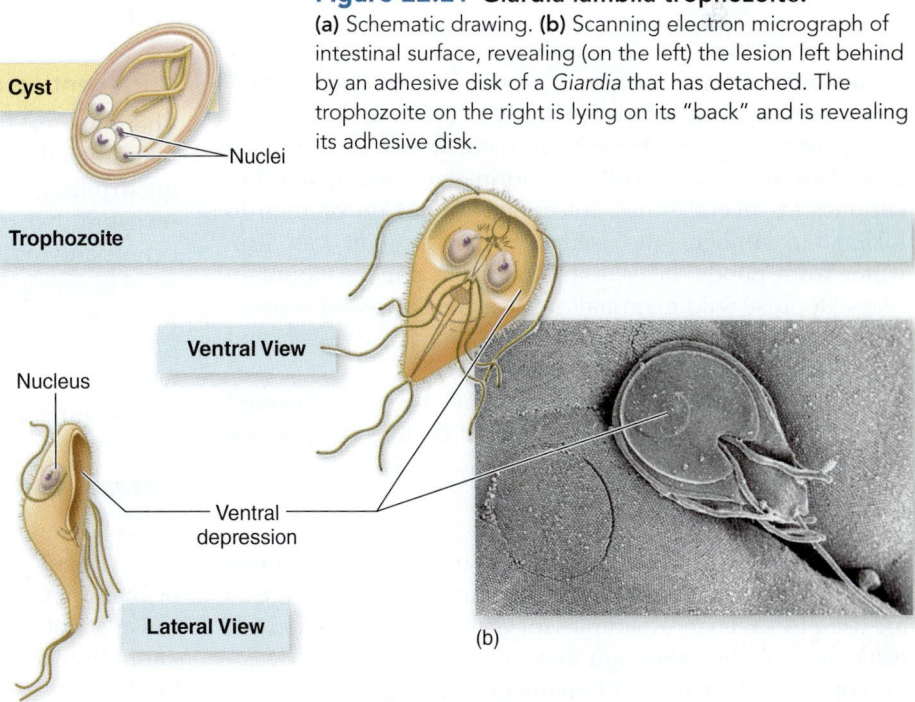

Figure 22.21 *Giardia lamblia* trophozoite.
(a) Schematic drawing. **(b)** Scanning electron micrograph of intestinal surface, revealing (on the left) the lesion left behind by an adhesive disk of a *Giardia* that has detached. The trophozoite on the right is lying on its "back" and is revealing its adhesive disk.

INSIGHT 22.3   A New Take on Number Two

You use one every day, probably several times. But what if you didn't have anywhere to, you know, *go*? In the industrialized world, toilets are a normal part of daily living, and we don't think twice about flushing away our business or what happens to it after we've deposited it. However, in developing countries, sometimes pooping in public is the only option—2.6 billion people have no access to a safe and affordable way to eliminate their waste. Consequently, human fecal material is deposited directly into lakes, rivers, and streams, causing 2.5 billion cases of diarrhea each year and 1.5 billion deaths in children. Chronic diarrhea inhibits the uptake of nutrients essential to childhood development and growth. Installing complex sewer systems requiring millions of gallons of water and electricity to treat the waste is expensive and unfeasible in developing countries.

To address this, the Bill and Melinda Gates Foundation launched the "Reinvent the Toilet Challenge" to incentivize scientists, researchers, and designers to develop a toilet that does not rely on a water source, sewer, or electrical connections and can produce something useful from the waste, all for under 5 cents per use. The winning design, garnering the $100,000 prize, went to the team at California Institute of Technology, who developed a solar-powered toilet capable of generating hydrogen and electricity. Second prize went to the team from Loughborough University in the United Kingdom, whose toilet was able to produce biological charcoal, minerals, and clean water from waste. All of the designs were showcased at the "Reinvent the Toilet Fair," where synthetic feces made from soy and rice were used to test the prototypes. The winning designs will still have to go through design modifications, field testing, and cost

In many countries, people don't have access to modern sewer systems and use primitive latrines, water sources, and outdoor toileting.

cutting before they can be used in developing nations; but an inexpensive and effective toilet system has the potential to save millions of lives by giving people somewhere sanitary "to go."

Source: 2012. www.wired.com/design/2012/08/gates-foundation-funds -better-toilet-design/all/

---

contact with infected people or contaminated objects. Infection can occur with a dose of only 10 to 100 cysts.

Giardia epidemics have been traced to water from fresh mountain streams as well as chlorinated municipal water supplies in several states. Infections are not uncommon in hikers and campers who used what they thought was clean water from ponds, lakes, and streams in remote mountain areas. Because wild mammals such as muskrats and beavers are intestinal carriers, they could account for cases associated with drinking water from these sources.

Cases of fecal-oral transmission have been documented in day care centers; food contaminated by infected persons has also transmitted the disease.

▶ Culture and Diagnosis

Diagnosis of giardiasis can be difficult because the organism is shed in feces only intermittently. Sometimes ELISA tests are used to screen fecal samples for *Giardia* antigens, and PCR tests are available, although they are mainly used for detection of the protozoan in environmental samples.

▶ Prevention and Treatment

There is a vaccine against *Giardia* that can be given to animals, including dogs. No human vaccine is available. Avoiding drinking from freshwater sources is the major preventive measure that can be taken. The agent is killed by boiling, ozone, and iodine; but unfortunately, the amount of chlorine used in municipal water supplies does not destroy the cysts. Because cysts can be present in treated municipal water supplies, water agencies have had to rethink their policies on water maintenance and testing.

Treatment is with tinidazole or nitazoxanide.

Entamoeba

Amoebas are widely distributed in aqueous habitats and are frequent parasites of animals, but only a small number of them have the necessary virulence to invade tissues and cause serious pathology. One of the most significant pathogenic amoebas is *Entamoeba histolytica* (en"-tah-mee'-bah his"-toh-lit'-ih-kuh). The relatively simple life cycle of this parasite alternates between a large trophozoite that is motile by means of pseudopods and a smaller, compact, nonmotile cyst **(figure 22.22, a–c).** The trophozoite lacks most of the organelles of other eukaryotes, and it has a large single nucleus that contains a prominent nucleolus called a *karyosome.* Amoebas from fresh specimens are often packed with food vacuoles containing host cells and bacteria. The mature cyst is encased in a thin yet tough wall and contains four nuclei as well as distinctive cigar-shaped bodies called *chromatoidal bodies,* which are actually dense clusters of ribosomes.

▶ Signs and Symptoms

As hinted to by its species name, tissue damage is one of the formidable characteristics of untreated *E. histolytica* infection. Clinical amoebiasis exists in intestinal and extraintestinal forms. The initial targets of intestinal amoebiasis are the cecum, appendix, colon, and rectum. The amoeba secretes enzymes that dissolve tissues, and it actively penetrates deeper layers of the mucosa, leaving erosive ulcerations **(figure 22.22d).** This phase is marked by dysentery (bloody, mucus-filled stools), abdominal pain, fever, diarrhea, and weight loss. The most life-threatening manifestations of intestinal infection are hemorrhage, perforation, appendicitis, and

tumorlike growths called amebomas. Lesions in the mucosa of the colon have a characteristic flasklike shape.

Extraintestinal infection occurs when amoebas invade the viscera of the peritoneal cavity. The most common site of invasion is the liver. Here, abscesses containing necrotic tissue and trophozoites develop and cause amoebic hepatitis. Another rarer complication is pulmonary amoebiasis. Other infrequent targets of infection are the spleen, adrenals, kidney, skin, and brain. Severe forms of the disease result in about a 10% fatality rate.

▶ Pathogenesis and Virulence Factors

Amoebiasis begins when viable cysts are swallowed and arrive in the small intestine, where the alkaline pH and digestive juices of this environment stimulate excystment. Each cyst releases four trophozoites, which are swept into the cecum and large intestine. There, the trophozoites attach by fine pseudopods **(figure 22.22e),** multiply, actively move about, and feed. In about 90% of patients, infection is asymptomatic or very mild, and the trophozoites do not invade beyond the most superficial layer. The severity of the infection can vary with the strain of the parasite, inoculum size, diet, and host resistance.

The secretion of lytic enzymes by the amoeba seems to induce apoptosis of host cells. This means that the host is contributing to the process by destroying its own tissues on cue from the protozoan. The invasiveness of the amoeba is also a clear contributor to its pathogenicity.

▶ Transmission and Epidemiology

Entamoeba is harbored by chronic carriers whose intestines favor the encystment stage of the life cycle. Cyst

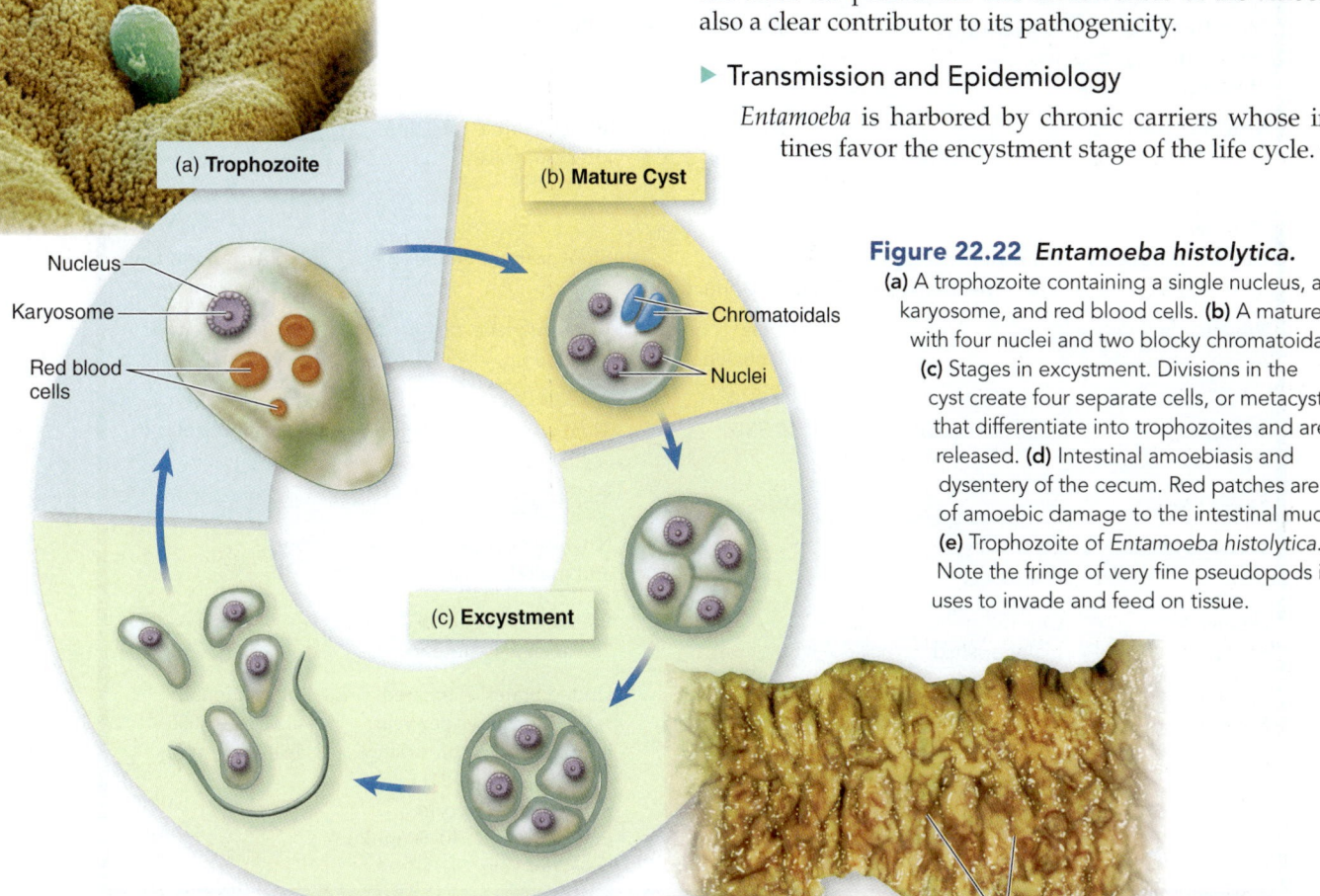

(e)

(a) **Trophozoite**

(b) **Mature Cyst**

Nucleus
Karyosome
Red blood cells

Chromatoidals
Nuclei

(c) **Excystment**

(d) **Erosion of the Intestine**   Ulcerations

Figure 22.22 *Entamoeba histolytica.*
(a) A trophozoite containing a single nucleus, a karyosome, and red blood cells. **(b)** A mature cyst with four nuclei and two blocky chromatoidals. **(c)** Stages in excystment. Divisions in the cyst create four separate cells, or metacysts, that differentiate into trophozoites and are released. **(d)** Intestinal amoebiasis and dysentery of the cecum. Red patches are sites of amoebic damage to the intestinal mucosa. **(e)** Trophozoite of *Entamoeba histolytica.* Note the fringe of very fine pseudopods it uses to invade and feed on tissue.

formation cannot occur in active dysentery because the feces are so rapidly flushed from the body; but after recuperation, cysts are continuously shed in feces.

Humans are the primary hosts of *E. histolytica*. Infection is usually acquired by ingesting food or drink contaminated with cysts released by an asymptomatic carrier. The amoeba is thought to be carried in the intestines of one-tenth of the world's population, and it kills up to 100,000 people a year. Its geographic distribution is partly due to local sewage disposal and fertilization practices. Occurrence is highest in tropical regions (Africa, Asia, and Latin America), where "night soil" (human excrement) or untreated sewage is used to fertilize crops and sanitation of water and food can be substandard. Although the prevalence of the disease is lower in the United States, as many as 10 million people could harbor the agent.

▶ Culture and Diagnosis

Diagnosis of this protozoal infection relies on a combination of tests, including microscopic examination of stool for the characteristic cysts or trophozoites, ELISA tests of stool for

E. histolytica antigens, and serological testing for the presence of antibodies to the pathogen. PCR testing is currently being refined. It is important to differentiate *E. histolytica* from the similar *Entamoeba coli* and *Entamoeba dispar*, which occur as normal biota.

▶ Prevention and Treatment

No vaccine yet exists for *E. histolytica*, although several are in development. Prevention of the disease therefore relies on purification of water. Because regular chlorination of water supplies does not kill cysts, more rigorous methods such as boiling or iodine are required.

Effective treatment usually involves the use of drugs such as metronidazole (Flagyl) or chloroquine. Dehydroemetine is used to control symptoms, but it will not cure the disease. Other drugs are given to relieve diarrhea and cramps, while lost fluid and electrolytes are replaced by oral or intravenous therapy. Infection with *E. histolytica* provokes antibody formation against several antigens, but permanent immunity is unlikely and reinfection can occur **(Disease Table 22.7)**.

Disease Table 22.7   Chronic Diarrhea

Causative Organism(s)	Enteroaggregative *E. coli* (EAEC)	*Cyclospora cayetanensis*	*Giardia lamblia*	*Entamoeba histolytica*
Most Common Modes of Transmission	Vehicle (food, water), fecal-oral	Fecal-oral, vehicle	Vehicle, fecal-oral, direct and indirect contact	Vehicle, fecal-oral
Virulence Factors	?	Invasiveness	Attachment to intestines alters mucosa	Lytic enzymes, induction of apoptosis, invasiveness
Culture/Diagnosis	Difficult to distinguish from other *E. coli*	Stool examination, PCR	Stool examination, ELISA	Stool examination, ELISA, serology
Prevention	?	Washing, cooking food, personal hygiene	Water hygiene, personal hygiene	Water hygiene, personal hygiene
Treatment	None, or ciprofloxacin	TMP-SMZ	Tinidazole, nitazoxanide	Metronidazole or chloroquine
Fever Present	No	Usually	Not usually	Yes
Blood in Stool	Sometimes, mucus also	No	No, mucus present (greasy and malodorous)	Yes
Distinctive Features	Chronic in the malnourished	–	Frequently occurs in backpackers, campers	–
Epidemiological Features	Developing countries: 87% of chronic diarrhea in children >2 years old	United States: estimated 16,000 cases per year; internationally: endemic in 27 countries, mostly tropical	United States: estimated 1.2 million cases per year; internationally: prevalence rates from 2% to 5% in industrialized world internationally: 40–50 million cases per year	Internationally: 40,000–100,000 deaths annually

Hepatitis

When certain viruses infect the liver, they cause **hepatitis,** an inflammatory disease marked by necrosis of hepatocytes and a response by mononuclear white blood cells that swells and disrupts the liver architecture. This pathologic change interferes with the liver's excretion of bile pigments such as bilirubin into the intestine. When bilirubin, a greenish-yellow pigment, accumulates in the blood and tissues, it causes **jaundice,** a yellow tinge in the skin and eyes. The condition can be caused by a variety of different viruses, including cytomegalovirus and Epstein-Barr virus. The others are all called "hepatitis viruses" but only because they all can cause this inflammatory condition in the liver. They are quite different from one another. While there are some recently discovered hepatitis viruses, they are not yet well characterized so we will cover the five that are well understood, named hepatitis A through hepatitis E.

Note that noninfectious conditions can also cause inflammation and disease in the liver, including some autoimmune conditions, drugs, and alcohol overuse.

Hepatitis A Virus

Hepatitis A virus (HAV) is a nonenveloped, single-stranded RNA enterovirus. It belongs to the family *Picornaviridae*. In general, HAV disease is far milder and shorter term than the other forms.

▶ Signs and Symptoms

Most infections by this virus are either subclinical or accompanied by vague, flulike symptoms. In more overt cases, the presenting symptoms may include jaundice and swollen liver. Darkened urine is often seen in this and other hepatitises. Jaundice is present in only about 10% of the cases. Hepatitis A occasionally occurs as a fulminating disease and causes liver damage, but this manifestation is quite rare. The virus is not oncogenic (cancer causing), and complete uncomplicated recovery results.

▶ Pathogenesis and Virulence Factors

The hepatitis A virus is generally of low virulence. Most of the pathogenic effects are thought to be the result of host response to the presence of virus in the liver.

▶ Transmission and Epidemiology

There is an important distinction between this virus and hepatitis B and C viruses: Hepatitis A virus is spread through the fecal-oral route (and is sometimes known as infectious hepatitis). In general, the disease is associated with deficient personal hygiene and lack of public health measures. In countries with inadequate sewage control, most outbreaks are associated with feces-contaminated water and food. Rates of infection in the United States have fallen nearly 90% in the past 20 years, though 20,000 cases still occur annually.

Most of these are a result of close institutional contact, unhygienic food handling, eating shellfish, sexual transmission, or travel to other countries. In 2003, the largest single hepatitis A outbreak to date in the United States was traced to contaminated green onions used in salsa dips at a Mexican restaurant. At least 600 people who had eaten at the restaurant fell ill with hepatitis A.

Hepatitis A occasionally can be spread by blood or blood products, but this is the exception rather than the rule. In developing countries, children are the most common victims, because exposure to the virus tends to occur early in life, whereas in North America and Europe, more cases appear in adults. Because the virus is not carried chronically, the principal reservoirs are asymptomatic, short-term carriers (often children) or people with clinical disease.

▶ Culture and Diagnosis

Diagnosis of the disease is aided by detection of anti-HAV IgM antibodies produced early in the infection and by tests to identify HA antigen or virus directly in stool samples.

▶ Prevention and Treatment

Prevention of hepatitis A is based primarily on immunization. Two inactivated viral vaccines (Havrix and VAQTA) are used in the United States today. Short-term protection can be conferred by passive immune globulin. This treatment is useful for people who have come in contact with HAV-infected individuals or who have eaten at a restaurant that was the source of a recent outbreak. It has also recently been discovered that administering Havrix after exposure can prevent symptoms. In the 2003 green onion outbreak, 9,000 patrons of the Mexican restaurant received passive immunization as a precaution. A combined hepatitis A/hepatitis B vaccine, called Twinrix, is recommended for people who may be at risk for both diseases, such as people with chronic liver dysfunction, intravenous drug users, and men who have sex with men. Travelers to areas with high rates of both diseases should obtain vaccine coverage as well. Development of active natural immunity toward hepatitis A virus leads to lifelong protection from reinfection.

No specific medication is available for hepatitis A once the symptoms begin. Drinking lots of fluids and avoiding liver irritants such as aspirin or alcohol will speed recovery. Patients who receive immune globulin early in the disease usually experience milder symptoms than patients who do not receive it.

Hepatitis B Virus

Hepatitis B virus (HBV) is an enveloped DNA virus in the family *Hepadnaviridae*. Intact viruses are often called Dane particles. An antigen of clinical and immunologic significance is the surface (or S) antigen. The genome is partly double-stranded and partly single-stranded.

▶ Signs and Symptoms

In addition to the direct damage to liver cells, the spectrum of hepatitis disease may include fever, chills, malaise, anorexia, abdominal discomfort, diarrhea, and nausea. Rashes may appear and arthritis may occur. Hepatitis B infection can be very serious, even life-threatening. A small number of patients develop glomerulonephritis and arterial inflammation. Complete liver regeneration and restored function occur in most patients; however, a small number of patients develop chronic liver disease in the form of necrosis or cirrhosis (permanent liver scarring and loss of tissue). In some cases, chronic HBV infection can lead to liver cancer.

Patients who become infected as children have significantly higher risks of long-term infection and disease. In fact, 90% of neonates infected at birth develop chronic infection, as do 30% of children infected between the ages of 1 and 5, but only 6% of persons infected after the age of 5. This finding is one of the major justifications for the routine vaccination of children. Also, infection becomes chronic more often in men than in women. The mortality rate is 15% to 25% for people with chronic infection.

HBV is known to be a cause of **hepatocellular carcinoma.** Investigators have found that mass vaccination against HBV in Taiwan, begun 18 years ago, has resulted in a significant decrease in liver cancer in that country. (Taiwan previously had one of the highest rates of this cancer.) It is speculated that cancer is probably a result of infection early in life and the long-term carrier state.

Some patients infected with hepatitis B are coinfected with a particle called the delta agent, sometimes also called a **hepatitis D** virus. This agent seems to be a defective RNA virus that cannot produce infection unless a cell is also infected with HBV. Hepatitis D virus invades host cells by "borrowing" the outer receptors of HBV. When HBV infection is accompanied by the delta agent, the disease becomes more severe and is more likely to progress to permanent liver damage.

▶ Pathogenesis and Virulence Factors

The hepatitis B virus enters the body through a break in the skin or mucous membrane or by injection into the bloodstream. Eventually, it reaches the liver cells (hepatocytes) where it multiplies and releases viruses into the blood during an incubation period of 4 to 24 weeks (7 weeks average). Surprisingly, the majority of those infected exhibit few overt symptoms and eventually develop an immunity to HBV, but some people experience the symptoms described earlier. The precise mechanisms of virulence are not clear. The ability of HBV to remain latent in some patients contributes to its pathogenesis. New research shows that a molecule dubbed "HBx" (hepatitis B X protein) plays a role in promoting viral replication in liver cells. Strangely, hepatitis B infection seems to be able to influence the gender of offspring. If one parent is a carrier, the child is more likely to be male than female.

▶ Transmission and Epidemiology

An important factor in the transmission pattern of hepatitis B virus is that it multiplies exclusively in the liver, which continuously seeds the blood with viruses. Electron microscopic studies have revealed up to 10^7 virions per milliliter of infected blood. Even a minute amount of blood (a *millionth* of a milliliter) can transmit infection. The abundance of circulating virions is so high and the minimal dose so low that such simple practices as sharing a toothbrush or a razor can transmit the infection. HBV has also been detected in semen and vaginal secretions, and it can be transmitted by these fluids. Growing concerns about virus spread through donated organs and tissue are prompting increased testing prior to surgery. Spread of the virus by means of close contact in families or institutions is also well documented. Vertical transmission is possible, and it predisposes the child to development of the carrier state and increased risk of liver cancer. The disease is sometimes known as *serum hepatitis.*

Hepatitis B is an ancient disease that has been found in all populations, although the incidence and risk are highest among people living under crowded conditions, drug addicts, the sexually promiscuous, and those in certain occupations, including people who conduct medical procedures involving blood or blood products.

This virus is one of the major infectious concerns for health care workers. Needlesticks can easily transmit the virus; therefore, most workers are required to have the full series of HBV vaccinations. Unlike the more notorious, but less resilient HIV, HBV remains infective for days in dried blood, for months when stored in serum at room temperature, and for decades if frozen. Although it is not inactivated after 4 hours of exposure to 60°C, boiling for the same period can destroy it. Disinfectants containing chlorine, iodine, and glutaraldehyde show potent anti–hepatitis B activity.

Cosmetic manipulation such as tattooing and ear or body piercing can expose a person to infection if the instruments are not properly sterilized. The only reliable method for destroying HBV on reusable instruments is autoclaving.

▶ Culture and Diagnosis

Serological tests can detect either virus antigen or antibodies. Radioimmunoassay and ELISA testing permit detection of the important surface antigen of HBV very early in infection. These same tests are essential for screening blood destined for transfusions, semen in sperm banks, and organs intended for transplant. Antibody tests are most valuable in patients who are negative for the antigen.

▶ Prevention and Treatment

Since 1981, the primary prevention for HBV infection is vaccination. The most widely used vaccines are recombinant, containing the pure surface antigen cloned in yeast cells. Vaccines are given in three doses over 18 months, with occasional boosters. Vaccination is a must for medical and dental workers and students, patients receiving multiple transfusions, immunodeficient persons, and cancer patients. The vaccine is also now strongly recommended for all newborns as part of a routine immunization schedule. As just mentioned, a combined vaccine for HAV/HBV may be appropriate for certain people.

Passive immunization with hepatitis B immune globulin (HBIG) gives significant immediate protection to people who have been exposed to the virus through needle puncture, broken blood containers, or skin and mucosal contact with blood. Another group for whom passive immunization is highly recommended is neonates born to infected mothers.

Mild cases of hepatitis B are managed by symptomatic treatment and supportive care. Chronic infection can be controlled with recombinant human interferon, tenofovir, or entecavir. Each of these can help to slow virus multiplication and prevent liver damage in many but not all patients. None of the drugs are considered curative. Different drug regimens are called for when a patient is coinfected with HBV and HIV.

Hepatitis C Virus

Hepatitis C is sometimes referred to as the "silent epidemic" because 4.1 million Americans are infected with the virus, but it takes many years to cause noticeable symptoms. In the United States, its incidence fell between 1992 and 2003, but no further decreases have been seen since then. Liver failure from hepatitis C is one of the most common reasons for liver transplants in this country. Hepatitis C is an RNA virus in the *Flaviviridae* family. It is closely related to viruses causing West Nile fever and yellow fever. It used to be known as "non-A non-B" virus. It is usually diagnosed with a blood test for antibodies to the virus.

▶ Signs and Symptoms

People have widely varying experiences with this infection. It shares many characteristics of hepatitis B disease, but it is much more likely to become chronic. Of those infected, 75% to 85% will remain infected indefinitely. (In contrast, only about 6% of persons who acquire hepatitis B after the age of 5 will be chronically infected.) With HCV infection, it is possible to have severe symptoms without permanent liver damage, but it is more common to have chronic liver disease even if there are no overt symptoms. Cancer may also result from chronic HCV infection. Worldwide, HBV infection is the most common cause of liver cancer, but in the United States it is more likely to be caused by HCV.

▶ Pathogenesis and Virulence Factors

The virus is so adept at establishing chronic infections that researchers are studying the ways that it evades immunologic detection and destruction. The virus's core protein seems to play a role in the suppression of cell-mediated immunity as well as in the production of various cytokines. The protein may also be responsible for altering mitochondrial activity in HCV-infected cells. Scientists recently identified that HCV enters liver cells using the same receptors utilized for cholesterol entry. This discovery has revealed a potential target for therapeutic drug development.

▶ Transmission and Epidemiology

This virus is acquired in similar ways to HBV. It is more commonly transmitted through blood contact (both "sanctioned," such as in blood transfusions, and "unsanctioned," such as needle sharing by injecting drug users) than through transfer of other body fluids. Vertical transmission is also possible. Before a test was available to test blood products for this virus, it seems to have been frequently transmitted through blood transfusions. Hemophiliacs who were treated with clotting factor prior to 1985 were infected with HCV at a high rate. Once blood began to be tested for HIV (in 1985) and screened for so-called "non-A non-B" hepatitis, the risk of contracting HCV from blood was greatly reduced. The current risk for transfusion-associated HCV is thought to be 1 in 100,000 units transfused. The risk of HCV transmission through contaminated donor tissues is still of concern today, however.

Because HCV was not recognized sooner, a relatively large percentage of the population is infected. Eighty percent of the roughly 3.2 million affected in this country are suspected to have no symptoms. In 2012, the CDC recommended that all baby boomers (those born between 1945 and 1965) be tested for HCV. It has a very high prevalence in parts of South America, Central Africa, and China.

▶ Prevention and Treatment

There is currently no vaccine for hepatitis C. Various treatment regimens have been attempted; most include the use of therapeutic interferon and a more effective derivative of interferon called pegylated interferon. Some clinicians also prescribe ribavirin to try to suppress viral multiplication. The addition of protease and polymerase inhibiting drugs to this combination treatment appear to be beneficial as well. The treatments are not curative, but they may prevent or lessen damage to the liver. In 2012, two companies reported drug

A Note About Hepatitis E Virus

Another RNA virus, called hepatitis E, causes a type of hepatitis very similar to that caused by hepatitis A. Like hepatitis A virus, it is a single-stranded nonenveloped RNA virus. However, its exact classification has still not been determined. The disease it causes is usually self-limiting and often goes undiagnosed. The jaundice that is typical of other forms of hepatitis is not seen in hepatitis E infection. In rare cases, hepatitis E disease can lead to acute liver failure. Pregnant women in their third trimester are at highest risk for this severe form of disease, and the fatality rate is nearly 20%. Overall, hepatitis E infection is more common in developing countries. A majority of the cases reported in the United States occur in people who have traveled to these endemic regions. Hepatitis E virus is transmitted by the fecal-oral route, mainly through contaminated water and food. The infection does not seem to be transmitted from person to person, though blood transfusions have been documented to transmit the pathogen to uninfected patients. There is currently no vaccine.

Disease Table 22.8   Hepatitis

Causative Organism(s)	Hepatitis A or E virus	Hepatitis B virus	Hepatitis C virus
Most Common Modes of Transmission	Fecal-oral, vehicle	Parenteral (blood contact), direct contact (especially sexual), vertical	Parenteral (blood contact), vertical
Virulence Factors	–	Latency	Core protein suppresses immune function?
Culture/Diagnosis	IgM serology	Serology (ELISA, radioimmunoassay)	Serology
Prevention	Hepatitis A vaccine or combined; HAV/HBV vaccine	HBV recombinant vaccine	–
Treatment	HAV: hepatitis A vaccine or immune globulin; HEV: immune globulin	Interferon, tenofovir, or entecavir	(Pegylated) interferon, with or without ribavirin
Incubation Period	2–7 weeks	1–6 months	2–8 weeks
Epidemiological Features	Hepatitis A, United States: 20,000 cases annually and 40% of adults show evidence of prior infection; internationally: 1.4 million cases per year; hepatitis E, internationally: 20 million infections per year; 60% in East and Southeast Asia	United States: prevalence rate 1.5 per 100,000; 800,000 to 1.4 million have chronic infection; internationally: 240 million	United States: estimated 17,000 new cases per year; 3.2 million with chronic HCV; internationally: 150 million chronically infected

regimens that cleared the virus, but they are both years away from widespread use **(Disease Table 22.8).**

22.3 Learning Outcomes—Assess Your Progress

5. List the possible causative agents for the following infectious gastrointestinal conditions: dental caries, periodontal diseases, mumps, and gastric ulcers.
6. Name eight bacterial and three nonbacterial causes of acute diarrhea, and identify the most common cause of food-borne illness in the United States.
7. Name one distinct feature for each of the acute diarrhea pathogens.
8. Differentiate between food poisoning and food-borne infection.
9. Identify three causative agents for chronic diarrhea.
10. Differentiate among the main types of hepatitis and discuss causative agents, modes of transmission, diagnostic techniques, prevention, and treatment of each.

22.4 Gastrointestinal Tract Diseases Caused by Helminths

Helminths that parasitize humans are amazingly diverse, ranging from barely visible roundworms (0.3 mm) to huge tapeworms (25 m long). In the introduction to these organisms

in chapter 5, we grouped them into three categories—nematodes (roundworms), trematodes (flukes), and cestodes (tapeworms)—and discussed basic characteristics of each group. You may wish to review those sections before continuing. In this section, we examine the intestinal diseases caused by helminths. Although they can cause symptoms that may be mistaken for some of the diseases discussed elsewhere in this chapter, helminthic diseases are usually accompanied by an additional set of symptoms that arise from the host response to helminths. Helminthic infection usually provokes an increase in granular leukocytes called eosinophils, which have a specialized capacity to destroy multicellular parasites. This increase, termed **eosinophilia,** is a hallmark of helminthic infection and is detectable in blood counts. If the following symptoms occur coupled with eosinophilia, helminthic infection should be suspected. Many of these infections are considered "neglected tropical diseases"—infections that cast a large burden of disease in the poorest countries of the world yet receive the least recognition and research funding today. Due to the efforts of dedicated tropical disease medicine specialists and organizations like both the Carter and Gates Foundations, some of these helminthic diseases are on the decline.

Helminthic infections may be acquired through the fecal-oral route or through penetration of the skin, but most of them spend part of their lives in the intestinal tract. **Figure 22.23** depicts the four different types of life cycles of the helminths. While the worms are in the intestines, they can

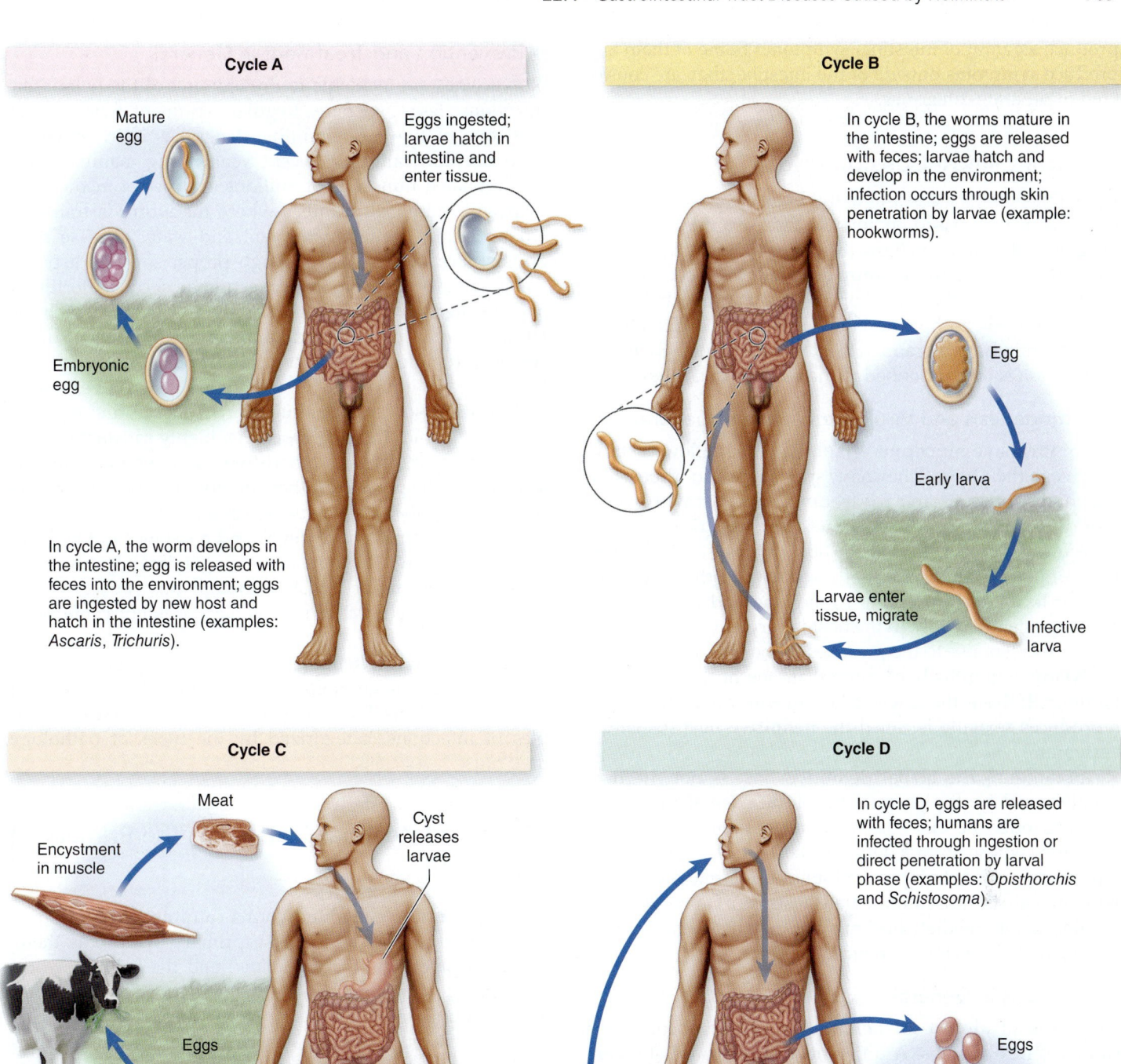

Cycle A

Mature egg

Eggs ingested; larvae hatch in intestine and enter tissue.

Embryonic egg

In cycle A, the worm develops in the intestine; egg is released with feces into the environment; eggs are ingested by new host and hatch in the intestine (examples: *Ascaris*, *Trichuris*).

Cycle B

In cycle B, the worms mature in the intestine; eggs are released with feces; larvae hatch and develop in the environment; infection occurs through skin penetration by larvae (example: hookworms).

Egg

Early larva

Larvae enter tissue, migrate

Infective larva

Cycle C

Meat

Encystment in muscle

Cyst releases larvae

Eggs

In cycle C, the adult matures in human intestine; eggs are released with feces into the environment; eggs are eaten by grazing animals; larval forms encyst in tissue; humans eating animal flesh are infected (example: *Taenia*).

Cycle D

In cycle D, eggs are released with feces; humans are infected through ingestion or direct penetration by larval phase (examples: *Opisthorchis* and *Schistosoma*).

Eggs

First larval stage

Animal flesh

Organ such as intestine or bladder

Second larval stage

Figure 22.23 Four basic helminth life and transmission cycles.

produce a gamut of intestinal symptoms. Some of them also produce symptoms outside of the intestine; they are considered in separate categories.

General Clinical Considerations

Up to this point, the diseases in this book have been arranged in the same way, based on how the disease appears in terms of signs and symptoms (how the patient appears upon presentation to the health care provider). But this section on helminthic diseases adopts a bit of a different approach. We talk about diagnosis, pathogenesis and prevention, and treatment of the helminths as a group in the next subsections. Each type of infection is then described in the sections that follow.

▶ Pathogenesis and Virulence Factors in General

Helminths have numerous adaptations that allow them to survive in their hosts. They have specialized mouthparts for attaching to tissues and for feeding, enzymes with which they liquefy and penetrate tissues, and a cuticle or other covering to protect them from host defenses. In addition, their organ systems are usually reduced to the essentials: getting food and processing it, moving, and reproducing. The damage they cause in the host is very often the result of the host's response to the presence of the invader.

Many helminths have more than one host during their lifetimes. If this is the case, the host in which the adult worm reproduces sexually is called the **definitive host** (usually a vertebrate).

Sometimes the actual definitive host is not the host usually used by the parasite but an accidental bystander. Humans often become the accidental definitive hosts for helminths whose normal definitive host is a cow, pig, or fish. Larval stages of helminths are found in intermediate hosts. Humans can serve as intermediate hosts, too. Helminths may require no intermediate host at all or may need one or more intermediate hosts for their entire life cycle.

▶ Diagnosis in General

Diagnosis of almost all helminthic infections follows a similar series of steps. A differential blood count showing eosinophilia and serological tests indicating sensitivity to helminthic antigens all provide indirect evidence of worm infection. A history of travel to the tropics or immigration from those regions is also helpful, even if it occurred years ago, because some flukes and nematodes persist for decades. The most definitive evidence, however, is the discovery of eggs, larvae, or adult worms in stools or in tissues. The worms are sufficiently distinct in morphology that positive identification can be based on any stage, including eggs. That said, not all of these diseases result in eggs or larval stages that can easily be found in stool.

▶ Prevention and Treatment in General

No vaccines are available to prevent any of the helminthic infections described here. Regular treatment twice a year with one of the antihelminthic drugs has been shown to keep people healthy. In recent years, drug manufacturers have donated hundreds of millions of doses of medicine to help with this goal. In areas where the worm is transmitted by fecally contaminated soil and water, disease rates are significantly reduced through proper sewage disposal, using sanitary latrines, avoiding human feces as fertilizer, and disinfection of the water supply. In cases in which the larvae invade through the skin, people should avoid direct contact with infested water and soil. Food-borne disease can be avoided by thoroughly washing and cooking vegetables and meats. Also, because adult worms, larvae, and eggs are sensitive to cold, freezing foods is a highly satisfactory preventive measure. These methods work best if humans are the sole host of the parasite; if they are not, control of reservoirs or vector populations may be necessary.

Some helminths have developed resistance to the drugs used to treat them. In some cases, surgery may be necessary to remove worms or larvae, although this procedure can be difficult if the parasite load is high or is not confined to one area.

The variety of helminthic infections is very large. A whole branch of microbiology, called parasitology, is devoted to these organisms. Following are some representative examples of infections, categorized by the types of pathology they cause.

Disease: Intestinal Distress as the Primary Symptom

Both tapeworms and roundworms can infect the intestinal tract in such a way as to cause primary symptoms there. The pork tapeworm (*Taenia solium*) and the fish tapeworm (*Diphyllobothrium latum*) are highlighted, as well as two nematodes (roundworms): the whipworm *Trichuris trichiura* and the pinworm *Enterobius vermicularis*. Both of the roundworms are deposited in the small intestine and migrate to the large intestine. We start with these.

Trichuris trichiura

The common name for this nematode—whipworm—refers to its likeness to a miniature buggy whip. Its life cycle and transmission is of the cycle A type (see figure 22.23). Humans are the sole host. Trichuriasis has its highest incidence in areas of the tropics and subtropics that have poor sanitation. Embryonic eggs deposited in the soil are not immediately infective and continue development for 3 to 6 weeks in this habitat. Ingested eggs hatch in the small intestine, where the larvae attach, penetrate the outer wall, and go through several molts. The mature adults move to the large intestine and gain a hold with their long, thin heads, while the thicker tail dangles free in the intestinal lumen. Following sexual maturation and fertilization, the females eventually lay 3,000 to 5,000 eggs daily into the bowel. The entire cycle requires about 90 days, and untreated infection can last up to 2 years.

Symptoms of this infection may include localized hemorrhage of the bowel caused by worms burrowing and piercing intestinal mucosa. This can also provide a portal of entry for secondary bacterial infection. Heavier infections can cause dysentery, loss of muscle tone, and rectal prolapse, which can prove fatal in children.

Enterobius vermicularis

This nematode is often called the pinworm, or seatworm. It is the most common worm disease of children in temperate zones. Some estimates put the prevalence of this infection in the United States at 5% to 15%, although most experts feel that this has declined in recent years. The transmission of this roundworm is of the cycle A type.

Freshly deposited eggs have a sticky coating that causes them to lodge beneath the fingernails and to adhere to fomites. Upon drying, the eggs become airborne and settle in house dust. Eggs are ingested from contaminated food or drink and from self-inoculation from one's own fingers. Eggs hatch in the small intestine and release larvae that migrate to the large intestine. There the larvae mature into adult worms and mate.

The hallmark symptom of this condition is pronounced anal itching when the mature female emerges from the anus and lays eggs. Although infection is not fatal and most cases are asymptomatic, the afflicted child can suffer from disrupted sleep and sometimes nausea, abdominal discomfort, and diarrhea. A simple rapid test can be performed by pressing a piece of transparent adhesive tape against the anal skin and then applying it to a slide for microscopic examination. When one member of the family is diagnosed, the entire family should be tested and/or treated because it is likely that multiple members are infected.

Taenia solium

In contrast to the last two helminths, this one is a tapeworm. Adult worms are usually around 5 meters long and have a scolex with hooklets and suckers to attach to the intestine **(figure 22.24)**. Disease caused by *T. solium* (the pig tapeworm) is distributed worldwide but is mainly concentrated in areas where humans live in close proximity with pigs or eat undercooked pork. In pigs, the eggs hatch in the small intestine and the released larvae migrate throughout the organs. Ultimately, they encyst in the muscles, becoming

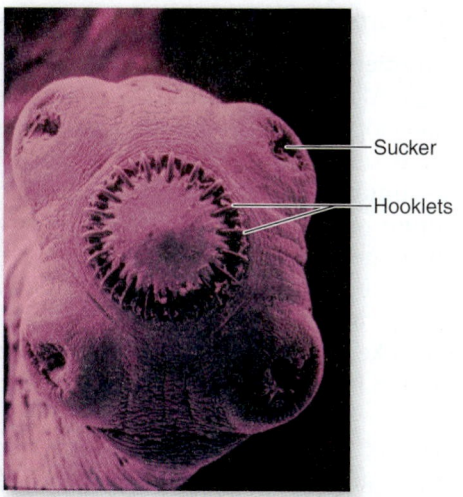

Figure 22.24
Tapeworm
characteristics.

Sucker

Hooklets

(a) Tapeworm scolex showing sucker and hooklets.

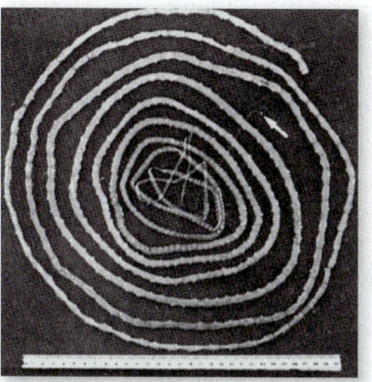

(b) Adult *Taenia saginata*. The arrow points to the scolex; the remainder of the tape, called the strobila, has a total length of 5 meters.

cysticerci, young tapeworms that are the infective stage for humans. When humans ingest a live cysticercus in pork, the coat is digested and the organism is flushed into the intestine, where it firmly attaches by the scolex and develops into an adult tapeworm. Infection with *T. solium* can take another form when humans ingest the tapeworm eggs rather than cysticerci. Although humans are not the usual intermediate hosts, the eggs can still hatch in the intestine, releasing tapeworm larvae that migrate to all tissues. They form bladderlike sacs throughout the body that can cause serious damage. This transmission and life cycle are shown in cycle C in figure 22.23. The pork tapeworm is not the same as the more commonly known pork helminthic infection, *trichinosis.* It is discussed in a later section.

For such a large organism, it is remarkable how few symptoms a tapeworm causes. Occasionally, a patient discovers proglottids in his or her stool, and some patients complain of vague abdominal pain and nausea.

Diphyllobothrium latum

This tapeworm has an intermediate host in fish. It is common in the Great Lakes, Alaska, and Canada. Humans are its definitive host. It develops in the intestine and can cause long-term symptoms. It can be transmitted in raw food such as sushi and sashimi made from salmon. (Reputable sushi restaurants employ authentic sushi chefs who are trained to carefully examine fish for larvae and other signs of infection.) It must also be recognized that transmission of this pathogen can occur in the United States through the consumption of undercooked or lightly smoked trout, perch or pike—all very common fish to the American freshwater sport fisherman.

As is the case with most tapeworms, symptoms are minor and usually vague and include possible abdominal discomfort or nausea. The tapeworm seems to have the ability to absorb and use the vitamin B_{12}, making it unavailable to its human host. Anemia is therefore sometimes reported with this infection. You should be aware that certain people of Scandinavian descent have a genetic predisposition for not adsorbing B_{12}. In these patients, *Diphyllobothrium latum* infection can be quite dangerous **(Disease Table 22.9).**

Disease: Intestinal Distress Accompanied by Migratory Symptoms

A diverse group of helminths enter the body as larvae or eggs, mature to the worm stage in the intestine, and then migrate into the circulatory and lymphatic systems, after which they travel to the heart and lungs, migrate up the respiratory tree to the throat, and are swallowed. This journey returns the mature worms to the intestinal tract where they then take up residence. All of these conditions, in addition to causing symptoms in the digestive tract, may induce inflammatory reactions along their migratory routes, resulting in eosinophilia and, during their lung stage, pneumonia. Two examples of this type of infection follow.

Ascaris lumbricoides

Ascaris lumbricoides is a giant intestinal roundworm (up to 300 mm—a foot or more—long) that probably accounts for the greatest number of worm infections worldwide (estimated at 1 billion cases). It was first identified and described by Linnaeus in 1758. Most reported cases in the United States occur in the southeastern states. *Ascaris* spends its larval and

Disease Table 22.9   Intestinal Distress

Causative Organism(s)	*Trichuris trichiura* (whipworm)	*Enterobius vermicularis* (pinworm)	*Taenia solium* (pork tapeworm)	*Diphyllobothrium latum* (fish tapeworm)
Most Common Modes of Transmission	Cycle A: vehicle (soil)—also fecal-oral	Cycle A: vehicle (food, water), fomites, self-inoculation	Cycle C: vehicle (pork)—also fecal-oral	Cycle C: vehicle (seafood)
Virulence Factors	Burrowing and invasiveness	–	–	Vitamin B_{12} usage
Culture/Diagnosis	Blood count, serology, egg or worm detection	Adhesive tape	Blood count, serology, egg or worm detection	Blood count, serology, egg or worm detection
Prevention	Hygiene, sanitation	Hygiene	Cook meat, avoid pig feces	Cook meat
Treatment	Mebendazole	Mebendazole, piperazine	Praziquantel	Praziquantel
Distinctive Features	Humans sole host	Common in United States	Tapeworm; intermediate host is pigs	Large tapeworm; anemia
Epidemiological Features	United States: prevalence >0.1%; internationally: prevalence as high as 80% in Southeast Asia, Africa, the Caribbean, and Central and South America	United States: prevalence in children 0.2%–20%	United States: 1,000 cases diagnosed per year; internationally: 50 million infected with *T. solium* or related tapeworm	Internationally: estimated 20 million infections worldwide

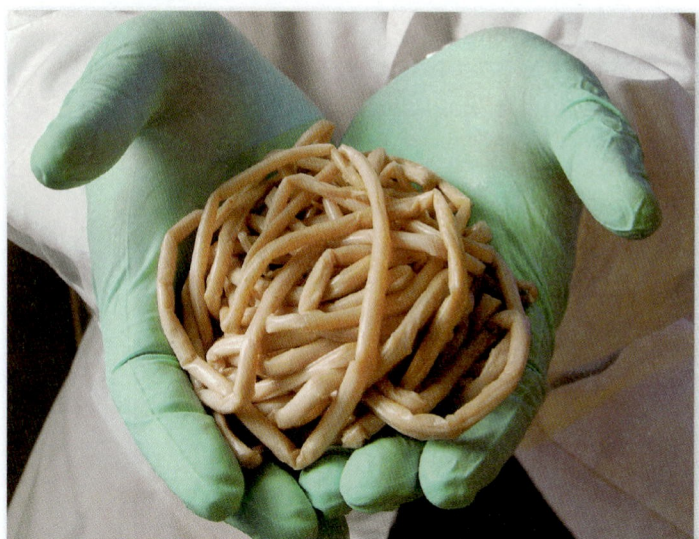

Figure 22.25 A mass of *Ascaris lumbricoides* worms.   These worms were passed by a child in Kenya in 2007.

adult stages in humans and releases embryonic eggs in feces, which are then spread to other humans through food, drink, or contaminated objects placed in the mouth. The eggs thrive in warm, moist soils and resist cold and chemical disinfectants, but they are sensitive to sunlight, high temperatures, and drying. After ingested eggs hatch in the human intestine, the larvae embark upon an odyssey in the tissues. First, they penetrate the intestinal wall and enter the lymphatic and circulatory systems. They are swept into the heart and eventually arrive at the capillaries of the lungs. From this point, the larvae migrate up the respiratory tree to the glottis. Worms entering the throat are swallowed and returned to the small intestine, where they reach adulthood and reproduce, producing up to 200,000 fertilized eggs a day.

Even as adults, male and female worms are not attached to the intestine and retain some of their exploratory ways. They are known to invade the biliary channels of the liver and gallbladder, and on occasion the worms emerge from the nose and mouth. Severe inflammatory reactions mark the migratory route; and allergic reactions such as bronchospasm, asthma, or skin rash can occur. Heavy worm loads can retard the physical and mental development of children **(figure 22.25)**. One factor that contributes to intestinal worm infections is self-reinoculation due to poor personal hygiene.

Necator americanus and *Ancylostoma duodenale*

These two different nematodes are called by the common name "hookworm." *Necator americanus* (nee-kay'-tor ah-mer"-ih-cah'-nus) is endemic to the New World, and *Ancylostoma duodenale* (an'-kih-los'-toh-mah doo-oh-den-ah'-lee) is endemic to the Old World, although the two species overlap in parts of Latin America. Otherwise, with respect to transmission, life cycle, and pathology, they are usually lumped together. The *hook* refers to the adult's oral cutting plates on its curved anterior end, by which it anchors to the intestinal villi **(figure 22.26)**. Nearly 800 million people suffer

from hookworm infections worldwide. *Ascaris*, hookworm, and whipworm infections are termed *soil-transmitted helminthic infections;* collectively, they account for a large percentage of the total disease burden in the world today.

Unlike other intestinal worms, hookworm larvae hatch outside the body and infect by penetrating the skin. Hookworm transmission is described by cycle B (see figure 22.23). Ordinarily, the parasite is present in soil contaminated with human feces. It enters sites on bare feet such as hair follicles, abrasions, or the soft skin between the toes, but cases have occurred via mud that was splattered on the ankles of people wearing shoes. Infection has even been reported in people handling soiled laundry.

On contact, the hookworm larvae actively burrow into the skin. After several hours, they reach the lymphatic or blood circulation and are immediately carried into the heart and lungs. The larvae proceed up the bronchi and trachea to the throat. Most of the larvae are swallowed with sputum and arrive in the small intestine, where they anchor, feed on blood, and mature. Eggs first appear in the stool about 6 weeks after the time of entry, and the untreated infection can last about 5 years.

Symptoms from these infections follow the progress of the worm in the body. A localized dermatitis called *ground itch* may be caused by the initial penetration of larvae. The transit of the larvae to the lungs is ordinarily brief, but it can cause symptoms of pneumonia and eosinophilia. The potential for injury is greatest during the intestinal phase, when heavy worm burdens can cause nausea, vomiting, cramps, and bloody diarrhea. Because blood loss is significant, iron-deficient anemia develops, and infants are especially susceptible to hemorrhagic shock. Chronic fatigue, listlessness, apathy, and anemia worsen with chronic and repeated infections.

Hookworm infections are treated with antihelminthic drugs, but frequent reinfection is a problem. In 2000, the Bill and Melinda Gates Foundation, recognizing the impact of

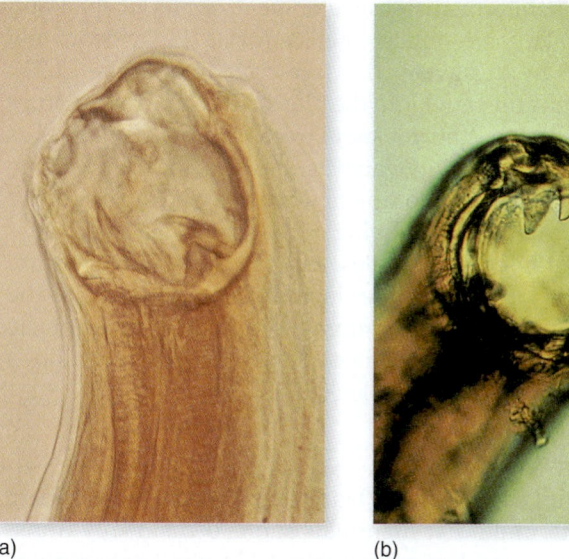

(a)   (b)

Figure 22.26 Cutting teeth on the mouths of hookworms.
(a) *Necator americanus.* **(b)** *Ancylostoma duodenale.*

Disease Table 22.10 Intestinal Distress Accompanied by Migratory Symptoms		
Causative Organism(s)	*Ascaris lumbricoides* (intestinal roundworm)	*Necator americanus* and *Ancylostoma duodenale* (hookworms)
Most Common Modes of Transmission	Cycle A: vehicle (soil/fecal-oral), fomites, self-inoculation	Cycle B: vehicle (soil), fomite
Virulence Factors	Induction of hypersensitivity, adult worm migration, and abdominal obstruction	Induction of hypersensitivity, adult worm migration, and abdominal obstruction
Culture/Diagnosis	Blood count, serology, egg or worm detection	Blood count, serology, egg or worm detection
Prevention	Hygiene	Sanitation
Treatment	Albendazole	Albendazole
Distinctive Features	Roundworm; 1 billion persons infected	Penetrates skin, serious intestinal symptoms
Epidemiological Features	Internationally: 25% prevalence; 80,000 to 100,000 deaths per year	United States: widespread in Southeast until early 20th century; internationally: 800 million infected

worldwide hookworm infections, contributed $18 million to the development of a hookworm vaccine; in 2011, they added another $12 million to this research (**Disease Table 22.10**).

Liver and Intestinal Disease

One group of worms that lands in the intestines has a particular affinity for the liver. Three of these worms are trematodes (flatworms), and they are categorized as liver flukes.

Opisthorchis sinensis and Clonorchis sinensis

Opisthorchis sinensis and *Clonorchis sinensis* are two worms known as Chinese liver flukes. They complete their sexual development in mammals such as humans, cats, dogs, and swine. Their intermediate development occurs in snail and fish hosts. Humans ingest metacercariae in inadequately cooked or raw freshwater fish (see cycle D in figure 22.23). Larvae hatch and crawl into the bile duct, where they mature and shed eggs into the intestinal tract. Feces containing eggs are passed into standing water that harbors the intermediate snail host. The cycle is complete when infected snails release cercariae that invade fish living in the same water.

Symptoms of *Opisthorchis* and *Clonorchis* infection are slow to develop but include thickening of the lining of the bile duct and possible granuloma formation in areas of the liver if eggs enter the stroma of the liver. If the infection is heavy, the bile duct can be blocked.

Fasciola hepatica

This liver fluke (**figure 22.27**) is a common parasite in sheep, cattle, goats, and other mammals and is occasionally transmitted to humans. Periodic outbreaks in temperate regions of Europe and South America are associated with eating wild watercress. The life cycle is very complex, involving the mammal as the definitive host, the release of eggs in the feces, the hatching of eggs in the water into *miracidia*, invasion of freshwater snails, development and release of cercariae, encystment of metacercariae on a water plant, and ingestion of the cyst by a mammalian host eating the plant. The cysts release young flukes into the intestine that wander to the liver, lodge in the gallbladder, and develop into adults. Humans develop symptoms of vomiting, diarrhea, hepatomegaly, and bile obstruction only if they are chronically infected by a large number of flukes (**Disease Table 22.11**).

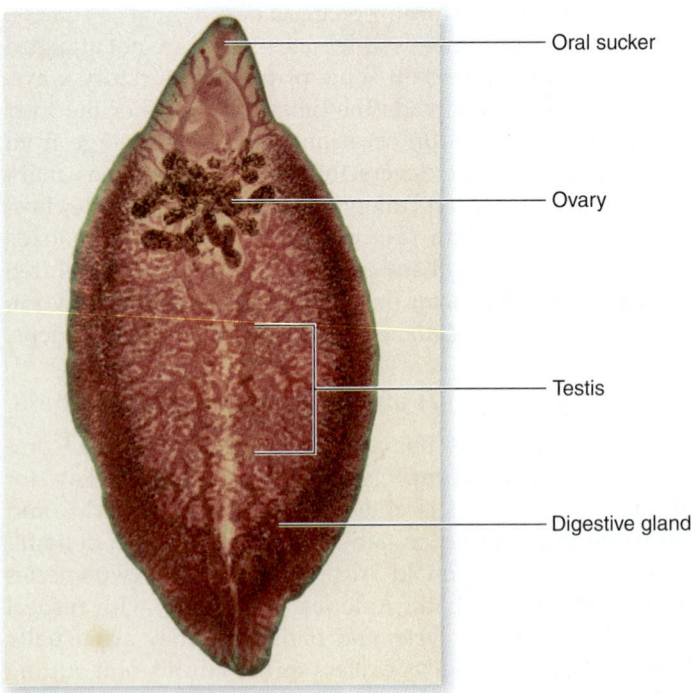

Oral sucker

Ovary

Testis

Digestive gland

Figure 22.27 *Fasciola hepatica*, the sheep liver fluke (2×).

Disease Table 22.11	Liver and Intestinal Disease	
Causative Organism(s)	*Opisthorchis sinensis, Clonorchis sinensis*	*Fasciola hepatica*
Most Common Modes of Transmission	Cycle D: vehicle (fish or crustaceans)	Cycle D: vehicle (water and water plants)
Virulence Factors	–	–
Culture/Diagnosis	Blood count, serology, egg or worm detection	Blood count, serology, egg or worm detection
Prevention	Cook food, sanitation of water	Sanitation of water
Treatment	Praziquantel	Triclabendazole
Distinctive Features	Live in liver	Live in liver and gallbladder
Epidemiological Features	United States: most cases imported; internationally: 56 million infected	

Disease: Muscle and Neurological Symptoms

Trichinosis is an infection transmitted by eating pork (and sometimes other wildlife) that have the cysts of *Trichinella* species embedded in the meat. The life cycle of this nematode is spent entirely within the body of a mammalian host such as a pig, bear, cat, dog, or rat. In nature, the parasite is maintained in an encapsulated (encysted) larval form **(figure 22.28)** in the muscles of these animal reservoirs and is transmitted when other animals prey upon them. The disease cannot be transmitted from one human to another except in the case of cannibalism.

The cyst envelope is digested in the stomach and small intestine, which liberates the larvae. After burrowing into the intestinal mucosa, the larvae reach adulthood and mate. The larvae that result from this union penetrate the intestine and enter the lymphatic channels and blood. All tissues are at risk for invasion, but final development occurs when the coiled larvae are encysted in the skeletal muscle. At maturity, the cyst is about 1 mm long and can be observed by careful inspection of meat. Although larvae can deteriorate over time, they have also been known to survive for years.

Symptoms may be unnoticeable or they could be life-threatening, depending on how many larvae were ingested in the tainted meat. The first symptoms, when present, mimic influenza or viral fevers, with diarrhea, nausea, abdominal pains, fever, and sweating. The second phase, brought on by the mass migration of larvae and their entrance into muscle, produces puffiness around the eyes, intense muscle and joint pain, shortness of breath, and pronounced eosinophilia. The most serious life-threatening manifestations are heart and brain involvement. Although the symptoms eventually subside, a cure is not available once the larvae have encysted in muscles.

The most effective preventive measures for trichinosis are to adequately store and cook pork and wild meats **(Disease Table 22.12)**.

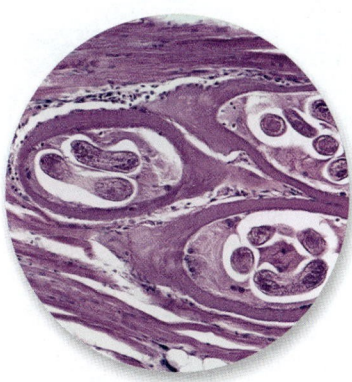

Figure 22.28
Trichinella cysts embedded in pork muscle.

Disease Table 22.12	Muscle and Neurological Symptoms
Causative Organism(s)	*Trichinella* species
Most Common Modes of Transmission	Vehicle (food)
Virulence Factors	–
Culture/Diagnosis	Serology combined with clinical picture; muscle biopsy
Prevention	Cook meat
Treatment	Mebendazole, steroids
Distinctive Features	Brain and heart involvement can be fatal
Epidemiological Features	United States: 20 cases per year; internationally: 10,000 cases per year

Liver Disease

When liver swelling or malfunction is accompanied by eosinophilia, **schistosomiasis** should be suspected. The disease is caused by the blood flukes *Schistosoma mansoni* and *S. japonicum,* species that are morphologically and geographically distinct but share similar life cycles, transmission methods, and general disease manifestations. Schistosomiasis is one of the few infectious agents that can invade intact skin.

▶ Signs and Symptoms

The first symptoms of infection are itchiness in the area where the worm enters the body, followed by fever, chills, diarrhea, and cough. The most severe consequences, associated with chronic infection, are hepatomegaly, liver disease, and splenomegaly. Other serious conditions caused by a different schistosome occur in the urinary tract—bladder obstruction and blood in the urine. This condition is discussed in chapter 23 (genitourinary tract diseases). Occasionally, eggs from the worms are carried into the central nervous system and heart and create a severe granulomatous response. Adult flukes can live for many years and, by eluding the immune defenses, cause a chronic affliction.

▶ Causative Agent

Schistosomes are trematodes, or flukes (see chapter 5), but they are more cylindrical than flat **(figure 22.29).** They are often called blood flukes. Flukes have digestive, excretory, neuromuscular, and reproductive systems, but they lack circulatory and respiratory systems. Humans are the definitive hosts for the blood fluke, and snails are the intermediate host.

▶ Pathogenesis and Virulence Factors

This parasite is clever indeed. Once inside the host, it coats its outer surface with proteins from the host's bloodstream, basically "cloaking" itself from the host defense system. This coat reduces its surface antigenicity and allows it to remain in the host indefinitely.

▶ Transmission and Epidemiology

The life cycle of the schistosome is of the D type and is very complex (as shown in figure 22.29). The cycle begins when infected humans release eggs into irrigated fields or ponds, either by deliberate fertilization with excreta or by defecating or urinating directly into the water. The egg hatches in the water and gives off an actively swimming ciliated larva called a **miracidium,** which instinctively swims to a snail and burrows into a vulnerable site, shedding its ciliated covering in the process. In the body of the snail, the miracidium multiplies into a larger,

fork-tailed swimming larva called a **cercaria.** Cercariae are given off by the thousands into the water by infected snails.

Upon contact with a human wading or bathing in water, cercariae attach themselves to the skin by ventral suckers and penetrate into hair follicles. They pass into small blood and lymphatic vessels and are carried to the liver. Here, the schistosomes achieve sexual maturity, and the male and female worms remain permanently entwined to facilitate mating see **figure 22.29.** In time, the pair migrates to and lodges in small blood vessels at specific sites. *Schistosoma mansoni* and *S. japonicum* end up in the mesenteric venules of the small intestine. While attached to these intravascular sites, the worms feed upon blood, and the female lays eggs that are eventually voided in feces or urine.

The disease is endemic to 74 countries located in Africa, South America, the Middle East, and the Far East. *S. mansoni* is found throughout these regions but not in the Far East.

S. japonicum has a much smaller geographic distribution than *S. mansoni,* only being found in the Far East. Schistosomiasis (including the urinary tract form) is the second most prominent parasitic disease after malaria, probably affecting 200 million people at any one time worldwide. Recent increases in its occurrence in Africa have been attributed to new dams on the Nile River, which have provided additional habitats for snail hosts.

▶ Culture and Diagnosis

Diagnosis depends on identifying the eggs in urine or feces. The clinical pictures of hepatomegaly, splenomegaly, or both also contribute to the diagnosis.

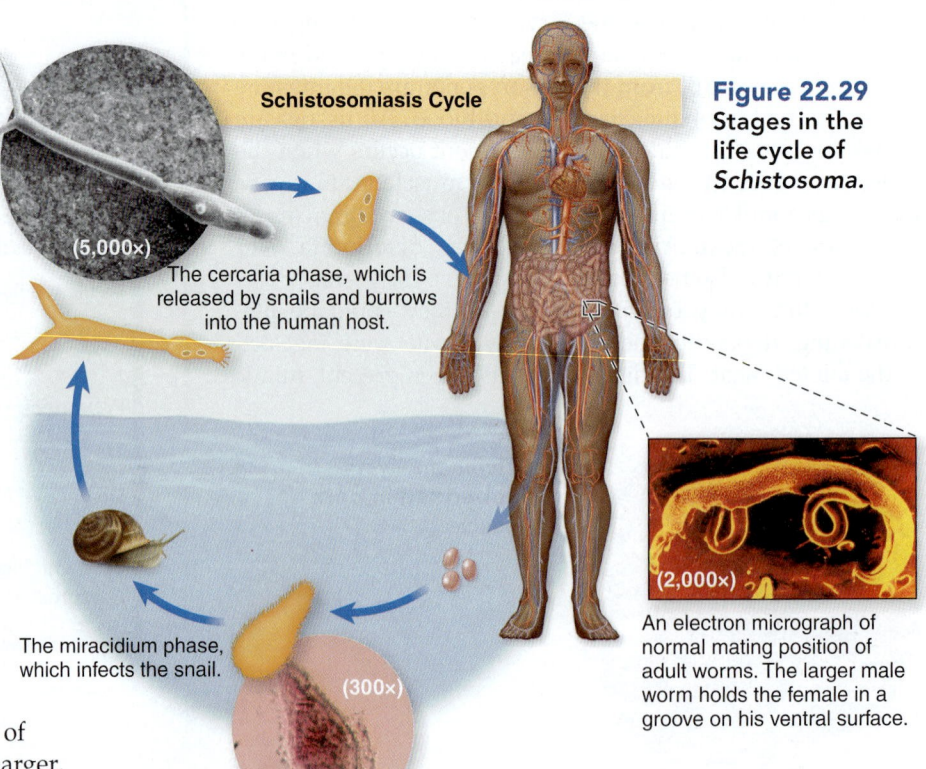

Schistosomiasis Cycle

(5,000×)

The cercaria phase, which is released by snails and burrows into the human host.

The miracidium phase, which infects the snail.

(300×)

Figure 22.29
Stages in the life cycle of *Schistosoma*.

(2,000×)

An electron micrograph of normal mating position of adult worms. The larger male worm holds the female in a groove on his ventral surface.

▶ **Prevention and Treatment**

The cycle of infection cannot be broken as long as people are exposed to untreated sewage in their environment. It is quite common for people to be cured and then to be reinfected because their village has no sewage treatment. A vaccine would provide widespread control of the disease, but so far none is licensed. More than one vaccine is in development, however.

Praziquantel is the drug treatment of choice. It works by crippling the worms, making them more antigenic and thereby allowing the host immune response to eliminate them. Because this is currently the only drug of choice in the treatment of disease today, epidemiologists are focusing their attention on the development of drug-resistant strains. Clinicians use an "egg-hatching test" to determine whether an infection is current and whether treatment is actually killing the eggs. Urine or feces containing eggs are placed in room-temperature water, and if miracidia emerge, the infection is still "active" **(Disease Table 22.13).**

Disease Table 22.13 Liver Disease

Causative Organism(s)	*Schistosoma mansoni, S. japonicum*
Most Common Modes of Transmission	Cycle D: vehicle (contaminated water)
Virulence Factors	Antigenic "cloaking"
Culture/Diagnosis	Identification of eggs in feces, scarring of intestines detected by endoscopy
Prevention	Avoiding contaminated vehicles
Treatment	Praziquantel
Distinctive Features	Penetrates skin, lodges in blood vessels of intestine, damages liver
Epidemiological Features	Internationally: 230 million new infections per year by these and the urinary schistosome

22.4 Learning Outcomes—Assess Your Progress

11. Describe some distinguishing characteristics and commonalities seen in helminthic infections.

12. List four helminths that cause primarily intestinal symptoms, and identify which life cycle each follows and one unique fact about each helminth.

13. List three helminths that cause intestinal symptoms that may be accompanied by migratory symptoms, identifying which life cycle each follows and one unique fact about each helminth.

14. List the modes of transmission for each of the helminthic infections resulting in liver and intestinal symptoms. These are infections caused by *Opisthorchis sinensis, Clonorchis sinensis,* and *Fasciola hepatica.*

15. Describe the type of disease caused by *Trichinella* species.

16. Diagram the life cycle of *Schistosoma mansoni* and *S. japonicum,* and describe the importance of these organisms in world health.

▶ Summing Up

Taxonomic Organization Microorganisms Causing Disease in the GI Tract

Microorganism	Disease	Location of Disease Table
Gram-positive endospore-forming bacteria		
Clostridium difficile	Antibiotic-associated diarrhea	Acute diarrhea, p. 692
Clostridium perfringens	Food poisoning	Acute diarrhea and/or vomiting, p. 694
Bacillus cereus	Food poisoning	Acute diarrhea and/or vomiting, p. 694
Gram-positive bacteria		
Streptococcus mutans	Dental caries	Dental caries, p. 675
Streptococcus sobrinus	Dental caries	Dental caries, p. 675
Staphylococcus aureus	Food poisoning	Acute diarrhea and/or vomiting, p. 694
Gram-negative bacteria		
Campylobacter jejuni	Acute diarrhea	Acute diarrhea, p. 692
Helicobacter pylori	Gastritis/gastric ulcers	Gastritis/gastric ulcers, p. 681
Escherichia coli O157:H7	Acute diarrhea plus hemolytic syndrome	Acute diarrhea, p. 692
Other *E. coli*	Acute or chronic diarrhea	Acute diarrhea, p. 692 Chronic diarrhea, p. 698
Salmonella	Acute diarrhea or typhoid fever	Acute diarrhea, p. 692
Shigella	Acute diarrhea and dysentery	Acute diarrhea, p. 692
Vibrio cholerae	Cholera	Acute diarrhea, p. 692
Tannerella forsythia, Aggregatibacter actinomycetemcomitans, Porphyromonas gingivalis, Treponema vincentii, Prevotella intermedia, Fusobacterium	Periodontal disease	Periodontal disease, p. 677
DNA viruses		
Hepatitis B virus	"Serum" hepatitis	Hepatitis, p. 702
RNA viruses		
Hepatitis A virus	"Infectious" hepatitis	Hepatitis, p. 702
Hepatitis C virus	"Serum" hepatitis	Hepatitis, p. 702
Hepatitis E virus	"Infectious" hepatitis	Hepatitis, p. 702
Mumps virus	Mumps	Mumps, p. 670
Norovirus	Acute diarrhea	Acute diarrhea, p. 692
Rotavirus	Acute diarrhea	Acute diarrhea, p. 692
Protozoa		
Entamoeba histolytica	Chronic diarrhea	Chronic diarrhea, p. 698
Cryptosporidium	Acute diarrhea	Acute diarrhea, p. 692
Cyclospora	Chronic diarrhea	Chronic diarrhea, p. 698
Giardia lamblia	Chronic diarrhea	Chronic diarrhea, p. 698
Helminths—nematodes		
Ascaris lumbricoides	Intestinal distress plus migratory symptoms	Intestinal distress plus migratory symptoms, p. 708
Enterobius vermicularis	Intestinal distress	Intestinal distress, p. 706
Trichuris trichiura	Intestinal distress	Intestinal distress, p. 706
Necator americanus and	Chronic diarrhea	Chronic diarrhea, p. 698
Ancylostoma duodenale	Intestinal distress plus migratory symptoms	Intestinal distress plus migratory symptoms, p. 708
Trichinella spp.	Muscle and neurological symptoms	Muscle and neurological symptoms, p. 710
Helminths—cestodes		
Diphyllobothrium latum	Intestinal distress	Intestinal distress, p. 706
Opisthorchis sinensis and		
Clonorchis sinensis	Liver and intestinal disease	Liver and intestinal disease, p. 709
Fasciola hepatica	Liver and intestinal disease	Liver and intestinal disease, p. 709
Helminths—trematodes		
Schistosoma mansoni, S. japonicum	Schistosomiasis	Helminthic liver disease, p. 711

INFECTIOUS DISEASES AFFECTING
The Gastrointestinal Tract

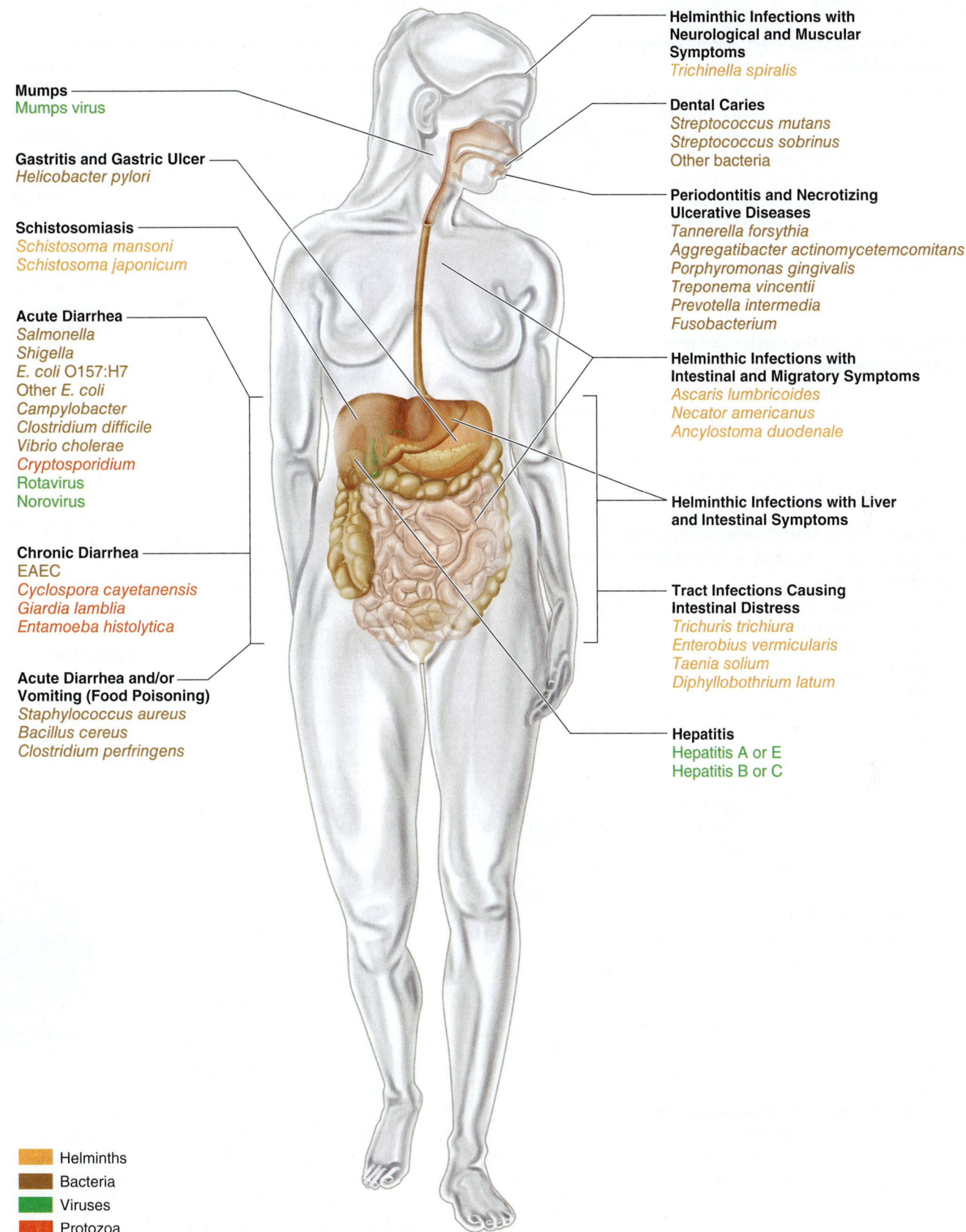

Mumps
Mumps virus

Gastritis and Gastric Ulcer
Helicobacter pylori

Schistosomiasis
Schistosoma mansoni
Schistosoma japonicum

Acute Diarrhea
Salmonella
Shigella
E. coli O157:H7
Other *E. coli*
Campylobacter
Clostridium difficile
Vibrio cholerae
Cryptosporidium
Rotavirus
Norovirus

Chronic Diarrhea
EAEC
Cyclospora cayetanensis
Giardia lamblia
Entamoeba histolytica

**Acute Diarrhea and/or
Vomiting (Food Poisoning)**
Staphylococcus aureus
Bacillus cereus
Clostridium perfringens

**Helminthic Infections with
Neurological and Muscular
Symptoms**
Trichinella spiralis

Dental Caries
Streptococcus mutans
Streptococcus sobrinus
Other bacteria

**Periodontitis and Necrotizing
Ulcerative Diseases**
Tannerella forsythia
Aggregatibacter actinomycetemcomitans
Porphyromonas gingivalis
Treponema vincentii
Prevotella intermedia
Fusobacterium

**Helminthic Infections with
Intestinal and Migratory Symptoms**
Ascaris lumbricoides
Necator americanus
Ancylostoma duodenale

**Helminthic Infections with Liver
and Intestinal Symptoms**

**Tract Infections Causing
Intestinal Distress**
Trichuris trichiura
Enterobius vermicularis
Taenia solium
Diphyllobothrium latum

Hepatitis
Hepatitis A or E
Hepatitis B or C

Helminths
Bacteria
Viruses
Protozoa

System Summary Figure 22.30

Chapter Summary

22.1 The Gastrointestinal Tract and Its Defenses (ASM Guidelines* 3.4, 5.4, 6.4)

- The gastrointestinal (GI) tract is composed of *eight* main sections—the mouth, pharynx, esophagus, stomach, small intestine, large intestine, rectum, and anus; and *four* accessory organs—the salivary glands, liver, gallbladder, and pancreas.
- The GI tract has a very heavy load of microorganisms, and it encounters millions of new ones every day. There are significant mechanical, chemical, and antimicrobial defenses to combat microbial invasion.

22.2 Normal Biota of the Gastrointestinal Tract (ASM Guidelines 3.4, 5.4, 6.4)

- Bacteria abound in all of the eight main sections of the gastrointestinal tract. Even the highly acidic stomach is colonized.
- Normal biota provide more functions in the GI tract than ever recognized before. A diverse biota seems to be necessary for overall health.

22.3 Gastrointestinal Tract Diseases Caused by Microorganisms (Nonhelminthic) (ASM Guidelines 5.3, 5.4, 6.4, 8.3)

- **Tooth and Gum Infections:** Alpha-hemolytic *Streptococcus mutans* and *Streptococcus sobrinus* are main causes of dental caries.
- **Dental Caries (Tooth Decay):** This is the most common infectious disease of human beings.
- **Periodontal Disease:** Some form of this disease affects nearly everyone.
  - Periodontitis: The anaerobic bacteria *Tannerella forsythia* (formerly *Bacteroides forsythus*), *Aggregatibacter actinomycetemcomitans*, *Porphyromonas*, *Fusobacterium*, and spirochete species are causative agents.
  - Necrotizing Ulcerative Gingivitis and Periodontitis: Necrotizing ulcerative gingivitis (NUG) and necrotizing ulcerative periodontitis (NUP) are synergistic infections involving *Treponema vincentii*, *Prevotella intermedia*, and *Fusobacterium* species.
- **Mumps:** Swelling of the salivary gland—a condition called parotitis. Mumps is caused by an enveloped, single-stranded RNA virus (mumps virus) from the genus *Paramyxovirus*.
- **Gastritis and Gastric Ulcers:** Gastritis: sharp or burning pain emanating from the abdomen. Gastric ulcers: actual lesions in the mucosa of the stomach (gastric ulcers) or in the uppermost portion of the small intestine (duodenal ulcer). *Helicobacter pylori*, a curved gram-negative rod, is causative agent.

- **Acute Diarrhea (With or Without Vomiting):** In the United States, a third of all acute diarrhea is transmitted by contaminated food.
  - *Salmonella: Salmonella enteritidis* is divided into many serotypes, based on major surface antigens. Animal and dairy products are often contaminated with the bacterium. Typhoid fever, caused by *S. enteritidis* variant Typhi is a progressive, invasive infection that can lead to septicemia.
  - *Shigella: Shigella* species give symptoms of frequent, watery, bloody stools; fever; and often intense abdominal pain. Diarrhea containing blood and mucus is also called dysentery. The bacterium *Shigella dysenteriae* produces a heat-labile exotoxin called shiga toxin.
  - Shiga-Toxin-Producing *E. coli* (STEC): Dozens of different strains of *E. coli* exist. *E. coli* O157:H7 and its close relatives are most virulent. This group of *E. coli* is referred to as shiga-toxin-producing *E. coli* (STEC). These *E. coli* are the agent of a spectrum of conditions, ranging from mild gastroenteritis with fever to bloody diarrhea. About 10% of patients develop hemolytic uremic syndrome (HUS), a severe hemolytic anemia that can cause kidney damage and failure. Virulence is due to shiga toxins.
  - Other *E. coli*: At least five other categories of *E. coli* cause diarrheal diseases. These are enterotoxigenic *E. coli* (traveler's diarrhea), enteroinvasive *E. coli*, enteropathogenic *E. coli*, enteroaggregative *E. coli*, and diffusely adherent *E. coli*.
  - *Campylobacter:* Symptoms are frequent watery stools, fever, vomiting, headaches, and severe abdominal pain. Infrequently, infection can lead to serious neuromuscular paralysis called *Guillain-Barré syndrome*.
  - *Clostridium difficile* causes a condition called pseudomembranous colitis (antibiotic-associated colitis), precipitated by therapy with broad-spectrum antibiotics.
  - *Vibrio cholerae:* Symptoms of secretory diarrhea and severe fluid loss can lead to death in less than 48 hours. Produces enterotoxin called cholera toxin (CT), which disrupts the normal physiology of intestinal cells.
  - *Cryptosporidium:* Intestinal waterborne protozoan that infects mammals, birds, and reptiles.
  - Rotavirus: Common cause of diarrhea that is most dangerous to children ages 6 to 24 months.
  - Norovirus: Most common cause of food-borne illness in the United States; also transmitted via fecal-oral route.
- **Acute Diarrhea with Vomiting Caused by Exotoxins (Food Poisoning):** Food poisoning refers to symptoms in the gut that are caused by a preformed toxin.

*Source: ASM Curriculum Guidelines (American Society for Microbiology, 2012). Complete guidelines in appendix B of this book.

- *Staphylococcus aureus* Exotoxin: Heat-stable enterotoxin requires 100°C for 30 minutes for inactivation. Ingested toxin acts on gastrointestinal epithelium and stimulates nerves; acute symptoms of cramping, nausea, vomiting, and diarrhea.
- *Bacillus cereus* Exotoxin: *B. cereus* is common resident on vegetables and soil. Produces two exotoxins; one causes a diarrheal-type disease, the other causes an emetic disease.
- *Clostridium perfringens* Exotoxin: The toxin initiates acute abdominal pain, diarrhea, and nausea in 8 to 16 hours.
- **Chronic Diarrhea:** Chronic diarrhea lasts longer and is generally less severe than acute diarrhea.
  - Enteroaggregative *E. coli* (EAEC): EAEC is particularly associated with chronic disease, especially in children. Transmission is through contaminated food and water.
  - *Cyclospora:* *C. cayetanensis* is a protozoan transmitted via the fecal-oral route; associated with fresh produce and water.
  - *Giardia:* *G. lamblia* is a protozoan that can cause diarrhea of long duration, abdominal pain, and flatulence. Freshwater is common vehicle of infection.
  - *Entamoeba:* *E. histolytica* is a freshwater protozoan that causes intestinal amoebiasis, targeting the cecum, appendix, colon, and rectum, leading to dysentery, abdominal pain, fever, diarrhea, and weight loss.
- **Hepatitis:** Inflammatory disease marked by necrosis of hepatocytes and a mononuclear response that swells and disrupts the liver, causing jaundice. Can be caused by a variety of different viruses.
  - Hepatitis A Virus (HAV): A nonenveloped, single-stranded RNA enterovirus of low virulence. Spread through fecal-oral route. Inactivated vaccine available.
  - Hepatitis B Virus (HBV): Enveloped DNA virus in the family *Hepadnaviridae*. Can be very serious, even life-threatening; some patients develop chronic liver disease in the form of necrosis or cirrhosis. Also associated with hepatocellular carcinoma. Some patients infected with hepatitis B are coinfected with the delta agent, sometimes also called hepatitis D Virus. HBV transmitted by blood and other bodily fluids. Virus is major infectious concern for health care workers.
  - Hepatitis C Virus: RNA virus in *Flaviviridae* family. Shares characteristics of hepatitis B disease, but is much more likely to become chronic. More commonly transmitted through blood contact than through other body fluids.

22.4 Gastrointestinal Tract Diseases Caused by Helminths (ASM Guidelines 5.3, 5.4, 6.4, 8.3)

- **General Clinical Considerations:** Helminths are low on virulence factors and high on the ability to persist in the host.

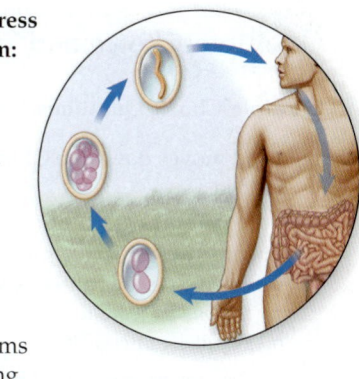

- **Disease: Intestinal Distress as the Primary Symptom:** Both tapeworms and roundworms can infect intestinal tract in such a way as to cause primary symptoms there.
  - *Trichuris trichiura:* Symptoms may include localized hemorrhage of the bowel, caused by worms burrowing and piercing intestinal mucosa.
  - *Enterobius vermicularis:* "Pinworm"; most common worm disease of children in temperate zones. Not fatal, and most cases are asymptomatic.
  - *Taenia solium:* This tapeworm transmitted to humans by raw or undercooked pork. Other tapeworms of the genus *Taenia,* such as the beef tapeworm *Taenia saginata,* infect humans.
  - *Diphyllobothrium latum:* The intermediate host is fish; can be transmitted in raw food such as sushi and sashimi made from salmon.
- **Disease: Intestinal Distress Accompanied by Migratory Symptoms:** These helminths damage tissues as they move from the intestines to the circulatory system tissues and back to the intestines.
  - *Ascaris lumbricoides:* Intestinal roundworm that releases eggs in feces; eggs then spread to other humans through fecal-oral routes.
  - *Necator americanus* and *Ancylostoma duodenale:* Both called by the common name "hookworm." Hookworm larvae hatch outside the body in soil contaminated with feces and infect by penetrating skin.
- **Liver and Intestinal Disease:** One group of worms has a particular affinity for the liver—liver flukes.
  - *Opisthorchis sinensis* and *Clonorchis sinensis:* Humans infected by eating inadequately cooked or raw freshwater fish and crustaceans.
  - *Fasciola hepatica:* Common parasite in sheep, cattle, goats, and other mammals. Humans develop symptoms only if chronically infected by a large number of flukes.
- **Disease: Muscle and Neurological Symptoms:** When this helminth infects the brain, serious neurological consequences ensue.
  - Trichinosis: Transmitted by eating undercooked pork that has cysts of *Trichinella* embedded in the meat.
- **Liver Disease:** Blood flukes cause disease in the digestive and urinary systems.
  - Schistosomiasis in intestines is caused by blood flukes *Schistosoma mansoni* and *S. japonicum.* Symptoms include fever, chills, diarrhea, liver and spleen disease.

Multiple-Choice and True-False Questions │ Bloom's Levels 1 and 2: Remember and Understand

Multiple-Choice Questions. Select the correct answer from the options provided.

1. Food moves down the GI tract through the action of
   a. cilia.
   b. peristalsis.
   c. gravity.
   d. microorganisms.

2. The microorganism(s) most associated with acute necrotizing ulcerative periodontitis (ANUP) is (are)
   a. *Treponema vincentii.*
   b. *Prevotella intermedia.*
   c. *Fusobacterium.*
   d. all of the above.

3. Gastric ulcers are caused by
   a. *Treponema vincentii.*
   b. *Prevotella intermedia.*
   c. *Helicobacter pylori.*
   d. all of the above.

4. Virus family *Paramyxoviridae* contains viruses that cause which of the following diseases?
   a. measles
   b. mumps
   c. influenza
   d. both a and b
   e. both b and c

5. Which of these microorganisms is considered the most common cause of diarrhea (not just food-borne) in the United States?
   a. *E. coli*
   b. *Salmonella*
   c. *Campylobacter*
   d. *Shigella*

6. Which of these microorganisms is associated with Guillain-Barré syndrome?
   a. *E. coli*
   b. *Salmonella*
   c. *Campylobacter*
   d. *Shigella*

7. This microorganism is commonly associated with fried rice and produces an emetic (vomiting) toxin.
   a. *Bacillus cereus*
   b. *Clostridium perfringens*
   c. *Shigella*
   d. *Staphylococcus aureus*

8. This endospore former contaminates meats as well as vegetables and is also the causative agent of gas gangrene.
   a. *Bacillus cereus*
   b. *Clostridium perfringens*
   c. *Shigella*
   d. *Staphylococcus aureus*

9. This hepatitis virus is an enveloped DNA virus.
   a. hepatitis A virus
   b. hepatitis B virus
   c. hepatitis C virus
   d. hepatitis E virus

10. In which helminth life cycle is a grazing animal involved?
    a. A
    b. B
    c. C
    d. D

True-False Questions. If the statement is true, leave as is. If it is false, correct it by rewriting the sentence.

11. Mumps is a disease that affects humans and several other species.

12. *Giardia lamblia* is a water-borne, flagellated protozoan often associated with chronic diarrhea.

13. Pseudomembranous colitis (or antibiotic-associated colitis) is caused by *Clostridium difficile.*

14. Poor oral health has been associated with heart disease.

15. *Enterobius vermicularis,* commonly known as the pinworm, is a common cause of anal itching in young children in the United States.

Critical Thinking Questions │ Bloom's Levels 3, 4, and 5: Apply, Analyze, and Evaluate

Critical thinking is the ability to reason and solve problems using facts and concepts. These questions can be approached from a number of angles and, in most cases, they do not have a single correct answer.

1. a. Summarize the current knowledge of normal biota in the human mouth. To what domain of life do most of these microorganisms belong?
   b. Explain why many patients undergoing dental surgery are placed on antibiotic therapy before the procedure.

2. List the microorganisms involved in tooth decay and discuss the sequence of events leading to the development of periodontitis.

3. Provide evidence in support of or refuting the following statement: Blood type can help predict your risk of developing ulcers.

4. a. Summarize the characteristics that differentiate food poisoning from other gastrointestinal diseases.
   b. Conduct additional research and discuss at least two new methods for detecting or eliminating microbial contaminants in the global food supply today.

5. a. In May of 2010, the CDC began to notice a sudden increase in the number of salmonellosis cases reported that month. That trend continued until July 2010, when the number of cases began to slowly decline. Explain what type of epidemic these data represent, and why the number of cases finally went down after 3 months.

   b. The above cases turned out to be linked to *Salmonella* contamination found at egg-producing facilities in Iowa. Nearly 2,000 people became ill in multiple states. Explain how proper cooking of the eggs involved in this outbreak did not ensure elimination of the food-borne pathogen.

6. a. An outbreak of cholera occurred in Haiti following a devastating earthquake in 2010. Based upon your knowledge of the bacterium involved, discuss the factors that may have allowed for the outbreak to develop within this country.
   b. Explain why many recreational water parks have chosen to use ultraviolet light filtration systems to effectively treat their water supply.

7. Create a chart comparing the modes of transmission and means of prevention for all types of hepatitis viruses.

8. a. Describe the methods used to definitively diagnose helminthic infections in humans.
   b. Explain why antihelminthic drugs are so difficult to develop, and list at least three therapeutic targets of successful drugs used today.

9. Regarding food safety,
   a. Explain whether there is a greater risk for *E. coli* O157:H7 infection when consuming a hamburger compared to consuming a steak.
   b. Discuss whether or not food poisoning can still occur after consuming a reheated pot of soup known to be contaminated with *Staphylococcus aureus*.

   c. Summarize the microbial risks associated with consuming raw seafood or shellfish today.
10. One hundred individuals present to a health care clinic complaining of gastrointestinal distress. As the head triage physician, please detail all steps you will take to definitively determine what disease the patients are suffering from and to identify the source of the apparent outbreak.

Concept Connections | Bloom's Levels 4 and 6: Analyze and Create

This activity ties together multiple concepts in the chapter.

You have recently finished your nursing degree and are working at a health care clinic. A patient presents with symptoms of hepatitis. He also reports that he is a health care worker.

1. What are the general symptoms of hepatitis?
2. As a clinician, what is your first assumption about the source of the infection?
3. What virus(es) is (are) potentially involved through this source?
4. What treatments, if any, are available?

The patient further reports that he has been fully vaccinated against hepatitis as a requirement for his employment and has had no needlestick injuries.

5. Knowing that the patient has been vaccinated, what hepatitis-causing virus can be eliminated as a cause of this patient's hepatitis?
6. What is the potential source of infection?
7. What treatments, if any, are available?
8. What other vaccination might you recommend to this patient?

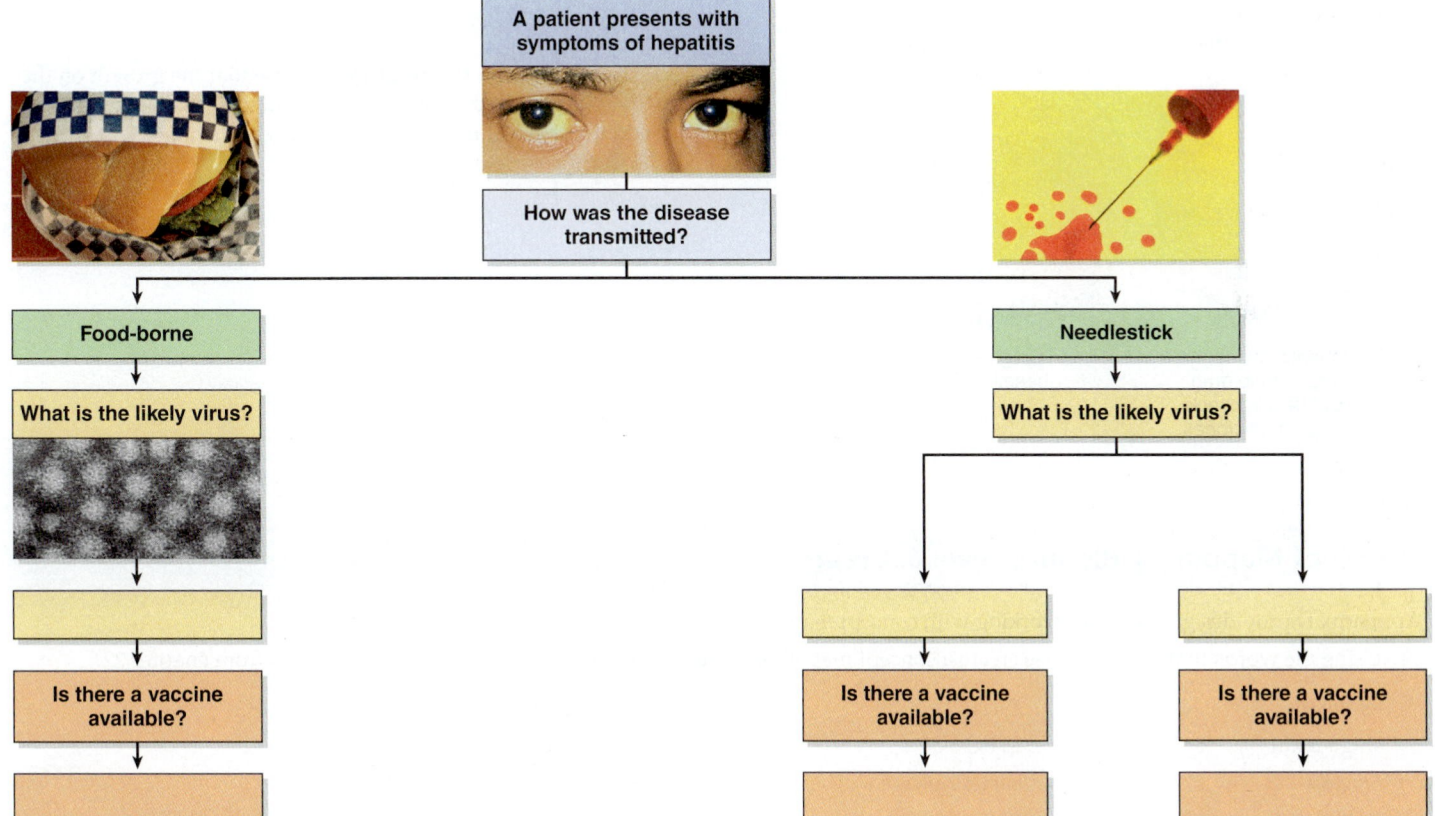

Visual Connections | Bloom's Level 5: Evaluate

These questions use visual images or previous content to make connections to this chapter's concepts.

1. **From chapter 13, figure 13.6b.** Imagine for a minute that the organism in this illustration is *E. coli* O157:H7. What would be one reason to *not* treat a patient having this infection with powerful antibiotics?

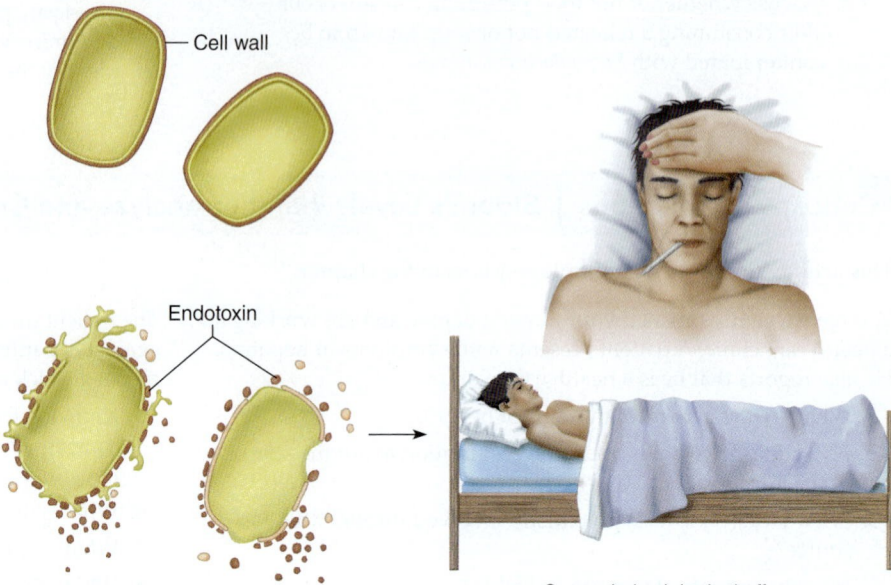

Cell wall

Endotoxin

General physiological effects—fever, malaise, aches, shock

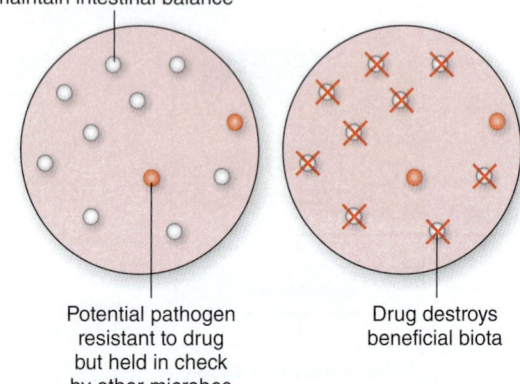

Normal biota important to maintain intestinal balance

Potential pathogen resistant to drug but held in check by other microbes

Drug destroys beneficial biota

2. **From chapter 12, figure 12.15.** Assume that the growth on the first plate represents normal intestinal microbiota. How could you use these illustrations to explain the development of *C. difficile*–associated colitis?

Concept Mapping | Bloom's Level 6: Create

Appendix D provides guidance for working with concept maps.

1. Using the words that follow, please create a concept map illustrating the relationships among these key terms from chapter 22.

exotoxins	*E. coli*	transduction	EAEC
shiga toxin	STEC	protein synthesis	
Shigella	bacteriophage	EIEC	

www.mcgrawhillconnect.com

Enhance your study of this chapter with study tools and practice tests. Also ask your instructor about the resources available through ConnectPlus, including the media-rich eBook, interactive learning tools, and animations.

Infectious Diseases Affecting the Genitourinary System

Case File 23

HPV Vaccine for All?

In 2006, the FDA licensed Gardasil, a vaccine that protects against four strains of the human papillomavirus (HPV) that cause cervical cancer in women. In 2009, another vaccine, Cervarix, was licensed that protects against two strains of HPV that cause cervical cancer. In 2013, the CDC released data showing that the HPV infection rate among girls ages 14 to 19 has been reduced by 56% in the years since the vaccine was introduced. This was true even though only one-third of girls in the recommended age group had received their full set of vaccinations. The protection of unvaccinated women represents a type of **herd immunity,** but according to guidelines from the CDC, all women ages 11 to 26 should receive both vaccines to protect against cervical cancer.

- The human papillomavirus is often associated with warts. How does it cause cervical cancer?
- Why should young women and preteens receive the HPV vaccines?
- What are the recommendations for males to receive the HPV vaccine(s)?

Continuing the Case appears on page 737.

Outline and Learning Outcomes

23.1 The Genitourinary Tract and Its Defenses
1. Draw or describe the anatomical features of the genitourinary tracts of both genders.
2. List the natural defenses present in the genitourinary tracts.

23.2 Normal Biota of the Genitourinary Tract
3. List the types of normal biota presently known to occupy the genitourinary tracts of both genders.
4. Summarize how the microbiome of the female reproductive tract changes over time.

23.1 The Genitourinary Tract and Its Defenses

As suggested by the name, the structures considered in this chapter are really two distinct organ systems. The *urinary tract* has the job of removing substances from the blood, regulating certain body processes, and forming urine and transporting it out of the body. The *genital system* has reproduction as its major function. It is also called the *reproductive system.*

The urinary tract includes the kidneys, ureters, bladder, and the urethra **(figure 23.1).** The kidneys remove metabolic wastes from the blood, acting as a sophisticated filtration system. Ureters are tubular organs extending from each kidney to the bladder. The bladder is a collapsible organ that stores urine and empties it into the urethra, which is the conduit of urine to the exterior of the body. In males, the urethra is also the terminal organ of the reproductive tract, but in females the urethra is separate from the vagina, which is the outermost organ of the reproductive tract.

The most obvious defensive mechanism is the flushing action of the urine flowing out of the system. The flow of urine also encourages the **desquamation** (shedding) of the epithelial cells lining the urinary tract. For example, each time a person urinates, he or she loses hundreds of thousands of epithelial cells! Any microorganisms attached to them are also shed, of course. Probably the most common microbial threat to the urinary tract is the group of microorganisms that comprise the normal biota in the gastrointestinal tract, because the two organ systems are in close proximity. But the cells of the epithelial lining of the urinary tract have different chemicals on their surfaces than do those lining the GI tract. For that reason, most bacteria that are adapted to adhere to the chemical structures in the GI tract cannot gain a foothold in the urinary tract.

Urine, in addition to being acidic, also contains two antibacterial proteins, lysozyme and lactoferrin. You may recall that lysozyme is an enzyme that breaks down peptidoglycan. Lactoferrin is an iron-binding protein that inhibits bacterial growth. Finally, secretory IgA specific for previously encountered microorganisms can be found in the urine.

The male reproductive system produces, maintains, and transports sperm cells and is the source of male sex hormones. It consists of the *testes,* which produce sperm cells and hormones, and the *epididymides,* which are coiled tubes leading out of the testes. Each epididymis terminates in a *vas deferens,* which combines with the seminal vesicle and terminates in the ejaculatory duct **(figure 23.2).** The contents of the ejaculatory duct empty into the urethra during

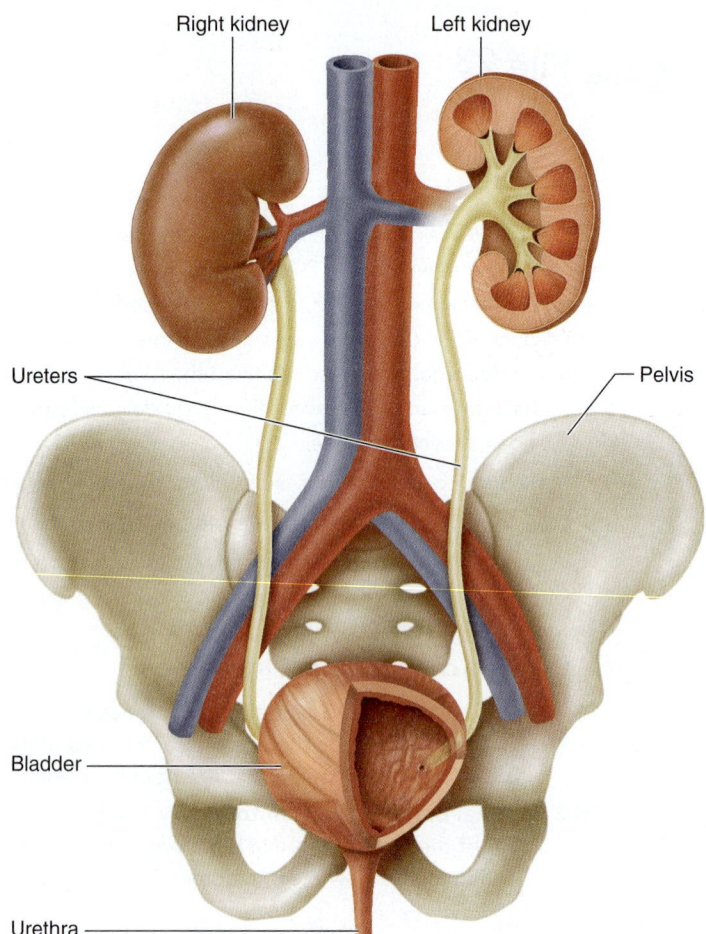

Figure 23.1 **The urinary system.**

Male Cross Section

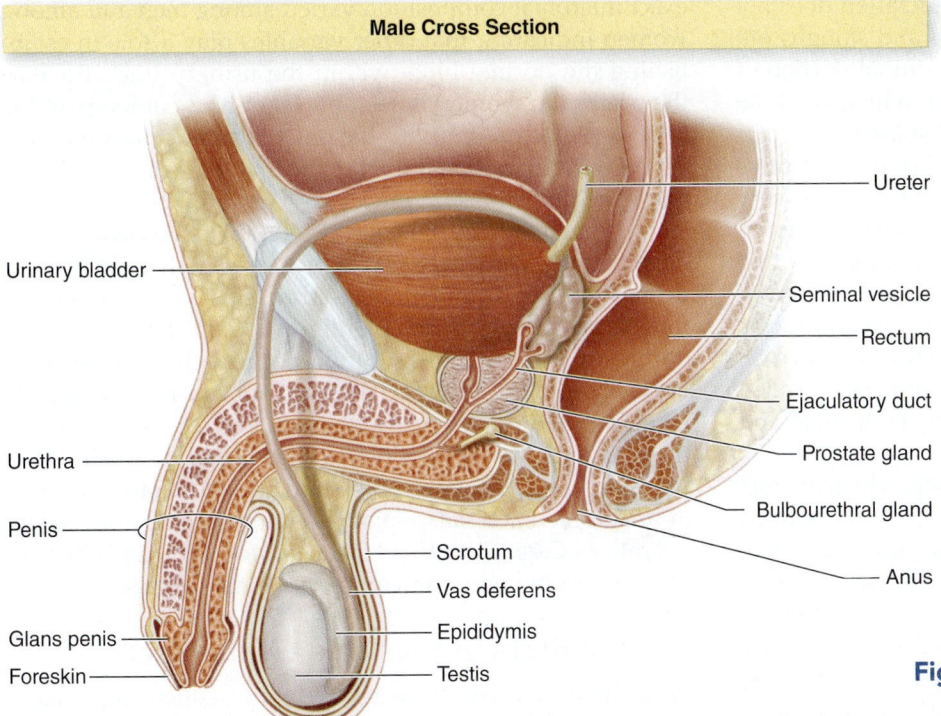

Ureter

Urinary bladder

Seminal vesicle

Rectum

Ejaculatory duct

Urethra

Prostate gland

Penis

Bulbourethral gland

Scrotum

Anus

Vas deferens

Glans penis

Epididymis

Foreskin

Testis

Figure 23.2 The male reproductive system.

Female Cross Section

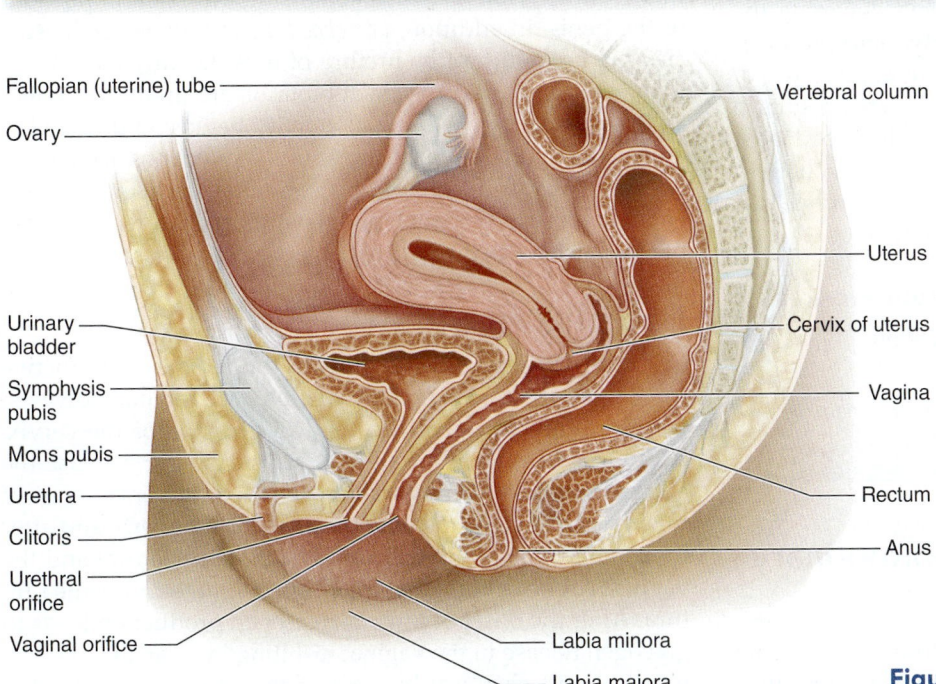

Fallopian (uterine) tube

Ovary

Vertebral column

Uterus

Urinary bladder

Cervix of uterus

Symphysis pubis

Vagina

Mons pubis

Urethra

Clitoris

Rectum

Urethral orifice

Anus

Vaginal orifice

Labia minora

Labia majora

Figure 23.3 The female reproductive system.

ejaculation. The *prostate gland* is a walnut-shaped structure at the base of the urethra. It also contributes to the released fluid (semen). The external organs are the scrotum, containing the testes, and the *penis,* a cylindrical organ that houses the urethra. As for its innate defenses, the male reproductive system also benefits from the flushing action of the urine, which helps move microorganisms out of the system.

The female reproductive system consists of the *uterus,* the *fallopian tubes* (also called uterine tubes), *ovaries,* and *vagina* **(figure 23.3).** During childbearing years, an egg is released from one of the ovaries approximately every 28 days. It enters the fallopian tubes, where fertilization by sperm may take place if sperm are present. The fertilized egg moves through the fallopian tubes to the uterus, where it is implanted in the uterine lining. If fertilization does not occur, the lining of the uterus degenerates and sloughs off; this is the process of menstruation. The terminal portion of the female reproductive tract is the vagina, which is a tube about 9 cm long. The vagina is the exit tube for fluids from the uterus, the channel for childbirth, and the receptive chamber for the penis during sexual intercourse. One very important tissue of the female reproductive tract is the *cervix*, which is the lower one-third of the uterus and the part that connects to the vagina. The opening of the uterus is part of the cervix. The cervix is a common site of infection in the female reproductive tract. Ongoing research has shown an innate protective effect against viral infection from so-called restriction factors produced by cells lining both male and female reproductive tracts.

The natural defenses of the female reproductive tract vary over the lifetime of the woman. The vagina is lined with mucous membranes and, thus, has the protective covering of secreted mucus. During childhood and after menopause, this mucus is the major nonspecific defense of this system. Secretory IgA antibodies specific for any previously encountered infections would be present on these surfaces. During a woman's reproductive years, a major portion of the defense is provided by changes in the pH of the vagina brought about by the release of estrogen. This hormone stimulates the vaginal mucosa to secrete glycogen, which certain bacteria can ferment into acid, lowering the pH of the vagina to about 4.5. Before puberty, a girl produces little estrogen and little glycogen and has a vaginal pH of about 7. The change in pH beginning in adolescence results in a vastly different normal biota in the vagina, described later. The biota of women in their childbearing years is thought to prevent the establishment and invasion of microbes that might have the potential to harm a developing fetus.

23.1 Learning Outcomes—Assess Your Progress

1. Draw or describe the anatomical features of the genitourinary tracts of both genders.
2. List the natural defenses present in the genitourinary tracts.

23.2 Normal Biota of the Genitourinary Tract

In both genders, the outer region of the urethra harbors some normal biota. The kidney, ureters, bladder, and upper urethra were previously thought to be sterile. However, recent data suggest that some of these areas may actually contain microbiota that are simply unculturable using currently available methods. Genomic analysis of aseptically obtained urine samples from women showed the presence of a variety of microorganisms. These included known residents of the urethra (nonhemolytic streptococci, staphylococci, corynebacteria, and some lactobacilli) and additionally *Prevotella*, *Veillonella*, and *Gardnerella* species. However, the exact microbial composition varied among men and among women indicating that other variables play a role in establishing the normal biota within the urinary tract. Because the urethra in women is so short (about 3.5 cm long) and is in such close proximity to the anus, it can act as a pipeline for bacteria from the GI tract to the bladder, resulting in urinary tract infections. In men, removal of the penile foreskin triggers a change in the composition of the known normal biota on the outer surface of the penis and perhaps in the urethra as well. In an uncircumcised penis, the area under the foreskin is colonized by several microbial species, the most dominant being anaerobic gram-negative bacteria such as *Prevotella* and *Clostridiales*. These microbes tend to stimulate an inflammatory response at the skin surface and may make these men more susceptible to infections, especially HIV, which infects immune cells. Circumcision leads to a decrease in anaerobic bacterial populations and microbial diversity overall, as *Corynebacterium*, *Pseudomonas*, and *Staphylococcus* species become predominant in these individuals.

Normal Biota of the Male Genital Tract

With the easy access to whole-genome sequencing, a close inspection of the male genital tract microbiome is now underway. Recent studies have revealed that the normal biota of the male genital tract (that is, in the urethra) is composed of many of the same residents colonizing the external portions of the penis. In addition, *Lactobacillus* and *Streptococcus* species can be found in the urethra of most healthy men. What is interesting is that the microbiota of the urethra in males apparently shifts once sexual activity begins, and microbes associated with sexually transmitted infections (STIs) can begin to take up residence in the genital tract. In fact, men engaging in vaginal, anal, or even oral intercourse can often harbor bacteria that produce bacterial vaginosis in females.

Normal Biota of the Female Genital Tract

In the uterus and other organs of the female reproductive tract above it, there is no normal biota. Microbiota, however, were recently identified on the exterior side of the cervix. Although a new finding, it was not surprising because the adjacent vaginal canal is colonized by a diverse array of microorganisms. As just mentioned, before puberty and after menopause, the pH of the vagina is close to neutral and the vagina harbors a biota that is similar to that found in the urethra. After the onset of puberty, estrogen production leads to glycogen release in the vagina, resulting in an acidic pH. The physical and chemical barriers of the vagina select for the growth of normal biota such as *Lactobacillus* species, which thrive in the acidic environment. But these microbes also contribute to the low pH environment, converting sugars to acid. Their predominance in the vagina, combined with the acidic environment, discourages the growth of many microorganisms and actually plays a major role in developing the overall composition of the vaginal biota. Even though the *Lactobacilli* dominate the normal biota of the vagina in most women, studies show that this is not the case in all women.

Genitourinary Tract Defenses and Normal Biota

	Defenses	Normal Biota
Urinary Tract (Both Genders)	Flushing action of urine; specific attachment sites not recognized by most nonnormal biota; shedding of urinary tract epithelial cells, secretory IgA, lysozyme, and lactoferrin in urine	Nonhemolytic *Streptococcus, Staphylococcus, Corynebacterium, Lactobacillus, Prevotella, Veillonella, Gardnerella*
Female Genital Tract (Childhood and Postmenopausal)	Mucus secretions, secretory IgA	Same as for urinary tract
Female Gential Tract (Childbearing Years)	Acidic pH, mucus secretions, secretory IgA	Predominantly *Lactobacillus* but also *Candida*
Male Genital Tract	Same as for urinary tract	Urethra: same as for urinary tract; outer surface of penis: *Pseudomonas* and *Staphylococcus*; sulcus of uncircumcised penis: anaerobic gram-negatives

Others show higher percentages of anaerobic bacteria such as *Prevotella, Sneathia,* or *Streptococcus species.* Scientists have shown that the microbial makeup can actually shift dramatically during the menstrual cycle and during pregnancy and changes can even occur over just a few days. All of this information reflects the fact that there is no "core" or common vaginal biota composition during childbearing years and this microbiome is not always stable. Future studies will provide a better understanding of how these microorganisms maintain a healthy, disease-free vaginal canal over time.

The estrogen-glycogen effect continues throughout the childbearing years until menopause. Genomic techniques have led to new findings about the normal biota in postmenopausal women. In contrast to women in their childbearing years, the normal biota composition in postmenopausal women appears to be stable over time. Although *Lactobacillus* and *Gardnerella* species are still common, there is a drop in other characteristic microbial species seen in premenopausal women. It was also noted that the number of *Lactobacilli* decreases as vaginal dryness increases, which opens a new door to investigate shifts in microbial composition in women suffering from this common symptom of menopause. Note that the very common fungus *Candida albicans* is also present at low levels in the healthy female reproductive tract.

23.2 Learning Outcomes—Assess Your Progress

3. List the types of normal biota presently known to occupy the genitourinary tracts of both genders.
4. Summarize how the microbiome of the female reproductive tract changes over time.

23.3 Urinary Tract Diseases Caused by Microorganisms

We consider two types of diseases in this section. **Urinary tract infections (UTIs)** result from invasion of the urinary system by bacteria or other microorganisms. Leptospirosis, by contrast, is a spirochete-caused disease transmitted by contact of broken skin or mucous membranes with contaminated animal urine.

Urinary Tract Infections (UTIs)

Even though the flushing action of urine helps to keep infections to a minimum in the urinary tract, urine itself is a good growth medium for many microorganisms. When urine flow is reduced or bacteria are accidentally introduced into the bladder, an infection of that organ (known as *cystitis*) can occur. Occasionally, the infection can also affect the kidneys, in which case it is called *pyelonephritis*. If an infection is limited to the urethra, it is called *urethritis*.

▶ Signs and Symptoms

Cystitis is a disease of sudden onset. Symptoms include pain in the pubic area, frequent urges to urinate even when the bladder is empty, and burning pain accompanying urination (called *dysuria*). The urine can be cloudy due to the presence of bacteria and white blood cells. It may have an orange tinge from the presence of red blood cells (*hematuria*). Low-grade fever and nausea are frequently present. If back pain is present and fever is high, it is an indication that the kidneys may also be involved (pyelonephritis). Pyelonephritis is a serious infection that can result in permanent damage to the kidneys if improperly or inadequately treated. If only the bladder is involved, the condition is sometimes called acute uncomplicated UTI.

▶ Causative Agents

In 95% of cystitis and pyelonephritis cases, the cause is bacteria that are either normal biota in the gastrointestinal tract or, as **Insight 23.1** suggests, antibiotic-resistant strains acquired by eating poultry. *Escherichia coli* is by far the most common cause, accounting for approximately 80% of urinary tract infections. *Staphylococcus saprophyticus* and *Enterococcus* are also common culprits. These last two are only referenced in Disease Table 23.1 following the discussion of *E. coli*.

INSIGHT 23.1 Is Your Chicken Salad Causing a UTI?

Chicken is the preferred low-fat, low-calorie protein source in many American diets, but an increasing body of research is linking chronic, drug-resistant urinary tract infections (UTIs) to poultry. A recent study published in the journal *Emerging Infectious Diseases* showed a genetic link between *Escherichia coli* found in retail poultry and the strain of "extraintestinal pathogenic *E. coli*" (ExPEC) that causes UTIs. Each year, 6 to 8 million UTIs are diagnosed in the United States and 130 to 175 million are diagnosed worldwide. The main means of transmission is considered to be "autoinoculation" or improper hygiene in which *E. coli* is transferred from the anus to the urethra. Women are more susceptible to UTIs than males due to the comparatively short length of the female urethra and its proximity to the anus. But why are the ExPEC bacteria in the gut becoming antibiotic resistant? Several studies have shown that the *E. coli* strain transmitted by retail chicken was antibiotic resistant due to overuse of antibiotics for growth promotion in poultry production. UTIs caused by these antibiotic-resistant strains can lead to recurrent UTIs in both men and women. UTIs can lead to more serious infections such as pyelonephritis (kidney infection) and sepsis if not treated.

Researchers estimate that the health care costs of UTIs in the United States are as high as 1 to 2 billion dollars per year, and these drug-resistant strains make treatment increasingly more difficult. There is also difficulty in establishing a direct link between retail poultry and UTIs, because doctors and clinicians don't consider the link between improper food handling and a UTI and don't ask patients about their dietary habits when diagnosing and treating a UTI. However, the genetic link between ExPEC, retail poultry, and UTIs strongly suggests that the chicken in your salad—rather than poor hygiene practices—may actually transmit recurrent UTIs.

Source: 2012. *Emerg. Infect. Dis.* vol. 18, no. 3, p 415.

The *E. coli* species that cause UTIs are ones that exist as normal biota in the gastrointestinal tract. These uropathogenic *E. coli* (UPEC) are called extraintestinal *E. coli* (ExPEC) and are different from the strains that cause diarrhea and other digestive tract diseases.

▶ Pathogenesis and Virulence Factors

Uropathogenic *E. coli* (UPEC) secure themselves in the gastrointestinal tract using specific adhesins on the ends of long fimbriae. They can reside there without causing disease but can also travel to other locations in the body where they may cause disease. UPEC bacteria can also use these adhesins to attach to slightly different chemicals present on the epithelial lining of the urinary tract. These microbes also have different fimbriae with adhesins that recognize chemicals only present on cells lining the ureters and kidney. UPEC strains exhibit motility that allows them to travel along mucosal surfaces, so they seem to be specially adapted to ascending the urinary system. Recently identified toxins produced by these pathogenic bacteria also appear to enhance their ability to cause disease. These chemicals induce an inflammatory response that we experience as symptoms and that may lead to scarring in the ureters and kidneys.

▶ Transmission and Epidemiology

Community-acquired UTIs are nearly always "transmitted" *not* from one *person* to another but from one *organ system* to another, namely from the GI tract to the urinary system. They are much more common in women than in men because of the shorter length of the female urethra and because of the nearness of the female urethral opening to the anus (see figure 23.3). Many women experience what have been referred to as "recurrent urinary tract infections," although it is now known that some UPEC strains can invade the deeper tissue of the urinary tract and therefore avoid being destroyed by antibiotics. They can emerge later to cause symptoms again. It is not clear how many "recurrent" infections are actually infections that reactivate in this way.

Note that urinary tract infections are also the most common of healthcare-associated infections. Patients of both sexes who have urinary catheters are susceptible to infections with a variety of microorganisms, not just the three mentioned here. These include strains of *E. coli* and *K. pneumoniae* known as carbapenem-resistant *Enterobacteriaceae* (CRE) or vancomycin-resistant enterococci (VRE).

▶ Prevention

There are currently two vaccines in clinical trials for recurrent infection, but progress is slow because the antibodies induced do not accumulate in significant amounts at the site of infection—the mucosal surfaces of the urinary tract. For now, prevention of all UTIs relies on more basic practices, such as emptying the bladder frequently and (for females) wiping from front to back after a bowel movement. People who are predisposed to UTIs often drink cranberry juice to prevent the disease. Scientists have found that there are multiple compounds in the juice that help to discourage the attachment of *E. coli* strains to urinary epithelium.

▶ Treatment

Sulfa drugs such as trimethoprim-sulfamethoxazole are most often used for UTIs of various etiologies. If there is a lot of resistance to this treatment in the local area, other drugs must be used. Often another nonantibiotic drug called phenazopyridine (Pyridium) is administered simultaneously. This drug relieves the very uncomfortable symptoms of burning and urgency. However, some physicians are reluctant to administer this medication for fear that it may mask worsening symptoms; when Pyridium is used, it should be used only for a maximum of 2 days. Pyridium is an azo dye and causes the urine to turn a dark orange to red color, and may also stain contact lenses if you wear them during treatment. A large percentage of UPEC strains are resistant to penicillin derivatives, so these should be avoided. Also, a new ExPEC strain of *E. coli* (ST131) has evolved that is highly virulent and, more troubling, resistant to multiple antibiotics. Medical professionals are ringing alarm bells about this strain saying that if it acquires resistance to one more class of antibiotics it will become virtually untreatable. More disturbing is the fact that it has become widespread geographically around the world today; it can even be transmitted among family members **(Disease Table 23.1).**

Leptospirosis

This infection is a zoonosis associated with wild animals and domesticated animals. It can affect the kidneys, liver, brain, and eyes. It is considered in this section because it can have its major effects on the kidneys and because its presence in animal urinary tracts causes it to be shed into the environment through animal urine.

▶ Signs and Symptoms

Leptospirosis has two phases. During the early—leptospiremic—phase, the pathogen appears in the blood and cerebrospinal fluid. Symptoms are sudden high fever, chills, headache, muscle aches, conjunctivitis, and vomiting. During the second—immune—phase, the blood infection is cleared by natural defenses. This period is marked by milder fever; headache due to leptospiral meningitis; and *Weil's syndrome*, a cluster of symptoms characterized by kidney invasion, hepatic disease, jaundice, anemia, and neurological disturbances. Long-term disability and even death can result from damage to the kidneys and liver, but they occur primarily with the most virulent strains and in elderly persons.

▶ Causative Agent

Leptospires are typical spirochete bacteria marked by tight, regular, individual coils with a bend or hook at one or both ends **(figure 23.4).** *Leptospira interrogans* (lep"-toh-spy'-rah in-terr'-oh-ganz) is the species that causes leptospirosis in humans and animals. There are nearly 200 different serotypes of this species distributed among various animal groups, which accounts for extreme variations in the disease manifestations in humans.

Disease Table 23.1	Urinary Tract Infections (Cystitis, Pyelonephritis)		
Causative Organism(s)	*Escherichia coli*	*Staphylococcus saprophyticus*	*Enterococcus*
Most Common Modes of Transmission	Endogenous transfer from GI tract (opportunism)	Opportunism	Opportunism
Virulence Factors	Adhesins, motility	–	–
Culture/Diagnosis	Often "bacterial infection" diagnosed on basis of increased white cells in urinalysis; if culture performed, bacteria may or may not be identified to species level		
Prevention	Vaccine may be available soon; hygiene practices	Hygiene practices	Hygiene practices
Treatment	If <20% of local *E. coli* are resistant to TMP-SMX, use this; if >20%, use nitrofurantoin or fosfomycin; Pyridium also given for symptom relief		
Distinctive Features	–	–	–
Epidemiological Features	Causes 90% of community-acquired UTIs and majority of healthcare-associated UTIs	Causes 5%–20% of community-acquired UTIs; not frequent cause of healthcare-associated UTIs	Frequent cause of healthcare-associated UTIs; frequent cause in children with urinary system abnormalities; vancomycin-resistant strains common

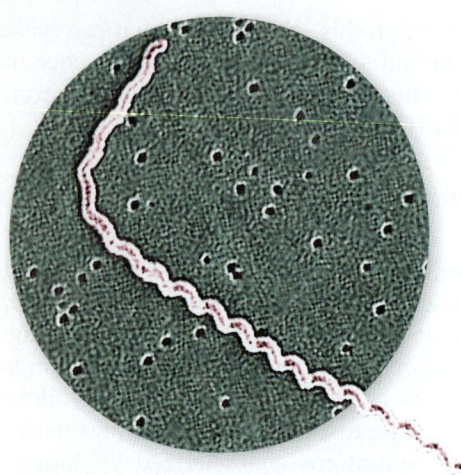

Figure 23.4 *Leptospira interrogans,*
the agent of leptospirosis.   Note the
curved hook at the ends of the spirochete.
The green disk is a piece of filter paper and
the holes are the pores of the filter.

Pathogenesis and Virulence Factors

In 2003, Chinese scientists sequenced the entire genome of
this bacterium and found a series of genes that code for virulence factors such as adhesins and invasion proteins. These
factors allow the pathogen to rapidly penetrate host cells and
enter into the bloodstream and enable the bacterium to cause
cell death in kidney tissue. Because it appears that the bacterium evolved from its close relatives, which are free-living
and cause no disease, finding out how the bacterium acquired
these genes will be useful in understanding its pathogenesis.

Transmission and Epidemiology

Leptospirosis is a zoonosis, affecting wild animals such as
rodents, skunks, raccoons, and foxes and some domesticated
animals—particularly horses, dogs, cattle, and pigs. It is
found throughout the world, although it is more common in
the tropics. It is an occupational hazard of people who work
with animals or in the outdoors. Leptospires that are shed
in the urine of an infected animal can survive for several
months in neutral or alkaline soil or water. Infection occurs
almost entirely through contact of skin abrasions or mucous
membranes with animal urine or some environmental source
containing urine. In 1998, dozens of athletes competing in the
swimming phase of a triathlon in Illinois contracted leptospirosis from the water. In late 2009, the Philippines experienced a major outbreak after a series of typhoons flooded the
country. At one point, 350 new cases a day were diagnosed.
Today, leptospirosis is becoming an increasingly significant
disease in urban slums around the world. The disease does
not appear to be easily transmissible from person to person.

Prevention

A preventive vaccine for humans is currently being investigated, though a protective vaccine for use in animals is
used to reduce the spread of the pathogen. For now, the best
prevention is to wear protective footwear and clothing and

to avoid swimming and wading in natural water sources that
are frequented by livestock. Anyone participating in aquatic
recreational activities should be aware of this infection, especially in more tropical regions of the world.

Treatment

Early treatment with doxycycline rapidly reduces symptoms
and shortens the course of disease, but delayed therapy is
less effective. In severe disease, penicillin G or ceftriaxone
should be used. Other spirochete diseases, such as syphilis
(described later), also exhibit this same pattern of reduced
antibiotic susceptibility over time **(Disease Table 23.2).**

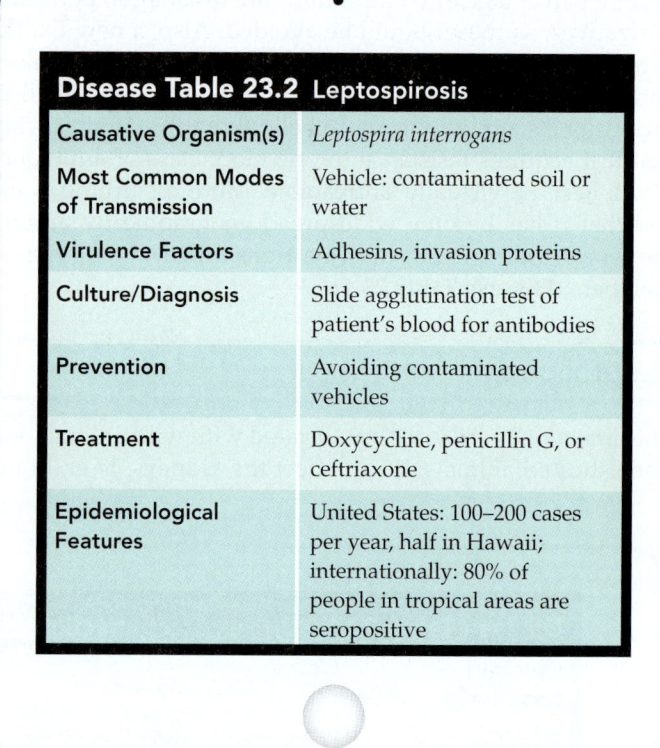

Disease Table 23.2 Leptospirosis	
Causative Organism(s)	*Leptospira interrogans*
Most Common Modes of Transmission	Vehicle: contaminated soil or water
Virulence Factors	Adhesins, invasion proteins
Culture/Diagnosis	Slide agglutination test of patient's blood for antibodies
Prevention	Avoiding contaminated vehicles
Treatment	Doxycycline, penicillin G, or ceftriaxone
Epidemiological Features	United States: 100–200 cases per year, half in Hawaii; internationally: 80% of people in tropical areas are seropositive

Urinary Schistosomiasis

In chapter 22, we talked about schistosomiasis, because one
of its two distinct disease manifestations occurs in the liver
and spleen, both parts of the digestive system. One particular species of this trematode (helminth) lodges in the blood
vessels of the bladder. This may or may not result in symptoms. Blood in the urine and, eventually, bladder obstruction
can occur.

Signs and Symptoms

As with the other forms of schistosomiasis, the first symptoms of infestation are itchiness in the area where the helminth enters the body, followed by fever, chills, diarrhea, and
cough. Urinary tract symptoms occur at a later date. Remember that adult flukes can live for many years and, by eluding
the immune defenses, cause chronic infection.

▶ **Causative Agent**

The urinary manifestations occur if a host is infected with a particular species of schistosome, *Schistosoma haematobium*. It is found throughout Africa, the Caribbean, and the Middle East. (*S. mansoni* and *S. japonicum* are the species responsible for liver manifestations.) *Schistosomes* are trematodes, or flukes (illustrated in figure 22.29). Humans are the definitive hosts for schistosomes, and snails are the intermediate hosts.

▶ **Pathogenesis and Virulence Factors**

Like the other species, *S. haematobium* is able to invade intact skin and attach to vascular endothelium. It engages in the same antigenic cloaking behavior as the other two species. The disease manifestations occur when the eggs in the bladder induce a massive granulomatous response that leads to leakage in the blood vessels and blood in the urine. Significant portions of the bladder eventually can be filled with granulomatous tissue and scar tissue. Function of the bladder is decreased or halted altogether. Chronic infection with *S. haematobium* can also lead to bladder cancer.

▶ **Transmission and Epidemiology**

The life cycle of the schistosome is described completely in chapter 22. After the helminths pass into small blood and lymphatic vessels, they are carried to the liver. Eventually *S. haematobium* enters the venous plexus of the bladder. While attached to these intravascular sites, the helminths feed upon blood, and the female lays eggs that are eventually voided in urine. The appropriate snail vector does not exist in the United States, so cases found here are virtually all imported.

▶ **Culture and Diagnosis**

Diagnosis depends on identifying the eggs in urine. Newly developed genotypic tests may prove to be more sensitive in the detection of disease.

▶ **Prevention and Treatment**

The cycle of infection cannot be broken as long as people are exposed to untreated sewage in their environment. It is quite common for people to be cured and then to be reinfected because their village has no sewage treatment. A vaccine would provide widespread control of the disease, but so far none is licensed. More than one vaccine is in development, however.

Praziquantel is the drug treatment of choice and is quite effective at eliminating the helminths, though drug resistance is developing in many areas of the world today (**Disease Table 23.3**).

23.3 Learning Outcomes—Assess Your Progress

5. List the possible causative agents for each type of urinary tract infection: cystitis/pyelonephritis, leptospirosis, and schistosomiasis.
6. Discuss the epidemiology of the three types of urinary tract infection.

Disease Table 23.3	Urinary Schistosomiasis
Causative Organism(s)	*Schistosoma haematobium*
Most Common Modes of Transmission	Vehicle: contaminated water
Virulence Factors	Antigenic "cloaking," induction of granulomatous response
Culture/Diagnosis	Identification of eggs in urine, PCR methods
Prevention	Avoiding contaminated vehicles
Treatment	Praziquantel
Epidemiological Features	Endemic in Africa, Middle East, India, and Turkey; in sub-Saharan Africa: 120 million infected

23.4 Reproductive Tract Diseases Caused by Microorganisms

We saw earlier that reproductive tract diseases in men almost always involve the urinary tract as well, and this is sometimes but not always the case with women. Not all reproductive tract diseases are sexually transmitted, though many are.

We begin this section with a discussion of infections that are symptomatic primarily in women: *vaginitis* and *vaginosis*. Men may also harbor similar infections with or without symptoms. We next consider three broad categories of sexually transmitted infections (STIs): *discharge diseases* in which increased fluid is released in male and female reproductive tracts; *ulcer diseases* in which microbes cause distinct open lesions; and the *wart diseases*. The discharge diseases are responsible for large numbers of infertility cases. Herpes and human papillomavirus (HPV) infections are incurable and therefore simply increase in their prevalence over time. The section concludes with a neonatal disease caused by group B *Streptococcus* colonization.

Vaginitis

▶ **Signs and Symptoms**

Vaginitis, an inflammation of the vagina, is a condition characterized by some degree of vaginal itching, depending on the etiologic agent. Symptoms may also include burning, and sometimes a discharge, which may take different forms as well.

▶ **Causative Agents**

The most common cause of vaginitis is *Candida albicans*. The vaginal condition caused by this fungus is known as a *yeast infection.* Most women experience this condition one or multiple times during their lives. Other bacteria—and even protozoa, such as *Trichomonas*—can also cause vaginal infections.

Candida albicans

Candida albicans is a dimorphic fungus that is normal biota in from 50% to 100% of humans, living in low numbers on many mucosal surfaces such as the mouth, gastrointestinal tract, vagina, and so on. The vaginal condition it causes is often called vulvovaginal candidiasis. The yeast is easily detectable on a wet prep or a Gram stain of material obtained during a pelvic exam **(figure 23.5).** The presence of pseudohyphae in the smear is a clear indication that the yeast is growing rapidly and causing a yeast infection.

▶ ### Pathogenesis and Virulence Factors

The fungus grows in thick curdlike colonies on the walls of the vagina. The colony debris contributes to a white vaginal discharge. In otherwise healthy people, the fungus is not invasive and limits itself to this surface infection. Please note, however, that *Candida* infections of the bloodstream do occur and they have high mortality rates. They do not normally stem from vaginal infections with the fungus, however, and are seen most frequently in hospitalized patients. AIDS patients are also at risk of developing systemic *Candida* infections.

▶ ### Transmission and Epidemiology

Vaginal infections with this organism are nearly always opportunistic. Disruptions of the normal bacterial biota or even minor damage to the mucosal epithelium in the vagina can lead to overgrowth by this fungus. Disruptions may be mechanical, such as wearing very tight pants, or they may be chemical, as when broad-spectrum antibiotics taken for some other purpose temporarily diminish the vaginal bacterial population. Diabetics and pregnant women are also predisposed to vaginal yeast overgrowths. Some women are prone to this condition during menstruation.

It is possible to transmit this microbe through sexual contact, especially if a woman is experiencing an overgrowth of the yeast. The recipient's immune system may well subdue the potential pathogen so that it acts as normal biota in them. But the yeast may be passed back to the original partner during further sexual contact after treatment. Because of this, it is recommended that a patient's sexual partner also be treated to short-circuit the possibility of retransmission. The important thing to remember is that *Candida* is an opportunistic fungus; women with HIV infection experience frequently recurring yeast infections. However, a small percentage of women with no underlying immune disease experience chronic or recurrent vaginal infection with *Candida* for reasons that are not clear.

▶ ### Prevention and Treatment

No vaccine is available for *C. albicans*. Topical and oral azole drugs are used to treat vaginal candidiasis, and some of them are now available over the counter. If infections recur frequently or fail to resolve, it is important to see a physician for evaluation.

Trichomonas vaginalis

Trichomonads are small, pear-shaped protozoa with four anterior flagella and an undulating membrane **(figure 23.6).** *Trichomonas vaginalis* seems to cause asymptomatic infections in approximately 50% of females and males, despite its species name. Trichomonads are considered asymptomatic infectious agents rather than normal biota because of evidence that some people experience long-term negative effects. Even though *Trichomonas* is a protozoan, it has no cyst form and it does not survive long outside of the host.

▶ ### Pathogenesis and Virulence Factors

Many cases are asymptomatic, and men seldom have symptoms. Women often have vaginitis symptoms, which can

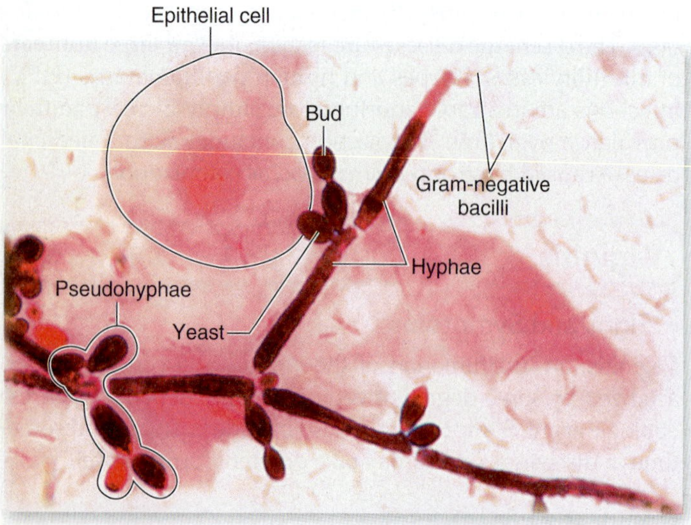

Epithelial cell
Bud
Gram-negative bacilli
Hyphae
Pseudohyphae
Yeast

Figure 23.5 Gram stain of *Candida albicans* in a vaginal smear.

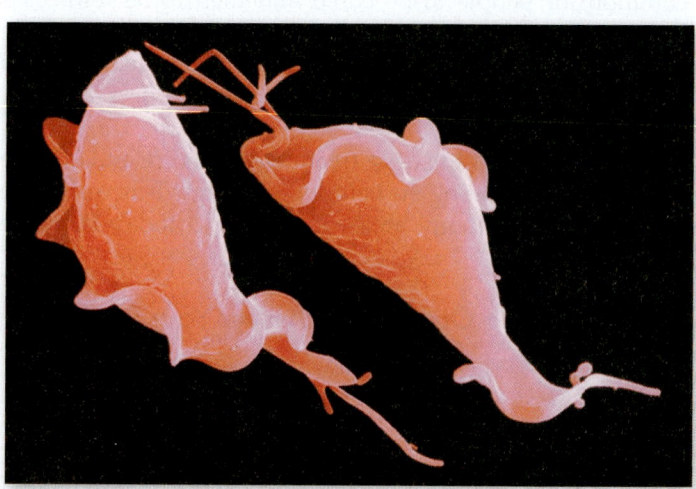

Figure 23.6 *Trichomonas vaginalis*.

include a white to green frothy discharge. Chronic infection can make a person more susceptible to other infections, including HIV. Also, women who become infected during pregnancy are predisposed to premature labor and low-birth-weight infants. Chronic infection may also lead to infertility. Recent research has suggested a link between *Trichomonas* and prostate cancer. Scientists have found that the protozoan can activate a set of proteins whose cascade of effects can increase prostate cancer risk.

▶ **Transmission and Epidemiology**

Because *Trichomonas* is common biota in so many people, it is easily transmitted through sexual contact. It has been called the most common nonviral sexually transmitted infection, and it is estimated that 10 million Americans have the infection. It does not appear to undergo opportunistic shifts within its host (that is, becoming symptomatic under certain conditions); rather, the protozoan causes symptoms when transmitted to a noncarrier. Some recent data suggest that the protozoan can be transmitted through communal bathing, public facilities, and from mother to child, but these types of transmission are rare in most populations.

▶ **Prevention and Treatment**

There is no vaccine for *Trichomonas*. The antiprotozoal drug metronidazole is the drug of choice, although some isolates are resistant to it **(Disease Table 23.4)**.

Vaginosis

There is a particularly common—and misunderstood—condition in women in their childbearing years. This condition is usually called vaginosis rather than vaginitis because it doesn't appear to induce inflammation in the vagina. It is

also known as BV, or bacterial vaginosis. Despite the absence of an inflammatory response, a vaginal discharge is associated with the condition. It is often characterized by a fishy odor, and itching is common. But it is also true that many women have this condition with no noticeable symptoms.

Vaginosis is most likely a result of a reduction in the number of "good bacteria" (lactobacilli) in the vagina. The growth of additional microbes plays a role in the development of vaginosis, and new research shows that the diversity of microbiota in cases of BV is much higher than in the healthy vagina. It appears now that this condition should be considered the result of a mixed infection. Many of these bacteria are normally found in low numbers in a healthy vagina, including *Gardnerella vaginalis*, a facultatively anaerobic bacterium, as well as *Atopobium*, which is an aerobe, and *Mobiluncus* species, which are anaerobic. The imbalance of the normal biota can leave the vagina open for infection by other opportunistic pathogens, including a newly recognized organism associated with preterm labor (*Leptotrichia amnionii*). There does not appear to be a common set of microbes associated with all cases of BV, and the often-mentioned fishy odor comes from the metabolic by-products produced by many of these anaerobic bacteria.

▶ **Pathogenesis and Virulence Factors**

The mechanism of damage in this disease is not well understood, but some of the outcomes are. Besides the symptoms just mentioned, vaginosis can lead to complications such as pelvic inflammatory disease (PID), to be discussed later in the chapter; infertility; and, more rarely, ectopic pregnancies. Babies born to some mothers with vaginosis have low birth weights.

▶ **Transmission and Epidemiology**

This mixed infection is not considered to be sexually transmitted, although women who have never had sex

Disease Table 23.4	Vaginitis	
Causative Organism(s)	*Candida albicans*	*Trichomonas vaginalis*
Most Common Modes of Transmission	Opportunism	Direct contact (STI)
Virulence Factors	–	–
Culture/Diagnosis	Wet prep or Gram stain	Protozoa seen on Pap smear or Gram stain
Prevention	–	Barrier use during intercourse
Treatment	Topical or oral azole drugs, some over-the-counter drugs	Metronidazole, tinidazole
Distinctive Features	White curdlike discharge	Discharge may be greenish
Epidemiological Features	United States: 20% of all vaginitis; 75% women reported to have had at least one infection in their lifetimes	7–8 million women infected per year

rarely develop the condition. It is very common in sexually active women. We do not know exactly what causes the off-kilter balance of biota in the vagina. The low pH typical of the vagina is usually higher in vaginosis, but it is not clear whether this causes or is caused by the change in bacterial biota.

▶ Culture and Diagnosis

The condition can be diagnosed by a variety of methods. Sometimes a simple stain of vaginal secretions is used to examine sloughed vaginal epithelial cells. In vaginosis, some cells will appear to be nearly covered with adherent bacteria. (In normal times, vaginal epithelial cells are sparsely covered with bacteria.) These cells are called clue cells and are a helpful diagnostic indicator **(figure 23.7)**. They can also be found on Pap smears. Due to the complex nature of the infection, genomic analysis of vaginal swabs is often necessary to diagnose disease.

▶ Prevention and Treatment

No known prevention exists. Asymptomatic cases are generally not treated. Women who find the condition uncomfortable or who are planning on becoming pregnant should be treated. Women who use intrauterine devices (IUDs) for contraception should also be treated because IUDs can provide a passageway for the bacteria to gain access to the upper reproductive tract. The usual treatment is oral or topical metronidazole or clindamycin **(Disease Table 23.5).**

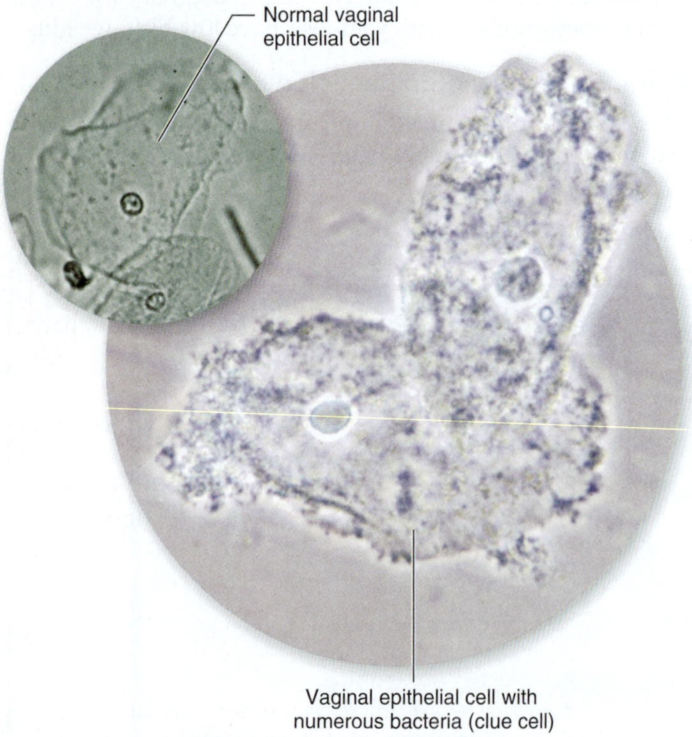

Disease Table 23.5 Vaginosis	
Causative Organism(s)	Mixed infection
Most Common Modes of Transmission	Opportunism or STI
Virulence Factors	–
Culture/Diagnosis	Visual exam of vagina, or clue cells seen in Pap smear or other smear
Prevention	–
Treatment	Metronidazole or clindamycin
Distinctive Features	Discharge may have fishy smell
Epidemiological Features	United States: Estimated 7.4 million new cases per year; internationally: prevalence rates vary by country from 20%–51%

Figure 23.7 Clue cell in bacterial vaginosis. These epithelial cells came from a pelvic exam. The cells in the large circle have an abundance of bacteria attached to them.

— Normal vaginal epithelial cell

Vaginal epithelial cell with numerous bacteria (clue cell)

Prostatitis

Prostatitis is an inflammation of the prostate gland (see figure 23.2). It can be acute or chronic. Acute prostatitis is virtually always caused by bacterial infection. The bacteria are usually normal biota from the intestinal tract or may have caused a previous urinary tract infection. Chronic prostatitis is also often caused by bacteria. Researchers have found that chronic prostatitis, often unresponsive to antibiotic treatment, can be caused by mixed biofilms of bacteria in the prostate. Some forms of chronic prostatitis have no known microbial cause, though many infectious disease specialists feel that one or more bacteria are involved that are simply not culturable with current techniques.

Symptoms may include pain in the groin and lower back, frequent urge to urinate, difficulty in urinating, blood in the urine, and painful ejaculation. Acute prostatitis is accompanied by fever and chills and flulike symptoms. Patients appear to be quite ill with the acute form of the disease.

Treatment involves broad-spectrum antibiotics. Also, muscle relaxers or drugs called alpha blockers, which relax the neck of the bladder, may be prescribed. Prostatitis is distinct from prostate cancer, although some of the symptoms may be similar **(Disease Table 23.6).**

Disease Table 23.6 Prostatitis

Causative Organism(s)	GI tract biota
Most Common Modes of Transmission	Endogenous transfer from GI tract; otherwise unknown
Virulence Factors	Various
Culture/Diagnosis	Digital rectal exam to examine prostate; culture of urine or semen
Prevention	None
Treatment	Antibiotics, muscle relaxers, alpha blockers
Distinctive Features	Pain in genital area and/or back, difficulty urinating
Epidemiological Features	United States: 50% of men experience during lifetime

Discharge Diseases with Major Manifestation in the Genitourinary Tract

Discharge diseases are those in which the infectious agent causes an increase in fluid discharge in the male and female reproductive tracts. Examples are trichomoniasis, gonorrhea, and *Chlamydia* infection. The causative agents are transferred to new hosts when the fluids in which they live contact the mucosal surfaces of the receiving partner. Trichomoniasis has been described in the preceding section because its main disease manifestation is considered to be vaginitis. In this section, we cover the other two major discharge diseases: gonorrhea and *Chlamydia* infection.

Gonorrhea

Gonorrhea has been known as a sexually transmitted disease since ancient times. Its name originated with the Greek physician Claudius Galen, who thought that it was caused by an excess flow of semen. For a fairly long period in history, gonorrhea was confused with syphilis. Later, microbiologists went on to cultivate *Neisseria gonorrhoeae,* also known as the **gonococcus,** and proved conclusively that it alone was the etiologic agent of gonorrhea.

▶ Signs and Symptoms

In the male, infection of the urethra elicits urethritis, painful urination and a yellowish discharge, although a relatively large number of cases are asymptomatic. In most cases, infection is limited to the distal urogenital tract, but it can occasionally spread from the urethra to the prostate gland and epididymis (refer to figure 23.2). Scar tissue formed in the spermatic ducts during healing of an invasive infection can render a man infertile. This outcome is becoming increasingly rare with improved diagnosis and treatment regimens.

In the female, it is likely that both the urinary and genital tracts will be infected during sexual intercourse. A mucopurulent (containing mucus and pus) or bloody vaginal discharge occurs in about half of the cases, along with painful urination if the urethra is affected. Major complications occur when the infection ascends from the vagina and cervix to higher reproductive structures such as the uterus and fallopian tubes **(figure 23.8).** One disease resulting from this progression is **salpingitis** (sal"-pin-jy'-tis). This inflammation of the fallopian tubes may be isolated, or it may also include inflammation of other parts of the upper reproductive tract, called pelvic inflammatory disease (PID). It is not unusual for the microbe that initiates PID to become involved in mixed infections with anaerobic bacteria. The buildup of scar tissue from PID can block the fallopian tubes, causing sterility or ectopic pregnancies **(Insight 23.2).**

Serious consequences of gonorrhea can occur outside of the reproductive tract. In a small number of cases, the gonococcus enters the bloodstream and is disseminated to the joints and skin. Involvement of the wrist and ankle can lead to chronic arthritis and a painful, sporadic, papular rash on the limbs. Rare complications of gonococcal bacteremia are meningitis and endocarditis.

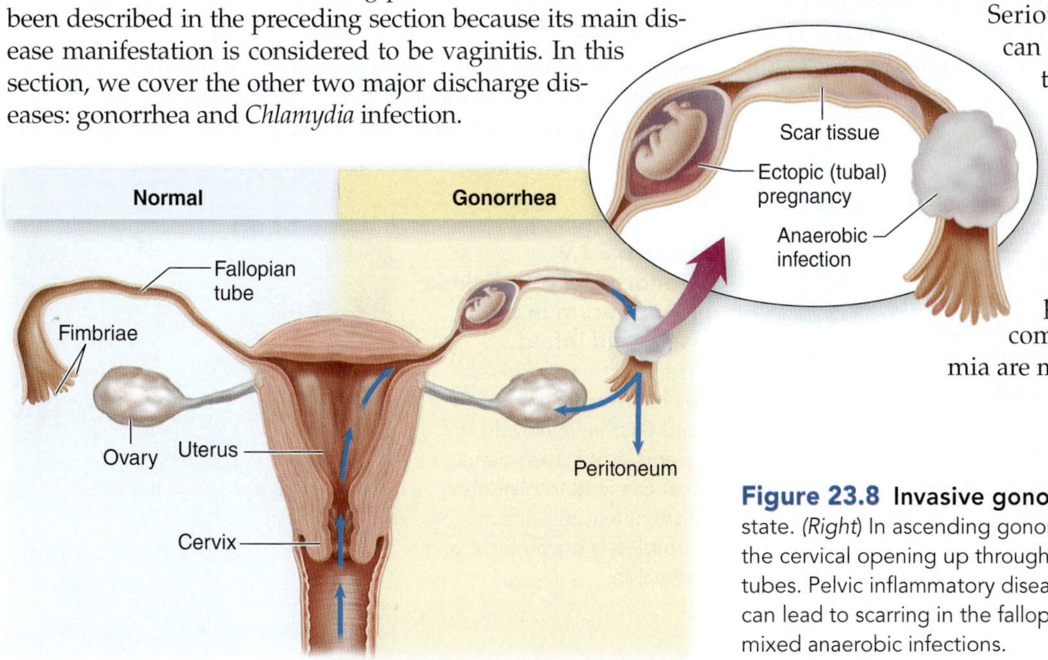

Figure 23.8 Invasive gonorrhea in women. (*Left*) Normal state. (*Right*) In ascending gonorrhea, the gonococcus is carried from the cervical opening up through the uterus and into the fallopian tubes. Pelvic inflammatory disease (PID) is a serious complication that can lead to scarring in the fallopian tubes, ectopic pregnancies, and mixed anaerobic infections.

Normal — Gonorrhea
Fallopian tube
Fimbriae
Ovary — Uterus
Cervix
Peritoneum
Scar tissue
Ectopic (tubal) pregnancy
Anaerobic infection

Pelvic inflammatory disease (PID) is a generalized term for infection of the upper reproductive structures of women most often caused by *Chlamydia trachomatis* or *Neisseria gonorrhoeae*. According to CDC statistics, 80% to 90% of all infections caused by *C. trachomatis* and 50% of infections caused by *N. gonorrhoeae* are asymptomatic, which can then progress to PID. Most often, the uterus, fallopian tubes, and the ovaries are involved. Because there is no normal biota in these organs, inflammation resulting from infection with these organisms can lead to scar tissue and pelvic adhesions that can cause pelvic pain, discharge, fever, nausea, diarrhea, painful urination, and pain during intercourse. There is often great variation in the symptoms, and often women are misdiagnosed or show no symptoms at all. PID can be treated with broad-spectrum antibiotics, but undiagnosed, subclinical, or recurrent infection can result in scarring in the uterus and fallopian tubes, which can lead to ectopic pregnancy and infertility.

The CDC reports that 47.4% of U.S. high-school students surveyed in 2011 have had sexual intercourse. The same report shows that nearly half of the 19 million new STIs each year are among young people ages 15 to 24 years. These statistics indicate that potentially millions of young teenage girls are at risk for *Chlamydia* and gonorrhea, which can lead to PID and infertility later in life. A study conducted by Johns Hopkins University showed that young teenage girls are unlikely to seek treatment for the early symptoms of PID on their own with their family doctor or at an outpatient

clinic. It is likely that they are afraid to tell their parents about their sexual activity or ask for help in seeking medical treatment for an STI. The study showed that more often, teenaged girls and young women are hospitalized with symptoms of PID, indicating that they wait until symptoms are so severe that either the parents notice or they are in so much pain that they are forced to ask for help. Unfortunately, costs for emergency room and hospital visits are 6 to 12 times higher than an outpatient visit, and the delay may result in greater damage to reproductive organs.

Source: 2012. www.hopkinsmedicine.org/healthlibrary/conditions/gynecological_health/pelvic_inflammatory_disease_pid_85,P01552/.

Children born to gonococcus carriers are also in danger of being infected as they pass through the birth canal. Because of the potential harm to the fetus, physicians usually screen pregnant mothers for its presence. Gonococcal eye infections are very serious and often result in keratitis, ophthalmia neonatorum, and even blindness (**figure 23.9**). A universal precaution to prevent such complications is the use of antibiotic eyedrops or ointments (usually erythromycin) for newborn babies. The pathogen may also infect the pharynx and respiratory tract of neonates. Finding gonorrhea in children other than neonates is strong evidence of sexual abuse by infected adults, and it calls for child welfare consultation along with thorough bacteriologic analysis.

▶ **Causative Agent**

N. gonorrhoeae is a pyogenic gram-negative diplococcus. It appears as pairs of kidney bean–shaped bacteria, with their flat sides touching (**figure 23.10**).

▶ **Pathogenesis and Virulence Factors**

Successful attachment is key to the organism's ability to cause disease. Gonococci use specific chemicals on the tips of fimbriae to anchor themselves to mucosal epithelial

A Note About HIV and Hepatitis B and C

This chapter is about diseases whose *major* (presenting) symptoms occur in the genitourinary tract. But some sexually transmitted infections do not have their major symptoms in this system. HIV and hepatitis B and C can all be transmitted in several ways, one of them being through sexual contact. HIV is considered in chapter 20 because its major symptoms occur in the cardiovascular and lymphatic systems. Because the major disease manifestations of hepatitis B and C occur in the gastrointestinal tract, these diseases are discussed in chapter 22. Anyone diagnosed with any sexually transmitted infection should also be tested for HIV.

Figure 23.9
Gonococcal ophthalmia neonatorum in a week-old infant.
The infection is marked by intense inflammation and edema; if allowed to progress, it causes damage that can lead to blindness. Fortunately, this infection is completely preventable and treatable.

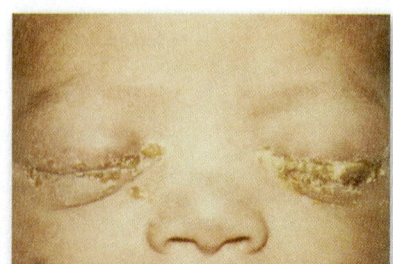

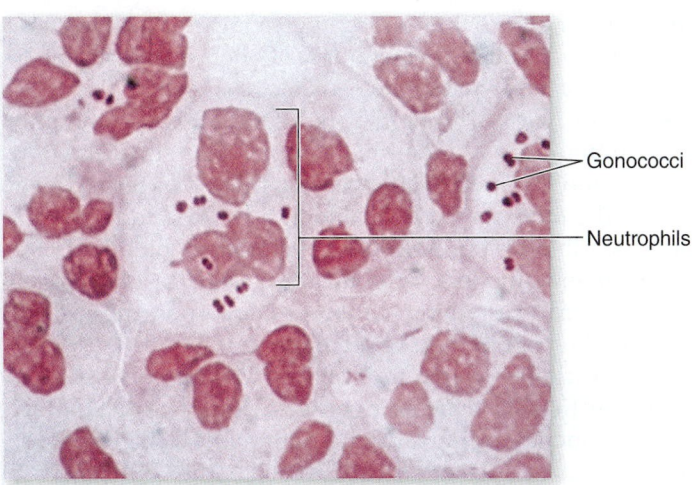

Figure 23.10 Gram stain of urethral pus from a male patient with gonorrhea (1,000×). Note the intracellular (phagocytosed) gram-negative diplococci (arranged side-to-side) in polymorphonuclear leukocytes (neutrophils).

cells. They only attach to nonciliated cells of the urethra and the cervix, for example. Once the bacterium attaches, it invades the cells and multiplies within the basement membrane.

The fimbriae may also play a role in slowing down effective immunity. The fimbrial proteins are controlled by genes that can be turned on or off, depending on the bacterium's situation. This phenotypic change is called phase variation.

In addition, the genes can rearrange themselves to put together fimbriae of different configurations. This antigenic variation confuses the body's immune system. Antibodies that previously recognized fimbrial proteins may not recognize them once they are rearranged.

The gonococcus also possesses an enzyme called IgA protease, which can cleave IgA molecules stationed for protection on mucosal surfaces. In addition, it pinches off pieces of its outer membrane. These "blebs," containing endotoxin, probably play a role in pathogenesis because they can stimulate portions of the nonspecific defense response, resulting in localized damage.

▶ **Transmission and Epidemiology**

N. gonorrhoeae does not survive more than 1 or 2 hours on fomites and is most infectious when transferred to a suitable mucous membrane. Except for neonatal infections, the gonococcus spreads through some form of sexual contact. The pathogen requires an appropriate portal of entry that is genital or extragenital (rectum, eye, or throat).

Gonorrhea is a strictly human infection that occurs worldwide and ranks among the most common sexually transmitted infections. Although about 350,000 cases are reported in the United States each year, it is estimated that the actual incidence is much higher—in the millions if one counts asymptomatic infections. However, this total number is on the decline since 2006. Please refer to **figure 23.11** on this page and also "A Note About STI Statistics" on the next page.

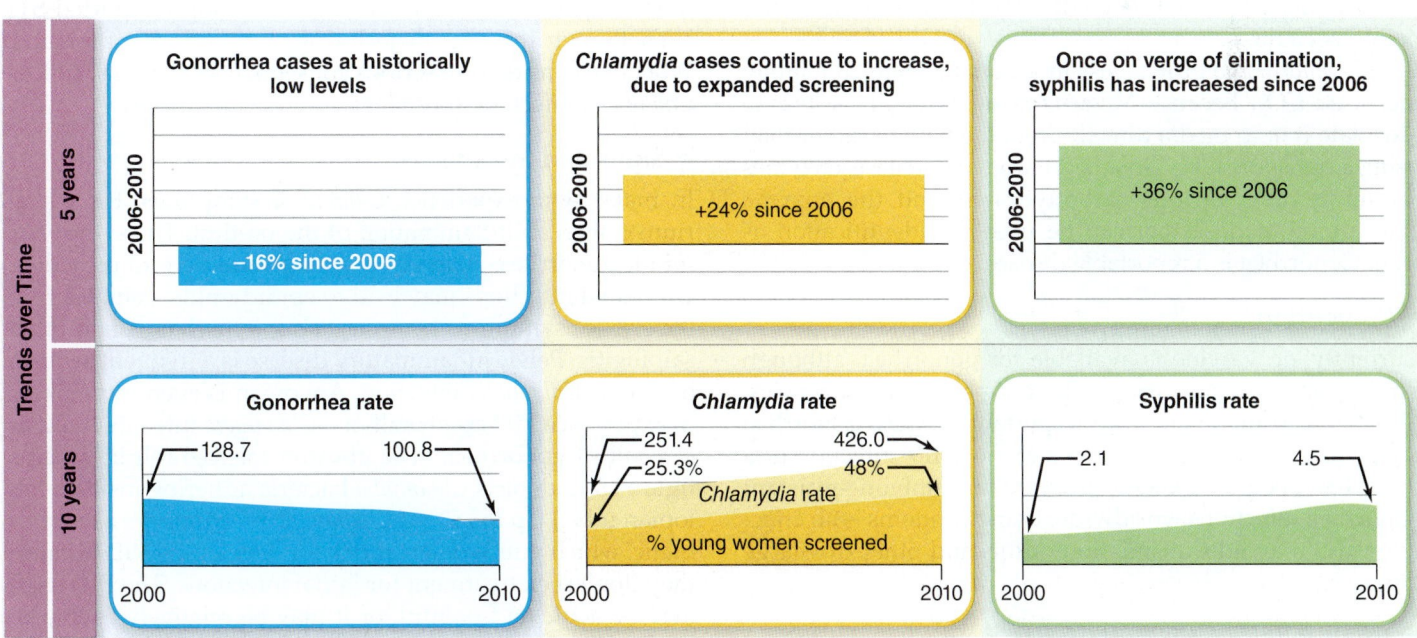

Figure 23.11 Snapshot of sexually transmitted diseases (infections) in the United States, 2010.
Data from the Centers for Disease Control and Prevention.

A Note About STI Statistics

It is difficult to compare the incidence of different STIs to one another, for several reasons. The first is that many, many infections are "silent"; therefore, infected people don't access the health care system and don't get counted. Of course, we know that many silent infections are actually causing damage that won't be noticed for years; when it is, the original causative organism is almost never sought out. The second reason is that only some STIs are officially reportable to health authorities. *Chlamydia* infection and gonorrhea are, for example, but herpes and HPV are not (see table 13.11). In each section, we will try to present accurate estimates of the prevalence and/or incidence of the diseases as we know them. Finally, figure 23.21 provides a visual representation of best estimates of new cases of each disease each year in the United States.

It is important to consider the reservoir of asymptomatic males and females when discussing the transmission of the infection. Because approximately 10% of infected males and 50% of infected females experience no symptoms, it is often spread unknowingly.

▶ Culture and Diagnosis

In males, it is easy to diagnose this disease; a Gram stain of urethral discharge is diagnostic. The normal biota of the male urethra is so sparse that it is easy to see the diplococcus inside of neutrophils (see figure 23.10). In females, other methods, such as ELISA or PCR tests, are called for. Alternatively, the bacterium can be cultured on Thayer-Martin agar, a rich chocolate agar base with added antibiotics that inhibit competing bacteria.

N. gonorrhoeae grows best in an atmosphere containing increased CO_2. Because *Neisseria* is so fragile, it is best to inoculate it onto media directly from the patient rather than using a transport tube. Gonococci produce catalase, enzymes for fermenting various carbohydrates, and the enzyme cytochrome oxidase that can be used for identification as well. Gonorrhea is a reportable disease.

▶ Prevention

Currently, no vaccine is available for gonorrhea, although finding one is a priority for government health agencies. This has become even more important because gonorrhea infection greatly enhances one's risk of HIV infection and there has been a dramatic increase in antibiotic-resistant gonorrhea infections worldwide. Using condoms is an effective way to avoid transmission of this and other discharge diseases.

▶ Treatment

The CDC runs a program called the Gonococcal Isolate Surveillance Project (GISP) to monitor the occurrence of antibiotic resistance in *N. gonorrhoeae*. Penicillin was traditionally the drug of choice, but a large percentage of isolates now are able to produce penicillinase. Others are resistant to tetracycline and quinolones (like ciprofloxacin). In 2011, a strain of *N. gonorrhoeae* that is resistant to all commonly used antibiotics was identified in Japan. This strain has now spread into Europe and may lead to the use of carbapenem drugs, the most potent available in the world today, as the only possible treatment. This development highlights the need for practitioners to be aware of local resistance patterns before prescribing antibiotics for gonorrhea today. The GISP provides this local data. Every month in 28 local STD (STI) clinics around the country, *N. gonorrhoeae* isolates from the first 25 males diagnosed with the infection are sent to regional testing labs, their antibiotic sensitivities are determined, and the data are provided to the GISP program at the CDC. Although the highly resistant strain has not yet been observed in the United States, the CDC has recently changed its overall recommendation for the treatment of gonorrhea worldwide. The Centers now advise the use of ceftriaxone + azithromycin or doxycycline. This is a major change in patient treatment, but it is hoped that making this switch will slow down the spread of the highly resistant strain.

Chlamydia Disease

Genital *Chlamydia* infection is the most common reportable infectious disease in the United States. Annually, more than 1 million cases are reported but the actual infection rate may be 5 to 7 times that number. The overall prevalence among sexually active young women ages 14 to 19 years is 6.8% according to the CDC. It is at least two to three times as common as gonorrhea and is the most commonly reported STI even though the vast majority of cases are asymptomatic. When we consider the serious consequences that may follow *Chlamydia* infection, those facts are very disturbing.

▶ Signs and Symptoms

In males who experience *Chlamydia* symptoms, the bacterium causes an inflammation of the urethra. The symptoms mimic gonorrhea—namely, discharge and painful urination. Untreated infections may lead to epididymitis. Females who experience symptoms have cervicitis, a discharge, and often salpingitis. Pelvic inflammatory disease is a frequent sequela of female *Chlamydia* infection. A woman is even more likely to experience PID as a result of a *Chlamydia* infection than as a result of gonorrhea. (The electron micrograph in process figure 23.12 depicts *Chlamydia* bacteria adhering inside a fallopian tube.) Up to 75% of *Chlamydia* infections are asymptomatic, which puts women at risk for developing PID because they don't seek treatment for initial infections. The PID itself may be acute and painful, or it may be relatively asymptomatic, allowing damage to the upper reproductive tract to continue unchecked.

Certain strains of *C. trachomatis* can invade the lymphatic tissues, resulting in another condition called lymphogranuloma venereum. This condition is accompanied by headache, fever, and muscle aches. The lymph nodes near the lesion begin to fill with granuloma cells and become enlarged and tender. These "nodes" can cause long-term lymphatic obstruction that leads to chronic, deforming edema of the genitalia or anus. The disease is endemic to South America, Africa, and Asia but occasionally occurs in other parts of the world. Its incidence in the United States is about 500 cases per year, with the number of cases increasing in men who have sex with men.

Babies born to mothers with *Chlamydia* infections can develop eye infections and also pneumonia if they become infected during passage through the birth canal. Infant conjunctivitis caused by contact with maternal *Chlamydia* infection is the most prevalent form of conjunctivitis in the United States (100,000 cases per year). Antibiotic drops or ointment applied to newborns' eyes are chosen to eliminate both *Chlamydia* and *N. gonorrhoeae*.

▶ Causative Agent

C. trachomatis is a very small gram-negative bacterium. It lives inside host cells as an obligate intracellular parasite. All *Chlamydia* species alternate between two distinct stages: (1) a small, metabolically inactive infectious form called the elementary body, which is released by the infected host cell; and (2) a larger, noninfectious, actively dividing form called the reticulate body, which grows within the host cell vacuoles **(process figure 23.12)**. Elementary bodies are tiny, dense spheres shielded by a rigid, impervious envelope that ensures survival outside the eukaryotic host cell. Studies of reticulate bodies indicate that they are "energy parasites," entirely lacking enzyme systems for synthesizing ATP, although they do possess ribosomes and mechanisms for synthesizing proteins, DNA, and RNA. Reticulate bodies ultimately become elementary bodies during their life cycle.

▶ Pathogenesis and Virulence Factors

Chlamydia's ability to grow intracellularly contributes to its virulence because it escapes certain aspects of the host's immune response. Also, the bacterium has a unique cell wall that apparently prevents the phagosome from fusing with the lysosome inside phagocytes. The presence of the bacteria inside cells causes the release of cytokines that provoke intense inflammation. This defensive response leads to most of the actual tissue damage in *Chlamydia* infection. Of course, the last step of inflammation is repair, which often results in scarring, as described in Insight 23.2. This can have disastrous effects on a narrow tube like the fallopian tube.

▶ Transmission and Epidemiology

The reservoir of pathogenic strains of *C. trachomatis* is the human body. The microbe shows an astoundingly broad distribution within the population. More alarming is the fact that *Chlamydia* infections have risen steadily over the past few years to reporting levels that have never been seen with any other CDC-notifiable disease. Adolescent women

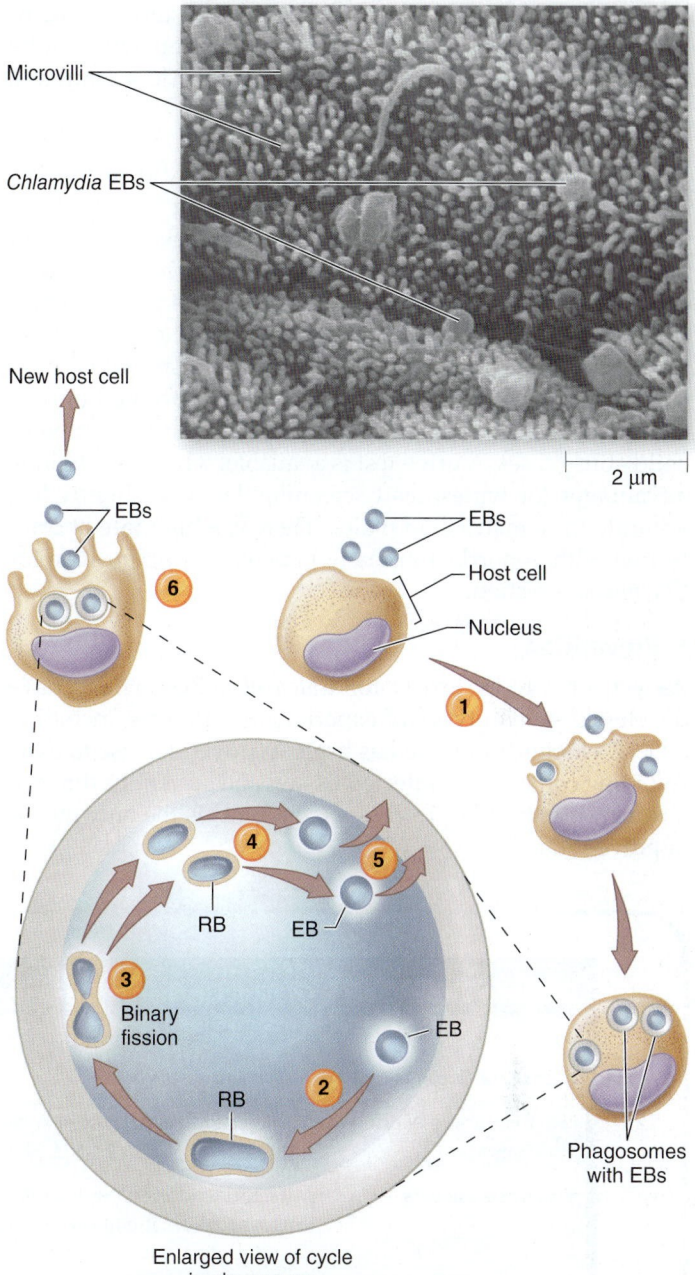

Process Figure 23.12 The life cycle of *Chlamydia*. The infectious stage, or elementary body (EB), is taken into phagocytic vesicles by the host cell. ① In the phagosome, each elementary body develops into a reticulate body (RB). ② Reticulate bodies multiply by regular binary fission. ③ and ④ mature RBs become reorganized into EBs. ⑤ Completed EBs are released from the host cell. Inset features a micrograph of *C. trachomatis* adhering to a fallopian tube (1,750×).

are more likely than older women to harbor the bacterium because it prefers to infect cells that are particularly prevalent on the adolescent cervix. This, along with increased screening rates in women, may in part explain why disease incidence is nearly 3 times higher in females than in males.

It is transmitted through sexual contact and also vertically. Fifty percent of babies born to infected mothers will acquire conjunctivitis (more common) or pneumonia (less common).

▶ **Culture and Diagnosis**

Infection with this microorganism is usually detected initially using a rapid technique such as PCR or ELISA. Direct fluorescent antibody detection is also used. Serology is not always reliable. In addition, antibody to *Chlamydia* is very common in adults and often indicates past, not present, infection. Isolating the bacterium and growing it in cell culture is the best method for detecting this bacterium, but because it is time-consuming and expensive, it is performed only in cases where 100% accuracy is required—such as in rape or child abuse cases. A urine test is available, which has definite advantages for widespread screening, but it is slightly less accurate for females than males. There is a high rate of coinfection with gonorrhea in many patients testing positive for *Chlamydia* infection.

▶ **Prevention**

As yet, no vaccine exists for *Chlamydia*. Researchers have developed several types of experimental vaccines, including a DNA vaccine, but none has been approved for use to date. Avoiding contact with infected tissues and secretions through abstinence or barrier protection (condoms) is the only means of prevention.

▶ **Treatment**

Treatment for this infection relies on being aware of it, so part of the guidelines issued by the CDC is a recommendation for annual screening of young women for presence of the bacterium. It is also recommended that older women with some risk factor (new sexual partner, for instance) also be screened. If infection is found, treatment is usually with doxycycline or azithromycin. Coinfection with gonorrhea should be assumed and treated similarly. Note that according to public health officials, many patients become reinfected soon after treatment; therefore, the recommendation is that patients be rechecked for *Chlamydia* infection 3 to 4 months after treatment. Treatment of all sexual partners of the patient is also recommended to prevent reinfection. Repeated infections with *Chlamydia* increase the likelihood of PID and other serious sequelae **(Disease Table 23.7).**

Genital Ulcer Diseases

Three common infectious conditions can result in lesions on a person's genitals: syphilis, chancroid, and genital herpes. In this section, we consider each of these. One very important fact to remember about the ulcer diseases is that having one of them increases the chances of infection with HIV because of the open lesions.

Disease Table 23.7 Genital Discharge Diseases (in Addition to Vaginitis/Vaginosis)

	Gonorrhea	Chlamydia
Causative Organism(s)	*Neisseria gonorrhoeae*	*Chlamydia trachomatis*
Most Common Modes of Transmission	Direct contact (STI), also vertical	Direct contact (STI), vertical
Virulence Factors	Fimbrial adhesins, antigenic variation, IgA protease, membrane blebs/endotoxin	Intracellular growth resulting in avoiding immune system and cytokine release, unusual cell wall preventing phagolysosome fusion
Culture/Diagnosis	Gram stain in males, rapid tests (PCR, ELISA) for females, culture on Thayer-Martin agar	PCR or ELISA, can be followed by cell culture
Prevention	Avoid contact; condom use	Avoid contact; condom use
Treatment	Coinfection by gonorrhea and *Chlamydia* should be assumed; treat with doxycycline or azithromycin	
Distinctive Features	Rare complications include arthritis, meningitis, endocarditis	More commonly asymptomatic than gonorrhea
Effects on Fetus	Eye infections, blindness	Eye infections, pneumonia
Epidemiological Features	United States: rates decreased 16% between 2006 and 2010; internationally: 26 million cases	United States: 2.8 million new infections per year; internationally: eye infection (trachoma) has 90% prevalence rate in developing world

Cervical cancer occurs as a result of long-standing infection with particular types of HPV. Cervical cancer can be insidious because often there are no outward signs or symptoms and the cancer can only be detected by a Papanicolaou (Pap) smear that examines cells scraped from the cervix for abnormalities.

Recent studies have shown that there has been a rise in oropharyngeal cancer linked to HPV in men. Usually oropharyngeal cancer is linked to alcohol and tobacco use, but researchers have recently found a 225% increase in oral cancer among men caused by HPV, mainly transmitted by oral sex. A CDC report showed that two-thirds of people ages 15 to 24 have engaged in oral sex before they have intercourse. Many young people consider oral sex a safe way to engage in sexual activity but may not realize the potential for transmission of disease through this activity. Studies have shown that oropharyngeal cancers caused by HPV in males are the same types that cause cervical cancer in females. For this reason and to reduce female exposure, the CDC recommends that boys and young men ages 11 to 21 also receive the HPV vaccine.

- What are the HPV vaccination rates in the United States?
- Why might parents be reluctant to vaccinate their children against HPV?

Syphilis

The origin of **syphilis**[1] is an obscure yet intriguing topic of speculation. The disease was first recognized at the close of the 15th century in Europe, a period coinciding with the return of Columbus from the West Indies. From this, some medical scholars have concluded that syphilis was introduced to Europe (i.e., the Old World) from the New World. However, recent analysis of data points to the fact that the predecessor of the spirochete that causes the disease actually traveled in reverse—from the Old World to the New World. This predecessor, which was a *non*–sexually transmitted pathogen, evolved in the Old World—through a combination of the immunologically naive population of Europe, the European wars, and sexual promiscuity—and set the stage for worldwide transmission of syphilis that continues to this day.

A disturbing chapter of syphilis history in the United States is worth noting here. Beginning in 1932, the U.S. government conducted a study called the Tuskegee Study of Untreated Syphilis in the Negro Male, which eventually involved 399 indigent African-American men living in the South. Infected men were recruited into the study, which

sought to document the natural progression of the disease. These men were never told that they had syphilis and were never treated for it, even after penicillin was shown to be an effective cure. The study ended in 1972 after it became public. Much later, in 1997, President Bill Clinton issued a public apology on behalf of the U.S. government, and the government has paid millions of dollars in compensation to the victims and their heirs.

▶ Signs and Symptoms

Untreated syphilis is marked by distinct clinical stages designated as *primary, secondary,* and *tertiary syphilis*. The disease also has latent periods of varying duration during which it is quiescent. The spirochete appears in the lesions and blood during the primary and secondary stages and, thus, is transmissible at these times. During the early latency period between secondary and tertiary syphilis, it is also transmissible. Syphilis is largely nontransmissible during the "late latent" and tertiary stages. Symptoms of each of these stages and congenital syphilis are briefly described here.

Primary Syphilis The earliest indication of syphilis infection is the appearance of a hard **chancre** (shang'-ker) at the site of entry of the pathogen (see Disease Table 23.8 for photos of all three types of genital lesions). A chancre appears after an incubation period that varies from 9 days to 3 months. The chancre begins as a small, red, hard bump that enlarges and breaks down, leaving a shallow crater with firm margins. The base of the chancre beneath the encrusted surface swarms with spirochetes. Most chancres appear on the internal and external genitalia, but about 20% occur on the lips, oral cavity, nipples, or fingers, or around the anus. Because these ulcers tend to be painless, they may escape notice, especially when they are on internal surfaces. Lymph nodes draining the affected region become enlarged and firm, but systemic symptoms are absent at this point. The chancre heals spontaneously without scarring in 3 to 6 weeks, but the healing is deceptive because the spirochete has escaped into the circulation and is entering a period of tremendous activity.

Secondary Syphilis About 3 weeks to 6 months after the chancre heals, the secondary stage appears. By then, many systems of the body have been invaded and the signs and symptoms are more profuse and intense. Initial symptoms are fever, headache, and sore throat, followed by lymphadenopathy and a peculiar red or brown rash that breaks out on all skin surfaces, including the palms of the hands and the soles of the feet **(figure 23.13)**. A person's hair often falls out. Like the chancre, the lesions contain viable spirochetes and disappear spontaneously in a few weeks. The major complications of this stage, occurring in the bones, hair follicles, joints, liver, eyes, and brain, can linger for months and years.

Latency and Tertiary Syphilis After resolution of secondary syphilis, about 30% of infections enter a highly varied latent period that can last for 20 years or longer. During latency, although antibodies to the bacterium are readily detected, the bacterium itself is not. The final stage of the

---

1. The term *syphilis* first appeared in a poem entitled "Syphilis sive Morbus Gallicus" by Fracastorius (1530), about a mythical shepherd whose name eventually became synonymous with the disease from which he suffered.

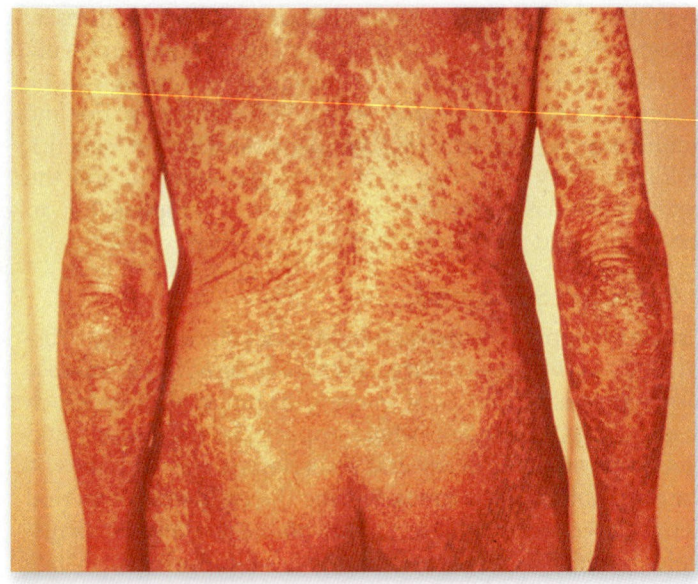

(a)

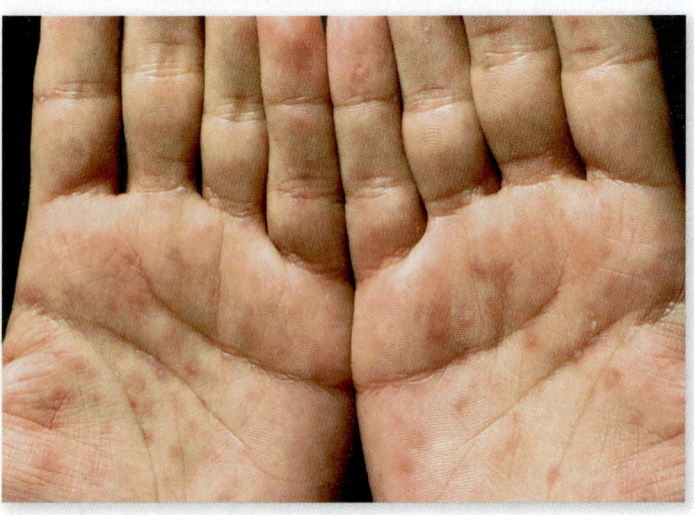

(b)

Figure 23.13 **Symptom of secondary syphilis.** The skin rash in secondary syphilis can form on the trunk, arms, and even palms and soles (these latter locations are particularly diagnostic). The rash does not hurt or itch and can persist for months.

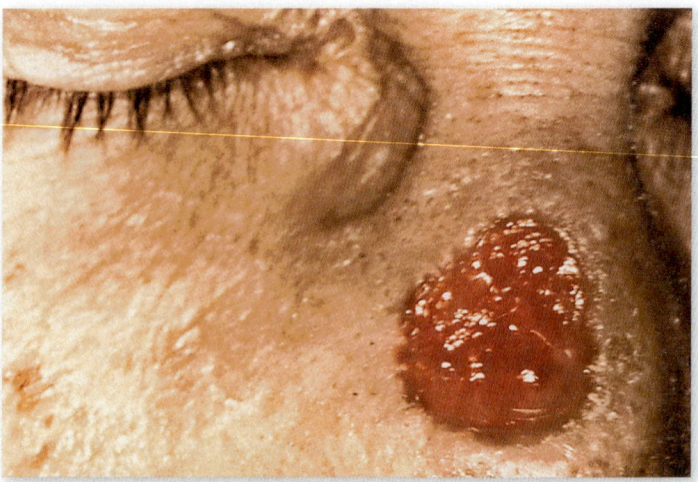

Figure 23.14 **The pathology of late, or tertiary, syphilis.** An ulcerating syphilis tumor, or gumma, appears on the nose of this patient. Other gummas can be internal.

of the nervous system, but it shows particular affinity for the blood vessels in the brain, cranial nerves, and dorsal roots of the spinal cord. The diverse results include severe headaches, convulsions, atrophy of the optic nerve, blindness, dementia, and a sign called the Argyll-Robertson pupil—a condition caused by adhesions along the inner edge of the iris that fix the pupil's position into a small irregular circle.

Congenital Syphilis   The syphilis bacterium can pass from a pregnant woman's circulation into the placenta and can be carried throughout the fetal tissues. An infection leading to congenital syphilis can occur in any of the three trimesters, but it is most common in the second and third. The pathogen inhibits fetal growth and disrupts critical periods of development with varied consequences, ranging from mild to the extremes of spontaneous miscarriage or stillbirth. Early congenital syphilis encompasses the period from birth to 2 years of age and is usually first detected 3 to 8 weeks after birth. Infants often demonstrate such signs as profuse nasal discharge **(figure 23.15a)**, skin eruptions, bone deformation, and nervous system abnormalities. The late form gives rise to an unusual assortment of problems in the bones, eyes, inner ear, and joints and causes the formation of Hutchinson's teeth **(figure 23.15b)**. The number of congenital syphilis cases is closely tied to the incidence in adults.

▶ Causative Agent

Treponema pallidum, a spirochete, is a thin, regularly coiled cell with a gram-negative cell wall. It is a strict parasite with complex growth requirements that necessitate cultivating it in living host cells. Most spirochete bacteria are nonpathogenic; *Treponema* and *Leptospira,* described earlier, are among the pathogens of this group.

Syphilis is a complicated disease to diagnose. Not only do the stages each mimic other diseases, but their appearance can also be so separated in time as to seem unrelated. The chancre and secondary lesions must be differentiated

disease, tertiary syphilis, is relatively rare today because of widespread use of antibiotics. But it is so damaging that it is important to recognize. By the time a patient reaches this phase, numerous pathologic complications occur in susceptible tissues and organs. Cardiovascular syphilis results from damage to the small arteries in the aortic wall. As the fibers in the wall weaken, the aorta is subject to distension and fatal rupture. The same pathologic process can damage the aortic valves, resulting in insufficiency and heart failure.

In one form of tertiary syphilis, painful swollen syphilitic tumors called **gummas** (goo-mahz′) develop in tissues such as the liver, skin, bone, and cartilage **(figure 23.14).** Gummas are usually benign and only occasionally lead to death, but they can impair function. Neurosyphilis can involve any part

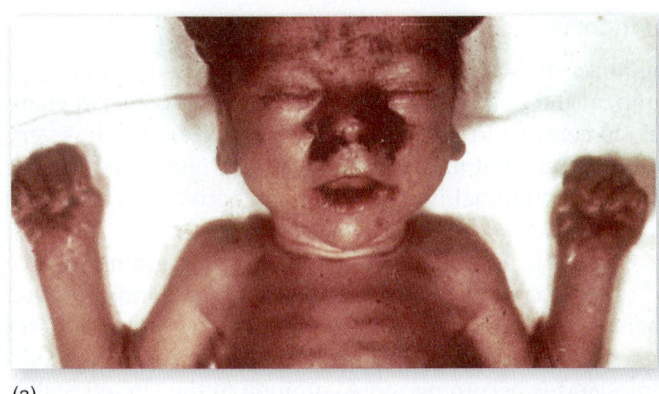

(a)

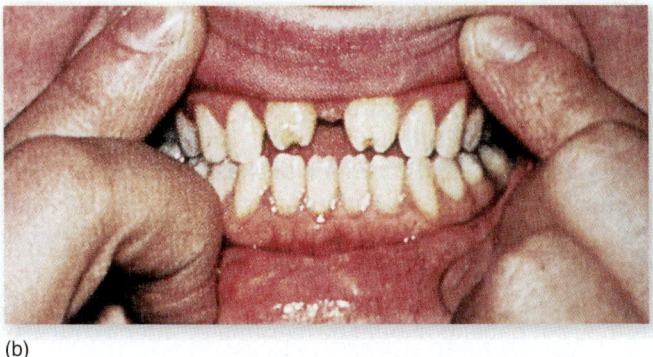

(b)

Figure 23.15 Congenital syphilis. (a) An early sign is snuffles, a profuse nasal discharge that obstructs breathing. (b) A common characteristic of late congenital syphilis is notched, barrel-shaped incisors (Hutchinson's teeth).

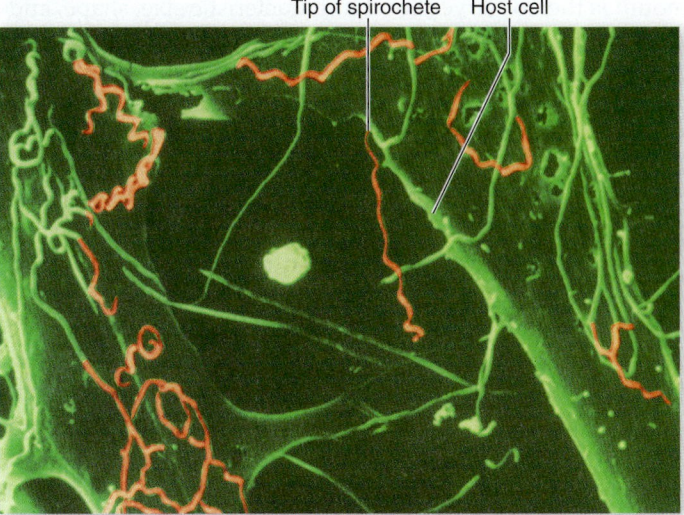

Tip of spirochete   Host cell

Figure 23.16 Electron micrograph of the syphilis spirochete attached to cells.

from bacterial, fungal, and parasitic infections; tumors; and even allergic reactions. Overlapping symptoms of sexually transmitted infections that the patient is concurrently experiencing, such as gonorrhea or *Chlamydia*, can further complicate diagnosis. The disease can be diagnosed using two different strategies: either by detecting the bacterium in patient lesions or by looking for antibodies in the patient's blood.

▶ **Pathogenesis and Virulence Factors**

Brought into direct contact with mucous membranes or abraded skin, *T. pallidum* binds avidly by its hooked tip to the epithelium **(figure 23.16)**. At the binding site, the spirochete multiplies and penetrates the capillaries nearby. Within a short time, it moves into the circulation, and the body is literally transformed into a large receptacle for incubating the pathogen. Virtually any tissue is a potential target.

The specific factor that accounts for the virulence of the syphilis spirochete appears to be outer membrane lipoproteins. These molecules appear to stimulate a strong inflammatory response, which is helpful in clearing the organism but can produce damage as well. *T. pallidum* produces no toxins and does not appear to kill cells directly. Studies have shown that, although phagocytes seem to act against it and several types of antitreponemal antibodies are formed, immune responses are unable to contain it. The primary

lesion occurs when the spirochetes invade the spaces around arteries and stimulate an inflammatory response. Organs are damaged when granulomas form at these sites and block circulation.

▶ **Transmission and Epidemiology**

Humans are evidently the sole natural hosts and source of *T. pallidum*. The bacterium is extremely fastidious and sensitive and cannot survive for long outside the host, being rapidly destroyed by heat, drying, disinfectants, soap, high oxygen tension, and pH changes. It survives a few minutes to hours when protected by body secretions and about 36 hours in stored blood. Research with human subjects has demonstrated that the risk of infection from an infected sexual partner is 12% to 30% per encounter. The bacterium can also be transmitted to the fetus in utero. Syphilis infection through blood transfusion or exposure to fomites is rare.

For centuries, syphilis was a common and devastating disease in the United States, so much so that major medical centers had "Departments of Syphilology." Its effect on social life was enormous. This effect diminished quickly when antibiotics were discovered. In the 20th and 21st centuries, syphilis, like other STIs, has experienced periodic increases during times of social disruption. Most cases tend to be concentrated in larger metropolitan areas among prostitutes, their contacts, and crack cocaine users. If you examine figure 23.11, you will see that syphilis has been increasing since 2006. Syphilis continues to be a serious problem worldwide, especially in Africa and Asia. As mentioned previously, persons with syphilis often suffer concurrent infections with other STIs, including higher risks of HIV infection.

▶ **Culture and Diagnosis**

Syphilis can be detected in patients most rapidly by using dark-field microscopy of a suspected lesion. The lesions are gently squeezed or scraped to extract clear fluid. A wet mount is then observed for the characteristic size, shape, and motility of *T. pallidum*. Another microscopic test for discerning the spirochete directly in samples is direct immunofluorescence staining with monoclonal antibodies.

Very commonly, blood tests are used for this diagnosis. These tests are based on detection of antibody formed in response to *T. pallidum* infection. The best test is one that specifically reacts with treponemal antigens. Additional specific tests are available when considered necessary. One of these is the indirect immunofluorescent method called the FTA-ABS (fluorescent treponemal antibody absorbance) test. The test serum is first allowed to react with treponemal cells and then reacted with antihuman globulin antibody labeled with fluorescent dyes. If antibodies to the treponeme are present, the fluorescently labeled antibody will bind to the human antibody bound to the treponemal cells. The result is highly visible with a fluorescence microscope. A PCR test is available for syphilis, but its accuracy is dependent on the type of tissue being tested.

▶ Prevention

The core of an effective prevention program depends on detection and treatment of the sexual contacts of syphilitic patients. Public health departments and physicians are charged with the task of questioning patients and tracing their contacts. All individuals identified as being at risk, even if they show no signs of infection, are given immediate prophylactic penicillin in a single long-acting dose.

The barrier effect of a condom provides excellent protection during the primary phase. Protective immunity apparently does arise in humans, allowing the prospect of an effective immunization program in the future, although no vaccine exists currently.

▶ Treatment

Throughout most of history, the treatment for syphilis was a dose of mercury or even a "mercurial rub" applied to external lesions. In 1906, Paul Ehrlich discovered that a derivative of arsenic called salvarsan could be very effective. The fact that toxic compounds like mercury and arsenic were used to treat syphilis gives some indication of how dreaded the disease was and to what lengths people would go to rid themselves of it.

Once penicillin became available, it replaced all other treatments, and penicillin G retains its status as a wonder drug in the treatment of acute-stage syphilis. It is given parenterally in large doses with benzathine or procaine. The goal is to maintain a blood level lethal to the spirochete for at least 7 days. Alternative drugs (tetracycline and doxycycline) are less effective, and they are indicated only if penicillin allergy has been documented. It is important that patients be monitored for successful clearance of the spirochete. Macrolide-resistant strains have been identified in China, a country where this disease was virtually eliminated before it recently resurfaced in epidemic proportions.

Chancroid

This ulcerative disease usually begins as a soft papule, or bump, at the point of contact. It develops into a "soft chancre" (in contrast to the hard syphilis chancre), which is very painful in men but may be unnoticed in women (see Disease Table 23.8). Inguinal lymph nodes can become very swollen and tender.

Chancroid is caused by a **pleomorphic** gram-negative rod called *Haemophilus ducreyi*. Recent research indicates that a hemolysin (exotoxin) is important in the pathogenesis of chancroid disease. It is very common in the tropics and subtropics and is becoming more common in the United States. Chancroid is transmitted exclusively through direct contact and is considered a sexually transmitted infection. This disease is associated with sex workers and poor hygiene; uncircumcised men seem to be more commonly infected than those who have been circumcised. People may carry this bacterium asymptomatically.

No vaccine exists. Prevention of chancroid is the same as for other sexually transmitted infections: Avoid contact with infected tissues, either by abstaining from sexual contact or by proper use of barrier protection.

Antibiotics such as azithromycin and ceftriaxone are effective, but patients should be reexamined after a course of treatment to ensure that the bacterium has been eliminated.

Genital Herpes

Virtually everyone becomes infected with a herpesvirus at some time, because this large family of viruses can infect a wide range of host tissues. (We studied three herpesviruses in chapter 21 alone.) Genital herpes is caused by herpes simplex viruses (HSVs). Two types of HSV have been identified, HSV-1 and HSV-2. Other members of the *Herpesviridae* family are herpes zoster (causing chickenpox and shingles), cytomegalovirus (associated with congenital disease and also with HIV-associated disease), Epstein-Barr virus (causing infection of the lymphoid tissue as in infectious mononucleosis), and more recently identified viruses (herpesvirus-6, -7, and -8).

Genital herpes is much more common than most people think.

▶ Signs and Symptoms

Genital herpes infection has multiple presentations. After initial infection, a person may notice no symptoms. Alternatively, herpes could cause the appearance of single or multiple vesicles on the genitalia, perineum, thigh, and buttocks. The vesicles are small and are filled with a clear fluid (see Disease Table 23.8). They are intensely painful to the touch. The appearance of lesions the first time you get them can be accompanied by malaise, anorexia, fever, and bilateral swelling and tenderness in the groin. Occasionally, central nervous system symptoms such as meningitis or encephalitis can develop. Thus, we see that initial infection can be either completely asymptomatic or serious enough to require hospitalization.

After recovery from initial infection, a person may have recurrent episodes of lesions. They are generally less severe than the original symptoms, although the whole gamut of possible severity is seen here as well. Some people never have recurrent lesions. Others have nearly constant outbreaks with little recovery time between them. On average, the number of recurrences is four or five a year. Their frequency tends to decrease over the course of years.

In most cases, patients remain asymptomatic or experience recurrent "surface" infections indefinitely. Very rarely, complications can occur. Every year, one or two persons per million with chronic herpes infections develop encephalitis. The virus disseminates along nerve pathways to the brain (although it can also infect the spinal cord). The effects on the central nervous system begin with headache and stiff neck and can progress to mental disturbances and coma. The fatality rate in untreated encephalitis cases is 70%, although treatment with acyclovir is effective. Patients with underlying immunodeficiency are more prone to severe, disseminated herpes infection than are immunocompetent patients. Of greatest concern are patients receiving organ grafts, cancer patients on immunosuppressive therapy, those with congenital immunodeficiencies, and AIDS patients. Recent data suggest that people with HSV-1 are more prone to Alzheimer's disease, particularly if they carry a particular variant of a particular gene. This is quite sobering when you think that approximately 80% of elderly people are HSV-1-positive, and up to 30% of them carry the gene variant. However, there is hope: Anti-herpes drugs may make a difference in Alzheimer's in these people.

Herpes of the Newborn   Although HSV infections in healthy adults are annoying and unpleasant, only rarely are they life-threatening. However, in the neonate and the fetus **(figure 23.17),** HSV infections are very destructive and can be fatal. Most cases occur when infants are contaminated by the mother's reproductive tract immediately before or during birth, but they have also been traced to hand transmission from the mother's lesions to the baby. Because HSV-2 is more often associated with genital infections, it is more frequently involved; however, HSV-1 infection has similar complications. In infants whose disease is confined to the mouth, skin, or eyes, the mortality rate is 30%, but disease affecting the central nervous system has a 50% to 80% mortality rate.

Because of the danger of herpes to fetuses and newborns and also because of the increase in the number of cases of genital herpes, it is now standard procedure to screen pregnant women for the herpesvirus early in their prenatal care. (Don't forget that most women who are infected do not even know it.) Pregnant women with a history of recurrent infections must be constantly monitored for any signs of viral shedding, especially in the last 4 weeks of pregnancy. If no evidence of recurrence is seen, vaginal birth is indicated, but any evidence of an outbreak at the time of delivery necessitates a cesarean section.

▶ Causative Agent

Both HSV-1 and HSV-2 can cause genital herpes if the virus contacts the genital epithelium, although HSV-1 is thought of as a virus that infects the oral mucosa, resulting in "cold sores" or "fever blisters" **(figure 23.18),** and HSV-2 is thought of as the genital virus. In reality, either virus can infect either region, depending on the type of contact that transmits the infectious agent.

HSV-1 and HSV-2 are DNA viruses with icosahedral capsids and envelopes containing glycoprotein spikes. Like other enveloped viruses, herpesviruses are prone to deactivation by organic solvents or detergents and are unstable outside the host's body.

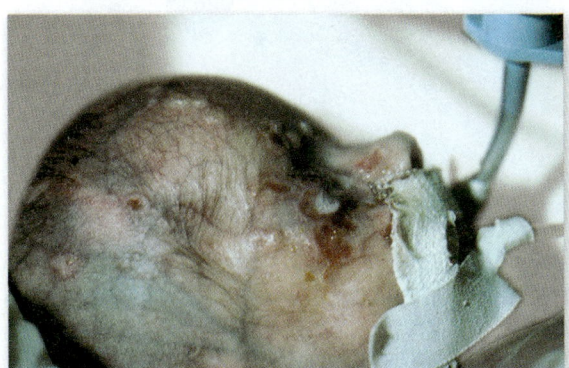

Figure 23.17 Neonatal herpes simplex.   This premature infant was born with the classic "cigarette burn" pattern of HSV infection. Babies can be born with the lesions or develop them 1 to 2 weeks after birth.

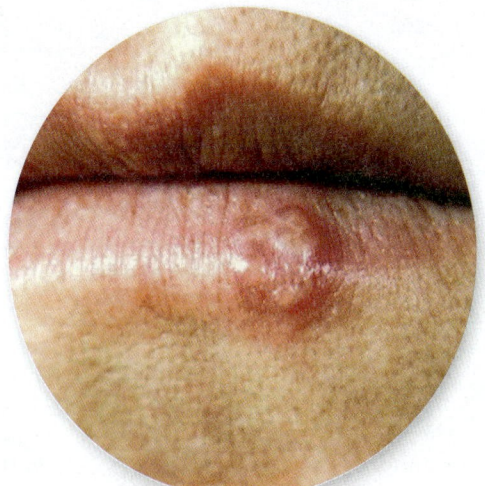

Figure 23.18 Oral herpes infection.   Tender itchy papules erupt around the mouth and progress to vesicles that burst, drain, and scab over. These sores and fluid are highly infectious and should not be touched.

▶ **Pathogenesis and Virulence Factors**

Herpesviruses have a tendency to become latent. During latency, some type of signal causes most of the HSV genome not to be transcribed. This allows the virus to be maintained within cells of the nervous system between episodes. Recent research has found that microRNAs are in part responsible for the latency of HSV-1. It is further suggested that in some peripheral cells, viral replication takes place at a constant, slow rate, resulting in constant low-level shedding of the virus without lesion production.

HSV-2 (or HSV-1, if it has infected the genital region) usually becomes latent in the ganglion of the lumbosacral spinal nerve trunk **(figure 23.19)**. Reactivation of the virus can be triggered by a variety of stimuli, including stress, UV radiation (sunlight), injury, menstruation, or another microbial infection. At that point, the virus begins manufacturing large numbers of entire virions, which cause new lesions on the surface of the body served by the neuron, usually in the same site as previous lesions.

HSV-1 (or HSV-2 if it is in the oral region) behaves in a similar way, but it becomes latent in the trigeminal nerve, which has extensive innervations in the oral region.

▶ **Transmission and Epidemiology**

Herpes simplex infection occurs globally in all seasons and among all age groups. Because these viruses are relatively sensitive to the environment, transmission is primarily through direct exposure to secretions containing the virus. People with active lesions are the most significant source of infection, but studies indicate that genital herpes can be transmitted even when no lesions are present.

It is estimated that about 20% of American adults have genital herpes. As many as 50% to 90% of people who are infected don't even know it, either because they have rare symptoms that they fail to recognize or because they have no symptoms at all.

▶ **Culture and Diagnosis**

These two viruses—HSV-1 and HSV-2—are sometimes diagnosed based on the characteristic lesions alone. PCR tests are available to test for these viruses directly from lesions. Alternatively, antibody to either of the viruses can be detected from blood samples. Detecting antibody to either HSV-1 or HSV-2 in blood does not necessarily indicate whether the infection is oral or genital or whether the infection is new or preexisting.

Herpes-infected mucosal cells display notable characteristics in a Pap smear **(figure 23.20)**. Laboratory culture and specific tests are essential for diagnosing severe or complicated herpes infections. They are also used when screening pregnant women for the presence of virus on the vaginal mucosa. A specimen of tissue or fluid is inoculated into a primary cell culture line and then observed for cytopathic effects that are characteristic for specific viruses.

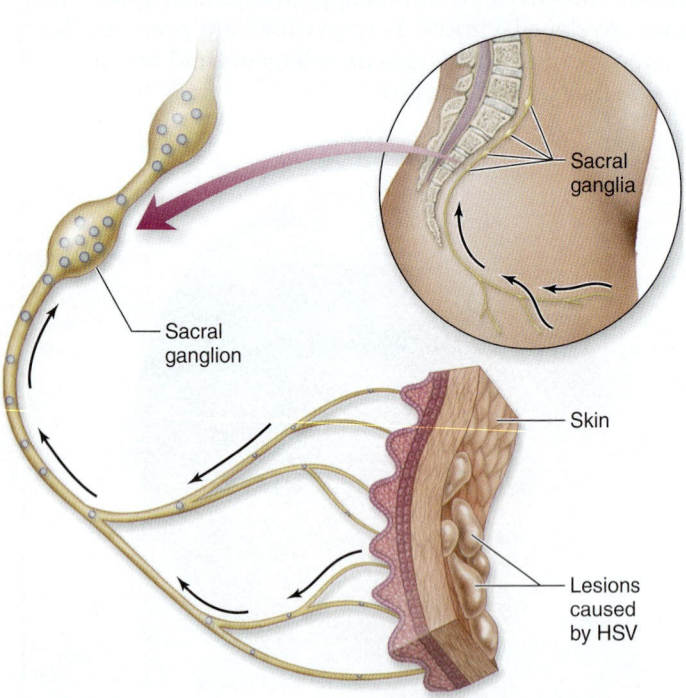

Figure 23.19 Latent HSV in lumbosacral ganglion. The ganglion is the nerve root near the base of the spine. When the virus is reactivated, it travels down the neuron to the body's surface.

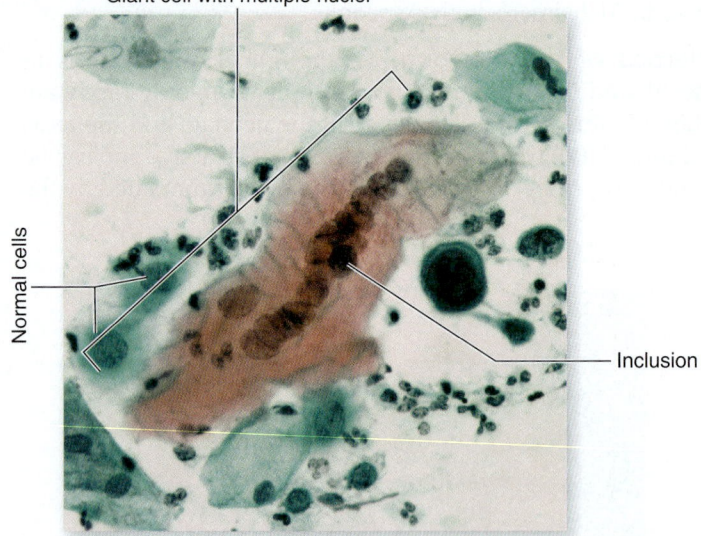

Figure 23.20 The appearance of herpesvirus infection in a Pap smear. A Pap smear of a cervical scraping shows enlarged (multinucleate giant) cells and intranuclear inclusions typical of herpes simplex, type 2. This appearance is not specific for the simplex form of herpesvirus, but most other herpesviruses do not infect the reproductive mucosa. This figure also highlights the fact that Pap smears, while intended primarily to detect cervical cancer, can also provide information about other infections.

▶ **Prevention**

No vaccine is currently licensed for HSV, but more than one is being tested in clinical trials, meaning that vaccines may become available very soon. In the meantime, avoiding contact with infected body surfaces is the only way to avoid HSV. Condoms provide good protection when they actually cover the site where the lesion is, but lesions can occur outside of the area covered by a condom. In general, people experiencing active lesions should avoid sex. Because the virus can be shed when no lesions are present, barrier protection should be practiced at all times by persons infected with HSV.

Mothers with cold sores should be careful in handling their newborns; they should never kiss their infants on the mouth. Some of the drugs used to "treat" genital herpes really function to prevent recurrences of lesions. In this way, they serve as protection for potential partners of people with herpes. It is important to remember that herpes infection is a lifetime infection.

▶ **Treatment**

Several agents are available for treatment. These agents often result in reduced viral shedding and a decrease in the frequency of lesion occurrence. They are not curative. Acyclovir and its derivatives (famciclovir or valacyclovir) are very effective. Topical formulations can be applied directly to lesions, and pills are available as well. Sometimes medicines are prescribed on an ongoing basis to decrease the frequency of recurrences, and sometimes they are prescribed to be taken at the beginning of a recurrence to shorten it **(Disease Table 23.8).**

Wart Diseases

In this section, we describe two viral STIs that cause wart-like growths. The more serious disease is caused by the *human papillomavirus (HPV)*; the other condition, called *molluscum contagiosum,* apparently has no serious effects outside of the growths themselves.

Disease Table 23.8	Genital Ulcer Diseases		
	Syphilis	**Chancroid**	**Herpes**
Causative Organism(s)	*Treponema pallidum*	*Haemophilus ducreyi*	Herpes simplex 1 and 2
Most Common Modes of Transmission	Direct contact and vertical	Direct contact (vertical transmission not documented)	Direct contact, vertical
Virulence Factors	Lipoproteins	Hemolysin (exotoxin)	Latency
Culture/Diagnosis	Direct tests (immunofluorescence, dark-field microscopy), blood tests for treponemal and nontreponemal antibodies, PCR	Culture from lesion	Clinical presentation, PCR, Ab tests, growth of virus in cell culture
Prevention	Antibiotic treatment of all possible contacts, avoiding contact	Avoiding contact	Avoiding contact, antivirals can reduce recurrences
Treatment	Penicillin G	Ceftriaxone or azithromycin	Acyclovir and derivatives
Distinctive Features	Three stages of disease plus latent period, possibly fatal	No systemic effects	Ranges from asymptomatic to frequent recurrences
Effects on Fetus	Congenital syphilis	None	Blindness, disseminated herpes infection
Appearance of Lesions			Vesicles
Epidemiological Features	United States: estimated 90,000 new cases per year; internationally: estimated 12 million new infections per year	United States: no more than 200 per year; internationally: estimated 7 million cases annually	United States: 20% prevalence in adults; internationally: estimated 536 million infected in 15–49 age group

Human Papillomaviruses

These viruses are the causative agents of genital warts. But an individual can be infected with these viruses without having any warts, while still risking serious consequences.

▶ Signs and Symptoms

Symptoms, if present, may manifest as warts—outgrowths of tissue on the genitals (see Disease Table 23.9). In females, these growths can occur on the vulva and in and around the vagina. In males, the warts can occur in or on the penis and the scrotum. In both sexes, the warts can appear in or on the anus and even on the skin around the groin, such as the area between the thigh and the pelvis. The warts themselves range from tiny, flat, inconspicuous bumps to extensively branching, cauliflower-like masses called **condyloma acuminata.** The warts are unsightly and can be obstructive, but they don't generally lead to more serious symptoms.

Other types of HPV can lead to more subtle symptoms. Certain types of the virus infect cells on the female cervix. This infection may be "silent," or it may lead to abnormal cell changes in the cervix. Some of these cell changes can eventually result in malignancies of the cervix. The vast majority of cervical cancers are caused by HPV infection. (It is possible that chronic infections with other microorganisms cause a very small percentage of cervical malignancies.) Approximately 4,000 women die each year in the United States from cervical cancer. In 2007, a link between oral sexual activity and an increased risk of throat cancer, presumably due to HPV, was established. In 2011, an even more alarming trend was documented in men— HPV-16, a strain associated with cervical cancer, was found in over 70% of throat cancer specimens. If the current infection rate continues, it is estimated that HPV will cause more cases of throat cancer than cervical cancer by the year 2020.

Males can also get genitourinary tract cancer from infection with these viruses. The sites most often affected are the penis and the anus. These cases are much less common than cervical cancer.

▶ Causative Agent

The human papillomaviruses are a group of nonenveloped DNA viruses belonging to the *Papovaviridae* family. There are more than 100 different types of HPV. Some types are specific for the mucous membranes; others invade the skin. Some of these viruses are the cause of plantar warts, which often occur on the soles of the feet. Other HPVs cause the common or "seed" warts and flat warts.

Among the HPVs that infect the genital tract, some are more likely to cause the appearance of warts. Ninety percent of genital warts are associated with HPV-6 and HPV-11 infection. Others that have a preference for growing on the cervix can lead to cancerous changes. Two types in particular, HPV-16, and HPV-18, are most closely associated with development of cervical cancer. Other types put you at higher risk for vulvar or penile cancer.

▶ Pathogenesis and Virulence Factors

The major virulence factors for cancer-causing HPVs are **oncogenes,** which code for proteins that interfere with normal host cell function, resulting in uncontrolled growth.

▶ Transmission and Epidemiology

Young women have the highest rate of HPV infections; 25% to 46% of women under the age of 25 are infected with genital HPV. It is estimated that 14% of female college students become infected with this incurable condition each year. The CDC estimates that 50% of sexually active adults will become infected with one of the HPVs in their lifetimes. It is probably safe to assume that any unprotected sex carries a good chance of encountering either HSV or HPV.

The mode of transmission is direct contact. Autoinoculation is also possible—meaning that the virus can be spread to other parts of the body by touching warts. Indirect transmission occurs but is more common for nongenital warts caused by HPV.

▶ Culture and Diagnosis

PCR-based screening tests can be used to test samples from a pelvic exam for the presence of dangerous HPV types. These tests are now recommended for women over the age of 30. A Greek-born physician named George Papanicolaou, developed what is now known as the "Pap smear" in the early part of the 20th century, in which changes in vaginal smears were evaluated for precancerous changes. The Pap smear is still the single best screening procedure available for cervical cancer and has caused a 74% decrease in the incidence of cervical cancer since 1955. The procedure is simple and relatively painless. During a pelvic exam, a sample of cells is taken from the cervix using a wooden spatula or small cervical brush. Then the sample is "smeared" onto a glass microscope slide and preserved with a fixative. In a newer method, the fluid is saved, and later it is automatically applied in a thin layer to a microscope slide called a "thin prep." Whether the slide is produced with a "smear" or a "thin prep," it is then viewed microscopically by a technician or by a computer so that abnormal cells can be detected. Nearly all cervical cancer can be prevented if women get Pap smears on the recommended schedule.

▶ Prevention

When discussing HPV prevention, we must consider two possibilities. One of these is infection with the viruses, which is prevented the same way other sexually transmitted infections are prevented—by avoiding direct, unprotected contact, but also by one of the two vaccines available for it. Gardasil protects against four of the most carcinogenic HPVs (6, 11, 16, and 18). Cervarix protects against the top two, 16 and 18. The three-dose vaccine regimen is recommended for both girls and boys at the age of 11 or 12. People as old as 26 who have not yet received all three doses are also encouraged to get vaccinations. Health officials emphasize that vaccinated women should still get Pap smears to screen for cervical cancer, because other strains of the HPV can also cause the condition.

The good news is that cervical cancer is slow in developing, so that even if a woman is infected with a malignant HPV type, regular screening of the cervix can detect abnormal changes early. Precancerous changes show up very early, and the development process can be stopped by removal of the affected tissue. Women should have their first Pap smear by age 21 or within 3 years of their first sexual activity, whichever comes first. Between the ages of 30 and 65, women may elect to get an HPV test at the time of their Pap smear. A negative HPV test could mean that you need not repeat the Pap smear for a period of 2 to 5 years, depending on your doctor's advice.

▶ Treatment

Genital warts can be removed through a variety of methods, some of which can be used at home. Many infections are eventually cleared by the immune system, but this is very unpredictable and may take up to 2 years. Current studies show that even when tests for viral DNA are negative, the HPV may still reside latently in an infected female—for up to 20 years or more.

Treatment of cancerous cell changes is an important part of HPV therapy, and it can only be instituted if the changes are detected through Pap smears. Again, the *results* of the infection are treated (cancerous cells removed), but the viral infection is not amenable to treatment and relies upon the activity of the host immune system.

Molluscum Contagiosum

An unclassified virus in the family *Poxviridae* can cause a condition called molluscum contagiosum. This disease can take the form of skin lesions, and it can also be transmitted sexually. The wart-like growths that result from this infection can be found on the mucous membranes or the skin of the genital area (see Disease Table 23.9). Few problems are associated with these growths beyond the warts themselves. In severely immunocompromised people, the disease can be more serious, resulting in extensive growth of warts.

The virus causing these growths can also be transmitted through fomites such as clothing or towels and through autoinoculation (**Disease Table 23.9**).

Disease Table 23.9	Wart Diseases	
	HPV	**Molluscum Contagiosum**
Causative Organism(s)	Human papillomaviruses	Poxvirus, sometimes called the molluscum contagiosum virus (MCV)
Most Common Modes of Transmission	Direct contact (STI), also autoinoculation, indirect contact	Direct contact (STI), also indirect and autoinoculation
Virulence Factors	Oncogenes (in the case of malignant types of HPV)	–
Culture/Diagnosis	PCR tests for certain HPV types, clinical diagnosis	Clinical diagnosis, also histology, PCR
Prevention	Vaccine available; avoid direct contact; prevent cancer by screening cervix	Avoid direct contact
Treatment	Warts or precancerous tissue can be removed; virus not treatable	Warts can be removed; virus not treatable
Distinguishing Features	Infection may or may not result in warts; infection may result in malignancy	Wart-like growths are only known consequence of infection
Effects on Fetus	May cause laryngeal warts	–
Appearance of Lesions		
Epidemiological Features	United States: estimated 6 million new infections per year; 12,000 new cases of HPV-associated cervical cancer	United States: affects 2%–10% of children annually

Group B *Streptococcus* "Colonization"—Neonatal Disease

Ten to forty percent of women in the United States are colonized, asymptomatically, by a beta-hemolytic *Streptococcus* in Lancefield group B (GBS). Nonpregnant women experience no ill effects from this colonization. But when these women become pregnant and give birth, about half of their infants become colonized by the bacterium during passage through the birth canal or by ascension of the bacteria through ruptured membranes; thus, this colonization is considered a reproductive tract disease.

A small percentage of infected infants experience life-threatening bloodstream infections, meningitis, or pneumonia. If they recover from these acute conditions, they may have permanent disabilities such as developmental disabilities, hearing loss, or impaired vision. In some cases, the mothers also experience disease, such as amniotic infection or subsequent stillbirths. Although GBS infections have declined in the United States, they remain a major threat to infant morbidity and mortality worldwide.

In 2002, the CDC recommended that all pregnant women be screened for group B *Streptococcus* colonization at 35 to 37 weeks of pregnancy. Because colonization has been associated with preterm birth, recommendations for earlier testing are sometimes warranted. Women positive for the bacterium should be treated with penicillin or ampicillin unless the bacterium is found to be resistant to these—and unless allergy to penicillin is present, in which case erythromycin may be used. Development of a protective vaccine is underway, but a major hurdle is the ability to design one vaccine that will protect all populations against the many serotypes of this bacterium (**Disease Table 23.10**).

We wrap up this chapter with a figure that helps you put the different sexually transmitted infections in perspective. **Figure 23.21** provides estimates of annual new infections, whether they result in frank symptoms or are asymptomatic.

Disease Table 23.10	Group B *Streptococcus* Colonization
Causative Organism(s)	Group B *Streptococcus*
Most Common Modes of Transmission	Vertical
Virulence Factors	–
Culture/Diagnosis	Culture of mother's genital tract
Prevention/Treatment	Treat mother with penicillin/ampicillin
Epidemiological Features	United States: vaginal carriage rates are 15%–45%; neonatal sepsis due to this occurs in 1.8–3.2 per 1,000 live births; internationally: vaginal carriage rates 12%–27%

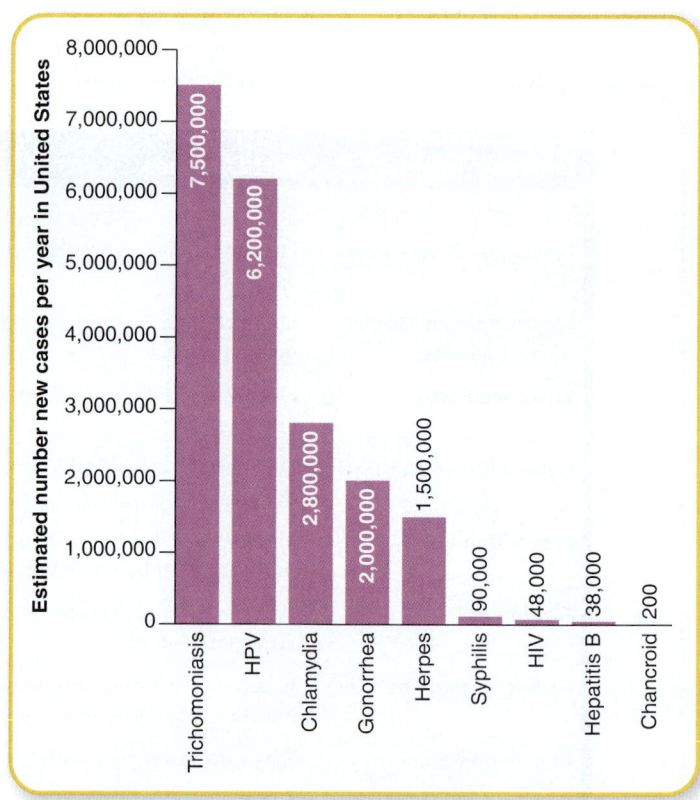

Figure 23.21 Estimates of number of new cases of various sexually transmitted infections each year in the United States. Data were compiled from multiple sources.

Case File 23   *Wrap-Up*

According to a recent CDC report, nearly one in two U.S. high-school students has had sexual intercourse. With the majority of individuals acquiring HPV soon after their first sexual experience, millions of teens are at risk. Because the HPV vaccine requires a series of three vaccinations, often teens and preteens don't get the full vaccination series even if they receive the first dose. According to the CDC, overall vaccination rates for other diseases such as tetanus, diphtheria, pertussis, hepatitis, and influenza are increasing but vaccination rates for HPV are between 15% and 56% depending on the state surveyed. As of early 2013, vaccination rates among males are only at about 8%, although the recommendation for males to receive the vaccine came out in October of 2011.

Parents may be reluctant to pursue the HPV vaccine for their preteens because they are not ready to think about their children having sex. Cancer caused by HPV may not manifest itself for decades after the initial infection and parents may not see the need to discuss the risks with their preteens, even though the infection could occur in the teenage years. Pediatricians can be effective advocates for the HPV vaccine with their preteen patients; good communication between parents, pediatricians, and preteens can increase awareness for the need for the HPV vaccine early in life.

23.4 Learning Outcomes—Assess Your Progress

7. List the possible causative agents for each of the following infectious reproductive tract conditions: vaginitis, vaginosis, prostatitis, genital discharge diseases, genital ulcer diseases, and wart diseases.
8. Identify which of the preceding conditions can cause disease through vertical transmission.
9. Distinguish between vaginitis and vaginosis.
10. Summarize important aspects of prostatitis.
11. Discuss pelvic inflammatory disease, and identify which organisms are most likely to cause it.
12. Provide some detail about HPV vaccination.
13. Identify the most important risk group for group B *Streptococcus* infection, and discuss why these infections are so dangerous in this population.

▶ Summing Up

Taxonomic Organization Microorganisms Causing Disease in the Genitourinary Tract

Microorganism	Disease	Location of Disease Table
Gram-positive bacteria		
Staphylococcus saprophyticus	Urinary tract infection	UTI, p. 725
Group B *Streptococcus*	Neonatal disease	Group B strep neonatal disease, p. 746
Gram-negative bacteria		
Enterococcus	Urinary tract infection	UTI, p. 725
Escherichia coli	Urinary tract infection	UTI, p. 725
Leptospira interrogans (spirochete)	Leptospirosis	Leptospirosis, p. 726
Neisseria gonorrhoeae	Gonorrhea	Discharge diseases, p. 736
Chlamydia trachomatis	*Chlamydia*	Discharge diseases, p. 736
Treponema pallidum (spirochete)	Syphilis	Genital ulcer diseases, p. 743
Haemophilus ducreyi	Chancroid	Genital ulcer diseases, p. 743
DNA viruses		
Herpes simplex viruses 1 and 2	Genital herpes	Genital ulcer diseases, p. 743
Human papillomaviruses	Genital warts, cervical carcinoma	Wart diseases, p. 745
Poxviruses	Molluscum contagiosum	Wart diseases, p. 745
Fungi		
Candida albicans	Vaginitis	Vaginitis, p. 729
Protozoa		
Trichomonas vaginalis	Trichomoniasis (vaginitis)	Vaginitis, p. 729
Helminth—trematode		
Schistosoma haematobium	Urinary schistosomiasis	Urinary schistosomiasis, p. 727

INFECTIOUS DISEASES AFFECTING
The Genitourinary Tract

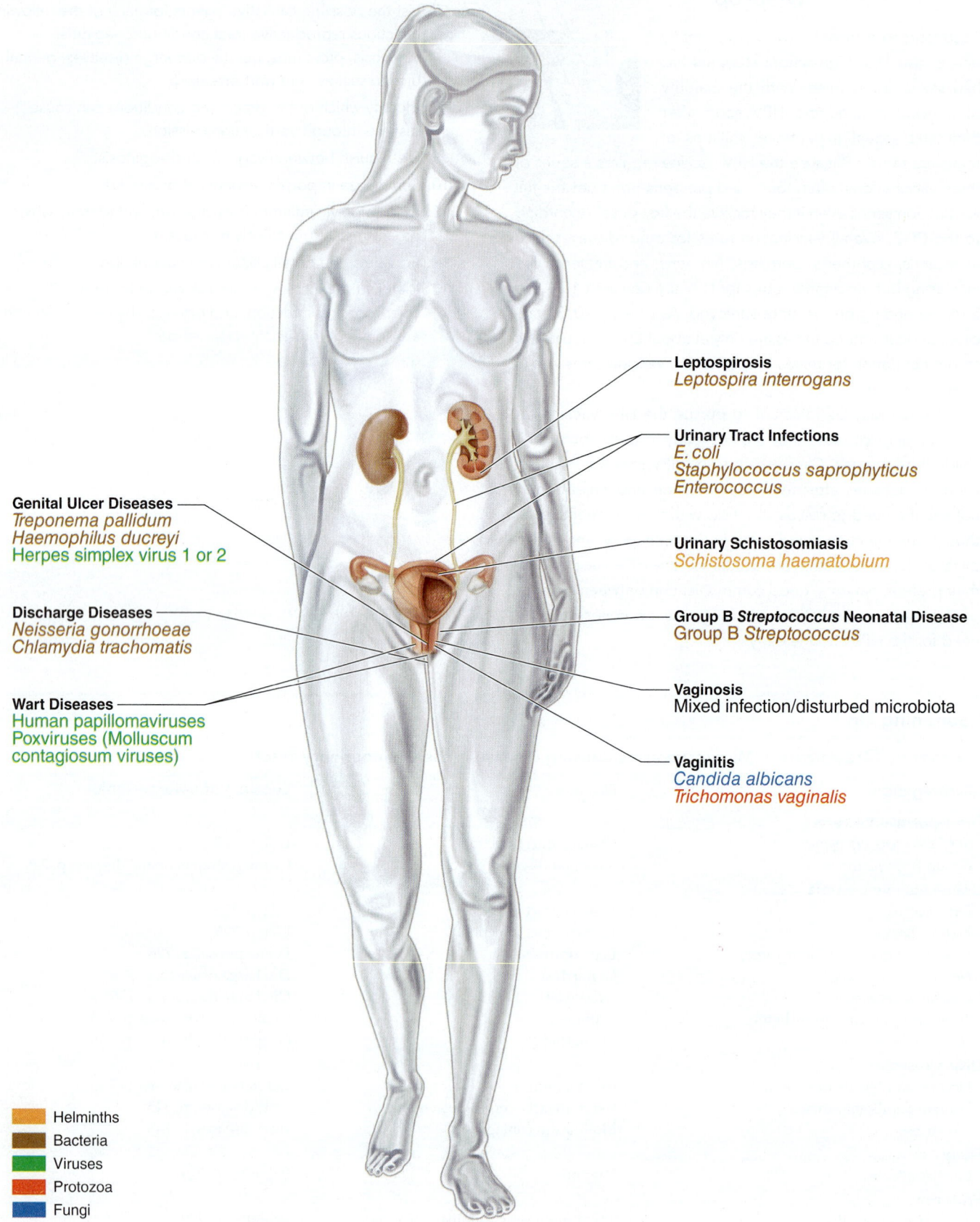

Leptospirosis
Leptospira interrogans

Urinary Tract Infections
E. coli
Staphylococcus saprophyticus
Enterococcus

Genital Ulcer Diseases
Treponema pallidum
Haemophilus ducreyi
Herpes simplex virus 1 or 2

Urinary Schistosomiasis
Schistosoma haematobium

Discharge Diseases
Neisseria gonorrhoeae
Chlamydia trachomatis

Group B *Streptococcus* Neonatal Disease
Group B *Streptococcus*

Vaginosis
Mixed infection/disturbed microbiota

Wart Diseases
Human papillomaviruses
Poxviruses (Molluscum
contagiosum viruses)

Vaginitis
Candida albicans
Trichomonas vaginalis

Helminths
Bacteria
Viruses
Protozoa
Fungi

System Summary Figure 23.22

INFECTIOUS DISEASES AFFECTING
The Genitourinary Tract

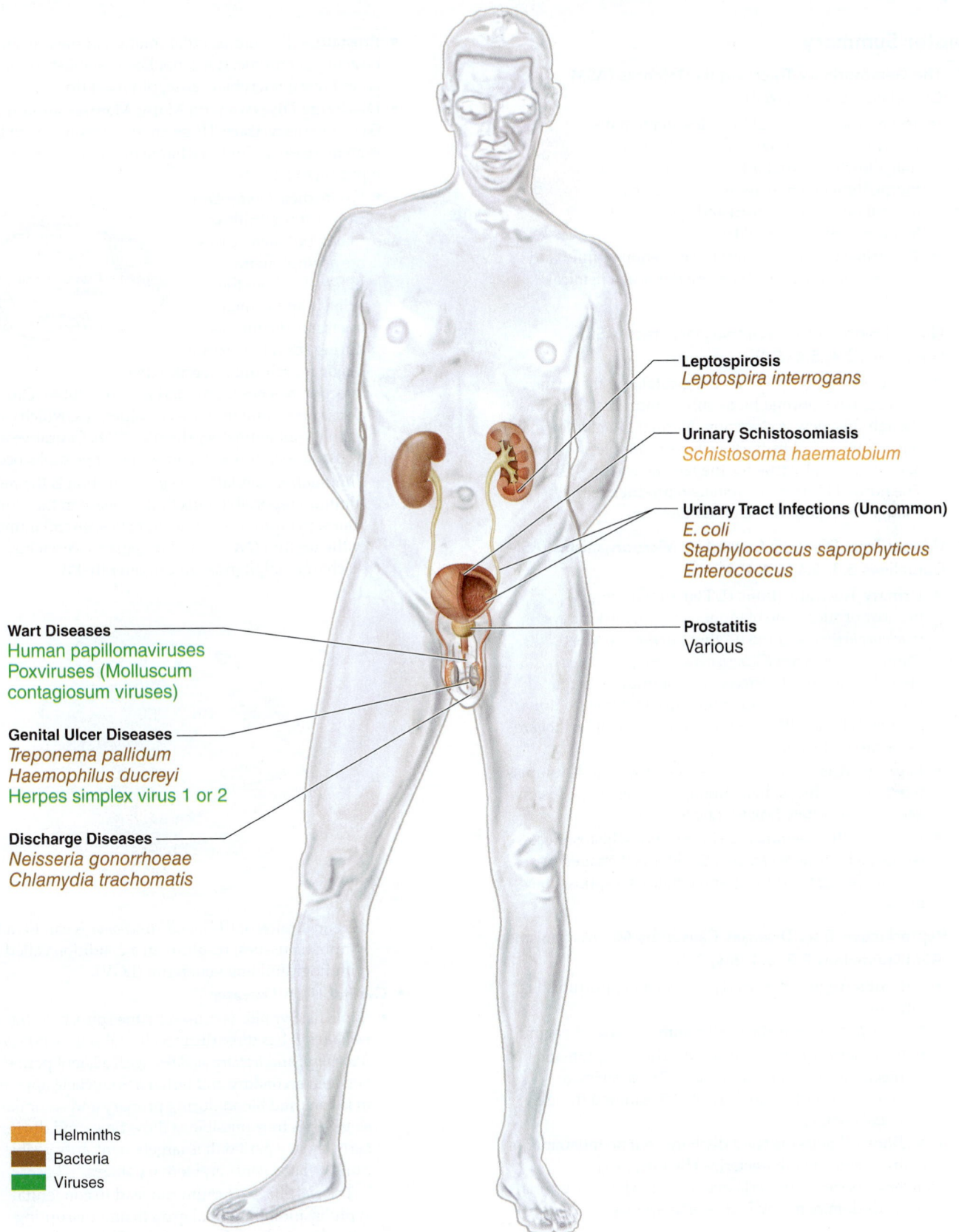

Leptospirosis
Leptospira interrogans

Urinary Schistosomiasis
Schistosoma haematobium

Urinary Tract Infections (Uncommon)
E. coli
Staphylococcus saprophyticus
Enterococcus

Prostatitis
Various

Wart Diseases
Human papillomaviruses
Poxviruses (Molluscum
contagiosum viruses)

Genital Ulcer Diseases
Treponema pallidum
Haemophilus ducreyi
Herpes simplex virus 1 or 2

Discharge Diseases
Neisseria gonorrhoeae
Chlamydia trachomatis

Helminths
Bacteria
Viruses

System Summary Figure 23.23

Chapter Summary

Chapter Summary

23.1 The Genitourinary Tract and Its Defenses (ASM Guidelines* 3.4, 5.4, 6.4)

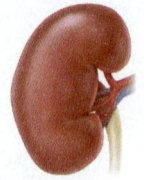

- The reproductive tract in males and females is composed of structures and substances that allow for sexual intercourse and the creation of a new fetus; protected by normal mucosal defenses and specialized features (such as low pH).
- The urinary system allows the excretion of fluid and wastes from the body. It has mechanical, chemical defense mechanisms.

23.2 Normal Biota of the Genitourinary Tract (ASM Guidelines 3.4, 5.4, 6.4)

- Current knowledge is that the genital and the urinary systems have normal biota only in most distal regions, though this view is changing. Normal biota in the male reproductive and urinary systems resemble skin biota. Same is generally true for the female urinary system. The normal biota in the female reproductive tract changes over the course of her lifetime.

23.3 Urinary Tract Diseases Caused by Microorganisms (ASM Guidelines 5.3, 5.4, 6.4, 8.3)

- **Urinary Tract Infections (UTIs):** Can occur at a number of sites—the bladder (*cystitis*), the kidneys (*pyelonephritis*), and the urethra (*urethritis*). Most common causes are *Escherichia coli*, *Staphylococcus saprophyticus*, and *Enterococcus*. Community-acquired UTIs are most often transmitted from GI tract to urinary system. UTIs are the most common cause of healthcare-associated infections.
- **Leptospirosis:** Zoonosis is associated with wild animals that affects kidneys, liver, brain, and eyes. The causative agent is spirochete *Leptospira interrogans*.
- **Urinary Schistosomiasis:** This form of schistosomiasis is caused by *S. haematobium*. Bladder is damaged by trematode eggs and the granulomatous response they induce.

23.4 Reproductive Tract Diseases Caused by Microorganisms (ASM Guidelines 5.3, 5.4, 6.4, 8.3)

- **Vaginitis:** Vaginitis is nearly always an opportunistic infection.
  - *Candida albicans* is the most common cause of vaginitis.
  - *Trichomonas vaginalis* causes mostly asymptomatic infections in females and males. *Trichomonas*, a flagellated protozoan, is easily transmitted through sexual contact.
- **Vaginosis:** Vaginosis has a discharge but no inflammation in the vagina. It is characterized by the presence of diverse species of bacteria instead of the healthy vagina, which is dominated by *Lactobacillus* species.

* Source: *ASM Curriculum Guidelines* (American Society for Microbiology, 2012). Complete guidelines in appendix B of this book.

- **Prostatitis:** Prostatitis, inflammation of the prostate, can be acute or chronic. It has not been established that all cases have a microbial cause, but most do.
- **Discharge Diseases with Major Manifestation in the Genitourinary Tract:** These are diseases in which there is an increase in fluid discharge in the male and female reproductive tracts.
  - **Gonorrhea:** Gonorrhea can elicit urethritis in males, but many cases are asymptomatic. In females, both the urinary and genital tracts may be infected during sexual intercourse.

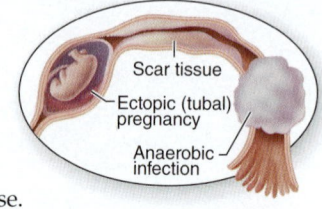

    Major complications occur when infection reaches uterus and fallopian tubes. One disease resulting from this is salpingitis, which can lead to pelvic inflammatory disease (PID). Causative agent, *Neisseria gonorrhoeae*, is a gram-negative diplococcus.
  - *Chlamydia:* Genital *Chlamydia* infection is the most common reportable infectious disease in the United States. Symptoms include, in males, an inflammation of the urethra (NGU), and in females, cervicitis, discharge, salpingitis, and frequently PID.

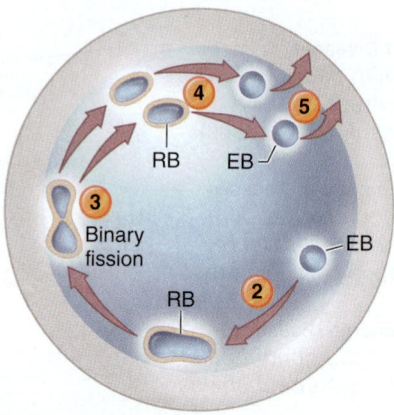

    Certain strains of *Chlamydia trachomatis* can invade lymphatic tissues, resulting in a condition called lymphogranuloma venereum (LGV).
- **Genital Ulcer Diseases**
  - **Syphilis**: Syphilis is caused by the spirochete *Treponema pallidum*. It has three distinct clinical stages: *primary*, *secondary*, and *tertiary syphilis*, with a latent period between secondary and tertiary. Spirochete appears in lesions and blood during primary and secondary stages; it is transmissible at these times and during early latency period. It is largely nontransmissible during "late latent" and tertiary stages.

    The syphilis bacterium can lead to congenital syphilis, inhibiting fetal growth and disrupting critical periods of development. This can lead to spontaneous miscarriage or stillbirth.
  - **Chancroid:** Chancroid is caused by *Haemophilus ducreyi*, a pleomorphic gram-negative rod, and is

transmitted exclusively through direct—mainly sexual—contact.

- **Genital Herpes:** Genital herpes is caused by herpes simplex viruses (HSVs). There are two types: HSV-1 and HSV-2. There may be no symptoms, or there may be fluid-filled, painful vesicles on genitalia, perineum, thigh, and buttocks. In severe cases, meningitis or encephalitis can develop. Patients remain asymptomatic or experience recurrent "surface" infections indefinitely. Infections in neonate and fetus can be fatal.
- **Wart Diseases**
  - **Human Papillomaviruses:** The causative agents of genital warts are human papillomaviruses. Certain types infect cells on the female cervix that eventually

result in malignancies of the cervix. Males can also get cancer from these viral types.

Genital warts can be removed, but the virus will remain. Treatment of cancerous cell changes—detected through Pap smears in females—is an important part of HPV therapy. Vaccine for several types of HPV is now available.

- **Molluscum Contagiosum:** Caused by a virus in the family *Poxviridae*, molluscum contagiosum can take the form of wart-like growths in the membranes of the genitalia, and it can also be transmitted sexually.
- **Group B *Streptococcus* "Colonization"—Neonatal Disease:** Asymptomatic colonization of women by a beta-hemolytic *Streptococcus* in Lancefield group B is very common. It can cause preterm delivery and infections in newborns.

Multiple-Choice and True-False Questions | Bloom's Levels 1 and 2: Remember and Understand

Multiple-Choice Questions. Select the correct answer from the options provided.

1. Cystitis is an infection of the
   a. bladder.
   c. kidney.
   b. urethra.
   d. vagina.

2. A form of vaginitis is caused by
   a. *Neisseria gonorrhoeae.*
   c. *Treponema pallidum.*
   b. *Chlamydia trachomatis.*
   d. *Trichomonas vaginalis.*

3. Leptospirosis transmission to humans is
   a. person to person.
   c. by mosquitoes.
   b. by fomites.
   d. by contaminated soil or water.

4. Syphilis is caused by
   a. *Treponema pallidum.*
   c. *Trichomonas vaginalis.*
   b. *Neisseria gonorrhoeae.*
   d. *Haemophilus ducreyi.*

5. "Yeast infections" are caused by
   a. *Candida albicans.*
   d. all of the above.
   b. group B *Streptococcus.*
   e. none of the above.
   c. *Trichomonas.*

6. This dimorphic fungus is a common cause of vaginitis.
   a. *Candida albicans*
   c. *Trichomonas*
   b. *Gardnerella*
   d. all of the above

7. There are estimates that approximately _____% of adult Americans have genital herpes.
   a. 2
   c. 20
   b. 10
   d. 50

8. Genital herpes transmission can be reduced or prevented by all of the following *except*
   a. a condom.
   c. the contraceptive pill.
   b. abstinence.
   d. a female condom.

9. The drug Flagyl can be used to treat the protozoan infection
   a. *Neisseria gonorrhoeae.*
   c. *Treponema pallidum.*
   b. *Chlamydia trachomatis.*
   d. *Trichomonas vaginalis.*

10. Which group(s) should be vaccinated for HPV infection?
    a. female college students who were not vaccinated at age 11 or 12
    b. male college students who were not vaccinated at age 11 or 12
    c. baby boomers who were not vaccinated at age 11 or 12
    d. two of the above

True-False Questions. If the statement is true, leave as is. If it is false, correct it by rewriting the sentence.

11. Genital herpes can be treated with acyclovir.

12. Chancroid is caused by a fungus.

13. The vast majority of cervical cancers are caused by human papillomavirus.

14. *Chlamydia* infection is the most common STI in the United States.

15. Group B *Streptococcus* infection is generally silent in adult females.

Critical Thinking Questions | Bloom's Levels 3, 4, and 5: Apply, Analyze, and Evaluate

Critical thinking is the ability to reason and solve problems using facts and concepts. These questions can be approached from a number of angles and, in most cases, they do not have a single correct answer.

1. Because the acidic vaginal pH is maintained in part by normal biota during reproductive years to prevent microbial infection, describe how sperm are capable of gaining entrance into the uterus on their way to fertilizing an ovum.

2. a. Urine in the bladder has always been thought to be a sterile fluid. Describe the diversity of microbes being identified in urine today, and explain why previous studies did not detect these microbes in the past.
   b. Describe how ExPEC strains of *E. coli* are able to cause disease at a site distant from the area of the body they normally colonize.

3. A 38-year-old male living in Hawaii tried to rescue some of the family's belongings as the basement filled with water from nearby stream flooding. Within a few days, he developed flulike symptoms, but these symptoms rapidly cleared on their own. Several weeks later, however, he developed a painful headache and jaundice; at this point, he immediately sought medical attention. Urinalysis revealed the signs of a distinct pathogen, and serology showed he had increasing levels of IgM antibody. Explain what disease you think this man was suffering from and describe the causative agent.

4. Summarize how a laboratory technologist would identify a case of vaginosis versus a case of vaginitis from a vaginal swab specimen.

5. a. Describe the main characteristics of PID, and list the two microorganisms that are most commonly associated with this disease.
   b. Conduct additional research on current drug-resistant strains of *N. gonorrhoeae* and create a map showing their worldwide distribution today.

6. a. Explain why microscopic analysis of a urine specimen is more accurate for *Chlamydia* screening in males than in females.

   b. Describe the life cycle of *Chlamydia,* and explain how it plays a direct role in the pathogen's ability to cause pelvic inflammatory disease.

7. a. A young man presents to his primary care physician with genital lesions and he is told that he has herpes. He refuses to believe this diagnosis because he is in a long-standing relationship with a woman who clearly has never shown signs of vaginal herpes lesions. Construct an informative response to this patient based upon the information in this chapter.
   b. Thinking about the previous question, explain why the actual number of people in the United States who have genital herpes may be a lot higher than official statistics depict.

8. Summarize the clinical stages of syphilis. Suggest reasons for the observed increase in the number of syphilis cases identified in recent years in the United States and abroad.

9. Provide evidence in support of or refuting the following statement: An HPV screening test is readily available for men today and involves analysis of a urine specimen.

10. Why are urinary tract infections such common healthcare-associated infections? Conduct additional research and provide an update on the significance of VRE, CRE, and *E. coli* ST131 infections today.

Concept Connections | Bloom's Levels 4 and 6: Analyze and Create

This activity ties together multiple concepts in the chapter.

1. What are the signs and symptoms of a urinary tract infection (UTI)?

2. What organisms are the likely causes of a UTI?

3. What are the likely treatments of a UTI?

4. If you are a male, what is another cause of painful urination in addition to a UTI?

5. What are the likely causative organisms and the subsequent infections if there is discharge associated with painful urination?

6. What are the possible treatments of these infections? What are the treatment options if an organism is resistant to antibiotics?

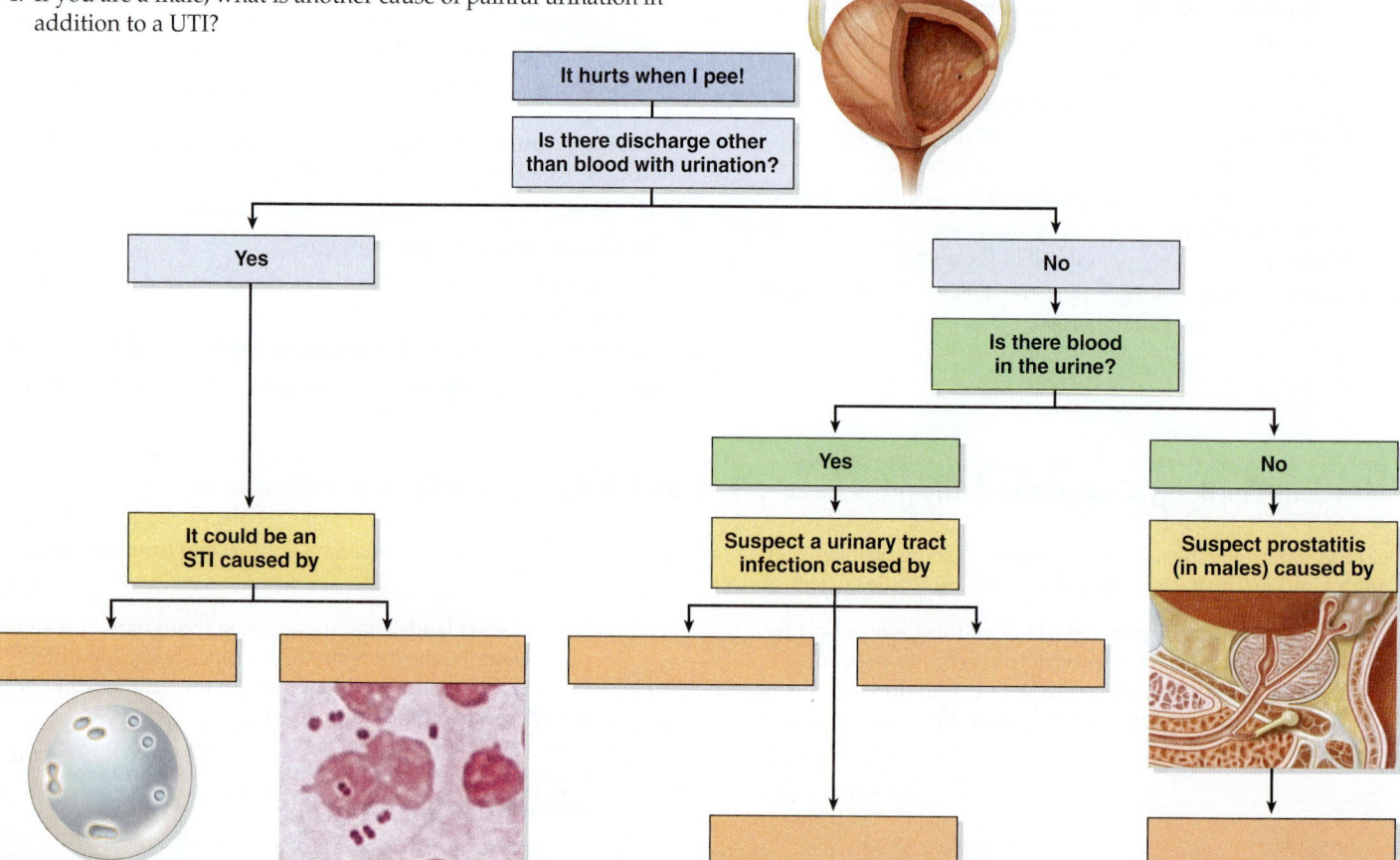

Visual Connections | Bloom's Level 5: Evaluate

This question uses visual images or previous content to make connections to this chapter's concepts.

1. a. **From chapters 20 and 23, figures 20.16 and 23.13a.**
   Compare these two rashes. What kind of information would help you determine the diagnosis in both cases?
   b. Now compare both of these to the rashes summarized in **Disease Table 18.7** (p. 532–533) . Which of the diseases

in Disease Table 18.7 most resembles the rashes in the preceding question, and how would you distinguish among it, the rash from figure 20.16, and the one from figure 23.13a?

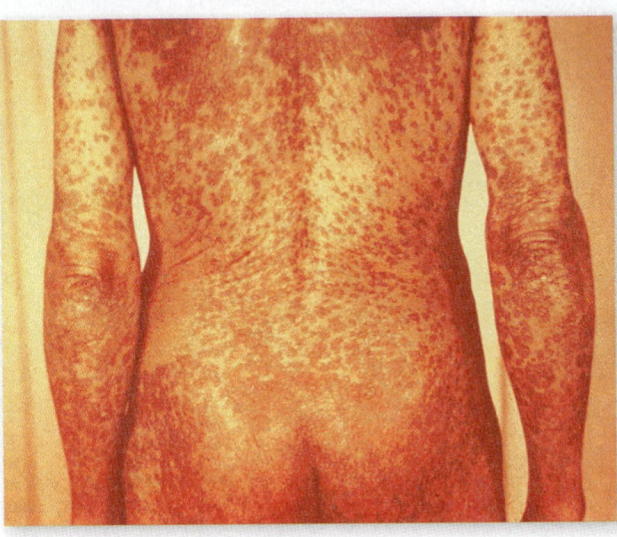

(a)

Concept Mapping | Bloom's Level 6: Create

Appendix D provides guidance for working with concept maps.

1. Using the words that follow, please create a concept map illustrating the relationships among these key terms from chapter 23.

genital warts	bacterium	ulcers	cancer
discharge	molluscum contagiosum	warts	
herpes	virus	syphilis	
chancroid	curable	incurable	

www.mcgrawhillconnect.com
Enhance your study of this chapter with study tools and practice tests. Also ask your instructor about the resources available through ConnectPlus, including the media-rich eBook, interactive learning tools, and animations.

24

The *Deepwater Horizon* oil spill in the Gulf of Mexico.

Microbes and the Environment

Case File 24

Where Did All the Methane Go?

The British Petroleum (BP) *Deepwater Horizon* oil spill in the Gulf of Mexico in April 2010 was the largest oil spill in history, releasing an estimated 4.9 million barrels of oil (205.8 million gallons), contaminating 665 miles of coastline, and having an untold impact on the environment. Among the many concerns about the impact that the massive spill would have on the Gulf of Mexico and its flora and fauna, scientists were concerned that the methane released by the spill would eventually make its way into the atmosphere and, as a greenhouse gas, would contribute to global warming. Others speculated that it would remain trapped in the ocean, suffocating all life and creating "dead zones." According to measurements by the Department of Energy (DOE) and BP, methane was proportionately the most abundant substance released in the spill as compared to crude oil and octane.

When researchers studied methane levels in the ocean shortly after the spill, they found a layer 110 to 220 yards thick about two-thirds of a mile below the surface at concentrations 10,000 to 100,000 times the normal levels—the highest concentration of methane ever measured in the ocean. As scientists continued to monitor the methane concentration in the Gulf, they made a surprising finding: The concentration of methane decreased to nearly zero within the space of a few months. What accounted for the dramatic drop in methane levels in the Gulf? You guessed it: bacteria.

- Where did the methane-consuming bacteria in the Gulf of Mexico come from?
- What microbial species were involved in the methane consumption?

Continuing the Case appears on page 758.

Outline and Learning Outcomes

24.1 Ecology: The Interconnecting Web of Life
1. Define *microbial ecology*.
2. Summarize why our view of the abundance of microbes on earth has changed in recent years.
3. Discuss the terms *ecosystem* and *community* in relation to one another.

4. Differentiate between habitat and niche.
5. Draw an example of an energy pyramid, labeling primary producers, consumers, and decomposers.
6. Define *bioremediation,* and provide one example of its use today.

24.2 The Natural Recycling of Bioelements

7. List five important elements of biogeochemical cycles.
8. Diagram a carbon cycle.
9. Explain the role of methanogens within the carbon cycle.
10. List the four reactions involved in the nitrogen cycle.
11. Describe the process of nitrogen fixation, and explain how microbes play a role in this biochemical reaction.
12. Briefly summarize both the sulfur and phosphorus cycles.

24.3 Microbes on Land and in Water

13. Outline the basic process used to perform metagenomic analysis of the environment.
14. List two important symbiotic partnerships that occur in the soil.
15. Diagram the hydrologic cycle.
16. Discuss how metagenomic sampling has changed our view of deep subsurface and oceanic microbiology.
17. List the stratified regions of large bodies of standing water, and describe how microbes are affected by this layering.
18. Define *eutrophication,* and explain how microbes are responsible for its impact on aquatic life.

This chapter emphasizes microbial activities that help maintain, sustain, and control the life-support systems on the earth. This subject is explored from the standpoint of the natural roles of microorganisms in the environment and their contributions to the ecological balance, including soil, water, and mineral cycles.

24.1 Ecology: The Interconnecting Web of Life

The study of microbes in their natural habitats is known as **microbial ecology;** the study of the practical uses of microbes in food processing, industrial production, and biotechnology is known as industrial or **applied microbiology** (see chapter 25). The two areas actually overlap to a considerable degree—largely because most natural habitats have been altered by human activities. Human intervention in natural settings has changed the earth's warming and cooling cycles, increased wastes in soil, polluted water, and altered some of the basic relationships between microbial, plant, and animal life. We know one thing for certain: Microbes—the most vast and powerful resource of all—will be silently working in nature.

In chapter 7, we first touched upon the widespread distribution of microorganisms and their adaptations to most habitats of the world, from extreme environments to more welcoming locations. Although we have known for a long time that geologic features on the earth, including coal and limestone, are formed in small or large part by microbes, it is only recently that we have come to understand the sheer mass of microbial life present on our planet. Remember that the vast majority of microbial life has not yet been cultured. With the development of genomic techniques that do not rely on cultivating bacteria, we have discovered abundant microbial life all over—and within and around—our planet **(figure 24.1)**. We are learning about the planet-shaping effects of bacteria deep in the earth's core and deep within

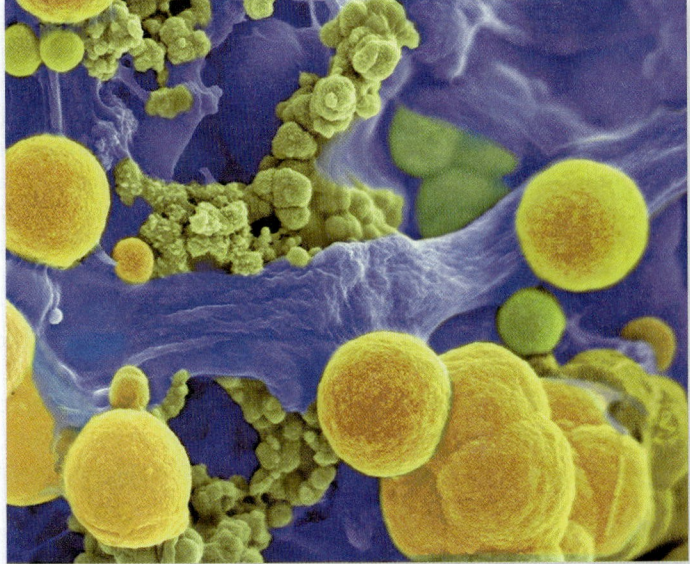

Figure 24.1 A sample of water from a deep cavern as imaged by scanning electron microscopy. This view shows a bacterial biofilm that actively forms mineral deposits of zinc and sulfate (light green and yellow). This single image brings focus to several themes of this chapter: (1) Microbes work together in mixed communities, (2) microbes can alter the chemistry of the nonliving environment, and (3) microbes can be used to control undesirable wastes created by humans.

glaciers. The sheer abundance of viral life in the oceans has been a huge surprise.

Ecological studies deal with both the biotic and the abiotic components of an organism's environment. **Biotic** factors are defined as any living or dead organisms[1] that occupy an organism's habitat. **Abiotic** factors include nonliving

---
1. Biologists make a distinction between nonliving and dead. A nonliving thing has never been alive, whereas a dead thing was once alive but no longer is.

components such as atmospheric gases, minerals, water, temperature, and light. You may recall these from chapters 7 and 11 as factors that affect microbial growth. A collection of organisms together with its surrounding physical and chemical factors is defined as an **ecosystem.**

The Organization of Ecosystems

The earth initially may seem like a random, chaotic place, but it is actually an incredibly organized, fine-tuned machine. Ecological relationships exist at several levels, ranging from the entire earth all the way down to a single organism **(figure 24.2).** The most all-encompassing of these levels, the **biosphere,** contains all physical locations on earth that support life, including the thin envelope of life that surrounds the earth's surface and extends several miles below. This global ecosystem is comprised of the **hydrosphere** (water), the **lithosphere** (a few miles into the soil), and the **atmosphere** (a few miles into the air). The biosphere maintains or creates the conditions of temperature, light, gases, moisture, and minerals required for life processes. The biosphere can be naturally subdivided into terrestrial and aquatic realms. The terrestrial realm is usually distributed into particular climatic regions called **biomes** (by'-ohmz), each of which is characterized by a dominant plant form, temperature, and precipitation. Particular biomes include grassland, desert, mountain, and tropical rain forest. The aquatic biosphere is generally divisible into freshwater and marine realms. The earth's crust also supports a vast and diverse number of life forms, estimated to be equal to or even greater than life as we know it in aquatic and terrestrial realms.

Biomes and aquatic ecosystems are generally composed of mixed assemblages of organisms that live together at the same place and time and that usually exhibit well-defined nutritional or behavioral interrelationships. These clustered associations are called **communities.** Although most communities are identified by their easily visualized dominant plants and animals, they also contain a complex assortment of bacteria, fungi, algae, protozoa, and viruses. The basic units of community structure are **populations,** groups of organisms of the same kind. For organisms with sexual reproduction, this level is the species. In contrast, bacteria are classified using taxonomic units such as "strain" The organizational unit of a population is the individual organism; each multicellular organism, in turn, has its own levels of organization (organs, tissues, cells). Within an ecosystem, each organism tends a recognizable habitat and niche. The **habitat** is the physical location and environment to which an organism has adapted. In the case of microorganisms, the habitat is frequently a *microenvironment*, where particular qualities of oxygen, light, or nutrient content are suitable for that microorganism. The **niche** is the overall role that a species (or population) serves in a community. This includes such activities as nutritional intake (what it eats), position in the community structure (what is eating it), and rate of population growth. A niche can be broad (such as scavengers that feed on nearly any organic food source) or narrow (microbes that decompose cellulose in forest litter).

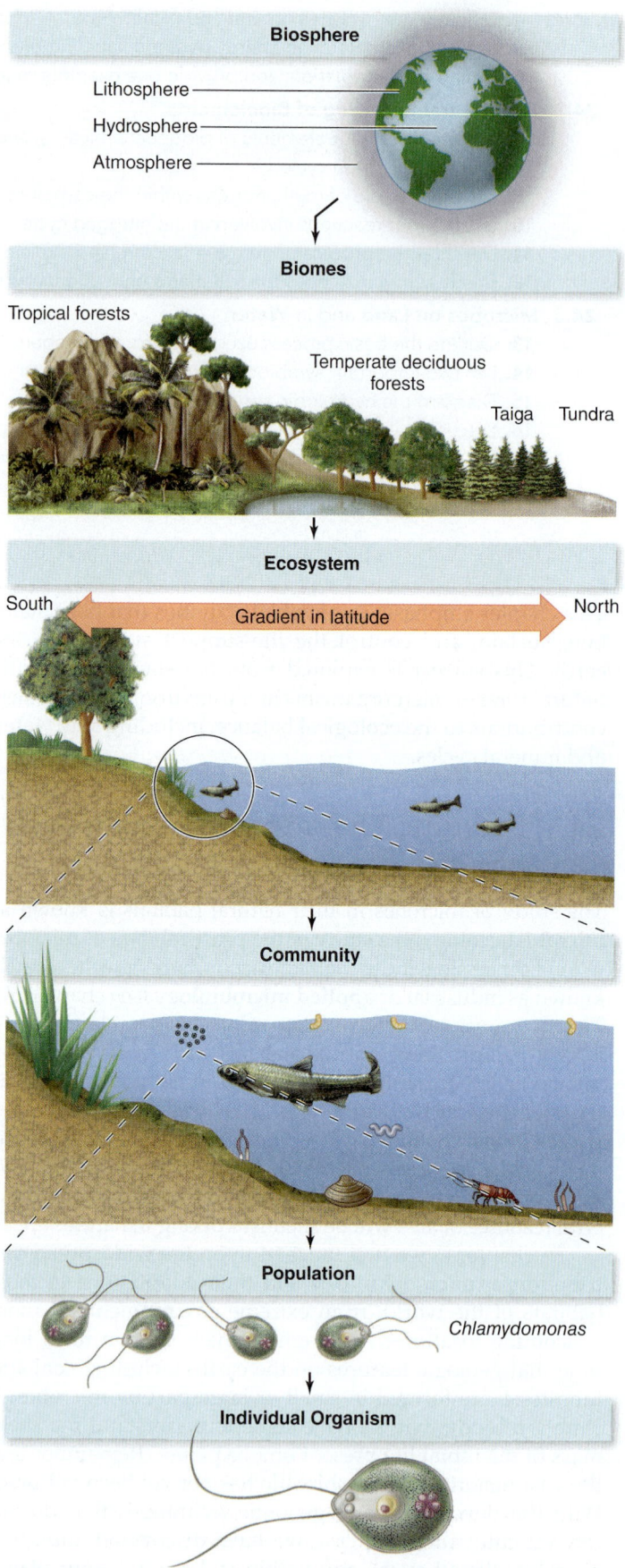

Figure 24.2 Levels of organization in an ecosystem, ranging from the biosphere to the individual organism.

INSIGHT 24.1

Colonizing Mars with Bacteria?

The Mars rover *Curiosity* has captured the attention of scientists, astronomers, and amateur stargazers alike with its fascinating pictures and analysis of the surface of Mars. One of its missions is to determine if there ever was water on the surface of the Red Planet. They did find evidence of past water, but after a recent controversy came to light, some NASA officials are hoping that it won't find water there in the present day. To protect the rover from any contamination, technicians worked in clean-room conditions, routinely wiped all surfaces with alcohol and sterilized equipment whenever possible. Six months before the launch of *Curiosity,* engineers became concerned about an important drill mechanism designed to dig into the surface of the planet in search of water. They decided to open the sterilized box holding the drill bit to do some maintenance to prevent any problems during the landing. When they did so, they exposed the drill bit to microbes from earth that could make their way to Mars. According to NASA's Planetary Protection Guidelines, the rover was required to carry no more than 300,000 bacterial endospores on any surface that might make contact with the Martian landscape. However, bacterial endospores are impervious to the disinfecting action of alcohol and are known to survive the vacuum and radiation of space; if *Curiosity* finds liquid water, any endospores contaminating the surface of the rover could germinate and reproduce. It was estimated that approximately 250,000 endospores may have survived the space flight as a result of the breach in sterilization when the drill bit box was opened.

Source: 2012. www.businessinsider.com/mars-rover-contaminated-with-earth-bacteria-2012-9.

The Mars rover *Curiosity.*

benefits of photosynthesis for the bacterium to use. The bacterium, named *Desulforudis audaxviator,* has to extract everything it needs from an abiotic environment. Apparently it garners energy from metabolizing the hydrogen and sulfate produced from the radioactive decay of uranium in the rocks, and it possesses genes that enable it to leach inorganic carbon and nitrogen from the environment. The interesting spin on this discovery is that it now suddenly makes the possibility of finding microbial life on other planets more plausible. As one researcher said of *Desulforudis:* "This is just the kind of organism that could survive on Mars." NASA's *Curiosity* rover continues to seek out evidence of these potential life forms today (see **Insight 24.1**).

Energy and Nutritional Flow in Ecosystems

All living things must obtain nutrients and a usable form of energy from their abiotic and biotic environments. The energy and nutritional relationships in ecosystems can be described in a number of convenient ways. A **food chain,** or **energy pyramid,** provides a simple summary of the general trophic (feeding) levels, designated as producers, consumers, and decomposers, and traces the flow and quantity of available energy from one level to another **(figure 24.3).** It is worth

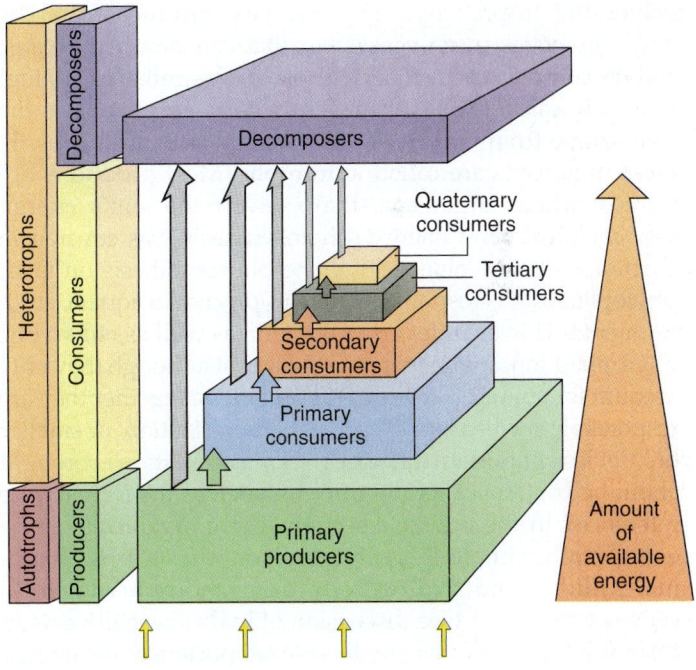

Figure 24.3 A trophic and energy pyramid. The relative size of the blocks indicates the number of individuals that exist at a given trophic level. The orange arrow on the right indicates the amount of usable energy from producers to top consumers. Both the number of organisms and the amount of usable energy decrease with each trophic level. Decomposers are an exception to this pattern but only because they can feed from all trophic levels (gray arrows). Blocks shown on the left indicate the general nutritional types and levels that correspond with the pyramid.

Microbes in natural ecosystems exhibit an amazing tendency to adapt to extreme environments. Pure cultures are seldom found anywhere in nature. One exception to this rule is particularly noteworthy. In 2008, researchers found a bacterium living completely alone, with no other life forms in its ecosystem. It was found in a South African gold mine, in fluid collected in cracks in the rock 2 miles below the surface. Obviously, there is no light there, and there are also no photosynthetic organisms (such as plants) to offer the indirect

Table 24.1 The Major Roles of Microorganisms in Ecosystems

Role	Description of Activity	Examples of Microorganisms Involved
Primary producers	Photosynthesis	Algae, bacteria, sulfur bacteria
	Chemosynthesis	Chemolithotrophic bacteria in thermal vents
Consumers	Predation	Free-living protozoa that feed on algae and bacteria; some fungi that prey upon nematodes
Decomposers	Degradation of plant and animal matter and wastes	Soil saprobes (primarily bacteria and fungi) that degrade cellulose, lignin, and other complex macromolecules
	Mineralization of organic nutrients	Soil bacteria that reduce organic compounds to inorganic compounds such as CO_2 and minerals
Cycling agents for biogeochemical cycles	Recycling compounds containing carbon, nitrogen, phosphorus, sulfur	Specialized bacteria that transform elements into different chemical compounds and keep them cycling from the biotic to the abiotic and back to the biotic phases of the biosphere
Parasites	Living and feeding on hosts	Viruses, bacteria, protozoa, fungi, and worms that play a role in population control

noting that microorganisms are the only living beings that exist at all three major trophic levels. The nutritional roles of microorganisms in ecosystems are summarized in **table 24.1.**

Life would not be possible without **primary producers,** because they provide the fundamental energy source that drives the trophic pyramid. Primary producers are the only organisms in an ecosystem that can produce organic carbon compounds such as glucose by assimilating (fixing) inorganic carbon (CO_2) from the atmosphere. If CO_2 is the sole source from which they can obtain carbon for growth, these organisms are called **autotrophs.** Most producers are photosynthetic organisms that convert the sun's energy into chemical bond energy (photosynthesis was covered in chapter 8). While plants dominate photosynthesis on land, phytoplankton is responsible for this process in aquatic environments. This includes cyanobacteria as well as eukaryotic phytoplankton, and recent data show that although they only account for about 2% of the earth's biomass, together they are responsible for nearly 50% of all carbon fixation. A smaller but not less important amount of CO_2 assimilation is brought about by bacteria called lithotrophs, such as the *Desulforudis* referenced in the previous section. These organisms derive energy from simple inorganic compounds such as ammonia, sulfides, and hydrogen by using redox reactions. In certain ecosystems (see discussion of hydrothermal vents in Insight 7.1), lithotrophs are the sole supporters of the energy pyramid as primary producers.

Consumers feed on other living organisms and obtain energy from bonds present in the organic substrates they contain. The category includes animals, protozoa, and a few bacteria and fungi. A pyramid usually has several levels of consumers, ranging from *primary consumers* (grazers or herbivores), which consume producers; to *secondary consumers* (carnivores), which feed on primary consumers; to *tertiary consumers,* which feed on secondary consumers; and up to *quaternary consumers* (usually the last level), which feed

on tertiary consumers. **Figures 24.4** and **24.5** show specific organisms at these levels.

Decomposers, primarily microbes inhabiting soil and water, break down and absorb the organic matter of dead organisms, including plants, animals, and other microorganisms. Because of their biological function, decomposers are active at all levels of the food pyramid. Without this important nutritional class of saprobes, the biosphere would stagnate and die. The work of decomposers is to reduce organic matter into inorganic minerals and gases

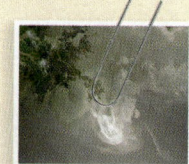

Case File 24   *Continuing the Case*

Methane-consuming microbes, also known as **methanotrophs,** exist naturally in the environment and cluster around methane-seeping vents in the ocean floor, living off of the small amounts of methane released there. However, when the *Deepwater Horizon* exploded, there was a massive increase in methane in the area—a banquet for methanogens. For 2 months, they gorged themselves on the abundance of methane and consumed nearly all of the methane released in the spill. Scientists estimate that these microbes consumed 200,000 tons of oil and natural gas within 5 months after the spill. They utilized metagenomic analysis to determine the identity of the methane consumers and found *Oceanospirillales, Colwellia, Methylococcaceae,* and *Cycloclasticus.* All of these organisms are also **psychrophilic** and grow optimally at 40°F, the temperature at the depths where the largest concentration of methane was found in the Gulf.

■ How did scientists measure the volume of methane that had been consumed by microbes?

Food Chain

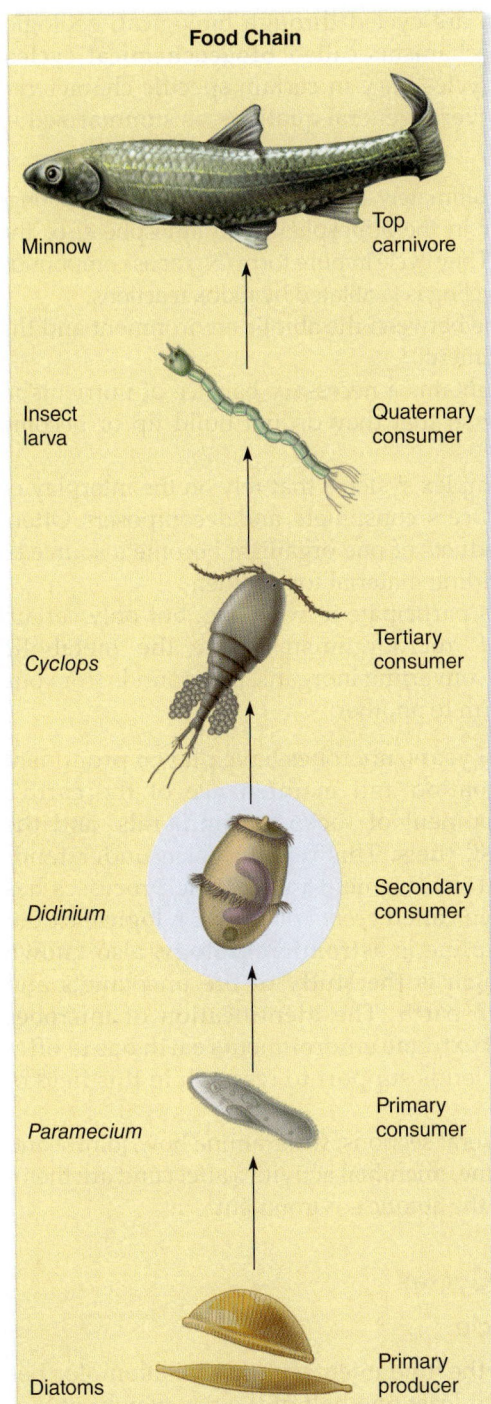

Food Web

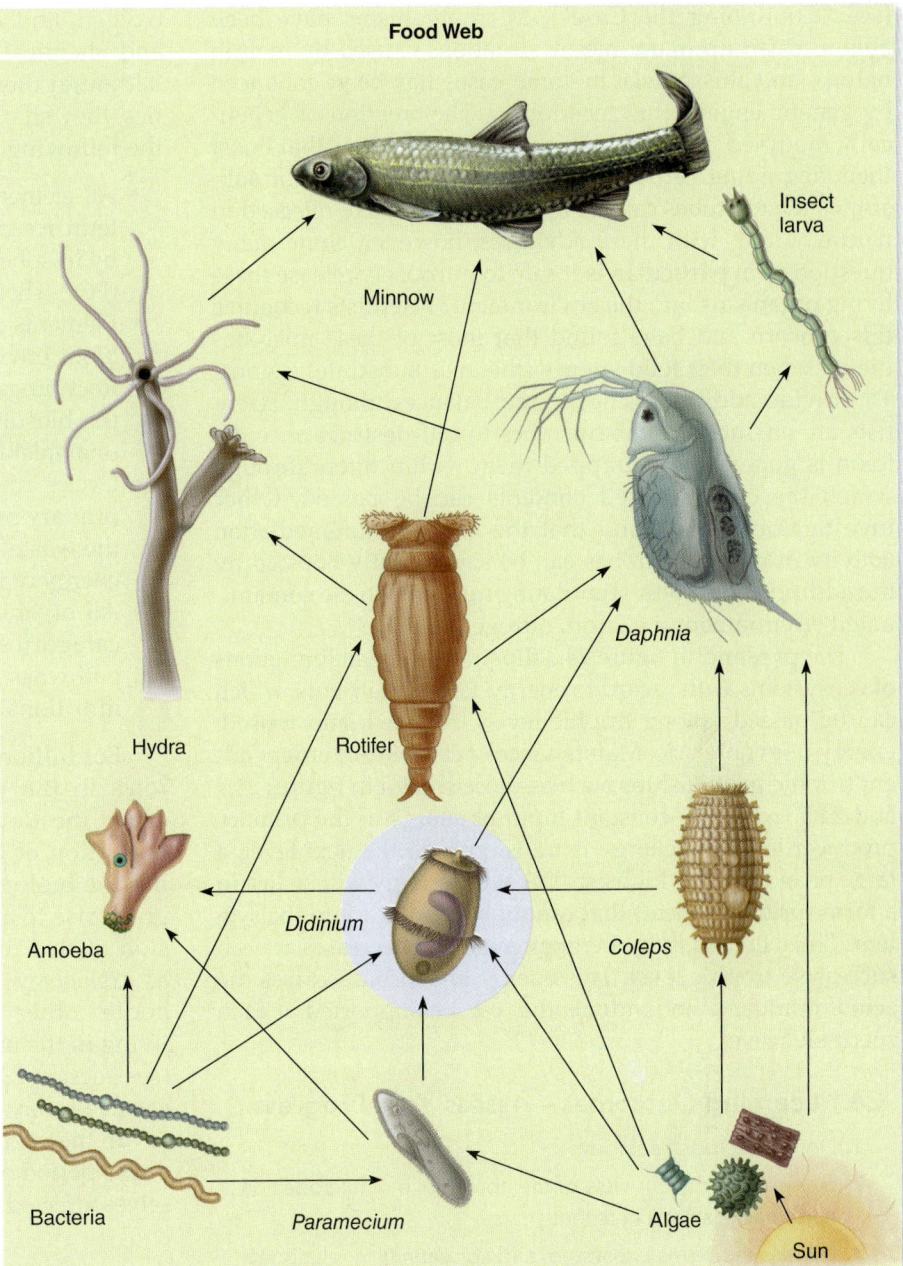

Figure 24.5 Food web. More complex trophic patterns are accurately depicted by a food web, which traces the multiple feeding options that exist for most organisms. *Note:* Arrows point toward the consumers. Compare this pattern of feeding with the chain in figure 24.4 (organisms not to scale).

Figure 24.4 Food chain. A food chain is the simplest way to present specific feeding relationships among organisms, but it may not reflect the total nutritional interactions in a community (figure not to scale).

that can be cycled back into the ecosystem, especially for the use of primary producers. This process, also called **mineralization,** is so efficient that almost all biological compounds can be reduced by some type of decomposer. Numerous microorganisms decompose cellulose and lignin, polysaccharides from plant cell walls that account for the vast bulk of detritus in soil and water. Surprisingly, decomposers can also break down most human-made compounds that are not naturally found on earth. This process is referred to as **bioremediation.**

A variety of microbes isolated from the environment have been found to degrade anything from radioactive waste to Styrofoam. Often, bioremediation involves more than one kind of microbe, and the collection of participating microbes in this process is known as a **consortium** (see "Continuing the Case"). Microorganisms have been able to clean up many messy situations caused by man in nature, and this power in some cases has been enhanced by genetic engineering technology. The creation of genetically modified microorganisms containing genes that boost their degrading capabilities will widen the range of substrates the microbes can act upon when they are released in nature. Along with these advances, however, come many questions—in particular, is it safe to purposely release these living organisms into the environment? Scientists recognize this concern and have found that most of these microbes die off when their food supply (the toxic substrate) is gone. To provide additional checks and balances, though, scientists are engineering the microbes to self-destruct once the toxin is gone or have trapped them within filters through which the contaminated contents can be passed. Other investigators have found that the natural bioremediation activity of native microbes can be significantly boosted by the addition of growth-stimulating nutrients to the contaminated environment (i.e., iron, nitrogen).

The pyramid in figure 24.3 illustrates several limitations of ecosystems with regard to energy. Unlike nutrients, which can be passed among trophic levels, recycled, and reused, energy does not cycle. Maintenance of complex interdependent trophic relationships such as those shown in figures 24.4 and 24.5 requires a constant input of energy at the primary producer level. As energy is transferred to the next level, a large proportion (as high as 90%) of the energy will be lost in a form (primarily heat) that cannot be fed back into the system. Thus, the amount of energy available decreases at each successive trophic level. This energy loss also decreases the actual number of individuals that can be supported at each successive level.

24.1 Learning Outcomes—Assess Your Progress

1. Define *microbial ecology.*
2. Summarize why our view of the abundance of microbes on earth has changed in recent years.
3. Discuss the terms *ecosystem* and *community* in relation to one another.
4. Differentiate between habitat and niche.
5. Draw an example of an energy pyramid, labeling producers, consumers, and decomposers.
6. Define *bioremediation,* and provide one example of its use today.

24.2 The Natural Recycling of Bioelements

Because of the finite supply of life's building blocks, the long-term sustenance of the biosphere results from continuous **recycling** of elements and nutrients. Essential elements such as carbon, nitrogen, sulfur, phosphorus,

oxygen, and iron are cycled through biological, geologic, and chemical mechanisms called **biogeochemical cycles.** Although these cycles vary in certain specific characteristics, they share several general qualities, as summarized in the following list:

- All elements ultimately originate from a nonliving, long-term reservoir in the atmosphere, the lithosphere, or the hydrosphere. They cycle in pure form (N_2) or as compounds (PO_4). Their cycling is facilitated by redox reactions.
- Elements cycle between the abiotic environment and the biotic environment.
- Recycling maintains a necessary balance of nutrients in the biosphere so that they do not build up or become unavailable.
- Cycles are complex systems that rely on the interplay of primary producers, consumers, and decomposers. Often, the waste products of one organism become a source of energy or building material for another.
- All organisms participate in recycling, but only certain categories of microorganisms have the metabolic pathways for converting inorganic compounds from one nutritional form to another.

For billions of years, microbes have played prominent roles in the formation and maintenance of the earth's crust, the development of rocks and minerals, and the formation of fossil fuels. This revolution in understanding the biological involvement in geologic processes has given rise to a field called *geomicrobiology.* A logical extension of this discipline is **astromicrobiology,** also known as *exobiology,* which is the study of life on planets and bodies other than earth. The identification of microbes living in the most extreme environments earth has to offer (see section 24.3) lends support to research in this field of science today.

In the next several sections, we examine how, jointly and over a period of time, microbial activities affect and are themselves affected by the abiotic environment.

Atmospheric Cycles

The Carbon Cycle

Because carbon is the fundamental atom in all biomolecules and accounts for at least one-half of the dry weight of biomass, the **carbon cycle** is more intimately associated with the energy transfers and trophic patterns in the biosphere than are other elements. Carbon exists predominantly in the mineral state and as an organic reservoir in the bodies of organisms. A much smaller amount of carbon also exists in the gaseous state as carbon dioxide (CO_2), carbon monoxide (CO), and methane (CH_4). In general, carbon is recycled through ecosystems via carbon fixation, respiration, or fermentation of organic molecules, limestone decomposition, and methane production. A convenient starting point from which to trace the movement of carbon is with carbon dioxide, which occupies a central position in the cycle and represents a large common pool that diffuses into all parts of the ecosystem **(figure 24.6).** As a general rule, the cycles of oxygen and hydrogen are closely allied to the carbon cycle.

The principal users of the atmospheric carbon dioxide pool are photosynthetic autotrophs (photoautotrophs) such as plants, algae, and bacteria. An estimated 165 billion tons of organic material per year are produced by terrestrial and aquatic photosynthesis, with phytoplankton contributing to nearly half of the earth's overall photosynthetic output. Although we don't yet know exactly how many autotrophs exist in the earth's crust, a small amount of CO_2 is used by bacteria (chemolithoautotrophs) that derive their energy from bonds in inorganic chemicals. A review of the general equation for photosynthesis in figure 8.24 reveals that phototrophs use energy from the sun to fix CO_2 into organic compounds such as glucose that can be used in synthesis. Photosynthesis is also the primary means by which the atmospheric supply of O_2 is regenerated.

While photosynthesis removes CO_2 from the atmosphere and converts solar energy into stored chemical energy, respiration and fermentation release CO_2 and convert stored chemical energy into kinetic energy and work. As you may recall from the discussion of aerobic respiration in chapter 8, in the presence of O_2, organic compounds such as glucose are degraded completely to CO_2 with the release of energy and the formation of H_2O. Carbon dioxide is also released by anaerobic respiration and by certain types of fermentation reactions.

A small but important phase of the carbon cycle involves certain limestone deposits composed primarily of calcium carbonate ($CaCO_3$). Limestone is produced when marine organisms such as mollusks, corals, protozoa, and algae form hardened shells by combining carbon dioxide and calcium ions from the surrounding water. When these organisms die, the durable skeletal components accumulate in marine deposits. As these immense deposits are gradually exposed by geologic upheavals or receding ocean levels, various decomposing agents liberate CO_2 and return it to the CO_2 pool of the water and atmosphere.

The complementary actions of photosynthesis and respiration, along with other natural CO_2-releasing processes such as limestone erosion and volcanic activity, have maintained a relatively stable atmospheric pool of carbon dioxide. Recent figures show that this balance is being disturbed as humans burn *fossil fuels* and other organic carbon sources. Fossil fuels, including coal, oil, and natural gas, were formed through millions of years of natural biological and geologic activities. Humans are so dependent upon this energy source that, within the past 25 years, the proportion of CO_2 in the atmosphere has steadily increased from 320 to 400 parts per million (ppm) **(figure 24.7).** Although this increase may seem insignificant, scientists now feel it has begun to disrupt the delicate temperature balance of the biosphere, leading to global warming.

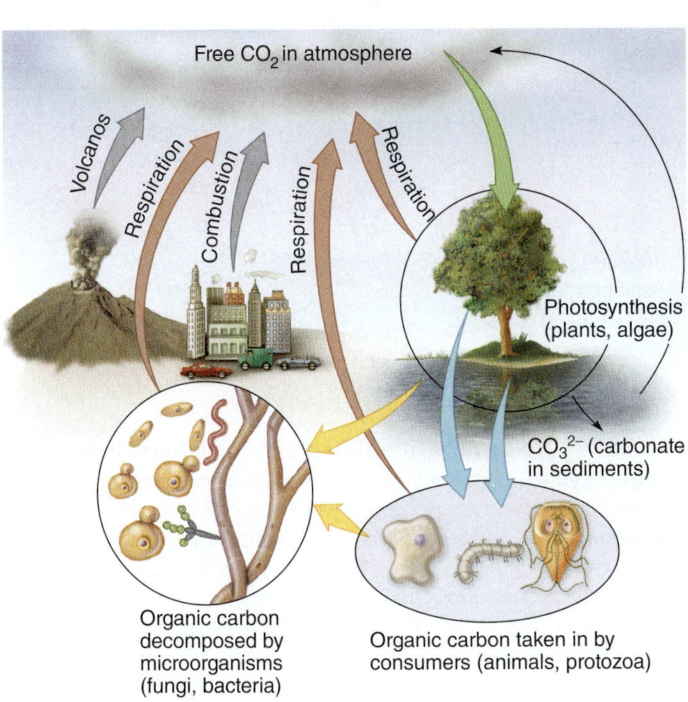

Figure 24.6 The carbon cycle. This cycle traces carbon from the CO_2 pool in the atmosphere to the primary producers (green) where it is fixed into protoplasm. Organic carbon compounds are taken in by consumers (blue) and decomposers (yellow) that produce CO_2 through respiration and return it to the atmosphere (pink). Combustion of fossil fuels and volcanic eruptions also add to the CO_2 pool. Some of the CO_2 is carried into inorganic sediments by organisms that synthesize carbonate (CO_3) skeletons. In time, natural processes acting on exposed carbonate skeletons can liberate CO_2.

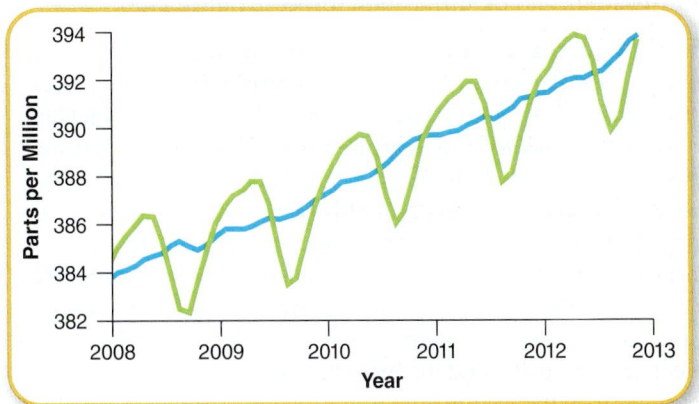

Figure 24.7 Global monthly mean CO_2 amounts since 2008. CO_2 is measured at a globally distributed network of air sampling sites above marine environments. The green line represents monthly mean values. The blue line uses the same data but is smoothed to remove seasonal variation.

Compared with carbon dioxide, methane gas (CH_4) plays a secondary part in the carbon cycle, though it can be a significant product in anaerobic ecosystems dominated by **methanogens** (methane producers). In general, when methanogens reduce CO_2 by means of various oxidizable substrates, they give off CH_4. The practical applications of methanogens are covered in chapter 25 in a section on sewage treatment.

Today methane is recognized as a more potent greenhouse gas than CO_2 as it can trap nearly 20 times more heat in the atmosphere. Much of this comes from gas—intestinal gas, that is. Methane released from the gastrointestinal tracts of ruminant animals such as cattle, goats, and sheep contributes to an estimated 20% of global methane production. Methanogens in the GI tract of humans as well as tiny termites also add to this value each year. This has led microbiologists to wonder, If we alter their diet, will cows produce less methane gas? The current data suggest that the answer is "yes," which may pave the way for a significant reduction in this aspect of global warming in the near future.

The Nitrogen Cycle

Nitrogen (N_2) gas is the most abundant component of the atmosphere, accounting for nearly 79% of air volume. As we will see, this extensive reservoir in the air is largely unavailable to most organisms. Only about 0.03% of the earth's nitrogen is combined (or fixed) in some other form such as nitrates (NO_3), nitrites (NO_2), ammonium ion (NH_4^+), and organic nitrogen compounds (proteins, nucleic acids).

The **nitrogen cycle** is more intricate than other cycles because it involves such a diversity of specialized microbes to maintain the flow of the cycle. In many ways, it is actually more of a nitrogen "web" because of the array of reactions that occur. Higher plants can utilize NO_3^- and NH_4^+; animals must receive nitrogen in organic form from plants or other animals; however, microorganisms can use all forms of nitrogen: NO_2^-, NO_3^-, NH_4^+, N_2, and organic nitrogen. The cycle includes four basic types of reactions: nitrogen fixation, ammonification, nitrification, and denitrification (**process figure 24.8**).

Root Nodules: Natural Fertilizer Factories   A significant symbiotic association occurs between **rhizobia** (ry-zoh'-bee-uh) (bacteria in genera such as *Rhizobium, Bradyrhizobium,* and *Azorhizobium*) and **legumes** (plants such as soybeans, peas, alfalfa, and clover that characteristically produce seeds in pods). The infection of legume roots by these gram-negative, motile, rod-shaped bacteria causes the formation of special nitrogen-fixing organs called **root nodules** (**figure 24.9**). Nodulation begins when rhizobia colonize specific sites on root hairs. From there, the bacteria invade deeper root cells and induce the cells to form tumorlike masses. The bacterium's enzyme system supplies a constant source of reduced nitrogen to the plant, and the plant furnishes nutrients and energy for the activities of the bacterium. The legume uses the NH_4^+ to aminate (add an amino group to) various carbohydrate intermediates and thereby synthesize amino acids and other nitrogenous compounds that are used in plant and animal biosynthesis.

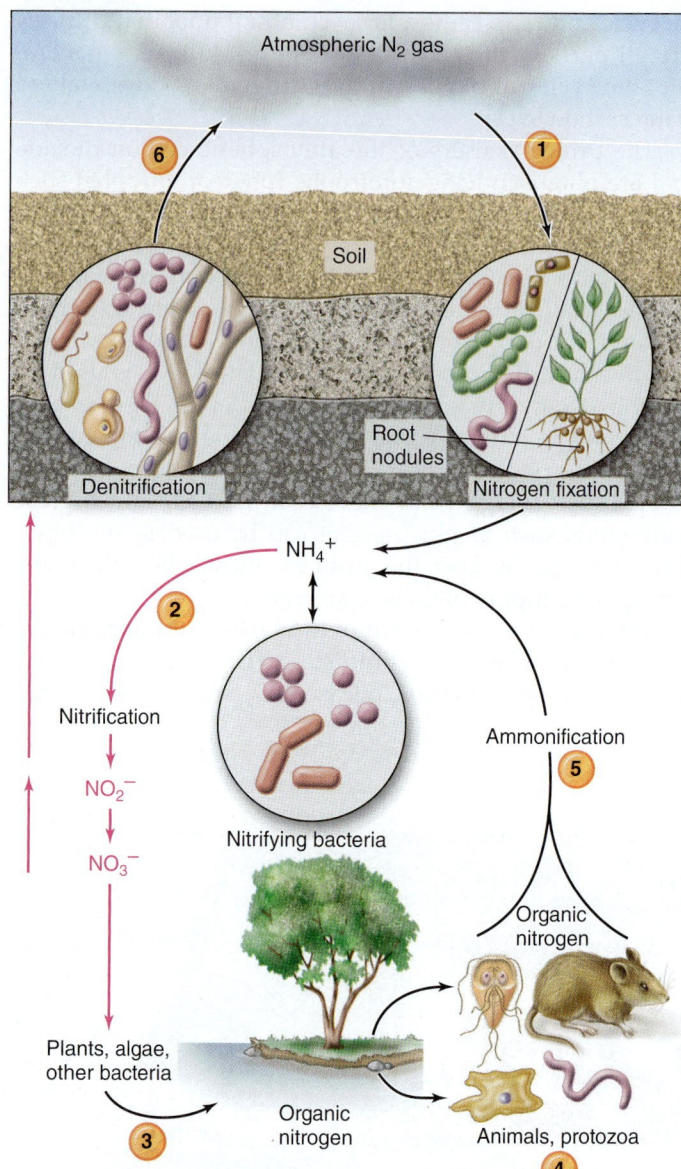

Process Figure 24.8 A simplified view of events in the nitrogen cycle.   **(1)** In nitrogen fixation, gaseous nitrogen (N_2) is acted on by nitrogen-fixing bacteria, which give off ammonia (NH_3). **(2)** Ammonia is converted to nitrite (NO_2^-) and nitrate (NO_3^-) by nitrifying bacteria in nitrification. **(3)** Plants, algae, and bacteria use nitrates to synthesize nitrogenous organic compounds (proteins, amino acids, nucleic acids). **(4)** Organic nitrogen compounds are used by animals and other consumers. **(5)** In ammonification, nitrogenous macromolecules from wastes and dead organisms are converted to NH_4^+ by ammonifying bacteria. NH_4^+ can be used directly by plants and algae, transformed into nitrates that can be used by plants and algae, or **(6)** returned to the atmospheric N_2 form by denitrifying bacteria (denitrification).

Plant–bacteria associations have great practical importance in agriculture, because an available and usable source of nitrogen is often a limiting factor in the growth of crops. The self-fertilizing nature of legumes makes them valuable food plants in areas with poor soils and in countries with limited resources. It has been shown that crop health and yields

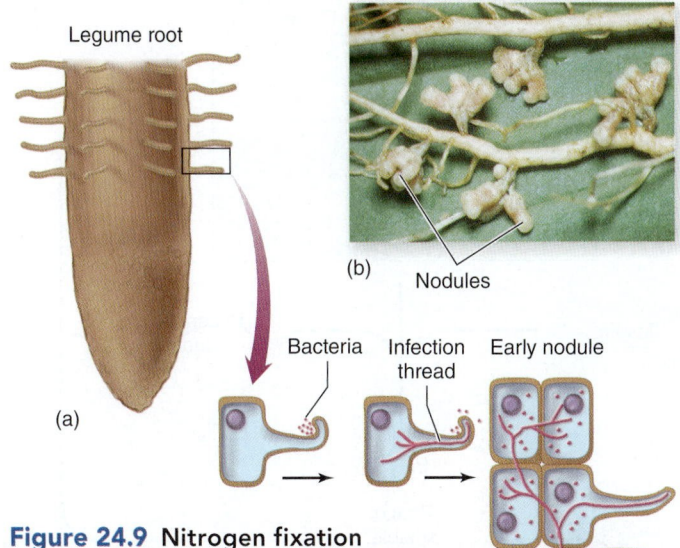

Legume root

(b) Nodules

(a)

Bacteria   Infection   Early nodule
thread

Figure 24.9 Nitrogen fixation through symbiosis. **(a)** Events leading to formation of root nodules. Cells of the bacterium *Rhizobium* attach to a legume root hair and cause it to curl. Invasion of the legume root proper by *Rhizobium* initiates the formation of an infection thread that spreads into numerous adjacent cells. The presence of bacteria in cells causes nodule formation. **(b)** Mature nodules that have developed in a sweet clover plant.

(a)   (b)

Figure 24.10 Inoculating legume seeds with *Rhizobium* bacteria increases the plant's access to nitrogen. **(a)** The legumes in this photo were inoculated and are healthy. **(b)** The poor growth and yellowish color of the uninoculated legumes in this photo indicate a lack of nitrogen.

can be improved by inoculating legume seeds with pure cultures of rhizobia, because the soil is often deficient in the proper strain of bacteria for forming nodules **(figure 24.10).**

Ammonification, Nitrification, and Denitrification In another part of the nitrogen cycle, nitrogen-containing organic matter is decomposed by various bacteria (i.e., *Clostridium, Proteus*) that live in the soil and water. Organic detritus consists of large amounts of protein and nucleic acids from dead organisms and nitrogenous animal wastes such as urea and uric acid. The decomposition of these substances splits off amino groups and produces NH_4^+. This process is thus known as **ammonification.** The ammonium released can be reused by certain plants or converted to other nitrogen compounds, as discussed next.

The oxidation of NH_4^+ to NO_2^- and NO_3^- is a process called **nitrification.** It is an essential conversion process for generating the most oxidized form of nitrogen (nitrate, NO_3). This reaction occurs in two phases and involves two different kinds of lithotrophic bacteria in soil and water. In the first phase, certain gram-negative genera such as *Nitrosomonas, Nitrosospira,* and *Nitrosococcus* oxidize NH_3 to NO_2^- as a means of generating energy. Nitrite is rapidly acted upon by a second group of nitrifiers, including *Nitrobacter, Nitrosospira,* and *Nitrococcus,* which perform the *final* oxidation of NO_2^- to NO_3^-. Nitrates can be assimilated through several routes by a variety of organisms (plants, fungi, and bacteria). Nitrate and nitrite are also important in anaerobic respiration where they serve as terminal electron acceptors; some bacteria use them as a source of oxygen as well.

The nitrogen cycle is complete when nitrogen compounds are returned to the reservoir in the air by a reaction series that converts NO_3^- through intermediate steps to atmospheric nitrogen. The first step, which involves the reduction of nitrate to nitrite, is so common that hundreds of different bacterial species can do it. Several genera such as *Bacillus, Pseudomonas, Spirillum,* and *Thiobacillus* can carry out this **denitrification process** to completion as follows:

$$NO_3^- \rightarrow NO_2^- \rightarrow NO \rightarrow N_2O \rightarrow N_2 \text{ (gas)}$$

When this process is not carried through to completion, the greenhouse gas nitrous oxide (N_2O) is added to the atmosphere.

Sedimentary Cycles

The Sulfur Cycle

The sulfur cycle resembles the carbon cycle more than the nitrogen cycle in that sulfur is mostly in solid form and originates from natural sedimentary deposits in rocks, oceans, lakes, and swamps rather than from the atmosphere. Sulfur exists in the elemental form (S) and as hydrogen sulfide gas (H_2S), sulfate (SO_4), and thiosulfate (S_2O_3). Most of the oxidations and reductions that convert one form of inorganic sulfur to another are accomplished by bacteria. Plants and many microorganisms can assimilate only SO_4, and animals must have an organic source. Organic sulfur occurs in the amino acids cystine, cysteine, and methionine, which contain sulfhydryl (—SH) groups and form disulfide (S—S) bonds that contribute to the stability and configuration of proteins.

One of the most remarkable contributors to the cycling of sulfur in the biosphere are the thiobacilli. These gram-negative, motile rods flourish in mud, sewage, bogs, mining drainage, and brackish springs that can be inhospitable to organisms that require complex organic nutrients. But the metabolism of these specialized lithotrophic bacteria is adapted to extracting energy by oxidizing elemental sulfur, sulfides, and thiosulfate. One species, *T. thiooxidans,* is so efficient at this process that it secretes large amounts of sulfuric acid into its environment. The marvel of this bacterium is its ability to create and survive in the most acidic habitats on the earth. It also plays an essential part in the phosphorus cycle, and its relative, *T. ferrooxidans*, participates in the cycling of iron. Other bacteria that can oxidize sulfur to sulfates are the photosynthetic sulfur bacteria mentioned in the section on photosynthesis.

The sulfates formed from oxidation of sulfurous compounds are assimilated into biomass by a wide variety of organisms. The sulfur cycle reaches completion when inorganic and organic sulfur compounds are reduced. Bacteria in the genera *Desulfovibrio* and *Desulfuromonas* anaerobically reduce sulfates to hydrogen sulfide or metal sulfide as the final step in electron transport. Sites in ocean sediments and mud where these bacteria live usually emanate a strong, rotten-egg stench from H_2S and may be blackened by the iron they contain.

The Phosphorus Cycle

Phosphorus is an integral component of DNA, RNA, and ATP, and all life depends upon a constant supply of it. It cycles between the abiotic and biotic environments almost exclusively as inorganic phosphate (PO_4) rather than its elemental form **(figure 24.11).** The chief inorganic reservoir is phosphate rock, which contains the insoluble compound fluorapatite, $Ca_5(PO_4)_3F$. Before it can enter biological systems, this mineral must be *phosphatized*—converted into more soluble PO_4^{3-} by the action of acid. Phosphate is released naturally when the sulfuric acid produced by *Thiobacillus* dissolves phosphate rock. Soluble phosphate in the soil and water is the principal source for autotrophs, which fix it onto organic molecules and pass it on to heterotrophs in this form. Organic phosphate is returned to the pool of soluble phosphate by decomposers, and it is finally cycled back to the mineral reservoir by slow geologic processes such as sedimentation. Because the low phosphate content of many soils can limit productivity, phosphate is added to soil to increase agricultural yields. The excess runoff of fertilizer into the hydrosphere is often responsible for overgrowth of aquatic microbes, which can lead to devastating effects on these environments (see eutrophication in a subsequent section on aquatic habitats).

Other Forms of Cycling

The involvement of microbes in cycling elements and compounds can be escalated by the introduction of toxic substances into the environment. Such toxic elements as arsenic, chromium, lead, and mercury, as well as hundreds of thousands of synthetic chemicals introduced into the environment over the past hundred years, are readily caught up in biodegradative cycles by microbial actions. Some of these chemicals will be converted into less harmful substances, but others, such as PCB and heavy metals, persist and flow along with nutrients into all levels of the biosphere. If such a pollutant accumulates in living tissue and is not excreted, it can be accumulated by living things through the natural trophic flow of the ecosystem. This process is known as bioaccumulation. Microscopic producers such as bacteria and algae begin the accumulation process. With each new level of the food chain, the consumers gather

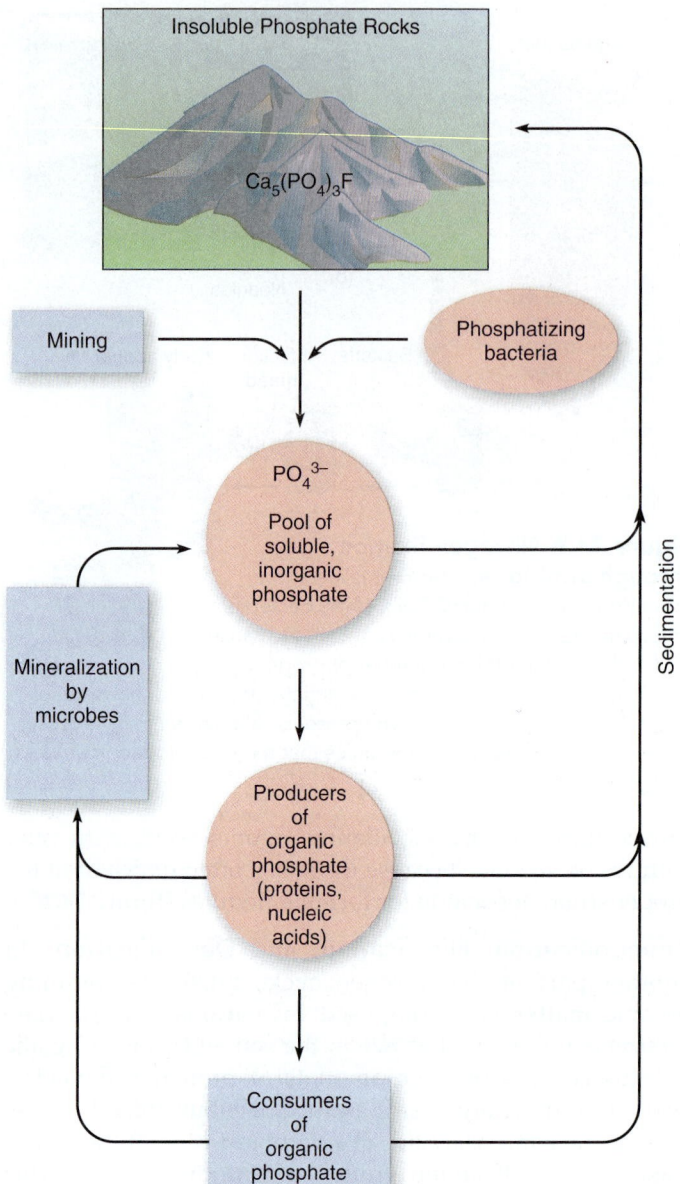

Figure 24.11 The phosphorus cycle. The pool of phosphate existing in sedimentary rocks is released into the ecosystem either naturally by erosion and microbial action or artificially by mining and the use of phosphate fertilizers. Soluble phosphate (PO_4^{3-}) is cycled through producers, consumers, and decomposers back into the soluble pool of phosphate, or it is returned to sediment in the aquatic biosphere.

an increasing amount of the chemical, until the top consumers can contain toxic levels.

Recent studies have disclosed increased mercury content in fish taken from oceans and freshwater lakes in North America and even in canned tuna, adding to the risk in consumption of these products. A similar trend was recently identified when researchers observed the concentration of arsenic in freshwater systems. Additionally, there are high levels of compounds such as PCB, DDT, and various insecticides in human breast milk, illustrating our role in the food chain.

24.2 Learning Outcomes—Assess Your Progress

7. List five important elements of biogeochemical cycles.

8. Diagram a carbon cycle.

9. Explain the role of methanogens within the carbon cycle.

10. List the four reactions involved in the nitrogen cycle.

11. Describe the process of nitrogen fixation, and explain how microbes play a role in this biochemical reaction.

12. Briefly summarize both the sulfur and phosphorus cycles.

1 Environmental sampling

24.3 Microbes on Land and in Water

As you have heard several times already in this book, until fairly recently, our understanding of which microbes inhabited a place, whether it was the human gut or your backyard pond, relied on culturing them. Just as there is a Human Microbiome Project, scientists have been busy using these techniques to identify the microbes living in the environment, which includes land, water, and air. The field of environmental genomics has revealed many surprising things, such as bacteria living in glaciers and deep under the seafloor—places once thought to be uninhabitable by any life forms.

Environmental Sampling in the Genomic Era

The methods for identifying bacteria and genes in the environment are evolving rapidly. When the genes of all microbes in a habitat are sampled, it is called metagenomics. We will discuss the basic principles here. The process always begins with an environmental sample, such as a gallon of seawater or a gram of soil. Techniques are available to extract the DNA from such samples. Fragments can then be cloned into plasmid vectors in the same way that we described in chapter 10. A library of DNA fragments can then be preserved and amplified. If specific gene sequences are sought (and known in advance), PCR can then be performed directly on the environmental sample to fish the "needle out of the haystack." This method is commonly used when seeking 16S rRNA molecules. Novel sequences are still found in this way as the "fishing" is done with conserved sequences found on the different molecules. Once the nucleotide fragments are retrieved using either method, they can be sequenced, and it can be determined whether they match known sequences or are new to us. The DNA sequences can also be cloned into expression vectors, which can then be screened for their functions, or their products. These processes are summarized in **process figure 24.12.**

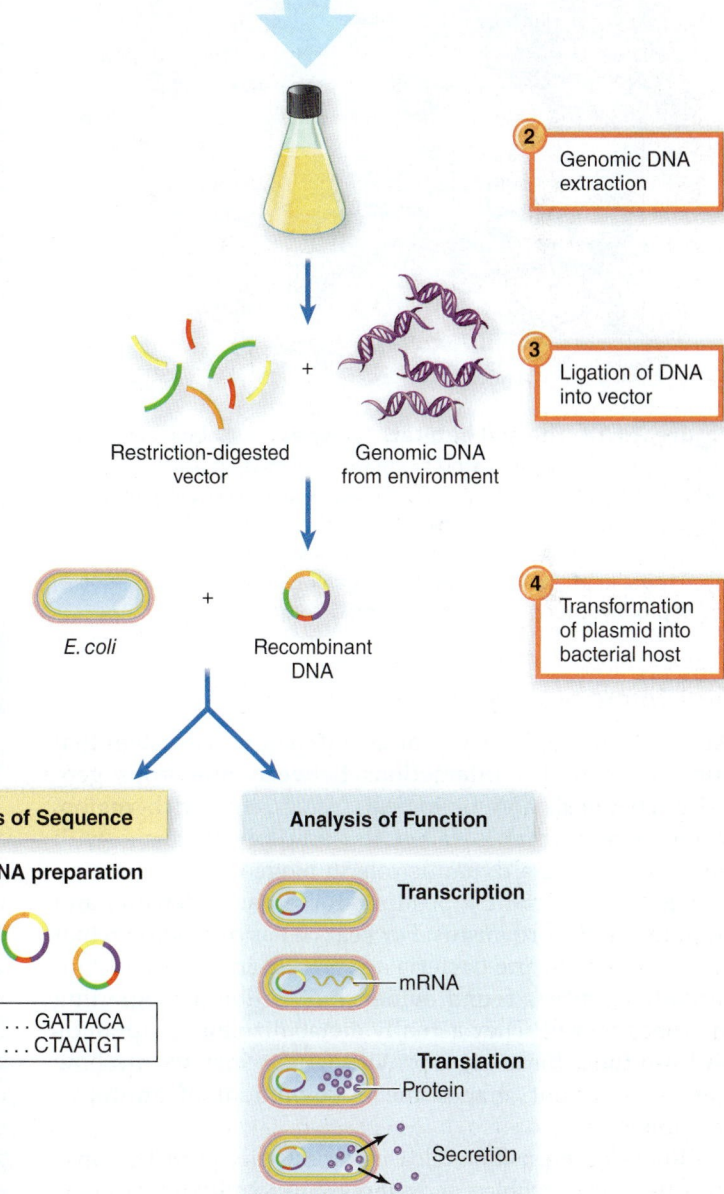

Process Figure 24.12 Construction and screening of genomic libraries directly from the environment.

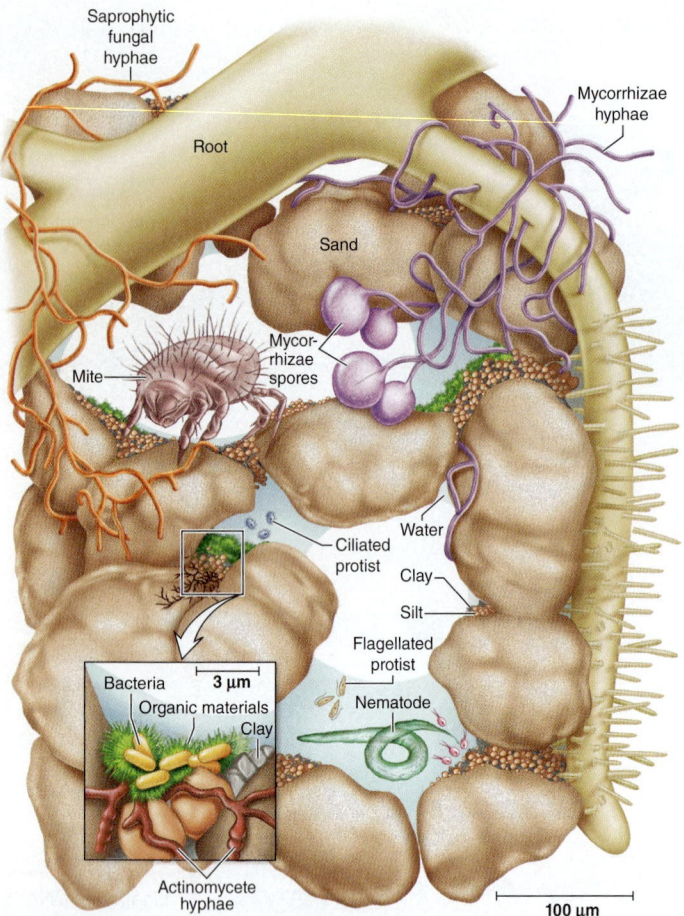

Figure 24.13 The soil habitat.   A typical soil habitat contains a mixture of clay, silt, and sand along with soil organic matter. Roots and animals (e.g., nematodes and mites), as well as protozoa and bacteria, consume oxygen, which rapidly diffuses into the soil pores where the microbes live. Note that two types of fungi are present: mycorrhizal fungi, which derive their organic carbon from plant roots; and saprophytic fungi, which help degrade organic material.

Soil Microbiology

At the microscopic level, soil is a dynamic ecosystem that supports complex interactions between numerous geologic, chemical, and biological factors. This rich region, found within what is called the lithosphere, teems with microbes, serves a dynamic role in biogeochemical cycles, and is an important repository for organic detritus and dead terrestrial organisms. For years it has been known that antibiotic-producing bacteria are present in these soils, but recently scientists found evidence of antibiotic-degrading microbes as well. They actually metabolize the compounds and use them for energy, providing new insight into how the environment shapes the development of antibiotic resistance.

Rock decomposition releases various-size particles, ranging from rocks, pebbles, and sand grains to microscopic morsels that lie in a loose aggregate **(figure 24.13).** The porous structure of soil creates various-size pockets or spaces that

provide numerous microhabitats. Some spaces trap moisture and form a liquid phase in which mineral ions and other nutrients are dissolved. Other spaces trap air that will provide gases to soil microbes, plants, and animals. Water-saturated soils contain less oxygen, and dry soils have more. Gas tensions in soil can also vary vertically. In general, the concentration of O_2 decreases and that of CO_2 increases with the depth of soil. Aerobic and facultative organisms tend to occupy looser, drier soils, whereas anaerobes are adapted to waterlogged, poorly aerated soils.

Within the superstructure of the soil are varying amounts of humus, the slowly decaying organic litter from plant and animal tissues. This soft, crumbly mixture holds water like a sponge. It is also an important habitat for microbes that decompose the complex litter and gradually recycle nutrients. The humus content varies with climate, temperature, moisture and mineral content, and microbial action. For instance, the moisture and warmth of the tropics promote rapid microbial decomposition and thereby reduce humus levels, and the high levels of precipitation wash away the nutrients mobilized by the microbes. These processes cause the moist tropics to have very poor soils for agriculture. Native tropical rainforests are adapted to these conditions and therefore thrive, whereas agricultural crops do poorly without inputs of fertilizer. On the other hand, the moderate climate of the temperate zone provides a balance of plant growth and microbial decomposition that causes accumulations of humus, and naturally fertile soils.

Disease Connection

The vast majority of microbes that are found in soil are not human pathogens. The biota in soil thrive in cool, dry conditions—quite a different environment than the human body. However, some pathogens can survive in soil. Bacteria that are capable of forming endospores survive in the endospore form for years in soil. These include the bacteria causing tetanus (*Corynebacterium diphtheriae*) and anthrax (*Bacillus anthracis*).

Living Activities in Soil

The rich culture medium of the soil supports a fantastic array of microorganisms (bacteria, fungi, algae, protozoa, and viruses). A gram of moist loam soil with high humus content can have a microbe count as high as 10 billion. Some of the most distinctive biological interactions occur in the **rhizosphere,** the zone of soil surrounding the roots of plants, which contains associated bacteria, fungi, and protozoa (see figure 24.13). Plants interact with soil microbes in a truly synergistic fashion. Studies have shown that a rich microbial community grows in a biofilm around the root hairs and other exposed surfaces. Their presence stimulates the plant to exude growth factors such as carbon dioxide, sugars, amino acids, and vitamins. These nutrients are released into fluid spaces, where they can be readily captured by microbes. Bacteria and fungi likewise contribute to plant survival by releasing

Figure 24.14 Mycorrhizae. These symbiotic associations between fungi and plant roots favor the absorption of water and minerals from the soil.

hormonelike growth factors and protective substances. They are also important in converting minerals into forms usable by plants. We saw numerous examples in the nitrogen, sulfur, and phosphorus cycles.

We previously observed that plants can form close symbiotic associations with microbes to fix nitrogen. Other mutualistic partnerships between plant roots and microbes are **mycorrhizae** (my"-koh-ry'-zee). These associations occur when various species of basidiomycetes, ascomycetes, or zygomycetes attach themselves to the roots of vascular plants **(figure 24.14).** The plant feeds the fungus through photosynthesis, and the fungus sustains the relationship in several ways. By extending its mycelium into the rhizosphere, it helps anchor the plant and increases the surface area for capturing water from dry soils and minerals from poor soils. Plants with mycorrhizae can inhabit severe habitats more successfully than plants without them.

The topsoil, which extends a few inches to a few feet from the surface, supports a host of burrowing animals such as nematodes, termites, and earthworms. Many of these animals are decomposer organisms that break down organic nutrients through digestion and also mechanically reduce or fragment the size of particles so that they are more readily mineralized by microbes. Aerobic bacteria initiate the digestion of organic matter into carbon dioxide and water and generate minerals such as sulfate, phosphate, and nitrate, which can be further degraded by anaerobic bacteria. Fungal enzymes increase the efficiency of soil decomposition by hydrolyzing complex natural substances such as cellulose, keratin, lignin, chitin, and paraffin.

The soil is also a repository for agricultural, industrial, and domestic wastes such as insecticides, herbicides, fungicides, manufacturing wastes, and household chemicals. Applied microbiologists, using expertise from engineering, biotechnology, and ecology, work to explore the feasibility of harnessing indigenous soil microbes to break down undesirable hydrocarbons and pesticides through bioremediation (see chapter 25).

Deep Subsurface Microbiology

For the past 30 years, scientists have been sampling the deep subsurface—2 miles and more below the surface. From the very beginning of these studies, the results have been astounding. With the advent of genomic sampling, the discoveries are piling up. The most fascinating information has been found in deep ocean sediments, where the microbial diversity is immense and expands to depths thought unimaginable. For example, bacteria have been found 30 meters beneath the seafloor in clay that was deposited there 86 million years ago. The clay contains infinitesimally small levels of nutrients, yet bacteria and archaea were found. These extreme environments around the globe, characterized by high pressures, cold temperatures, and the complete absence of light, may provide new insights into biogeochemical processes and more. Scientists search to determine how bacteria survive in very nearly abiotic environments and along the way are discovering new metabolic capabilities that may turn out to provide clues to the very origin of life.

Aquatic Microbiology

Water occupies nearly three-fourths of the earth's surface. The **hydrologic cycle (figure 24.15)** begins when surface water (lakes, oceans, rivers) exposed to the sun and wind evaporates and enters the vapor phase of the atmosphere. Living beings contribute to this reservoir by various activities. Most of the water that falls on land is first returned to the atmosphere by plants, through movement of soil water through plants, ending as evaporation through leaves. All aerobic organisms give off water during respiration. Airborne moisture accumulates in the atmosphere, most conspicuously as clouds.

Water is returned to the earth through condensation or precipitation (rain, snow), a process influenced by bacteria (see chapter 1 Case File). The largest proportion of precipitation falls back into surface waters, where it circulates rapidly between running water and standing water. Only about 2% of water seeps into the earth or is bound in ice, but these are very important reservoirs. **Figure 24.16** shows how water is distributed on the earth. Surface water collects in extensive subterranean pockets produced by the underlying layers of rock, gravel, and sand. This process forms a deep groundwater source called an **aquifer.** The water in aquifers circulates very slowly and is an important replenishing source for surface water. It can resurface through springs, geysers, and hot vents; it is also tapped as the primary supply for one-fourth of all water used by humans.

Although the total amount of water in the hydrologic cycle has not changed over millions of years, its distribution and quality have been greatly altered by human activities.

Figure 24.15 The hydrologic cycle. The largest proportion of water cycles through evaporation, transpiration, and precipitation between the hydrosphere and the atmosphere. Other reservoirs of water exist in the groundwater or deep storage aquifers in sedimentary rocks. Plants add to this cycle by releasing water through transpiration, and heterotrophs release it through respiration.

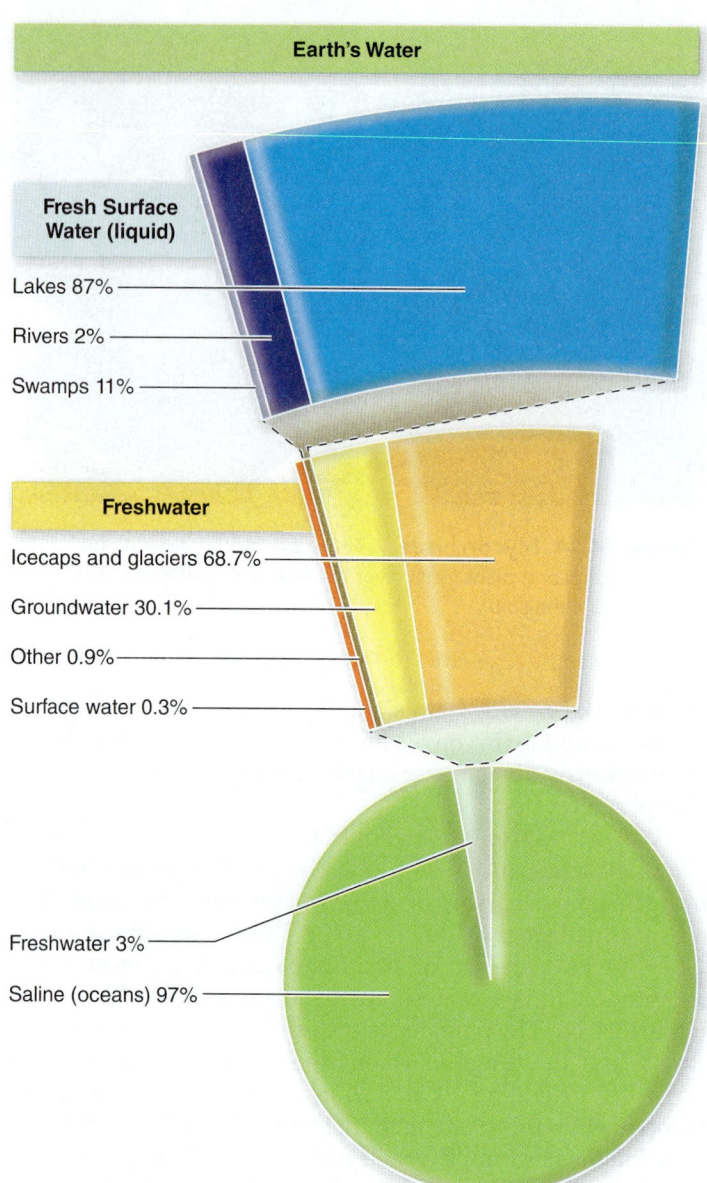

Figure 24.16 The distribution of water on earth.

Two serious problems have arisen with aquifers. First, as a result of increased well drilling, land development, and persistent local droughts, the aquifers in many areas have not been replenished as rapidly as they have been depleted. As these reservoirs have been used up in many places around the world, including parts of the United States, humans have had to rely on other delivery systems such as pipelines, dams, and reservoirs, which can further disrupt the cycling of water. Second, because water picks up materials when falling through air or percolating through the ground, aquifers are also important collection points for pollutants. As we will see in chapter 25, the proper management of water resources is one of the greatest challenges of this century.

Marine Environments

The ocean exhibits extreme variations in salinity, depth, temperature, hydrostatic pressure, and mixing. Even so, it supports a great abundance of bacteria and viruses, the extent of which has only been appreciated in very recent years. In 2004,

J. Craig Venter (the same Venter from the Human Genome Project, chapter 10) set sail on a 100 ft yacht to the Sargasso Sea to get the DNA profile of an entire ecosystem. The Sargasso Sea was thought to be relatively sparsely populated by life forms, as it was nutrient-poor. His team found a rich variety of life and discovered 1,800 new species and more than 1.2 million new genes. His group widened the search over the next 2 years and sailed around the world, collecting ocean samples all along the way. They eventually discovered 6 million new genes and thousands of new proteins—essentially doubling the number of known proteins. Proteins, of course, are responsible for nearly all of the activities of cells. The capabilities of microbes have been greatly underappreciated, in other words, because we have not been able to cultivate the vast majority of them in the laboratory. Indeed, a tiny

Novel Hot Spring Viruses Migrate on Water Droplets

Hot springs represent a unique environment—high temperatures and the low pH of the water support the growth of a wide variety of archaea including *Acidianus, Metallosphaera, Stygiolobus, Sulfolobus, Sulfurococcus,* and *Sulphurisphaera.* Until recently, little was known about the viruses that infect these unique organisms. Researchers from Montana State University at Bozeman studied the viral populations that infect *Sulfolobus* in the hot springs in Yellowstone National Park and in hot springs in Japan and Iceland. The

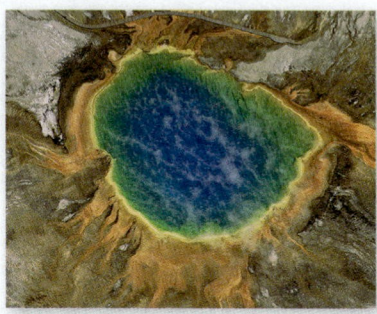

viruses were from the family *Fueslloviridae* and had a unique spindle or "lemon" shape that is only found in viruses infecting archaea. The researchers found migratory patterns of these viruses between hot springs up to 20 miles apart, postulating that the viruses traveled on water droplets in the steam from the hot springs. They hypothesized that airborne migration of viruses may explain how nearly identical viruses found in the hot springs of Russia, Iceland, and Japan turn up in Yellowstone Park.

As MSU researchers studied these flighty viruses infecting archaea, they made another startling discovery. Metagenomic analysis of hot spring viruses revealed positive-strand RNA viruses of archaea completely unrelated to all known RNA viruses. Until now, the only viruses infecting archaea that had been discovered were dsDNA viruses. The discovery of these novel positive-strand RNA viruses in archaea sheds light on the possible origin of RNA viruses that infect eukaryotic cells. Scientists still need to isolate the archaea and their viral parasites to study them in more detail in the laboratory to find the link between archaeal and eukaryotic RNA viruses.

Source: 2012. J. Virol. vol. 86, no. 10, p. 5562.

and therefore pose no danger to humans, but as parasites of bacteria, they appear to be a natural control mechanism for these populations. Plus, their lysis of bacteria plays an important role in the turnover of nutrients in the ocean. From an evolutionary standpoint, however, an interesting discovery has been made: The oceans represent the most extensive gene pool known currently to humankind, a virtual toolbox of nature-engineered genes. Because bacteriophages can swap and exchange any number of these microbial genes, their ability to acquire new traits that increase their survival plays a major role in their biology and their evolution. In one example, scientists found that cyanophages contain genes that can alter the light-harvesting abilities of their cyanobacterial hosts. This means that the bacteriophages can use the genes they acquire from microbes in their environment to boost the photosynthetic rate of their hosts. But why? It's all about survival and, more specifically, energy; experiments show that this boost in photosynthesis provides the energy the phages require to complete their life cycle. It will take many years to fully uncover the intricacies of the oceanic virome, but these studies are sure to reveal even more exciting information along the way.

Some microbial inhabitants of the ocean produce the periodic emergence of *red tides* around ocean coastlines **(figure 24.17)**. Scientists call these "harmful algal blooms" (HABs). Environmental factors cause an increase in the number of these algae, leading to an increase in food for organisms farther up the food chain but also resulting in toxin production that can harm fish, shellfish, and even humans who may ingest the seafood or swim in the water. These algae produce a potent muscle toxin that can be concentrated by shellfish through filtration feeding. When humans eat clams, mussels, or oysters that contain the toxin, they develop paralytic shellfish poisoning. People living in coastal areas are cautioned to not eat shellfish during those months of the year associated with red tides (which varies from one area to another).

Freshwater Communities

The freshwater environment is a site of tremendous microbiological activity. Microbial distribution is associated with sunlight, temperature, oxygen levels, and nutrient availability. The uppermost portion is the most productive self-sustaining region because it contains large amounts of **plankton,** a floating microbial community that drifts with wave action and currents. A major member of this assemblage is the phytoplankton, containing a variety of photosynthetic algae and cyanobacteria. The phytoplankton provides nutrition for **zooplankton,** composed of microscopic consumers such as protozoa and invertebrates that filter, feed, prey, or scavenge. The plankton supports numerous other trophic levels such as larger invertebrates and fish. With its high nutrient content, the deeper regions also support an extensive variety and concentration of organisms, including aquatic plants, aerobic bacteria, and anaerobic bacteria actively involved in recycling organic detritus.

microbe called *Prochlorococcus*, which is responsible for the majority of the photosynthesis occurring in the ocean (which in itself accounts for 50% of global photosynthesis), was only discovered in 1986, even though there are a trillion trillion of them in the planet's oceans.

We know that high salinity is an effective physical measure to control microbial growth. The Dead Sea is a hypersaline body of water whose salinity is nearly 35%, which led microbiologists to believe that it would be devoid of all microbial life. It was very shocking then to find that it was teeming with microbes thriving in the very water that should prevent their survival. Most of the organisms were identified as salt-tolerant archaea. Oceans also contain an estimated 10 million viruses per milliliter. Most of these viruses are bacteriophages

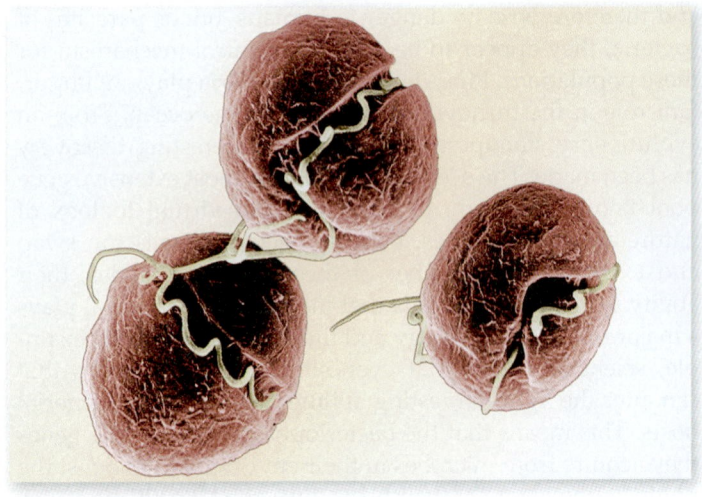
(a)

(b)

Figure 24.17 Red tides. (a) Single-celled red algae called dinoflagellates (*Gymnodinium* shown here) bloom in high-nutrient, warm seawater and impart a noticeable red color to it. (b) An aerial view of California coastline in the midst of a red tide. (c) Fish washed ashore during a red tide bloom.

Larger bodies of standing water develop gradients in temperature or thermal stratification, especially during the summer **(figure 24.18).** The upper region, called the *epilimnion,* is warmest; the deeper *hypolimnion* is cooler. Between these is a buffer zone, the **thermocline,** that ordinarily prevents the mixing of the two. Twice a year, during the warming cycle of spring and the cooling cycle of fall, temperature changes in the water column break down the thermocline and cause the water from the two strata to mix. Mixing disrupts the stratification and creates currents that bring nutrients up from the sediments. This process, called *upwelling,* is associated with increased activity by certain groups of microbes. Because oxygen is not very soluble in water and is rapidly used up by the plankton, its concentration forms

(c)

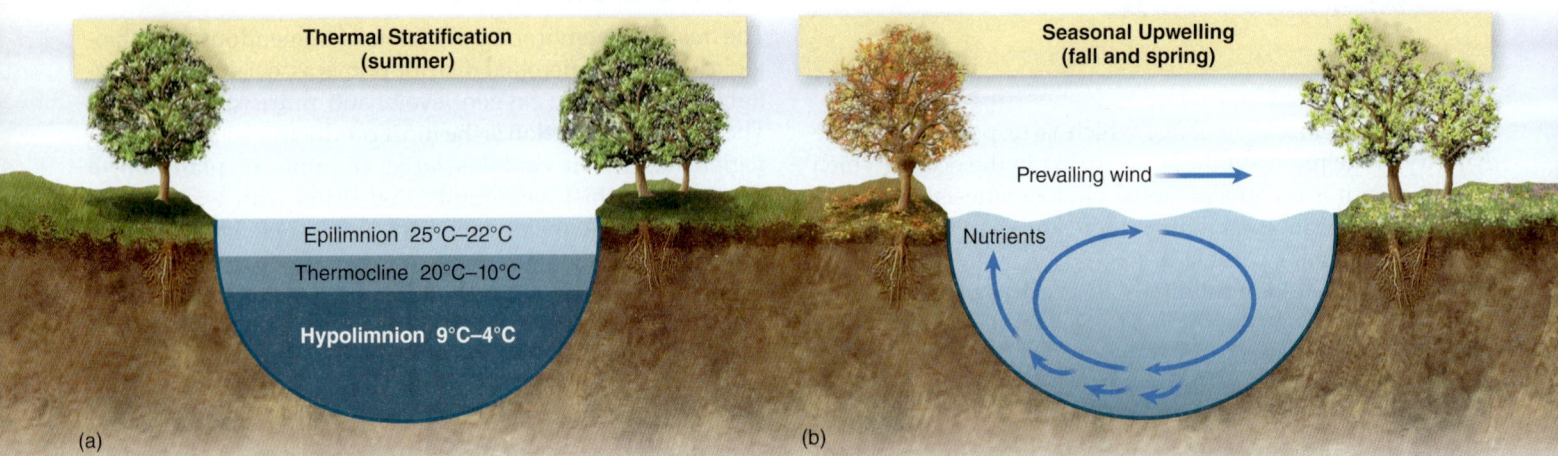
(a)   (b)

Figure 24.18 Profiles of a lake.   (a) During summer, a lake becomes stabilized into three major temperature strata. (b) During fall and spring, cooling or heating of the water disrupts the temperature strata and causes upwelling of nutrients from the bottom sediments.

a gradient from highest in the epilimnion to lowest at the bottom. In general, the amount of oxygen that can be dissolved is dependent on temperature. Warmer strata on the surface tend to carry lower levels of this gas. But of all the characteristics of water, the greatest range occurs in nutrient levels. Nutrient-deficient aquatic ecosystems are called **oligotrophic** (ahl″-ih-goh-trof′-ik). Species that can make a living on such starvation rations are *Hyphomicrobium* and *Caulobacter*. These bacteria have special stalks that capture even minuscule amounts of hydrocarbons present in oligotrophic habitats. The addition of excess quantities of nutrients to aquatic ecosystems, called **eutrophication,** often wreaks havoc on the communities involved. The sudden influx of abundant nutrients along with warm temperatures encourages a heavy surface growth of cyanobacteria and algae similar to red tides in oceans **(figure 24.19)**. This heavy mat of biomass effectively shuts off the oxygen supply to the lake below. The oxygen content below the surface is further depleted by aerobic heterotrophs that actively decompose the organic matter. The lack of oxygen greatly disturbs the ecological balance of the community. It causes massive die-offs of strict aerobes (fish, invertebrates), and only anaerobic or facultative microbes will survive. This effect can be triggered by the addition of industrial wastes, detergents in household wastewater, or runoff from manure and fertilizer-rich fields and can be remediated over long periods of time.

Figure 24.19 Heavy surface growth of algae and cyanobacteria in a eutrophic pond.

24.3 Learning Outcomes—Assess Your Progress

13. Outline the basic process used to perform metagenomic analysis of the environment.

14. List two important symbiotic partnerships that occur in the soil.

15. Diagram the hydrologic cycle.

16. Discuss how metagenomic sampling has changed our view of deep subsurface and oceanic microbiology.

17. List the stratified regions of large bodies of standing water, and describe how microbes are affected by this layering.

18. Define *eutrophication*, and explain how microbes are responsible for its impact on aquatic life.

Case File 24   *Wrap-Up*

Two methods were used to determine the volume of methane and oil consumed by microbes. When methanogens metabolize methane, they utilize oxygen, so scientists were able to correlate the depletion of oxygen in the spill area to the amount of methane consumed. Researchers noted that the rate of methane consumption was aided by the addition of dispersants (chemicals designed to break up the oil into smaller particles) at the source of the spill. The smaller particles of oil, gas, and methane produced by the dispersants facilitated metabolism of these materials by microbes. Researchers found depleted oxygen levels in the areas where they had first measured methane; they also found that the amount of oxygen consumed by microbes corresponded to the amount of methane that was consumed.

The second type of analysis used radioactively labeled methane, ethane, and propane introduced to samples of unpolluted water from the Gulf. As microbes naturally occurring in the water utilized the radioactive methane, ethane, and propane, they incorporated radioactive carbon atoms into their DNA. Scientists then analyzed the radioactive DNA to identify which organisms consumed each molecule. They found that *Colwellia* consumed ethane and propane and *Methylococcaceae* were the predominant consumers of methane.

There is still a great deal of work to be done to overcome the devastating effects of the *Deepwater Horizon* oil spill. However, these insights into how microbes consume methane and other oil products may aid in future remediation efforts. They remind us that "simple" microbes have dealt with all kinds of environments and have adapted to thriving in them.

Source: 2012. Env. Microb. vol. 14, no. 9, p. 2405.

Chapter Summary

24.1 Ecology: The Interconnecting Web of Life (ASM Guidelines* 3.1, 3.2, 3.3, 5.1, 5.3, 5.4, 6.1, 6.2, 6.3, 6.4)

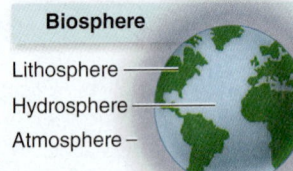

- The study of ecology includes both living (biotic) and nonliving (abiotic) components of the earth.
- Ecosystems are organizations of living populations in specific habitats. Environmental ecosystems require a continuous outside source of energy for survival and a nonliving habitat consisting of soil, water, and air.
- A living community is composed of populations that show a pattern of energy and nutritional relationships called a food web. Microorganisms are essential producers and decomposers in any ecosystem.
- Bioremediation is the use of naturally occurring or genetically engineered microorganisms to remove pollutants within a contaminated environment.

24.2 The Natural Recycling of Bioelements (ASM Guidelines 1.1, 3.1, 3.2, 3.3, 4.5, 5.1, 5.3, 5.4, 6.1, 6.2, 6.3, 6.4)

- Nutrients and minerals necessary to communities and ecosystems must be continuously recycled. These biogeochemical cycles involve transformation of elements from inorganic to organic forms and back again. Specific types of microorganisms are needed to convert many nutrients from one form to another.
- Elements of critical importance to all ecosystems that cycle through various forms are carbon, nitrogen, sulfur,

*Source: *ASM Curriculum Guidelines* (American Society for Microbiology, 2012). Complete guidelines in appendix B of this book.

phosphorus, and water. Carbon and nitrogen are part of the atmospheric cycle. Sulfur and phosphorus are part of the sedimentary cycling of nutrients.

24.3 Microbes on Land and in Water (ASM Guidelines 1.1, 3.1, 3.2, 3.3, 4.5, 5.1, 5.3, 5.4, 6.1, 6.2, 6.3, 6.4)

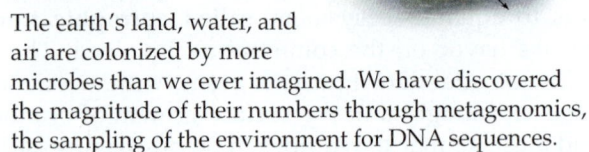

- The earth's land, water, and air are colonized by more microbes than we ever imagined. We have discovered the magnitude of their numbers through metagenomics, the sampling of the environment for DNA sequences.
- The lithosphere is an ecosystem in which mineral-rich rocks are decomposed to organic humus, the base for the soil community.
- The deep subsurface, below land and sea, is colonized by a rich array of microbes that have a wide variety of metabolic capabilities.
- The food web of the aquatic community is built on phytoplankton and zooplankton. The nature of the aquatic community varies with the temperature, depth, minerals, and amount of light present in each zone.

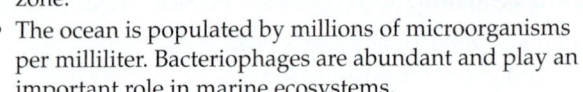

- The ocean is populated by millions of microorganisms per milliliter. Bacteriophages are abundant and play an important role in marine ecosystems.
- Eutrophication of freshwater and marine systems is caused by the addition of excess nutrients. It causes major disruptions in the ecology of these systems.

Multiple-Choice and True-False Questions | Bloom's Levels 1 and 2: Remember and Understand

Multiple-Choice Questions. Select the correct answer from the options provided.

1. Which of the following is *not* a major subdivision of the biosphere?
   a. hydrosphere
   b. lithosphere
   c. stratosphere
   d. atmosphere

2. A/an _____ is defined as a collection of populations sharing a given habitat.
   a. biosphere
   b. community
   c. biome
   d. ecosystem

3. The quantity of available nutrients _____ from the lower levels of the energy pyramid to the higher ones.
   a. increases
   b. decreases
   c. remains stable
   d. cycles

4. Which of the following is considered a greenhouse gas?
   a. CO_2
   b. CH_4
   c. N_2O
   d. all of these

5. Root nodules contain _____, which can _____.
   a. *Azotobacter*, fix N_2
   b. *Nitrosomonas*, nitrify NH_3^-
   c. rhizobia, fix N_2
   d. *Bacillus*, denitrify NO_3^-

6. Which element(s) has/have an inorganic reservoir that exists primarily in sedimentary deposits?
   a. nitrogen
   b. phosphorus
   c. sulfur
   d. both b and c

7. What percentage of the earth's biomass is made of microbes?
   a. a small fraction
   b. at least half of it
   c. all of it
   d. 25% of it

8. Genomic analysis of the land, sea, and air has shown us that
   a. there are many more animals than we expected.
   b. there are many fewer microbes than we expected.
   c. seawater is much more sterile than we expected.
   d. microbes colonize places we never imagined.

9. Microbes in the environment are likely to be
   a. living in biofilms on surfaces.
   b. living solitary and planktonic lives.
   c. nonculturable in the lab.
   d. two of the above.

10. Recent studies reveal that
    a. 100% of photosynthesis is accomplished by plants.
    b. viruses may well augment bacterial photosynthesis.
    c. the sun is not the only source of energy for photosynthesis.
    d. none of the above is true.

True-False Questions. If the statement is true, leave as is. If it is false, correct it by rewriting the sentence.

11. Pure cultures are very common in the biosphere.

12. Bioremediation usually involves more than one type of microorganism.

13. The production of all nitrogenous compounds begins with the process called nitrogen fixation.

14. The high mercury content found in some fish is the result of a process called bioaccumulation.

15. As far as we know, all microorganisms exist in multiple-species communities.

Critical Thinking Questions | Bloom's Levels 3, 4, and 5: Apply, Analyze, and Evaluate

Critical thinking is the ability to reason and solve problems using facts and concepts. These questions can be approached from a number of angles and, in most cases, they do not have a single correct answer.

1. a. List examples of biotic and abiotic factors that contribute to a microbe's ability to survive within a habitat.
   b. Define the term *niche*, and describe the many roles microbes fulfill in an ecosystem.

2. Using the chapter 24 Case File as an example, construct a paragraph illustrating how a consortium of microbes is often required for the process of bioremediation.

3. a. Outline the general characteristics of a biogeochemical cycle.
   b. Conduct additional research, and discuss two recent discoveries in the field of geomicrobiology.

4. Provide evidence in support of or refuting the following statement: Microbial life was recently identified on Mars.

5. Summarize the role microbes play in the cycling of carbon, and discuss their possible influence on global warming.

6. a. Looking at the photo of the three plants, explain a symbiotic relationship within the rhizosphere that could account for the observed difference in height among these plants.

   b. Many people use animal manure to fertilize their garden crops. Discuss how this application benefits the growing plants, and discuss whether or not it poses any risks to human or environmental health.

7. a. Outline the modes of cycling water through the lithosphere, hydrosphere, and atmosphere.
   b. In the chapter 1 Case File, you learned about the role of microbes in the formation of precipitation. Expanding on this idea, could changing weather patterns alter the normal habitat range of environmental microbes today? Explain your answer.

8. a. What causes the formation of the epilimnion, hypolimnion, and thermocline? Explain why many largely populated fisheries are located in areas of the ocean where these layers are mixed by strong currents.
   b. In the 1970s, Lake Erie became known as the "Dead Sea of North America." It was found that local industries, farms, and households all contributed to the loss of life in this Great Lake. Explain the microbial process that occurred in Lake Erie. What biogeochemical cycle was involved and how was this lake brought back to life?

9. Discuss the impact of global warming on the growth of microbes in the ocean today. How will this affect food webs within marine ecosystems?

10. PCBs are human-made pollutants that are not synthesized in nature. However, a remote lake in Alaska was found to contain PCBs even though humans had never set foot near this body of water. Based upon what you have learned in this chapter, develop a hypothesis that could explain this finding.

Concept Connections | Bloom's Levels 4 and 6: Analyze and Create

This activity ties together multiple concepts from the chapter.

1. What are the roles of microbes in the carbon cycle?
2. How are microbes involved in photosynthesis? Give at least one example.
3. How do microbes act as consumers? Give at least one example.
4. How do microbes act as decomposers? Give at least one example.

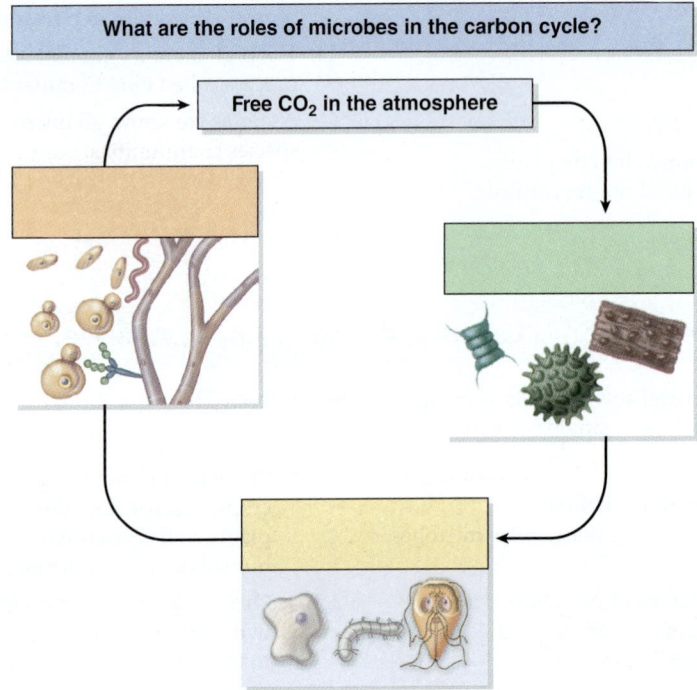

Visual Connections | Bloom's Level 5: Evaluate

These questions use visual images or previous content to make connections to this chapter's concepts.

1. **From chapter 6, figure 6.19.** We suggested that bacteriophages in the ocean appear to play a role in photosynthesis and the turnover of nutrients. Which of these two activities is more likely to be accomplished when the bacteriophage is in the lysogenic state?

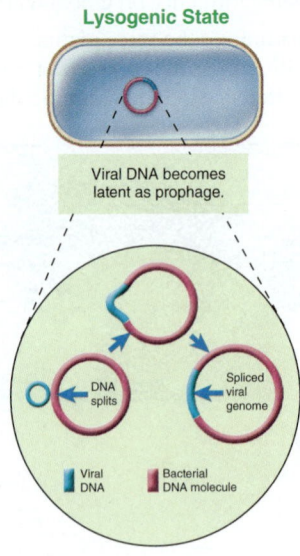

2. **From chapter 8, figure 8.24.** What process does this represent? How does it link to the biogeochemical cycles from this chapter?

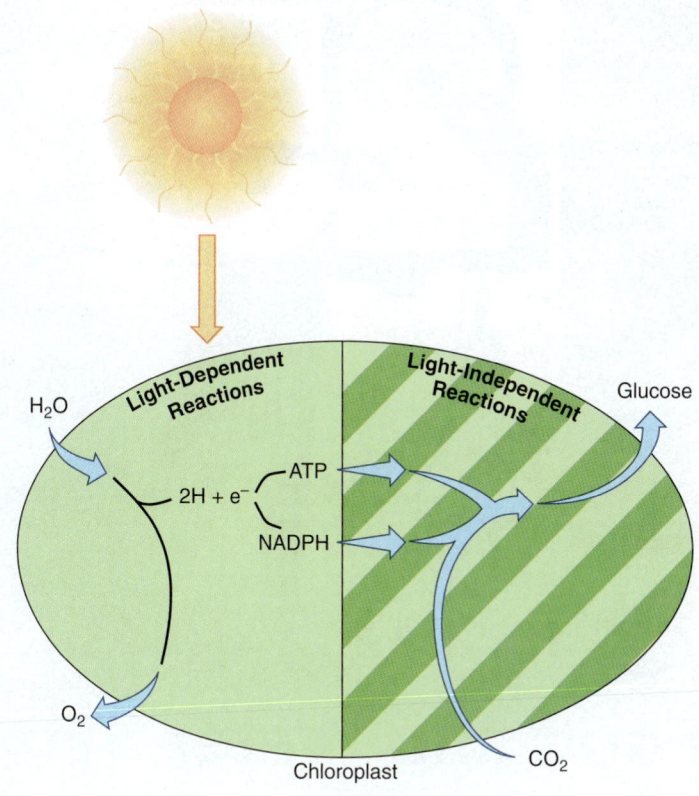

Concept Mapping | Bloom's Level 6: Create

Appendix D provides guidance for working with concept maps.

1. Using the words that follow, please create a concept map illustrating the relationships among these key terms from chapter 24.

lithosphere	oceans and lakes	phytoplankton	eutrophication
rhizosphere	bioremediation	zooplankton	
mycorrhizae	genomic sampling	oligotrophic	

25

Applied Microbiology and Food and Water Safety

Case File 25

Raw Milk: Worth the Risk?

In 2011, federal officials arrested three members of Rawesome Foods, an organic food collective in Ventura County, California, after a yearlong investigation. Their crime: selling raw or unpasteurized milk. Sale and consumption of raw milk is legal in California and 28 other states, but the milk is subject to inspection and regulation and, in some states, may not be sold in grocery stores. Other small farmers in California and other states have been subject to similar raids and criminal charges for distributing raw milk. Passionate proponents of raw milk believe that there is a higher nutritional content in unpasteurized milk and that the pasteurization process kills helpful bacteria, denatures enzymes, and inactivates beneficial compounds in milk. They believe that lactose intolerance, food allergies, and other maladies can be traced to the consumption of pasteurized milk. In addition, they claim that the flavor and texture of unpasteurized and non-homogenized milk far exceed what can be bought in the grocery store.

- Why is most milk in the United States pasteurized?
- Why is raw or unpasteurized milk a health concern?

Continuing the Case appears on page 789.

Outline and Learning Outcomes

25.1 Applied Microbiology and Biotechnology
    1. Compose a sentence about the history of applied microbiology.
    2. Define *biotechnology*, and explain how its uses have changed in modern times.

25.2 Microorganisms in Water and Wastewater Treatment
    3. Outline the steps in water purification.
    4. Differentiate water purification from sewage treatment.

5. Describe the primary and secondary phases of sewage treatment.
6. Discuss how wastes can be converted into usable forms of energy.
7. List five important pathogens of drinking water.
8. Provide examples of indicator bacteria, and describe their role in the survey of water quality.
9. Summarize methods for identifying and quantifying microbial contaminants in water supplies.

25.3 Microorganisms Making Food and Spoiling Food

10. Name five foods and/or beverages that are produced using microbial fermentation.
11. Summarize the microbial process that leads to leavening in bread.
12. Write the equation showing how yeasts convert sugar into alcoholic beverages.
13. Discuss how microorganisms themselves are useful as food products today.
14. Provide background on current HACCP guidelines, and describe how they are used to maintain food safety.
15. Report 10-year trends in the incidence of food-borne illness within the United States.
16. Outline basic principles of using temperature to preserve food.
17. List methods other than temperature currently used for preservation in the food industry.

25.4 Using Microorganisms to Make Things We Need

18. State the general aim(s) of industrial microbiology.
19. Distinguish between primary and secondary metabolites.
20. List the four steps of industrial product production from microbes.
21. Identify five industrial products made by microorganisms, and describe their applications.

25.1 Applied Microbiology and Biotechnology

This chapter emphasizes the artificial applications of microbes in communal waste remediation; water treatment; and the manufacture of food, medical, biochemical, drug, and agricultural products. Microbes have evolved by responding to functional pressures, such as when nutrients are limited or unevenly available, or when other organisms are competing for the nutrients. Applied and industrial microbiologists have learned from microbes' own survival mechanisms and have devised ways to manipulate them for use by people.

The profound and sweeping involvement of microbes in the natural world is inescapable. Although our daily encounters with them usually go unnoticed, human and microbial life are clearly intertwined on many levels. It is no wonder that long ago humans realized the power of microbes and harnessed them for specific metabolic tasks. The practical applications of microorganisms in manufacturing products or carrying out a particular decomposition process belong to the large and diverse area of **biotechnology.** Biotechnology has an ancient history, dating back nearly 6,000 years to those first observant humans who discovered that grape juice left to sit resulted in wine or that bread dough properly infused with a starter would rise. Modern biotechnology includes the use of genetic engineering methods to boost or augment the naturally occurring abilities of microbes. The field of biotechnology is providing hundreds of applications in industry, medicine, agriculture, food sciences, and environmental protection. Most biotechnological systems involve the actions of bacteria, yeasts, molds, and algae that have been selected or altered to synthesize a certain food, drug, organic acid, alcohol, or vitamin. Many such food and industrial end products are obtained through **fermentation**—a general term used here to refer to the mass, controlled culture of microbes to produce desired organic compounds. It also includes the use of microbes in sewage control, pollution control, metal mining, and bioremediation, which was introduced in chapter 24 **(Insight 25.1).**

25.1 Learning Outcomes—Assess Your Progress

1. Compose a sentence about the history of applied microbiology.
2. Define *biotechnology*, and explain how its uses have changed in modern times.

25.2 Microorganisms in Water and Wastewater Treatment

Most drinking water comes from rivers, aquifers, and springs. Only in remote, undeveloped, or high mountain areas is this water used in its natural form. In most cities, it must be treated before it is supplied to consumers. Water supplies such as deep wells that are relatively clean and free of contaminants require less treatment than those from surface sources laden with wastes. The stepwise process in water purification as carried out by most cities is shown in **process figure 25.1.** Steps ① to ④ of the figure outline what happens to water between its natural source and the point at which it flows through your faucet at home. It involves filtration and chemical disinfection processes that make the water safe to drink.

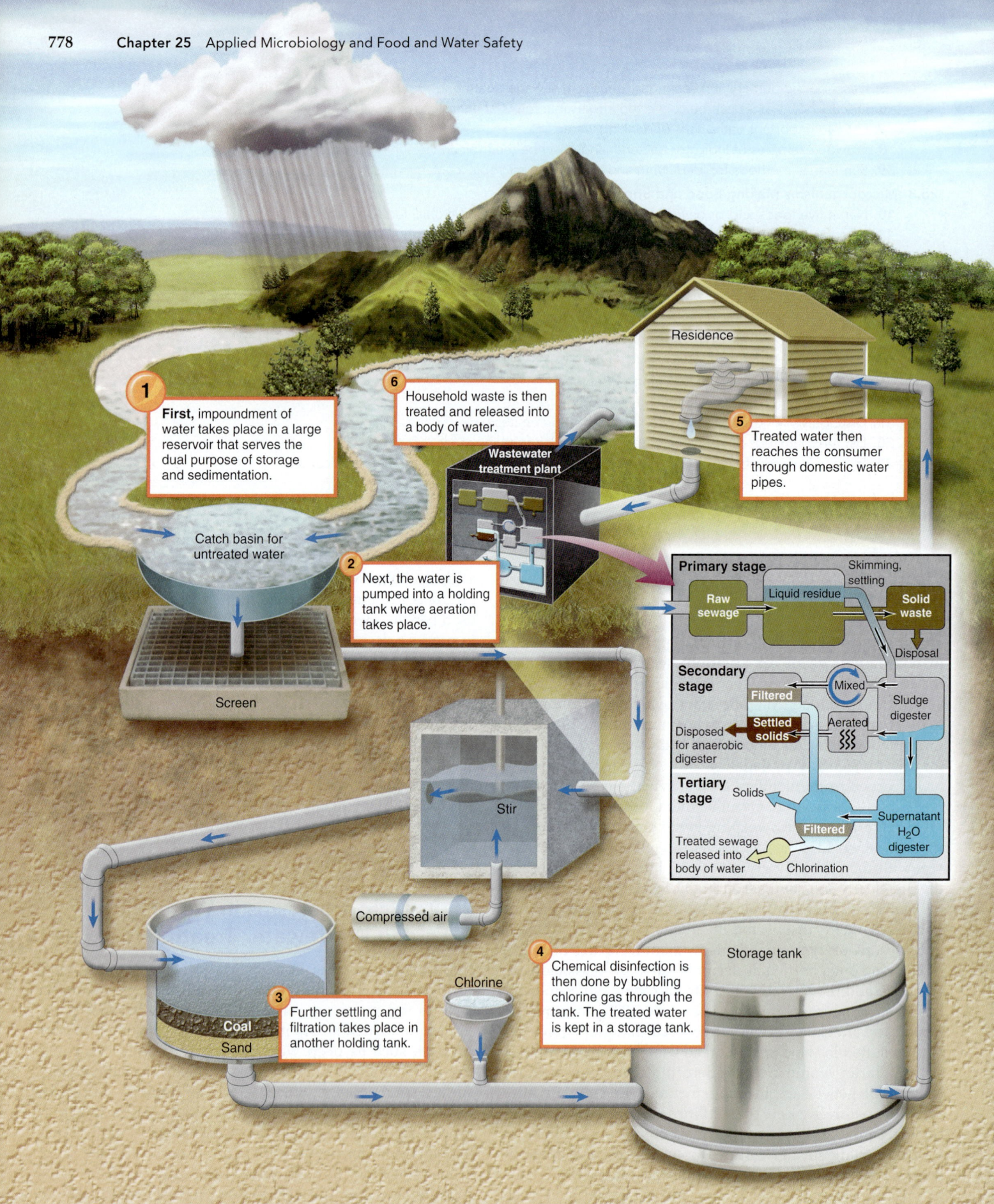

1 **First,** impoundment of water takes place in a large reservoir that serves the dual purpose of storage and sedimentation.

6 Household waste is then treated and released into a body of water.

Residence

5 Treated water then reaches the consumer through domestic water pipes.

Catch basin for untreated water

Wastewater treatment plant

2 Next, the water is pumped into a holding tank where aeration takes place.

Screen

Primary stage

Skimming, settling

Raw sewage   Liquid residue   Solid waste

Disposal

Secondary stage

Mixed

Filtered   Sludge digester

Disposed for anaerobic digester   Settled solids   Aerated

Tertiary stage   Solids

Treated sewage released into body of water   Filtered   Supernatant H₂O digester

Chlorination

Stir

Compressed air

Chlorine

4 Chemical disinfection is then done by bubbling chlorine gas through the tank. The treated water is kept in a storage tank.

Storage tank

3 Further settling and filtration takes place in another holding tank.

Coal
Sand

Process Figure 25.1 **The major steps in water purification and sewage treatment.**

INSIGHT 25.1 Trash to Treasure: Using Bioremediation to Clean Up a Garbage Dump

From 1972 to 1984, "El Morro" was a landfill in the Moravia Hill neighborhood in Medellín, Colombia. Impoverished people lived in the dump and picked through the trash to find food, water, clothing, and shelter. At one point, up to 50,000 people lived there, growing vegetables from the contaminated water that accumulated in the landfill. Recently, the Colombian government relocated the inhabitants of El Morro to better living conditions, but the landfill remained. Jerry Sims from the University of Illinois, along with Andres Gomez from the National University of Colombia in Medellín, conducted research on El Morro to determine if it could be cleaned up through bioremediation. There were no records of what had been dumped there, and digging up the landfill was economically unfeasible to determine what substances were there. Gomez and Sims, along with a team from the University of Colombia, analyzed samples of bacteria in the landfill and found that these microbes were already performing bioremediation on the contaminants in the landfill. Different communities of microbes at different depths of El Morro were found. Based on these results, the Colombian government gave Sims, Gomez, and their colleagues at the University of Colombia permission to further bioremediate the landfill by giving the indigenous microbes extra nutrients to speed up their metabolic processes and spur on bioremediation of the contaminants in the landfill. The

hope is that the El Morro landfill can one day be transformed from a contaminated dump that once was the home of impoverished people into a public park that can benefit the community.

Source: 2011. Soil Biol. Biochem. vol. 43, no. 6, p. 1275.

In many parts of the world, the same water that serves as a source of drinking water is also used as a dump for solid and liquid wastes **(figure 25.2)**. Continued pressure on the earth's finite water resources may require reclaiming and recycling of contaminated water such as sewage. Sewage is the used wastewater draining out of homes and industries that contains a wide variety of chemicals, debris, and microorganisms. Sewage contains large amounts of solid wastes, dissolved organic matter, and toxic chemicals that pose a health risk. To remove all potential health hazards, treatment typically requires three phases: The *primary stage* separates out large matter; the *secondary stage* reduces remaining matter and can remove some toxic substances; and the *tertiary stage* completes the purification of the water (see the inset in process figure 25.1). Microbial activity is an integral part of the overall process. The newest systems use *membrane bioreactors,* which are combinations of microbial communities and high-efficiency membranes that are much more effective at removing contaminants. The systems for sewage treatment are massive engineering marvels.

In the primary phase of treatment, floating bulkier materials such as paper, plastic waste, and bottles are skimmed off. The remaining smaller, suspended particulates are allowed to settle. Sedimentation in settling tanks usually takes 2 to 10 hours and leaves a mixture rich in organic matter. This aqueous portion is carried into a secondary phase of active microbial decomposition, or biodegradation. In this phase, a diverse community of natural bioremediators (bacteria, algae, and protozoa) aerobically decomposes the remaining particles of wood, paper, fabrics, petroleum, and organic molecules

Figure 25.2 Water: one source, many uses.

inside a large digester tank **(figure 25.3)**. This forms a suspension of material called *sludge* that tends to settle out and slow the process. To hasten aerobic decomposition of the sludge, most processing plants have systems to *activate* it by injecting air, mechanically stirring it, and recirculating it. A large amount of organic matter is mineralized into sulfates, nitrates, phosphates, carbon dioxide, and water. Certain volatile gases such as hydrogen sulfide, ammonia, nitrogen, and methane may also be released. Water from this process is siphoned off and carried to the tertiary phase, which involves further filtering and chlorinating prior to discharge. Such reclaimed sewage water is usually used to water golf courses and parks for use in aquaculture, or it is gradually released into large bodies of water. The safety of this practice has recently come into question, due to the persistence of antibiotics in the released water and the risk of spreading antibiotic-resistant bacteria that have managed to survive the process.

(a)

(b)

Figure 25.3 Treatment of sewage and wastewater.
(a) Digester tanks used in the primary phase of treatment; each tank can process several million gallons of raw sewage a day. **(b)** View inside the secondary reactor shows the large stirring paddle that mixes the sludge to aerate it to encourage microbial decomposition.

In some cases, the solid waste that remains after aerobic decomposition is harvested and reused. Its rich content of nitrogen, potassium, and phosphorus makes it a useful fertilizer. It is estimated that 50% of the 8 million tons of sludge (also called "biosolids") made in the United States annually is recycled and applied to land. This has been viewed as a "green" alternative to burying or burning the sludge. However, scientists are now raising concerns that hundreds of thousands of pounds of potent antimicrobial substances such as triclosan are also being spread on the ground, because these chemicals accumulate in the sludge and are not degraded by the typical process of wastewater treatment.

Recently, scientists found a way to harness the bacteria found in sewage to construct a microbial fuel cell to produce usable energy **(Insight 25.2)**. In these experiments, wastewater bacteria form biofilms on special rods inserted in the sewage that is being treated. These biofilms generate electrons that are transferred via copper wires to cathodes, producing electricity. The development of anaerobic or methane digesters has also provided an additional method for converting waste products into energy. These structures are built to house a variety of microbes that can metabolize compounds received from the input of agricultural (manure) or industrial (spent mash from beer brewing) waste products. Methane produced and captured from this process can then be used to power the energy needs of these farms and breweries—all from substances that in many cases the businesses would have to pay to dispose of. Considering the mounting waste disposal and energy shortage problems, these technologies are gaining more momentum every day.

Water Monitoring to Prevent Disease
Microbiology of Drinking Water Supplies

We do not have to look far for overwhelming reminders of the importance of safe water. Worldwide epidemics of cholera have killed thousands of people, and an outbreak of *Cryptosporidium* in Wisconsin in the 1990s that affected 370,000 people was traced to a contaminated municipal water supply. In a large segment of the world's population, the lack of sanitary water is responsible for billions of cases of diarrheal illness that kill 3 million children each year (see chapter 22). In the United States, millions of people develop waterborne illness every year.

Good health is dependent on a clean, potable (drinkable) water supply. This means the water must be free of pathogens; dissolved toxins; and disagreeable turbidity, odor, color, and taste. As we shall see, water of high quality does not come easily, and we must look to microbes as part of the problem *and* part of the solution.

Through ordinary exposure to air, soil, and effluents, surface waters usually acquire harmless, saprobic microorganisms. But along its course, water can also pick up pathogenic contaminants. Among the most prominent waterborne pathogens of recent times are the protozoa *Giardia* and *Cryptosporidium;* the bacteria *Campylobacter, Salmonella,*

| INSIGHT 25.2 | **Microbial Fuel Cells: Turning Your Poop into Electricity** |

Is the future of energy production being flushed down your toilet? Since 1911, scientists have known that bacteria can produce electrical energy when they degrade organic compounds in batteries called microbial fuel cells (MFCs). The idea is simple: A wire connects two electrodes with bacteria growing in a biofilm around the anode (negative electrode) in an anaerobic environment. As the bacteria metabolize, they produce carbon dioxide, electrons, and protons (hydrogen ions). The electrons travel through a wire connecting the anode and the cathode (positive electrode), and hydrogen ions pass through a semipermeable membrane to produce electricity and pure water when they reach the cathode. Nutrients can be provided to the bacteria surrounding the anode in the form of wastewater and sewage.

Unfortunately, it has been difficult to develop an efficient MFC. One of the main hurdles is that it must operate at a neutral pH to promote the growth of bacteria at the anode. Hydroxide (OH⁻) ions accumulate at the cathode and raise the pH, causing a decrease in energy production. Recent research efforts have focused on improving MFC efficiency. A group at Arizona State University's Biodesign Institute studied experimental polymers that more efficiently bind OH⁻ ions at the cathode to improve energy production. Scientists at Penn State found that combining a reverse electrodialysis system with MFCs can generate electricity in a wastewater treatment facility. Other engineers have found a catalyst that is 5% of the cost of platinum catalysts currently used in MFCs.

MFC technology has many useful applications, not the least of which is sewage treatment. Currently, electricity is used to pretreat sewage to remove toxins and nonbiodegradable materials. Energy produced by MFCs can offset the cost of electricity used in pretreatment, and fewer pollutants will be dumped into rivers and oceans as a result. The same MFC technology can be used in treating wastewater from breweries and other production plants, in the desalination of seawater, and in hydrogen production.

Source: 2012. www.sciencedaily.com/releases/2012/07/120710133100.htm.

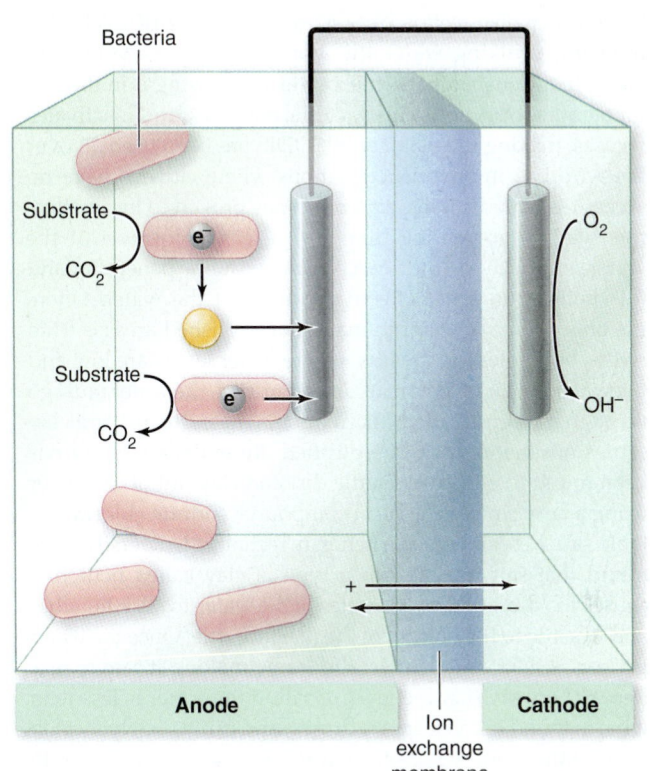

In a microbial fuel cell, bacteria (depicted as pink ovals) in the anode oxidize an organic substrate (such as wastewater) to CO_2 and transfer metabolic electrons to an electrode either through direct contact or a soluble electron shuttle (e.g., riboflavin; depicted as a yellow circle). Electrons travel through an external circuit to the cathode, where oxygen is reduced to hydroxide ions by removing protons from water. Charge balance is maintained by cations moving toward the anode and/or anions moving toward the cathode through an ion exchange membrane.

Source: Michaela A. TerAvest and Largus T. Angenent, Cornell University.

Shigella, Vibrio, and *Mycobacterium;* and hepatitis A and Norwalk viruses. Some of these agents (especially encysted protozoa) can survive in natural waters for long periods without a human host, whereas others are present only transiently and are rapidly lost. The microbial content of drinking water must be continuously monitored to ensure that the water is free of infectious agents.

Attempting to survey water for specific pathogens can be very difficult and time-consuming, so most assays of water purity are more focused on detecting fecal contamination. High fecal levels can mean the water contains pathogens and is consequently unsafe to drink. Thus, wells, reservoirs, and other water sources can be analyzed for the presence of various **indicator bacteria.** These species are intestinal residents of birds and mammals, and they are readily identified using routine lab procedures.

In the late 1800s, it was suggested that a good way to determine if water or its products had been exposed to feces was to test for *E. coli.* Although most *E. coli* strains are not pathogenic, they almost always come from a mammal's intestinal tract, so their presence in a sample is a clear indicator of fecal contamination. Because at the time it was too difficult to differentiate *E. coli* from the closely related species of *Citrobacter, Klebsiella,* and *Enterobacter,* laboratories instead simply reported whether a sample contained one of these isolates. (All of these bacteria ferment lactose and are phenotypically similar.) The terminology adopted was *coliform-positive* or *coliform-negative* (*coliform* means "*E. coli*–like"). Coliforms, then, are gram-negative, lactose-fermenting, gas-producing bacteria such as those previously mentioned.

The use of this coliform assay has been the standard procedure since 1914, and it is still in widespread use. Pick up a

Cleaning Up Muddy Water

Solar water disinfection or SODIS is a method of safely disinfecting drinking water by simply placing contaminated water in a transparent plastic bottle and leaving it in the sun for 6 hours. UVA light kills bacteria, parasites, and inactivates viruses, making the water safe. SODIS has been used all over the world in impoverished nations where citizens have no access to clean drinking water and has proven to be an effective way of preventing diarrheal disease. However, if the water is muddy or murky, UVA rays cannot penetrate, and the method does not effectively disinfect the water. Often, the only source of water in many impoverished areas is from rivers, boreholes, or streams, where water levels are low and the water is murky. Before it can be purified, the water must go through a process called flocculation, where the dirt and clay settles out before it can be purified. Recently, a method has been found to effectively settle dirt and clay out of the water using a very simple chemical compound: sodium chloride, or table salt. Researchers at Michigan Technical University have found that salt mixed with a type of clay called bentonite causes the dirt and clay particles in the water to stick together and to settle to the bottom of the plastic bottle. Once the water is clear, SODIS can disinfect the water, making it safe to use. Researchers say that the level of salt in the water is less than what is found in most sports drinks, and can be safely used in areas of the world where clean water is desperately needed to prevent diarrheal disease in children.

Source: 2012.www.sciencedaily.com/releases/2012/05/120501134315.htm.

is that the three other bacterial species already mentioned, among others, are commonly found growing in fecal-free environments such as freshwater and plants that eventually become food. In other words, if you're not looking specifically for *E. coli,* you can't be sure you're looking for feces.

In 1995, there was a minor panic when media outlets reported that iced tea from restaurants contained significant numbers of "fecal coliforms." The public was outraged. One headline read, "Iced Tea Worse Than River Water." Restaurants were named, and their reputations were damaged. When scientists did more detailed testing, they found that the predominant species found were *Klebsiella* and *Enterobacter,* both of which are normal colonizers of plants, such as tea leaves. Furthermore, despite the reports of widespread contamination with large numbers of "fecal coliforms," no one became sick from drinking the iced tea.

Microbiologists are now advocating that *E. coli* alone be used as an indicator of fecal contamination. Newer identification techniques make this as simple as, if not simpler than, the standard coliform tests. Other tests, including genotypic microarray assays, can now rapidly detect and identify multiple microbes in a single sample. But old habits die hard, and regulatory and public laboratories are proving slow to convert to the *E. coli* standard. For that reason, we present some of the older (and commonly used) methods in this section.

Water Quality Assays   A rapid method for testing the total bacterial levels in water is the standard plate count. In this technique, a small sample of water is spread over the surface of a solid medium. The numbers of colonies that develop provide an estimate of the total viable population without differentiating coliforms from other microbial species. This information is particularly helpful in evaluating the effectiveness of various water purification stages. Another general indicator of water quality is the level of dissolved oxygen it contains. It is established that water containing high levels of organic matter and bacteria will have a significantly lower concentration of oxygen due to the metabolism of aerobic microorganisms.

Coliform Enumeration   Water quality departments employ some standard assays for routine detection and quantification of coliforms. The techniques available are

- simple tests that detect the presence of coliforms but do not quantify them,
- rapid tests that isolate coliform colonies and quantify the coliforms present, and
- rapid tests that identify specific coliforms and determine numbers within a sample.

In many circumstances (drinking water, for example), it is important to *differentiate* between facultative coliforms (*Enterobacter*) that are often found in other habitats (soil, water) *and* true **fecal coliforms** that live mainly in human and animal intestines.

The membrane filter method is a widely used rapid method that can be used in the field or lab to process and test larger quantities of water (100 to 200 mL). This method

newspaper in the summer, and you will likely find a report about a swimming pool or a river with a high coliform count. Coliform counts are also used to regulate food production and to trace the causes of food-borne outbreaks. Recently, microbiologists have noted serious problems with the use of coliforms to indicate fecal contamination. The main issue

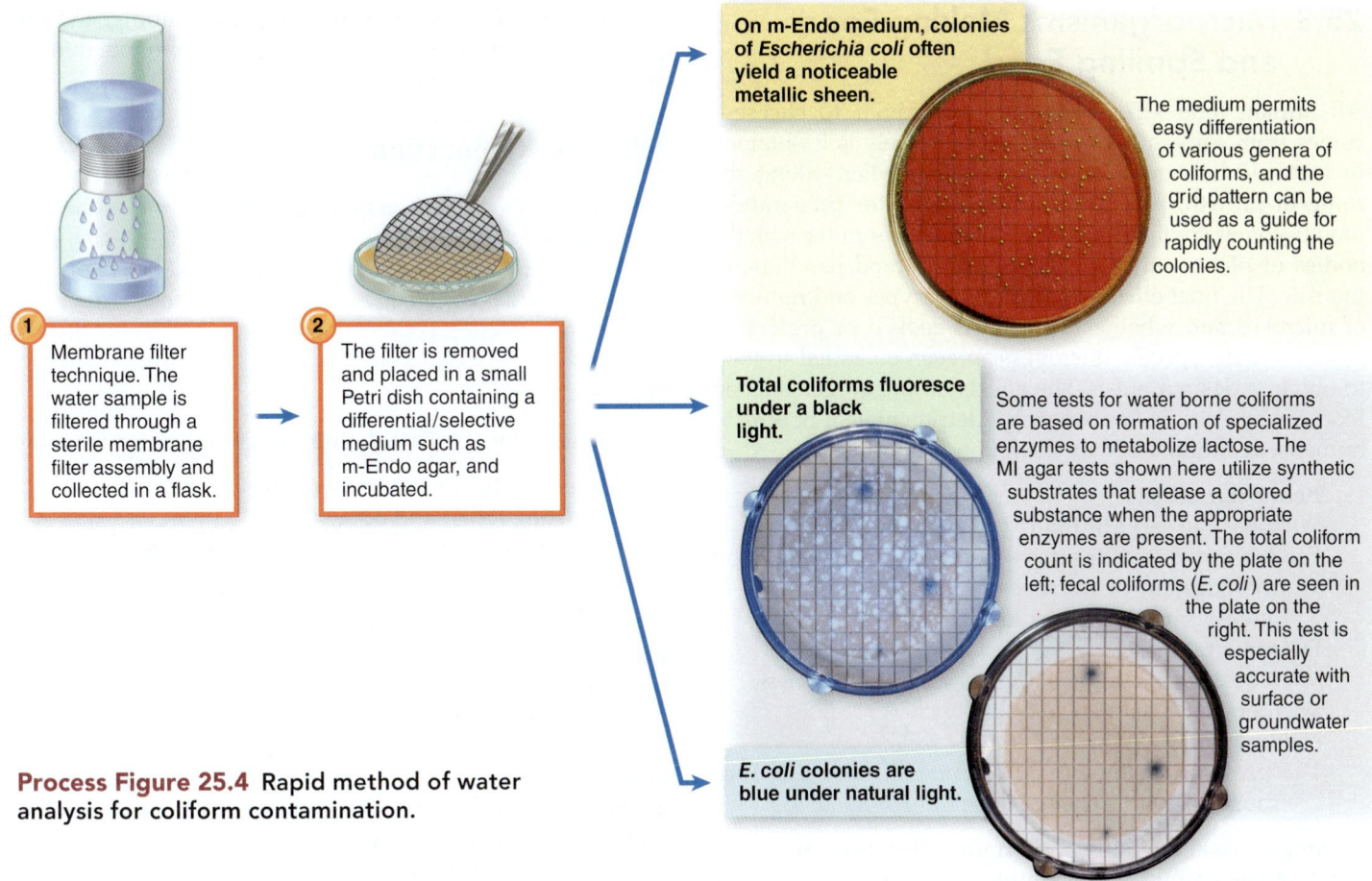

Process Figure 25.4 **Rapid method of water analysis for coliform contamination.**

On m-Endo medium, colonies of *Escherichia coli* often yield a noticeable metallic sheen.

The medium permits easy differentiation of various genera of coliforms, and the grid pattern can be used as a guide for rapidly counting the colonies.

Total coliforms fluoresce under a black light.

Some tests for water borne coliforms are based on formation of specialized enzymes to metabolize lactose. The MI agar tests shown here utilize synthetic substrates that release a colored substance when the appropriate enzymes are present. The total coliform count is indicated by the plate on the left; fecal coliforms (*E. coli*) are seen in the plate on the right. This test is especially accurate with surface or groundwater samples.

E. coli colonies are blue under natural light.

1. Membrane filter technique. The water sample is filtered through a sterile membrane filter assembly and collected in a flask.

2. The filter is removed and placed in a small Petri dish containing a differential/selective medium such as m-Endo agar, and incubated.

is more suitable for dilute fluids, such as drinking water, that are relatively free of particulate matter; and it is less suitable for water containing heavy microbial growth or debris. This technique is related to the method described in chapter 11 for sterilizing fluids by filtering out microbial contaminants, except that in this system, the filter containing the trapped microbes is the desired end product. The steps in membrane filtration are diagrammed in **process figure 25.4.** After filtration, the membrane filter is placed in a Petri dish containing selective medium. After incubation, both nonfecal and fecal coliform colonies can be counted and often presumptively identified by their distinctive characteristics on these media.

Another more time-consuming but useful technique is the **most probable number (MPN)** procedure, which detects coliforms by a series of *presumptive, confirmatory,* and *completed* tests. The presumptive test involves three subsets of fermentation tubes, each containing different amounts of lactose or lauryl tryptose broth. The three subsets are inoculated with various-size water samples. After 24 hours of incubation, the tubes are evaluated for gas production. A positive test for gas formation is presumptive evidence of coliforms; negative for gas means no coliforms. The number of positive tubes in each subset is tallied, and this set of numbers is applied to a statistical table to estimate the most likely or probable concentration of coliforms.

When a test is negative for coliforms, the water is considered generally fit for human consumption, but even slight coliform levels are allowable under some circumstances. For example, municipal waters can have a maximum of 4 coliforms per 100 mL; private wells can have an even higher count. There is no acceptable level for fecal coliforms, enterococci, viruses, or pathogenic protozoa in drinking water. Waters that will not be consumed but are used for fishing or swimming are permitted to have counts of 70 to 200 coliforms per 100 mL. If the coliform level of recreational water reaches 1,000 coliforms per 100 mL, health departments usually bar its usage.

25.2 Learning Outcomes—Assess Your Progress

3. Outline the steps in water purification.
4. Differentiate water purification from sewage treatment.
5. Describe the primary and secondary phases of sewage treatment.
6. Discuss how wastes can be converted into usable forms of energy.
7. List five important pathogens of drinking water.
8. Provide examples of indicator bacteria, and describe their role in the survey of water quality.
9. Summarize methods for identifying and quantifying microbial contaminants in water supplies.

25.3 Microorganisms Making Food and Spoiling Food

All human food—from vegetables to caviar to cheese—comes from some other organism, and rarely is it obtained in a sterile, uncontaminated state. Somewhere along the route of growth, procurement, processing, or preparation, food becomes contaminated with microbes from the soil, the bodies of plants and animals, water, air, food handlers, or utensils. The final effects depend on the types and numbers of microbes and whether the food is cooked or preserved. In some cases, specific microbes can even be added to food to obtain a desired effect. The effects of microorganisms on food can be classified as beneficial, detrimental, or neutral to humans, as summarized by the following outline:

Beneficial Effects
Microbes can serve as food.
Food is fermented or otherwise chemically changed by the addition of microbes or microbial products to alter or improve flavor, taste, or texture.
Detrimental Effects
Microbes cause food poisoning or food-borne illness.
Microbes spoil food.
Neutral Effects
The presence or growth of certain microbes does not cause disease or change the nature of the food.

As long as food contains no harmful substances or organisms, its suitability for consumption is largely a matter of taste. But what tastes like rich flavor to some may seem like decay to others. The test of whether certain foods are edible is guided by culture, experience, and preference. The flavors, colors, textures, and aromas of many cultural delicacies are supplied by bacteria and fungi. Poi, pickled cabbage, Norwegian fermented fish, and Limburger cheese are notable examples. If you examine the foods of most cultures, you will find some foods that derive their delicious and at times very unique flavor from microbes.

Microbial Fermentations in Food Products from Plants

In contrast to methods that destroy or keep out unwanted microbes, many culinary procedures deliberately add microorganisms and encourage them to grow. Common substances such as bread, cheese, beer, wine, yogurt, and pickles are the result of **food fermentations.** These reactions actively encourage biochemical activities that impart a particular taste, smell, or appearance to food. The microbe or microbes can occur naturally on the food substrate, as in sauerkraut, or they can be added as pure or mixed samples of known bacteria, molds, or yeasts called **starter cultures.** Many food fermentations are synergistic, with a series of microbes acting in concert to convert a starting substrate to the desired end product. Because large-scale production of fermented milk, cheese, bread, alcoholic brews, and vinegar depends upon inoculation with starter cultures,

considerable effort is spent selecting, maintaining, and preparing these cultures and excluding contaminants that can spoil the fermentation.

Disease Connection

Recently, a food-borne outbreak was linked to a Maryland company's use of a contaminated starter culture for making a fermented soy product called tempeh. Eighty-eight people were sickened by a nontyphoidal strain of *Salmonella*.

Bread

Microorganisms accomplish three functions in bread making:

1. leavening the flour-based dough,
2. imparting flavor and odor, and
3. conditioning the dough to make it workable.

Leavening is achieved primarily through the release of gas to produce a porous and spongy product. Without leavening, bread dough remains dense, flat, and hard. Although various microbes and leavening agents can be used, the most common ones are various strains of the baker's yeast *Saccharomyces cerevisiae.* Other gas-forming microbes such as coliform bacteria, certain *Clostridium* species, heterofermentative lactic acid bacteria, and wild yeasts can be employed, depending on the type of bread desired.

Yeast metabolism requires a source of fermentable sugar such as maltose or glucose. Because the yeast respires aerobically in bread dough, the chief products of maltose fermentation are carbon dioxide and water rather than alcohol (the main product in beer and wine). Other contributions to bread texture come from kneading, which incorporates air into the dough, and from microbial enzymes, which break down flour proteins (gluten) and give the dough elasticity.

Besides carbon dioxide production, bread fermentation generates other volatile organic acids and alcohols that impart delicate flavors and aromas. These are especially well developed in handmade bread, which is leavened more slowly than commercial bread. Yeasts and bacteria can also impart unique flavors, depending upon the culture mixture and baking techniques used. The pungent flavor of rye bread, for example, comes in part from starter cultures of lactic acid bacteria such as *Lactobacillus plantarum*, *L. brevis*, *L. bulgaricus*, *Leuconostoc mesenteroides*, and *Streptococcus thermophilus*. Sourdough bread gets its unique tang from *Lactobacillus sanfranciscensis*.

Beer

The production of alcoholic beverages takes advantage of another useful property of yeasts. By fermenting carbohydrates in fruits or grains anaerobically, they produce ethyl alcohol, as shown by this equation:

$$C_6H_{12}O_6 \rightarrow 2C_2H_5OH + 2CO_2$$

Yeast + Sugar = Ethanol + Carbon dioxide

Depending on the starting materials, the type of yeast used, and the processing method, alcoholic beverages vary in alcohol content and flavor. The principal types of fermented beverages are beers, wines, and spirit liquors.

The earliest evidence of beer brewing appears in ancient tablets by the Sumerians and Babylonians around 6000 BC. The starting ingredients for both ancient and present-day versions of beer, ale, stout, porter, and other variations are water, malt (barley grain), hops, and special strains of yeasts. The steps in brewing include malting, mashing, adding hops, fermenting, aging, and finishing.

For brewer's yeast to convert the carbohydrates in grain into ethyl alcohol, the barley must first be sprouted and softened to make its complex nutrients available to yeasts. This process, called **malting,** releases amylases that convert starch to dextrins and maltose, and proteases that digest proteins. Other sugar and starch supplements added in some forms of beer are corn, rice, wheat, soybeans, potatoes, and sorghum. After the sprouts have been separated, the remaining malt grain is dried and stored in preparation for mashing.

The malt grain is soaked in warm water and ground up to prepare a **mash.** Sugar and starch supplements are then introduced to the mash mixture, which is heated to a temperature of about 65°C to 70°C. During this step, the starch is hydrolyzed by amylase and simple sugars are released. Heating this mixture to 75°C stops the activity of the enzymes. Solid particles are next removed by settling and filtering (the result is what can then be added to an anaerobic digester). **Wort,** the clear fluid that comes off, is rich in dissolved carbohydrates. It is boiled for about 2.5 hours with **hops,** the dried scales of the female flower of *Humulus lupulus* **(figure 25.5),** to extract the bitter acids and resins that give aroma and flavor to the finished product. Boiling also caramelizes the sugar and imparts a golden or brown color, destroys any bacterial contaminants that can destroy flavor, and concentrates the mixture. The filtered and cooled supernatant is then ready for the addition of yeasts and fermentation.

Fermentation begins when wort is inoculated with a species of *Saccharomyces* that has been specially developed for beer making. Top yeasts such as *Saccharomyces cerevisiae* function at the surface and are used to produce the higher alcohol content of *ales*. Bottom yeasts such as *S. uvarum* (*carlsbergensis*) function deep in the fermentation vat and are used to make other beers. In both cases, the initial inoculum of yeast starter is aerated briefly to promote rapid growth and increase the load of yeast cells. Shortly thereafter, an insulating blanket of foam and carbon dioxide develops on the surface of the vat and promotes anaerobic conditions **(figure 25.6).** During 8 to 14 days of fermentation, the wort sugar is converted chiefly to ethanol and carbon dioxide. The diversity of flavors in the finished product is partly due to the release of small amounts of glycerol, acetic acid, and esters. Fermentation is self-limited, and it essentially ceases when a concentration of 3% to 6% ethyl alcohol is reached.

Freshly fermented, or "green," beer is **lagered,** meaning it is held for several weeks to months in vats near 0°C. During this maturation period, yeast, proteins, resin, and other materials settle, leaving behind a clear, mellow fluid. Lager beer is subjected to a final filtration step to remove any residual yeasts that could spoil it. Finally, it is carbonated with carbon dioxide collected during fermentation and packaged in kegs, bottles, or cans.

Scientists are manipulating microbes to produce higher yields of alcohol from additional substrates. For example, a genetically engineered strain of *E. coli* has recently been created that produces ethanol from brown seaweed.

Wine and Liquors

Wine is traditionally considered any alcoholic beverage arising from the fermentation of grape juice, but practically any fruit can be rendered into wine. The essential starting point is the preparation of **must,** the juice given off by crushed fruit that is used as a substrate for fermentation. In general, grape wines are either white or red. The color comes from the skins of the grapes, so white wine is prepared either from white-skinned

Figure 25.6 Anaerobic conditions in homemade beer production. A layer of carbon dioxide foam keeps oxygen out.

Figure 25.5 Hops. Female flowers of hops, the herb that gives beer some of its flavor and aroma.

grapes or from red-skinned grapes that have had the skin removed. Red wine comes from the red- or purple-skinned varieties. Major steps in making wine include must preparation (crushing), fermentation, storage, and aging **(figure 25.7).**

For proper fermentation, the must should contain 12% to 25% glucose or fructose, so the art of wine making begins in the vineyard. Grapes are harvested when their sugar content reaches 15% to 25%, depending on the type of wine to be made. Grapes from the field carry a mixed biofilm on their surface called the *bloom* that can serve as a source of wild yeasts. Some winemakers allow these natural yeasts to dominate, but many wineries inoculate the must with a special strain of *Saccharomyces cerevisiae,* variety *ellipsoideus.* To discourage yeast and bacterial spoilage agents, winemakers sometimes treat grapes with sulfur dioxide or potassium metabisulfite. The inoculated must is thoroughly aerated and mixed to promote rapid aerobic growth of yeasts, but when the desired level of yeast growth is achieved, anaerobic alcoholic fermentation is begun.

The temperature of the vat during fermentation must be carefully controlled to facilitate alcohol production. The length of fermentation varies from 3 to 5 days in red wines and from 7 to 14 days in white wines. The initial fermentation yields ethanol concentrations reaching 7% to 15% by volume, depending on the type of yeast, the source of the juice, and ambient conditions. The fermented juice (raw wine) is decanted and transferred to large vats to settle and clarify. Before the final aging process, it is flash-pasteurized to kill microorganisms and filtered to remove any remaining yeasts and sediments. Wine is aged in wooden casks for varying time periods (months to years), after which it is bottled and stored for further aging. During aging, nonmicrobial changes produce aromas and flavors (the bouquet) characteristic of a particular wine.

The fermentation processes discussed thus far can only achieve a maximum alcoholic content of 17%, because concentrations above this level inhibit the metabolism of the yeast. The fermentation product must be distilled to obtain higher concentrations such as those found in liquors. During distillation, heating the liquor separates the more volatile alcohol from the less volatile aqueous phase. The alcohol is

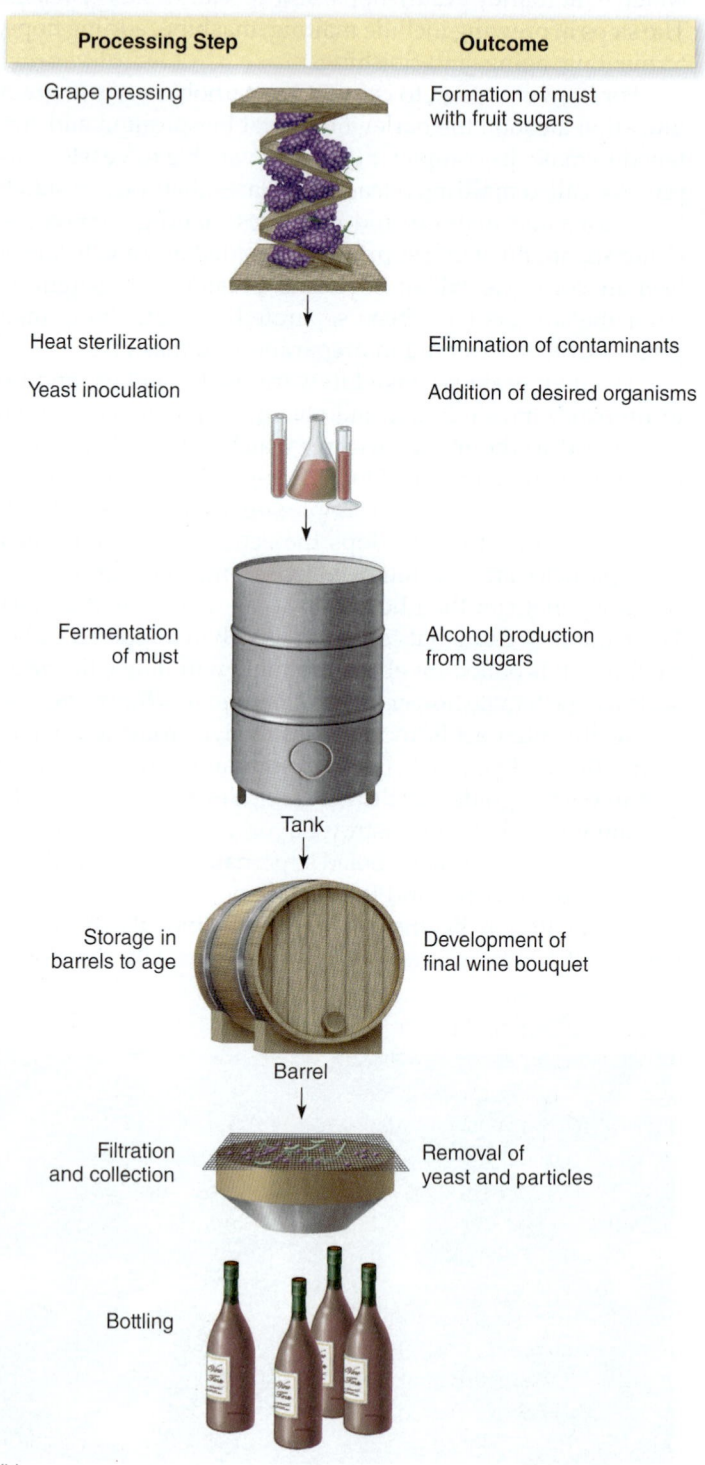

Processing Step		Outcome
Grape pressing		Formation of must with fruit sugars
Heat sterilization		Elimination of contaminants
Yeast inoculation		Addition of desired organisms
Fermentation of must		Alcohol production from sugars
	Tank	
Storage in barrels to age		Development of final wine bouquet
	Barrel	
Filtration and collection		Removal of yeast and particles
Bottling		

(a)

(b)

Figure 25.7 Wine making.   (a) Wine fermentation vats in a large commercial winery. (b) General steps in wine making.

then condensed and collected. The alcohol content of distilled liquors is rated by *proof,* a measurement that is usually two times the alcohol content. Thus, 80 proof vodka contains 40% ethyl alcohol.

Distilled liquors originate through a process similar to wine making, although the starting substrates can be extremely diverse. In addition to distillation, liquors can be subjected to special treatments such as aging to provide unique flavor or color. Vodka, a colorless liquor, is usually prepared from fermented potatoes, and rum is distilled from fermented sugarcane. Assorted whiskeys are derived from fermented grain mashes; rye whiskey is produced from rye mash, and bourbon from corn mash. Brandy is distilled grape, peach, or apricot wine.

The resident microbiologist at any brewery, winery, or distillery plays a vital role in the production of a quality product. Constant monitoring at every step of the process is required to make sure the yeasts are thriving and that uninvited microbial guests have not joined the fermentation process.

Other Fermented Plant Products

Fermentation provides an effective way of preserving vegetables, as well as enhancing flavor with lactic acid and salt. During pickling fermentations, vegetables are immersed in an anaerobic salty solution (brine) to extract sugar and nutrient-laden juices. The salt also disperses bacterial clumps, and its high osmotic pressure inhibits proteolytic bacteria and sporeformers that can spoil the product.

Sauerkraut is a fermentation product of cabbage. Cabbage is washed, wilted, shredded, salted, and packed tightly into a fermentation vat. Weights cover the cabbage mass and squeeze out its juices. The fermentation is achieved by natural cabbage microbiota or by an added culture. The initial agent of fermentation is *Leuconostoc mesenteroides,* which grows rapidly in the brine and produces lactic acid. It is followed by *Lactobacillus plantarum,* which continues to raise the acid content to as high as 2% (pH 3.5) by the end of fermentation. The high acid content restricts the growth of spoilage microbes.

Fermented cucumber pickles come chiefly in salt and dill varieties. Salt pickles are prepared by washing immature cucumbers, placing them in barrels of brine, and allowing them to ferment for 6 to 9 weeks. The brine can be inoculated with *Pediococcus cerevisiae* and *Lactobacillus plantarum* to avoid unfavorable qualities caused by natural microbiota and to achieve a more consistent product. Fermented dill pickles are prepared in a somewhat more elaborate fashion, with the addition of dill herb, spices, garlic, onion, and vinegar.

Natural vinegar is produced when the alcohol in fermented plant juice is oxidized to acetic acid, which is responsible for the pungent odor and sour taste. Although a reasonable facsimile of vinegar could be made by mixing about 4% acetic acid and a dash of sugar in water, this preparation would lack the traces of various esters, alcohol, glycerin, and volatile oils that give natural vinegar its pleasant character. Vinegar is actually produced in two stages. The first stage is similar to wine or beer making, in which a plant juice is fermented to alcohol by *Saccharomyces.* The second

stage involves an aerobic fermentation carried out by acetic acid bacteria in the genera *Acetobacter* and *Gluconobacter.* These bacteria oxidize the ethanol in a two-step process, as shown here:

$$2C_2H_5OH + \tfrac{1}{2} O_2 \rightarrow CH_3CHO + H_2O$$
Ethanol $\qquad\qquad$ Acetaldehyde

$$CH_3CHO + \tfrac{1}{2} O_2 \rightarrow CH_3COOH$$
Acetaldehyde $\qquad\qquad$ Acetic acid

The abundance of oxygen necessary in commercial vinegar making is furnished by exposing inoculated raw material to air by arranging it in thin layers in open trays, allowing it to trickle over loosely packed beechwood twigs and shavings, or aerating it in a large vat. Different types of vinegar are derived from substrates such as apple cider (cider vinegar), malted grains (malt vinegar), and grape juice (wine vinegar).

Microbes in Milk and Other Dairy Products

Milk has a highly nutritious composition. It contains an abundance of water and is rich in minerals, protein (chiefly casein), butterfat, sugar (especially lactose), and vitamins. Previously it was thought to be sterile in the udder, but human milk was recently found to contain a rich microbiota, suggesting that other mammals may follow suit. Of course, as it passes out of the teat, it is inoculated by more biota. Other microbes can be introduced by milking utensils. Because milk is a nearly perfect culture medium, it is highly susceptible to microbial growth. When raw milk is left at room temperature, a series of bacteria ferment the lactose, produce acid, and alter the milk's content and texture **(figure 25.8).** This progression can occur naturally, or it can be induced, as in the production of cheese and yogurt.

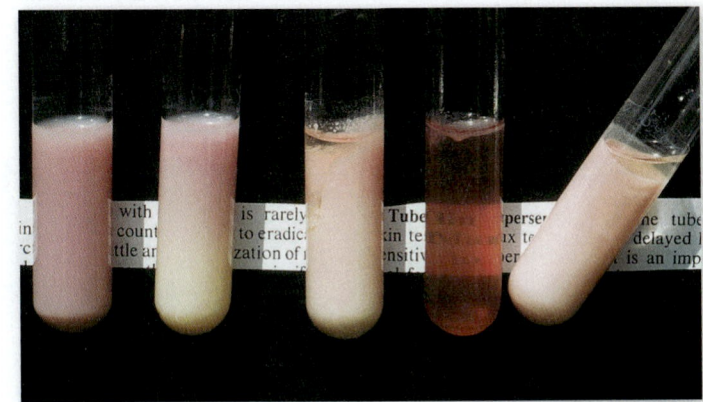

Figure 25.8 Microbes at work in milk products. Litmus milk is a medium used to indicate pH and consistency changes in milk resulting from microbial action. The first tube is an uninoculated, unchanged control (a purplish color). The second tube has a white, decolorized zone indicative of litmus reduction. The third tube has become acidified (pink), and its proteins have formed a loose curd. In the fourth tube, digestion of milk proteins has caused complete clarification or peptonization of the milk. The fifth tube shows a well-developed solid curd overlaid by a clear fluid, the whey.

In the initial stages of milk fermentation, lactose is rapidly attacked by *Streptococcus lactis* and *Lactobacillus* species. The resultant lactic acid accumulation and lowered pH cause the milk proteins to coagulate into a solid mass called the **curd.** Curdling also causes the separation of a watery liquid called **whey** on the surface. Curd can be produced by microbial action or by an enzyme, **rennin** (casein coagulase). Although this enzyme was traditionally isolated from the stomach of unweaned calves, today it is produced in large quantities by genetically engineered microbes. In fact, this was the first product made through genetic technology that was FDA-approved for human consumption.

Cheese

Since 5000 BC, various forms of cheese have been produced by spontaneous fermentation of cow, goat, or sheep milk. Present-day, large-scale cheese production is carefully controlled and uses only freeze-dried samples of pure cultures. These are first inoculated into a small quantity of pasteurized milk to form an active starter culture. This amplified culture is subsequently inoculated into a large vat of milk, where rapid curd development takes place. Such rapid growth is desired because it promotes the overgrowth of the desired inoculum and prevents the activities of undesirable contaminants. Rennin is usually added to increase the rate of curd formation.

After its separation from whey, the curd is rendered to produce soft, semisoft, or hard cheese **(figure 25.9).** The composition of cheese is varied by adjusting water, fat, acid, and salt content. Cottage and cream cheese are examples of the soft, more perishable variety. After light salting and the optional addition of cream, they are ready for consumption without further processing. Other cheeses acquire their character from "ripening," a complex curing process involving bacterial, mold, and enzyme reactions that develop the final flavor, aroma, and other features characteristic of particular cheeses.

Figure 25.9 Cheese making. The curd-cutting stage in the making of cheddar cheese.

The distinctive traits of soft cheeses such as Limburger, Camembert, and Liederkranz are acquired by ripening with a reddish-brown mucoid coating of yeasts, micrococci, and molds. The microbial enzymes permeate the curd and ferment lipids, proteins, carbohydrates, and other substrates. This process leaves assorted acids and other by-products that give the finished cheese powerful aromas and delicate flavors. Semisoft varieties of cheese such as Roquefort, bleu, or Gorgonzola are infused and aged with a strain of *Penicillium roqueforti* mold. Hard cheeses such as Swiss, cheddar, and Parmesan develop a sharper flavor by aging with selected bacteria. The pockets in Swiss cheese come from entrapped carbon dioxide formed by *Propionibacterium*, which is also responsible for its bittersweet taste.

Other Fermented Milk Products

Yogurt is formed by the fermentation of milk by *Lactobacillus bulgaricus* and *Streptococcus thermophilus.* These organisms produce organic acids and other flavor components and can grow in such numbers that a gram of yogurt regularly contains 100 million bacteria. Live cultures of *Lactobacillus acidophilus* are an important additive to acidophilus milk, which is said to benefit digestion and to help maintain the normal biota of the intestine. This bacterium, along with *Bifidobacterium bifidum*, is also used in yogurt products specifically marketed to regulate digestion. Fermented milks such as kefir, koumiss, and buttermilk are a basic food source in many cultures.

Microorganisms as Food

At first, the thought of eating bacteria, molds, algae, and yeasts may seem odd or unappetizing. We do eat their macroscopic relatives, such as mushrooms, truffles, and seaweed, but we are used to thinking of the microscopic forms as agents of decay and disease or, at most, as food flavorings. The consumption of microorganisms is not a new concept. In Germany during World War II, it became necessary to supplement the diets of undernourished citizens by adding yeasts and molds to foods. Several countries now commercially mass-produce food yeasts, bacteria, and in a few cases, algae. Although eating microbes has yet to win total public acceptance, their use as feed supplements for livestock is increasing. In England, an animal feed called Pruteen is produced by mass culture of the bacterium *Methylophilus methylotrophus*. Mycoprotein, a product made from the fungus *Fusarium graminearum*, is also sold there. Recently, it was found that certain types of algae produce nearly 40% more protein than soybeans. Algae, thus, have turned into potentially marketable foods for use in aquaculture and other animal populations.

Health food stores carry bottles of dark green pellets or powder made up of a spiral-shaped cyanobacterium called *Spirulina*. This microbe is harvested from the surface of lakes and ponds, where it grows in great mats. In some parts of Africa and Mexico, *Spirulina* has become a viable alternative to green plants as a primary nutrient source. It can be eaten in its natural form or added to other foods and beverages.

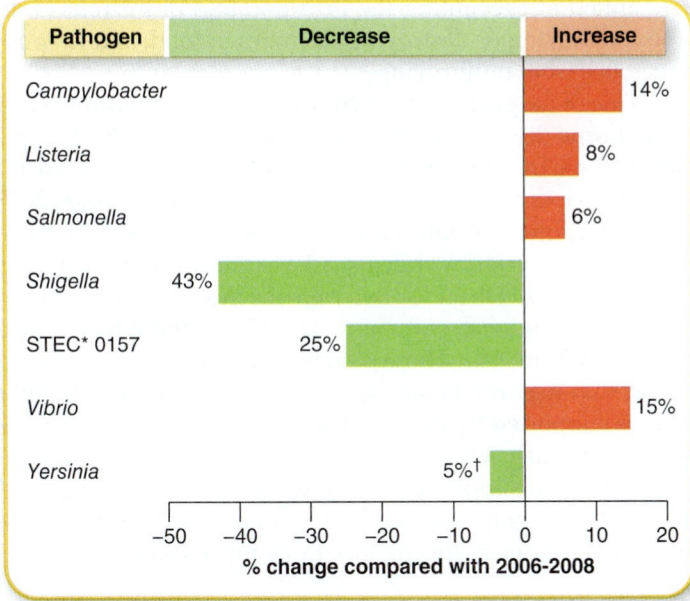

| Pathogen | Decrease | Increase |

*Shiga-toxin-producing *Escherichia coli*
[†]Not statistically significant

Figure 25.10 Changes in incidence of laboratory-confirmed bacterial infections in the United States, 2011.

Microbial Involvement in Food-Borne Diseases

In the United States, 48 million people suffer each year from some form of food infection (see chapter 22). Until very recently, reports of food poisoning were escalating rapidly in the United States and worldwide. Outbreaks attributed to common pathogens (*Salmonella, E. coli, Vibrio,* hepatitis A, *Listeria, Campylobacter,* and various protozoa) had approximately doubled between 1980 and 2000. A major factor in the escalation was the mass production and distribution of processed food such as raw vegetables, fruits, and meats. Improper handling can lead to gross contamination of these products with soil or animal wastes.

Growing concerns about food safety led to a new approach to regulating the food industry. Hazard Analysis and Critical Control Point (HACCP) is a management system that is used to assess safety risks in the growth, harvesting, processing, and distribution of food items. It involves the identification, evaluation, control, and prevention of hazards at all stages of the food production process. Since 1998, HACCP principles have been accepted by the U.S. Department of Agriculture and by the Food and Drug Administration, as well as by other global agencies. The success of the current HACCP system depends on all levels of food industry adhering to the established guidelines. For 10 years after its implementation, illnesses caused by most of the food-borne organisms had significantly decreased. However, 2011 and 2012 showed increases (**figure 25.10** shows 2011 data). Keep in mind that many reported food-poisoning outbreaks occur where contaminated food has been served to large groups of

people,[1] but most cases probably occur in the home and are not reported.

In 2011, President Obama signed a sweeping new law called the Food Safety Modernization Act (FSMA). Its implementation was delayed, but in 2013 the Food and Drug

1. One-third of all reported cases result from eating restaurant food.

Administration started releasing the proposed rules for public commentary. The FDA says it is the most sweeping reform of food safety in 70 years. It changes FDA's focus to prevention of food-borne outbreaks, as opposed to investigation after the fact.

Prevention Measures for Food Poisoning and Spoilage

It will never be possible to avoid all types of food-borne illness because of the ubiquity of microbes in air, water, food, and the human body. But most types of food poisoning require the growth of microbes in the food. In the case of food infections, an infectious dose (sufficient cells to initiate infection) must be present; and in food intoxication, enough cells to produce the toxin must be present. Thus, food poisoning or spoilage can be prevented by proper food handling, preparation, and storage. The methods shown in **figure 25.11**

are aimed at preventing the incorporation of microbes into food, removing or destroying microbes in food, and keeping microbes from multiplying.

One of the greatest concerns today is the increasing risk of food-borne disease brought about by imported food. According to the CDC, 39 outbreaks and 2,348 illnesses were attributed to imported food from 15 countries from the years 2005 to 2010. More alarming is the fact that nearly 50% of these outbreaks occurred within just a 1-year time frame—between 2009 and 2010. Fish and spices were the most common sources of infection in these outbreaks, and 45% of these foods were imported from Asia. These data indicate a possible escalating trend that may not be going away any time soon, a thought that is supported by the fact that the United States continues to import nearly 90% of its seafood and up to 60% of the country's produce supply. The tracking and testing of imported foods are currently substandard, but the FSMA contains provisions to significantly enhance those procedures.

Preventing the Incorporation of Microbes into Food

Most agricultural products such as fruits, vegetables, grains, meats, eggs, and milk are naturally exposed to microbes. Vigorous washing reduces the levels of contaminants in fruits and vegetables, whereas meat, eggs, and milk must be taken from their animal source as aseptically as possible. Aseptic techniques are also essential in the kitchen. Contamination of foods by fingers can be easily remedied by hand washing and proper hygiene, and contamination by flies or other insects can be stopped by covering foods or eliminating pests from the kitchen. Care and common sense also apply in managing utensils. It is important to avoid cross-contaminating food by, for example, using the same cutting board for meat and vegetables without disinfecting it between uses.

Preventing the Survival or Multiplication of Microbes in Food

Because it is not possible to eliminate all microbes from certain types of food by clean techniques alone, a more efficient approach is to preserve the food by physical or chemical methods. Hygienically preserving foods is especially important for large commercial companies that process and sell bulk foods and must ensure that products are free from harmful

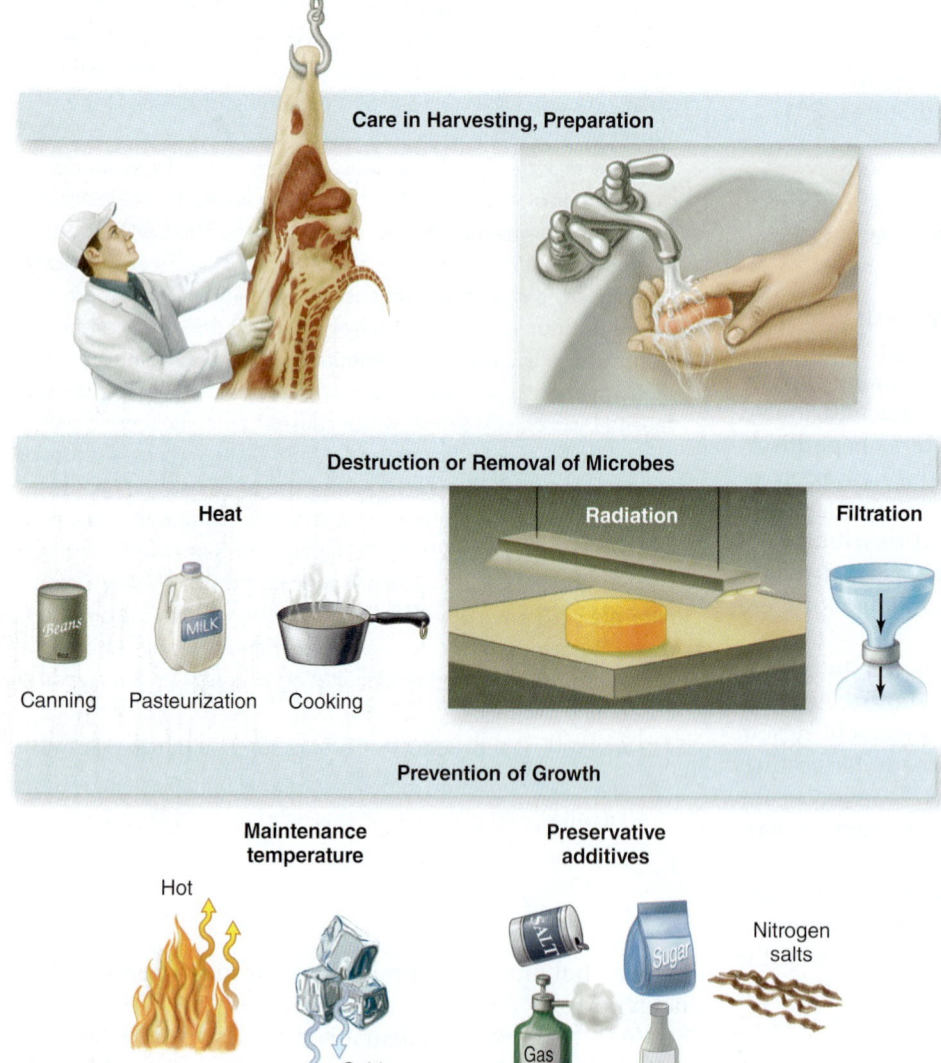

Figure 25.11 The primary methods of preventing food poisoning and food spoilage.

contaminants. Regulations and standards for food processing are administered by two federal agencies: the Food and Drug Administration (FDA) and the U.S. Department of Agriculture (USDA).

Temperature and Food Preservation

Heat is a common way to destroy microbial contaminants or to reduce the load of microorganisms. Commercial canneries preserve food in hermetically sealed containers that have been exposed to high temperatures over a specified time period. The temperature used depends on the type of food, and it can range from 60°C to 121°C, with exposure times ranging from 20 minutes to 115 minutes. The food is usually processed at a thermal death time (TDT; see chapter 11) that will destroy the main spoilage organisms and pathogens but will not alter the nutrient value or flavor of the food. For example, tomato juice must be heated to between 121°C and 132°C for 20 minutes to ensure destruction of the spoilage agent *Bacillus coagulans.* Most canning methods are rigorous enough to sterilize the food completely, but some only render the food "commercially sterile," which means it contains live bacteria that are unable to grow under normal conditions of storage.

Another use of heat is **pasteurization,** usually defined as the application of heat below 100°C to destroy nonresistant bacteria and yeasts in liquids such as milk, wine, and fruit juices. The heat is applied in the form of steam, hot water, or even electrical current. The most prevalent technology is the *high-temperature short-time (HTST),* or flash method, using extensive networks of tubes that expose the liquid to 72°C for 15 seconds **(figure 25.12).** An alternative method, ultra-high-temperature (UHT) pasteurization, steams the product until it reaches a temperature of 134°C for at least 1 second. Although milk processed this way is not actually sterile, it is often marketed as sterile, with a shelf life of up to 3 months. Older methods involve large bulk tanks that hold the fluid at a lower temperature for a longer time, usually 62.3°C for 30 minutes.

Cooking temperatures used to boil, roast, or fry foods can render them free or relatively free of living microbes if carried out for sufficient time to destroy any potential pathogens. A quick warming of chicken or an egg is inadequate to kill microbes such as *Salmonella.* In fact, any meat is a potential source of infectious agents and should be adequately cooked. Because most meat-associated food poisoning is caused by non-endospore-forming bacteria, heating the center of meat to at least 80°C and holding it there for 30 minutes is usually sufficient to kill pathogens. Roasting or frying food at temperatures of at least 200°C or boiling it will reduce microbial contamination to a safe level for consumption.

Any perishable raw or cooked food that could serve as a growth medium must be stored to prevent the multiplication of microorganisms that have survived during processing or handling. Because most food-borne bacteria and molds that are agents of spoilage or infection can multiply at room temperature, manipulation of the holding temperature is a useful

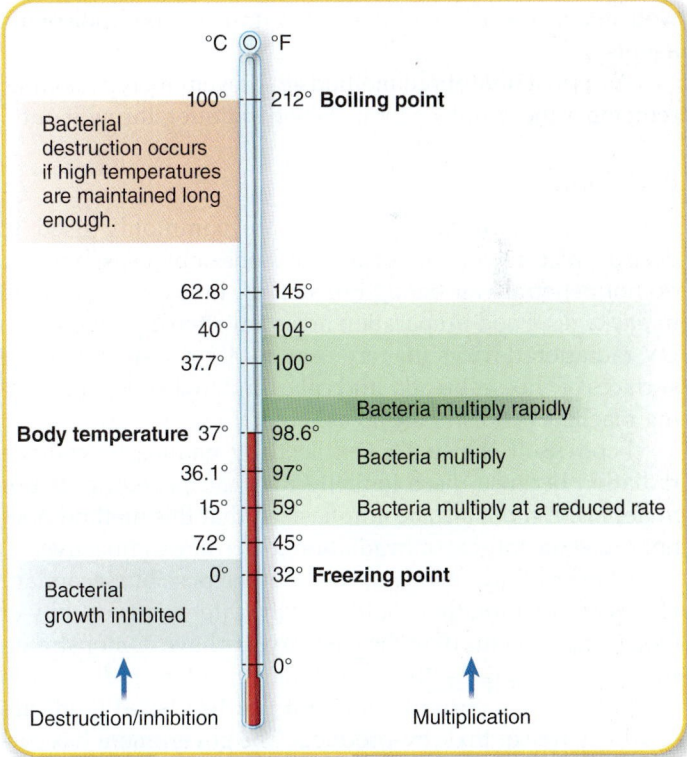

Figure 25.13 Temperatures favoring and inhibiting the growth of microbes in food. Most microbial agents of disease or spoilage grow in the temperature range of 15°C to 40°C. Preventing unwanted growth in foods in long-term storage is best achieved by refrigeration or freezing (4°C or lower). Preventing microbial growth in foods intended to be consumed warm in a few minutes or hours requires maintaining the foods above 60°C.

Figure 25.12 A modern flash pasteurizer, a system used in dairies for high-temperature short-time (HTST) pasteurization.

preservation method **(figure 25.13).** A good general directive is to store foods at temperatures below 4°C or above 60°C.

Regular refrigeration reduces the growth rate of most mesophilic bacteria by 10 times, although some psychrotrophic microbes can continue to grow at a rate that causes spoilage. This factor limits the shelf life of milk, because even at 7°C, a population could go from a few cells to a billion in 10 days. Pathogens such as *Listeria monocytogenes* and *Salmonella* can also continue to grow in refrigerated foods. Freezing is a longer-term method for cold preservation. Foods can be either slow-frozen for 3 to 72 hours at −15°C to −23°C or rapidly frozen for 30 minutes at −17°C to −34°C. Freezing is not a microbicidal method; because it cannot be counted upon to kill microbes, rancid, spoiled, or infectious foods will still be unfit to eat after freezing and defrosting. *Salmonella* is known to survive several months in frozen chicken and ice cream, and *Vibrio parahaemolyticus* can survive in frozen shellfish. For this reason, frozen foods should be defrosted rapidly and immediately cooked or reheated. However, even this practice will not prevent staphylococcal intoxication if the toxin is already present in the food before it is heated.

Foods such as soups, stews, gravies, meats, and vegetables that are generally eaten hot should not be maintained at warm or room temperatures, especially in settings such as cafeterias, banquets, and picnics. The use of a hot plate, chafing dish, or hot water bath will maintain foods above 60°C, well above the incubation temperature of food-poisoning agents.

As a final note about methods to prevent food poisoning, remember the simple axiom: "When in doubt, throw it out."

Radiation

Ultraviolet (nonionizing) lamps are commonly used to destroy microbes on the surfaces of foods or utensils, but they do not penetrate far enough to sterilize bulky foods or food in packages. Food preparation areas are often equipped with UV radiation devices that are used to destroy spores on the surfaces of cheese, breads, and cakes and to disinfect packaging machines and storage areas.

Food itself is usually sterilized by gamma or cathode radiation because these ionizing rays can penetrate denser materials. It must also be emphasized that this method does not cause the targets of irradiation to become radioactive.

Concerns have been raised about the possible secondary effects of radiation that could alter the safety and edibility of foods. Experiments over the past 30 years have demonstrated some side reactions that affect flavor, odor, and vitamin content; but it is currently thought that irradiated foods are relatively free of toxic by-products. The government has currently approved the use of radiation in sterilizing beef, pork, poultry, fish, spices, grain, shell eggs, and some fruits and vegetables. The FDA has mandated that food labels contain the Radura symbol (the international symbol for irradiation) and the statement "treated with radiation" or "treated by irradiation." Radiation also increases the shelf life of perishable foods, thus lowering their cost.

Other Forms of Preservation

The addition of chemical preservatives to many foods can prevent the growth of microorganisms that could cause spoilage or disease. Preservatives include natural chemicals such as salt (NaCl) or table sugar and artificial substances such as ethylene oxide. The main classes of preservatives are organic acids, nitrogen salts, sulfur compounds, oxides, salt, and sugar.

Organic acids, including lactic, benzoic, and propionic acids, are among the most widely used preservatives. They are added to baked goods, cheeses, pickles, carbonated beverages, jams, jellies, and dried fruits to reduce spoilage from molds and some bacteria. Nitrites and nitrates are used primarily to maintain the red color of cured meats (hams, bacon, and sausage). By inhibiting the germination of *Clostridium botulinum* spores, they also prevent botulism intoxication, but their effects against other microorganisms are limited. Sulfite prevents the growth of undesirable molds in dried fruits, juices, and wines and retards discoloration in various foodstuffs. Ethylene and propylene oxide gases disinfect various dried foodstuffs. Their use is restricted to fruit, cereals, spices, nuts, and cocoa.

The high osmotic pressure contributed by hypertonic levels of salt plasmolyzes bacteria and fungi and removes moisture from food, thereby inhibiting microbial growth. Salt is commonly added to brines, pickled foods, meats, and fish. However, it does not retard the growth of pathogenic halophiles such as *Staphylococcus aureus*, which grows readily even in 7.5% salt solutions. The high sugar concentrations of candies, jellies, and canned fruits also exert an osmotic preservative effect, though some molds can still grow in this environment, leading to food spoilage and the potential for food-borne disease. Other chemical additives that function in preservation are alcohols and antibiotics. Alcohol is added to flavoring extracts, and antibiotics are approved for treating the carcasses of chickens, fish, and shrimp.

Food can also be preserved by **desiccation,** a process that removes moisture needed by microbes for growth by exposing the food to dry, warm air. Solar drying was traditionally used for fruits and vegetables, but modern commercial dehydration is carried out in rapid-evaporation mechanical devices. Drying is not a reliable microbicidal method, however. Numerous resistant microbes such as micrococci, coliforms, staphylococci, salmonellae, and fungi survive in dried milk and eggs, which can subsequently serve as agents of spoilage and infections.

In 2006, the Food and Drug Administration approved the spraying of bacteriophages onto ready-to-eat meat products. The bacteriophages are specific for *Listeria* and will act to kill the bacteria that would not otherwise be killed because the cold cuts and poultry are usually not cooked before consumption. Researchers are also testing bacteriophage sprays to rid chicken carcasses of *Salmonella* before processing.

25.3 Learning Outcomes—Assess Your Progress

10. Name five foods and/or beverages that are produced using microbial fermentation.
11. Summarize the microbial process that leads to leavening in bread.
12. Write the equation showing how yeasts convert sugar into alcoholic beverages.
13. Discuss how microorganisms themselves are useful as food products today.
14. Provide background on current HACCP guidelines, and describe how they are used to maintain food safety.
15. Report 10-year trends in the incidence of food-borne illness within the United States.

16. Outline basic principles of using temperature to preserve food.
17. List methods other than temperature currently used for preservation in the food industry.

25.4 Using Microorganisms to Make Things We Need

Virtually any large-scale commercial enterprise that enlists microorganisms to manufacture consumable materials is part of the realm of industrial microbiology. Traditionally, the name pertains primarily to bulk production of organic compounds such as antibiotics, hormones, vitamins, acids, solvents (table 25.1), and enzymes (table 25.2). There are also interesting attempts to place microbes in strategic places outside of the lab to work their magic at improving

Table 25.1 Industrial Products of Microorganisms

Chemical	Microbial Source	Substrate	Applications
Pharmaceuticals			
Cephalosporins	*Cephalosporium*	Glucose	Antibacterial antibiotics, broad-spectrum
Penicillins	*Penicillium chrysogenum*	Lactose	Antibacterial antibiotics, broad- and narrow-spectrum
Vitamin B_{12}	*Pseudomonas*	Molasses	Dietary supplement
Steroids (hydrocortisone)	*Rhizopus, Cunninghamella*	Deoxycholic acid, stigmasterol	Treatment of inflammation, allergy; hormone replacement therapy
Insulin	Original source is human; manufactured in *E. coli*	Nutrient broths	Diabetic therapy
Food additives and amino acids			
Citric acid	*Aspergillus, Candida*	Molasses	Acidifier in soft drinks; used to set jam; candy additive; fish preservative; retards discoloration of crabmeat; delays browning of sliced peaches
Xanthan	*Xanthomonas*	Glucose medium	Food stabilizer; not digested by humans
Acetic acid	*Acetobacter*	Any ethylene source, ethanol	Food acidifier; used in industrial processes
Miscellaneous			
Ethanol	*Saccharomyces*	Beets, cane, grains, wood, wastes	Additive to gasoline (gasohol)
Acetone	*Clostridium*	Molasses, starch	Solvent for lacquers, resins, rubber, fat, oil
Glycerol	Yeast	By-product of alcohol fermentation	Explosive (nitroglycerine)
Dextran	*Klebsiella, Acetobacter, Leuconostoc*	Glucose, molasses, sucrose	Polymer of glucose used as adsorbents, blood expanders, and in burn treatment; a plasma extender; used to stabilize ice cream, sugary syrup, candies

Table 25.2 Industrial Enzymes and Their Uses

Enzyme	Source	Application
Amylase	*Aspergillus, Bacillus, Rhizopus*	Flour supplement, desizing textiles, mash preparation, syrup manufacture, digestive aid, precooked foods, spot remover in dry cleaning
Cellulase	*Aspergillus, Trichoderma*	Denim finishing ("stone-washing"), digestive aid, increase digestibility of animal feed, degradation of wood or wood by-products
Hyaluronidase	Various bacteria	Medical use in wound cleansing, preventing surgical adhesions
Keratinase	*Streptomyces*	Hair removal from hides in leather preparation
Pectinase	*Aspergillus, Sclerotina*	Clarifies wine, vinegar, syrups, and fruit juices by degrading pectin, a gelatinous substance; used in concentrating coffee
Proteases	*Aspergillus, Bacillus, Streptomyces*	To clear and flavor rice wines, process animal feed, remove gelatin from photographic film, recover silver, tenderize meat, unravel silkworm cocoon, remove spots
Streptokinase	*Streptococcus*	Medical use in clot digestion, as a blood thinner

materials and processes **(Insight 25.4).** Many of the processing steps involve fermentations similar to those described in food technology, but industrial processes usually occur on a much larger scale, produce a specific compound, and involve numerous complex stages. The aim of industrial microbiology is to produce chemicals that can be purified and packaged for sale or for use in other commercial processes. Thousands of tons of organic chemicals worth several billion dollars are produced by this industry every year. To create one of these products, an industry must determine which microbes, starting compounds, and growth conditions work best. The research and development involved usually require an investment of 10 to 15 years and billions of dollars. However, in many cases, it has been worth the investment—financially and environmentally.

One of the most active areas of research in industrial microbiology is the use of cyanobacteria and algal species to produce biofuels. Although the U.S. government envisions microbial biofuels replacing a significant percentage of fossil fuel use in transportation, there are still significant stumbling blocks in meeting this goal. You may recall that original attempts to produce biofuels involved plants such as corn and soybeans. This proved to be unpopular because the plants, acreage, and water used to grow them could have been used for food. Scientists quickly realized that microorganisms—specifically, algae—could produce oil when exposed to sunlight within a photobioreactor **(figure 25.14).** This is a closed system that provides optimal photosynthetic growth conditions for the microbes, typically on a large scale. These systems use photosynthesis to manufacture O_2 and

INSIGHT 25.4   Bacteria Heal Concrete

Concrete, which is a mixture of cement, aggregate substances (such as gravel), and water, is the most widely used construction material in the world. It was used in the Roman Empire and has a variety of desirable qualities that make it popular. However, it has one drawback: It develops tiny cracks over time, and water seeps into those cracks, leading to widened cracks and weakening the concrete. For this reason, concrete structures have to be reinforced with steel, such as reinforcing bars ("rebar") incorporated in the concrete.

However, scientists in the Netherlands have recruited bacteria to do the work of reinforcement. They are using a bacterium that is capable of producing limestone, a strong rock substance consisting of calcium carbonate. The bacteria are *Bacillus* species that form endospores and are, therefore, very hardy; with some extra help, they can survive the mixing and hardening process. The scientists add the endospores and nutrients into the concrete mixture. Later, when water seeps into the tiny cracks, the nutrients and water cause the bacteria to germinate and to start pro-

ducing limestone. The researchers predict that this innovation can eliminate the need for steel reinforcement of many concrete structures.

Source: 2010. New Scient. vol. 15, p. 45.

Figure 25.14. Algal bioreactor. The photobioreactor contains algae, water, and trace elements.

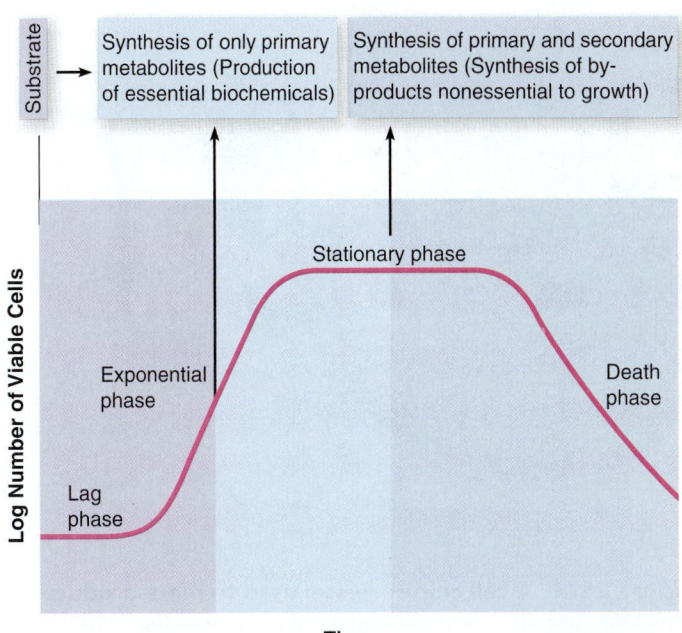

Figure 25.15 The origins of primary and secondary microbial metabolites harvested by industrial processes.

biomass—namely, lipids or oils. Early experiments showed that these microbes were easily grown and could produce vast quantities of easily biodegradable oil in a controlled manner. Many government and private industries around the world began investing in massive research projects to bring this technology to a scale that would allow the offset of fossil fuel use.

So why aren't we there yet? The main reason appears to be the demand on resources. Studies have shown that the use of current technologies for large-scale production of algal and cyanobacterial biofuels is not sustainable, meaning that the requirements for energy, water, and nutrients will have an overall negative impact on the environment. These data in no way signal the death of microbial biofuel production; instead, they have motivated scientists to find new solutions to this problem. Researchers have looked to the genetics of microorganisms such as *Synechococcus*, a cyanobacterium, to see how biomass production could be enhanced. Along the way, they uncovered a pathway of the Krebs cycle thought to be nonexistent in these microbes, information that may lead to the creation of more effective and efficient biofuel-producing microbes. Several research teams, including those from NASA, the U.S. military, and academia, are now forging ahead with the next phase of biofuel research.

Very often, the microbes used by biotechnology and fermentation industries are mutant strains of fungi or bacteria that selectively synthesize large amounts of various metabolic intermediates, or **metabolites.** Two basic kinds of metabolic products are harvested by industrial processes: (1) *Primary metabolites* are produced during the major metabolic pathways and are essential to the microbe's function. (2) *Secondary metabolites* are by-products of metabolism that may not be critical to the microbe's function **(figure 25.15).** In general, primary products are compounds such as amino acids and organic acids synthesized during the logarithmic phase of microbial growth, and secondary products are compounds such as vitamins, antibiotics, and steroids synthesized during the stationary phase (see chapter 7). Most strains of industrial microorganisms have been chosen for

their high production of a particular primary or secondary metabolite. The use of genetic engineering technology and now synthetic biology has enhanced not only the overall output of these microorganisms but the diversity of metabolites being synthesized as well.

Industrial microbiologists have several tricks to increase the amount of the chosen end product. First, they can manipulate the growth environment to increase the synthesis of a metabolite. For instance, adding lactose instead of glucose as the fermentation substrate increases the production of penicillin by *Penicillium.* Another strategy is to select microbial strains that genetically lack a feedback system to regulate the formation of end products, thus encouraging mass accumulation of this product. Many syntheses occur in sequential fashion, wherein the waste products of one organism become the building blocks of the next. During these *biotransformations*, the substrate undergoes a series of slight modifications, each of which gives off a different by-product. The production of an antibiotic such as tetracycline requires several microorganisms and 72 separate metabolic steps.

From Microbial Factories to Industrial Factories

Industrial fermentations begin with microbial cells acting as living factories. When exposed to optimum conditions, they multiply in massive numbers and synthesize large volumes of a desired product. Producing appropriate levels of growth and fermentation requires cultivation of the microbes in a carefully controlled environment **(figure 25.16).** This process is basically similar to culturing bacteria in a test tube of nutrient broth. It requires a sterile medium containing appropriate nutrients, protection from contamination, provisions for

Figure 25.16 A cell culture vessel used to mass-produce pharmaceuticals. Such elaborate systems require the highest levels of sterility and clean techniques.

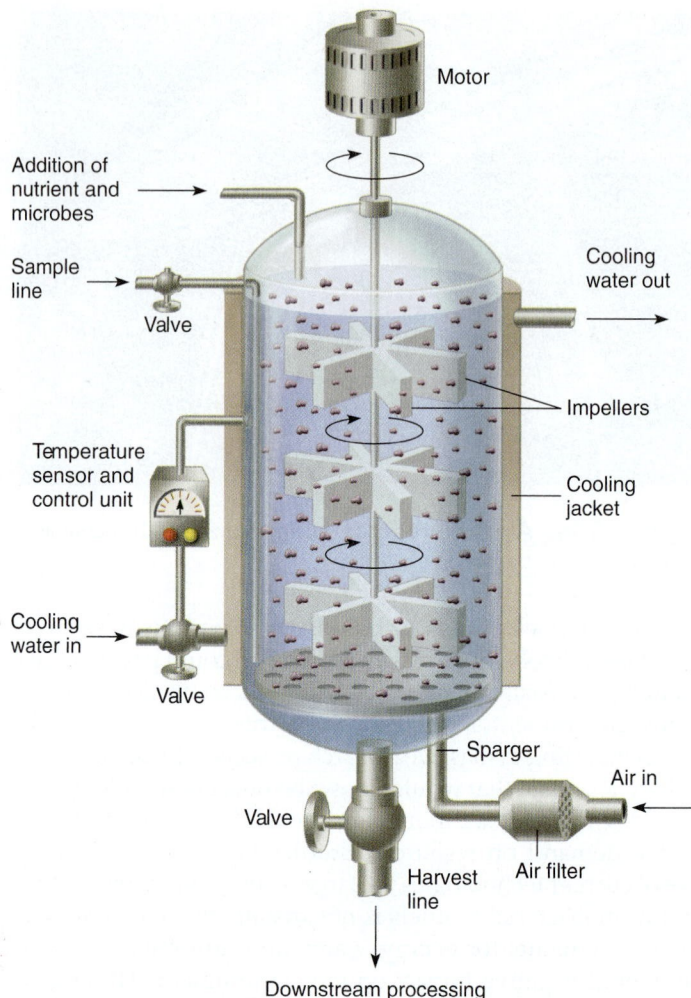

Figure 25.17 A schematic diagram of an industrial fermentor for mass culture of microorganisms. Such instruments are equipped to add nutrients and cultures; to remove product under sterile or aseptic conditions; and to aerate, stir, and cool the mixture automatically.

introduction of sterile air or total exclusion of air, and a suitable temperature and pH.

Commercial fermentation processes are worked out on a small scale in a lab and then *scaled up* to a large commercial venture. An essential component for scaling up is a **fermentor,** a device in which mass cultures are grown, reactions take place, and product develops (similar to the photobioreactors used for biofuel production). Some fermentors are large tubes, flasks, or vats; but most industrial types are metal cylinders with built-in mechanisms for stirring, cooling, monitoring, and harvesting product **(figure 25.17).** Fermentors are made of materials that can withstand pressure and are rustproof, nontoxic, and leakproof. They range in holding capacity from small, 5-gallon systems used in research labs to larger, 5,000- to 100,000-gallon vessels and, in some industries, to tanks of 250 million to 500 million gallons. For optimum yield, a fermentor must duplicate the actions occurring in a tiny volume (a test tube) on a massive scale. Most microbes performing fermentations have an aerobic metabolism, and the large volumes make it difficult to provide adequate oxygen. Fermentors have a built-in device called a *sparger* that aerates the medium to promote aerobic growth. Paddles (*impellers*) located in the central part of the fermentor increase the contact between the microbe and the nutrients by vigorously stirring the fermentation mixture. Their action also maintains the mixture's uniformity.

Substance Production

The general steps in mass production of organic substances in a fermentor are illustrated in **figure 25.18.** These can be summarized as

1. introduction of microbes and sterile media into the reaction chamber;
2. fermentation;
3. *downstream processing* (recovery, purification, and packaging of product); and
4. removal of waste.

All phases of production must be carried out aseptically and monitored (usually by computer) for rate of flow and quality of product. The starting raw substrates include crude plant residues, molasses, sugars, fish and meat meals, and whey. Additional chemicals can be added to control pH or to increase the yield. In *batch fermentation*, the substrate is added to the system all at once and taken through a limited run until product is harvested. In *continuous feed systems*, nutrients are continuously fed into the reactor and the product is siphoned off throughout the run.

Table 25.1 itemizes some of the major pharmaceutical substances, food additives, and solvents produced by microorganisms. Some newer technologies employ extremophilic archaea and their enzymes to run the processes at high or low temperatures or in high-salt conditions. Hyperthermophiles

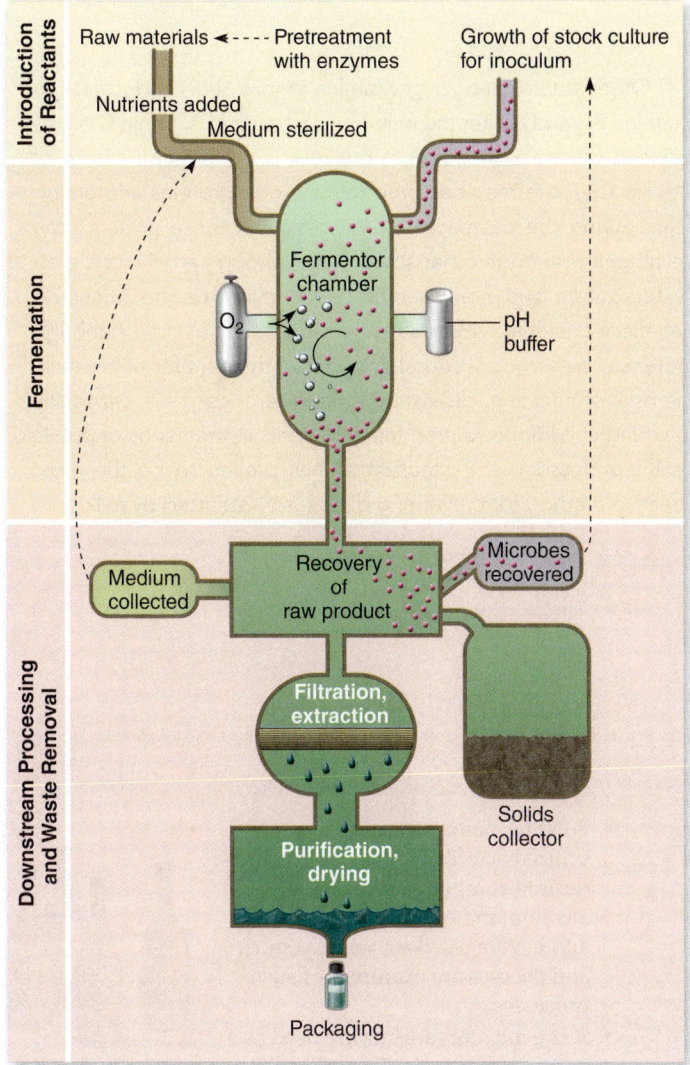

Figure 25.18 The general layout of a fermentation plant. These general steps are followed for industrial production of drugs, enzymes, fuels, vitamins, and amino acids.

to the fermentation vessel during the most appropriate phase of growth. These experiences with penicillin have provided an important model for the manufacture of other antibiotics.

Several steroid hormones used in therapy are produced industrially. Corticosteroids of the adrenal cortex, cortisone and cortisol (hydrocortisone), are invaluable for treating inflammatory and allergic disorders; and female hormones such as progesterone or estrogens are the active ingredients in birth control pills. For years, the production of these hormones was tedious and expensive because it involved purifying them from slaughterhouse animal glands or chemical syntheses. In time, it was shown that, through biotransformation, various molds could convert a precursor compound called diosgenin into cortisone. By the same means, stigmasterol from soybean oil could be transformed into progesterone.

Modern biotechnology has created a more straightforward method of pharmaceutical-grade drug production. Insulin is a classic example of how genetic manipulation was utilized to create a bacterium that could rapidly produce mass quantities of human insulin. Humulin was first approved by the FDA for human use in 1982 and has paved the way for the development of similar methods to produce therapeutic substances including tissue plasminogen activator (tPA) and human growth hormone (HGH). An important point to remember in all of the examples mentioned in this section, though, is the high need for safety testing of each product. The presence of contaminating microbes or microbial toxins—namely, lipopolysaccharides—can be devastating to the health of the individual using the very drug that should be treating his or her illness.

Miscellaneous Products

Enzymes are critical to chemical manufacturing, the agriculture and food industries, textile and paper processing, and even laundry and dry cleaning. The advantage of enzymes is that they are very specific in their activity and are readily produced and released by microbes. Mass quantities of proteases, amylases, lipases, oxidases, and cellulases are produced by fermentation technology (see table 25.2). The wave of the future appears to be custom designing enzymes to perform a specific task by altering their amino acid content. Other compounds of interest that can be mass-produced by microorganisms are amino acids, organic acids, solvents, and natural flavor compounds to be used in air fresheners and foods.

have been adapted for high-temperature detergent and enzyme production. Psychrophiles are used for cold processing of reagents for molecular biology and medical tests. Halophiles are effective for processing of salted foods and dietary supplements.

Pharmaceutical Products

Health care products derived from microbial biosynthesis include antibiotics, hormones, vitamins, and vaccines. The first mass-produced antimicrobial was penicillin, which came from *Penicillium chrysogenum,* a mold first isolated from a cantaloupe in Wisconsin. The current strain of this species has gone through decades of selective mutation and screening to increase its yield. (The original wild *P. chrysogenum* synthesized 60 mg/mL of medium, and later isolates yielded 85,000 mg/mL.) The semisynthetic penicillin derivatives are produced by introducing the assorted side-chain precursors

25.4 Learning Outcomes—Assess Your Progress

18. State the general aim(s) of industrial microbiology.

19. Distinguish between primary and secondary metabolites.

20. List the four steps of industrial product production from microbes.

21. Identify five industrial products made by microorganisms, and describe their applications.

Case File 25   *Wrap-Up*

Proponents of the benefits of raw milk point to a number of studies published in Europe showing a reduction in asthma and allergies in children who consumed raw milk. These studies analyzed IgE levels in children in Germany, Austria, the Netherlands, Sweden, and Switzerland. They found a significant reduction in allergies, asthma, eczema, and other atopic allergies among children who consumed raw milk as compared to children who drank pasteurized milk. However, the researchers involved in this study admit that the children consuming raw milk were raised on a farm and drank the milk of their own cows. Numerous studies have shown that children raised on farms have lowered incidences of asthma and allergies because of their constant exposure to animals and the less hygienic environment—not necessarily because they consumed raw milk.

Other studies analyzing vitamins in milk show a decrease in vitamins B_{12} and E after the milk was pasteurized. Vitamin C is also reduced during pasteurization, but milk is not a major source of vitamin C. There may be some reduction in proteins and amino acids during the pasteurization process, but these proteins can be obtained through other sources and through a balanced diet. Pasteurization and homogenization of milk have no effect on beneficial bacterial enzymes or the digestibility of milk. Although there may be some anecdotal evidence of the benefits of raw milk, the risks of infection, disease, and long-term damage, especially to young children, outweigh those benefits in the minds of public health professionals. Pasteurization has proven to be the most effective method for preventing diseases transmitted by milk.

Source: 2012. Emerg. Infect. Dis. 18, no. (3): 385–91.

Chapter Summary

25.1   Applied Microbiology and Biotechnology (ASM Guidelines* 1.2, 3.1, 3.2, 4.5, 5.3, 6.1, 6.2, 6.3)

- The use of microorganisms for practical purposes to benefit humans is called biotechnology.

25.2   Microorganisms in Water and Wastewater Treatment (ASM Guidelines 1.2, 3.1, 3.2, 4.5, 5.3, 6.1, 6.2, 6.3, 8.2, 8.4)

- Wastewater or sewage is treated in three stages to remove organic material, microorganisms, and chemical pollutants. The primary phase removes physical objects from the wastewater. The secondary phase removes the organic matter by biodegradation. The tertiary phase disinfects the water and removes chemical pollutants.

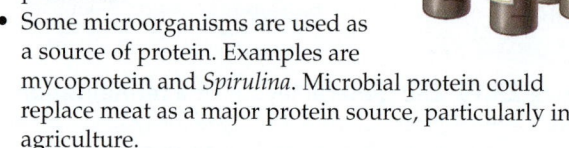

- Significant waterborne pathogens include protozoa, bacteria, and viruses.
- Water quality assays screen for coliforms as indicator organisms or may assess the most probable number of microorganisms. As these results may be misleading, more emphasis is being placed on identifying *E. coli* to indicate fecal contamination.

25.3   Microorganisms Making Food and Spoiling Food (ASM Guidelines 1.2, 3.1, 3.2, 4.5, 5.3, 6.1, 6.2, 6.3)

- The presence of microorganisms in food can be beneficial, detrimental, or of neutral consequence to human consumers.

- Food fermentation processes utilize bacteria or yeasts to produce desired components such as alcohols and organic acids in foods and beverages. Beer, wine, yogurt, and cheeses are examples of such processes.

- Some microorganisms are used as a source of protein. Examples are mycoprotein and *Spirulina*. Microbial protein could replace meat as a major protein source, particularly in agriculture.

- Food-borne disease can be an intoxication caused by microbial toxins produced as by-products of microbial decomposition of food, or it can be a food infection when pathogenic microorganisms in the food attack the human host after being consumed.

- Heat, radiation, chemicals, and drying are methods used to limit numbers of microorganisms in food. The type of method used depends on the nature of the food and the type of pathogens or spoilage agents it contains.

25.4   Using Microorganisms to Make Things We Need (ASM Guidelines 1.2, 3.1, 3.2, 4.5, 5.3, 6.1, 6.2, 6.3, 6.4)

- *Industrial microbiology* refers to the bulk production of any organic compound derived from microorganisms.
- Scientists are using cyanobacteria and algae to create biofuels, hoping to replace large portions of fossil fuels currently being utilized.
- Industrial processes produce antibiotics, hormones, vitamins, acids, solvents, vaccines, and enzymes from microbes.

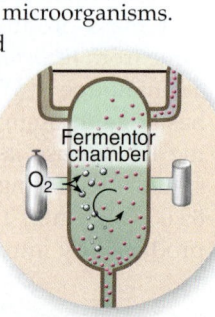

Source: ASM Curriculum Guidelines (American Society for Microbiology, 2012). Complete guidelines in appendix B of this book.

Multiple-Choice and True-False Questions | Bloom's Levels 1 and 2: Remember and Understand

Multiple-Choice Questions. Select the correct answer from the options provided.

1. Drinking water utilities monitor their production system for the occurrence of
   a. methanogens.   c. nematodes.
   b. coliform bacteria.   d. yeasts.

2. Milk is usually pasteurized by
   a. the high-temperature short-time method.
   b. ultrapasteurization.
   c. the batch method.
   d. electrical currents.

3. During sewage treatment, microbial action on a large scale first takes place in the
   a. primary phase.
   b. secondary phase.
   c. Microbial action is not a part of sewage treatment.
   d. Microbial action takes place after the secondary phase.

4. Which of the following is unlikely to be a waterborne pathogen?
   a. *Giardia lamblia*   c. *Vibrio*
   b. *Salmonella*   d. *Staphylococcus*

5. The "bloom" in wine making refers to
   a. the flowering of the grape plant.
   b. the biofilm on the skin of the grapes.
   c. the fermentation taking place in vats.
   d. none of the above.

6. When algae produce biofuels, what is the other significant by-product of photosynthesis?
   a. CO_2 c. waste
   b. energy   d. O_2

7. Secondary metabolites of microbes are formed during the _____ phase of growth.
   a. exponential   c. trophophase
   b. stationary   d. idiophase

8. In industrial fermentation, which step precedes *downstream processing*?
   a. removal of waste
   b. introduction of microbes into chamber
   c. packaging of product
   d. fermentation

9. Which of the following is/are currently being produced through biotechnology?
   a. glycerol
   b. vitamins
   c. steroids
   d. all of the above

10. In biotechnology, fermentation refers to
    a. the anaerobic metabolism of microorganisms.
    b. the creation of alcoholic beverages.
    c. the mass culturing of microorganisms to yield large quantities of products.
    d. all of the above.

True-False Questions. If the statement is true, leave as is. If it is false, correct it by rewriting the sentence.

11. Raw sewage is still being dumped into the aquatic environment in many places around the world.

12. Food products should always be kept completely free of microorganisms.

13. Alcoholic beverages are produced by the fermentation of sugar to ethanol and carbon dioxide.

14. The incidence of many food-borne illnesses declined in the period between 1998 and 2008.

15. Refrigerating food prevents the growth of all bacteria.

Critical Thinking Questions | Bloom's Levels 3, 4, and 5: Apply, Analyze, and Evaluate

Critical thinking is the ability to reason and solve problems using facts and concepts. These questions can be approached from a number of angles and, in most cases, they do not have a single correct answer.

1. a. Summarize one beneficial use of sludge today, and discuss one potential environmental risk in the use of this substance.
   b. Explain how waste is handled in the United States when the household is not connected to a sewage treatment facility.

2. Every year, supposedly safe municipal water supplies cause outbreaks of enteric illness.
   a. Using the Wisconsin outbreak example from the 1990s, explain how pathogens can slip through the processes of water analysis and treatment undetected and untreated.
   b. Looking at process figure 25.1, discuss how bioterrorists could target municipal water supplies and what pathogens would be utilized for this type of dispersal.

3. Provide evidence in support of or refuting the following statement: Humans consume microbes every day with little health risk.

4. a. List examples of microbes used in the production of fermented products such as beer, cheese, and pickles.
   b. Explain why only specific microbes are selected to produce each of these distinct fermented products.

5. a. If fermentation of sugars to produce alcohol in wine is anaerobic, why do winemakers make sure that the early phase of yeast growth is aerobic?
   b. In many cases, the identification of bacteria in a tank of beer signals trouble for the brewing process. Explain biochemically why this is so.

6. a. Predict the differences in the outcome if raw milk is incubated for 48 hours versus pasteurized milk being incubated for the same length of time.
   b. Many soldiers are deployed to areas of the world where there is no refrigeration. Explain how these individuals can still have safe dairy products to consume each day.

7. You are working for the county health department this summer, and your job is to regularly survey the local beaches for microbial contamination.
   a. Describe the tests you may use to analyze the water quality in these recreational areas.
   b. You identify high levels of *Enterobacter* species in a sample one day. Summarize how you will convey this information to the beach supervisor who asks, "Are we shutting down today?"
   c. Explain why there is little tolerance for fecal coliforms in drinking or recreational water.

8. Further investigate the HACCP system, and summarize two methods currently used to detect unsafe food industry practices. Provide examples illustrating how these observations have changed how food is harvested, manufactured, and dispersed in the United States today.

9. Conduct additional research and discuss current studies supporting the feasibility of using applied microbiology to offset the global energy challenges that exist today.

10. Summarize examples of how modern biotechnology is aiding humans medically, nutritionally, and environmentally.

Concept Connections | Bloom's Levels 4 and 6: Analyze and Create

This activity ties together multiple concepts in the chapter.

1. What precautions are taken at the harvesting step of food production to prevent food-borne diseases?

2. What processes do liquid foods go through to remove or destroy microbes?

3. How can the growth of microbes in solid processed foods be prevented?

4. What precautions should be taken with raw foods to prevent the growth of microbes and prevent food-borne disease?

5. What precautions should be taken with food cooked at home or in a restaurant?

6. What single process is fundamental in preventing food-borne disease?

7. Are there any circumstances in which food should be sterilized?

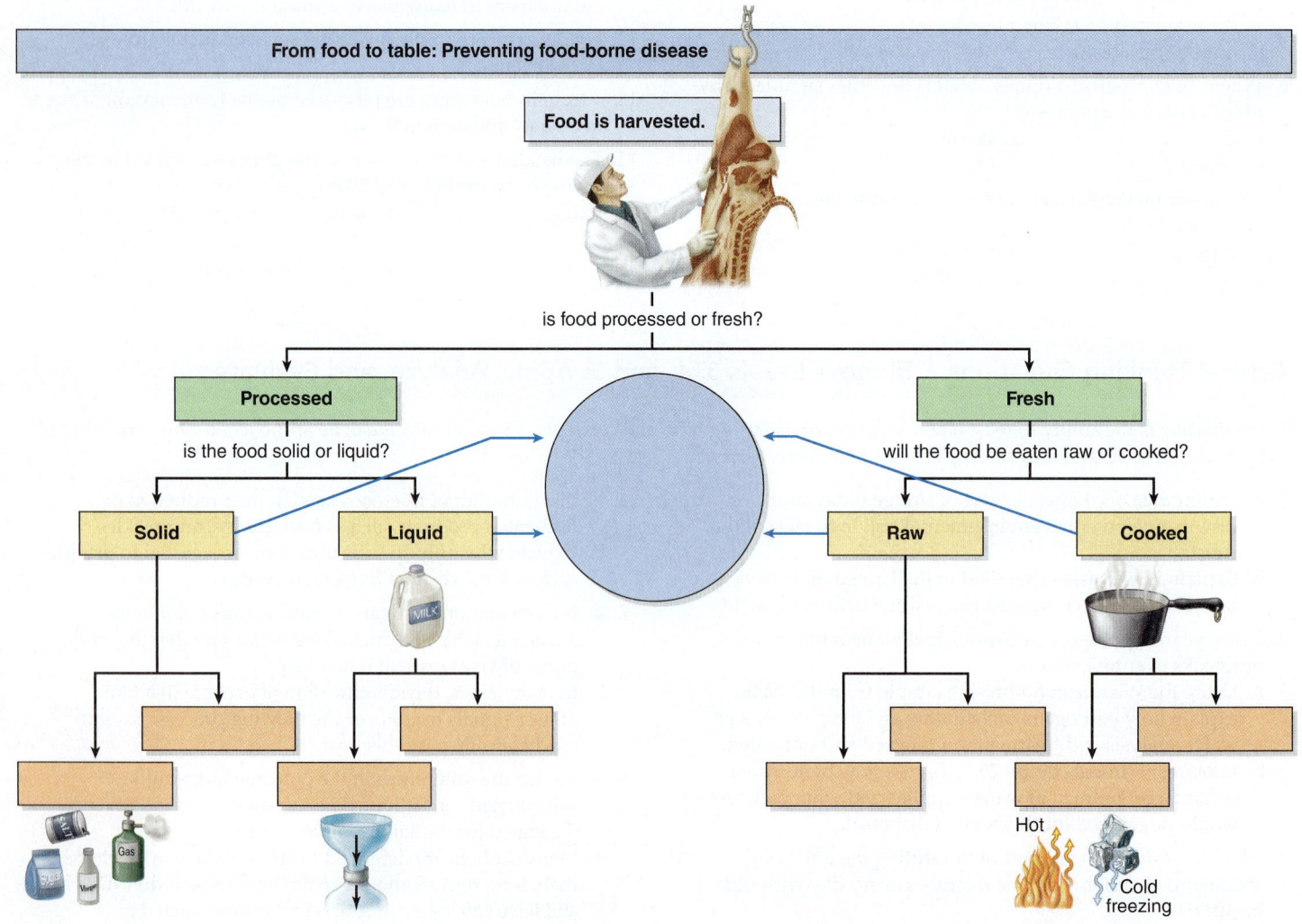

Visual Connections | Bloom's Level 5: Evaluate

These questions use visual images or previous content to make connections to this chapter's concepts.

1. **From chapter 3, figure 3.5*b*.** If this MacConkey agar plate was inoculated with well water, would you report that coliforms were present in the water?

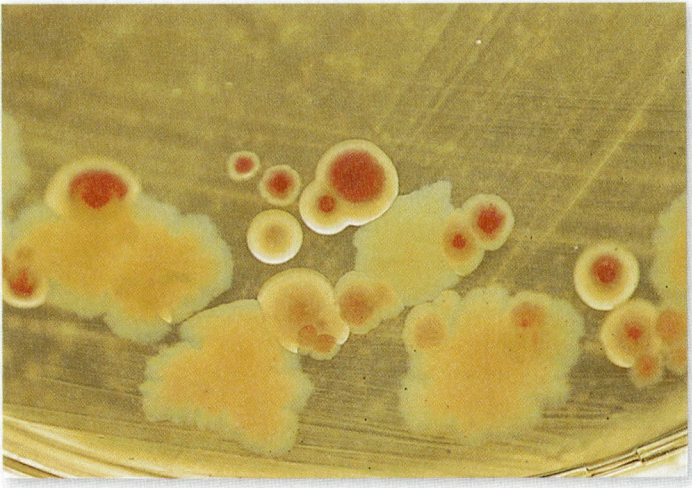

2. **From chapter 3, selective and differential media.** This is a plate with blood agar on the top and MacConkey agar on the bottom. (You studied both of these agars in chapter 3.) It has been inoculated with samples from a wooden cutting board that had been exposed to a raw chicken carcass. Knowing what you know about the properties of these two agars, say as much as you can about the bacteria growing on them.

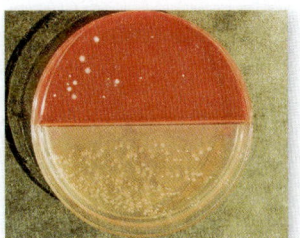

Concept Mapping | Bloom's Level 6: Create

Appendix D provides guidance for working with concept maps.

1. Using the words that follow, please create a concept map illustrating the relationships among these key terms from chapter 25.

primary metabolites	fermentation	downstream processing	pH
secondary metabolites	microbes	substrate	biotransformations

Exponents

Dealing with concepts such as microbial growth often requires working with numbers in the billions, trillions, and even greater. A mathematical shorthand for expressing such numbers is with exponents. The exponent of a number indicates how many times (designated by a superscript) that number is multiplied by itself. These exponents are also called common *logarithms,* or logs. The following chart, based on multiples of 10, summarizes this system.

Exponential Notation for Base 10				
Number	**Quantity**	**Exponential Notation***	**Number Arrived at By:**	**One Followed By:**
1	One	10^0	Numbers raised to zero power are equal to one	No zeros
10	Ten	10^{1}**	10×1	One zero
100	Hundred	10^2	10×10	Two zeros
1,000	Thousand	10^3	$10 \times 10 \times 10$	Three zeros
10,000	Ten thousand	10^4	$10 \times 10 \times 10 \times 10$	Four zeros
100,000	Hundred thousand	10^5	$10 \times 10 \times 10 \times 10 \times 10$	Five zeros
1,000,000	Million	10^6	10 times itself 6 times	Six zeros
1,000,000,000	Billion	10^9	10 times itself 9 times	Nine zeros
1,000,000,000,000	Trillion	10^{12}	10 times itself 12 times	Twelve zeros
1,000,000,000,000,000	Quadrillion	10^{15}	10 times itself 15 times	Fifteen zeros
1,000,000,000,000,000,000	Quintillion	10^{18}	10 times itself 18 times	Eighteen zeros

Other large numbers are sextillion (10^{21}), septillion (10^{24}), and octillion (10^{27}).

*The proper way to say the numbers in this column is 10 raised to the nth power, where n is the exponent. The numbers in this column can also be represented as 1×10^n, but for brevity, the $1 \times$ can be omitted.

**The exponent 1 is usually omitted.

Converting Numbers to Exponent Form

As the chart shows, using exponents to express numbers can be very economical. When simple multiples of 10 are used, the exponent is always equal to the number of zeros that follow the 1, but this rule will not work with numbers that are more varied. Other large whole numbers can be converted to exponent form by the following operation: First, move the decimal (which we assume to be at the end of the number) to the left until it sits just behind the first number in the series (example: 3568. = 3.568). Then count the number of spaces (digits) the decimal has moved; that number will be the exponent. (The decimal has moved from 8. to 3., or 3 spaces.) In final notation, the converted number is multiplied by 10 with its appropriate exponent: 3568 is now 3.568×10^3.

Rounding Off Numbers

The notation in the previous example has not actually been shortened, but it can be reduced further by rounding off the decimal fraction to the nearest thousandth (three digits), hundredth (two digits), or tenth (one digit). To round off a number, drop its last digit and either increase the one next to it or leave it as it is. If the number dropped is 5, 6, 7, 8, or 9, the subsequent digit is increased by one (rounded up); if it is 0, 1, 2, 3, or 4, the subsequent digit remains as is. Using the example of 3.528, removing the 8 rounds off the 2 to a 3 and produces 3.53 (two digits). If further rounding is desired, the same rule of thumb applies, and the number becomes 3.5 (one digit). Other examples of exponential conversions follow.

Number	Is the Same As	Rounded Off, Placed in Exponent Form
16,825.	$1.6825 \times 10 \times 10 \times 10 \times 10$	1.7×10^4
957,654.	$9.57654 \times 10 \times 10 \times 10 \times 10 \times 10$	9.58×10^5
2,855,000.	$2.855000 \times 10 \times 10 \times 10 \times 10$ $\times 10 \times 10$	2.86×10^6

Negative Exponents

The numbers we have been using so far are greater than 1 and are represented by positive exponents. But the correct notation for numbers less than 1 involves negative exponents (10 raised to a negative power, or 10^{-n}). A negative exponent says that the number has been divided by a certain power of 10 (10, 100, 1,000). This usage is handy when working with concepts such as pH that are based on very small numbers otherwise needing to be represented by large decimal fractions—for example, 0.003528. Converting this and other such numbers to exponential notation is basically similar to converting positive numbers, except that you work from left to right and the exponent is negative. Using the example of 0.003528, first convert the number to a whole integer followed by a decimal fraction and keep track of the number of spaces the decimal point moves (example: 0.003528 = 3.528). The decimal has moved three spaces from its original position, so the finished product is 3.528×10^{-3}. Other examples follow.

Number	Is the Same As	Rounded Off, Expressed with Exponents
0.0005923	$\dfrac{5.923}{10 \times 10 \times 10 \times 10}$	5.92×10^{-4}
0.00007295	$\dfrac{7.295}{10 \times 10 \times 10 \times 10 \times 10}$	7.3×10^{-5}

ASM Curriculum Guidelines for Undergraduate Microbiology

Guidelines were created by microbiology educators and adopted in 2011 by the American Society for Microbiology. They are divided into two parts. The first part contains concepts and statements, and the second part focuses on skills and competencies.

Part 1: Concepts and Statements

Evolution

1.1 Cells, organelles (e.g., mitochondria and chloroplasts) and all major metabolic pathways evolved from early prokaryotic cells.

1.2 Mutations and horizontal gene transfer, with the immense variety of microenvironments, have selected for a huge diversity of microorganisms.

1.3 Human impact on the environment influences the evolution of microorganisms (e.g., emerging diseases and the selection of antibiotic resistance).

1.4 The traditional concept of species is not readily applicable to microbes due to asexual reproduction and the frequent occurrence of horizontal gene transfer.

1.5 The evolutionary relatedness of organisms is best reflected in phylogenetic trees.

Cell Structure and Function

2.1 The structure and function of microorganisms have been revealed by the use of microscopy (including bright field, phase contrast, fluorescent, and electron).

2.2 Bacteria have unique cell structures that can be targets for antibiotics, immunity, and phage infection.

2.3 Bacteria and Archaea have specialized structures (e.g., flagella, endospores, and pili) that often confer critical capabilities.

2.4 While microscopic eukaryotes (for example, fungi, protozoa, and algae) carry out some of the same processes as bacteria, many of the cellular properties are fundamentally different.

2.5 The replication cycles of viruses (lytic and lysogenic) differ among viruses and are determined by their unique structures and genomes.

Metabolic Pathways

3.1 Bacteria and Archaea exhibit extensive, and often unique, metabolic diversity (e.g., nitrogen fixation, methane production, anoxygenic photosynthesis).

3.2 The interactions of microorganisms among themselves and with their environment are determined by their metabolic abilities (e.g., quorum sensing, oxygen consumption, nitrogen transformations).

3.3 The survival and growth of any microorganism in a given environment depends on its metabolic characteristics.

3.4 The growth of microorganisms can be controlled by physical, chemical, mechanical, or biological means.

Information Flow and Genetics

4.1 Genetic variations can impact microbial functions (e.g., in biofilm formation, pathogenicity, and drug resistance).

4.2 Although the central dogma is universal in all cells, the processes of replication, transcription, and translation differ in Bacteria, Archaea, and Eukaryotes.

4.3 The regulation of gene expression is influenced by external and internal molecular cues and/or signals.

4.4 The synthesis of viral genetic material and proteins is dependent on host cells.

4.5 Cell genomes can be manipulated to alter cell function.

Microbial Systems

5.1 Microorganisms are ubiquitous and live in diverse and dynamic ecosystems.
5.2 Most bacteria in nature live in biofilm communities.
5.3 Microorganisms and their environment interact with and modify each other.
5.4 Microorganisms, cellular and viral, can interact with both human and nonhuman hosts in beneficial, neutral, or detrimental ways.

Impact of Microorganisms

6.1 Microbes are essential for life as we know it and the processes that support life (e.g., in biogeochemical cycles and plant and/or animal microflora).
6.2 Microorganisms provide essential models that give us fundamental knowledge about life.
6.3 Humans utilize and harness microorganisms and their products.
6.4 Because the true diversity of microbial life is largely unknown, its effects and potential benefits have not been fully explored.

Part 2: Skills and Competencies

Scientific Thinking

7.1 Ability to apply the process of science
- Demonstrate an ability to formulate hypotheses and design experiments based on the scientific method.
- Analyze and interpret results from a variety of microbiological methods and apply these methods to analogous situations.

7.2 Ability to use quantitative reasoning
- Use mathematical reasoning and graphing skills to solve problems in microbiology.

7.3 Ability to communicate and collaborate with other disciplines
- Effectively communicate fundamental concepts of microbiology in written and oral format.
- Identify credible scientific sources and interpret and evaluate the information therein.

7.4 Ability to understand the relationship between science and society
- Identify and discuss ethical issues in microbiology.

Microbiology Laboratory Skills

8.1 Properly prepare and view specimens for examination using microscopy (bright field and, if possible, phase contrast).
8.2 Use pure culture and selective techniques to enrich for and isolate microorganisms.
8.3 Use appropriate methods to identify microorganisms (media-based, molecular, and serological).
8.4 Estimate the number of microorganisms in a sample (using, for example, direct count, viable plate count, and spectrophotometric methods).
8.5 Use appropriate microbiological and molecular lab equipment and methods.
8.6 Practice safe microbiology, using appropriate protective and emergency procedures.
8.7 Document and report on experimental protocols, results and conclusions.

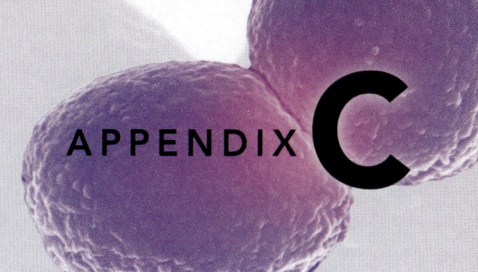

Answers to Multiple-Choice, True-False, and Matching Questions

Chapter 1

1. d
2. d
3. d
4. a
5. c
6. c
7. c
8. d
9. 1st col:
   3, 7, 4, 2
   2nd col:
   8, 5, 6, 1
10. c
11. F: Organisms in the same family are more closely related than those in the same order.
12. F: Eukaryotes and prokaryotes emerged independently.
13. T
14. T
15. T

Chapter 2

1. c
2. c
3. b
4. d
5. c
6. a
7. b
8. c
9. a
10. b
11. T
12. F: Covalent bonds are those formed when two elements share electrons.
13. F: A compound is called "organic" if it contains both carbon and hydrogen bonded

together in various combinations.
14. T
15. F: Membranes are mainly composed of macro-molecules called phospholipids.

Chapter 3

1. b
2. c
3. b
4. b
5. c
6. c
7. a
8. b
9. ab, df, abf, ef, af, bc, bf, bf
10. c or d
11. F: Agar is not easily decomposed by micro-organisms (although gelatin can be).
12. T
13. F: The factor that most limits the clarity of an image in a microscope is the *resolution*.
14. F: Living specimens can be examined with phase-contrast or differential interference microscopy.
15. F: The best stain to use to visualize a microorganism with a large capsule is a negative stain.

Chapter 4

1. d
2. c
3. a
4. c
5. b
6. d
7. b
8. d
9. b
10. c
11. F: One major difference in the envelope structure between gram-positive bacteria and gram-negative bacteria is the presence or absence of an outer membrane.
12. F: A research microbiologist looking at evolutionary relatedness between two bacterial species is more likely to use *Bergey's Manual of Systematic Bacteriology*.
13. T
14. T
15. T

Chapter 5

1. b
2. d or e
3. d
4. b
5. b
6. d
7. d
8. b
9. c
10. a
11. T
12. T
13. F: Both the trophozoite

and the cyst stages of protozoan life cycles can be infective.
14. F: In humans, fungi can infect skin, mucous membranes, lungs, and other areas.
15. F: Fungi generally derive nutrients by digesting organic substrates.

Chapter 6

1. c
2. d
3. d
4. a
5. a
6. b
7. d
8. a
9. See figure 6.4, page 147
10. rabies, cold sores, genital warts, mumps, rubella
11. T
12. F: A viral capsid is composed of subunits called capsomeres.
13. F: The envelope of an animal virus is derived from the cell membrane or nuclear membrane of the host cell.
14. F: The nucleic acid of animal viruses enters the cell through a process called penetration.

Chapter 7

1. a
2. a
3. c
4. b
5. a
6. a
7. b
8. c
9. c
10. b
11. F: Active transport of a substance across a membrane requires energy.
12. T
13. T
14. F: An obligate halophile is an organism that requires high salt concentration to grow.
15. F: A facultative anaerobe can grow with or without oxygen.

Chapter 8

1. a
2. d
3. d
4. b
5. b
6. b
7. a
8. b
9. c
10. c
11. T
12. T
13. F: One cycle of fermentation yields much less energy than one cycle of aerobic respiration.
14. T

15. T

Chapter 9

1. b
2. b
3. b
4. c
5. b
6. a
7. b
8. d
9. b
10. c
11. F: The DNA base pairs are held together primarily by hydrogen bonds.
12. F: Although some mutations are harmful, many are neutral or helpful.
13. T
14. F: Messenger RNA is formed by transcription of a gene on the DNA template strand.
15. T

Chapter 10

1. c
2. d
3. a
4. d
5. b
6. c
7. c
8. d
9. a
10. d
11. F: The synthetic unit of the polymerase chain reaction is the amplicon.
12. T

15. F: Exoenzymes are produced inside a cell then released to the outside.

13. F: A DNA fragment with 450 bp will migrate farther toward the positive pole (away from the origin) than one with 2,500 bp.
14. T
15. F: Plasmids and bacteriophages are commonly used as cloning vectors.

Chapter 11

1. c
2. a
3. c
4. b
5. d
6. b
7. d
8. d
9. a
10. b
11. T
12. F: The acceptable temperature-pressure combination for an autoclave is 121°C and 15 psi.
13. F: Ionizing radiation dislodges electrons from atoms.
14. T
15. F: Prions are highly resistant to denaturation by heat.

Chapter 12

1. b
2. c
3. b
4. d
5. b

6. d
7. a or b
8. c
9. c
10. b
11. F: Most antiviral agents work by blocking an essential viral activity.
12. F: Sulfonamide drugs work by disrupting folic acid synthesis.
13. T
14. T
15. F: Drug resistance can occur when a bacterium stops being susceptible to an antibiotic.

Chapter 13

1. a
2. b
3. d
4. c
5. d
6. b
7. c
8. d
9. a
10. a
11. F: The presence of a few bacteria in the blood is called bacteremia.
12. F: Resident microbiota are not commonly found in the kidney.
13. F: A health-care associated infection is one that is acquired in a hospital or medical facility.
14. F: The general term that describes a decrease in the number of white blood cells is leukopenia.
15. T

Chapter 14

1. b
2. b
3. d
4. b
5. b
6. c
7. d
8. a
9. d
10. c
11. F: The liquid component of unclotted blood is called plasma.
12. F: Pyrogenic bacteria are commonly associated with fever.
13. T
14. T
15. F: The immune system uses markers on the surface of cells to distinguish self from nonself.

Chapter 15

1. a
2. c
3. a
4. c
5. b
6. a
7. e
8. c
9. d
10. d
11. T
12. F: Antibodies are secreted by plasma cells.
13. F: Vaccination is artificial active immunity.
14. F: IgA antibodies are found in body secretions.
15. F: The process of reducing the virulence of microbes so they can be used in vaccines is called attenuation.

Chapter 16

1. d
2. d
3. c
4. c
5. b
6. a
7. d
8. d
9. d
10. d
11. T
12. F: A positive tuberculin skin test is an example of delayed hyper-sensitivity.
13. F: Contact dermatitis can be caused by chemicals absorbed through the skin.
14. T
15. T

Chapter 17

1. b
2. a
3. b
4. c
5. c
6. b
7. a
8. a
9. d
10. c
11. T
12. T
13. T
14. F: Microorgan-isms that are grown from clinical sam-ples should be evaluated to determine their significance.
15. T

Chapter 18

1. b
2. a
3. d
4. e
5. c
6. b
7. d
8. d
9. b
10. c
11. F: The enzyme coagulase is associated with pathogenic strains of *Staphylococcus aureus.*
12. F: Fifth disease has no vaccine and no treatment.
13. T
14. F: The blistering and peeling of the skin in SSS are due to the ability of *S. aureus* to produce exfoliative toxins.
15. F: Although the Human Microbiome Project has identified five major taxa represented in the normal skin biota, there are large differences among people with respect to types of microbes found on various skin sites.

Chapter 19

1. d
2. a
3. c
4. d
5. b
6. d
7. b
8. a
9. b
10. a
11. F: *Toxoplasma gondii* is a protozoan.
12. F: Oral fluconazole is the first line treatment for people with disseminated disease.
13. T
14. F: In the United States, wild animals are a common reservoir for rabies.
15. T

Chapter 20

1. c
2. d
3. b
4. b
5. a
6. b
7. d
8. d
9. b
10. d
11. T
12. F: Respiratory tract infection with *Pneumocystis jirovecii* is an ADI.
13. F: Lyme disease is caused by *Borrelia burgdorferi.*
14. F: Yellow fever is caused by a virus transmitted by mosquitoes.
15. T

Chapter 21

1. d
2. d
3. b
4. b
5. d
6. c
7. d
8. c
9. a
10. c
11. T
12. T
13. F: BCG vaccine is used in other countries to prevent TB.
14. F: RSV is a respiratory infection associated with infants.
15. F: The "flu shot" is an inactivated virus and cannot cause influenza.

Chapter 22

1. b
2. d
3. c
4. d
5. c
6. c
7. a
8. b
9. b
10. c
11. F: Humans are the only natural host for the mumps virus.
12. T
13. T
14. T
15. T

Chapter 23

1. a
2. d
3. d
4. a
5. a
6. a
7. c
8. c
9. d
10. d
11. T
12. F: Chancroid is caused by a bacterium.
13. T
14. F: *Chlamydia* is the most common reportable infectious disease in the U.S.
15. T

Chapter 24

1. c
2. b
3. b
4. d
5. c
6. d
7. b
8. d
9. d
10. b
11. F: Pure cultures are very rare in the biosphere.
12. T
13. T
14. T
15. F: One microbe has been found to live in a single-organism community.

Chapter 25

1. b
2. a
3. b
4. d
5. b
6. d
7. b
8. d
9. d
10. c
11. T
12. F: Food products will usually be colonized by harmless micro-organisms.
13. T
14. T
15. F: Refrigerating food prevents the growth of many bacteria, but some pathogens, such as *Listeria* and *Salmonella,* can continue to grow at low temperatures.

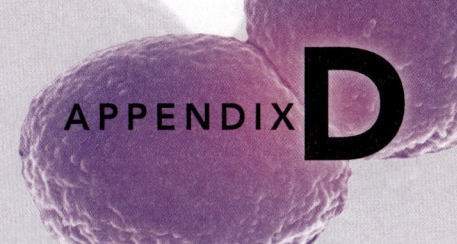

An Introduction to Concept Mapping

Concept maps are visual tools for presenting and organizing what you have learned. They can take the place of an outline, though for most people they contain much more meaning and can illustrate connections and interconnections in ways that ordinary outlines cannot. They are also very flexible. If you are creating a concept map, there is a nearly infinite number of ways that it can be put together and still be "correct." Concept maps are also a way to incorporate and exploit your own creative impulses, so that you are not stuck inside a rigid framework but can express your understanding of concepts and their connections in ways that make sense to you.

This is an example of a relatively large concept map:

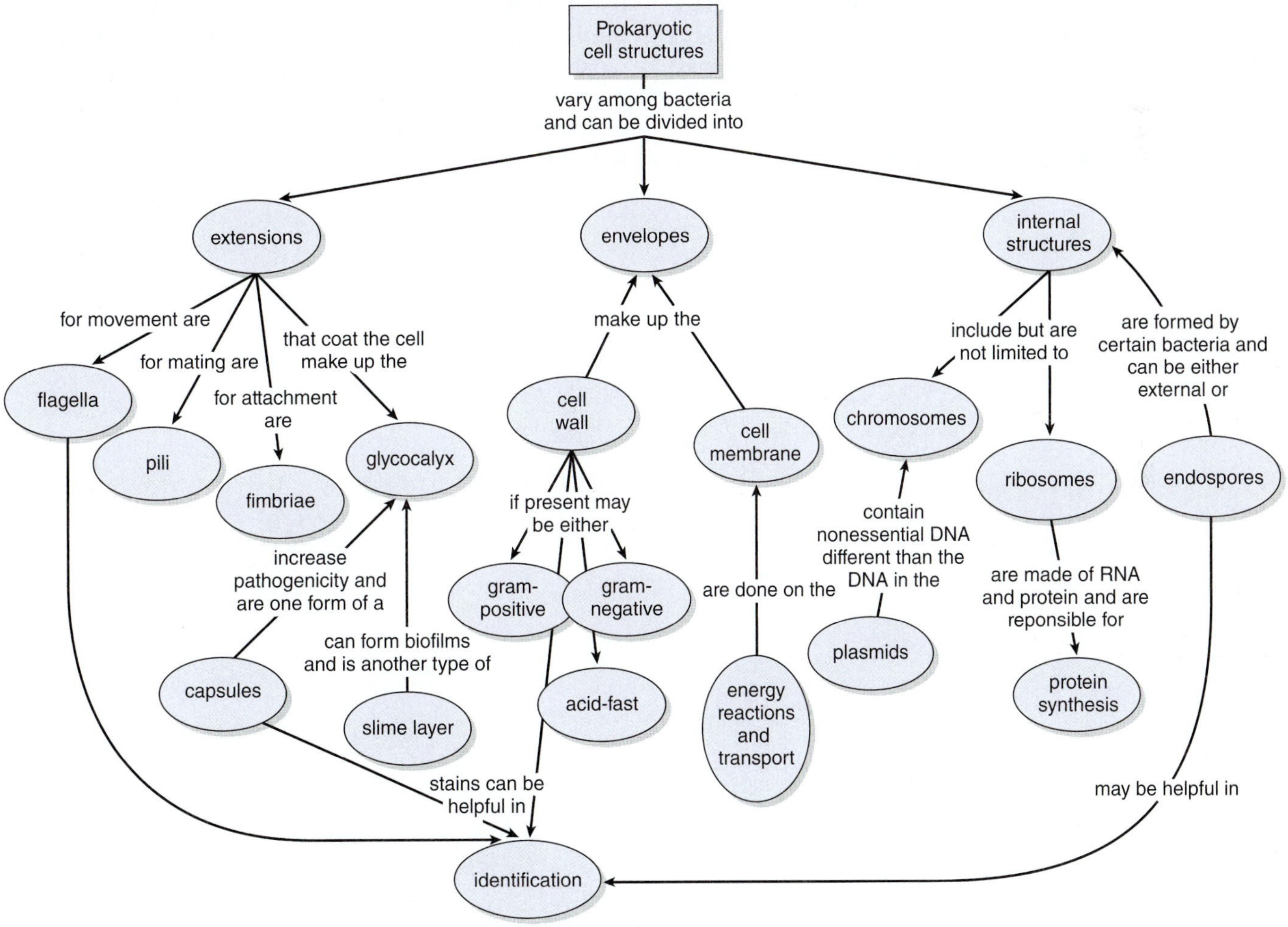

Prokaryotic cell structures

vary among bacteria and can be divided into

extensions

envelopes

internal structures

for movement are

for mating are

that coat the cell make up the

flagella

for attachment are

pili

fimbriae

glycocalyx

make up the

cell wall

cell membrane

include but are not limited to

are formed by certain bacteria and can be either external or

chromosomes

ribosomes

endospores

if present may be either

increase pathogenicity and are one form of a

contain nonessential DNA different than the DNA in the

are made of RNA and protein and are reponsible for

capsules

can form biofilms and is another type of

gram-positive

gram-negative

are done on the

plasmids

protein synthesis

slime layer

acid-fast

energy reactions and transport

stains can be helpful in

may be helpful in

identification

There is a wide variety of different ways to work with concept maps, such as using them as an introductory overview of material or using them as an evaluation tool. There are even software programs that enable concept mappers to create elaborate maps, complete with sound bytes and photos. Some of these will even convert an outline into a concept map for you.

All concept maps are made of two basic components:

1. Boxes or circles, each containing a single *concept*, which is most often a noun. The boxes are arranged on the page in vertical, horizontal, or diagonal rows or arrangements. They may also be arranged in a more free-form manner.
2. Connecting lines that join each concept box to at least one other box. Each connecting line has a word or a phrase associated with it—a linking word. These words/phrases are almost never nouns—but are verbs (like "requires") or adjectives or adverbs (like "underneath").

In the end, a picture is created that maps what you know about a subject. It illustrates which concepts are bigger and which are details. It illustrates that multiple concepts may be connected. Experts say that concept maps almost always lead us to conclude that all concepts in a subject can be connected in some way. This is true! And nowhere is it truer than in biology. The trick is to get used to finding the right connecting word to show how two concepts are, indeed, related. When you succeed, you will know the material in a deeper way than is possible by simply answering a single question or even a series of questions.

The kind of concept map you will see in this book is one in which you will be provided a list of words to be used as concepts. You will be asked to draw the boxes and put the words in them in some way that makes sense. Here, there will be a lot of variability based on your view of how the concepts might relate to each other. After you put the concepts in your own boxes, you will need to add linking words/phrases. By the time you have drawn your boxes and added the concepts, you will have many ideas about what kind of linkers you want.

Many students report that their first experiences with concept mapping can be frustrating. But when they have invested some time in their first few concept maps, many of them find they can never "go back" to organizing information in linear ways. Maps can make the time fly when you're studying. And creating concept maps with a partner or a group is also a great way to review material in a meaningful way. Give concept maps a try. Let your creative side show!

Glossary

A

abiogenesis   The belief in spontaneous generation as a source of life.

abiotic   Nonliving factors such as soil, water, temperature, and light that are studied when looking at an ecosystem.

ABO blood group system   Developed by Karl Landsteiner in 1904; the identification of different blood groups based on differing isoantigen markers characteristic of each blood type.

abscess   An inflamed, fibrous lesion enclosing a core of pus.

acellular vaccine   A vaccine preparation that contains specific antigens such as the capsule or toxin from a pathogen and not the whole microbe. Acellular (without a cell).

acid-fast   A term referring to the property of mycobacteria to retain carbol fuchsin even in the presence of acid alcohol. The staining procedure is used to diagnose tuberculosis.

acid-fast stain   A solution containing carbol fuchsin, which, when bound to lipids in the envelopes of *Mycobacterium* species, cannot be removed with an acid wash.

acidic   A solution with a pH value below 7 on the pH scale.

acidic fermentation   An anaerobic degradation of pyruvic acid that results in organic acid production.

acidophilic   An organism whose optimal growth pH is 2.0 or lower.

acquired immunodeficiency syndrome   See *AIDS*.

actin   Protein component of long filaments of protein arranged under the cell membrane of bacteria; contributes to cell shape and division.

actin filaments   Long, thin, protein strands found throughout a eukaryotic cell—but mainly concentrated just inside the cell membrane.

actinomycetes   A group of filamentous, funguslike bacteria.

active immunity   Immunity acquired through direct stimulation of the immune system by antigen.

active site   The specific region on an apoenzyme that binds substrate. The site for reaction catalysis.

active transport   Nutrient transport method that requires carrier proteins in the membranes of the living cells and the expenditure of energy.

acute   Characterized by rapid onset and short duration.

acyclovir   A synthetic purine analog that blocks DNA synthesis in certain viruses, particularly the herpes simplex viruses.

adenine (A)   One of the nitrogen bases found in DNA and RNA, with a purine form.

adenosine deaminase (ADA) deficiency   An immunodeficiency disorder and one type of SCIDs that is caused by an inborn error in the metabolism of adenine. The accumulation of adenine destroys both B and T lymphocytes.

adenosine triphosphate (ATP)   A nucleotide that is the primary source of energy to cells.

adhesion   The process by which microbes gain a more stable foothold at the portal of entry; often involves a specific interaction between the molecules on the microbial surface and the receptors on the host cell.

adjuvant   In immunology, a chemical vehicle that enhances antigenicity, presumably by prolonging antigen retention at the injection site.

adsorption   A process of adhering one molecule onto the surface of another molecule.

aerobe   A microorganism that lives and grows in the presence of free gaseous oxygen (O_2).

aerobic respiration   Respiration in which the final electron acceptor in the electron transport chain is oxygen (O_2).

aerosols   Suspensions of fine dust or moisture particles in the air that contain live pathogens.

aerotolerant   The state of not utilizing oxygen but not being harmed by it.

agammaglobulinemia   Also called *hypogammaglobulinemia*. The absence of or severely reduced levels of antibodies in serum.

agar   A polysaccharide found in seaweed and commonly used to prepare solid culture media.

agglutination   The aggregation by antibodies of suspended cells or similar-size particles (agglutinogens) into clumps that settle.

agranulocyte   One form of leukocyte (white blood cell) having globular, nonlobed nuclei and lacking prominent cytoplasmic granules.

AIDS   Acquired immunodeficiency syndrome. The complex of signs and symptoms characteristic of the late phase of human immunodeficiency virus (HIV) infection.

alcoholic fermentation   An anaerobic degradation of pyruvic acid that results in alcohol production.

algae   Photosynthetic, plantlike organisms that generally lack the complex structure of plants; they may be single-celled or multicellular and inhabit diverse habitats such as marine and freshwater environments, glaciers, and hot springs.

allele   A gene that occupies the same location as other alternative (allelic) genes on paired chromosomes.

allergen   A substance that provokes an allergic response.

allergy   The altered, usually exaggerated, immune response to an allergen. Also called *hypersensitivity*.

alloantigen   An antigen that is present in some but not all members of the same species.

allograft   Relatively compatible tissue exchange between nonidentical members of the same species. Also called *homograft*.

allosteric   Pertaining to the altered activity of an enzyme due to the binding of a molecule to a region other than the enzyme's active site.

Ames test   A method for detecting mutagenic and potentially carcinogenic agents based upon the genetic alteration of nutritionally defective bacteria.

amination   The addition of an amine ($-NH_2$) group to a molecule.

amino acids   The building blocks of protein. Amino acids exist in 20 naturally occurring forms that impart different characteristics to the various proteins they compose.

aminoglycoside   A complex group of drugs derived from soil actinomycetes that impairs ribosome function and has antibiotic potential. Example: streptomycin.

ammonification   Phase of the nitrogen cycle in which ammonia is released from decomposing organic material.

amphibolism   Pertaining to the metabolic pathways that serve multiple functions in the breakdown, synthesis, and conversion of metabolites.

amphipathic   Relating to a compound that has contrasting characteristics, such as hydrophilic-hydrophobic or acid-base.

amphitrichous   Having a single flagellum or a tuft of flagella at opposite poles of a microbial cell.

amplicon   DNA strand that has been primed for replication during polymerase chain reaction.

anabolism   The energy-consuming process of incorporating nutrients into protoplasm through biosynthesis.

anaerobe   A microorganism that grows best, or exclusively, in the absence of oxygen.

anaerobic respiration   Respiration in which the final electron acceptor in the electron transport chain is an inorganic molecule containing sulfate, nitrate, nitrite, carbonate, and so on.

anamnestic response   In immunology, an augmented response or memory related to a prior stimulation of the immune system by antigen. It boosts the levels of immune substances.

anaphylaxis   The unusual or exaggerated allergic reaction to antigen that leads to severe respiratory and cardiac complications.

anion   A negatively charged ion.

annotating   In the context of genome sequencing, it is the process of assigning biological function to genetic sequence.

anoxygenic   Non-oxygen-producing.

antagonism   Relationship in which microorganisms compete for survival in a common environment by taking actions that inhibit or destroy another organism.

antibiotic   A chemical substance from one microorganism that can inhibit or kill another microbe even in minute amounts.

antibody   A large protein molecule evoked in response to an antigen that interacts specifically with that antigen.

antibody-mediated immunity   Specific protection from disease provided by the products of B cells.

anticodon   The trinucleotide sequence of transfer RNA that is complementary to the trinucleotide sequence of messenger RNA (the codon).

antigen (Ag)   Any cell, particle, or chemical that induces a specific immune response by B cells or T cells and can stimulate resistance to an infection or a toxin. See *immunogen*.

antigen binding site   Specific region at the ends of the antibody molecule that recognizes specific antigens. These sites have numerous shapes to fit a wide variety of antigens.

antigenic drift   Minor antigenic changes in the influenza A virus due to mutations in the spikes' genes.

antigenic shift   Major changes in the influenza A virus due to recombination of viral strains from two different host species.

antigenicity   The property of a substance to stimulate a specific immune response such as antibody formation.

antigen-presenting cells   Cells of the immune system that digest foreign cells and particles and place pieces of them on their own surfaces in such a way that other cells of the immune system recognize them.

antihistamine   A drug that counters the action of histamine and is useful in allergy treatment.

antimicrobial   A special class of compounds capable of destroying or inhibiting microorganisms.

antimicrobial peptides   Short protein molecules found in epithelial cells; have the ability to kill bacteria.

antiparallel   A description of the two strands of DNA, which are parallel to each other, but the orientation of the deoxyribose and phosphate groups run in the opposite directions, with the 5′ carbon at the top of the leading strand and the 3′ carbon at the top of the lagging strand.

antisense DNA   A DNA oligonucleotide that binds to a specific piece of RNA, thereby inhibiting translation; used in gene therapy.

antisense RNA   An RNA oligonucleotide that binds to a specific piece of RNA, thereby inhibiting translation; used in gene therapy.

antisepsis   Chemical treatments to kill or inhibit the growth of all vegetative microorganisms on body surfaces.

antiseptic   A growth-inhibiting agent used on tissues to prevent infection.

antiserum   Antibody-rich serum derived from the blood of animals (deliberately immunized against infectious or toxic antigen) or from people who have recovered from specific infections.

antitoxin   Globulin fraction of serum that neutralizes a specific toxin. Also refers to the specific antitoxin antibody itself.

apicomplexans   A group of protozoans that lack locomotion in the mature state.

apoenzyme   The protein part of an enzyme, as opposed to the nonprotein or inorganic cofactors.

apoptosis   The genetically programmed death of cells that is both a natural process of development and the body's means of destroying abnormal or infected cells.

appendages   Accessory structures that sprout from the surface of bacteria. They can be divided into two major groups: those that provide motility and those that enable adhesion.

applied microbiology   The study of the practical uses of microorganisms.

aqueous   Referring to solutions in which water is used as the solvent.

aquifer   A subterranean water-bearing stratum of permeable rock, sand, or gravel.

archaea   Prokaryotic single-celled organisms of primitive origin that have unusual anatomy, physiology, and genetics and live in harsh habitats; when capitalized **(Archaea)**, the term refers to one of the three domains of living organisms as proposed by Woese.

arthroconidia   Reproductive body of *Coccidioides immitis*; also *arthrospore*.

Arthus reaction   An immune complex phenomenon that develops after repeat injection. This localized inflammation results from aggregates of antigen and antibody that bind, complement, and attract neutrophils.

artificial immunity   Immunity that is induced as a medical intervention, either by exposing an individual to an antigen or administering immune substances to him or her.

ascospore   A spore formed within a saclike cell (ascus) of Ascomycota following nuclear fusion and meiosis.

ascus   Special fungal sac in which haploid spores are created.

asepsis   A condition free of viable pathogenic microorganisms.

aseptic technique   Methods of handling microbial cultures, patient specimens, and other sources of microbes in a way that prevents infection of the handler and others who may be exposed.

assay medium   Microbiological medium used to test the effects of specific treatments to bacteria, such as antibiotic or disinfectant treatment.

assembly (viral)   The step in viral multiplication in which capsids and genetic material are packaged into virions.

asthma   A type of chronic local allergy in which the airways become constricted and produce excess mucus in reaction to allergens, exercise, stress, or cold temperatures.

astromicrobiology   A branch of microbiology that studies the potential for and the possible role of microorganisms in space and on other planets.

asymptomatic   An infection that produces no noticeable symptoms even though the microbe is active in the host tissue.

asymptomatic carrier   A person with an inapparent infection who shows no symptoms of being infected yet is able to pass the disease agent on to others.

atmosphere   That part of the biosphere that includes the gaseous envelope up to 14 miles above the earth's surface. It contains gases such as carbon dioxide, nitrogen, and oxygen.

atom   The smallest particle of an element to retain all the properties of that element.

atopy   Allergic reaction classified as type I, with a strong familial relationship; caused by allergens such as pollen, insect venom, food, and dander; involves IgE antibody; includes symptoms of hay fever, asthma, and skin rash.

ATP synthase   A unique enzyme located in the mitochondrial cristae and chloroplast grana that harnesses the flux of hydrogen ions to the synthesis of ATP.

attenuate   To reduce the virulence of a pathogenic bacterium or virus by passing it through a nonnative host or by long-term subculture.

AUG (start codon)   The codon that signals the point at which translation of a messenger RNA molecule is to begin.

autoantibody   An "anti-self" antibody having an affinity for tissue antigens of the subject in which it is formed.

autoclave   A sterilization chamber that allows the use of steam under pressure to sterilize materials. The most common temperature/pressure combination for an autoclave is 121°C and 15 psi.

autograft   Tissue or organ surgically transplanted to another site on the same subject.

autoimmune disease   The pathologic condition arising from the production of antibodies against autoantigens. Example: rheumatoid arthritis. Also called *autoimmunity.*

autoimmune regulator (AIRE)   A protein that regulates the transcription of self antigens in the thymus; defects in AIRE can lead to inappropriate responses to self antigens.

autoinducer   A chemical produced when bacteria have reached a specific concentration of cells, or quorum, causing the bacteria to behave as a group in various physiological activities including bioluminescence and biofilm formation.

autotroph   A microorganism that requires only inorganic nutrients and whose sole source of carbon is carbon dioxide.

axenic   A sterile state such as a pure culture. An axenic animal is born and raised in a germ-free environment. See *gnotobiotic.*

axial filament   A type of flagellum (called an *endoflagellum*) that lies in the periplasmic space of spirochetes and is responsible for locomotion. Also called *periplasmic flagellum.*

azole   Five-membered heterocyclic compounds typical of histidine, which are used in antifungal therapy.

B

B lymphocyte (B cell)   A white blood cell that gives rise to plasma cells and antibodies.

bacillus   Bacterial cell shape that is cylindrical (longer than it is wide).

bacitracin   Antibiotic that targets the bacterial cell wall; component of over-the-counter topical antimicrobial ointments.

back-mutation   A mutation that counteracts an earlier mutation, resulting in the restoration of the original DNA sequence.

bacteremia   The presence of viable bacteria in circulating blood.

Bacteria   When capitalized can refer to one of the three domains of living organisms proposed by Woese, containing all nonarchaea prokaryotes.

bacteria   (singular, *bacterium*) Category of prokaryotes with peptidoglycan in their cell walls and circular chromosome(s). This group of small cells is widely distributed in the earth's habitats.

bacterial chromosome   A circular body in bacteria that contains the primary genetic material. Also called *nucleoid.*

bactericide   An agent that kills bacteria.

bacteriophage   A virus that specifically infects bacteria.

bacteristatic   Any process or agent that inhibits bacterial growth.

bacterium   A tiny unicellular prokaryotic organism that usually reproduces by binary fission and usually has a peptidoglycan cell wall, has various shapes, and can be found in virtually any environment.

barophile   A microorganism that thrives under high (usually hydrostatic) pressure.

basement membrane   A thin layer (1–6 μm) of protein and polysaccharide found at the base of epithelial tissues.

basic   A solution with a pH value above 7 on the pH scale.

basidiospore   A sexual spore that arises from a basidium. Found in Basidiomycota fungi.

basophil   A motile polymorphonuclear leukocyte that binds IgE. The basophilic cytoplasmic granules contain mediators of anaphylaxis and atopy.

beta oxidation   The degradation of long-chain fatty acids. Two-carbon fragments are formed as a result of enzymatic attack directed against the second or beta carbon of the hydrocarbon chain. Aided by coenzyme A, the fragments enter the Krebs cycle and are processed for ATP synthesis.

beta-lactamase   An enzyme secreted by certain bacteria that cleaves the beta-lactam ring of penicillin and cephalosporin and thus provides for resistance against the antibiotic. See *penicillinase.*

binary fission   The formation of two new cells of approximately equal size as the result of parent cell division.

binomial system   Scientific method of assigning names to organisms that employs two names to identify every organism—genus name plus species name.

biochemistry   The study of organic compounds produced by (or components of) living things. The four main categories of biochemicals are carbohydrates, lipids, proteins, and nucleic acids.

biofilm   A complex association that arises from a mixture of microorganisms growing together on the surface of a habitat.

biogenesis   Belief that living things can only arise from others of the same kind.

biogeochemical cycle   A process by which matter is converted from organic to inorganic form and returned to various nonliving reservoirs on earth (air, rocks, and water) where it becomes available for reuse by living things. Elements such as carbon, nitrogen, and phosphorus are constantly cycled in this manner.

bioinformatics   The use of computer software to determine the function of genes through analysis of the DNA and protein sequences.

biological vector   An animal that not only transports an infectious agent but plays a role in the life cycle of the pathogen, serving as a site in which it can multiply or complete its life cycle. It is usually an alternate host to the pathogen.

bioluminescence   The production of light by various species of bacteria, fish, insects, and some animals through the conversion of chemical energy into light.

biomarkers   Proteins, chemicals, or other substances that can be used as indicators of normal biological processes, disease, exposure to an environmental substance, or a reaction to a drug; measured in various bodily substances such as saliva, blood, urine, and hair—and even in the breath.

biomes   Particular climate regions in a terrestrial realm.

bioremediation   Decomposition of harmful chemicals by microbes or consortia of microbes.

biosphere   Habitable regions comprising the aquatic (hydrospheric), soil-rock (lithospheric), and air (atmospheric) environments.

biosurfactants   Surface-acting agents such as soaps and cleaning agents derived from bacteria and fungi rather than fossil fuels. See *surfactant.*

biota   Beneficial or harmless resident bacteria commonly found on and/or in the human body.

biotechnology   The use of microbes or their products in the commercial or industrial realm.

biotic   Living factors such as parasites, food substrates, or other living or once-living organisms that are studied when looking at an ecosystem.

blocking antibody   The IgG class of immunoglobulins that competes with IgE antibody for allergens, thus blocking the degranulation of basophils and mast cells.

blood cells   Cellular components of the blood consisting of red blood cells, primarily responsible for the transport of oxygen and carbon dioxide, and white blood cells, primarily responsible for host defense and immune reactions.

blood-brain barrier   Decreased permeability of the walls of blood vessels in the brain, restricting access to that compartment.

botulinum   *Clostridium botulinum* toxin. Ingestion of this potent exotoxin leads to flaccid paralysis.

bradykinin   An active polypeptide that is a potent vasodilator released from IgE-coated mast cells during anaphylaxis.

broad-spectrum   Denotes drugs that have an effect on a wide variety of microorganisms.

bubo   The swelling of one or more lymph nodes due to inflammation.

bubonic plague   The form of plague in which bacterial growth is primarily restricted to the lymph and is characterized by the appearance of a swollen lymph node referred to as a *bubo.*

budding   See *exocytosis.*

bulbar poliomyelitis   Complication of polio infection in which the brain stem, medulla, or cranial nerves are affected. Leads to loss of respiratory control and paralysis of the trunk and limbs.

bullous   Consisting of fluid-filled blisters.

bursa of Fabricius   A lymphatic gland in the cloaca in birds in which antibody-producing cells were first demonstrated and described as "bursa-derived cells," giving these types of cells the name *B cells.*

C

calculus   Dental deposit formed when plaque becomes mineralized with calcium and phosphate crystals. Also called *tartar*.

Calvin cycle   A series of reactions in the second phase of photosynthesis that generates glucose.

cancer   Any malignant neoplasm that invades surrounding tissue and can metastasize to other locations. A carcinoma is derived from epithelial tissue, and a sarcoma arises from proliferating mesodermal cells of connective tissue.

capsid   The protein covering of a virus's nucleic acid core. Capsids exhibit symmetry due to the regular arrangement of subunits called *capsomers*. See *icosahedron*.

capsomer   A subunit of the virus capsid shaped as a triangle or disc.

capsule   In bacteria, the loose, gel-like covering or slime made chiefly of polysaccharides. This layer is protective and can be associated with virulence.

capsule staining   Any staining method that highlights the outermost polysaccharide and/or protein structure on a bacterial, fungal, or protozoal cell.

carbohydrate   A compound containing primarily carbon, hydrogen, and oxygen in a 1:2:1 ratio.

carbohydrate fermentation medium   A growth medium that contains sugars that are converted to acids through fermentation. Usually contains a pH indicator to detect acid protection.

carbon cycle   That pathway taken by carbon from its abiotic source to its use by producers to form organic compounds (biotic), followed by the breakdown of biotic compounds and their release to a nonliving reservoir in the environment (mostly carbon dioxide in the atmosphere).

carbon fixation   Reactions in photosynthesis that incorporate inorganic carbon dioxide into organic compounds such as sugars. This occurs during the Calvin cycle and uses energy generated by the light reactions. This process is the source of all production on earth.

carbuncle   A deep staphylococcal abscess joining several neighboring hair follicles.

carotenoid   Yellow, orange, or red photosynthetic pigments.

carrier   A person who harbors infections and inconspicuously spreads them to others. Also, a chemical agent that can accept an atom, chemical radical, or subatomic particle from one compound and pass it on to another.

catabolism   The chemical breakdown of complex compounds into simpler units to be used in cell metabolism.

catalyst   A substance that alters the rate of a reaction without being consumed or permanently changed by it. In cells, enzymes are catalysts.

catalytic site   The niche in an enzyme where the substrate is converted to the product (also *active site*).

catarrhal   A term referring to the secretion of mucus or fluids; term for the first stage of pertussis.

cation   A positively charged ion.

cell   An individual membrane-bound living entity; the smallest unit capable of an independent existence.

cell wall   In bacteria, a rigid structure made of peptidoglycan that lies just outside the cytoplasmic membrane; eukaryotes also have a cell wall, but it may be composed of a variety of materials.

cell-mediated immunity   The type of immune responses brought about by T cells, such as cytotoxic and helper effects.

cellulitis   The spread of bacteria within necrotic tissue.

cellulose   A long, fibrous polymer composed of β-glucose; one of the most common substances on earth.

cephalosporins   A group of broad-spectrum antibiotics isolated from the fungus *Cephalosporium*.

cercaria   The free-swimming larva of the schistosome trematode that emerges from the snail host and can penetrate human skin, causing schistosomiasis.

cestode   The common name for tapeworms that parasitize humans and domestic animals.

chancre   The primary sore of syphilis that forms at the site of penetration by *Treponema pallidum*. It begins as a hard, dull red, painless papule that erodes from the center.

chancroid   A lesion that resembles a chancre but is soft and is caused by *Haemophilus ducreyi*.

chemical bond   A link formed between molecules when two or more atoms share, donate, or accept electrons.

chemical mediators   Small molecules that are released during inflammation and specific immune reactions that allow communication between the cells of the immune system and facilitate surveillance, recognition, and attack.

chemiosmosis   The generation of a concentration gradient of hydrogen ions (called the *proton motive force*) by the pumping of hydrogen ions to the outer side of the membrane during electron transport.

chemoautotroph   An organism that relies upon inorganic chemicals for its energy and carbon dioxide for its carbon. Also called a *chemolithotroph*.

chemoheterotroph   Microorganisms that derive their nutritional needs from organic compounds.

chemokines   Chemical mediators (cytokines) that stimulate the movement and migration of white blood cells.

chemostat   A growth chamber with an outflow that is equal to the continuous inflow of nutrient media. This steady-state growth device is used to study such events

as cell division, mutation rates, and enzyme regulation.

chemotactic factors   Chemical mediators that stimulate the movement of white blood cells. See *chemokines*.

chemotaxis   The tendency of organisms to move in response to a chemical gradient (toward an attractant or to avoid adverse stimuli).

chemotherapy   The use of chemical substances or drugs to treat or prevent disease.

chemotroph   Organism that oxidizes compounds to feed on nutrients.

chitin   A polysaccharide similar to cellulose in chemical structure. This polymer makes up the horny substance of the exoskeletons of arthropods and certain fungi.

chlorophyll   A group of mostly green pigments that are used by photosynthetic eukaryotic organisms and cyanobacteria to trap light energy to use in making chemical bonds.

chloroplast   An organelle containing chlorophyll that is found in photosynthetic eukaryotes.

cholesterol   Best-known member of a group of lipids called *steroids*. Cholesterol is commonly found in cell membranes and animal hormones.

chromatin   The genetic material of the nucleus. Chromatin is made up of nucleic acid and stains readily with certain dyes.

chromosome   The tightly coiled bodies in cells that are the primary sites of genes.

chronic   Any process or disease that persists over a long duration.

chronic carrier   An individual who has recovered from an initial infection but continues to harbor and shed infectious agents for a long period of time.

cilium   (plural, *cilia*) Eukaryotic structure similar to a flagellum that propels a protozoan through the environment.

class   In the levels of classification, the division of organisms that follows phylum.

clonal selection theory   A conceptual explanation for the development of lymphocyte specificity and variety during immune maturation.

clone   A colony of cells (or group of organisms) derived from a single cell (or single organism) by asexual reproduction. All units share identical characteristics. *Clone* is also used as a verb to refer to the process of producing a genetically identical population of cells or genes.

cloning host   An organism such as a bacterium or a yeast that receives and replicates a foreign piece of DNA inserted during a genetic engineering experiment.

coagulase   A plasma-clotting enzyme secreted by *Staphylococcus aureus*. It contributes to virulence and is involved in forming a fibrin wall that surrounds staphylococcal lesions.

coccobacillus   An elongated coccus; a short, thick, oval-shaped bacterial rod.

coccus A spherical-shaped bacterial cell.

codon A specific sequence of three nucleotides in mRNA (or the sense strand of DNA) that constitutes the genetic code for a particular amino acid.

coenzyme A complex organic molecule, several of which are derived from vitamins (e.g., nicotinamide, riboflavin). A coenzyme operates in conjunction with an enzyme. Coenzymes serve as transient carriers of specific atoms or functional groups during metabolic reactions.

coevolution A biological process whereby a change in the genetic composition in one organism leads to a change in the genetics of another organism.

cofactor An enzyme accessory. It can be organic, such as coenzymes, or inorganic, such as Fe^{2+}, Mn^{2+}, or Zn^{2+} ions.

cold sterilization The use of nonheating methods such as radiation or filtration to sterilize materials.

coliform A collective term that includes normal enteric bacteria that are gram-negative and lactose-fermenting.

colony A macroscopic cluster of cells appearing on a solid medium, each arising from the multiplication of a single cell.

comedo A skin lesion commonly associated with acne that develops over the pore leading out of a hair follicle; a "whitehead" composed of sebum, cellular debris, and bacteria, which blocks the pore.

commensalism An unequal relationship in which one species derives benefit without harming the other.

common-source epidemic An outbreak of disease in which all affected individuals were exposed to a single source of the pathogen, even if they were exposed at different times.

communicable infection Capable of being transmitted from one individual to another.

community The interacting mixture of populations in a given habitat.

competent Referring to bacterial cells that are capable of absorbing free DNA in their environment either naturally or through induction by exposure to chemicals or electrical currents.

competitive inhibition Control process that relies on the ability of metabolic analogs to control microbial growth by successfully competing with a necessary enzyme to halt the growth of bacterial cells.

complement In immunology, serum protein components that act in a definite sequence when set in motion either by an antigen-antibody complex or by factors of the alternative (properdin) pathway.

complementary DNA (cDNA) DNA created by using reverse transcriptase to synthesize DNA from RNA templates.

compounds Molecules that are a combination of two or more different elements.

concentration The expression of the amount of a solute dissolved in a certain amount of solvent. It may be defined by weight, volume, or percentage.

condyloma acuminata Extensive, branched masses of genital warts caused by infection with human papillomavirus.

congenital Transmission of an infection from mother to fetus.

congenital rubella Transmission of the rubella virus to a fetus *in utero*. Injury to the fetus is generally much more serious than it is to the mother.

conidia Asexual fungal spores shed as free units from the tips of fertile hyphae.

conidiospore A type of asexual spore in fungi; not enclosed in a sac.

conjugated vaccines Subunit vaccines combined with carrier proteins, often from other microbes, to make them more immunogenic.

conjugation In bacteria, the contact between donor and recipient cells associated with the transfer of genetic material such as plasmids. Can involve special (sex) pili. Also a form of sexual recombination in ciliated protozoans.

conjunctiva The thin fluid-secreting tissue that covers the eye and lines the eyelid.

consortium A group of microbes that includes more than one species.

constitutive enzyme An enzyme present in bacterial cells in constant amounts, regardless of the presence of substrate. Enzymes of the central catabolic pathways are typical examples.

consumer An organism that feeds on producers or other consumers. It gets all nutrients and energy from other organisms (also called *heterotroph*). May exist at several levels, such as primary (feeds on producers) and secondary (feeds on primary consumers).

contagious Communicable; transmissible by direct contact with infected people and their fresh secretions or excretions.

contaminant An impurity; any undesirable material or organism.

contaminated culture A medium that once held a pure (single or mixed) culture but now contains unwanted microorganisms.

contigs Contiguous sets of overlapping nucleotide sequences determined by sequencing fragments of a genome in a genomic library.

convalescence Recovery; the period between the end of a disease and the complete restoration of health in a patient.

convalescent period The period after the period of invasion in which the patient's immune system responds to the infection, signs and symptoms gradually decline, and the patient's health gradually returns.

corepressor A molecule that combines with inactive repressor to form active repressor, which attaches to the operator gene site and inhibits the activity of structural genes subordinate to the operator.

cornea The transparent, dome-shaped tissue covering the iris, pupil, and anterior chamber of the eye composed of five to six layers of quickly regenerating epithelial cells.

covalent A type of chemical bond that involves the sharing of electrons between two atoms.

covalent bond A chemical bond formed by the sharing of electrons between two atoms.

Creutzfeldt-Jakob disease (CJD) A spongiform encephalopathy caused by infection with a prion. The disease is marked by dementia, impaired senses, and uncontrollable muscle contractions.

crista The infolded inner membrane of a mitochondrion that is the site of the respiratory chain and oxidative phosphorylation.

culture The visible accumulation of microorganisms in or on a nutrient medium. Also, the propagation of microorganisms with various media.

curd The coagulated milk protein used in cheese making.

cutaneous Second level of skin, including the stratum corneum and occasionally the upper dermis.

cyst The resistant, dormant but infectious form of protozoans. Can be important in spread of infectious agents such as *Entamoeba histolytica* and *Giardia lamblia*.

cysteine A sulfide-containing amino acid that usually produces covalent disulfide bonds in an amino acid sequence, contributing to the tertiary structure of the protein.

cytochrome A group of heme protein compounds whose chief role is in electron and/or hydrogen transport occurring in the last phase of aerobic respiration.

cytokine A chemical substance produced by white blood cells and tissue cells that regulates development, inflammation, and immunity.

cytopathic effects (CPEs) The degenerative changes in cells associated with virus infection.

cytoplasm Dense fluid encased by the cell membrane; the site of many of the cell's biochemical and synthetic activities.

cytoplasmic membrane Lipid bilayer that encloses the cytoplasm of bacterial cells.

cytosine (C) One of the nitrogen bases found in DNA and RNA, with a pyrimidine form.

cytotoxicity The ability to kill cells; in immunology, certain T cells are called *cytotoxic T cells* because they kill other cells.

D

daptomycin A lipopetide antibiotic that disrupts the cytoplasmic membrane.

deamination The removal of an amino group from an amino acid.

death phase End of the cell growth due to lack of nutrition, depletion of environment, and accumulation of wastes. Population of cells begins to die.

debridement Trimming away devitalized tissue and foreign matter from a wound.

decomposer   A consumer that feeds on organic matter from the bodies of dead organisms. These microorganisms feed from all levels of the food pyramid and are responsible for recycling elements (also called *saprobes*).

decomposition   The breakdown of dead matter and wastes into simple compounds that can be directed back into the natural cycle of living things.

decontamination   The removal or neutralization of an infectious, poisonous, or injurious agent from a site.

definitive host   The organism in which a parasite develops into its adult or sexually mature stage. Also called the *final host*.

degerm   To physically remove surface oils, debris, and soil from skin to reduce the microbial load.

dehydration synthesis   During the formation of a carbohydrate bond, the step in which one carbon molecule gives up its OH group and the other loses the H from its OH group, thereby producing a water molecule. This process is common to all polymerization reactions.

denaturation   The loss of normal characteristics resulting from some molecular alteration. Usually in reference to the action of heat or chemicals on proteins whose function depends upon an unaltered tertiary structure.

dendritic cell   A large, antigen-processing cell characterized by long, branchlike extensions of the cell membrane.

denitrification   The end of the nitrogen cycle when nitrogen compounds are returned to the reservoir in the air.

deoxyribonucleic acid (DNA)   The nucleic acid often referred to as the "double helix." DNA carries the master plan for an organism's heredity.

deoxyribose   A 5-carbon sugar that is an important component of DNA.

dermatophytes   A group of fungi that cause infections of the skin and other integument components. They survive by metabolizing keratin.

dermolytic   Capable of damaging the skin.

desensitization   See *hyposensitization*.

desiccation   To dry thoroughly. To preserve by drying.

desquamate   To shed the cuticle in scales; to peel off the outer layer of a surface.

diabetes mellitus   A disease involving compromise in insulin function. In one form, the pancreatic cells that produce insulin are destroyed by autoantibodies; in another, the pancreas does not produce sufficient insulin.

diapedesis   The migration of intact blood cells between endothelial cells of a blood vessel such as a venule.

dichotomous keys   Flow charts that offer two choices or pathways at each level.

differential medium   A single substrate that discriminates between groups of microorganisms on the basis of differences in their appearance due to different chemical reactions.

differential stain   A technique that utilizes two dyes to distinguish between different microbial groups or cell parts by color reaction.

diffusion   The dispersal of molecules, ions, or microscopic particles propelled down a concentration gradient by spontaneous random motion to achieve a uniform distribution.

DiGeorge syndrome   A birth defect usually caused by a missing or incomplete thymus that results in abnormally low or absent T cells and other developmental abnormalities.

dimorphic   In mycology, the tendency of some pathogens to alter their growth form from mold to yeast in response to rising temperature.

dipicolinic acid   An organic acid found in the walls of endospores; contributes to their extreme resistance to chemicals, drying, and heat.

diplococci   Spherical or oval-shaped bacteria, typically found in pairs.

direct (total) cell count   1. Counting total numbers of individual cells being viewed with magnification. 2. Counting isolated colonies of organisms growing on a plate of media as a way to determine population size.

disaccharide   A sugar containing two monosaccharides. Example: sucrose (fructose + glucose).

disease   Any deviation from health, as when the effects of microbial infection damage or disrupt tissues and organs.

disinfection   The destruction of pathogenic nonsporulating microbes or their toxins, usually on inanimate surfaces.

division   In the levels of classification, an alternate term for phylum.

DNA   See *deoxyribonucleic acid*.

DNA polymerase   Enzyme responsible for the replication of DNA. Several versions of the enzyme exist, each completing a unique portion of the replication process.

DNA profiling   A pattern of restriction enzyme fragments that is unique for an individual organism.

DNA vaccine   A newer vaccine preparation based on inserting DNA from pathogens into host cells to encourage them to express the foreign protein and stimulate immunity.

domain   In the levels of classification, the broadest general category to which an organism is assigned. Members of a domain share only one or a few general characteristics.

doubling time   Time required for a complete fission cycle—from parent cell to two new daughter cells. Also called *generation time*.

droplet nuclei   The dried residue of fine droplets produced by mucus and saliva sprayed while sneezing and coughing. Droplet nuclei are less than 5 μm in diameter (large enough to bear a single bacterium and small enough to remain airborne for a long time) and can be carried by air currents. Droplet nuclei are drawn deep into the air passages.

drug resistance   An adaptive response in which microorganisms begin to tolerate an amount of drug that would ordinarily be inhibitory.

dysentery   Diarrheal illness in which stools contain blood and/or mucus.

dyspnea   Difficulty in breathing.

E

ecosystem   A collection of organisms together with its surrounding physical and chemical factors.

ectoplasm   The outer, more viscous region of the cytoplasm of a phagocytic cell such as an amoeba. It contains microtubules, but not granules or organelles.

eczema   An acute or chronic allergy of the skin associated with itching and burning sensations. Typically, red, edematous, vesicular lesions erupt, leaving the skin scaly and sometimes hyperpigmented.

edema   The accumulation of excess fluid in cells, tissues, or serous cavities. Also called *swelling*.

electrolyte   Any compound that ionizes in solution and conducts current in an electrical field.

electron   A negatively charged subatomic particle that is distributed around the nucleus in an atom.

electrophoresis   The separation of molecules by size and charge through exposure to an electrical current.

element   A substance comprising only one kind of atom that cannot be degraded into two or more substances without losing its chemical characteristics.

ELISA   Abbreviation for **e**nzyme-linked **i**mmuno**s**orbent **a**ssay, a very sensitive serological test used to detect antibodies in diseases such as AIDS.

emetic   Inducing to vomit.

encephalitis   An inflammation of the brain, usually caused by infection.

endemic disease   A native disease that prevails continuously in a geographic region.

endergonic reaction   A chemical reaction that occurs with the absorption and storage of surrounding energy. Antonym: *exergonic*.

endocytosis   The process whereby solid and liquid materials are taken into the cell through membrane invagination and engulfment into a vesicle.

endoenzyme   An intracellular enzyme, as opposed to enzymes that are secreted.

endogenous   Originating or produced within an organism or one of its parts.

endoplasm   The granular inner region of the cytoplasm of a eukaryotic cell that contains the nucleus, mitochondria, and vacuoles.

endoplasmic reticulum (ER)　An intracellular network of flattened sacs or tubules with or without ribosomes on their surfaces.

endospore　A small, dormant, resistant derivative of a bacterial cell that germinates under favorable growth conditions into a vegetative cell. The bacterial genera *Bacillus* and *Clostridium* are typical sporeformers.

endosymbiosis　Relationship in which a microorganism resides within a host cell and provides a benefit to the host cell.

endotoxic shock　A massive drop in blood pressure caused by the release of endotoxin from gram-negative bacteria multiplying in the bloodstream.

endotoxin　A bacterial toxin that is not ordinarily released (as is exotoxin). Endotoxin is composed of a phospholipid-polysaccharide complex that is an integral part of gram-negative bacterial cell walls. Endotoxins can cause severe shock and fever.

energy of activation　The minimum energy input necessary for reactants to form products in a chemical reaction.

energy pyramid　An ecological model that shows the energy flow among the organisms in a community. It is structured like the food pyramid but shows how energy is reduced from one trophic level to another.

enriched medium　A nutrient medium supplemented with blood, serum, or some growth factor to promote the multiplication of fastidious microorganisms.

enteric　Pertaining to the intestine.

enterotoxin　A bacterial toxin that specifically targets intestinal mucous membrane cells. Enterotoxigenic strains of *Escherichia coli* and *Staphylococcus aureus* are typical sources.

enumeration medium　Microbiological medium that does not encourage growth and allows for the counting of microbes in food, water, or environmental samples.

enzyme　A protein biocatalyst that facilitates metabolic reactions.

enzyme induction　One of the controls on enzyme synthesis. This occurs when enzymes appear only when suitable substrates are present.

enzyme repression　The inhibition of enzyme synthesis by the end product of a catabolic pathway.

eosinophil　A leukocyte whose cytoplasmic granules readily stain with red eosin dye.

epidemic　A sudden and simultaneous outbreak or increase in the number of cases of disease in a community.

epidemiology　The study of the factors affecting the prevalence and spread of disease within a community.

epitope　The precise molecular group of an antigen that defines its specificity and triggers the immune response.

Epstein-Barr virus (EBV)　Herpesvirus linked to infectious mononucleosis, Burkitt's lymphoma, and nasopharyngeal carcinoma.

erysipelas　An acute, sharply defined inflammatory disease specifically caused by hemolytic *Streptococcus*. The eruption is limited to the skin but can be complicated by serious systemic symptoms.

erythrocytes　Red blood cells; involved in the transport of oxygen and carbon dioxide.

erythrogenic toxin　An exotoxin produced by lysogenized group A strains of β-hemolytic streptococci that is responsible for the severe fever and rash of scarlet fever in the nonimmune individual. Also called a *pyrogenic toxin*.

eschar　A dark, sloughing scab that is the lesion of anthrax and certain rickettsioses.

essential nutrient　Any ingredient such as a certain amino acid, fatty acid, vitamin, or mineral that cannot be formed by an organism and must be supplied in the diet. A growth factor.

ethylene oxide　A potent, highly water-soluble gas invaluable for gaseous sterilization of heat-sensitive objects such as plastics, surgical and diagnostic appliances, and spices.

etiologic agent　The microbial cause of disease; the pathogen.

eubacteria　Term sometimes used for nonarchaea prokaryotes, means "true bacteria."

Eukarya　One of the three domains (sometimes called *superkingdoms*) of living organisms, as proposed by Woese; contains all eukaryotic organisms.

eukaryote　A member of the domain Eukarya whose cells have a well-defined nucleus and membrane-bound organelles; includes plants, animals, fungi, protozoa, and algae.

eukaryotic cell　A cell that differs from a prokaryotic cell chiefly by having a nuclear membrane (a well-defined nucleus), membrane-bounded subcellular organelles, and mitotic cell division.

eutrophication　The process whereby dissolved nutrients resulting from natural seasonal enrichment or industrial pollution of water cause overgrowth of algae and cyanobacteria to the detriment of fish and other large aquatic inhabitants.

evolution　Scientific principle that states that living things change gradually through hundreds of millions of years, and these changes are expressed in structural and functional adaptations in each organism. Evolution presumes that those traits that favor survival are preserved and passed on to following generations, and those traits that do not favor survival are lost.

exanthem　An eruption or rash of the skin.

exergonic　A chemical reaction associated with the release of energy to the surroundings. Antonym: *endergonic*.

exfoliative toxin　A poisonous substance that causes superficial cells of an epithelium to detach and be shed. Example: staphylococcal exfoliatin. Also called an *epidermolytic toxin*.

exocytosis　The process that releases enveloped viruses from the membrane of the host's cytoplasm.

exoenzyme　An extracellular enzyme chiefly for hydrolysis of nutrient macromolecules that are otherwise impervious to the cell membrane. It functions in saprobic decomposition of organic debris and can be a factor in invasiveness of pathogens.

exogenous　Originating outside the body.

exon　A stretch of eukaryotic DNA coding for a corresponding portion of mRNA that is translated into peptides. Intervening stretches of DNA that are not expressed are called *introns*. During transcription, exons are separated from introns and are spliced together into a continuous mRNA transcript.

exotoxin　A toxin (usually protein) that is secreted and acts upon a specific cellular target. Examples: botulin, tetanospasmin, diphtheria toxin, and erythrogenic toxin.

exponential　Pertaining to the use of exponents, numbers that are typically written as a superscript to indicate how many times a factor is to be multiplied. Exponents are used in scientific notation to render large, cumbersome numbers into small workable quantities.

exponential growth phase　The period of maximum growth rate in a growth curve. Cell population increases logarithmically.

extrapulmonary tuberculosis　A condition in which tuberculosis bacilli have spread to organs other than the lungs.

extremophiles　Organisms capable of living in harsh environments, such as extreme heat or cold.

exudate　Fluid that escapes cells into the extracellular spaces during the inflammatory response.

F

facilitated diffusion　The passive movement of a substance across a plasma membrane from an area of higher concentration to an area of lower concentration utilizing specialized carrier proteins.

facultative　Pertaining to the capacity of microbes to adapt or adjust to variations; not obligate. Example: the presence of oxygen is not obligatory for a facultative anaerobe to grow. See *obligate*.

family　In the levels of classification, a midlevel division of organisms that groups more closely related organisms than previous levels. An order is divided into families.

fastidious　Requiring special nutritional or environmental conditions for growth. Said of bacteria.

fecal coliforms　Any species of gram-negative lactose-positive bacteria (primarily *Escherichia coli*) that live primarily in the intestinal tract and not the environment. Finding evidence of these bacteria in a water or food sample is substantial evidence of fecal contamination and potential for infection (see *coliform*).

fermentation　The extraction of energy through anaerobic degradation of substrates into simpler, reduced metabolites. In large industrial processes, *fermentation* can mean any use of microbial metabolism to manufacture organic chemicals or other products.

fermentor   A large tank used in industrial microbiology to grow mass quantities of microbes that can synthesize desired products. These devices are equipped with means to stir, monitor, and harvest products such as drugs, enzymes, and proteins in very large quantities.

ferritin   An iron-binding protein found in all cell types.

fertility (F) factor   Donor plasmid that allows synthesis of a pilus in bacterial conjugation. Presence of the factor is indicated by F^+, and lack of the factor is indicated by F^-.

filament   A helical structure composed of proteins that is part of bacterial flagella.

fimbria   A short, numerous-surface appendage on some bacteria that provides adhesion but not locomotion.

Firmicutes   Taxonomic category of bacteria that have gram-positive cell envelopes.

flagellar staining   A staining method that highlights the flagellum of a bacterium.

flagellum   A structure that is used to propel the organism through a fluid environment.

fluid mosaic model   A conceptualization of the molecular architecture of cellular membranes as a bilipid layer containing proteins. Membrane proteins are embedded to some degree in this bilayer, where they float freely about.

fluorescence   The property possessed by certain minerals and dyes to emit visible light when excited by ultraviolet radiation. A fluorescent dye combined with specific antibody provides a sensitive test for the presence of antigen.

fluorescent *in situ* hybridization (FISH)   A technique in which a fluorescently labeled DNA or RNA probe is used to locate a specific sequence of DNA in an organism without removing it from its natural environment.

fluoroquinolones   Synthetic antimicrobial drugs chemically related to quinine. They are broad-spectrum and easily adsorbed from the intestine.

focal infection   Occurs when an infectious agent breaks loose from a localized infection and is carried by the circulation to other tissues.

folliculitis   An inflammatory reaction involving the formation of papules or pustules in clusters of hair follicles.

fomite   Virtually any inanimate object an infected individual has contact with that can serve as a vehicle for the spread of disease.

food chain   A simple straight-line feeding sequence among organisms in a community.

food fermentations   Addition to and growth of known cultures of microorganisms in foods to produce desirable flavors, smells, or textures. Includes cheeses, breads, alcoholic beverages, and pickles.

food poisoning   Symptoms in the intestines (which may include vomiting) induced by preformed exotoxin from bacteria.

food web   A complex network that traces all feeding interactions among organisms in a community (see *food chain*). This is considered to be a more accurate picture of food relationships in a community than a food chain.

formalin   A 37% aqueous solution of formaldehyde gas; a potent chemical fixative and microbicide.

frameshift mutation   An insertion or deletion mutation that changes the codon reading frame from the point of the mutation to the final codon. Almost always leads to a nonfunctional protein.

free energy   Energy in a chemical system that can be used to do work.

fructose   One of the carbohydrates commonly referred to as sugars. Fructose is commonly fruit sugars.

functional group   In chemistry, a particular molecular combination that reacts in predictable ways and confers particular properties on a compound. Examples: —COOH, —OH, —CHO.

fungemia   The condition of fungi multiplying in the bloodstream.

fungi   (singular, *fungus*) Macroscopic and microscopic heterotrophic eukaryotic organisms that can be uni- or multicellular.

furuncle   A boil; a localized pyogenic infection arising from a hair follicle.

G

gamma globulin   The fraction of plasma proteins high in immunoglobulins (antibodies). Preparations from pooled human plasma containing normal antibodies make useful passive immunizing agents against pertussis, polio, measles, and several other diseases.

gas gangrene   Disease caused by a clostridial infection of soft tissue or wound. The name refers to the gas produced by the bacteria growing in the tissue. Unless treated early, it is fatal. Also called *myonecrosis*.

gastritis   Pain and/or nausea, usually experienced after eating; result of inflammation of the lining of the stomach.

gel electrophoresis   A laboratory technique for separating DNA fragments according to length by employing electricity to force the DNA through a gel-like matrix typically made of agarose. Smaller DNA fragments move more quickly through the gel, thereby moving farther than larger fragments during the same period of time.

gene   A site on a chromosome that provides information for a certain cell function. A specific segment of DNA that contains the necessary code to make a protein or RNA molecule.

gene probe   Short strands of single-stranded nucleic acid that hybridize specifically with complementary stretches of nucleotides on test samples and thereby serve as a tagging and identification device.

gene therapy   The introduction of normal functional genes into people with genetic diseases such as sickle-cell anemia and cystic fibrosis. This is usually accomplished by a virus vector.

generation time   Time required for a complete fission cycle—from parent cell to two new daughter cells. Also called *doubling time*.

genetic engineering   A field involving deliberate alterations (recombinations) of the genomes of microbes, plants, and animals through special technological processes.

genetics   The science of heredity.

genome   The complete set of chromosomes and genes in an organism.

genomic libraries   Collections of DNA fragments representing the entire genome of an organism inserted into plasmids and stored in vectors such as bacteria or yeast.

genomics   The systematic study of an organism's genes and their functions.

genotype   The genetic makeup of an organism. The genotype is ultimately responsible for an organism's phenotype, or expressed characteristics.

genus   In the levels of classification, the second most specific level. A family is divided into several genera.

geomicrobiology   A branch of microbiology that studies the role of microorganisms in the earth's crust.

germ free   See *axenic*.

germ theory of disease   A theory first originating in the 1800s that proposed that microorganisms can be the cause of diseases. The concept is actually so well established in the present time that it is considered a fact.

germicide   An agent lethal to non-endospore-forming pathogens.

gingivitis   Inflammation of the gum tissue in contact with the roots of the teeth.

glomerulonephritis   A complication of an infection by *Streptococcus pyogenes* in which antigen-antibody complexes become deposited in the glomeruli of the kidneys; characterized by nephritis, blood in the urine, increased blood pressure, and occasionally heart failure or permanent kidney damage.

gluconeogenesis   The formation of glucose (or glycogen) from noncarbohydrate sources such as protein or fat. Also called *glyconeogenesis*.

glucose   One of the carbohydrates commonly referred to as sugars. Glucose is characterized by its 6-carbon structure.

glycan   A type of carbohydrate or polysaccharide that is combined with another organic molecule such as a lipid or protein; examples include peptidoglycan and glycocalyx.

glycerol   A 3-carbon alcohol, with three OH groups that serve as binding sites.

glycocalyx   A filamentous network of carbohydrate-rich molecules that coats cells.

glycogen   A glucose polymer stored by cells.

glycolysis   The energy-yielding breakdown (fermentation) of glucose to pyruvic or lactic acid. It is often called *anaerobic glycolysis* because no molecular oxygen is consumed in the degradation.

glycosidic bond   A bond that joins monosaccharides to form disaccharides and polymers.

gnotobiotic   Referring to experiments performed on germ-free animals.

Golgi apparatus   An organelle of eukaryotes that participates in packaging and secretion of molecules.

gonococcus   Common name for *Neisseria gonorrhoeae*, the agent of gonorrhea.

Gracilicutes   Taxonomic category of bacteria that have gram-negative envelopes.

graft versus host disease (GVHD)   A condition associated with a bone marrow transplant in which T cells in the transplanted tissue mount an immune response against the recipient's (host) normal tissues.

Gram stain   A differential stain for bacteria useful in identification and taxonomy. Gram-positive organisms appear purple from crystal violet mordant retention, whereas gram-negative organisms appear red after loss of crystal violet and absorbance of the safranin counterstain.

gram-negative   A category of bacterial cells that describes bacteria with an outer membrane, a cytoplasmic membrane, and a thin cell wall.

gram-positive   A category of bacterial cells that describes bacteria with a thick cell wall and no outer membrane.

grana   Discrete stacks of chlorophyll-containing thylakoids within chloroplasts.

granulocyte   A mature leukocyte that contains noticeable granules in a Wright stain. Examples: neutrophils, eosinophils, and basophils.

granuloma   A solid mass or nodule of inflammatory tissue containing modified macrophages and lymphocytes. Usually a chronic pathologic process of diseases such as tuberculosis or syphilis.

granzymes   Enzymes secreted by cytotoxic T cells that damage proteins of target cells.

Graves' disease   A malfunction of the thyroid gland in which autoantibodies directed at thyroid cells stimulate an overproduction of thyroid hormone (hyperthyroidism).

group translocation   A form of active transport in which the substance being transported is altered during transfer across a plasma membrane.

growth curve   A graphical representation of the change in population size over time. This graph has four periods known as lag phase, exponential or log phase, stationary phase, and death phase.

growth factor   An organic compound such as a vitamin or amino acid that must be provided in the diet to facilitate growth. An essential nutrient.

guanine (G)   One of the nitrogen bases found in DNA and RNA in the purine form.

Guillain-Barré syndrome   A neurological complication of infection or vaccination.

gumma   A nodular, infectious granuloma characteristic of tertiary syphilis.

gut-associated lymphoid tissue (GALT)   A collection of lymphoid tissue in the gastrointestinal tract that includes the appendix, the lacteals, and Peyer's patches.

gyrase   The enzyme responsible for supercoiling DNA into tight bundles; a type of topoisomerase.

H

habitat   The environment to which an organism is adapted.

halogens   A group of related chemicals with antimicrobial applications. The halogens most often used in disinfectants and antiseptics are chlorine and iodine.

halophile   A microbe whose growth is either stimulated by salt or requires a high concentration of salt for growth.

hapten   An incomplete or partial antigen. Although it constitutes the determinative group and can bind antigen, hapten cannot stimulate a full immune response without being carried by a larger protein molecule.

hay fever   A form of atopic allergy marked by seasonal acute inflammation of the conjunctiva and mucous membranes of the respiratory passages. Symptoms are irritative itching and rhinitis.

healthcare-associated infection   Formerly referred to as "nosocomial infection," any infection acquired as a direct result of a patient's presence in a hospital or health care setting.

helical   Having a spiral or coiled shape. Said of certain virus capsids and bacteria.

helminth   A term that designates all parasitic worms.

helper T cell   A class of thymus-stimulated lymphocytes that facilitate various immune activities such as assisting B cells and macrophages. Also called a *T helper cell.*

hemagglutinin   A molecule that causes red blood cells to clump or agglutinate. Often found on the surfaces of viruses.

hematopoiesis   The process by which the various types of blood cells are formed, such as in the bone marrow.

hemoglobin   A protein in red blood cells that carries iron.

hemolysin   Any biological agent that is capable of destroying red blood cells and causing the release of hemoglobin. Many bacterial pathogens produce exotoxins that act as hemolysins.

hemolytic disease   Incompatible Rh factor between mother and fetus causes maternal antibodies to attack the fetus and trigger complement-mediated lysis in the fetus.

hemolytic uremic syndrome (HUS)   Severe hemolytic anemia leading to kidney damage or failure; can accompany *E. coli* O157:H7 intestinal infection.

hemolyze   When red blood cells burst and release hemoglobin pigment.

hepatitis   Inflammation and necrosis of the liver, often the result of viral infection.

hepatitis A virus (HAV)   Enterovirus spread by contaminated food responsible for short-term (infectious) hepatitis.

hepatitis B virus (HBV)   DNA virus that is the causative agent of serum hepatitis.

hepatitis D   The delta agent; a defective RNA virus that cannot reproduce on its own unless a cell is also infected with the hepatitis B virus.

hepatocellular carcinoma   A liver cancer associated with infection with hepatitis B virus.

herd immunity   The status of collective acquired immunity in a population that reduces the likelihood that nonimmune individuals will contract and spread infection. One aim of vaccination is to induce herd immunity.

heredity   Genetic inheritance.

hermaphroditic   Containing the sex organs for both male and female in one individual.

herpes zoster   A recurrent infection caused by latent chickenpox virus. Its manifestation on the skin tends to correspond to dermatomes and to occur in patches that "girdle" the trunk. Also called *shingles.*

heterotroph   An organism that relies upon organic compounds for its carbon and energy needs.

hexose   A 6-carbon sugar such as glucose and fructose.

hierarchies   Levels of power. Arrangement in order of rank.

histamine   A cytokine released when mast cells and basophils release their granules. An important mediator of allergy, its effects include smooth muscle contraction, increased vascular permeability, and increased mucus secretion.

histiocyte   Another term for *macrophage.*

histone   Proteins associated with eukaryotic DNA. These simple proteins serve as winding spools to compact and condense the chromosomes.

holoenzyme   An enzyme complete with its apoenzyme and cofactors.

hops   The ripe, dried fruits of the hop vine (*Humulus lupulus*) that are added to beer wort for flavoring.

horizontal gene transfer   Transmission of genetic material from one cell to another through nonreproductive mechanisms, such as from one organism to another living in the same habitat.

host   Organism in which smaller organisms or viruses live, feed, and reproduce.

host range   The limitation imposed by the characteristics of the host cell on the type of virus that can successfully invade it.

human immunodeficiency virus (HIV)   A retrovirus that causes acquired immunodeficiency syndrome (AIDS).

Human Microbiome Project   A project of the National Institutes of Health to identify microbial inhabitants of the human body and their role in health and disease; uses metagenomic techniques instead of culturing.

hybridization   A process that matches complementary strands of nucleic acid (DNA-DNA, RNA-DNA, RNA-RNA). Used for locating specific sites or types of nucleic acids.

hydration   The addition of water as in the coating of ions with water molecules as ions enter into aqueous solution.

hydrogen bond   A weak chemical bond formed by the attraction of forces between molecules or atoms—in this case, hydrogen and either oxygen or nitrogen. In this type of bond, electrons are not shared, lost, or gained.

hydrologic cycle   The continual circulation of water between hydrosphere, atmosphere, and lithosphere.

hydrolysis   A process in which water is used to break bonds in molecules. Usually occurs in conjunction with an enzyme.

hydrophilic   The property of attracting water. Molecules that attract water to their surface are called *hydrophilic*.

hydrophobic   The property of repelling water. Molecules that repel water are called *hydrophobic*.

hydrosphere   That part of the biosphere that encompasses water-containing environments such as oceans, lakes, rivers.

hyperthermophile   An organism whose optimal growth temperature is above 80°C (176°F), with a temperature range from 60°C to 113°C (140°F to 235°F).

hypertonic   Having a greater osmotic pressure than a reference solution.

hyphae   The tubular threads that make up filamentous fungi (molds). This web of branched and intertwining fibers is called a *mycelium*.

hypogammaglobulinemia   An inborn disease in which the gamma globulin (antibody) fraction of serum is greatly reduced. The condition is associated with a high susceptibility to pyogenic infections.

hyposensitivity diseases   Diseases in which there is a diminished or lack of immune reaction to pathogens due to incomplete immune system development, immune suppression, or destruction of the immune system.

hyposensitization   A therapeutic exposure to known allergens designed to build tolerance and eventually prevent allergic reaction.

hypothesis   A tentative explanation of what has been observed or measured.

hypotonic   Having a lower osmotic pressure than a reference solution.

I

icosahedron   A regular geometric figure having 20 surfaces that meet to form 12 corners. Some virions have capsids that resemble icosahedral crystals.

immune complex reaction   Type III hypersensitivity of the immune system. It is characterized by the reaction of soluble antigen with antibody, and the deposition of the resulting complexes in basement membranes of epithelial tissue.

immune privilege   The restriction or reduction of immune response in certain areas of the body that reduces the potential damage to tissues that a normal inflammatory response could cause.

immune tolerance   Tolerance to self; the inability of one's immune system to react to self proteins or antigens.

immunity   An acquired resistance to an infectious agent due to prior contact with that agent.

immunocompetence   The ability of the body to recognize and react with multiple foreign substances.

immunodeficiency   Immune function is incompletely developed, suppressed, or destroyed.

immunogen   Any substance that induces a state of sensitivity or resistance after processing by the immune system of the body.

immunoglobulin (Ig)   The chemical class of proteins to which antibodies belong.

immunology   The study of the system of body defenses that protect against infection.

immunopathology   The study of disease states associated with overreactivity or underreactivity of the immune response.

in utero   Literally means "in the uterus"; pertains to events or developments occurring before birth.

in vitro   Literally means "in glass," signifying a process or reaction occurring in an artificial environment, as in a test tube or culture medium.

in vivo   Literally means "in a living being," signifying a process or reaction occurring in a living thing.

inapparent   Referring to an infection in which infectious agents have entered the body and some signs of infection are present but no disease symptoms are manifest; also described as *subclinical* or *asymptomatic*.

incidence   In epidemiology, the number of new cases of a disease occurring during a period.

inclusion   A relatively inert body in the cytoplasm such as storage granules, glycogen, fat, or some other aggregated metabolic product.

inclusion body   One of a variety of different storage compartments in bacterial cells.

incubate   To isolate a sample culture in a temperature-controlled environment to encourage growth.

incubation period   The period from the initial contact with an infectious agent to the appearance of the first symptoms.

index case   The first case of a disease identified in an outbreak or epidemic.

indicator bacteria   In water analysis, any easily cultured bacteria that may be found in the intestine and can be used as an index of fecal contamination. The category includes coliforms and enterococci. Discovery of these bacteria in a sample means that pathogens may also be present.

induced mutation   Any alteration in DNA that occurs as a consequence of exposure to chemical or physical mutagens.

inducer   A molecule in an inducible operon responsible for initiating transcription of the operon by removing the repressor from the operator section of the DNA, allowing transcription to proceed.

inducible enzyme   An enzyme that increases in amount in direct proportion to the amount of substrate present.

inducible operon   An operon that under normal circumstances is not transcribed. The presence of a specific inducer molecule can cause transcription of the operon to begin.

induction   The process whereby a bacteriophage in the prophage state is activated and begins replication and enters the lytic cycle.

induration   Area of hardened, reddened tissue associated with the tuberculin test.

infection   The entry, establishment, and multiplication of pathogenic organisms within a host.

infectious disease   The state of damage or toxicity in the body caused by an infectious agent.

inflammation   A natural, nonspecific response to tissue injury that protects the host from further damage. It stimulates immune reactivity and blocks the spread of an infectious agent.

inoculating loop   A tool used in the microbiology laboratory sometimes comprised of a platinum or nichrome wire loop attached to a heat-proof handle.

inoculation   The implantation of microorganisms into or upon culture media.

inorganic chemicals   Molecules that lack the basic framework of the elements of carbon and hydrogen.

insertion elements   The smallest transposable elements, consisting only of tandem repeats that are capable of inserting themselves into DNA but do not carry any genes.

integument   The outer surfaces of the body: skin, hair, nails, sweat glands, and oil glands.

interferon (IFN)   Natural human chemical that inhibits viral replication; used therapeutically to combat viral infections and cancer.

interleukins   A class of chemicals released from host cells that have potent effects on immunity.

intermediate filament   Proteinaceous fibers in eukaryotic cells that help provide support to the cells and their organelles.

intoxication   Poisoning that results from the introduction of a toxin into body tissues through ingestion or injection.

intron   The segments on split genes of eukaryotes that do not code for polypeptide.

They can have regulatory functions. See *exon*.

iodophor A combination of iodine and an organic carrier that is a moderate-level disinfectant and antiseptic.

ion An unattached, charged particle.

ionic bond A chemical bond in which electrons are transferred and not shared between atoms.

ionization The aqueous dissociation of an electrolyte into ions.

ionizing radiation Radiant energy consisting of short-wave electromagnetic rays (X ray) or high-speed electrons that cause dislodgment of electrons on target molecules and create ions.

irradiation The application of radiant energy for diagnosis, therapy, disinfection, or sterilization.

isograft Transplanted tissue from one monozygotic twin to the other; transplants between highly inbred animals that are genetically identical.

isolation The separation of microbial cells by serial dilution or mechanical dispersion on solid media to create discrete colonies.

isoniazid Older drug that targets the bacterial cell wall; used against *M. tuberculosis*.

isotonic Two solutions having the same osmotic pressure such that, when separated by a semipermeable membrane, there is no net movement of solvent in either direction.

isotope A version of an element that is virtually identical in all chemical properties to another version except that their atoms have slightly different atomic masses.

J

jaundice The yellowish pigmentation of skin, mucous membranes, sclera, deeper tissues, and excretions due to abnormal deposition of bile pigments. Jaundice is associated with liver infection, as with hepatitis B virus and leptospirosis.

JC virus (JCV) Causes a form of encephalitis (progressive multifocal leukoencephalopathy), especially in AIDS patients.

K

keratin Protein produced by outermost skin cells that provide protection from trauma and moisture.

killed or inactivated vaccine A whole-cell or intact virus preparation in which the microbes are dead or preserved and cannot multiply but are still capable of conferring immunity.

killer T cells A T lymphocyte programmed to directly affix cells and kill them. See *cytotoxicity*.

kingdom In the levels of classification, the second division from more general to more specific. Each domain is divided into kingdoms.

Koch's postulates A procedure to establish the specific cause of disease. In all cases of infection, (1) the agent must be found; (2) inoculations of a pure culture must reproduce the same disease in animals; (3) the agent must again be present in the experimental animal; and (4) a pure culture must again be obtained.

Koplik's spots Tiny red blisters with central white specks on the mucosal lining of the cheeks. Symptomatic of measles.

L

labile In chemistry, molecules, or compounds that are chemically unstable in the presence of environmental changes.

lactoferrin A protein in mucosal secretions, tears, and milk that contains iron molecules and has antimicrobial activity.

lactose One of the carbohydrates commonly referred to as sugars. Lactose is commonly found in milk.

lactose (*lac*) operon Control system that manages the regulation of lactose metabolism. It is composed of three DNA segments, including a regulator, a control locus, and a structural locus.

lag phase The early phase of population growth during which no signs of growth occur.

lager The maturation process of beer, which is allowed to take place in large vats at a reduced temperature.

lagging strand The newly forming 5' DNA strand that is discontinuously replicated in segments (Okazaki fragments).

latency The state of being inactive. Example: a latent virus or latent infection.

leading strand The newly forming 3' DNA strand that is replicated in a continuous fashion without segments.

leaven To lighten food material by entrapping gas generated within it. Example: the rising of bread from the CO_2 produced by yeast or baking powder.

legumes Plants that produce seeds in pods. Examples include soybeans and peas.

lesion A wound, injury, or some other pathologic change in tissues.

leukocidin A heat-labile substance formed by some pyogenic cocci that impairs and sometimes lyses leukocytes.

leukocytes White blood cells. The primary infection-fighting blood cells.

leukocytosis An abnormally large number of leukocytes in the blood, which can be indicative of acute infection.

leukopenia A lower-than-normal leukocyte count in the blood that can be indicative of blood infection or disease.

leukotriene An unsaturated fatty acid derivative of arachidonic acid. Leukotriene functions in chemotactic activity, smooth muscle contractility, mucus secretion, and capillary permeability.

library In biotechnology, a collection of DNA fragments obtained by exposing the DNA to restriction enzymes, separating the fragments through gel electrophoresis, and inserting the fragments into plasmids.

ligase An enzyme required to seal the sticky ends of DNA pieces after splicing.

light-dependent reactions The series of reactions in photosynthesis that are driven by the light energy (photons) absorbed by chlorophyll. They involve splitting of water into hydrogens and oxygen, transport of electrons by NADP, and ATP synthesis.

light-independent reactions The series of reactions in photosynthesis that can proceed with or without light. It is a cyclic system that uses ATP from the light reactions to incorporate or fix carbon dioxide into organic compounds, leading to the production of glucose and other carbohydrates (also called the *Calvin cycle*).

lipase A fat-splitting enzyme. Example: triacylglycerol lipase separates the fatty acid chains from the glycerol backbone of triglycerides.

lipid A term used to describe a variety of substances that are not soluble in polar solvents such as water but will dissolve in nonpolar solvents such as benzene and chloroform. Lipids include triglycerides, phospholipids, steroids, and waxes.

lipopolysaccharide A molecular complex of lipid and carbohydrate found in the bacterial cell wall. The lipopolysaccharide (LPS) of gram-negative bacteria is an endotoxin with generalized pathologic effects such as fever.

lipoteichoic acid Anionic polymers containing glycerol that are anchored in the cytoplasmic membranes of gram-positive bacteria.

liquid media Growth-supporting substance in fluid form.

lithoautotrophs Bacteria that rely on inorganic minerals to supply their nutritional needs. Sometimes referred to as *chemoautotrophs*.

lithosphere That part of the biosphere that encompasses the earth's crust, including rocks and minerals.

lithotroph An autotrophic microbe that derives energy from reduced inorganic compounds such as N_2S.

live attenuated vaccines Vaccines composed of living organisms that have been weakened and cannot cause disease.

localized infection Occurs when a microbe enters a specific tissue, infects it, and remains confined there.

log phase Maximum rate of cell division during which growth is geometric in its rate of increase. Also called *exponential growth phase*.

lophotrichous Describing bacteria having a tuft of flagella at one or both poles.

luciferase An enzyme involved in light production in bioluminescent organisms.

lumen The cavity within a tubular organ.

lymphadenitis Inflammation of one or more lymph nodes. Also called *lymphadenopathy*.

lymphatic system A system of vessels and organs that serve as sites for development of immune cells and immune reactions. It includes the spleen, thymus, lymph nodes, and GALT.

lymphocyte The second most common form of white blood cells.

lyophilization A method for preserving microorganisms (and other substances) by freezing and then drying them directly from the frozen state.

lyse To burst.

lysis The physical rupture or deterioration of a cell.

lysogenic conversion A bacterium acquires a new genetic trait due to the presence of genetic material from an infecting phage.

lysogeny The indefinite persistence of bacteriophage DNA in a host without bringing about the production of virions.

lysosome A cytoplasmic organelle containing lysozyme and other hydrolytic enzymes.

lysozyme An enzyme found in sweat, tears, and saliva that breaks down bacterial peptidoglycan.

M

macromolecules Large, molecular compounds assembled from smaller subunits, most notably biochemicals.

macronutrient A chemical substance required in large quantities (phosphate, for example).

macrophage A white blood cell derived from a monocyte that leaves the circulation and enters tissues. These cells are important in nonspecific phagocytosis and in regulating, stimulating, and cleaning up after immune responses.

macroscopic Visible to the naked eye.

major histocompatibility complex (MHC) A set of genes in mammals that produces molecules on surfaces of cells that differentiate among different individuals in the species.

malt The grain, usually barley, that is sprouted to obtain digestive enzymes and dried for making beer.

maltose One of the carbohydrates referred to as sugars. A fermentable sugar formed from starch.

Mantoux test An intradermal screening test for tuberculin hypersensitivity. A red, firm patch of skin at the injection site greater than 10 mm in diameter after 48 hours is a positive result that indicates current or prior exposure to the TB bacillus.

marker Any trait or factor of a cell, virus, or molecule that makes it distinct and recognizable. Example: a genetic marker.

mash In making beer, the malt grain is steeped in warm water, ground up, and fortified with carbohydrates to form mash.

mast cell A nonmotile connective tissue cell implanted along capillaries, especially in the lungs, skin, gastrointestinal tract, and genitourinary tract. Like a basophil, its granules store mediators of allergy.

matrix The dense ground substance between the cristae of a mitochondrion that serves as a site for metabolic reactions.

matter All tangible materials that occupy space and have mass.

maximum temperature The highest temperature at which an organism will grow.

MDR-TB Multidrug-resistant tuberculosis.

mechanical vector An animal that transports an infectious agent but is not infected by it, such as houseflies whose feet become contaminated with feces.

medium (plural, *media*) A nutrient used to grow organisms outside of their natural habitats.

meiosis The type of cell division necessary for producing gametes in diploid organisms. Two nuclear divisions in rapid succession produce four gametocytes, each containing a haploid number of chromosomes.

membrane In a single cell, a thin double-layered sheet composed of lipids such as phospholipids and sterols and proteins.

memory (immunologic memory) The capacity of the immune system to recognize and act against an antigen upon second and subsequent encounters.

memory cell The long-lived progeny of a sensitized lymphocyte that remains in circulation and is genetically programmed to react rapidly with its antigen.

Mendosicutes Taxonomic category of bacteria that have unusual cell walls; archaea.

meninges The tough tri-layer membrane covering the brain and spinal cord. Consists of the dura mater, arachnoid mater, and pia mater.

meningitis An inflammation of the membranes (meninges) that surround and protect the brain. It is often caused by bacteria such as *Neisseria meningitidis* (the meningococcus) and *Haemophilus influenzae*.

mesophile Microorganisms that grow at intermediate temperatures.

messenger RNA (mRNA) A single-stranded transcript that is a copy of the DNA template that corresponds to a gene.

metabolism A general term for the totality of chemical and physical processes occurring in a cell.

metabolites Small organic molecules that are intermediates in the stepwise biosynthesis or breakdown of macromolecules.

metabolomics The study of the complete complement of small chemicals present in a cell at any given time.

metachromatic Exhibiting a color other than that of the dye used to stain it.

metachromatic granules A type of inclusion in storage compartments of some bacteria that stain a contrasting color when treated with colored dyes.

metagenomics The study of all the genomes in a particular ecological niche, as opposed to individual genomes from single species.

methanogens Methane producers.

methanotrophs Certain species of bacteria that derive carbon and energy through the oxidation of methane.

MHC Major histocompatibility complex.

MIC Abbreviation for **m**inimum **i**nhibitory **c**oncentration. The lowest concentration of antibiotic needed to inhibit bacterial growth in a test system.

micelles Small droplets of oil or other hydrophobic material coated by a surfactant dispersed in water due to reduced surface tension.

micro RNA Short sequences of RNA that are capable of binding to mRNA, ultimately repressing the production of a particular protein; found in all eukaryotes, viruses, and many bacteria.

microaerophile An aerobic bacterium that requires oxygen at a concentration less than that in the atmosphere.

microarray A glass, silicon, or nylon chip that contains sequences from tens of thousands of different genes in cDNA form that fluoresce when complementary DNA binds to them, indicating what mRNA molecules are present in a cell under varying conditions.

microbe See *microorganism*.

microbial antagonism Relationship in which microorganisms compete for survival in a common environment by taking actions that inhibit or destroy another organism.

microbial ecology The study of microbes in their natural habitats.

microbicides Chemicals that kill microorganisms.

microbiology A specialized area of biology that deals with living things ordinarily too small to be seen without magnification, including bacteria, archaea, fungi, protozoa, and viruses.

microbistatic The quality of inhibiting the growth of microbes.

micronutrient A chemical substance required in small quantities (trace metals, for example).

microorganism A living thing ordinarily too small to be seen without magnification; an organism of microscopic size.

microscopic Invisible to the naked eye.

microscopy Science that studies structure, magnification, lenses, and techniques related to use of a microscope.

microtubules Long hollow tubes in eukaryotic cells; maintain the shape of the cell and transport substances from one part of cell to another; involved in separating chromosomes in mitosis.

mineralization The process by which decomposers (bacteria and fungi) convert organic debris into inorganic and elemental form. It is part of the recycling process.

minimum inhibitory concentration (MIC) The smallest concentration of drug needed to visibly control microbial growth.

minimum temperature The lowest temperature at which an organism will grow.

miracidium The ciliated first-stage larva of a trematode. This form is infective for a corresponding intermediate host snail.

missense mutation A mutation in which a change in the DNA sequence results in a different amino acid being incorporated into a protein, with varying results.

mitochondrion A double-membrane organelle of eukaryotes that is the main site for aerobic respiration.

mitosis Somatic cell division that preserves the somatic chromosome number.

mixed acid fermentation An anaerobic degradation of pyruvic acid that results in more than one organic acid being produced (e.g., acetic acid, lactic acid, succinic acid).

mixed culture A container growing two or more different, known species of microbes.

mixed infection Occurs when several different pathogens interact simultaneously to produce an infection. Also called a *synergistic infection.*

molecule A distinct chemical substance that results from the combination of two or more atoms.

molluscum contagiosum Poxvirus-caused disease that manifests itself by the appearance of small lesions on the face, trunk, and limbs. Can be associated with sexual transmission.

monocyte A large mononuclear leukocyte normally found in the lymph nodes, spleen, bone marrow, and loose connective tissue. This type of cell makes up 3% to 7% of circulating leukocytes.

monomer A simple molecule that can be linked by chemical bonds to form larger molecules.

mononuclear phagocyte system A collection of monocytes and macrophages scattered throughout the extracellular spaces that function to engulf and degrade foreign molecules.

monosaccharide A simple sugar such as glucose that is a basic building block for more complex carbohydrates.

monotrichous Describing a microorganism that bears a single flagellum.

morbidity A diseased condition.

morbidity rate The number of persons afflicted with an illness under question or with illness in general, expressed as a numerator, with the denominator being some unit of population (as in $x/100,000$).

mordant A chemical that fixes a dye in or on cells by forming an insoluble compound and thereby promoting retention of that dye. Example: Gram's iodine in the Gram stain.

morphology The study of organismic structure.

mortality rate The number of persons who have died as the result of a particular cause or due to all causes, expressed as a numerator, with the denominator being some unit of population (as in $x/100,000$).

most probable number (MPN) Test used to detect the concentration of contaminants in water and other fluids.

motility Self-propulsion.

must Juices expressed from crushed fruits that are used in fermentation for wine.

mutagen Any agent that induces genetic mutation. Examples: certain chemical substances, ultraviolet light, radioactivity.

mutant strain A subspecies of microorganism that has undergone a mutation, causing expression of a trait that differs from other members of that species.

mutation A permanent inheritable alteration in the DNA sequence or content of a cell.

mutualism Organisms living in an obligatory but mutually beneficial relationship.

mycelium The filamentous mass that makes up a mold. Composed of hyphae.

mycolic acid A thick, waxy, long-chain fatty acid found in the cell wall of *Mycobacterium* and *Nocardia* that confers resistance to chemicals and dyes.

Mycoplasma A genus of bacteria; contain no peptidoglycan/cell wall, but the cytoplasmic membrane is stabilized by sterols.

mycorrhizae Various species of fungi adapted in an intimate, mutualistic relationship to plant roots.

mycosis Any disease caused by a fungus.

myonecrosis Death of muscle tissue.

N

NAD/NADH Abbreviations for the oxidized/reduced forms of nicotinamide adenine dinucleotide, an electron carrier. Also known as the vitamin niacin.

narrow-spectrum Denotes drugs that are selective and limited in their effects. For example, they inhibit either gram-negative or gram-positive bacteria but not both.

natural immunity Any immunity that arises naturally in an organism via previous experience with the antigen.

natural killer (NK) cells Cells that are derived directly from lymphoid stem cells that do not have specific antigen receptors and directly attack and kill virus-infected and cancer cells.

natural selection A process in which the environment places pressure on organisms to adapt and survive changing conditions. Only the survivors will be around to continue the life cycle and contribute their genes to future generations. This is considered a major factor in evolution of species.

necrosis A pathologic process in which cells and tissues die and disintegrate.

negative stain A staining technique that renders the background opaque or colored and leaves the object unstained so that it is outlined as a colorless area.

negative-sense RNA Single-stranded viral RNA that is complementary to positive-sense RNA and must be converted into positive-sense RNA before it can be translated.

nematode A common name for helminths called *roundworms.*

nephritis Inflammation of the kidney.

neurons Cells that make up the tissues of the brain and spinal cord that receive and transmit signals to and from the peripheral nervous system and central nervous system.

neurotropic Having an affinity for the nervous system. Most likely to affect the spinal cord.

neutralization The process of combining an acid and a base until they reach a balanced proportion, with a pH value close to 7.

neutron An electrically neutral particle in the nuclei of all atoms except hydrogen.

neutrophil A mature granulocyte present in peripheral circulation, exhibiting a multilobular nucleus and numerous cytoplasmic granules that retain a neutral stain. The neutrophil is an active phagocytic cell in bacterial infection.

niche In ecology, an organism's biological role in or contribution to its community.

nitrification Phase of the nitrogen cycle in which ammonium is oxidized.

nitrogen base A ringed compound of which pyrimidines and purines are types.

nitrogen cycle The pathway followed by the element nitrogen as it circulates from inorganic sources in the nonliving environment to living things and back to the nonliving environment. The longtime reservoir is nitrogen gas in the atmosphere.

nitrogen fixation A process occurring in certain bacteria in which atmospheric N_2 gas is converted to a form (NH_4) usable by plants.

nitrogenous base A nitrogen-containing molecule found in DNA and RNA that provides the basis for the genetic code. Adenine, guanine, and cytosine are found in both DNA and RNA, while thymine is found exclusively in DNA and uracil is found exclusively in RNA.

nomenclature A set system for scientifically naming organisms, enzymes, anatomical structures, and so on.

noncommunicable An infectious disease that does not arrive through transmission of an infectious agent from host to host.

noncompetitive inhibition Form of enzyme inhibition that involves binding of a regulatory molecule to a site other than the active site.

nonionizing radiation Method of microbial control, best exemplified by ultraviolet light, that causes the formation of abnormal bonds within the DNA of microbes, increasing the rate of mutation. The primary limitation of nonionizing radiation is its inability to penetrate beyond the surface of an object.

nonpolar A term used to describe an electrically neutral molecule formed by covalent bonds between atoms that have the same or similar electronegativity.

nonself Molecules recognized by the immune system as containing foreign markers, indicating a need for immune response.

nonsense codon A triplet of mRNA bases that does not specify an amino acid but signals the end of a polypeptide chain.

nonsense mutation   A mutation that changes an amino-acid-producing codon into a stop codon, leading to premature termination of a protein.

normal biota (also **normal microbiota**)   The native microbial forms that an individual harbors.

nosocomial infection   An infection not present upon admission to a hospital but incurred while being treated there.

notifiable disease   A disease that once diagnosed by a doctor is required to be reported to local, state, or national health authorities in order to prevent and control the disease.

nucleocapsid   In viruses, the close physical combination of the nucleic acid with its protective covering.

nucleoid   The basophilic nuclear region or nuclear body that contains the bacterial chromosome.

nucleolus   A granular mass containing RNA that is contained within the nucleus of a eukaryotic cell.

nucleotide   The basic structural unit of DNA and RNA; each nucleotide consists of a phosphate, a sugar (ribose in RNA, deoxyribose in DNA), and a nitrogenous base such as adenine, guanine, cytosine, thymine (DNA only), or uracil (RNA only).

null cells   Lymphocytes derived from lymphoid stem cells; primarily natural killer (NK) cells that can act together with other parts of the immune response or independently.

nutrient   Any chemical substance that must be provided to a cell for normal metabolism and growth. Macronutrients are required in large amounts, and micronutrients in small amounts.

nutrition   The acquisition of chemical substances by a cell or organism for use as an energy source or as building blocks of cellular structures.

O

obligate   Without alternative; restricted to a particular characteristic. Example: an obligate parasite survives and grows only in a host; an obligate aerobe must have oxygen to grow; an obligate anaerobe is destroyed by oxygen.

Okazaki fragment   In replication of DNA, a segment formed on the lagging strand in which biosynthesis is conducted in a discontinuous manner dictated by the $5' \rightarrow 3'$ DNA polymerase orientation.

oligodynamic action   A chemical having antimicrobial activity in minuscule amounts. Example: Certain heavy metals are effective in a few parts per billion.

oligonucleotides   Short pieces of DNA or RNA that are easier to handle than long segments.

oligotrophic   Nutrient-deficient ecosystem.

oncogene   A naturally occurring type of gene that when activated can transform a normal cell into a cancer cell.

oncovirus   Mammalian virus capable of causing malignant tumors.

operator   In an operon sequence, the DNA segment where transcription of structural genes is initiated.

operon   A genetic operational unit that regulates metabolism by controlling mRNA production. In sequence, the unit consists of a regulatory gene, inducer or repressor control sites, and structural genes.

opportunistic   In infection, ordinarily nonpathogenic or weakly pathogenic microbes that cause disease primarily in an immunologically compromised host.

opsonization   The process of stimulating phagocytosis by affixing molecules (opsonins such as antibodies and complement) to the surfaces of foreign cells or particles.

optimum temperature   The temperature at which a species shows the most rapid growth rate.

orbitals   The pathways of electrons as they rotate around the nucleus of an atom.

order   In the levels of classification, the division of organisms that follows class. Increasing similarity may be noticed among organisms assigned to the same order.

organelle   A small component of eukaryotic cells that is bounded by a membrane and specialized in function.

organic chemicals   Molecules that contain the basic framework of the elements carbon and hydrogen.

osmophile   A microorganism that thrives in a medium having high osmotic pressure.

osmosis   The diffusion of water across a selectively permeable membrane in the direction of lower water concentration.

outer membrane   An additional membrane possessed by gram-negative bacteria; a lipid bilayer containing specialized proteins and polysaccharides. It lies outside of the cell wall.

oxidation   In chemical reactions, the loss of electrons by one reactant.

oxidation-reduction   Redox reactions, in which paired sets of molecules participate in electron transfers.

oxidative phosphorylation   The synthesis of ATP using energy given off during the electron transport phase of respiration.

oxygenic   Any reaction that gives off oxygen; usually in reference to the result of photosynthesis in eukaryotes and cyanobacteria.

P

palindrome   A word, verse, number, or sentence that reads the same forward or backward. Palindromes of nitrogen bases in DNA have genetic significance as transposable elements, as regulatory protein targets, and in DNA splicing.

palisades   The characteristic arrangement of *Corynebacterium* cells resembling a row of fence posts and created by snapping.

PAMPs   Pathogen-associated molecular patterns. Chemical signatures present on many different microorganisms but not on host that are recognized by host as foreign.

pandemic   A disease afflicting an increased proportion of the population over a wide geographic area (often worldwide).

papilloma   Benign, squamous epithelial growth commonly referred to as a *wart*.

parasite   An organism that lives on or within another organism (the host), from which it obtains nutrients and enjoys protection. The parasite produces some degree of harm in the host.

parasitic   The relationship between a parasite and its host in which the parasite lives on or within the host and damages the host in some way; characteristic of an organism considered to be a parasite.

parasitism   A relationship between two organisms in which the host is harmed in some way, while the colonizer benefits.

parenteral   Administering a substance into a body compartment other than through the gastrointestinal tract, such as via intravenous, subcutaneous, intramuscular, or intramedullary injection.

paroxysmal   Events characterized by sharp spasms or convulsions; sudden onset of a symptom such as fever and chills.

passive carrier   Persons who mechanically transfer a pathogen without ever being infected by it, for example, a health care worker who doesn't wash his or her hands adequately between patients.

passive immunity   Specific resistance that is acquired indirectly by donation of preformed immune substances (antibodies) produced in the body of another individual.

passive transport   Nutrient transport method that follows basic physical laws and does not require direct energy input from the cell.

pasteurization   Heat treatment of perishable fluids such as milk, fruit juices, or wine to destroy heat-sensitive vegetative cells, followed by rapid chilling to inhibit growth of survivors and germination of spores. It prevents infection and spoilage.

pathogen   Any agent (usually a virus, bacterium, fungus, protozoan, or helminth) that causes disease.

pathogen-associated molecular patterns (PAMPs)   Molecules on the surfaces of many types of microbes that are not present on host cells that mark the microbes as foreign.

pathogenicity   The capacity of microbes to cause disease.

pathogenicity islands   Areas of the genome containing multiple genes that contribute to a new trait for the organism that increases its ability to cause disease.

pathognomic   Distinctive and particular to a single disease; suggestive of a diagnosis.

pathologic   Capable of inducing physical damage on the host.

pathology   The structural and physiological effects of disease on the body.

pattern recognition receptors (PRRs) Molecules on the surface of host defense cells that recognize pathogen-associated molecular patterns on microbes.

pellicle A membranous cover; a thin skin, film, or scum on a liquid surface; a thin film of salivary glycoproteins that forms over newly cleaned tooth enamel when exposed to saliva.

pelvic inflammatory disease (PID) An infection of the uterus and fallopian tubes that has ascended from the lower reproductive tract. Caused by gonococci and chlamydias.

penetration (viral) The step in viral multiplication in which virus enters the host cell.

penicillinase An enzyme that hydrolyzes penicillin; found in penicillin-resistant strains of bacteria.

penicillins A large group of naturally occurring and synthetic antibiotics produced by *Penicillium* mold and active against the cell wall of bacteria.

pentose A monosaccharide with five carbon atoms per molecule. Examples: arabinose, ribose, xylose.

peptide Molecule composed of short chains of amino acids, such as a dipeptide (two amino acids), a tripeptide (three), and a tetrapeptide (four).

peptide bond The covalent union between two amino acids that forms between the amine group of one and the carboxyl group of the other. The basic bond of proteins.

peptidoglycan A network of polysaccharide chains cross-linked by short peptides that forms the rigid part of bacterial cell walls. Gram-negative bacteria have a smaller amount of this rigid structure than do gram-positive bacteria.

perforin Proteins released by cytotoxic T cells that produce pores in target cells.

perinatal In childbirth, occurring before, during, or after delivery.

period of invasion The period during a clinical infection when the infectious agent multiplies at high levels, exhibits its greatest toxicity, and becomes well established in the target tissues.

periplasmic space The region between the cell wall and cell membrane of the cell envelopes of gram-negative bacteria.

peritrichous In bacterial morphology, having flagella distributed over the entire cell.

petechiae Minute hemorrhagic spots in the skin that range from pinpoint- to pinhead-size.

Peyer's patches Oblong lymphoid aggregates of the gut located chiefly in the wall of the terminal and small intestine. Along with the tonsils and appendix, Peyer's patches make up the gut-associated lymphoid tissue that responds to local invasion by infectious agents.

pH The symbol for the negative logarithm of the H ion concentration; p (power) or $[H^+]_{10}$. A system for rating acidity and alkalinity.

phage A bacteriophage; a virus that specifically parasitizes bacteria.

phagocyte A class of white blood cells capable of engulfing other cells and particles.

phagocytosis A type of endocytosis in which the cell membrane actively engulfs large particles or cells into vesicles.

phagolysosome A body formed in a phagocyte, consisting of a union between a vesicle containing the ingested particle (the phagosome) and a vacuole of hydrolytic enzymes (the lysosome).

phagosome A vacuole formed within a phagocytic cell when it extends its pseudopods to enclose a cell or particle.

phase variation The process of bacteria turning on or off a group of genes that changes its phenotype in a heritable manner.

phenotype The observable characteristics of an organism produced by the interaction between its genetic potential (genotype) and the environment.

phosphate An acidic salt containing phosphorus and oxygen that is an essential inorganic component of DNA, RNA, and ATP.

phospholipid A class of lipids that compose a major structural component of cell membranes.

phosphorylation Process in which inorganic phosphate is added to a compound.

photoactivation (light repair) A mechanism for repairing DNA with ultraviolet-light-induced mutations using an enzyme (photolyase) that is activated by visible light.

photoautotroph An organism that utilizes light for its energy and carbon dioxide chiefly for its carbon needs.

photolysis The splitting of water into hydrogen and oxygen during photosynthesis.

photon A subatomic particle released by electromagnetic sources such as radiant energy (sunlight). Photons are the ultimate source of energy for photosynthesis.

photophosphorylation The process of electron transport during photosynthesis that results in the synthesis of ATP from ADP.

photosynthesis A process occurring in plants, algae, and some bacteria that traps the sun's energy and converts it to ATP in the cell. This energy is used to fix CO_2 into organic compounds.

phototaxis The movement of organisms in response to light.

phototrophs Microbes that use photosynthesis to feed.

phycobilin Red or blue-green pigments that absorb light during photosynthesis.

phylum In the levels of classification, the third level of classification from general to more specific. Each kingdom is divided into numerous phyla. Sometimes referred to as a *division*.

pili (singular, *pilus*) Long, tubular structures made of pilin protein produced by gram-negative bacteria and used for conjugation.

pinocytosis The engulfment, or endocytosis, of liquids by extensions of the cell membrane.

plankton Minute animals (zooplankton) or plants (phytoplankton) that float and drift in the limnetic zone of bodies of water.

plantar warts Deep, painful warts on the soles of the feet as a result of infection by human papillomavirus.

plaque In virus propagation methods, the clear zone of lysed cells in tissue culture or chick embryo membrane that corresponds to the area containing viruses. In dental application, the filamentous mass of microbes that adheres tenaciously to the tooth and predisposes to caries, calculus, or inflammation.

plasma The carrier fluid element of blood.

plasma cell A progeny of an activated B cell that actively produces and secretes antibodies.

plasmids Extrachromosomal genetic units characterized by several features. A plasmid is a double-stranded DNA that is smaller than and replicates independently of the cell chromosome; it bears genes that are not essential for cell growth; it can bear genes that code for adaptive traits; and it is transmissible to other bacteria.

platelets Formed elements in the blood that develop when megakaryocytes disintegrate. Platelets are involved in hemostasis and blood clotting.

pleomorphism Normal variability of cell shapes in a single species.

pluripotential Stem cells having the developmental plasticity to give rise to more than one type. Example: undifferentiated blood cells in the bone marrow.

pneumococcus Common name for *Streptococcus pneumoniae,* the major cause of bacterial pneumonia.

pneumonia An inflammation of the lung leading to accumulation of fluid and respiratory compromise.

pneumonic plague The acute, frequently fatal form of pneumonia caused by *Yersinia pestis.*

point mutation A change that involves the loss, substitution, or addition of one or a few nucleotides.

point-source epidemic An outbreak of disease in which all affected individuals were exposed to a single source of the pathogen at a single point in time.

polar Term to describe a molecule with an asymmetrical distribution of charges. Such a molecule has a negative pole and a positive pole.

poliomyelitis An acute enteroviral infection of the spinal cord that can cause neuromuscular paralysis.

polymer A macromolecule made up of a chain of repeating units. Examples: starch, protein, DNA.

polymerase An enzyme that produces polymers through catalyzing bond formation between building blocks (polymerization).

polymerase chain reaction (PCR)   A technique that amplifies segments of DNA for testing. Using denaturation, primers, and heat-resistant DNA polymerase, the number can be increased several-million-fold.

polymicrobial   Involving multiple distinct microorganisms.

polymorphonuclear leukocytes (PMNLs)   White blood cells with variously shaped nuclei. Although this term commonly denotes all granulocytes, it is used especially for the neutrophils.

polymyxin   A mixture of antibiotic polypeptides from *Bacillus polymyxa* that are particularly effective against gram-negative bacteria.

polypeptide   A relatively large chain of amino acids linked by peptide bonds.

polyribosomal complex   An assembly line for mass production of proteins composed of a chain of ribosomes involved in mRNA transcription.

polysaccharide   A carbohydrate that can be hydrolyzed into a number of monosaccharides. Examples: cellulose, starch, glycogen.

population   A group of organisms of the same species living simultaneously in the same habitat. A group of different populations living together constitutes the community level.

porin   Transmembrane protein of the outer membrane of gram-negative cells that permits transport of small molecules into the periplasmic space but bars the penetration of larger molecules.

portal of entry   Route of entry for an infectious agent; typically a cutaneous or membranous route.

portal of exit   Route through which a pathogen departs from the host organism.

positive stain   A method for coloring microbial specimens that involves a chemical that sticks to the specimen to give it color.

positive-sense RNA   Single-stranded viral RNA that has the same sequence as host cell mRNA and can be directly translated into viral proteins.

posttranslational   Referring to modifications to the protein structure that occur after protein synthesis is complete, including removal of formyl methionine, further folding of the protein, addition of functional groups, or addition of the protein to a quaternary structure.

prebiotics   Nutrients used to stimulate the growth of favorable biota in the intestine.

precipitation   An immune testing reaction in which the antigen to be examined is a soluble molecule that is made insoluble and visible to the naked eye by the addition of an antibody.

prevalence   The total number of cases of a disease in a certain area and time period.

primary infection   An initial infection in a previously healthy individual that is later complicated by an additional (secondary) infection.

primary response   The first response of the immune system when exposed to an antigen.

primary structure   Initial protein organization described by type, number, and order of amino acids in the chain. The primary structure varies extensively from protein to protein.

primers   Synthetic oligonucleotides of known sequence that serve as landmarks to indicate where DNA amplification will begin.

prion   A concocted word to denote "proteinaceous infectious agent"; a cytopathic protein associated with the slow-virus spongiform encephalopathies of humans and animals.

probes   Small fragments of single-stranded DNA (RNA) that are known to be complementary to the specific sequence of DNA being studied.

probiotics   Preparations of live microbes used as a preventive or therapeutic measure to displace or compete with potential pathogens.

prodromal stage   A short period of mild symptoms occurring at the end of the period of incubation. It indicates the onset of disease.

producer   An organism that synthesizes complex organic compounds from simple inorganic molecules. Examples would be photosynthetic microbes and plants. These organisms are solely responsible for originating food pyramids and are the basis for life on earth (also called *autotroph*).

product(s)   In a chemical reaction, the substance(s) that is(are) left after a reaction is completed.

proglottid   The egg-generating segment of a tapeworm that contains both male and female organs.

progressive multifocal leukoencephalopathy (PML)   An uncommon, fatal complication of infection with JC virus (polyomavirus).

prokaryote   A single-celled organism that does not have special structures such as a nucleus or membrane-bound organelles; includes bacteria and archaea.

promoter   Part of an operon sequence. The DNA segment that is recognized by RNA polymerase as the starting site for transcription.

promoter region   The site composed of a short signaling DNA sequence that RNA polymerase recognizes and binds to commence transcription.

propagated epidemic   An outbreak of disease in which the causative agent is passed from affected persons to new persons over the course of time.

prophage   A lysogenized bacteriophage; a phage that is latently incorporated into the host chromosome instead of undergoing viral replication and lysis.

prophylactic   Any device, method, or substance used to prevent disease.

prostaglandin   A hormonelike substance that regulates many body functions. Prostaglandin comes from a family of organic acids containing 5-carbon rings that are essential to the human diet.

protease   Enzymes that act on proteins, breaking them down into component parts.

protease inhibitors   Drugs that act to prevent the assembly of functioning viral particles.

protein   Predominant organic molecule in cells, formed by long chains of amino acids.

proteomics   The study of an organism's complement of proteins (its *proteome*) and functions mediated by the proteins.

proton   An elementary particle that carries a positive charge. It is identical to the nucleus of the hydrogen atom.

protoplast   A bacterial cell whose cell wall is completely lacking and that is vulnerable to osmotic lysis.

protozoa   A group of single-celled, eukaryotic organisms.

provirus   The genome of a virus when it is integrated into a host cell's DNA.

PRRs   Pattern recognition receptors. Molecules on the surface of host cells that recognize pathogen-associated molecular patterns (PAMPs) on microbial cells.

pseudohypha   A chain of easily separated, spherical to sausage-shaped yeast cells partitioned by constrictions rather than by septa.

pseudomembrane   A tenacious, noncellular mucous exudate containing cellular debris that tightly blankets the mucosal surface in infections such as diphtheria and pseudomembranous enterocolitis.

pseudopodium   A temporary extension of the protoplasm of an amoeboid cell. It serves both in amoeboid motion and for food gathering (phagocytosis).

pseudopods   Protozoan appendage responsible for motility. Also called "false feet."

psychrophile   A microorganism that thrives at low temperature (0°C–20°C), with a temperature optimum of 0°C–15°C.

pure culture   A container growing a single species of microbe whose identity is known.

purine   A nitrogen base that is an important encoding component of DNA and RNA. The two most common purines are adenine and guanine.

pus   The viscous, opaque, usually yellowish matter formed by an inflammatory infection. It consists of serum exudate, tissue debris, leukocytes, and microorganisms.

pyogenic   Pertains to pus formers, especially the pyogenic cocci: pneumococci, streptococci, staphylococci, and neisseriae.

pyrimidine   Nitrogen bases that help form the genetic code on DNA and RNA. Uracil, thymine, and cytosine are the most important pyrimidines.

pyrimidine dimer   The union of two adjacent pyrimidines on the same DNA strand, brought about by exposure to ultraviolet light. It is a form of mutation.

pyrogen   A substance that causes a rise in body temperature. It can come from pyrogenic microorganisms or from polymorphonuclear leukocytes (endogenous pyrogens).

Q

quaternary structure   Most complex protein structure characterized by the formation of large, multiunit proteins by more than one of the polypeptides. This structure is typical of antibodies and some enzymes that act in cell synthesis.

quats   A word that pertains to a family of surfactants called *quaternary ammonium compounds*. These detergents are only weakly microbicidal and are used as sanitizers and preservatives.

quinine   A substance derived from cinchona trees that was used as an antimalarial treatment; has been replaced by synthetic derivatives.

quorum sensing   The ability of bacteria to regulate their gene expression in response to sensing bacterial density.

R

rabies   The only rhabdovirus that infects humans. Zoonotic disease characterized by fatal meningoencephalitis.

radiation   Electromagnetic waves or rays, such as those of light given off from an energy source.

rales   Sounds in the lung, ranging from clicking to rattling; indicate respiratory illness.

random amplified polymorphic DNA (RAPD)   A type of PCR that utilizes primers with random sequences in order to identify microbial populations that are relatively unknown.

reactants   Molecules entering or starting a chemical reaction.

real image   An image formed at the focal plane of a convex lens. In the compound light microscope, it is the image created by the objective lens.

receptor   Cell surface molecules involved in recognition, binding, and intracellular signaling.

recombinant   An organism that contains genes that originated in another organism, whether through deliberate laboratory manipulation or natural processes.

recombinant DNA technology   A technology, also known as genetic engineering, that deliberately modifies the genetic structure of an organism to create novel products, microbes, animals, plants, and viruses.

recombination   A type of genetic transfer in which DNA from one organism is donated to another.

recycling   A process that converts unusable organic matter from dead organisms back into their essential inorganic elements and returns them to their nonliving reservoirs

to make them available again for living organisms. This is a common term that means the same as mineralization and decomposition.

redox   Denoting an oxidation-reduction reaction.

reducing medium   A growth medium that absorbs oxygen and allows anaerobic bacteria to grow.

redundancy   The property of the genetic code that allows an amino acid to be specified by several different codons.

refraction   In optics, the bending of light as it passes from one medium to another with a different index of refraction.

refractive index   The measurement of the degree of light that is bent, or refracted, as it passes between two substances such as air, water, or glass.

regulated enzymes   Enzymes whose extent of transcription or translation is influenced by changes in the environment.

regulator   DNA segment that codes for a protein capable of repressing an operon.

regulatory B cells (B_{reg} cells)   A type of activated B cell that controls the immune response.

release   The final step in the multiplication cycle of viruses in which the assembled viral particle exits the host cell and moves on to infect another cell.

rennin   The enzyme casein coagulase, which is used to produce curd in the processing of milk and cheese.

replication   In DNA synthesis, the semiconservative mechanisms that ensure precise duplication of the parent DNA strands.

replication fork   The Y-shaped point on a replicating DNA molecule where the DNA polymerase is synthesizing new strands of DNA.

reportable disease   Those diseases that must be reported to health authorities by law.

repressible operon   An operon that under normal circumstances is transcribed. The buildup of the operon's amino acid product causes transcription of the operon to stop.

repressor   The protein product of a repressor gene that combines with the operator and arrests the transcription and translation of structural genes.

reservoir   In disease communication, the natural host or habitat of a pathogen.

resident biota   The deeper, more stable microbiota that inhabit the skin and exposed mucous membranes, as opposed to the superficial, variable, transient population.

resistance (R) factor   Plasmids, typically shared among bacteria by conjugation, that provide resistance to the effects of antibiotics.

resolving power   The capacity of a microscope lens system to accurately distinguish between two separate entities that lie close to each other. Also called *resolution*.

respiratory chain   A series of enzymes that transfer electrons from one to another,

resulting in the formation of ATP. It is also known as the electron transport chain. The chain is located in the cell membrane of bacteria and in the inner mitochondrial membrane of eukaryotes.

restriction endonuclease   An enzyme present naturally in cells that cleaves specific locations on DNA. It is an important means of inactivating viral genomes, and it is also used to splice genes in genetic engineering.

restriction fragment length polymorphisms (RFLPs)   Variations in the lengths of DNA fragments produced when a specific restriction endonuclease acts on different DNA sequences.

restriction fragments   Short pieces of DNA produced when DNA is exposed to restriction endonucleases.

reticuloendothelial system   Also known as the *mononuclear phagocyte system*, it pertains to a network of fibers and phagocytic cells (macrophages) that permeates the tissues of all organs. Examples: Kupffer cells in liver sinusoids, alveolar phagocytes in the lung, microglia in nervous tissue.

retrotransposon   A transmissible element capable of translating itself from DNA to RNA to make many copies of itself and then translating itself back into DNA in order to insert itself into a new location on the chromosome.

retrovirus   A group of RNA viruses (including HIV) that have the mechanisms for converting their genome into a double strand of DNA that can be inserted on a host's chromosome.

reverse transcriptase (RT)   The enzyme possessed by retroviruses that carries out the reversion of RNA to DNA—a form of reverse transcription.

Rh factor   An isoantigen that can trigger hemolytic disease in newborns due to incompatibility between maternal and infant blood factors.

rhizobia   Bacteria that live in plant roots and supply supplemental nitrogen that boosts plant growth.

rhizosphere   The zone of soil, complete with microbial inhabitants, in the immediate vicinity of plant roots.

ribonucleic acid (RNA)   The nucleic acid responsible for carrying out the hereditary program transmitted by an organism's DNA.

ribose   A 5-carbon monosaccharide found in RNA.

ribosomal RNA (rRNA)   A single-stranded transcript that is a copy of part of the DNA template.

ribosome   A bilobed macromolecular complex of ribonucleoprotein that coordinates the codons of mRNA with tRNA anticodons and, in so doing, constitutes the peptide assembly site.

ribozyme   A part of an RNA-containing enzyme in eukaryotes that removes intervening sequences of RNA called *introns* and splices together the true coding sequences (exons) to form a mature messenger RNA.

RNA polymerase   Enzyme process that translates the code of DNA to RNA.

root nodules   Small growths on the roots of legume plants that arise from a symbiotic association between the plant tissues and bacteria (rhizobia). This association allows fixation of nitrogen gas from the air into a usable nitrogen source for the plant.

rough endoplasmic reticulum (RER)
Microscopic series of tunnels that originates in the outer membrane of the nuclear envelope and is used in transport and storage. Large numbers of ribosomes, partly attached to the membrane, give the rough appearance.

rubeola (red measles)   Acute disease caused by infection with Morbillivirus.

S

S layer   Single layer of thousands of copies of a single type of protein linked together on the surface of a bacterial cell that is produced when the cell is in a hostile environment.

saccharide   Scientific term for *sugar*. Refers to a simple carbohydrate with a sweet taste.

salpingitis   Inflammation of the fallopian tubes.

sanitize   To clean inanimate objects using soap and degerming agents so that they are safe and free of high levels of microorganisms.

saprobe   A microbe that decomposes organic remains from dead organisms. Also known as a *saprophyte* or *saprotroph*.

sarcina   A cubical packet of 8, 16, or more cells; the cellular arrangement of the genus *Sarcina* in the family *Micrococcaceae*.

satellitism   A commensal interaction between two microbes in which one can grow in the vicinity of the other due to nutrients or protective factors released by that microbe.

saturation   The complete occupation of the active site of a carrier protein or enzyme by the substrate.

schistosomiasis   Infection by blood fluke, often as a result of contact with contaminated water in rivers and streams. Symptoms appear in liver, spleen, or urinary system depending on species of *Schistosoma*. Infection may be chronic.

schizogony   A process of multiple fission whereby first the nucleus divides several times, and subsequently the cytoplasm is subdivided for each new nucleus during cell division.

scientific method   Principles and procedures for the systematic pursuit of knowledge, involving the recognition and formulation of a problem, the collection of data through observation and experimentation, and the formulation and testing of a hypothesis.

scolex   The anterior end of a tapeworm characterized by hooks and/or suckers for attachment to the host.

sebaceous glands   The sebum- (oily, fatty) secreting glands of the skin.

sebum   Low pH, oil-based secretion of the sebaceous glands.

secondary infection   An infection that compounds a preexisting one.

secondary response   The rapid rise in antibody titer following a repeat exposure to an antigen that has been recognized from a previous exposure. This response is brought about by memory cells produced as a result of the primary exposure.

secondary structure   Protein structure that occurs when the functional groups on the outer surface of the molecule interact by forming hydrogen bonds. These bonds cause the amino acid chain to either twist, forming a helix, or to pleat into an accordion pattern called a β-*pleated sheet.*

secretory antibody   The immunoglobulin (IgA) that is found in secretions of mucous membranes and serves as a local immediate protection against infection.

selective media   Nutrient media designed to favor the growth of certain microbes and to inhibit undesirable competitors.

selectively permeable   Describes a property of cell membranes in which certain substances are able to pass through the membrane, while other substances cannot pass through unaided and require special carrier proteins in order to enter or exit the cell.

selectively toxic   Property of an antimicrobial agent to be highly toxic against its target microbe, while being far less toxic to other cells, particularly those of the host organism.

self   Natural markers of the body that are recognized by the immune system.

self-limited   Applies to an infection that runs its course without disease or residual effects.

semiconservative replication   In DNA replication, the synthesis of paired daughter strands, each retaining a parent strand template.

semisolid media   Nutrient media with a firmness midway between that of a broth (a liquid medium) and an ordinary solid medium; motility media.

semisynthetic   Drugs that, after being naturally produced by bacteria, fungi, or other living sources, are chemically modified in the laboratory. Compare to *synthetic.*

sepsis   The state of putrefaction; the presence of pathogenic organisms or their toxins in tissue or blood.

septic shock   Blood infection resulting in a pathologic state of low blood pressure accompanied by a reduced amount of blood circulating to vital organs. Endotoxins of all gram-negative bacteria can cause shock, but most clinical cases are due to gram-negative enteric rods.

septicemia   Systemic infection associated with microorganisms multiplying in circulating blood.

septicemic plague   A form of infection with *Yersinia pestis* occurring mainly in the bloodstream and leading to high mortality rates.

septum   A partition or cellular cross wall, as in certain fungal hyphae.

sequela   A morbid complication that follows a disease.

sequence map   A map that shows the exact order of DNA bases in an organism determined by whole-genome shotgun sequencing.

sequencing   Determining the actual order and types of bases in a segment of DNA.

serology   The branch of immunology that deals with *in vitro* diagnostic testing of serum.

serotonin   A vasoconstrictor that inhibits gastric secretion and stimulates smooth muscle.

serotyping   The subdivision of a species or subspecies into an immunologic type, based upon antigenic characteristics.

serous   Referring to serum, the clear fluid that escapes cells during the inflammatory response.

serum   The clear fluid expressed from clotted blood that contains dissolved nutrients, antibodies, and hormones but not cells or clotting factors.

serum sickness   A type of immune complex disease in which immune complexes enter circulation, are carried throughout the body, and are deposited in the blood vessels of the kidney, heart, skin, and joints. The condition may become chronic.

severe acute respiratory syndrome (SARS)
A severe respiratory disease caused by infection with a newly described coronavirus.

severe combined immunodeficiencies
A collection of syndromes occurring in newborns caused by a genetic defect that knocks out both B- and T-cell types of immunity. There are several versions of this disease, termed *SCIDs* for short.

"sex" pilus   A conjugative pilus.

sexually transmitted disease (STD)
Infection resulting from pathogens that enter the body via sexual intercourse or intimate, direct contact; also known as *sexually transmitted infection (STI).*

shiga toxin   Heat-labile exotoxin released by some *Shigella* species and by *E. coli* O157:H7; responsible for worst symptoms of these infections.

shiga-toxin-producing E. coli (STEC)   A strain of *E. coli* that produces the shiga toxin.

shingles   Lesions produced by reactivated human herpesvirus 3 (chickenpox) infection; also known as herpes zoster.

siderophores   Low-molecular-weight molecules produced by many microorganisms that can bind iron very tightly.

sign   Any abnormality uncovered upon physical diagnosis that indicates the presence of disease. A *sign* is an objective assessment of disease, as opposed to a *symptom*, which is the subjective assessment perceived by the patient.

silent mutation A mutation that, because of the degeneracy of the genetic code, results in a nucleotide change in both the DNA and mRNA but not the resultant amino acid and thus, not the protein.

simple stain Type of positive staining technique that uses a single dye to add color to cells so that they are easier to see. This technique tends to color all cells the same color.

single nucleotide polymorphism (SNP) An alteration in a gene sequence in which a single nucleotide base is altered.

slime layer A diffuse, unorganized layer of polysaccharides and/or proteins on the outside of some bacteria.

smooth endoplasmic reticulum (SER) A microscopic series of tunnels lacking ribosomes that functions in the nutrient processing function of a cell.

solute A substance that is uniformly dispersed in a dissolving medium or solvent.

solution A mixture of one or more substances (*solutes*) that cannot be separated by filtration or ordinary settling.

solvent A dissolving medium.

source The person or item from which an infection is directly acquired. See *reservoir*.

species In the levels of classification, the most specific level of organization.

specificity In immunity, the concept that some parts of the immune system only react with antigens that originally activated them.

spike A receptor on the surface of certain enveloped viruses that facilitates specific attachment to the host cell.

spirillum A type of bacterial cell with a rigid spiral shape and external flagella.

spirochete A coiled, spiral-shaped bacterium that has endoflagella and flexes as it moves.

spliceosome A molecule composed of RNA and protein that removes introns from eukaryotic mRNA before it is translated by forming a loop in the intron, cutting it from the mRNA, and joining exons together.

spontaneous generation Early belief that living things arose from vital forces present in nonliving, or decomposing, matter.

spontaneous mutation A mutation in DNA caused by random mistakes in replication and not known to be influenced by any mutagenic agent. These mutations give rise to an organism's natural, or background, rate of mutation.

sporadic Description of a disease that exhibits new cases at irregular intervals in unpredictable geographic locales.

sporangiospore A form of asexual spore in fungi; enclosed in a sac.

sporangium A fungal cell in which asexual spores are formed by multiple cell cleavage.

spore A differentiated, specialized cell form that can be used for dissemination, for survival in times of adverse conditions, and/or for reproduction. Spores are usually unicellular and may develop into gametes or vegetative organisms.

sporozoite One of many minute elongated bodies generated by multiple division of the oocyst. It is the infectious form of the malarial parasite that is harbored in the salivary gland of the mosquito and inoculated into the victim during feeding.

sporulation The process of spore formation.

start codon The nucleotide triplet AUG that codes for the first amino acid in protein sequences.

starter culture The sizable inoculation of pure bacterial, mold, or yeast sample for bulk processing, as in the preparation of fermented foods, beverages, and pharmaceuticals.

stationary growth phase Survival mode in which cells either stop growing or grow very slowly.

stem cells Pluripotent, undifferentiated cells.

sterile Completely free of all life forms, including spores and viruses.

sterilization Any process that completely removes or destroys all viable microorganisms, including viruses, from an object or habitat. Material so treated is *sterile*.

stop codon One of the codons UAA, UAG, and UGA, which have no corresponding tRNA and thus signal the end of transcription; also known as a nonsense codon.

strain In microbiology, a set of descendants cloned from a common ancestor that retain the original characteristics. Any deviation from the original is a different strain.

strict or obligate anaerobe An organism that does not use oxygen gas in metabolism and cannot survive in oxygen's presence.

stroma The matrix of the chloroplast that is the site of the dark reactions.

structural gene A gene that codes for the amino acid sequence (peptide structure) of a protein.

subacute Indicates an intermediate status between acute and chronic disease.

subacute sclerosing panencephalitis (SSPE) A complication of measles infection in which progressive neurological degeneration of the cerebral cortex invariably leads to coma and death.

subclinical A period of inapparent manifestations that occurs before symptoms and signs of disease appear.

subculture To make a second-generation culture from a well-established colony of organisms.

subcutaneous The deepest level of the skin structure.

subspecies Bacteria of the same species that have differing characteristics; also known as a bacterial *type*.

substrate The specific molecule upon which an enzyme acts.

subunit vaccine A vaccine preparation that contains only antigenic fragments such as surface receptors from the microbe. Usually in reference to virus vaccines.

sucrose One of the carbohydrates commonly referred to as sugars. Common table or cane sugar.

sulfonamide Antimicrobial drugs that interfere with the essential metabolic process of bacteria and some fungi.

superantigens Bacterial toxins that are potent stimuli for T cells and can be a factor in diseases such as toxic shock.

superficial mycosis A fungal infection located in hair, nails, and the epidermis of the skin.

superinfection An infection occurring during antimicrobial therapy that is caused by an overgrowth of drug-resistant microorganisms.

superoxide A toxic derivative of oxygen; (O_2^-).

surfactant A surface-active agent that forms a water-soluble interface. Examples: detergents, wetting agents, dispersing agents, and surface tension depressants.

symbiosis An intimate association between individuals from two species; used as a synonym for *mutualism*.

symptom The subjective evidence of infection and disease as perceived by the patient.

syncytium A multinucleated protoplasmic mass formed by consolidation of individual cells.

syndrome The collection of signs and symptoms that, taken together, paint a portrait of the disease.

synergism The coordinated or correlated action by two or more drugs or microbes that results in a heightened response or greater activity.

synthesis (viral) The step in viral multiplication in which viral genetic material and proteins are made through replication and transcription/translation.

synthetic Referring to a chemotherapeutic agent manufactured entirely through chemical processes in the laboratory that mimics the actions of antibiotics. Compare to *semisynthetic*.

syphilis A sexually transmitted bacterial disease caused by the spirochete *Treponema pallidum*.

systemic Occurring throughout the body; said of infections that invade many compartments and organs via the circulation.

systemic infection An infection spread by the blood to multiple sites in the body at some distance from the site of initial infection.

T

T lymphocyte (T cell) A white blood cell that is processed in the thymus and is involved in cell-mediated immunity.

Taq polymerase DNA polymerase from the thermophilic bacterium *Thermus aquaticus* that enables high-temperature replication of DNA required for the polymerase chain reaction.

tartar   See *calculus*.

taxa   Taxonomic categories.

taxonomy   The formal system for organizing, classifying, and naming living things.

teichoic acid   Anionic polymers containing glycerol that appear in the walls of gram-positive bacteria.

telomeres   Areas of repeating DNA sequences at the ends of a linear chromosome that protect the chromosome from being deteriorated during rounds of DNA replication.

temperate phage   A bacteriophage that enters into a less virulent state by becoming incorporated into the host genome as a prophage instead of in the vegetative or lytic form that eventually destroys the cell.

template strand   The strand in a double-stranded DNA molecule that is used as a model to synthesize a complementary strand of DNA or RNA during replication or transcription.

Tenericutes   Taxonomic category of bacteria that lack cell walls.

teratogenic   Causing abnormal fetal development.

tertiary structure   Protein structure that results from additional bonds forming between functional groups in a secondary structure, creating a three-dimensional mass.

tetanospasmin   The neurotoxin of *Clostridium tetani,* the agent of tetanus. Its chief action is directed upon the inhibitory synapses of the anterior horn motor neurons.

tetracyclines   A group of broad-spectrum antibiotics with a complex 4-ring structure.

tetrads   Groups of four.

theory   A collection of statements, propositions, or concepts that explains or accounts for a natural event.

theory of evolution   The evidence cited to explain how evolution occurs.

therapeutic index   The ratio of the toxic dose to the effective therapeutic dose that is used to assess the safety and reliability of the drug.

thermal death point   The lowest temperature that achieves sterilization in a given quantity of broth culture upon a 10-minute exposure. Examples: 55°C for *Escherichia coli,* 60°C for *Mycobacterium tuberculosis,* and 120°C for spores.

thermal death time   The least time required to kill all cells of a culture at a specified temperature.

thermocline   A temperature buffer zone in a large body of water that separates the warmer water (the epilimnion) from the colder water (the hypolimnion).

thermophile   A microorganism that thrives at a temperature of 50°C or higher.

thylakoid   Vesicles of a chloroplast formed by elaborate folding of the inner membrane to form "discs." Solar energy trapped in the thylakoids is used in photosynthesis.

thymine (T)   One of the nitrogen bases found in DNA but not in RNA. Thymine is in a pyrimidine form.

thymus   Butterfly-shaped organ near the tip of the sternum that is the site of T-cell maturation.

tincture   A medicinal substance dissolved in an alcoholic solvent.

tinea   Ringworm; a fungal infection of the hair, skin, or nails.

tinea versicolor   A condition of the skin appearing as mottled and discolored skin pigmentation as a result of infection by the yeast *Malassezia furfur.*

titer   In immunochemistry, a measure of antibody level in a patient, determined by agglutination methods.

toll-like receptors (TLRs)   A category of pattern recognition receptors that binds to pathogen-associated molecular patterns on microbes.

tonsils   A ring of lymphoid tissue in the pharynx that acts as a repository for lymphocytes.

topoisomerases   Enzymes that can add or remove DNA twists and thus regulate the degree of supercoiling.

toxemia   Condition in which a toxin (microbial or otherwise) is spread throughout the bloodstream.

toxigenicity   The tendency for a pathogen to produce toxins. It is an important factor in bacterial virulence.

toxin   A specific chemical product of microbes, plants, and some animals that is poisonous to other organisms.

toxinosis   Disease whose adverse effects are primarily due to the production and release of toxins.

toxoid   A toxin that has been rendered nontoxic but is still capable of eliciting the formation of protective antitoxin antibodies; used in vaccines.

trace elements   Micronutrients (zinc, nickel, and manganese) that occur in small amounts and are involved in enzyme function and maintenance of protein structure.

transamination   The transfer of an amino group from an amino acid to a carbohydrate fragment.

transcript   A newly transcribed RNA molecule.

transcription   mRNA synthesis; the process by which a strand of RNA is produced against a DNA template.

transcriptome   The genomic analysis of the entire complement of RNA molecules produced by a cell.

transduction   The transfer of genetic material from one bacterium to another by means of a bacteriophage vector.

transfection   The introduction of DNA into eukaryotic cells from the environment by exposing cells to chemicals or electrical currents; similar to transformation.

transferrin   A protein in the plasma fraction of blood that transports iron.

transfer RNA (tRNA)   A transcript of DNA that specializes in converting RNA language into protein language.

transformation   In microbial genetics, the transfer of genetic material contained in "naked" DNA fragments from a donor cell to a competent recipient cell.

transgenic   Referring to genetically modified organisms that contain foreign genes inserted into their genome through recombinant DNA technologies.

translation   Protein synthesis; the process of decoding the messenger RNA code into a polypeptide.

translocation   The movement of the ribosome from one codon to the next during translation of the mRNA sequence after the peptide bond has been formed between amino acids.

transmissible spongiform encephalopathies (TSEs)   Diseases caused by proteinaceous infectious particles (also known as *prions*).

transposon   A DNA segment with an insertion sequence at each end, enabling it to migrate to another plasmid, to the bacterial chromosome, or to a bacteriophage.

transport medium   Microbiological medium that is used to transport specimens.

trematode   A category of helminth; also known as flatworm or fluke.

trichinosis   Infection by the *Trichinella spiralis* parasite, usually caused by eating the meat of an infected animal. Early symptoms include fever, diarrhea, nausea, and abdominal pain that progress to intense muscle and joint pain and shortness of breath. In the final stages, heart and brain function are at risk, and death is possible.

triglyceride   A type of lipid composed of a glycerol molecule bound to three fatty acids.

triplet   See *codon*.

trophozoite   A vegetative protozoan (feeding form) as opposed to a resting (cyst) form.

true pathogen   A microbe capable of causing infection and disease in healthy persons with normal immune defenses.

tubercle   In tuberculosis, the granulomatous well-defined lung lesion that can serve as a focus for latent infection.

tuberculin   A glycerinated broth culture of *Mycobacterium tuberculosis* that is evaporated and filtered. Formerly used to treat tuberculosis, tuberculin is now used chiefly for diagnostic tests.

tuberculin reaction   A diagnostic test in which PPD, or purified protein derivative (of *M. tuberculosis*), is injected superficially under the skin and the area of reaction measured; also called the *Mantoux test.*

tubulin   Protein component of long filaments of protein arranged under the cell membrane of bacteria; contribute to cell shape and division.

turbid   Cloudy appearance of nutrient solution in a test tube due to growth of microbe population.

tyndallization   Fractional (discontinuous, intermittent) sterilization designed to destroy spores indirectly. A preparation is exposed to flowing steam for an hour, and then the mineral is allowed to incubate to permit spore germination. The resultant vegetative cells are destroyed by repeated steaming and incubation.

type   Bacteria of the same species that have differing characteristics; also known as a bacterial *subspecies.*

typhoid fever   Form of salmonellosis. It is highly contagious. Primary symptoms include fever, diarrhea, and abdominal pain. Typhoid fever can be fatal if untreated.

U

ubiquitous   Present everywhere at the same time.

ultraviolet (UV) radiation   Radiation with an effective wavelength from 240 nm to 260 nm. UV radiation induces mutations readily but has very poor penetrating power.

uncoating   The process of removal of the viral coat and release of the viral genome by its newly invaded host cell.

universal donor   In blood grouping and transfusion, a group O individual whose erythrocytes bear neither agglutinogen A nor B.

universal precautions (UPs)   Centers for Disease Control and Prevention guidelines for health care workers regarding the prevention of disease transmission when handling patients and body substances.

uracil (U)   One of the nitrogen bases in RNA but not in DNA. Uracil is in a pyrimidine form.

urinary tract infection (UTI)   Invasion and infection of the urethra and bladder by bacterial residents, most often *E. coli.*

V

vaccination   Exposing a person to the antigenic components of a microbe without its pathogenic effects for the purpose of inducing a future protective response.

vaccine   Originally used in reference to inoculation with the cowpox or vaccinia virus to protect against smallpox. In general, the term now pertains to injection of whole microbes (killed or attenuated), toxoids, or parts of microbes as a prevention or cure for disease.

vacuoles   In the cell, membrane-bounded sacs containing fluids or solid particles to be digested, excreted, or stored.

valence   The combining power of an atom based upon the number of electrons it can either take on or give up.

van der Waals forces   Weak attractive interactions between molecules of low polarity.

vancomycin   Antibiotic that targets the bacterial cell wall; used often in antibiotic-resistant infections.

variable region   The antigen binding fragment of an immunoglobulin molecule, consisting of a combination of heavy and light chains whose molecular conformation is specific for the antigen.

varicella   Informal name for virus responsible for chickenpox as well as shingles; also known as *human herpesvirus 3* (*HHV-3*).

variolation   A hazardous, outmoded process of deliberately introducing smallpox material scraped from a victim into the nonimmune subject in the hope of inducing resistance.

vasoactive   Referring to chemical mediators involved in the immune response that act on endothelial cells or the smooth muscle of blood vessels causing them to either constrict or relax.

vector   An animal that transmits infectious agents from one host to another, usually a biting or piercing arthropod like the tick, mosquito, or fly. Infectious agents can be conveyed mechanically by simple contact or biologically whereby the parasite develops in the vector. A genetic element such as a plasmid or a bacteriophage used to introduce genetic material into a cloning host during recombinant DNA experiments.

vegetative   In describing microbial developmental stages, a metabolically active feeding and dividing form, as opposed to a dormant, seemingly inert, nondividing form. Examples: a bacterial cell versus its spore; a protozoan trophozoite versus its cyst.

vehicle   An inanimate material (solid object, liquid, or air) that serves as a transmission agent for pathogens.

vesicle   A blister characterized by a thin-skinned, elevated, superficial pocket filled with serum.

viable nonculturable (VNC)   Describes microbes that cannot be cultivated in the laboratory but that maintain metabolic activity (i.e., are alive).

vibrio   A curved, rod-shaped bacterial cell.

viremia   The presence of viruses in the bloodstream.

virion   An elementary virus particle in its complete morphological and thus infectious form. A virion consists of the nucleic acid core surrounded by a capsid, which can be enclosed in an envelope.

viroid   An infectious agent that, unlike a virion, lacks a capsid and consists of a closed circular RNA molecule. Although known viroids are all plant pathogens, it is conceivable that animal versions exist.

virome   The genome sequences of the entire complement of viruses that live on or in an organism.

virtual image   In optics, an image formed by diverging light rays; in the compound light microscope, the second, magnified visual impression formed by the ocular from the real image formed by the objective.

virulence   In infection, the relative capacity of a pathogen to invade and harm host cells.

virulence factors   A microbe's structures or capabilities that allow it to establish itself in a host and cause damage.

virus   Microscopic, acellular agent composed of nucleic acid surrounded by a protein coat.

virus particle   A more specific name for a virus when it is outside of its host cells.

vitamins   A component of coenzymes critical to nutrition and the metabolic function of coenzyme complexes.

W

Western blot test   A procedure for separating and identifying antigen or antibody mixtures by two-dimensional electrophoresis in polyacrylamide gel, followed by immune labeling.

wheal   A welt; a marked, slightly red, usually itchy area of the skin that changes in size and shape as it extends to adjacent area. The reaction is triggered by cutaneous contact or intradermal injection of allergens in sensitive individuals.

whey   The residual fluid from milk coagulation that separates from the solidified curd.

white blood cells   In contrast to erythrocytes, or red blood cells, white blood cells are clear or colorless and include granulocytes (neutrophils, eosinophils, and basophils) and agranulocytes (lymphocytes and monocytes).

whole blood   A liquid connective tissue consisting of blood cells suspended in plasma.

wild type   The natural, nonmutated form of a genetic trait.

wobble   A characteristic of amino acid codons in which the third base of a codon can be altered without changing the code for the amino acid.

wort   The clear fluid derived from soaked mash that is fermented for beer.

X

XDR-TB   Extensively drug-resistant tuberculosis (worse than multidrug-resistant tuberculosis).

xenograft   The transfer of a tissue or an organ from an animal of one species to a recipient of another species.

Z

zoonosis   An infectious disease indigenous to animals that humans can acquire through direct or indirect contact with infected animals.

zooplankton   The collection of nonphotosynthetic microorganisms (protozoa, tiny animals) that float in the upper regions of aquatic habitat and together with phytoplankton comprise the plankton.

Credits

Photo Credits

Visuals Unlimited; p. 547 (conjunctivitis): ©Science VU-Bascom Palmer Institute/Visuals Unlimited.

Chapter 19

Opener: ©Jean-Marc Giboux/Getty Images; 19.3: ©Kwangshin Kim/Science Source; 19.5: ©Dr. Ken Greer/Visuals Unlimited; p. 556 (map): CDC; 19.6: ©Kathy Park Talaro/Visuals Unlimited; 19.7: ©Dr. Gary Gaugler/Visuals Unlimited; 19.8: ©Gordon Love, M.D. VA, North CA Healthcare System, Martinez, CA; 19.9a: CDC/Dr. Hardin; 19.9b: CDC/Mercy Hosp Toledo OH; Brian J. Harrington; 19.11: CDC; p. 565: ©Transcendental Graphics/Archive Photos/Getty Images; 19.14a: ©M. Abbey/Science Source; 19.14b: ©Pr. J.J. Hauw/ISM/Phototake; p. 571 (farmer): ©Echo/Getty Images RF; 19.15a: ©Tektoff RM/CNRI/SPL/Custom Medical Stock Photo; p. 572: ©Geostock/Getty Images RF; 19.16 (bat): ©PhotoLink/Getty Images RF; 19.16 (skunk): ©Stockbyte/Getty Images RF; 19.16 (raccoon): ©Corbis RF; 19.16 (tx fox): ©Ingram Publishing RF; 19.16 (az fox): ©Corbis RF; p. 574: Courtesy Sharon Ellis Pratt and Randle Ellis; 19.17b: ©Science VU-CDC/Visuals Unlimited; 19.19: CDC; 19.20: ©John D. Cunningham/Visuals Unlimited; 19.21: Courtesy Dr. T. F. Sellers, Jr; 19.22c: CDC/Dr. Thomas F. Sellers; 19.25: ©Biophoto Associates/Science Source; p. 588 (spherules): ©Gordon Love, M.D. VA, North CA Healthcare System, Martinez, CA; p. 588 (cocci): ©Kwangshin Kim/Science Source; p. 588 (rods): ©Dr. Gary Gaugler/Visuals Unlimited; p. 589: ©A.M. Siegelman/Visuals Unlimited.

Chapter 20

Opener, 20.4a: CDC; 20.4b: Thomas V. Inglesby, et al., "Plague as a Biological Weapon," *JAMA*, Vol. 283 (17), May 3, 2000, pp. 2281-2290. Used with permission; p. 597 (art): ©Corbis; 20.5: CDC and Prevention, National Center for Emerging and Zoonotic Infectious Diseases (NCEZID), Division of Vector-Borne Diseases (DVBD); 20.7 (top): CDC; 20.7 (bottom): CDC/James Gathany; 20.8: ©Science VU-Charles W. Stratton/Visuals Unlimited; 20.9: ©Scott Camazine/Science Source; 20.10, p. 602 (mouse): CDC; 20.11a: ©Dr. John Cunningham/Visuals Unlimited; 20.11b: Originally published in Blood. Peter Maslak and Susan McKenzie, "Infectious Mononucleosis" *Blood*. 2004. ©American Society of Hematology. Used with permission; 20.12: ©A.M. Siegelman/Visuals Unlimited; 20.14: McCaul and Williams, "Development Cycle of C. Burnetii," *Journal of Bacteriology*, 147:1063, 1981. Reprinted with permission of American Society for Microbiology; 20.15: ©Kenneth E. Greer/Visuals Unlimited; 20.16: CDC; 20.16 (tick): ©PhotoLink/Getty Images RF; 20.17: ©The Natural History Museum/The Image Works; 20.18: ©Dr. David Phillips/Visuals Unlimited/Corbis; 20.19: Courtesy Stephen B. Aley, PhD., University of Texas at El Paso; 20.22: ©Anthony Asael/Art in All of Us/Corbis; p. 620: ©SPL/Science Source; p. 630 (lice): CDC/Alexander J. da Silva & PhD & Melanie Moser; p. 630 (Black Death): ©Corbis; p. 630 (Triatomine): ©The Natural History Museum/The Image Works; p. 631: Courtesy Steve Kunkel.

Chapter 21

Opener: ©McGraw-Hill Education/Gary He; 21.1: ©Ellen R. Dirksen/Visuals Unlimited; 21.3: ©Pulse Picture Library/CMP Images/PhotoTake; 21.4a: Multimedia Library, Congenital Heart Disease, Children's Hospital, Boston. Editor: Robert Geggel, MD. Www.childrenshospital.org/mml/cvp/; 21.4b: ©Dr. E. Walker/Science Source; 21.6a: ©McGraw-Hill Education; 21.6b: ©Science Source; 21.7: Federal Agriculture Research Centre, Germany; 21.8-9: CDC; p. 648 (WWI): ©Ingram Publishing RF; p. 648

(MLK): Library of Congress Prints and Photographs Division (LC-USZ62-111158); p. 648 (scanner): © imago stock&people/Newscom; 21.13: ©John D. Cunningham/Visuals Unlimited; 21.15: ©CDC/PhotoTake; 21.16: ©Dr. Leonid Heifets, National Jewish Medical Research Center; 21.17: ©SIU Bio Med/Custom Medical Stock Photo; 21.19a: Nester et al. Microbiology: A Human Perspective, 4th ed. ©Evans Roberts; 21.19b: ©L.M. Pope and D.R. Grote/Biological Photo Service; p. 659 (hasmat): ©Kenneth Lambert/AP Images; p. 659 (Y. pestis, anthrax): CDC; 21.21: ©Tom Volk; 21.22: CDC/,Viral Special Pathogens Branch; p. 663 (mouse): CDC/James Gathany; p. 668: ©Tom Volk; p. 669: ©Jack Bostrack/Visuals Unlimited.

Chapter 22

Opener: ©Jupiterimages/ImageSource RF; 22.4a: ©BSIP/Phototake; 22.4b: ©Stanley Flegler/Visuals Unlimited; p. 676 (woman): ©Getty Images/SW Productions RF; 22.6: ©Science VU-Max A. Listgarten/Visuals Unlimited; 22.7: CDC; 22.8a: Exeen M. Morgan and Fred Rapp, "Measles virus and its associated diseases," *Bacteriological Reviews*, Vol. 41(3), Sept. 1977, pp. 636-666. Reprinted by permission of American Society for Microbiology; 22.9a: ©PhotoTake; 22.12 (healthy): ©Gastrolab/Science Source; 22.12 (diseased): CDC; 22.13: R.R. Colwell and D.M. Rollins, "Viable but Nonculturable Stage of Campylobacter jejuni and its Role in Survival in the Natural Aquatic Environment," *Applied and Environmental Microbiology*, Vol. 52(3), Sept. 1986, pp. 531-538. Reprinted with permission of American Society for Microbiology; 22.14a: ©David Musher/Science Source; 22.14b: Courtesy Fred Pittman; 22.14c: ©Benjamin/Custom Medical Stock Photo; 22.15: CDC/Janice Haney Carr; 22.16: ©Moredun Animal Health Ltd./Science Source; p. 689 (patient): CDC; 22.17: ©Custom Medical Stock Photo; 22.18: ©K.G. Murti/Visuals Unlimited; 22.19: ©Iruka Okeke; 22.20: ©Ynes R. Ortega; 22.21: CDC/Dr. Stan Erlandsen; p. 696: ©Dr. Parvinder Sethi RF; 22.22d: ©Science VU-Charles W. Stratton/Visuals Unlimited; 22.22e: ©Eye of Science/Science Source; 22.24a: ©Stanley Flegler/Visuals Unlimited; 22.24b: Katz et al., "Parasitic Diseases," 1982 ©Springer-Verlag. With kind permission of Springer Science and Business Media; 22.25: CDC/Henry Bishop; 22.26a: ©R. Calentine/Visuals Unlimited; 22.26b: ©Science VU-Fred Marsik/Visuals Unlimited; 22.27: ©A.M. Siegelman/Visuals Unlimited; 22.28: ©Ed Reschke/Photolibrary/Getty Images; 22.29 (miracidium): ©Cabisco/Visuals Unlimited; 22.29 (cercaria): Courtesy Harvey Blankespoor; 22.29 (mating): ©Science VU/Visuals Unlimited; p. 714: CDC/Janice Haney Carr; p. 717 (food): ©Spike Mafford/Getty Images RF; p. 717 (eyes): CDC; p. 717 (needle): ©Stockbyte/PunchStock RF.

Chapter 23

Opener: CDC/James Gathany; p. 724: ©Pixtal/AGE Fotostock RF; 23.4: CDC; 23.5: Courtesy Danny L. Wiedbrauk; 23.6: ©David M. Phillips/Science Source; 23.7 (both): CDC; p. 732 (women): ©Design Pics/Don Hammond RF; 23.9: CDC/J. Pledger; 23.10: ©George J. Wilder/Visuals Unlimited; 23.11: CDC; 23.12: Courtesy Morris D. Cooper, Ph.D., Professor of Medical Microbiology, Southern Illinois University School of Medicine, Springfield, IL; 23.13a: ©Science VU/Visuals Unlimited; 23.13b: ©Martin M. Rotke/Science Source; 23.14: CDC/J. Pledger; 23.15a-b: ©Science VU/CDC/Visuals Unlimited; 23.16: ©Custom Medical Stock Photo, Inc; 23.17: ©Kenneth E. Greer/Visuals Unlimited; 23.20: ©Dr. E. Walker/Science Source; p. 743 (syphilis): ©Biophoto Associates/Science Source; p. 743 (chancroid): ©Dr. M.A. Ansary/Science Source; p. 743 (herpes): ©Dr.

p. Marazzi/Science Source; p. 745 (HPV): ©Kenneth E. Greer/Visuals Unlimited; p. 745 (molluscum): ©Dr. p. Marazzi/Science Source; p. 752: ©George J. Wilder/Visuals Unlimited; p. 753 (hand): CDC; p. 753 (tick): ©PhotoLink/Getty Images RF; p. 753 (back): ©Science VU/Visuals Unlimited.

Chapter 24

Opener: Jeff Schmaltz/NASA; MODIS Rapid Response Team; 24.1: Reprinted Cover image from December 1, 2000 Science with permission from Jillian Banfield, Vol 290, 12/1/2000. ©2000 American Association for the Advancement of Science; Image Courtesy Jillian Banfield; p. 757: NASA/JPL-Caltech; 24.9b: ©John D. Cunningham/Visuals Unlimited; 24.10a-b: ©Sylvan Wittwer/Visuals Unlimited; 24.14: ©John D. Cunningham/Visuals Unlimited; p. 769: ©Creatas RF/Punchstock RF; 24.17a: ©Dr. David Phillips/Visuals Unlimited/Getty Images; 24.17b: ©Carleton Ray/Science Source; 24.17c: ©Phillip Slattery/Visuals Unlimited; 24.19: ©John D. Cunningham/Visuals Unlimited; p. 773: ©Science VU/R.Roncadori/Visuals Unlimited.

Chapter 25

Opener: ©Apply Pictures/Alamy RF; p. 779 (dump): ©D. Falconer/PhotoLink/Getty Images RF; 25.2: ©Paula Bronstein/Getty Images; 25.3a-b: Courtesy Sanitation Districts of Los Angeles County; p. 782: ©Khalil Senosi/AP Images; 25.4 (Endo): ©Kathy Park Talaro; 25.4 (black light): Reprinted from EPA Method 1604 (EPA-821-R-02-024) Courtesy Dr. Kristen Brenner from the Microbial Exposure Research Branch, Microbiological and Chemical Exposure Assessment Research Division, National Exposure Research Laboratory, Office of Research and Development, U.S Environmental Protection Agency; 25.4 (natural light): Reprinted from EPA Method 1604 (EPA-821-R-02-024) Courtesy Dr. Kristen Brenner from the Microbial Exposure Research Branch, Microbiological and Chemical Exposure Assessment Research Division, National Exposure Research Laboratory, Office of Research and Development, U.S Environmental Protection Agency; 25.5: ©John D. Cunningham/Visuals Unlimited; 25.6: ©Ryan Hatch; 25.7a: ©Javier Larrea/age fotostock/Getty Images; 25.8: ©Kathy Park Talaro/Visuals Unlimited; 25.9: ©Joe Munroe/Science Source; 25.12: ©Kathy Park Talaro; p. 794: ©David Papazian/Getty Images RF; 25.14: ©Volker Steger/Science Source; 25.16: ©J.T. MacMillan; p. 801 (both): ©Kathy Park Talaro.

Line Art

Chapter 6

F6.9b, p 149: Source: "Studies on the Structure of Vaccinia Virus" by J.C.N. Westwood, W.J. Harris, H.T. Zwartouw, D.H.J. Titmuss and G. Appleyard from *J. gen. Microbiol.* (1964), 34, 67-78.

Chapter 8

F8.17, p 217: Adapted from LIFE: *The Science of Biology*, 7/e, by Purves, Sadava, Orioans & Heller. © Sinauer Associates, Inc. Reprinted with permission.

Chapter 10

F10.13, p 286: "Gene predictions tell an ever-changing story" by Peter Aldhous from *New Scientist*, 29 July 2009. © 2009 Reed Business Information – UK. All rights reserved. Distributed by Tribune Media Services.

Chapter 13

F13.18b, p 384: Source: Centers for Disease Control and Prevention. F13.19a, p 385: Source: Centers for Disease

Control and Prevention. F13.19b, p 385: Source: Centers for Disease Control and Prevention. F13.19c, p 385: Source: Centers for Disease Control and Prevention. F13.20a, p 386: Source: Centers for Disease Control and Prevention. F13.20b, p 386: Source Centers for Disease Control and Prevention.

Chapter 15

T15.A, p 449: Source: https://www.wp.dh.gov.uk/immunisation/files/2012/12/8252_VU_195_Nov_2012_05_accessible.pdf. TA15.8a, p 450: Source: Centers for Disease Control and Prevention TA15.8b, p 450: Source: Centers for Disease Control and Prevention. TA15.9, p 451: Source Centers for Disease Control and Prevention.

Chapter 21

F21.18, p 655: Source: Reproduced, with the permission of the publisher, from http://www.who.int/tb/challenges/mdr/drs_maps_oct2011.pdf by permission of World Health Organization.

Chapter 22

F22.11, p 682: Source: http://www.cdc.gov/eid/article/17/1/p1-1101_article.htm. Data from Scallan, Elaine, et al., Foodborne Illness Acquired in the United States-Major Pathogens, Centers for Disease Control and Prevention.

Chapter 24

F24.7, p 761: Source: http://www.esrl.noaa.gov/gmd/ccgg/trends/global.html. F24.7, p 761: Source: http://ga.water.usgs.gov/edu/graphics/earthwheredistribution.gif

Chapter 25

F25.10, p 789: Source: http://www.cdc.gov/features/dsfoodnet/

Endpapers

End Papers (4 graphics): Source: Centers for Disease Control and Prevention.

Index

Note: In this index page numbers set in **boldface** refer to boxed material, page numbers followed by an f refer to figures, page numbers set in *italics* refer to definitions or introductory discussions, page numbers followed by a t refer to tables, and page numbers beginning with an A refer to the appendix.

Displaying Disease Statistics

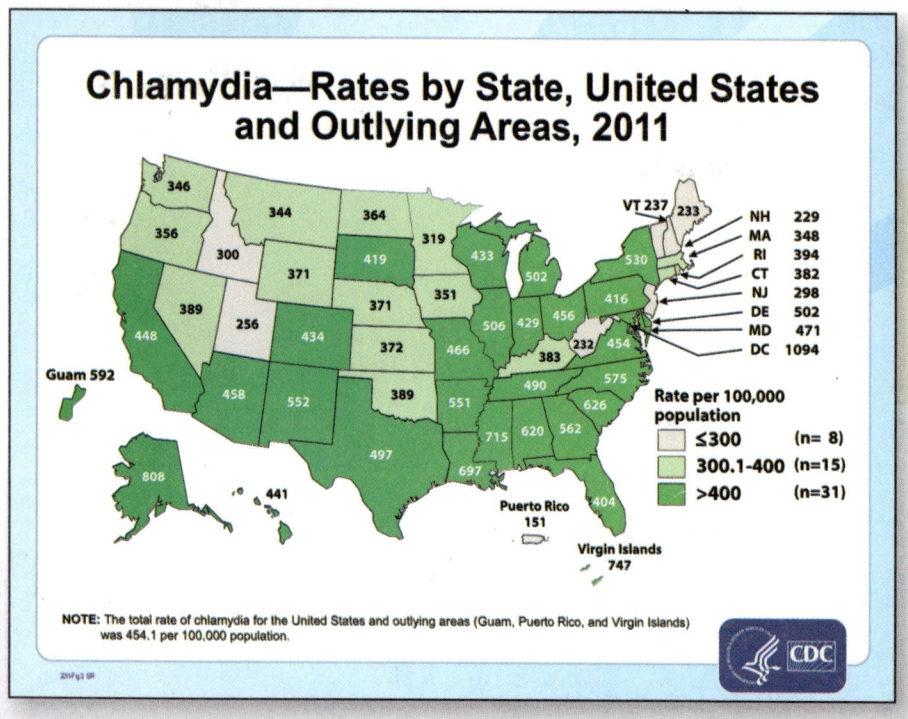

Chlamydia—Rates by State, United States and Outlying Areas, 2011

NH	229
MA	348
RI	394
CT	382
NJ	298
DE	502
MD	471
DC	1094

Rate per 100,000 population
- ≤300 (n= 8)
- 300.1-400 (n=15)
- >400 (n=31)

Guam 592
Puerto Rico 151
Virgin Islands 747

NOTE: The total rate of chlamydia for the United States and outlying areas (Guam, Puerto Rico, and Virgin Islands) was 454.1 per 100,000 population.

CDC

> Infectious disease specialists use a number of different methods to visually represent the numbers of disease cases or deaths.

> Some methods emphasize the geographical distribution of disease.

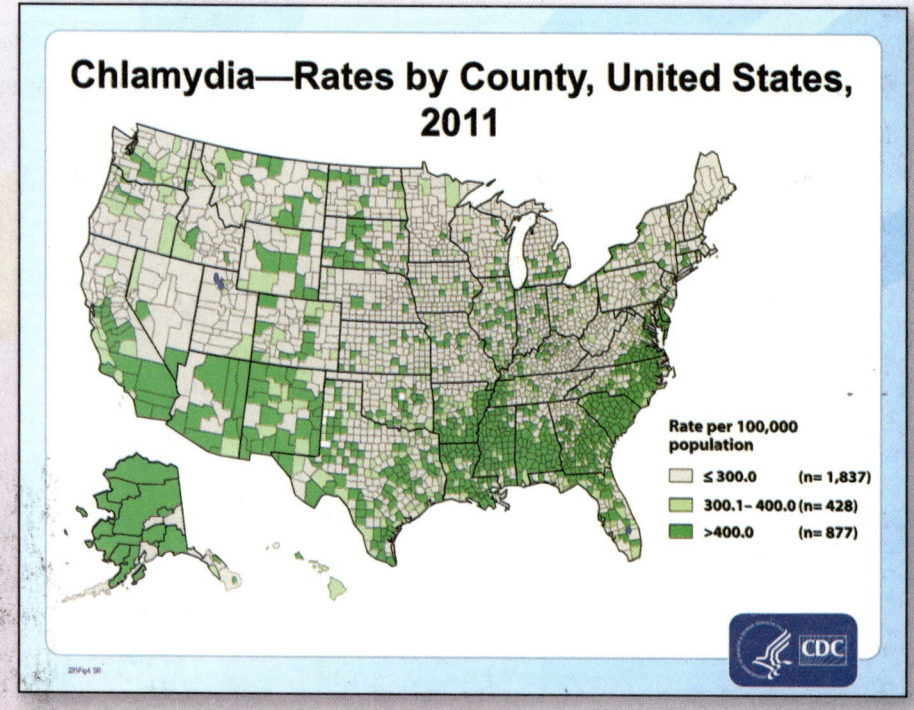

Chlamydia—Rates by County, United States, 2011

Rate per 100,000 population
- ≤ 300.0 (n= 1,837)
- 300.1– 400.0 (n= 428)
- >400.0 (n= 877)

CDC